Numerical Modelling in Applied Geodynamics

Numerical Modelling in Applied Geodynamics

Jiří Nedoma
Institute of Computer Science
Prague, Czech Republic

JOHN WILEY & SONS
Chichester • New York • Weinheim • Brisbane • Singapore • Toronto

National 01243 779777
International (+44) 1243 779777
e-mail (for orders and customer service enquiries): cs-books@wiley.co.uk
Visit our Home Page on http://www.wiley.co.uk
or
http://www.wiley.com

Other Wiley Editorial Offices

John Wiley & Sons, Inc., 605 Third Avenue,
New York, NY 10158-0012, USA

Wiley VCH Verlag GmbH, Pappelallee 3,
D-69469 Weinheim, Germany

Jacaranda Wiley Ltd, 33 Park Road, Milton,
Queensland 4064, Australia

John Wiley & Sons (Asia) Pte Ltd, 2 Clementi Loop #02-01,
Jin Xing Distripark, Singapore 0512

John Wiley & Sons (Canada) Ltd, 22 Worcester Road,
Rexdale, Ontario M9W 1L1, Canada

Library of Congress Cataloging-in-Publication Data

Nedoma, J. (Jiří)
Nedoma.
Numerical modelling in applied geodynamics / J.
p. cm.
Includes bibliographical references and index.
ISBN 0-471-97461-7 (hardcover)
1. Geodynamics — Mathematical models. I. Title.
QE501.4.M38N43 1997
551.1′015194 — DC21 97-20322
CIP

British Library Cataloguing in Publication Data

A catalogue record for this book is available from the British Library

ISBN 0 471 97461 7

Typeset in 10/12pt Times by Laser Words, Madras, India
Printed and bound in Great Britain by Bookcraft (Bath) Ltd
This book is printed on acid-free paper responsibly manufactured from sustainable forestry, in which at least two trees are planted for each one used for paper production.

To my parents, sister and daughter

Contents

Preface

The fundamental principle of ethics is reverence for life. Progress in knowledge and abilities promote mankind only in trivialities. The essence is that we become better and more profound human beings.

Dr. Albert Schweizer
Die Lehre der Ehrfurcht vor dem Leben

This monograph, based on the author's results during the last twenty five years, discusses a global geodynamical model of the Earth as well as models of rift and subduction (collision) zones and the physical and mathematical theories and algorithms, solving such problems.

The model is based on the hypothesis of tectonics of lithospheric plates (plate tectonics), on the assumption of gradual density differentiation on the masses of the Earth, on their phase transitions under high pressures and temperatures and on the theory of thermo-elasto-visco-plasticity, and takes into account the theoretical and experimental results pertinent to all geophysical fields.

Geological and geophysical facts concerning the structure of the Earth as a whole, as well as its separate regions, based on geological and geophysical observations *in situ*, as well as on theoretical results, have in the past been mis-interpreted. The primary purpose for any new interpretation is to re-analyse old conclusions and only adopt those which will prove to be an asset for further study.

The presented geodynamic model explains the origination of the primary core and the processes taking place, even in the present, at the core-mantle boundary by means of the density differentiation of molten masses of the lower mantle. The layer of light materials, enriched with Fe, Mg and other elements generated during this density differentiation, gives rise to an unstable layer at the outer core-mantle boundary, which displays a tendency to adopt a new and more stable state. As a result of this development, the materials forming it flow through GTLV-channels, supported by convective flow partly in the mantle and partly in the core, into the upper parts of the Earth, where rift zones are generated. Rift zones are regions where a new oceanic lithosphere and, consequently, regions of the primary strain on the lithosphere originate. After physical limits have been exceeded, the strained lithosphere breaks up into separate lithospheric megaplates which collide with one another. The model interprets the mechanism of generation of rift zones as well as the mechanism of collision of lithospheric blocks and plates on the basis of the mechanism of dynamic contact of elasto-visco-plastic bodies subjected to considerable temperatures, derived from the theory of thermo-elasto-visco-plasticity, taking into account the results of studies of all geophysical fields. It also explains the origin of the primary geomagnetic field and its variations, as well as its relation to the geodynamic processes within the Earth. It expounds the possible origin of the primary continent of Pangaea, the foundations of the original ocean bed, of the present ocean bed in areas of inland and smaller seas, as well as in smaller areas of continent in regions of subduction, the existence and mechanism of recent movements, the existence of unfolded sediments in deep-sea trenches, of the Conrad, Mohorovičić and other transitional zones (and, consequently, the non-existence of well-defined boundaries as originally interpreted), which has also been proved by the results derived from available materials of deep drilling (around 10 km).

To conclude, it may be surmised that the presented geodynamic model is probably broad enough for it to be used to describe all sufficiently large bodies of our planetary system (i.e. planets and some large moons). The existence of hitherto observed manifestations of volcanic activity on some planets and moons renders these deliberations quite probable.

The main goals of the book, are above all to present the mathematical models of these geophysical problems and methods of their solutions based on the modern mathematical analyses.

The book is divided into five parts, dealing with a summary of functional and numerical analyses, numerical models of deformation of geological bodies, numerical models of plate tectonics and mathematical problems of geophysical fields, respectively.

Chapter 1 gives an introduction to the geodynamics of the Earth and discusses the global geodynamic model of the Earth.

Chapter 2 summarizes the essentials of the theory of functional analysis, theory of spaces, convex analysis, variational equalities and inequalities, finite element methods as well as methods of linear and nonlinear algebra.

Chapter 3 deals with the deformation of geological bodies, introducing the mathematical theory of continuum mechanics and applications in geological practice.

Chapter 4 develops the physical and mathematical theory of plate tectonics and summarizes the rheology of the lithosphere and the upper mantle. The mathematical models of geodynamic processes in upper parts of the Earth are given and discussed.

Chapter 5 analyses the mathematical models of the rift zones. The models are based on the thermo-Bingham rheology. Numerical analyses of the model problems are given.

Chapters 6 and 7 deal with the simple and realistic mathematical models of collision and subducted zones. The simple models of collision and subducted zones are based on the basic boundary value problems of elasticity. variational formulations and weak solutions of basic boundary value problems of linear elasticity formulated in displacements and stresses are given and analysed. The 3D models of periodic structures, based on the homogenisation method and the finite element method, are numerically analysed. Models based on the basic boundary value problems of visco-elasticity, plasticity as well as thermo-elasticity are also analysed.

Realistic mathematical models of collision and subducted zones are based on contact problems with and without friction in elasticity and thermo-elasticity, which solve numerically variational inequalities. These include moreover, mathematical models with recrystallization based on coupled contact-two phase Stefan problems of linear thermoelasticity and the deformation theory of thermoplasticity. Numerical verification of geodynamic processes in the lithosphere of the West Carpathians, in the Central Aleutians as well as those of the Himalayas and the Andaman-Nicobar area are given and numerical results are discussed. The theory of plate tectonics presented in this monograph is also applied to the Mediterranean area, Japan arc system and to the subduction zones in areas of the North and South American plates as well as to the hot-spot areas.

In Chapter 8, algorithms for numerical solutions of plate tectonic problems, based on the variational inequalities, are developed and presented.

Chapter 9 describes the geothermal models based on linear and nonlinear boundary value problems and their numerical finite element approximations.

Chapter 10 introduces the magneto-telluric theory and formulations of magneto-telluric models of the Earth in the frequency and time domains and their numerical solutions based on the author's results from the end of the 1960's and the beginning of the 1970.

Chapter 11 deals with some problems of propagation of seismic (elastic) waves in isotropic and anisotropic rocks of the Earth. Practical applications are also discussed.

In Chapter 12, global geodynamic, geomagnetic and gravity models of the Earth and planets are presented and analysed. Firstly the geodynamic processes in the mantle are discussed. Then boundary value problems are formulated and numerically analysed. The coupled-Stefan-like-Bingham model problems of geodynamic processes in the mantle and their numerical approximations are also analysed and discussed. Mathematical analyses of geodynamic processes in the Earth's and planets's interiors as well as thermo-magnetodynamic problems, ionization and metallization of rocks are formulated and analysed. This chapter develops a global gravity model and shows how the gravity field depends on the evolution of the Earth. The theory developed in this monograph is generalized to the magnetic anomaly observed on the planet Uranus, based on the second change transition theory.

This book is addressed to graduate students as well as to researchers and specialists working in the fields of numerical simulation of geophysical and geological processes inside the Earth, planets and moons as well as specialists in the field of numerical simulation of partial differential equations. As a graduate text for the PhD courses in mathematical geophysics, geology and applied and numerical mathematics it, can be used in its entirety. For easier study, an introduction to modern mathematical methods has been given. The book is also addressed to geologists and geophysicists to help simulate geodynamical processes in the Earth and verify their scientific ideas. It is evident that for the verification of geological ideas, the correct data must be used, as incorrect data gives incorrect models and therefore, incorrect results. The evolution models of the subducted and collision zones depend also on a time scale. It is evident that for different time periods we study the different evolution models, i.e. models which enable us to simulate different geological situations. By comparing the obtained numerical results with the observed data we get material for the better understanding and interpretation of the real geological situation. Therefore, the methods discussed in the book facilitate the simulation of the investigated processes inside the Earth and after analysis of obtained data should give a more accurate interpretation of the studied geological situations. However in every case the specialists must take into account the fact that all steps preceding the mathematical models are intuitive only, and therefore, cannot be proved. Besides these facts, stable algorithms must also be used, to avoid incorrect interpretation of data and therefore, incorrect scientific conclusions.

Acknowledgements

I gratefully acknowledge the help of all who contributed to the completion of this work. My special thanks go out to my institute for their positive attitude to my work concerned with the solutions of the COPERNICUS project No CP 94 0820-HIPERGEOS "High Performance Computations in Geosciences: Safety of Construction with Respect to Rock Deformations and Movements" under the leadership of Prof. Owe Axelsson. The author's special thanks go out to the Dept. of Mathematics of the Faculty of Applied Sciences, University of West Bohemia, Pilsen for their positive offer to read parts of the book. I gratefully acknowledge their numerous useful suggestions and discussions in dealing with the geodynamic problems. I wish to thank to Dr. Jana Kotkvá PhD and Prof. Dr. Karel Segeth PhD for their excellent translation work and for correcting and reading some parts of the book. Of great importance to the completion of the present book was the help of Prof. Dr. Karel Najzar PhD, Prof. Dr. Igor Bock PhD, Dr. A.K. Singh PhD, who read some of parts of the book as well as of my PhD students Mgr. Zdeněk Kestřánek, Mgr. Alena Baštová, Mgr. Marek Novický and Mgr. Jan Dvořák, who computed some numerical models presented in the book and who also read some of parts of the book. I wish to thank Prof. R.K. Verma, Prof. J.G. Negi, Prof. V.C. Thakur, Dr. K.K. Sharma, Prof. M. Mukhopadhyay for his useful discussions concerning analyses of the Himalayas and the Andaman-Nicobar Island Arc system. Finally my thanks go out to the Kartografia a.s., Prague, who provided the map of the divided lithosphere into the sub-blocks for publication in the book.

List of Notation

a.e.	almost everywhere
$\Rightarrow$	implication, e.g. $A \Rightarrow B$ means that A implies B
R^n	n-dimensional space over the field of real numbers
$\Omega C R^n$	domain in R^n an open connected set
$\partial\Omega$, Γ, Γ_u, Γ_τ	boundary of the domain Ω and its parts
$\overline{\Omega}$	closure of the domain Ω, $\overline{\Omega} = \Omega \cup \partial\Omega$
$\forall$	for all
$\exists$	there exists
$F : R^n \to R^m$	mapping F from R^n into R^m
$f_n \to f$	strong convergence of the sequence $\{f_n\}$
$f_n \longrightarrow f$, $f_n \to f$ weakly	weak convergence of the sequence $\{f_n\}$
$\boldsymbol{x} = (x_i)$, $i = 1, \ldots, n$	point in R^n
$\lvert\boldsymbol{x} - \boldsymbol{y}\rvert = \left(\sum_{i=1}^{n}(x_i - y_i)^2\right)^{1/2}$	distance between two points $\boldsymbol{x}, \boldsymbol{y} \in R^n$
$\boldsymbol{u}(\vec{u}) \in V(K)$	vector $\boldsymbol{u}$ belongs to the space V (to the set K)
(a, b)	an open interval, $a < x < b$
$\langle a, b\rangle$	closed interval, i.e. the set of $a \leq x \leq b$
supp f	support of the function f, i.e. the closure of the set of those points at which $f \neq 0$
$H^k(\Omega \equiv W^{k,p}(\Omega)$	the Sobolev space
$H_0^1(\Omega \equiv W_0^{1,2}(\Omega)$	
$H^{k,p}(\Omega)$	the space of functions from $L^p(\Omega)$ having generalized derivations up to the order k from $L^p(\Omega)$
$\partial(\Omega)$	space of functions with compact support
$C^{(k)}(\Omega)$	space of functions continuous in Ω whose derivatives up to order k are continuous in Ω
$(u, v)_k$	inner (scalar) product in $C^{(k)}(\Omega)$
$L^2(\Omega)$	space of functions square integrable in Ω
$\langle u, v\rangle$	dual pairing
$L^p(\Omega) = W^{o,p}(\Omega)$	
$\lvert\alpha\rvert$	$= \alpha_1 + \cdots + \alpha_n$ - length of multi-index α
$f_i x \to f(x)$	function correspondence
$H^1(\Omega) \equiv W^{1,2}(\Omega)$	
$\lVert\cdot\rVert_{k,p}$	the norm of the space $H^{k,p}(\Omega)$
$W_{\text{per}}^{k,p}(\gamma)$	the space of functions from $W^{k,p}(\gamma)$, which are periodic on the interval γ

$X_1 + \cdots + X_n = \prod_{i=1}^{n} X_i$	product of spaces
X', X''	dual and second dual spaces of X
$f\mid_\Omega$	restriction of the function f on the set Ω
$D^\alpha u$	derivative of the order $\lvert\alpha\rvert$
$R_\varepsilon u$	mollifier of the function u
diam (Ω) (or diam Ω)	diameter of the set Ω
$\boldsymbol{n} = (n_i),\ i = 1, \ldots, N$	outer normal
$\hookrightarrow$	imbedding
$\hookrightarrow\hookrightarrow$	compact imbedding
C_v, C_p, c_v, c_p	specific heat at constant volume and pressure, the second pair per unit mass

$u_{i,j} = \partial u_i/\partial x_j,\ u_{i,t} = \partial u_i/\partial t,$
$u_{i,tt} = \partial^2 u_i/\partial t^2,$
$\partial u_{i,jk} = \partial^2 u_i \partial x_j \partial x_k$

$\boldsymbol{n}$ - the unit unusual vector
Div $\boldsymbol{A} = (\boldsymbol{n}, \boldsymbol{A}_1 - \boldsymbol{A}_2)$ - the surface divergence of the vector $\boldsymbol{A}$
Rot $\boldsymbol{A} = [\boldsymbol{n}, \boldsymbol{A}_1 - \boldsymbol{A}_2]$ - the surface rotation of the vector $\boldsymbol{A}$
div $\boldsymbol{A} = \nabla \cdot \boldsymbol{A} = \partial A_i/\partial x_i = A_{i,i}$

$$\text{rot } \boldsymbol{A} = \nabla \times \boldsymbol{A} = [\nabla, \boldsymbol{A}] = \begin{pmatrix} \boldsymbol{i}_1 & \boldsymbol{i}_2 & \boldsymbol{i}_3 \\ \partial/\partial x_1 & \partial/\partial x_2 & \partial/\partial x_3 \\ A_1 & A_2 & A_3 \end{pmatrix}$$

$u_{i,j} = \partial u_i/\partial x_j$
$u_{i,t} = \partial u_i/\partial t$
$u_{i,tt} = \partial^2 u_i/\partial t^2$
$\tau_{ij,j} = \partial \tau_{ij}/\partial x_j$
$f_{,i} = \partial f/\partial x_i$

PART I

Geodynamics: An Introduction

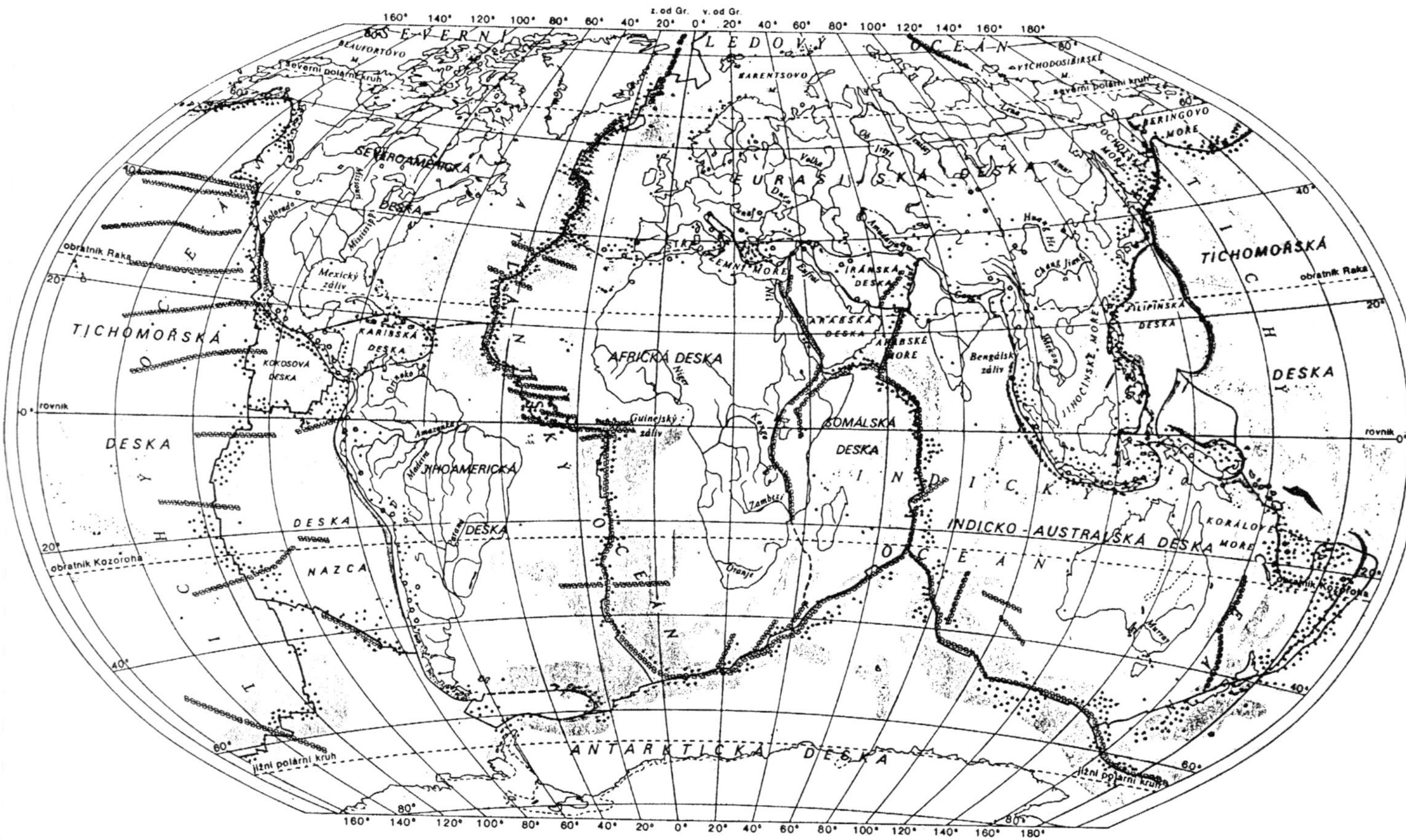

Figure 1.1. Principal lithospheric plates and geological blocks of the Earth

1

Geodynamics of the Earth: General Introduction

1.1. INTRODUCTION

The study of the geodynamic processes taking place within the Earth, which has been carried out over the last six to nine decades, is founded on various, sometimes even contradictory, assumptions and hypotheses, e.g. on the plutonic, expansional, contractional, isostatic, pulsational, oscillational and rationally dynamic hypotheses, and on the hypotheses of extraterrestrial influences, convective currents, continental drift and, more recently, the tectonics of lithospheric plates — plate tectonics.

If we wish to describe physically and mathematically the geodynamic processes taking place in the Earth's interior, we must adopt one of the hypotheses mentioned above and, on its basis, develop our own theory. The hypothesis which is most probable at this time is that of the tectonics of lithospheric places — plate tectonics. This will be used as a basis for developing a theory which will enable the geodynamic model to describe the intra-terrestrial geodynamic processes. The geodynamic model of the Earth will be developed on the hypothesis of the tectonics of lithospheric plates, on the assumption of density differentiation of the Earth's masses and their phase transitions under high pressures and temperatures, and on the theory of thermo-elasto-plasticity, taking into account all the available data, both experimental and theoretical, pertaining to all geophysical fields.

Plate tectonics is a hypothesis explaining tectonic and seismic activity in the upper parts of the Earth by operation of several large and several tens of smaller geological bodies — lithospheric plates and geological blocks, into which the upper part of the Earth is broken, see Figure 1.1. The term lithosphere relates to the outer solid upper part of the Earth, broken into blocks, moving on the underlying asthenosphere. The thickness of the lithosphere is about 100 km on average. The asthenosphere is a layer below the lithosphere, characterized by lower velocities of seismic waves and lower viscosity (of 10^{17} to 10^{20} Pa). Boundaries between geological blocks and plates are situated within seismic zones, where most of the mechanical energy is being consumed. These zones are generally very narrow, and they represent sites where the differential movements between geological bodies proceed. Figure 1.1 shows that the plate boundaries are located mainly in the oceans (about 120 000 km), from which only two plates, Nazca and Cocos, are purely oceanic plates; the remaining approximately 25 000 km of the plate boundaries are continental. Continental boundaries have a diffusive character, represented by seismic zones of several hundred kilometres width. Only the Eurasian plate is of a purely continental character. Plate tectonics is characterised by the fact that each of two geological bodies move against each other. Boundaries between geological bodies are of several types:

(i) divergent boundaries — plates in the vicinity of boundary accretion (rift zones, mid-ocean ridges); sign R,

(ii) convergent boundaries — volume of the plates decreases, oceanic part of the plates is consumed (subduction zones), continental parts of the plates cannot be consumed because of their considerable thickness (collision zones); sign T,

(iii) transform boundaries — plates move against one another by translation movement, their volume does not change; sign F.

Three plates then associate in triple points — RRR, RRF, RFR, RFT, TTT, TFF, FFF, RTT, TTF.

We distinguish 10 major lithospheric plates: African, Antarctic, Arabian, Eurasian, Indian, South American, Cocos, Nazca, Pacific and North American.

According to Holmes (1931) and more recent works, e.g. Davies (1980), plate tectonics results from heat convection in the mantle. It is not completely clear, however, how large a part of the Earth's mantle is involved in the convection and what is the character of the mass flow in the mantle. The present studies show that the only geophysical quantities which enable us to study flow in the mantle are global gravity anomalies, i.e. Faye gravity anomalies and geoid heights. However, it is very difficult to find a corresponding mathematical model for extreme mechanical and heat conditions in the mantle. The present mathematical models are based on models of Newtonian liquid in the Boussinesq's approximation. These models are characterized by Prandtl, Reynolds and Rayleigh dimensionless numbers. The Prandtl number is an expression of the ratio between the kinematic viscosity and the heat conductivity. The Reynolds number represents the ratio between the force causing acceleration of the liquid and the cohesion force impeding its motion. The Rayleigh number is proportional to the temperature difference between the upper and lower boundary of the moving liquid, therefore it expresses the ratio between the time necessary for heating up of the liquid layer by heat conduction and the time which is necessary for one run around one convection cell. In the mantle, the Prandtl number is very high, whilst the Reynolds number is very small (of about 10^{-10}), which in practice means that viscous forces prevail over the inertia. The Rayleigh number changes its value in a very narrow interval (from 10^5 to 10^9), which indicates that temperature gradients are very high and increase towards the Earth's surface (see Peltier (1989)).

The existence of convection currents is proved indirectly by gravimetric and geothermic methods — changes of gravitational and heat fields of the Earth.

Convection currents are situated very deep under the Earth's surface and they are, moreover, very slow (several cms per year). McKenzie *et al.* (1980) and others have demonstrated that the flow in the mantle is not represented by regular convection cells and that it is not parallel to the plate movement. It therefore has the character of three-dimensional circulation. The question then arises, what is the maximum depth of the mantle convection directly associated with the movement of lithospheric plates? There are two current models postulated by geophysicists dealing with mantle convection research: whole-mantle convection, formulated in 1931 by Holmes, and the upper-mantle model formulated in 1971 by Isacks and Molnar see Figure 2a. The latter authors assumed that the viscosity of the lower mantle is several orders of magnitude higher than in the upper mantle and therefore the subducting lithosphere cannot penetrate to depths greater than 700 km. According to Richter (1979) and McKenzie (1983), this is related to the change of the chemical composition of the mantle at depths of 650-700 km and the fact that there are no deep earthquakes below these depths. Based on these hypotheses, upper and lower mantle convection are more or less independent of each other see Figure 2b. On the contrary, many geophysicists assume that the viscosity of the upper and lower mantle are practically the same ($\sim 10^{21}$ Pa; see Cathles (1975), Peltier and Andrews (1976), O'Connell (1977), Elsasser, Olson and Marsh (1979), and Hager and O'Connell (1979)). This assumption implies whole-mantle convection, however it does not explain generation of deep earthquakes. According to Davies (1980), there is a high internal friction which prevents the subducting plate from penetrating great depths and into the lower mantle. Advocates of whole-mantle convection assume that there is a viscous link between asthenosphere and lithosphere, and that convection is the driving force of the motion of lithospheric plates and blocks. These simply drift because of the whole-mantle convection. High friction in the contact zone of two plates can then only be overcome by shear force in the lower part of the lithosphere, linked by viscosity with the mantle convection. In the present work we will try to identify and explain some imperfections of these hypotheses, and to find a mathematical apparatus which would, with higher precision and better understanding of rheology of the lithosphere and the Earth's mantle, help to find a more appropriate model.

Divergent boundaries are characterized by shallow earthquakes, high heat flow and intense volcanism, as well as by anomalous magnetic fields, almost parallel to mid-ocean ridges. Almost regular alternation of 10-30 km wide zones with positive and negative anomalies of amplitudes 200-400 nT is being explained as a result of sea-floor spreading, during which the natural magnetic polarization of basalts and their intrusive equivalents retain polarity of the magnetic field at the time of their cooling and transition over the Curie point. Seismicity in these areas is two orders of magnitude lower than in the areas of convergent boundaries. At the same time, brittle behaviour of the rocks reaches a depth of only a few kilometres. The majority of differential movements in rocks in these areas proceeds by creep (aseismic flow). The direction during earthquakes is strike slip. Shakes last several hours to several days without any dominant shock.

Systems of faults — almost perpendicular to ridges in the case of quick ridges and slightly oblique in the case of slow ridges (5-40°) — are typical of transform boundaries.

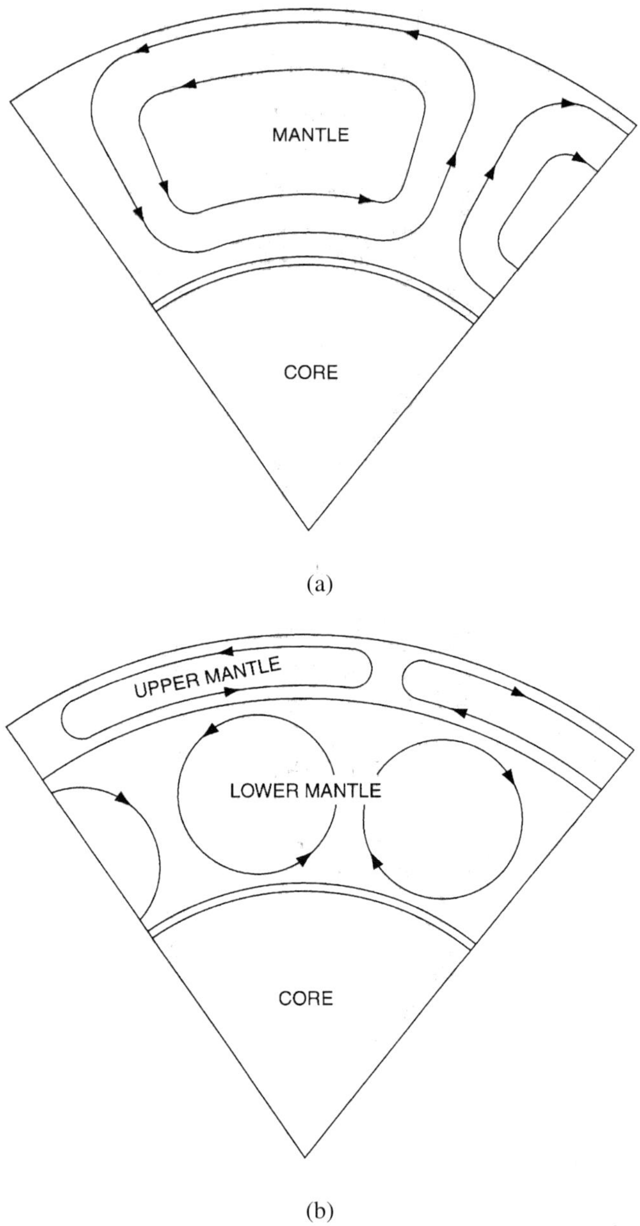

Figure 1.2. Two Types of Mantle Convection-Schematic Cross-Sections

On transform faults, wide V shape valleys with stair-like faults are formed. The deep San Andreas Fault in California, USA, is one of the most well known transform faults. In its southern part, about 300 km in length, earthquakes are common. Its central part, 250 km long, is characterized by compressive tectonics and catastrophic earthquakes. It is followed by a 120 km long aseismic part, and the more than 250 km long San Andreas proper, with compressive tectonics, oblique slip and catastrophic earthquakes.

Both parts with catastrophic earthquakes are seismically quiet in between the earthquakes. Due to the penetrating rift, the transform fault is already situated on a continent. The majority of transform faults are being explained as a natural consequence of the formation of a new oceanic lithosphere and magma cooling.

Convergent boundaries occur in sites of oceanic lithospheric consumption and in those of continental lithospheric collision. The sites where the oceanic lithosphere is consumed are called "subduction zones". Deep ocean trenches associated with Faye gravity anomalies represent the surface manifestation of these zones. Subduction zones are accompanied by intense seismicity in obliquely inclined layers, the so-called Wadati-Benioff zones, 10–30 km in thickness, in which more than 85% of seismic energy is consumed. The majority of Wadati-Benioff zones are slightly curved with a relatively cold elastic part of subducting lithospheric plate. They reach to depths of as much as 650 km; their geometry is more or less curved to 200–300 km depth, below these depths occur more irregularities associated with aseismic zones. At great depths in areas of subduction zones, relics of lithosphere of older orogenic cycles have been observed. Another type of convergent boundary besides subduction zones are collision zones. Continental collisions represent a final stage of the evolution of convergent plate boundaries, at present characterized by the Alpides (Alpine-Himalayan mountain belt). In some areas, continental collision passes into subduction (Brahmaputra River valley, eastern Himalaya, and Burma plate).

In the next section we shall introduce the global geodynamic model of the Earth which, instead of geological observations, also fulfils the theory of thermo-visco-plasticity and properties of several geological bodies being in contact, as well as properties of melted, solidified, recrystallized, metallized and ionized rocks under very high temperatures and pressures. Moreover, the standpoint of evolution of the planetary systems must also be assumed.

1.2. GLOBAL MODEL OF THE EARTH

1.2.1. Geodynamic Processes in the Earth's Core and Mantle, their Causes

The earth was most probably created by an accumulation of interstellar matter on material condensation nuclei. After a sufficient amount of interstellar matter had been accumulated, the temperature within the original Earth increased, due to the effect of gravity forces and the inherent radioactivity of rocks of the original interstellar matter, until it reached the melting point of the rocks. Rocks of interstellar matter in the neighbourhood of the centre of gravity of the original Earth melted and the first density differentiation of the molten matter occurred. The heavier matter remained in the vicinity of the centre of gravity, whilst the lighter substances and the more liquid substances of easily molten rocks had a tendency to rise. The Earth's core began to form. The matter of the core was thus differentiated, and the initial inner and outer cores originated.

As thc core increased in size, its temperature and the pressure began to increase, and the first phase transitions occured within the core. The temperature in the neighbourhood of the core also increased. The matter of the initial Earth outside its core was thus subjected to considerable temperatures and pressures, and the Earth's mantle began to form.

The materials of the mantle at the core-mantle boundary were and are still being melted. The heavier substances, attracted by the matter of the inner core, descend and increase the size of the inner core, whereas the lighter substances and the rocks easier to melt gradually amalgamate with the substances of the outer core formed by molten rocks in liquid state and also containing various gaseous components. These, moreover, are strongly electrically and heat conducting. Since the surface of the outer liquid (strongly visco-plastic) core was being cooled from the mantle side and heated from the inner-core side, heat convection was generated in the outer liquid core, to which contributions are made by the transport of the heavier substances towards the inner core, the effects of different dynamic ellipticities of the core and mantle, the precision of the Earth's axis of rotation and, in particular, of the Coriolis force (Hide (1953)). As a result of all these effects, thermo-electric forces and currents are generated (Nedoma (1984), (1986a, b)).

In the deeper parts of the outer liquid (strongly visco-plastic) core, the molten liquid matter has a tendency to rotate faster on average than its upper parts, where the rotation is slower due to the viscosity of the medium and in order to preserve angular momentum. These effects result in a kind of stratification of the outer core. The velocity of the convective flow and thermo-magneto-hydrodynamic turbulence is relatively high, about 20 km/year. The matter in the outer core is strongly heated and exposed to pressures, and it is more electrically conductive as a result of phase transitions and metallizations as well as ionizations, and also less viscous (Artyushkov (1979), Cathles (1975)) with a viscosity coefficient $\eta \sim 10^2$–10^9 poise and density 9.8–12.5 gcm^{-3}. This convective flow and thermo-magneto-hydrodynamic turbulence then generate a primary geomagnetic field in the outer core; the field's mechanism acts as a double or multiple hydrodynamic dynamo (Rikitake (1958), Bucha (1975)), just as a multiple thermo-hydrodynamic dynamo.

Moreover, since the viscous and Coulomb friction at the mantle-outer core boundary is small, the mantle is not couple with the core by large mechanical forces, and, consequently, rotates faster than the core. This is responsible for the observed westward drift of the geomagnetic field. The mantle-outer core boundary, i.e. the boundary at which the density differentiation of the molten material takes place, is itself uneven due to the non-uniform melting of the mantle rocks (caused by the non-uniformity in the temperature distribution in the upper parts of the outer core and mantle, by the different degrees of fusibility of the mantle rocks as well as the property of "mushy zone"), or to possible recrystallization processes in accumulated light substances and the possible consequences of the non-uniform shape of the initial core-mantle boundary. This may in turn be responsible for temporal changes in the rotation of the mantle and of the whole Earth, and may have a feedback effect on the primary geomagnetic field (time variations of the latter) and on the geodynamic processes in the Earth's core and the mantle. It is evident that the main geomagnetic field and its variations are generated by the co-action of the non-uniform Earth's rotation and the high electric conductivity of the core matter, as well as the thermo-hydromagnetic forces generated by the motion of matter in the boundary regions of the outer core and mantle.

Deflection of the axis of the magnetic field from the axis of rotation (by approximately 10°) is probably due to the asymmetrically distributed convective cells and inhomogeneous distribution of masses in the outer core and the mantle. This inhomogeneity and the non-uniformity in the rotation of the Earth are responsible for the irregularity in the action of the Coriolis and Poincaré forces, which in turn is reflected in the effect on thermo-magnetohydrodynamic turbulence in the outer core.

The inhomogeneities in the composition of the molten liquid mass may be generated not only by the processes associated with the density differentiation and rotation of the Earth, but also as a result of the pressure differences in the upper and lower parts of convective vortices. Since the generated electric currents and convective motions caused by the rotating mass of the outer core heated from below and cooled from above, the axis of these motions in the ideal case is the Earth's axis of rotation. In addition the direction of the axis of the generated magnetic field will be close to the Earth's axis of rotation. These changes in the motion of masses of the outer core and their irregularities may cause inhomogeneities in the regeneration of the magnetic field and, consequently, a reversal of the Earth's magnetic field.

One must bear in mind that the geodynamic processes within the Earth are closely related to thermo-magnetohydrodynamic processes and that they affect one another. To be able to explain all these processes, it is necessary to consider all of the above effects simultaneously and, moreover, to also consider the expansion of the Earth's masses due to the temperature rising within the Earth (although this effect will probably not be as large as presented by some authors, the advocates of the theory of the Earth's expansion).

The inner core is most probably formed by a stratified polycrystalline structure in thermo-visco-plastic state (S-waves propagate through it), characterized by a high density, $\rho \sim 13\text{-}14\ \text{gcm}^{-3}$, originated by and, at the same time, subject to high pressures and temperatures, as well as to the Coriolis force. In the outer part of the inner core, i.e. at depths of between 4703 and 5154 km, high pressures and temperatures give rise to phase transition of the heavier materials of the outer core, which have reached the boundary of the inner core as a result of density differentiation and motions of masses in the outer core. The material of the inner core will probably consist of oxides, mainly FeO and MgO but also SiO_2, and possibly of other oxides, Fe alloys and other elements in a metallized state. Due to the phase transitions, endogeneous energy is released in the outer part of the inner core, and is transferred by thermo-magnetohydrodynamic turbulence into higher parts of the core, where it is transformed into thermal energy; the latter is enhanced by that part of the thermal energy generated by conversion of part of the energy of the electromagnetic field into heat.

After the density differentiation of the materials of the inner and outer core, the remaining light materials from the molten mixture at the outer core-mantle boundary, enriched with heavier elements, Fe, Mn, U, Pb, Zn and others (also in gaseous state), accumulate under the mantle just below its boundary with the external core. Since the mantle consists of relatively heavier material, this layer of light molten materials becomes unstable, forming blocks of considerable sizes (several hundreds of kilometres). As proved by laboratory experiments in the years 1971–1975 (Kawai and Mochizuki (1971), Kawai and Nishiyama (1974), (1975)), under a load of several megabars silicate rocks (Fe_2SiO_4, etc.) decompose into oxides (SiO_2, FeO, MgO, etc.). It is probable that similar processes also take place at the mantle-core boundary. Moreover, they are accompanied by phase transition of oxides into metallized state. FeO and Fe alloys and possibly also other metallized oxides and alloys become part of the core, and the unmetallized oxides SiO_2, MgO and oxides of other light elements, enriched with oxides of heavier elements, form the light materials which accumulate below the mantle. Their behaviour is similar to that of a viscous liquid with high viscosity (or of a strongly visco-plastic to visco-liquid molten mass).

The lower mantle can be characterized as a thermo-visco-plastic medium, strongly heated from below, i.e. from the side of the outer core, and cooled from above. This generates convective heat flow and so-called convective cells and channels of increased temperature and decreased viscosity (GTLV-channels) between them, see Figure 1.3. The light, strongly heated viscous material from the core-mantle

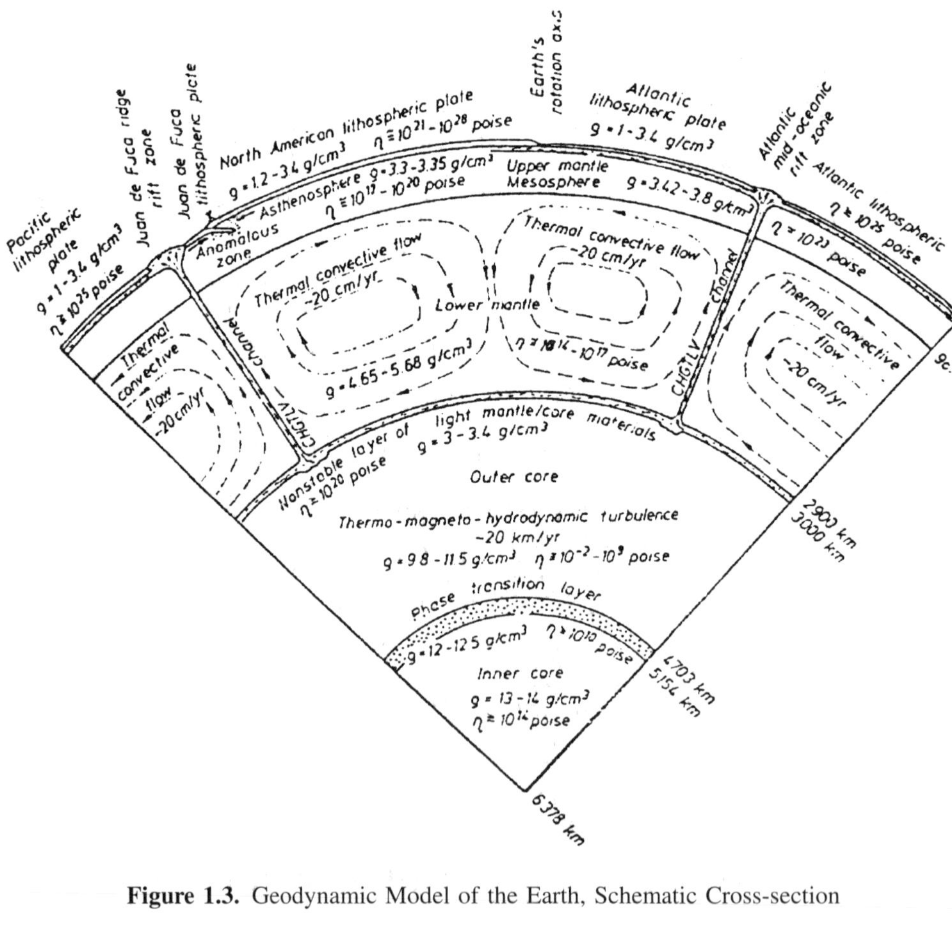

Figure 1.3. Geodynamic Model of the Earth, Schematic Cross-section

boundary, which was accumulated here in large amounts during the first density differentiation in the core of the primary Earth, represents a very unstable layer. The first density differentiation resulted in spatially non-uniform accumulation of light materials.

Also the origination of convective cells was spatially non-uniform, because the masses of the primary Earth's mantle were still considerably inhomogeneous when compared with the present values. This caused the effusion of light masses in the place where the GTLV-channel was largest and the mantle of the primary Earth thinnest on the Earth's surface. After spreading over the surface and solidifying, it initiated the primary continent of Pangaea. A similar phenomenon has been observed by the author in experimental practice if a visco-plastic medium exhibiting a strong inhomogeneity is heated on one side and cooled on the opposite. The more heated and, therefore, lighter and less viscous liquid effuses to the surface, initially in one place located close to the inhomogeneity, this mass effusion being large, and later at other places but with smaller mass effusion.

1.2.2. Geodynamic Processes in the Upper Parts of the Earth

In the further geological evolution of the Earth, the density differentiation at the core-mantle boundary became continuous and has endured until the present time. In order to understand the processes taking place in the upper parts of the Earth, we shall describe the processes associated with this density differentiation at the core-mantle boundary in greater detail.

Origin and development of the rift

Light materials, which after accumulation below the core-mantle boundary form an unstable layer, display a tendency to produce a stable situation by escaping through GTLV-channels into the upper parts of the Earth and even to the Earth's surface under the effective aid of convective flow in the lower mantle and of the Coriolis force. Under this type of convective flow in the mantle, the heated light material moves along the boundaries of convective cells through the GTLV-channels. The spatial distribution of these cells depends on the light matter accumulated below, the non-uniformities of the temperature distribution in the light matter, the irregularities in rotation around the Earth's axis of rotation, the heat convection in the outer core and, particularly, the non-uniformities both of the heating of the lower mantle boundary and of the inhomogeneities distributed within the mantle.

By passing through the GTLV-channel, the light material is heated to temperatures of about 1000–1300°C and rises to the lower boundary of the lithosphere which it heats and melts on the one hand, and strains mechanically in the vertical direction, on the other. After effusing below the lithosphere, it also strains it horizontally. The light material penetrates below the lithosphere, where because the light material has become heated by passing through the GTLV-channels, i.e. through the gravity field, and consequently its viscosity has decreased, it forms a layer of low viscosity — the asthenosphere, where seismic waves propagate at a lower velocity and are attenuated

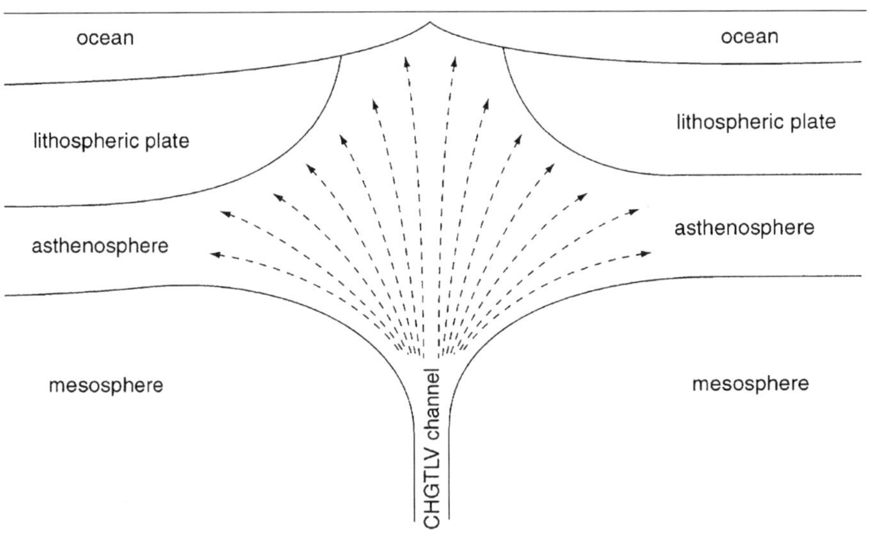

Figure 1.4. The Rift Zone, Schematic Cross-section

more rapidly see Figure 1.4. The non-uniformity and the velocities of the masses moving through the asthenosphere is then the result of all the processes acting in the generation of convective thermal flow and effusion of the heated material below the lithosphere.

The zone below the lithosphere is thus characterized by a low-velocity channel of seismic waves, which differs for P-waves (Press (1970)) (less thick down to depths of about 70 to 120 km, maximum to 150 km) and for S-waves (from 70 to 250–350 km). At depths exceeding 250 km the density and velocity of seismic waves increase again. At depths of 350–400 km and 620–650 km there are layers with a rapid increase in seismic velocity and density (LePichon, Francheteau and Bonnin (1973)). The zone between 350 and 850 km, or 984 km, is referred to as the transitional zone, characterized by a layered structure and higher viscosity. It is probable that it was and is being formed gradually by the density differentiation of the materials of the asthenosphere and of lithospheric plates, which have penetrated it in the subduction zones. This transitional zone is referred to as the mesosphere. Its material is assumed to be composed of ultrabasic ferromagnesite silicates ($<45\%$ SiO_2), i.e. of olivine, pyroxene, garnet, etc., and of their phase transitions (under relatively low temperatures and pressures) (LePichon, Francheteau and Bonnin (1973), Ringwood (1969) and others). No convective flow has so far been traced in it, as the archaic subducted lithosphere is observed near the new subduction zones only.

After formation of the asthenosphere, the lithosphere is further strained by horizontal forces under friction (viscous and Coulombian), namely near the rift zones. The mechanical strain in the vertical and horizontal directions, together with its continual melting, results in the opening of a rift. As proved by paleomagnetic studies, rifts appeared in various places and at different rates. In the first stage of rifting, the vertical components are predominant (e.g. the Baykal rift zone, the East African rift, etc.), in the second stage, horizontal components prevail (Rhine Graben, Red Sea and the Bay of Aden). These new rift zones appeared in the immediate vicinity of old rifts, which is the result of the generation of convective cells that have not changed their locations very much in the course of geological epochs. Only the intensity of the effusions of light materials under the lithosphere changed, so that the brief opening of the rift was probably followed by its closure and then re-opening. This process may have happened several times. Older and younger sediments accumulated in the open rift and, as a result of a substantial effusion of light materials taking place over a geologically relatively long interval, both lithospheric plates became separated and moved away from each other (divergent motion). Thus, older sediments can be found along the edge of one of the lithospheric plates (e.g. at the eastern edge of the American plate the sediments are about 155 million years old), whereas the sediments near the other are considerably younger (e.g., close to the north-western edge of the African plate the sediments has an age of about 110 million years).

The courses taken by the rifts (N-S, E-W), depend on the shape and location of the convective cells; apart from temperature, their origin is also affected substantially by the rotational dynamics of the Earth. Moreover, the convective cells will adopt spatial formations, optimal in shape in a certain sense with respect to all of these effects.

Due to viscous and Coulombian friction, the opening of the rift represented the beginning of the action of oblique and horizontal forces on a larger scale (Nedoma (1986a, b)). The immediate consequences of these horizontal motions are transverse fault zones along which parts of the ridge in the rift region are displaced with respect to one another (e.g. the mid-ocean ridges in the Atlantic, Pacific and Indian Oceans). The forming of these transverse fault zones is also related to the inhomogeneous influence of friction. The lithospheric plate behaves like a constrained plate on which non-uniformly distributed horizontal forces act in the horizontal direction. The absolute magnitude of the individual components of the horizontal forces depends on the effusion of light masses and their spreading below the lithosphere. Compressive and tensile stresses are thus generated, and these are responsible (due to shear) to the rift zone, i.e. the mid-ocean ridge. The originally compact lithosphere (of the oceanic or continental type) separates in the neighbourhood of the rift into sub-blocks, which rotate due to the action of horizontal and vertical forces, as well as the collisions between the sub-blocks themselves.

Origin and development of oceanic and continental lithosphere

The light material enriched with FeO, MgO, etc. (as can be observed for example in the rocks from the bottom of the Red Sea) from the outercore-mantle boundary, which is again density differentiated, is gradually pressed into the open rift. The lightest materials rise to the Earth's surface, where they solidify and give rise to the granitic layer (second oceanic layer). The heavier materials remain in the deeper parts, generating the basaltic layer (third oceanic layer) formed by rocks of the type amphibolitic gabbro, amphibolites, serpentine, green shales, etc. Even heavier materials do not reach the lower boundary of the third oceanic layer (the Mohorovičić transitional zone), where, under higher pressures and temperatures, they form the lithospheric part of the upper mantle (the lower part of the lithosphere; by which we understand the visco-elastic to visco-plastic upper part of the Earth, whose lower boundary is identical with the upper boundary of the low-velocity channel of seismic waves and of the low-viscosity channel), characterized by the quasi-anisotropy of the materials forming it (the possible physical consequence of density differentiation under specific conditions at these depths). As density differentiation is again involved, the transitions between the separated layers will not be well defined but will constitute transitional zones. That is why we speak of Conrad's and Mohorovičić's transitional zones. Results of deep boreholes (around 10 km) seem to corroborate these assumptions.

The continental lithosphere originated in very much the same way, i.e. by density differentiation of the primary light material generated by the initial density differentiation in the primary core which effused non-symmetrically onto the Earth's surface to form Pangaea. The lightest materials formed the uppermost layer, a granite layer 15-20 km thick, separated by the Conrad transition zone from the heavier materials constituting the basaltic layer. The latter is separated by the Mohorovičić transition zone (at a depth of 25-60 km) for the layer of the lower lithosphere, containing relatively heavy materials of the effused light mass and, as in the case of the oceanic lithosphere, characterized by the quasi-anisotropy of the materials constituting it. The bottom of the primary ocean was formed of rocks of the original terrestrial primary mass, penetrated gradually in places by the light materials from the density differentiation in the primary core, which penetrated to the primary surface of the Earth and basified in the same way as it actually does in internal seas. Once the outer core was formed, density differentiation occurred at the outer core-mantle boundary, convective cells were formed and the processes described above took place; and these were repeated with greater intensity at intervals of 200-220 million years. In the course of geological epochs, the original bottom of the primary ocean changed substantially as a result of the generation of rifts and the related tectonics, as we shall see later. As the origin of the lithosphere is the consequence of density differentiation of material, the individual layers forming it are not separated from one another by surfaces of discontinuity but by transitional zones. We are, therefore, justified in speaking of surfaces of discontinuity of the n-th order ($n \geq 1$) as we commonly do in seismology.

Subduction and obduction of lithospheric plates

As a result of convective flow in the mantle, rift zones open at various locations on the Earth's surface. Consequently, the lithospheric megaplate is strained by horizontal forces acting from various sides. In the plate, as a stress guide, compressive and tensile stresses are generated, and these are the cause of abyssal faults in the lithospheric megaplate once the limiting strain values of the plate have been exceeded. During the subsequent stage of development, the lithospheric plates collide with the blocks into which the original lithospheric megaplate has broken up (Uyeda (1982)).

In the region of the deep fault, the two lithospheric plates come into contact (Nedoma (1984), (1986a, b), (1990a, b). In the first phase, they are pressed against each other. Due to horizontal forces, the lower part of the lithosphere is pressed more strongly. The near-surface part of the fault thus begins to open up, the extent of the contact zone in the deep fault increases even more due to the deformation of the lithospheric plates, with the bottom part of the contact zone between the lithospheric plates being deformed most. The lithospheric plate which is acted upon by absolutely smaller horizontal forces, or which is less thick, begins to be subducted (descend) below the other, thicker, collisional lithospheric plate. If the plates are of the same thickness, the forces acting on both of them and the deep fault determine which of the plates will be subducted and which will override it. Convergent motion of lithospheric plates thus occurs. A moving lithospheric plate is referred to as an invasion lithospheric plate.

It may also happen that the closing of rifts relieves both plates, the contact surfaces will be "smoothed" due to relatively small mutual displacements, the plates having contact in a relatively short zone (see Figure 1.5). While the opened part of the contact zone is filled with sediments at the surface, the opened part at the bottom of the lithosphere is filled the molten asthenospheric rocks. When the rift opens up again, the contact zone first begins to increase in the course of the deformation process over a relatively short geological interval. At the same time, it is still being filled with sediments at the surface and with asthenospheric material at the bottom part of the lithospheric. Further opening of the rift causes the lithospheric plates to move. The collision zone begins to dimish and, as a result of the horizontal forces acting at the bottom part of the lithospheric plates, the asthenospheric materials from the opened bottom part of the colliding plates begins to be pressed out. The upper opened part of the contact zone at the surface of the lithosphere continues to open up, to the detriment of one of the plates, until one plate begins to descend below the other. On the other hand, this invading plate begins to be thrust over the down-going plate. We speak of subduction and obduction of lithospheric plates.

The heads of both of the collisional plates are crushed and broken up; they rotate with respect to one another. The heads of the lithospheric continental blocks and the sedimentary fillings of the contact zone and the basins between continental blocks are folded.

At the same time, larger and smaller blocks of the broken-up lithosphere of the continental, quasi-oceanic or purely oceanic types collide, depending on the region of the Earth's lithosphere in which the collision is taking place. However, for subduction

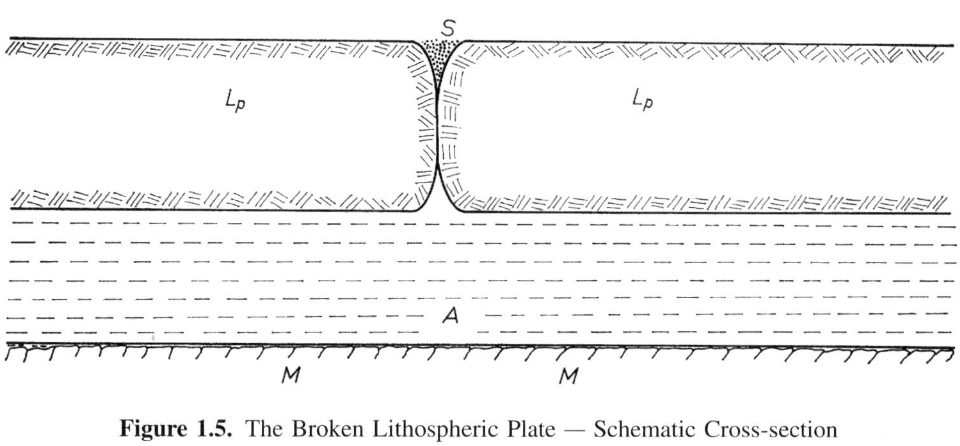

Figure 1.5. The Broken Lithospheric Plate — Schematic Cross-section

to occur, the invasion plates must move at a greater speed, about 5-8 cm/year or more. If the speed of the collisional lithospheric plates is 5-8 cm/year and if at least one of them is of the oceanic type, deep-sea trenches are generated in places where the oceanic lithospheric plate descends below the continental (or oceanic) plate. At the head of the plate which is colliding with a continental or oceanic lithospheric plate, sedimentary fillings of basins are folded and even whole mountain ranges are generated (e.g. the Andes, Cordilleras, Himalayas, Alps).

If the speed of both collisional plates or blocks is low, the collision will not be reflected in the subduction of the oceanic lithospheric plate, but both blocks will be deformed, their thickness will increase and, possibly, folded sediments will be thrust over them (e.g. the Carpathians, the African Plate). At the same time, the centre of the deformed lithospheric block or megaplate will bulge out at the bottom and grow thinner (Pannonian Basin, African Plate), see Figure 1.6. The space where the lithosphere has bulged out is immediately filled with molten asthenospheric material. If the lithospheric bulge is large (e.g. in the Pannonian Basin, in the Central Africa), the centre of the block or plate will display an anomalous heat flow, the latter being closely related to tectonic processes. This heat flow has anomalous value in the regions of rift zones, in regions of enhanced volcanic activity which accompany rift zones and subduction zones, in collision zones of lithospheric plates, and in regions where lithospheric plates or blocks bulge out as in the case of the Pannonian Basin or the Central African plate.

The collision of lithospheric plates itself, i.e. their mutual subduction or obduction, is the actual cause of the large stresses in the lithosphere. The lithosphere behaves as a stress guide. To appreciate the part this process plays in tectonogenesis, we shall investigate this process in greater detail. As indicated by numerical results, discussed in later chapters, in the obducting invasion plate there will be a internal transfer of mass to the place where it has grown thicker and which is not very far from the head of the plate. In the case of an oceanic plate, this internal mass transfer gives rise to island arcs such as the Andaman-Nicobar or the Aleutian island arcs, in the case of a continental plate it produces mountain ranges (the Andes, Himalayas, Alps, etc), and, moreover, the broken-up and folded sediments are also thrust over the "thickened" mass of the continental lithosphere. Both lithospheric plates become deformed. Their deformation is accompanied by enormous stresses in both plates.

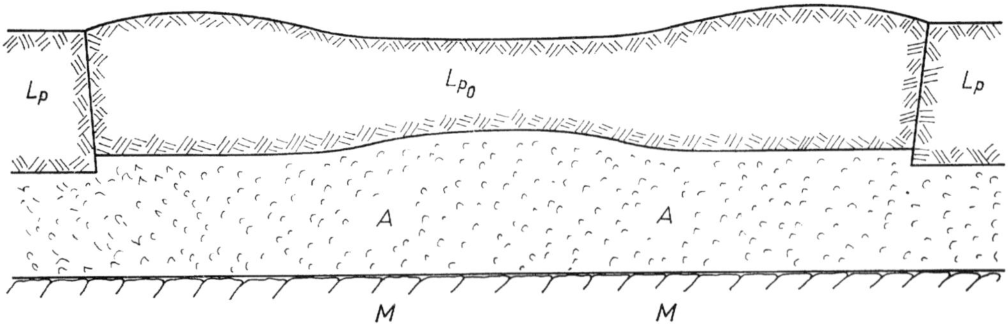

Figure 1.6. Collision of Three Lithospheric Plates without Subduction: Schematic Cross-Section

At the contact of the overriding plate with downgoing one, the normal component of the stress is characterized particularly by a pressure maximum closer to the lower edge of the lithosphere. A zone of tensile stress is created distance on either side of the compressive stress maximum (the numerical results will be discussed in detail later in connection with numerical simulation of the here discussed plate tectonic problems). As a result of the compressive and tensile stresses, faults appear, often reaching up to the Earth's surface. These faults may, and probably do, have continuations in the subducted lithospheric plate so that they may reach considerable depths (even several tens of kilometres). Coulombian friction acts at the contact of two colliding plates (Nedoma (1982), (1984), (1986a, b), (1990a, b), (1994)). Due to friction in the contact zone, both plates are heated and their material melted, and in more deeply subducted plates the masses of the asthenosphere and mesosphere are also heated.

The contact zone is characterized also by the zone of deep earthquake foci — the Wadati-Benioff zone. As the results of a number of authors (e.g. Barazangi and Issacks (1971) indicate, the Wadati-Benioff zone is divided by aseismic faults into two or more seismically active zones.

In the contact zone, the molten material forms magmatic deposits, the so-called volcanic chambers, from which the magma, together with gases, ascends to the Earth's surface, where it forms volcanoes (see Nedoma (1984), (1986a, b), (1990a, b), (1994)). The activity of volcanoes is accompanied by earthquakes. There are relations between rock composition and the depth of earthquake epicentres between low volcanoes. In nature there are also such volcanic formations (i.e. volcanoes and volcanic chambers) Which are not accompanied by earthquakes (e.g. the volcanic crater about 7 km from the centre of the town of Honiara in the Solomon Islands). The speed of the ascending magma varies considerably, from several hundreds of metres to two kilometres per day. A number of volcanoes whose earthquake epicentres are at larger depths display an enhanced content of potassium (K_2O), i.e. volcanoes with deeper supply channels and magma sources. Thus, the potassium (K_2O) content increases from 0.80 to 2.20% with the depth of the magma source increasing from 100 to 200 km. There also appears to be a relation between the zonal distribution of volcanic products and the distance from the collision zone. The alkaline magmas of oceanic composition occur on the ocean bed. Products with a higher content of silica (SiO_2) appear in oceans close to island arcs. Closer to the continent alkaline-calcareous andensites join them and, with increasing distance from the archipelago, they contain more potassium and sodium (K_2O, Na_2O). One can thus see that the mineral and chemical composition of volcanites depends on the distance from archipelagos and, therefore, on the supply of the supply channels (deep faults) and chambers of magma in the bulged out domains. Their relation to density differentiation is evident (Nedoma (1984), (1986a, b), (1990a, b)). Hence the volcanoes can be of eruptive and non-eruptive (basaltic) type.

On the Earth's surface, we even observe cases in which the rift zone and subduction zone are located closer to or further away from each other (San Andreas, Subduction of the Juan de Fuca and Gorda plates beneath the North American Plate). Figure 1.7 shows a schematic section of the subduction zone of the Juan de Fuca plate below the North American lithospheric plate, whose modification corresponds to the concept of the presented geodynamic model.

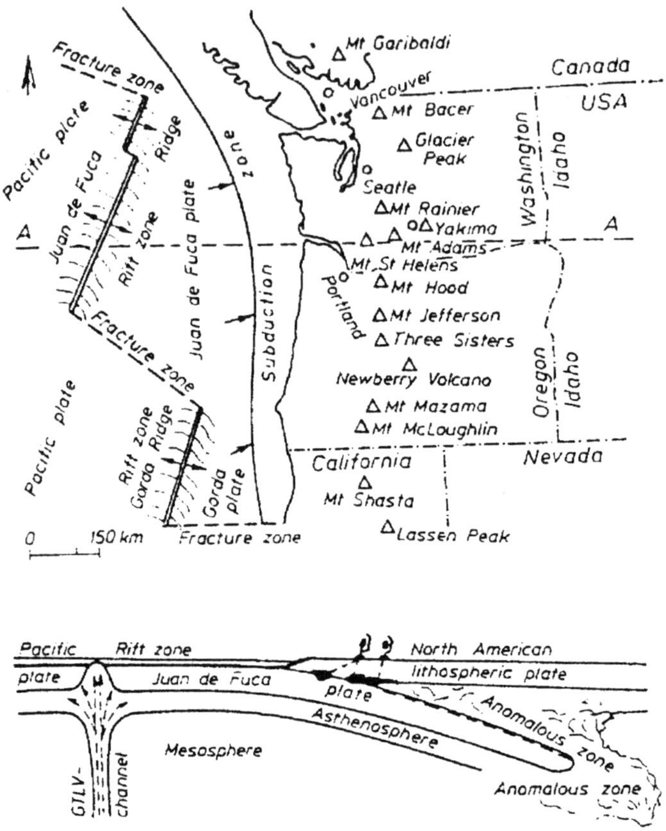

Figure 1.7. The Rift and Subduction Zones of the Juan de Fuca and Gorda Plates (modified after Atwater (1970), Forworthy and Hill (1982), Nedoma (1986b), (1990b))

Some consequences of the subduction and obduction of lithospheric plates

Seismic sounding in areas of deep-sea trenches (Aleutians, the Peruvian-Chilean trench) has dislocated great thicknesses of un folder or otherwise perturbed sediments. As was shown in the author's papers, due to compressive and shear stresses caused by the invasion plate, the subducting plate is broken up by longitudinal faults into separate sub-blocks. The sub-block carrying the horizontally deposited sediments is displaced upwards relatively, at roughly the same speed as the end of the overriding plate, whereas the descending plate is displaced along the oblique faults of the block carrying the sediments. Within a particular interval of time, therefore, the sediments are unable to fold (which is probably the case of the Aleutians and of the Peruvian-Chilean trench). The overriding plate exerts pressure on the downgoing one. That part of the descending plate which is under direct pressure from the overgoing one is pressed into the asthenosphere along the fault in the latter plate (more accurately, along the faults which create the sub-block carrying the unfolded sediments). In this way, the overriding plate enables the lower part of the submerged lithosphere to more away from the main part of the downgoing plate in some cases, see Figure 1.8. (This

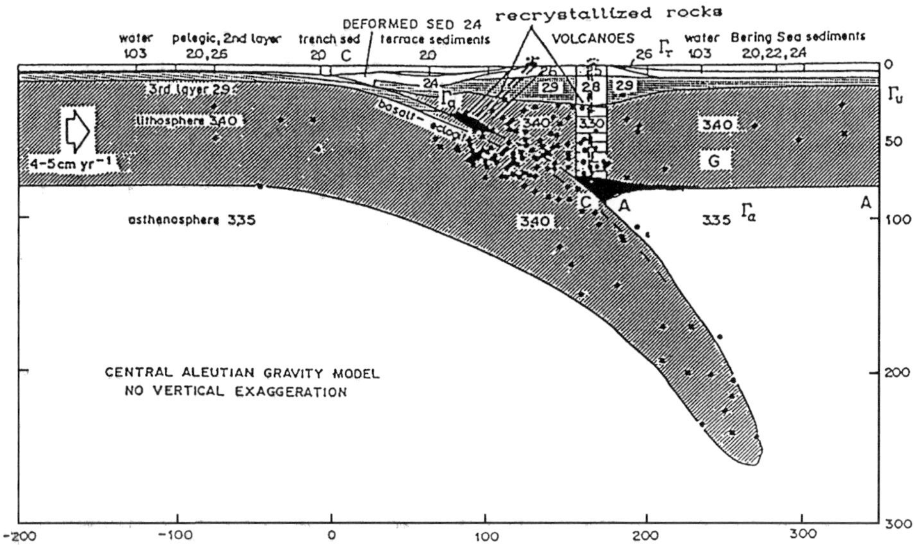

(a)

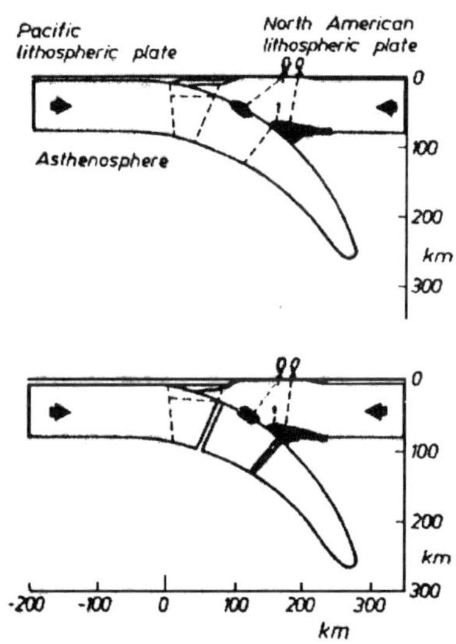

(b)

Figure 1.8. Schematic Cross-section of the Central Aleutian Arc, a) Contemporary state, b) probable further evolutionary state

figure has been modified after the results of Grow and Atwater (1973), Grow (1973), LePichon *et al.* (1973), Nedoma *et al.* (1989) and Nedoma (1990b).)

These fragments of the submerged oceanic lithosphere were identified in the Pacific in the region of the Tonga and Nasca plates. The position of the fragments of the oceanic submerged lithosphere depends on the rate of displacement of both colliding lithospheric plates (see Barazangi and Issacks (1971) and Nedoma (1990b)).

In the next stage, probably due to the displacement of the oceanic lithospheric plate, the head of the overriding plate and of the sub-block carrying the hitherto unfolded sediments is folded and broken up. The speeds of the two colliding plates, the magnitude of the forces acting in the collision zone and of the Coulombian friction at the contact fault planes and surfaces of the overgoing and descending lithospheric plates, according to the author's theory, determine whether or not displacement occurs at the contact surfaces of the faults and in which way the subducting lithospheric plate will move. The conditions describing these processes are called contact conditions; they will be derived and discussed in the next sections. Simpler and more complicated models from the physical as well as mathematical point of view will be discussed in chapters 4-7 and algorithms which solve these problems in chapter 7. It can be noted that more realistic models will be very complicated due to partially melted rocks at the contact boundaries of colliding plates and geological blocks in lower parts of the lithosphere, and also due to brittle properties of rocks near the Earth's surface.

If two continental megaplates or a continental and an oceanic megaplate collide, we observe basification of the continental lithosphere in marginal and island seas, or at the edges of continental plates. The basification is due to the breaking up of the lithosphere at the head of the two colliding plates into separate sub-blocks by a great number of large transverse dislocations, due to their displacement along the contact surfaces between them, and by the penetration of basic rocks from asthenosphere to the surface. This process gives rise to basified penetrations of continental and oceanic lithosphere (Bay of California, Gulf of Mexico, Caribbean sea, Mediterranean and Black Seas, Caspian Sea, marginal seas in the pacific area, etc.) With a transitional structure between the continental and oceanic structures. In the course of these penetrations of basic rocks, a physico-chemical restructuring of granitoid rocks of the original continental lithosphere also occurs.

REFERENCES

Artyushkov, E.V. (1979), *Geodinamika*, Nauka, Moscow (in Russian).

Atwater, T. (1970), "Implications of Plate Tectonics for the Cenozoic Evolution of Western North America", *Geol. Soc. Am. Bull.*, **81**, 2513-3536.

Barazangi, M. and Issacks, B. (1971), "Lateral Variations of Seismic Wave Attenuation in the Upper Mantle Above the Inclined Earthquake Zone at the Tonga Island Arc", *J. Geophys. Res.*, **76**, 8493-8516.

Bucha, V. (1975), *The Geomagnetic Field and Its Contribution to Clearing up the Evolution of the Earth* (in Czech), Academia, Prague.

Cathles, L.M. (1975), *The Viscosity of the Earth's Mantle*, Princeton University Press, Princeton, N.J.

Davies, G.F. (1980), "Mechanism of Subducted Lithosphere", *J. Geophys. Res.*, **85**, 6304-6318.

Elsasser, W.M., Olson, P. and Marsh, B.M. (1979), "The Depth of Mantle Convection", *J. Geophys. Res.*, **4**, 147-155.

Forworthy, B.L. and Hill, M. (1982), "Volcanic Eruptions of 1980 at Mt. St. Helens — The first 100 Days", *Geol. Survey Professional Paper*, **1249**, US Gov. Printing Off., Washington.

Grow, J.A. (1973), "Crustal and Upper Mantle Structure of the Aleutian Arc", *Geol. Soc. Am. Bull.*, **84**, 2169-2192.
Grow, J.A. and Atwater, T. (1973), "Mid-Tertiary Tectonic Transition in the Aleutian Arc", *Geol. Soc. Am. Bull.*, **81**, 3715-3721.
Hager, B.H. and O'Connell, R.J. (1979), "Kinematic Models of Large-Scale Flow in the Earth's Mantle", *J. Geophys. Res.*, **84**, 1031-1048.
Hide, R. (1953), "Some Experiments of Thermal Convection in a Rotating Liquid", *Quart. J. Roy. Meteor. Soc.*, **79**, 161.
Holmes, A. (1931), "Radioactivity and Earth's Movements", *Trans. Geol. Soc. Glasgow*, **18**, 559-606.
Kawai, N. and Mochizuki, S. (1971), "Metallic States in the Three Third Transition Metal Oxides FeO_3, Cr_2O_3 and TiO_3 Under Static High Pressure", *Phys. Rev. Lett.*, **A36**, 1.
Kawai, N. and Nishiyama, S. (1974), "Conductive SiO_2 at High Pressure". *Proc. Jap. Acad.*, **50**.
Kawai, N. and Nishiyama, S. (1975), "Insulator-Metal Transition in $Si0_2$, FeO and MgO", *Rev. Phys. and Chem. Japan*, Spec. Issue 1975.
LePichon, X., Francheteau, J. and Bonnin, J. (1973), *Plate Tectonics (Developments in Geophysics, 6)*, Elsevier, Amsterdam, London, New York.
McKenzie, D.P. (1983), "The Earth's Mantle", *Sci Am.*, **249**, 67-78.
McKenzie, D.P. *et al.* (1980), "Platform of Mantle Convection beneath the Pacific Ocean", *Nature*, **288**, 442-446.
Nedoma, J. (1982), "Thermo-Elastic Stress-Strain Analysis of the Geodynamic Mechanism", *Gerlands Beitr. Geophysik*, **91**, 75-89.
Nedoma, J. (1984), "On the Global Geodynamic Model of the Earth", *Gerlands Beitr. Geophysik 93*, **4**, 241-260.
Nedoma, J. (1986a, b), "On the Geodynamic Model of the Earth and the Plate Tectonic Hypothesis I-III", *Gerlands Beitr. Geophysik 95*, **1**, 36-62; **2**, 89-105.
Nedoma, J. (1990a, b). "Tectonic Evolution of Collision Zones I-II", *Gerlands Beitr. Geophysik 99*, **2**, 97-108, 109-121.
Nedoma, J. (1994), "Numerical Modelling of Tectonic Evolution of Collision Zones", *20th Int. Conference on Mathematical Geophysics*, La Citadelle, V'lle Franche Sur Mer, France, June 19-24, 1994.
Nedoma, J., Haslinger, J., and Hlaváček, I. (1989), "The Problem of an Obducting Lithospheric plate in the Aleutian Arc System: A Finite Element Analysis in the Frictionless Case", *Math. Comput. Modelling* **12**, **1**, 61-75.
O'Connell, R.J. (1977), "On the Scale of Mantle Convection", *Tectophysics*, **38**, 119-136, Amsterdam.
Peltier, W.R. (Ed.) (1989), *Mantle Convection, Plate Tectonics and Global Dynamics*, Gordon and Breach Science Publishers, New York, London, Paris, Montreux, Tokyo, Melbourne.
Peltier, W.R. and Andrews, J.T. (1976), "Glacial-isostatic Adjustment: 1, The Forward Problem", *Geophys. J. Roy. Astr. Soc.*, **46**, 605-646, Oxford.
Press, F. (1970), "Regionalized Earth's Model", *J. Geophys. Res.*, **75**, 6575-6581.
Richter, F.M. (1979), "Focal Mechanism and Seismic Energy Release of Deep and Intermediate Earthquakes in the Tonga-Kermandec Region and Their Bearing on the Depth Extent of Mantle Flow". *J. Geophys. Res.*, **84**, 6783-6795.
Rikitake, T. (1958), "Oscillations of a System of Disc Dynamo", *Proc. Cambridge Philos. Soc.*, **54**, 89.
Ringwood, A.E. (1969), "Composition and Evolution of the Upper Mantle", in Hart, P.J., "The Earth's Crust and Upper Mantle", *Amer. Geophys. Un. Monogr. 13* (1969), **1**, 17.
Uyeda, S. (1982), "Subduction Zones: An Introduction to Comparative Subductology", *Tectonophysics*, **81**, 133-159.

PART II

Tools to Functional and Numerical Analyses

2

An Introduction to Variational Methods and Related Problems

2.1. A BRIEF SURVEY OF FUNCTIONAL ANALYSIS

Since functional analysis is one of the important mathematical disciplines, having an application in mathematical analyses of the model problem and their numerical approximations, as well as in the analysis of algebraic problems, we give a short survey of the main ideas and results from functional analysis in their generalized form. All results will be presented without proofs which, if required, can be found in special monographs of functional analysis (e.g. Sobolev (1963), Taylor (1967), Yosida (1974), Ljusternik and Sobolev (1974)).

2.1.1. Vector and Metric Spaces

Let us denote by X a set of elements u_i $(i = 1, \ldots)$, u, v, w, x, y, z. Then addition and multiplication in X are defined as follows: let $u_i \in X$, $i = 1, 2, \ldots, n$, then the element $u \in X$, the so-called sum of u_i, is defined by $u = \Sigma_{i=1}^{n} u_i$ and the element $v \in X$, the so-called λ-multiple of u, is defined by $v = \lambda u = \lambda \Sigma_{i=1}^{n} u_i$.

The set X with addition and multiplication by scalars, defined above, is called a real or complex vector space if for every real or complex elements $u, y, z \in X$:

a) $z + u = u + z$,

b) $u + (y + z) = (u + y) + z$,

c) in X there exists a uniquely determined element Θ, the so-called zero element, defined by $z + \Theta = z$ for every $z \in X$,

d) for each $u \in X$ there exists a uniquely determined element $-u \in X$ defined by $u + (-u) = \Theta$,

e) $\lambda(u+z) = \lambda u + \lambda z$, $\lambda \in R$ in the case of a real vector space and $\lambda \in C$ in the case of a complex vector space, where $R = R^1$ denotes a real line and $C = C^1$ denotes a complex plane,

f) $(\lambda + \mu)u = \lambda u + \mu u$, $\lambda, \mu \in R$ in a real vector space case and $\lambda, \mu \in C$ in a complex vector space case,

g) $\lambda(\mu u) = (\lambda\mu)u$, $\lambda, \mu \in R$ in a real vector space case and $\lambda, \mu \in C$ in a complex vector space case,

h) $1 \cdot u = u$,

i) $\Theta \cdot u = \Theta$

Let $y, z \in X$. Then the difference $y - z$ of two elements $y \in X$ and $z \in X$ is defined by $y - z = y + (-z)$ and is called subtraction.

Remark: Instead of a vector space we speak about a linear space or linear set (real or complex).

Let M be a subset of a vector space X and let Y denote a vector space contained in X and containing the set M. Then $[M] = \cap Y$, where the intersection is taken over all vector spaces Y, is called the linear hull of M. It is evident that $[M]$ is also a vector space.

A vector space is called an algebra if for every ordered pair $y \in X$, $z \in X$, a product $y \cdot z$ is defined as follows:

a) $(y \cdot z) \cdot w = y \cdot (z \cdot w)$,

b) $y \cdot (z + w) = y \cdot z + y \cdot w$,

c) $(y + z) \cdot w = y \cdot w + z \cdot w$,

d) $(\lambda y) \cdot (\mu z) = (\lambda\mu)(y \cdot z)$.

Let X be a set of elements $w, y, z, \ldots$. A nonnegative function ρ defined on $X \times X$ is called a metric if it satisfies:

a) $\rho(y, z) = 0$ if and only if $y = z$,

b) $\rho(y, z) = \rho(z, y)$,

c) $\rho(y, z) \leq \rho(y, w) + \rho(w, z)$.

Then a set X with a metric ρ is called a metric space.

Let X be a vector space. A nonnegative function $\|y\|_X \equiv \|y\|$, defined on X, is called a norm on X if it satisfies

a) $\|y\| = 0$ if and only if $y = 0$,

b) $\|\lambda y\| = |\lambda| \, \|y\|$, the so-called homogeneity axiom,

c) $\|y + z\| \leq \|y\| + \|z\|$, the so-called triangle inequality.

A vector space X with a norm $\|y\|$ is called a normed linear (or vector) space and the number $\|y\|$ is called the norm of $y \in X$. Some authors use the notation $(X, \|w\|_X)$ for the normed linear space X. If the function $\|\cdot\|$ satisfies only axioms b) and c) then it is called a seminorm. Every normed linear space is a metric space with metric $\rho(y, z) = \|y - z\|$.

Let X be a normed linear space, let $y \in X$, $r > 0$. Then the set

$$B(y, r) = \{z \in X; \|y - z\| < r\}$$

is called an open ball with its centre in y and radius r. A subset $M \subset X$ is called an open set in X if for every $y \in M$ there exists an $r = r(y) > 0$ such that $B(y, r) \subset$M. A subset $M \subset X$ is called a closed set in X if $X - M$ is an open set in X.

2.1.2. Inner Product, Banach and Hilbert Spaces

Let X be a real or complex vector space. Then a real or complex, respectively, function defined on $X \times X$ is called an inner (scalar) product of $y \in X$, $z \in X$ and is denoted by (y, z), $y, z \in X$, if it satisfies

a) $(y + z, w) = (y, w) + (z, w)$,

b) $(\lambda y, z) = \lambda(y, z)$ with $\lambda \in R$ in the real vector space case and $\lambda \in C$ in the complex vector space case,

c) $\overline{(y, z)} = (z, y)$,

d) $(y, y) > 0$ if $y \neq 0$.

A vector space X with an inner product is called an inner product space or a pre-Hilbert space or unitary space.

For $y \in X$ we define $\|y\| = (y, y)^{1/2}$, a norm on X generated by the inner product. Moreover, the inequalities $|(y, z)| \leq \|y\| \, \|z\|$, the so-called Schwarz inequality, and $\|y + z\| \leq \|y\| + \|z\|$ hold. Then X is a normed linear space.

Let X be a normed linear space and $\{y_n\}_{n=1}^{\infty} a$ sequence in X. Then the sequence y_n converges to $y \in X$ in X, and we denote it by $y_n \to y$ in X, if $\lim_{n\to\infty} \|y_n - y\| = 0$, i.e. if for every $\varepsilon > 0$ there exists an $n_o = n_o(\varepsilon) \in N$ such that $\|y_n - y\| < \varepsilon$ holds for all $n > n_o$. We say that the sequence $\{y_n\}_{n=1}^{\infty}$ converges strongly in X or in the norm of X. If the sequence $\{y_n\}_{n=1}^{\infty}$ converges strongly in X then the sequence $\{y_n\}_{n=1}^{\infty}$ is to be convergent in X and y is called its limit.

A sequence $\{y_n\}_{n=1}^{\infty}$ in X is called a fundamental or Cauchy sequence if $\lim_{m,n\to\infty} \|y_m - y_n\| = 0$, i.e. if for every $\varepsilon > 0$ there exists an $n_o = n_o(\varepsilon) \in N$ such that $\|y_m - y_n\| < \varepsilon$ holds for all $m, n > n_o$. It is evident that every convergent sequence is fundamental, but the converse assertion is not generally valid.

Let M be a subset of a normed linear space X. Then the closure of M in X, i.e. $\overline{M}$, is defined as the set of all elements $y \in X$ such that there exists a sequence $\{y_n\}_{n=1}^{\infty}$ in M converging strongly in M, i.e. $y_n \to y$ in X. It is evident that $M \subset \overline{M}$. The set M is closed in X if and only if $M = \overline{M}$. A subset $M \subset X$ is said to be dense in X if $\overline{M} = X$. A subset of a normed linear space X is called a subspace of X if it is a linear set which is closed in X. It is evident that linear subsets need not be closed in X while linear subspaces due to the definition must be closed in X. A subspace M of a normed linear space $(X, \|\cdot\|_X)$ is also a normed linear space with the norm $\|y\|_M \equiv \|y\|_X$, $y \in M$.

Let $n \in \boldsymbol{N}$. Let X_i, $i = 1, \ldots, n$ be normed linear spaces and let $X = \Pi_{i=1}^n X_i$ be the Cartesian product of X_i, $i = 1, \ldots, n$, i.e. a set of all $y = (y_1, \ldots, y_n)$ such that $y_i \in X_i$, $i = 1, \ldots, n$. Then X is also a normed linear space with the norm

$$\|y\|_X = \left(\sum_{i=1}^n \|y_i\|_{X_i}^p\right)^{1/p}, 1 \leq p < \infty, \text{ or } \|y\|_X = \max_{i=1,\ldots,n} \|y_i\|_{X_i}.$$

A normed linear space X is called separable if there exists a countable set $M \subset X$ which is dense in X. If M is an uncountable subset of a normed linear space X, then a normed linear space X is not separable, i.e. if there exists a $\delta > 0$ such that $\|y - z\| > \delta$ for every pair $y, z \in M$, $y \neq z$. A subspace of a separable normed linear space X is also a separable normed linear space. Let X_i, $i = 1, \ldots, n$, $n \in \boldsymbol{N}$, be separable normed linear spaces. Then the product $X = \Pi_{i=1}^n X_i$ is also a separable normed space.

A normed linear space X is said to be complete if every fundamental sequence $\{y_n\}_{n=1}^\infty$ is convergent in X, which means that it has a limit $y \in$X. A complete normed linear space is called a Banach space.

Let X be an inner product space with inner product (y, z) and let X be complete with respect to the norm $\|y\| = (y, y)^{1/2}$, then X is the so-called Hilbert space. A subspace of a Banach and/or of a Hilbert space is also a Banach or Hilbert, respectively, space. Let X_i, $i = 1, \ldots, n$, $n \in \boldsymbol{N}$, be Banach or Hilbert spaces. Then the Cartesian product $X = \Pi_{i=1}^n X_i$ is also a Banach or Hilbert, respectively, space.

Let X be a Banach space. We say that a sequence $\{y_n\}_{n=1}^\infty$ in X is the so-called Schauder basis of X if every $y \in X$ has a unique representation $y = \Sigma_{i=1}^\infty a_i y_i$, i.e. $\lim_{n\to\infty} \left\| y - \Sigma_{i=1}^n a_i y_i \right\| = 0$, where a_i, $i \in \boldsymbol{N}$, are constants and are called the coefficients of y with respect to the Schauder basis $\{y_i\}_{i=1}^\infty$. It is evident that every Banach space with a Schauder basis is separable.

A set M in a normed linear space X is said to be relatively (conditionally) compact or precompact, respectively, if every sequence in M contains a convergent subsequence. If M is closed and relatively compact then M is said to be compact. Every relatively compact set M is bounded. It is evident that a closed subset of a compact set is compact.

Let M_i $(i = 1, \ldots, n)$ be relatively compact subsets of Banach spaces M_i. Then $\Pi_{i=1}^n M_i$ is a relatively compact subset of $\Pi_{i=1}^n X_i$.

Let $\varepsilon > 0$, let M be a set in a normed linear space X, then a set E is called the ε-net of M if for every $y \in M$ there exists a $y_\varepsilon \in E$ such that $\|y - y_\varepsilon\|_X < \varepsilon$. A set M in a Banach space X is relatively compact if and only if for every $\varepsilon > 0$ there exists a finite ε-net of M. If M_i $(i = 1, \ldots, n)$ are relatively compact subsets of Banach spaces X_i $(i = 1, \ldots, n)$, then $\Pi_{i=1}^n M_i$ is a relatively compact subset of $\Pi_{i=1}^n X_i$.

2.1.3. Operators and Functionals

Let X, Y be normed linear spaces (real or complex) and let M be a set in X. Let us suppose that for every $y \in M$ there exists a uniquely determined $z \in Y$ such that $z = Ay$, where A is the so-called operator (or mapping function or abstract function) defined on M and M is the so-called domain of the operator A and is denoted by $\mathscr{D}(A)$. We also say that the operator A is an operator from X into Y which maps M into Y. The set $\mathscr{R}(A) = \{w \in Y; w = Ay \text{ with } y \in M\}$ is called the range of the operator A. If for all $x, y \in M$ we have $Ax \not\equiv Ay$ provided $x \not\equiv y$, then every $w \in \mathscr{R}(A)$ is assigned a uniquely determined element $y \in M$ by $Ay = w$. Then $y = A^{-1}w$ and A^{-1} is the inverse operator to A. Moreover, $\mathscr{D}(A^{-1}) = \mathscr{R}(A)$, $\mathscr{R}(A^{-1}) = \mathscr{D}(A) = M$. If $\mathscr{R}(A) = Y$, we say that the operator A maps M onto Y.

Let X, Y be normed linear (real or complex) spaces. An operator A from X into Y is called a linear operator if $\mathscr{D}(A)$ is a linear set and if:

a) $A(x + y) = Ax + Ay$,

b) $A(\lambda x) = \lambda Ax$,

for all $x, y \in \mathscr{D}(A)$ and for every scalar λ.

An operator A from X into Y is said to be continuous if $y_i \to y$ in X implies $Ay_i \to Ay$ in Y for any sequence $\{y_i\}_{i=1}^{\infty}$ provided y_i, $y \in \mathscr{D}(A)$.

A linear operator A from X into Y is said to be bounded if $\sup \|A\|_Y < \infty$, where the supremum is taken over all $y \in \mathscr{D}(A)$ such that $\|y\|_X \leq 1$. We see that a linear operator is bounded if and only if it is continuous.

Let A be a continuous linear operator from X into Y. The norm of the operator A (the operator norm) is defined by $\|A\| = \sup \|Ax\|_Y$, where the supremum is taken over all $y \in \mathscr{D}(A)$ such that $\|x\|_X \leq 1$ or by

$$\|A\| = \sup_{\substack{x \in \mathscr{D}(A) \\ \|x\|_X = 1}} \|Ax\|_Y = \sup_{\substack{x \in \mathscr{D}(A) \\ x \neq 0}} \|Ax\|_Y \,/\, \|x\|_X \,.$$

Theorem 2.1 (Banach): Let X, Y be Banach spaces, A a linear operator from X onto Y with $\mathscr{D}(A) = X$ and $\mathscr{R}(A) = Y$. Suppose that A is continuous and that A^{-1} exists. Then the inverse operator A^{-1} is continuous.

Let A be a linear operator from a normed linear space X into a normed linear space Y with $\mathscr{D}(A)$=X. The operator A is said to be completely continuous (compact) if it maps every bounded set in X onto a relatively compact set in Y. We see that every completely continuous linear operator is continuous.

Theorem 2.2 (Banach-Steinhaus): Let X be a Banach space, Y a normed linear space. Then the sequence $\{A_i\}_{i=1}^{\infty}$ of bounded linear operators from X into Y with $\mathscr{D}(A) = X$ satisfies $A_i x \to Ax$ in Y for every $x \in X$ if and only if: a) the sequence $\{\|A_i\|\}_{i=1}^{\infty}$ is bounded, and b) $A_i x \to Ax$ in Y for every $x \in M$, where M is a dense subset of X.

Two normed linear (real or complex) spaces are said to be isomorphic if there exists a continuous linear operator A such that $\mathscr{D}(A) = X$, $\mathscr{R}(A) = Y$, and A^{-1} exists and is continuous. The operator A is then called an isomorphism between X and Y or an isomorphism mapping.

Two metric spaces X with metric ρ_1 and Y with metric ρ_2 are said to be isometric if there exists a mapping B such that $\mathscr{D}(B) = X$, $\mathscr{R}(B) = Y$ and $\rho_1(x, y) = \rho_2(Bx, By)$ for every $x, y \in X$.

Two normed linear spaces X, Y are said to be isometrically isomorphic if there exists a linear operator A such that $\mathscr{D}(A) = X$, $\mathscr{R}(A) = Y$ and $\|x - y\|_X = \|Ax - Ay\|_Y$ for every $x, y \in X$.

Let X, Y be two isomorphic normed linear spaces. Then a) if X is separable then Y is also separable, b) if X is complete then Y is also complete, c) if X has a Schauder basis then Y also has a Schauder basis.

Let X, Y be two normed linear spaces and let $X \subset Y$. Then the identity operator I from X into Y with $\mathscr{D}(I) = \mathscr{R}(I) = X$ is defined as the operator which maps every element $x \in X$ onto itself, i.e. $Ix = x$, regarded as an element of Y. The operator I is a linear operator. If it is also continuous, then it is called the imbedding operator or the imbedding from X into Y, and we write $X \hookrightarrow Y$. While the elements in X and in $\mathscr{R}(I) \subset Y$ are the same, the norms of x in X and in Y are different, i.e. $\|x\|_X \neq \|x\|_Y$. Continuity of the imbedding implies that there exists a constant $c > 0$ such that $\|x\|_Y \leq c\,\|x\|_X$ for every $x \in$X. If the imbedding operator is completely continuous then $X \hookrightarrow\hookrightarrow Y$.

Let X be a vector space and suppose that $\|\cdot\|_1$ and $\|\cdot\|_2$ are two norms on X. Then these norms are equivalent if there exist constants $c_1 > 0$, $c_2 > 0$ such that $c_1\,\|x\|_1 \leq \|x\|_2 \leq c_2\,\|x\|_1$ for all $x \in X$. Then the imbedding from $(X, \|\cdot\|_1)$ into $(X, \|\cdot\|_2)$ is an isomorphism.

Let X be a real or complex normed space, and Y the real line R or the complex plane C, respectively. A linear operator from X into Y is called a real or complex, respectively, linear functional and it will be denoted by φ, ψ, etc. and its value at $x \in X$ by $\varphi(x)$. The same results given above for linear continuous operators are valid also in the case of continuous linear functionals. A norm of a linear functional φ is defined by $\|\varphi\| = \sup|\varphi(x)|$, where the supremum is taken over all $x \in \mathscr{D}(\varphi)$ such that $\|x\|_X \leq 1$.

Continuous linear functionals defined on linear subsets of X can be extended to the entire space X as follows: let φ be a continuous linear functional defined on a linear set M which is dense in X. Then there exists a uniquely determined continuous linear functional ψ defined on X such that $\psi(x) = \varphi(x)$ for all $x \in M$.

Theorem 2.3 (Hahn-Banach): Let φ be a continuous linear functional defined on a linear subset M of a normed linear space X. Then there exists a continuous linear functional ψ defined on X such that $\psi(x) = \varphi(x)$ for $x \in M$ and $\|\psi\| = \|\varphi\|$.

2.1.4. Dual Spaces, Weak Convergence and Reflexivity

Let G be the set of all continuous linear functionals defined on X. Then G is a vector space as the addition of functionals and the multiplication of a functional by a scalar can be defined, i.e. $(\varphi + \psi)(x) = \varphi(x) + \psi(x)$ and $(\lambda\varphi)(x) = \lambda \cdot \varphi(x)$, where $\varphi, \psi \in G$, and λ is a scalar. Moreover, G is a normed linear space with the norm defined by $\|\varphi\| = \sup|\varphi(x)|$ and is called the dual (adjoint, conjugate) space of X and denoted by X'.

Let X be a normed linear space. Then the dual space X' is a Banach space.

Let X be a normed linear space, and $\{x_i\}_{i=1}^{\infty}$ a sequence in X. We say that x_i converges weakly to $x \in X$, i.e. $x_i \to x$ weakly, if $\lim_{i\to\infty} \varphi(x_i) = \varphi(x)$ for every $\varphi \in X'$. It is known that every weakly convergent sequence is bounded.

Theorem 2.4 (modified Banach-Steinhaus theorem): Let X be a Banach space. The sequence $\{x_i\}_{i=1}^{\infty}$ converges to $x \in X$ weakly if and only if: a) the sequence $\{\|x_i\|_X\}_{i=1}^{\infty}$ is bounded, and b) $\lim_{i\to\infty} \varphi(x_i) = \varphi(x)$ for all $\varphi \in G_{o'}$, where G_o is a dense subset of the dual space X'.

Let X be a Banach space, and X' its dual space. Then we define the dual of X' by $X'' = (X')'$.

Let $x'', y'', \ldots \in X''$. The operator J from X into X'', with $\mathscr{D}(J) = X$ defined by $(Jx)(\varphi) = \varphi(x)$ for $\varphi \in X'$, and $x \in X$, is the so-called canonical mapping from X into X'' and it holds $x''(\varphi) = \varphi(x)$. Let us denote by $J(X)$ the image of X in the canonical mapping J. Then J is an isometric isomorphism between X and $J(X)$.

A Banach space X is said to be reflexive if $J(X) = X''$. Let X be a reflexive Banach space. Then every subspace of X is also reflexive. Furthermore, the product of a finite number of reflexive Banach spaces is a reflexive Banach space. If X is a separable reflexive space, then its dual X' is separable.

Let X, Y be two isomorphic normed linear spaces and let X be a reflexive Banach space. Then Y is also a reflexive Banach space. A sequence $\{x_i\}_{i=1}^{\infty}$ of elements of a Banach space X is said to be weakly fundamental if for every $x' \in X'$ the sequence $\{x'(x_i)\}_{i=1}^{\infty}$ is a fundamental sequence of real numbers. We say that a Banach space is weakly complete if every weakly fundamental sequence is weakly convergent. Every reflexive Banach space is weakly complete (see Dunford and Schwartz (1958)).

2.1.5. Physical Significance of the Inner Products and the Norms

In many cases in mathematics as well as in physics we use abstractions; by which we mean to extract the common property from otherwise entirely different concepts. Such abstractions in physics are for example the concepts of energy or stresses (elastic, plastic, thermal, Maxwell's, etc.), and in functional analysis examples are vectors, tensors, inner (scalar) products, norms, operators, etc. In physics it is known for example that energy cannot be measured by numbers but by numbers times units; and that its unit, the product of the force unit with the length unit, reflects the range on which this abstract entity is measured. Similarly in the case of stresses. It is clear that we have to go through such a function space to find quantitative results, and then we can use functional analysis as a tool of physical and geophysical analyses. Hence for a norm we also need a physical quantity. It should be a real scalar, as real scalars are real numbers with a unit and so have a one-to-one correspondence to real numbers, e.g. for linear elastic systems we use the root of the double elastic energy.

Now let us assume fixed linear elastic systems. By a linear elastic system is meant a linear connection between static and geometric quantities, and the existence of an elastic energy, which means that work done in deforming an elastic system can always be retrieved in allowing the relaxing system to do work on some other body. The linearity of the system yields that the double elastic energy of a state of a given body, denoted by f (or g), must be a positive quadratic form, which we denote by (f, f). Let f and g be two states of the system, then for their linear superposition $f + g$ we find

$$(f + g, f + g) = (f, f) + (g, g) + (f, g) + (g, f),$$

i.e, the quadratic form is distributive and homogeneous. The existence of a unique double energy implies

$$(f, g) = (g, f),$$

i.e. the quadratic form is commutative. The interaction energy (f, g) has all necessary properties for a scalar (inner) product defining a Hilbert space $\mathscr{H}$, the space of state of the given elastic system.

Next we shall define the load stress state, where the load stress state f is generated by forces $\boldsymbol{P}$ acting on the elastic system, in which the system of displacements $\boldsymbol{u}$ are generated. The double work done by $\boldsymbol{P}$ can be written as

$$(\boldsymbol{P}, \boldsymbol{u}) = (f, f).$$

The inner (scalar) products and the corresponding norms will be defined later in more details for special cases of investigated geophysical fields.

2.2. INTRODUCTION TO THE THEORY OF SPACES

2.2.1. Hilbert Spaces

The symbol Ω denotes an N-dimensional domain, i.e. an open connected set in the Euclidean space R^N. Moreover, it is assumed that $\Omega = \cup_{\iota=1}^{S} \Omega^{\iota}$. The boundary of the domain Ω will be denoted by $\partial\Omega$ and its parts by Γ_u, Γ_τ, Γ_c, or Γ_i, $i = 1, 2, \ldots, n$, respectively. The coordinates of a point of R^N will be denoted by $\boldsymbol{x} = (x_i)$, $i = 1, \ldots, N$.

The set M, elements of which are functions on some set M_1, is called the linear set if with every two functions $u(\boldsymbol{x})$ and $v(\boldsymbol{x})$ from M the function $au(\boldsymbol{x}) + bv(\boldsymbol{x})$ also belongs to M, where a and b are arbitrary (real) constants.

By the inner (scalar) product of functions $u(\boldsymbol{x})$ and $v(\boldsymbol{x})$ from the linear set M we understand

$$(u, v) = \int_\Omega u(\boldsymbol{x})v(\boldsymbol{x})\mathrm{d}\boldsymbol{x}.$$

The inner product has the following properties:

$$(u, v) = (v, u),$$

$$(a_1u_1 + a_2u_2, v) = a_1(u_1, v) + a_2(u_2, v),$$

$$(u, u) \geq 0,$$

$$(u, u) = 0 \Leftrightarrow u(\boldsymbol{x}) = 0 \text{ in } \Omega.$$

By the norm of a function $u(\boldsymbol{x})$ from the linear set M we understand the non-negative number

$$\|u\| = (u, u)^{1/2} = \left(\int_{\Omega} u^2(\boldsymbol{x}) \mathrm{d}\boldsymbol{x} \right)^{1/2} .$$

The norm has the following properties:

$$\|u\| \geq 0, \quad \|u\| = 0 \Leftrightarrow u(\boldsymbol{x}) = 0 \text{ in } \Omega,$$

$$\|au\| = |a| \, \|u\| , \quad |(u, v)| \leq \|u\| \, \|v\| ,$$

$$\|u + v\| \leq \|u\| + \|v\| \quad \text{(the so-called triangle inequality)}$$

$$| \, \|u\| - \|v\| \, | \leq \|u - v\| .$$

The distance of two functions $u(\boldsymbol{x})$ and $v(\boldsymbol{x})$ from the linear set M is represented by the norm of the difference of these two functions, i.e. $\rho(u, v) = \|u(\boldsymbol{x}) - v(\boldsymbol{x})\|$, for which

$$\rho(u, v) \geq 0, \quad \rho(u, v) = 0 \Leftrightarrow u(\boldsymbol{x}) \equiv v(\boldsymbol{x}),$$

$$\rho(u, v) = \rho(v, u), \quad \rho(u, z) \leq \rho(u, v) + \rho(v, z).$$

We say that the linear set M is a metric space if for every pair of functions $u \in M$ and $v \in M$ there is defined a number $\rho(u, v) = \|u(\boldsymbol{x}) - v(\boldsymbol{x})\|$, the so-called distance of elements $u(\boldsymbol{x})$ and $v(\boldsymbol{x})$, with properties given above. In the literature it is known as the axiom of the metric.

We say that a sequence $\{u_m\}$ of elements u_m of a metric space M converges in that space to the element $u \in M$ if $\lim_{m\to\infty} \rho(u_m, u) = 0$ and we write $\lim_{m\to\infty} u_m = u$ in M, where u is called the limit element of the sequence $\{u_m\}$.

The set N is called dense in the space M if $\overline{N} = M$. If there exists an at most countable set dense in M, then the space M is called separable. A metric space M is called complete if every Cauchy sequence $\{u_m\}$ converges in this space to an element $u \in M$.

The linear set M is a linear normed space if on this set a norm, with axioms of the norm, is defined.

A linear normed metric space M, complete in the metric $\rho(u, v) = \|u - v\|$, is a Banach space.

Let M be a linear set with the metric $\rho(u, v) = \|u - v\|$, where the norm is defined by the inner product (u, u), i.e. $\|u\| = (u, u)^{1/2}$. If M is an infinite dimensional and complete space (i.e. a complete real or complex space with the inner product) then it is called a Hilbert space. If M is a finite dimensional space with the inner product then it is called a Euclidean space. The Euclidean space is denoted by the symbol R or E. The Hilbert space is denoted by the symbol H.

We say that $u \in H$ is a linear combination, in H, of the elements $v_1, \ldots, v_n \in H$, if $u = \Sigma_{i=1}^{n} a_i v_i$, where a_i, $i = 1, \ldots, n$ are suitable real constants. We say that the elements $u_i \in H$, $i = 1, \ldots, n$ are linearly dependent in H if at least one of them is a linear combination of the others, and are linearly independent in H if none of them is a linear combination of the others. The elements $u_i \in H$, $i = 1, \ldots, n$ are linearly dependent in H if and only if their Gramm's determinant

$$J(u_1, \ldots, u_n) = \begin{vmatrix} (u_1, u_1), (u_1, u_2), \ldots, (u_1, u_n) \\ (u_2, u_1), (u_2, u_2), \ldots, (u_2, u_n) \\ \ldots\ldots\ldots\ldots\ldots\ldots\ldots\ldots\ldots \\ (u_n, u_1), (u_n, u_2), \ldots, (u_n, u_n) \end{vmatrix} = 0,$$

and are linearly independent in H if and only if $J \neq 0$. The elements $u, v \in H$ are called orthogonal in H if $(u, v) = 0$, i.e. $u \perp v$. If every two distinct elements of a system $u_1, u_2, \ldots, u_n, \ldots, u_i \in H, i = 1, 2, \ldots$ are orthogonal in H, i.e. if $(u_i, u_j) = 0$ for $i \neq j$, then the system u_i, $i = 1, 2, \ldots$, is called orthogonal in H; if, moreover, its elements u_i are normed in H, i.e. $\|u_i\| = 1$, then the system is called orthonormal in H.

Let $u_1, u_2, \ldots, u_n, \ldots, u_i \in H$, $i = 1, 2, \ldots$ be an orthonormal system in a Hilbert space H. Then the series $\Sigma_{i=1}^{\infty} a_i u_i$, where a_i are real constants, is convergent in H, if and only if the series $\Sigma_{i=1}^{\infty} a_i$ is convergent.

The system $u_1, u_2, \ldots, u_n, \ldots, u_i \in H$, $i = 1, 2, \ldots$ is called closed in the space H if the only element which is orthogonal to all the elements of this system is the null-element of H. The orthogonal system $u_1, u_2, \ldots, u_n, \ldots, u_i \in H$, $i = 1, 2, \ldots$ is complete in a Hilbert space if and only if it is closed in this space.

The system $\varphi_1, \varphi_2, \ldots, \varphi_n, \ldots$ (not necessarily orthonormal) of elements of a Hilbert space is called complete in this space if the set of all linear combinations of its elements is dense in H (i.e. if to every element $u \in H$ and to every $\varepsilon > 0$ it is possible to find a positive integer n and the numbers $a_i^{(n)}$, $i = 1, \ldots, n$, so that $\rho(u, \Sigma_{i=1}^{n} a_i^{(n)} \varphi_i) < \varepsilon$). If, moreover, the system $\{\varphi_i\}$, $i = 1, 2, \ldots$ is linearly independent in H, then this system constitutes a base in H. If this system is orthonormal, then the base of the space H is orthonormal too. Every separable Hilbert space has an orthonormal base.

A linear set M of elements of a Hilbert space H with the same metric as in H is also a metric space. If M is a complete space then it is a subspace of H. Then the space M is a linear subspace of H if and only if the set M is closed in H.

Let M be a linear subspace of a Hilbert space H. Then every element $u \in H$ can be uniquely decomposed into the sum $u = v + w$, where $v \in M$ and $w \perp M$, $w \in K$. Then we say that the element $v \in M$ is the orthogonal projection of u into M. Moreover, the set K of all elements $w \in H$ orthogonal to the linear subspace M of a Hilbert space is also a linear subspace. Then K is orthogonal to M and we write $K \perp M$, so that $H = M \oplus K$, and we say that H is an orthogonal sum of the subspaces M and K. The space K is called the orthogonal complement of the subspace M in H, i.e. $K = H \ominus M$. Let the sequence $\{u_n\}$ converges in H to the element u, and the sequence $\{v_n\}$ to v. Then the sequence of inner products converges to the number (u, v), i.e. if $u_n \to u$, $v_n \to v$ in H then $(u_n, v_n) \to (u, v)$, and if $u_n \to u$, then $(u_n, v) \to (u, v)$ for every $v \in$H. Furthermore, if $u_n \to u$, then $\|u_n\| \to \|u\|$.

2.2.2. Operators and Functionals in Hilbert Spaces

Let two sets M and N be given. We say that on the set M there is defined an operator A mapping the set M into the set N if a rule is given by which for every element $u \in M$ there is a uniquely determined element $v \in N$. Then we write $v = Au$.

M is called the domain of the operator A and is also denoted by D_A. The set $\underline{N}$ of all $v \in N$ which are obtained by $v = Au$ for all $u \in M$ is called the range of the operator A and is denoted by R_A.

We say that the operator A maps the set M onto the set N if it maps M into N and if for every element $v \in N$ there exists at least one element $u \in M$ such that $Au = v$.

The operator B defined on N, which assigns to every element $v \in N$ just that element $u \in M$ for which $v = Au$ is called the inverse operator to the operator A and we write $B = A^{-1}$. For all $u \in M$ and for all $v \in N$ we have $A^{-1}Au = u$, $AA^{-1}v = v$.

An operator A is called linear if its domain D_A is a linear set and if for arbitrary elements $u_i \in D_A$, $i = 1, \ldots, n$, and for arbitrary (real) numbers a_i, $i = 1, \ldots, n$, the relation

$$A(a_1u_1 + \cdots + a_nu_n) = a_1Au_1 + \cdots + a_nAu_n$$

holds.

Let P be a metric space with the metric ρ. An operator A mapping this space into the space P again is called contractive in this space if there exists a number α, $0 < \alpha < 1$, such that for every pair of elements $x, y \in P$ we have $\rho(Ax, Ay) \leq \alpha\rho(x, y)$.

Theorem 2.5 (The Banach fixed point theorem): Let A be a contractive operator in a complete metric space P. Then the equation $x = Ax$ has one and only one solution in this space, i.e. there exists exactly one element $u \in P$ for which $u = Au$ holds. This element can be obtained as the limit of a sequence of elements $u_n \in P$, $u = \lim_{n\to 0} u_n$, where $u_{n+1} = Au_n$, $n = 1, 2, \ldots$, while the element $u_1 \in P$ may be chosen arbitrarily.

An operator I is called the identity (unit) operator if it assigns to every element $u \in D_I$ the same element u again. An operator O is called the null operator if it assigns to every $u \in D_O$ the null-element of the space H.

An operator A is called continuous at the point $u_o \in D_A$ if for every sequence of elements $u_i \in D_A$ for which $\lim_{i\to\infty} u_i = u_o$ in H the relation $\lim_{i\to\infty} Au_i = Au_o$ in H, i.e. $\lim_{i\to\infty} \|Au_i - Au_o\| = 0$. We say that the operator A is continuous in D_A if it is continuous at every point of its domain D_A. An operator A is called bounded in its domain D_A if it is possible to find a number $c \geq 0$ such that for all $u \in D_A$ we have

$$\|Au\| \leq c\,\|u\| \quad (\|Au\| \leq \|A\|\,\|u\| \quad \text{holds for every } u \in D_A).$$

The least of the numbers c for which the relation $\|Au\| \leq c\,\|u\|$ is satisfied is called the norm of the operator A and $c = \|A\|$.

If the linear operator A (in a Hilbert space H) is bounded in D_A, then it is also continuous in D_A.

Let D_A be a linear set dense in H. An operator A, linear in D_A, is called symmetric in D_A if for every pair of elements u, v from D_A we have $(Au, v) = (u, Av)$.

An operator A is called positive in D_A if it is symmetric and if, for all $u \in D_A$, the relations $(Au, u) \geq 0$ and $(Au, u) = 0 \Rightarrow u = 0$ in D_A hold. If, moreover, there exists a constant $c > 0$ such that for all $u \in D_A$ the relation $(Au, u) \geq c^2\,\|u\|^2$ holds, then the operator A is called positive definite.

An operator F which maps its domain D_F into the set of real (complex) numbers is called a functional (real, complex) and is denoted by $F(u)$. It is evident that, as a functional is a special case of an operator, all concepts, definitions and results given above for an operator are valid in the case of a functional $F(u)$.

Theorem 2.6 (Riesz): Every linear bounded functional $F(u)$ in a Hilbert space H can be expressed in the form $F(u) = (u, v)$, where v is an element of H uniquely determined by the functional F. Moreover, $\|v\| = \|F\|$.

(For the proof see Rektorys (1980).)

2.2.3. Sobolev Spaces

Firstly a Lipschitz continuous function and a Lipschitz boundary will be defined. We say that a function $f(\boldsymbol{x})$ is Lipschitz continuous, or satisfies the Lipschitz condition, if there exists a constant $c_o > 0$ such that

$$|f(\boldsymbol{x}) - f(\boldsymbol{y})| < c_o |\boldsymbol{x} - \boldsymbol{y}| \tag{2.1}$$

for every $\boldsymbol{x} = (x_1, \ldots, x_n) \in K_n$, $\boldsymbol{y} = (y_1, \ldots, y_n) \in K_n$, where K_n is an n-dimensional open cube defined by

$$K_n = \{\boldsymbol{x} \mid |x_i| < \alpha = \text{constant} > 0 \text{ for } i = 1, \ldots, n\}. \tag{2.2}$$

A function f defined on Ω is said to satisfy the Hölder condition with exponent k, $0 \leq k \leq 1$, if there exists a nonnegative constant $c = c(f)$ such that

$$|f(\boldsymbol{x}) - f(\boldsymbol{y})| \leq c\,|\boldsymbol{x} - \boldsymbol{y}|^k \text{ for all } \boldsymbol{x}, \boldsymbol{y} \in \Omega.$$

For $k = 1$, the condition is known as the Lipschitz condition.

Let Ω be a domain with a boundary $\partial\Omega$. A boundary $\partial\Omega$ is said to be a Lipschitz boundary if there exist positive real numbers $\alpha > 0$, $\beta > 0$ such that for each $\boldsymbol{x} \in \partial\Omega$, the Cartesian coordinate system can be rotated and translated to $\boldsymbol{x}$ in such a way that it holds that:

There exists a function $g(x_1, \ldots, x_{n-1})$ defined on K_{n-1}, where K_{n-1} is an $(n-1)$-dimensional open cube defined by Equation 2.2 such that

$$g(x_1, \ldots, x_{n-1}) = x_n, \quad (x_1, \ldots, x_{n-1}) \in K_{n-1}$$

and satisfying the Lipschitz condition

$$|g(x_1, \ldots, x_{n-1}) - g(y_1, \ldots, y_{n-1})| < c_o\,|(x_1, \ldots, x_{n-1}) - (y_1, \ldots, y_{n-1})|$$

for every $(x_1, \ldots, x_{n-1}) \in K_{n-1}$ and $(y_1, \ldots, y_{n-1}) \in K_{n-1}$.

Moreover, all the points $\boldsymbol{x} = (x_1, \ldots, x_{n-1}, x_n)$ such that $(x_1, \ldots, x_{n-1}) \in K_{n-1}$ and $g(x_1, \ldots, x_{n-1}) < x_n < g(x_1, \ldots, x_{n-1}) + \beta$ lie inside Ω while all the points $\boldsymbol{x} = (x_1, \ldots, x_{n-1}, x_n)$, $(x_1, \ldots, x_{n-1}) \in K_{n-1}$, $g(x_1, \ldots, x_{n-1}) - \beta < x_n < g(x_1, \ldots, x_{n-1})$ lie outside $\overline{\Omega}$.

Example: For the 2D-case

$$(x_1, \ldots, x_{n-1}) \equiv x_1, \quad x_n \equiv x_2.$$

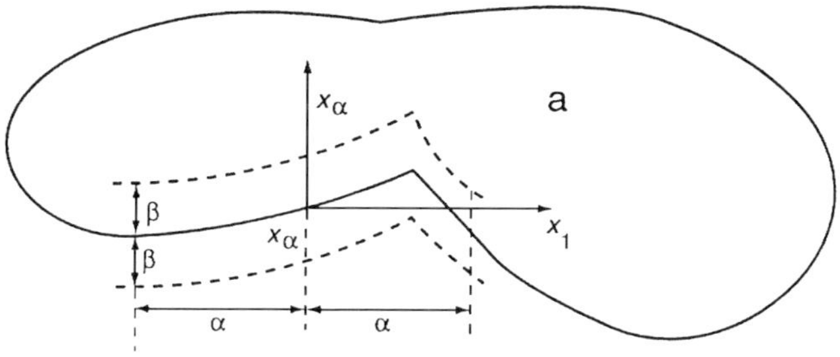

Figure 2.1. Domain with a Lipschitz Boundary

For illustration see Figure 2.1.

A real function f is said to be integrable in the Lebesgue sense if it is measurable on Ω and the Lebesgue integral $\int_\Omega f(\boldsymbol{x})\mathrm{d}\boldsymbol{x}$ exists, but may be $\pm\infty$. The Lebesgue measure of a measurable subset $M \subset R^N$ will be denoted by $\mu(M)$.

Theorem 2.7 (Levi)**:** Let $\{f_i\}_{i=1}^{\infty}$ be a sequence of Lebesgue integrable functions on a measurable set $\Omega \subset R^N$ such that $f_1(\boldsymbol{x}) \leq f_2(\boldsymbol{x}) \leq \ldots \leq f_i(\boldsymbol{x}) \leq f_{i+1}(\boldsymbol{x}) \leq \ldots$ for almost all $\boldsymbol{x} \in \Omega$, and that

$$\int_\Omega f_1(\boldsymbol{x})\mathrm{d}\boldsymbol{x} > -\infty.$$

Then for almost all $\boldsymbol{x} \in \Omega$ the limit $f(\boldsymbol{x}) = \lim_{i\to\infty} f_i(x)$ exists, the function f is integrable and

$$\int_\Omega \lim_{i\to\infty} f_i(\boldsymbol{x})\mathrm{d}\boldsymbol{x} = \int_\Omega f(\boldsymbol{x})\mathrm{d}\boldsymbol{x} = \lim_{i\to\infty} \int_\Omega f_i(\boldsymbol{x})\mathrm{d}\boldsymbol{x}.$$

Theorem 2.8: Let $\{f_i\}_{i=1}^{\infty}$ be a sequence of measurable functions on a measurable set $\Omega \subset R^N$ which converges for almost all $\boldsymbol{x} \in \Omega$ to $f(\boldsymbol{x})$.

Suppose that there exists a function g with finite Lebesgue integral over Ω such that $|f_i(\boldsymbol{x})| \leq g(\boldsymbol{x})$ for all $i \in N$ (the set of all positive integers) and almost all $x \in \Omega$.

Then f_i and $f (i = 1, 2, \ldots)$ have finite integrals and

$$\int_\Omega f(\boldsymbol{x})\mathrm{d}\boldsymbol{x} = \lim_{i\to\infty} \int_\Omega f_i(\boldsymbol{x})\mathrm{d}\boldsymbol{x}.$$

The theorem is known as the Lebesgue's convergence theorem.

Theorem 2.9 (Vitali)**:** Let $\{f_i\}_{i=1}^{\infty}$ be a sequence of functions with finite integrals over a measurable set $\Omega \subset R^N$. Suppose that $\lim_{i\to\infty} f_i(\boldsymbol{x}) = f(\boldsymbol{x})$ for almost all $\boldsymbol{x} \in \Omega$ and let f be an almost everywhere finite function. Suppose that the following condition is satisfied:

($\mathscr{P}$): For every $\varepsilon > 0$ there exists a $\delta > 0$ with the property: if $B \subset \Omega$, and $\mu(B) < \delta$, then $\int_B |f_i(\boldsymbol{x})|\,\mathrm{d}\boldsymbol{x} < \varepsilon$ for all $i \in N$ (the set of all positive integers).

Then the function f has a finite integral over Ω and

$$\lim_{i\to\infty} \int_\Omega f_i(\boldsymbol{x})\mathrm{d}\boldsymbol{x} = \int_\Omega f(\boldsymbol{x})\mathrm{d}\boldsymbol{x}.$$

(For the proofs see, for example, Hewitt and Stromberg (1969), Dunford and Schwartz (1958).)

Let $p \in \langle 0, \infty)$. Let Ω be a (Lebesque) measurable subset of Euclidean space R^N. Denote by $L^p(\Omega)$ the set of all real-valued or complex-valued measurable functions $f(\boldsymbol{x})$ defined almost everywhere on Ω and such that the Lebesque integral $\int_\Omega |f(\boldsymbol{x})|^p \,\mathrm{d}\boldsymbol{x}$ is finite. The seminorm will be defined by

$$\|f\|_{L^p(\Omega)} \equiv \|f\|_p = \left(\int_\Omega |f(\boldsymbol{x})|^p \,\mathrm{d}\boldsymbol{x}\right)^{1/p}.$$

Remark: The notation $L_p(\Omega)$ is also used.

Remark: Similarly, a real (complex) function $f(\boldsymbol{x})$ is said to be square integrable in the domain Ω if the integrals $\int_\Omega f(\boldsymbol{x})\mathrm{d}\boldsymbol{x}$ and $\int_\Omega f^2(\boldsymbol{x})\mathrm{d}\boldsymbol{x}$ are convergent, if they exist in the Lebesgue sense and are finite.

A set M is said to be open in $\partial\Omega$ if for any $\boldsymbol{x} \in M$ we can find an open sphere B containing $\boldsymbol{x}$ such that $B \cap \partial\Omega \subset M$.

Let Ω be a domain with a Lipschitz boundary $\partial\Omega$. Then we denote by $C(\Omega)$ the space of all functions defined and continuous on Ω and by $C(\overline{\Omega})$ the space of all functions continuous in Ω and continuously extendable to $\overline{\Omega} = \Omega \cup \partial\Omega$. By $C^{(k)}(\Omega)$, $1 \leq k < \infty$, we denote the space of all functions continuous in Ω whose derivatives up to order k are continuous in Ω, i.e. the space of all functions $u \in C^{(k)}(\Omega)$ such that $D^\alpha u \in C(\overline{\Omega})$ for all $\alpha \in M_{N,k}$, where $M_{N,k}$ is the set of all N-dimensional multi-indices of length less than or equal to k. Similarly by $C^{(k)}(\overline{\Omega})$ we denote the space of all functions continuous in $\overline{\Omega}$ whose derivatives up to order k are continuous in $\overline{\Omega}$, i.e. continuous in Ω and continuously extendable to $\overline{\Omega} = \Omega \cup \partial\Omega$ whose derivatives up to order k are continuous in Ω and continuously extendable to $\overline{\Omega} = \Omega \cup \partial\Omega$. In $C^{(k)}(\Omega)$ we define an inner (scalar) product by

$$(u, v)_k = \int_\Omega \sum_{|\alpha| \leq k} D^\alpha u D^\alpha v \mathrm{d}\boldsymbol{x},$$

where

$$D^\alpha u = \frac{\partial^{|\alpha|} u}{\partial x_1^{\alpha_1} \partial x_2^{\alpha_2} \partial x_3^{\alpha_3}}, \qquad |\alpha| = \sum_{i=1}^{3} \alpha_i.$$

By this scalar product the norm in $C^{(k)}(\Omega)$

$$\|u\|_{k,2} = (u, u)^{1/2},$$

is defined.

By completion of the space $C^{(k)}(\overline{\Omega})$ in the norm $\|u\|_{k,2}$, we obtain the Sobolev space $H^k(\Omega) \equiv W^{k,2}(\Omega)$. For $k = 0$ $H^o(\Omega) \equiv L^2(\Omega)$, we have the space of functions square integrable in the domain Ω. The Sobolev space as well as the space $L^2(\Omega)$ are Hilbert spaces.

We say that supp f is the support of the function $f(\boldsymbol{x})$, if supp $f(\boldsymbol{x}) = \overline{M}$, $M = \{\boldsymbol{x} \in \Omega \,|\, f(\boldsymbol{x})| \neq 0\}$ and $\overline{M}$ denotes its closure.

We denote by $D(\overline{\Omega})$ the space of all functions having derivatives of all orders continuous in Ω and continuously extendable to $\overline{\Omega}$. The space $D(\overline{\Omega})$ is dense in $H^k(\Omega)$ if Ω is a domain with a Lipschitz boundary. By $\mathscr{D}(\Omega)$ we denote the space of functions with compact support in Ω, i.e.

$$\mathscr{D}(\Omega) = \{v \mid v \in D(\overline{\Omega}), \qquad \text{supp } v \subseteq \Omega\}.$$

By $C_o^{(k)}(\overline{\Omega})$ we denote the space of all functions $v \in C^{(k)}(\Omega)$ satisfying $D^\alpha v = 0$ on $\partial\Omega$ for which $|\alpha| \leq k - 1$, i.e.

$$C_o^{(k)}(\overline{\Omega}) = \{v \mid v \in C^{(k)}(\overline{\Omega}), D^\alpha v = 0 \text{ on } \partial\Omega, |\alpha| \leq k - 1\},$$

the closure of which in $H^k(\Omega)$ is denoted by $H_o^k(\Omega)$.

By $C^{(\infty)}(\Omega)$ we denote the set of all functions defined on Ω which have derivatives of any order on Ω, i.e. $C^{(\infty)}(\Omega) = \cap_{k=0}^{\infty} C^{(k)}(\Omega)$. Likewise we denote $C^{(\infty)}(\overline{\Omega}) = \cap_{k=0}^{\infty} C^{(k)}(\overline{\Omega})$. The functions in $C_o^{(\infty)}(\Omega)$ are also called finite functions. (Instead of $C_o^{(\infty)}(\Omega)$, the symbol $\mathscr{D}(\Omega)$ is used, especially in distribution theory.)

Let $p \in \langle 1, \infty)$ and let Ω be a nonempty open subset of R^N. Denote by $L^p_{\text{loc}}(\Omega)$ (or $L_{p,\text{loc}}(\Omega)$) the set of all functions u defined almost everywhere on Ω such that $u \in L^p(\Omega_1)$ for every bounded set Ω_1 for which $\overline{\Omega}_1 \subset \Omega$.

Theorem 2.10: The spaces $C(\overline{\Omega})$ and $C^{(k)}(\overline{\Omega})$, $k \in N$, are separable.

(For the proof see for example, Kufner, John and Fučík. (1977).)

Theorem 2.11: The space $C^{(k)}(\overline{\Omega})$, $k \in N$, has a Schauder basis.

(For the proof see for example, Kufner, John and Fučík. (1977).)

Theorem 2.12: Let Ω be a nonempty bounded open subset of R^N. Then the set $C_o^{(\infty)}(\Omega)$ is dense in $L^p(\Omega)$ for arbitrary $p \in \langle 1, \infty)$.

(For proof see, for example Kufner, John and Fučík. (1977).)

Theorem 2.13: Let Ω be a nonempty bounded open subset of R^N. Let $p \in \langle 0, \infty)$. Then the normed linear space $L^p(\Omega)$ is separable.

Theorem 2.14: Let $\dfrac{1}{p} + \dfrac{1}{p'} = 1$ and let $g \in L^{p'}(\Omega)$ and denote

$$\varphi_g(f) = \int_\Omega f(\boldsymbol{x}) g(\boldsymbol{x}) \mathrm{d}\boldsymbol{x}$$

for $f \in L^p(\Omega)$. Then $\varphi_g \in (L^p(\Omega))'$ and $\|\varphi_g\| = \|g\|_{p'}$ where $(L^p(\Omega))'$ is the dual space of $L^p(\Omega)$.

(For the proof see, for example, Kufner, John and Fučík. (1977).)

Theorem 2.15: Let $p \in \langle 1, \infty)$. Then the spaces $L^p(\Omega)$ are Banach spaces and reflexive.

Theorem 2.16: Let Ω be a nonempty bounded open subset of R^N. Then the set $C_o^{(\infty)}(\Omega)$ is dense in $L^p(\Omega)$ for arbitrary $p \in \langle 1, \infty)$.

Let $L^\infty(\Omega)$ denote the set of all (Lebesgue) measurable functions f defined almost everywhere on Ω and such that there exists a constant $c > 0$ and a set $\Omega' \subset \Omega$, both depending on f, and with Lebesgue measure $\mu(\Omega') = 0$ and with property $|f(\boldsymbol{x})| \leq c$ for all $\boldsymbol{x} \in \Omega - \Omega'$. The norm $\|f\|_\infty$ is defined by

$$\|f\|_{L^\infty(\Omega)} \equiv \|f\|_\infty = \text{vrai max}\,|f(\boldsymbol{x})| \equiv \inf_{\substack{\Omega' \\ \mu(\Omega')=0}} \{ \sup_{x \in \Omega - \Omega'} |f(\boldsymbol{x})|\}.$$

The norm is sometimes denoted by ess $\sup|f(\boldsymbol{x})|$ or vrai $\sup|f(\boldsymbol{x})|$. The space $L^\infty(\Omega)$ is a Banach space.

Theorem 2.17: Let Ω be a nonempty open set of R^N. Then $L^\infty(\Omega)$ is not separable.

Theorem 2.18: Let Ω be a nonempty bounded open set of R^N. Then $L^\infty(\Omega)$ is not reflexive. (For the proofs see, for example Kufner, John and Fučík (1977).)

Let f_i, $i = 1, 2, \ldots, f$ be functions defined on a nonempty bounded subset Ω of R^N. Then:

i) the sequence $\{f_i\}_{i=1}^\infty$ converges to f pointwise on Ω if

$$\lim_{i \to \infty} f_i(\boldsymbol{x}) = f(\boldsymbol{x}) \quad \text{for every } \boldsymbol{x} \in \Omega,$$

ii) the sequence $\{f_i\}_{i=1}^\infty$ converges to f uniformly on Ω if for every $\varepsilon > 0$ there exists an $n_o > 0$ such that for every $n \geq n_o$ and every $x \in \Omega$ we have $|f_i(\boldsymbol{x}) - f(\boldsymbol{x})| < \varepsilon$,

iii) the sequence $\{f_i\}_{i=1}^\infty$ converges to f locally uniformly on Ω if $\{f_i\}_{i=1}^\infty$ converges to f uniformly on each compact subset $\Omega' \subset \Omega$,

iv) the sequence $\{f_i\}_{i=1}^\infty$ converges to f uniformly up to small sets on Ω if for every $\varepsilon > 0$ there exists an $M \subset \Omega$, $\mu(M) < \varepsilon$, such that $\{f_i\}_{i=1}^\infty$ converges uniformly to f on $\Omega - M$,

v) the sequence $\{f_i\}_{i=1}^\infty$ converges to f almost everywhere on Ω if there exists a set $M \subset \Omega$ of measure zero such that $\{f_i\}_{i=1}^\infty$ converges pointwise to f on $\Omega - M$.

Theorem 2.19: Let $p \in (1, \infty)$. The sequence $\{f_i\}_{i=1}^\infty \subset L^p(0, 1)$ converges weakly to $f \in L^p(0, 1)$ if and only if the following conditions are satisfied:

i) the sequence $\{\|f_i\|_p\}_{i=1}^\infty$ is bounded,

ii) $\lim_{i\to\infty} \int_0^\tau f_i(t)\mathrm{d}t = \int_0^\tau f(t)\mathrm{d}t$ for every $\tau \in (0, 1)$.

Theorem 2.20: Let Ω be a nonempty bounded open subset of R^N. Then there exist Schauder bases in the space $L^p(\Omega)$, $1 \leq p < \infty$.

Theorem 2.21: The spaces $L^p(\Omega)$, $1 < p < \infty$, are weakly complete.

(For the proofs see, for example, Kufner, John and Fučík (1977).)

Theorem 2.22: The Sobolev space $H^{(k)}(\Omega)$ is separable.

Theorem 2.23: The space $C^{(\infty)}(\Omega)$ is dense in the Sobolev space $H^{(k)}(\Omega)$.

(For the proof see, for example, Nečas (1967).)

Remark: The elements of $H^{(k)}(\Omega)$ are, on one hand, the functions of $C^{(\infty)}(\overline{\Omega})$, and, on the other hand, functions of $L^2(\Omega)$ which have generalized derivatives $D^\alpha u$ and which can be approximated with arbitrary accuracy, in the metric determined by the inner product (u, v), by functions from $C^{(\infty)}(\overline{\Omega})$. So that to $H^{(k)}(\Omega)$ there belong functions which have continuous in $\overline{\Omega}$ derivatives up to the order $k - 1$ and continuous or piecewise continuous derivatives of order k. For $u \in C^{(\infty)}(\overline{\Omega})$, the generalized derivatives are derivatives in the classical sense. Of course, if $u \in H^{(k)}(\Omega)$ only, this need not be the case.

Furthermore, we shall introduce the Sobolev spaces $H^{(k,p)}(\Omega)$. Let $p \in \langle 1, \infty)$ and let $1 \leq k < \infty$. Denote by $H^{(k,p)}(\Omega)$ the set of all functions $u \in L^p(\Omega)$ whose generalized derivatives $D^\alpha u$ for $\alpha \in M_{N,k}$ are elements of the space $L^p(\Omega)$. Therefore, $H^{(k,p)}(\Omega)$ is a linear subset of $L^p(\Omega)$. The norm is defined by

$$\|u\|_{H^{(k,p)}(\Omega)} = \sum_{|\alpha| \leq k} \|D^\alpha u\|_{L^p(\Omega)}$$

for $u \in H^{(k,p)}(\Omega)$(where the equality on the space $H^{(k,p)}(\Omega)$ is taken in the sense of almost everywhere).

Theorem 2.24:

(i) the space $H^{(k,p)}(\Omega)$ is a Banach space,

(ii) the space $H^{(k,p)}(\Omega)$ is reflexive if $p \in \langle 1, \infty)$,

(iii) the space $H^{(k,p)}(\Omega)$ is separable if $p \in \langle 1, \infty)$,

(iv) $C_o^{(\infty)}(\Omega) \subset H^{(k,p)}(\Omega)$.

(v) if Ω is bounded then $C_o^{(\infty)}(\overline{\Omega}) \subset H^{(k,p)}(\Omega)$.

(For the proof see Kufner, John and Fučík (1977).)

Denote by $H_o^{k,p}(\Omega)$ the closure of $C_o^{(\infty)}(\Omega)$ in the norm of the space $H^{(k,p)}(\Omega)$. The space $H_o^{k,p}(\Omega)$ is a subspace of $H^{k,p}(\Omega)$ and its norm is the same as on $H^{k,p}(\Omega)$.

Let Ω be a domain in R^N and X a normed linear space with the norm $\|\cdot\|_X$. A mapping u from Ω or from $\overline{\Omega}$ into X is said to be an abstract (a function with values in X) defined on Ω or on $\overline{\Omega}$, respectively. An abstract function u is said to be continuous at the point $\boldsymbol{x}_o \in \Omega$ if $\lim_{\boldsymbol{x} \to \boldsymbol{x}_o} \|u(\boldsymbol{x}) - u(\boldsymbol{x}_o)\|_X = 0$, $\boldsymbol{x} \in X$ and the abstract function v_i defined on Ω is called the partial derivative of the abstract function u with respect to the variable x_i and defined by $\partial u/\partial x_i$, if

$$\lim_{h\to 0} \|(u(x_1, \dots, x_{i-1}, x_i + h, x_{i+1}, \dots, x_N) - u(x))/h - v_i(x)\|_X = 0.$$

For derivatives of higher orders of abstract functions we shall use the symbols D^αu.

Let X be a Banach space. The set of all abstract functions u defined and continuous on Ω with values in X will be denoted by $C(\Omega; X)$. Analogously as above we define the spaces $C^{(k)}(\Omega; X), \dots, C_o^\infty(\Omega; X)$.

Theorem 2.25: Let X be a Banach space. Then $C^{(k)}(\overline{\Omega}; X)$ is a Banach space.

Theorem 2.26: The space $C(\overline{\Omega}; X)$ is separable if and only if the Banach space X is separable.

(For the proof see Kufner, John and Fučík (1977).)

Theorem 2.27: Let X be a separable Banach space. Then the space $C^{(k)}(\overline{\Omega}; X)$ is separable.

Let $\{u_i\}_{i=1}^\infty$ be a sequence in $C^{(k)}(\overline{\Omega})$. We say that the sequence $\{u_i\}_{i=1}^\infty$ converges to $u \in C^{(k)}(\overline{\Omega})$ weakly if and only if:

i) the set $\{u_i\}_{i=1}^\infty$ is bounded in $C^{(k)}(\overline{\Omega})$,

ii) $D^\alpha u_i$ converges to $D^\alpha u$ pointwise in $\overline{\Omega}$, i.e. $D^\alpha u(\boldsymbol{x}) = \lim_{i\to\infty} D^\alpha u_i(\boldsymbol{x})$ for all $\boldsymbol{x} \in \overline{\Omega}$ and $|\alpha| \le k$.

Let $\Omega \in R^N$ be a nonempty bounded open set. Let us consider functions, the so-called abstract functions, defined on Ω whose values are in a general Banach space X with norm $\|\cdot\|$ and zero element Θ.

A mapping $f : \Omega \to X$ is said to be a simple function if there exist $c_i \in X$, $i = 1, \dots, n$, and measurable subsets B_i of Ω, $i = 1, \dots, n$, $B_i \cap B_j = 0$ for $i \neq j$, such that $f(\boldsymbol{x}) = \Sigma_{i=1}^n \chi_{B_i}(\boldsymbol{x}) c_i$, where χ_{B_i} denotes the characteristic function of the set B_i.

A function $f : \Omega \to X$ is said to be a strongly measurable function if there exists a sequence $\{f_i\}_{i=1}^\infty$ of simple functions such that $\lim_{i\to\infty} \|f_i(\boldsymbol{x}) - f(\boldsymbol{x})\| = 0$ for almost all $\boldsymbol{x} \in \Omega$.

Lemma 2.1: Let $f : \Omega \to X$ be a strongly measurable function. Then the function $\|f(\boldsymbol{x})\| : \Omega \to R$ is measurable in the usual Lebesgue sense.

(For the proof see, for example, Kufner, John and Fučík (1977).)

Let $f : \Omega \to X$ be a simple function, i.e. $f(\boldsymbol{x}) = \Sigma_{i=1}^n \chi_{B_i}(\boldsymbol{x}) c_i$, then the Bochner integral of f is defined by

$$\int_\Omega f(\boldsymbol{x}) \mathrm{d}\boldsymbol{x} = \sum_{i=1}^n \mu(B_i) c_i.$$

Lemma 2.2:

(i) Let $f : \Omega \to X$ be a simple function, i.e.

$$f(\boldsymbol{x}) = \sum_{i=1}^{n} \chi_{B_i}(\boldsymbol{x}) c_i .$$

Then $$\left\| \int_\Omega f(\boldsymbol{x}) \mathrm{d}\boldsymbol{x} \right\| \leq \int_\Omega \| f(\boldsymbol{x}) \| \, \mathrm{d}\boldsymbol{x} .$$

(ii) Let f, g be two simple functions, and let $\alpha, \beta \in$R. Then $\alpha f + \beta g$ is also simple function and

$$\int_\Omega (\alpha f(\boldsymbol{x}) + \beta g(\boldsymbol{x})) \mathrm{d}\boldsymbol{x} = \alpha \int_\Omega f(\boldsymbol{x}) \mathrm{d}\boldsymbol{x} + \beta \int_\Omega g(\boldsymbol{x}) \mathrm{d}\boldsymbol{x} .$$

The proof is obvious.

A function $f : \Omega \to X$ is said to be Bochner integrable if there exists a sequence $\{f_i\}_{i=1}^{\infty}$ of simple functions such that

$$\lim_{i \to \infty} \| f(\boldsymbol{x}) - f_i(\boldsymbol{x}) \| = 0 \tag{2.3}$$

for almost all $\boldsymbol{x} \in \Omega$ and if

$$\lim_{i \to \infty} \int_\Omega \| f(\boldsymbol{x}) - f_i(\boldsymbol{x}) \| \, \mathrm{d}\boldsymbol{x} = 0. \tag{2.4}$$

For measurable $B \subset \Omega$ and for strongly measurable and Bochner integrable function f we define the Bochner integral by

$$\int_B f(\boldsymbol{x}) \mathrm{d}\boldsymbol{x} = \lim_{i \to \infty} \int_\Omega \chi_B(\boldsymbol{x}) f_i(\boldsymbol{x}) \mathrm{d}\boldsymbol{x} ,$$

where the sequence $\{f_i\}_{i=1}^{\infty}$ satisfies equations 2.3 and 2.4, and the integral on the right hand side exists and its value is independent of the sequence $\{f_i\}_{i=1}^{\infty}$ of simple functions with properties given by equations 2.3 and 2.4.

(For the proof see, for example, Kufner, John and Fučík (1977).)

Theorem 2.28 (Bochner's theorem)**:** A strongly measurable function $f : \Omega \to X$ is Bochner integrable if and only if the real-valued function $\| f(\boldsymbol{x}) \|$ has a finite Lebesgue integral. Moreover, for the Bochner integrable function $f : \Omega \to X$

$$\left\| \int_\Omega f(\boldsymbol{x}) \mathrm{d}\boldsymbol{x} \right\| \leq \int_\Omega \| f(\boldsymbol{x}) \| \, \mathrm{d}\boldsymbol{x}$$

holds.

Corollary: Let $T : X \to Y$ be a bounded linear mapping from a Banach space X into another Banach space Y. If the function $f : \Omega \to X$ is Bochner integrable then the function $Tf : \Omega \to Y$ is also Bochner integrable and

$$\int_\Omega Tf(\boldsymbol{x}) \mathrm{d}\boldsymbol{x} = T \left(\int_\Omega f(\boldsymbol{x}) \mathrm{d}\boldsymbol{x} \right) .$$

Corollary: Let $f : \Omega \to X$ be Bochner integrable. Then for almost all $\boldsymbol{x}_o \in \Omega$ we have that

$$\lim_{\delta \to 0+} (2\delta)^{-1} \int_{K_\delta(\boldsymbol{x}_o)} f(\boldsymbol{x}) \mathrm{d}\boldsymbol{x} = f(\boldsymbol{x}_o),$$

where $K_\delta(\boldsymbol{x}_o)$ is a cube in Ω with the centre at $\boldsymbol{x}_o$ and the length of its edge equal to 2δ.

(For the proof see Hewitt and Stromberg (1969).)

Moreover, it holds

$$\left\| (2\delta)^{-N} \int_{K_\delta(\boldsymbol{x}_o)} f(\boldsymbol{x}) \mathrm{d}\boldsymbol{x} - f(\boldsymbol{x}_o) \right\| \le (2\delta)^{-N} \int_{K_\delta(\boldsymbol{x}_o)} \| f(\boldsymbol{x}) - f(\boldsymbol{x}_o) \| \, \mathrm{d}\boldsymbol{x} .$$

Let $1 \le p < \infty$, let X be a Banach space with the norm $\|\cdot\|$ and let Ω be a nonempty bounded open subset of R^N. Denote by $L^p(\Omega; X)$ the set of all mappings $f : \Omega \to X$ which are strongly measurable and such that the Lebesgue integrals $\int_\Omega \|f(\boldsymbol{x})\|^p \, \mathrm{d}\boldsymbol{x}$ are finite. We shall say that two abstract functions $f_1, f_2 \in L^p(\Omega; X)$ are equal if and only if $f_1(\boldsymbol{x}) = f_2(\boldsymbol{x})$ for almost $\boldsymbol{x} \in \Omega$.

Theorem 2.29: It holds that

$$L^p(\Omega; X) \subset L^1(\Omega; X).$$

From the properties of Bochner's integral and Theorem 2.29 it is easy to show that every $f \in L^p(\Omega; X)$ is Bochner integrable and that the inequality $\left\| \int_\Omega f(\boldsymbol{x}) \mathrm{d}\boldsymbol{x} \right\| \le \int_\Omega \|f(\boldsymbol{x})\| \, \mathrm{d}\boldsymbol{x}$ holds. Furthermore, it is easy to show that $L^p(\Omega; X)$ is a vector space with a norm

$$\|f(\boldsymbol{x})\|_{L^p(\Omega;X)} \equiv \| \, \|f(\boldsymbol{x})\| \, \|_p = \left(\int_\Omega \|f(\boldsymbol{x})\|^p \, \mathrm{d}\boldsymbol{x} \right)^{1/p} .$$

Theorem 2.30: The spaces $L^p(\Omega; X)$, $1 \le p < \infty$, are Banach spaces.

(For the proof see, for example Kufner, John and Fučík (1977).)

Let us denote by $L^\infty(\Omega; X)$ the set of all strongly measurable mappings $f : \Omega \to X$ such that $\operatorname{ess\,sup}_{x \in \Omega} \|f(\boldsymbol{x})\| \equiv \text{vrai max} \|f(\boldsymbol{x})\| < \infty$. Its norm is defined by $\| \, \|f(\boldsymbol{x})\| \, \|_p = \text{vrai max} \|f(\boldsymbol{x})\|$.

Theorem 2.31: The space $L^\infty(\Omega; X)$ is complete.

(For the proof see, for example, Kufner, John and Fučík (1977).)

2.2.4. Survey of the Theory of Distribution

We say that supp f is the support of the function $f(\boldsymbol{x})$, if supp $f(\boldsymbol{x}) = \overline{M}$, $M = \{\boldsymbol{x} \in \Omega \mid f(\boldsymbol{x}) \ne 0\}$ and $\overline{M}$ denotes its closure.

Let Ω be a nonempty domain in R^N. Let $\mathscr{D}(\Omega)$ be a space of all complex (real) functions having derivatives of all orders in R^N with bounded supports (in the theory of spaces this space will be identical with the space $C_o^{(\infty)}(\Omega)$).

Theorem 2.32 (approximation theorem): Let $\varepsilon > 0$ be arbitrary. Then every continuous function f with bounded support can be uniformly approximated with an error at most ε by a function $\varphi \in \mathscr{D}(\Omega)$. A function φ can be found such that its support is a subset of an arbitrary neighbourhood of a support K of the function f.

(For the proof see Schwartz (1954).

Let $\{\varphi_n\}_{n=1}^{\infty} \subset \mathscr{D}(\Omega)$. Then $\varphi_n \to 0$ in $\mathscr{D}(\Omega)$ if there exists a compact set $K \subset \mathscr{D}(\Omega)$ such that:

(i) $\operatorname{supp} \varphi_n \subset K$ for all n from the set of all positive integers N,
(ii) $\lim_{n\to\infty} D^{\alpha}\varphi_n = 0$ for all arbitrary multi-index α uniformly on Ω.

Definition 2.1: By the distribution T we call a continuous linear functional on a vector space $\mathscr{D}(\Omega)$.

It means that the mapping $T : \mathscr{D}(\Omega) \to R$ adjoins to every function $\varphi \in \mathscr{D}(\Omega)$ a complex (real) number $T(\varphi) \equiv \langle T, \varphi\rangle$ satisfying:

(i) $T(\alpha\varphi + \beta\psi) = \alpha T(\varphi) + \beta T(\psi)$ for all $\alpha, \beta \in R$, $\varphi, \psi \in \mathscr{D}(\Omega)$,
(ii) $\lim_{n\to\infty} T(\varphi_n) = 0$ for an arbitrary sequence $\{\varphi_n\}_{n=1}^{\infty} \subset \mathscr{D}(\Omega)$ such that $\varphi_n \to 0$ in $\mathscr{D}(\Omega)$.

The set of all distributions is also a vector space and is denoted by $\mathscr{D}'(\Omega)$.

Definition 2.2: We call $\langle .\ .\rangle$ a duality pairing, i.e a bilinear map of $\mathscr{D}'(\Omega) \times \mathscr{D}(\Omega)$ into R.

Example: Let $\boldsymbol{x} \in \Omega$. We define a mapping $\delta(\boldsymbol{x})$ by $\delta(\boldsymbol{x}) \equiv \delta(\varphi(\mathbf{x})) = \varphi(\boldsymbol{x})$, $\varphi \in \mathscr{D}(\Omega)$. It is evident that $\delta(\boldsymbol{x}) \in \mathscr{D}'(\Omega)$. The distribution $\delta(\boldsymbol{x})$ is known as the Dirac function.

Example: Let f be a locally integrable function, i.e. a function which is integrable on every bounded set (in the theory of spaces we shall see that $f \in L^1_{\text{loc}}(\Omega)$). We define the mapping T by

$$T(\varphi) = \int_{\Omega} f(\boldsymbol{x})\varphi(\boldsymbol{x})\mathrm{d}\mathbf{x}, \qquad \varphi \in \mathscr{D}(\Omega).$$

Then $T \in \mathscr{D}'(\Omega)$. The integral is correct as we integrate not over R^N but only over the support of the function φ. The function f is continuous on the supp φ, the function φ is a continuous function, so the function $f(\mathbf{x})\varphi(\boldsymbol{x})$ is integrable. It is evident that the integral is a linear functional of the variable φ and is continuous on $\mathscr{D}(\Omega)$.

(For the proof see Schwartz (1957).)

Theorem 2.33: Functions f and g define the same functional T if and only if they are equal almost everywhere.

(For the proof see Schwartz (1957).)

We say that the distribution T is equal to zero in an open set $\Omega \subset R^N$, if $\langle T, \varphi\rangle = 0$ for every function $\varphi \in \mathscr{D}(\Omega)$, support of which is a subset of Ω.

Let us define derivatives of distributions $\partial T/\partial x_i$ in R^N. The derivative $\partial T/\partial x_i$ of a distribution T is defined by the relation

$$\partial T(\varphi)/\partial x_i = -T(\partial\varphi/\partial x_i), \quad \varphi \in \mathscr{D}(\Omega).$$

It is evident that $\partial T/\partial x_i$ is a distribution as for a continuously differentiable function f it holds

$$\langle \partial f/\partial x_i, \varphi\rangle = \int_{R^N} \partial f/\partial x_i \varphi \mathrm{d}\boldsymbol{x}.$$

Applying the Fubini's theorem (see below, Theorem 2.38) we have

$$\int_{x_1}\int_{x_2}\cdots\int_{x_{i-1}}\int_{x_{i+1}}\cdots\int_{x_n} \mathrm{d}x_1\ldots\mathrm{d}x_{i-1}\mathrm{d}x_{i+1}\ldots\mathrm{d}x_n \int_{-\infty}^{+\infty} \partial f/\partial x_i\varphi \mathrm{d}x_i$$

and then by integration by parts, and since

$$\int_{-\infty}^{+\infty} \partial f/\partial x_i\varphi \mathrm{d}x_i = [f\varphi]_{-\infty}^{+\infty} - \int_{-\infty}^{+\infty} f\partial\varphi/\partial x_i \mathrm{d}x_i$$

and since $[f\varphi]_{-\infty}^{+\infty} = 0$ due to $\varphi = 0$ outside a bounded set, we obtain

$$\langle \partial f/\partial x_i, \varphi\rangle = \int_{R^N} \partial f/\partial x_i\varphi \mathrm{d}\boldsymbol{x} = -\int_{R^N} f\partial\varphi/\partial x_i \mathrm{d}\boldsymbol{x} = -\langle f, \partial\varphi/\partial x_i\rangle.$$

Similarly we define $D^\alpha T$, i.e.

$$\frac{\partial^{|\alpha|}T}{\partial \boldsymbol{x}_1^{\alpha_1}\ldots\partial \boldsymbol{x}_N^{\alpha_N}}$$

by $D^\alpha T(\varphi) = (-1)^{|\alpha|}T(D^\alpha\varphi)$, $\varphi \in \mathscr{D}(\Omega)$. We call $D^\alpha T$ the α-th distributional derivatives of T.

Two arbitrary distributions T and S cannot be multiplied, as, for example, for locally integrable function f which determines a distribution T, a function f^2 is not locally integrable and therefore does not determine a distribution. For example, such a function is $f(\boldsymbol{x}) = 1/\mathrm{sqrt}(|\boldsymbol{x}|)$, $f^2(\mathbf{x}) = 1/|\boldsymbol{x}|$, $\boldsymbol{x} \in R$. It is evident that if T is an irregular function then S must be a "very" regular function, i.e. it must be defined by a function which has derivatives of any order on Ω. Let $T \in \mathscr{D}'(\Omega)$ be an arbitrary distribution and ψ be a function defined on $\overline{\Omega}$ which have derivatives up to the order $k = \infty$ on $\overline{\Omega}$. Then

$$S(\varphi) = T(\psi\varphi)$$

is an element of $\mathscr{D}'(\Omega)$ which is usually denoted by ψT and is called "the product of T and the scalar function ψ".

Theorem 2.34: Let T be an arbitrary distribution, $T \in \mathscr{D}'(\Omega)$, $\psi \in C^{(\infty)}(\overline{\Omega})$. Then there exists a distribution ψT for which

$$\langle\psi T, \varphi\rangle = \langle T, \psi\varphi\rangle.$$

Moreover, for its derivation it holds

$$\frac{\partial}{\partial \boldsymbol{x}_i}(\psi T) = \frac{\partial \psi}{\partial \boldsymbol{x}_i} T + \psi \frac{\partial T}{\partial \boldsymbol{x}_i}$$

(For the proof see, for example Schwartz (1957).)

We say that a sequence of distributions T_i converges for $i \to \infty$ to a distribution T if for every function $\varphi \in \mathscr{D}(\Omega)$ a sequence of complex numbers $\langle T_i, \varphi \rangle$ converges for $i \to \infty$ to a complex number $\langle T, \varphi \rangle$. The convergence is taken in the sense of weak topology of the space $\mathscr{D}'(\Omega)$

2.2.5. Some Fundamental Theorems

Theorem 2.35 (Green theorem): Let Ω be a domain with Lipschitz boundary $\partial\Omega$ and let $f, g \in C^1(\Omega)$, $\boldsymbol{u}, \boldsymbol{v} \in [C^1(\Omega)]^N$, $N = 2, 3$ be scalar or vector functions, respectively. Then

$$\int_\Omega \partial f / \partial x_i g \mathrm{d}\boldsymbol{x} = \int_{\partial\Omega} f g n_i \mathrm{d}s - \int_\Omega f \partial g / \partial x_i \mathrm{d}\boldsymbol{x} \tag{2.5}$$

and

$$\int_\Omega \partial u_i / \partial x_j v_i \mathrm{d}\boldsymbol{x} = \int_{\partial\Omega} u_i v_i n_j \mathrm{d}s - \int_\Omega u_i \partial v_i / \partial x_j \mathrm{d}\boldsymbol{x}, \tag{2.6}$$

where n_j is the j-th component of the unit outward normal to $\partial\Omega$. Moreover, it can be shown that if the boundary $\partial\Omega$ of the domain Ω is a Lipschitz boundary, then the vector of the outward normal exists almost everywhere on $\partial\Omega$, and the integral over $\partial\Omega$ in Equations 2.5 and 2.6 exist and are convergent.

(For the proof see Rektorys (1980).)

Theorem 2.36 (trace theorem): There exists a continuous linear mapping $T : H^1(\Omega) \to L^2(\partial\Omega)$ such that $Tu = u$ for $u \in C^1(\overline{\Omega})$, where Tu is the so-called trace of the function u.

(For the proof see Oden and Reddy (1976), Nečas and Hlaváček (1981).)

In the next chapters we shall prove that if the coefficient of the equation considered is sufficiently smooth, and if the weak solution $u(\boldsymbol{x})$ is sufficiently smooth, then it is the classical solution of the investigated problem. It is now a question whether in the case of sufficiently smooth data it is necessary to assume the smoothness of the weak solution, or whether this smoothness is itself already a consequence of the smoothness of the given data. The answer is based on the so-called embedding theorems. The embedding is defined as the identical mapping. A completely continuous mapping in a metric space is a continuous linear mapping that maps any bounded sequence onto a sequence from which a convergent subsequence can be extracted.

Theorem 2.37 (Rellich): The embedding of $H^1(\Omega)$ in $L^2(\Omega)$ is completely continuous.

(For the proof see Nečas and Hlaváček (1981).)

Lemma 2.3: Let Ω be a domain with a Lipschitz boundary in an N-dimensional Euclidean space R^N, let $2k > N$. Then every function $u \in H^k(\Omega)$ is continuous in $\overline{\Omega}$,

$$u \in H^k(\Omega), \qquad 2k > N \Rightarrow u \in C(\overline{\Omega}).$$

(For the proof see Nečas (1967).)

Lemma 2.4 (Fatou's lemma): Let $\{f_i\}_{i=1}^{\infty}$ be a sequence of measurable functions which are nonnegative almost everywhere on Ω. Then $\lim_{i\to\infty} \inf f_i(\boldsymbol{x})$ is integrable and

$$\int_\Omega \liminf_{i\to\infty} f_i(x)\mathrm{d}x \le \liminf_{i\to\infty} \int_\Omega f_i(x)\mathrm{d}x.$$

(For the proof see Jarník (1955).)

Theorem 2.38 (Fubini's theorem): Let $\Omega_i \subset R^N$ $(i = 1, 2)$ be measurable and put $\Omega = \Omega_1 \times \Omega_2$. Let $f(\boldsymbol{x}, \boldsymbol{y})$ be integrable over Ω. Then for almost all $\boldsymbol{x} \in \Omega_1$ and $\boldsymbol{y} \in \Omega_2$ the integrals

$$\int_{\Omega_1} f(\boldsymbol{x}, \boldsymbol{y})\mathrm{d}\boldsymbol{x} \text{ and } \int_{\Omega_2} f(\boldsymbol{x}, \boldsymbol{y})\mathrm{d}\boldsymbol{y}$$

exist. Moreover

$$\int_\Omega f(\boldsymbol{x}, \boldsymbol{y})\mathrm{d}\boldsymbol{x}\,\mathrm{d}\boldsymbol{y} = \int_{\Omega_1}\left(\int_{\Omega_2} f(\boldsymbol{x}, \boldsymbol{y})\mathrm{d}\boldsymbol{y}\right)\mathrm{d}\boldsymbol{x} = \int_{\Omega_2}\left(\int_{\Omega_1} f(\boldsymbol{x}, \boldsymbol{y})\mathrm{d}\boldsymbol{x}\right)\mathrm{d}\boldsymbol{y}.$$

(For the proof see Jarník (1955).)

Lemma 2.5 (Young's inequality): Let f be a continuous real, strictly increasing function defined on $\langle 0, \infty)$ such that

$$\lim_{u\to\infty} f(u) = \infty \text{ and } f(0) = 0.$$

Let $g = f^{-1}$. For all $x \in \langle 0, \infty)$, define $\varphi(x) = \int_0^x f(u)\mathrm{d}u$, $\psi(x) = \int_0^x g(v)\mathrm{d}v$. Then for all $a, b \in \langle 0, \infty)$ we have $ab \le \varphi(a) + \psi(b)$ and the equality occurs if and only if $b = \varphi(a)$.

(For the proof see (Kufner, John and Fučík (1977).)

Corollary: Given $p > 1$ and nonnegative real numbers a and b, we have

$$ab \le \frac{a^p}{p} + \frac{b^{p'}}{p'}, \qquad \text{where } p' = \frac{p}{p-1}, \qquad \text{i.e. } \frac{1}{p} + \frac{1}{p'} = 1$$

The equality holds if and only if $a^p = b^{p'}$.

Lemma 2.6 (Hölder's inequality): Let $f \in L^p(\Omega)$ and $g \in L^{p'}(\Omega)$, $p > 1$. Then $f \cdot g \in L^1(\Omega)$ and

$$\left|\int_\Omega f(\boldsymbol{x})g(\boldsymbol{x})\mathrm{d}\boldsymbol{x}\right| \le \int_\Omega |f(\boldsymbol{x})g(\boldsymbol{x})|\,\mathrm{d}\boldsymbol{x} \le \|f\|_p \|g\|_{p'}.$$

The equality $\int_\Omega |f(\boldsymbol{x})g(\boldsymbol{x})|\,\mathrm{d}\mathbf{x} = \|f\|_p \|g\|_{p'}$ holds if and only if there exist nonnegative real numbers a and b, not both simultaneously equal zero, such that $a\,|f(\boldsymbol{x})|^p = b\,|g(\boldsymbol{x})|^{p'}$ almost everywhere on Ω.

(For the proof see Kufner, John and Fučík (1977).)

Lemma 2.7 (Minkowski's inequality): For $1 \le p < \infty$ and $f, g \in L^p(\Omega)$ we have $f + g \in L^p(\Omega)$ and

$$\|f + g\|_p \le \|f\|_p + \|g\|_p.$$

(For the proof see Kufner, John and Fučík (1977).)

Denote by S the set of all functions φ satisfying: $\varphi \in C_o^\infty(R^N)$, $\varphi(\boldsymbol{x}) \geq 0$ for all $\boldsymbol{x} \in R^N$, $\int_{R^N} \varphi(\boldsymbol{x})\mathrm{d}\boldsymbol{x} = 1$, supp $\varphi = \{\boldsymbol{x} \in R^N; |\boldsymbol{x}| \leq 1\}$.

Let $\varepsilon > 0$ and let $\varphi \in S$. For $u \in L^1(\Omega)$ put

$$(R_\varepsilon u)(\boldsymbol{x}) = \varepsilon^{-N} \int_\Omega \varphi(\varepsilon^{-1}(\boldsymbol{x} - \boldsymbol{y}))u(\boldsymbol{y})\mathrm{d}\boldsymbol{y}.$$

Then the mapping R_ε is the so-called mollifier.

Theorem 2.39: Let Ω be a nonempty bounded open subset of R^N. Let $u \in L^p(\Omega)$. Then

(i) $R_\varepsilon u \in C^\infty(R^N)$,
(ii) $\lim_{\varepsilon \to 0_+} \|R_\varepsilon u - u\|_p = 0$.

(For the proof see Kufner, John and Fučík (1977).)

Due to the definition of distribution we can prove the following theorems:

Theorem 2.40: Let $u \in L^p_{\text{loc}}(\Omega)$ and let $D^\alpha u \in L^p_{\text{loc}}(\Omega)$, $p \geq 1$. Then for $\varepsilon > 0$ and $\boldsymbol{x} \in \Omega$

$$(D^\alpha R_\varepsilon u)(\boldsymbol{x}) = (R_\varepsilon D^\alpha u)(\boldsymbol{x})$$

provided dist $(\boldsymbol{x}, \partial\Omega) > \varepsilon$.
(For the proof see Kufner, John and Fučík (1977).)

Theorem 2.41: Let Ω, Ω^* be open sets in R^N, with Ω^* bounded and such that $\overline{\Omega}^* \subset \Omega$. Let $u \in L^p_{\text{loc}}(\Omega)$ and $D^\alpha u \in L^p_{\text{loc}}(\Omega)$, $p \geq 1$. Then

$$\lim_{\varepsilon \to \infty} \|D^\alpha(R_\varepsilon u) - D^\alpha u\|_{L^p(\Omega^*)} = 0.$$

(For the proof see Kufner, John and Fučík (1977).)

Lemma 2.8 (Dubois-Reymond): Suppose that

$$\int_\Omega f(\boldsymbol{x})\mathrm{d}\boldsymbol{x} = 0$$

for every region Ω contained in a domain D. If $f(\boldsymbol{x})$ is continuous for $\boldsymbol{x}$ in D, then $f(\boldsymbol{x}) \equiv 0$ for $\boldsymbol{x}$ in D.

Theorem 2.42 (Reynolds transport theorem): For any sufficiently smooth function g and region $\Omega(t)$,

$$\frac{\mathrm{d}}{\mathrm{d}t} \int_{\Omega(t)} g\mathrm{d}\boldsymbol{x} = \int_{\Omega(t)} \frac{\partial g}{\partial t}\mathrm{d}\boldsymbol{x} + \int_{\partial\Omega(t)} g\boldsymbol{v} \cdot \boldsymbol{n}\mathrm{d}s$$

where $\boldsymbol{n}$ is the unit exterior normal to the boundary $\partial\Omega$ of the region Ω, $\boldsymbol{v}$ is a velocity vector, t is time, $\mathrm{d}\boldsymbol{x}$ is a volume element and $\mathrm{d}s$ is a surface element.

Theorem 2.43 (Poincaré inequality): Assume that Ω is a bounded connected open set of R^N, $N = 2, 3$. Then there exists a constant $c > 0$ such that

$$|v|_{H^{k,p}(\Omega)} \le c \left(\sum_{|\alpha|=k} \|D^\alpha v\|^p_{L^p(\Omega)} + \sum_{|\alpha|\le k-1} \left| \int_\Omega D^\alpha v(\boldsymbol{x}) \mathrm{d}\boldsymbol{x} \right|^p \right)^{1/p}$$

for all $v \in H^{k,p}(\Omega)$.

(For the proof see Kufner, John and Fučík (1977).)

Lemma 2.9 (Gronwall lemma):

(i) Let $f \in L^1(t_o, t_1)$ be a non-negative function, and g and φ be continuous functions on $\langle t_o, t_1 \rangle$. If φ satisfies

$$\varphi(t) \le g(t) + \int_{t_o}^t f(\tau)\varphi(\tau)\mathrm{d}\tau \quad \forall t \in \langle t_o, t_1 \rangle,$$

then

$$\varphi(t) \le g(t) + \int_{t_o}^t f(s)g(s)\exp\left(\int_s^t f(\tau)\mathrm{d}\tau\right)\mathrm{d}s \quad \forall t \in \langle t_o, t_1 \rangle.$$

Moreover if g is non-decreasing, then

$$\varphi(t) \le g(t)\exp\left(\int_{t_o}^t f(\tau)\mathrm{d}\tau\right) \qquad \forall t \in \langle t_o, t_1 \rangle.$$

(ii) Discrete version: Assume that k_n is a non-negative sequence, and that the sequence φ_n satisfies

$$\varphi_o \le g_o,$$

$$\varphi_n \le g_o + \sum_{i=0}^{n-1} p_i + \sum_{i=0}^{n-1} k_i\varphi_i, \qquad n \ge 1.$$

Then φ_n satisfies

$$\varphi_1 \le g_o(1 + k_o) + p_o,$$

$$\varphi_n \le g_o \prod_{i=0}^{n-1}(1 + k_i) + \sum_{i=0}^{n-2} p_i \prod_{j=i+1}^{n-1}(1 + k_j) + p_{n-1}, \qquad n \ge 2.$$

Moreover, if $g_o \ge 0$ and $p_n \ge 0$ for $n \ge 0$, then it follows

$$\varphi_n \le \left(g_o + \sum_{i=0}^{n-1} p_i\right)\exp\left(\sum_{i=0}^{n-1} k_i\right), \qquad n \ge 1.$$

(For the proof see Quarteroni and Valli (1994), Lions and Magenes (1968).)

Definition 2.3 (Fourier transform): Let $u(\boldsymbol{x}) \in L^2(R^N)$ be a complex function. Then the Fourier transform of u in $L^2(R^N)$ is the function $\tilde{u}$ defined by

$$\tilde{u}(\boldsymbol{y}) = \mathscr{F}(u) = \int_{R^N} u(\boldsymbol{x})\exp(-i\boldsymbol{x}\boldsymbol{y})\mathrm{d}\boldsymbol{x},$$

where $\boldsymbol{x} = (x_i)$, $\boldsymbol{y} = (y_i)$, $i = 1, \ldots, N$ are points of R^N, $i = (-1)^{1/2}$ and $\boldsymbol{xy} = \Sigma_{i=1}^N x_i y_i$, $\mathrm{d}\boldsymbol{x} = \mathrm{d}x_1 \ldots \mathrm{d}x_N$. The Fourier transform $\mathscr{F}$ is an isomorphism of $L^2(R^N)$ onto itself, and the inverse Fourier transform is defined almost everywhere by

$$u(\boldsymbol{x}) = \mathscr{F}^{-1}(\tilde{u}) = (2\pi)^{-N} \int_{R^N} \tilde{u}(\boldsymbol{y}) \exp(i\boldsymbol{xy}) \mathrm{d}\boldsymbol{y}.$$

Theorem 2.44: The Fourier transform has the following properties:
Let us define

$$S(R^N) = \{v \mid \boldsymbol{x}^\alpha D^\beta v(\boldsymbol{x}) \in L^2(R^N), \quad \forall \alpha, \beta\}.$$

Clearly, every $v \in S(R^N)$ is a function infinitesimally differentiable on R^N and, since $\boldsymbol{x}^\alpha D^\beta v \in L^2(R^N)$ for every α, β then

$$\left|\boldsymbol{x}^\alpha D^\beta v(\boldsymbol{x})\right| \to 0 \text{ if } |\boldsymbol{x}| \to \infty; \quad \forall \alpha, \beta \text{ and ess } \sup_{\boldsymbol{x} \in R^N} \left|\boldsymbol{x}^\alpha D^\beta v(\boldsymbol{x})\right| < \infty \quad \forall \alpha, \beta.$$

For the convergence property in $S(R^N)$ it follows

$$\lim_{m \to \infty} \text{ ess } \sup_{\boldsymbol{x} \in R^N} \left|\boldsymbol{x}^\alpha D^\beta v_m(\boldsymbol{x})\right| = 0 \quad \text{for any } \alpha, \beta.$$

Then

(i) $u \mapsto \tilde{u}$ is a linear mapping from $S(R^N)$ to $S(R^N)$,

(ii) $\partial \tilde{u}(\boldsymbol{y})/\partial x_j = i y_j \tilde{u}(\boldsymbol{y})$,

(iii) $u(\boldsymbol{x}) = (2\pi)^{-N} \int_{R^N} \tilde{u}(\boldsymbol{y}) \exp(i\boldsymbol{xy}) \mathrm{d}\boldsymbol{y}$

(iv) $\int_{R^N} u(\boldsymbol{x}) \overline{v(\boldsymbol{x})} \mathrm{d}\mathbf{x} = (2\pi)^{-N} \int_{R^N} \tilde{u}(\boldsymbol{y}) \overline{\tilde{v}(\boldsymbol{y})} \mathrm{d}\boldsymbol{y}$, the so-called Parseval's identity. If $u(\boldsymbol{x}) = v(\boldsymbol{x})$ then the Fourier transform is a bounded linear mapping from $S(R^N)$ to $S(R^N)$.

Let $\overline{S} \equiv H^1(R^N)$, then

(v)
$$u \in H^1(R^N) \Leftrightarrow \int_{R^N} (1 + |\boldsymbol{y}|^2) \, |\tilde{u}(\boldsymbol{y})|^2 \, \mathrm{d}\boldsymbol{y} < \infty,$$

$$\|u\|_{1,2} = (2\pi)^{-(1/2)N} \left(\int_{R^N} (1 + |\boldsymbol{y}|^2) \, |\tilde{u}(\boldsymbol{y})|^2 \, \mathrm{d}\boldsymbol{y} \right)^{1/2}.$$

Theorem 2.45: Let $\Omega \subset R^N$ be a domain with a Lipschitz boundary. Then there exist constants $c_i > 0$, $i = 1, 2$, such that

$$c_1 \|u\|_{L^2(\Omega)} \le \|u\|_{-1,2} + \sum_{i=1}^N \|\partial u/\partial x_i\|_{-1,2} \le c_2 \|u\|_{L^2(\Omega)}$$

holds for every $u \in L^2(\Omega)$.

(For the proof see Nečas and Hlaváček (1981).)

As a special case we have the following theorem.

Theorem 2.46: Let $u \in L^2(R^N)$, then

$$\|u\|_{0,2}^2 = \|u\|_{-1,2}^2 + \sum_{i=1}^{N} \|\partial u/\partial x_i\|_{-1,2}^2 .$$

(For the proof see Nečas and Hlaváček (1981).)

2.3. FUNDAMENTALS OF CONVEX ANALYSIS

2.3.1. Convex Sets and Convex Functions

Let V be a scalar or a vector space over R. The extended real line is defined by $\overline{R} = R \cup \{\pm\infty\}$. If u and v are two points of V, then u and v are the so-called end points of the line-segment denoted by $[u, v] = \{\lambda u + (1-\lambda)v \mid 0 \leq \lambda \leq 1\}$.

Definition 2.4: A set $W \subset V$ is said to be convex if and only if for every pair of elements (u, v) of W the segment $[u, v]$ is contained in V.

It is known that a set $W \subset V$ is convex if and only if for every finite subset of elements $\{u_i\}_{i=1}^n$ of W, and for every family of real positive numbers $\lambda_1, \cdots, \lambda_n$ with $\lambda_1 + \cdots + \lambda_n = 1$ the linear combination $\Sigma_{i=1}^n \lambda_i u_i \in W$.

Let V be a real scalar or vector space and let us consider mappings L of W into R, that is the values $+\infty$ and $-\infty$ are allowed to the functions under consideration.

Definition 2.5: Let W be a convex subspace of V, and let L be a mapping of W into R. Then a function L is said to be convex if, for every u and v in W, it holds that

$$L(\lambda u + (1-\lambda)v) \leq \lambda L(u) + (1-\lambda)L(v) \quad \forall\lambda \in \langle 0, 1\rangle, \tag{2.7}$$

whenever the right-hand side is defined.

A function L is said to be strictly convex if it is convex and the strict inequality of Equation 2.7 holds, $\forall u, v \in W$, $u \neq v$ and $\forall\lambda \in (0, 1)$.

The inequality of Equation 2.7 must therefore be valid unless $L(u) = -L(v) = \pm\infty$.

2.3.2. Minimization of Functionals

Let V be a reflexive real Banach space with norm $\|.\|$, let W be a non-empty closed convex subset of V and let $L : W \to R$ be a real function defined on W. Then the minimization problem for the function L relative to the set W is as follows find $u \in W$ such that:

$$L(u) = \inf_{v \in W} L(v).$$

The principal issue is to give sufficient conditions under which the minimization problem has a solution.

Definition 2.6: A functional $L : W \to R$ is said to be weakly lower semicontinuous on W if, for any sequence $\{u_k\} \subset W$ with the property that $\{u_k\}$ converges weakly to $u \in W$, we have

$$\lim_{k\to\infty} \inf L(u_k) \geq L(u), \tag{2.8}$$

and is weakly upper semicontinuous if

$$\lim_{k\to\infty} \sup L(u_k) \leq L(u). \tag{2.9}$$

If $u_k \to u$ strongly in W and Equation 2.8 holds, then L is said to be lower semicontinuous. Upper semicontinuity is defined in an analogous way.

Remark: Recall that $\{u_k\}$ converges strongly to u if and only if $\lim_{k\to\infty} \|u_k - u\| = 0$, and $\{u_k\}$ converges weakly to u whenever $\lim_{k\to\infty} L(u_k) = L(u) \, \forall L \in W'$, where W' is the dual of W.

The term "lower semicontinuity" has the following meaning. The continuity is defined as follows

$$\forall \varepsilon > 0 \quad \exists n_o \quad \forall k > n_o \Rightarrow L(u) - \varepsilon < L(u_k) < L(u) + \varepsilon.$$

If we consider the left-hand side of the last inequalities only, then we obtain lower semicontinuity.

Definition 2.7: A functional $L : W \to R$ is G-differentiable (Gâteaux differentiable) at a point $u \in W$ if there exists a linear functional $DL(u) \in V'$ such that, for every $v \in W$,

$$\lim_{t\to 0} \frac{\partial}{\partial t} L(u + tv) = \langle DL(u), v\rangle \equiv DL(u, v), \tag{2.10}$$

where t is an arbitrary positive number and $\langle .,. \rangle$ denotes duality pairing on $V' \times V$, which means that if $f \in V'$ then we write $g(v) = \langle g, v\rangle$.

Higher order derivatives are defined in an analogous way. We call $DL(u)$ the gradient of L at u and $DL(u, v)$ the Gâteaux derivative of L at u in the direction v.

Definition 2.8: The functional L is said to be coercive if

$$\lim L(u) = +\infty \text{ as } |u| \to \infty. \tag{2.11}$$

Theorem 2.47: Let L be a functional that maps V into R^1. Let L be coercive and weakly lower semicontinuous. Then L attains its minimum in V.
(For the proof see Nečas and Hlaváček (1981).)

Theorem 2.48 (generalized Weierstrass minimization theorem)**:** Let L be a real functional defined on a non-empty closed convex subset W of a reflexive Banach space V. Moreover, let the following conditions hold:

(i) L is weakly lower semicontinuous on W and W is bounded, or
(ii) L is weakly lower semicontinuous and coercive on W.

Then L attains its minimum on W, i.e. there exists at least one $u \in W$ such that

$$L(u) = \inf_{v\in W} L(v) \tag{2.12}$$

or

$$L(u) \leq L(v) \quad \forall v \in W, \tag{2.13}$$

holds, where Equation 2.13 represents the equivalent form to Equation 2.12.

Theorem 2.49: Let $L : V \to R$ satisfy the following conditions:

(i) L is convex,

(ii) L is G-differentiable on V.

Then L is weakly lower semicontinuous on V.

Theorem 2.50: Let $L : W \to R$ be G-differentiable at every point $u \in W$, i.e. $DL(u, v)$ exists for all $u \in$W. Suppose that:

(i) $DL(u + w, w) - DL(u, w) \geq 0$ for all $u, w \in W$,

(ii) $DL(u, w)$ is continuous on any line segment in the space W in the variable u, i.e. $L(t) = DL(u + tv, w)$ is a continuous function of the variable t for all elements $u, v, w \in W$.

Then L is weakly lower semi-continuous.

Theorem 2.51: Let $L : W \to R$ be G-differentiable. Then the following conditions are equivalent:

(i) L is convex,

(ii) $\langle DL(u) - DL(v), u - v\rangle \equiv DL(u - v, u - v) \geq 0\,\forall u, v \in W$,

(iii) $L(v) - L(u) \geq \langle DL(u), v - u\rangle \equiv DL(u, v - u)\,\forall u, v \in W$.
Moreover, if L is twice G-differentiable, then the above conditions are also equivalent to

(iv) $\langle D^2L(u)(v), v\rangle \equiv D^2L(u; v, v) \equiv \lim_{t\to 0} \frac{\partial}{\partial t} DL(u + tv, v) \geq 0\,\forall v \in W$.

From all these results we obtain the following theorem:

Theorem 2.52: Let L be a real, coercive, convex, G-differentiable functional defined on a possibly unbounded, nonempty, closed convex subset W of a reflexive Banach space V. Then L attains its minimum on W. If, in addition, L is strictly convex then the minimizer is unique.

Remark: If $L : W \to R$ is G-differentiable, then its gradient $DL(.)$ is a map from V into V'. If $DL(.) : V \to V'$ is the gradient of the functional L, then $DL(.)$ is said to be potential. The second derivative $D^2L(u; v, v)$ is bilinear on V and $D^2L : V \times V \to$V'. If DL satisfies condition (ii) of Theorem 2.51, then DL is a monotone operator and hence the gradient of any convex G-differentiable functional is monotone; if moreover the strict inequality in (ii) holds for $u \neq v$, then DL is strictly monotone. Similarly, if L is strictly convex and G-differentiable, then DL is strictly monotone. Suppose $L : V \to \overline{R}$ is convex but not G-differentiable on all of V. Then many of the ideas concerning the minimization of G-differentiable functions can be extended if L is sub-differentiable at points in V. The sub-differential of L is defined as follows.

Definition 2.9: The set $\partial L(u) \subset V'$ of all linear functionals f such that

$$L(v) - L(u) \geq \langle f, v - u\rangle \quad \forall v \in V, \tag{2.14}$$

with $|L(u)| < \infty$, is called the sub-differential of L at u and any $f \in \partial L(u)$ is a sub-gradient of L at u.

Theorem 2.53: Let W be a subset of a normed linear space V and let L be a G-differentiable functional mapping W into R. If u is a minimizer of L in W, then u may be characterized in one of the following ways:

(i) if W is a non-empty closed convex subset of V, then

$$\langle DL(u), v-u\rangle \equiv DL(u, v-u) \geq 0 \quad \forall v \in W, \tag{2.15}$$

(ii) if W is a non-empty closed convex subset of V and $u \in \text{int } W$ (where int $W \equiv$ interior of W), then

$$\langle DL(u), v\rangle \equiv DL(u, v) = 0 \quad \forall v \in W, \tag{2.16}$$

(iii) if W is a non-empty closed convex cone with vertex w, then

$$\langle DL(u), u-w\rangle \equiv DL(u, u-w) = 0 \text{ and } \langle DL(u), v-w\rangle \equiv DL(u, v-w) \geq 0 \quad \forall v \in W, \tag{2.17}$$

(iv) if W is a linear subspace translated by a vector (or scalar) w with respect to the origin (i.e. the so-called variety with respect to w), then

$$\langle DL(u), v\rangle \equiv DL(u, v) = 0 \quad \forall v \in V \text{ such that } v-w \in W, \tag{2.18}$$

(v) if W is a linear subspace of V, then

$$\langle DL(u), v\rangle \equiv DL(u, v) = 0 \quad \forall v \in W. \tag{2.19}$$

Remark: The set W is called the convex cone with vertex at the origin, if it is closed and moreover if elements $u, v \in W$ then also elements $tu + su \in W$, $t \geq 0$, $s \geq 0$.

Remark: The inequality of Equation 2.15 represents a variational inequality and it characterizes a minimizer of any G-differentiable function defined on a non-empty closed convex set W. The convexity of W is essential.

Theorem 2.54: Let W be non-empty, closed, and convex and suppose that $L : W \subset V \to R$ is of the form $L = L_o + \Phi$, where L_o and Φ are convex and lower semi-continuous and L_o is G-differentiable on K. Then, if u is a minimizer on W

$$\begin{aligned} &\langle DL_o(u), v-u\rangle + \Phi(v) - \Phi(u) \equiv DL_o(u, v-u) + \Phi(v) - \Phi(u) \geq 0 \quad \forall v \in W, \\ &\langle DL_o(v), v-u\rangle + \Phi(v) - \Phi(u) \equiv DL_o(u, v-u) + \Phi(v) - \Phi(u) \geq 0 \quad \forall v \in W, \end{aligned} \tag{2.20}$$

Conversely, if either Equation 2.20 holds for $u \in W$, then u is a minimizer of L.

Remark: An example of a function Φ is the function $\Phi : H^1(\Omega) \to R$ defined by

$$\Phi(v) = \int_\Omega g\,|v|\,\mathrm{d}x,$$

where g is a given smooth function. The function Φ is convex and lower semi-continuous, but not G-differentiable.

Remark: Proofs of the theorems given above can be found in Céa (1971), Ekeland and Temam (1976).

2.3.3. Minimization of Quadratic Functionals

Special cases of the theory presented in the previous section are abstract elliptic problems.

Let V be a real Hilbert space with scalar product $(.,.)$ and with norm $\|.\|^2 = (.,.)$. Let $W \subseteq V$ be a nonempty, convex, bounded subset. Let V' be a dual space of V. Let $f \in V'$, $u \in V$ then

$$\|f\|_* = \sup_{\substack{v \neq 0 \\ v \in V}} \frac{|(f, v)|}{\|v\|}$$

is the norm of dual space V' of V. Let $a(.,.)$ be a bilinear form, $a : W \times W \to R$.

Definition 2.10: A bilinear form $a : W \times W \to R$ is said to be symmetric on a subset W if

$$a(u, v) = a(v, u) \quad \forall u, v \in W,$$

and is said to be bounded on W if there exists constant $c_o > 0$ such that

$$|a(u, v)| \leq c_o \|u\| \|v\| \quad \forall u, v \in W,$$

and is said to be V-elliptic on W, if there exists constant $c > 0$ such that

$$a(v, v) \geq c \|v\|^2 \quad \forall v \in W.$$

Definition 2.11: Let $W \subseteq V$ be a nonempty, convex and bounded subset of V, let $a : W \times W \to R$ be a bilinear form and let $f \in V'$. Then $u \in W$ is said to be a solution of an elliptic problem if

$$a(u, v - u) \geq (f, v - u) \quad \forall v \in W \tag{2.21}$$

If $W \not\subseteq V$ and W is not a linear space, then Equation 2.21 represents an abstract variational inequality. If $W = V$, then the equation represents an abstract variational equality. In this last case, Equation 2.21 can be rewritten in the form

$$a(u, v) = (f, v) \quad \forall v \in W. \tag{2.22}$$

Theorem 2.55 (Lax-Milgram)**:** Let $a : W \times W \to R$ be a V-elliptic and bounded on W bilinear form, and let $f \in V'$. Then there exists one and only one solution of Equation 2.21. Moreover,

$$\|u\| \leq c^{-1} \|f\|_* .$$

(For the proof see Glowinski, Lions and Trémoliéres (1976).)

If the bilinear form is symmetric, bounded and V-elliptic, then Equation 2.21 is equivalent to the following problem.

Find $u \in W$ such that

$$L(u) \leq L(v) \quad \forall v \in W, \tag{2.23}$$

where

$$L(v) = \tfrac{1}{2} a(v, v) - (f, v). \tag{2.24}$$

The problem of Equation 2.23 is equivalent to the following problem:

Problem ($\mathscr{P}$): find $u \in W$ such that

$$L(u) = \min_{v \in W} L(v). \tag{2.25}$$

A direct application of the results of the previous section will be given in the next theorem.

Let L be a functional defined by Equation 2.24 $\forall v \in V$, where V is a Hilbert space, $a(.,.)$ a symmetric, continuous, V-elliptic bilinear form from $V \times V$ into R and f is a bounded linear functional on V, i.e. for every $u, v \in V$

$$a(u, v) = a(v, u), \quad |a(u, v)| \le c_o \|u\| \|v\|, \quad a(u, v) \ge c \|u\|^2, \tag{2.26}$$

where c_o and c are positive constants. Let Q be another Hilbert space and let B be a continuous linear operator from V onto Q such that

$$B \in \mathscr{M}(V, Q), \quad \text{range}(B) = Q, \quad |Bv| \le C \|v\| \quad \forall v \in V, \tag{2.27}$$

where $|.|$ denotes the norm on Q, and C is a constant. The transpose $B^* (\equiv B^T)$ of B is defined by

$$B^* \in \mathscr{M}(Q', V'), \quad [w, Bv] = \langle B^* w, v\rangle \quad \forall w \in Q', \quad \forall v \in V', \tag{2.28}$$

where $[.,.]$ and $\langle .,.\rangle$ denote duality pairing on $Q' \times Q'$ and $V' \times V'$, respectively. The classical closed range theorem (see For example Yosida (1974), p. 205-208) and Equation 2.27 give

$$\exists \alpha \in R, \quad \alpha > 0$$

such that
$$\sup_{v \in V - \{0\}} \frac{[w, Bv]}{\|v\|} \ge \alpha |w|_* \quad \forall w \in Q', \tag{2.29}$$

where $|.|_*$ denotes the norm on the dual Q' of Q, i.e. B^* is bounded below as a linear operator from Q' into V'.

Theorem 2.56: Let Equations 2.24 to 2.27 hold and let W be a non-empty closed and convex set in V. Then there exists a unique solution u to the minimization problem of Equation 2.23. Moreover, this minimizer is the unique solution of the variational inequality

$$u \in W : a(u, v - u) \ge (f, v - u) \quad \forall v \in W. \tag{2.30}$$

Proof: Since a(.,.) is bilinear form, then

$$(1 - \Theta)L(u) + \Theta L(v) = L((1 - \Theta)u + \Theta v) + \tfrac{1}{2}(1 - \Theta)\Theta a(u - v, u - v) \tag{2.31}$$

for every $u, v \in V$ and $0 \le \Theta \le 1$. Due to Equation 2.26, L is strictly convex and Gâteaux differentiable, so that for every $u, v \in V$

$$\lim_{\varepsilon \to 0} \frac{\partial}{\partial \varepsilon} L(u + \varepsilon v) = DL(u, v) = a(u, v) - f(v) \le (c_o \|u\| + \|f\|_*) \|v\|.$$

The assertions of this theorem thus follow immediately from Theorems 2.52 and 2.54.

It can be noted that Equation 2.31 implies the relation

$$L(v) - L(u) = DL(u, v - u) + \tfrac{1}{2}a(u - v, u - v). \tag{2.32}$$

Remark: By the operator B in Equation 2.27 we can introduce an abstract form of an inequality constraint on minimizers of the functional L. Let B be defined by Equation 2.27 and $b \in Q$, then we define the set $W \subset V$ by

$$W = \{v \mid v \in V, \quad Bv - b \leq 0 \text{ in } Q\}.$$

Since B is linear and continuous, and since the positive cone in Q is closed, then W is a closed convex set in V. We say that W represents the constraint set in the minimization problem of Equation 2.23.

If the bilinear form a(.,.) is nonsymmetric, then instead of Equation 2.25 we have

$$|a(u, v)| \leq c_o \|u\| \|v\|, \quad a(u, v) \geq c \|u\|^2, \quad \forall u, v \in V. \tag{2.33}$$

Then we have the following theorem:

Theorem 2.57: Let Equations 2.24 and 2.33 hold, and let W be a non-empty closed and convex set in V. Then there exists a unique solution u to the variational inequality

$$u \in W: \quad a(u, v - u) \geq (f, v - u) \quad \forall v \in W. \tag{2.34}$$

(For the proof see Kinderlehrer and Stampacchia (1980), Kikuchi and Oden (1988).)

2.3.4 Theory of Duality, Saddle Points and Lagrange Multipliers

Variational problems discussed in the previous section arise in constrained minimization problems, i.e. problems where we need to find a minimum of a functional L subject to the constraint that the minimizers lie in a convex set W. From the convex theory it is known that such problems can be formulated as saddle point problems using the method of Lagrange multipliers. Sufficient conditions for the existence of a saddle point can be found in Céa (1971) or in Ekeland and Temam (1976).

Let V and Q' be Banach spaces with norms $\|.\|$ and $|.|_*$, respectively, and W and Λ be nonempty convex subsets of V and Q', respectively. Let $\mathscr{L} : W \times \Lambda \to R$ be a real functional defined on $W \times \Lambda$.

Definition 2.12: A point $(u_s, \lambda) \in W \times \Lambda$ is said to be a saddle point of a functional $\mathscr{L}(v) \in W \times \Lambda \to R^1$ if and only if

$$\mathscr{L}(u_s, \mu) \leq \mathscr{L}(u_s, \lambda) \leq \mathscr{L}(v, \lambda) \quad \forall (v, \mu) \in W \times \Lambda. \tag{2.35}$$

Theorem 2.58: The functional $\mathscr{L}$ defined on $W \times \Lambda$ with real values possesses a saddle point (u_s, λ) on $W \times \Lambda$ if and only if

$$\max_{\mu \in \Lambda} \inf_{v \in W} \mathscr{L}(v, \mu) = \min_{v \in W} \sup_{\mu \in \Lambda} \mathscr{L}(v, \mu) \tag{2.36}$$

and this number is equal to $\mathscr{L}(u_s, \lambda)$.

(For the proof see Céa (1971), Ekeland and Temam (1976).)

The problem investigated by Equation 2.25 can be rewritten in the following equivalent form.

Problem ($\mathscr{P}_{\mathscr{L}}$): Find $u \in K$ such that

$$L(u) = \min_{v \in W} \sup_{\mu \in \Lambda} \mathscr{L}(v, \mu). \tag{2.37}$$

Theorem 2.59: Let V and Z be two reflexive Banach spaces. Assume

(i) $W \subset V$ is convex and nonempty,

(ii) $\Lambda \subset Z$ is convex, closed and nonempty,

(iii) $\forall v \in W,\ \mu \to \mathscr{L}(v, \mu)$ is concave and upper semicontinuous (u.s.c),

(iv) $\forall \mu \in \Lambda, v \to \mathscr{L}(v, \mu)$ is convex and lower semicontinuous (l.s.c),

(v) $\exists \mu_o \in \Lambda$ such that

$$\lim_{\substack{v \in W \\ \|v\| \to \infty}} \mathscr{L}(v, \mu_o) = +\infty,$$

(vi) $\exists u_o \in W$ such that

$$\lim_{\substack{\mu \in \Lambda \\ \|\mu\| \to \infty}} \mathscr{L}(u_o, \mu) = -\infty.$$

Then $\mathscr{L}$ possesses at least one saddle point of $\mathscr{L}$ and

$$\mathscr{L}(u_s, \lambda) = \min_{v \in W} \sup_{\mu \in \Lambda} \mathscr{L}(v, \mu) = \max_{\mu \in \Lambda} \inf_{v \in W} \mathscr{L}(v, \mu).$$

(For the proof see Céa (1971), Ekeland and Temam (1976).)

Theorem 2.60: We assume, in addition to (i)-(iv) of the previous theorem, that

(i) $\forall v \in W,\ \mu \to \mathscr{L}(v, \mu)$ is Gateaux-differentiable,

(ii) $\forall \mu \in \Lambda,\ v \to \mathscr{L}(v, \mu)$ is Gateaux-differentiable.

Then $(u_s, \lambda) \in W \times \Lambda$ is the saddle point of $\mathscr{L}$ if and only if

$$\langle \partial \mathscr{L}(u_s, \lambda)/\partial v, v - u_s \rangle \geq 0 \quad \forall v \in W,$$

$$\langle \partial \mathscr{L}(u_s, \lambda)/\partial \lambda, \mu - \lambda \rangle \leq 0 \quad \forall \mu \in \Lambda,$$

where $\langle .,. \rangle$ denotes a bilinear pairing in the usual sense.

(For the proof see Céa(1971), Ekeland and Temam (1976).)

Lemma 2.10: Let V be a real Hilbert space, and let $M \subset V$ be convex, closed and nonempty. Let an operator P be a projection V onto M and let $u \in V$ be arbitrary. Then the following presumptions

(i) $w = Pu$,

(ii) $((u - w, v - w)) \leq 0 \qquad \forall v \in M$,

where $((.,.))$ denotes a scalar product in V, are equivalent.

(For the proof see Nečas (1967).)

2.3.5 Penalty Methods

Penalty techniques provide an alternative approach to constrained optimization problems. Constraints are evoked by an abstraction of constraints on the admissible displacements, such as those restricting the values $v \to v \cdot n$ of the normal displacement in contact problems or the change in volume (i.e. div $v = 0$) in problems involving incompressible materials, as well as in problems involving a magnetic field (i.e. div $B = 0$). The idea of the penalty technique is as follows.

Let L be the functional $L : V \to R$. We minimize L subject to the constraint that the minimizers u belong to a convex set $W \subset V$. We append to the functional L a "penalty functional" P which increases in magnitude according to how severely the constraint is violated.

Let L be a coercive weakly lower semicontinuous functional defined on reflexive Banach space V. Let W be a non-empty, closed, convex subset of V, $W \subset V$. The penalty technique consists of introducing a new functional L_ε, depending on a positive real parameter ε, such that

$$L_\varepsilon(v) = L(v) + \varepsilon^{-1}P(v), \quad \varepsilon > 0, \tag{2.38}$$

where $P : V \to R$ is a penalty functional satisfying:

$$\begin{aligned} &\text{(i)} \quad P : V \to R \text{ is weakly lower semicontinuous,} \\ &\text{(ii)} \quad P(v) \geq 0,\ P(v) = 0 \text{ if and only if } v \in W, \\ &\text{(iii)} \quad P : V \to R \text{ is } G-\text{differentiable on } V. \end{aligned} \tag{2.39}$$

Due to the properties of L and P, the functional $L_\varepsilon(v)$ is coercive and weakly lower semicontinuous. Thus, due to the generalized Weierstrass theorem (see Theorem 2.48), for each $\varepsilon > 0$ there exists a $u_\varepsilon \in V$ which minimizes the functional L_ε:

$$F_\varepsilon(u) = \inf_{v \in V} L_\varepsilon(v). \tag{2.40}$$

Moreover, if L and P are G-differentiable, then u_ε is defined by

$$\langle DL(u_\varepsilon), v\rangle + \varepsilon^{-1}\langle DP(u_\varepsilon), v\rangle \equiv DL(u_\varepsilon, v) + \varepsilon^{-1}DP(u_\varepsilon, v) = 0 \quad \forall v \in V. \tag{2.41}$$

It must be proved that the solutions of Equations 2.40 or 2.41 generate a sequence u_ε converging to u of the original minimization problem in Equation 2.25. Since u_ε minimizes L_ε, then

$$L(u_\varepsilon) + \varepsilon^{-1}P(u_\varepsilon) \leq L(v) + \varepsilon^{-1}P(v) \quad \forall v \in V.$$

If $v \in W$, $P(v) = 0$ and since $\varepsilon^{-1}P(u_\varepsilon) \geq 0$, then

$$L(u_\varepsilon) \leq L(v) \quad \forall v \in W. \tag{2.42}$$

Let $\{u_\varepsilon\}$ be a sequence of solutions of Equation (2.40) for $\varepsilon \to 0$. As L is coercive, this implies that there exists a constant $c > 0$, independent of ε, such that $\|u_\varepsilon\| < c$. Since V is reflexive, then from the sequence $\{u_\varepsilon\}$ the subsequence, denoted also by $\{u_\varepsilon\}$, can be taken such that $u_\varepsilon \to u$ weakly in V, $u \in$V.

If P is nonnegative, then due to the lower semicontinuity of L we obtain

$$\liminf_{\varepsilon\to 0} L_\varepsilon(u_\varepsilon) \geq \lim_{\varepsilon\to 0} L(u_\varepsilon) \geq L(u).$$

We shall now show that the weak limit u lies in W. For this, let v_o be a fixed element of W. Then

$$L(u_\varepsilon) + \varepsilon^{-1}P(u_\varepsilon) \leq L(v_o) + \varepsilon^{-1}P(v_o) = L(v_o).$$

Hence

$$P(u_\varepsilon) \leq \varepsilon(L(v_o) - L(u_\varepsilon)) \leq c\varepsilon,$$

where c is a constant. Hence

$$P(u) \leq \liminf_{\varepsilon\to 0} P(u_\varepsilon) \leq 0$$

which, according to condition (ii) of Equation 2.39 implies that $u \in W$. Thus we proved the following theorem.

Theorem 2.61: Let $L : V \to R$ be a coercive and weakly lower semicontinuous functional, and let W be a non-empty, closed, convex subset of the reflexive Banach space V. Let $P : V \to R$ be a penalty functional satisfying Equation 2.39 (i)-(ii). Then, for every $\varepsilon > 0$, there exists a solution u_ε to Equation 2.40 obtained as $\varepsilon \to 0$. Moreover, there exists a subsequence $\{u_\varepsilon\}$ of solutions of Equation 2.40 with such solutions converging weakly to $u \in V$, where

$$u \in W : L(u) = \inf_{v\in W} L(v). \tag{2.43}$$

Moreover, if L and P are G-differentiable, the u_ε satisfies Equation 2.41 for each $\varepsilon > 0$ and the limit u satisfies the variational inequality

$$\langle DL(u), v - u\rangle \equiv DL(u, v - u) = 0 \quad \forall v \in W. \tag{2.44}$$

2.4. FUNDAMENTALS OF VARIATIONAL EQUALITIES AND INEQUALITIES AND THEIR NUMERICAL APPROXIMATIONS

2.4.1. Elliptic Variational Inequalities and Equalities the Continuous Problem

Let us denote by V a real Hilbert space with inner product $(.,.)$ and associated norm $\|.\|$, by V' the dual space of V and by K a closed convex non-empty subset of V. Furthermore, by $a(.,.) : V \times V \to R$ we denote a bilinear continuous and V-elliptic form on $V \times V$, by $S : V \to R$ a continuous linear functional and by $j(.) : V \to R \cup \{+\infty\}$ a convex lower semi-continuous (l.s.c.) and proper functional. We say that the functional $j(.) : V \to R \cup \{+\infty\}$ is proper if $j(v) > -\infty$, $\forall v \in V$ and $j \neq +\infty$.

We shall investigate two variational problems:

Problem $(\mathscr{P}_1)_V$: Find $u \in V$ such that u is a solution of the variational inequality

$$a(u, v - u) \geq S(v - u) \quad \forall u, v \in K. \tag{2.45}$$

Problem $(\mathscr{P}_2)_v$: Find $u \in V$ such that u is a solution of the problem

$$a(u, v-u) + j(v) - j(u) \geq S(v-u) \quad \forall u, v \in V. \tag{2.46}$$

Remark: If $K = V$ and $j(v) \equiv 0$, then problems $(\mathscr{P}_1)_v$ and $(\mathscr{P}_2)_v$ represent classical variational elliptic problems, i.e. classical variational equalities

$$a(u, v) = S(v) \quad \forall u, v \in V. \tag{2.47}$$

From the theory of variational inequalities it is known that the problem $(\mathscr{P}_1)_v$ is a particular case of the problem $(\mathscr{P}_2)_v$ if the functional $j(.)$ in $(\mathscr{P}_2)_v$ is replaced by the indicator functional I_k of K defined by $I_k(v) = 0$ for $v \in K$, and $I_k = +\infty$ for $v \notin K$. The indicator functional I_k is a convex l.s.c. and proper functional. Then the problem $(\mathscr{P}_1)_v$ is equivalent to the problem of finding $u \in V$ such that

$$a(u, v-u) + I_k(v) - I_k(u) \geq S(u-v) \quad \forall v \in V. \tag{2.48}$$

In the next section we shall prove the existence theorems based on the results of Lions and Stampacchia (1967), Ciarlet (1978), Glowinski *et al.* (1976), Glowinski (1984).

Theorem 2.62: The problem $(\mathscr{P}_1)_v$ has a unique solution.

Proof: (a) Existence: From the Riesz theorem there exist an operator $A \in \mathscr{L}(v, v)$ and the functional $f \in V$ such that $a(u, v) = (Au, v) \forall u, v \in V$ and $S(v) = (f, v) \forall v \in$V. Then the problem $(\mathscr{P}_1)_v$ is equivalent to finding $u \in V$ such that

$$(u - \rho(Au - f) - u, v - u) \leq 0 \quad \forall u, v \in K, \rho > 0,$$

which then is equivalent to finding u such that

$$u = P_k(u - \rho(Au - f)), \qquad \rho > 0,$$

where P_k is the projection operator, $P_k : V \to K$ in the $\|.\|$ norm, i.e. the operator P_k is a contraction. Let us define

$$z(v) = P_k(v - \rho(Av - f)).$$

Let $v_1, v_2 \in V$. Then

$$\begin{aligned}\|z(v_1) - z(v_2)\|^2 &\leq \|v_2 - v_1\|^2 + \rho^2 \|A(v_2 - v_1)\|^2 - 2\rho a(v_2 - v_1, v_2 - v_1)\\ &\leq (1 - 2\alpha\rho + \rho^2 \|A\|^2) \|v_2 - v_1\|^2 .\end{aligned}$$

Hence z is a strict contraction mapping if $0 < \rho < 2\alpha \|A\|^{-2}$. For $\rho \in \langle 0, 2\alpha \|A\|^{-2}\rangle$ the fixed-point problem has a unique solution, which completes the proof of the existence of the problem $(\mathscr{P}_1)_v$.

(b) Uniqueness: Let u_1, u_2 be two solutions of the problem $(\mathscr{P}_1)_v$. Then

$$a(u_1, v - u_1) \geq S(v - u_1) \quad \forall v, u_1 \in K, \tag{2.49}$$

$$a(u_2, v - u_2) \geq S(v - u_2) \quad \forall v, u_2 \in K. \tag{2.50}$$

Let us put $v = u_2$ in Equation 2.49 and $v = u_1$ in Equation 2.50 and then add both inequalities. Then, due to the V-ellipticity of the bilinear form $a(u, v)$, we have

$$\alpha \|u_2 - u_1\|^2 \leq a(u_2 - u_1, u_2 - u_1) \leq 0.$$

Hence $u_1 = u_2$, which completes the proof.

Remark: The algorithm for the numerical solution of the problem $(\mathscr{P}_1)_v$ will be based on the proof of the previous theorem, i.e. let $u^o \in V$ be an arbitrary. Then for $n \geq 0$, assuming that u^n is known, define u^{n+1} by

$$u^{n+1} = P_k(u^n - \rho(Au^n - f)). \tag{2.51}$$

Moreover, $u^n \to u$ strongly in V, where u is the solution of the problem $(\mathscr{P}_1)_v$.

If the bilinear form $a(u, v)$ is symmetric, then we have the following theorem.

Theorem 2.63: Let the bilinear form $a(u, v)$ in Equation 2.45 be symmetric. Then the problem $(\mathscr{P}_1)_v$ has a unique solution.

Proof: Let $L_o : V \to R$ be the functional defined by

$$L_o(v) = \tfrac{1}{2}a(v, v) - S(v).$$

Then

(i) the functional $L_o(v)$ is coercive, i.e.

$$\lim_{\|v\|\to+\infty} L_o = +\infty \text{ as } L_o(v) = \tfrac{1}{2}a(v, v) - S(v) \geq \tfrac{1}{2}\alpha \|v\|^2 - \|f\| \|v\|,$$

(ii) the functional $L_o(v)$ is strictly convex.

As $S(v)$ is linear, it is sufficient to prove that the functional $v \to a(v, v)$ is strictly convex. We have $0 < a(v - u, v - u) = a(u, u) + a(v, v) - 2a(u, v)$, $u, v \in$V. Hence $2a(u, v) < a(u, u) + a(v, v)$. Let $0 < t < 1$, $u, v \in V$. Then $a(tu + (1 - t)v, tu + (1 - t)v) = t^2a(u, u) + 2t(1 - t)a(u, v) + (1 - t^2)$ $a(v, v) < ta(u, u) + (1 - t)a(v, v)$. Hence the functional $v \to a(v, v)$ is strictly convex.

Due to the convex analysis (see Section 2.3 or Ekeland and Temam (1976), Céa (1971)) the minimization problem of finding u such that

$$L_o(u) \leq L_o(v) \quad \forall u, v \in K \tag{2.52}$$

has a unique solution, which completes the proof.

Remark: The problem of Equation 2.51 is equivalent to the problem of finding u such that

$$(L'_o(u), v - u) \geq 0 \quad \forall u, v \in K,$$

where $L'_o(u)$ is the Gâteaux derivative of L_o at u. Since for $a(u, v)$ symmetric $(L'_o(u), v) = a(u, v) - S(v)$ then $L'_o(u) = Au - f\, \forall u \in V$. Then Equation 2.51 will be of the following form, known as the so-called gradient-projection method with ρ being constant:

$$u^{n+1} = P_k(u^n - \rho(L'_o(u^n))).$$

Theorem 2.64: Problem $(\mathscr{P}_2)_v$ with symmetric continuous bilinear form has a unique solution. Moreover, the variational formulation in Equation 2.46 is equivalent to the formulation

$$L(u) \leq L(v) \quad \forall u, v \in V, \tag{2.53}$$

where $L(v) = L_o(v) + j(v)$, $L_o(v) = \frac{1}{2}a(v, v) - S(v)$.

Proof: Let $a : V \times V \to R$ be a symmetric continuous bilinear V-elliptic form, and $\alpha > 0$ be a V-ellipticity constant. Let $f \in V'$ and $j : V \to R$ be a convex, l.s.c. proper functional. Let $L(v) =$

$\frac{1}{2}a(v, v) + j(v) - S(v)$. Then the problem $(\mathscr{P}_2)_v$ is equivalent to finding u such that

$$L(u) \leq L(v) \quad \forall u, v \in V,$$

where $L(v) = L_o(v) + j(v)$, $L_o(v) = \frac{1}{2}a(v, v) - S(v)$.

Let $0 < t \leq 1$ and let u be a solution of Equation 2.53. Then for $\forall v \in V$ we have

$$L(u) \leq L(u + t(v - u)).$$

Then

$$\begin{aligned} 0 &\leq L_o(u + t(v - u)) - L_o(u) + j(u + t(v - u)) - j(u) \\ &\leq L_o(u + t(v - u)) - L_o(u) + t(j(v) - j(u)) \quad \forall v \in V, \end{aligned}$$

as the functional $j(v)$ is convex. Hence, dividing by t and taking the limit for $t \to 0$, we obtain

$$0 \leq (L_o'(u), v - u) + j(v) - j(u) \quad \forall v \in V.$$

As

$$(L_o'(u), v - u) = a(u, v) - S(v) \quad \forall u, v \in V,$$

then

$$a(u, v - u) + j(v) - j(u) \geq S(v - u) \quad \forall v \in V,$$

which proves the necessity that the problem $(\mathscr{P}_2)_v$ is equivalent to Equation 2.46.

To prove sufficiency of Equation 2.46 we shall assume that u is a solution of the equation. Then for $v \in V$

$$L(v) - L(u) = \tfrac{1}{2}[a(v, v) - a(u, u)] + j(v) - j(u) - S(v - u).$$

But

$$a(v, v) - a(u + v - u, u + v - u) = a(u, u) + 2a(u, v - u) + a(u - v, u - v).$$

Hence

$$L(v) - L(u) = a(u, v - u) + j(v) - j(u) - S(v - u) + \tfrac{1}{2}a(v - u, v - u).$$

Since u is a solution of Equation 2.46 and since $a(v - u, v - u) \geq 0$, then we obtain that $L(v) - L(u) \geq 0$ and therefore u is a solution of Equation 2.53.

To prove the existence and uniqueness of the solution u, we see that since $a(v, v)$ is strictly convex, the functional $j(v)$ is convex and $S(v)$ is linear, the functional $L(v)$ is strictly convex and l.s.c., $j(v)$ is l.s.c. and $S(v)$ is continuous. Moreover (see Ekeland and Temam (1976)), since $j(v)$ is convex, l.s.c., and proper, there exists $\lambda \in V'$, $\mu \in R$ such that

$$j(v) = \lambda(v) + \mu,$$

then

$$\begin{aligned} L(v) &= \tfrac{1}{2}\alpha \|v\|^2 - \|\lambda\| \|v\| - \|f\| \|v\| + \mu \\ &= \left[\left(\tfrac{1}{2}\alpha\right)^{1/2} \|v\| - 2^{-(1/2)}\alpha^{-(1/2)}(\|\lambda\| + \|f\|)\right]^2 + \mu - (2\alpha)^{-1}(\|\lambda\| + \|f\|)^2. \end{aligned}$$

Hence in the limit $\|v\| \to +\infty$, we obtain

$$L(v) \to +\infty \text{ as } \|v\| \to +\infty.$$

Due to Ekeland and Temam (1976), Céa (1971) there exists a unique solution of the problem of Equation 2.53 and therefore of Equation 2.46 also.

2.4.2. Elliptic Variational Inequalities and Equalities. The Approximate Problem

We shall approximate the space V by its FEM approximation V_h or generally by a family $\{V_h\}_h$ of closed subspaces of V, with V_h a closed subspace of V and the set K by its FEM approximation $K_h \subset V_h$, $\forall h$ or generally by a family $\{K_h\}_h$ of closed convex nonempty subspaces of V with $K_h \subset V_h$, $\forall h$, but in general $K_h \not\subset K$. Moreover, K_h satisfies:

(i) if $\{v_h\}_h$ is such that $v_h \in K_h$, $\forall h$ and $\{v_h\}_h$ is bounded in V, then the weak cluster point of $\{v_h\}_h$ belongs to K;

(ii) there exists $X \subset V$, $\overline{X} = K$ and $r_h : X \to K_h$ such that $\lim_{h\to 0} r_h v = v$ strongly in V, $\forall v \in X$.

It is evident that if $K_h \subset K$, $\forall h$, then (i) is satisfied because K is weakly closed.

Problem $(\mathscr{P}_1)_{vh}$: Find $u_h \in K_h$ such that u_h is a solution of the problem

$$a(u_h, v_h - u_h) \geq S(v_h - u_h) \quad \forall v_h \in K_h. \tag{2.54}$$

Problem $(\mathscr{P}_2)_{vh}$: Find $u_h \in V_h$ such that u_h is a solution of the problem

$$a(u_h, v_h - u_h) + j_h(v_h) - j_h(u_h) \geq S(v_h - u_h) \quad \forall v_h \in V_h. \tag{2.55}$$

Theorem 2.65: The problem $(\mathscr{P}_1)_{vh}$ has a unique solution.

Proof: The proof is similar of that of Theorem 2.62 taking V to be V_h and K to be K_h.

The next theorem gives the weak and strong convergences and a priori estimates for $\{u_h\}_h$. The convergence results are based on the papers of Falk (1974), Mosco and Strang (1974), Ciarlet (1978), Glowinski (1984), Glowinski *et al.* (1976), Brezzi *et al.* (1977).

Theorem 2.66: Let K, K_h be closed convex nonempty subsets of V or V_h, defined above. Let u be a solution of Equation 2.45 and u_h be a solution of 2.54. Then

$$\lim_{h\to 0} \|u_h - u\|_V = 0.$$

Proof: (a) A priori estimates of u_h: Since u_h is the solution of Equation 2.54, then

$$a(u_h, v_h - u_h) \geq S(v_h - u_h) \quad \forall v_h \in K_h.$$

Hence

$$a(u_h, u_h) \leq a(u_h, v_h) - S(v_h - u_h).$$

Due to the V-ellipticity of a(.,.), we have

$$\alpha \|u_h\|^2 \leq \|A\| \, \|u_h\| \, \|v_h\| + \|f\| \, (\|v_h\| + \|u_h\|) \quad \forall v_h \in K_h. \tag{2.56}$$

Let $v_o \in X$, $v_h = r_h v_o \in K_h$. Due to the condition (ii) on K_h then $r_h v_o \to v_o$ strongly in V and therefore $\|v_h\|$ is uniformly bounded. Then from Equation 2.56 we obtain

$$\|u_h\|^2 \leq \alpha^{-1}\{(c\,\|A\| + \|f\|)\,\|u_h\| + c\,\|f\|\} = c_1\,\|u_h\| + c_2,$$

where $c_1 = \alpha^{-1}(c\,\|A\| + \|f\|)$, $c_2 = \alpha^{-1}c\,\|f\|$. Hence

$$\|u_h\| \leq c, \quad \forall h. \tag{2.57}$$

(b) Weak convergence: From Equation 2.57 we see that u_h is uniformly bounded. Then there exists a subsequence $\{u_{hi}\}$ such that u_{hi} converges to $\underline{u}$ weakly in V. Due to the condition (i) on $\{K_h\}_h$ then $\underline{u} \in K$. We shall prove that $\underline{u}$ is a solution of the problem $(\mathscr{P}_1)_v$.

Let $v \in X$, $v_{hi} = r_{hi}v$. Since

$$a(u_{hi}, u_{hi}) \leq a(u_{hi}, v_{hi}) - S(v_{hi} - u_{hi}) \quad \forall v_{hi} \in K_{hi},$$

then

$$a(u_{hi}, u_{hi}) \leq a(u_{hi}, r_{hi}v) - S(r_{hi}v - u_{hi}).$$

But $r_{hi}v \to v$ strongly and $u_{hi} \to \underline{u}$ weakly as $h_i \to 0$. Then

$$\liminf_{h_i \to 0} a(u_{hi}, u_{hi}) \leq a(\underline{u}, v) - S(v - \underline{u}) \quad \forall v \in X. \tag{2.58}$$

Since

$$0 \leq a(u_{hi} - \underline{u}, u_{hi} - \underline{u}) \leq a(u_{hi}, u_{hi}) - a(u_{hi}, \underline{u}) - a(\underline{u}, u_{hi}) + a(\underline{u}, \underline{u})$$

then

$$a(u_{hi}, \underline{u}) + a(\underline{u}, u_{hi}) - a(\underline{u}, \underline{u}) \leq a(u_{hi}, u_{hi}).$$

In the limit $h_i \to 0$ we have

$$a(\underline{u}, \underline{u}) \leq \liminf_{h_i \to 0} a(u_{hi}, u_{hi}).$$

Hence using Equation 2.58 we obtain

$$a(\underline{u}, \underline{u}) \leq \liminf_{h_i \to 0} a(u_{hi}, u_{hi}) \leq a(\underline{u}, \underline{u}) - S(v - \underline{u}) \quad \forall v \in X.$$

Hence

$$a(\underline{u}, v - \underline{u}) \geq S(v - \underline{u}) \quad \forall v \in X, \underline{u} \in K.$$

But X is dense in K, and $a(.,.)$, $S(.)$ are continuous then

$$a(\underline{u}, v - \underline{u}) \geq S(v - \underline{u}) \quad \forall v \in K, \underline{u} \in K.$$

Hence $\underline{u}$ is a solution of the problem $(\mathscr{P}_1)_v$ and, due to Theorem 62 is unique. Thus $\underline{u} = u$ and hence u is the cluster point of $\{u_h\}_h$ in the weak topology of V and therefore the whole $\{u_h\}_h$ converges to u weakly.

(c) Strong convergence: The bilinear form $a(.,.)$ is V-elliptic. Hence

$$0 \leq \alpha\,\|u_h - u\|^2 \leq a(u_h - u, u_h - u) = a(u_h, u_h) - a(u_h, u) - a(u, u_h) + a(u, u), \tag{2.59}$$

where u, u_h are the solutions of problems $(\mathscr{P}_1)_v$ and $(\mathscr{P}_1)_{vh}$, respectively. Since u_h is the solution of the problem $(\mathscr{P}_1)_{vh}$ and $r_h v \in K_h$ for any $v \in X$, from $(\mathscr{P}_1)_{vh}$, then

$$a(u_h, u_h) \leq a(u_h, r_h v) - S(r_h v - u_h) \quad \forall v \in X. \tag{2.60}$$

Since $\lim_{h\to 0} u_h = u$ weakly in V, and due to the condition (ii), $\lim_{h\to 0} r_h v = v$ strongly in V, then Equation 2.60 follows from Equation 2.59 and in limit $\forall v \in X$, we obtain

$$0 \le \alpha \liminf_{h\to 0} \|u_h - u\|^2 \le \alpha \limsup_{h\to 0} \|u_h - u\|^2 \le a(u, v-u) - S(v-u), \tag{2.61}$$

which due to the density and continuity also holds for $\forall v \in K$. Putting $v = u$ in Equation 2.61 we have

$$\lim_{h\to 0} \|u_h - u\|^2 = 0,$$

i.e. strong convergence.

To find a priori estimates and to prove the weak and strong convergences of the problem $(\mathscr{P}_2)_{vh}$, we shall give some fundamental preliminary remarks.

Let $h \to 0$ be given, let $\{V_h\}_h$ be a family of closed subspaces of V. We suppose that the family $\{V_h\}_h$ of closed subspaces of V satisfies:

(i) there exists $U \subset V$ such that $\overline{U} = V$, and for $\forall h$ a mapping $r_h : U \to V_h$ such that $\lim_{h\to 0} r_h v = v$ strongly in V, $\forall v \in U$.

We approximate the functional $j(v)$ by $j_h(v_h)\,\forall h$ with properties: $j_h : V_h \to \overline{R}$, j_h is convex, l.s.c., and uniformly proper in h. We say that the family $\{j_h\}_h$ is uniformly proper in h if there exist $\lambda \in V'$, $\mu \in R$ such that

$$j_h(v_h) \ge \lambda(v_h) + \mu \quad \forall v_h \in V_h, \quad \forall h, \tag{2.62}$$

(ii) if $v_h \to v$ weakly in V then $\lim_{h\to 0}\inf j_h(v_h) \ge j(v)$,

(iii) $\lim_{h\to 0} j_h(r_h v) = j(v)\,\forall v \in U$, where U is defined above.

(iv) if $j(v)$ is a continuous functional, then $j_h(v_h)$, satisfying (ii) and (iii), is also continuous.

Theorem 2.67: Problem $(\mathscr{P}_2)_{vh}$ has a unique solution.

Proof: The proof is similar of that of Theorem 2.63 taking V to V_h and $j(v)$ to $j_h(v_h)$.

Theorem 2.68: Let $\{V_h\}_h$, $\{j_h\}_h$ be a family of closed subspaces of V and a family of approximate functionals of the functional $j(v)$ defined above. Then

$$\lim_{h\to 0} \|u_h - u\| = 0, \quad \lim_{h\to 0} j_h(u_h) = j(u).$$

Proof: The proof is similar of that of Theorem 2.66, as follows.
(a) A priori estimates of u_h: We shall show also that there exist constants $c_1 > 0$, $c_2 > 0$ independent of h such that

$$\|u_h\|^2 \le c_1 \|u_h\| + c_2 \quad \forall h.$$

Since u_h is the solution of the problem $(\mathscr{P}_2)_{vh}$, then

$$a(u_h, u_h) + j_h(u_h) \le a(u_h, v_h) + j_h(v_h) - S(v_h - u_h) \quad \forall v_h \in V_h$$

and then due to Equation 2.62 we have

$$\alpha \|u_h\|^2 \le \|\lambda\| \|u_h\| + |\mu| + \|A\| \|u_h\| \|v_h\| + |j_h(v_h)| + \|f\| (\|v_h\| + \|u_h\|).$$

Let $v_o \in U$, $v_h = r_h v_o$. Due to (i) and (iii), there exists a constant c, independent of h, such that $\|v_h\| \le c$, $|j_h(v_h)| \le c$. Then

$$\|u_h\|^2 \le \alpha^{-1}(\|\lambda\| + c\,\|A\| + \|f\|)\,\|u_h\| + c\alpha^{-1}(1 + \|f\|) + \alpha^{-1}\,|\mu| = c_1\,\|u_h\| + c_2,$$

where $c_1 = \alpha^{-1}(\|\lambda\| + c\,\|A\| + \|f\|)$ and $c_2 = c\alpha^{-1}(1 + \|f\|) + \alpha^{-1}\,|\mu|$. Hence

$$\|u_h\| \le c \quad \forall h, \qquad c \text{ is a constant.}$$

(b) Weak convergence: The last estimate $\|u_h\| \le c\,\forall h$ implies that there exists a subsequence $\{u_{hi}\}_{hi}$ such that $u_{hi} \to u_h$ weakly in V. As u_h is the solution of the problem $(\mathscr{P}_2)_{vh}$, $r_h v \in V_h$, $\forall h$ and $v \in U$, then

$$a(u_{hi}, u_{hi}) + j_{hi}(u_{hi}) \le a(u_{hi}, r_{hi}v) + j_{hi}(r_{hi}v) - S(r_{hi}v - u_{hi}). \tag{2.63}$$

From the weak convergence of $\{u_{hi}\}$ and (iii), we have

$$\lim_{h_i \to 0} \inf(a(u_{hi}, u_{hi}) + j_{hi}(u_{hi})) \le a(\underline{u}, v) + j(v) - S(v - \underline{u}) \quad \forall v \in U.$$

By taking the limit and (ii), then similarly as above

$$a(\underline{u}, \underline{u}) + j(\underline{u}) \le \lim_{h_i \to 0} \inf(a(u_{hi}, u_{hi}) + j_{hi}(u_{hi})).$$

Hence using Equation 2.63 and due to the density of U, we have

$$a(\underline{u}, v - \underline{u}) + j(v) - j(\underline{u}) \le S(v - \underline{u}) \quad \forall v \in V, \underline{u} \in V.$$

Hence u is a solution of the problem $(\mathscr{P}_2)_v$ and therefore $\underline{u} = u$ is a unique solution of the problem $(\mathscr{P}_2)_v$ and therefore $\{u_h\}_h$ converges to u weakly.

(c) Strong convergence: As the bilinear form $a(u, v)$ is V-elliptic, then from Equation 2.55 it follows

$$\begin{aligned}\alpha\,\|u_h - u\|^2 + j_h(u_h) &\le a(u_h - u, u_h - u) + j_h(u_h)\\ &= a(u_h, u_h) - a(u, u_h) - a(u_h, u) + a(u, u) + j_h(u_h)\\ &\le a(u_h, r_h v) + j_h(r_h v) - S(r_h v - u_h) - a(u, u_h)\\ &\quad - a(u_h, u) + a(u, u) \quad \forall v \in U.\end{aligned}$$

The right hand side converges to $a(u, v - u) + j(v) - S(v - u)$ for $h \to 0$, $\forall v \in$U. Hence

$$\begin{aligned}\lim_{h\to 0}\inf j_h(u_h) &\le \lim_{h\to 0}\inf(\alpha\,\|u_h - u\|^2 + j_h(u_h))\\ &\le \lim_{h\to 0}\sup(\alpha\,\|u_h - u\|^2 + j_h(u_h)) \le a(u, v - u) + j(v) - S(v - u) \quad \forall v \in U.\end{aligned}$$

Due to the density of U, the last inequality holds for $\forall v \in V$. Therefore, replacing v by u and using (ii), we obtain

$$j(u) \le \lim_{h\to 0}\inf j_h(u_h) \le \lim_{h\to 0}\sup(\alpha\,\|u - u_h\|^2 + j_h(u_h)) \le j(u),$$

so that

$$\lim_{h\to 0} j_h(u_h) = j(u) \text{ and } \lim_{h\to 0}\|u_h - u\| = 0,$$

which completes the proof.

2.4.3. Application of Penalty Methods to Variational Inequalities

To solve both problems numerically the penalty technique can be used. But in our case the penalty method will be applied for the problem $(\mathscr{P}_1)_v$ only.

For the functional $j(v)$ we can suppose that it has the following properties:

$$j : V \to \overline{R} \text{ is convex, proper and l.s.c.,}$$

$$j(v) = 0 \Leftrightarrow v \in K,$$

$$j(v) \geq 0 \quad \forall v \in V.$$

Let $\varepsilon > 0$ be given. Then we define $j_\varepsilon : V \to \overline{R}$ by

$$j_\varepsilon(v) = \varepsilon^{-1} j(v).$$

Then the penalized problem $(\mathscr{P}_1)_{vp}$ associated with $j(v)$ is as follows:

Problem $(\mathscr{P}_1)_{vp}$: find $u_\varepsilon \in V$ such that

$$a(u_\varepsilon, v - u_\varepsilon) + j_\varepsilon(v) - j_\varepsilon(u_\varepsilon) \geq S(v - u_\varepsilon) \quad \forall v \in V. \tag{2.64}$$

If j_ε is differentiable then the solution u_ε of Equation 2.64 represents a solution of a nonlinear variational equation

$$a(u_\varepsilon, v) + \langle j'_\varepsilon(u_\varepsilon), v\rangle = S(v) \quad \forall v \in V, u_\varepsilon \in V, \tag{2.65}$$

where $j'_\varepsilon(v) \in V'$ (V' is the dual space of V) represents the differential of j_ε at v and $\langle .,. \rangle$ is the duality pairing between V' and V. If $a(.,.)$ is symmetric, then the penalized problem $(\mathscr{P}_1)_{vp}$ is equivalent to the minimization problem given by:

Problem $(\mathscr{P}_1)_{\mathrm{vps}}$: find $u_\varepsilon \in V$ such that

$$L_\varepsilon(u_\varepsilon) \leq L_\varepsilon(v) \quad \forall v \in V, \tag{2.66}$$

where $L_\varepsilon(v) = \frac{1}{2}a(v, v) - S(v) + j_\varepsilon(v)$.

Theorem 2.69: Let V, K, $a(.,.)$, $S(.)$, $j(.)$ be as defined above. Then

$$\lim_{\varepsilon \to 0} \|u_\varepsilon - u\| = 0, \quad \lim_{\varepsilon \to 0} j_\varepsilon(u_\varepsilon) = 0,$$

where u, u_ε are solutions of the Problem $(\mathscr{P}_1)_v$ or $(\mathscr{P}_1)_{vp}$, respectively.

Proof: The proof is similar of that of Theorem 68.
(a) A priori estimates: From Equation 2.64 we have

$$a(u_\varepsilon, u_\varepsilon) + j_\varepsilon(u_\varepsilon) \leq a(u_\varepsilon, v) + j_\varepsilon(v) - S(v - u_\varepsilon) \quad \forall v \in V. \tag{2.67}$$

Since $j_\varepsilon(v) = \varepsilon^{-1} j(v) = 0 \, \forall v \in K$(as $j(v) = 0 \Leftrightarrow v \in K$), then

$$a(u_\varepsilon, u_\varepsilon) + j_\varepsilon(u_\varepsilon) \leq a(u_\varepsilon, v) - S(v - u_\varepsilon) \quad \forall v \in K.$$

Let $v_o \in K$. Substitute $v = v_o$ in the last inequality. Then due to V-ellipticity and as $j(v) \geq 0 \, \forall v \in V$, $j_\varepsilon(v) = \varepsilon^{-1} j(v)$, we have

$$\alpha \|u_\varepsilon\|^2 \leq \|A\| \, \|u_\varepsilon\| \, \|v_o\| + \|f\| \, (\|u_\varepsilon\| + \|v_o\|), \tag{2.68}$$

$$0 \leq j_\varepsilon(u_\varepsilon) \leq \varepsilon(\|A\| \, \|u_\varepsilon\| \, \|v_o\| + \|f\| \, (\|u_\varepsilon\| + \|v_o\|)). \tag{2.69}$$

Similarly as above we obtain

$$\|u_\varepsilon\| \le c_1, \quad \forall \varepsilon > 0 \tag{2.70}$$

and from Equation 2.69 and the last estimate, we obtain

$$0 \le j_\varepsilon(u_\varepsilon) \le c_2, \tag{2.71}$$

where c_1 and c_2 are constants independent of ε.

(b) Weak convergence: As $\|u_\varepsilon\| \le c_1$ then there exists a sequence $\{u_\varepsilon\}_\varepsilon$ such that

$$\lim_{\varepsilon \to 0} u_\varepsilon = \underline{u} \quad \text{weakly in } V, \quad \underline{u} \in V. \tag{2.72}$$

Due to Equation 2.71 and 2.72, and the weak lower semicontinuity of $j(.)$ then

$$j(\underline{u}) = 0 \Rightarrow \underline{u} \in K (\text{as } j(v) = 0 \Leftrightarrow v \in K). \tag{2.73}$$

From Equation 2.67 it follow that

$$a(u_\varepsilon, u_\varepsilon) \le a(u_\varepsilon, v) - S(v - u_\varepsilon) \quad \forall v \in K.$$

In limit $\varepsilon \to 0$ we obtain

$$a(\underline{u}, \underline{u}) \le \liminf_{\varepsilon \to 0} a(u_\varepsilon, u_\varepsilon) \le a(\underline{u}, v) - S(v - \underline{u}) \quad \forall v \in K.$$

Hence, and using Equation 2.73, we obtain

$$a(\underline{u}, v - \underline{u}) \ge S(v - \underline{u}) \quad \forall v \in K, \underline{u} \in K.$$

Similarly as above $\underline{u} = u$ and the whole $\{u_\varepsilon\}_\varepsilon$ converges weakly to u.

(c) Strong convergence: Due to $j(v) = 0 \Leftrightarrow v \in K$, $j(v) \ge 0 \, \forall v \in K$ and from Equation 2.64 we obtain

$$\begin{aligned} 0 &\le \alpha \|u_\varepsilon - u\|^2 + j_\varepsilon(u_\varepsilon) \\ &\le a(u_\varepsilon - u, u_\varepsilon - u) + j_\varepsilon(u_\varepsilon) \\ &\le a(u_\varepsilon, u_\varepsilon) + j_\varepsilon(u_\varepsilon) - a(u, u_\varepsilon) - a(u_\varepsilon, u) + a(u, u) \\ &\le a(u_\varepsilon, v) - S(v - u_\varepsilon) - a(u, u_\varepsilon) - a(u_\varepsilon, u) + a(u, u) \quad \forall v \in K. \end{aligned}$$

Hence, since $\{u_\varepsilon\}_\varepsilon \to u$ weakly, we have

$$\begin{aligned} 0 \le \liminf_{\varepsilon \to 0} (\alpha \|u_\varepsilon - u\|^2 + j_\varepsilon(u_\varepsilon)) &\le \limsup_{\varepsilon \to 0} (\alpha \|u_\varepsilon - u\|^2 + j_\varepsilon(u_\varepsilon)) \\ &\le a(u, v - u) - S(v - u) \quad \forall v \in K. \end{aligned}$$

Hence, setting $v = u$, we obtain

$$\lim_{\varepsilon \to 0} (\alpha \|u_\varepsilon - u\|^2 + j_\varepsilon(u_\varepsilon)) = 0,$$

hence $\lim_{\varepsilon \to 0} \|u_\varepsilon - u\|^2 = 0$, $\lim_{\varepsilon \to 0} j_\varepsilon(u_\varepsilon) = 0$, which completes the proof.

2.5. QUADRATIC FUNCTIONALS IN R^N

2.5.1. Tools from the Quadratic Programming

In this section we use the notation given above: the transpose of a matrix C will be denoted by C^T. A square matrix C is nonsingular if its determinant $\det C \ne 0$. The inverse of the matrix C, if it exists, will be denoted by C^{-1}. If $C^T = C$, then C is said to be a symmetric matrix. A symmetric matrix C is nonnegative if the inequality $\boldsymbol{x}^T C\boldsymbol{x} \geq 0$ is valid for all vectors $\boldsymbol{x}$ (we denote it also as $\forall \boldsymbol{x}$), and is positive definite if $\boldsymbol{x}^T C\boldsymbol{x} > 0$ whenever $\boldsymbol{x} \neq 0$ and is positive semi-definite if $\boldsymbol{x}^T C\boldsymbol{x} \geq 0$ whenever $\boldsymbol{x} \neq 0$. A matrix C is nonpositive if $-C$ is nonnegative and is negative definite (semi-definite) if $-C$ is positive definite (semi-definite). The function $f(\boldsymbol{x}) = \boldsymbol{x}^T C\boldsymbol{x}$ is called a quadratic form and is nonnegative, nonpositive, positive definite, positive semi-definite, negative definite, or negative semi-definite having the same property as the matrix C.

Let S be a set and let X be a subset of R^n. Let R be a set of real numbers. A functional f on X is defined as a mapping $f : X \subseteq R^n \to R^1$ and its range by

$$R_f = \{k \in R \mid f(\boldsymbol{x}) = k \text{ for some } \boldsymbol{x} \in X\}.$$

For $\boldsymbol{x} = (x_1, x_2, \ldots, x_n)^T$, $x_i \in R$, then $S = R^n$ and f represents a function of n real variables $x_1, x_2, \ldots, x_n$ and we shall write $f(\boldsymbol{x}) = f(x_1, x_2, \ldots, x_n)$. Further, we shall define the neighbourhood of $\boldsymbol{x} \in R^n$ of radius $\varepsilon > 0$ by

$$S(\boldsymbol{x}, \varepsilon) = \{z \in R^n \mid 0 \leq \|\boldsymbol{x} - z\| < \varepsilon\},$$

where $\|.\|$ denotes the Euclidean vector norm, i.e. $\|\boldsymbol{x}\| = (\Sigma_{i=1}^n x_i^2)^{1/2}\, \forall \boldsymbol{x} \in R^n$ and we shall denote it by $S(\boldsymbol{x}, \varepsilon)$.

Furthermore, we shall assume that X will denote an open subset R^n, i.e. for every $\boldsymbol{x} \in X$ there exists $\varepsilon > 0$ such that $S(\boldsymbol{x}, \varepsilon) \subset X$.

In the following, the problem of finding a local solution to the problem:

$$\text{minimize } f(\boldsymbol{x}), \quad \boldsymbol{x} \in R^n \tag{2.74}$$

will be considered. The minimizing point or minimizer of Equation 2.74 (see Definitions 13 and 14, below) will be denoted by $\underline{\boldsymbol{x}}$.

Consider next an arbitrary real-valued continuous function f on an open set $X \subseteq R^n$ possessing continuous first- and second-order partial derivatives on X. Then a function f is said to be of class $C^{(2)}$ on X if it is continuous and has continuous derivatives of all orders ≤ 2 on X.

Definition 2.13: Let us assume a function $f : X \subseteq R^n \to R^1$. We say that $\underline{\boldsymbol{x}} \in X$ is a local minimizer of f if there exists an $\varepsilon > 0$ such that $f(\underline{\boldsymbol{x}}) \leq f(\boldsymbol{x})\, \forall \boldsymbol{x} \in S(\underline{\boldsymbol{x}}, \varepsilon) \cap X, \boldsymbol{x} \neq \underline{\boldsymbol{x}}$. If $f(\underline{\boldsymbol{x}}) < f(\boldsymbol{x})\, \forall \boldsymbol{x} \in S(\underline{\boldsymbol{x}}, \varepsilon) \cap X$, $\boldsymbol{x} \neq \underline{\boldsymbol{x}}$, then $\underline{\boldsymbol{x}}$ is a strong local minimizer of f. If $f(\underline{\boldsymbol{x}}) \leq f(\boldsymbol{x})\, \forall \boldsymbol{x} \in X$, then $\underline{\boldsymbol{x}}$ is a global minimizer of f. If $f(\underline{\boldsymbol{x}}) < f(\boldsymbol{x})\, \forall \boldsymbol{x} \in X$, $\boldsymbol{x} \neq \underline{\boldsymbol{x}}$, then $\underline{\boldsymbol{x}}$ is a strong global minimizer of f.

Remark: In a similar way we define local, strong local, global, and strong global maximizer. It is evident that a maximizer (local,etc.) of a functional f is a minimizer (local, etc.) of the functional $-f$.

Definition 2.14: Let $f \in C^{(1)}(X)$, then the gradient of f at $\boldsymbol{x} \in X$ is the vector

$$f^{(1)}(\boldsymbol{x}) \equiv f'(\boldsymbol{x}) = (\partial f(\boldsymbol{x})/\partial x_i)^T, \quad i = 1, \ldots, n$$

and is denoted by grad $f(\boldsymbol{x})$ or $\nabla f(\boldsymbol{x})$ as well as $\boldsymbol{g}(\boldsymbol{x})$ [note that mapping $\boldsymbol{g} : X \subset R^n \to R^n$ is a gradient at $\boldsymbol{x} \in X$ if there exists a G-differentiable functional $f : X \subset R^n \to R^1$ such that $\boldsymbol{g}(\boldsymbol{x}) = (f'(\boldsymbol{x}))^T \; \forall \boldsymbol{x} \in X$.]

Let $f \in C^{(2)}(X)$, then the Hessian of f at $\boldsymbol{x} \in X$ is the symmetric matrix (i.e. the linear mapping $L(R^n)$ from $R^n \to R^n$)

$$C(\boldsymbol{x}) \equiv f^{(2)}(\boldsymbol{x}) = (c_{ij})^n_{i,j=1},$$

where $f^{(2)}(\boldsymbol{x}) \equiv f''(\boldsymbol{x}) = (\partial^2 f(\boldsymbol{x})/\partial x_i \partial x_j)$, $i, j = 1, \ldots, n$ and $c_{ij} = \partial^2 f(\boldsymbol{x})/\partial x_i \partial x_j$. The Hessian of f at $\boldsymbol{x}$ is also frequently denoted in the literature by $H(\boldsymbol{x})$ or $A(\boldsymbol{x})$. At each point $\boldsymbol{x} \in X$ the Taylor expansion of $f \in C^{(1)}(X)$ is defined by

$$\begin{aligned} f(\boldsymbol{x} + \boldsymbol{v}) &= f(\boldsymbol{x}) + \sum\nolimits_{i=1}^{n} \partial f(\boldsymbol{x})/\partial x_i v_i + \tfrac{1}{2} \sum\nolimits_{i=1}^{n} \partial^2 f(\boldsymbol{x})/\partial x_i \partial x_j v_i v_j + R(\boldsymbol{x}, \boldsymbol{v}) \\ &\equiv f(\boldsymbol{x}) + \boldsymbol{g}^T(\boldsymbol{x})\boldsymbol{v} + o(\|\boldsymbol{v}\|), \end{aligned} \tag{2.75}$$

where $R(\boldsymbol{x}, \boldsymbol{v})$ is the so-called remainder and it satisfies

$$R(\boldsymbol{x}, \boldsymbol{v}) = \tfrac{1}{2} \sum\nolimits_{i=1}^{n} v_i v_j [\partial^2 f(\boldsymbol{x} + \Theta \boldsymbol{v})/\partial x_i \partial x_j - \partial^2 f(\boldsymbol{x})/\partial x_i \partial x_j], \qquad \Theta \in (0, 1),$$

$$\lim_{\boldsymbol{v} \to 0} R(\boldsymbol{x}, \boldsymbol{v}) = 0, \qquad \text{i.e.} \quad R(\boldsymbol{x}, \boldsymbol{v}) \cong o(\|\boldsymbol{v}\|^2).$$

Similarly, the Taylor expansion of $f \in C^{(2)}(X)$ for $\boldsymbol{x} \in X$ is defined by

$$f(\boldsymbol{x} + \boldsymbol{v}) = f(\boldsymbol{x}) + \boldsymbol{g}^T(\boldsymbol{x})\boldsymbol{v} + \tfrac{1}{2}\boldsymbol{v}^T C(\boldsymbol{x})\boldsymbol{v} + o(\|\boldsymbol{v}\|^2). \tag{2.76}$$

Let $f \in C^{(1)}(X)$. Then $\underline{\boldsymbol{x}} \in X$ is a stationary (critical, optimal, minimum) point of f if $\boldsymbol{g}(\underline{\boldsymbol{x}}) = 0$.

Definition 2.15: Let $f \in C^{(k)}(X)$, $\boldsymbol{x} \in X \subseteq R^n$. Let $\boldsymbol{y} \in R^n$, where $\|\boldsymbol{y}\| = 1$. Then, the m-th-order directional derivative of f at $\boldsymbol{x}$ in the direction $\boldsymbol{y}$ is defined by

$$f^{(k)}(\boldsymbol{x}; \boldsymbol{y}) = d^k f(\boldsymbol{x} + t\boldsymbol{y})/\mathrm{d}t^k \big|_{t=0}.$$

From this definition for $f \in C^{(1)}(X)$ we obtain

$$\begin{aligned} f^{(1)}(\boldsymbol{x}; \boldsymbol{y}) &= \mathrm{d} f(x_1 + t y_1, \ldots, x_n + t y_n)/\mathrm{d}t \big|_{t=0} \\ &= \sum\nolimits_{i=1}^{n} \partial f(x_1, \ldots, x_n)/\partial x_i y_i = \boldsymbol{g}^T(\boldsymbol{x})\boldsymbol{y}, \end{aligned} \tag{2.77}$$

and for the case of $f \in C^{(2)}(X)$, then by further differentiation of Equation 2.77, we obtain

$$f^{(2)}(\boldsymbol{x}; \boldsymbol{y}) = \boldsymbol{y}^T C(\boldsymbol{x})\boldsymbol{y}. \tag{2.78}$$

Theorem 2.70: The necessary and sufficient conditions for $\underline{\boldsymbol{x}}$ to be a local minimizer are the following:

(i) $\underline{\boldsymbol{x}}$ is a stationary point of f,

(ii) $\underline{\boldsymbol{x}}$ is a strong minimizer of f.

Proof: To prove (i) we shall assume that $f \in C^{(1)}(X)$. The Taylor expansion Equation 2.75 of f for $\boldsymbol{x} = \underline{\boldsymbol{x}}$ and $\boldsymbol{v}(\underline{\boldsymbol{x}}) = -v\boldsymbol{g}(\underline{\boldsymbol{x}})$, $0 < v < v_o$, v being a real variable, gives

$$f(\underline{\boldsymbol{x}} + \boldsymbol{v}) = f(\underline{\boldsymbol{x}}) - v\,\|\boldsymbol{g}(\underline{\boldsymbol{x}})\|^2 + o(v).$$

On the contrary we shall assume that $\underline{\boldsymbol{x}}$ is not a stationary point of f; then $\boldsymbol{g}(\underline{\boldsymbol{x}}) \neq 0$. Then $f(\underline{\boldsymbol{x}} + \boldsymbol{v}) < f(\underline{\boldsymbol{x}})$ as $+v\,\|\boldsymbol{g}(\underline{\boldsymbol{x}})\|^2 > o(v)$ and, therefore, in this case $\underline{\boldsymbol{x}}$ is not a local minimizer.

To prove (ii) we shall assume that $f \in C^{(2)}(X)$ and that (i) holds. Then the Taylor expansion Equation 2.76 of f for $\boldsymbol{x} = \underline{\boldsymbol{x}}$ and $\boldsymbol{g}(\underline{\boldsymbol{x}}) = 0$ gives

$$f(\underline{\boldsymbol{x}} + \boldsymbol{v}) = f(\underline{\boldsymbol{x}}) + \tfrac{1}{2}\boldsymbol{v}^T C(\underline{\boldsymbol{x}})\boldsymbol{v} + o(\|\boldsymbol{v}\|^2).$$

Hence

$$f(\underline{\boldsymbol{x}} + \boldsymbol{v}) - f(\underline{\boldsymbol{x}}) \geq \tfrac{1}{2}\lambda_1\,\|\boldsymbol{v}\|^2 + o(\|\boldsymbol{v}\|^2), \tag{2.79}$$

as $\boldsymbol{v}^T C(\underline{\boldsymbol{x}})\boldsymbol{v} \geq \lambda_1\,\|\boldsymbol{v}\|^2\ \forall \boldsymbol{v} \in R^n$, where $\lambda_1 > 0$ is the smaller eigenvalue of the positive definite matrix $C(\underline{\boldsymbol{x}})$. Since the right hand side of Equation 2.79 for $\|\boldsymbol{v}\|$ sufficiently small is positive, then $f(\underline{\boldsymbol{x}} + \boldsymbol{v}) > f(\underline{\boldsymbol{x}})$ in some neighbourhood of $\underline{\boldsymbol{x}}$, so that $\underline{\boldsymbol{x}}$ is a strong local minimizer of the function f.

Remark: The case of a local maximizer of f can be investigated in a similar way. If $C(\underline{\boldsymbol{x}})$ is negative definite, then $\underline{\boldsymbol{x}}$ is a strong local maximizer. If $C(\underline{\boldsymbol{x}})$ is positive (negative) semi-definite, then a stationary point $\underline{\boldsymbol{x}}$ may or may not be a local minimizer (maximizer).

Theorem 2.71: The necessary and sufficient conditions for $\underline{\boldsymbol{x}}$ to be a local minimizer are

(i) $f^{(1)}(\underline{\boldsymbol{x}};\boldsymbol{y}) = 0$ for all directions $\boldsymbol{y}$,

(ii) $f^{(2)}(\underline{\boldsymbol{x}};\boldsymbol{y}) > 0$ for all directions $\boldsymbol{y}$.

Proof: Since $f^{(1)}(\underline{\boldsymbol{x}};\boldsymbol{y}) = \boldsymbol{g}^T(\underline{\boldsymbol{x}})\boldsymbol{y}$, then due to the Cauchy-Schwarz inequality $\left(|\boldsymbol{x}^T\boldsymbol{y}| \leq \|\boldsymbol{x}\|\,\|\boldsymbol{y}\|\ \forall \boldsymbol{x}, \boldsymbol{y} \in R^n\right)$, we have

$$\max_{\|\boldsymbol{y}\|=1} \left|f^{(1)}(\underline{\boldsymbol{x}};\boldsymbol{y})\right| \leq \left|f^{(1)}(\underline{\boldsymbol{x}};\underline{\boldsymbol{y}})\right| = \|\boldsymbol{g}(\boldsymbol{x})\|,$$

where $\underline{\boldsymbol{y}} \equiv g(\underline{\boldsymbol{x}})/\,\|\boldsymbol{g}(\underline{\boldsymbol{x}})\|$. Hence $f^{(1)}(\underline{\boldsymbol{x}};\boldsymbol{y}) = 0$ for all directions $\boldsymbol{y}$ if and only if $\boldsymbol{g}(\underline{\boldsymbol{x}}) = 0$, i.e. (according to Definition 2.14) if and only if $\boldsymbol{x}$ is a critical (stationary) point of f. Furthermore, it follows from Equation 2.78 that $f^{(2)}(\underline{\boldsymbol{x}};\boldsymbol{y}) > 0$ for all directions $\boldsymbol{y}$ if and only if $C(\underline{\boldsymbol{x}})$ is positive definite.

So we see that for $f^{(1)}(\underline{\boldsymbol{x}};\boldsymbol{y}) = 0$, $f^{(2)}(\underline{\boldsymbol{x}};\boldsymbol{y}) \geq 0$ for all $\boldsymbol{y} \neq 0$ we have a local minimum point $\underline{\boldsymbol{x}}$; for $f^{(1)}(\underline{\boldsymbol{x}};\boldsymbol{y}) = 0$, $f^{(2)}(\underline{\boldsymbol{x}};\boldsymbol{y}) > 0$ for all $\boldsymbol{y} \neq 0$ the point $\underline{\boldsymbol{x}}$ afforts a strict minimum to f. If these inequalities are reversed, then we obtain criteria for local maxima. The point $\underline{\boldsymbol{x}}$ at which $f^{(1)}(\underline{\boldsymbol{x}};\boldsymbol{y}) = 0$ is frequently called a critical (stationary) point of f, and $f(\underline{\boldsymbol{x}})$ is called the corresponding critical value of f. [Note that a point $\underline{\boldsymbol{x}} \in \mathrm{int}(X)$ is a critical point of $f : X \subseteq R^n \to R^1$ if f is G-differentiable at $\underline{\boldsymbol{x}}$ and $f^{(1)}(\underline{\boldsymbol{x}}) = 0$. Let the second G-derivative of f exist. If $\underline{\boldsymbol{x}}$ is a critical point of f and $f^{(2)}(\underline{\boldsymbol{x}})$ is positive definite, then $\underline{\boldsymbol{x}}$ is a minimum point of f (or a proper local minimizer of f.] The critical (stationary) point $\underline{\boldsymbol{x}}$ is said to be nondegenerate if $\det f^{(2)}(\underline{\boldsymbol{x}};\boldsymbol{y}) = \det(C) \neq 0$. Locally, a nondegenerate critical point $\underline{\boldsymbol{x}}$ of f is a minimum point if $f^{(2)}(\underline{\boldsymbol{x}})$ is positive definite, is a maximum point if $f^{(2)}(\underline{\boldsymbol{x}})$ is negative definite, and is a saddle point if $f^{(2)}(\underline{\boldsymbol{x}})$ is indefinite.

Remark: Numerical methods for finding a strong local minimizer of a functional f make full use of properties of functionals, especially their geometrical properties.

Definition 2.16: Let $f : X \subseteq R^n \to R^1$ and let R_f be its range. Furthermore, let $k \in R_f$, then the set

$$S_k = \{\boldsymbol{x} \in X \mid f(\boldsymbol{x}) = k\}$$

represents the level surface of the functional f for the value k.

The idea of using numerical methods to find a strong local minimizer of a functional is as follows: let $\underline{\boldsymbol{x}} \in X$ be a strong local minimizer of a functional f and let $f(\underline{\boldsymbol{x}}) = \underline{k}$. Due to Definition 13 the intersection of $S_{\underline{k}}$ and the neighbourhood $S(\underline{\boldsymbol{x}}, \varepsilon)$, ε being small, is composed of the single point $\underline{\boldsymbol{x}}$. In the case when $k > \underline{k}$, then the level surfaces surround $\underline{\boldsymbol{x}}$. Let $\boldsymbol{x}$ belong to S_k, then the gradient vector $\boldsymbol{g}(\boldsymbol{x})$ is perpendicular to S_k at $\boldsymbol{x}$ and its direction is in the direction of most rapid increase of the functional. Therefore, gradients near a strong minimizer direct outward, but inward near a strong local maximizer.

Let $f : R^n \to R^1$ be a quadratic functional

$$f(\boldsymbol{x}) = \tfrac{1}{2}(\boldsymbol{x}, C\boldsymbol{x}) - (\boldsymbol{b}, \boldsymbol{x}) \equiv \tfrac{1}{2}\boldsymbol{x}^T C\boldsymbol{x} - \boldsymbol{b}^T\boldsymbol{x}, \tag{2.80}$$

where $\boldsymbol{x} \in R^n$, C is an $n \times n$ symmetric matrix (Hessian), and $\boldsymbol{b} \in R^n$. Then the gradient $\boldsymbol{g}(\boldsymbol{x}) = C\boldsymbol{x} - \boldsymbol{b}$ and the Hessian $C(\boldsymbol{x}) = C$, i.e. its Hessian is constant. A point $\underline{\boldsymbol{x}}$ is a critical (stationary) point of f if the gradient $\boldsymbol{g}(\boldsymbol{x})$ at $\underline{\boldsymbol{x}}$ vanishes, i.e. if $C\underline{\boldsymbol{x}} - \boldsymbol{b} = 0$. If C is nonsingular, then $\underline{\boldsymbol{x}} = C^{-1}\boldsymbol{b}$. Hence

$$f(\boldsymbol{x}) = \tfrac{1}{2}(\boldsymbol{x} - \underline{\boldsymbol{x}})^T C(\boldsymbol{x} - \underline{\boldsymbol{x}}) - \tfrac{1}{2}\boldsymbol{b}^T\underline{\boldsymbol{x}}. \tag{2.81}$$

Next we shall assume that C is a positive definite matrix.

A critical (stationary) point of f is a point $\underline{\boldsymbol{x}}$ such that $f^{(1)}(\underline{\boldsymbol{x}}) = 0$. Hence $\underline{\boldsymbol{x}}$ is a critical point of the functional f if and only if it is a solution of the system of linear algebraic equations

$$C\boldsymbol{x} = \boldsymbol{b}. \tag{2.82}$$

If C is nonsingular then there is exactly one solution

$$\underline{\boldsymbol{x}} = C^{-1}\boldsymbol{b}, \tag{2.83}$$

where $\underline{\boldsymbol{x}}$ is the unique critical (stationary) point. In the case when C is singular and $\underline{\boldsymbol{x}}$ is a solution of Equation 2.80, then every solution of that equation can be found in the form $\underline{\underline{\boldsymbol{x}}} = \underline{\boldsymbol{x}} + \boldsymbol{w}$, where $\boldsymbol{w}$ is a null vector of C, i.e. $\boldsymbol{w}$ satisfies $C\boldsymbol{w} = 0$. It holds that

$$f(\boldsymbol{x} + \boldsymbol{v}) = f(\boldsymbol{x}) + \boldsymbol{v}^T(C\boldsymbol{x} - \boldsymbol{b}) + \tfrac{1}{2}\boldsymbol{v}^T C\boldsymbol{v}, \tag{2.84}$$

which follows from Equation 2.80 replacing $\boldsymbol{x}$ with $\boldsymbol{x} + \boldsymbol{v}$. Applying Taylor's expansion to Equation 2.84 at point $\boldsymbol{x}$ gives

$$f(\boldsymbol{x} + \boldsymbol{v}) = f(\boldsymbol{x}) + \boldsymbol{v}^T\boldsymbol{g}(\boldsymbol{x}) + \tfrac{1}{2}\boldsymbol{v}^T C(\boldsymbol{x})\boldsymbol{v} + o(\|\boldsymbol{v}\|^2). \tag{2.85}$$

If $\underline{x}$ is a critical (stationary) point of f, i.e. $f^{(1)}(\underline{x}) = 0$, then, by replacing $\boldsymbol{x}$ by $\underline{\boldsymbol{x}}$ and $\boldsymbol{v}$ by $\boldsymbol{x} - \underline{\boldsymbol{x}}$ in Equation 2.85, we have

$$f(\boldsymbol{x}) = f(\underline{\boldsymbol{x}}) + \tfrac{1}{2}(\boldsymbol{x} - \underline{\boldsymbol{x}})^T C(\boldsymbol{x} - \underline{\boldsymbol{x}}) \tag{2.86}$$

for f relative to a critical (stationary) point of f. From the geometrical point of view and for C being nonsingular, the critical (stationary) point $\underline{\boldsymbol{x}} = C^{-1}\boldsymbol{b}$ of f is the centre of the quadratic surface

$$f(\boldsymbol{x}) = k. \tag{2.87}$$

The level surfaces given by Equation 2.87 for a positive definite quadratic function f represent ellipsoids having $\underline{\boldsymbol{x}} = C^{-1}\boldsymbol{b}$ as their common centre. Hence, the problem of minimizing f is equivalent to the geometrical problem of finding the centre of an ellipsoid.

Further, we denote by $\{\lambda_i, \boldsymbol{w}_i\}_{i=1}^n$ the eigensolutions the eigenvalues λ_i and eigenvectors $\boldsymbol{w}_i$ of the symmetric matrix C, i.e.

$$C\boldsymbol{w}_i = \lambda_i \boldsymbol{w}_i, \quad i = 1, 2, \ldots, n. \tag{2.88}$$

Then the eigenvalues λ_i are real and $\lambda_1 \leq \lambda_2 \leq \ldots \leq \lambda_n$, and the corresponding eigenvectors satisfy the orthonormality condition $\boldsymbol{w}_i^T \boldsymbol{w}_j = \delta_{ij}$, $i, j, = 1, 2, \ldots, n$, where δ_{ij} is the Kronecker symbol. Let us denote $\Lambda = \text{diag}\ (\lambda_i)_{i=1}^n$, $W = (\boldsymbol{w}_i)_{i=1}^n$ matrices of eigenvalues and eigenvectors. The matrix W is, due to an orthogonality condition an orthogonal matrix, i.e. $W^{-1} = W^T$. Then Equation 2.88 can be rewritten in the form

$$CW = W\Lambda. \tag{2.89}$$

Let $z = W^T(\boldsymbol{x} - \underline{\boldsymbol{x}})$ then Equation 2.80 can be rewritten as follows

$$\underline{f}(z) \equiv f(Wz + \underline{\boldsymbol{x}}) = \tfrac{1}{2}z^T W^T C W z - \tfrac{1}{2}\boldsymbol{b}^T \underline{\boldsymbol{x}} = \tfrac{1}{2}z^T \Lambda z - \underline{b} = \tfrac{1}{2}\sum_{i=1}^n \lambda_i z_i^2 - \underline{b}, \tag{2.90}$$

where $\underline{b} = \frac{1}{2}\boldsymbol{b}^T \underline{\boldsymbol{x}}$. We see that $z = 0$ is the strong global minimizer of $\underline{f}(z)$. If C is a negative definite matrix, then $z = 0$ is the strong global maximizer, and if its eigenvalues are negative the range of $\underline{f}$ is $(-\infty, -\underline{b}\rangle$. If C has positive and negative eigenvalues, then the range of $\underline{f}$ is a $(-\infty, \infty)$, but though f is stationary at $z = 0$ it does not possess minimizer or maximizer there.

For $k > \underline{b}$ the level surface S_k is ellipsoid defined by

$$\sum_{i=1}^n \lambda_i z_i^2 = 2(k - \underline{b}) > 0.$$

The quantity

$$D_k = \inf_{\boldsymbol{y} \in \Sigma_k} \{ \sup_{\boldsymbol{x} \in S_k} \|\boldsymbol{x} - \boldsymbol{y}\| \,/\, \inf_{\boldsymbol{x} \in S_k} \|\boldsymbol{x} - \boldsymbol{y}\| \} \geq 1,$$

where Σ_k represents the set of all interior points of S_k, is the so-called measure of the distortion of S_k. In the case of a quadratic functional it has the value

$$k(C) = \lambda_n / \lambda_1 \tag{2.91}$$

and is called the spectral condition number of the Hessian C. It plays a critical role in the analysis of the methods of steepest descent and the conjugate gradient methods.

Let a system of linear algebraic equations

$$Cx = b \tag{2.92}$$

be given. Then the residual of the system at x is the vector

$$r = r(x) = b - Cx, \tag{2.93}$$

as r depends also on x. As C is symmetric, then the vector r is a negative gradient of f, i.e.

$$r = -g(x) = b - Cx. \tag{2.94}$$

Hence, we see that the residual $r(x)$ is in the direction of steepest descent of f at the point x. The methods based on this idea are known as methods of steepest descent or gradient methods.

2.5.2 Modern Iterative Methods

Classical iterative methods were based on an eigenvalue analysis, while modern iterative methods and their acceleration parameters are computed by a closed formula depending upon iterates. Above all methods of gradient and the conjugate gradient method will be given. For more details see, For example, Hestenes (1980), Golub and Van Loan (1989), Axelsson and Barker (1984), Axelsson (1994).

2.5.2.1. Method of the steepest descent or the gradient method

The method of steepest descent is also known in the literature as the gradient method or the Richardson method.

Let us consider the following problem

$$\text{minimize } f(x) = \tfrac{1}{2}(x, Cx) - (b, x) \equiv \tfrac{1}{2}x^T Cx - b^T x. \tag{2.95}$$

To solve this equation the iterative methods with advantages will be used. Then, for finding a local minimizer of a functional $f \in R^n$ the following iterative processes

$$x^{k+1} = x^k + \alpha_k p_k \tag{2.96}$$

will be used, where p_k is a vector determining the direction of motion from point x^k, and α_k is a numerical factor whose value determines the length of the step in the direction p_k. As it was shown for example in Axelsson and Barker (1984) in the method of steepest descent (the gradient method), a vector $p_k = -g(x^k) \equiv \nabla_k x$ and

$$\alpha_k = -p_k^T g(x^k)/p_k^T C p_k = g^T(x^k) g(x^k)/g^T(x^k) C g(x^k), \tag{2.97}$$

and then Equation 2.96 can be written in the following form

$$x^{k+1} = x^k - \alpha_k g(x^k). \tag{2.98}$$

Hence the method of steepest descent for the quadratic function of Equation 2.95 leads to the following algorithm:

Algorithm SD1: for $k = 0, 1, 2, \ldots$

$$\begin{aligned} \boldsymbol{g}^k &\equiv \boldsymbol{g}(\boldsymbol{x}^k) = C\boldsymbol{x}^k - \boldsymbol{b}, \\ \alpha_k &= \boldsymbol{g}^T(\boldsymbol{x}^k)\boldsymbol{g}(\boldsymbol{x}^k)/\boldsymbol{g}^T(\boldsymbol{x}^k)C\boldsymbol{g}(\boldsymbol{x}^k) = \boldsymbol{r}_k^T\boldsymbol{r}_k/\boldsymbol{r}_k^T C\boldsymbol{r}_k = |\boldsymbol{r}_k|^2/\boldsymbol{r}_k^T C\boldsymbol{r}_k, \\ \boldsymbol{x}^{k+1} &= \boldsymbol{x}^k - \alpha_k\boldsymbol{g}(\boldsymbol{x}^k) = \boldsymbol{x}^k + |\boldsymbol{r}_k|^2/\boldsymbol{r}_k^T C\boldsymbol{r}_k, \end{aligned} \tag{2.99}$$

where $\boldsymbol{r}_k = -\boldsymbol{g}(\boldsymbol{x}^k)$ and

$$\lim_{k\to\infty} \boldsymbol{x}^k = \underline{\boldsymbol{x}}^k = C^{-1}\boldsymbol{b}.$$

Since due to Equation 2.99

$$\boldsymbol{g}(\boldsymbol{x}^{k+1}) = \boldsymbol{g}(\boldsymbol{x}^k) - \alpha_k C\boldsymbol{g}(\boldsymbol{x}^k), \tag{2.100}$$

then the algorithm of Equation 2.99 can be rewritten as follows:

Algorithm SD2: $\boldsymbol{g}(\boldsymbol{x}^o) = C\boldsymbol{x}^o - \boldsymbol{b} = -\boldsymbol{r}_o.$

For $k = 0, 1, 2, \ldots$

$$\begin{aligned} \alpha_k &= \boldsymbol{g}^T(\boldsymbol{x}^k)\boldsymbol{g}(\boldsymbol{x}^k)/\boldsymbol{g}^T(\boldsymbol{x}^k)C\boldsymbol{g}(\boldsymbol{x}^k) = \boldsymbol{r}_k^T\boldsymbol{r}_k/\boldsymbol{r}_k^T C\boldsymbol{r}_k = |\boldsymbol{r}_k|^2/\boldsymbol{r}_k^T C\boldsymbol{r}_k, \\ \boldsymbol{x}^{k+1} &= \boldsymbol{x}^k - \alpha_k\boldsymbol{g}(\boldsymbol{x}^k) = \boldsymbol{x}^k + \alpha_k\boldsymbol{r}_k, \\ \boldsymbol{g}(\boldsymbol{x}^{k+1}) &= \boldsymbol{g}(\boldsymbol{x}^k) - \alpha_k C\boldsymbol{g}(\boldsymbol{x}^k), \quad (\text{or } \boldsymbol{r}_{k+1} = \boldsymbol{r}_k - \alpha_k C\boldsymbol{r}_k) \end{aligned} \tag{2.101}$$

end k;

and

$$\lim_{k\to\infty} \boldsymbol{x}^k = \underline{\boldsymbol{x}} = C^{-1}\boldsymbol{b}.$$

We shall estimate the rate of convergence of the method of steepest descent, i.e. we shall estimate $\|\boldsymbol{x} - \underline{\boldsymbol{x}}\|$, the Euclidean norm of the error vector $\boldsymbol{x} - \underline{\boldsymbol{x}}$, representing the distance between $\boldsymbol{x}$ and $\underline{\boldsymbol{x}}$. An alternative idea for analysation of the convergence of the method of steepest descent is to use $f(\boldsymbol{x}) - f(\underline{\boldsymbol{x}})$ as a measure of error.

To analyse the convergence of the method of steepest descent, we shall rewrite the functional of Equation 2.95 into the "energy" form

$$f(\boldsymbol{x}) = \tfrac{1}{2}(\boldsymbol{x} - \underline{\boldsymbol{x}})^T C(\boldsymbol{x} - \underline{\boldsymbol{x}}) + f(\underline{\boldsymbol{x}}), \tag{2.102}$$

where $\underline{\boldsymbol{x}} = C^{-1}\boldsymbol{b}$ is the minimizer of Equation 2.95.

Let $(\boldsymbol{x}, \boldsymbol{y}) = \boldsymbol{x}^T C\boldsymbol{y}$ and $\|\boldsymbol{x}\| = (\boldsymbol{x}, \boldsymbol{x})^{1/2} = (\boldsymbol{x}^T C\boldsymbol{x})^{1/2}$ be the energy inner product and energy norm corresponding to a positive definite matrix C. Then from Equation 2.102 it follows that

$$(f(\boldsymbol{x}) - f(\underline{\boldsymbol{x}})) = \tfrac{1}{2}\|\boldsymbol{x} - \underline{\boldsymbol{x}}\|^2 \tag{2.103}$$

and hence

$$\|\boldsymbol{x} - \underline{\boldsymbol{x}}\| = (2(f(\boldsymbol{x}) - f(\underline{\boldsymbol{x}})))^{1/2}. \tag{2.104}$$

Then the rate of convergence of the method of steepest descent in the energy norm is solved in the next theorem.

Theorem 2.72: It holds that

$$\left\|x^k - \underline{x}\right\| \leq [(\kappa(C) - 1)/(\kappa(C) + 1)]^k \left\|x^o - \underline{x}\right\|, \tag{2.105}$$

where $\kappa(C)$ is the spectral condition number of the matrix C.

Proof: From Equation 2.101 and using the relation $x - \underline{x} = C^{-1}g(x)$ we have

$$(x - \underline{x})^T C(x - \underline{x}) = (C^{-1}g(x))^T CC^{-1}g(x) = g^T(x)C^{-1}g(x). \tag{2.106}$$

Since C is a positive definite matrix it has positive eigenvalues $0 < \lambda_1 \leq \lambda_2 \leq \ldots \leq \lambda_n$ and its spectral condition number $\kappa(C) = \lambda_n/\lambda_1 \geq 1$. Then from Axelsson and Barker (1984)

$$1 \leq (x^T Cx)(x^T C^{-1}x)(x^T x)^{-2} \leq (\lambda_1 + \lambda_n)^2(4\lambda_1\lambda_n)^{-1} \quad \forall x \in R^n. \tag{2.107}$$

Then using Equations 2.106, 2.100 and 2.107 we have

$$\begin{aligned}
&(x^{k+1} - \underline{x})^T C(x^{k+1} - \underline{x})g^T(x^{k+1})C^{-1}g(x^{k+1}) \\
&= g^T(x^k)(I - \alpha_k C)C^{-1}(I - \alpha_k C)g(x^k) \\
&= g^T(x^k)(I - \alpha_k C)^2 C^{-1}g(x^k) \\
&= \{1 - [g^T(x^k)g(x^k)]^2/[(g^T(x^k)Cg(x^k))(g^T(x^k)C^{-1}g(x^k)]\}(x^k - \underline{x})^T C(x^k - \underline{x}) \\
&\leq (1 - 4\lambda_1\lambda_n/(\lambda_1 + \lambda_n)^2)(x^k - \underline{x})^T C(x^k - \underline{x}).
\end{aligned}$$

Hence

$$(x^{k+1} - \underline{x})^T C(x^{k+1} - \underline{x}) = (\kappa(C) - 1)^2(\kappa(C) + 1)^{-2}(x^k - \underline{x})^T C(x^k - \underline{x}),$$

and hence

$$\left\|x^{k+1} - \underline{x}\right\| \leq (\kappa(C) - 1)/(\kappa(C) + 1)\left\|x^k - \underline{x}\right\|.$$

From here Equation 2.105 follows immediately.

Preconditioning

To obtain better convergence properties of this method the preconditioning technique will be used. The idea of the preconditioning technique is as follows:

Let W be a positive definite $n \times n$ matrix of the form $W = EE^T$, let C be a positive definite matrix, and let us introduce the transformation $y = E^T x$. Then the quadratic functional $f(x)$ can be expressed in terms of a new variable y in the form

$$\underline{f}(y) = f(E^{-T}y) = \tfrac{1}{2}(y, \overline{C}y) - (\underline{b}, y) \equiv \tfrac{1}{2}y^T\overline{C}y - \underline{b}^T y, \tag{2.108}$$

where

$$\overline{C} = E^{-1}CE^{-T}, \quad \underline{b} = E^{-1}b. \tag{2.109}$$

It is evident that $\overline{C}$ is also symmetric. Since $y^T\overline{C}y = x^T Cx$ and $x^T Cx > 0 \, \forall x \neq 0$ and since C is assumed to be positive definite, then $y^T\overline{C}y > 0 \, \forall y \neq 0$, and hence $\overline{C}$ is also positive definite. Moreover, since $E^{-T}\overline{C}E^T = E^{-T}E^{-1}C = W^{-1}C$ then $\overline{C}$ and

$W^{-1}C$ have the same eigenvalues $0 < \overline{\lambda}_1 \leq \overline{\lambda}_2 \leq \ldots \leq \overline{\lambda}_n$, and therefore the spectral condition number $\kappa(\overline{C}) = \overline{\lambda}_n/\overline{\lambda}_1$ is determined by W and C.

The algorithm of the preconditioned steepest descent method is then the following:

Algorithm PSD1: Let $\boldsymbol{y}^o$ be arbitrary. For $k = 0, 1, 2, \ldots$

$$\begin{aligned} &\underline{\boldsymbol{g}}(\boldsymbol{y}^k) = \overline{C}\boldsymbol{y}^k - \underline{\boldsymbol{b}}, \\ &\underline{\alpha}_k = \underline{\boldsymbol{g}}^T(\boldsymbol{y}^k)\underline{\boldsymbol{g}}(\boldsymbol{y}^k)/\underline{\boldsymbol{g}}^T(\boldsymbol{y}^k)\overline{C}\underline{\boldsymbol{g}}(\boldsymbol{y}^k), \quad \boldsymbol{y}^{k+1} = \boldsymbol{y}^k - \underline{\alpha}_k\underline{\boldsymbol{g}}(\boldsymbol{y}^k), \\ &\text{end } k; \end{aligned} \tag{2.110}$$

and $\lim_{k\to\infty} \boldsymbol{y}^k = \underline{\boldsymbol{y}} = (\overline{C})^{-1}\underline{\boldsymbol{b}}$.

Since $\boldsymbol{x}^k = E^{-T}\boldsymbol{y}^k$, $\boldsymbol{g}(\boldsymbol{x}^k) = C\boldsymbol{x}^k - \boldsymbol{b}$ for $k = 0, 1, 2, \ldots$, and since Equation 2.109 holds, then using Equation 2.110 we can find

$$\begin{aligned} \underline{\boldsymbol{g}}(\boldsymbol{y}^k) &= E^{-1}\boldsymbol{g}(\boldsymbol{x}^k), \\ \underline{\alpha}_k &= \boldsymbol{g}^T(\boldsymbol{x}^k)W^{-1}\boldsymbol{g}(\boldsymbol{x}^k)/(W^{-1}\boldsymbol{g}(\boldsymbol{x}^k))^T C(W^{-1}\boldsymbol{g}(\boldsymbol{x}^k)) \\ &= \boldsymbol{g}^T(\boldsymbol{x}^k)\boldsymbol{h}(\boldsymbol{x}^k)/(\boldsymbol{h}^T(\boldsymbol{x}^k)C\boldsymbol{h}(\boldsymbol{x}^k)), \end{aligned}$$

and

$$\boldsymbol{y}^{k+1} = E^T(\boldsymbol{x}^k - \underline{\alpha}_k\boldsymbol{h}(\boldsymbol{x}^k)),$$

where $\boldsymbol{h}(\boldsymbol{x}^k) = W^{-1}\boldsymbol{g}(\boldsymbol{x}^k)$.

Since $\boldsymbol{y}^{k+1} = E^T\boldsymbol{x}^{k+1}$ then we can find the following algorithm.

Algorithm PSD2: Let $\boldsymbol{x}^o$ be arbitrary. For $k = 0, 1, 2, \ldots$.

$$\begin{aligned} &\boldsymbol{g}(\boldsymbol{x}^k) = C\boldsymbol{x}^k - \boldsymbol{b}, \\ &\boldsymbol{h}(\boldsymbol{x}^k) = W^{-1}\boldsymbol{g}(\boldsymbol{x}^k), \\ &\alpha_k = \boldsymbol{g}^T(\boldsymbol{x}^k)\boldsymbol{h}(\boldsymbol{x}^k)/\boldsymbol{h}^T(\boldsymbol{x}^k)C\boldsymbol{h}(\boldsymbol{x}^k), \\ &\boldsymbol{x}^{k+1} = \boldsymbol{x}^k - \alpha_k\boldsymbol{h}(\boldsymbol{x}^k), \\ &\text{end } k; \end{aligned} \tag{2.111}$$

and $\lim_{k\to\infty} \boldsymbol{x}^k = \underline{\boldsymbol{x}} = C^{-1}\boldsymbol{b}$ and since $\underline{\boldsymbol{y}} = E^T\underline{\boldsymbol{x}}$ then $\underline{\boldsymbol{y}} = (\overline{C})^{-1}\underline{\boldsymbol{b}}$, so that $\underline{\boldsymbol{x}}$ and $\underline{\boldsymbol{y}}$ are minimizers of $f(\boldsymbol{x})$ and $\underline{f}(\boldsymbol{y})$, respectively.

Moreover, since

$$\begin{aligned} \boldsymbol{y}^k - \underline{\boldsymbol{y}} &= E^T(\boldsymbol{x}^k - \underline{\boldsymbol{x}}), \\ \left\|\boldsymbol{y}^k - \underline{\boldsymbol{y}}\right\|^2 &= [E^T(\boldsymbol{x}^k - \underline{\boldsymbol{x}})]^T E^{-1}C(E^{-1})^T[E^T(\boldsymbol{x}^k - \underline{\boldsymbol{x}})] \\ &= (\boldsymbol{x}^k - \underline{\boldsymbol{x}})^T C(\boldsymbol{x}^k - \underline{\boldsymbol{x}}) = \left\|\boldsymbol{x}^k - \underline{\boldsymbol{x}}\right\|^2, \end{aligned}$$

then due to Theorem 2.72

$$\left\|\boldsymbol{y}^k - \underline{\boldsymbol{y}}\right\| \leq (\kappa(\overline{C}) - 1)^k(\kappa(\overline{C}) + 1)^{-k}\left\|\boldsymbol{y}^o - \underline{\boldsymbol{y}}\right\|$$

and $\left\|\boldsymbol{x}^k - \underline{\boldsymbol{x}}\right\| \leq (\kappa(C) - 1)^k(\kappa(C) + 1)^{-k}\left\|\boldsymbol{x}^o - \underline{\boldsymbol{x}}\right\|$.

If W can be found such that $\kappa(\overline{C}) < \kappa(C)$ then the rate of convergence of Equation 2.111, i.e. of the algorithm PSD2, will be better than that of Equation 2.101; i.e. the algorithm SD2.

Remark: Since the matrix C can be split as $C = P - N$, where P is a non-singular and easy to invert matrix, then the sequence $\{\boldsymbol{x}^k\}$ is generated by

$$P\boldsymbol{x}^{k+1} = N\boldsymbol{x}^k + \boldsymbol{b}, \quad k \geq 0, \quad \boldsymbol{x}^o \text{ is given,} \tag{2.112}$$

and since the iteration above is equivalent to

$$P(\boldsymbol{x}^{k+1} - \boldsymbol{x}^k) = \boldsymbol{r}^k, \quad k \geq 0, \tag{2.113}$$

then the Jacobi, Gauss-Seidel, S.O.R., S.S.O.R. methods can be interpreted as stationary preconditioned steepest descent methods, with the acceleration parameter $\alpha = 1$ and the preconditioner $P = D$ or $P = D + L$ or $P = \omega^{-1}D + L$ or $P = (\omega(2-\omega))^{-1}(D + \omega L)D^{-1}(D + \omega U)$, respectively, where D is diagonal, L and U are lower and upper triangular matrices of the split matrix C, i.e. $C = D + L + U$.

2.5.2.2. *The conjugate gradient method or the CG-method*

For a complete derivation of the CG-method we refer for example to Barker and Axelsson (1984), Golub and Van Loan (1989), Axelsson (1994).

Assume that C is an $n \times n$ symmetric and positive definite matrix. To minimize the quadratic functional

$$f(\boldsymbol{x}) = \tfrac{1}{2}(\boldsymbol{x}, C\boldsymbol{x}) - (\boldsymbol{b}, \boldsymbol{x}) \equiv \tfrac{1}{2}\boldsymbol{x}^T C\boldsymbol{x} - \boldsymbol{b}^T\boldsymbol{x}, \tag{2.114}$$

we are concerned with finding its minimizer $\underline{\boldsymbol{x}} = C^{-1}\boldsymbol{b}$. We also notice that

$$\boldsymbol{r}^k = -\boldsymbol{g}(\boldsymbol{x}^k) = \boldsymbol{b} - C\boldsymbol{x}^k \tag{2.115}$$

is the residual vector and it represents the negative gradient of f at $\boldsymbol{x}^k$.

The method of steepest descent (or the gradient method or the Richardson method) was characterized by incrementing iterates in the direction of the negative gradient of f:

$$\boldsymbol{x}^{k+1} = \boldsymbol{x}^k + \alpha_k\boldsymbol{r}^k = \boldsymbol{x}^k - \alpha_k\boldsymbol{g}(\boldsymbol{x}^k). \tag{2.116}$$

In the conjugate direction methods, the iteration process is obtained from Equation 2.116 by replacing $\boldsymbol{r}^k$ with a new direction $\boldsymbol{p}_k$ that is not parallel to that of the gradient (Hestenes and Stiefel (1952), Hestenes (1980)). In the conjugate gradient method (CG-method), the directions $\{\boldsymbol{p}_k\}$ satisfy the orthogonality property $\boldsymbol{p}_j^T C\boldsymbol{p}_k = 0$ for $j \neq$k. The iteration process is of the type

$$\boldsymbol{x}^{k+1} = \boldsymbol{x}^k + \alpha_k\boldsymbol{p}_k, \quad k \geq 0, \tag{2.117}$$

where $\boldsymbol{p}_k$ is a vector determining the new direction of motion from point $\boldsymbol{x}^k$, and α_k represents the length of the step in the new direction $\boldsymbol{p}_k$. We shall take α_k in the form

$$\alpha_k = -\boldsymbol{p}_k^T\boldsymbol{g}(\boldsymbol{x}^k)/\boldsymbol{p}_k^T C\boldsymbol{p}_k = \boldsymbol{p}_k^T\boldsymbol{r}^k/\boldsymbol{p}_k^T C\boldsymbol{p}_k, \tag{2.118}$$

where $\boldsymbol{r}^k = -\boldsymbol{g}(\boldsymbol{x}^k) = \boldsymbol{b} - C\boldsymbol{x}^k$. Multiplying Equation 2.117 by C and subtracting vector $\boldsymbol{b}$ we obtain

$$C\boldsymbol{x}^{k+1} - \boldsymbol{b} = C\boldsymbol{x}^k - \boldsymbol{b} + \alpha_k C\boldsymbol{p}_k, \quad k \geq 0,$$

and hence

$$\boldsymbol{g}(\boldsymbol{x}^{k+1}) = \boldsymbol{g}(\boldsymbol{x}^k) + \alpha_k C\boldsymbol{p}_k. \tag{2.119}$$

Multiplying this equation by $\boldsymbol{p}_k^T$ and using Equation 2.118 we have

$$\boldsymbol{p}_k^T\boldsymbol{g}(\boldsymbol{x}^{k+1}) = \boldsymbol{p}_k^T\boldsymbol{g}(\boldsymbol{x}^k) + \alpha_k\boldsymbol{p}_k^T C\boldsymbol{p}_k = 0.$$

The "new search direction" $\boldsymbol{p}_{k+1}$ will be determined by an iteration process

$$\boldsymbol{p}_{k+1} = -\boldsymbol{g}(\boldsymbol{x}^{k+1}) + \beta_k\boldsymbol{p}_k, \quad k \geq 0, \tag{2.120}$$

where $\boldsymbol{p}_o = -\boldsymbol{g}(\boldsymbol{x}^o)$ and $\beta_o, \beta_1 \ldots$ remain to be determined.

Remark: In the method of steepest descent, $\beta_k = 0$ and the new search direction is found in the plane spanned by the gradient at the most recent point and the previous search direction.

To determine $\{\boldsymbol{r}^k\}$ we minimize in each iteration the error $\|\boldsymbol{x} - \underline{\boldsymbol{x}}\|$ over a certain subset of R^n. Axelsson and Barker (1984) show that β_k in the iteration process is determined by

$$\beta_k = \boldsymbol{g}^T(\boldsymbol{x}^{k+1})C\boldsymbol{p}_k/\boldsymbol{p}_k^T C\boldsymbol{p}_k. \tag{2.121}$$

Since the gradient can be determined directly by $\boldsymbol{g}(\boldsymbol{x}^{k+1}) = C\boldsymbol{x}^{k+1} - \boldsymbol{b}$ or recursively from Equation 2.119, then β_k and α_k can be determined in several ways. Moreover, since due to Equation 2.120

$$\begin{aligned}\boldsymbol{p}_k &= -\boldsymbol{g}(\boldsymbol{x}^k) + \beta_{k-1}\boldsymbol{p}_{k-1} = -\boldsymbol{g}(\boldsymbol{x}^k) + \beta_{k-1}(-\boldsymbol{g}(\boldsymbol{x}^{k-1}) + \beta_{k-2}\boldsymbol{p}_{k-2}) \\ &= -\boldsymbol{g}(\boldsymbol{x}^k) - \beta_{k-1}\boldsymbol{g}(\boldsymbol{x}^{k-1}) + \beta_{k-1}\beta_{k-2}(-\boldsymbol{g}(\boldsymbol{x}^{k-2}) + \beta_{k-3}\boldsymbol{p}_{k-3}) = \ldots,\end{aligned}$$

and since due to Equation 2.119

$$C\boldsymbol{p}_k = \alpha_k^{-1}(\boldsymbol{g}(\boldsymbol{x}^{k-1}) - \boldsymbol{g}(\boldsymbol{x}^k)),$$

then from Equation (2.121) we can find

$$\beta_k = \boldsymbol{g}^T(\boldsymbol{x}^{k+1})\boldsymbol{g}(\boldsymbol{x}^{k+1})/(\boldsymbol{g}^T(\boldsymbol{x}^k)\boldsymbol{g}(\boldsymbol{x}^k)), \tag{2.122}$$

i.e. β_k does not depend on the Hessian. Moreover, for $\boldsymbol{p}_k$ we find that

$$\boldsymbol{p}_k\boldsymbol{g}(\boldsymbol{x}^k) = -\boldsymbol{g}^T(\boldsymbol{x}^k)\boldsymbol{g}(\boldsymbol{x}^k),$$

and then

$$\alpha_k = \boldsymbol{g}^T(\boldsymbol{x}^k)\boldsymbol{g}(\boldsymbol{x}^k)/\boldsymbol{p}_k^T C\boldsymbol{p}_k. \tag{2.123}$$

Such type of algorithm is very important for computer implementation and is also useful for nonquadratic functionals.

Therefore, the CG-method can be formulated as follows.

Algorithm CG1: Let $\boldsymbol{x}^o$ be an initial guess, $\boldsymbol{r}^o := \boldsymbol{b} - C\boldsymbol{x}^o$ and set $\boldsymbol{p}_o := \boldsymbol{r}^o = -\boldsymbol{g}(\boldsymbol{x}^o)$. For each $k \in N$, the k-th iteration can be determined by the formulae

$$\begin{aligned}
\alpha_k &= \boldsymbol{g}^T(\boldsymbol{x}^k)\boldsymbol{g}(\boldsymbol{x}^k)/\boldsymbol{p}_k^T C\boldsymbol{p}_k = \boldsymbol{r}_k^T\boldsymbol{r}_k/\boldsymbol{p}_k^T C\boldsymbol{p}_k = |\boldsymbol{r}_k|^2/\boldsymbol{p}_k^T C\boldsymbol{p}_k,\\
\boldsymbol{x}^{k+1} &= \boldsymbol{x}^k + \alpha_k\boldsymbol{p}_k,\\
\boldsymbol{g}(\boldsymbol{x}^{k+1}) &= \boldsymbol{g}(\boldsymbol{x}^k) + \alpha_k C\boldsymbol{p}_k = -\boldsymbol{r}_k + \alpha_k C\boldsymbol{p}_k = -\boldsymbol{r}_{k+1},\\
\beta_k &= \boldsymbol{g}^T(\boldsymbol{x}^{k+1})\boldsymbol{g}(\boldsymbol{x}^{k+1})/(\boldsymbol{g}^T(\boldsymbol{x}^k)\boldsymbol{g}(\boldsymbol{x}^k)) = \boldsymbol{r}_{k+1}^T\boldsymbol{r}_{k+1}/\boldsymbol{r}_k^T\boldsymbol{r}_k = |\boldsymbol{r}_{k+1}|^2/|\boldsymbol{r}_k|^2,\\
\boldsymbol{p}_{k+1} &= -\boldsymbol{g}(\boldsymbol{x}^{k+1}) + \beta_k\boldsymbol{p}_k = \boldsymbol{r}_{k+1} + \beta_k\boldsymbol{p}_k.
\end{aligned} \tag{2.124}$$

The computer implementation of the above algorithm can be written as follows:

begin ε be given; $\boldsymbol{x}^o$ and $\boldsymbol{b}$ are given input data;

$\boldsymbol{x} := \boldsymbol{x}^o$; $\boldsymbol{g} := \boldsymbol{b}$; $\boldsymbol{g} := C\boldsymbol{x} - \boldsymbol{g}$; $\Theta := \boldsymbol{g}^T\boldsymbol{g}$;

if $\Theta \le \varepsilon$ then stop;

$\boldsymbol{p} := -\boldsymbol{g}$;

A: $\boldsymbol{h} := C\boldsymbol{p}$; $\alpha := \Theta/\boldsymbol{p}^T\boldsymbol{h}$: $\boldsymbol{x} := \boldsymbol{x} + \alpha\boldsymbol{p}$;

$\boldsymbol{g} := \boldsymbol{g} + \alpha\boldsymbol{h}$; $\vartheta := \boldsymbol{g}^T\boldsymbol{g}$;

if $\vartheta \le \varepsilon$ then stop;

$\beta := \vartheta/\Theta$; $\Theta := \vartheta$; $\boldsymbol{p} := -\boldsymbol{g} + \beta\boldsymbol{p}$; goto A;

end;

The iterate $\boldsymbol{x}^{k+1}$ is equal to the previous $\boldsymbol{x}^k$ if and only if $\boldsymbol{p}_k = 0$ or $\alpha_k = 0$, which again means that $\boldsymbol{x}^k = C^{-1}\boldsymbol{b}$.

Convergence analysis of CG-method

Let us denote $\boldsymbol{e}_k \equiv \boldsymbol{x}_k - \underline{\boldsymbol{x}}$, and for any fixed direction $\boldsymbol{p}_k$ let us put

$$\boldsymbol{e}_{k+1}(\alpha) = \boldsymbol{x}^{k+1} - \underline{\boldsymbol{x}} = (\boldsymbol{x}^k + \alpha\boldsymbol{p}_k) - \underline{\boldsymbol{x}}.$$

The values α_k of the CG1 algorithm, given in Equation 2.124, are such that for one of them

$$(\mathrm{d}/\mathrm{d}\alpha\, \|\boldsymbol{e}_{k+1}(\alpha)\|^2)\Big|_{\alpha=\alpha_k} = 0, \tag{2.125}$$

i.e. α_k is the optimal parameter minimizing the functional $f(\boldsymbol{x}^k + \alpha\boldsymbol{p}_k)$ among all possible $\alpha \in R^1$. Moreover, β_k from Equation 2.124 is such that $(\boldsymbol{p}_k, C\boldsymbol{p}_k) \equiv \boldsymbol{p}_k^T C\boldsymbol{p}_k = 0$ for all $k \in N$.

The Krylov space of the order k, generated by $\boldsymbol{r}_o$, will be defined by

$$K_k(\boldsymbol{r}_o) = \mathrm{span}\{\boldsymbol{r}_o, C\boldsymbol{r}_o, \ldots, C^{k-1}\boldsymbol{r}_o\}. \tag{2.126}$$

Lemma 2.11: For $k \geq 0$ it holds that

$$\text{span}\,\{\boldsymbol{p}_o, \boldsymbol{p}_1, \ldots, \boldsymbol{p}_k\} = \text{span}\{\boldsymbol{r}_o, C\boldsymbol{r}_o, \ldots, C^k\boldsymbol{r}_o\} \equiv K_{k+1}(\boldsymbol{r}_o). \tag{2.127}$$

Proof: From Equation 2.124, i.e. $\boldsymbol{p}_{k+1} = \boldsymbol{r}_{k+1} + \beta_k \boldsymbol{p}_k$ and $\boldsymbol{p}_o = \boldsymbol{r}_o$, it follows that

$$\text{span}\,\{\boldsymbol{p}_o, \boldsymbol{p}_1, \ldots, \boldsymbol{p}_k\} = \text{span}\{\boldsymbol{r}_o, \boldsymbol{r}_1, \ldots, \boldsymbol{r}_k\}, \quad k \geq 0.$$

Assume that span $\{\boldsymbol{p}_o, \boldsymbol{p}_1, \ldots, \boldsymbol{p}_k\} = K_{k+1}(\boldsymbol{r}_o)$. Then $C\boldsymbol{p}_k \in K_{k+2}(\boldsymbol{r}_o)$, and since $\boldsymbol{r}_{k+1} = \boldsymbol{r}_k - \alpha_k C\boldsymbol{p}_k$ then $\boldsymbol{r}_{k+1} \in K_{k+2}(\boldsymbol{r}_o)$, i.e. span $\{\boldsymbol{p}_o, \boldsymbol{p}_1, \ldots, \boldsymbol{p}_{k+1}\} \in K_{k+2}(\boldsymbol{r}_o)$. On the other hand, $C^k\boldsymbol{r}_o \in$ span $\{\boldsymbol{p}_o, \boldsymbol{p}_1, \ldots, \boldsymbol{p}_k\}$ by the induction hypothesis and consequently $C^{k+1}\boldsymbol{r}_o \in$ span $\{C\boldsymbol{p}_o, C\boldsymbol{p}_1, \ldots, C\boldsymbol{p}_k\}$. Thus, due to $\boldsymbol{r}_{k+1} = \boldsymbol{r}_k - \alpha_k C\boldsymbol{p}_k$, we have $C^{k+1}\boldsymbol{r}_o \in$ span $\{\boldsymbol{r}_o, \boldsymbol{r}_1, \ldots, \boldsymbol{r}_{k+1}\}$ and hence Equation 2.127 follows.

In the Krylov space $K_k(\boldsymbol{r}_o)$ for each $k \geq 1$, $\boldsymbol{x}^k$ satisfies

$$f(\boldsymbol{x}^k) \leq f(\boldsymbol{v}) \quad \forall \boldsymbol{v} \in \boldsymbol{x}^o + K_k(\boldsymbol{r}_o). \tag{2.128}$$

Lemma 2.12: It holds that where

$$\left.\begin{aligned} (\boldsymbol{r}_k, \boldsymbol{r}_l) &= 0 \qquad (2.129) \\ (\boldsymbol{p}_k, C\boldsymbol{p}_l) &= 0 \qquad (2.130) \end{aligned}\right\} \text{ for } k \neq 1.$$

Proof: The induction technique on l will be used. Firstly, it holds that $(\boldsymbol{r}_1, \boldsymbol{r}_o) = 0$, $(C\boldsymbol{p}_1, \boldsymbol{p}_o) = 0$. Let us assume that

$$(\boldsymbol{r}_i, \boldsymbol{r}_j) = 0, (C\boldsymbol{p}_i, \boldsymbol{p}_j) = 0 \quad \text{for } 0 \leq j < i \leq l$$

and we shall prove that these relations are valid also for $0 \leq j \leq i \leq l+1$. Putting $j = l$, $i = l+1$ then

$$(\boldsymbol{p}_l, \boldsymbol{r}_{l+1}) = 0, \quad (p_l, C\boldsymbol{p}_l) = (\boldsymbol{r}_l, C\boldsymbol{p}_l), \quad (\boldsymbol{r}_{l+1}, \boldsymbol{r}_l) = 0.$$

Let $1 \leq j < l$, $i = l+1$, then

$$(\boldsymbol{r}_{l+1}, \boldsymbol{r}_j) = (\boldsymbol{r}_l, \boldsymbol{r}_i) - \alpha_l(C\boldsymbol{p}_l, \boldsymbol{r}_j) = -\alpha_l(C\boldsymbol{p}_l, \boldsymbol{p}_j - \beta_{j-1}\boldsymbol{p}_{j-1}) = 0.$$

Since $\boldsymbol{r}_o = \boldsymbol{p}_o$, the same is valid for $j = 0$. Furthermore, since C is symmetric, then for $0 \leq j < l$ we have

$$(\boldsymbol{p}_{l+1}, C\boldsymbol{p}_j) = (\boldsymbol{r}_{l+1}, C\boldsymbol{p}_j) + \beta_l(C\boldsymbol{p}_l, \boldsymbol{p}_j) = (\boldsymbol{r}_{l+1}, \boldsymbol{r}_j - \boldsymbol{r}_{j+1})\alpha_j^{-1} = 0,$$

when $\alpha_j \neq 0$. For the case when $\alpha_j = 0$, then $\boldsymbol{r}_j = \boldsymbol{p}_j = 0$, which completes the proof.

From this lemma it follows that $\boldsymbol{r}_k = 0$ (i.e. from Equation 2.129 for some $k \leq n$, where n is the order of the matrix and the conjugate gradient method converges to the exact solution in a number of iterations not greater than n. Moreover, if the matrix C has l different eigenvalues, where $l < n$, then the method is converged after l iterations.

As a result we shall prove the fundamental convergence theorem of the conjugate gradient method.

Theorem 2.73: The following estimate holds:

$$\begin{aligned} \|\boldsymbol{e}_k\| = \left\|\boldsymbol{x}^k - \underline{\boldsymbol{x}}\right\| &\leq 2\{([\kappa(C)]^{1/2} - 1)/([\kappa(C)]^{1/2} + 1)\}^k \|\boldsymbol{x}^o - \underline{\boldsymbol{x}}\| \\ &= 2\{([\kappa(C)]^{1/2} - 1)/([\kappa(C)]^{1/2} + 1)\}^k \|\boldsymbol{e}_o\|, \quad k \in N, \end{aligned} \tag{2.131}$$

where $\kappa(C)$ is the spectral condition number of the matrix C.

Proof: From Equation 2.124, i.e. $\boldsymbol{x}^{k+1} = \boldsymbol{x}^k + \alpha_k \boldsymbol{p}_k,\ k \geq 0$, we have

$$\boldsymbol{x}^{k+1} - \boldsymbol{x}^o = \sum_{i=0}^{k} \alpha_i \boldsymbol{p}_i.$$

Then, taking $0 \leq j \leq k$ and putting $\underline{\boldsymbol{x}} = C^{-1}\boldsymbol{b}$, and using Lemma 2.11, we obtain

$$(C\boldsymbol{p}_j, \boldsymbol{x}^{k+1} - \boldsymbol{x}^o) = \alpha_j (C\boldsymbol{p}_j, \boldsymbol{p}_j) = (\boldsymbol{p}_j, \boldsymbol{r}_j) = (C\boldsymbol{p}_j, \underline{\boldsymbol{x}} - \boldsymbol{x}^j) = (C\boldsymbol{p}_j, \underline{\boldsymbol{x}} - \boldsymbol{x}^o).$$

Hence, $\boldsymbol{x}^{k-1} - \boldsymbol{x}^o$ is an orthogonal projection of $\underline{\boldsymbol{x}} - \boldsymbol{x}^o$ onto the space $K_k(\boldsymbol{r}_o) = \text{span}\{\boldsymbol{p}_o, \boldsymbol{p}_1, \ldots, \boldsymbol{p}_k\}$ with respect to the scalar product, defined by $(\boldsymbol{u}, \boldsymbol{v}) = (C\boldsymbol{u}, \boldsymbol{v})$.

Thus $\boldsymbol{x}^{k+1} - \boldsymbol{x}^o$ is the projection of $\underline{\boldsymbol{x}} - \boldsymbol{x}^o$ onto $K_{k+1}(\boldsymbol{r}_o)$ and

$$\left\|\underline{\boldsymbol{x}} - \boldsymbol{x}^{k+1}\right\| = \left\|\underline{\boldsymbol{x}} - \boldsymbol{x}^o - (\boldsymbol{x}^{k+1} - \boldsymbol{x}^o)\right\| = \min_{\boldsymbol{w} \in K_{k+1}(\boldsymbol{r}_o)} \|\underline{\boldsymbol{x}} - \boldsymbol{x}_o - \boldsymbol{w}\|.$$

Since $\boldsymbol{r}_o = C(\underline{\boldsymbol{x}} - \boldsymbol{x}^o)$, any $\boldsymbol{w} \in K_{k+1}(\boldsymbol{r}_o)$ can be written as

$$\boldsymbol{w} = \sum_{j=1}^{k+1} \gamma_j C_j (\underline{\boldsymbol{x}} - \boldsymbol{x}^o), \quad r_j \in R$$

and consequently

$$\left\|\underline{\boldsymbol{x}} - \boldsymbol{x}^{k+1}\right\| = \min_{p \in P^*_{k+1}} \|p(C)(\underline{\boldsymbol{x}} - \boldsymbol{x}^o)\|,$$

where P^*_{k+1} represents the set of polynomials $p : R \to R$ of degree less than or equal to $k+1$ and such that $p(0) = 1$.

Since $C = C^T > 0$ then there exists an orthogonal basis given by the eigenvectors of C such that the eigenvalues λ_j of C are strictly positive. By expanding $\underline{\boldsymbol{x}} - \boldsymbol{x}^o$ with respect to these eigenvectors then

$$\|p(C)(\underline{\boldsymbol{x}} - \boldsymbol{x}^o)\| \leq \max_{1 \leq j \leq n} |p(\lambda_j)| \, \|(\underline{\boldsymbol{x}} - \boldsymbol{x}^o)\|.$$

One can find

$$p(x) = \frac{T_{k+1}((\lambda_{\max} + \lambda_{\min} - 2x)/(\lambda_{\max} - \lambda_{\min}))}{T_{k+1}((\lambda_{\max} + \lambda_{\min})/(\lambda_{\max} - \lambda_{\min}))}$$

where we denote by $T_{k+1}(x)$ the Chebyshev polynomial of degree $k+1$. Since $|T_{k+1}(w)| \leq 1$ for $|w| \leq 1$, then it follows that

$$\max_{x \in [\lambda_{\max}, \lambda_{\min}]} |p(x)| \leq [T_{k+1}((\lambda_{\max} + \lambda_{\min})/(\lambda_{\max} - \lambda_{\min}))]^{-1}.$$

Hence

$$\max_{x \in [\lambda_{\max}, \lambda_{\min}]} |p(x)| \leq 2\{1 + [(\lambda_{\max}^{1/2} - \lambda_{\min}^{1/2})/(\lambda_{\max}^{1/2} + \lambda_{\min}^{1/2})]^{2k+2}\}^{-1} [(\lambda_{\max}^{1/2} - \lambda_{\min}^{1/2})/(\lambda_{\max}^{1/2} + \lambda_{\min}^{1/2})]^{k+1}$$

and hence the assertion of the theorem holds.

2.5.2.3. *The preconditioned conjugate gradient method or the PCG-method*

In the case when the spectral condition number $\kappa(C)$ is large, then the preconditioning technique will be used.

The initial problem of minimizing the functional

$$f(\boldsymbol{x}) = \tfrac{1}{2}(\boldsymbol{x}, C\boldsymbol{x}) - (\boldsymbol{b}, \boldsymbol{x}) \equiv \tfrac{1}{2}\boldsymbol{x}^T C\boldsymbol{x} - \boldsymbol{b}^T\boldsymbol{x}, \quad \boldsymbol{x} \in R^n,$$

will be reduced to that of minimizing the related functional

$$\underline{f}(y) = \tfrac{1}{2}(y, \overline{C}y) - (\underline{b}, y) \equiv \tfrac{1}{2}y^T\overline{C}y - \underline{b}^T y, \quad y \in R^n,$$

where $\overline{C} = E^{-1}CE^{-T}$, $\underline{b} = E^{-1}b$, E is some nonsingular matrix, $E^{-T} \equiv (E^T)^{-1}$ and then $W = EE^T$ is positive definite. The idea of the preconditioning is similar to the case of the method of steepest descent (SD-method). Since the rate of convergence of the conjugate gradient method depends on the spectral condition number of the Hessian, the idea of preconditioning of the CG-method is then similar of that of the SD-method. Since $W = EE^T$ and the matrix E makes $\kappa(\overline{C}) < \kappa(C)$, the rate of convergence of the PCG-method is faster than the rate of convergence of the CG-method, and $\kappa(\overline{C}) = \overline{\lambda}_n/\overline{\lambda}_1$, where $\overline{\lambda}_1$ and $\overline{\lambda}_n$ are the smallest and largest eigenvalues, respectively, of W^{-1}C. The preconditioned conjugate gradient method (PCG-method) can be defined as follows:

Algorithm PCG1: Let x^o be arbitrary, $g(x^o) = Cx^o - b$, $h(x^o) = W^{-1}g(x^o)$, $p_o = -h(x^o)$.

$$\begin{aligned}
&\text{For } k = 0, 1, 2, \ldots\\
&\alpha_k = g^T(x^k)h(x^k)/p_k^T Cp_k;\\
&x^{k+1} = x^k + \alpha_k p_k;\\
&g(x^{k+1}) = g(x^k) + \alpha_k Cp_k;\\
&h(x^{k+1}) = W^{-1}g(x^{k+1});\\
&\beta_k = g^T(x^{k+1})h(x^{k+1})/g^T(x^k)h(x^k);\\
&p_{k+1} = -h(x^{k+1}) + \beta_k p_k;\\
&\text{end } k;
\end{aligned} \tag{2.132}$$

where $W^{-1} = (EE^T)^{-1}$, and $h(x^{k+1}) = W^{-1}g(x^{k+1})$ represents the solution of a system of equations with matrix $W = EE^T$.

Theorem 2.74: The following estimate holds

$$\begin{aligned}
\|e_k\| = \|x^k - \underline{x}\| &\le 2\{[(\kappa(\overline{C}))^{1/2} - 1]/[(\kappa(\overline{C}))^{1/2} + 1]\}^k \|x^o - \underline{x}\|\\
&= 2\{[(\kappa(\overline{C}))^{1/2} - 1]/[(\kappa(\overline{C}))^{1/2} + 1]\}^k \|e_o\|,
\end{aligned}$$

where $\kappa(\overline{C})$ is the spectral condition number of $\overline{C}$.

Proof: To prove the theorem, a similar technique as for the non-preconditioned case is used, but the scalar product is defined in the form induced by the preconditioned matrix $\overline{C}$.

2.5.2.4. *Conjugate gradient methods for non-symmetric problems*

Some geophysical problems, such as propagation of seismic waves in inhomogeneous media, and magnetotelluric problems, etc., lead to systems of linear algebraic equations with non-symmetric matrices. In this section we shall give a short outline of methods of solving such types of problems.

Let C be a non-singular and non-symmetric $n \times n$ matrix and let

$$C = C_s + C_{ss}, \quad C_s = \tfrac{1}{2}(C + C^T), \quad C_{ss} = \tfrac{1}{2}(C - C^T), \tag{2.133}$$

where C_s is the symmetric part of C and C_{ss} is the skew-symmetric part of C. If C is positive then C_s is also definite.

Conjugate gradient normal residual (CGNR-) method

The CGNR-method is based on the CG-method for symmetric systems of algebraic equations. Then the non-singular system of algebraic equations $\boldsymbol{Cx} = \boldsymbol{b}$ can be converted into the symmetric and positive definite one i.e. into the so-called system of normal equations

$$C^T C\boldsymbol{x} = C^T\boldsymbol{b}, \tag{2.134}$$

also known from the theory of least squares method and then applying the conjugate gradient method. The algorithm is then the following:

Algorithm CGNR: Let $\boldsymbol{x}^o$ be given, let $\boldsymbol{r}_o = \boldsymbol{b} - C\boldsymbol{x}^o$, $\boldsymbol{p}_o = C^T\boldsymbol{r}_o$.

$$\begin{aligned}
&\text{For } k = 0, 1, 2, \ldots\\
&\alpha_k = (C^T\boldsymbol{r}_k, C^T\boldsymbol{r}_k)/(C\boldsymbol{p}_k, C\boldsymbol{p}_k) = \left|C^T\boldsymbol{r}_k\right|^2 / \left|C\boldsymbol{p}_k\right|^2;\\
&\boldsymbol{x}^{k+1} = \boldsymbol{x}^k + \alpha_k\boldsymbol{p}_k;\\
&\boldsymbol{r}_{k+1} = \boldsymbol{r}_k - \alpha_k C\boldsymbol{p}_k;\\
&\beta_{k+1} = (C^T\boldsymbol{r}_{k+1}, C^T\boldsymbol{r}_{k+1})/(C^T\boldsymbol{r}_k, C^T\boldsymbol{r}_k) = \left|C^T\boldsymbol{r}_{k+1}\right|^2 / \left|C\boldsymbol{r}_k^T\right|;\\
&\boldsymbol{p}_{k+1} = C^T\boldsymbol{r}_{k+1} + \beta_{k+1}\boldsymbol{p}_k;\\
&\text{end } k;
\end{aligned} \tag{2.135}$$

This algorithm is convergent and theoretically it converges in no more than n iterations like the CG-method.

Generalized conjugate residual (GCR-) method

This method is a special case of the CG-method that minimizes the residual $|\boldsymbol{r}_k|$ at each iteration and chooses direction vectors $\{\boldsymbol{p}_k\}$ that are $C^T C$-conjugate, i.e. they satisfy the orthogonality conditions

$$(\boldsymbol{p}_i, C^T C\boldsymbol{p}_j) = 0 \text{ for } i \neq j.$$

Let C be a non-symmetric matrix, and let

$$K_j(\boldsymbol{r}_o) = \operatorname{span}\{\boldsymbol{r}_o, C\boldsymbol{r}_o, C^2\boldsymbol{r}_o, \ldots, C^{j-1}\boldsymbol{r}_o\},$$

be the Krylov space of order j generated by $\boldsymbol{r}_o$, where j is a positive integer. Then we have the following algorithm:

Algorithm GCR: Let x^o be given, let $r_o = b - Cx^o$, $p_o = r_o$.

For $k = 0, 1, 2, \ldots$

$$\alpha_k = (r_k, Cp_k)/(Cr_k, Cr_k) = (r_k, Cp_k)/\,|Cr_k|^2\,;$$

$$x^{k+1} = x^k + \alpha_k p_k;$$

$$r_{k+1} = r_k - \alpha_k C p_k; \tag{2.136}$$

$$\beta_{kj} = -(Cr_{k+1}, Cp_j)/(Cp_j, Cp_j) = -(Cr_{k+1}, Cp_j)/\left|Cp_j\right|^2; k - k_o + 1 \le j \le k,$$

$$p_{k+1} = r_{k+1} + \sum\nolimits_{j=k-k_{o+1}}^{k} \beta_{kj} p_j;$$

$$Cp_{k+1} = Cr_{k+1} + \sum\nolimits_{j=k-k_{o+1}}^{k} \beta_{kj} Cp_j$$

for $k_o = k + 1$;

end k;

The GCR-method converges if C is positive definite and

$$\|r_k\| \le \{1 - (\lambda_{\min}(C_s))^2/(\lambda_{\min}(C^T C))\}^{k/2}\, \|r_o\|\,.$$

If C is not positive definite it breaks down.

Bi-conjugate gradient (Bi-CG-) method

The Bi-CG-method is a variant of the CG-method for non-symmetric matrices and was introduced by Fletcher (1976). We construct a residual r_k orthogonal with respect to another row of vectors $\underline{r}_o, \underline{r}_1, \ldots, \underline{r}_{k-1}$. The algorithm of the Bi-CG-method is the following:

Algorithm Bi-CG: Let x^o be given and let $r_o = b - Cx^o$, $p_o = r_o$. Further, let $\underline{r}_o$ be a vector such that $\rho_o = (\underline{r}_o, r_o) \ne 0$, e.g. $\underline{r}_o = r_o$ and let us put $\underline{p}_o = \underline{r}_o$.

For $k = 0, 1, 2, \ldots$

$$v_k = Cp_k;$$

$$\alpha_k = \rho_k/(v_k, \underline{p}_k);$$

$$x^{k+1} = x^k + \alpha_k p_k;$$

$$r_{k+1} = r_k - \alpha_k v_k; \quad \underline{r}_{k+1} = \underline{r}_k - \alpha_k C^T \underline{p}_k; \tag{2.137}$$

$$\rho_{k+1} = (r_{k+1}, \underline{r}_{k+1});$$

$$\beta_{k+1} = \rho_{k+1}/\rho_k;$$

$$p_{k+1} = r_{k+1} + \beta_{k+1} p_k; \quad \underline{p}_{k+1} = \underline{r}_{k+1} + \beta_{k+1} \underline{p}_k;$$

end k;

The Bi-CG-method converges after no more than n iterations. For the symmetric matrices this method also converges, but it does require twice as many operations as the CG-method.

2.5.3 The Multi-Grid Method

In (Axelsson and Barker (1984)) the multi-grid method was derived from the classical point of view, and based on the analysis of variational problems and their finite element approximations. In this section we shall briefly discuss the multi-grid method from the algebraic point of view.

The basic idea of the method depends on the change of the grid, as the linear system arose from a discretization of partial differential equations, like finite difference or finite element approximations, over a certain mesh. Such discretizations of PDEs lead to a requirement to solve a system of linear algebraic equations. Moreover, we see in the theory about classical iterative methods that these methods are effective for the spectral radius of the iterative matrix B to be close to one (but $\rho(B) < 1$). A Fourier analysis though shows that large eigenvalues are associated with errors of low frequency, on which depend the convergence of the iterative process, while the high frequency errors are rapidly damped and therefore the convergence does not depend on them.

Discretization of the variational problems discussed required the solution of a family of linear problems, which in the operator form can be written as

$$C_k \boldsymbol{x}_k = \boldsymbol{b}_k, \quad k = 0, 1, 2, \ldots, N, \tag{2.138}$$

where C_k is an invertible linear operator (matrix) on a finite dimensional space V_k, dim $V_k < V_{k+1}$, $k = 0, 1, 2, \ldots, N-1$, and $\boldsymbol{b}_k$ is a given right hand side. Our aim is to solve Equation 2.138 for $k = N$. Let us put $V = V_N$, $C = C_N$, $\boldsymbol{b} = \boldsymbol{b}_N$, $\boldsymbol{x} = \boldsymbol{x}_N$, and let C_k, $k < N$ be an auxiliary. Let h_k be a step of a used grid and let $h_{k-1} = 2h_k$. Then Equation 2.138 creates a family of discretizations with a factor 2 (generally with a factor s), and to solve each of these we have to solve an iterative process

$$\boldsymbol{x}_k^{i+1} = \boldsymbol{x}_k^i + P_k(\boldsymbol{b}_k - C_k \boldsymbol{x}_k^i), \tag{2.139}$$

where P_k denotes a preconditioner for C_k. Let us denote the error vector by $\boldsymbol{e}_k^i = \boldsymbol{x}_k^i - \boldsymbol{x}_k$ and let

$$\boldsymbol{e}_k^{i+1} = B_k e_k^i, \quad B_k = I - P_k C_k. \tag{2.140}$$

Then we have the multi-grid algorithm consisting of several steps combined into three main steps the pre-smoothing on the fine grid, the coarse grid correction, and the post-smoothing on the fine grid and representing an acceleration algorithm of the type Equation 2.139 with $k = N$.

1. Pre-smoothing on the fine grid:
   (i) Let $\boldsymbol{x}_k^o$ be given, then we execute the iteration process

$$\boldsymbol{x}_k^i = \boldsymbol{x}_k^{i-1} + P_k(\boldsymbol{b}_k - C_k \boldsymbol{x}_k^{i-1}), \quad i = 1, \ldots, m. \tag{2.141}$$

2. Coarse grid correction:
   (ii) Compute the residual vector

$$\boldsymbol{r}_k = \boldsymbol{b}_k - C_k \boldsymbol{x}_k^m. \tag{2.142}$$

(iii) Set $\boldsymbol{r}_{k-1} = R_k^{k-1}\boldsymbol{r}_k$, representing a restriction of the residual vector on the coarse grid, where R_k^{k-1} is a restriction operator, $R_k^{k-1} : V_k \to V_{k-1}$.

(iv) Solve the residual (defect) problem

$$C_{k-1}\boldsymbol{x}_{k-1} = \boldsymbol{r}_{k-1}. \tag{2.143}$$

(v) Set $\underline{\boldsymbol{x}}_k = \boldsymbol{x}_k^m + S_{k-1}^k \boldsymbol{x}_{k-1}$, representing the correction of the solution in V_k, where S_{k-1}^k denotes a prolongation operator, i.e. the operator $S_{k-1}^k : V_{k-1} \to V_k$.

3. Post-smoothing on the fine grid:
   (vi) Let us set $\underline{\underline{x}}_k^o = \underline{\boldsymbol{x}}_k$, then we execute the iterative process

$$\underline{\underline{\boldsymbol{x}}}_k^i = \underline{\underline{\boldsymbol{x}}}_k^{i-1} + P_k(\boldsymbol{b}_k - C_k \underline{\underline{\boldsymbol{x}}}_k^{i-1}), \quad i = 1, 2, \ldots, M. \tag{2.144}$$

One of the numbers of pre- and post-smoothing steps m and M can be equal to zero, but in the general case they depend on k and may be determined adaptively.

Concerning the problem in Equation 2.143, if the dimension V_{k-1} is low it can be solved exactly by a direct method and in this case we have a two-grid method (see Figure 2.2). If the coarse grid problem of Equation 2.143 is solved recursively by γ applications of the multi-grid algorithm we speak of basic multi-grid cycles, for $\gamma = 1$ of V-cycles, for $\gamma = 2$ of W-cycles. The structure of one cycle of the multi-grid method is given in Figure 2.2, where symbols ○, □ and / or \ denote smoothing, solving exactly the fine-to-coarse and coarse-to-fine transfer, respectively.

If we expand the errors at the k-th step into a discrete Fourier series, we can deduce that a relaxation method is very efficient in smoothing the errors, which represents the high frequency error components, whereas for low frequencies the amplitude reduction is very slight (see Figure 2.3). The smoothing properties of an iterative algorithm are measured by a smoothing factor, which is smaller than $\frac{1}{2}$, independently of the step of the grid h_k. When passing to the coarse grid, for which $h_{k-1} = 2h_k$, the high frequency components on the fine grid are not visible on the coarse one, due to aliasing, and therefore they cannot be modified by use of this grid. Defect (residual) corrections smooth the error components that are associated with low frequencies on the fine grid.

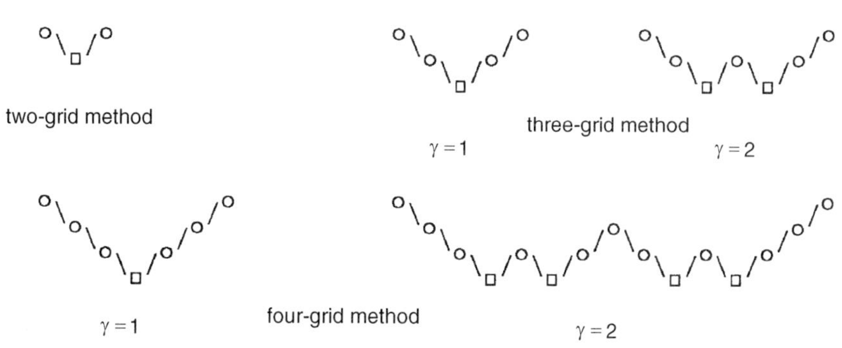

Figure 2.2. Multi-grid Cycles

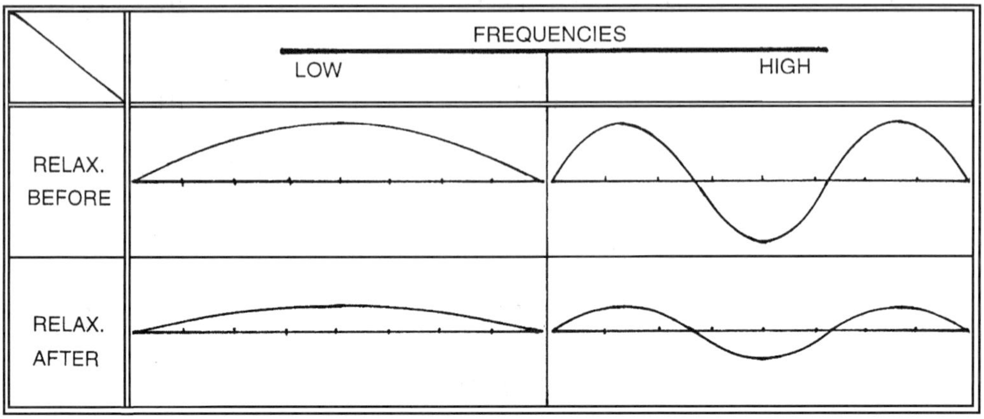

Figure 2.3. Error Smoothing Behaviour of Relaxation Method before and after Relaxation

The convergence of the multi-grid method is studied in detail in very many of papers and books (e.g. Hackbusch (1985), Mandel, McCormick and Bank (1987), Golub and Van Loan (1983), Golub and Ortega (1993), Bramble (1993)). We shall give here the idea for the two-grid cycle only.

Let the error be transformed as

$$\boldsymbol{e}_k^n = M_k \boldsymbol{e}_k^o, \tag{2.145}$$

where $\boldsymbol{e}_k^n$ and $\boldsymbol{e}_k^o$ represent the new and oldest errors respectively, and M_k is the two-grid iteration operator defined as

$$M_k = B_k^M T_k B_k^m, \tag{2.146}$$

where T_k is the coarse grid correction operator defined by

$$T_k = I - S_{k-1}^k C_{k-1}^{-1} R_k^{k-1} C_k. \tag{2.147}$$

For the convergence, $\rho_k(M_k) = \|M_k\| \le c < 1$, for each $k = 0, 1, \ldots, r$ for a suitable matrix norm, holds, where ρ_k is a two-grid convergence factor. To estimate the convergence factor for the multi-grid V-cycle and W-cycle we define it by

$$\varepsilon_k = \max_{\boldsymbol{e}_k^S \neq 0} \left\{ \left| \boldsymbol{e}_k^{s+1} \right| \left| \boldsymbol{e}_k^s \right|^{-1} \right\}, \tag{2.148}$$

where e_k^s and $\boldsymbol{e}_k^{s+1}$ are the errors before and after one step of the multi-grid iteration. For example, for the W-cycle

$$\varepsilon_k \le \rho_k + c\varepsilon_{k-1}^2, \quad k = 1, 2, \ldots, r,$$

provided that the restriction and prolongation operators are bounded in suitable norms. If ρ_k is bounded by a small enough constant c and, moreover, independent of k, then $\varepsilon_r \le c < 1$ and hence it follows that the multi-grid convergence factor is also bounded away from 1 independently of the number of levels r. Let us assume that the cost of smoothing on V_k is proportional to the dimension of a finite dimensional space V_k (we denote it by N_k) then the cost of the V-cycle and W-cycle are also proportional to N_k, i.e. the multi-grid method behaves with optimal asymptotical computational complexity.

2.6. QUADRATIC FUNCTIONALS WITH CONSTRAINTS IN R^N

2.6.1. Introduction

Quadratic functionals with constraints play an important role in optimization theory and its application in structural mechanics as well as in geodynamics. These types of problems lead to the minimizing of quadratic functions with the constraint functions, which are linear. During the last 25 years a number of methods have been developed. The basic methods are given in Hestenes (1980). Additional algorithms can be found in various books and journals concerned with optimization theory and with numerical analysis (see Fiaco and Cormick (1968), Pshenichny and Danilin (1978), Hestenes (1980), Fletcher (1981), Gill, Murray and Wright (1981), Pshenichny (1983) and many other papers).

2.6.2. Problems of Quadratic Programming

Problems of quadratic programming consist of finding a solution $\underline{x}$ to a problem (which we formulate in both use notation)

Problem ($\mathscr{P}$):

$$\min_{x} f(\boldsymbol{x}) = \tfrac{1}{2}(\boldsymbol{x}, C\boldsymbol{x}) + (\boldsymbol{b}, \boldsymbol{x}) = \tfrac{1}{2}\boldsymbol{x}^T C\boldsymbol{x} + \boldsymbol{b}^T\boldsymbol{x}$$

with constraints

$$A\boldsymbol{x} = \boldsymbol{d} \quad \text{i.e.} \quad (\boldsymbol{a}_i, \boldsymbol{x}) - d_i = \boldsymbol{a}_i^T\boldsymbol{x} - d_i = 0, \quad i \in J = \{1, \dots, m\}, \tag{2.149}$$

$$A\boldsymbol{x} \le \boldsymbol{d} \quad \text{i.e.} \quad (\boldsymbol{a}_i, \boldsymbol{x}) - d_i = \boldsymbol{a}_i^T\boldsymbol{x} - d_i \le 0, \quad i \in J = \{1, \dots, m\}, \tag{2.150}$$

$$A\boldsymbol{x} \le \boldsymbol{d} \quad \text{i.e.} \quad (\boldsymbol{a}_i, \boldsymbol{x}) - d_i = \boldsymbol{a}_i^T\boldsymbol{x} - d_i \le 0, \quad i \in J^- = \{1, \dots, k\}, \tag{2.151}$$

$$(\boldsymbol{a}_i, \boldsymbol{x}) - d_i = \boldsymbol{a}_i^T\boldsymbol{x} - d_i = 0, \quad i \in J^o = \{k+1, \dots, m\},$$

where C is an $n \times n$ symmetric positive (semi-) definite matrix, the so-called Hessian matrix, $\boldsymbol{b} \in R^n$, $\boldsymbol{x} \in R^n$, $i \in J \equiv J^- \cup J^o = \{1, \dots, m\}$, where J^- and J^o are finite sets of indices and d_i are corresponding numbers.

If the Hessian matrix C is positive semi-definite, i.e. $(\boldsymbol{x}, C\boldsymbol{x}) \ge 0$ for all $\boldsymbol{x}$, then $\boldsymbol{x}$ is a global solution, and if C is positive definite, i.e. $(\boldsymbol{x}, C\boldsymbol{x}) > 0$ for all $\boldsymbol{x}$, then $\boldsymbol{x}$ is unique. If the Hessian C is infinite then local solutions which are not global can occur. These results follow from the convexity of the function $f(\boldsymbol{x})$.

It is known that the quadratic function $f(\boldsymbol{x}) = \frac{1}{2}(\boldsymbol{x}, C\boldsymbol{x}) + (\boldsymbol{b}, \boldsymbol{x})$ is convex if and only if matrix C is positive definite. The problem with the constraints given in Equation 2.149 is known as a problem of fixed extreme. The problems with the constraints given in Equation 2.150 and 2.151 can be transformed to the problem with constraint of Equation 2.149 by using slack (complementary) variables z_i, so that

$$\min_{\boldsymbol{x}} f(\boldsymbol{x}) = \tfrac{1}{2}(\boldsymbol{x}, C\boldsymbol{x}) + (\boldsymbol{b}, \boldsymbol{x}) \equiv \tfrac{1}{2}\boldsymbol{x}^T C\boldsymbol{x} + \boldsymbol{b}^T\boldsymbol{x} \tag{2.152}$$

$$\text{with constraints } (\boldsymbol{a}_i, \boldsymbol{x}) + z_i^2 - d_i = 0, \quad i \in J,$$

$$\text{or} \quad \text{with constraints } (\boldsymbol{a}_i, \boldsymbol{x}) + z_i^2 - d_i = 0, \quad i \in J^- \cup J^o.$$

Remark: In the theory of optimization it is known that several types of constraints can be mutually transformed. Generally the constraints are often written in the form of concave functions $f_i(\boldsymbol{x})$. In our investigations functions $f_i(\boldsymbol{x})$ are linear functions only.

It is clear that the constraints

$$-\boldsymbol{x}_i \le 0, \quad i \in J \tag{2.153}$$

are of the form of those in Equation 2.150. Similarly, the constraints

$$c_i \le \boldsymbol{x}_i \le d_i, \quad i \in J = J^- \cup J^o \tag{2.154}$$

are of the same form, because these constraints can be written in the form

$$-\boldsymbol{x}_i + c_i \le 0, \quad \boldsymbol{x}_i - d_i \le 0, \quad i \in J. \tag{2.155}$$

Frequently, the constraints defined by the inequality and equality constraints can be written as

$$f_i(\boldsymbol{x}) \le 0, \quad i \in J^-, \quad f_i(\boldsymbol{x}) = 0, \quad i \in J^o. \tag{2.156}$$

These constraints are equivalent to the constraints

$$f_i(\boldsymbol{x}) \le 0, \quad i \in J^-, \quad f_i(\boldsymbol{x}) \le 0, \quad -f_i(\boldsymbol{x}) \le 0, \quad i \in J^o. \tag{2.157}$$

In many cases the constraints are considered in the form

$$f_i(\boldsymbol{x}) = 0, \quad i \in J, \quad \boldsymbol{x}^j \ge 0, \quad j = 1, 2, \ldots, m \le n, \tag{2.158}$$

where $\boldsymbol{x}^j$ is the j-th coordinate of $\boldsymbol{x}$. Then the constraints are obviously of the form of Equation 2.156 and hence can be rewritten in the form of Equation 2.150. By the introduction of slack variables $y^1, \ldots, y^m$, then the constraints of Equation 2.149 can be rewritten as

$$f_i(\boldsymbol{x}) + y^i = 0, \quad y^i \ge 0, \quad i = 1, \ldots, m. \tag{2.159}$$

In this case the problem of minimizing $f(\boldsymbol{x})$ subject to the constraints given in Equation 2.149 is therefore equivalent to minimizing $\underline{f}(\boldsymbol{x}, \boldsymbol{y}) = f(\boldsymbol{x})$ subject to the constraints of Equation 2.159. Furthermore, by introducing slack variables $y^1, \ldots, y^m$ the minimum problem of Equation 2.151, i.e. the problem

$$\min_{\boldsymbol{x}} f(\boldsymbol{x}), \quad f_i(\boldsymbol{x}) \le 0, \quad i \in J^-, \quad f_i(\boldsymbol{x}) = 0, \quad i \in J^o \tag{2.160}$$

is equivalent to the equality constrained minimum problem

$$\min_{x} f(\boldsymbol{x}), \quad f_i(\boldsymbol{x}) + (y^i)^2 = 0, \quad i \in J^-, \quad f_i(\boldsymbol{x}) = 0, \quad i \in J^o. \tag{2.161}$$

It is evident that the problem in Equation 2.160 can be solved by applying equality constrained methods to the problem in Equation 2.161. This procedure is particularly effective when the number of inequality constraints is small, i.e. when the set J^- is small.

Definition 2.17: Let $(\boldsymbol{a}_i, \boldsymbol{x})$ have a continuous first partial differential at the neighbourhood of a point $\boldsymbol{x}_o \in K_o$, where $K_o = \{(\boldsymbol{a}_i, \boldsymbol{x}) = \boldsymbol{d}_i, i \in J\}$ and the rank of a Jacobi matrix $(\partial(\boldsymbol{a}_i, \boldsymbol{x}_o)/\partial x_j)$ at the point $\boldsymbol{x}_o$ is m. Then the point $\boldsymbol{x}_o \in K_o$ is called $r1$-regular.

Definition 2.18: Let $(\boldsymbol{a}_i, \boldsymbol{x})$ have a continuous first partial diferential at the neighbourhood of a point $\boldsymbol{x}_o$. Let denote $J^- = \{i \mid (\boldsymbol{a}_i, \boldsymbol{x}_o) - d_i < 0\}$, $J^o = \{i \mid (\boldsymbol{a}_i, \boldsymbol{x}) - d_i = 0\}$. A point $\boldsymbol{x}_o \in K_o$ is called $r2$-regular according to the set K_o, if $\boldsymbol{x}_o$ is an $r1$-regular point of the set $K_o = \{(\boldsymbol{a}_i, \boldsymbol{x}) = \boldsymbol{d}_i, i \in J\}$.

Definition 2.19: Let the convex quadratic functional $f(\boldsymbol{x})$ be defined at point $\boldsymbol{x}_o$ and have a finite value. Then the subgradient $\boldsymbol{g}$ of $f(\boldsymbol{x})$ at point $\boldsymbol{x}_o$ is defined by

$$f(\boldsymbol{x}) - f(\boldsymbol{x}_o) \geq (\boldsymbol{g}, \boldsymbol{x} - \boldsymbol{x}_o) \text{ for any } \boldsymbol{x}.$$

It can be shown that if $f(\boldsymbol{x})$ is continuous at point $\boldsymbol{x}_o$ then at this point there exist subgradients, and the set of these subgradients is convex, closed and bounded. If $f(\boldsymbol{x})$ is differentiable, then $f'(\boldsymbol{x}_o)$ is a subgradient of $f(\boldsymbol{x})$ at point $\boldsymbol{x}_o$. If $\boldsymbol{g}_1, \ldots, \boldsymbol{g}_k$ are subgradients of convex functions $f_1(\boldsymbol{x}), \ldots, f_k(\boldsymbol{x})$ at point $\boldsymbol{x}_o$, then $\Sigma_{i=1}^k \alpha_i \boldsymbol{g}_i$ is a subgradient of function $\Sigma_{i=1}^k \alpha_i f_i(\boldsymbol{x})$, $\alpha_i \geq 0$, $i = 1, \ldots, k$.

Definition 2.20: If for any $\boldsymbol{x}_1, \boldsymbol{x}_2 \in R^n$ and any $0 \leq \lambda \leq 1$ the inequality

$$f(\lambda \boldsymbol{x}_1 + (1 - \lambda)\boldsymbol{x}_2) \geq \lambda f(\boldsymbol{x}_1) + (1 - \lambda) f(\boldsymbol{x}_2)$$

is satisfied, then the function $f(\boldsymbol{x})$ is concave. Moreover, the function $-f(\boldsymbol{x})$ is a convex function.

Definition 2.21: A set K is called a convex cone if the set is convex and together with every point $\boldsymbol{x} \in K$ it contains all points $\lambda \boldsymbol{x}$ with $\lambda \geq 0$.

Further, let K be a convex cone. Then the set of all vectors $\boldsymbol{y} \in R^n$ satisfying for any $\boldsymbol{x} \in K$ the inequality $(\boldsymbol{x}, \boldsymbol{y}) \geq 0$ is the so-called conjugate cone K^*.

It can be shown that the conjugate cone K^* is a closed convex cone and that $(K^*)^* = K$.

Lemma 2.13: Let K be a convex cone. Then $\boldsymbol{x}_o \in \overline{K}$ if and only if $(\boldsymbol{x}_o, \boldsymbol{y}) \geq 0$ for any $\boldsymbol{y} \in K^*$.

To prove the assertion of this lemma the controversy technique can be used. (For the details see Pshenichny and Danilin (1978).)

Definition 2.22: A cone K is called polyhedral if there exists a finite set of n-dimensional vectors $\boldsymbol{a}_i$, $i = 1, \ldots, m$ such that with $\boldsymbol{x} \in K$

$$\boldsymbol{x} = \sum_{i=1}^{m} \lambda_i \boldsymbol{a}_i, \quad \lambda_i \geq 0, \quad i = 1, 2, \ldots, m, \tag{2.162}$$

is valid, and conversely Equation 2.162 implies that $\boldsymbol{x} \in K$.

It can be shown that a polyhedral cone is closed.

Lemma 2.14: Let the cone K be defined by a system of linear inequalities

$$(\boldsymbol{a}_i, \boldsymbol{x}) \geq 0, \quad i = 1, \ldots, m,$$

where $\boldsymbol{a}_i \in R^n$. Then the conjugate cone K^* is a polyhedral cone and is created from points $\boldsymbol{y} = \Sigma_{i=1}^{m} \lambda_i \boldsymbol{a}_i$, $\lambda_i \geq 0$, $i = 1, 2, \ldots, m$.

Proof: Let the cone $\underline{K}$ be a set $\underline{K} = \{\boldsymbol{y} \mid \boldsymbol{y} = \Sigma_{i=1}^{m} \lambda_i \boldsymbol{a}_i, \lambda_i \geq 0, i = 1, 2, \ldots, m\}$. Then $\underline{K}^*$ is a set of $\boldsymbol{x}$, for which $(\boldsymbol{x}, \boldsymbol{y}) \geq 0$, $\boldsymbol{y} \in \underline{K}$. Then

$$\left(\boldsymbol{x}, \sum_{i=1}^{m} \lambda_i \boldsymbol{a}_i\right) = \sum_{i=1}^{m} \lambda_i (\boldsymbol{x}, \boldsymbol{a}_i) \geq 0, \quad \lambda_i \geq 0,$$

which is valid for any $\lambda_i \geq 0$ only if $(\boldsymbol{a}_i, \boldsymbol{x}) \geq 0$, $i = 1, \ldots, m$, i.e. if $\boldsymbol{x} \in K$. Hence $\underline{K}^* = K$. Since $\underline{K}$ is a polyhedral cone, then $\underline{K}$ is closed and $(\underline{K}^*)^* = \underline{K}$. Hence $K^* = \underline{K}$.

Definition 2.23: Let K be a convex set. Let $\boldsymbol{x}_o \in K$. $K(\boldsymbol{x}_o) \equiv K^o$ denotes a set of vectors $\boldsymbol{y}$ such that $\boldsymbol{y} \in K^o$ if and only if there is an $\alpha > 0$ such that $\boldsymbol{x}_o + \alpha \boldsymbol{y} \in K$. The set K^o is called the cone of admissible directions for K at point $\boldsymbol{x}_o$.

It can be shown that if K^o is a convex cone and if $\boldsymbol{y} \in K^o$ and $\boldsymbol{x}_o + \alpha_o \boldsymbol{y} \in K$ then $\boldsymbol{x}_o + \alpha \boldsymbol{y} \in K$ for any $0 \leq \alpha \leq \alpha_o$.

Lemma 2.15: A point $\underline{\boldsymbol{x}}$ is the minimum (critical) point, of $f(\boldsymbol{x})$ in K if and only if the inequality

$$(f'(\underline{\boldsymbol{x}}), \boldsymbol{x} - \underline{\boldsymbol{x}}) \geq 0, \quad \boldsymbol{x} \in K \tag{2.163}$$

is satisfied.
(For the proof see, For example, Pshenichny and Danilin (1978).)

Let $\boldsymbol{a}_i \in R^n$, $i \in J = J^- \cup J^o$, where J^- and J^o are finite sets of indices defined above (see Definition 2.18). Let the domain K be defined by the system of inequalities and equalities

$$(\boldsymbol{a}_i, \boldsymbol{x}) - d_i \leq 0, \quad i \in J^-, \quad (\boldsymbol{a}_i, \boldsymbol{x}) - d_i = 0, \quad i \in J^o. \tag{2.164}$$

Let $0 \leq \alpha \leq \alpha_o$. Then $\boldsymbol{y} \in K^o$ if $\boldsymbol{x}_o + \alpha \boldsymbol{y} \in K$. Since $\boldsymbol{x}_o + \alpha \boldsymbol{y} \in K$ then $\boldsymbol{x}_o + \alpha \boldsymbol{y}$ satisfies Equation 2.164 if

$$(\boldsymbol{a}_i, \boldsymbol{y}) \leq 0, \quad i \in J^-(\boldsymbol{x}_o), \quad (\boldsymbol{a}_i, \boldsymbol{y}) = 0, \quad i \in J^o. \tag{2.165}$$

Hence the cone K^o is defined by Equation 2.165, and similarly by

$$(-\boldsymbol{a}_i, \boldsymbol{y}) \geq 0, \quad i \in J^-(\boldsymbol{x}_o), \quad (\boldsymbol{a}_i, \boldsymbol{y}) \geq 0, \quad i \in J^o, \quad (-\boldsymbol{a}_i, \boldsymbol{y}) \geq 0, \quad i \in J^o.$$

Then due to Lemma 2.14 the vector $\boldsymbol{y} \in (K^o)^*$ can be found as

$$\boldsymbol{y} = -\sum_{i\in J^-(\boldsymbol{x}_o)} u_i \boldsymbol{a}_i - \sum_{i\in J^o} u_i \boldsymbol{a}_i, \quad u_i \geq 0, \quad i \in J^-(\boldsymbol{x}_o). \tag{2.166}$$

Then we have the following theorem.

Theorem 2.75: Let $f(\boldsymbol{x}) = \frac{1}{2}(\boldsymbol{x}, C\boldsymbol{x}) + (\boldsymbol{b}, \boldsymbol{x}) = \frac{1}{2}\boldsymbol{x}^T C\boldsymbol{x} + \boldsymbol{b}^T\boldsymbol{x}$ be a convex differentiable quadratic function, and K be the set defined by $(\boldsymbol{a}_i, \boldsymbol{x}) - d_i = \boldsymbol{a}_i^T\boldsymbol{x} - d_i \leq 0$, $i \in J^-$, $(\boldsymbol{a}_i, \boldsymbol{x}) - d_i = \boldsymbol{a}_i^T\boldsymbol{x} - d_i = 0$, $i \in J^o$. Then the point $\underline{\boldsymbol{x}}$ is the critical point of $f(\boldsymbol{x})$ in K if the necessary and sufficient conditions are valid, i.e. if there exist numbers u_i, $i \in J^- \cup J^o$ such that

$$f'(\boldsymbol{x}) + \sum_{i\in J^-\cup J^o} u_i \boldsymbol{a}_i = 0, \quad u_i \geq 0, i \in J^-, \quad u_i = 0 \text{ if } (\boldsymbol{a}_i, \underline{\boldsymbol{x}}) - d_i < 0, \quad i \in J^-, \tag{2.167}$$

where u_i are Lagrange multipliers.

Theorem 2.76: A point $\underline{\boldsymbol{x}}$ is the minimum point of the convex differentiable quadratic function over the whole space if the following necessary and sufficient condition

$$f'(\underline{\boldsymbol{x}}) = 0 \tag{2.168}$$

is satisfied.
The proof follows from Theorem 2.75.

Theorem 2.77: A point $\underline{\boldsymbol{x}}$ is the minimum point of the convex differentiable quadratic function in the set $\boldsymbol{x}_i \geq 0$, $i \in J$, if the following necessary and sufficient conditions

$$\partial f(\underline{\boldsymbol{x}})/\partial \boldsymbol{x}_i \geq 0 \text{ if } \boldsymbol{x}_i = 0, \quad i \in J, \tag{2.169}$$

$$\partial f(\underline{\boldsymbol{x}})/\partial \boldsymbol{x}_i = 0 \text{ if } \boldsymbol{x}_i \neq 0. \quad i \notin J. \tag{2.170}$$

are valid.

Lemma 2.16: Let the function $f(\boldsymbol{x})$ have all first derivatives. Then $f(\boldsymbol{x})$ is convex if and only if for all pairs of points $\boldsymbol{x}, \boldsymbol{y} \in R^n$, $\boldsymbol{x} \neq y$ it holds that

$$f'(y)(y - \boldsymbol{x}) \geq f(y) - f(\boldsymbol{x}) \text{ or } f(y) - f(\boldsymbol{x}) \geq f'(\boldsymbol{x})(y - \boldsymbol{x}),$$

and $f(\boldsymbol{x})$ is pure convex if and only if

$$f'(y)(y - \boldsymbol{x}) > f(y) - f(\boldsymbol{x}) \text{ or } f(y) - f(\boldsymbol{x}) > f'(\boldsymbol{x})(y - \boldsymbol{x}).$$

To give the necessary and sufficient conditions for a minimum of a function with constraints we state the well-known Kuhn-Tucker theorem:

Theorem 2.78 (Kuhn-Tucker)**:** Let convex functions $f_i(\boldsymbol{x})$, $i = 1, \ldots, m$ and a convex set K be given. Let $\underline{\boldsymbol{x}}$ be the critical (stationary, minimum) point of a convex quadratic function

$$f(\boldsymbol{x}) = \frac{1}{2}(\boldsymbol{x}, C\boldsymbol{x}) + (\boldsymbol{b}, \boldsymbol{x}) = \frac{1}{2}\boldsymbol{x}^T C\boldsymbol{x} + \boldsymbol{b}^T\boldsymbol{x} \tag{2.171}$$

with constraints

$$f_i(\boldsymbol{x}) \leq 0, \quad i = 1, \ldots, m, \quad \boldsymbol{x} \in K \tag{2.172}$$

and let there exist a point $\boldsymbol{x}_1 \in K$ such that

$$f_i(\boldsymbol{x}_1) < 0, \quad i = 1, \ldots, m.$$

Then there exist numbers $u_i \geq 0$, $i = 1, \ldots, m$ such that

$$f(\underline{\boldsymbol{x}}) + \sum_{i=1}^{m} u_i f_i(\underline{\boldsymbol{x}}) \leq f(\boldsymbol{x}) + \sum_{i=1}^{m} u_i f_i(\boldsymbol{x}), \quad \boldsymbol{x} \in K, \tag{2.173}$$

$$u_i f_i(\underline{\boldsymbol{x}}) = 0, \quad i = 1, \ldots, m. \tag{2.174}$$

The conditions given in Equations 2.173 and 2.174 are necessary and sufficient conditions for the minimum of $f(\boldsymbol{x})$.

The other formulation of the Kuhn-Tucker theorem is in the Arrow-Enthoven version, which is also valid for $f_i(\boldsymbol{x})$ being quasi-convex.

Theorem 2.79 (Arrow-Enthoven): Sufficient conditions for $\underline{\boldsymbol{x}} \in R^n$ to be an optimal solution of minimization of Equations 2.171 and 2.172 is satisfaction of Kuhn-Tucker's conditions, i.e. for the problem

$$\min_{\boldsymbol{x}} f(\boldsymbol{x}) = \tfrac{1}{2}(\boldsymbol{x}, C\boldsymbol{x}) + (\boldsymbol{b}, \boldsymbol{x}) = \tfrac{1}{2}\boldsymbol{x}^T C\boldsymbol{x} + \boldsymbol{b}^T\boldsymbol{x} \tag{2.175}$$

subject to

$$f_i(\boldsymbol{x}) \leq 0, \quad i = 1, \ldots, m, \quad \boldsymbol{x} \in K, \tag{2.176}$$

there exist u_i such that

$$\left.\begin{aligned} & f'(\underline{\boldsymbol{x}}) + \sum_{i=1}^{m} u_i f_i(\underline{\boldsymbol{x}}) = 0, \\ & f_i(\underline{\boldsymbol{x}}) \leq 0, \quad i = 1, \ldots, m, \\ & u_i f_i(\underline{\boldsymbol{x}}) = 0, \quad i = 1, \ldots, m, \\ & u_i \geq 0, \quad i = 1, \ldots, m, \end{aligned}\right\} \tag{2.177}$$

then $\underline{\boldsymbol{x}}$ is an optimal solution of Equations 2.175 and 2.176.

Proof: As f($\boldsymbol{x}$), $f_i(\boldsymbol{x})$, $i - 1, 2, \ldots, m$, are assumed to be convex functions, then due to Lemma 2.16

$$\begin{aligned} f(\boldsymbol{x}) - f(\underline{\boldsymbol{x}}) &\geq f'(\underline{\boldsymbol{x}})(\boldsymbol{x} - \underline{\boldsymbol{x}}), \\ f_i(\boldsymbol{x}) - f_i(\underline{\boldsymbol{x}}) &\geq f_i'(\underline{\boldsymbol{x}})(\boldsymbol{x} - \underline{\boldsymbol{x}}). \end{aligned} \tag{2.178}$$

Let $\boldsymbol{x} \in K$ be an arbitrary admissible solution, then applying Equations 2.178 and 2.177 we have

$$\begin{aligned} f(\boldsymbol{x}) - f(\underline{\boldsymbol{x}}) \geq f'(\underline{\boldsymbol{x}})(\boldsymbol{x} - \underline{\boldsymbol{x}}) &= -\sum_{i=1}^{m} u_i f_i'(\underline{\boldsymbol{x}})(\boldsymbol{x} - \underline{\boldsymbol{x}}) \geq -\sum_{i=1}^{m} u_i(f_i(\boldsymbol{x}) - f_i(\underline{\boldsymbol{x}})) \\ &= -\sum_{i=1}^{m} u_i f_i(\boldsymbol{x}) \geq 0. \end{aligned}$$

Hence for an arbitrary $\boldsymbol{x} \in K$ follows that $f(\boldsymbol{x}) \geq f(\underline{\boldsymbol{x}}) \geq 0$ and therefore $\underline{\boldsymbol{x}}$ is an optimal solution of Equations 2.175 and 2.176.

In our special case,

$$\min_{\boldsymbol{x}} f(\boldsymbol{x}) = \tfrac{1}{2}(\boldsymbol{x}, C\boldsymbol{x}) + (\boldsymbol{b}, \boldsymbol{x}) = \tfrac{1}{2}\boldsymbol{x}^T C\boldsymbol{x} + \boldsymbol{b}^T\boldsymbol{x}$$

subject to $A\boldsymbol{x} \leq d$

the Kuhn-Tucker conditions are as follows

$$-2C\boldsymbol{x} - A^T u = \boldsymbol{b},$$

$$A\boldsymbol{x} + y = \boldsymbol{d},$$

$$u \geq 0, \quad \boldsymbol{y} \geq 0,$$

$$u_i y_i = 0, \quad i = 1, 2, \ldots, m.$$

2.6.3. Lagrange Multipliers and Dual Problem

The most important concept in the theory of constrained optimization represents the way in which the so-called Lagrange multipliers are introduced. In the previous section concerning the unconstrained optimization we saw that the relation $\boldsymbol{g}(\boldsymbol{x}) = f'(\boldsymbol{x}) = 0$ was a necessary condition for a minimum (critical, stationary) point. But it can be appreciated that the concept of a critical point is fundamental to the subject of unconstrained optimization, and will also be a necessary condition for a local minimizer. Lagrange multipliers arise when similar necessary conditions are introduced for the solution $\underline{\boldsymbol{x}}$ of the constrained minimization problem

$$\begin{aligned}
&\min_{\boldsymbol{x}} f(\boldsymbol{x}) = \tfrac{1}{2}(\boldsymbol{x}, C\boldsymbol{x}) + (\boldsymbol{b}, \boldsymbol{x}) = \tfrac{1}{2}\boldsymbol{x}^T C\boldsymbol{x} + \boldsymbol{b}^T\boldsymbol{x} \\
&\text{subject to } A\boldsymbol{x} \leq d, \quad \text{i.e.} \qquad\qquad (2.179) \\
&(\boldsymbol{a}_i, \boldsymbol{x}) - d_i = \boldsymbol{a}_i^T\boldsymbol{x} - d_i \leq 0, \quad i \in J^-, \quad \text{i.e. } f_i(\boldsymbol{x}) \leq 0, \quad i \in J^-, \\
&(\boldsymbol{a}_i, \boldsymbol{x}) - d_i = \boldsymbol{a}_i^T\boldsymbol{x} - d_i = 0, \quad i \in J^o, \quad \text{i.e. } f_i(\boldsymbol{x}) = 0, \quad i \in J^o.
\end{aligned}$$

A local minimizer must be a feasible point, and other necessary conditions illustrate the need for there to be no feasible descent directions at $\underline{\boldsymbol{x}}$. Suppose that a feasible incremental step $\boldsymbol{v}$ is taken from the minimizing point $\underline{\boldsymbol{x}}$. Taylor's expansion gives

$$f_i(\underline{\boldsymbol{x}} + \boldsymbol{v}) = f_i(\underline{\boldsymbol{x}}) + \boldsymbol{v}^T f_i'(\underline{\boldsymbol{x}}) + o(\|\boldsymbol{v}\|). \tag{2.180}$$

In the case when $f_i(\underline{\boldsymbol{x}} + \boldsymbol{v}) = f_i(\underline{\boldsymbol{x}}) = 0$, then any feasible incremental step lies along a feasible direction $\boldsymbol{p}$ and satisfies

$$\boldsymbol{p}^T f_i'(\underline{\boldsymbol{x}}) = 0, \quad i \in J^o. \tag{2.181}$$

In a real case a feasible incremental step $\boldsymbol{v}$ can be constructed given any such $\boldsymbol{p}$. If in addition $f(\boldsymbol{x})$ has negative slope along $\boldsymbol{p}$, that is

$$\boldsymbol{p}^T f_i'(\underline{\mathbf{x}}) < 0, \tag{2.182}$$

then the direction $\boldsymbol{p}$ is a descent direction and the feasible incremental step along $\boldsymbol{p}$ reduces $f(\boldsymbol{x})$. But it cannot be done because $\underline{\boldsymbol{x}}$ is a local minimizer. Then there cannot be only a linear combination of f_i', $i \in J^o$, i.e.

$$\boldsymbol{g}(\boldsymbol{x}) = \sum\nolimits_{i \in J^o} f_i'(\underline{\boldsymbol{x}})\lambda_i = \underline{A\vec{\lambda}} \tag{2.183}$$

where $\{\underline{\lambda}_i\} = \vec{\underline{\lambda}}$ are Lagrange multipliers, associated with the point $\underline{x}$, and $\underline{A}$ denotes a matrix with columns $f_i'(\underline{x})$, $i \in J^o$. The last condition (Equation 2.183) is a necessary condition for a local minimizer. From this equation $\vec{\underline{\lambda}}$ is defined uniquely if $\underline{A}$ has full rank by

$$\vec{\underline{\lambda}} = \underline{A}^+ \boldsymbol{g}(\underline{x}), \tag{2.184}$$

where $\underline{A}^+ = (\underline{A}^T\underline{A})^{-1}\underline{A}^T$ is the generalized inverse of $\underline{A}$. From Equation 2.183 it is seen that if there are no constraints then $\boldsymbol{g}(\boldsymbol{x}) = 0$, i.e. the equation reduces to the condition for a stationary (critical) point.

These conditions give rise to an idea for solving equality constraint problems. This method is known as the classical method of Lagrange multipliers and consists of finding vectors $\boldsymbol{x}$ and $\vec{\underline{\lambda}}$, which solve equations arising from Equation 2.183 and feasibility, i.e.

$$\begin{aligned} \boldsymbol{g}(\boldsymbol{x}) &= \sum\nolimits_{i \in J^o} f_i'(\boldsymbol{x})\lambda_i, \\ f_i(\boldsymbol{x}) &= 0, \quad i \in J^o. \end{aligned} \tag{2.185}$$

This system is well determined, as for m equality constraints we have $n + m$ equations and $n + m$ unknowns $\boldsymbol{x}$ and $\vec{\lambda}$.

Let us introduce the Lagrangian function

$$\mathscr{L}(\boldsymbol{x}, \vec{\lambda}) = f(\boldsymbol{x}) - \sum\nolimits_{i=1}^{m} \lambda_i f_i(\boldsymbol{x}). \tag{2.186}$$

Hence Equation 2.185 is equivalent to the expression

$$\nabla\mathscr{L}(\underline{x}, \vec{\underline{\lambda}}) = 0, \tag{2.187}$$

where $\nabla = (\nabla_x, \nabla_\lambda)^T = (\partial/\partial x_i, \partial/\partial\lambda)^T$.

Theorem 2.80: Let $f(\boldsymbol{x}) = \frac{1}{2}(\boldsymbol{x}, C\boldsymbol{x}) + (\boldsymbol{b}, \boldsymbol{x}) = \frac{1}{2}\boldsymbol{x}^T C\boldsymbol{x} + \boldsymbol{b}^T\boldsymbol{x}$ be a quadratic function and let $f_i(\boldsymbol{x})$ be constraint functions. Then the point $\underline{x}$ is a minimum (stationary or critical) point of the problem:

$$\begin{aligned} &\min_{x} f(\boldsymbol{x}) = \tfrac{1}{2}(\boldsymbol{x}, C\boldsymbol{x}) + (\boldsymbol{b}, \boldsymbol{x}) = \tfrac{1}{2}\boldsymbol{x}^T C\boldsymbol{x} + \boldsymbol{b}^T\boldsymbol{x} \\ &\text{subject to } f_i(\boldsymbol{x}) = 0, \quad i \in J^o, \end{aligned} \tag{2.188}$$

if Equation 2.187 is valid. That is if

$$\nabla\mathscr{L}(\underline{x}, \vec{\underline{\lambda}}) = 0,$$

where

$$\mathscr{L}(\underline{x}, \vec{\underline{\lambda}}) = f(\boldsymbol{x}) - \sum\nolimits_{i \in J} \lambda_i f_i(\boldsymbol{x}). \tag{2.189}$$

Proof: The method of Lagrange multipliers leads to the solution of Equation 2.187 or equivalently Equation 2.185. Since $\nabla_\lambda\mathscr{L} \equiv \partial/\partial\lambda\mathscr{L}(\underline{x}, \vec{\underline{\lambda}}) = 0$ then $\underline{x}$ is feasible. Since $\nabla_x\mathscr{L}(\underline{x}, \vec{\underline{\lambda}}) = \partial/\partial x_i\mathscr{L}(\underline{x}, \vec{\underline{\lambda}}) = 0$ then $\mathscr{L}(\boldsymbol{x}, \vec{\underline{\lambda}})$ is stationary at $\underline{x}$ for all changes $\boldsymbol{v}$, and therefore, for all feasible changes. If $\boldsymbol{v}$ is a feasible change, then $\mathscr{L}(\underline{x} + \boldsymbol{v}, \vec{\underline{\lambda}}) = f(\underline{x} + \boldsymbol{v})$. Hence $f(\boldsymbol{x})$ is stationary at the point $\underline{x}$ for all feasible changes $\boldsymbol{v}$. If Equation 2.187 can be solved, then this solution $\underline{x}$ is a constrained stationary point of $f(\boldsymbol{x})$. The opposite assumptions are not correct as it is possible that $\underline{x}$ can be a minimizer but it does not satisfy Equation 2.187. [Note that an example which proves this assertion is the function $f(x) = \pm(x - 1)^3 + y$ and with the same constraints. The point $\underline{x} = (0, 1)$ can be a constrained stationary point but it does not satisfy Equation 2.187.]

To find the Kuhn-Tucker condition of the problem

$$\min_{x} f(\boldsymbol{x}) = \tfrac{1}{2}(\boldsymbol{x}, C\boldsymbol{x}) + (\boldsymbol{b}, \boldsymbol{x}) = \tfrac{1}{2}\boldsymbol{x}^T C\boldsymbol{x} + \boldsymbol{b}^T\boldsymbol{x}$$
$$\text{subject to } f_i(\boldsymbol{x}) = A\boldsymbol{x} - \boldsymbol{d} \leq 0, \quad i \in J,$$

we have

$$\mathscr{L}(\boldsymbol{x}, \vec{\lambda}) = \tfrac{1}{2}\boldsymbol{x}^T C\boldsymbol{x} + \boldsymbol{b}^T\boldsymbol{x} + \vec{\lambda}(A\boldsymbol{x} - d).$$

Hence

$$\partial\mathscr{L}/\partial x = C\boldsymbol{x} + b + A^T\vec{\lambda} = \boldsymbol{v}, \quad \partial\mathscr{L}/\partial\lambda = A\boldsymbol{x} - d = -\boldsymbol{y},$$

where $\boldsymbol{v}$ and $\boldsymbol{y}$ denote $\partial\mathscr{L}/\partial x$ and $\partial\mathscr{L}/\partial\lambda$, respectively. Hence the Kuhn-Tucker conditions are the following:

$$C\boldsymbol{x} + \boldsymbol{b} + A^T\vec{\lambda} \geq 0, \quad A\boldsymbol{x} - \boldsymbol{d} \leq 0,$$
$$\boldsymbol{x}(C\boldsymbol{x} + \boldsymbol{b} + A^T\vec{\lambda}) = 0, \quad \vec{\lambda}(A\boldsymbol{x} - \boldsymbol{d}) = 0,$$
$$\vec{\lambda} \geq 0$$

or in the other form

$$-C\boldsymbol{x} - A^T\vec{\lambda} + \boldsymbol{v} = \boldsymbol{b},$$
$$A\boldsymbol{x} + \boldsymbol{y} = \boldsymbol{d},$$
$$\vec{\lambda} \geq 0, \quad \boldsymbol{v} \geq 0, \quad \boldsymbol{y} \geq 0,$$
$$x_i v_i = 0, \quad i = 1, 2, \ldots, n,$$
$$\lambda_j \lambda_j = 0, \quad j = 1, 2, \ldots, m.$$

The dual formulation is based on the Lagrangian function.

Theorem 2.81: Let $\underline{x}$ be the minimal (critical) point of the primal convex quadratic problem:

$$\min_{x} f(\boldsymbol{x}) = \tfrac{1}{2}(\boldsymbol{x}, C\boldsymbol{x}) + (\boldsymbol{b}, \boldsymbol{x}) = \tfrac{1}{2}\boldsymbol{x}^T C\boldsymbol{x} + \boldsymbol{b}^T\boldsymbol{x}$$
$$\text{subject to } f_i(\boldsymbol{x}) = A\boldsymbol{x} - \boldsymbol{d} \leq 0, \quad i \in J,$$

where $f(\boldsymbol{x}), f_i(\boldsymbol{x}) \in C^1$. Then $(\underline{x}, \underline{\vec{\lambda}})$ solves the dual problem:

$$\text{maximize } \mathscr{L}(\boldsymbol{x}, \vec{\lambda}),$$
$$\text{subject to } \partial/\partial x_i \mathscr{L}(\boldsymbol{x}, \vec{\lambda}) = 0, \quad \vec{\lambda} \geq 0.$$

Moreover, the minimum primal and maximum dual function values are equal, i.e.

$$f(\underline{x}) = \mathscr{L}(\underline{x}, \underline{\vec{\lambda}}).$$

Proof: Due to the Kuhn-Tucker theorem there exist

$$\partial/\partial x_i \mathscr{L}(\underline{x}, \vec{\underline{\lambda}}) = 0 \text{ (dual feasibility)},$$

$$f_i(\underline{x}) = 0, \quad i \in J^o, \quad f_i(\underline{x}) \leq 0, \quad i \in J^-, \quad \lambda_i \geq 0, \quad i \in J^-,$$

$$\underline{\lambda}_i f_i(\underline{x}) = 0, \quad i \in J = J^- \cup J^o.$$

Hence $f(\underline{x}) = \mathscr{L}(\underline{x}, \vec{\underline{\lambda}})$. Further, let x, λ be dual feasible. Then using $\vec{\lambda} \geq 0$, the convexity of $\mathscr{L}$ and since $\partial/\partial x_i \mathscr{L}(x, \vec{\lambda}) = 0$,

$$\mathscr{L}(\underline{x}, \vec{\underline{\lambda}}) = f(\underline{x}) \geq f(\underline{x}) - \sum_{i=1}^{m} \lambda_i f_i(\underline{x}) = \mathscr{L}(\underline{x}, \vec{\lambda}) \geq \text{convexity}$$

$$\geq \mathscr{L}(x, \vec{\lambda}) + (\underline{x} - x)^T \partial/\partial x_i \mathscr{L}(x, \vec{\lambda}) = \mathscr{L}(x, \vec{\lambda}).$$

From here $\mathscr{L}(\underline{x}, \vec{\lambda}) \geq \mathscr{L}(x, \vec{\lambda})$ and thus $\underline{x}$, $\underline{\lambda}$ solve the dual problem.

2.6.4. The Generalized Gradient Method and the Gradient Projection Method

Let the following problem be solved:

$$\min_{x} f(x) = \tfrac{1}{2}(x, Cx) + (b, x) = \tfrac{1}{2}x^T Cx + b^T x \tag{2.190}$$

$$\text{subject to } f_i(x) \leq 0, \quad i \in J,$$

i.e.

$$\min_{x} f(x) = \tfrac{1}{2}(x, Cx) + (b, x) = \tfrac{1}{2}x^T Cx + b^T x \tag{2.191}$$

$$\text{subject to } f_i(x) \leq 0, \quad i \in J^-, \quad \text{i.e. } (a_i, x) - d_i = a_i^T x - d_i \leq 0, \quad i \in J^-,$$

$$f_i(x) = 0, \quad i \in J^o, \quad \text{i.e. } (a_i, x) - d_i = a_i^T x - d_i = 0, \quad i \in J^o.$$

Since the constraints are assumed to be linear, then after the transformation given above $f_i(x) = 0$, $i \in J$ or $Ax - d = 0$, where A is an $m \times n$-dimensional matrix of rank m. The solution of $f_i(x) = 0$ can be find in the form

$$x = x_1 + u_1 y_1 + \cdots + u_m y_m = x_1 + Uy, \tag{2.192}$$

where $u_1, \ldots, u_m$ are $m_1 = n - m$ linearly independent solutions of $f_i(x) = 0$ and are the column vectors of the matrix U. Let us put

$$\underline{f}(y) = f(x_1 + Uy). \tag{2.193}$$

Hence the gradient is given by

$$\underline{f}'(y) = U^T f'(x_1 + Uy) = U^T f'(x). \tag{2.194}$$

Then the iterative process can be written in the usual way

$$y_{k+1} = y_k - C_k f'(y_k) \tag{2.195}$$

and as $\boldsymbol{x}_k = \boldsymbol{x}_1 U \boldsymbol{y}_k$, then

$$\boldsymbol{x}_{k+1} = \boldsymbol{x}_k - UC_k U^T f'(\boldsymbol{x}_k) \text{ with } f_k(\boldsymbol{x}_1) = 0, \tag{2.196}$$

representing a generalized gradient type method. Let us put $C_k = \alpha_k I$, then Equation 2.196 can be written in the form

$$\boldsymbol{x}_{k+1} = \boldsymbol{x}_k - \alpha_k C f'(\boldsymbol{x}_k) = \boldsymbol{x}_k + \alpha_k \boldsymbol{p}_k, \tag{2.197}$$

where $\boldsymbol{p}_k = Cf'(\boldsymbol{x}_k)$, $C = UU^T$ and where α_k can be determined in a similar way as earlier in Section 2.6.2. Since the gradient $f'(\underline{\boldsymbol{x}})$ is orthogonal to the manifold $f_i(\underline{\boldsymbol{x}}) = 0$ at $\underline{\boldsymbol{x}}$, then $Cf'(\underline{\boldsymbol{x}}) = 0$. When $\boldsymbol{p}_k$ is small we finish the iterative process by a suitable estimate of $\underline{\boldsymbol{x}}$.

Let $\boldsymbol{x}_1 = A^T(AA^T)^{-1}\boldsymbol{d} = A^{-1}\boldsymbol{d}$, and $C = I - A^T(AA^T)^{-1}A = I - A^{-T}A$, where $A^{-1} = A^T(AA^T)^{-1}$ is the pseudo-inverse of A. Then Equation 2.197 can be rewritten in the form:

Let $\boldsymbol{x}_1$ be an initial guess. Then we compute

$$\begin{aligned} &A_k, A_k^{-1} \equiv A_k^T(A_k A_k^T)^{-1}, C_k = I - A_k^{-1}A_k, \\ &\tilde{\boldsymbol{x}}_{k+1} = \boldsymbol{x}_k - \alpha_k C_k f'(\boldsymbol{x}_k), \\ &\boldsymbol{x}_{k+1} = \tilde{\boldsymbol{x}}_{k+1} - A_k^{-1} g(\boldsymbol{x}_k), \end{aligned} \tag{2.198}$$

where α_k can be determined as in Section 2.6.2. This method is known as the gradient projection method and can also be used for nonlinear constraints.

2.6.5. The Active Set Method

In this section we shall limit ourselves to the positive definite case only. But this method can be extended for positive and negative semi-definite cases, as well as for the indefinite case (see for example Fletcher (1981)).

In the method of active set, certain constraints are considered as equalities and are indexed by the set J, the so-called active set. Since we shall investigate quadratic functions only, there may be $m(0 \le m \le n)$ active constraints. Active constraints at any point $\boldsymbol{x}$ are defined by the index set, the so-called active set

$$J = \{i \mid f_i(\boldsymbol{x}) = 0\}. \tag{2.199}$$

For iteration k a feasible point $\boldsymbol{x}_k$ is known and satisfies the active equality constraints $(\boldsymbol{a}_i, \boldsymbol{x}_k) - d_i = 0,\ i \in J$. In the case for which $(\boldsymbol{a}_i, \boldsymbol{x}_k) - d_i > 0,\ i \in J$, the current active set J is equivalent to the set of active constraints J_k defined by Equation 2.199. Each iterate $\boldsymbol{x}_k$ is a feasible point and each iteration attempts to locate the solution to an equality problem with only the active constraints, i.e.

$$\min_{\boldsymbol{v}} f(\boldsymbol{v}) = \tfrac{1}{2}(\boldsymbol{v}, C\boldsymbol{v}) + (\boldsymbol{b}_k, \boldsymbol{v}) = \tfrac{1}{2}\boldsymbol{v}^T C\boldsymbol{v} + \boldsymbol{b}_k^T \boldsymbol{v}, \tag{2.200}$$

subject to $f_i(\boldsymbol{v}) = 0, \quad i \in J$, i.e. $(\boldsymbol{a}_i, \boldsymbol{v}) - d_i = \boldsymbol{a}_i^T\boldsymbol{v} - d_i = 0, \quad i \in J,$

obtained by shifting the origin to the point $\boldsymbol{x}_k$, where $\boldsymbol{v}$ represents a correction, $\boldsymbol{b}_k = \boldsymbol{b} + C\boldsymbol{x}_k \equiv \boldsymbol{g}(\boldsymbol{x}_k)$, and $\boldsymbol{g}(\boldsymbol{x}_k)$ is the gradient of f at $\boldsymbol{x}_k$. The problem in Equation 2.200 is equivalent to the problem

$$\min f(\boldsymbol{v}) = \tfrac{1}{2}(\boldsymbol{v}, C\boldsymbol{v}) + (\boldsymbol{b}, \boldsymbol{v}) = \tfrac{1}{2}\boldsymbol{v}^T C\boldsymbol{v} + \boldsymbol{b}^T\boldsymbol{v}, \tag{2.201}$$

$$\text{subject to } A\boldsymbol{v} = d, \quad \text{i.e. } (\boldsymbol{a}_i, \boldsymbol{v}) - d_i \equiv \boldsymbol{a}_i^T\boldsymbol{v} - d_i = 0, \quad i \in J,$$

and therefore can be solved, by the generalized gradient method, etc. If $\boldsymbol{v}_k$ is feasible corresponding to the constraints, which are not in J, then $\boldsymbol{x}_{k+1} = \boldsymbol{x}_k + \boldsymbol{v}_k$. In the opposite case a line search is made in the direction of $\boldsymbol{v}_k$ to find the best feasible point as a solution of Equation 2.200 in a search direction $\boldsymbol{p}_k$, and we find α_k by solving

$$\alpha = \min\left(1, \min_{\substack{i \notin J \\ (\boldsymbol{a}_i, \boldsymbol{p}_k) < 0}} \frac{d_i - (\boldsymbol{a}_i, \boldsymbol{x}_k)}{(\boldsymbol{a}_i, \boldsymbol{p}_k)}\right), \tag{2.202}$$

and therefore $\boldsymbol{x}_{k+1} = \boldsymbol{x}_k + \alpha_k\boldsymbol{p}_k$. If $\alpha_k < 1$ in Equation 2.202 then new constraints become active, defined by the index j, which achieve the minimum in the equation, and this index is added to the active set J.

If $\boldsymbol{x}_k$ solves the current equality problem for which $\boldsymbol{v} = 0$, then it is possible to compute multipliers u_i for the active constraints. Then $\boldsymbol{x}_k$ and $\vec{\lambda}_k$ satisfy the Kuhn-Tucker conditions for the inequality constraint problem with $u_i \geq 0$, $\forall i \in J^-$. Now test whether $u_i^k \geq 0 \, \forall i \in J_k \cap J^-$; if so then the Kuhn-Tucker conditions are satisfied. According to the Kuhn-Tucker theorem and since $f(\boldsymbol{x})$ is convex, we have a global solution. Otherwise, there exists an index $j \in J_k \cap J^-$ such that $u_j^k < 0$. Then we reduce $f(\boldsymbol{x})$ with the constraint j to be inactive. Then $J_i = J' - \{j\}$ and the algorithm continues as before by solving the corresponding equality problem. In many cases there are several j for which $u_j^k < 0$, and then we select index j successfully based on solving

$$\min_{i \in J \cap J^-} u_i^k. \tag{2.203}$$

The conjugate gradient methods with constraints based on this active set method will be used for solving contact problems of plate tectonics. These methods will be discussed in connection with the mentioned plate tectonic problems.

2.7. NONQUADRATIC FUNCTIONALS IN R^N AND ITERATIVE METHODS FOR THE SOLUTION OF SYSTEMS OF NON-LINEAR EQUATIONS

2.7.1. Introduction

A great number of geophysical problems, such as the problem of geothermal and geomagnetic fields, geodynamic problems, etc., are described by the non-linear boundary value problems. In such problems physical parameters depend on the temperature, intensity of magnetic field, etc., and eventually on their derivatives as well. In some geodynamic problems we approximate a geological body by an elasto-plastic plane body of shape Ω. Then we solve

$$-\frac{\partial}{\partial x}\left(m\left(|\text{grad}u|^2\frac{\partial u}{\partial x}\right)\right)-\frac{\partial}{\partial y}\left(m\left(|\text{grad}u|^2\frac{\partial u}{\partial y}\right)\right)=f(x,y)\text{ in }\Omega,$$

where the function m characterizes the property of the rock and represents the load. In such type of problems the boundary conditions can be non-linear, e.g.

$$u|_{\partial\Omega_1}=g,\ m\left(|\text{grad}u|^2\frac{\partial u}{\partial \nu}\right)\bigg|_{\partial\Omega_2}=h,$$

where $g=g(x,y)$ and $h=h(x,y)$ are given functions defined on $\partial\Omega_1$ and $\partial\Omega_2$, respectively, and ν is the outer normal to $\partial\Omega_2$.

In some geodynamic problems the deformation of the lithosphere or geological blocks is described by the boundary value problem for the von Karmán equations, for the Navier-Stokes equations or for the equations of motion in the Bingham visco-plastic rheology. In the section of the mathematical simulation of geodynamic processes in the rift zones, we shall analyse the boundary value problem for the visco-plastic fluid of the Bingham type. Other types of non-linear problems, e.g. problems of plate tectonics, lead us to solve contact problems of Signorini type. Since materials in the Earth's interior are melted, recrystallized or solidified, then such types of problem lead us to solve free boundary problems, the so-called phase transition problems or the Stefan problems. In geodynamics many problems are coupled, and then they can be described by the non-linear or by the linear and non-linear rheologies of the media.

The differential equations describing the geophysical and geological problems represent certain physical phenomena. However, the mathematical description of physical phenomena necessarily entails some simplification; were all the factors to be included in the representations, mathematically unsolvable problems would frequently arise. Consequently, the mathematical description of the geological reality is actually no more than a first approximation of the physical reality. A more exact representation of the rheology of the Earth would lead to non-linear problems. Their numerical solutions lead to solving a system of non-linear algebraic equations. Numerical solutions of non-linear boundary value problems based on the variational formulations lead to minimizing a nonquadratic functional $F : X \subseteq R^n \to R^1$. The problem of solving a system of non-linear equations may be replaced by an equivalent problem of minimizing a functional F, and conversely the problem of minimizing a functional F may be replaced by an equivalent problem of solving a system of non-linear equations.

2.7.2. The Newton Method and its Modifications

Let F be an n-dimensional function $F : X \subseteq R^n \to R^n$, let $\alpha \neq 0$ be a constant and C an $n \times n$ nonsingular matrix. Then the iteration process in a single variable x is defined by

$$x_{k+1}=x_k-\alpha^{-1}F(x_k),\quad k=0,1,2,\ldots,n-1. \tag{2.204}$$

The geometric representation consists of replacing F at some approximation $x_k (k = 0, 1, 2, \ldots)$ of a root $\underline{x}$ by a linear function with a suitable (fixed) slope $\alpha \neq 0$, and then in taking the root x_{k+1} of this linear function as a new approximation of x.

The iteration process in n variables is then defined by

$$\boldsymbol{x}_{k+1} = \boldsymbol{x}_k - C^{-1} F(\boldsymbol{x}_k), \quad k = 0, 1, 2, \ldots, \tag{2.205}$$

which geometrically means that $\boldsymbol{x}_{k+1}$ is the intersection of the n hyperplanes

$$\sum\nolimits_{j=1}^{n} c_{ij}(x_j - x_{j,k}) + F_i(\boldsymbol{x}_k) = 0, \quad i = 1, 2, \ldots, n$$

in R^{n+1} with hyperplane $\boldsymbol{x} = 0$.

If $C = \alpha I, \alpha \neq 0$, where I is the identity matrix, then the iteration process of Equation 2.204 is applied to each component F_i of F separately. If $C = F'(x_o)$, where $F'(\boldsymbol{x})$ denotes the G-derivative at $\boldsymbol{x}$, then

$$\boldsymbol{x}_{k+1} = \boldsymbol{x}_k - (F'(\boldsymbol{x}_o))^{-1} F(\boldsymbol{x}_k), \quad k = 0, 1, 2, \ldots \tag{2.206}$$

and we have the so-called simplified Newton method. It is well known that a special form of F leads to a special choice of C. Thus for a decomposition

$$F(\boldsymbol{x}) = C\boldsymbol{x} - G(\boldsymbol{x}), \tag{2.207}$$

where C is a non-singular matrix and G is non-linear, $G : X \subseteq R^n \to R^n$, we have

$$C^{-1}G(\boldsymbol{x}) = \boldsymbol{x} - C^{-1}F(\boldsymbol{x})$$

and therefore from the so-called natural Picard iteration we obtain

$$\boldsymbol{x}_{k+1} = C^{-1}G(\boldsymbol{x}_k) = \boldsymbol{x}_k - C^{-1}F(\boldsymbol{x}_k), \quad k = 0, 1, 2, \ldots. \tag{2.208}$$

Now let us assume that the n-dimensional mapping F, defined above, with the G-derivative $F'(\boldsymbol{x}_k)$ of F at $\boldsymbol{x}_k$ gives the n-dimensional Newton method in the form

$$\boldsymbol{x}_{k+1} = \boldsymbol{x}_k - (F'(\boldsymbol{x}_k))^{-1} F(\boldsymbol{x}_k), \quad k = 0, 1, 2, \ldots. \tag{2.209}$$

Geometrically the step from $\boldsymbol{x}_k$ to $\boldsymbol{x}_{k+1}$ in this equation may be interpreted as the intersection of the n hyperplanes $F_i'(\boldsymbol{x}_k)(\boldsymbol{x} - \boldsymbol{x}_k) + F_i(\boldsymbol{x}_k)$ (which is the affine function of each component F_i of F) in R^{n+1} with the hyperplane $\boldsymbol{x} = 0$.

These hyperplanes represent the tangent hyperplanes of F_i at $\boldsymbol{x}_k$. We set the next approximation $\boldsymbol{x}_{k+1}$ equal to the unique solution of

$$\sum_{j=1}^{n} F_{i,j}'(x_k)(x_j - x_{j,k}) + F_i(\boldsymbol{x}_k) = 0, \quad i = 1, 2, \ldots, n.$$

Let $G : X \subseteq R^n \to R^n$ be an arbitrary mapping, then

$$\boldsymbol{x}_{k+1} = \boldsymbol{x}_k - (F'(\boldsymbol{x}_k))^{-1} F(\boldsymbol{x}_k) + (n-1)G(\boldsymbol{x}_k), \quad k = 0, 1, \ldots. \tag{2.210}$$

There are numerous other modifications of Newton's method. We introduce three such modifications:

$$\boldsymbol{x}_{k+1} = \boldsymbol{x}_k - \omega_k (F'(\boldsymbol{x}_k))^{-1} F(\boldsymbol{x}_k), \quad k = 0, 1, 2, \ldots, \tag{2.211}$$

$$\boldsymbol{x}_{k+1} = \boldsymbol{x}_k - (F'(\boldsymbol{x}_k) + \lambda_k I)^{-1} F(\boldsymbol{x}_k), \quad k = 0, 1, 2, \ldots, \tag{2.212}$$

$$\boldsymbol{x}_{k+1} = \boldsymbol{x}_k - (F'(\boldsymbol{x}_k))^{-1} [F(\boldsymbol{x}_k) + F(\boldsymbol{x}_k - (F'(\boldsymbol{x}_k))^{-1} F(\boldsymbol{x}_k))], \quad k = 0, 1, 2, \ldots, \tag{2.213}$$

where ω_k are chosen in such a way that the norm-reducing

$$\|F(\boldsymbol{x}_{k+1})\| \leq \|F(\boldsymbol{x}_k)\|, \quad k = 0, 1, 2, \ldots \tag{2.214}$$

holds, and λ_k are selected in such a way that $F'(\boldsymbol{x}_k) + \lambda_k I$ is non-singular, if $F'(\boldsymbol{x}_k)$ itself is singular and Equation 2.214 holds.

From the Newton method, using the difference approximation of $F'(\boldsymbol{x}_k)$, we obtain the discretized Newton methods. It holds that

$$\partial_k F_i(\boldsymbol{x}) = h_{ik}^{-1} (F_i(\boldsymbol{x} + h_{ik} \boldsymbol{e}_k) - F_i(\boldsymbol{x})),$$

where h_{ik} are given discretization parameters and $\boldsymbol{e}_k$ is the k-th co-ordinate vector. Let $\Delta_{ik}(\boldsymbol{x}, h)$ denote any difference approximation of $\partial_k F_i(\boldsymbol{x})$ with the property that if $\partial_k F_i(\boldsymbol{x})$ exists, then $\lim_{h \to 0} \Delta_{ik}(\boldsymbol{x}, h) = \partial_k F_i(\boldsymbol{x})$, $i, k = 1, \ldots, n$. Let us denote by $J(\boldsymbol{x}, h)$ the difference matrix, defined by $J(\boldsymbol{x}, h) = (\Delta_{ik}(\boldsymbol{x}, h))$. The simplest choice of the parameters is $h \equiv h_k$, i.e. the differencing parameters are held constant throughout the iteration. Then we find

$$\boldsymbol{x}_{k+1} = \boldsymbol{x}_k - J(\boldsymbol{x}_k, h_k)^{-1} F(\boldsymbol{x}_k), \quad k = 0, 1, \ldots. \tag{2.215}$$

In the special case of the one-dimensional discretized Newton method we find

$$x_{k+1} = x_k - [(x_{k-1} - x_k)^{-1} (F(x_{k-1}) - F(x_k))]^{-1} F(x_k), \quad k = 0, 1, 2, \ldots, \tag{2.216}$$

where $h_k = x_{k-1} - x_k$, which is known as the secant method.

Remark: The special well-known one-dimensional case of the discretized Newton method is the so-called regula falsi iteration method.

2.7.3. Generalized Linear Methods

In the next section we refer to generalized linear methods. By this we mean the non-linear Gauss-Seidel or SOR methods, a composite Newton-SOR method (with Newton's method as the primary and SOR as the secondary iterations), one-step Newton-SOR or one-step Newton-Gauss-Seidel or one-step Newton-Jacobi methods, the l-step Newton-Jacobi method and the one-step Newton-ADI method.

To derive the generalized linear methods for the non-linear equation $F(\boldsymbol{x}) = 0$ we assume that $\boldsymbol{x}_k$ has been determined. The next Newton iterate will be found as the solution of the linear system

$$\sum_{j=1}^{n} F'_{ij}(x_k)(x_j - x_{j,k}) + F_i(\boldsymbol{x}_k) = 0, \quad i = 1, \ldots, n. \tag{2.217}$$

Let us derive the Newton-SOR iteration method. Assume that $\boldsymbol{x}_k$ has been determined. Then the next Newton iterate will be the solution of the linear system in Equation 2.2.17, which will be approximated by the SOR method.

Let the matrix $F'(\boldsymbol{x}_k)$ be decomposed as

$$C_k \equiv F'(\boldsymbol{x}_k) = P_k - N_k = D_k + L_k + U_k, \tag{2.218}$$

where D_k, L_k, and U_k are diagonal, strictly lower triangular and strictly upper triangular matrices, respectively. Let us assume that D_k is non-singular and let us define the matrix $\mathscr{L}_{\omega k}$ (see Axelsson (1994)) as

$$\mathscr{L}_{\omega k} = (D_k + \omega_k L_k)^{-1}[(1 - \omega_k)D_k - \omega_k U_k] = P_k^{-1} N_k, \tag{2.219}$$

where ω_k represents some relaxation parameter, $0 < \omega_k < 2$, and where $P_k = \omega_k^{-1}(D_k + \omega_k L_k)$, and $N_k = \omega_k^{-1}[(1 - \omega_k)D_k - \omega_k U_k]$.

The system in Equation 2.217 will be solved by using the SOR method. Let us denote the SOR iterates by $\boldsymbol{x}_{k,l}$, where the index k denotes the Newton iteration and l denotes the SOR iterations. Since $P_k^{-1}C_k = P_k^{-1}(P_k - N_k) = I - \mathscr{L}_{\omega k}$, and

$$(I + \cdots + \mathscr{L}_{\omega k}^{l})(I - \mathscr{L}_{\omega k}) = I - \mathscr{L}_{\omega k}^{l+1} \tag{2.220}$$

and since the $(l + 1)$-th SOR iteration may be written in the form

$$\begin{aligned} \boldsymbol{x}_{k,l+1} &= (D_k + \omega_k L_k)^{-1}[(1 - \omega_k)D_k - \omega_k U_k]\boldsymbol{x}_{k,l} + \omega_k(D_k + \omega_k L_k)^{-1}\boldsymbol{b}_k \\ &= \boldsymbol{x}_{k,l} - \omega_k(D_k + \omega_k L_k)^{-1}(C_k \boldsymbol{x}_{k,l} - \boldsymbol{b}_k), \quad l = 0, 1, 2, \ldots, \end{aligned} \tag{2.221}$$

then

$$\boldsymbol{x}_{k,l+1} = \boldsymbol{x}_{k,o} - \omega_k(I + \cdots + \mathscr{L}_{\omega k}^{l-1})(D_k + \omega_k L_k)^{-1}[F'(\boldsymbol{x}_k)(\boldsymbol{x}_{k,o} - \boldsymbol{x}_k) + F(\boldsymbol{x}_k)]. \tag{2.222}$$

Let us set $\boldsymbol{x}_{k,o} \equiv \boldsymbol{x}_k, \boldsymbol{x}_{k,l_k} \equiv \boldsymbol{x}_{k+1}$, where l_k is the total number of SOR iterations at the k-th Newtonian step, then Equation 2.222 yields

$$\boldsymbol{x}_{k+1} = \boldsymbol{x}_k - \omega_k(I + \cdots + \mathscr{L}_k^{l_k - 1})(D_k + \omega_k L_k)^{-1}F(\boldsymbol{x}_k), \tag{2.223}$$

which is called the generalized Newton-SOR method.

For $l_k = 1$, $k = 0, 1, 2, \ldots$ in Equation 2.223 we obtain the one-step Newton-SOR method

$$\boldsymbol{x}_{k+1} = \boldsymbol{x}_k - \omega_k(D_k + \omega_k L_k)^{-1}F(\boldsymbol{x}_k), \quad k = 0, 1, 2, \ldots. \tag{2.224}$$

For $l_k = l$, $k = 0, 1, 2, \ldots.$ in Equation 2.223 we obtain the l-step Newton-SOR method. All these methods of the Newton-SOR type are only a special case of the more general principle of combining non-linear iterative methods of Newtonian type with linear iterative methods of SOR type in order to form composite methods. This is the principle for the other types of composite methods based on the Newtonian or secant methods, and on the Jacobi, Gauss-Seidel, SOR or ADI types of methods. For more details see for example Ortega and Rheinboldt (1970), Daniel (1971). For information we shall introduce the following methods:

(i) The l-step Newton-Jacobi method

$$\boldsymbol{x}_{k+1} = \boldsymbol{x}_k - [I + \cdots + B_k^{l-1}]D_k^{-1}F(\boldsymbol{x}_k), \quad k = 0, 1, 2, \ldots, \tag{2.225}$$

where $F'(\boldsymbol{x}_k) = P_k - N_k$, and where $P_k = D_k, N_k = -(L_k + U_k)$ so that $B_k = D_k^{-1}(L_k + U_k)$.

(ii) The l-step Newton-Gauss-Seidel method

$$\boldsymbol{x}_{k+1} = \boldsymbol{x}_k - [I + \cdots + \mathscr{L}_1^{l-1}](D_k + L_k)^{-1}F(\boldsymbol{x}_k), \quad k = 0, 1, 2, \ldots, \tag{2.226}$$

where $F'(\boldsymbol{x}_k) = P_k - N_k, P_k = D_k + L_k, N_k = -U_k$.

(iii) The one-step Newton-Peaceman-Rachford method

$$\begin{aligned}\boldsymbol{x}_{k+1} &= \boldsymbol{x}_k - (\alpha_k I + C_{1k})^{-1}(\alpha_k I + C_{2k})^{-1}(\alpha_k I - C_{1k})^{-1}(\alpha_k I - C_{2k})^{-1}F(\boldsymbol{x}_k) \\ &= \boldsymbol{x}_k - 2\alpha_k(\alpha_k I + C_{1k})^{-1}(\alpha_k I + C_{2k})^{-1}F(\boldsymbol{x}_k), \quad k = 0, 1, 2, \ldots,\end{aligned} \tag{2.227}$$

where $F'(\boldsymbol{x}_k) = C_{1k} + C_{2k}$. The later iteration means that the linear system $C\boldsymbol{x} - \boldsymbol{b} = F'(\boldsymbol{x}_k)(\boldsymbol{x} - \boldsymbol{x}_k) + F(\boldsymbol{x}_k) = 0$ is solved.

2.7.4. Non-Linear Jacobi-, Gauss-Seidel-, SOR-, Peaceman-Rachford-Newton Methods

The goal of this section is to give non-linear solutions of Jacobi, Gauss-Seidel and SOR-Newton types. We shall solve numerically the non-linear equation

$$F(\boldsymbol{x}) = 0. \tag{2.228}$$

Then we solve the i-th equation

$$F_i(x_{1,k+1}, \ldots, x_{i-1,k+1}, x_i, x_{i+1,k}, \ldots, x_{n,k}) = 0 \tag{2.229}$$

for x_i, where F_i are components of F, and where we set $x_{i,k+1} = x_i$. Thus, in order to obtain $\boldsymbol{x}_{k+1}$ from $\boldsymbol{x}_k$, we must solve the n one-dimensional non-linear Equations 2.229 for $i = 1, \ldots, n$. This algorithm is the well-known non-linear Gauss-Seidel iterative method. If we put $x_{i,k+1} = x_{i,k} + \omega_k(x_i - x_{i,k})$, then we obtain a non-linear SOR method.

The non-linear Jacobi method follows from the fact that the k-th stage may be defined by solving the equations

$$F_i(x_{1,k}, \ldots, x_{i-1,k}, x_i, x_{i+1,k}, \ldots, x_{n,k}) = 0, \quad i = 1, 2, \ldots, n, \tag{2.230}$$

for x_i, and then setting $x_{i,i+1} = x_i$, $i = 1, \ldots, n$.

To obtain explicit formulae analogous to a general SOR-Newton iteration then $l_{k,i}$ Newton steps are applied to Equation 2.229. Then we find an approximate solution $\tilde{x}_i$ and we set $x_{i,k+1} = x_{i,k} + \omega(\tilde{x}_i - x_{i,k})$. If $l_{k,i} = 1$ then we have the one-step SOR-Newton method; if $l_{k,i} = 1$ and $\omega = 1$ then we have the one-step Gauss-Seidel-Newton method. Hence,

$$x_{i,k+1} = x_{i,k} - \omega F_i(x^{k,i})(\partial_i F_i(x^{k,i}))^{-1}, \quad i = 1, 2, \ldots, n, \quad k = 0, 1, 2, \ldots, \tag{2.231}$$

where we set

$$x^{k,i} = (x_{1,k+1}, \ldots, x_{i-1,k+1}, x_{i,k}, \ldots x_{n,k})^T. \tag{2.232}$$

For $\omega = 1$ in Equation 2.231 we have the one-step Gauss-Seidel-Newton iterative method.

Moreover, the one-step Jacobi-Newton iterative method is defined as follows

$$x_{i,k+1} = x_{i,k} - F_i(\boldsymbol{x}_k)(\partial_i F_i(\boldsymbol{x}_k))^{-1}, \quad k = 0, 1, 2, \ldots, \quad i = 1, \ldots, n. \tag{2.233}$$

Both algorithms (the one-step Newton-Jacobi and one-step Jacobi- Newton iterative methods) are identical, but the l-step methods are different. The one-step SOR-Newton and one-step Newton-SOR iterative methods are different, except for the case when $A\boldsymbol{x} = \varphi\boldsymbol{x}$, $A \in L(R^n)$ and $\varphi'(\boldsymbol{x})$ is diagonal, then both methods are identical.

In the case when the Newton method in Equation 2.231 or 2.233 is replaced by an other method, e.g. by the secant method, we obtain the one-step SOR-secant method in the form

$$x_{i,k+1} = x_{i,k} - \omega[F_i(x^{k,i}) - F_i(x^{k,i} + (x_{i,k+1} - x_{i,k})e_i)(x_{i,k} - x_{i,k-1})^{-1}]^{-1} F_i(x^{k,i}), \tag{2.234}$$

where $x^{k,i}$ is defined by Equation 2.232, and e_i are the coordinate vectors in R^n.

Now let us decompose $F = F_1 + F_2$, then the non-linear Peaceman-Rachford method can be found as

$$(\alpha I + F_1)\boldsymbol{x}_{k+(1/2)} = (\alpha I - F_2)\boldsymbol{x}_k, \tag{2.235}$$

$$(\alpha I + F_2)\boldsymbol{x}_{k+1} = (\alpha I - F_1)\boldsymbol{x}_{k+(1/2)}, \quad k = 0, 1, 2, \ldots . \tag{2.236}$$

Let $F(\boldsymbol{x}) = C\boldsymbol{x} - \boldsymbol{b}$ represent the linearized case and let us decompose C as $C = H + V$. Then $F_1(\boldsymbol{x}) = H\boldsymbol{x} - \frac{1}{2}\boldsymbol{b}$, $F_2(\boldsymbol{x}) = V\boldsymbol{x} - \frac{1}{2}\boldsymbol{b}$, and Equation 2.235 and 2.236 reduce to the form similar to the Peaceman-Rachford method. Both F_1 and F_2 are non-linear and thus one step of the non-linear Peaceman-Rachford method requires the solution of the two non-linear systems in Equations 2.235 and 2.236 for $\boldsymbol{x}_{k+(1/2)}$ and $\boldsymbol{x}_{k+1}$. We solve these non-linear equations by Newton's method. Thus, we obtain a general Peaceman-Rachford-Newton method in which l_k Newton steps are taken in the first equation, starting from $\boldsymbol{x}_k$; hence we obtain an approximate solution $\tilde{\boldsymbol{x}}_{k+(1/2)}$ to $\boldsymbol{x}_{k+(1/2)}$, and then n_k Newton steps are taken in the second equation, starting from $\tilde{\boldsymbol{x}}_{k+(1/2)}$, hence we obtain the iterate solution $\boldsymbol{x}_{k+1}$ of the composite method. If we put $l_k = n_k = 1$, we have the one-step Peaceman-Rachford-Newton method in the form

$$\boldsymbol{x}_{k+(1/2)} = \boldsymbol{x}_k - (\alpha I + F_1'(\boldsymbol{x}_k))^{-1} F(\boldsymbol{x}), \tag{2.237}$$

$$x_{k+1} = x_{k+(1/2)} - (\alpha I + F_2'(x_{k+(1/2)}))^{-1} F(x_{k+(1/2)}), \quad k = 0, 1, 2, \ldots, \tag{2.238}$$

where α is a parameter.

2.7.5. Modern Non-Linear Iterative Methods

2.7.5.1. Introduction

A variational solution of many non-linear geophysical problems has the form: minimize a given functional L defined on some function space X over a certain subset X_o, i.e.

$$\text{find } \boldsymbol{x} \in X \text{ such that } L(\underline{\boldsymbol{x}}) = \inf_{\boldsymbol{x} \in X_o} L(\boldsymbol{x}). \tag{2.239}$$

Numerical solutions then lead to solving related finite-dimensional minimization problems, which approximate the (infinite-dimensional) problems of Equation 2.239, representing the problems of solving a system of non-linear equations. Then these problems may be replaced by an equivalent problem of minimizing a functional $F : X \subseteq R^n \to R^1$. Let us consider a quadratic functional F in the form

$$F : R^n \to R^1,\ F(\boldsymbol{x}) = \tfrac{1}{2}\boldsymbol{x}^T C\boldsymbol{x} - \boldsymbol{b}^T\boldsymbol{x} + \boldsymbol{c}. \tag{2.240}$$

If C is an $n \times n$ symmetric, positive definite matrix (Hessian), then F has a unique global minimizer $\underline{\boldsymbol{x}}$ which is the solution of the linear system

$$(F'(\boldsymbol{x}))^T = C\boldsymbol{x} - \boldsymbol{b} = 0. \tag{2.241}$$

However we shall give various methods for minimizing a nonquadratic functional $F : R^n \to R^1$ by replacing F at the k-th step of the iteration by a quadratic functional F_k which approximates F in a neighbourhood of the k-th iterate $\boldsymbol{x}_k$. Then the minimizer of F_k we take as the next iterate $\boldsymbol{x}_{k+1}$.

Let us assume that $F : X \subseteq R^n \to R^1$ is twice F-differentiable at $\boldsymbol{x}_k \in X$, and let $H(\boldsymbol{x})$ be the Hessian of F. [Note that the mapping $F : X \subseteq R^n \to R^m$ is F- (Frechet-) differentiable at $x \in$int (X) if there is an $A \in L(R^n, R^m)$ such that $\lim_{h \to O} h^{-1}\, ||F(x+h) - F(x) - Ah|| = 0$. The linear operator $A \equiv F'(x)$ is called the F-derivative of F at x.] Applying the Taylor's expansion on F, for every $\boldsymbol{x} \in X$ we find

$$F(\boldsymbol{x}) = F(\boldsymbol{x}_k) + F'(\boldsymbol{x}_k)(\boldsymbol{x} - \boldsymbol{x}_k) + \tfrac{1}{2}(\boldsymbol{x} - \boldsymbol{x}_k)^T H(\boldsymbol{x}_k)(\boldsymbol{x} - \boldsymbol{x}_k) + R(\boldsymbol{x} - \boldsymbol{x}_k), \tag{2.242}$$

where $R(.)$ denotes the rest. Since $\lim_{h \to O} R(\boldsymbol{h})\, \|\boldsymbol{h}\|^{-2} = 0$, then for $\boldsymbol{x}$ sufficiently close to $\boldsymbol{x}_k$ the quadratic functional $F_k(\boldsymbol{x})$ defined by

$$F_k(\boldsymbol{x}) = F(\boldsymbol{x}_k) + F'(\boldsymbol{x}_k)(\boldsymbol{x} - \boldsymbol{x}_k) + \tfrac{1}{2}(\boldsymbol{x} - \boldsymbol{x}_k)^T H(\boldsymbol{x}_k)(\boldsymbol{x} - \boldsymbol{x}_k). \tag{2.243}$$

approximates F near $\boldsymbol{x}_k$. Since F, due to our assumption, has a second F-derivative at $\boldsymbol{x}$, then $F''(\boldsymbol{x})$, i.e. the Hessian matrix $H(\boldsymbol{x})$, is symmetric. In the case when $H(\boldsymbol{x})$ is positive definite, the unique global minimizer of F_k is the solution of the linear system

$$H(\boldsymbol{x}_k)(\boldsymbol{x} - \boldsymbol{x}_k) = (F'(\boldsymbol{x}_k))^T \tag{2.244}$$

and then

$$x_{k+1} = x_k - (H(x_k))^{-1}(F'(x_k))^T \tag{2.245}$$

is the next Newton iterate. In the case, when $H(x_k)$ is non-singular but not positive definite, then a Newton iterate of the form of Equation 2.245 corresponds to finding the unique critical point of the quadratic functional F_k. On the other hand, if $H(x_k)$ is singular, or indefinite, then instead of Equation 2.243 we consider the quadratic functional

$$F_k(x) = F(x_k) + F'(x_k)(x - x_k) + \tfrac{1}{2}(x - x_k)^T[H(x_k) + \alpha_k I](x - x_k), \tag{2.246}$$

where α_k is a scalar such that $H(x_k) + \alpha_k I$ is positive definite. Then F_k has a unique global minimizer x_{k+1} and

$$x_{k+1} = x_k - (F'(x_k) + \alpha_k I)^{-1}F(x_k), \quad k = 0, 1, 2, \dots . \tag{2.247}$$

2.7.5.2. *Descent methods*

These methods are based on the idea that the iterates decrease the function value at each stage, i.e. that

$$F(x_{k+1}) \leq F(x_k), \quad k = 0, 1, 2, \dots . \tag{2.248}$$

To derive these types of methods we shall consider the Newton iteration of Equation 2.245 with an added parameter α_k, which may be determined in the form

$$x_{k+1} = x_k - \alpha_k(F'(x_k))^{-1}F(x_k), \quad k = 0, 1, 2, \dots . \tag{2.249}$$

Then the parameter α_k, under the assumption $F(x_k) \neq 0$, may be found in such a way that Equation 2.248 holds. For Newton's method we put

$$p_k = (F'(x_k))^{-1}F(x_k). \tag{2.250}$$

Move then the next lemma proves our above assertion:

Lemma 2.17: Let $F : X \subseteq R^n \to R^1$ be G-differentiable at $x \in \text{int}(X)$ and let, for some $p \in R^n$, $F'(x)p > 0$. Then there exists a $\delta > 0$ such that

$$F(x - \alpha p) < F(x) \quad \forall \alpha \in (0, \delta).$$

Proof: Since

$$\lim_{\alpha \to 0} \alpha^{-1}(F(x - \alpha p) - F(x)) + F'(x)p = 0, \; F'(x)p > 0,$$

then there exists a $\delta > 0$ such that $x - \alpha p \in X \; \forall \alpha \in (0, \delta)$ and

$$\alpha^{-1}(F(x - \alpha p) - F(x)) + F'(x)p < F'(x)p.$$

Hence, the assertion follows immediately.

Since for the Newton method p_k is defined by Equation 2.250, then if the Hessian $H(x_k) = F'(x_k)$ is positive definite, then $(F'(x_k))^{-1}$ is also positive definite and

$$F'(x_k)p_k = (F(x_k))^T(F'(x_k))^{-1}F(x_k) > 0$$

provided that $F(\boldsymbol{x}_k) \neq 0$. Due to Lemma 17, $\alpha_k > 0$ in Equation 2.249 may be chosen so that Equation 2.248 is satisfied. For small α_k we have the so-called damped Newton method.

Due to Equation 2.250, the Newton iteration of Equation 2.249 can be written as

$$\boldsymbol{x}_{k+1} = \boldsymbol{x}_k - \alpha_k \boldsymbol{p}_k, \quad k = 0, 1, 2, \ldots, \tag{2.251}$$

where the vector $-\boldsymbol{p}_k$ has the meaning of a direction along which a new iterate $\boldsymbol{x}_{k+1}$ will be chosen, and α_k is a scalar which has the meaning of a step-length from $\boldsymbol{x}_k$ to $\boldsymbol{x}_{k+1}$.

If F is differentiable at $\boldsymbol{x}_k$ then $-\boldsymbol{p}_k$ may be chosen as the direction of maximal local decrease of F. Then this direction $\boldsymbol{p}_k$ is a best direction if

$$\left\| F'(\boldsymbol{x}_k) \right\| = F'(\boldsymbol{x}_k)\boldsymbol{p}_k \, \|\boldsymbol{p}_k\|^{-1},$$

and is called a direction of steepest descent, and the method defined by Equation 2.251 is called a method of steepest descent.

Lemma 2.18: Let C be an $n \times n$ symmetric, positive definite matrix and let F be G-differentiable at $\boldsymbol{x}$, then the direction of steepest descent of F at $\boldsymbol{x}$ in the norm $\|\boldsymbol{x}\|^2 = \boldsymbol{x}^T C \boldsymbol{x}$ is given by $-C^{-1}(F'(\boldsymbol{x}))^T$.

(For the proof see Ortega and Rheinboldt (1970), p. 245.)

As a consequence of this lemma we then derive the following method:

$$\boldsymbol{x}_{k+1} = \boldsymbol{x}_k - \alpha_k C_k^{-1}(F'(\boldsymbol{x}_k))^T, \quad k = 0, 1, 2, \ldots, \tag{2.252}$$

where C_k is a matrix which may change from step to step, but in many cases it is symmetric and positive definite. If the Hessian matrix $C_k = H(\boldsymbol{x}_k)$ is positive definite for each $\boldsymbol{x}_k$ then we have the damped Newton iterative method of Equation 2.249.

Next we introduce the method defined by

$$\boldsymbol{x}_{k+1} = \boldsymbol{x}_k - \alpha_k C_k (F'(\boldsymbol{x}_k))^T, \quad k = 0, 1, 2, \ldots, \tag{2.253}$$

where the matrices C_k change from step to step and are defined recursively as

$$\begin{aligned} C_{k+1} = C_k &+ \frac{(\boldsymbol{x}_{k+1} - \boldsymbol{x}_k)(\boldsymbol{x}_{k+1} - \boldsymbol{x}_k)^T}{(\boldsymbol{x}_{k+1} - \boldsymbol{x}_k)^T[(F'(x_{k+1}))^T - (F'(x_k))^T]} \\ &- \frac{C_k[(F'(x_{k+1}))^T - ((F'(x_k))^T](C_k[(F'(x_{k+1}))^T - (F'(x_k))^T])^T}{[(F'(X_{k+1}))^T - (F'(x_k))^T]^T C_k[(F'(x_{k+1}))^T - (F'(x))^T]}, \\ &\qquad k = 0, 1, 2, \ldots, \end{aligned}$$

where C_o is an arbitrary, symmetric, positive definite matrix (e.g. the identity matrix), and hence C_k are positive definite matrices. The method is known as the Davidson-Fletcher-Powell method.

2.7.5.3. *Conjugate gradient method*

In the linear case of this method we saw that any two non-zero vectors $\boldsymbol{p}$ and $\boldsymbol{q} \in R^n$ are conjugate with respect to a matrix C, if $\boldsymbol{p}^T C\boldsymbol{q} = 0$. A set of non-zero vectors $\{\boldsymbol{p}_l\}_{l=1}^n$ is a conjugate basis if $\boldsymbol{p}_i$ and $\boldsymbol{p}_j$ are conjugate for all $i \neq j$.

The main problem is to determine vectors $\boldsymbol{p}_j$, which are generated simultaneously with the $\boldsymbol{x}_j$. Let $F : X \subset R^n \to R^1$ be twice F-differentiable. Then the algorithm is as follows:

$$\begin{aligned}
\boldsymbol{x}_{k+1} &= \boldsymbol{x}_k - \alpha_k \boldsymbol{p}_k, \\
F(\boldsymbol{x}_{k+1}) &= \min\{F(\boldsymbol{x}_k - \alpha\boldsymbol{p}_k) \mid \boldsymbol{x}_k - \alpha\boldsymbol{p}_k \in X\}, \\
\boldsymbol{p}_o &= (F'(\boldsymbol{x}_o))^T, \\
\beta_k &= F''(\boldsymbol{x}_{k+1})(F'(\boldsymbol{x}_k))^T \boldsymbol{p}_k [F''(\boldsymbol{x}_{k+1})\boldsymbol{p}_k\boldsymbol{p}_k]^{-1}, \\
\boldsymbol{p}_{k+1} &= (F'(\boldsymbol{x}_{k+1}))^T - \beta_k \boldsymbol{p}_k,
\end{aligned} \tag{2.254}$$

where α_k is determined by the minimization principle in the second equation. The above algorithm of the CG method is known in the literature as the Daniel algorithm.

For $F(\boldsymbol{x}) = \frac{1}{2}\boldsymbol{x}^T C\boldsymbol{x} - \boldsymbol{b}^T\boldsymbol{x} + \boldsymbol{c}$, $F'(\boldsymbol{x}) = C\boldsymbol{x} - \boldsymbol{b}$ and $F''(\boldsymbol{x})\boldsymbol{hk} = \boldsymbol{k}^T C\boldsymbol{h}$, and Equation 2.254 reduces to the classical conjugate gradient method.

The further CG-algorithm is known as the Fletcher-Reeves algorithm and can be written as follows

$$\begin{aligned}
\boldsymbol{x}_{k+1} &= \boldsymbol{x}_k - \alpha_k \boldsymbol{p}_k, \\
F(\boldsymbol{x}_{k+1}) &= \min\{F(\boldsymbol{x}_k - \alpha\boldsymbol{p}_k) \mid \boldsymbol{x}_k - \alpha\boldsymbol{p}_k \in X\}, \\
\boldsymbol{p}_o &= (F'(\boldsymbol{x}_o))^T, \\
\beta_k &= -F'(\boldsymbol{x}_{k+1})(F'(\boldsymbol{x}_{k+1}))^T (F'(\boldsymbol{x}_k)((F'(\boldsymbol{x}_k))^T)^{-1}, \\
\boldsymbol{p}_{k+1} &= (F'(\boldsymbol{x}_{k+1}))^T - \beta_k \boldsymbol{p}_k,
\end{aligned} \tag{2.255}$$

where α_k is determined by the minimization principle of the second equation.

The conjugate gradient method — the Daniel algorithm for any $\boldsymbol{x}_o \in R^n$ is well defined, and converges to the unique critical point $\underline{\boldsymbol{x}}$ of F in R^n.

2.7.5.4. *The Gauss-Newton method*

Let mappings $F : X \subset R^n \to R^m$, $f : R^m \to R^1$ be given and let $G : X \subset R^n \to R^1$ be defined by

$$G(\boldsymbol{x}) = f(F(\boldsymbol{x})) \text{ for all } \boldsymbol{x} \in X. \tag{2.256}$$

If $f = \frac{1}{2}\boldsymbol{x}^T\boldsymbol{x}$ is the quadratic functional, then the minimizer of $G(\boldsymbol{x}) = \frac{1}{2}(F(\boldsymbol{x}))^T$ $F(\boldsymbol{x})$ is a least-squares solution of $F(\boldsymbol{x}) = 0$.

Further, let us assume that the mapping F is F-differentiable on X and let $\boldsymbol{x}_k$ denote a k-th approximation of a minimizer of Equation 2.256. To compute the next approximation $\boldsymbol{x}_{k+1}$ we minimize the functional

$$G_k(\boldsymbol{x}) = f(F(\boldsymbol{x}_k) + F'(\boldsymbol{x}_k)(\boldsymbol{x} - \boldsymbol{x}_k)), \tag{2.257}$$

obtained by a linearization of F about $\boldsymbol{x}_k$. Assume f is twice F-differentiable, then a minimizer of Equation 2.257 can be approximated by one-step of the Newton iteration starting from $\boldsymbol{x}_k$. Then

$$\begin{aligned} \boldsymbol{x}_{k+1} &= \boldsymbol{x}_k - (H_{Gk}(\boldsymbol{x}_k))^{-1}(G_k'(\boldsymbol{x}_k))^T \\ &= \boldsymbol{x}_k - [(F'(\boldsymbol{x}_k))^T H_f(F(\boldsymbol{x}_k))F'(\boldsymbol{x}_k)]^{-1}(F'(\boldsymbol{x}_k))^T(f'(F(\boldsymbol{x}_k)))^T, \end{aligned} \tag{2.258}$$

where

$$G_k'(\boldsymbol{x}) = f'(F(\boldsymbol{x}_k) + F'(\boldsymbol{x}_k)(\boldsymbol{x} - \boldsymbol{x}_k))F'(\boldsymbol{x}_k)$$

and the Hessian

$$H_{Gk}(\boldsymbol{x}) = (F'(\boldsymbol{x}_k))^T H_f(F(\boldsymbol{x}_k) + F'(\boldsymbol{x}_k)(\boldsymbol{x} - \boldsymbol{x}_k))F'(\boldsymbol{x}_k),$$

under the assumptions that the above inverse exists. The algorithm of Equation 2.258 is known as a generalized Gauss-Newton method.

Furthermore, we may modify Equation 2.258 with parameters α_k and λ_k and with $f = \frac{1}{2}\boldsymbol{x}^T\boldsymbol{x}$ as

$$\boldsymbol{x}_{k+1} = \boldsymbol{x}_k - \alpha_k[(F'(\boldsymbol{x}_k))^T F'(\boldsymbol{x}_k) + \lambda_k I]^{-1}(F'(\boldsymbol{x}_k))^T F(\boldsymbol{x}_k). \tag{2.259}$$

The matrix $(F'(\boldsymbol{x}_k))^T F'(\boldsymbol{x}_k)$ is symmetric, positive semi-definite, and then the inverse in Equation 2.259 exists provided that $\lambda_k > 0$. The parameter α_k can be found such that $G(\boldsymbol{x}_{k+1}) \leq G(\boldsymbol{x}_k)$. Moreover, a suitable selection of λ_k may ensure that $G(\boldsymbol{x}_{k+1}) \leq G(\boldsymbol{x}_k)$.

REFERENCES

Axelsson, O. (1994), *Iterative Solution Methods*, Cambridge University Press, Cambridge.

Axelsson, O. and Barker, V.A. (1984), *Finite Eelement Solutions of Boundary Value Problems*, Academic Press, Orlando.

Bramble, J.H. (1993), *Multi-grid Methods*, Longman, Harlow.

Brezzi, F., Hager, W. and Raviart, P.A. (1977), "Error Estimates for the Finite Element Solution of Variational Inequalities I", *Numer. Math.*, **28**, 431-443.

Céa, J. (1971), *Optimisation, Théorie et Algorithmes*, Dunod, Paris.

Ciarlet, P.G. (1978), *The Finite Element Method for Elliptic Problems*, North-Holland Publishing Company, Amsterdam, New York, Oxford.

Daniel, J.W. (1971), *The Approximate Minimization of Functionals*, Prentice-Hall Inc., Englewood Cliffs, N.J.

Dunford, N. and Schwartz, J.T. (1958), *Linear Operators. 1. General Theory*, Interscience Publishers Inc., New York, London.

Ekeland, I. and Temam, R.(1976), *Convex Analysis and Variational Problems*, North-Holland Publishing Company, Amsterdam, Oxford.

Falk, R.S. (1974), "Error Estimates for the Approximation of a Class of Variational Inequalities", *Math. Comput.*, **28**, 963-971.

Fiaco, A.V. and Cormick, G.P. (1968), *Nonlinear Programming: Sequential Unconstrained Minimization Techniques*, J. Wiley & Sons, New York.

Fletcher, R. (1976), "Conjugate Gradient Methods for Indefinite Systems", in *Numerical Analysis* (Watson, G.A. (Ed.)), Spinger Vlg., Berlin.

Gill, P.E., Murray, W. and Wright, M.H. (1981), *Practical Optimization*, Academic Press, New York and London.

Glowinski, R. (1984), *Numerical Methods for Non-linear Variational Problems*, Springer Verlag, New York, Berlin, Heidelberg, Tokyo.

Glowinski, R., Lions, J.L. and Trémoliéres, R. (1976), *Numerique des Inéqualitions Variationnelles*, Dunod, Paris.

Golub, G.H. and Ortega, J.M. (1993), *Scientific Computing: An Introduction with Parallel Computing*, Academic Press, San Diego.

Golub, G.H. and Van Loan, C.F. (1983) and (1989), *Matrix Computation, 1st and 2nd Editions*, The John Hapkins University Press, Baltimore.

Hackbusch, W. (1985), *Multi-grid Methods and Applications*, Springer Vlg., Berlin.

Hestenes, M.R. (1980), *Conjugate Direction Methods in Optimization*, Springer Vlg., New York.

Hestenes, M.R. and Stiefel, E. (1952), "Methods of Conjugate Gradients for Solving Linear Systems", *J. Res. Nat. Bus. Stand.*, **49**, 409-436.

Hewitt, E. and Stromberg, K. (1969), *Real and Abstract Analysis*, Springer Verlag, New York, Berlin.

Jarník, V. (1955), *Integral Calculation*, Academia, Prague (in Czech).

Kikuchi, N. and Oden, J.T. (1988), *Contact Problems in Elasticity. A Study of Variational Inequalities and Finite Element Methods*, SIAM-Studies in Applied Mathematics, Philadelphia.

Kinderlehrer, D. and Stampacchia, G. (1980), *An Introduction to Variational Inequalities and their Application*, Academic Press, New York.

Kufner, A., John, O. and Fučík, S. (1977), *Function Spaces*, Academia, Prague.

Lions, J.L. and Magenes, E. (1968), *Problémes aux Limites Non-Homogénes et Applications*, Dunod, Paris.

Lions, J.L. and Stampacchia, G. (1967), "Variational Inequalities", Comm. Pure Appl. Math., **20**, 493-519.

Ljusternik, L.A. and Sobolev, V.I. (1974), *Elements of Functional Analysis*, Hindustan Publishing Corp., Delhi; Halstadt Press, New York.

Mandel, J. McCornick, S.F. and Bank, R.E. (1987), "Variational Multi-grid Theory", in *Multi-grid Methods* (McCornick, S.F. (Ed)), SIAM, Philadelphia, 131-177.

Mosco, V. and Strang, G. (1974), "One-Sided Approximation and Variational Inequalities", Bull. Am. Math. Soc, **80**, 308-312.

Nečas, J. (1967), *Les Méthodes Directes en Théorie des Équations Elliptiques*, (Masson et Cie, Editeurs), Academia, Prague.

Nečas, J. and Hlaváček, I. (1981), *Mathematical Theory of Elastic and Elasto-Plastic Bodies: An Introduction*, Elsevier, Amsterdam, Oxford, New York.

Oden, J.T. and Reddy, J.N. (1976), *An Introduction to the Mathematical Theory of Finite Elements*, J. Wiley & Sons, New York, London, Sydney, Toronto.

Ortega, J.M. and Rheinboldt, W.C. (1970), *Iterative Solution of Nonlinear Equations in Several Variables*, Academic Press New York and London.

Pshenichny, B.N. (1983), *The Linearization Method for Constrained Optimization*, Springer Vlg., Berlin.

Pshenichny, B.N. and Danilin, Y.M. (1978), *Numerical Methods in Extremal Problems*, Mir Publishers, Moscow.

Quarteroni, A. and Valli, A. (1994), *Numerical Approximation of Partial Differential Equations*, Springer Verlag, Berlin, Heidelberg.

Rektorys, K. (1980), *Variational Methods in Mathematics, Science and Engineering*, D. Reidel Publishing Company, Dordrecht-Holland, Boston U.S.A.

Schwartz, L. (1957), *Theorie des Distributions*, Herman, Paris.

Sobolev, S.L. (1963). "Applications of Functional Analysis in Mathematical Physics", *American Mathematical Society*, **MR 29**, 2624 Providence, R.I..

Taylor, A.E. (1967), *Introduction to Functional Analysis*, J. Wiley & Sons, New York.

Yosida, K. (1974), *Functional Analysis*, Springer Verlag, New York, Heidelberg.

PART III

Numerical Models of Deformation of Geological Bodies

3

Deformation of Geological Bodies and Stress-Strain Measurements in Situ

3.1. INTRODUCTION TO THE MATHEMATICAL THEORY OF CONTINUUM MECHANICS

Classical continuum mechanics has its triumphs — the problems of the modern thermo-elasto-visco-plasticity and the problems of the modern supersonic jet; and its enduring challenges — the problems of turbulence, the problems of geodynamic processes, and earthquake models. But there are also modern aspects of continuum mechanics that have emerged in the last decades, and which are characterized by the modern mathematical theory of continuum mechanics. The present study is designed to serve as an introduction to both classical and modern continuum theory (See Brdička (1959), Brdička and Hladík (1987), Desloge (1982), Landau and Lifshitz (1973), Lanczos (1962), Serrin (1959), Levin (1977), Jaunzemis (1967), Washizu (1968), Duvaut and Lions (1972), Fichera (1972), Nečas and Hlaváček (1981), and many other monographs and papers).

Since the modern theory of continuum mechanics is closely connected with thermodynamics, we also give a brief survey to mathematical thermodynamics. Our thermodynamic studies will be based on the results of modern mathematical and rational thermo-dynamics given in Hill (1950), Callen (1960), Hasse (1968), Zemansky (1968), Gladsdorff and Prigogine (1971), Gurtin (1974), Woods (1975), Serrin (1979), Truesdell (1969), (1984), Owen (1984), Samohýl (1982).

3.1.1. Brief Survey of Mathematical Thermodynamics

Thermodynamics is one part of theoretical physics studying general properties of physical systems in equilibrium, as well as general (representative) regularities holding in their steady state. Much of mathematical thermodynamics concerns thermal and mechanical interactions between a physical system and its environment — its surroundings. In thermodynamics we firstly define the physical systems as the portions which are set aside and on which the attention is focussed. Everything outside the system which has a direct bearing on its behaviour is known as the surroundings. When a physical system has been chosen, the next step is to describe it in terms of physical quantities describing the behaviour of the system or its interactions with the surroundings, or both. A statement of the relative amounts of substances creating the system represents a description of the composition of the system. At any moment, the system occupies a certain volume, denoted by V. The properties of macroscopic bodies occupying the volume V and creating the physical system causally depend on the motion, the temperature T, the density of particles of the body, the pressure in the past as well as in the present time. One can say that by this the deterministic principle is motivated. These quantities refer to the gross characteristics, or large-scale properties, of the physical system and provide a macroscopic description. They are the so-called macroscopic or thermodynamic coordinates. From thermodynamic theory it is known that in general two points of view are used in thermodynamic analyses. Here the macroscopic point of view will be applied. A system that may be described in terms of thermodynamic co-ordinates is called a thermodynamic system. In this section the physical systems will be represented by the liquid and the rigid bodies and their deformation will be investigated.

3.1.1.1. The first law of thermodynamics

The state of a system in which a pair of independent coordinates, denoted by X_1 and X_2, have definite values which remain constant so long as the external conditions are unchanged is called an equilibrium state. The existence of an equilibrium state in one system depends on the proximity of the other systems and on the nature of the wall separating them. These walls can be adiabatic and diathermic. If a wall is adiabatic, states X_2, X_1 for the system denoted by A_1, and X_2', X_1' for the system A_2, may co-exist as equilibrium states for any attainable values of the four quantities, provided only that the wall is able to withstand the stress associated with the difference between the two sets of co-ordinates (e.g., the adiabatic wall can be demonstrated by an asbestos layer, etc.). For a diathermic wall the values of X_2, X_1 and X_2', X_1' will change spontaneously until an equilibrium state of the combined system is attained (e.g., the diathermic wall can be demonstrated by a thin metallic layer). The two systems are then said to be in thermal equilibrium with each other. Thermal equilibrium is the state achieved by two or more systems, characterized by restricted values of the co-ordinates of the systems, after they have been in communication with one another through a diathermic wall.

To describe the state of the investigated deformable body we shall start from a number of conservation laws, which lead into a number of conservation law equations. But these conservation equations, such as the continuity equation, the equation of motion and the energy equation, do not amount to the number of unknown physical quantities, e.g. the velocity $v(x, t)$, the density $\rho(x, t)$, the displacement vector $u(x, t)$, the temperature $T(x, t)$, the pressure $p(x, t)$, intensities of electric and magnetic fields $E(x, t)$ and $H(x, t)$ respectively, and the heat flow $q(x, t)$. In order to complete the conservation law system, additional thermodynamics relations (constitutive equations) must be added. In our considerations we shall use the results of phenomenological thermodynamics based on rational thermodynamics. Constitutive equations represent macroscopic mathematical models of a real material, and accentuate its characteristic features and properties which are suitable for our applications. In many cases constitutive equations can be represented by very complicated functional dependences between all physical quantities. In several cases the constitutive equations can describe materials with memories of various types. Materials without memory are for example classical elastic materials, whilst materials with memory are for example visco-plastic, visco-thermo-elastic or visco-thermo-plastic materials.

When a system is completely surrounded by an adiabatic envelope, it may still be coupled to the outside so that work may be done. It is an important fact of experience that the state of a system may be caused to change from a given initial state to a suitable final state by the performance of adiabatic work only. The generalization of this result is known as the first law of thermodynamics:

> If a system is caused to change from an initial state to a final state by adiabatic means only, the work done is the same for all adiabatic paths connecting the two states.

It was concluded from classical mechanics that there exists a function of the space co-ordinates of the body whose final value minus its initial value is equal to the work done. This function is called the potential-energy function. Similarly in the electromagnetic field there exists the electric-potential function. It therefore follows from the first law of thermodynamics that there exists a function of the co-ordinates of a thermodynamic system whose value at the final state minus its value at the initial state is equal to the adiabatic work done in going from one state to the other. This function is known as the internal-energy function and is denoted by U. The physical interpretation of the difference between the values corresponding to the final and the initial states is the increase in energy of the system. Therefore, the equality of the increase of energy and the adiabatic work expresses the principle of the conservation of energy. The internal energy is a function of as many thermodynamic co-ordinates as are necessary to specify the state of a system.

If the system is in an equilibrium state and if none of the external force fields operate on it, then the total energy of the system is identical to its internal energy. But in the general case of a nonadiabatic system, the nonadiabatic work is not equal to the difference of the final and the initial values of the internal energy. Thus, when a system whose surroundings are at a different temperature and on which work may be done undergoes a process, the energy transferred by nonmechanical means, equal to the difference between the internal-energy change and the work done, is called heat and is denoted by Q.

Then the first law of thermodynamics in its differential form can be written as follows

$$\mathrm{d}Q = \mathrm{d}U + \mathrm{d}W. \tag{3.1}$$

But one can say that a state of the investigated system (e.g. a volume of a continuum or a particle) is given if any parameters a_i, which completely define all characteristics of the system, are given. These parameters create a complete system (basis) of defining parameters. The parameters a_i are called parameters of a state. Then the work of the system upon passing from one state to a second state can be written in the integral from

$$W = \int_1^2 \sum_{i=1}^n A_i \mathrm{d}a_i, \tag{3.2}$$

where A_i is a function of state of the system (a state function), i.e. a function of external parameters a_i, of temperature Θ and of internal parameters ξ_i. From the mathematical point of view the elementary work $\mathrm{d}W = \Sigma A_i \mathrm{d}a_i$ is a total differential of any function of state of the system. In the adiabatic system $\mathrm{d}W = -\mathrm{d}U$, therefore $\mathrm{d}U = \mathrm{d}U(a_i, \Theta, \xi_i)$. Hence, and using Equations 3.1 and 3.2,

$$\mathrm{d}Q = \mathrm{d}U(a_i, \Theta, \xi_i) + \sum_{i=1}^n A_i \mathrm{d}a_i. \tag{3.3}$$

Since the integral $\int \mathrm{d}Q$ depends on the integral path, then $\mathrm{d}Q$ is not a total differential. Hence the quantity Q represents thermal energy and is called heat. For adiabatic processes $\mathrm{d}Q = 0$, for isothermic processes $\mathrm{d}\Theta/\mathrm{d}t = 0$. Then $\Theta =$ constant or grad $\Theta = 0$, and/or $\Theta = f(t)$.

By the state of equilibrium we mean the state which will occur in the system during the time, earlier or later, that the system is under the specific external conditions characterized by specific constant values of external parameters, and by specific constant temperature of neighbourhood bodies. In thermodynamic equilibrium all internal parameters of the system are functions of external parameters and the temperature of the system.

The temperature of a system is a property that determines whether or not a system is in thermal equilibrium with other systems. The temperature Θ is the objective scalar which determines how hot is the particle being situated in a given place of an actual configuration ($\Theta \geq 0$ always).

When heat is absorbed by a system, a change of temperature may or may not take place, depending on the process. If a system undergoes a change of temperature from Θ_1 to Θ_2 during the transfer of Q units of heat, the average heat capacity $\underline{C}$ of the system is defined as $\underline{C} = Q/(\Theta_2 - \Theta_1)$. In the limit as $\Theta_2 \to \Theta_1$ the heat capacity $\underline{C} = \mathrm{d}Q/\mathrm{d}\Theta$, i.e. it is a ratio of heat increment to temperature increment in the system. The heat capacity of a system per unit mass is called the specific heat capacity or specific heat and is defined by

$$c = m^{-1}\mathrm{d}Q/\mathrm{d}\Theta = \mathrm{d}q/\mathrm{d}\Theta, \quad q = Q/m. \tag{3.4}$$

The heat capacity may be negative, positive, zero or infinite, depending on the process the system undergoes during the heat transfer. It has a definite value only for a definite process, for example in the case when the pressure, the volume, the area, the surface tension, the electromagnetic fields, the charge or the magnetization, etc. are constant. Then we speak about the heat capacity at constant pressure $c_p = (\partial Q/\partial\Theta)_p$, the heat capacity at constant volume $c_v = (\partial Q/\partial\Theta)_V$, the heat capacity at constant electromagnetic field $c_E = (\partial Q/\partial\Theta)_E$, the heat capacity at constant charge $c_\eta = (\partial Q/\partial\Theta)_\eta$, the heat capacity at constant magnetic field $c_H = (\partial Q/\partial\Theta)_H$, the heat capacity at constant magnetization $c_M = (\partial Q/\partial\Theta)_M$, etc.. Each heat capacity is a function of two variables, and practically may be regarded as constant. Very often, one heat capacity is without much error relative to another; so that the c_H of a paramagnetic solid material is at times practically equal to c_p. For the molar heat capacity at constant magnetic field of a paramagnetic solid at low temperatures, $c_H = (b + cH^2)/\Theta^2 + D\Theta^3$, where b, c, D (b is the Stefan-Boltzman constant, D is the Debue function) are constants, and for the molar heat capacity at constant pressure of a gas, $c_p = a + b\Theta - c\Theta^{-2}$, where a, b, c are constants; and for a paramagnetic solid obeying Curie's equation $c_M = (\partial U/\partial\Theta)_M$, $c_H = (\partial U/\partial\Theta)_H + M^2/c_o$, where c_o is a Curie's constant. If the infinitesimal process is quasi-static, then $\mathrm{d}U$ and $\mathrm{d}W$ can be expressed in terms of thermodynamic co-ordinates only, and the first law of thermodynamics becomes

$$\mathrm{d}Q = \mathrm{d}U + p\mathrm{d}V, \tag{3.5}$$

where U is a function of any two of the three thermodynamic co-ordinates and p is a function of V and Θ.

Let a state of the system be determinated by a volume V and by a temperature Θ. The energy of the system in an equilibrium state is $U = U(V, \Theta)$, i.e. it is a function of a volume V and a temperature Θ. A quantity of heat obtained by the system over the quasistatic process is given by

$$\begin{aligned}\mathrm{d}Q = \mathrm{d}U + p\mathrm{d}V &= (\partial U/\partial\Theta)_V\mathrm{d}\Theta + (\partial U/\partial V)_\Theta\mathrm{d}V + p\mathrm{d}V\\ &= (\partial U/\partial\Theta)_V\mathrm{d}\Theta + [(\partial U/\partial V)_\Theta + p]\mathrm{d}V.\end{aligned} \tag{3.6}$$

If the volume is constant, then $\mathrm{d}V = 0$ and

$$\mathrm{d}Q = (\partial U/\partial\Theta)_V\mathrm{d}\Theta, \quad C_v = (\partial U/\partial\Theta)_V.$$

Let the body be heated at constant pressure, then

$$C_p = \mathrm{d}Q/\mathrm{d}\Theta = (\partial U/\partial\Theta)_V + [(\partial U/\partial V)_\Theta + p](\partial V/\partial\Theta)_p \tag{3.7}$$

and hence

$$C_p = C_v + [(\partial U/\partial V)_\Theta + p](\partial V/\partial\Theta)_p,$$

where the term $[(\partial U/\partial V)_\Theta + p](\partial V/\partial\Theta)_p$ represents the quantity of heat needed for executing the work $(p\partial V/\partial\Theta)$ and for a change of energy $[(\partial U/\partial V)_\Theta(\partial V/\partial\Theta)_p]$ connected with a change of volume. The unit of the capacity of heat is the unit of work per one degree of temperature in the chosen scale. As the inner energy of the body is proportional to the body mass we can define the specific heat as a capacity of heat of the unit mass of the body. The capacity of heat c_v and c_p are positive values as the temperature increases if the body gets heat. For the ideal gas the state equation is given by $pV = R\Theta$, where R is the gas constant. Due to Joule's law $(\partial U/\partial V)_\Theta = 0$, then $(\partial V/\partial\Theta)_p = R/p$ and then $C_p = C_v + R$ is the so-called Mayer's relation.

Then the first law of thermodynamics can be written in special cases as

$$\mathrm{d}Q = \mathrm{d}U + p\mathrm{d}V - \boldsymbol{E}\mathrm{d}\eta, \tag{3.8}$$

where U is a function of a pressure p, a volume V, a temperature Θ, an electric field intensity $\boldsymbol{E}$, and an electric charge η, and in the case of a paramagnetic gas as

$$\mathrm{d}Q = \mathrm{d}U + p\mathrm{d}V - \boldsymbol{H}\mathrm{d}\boldsymbol{M}, \tag{3.9}$$

where U is a function of three of p, V, Θ, $\boldsymbol{H}$ and $\boldsymbol{M}$; i.e. a pressure p, a volume V, a temperature Θ, an intensity of magnetic field $\boldsymbol{H}$, and a magnetization $\boldsymbol{M}$.

In many cases, instead of volume V and temperature Θ as fundamental variables of the system, it is better to use the pressure p and temperature Θ. Then

$$\mathrm{d}Q = \mathrm{d}U + p\mathrm{d}V = \mathrm{d}(U + pV) - V\mathrm{d}p = \mathrm{d}H - V\mathrm{d}p, \tag{3.10}$$

where $H = U + pV$ is the so-called enthalpy. Then for the capacity of heat at constant pressure we obtain

$$\mathrm{d}Q = (\partial H/\partial\Theta)_p\mathrm{d}\Theta + [(\partial H/\partial p)_\Theta - V]\mathrm{d}p, \tag{3.11}$$

$$C_p = (\mathrm{d}Q/\mathrm{d}\Theta)_p = (\partial H/\partial\Theta)_p. \tag{3.12}$$

Hence $C_p \neq (\partial U/\partial\Theta)_p$. Let us assume the reversible adiabatic expansion or compression of a body. Then $\mathrm{d}Q = 0$, hence

$$(\partial U/\partial\Theta)_V\mathrm{d}\Theta + [(\partial U/\partial V)_\Theta + p]\mathrm{d}V = 0. \tag{3.13}$$

Since $C_v = (\partial U/\partial\Theta)_V$, $C_p = (\partial U/\partial\Theta)_V + [(\partial U/\partial V)_\Theta + p](\partial V/\partial\Theta)_p$, then

$$C_v\mathrm{d}\Theta + (C_p - C_v)/(\partial U/\partial\Theta)_p\mathrm{d}V = \mathrm{d}\Theta + (\gamma - 1)/(\partial U/\partial\Theta)_p\mathrm{d}V = 0, \tag{3.14}$$

where $\gamma = C_p/C_v$ and the last equation is valid for the gas, the liquid or the solid body.

3.1.1.2 Principle of Carathéodory and formulation of the second law of thermodynamics

In 1824 the young French engineer Nicolas Lénard Sudi Carnot described an ideal engine operating in a particularly simple cycle known today as the Carnot cycle. Analyses of the Carnot cycles of a real or ideal gas, of a mixture of a liquid and vapour, of a reversible electric cell, of a paramagnetic substance, of a refrigerator, etc., lead to the formulation of the second law of thermodynamics. Firstly the Carnot idea was elaborated by the physicists Kelvin and Clausius. In 1909, the mathematician Carathéodory set himself the problem of finding a statement of the second law of thermodynamics, without the Carnot idea of engines and refrigerators, but only by mathematical deduction, which would lead to the existence of an entropy function. His formulation of the second law of thermodynamics was based on the mathematical theorem, proved by him, which we present as follows.

Theorem 3.1 (Carathéodory): Let x_i, $i = 1, 2, 3$ be the rectangular co-ordinates. In the neighbourhood of any arbitrary point x_o there are points which are not accessible from x_o along solution curves of the equation

$$\sum_{i=1}^{3} A_i(x_1, x_2, x_3)\mathrm{d}x_i = 0 \tag{3.15}$$

if, and only if, the equation is integrable. The equation is said to be integrable if there exist functions $\lambda(x_1, x_2, x_3)$ and $F(x_1, x_2, x_3)$ such that

$$\sum_{i=1}^{3} A_i(x_1, x_2, x_3)\mathrm{d}x_i = \lambda(x_1, x_2, x_3)\mathrm{d}F(x_1, x_2, x_3). \tag{3.16}$$

(For the proof see Carathéodory (1909), Buchdahl (1949).)

The theorem is a special case of the generalized Carathéodory theorem about the holonomian Pfaff's forms, i.e. forms having an integrable factor or being a total differential.

Theorem 3.2 (Generalized Carathéodory): Let $\mathrm{d}\omega_k$ be a linear form in total differentials $\mathrm{d}x_1, \ldots, \mathrm{d}x_n$ of the independent variable $x_1, \ldots, x_n$, the so-called Pfaff's form, defined by

$$\mathrm{d}\omega_n(x_1, \ldots, x_n) = \sum_{i=1}^{n} A_i(x_1, \ldots, x_n)\mathrm{d}x_i. \tag{3.17}$$

Let coefficients $A_i = A_i(x_1, \ldots, x_n)$ be defined in a simply connected space $\{x_1, \ldots, x_n\}$. The necessary and sufficient condition for the existence of the integrable factor of the Pfaff form $\mathrm{d}\omega_n$ is that in the neighbourhood of any arbitrary point $x_o \equiv (x_1, \ldots, x_n)$ there are points which are not accessible from the point x_o along the path $\mathrm{d}\omega_n = 0$. Then $\mathrm{d}\sigma_n = \mu\mathrm{d}\omega_n$ is the total differential of the unambiguous function $\sigma_n(x_1, \ldots, x_n)$, where $\mu(x_1, \ldots, x_n)$ is an integrable factor.

(For the proof see Stepanov (1945).)

The process is said to be quasi-static in equilibrium if all parameters of the system change from the physical point of view very slowly, so that the system is all the time in an equilibrium.

In the general case the work W is defined by

$$W = \int \sum_{i=1}^{n} A_i \mathrm{d}a_i, \tag{3.18}$$

where $A_i = A_i(a_1, \ldots, a_n, \Theta)$. Since the work for a closed path is equal to zero then from the theory of circle integral the integrand is a total differential of a certain unambiguous function of parameters a_i, $i = 1, \ldots, n$, which depends on the temperature Θ like a parameter and therefore can be written in the form

$$\mathrm{d}W = \sum_{i=1}^{n} A_i \mathrm{d}a_i = -\mathrm{d}\Psi(a_1, \ldots, a_n, \Theta) = -\sum_{i=1}^{n} \partial\Psi/\partial a_i \mathrm{d}a_i. \tag{3.19}$$

This equality is valid at a constant temperature so that the differential $\mathrm{d}\Psi$ is taken at $\Theta =$ constant. The state function $\Psi(a_1, \ldots, a_n)$ is called the free energy. It is evident that for isothermic processes a free energy has the same meaning as a potential energy in mechanics. From Equation 3.19 it is evident that the free energy was determined with an accuracy till an additive constant. From Equation 3.19, since a_i are independent variables, it follows that

$$A_i = -\partial\Psi/\partial a_i, \quad i = 1, \ldots, n \tag{3.20}$$

holds for an arbitrary state of equilibrium corresponding to the determinated values of a_i and Θ. Hence

$$\partial A_i/\partial a_j = \partial A_j/\partial a_i. \tag{3.21}$$

The second law of thermodynamics results:

Theorem 3.3: (a) The Quantity of heat, obtained by the system by an arbitrary reversible process, always has an integrable factor.

(b) Between integrable factors of the expression $\mathrm{d}Q$ there exists a factor depending on the temperature of the system only.

Let us denote by $T = T(\Theta)$ the reciprocal value of this factor, and then

$$\mathrm{d}Q = T(\Theta)\mathrm{d}S(a_1, \ldots, a_n, \Theta), \tag{3.22}$$

where $S = S(a_1, \ldots, a_n, \Theta)$ is a certain state function of the system and is called the entropy of the body. The quantity T is called the absolute temperature or the Kelvin temperature.

Proof: The theorem can be proved by applying the generalized Carathéodory's theorem or by his three component version. We use the second case.

1. It holds that

$$\mathrm{d}Q = \mathrm{d}U + \mathrm{d}W = \mathrm{d}U + \sum_{i=1}^{n} A_i \mathrm{d}a_i.$$

Hence

$$\mathrm{d}Q = \sum_{i=1}^{n} (\partial U/\partial a_i + A_i)\mathrm{d}a_i + \partial U/\partial\Theta \mathrm{d}\Theta. \tag{3.23}$$

Since from Equation 3.19, $\mathrm{d}W = -\Sigma\partial\Psi/\partial a_i \mathrm{d}a_i$, is valid for any arbitrary reversible process, the work for an arbitrary quasi-static process can be written as

$$\mathrm{d}W = -\mathrm{d}\Psi + \partial\Psi/\partial\Theta\mathrm{d}\Theta, \tag{3.24}$$

where $\mathrm{d}\Psi = \Sigma\partial\Psi/\partial a_i \mathrm{d}a_i + \partial\Psi/\partial\Theta\mathrm{d}\Theta$ is a total differential of a function $\Psi(a_i, \Theta)$, $i = 1, \ldots, n$. Then

$$\mathrm{d}Q = \mathrm{d}U + \mathrm{d}W = \mathrm{d}(U - \Psi) + \partial\Psi/\partial\Theta\mathrm{d}\Theta = \mathrm{d}G - \sigma\mathrm{d}\Theta, \tag{3.25}$$

where $G = U - \Psi$ is the so-called fixed energy or the Gibb's function, and $\sigma = -\partial\Psi/\partial\Theta$. By this technique we transform Equation 3.23 with $n + 1$ variables a_i, Θ $(i = 1, \ldots, n)$ to Equation 3.25 with three variables G, Θ, σ. Due to the Carathéodory theorem

$$\mathrm{d}Q = \mathrm{d}G - \sigma\mathrm{d}\Theta = 0. \tag{3.26}$$

Then there are two cases:

(a) G, σ, Θ are independent variables. Then the integral of the equation $\mathrm{d}Q = 0$ is a curve. Setting $G = g(\Theta)$ and using Equation 3.26, $\sigma = g'(\Theta)$, which is the equation of the curve in the 3D space of variables G, σ, Θ. It can be shown that since g is an arbitrary function in the plane (G, Θ), a curve can be found going through any arbitrary points in the given directions. From this assertion it follows that from an arbitrary state (G, Θ, σ) we shall pass to the other state along the adiabatic path defined by the equation $\mathrm{d}Q = 0$.

(b) Let one of the variables be a function of the remaining two variables, e.g. $G = g(\sigma, \Theta)$. Then

$$\mathrm{d}Q = \partial g(\sigma, \Theta)/\partial\sigma\mathrm{d}\sigma + [\partial g(\sigma, \Theta)/\partial\Theta - \sigma]\mathrm{d}\Theta.$$

Then to the Carathéodory theorem for two independent variables, there exists an integrable factor $\mu = 1/\Lambda$ such that $\mu\mathrm{d}Q = \mathrm{d}\eta$ or $\mathrm{d}Q = \Lambda\mathrm{d}\eta$, where $\eta = \eta(\sigma, \Theta)$, $\Lambda = \Lambda(\sigma, \Theta)$. Integrating Equation 3.26 we obtain

$$\eta(\sigma, \Theta) = c, \quad \text{where } c \text{ is a constant.} \tag{3.27}$$

From there we see that from an arbitrary state (G, σ, Θ) we can pass to such a state which satisfies the condition of Equation 3.27 with the same value of c. In this case there exist adiabatically unattainable states. So we proved the existence of the integrable factor and that $G = g(\sigma, \Theta)$, which implies $G = U - \Psi = g(-\partial\Psi/\partial\Theta, \Theta)$.

2. To prove the second assertion we shall prove that between integrable factors there exists one factor depending on the temperature only and being a universal function of temperature.

Let the total energy $U = U_1 + U_2$. Then the free energy $\Psi = \Psi_1 + \Psi_2$. Differentiating over Θ, we obtain

$$\sigma = \partial\Psi/\partial\Theta = \partial\Psi_1/\partial\Theta + \partial\Psi_2/\partial\Theta = \sigma_1 + \sigma_2.$$

Moreover, $G = U - \Psi = U_1 + U_2 - (\Psi_1 + \Psi_2) = G_1 + G_2$. Since $G = g(\sigma, \Theta)$ then

$$G_1 = g_1(\sigma_1, \Theta), \quad G_2 = g_2(\sigma_2, \Theta).$$

Hence and since $G = G_1 + G_2$, $\sigma = \sigma_1 + \sigma_2$ we have

$$g(\sigma_1 + \sigma_2, \Theta) = g_1(\sigma_1, \Theta) + g_2(\sigma_2, \Theta).$$

Differentiating over σ_2 and σ_1, where Θ is constant so for simplicity we omit it, we have

$$g'(\sigma_1 + \sigma_2) = g_2'(\sigma_2), \quad g'(\sigma_1 + \sigma_2) = g_1'(\sigma_1).$$

Hence $g_2'(\sigma_2) = g_1'(\sigma_1)$. This equality is valid if g_1' and g_2' are independent of σ_1 and σ_2, and therefore

$$g_1'(\sigma_1) = g_2'(\sigma_2) = g'(\sigma_1 + \sigma_2) = \alpha(\Theta). \tag{3.28}$$

Since this equation is valid for any arbitrary system, $\alpha(\Theta)$ is a universal function of Θ. From Equation 3.28 it follows that

$$\begin{aligned} g_1(\sigma_1) &= \alpha(\Theta)\sigma_1 + \beta_1(\Theta), \\ g_2(\sigma_2) &= \alpha(\Theta)\sigma_2 + \beta_2(\Theta), \\ g(\sigma) &= \alpha(\Theta)\sigma + \beta(\Theta). \end{aligned} \tag{3.29}$$

Since $g = g_1 + g_2$, then

$$\alpha(\Theta)(\sigma_1 + \sigma_2) + \beta(\Theta) = \alpha(\Theta)(\sigma_1 + \sigma_2) + \beta_1(\Theta) + \beta_2(\Theta).$$

Thus $\beta(\Theta) = \beta_1(\Theta) + \beta_2(\Theta)$.

From the last part of Equation (3.29), and again including the variable Θ, we have

$$\sigma = (g(\sigma, \Theta) - \beta(\Theta))/\alpha(\Theta) = (G - \beta)/\alpha.$$

Hence and using Equation 3.25 we have

$$\mathrm{d}Q = \mathrm{d}G - \sigma\mathrm{d}\Theta = \mathrm{d}G - \alpha^{-1}(G - \beta)\mathrm{d}\Theta.$$

Due to the Carathéodary theorem this expression has an integrable factor, which we denote by $1/T$. Then

$$T^{-1}(\mathrm{d}G - \alpha^{-1}(G - \beta)\mathrm{d}\Theta) = \mathrm{d}S,$$

where $\mathrm{d}S$ is a total differential. Let $T = T(\Theta)$, then $\mathrm{d}T/T = \mathrm{d}\Theta/\alpha(\Theta)$ and

$$\mathrm{d}S = \mathrm{d}G/T - T^{-1}(G - \beta)\mathrm{d}T/T = \mathrm{d}G/T - T^{-2}G\mathrm{d}T + T^{-2}\beta(\Theta)\mathrm{d}T$$

and

$$S = G/T + \int T^{-2}\beta(\Theta)\mathrm{d}T. \tag{3.30}$$

Since α is a universal function of the temperature T, then also T is a universal function of the function Θ, being the same for all bodies. The quantity T represents the absolute temperature. Thus we have proved Equation 3.22, $\mathrm{d}Q = T\mathrm{d}S$, where $S = S(\sigma, \Theta) = S(\sigma, -\partial\Psi/\partial\Theta)$. And since Ψ is a function of a_i, Θ, $i = 1, \ldots, n$ then $S = S(a_1, \ldots, a_n, \Theta)$, which completes the proof.

Physically Equation 3.22 has the following meaning:

At the reversible process the quantity of heat obtained by the system is equal to the absolute temperature T multiplied by the differential of entropy (change of entropy). Equation 3.22 can be rewritten in the following form

$$T\mathrm{d}S = \mathrm{d}U + \sum\nolimits_{i=1}^{n} A_i \mathrm{d}a_i, \tag{3.31}$$

representing the general form of the second law of thermodynamics for reversible processes. Integrating this equation, we obtain

$$S - S_o = \int T^{-1}\mathrm{d}Q = \int T^{-1}\left(\mathrm{d}U + \sum\nolimits_{i=1}^{n} A_i \mathrm{d}a_i\right).$$

Since the constant of integration S_o has no physical meaning and depends on the choice of an initial state, then the entropy is determined till the additive constant.

Because the energy U and external forces A_i are, for all physically admissible states, unambiguously continuous and bounded state functions, then for all states for which $T \neq 0$ the entropy S is an unambiguous function of a_i, $T (i = 1, \ldots, n)$. Therefore,

$$\oint T^{-1}\left(\mathrm{d}U + \sum\nolimits_{i=1}^{n} A_i \mathrm{d}a_i\right) = 0,$$

as the integrand represents the total differential of an unambiguous function, and therefore for a closed reversible process

$$\oint T^{-1}\mathrm{d}Q = 0. \tag{3.32}$$

The last equation is the so-called Clausius' equation.

From Equation 3.31 it follows that

$$\mathrm{d}S = T^{-1}\left(\sum\nolimits_{i=1}^{n} (\partial U/\partial a_i + A_i)\mathrm{d}a_i + \partial U/\partial T \mathrm{d}T\right) = \sum\nolimits_{i=1}^{n} X_i \mathrm{d}a_i + Y\mathrm{d}T, \tag{3.33}$$

where we denote $X_i = T^{-1}(\partial U/\partial a_i + A_i)$, $Y = T^{-1}\partial U/\partial T$. In order that this equation will be a total differential, we have

$$\begin{aligned} \partial X_i/\partial T &= \partial Y/\partial a_i, \quad i = 1, \ldots, n \quad \text{and} \\ \partial X_i/\partial a_j &= \partial X_j/\partial a_i, \quad i \neq j, \quad i, j = 1, \ldots, n. \end{aligned} \tag{3.34}$$

Then

$$(\partial U/\partial a_i)_T + A_i = T(\partial A_i/\partial T) \text{ and } \partial A_i/\partial a_j = \partial A_j/\partial a_i,$$

where the last system of equations was derived above.

From Equation 3.30 we have

$$S = T^{-1}G + \int T^{-2}\beta(\Theta)\mathrm{d}T = T^{-1}(U - \Psi)\int T^{-2}\beta(\Theta)\mathrm{d}T$$

and hence

$$\Psi = U - TS + T\int T^{-2}\beta(\Theta)\mathrm{d}T. \tag{3.35}$$

We saw above that the free energy Ψ was determined till the arbitrary additive function of temperature. If we put $\beta(\Theta)$ then the last term in the above equation is equal to zero. Then

$$\Psi = U - TS \tag{3.36}$$

or, in the diferential form,

$$\mathrm{d}\Psi = \mathrm{d}U - T\mathrm{d}S - S\mathrm{d}T = -\sum\nolimits_{i=1}^{n} A_i \mathrm{d}a_i - S\mathrm{d}T. \tag{3.37}$$

Since $\mathrm{d}\Psi = \partial\Psi/\partial a_i + \partial\Psi/\partial T\mathrm{d}T$, then $A_i = -\partial\Psi/\partial a_i$, $S = -\partial\Psi/\partial T$ (at constant T in the first case and at constant a_i in the second case). Hence and using Equation 3.36 we obtain

$$U = \Psi - T\partial\Psi/\partial T,$$

which is called the Gibbs-Helmholtz equation. From here we see that the free energy, if it is independent of the temperature T, is equal to the total energy of the system.

To characterize properties of a body, whose state is determined by a volume V and by a temperature T at a thermodynamic equilibrium, instead of its state equation (representing the relation between a pressure p, a volume V and a temperature T) we must know a second equation, representing e.g. its energy as a function of a volume V and a temperature T.

The second law of thermodynamics reduces the number of these equations, which are needed for characterization of the body, to one equation only. For this fact we shall use the characteristic functions. To derive characteristic functions we shall use Equation 3.31, i.e.

$$T\mathrm{d}S = \mathrm{d}U + \sum\nolimits_{i=1}^{n} A_i \mathrm{d}a_i.$$

Let the function $S = S(a_i, U)$ be known, then

$$\partial S(a_i, U)/\partial U = T^{-1}, \quad \partial S(a_i, U)/\partial a_j = T^{-1}A_j.$$

Hence, solving the first equation for U, we shall find $U = U(a_i, T)$ and these equations determine the external forces. Since

$$\mathrm{d}\Psi = -S\mathrm{d}T - \sum\nolimits_{i=1}^{n} A_i \mathrm{d}A_i$$

we have

$$\partial\Psi(a_i, T)/\partial T = -S, \quad \partial\Psi(a_i, T)/\partial a_i = -A_i, \quad U = \Psi - T(\partial\Psi/\partial T)_a.$$

If the function $\Psi(a_i, T)$ is known, then we can determine all fundamental quantities characterizing the system being investigated.

Let the volume V be an external parameter. Then, in this special case, we denote the free energy Ψ by $F(U, T)$, and it holds that:

$$\mathrm{d}F = -S\mathrm{d}T - p\mathrm{d}V, \tag{3.38}$$

$$p = -\partial F/\partial V, \tag{3.39}$$

$$S = -\partial F/\partial T, \tag{3.40}$$

$$U = F - T\partial F/\partial T. \tag{3.41}$$

Equation 3.39 represents the state equation and Equation 3.41 determines the energy as a function of a temperature and a volume. Now let a pressure and temperature T be external parameters, then the characteristic function can be determined from Equation 3.38. We have

$$\mathrm{d}(F + p\mathrm{d}V) = -S\mathrm{d}T + V\mathrm{d}p$$

and we put

$$\Phi = \Phi(p, T) = F + pV = U - TS + pV = H - TS. \tag{3.42}$$

Hence

$$\mathrm{d}\Phi = -S\mathrm{d}T + V\mathrm{d}p \tag{3.43}$$

and thus

$$V = \partial\Phi/\partial p, \quad S = -\partial\Phi/\partial T. \tag{3.44}$$

Then for the enthalpy H, using Equations 3.42 and 3.44, we obtain

$$H = \Phi - T\partial\Phi/\partial T. \tag{3.45}$$

The function Φ is the so-called thermodynamic potential or the free enthalpy, and $\Phi = \Phi(p, T)$.

3.1.1.3. The reversible and irreversible states

In thermodynamics, work is a macroscopic phenomenon. The performance of work may always be described in terms of the raising or lowering of a weight, or the winding or unwinding of a spring, i.e. by the operation of a device that serves to increase or decrease the potential energy of a mechanical system. Now let us suppose that a process occurs in which:

a) the system proceeds from an initial state (1) to a final state (2),
b) the suspended weight is lowered to an extent that W units of work are performed,
c) a transfer of heat Q takes place from the system to the series of reservoirs.

If, at the conclusion of this process, the system may be restored to its initial state (1), the weight lifted to its former level, and the reservoirs caused to part with the same amount of heat Q, without producing any changes in any other mechanical device or reservoir in the universe (here universe is meant in its local technical sense), the original process is said to be reversible. Therefore, a reversible process is one that is performed in such a way that, at the conclusion of the process, both the system and the local surroundings may be restored to their initial states, without producing any changes in the rest of the universe. A process that does not fulfill these hard conditions is said to be irreversible. The question immediately arises as to whether natural processes are reversible or not. The answer to this question gives the second law of thermodynamics. There are several types of irreversibility:

a) external mechanical irreversibility, represented by

   (i) processes involving the isothermal transformation of work through a system into internal energy of a reservoir, e.g. inelastic deformation of a solid in contact with a reservoir or magnetic hysteresis of a material in contact with a reservoir, or transfer of electricity through a resistor in contact with a reservoir, etc.,

   (ii) processes involving the adiabatic transformation of work into internal energy of a system, i.e. inelastic deformation of a thermally insulated solid, magnetic hysteresis of a thermally insulated material, or transfer of electricity through a thermally insulated resistor, etc.

   This type of process is accompanied by a rise in temperature of the system from temperature Θ_1 to Θ_2. The internal energy of the system would have to be decreased by extracting $U_2 - U_1$ units of heat, thus lowering the temperature from Θ_2 to Θ_1, and this heat would have to be completely converted into work. The transformation of work into internal energy either of a system or of a reservoir is seen to take place through the agency of such phenomena as viscosity, friction, inelasticity, electric resistance, magnetic hysteresis, etc.. These effects are known as dissipative effects and the work is said to be dissipated.

b) external and internal thermal irreversibility — represented by processes involving a transfer of heat between a system and a reservoir by virtue of a finite temperature difference, e.g. conduction or radiation of heat from a system to a cooler reservoir, as well as conduction or radiation of heat through a system from a hot reservoir to a cooler one.

c) internal mechanical irreversibility — represented by processes involving the transformation of internal energy of a system into mechanical energy and then back into internal energy again, e.g. an ideal gas rushing into a vacuum, known as a free expansion, or gas seeping through a porous plug, etc..

d) chemical irreversibility — represented by processes that go on in nature involving a spontaneous change of internal structure, chemical composition, density or crystal form, etc., e.g. all chemical reactions, solution of solid in liquid, etc..

One can say that all processes are irreversible, as the equilibrium state of the system under a change of external parameters or temperature of neighbouring bodies is violated, and in the system investigated irreversible processes occur. In the course of time the system comes into a new equilibrium state, which is represented by new external conditions. It is evident that the reversible state is a special case of an irreversible state, where the transition from one equilibrium state into the other is quasi-static (i.e. very slow).

Let us assume an adiabatic irreversible transition of a system from one equilibrium state into another. It can be shown that at this transition the entropy of a system increases. Then we have the following theorem:

Theorem 3.4: Let a system in an equilibrium state be described by parameters a_i, $i = 1, \dots, n$ and by an energy U. Let the transition of a system (a_i, U) when passing from one equilibrium state into another be adiabatic and irreversible. Then the entropy of the system is increased.

Proof: Let the system be in an equilibrium state (a_{oi}, U_o), the entropy of which is $S_o = S(a_{oi}, U_o)$. Let external parameters a_i be quickly adiabatically changed into a_i. Thus the equilibrium of the system is violated, but in the course of time the system reaches a new equilibrium state (a_i', U'). The entropy of this state will be $S' = S(a_i', U')$. Generally $S' \neq S_o$. Then by a quasi-static adiabatic change the external parameters a_i' are transferred into the initial ones a_{oi}. In this case the energy of the system is changed and has a value U'', but the entropy of the system is the same as above, i.e. S', so that $S(a_{oi}, U'') = S(a_o', U') = S'$ and the system is now in a state (a_{oi}, U''). We see that $U'' \neq U_o$ and therefore the system takes work $W = U_o - U''$. Due to the second law of thermodynamics (in the version where there does not exist a perpetum mobile of the second order), this work $W \leq 0$ and therefore $U_o \leq U''$. But as $(\partial S/\partial U) = T^{-1}$, then S is a non-decreasing function of U, and as

$$S' - S_o = S(a_{oi}, U'') - S(a_{oi}, U_o) \geq 0$$

then

$$S' \geq S_o,$$

which completes the proof.

Hence follows the second law of thermodynamics for irreversible processes, which will be formulated in the next theorem:

Theorem 3.5 (the second law of thermodynamics): For an irreversible process

$$\mathrm{d}Q \leq T\mathrm{d}S \quad \text{i.e. } \mathrm{d}S \geq T^{-1}\mathrm{d}Q, \tag{3.46}$$

where

$$T\mathrm{d}S = \mathrm{d}U + \sum\nolimits_{i=1}^{n} A_i \mathrm{d}a_i, \tag{3.47}$$

represents the generalized second law of thermodynamics.

The entropy was determined from Equation 3.47 representing the second law of thermodynamics for quasi-static processes, which are expressed by a sequence of equilibrium states. Since a state of a system at equilibrium is determined by the given external parameters and the temperature (or by external parameters and the energy), the entropy as the state function is determined as a function of these quantities. Also the free energy Ψ expressed by the relation $\Psi = U - TS$ is determined for equilibrium states only.

In an unequilibrium state, internal parameters of the system are not a function of external parameters and the temperature. Therefore, the unequilibrium state of the system will be characterized, in addition to the external parameters and the temperature (the energy), also by one or several internal parameters. The entropy and the free energy must in an unequilibrium state be a function of a state, i.e. a function of external parameters and the temperature and also of internal parameters characterizing the investigated equilibrium state. The entropy of an unequilibrium state is defined as an entropy of an equilibrium state (which is balanced under a function of a complement force field), characterized by the same values of the considered internal parameters. Thus the system is characterized by external parameters a_i, $i = 1, \ldots, n$ (characterizing the state of the system), and by external parameters x_i, $i = 1, \ldots, m$ (characterizing the complement force field, e.g. the gravity field, which is characterized by the acceleration of gravity g, and then $x_1 = g$).

In the case of quasi-static processes under a function of a complement force field, the change of entropy is given by

$$T\mathrm{d}S = \mathrm{d}U^* + \sum_{i=1}^{n} A_i \mathrm{d}a_i + \sum_{j=1}^{m} \xi_j \mathrm{d}x_j, \tag{3.48}$$

where U^* represents the energy of the system also under a function of a complement force field and ξ_j, $j = 1, \ldots, m$, represent generalized external forces corresponding to external parameters X_j. Therefore, ξ_j represent internal parameters (e.g. a pressure is an internal parameter, the volume of a body is an external parameter and $m\xi$, where m is a mass of a body and ξ a vertical co-ordinate of the mass centre of a body, is the generalized external force. Hence work under the function of the gravity field is $m\xi dg$). Then the unequilibrium state, if a complement force field was omitted, will be characterized by the same values of ξ_j and a_i, but its energy U will be different from U^* by about a potential energy $\Sigma_{j=1}^{m} x_j \xi_j$. Thus

$$U = U^* + \sum_{j=1}^{m} x_j \xi_j \tag{3.49}$$

and hence

$$U^* = U - \sum_{j=1}^{m} x_j \xi_j. \tag{3.50}$$

Then from Equation 3.48 and 3.50 we obtain

$$T\mathrm{d}S = \mathrm{d}U + \sum_{i=1}^{n} A_i \mathrm{d}a_i - \sum_{j=1}^{m} \xi_j \mathrm{d}x_j, \tag{3.51}$$

which is correct for transitions between equilibrium states (e.g. for irreducible processes being connected with a change of internal parameters). At the same time generalized external forces A_i are functions of an energy U and external parameters a_i and also of internal parameters ξ_j; and the parameters x_j are also functions of ξ_j. Therefore, from Equation 3.51 it follows that $x_j = -T\partial S(U, a_i, \xi_j)/\partial \xi_j$.

For the entropy of the thermally non-homogeneous system we have

$$\mathrm{d}S = \sum_k T_k^{-1} \left(\mathrm{d}U_k + \sum_{i=1}^{n} A_{ki} \mathrm{d}a_{ki} \right), \tag{3.52}$$

where T_k, U_k are temperatures and energies of different parts of the system and $\Sigma_{i=1}^{n} A_{ki} \mathrm{d}a_{ki}$ are their executed works.

To determine a free energy of an unequilibrium state we have

$$\Psi = U - TS = U^* + \sum_{j=1}^{m} x_j\xi_j - TS' = \Psi^* + \sum_{j=1}^{m} x_j\xi_j, \tag{3.53}$$

where $\Pi = -\Sigma_{j=1}^{m} x_j\xi_j$, represents the potential energy of the force field. Since $\Psi^* = U^* - TS$, then

$$\mathrm{d}\Psi^* = -S\mathrm{d}T - \sum_{i=1}^{n} A_i\mathrm{d}a_i - \sum_{j=1}^{m} \xi_j\mathrm{d}x_j. \tag{3.54}$$

Hence and using Equation 3.53 we obtain

$$\mathrm{d}\Psi(T, a_i, \xi_j) = -S\mathrm{d}T - \sum_{i=1}^{n} A_i\mathrm{d}a_i + \sum_{j=1}^{m} x_j\mathrm{d}\xi_j, \tag{3.55}$$

where S, A_i and x_j are functions of T, a_i and ξ_j.

As a generalized case of Theorem 3.4 for irreversible processes we have the following theorem.

Theorem 3.6: At every adiabatic process the entropy of the system increases or it does not change.

Proof: (a) Let the system investigated be thermally homogeneous. We say that the system at first is in an unequilibrium state, characterized by certain values of the energy U_1 and external and internal parameters a_{1i}, ξ_{1j}. Then the entropy of this state is $S_1 = S(U_1, a_{1i}, \xi_{1j})$. Then at the adiabatic insulation the external parameters a_i change. In the system an unequilibrium state is obtained and the system translates into another, generally into an unequilibrium state which is characterized by U_2, a_{2i}, ξ_{2j} with entropy $S_2 = S(U_2, a_{2i}, \xi_{2j})$. At this moment we introduce a complement force field characterized by x_{2j} such that the values of the internal parameters ξ_j catch values corresponding to an equilibrium state of this force field. For equilibrium, the entropy is equal to S_2 and $a_i = a_{2i}$, $\xi_j = \xi_{2j}$, but the energy $U_2^* \neq U_2$ and $U_2^* = U_2 - \Sigma_{j=1}^{m} x_{2j}\xi_{2j}$, where $-\Sigma_{j=1}^{m} x_{2j}\xi_{2j} = \Pi_2$ is the potential energy of this force field. The force field acts immediately and the work of the system after introduction of a complement force field is equal to $\Sigma_{j=1}^{m} x_{2j}\xi_{2j}$.

By the quasi-static adiabatic change of a_i and x_j the system, which is in equilibrium, is transfered into the state characterized by $a_i = a_{1i}$ and $x_j = x_{1j}$. Then the entropy of the system does not change and is equal to S_2. Then the force field will act. At the same time the system makes the work and the system will be in an unequilibrium state, for which $a_i = a_{1i}$, $\xi_j = \xi_{1j}$ and the energy $U = U_3$. The entropy then is $S_2 = S(U_3, a_{1i}, \xi_{1j})$.

The work of the system from the beginning to this moment is equal to

$$W = U_1 - U_3.$$

Due to the second law of thermodynamics (i.e. in the version where there does not exist a perpetum mobile of the second order) this work is ≤ 0 and therefore $U_1 \leq U_3$. But since $(\partial S/\partial U) = T^{-1} > 0$, then

$$S_2 = S(U_3, a_{1i}, \xi_{1j}) \geq S(U_1, a_{1i}, \xi_{1j}) = S_1,$$

which completes the first part of the proof.

(b) Let the system investigated be thermally non-homogeneous. Let us denote by a_{ki}, ξ_{kj}, U_k, S_k, T_k the external and internal parameters, an energy, an entropy, and a temperature in every thermally homogeneous part of the system. Then

$$\mathrm{d}S_k = T_k^{-1}\left(\mathrm{d}U_k + \sum_{i=1}^{n} A_{ki}\mathrm{d}a_{ki}\right) + (\mathrm{d}S_k)_{a,U} = T_k^{-1}\mathrm{d}Q_k + (\mathrm{d}S_k)_{a,U}, \tag{3.56}$$

where $\mathrm{d}Q_k = \Sigma_l \mathrm{d}Q_{kl}$, and where $\mathrm{d}Q_{kl} = -\mathrm{d}Q_{lk}$ is a quantity of heat, which obtains the k-th part of the system from the l-th part. The change of entropy is

$$\begin{aligned}\mathrm{d}S &= \sum\nolimits_k \mathrm{d}S_k = \sum\nolimits_{k,l} T_k^{-1}\mathrm{d}Q_{kl} + \sum\nolimits_k (\mathrm{d}S_k)_{a,U}\\ &= \tfrac{1}{2}\sum\nolimits_{k,l} \mathrm{d}Q_{kl}(T_k^{-1} - T_l^{-1}) + \sum\nolimits_k (\mathrm{d}S_k)_{a,U}. \qquad (3.57)\end{aligned}$$

We have $(\mathrm{d}S_k)_{a,U} \geq 0$ and $\mathrm{d}Q_{kl}$ has the same sign as $T_k^{-1} - T_l^{-1}$ as the heat is passing from the more hot body to the less hot one. Therefore $\mathrm{d}S \geq 0$, which completes the proof.

Remark: At the adiabatic process $\mathrm{d}Q = \mathrm{d}U + \Sigma_i A_i \mathrm{d}a_i = 0$. Then the change of entropy is given by

$$\sum\nolimits_{j=1}^{m} (\partial S/\partial \xi_j)_{a,U} d\xi_j/dt \geq 0. \qquad (3.58)$$

Theorem 3.7: For the system, which at an isothermic irreversible process does not make the work, the free energy decreases or does not change.

Proof: The idea of the proof is similar of that of Theorem 3.6. The system made the circle isothermic process, in which the work (due to the second law of thermodynamics) cannot be positive. This work is the sum of the work $\Sigma_{j=1}^m x_{2j}\xi_{2j}$ (needed for a starting force field), the work at the quasi-static process equals to $\Psi_2^* - \Psi_1^*$, and the work $-\Sigma_{j=1}^m x_{1j}\xi_{1j}$ (needed for release of a force field) so that

$$W = \Psi_2^* - \Psi_1^* + \sum\nolimits_{j=1}^{m} (x_{2j}\xi_{2j} - x_{1j}\xi_{1j}) = \Psi_2 - \Psi_1 \leq 0,$$

which completes the proof.

Remark: It holds

$$\sum\nolimits_{j=1}^{m} (\partial \Psi/\partial \xi_j)_{a_i,T} \mathrm{d}\xi_j/\mathrm{d}t \leq 0. \qquad (3.59)$$

For the system the following equilibrium conditions hold:

Theorem 3.8: It holds

$$S(U, a_i, \xi_j) = \max, \qquad (3.60)$$

if $U =$ constant, $a_i =$ constant and the internal parameters ξ_j do not change, and

$$\Psi(T, a_i, \xi_j) = \min, \qquad (3.61)$$

if $T =$ constant, $a_i =$ constant and the internal parameters ξ_j do not change. Moreover, the entropy S and the free energy Ψ can have several maxima and several minima, respectively.

Proof: Let the system be defined by a_i and U (adiabatic isolation). In this case the condition of equilibrium is that the entropy of the system $S(U, a_i, \xi_j)$ gains its maximum for internal parameters ξ_j, corresponding to the equilibrium, and for the given U and a_i. Since $\Sigma_j(\partial S/\partial \xi_j)_{a,U}\mathrm{d}\xi_j/\mathrm{d}t \geq 0$, then $S(U, a_i, \xi_j) = \max$ is the sufficient condition for the equilibrium of the system (parameters ξ_j change) as if the entropy for the given U and a_i and certain $\xi_j = \xi_{oj}$ has a maximum, then for all changes of ξ_j the entropy may decrease only. But this is not correct as the entropy for adiabatic processes $\mathrm{d}S \geq 0$. It is evident that the conditions of the equilibrium will be fulfilled, if $(\partial S/\partial \xi_i)_{U,a_i} = 0$, $i = 1, 2, \ldots$ and $(\mathrm{d}^2 S)_{U,a_i} = \Sigma_{jk}(\partial^2 S/\partial \xi_j \partial \xi_k)\mathrm{d}\xi_j \mathrm{d}\xi_k < 0$, where ξ_j are assumed to be independent. If the differential of the second order $(\mathrm{d}^2 S)_{U,a_i} > 0$ the equilibrium does not exist; if it is equal to zero, then the differential of higher order must be investigated.

Similarly, if the temperature of the system T and external parameters a_i are given, then the equilibrium of the system is guaranteed if the free energy $\Psi(T, a_i, \xi_j) = \min$, for $T = \text{constant}$, $a_i = \text{constant}$ and ξ_j do not change. Moreover, the equilibrium of the system is guaranteed if $(\partial\Psi/\partial\xi_j)_{T,a_i} = 0$, $j = 1, 2, \ldots$ and $(\mathrm{d}^2\Psi)_{T,a_i} > 0$. It is evident that the maximum in the first case and the minimum in the second one are not unique, so that the system has several equilibrium states, which complete the proof.

Remark: We say that the system is absolutely stable if the maximum of $S(U, a_i, \xi_j)$ is the greatest of it or if the minimum of $\Psi(T, a_i, \xi_j)$ is the smallest of it. All other equilibrium states are called the metastable equilibrium.

Theorem 3.9: For an arbitrary internal parameter ξ_i the relation

$$T(\partial S(U, a_i, \xi_j)/\partial\xi_j)_{U,a_i} = -(\partial\Psi(T, a_i, \xi_j)/\partial\xi_j)_{T,a_i} \tag{3.62}$$

is valid.

Proof: Since $\Psi = U - TS$, then for $a_i = \text{constant}$ we have

$$(\partial\Psi/\partial\xi_j)_{T,a_i}\mathrm{d}\xi_j + (\partial\Psi/\partial T)_{\xi_j,a_i}\mathrm{d}T = \mathrm{d}U - S\mathrm{d}T - T(\partial S/\partial U)_{\xi_j,a_i}\mathrm{d}U - T(\partial S/\partial\xi_j)_{U,a_i}\mathrm{d}\xi_j.$$

Since

$$T\mathrm{d}S = \mathrm{d}U + \sum\nolimits_{i=1}^{n} A_i \mathrm{d}a_i - \sum\nolimits_{j=1}^{m} x_j \mathrm{d}\xi_j,$$

$$\mathrm{d}\Psi(T, a_i, \xi_j) = -S\mathrm{d}T - \sum\nolimits_{i=1}^{n} A_i \mathrm{d}a_i + \sum\nolimits_{j=1}^{m} x_j \mathrm{d}\xi_j,$$

then the assertion follows immediately, which completes the proof.

3.1.1.4 Some theorems

Theorem 3.10: It holds

$$T\mathrm{d}S = C_v\mathrm{d}T + T(\partial p/\partial T)_V\mathrm{d}V. \tag{3.63}$$

Proof: Since $\mathrm{d}U = T\mathrm{d}S - p\mathrm{d}V$, $T\mathrm{d}S = C_v\mathrm{d}T + T(\partial p/\partial T)_V\mathrm{d}V$ and $C_V = (\partial U/\partial T)_V$, then $\mathrm{d}U = C_V\mathrm{d}T + (T(\partial p/\partial T)_V - p)\mathrm{d}V = (\partial U/\partial T)_V\mathrm{d}T + (T(\partial p/\partial T)_V - p)\mathrm{d}V$. But $\mathrm{d}U = (\partial U/\partial T)_V\mathrm{d}T + (\partial U/\partial V)_T\mathrm{d}V$, hence the assertion follows immediately, which completes the proof.

Theorem 3.11: It holds

$$T\mathrm{d}S = C_p\mathrm{d}T - T(\partial V/\partial T)_p\mathrm{d}p. \tag{3.64}$$

Proof: Since $\mathrm{d}S = (\partial S/\partial T)_p\mathrm{d}T + (\partial S/\partial p)_T\mathrm{d}p$, $T(\partial S/\partial T)_p = C_p$ and $(\partial S/\partial p)_T = -(\partial V/\partial T)_p$ then

$$T\mathrm{d}S = T(\partial S/\partial T)_p\mathrm{d}T + T(\partial S/\partial p)_T\mathrm{d}p = C_p\mathrm{d}T - T(\partial V/\partial T)_p\mathrm{d}p.$$

Theorem 3.12: It holds

$$(\partial U/\partial T)_T = T(\partial p/\partial T)_V - p. \tag{3.65}$$

Proof: Since $\mathrm{d}U = T\mathrm{d}S - p\mathrm{d}V$, $T\mathrm{d}S = C_v\mathrm{d}T + T(\partial p/\partial T)_V\mathrm{d}V$ and $C_v = (\partial U/\partial T)_V$, then $\mathrm{d}U = C_v\mathrm{d}T + (T(\partial p/\partial T)_V - p)\mathrm{d}V = (\partial U/\partial T)_V\mathrm{d}T + (T(\partial p/\partial T)_V - p)\mathrm{d}V$. But $\mathrm{d}U = (\partial U/\partial T)_V\mathrm{d}T + (\partial U/\partial V)_T\mathrm{d}V$, hence the assertion of the proof follows immediately.

Theorem 3.13: It holds that

$$C_p - C_v = -T(\partial V/\partial T)_p^2(\partial p/\partial V)_T^{-1}, \tag{3.66}$$

$$C_p - C_v = k^{-1}TV\beta^2. \tag{3.67}$$

Proof: From Theorems 3.10 and 3.11 we obtain

$$C_p\mathrm{d}T - T(\partial V/\partial T)_p\mathrm{d}p = C_v\mathrm{d}T + T(\partial p/\partial T)_V\mathrm{d}V. \quad \text{Hence}$$

$$\mathrm{d}T = (C_p - C_v)^{-1}[T(\partial p/\partial T)_V\mathrm{d}V + T(\partial U/\partial T)_p\mathrm{d}p] = (\partial T/\partial V)_p\mathrm{d}V + (\partial T/\partial p)_V\mathrm{d}p.$$

Therefore

$$(\partial T/\partial V)_p = (C_p - C_v)^{-1}T(\partial p/\partial T)_V, \quad (\partial T/\partial p)_V = (C_p - C_v)^{-1}T(\partial V/\partial T)_p.$$

Hence

$$C_p - C_v = T(\partial U/\partial T)_p(\partial p/\partial T)_V^{-1}.$$

Since for the differential from the theorem of partial differential calculus*

$$(\partial p/\partial T)_V = -(\partial V/\partial T)_p(\partial p/\partial V)_T$$

then

$$C_p - C_v = -T(\partial V/\partial T)_p^2(\partial p/\partial V)_T,$$

which completes the first part of the proof.

Since $\beta = V^{-1}(\partial V/\partial T)_p$, and $k = -V^{-1}(\partial V/\partial p)_V$ then

$$C_p - C_v = (-V^{-1}(\partial V/\partial p)_T)^{-1}TV[V^{-1}(\partial V/\partial T)_p]^2 = k^{-1}TV\beta^2,$$

which completes the proof.

Remark: Equation 3.66 is one of the most important equations of thermodynamics. From this equation we see that for $T \to 0$ then $C_p \to C_v$, and at the absolute zero temperature both heat capacities are equal. Moreover, $C_p = C_v$ when $(\partial V/\partial T)_p = 0$, and as $(\partial p/\partial V)_T$ is always negative then $(\partial p/\partial V)_T^2$ is positive, and then $C_p - C_v$ can be negative, or C_p can never be less than C_v.

Theorem 3.14: It holds

$$\gamma = C_p/C_v = (\partial p/\partial V)_S/(\partial p/\partial V)_T, \tag{3.68}$$

where γ is the so-called ratio of heat capacities, and the adiabatic compressibility is defined as the reciprocal of the adiabatic bulk modulus, and $\gamma = k/k_s$, where $k_s = -V^{-1}(\partial V/\partial p)_S$, $k = -V^{-1}(\partial V/\partial p)_T$.

Proof: From Theorems 3.10 and 3.11 we have

$$T\mathrm{d}S = C_p\mathrm{d}T - T(\partial V/\partial T)_p = C_v\mathrm{d}T + T(\partial p/\partial T)_V\mathrm{d}V.$$

Since at constant S

$$C_p\mathrm{d}T_S = T(\partial V/\partial T)_p\mathrm{d}p_S, \quad C_v\mathrm{d}T_S = -T(\partial p/\partial T)_V\mathrm{d}V_S,$$

* For differentials $\mathrm{d}x$ and $\mathrm{d}y$ we have $\mathrm{d}x = (\partial x/\partial y)_z\mathrm{d}y + (\partial x/\partial z)_y$, $\mathrm{d}y = (\partial y/\partial x)_z + (\partial y/\partial z)_x\mathrm{d}z$. Hence $\mathrm{d}x = (\partial x/\partial y)_z(\partial y/\partial x)_z\mathrm{d}x + [(\partial x/\partial y)_z(\partial y/\partial z)_x + (\partial x/\partial z)_z]\mathrm{d}z$. For $\mathrm{d}x = 0$ and $\mathrm{d}z \neq 0$ it follows that $(\partial x/\partial y)_z(\partial y/\partial x)_z + (\partial x/\partial z)_y = 0$.

we have

$$C_p/C_v = -((\partial V/\partial T)_p(\partial p/\partial T)_V^{-1})(\partial p/\partial V)_S = (\partial V/\partial p)_T(\partial p/\partial V)_S.$$

As $k_s = -V^{-1}(\partial V/\partial p)_S$, $k = -V^{-1}(\partial V/\partial p)_T$, the assertion of the theorem follows immediately.

Remark: If we know the ratio of heat capacities and k then k_s can be calculated. Moreover, since $c_p/c_v = k/k_s$ and $c_p - c_v = k^{-1}Tv\beta^2$, we get $k - k_s = c_p^{-1}Tv\beta^2$, and as $k_s = (\rho v_p^2)^{-1}$, $k = \rho^{-1}(v_p^{-2} + c_p^{-1}\beta^2 T)$, where $v_p = (\rho^{-1}b_s)^{-1}$ (in m/sec) is a velocity of a longitudinal wave, or $v_p = (\rho^{-1}b_s)^{-1} = ((\partial p/\partial \rho)_S)^{1/2} = (\rho k_s)^{-1/2}$.

Remark: In the case of an infinitesimal reversible process of a stretched bar, the two laws of thermodynamics yield the equation

$$TdS = dU - \tau dl = c_\tau dT + T(\partial l/\partial T)_\tau d\tau, \tag{3.69}$$

where $c_\tau = (\partial U/\partial T)_\tau$.

Let $V = V(T, p)$, $p = p(T, V)$, $T = T(p, V)$ be the state equations, then

$$\begin{aligned} dV &= (\partial V/\partial T)_p dT + (\partial V/\partial p)_T, \\ dp &= (\partial p/\partial T)_V + (\partial p/\partial V)_T dV, \\ dT &= (\partial T/\partial V)_V + (\partial T/\partial V)_p dV. \end{aligned} \tag{3.70}$$

Then we define
(i) the volume expansivity β by

$$\beta = V^{-1}(\partial V/\partial T)_p. \tag{3.71}$$

Strictly speaking, β is a function of T and p.
(ii) the isothermal bulk modulus b and the adiabatic bulk modulus b_s by

$$b = -V(\partial p/\partial V)_T, \quad b_s = -V(\partial p/\partial V)_S, \tag{3.72}$$

where S is the entropy. It is clear that the isothermal bulk modulus has the same units as a pressure (i.e. dyn cm^{-2}) and that it is a function of T and p.
(iii) the isothermal compressibility k and the adiabatic compressibility k_s by

$$k = -V^{-1}(\partial V/\partial p)_T, \quad k_s = -V^{-1}(\partial V/\partial p)_S, \tag{3.73}$$

as the reciprocal of the isothermal or adiabatic bulk modulus, respectively, which are functions of T and p.

Remark: At constant volume $dp = \beta b dT$.

The states of thermodynamic equilibrium are connected by an equation of state that as a rule cannot be expressed by a simple equation. Let a bar at constant temperature T within the limit of elasticity and constant cross-section s be subject to the external force P. Then Hooke's law holds

$$\tau = \text{const.}(l - l_o), \tag{3.74}$$

where τ is the tension in the bar (measured in dynes), l is the length of the bar (measured in cm), and l_o is the length at zero tension. If the tension is not balanced by an external force, the bar will describe some sort of accelerated motion such as a vibration, and will pass through states that cannot be described in thermodynamic co-ordinates referring to the bar as a whole.

If a bar undergoes an infinitesimal change from one state of equilibrium to another, then the infinitesimal change of length is an exact differential of the form

$$\mathrm{d}l = (\partial l/\partial T)_\tau \mathrm{d}T + (\partial l/\partial \tau)_T \mathrm{d}\tau,$$

where both partial derivatives are functions of T and τ, and are connected with important physical quantities. Let the change of temperature be infinitesimal, then

$$\alpha = l^{-1}(\partial l/\partial T)_\tau \tag{3.75}$$

is the linear expansivity. Measurements of α show that it depends only slightly on T and varies mostly with temperature T, but in a small temperature range it is practically constant. Furthermore, let the change of tension be infinitesimal, then

$$E = s^{-1}l(\partial \tau/\partial l)_T \tag{3.76}$$

is the isothermal Young's modulus. Experiments show that E depends little on τ and mostly on T, but for a small temperature range it is practically constant. Moreover,

$$(\partial \tau/\partial T)_l = -(\partial \tau/\partial l)_T(\partial l/\partial T)_\tau = -\alpha s E, \tag{3.77}$$

which is a corollary of theorems in partial differential calculus. If a bar undergoes an infinitesimal change from an initial equilibrium state to a final equilibrium state then the change in tension is equal to

$$\mathrm{d}\tau = (\partial \tau/\partial T)_l \mathrm{d}T + (\partial \tau/\partial l)_T \mathrm{d}l = -\alpha s E \mathrm{d}T + sEl^{-1}\mathrm{d}l, \tag{3.78}$$

from which for constant temperature the Hooke's law follows immediately.

Remark: The equation of state of an ideal elastic material is

$$\tau = cT(l_o^{-1}l - l^{-2}l_o^2), \tag{3.79}$$

where c is a constant, and l_o is the length at zero tension and is a function of the temperature only. Then the isothermal Young's modulus is given by

$$E = s^{-1}\tau + 3s^{-1}l^{-2}cTl_o^2 = s^{-1}\tau + E_o l^{-2}l_o^2, \tag{3.80}$$

where $E_o = 3s^{-1}cT$ is the isothermal Young's modulus at zero tension. The linear expansivity is given by

$$\alpha = \alpha_o - (sET)^{-1}\tau, \tag{3.81}$$

where $\alpha_o = l_o^{-1}\mathrm{d}l_o/\mathrm{d}T$ is the linear expansivity of zero tension. Let the tension in a bar be increased quasi-statically and isothermally from τ_1 to τ_2. Let the length l, cross-sectional area s, and isothermal Young's modulus of the bar be constant, then the work

$$W = -\tfrac{1}{2}(sE)^{-1}l(\tau_2^2 - \tau_1^2). \tag{3.82}$$

Remark: For a rectangular bar of volume $V = l_1 l_2 l_3$, then

$$\beta = (V^{-1}\partial V/\partial T)_p = (l_1^{-1}\partial l_1/\partial T)_p + (l_2^{-1}\partial l_2/\partial T)_p + (l_3^{-1}\partial l_3/\partial T)$$
$$\cong \alpha_1 + \alpha_2 + \alpha_3,$$

and also

$$\beta = -\rho^{-1}\partial\rho/\partial T,$$

as the specific volume v is the reciprocal value of the density ρ. For an isotropic solid, then $\alpha_1 = \alpha_2 = \alpha_3 = \alpha$ and $\beta = 3\alpha$. Similarly, for isothermal compressibility

$$k = -V^{-1}(\partial V/\partial p)_T = \delta_1 + \delta_2 + \delta_3,$$

where $\delta_i = -l_i^{-1}\partial l_i/\partial p$ $(i = 1, 2, 3)$ is the linear compressibility and $V = l_1 l_2 l_3$, $\partial V/\partial p = l_2 l_3 \partial l_1/\partial p + l_1 l_3 \partial l_2/\partial p + l_1 l_2 \partial l_3/\partial p$. For an isotropic solid $\delta_1 = \delta_2 = \delta_3 = \delta$ and $k = 3\delta$.

3.1.1.5 Phase changes of the first and second orders

In thermodynamics every homogeneous system is called a phase. Such a system in all parts and therefore in all points has the same physical properties. But the phase can be constituted from several components, e.g. mixtures. On the other hand different crystaline modifications are also different phases.

Let the system created by one homogeneous phase (crystalline or fluid) without chemical reactions in this system be assumed. Assume external forces do not act on the system. Let us use the equilibrium conditions and then find the stability conditions. As an external parameter we choose a pressure, and then a temperature will be given, when a specific volume v will be assumed as an internal parameter.

To study the equilibrium we shall construct the free energy of an irreversible state as a function of p, T and the internal parameter v. As the external parameter is a pressure p, then the thermodynamic potential will be used. Now by introducing additional forces we transform the system by isothermic quasi-static means into the state with relevant values of the internal parameter. Then the additional forces (e.g. external pressure, denoted by $p(v)$) are cut off and the exerted work will be determined. This work is equal to the difference of free energies, i.e. this work is equal to the difference of initial and final values of thermodynamic potential $F + pV$. Let $F = fm$, where f is a specific free energy of the phase investigated and m is its mass. Then this work can be written in the form

$$m[f(v, T) + p(v)v - f_o - p_o v] = m[f(v, T) + p(v)v] - \Phi_o,$$

where v is a specific volume, $V = mv$ and $\Phi_o = m(f_o + p_o v_o)$ is the thermodynamic potential of the initial state. After repeating additional forces (which are assumed to be very quick so that the volume v is not changed) the work is equal to $m(p(v) - p)v$. Then the difference

$$m[f(v, T) + p(v)v] - \Phi_o - m[p(v) - p]v = m[f(v, T) + pv] = \Phi(p, T, v)$$

represents the difference of free energies (thermodynamic potentials) of an arbitrary equilibrium in an initial state Φ_o and in a non-equilibrium state (p, T, v). It is known that the phase is in an equilibrium if the thermodynamic potential $\Phi = m\varphi$, where $\varphi = \varphi(p, T, v) + pv$ is a specific thermodynamic potential, acquires its minimum for varying internal parameters, e.g. v, and constant p and T. Therefore, $(\partial\varphi/\partial v)_{p,T} = 0$ and hence $\partial f/\partial v + p = 0$ or $p = -\partial f/\partial v$, which represents the equation of state. The second derivative gives $(\partial^2\varphi/\partial v^2)_{p,T} = \partial^2 f/\partial v^2 = -\partial p/\partial v$. If $\partial p/\partial v < 0$, then if the volume increases the pressure will decrease and the state is stable. If $\partial p/\partial v > 0$, then the equilibrium is unstable and such states are not in an equilibrium. The case $\partial p/\partial v = 0$ is connected with the critical point.

In the familiar phase transitions — melting, solidification, recrystallization, vaporization, sublimation — the temperature and pressure remain constant while the entropy and volume change. The different phases will be indicated by superscripts in order to reserve subscripts for specifying different states of the same phase or different substances.

Theorem 3.15: Let us assume two phases described by (p_1, T_1, v_1, m_1) and (p_2, T_2, v_2, m_2), where p_i, T_i, v_i, m_i, $i = 1, 2$ are pressure, temperature, specific volume and mass in both phases. Then the thermodynamic potential Φ is defined by

$$\Phi(p, T, v_1, v_2, m_1) = m_1(f_1 + pv_1) + m_2(f_2 + pv_2).$$

Then the conditions of equilibrium are

$$p_1 = p_2 = p, \quad \varphi_1(p, T) = \varphi_2(p, T). \tag{3.83}$$

Moreover, $p = p_s(T)$ is the curve of equilibrium of two phases.

Proof: Let us assume an equilibrium of the system created by two phases, which can transform one to another. Let the state of the system be determined by the pressure p and the temperature T, and by three independent internal parameters — by specific volumes v_1, v_2 of both phases and the mass m_1 of one phase as $m_2 = m - m_1$, where m is the mass of both phases.

In the state of equilibrium the thermodynamic potential of the system is equal to

$$m_1[f_1(v_1, T) + p_1(v_1)v_1] + m_2[f_2(v_2, T) + p_2(v_2)v_2].$$

When the pressure of the first phase changes from $p_1(v_1)$ to p, then the work of the first phase is $m_1v_1(p_1(v_1) - p)$ and the work of the second phase at the change of the pressure from $p_2(v_2)$ to p is $m_2v_2(p_2(v_2) - p)$. Then the thermodynamic potential of the investigated state is

$$\begin{aligned}\Phi(p, T, v_1, v_2, m_1) &= m_1[f_1(v_1, T) + p_1(v_1)v_1] + m_2[f_2(v_2, T) + p_2(v_2)v_2] - m_1v_1(p_1(v_1) - p)\\ &\quad - m_2v_2(p_2(v_2) - p)\\ &= m_1(f_1(v_1, T) + pv_1) + m_2(f_2(v_2, T) + pv_2).\end{aligned}$$

To determine the condition of the equilibrium we find the minimum of Φ when p and T are constant. Then $\mathrm{d}\Phi = 0$ and hence

$$\mathrm{d}\Phi = m_1(\partial f_1/\partial v_1 + p)\mathrm{d}v_1 + m_2(\partial f_2/\partial v_2 + p)\mathrm{d}v_2 + (f_1 + pv_1)\mathrm{d}m_1 + (f_2 + pv_2)\mathrm{d}m_2 = 0.$$

Since $\mathrm{d}m_2 = \mathrm{d}(m - m_1) = -\mathrm{d}m_1$, and v_1, v_2 and m_1 are independent variables, we obtain

$$\partial f_1/\partial v_1 = -p, \quad \partial f_2/\partial v_2 = -p, \quad f_1 + pv_1 = f_2 + pv_2. \tag{3.84}$$

Since $\partial f_1/\partial v_1 = p_1$, and $\partial f_2/\partial v_2 = p_2$ and $\varphi_1 = f_1 + pv_1$, $\varphi_2 = f_2 + pv_2$ then the above equation gives

$$p_1 = p_2 = p \text{ and } \varphi_1 = \varphi_2. \tag{3.85}$$

As in the equilibrium equations the mass does not occur, and then at the change of phases the equilibrium is not violated when the density of phases are not changed. Then Equation 3.84 gives three equations for v_1, v_2, p, T. From the last of these equations we then obtain

$$\varphi_1(p, T) = \varphi_2(p, T), \tag{3.86}$$

where v_1, v_2 were eliminated. Solving Equation 3.86 for p we obtain

$$p = p_s(T) \tag{3.87}$$

representing in the (p, T) plane the curve of equilibrium of two phases, which completes the proof.

Due to the condition in Equation 3.86, the specific thermodynamic potential φ is in the point of the phase change boundary conditions, and its first derivatives $\partial\varphi/\partial p = v$ and $\partial\varphi/\partial T = -s$, where v and s are specific volume ($V = mv$) and specific entropy ($S = ms$), are changed with jump. Such type of phase changes are known as phase transitions of the first order. The jump of entropy is equal to the latent heat divided by an absolute temperature. At the transition of the masses $\mathrm{d}m_2$ from the first to the second phases, the heat $\mathrm{d}Q = l_{12}\mathrm{d}m_2$ is absorbed, where l_{12} is a latent heat of the unity of mass.

Let us consider the reversible isothermal isobaric change of 1 mole of substance from phase 1 to phase 2. Using Theorem 3.10 for the specific entropy case we obtain

$$T\mathrm{d}s = c_v\mathrm{d}T + T(\partial p/\partial T)_v\mathrm{d}v. \tag{3.88}$$

Integrating over the whole change of phase, and remembering that $(\partial p/\partial T)_v$ is independent of v, we obtain

$$T(s_2 - s_1) = T\mathrm{d}p/\mathrm{d}T(v_2 - v_1).$$

Hence using Equation 3.87 and as $s_2 - s_1 = T^{-1}l_{12}$ we have

$$\mathrm{d}p_s/\mathrm{d}T = l_{12}(T(v_2 - v_1))^{-1}. \tag{3.89}$$

This equation is the well-known Clapeyron-Clausius' equation and it applies to any first order phase change (or transition) that takes place at constant T and p.

In the case when the specific thermodynamic potential φ is, in the point of the phase change boundary, continuous including its first derivatives $\partial\varphi/\partial p = v$, $\partial\varphi/\partial T = -s$, but its second derivatives

$$\partial^2\varphi/\partial p^2 = \partial v/\partial p, \quad \partial^2\varphi/\partial p\partial T = \partial v/\partial T, \quad \partial^2\varphi/\partial T^2 = T^{-1}c_p \tag{3.90}$$

are changed by a jump, we speak about the phase transitions of the second order. The point of the second order phase change boundary is known as the Curie point.

Remark: We cannot investigate where the pressure and temperature at which all three phases (i.e. liquid, solid and vapour) of a pure substance coexist. For information the triple point is merely the point of intersection of the sublimation and vaporization curves.

3.1.2. Introduction to the Tensor Analysis

Consider two Cartesian systems, unprimed and primed, with common origin O. Denote unit vectors in these two systems by $\boldsymbol{e}_i$ and $\boldsymbol{e}'_j$, respectively. Let the components of a vector $\mathbf{x}$ be denoted by x_i and x'_j, so that

$$\boldsymbol{x} = \sum_{i=1}^{3} x_i \boldsymbol{e}_i = \sum_{j=1}^{3} x'_j \boldsymbol{e}'_j. \tag{3.91}$$

To find the relation between the two sets of components we take the scalar product of Equation 3.91 with $\boldsymbol{e}_k$, so that

$$\sum_{i=1}^{3} x_i(\boldsymbol{e}_i.\boldsymbol{e}_k) = \sum_{j=1}^{3} x'_j(\boldsymbol{e}'_j.\boldsymbol{e}_k).$$

Set $\boldsymbol{e}_i.\boldsymbol{e}_k = \delta_{ik}$, where δ_{ik} is the Kronecker delta, and define

$$a_{jk} \equiv \boldsymbol{e}'_j.\boldsymbol{e}_k, \tag{3.92}$$

and we obtain

$$x_k = \sum_{j=1}^{3} a_{jk}x'_j, \quad k = 1, 2, 3. \tag{3.93}$$

Similarly,

$$x'_j = \sum_{k=1}^{3} a_{jk}x_k, \quad j = 1, 2, 3.$$

Now we may generalize our results for N-dimensional Euclidean space R^N. Now, let us consider an N-dimensional Euclidean space R^N and an orthogonal transformation from $R^N \to R^N$, $\boldsymbol{x} = (x_i)_{i=1}^N \to \mathbf{x}' = (x'_j)_{j=1}^N$ and

$$x'_j = a_{jk}x_k + c'_j, \quad j = 1, \dots, N, \tag{3.94}$$

where the matrix (a_{jk}) is orthogonal, so that

$$a_{ik}a_{jk} = \delta_{ij}, \quad i, j = 1, \dots, N, \tag{3.95}$$

where δ_{ij} is the Kronecker delta and where the Einstein summation convention (about the repeated indices) is used. For the inverse transformation we find

$$x_k = a_{ik}x'_i + c_k, \tag{3.96}$$

where $c_k = -a_{ik}c'_i$.

Definition 3.1: A zeroth order tensor is a scalar "s" such that

(i) in any Cartesian coordinate system there is a rule for associating s with a real number;

(ii) this real number is the same no matter what Cartesian co-ordinate system is used.

An n-th order tensor $\underline{\boldsymbol{T}}$, $n = 1, 2, 3, \dots$ is such that

(i) in any Cartesian co-ordinate system there is a rule for associating the tensor $\underline{\boldsymbol{T}}'$ with a unique order set of 3^n quantities $T_{i_1 \dots i_n}$, the so-called components of tensor $\underline{\boldsymbol{T}}$;

(ii) if $T_{i_1 \ldots i_n}$ and $T'_{j_1 \ldots j_n}$ are components of $\underline{\boldsymbol{T}}$ in two different Cartesian systems then

$$T_{i_1 \ldots i_n} = a_{i_1 j_1} \ldots \ldots a_{i_n j_n} T'_{j_1 \ldots j_n} \tag{3.97}$$

under the transformation of Equation 3.94.

For a second order tensor, the last equation leads to $T_{ij} = a_{ik} a_{jl} T_{kl}$. Let $\underline{\boldsymbol{I}}$ be an entity with nine components given by $[\underline{\boldsymbol{I}}]_{ij} = \delta_{ij}$. Then $\underline{\boldsymbol{I}}$ is a second order tensor, the so-called unit tensor, for which $\delta_{ij} = a_{ik} a_{jl} \delta_{kl} = a_{il} a_{jl} = \delta_{ij}$. Let $\underline{\boldsymbol{E}}$ be an entity with 27 components $[\underline{\boldsymbol{E}}]_{ijk} = \varepsilon_{ijk}$. Then $\underline{\boldsymbol{E}}$ is a third order tensor, the so-called Levi-Cività skew-symmetric tensor (or alternative or permutation tensor) such that transformations between co-ordinate systems are restricted to proper rotations, i.e.

$$\varepsilon_{ijk} = a_{il} a_{jm} a_{kn} \varepsilon_{lmn},$$

and moreover,

$$\begin{aligned} \varepsilon_{ijk} &= 0 \text{ if two of the integers } i, j, k \text{ are equal or if all three are equal;} \\ \varepsilon_{ijk} &= 1 \text{ if } (ijk) \text{ is an even permutation of } (123); \\ \varepsilon_{ijk} &= -1 \text{ if } (ijk) \text{ is an odd permutation of } (123). \end{aligned} \tag{3.98}$$

Definition 3.2: The determinant $|A|$ of a three-by-three matrix $A = (A_{ij})$ is given by

$$|A| = \varepsilon_{ijk} A_{1i} A_{2j} A_{3k}.$$

Theorem 3.16: It holds that

(i)
$$\varepsilon_{lmn} |A| = \varepsilon_{ijk} A_{il} A_{jm} A_{kn} = \varepsilon_{ijk} A_{li} A_{mj} A_{nk},$$

(ii)
$$|AB| = |A|\,|B|,$$

(iii)
$$\begin{vmatrix} A_{il} & A_{im} & A_{in} \\ A_{jl} & A_{jm} & A_{jn} \\ A_{kl} & A_{km} & A_{kn} \end{vmatrix} = \varepsilon_{ijk} \varepsilon_{lmn} |A|, \tag{3.99}$$

(iv)
$$\varepsilon_{ijk} \varepsilon_{lmk} = \delta_{il} \delta_{jm} - \delta_{im} \delta_{jl}.$$

Definition 3.3: Let $\underline{\boldsymbol{A}} = (A_{ij})$ and $\underline{\boldsymbol{B}} = (B_{ij})$ be second order tensors, and let c be a scalar. Then the sum, difference and scalar multiple are defined as

$$[\underline{\boldsymbol{A}} \pm \underline{\boldsymbol{B}}]_{ij} = A_{ij} \pm B_{ij}; \quad [c\underline{\boldsymbol{A}}]_{ij} = cA_{ij}, \tag{3.100}$$

and similarly for n-th order tensors.

The zero tensor $\underline{\boldsymbol{O}}$ is an entity whose components in any Cartesian co-ordinate system are all equal to zero. Two tensors $\underline{\boldsymbol{A}}$ and $\underline{\boldsymbol{B}}$ are equal if their difference is the zero tensor, i.e. $\underline{\mathbf{A}} = \underline{\mathbf{B}} \Leftrightarrow \underline{\boldsymbol{A}} - \underline{\boldsymbol{B}} = \underline{\mathbf{O}}$.

Theorem 3.17: Tensors of the same order form a vector space under addition and scalar multiplication.

Definition 3.4: A contraction of an n-th order tensor, $n \geq 2$, is formed by setting two of its indices equal and performing the resulting sum. The tensor $\underline{\mathbf{A}} = (A_{ij})$ of order 2 is said to be symmetric if $A_{ij} = A_{ji}$, i.e. $\underline{\mathbf{A}} = \underline{\mathbf{A}}^T$, and is skew-symmetric if $A_{ij} = -A_{ji}$, i.e. $\underline{\mathbf{A}} = -\underline{\mathbf{A}}^T$.

Definition 3.5: Let $\underline{\mathbf{A}} = (A_{ij})$ and $\underline{\boldsymbol{B}} = (B_{kl})$ be second order tensors. Then the contraction product $\underline{\boldsymbol{A}}.\underline{\boldsymbol{B}}$ is defined by

$$[\underline{\boldsymbol{A}}.\underline{\boldsymbol{B}}]_{ik} = A_{ij}B_{jk}. \tag{3.101}$$

Definition 3.6: The inner product of two vectors $\boldsymbol{A}$ and $\mathbf{B}$ in R^N is defined by

$$\boldsymbol{A} \cdot \boldsymbol{B} = A_i B_i. \tag{3.102}$$

The vector product of two vectors $\boldsymbol{A}$ and $\boldsymbol{B}$ in R^N is defined by

$$\boldsymbol{A} \times \boldsymbol{B} \equiv [\boldsymbol{A}, \boldsymbol{B}] = \boldsymbol{C}, \tag{3.103}$$

where $C = (c_i)$, $c_i = \varepsilon_{ijk} A_j B_k$.

Theorem 3.18: Let $\underline{\boldsymbol{T}}$ be a second order tensor. Then, uniquely it holds that

$$\underline{\boldsymbol{T}} = \underline{\boldsymbol{S}} + \underline{\boldsymbol{A}}, \tag{3.104}$$

where $\underline{\boldsymbol{S}}$ is a symmetric second order tensor and $\underline{\boldsymbol{A}}$ is an anti-symmetric second order tensor.

Proof: Since $\underline{\mathbf{S}} = \frac{1}{2}(\underline{\mathbf{T}} + \underline{\boldsymbol{T}}^T)$, $\underline{\boldsymbol{A}} = \frac{1}{2}(\underline{\boldsymbol{T}} - \underline{\boldsymbol{T}}^T)$, then Equation 3.104 follows immediately.

Definition 3.7: Tensors whose components are given by the same set of numbers in all co-ordinate systems are called isotropic.

Theorem 3.19:

(i) There are no nontrivial isotropic vectors of order one

(ii) Any isotropic tensor of order 2 can be written $\lambda\underline{\boldsymbol{I}}$ for some scalar λ

(iii) Any isotropic tensor of order 3 can be written $\lambda\underline{\boldsymbol{E}}$ for some scalar λ;

(iv) If $\underline{\boldsymbol{T}}$ is an isotropic tensor of order 4 then

$$[\underline{\boldsymbol{T}}]_{ijkl} = \lambda\delta_{ij}\delta_{kl} + \mu(\delta_{ik}\delta_{jl} + \delta_{il}\delta_{jk}) + \kappa(\delta_{ik}\delta_{jl} - \delta_{il}\delta_{jk})$$

for some scalars λ, μ, κ.

Proof: The idea of the proof consists of the following: let $\boldsymbol{v}$ be an isotropic vector, and let its components be v_i relative to a certain basis $\boldsymbol{e}_i$. Let there be another set of basis vectors satisfying $\boldsymbol{e}'_1 = \boldsymbol{e}_2$, $\mathbf{e}'_2 = \boldsymbol{e}_3$, $\boldsymbol{e}'_3 = \boldsymbol{e}_1$, and then

$$v'_1 = v_2, \quad v'_2 = v_3, \quad v'_3 = v_1. \tag{3.105}$$

As we assume that $\boldsymbol{v}$ is isotropic then $v'_1 = v_1$, $v'_2 = v_2$, $v'_3 = v_3$. Hence using the last equation it follows that

$$v'_1 = v_2 = v'_2 = v_3 = v'_3 = v_1. \tag{3.106}$$

Let there be a third set of basis vectors $\boldsymbol{e}''_i$ such that

$$\boldsymbol{e}''_1 = \boldsymbol{e}_2, \quad \boldsymbol{e}''_2 = -\boldsymbol{e}_1, \quad \boldsymbol{e}''_3 = \boldsymbol{e}_3.$$

Then $v_1'' = v_2$, $v_2'' = -v_1$, $v_3'' = v_3$. But as $\mathbf{v}$ is an isotropic vector then $v_2'' = v_2$, so that $v_2 = -v_1$. This is consistent with Equation 3.106 if and only if $v_1 = v_2 = v_3 = 0$. The only isotropic vector is thus the zero vector, which proves (i). Similarly (ii)-(iv) can be proved.

Definition 3.8: The concepts of limit, continuity and derivative are defined for tensor functions in exact analogy with their definition for scalar functions. Moreover, we define

(i) $[\nabla]_i \equiv \partial_i \equiv \partial/\partial x_i$.

Let $\underline{T}$ be a second order tensor, then

(ii) grad $\underline{T} \equiv [\nabla \underline{\mathbf{T}}]_{kij} \equiv \partial_k T_{ij} \equiv T_{ij,k}$,

(iii) $[\text{div } \underline{T}]_j \equiv [\nabla.\underline{\mathbf{T}}]_j \equiv \partial_i T_{ij} \equiv T_{ij,i}$,

(iv) $[\text{curl } \underline{T}]_{il} \equiv [\nabla \times \underline{T}]_{il} \equiv \varepsilon_{ijk}\partial_j T_{kl} \equiv \varepsilon_{ijk} T_{kl,j}$.

Theorem 3.20: Let $\underline{T}$ be a tensor field of order k, then

(i) $\nabla \underline{T}$ is a tensor field of order $k+1$,

(ii) $\nabla \cdot \underline{T}$ is a tensor field of order $k-1$,

(iii) $\nabla \times \underline{T}$ is a tensor field of order k.

Theorem 3.21: Let f, g be scalar functions and $\boldsymbol{A}, \boldsymbol{B}$ be vector functions. Then

(i) $\nabla(fg) = f\nabla g + g\nabla f$,

(ii) $\nabla(\boldsymbol{AB}) = (\boldsymbol{B}\nabla)\boldsymbol{A} + (\boldsymbol{A}\nabla)\boldsymbol{B} + \boldsymbol{B} \times \text{rot } \boldsymbol{A} + \boldsymbol{A} \times \text{rot } \boldsymbol{B}$,

(iii) $\text{div } (f\boldsymbol{A}) = f \text{div } \boldsymbol{A} + \boldsymbol{A}\nabla f$,

(iv) $\text{div } [\boldsymbol{A} \times \boldsymbol{B}] = \boldsymbol{B} \text{ rot } \boldsymbol{A} - \boldsymbol{A} \text{ rot } \boldsymbol{B}$,

(v) $\text{rot } (f\boldsymbol{A}) = f \text{ rot } \boldsymbol{A} - \mathbf{A} \times \nabla f$,

(vi) $\text{rot } [\boldsymbol{A} \times \boldsymbol{B}] = (\boldsymbol{B}\nabla)\boldsymbol{A} - (\boldsymbol{A}\nabla)\boldsymbol{B} + \boldsymbol{A} \text{ div } \boldsymbol{B} - \boldsymbol{B} \text{ div } \boldsymbol{A}$,

(vii) $\text{rot rot } \boldsymbol{A} = \nabla \text{ div } \boldsymbol{A} - \Delta \boldsymbol{A}$.

Remark: The obtained results of the tensor algebra and tensor analysis for a Cartesian co-ordinate system can also be extended for orthogonal curvilinear co-ordinates. For more details see for example, Brdička (1959), Nečas and Hlaváček (1981).

3.1.3. Field Equations of Continuum Mechanics

3.1.3.1. Kinematics of deformable continuum

Let Ω be a domain in R^N, $N = 2, 3$ occupied by a continuum medium. Let us denote its boundary by $\partial\Omega$. The continuum occupying a domain Ω will be under the effects of acting forces. Forces acting onto a continuum produce motion. The study of the motion of a continuum without regard to the acting forces that produce the motion represents kinematics as opposed to dynamics — the study of the influence of dynamic forces on motion — and to statics — the study of the influence of acting static forces on deformation of a continuum (which are independent of time).

As a continuum moves, it is natural to determine the position of each point as a function of the time $t \in \bar{I} \equiv \langle t_o, t_1 \rangle$, where t_o is an initial time and t_1 a final time of the time interval I. Firstly we identify its position $\boldsymbol{x}_o$ at a certain initial time t_o. If $\boldsymbol{x}$ is the position vector of this point at t, we describe the motion of the medium by

$$\boldsymbol{x} = \boldsymbol{x}(\boldsymbol{x}_o, t), \quad \boldsymbol{x}(\boldsymbol{x}_o, t_o) \equiv \boldsymbol{x}_o. \tag{3.107}$$

This equation means that $\boldsymbol{x}$ is the position at time t of a point that was initially, i.e. at time $t = t_o$, located at a point $\boldsymbol{x}_o$. For fixed $\boldsymbol{x}_o$ the curve $\boldsymbol{x} = \boldsymbol{x}(\boldsymbol{x}_o, t)$ for $t \geq t_o$ is called a particle path. Let $\boldsymbol{x}_o(\boldsymbol{x}, t)$ be the function of $\boldsymbol{x}$ and t obtained by solving Equation 3.107 for $\boldsymbol{x}_o$. The result $\boldsymbol{x}_o = \boldsymbol{x}_o(\boldsymbol{x}, t)$ states that $\boldsymbol{x}_o$ is the initial position of the particle at time t located at $\boldsymbol{x}$. It is evident that $\boldsymbol{x}_o(\boldsymbol{x}, t)$ and $\boldsymbol{x}(\boldsymbol{x}_o, t)$ are a pair of inverse functions, i.e. that $\boldsymbol{x}_o(\boldsymbol{x}(\boldsymbol{x}_o, t), t) \equiv \boldsymbol{x}_o$ and $\boldsymbol{x}(\boldsymbol{x}_o(\boldsymbol{x}, t), t) \equiv \boldsymbol{x}$.

Further, we shall postulate (i) impenetrability of matter and (ii) smoothness of motion by the following speculations:

(i) $\boldsymbol{x}_o(\boldsymbol{x}, t)$ and $\boldsymbol{x}(\boldsymbol{x}_o, t)$ will be assumed to be single-valued invertible functions so that, at time t, for each $\boldsymbol{x}_o$ there exists a unique $\boldsymbol{x}$ and vice versa, and

(ii) $\boldsymbol{x}_o(\boldsymbol{x}, t)$ and $\boldsymbol{x}(\boldsymbol{x}_o, t)$ will be assumed to have as many continuous partial derivatives as required for the various operations we shall perform on these functions.

To demonstrate this speculation we shall consider the density. Similarly this speculation holds for all physical quantities of the continuum. Let $\rho_o(\boldsymbol{x}_o, t)$ denote the mass per unit volume at time t of the particle initially located at $\boldsymbol{x}_o$. Let $\rho(\boldsymbol{x}, t)$ denote the mass per unit volume at time t, at fixed point $\boldsymbol{x}$. Then

$$\rho(\boldsymbol{x}, t) = \rho_o(\boldsymbol{x}_o(\boldsymbol{x}, t), t). \tag{3.108}$$

The expression $\rho_o(\boldsymbol{x}_o(\boldsymbol{x}, t), t)$ describes the density now associated with a particle that was initially located at the initial location of the particle now at $\boldsymbol{x}$, i.e. at time t. We see that $\rho_o(\boldsymbol{x}_o(\boldsymbol{x}, t), t)$ is the density associated with a particle now at $\boldsymbol{x}$. Substituting $\boldsymbol{x} = \boldsymbol{x}(\boldsymbol{x}_o, t)$ into Equation 3.108, and since $\boldsymbol{x}_o(\boldsymbol{x}(\boldsymbol{x}_o, t), t) \equiv \boldsymbol{x}_o$, we find that

$$\rho(\boldsymbol{x}(\boldsymbol{x}_o, t), t) = \rho_o(\boldsymbol{x}_o, t). \tag{3.109}$$

From this speculation we see that Equations 3.108 and 3.109 describe the transformation between the spatial description (i.e. $\boldsymbol{x}, t$ independent variables) and material description (i.e. $\boldsymbol{x}_o, t$ independent variables).

In the case of the velocity we define the velocity of a particle by the rate of change of the particle's position with time as

$$\boldsymbol{v}_o(\boldsymbol{x}_o, t) \equiv \partial \boldsymbol{x}(\boldsymbol{x}_o, t)/\partial t. \tag{3.110}$$

Further, we define $\boldsymbol{v}(\boldsymbol{x}, t)$, the velocity in spatial coordinates, by

$$\boldsymbol{v}(\boldsymbol{x}, t) = \boldsymbol{v}_o(\boldsymbol{x}_o(\boldsymbol{x}, t), t). \tag{3.111}$$

Hence, and using the fact that $\boldsymbol{x}_o(\boldsymbol{x}(\boldsymbol{x}_o, t), t) \equiv \boldsymbol{x}_o$, we obtain

$$\boldsymbol{v}(\boldsymbol{x}(\boldsymbol{x}_o, t), t) = \boldsymbol{v}_o(\boldsymbol{x}_o, t). \tag{3.112}$$

To derive their material derivatives:

(a) in the case of the density, firstly we compute a partial derivative of ρ_o with respect to t and we regard the initial position $\boldsymbol{x}_o$ as fixed, and secondly we compute a partial derivative of $\rho(\boldsymbol{x}, t)$, the density described in spatial variables, and we regard the spatial position $\boldsymbol{x}$ as fixed. Between $\boldsymbol{x}_o$ and $\boldsymbol{x}$ there exist mutual relations, and then will $N = 3$:

$$\frac{\partial \rho_o(\boldsymbol{x}_o, t)}{\partial t} = \left.\frac{\partial \rho}{\partial x_1}\right|_{\boldsymbol{x}=\boldsymbol{x}(\boldsymbol{x}_o,t)} \frac{\partial x_1}{\partial t} + \left.\frac{\partial \rho}{\partial x_2}\right|_{\boldsymbol{x}=\boldsymbol{x}(\boldsymbol{x}_o,t)} \frac{\partial x_2}{\partial t} + \left.\frac{\partial \rho}{\partial x_3}\right|_{\boldsymbol{x}=\boldsymbol{x}(\boldsymbol{x}_o,t)} \frac{\partial x_3}{\partial t} + \left.\frac{\partial \rho}{\partial t}\right|_{\boldsymbol{x}=\boldsymbol{x}(\boldsymbol{x}_o,t)}. \tag{3.113}$$

Now $\partial x_i/\partial t$ evaluated at $(\boldsymbol{x}_o, t)$ is $v_{oi}(\boldsymbol{x}_o, t)$, $i = 1, \ldots, N$, representing the i-th velocity component expressed in material variables. But our goal is to give the expression for $\partial \rho_o(\boldsymbol{x}_o, t)/\partial t$ in Equation 3.113 in spatial co-ordinates. Then after the substitution $\boldsymbol{x}_o = \boldsymbol{x}_o(\boldsymbol{x}, t)$, we firstly put

$$v_{oi}(\boldsymbol{x}_o, t)|_{\boldsymbol{x}_o=\boldsymbol{x}_o(\boldsymbol{x},t)} = v_i(\boldsymbol{x}, t), \quad i = 1, \ldots, N, \tag{3.114}$$

then we have

$$\left.\frac{\partial \rho_o(\boldsymbol{x}_o, t)}{\partial t}\right|_{\boldsymbol{x}_o=\boldsymbol{x}_o(\boldsymbol{x},t)} = \frac{\partial \rho(\boldsymbol{x}, t)}{\partial t} + \sum_{i=1}^{N} v_i(\boldsymbol{x}, t) \frac{\partial \rho(\boldsymbol{x}, t)}{\partial x_i}. \tag{3.115}$$

Frequently the notation d/dt for material or substantial derivative is used, i.e. $\mathrm{d}\rho(\boldsymbol{x}, t)/\mathrm{d}t \equiv \partial \rho_o(\boldsymbol{x}_o, t)/\partial t|_{\boldsymbol{x}_o=\boldsymbol{x}_o(\boldsymbol{x},t)}$. Then

$$\begin{aligned}\frac{\mathrm{d}\rho(\boldsymbol{x}, t)}{\mathrm{d}t} &= \frac{\partial \rho(\boldsymbol{x}, t)}{\partial t} + \boldsymbol{v}(\boldsymbol{x}, t).\nabla\rho(\boldsymbol{x}, t)\\ &= \frac{\partial \rho(\boldsymbol{x}, t)}{\partial t} + \boldsymbol{v}(\boldsymbol{x}, t).\mathrm{grad}\ \rho(\boldsymbol{x}, t).\end{aligned} \tag{3.116}$$

The last expression, the material derivative of density, describes the time rate of change of the density for a fixed particle.

(b) in the case of the velocity of the particle we use the following consideration: if $\boldsymbol{v}(\boldsymbol{x}, t)$ represents the velocity of the particle at position $\boldsymbol{x}$ and time t, then $\mathrm{d}\boldsymbol{v}(\boldsymbol{x}, t)/\mathrm{d}t$ represents the acceleration of the particle located at position $\boldsymbol{x}$ and time t. Then

$$\left.\frac{\partial \boldsymbol{v}_o(\boldsymbol{x}_o, t)}{\partial t}\right|_{\boldsymbol{x}_o=\boldsymbol{x}_o(\boldsymbol{x},t)} = \frac{\mathrm{d}\boldsymbol{v}(\boldsymbol{x}, t)}{\mathrm{d}t},$$

$$\begin{aligned}\frac{\mathrm{d}\boldsymbol{v}(\boldsymbol{x}, t)}{\mathrm{d}t} &= \frac{\partial \boldsymbol{v}(\boldsymbol{x}, t)}{\partial t} + (\boldsymbol{v}(\boldsymbol{x}, t).\nabla)\boldsymbol{v}(\boldsymbol{x}, t)\\ &= \frac{\partial \boldsymbol{v}(\boldsymbol{x}, t)}{\partial t} + (\boldsymbol{v}(\boldsymbol{x}, t).\mathrm{grad})\boldsymbol{v}(\boldsymbol{x}, t),\end{aligned} \tag{3.117}$$

where $\partial \boldsymbol{v}/\partial t$ and $(\boldsymbol{v}.\nabla)\boldsymbol{v}$ (or $(\boldsymbol{v}.\mathrm{grad})\boldsymbol{v}$) represent the local and convective derivatives. Furthermore, for the i-th acceleration component, $\mathrm{d}\boldsymbol{v}/\mathrm{d}t$ can be written in the form

$$\frac{\mathrm{d}v_i(\boldsymbol{x}, t)}{\mathrm{d}t} = \frac{\partial v_i(\boldsymbol{x}, t)}{\partial t} + \sum_{j=i}^{N} v_j(\boldsymbol{x}, t) \frac{\partial v_i(\boldsymbol{x}, t)}{\partial x_j}. \tag{3.118}$$

The acceleration of the particle at $(\boldsymbol{x}, t)$ can be written in the form

$$\lim_{\Delta t\to 0} \frac{\boldsymbol{v}(\boldsymbol{x}+\Delta \boldsymbol{x}, t+\Delta t)-\boldsymbol{v}(\boldsymbol{x}, t)}{\Delta t}$$

where $\Delta x \cong \boldsymbol{v}(\boldsymbol{x}, t)\Delta t$ and the numerator $\cong \nabla \boldsymbol{v}\cdot \Delta x + \boldsymbol{v}_{,t}\Delta t = (\nabla \boldsymbol{v}\cdot \boldsymbol{v} + \boldsymbol{v}_{,t})\Delta t$. Hence the result follows immediately. The term $\boldsymbol{v}_{,t} \equiv \partial \boldsymbol{v}/\partial t$ is the so-called local acceleration, and the term $\boldsymbol{v}\cdot\nabla\boldsymbol{v}$ is the convective acceleration. In general, we speak about the local part $\partial/\partial t$ and the convective part $\boldsymbol{v}\cdot\nabla$ of the derivative $\mathrm{d}/\mathrm{d}t$.

Let $\Omega(t)$ denote the domain occupied by the moving continuum at time t and let $\Omega(0)$ denote the domain occupied by this continuum at initial time $t = t_o (\equiv 0)$. Let the subsequent motion of any point $\boldsymbol{x}$ be described by $\boldsymbol{x} = \boldsymbol{x}(\boldsymbol{x}_o, t)$. Then the domain $\Omega(t)$ occupied by the designated continuum at time t is the image of $\Omega(0)$ at time t under the mapping $\boldsymbol{x} = \boldsymbol{x}(\boldsymbol{x}_o, t)$. Let $V(0)$ denote the volume of $\Omega(t)$, then

$$V(t) = \int_{\Omega(0)} \mathrm{d}\boldsymbol{x}_o \quad \text{and} \quad V(t) = \int_{\Omega(t)} \mathrm{d}\boldsymbol{x} = \int_{\Omega(0)} J(\boldsymbol{x}_o, t)\mathrm{d}\boldsymbol{x}_o \tag{3.119}$$

where in the second integral the change of the variables of integration to x_{oi} was made, and where $J(\boldsymbol{x}_o, t) \equiv \partial(x_i)/\partial(x_{oi})$ is the Jacobian of this transformation of variables, where $i = 1, \ldots, N$. It is evident that no absolute value sign is needed on J as the components of $\boldsymbol{x}$ and $\boldsymbol{x}_o$ are both determined in right-handed co-ordinate systems, and further we see that if $V(0)$ is the volume of a small domain centred about $\boldsymbol{x}_o$ then $V(t)/V(0) \cong J$. The Jacobian at point $\boldsymbol{x}_o$ and time t represents the dilatation of an infinitesimal volume initially at $\boldsymbol{x}_o$, where the dilatation is the ratio of volume occupied by the infinitesimal material region at time t to its initial volume.

For the time derivative of the Jacobian we have

$$\frac{\partial J(\boldsymbol{x}_o, t)}{\partial t} = \sum_{i=1}^{N} \left.\frac{\partial v_i}{\partial x_i}\right|_{\boldsymbol{x}=\boldsymbol{x}(\boldsymbol{x}_o,t)} J(\boldsymbol{x}_o, t). \tag{3.120}$$

If we write this equation in spatial coordinates, we use the substitution $\boldsymbol{x}_o = \boldsymbol{x}_o(\boldsymbol{x}, t)$ and use the definition of $\nabla\cdot\boldsymbol{v}$, and then we obtain the so-called Euler expansion formula

$$\frac{\mathrm{d}J(\boldsymbol{x}_o(\boldsymbol{x}, t), t)}{\mathrm{d}t} = (\nabla\cdot\boldsymbol{v})J(\boldsymbol{x}_o(\boldsymbol{x}, t), t). \tag{3.121}$$

We see that in material variables $\partial J/\partial t = ((\nabla\cdot\boldsymbol{v})|_{\boldsymbol{x}=\boldsymbol{x}(\boldsymbol{x}_o,t)})J$ and in spatial variables $\mathrm{d}J/\mathrm{d}t = (\nabla\cdot\boldsymbol{v})J$.

3.1.3.2. Conservation laws in their integral and differential forms

In the previous section we gave a characterization of a specified portion of continuum occupying the region $\Omega(t)$ at time t. From this it is evident that the mass of a portion of continuum does not vary as time increases is one way to state what is meant by conservation of mass. Hence we can assume that

$$\frac{\mathrm{d}}{\mathrm{d}t}\int_{\Omega(t)} \rho(\boldsymbol{x}, t)\mathrm{d}\boldsymbol{x} = 0, \tag{3.122}$$

where $\Omega(t)$ is arbitrary and $\rho(\boldsymbol{x}, t)$ denotes the density at point $\boldsymbol{x}$ and time t, represents the law of conservation of mass in its integral form. Since $\Omega(t)$ is arbitrary then due to the Dubois-Reymond lemma we shall be able to transform the integral formulation of this law to its differential form. Take the derivative inside the integral in the above equation, and then by using the transformation $\boldsymbol{x} = \boldsymbol{x}(\boldsymbol{x}_o, t)$ we shall transform it to initial coordinates $\boldsymbol{x}_o$. Hence

$$\begin{aligned}\frac{\mathrm{d}}{\mathrm{d}t}\int_{\Omega(t)}\rho(\boldsymbol{x}, t)\mathrm{d}\boldsymbol{x} &= \frac{\mathrm{d}}{\mathrm{d}t}\int_{\Omega(0)}\rho(\boldsymbol{x}(\boldsymbol{x}_o, t), t)J(\boldsymbol{x}_o, t)\mathrm{d}\boldsymbol{x}_o \\ &= \frac{\mathrm{d}}{\mathrm{d}t}\int_{\Omega(0)}\rho_o(\boldsymbol{x}_o, t)J(\boldsymbol{x}_o, t)\mathrm{d}\boldsymbol{x}_o \\ &= \int_{\Omega(0)}\left(\frac{\partial\rho_o(\boldsymbol{x}_o, t)}{\partial t}J(\boldsymbol{x}_o, t) + \rho_o(\boldsymbol{x}_o, t)\frac{\partial J(\boldsymbol{x}_o, t)}{\partial t}\right)\mathrm{d}\boldsymbol{x}_o \\ &= \int_{\Omega(0)}\left(\frac{\partial\rho_o(\boldsymbol{x}_o, t)}{\partial t} + \rho_o(\boldsymbol{x}_o, t)(\nabla\cdot\boldsymbol{v})|_{\boldsymbol{x}=\boldsymbol{x}(\boldsymbol{x}_o, t)}\right)J(\boldsymbol{x}_o, t)\mathrm{d}\boldsymbol{x}_o.\end{aligned} \tag{3.123}$$

Hence, making the variable change $\boldsymbol{x}_o = \boldsymbol{x}_o(\boldsymbol{x}, t)$ and using the results of the previous section, we obtain

$$\frac{\mathrm{d}}{\mathrm{d}t}\int_{\Omega(t)}\rho(\boldsymbol{x}, t)\mathrm{d}\boldsymbol{x} = \int_{\Omega(t)}\left\{\frac{\mathrm{d}\rho(\boldsymbol{x}, t)}{\mathrm{d}t} + \rho(\boldsymbol{x}, t)[\nabla\cdot\boldsymbol{v}(\boldsymbol{x}, t)]\right\}\mathrm{d}\boldsymbol{x}. \tag{3.124}$$

Due to the Dubois-Reymond lemma

$$\frac{\mathrm{d}\rho(\boldsymbol{x}, t)}{\partial t} + \rho(\boldsymbol{x}, t)[\nabla\cdot\boldsymbol{v}(\boldsymbol{x}, t)] = 0. \tag{3.125}$$

The last equation is also known as the continuity equation. Since $\mathrm{d}\rho/\mathrm{d}t = \partial\rho/\partial t + (\boldsymbol{v}\cdot\nabla)\rho$ then Equation 3.125 can be rewritten in the following form

$$\frac{\partial\rho(\boldsymbol{x}, t)}{\partial t} + \nabla\cdot(\rho(\boldsymbol{x}, t)\boldsymbol{v}(\boldsymbol{x}, t)) = 0. \tag{3.126}$$

This equation can be derived in another way. We shall start from the fact that conservation of mass requires that for a region fixed in space the rate of increase of the mass contained in this region must equal the net mass flux into the region. Then only flow of mass across the boundary can cause a change in the mass of the material contained in a fixed region, since mass is neither created nor destroyed. Therefore, we have

$$\frac{\mathrm{d}}{\mathrm{d}t}\int_{\Omega}\rho(\boldsymbol{x}, t)\mathrm{d}\boldsymbol{x} = -\int_{\partial\Omega}\rho(\boldsymbol{x}, t)\boldsymbol{v}(\boldsymbol{x}, t)\boldsymbol{n}(\boldsymbol{x}, t)\mathrm{d}s, \tag{3.127}$$

where Ω is an arbitrary region fixed in space, $\partial\Omega$ its boundary, $\boldsymbol{n}$ a unit exterior normal to $\partial\Omega$, and $\boldsymbol{v}(\boldsymbol{x}, t)$ a velocity of a mass particle. Then applying the Green's theorem we obtain

$$\int_{\Omega}\left\{\frac{\partial\rho(\boldsymbol{x}, t)}{\partial t} + \nabla\cdot[\boldsymbol{v}(\boldsymbol{x}, t)\rho(\boldsymbol{x}, t)]\right\}\mathrm{d}\boldsymbol{x} = 0. \tag{3.128}$$

Since the region Ω is arbitrary and as we assume that the integrand is continuous then, due to the Dubois-Reymond lemma, Equation 3.126 follows immediately.

To derive the laws of balance of linear and angular momentum (or moment of momentum) we shall start from ideas of the particle mechanics, where a particle M, which has mass m and moves with velocity $\boldsymbol{v}$, possesses linear momentum m$\boldsymbol{v}$. Due to the Newtonian second law ($m\boldsymbol{a} = \boldsymbol{F}$), the rate of change of the linear momentum of a particle M is equal to the forces acting on the particle. In the case of continuum mechanics such forces are body forces (from which the best known is the gravity force) and surface forces. To generalize the Newtonian second law to continuum mechanics, we define the linear momentum at time t possessed by a continuum of density $\rho(\boldsymbol{x}, t)$ that occupies a region $\Omega(t)$ by $\int_{\Omega(t)} \rho(\boldsymbol{x}, t)\boldsymbol{v}(\boldsymbol{x}, t)\mathrm{d}\boldsymbol{x}$, where $\boldsymbol{v}$ is the velocity vector of a moving continuum. Moreover, the time rate of change of the last integral is assumed to be equal to the forces acting on the continuum occupying $\Omega(t)$. We thus consider the body and surface forces acting on the continuum, which are defined by $\int_{\Omega(t)} \rho(\boldsymbol{x}, t)\boldsymbol{f}(\boldsymbol{x}, t)\mathrm{d}\boldsymbol{x}$ in the case of body forces, where $\boldsymbol{f}(\boldsymbol{x}, t)$ is the body force vector per unit mass, and $\int_{\partial\Omega(t)} \vec{\tau}(\boldsymbol{x}, t, \boldsymbol{n})\mathrm{d}s$ in the case of surface forces, where $\vec{\tau}(\boldsymbol{x}, t, \boldsymbol{n})$ represents forces per unit area exerted at point $\boldsymbol{x}$ of $\partial\Omega$ by the material exterior to Ω on the material interior to Ω, and $\boldsymbol{x}$ is the position vector of the surface element, t is time, and $\boldsymbol{n}$ is the unit exterior normal to $\partial\Omega$. The surface forces have the meaning of surface stress vectors. In $\int_{\partial\Omega(t)} \vec{\tau}(\boldsymbol{x}, t, \boldsymbol{n})\mathrm{d}s$ we have implicitly assumed that force per unit area (stress) approaches a limit as the area approaches zero.

Then the rate of change of linear momentum in a material region $\partial\Omega$ is equal to the sum of body and surface forces, i.e.

$$\frac{\mathrm{d}}{\mathrm{d}t}\int_{\Omega(t)} \rho\boldsymbol{v}\,\mathrm{d}\boldsymbol{x} = \int_{\Omega(t)} \rho\boldsymbol{f}\,\mathrm{d}\boldsymbol{x} + \int_{\partial\Omega(t)} \vec{\tau}\mathrm{d}s. \tag{3.129}$$

Now for a sufficiently smooth function $\boldsymbol{g}$ and a region occupied by a continuum $\Omega(t)$, assuming conservation of mass, we have

$$\frac{\mathrm{d}}{\mathrm{d}t}\int_{\Omega(t)} \rho\boldsymbol{g}\,\mathrm{d}\boldsymbol{x} = \int_{\Omega(t)} \rho\left(\frac{\partial\boldsymbol{g}}{\partial t} + (\boldsymbol{v}\cdot\nabla)\boldsymbol{g}\right)\mathrm{d}\boldsymbol{x},$$

and then

$$\int_{\Omega(t)} \rho\left(\frac{\partial\boldsymbol{v}}{\partial t} + (\boldsymbol{v}\cdot\nabla)\boldsymbol{v}\right)\mathrm{d}\boldsymbol{x} - \int_{\Omega(t)} \rho\boldsymbol{f}\mathrm{d}\boldsymbol{x} = \int_{\partial\Omega(t)} \vec{\tau}\mathrm{d}s, \tag{3.130}$$

which represents the integral form of the momentum balance law. Now $\vec{\tau}(\boldsymbol{x}, t, \boldsymbol{n}) = \boldsymbol{n}\cdot\vec{\tau}(\boldsymbol{x}, t)$, is the stress vector (which will be discussed in detail in the next section), with components $\tau_i(\boldsymbol{x}, t, \boldsymbol{n}) = \tau_{ij}(\boldsymbol{x}, t)n_j$, where τ_{ij} is the so-called stress tensor, and then

$$\int_{\Omega(t)} \rho\left(\frac{\partial\boldsymbol{v}}{\partial t} + (\boldsymbol{v}\cdot\nabla)\boldsymbol{v}\right)\mathrm{d}\boldsymbol{x} - \int_{\Omega(t)} \rho\boldsymbol{f}\mathrm{d}\boldsymbol{x} = \int_{\partial\Omega(t)} \boldsymbol{n}\cdot\vec{\tau}\mathrm{d}s. \tag{3.131}$$

Using the divergence theorem we obtain

$$\int_{\Omega(t)} \rho\left(\frac{\partial\boldsymbol{v}}{\partial t} + (\boldsymbol{v}\cdot\nabla)\boldsymbol{v}\right)\mathrm{d}\boldsymbol{x} - \int_{\Omega(t)} \rho\boldsymbol{f}\mathrm{d}\boldsymbol{x} = \int_{\Omega(t)} \nabla\cdot\vec{\tau}\mathrm{d}s. \tag{3.132}$$

Since $\Omega(t)$ is arbitrary and if we assume that the integrand is continuous then

$$\rho\left(\frac{\partial \boldsymbol{v}}{\partial t}+(\boldsymbol{v}\cdot\nabla)\boldsymbol{v}\right)-\rho\boldsymbol{f}=\operatorname{div}\vec{\tau}\quad(\equiv\nabla\cdot\vec{\tau}) \tag{3.133}$$

which for the i-th component of $\boldsymbol{v}$ can be written in the following form

$$\rho\left(\frac{\partial v_i}{\partial t}+v_j\frac{\partial v_i}{\partial x_j}\right)-\rho f_i=\frac{\partial \tau_{ij}}{\partial x_j}. \tag{3.134}$$

The relationships of Equation 3.133 in the vector form and of Equation 3.134 in the component form are the so-called Cauchy's differential equations expressing for any continuum the balance of linear momentum or the so-called equations of motion of a moving continuum. These equations represent initial equations for the description of dynamical processes in a moving continuum. The first term represents inertial forces, which for a uniform motion and for a static case are equal to zero. The second represents body forces, and the third internal forces in a moving continuum.

To derive the law of angular momentum (i.e. moment of momentum) we retain the idea from particle mechanics. If a mass point at the point $\boldsymbol{x}$ possesses momentum $\boldsymbol{M}$, then its moment of momentum about O is defined as $\boldsymbol{x}\times\boldsymbol{M}\equiv[\boldsymbol{x},\boldsymbol{M}]$. Let us suppose that a system of mass points is subject to mutual central forces (e.g. gravity forces) which are directed along the lines joining the interacting particles. Then the rate of change of each particle's moment of momentum is equal to the sum of the torques acting on the particle. In the case of continuum mechanics, in addition to centrally acting body forces like gravity, contact forces that differ from material to material must also be assumed. Then in an inertial or non-moving co-ordinate system, the rate of change of the moment of momentum is equal to the torques exerted by body forces and surface forces. To formulate this assumption more precisely, we generalize the concepts of moment of momentum and torque used in particle mechanics. Hence we postulate

$$\frac{\mathrm{d}}{\mathrm{d}t}\int_{\Omega(t)}(\boldsymbol{x}\times\rho\boldsymbol{v})\mathrm{d}\boldsymbol{x}=\int_{\Omega(t)}(\boldsymbol{x}\times\rho\boldsymbol{f})\mathrm{d}\boldsymbol{x}+\int_{\partial\Omega(t)}(\boldsymbol{x}\times\vec{\tau})\mathrm{d}\boldsymbol{x}. \tag{3.135}$$

Since $\mathrm{d}\boldsymbol{x}/\mathrm{d}t=\boldsymbol{v}$ and since

$$\frac{\mathrm{d}}{\mathrm{d}t}\int_{\Omega(t)}\rho\boldsymbol{f}\,\mathrm{d}\boldsymbol{x}=\int_{\Omega(t)}\rho\left(\frac{\partial\boldsymbol{f}}{\partial t}+(\boldsymbol{v}\cdot\nabla)\boldsymbol{f}\right)\mathrm{d}\boldsymbol{x}\quad\left(\equiv\int_{\Omega(t)}\rho\frac{\mathrm{d}\boldsymbol{f}}{\mathrm{d}t}\mathrm{d}\boldsymbol{x}\right),$$

and using the divergence theorem, we obtain

$$\begin{aligned}\int_{\Omega(t)}\left(\frac{\partial}{\partial t}+(\boldsymbol{v}\cdot\nabla)\right)(\boldsymbol{x}\times\rho\boldsymbol{v})\mathrm{d}\boldsymbol{x}&\equiv\int_{\Omega(t)}\left(\boldsymbol{x}\times\rho\left(\frac{\partial\boldsymbol{v}}{\partial t}+(\boldsymbol{v}\cdot\nabla)\boldsymbol{v}\right)\right)\mathrm{d}\boldsymbol{x}\\&=\int_{\Omega(t)}(\boldsymbol{x}\times\rho\boldsymbol{f})\mathrm{d}\boldsymbol{x}+\int_{\Omega(t)}\nabla\cdot\vec{\tau}\mathrm{d}\boldsymbol{x}-\int_{\Omega(t)}\tau_s\mathrm{d}\boldsymbol{x}.\end{aligned} \tag{3.136}$$

Hence

$$\int_{\Omega(t)} \left(\boldsymbol{x} \times \left(\rho \left(\frac{\partial \boldsymbol{v}}{\partial t} + (\boldsymbol{v} \cdot \nabla)\boldsymbol{v} \right) - \rho \boldsymbol{f} - \nabla \cdot \vec{\tau} \right) \right) \mathrm{d}\boldsymbol{x} = \int_{\Omega(t)} \vec{\tau}_s \mathrm{d}\boldsymbol{x}, \qquad (3.137)$$

where $(\vec{\tau}_s)_i = \varepsilon_{ijk}\tau_{jk}$. Due to the momentum balance law the left hand side vanishes and since $\Omega(t)$ is arbitrary, then $\vec{\tau}_s = 0$, where

$$(\vec{\tau}_s)_i = \varepsilon_{ijk}\tau_{jk} = (\tau_{23} - \tau_{32}, \tau_{31} - \tau_{13}, \tau_{12} - \tau_{21}) = 0, \text{ i.e } \tau_{ij} = \tau_{ji}.$$

In the component form Equation 3.135 is of the form

$$\frac{\mathrm{d}}{\mathrm{d}t} \int_{\Omega(t)} \varepsilon_{ijk} x_j \rho v_k \mathrm{d}\boldsymbol{x} = \int_{\Omega(t)} \varepsilon_{ijk} x_j f_k \mathrm{d}\boldsymbol{x} + \int_{\partial\Omega(t)} \varepsilon_{ijk} x_j \tau_{kl} n_l \mathrm{d}s, \; i = 1, \ldots N, \qquad (3.138)$$

where ε_{ijk} is the Levi-Civita tensor. Similarly as above we have

$$\begin{aligned}
\int_{\partial\Omega(t)} \varepsilon_{ijk} x_j \tau_k \mathrm{d}s &= \int_{\partial\Omega(t)} \varepsilon_{ijk} x_j \tau_{kl} n_l \mathrm{d}s \\
&= \int_{\partial\Omega(t)} \varepsilon_{ijk} \frac{\partial}{\partial x_1} (x_j \tau_{kl}) \mathrm{d}s \\
&= \int_{\Omega(t)} \varepsilon_{ijk} \left(\delta_{lj} \tau_{lk} + x_j \frac{\partial}{\partial x_l} \tau_{kl} \right) \mathrm{d}x.
\end{aligned}$$

Hence

$$\frac{\mathrm{d}}{\mathrm{d}t} \int_{\Omega(t)} \varepsilon_{ijk} x_j \rho v_k \mathrm{d}\boldsymbol{x} = \int_{\Omega(t)} \varepsilon_{\mathrm{ijk}} x_j f_k \mathrm{d}\boldsymbol{x} + \int_{\Omega(t)} \varepsilon_{ijk} \left(\delta_{lj} \tau_{lk} + x_j \frac{\partial}{\partial x_l} \tau_{kl} \right) \mathrm{d}\boldsymbol{x}, \qquad (3.139)$$

from which $\varepsilon_{ijk}\tau_{jk} = 0$ follows immediately and thus we have proved that

$$\tau_{ij} = \tau_{ji}, \qquad (3.140)$$

i.e. the symmetry of the stress tensor τ_{ij}.

To derive the law of energy conservation from thermodynamics it is known that the rate of the total energy of the system (i.e. the sum of kinetic and intrinsic energy) is equal to the sum of the effect of external forces acting on the system and of the energy influx per unit time. Thus

$$\frac{\mathrm{d}}{\mathrm{d}t} \int_{\Omega(t)} \rho \left(\frac{1}{2} \boldsymbol{v}^2 + e \right) \mathrm{d}\boldsymbol{x} = \int_{\Omega(t)} \boldsymbol{f}\boldsymbol{v} \,\mathrm{d}\boldsymbol{x} + \int_{\partial\Omega(t)} \vec{\tau} \cdot \boldsymbol{v} \,\mathrm{d}s + \int_{\Omega(t)} \rho w \mathrm{d}\boldsymbol{x} - \int_{\partial\Omega(t)} \boldsymbol{q} \cdot \boldsymbol{n} \,\mathrm{d}s, \qquad (3.141)$$

which can be written or in the form

$$\frac{\mathrm{d}}{\mathrm{d}t} \int_{\Omega(t)} \rho \left(\frac{1}{2} \boldsymbol{v}^2 + e \right) \mathrm{d}\boldsymbol{x} = \int_{\Omega(t)} f_i v_i \mathrm{d}\boldsymbol{x} + \int_{\partial\Omega(t)} \tau_{ij} n_j v_i \mathrm{d}s + \int_{\Omega(t)} \rho w \mathrm{d}\boldsymbol{x} - \int_{\partial\Omega(t)} q_i n_i \mathrm{d}s, \qquad (3.142)$$

where the scalar e represents an intrinsic energy of the medium, w represents an influx of energy and the vector $\boldsymbol{q}$ represents a transport of energy. Hence applying

Equations 3.126, 3.133 or 3.134, using Equation 3.140 and putting

$$\mathrm{D}_{ij} = \frac{1}{2}\left(\frac{\partial v_i}{\partial x_j} + \frac{\partial v_j}{\partial x_i}\right), \tag{3.143}$$

where D_{ij} is the strain rate tensor, which will be discussed in detail in Section 3.1.3.4, we obtain

$$\int_{\Omega(t)} \rho \left(\frac{\mathrm{d}e}{\mathrm{d}t}\right) \mathrm{d}\boldsymbol{x} = \int_{\Omega(t)} \tau_{ij} D_{ij} d\boldsymbol{x} + \int_{\Omega(t)} \left(\rho w - \frac{\partial q_i}{\partial x_i}\right) \mathrm{d}\boldsymbol{x}. \tag{3.144}$$

Since $\Omega(t)$ is arbitrary and if we assume as above that the integrand is continuous, then

$$\rho \frac{\mathrm{d}e}{\mathrm{d}t} = \tau_{ij} D_{ij} + \rho w - \frac{\partial q_i}{\partial x_i}, \tag{3.145}$$

representing the differential form of the law of energy conservation.

3.1.3.3. Stress tensor

At time t, consider a domain $\Omega(t) \subset R^3$ with boundary $\partial\Omega(t)$. Let $\boldsymbol{f}(\boldsymbol{x}, t) = (f_i(\boldsymbol{x}, t))$, $i = 1, 2, 3$, be a density of forces acting on every volume element $G(t)$ and such that $\overline{G}(t) \subset \Omega(t)$. We shall assume that $G(t)$ has a Lipschitz boundary $\partial G(t)$. Furthermore, we shall assume that $\boldsymbol{f} \in [C(\Omega)]^3$ for every $t \in \overline{I} \equiv \langle t_o, t_1 \rangle$. Let $\boldsymbol{x} \in \partial G(t)$ and let us denote by $\boldsymbol{n}$ the unit outer normal to $\partial G(t)$ at the point $\boldsymbol{x}$. Then for fixed time t, $\vec{\tau}(\boldsymbol{x}, t, \boldsymbol{n})$ is the force per unit area exerted at the point $\boldsymbol{x} \in \partial G$ by the material exterior to G on the material interior to G. It depends on the point $\boldsymbol{x}$ and on the direction of the normal $\boldsymbol{n}$. We assume that $\tau_i \in C(\overline{\Omega} \times \partial S)$, $i = 1, 2, 3$, where S is the unit sphere. The dimensions are $[\tau_i] = \mathrm{kp/m^2}$ and $[f_i] = \mathrm{kp/m^3}$.

From the momentum balance law

$$\int_{\Omega(t)} \rho \left(\frac{\partial \boldsymbol{v}}{\partial t} + (\boldsymbol{v} \cdot \nabla)\boldsymbol{v}\right) \mathrm{d}\boldsymbol{x} - \int_{\Omega(t)} \rho \boldsymbol{f} \,\mathrm{d}\boldsymbol{x} = \int_{\partial\Omega(t)} \vec{\tau} \mathrm{d}s, \tag{3.146}$$

we deduce an important result about the stress vector. Let us consider a family of similar domains G_h, characterized by the length h. Such domains are e.g. rectangular parallelepipeds with sides of lengths ph, rh, sh. These domains G_h have the same shape and their volume is $\mathscr{V}_h = ah^3$ and area $S_h = bh^2$, where constants a and b depend only on the shape of domains. Let $g(\boldsymbol{x})$ be any continuous function, then from the integral mean value theorem we obtain

$$\int_{G_h(t)} g(\boldsymbol{x}) \mathrm{d}\boldsymbol{x} = ah^3 g(\overline{\boldsymbol{x}}), \quad \overline{\boldsymbol{x}} \in G_h. \tag{3.147}$$

Applying this equation to the right hand side of Equation 3.146 and then dividing by h^2 and in the limit for $h \to 0$ we obtain

$$\lim_{h \to 0} h^{-2} \int_{\partial G_h(t)} \vec{\tau} \mathrm{d}\boldsymbol{x} = 0, \tag{3.148}$$

as both integrals are proportional to h^3. To interpret this equation we see that if the material in the domain G_h were subject to surface stresses and to no other forces, and if this material were in equilibrium, then the total force on the material would have to vanish, i.e. $\int_{\partial G_h(t)} \vec{\tau} \mathrm{d}\boldsymbol{x} = 0$. But each domain G_h is moving and is subject to body forces. But as $h \to 0$, the effect of the surface stress term becomes more and more dominant compared to the volume integrals $\int_{G_h(t)} \rho(\partial \boldsymbol{v}/\partial t + (\boldsymbol{v} \cdot \nabla)\boldsymbol{v})\mathrm{d}\boldsymbol{x}$ and $\int_{G_h(t)} \rho \boldsymbol{f} \,\mathrm{d}\boldsymbol{x}$, representing the effects of inertia and body forces. Equation 3.148 is known as the principle of local stress equilibrium.

A consequence of the principle of local stress equilibrium is Newton's third law. To derive it we shall consider a rectangular parallelepiped G_h of dimensions $\varepsilon h \times h \times h$, $\varepsilon \ll 1$, containing a fixed point $\boldsymbol{x}$ on one corner. Let $\boldsymbol{n}$, $-\boldsymbol{n}$, $\boldsymbol{n}_t$, $-\boldsymbol{n}_t$, $\boldsymbol{n}_G$, $-\boldsymbol{n}_G$ be the unit exterior normals to the front, rear, top, bottom, right, and left faces of G_h (in points x_i, $i = 1, \ldots, 6$), respectively. By the integral mean value theorem for the i-th component of the surface integral in Equation 3.148 we have

$$\lim_{h \to 0} (h^{-2}([\tau_i(\boldsymbol{x}_1, t, \boldsymbol{n}) + \tau_i(\boldsymbol{x}_2, t, -\boldsymbol{n}) + \varepsilon B_i]) = 0, \quad i = 1, 2, 3, \tag{3.149}$$

where

$$B_i = \tau_i(\boldsymbol{x}_3, t, \boldsymbol{n}_t) + \tau_i(\boldsymbol{x}_4, t, -\boldsymbol{n}_t) + \tau_i(\boldsymbol{x}_5, t, \boldsymbol{n}_G) + \tau_i(\boldsymbol{x}_6, t, -\boldsymbol{n}_G).$$

Then Equations 3.148 and 3.149 yield

$$\tau_i(\boldsymbol{x}, t, \boldsymbol{n}) + \tau_i(\boldsymbol{x}, t, -\boldsymbol{n}) + \varepsilon B_i = 0, \quad i = 1, 2, 3. \tag{3.150}$$

In the limit $h \to 0$ in each τ_i, all x_i $(i = 1, \ldots, 6)$ approach $\boldsymbol{x}$, as we assume that $\vec{\tau}$ is continuous in a domain containing $\boldsymbol{x}$. Using the continuity and hence the boundedness of B_i, we see that εB_i can be made arbitrarily small so that

$$\tau_i(\boldsymbol{x}, t, \boldsymbol{n}) + \tau_i(\boldsymbol{x}, t, -\boldsymbol{n}) = 0 \text{ or } \vec{\tau}(\boldsymbol{x}, t, \boldsymbol{n}) + \vec{\tau}(\boldsymbol{x}, t, -\boldsymbol{n}) = 0, \tag{3.151}$$

which represents the principle of action and reaction (the so-called Newton's third law for continua).

The stress vector $\vec{\tau}(\boldsymbol{x}, t, \boldsymbol{n})$ can be expressed as a linear combination of $\vec{\tau}(\boldsymbol{x}, t, \boldsymbol{e}_1)$, $\vec{\tau}(\boldsymbol{x}, t, \boldsymbol{e}_2)$, $\vec{\tau}(\boldsymbol{x}, t, \boldsymbol{e}_3)$, where $\boldsymbol{e}_1 = (1, 0, 0)$, $\boldsymbol{e}_2 = (0, 1, 0)$, $\boldsymbol{e}_3 = (0, 0, 1)$. Let us consider for fixed t a tetrahedron $G_h(t) \subset \Omega(t)$ defined by OABC, three edges of which are parallel to the co-ordinate system, defined by the orthogonal basis vectors $\boldsymbol{e}_i$, $i = 1, 2, 3$. The domain OABC has a Lipschitz boundary and the outer normal $\boldsymbol{n}$ to ABC exists. The mutually perpendicular faces of area S_i are normal to the orthonormal basis vectors $\boldsymbol{e}_i$ and meet at the point $O \equiv \boldsymbol{x}$. The surface ABC has area S and we define $h \equiv S^{1/2}$. Then h is the distance of the point $O \equiv \boldsymbol{x}$ from the plane ABC and the volume is $|G_h| = \frac{1}{3}hS$. Let us apply Equation 3.148 to the tetrahedron OABC.

By using the integral mean theorem, then for the j-th component of the force $\boldsymbol{F}$ acting on the face of area S_i we obtain

$$F^i_j = S_i \tau_j(\boldsymbol{x}^i, t, -\boldsymbol{e}_i), \quad i, j = 1, 2, 3,$$

where $\boldsymbol{x}^i$ is a point in this face. Let $\boldsymbol{x}^o \in ABC$, then similarly as above for the j-th component of the force acting on ABC we obtain

$$F_j = S\tau_j(\boldsymbol{x}^o, t, \boldsymbol{n}), \quad j = 1, 2, 3.$$

From the analytical geometry it is known that S_i is a projection of S on the i-th co-ordinate plane, so that

$$S_i = n_i S = [\boldsymbol{n} \cdot \boldsymbol{e}_i]S,$$

where n_i is the i-th component of the normal $\boldsymbol{n}$. Hence using Equation 3.148, and as $h = S^{\frac{1}{2}}$ we have

$$\sum_{i=1}^{3} n_i \tau_j(\boldsymbol{x}, t, -\boldsymbol{e}_i) + \tau_j(\boldsymbol{x}, t, \boldsymbol{e}) = 0.$$

From the principle of action and reaction (Equation 3.151) we have

$$\tau_j(\boldsymbol{x}, t, \boldsymbol{n}) = \sum_{i=1}^{3} n_i \tau_j(\boldsymbol{x}, t, \boldsymbol{e}_i), \quad j = 1, 2, 3.$$

Since the point $\boldsymbol{x}$ is the general point of the domain $\Omega(t)$ we can define τ_{ij} by

$$\tau_{ij}(\boldsymbol{x}, t) \equiv \tau_j(\boldsymbol{x}, t, \boldsymbol{e}_i), \quad i, j = 1, 2, 3,$$

and then

$$\tau_j(\boldsymbol{x}, t, \boldsymbol{n}) = \sum_{i=1}^{3} n_i \tau_{ij}(\boldsymbol{x}, t), \quad j = 1, 2, 3, \tag{3.152}$$

or, using the well known Einstein's summation notation,

$$\tau_j(\boldsymbol{x}, t, \boldsymbol{n}) = n_i \tau_{ij}(\boldsymbol{x}, t), \quad j = 1, 2, 3, \tag{3.153}$$

and in the vector form

$$\vec{\tau}(\boldsymbol{x}, t, \boldsymbol{n}) = \boldsymbol{n} \cdot \vec{\tau}(\boldsymbol{x}, t). \tag{3.154}$$

The matrix $\tau_{ij}(\boldsymbol{x}, t)$ represents the stress tensor at the point $\boldsymbol{x}$ and time t. Equation 3.152 (or 3.153) shows the dependence of the stress vector $\vec{\tau}$ on the exterior unit vector $\boldsymbol{n}$. From here we see that this dependence is linear. The stress tensor τ_{ij} is a tensor of order two.

Let us show that the stress tensor has a character of tensor of order two. Let us consider a new cartesian co-ordinate system with mutually perpendicular unit base vector $\boldsymbol{e}'_i$. The stress tensor $\tau'_{ij}(\boldsymbol{x}, t)$ can be derived in the same way as above, so that

$$\tau'_j(\boldsymbol{x}, t, n) = n'_i \tau'_{ij}(\boldsymbol{x}, t), \quad j = 1, 2, 3, \tag{3.155}$$

where τ'_{ij} is the j-th component of the stress vector acting on the surface element whose exterior normal points are in the direction of $\boldsymbol{e}'_i$. Let us assume that Equation 3.153 and 3.155 consistent. Let us determine the relation between τ_{ij} and τ'_{ij}. From the tensor analysis it is known (see e.g. Brdička (1959)) that between both systems the transformations $x_i = a_{ij}x'_j$ and $x'_j = a_{ij}x_i$ are valid. Then using this transformation for $\vec{\tau}'(\boldsymbol{x}, t)$ and $\boldsymbol{n}'(\boldsymbol{x}, t)$ we obtain

$$\tau'_j(\boldsymbol{x}, t) = n'_i \tau'_{ij}(\boldsymbol{x}, t) = a_{li} n_l \tau'_{ij}(\boldsymbol{x}, t),$$

$$\tau'_j(\boldsymbol{x}, t) = a_{kj} \tau_k(\boldsymbol{x}, t) = a_{li} n_l \tau'_{ij}(\boldsymbol{x}, t).$$

Hence substituting for $\tau_j(\boldsymbol{x}, t)$ by using k instead of j as a free subscript in Equation 3.153

$$a_{kj} n_i \tau_{ik}(\boldsymbol{x}, t) = a_{li} n_l \tau'_{ij}(\boldsymbol{x}, t).$$

Multiplying both sides of the last equation by a_{nj} we have

$$a_{nj} a_{kj} n_i \tau_{ik}(\boldsymbol{x}, t) = a_{nj} a_{li} n_l \tau'_{ij}(\boldsymbol{x}, t).$$

Since $a_{nj} a_{kj} = \delta_{nk}$, then

$$n_i \tau_{in}(\boldsymbol{x}, t) = a_{nj} a_{li} n_l \tau'_{ij}(\boldsymbol{x}, t). \tag{3.156}$$

But Equation 3.153 is true for any unit vector $\boldsymbol{n}$. Let $\boldsymbol{n} = (1, 0, 0)$, then from Equation 3.156 it follows that

$$\tau_{1n}(\boldsymbol{x}, t) = a_{1i} a_{nj} \tau'_{ij}(\boldsymbol{x}, t). \tag{3.157}$$

Similarly for $\boldsymbol{n} = (0, 1, 0)$ and $\boldsymbol{n} = (0, 0, 1)$. Hence,

$$\tau_{kn}(\boldsymbol{x}, t) = a_{ki} a_{nj} \tau'_{ij}(\boldsymbol{x}, t), \tag{3.158}$$

which proves that the stress tensor has a character of tensor of order two. An expression that has the same value in all co-ordinate systems is known as an invariant. It can be shown that $\Theta \equiv \tau_{ii}$ is an invariant. Really, $\Theta = \tau_{ii} = \tau_{11} + \tau_{22} + \tau_{33}$ is an invariant as

$$\tau'_{ii}(\boldsymbol{x}, t) = a_{ij} a_{ik} \tau_{jk}(\boldsymbol{x}, t) = \delta_{jk} \tau_{jk}(\boldsymbol{x}, t) = \tau_{jj}(\boldsymbol{x}, t).$$

We define the normal stress $N(\boldsymbol{x}, t)$ as the projection of the stress vector $\vec{\tau}(\boldsymbol{x}, t, \boldsymbol{n})$ onto the normal. Thus

$$N(\boldsymbol{x}, t, \boldsymbol{n}) = \vec{\tau}(\boldsymbol{x}, t, \boldsymbol{n}) \cdot \boldsymbol{n} = \tau_i(\boldsymbol{x}, t) n_i = \tau_{ij} n_i n_j, \tag{3.159}$$

which represents a quadratic form in the components of $\boldsymbol{n}$. For $N > 0$ we speak of tension, for $N < 0$ of pressure. The difference

$$\vec{\Sigma}(\boldsymbol{x}, t, \boldsymbol{n}) = \vec{\tau}(\boldsymbol{x}, t, \boldsymbol{n}) - N(\boldsymbol{x}, t, \boldsymbol{n}) \cdot \boldsymbol{n} \tag{3.160}$$

is called the shear or tangential stress vector.

To define the principal stresses and principal directions of stress we find a direction $\boldsymbol{n}$ for which the tangential stresses vanish. For such a direction $\boldsymbol{n}$ it holds

$$\tau_{ij}(\boldsymbol{x}, t) n_j = \tau(\boldsymbol{x}, t) n_i, \quad i = 1, 2, 3,$$

representing a system of three homogeneous linear equations for normal components n_i, where τ is a parameter of proportionality. The necessary condition for the existence of a non-zero solution $\boldsymbol{n}$ is

$$\det(\tau_{ij} - \tau \delta_{ij}) = 0 \tag{3.161}$$

or in equivalent form

$$-\tau^3 + \Theta_1 \tau^2 - \Theta_2 \tau + \Theta_3 = 0, \tag{3.162}$$

where Θ_i, $i = 1, 2, 3$, are invariants and

$$\Theta_1 = \Theta \equiv \tau_{11} + \tau_{22} + \tau_{33} = \tau_{ii},$$

$$\Theta_2 = \begin{vmatrix} \tau_{11} & \tau_{12} \\ \tau_{21} & \tau_{22} \end{vmatrix} + \begin{vmatrix} \tau_{11} & \tau_{13} \\ \tau_{31} & \tau_{33} \end{vmatrix} + \begin{vmatrix} \tau_{22} & \tau_{23} \\ \tau_{32} & \tau_{33} \end{vmatrix} = \frac{1}{2!}\varepsilon_{ijk}\varepsilon_{imn}\tau_{jm}\tau_{kn},$$

$$\Theta_3 = \det(\tau_{ij}) = \tfrac{1}{3!}\varepsilon_{ijk}\varepsilon_{lmn}\tau_{il}\tau_{jm}\tau_{kn}.$$

As the stress tensor is symmetric, then the solution of Equation 3.161 is represented by three real roots τ_1, τ_2, τ_3, which are independent of the co-ordinate system since the characteristic polynomial of the linear mapping, characterized by the matrix (τ_{ij}), is independent of the choice of the co-ordinate system.

Definition 3.9: The eigenvalues τ_i of the linear mapping given by the matrix (τ_{ij}) are called the principal stresses, and the eigenvectors of this linear mapping, the matrix (τ_{ij}), are called the principal directions of stress. The planes perpendicular to the principal directions of stress are designated principal stress planes (and the stress acting within this plane is the principal stress).

If $\tau_1 \geq \tau_2 \geq \tau_3$ then eigenvectors n_i corresponding to these eigenvalues can be taken in such a way that they are mutually orthogonal unit vectors. In the case when $\tau_1 = \tau_2 = \tau_3$, then three arbitrary mutually orthogonal unit vectors for the eigenvectors can be taken. For a principal direction n_i it holds

$$|\vec{\tau}(\boldsymbol{x}, t, n_i)| = |\tau_i(\boldsymbol{x}, t)| \cdot |\boldsymbol{n}| = |\tau_i(\boldsymbol{x}, t)| \quad \forall t.$$

In a new co-ordinate system whose axes coincide with the principal directions we have $\tau'_{ii}(\boldsymbol{x}, t) = \tau_i(\boldsymbol{x}, t)$ (i not summed), $\tau'_{ij}(\boldsymbol{x}, t) = 0 (i \neq j)$. Then we can define the Cauchy quadric of stress.

Definition 3.10: Let $k \neq 0$ be an arbitrary real number. Let t be fixed. The quadratic surface defined by

$$\tau_{ij}\xi_i\xi_j = \pm k^2$$

is called the Cauchy quadric of stress.

If $\tau_1 = \tau_2$, there is a symmetry to the axis ξ_3. If $\tau_1 = \tau_2 = \tau_3$, there is a symmetry with respect to the origin.

Let $\tau_1 \geq \tau_2 \geq \tau_3 > 0$, then for $+k^2$ we have a general ellipsoid, and for $-k^2$ the empty set. In this case there is tension in every direction (see for example Brdička (1959) or Nečas, and Hlaváček (1981)).

Let $0 > \tau_1 \geq \tau_2 \geq \tau_3$, then for $+k^2$ we have the empty set and for $-k^2$ an ellipsoid. In this case there is pressure in every direction.

Let $\tau_1 \geq \tau_2 > 0 > \tau_3$, then for $+k^2$ we have a hyperboloid of one sheet, and for $-k^2$ we have a hyperboloid of two sheet.

In the term of mean normal stress in introduced for the expression $\Theta_s = \frac{1}{3}\Theta = \frac{1}{3}\tau_{ii}$ the volume stress tensor is defined by

$$\tau_s = \begin{Vmatrix} \Theta_s & 0 & 0 \\ 0 & \Theta_s & 0 \\ 0 & 0 & \Theta_s \end{Vmatrix},$$

the components of which change the volume of the body, t being fixed. As shear stresses $\vec{\Sigma}$ do not affect the body's volume, but alter its form, another matrix will be introduced by

$$D_s = \begin{Vmatrix} \tau_{11} - \Theta_s & \tau_{12} & \tau_{13} \\ \tau_{21} & \tau_{22} - \Theta_s & \tau_{23} \\ \tau_{31} & \tau_{32} & \tau_{33} - \Theta_s \end{Vmatrix} = (\tau_{ij} - \tfrac{1}{3}\Theta\delta_{ij}),$$

which is called the deviator of the stress tensor or the stress deviator and is also denoted as τ_{ij}^D. Generally, it holds $\tau_{ij} = \tau_s + D_s$.

To find the maximum shear stresses at a given point $\boldsymbol{x}$ and at a given t, firstly we have $|\vec{\tau}(\boldsymbol{x}, t, \boldsymbol{n})|^2 = (N(\boldsymbol{x}, t, \boldsymbol{n}))^2 + \left|\vec{\Sigma}(\boldsymbol{x}, t, \boldsymbol{n})\right|^2$. Then transforming the co-ordinate system into the system with co-ordinates identical with principal directions, and since $\tau_i = \tau_i n_i$, we have

$$N^2 = (\tau_1 n_1^2 + \tau_2 n_2^2 + \tau_3 n_3^2)^2, \quad |\vec{\tau}|^2 = \sum\nolimits_{i=1}^{3} (\tau_i n_i)^2,$$

so that

$$\left|\vec{\Sigma}\right|^2 = \tau_1^2 n_1^2 + \tau_2^2 n_2^2 + \tau_3^2 n_3^2 - (\tau_1 n_1^2 + \tau_2 n_2^2 + \tau_2 n_2^2)^2. \tag{3.163}$$

Moreover, it holds that $\sum_{i=1}^{3} n_i^2 = 1$. This equation is a subsidiary condition. To determine the external value of Equation 3.163 the technique of Lagrange multipliers can be used. This well-known technique searchs for the maximum value of the function f under the constraint condition g as

$$\frac{\partial f}{\partial n_i} + \lambda \frac{\partial g}{\partial n_i} = 0, \quad i = 1, 2, 3, \quad \lambda \in R^1,$$

where in our case $f \equiv \left|\vec{\Sigma}(\boldsymbol{x}, t, \boldsymbol{n})\right|^2$ and $g \equiv \sum_{i=1}^{3} n_i^2 = 1$. Hence

$$(\lambda + \tau_1^2 - 2\tau_1(\tau_1 n_1^2 + \tau_2 n_2^2 + \tau_3 n_3^2))n_1 = (\lambda + \tau_1^2 - 2\tau_1 N)n_1 = 0,$$
$$(\lambda + \tau_2^2 - 2\tau_2(\tau_1 n_1^2 + \tau_2 n_2^2 + \tau_3 n_3^2))n_2 = (\lambda + \tau_2^2 - 2\tau_2 N)n_2 = 0,$$
$$(\lambda + \tau_3^2 - 2\tau_3(\tau_1 n_1^2 + \tau_2 n_2^2 + \tau_3 n_3^2))n_3 = (\lambda + \tau_3^2 - 2\tau_3 N)n_3 = 0.$$

The necessary condition for a non-trivial solution of this system is the condition that at least one of the expressions in brackets is zero. But this condition is not sufficient as we would find a principal direction, and so we solve the system for two unknowns λ and N. Then the second condition is the vanishing of the Vandermonde determinant

$$\begin{vmatrix} 1 & \tau_1 & \tau_1^2 \\ 1 & \tau_2 & \tau_2^2 \\ 1 & \tau_3 & \tau_3^2 \end{vmatrix} = 0.$$

Analysis of this system of linear equations shows that the maximum shear stress satisfies one of the following relations:

$$(0, \pm\tfrac{1}{2}\sqrt{2}, \pm\tfrac{1}{2}\sqrt{2} \Rightarrow |\Sigma_1| = \tfrac{1}{2}\,|\tau_2 - \tau_3|$$
$$(\pm\tfrac{1}{2}\sqrt{2}, 0, \pm\tfrac{1}{2}\sqrt{2}) \Rightarrow |\Sigma_2| = \tfrac{1}{2}\,|\tau_3 - \tau_1|\,,$$
$$(\pm\tfrac{1}{2}\sqrt{2}, \pm\tfrac{1}{2}\sqrt{2}, 0) \Rightarrow |\Sigma_3| = \tfrac{1}{2}\,|\tau_1 - \tau_2|\,.$$

These shear stresses are called principal shear stresses. If we assume only sign "+", then $\Sigma_1 + \Sigma_2 + \Sigma_3 = 0$.

Definition 3.11: The following expression

$$\tilde{\Sigma} = \left\{1/6[(\tau_1 - \tau_2)^2 + (\tau_2 - \tau_3)^2 + (\tau_1 - \tau_3)^2]\right\}^{\frac{1}{2}}$$
$$= \left\{1/6[(\tau_{11} - \tau_{22})^2 + (\tau_{22} - \tau_{33})^2 + (\tau_{33} - \tau_{11})^2] + \tau_{12}^2 + \tau_{23}^2 + \tau_{31}^2\right\}^{\frac{1}{2}} \quad (3.164)$$

represents the so-called intensity of shear stress, which is an invariant.

Remark: Its crucial meaning is in the theory of inelasticity, namely in the theory of plasticity.

3.1.3.4. Strain and strain rate tensors

In the real world practically all bodies are strained (deformed) by the action of loaded forces, which in the body evoke stresses, both normal and shear stresses. Special attention will be paid to elastic strain, i.e. one which disappears completely after de-loading (relieving) the body. An important simplifying feature of two large classes of problems in elasticity is that displacements in the first case are great and in the second are small. In the patter case this means that the distinction between material and spatial co-ordinates can be ignored to first approximations. Deformation of the body consists of a change of volume and an alteration of its form. It will be shown that the shear does not cause any volume reduction, but only a change of form (shape). Such a change of form (distortion) is produced by an alteration of right angles of the elementary cube.

At time $t \in I \subset R^1$ let us consider the domain $\Omega(t) \in R^3$, which is occupied by the moving medium at time $t \in$I. The domain $\Omega(t)$ need not represent the whole region occupied by the moving medium at time t. In many cases in geodynamics it can be a small part of the Earth body, which is our present interest, e.g. a $2D$ model of the Earth mantle, approximated by the infinite layer, from which we study only a small part. As the medium moves, then for two different times t_1 and t_2, $t_1 \neq t_2$, the domains $\Omega(t_1)$ and $\Omega(t_2)$ are formed by different media particles, which can enter or leave the investigated domains through their boundaries $\partial\Omega(t_1)$ and $\partial\Omega(t_2)$. The kinematics of a moving medium was discussed in Section 3.1.3.1. We saw that we have two different possibilities of how to describe the motion of the medium in the region investigated, which closely connects with the description of the deformation of the medium.

The first possibility is known as the *Lagrangian description* of the moving medium. It considers the motion of each particular medium particle at time t. Its position at time t is generally determined by curvilinear co-ordinate X, so that

$$\boldsymbol{x} = \boldsymbol{x}(\boldsymbol{X}, t) \quad (\text{i.e. } x_i = x_i(X_i, t), \quad i = 1, 2, 3), \tag{3.165}$$

where X_i are the so-called Lagrangian co-ordinates. It is evident that four quantities (X_i and t, $i = 1, 2, 3$) determine at time t the position of the particle of the medium. As the medium is continuous then the functions $\boldsymbol{x}(\boldsymbol{X}, t)$ are also continuous functions. Then the velocity and acceleration of the medium particle given by the reference (X_i) are defined by partial derivatives

$$\underline{\boldsymbol{v}}(\boldsymbol{X}, t) = \frac{\partial \boldsymbol{x}(\boldsymbol{X}, \boldsymbol{t})}{\partial t}, \tag{3.166}$$

$$\underline{\boldsymbol{a}}(\boldsymbol{X}, t) \equiv \underline{\boldsymbol{v}}_{,t}(\boldsymbol{X}, t) = \frac{\partial \underline{\boldsymbol{v}}(\boldsymbol{X}, t)}{\partial t} = \frac{\partial^2 \boldsymbol{x}(\boldsymbol{X}, t)}{\partial^2 t}. \tag{3.167}$$

For example, in the MKS (Giorgi) system the dimension of $\boldsymbol{x}$ is $[\boldsymbol{x}] \equiv m$, and the dimensions of velocity $[\underline{\boldsymbol{v}}]$ and acceleration $[\underline{\boldsymbol{a}}]$ are ms^{-1} and ms^{-2}, respectively.

But in more investigated cases we want to know the state of the medium and its changes with dependence on time. For such type of investigations we use the second possibility of description of the moving medium, known as the *Eulerian description*. It is based on the determination of the velocity $\boldsymbol{v}(\boldsymbol{x}, t)$ of the moving medium particle passing through the point at time t. Since $\boldsymbol{x} = \boldsymbol{x}(\boldsymbol{X}, t)$ then $\boldsymbol{v}(\boldsymbol{x}, t) = \underline{\boldsymbol{v}}(\boldsymbol{X}, t) = \dfrac{\partial \boldsymbol{x}(\boldsymbol{X}, t)}{\partial t}$, and thus as we see above

$$\begin{aligned} \frac{\mathrm{d}\boldsymbol{v}(\boldsymbol{x}, t)}{\mathrm{d}t} &= \frac{\partial \boldsymbol{v}(\boldsymbol{x}, t)}{\partial t} + (\boldsymbol{v}(\boldsymbol{x}, t).\nabla)\boldsymbol{v}(\boldsymbol{x}, t) \\ &= \frac{\partial \boldsymbol{v}(\boldsymbol{x}, t)}{\partial t} + (\boldsymbol{v}(\boldsymbol{x}, t).\mathrm{grad})\boldsymbol{v}(\boldsymbol{x}, t), \end{aligned} \tag{3.168}$$

where $\partial \boldsymbol{v}/\partial t$ and $(\boldsymbol{v} \cdot \nabla)\boldsymbol{v}$ (or $(\boldsymbol{v}.\mathrm{grad})\boldsymbol{v}$) represent the local and convective derivatives. Furthermore, for the i-th acceleration component, $\mathrm{d}\boldsymbol{v}/\mathrm{d}t$ can be written in the form

$$\frac{\mathrm{d}v_i(\boldsymbol{x}, t)}{\mathrm{d}t} = \frac{\partial v_i(\boldsymbol{x}, t)}{\partial t} + \sum_{j=i}^{N} \frac{\partial v_i(\boldsymbol{x}, t)}{\partial x_j} v_j(\boldsymbol{x}, t). \tag{3.169}$$

If we use the Eulerian description, then the investigated quantity is represented by a function dependent on $\boldsymbol{x}$ and t, as we saw above in the case of the density $\rho(\boldsymbol{x}, t)$. Hence the local derivative (i.e. $\partial/\partial t$) results from the dependence of the investigated function (e.g. ρ) on time, and the convective derivative $\boldsymbol{v} \cdot \mathrm{grad}\rho$ is a consequence of the transport of the investigated quantity (e.g. ρ) by the moving medium, as we saw above.

The displacement vector $\boldsymbol{u}(\boldsymbol{X}, t)$ is defined as the difference between a particle's present position $\boldsymbol{x}(\boldsymbol{X}, t)$ and its initial position $\boldsymbol{X}$. That is

$$\boldsymbol{u}(\boldsymbol{X}, t) = \boldsymbol{x}(\boldsymbol{X}, t) - \boldsymbol{X} \,(\text{ or } u_i(\boldsymbol{X}, t) = x_i(\boldsymbol{X}, t) - X_i).$$

Let us consider the position of nearby points after displacement. Then we have

$$x_i(\boldsymbol{X}_2, t) - x_i(\boldsymbol{X}_1, t) = X_{2i} - X_{1i} + \partial u_i(\boldsymbol{X}, t)/\partial x_j(X_{2j} - X_{1j}) + \cdots,$$

where $\partial u_i/\partial x_j \equiv u_{i,j}$ denotes the displacement gradient tensor.

Let $\Omega(t)$ be a domain occupied by an elastic body at time t and let us consider a point $\boldsymbol{x} \in \Omega(t)$. Let us assume that the body is loaded. Then after its deformation, the body is transformed into a body $\Omega'(t)$ and the point $\boldsymbol{x}$ into a point $\boldsymbol{y}$, so that

$$\boldsymbol{y}(\boldsymbol{x}, t) = \boldsymbol{x}(t) + \boldsymbol{u}(\boldsymbol{x}, t),$$

where $\boldsymbol{u}(\boldsymbol{x}, t)$ is the displacement vector, defined as the difference between a particle's present position and its initial position.

Function $\boldsymbol{u}(\boldsymbol{x}, t)$ is assumed to belong to $[C^{(3)}(\Omega)]^3$, t being fixed. Let P, Q be two points in a non-deformed elastic body such that $\overline{PQ} = \mathrm{d}\boldsymbol{x}$, and such that their co-ordinates are x_i or $x_i + \mathrm{d}x_i$, respectively. After deformation let the body points P, Q be transformed into points P', Q', let $\overline{P'Q'} = \mathrm{d}\boldsymbol{y}$ and let their co-ordinates be y_i and $y_i + \mathrm{d}y_i$. Then from the Taylor expansion of the function $u_i(x_j + \mathrm{d}x_j, t)$, t being fixed, where we restrict ourselves to the first degree only, we have

$$\mathrm{d}y_i = \mathrm{d}x_i + \frac{\partial u_i}{\partial x_j}\mathrm{d}x_j = \left(\delta_{ij} + \frac{\partial u_i}{\partial x_j}\right)\mathrm{d}x_j,$$

where δ_{ij} is the Kroneker symbol. Since $|\mathrm{d}\boldsymbol{x}|^2 = \mathrm{d}x_i\mathrm{d}x_i = \delta_{jk}\mathrm{d}x_j\mathrm{d}x_k$, and $|\mathrm{d}\boldsymbol{y}|^2 = \mathrm{d}y_i\mathrm{d}y_i$, then

$$\begin{aligned}|\mathrm{d}\boldsymbol{y}|^2 - |\mathrm{d}\boldsymbol{x}|^2 &= \left(\left(\delta_{ij} + \frac{\partial u_i}{\partial x_j}\right)\left(\delta_{ik} + \frac{\partial u_i}{\partial x_k}\right) - \delta_{jk}\right)\mathrm{d}x_j\mathrm{d}x_k \\ &= 2\varepsilon_{jk}\mathrm{d}x_j\mathrm{d}x_k,\end{aligned}$$

where

$$\varepsilon_{jk} = \frac{1}{2}\left(\frac{\partial u_j}{\partial x_k} + \frac{\partial u_k}{\partial x_j} + \frac{\partial u_i}{\partial x_j}\frac{\partial u_i}{\partial x_k}\right)$$

is the so-called finite strain tensor, which is a symmetric tensor. Similarly as for the case of the stress tensor, the finite strain tensor is represented by the 3×3 symmetric matrix (ε_{ij}). The tensor character of ε_{ij} follows from its definition. Let $X_i = a_{ij}x_j + C_i$, then it implies that $x_j = a_{ij}X_i + c_j$. Let $\boldsymbol{u}$ be the displacement vector. Then $\underline{u}_i(\boldsymbol{X}, t) = a_{ij}u_j(\boldsymbol{X}, t)$. Hence

$$\begin{aligned}\frac{\partial \underline{u}_i}{\partial x_j} &= a_{ik}\frac{\partial u_k}{\partial x_l}\frac{\partial u_l}{\partial x_j} = a_{ik}a_{jl}\frac{\partial u_k}{\partial x_l}, \\ \frac{\partial \underline{u}_i}{\partial x_j}\frac{\partial \underline{u}_k}{\partial x_i} &= a_{kl}a_{jm}\frac{\partial u_l}{\partial x_m}a_{kr}a_{is}\frac{\partial u_r}{\partial x_s} = a_{jm}a_{is}\frac{\partial u_r}{\partial x_m}\frac{\partial u_r}{\partial x_s}.\end{aligned}$$

From here we see that the linear as well as the quadratic terms are transformed as tensors of the second order, so that the tensor character of the finite strain tensor is proved. Hence the finite strain tensor is transformed from a Cartesian co-ordinate system into the other one by

$$\underline{\varepsilon}_{ij} = a_{ik} a_{jl} \varepsilon_{kl}.$$

The physical meaning of the components of the finite strain tensor follows from the fact that

$$\frac{|\mathrm{d}\boldsymbol{y}|^2 - |\mathrm{d}\boldsymbol{x}|^2}{|\mathrm{d}\boldsymbol{x}|} = \varepsilon_{PQ}$$

represents a relative extension in the point P in the direction $\overline{PQ} = \overline{\boldsymbol{x}, \boldsymbol{x} + \Delta\boldsymbol{x}} \equiv \mathrm{d}\boldsymbol{x}$ before and after deformation. For the sign "+" we have an extension, for the sign "−" we have an abbreviation (reduction). It holds that $|\mathrm{d}\boldsymbol{y}|^2 - |\mathrm{d}\boldsymbol{x}|^2 = 2\varepsilon_{ij}\mathrm{d}x_i\mathrm{d}x_j = \varepsilon_{PQ}\,|\mathrm{d}\boldsymbol{x}|^2$.

The linear part of the finite strain tensor ε_{ij} defines the so-called small strain tensor e_{ij}, i.e.

$$e_{ij}(\boldsymbol{u}) = \tfrac{1}{2}(\partial u_i/\partial x_j + \partial u_j/\partial x_i), \quad e_{ij}(\boldsymbol{u}) = e_{ji}(\boldsymbol{u}), \quad i, j = 1, ..N. \tag{3.170}$$

Since $\mathrm{d}y_i - \mathrm{d}x_i = \delta(\mathrm{d}x_i) = \partial u_i/\partial x_j \mathrm{d}x_j$ and its differentiation $\delta(\mathrm{d}v_i) = \partial v_i/\partial x_j \mathrm{d}x_j$ and since

$$\partial u_i/\partial x_j = \tfrac{1}{2}(\partial u_i/\partial x_j + \partial u_j/\partial x_i) + \tfrac{1}{2}(\partial u_i/\partial x_j - \partial u_j/\partial x_i) = e_{ij}(\boldsymbol{u}) + \omega_{ij}(\boldsymbol{u}),$$

$$\partial v_i/\partial x_j = \tfrac{1}{2}(\partial v_i/\partial x_j + \partial v_j/\partial x_i) + \tfrac{1}{2}(\partial v_i/\partial x_j - \partial v_j/\partial x_i) = D_{ij}(\boldsymbol{v}) + W_{ij}(\boldsymbol{v}),$$

then

$$\delta(\mathrm{d}x_i) = e_{ij}(\boldsymbol{u})\mathrm{d}x_j + \omega_{ij}(\boldsymbol{u})\mathrm{d}x_j, \quad \delta(\mathrm{d}v_i) = D_{ij}(\boldsymbol{v})\mathrm{d}x_j + W_{ij}(\boldsymbol{v})\mathrm{d}x_j,$$

where $D_{ij}(\boldsymbol{v})$ is the strain rate tensor, $i, j = 1, .., N$ and $\omega_{ij}(\boldsymbol{u})$ and $W_{ij}(\boldsymbol{v})$ are antisymmetric (skew-symmetric) tensors connected with a small rotation or with a small rate rotation of the considered rigid element and where

$$\begin{aligned} &\omega_{ij}(\boldsymbol{u}) = \tfrac{1}{2}(\partial u_i/\partial x_j - \partial u_j/\partial x_i), \quad \omega_{ij}(\boldsymbol{u}) = -\omega_{ji}(\boldsymbol{u}), \\ &D_{ij}(\boldsymbol{v}) = \tfrac{1}{2}(\partial v_i/\partial x_j - \partial v_j/\partial x_i), \quad D_{ij}(\boldsymbol{v}) = D_{ji}(\boldsymbol{v}), \\ &W_{ij}(\boldsymbol{v}) = \tfrac{1}{2}(\partial v_i/\partial x_j - \partial v_j/\partial x_i), \quad W_{ij}(\boldsymbol{v}) = -W_{ji}(\boldsymbol{v}), \quad i, j = 1, .., N. \end{aligned} \tag{3.171}$$

Similarly as in the stress tensor case we can define the principal strains, principal directions of strains as well as invariants of strain tensor, from which the first one represents the cubical dilatation. The similar relations can be also defined for the strain rate tensor. Moreover, because of the symmetry of D_{ij} and the antisymmetry of W_{ij} then the motion of the considered rigid element can be regarded as the superposition of a uniform translation with velocity $\boldsymbol{v}(\boldsymbol{x}, t)$ and a solid body rotation with angular velocity vector $\tfrac{1}{2}\mathrm{rot}\,\boldsymbol{v}(\boldsymbol{x}, t)$.

Solving the system of partial differential equations

$$\partial u_i/\partial x_j + \partial u_j/\partial x_i = 2e_{ij}(\boldsymbol{u}),$$

for unknown functions u_i, $i = 1, \ldots, N$, where $e_{ij}(\boldsymbol{u}) \in C^2(\Omega)$, then the compatibility equations (the so-called Saint-Venant equations of compatibility)

$$\varepsilon_{\mathrm{rim}}\varepsilon_{\mathrm{sjn}}\partial^2 e_{ij}/\partial x_m \partial x_n = 0$$

must be satisfied. For more details see e.g. Brdička (1959), Nečas and Hlaváček (1981).

3.1.4. Constitutive Laws in the Theory of Elasticity and Visco-Elasticity

3.1.4.1. Constitutive laws of Hookean elasticity

In this section we shall formulate the basic constitutive equations for classical solid mechanics. We begin with physical observations. An important simplifying feature of a large class of problems in elasticity is that displacements are very small. As we saw above this means that the distinction between material and spatial co-ordinates can be ignored to a first approximation. Then the strain tensor $e_{ij}(\boldsymbol{u})$ is defined by

$$e_{ij}(\boldsymbol{u}) = \tfrac{1}{2}(\partial u_i(\boldsymbol{x}, t)/\partial x_j + \partial u_j(\boldsymbol{x}, t)/\partial x_i). \qquad (3.171)$$

The well-known experiment of loading a bar of length l and constant cross-section q gives us the relation between the relative extension $e = \Delta l/l$ and the stress $\tau = P/q$, where P is a loading force (see Figure 3.1). The part of the diagram between the points O and A is linear and practically corresponds to the elastic behaviour of the rocks. The stress τ_A corresponding to the point A is the so-called proportional limit.

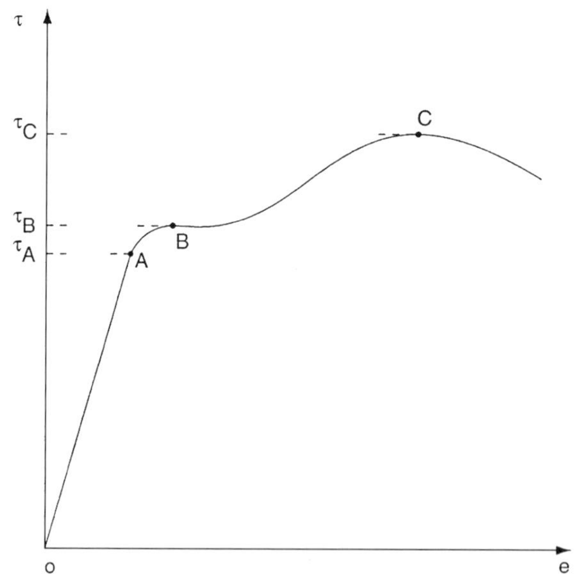

Figure 3.1. The Tension Test Diagram

The part between the points A and B is curved and the extension increases suddenly at the point B without any change in the stress. The value τ_B is the so-called yield point. If the stress further increases, the extension also increases up to the point C. The value τ_c is called the strength limit. Further stretching leads to breaking of the bar.

Now let us stop the loading of the bar and make the loading force decrease to zero. If we come back to the origin along the same curve, then the maximum of the corresponding stresses is called the elastic limit τ_E and it satisfies $\tau_E \geq \tau_A$. The deformation corresponding to stresses $\tau \leq \tau_E$ is called elastic. If the loading force was removed after stress greater than τ_E, part of the extension vanishes and part remains permanently. This part of the deformation is called plastic. The magnitude of the permanent deformation depends on the physical properties of rocks, such as viscosity, temperature, pressure, etc. Elastic deformations are studied in the theories of elasticity and visco-elasticity; the plastic deformations are studied the in theory of plasticity. In the case of linear theory of elasticity, $\tau < \tau_A$ and the relative extension e is proportional to the stress τ as $\tau = Ee$, where E is the so-called Young's modulus of elasticity. The relation $\tau = Ee$ is known as the Hooke's law.

Let us generalize the Hooke's law. Let us assume that at the point $\boldsymbol{x} \in \Omega$ the relations between stress and strain tensors are defined by

$$\tau_{ij} = c_{ijkl}e_{kl}, \quad i, j, k, l = 1, \ldots, N. \tag{3.172}$$

The coefficients c_{ijkl} are symmetric and Lipschitz, so that

$$c_{ijkl} = c_{jikl} = c_{ijlk} = c_{klij}, \quad c_{ijkl}e_{ij}e_{kl} \geq c_o e_{ij}e_{ij}, \quad c_o > 0. \tag{3.173}$$

Definition 3.12: Assume the elastic coefficients $c_{ijkl}(\boldsymbol{x})$ do not depend on $\boldsymbol{x} \in \Omega$, then the rock material is said to be homogeneous; and if they do depend on $\boldsymbol{x} \in \Omega$, then it is non-homogeneous. If the coefficients $c_{ijkl}(\boldsymbol{x})$ do not depend on the choice of co-ordinate system, then the rock material is said to be isotropic at $\boldsymbol{x} \in \Omega$, whilst if the coefficients $c_{ijkl}(\boldsymbol{x})$ do depend on the choice of co-ordinate system, it is anisotropic at $\boldsymbol{x} \in \Omega$. For isotropic rocks

$$c_{ijkl} = \lambda\delta_{ij}\delta_{kl} + \mu(\delta_{ik}\delta_{jl} + \delta_{il}\delta_{jk}), \tag{3.174}$$

where $\lambda = c_{1122}$ and $\mu = c_{1212}$ are the so-called Lamé coefficients.

Note that

$$\tau_{kk} = (3\lambda + 2\mu)e_{kk}, \tag{3.175}$$

then Equation 3.172 and 3.174 yield

$$e_{ij} = (2\mu)^{-1}\tau_{ij} - [2\mu(3\lambda + 2\mu)]^{-1}\lambda\tau_{kk}\delta_{ij}, \tag{3.176}$$

or in general

$$e_{ij} = s_{ijkl}\tau_{kl}, \tag{3.177}$$

where

$$s_{ijkl} = s_{jikl} = s_{ijlk} = s_{klij}, \quad s_{ijkl}\tau_{ij}\tau_{kl} \geq s_o\tau_{ij}\tau_{ij}, s_o > 0, \tag{3.178}$$

represent the inverse Hooke's laws for isotropic and anisotropic rocks, respectively.

In the tension test under consideration, the ratio of axial stress $N = \boldsymbol{P}/q$ to axial strain e_{11} is called Young's modulus E. The ratio of lateral contraction to longitudinal extension is designated by Poisson's ration $\nu = |e_{22}/e_{11}| = |e_{33}/e_{11}|$. One can find that

$$\begin{aligned} E &= \mu(3\lambda + 2\mu)(\lambda + \mu)^{-1}, \quad & \nu &= \tfrac{1}{2}\lambda(\lambda + \mu)^{-1}, \\ \lambda &= E\nu(1 + \nu)^{-1}(1 - 2\nu)^{-1}, \quad & \mu &= \tfrac{1}{2}E(1 + \nu)^{-1}. \end{aligned} \tag{3.179}$$

Now let us assume the homogeneous cube. Suppose that the size of the cube may be neglected in comparison with depth. Then $P_i = \tau_{ij}n_j = -pn_i$, $p = \text{constant} > 0$, $i = 1, \ldots, N$. Since the Cauchy quadric of stress is a sphere, then on every area only the normal stress acts; as $p > 0$ then p is a pressure. Then $\tau_{ij} = -p\delta_{ij}$. Consider the first invariant of the stress and strain tensors. Then $\Theta = \tau_{ii} = -3p$. Then summing the first three of Equations 3.172 and 3.174, i.e. of

$$\tau_{ij}(\boldsymbol{x}) = \lambda(\boldsymbol{x})\delta_{ij}\vartheta(\boldsymbol{x}) + 2\mu(\boldsymbol{x})e_{ij}(\boldsymbol{x}), \quad \text{where } \vartheta(\boldsymbol{x}) = e_{kk}(\boldsymbol{u}(\boldsymbol{x})), \tag{3.180}$$

we find

$$\Theta = \tau_{kk} = (3\lambda + 2\mu)\vartheta.$$

Hence, and using $\Theta = -3p$, we obtain

$$\vartheta = -3p(3\lambda + 2\mu)^{-1} = -p(\lambda + \tfrac{2}{3}\mu) = -p/k,$$

where

$$k = \lambda + \tfrac{2}{3}\mu > 0 \tag{3.181}$$

is the so-called bulk modulus. The sign "minus" denotes that the volume is compressed.

Let us consider a cube loaded by a constant shear $P_i = P > 0$ on the four faces in the direction of the x_1 and x_2 axes only, and for the remaining two faces we put $P_i = 0$. Then

$$\tau_{ij} = P(\delta_{i1}\delta_{j2} + \delta_{i2}\delta_{j1}).$$

The cubical dilatation in such a case is $\vartheta = \Theta(3\lambda + 2\mu)^{-1} = 0$ since $\Theta = 0$. Thus $e_{ij} = (2\mu)^{-1}\tau_{ij}$.

The off-diagonal components of the strain tensor, i.e. components $e_{12}, e_{13}, e_{23}, e_{21}, e_{31}, e_{32}$, are called relative shears, and the limit $2e_{12} \doteq \lim_{t\to 0^+} \alpha_{12}(t) > 0$ represents the angle of shear (similarly for e_{13} and e_{23}). Then

$$2e_{ij} = \mu^{-1}\tau_{ij} = \mu^{-1}P > 0, \quad i \neq j,$$

where we then conclude that $\mu > 0$. Moreover, since $k > 0$ then $\lambda > 0$.

3.1.4.2. *Constitutive laws of linear visco-elasticity*

Visco-elastic properties of rocks characterize rocks with memory in the sense that the state of stress at the instant t depends on all deformations undergone by the rocks in previous time. The rocks are called linear if the constitutive law establishes a linear relation between the stress tensor and the strain tensor.

It is known that the rocks are endowed with a memory. One type of rocks are endowed with a "short memory", while the others are endowed with a "long memory". An important class of visco-elastic constitutive laws concerns rocks for which a linear relationship exists between the derivatives with respect to time of the strain (i.e. of the strain rate) tensor and the stress tensor. For the rocks with short memory the constitutive law is as follows

$$\tau_{ij}(t) = c^{(o)}_{ijkl}(\boldsymbol{x})e_{kl}(\boldsymbol{u}(\boldsymbol{x},t)) + c^{(1)}_{ijkl}(\boldsymbol{x})\frac{\partial}{\partial t}e_{kl}(\boldsymbol{u}(\boldsymbol{x},t)), \tag{3.182}$$

where the derivatives with respect to t are to be understood in the sense of distribution, and tensors τ_{ij} and e_{ij} are equal to zero for $t < 0$. The state of stress at the moment t depends only on the deformation at time t and at the immediately preceeding time. The coefficients $c^{(o)}_{ijkl}(\boldsymbol{x})$ and $c^{(1)}_{ijkl}(\boldsymbol{x})$ have a meaning of coefficients of elasticity and of viscosity, respectively, and satisfy the symmetry and Lipschitz conditions

$$c^{(r)}_{ijkl} = c^{(r)}_{jikl} = c^{(r)}_{ijlk} = c^{(r)}_{klij},\ c^{(r)}_{ijkl}e_{ij}e_{kl} \geq c^{(r)}_{o}e_{ij}e_{ij},\ c^{(r)}_{o} > 0,\ r = 0, 1. \tag{3.183}$$

The coefficients $c^{(1)}_{ijkl}$ are constructed for the field of velocity vectors, hence are coefficients of viscosity. The scalar $c^{(r)}_{ijkl}e_{ij}e_{kl}$ has a meaning of dissipation of energy for $r = 1$ and is positive, which follows from the condition in Equation 3.183 for $r = 1$; and has a meaning of a rate deformation energy for $r = 0$ and is also positive.

For the rocks with long memory, the constitutive law is as follows

$$\tau_{ij}(t) = c_{ijkl}(\boldsymbol{x})e_{kl}(\boldsymbol{u}(\boldsymbol{x},t)) + \int_{t_0}^{t} b_{ijkl}(t-s)e_{kl}(s)\mathrm{d}s, \tag{3.184}$$

where tensors τ_{ij} and e_{ij} are equal to zero for $t < 0$. This constitutive law describes the rocks with long memory for which the state of stress at time t depends on the deformation at time t and also on the deformations at the times preceeding t. The coefficients $c_{ijkl}(\boldsymbol{x})$ are bounded in $\boldsymbol{x}$ and satisfy

$$c_{ijkl} = c_{jikl} = c_{ijlk} = c_{klij}, \quad c_{ijkl}e_{ij}e_{kl} \geq c_o e_{ij}e_{ij}, \quad c_o > 0, \tag{3.185}$$

and the coefficient $b_{ijkl}(t)$, representing the effect of the memory of the rocks, depend on $\boldsymbol{x}$ and t, are bounded in $\boldsymbol{x}$ and t and satisfy

$$b_{ijkl}(\boldsymbol{x},t) = b_{jikl}(\boldsymbol{x},t) \tag{3.186}$$

and the regularity assumption

$$b_{ijkl},\ \frac{\partial}{\partial t}b_{ijkl},\ \frac{\partial^2}{\partial t^2}b_{ijkl} \in L^{\infty}(\Omega \times (t_o, t_1)). \tag{3.187}$$

3.1.5. Constitutive Laws in the Theory of Plasticity

We saw above that for $\tau \geq \tau_E$, where the permanent part of deformation remains, the rocks behave as though plastic. But the notation "plastic rocks" covers rocks with various types of behaviour, such as elasto-visco-plastic, rigid-perfectly-plastic, plastic with work hardening, non-linear behaviour rocks, elastic-inelastic rocks, etc.. In all these types of approximations of rocks there exists a threshold separating two types of behaviour — elastic and plastic. This threshold of plasticity can be fixed, or it can be dependent on the history of the previous deformations. In the second case we speak about rocks undergoing work hardening.

To formulate the constitutive law of plasticity (see Figure 3.2), we see that the segment $\overline{OA}$ characterizes the elastic property of the rocks. Now let us place ourselves at a point P and let e decrease. Then we observe that the representative point (e, τ) describes a line starting from P, parallel to $\overline{OA}$ and crossing the e-axis at a point Q. At the point P, the behaviour of the rocks is then no longer reversible and therefore represents a plastic region as the rest of the deformation remains. Moreover, $\overline{PQ} > \overline{OA}$ as long as the arc $\overline{Az}$ is not a half-line parallel to $\overline{Oe}$: this effect represents the phenomenon of work hardening. In the case when the segment $\overline{Az}$ is parallel to $\overline{Oe}$, we denote such type of rocks as perfectly plastic.

To formulate the constitutive law for such perfectly plastic rocks we see that for every point (e, τ) of the region between the Oe-axis and the graph OAz, a constitutive law can be found in the form e being a function of τ or τ being a function of e. Indeed, if starting from a state (e, τ), $\tau \leq g$, we increase e by a small value, we denote it as de, then τ changes by a small value dτ such that

$$\mathrm{d}e = c\mathrm{d}\tau + \tilde{\lambda}, \tag{3.188}$$

where

$$\tilde{\lambda} = 0 \quad \text{when } \tau < g \text{ or when } \tau = g, \quad \mathrm{d}\tau < 0,$$

$$\tilde{\lambda} \geq 0 \quad \text{when } \tau = g \text{ and } \mathrm{d}\tau = 0.$$

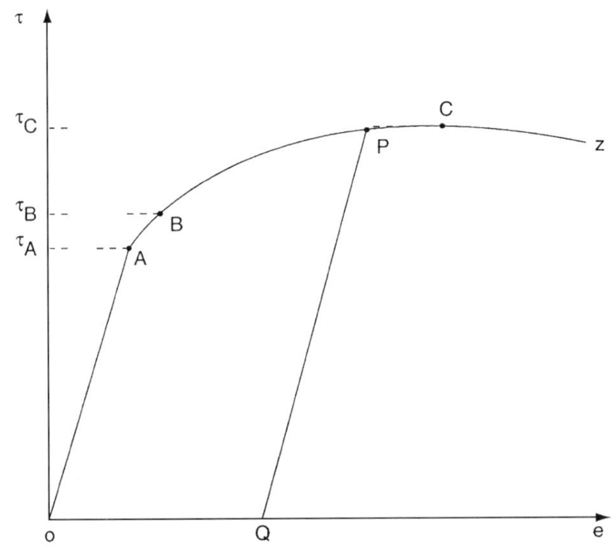

Figure 3.2. Tension Test Diagram. Derivation of the Constitutive Law of Plasticity

The value 1/c denotes the slope of the segment $\overline{OA}$.

If the increase de takes place during the time dt, then Equation 3.188 can be written as follows

$$\dot{e} = c\dot{\tau} + \lambda, \quad \dot{e} = e(\dot{\boldsymbol{u}}) \equiv \boldsymbol{D}, \tag{3.189}$$

where

$$\begin{aligned} \lambda = 0 &\quad \text{when } \tau < g \text{ or when } \tau = g, \dot{\tau} < 0, \\ \lambda \geq 0 &\quad \text{when } \tau = g \text{ and } \dot{\tau} = 0, \end{aligned} \tag{3.190}$$

and where $\dot{e} \equiv \boldsymbol{D}$ is the strain rate and $\dot{\tau}$ is the stress rate. The relationships in Equation 3.190 can be rewritten as

$$\begin{aligned} \lambda(\tau - \sigma) \leq 0 \quad \forall \tau \leq g, \\ \lambda \dot{\tau} = 0. \end{aligned} \tag{3.191}$$

If $c \to 0$, then $\dot{e} = \lambda$ and we speak about the constitutive law for rigid perfectly plastic rocks.

The constitutive law in Equation 3.188 can be generalized to N-dim ($N = 2, 3$) perfectly plastic rocks. We shall assume that the elastic region is defined by

$$\mathscr{F}(\tau_{ij}) < 0, \tag{3.192}$$

and the plastic region by

$$\mathscr{F}(\tau_{ij}) = 0, \tag{3.193}$$

where τ_{ij} are components of the stress tensor and satisfy $\tau_{ij} = \tau_{ji}$. The function $\mathscr{F}$ is convex and continuous with respect to τ_{ij}, so that the domain of R^6, where $\mathscr{F}(\tau_{ij}) \leq 0$, is a closed convex set. The function $\mathscr{F}$ can be defined by using the deviator τ_{ij}^D of the tensor τ_{ij} as:

(i) the Von Mises model:

$$\mathscr{F}(\tau_{ij}) = \tfrac{1}{2}\tau_{ij}^D \tau_{ij}^D - k^2, \tag{3.194}$$

where k is a given constant,

(ii) the Tresca's model:

$$\mathscr{F}(\tau_{ij}) = \sup_{I,J;I \neq J} |\tau_I - \tau_J| - g, \tag{3.195}$$

where the τ_I are the eigenvalues of the stress tensor τ_{ij} and where g is a positive constant.

Since in both cases given by Equations 3.194 and 3.195 the function $\mathscr{F}$ is independent of the deformation, then both cases describe perfectly plastic rocks.

The constitutive law of Equation 3.188 can be generalized as

$$\begin{aligned} e_{ij}(\dot{\boldsymbol{u}}) &= s_{ijkl}\dot{\tau}_{kl} + \lambda_{ij}, \\ \lambda_{ij}(\tau_{ij} - \sigma_{ij}) &\leq 0 \quad \forall \tau_{ij} \text{ with } \mathscr{F}(\tau_{ij}) \leq 0, \\ \lambda_{ij}\dot{\tau}_{ij} &= 0, \end{aligned} \tag{3.196}$$

where the coefficients S_{ijkl} are the coefficients of elasticity and satisfy

$$s_{ijkl} = s_{jikl} = s_{ijlk} = s_{klij}, \quad s_{ijkl}\tau_{ij}\tau_{kl} \geq s_o\tau_{ij}\tau_{ij}, \quad s_o > 0. \tag{3.197}$$

The quantity $s_{ijkl}\tau_{kl}$ represents the elastic part of the strain rate tensor $\dot{e}_{ij}(\boldsymbol{u}) \equiv e_{ij}(\dot{\boldsymbol{u}})$, and λ_{ij} represents the velocity of plastic deformation.

Remark: The constitutive law for an elastic perfectly plastic material is also known as the Prandtl-Reuss constitutive law.

The constitutive law for elastic perfectly plastic rocks can be generalized for elasto-visco-plastic rocks. Then we will describe a model of a law of elasto-visco-plastic behaviour of rocks, where the rock will behave elastically below the threshold of plasticity and visco-plastically beyond.

For this reason we introduce the closed convex set

$$K = \{\tau \mid \tau = (\tau_{ij}) \in R^6, \quad \tau_{ij} = \tau_{ji}, \quad \mathscr{F}(\tau_{ij}) \leq 0\},$$

where $\mathscr{F}$ is a convex continuous function of the six scalar variables τ_{ij} and the mapping $\tau \rightarrow P_K(\tau)$, which is an orthogonal projection, is the usual Euclidean structure of $\tau \in R^6$ on the convex set K. Then the constitutive law can be written as

$$\begin{aligned} e_{ij}(\dot{\boldsymbol{u}}) &= s_{ijkl}\dot{\tau}_{kl} + \lambda_{ij}, \\ \{\lambda_{ij}\} &= 0 \text{ if } \mathscr{F}(\tau_{ij}) < 0, \\ \lambda_{ij} &= (2\hat{\mu})^{-1}(\tau_{ij} - (P_K\tau)_{ij}) \text{ if } \mathscr{F}(\tau_{ij}) \geq 0, \end{aligned} \tag{3.198}$$

where $\hat{\mu}$ is a positive scalar, having the meaning of a viscosity coefficient.

The special case when

$$\begin{aligned} & e_{ij}(\boldsymbol{u}) = s_{ijkl}\tau_{kl} + \lambda_{ij}, \\ \text{with} \quad & \mathscr{F}(\tau_{ij}) \leq 0, \text{ and} \\ & \lambda_{ij} = (\sigma_{ij} - \tau_{ij}) \leq 0 \quad \forall\sigma_{ij}, \quad \mathscr{F}(\sigma_{ij}) \leq 0, \end{aligned} \tag{3.199}$$

is known as the Hencky's law of plasticity.

A special case of the non-linear Hooke's law is the stress-strain relation for elasto-plastic rocks in the deformation theory of plasticity. To derive it we shall assume that the Lamé coefficients depend on the first two invariants ϑ_1 and ϑ_2 of the small strain tensor. Moreover, we shall also assume that the relation between the first invariants of stress and strain is linear. Then

$$\tau_{ij} = [k(\boldsymbol{x}) - \tfrac{2}{3}\mu(\boldsymbol{x}, \gamma)]\vartheta_1\delta_{ij} + 2\mu(\boldsymbol{x}, \gamma)e_{ij},$$

where $k(\boldsymbol{x})$ is the bulk modulus and γ is defined by

$$\begin{aligned} \gamma &= \{\tfrac{1}{3}[(e_{11} - e_{22})^2 + (e_{22} - e_{33})^2 + (e_{11} - e_{33})^2 + 6(e_{12}^2 + e_{23}^2 + e_{13}^2)]\}^{1/2} \\ &= (-\tfrac{1}{3}\vartheta_1^2 + e_{ij}e_{ij}))^{1/2} \\ &= (\tfrac{2}{9}(\vartheta_1^2 - 3\vartheta_2))^{1/2}. \end{aligned}$$

Putting $S^2 = \tau_{ij}^D \tau_{ij}^D$, where τ_{ij}^D is the stress deviator, then

$$S = 2\mu\gamma$$

is the so-called Hencky's relation.

3.1.6. Basic Boundary Value Problems of the Linear Theory of Elasticity

Let the domain Ω be occupied by the deformed geological body, and let $\partial\Omega$ be its Lipschitz boundary. Let $\partial\Omega = \Gamma_u \cup \Gamma_\tau \cup R$, where Γ_u represents the part of the boundary $\partial\Omega$ where a displacement is prescribed, Γ_τ represents the part of the boundary $\partial\Omega$ where a loading is prescribed and R is the set of zero surface measure. Let us assume that the generalized Hooke's law with $c_{ijkl} \in C^1(\overline{\Omega})$ holds. Let the displacement $\boldsymbol{u} \in [C^2(\Omega)]^N$ and let the equilibrium equation

$$\partial\tau_{ij}/\partial x_j + f_i = 0, \tag{3.200}$$

with $\boldsymbol{f}(\boldsymbol{x}) \in [C^1(\overline{\Omega})]^N$ in Ω, be satisfied.

If to agree with $\Gamma_u = 0$ then we speak about the first basic boundary value problem of elasticity, if $\Gamma_\tau = 0$ then we speak about the second basic boundary value problem of elasticity, and if $\Gamma_\tau \neq 0$, $\Gamma_u \neq 0$ then we speak about the mixed boundary value problem of elasticity.

First Basic Boundary Value Problem: We seek the displacement $\boldsymbol{u} \in [C^1(\overline{\Omega}) \cap C^2(\Omega)]^N$, which satisfies the equilibrium equation (Equation 3.200) and

$$\tau_{ij}(\boldsymbol{x}) n_j(\boldsymbol{x}) = P_i(\boldsymbol{x}), \quad \boldsymbol{x} \in \partial\Omega, \tag{3.201}$$

where $\boldsymbol{P}(\boldsymbol{x}) \in [C(\partial\Omega)]^N$ is prescribed on the boundary $\partial\Omega$ and where $\boldsymbol{n}$ denotes the outer normal to $\partial\Omega$.

Second Basic Boundary Value Problem: We seek the displacement $\boldsymbol{u} \in [C(\overline{\Omega}) \cap C^2(\Omega)]^N$, which satisfies the equilibrium equation (Equation 3.200) and

$$\boldsymbol{u}(\boldsymbol{x}) = \boldsymbol{u}_o(\boldsymbol{x}), \quad \boldsymbol{x} \in \partial\Omega, \tag{3.202}$$

where $\boldsymbol{u}_o(\boldsymbol{x}) \in [C(\partial\Omega)]^N$ is prescribed on the boundary $\partial\Omega$.

Mixed Basic Boundary Value Problem: We seek the displacement $\boldsymbol{u} \in [C^1(\Omega) \cup \Gamma_\tau]^N \cap [C(\Omega \cup \Gamma_u)]^N \cap [C^2(\Omega)]^N$, which satisfies the equilibrium equation (Equation 3.200) and the boundary conditions in Equations 3.201 and 3.202, where $\boldsymbol{P}(\boldsymbol{x}) \in [C(\Gamma_\tau)]^N$, $\boldsymbol{u}_o \in [C(\Gamma_u)]^N$ are prescribed on the boundary Γ_τ or on the boundary Γ_u, respectively.

3.1.7. Variational Principles in the Theory of Elasticity

Let us consider the domain Ω being occupied by the deformed geological body, and let $\partial\Omega$ be its Lipschitz boundary. Let $\partial\Omega = \Gamma_u \cup \Gamma_\tau \cup R$, where Γ_u represents the part of the boundary $\partial\Omega$ where a displacement is prescribed, Γ_τ represents the part of the boundary $\partial\Omega$, where a loading is prescribed and R is the set of zero surface measure. We shall assume that all the functions of displacement and stress are to be smooth enough.

Next we shall define the basic variational principles of the deformation theory of elasticity.

Definition 3.13: A statically admissible stress field is a tensor function $\tau(\boldsymbol{x})$, being symmetric, i.e. $\tau_{ij} = \tau_{ji}$ and its components τ_{ij} satisfy the equations of equilibrium

$$\frac{\partial \tau_{ij}}{\partial x_j} + f_i = 0 \quad \text{in } \Omega \tag{3.203}$$

and the boundary conditions

$$\tau_{ij} n_j = P_i \quad \text{on } \Gamma_\tau. \tag{3.204}$$

A geometrically admissible displacement field is a vector function $\boldsymbol{u}(\boldsymbol{x})$ satisfying the boundary conditions

$$\boldsymbol{u} = \boldsymbol{u}_o \quad \text{on } \Gamma_u. \tag{3.205}$$

Theorem 3.22 (Principle of Virtual Work)**:** Let $\tau(\boldsymbol{x})$ be a statically admissible stress field, and let $\boldsymbol{u}(\boldsymbol{x})$ be a geometrically admissible displacement field. Let $e_{ij}(\boldsymbol{u}) = \frac{1}{2}(\partial u_i/\partial x_j + \partial u_j/\partial x_i)$ be a small strain tensor. Then we find

$$\int_\Omega \tau_{ij} e_{ij}(\boldsymbol{u}) \mathrm{d}\boldsymbol{x} = \int_\Omega f_i u_i \mathrm{d}\boldsymbol{x} + \int_{\Gamma_\tau} P_i u_i \mathrm{d}s + \int_{\Gamma_u} \tau_{ij} n_j u_{oi} \mathrm{d}s. \tag{3.206}$$

Proof: The proof follows from the equations of equilibrium (Equation 3.203), the boundary conditions of Equations 3.204 and 3.205 and the Green's theorem.

Theorem 3.23 (Principle of Virtual Displacements)**:** Let $\underline{\boldsymbol{u}}_o$ and $\underline{\boldsymbol{u}}_o + \delta\boldsymbol{u}$ be geometrically admissible displacement fields. Let the actual stress field τ_o be statically admissible. Then

$$\int_\Omega \tau_{oij} \delta e_{ij} \mathrm{d}\boldsymbol{x} = \int_\Omega f_i \delta u_i \mathrm{d}\boldsymbol{x} + \int_{\Gamma_\tau} P_i \delta u_i \mathrm{d}s, \tag{3.207}$$

where $\delta e_{ij} = e_{ij}(\delta\boldsymbol{u})$.

Proof: We substitute τ_o into the principle of virtual work. Furthermore, we substitute the actual displacement fields $\underline{\boldsymbol{u}}_o$ and $\underline{\boldsymbol{u}}_o + \delta\boldsymbol{u}$. Then, subtracting the resultant equalities we find Equation 3.207.

Remark: The quantity $\delta\boldsymbol{u}$ is called the "virtual displacement" or "variation" of $\boldsymbol{u}$. Due to Definition 3.13 we see that $\delta u_i = 0$ on Γ_u. Generally, the principle of displacements holds for any type of stress-strain relations (Hooke's laws).

Theorem 3.24 (Principle of Virtual Stresses)**:** Let $\underline{\boldsymbol{u}}_o$ be the actual displacement field, and let τ_o be the actual stress field and $\tau_o + \delta\tau$ be an arbitrary stress field, which are both statically admissible stress fields. Then

$$\int_\Omega e_{ij}(\underline{\boldsymbol{u}}_o)\delta\tau_{ij}\mathrm{d}\boldsymbol{x} = \int_{\Gamma_u} \underline{u}_{oi}\delta\tau_{ij}n_j\mathrm{d}s. \tag{3.208}$$

Proof: We substitute the actual displacement field $\underline{\boldsymbol{u}}_o$ into the principle of virtual work. Furthermore, we substitute the actual stress field τ_o on the one hand and, on the other hand, an arbitrary field $\tau_o + \delta\tau$. Subtracting the resultant equalities, we obtain Equation 3.208.

Remark: The $\delta\tau_{ij}$ are components of the virtual stress or variation of stress. Due to Definition 3.13, $\delta\tau_{ij} = \delta\tau_{ji}$ and $\partial\delta\tau_{ij}/\partial x_j = 0$, $\delta\tau_{ij}n_j = 0$ on Γ_τ. Then Equation 3.208 is valid for an arbitrary Hooke's law.

Theorem 3.25 (Principle of Minimum Potential Energy): Let the generalized Hooke's law

$$\tau_{ij} = c_{ijkl}e_{kl}(\boldsymbol{u}) \tag{3.209}$$

be given. Let

$$c_{ijkl} = c_{jikl} = c_{ijlk} = c_{klij},$$

$$2W(e) = c_{ijkl}e_{ij}e_{kl} \geq c_o e_{ij}e_{ij},\ c_o = \text{ constant} > 0 \tag{3.210}$$

be valid for all symmetric tensors $e_{ij}(\boldsymbol{u})$ and almost every $\boldsymbol{x} \in \Omega$. Let τ_o be statically admissible. Then the solution $\underline{\boldsymbol{u}}_o$ of the mixed problem of Equation 3.203–3.205 gives the functional of potential energy

$$J(\boldsymbol{u}) = \tfrac{1}{2}\int_\Omega c_{ijkl}(\boldsymbol{x})e_{ij}(\boldsymbol{u})e_{kl}(\boldsymbol{u})\mathrm{d}\boldsymbol{x} - \int_\Omega f_i u_i \mathrm{d}\boldsymbol{x} - \int_{\Gamma_\tau} P_i u_i \mathrm{d}s \tag{3.211}$$

its minimum value over the set of geometrically admissible fields. Moreover, from the principle of minimum potential energy the principle of virtual displacements follows immediately.

Proof: Substitute the Hooke's law (Equation 3.209) into the principle of virtual displacements, then we obtain

$$\int_\Omega c_{ijkl}e_{kl}(\underline{\boldsymbol{u}}_o)\delta e_{ij}\mathrm{d}\boldsymbol{x} - \int_\Omega f_i\delta u_i\mathrm{d}\boldsymbol{x} - \int_{\Gamma_\tau} P_i\delta u_i\mathrm{d}s = 0. \tag{3.212}$$

But

$$\delta W(\boldsymbol{e}(\underline{\boldsymbol{u}}_o)) = \frac{\mathrm{d}}{\mathrm{d}t}W[\boldsymbol{e}(\underline{\boldsymbol{u}}_o + t\delta\boldsymbol{u})]\bigg|_{t=0} = c_{ijkl}e_{kl}(\underline{\boldsymbol{u}}_o)e_{ij}(\delta\boldsymbol{u}).$$

Hence, Equation 3.212 yields

$$\delta\left\{\int_\Omega W(\boldsymbol{e}(\underline{\boldsymbol{u}}_o))\mathrm{d}\boldsymbol{x} = \int_\Omega f_i\underline{u}_{oi}\mathrm{d}\boldsymbol{x} - \int_{\Gamma_\tau} P_i\underline{u}_{oi}\mathrm{d}s\right\} \equiv \delta J(\boldsymbol{u}_o) = 0. \tag{3.213}$$

Then $\underline{\boldsymbol{u}}_o$ is a critical (minimum, stationary) point of the functional $J(\boldsymbol{u})$. For a geometrically admissible displacement field $\underline{\boldsymbol{u}}_o + \delta\boldsymbol{u}$, we find

$$\begin{aligned} J(\underline{\boldsymbol{u}}_o + \delta\boldsymbol{u}) - J(\underline{\boldsymbol{u}}_o) &= \int_\Omega [W(\boldsymbol{e}(\underline{\boldsymbol{u}}_o + \delta\boldsymbol{u})) - W(\boldsymbol{e}(\underline{\boldsymbol{u}}_o))]\mathrm{d}\boldsymbol{x} - \int_\Omega f_i\delta u_i d\boldsymbol{x} - \int_{\Gamma_\tau} P_i\delta u_i\mathrm{d}s \\ &= \int_\Omega c_{ijkl}e_{ij}(\underline{\boldsymbol{u}}_o)\delta e_{kl}\mathrm{d}\boldsymbol{x} + \tfrac{1}{2}\int_\Omega c_{ijkl}\delta e_{ij}\delta e_{kl}\mathrm{d}\boldsymbol{x} - \int_\Omega f_i\delta u_i\mathrm{d}\boldsymbol{x} - \int_{\Gamma_\tau} P_i\delta u_i\mathrm{d}s \\ &= \tfrac{1}{2}\int_\Omega c_{ijkl}\delta e_{ij}\delta e_{kl}\mathrm{d}\boldsymbol{x} \geq \tfrac{1}{2}c_o\int_\Omega \delta e_{ij}\delta e_{ij}\mathrm{d}\boldsymbol{x} \geq 0. \end{aligned} \tag{3.214}$$

From which the existence of a minimum of $J(\boldsymbol{u})$ follows.

The last assertion follows from the assumption that the Hooke's law (Equation 3.209) holds and that $\delta J(\underline{u}_o) = 0$, i.e. that

$$\delta\{\int_\Omega W(e(\underline{u}_o))\mathrm{d}x - \int_\Omega f_i \underline{u}_{oi}\mathrm{d}x - \int_{\Gamma_\tau} P_i \underline{u}_{oi}\mathrm{d}s\} = 0.$$

Hence,

$$\int_\Omega \tau_{oij}\delta e_{ij}\mathrm{d}x - \int_\Omega f_i \underline{u}_{oi}\mathrm{d}x - \int_{\Gamma_\tau} P_i \underline{u}_{oi}\mathrm{d}s = 0,$$

which finishes the proof.

Theorem 3.26 (Principle of Minimum Complementary Energy): Let τ_o the actual stress field of the mixed problem of Equations 3.203-3.205, be statically admissible. Then the functional of complementary energy

$$S(\tau) = \tfrac{1}{2}\int_\Omega s_{ijkl}\tau_{ij}\tau_{kl}\mathrm{d}x - \int_{\Gamma_u} u_{oi}\tau_{ij}n_j\mathrm{d}s \tag{3.215}$$

makes a minimum over the set of all statically admissible stress fields.

Proof: Firstly, we shall derive the functional of the complementary energy. We shall start with the principle of minimum potential energy (Equation 3.211). Let us regard the strain-displacement relation, i.e. $e_{ij} = \frac{1}{2}(\partial u_i/\partial x_j + \partial u_j/\partial x_i)$, and the boundary condition, Equation 3.205, as subsidiary conditions. Then the problem leads to our finding a constrained extremum. Such a problem is solved by using the Lagrange multipliers. Let us introduce the functional

$$\begin{aligned} H(e, u, \lambda, \mu) = &\int_\Omega \{\tfrac{1}{2}c_{ijkl}e_{ij}e_{kl} + \lambda_{ij}[\tfrac{1}{2}(\partial u_i/\partial x_j + \partial u_j/\partial x_i) - e_{ij}]\}\mathrm{d}x \\ &- \int_\Omega f_i u_i \mathrm{d}x - \int_{\Gamma_\tau} P_i u_i \mathrm{d}s + \int_{\Gamma_u} \mu_i(u_{oi} - u_i)\mathrm{d}s, \end{aligned} \tag{3.216}$$

where $\lambda_{ij}(x) = \lambda_{ji}(x)$ and μ_{ij} are Lagrange multipliers.

Similarly as in the proof of Theorem 3.25, we seek the stationary value of the functional H without constraints. Therefore, we apply the variation and integration by parts, and then

$$\begin{aligned} \delta H = &\int_\Omega \{(c_{ijkl}e_{kl} - \lambda_{ij})\delta e_{ij} - (\partial\lambda_{ik}/\partial x_k + f_i)\delta u_i + [\tfrac{1}{2}(\partial u_i/\partial x_j + \partial u_j/\partial x_i) - e_{ij}]\delta\lambda_{ij}\}\mathrm{d}x \\ &+ \int_{\Gamma_u} \{(\lambda_{ij}n_j - \mu_i)\delta u_i + (u_{oi} - u_i)\delta\mu_i\}\mathrm{d}s - \int_{\Gamma_\tau} (P_i - \lambda_{ij}n_j)\delta u_i \mathrm{d}s = 0. \end{aligned}$$

Hence,

$$\lambda_{ij} = c_{ijkl}e_{kl} \quad \text{in } \Omega, \tag{3.217}$$

$$\frac{\partial\lambda_{ik}}{\partial x_k} + f_i = 0 \quad \text{in } \Omega, \tag{3.218}$$

$$\mu_i = \lambda_{ij}n_j \quad \text{on } \Gamma_u, \tag{3.219}$$

$$\lambda_{ij}n_j = P_i \quad \text{on } \Gamma_\tau. \tag{3.220}$$

Hence, we see that the function λ_{ij} has the meaning of a stress tensor, and the μ_i have the meaning of surface force components. Equation 3.218 has the meaning of the equation of equilibrium of the elastic body, Equation 3.220 has the meaning of surface loading, while Equation 3.217 has the meaning of Hooke's law. From Equation 3.217 we find

$$e_{ij} = s_{ijkl}\lambda_{kl}, \tag{3.221}$$

where

$$s_{ijkl} = s_{jikl} = s_{ijlk} = s_{klij}. \tag{3.222}$$

The corresponding quadratic form in τ_{ij} is uniformly positive definite in Ω.

Since by integrating by parts we obtain

$$\tfrac{1}{2}\int_{\Omega} \lambda_{ij}(\partial u_i/\partial x_j + \partial u_j/\partial x_i)\mathrm{d}\boldsymbol{x} = -\int_{\Omega} \partial\lambda_{ij}/\partial x_j u_i \mathrm{d}\boldsymbol{x} + \int_{\partial\Omega} \lambda_{ij} n_j u_i \mathrm{d}s,$$

using Equations 3.218-3.220, then Equation 3.216 yields

$$S_1(\tau) = -\int_{\Omega} \tfrac{1}{2} s_{ijkl}\tau_{ij}\tau_{kl}\mathrm{d}\boldsymbol{x} + \int_{\Gamma_u} u_{oi}\tau_{ij} n_j \mathrm{d}s,$$

where instead of λ_{ij} we write τ_{ij}.

Due to Lemma 3.1 of Nečas and Hlaváček (1981), if $J(\underline{\boldsymbol{u}}_o) = \min J(\boldsymbol{u})$ on the set of geometrically admissible displacements holds for some $\underline{\boldsymbol{u}}_o$, then $S_1(\tau_o) = \max S_1(\tau)$ is valid for some τ_o on the set of statically admissible stress, so that $S_1(\tau_o) = J(\underline{\boldsymbol{u}}_o)$.

Remark: For a statically admissible stress field, the principle of minimum complementary energy, also known as the Castigliano-Menabrea principle, can also be derived from the principle of virtual stresses by using the inverse Hooke's law.

In practice as well as in the theory of elasticity the hybrid principle, the so-called Hellinger-Reissner principle and the Hu Hai Chand and Washizu principle are often used. (For more details see e.g. Washizu (1955), (1968), Nečas and Hlaváček (1981).) All these above mentioned principles are the initial principles for the numerical solution of problems in the structural geology. But in the next section we mainly limited ourselves to the principle of minimum potential energy only.

3.1.8. The Space of Functions with Finite Energy and Some Theorems

Let Ω be a domain with a Lipschitz boundary $\partial\Omega$. Let $\boldsymbol{u}$ denote the displacement vector, and e_{ij} the small strain tensor. We shall define the so-called strain energy, which occurs in the functional of potential energy $J(\boldsymbol{u})$ defined in the previous section, by the integral $\frac{1}{2}\int_{\Omega} c_{ijkl} e_{ij}(\boldsymbol{u}) e_{kl}(\boldsymbol{u})\mathrm{d}\boldsymbol{x}$. Let $\boldsymbol{u} \in [H^1(\Omega)]^N$, $N = 2, 3$, then the strain energy is finite since the coefficients of elasticity c_{ijkl} are assumed to be bounded and measurable functions. The inner (scalar) product will be defined in the usual way, i.e. $(\boldsymbol{u}, \boldsymbol{v}) = \int_{\Omega} \Sigma_{|\alpha| \le k} D^{\alpha}\boldsymbol{u} D^{\alpha}\boldsymbol{v}\,\mathrm{d}\boldsymbol{x}$. This inner product induces the norm $\|\boldsymbol{u}\| = (\boldsymbol{u}, \boldsymbol{u})^{1/2}$. Moreover, we shall write $\mathrm{W}^o = [H_o^1(\Omega)]^N (\equiv [W_o^{1,2}(\Omega)]^N)$.

Next we shall introduce some definitions and theorems.

Theorem 3.26: For $g \in \mathrm{W}^o = H_o^1(\Omega)$ the norms $\|g\|_{1,2}$ and the norm $\|g\|'$, defined by $\|g\|' \equiv (\int_{\Omega} \Sigma_{i=1}^N (\partial g/\partial x_i)^2 \mathrm{d}\boldsymbol{x})^{1/2}$, are equivalent. (For the proof see Nečas and Hlaváček (1981), p. 74.)

Later we shall use the well-known Korn's inequality. Firstly we introduce the following theorem:

Theorem 3.27: Let Ω be a domain in R^N, $N = 2, 3$, and let $\boldsymbol{u} \in [H_o^1(\Omega)]^N$. Then

$$\int_\Omega e_{ij}(\boldsymbol{u})e_{ij}(\boldsymbol{u})\mathrm{d}\boldsymbol{x} \geq \tfrac{1}{2}\int_\Omega \frac{\partial u_i \partial u_i}{\partial x_j \partial x_j}\mathrm{d}\boldsymbol{x}. \tag{3.223}$$

(For the proof see Nečas and Hlaváček (1981), p. 77.)

During the deformation of the geological body a displacement $\boldsymbol{u}$ consists of rigid-body translation and rotation. Intuitively in a non-deformed geological body a displacement $\boldsymbol{u}$ also consists of rigid-body translation and rotation. In the non-deformed state $e_{ij} = 0$.

Theorem 3.28: Let Ω be a domain in $R^N, N = 2, 3$ and let $\boldsymbol{u} \in [H^1(\Omega)]^N$. Then

$$e_{ij}(\boldsymbol{u}) = 0, \quad i, j = 1, \ldots, N \Rightarrow \boldsymbol{u} = \boldsymbol{a} + \boldsymbol{b} \times \boldsymbol{x}. \tag{3.224}$$

(For the proof see Nečas and Hlaváček (1981), p.78.)

Let us define the set

$$P \equiv \{\boldsymbol{v} \mid \boldsymbol{v} \in [H^1(\Omega)]^N; \quad \boldsymbol{v} = \boldsymbol{a} + \boldsymbol{b} \times \boldsymbol{x}\}, \tag{3.225}$$

which is a finite-dimensional linear subset of $W = [H^1(\Omega)]^N$, $N = 2, 3$ and therefore is closed in W.

Definition 3.14: Let V be a closed subspace of W such that $W^o \subset V \subset W$. Let $P_V = P \cap V$ and let Q_V be the orthogonal complement of P_V in V, i.e. $V = P_V \oplus Q_V$. Then the inequality

$$\int_\Omega e_{ij}(\boldsymbol{u})e_{ij}(\boldsymbol{u})\mathrm{d}\boldsymbol{x} \geq \|\boldsymbol{v}\|_{1,2}^2 \qquad \text{for any } \boldsymbol{v} \in Q_V, \tag{3.226}$$

is the so-called Korn's inequality.

Definition 3.15: We say that strains are coercive if

$$\int_\Omega (e_{ij}(\boldsymbol{v})e_{ij}(\boldsymbol{v}) + v_i v_i)\mathrm{d}\boldsymbol{x} \geq c\,\|\boldsymbol{v}\|_{1,2}^2 \tag{3.227}$$

for any $\boldsymbol{v} \in W = [H^1(\Omega)]^N, N = 2, 3$, is valid.

Theorem 3.29: Let $\Omega \subset R^N$, $N = 2, 3$, be a domain with a Lipschitz boundary $\partial\Omega$. Then the coerciveness of strains, given by Equation 3.227, holds. (For the proof see Nečas and Hlaváček (1981), p. 84.)

Theorem 3.30: Let $\Omega \subset R^N$, $N = 2, 3$, be a domain with a Lipschitz boundary $\partial\Omega$. Let V be a closed subspace of W such that $W^o \subset V \subset W$. Let $P_V = P \cap V$ and let Q_V be the orthogonal complement of P_V in V, i.e. $V = P_V \oplus Q_V$. Then the Korn's inequality

$$\int_\Omega e_{ij}(\boldsymbol{u})e_{ij}(\boldsymbol{u})\mathrm{d}\boldsymbol{x} \geq \|\boldsymbol{v}\|_{1,2}^2 \qquad \text{for any } \boldsymbol{v} \in Q_V \tag{3.228}$$

holds.
(For the proof see Nečas and Hlaváček (1981), p. 85.)

3.2. DEFORMATION OF GEOLOGICAL BLOCKS AND THEIR LOCAL REGIONS: APPLICATIONS IN PRACTICE

3.2.1. Simple Models of Deformation of Geological Blocks based on Boundary Value Problems of Linear Elasticity

In this section we shall formulate the model problem of the deformation of an elastic geological body which can be described by the mixed boundary value problem of linear elasticity, as such types of problems are typical in structural geology.

Let Ω be a domain occupied by an N-dimensional geological body, $(N = 2, 3)$ and let $\partial\Omega$ be its Lipschitz boundary. Let the boundary $\partial\Omega$ be divided into disjoint parts Γ_τ, Γ_u and $\mathscr{R}$ such that $\partial\Omega = \Gamma_\tau \cup \Gamma_u \cup \mathscr{R}$, where Γ_τ and Γ_u are open in $\partial\Omega$ and the surface measure of $\mathscr{R}$ is zero. Let the elastic coefficients c_{ijkl} are bounded and measurable so that $c_{ijkl} \in L^\infty(\Omega)$. Then our problem will be defined as follows:

Problem ($\mathscr{P}$): Let Γ_τ and Γ_u be two disjoint sets open in $\partial\Omega$, let $\mathscr{R}$ be the set of zero surface measure, and $\partial\Omega = \Gamma_\tau \cup \Gamma_u \cup \mathscr{R}$. Find a displacement vector $\boldsymbol{u}$ satisfying

$$\partial/\partial x_i(c_{ijkl}(\boldsymbol{x})e_{kl}(\boldsymbol{u})) + f_i = 0 \quad \text{in } \Omega \tag{3.229}$$

and boundary conditions

$$\tau_{ij}(\boldsymbol{x})n_j(\boldsymbol{x}) = P_i(\boldsymbol{x}), \quad \boldsymbol{x} \in \Gamma_\tau, \tag{3.230}$$

$$\boldsymbol{u}(\boldsymbol{x}) = \boldsymbol{u}_o(\boldsymbol{x}), \quad \boldsymbol{x} \in \Gamma_u, \tag{3.231}$$

where $\boldsymbol{f}(\boldsymbol{x}) = (f_i(\boldsymbol{x})), \boldsymbol{f} \in [L^2(\Omega)]^N$, $\boldsymbol{P}(\boldsymbol{x}) = (P_i(\boldsymbol{x})), \boldsymbol{P} \in [L^2(\Omega)]^N$, $\boldsymbol{u}_o(\mathbf{x}) = (u_{oi}(\boldsymbol{x}))$, $i = 1, \dots, N$, $\boldsymbol{u}_o \in W$, determine the displacement $\boldsymbol{u}$ of Γ_u in the sense of traces.

For the isotropic case the coefficients of elasticity are defined by $c_{ijkl} = \lambda\delta_{ij}\delta_{kl} + \mu(\delta_{ik}\delta_{jl} + \delta_{il}\delta_{jk})$, where λ and μ are the well-known Lamé coefficients.

If $\Gamma_u \equiv 0$ then the problem leads to the first basic boundary value problem of elasticity, and for $\Gamma_\tau \equiv 0$ to the second basic boundary value problem of elasticity.

Let us define the space $V = [\bar{v}]^N$, $N = 2, 3$, i.e. as the closure of the linear set v in the topology of the Sobolev space $H^1(\Omega)$, where the linear set v is defined by

$$\boldsymbol{v} = \{v \mid v \in C^1(\Omega), \ \text{supp } v \subset \Omega \cup \Gamma_\tau\}$$

Definition 3.16: We say that a function $\boldsymbol{u} \in W$ is the weak (or generalized) solution of the problem ($\mathscr{P}$) if

$$\boldsymbol{u} - \boldsymbol{u}_o \in V, \int_\Omega c_{ijkl}(\boldsymbol{x})e_{ij}(\boldsymbol{u})e_{kl}(\boldsymbol{v})\mathrm{d}\boldsymbol{x} = \int_\Omega f_i v_i \mathrm{d}\boldsymbol{x} + \int_{\Gamma_\tau} P_i v_i \mathrm{d}s \quad \forall \boldsymbol{v} \in \boldsymbol{V}. \tag{3.232}$$

Now we shall show that the problem ($\mathscr{P}$) is well-posed. By well-posedness we mean the existence, uniqueness, and continuous dependence of the solution on the given data $(\boldsymbol{f}, \boldsymbol{P}, \boldsymbol{u}_o)$, i.e. if we change only a little the given data of body forces $\boldsymbol{f}$ in Ω, of surface forces $\boldsymbol{P}$ on Γ_τ and of the given displacement $\boldsymbol{u}_o$ on Γ_u, then the solution of our problem ($\mathscr{P}$) will change also a little. In the opposite case we speak about ill-posedness.

Theorem 3.31: Let $\boldsymbol{u}_o \in W$, $\boldsymbol{f} \in [L^2(\Omega)]^N$, $\boldsymbol{P} \in [L^2(\Gamma_\tau)]^N$ and let $\Gamma_u \neq 0$. Let there exist $c_o > 0$ such that $c_{ijkl}(\boldsymbol{x}) \geq c_o$ for every $\boldsymbol{x} \in \overline{\Omega}$ (or for almost every $\boldsymbol{x} \in \Omega$). Then there exists one and only one weak solution $\boldsymbol{u}$ of the problem $(\mathscr{P})$ and

$$\|\boldsymbol{u}\|_W \leq c(\|\boldsymbol{f}\|_{[L^2(\Omega)]^N} + \|\boldsymbol{P}\|_{[L^2(\Gamma_\tau)]^N} + \|\boldsymbol{u}_o\|_W). \tag{3.233}$$

Proof: We shall prove that the minimum of the potential energy $J(\boldsymbol{u})$ is the weak solution. We have

$$\begin{aligned} J(\boldsymbol{u}) &= \tfrac{1}{2}\int_\Omega c_{ijkl}(\boldsymbol{x})e_{ij}(\boldsymbol{u})e_{kl}(\boldsymbol{u})\mathrm{d}\boldsymbol{x} - \int_\Omega f_i u_i \mathrm{d}\boldsymbol{x} - \int_{\Gamma_\tau} P_i u_i \mathrm{d}s \\ &= \tfrac{1}{2}a(\boldsymbol{u},\boldsymbol{v}) - S(\boldsymbol{v}), \end{aligned}$$

where $a(\boldsymbol{u},\boldsymbol{v})$ is the bi-linear form and $S(\boldsymbol{v})$ the linear functional defined by

$$a(\boldsymbol{u},\boldsymbol{v}) = \int_\Omega c_{ijkl}(\boldsymbol{x})e_{ij}(\boldsymbol{u})e_{kl}(\boldsymbol{v})\mathrm{d}\boldsymbol{x},\ S(\boldsymbol{v}) = \int_\Omega f_i v_i \mathrm{d}\boldsymbol{x} + \int_{\Gamma_\tau} P_i v_i \mathrm{d}s.$$

The bi-linear form $a(\boldsymbol{u},\boldsymbol{v})$ is bounded, V-elliptic and symmetric on V whilst $S(\boldsymbol{v})$ is linear and continuous on W.

We see that the functional $J(\boldsymbol{u})$ is Gâteaux differentiable at every point $\boldsymbol{u} \in W$ and that

$$DJ(\boldsymbol{u},\boldsymbol{v}) = a(\boldsymbol{u},\boldsymbol{v}) - S(\boldsymbol{v}). \tag{3.234}$$

Really, for $\boldsymbol{u},\boldsymbol{v} \in W, t \in R^1$ we have

$$\lim_{t\to 0} t^{-1}[\tfrac{1}{2}a(\boldsymbol{u}+t\boldsymbol{v},\boldsymbol{u}+t\boldsymbol{v}) - S(\boldsymbol{u}+t\boldsymbol{v}) - \tfrac{1}{2}a(\boldsymbol{u},\boldsymbol{u}) + S(\boldsymbol{u})] = a(\boldsymbol{u},\boldsymbol{v}) - S(\boldsymbol{v}).$$

From the convex analysis it is known that if a differentiable functional $\Phi(\boldsymbol{w})$ attains its minimum at the point $\boldsymbol{w}^o$ in a Hilbert space then

$$D\Phi(\boldsymbol{w}^o,\boldsymbol{v}) = 0 \text{ for all } \boldsymbol{v} \in H. \tag{3.235}$$

Let us put $\boldsymbol{u} = \boldsymbol{u}_o + \boldsymbol{w}$ for $\boldsymbol{w} \in V$, and put $H = V$. Let us put

$$J(\boldsymbol{u}) = J(\boldsymbol{u}_o + \boldsymbol{w}) = \Phi(\boldsymbol{w}). \tag{3.236}$$

Then the functional J attains its minimum at the point $\boldsymbol{u}^o = \boldsymbol{u}_o + \boldsymbol{w}^o$ if and only if Φ attains its minimum at the point $\boldsymbol{w}^o$ in V. Hence $\boldsymbol{u}^o = \boldsymbol{u}_o + \boldsymbol{w}^o$ is a minimum (critical, stationary) point of J. Since

$$D\Phi(\boldsymbol{w},\boldsymbol{v}) = DJ(\boldsymbol{u}_o + \boldsymbol{w},\boldsymbol{v}) = DJ(\boldsymbol{u},\boldsymbol{v}) \quad \forall \boldsymbol{w},\boldsymbol{v} \in V, \tag{3.237}$$

then due to Equation 3.235, $DJ(\boldsymbol{u}^o,\boldsymbol{v}) = 0 \forall \boldsymbol{v} \in$V. From Equations 3.234 and 3.232 we see that $\boldsymbol{u}^o$ is a weak solution in the sense of Definition 3.16.

The existence of a weak solution $\boldsymbol{u}^o$ follows from the existence of the minimum of $\Phi(\boldsymbol{w})$ in V. From convex analysis it is known that the functional Φ attains its minimum in V if Φ is coercive, i.e. if $\lim_{\|\boldsymbol{w}\|\to\infty, \boldsymbol{w}\in V} \Phi(\boldsymbol{w}) = +\infty$, and weakly lower semi-continuous, i.e. if $\Phi(\boldsymbol{w})$ is G-differentiable at every $\boldsymbol{w} \in V$ and $D\Phi(\boldsymbol{w}+\boldsymbol{v},\boldsymbol{v}) - D\Phi(\boldsymbol{w},\boldsymbol{v}) \geq 0 \forall \boldsymbol{w},\boldsymbol{v} \in V$ and $D\Phi(\boldsymbol{w},\boldsymbol{v})$ is continuous on any segment in the space V.

To prove the coerciveness we see that the functional $S(\boldsymbol{w})$ is linear and continuous on W; therefore there exists $c_1 > 0$ such that $S(\boldsymbol{w}) \leq c_1 \|\boldsymbol{w}\|_{1,2}$ for every $\boldsymbol{w} \in$W. From the definition of Φ and from V-ellipticity of the bi-linear form $a(\boldsymbol{w},\boldsymbol{w})$, then for $\boldsymbol{w} \in V$ we have

$$\Phi(\boldsymbol{w}) = \tfrac{1}{2}a(\boldsymbol{w},\boldsymbol{w}) - S(\boldsymbol{w}) \geq \tfrac{1}{2}c\|\boldsymbol{w}\|_{1,2}^2 - c_1\|\boldsymbol{w}\|_{1,2},$$

from which the coerciveness follows immediately.

To prove the weak lower semi-continuity we see that the functional Φ is G-differentiable at every point $\boldsymbol{w} \in$V. Then

$$D\Phi(\boldsymbol{w}+\boldsymbol{v},\boldsymbol{v}) - D\Phi(\boldsymbol{w},\boldsymbol{v}) = \int_\Omega c_{ijkl}(\boldsymbol{x})e_{ij}(\boldsymbol{v})e_{kl}(\boldsymbol{v})\mathrm{d}\boldsymbol{x} \geq c\|\boldsymbol{v}\|_W^2 \quad \forall \boldsymbol{v} \in V,$$

due to Equations 3.234 and 3.237, the boundedness of elastic coefficients and the Korn inequality. Moreover, the functional $D\Phi(\boldsymbol{w}, \boldsymbol{v})$ is continuous on any line segment in the variable $\boldsymbol{w}$, Therefore, it results that it is weakly lower semi-continuous.

From the convex analysis it is known that if the functional is coercive and weakly lower semi-continuous then it attains its minimum. Hence the existence of the weak solution follows immediately. The uniqueness of the weak solution can be proved by the usual way. Let us assume that $\boldsymbol{u}_1$ and $\boldsymbol{u}_2$ are two weak solutions of our problem. Then $\boldsymbol{U} = u_1 - \boldsymbol{u}_2$, $\boldsymbol{U} \in V$. From Equation 3.232 it follows that $a(\boldsymbol{U}, \boldsymbol{v}) = 0$ for all $\boldsymbol{v} \in V$. Setting $\boldsymbol{v} = \boldsymbol{U}$, then $\|\boldsymbol{U}\|_{1,2} = 0$, hence $\boldsymbol{u}_1 = \boldsymbol{u}_2$.

To prove Equation 3.233, we put $\boldsymbol{u} = \boldsymbol{u}_o + \boldsymbol{w}$ into Equation 3.232. Hence

$$a(\boldsymbol{w}, \boldsymbol{v}) = -a(\boldsymbol{u}_o, \boldsymbol{v}) + S(\boldsymbol{v}) \quad \forall \boldsymbol{v} \in V.$$

Putting $\boldsymbol{v} = \boldsymbol{w}$ and applying the Korn inequality, we obtain

$$c\,\|\boldsymbol{w}\|_W^2 \le a(\boldsymbol{w}, \boldsymbol{w}) \le c\,\|\boldsymbol{w}\|_W\,(\|\boldsymbol{f}\|_{[L^2(\Omega)]^N} + \|\boldsymbol{P}\|_{[L^2(\Gamma_\tau)]^N} + \|\boldsymbol{u}_o\|_W).$$

Hence, and since $\|\boldsymbol{u}\|_W \le \|\boldsymbol{u}_o\|_W + \|\boldsymbol{w}\|_W$, then Equation 3.233 follows immediately.

In the case when $\Gamma_u = 0$ we have the first basic boundary value problem of elasticity. To prove the existence theorem the following necessary conditions must be satisfied.

Theorem 3.32: Necessary conditions for the existence of the solution of the first basic boundary value problem of elasticity are the conditions of total equilibrium:

$$\int_\Omega f_i \mathrm{d}\boldsymbol{x} + \int_{\partial\Omega} P_i \mathrm{d}s = 0, \quad i = 1, \ldots, N,$$

$$\int_\Omega (\boldsymbol{x} \times \boldsymbol{f})_i \mathrm{d}\boldsymbol{x} + \int_{\partial\Omega} (\boldsymbol{x} \times \boldsymbol{P})_i \mathrm{d}s = 0, \quad i = 1, \ldots, N. \tag{3.238}$$

(For the proof see Nečas and Hlaváček (1981), p. 95.)

Theorem 3.33: Let the necessary conditions of total equilibrium (Equation 3.238) be satisfied. Then there exists a unique solution of the first basic boundary value problem of elasticity and

$$\|\boldsymbol{u}\|_W \le c(\|\boldsymbol{f}\|_{[L^2(\Omega)]^N} + \|\boldsymbol{P}\|_{[L^2(\partial\Omega)]^N}). \tag{3.239}$$

(For the details of the proof see Nečas and Hlaváček (1981), p. 96.)

3.2.2. Simple Models of Deformation of Geological Blocks Based on Boundary Value Problems in the Deformation Theory of Thermo-plasticity

3.2.2.1. Formulation of the problem

Let us denote by Ω the region occupied by the rock massif. We shall assume that the region Ω with Lipschitz boundary $\partial\Omega = \Gamma_u \cup \Gamma_\tau \cup \mathscr{R}$ occupies the rock massif including a high level radioactive waste repository, or another great structure, where $\Gamma_u = {}^1\Gamma_u \cup {}^2\Gamma_u$, $\Gamma_\tau = {}^1\Gamma_\tau \cup {}^2\Gamma_\tau$ and $\mathscr{R}$ is the boundary of the zero measure.

The fundamental equations for the thermo-plastic media follow from the general conservation law of energy and the law of momentum conservation, which in differential form are as follows

$$-\frac{\partial}{\partial x_i}\left(k_{ij}(\boldsymbol{x})\frac{\partial T(\boldsymbol{x})}{\partial x_j}\right) = Q(\boldsymbol{x}), \quad i, j = 1, \ldots, N, \quad \boldsymbol{x} \in \Omega, \tag{3.240}$$

$$\frac{\partial \tau_{ij}}{\partial x_j} + F_i = 0, \quad i, j = 1, \ldots, N, \tag{3.241}$$

where ρ is a density, T is a temperature, Q represents thermal sources, κ_{ij} are coefficients of thermal conductivity and $\boldsymbol{x} = (x_i)$ are the Cartesian co-ordinates. Furthermore, $\boldsymbol{u} = (u_i)_{i=1}^N$ is a displacement vector, τ_{ij} is the stress tensor of the plastic body defined by the constitutional stress-strain relation, and F_i represents body forces.

We shall assume that the Lamé coefficient is a function of the first two invariants of the small strain tensor $e_{ij}(\boldsymbol{u})$, defined by

$$e_{ij}(\boldsymbol{u}) = \tfrac{1}{2}(\partial u_i/\partial x_j + \partial u_j/\partial x_i), \quad i, j = 1, \ldots, N. \tag{3.242}$$

Furthermore, we shall assume that the relation between the first invariants of stress and strain is linear. Then the stress-strain relation is defined by the so-called Hencky-Mises relation

$$\tau_{ij} = [k(\boldsymbol{x}) - \tfrac{2}{3}\mu(\boldsymbol{x}, \gamma^2(\boldsymbol{u}))]\vartheta(\boldsymbol{u})\delta_{ij} + 2\mu(\boldsymbol{x}, \gamma^2(\boldsymbol{u}))e_{ij}(\boldsymbol{u}) - \beta_{ij}(T - T_o),$$
$$i, j = 1, \ldots, N, \tag{3.243}$$

where $k(\boldsymbol{x}) = \lambda(\boldsymbol{x}) + \frac{2}{3}\mu(\boldsymbol{x})$ is the bulk modulus and where $\gamma(\boldsymbol{u}, \boldsymbol{v}) = -\frac{2}{3}\vartheta(\boldsymbol{u})\vartheta(\boldsymbol{v}) + 2e_{ij}(\boldsymbol{u})e_{ij}(\boldsymbol{v})$, and $\gamma(\boldsymbol{u}, \boldsymbol{u}) = \gamma^2(e_{ij}(\boldsymbol{u})) = \gamma^2(\boldsymbol{u}) = \gamma^2$, $\gamma^2(\boldsymbol{u}) = e_{ij}^D e_{ij}^D$, where $e_{ij}^D = e_{ij} - e_{kk}\delta_{ij}$ is the deviator of the strain tensor, $\vartheta = \vartheta_1 = e_{kk}$, $\tau_{ij}^D = 2\mu e_{ij}^D$, is the stress deviator, and the invariant $s = \tau_{ij}^D\tau_{ij}^D$ represents the intensity of shear stress. The relation $s = (2\mu\gamma)^{1/2}$ is the so-called Hencky's relation. Let us suppose the existence $k_o > 0$, $\mu_o > 0$ such that

$$0 < k_o \le k(\boldsymbol{x}) \le k_1 < +\infty, \quad 0 < \mu_o \le \mu(\boldsymbol{x}, s) \le 3/2k(\boldsymbol{x}) \quad \forall \boldsymbol{x} \in \overline{\Omega}, \quad s \ge 0, \tag{3.244}$$

where the functions $k(\boldsymbol{x})$, $\mu(\boldsymbol{x}, s)$ are continuous and $\mu(\boldsymbol{x}, s)$ is also continuously differentiable with respect to s and

$$m_1 \ge \mu(\boldsymbol{x}, s) + 2s\frac{\partial \mu(x, s)}{\partial s} \ge 0 \quad \forall \boldsymbol{x} \in \overline{\Omega}, \quad s \ge 0. \tag{3.245}$$

Then we shall deal with the quasi steady-state problem ($\mathscr{P}$) defined by Equations 3.240, 3.241 and 3.243, with the following boundary conditions

$$\kappa_{ij}\frac{\partial T(x)}{\partial x_j}n_i = q(\boldsymbol{x}), \boldsymbol{u}(\boldsymbol{x}) = \boldsymbol{u}_o(\boldsymbol{x}) \quad \text{on } \Gamma_u, \quad i, j = 1, \ldots, N,$$

$$T(\boldsymbol{x}) = T_1(\boldsymbol{x}), \tau_{ij}n_j = P_i \quad \text{on } \Gamma_\tau, \quad i, j = 1, \ldots, N, \tag{3.246}$$

where $\boldsymbol{n} = (n_i)_{i=1}^N$ is the outerward normal to $\partial\Omega$, and $q(\boldsymbol{x})$ is the heat flow.

Let us transform the heat problem to one with homogeneous boundary conditions. In Equation 3.240 let us replace T by $T + w$, where w is a sufficiently smooth function satisfying Equation 3.246 on Γ_τ. This transformation changes the functions in the boundary conditions. It can be shown that the heat sources and heat flow $q(\boldsymbol{x})$ are given by

$$Q(\boldsymbol{x}) \to Q(\boldsymbol{x}) + \frac{\partial}{\partial x_i}\left(\kappa_{ij}(\boldsymbol{x})\frac{\partial w(x)}{\partial x_j}\right) \in L^2(\overline{\Omega}),$$

$$q(\boldsymbol{x}) \to q(\boldsymbol{x}) + \kappa_{ij}(\boldsymbol{x})\frac{\partial W(x)}{\partial x_j} n_i \in L^2(\Gamma_u).$$

Then due to this above transformation

$$T(\boldsymbol{x}) = 0 \quad \text{on } \Gamma_\tau, \tag{3.247}$$

and due to the definition of τ_{ij} in Equation 3.243, the boundary condition of Equation 3.246 is non-linear, i.e.

$$\tau_{ij}n_j = [k(\boldsymbol{x}) - \tfrac{2}{3}\mu(\boldsymbol{x}, \gamma^2)]\vartheta(\boldsymbol{u})n_i + 2\mu(\boldsymbol{x}, \gamma^2)e_{ij}(\boldsymbol{u})n_j = P_i \quad \text{on } \Gamma_\tau, \quad i = 1, \ldots, N. \tag{3.248}$$

3.2.2.2. *Variational formulation of the problem*

Let $\Omega \subset R^N$, $N = 2, 3$, be the region with a Lipschitz boundary $\partial\Omega = \overline{\Gamma}_\tau \cup \overline{\Gamma}_u \cup \mathscr{R}$, and measure $\mathscr{R} = 0$, occupied by a plastic geological body. Let $\boldsymbol{x} = (x_i)_{i=1}^N$ be Cartesian co-ordinates and let $\boldsymbol{n} = (n_i)_{i=1}^N$ be the outward normal to $\partial\Omega$. Let us denote by (.,.) the scalar product in $L^2(\Omega)$ and $[L^2(\Omega)]^N$, by $\|.\|_k$ the norm in $H^k(\Omega)$ and $[H^k(\Omega)]^N$, k being an integer, $k \in R^1$, where $H^k(\Omega)$ denotes the Sobolev space in the usual sense. Furthermore, let us denote by

$${}^1V = \{z \mid z \in H^1(\Omega), z = 0 \text{ on } \Gamma_\tau \text{ in the sense of traces}\}$$

the space of virtual temperatures, and by

$$V = \{\boldsymbol{v} \mid \boldsymbol{v} \in W = [H^1(\Omega)]^N, \quad \boldsymbol{v} = 0 \text{ on } \Gamma_u \text{ in the sense of traces}\}$$

the space of virtual displacements.

We suppose that the κ_{ij} are symmetric and Lipschitz on Ω, i.e.

$$\kappa_{ij} = \kappa_{ji}, \quad 0 < \kappa_o \leq (\kappa_{ij}(\boldsymbol{x}, T)\xi_i\xi_j\,|\xi|^2 \leq \kappa_1 < +\infty \quad \forall \boldsymbol{x} \in \Omega, \quad \xi = (\xi_i) \in R^N, \tag{3.249}$$

where $|.|$ denote the Euclidean norm and

$$0 < k_o \leq k(\boldsymbol{x}) \leq k_1 < +\infty, \quad 0 < \mu_o \leq \mu(\boldsymbol{x}, s) \leq 3/2k(\boldsymbol{x}) \quad \forall \boldsymbol{x} \in \overline{\Omega}, \quad s \geq 0. \tag{3.250}$$

Definition 3.17: By a weak (variational,generalized) solution of the problem ($\mathscr{P}$) we mean a pair of functions $(T, \boldsymbol{u})$, $T \in {}^1V, \boldsymbol{u} \in V$ satisfying the following conditions:

$$T \in {}^1V, \quad \boldsymbol{u} - \boldsymbol{u}_o \in V \text{ such that}$$

$$a_T(T, z) = s(z) \quad \forall z \in {}^1V, \tag{3.251}$$

$$a(\boldsymbol{w}; \boldsymbol{u}, \boldsymbol{v}) + b_s(T, \boldsymbol{v}) = S(\boldsymbol{v}) \quad \forall \boldsymbol{v} \in \boldsymbol{V}, \tag{3.252}$$

where

$$a_T(T, z) = \int_\Omega \kappa_{ij} \frac{\partial T}{\partial x_j} \frac{\partial z}{\partial x_i} \mathrm{d}\boldsymbol{x}, \tag{3.253}$$

$$s(z) = \int_\Omega Qz \mathrm{d}\boldsymbol{x}, \tag{3.254}$$

$$a(\boldsymbol{u}; \boldsymbol{u}, \boldsymbol{v}) = \int_\Omega [k(\boldsymbol{x}) - \tfrac{2}{3}\mu(\gamma^2(\boldsymbol{u}))\vartheta(\boldsymbol{u})\vartheta(\boldsymbol{v}) + 2\mu(\gamma^2(\boldsymbol{u}))e_{ij}(\boldsymbol{u})e_{ij}(\boldsymbol{v})]\mathrm{d}\boldsymbol{x}, \tag{3.255}$$

$$S(\boldsymbol{v}) = \int_\Omega F_i v_i \mathrm{d}\boldsymbol{x} + \int_{\Gamma_\tau} P_i v_i \mathrm{d}s, \; b_s(T, \boldsymbol{v}) = \int_\Omega \frac{\partial}{\partial x_j}(\beta_{ij}(T - T_o)v_i \mathrm{d}\boldsymbol{x}. \tag{3.256}$$

Remark: To obtain the weak solution of the problem, we integrated Equation 3.249 and 3.250 by an arbitrary $z \in H^1(\Omega)$, $\boldsymbol{v} \in V$ and applied the Green's theorem in both cases and then we found the weak formulation in the form of Equations 3.251 and 3.252 with Equations 3.253-3.256.

It can be shown that any classical solution of the problem $(\mathscr{P})$ is also a weak solution. On the other hand, if the solution is smooth enough, then it is also a classical solution of our problem. Details for the case of the thermal part of the problem will be given in Chapter 9. In the plastic part of the problem we integrate Equation 3.252 with Equation 3.255 by parts. Then we find

$$\int_\Omega \partial/\partial x_j[(k(\boldsymbol{x}) - \tfrac{2}{3}\mu(\gamma^2(\boldsymbol{u}))\vartheta(\boldsymbol{u})\delta_{ij} + 2\mu(\gamma^2(\boldsymbol{u}))e_{ij}(\boldsymbol{u}) - \beta_{ij}(T - T_o) - F_i]v_i \mathrm{d}\boldsymbol{x}$$
$$+ \int_{\Gamma_u} [k(\boldsymbol{x}) - \tfrac{2}{3}\mu(\gamma^2(\boldsymbol{u}))\vartheta(\boldsymbol{u})n_i + 2\mu(\gamma^2(\boldsymbol{u}))e_{ij}(\boldsymbol{u})n_j - \beta_{ij}(T - T_o)n_j]v_i \mathrm{d}s$$
$$+ \int_{\Gamma_\tau} [(k(\boldsymbol{x}) - \tfrac{2}{3}\mu(\gamma^2(\boldsymbol{u}))\vartheta(\boldsymbol{u})n_i + 2\mu(\gamma^2(\boldsymbol{u}))e_{ij}(\boldsymbol{u})n_j$$
$$- \beta_{ij}(T - T_o)n_j - P_i]v_i \mathrm{d}s = 0 \quad \boldsymbol{v} \in V.$$

Hence the equation of equilibrium and the boundary conditions are satisfied, i.e. $\partial\tau_{ij}/\partial x_j - F_i = 0$ in Ω and $\tau_{ij}(\boldsymbol{u})n_j = P_i$, $i = 1, \ldots, N, N = 2, 3$ on Γ_τ, and $\boldsymbol{u} = \boldsymbol{u}_o$ on Γ_u.

Since the problem is assumed to be quasi-coupled then both problems can be analysed separately. Since the problem of the thermal part of the problem studied is a special case of the problem discussed in Chapter 9, we shall give the results only.

The bi-linear form $a_T(.,.)$ is symmetric, i.e.

$$a_T(T, z) = a_T(z, T), \quad \forall T, z \in {}^1V,$$

bounded, i.e. there exists $c_o > 0$ such that

$$|a_T(T, z)| = \left| \int_\Omega \kappa_{ij} \frac{\partial T}{\partial x_j} \frac{\partial z}{\partial x_i} \mathrm{d}\boldsymbol{x} \right| \leq c_o \|T\|_{1_V} \|z\|_{1_V} \quad \forall T, z \in {}^1V,$$

and ${}^1V$-elliptic, i.e. there exists $c > 0$ such that

$$a_T(T, T) = \int_\Omega \kappa_{ij} \frac{\partial T}{\partial x_j} \frac{\partial T}{\partial x_i} \mathrm{d}\boldsymbol{x} \geq c \|T\|^2_{1_V} \quad \forall T \in {}^1V.$$

Moreover, the linear form $s(z)$ is bounded on ${}^1V$, i.e. there exists a constant $c_2 > 0$ such that

$$|s(z)| \leq c_2 \|z\|_1 \quad \forall z \in {}^1V.$$

Then we have proved the following lemma:

Lemma 3.1: If the bi-linear form $a_T(T, z)$ is bounded on $H^1(\Omega) \times H^1(\Omega)$ and ${}^1V$-elliptic and if the linear form $s(z)$ is bounded on ${}^1V$, then the problem in Equation 3.251 i.e. the problem find $T \in {}^1V$ such that

$$a_T(T, z) = s(z) \quad \forall z \in {}^1V,$$

has one and only one solution.

Lemma 3.2: Let the assumptions of Lemma 3.1 be satisfied. Then the problem in Equation 3.251 is equivalent to the problem: find $T \in {}^1V$ such that

$$l(T) = \inf_{T \in {}^1V} l(z), \tag{3.257}$$

where

$$l(z) = \tfrac{1}{2} a_T(z, z) - s(z) \tag{3.258}$$

Proof: Let $T \in {}^1V$ be the solution of Equation 3.251. Then, since $a_T(.,.)$ is a symmetric bi-linear form, we have

$$l(z) = \tfrac{1}{2} a_T(z, z) - a_T(T, z) = \tfrac{1}{2} a_T(z - T, z - T) - a_T(T, T).$$

Then the functional $l(z)$ attains its minimum just for $z = T$ as the term $\frac{1}{2} a_T(T, T)$ is constant.

On the other hand, since the bi-linear form $a_T(z, z)$ is ${}^1V$-elliptic, then from the foregoing definition of $l(z)$ we find that just the element T minimizes $l(z)$, where T represents the solution of Equation 3.251. Therefore, both problems in Equation 3.251 and Equations 3.257 and 3.258 are equivalent, which completes the proof.

Now we shall investigate the plastic part of the problem. Firstly, we note that since $\beta_{ij} \in C^1(\overline{\Omega})$, $T, T_o \in H^1(\Omega)$, then the term $\partial/\partial x_j(\beta_{ij}(T - T_o)) \in L^2(\Omega)$ and, therefore, it can be comprehended into the body forces. Therefore, we put $F_i(\boldsymbol{x}) \equiv F_i(\boldsymbol{x}) - \partial/\partial x_j(\beta_{ij}(T - T_o))$.

Now we shall derive a functional of potential energy. Let $L(\boldsymbol{u})$ denote such a functional of potential energy. Similarly as in an elastic case we know that the solution $\boldsymbol{u}$ is a critical point of the functional $L(\boldsymbol{u})$. Therefore, $DL(\boldsymbol{u}, \boldsymbol{v}) = 0$ for all $\boldsymbol{v} \in V$.

Let us assume that the functional $L(\boldsymbol{u})$ exists and let us assume that $\boldsymbol{u}, \boldsymbol{v} \in [C^1(\overline{\Omega})]^N$. Let us assume that the functional $L(\boldsymbol{v})$ is Gâteaux differential continuous on every line segment in the variable $\boldsymbol{u}$. Then

$$L(\boldsymbol{u}) - L(\boldsymbol{0}) = \int_0^1 DL(t\boldsymbol{u}, \boldsymbol{u})\mathrm{d}t.$$

The left hand side of Equations 3.252 with Equation 3.255 gives

$$\int_0^1 \int_\Omega [(tk(\boldsymbol{x}) - \tfrac{2}{3}\mu(\gamma^2(t\boldsymbol{u}))\vartheta^2(\boldsymbol{u}) + 2t\mu(\gamma^2(t\boldsymbol{u}))e_{ij}(\boldsymbol{u})e_{ij}(\boldsymbol{u})]\mathrm{d}\boldsymbol{x}\,\mathrm{d}t$$

$$= \int_\Omega \int_0^1 [(tk(\boldsymbol{x}) - \tfrac{2}{3}\mu(\gamma^2(t\boldsymbol{u}))\vartheta^2(\boldsymbol{u}) + 2t\mu(\gamma^2(t\boldsymbol{u}))e_{ij}(\boldsymbol{u})e_{ij}(\boldsymbol{u})]\mathrm{d}t\mathrm{d}\boldsymbol{x}$$

$$= \int_\Omega \left[\tfrac{1}{2}k\vartheta^2(\boldsymbol{u}) + \gamma^2(\boldsymbol{u})\int_0^1 \mu(t^2\gamma^2(\boldsymbol{u}))t\mathrm{d}t\right]\mathrm{d}\boldsymbol{x}$$

as for $\boldsymbol{u}, \boldsymbol{v}$, which are smooth enough, the order of integration may be changed, and $\gamma(\boldsymbol{u}, \boldsymbol{v}) = -\frac{2}{3}\vartheta(\boldsymbol{u})\vartheta(\boldsymbol{v}) + 2e_{ij}(\boldsymbol{u})e_{ij}(\boldsymbol{v})$ and $\gamma^2(t\boldsymbol{u}) = t^2\gamma^2(\boldsymbol{u})$ may be used. Now let $\gamma(\boldsymbol{u}) = 0$ then $\int_0^1 \mu(0)t\mathrm{d}t = \frac{1}{2}\mu(0)$. Further, let $\gamma(\boldsymbol{u}) \neq 0$, then using the substitution $\tau = t\gamma^2(\boldsymbol{u})$ we obtain

$$\int_0^1 \mu(t^2\gamma^2(\boldsymbol{u})t\mathrm{d}t = \tfrac{1}{2}\int_0^1 \mu(\tau\gamma^2(\boldsymbol{u}))\mathrm{d}\tau = (2\gamma^2(\boldsymbol{u}))^{-1}\int_0^{\gamma^2(\boldsymbol{u})} \mu(\sigma)\mathrm{d}\sigma,$$

where $\sigma = \tau\gamma^2(\boldsymbol{u})$. Then we find

$$L(\boldsymbol{u}) = \tfrac{1}{2}\int_\Omega [k(\boldsymbol{x})\vartheta^2(\boldsymbol{u}) + \int_0^{\gamma(\boldsymbol{u},\boldsymbol{u})} \mu_(\sigma)\mathrm{d}\sigma]\mathrm{d}\boldsymbol{x} - \int_\Omega F_i u_i \mathrm{d}\boldsymbol{x} - \int_{\Gamma_\tau} P_i u_i ds, \quad (3.259)$$

representing the functional of potential energy of a plastic body.

Let $\boldsymbol{u} \in W = [H^1(\Omega)]^N$, $\boldsymbol{v} \in V$, then due to the definition of the G-differential of L and since $\mathrm{d}/\mathrm{d}t\gamma^2(\boldsymbol{u} + t\boldsymbol{v})\big|_{t=0} = 2\gamma(\boldsymbol{u}, \boldsymbol{v})$, we find

$$\frac{\mathrm{d}}{\mathrm{d}t}L(\boldsymbol{u} + t\boldsymbol{v})\bigg|_{t=0} = DL(\boldsymbol{u}, \boldsymbol{v})$$

$$= \int_\Omega [k(\boldsymbol{x})\vartheta(\boldsymbol{u})\vartheta(\boldsymbol{v}) + \mu(\gamma^2(\boldsymbol{u}))\gamma(\boldsymbol{u}, \boldsymbol{v})]\mathrm{d}\boldsymbol{x} - \int_\Omega F_i v_i \mathrm{d}\boldsymbol{x} - \int_{\Gamma_\tau} P_i v_i \mathrm{d}s. \quad (3.260)$$

From the convex theory (see Chapter 2) it is known that the functional $L : V \to R^1$, which is coercive and weakly lower semi-continuous and where V is a Hilbert space, attains its minimum in V. To prove this fact we prove that the functional L is coercive, i.e. $\lim L(\boldsymbol{u}) = +\infty$ as $\|\boldsymbol{u}\| \to \infty$, and is weakly lower semi-continuous, i.e. $\boldsymbol{u}_n \to \boldsymbol{u} \Rightarrow$ lin inf $L(\boldsymbol{u}_n) \geq L(\boldsymbol{u})$, where $\boldsymbol{u}_n \to \boldsymbol{u}$ weakly.

Firstly, we shall prove that the functional L, defined on a Hilbert space V, has a continuous differential, which is strictly monotone, i.e.

$$DL(\boldsymbol{u} + \boldsymbol{v}, \boldsymbol{v}) - DL(\boldsymbol{u}, \boldsymbol{v}) \geq c\,\|\boldsymbol{v}\|^2, \quad c > 0.$$

Let $\boldsymbol{u}, \boldsymbol{v} \in W$. The second Gâteaux differential is defined by

$$D^2L(\boldsymbol{u}; \boldsymbol{v}, \boldsymbol{w}) = \frac{\mathrm{d}}{\mathrm{d}t}DL(\boldsymbol{u} + t\boldsymbol{w}, \boldsymbol{v})\bigg|_{t=0}$$

$$= \int_\Omega [k(\boldsymbol{x})\vartheta(\boldsymbol{w})\vartheta(\boldsymbol{v}) + 2\mu'(\gamma^2(\boldsymbol{u}))\gamma(\boldsymbol{u}, \boldsymbol{w})\gamma(\boldsymbol{u}, \boldsymbol{v}) + \mu(\gamma^2(\boldsymbol{u}))\gamma(\boldsymbol{w}, \boldsymbol{v})]\mathrm{d}\boldsymbol{x}, \quad (3.261)$$

where the theorem about differentiating under the integral sign was used.

Let $\Omega = \Omega_1 + \Omega_2 = \{x \in \Omega \mid \mu'(\gamma^2(u))(x) \geq 0\} \cup \{x \in \Omega \mid \mu'(\gamma^2(u))(x) < 0\}$. Then

$$DL(u+v,v) - DL(u,v) = \int_0^1 D^2L(u+tv;v,v)\mathrm{d}t$$

$$= \int_0^1 \int_{\Omega_1 \cup \Omega_2} [k(x)\vartheta(v)\vartheta(v) + 2\mu'(\gamma^2(u+tv))\gamma(u+tv,v)\gamma(u+tv,v)$$

$$+ \mu(\gamma^2(u+tv))\gamma(v,v)]\mathrm{d}x\,\mathrm{d}t$$

$$\geq \mu_o \int_0^1 \int_{\Omega_1} \gamma(v,v)\mathrm{d}x\,\mathrm{d}t + \tfrac{2}{3}\mu_o \int_0^1 \int_{\Omega_1} \vartheta^2(v)\mathrm{d}x\,\mathrm{d}t + \tfrac{2}{3}\mu_o \int_0^1 \int_{\Omega_2} \vartheta^2(v)\mathrm{d}x\,\mathrm{d}t$$

$$+ \int_0^1 \int_{\Omega_2} [\mu(\gamma^2(u))\gamma(v,v) - 2\left|\mu'(\gamma^2(u))\right|(\gamma^2(u,v))^2]\mathrm{d}x\,\mathrm{d}t$$

$$\geq 2\mu_o \int_0^1 \int_{\Omega_1} e_{ij}(v)e_{ij}(v)\mathrm{d}x + \tfrac{2}{3}\mu_o \int_0^1 \int_{\Omega_2} \vartheta^2(v)\mathrm{d}x\,\mathrm{d}t$$

$$+ \int_0^1 \int_{\Omega_2} [\mu(\gamma^2(u))\gamma(v,v) + 2\mu'(\gamma^2(u))\gamma(u,u)\gamma(v,v)]\mathrm{d}x\,\mathrm{d}t$$

$$\geq 2\mu_o \int_0^1 \int_{\Omega_1} e_{ij}(v)e_{ij}(v)\mathrm{d}x + \tfrac{2}{3}\mu_o \int_0^1 \int_{\Omega_2} \vartheta^2(v)\mathrm{d}x\,\mathrm{d}t + \kappa \int_0^1 \int_{\Omega_2} \gamma(v,v)\mathrm{d}x\,\mathrm{d}t$$

$$\geq 2\mu_o \int_0^1 \int_{\Omega_1} e_{ij}(v)e_{ij}(v)\mathrm{d}x\,\mathrm{d}t + \tfrac{2}{3}(\mu_o - \kappa) \int_0^1 \int_{\Omega_2} \vartheta^2(v)\mathrm{d}x\,\mathrm{d}t$$

$$+ 2\kappa \int_0^1 \int_{\Omega_2} e_{ij}(v)e_{ij}(v)\mathrm{d}x\,\mathrm{d}t$$

$$\geq c \int_0^1 \int_{\Omega_1} e_{ij}(v)e_{ij}(v)\mathrm{d}x\,\mathrm{d}t,$$

where $(\gamma^2(u,u))^2 \leq \gamma(u,u)\gamma(v,v)$ (due to the Schwarz inequality) and $\mu_o \geq \kappa$ were used. Applying the Korn inequality, we have

$$\int_\Omega e_{ij}(v)e_{ij}(v)dx \geq c\,\|v\|^2_{[H^1(\Omega)]^N}. \tag{3.262}$$

Thus we find

$$DL(u+v,v) - DL(u,v) \geq c \int_\Omega e_{ij}(v)e_{ij}(v)\mathrm{d}x \geq c\,\|v\|^2_{[H^1(\Omega)]^N}. \tag{3.263}$$

To prove the coerciveness of $L(u)$, $u \in V$, we see that

$$L(u) = \int_0^1 DL(tu,u)\mathrm{d}t = \int_0^1 t^{-1}DL(tu,tu)\mathrm{d}t. \tag{3.264}$$

From Equation 3.263, putting $\boldsymbol{u} = 0$, $\boldsymbol{v} = t\boldsymbol{u}$, we find

$$DL(t\boldsymbol{u}, t\boldsymbol{u}) - DL(\boldsymbol{0}, t\boldsymbol{u}) \geq c \int_\Omega e_{ij}(t\boldsymbol{u})e_{ij}(t\boldsymbol{u})\mathrm{d}\boldsymbol{x}.$$

Thus

$$DL(t\boldsymbol{u}, t\boldsymbol{u}) \geq ct^2 \int_\Omega e_{ij}(\boldsymbol{u})e_{ij}(\boldsymbol{u})\mathrm{d}\boldsymbol{x} + tDL(\boldsymbol{0}, \boldsymbol{u}).$$

Hence, and using Equation 3.263, yields

$$L(\boldsymbol{u}) = \int_0^1 t^{-1}DL(t\boldsymbol{u}, t\boldsymbol{u})dt \geq \int_0^1 ct \int_\Omega e_{ij}(\boldsymbol{u})e_{ij}(\boldsymbol{u})\mathrm{d}\boldsymbol{x}\,\mathrm{d}t + \int_0^1 DL(\boldsymbol{0}, \boldsymbol{u})\mathrm{d}t.$$

Using the Korn inequality, then for any $\boldsymbol{u} \in V$

$$\begin{aligned} L(\boldsymbol{u}) &\geq c\,\|\boldsymbol{u}\|^2_{[H^1(\Omega)]^N} - c_1\,\|\boldsymbol{F}\|_{[L^2(\Omega)]^N}\,\|\boldsymbol{u}\|_{[H^1(\Omega)]^N} - c_2\,\|\boldsymbol{P}\|_{[L^2(\Gamma_\tau)]^N}\,\|\boldsymbol{u}\|_{[H^1(\Omega)]^N} \\ &\geq c\,\|\boldsymbol{u}\|^2_{[H^1(\Omega)]^N} - c\,\|\boldsymbol{u}\|_{[H^1(\Omega)]^N} \end{aligned} \tag{3.265}$$

holds.

Since the functional L possesses a Gâteaux differential at any point, then the function $\Phi(t) = L(\boldsymbol{u} + t(\boldsymbol{u}_n - \boldsymbol{u}))$ is differentiable. Thus

$$\Phi'(t) = DL(\boldsymbol{u} + t(\boldsymbol{u}_n - \boldsymbol{u}), \boldsymbol{u}_n - \boldsymbol{u}), \quad t \in \langle 0, 1\rangle,$$

which is continuous on the line segment in the variable $\boldsymbol{u}$. Then, if we integrate $\Phi'(t)$ for $t \in \langle 0, 1\rangle$ we have

$$\Phi(1) - \Phi(0) = L(\boldsymbol{u}_n) - L(\boldsymbol{u}) = \int_0^1 DL(\boldsymbol{u} + t(\boldsymbol{u}_n - \boldsymbol{u}), \boldsymbol{u}_n - \boldsymbol{u})\mathrm{d}t.$$

Since

$$\begin{aligned} L(\boldsymbol{u}_n) - L(\boldsymbol{u}) - DL(\boldsymbol{u}, \boldsymbol{u}_n - \boldsymbol{u}) &= \int_0^1 [DL(\boldsymbol{u} + t(\boldsymbol{u}_n - \boldsymbol{u}), \boldsymbol{u}_n - \boldsymbol{u}) - DL(\boldsymbol{u}, \boldsymbol{u}_n - \boldsymbol{u})]\mathrm{d}t \\ &= \int_0^1 t^{-1}[DL(\boldsymbol{u} + t(\boldsymbol{u}_n - \boldsymbol{u}), t(\boldsymbol{u}_n - \boldsymbol{u})) - DL(\boldsymbol{u}, t(\boldsymbol{u}_n - \boldsymbol{u})]\mathrm{d}t \geq 0, \end{aligned}$$

then

$$\liminf\,[L(\boldsymbol{u}_n) - L(\boldsymbol{u}) - DL(\boldsymbol{u}, \boldsymbol{u}_n - \boldsymbol{u})] \geq 0,$$

and since $DL(\boldsymbol{u}, \boldsymbol{u}_n - \boldsymbol{u}) \to 0$ as $\boldsymbol{u}_n \to \boldsymbol{u}$ weakly, then

$$\liminf L(\boldsymbol{u}_n) - L(\boldsymbol{u}) \geq 0, \text{ as } \boldsymbol{u}_n \to \boldsymbol{u} \text{ weakly,} \tag{3.266}$$

which proves that $DL(\boldsymbol{u}, \boldsymbol{v})$ is continuous on any line segment in the space V.

Since we proved Equation 3.263, then

$$DL(\boldsymbol{u} + \boldsymbol{v}, \boldsymbol{v}) - DL(\boldsymbol{u}, \boldsymbol{v}) \geq 0 \quad \text{for } \boldsymbol{u}, \boldsymbol{v} \in V. \tag{3.267}$$

Hence, and using the fact that $DL(\boldsymbol{u}, \boldsymbol{v})$ is continuous on any line segment in V, then L is weakly lower semi-continuous.

Let us put $\boldsymbol{u} = \boldsymbol{u}_o + \boldsymbol{w}$ and let $\varphi(\boldsymbol{w}) = L(\boldsymbol{u}_o + \boldsymbol{w})$. Then the functional $\varphi(\boldsymbol{v})$ is coercive on V and lower semi-continuous. Therefore, the functional φ attains its minimum in V (see Chapter 2). Then, there exists an element $\boldsymbol{w} \in V$, which is a minimum of φ in V. For this $\boldsymbol{w}$ and any $\boldsymbol{v} \in V$ we have $DL(\boldsymbol{u}_o + \boldsymbol{w}, \boldsymbol{v}) = 0$ representing the definition of the weak solution (Equation 3.252), so that $\boldsymbol{u}_o + \boldsymbol{w} = \boldsymbol{u}$ is a weak solution of the problem.

To prove the uniqueness of the solution we shall assume that $\boldsymbol{u}_1$ and $\boldsymbol{u}_2$ are two weak solutions. Then Equation 3.263 yields

$$0 = DL(\boldsymbol{u}_2, \boldsymbol{u}_2 - \boldsymbol{u}_1) - DL(\boldsymbol{u}_1, \boldsymbol{u}_2 - \boldsymbol{u}_1) = DL(\boldsymbol{u}_2 - \boldsymbol{u}_1, \boldsymbol{u}_2 - \boldsymbol{u}_1) \geq c \, \|\boldsymbol{u}_2 - \boldsymbol{u}_1\|^2_{1,N'},$$

since $\boldsymbol{u}_2 - \boldsymbol{u}_1 \in V$ as $\boldsymbol{u} - \boldsymbol{u}_o \in V$. Hence $\boldsymbol{u}_1 = \boldsymbol{u}_2$.

Then, using Lemma 3.1 and the last results we prove the following theorem:

Theorem 3.34: Let $\Gamma_u \neq 0$, $\Gamma_\tau \neq 0$ and $\Omega \in C^{0,1}$. Let $Q \in L^2(\overline{\Omega})$, $q \in L^2(\Gamma_u)$, $\kappa_{ij} \in L^\infty(\overline{\Omega})$, $\boldsymbol{u}_o \in [H^1(\Omega)]^N$, $\boldsymbol{F} \in [L^2(\overline{\Omega})]^N$, $\boldsymbol{P} \in [L^2(\Gamma_\tau)]^N$. Let the coefficient $k(\boldsymbol{x})$ and $\mu(\boldsymbol{x})$ satisfy Equation 3.250. Then there exists a unique weak solution $(T, \boldsymbol{u})$ of the problem $(\mathscr{P})_v$. This solution minimizes the functional $l(z)$ of Equation 3.258 and the functional of potential energy $L(\boldsymbol{v})$ of Equation 3.259 over the spaces ${}^1V$ and $\boldsymbol{u}_o + V$.

In the case when the part Γ_u of the boundary $\partial\Omega$ is equal to zero, i.e. $\Gamma_u \equiv 0$, then we have the first boundary value problem in thermo-plasticity. Similarly as in the elastic case the necessary conditions for the existence of the solution are the conditions of total equilibrium.

Consider the space V and let $P = \{\boldsymbol{v} \mid v = \boldsymbol{a} + \boldsymbol{b} \times \boldsymbol{x}\}$ be the space of all translations and rotations. Let Q_p be the orthogonal decomposition of P in W, i.e. $W = P \oplus Q_p$. Then we have the following theorem:

Theorem 3.35: Let $\Gamma_u \equiv 0$, then $\partial\Omega = \Gamma_\tau \cup \mathscr{R}$. Then the necessary conditions for the existence of the solution of the first boundary value problem in the deformation theory of plasticity are the conditions of total equilibrium:

$$\int_\Omega F_i \mathrm{d}\boldsymbol{x} + \int_{\Gamma_\tau} P_i \mathrm{d}s = 0, \quad i = 1, \ldots, N, \tag{3.266}$$

$$\int_\Omega (\boldsymbol{x} \times \boldsymbol{F})_i \mathrm{d}\boldsymbol{x} + \int_{\Gamma_\tau} (\boldsymbol{x} \times \boldsymbol{P})_i \mathrm{d}s = 0, \quad i = 1, \ldots, N. \tag{3.267}$$

Proof: Let $\boldsymbol{v} \in P$. Then substitute $\boldsymbol{v} \in P$ into Equation 3.252 of Definition 3.17. Then $e_{ij}(\boldsymbol{v}) = 0$. Hence $s(\boldsymbol{v}) = 0$, i.e.

$$\int_\Omega F_i v_i \mathrm{d}\boldsymbol{x} + \int_{\Gamma_\tau} P_i v_i \mathrm{d}s = 0 \quad \text{for all } \boldsymbol{v} \in P. \tag{3.268}$$

Since $\boldsymbol{v} = \boldsymbol{a} + \boldsymbol{b} \times \boldsymbol{x}$ with arbitrary vectors $\boldsymbol{a}$ and $\boldsymbol{b}$, then putting $\boldsymbol{b} = \boldsymbol{0}$, we obtain $\int_\Omega F_i a_i \mathrm{d}\boldsymbol{x} + \int_{\Gamma_\tau} P_i a_i \mathrm{d}s = 0$ for all $\boldsymbol{a} \in R^N$. Hence Equation 3.266 follows immediately.

Let us put $\boldsymbol{a} = \boldsymbol{0}$. Then from Equation 3.268 we obtain

$$\int_\Omega \boldsymbol{F}(\boldsymbol{b} \times \boldsymbol{x}) \mathrm{d}\boldsymbol{x} + \int_{\Gamma_\tau} \boldsymbol{P}(\boldsymbol{b} \times \boldsymbol{x}) \mathrm{d}s = 0 \quad \text{for all } \boldsymbol{b} \in R^N.$$

As $\boldsymbol{F}(\boldsymbol{b} \times \boldsymbol{x}) = \boldsymbol{b}(\boldsymbol{x} \times \boldsymbol{F})$ and $\boldsymbol{P}(\boldsymbol{b} \times \boldsymbol{x}) = \boldsymbol{b}(\boldsymbol{x} \times \boldsymbol{P})$, then Equation 3.267 follows immediately.

Theorem 3.36: Let Q_p be the orthogonal decomposition of P in W. Then for any $\boldsymbol{u} \in Q_p$

$$L(\boldsymbol{u}) \geq c_o \|\boldsymbol{u}\|_{1,2}^2 - c_1 \|\boldsymbol{u}\|_{1,2} . \tag{3.269}$$

Proof: The proof is similar to that above. Since

$$L(\boldsymbol{u}) = \int_0^1 DL(t\boldsymbol{u}, \boldsymbol{u})\mathrm{d}t = \int_0^1 DL(t\boldsymbol{u}, t\boldsymbol{u})t^{-1}\mathrm{d}t \tag{3.270}$$

and since $DL(\boldsymbol{u} + \boldsymbol{v}, \boldsymbol{v}) - DL(\boldsymbol{u}, \boldsymbol{v}) \geq c \int_\Omega e_{ij}(\boldsymbol{v})e_{ij}(\boldsymbol{v})\mathrm{d}\boldsymbol{x}$, then putting $\boldsymbol{u} = \boldsymbol{0}, \boldsymbol{v} = t\boldsymbol{u}$, we obtain

$$DL(t\boldsymbol{u}, t\boldsymbol{u}) \geq ct^2 \int_\Omega e_{ij}(\boldsymbol{u})e_{ij}(\boldsymbol{u})\mathrm{d}\boldsymbol{x} + tDL(\boldsymbol{0}, \boldsymbol{u}).$$

Hence, using Equation 3.270 and the Korn inequality we prove Equation 3.269.

Theorem 3.37: Let the necessary conditions of total equilibrium Equations 3.266 and 3.267 be satisfied. Then there exists a unique solution $(T, \boldsymbol{u})$ of the problem $(\mathscr{P})$ in Q_p. This solution minimizes the functional $l(z)$ of Equation 3.258 and the functional of potential energy L($\boldsymbol{u}$) of Equation 3.259 over the set Q_p.

Proof: The first part of the proof is similar to the coercive case. To prove the second part of the proof we see that the functional $L(\boldsymbol{v})$ is also coercive and weakly lower semi-continuous on Q_p which follows from the fact that

$$DL(\boldsymbol{u} + \boldsymbol{v}, \boldsymbol{v}) - DL(\boldsymbol{u}, \boldsymbol{v}) \geq c \int_\Omega e_{ij}(\boldsymbol{v})e_{ij}(\boldsymbol{v})\mathrm{d}\boldsymbol{x} \quad \forall u, \boldsymbol{v} \in W,$$

as well as from Theorem 3.36 and the Korn inequality.

Remark: It can also be proved that the weak solution depends continuously on the given data $\boldsymbol{F}$ and $\boldsymbol{P}$, i.e.

$$\|\boldsymbol{u}\|_1 \leq c \left\{ \|\boldsymbol{F}\|_{[L^2(\Omega)]^N} + \|\boldsymbol{P}\|_{[L^2(\Gamma_\tau)]^N} \right\} .$$

3.2.2.3. Numerical solution

To solve the problem $(\mathscr{P})_v$ numerically several methods based on the FEM and finite difference techniques in the thermal part of the problem and several methods based on the linearization of the plastic part of the problem can be used. In this section we use the FEM method (see for example Zlámal(1968), Babuška and Aziz (1972), Oden and Reddy (1976), Ciarlet (1978), Oganesian and Ruhovec (1979), Axelsson and Barker (1984), Křížek and Neittaanmäki (1990), Quarteroni and Valli (1994), Hackbusch (1992), etc.). For the case of the plastic part of the problem, methods which can be used are the Galerkin method, FEM, gradient methods and methods of steepest descent, Newton's method, method of contraction as well as secant modules method and method of successive approximations (see Mihlin (1966), Ortega and Rheinboldt (1970), Céa (1971), Ekeland and Temam (1976), Fučík and Kufner (1980), Nečas and Hlaváček (1981), Nečas (1983), etc.). The Newton's method is not convergent in the space V in general and it may be numerically unstable. Therefore, this method requires a regular solution only (for details see Mihlin (1966)). The secant modules method is rapidly convergent in practice and therefore will be discussed in detail in this section.

Since the problem is quasi-coupled we shall discuss numerical solutions of both problems separately. Firstly we shall discuss the thermal part of the problem.

Let $\Omega \subset R^N$ with boundary $\partial\Omega$, defined above, be triangulated, i.e. $\overline{\Omega}$ is covered for example by the system of triangles T_h in the 2D case and by the system of tetrahedra T_h in the 3D case, generating a triangulation $\mathscr{T}_h$. Denote by h the maximal side of all triangles in the 2D case and the maximal edge of all tetrahedra in the 3D case. Let $\{\mathscr{T}_h\}$ be a system of regular triangulations.

We define the finite element approximations of the space ${}^1V$ as

$$ {}^1V_h = \{z \in C(\Omega), z\,|_{T_h} \in P_1, z = 0 \text{ on } \Gamma_\tau, \quad \forall T_h \in \mathscr{T}_h\}. $$

The approximate solution of the problem $(\mathscr{P})_v$ will be defined as follows:

Definition 3.18: We say that a function $T_h \in {}^1V_h$ is a finite element approximation of the thermal part of the problem, if

$$ a_T(T_h, z) = s(z) \quad \forall z \in {}^1V_h. \tag{3.271} $$

Theorem 3.38: There exists a unique solution of the finite element approximation of Equation 3.271 for all h.

Proof: As the problem is quasi-coupled, it is sufficient to investigate the thermal part of the problem $(\mathscr{P})_v$ separately. Since this part is the standard FEM analysis, we refer to Chapter 9 and the FEM literature given above.

The uniqueness can be proved by the standard technique. Let $T_h^1, T_h^2 \in {}^1V_h$ be two approximate FEM solutions of Equation 3.271. Then

$$ c\,\|T_h^1 - T_h^2\|^2 \le a_T(T_h^1 - T_h^2, T_h^1 - T_h^2) \le 0. $$

This implies that $T_h^1 = T_h^2$ a.e. in Ω, which completes the proof.

Secondly, we shall discuss the numerical solution of the plastic part of the problem. For its solution several methods can be used; we shall discuss one type of gradient method and the method of secant modules.

Gradient Methods

In this section we shall limit ourselves to the method of steepest descent only. For more details about gradient methods see for example Céa (1971) and Chapter 2, Section 2.5.2.1. We shall solve the following problem $(\mathscr{P})$:

We seek $\boldsymbol{u} \in [H^1(\Omega)]^N$ such that

$$ \int_\Omega [(k(\boldsymbol{x}) - \tfrac{2}{3}\mu(\gamma^2(\boldsymbol{u}))\vartheta(\boldsymbol{u})\vartheta(\boldsymbol{v}) + 2\mu(\gamma^2(\boldsymbol{u}))e_{ij}(\boldsymbol{u})e_{ij}(\boldsymbol{v})]\mathrm{d}\boldsymbol{x} $$

$$ - \int_\Omega F_i v_i \mathrm{d}\boldsymbol{x} - \int_{\Gamma_\tau} P_i v_i \mathrm{d}s = 0 \tag{3.272} $$

holds for any $\boldsymbol{v} \in [H^1(\Omega)]^N$.

According to Equation 3.259 the functional of total potential energy is of the following form

$$L(\boldsymbol{u}) = \tfrac{1}{2}\int_{\Omega}[k(\boldsymbol{x})\vartheta^2(\boldsymbol{u}) + \int_0^{\gamma(\boldsymbol{u},\boldsymbol{u})}\mu(t)\mathrm{d}t]\mathrm{d}\boldsymbol{x} - \int_{\Omega}F_i u_i \mathrm{d}\boldsymbol{x} - \int_{\Gamma_\tau}P_i u_i \mathrm{d}s \qquad (3.273)$$

The method of steepest descent is as follows.

The functional L assumes, for every $\boldsymbol{u} \in [H^1(\Omega)]^N$, a certain value and therefore, geometrically we can imagine the functional L as a certain surface $L(\boldsymbol{u}) =$ constant > 0 above the space $[H^1(\Omega)]^N$. Then starting from an initial point, we desire to approach, as rapidly as possible, the next approximate point at which the functional $L(\boldsymbol{u})$ attains its minimum value on the half-line of steepest descent. This new approximate point lies on the level surface $L(\boldsymbol{u}) = c_1$, $c_1 =$ constant > 0. We construct the normal at this point again and continue with the above procedure. Then we obtain a new approximate point, etc..

Let us define the gradient of L at the point $\boldsymbol{u}$ as

$$DL(\boldsymbol{u}, \boldsymbol{v}) = (g(\boldsymbol{u}), \boldsymbol{v}),$$

and for the n-th approximation of $\boldsymbol{u}$, denoted by $\boldsymbol{u}_n$, let us put $g(\boldsymbol{u}^n) \equiv \boldsymbol{w}^n$. As a result we shall prove the following theorem:

Theorem 3.39: Let the functional $L(\boldsymbol{v})$ be defined by Equation 3.273, which has first and second Gâteaux differentials $DL(\boldsymbol{u}, \boldsymbol{v}) = (g(\boldsymbol{u}), \boldsymbol{v})$ and $D^2L(\boldsymbol{u}; \boldsymbol{v}, \boldsymbol{w})$. Let the second differential $D^2L(\boldsymbol{u}; \boldsymbol{v}, \boldsymbol{w})$ be continuous in $\boldsymbol{u}$ for fixed $\boldsymbol{v}$ and $\boldsymbol{w}$ and let

$$\left|D^2L(\boldsymbol{u}; \boldsymbol{v}, \boldsymbol{w})\right| \le c_1 \|\boldsymbol{v}\| \, \|w\|, \qquad (3.274)$$

$$D^2L(\boldsymbol{u}; \boldsymbol{v}, \boldsymbol{v}) \ge c_2 \|\boldsymbol{v}\|^2. \qquad (3.275)$$

Suppose that the n-th approximation $\boldsymbol{u}^n$ of $\boldsymbol{u}$ is known and let $\boldsymbol{w}^n = g(\boldsymbol{u}^n)$. If $\boldsymbol{w}^n = \boldsymbol{0}$ then the iterate process finishes; if $\boldsymbol{w}^n \neq \boldsymbol{0}$ then the next $(n+1)$-th approximation of $\boldsymbol{u}$ is given as follows: set

$$\boldsymbol{u}^{n+1} = \boldsymbol{u}^n - c_1^{-1}\boldsymbol{w}^n, \qquad (3.276)$$

then let $\rho_n \in \langle 0, \infty)$ be such that $L(\boldsymbol{u}^n - \rho_n \boldsymbol{w}^n)$ is a minimum of the potential functional L on the half-line $\boldsymbol{u}^n - \rho\boldsymbol{w}^n$ and set

$$\boldsymbol{u}^{n+1} = \boldsymbol{u}^n - \rho_n\boldsymbol{w}^n. \qquad (3.277)$$

Then $\boldsymbol{u}^n \to \boldsymbol{u}$, where $\boldsymbol{u}$ is the unique minimum of the functional of potential energy L.

Proof: The gradient $\boldsymbol{w}^n$ is determined from the relation

$$\int_{\Omega}[(k(\boldsymbol{x}) - \tfrac{2}{3}\mu(\gamma(\boldsymbol{u}_n, \boldsymbol{u}_n))\vartheta(\boldsymbol{u}_n)\vartheta(\boldsymbol{v}) + 2\mu(\gamma(\boldsymbol{u}_n, \boldsymbol{u}_n))e_{ij}(\boldsymbol{u}_n)e_{ij}(\boldsymbol{v})]\mathrm{d}\boldsymbol{x}$$

$$- \int_{\Omega}F_i v_i \mathrm{d}\boldsymbol{x} - \int_{\Gamma_\tau}P_i v_i \mathrm{d}s = \int_{\Omega}\frac{\partial w_i^n}{\partial x_j}\frac{\partial v_i}{\partial x_j}\mathrm{d}\boldsymbol{x},$$

where $(g(\boldsymbol{u}), \boldsymbol{v}) = \int_{\Omega}\frac{\partial w_i^n}{\partial v_i}\partial x_j \mathrm{d}\boldsymbol{x}$, and $\boldsymbol{w} = g(\boldsymbol{u})$. Since μ is supposed to be continuously differentiable, and

$$0 < \mu_o \le \mu(\gamma) \le \tfrac{3}{2}k(\boldsymbol{x}), \quad 0 < \alpha_o \le \mu(\gamma) + 2\gamma\mu'(\gamma) \le \alpha_1,$$

$$\gamma(\boldsymbol{u}, \boldsymbol{v}) = -\tfrac{2}{3}\vartheta(\boldsymbol{u})\vartheta(\boldsymbol{v}) + 2e_{ij}(\boldsymbol{u})e_{ij}(\boldsymbol{v}), \qquad (3.278)$$

then

$$DL^2(\boldsymbol{u};\boldsymbol{v},\boldsymbol{v}) = \int_\Omega [k(\boldsymbol{x})\vartheta(\boldsymbol{v})\vartheta(\boldsymbol{v}) + \mu(\gamma(\boldsymbol{u},\boldsymbol{u}))\gamma(\boldsymbol{v},\boldsymbol{v}) + 2\mu(\gamma(\boldsymbol{u},\boldsymbol{u}))(\gamma(\boldsymbol{u},\boldsymbol{v}))^2]\mathrm{d}\boldsymbol{x}$$

and hence, using the Korn inequality, we find Equation 3.274 and 3.275. Thus the functional L is coercive and weakly lower semi-continuous.

Now let us investigate

$$\begin{aligned} L(\boldsymbol{u}^n - \rho\boldsymbol{w}^n) &= L(\boldsymbol{u}^n) - \rho DL(\boldsymbol{u}^n, \boldsymbol{w}^n) + \tfrac{1}{2}\rho^2 D^2L(\boldsymbol{u}^n - r_n\rho\boldsymbol{w}^n; \boldsymbol{w}^n, \boldsymbol{w}^n) \\ &\le L(\boldsymbol{u}^n) - \rho(g(\boldsymbol{u}^n), \boldsymbol{w}^n) + \tfrac{1}{2}\rho^2 c_1 \|\boldsymbol{w}^n\|^2, \quad 0 < r_n < 1. \end{aligned} \tag{3.279}$$

The last expression on the right hand side of the inequality is the quadratic polynomial, which attains its minimum at $\rho = c_1^{-1}$. Since we construct $\boldsymbol{u}^{n+1} = \boldsymbol{u}^n - c_1^{-1}\boldsymbol{w}^n$ then Equation 3.279 yields

$$L(\boldsymbol{u}^n) - L(\boldsymbol{u}^{n+1}) \ge (2c_1)^{-1} \|\boldsymbol{w}^n\|^2. \tag{3.280}$$

Now let $\boldsymbol{u}^{n+1}$ be constructed by Equation 3.277, and denote it by $\underline{\boldsymbol{u}}^{n+1}$, and then Equation 3.280 yields

$$L(\boldsymbol{u}^n) - L(\underline{\boldsymbol{u}}^{n+1}) \ge (2c_1)^{-1} \|\boldsymbol{w}^n\|^2 \tag{3.281}$$

so that the estimation of equation 3.280 holds for $\boldsymbol{u}^{n+1}$ being constructed by Equations 3.276 and 3.277. The functional L satisfying Equation 3.275 is bounded from below since

$$\begin{aligned} L(\boldsymbol{u}) &= L(\boldsymbol{u}_o) + (g(\boldsymbol{u}_o), \boldsymbol{u} - \boldsymbol{u}_o) + \tfrac{1}{2}D^2L(\boldsymbol{u}_o + r(\boldsymbol{u} - \boldsymbol{u}_o); \boldsymbol{u} - \boldsymbol{u}_o, \boldsymbol{u} - \boldsymbol{u}_o) \\ &\ge \tfrac{1}{2}c_2 \|\boldsymbol{u} - \boldsymbol{u}_o\|^2 - c_o - c\|\boldsymbol{u} - \boldsymbol{u}_o\|. \end{aligned}$$

Then according to the estimate in Equation 3.280, $L(\boldsymbol{u}^n) \to c$ and $g(\boldsymbol{u}^n) \to 0$. Since L is coercive and weakly lower semi-continuous then there exists the unique minimum point of L on V. Then $g(\boldsymbol{u}) = 0$ and

$$c_2 \|\boldsymbol{u}^n - \boldsymbol{u}\|^2 \le (g(\boldsymbol{u}^n) - g(\boldsymbol{u}), \boldsymbol{u}^n - \boldsymbol{u}) \le \|g(\boldsymbol{u}^n)\| \, \|\boldsymbol{u}^n - \boldsymbol{u}\|.$$

Hence

$$c_2 \|\boldsymbol{u}^n - \boldsymbol{u}\| \le \|g(\boldsymbol{u}^n)\| \to 0. \tag{3.282}$$

Hence $\boldsymbol{u}^n \to \boldsymbol{u}$.

Since

$$c_2 \|\boldsymbol{u}^n - \boldsymbol{u}\|^2 \le \|g(\boldsymbol{u}^n)\| \le (2c_1)^{\frac{1}{2}}(L(\boldsymbol{u}^n) - L(\boldsymbol{u}^{n+1}))^{1/2},$$

we find the estimate of the error bound which is useful for convergent analysis.

Method of Secant Modules

Here the method of secant modules will be discussed. We shall assume that physical coefficients as well as the thermal conductivity κ_{ij} and thermal sources Q depend on the temperature and its gradient ∇T. Since the problem of the thermal part of the problem will be solved in detail in Chapter 9 we present here the idea only.

To solve the thermal part of the problem it is sometimes convenient to transform the problem into a sequence of linearized problems.

Let $Q \in C(\overline{\Omega})$ and $(\kappa_{ij}) \in (C^1(\overline{\Omega}))^{N\times N}$, moreover,

$$\kappa_{ij} = \kappa_{ji}, \quad i, j = 1, \ldots, N,$$

$$\kappa_{ij}\xi_i\xi_j |\xi|^{-2} \ge k_o \quad \forall(\xi_1, \ldots, \xi_N)^T \in R^N, \quad \boldsymbol{x} \in \overline{\Omega}.$$

For any fixed $z \in H^1(\Omega)$, we denote by $a_T(z; .,.)$, the symmetric bilinear form

$$a_T(w; z, v) = \int_\Omega \kappa_{ij}(\boldsymbol{x}, w, \nabla w) \frac{\partial z}{\partial x_i} \frac{\partial v}{\partial x_j} \mathrm{d}\boldsymbol{x}, \quad \forall z, v \in H^1(\Omega), \tag{3.283}$$

$$s(z) = \int_{\Gamma_\tau} qz ds + \int_\Omega Qz \mathrm{d}s = (q, z)_{0,\partial\Omega} + (Q, z)_{0,\Omega} \quad \forall z, v \in H^1(\Omega). \tag{3.284}$$

The variational formulation of the problem $(\mathscr{P})$ consists of finding $T \in {}^1V = H_o^1(\Omega)$ such that

$$a_T(T; T, z) = s(z) \quad \forall z \in {}^1V. \tag{3.285}$$

The corresponding discrete analogous reads.

We introduce the method of secant modules to approximate T_h. Let $T_h^o, \ldots, T_h^n \in {}^1V_h$ be already known. We define $T_h^{n+1} \in {}^1V_h$ by using the linearized problem

$$a_T(T_h^n; T_h^{n+1}, z_h) = s(z_h) \quad \forall z_h \in {}^1V_h. \tag{3.286}$$

Under some assumptions on $a_T(w; .,.)$ it may be proved that

$$\left\| T_h - T_n^n \right\|_{1,\Omega} \to 0 \qquad \text{as } n \to \infty.$$

The proof is based on Theorem 5.1. of Nečas and Hlaváček (1981), p. 263.

The method of secant modules for the plastic part of the problem is as follows:

Let $\boldsymbol{u}^n$ be the n-th approximation of the solution of the problem $(\mathscr{P})_v$. We look out for the $(n+1)$-th approximation of the weak solution, $\boldsymbol{u}^{n+1}$ of the linearized problem $(\mathscr{P}_\ell)_v$:

Seek $\boldsymbol{u}^{n+1}$ as the solution of the linearized problem for the linear Hooke's law

$$\tau_{ij} = [k(\boldsymbol{x}) - \tfrac{2}{3}\mu(\boldsymbol{x}, \gamma^2(\boldsymbol{u}^n))]\vartheta\delta_{ij} + 2\mu(\boldsymbol{x}, \gamma^2(\boldsymbol{u}^n)e_{ij}, \quad i, j = 1, \ldots, N,$$

i.e. for the weak solution of the linear problem

$$a(\boldsymbol{u}^n; \boldsymbol{u}^{n+1}, \boldsymbol{v}) = S(\boldsymbol{v}) \quad \forall \boldsymbol{v} \in V, \tag{3.289}$$

where

$$\begin{aligned} a(\boldsymbol{u}^n; \boldsymbol{u}^{n+1}, \boldsymbol{v}) = \int_\Omega &((k(\boldsymbol{x}) - \tfrac{2}{3}\mu(\gamma^2(\boldsymbol{u}^n))\vartheta(\boldsymbol{u}^{n+1})\vartheta(\boldsymbol{v}) \\ &+ 2\mu(\gamma^2(\boldsymbol{u}^n))e_{ij}(\boldsymbol{u}^{n+1})e_{ij}(\boldsymbol{v})]\mathrm{d}\boldsymbol{x}, \end{aligned} \tag{3.290}$$

$$S(\boldsymbol{v}) = \int_\Omega F_i v_i \mathrm{d}\boldsymbol{x} + \int_{\Gamma_\tau} P_i v_i \mathrm{d}s. \tag{3.291}$$

The main result is the next theorem:

Theorem 3.40: Let the functional $L(\boldsymbol{v})$ defined by Equation 3.259 be defined on a Hilbert space V. Let the functional $L(\boldsymbol{v})$ have a continuous differential which is strictly monotone, i.e.

$$DL(\boldsymbol{u} + \boldsymbol{v}, \boldsymbol{v}) - DL(\boldsymbol{u}, \boldsymbol{v}) \geq c \|\boldsymbol{v}\|^2, \quad c > 0. \tag{3.292}$$

Let the bi-linear form $a(\boldsymbol{w};\boldsymbol{u},\boldsymbol{v})$ be symmetric in $\boldsymbol{u},\boldsymbol{v}$ and such that

$$a(\boldsymbol{u};\boldsymbol{v},\boldsymbol{v}) \geq c\,\|\boldsymbol{v}\|^2, \quad c > 0, \tag{3.293}$$

$$|a(\boldsymbol{w};\boldsymbol{u},\boldsymbol{v})| \leq c\,\|\boldsymbol{u}\|\,\|\boldsymbol{v}\|, \tag{3.294}$$

$$a(\boldsymbol{u};\boldsymbol{u},\boldsymbol{v}) = DL(\boldsymbol{u},\boldsymbol{v}), \tag{3.295}$$

$$\tfrac{1}{2}a(\boldsymbol{u};\boldsymbol{v},\boldsymbol{v}) - \tfrac{1}{2}a(\boldsymbol{u};\boldsymbol{u},\boldsymbol{u}) - L(\boldsymbol{v}) + L(\boldsymbol{u}) \geq 0 \tag{3.296}$$

hold. Let $\boldsymbol{f} \in V$. Let $\boldsymbol{u}$ be the solution of the problem

$$DL(\boldsymbol{u},\boldsymbol{v}) = (\boldsymbol{f},\boldsymbol{v}) \quad \forall \boldsymbol{v} \in V, \tag{3.297}$$

which exists and is unique, and where we denoted by

$$(\boldsymbol{f},\boldsymbol{v}) = \int_\Omega F_i v_i d\boldsymbol{x} + \int_{\Gamma_\tau} P_i v_i \mathrm{d}s.$$

Let us denote by $\boldsymbol{u}^n$ the n-th approximation of the solution of the problem $(\mathscr{P})_v$ and let $\boldsymbol{u}^{n+1}$ be the solution of the problem

$$a(\boldsymbol{u}^n;\boldsymbol{u}^{n+1},\boldsymbol{v}) = (\boldsymbol{f},\boldsymbol{v}) \quad \forall \boldsymbol{v} \in V, \tag{3.298}$$

which exists and is unique. Then

$$\boldsymbol{u}^n \to \boldsymbol{u} \quad \text{in } V.$$

Proof: The condition in Equation 3.292 of this theorem was proved above. Conditions in Equations 3.294 and 3.298 are evident. Since $0 < \mu_o \leq \mu(\boldsymbol{x},\gamma) \leq \mu_1 \leq +\infty$ and

$$\int_\Omega e_{ij}(\boldsymbol{v})e_{ij}(\boldsymbol{v})\mathrm{d}\boldsymbol{x} \geq \tfrac{1}{2}\int_\Omega (\partial v_i/\partial x_i)^2 \mathrm{d}\boldsymbol{x},$$

then Equation 3.293 holds. The last condition (Equation 3.296) follows from the following consideration. Let us put

$$\mathscr{M}(\gamma) = \int_0^\gamma \mu(t)\mathrm{d}t.$$

Then Equation 3.296 yields

$$\int_\Omega [\mu(\gamma(\boldsymbol{u}))(\gamma(\boldsymbol{v}) - \gamma(\boldsymbol{u})) - (\mathscr{M}(\gamma(\boldsymbol{v})) - (\gamma(\boldsymbol{u})))]\mathrm{d}\boldsymbol{x} \geq \boldsymbol{0}.$$

$\mathscr{M}(\gamma)$ is a concave function if $\mathscr{M}'' = \mu' \leq 0$ and Equation 3.296 is satisfied.

Now we shall prove $\boldsymbol{u}^n \to \boldsymbol{u}$ in V. Let us put

$$\mathscr{L}_n(\boldsymbol{v}) = L(\boldsymbol{u}^n) - (\boldsymbol{f},\boldsymbol{v}) + \tfrac{1}{2}a(\boldsymbol{u}^n;\boldsymbol{v},\boldsymbol{v}) - \tfrac{1}{2}a(\boldsymbol{u}^n;\boldsymbol{u}^n,\boldsymbol{u}^n)$$

and

$$\varphi(\boldsymbol{v}) = L(\boldsymbol{v}) - (\boldsymbol{f},\boldsymbol{v}).$$

Due to Equation 3.296

$$\tfrac{1}{2}a(\boldsymbol{u}^n;\boldsymbol{u}^{n+1},\boldsymbol{u}^{n+1}) - \tfrac{1}{2}a(\boldsymbol{u}^n;\boldsymbol{u}^n,\boldsymbol{u}^n) - L(\boldsymbol{u}^{n+1}) + L(\boldsymbol{u}^n) \geq 0,$$

hence

$$\varphi(\boldsymbol{u}^{n+1}) \leq \mathscr{L}_n(\boldsymbol{u}^{n+1}) \leq \mathscr{L}(\boldsymbol{u}^n) = \varphi(\boldsymbol{u}^n),$$

as

$$\min_{v \in V} \mathscr{L}(v) = \mathscr{L}_n(\boldsymbol{u}^{n+1}).$$

From Equation 3.293 it follows

$$\begin{aligned} c\left\|\boldsymbol{u}^{n+1}-\boldsymbol{u}^{n}\right\|^{2} &\leq a(\boldsymbol{u}^{n};\boldsymbol{u}^{n+1}-\boldsymbol{u}^{n},\boldsymbol{u}^{n+1}-\boldsymbol{u}^{n}) \\ &= (\boldsymbol{f},\boldsymbol{u}^{n+1}-2\boldsymbol{u}^{n})+a(\boldsymbol{u}^{n};\boldsymbol{u}^{n},\boldsymbol{u}^{n}) \\ &= 2(\varphi(\boldsymbol{u}^{n})-L_{n}(\boldsymbol{u}^{n+1})) \\ &\leq 2(\varphi(\boldsymbol{u}^{n})-\varphi(\boldsymbol{u}^{n+1})). \end{aligned}$$

Since $\varphi(\boldsymbol{u})$ is bounded from below then

$$\lim_{n\to\infty}\left\|\boldsymbol{u}^{n+1}-\boldsymbol{u}^{n}\right\|=0.$$

Since the assumptions of Equations 3.292, 3.294, 3.295, 2.397 and 3.298 are satisfied then

$$\begin{aligned} c_{1}\left\|\boldsymbol{u}^{n}-\boldsymbol{u}\right\|^{2} &\leq DL(\boldsymbol{u}^{n}-\boldsymbol{u},\boldsymbol{u}^{n}-\boldsymbol{u}) \\ &= DL(\boldsymbol{u}^{n},\boldsymbol{u}^{n}-\boldsymbol{u})-DL(\boldsymbol{u},\boldsymbol{u}^{n}-\boldsymbol{u}) \\ &= a(\boldsymbol{u}^{n};\boldsymbol{u}^{n},\boldsymbol{u}^{n}-\boldsymbol{u})-(\boldsymbol{f},\boldsymbol{u}^{n}-\boldsymbol{u}) \\ &= a(\boldsymbol{u}^{n};\boldsymbol{u}^{n+1},\boldsymbol{u}^{n}-\boldsymbol{u})-(\boldsymbol{f},\boldsymbol{u}^{n}-\boldsymbol{u})+a(\boldsymbol{u}^{n};\boldsymbol{u}^{n}-\boldsymbol{u}^{n+1},\boldsymbol{u}^{n}-\boldsymbol{u}) \\ &= a(\boldsymbol{u}^{n};\boldsymbol{u}^{n}-\boldsymbol{u}^{n+1},\boldsymbol{u}^{n}-\boldsymbol{u}) \\ &\leq c_{2}\left\|u^{n}-\boldsymbol{u}^{n+1}\right\|\left\|\boldsymbol{u}^{n}-\boldsymbol{u}\right\|. \end{aligned}$$

Hence, in the limit $n\to\infty$ we find

$$\lim_{n\to\infty}\boldsymbol{u}^{n}=\boldsymbol{u},$$

which completes the proof.

Since the secant modules method converges very rapidly to the exact solution of the problem given, it is a very useful method for computation in such types of geomechanical problems. In this case we compute at every iteration the linearized problem $(\mathscr{P}_{\ell})_{v}$ (Equation 3.289). The linearized problems then are solved by the FEM technique.

Remark: It can be shown that the method of contraction is a special case of the secant modules method. For details see Céa (1971), Nečas and Hlaváček (1981).

3.3 STRESS-STRAIN MEASUREMENT IN SITU AND ANALYSIS OF THE MEASURED DATA

3.3.1. Problems, Approaches, and a Short Survey of Instrumentation

Knowledge of the state of stress in the Earth is fundamental for understanding virtually all tectonic processess. It is also of importance to the design of large structures like reservoir dams and large underground structures like underground mines, or radioactive and toxic industrial waste repositories.

Rock parameters which are to be measured or monitored by the field instrumentation in connection with the problems related to the project of in-situ stress-strain determination can be broadly classified as:

(i) rock mechanics site constants like in-situ stress, in-situ strain and in-situ strength of rock mass,

(ii) variable geodynamical and geomechanical parameters like ground movements and variation in stresses and strains,

(iii) parameters to evaluate performace of different strata control systems like fill pressure monitoring, etc..

Depending on use, the field instruments may be distinguished as:

Capital instruments, i.e. those which can be employed repeatedly like the hydrofracturing in-situ stress measuring equipment, overcoring equipment for in-situ stress measurements, borehole shear testers, telescopic and tape closuremeters, sliding micrometers, borehole inclinometer probes, the TRIVEC borehole probe, extensodefectometers, the C.S.I.R.-triaxial strain cell, rod, wire, laser, and borehole hydraulic strainmeters, tiltmeters, etc..

Consumable instruments, i.e. those which are irrecoverable or lost in stoping areas like stressmeters, borehole extensometers, instrumented bolts, load cells of hydraulic or remote indicating type and remote indicating closuremeters, doorstopper techniques based on resistant tensometry as well as probes for in-situ stress-strain monitoring.

To determine the present values of stress-strain fields at different locations in boreholes, the hydrofracturing technique and the overcoring method can be used. These types of methods will be discussed further.

The hydrofracturing technique was originally developed for fracture reopenings for oil production and then for measurements in deep mines where breakdown pressures are high, and subsequently also in geodynamical studies for the determination of the stress field in the lithosphere at large depths down to several kilometres. Stress determination requires the measurement of pressure data during fracturing and refracturing tests, and the consideration of the induced fracture plane orientation. The pressure data includes the breakdown pressure during fracture induction, the refracture pressure for fracture reopening, and the shut-in pressure to keep the induced fracture open against the acting normal stress. But the major difficulty in such measurements is that the packers have to withstand enormous differential pressures to seal the injection interval, since in the mines no water is, in general, present in the borehole to decrease the differential pressure. The hydraulic pressure is generated by an oil pump or by an air-driven water pump. The fracture orientation is generally obtained by an impression packer in connection with a magnetic borehole orientation tool. Unlike stress relief methods, it does not depend on the knowledge of rock moduli and is not unduly influenced by small scale heterogeneities in the rock properties or stress field. However, there are still certain complications in the use of hydrofracturing as an in-situ stress measurement technique, particularly in connection with the interpretation of the pressure time data and the fracture orientation for an inclined borehole. The great disadvantage of this method consists in the violation of rock in the place of the measurement so that here the stress-strain monitoring cannot be applied (Zoback and Haimson (1983), Warren and Smith (1985)).

The overcoring technique using borehole deformation gauges of several types was used for determination of the present stress-strain state in rock massif.

The stress relief methods are used extensively due to their convenience. Because of the necessity to overcore, the borehole length is usually limited to tens of metres. In recent practice of the determination of the present stresses by overcoring methods, two main techniques- the doorstopper and the U.S.B.M. borehole deformation technique- are used. In the doorstopper technique, strain gauge rosettes are attached to the base of a borehole. The radial strains are monitored as the tectonic stresses are relieved by overcoring. A component of the radial strain when plotted against the overcore depth typically peaks off at a depth roughly equal to the borehole radius, and then decays to a residual value (Leeman (1967), (1969), (1971), Engelder (1984), Borsetto, Mantinetti and Ribacchi (1984), Rahn (1984)). In the U.S.B.M. (United States Bureau of Mines) borehole deformation technique, a borehole of relatively large diameter is drilled and then a pilot hole of relatively small diameter is drilled further into the rock massif along the same axis. The instrument consists of six stainless steel buttons each of which rests on a strain gauge cantilever of 33 mm length, all encased in an air-tight stainless steel body. The strain gauges are placed to read bending strain of the cantilevers and are connected through an 8-wire cable to a switching unit and a strain indicator. Deformation of the borehole causes the buttons to move and thus increase or reduce the bending strain on the cantilevers. This bending strain is transformed into a diametral strain through a microprocessor in the strain indicator. The strain indicator is then balanced and the diametral strain is read out. Moreover, the switching unit helps record the deformation in the three co-planar directions.

Both methods are applied in at least three non-parallel, preferably orthogonal boreholes. Then from the secondary principal stresses and their directions in each borehole, the average ground stress components and also the principal stresses and their directions are computed (Wong and Walsh (1985), Wang and Wong (1987), Amadei and Goodman (1982)).

We saw that the stress-strain analysis can be based on several types of methods. We also saw that because of the necessity to overcore, the borehole length is usually limited to tens of metres. The author and his co-workers have found such types of methods which can also be useful for very deep boreholes. They have investigated several types of techniques for measuring the stress-strain field in situ. In fact, the author studied the following types of methods based on these principles:

(i) *The laser technique* — the measuring probe consisted of a laser of a semiconductor type, an interferometer and a postprocessing electronic device. It is the best method for measuring small displacements. Unfortunately, the types of lasers used strongly depend on temperature. Nevertheless, it is a method for future application.

(ii) *The resistential tensometric technique* — this method is often used for tensometric measurements in boreholes. The sensitivity of this technique is not very high.

(iii) *The semiconductor resistential tensometric technique* — the sensitivity is twice as high as in the previous case but it is strongly dependent on temperature.

(iv) *The Moiré technique* — the method is based on the analysis of interferential figures. The sensitivity of this method is not very high and the construction of probes in the boreholes is very difficult. The method is used for measurements of cracks and faults in accessible places in mines.

(v) *Hall's probe technique* — this method is not usable for tensometric probes in boreholes as the measuring elements are strongly dependent on the temperature and are very unstable.

(vi) *The induction measuring technique* — the method is based on the principle of large changes of electrical parameters with small displacements. The sensitivity of the method is very high, and the construction of the probe is not very difficult. The method is useful for measurements in boreholes and has been thus used.

(vii) *String tensometry* — the principle of this method is based on the relation between the tension of a steel measuring string and the frequency of its own natural transversal oscillations. The construction of the probe for the overcoring method as well as for the stress-strain in-situ monitoring is relatively simple. The transportation of measured data and its postprocessing are also relatively simple. The sensitivity is very high and is about half an order higher than in the case of the induction method. These are the reasons why this method has been investigated and why it will be discussed in this section in more detail.

Recent overcoring methods are useful only near the surface; because of the necessity to overcore, the borehole length is usually limited to tens of metres. The goal of the author's investigations was to find methods and techniques rendering application of the overcoring method to greater depths in the lithosphere, and consequently the stress-strain monitoring in the same places in the boreholes, where the overcoring method was applied. As a result of these investigations, the author showed that 3D string tensometry can be a useful technique for overcoring in boreholes of tens or hundreds of metres or several kilometres in depth, as well as for the stress-strain monitoring in places where the overcoring method was applied. Moreover, the author and Dr. Mantič showed that the technique of drilling at least two or more non-parallel holes is not so correct, and they presented a new approach for the postprocessing of measured stress-strain data. These problems will be solved in the next sections.

3.3.2. On Sensibility of the String Tensometric Method for the Stress-Strain Monitoring In Situ and the Overcoring Method

3.3.2.1. Introduction

We stated above that the overcoring method can be applied in very deep boreholes of about several kilometre depths as well as to the stress-strain monitoring in places where the overcoring method was applied. One of the best methods for such studies is based on string tensometry. A strain rosette consists of three (for plane elasticity) or more (in the case of spatial elasticity) straight surface lines (small pipes) which pass through or surround a point and which represent the axes of strain gauges employed to determine stress-strain state at that point. It has to be said that the stress-strain analysis is the art of arranging strain rosettes (measuring elements) at a number of points in the rock massif to be investigated, i.e. in shallow or deep boreholes as well as on walls of objects (i.e. of mines, nuclear power plants, high level radioactive waste repositories, etc.), of carrying out the measurements and computing the state of surface stress at these points. Stress-strain rosette analysis is usually based on the assumptions of isotropic and homogeneous material and of strain gradients so small that the strains can be considered substantially uniform over the area covered by the rosette gauges. These assumptions are not correct in rock massif, but in practice a simplification has to be implemented. For stress- strain measurements, the sensitivity of the rosettes (measuring elements) is the fundamental information. The probe for stress- strain analysis of rock massif discussed can also be constructed with the help of string tensometry (see Petřík, Kratochvíl and Záruba (1975), Nedoma *et al.* (1985a,b), (1986), Nedoma (1990c), Nedoma (1998)). As was shown in Nedoma (1990c), Nedoma (1998) the string tensometric method is a very important technique for stress-strain measurements in deep boreholes - for the overcoring method as well as for stress-strain monitoring as the measurement is relatively simple due to easy digitization of measured frequencies of oscillating strings.

3.3.2.2. Principle of the string tensometric method, and sensitivity of the method

The main principle of string tensometry (Petřík (1953), Nedoma (1990c)) is the relation between the tension of the steel measuring string and the frequency of its own natural transversal oscillations. The reading of the measured quantity can be carried out by (a) direct determination of the string frequency or the length of its period, or (b) using the compensation method to compare the frequency of the string with the frequency of a continuous tune comparable oscillator. The first technique is much better for further processing of obtained data and, therefore, was applied in our developed probes.

Frequency of the measuring steel string f ([c/s] or [Hz]) stretched by the force P ([dyn] or [N]), is given by

$$f = n(2l)^{-1}P^{1/2}(\rho S)^{-1/2}, \tag{3.299}$$

so that the relative extension (strain) ε is defined by

$$\varepsilon = \sigma/E = 4l^2\rho(nE)^{-1}f^2, \tag{3.300}$$

where l ([cm] or [m]) is the length of the string, n the order of harmonics, ρ ([gcm^{-3}] or [kgm^{-3}]) the density, ε, σ are strains and stresses with $\varepsilon = E^{-1}\sigma$, $\sigma = P/S$ ([Nmm^2] or [MPa]), S= $\frac{1}{4}\pi$d^2 is the area of the transverse cross-section of the string of diameter d ([cm] or [m]), and E is Young's modulus ([Nmm^{-2}]) of steel of the string. Further, $\sigma/\rho = c^2$, where c physically represents the speed of oscillations of the measuring string. For example, the stretching force $P = \sigma S$ for the measuring string of diameter d = 0.22 mm, frequency $f = 1500$ Hz, and length 35 mm is $P = \sigma S = \rho 4 l^2 f^2 \pi \frac{1}{4} \mathrm{d}^2 = \rho l^2 f^2 \pi \mathrm{d}^2 = 3.26781.10^6$ dyn = 3.3332 kg*.

Then the relative (comparative) strain Dε of the string is given by

$$D\varepsilon = \varepsilon_2 - \varepsilon_1 = 4l^2\rho(nE)^{-1}(f_2^2 - f_1^2), \tag{3.301}$$

where f_1, f_2, ε_1, ε_2 are two frequencies of the string and the corresponding strains. The equation represents the fundamental relation between the strain and frequency of the string. For the relative error of determination of the extension of the string from Equation 3.300, we obtain

$$2 \log f = \log \varepsilon + \log \mathrm{En}(4l^2\rho)^{-1},$$

hence

$$\mathrm{d}\varepsilon/\varepsilon = 2\mathrm{d}f/f, \tag{3.302}$$

which means that the relative error of the extension is equal to double the value of the relative error of the frequency and it is independent of the length of the string. The absolute error of the relative (comparative) strain $\delta\varepsilon$ due to the error of frequency measurement is given by

$$\delta\varepsilon = 2\varepsilon\delta f/f = 8\rho(\mathrm{nE})^{-1}l^2 f\delta f = 4(\rho\varepsilon)^{1/2}l(\mathrm{nE})^{-1/2}\delta f, \tag{3.303}$$

which means that the error of the measurement of $\delta\varepsilon$ due to δf grows with frequency and is independent of the length of string. From the third relation, we see that with the same relative strain ε the error of measurement increases with the string length. Therefore, short strings with low pre-tensions will be more sensitive.

These relations were derived under the assumptions of the existence of an ideal string and, ideal outside conditions, and the assumption of a small amplitude of an ideal string when the tension of the string is always constant. But unfortunately there exist many phenomena having negative influence on the string, such as:

(i) influence of the fixation,
(ii) influence of the temperature,
(iii) influence due to the deviation from Hooke's law,
(iv) influence of elongation and contraction of the measuring string,
(v) influence of rigidity of the measuring string,
(vi) influence of magnitude of the string amplitude,
(vii) influence of damping and outside media influences of non-linearity and damping,
(viii) influence of excitation of non-damped oscillations,
(ix) influence of plastic deformations and/or of creep,
(x) bounds of usefulness of the string tensometric method.

From all the facts further discussed, there follow criteria for the developed electronic circuit, as well as for construction of the string tensometric elements and probes.

3.3.2.3. Influences acting on the sensitivity of the string

In this section we shall investigate the main influences having some effects on the sensitivity of measurements in situ. To investigate these influences, it is necessary to possess knowledge about the measurement accuracy. Here we give the main ideas only. For more detail see Nedoma (1990c).

Influence of the Fixation, Temperature, Contraction, and Elongation on the Amplitude of Oscillations and on the Rigidity of the Measuring String

The fixation of the measuring string depends on the fixation force developing maximal clutching press of ties onto the measuring string. Local plastic deformations will be decreased by the use of indurated ties from steel of high rigidity and of special shape. Sticking, soldering, etc. is not useful as it results in new influences on the system given by physical properties of the paste or solder. In practice, the fixation of the string is not a point fixation as is assumed in the ideal case. The influence of fixation can be considered as a part of non-linearity, which will be investigated later.

The influence of temperature is one of the fundamental ones. The relative change of the string length evoked by the temperature change $\mathrm{d}t(^{o}\mathrm{C})$ is

$$\mathrm{d}\varepsilon_t = (\alpha_r - \alpha_s)\mathrm{d}t, \tag{3.304}$$

where α_r and α_s are thermal expansions of the rock and the measuring string. One of the thermal sources is the heat produced by the oscillation of the string ($\sim \rho\beta_{ij}T_o e_{ij}(\dot{\boldsymbol{u}})$, where ρ is the density of the string, β_{ij} is the coefficient of thermal expansion, T_o is the initial temperature and $e_{ij}(\dot{\boldsymbol{u}})$ is the strain rate tensor of the deformed string) or by Joule's heat. For this reason, the construction of electromagnets has to be done in such a way that the originating Joule's heat would be minimal. Due to damping, losses of the string energy are observed. The energy losted has to be compensated by energy delivered. For this reason, the amplitude of oscillation of the measuring string has to be minimal and the time of oscillation also has to be minimal. These conditions must be realized, and they have been realized by the author and his co-workers during the construction of string tensometers and tensometric probes.

In the course of derivation of the relation between the relative extension and the frequency, i.e. Equation 3.300, it was assumed that the length of the string during its loading is constant, and that in the case where it changes, its elongation is so small that it can be neglected. Unfortunately, these assumptions are not correct in practice as the contractions and elongations of measuring strings are observed. To derive the influence of contraction and elongation of the measuring string we will assume that during loading of the string the prolongation is given by the relative extension ε, so that the initial length l_o changes to $l = l_o(1+\varepsilon)$ and the initial contraction S_o to $S = S_o(1-\nu\varepsilon)$, where $\nu = \rho S$ is the mass of unit length of the measuring string, and also that the initial density ρ_o changes to $\rho = \rho_o(1+\varepsilon)^{-1}(1-\nu\varepsilon)^{-2}$. Then we find for the frequency of the string f that

$$f^2 = n^2(4l_o^2(1+\varepsilon)^2\rho_o)^{-1}E\varepsilon(1+\varepsilon)(1-\nu\varepsilon)^2. \tag{3.305}$$

Hence, after some algebra and assuming that $\nu = 0.3$ for steel, we find

$$\varepsilon = 4l_o^2\rho_o E^{-1} f^2 + \delta\varepsilon, \tag{3.306}$$

where E is Young's modulus, and $\delta\varepsilon \sim 1.6\varepsilon^2$ represents the error of measured relative extension. For changes of relative extensions less than 0.75%, the term $\delta\varepsilon \sim 1.6\varepsilon^2 < 10^{-6}$ and, therefore, can be neglected.

In the previous investigations, we also assumed that the string oscillates with small amplitude and that the tension of the string is constant all the time. In practice, we have another situation. The amplitude of the string in the course of maximal deviation has a non-zero value and the tension of the measuring string increases. Moreover, the string does not oscillate harmonically and, as a result, its frequency is higher, about

$$\mathrm{d}f = \pi^2 E(32\rho l^4 f)^{-1}a^2, \tag{3.307}$$

where a is the amplitude, f frequency, ρ density, E Young's modulus and l length of the string. Then for the error of the relative extension (relative strain), we obtain

$$\delta\varepsilon = 2\varepsilon\delta f/f = 2\varepsilon \mathrm{d}f/f = 2\varepsilon\pi^2 E(32\rho l^4 f^2)^{-1}a^2 = \pi^2 a^2(4nl^2)^{-1}, \tag{3.308}$$

as $f^2 = (4l^2\rho)^{-1}En\varepsilon$. For small amplitudes, the error $\delta\varepsilon$ in the case of the influence of magnitude of the amplitude of the measuring string and the amplitude given is constant. In practice, the amplitude at lower frequencies is higher than that at higher frequencies. Therefore, in order to suppress the error greater than $\delta\varepsilon$, the increase of the amplitude δa has to be, in the whole region of measurement, less than the value

$$\mathrm{d}a = 2\pi^{-1}n^{1/2}l(\delta\varepsilon)^{1/2}. \tag{3.309}$$

The experimental practice analyses of a great number of measurements in time imply that in a great range of measurements and long-term, relatively high loading of the measuring string, the dependence between the stress and strain is not yet linear. We will assume that it is quadratic, i.e. of the form

$$\sigma = E\varepsilon + E_1\varepsilon^2 = Ee(1+e\varepsilon), \tag{3.310}$$

where $e = E_1E^{-1}$, E_1 is an empiric constant, E is Young's modulus, and σ and ε are the stress and strain. The linear part of Equation 3.310 prevails in the domain of small loading only, i.e. in the domain of validity of Hooke's law. These speculations are not based on the theory of elasticity as they are speculations of the experimental practice only. One can show (see Petřík, Kratochvíl and Záruba (1975), Nedoma (1990c)) that if two strings of lengths l_1 and l_2 oscillate with two successive frequencies f and f', then the relation between the corresponding relative strains ε and ε' is of the form

$$\varepsilon = k\varepsilon' + \delta\varepsilon, \tag{3.311}$$

where $\delta\varepsilon$ is the error of the measurement of a relative strain due to the influence of the deviation of contraction and length of the measuring string and is of the form

$$\delta\varepsilon = k(1-k)e(1+2\varepsilon_o'/\varepsilon')(\varepsilon')^2, \tag{3.312}$$

where $k = l_1/l_2$ and l_1, l_2 are lengths of the strings oscillating with the frequencies f and f'. This error is of the second order due to ε and depends on pre-stresses, on the ratio $k = l_1/l_2$, and on the constant $e = E_1E^{-1}$.

Finally, the influence of rigidity of the measuring string will be discussed. The differential equation describing the motion of the measuring string in the case where its rigidity is assumed is (see Nedoma (1990c))

$$\frac{\mathrm{d}^2 y}{\mathrm{d}t^2} = -\frac{EJ}{\rho S}\frac{\mathrm{d}^4 y}{\mathrm{d}x^4} + \frac{\sigma}{\rho}\frac{\mathrm{d}^2 y}{\mathrm{d}x^2}, \tag{3.313}$$

where E is Young's modulus, J is the momentum of inertia of the string cross section with respect to the string axis, $\sigma = E\varepsilon$ represents Hooke's law, ρ is the density of the string, and S is the area of the string cross section. For $EJ = 0$, we obtain the wave equation for the measuring string without rigidity. The fixation of the string is realized by its compression in the ties so that the boundary conditions are of the form: $y = 0$, $\mathrm{d}y/\mathrm{d}x = 0$ for $x = 0$, $x = l$, l being the string length, and, moreover, the initial condition for $t = 0$ in the form $y = 0$ is given. An analytical solution does not exist and, therefore, it must be found numerically only.

In the case when the rigidity of the string is small then the frequency of the string can be found in the form

$$f = (2l)^{-1}(\sigma/\rho)^{1/2} + \delta f, \tag{3.314}$$

where $\delta f = \mathrm{d}(4l^2)^{-1}(E/\rho)^{1/2}$. The equation represents the influence of the rigidity of an oscillating string on its own natural frequency. The change of frequency of the string is inversely proportional to the square of the string length and is independent of the prestress of the string.

Influence of the Excitation of Damped and Non-Damped Oscillations

Our previous discussions were valid under the assumption that the measuring string oscillates in a vacuum. But in practice, the measuring string oscillates in the air which damps its oscillation. We will assume that the resistance of the medium is proportional to the instantaneous speed and that it has the opposite direction, i.e. $R = -c\mathrm{d}y/\mathrm{d}t$, where c is a constant of proportionality between the resistance and the speed of the oscillating string. The harmonic motion of the measuring string is caused by the force $P = -k^2y$. Then the equation of motion of the string is of the form $m\mathrm{d}v/\mathrm{d}t = R + P$ or of the form

$$\mathrm{d}^2y/\mathrm{d}t^2 + 2b\mathrm{d}y/\mathrm{d}t + \omega^2 y = 0, \tag{3.315}$$

where $v = \mathrm{d}y/\mathrm{d}t$, $\omega = k^2/m$ is the circular velocity and $b = -c(2m)^{-1}$ is the constant of damping. The solution of this equation is of the form

$$y = a_o \exp(-bt)\cos(\omega_o^2 - b^2)^{\frac{1}{2}}t = a\cos\omega t, \tag{3.316}$$

where $a = a_o \exp(-bt)$ is the amplitude of the damped oscillation, $\omega = (\omega_o^2 - b^2)^{1/2}$ is the circular frequency of the damped oscillation, ω_o is the circular frequency of the non-damped oscillations, and γ is the constant of integration. We see that the circular frequency of damped oscillations is less than that of non-damped oscillations. The period of oscillation $T = 2\pi/\omega = 2\pi/(\omega_o - b^2)^{1/2}$ increases with the increase of damping. After the lapse of one cycle $t = T$, we have $y = a_o \exp(-bt)$. The ratio of two successive deviations is constant. This ratio is called the damping λ, $\lambda = \exp(bt)$, and its natural logarithm $\delta = \ln\lambda = bT$ is called the logarithmic decrement. Hence for the damping constant b, we obtain $b = \delta T^{-1} = \delta\mathrm{f}$. The damping of the measuring string is the consequence of its own inner friction of mass, the air resistance, and the damping in the magnetic field of the device, as well as the consequence of the energy which is damped in the fixation of the string. The damping depends on the amplitude and frequency of the string. Unfortunately, single components cannot be determined theoretically. In practice, we know that for a given measuring string and electromechanical transducer, the damping constant b is, in a sufficiently large range, practically independent of the frequency.

Decrease of the natural frequency of the measuring string as a consequence of damping is indirectly proportional to the frequency. From $\omega = (\omega_o^2 - b^2)^{1/2}$ we find that $b = \omega_o^2 - \omega^2 = \frac{1}{2}(\omega_o + \omega)2(\omega_o - \omega)$. For a small damping, $\frac{1}{2}(\omega_o + \omega) \sim \omega_o$ and if we put $\mathrm{d}\omega_o = \omega_o - \omega$ then $\mathrm{d}\omega_o = b^2/(2\omega)$. Since $\omega = 2\pi f$ then $\mathrm{d}f = b^2(8\pi^2 f)^{-1}$ and thus

$$\mathrm{d}f/f \sim \tfrac{1}{2}\mathrm{d}\varepsilon/\varepsilon = b^2(8\pi^2 f^2)^{-1} = b^2 l^2 \rho(2\pi^2 E\varepsilon)^{-1}. \tag{3.317}$$

Hence $\mathrm{d}\varepsilon = b^2 l^2 \rho\pi^{-2} E^{-1}$ and we see that the error in determining the relative strain due to the damping is constant for the given measuring string, and is independent of frequency and elongation and, therefore, it has no great importance during measuring. Moreover, it was proved experimentally that the influence of electrical circuits is very low so that it can be neglected.

The direct influence of the outside medium onto the natural frequency of the measuring string is caused by a layer of molecules of the outside medium attached to the surface of the measuring string (i.e. by adhesion), which results in an increase of mass of the length unit of the measuring string. The frequency of the measuring string is given by the relation

$$f = f_o(1 - \tfrac{1}{2}\rho_p\rho^{-1}), \tag{3.318}$$

where f_o is the frequency of the measuring string in vacuum, ρ_p is the density of the medium, and ρ is the density of the measuring string. For the air $f \sim f_o$, as the error of measurement is less than 0.2% and can be neglected.

We saw that due to the damping the measuring string oscillates with damped oscillations. Then in the case of non-damped oscillations, we have to compensate the energy lost through the damping by the excitation of the string. Let the exciting force be of a harmonic type, i.e. $P_v = P_{vo}\sin\omega_o t$, where ω_o is the circular frequency of non-damped oscillations of the measuring string. Then an outer harmonic force with frequency ω_o operates on the initial damped oscillator — the measuring string, e.g. $P_v = P_{vo}\sin(\omega_o t + \varphi)$, where φ represents the effect of instability and different properties of a feedback loop of string oscillator. Then we put

$$P_v = P_{vo}\sin(\omega_o t + \varphi). \tag{3.319}$$

The equation of motion of the string oscillator is of the form

$$\mathrm{d}^2y/\mathrm{d}t^2 + 2b\mathrm{d}y/\mathrm{d}t + \omega^2 y = P_v \equiv P_{vo}\sin(\omega_o t + \varphi). \tag{3.320}$$

Its generalized solution is then

$$y = a_o\exp(-bt)\sin((\omega_o^2 - b^2)^{1/2}t + \psi) + a_v\sin(\omega_o t - \varphi). \tag{3.321}$$

We see that the motion of the string is determined by two components, i.e. by damped oscillations and non-damped oscillations. Thus the measuring string, instead of its own natural damped oscillations with frequency ω, oscillates with non-damped screwing oscillations with frequency $\omega_o = \omega$. Then during a short time, the damped oscillations disappear and the string oscillates with non-damped forced oscillations. Their amplitude and phase delay are given by

$$y = a_v\sin(\omega_o t - \varphi). \tag{3.322}$$

But since they have to satisfy the equation of motion we find

$$a_v = P_{vo}[((\omega^2 - \omega_o^2)^2 + 4b^2\omega_o^2)^2]^{-1}, \quad tg\varphi = 2b\omega_o(\omega^2 - \omega_o^2)^{-1},$$

and since $b = \delta f = \delta\omega_o(2\pi)^{-1}$,

$$tg\varphi = 2b\omega_o(\omega^2 - \omega_o^2)^{-1} = \delta\pi^{-1}(\omega^2/\omega_o^2 - 1)^{-1}, \tag{3.323}$$

hence

$$\omega/\omega_o = (1 + \delta(\pi tg\varphi)^{-1})^{1/2},$$

where δ is the logarithmic decrement of damped oscillations. Thus

$$\mathrm{d}\omega_o/\omega \sim -\tfrac{1}{2}\pi^{-1}\delta tg^{-1}\varphi. \tag{3.324}$$

This condition determines the choice of gain of the oscillator. The minimal error of decrease of the frequency of the measuring string caused by the excitation of non-damped oscillations is bounded by the difference of frequencies of the string oscillator in these limit states. The error is so small that it can be neglected. In practice, this type of oscillator is not very good. Similarly, the solution in the case when the dissipated energy of the measuring string is compensated by an exciting force, which is represented by a periodically repeated series of pulses, has no applications. These circuits are unstable, and excited higher harmonics were also observed. Therefore, the author and his co-workers developed a new string oscillator, the idea of which is the following: The excitation of the measuring string is carried out in such a way that a microprocessor emits exciting pulses and successively changes the frequency long as it finds the frequency which corresponds to the own natural frequency of the measuring string (see Figure 3.3). The measuring string starts oscillating. It can be excited by the same frequency as its own frequency or, which is better, by a sub-harmonic frequency. When a lower sub-harmonic is used then the maximal step of its period ΔT can be determined. But the energy needed is then delivered very slowly

Figure 3.3. Excitation of the Measuring String

and the interval of periods, which have to be acquired, is extensive. Our experiments and theory show that for the string with its own natural frequencies around 1 kHz, the third sub-harmonic is best. In fact, from the analysis of electric signals, it is known that their spectrum is defined by

$$S(n\omega_o) = T^{-1} \int_{(-1/2)\pi}^{(1/2)\pi} f(t) \exp(-in\omega_o t) \mathrm{d}t, \tag{3.325}$$

where n is the number of sub-harmonics, T is its period, τ is the length of the pulse, and ω_o is the circular frequency. In our case

$$f(t) = \begin{cases} 1 \cdots & \text{for } -\frac{1}{2}\pi < t < \frac{1}{2}\pi \\ 0 \cdots & \text{for } \frac{1}{2}\pi < t < T - \frac{1}{2}\pi, \end{cases}$$

and, therefore,

$$S = S(n\omega_o) = T^{-1} \int_{(-1/2)\tau}^{(1/2)\tau} f(t) \exp(-in\omega_o t) \mathrm{d}t = \tau T^{-1} \sin n\omega_o \tfrac{1}{2}\tau (n\omega_o \tfrac{1}{2}\tau)^{-1}. \tag{3.326}$$

It can be shown that the energy excited by harmonic components decreases with the increasing ratio τ/T and then, in the case when τ/T is not an entire fraction, all harmonic components are present. Therefore, all sub-harmonics can be used, although the intensity of excitation decreases with the lapse of sub-harmonics. However, the best ratio is $\tau/T \sim 0.2$ when the choice of 2 to 4 sub-harmonics is concerned. The first sub-harmonic is unsuitable as the exciting pulses can overlap the signal of rising oscillations of the measuring string and, therefore, it cannot be detected. The best one is the third sub-harmonic.

Influences of Non-Linearity and Damping

The oscillating measuring string is, in general, an oscillator, actually a non-linear oscillator with quadratic non-linearity. The equation of motion of a weakly non-linear system is of the form

$$\frac{\mathrm{d}^2 y}{\mathrm{d}t^2} + 2b\frac{\mathrm{d}y}{\mathrm{d}t} + \omega^2(1 + \alpha y^2)y = \beta P_v(y, t), \tag{3.327}$$

where y represents the deviation of the string, $b = c(2m)^{-1}$ the damping of the string, c being a proportionality constant between the resistance and the velocity of the oscillating string, m the mass of the string, ω the circular frequency, α the non-linearity coefficient of the system, caused by the rigidity of the measuring string, by inaccuracies in its fixation, by the non-linearity of the electronic circuit, etc., β is a parameter, and $P_v(y, t)$ represents the exciting force.

First, let us assume that $\alpha = 0$, $P_v = 0$. Then we have the case of as oscillating string with damping.

Second, for β being small and $\alpha = 0$, $P_v(y, t) = P_{vo}\sin(\omega_o t + \varphi)$, and we have the case of a string excited by non-damped oscillations, where ω_o is a circular frequency of non-damped oscillations. The measuring string oscillates with damped oscillations.

Third, let us assume that $\alpha y^2 \ll 1$, $b = 0$ (as the error in determination of the relative strain caused by damping can be neglected) and that the exciting force P_v is given by a periodic series of Dirac functions with period T, by which we approximate rectangular pulses, i.e.

$$P_v(y, t) = P_o(t)y = \omega y \sum_{k=-\infty}^{\infty} \delta(t - kT), \tag{3.328}$$

where $T = 2\pi/\omega_o$, $P_o(T) = y^{-1}P_v(y, t)$, and we let $\omega_o \ll \omega$. Then the equation of motion is of the form

$$\mathrm{d}^2y/\mathrm{d}t^2 + 2b\,\mathrm{d}y/\mathrm{d}t + \omega^2(1 + \alpha y^2)y - \beta\omega y \sum\nolimits_{k=-\infty}^{\infty} \delta(t - kT) = 0. \tag{3.329}$$

Thus for sufficiently large disturbances β and a sufficiently large non-linearity α, the system in Equation 3.329 has a periodic character. At the same time, the change of the string oscillator has a character which is close to a random character, and the energy of the string oscillator increases on average with time. Therefore, the system will be stochastically unstable. The solution of Equation 3.329 is

$$y = a_o \cos((\omega + D\omega)t + \varphi), \tag{3.330}$$

where $D\omega = \frac{3}{8}\alpha a_o^2\omega$. The author showed (Nedoma (1990c)), that the condition for instability is of the form

$$\beta\frac{D\omega}{\omega_o} \gg 1, \tag{3.331}$$

which represents the condition on which the phase of the string oscillator behaves chaotically. Since $I = \omega a_o^2$, I represents an adiabatic invariant of the non-linear string oscillator and Equation 3.331 can be rewritten as

$$\beta I\frac{\mathrm{d}D\omega(I)}{\mathrm{d}I}\omega_o^{-1} \gg 1, \quad D\omega = \tfrac{3}{8}\alpha I, \tag{3.332}$$

where DI represents the change of the adiabatic invariant system caused by the exciting force $P_o(t)$ during one period. The relation in Equation 3.332 holds for an arbitrary non-linearity.

If the exciting force $P_v(y, t) = P_o(t)y = a_o y\cos\omega_o t$, $\omega_o \ll \omega$, then

$$a_o \cong \omega^2(1 + \alpha y)^2, \quad \beta \cong \exp(-\omega/\omega_o)$$

and the condition of Equation 3.331 is never realized. Thus the system is stable.

If the exciting force is caused by the δ function then its frequency spectrum can be determined. The frequency spectrum of

$$P_o(t) = \omega \sum_{k=-\infty}^{\infty} \delta(t - kT), \quad T = 2\pi/\omega_o,$$

then has an infinite width.

The exciting forces can also be considered with a limited frequency spectrum. This is the case when the width of the frequency spectrum is sufficiently large, but such that the effective width of the pulse is much smaller than the period of oscillations of the measuring string (e.g. the case of rectangular pulses).

In the case when the exiting force is caused by a series of n pulses of width τ/n then the spectral representation of the function $P_o(t)$ is characterized by a great number of harmonics. This is the reason why the electronic circuit based on this type of exciting force is unstable. Such an electronic circuit has been verified experimentally with the same result as follows from the theory.

Therefore, as a result of all the above studies, the author and his co-worker (see Nedoma and Hiršl (1990)) developed an electronic circuit — a stable exciter for the excitation of oscillations of the measuring string. The excitation of oscillations of the measuring string is based on the idea that a microprocessor emits exciting pulses and then the frequency of the string is changed up to the corresponding natural frequency of the measuring string. At this moment, the measuring string begins to oscillate. To obtain reliable results, a statistical method applied to the measured data has to be used. Then the main value of the measured data is taken for the frequency of the measuring string, and the standard deviation informs us about the measurement in situ.

3.3.2.4. *Exciting force with a random phase*

We saw above that the exciting force can be a harmonic force $P_v = P_{vo} \sin \omega_o t$. In general, the exciting force can be assumed to be a random variable $\eta = P_v$. Then to every value of $P_v (-P_{vo} \le P_v \le P_{vo})$, there correspond infinitely many values

$$t_k = \pi k \omega_o^{-1} + (-1)^k \omega^{-1} \arcsin(P_v/P_{vo}), \quad k = 0, \pm 1, \pm 2, \ldots .$$

Then the probability density $W(y)$ is found, under the assumption that the random variable t has a uniform distribution on the interval $-\pi\omega^{-1} \le x \le \pi\omega^{-1}$ (i.e. $w(x) = \omega/(2\pi)$ for $|x| \le \pi/\omega$ and $w(x) = 0$ for $|x| > \pi/\omega$), in the form (see Nedoma (1990c))

$$W(y) = (\pi P_{vo}(1 - (y/P_{vo})^2)^{-1/2} \quad \text{for } |y| < P_{vo}, \tag{3.333}$$
$$W(y) = 0 \quad \text{for } |y| > P_{vo}.$$

This density depends on the amplitude of the exciting force P_{vo} and does not depend on the period $T = 2\pi/\omega$. The distribution function $F(y)$ is given by

$$F(y) = \tfrac{1}{2} + \pi^{-1} \arcsin y/P_{vo} \quad \text{for } |y| \le P_{vo}, \tag{3.334}$$
$$F(y) = 0 \quad \text{for } |y| > P_{vo}.$$

In the case when the exciting force, due to the fixation of the string and due to the location of the electromagnetic transducer out of the centre of the measuring string, etc., does not have quite a sinusoidal form, then, by expanding into a Fourier series and omitting the even components, the exciting force can be written in the form

$$P_v = \sum\nolimits_{k=1}^{n} P_{vok} \sin \omega_k t_k. \tag{3.335}$$

The density of every term of the series in this Equation is given by Equation 3.333. It depends on the amplitude P_{vok} only, and does not depend on the frequency ω_{ok}.

3.3.2.5. Problem of the reliability of measuring elements

As follows from the analysis presented above, the basic factor of the reliability of measuring elements is their accurate construction and their accurate assembly, as well as the maximal elimination of creep of the measuring strings. This last factor can be influenced by the choice of a suitable sort of steel (namely in the case of the construction of the probe for an aggressive environment) as well as by a suitable technology of turning the measuring string old. The long-term ability of the measuring element to work without disorders depends on several random factors in the mechanical part of the element as well as in the electronic circuit of the string oscillator. In the case of spatial elasticity of rock massif, we need to know six components of the stress tensor, which, for the construction of tensometric probes, means six independent measuring string elements. To obtain correct values of the stress-strain analysis of rock massif, rosettes with six, seven, eight, or nine measuring elements (and, better, with a greater number of measuring elements obtained by the enlargement, or at least doubling, of measuring string elements, which represents no great problem for the designer in string tensometry technique) are needed. It is the reason why the tensometric probes developed were constructed in such a way that the number of independent as well as dependent measuring string tensometers was sufficiently great. In the case of a defect of one measuring string element, this defective element is automatically disqualified and the measurement can continue further. For the reliability of an electronic circuit, all the results of the theory of electronic circuit reliability can be applied and the requirement of high quality of the hybrid integrated string oscillator circuit has to be imposed; in particular, in parallel connection and also in possible military specifications. For measuring string elements and probes constructed in such a way, there are no problems in proposing hybrid integrated string oscillators also with pre-processed measured data.

3.3.2.6. In-situ stress measuring system based on string tensometry

String Tensometers and Tensometric Rosettes

A string tensometer consists of cylindrical heads endowed with recesses and mutually connected by a soldered-in non-magnetic (e.g. brass or non-magnetic steel) thin-walled tube Ø 3/2.6 mm. The measuring string is anchored in the collet. Transversal oscillations of the measuring string are induced through the wall of the tube with the help of electromechanical transducers. The gauge base corresponds to the customer's requirement, and the sensitivity is $d\varepsilon[\mu m/m]/df$ [Hz] $\sim 0.5 \times 10^{-6}$ to $2. \times 10^{-6}$/Hz in dependence on the length of the gauge base. The frequency range is 600 Hz to 2.5 kHz with its centre at 1 kHz.

A 60° tensometric rosette consists of three heads with recesses mutually connected by tubes. A 45° tensometric rosette consists of four heads with recesses pairwise connected by tubes. Measuring strings are anchored in collets in the heads. Tensometric rosettes are suitable also for the overcoring method. The gauge base is 38 mm for the borehole of Ø 76 mm, 71 mm for that of Ø 112 mm, or it corresponds to the customer's requirement. The sensitivity is $d\varepsilon[\mu m/m]/df$ [Hz] ~ 0.38 to 0.5×10^{-6}/Hz.

Probes for the Overcoring Method and for the Stress-Strain In-Situ Monitoring

The probes employ measuring elements (sensors) of different shape according to the mode of use. In the overcoring method, a 60° rosette is employed for 2D elasticity and two 60° rosettes mutually turned and connected by measuring strings are employed for 3D elasticity (Figure 3.4). The relative strain measured is transferred to gauge tensometers via hydraulically stretched contacts (*3*) that ensure contact with the borehole wall. The relative strain caused by relieving after drilling out is measured by a 2D and/or 3D rosette of string tensometers consisting of measuring strings (*1*) anchored in anchoring heads and forced to oscillate with the help of electromechanical transducers (*2*). The strain measured is transferred to tensometers via contacts (*3*) stretched hydraulically in the pressure area in front of pistons. The contacts (*3*) are connected through sliding joints with the anchor stones of the string tensometers (*1*) and endowed with a blocking device (*5*). The water-tight cover of the probe consists of the casing with a separable part of the probe. The contacts (*3*) are sealed with rubber O-rings. The inside of the probe is filled with compressed inert gas to balance the pressure of borehole watering on the O-rings. The electronic part of the probe consists of a string oscillator and, moreover, a control and evaluation electronic device (in fact, a microcomputer). In the 3D sensors the 60° rosette are mutually turned about several degrees.

The measuring system consists of the measuring string (*1*) fixed in supporting heads (stones). The single-coil electromechanical transducer (*2*) is placed in the reach of the measuring string (*1*). Its input as well as output is connected via the amplifiers to the microcomputer. The microcomputer is further connected to the communication system and evaluation unit. The electromechanical transducer is designed as a single-coil system with a coil including a magnetic core and constant magnetic biassing. It is employed to force oscillations and, at the same time, to detect the frequency of natural oscillations of the steel string. The measurement of the oscillation of the gauge string is done repeatedly and the results are processed by statistical methods in the microcomputer. The output from the microcomputer is the mean value, standard deviation, and, possibly, further statistical data that characterizes the quality of measurement and is stored in the device memory. After removing the probe from the borehole, the data from the microcomputer is transferred by the communication system to the evaluation device where it is processed by methods described in the following paragraph.

To monitor stress, gauge units are employed in the form of a tetrahedron, two 60° rosettes mutually inclined and connected, or two 60° rosettes mutually turned and connected. These systems can then be doubled, tripled, etc.. The system of forcing the oscillations and measuring the frequency of the gauge string is a modification of the system employed in the overcoring method, with the only difference consisting in the

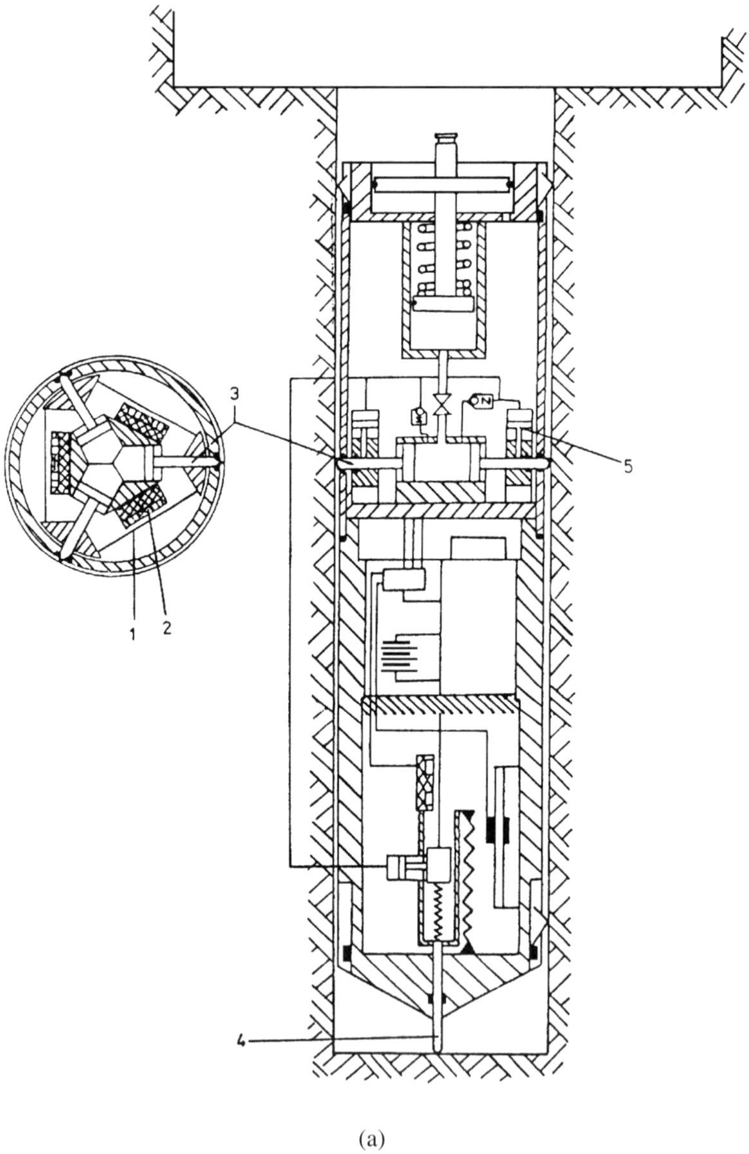

(a)

Figure 3.4. The Tensometric Probe for the Overcoring Method

(b)

Figure 3.4. *(continued)*

fact that the quantities measured are continually transferred by the communication system to the evaluation unit (a workstation) for further processing. Several basic types are shown in Figure 3.5.

The frequency range is 600 Hz to 2.5 kHz, the sensitivity of both the devices is similar to that of 60° rosettes, i.e. $d\varepsilon[\mu m/m]/df[Hz] \sim 0.5 \times 10^{-6}$ to $2. \times 10^{-6}$/Hz depending on the length of the gauge base. Probes for the 2D and 3D overcoring method have been designed for Ø 59 mm and Ø 76 mm, those for the monitoring in a series of types from Ø 76 mm to Ø 112 mm (or more) in dependence on the diameter used for the overcoring method.

3.3.3. Mathematical Models of In-Situ Stress-Strain Measurements based on the Method of Homogenization

3.3.3.1. Introduction

Geodynamical processes passing in the upper parts of the Earth are the results of the processes in deeper parts of the Earth's interior. For better understanding of these geodynamic processes and their consequences — the earthquakes — only complete studies of stress-strain fields in the investigated global and local regions can yield new data for precise analysis of geodynamical processes in collision zones, as well as

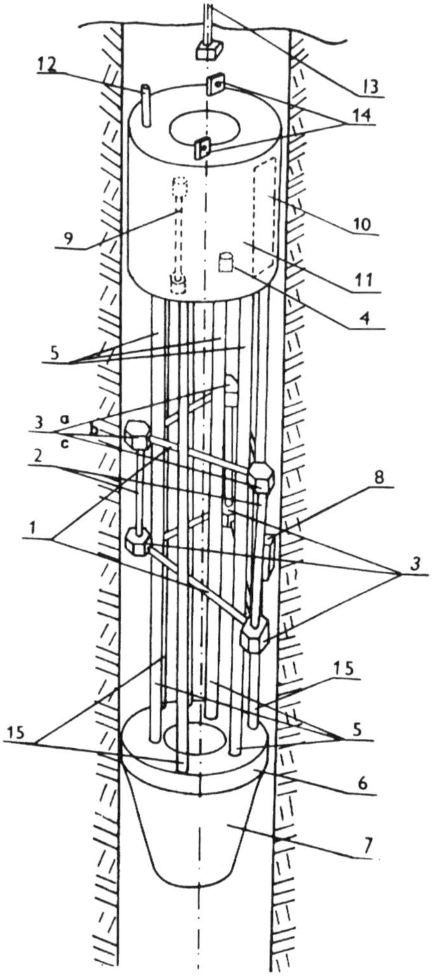

1. array of string tensometers,
2. string sensor,
3. fastening stones of string tensometers and the measured contact,
4. geophon,
5, 15. guides for the tensometric probe elements
6, 7. front of probe,
8. electromechanic transducers,
9. thermometer,
10. electronic equipment for the electromechanic transducers and transmission of the measured data onto the Earth's surface,
11. body of the probe for the electronic equipment,
12. cable for the transmission of the measured data,
13. descending device,
14. attachment for the descending device

(a)

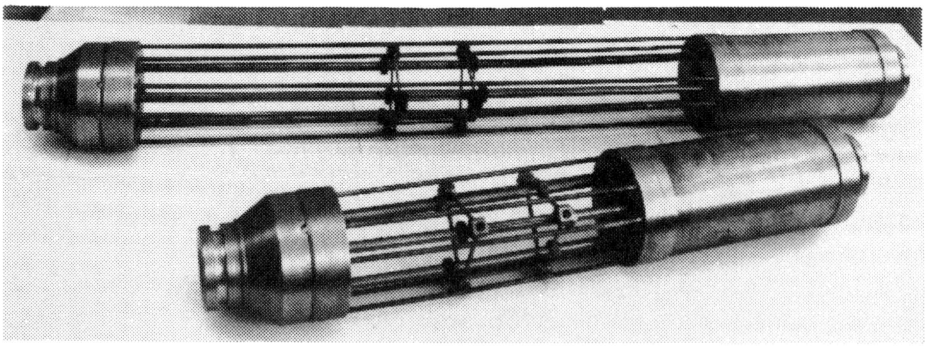

(b)

Figure 3.5. The Tensometric Probe for Stress-Strain Monitoring (a) schematic description of the probe: (b) the probe for deep boreholes, (c) the probe for shallow boreholes

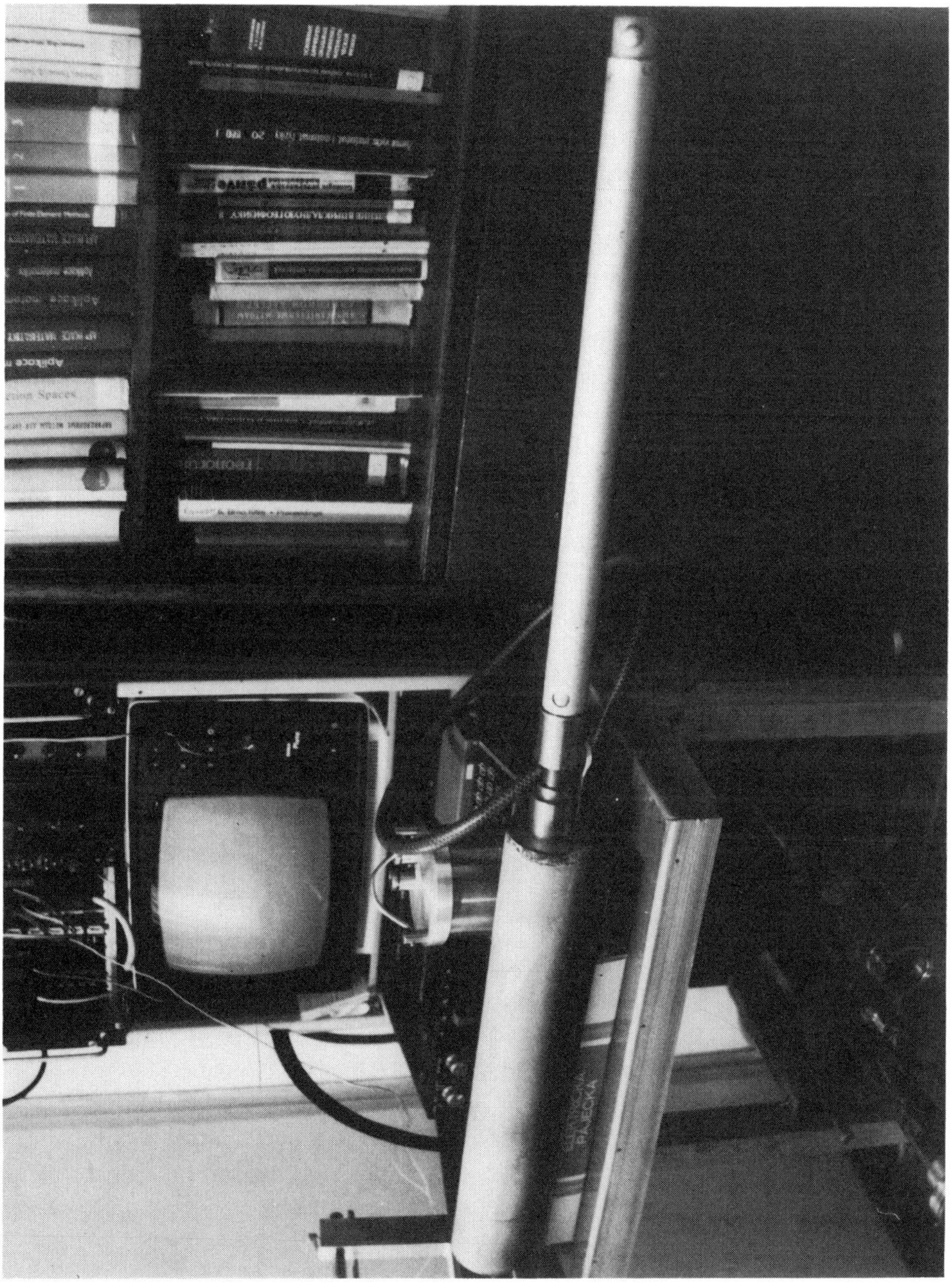

(c)

Figure 3.5. *(continued)*

for a better earthquake prediction. Such studies have to be based on the determination of stresses and strains, microcracking, and seismicity in deep boreholes located in places where the deformation energy increases and which are of great importance from the geodynamical point of view. The determination of in-situ stress-strain fields can be made by the overcoring method and by the monitoring of stress-strain fields in situ. In what follows, we will analyze models of the overcoring method and in-situ stress-strain monitoring.

3.3.3.2. Formulation of the problem and the main results

In geological as well as geophysical practice, real lithospheric rocks and geological structures are approximated by isotropic materials. Lithospheric rocks, however, are characterized by very complicated structures with anisotropic properties. Therefore, such isotropic approximations are very trivial and are not adequate for real geological situations. Nevertheless, lithospheric rocks are loaded with a very large load resulting in strongly anisotropic stress fields. To obtain some information about this stress-strain field in the lithosphere, stress-strain measurements based on the overcoring method and in-situ stress-strain monitoring in time have to be carried out.

We can assume that real rocks are formed of materials with periodically heterogeneous (or stochastic) structures which correspond to composite materials. In these cases, simulations of lithospheric rocks would be nontrivial and, therefore, impracticable. The homogenization technique is an important method to solve such problems and simulate complicated anisotropic (in general, stochastic) properties of rocks. From the physical point of view, the homogenization means replacing periodically heterogeneous (or stochastic) materials by equivalent homogeneous ones. From the mathematical point of view, in the case of periodically heterogeneous structures of rocks, it means that the equation with highly oscillating periodic coefficients is approximated by an equation with constant coefficients, which describes the anisotropic behaviour of rocks. The method was introduced by Babuška (1976) and then developed by many authors, such as Lions, Biroli, Bensoussan, Bachvalov, Skripnik, etc.

The mathematical theory of the evaluation of stress-strain data, obtained from stress-strain measurements in the lithosphere by applying the overcoring method and then by monitoring the stress-strain field in shallow or deep boreholes, was investigated for example by Wang and Wong (1987), Wong and Walsch (1987), Mantič (1988), Mantič and Nedoma (1989a,b), Nedoma, Bašlová and Mantič (1998).

Now we will describe a new approach to the postprocessing of measured stress-strain data introduced by Mantič and Nedoma. We will assume that, in a sufficiently large neighbourhood of the place where the stress-strain field will be measured and/or monitored, the stress-strain field before applying the borehole was inhomogeneous. We say that the stress-strain fields are inhomogeneous if they both depend on spatial co-ordinates in the investigated region. Moreover, by applying the borehole, the stress-strain fields are violated. Furthermore, we will assume that at larger distances from the borehole, the stress-strain fields are non-violated. Consider the in-situ stress-strain measurements by the overcoring method and/or by the monitoring, i.e. by long term observation of the stress-strain state in time. Then it is advantageous to use the string tensometric technique of measurements. The principle of string tensometry is the relation between the tension of the steel measuring string and the frequency of its natural transversal oscillations (in the range of kHz) which can be relatively easily digitized with sufficient accuracy and then further evaluated. In the case of stress-strain monitoring, after the borehole with the tensometric probe is filled up with concrete and after its hardening, the stresses are transferred to the concrete watering and the stress-strain field is then inhomogeneous in the neighbourhood of the measuring sensor (probe). This assumption is sufficient for all cases in which tensometric probes are situated at different geological levels (i.e. depths) in the borehole, placed at concrete benches of sufficient thickness so that the measurement is not influenced by the borehole drilling fluid. In the case of the overcoring method, we request that the wall of the borehole is sufficiently smooth only. On the assumption that the rock has constant Young's modulus E, Poisson's ratio ν, and density ρ, we can investigate the problem as a plane problem on a circle. From the analytical solution of such a problem, it follows (Timoshenko and Gudjev (1975)) that the relative error δ between the exact value of the stress tensor at distance R from the centre of the circle and the constant stress tensor at distance r is assumed to be less than the given value, e.g. 0.03 or smaller ($\delta \leq 0.03$). Then the relation $R/r \geq (2.5/\delta)^{1/2}$ has to be satisfied. Under the above assumptions, the minimal thickness of bench is approximately 1.2 m for the diameter of the borehole d $\sim$ 112 mm, or 70 cm for the diameter d $\sim$ 76 mm, or 2 m for d $\sim$ 216 mm. By a similar argument, we can restrict the investigated region at larger distances from the borehole, i.e., the above relation $R \geq (2.5/\delta)^{1/2} r$ is also valid, as follows from the analytical solution for a plate with a circular hole.

Our analysis of the measured data will be based on the idea that the expected stress and strain tensors can be found as linear combinations of bases of the spaces of all virtual stress and/or strain tensors. To construct these bases, a few boundary value problems for linear (quasi-)coupled thermo-elasticity and the 3D finite element method (FEM) for numerical simulations will be used.

Consider in-situ stress-strain measurements based on the overcoring method and the long-term observation of stress-strain changes in time. We will assume that in a sufficiently large domain in the neighbourhood of the place of the measurements, denoted by Ω, the rock massif is homogeneous and the stress-strain fields and their changes are inhomogeneous owing to the concentrations of stresses corresponding to the borehole. It is convenient to choose the domain so that the ratio of its radius R and the distances h and v to the radius of the borehole r is sufficiently large so that the stress-strain fields or their changes in time are, with a sufficient accuracy, homogeneous on the boundary. We denote the boundary by Γ_1^e. In what follows, we will investigate the overcoring method and the monitoring technique which can be applied in shallow and/or deep boreholes.

From the theory of spaces as well as from the theory of elasticity, it is known that a stress and/or strain tensor can be expressed as a linear combination of the basis of the space of the corresponding tensors. Since the stress and strain tensors are symmetric the bases of the corresponding tensors are of dimension six. To construct a basis of such spaces, we will simulate this basis by solving six different boundary value problems. These simulations can be based on:

(i) 3D boundary value problems of linear elasticity for the case of isotropic rocks, which were given by Mantič (1988), Mantič and Nedoma (1989a,b) and Nedoma (1990c),

(ii) 3D boundary value problems for the case of anisotropic rocks, which can also be simulated by the homogenization method, which will be discussed in this section (for details see Nedoma, Bašlová and Mantič (1998)).

The previous case of isotropic rocks is then a special instance of the anisotropic case and, therefore, it will not be investigated. In what follows, the two cases — the overcoring method and monitoring technique — will be investigated.

Overcoring Method

During the overcoring operation, we pass through the following situations:

(i) After the application of the borehole into rock massif, a tensometric probe with s measuring elements — tensometers — is located and fixed in the calibration borehole and adjusted in an initial state of the measurement. The original inhomogeneous tectonic stress-strain fields characterized by the stress and strain tensors τ_{ij}^* and e_{ij}^* $(i, j = 1, 2, 3)$ are changed due to the presence of the borehole, but at larger distances from the borehole they remain, with a sufficient accuracy, homogeneous.

(ii) After overcoring the calibration borehole (i.e. the borehole with smaller diameter), rock starts to be unloaded and then it deforms. The difference between the measurements of this state and of the initial state represents the original rock deformation caused by geotectonic processes.

In the course of drilling, the temperature of the rock core grows by the value Θ and thus the rock is also deformed by thermal stresses. Moreover, in both situations, the rock is deformed by the pressure p caused by the weight of the drilling fluid in the borehole.

Satisfying all the conditions discussed above, let us denote the investigated domain in the neighbourhood of the borehole by Ω in the case before applying the overcoring method and by Ω_o in the case after applying the overcoring method. Let us denote its boundary by $\partial\Omega = \Gamma_1^e \cup \Gamma_2^e \cup \Gamma^{\text{in}}$ in the case before applying the overcoring method and by $\partial\Omega_o = \Gamma_1^e \cup \Gamma_2^e \cup \Gamma_o^{\text{in}}$ in the case after applying the overcoring method. Let us denote by $\tau_{ij} = \tau_{ij}(\boldsymbol{x})$ the stress tensor of an unknown inhomogeneous stress field in the rock massif in the neighbourhood of the borehole. All these tensors create the space that we denote by S, and the space of all tensors τ_{ij}^*, corresponding to the rock massif violated due to applying the borehole, will be denoted by S^*.

Real situations according to items (i) and (ii) can be approximately described by boundary value problems of elasticity or (quasi-)coupled thermoelasticity in the domains Ω and Ω_o with the following boundary conditions:

Problem A1:

$$P_i(\boldsymbol{x}) = \tau_{ij}^* n_j(\boldsymbol{x}), \quad \boldsymbol{x} \in \Gamma_1^e, \tag{3.336}$$

$$P_i(\boldsymbol{x}) = -pn_i(\boldsymbol{x}), \quad \boldsymbol{x} \in \Gamma^{\text{in}}, \tag{3.337}$$

where $n_j(\boldsymbol{x})$, $j = 1, 2, 3$, is the outward normal to the boundary $\partial\Omega$, $\boldsymbol{P} = (P_i)$ represents the surface force and p the hydrostatic pressure caused by the borehole drilling fluid. On the boundary part Γ_2^e, the surface forces due to the inhomogeneity of the stress field are unknown (the lithostatic pressures can also be used). Moreover, their total moment and resultant force will be equal to the moment and resultant force of the surface forces on the fixed boundary Γ_2^e. In the vicinity of the calibration borehole the calculated stress field will be practically unaffected by the inaccurate boundary condition

$$u_i(\boldsymbol{x}) = 0, \qquad \boldsymbol{x} \in \Gamma_2^e, \tag{3.338}$$

if the distance H of the calibration borehole from the boundary Γ_2^e is sufficiently large.

Problem A2:

$$P_i(\boldsymbol{x}) = \tau_{ij}^* n_j(\boldsymbol{x}), \quad \boldsymbol{x} \in \Gamma_1^e, \tag{3.339}$$

$$P_i(\boldsymbol{x}) = -pn_i(\boldsymbol{x}), \quad \boldsymbol{x} \in \Gamma_o^{\text{in}}, \tag{3.340}$$

$$u_i(\boldsymbol{x}) = 0, \quad \boldsymbol{x} \in \Gamma_2^e. \tag{3.341}$$

Problem A3: In the case when the lithostatic pressure is considered we have the following boundary conditions:

$$P_i(\boldsymbol{x}) = \tau_{ij}^* n_j(\boldsymbol{x}), \quad \boldsymbol{x} \in \Gamma_1^e \cup \Gamma_2^e, \tag{3.342}$$

$$P_i(\boldsymbol{x}) = -pn_i(\boldsymbol{x}), \quad \boldsymbol{x} \in \Gamma_o^{\text{in}}. \tag{3.343}$$

During the application of the overcoring technique, the rock core is heated. The increase in temperature of the rock core by Θ is converted to kinematically equivalent fields of volume forces in the domain Ω_o and those of surface forces on Γ_o^{in}.

The stress field in the investigated domain satisfies the condition of equilibrium

$$\partial\tau_{ij}/\partial x_j + f_i = 0. \tag{3.344}$$

The stress-strain relation in the anisotropic case is defined by Hooke's law in linear elasticity,

$$\tau_{ij}(\boldsymbol{x}) = c_{ijkl}(\boldsymbol{x})e_{kl}(\boldsymbol{u}), \quad i, j = 1, 2, 3, \tag{3.345}$$

where the coefficients of elasticity $c_{ijkl}(\boldsymbol{x})$ satisfy the conditions of symmetry and the Lipschitz conditions

$$c_{ijkl}(\boldsymbol{x}) = c_{jikl}(\boldsymbol{x}) = c_{klij}(\boldsymbol{x}) = c_{ijlk}(\boldsymbol{x}), \tag{3.346}$$

$$c_o\,|\xi|^2 \le c_{ijkl}\xi_{ij}\xi_{kl} \le C_o\,|\xi|^2 \quad \text{for } \boldsymbol{x} \in \Omega, \tag{3.347}$$

where $c_o = 2\min\{\mu(\boldsymbol{x}), \boldsymbol{x} \in \Omega\}$ and $C_o = \max\{2\mu(\boldsymbol{x}) + 3\lambda(\boldsymbol{x}), \boldsymbol{x} \in \Omega\}$.

The stress-strain relation is also defined by Duhamel-Neumann's law in linear thermoelasticity,

$$\tau_{ij}(\boldsymbol{x}) = c_{ijkl}(\boldsymbol{x})e_{kl}(\boldsymbol{u}) - \beta_{ij}(\boldsymbol{x})(T(\boldsymbol{x}) - T_o(\boldsymbol{x})), \quad i, j = 1, 2, 3, \tag{3.348}$$

where the coefficients of elasticity $c_{ijkl}(\boldsymbol{x})$ satisfy Equations 3.346 and 3.347, $\beta_{ij}(\boldsymbol{x})$ represent coefficients of thermal expansion, $T(\boldsymbol{x})$ and $T_o(\boldsymbol{x})$ are the temperature and initial temperature at the point $\boldsymbol{x}$ of Ω, and $e_{ij}(\boldsymbol{u})$ is the small strain tensor defined by

$$e_{ij}(\boldsymbol{u}) = \tfrac{1}{2}(\partial u_i(\boldsymbol{x})/\partial x_j + \partial u_j(\boldsymbol{x})/\partial x_i), \quad i, j = 1, 2, 3. \tag{3.349}$$

For the case of isotropic rocks,

$$c_{ijkl} = \lambda(\boldsymbol{x})\delta_{ij}\delta_{kl} + \mu(\boldsymbol{x})(\delta_{ik}\delta_{jl} + \delta_{il}\delta_{jk}), \tag{3.350}$$

$$\beta_{ij}(\boldsymbol{x}) = \beta(\boldsymbol{x})\delta_{ij}, \tag{3.351}$$

where $\lambda(\boldsymbol{x})$ and $\mu(\boldsymbol{x})$ are the Lamé constants.

The change of temperature caused by the temperature of the rock massif and by applying the overcoring technique is described by the heat equation

$$\rho(\boldsymbol{x})c_e(\boldsymbol{x})\partial T(\boldsymbol{x}, t)/\partial t + \rho(\boldsymbol{x})\beta_{ij}(\boldsymbol{x})T_o(\boldsymbol{x})e_{ij}(\dot{\boldsymbol{u}}(\boldsymbol{x}, t)) - \partial/\partial x_i(\kappa_{ij}(\boldsymbol{x})\partial T(\boldsymbol{x}, t)/\partial x_j)$$
$$= W(\boldsymbol{x}, t) \quad \text{for a.e. } (\boldsymbol{x}, t) \in \Omega \times (t_o, t_1), \quad i, j = 1, 2, 3, \tag{3.352}$$

where $\rho(\boldsymbol{x})$ is density, $W(\boldsymbol{x}, t)$ thermal sources, and $\kappa_{ij}(\boldsymbol{x})$ coefficients of thermal conductivity satisfying the symmetry and Lipschitz conditions

$$\kappa_{ij}(\boldsymbol{x}) = \kappa_{ji}(\boldsymbol{x}), \tag{3.353}$$

$$\kappa_{ij}(\boldsymbol{x})\zeta_i\zeta_j \geq \kappa_o |\zeta|^2, \quad \kappa_o = \text{const} > 0. \tag{3.354}$$

Moreover, the term $\rho(\boldsymbol{x})\beta_{ij}(\boldsymbol{x})T_o(\boldsymbol{x})e_{ij}(\dot{\boldsymbol{u}}(\boldsymbol{x}, t))$ represents the dissipative energy evoked by the deformation of the rock massif and changed to heat.

For the isotropic case, $\kappa_{ij}(\boldsymbol{x}) = \kappa(\boldsymbol{x})\delta_{ij}$, $\beta_{ij}(\boldsymbol{x}) = \beta(\boldsymbol{x})\delta_{ij}$. But for simplicity, it is sufficient to assume the steady state case, i.e.,

$$\rho(\boldsymbol{x})\beta_{ij}(\boldsymbol{x})T_o(\boldsymbol{x})e_{ij}(\dot{u}(\boldsymbol{x})) - \partial/\partial x_i(\kappa_{ij}\partial T(\boldsymbol{x})/\partial x_j) = W(x) \text{ for a.e. } (\boldsymbol{x}) \in \Omega,$$
$$i, j = 1, 2, 3, \tag{3.355}$$

where the term $\rho(\boldsymbol{x})\beta_{ij}(\boldsymbol{x})T_o(\boldsymbol{x})e_{ij}(\dot{u}(\boldsymbol{x}))$ has a meaning of thermal sources.

We see that from the mathematical point of view, Problems A1 and A2 are mixed problems, and Problem A3 is the first problem of elasticity. From the mathematical theory of linear elasticity, it is known that the necessary conditions for the existence of the solution of the first problem of elasticity are the conditions of total equilibrium, i.e., the resultants of forces as well as of moments vanish:

$$\int_\Omega f_i \mathrm{d}\boldsymbol{x} + \int_{\partial\Omega} P_i \mathrm{d}s = 0, \tag{3.356}$$

$$\int_\Omega [\boldsymbol{x}, \boldsymbol{f}]_i \mathrm{d}\boldsymbol{x} + \int_{\partial\Omega} [\boldsymbol{x}, \boldsymbol{P}]_i \mathrm{d}s = 0, \quad i = 1, 2, 3. \tag{3.357}$$

Let the displacement fields $\boldsymbol{u}(\boldsymbol{x}), \boldsymbol{x} \in \Omega$, and $\boldsymbol{u}_o(\boldsymbol{x}), \boldsymbol{x} \in \Omega_o$, denote the solutions of Problems A1, A2, and A3. It is apparent that these solutions correspond, with a sufficient accuracy, to real stress-strain fields in the place of measurements in the borehole.

The dependence of τ_{ij} on τ_{ij}^* is linear, i.e. $\tau_{ij} = \mathscr{L}(\tau_{ij}^*)$ or $\tau_{ij} = \mathscr{L}(\tau_{ij}^*, p)$ (this case will be discussed in more detail in connection with the monitoring technique), where τ_{ij} is an element of the linear space S of all inhomogeneous stress tensors in the region Ω and the operator $\mathscr{L}$ is the linear operator $\mathscr{L} : S^* \to S$ (or $\mathscr{L} : S^* \times R \to S$ in the second case). This operator is one-to-one (single) as there exists only one solution to the given boundary conditions. The dimension of the space S is six. We can construct its natural basis as the solution of the above boundary value problem (one type of them), where we take the unit basis $\tau_{ij}^{(\iota)}$, $\iota = 1, \ldots, 6$ for the stress tensors τ_{ij}^*. We denote by $\tau_{ij}(\boldsymbol{x})$, $\boldsymbol{x} \in \Omega$, solutions of these problems. Then, before the violation of rock massif by the borehole, the stress tensor τ_{ij}^* can be written as a linear combination of the basis, i.e. in the form

$$\tau_{ij}^* = \sum_{\iota=1}^{6} c^{(\iota)} \tau_{ij}^{(\iota)}, \tag{3.358}$$

and the displacement vector as

$$\boldsymbol{u}^*(\boldsymbol{x}) = \sum_{\iota=1}^{6} c^{(\iota)} \boldsymbol{u}^{(\iota)}, \tag{3.359}$$

where $\boldsymbol{u}^{(\iota)}$ corresponds to the stress tensor $\tau_{ij}^{(\iota)}$.

Remark: In our further analysis, the last problem A3 will not be investigated.

Remark: In the formulation of these problems in thermo-elasticity, we replace the body force f_i by $f_i - \partial/\partial x_j(\beta_{ij}(T - T_o))$ and the surface force P_i by $P_i + \beta_{ij}(T - T_o)n_j$ in the anisotropic case, and by $f_i - \partial/\partial x_i(\beta(T - T_o))$ and $P_i + \beta(T - T_o)n_i$, where $\beta = (3\lambda + 2\mu)\alpha_t$, α_t is a coefficient of linear thermal expansion, in the isotropic case.

Remark: It is apparent that we have no chance of determining the exact value of the stress-strain field in the rock massif, anyway. We can only determine the stress-strain field in the rock massif which was violated by the borehole. Namely, every effect applied to rock massif violates the exact stress-strain field in this rock massif.

To compute the stress-strain field observed in situ by the overcoring method, we first analyse in more detail the strains on the individual measuring elements. The strains on the individual measuring elements (string tensometers) of the probe are observed during the in-situ measurement between the states before and after drilling. For the measured relative elongation defined as

$$\varepsilon^{(q,M)} = \Delta l(\boldsymbol{x}^{(q)}, \boldsymbol{x}^{(q)})/l(\boldsymbol{x}^{(q)}, \boldsymbol{x}^{(q)}), \tag{3.360}$$

it holds that (Mantič (1988))

$$\varepsilon^{(q,M)} = (\Delta \boldsymbol{u}(\boldsymbol{x}_1^{(q)}) - \Delta \boldsymbol{u}(\boldsymbol{x}_2^{(q)}))(\boldsymbol{x}_1^{(q)} - \boldsymbol{x}_2^{(q)}) \Big/ \left\| \boldsymbol{x}_1^{(q)} - \boldsymbol{x}_2^{(q)} \right\|^2, \tag{3.361}$$

where

$$\Delta \boldsymbol{u}(\boldsymbol{x}) = \boldsymbol{u}_o(\boldsymbol{x}) - \boldsymbol{u}(\boldsymbol{x}), \quad \boldsymbol{x} \in \Gamma_{\text{calib}}, \tag{3.362}$$

and where $\boldsymbol{x}_1^{(q)}$ and $\boldsymbol{x}_2^{(q)}$ are the position vectors of both ends of the q-th measuring element, Δl and l are the change of the length and the length of this element, and Γ_{calib} denotes the boundary of the borehole of small diameter (calibration borehole) in which the probe is located.

The computation of a relation between the unknown tectonic stress tensor τ_{ij}^* and the measured relative elongation (comparative, relative strain) of tensometers of the probe $\varepsilon^{(q,M)}$ is based on numerical solutions of some special cases of Problems A1 and A2. Let $\tau_{ij}^{(\iota)}$ ($\iota = 1, \ldots, 6$) be any basis of the Hilbert space of stress tensors $\tau^{3,3}$. We denote its elements by τ and σ. The scalar product is defined by

$$\sigma.\tau = \sigma_{ij}.\tau_{ij}.$$

Let $s_{ij}^{(\iota)}$ be the basis dual to $\sigma_{ij}^{(\iota)}$ in the sense that

$$\sigma^{(\iota)}.s^{(\upsilon)} = \delta_{\iota\upsilon}, \quad \delta_{\iota\upsilon} \text{ is the Kronecker symbol.}$$

Denote the displacement fields by

$$\boldsymbol{u}^{(\iota)}(\boldsymbol{x}), \boldsymbol{u}^{(\pi)}(\boldsymbol{x}), \quad \boldsymbol{x} \in \Omega,$$

and

$$\boldsymbol{u}_o^{(\iota)}(\boldsymbol{x}), \boldsymbol{u}_o^{(\pi)}(\boldsymbol{x}), \boldsymbol{u}_o^{(\Theta)}(\boldsymbol{x}), \quad \boldsymbol{x} \in \Omega_o,$$

i.e., the solutions of the special cases of Problems A1 and A2 changed by substitutions are

$$\tau_{ij}^* = \tau_{ij}^{(\iota)}, p = 0; \quad \tau_{ij}^* = 0, p = 1 \quad \text{in Equations 3.336 and 3.337,}$$

and

$$\tau_{ij}^* = \tau_{ij}^{(\iota)}, p = 0, \Theta = 0; \quad \tau_{ij}^* = 0, p = 1, \Theta = 0; \quad \tau_{ij}^* = 0, p = 0, \Theta = 1$$
$$\text{in Equations 3.339 to 3.341.}$$

Theoretically, we can express relative elongations (relative strains) of the q-th measuring elements of the probe between the states before and after drilling that are caused by tectonic stresses $\tau_{ij}^{(\iota)}$, $\iota = 1, \ldots, 6$, by the borehole drilling fluid pressure $p = 1$, and by the heating of the rock core by temperature $\Theta = 1$, as

$$\varepsilon^{(q,\iota)} = (\Delta\boldsymbol{u}^{(\iota)}(\boldsymbol{x}_1^{(q)}) - \Delta\boldsymbol{u}^{(\iota)}(\boldsymbol{x}_2^{(q)}))(\boldsymbol{x}_1^{(q)} - \boldsymbol{x}_2^{(q)}) \Big/ \left\|\boldsymbol{x}_1^{(q)} - \boldsymbol{x}_2^{(q)}\right\|^2, \quad (3.363)$$

$$\varepsilon^{(q,\pi)} = (\Delta\boldsymbol{u}^{(\pi)}(\boldsymbol{x}_1^{(q)}) - \Delta\boldsymbol{u}^{(\pi)}(\boldsymbol{x}_2^{(q)}))(\boldsymbol{x}_1^{(q)} - \boldsymbol{x}_2^{(q)}) \Big/ \left\|\boldsymbol{x}_1^{(q)} - \boldsymbol{x}_2^{(q)}\right\|^2, \quad (3.364)$$

$$\varepsilon^{(q,\Theta)} = (\Delta\boldsymbol{u}_o^{(\Theta)}(\boldsymbol{x}_1^{(q)}) - \Delta\boldsymbol{u}_o^{(\Theta)}(\boldsymbol{x}_2^{(q)}))(\boldsymbol{x}_1^{(q)} - \boldsymbol{x}_2^{(q)}) \Big/ \left\|\boldsymbol{x}_1^{(q)} - \boldsymbol{x}_2^{(q)}\right\|^2. \quad (3.365)$$

Let the unknown tectonic stress tensor τ_{ij}^* be the linear combination of elements of the basis $\tau_{ij}^{(\iota)}$, i.e.

$$\tau_{ij}^* = \tau_{ij}^{(\iota)}.c^{(\iota)}.$$

Then

$$c^{(\iota)} = s_{ij}^{(\iota)}.\tau_{ij}^{*}.$$

The difference $\Delta\boldsymbol{u}(\boldsymbol{x})$ of the displacement fields $\boldsymbol{u}_o(\boldsymbol{x})$ and $\boldsymbol{u}(\boldsymbol{x})$ on the boundary of the calibration borehole Γ_{calib} is easily seen to be

$$\Delta\boldsymbol{u}(\boldsymbol{x}) = c^{(\iota)}(\boldsymbol{u}_o^{(\iota)}(\boldsymbol{x}) - \boldsymbol{u}^{(\iota)}(\boldsymbol{x})) + p.(\boldsymbol{u}_o^{(\pi)}(\boldsymbol{x}) - \boldsymbol{u}^{(\pi)}(\boldsymbol{x})) + \Theta\boldsymbol{u}_o^{(\Theta)}(\boldsymbol{x}), \quad \boldsymbol{x} \in \Gamma_{\text{calib}}.$$

Then the measured relative elongations (relative strains) are given by

$$\varepsilon^{(q,\iota)} = c^{(\iota)}\varepsilon^{(q,\iota)} + p\varepsilon^{(q,\pi)} + \Theta\varepsilon^{(q,\Theta)}.$$

Putting

$$E_{ij}^{(q)} = \varepsilon_{ij}^{(q,\iota)}.s_{ij}^{(\iota)},\ \varepsilon^{(q,C)} = \varepsilon^{(q,M)} - p\varepsilon^{(q,\pi)} - \Theta\varepsilon^{(q,\Theta)},$$

where the superscript C denotes "correction", we obtain

$$E_{ij}^{(q)}.\tau_{ij}^{*} = \varepsilon^{(q,C)}. \tag{3.366}$$

After ordering the components of the stress tensor τ_{ij}^{*} into a column vector

$$\vec{\tau}^{*} = (\tau_{11}^{*}, \tau_{22}^{*}, \tau_{33}^{*}, \tau_{12}^{*}, \tau_{13,}^{*} \tau_{23}^{*})^{T} \tag{3.367}$$

and the components $E_{ij}^{(q)}$ into an $(s, 6)$ matrix $\boldsymbol{E}$, where s denotes the number of measuring elements, then the q-th row is

$$(\boldsymbol{E})_q = (\boldsymbol{E}_{11}^{(q)}, \boldsymbol{E}_{22}^{(q)}, \boldsymbol{E}_{33}^{(q)}, 2\boldsymbol{E}_{12}^{(q)}, 2\boldsymbol{E}_{13}^{(q)}, 2\boldsymbol{E}_{23}^{(q)}). \tag{3.368}$$

Writing the relative elongations (relative strains) $\varepsilon^{(q,C)}$ into the column vector

$$\vec{\varepsilon}^{(C)} = (\varepsilon^{(1,C)}, \varepsilon^{(2,C)}, \ldots, \varepsilon^{(s,C)})^{T}, \tag{3.369}$$

the relation in Equation 3.366 is equivalent to the system of linear equations

$$\boldsymbol{E} \cdot \vec{\tau}^{*} = \vec{\varepsilon}^{(C)}. \tag{3.370}$$

Note that the entries of the matrix $\boldsymbol{E}$ do not depend on the choice of the basis $\tau_{ij}^{(\iota)}$. To solve Equation 3.370 uniquely, the number s of the measuring elements has to be greater than or equal to six and the columns have to be linearly independent (which depends on the spatial arrangement of tensometers in the sensor of the probe). The solution of the generally "overdetermined" system in Equation 3.370 is a vector $\vec{\tau}^{*}$ minimizing the Euclidean norm of the residual $\left\|\varepsilon^{(C)} - \boldsymbol{E}\vec{\tau}\right\|$. If $\boldsymbol{E}^{+}$ denotes the Moore-Penrose generalized inverse of $\boldsymbol{E}$ (see Natterer (1986)), then it holds that

$$\vec{\tau}^{*} = \boldsymbol{E}^{+}\vec{\varepsilon}^{(C)}. \tag{3.371}$$

To calculate the entries $\boldsymbol{E}_{ij}^{(q)}$ in Equation 3.370 for a given type of tensometric probe located in the borehole, we have to solve the above $3D$ boundary value problems by the finite element method. If rocks are of periodic or stochastic structures then the homogenization method can be used with great success. It is apparent that the overcoring method discussed in this section can be applied to very deep watering and non-watering boreholes.

Stress-Strain Monitoring In Situ

Analysis of the stress-strain in-situ monitoring is analogous to the above and, therefore, many details can be omitted. It can be noted that the probe for the stress-strain in-situ monitoring can be located in places where the overcoring method was applied as, by the application of the overcoring method, the wall of the borehole is not destructed as in the case of the hydrofracturing technique.

Let us denote the investigated region in the neighbourhood of the borehole by Ω, and its boundary by $\partial\Omega$. Let $\Omega = \Omega_1 \cup \Omega_2$ where Ω_1 denotes the region created by rocks in the neighbourhood of the borehole and Ω_2 denotes the region occupied by concrete watering in the place of location of the probe in the borehole. The boundary of the investigated domain Ω will be denoted by $\partial\Omega = \Gamma_1^e \cup \Gamma_2^e \cup \Gamma_3^e \cup \Gamma_m^{\text{in}}$, where Γ_1^e, Γ_2^e, Γ_3^e denote the "outer part" of the boundary of the region Ω and Γ_m^{in} represents the "inner" boundary of the borehole.

Let us denote by $\tau_{ij}^* = \tau_{ij}^*(\boldsymbol{x})$ the unknown stress tensor of the initial inhomogeneous stress field in rock massif in the neighbourhood of the borehole. Such tensors create the space denoted by S^*. By opening the borehole as well as by erecting the concrete bench, the initial stress-strain fields are violated. Let us denote the resulting stress field by $\tau'_{ij} = \tau'_{ij}(\boldsymbol{x})$, $\boldsymbol{x} \in \Omega$. Let $R \gg r$ and H be sufficiently large ($H = 2R + h_s$, where h_s is the vertical dimension of the measuring sensor, r, R are diameters of boreholes and h_3 is the distance of the probe from Γ_3^e). First let

$$\tau'_j(\boldsymbol{x}) = \tau_{ij}^*(\boldsymbol{x}), \quad \boldsymbol{x} \in \Gamma_1^e \cup \Gamma_2^e \cup \Gamma_3^e, \quad i, j = 1, 2, 3. \tag{3.372}$$

Then for the surface forces on $\partial\Omega$ we find that

$$\tau'_{ij}(\boldsymbol{x}) n_j(\boldsymbol{x}) = \tau_{ij}^* n_j(\boldsymbol{x}) = P_i(\boldsymbol{x}), \quad \boldsymbol{x} \in \partial\Omega = \Gamma_1^e \cup \Gamma_2^e \cup \Gamma_3^e, \quad i, j = 1, 2, 3. \tag{3.373}$$

Second, let

$$\tau'_{ij}(\boldsymbol{x}) = \tau_{ij}^*(\boldsymbol{x}), \quad \boldsymbol{x} \in \Gamma_1^e \cup \Gamma_2^e, \quad i, j = 1, 2, 3. \tag{3.374}$$

We find for the surface forces on this part that

$$\tau'_{ij}(\boldsymbol{x}) n_j(\boldsymbol{x}) = \tau_{ij}^* n_j(\boldsymbol{x}) = P_i(\boldsymbol{x}), \quad \boldsymbol{x} \in \Gamma_1^e \cup \Gamma_2^e, \quad i, j = 1, 2, 3. \tag{3.375}$$

The surface forces on Γ_3^e are unknown in this case but they can be written in the form

$$\tau'_{ij}(\boldsymbol{x}) n_j(\boldsymbol{x}) = \tau_{ij}^* n_j(\boldsymbol{x}) = P_i(\boldsymbol{x}), \quad \boldsymbol{x} \in \Gamma_3^e, \quad i = 1, 2, 3. \tag{3.376}$$

The surface forces are equal to zero on the boundary Γ_m^{in} if the drilling watering is not present in the borehole, i.e., if

$$\tau'_{ij}(\boldsymbol{x}) n_j(\boldsymbol{x}) = 0, \quad \boldsymbol{x} \in \Gamma_m^{\text{in}}, \quad i, j = 1, 2, 3,$$

or in the case when the drilling watering is present, then

$$\tau'_{ij}(\boldsymbol{x})n_j(\boldsymbol{x}) = P_i(\boldsymbol{x}) = -pn_i(\boldsymbol{x}), \quad \boldsymbol{x} \in \Gamma_m^{\text{in}}, \quad i = 1, 2, 3, \tag{3.377}$$

where p is the hydrostatic pressure caused by the drilling watering in the place $\boldsymbol{x} \in \Gamma_m^{\text{in}}$. The stress-strain relation is given by Hooke's law in the case of elasticity and by Duhamel-Neumann's law in thermo-elasticity. The thermal field is determined by the solution of the boundary value problem for the heat equation similarly as above.

Remark: The hydrostatic pressure of watering p corresponds to lithostatic stresses in rock massif. Its physical meaning consists of the elimination of the closing of the borehole.

The analysis of the stress-strain fields in situ leads to the following boundary value problems in elasticity and/or thermo-elasticity.

Problem B1: Let the surface forces $P_i(\boldsymbol{x})$ be given by

$$\tau'_{ij}(\boldsymbol{x})n_j(\boldsymbol{x}) = P_i(\boldsymbol{x}), \quad \boldsymbol{x} \in \partial\Omega = \Gamma_1^e \cup \Gamma_2^e \cup \Gamma_3^e, \quad i = 1, 2, 3, \tag{3.378}$$

due to Equation 3.373 on the boundary $\partial\Omega$ of the region Ω.

The stress tensor τ in the region Ω satisfies the equation of equilibrium

$$\partial\tau'_{ij}/\partial x_j + \rho f_i = 0, \quad i, j = 1, 2, 3, \tag{3.379}$$

where the stress-strain relation is defined by Hooke's law in the elastic case and/or by Duhamel-Neumann's law in the thermo-elastic case.

From the mathematical theory of elasticity viewpoint, it is known that the necessary condition for the existence of the solution of the first problem of elasticity is the condition of total equilibrium, i.e., the resultants of forces as well as of moments vanish:

$$\int_\Omega f_i \mathrm{d}\boldsymbol{x} + \int_{\partial\Omega} P_i \mathrm{d}s = 0, \tag{3.380}$$

$$\int_\Omega [\boldsymbol{x}, \boldsymbol{f}]_i \mathrm{d}\boldsymbol{x} + \int_{\partial\Omega} [\boldsymbol{x}, \boldsymbol{P}]_i \mathrm{d}s = 0, \quad i = 1, 2, 3. \tag{3.381}$$

The analysis of this problem is discussed above in this chapter, and the detailed analysis is proved in (Nečas and Hlaváček (1981)). The solution of this problem represents the displacement field corresponding to the real stress field $\tau^*_{ij}(\boldsymbol{x})$ in the place of the borehole. The dependence between $\tau'_{ij}(\boldsymbol{x})$, $\boldsymbol{x} \in \Omega$, and τ^*_{ij} is linear, i.e. $\tau'_j = \mathscr{L}(\tau^*_{ij})$, where τ'_{ij} is an element of the space S' of all inhomogeneous stress tensors in Ω and the operator $\mathscr{L}$ is a linear operator, i.e. $\mathscr{L} : S^* \to S'$. This operator is unique, as in the overcoring method. The dimension of the space S' is six. As in the overcoring case, we construct its natural basis as the solution of the above boundary value problem, where the stress tensor τ^*_{ij} is taken as the unit basis $\tau^{(\iota)}_{ij}(\boldsymbol{x})$, $\iota = 1, \ldots, 6$. We denote this by $\tau^{(\iota)}_{ij}(\boldsymbol{x})$, $\boldsymbol{x} \in \Omega$. Then, before the violation of rock massif by the borehole and the construction of the concrete bench, the stress tensor can be written in the form

$$\tau^*_{ij} = \sum\nolimits_{\iota=1}^{6} c^{(\iota)}\tau^{(\iota)}_{ij} \tag{3.382}$$

and the displacement field in the form

$$\boldsymbol{u}^*(\boldsymbol{x}) = \sum\nolimits_{\iota=1}^{6} c^{(\iota)}\boldsymbol{u}^{(\iota)}(\boldsymbol{x}). \tag{3.383}$$

Problem B2 (Mixed problem of elasticity): Let the boundary conditions

$$\tau'_{ij}(\boldsymbol{x})n_j(\boldsymbol{x}) = P_i(\boldsymbol{x}), \quad \boldsymbol{x} \in \partial\Omega = \Gamma_1^e \cup \Gamma_2^e, \quad i = 1, 2, 3, \tag{3.384}$$

$$\tau'_{ij}(\boldsymbol{x})n_j(\boldsymbol{x}) = -pn_i(\boldsymbol{x}), \quad \boldsymbol{x} \in \Gamma_m^{\text{in}}, \quad i = 1, 2, 3, \tag{3.385}$$

be given on the boundary $\partial\Omega = \Gamma_1^e \cup \Gamma_2^e \cup \Gamma_m^{\text{in}} \cup R$ of the region Ω due to Equations 3.375 and 3.377.

Now we have to determine the condition on the boundary Γ_3^e. To determine this boundary condition, Saint-Venant's principle will be used. To establish the stress tensor τ_{ij}^*, the displacements on both ends of the string or other tensometer are essential. Let us denote by $\boldsymbol{x}_{q1}, \boldsymbol{x}_{q2}$ the position vectors of the ends of the q-th measuring element of the sensor (probe), by $l(\boldsymbol{x}_{q1}, \boldsymbol{x}_{q2})$ the length of the q-th measuring element in state 1 and by $\Delta l(\boldsymbol{x}_{q1}, \boldsymbol{x}_{q2})$ the change of the length of the measuring element between state 1 and state 2.

The influence of the inaccuracy of the given boundary condition — the surface force — on Γ_3^e with the increasing distance h_3 of the boundary Γ_3^e from the measuring element decreases under the assumption that the region Ω is the same and under the other conditions, provided that the surface forces on Γ_3^e are statically equivalent to the actual acting forces. On Γ_3^e, we can then prescribe the condition

$$\tau'_{ij}(\boldsymbol{x})n_j(\boldsymbol{x}) = P_i(\boldsymbol{x}), \quad \boldsymbol{x} \in \Gamma_3^e, \quad i, j = 1, 2, 3. \tag{3.386}$$

If the distance h_3 of the measuring sensor from the boundary Γ_3^e is sufficiently large we can find on the boundary Γ_3^e that

$$\boldsymbol{u}'(\boldsymbol{x}) = 0, \quad \boldsymbol{x} \in \Gamma_3^e, \tag{3.387}$$

which is equivalent, in the sense of statical equivalence, to the exact application of acting surface forces on Γ_3^e, as the resulting surface force and its moment (see Equations 3.356 and 3.357) acting on the whole region Ω have to be equal to zero and the boundary conditions on the other parts of the boundary are the same. It can be shown that for the distance h_3 the relation $h_3 > 12r$ is valid, where r is the diameter of the small borehole.

Problem B3: This represents three problems of the first boundary value problem of elasticity. The first one is the first boundary value problem of elasticity, where $\partial\Omega = \Gamma_1^e \cup \Gamma_2^e \cup \Gamma_3^e \cup \Gamma_m^{\text{in}} \cup R$, and the surface force $P_i(\boldsymbol{x}) = 0$ on Γ_m^{in}, i.e. the effect of the drilling watering is not considered.

The second one is the first boundary value problem of elasticity where we prescribe $P_i(\boldsymbol{x}) = -pn_i(\boldsymbol{x})$, $\boldsymbol{x} \in \Gamma_m^{\text{in}}$, $i = 1, 2, 3$, on Γ_m^{in} in the case when the drilling watering is considered.

The third one differs from the second problem in the fact that the condition $\Delta\boldsymbol{u}'(\boldsymbol{x}) = 0$ is prescribed on the boundary Γ_3^e.

Practically, in all the cases, the stress tensor and the surface forces have the meaning of the change of the stress tensor $\Delta\tau_{ij}(\boldsymbol{x})$ and of $\Delta P_i(\boldsymbol{x})$. Therefore, the displacement field $\boldsymbol{u}(\boldsymbol{x})$, as the solution of these problems, has the character of the change of the displacement field in the neighbourhood of the borehole.

Solutions of these problems of sufficient accuracy correspond to the stress field of the rock $\tau_{ij}^*(\boldsymbol{x})$, $\boldsymbol{x} \in \Omega$, in the neighbourhood of the borehole. The other considerations are similar to those in the case of the overcoring method.

In a similar way to the case of the overcoring technique, the solution of the above problems leads to solving the system of algebraic equations

$$\boldsymbol{Ec} = \vec{\varepsilon}^p, \tag{3.388}$$

where $\boldsymbol{E}$ is a $s \times 6$ matrix, $\boldsymbol{c}$ is of dimension 6 and $\vec{\varepsilon}^p$ is of dimension s.

That is, to solving

$$\sum_{\iota=1}^{6} c^{(\iota)}\varepsilon_q^{(\iota)} = -\varepsilon_q + p(\varepsilon_{qJ}^{(p)} - \varepsilon_q^{(p)} + \Theta\varepsilon_q^{(\Theta)}) \equiv \varepsilon_q^p, \quad q = 1, \ldots, s, \tag{3.389}$$

where ε_q^p represent deformation of the individual measuring elements (string tensometers), ε_q, $q = 1, \ldots, s$, represent a result of the self deformation of the rock, $\varepsilon_{qJ}^{(p)}$ are deformations caused by the hydrostatic pressure of the drilling watering p, and $\varepsilon_q^{(\Theta)}$, $q = 1, \ldots, s$ are deformations caused by thermal changes in rock (i.e. due to thermal stresses). All these quantities are determined practically by sensors in the probe. The quantities $\varepsilon_q^{(\iota)}$, $\varepsilon_q^{(\Theta)}$, and $\varepsilon_q^{(p)}$, $q = 1, \ldots, s$, $\iota = 1, \ldots, 6$, are found by the numerical simulation technique described above. The quantity $\varepsilon_{qJ}^{(p)} = -(1 - 2\nu)E^{-1}$, $q = 1, \ldots, s$, is known, as follows from the analytical solution of the first boundary value problem of linear elasticity with surface forces corresponding to the unit hydrostatic pressure where E denotes the Young's modulus.

For $s > 6$, the system cannot be solvable, but the vector $\boldsymbol{c}$ can always be found in such a way that $\|\boldsymbol{Ec} - \vec{\varepsilon}^p\|^2$, where $\|\cdot\|$ is the Euclidean norm, is minimized in the sense of the least squares method. From the analysis of this method, it is known that the problem leads to solving a system of linear algebraic equations with a rectangular matrix, which is optimal in the sense of the least squares method. Then the solution of this system is the vector $\vec{\varepsilon}_o = \boldsymbol{E}^+\vec{\varepsilon}^p$. Therefore, the pseudo-inverse $\boldsymbol{E}^+$, corresponding to the matrix $\boldsymbol{E}$, exists and is unique, and $\boldsymbol{E}^+ = \boldsymbol{E}^T(\boldsymbol{EE}^T)^{-1}$ for $s < 6$, $h(E) = s$, and $\boldsymbol{E}^+ = (\boldsymbol{E}^T\boldsymbol{E})^{-1}\boldsymbol{E}^T$ for $s \geq 6$, $h(\boldsymbol{E}) = 6$, where $h(\boldsymbol{E})$ is the rank of the matrix $\boldsymbol{E}$. If $s \leq 6$ then Equation 3.388 does not represent spatial elasticity in the place of the measurement. For $s \geq 6$, the solution of Equation 3.388 represents the information on the spatial (3D) stress-strain field in the place of measurement.

A Criterion for the Optimal Design of a Tensometric Sensor

The theory presented in this part of the section is valid for all types of stress-strain measurements, i.e. for all measurement methods.

To obtain the tectonic stress tensor τ_{ij}^* or the change $\Delta\tau_{ij}^*$ with maximum accuracy, it is necessary to optimize the space arrangement of measuring elements in the tensometric probe with regard to its application.

The entries of the matrix $\boldsymbol{E}$ and the components of the vectors $\vec{\varepsilon}^{(p)}$ and $\vec{\varepsilon}$ were found (computed and measured) inaccurately as we do not know accurate physical parameters of the rock (Young's modulus, Poisson's constant), the exact length of measuring elements, as well as the natural oscillations of the measuring string in the case of the string tensometry, etc.. Moreover, numerical results are influenced by the error of numerical approximation of the problem and by the error of numerical computation. Equation 3.388 is an overdetermined system of linear algebraic equations (according to the number of the measuring elements) with perturbed entries of the matrix $\boldsymbol{E}$ and components of the right hand side vector $\vec{\varepsilon}^p$. Due to Natterer (1986), it follows that the relative error of measurement $\Delta\vec{\tau}$ is proportional to the condition number $k(\boldsymbol{E})$ of the matrix $\boldsymbol{E}$,

$$k(\boldsymbol{E}) = \|\boldsymbol{E}\| \left\|\boldsymbol{E}^+\right\| .$$

As the columns of the matrix $\boldsymbol{E}$ are more independent the condition number $k(\boldsymbol{E})$ decreases. Our numerical experiments proved that this fact closely depends on the orientation and construction of the sensor. The analysis implies that the condition number $k(\boldsymbol{E})$ depends, in general, on the orientation of the co-ordinate system with regard to the measuring element. Detailed analyses show that the condition number $k(\hat{\boldsymbol{E}})$ of the matrix $\hat{\boldsymbol{E}}$ originated from the matrix $\boldsymbol{E}$ is invariant with respect to the choice of the co-ordinate system if its last three columns are multiplied by the coefficient $(2)^{-1/2}$. Numerical experiments proved this assertion (see Mantič (1988)).

For illustration, two simple tensometric sensors for the stress-strain monitoring in situ as well as for the overcoring method were analysed (see Figures 3.6 and 3.7). The rock in the neighbourhood of the probes and the concrete in the boreholes are considered isotropic with equal material parameters (E is Young's modulus, ν Poisson's ratio). The triangular tensometric sensor — tensometric rosette (see Figure 3.6) — is used in conditions of the plane stress state in rock massif (i.e. in the case of the string tensometry, the U.S.B.M. borehole deformation gauge technique, the strain indicator with switching unit technique, the C.S.I.R. doorstopper method, etc.). From the analysis of the dependence of $k(\hat{\boldsymbol{E}})$ on the sensor shape, it follows that:

(i) The optimal shape of the sensor is a 60° rosette, which is a result known from practice,

(ii) The optimal shape of the oblique seven-tensometer prism is the prism with ratio $z/\zeta = 0.943$, where ζ is the radius of the borehole, and z is the height of the triangular baseline of the prismatic sensor defining the slope of this baseline with respect to the plane perpendicular to the axis of the borehole. The optimal ratio z/ζ is $z/\zeta \sim 0.943$. For a tetrahedron, this ratio is $z/\zeta \sim -0.943$.

(iii) The condition number strongly increases if the baselines of the prism tend to be parallel, and grows to infinity in the case when the baselines are not connected by the tensometer, i.e., in the case when both baselines represent two different rosettes in space (as in the case of the stress relief method). This result proves that the theory of resulting plane measurements from two or more boreholes with non-parallel axes (e.g. with oblique axes) is not correct as the condition number tends to infinity in such cases.

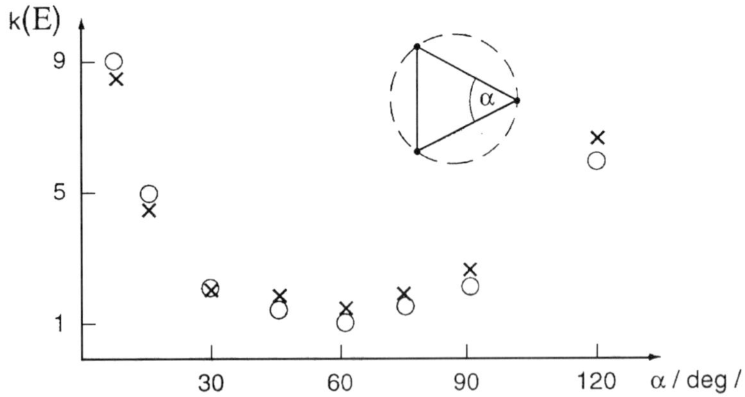

Figure 3.6. 60° rosette tensometric sensor-dependence of the condition number $\boldsymbol{k}(E)$ on the angle α, where X corresponds to Poisson's constant $\nu = 0.4$, while O to $\nu = 0.2$

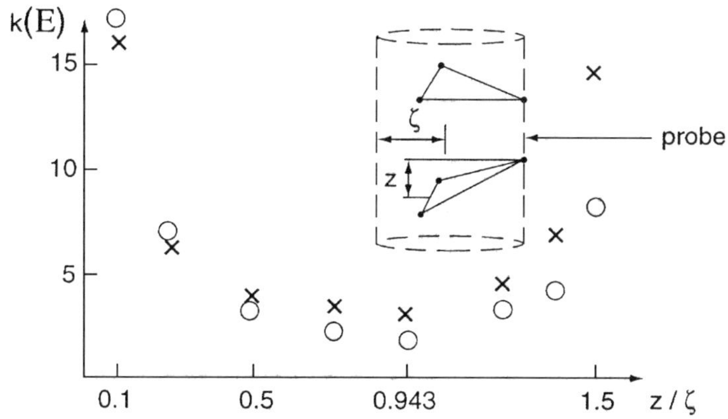

Figure 3.7. Seven-Tensometer Prism 3*D* tensometric sensor(seven independent tensomers, one being parallel with the borehole axis) — dependence of the condition number $k(E)$ on the parameter z/ζ, where *X* corresponds to Poisson's constant $\nu = 0.3321$, while *O* to $\nu = 0.1$

(iv) The shape of the sensor is independent of the physical parameters of rocks and of concrete waterings so that the shape of the sensor is universal for an arbitrary type of rock.

The optimal location of sensors and probes in the borehole in the region of our interest depends on the distribution of energy of deformation in this region. Analyses of these problems are based on the mathematical simulation of geodynamical processes discussed in Chapters 6 and 7 (see also Nedoma (1989), (1990a,b)) and on the investigations of earthquakes and mine shocks, namely on the time-dependent initial boundary value problems describing these types of problems. The prediction of earthquakes and mine shocks is based on the consistent monitoring of the stress-strain field in global as well as local regions of interest and on the mathematical modelling of geodynamical processes in these regions. In Nedoma (1989), it was shown that the geodynamical processes can be simulated, and consequently the earthquakes can be numerically "predicted"; more exactly, we can give better information for earthquake predictions only. But such a "numerical prediction of earthquakes" will only be a very simple approximation, while our simulation technique facilitates the simulations of conditions on the contact boundaries between colliding geological blocks as well as fractures in all colliding blocks. In this case we repeat the idea of the algorithm for the stress-strain borehole measurements with the difference that the bases of the spaces of stress (strain) tensors are found not from boundary value problems of elasticity (thermo-elasticity, thermo-plasticity) but from contact problems in elasticity (thermo-elasticity, thermo-plasticity) with friction, which will be discussed in Chapter 7. In case when the "numerical prediction" is connected with the in-situ measurement of the stress-strain field, the earthquake prediction has a physical basis. Further, we will use the following idea.

We analyse numerical data. Analysis of numerical results shows that geotectonic processes in the lithosphere are characterized by larger compressive and tensile stresses in several places there. The zones of tensile stresses are located at certain distances from the compressive stress maximum. It can be deduced that, as a result of compressive and tensile stresses, deep faults appear and extend to the Earth's surface in places in the obducted plate where the volcanoes occur. These deep faults originate from shear ruptures during longer time and thus the earthquake sources are of the shear fracture type, and, due to numerical results, similar implications are true for other types of deep faults. For shorter time, the deformation energy accumulates in the collision zone and can be released in the case when the absolute value of the tangential stress component $\tau_t^{kl}(\boldsymbol{x}, t)$ (where k, l are superscripts denoting two different lithospheric plates or geological blocks in contact) reaches the value of the frictional force $g_c^{kl}(\boldsymbol{x}, t)$ (for details, see Chapter 7). Thus the plates are mutually shifted, and one part of the deformation energy is released in the form of an earthquake. Such places discussed in the first and second cases can be determined numerically as the places where the deformation energy is accumulated. To identify these local regions, we use the finite element approximation. If the density of deformation energy in a local region is, with a certain coefficient, greater than the average density of deformation energy then such a region is suitable for the stress-strain monitoring. The second criterion is based on the analysis of conditions on the contact boundaries Γ_c^{kl} between lithospheric plates and blocks.

Homogenization Technique

In this section we shall discuss the model problems, which simulate the anisotropic properties of the rock massif as well as of the concrete watering, for generation of the bases of spaces of stress (strain) tensors, in connection with post-processing of the stress-strain measurements in situ.

Let Ω be a domain occupied by the rocks with periodic structure with a Lipschitz boundary $\partial\Omega = \Gamma_\tau \cup \Gamma_u \cup \mathscr{R}$. Let us assume that $\boldsymbol{f} \in [L^2(\Omega)]^3$, $\boldsymbol{P} \in [L^2(\Gamma_\tau)]^3$. The relation between the stress and strain tensors is defined by the linear Hooke's law for the periodic structure of the material defined by

$$\tau_{ij}^{\varepsilon}(\boldsymbol{x}) = c_{ijkl}^{\varepsilon} e_{kl}(\boldsymbol{u}^{\varepsilon}), \tag{3.390}$$

where the coefficients $c_{ijkl}^{\varepsilon}(\boldsymbol{x}) = c_{ijkl}(\boldsymbol{x}/\varepsilon)$ (where c_{ijkl} are Y-periodic* in all variables) form a matrix of the type $(3 \times 3 \times 3 \times 3)$ and satisfy

$$a_o^{\varepsilon} \le c_{ijkl}^{\varepsilon}(\boldsymbol{x})\xi_{ij}\xi_{kl}\,|\xi|^{-2} \le A_o^{\varepsilon} < +\infty \quad \forall \boldsymbol{x} \in \Omega,\ \xi = (\xi_{ij}) \in R^9, \tag{3.391}$$

$$c_{ijkl}^{\varepsilon}(\boldsymbol{x}) = c_{jikl}^{\varepsilon}(\boldsymbol{x}) = c_{klij}^{\varepsilon}(\boldsymbol{x}) = c_{ijlk}^{\varepsilon}(\boldsymbol{x}), \tag{3.392}$$

* Let $Y = [0, \underline{y}_1] \times [0, \underline{y}_2] \times [0, \underline{y}_3] (\underline{y}_i > 0)$ be the unit period in the variable y. A function $f(x, y, z)$ is said to be Y-periodic in y if $f(x, y_1 + k_1\underline{y}_1, y_2 + k_2\underline{y}_2, y_3 + k_3\underline{y}_3, z) = f(x, y_1, y_2, y_3, z)\, \forall k_i$, $i = 1, 2, 3$ where k_i integer. The integral average in y is denoted by

$$\mathscr{M}(f) = (\text{meas}(Y))^{-1} \int_Y f(x, y, z)\text{d}y.$$

If $f(x, y, z)$ is a Y-periodic function, then $\mathscr{M}(f_{,y_j}) = 0$.

where a_o^ε, A_o^ε are constants independent of $\boldsymbol{x} \in \Omega$ and $\xi \in R^9$ and $a_o^\varepsilon = 2\min\{\mu^\varepsilon(\boldsymbol{x}); \boldsymbol{x} \in \Omega\}$, $A_o^\varepsilon = \max\{2\mu^\varepsilon(\boldsymbol{x}) + 3\lambda^\varepsilon(\boldsymbol{x}); \boldsymbol{x} \in \Omega\}$.

Particularly, for an isotropic material with Lamé's constants λ^ε and μ^ε we have

$$c_{ijkl}^\varepsilon = \lambda^\varepsilon \delta_{ij}\delta_{kl} + \mu^\varepsilon(\delta_{ik}\delta_{jl} + \delta_{il}\delta_{jk}). \tag{3.393}$$

The stress tensor satisfies the equilibrium conditions

$$\tau_{ij,j}^\varepsilon(\boldsymbol{x}) + f_i(\boldsymbol{x}) = 0 \quad \forall \boldsymbol{x} \in \Omega. \tag{3.394}$$

For our model problem we use the following boundary conditions:

$$\tau_{ij}^\varepsilon(\boldsymbol{x})n_j = P_i(\boldsymbol{x}) \quad \text{on } \Gamma_\tau, \tag{3.395}$$

$$u^\varepsilon(\boldsymbol{x}) = u_o^\varepsilon(\boldsymbol{x}) \quad \text{on } \Gamma_u. \tag{3.396}$$

Let us investigate the case of $\varepsilon \to 0_+$. Coefficients $c_{ijkl}^\varepsilon(\boldsymbol{x})$ strongly oscillate in the interval Y, therefore the function $\boldsymbol{u}^\varepsilon(\boldsymbol{x})$ also strongly changes, but it is not a periodic function. We look for the asymptotic expansion of $\boldsymbol{u}^\varepsilon$ and τ^ε in the form

$$\boldsymbol{u}^\varepsilon(\boldsymbol{x}) = \boldsymbol{u}^o(\boldsymbol{x}, \boldsymbol{x}/\varepsilon) + \varepsilon \boldsymbol{u}^1(\boldsymbol{x}, \boldsymbol{x}/\varepsilon) + \varepsilon^2 \boldsymbol{u}^2(\boldsymbol{x}, \boldsymbol{x}/\varepsilon) + \cdots. \tag{3.397}$$

$$\tau^\varepsilon(\boldsymbol{x}) = \tau^o(\boldsymbol{x}, \boldsymbol{x}/\varepsilon) + \varepsilon \tau^1(\boldsymbol{x}, \boldsymbol{x}/\varepsilon) + \varepsilon^2 \tau^2(\boldsymbol{x}, \boldsymbol{x}/\varepsilon) + \cdots, \tag{3.398}$$

where the functions $\boldsymbol{u}^i(\boldsymbol{x}, \boldsymbol{y})$ are functions independent of ε and periodic in the variable $\boldsymbol{y}$. It holds

$$\frac{\mathrm{d}\boldsymbol{v}(\boldsymbol{x}, \boldsymbol{x}/\varepsilon)}{\mathrm{d}x_i} = \frac{\partial \boldsymbol{v}(\boldsymbol{x}, \boldsymbol{x}/\varepsilon)}{\partial x_i} + \varepsilon^{-1} \left.\frac{\partial \boldsymbol{v}(\boldsymbol{x}, \boldsymbol{x}/\varepsilon)}{\partial y_i}\right|_{\boldsymbol{y}=\boldsymbol{x}/\varepsilon}.$$

Let us put $\boldsymbol{u}^\varepsilon(\boldsymbol{x}) = \boldsymbol{u}^\varepsilon(\boldsymbol{x}, \boldsymbol{x}/\varepsilon)$ into Equation 3.394, and with Equation 3.390 we obtain

$$\begin{aligned} &\varepsilon^{-2}(c_{ijkl}u_{k,y_l}^\varepsilon)_{,y_j} - \varepsilon^{-1}[(c_{ijkl}u_{k,l}^\varepsilon)_{,y_j} + (c_{ijkl}u_{k,l}^\varepsilon)_{,j}] \\ &\quad -(c_{ijkl}u_{k,l}^\varepsilon)_{,j} = f_i \quad (i, j = 1, 2, 3) \quad \text{in } \Omega. \end{aligned} \tag{3.399}$$

Hence

$$-(c_{ijkl}^\varepsilon u_{k,l}^\varepsilon)_{,j} = (A^\varepsilon \boldsymbol{u}^\varepsilon)_i = ((\varepsilon^{-2}A^o + \varepsilon^{-1}A^1 + A^o)\boldsymbol{u}^\varepsilon)_i = f_i. \tag{3.400}$$

where we denote

$$\begin{aligned} (A^\varepsilon \boldsymbol{u}^\varepsilon)_i &= -(c_{ijkl}u_{k,l}^\varepsilon)_{,j}, \\ (A^o \boldsymbol{u}^\varepsilon)_i &= -(c_{ijkl}u_{k,y_l}^\varepsilon)_{,y_j}, \\ (A^1 \boldsymbol{u}^\varepsilon)_i &= -(c_{ijkl}u_{k,l}^\varepsilon)_{,y_j} - (c_{ijkl}u_{k,y_l}^\varepsilon)_{,j}, \\ (A^2 \boldsymbol{u}^\varepsilon)_i &= -(c_{ijkl}u_{k,l}^\varepsilon)_{,j}. \end{aligned}$$

Inserting Equations 3.397 and 3.398 into Equation 3.400, 3.395, and 3.396 and comparing the coefficients at the corresponding powers of ε (for all integers $k \geq -2$), we obtain

$$A^o \boldsymbol{u}^o = 0 \quad \text{in } \Omega \times Y, \tag{3.401}$$

$$A^o \boldsymbol{u}^1 + A^1 \boldsymbol{u}^o = 0 \quad \text{in } \Omega \times Y, \tag{3.402}$$

$$c_{ijkl}(\boldsymbol{y}) u^o_{k,y_l}(\boldsymbol{x},\boldsymbol{y}) n_j(\boldsymbol{x}) = 0 \quad \text{on } \Gamma_\tau \times Y, \tag{3.403}$$

$$A^o \boldsymbol{u}^2 + A^1 \boldsymbol{u}^1 + A^2 \boldsymbol{u}^o = f \quad \text{in } \Omega \times Y, \tag{3.404}$$

$$c_{ijkl}(\boldsymbol{y})(u^1_{k,y_l} + u^o_{k,l}) n_j(\boldsymbol{x}) = P_i(\boldsymbol{x}) \quad \text{on } \Gamma_\tau \times Y, \tag{3.405}$$

$$u^o_i(\boldsymbol{x}) = u_i(\boldsymbol{x}) \quad \text{on } \Gamma_u \times Y, \tag{3.406}$$

$$A^o \boldsymbol{u}^3 + A^1 \boldsymbol{u}^2 + A^2 \boldsymbol{u}^1 = 0 \quad \text{in } \Omega \times Y, \tag{3.407}$$

$$c_{ijkl}(\boldsymbol{y})(u^2_{k,y_l} + u^1_{k,l}) n_j(\boldsymbol{x}) = 0 \quad \text{on } \Gamma_\tau \times Y, \tag{3.408}$$

$$u^1_i(\boldsymbol{x}) = 0 \quad \text{on } \Gamma_u \times Y. \tag{3.409}$$

Lemma 3.3: Let $\boldsymbol{f} \in [L^2_{\text{per}}(Y)]^3$. $A^o \boldsymbol{u} = \boldsymbol{f}$ has a solution $\boldsymbol{u} \in [H^1_{\text{per}}(Y)]^3$ if and only if $\mathscr{M}(\mathbf{f}) = 0$. This solution is unique up to an additive constant.

Proof: The proof is based on the properties of Y-periodic functions and the Lax-Milgram theorem. To prove the lemma we have two cases:

(A) Let $\mathscr{M}(\boldsymbol{f}) = 0$. Let the range of the operator A^o, denoted by $D(A^o)$, be the set

$$D(A^o) = \{\boldsymbol{v} \mid \boldsymbol{v} \in \{H^1_{\text{per}}(Y)]^3 \cap A^o \boldsymbol{v} \in [L^2(Y)]^3\}.$$

For an arbitrary $\vec{\phi}$, $\vec{\psi} \in D(A^o)$ let

$$a(\vec{\phi}, \vec{\psi}) = (A^o \vec{\phi}, \vec{\psi}) = -\int_Y c_{ijkl} \phi_{k,y_l} \psi_{i,y_j} \mathrm{d}\boldsymbol{x},$$

$$f(\vec{\psi}) = (\boldsymbol{f}, \vec{\psi}) = \int_Y \boldsymbol{f} \vec{\psi} \mathrm{d}\boldsymbol{x}.$$

The problem $A^o \boldsymbol{u} = \boldsymbol{f}$ is equivalent to the problem

$$(A^o \boldsymbol{u}, \vec{\psi}) = (\boldsymbol{f}, \vec{\psi}) \quad \forall \vec{\psi} \in [H^1_{\text{per}}(Y)]^3,$$

or

$$a(\boldsymbol{u}, \vec{\psi}) = (\boldsymbol{f}, \vec{\psi}) \quad \forall \vec{\psi} \in [H^1_{\text{per}}(Y)]^3.$$

The bi-linear form a(.,.) is bounded on $D(A^o)$, and $\boldsymbol{f}$ is the functional continuous on $D(A^o)$. Let us denote by $H_{\text{per}} = [H^1_{\text{per}}(Y)]^3 \backslash R^3$ the space of all functions, whose components are different by up to any constant, i.e. $\vec{\psi} + \boldsymbol{c}$. Since $\mathscr{M}(\boldsymbol{f}) = 0$, then for $\boldsymbol{c} \in R^3$ it holds $(\boldsymbol{f}, \vec{\psi} + \boldsymbol{c}) = (\boldsymbol{f}, \vec{\psi})$ and therefore $\boldsymbol{f}$ is a continuous linear functional also on the space H_{per}. Moreover, for an arbitrary $\vec{\psi} \in H_{\text{per}}$

$$a(\vec{\psi}, \vec{\psi}) = \int_Y c_{ijkl} e_{ij}(\vec{\psi}) e_{kl}(\vec{\psi}) \mathrm{d}\mathbf{x} \geq c_o \int_Y e_{ij}(\vec{\psi}) e_{ij}(\vec{\psi}) \mathrm{d}\mathbf{x} \geq \alpha \left\| \vec{\psi} \right\|^2_{1,2}, c_o > 0.$$

Hence, the bi-linear form a($\vec{\psi}$, $\vec{\psi}$) is, on H_{per}, V-elliptic and due to the Lax-Milgram theorem we have proved on H_{per} the existence and uni-queness of the problem.

(B) For $\boldsymbol{u} \in [H^1_{\text{per}}(Y)]^3$ let $A^o\boldsymbol{u} = \boldsymbol{f}$, i.e. $-(c_{ijkl}u_{k,l})_{,y_j} = f_i$. Integrating over Y then $(\text{meas}(Y))^{-1}\int_Y -(c_{ijkl}u_{k,l})_{,y_j}\text{d}\boldsymbol{y} = 0$ as c_{ijkl} and $\boldsymbol{u}$ are Y-periodic functions, therefore $(\text{meas}(Y))^{-1}\int_Y f_i\text{d}\boldsymbol{y} = 0$. Hence $\mathscr{M}(\boldsymbol{f}) = 0$.

Remark: Index "per" implies periodic functions.

From (Equation 3.401) and Lemma 3.3 it follows that the function $\boldsymbol{u}^o$ does not depend on the variable $\boldsymbol{y}$, so that $\boldsymbol{u}^o = \boldsymbol{u}^o(\boldsymbol{x})$. Then from (Equation 3.402) we obtain

$$A^o\boldsymbol{u}^1 = -A^1\boldsymbol{u}^o = (c_{ijkl}(\boldsymbol{y})u^o_{k,l}(\boldsymbol{x}))_{,y_j}. \tag{3.410}$$

Due to Lemma 3.3 the above problem has a unique solution at every $\boldsymbol{x}$ from $[H^1_{\text{per}}(Y)]^3$. We shall seek such a solution in the form

$$u^1_g(\boldsymbol{x},\boldsymbol{y}) = -\chi^{kl}_g(\boldsymbol{y})u^o_{k,l}(\boldsymbol{x}) + \underline{u}_g(\boldsymbol{x}), \tag{3.411}$$

where $\chi \in [H^1_{\text{per}}(Y)]^3$. Therefore,

$$A^o(-\chi^{kl}_g(\boldsymbol{y})u^o_{k,l}(\boldsymbol{x}) + \underline{u}_g(\boldsymbol{x}))_{y_j} = (c_{ijkl}(\boldsymbol{y})u^o_{k,l}(\boldsymbol{x}))_{,y_j},$$

hence

$$(c_{ijgh}(\boldsymbol{y})(\chi^{kl}_{g,y_h}(\boldsymbol{y}) - \delta_{gk}\delta_{hl}))_{,y_j} = 0. \tag{3.412}$$

Equation 3.403 is fulfilled as $\boldsymbol{u}^o$ is independent on a variable $\boldsymbol{y}$. Equation 3.404, due to Lemma 3.3, has a unique solution from the space $[H^1_{\text{per}}(Y)]^3$ just when $\mathscr{M}(A^1\boldsymbol{u}^1 + A^2\boldsymbol{u}^o - \boldsymbol{f}) = 0$(as $\boldsymbol{u}^2 = 0$). Hence and using Equation 3.411 we obtain

$$\mathscr{M}((c_{ijgh}\chi^{kl}_{g,y_h}u^o_{k,l})_{,j} - (c_{ijkl}u^o_{k,l})_{,j} - f_i) = 0. \tag{3.413}$$

Using the notation

$$c^o_{ijkl} = \mathscr{M}(c_{ijgh}(\boldsymbol{y})(\delta_{gh}\delta_{hl} - \chi^{kl}_{g,y_h}(\boldsymbol{y}))), \tag{3.414}$$

then from Equation 3.413 we obtain

$$-(c^o_{ijkl}u^o_{k,l})_{,j} = f_i, \tag{3.415}$$

and Equation 3.405 yields

$$c^o_{ijkl}u^o_{k,l}n_j = P_i \quad \text{on } \Gamma_\tau, \tag{3.416}$$

and Equation 3.406 does not change.

Definition 3.19: The problem

$$-(c^o_{ijkl}u^o_{k,l})_{,j} = f_i \quad \text{in } \Omega, \tag{3.417}$$

$$c^o_{\text{ijkl}}u^o_{k,l}n_j = P_i \quad \text{on } \Gamma_\tau, \tag{3.418}$$

$$u^o_i = u_{oi} \quad \text{on } \Gamma_u \tag{3.419}$$

$i, j = 1, 2, 3$, with c^o_{ijkl} defined by Equation 3.414, where $\chi \in [H^1_{\text{per}}(Y)]^3$ satisfies Equation 3.412, is the so-called homogenization problem with regard to the initial problem. Constants c^o_{ijkl} are called the homogenization coefficients with regard to the initial one.

Remark: When c_{ijkl} are in the subdomain $\tilde{Y} \subset Y$ from the space $C^1(\tilde{Y})$ then the functions χ^{kl} are from $C^2(\tilde{Y})$.

Lemma 3.4: For homogenization coefficients it holds that

$$c^o_o \leq c^o_{ijkl}\xi_{ij}\xi_{kl}\,|\xi|^{-2} \leq C^o_o < +\infty \quad \forall \boldsymbol{x} \in \Omega, \quad \xi = (\xi_{ij}) \in R^9, \tag{3.420}$$

where c^o_o, C^o_o are constants independent of $\boldsymbol{x} \in \Omega$ and $\xi \in R^9$,
and

$$c^o_{ijkl} = c^o_{jikl} = c^o_{klij} = c^o_{ijlk}, \tag{3.421}$$

where $c^o_o = 2\min\{\mu^o(\boldsymbol{x}); \boldsymbol{x} \in \Omega\}$, $C^o_o = \max\{2\mu^o(\boldsymbol{x}) + 3\lambda^o(\boldsymbol{x}); \boldsymbol{x} \in \Omega\}$.

Proof: The assertions of the proof follow immediately from Equations 3.391 and 3.392

If a function f is a Y-periodic function then $\mathcal{M}(f_{,y_j}) = 0$. Therefore, $\mathcal{M}(-c_{ijkl,y_j}) = 0$ and thus the functions χ^{kl} exist and are unique if we add the condition $\mathcal{M}(\chi^{kl}) = 0$. According to Equation 3.392 we have

$$\chi^{kl} = \chi^{lk}. \tag{3.422}$$

For a homogeneous material $c^o_{ijkl} = c_{ijkl}$.

We derive the formula in Equation 3.414, i.e.

$$c^o_{ijkl} = \mathcal{M}(c_{ijgh}(\boldsymbol{y})(\delta_{gh}\delta_{hl} - \chi^{kl}_{g,y_h}(\boldsymbol{y}))),$$

where the functions $\chi^{kl} \in H_{\text{per}}$ are solutions of Equation 3.412, which can be rewritten in the form

$$\mathcal{M}(c_{ijgh}(\boldsymbol{y})(\chi^{kl}_{g,y_h}(\boldsymbol{y}) - \delta_{gk}\delta_{hl})_{,y_j}\varphi_i) = 0 \quad \forall \varphi \in H_{\text{per}}. \tag{3.423}$$

The solution exists and is unique up to additive constants which do not influence the value of c^o_{ijkl}. Choosing $\chi^{kl} \in H_{\text{per}}$ for the test function in Equation 3.423 and subtracting the equation from Equation 3.414 we obtain a symmetric formula

$$c^o_{ijkl} = \mathcal{M}(c_{pqgh}(\delta_{pi}\delta_{qj} - \chi^{ij}_g)(\delta_{gk}\delta_{hl} - \chi^{kl}_{g,y_h})). \tag{3.424}$$

Theorem 3.41: The solution $\boldsymbol{u}^\varepsilon$ of the problem in Equations 3.394-3.396 and 3.390 converges weakly to the solution $\boldsymbol{u}^o$ of the problem in Equations 3.417-3.419 in the space $[H^1(\Omega)]^3$ for $\varepsilon \to 0_+$.

The proof is similar to that of Birolli (1977), Bensounssan, Lions and Papanicolaou (1978).

To compute the derivative of χ^{kl} we start from the following information:

$$(c_{ijgh}(\boldsymbol{y})(\chi^{kl}_{g,y_h}(\boldsymbol{y}) - \delta_{gk}\delta_{hl}))_{,y_j} = 0. \tag{3.425}$$

Due to Equation 3.423 χ^{kl} are, in $[H^1_{\text{per}}(Y)]^3$, uniquely defined up to the additive constant.

The relation of Equation 3.425 can be rewritten in the following form

$$\int_Y (c_{ijgh}(\mathbf{y})(\chi^{kl}_{g,y_h}(\mathbf{y}) - \delta_{gk}\delta_{hl}))_{,y_j}\varphi_i \mathrm{d}\mathbf{x} = 0 \quad \forall \vec{\varphi} \in [H^1_{\mathrm{per}}(Y)]^3.$$

Let Y_i be subdomains of Y where the elastic coefficients are given. On Y_n let coefficients c^n_{ijkl} be given. Then

$$\int_Y (c_{ijgh}(\mathbf{y})(\chi^{kl}_{g,y_h}(\mathbf{y}) - \delta_{gk}\delta_{hl}))_{,y_j}\varphi_i \mathrm{d}\mathbf{x}$$

$$= \sum_{m=1}^{n} \int_{Y_m} (c^m_{ijgh}(\mathbf{y})(\chi^{kl}_{g,y_h}(\mathbf{y}) - \delta_{gk}\delta_{hl}))_{,y_j}\varphi_i \mathrm{d}\mathbf{x}$$

and hence

$$\int_{Y_n} (c^n_{ijgh}(\mathbf{y})(\chi^{kl}_{g,y_h}(\mathbf{y}) - \delta_{gk}\delta_{hl}))_{,y_j}\varphi_i \mathrm{d}\mathbf{x} = -\int_{Y_n} (c^n_{ijgh}(\mathbf{y})(\chi^{kl}_{g,y_h}(\mathbf{y}) - \delta_{gk}\delta_{hl}))\varphi_{i,y_j} \mathrm{d}\mathbf{x}$$

$$+ \int_{\partial Y_n} (c^n_{ijgh}(\mathbf{y})(\chi^{kl}_{g,y_h}(\mathbf{y}) - \delta_{gk}\delta_{hl}))\varphi_i n_j \mathrm{d}s.$$

Since in the contact point of two subdomains Y_i and Y_j the absolute value of the outward normals are the same (only their directions are opposite) then the relation of Equation 3.425 can be rewritten in the form

$$\int_Y c_{ijgh}(\chi^{kl}_{g,y_h}(\mathbf{y}) - \delta_{gk}\delta_{hl}))\varphi_{i,y_j} \mathrm{d}\mathbf{x} = 0 \quad \forall \vec{\varphi} \in [H^1_{\mathrm{per}}(Y)]^3, \tag{3.426}$$

$$c^m_{ijgh}(\chi^{kl}_{g,y_h}(\mathbf{y}) - \delta_{gk}\delta_{hl})) = c^n_{ijgh}(\chi^{kl}_{g,y_h}(\mathbf{y}) - \delta_{gk}\delta_{hl})) \tag{3.427}$$

for all subdomains Y_m and Y_n and in all points of $\overline{Y}_m \cap \overline{Y}_n$.

Such a formulation is not good for the numerical approximation. We shall show the Milka's idea (Milka (1989)). Equation 3.427 holds in clearly defined points of the mesh used. According to this fact the obtained matrix need not be a square matrix. Moreover, we do not know whether the obtained matrix is regular. Therefore, we shall solve another problem. Instead of the problem in Equation 3.425 we shall solve the following problem

$$\int_Y (c_{ijgh}(\overline{\chi}^{kl}_{g,y_h} - \delta_{gk}\delta_{hl}))_{,y_j}\varphi_i \mathrm{d}\mathbf{x} = 0 \quad \forall \vec{\varphi} \in [H^1_o(Y)]^3. \tag{3.428}$$

From the Green's theorem we obtain

$$\int_Y (c_{ijgh}(\overline{\chi}^{kl}_{g,y_h} - \delta_{gk}\delta_{hl}))\varphi_{i,y_j} \mathrm{d}\mathbf{x} = 0 \quad \forall \vec{\varphi} \in [H^1_o(Y)]^3. \tag{3.429}$$

This problem has just one solution $\overline{\chi}^{kl} \in [H^1_o(Y)]^3$. Let us prove this assertion.

Let $\boldsymbol{u}, \boldsymbol{v} \in [H^1_o(Y)]^3$ and let

$$a(\boldsymbol{u}, \boldsymbol{v}) = \int_Y c_{ijgh} u_{i,y_j} v_{g,y_h} \mathrm{d}\mathbf{x}, \tag{3.430}$$

$$f(\boldsymbol{v}) = \int_Y c_{ijkl} v_{i,y_j} \mathrm{d}\mathbf{x}, \tag{3.431}$$

k, l being fixed. But Equation 3.428 can be rewritten in the form

$$a(\overline{\chi}^{kl}, \vec{\varphi}) = f(\vec{\varphi}) \quad \forall \vec{\varphi} \in [H_o^1(Y)]^3. \tag{3.432}$$

The proof of the existence and uniqueness is similar to that of Lemma 3.3. It can be shown that the Lax-Milgram theorem can be used.

Let us investigate the relation between χ^{kl} and $\overline{\chi}^{kl}$. Let us denote

$$\mathscr{A}^{kl}\boldsymbol{u} = (c_{ijgh}(u_{g,y_h} - \delta_{gh}\delta_{hl}))_{,y_j}.$$

Then the function χ^{kl} is the solution of the problem

$$(\mathscr{A}^{kl}x^{kl}, \vec{\varphi}) = 0 \quad \forall \vec{\varphi} \in [H_{\text{per}}^1(Y)]^3$$

and the function $\overline{\chi}^{kl}$ is the solution of the problem

$$(\mathscr{A}^{kl}\overline{\chi}^{kl}, \vec{\varphi}) = 0 \quad \forall \vec{\varphi} \in [H_o^1(Y)]^3.$$

But every function from $[H_{\text{per}}^1(Y)]^3$ can be written as the sum of a function from $[H_o^1(Y)]^3$ and from $[H_{\text{per}}^1(Y)]^3$, which is non-zero in the small strip near the boundary of Y only. Therefore, the function $\mathscr{A}^{kl}\overline{\chi}^{kl}$ is non-zero on this strip only and therefore, we can approximate χ^{kl} by $\overline{\chi}^{kl}$ a.e. in Y. Near the boundary of Y the difference between χ^{kl} and $\overline{\chi}^{kl}$ will be substantial.

To compute $\overline{\chi}^{kl}$ we assume the problem

$$\int_Y (c_{ijgh}(\overline{\chi}_{g,y_h}^{kl} - \delta_{gk}\delta_{hl}))\varphi_{i,y_j}^n \mathrm{d}\boldsymbol{x} = 0 \quad \forall \vec{\varphi}^n \in \mathscr{B}, \tag{3.433}$$

where $\mathscr{B}$ is created by an arbitrary basis of $[H_o^1(Y)]^3$. Let us seek $\overline{\chi}^{kl}$ in the form $\overline{\chi}^{kl} = \Sigma_{\vec{\varphi}^i \in \mathscr{B}} \alpha_i^{kl} \vec{\varphi}^i$. Then we obtain a system of linear algebraic equations $A\vec{\alpha}^{kl} = \vec{b}^{kl}$, where $\vec{\alpha}^{kl}$ is the vector of unknown coefficients, $(b^{kl})_n = \int_Y c_{ijkl}\varphi_{i,y_j}^n \mathrm{d}\boldsymbol{x}$, and $(A)_{m,n} = \int_Y c_{ijgh}\varphi_{g,y_h}^m \varphi_{i,y_j}^n \mathrm{d}\boldsymbol{x}$ which is independent of k and l. The matrix A is regular and, more over, positive definite. The billinear form in Equation 3.430 is V-elliptic on $[H_o^1(Y)]^3$ and, due to the symmetry conditions, is symmetric. For $\boldsymbol{u} = 0$ we have $a(\boldsymbol{u}, \boldsymbol{u}) = 0$ and therefore it has all the properties of the inner product. Thus matrix A is the Gramm matrix and therefore symmetric and positive definite.

To define the space $\mathscr{B}$ a regular triangulation of Y and linear finite elements will be used. Let us denote by $\{V_k \mid k \in S\}$ the system of obtained tetrahedrons, and let us denote by h the diameter of the largest element in the finite element mesh. Let us assume that on every tetrahedron V_k, $k \in S$, coefficients c_{ijkl} are constant. Then the matrix A and the vector $\boldsymbol{b}$ can be constructed by using the algorithm described below.

Let Y be created from the finite number of subdomain Y_i with Lipschitz boundary. On every Y_i the homogeneous isotropic material with constant coefficients c_{ijkl} is defined. Moreover, the material is Y-periodic.

Due to the idea discussed above, in Y we define the mesh and we find the corresponding space of test functions $\mathscr{B}$ and then we find approximate values of $\overline{\chi}^{kl}$ in the mesh points of the given mesh.

On the elements laying on the boundary between Y_m and Y_n, the functions $\overline{\chi}^{kl}$ are in general discontinuous. Let M_1 be the set of all such elements. Its measure is of order h. Let M_2 be the set of all elements which have at least one side imbeded in Y, as there the functions $\overline{\chi}^{kl}$ and χ^{kl} are different. Its measure is also of the order h. On the set $Y \backslash (M_1 \cup M_2)$ the functions $\overline{\chi}^{kl}$ have continuous second derivatives. Since the bilinear form from Equation 3.430 is bounded and V-elliptic then $\|\boldsymbol{u} - \boldsymbol{u}_h\| \leq (M/\alpha)$, $\inf_{\boldsymbol{v} \in v_h} \|\boldsymbol{u} - \boldsymbol{v}\|$, and therefore the first derivative of $\overline{\chi}^{kl}$ can be approximated by linear test functions from $\mathscr{B}$ with an error of the order h.

The homogenized coefficients c^h_{ijkl} are found by the numerical integration

$$c^h_{ijkl} = \int_{Y \backslash (M_1 \cup M_2)} c_{ijgh}(\delta_{gk}\delta_{hl} - \overline{\chi}^{kl}_{g,y_h})\mathrm{d}\boldsymbol{x}\,.$$

Because the integrated function is computed with an error of the order h and the volume measure of $(M_1 \cup M_2)$ is also of the order h, then the computed coefficients c^h_{ijkl} converge to c^o_{ijkl} with an error also of the order h.

Remark: For rocks whose elastic coefficients c_{ijkl} are continuous together with their first derivatives on every subdomains Y_i, the above technique can also be used. In this case the coefficients c_{ijkl} are approximated by a constant on every tetrahedron only.

3.3.2.5. Finite Element Solution of the Problem

To solve the problem with homogenized coefficients numerically the finite element technique will be used.

Let a domain occupied by an investigated part of the rock massif be approximated by a 3D domain Ω, the boundary of which, denoted by $\partial\Omega$, is a polyhedron. Let the domain Ω be divided into a finite number of tetrahedrons V_k, $k = 1, \ldots, n$, such that two arbitrary tetrahedrons have either no common point or have a common vertex, a common edge or a common side. We restrict our considerations to the space V_h of all continuous functions which are linear on each tetrahedron of the given division. Let h be the largest diameter of all the tetrahedrons of the given division. We shall assume that the given division is regular. Let V_h be the finite element approximation of V consisting of all functions from V which on the tetrahedron V_k are equal to interpolation polynomials of the first degree. Then the problem leads to the following problem:

Problem ($\mathscr{P}_h$): Find $\boldsymbol{u}_h \in \boldsymbol{u}_o + V_h$ minimizing the functional

$$J(\boldsymbol{u}) = \tfrac{1}{2}a(\boldsymbol{u}, \boldsymbol{u}) - \int_\Omega f_i u_i \mathrm{d}\boldsymbol{x} - \int_{\Gamma_\tau} P_i u_i \mathrm{d}s \qquad (3.434)$$

on the set V_h, where $a(\boldsymbol{u}, \boldsymbol{v})$ is the bi-linear form defined by

$$a(\boldsymbol{u}, \boldsymbol{v}) = \int_\Omega c_{ijkl} e_{ij}(\boldsymbol{u}) e_{kl}(\boldsymbol{v}) \mathrm{d}\boldsymbol{x}\,. \qquad (3.435)$$

Let us denote by $\varepsilon(h) = \|\boldsymbol{u} - \boldsymbol{u}_h\|$ the error between the exact and the approximate solutions, where $\boldsymbol{u}_h \in V_h$. We shall prove that $\varepsilon(h) \to 0$ for $h \to 0_+$.

Theorem 3.42: Let $a : V \times V \to R_1$ be a symmetric, bounded and V-elliptic billinear form on V. Let V_h be the finite element approximation of V defined above with the following properties:

$$\forall \boldsymbol{v} \in \boldsymbol{u}_o + V \quad \exists \boldsymbol{v}_h \in \boldsymbol{u}_o + V_h : \boldsymbol{v}_h \to \boldsymbol{v}, \quad h \to 0_+ \tag{3.436}$$

$$\text{from } \boldsymbol{v}_h \to \boldsymbol{v} \quad \text{weakly}, \quad \boldsymbol{v}_h \in \boldsymbol{u}_o + V_h \Leftrightarrow \boldsymbol{v} \in \boldsymbol{u}_o + V. \tag{3.437}$$

Then

$$\|\boldsymbol{u} - \boldsymbol{u}_h\| \to 0 \text{ for } h \to 0_+. \tag{3.438}$$

Proof: At first

$$\lim_{\|\boldsymbol{v}_h\| \to \infty} J(\boldsymbol{v}_h) = +\infty. \tag{3.439}$$

Due to Equation 3.436 the sequence $\{\boldsymbol{v}_h\}$ exists, $\mathbf{v}_h \in \boldsymbol{u}_o + V_h$, such that

$$\boldsymbol{v}_h \to \boldsymbol{u}, \quad h \to 0_+. \tag{3.440}$$

Since J is continuous on $\boldsymbol{u}_o + V$ then we have

$$J(\boldsymbol{u}_h) \le J(\boldsymbol{v}_h) \to J(\boldsymbol{u}) \quad h \to 0_+. \tag{3.441}$$

The sequence $\{J(\boldsymbol{u}_h)\}$ is bounded, and from Equation (3.439) we see that it is bounded also in $\boldsymbol{u}_o + V$. Therefore there exists the subsequence $\{\boldsymbol{u}_{\not h}\}$ and the element $\boldsymbol{u}^*$ such that $\boldsymbol{u}_{\not h} \to \boldsymbol{u}^*$ weakly in $\boldsymbol{u}_o + V_h$, $\not h \to 0_+$. From Equation 3.437 and since the functional of potential energy J is weakly lower semi-continuous on $\boldsymbol{u}_o + V$ we have

$$J(\boldsymbol{u}^*) \le \lim_{h \to 0_+} \inf J(\boldsymbol{u}_{\not h}) \le \lim_{h \to 0_+} \inf J(\boldsymbol{v}_{\not h}) \le \lim_{h \to 0_+} \inf J(\boldsymbol{v}_h) = J(\boldsymbol{u}).$$

Hence $J(\boldsymbol{u}^*) = J(\boldsymbol{u})$ and therefore $\boldsymbol{u}^* \in \boldsymbol{u}_o + V$. Hence, we see that $\{\boldsymbol{u}_h\}$ converges weakly to $\boldsymbol{u}$. It can be shown that $\boldsymbol{u}_h \to \boldsymbol{u}$ in the norm of $\boldsymbol{u}_o + V$. Let $\{\boldsymbol{v}_h\}$ be a sequence such that Equation 3.440 holds. Then

$$\begin{aligned} J(\boldsymbol{v}_h) \ge J(\boldsymbol{u}_h) &= J(\boldsymbol{u}) + a(\boldsymbol{u}, \boldsymbol{u}_h - \boldsymbol{u}) - S(\boldsymbol{u}_h - \boldsymbol{u}) + \tfrac{1}{2} a(\boldsymbol{u} - \boldsymbol{u}_h, \boldsymbol{u} - \boldsymbol{u}_h) \\ &\ge J(\boldsymbol{u}) + a(\boldsymbol{u}, \boldsymbol{u}_h - \boldsymbol{u}) - S(\boldsymbol{u}_h - \boldsymbol{u}) + \tfrac{1}{2} c \|\boldsymbol{u} - \boldsymbol{u}_h\|^2, \end{aligned}$$

where $S(\boldsymbol{v}) = \int_\Omega f_i v_i \mathrm{d}\boldsymbol{x} + \int_{\Gamma_\tau} P_i v_i \mathrm{d}s$. Hence

$$\tfrac{1}{2} c \|\boldsymbol{u} - \boldsymbol{u}_h\|^2 \le J(\boldsymbol{v}_h) - J(\boldsymbol{u}) - a(\boldsymbol{u}, \boldsymbol{u}_h - \boldsymbol{u}) + S(\boldsymbol{u}_h - \boldsymbol{u}).$$

Hence, and from 3.441, $\|\boldsymbol{u} - \boldsymbol{u}_h\| \to 0$ for $h \to 0_+$.

The problem leads to finding the minimum of the potential energy

$$J(\boldsymbol{v}_h) = \tfrac{1}{2} a(\boldsymbol{v}_h, \boldsymbol{v}_h) - S(\boldsymbol{v}_h) \text{ on } V_h.$$

As $\boldsymbol{v}_h(\boldsymbol{x}) = \Sigma_{k=1}^n v_k \vec{\varphi}_k(\boldsymbol{x})$, then $a(\boldsymbol{v}_h, \boldsymbol{v}_h)$ is a quadratic form and $S(\boldsymbol{v}_h)$ is a linear form in the co-ordinates of the function $\boldsymbol{v}_h$ with respect to the basis $\{\vec{\varphi}_k\}_{k=1}^n$ of V_h, i.e.

$$J(\boldsymbol{v}_h) = \tfrac{1}{2} \sum_{k=1}^n \sum_{j=1}^n a(\vec{\varphi}_k, \vec{\varphi}_j) v_k v_j - \sum_{k=1}^n S(\vec{\varphi}_k) v_k.$$

Defining the elements K_{ij} of the symmetric stiff matrix K by the relations

$$K_{ij} = a(\vec{\varphi}_i, \vec{\varphi}_j), \quad i, j = 1, 2, \ldots, n \tag{3.442}$$

and the components of the vector F by

$$F_i = S(\vec{\varphi}_i), \quad i = 1, 2, \ldots, n, \tag{3.443}$$

we see that the matrix K is positive definite and therefore the function $\boldsymbol{u}_h \in V_h$ minimizes the functional $J(\boldsymbol{v}_h)$ on V_h if and only if the co-ordinates v_k, $k = 1, \ldots, n$, are the solutions of the system of linear equations

$$\frac{\partial J(\boldsymbol{v}_h)}{\partial \Delta} = 0, \tag{3.444}$$

where $\vec{\Delta} = (v_1, \ldots, v_n)^T$, i.e.

$$K\vec{\Delta} = \boldsymbol{F}. \tag{3.445}$$

The algorithm for computing K and $\boldsymbol{F}$ is a modification of the algorithm given in the next section for polynomials of the first order. To obtain a better approximation of our problem discussed here, polynomials of higher order and adaptive FEM can be used (see for example Kolář, Kratochvíl, Leitner and Ženíšek. (1979), Axelsson and Barker (1984), Babuška and Reinboldt (1980),(1984), etc.).

Algorithm

Next we shall discuss the algorithm of the finite element method in the generalized form, which can be used for numerical computation of the homogenized coefficients c^o_{ijkl}, and/or for numerical generations of the bases of stress (strain) spaces for overcoring and stress-strain monitoring post-processing models.

Let a 3D domain Ω, the boundary of which is a polyhedron $\partial\Omega$, be divided into a finite number of tetrahedrons V_k, $k = 1, \ldots, n$, such that two arbitrary tetrahedrons have either no common point, or have a common vertex, a common edge or a common side. We restrict our considerations to the space V_h of all continuous functions which are linear on each tetrahedron of the given division. The polynomial of the first degree, denoted by $u_r^{(k)}$, $r = 1, 2, 3$, $k = 1, 2, \ldots, n$, is uniquely determined on the tetrahedron V_k by the function values at the vertices of the tetrahedron. Let us denote the vertices of the tetrahedron by $P_i(x_i, y_i, z_i)$, $i = 1, \ldots, 4$ and let $\underline{x}_i = x_i - x_1$, $\underline{y}_i = y_i - y_1$, $\underline{z}_i = z_i - z_1$, $i = 2, 3, 4$.

Let V_o be the tetrahedron in the Cartesian co-ordinate system (ξ, η, ζ) with vertices $R_1(0, 0, 0)$, $R_2(1, 0, 0)$, $R_3(0, 1, 0)$, $R_4(0, 0, 1)$. It is well-known that the transformation

$$\begin{aligned} x &= x(\xi, \eta, \zeta) = x_1 + \underline{x}_2\xi + \underline{x}_3\eta + \underline{x}_4\zeta, \\ y &= y(\xi, \eta, \zeta) = y_1 + \underline{y}_2\xi + \underline{y}_3\eta + \underline{y}_4\zeta, \\ z &= z(\xi, \eta, \zeta) = z_1 + \underline{z}_2\xi + \underline{z}_3\eta + \underline{z}_4\zeta, \end{aligned} \tag{3.446}$$

transforms vertices P_i onto R_i, $i = 1, \ldots, 4$. Then from the linearity of this transformation it follows that the function

$$\underline{\boldsymbol{u}}_r^{(k)}(\xi, \eta, \zeta) = \boldsymbol{u}_r^{(k)}(x(\xi, \eta, \zeta), y(\xi, \eta, \zeta), z(\xi, \eta, \zeta)) = \boldsymbol{G}^T \underline{\boldsymbol{a}}_r \tag{3.447}$$

is a polynomial of the first order in the variables ξ, η, ζ, where

$$\boldsymbol{G} = (1, \xi, \eta, \zeta)^T \tag{3.448}$$

and

$$\underline{\boldsymbol{a}}_r = (a_1^{(r)}, a_2^{(r)}, a_3^{(r)}, a_4^{(r)})^T \tag{3.449}$$

denote the vector of the coefficients of the polynomial $\underline{\boldsymbol{u}}_r^{(k)}$. Moreover,

$$\underline{\boldsymbol{u}}_r^{(k)}(\xi(x, y, z), \eta(x, y, z), \zeta(x, y, z)) = \boldsymbol{u}_r^{(k)}(x, y, z) \tag{3.450}$$

and therefore

$$\frac{\partial \boldsymbol{u}_r^{(k)}}{\partial x} = \frac{\partial \boldsymbol{u}_r^{(k)}}{\partial \xi}\frac{\partial \xi}{\partial x} + \frac{\partial \boldsymbol{u}_r^{(k)}}{\partial \eta}\frac{\partial \eta}{\partial x} + \frac{\partial \boldsymbol{u}_r^{(k)}}{\partial \zeta}\frac{\partial \zeta}{\partial x}, \quad \text{etc.} \tag{3.451}$$

Let us denote

$$S = \begin{bmatrix} 1 & 0 & 0 & 1 \\ 0 & 1 & 0 & 1 \\ 0 & 0 & 1 & 1 \\ 0 & 0 & 0 & 1 \end{bmatrix} \tag{3.452}$$

$$\vec{\Delta}_r = (\underline{\boldsymbol{u}}_r^{(k)}(R_1), \underline{\boldsymbol{u}}_r^{(k)}(R_2), \underline{\boldsymbol{u}}_r^{(k)}(R_2))^T, \tag{3.453}$$

then

$$\vec{\Delta}_r = S\underline{\boldsymbol{a}}_r. \tag{3.454}$$

Hence

$$\underline{\boldsymbol{a}}_r = S^{-1}\vec{\Delta}_r. \tag{3.455}$$

Let us denote

$$T = \begin{bmatrix} \underline{x}_2 & \underline{x}_3 & \underline{x}_4 \\ \underline{y}_2 & \underline{y}_3 & \underline{y}_4 \\ \underline{z}_2 & \underline{z}_3 & \underline{z}_4 \end{bmatrix} \quad \text{and } \boldsymbol{J} = \det(T). \tag{3.456}$$

According to the substitution theorem we have

$$\int_{V_k} f(x, y, z)\mathrm{d}x\mathrm{d}y\mathrm{d}z = \int_{V_o} |\boldsymbol{J}|\, \underline{f}(\xi, \eta, \zeta)\mathrm{d}\xi\mathrm{d}\eta\mathrm{d}\zeta.$$

Let us denote $\boldsymbol{s} = (x, y, z) = (s_1, s_2, s_3)$ and

$$\boldsymbol{t} = (\xi, \eta, \zeta) = (t_1, t_2, t_3).$$

Then

$$\int_{V_k} c_{ijgh}\varphi_{g,h}^m \varphi_{i,j}^n \mathrm{d}\boldsymbol{x} = |\boldsymbol{J}|\, c_{ijgh} \int_{V_o} (S^{-1}\vec{\varphi}_g^m)^T \frac{\partial \boldsymbol{G}}{\partial t_k}\frac{\partial t_k}{\partial s_h}(S^{-1}\vec{\varphi}_i^n)\frac{\partial \boldsymbol{G}}{\partial t_l}\frac{\partial t_l}{\partial s_j}\mathrm{d}\boldsymbol{t}$$

$$= |\boldsymbol{J}|\, c_{ijgh}(\vec{\varphi}_g^m)^T (S^{-1})^T \int_{V_o} \frac{\partial \boldsymbol{G}}{\partial t_k}\frac{\partial \boldsymbol{G}^T}{\partial t_l}\frac{\partial t_k}{\partial s_h}\frac{\partial t_l}{\partial s_j}\mathrm{d}\boldsymbol{t}\, S^{-1}\vec{\varphi}_i^n.$$

Since

$$\frac{\partial t_i}{\partial s_j} = T_{ij}^{-1}, \quad \frac{\partial \boldsymbol{G}}{\partial t_i} = (0, \ldots, 0, 1, 0, \ldots 0),$$

where 1 is in the i-th position, and since

$$\int_{V_o} \xi^a \eta^b \zeta^c \mathrm{d}\boldsymbol{t} = \frac{a!b!c!}{(a+b+c+3)!}, \tag{3.457}$$

then

$$\int_{V_o} \frac{\partial \boldsymbol{G}}{\partial t_k} \frac{\partial \boldsymbol{G}^T}{\partial t_l} \frac{\partial t_k}{\partial s_h} \frac{\partial t_l}{\partial s_j} \mathrm{d}\boldsymbol{t} = \frac{1}{6} B_{hj},$$

where

$$B_{hj} = \begin{bmatrix} T_{1h}^{-1} T_{1j}^{-1} & T_{1h}^{-1} T_{2j}^{-1} & T_{1h}^{-1} T_{3j}^{-1} \\ T_{2h}^{-1} T_{1j}^{-1} & T_{2h}^{-1} T_{2j}^{-1} & T_{2h}^{-1} T_{3j}^{-1} \\ T_{3h}^{-1} T_{1j}^{-1} & T_{3h}^{-1} T_{2j}^{-1} & T_{3h}^{-1} T_{3j}^{-1} \\ 0 & 0 & 0 \end{bmatrix}$$

Let us denote

$$\tilde{K}_{hj} = \frac{|\boldsymbol{J}|}{6} (S^{-1})^T B_{hj} S^{-1}, \quad \vec{\varphi}^m = ((\varphi_1^m)^T, (\varphi_2^m)^T, (\varphi_3^m)^T)^T,$$

then

$$(A)_{m,n} = \int_{V_o} c_{ijgh} \varphi_{g,h}^m \varphi_{i,j}^n \mathrm{d}\boldsymbol{t} = (\vec{\varphi}^m)^T \tilde{K} \vec{\varphi}^n,$$

where $\tilde{K}$ is the 12×12 dimensional matrix of the form

$$\tilde{K} = \begin{bmatrix} a_{1j1h} \tilde{K}_{hj} & a_{1j2h} \tilde{K}_{hj} & a_{1j3h} \tilde{K}_{hj} \\ a_{2j1h} \tilde{K}_{hj} & a_{2j2h} \tilde{K}_{hj} & a_{2j3h} \tilde{K}_{hj} \\ a_{3j1h} \tilde{K}_{hj} & a_{3j2h} \tilde{K}_{hj} & a_{3j3h} \tilde{K}_{hj} \end{bmatrix} \tag{3.458}$$

If $\vec{\varphi}^m = 0$ or $\vec{\varphi}^n = 0$ in V_k, then $(A)_{m,n} = 0$. If $\vec{\varphi}^m \neq 0$ or $\vec{\varphi}^n \neq 0$ in V_k, then vectors $\vec{\varphi}^m, \vec{\varphi}^n$ have on just one position number 1 and elsewhere 0. We denote these positions by i_m and i_n. Then $(A)_{m,n} = (\tilde{K})_{i_m, i_n}$. The Jacobian as well as the matrix $\tilde{K}$ are computed for every tetrahedron V_k.

To set up the matrix A we use the following algorithm: let $A_{ij} = 0$, $i, j = 1, \ldots, n$. We compute the elementary matrix $\tilde{K}$ for all tetrahedrons V_k. We find all pairs of functions $\vec{\varphi}^m$ and $\vec{\varphi}^n$, which are non-zero on the given tetrahedron. Then for such pairs of functions we find indices i_m and i_n and we put

$$(A)_{i_m, i_n} = (A)_{i_m, i_n} + (\tilde{K})_{m,n}.$$

To compute $(b^{kl})_n = \int_Y c_{ijkl} \varphi_{i,y_j}^n \mathrm{d}\boldsymbol{x}$ we use the technique which is simple modification of that discussed above. We have

$$(b^{kl})_n = \int_{V_k} c_{ijkl} \varphi_{i,y_j}^n \mathrm{d}\boldsymbol{x}$$

$$= |\boldsymbol{J}|\, c_{ijkl} \int_{V_o} (S^{-1}\vec{\varphi}_i^n)^T \frac{\partial \boldsymbol{G}}{\partial t_l} \frac{\partial t_l}{\partial s_j} \mathrm{d}\boldsymbol{t}$$

$$= |\boldsymbol{J}|\, c_{ijkl} \int_{V_o} (\vec{\varphi}_i^n)^T (S^{-1})^T (T_{1j}^{-1}, T_{2j}^{-1}, T_{3j}^{-1}, 0)^T \mathrm{d}\boldsymbol{t}$$

$$= (\tilde{\varphi}^n)^T\, |\boldsymbol{J}| \int_{V_o} ((c_{1jkl}v_j)^T, (c_{2jkl}v_j)^T, (c_{3jkl}v_j)^T)^T \mathrm{d}\boldsymbol{t}$$

$$= \tfrac{1}{6}(\vec{\varphi}^n)^T\, |\boldsymbol{J}|\, v^{kl}, \qquad (3.459)$$

where $\underline{\boldsymbol{v}}_j = (S^{-1})^T (T_{1j}^{-1}, T_{2j}^{-1}, T_{3j}^{-1}, 0)^T$, and where Equation 3.457 was used. For $\vec{\varphi}^n \neq 0$ in V_k there exists just one number i such that $(\vec{\varphi}^n)_i = 1$. Then

$$(b^{kl})_n = (b^{kl})_n + \tfrac{1}{6}\, |\boldsymbol{J}|\, (\underline{\boldsymbol{v}}^{kl})_i.$$

Remark: The above FEM approximation of the 1st order, which was used for illustration only, does not estimate conditions between two different materials too accurately. Therefore, for a practical use the functions generated by the polynomials of the higher order must be used. Our numerical experiments presented later are based on the cubic 3D finite elements.

We saw (see Equation 3.384) that the problems discussed above lead to the following problem.

Find $\boldsymbol{u}_h \in \boldsymbol{u}_o + V_h$ minimizing the functional

$$J(\boldsymbol{u}) = \tfrac{1}{2}a(\boldsymbol{u}, \boldsymbol{u}) - \int_{\Omega} f_i u_i \mathrm{d}\boldsymbol{x} - \int_{\Gamma_\tau} P_i u_i \mathrm{d}s \qquad (3.460)$$

on the set V_h, where $a(\boldsymbol{u}, \boldsymbol{u})$ is a symmetric, positive definite quadratic form.

We rewrite Equation 3.460 into the equivalent algebraic formulation

$$J(\boldsymbol{u}) = \tfrac{1}{2}\boldsymbol{u}^T K \boldsymbol{u} - \boldsymbol{b}^T \boldsymbol{u}, \quad \boldsymbol{u} \in R^n, \qquad (3.461)$$

where K is the stiffness matrix, which is due to the properties of $a(\boldsymbol{u}, \boldsymbol{u})$, a symmetric and positive (semi-)definite matrix. To minimize the functional $J(\boldsymbol{u})$ from the above equation the Jacobi, Gauss-Seidel, SSOR or the conjugate gradient methods can be used.

Next we shall give some ideas of the preconditioned conjugate gradient method (PCGM) and of the preconditioned SSOR method (PSSORM). The preconditioned Jacobi and Gauss-Seidel methods are a special case of PSSORM. Firstly, we give the main idea of preconditioning for Equation 3.461 (see Axelsson and Barker (1984)).

Let C be a positive definite matrix which can be factored in the form $C = EE^T$, where E is a non-singular matrix (or in the more general form $C = EG^{-1}E^T$, where E is a sparse and lower triangular matrix and G is a diagonal one). By using the transformation $\boldsymbol{w} = E^T\boldsymbol{u}$ we transform Equation 3.461 into

$$\underline{J}(\boldsymbol{w}) = \tfrac{1}{2}\boldsymbol{w}^T \underline{K} \boldsymbol{w} - \underline{\boldsymbol{b}}^T w, \qquad (3.462)$$

where $\underline{K} = E^{-1}KE^{-T}$, $\underline{\boldsymbol{b}} = E^{-1}\boldsymbol{b}$ and where $E^{-T} = (E^{-1})^T$. The matrix $\underline{K}$ is also symmetric and since $\boldsymbol{w}^T\underline{K}\boldsymbol{w} = \boldsymbol{u}^TK\boldsymbol{u}$ and since $\boldsymbol{u}^TK\boldsymbol{u} > 0\,\forall\boldsymbol{u} \neq 0$, K being positive (semi-)definite, then $\boldsymbol{w}^T\underline{K}\boldsymbol{w} > 0\,\forall\boldsymbol{w} \neq 0$ and $\underline{K}$ is also positive (semi-)definite. Moreover, since $E^{-T}\underline{K}E^T = E^{-T}E^{-1}K = C^{-1}K$ we see that the matrices $\underline{K}$ and $C^{-1}K$ have the same eigenvalues $\underline{\lambda}_i$ and, therefore, also have the same spectral condition number $\underline{k}(\underline{K}) = \underline{\lambda}_N/\underline{\lambda}_1$. The rate of convergence of iterative methods depends on the spectral condition number. If C can be found such that $\underline{k}(\underline{K}) < \underline{k}(K)$, then the method with preconditioning converges faster than the initial one as the rate of convergence is better than that of the initial method.

(a) The Preconditioned Conjugate Gradient Method Due to our assumptions, matrix E is non-singular and C is positive definite. Then the sequence $\boldsymbol{w}_o, \boldsymbol{w}_1, \boldsymbol{w}_2, \ldots$, where $\boldsymbol{w}_k = E^T\boldsymbol{u}^k$, can be found by applying the ordinary conjugate gradient method to the quadratic functional $\underline{J}(w)$ of Equation 3.462. Thus $\lim_{k\to\infty}\boldsymbol{w}_k \to \underline{\boldsymbol{w}}$, where $\underline{\boldsymbol{w}}$ is the minimizer of the functional $\underline{J}(\boldsymbol{w})$.

Then the algorithm of the preconditioned conjugate gradient method is defined as follows (see Axelsson and Barker (1984)).

Let K be positive (semi-)definite. Let $C = EE^T$ be positive definite and E be non-singular. Let $\boldsymbol{u} = E^{-T}\boldsymbol{w}$. Let $\boldsymbol{g} = K\boldsymbol{u} - \boldsymbol{b}$ be the gradient of the functional of Equation 3.461. Then we find the minimizer of the functional $J(\boldsymbol{u})$ of Equation 3.461 by performing the iteration process defined by:

Start with $\boldsymbol{u}_o$ and compute $\boldsymbol{g}_o = \boldsymbol{g}(\boldsymbol{u}_o) = K\boldsymbol{u}_o - \boldsymbol{b}$, $\boldsymbol{h}_o = C^{-1}\boldsymbol{g}_o$, $\boldsymbol{d}_o = -\boldsymbol{h}_o$. Then we find

$$t_k = \boldsymbol{g}_k^T\boldsymbol{h}_k(\boldsymbol{d}_k^TK\boldsymbol{d}_k)^{-1}, \tag{3.463}$$

$$\boldsymbol{u}_{k+1} = \boldsymbol{u}_k + t_k\boldsymbol{d}_k, \tag{3.464}$$

$$\boldsymbol{g}_{k+1} = \boldsymbol{g}_k + t_kK\boldsymbol{d}_k, \tag{3.465}$$

$$\boldsymbol{h}_{k+1} = C^{-1}\boldsymbol{g}_{k+1}, \tag{3.466}$$

$$\alpha_k = \boldsymbol{g}_{k+1}^T\boldsymbol{h}_{k+1}(\boldsymbol{g}_k^T\boldsymbol{h}_k)^{-1}, \tag{3.467}$$

$$\boldsymbol{d}_{k+1} = -\boldsymbol{h}_{k+1} + \alpha_k\boldsymbol{d}_k, \tag{3.468}$$

where $k = 0, 1, 2, \ldots$.

Since $\lim_{k\to\infty}\boldsymbol{w}_k \to \underline{\boldsymbol{w}}$, where $\underline{\boldsymbol{w}}$ is the minimizer of the functional $\underline{J}(\boldsymbol{w})$, and since the relations $\boldsymbol{w}_k = E^T\boldsymbol{u}^k$, $k = 0, 1, 2, \ldots$ and $\underline{\boldsymbol{w}} = E^T\underline{\boldsymbol{u}}$, where $\underline{\boldsymbol{u}}$ is the minimizer of Equation 3.461, imply

$$\left\|\boldsymbol{w}_k - \underline{\boldsymbol{w}}_k\right\|_{\underline{K}} = \left\|\boldsymbol{u}_k - \underline{\boldsymbol{u}}_k\right\|_K,$$

then also $\lim_{k\to\infty}\boldsymbol{u}_k = \underline{\boldsymbol{u}}$. Due to Theorem 1.12 of Axelsson, and Barker (1984), p. 26, we have

$$\left\|\boldsymbol{w}_k - \underline{\boldsymbol{w}}_k\right\|_{\underline{K}} = \left\|\boldsymbol{u}_k - \underline{\boldsymbol{u}}_k\right\|_K \leq \varepsilon\left\|\boldsymbol{u}_o - \underline{\boldsymbol{u}}\right\|_K \quad \forall\boldsymbol{u}_o \in R^N.$$

Since the rate of convergence of the $\{\boldsymbol{w}_k\}$ depends on $\underline{k}(\underline{K})$ then the smallest number of iterations $k(\varepsilon)$ satisfies

$$k(\varepsilon) \leq \tfrac{1}{2}(\underline{k}(\underline{K}))^{1/2}\ln(2/\varepsilon) + 1. \tag{3.469}$$

It is evident (see Theorem 1.10 of Axelsson and Barker (1984), p. 24) that $k(\varepsilon) < N$, where N is the order of $\underline{K}$. Thus we have proved the following theorem.

Theorem 3.43: For the preconditioned conjugate gradient method the error estimate

$$\left\|\boldsymbol{w}_k - \underline{\boldsymbol{w}}_k\right\|_{\underline{K}} = \left\|\boldsymbol{u}_k - \underline{\boldsymbol{u}}_k\right\|_K \leq \varepsilon \left\|\boldsymbol{u}_o - \underline{\boldsymbol{u}}\right\|_K, \varepsilon > 0 \quad \forall \boldsymbol{u}_o \in R^N$$

holds, and the smallest number of iterations $k(\varepsilon)$ satisfies Equation 3.469

$$k(\varepsilon) \leq \tfrac{1}{2}(k(\underline{K}))^{1/2} \ln(2/\varepsilon) + 1.$$

Numerical experiments show that the conjugate gradient method is a numerically unstable method due to the cumulation round-off errors. Cumulation round-off errors after some first iterations are due to the orthogonalization of vectors in Equations 3.463-3.468. But by a suitable choice of preconditioning matrices C the method of preconditioned conjugate gradient will be useful for numerical calculations. Next we shall give one useful algorithm.

Matrix $\underline{K}$ is constructed from matrix K by the algorithm discussed above. Matrix K was constructed in such a way that at the place (i, j) of the matrix a value of a certain scalar product from basis functions $\vec{\varphi}_i$ and $\vec{\varphi}_j$ is given. But these functions are vector functions with three components. We shall number them in such a way that functions with non-zero first components have the lowest indices. Then we number functions with a non-zero second component and then functions with a non-zero third component. Then the obtained matrix can be written in the following form

$$K = \begin{bmatrix} K_{11} & K_{12} & K_{13} \\ K_{21} & K_{22} & K_{23} \\ K_{31} & K_{32} & K_{33} \end{bmatrix} \tag{3.470}$$

where K_{ij} is the square matrix of the scalar products of functions with non-zero i-th and j-th components. We define the matrices

$$K^s = \text{diag}(K_{11}, K_{22}, K_{33}), \tag{3.471}$$

which is denoted as the separate displacement component (SDC) part of the stiffness matrix K, and the matrix K^i defined by

$$K^i = \text{diag}(B, B, B), \quad B = K_{11} + K_{22} + K_{33}, \tag{3.472}$$

denoted in the literature as the isotropic approximation (ISO) to the stiffness matrix K. It can be shown that both matrices are symmetric and positive definite, i.e. there exist constants $c_1, \ldots, c_4$ such that

$$c_1(K^s\boldsymbol{u}, \boldsymbol{u}) \leq (K\boldsymbol{u}, \boldsymbol{u}) \leq c_2(K^s\boldsymbol{u}, \boldsymbol{u}) \quad \forall \boldsymbol{u} \in R^N,$$

$$c_1(K^i\boldsymbol{u}, \boldsymbol{u}) \leq (K\boldsymbol{u}, \boldsymbol{u}) \leq c_2(K^i\boldsymbol{u}, \boldsymbol{u}) \quad \forall \boldsymbol{u} \in R^N.$$

Matrices $(K^s)^{-1}$ and $(K^i)^{-1}$ are suitable for preconditioning of the matrix K.

Then the algorithm of PCGM from Equations 3.463-3.468 can be rewritten as follows.

We start with $\boldsymbol{u}_o$ and compute $\boldsymbol{g}_o = \boldsymbol{g}(\boldsymbol{u}_o) = K\boldsymbol{u}_o - \boldsymbol{b}$, $\boldsymbol{d}_o = (K^s)^{-1}\boldsymbol{g}_o$, $\gamma_o = \boldsymbol{g}_o^T(K^s)^{-1}\boldsymbol{g}_o$. Then we find

$$t_k = \gamma_o(\boldsymbol{d}_k^T K \boldsymbol{d}_k)^{-1}, \tag{3.473}$$

$$\boldsymbol{u}_{k+1} = \boldsymbol{u}_k + t_k \boldsymbol{d}_k, \tag{3.474}$$

$$\boldsymbol{g}_{k+1} = \boldsymbol{g}_k + t_k K \boldsymbol{d}_k, \tag{3.475}$$

$$\gamma_1 = g_k^T (K^s)^{-1} g_k. \tag{3.476}$$

If $\gamma_1 < \varepsilon^2$ then the iterative process is finished. In the opposite case we compute

$$\alpha_k = \gamma_o/\gamma_1, \gamma_o = \gamma_1, \tag{3.477}$$

$$\boldsymbol{d}_{k+1} = -(K^s)^{-1}\boldsymbol{g}_k + \alpha_k \boldsymbol{d}_k, \tag{3.478}$$

where $k = 0, 1, 2, \ldots$.

For this algorithm one can prove:

Theorem 3.44: Let

$$c = \left\|(K^s)^{-1}\right\| \cdot \|K\|.$$

Then at most $m = \mathrm{INT}(\frac{1}{2}c^{1/2}\ln(2/\varepsilon) + 1)$ iterations can be made. Moreover, after m iterations

$$(K\boldsymbol{g}_k, \boldsymbol{g}_k)^{1/2} < \varepsilon.$$

(For the proof see Blaheta (1987).)

Remark: The matrix $(K^s)^{-1}$ is not computed. The vector $\boldsymbol{w} = (K^s)^{-1}\boldsymbol{v}$ is obtained as a solution of $K^s\boldsymbol{w} = \boldsymbol{v}$. Since, in addition, the matrix K^s is symmetric and positive definite, we compute its triangle decomposition once and then for the other vectors $\boldsymbol{v}$ we find $\boldsymbol{w}$ by using a small number of arithmetic operations.

Remark: We can also write a similar algorithm for the matrix K^i, for which we obtain similar results as above.

In a brief survey we shall give some results obtained in the literature for the PSSOR method.

(b) The Preconditioned SSOR Method Let matrix K be decomposed as

$$K = D - L - L^T, \tag{3.479}$$

where D is the diagonal matrix and $(-L)$ is the strictly lower triangular one.

Let $0 < \omega < 2$ be the relaxation parameter. Let

$$C = (\omega^{-1}D - L)(\omega^{-1}D)^{-1}(\omega^{-1}D - L)^T \tag{3.480}$$

be the SSOR preconditioning matrix in the conjugate gradient algorithm of Equations 3.463-3.468. It is evident that matrix C is positive definite if and only if $0 < \omega < 2$, and is of the general form $C = EG^{-1}E^T$. Instead of Equation 3.480,

$$C = (2 - \omega)^{-1}(\omega^{-1}D - L)(\omega^{-1}D)^{-1}(\omega^{-1}D - L)^T \tag{3.481}$$

can be used for the preconditioning (see Axelsson and Barker (1984)). The analysis of the rate convergence with optimal SSOR preconditioning, as well as analysis of incomplete factorization and optimal storing in the computer memory are made in Axelsson and Barker (1984). All these problems are beyond the goal of this section.

Remark: In the case of the homogenization problem we must determine $\overline{\chi}^{kl}$ from the problem of Equation 3.456, i.e. from Equation 3.433:

$$\int_Y (c_{ijgh}(\overline{\chi}^{kl}_{g,y_h} - \delta_{gk}\delta_{hl}))\varphi^n_{i,y_j}\,\mathrm{d}\boldsymbol{x} = 0 \quad \forall \vec{\varphi}^n \in \mathscr{B},$$

where $\mathscr{B}$ is created by an arbitrary basis of $[H^1_o(Y)]^3$. We seek $\overline{\chi}^{kl}$ in the form $\overline{\chi}^{kl} = \Sigma_{\varphi^i \in \mathscr{B}} \alpha^{kl}_i \varphi^i$, where $\{\varphi^i\}$ is the basis of $\mathscr{B}$. Then we obtain a system of linear algebraic equations $A\vec{\alpha}^{kl} = \boldsymbol{b}^{kl}$, where $\vec{\alpha}^{kl}$ is the vector of unknown coefficients, $(b^{kl})_n = \int_Y c_{ijkl}\varphi^n_{i,y_j}\mathrm{d}\boldsymbol{x}$ and $(A)_{m,n} = \int_Y c_{ijgh}\varphi^m_{g,y_h}\varphi^n_{i,y_j}\mathrm{d}\boldsymbol{x}$ being independent of k and l. The matrix A is regular, positive definite and, moreover, very sparse. To solve this system of algebraic equations, the method of conjugate gradients can be used. Since the matrix A is very sparse, the round-off errors are not accumulated and therefore the method of conjugate gradients can be used without any difficulties.

Numerical Experiments

Composite materials play an important role in geology as practically all rock materials are composites. The structure of rocks is a composition of several types of crystals and, in general, is stochastic. In the first approximation we shall assume that the structure of the rocks is periodic and, therefore, can be transformed by the homogenization method to the anisotropic one. Next we shall present several types of rock materials, which simulate several types of anisotropic rocks. In our numerical experiments we approximated the rock material by its periodical structure simulated by a system of small cubes of different densities and different elastic parameters. We assumed these small cubes to be homogeneous and isotropic. Their number in the horizontal as well as vertical directions change from case to case. Then the values of elastic parameters also change from case to case as is shown from our numerical experiments, from which we shall give several numerical results.

In our numerical experiments we designate the type of approximation of rock materials "Type Number", the number of small cubes in a horizontal direction "Number of Cubes", about which the rock material is approximated, and the accuracy of computation of homogenization "Accuracy eps", and we also give the indices of elastic coefficients and their numerical values.

Numerical experiments:

1) Type Number: 1
   Elastic Coefficients: $\lambda_1 = 2.10^{10}$ Pa, $\mu_1 = 1.10^{10}$ Pa
   $\lambda_2 = 3.10^{10}$ Pa, $\mu_2 = 2.10^{10}$ Pa
   Number of Cubes: 2
   Accuracy eps: 10^{-7}

i	*j*	*k*	*l*	*numerical value*
1	1	1	1	5.50 e + 11
1	1	2	2	2.50 e + 11
1	1	3	3	2.50 e + 11
1	1	1	2	0.00 e + 11

i j k l	*numerical value*
1 1 2 3	0.00 e + 11
1 1 1 3	0.00 e + 11
2 2 2 2	5.50 e + 11
2 2 3 2	2.50 e + 11
2 2 1 2	0.00 e + 11
2 2 2 3	0.00 e + 11
2 2 1 3	0.00 e + 11
3 3 3 3	5.50 e + 11
3 3 1 2	0.00 e + 11
3 3 2 3	0.00 e + 11
3 3 1 3	0.00 e + 11
1 2 1 2	1.50 e + 11
1 2 2 3	0.00 e + 11
1 2 1 3	0.00 e + 11
2 3 2 3	1.50 e + 11
2 3 1 3	0.00 e + 11
1 3 1 3	1.50 e + 11

2) Type Number: 1
Elastic Coefficients: $\lambda_1 = 2.10^{10}$ Pa, $\mu_1 = 1.10^{10}$ Pa
$\lambda_2 = 3.10^{10}$ Pa, $\mu_2 = 2.10^{10}$ Pa
Number of Cubes: 4
Accuracy eps: 10^{-7}

This model gives the same values as the previous one.

3) Type Number: 2
Elastic Coefficients: $\lambda_1 = 2.10^{10}$ Pa, $\mu_1 = 1.10^{10}$ Pa
$\lambda_2 = 3.10^{10}$ Pa, $\mu_2 = 2.10^{10}$ Pa
Number of Cubes: 3
Accuracy eps: 10^{-7}

i j k l	*numerical value*
1 1 1 1	5.385117 e + 10
1 1 2 2	2.443377 e + 10
1 1 3 3	2.443377 e + 10
1 1 1 2	−5.163028 e + 02
1 1 2 3	2.067943 e + 06
1 1 1 3	−4.472232 e + 00
2 2 2 2	5.363613 e + 10
2 2 3 2	2.441576 e + 10
2 2 1 2	−2.252292 e + 01
2 2 2 3	−4.681956 e + 05

i j k l	*numerical value*
2 2 1 3	−8.864648 e + 00
3 3 3 3	5.363608 e + 10
3 3 1 2	1.056546 e + 01
3 3 2 3	−4.676974 e + 05
3 3 1 3	−3.113459 e + 01
1 2 1 2	1.462295 e + 10
1 2 2 3	−2.337181 e + 00
1 2 1 3	7.379588 e + 05
2 3 2 3	1.457190 e + 10
2 3 1 3	−2.121354 e + 00
1 3 1 3	1.462297 e + 10

4) Type Number: 2
Elastic Coefficients: $\lambda_1 = 2.10^{10}$ Pa, $\mu_1 = 1.10^{10}$ Pa
$\lambda_2 = 3.10^{10}$ Pa, $\mu_2 = 2.10^{10}$ Pa
Number of Cubes: 4
Accuracy eps: 10^{-7}

i j k l	*numerical value*
1 1 1 1	5.454546 e + 10
1 1 2 2	2.454546 e + 10
1 1 3 3	2.363637 e + 10
1 1 1 2	0.
1 1 2 3	−1.631580 e − 11
1 1 1 3	−2.204827 e − 12
2 2 2 2	5.454546 e + 10
2 2 3 2	2.363637 e + 10
2 2 1 2	0.
2 2 2 3	1.437553 e − 10
2 2 1 3	−2.204826 e − 12
3 3 3 3	5.090910 e + 10
3 3 1 2	0.
3 3 2 3	−1.984357 e − 11
3 3 1 3	−6.614477 e − 11
1 2 1 2	1.500000 e + 10
1 2 2 3	0.
1 2 1 3	2.182787 e − 01
2 3 2 3	1.333334 e + 10
2 3 1 3	4.849477 e − 12
1 3 1 3	1.333334 e + 10

REFERENCES

Amadei, B. and Goodman, R.E. (1982), "The Influence of Rock Anisotropy on Stress Measurement by Overcoring Techniques", *Roc. Mech.*, **15**, 167-180.

Axelsson, O. and Barker, V.A. (1984), *Finite Element Solution of Boundary Value Problems: Theory and Computation*, Academic Press, New York.

Babuška, I. (1976), "Solution of Interface Problems by Homogenization: I, II, III", *SIAM J. Math. Anal.*, **7** (1976), 603-634 (I), 635-645 (II), and **8** (1977), 923-937, (III).

Babuška, I. and Aziz, A.K. (1972), "Survey Lectures on the Mathematical Foundations of the Finite Element Method", in *The Mathematical Foundations of the Finite Element Method with Applications to Partial Differential Equations* (Aziz, A.K. (Ed.)) Academic Press, New York, 5-359.

Babuška, I. and Reinboldt, W.C. (1980), "Reliable Error Estimation and Mesh Adaptation for the Finite Element Method", in *Computational Methods in Nonlinear Mechanics* (Oden, J.T. (Ed.)), North-Holland.

Babuška, I. and Reinboldt, W.C. (1984), *Adaptive Finite Element Process in Structural Mechanics: Elliptic Problem Solvers*, Academic Press, 345-377.

Bensoussan, A., Lions, J.L. and Papanicolaou, G. (1978), *Asymptotic Analysis for Periodic Structures*, North-Holland Publ. Co., Amsterdam.

Biroli, M. (1977), "G-convergence for Elliptic Equations, Variational Inequalities and Quasi-variational Inequalities", *Rend. Sem. Mat. Fis.*, Milano, **47**, 269-328.

Blaheta, R. (1987), *Iterative Methods for Numerical Solution of Boundary Value Problems* (Thesis), Faculty of Mathematics and Physics, Charles University, Prague (in Czech).

Borsetto, M., Martinetti, S. and Ribacchi, R. (1984), "Interpretation of In-Situ Stress Measurements in Anisotropic Rocks with the Doorstopper Method", *Rock Mech. Rock Engmg.*, **17**, 167-182.

Brdička, M. (1959), *Continuum Mechanics*, Academia, Prague (in Czech).

Brdička, M. and Hladík, A. (1987), *Theoretical Mechanics*, Academia, Prague (in Czech).

Buchdahl, H.A. (1949), *American Journal of Physics*, January 1949.

Callen, H.B. (1960), *Thermodynamics*, J. Wiley & Sons, New York.

Carathéodory, K. (1909), *Math. Ann*, **67**, 355.

Céa, J. (1971), *Optimisation: Théorie et Algorithms*, Dunod, Paris.

Ciarlet, P.G. (1978), *The Finite Element Method for Elliptic Problems*, North-Holland Publ. Co., Amsterdam.

Desloge, E.A. (1982), *Classical Mechanics*, J. Wiley & Sons, New York.

Duvaut, G. and Lions, J.L. (1972), *Les Inéquations en Mécanique et en Physique*, Dunod, Paris.

Ekeland, I. and Temam, R. (1976), *Convex Analysis and Variational Problems*, North-Holland Publ. Co., Amsterdam.

Engelder, T. (1984), "The Time-Dependent Strain Relaxation of Algerie Granite", *Int. J. Rock Mech. Min. Sci. & Geomech. Abstr.*, **21**, 63-73.

Fichera, G. (1972), "Boundary Value Problems of Elasticity with Uni-lateral Constraints, in *Encyclopedia of Physics* (Flügge S. (Ed.)), **Vol. VIa/2**, Springer Verlag, Berlin.

Fučík, S. and Kufner, A. (1980), *Non-linear Differential Equations*, Elsevier, Amsterdam, Oxford, New York.

Fournier d'Albe, E.M. (1978), The UNESCO Programme on the Assessment and Mitigation of Earthquake Risk. In: Kárník, V. and Schenková, Z. (Eds). Proc. of the Symposium on the Analysis of Seismicity and on Seismic Risk, Academia, Prague, 15-24.

Glansdorff, P. and Prigogine, I. (1971), *Thermodynamic Theory of Structure, Stability and Fluctuations*, Wiley-Interscience, New York.

Gurtin, M.E. (1974), *Modern Continuum Thermodynamics, Mechanics Today*, **Vol. 1**, Pergamon Press, New York.

Hackbusch, W. (1992), *Elliptic Differential Equations*, Springer Verlag, Berlin, Heidelberg, New York, London, Paris, Tokyo, Hong Kong, Barcelona, Budapest.

Hasse, R. (1968), *Thermodynamics of Irreversible Processes*, Addison Wesley, New York.

Hill, R. (1950), *The Mathematical Theory of Plasticity*, Oxford.

Jaunzemis, W. (1967), *Continuum Mechanics*, Macmillan, New York.

Kolář, V., Kratochvíl, J., Leitner, F. and Ženíšek, A. (1972), (1979), *Calculation of Surface and Space Constructions by Finite Element Method*, SNTL, Prague (in Czech).

Křížek, M. and Neittaanmäki, P. (1990), *Finite Element Approximation of Variational Problems and Applications*, Longman Scientific & Technical, Harlow.

Lanczos, C. (1962), *The Variational Principles of Mechanics*, University of Toronto Press, Toronto.

Landau, L.D. and Lifshitz, E.M. (1973), *Mechanics*, Nauka, Moscow (in Russian).

Leeman, E.R. (1967), "The Borehole Deformation Type of Rock Stress Measuring Instrument", *Int. J. Rock Mech. & Min. Sci.*, **4**, 23-44.

Leeman, E.R. (1969), "The "Doorstopper" and Triaxial Rock Stress Measuring Instruments Developed by the C.S.I.R.", *Jl. S. Afr. Inst. Min. Metall.*, **69**, 305-319.

Leeman, E.R. (1971), "The CSIR "Doorstopper" and Triaxial Rock Stress Measuring Instruments", *Rock Mech.*, **3**, 25-50.

Levin, S. (Ed.) (1977), *Mathematics Applied to Continuum Mechanics*, Macmillan, New York.

Mantič, V. (1988), "Computation of Rock Stress State in the Neighbourhood of the Tensometric Probe by the Finite Element Method", *Research Report*, **V-370**, Institute of Computer Science AS CR, Prague (in Czech).

Mantič, V. and Nedoma, J. (1989a), "Optimal Design of Tensometric Probes for In-Situ Stress-Strain Measurements by Finite Element Analysis, and Their Optimal Dislocation in the Collision Region", in *Proc. of the 4th Int. Symposium on the Analysis of Seismicity and Seismic Risk* (Schenková, Z. (Ed.)), Geophysical Institute AS CR, Prague, 257-264.

Mantič, V. and Nedoma, J. (1989b), "Optimal Design of Tensometric Probes for Stress-Strain Measurements in Rock Massif Based on Finite Element Analysis, and Their Application for the Geomechanic Monitoring", *Maining Geomechanics*, ČSTV Ostrava, 94-102 (in Czech).

Mihlin, S.G. (1966), *The Numerical Implementation of Variational Methods*, Moscow (in Russian).

Milka, Z. (1989), *Numerical Solution of the Model Describing Geodynamic Processes in Medium with Periodic Structure* (Thesis), Faculty of Mathematics and Physics, Charles University, Prague (in Czech).

Natterer, F.(1986), "Numerical Treatment of Ill-Posed Problems", in *Inverse Problems* (Talenti, G. (Ed.)), Montecatini Terme, LNM 1225, 142-167.

Nečas, J. (1983), "Introduction to the Theory of Non-linear Elliptic Equations", *Teubner-Texte zur Mathematik*, Teubner Verlaggesellschaft, Leipzig.

Nečas, J. and Hlaváček, I. (1981), *Mathematical Theory of Elastic and Elasto-Plastic Bodies: An Introduction*, Elsevier, Amsterdam, Oxford, New York.

Nedoma, J. (1989), "On a New Theory of Earthquake Mechanism", in *Proc. of the 4th Int. Symposium on the Analysis of Seismicity and Seismic Risk* (Schenková, Z. (Ed.)), Geophysical Institute AS CR, Prague, 339-346.

Nedoma, J. (1990a), "Tectonic Evolution of Collision Zones: I. Mechanics of Tectonic Processes; II. Subducted Lithosphere", *Gerlands Beitr. Geophysik*, **99**, **2**, 97-108; 109-121.

Nedoma, J. (1990b), "Plate tectonics and the Analysis of Fracture Mechanics in Collision Zone: I., II.", *Contr. Geophys. Inst. Slov. Acad. Sci.*, **20**, 53-69; **21**, 38-58.

Nedoma, J. (1990c), "On the Stress-Strain Monitoring Methods In Situ", *Technical Report*, **V-477**, Institute of Computer Science AS CR, Prague (in Czech).

Nedoma, J. (1998), On the Stress-Strain Monitoring in Site and the Assessment of Seismic and Volcanic Hazards, Int. IDNDR Conference on "Modern Preparation and Response Systems for Earthquake", Tsunami and Volcanic Hazards, Santiago-Chile, April 27-30, 1998.

Nedoma, J. and Hiršl. P (1992), "Insertion of the Oscillation of the Measuring String", *Czechoslovakian Patent No. AO275448*, Prague.

Nedoma, J., Záruba, J., Novotná, E., Kohout, J., Vobořil, J., Kačerovský, J. and Holý, P. (1985a), "The Probe for Determing Stress-Strains in The Rocks by Using the Overcoring Method", *Czechoslovakian Patent No. AO226254*, Prague.

Nedoma, J., Záruba, J., Novotná, E., Hiršl, P. and Švanda, J. (1985b), "The Probe for Long-Term Observation of the Spatial Stress Changes in the Rock Massif", *Czechoslovakian Patent No. AO251876*, Prague.

Nedoma, J., Záruba, J., Novotná, E., Hiršl, P. and Švanda, J. (1986), "Tensometric Probes for Determing Stress-Strain in Boreholes", *Technical Report*, **V-220**, Institute of Computer Science AS CR, Prague (in Czech).

Nedoma, J. Bašlová, A. and Mantič, V. (1998). Post-processing of the Stress-Strain Monitoring Data for the Case of Seismic and Volcanic Hazards. Int. IDNDR Conference on "Modern Preparation and Response Systems for Earthquake", Tsunami and Volcanic Hazards, Santiago-Chile, April 27-30, 1998.

Oden, J.T. and Reddy, J.N. (1976), *An Introduction to the Mathematical Theory of Finite Elements*, J. Wiley & Sons, New York, London, Sydney, Toronto.

Oganesjan, L.A. and Rukhovec, L.A. (1979), *Variational Difference Methods for the Solution of Elliptic Problems*, Izd. Akad. Nauk ASSR, Jerevan (in Russian).

Ortega, J.M. and Rheinboldt, W.C. (1970), *Iterative Solution of Non-linear Equations in Several Variables*, Academic Press, New York, London.

Owen, D.R. (1984), *A First Course in the Mathematical Foundations of Thermodynamics*, Springer Verlag, New York, Berlin, Heidelberg, Tokyo.

Petřík, M. (1953), *String Tensometry*, SNTL, Prague (in Czech).

Petřík, M., Kratochvíl, J. and Záruba, J. (1975), "Development of the String and Semi-Conductor Tensometry for Application in Building", *Technical Report*, **335/75**, Technical University, Prague (in Czech).

Quarteroni, A. and Valli, A. (1994), *Numerical Approximation of Partial Differential Equations*, Springer Verlag, Berlin, Heidelberg, New York, London, Paris, Tokyo, Hong Kong, Barcelona, Budapest.

Rahn, W. (1984), "Stress Concentration Factors for the Interpretation of "Doorstopper" Stress Measurements in Anisotropic Rocks", *Int. J. Rock Mech. Min. Sci. & Geomech. Abstr.*, **21**, 313-326.

Samohýl, I. (1982), *Rational Thermodynamics*, Academia, Prague (in Czech).

Serrin Jr., J.B. (1959), "Mathematical Principles of Classical Mechanics", *Handbuch der Physik*, Band **8/1**, Strömungsmechanik 1, Springer Verlag, Berlin.

Stepanov, V.V. (1945), *The Course of Differential Equations*, Nauka, Moscow.

Timoshenko, S.P. and Gudjajev, D. (1975), *Continuum Mechanics*, Nauka, Moscow (in Russian).

Truesdell, C. (1969), *Rational Thermodynamics*, McGraw-Hill, New York.

Truesdell, C. (Ed.) (1984), *Rational Thermodynamics*, Springer Verlag, New York, Berlin, Heidelberg, Tokyo.

Wang, Y. and Wong, T.F. (1987), "Finite Element Analysis of Two Overcoring Techniques for In-Situ Stress Measurements", *Int. J. Rock Mech. Min. Sci. & Geomech. Abstr.*, **24**, 41-52.

Warren, W.E. and Schmith, C.W. (1985), "In-Situ Stress Estimates from Hydraulic Fracturing and Direct Observation of Crack Orientation", *J. Geophys. Res.*, **90**, 6829-6839.

Washizu, K. (1968), *Variational Methods in Elasticity and Plasticity*, Pergamon Press, Oxford, New York, Toronto, Sydney, Paris, Frankfurt.

Wong, T.F. and Walsch, J.B. (1985), "A Theoretical Analysis of Tectonic Stress Relief During Overcoring", *Int. J. Rock Mech. Min. Sci & Geomech. Abstr.*, **22**, 163-171.

Woods, L.C. (1975), *The Thermodynamics of Fluid Systems*, Clarendon Press, Oxford.
Zemansky, M.W. (1957), *Heat and Thermodynamics*, McGraw-Hill Company, New York, Toronto, London.
Zlámal, M. (1968), "On the Finite Element Method", *Num. Math.*, **12**, 394-409.
Zoback, M.D. and Haimson, B.C. (Eds.), (1983), *Hydraulic Fracturing Stress Measurements*, National Academic Press, Washingthon.

PART IV

Numerical Models of Plate Tectonics

4

Mathematical Models of Geodynamic Processes in the Lithosphere and of Plate Tectonics

4.1. INTRODUCTION

In this section the geodynamic processes in the domains of the subducted lithosphere will be studied and their mechanics will be discussed. The obtained results are intended to demonstrate and give a preliminary quantitative evaluation of the intimate relations between plate-driving forces and the geodynamic mechanics of the subducted lithosphere. Generally, the model problem discussed is so complicated according to the rheology of the Earth (see Scheidegger (1982), Ranalli (1987), Turcotte and Schubert (1982), Peltier (1989)) that at present it cannot be solved mathematically in its complete version. During the last twenty years, the first mathematical results for elastic, visco-elastic, thermo-elastic, thermo-plastic and thermo-elasto-visco-plastic media were obtained for classical and unilateral boundary value problems (see Duvaut and Lions (1972),(1976), Hlaváček and Lovíšek (1977), Hlaváček, Haslinger, Nečas and Lovíšek (1982), Nečas and Hlaváček (1981), Jarušek (1983), Nedoma (1983), (1987), (1990a,b), (1993), (1994a,b), Nedoma and Dvořák (1995), Kuczma and Rakowski (1994), Gioda (1990), Desai and Gioda (1990), Kuczma and Whiteman (1995), Panagiotopoulos (1985),(1993), Ascione and Grimaldi (1984), Haslinger and Hlaváček (1980),(1981), Kikuchi and Oden (1988), Coppetti and Elliott (1993), Elliott and Tang Qi (1994), Shi, Shillor and Zou (1991), Gilbert, Shi and Shillor (1990), Shi and Shillor (1990), etc.), and these can be applied to geodynamic problems. Contemporary geodynamic models lead to problems with classical boundary conditions (of the Dirichlet and Neumann type) for elastic, thermo-elastic, thermo-visco-elastic and plastic media (see Vanpé (1984), Andrews and Sleep (1974), Schmeling and Jacoby (1981), Toksösz and Hsui (1978), Bird and Baumgardner (1984), Bird (1978), McAdoo, Caldwell and Turcotte (1978), Park and Westbrook (1983), Turcotte, McAdoo and Caldwell (1978), Melosh and Raefsky

(1980), Zhao and Morgan (1987), Sato, Bhatia and Gupta (1996), etc.). Such model problems are only first approximations of the physical reality as they do not do justice to the actual geodynamic processes in the lithosphere. In this sense the model problem discussed in this section is a much better approximation of physical reality as this geodynamic model is closer to real physical reality in regions of the subducted lithosphere, and above all because it makes it possible to study the mutual movements of both colliding plates due to the function of tectonic and gravity forces. Since mathematical analysis is much more difficult, only relative simple model problems based on elasticity, visco-elasticity, thermo-elasticity and thermo-plasticity have been studied (Nedoma, Haslinger and Hlaváček (1989), Nedoma (1983), (1984), (1987), (1990a), (1994 b), Nedoma and Dvořák (1995)). For complicated model problems, mathematical as well as numerical analyses are not known and in this book will be indicated up to currently unsolved problems.

At present, the geodynamic problems are only very sparsly studied by complex research projects at geophysical or geological institutes by uniform types of methods. Hence no complex research exists in the whole branches of interest governed by a uniform methodology in the whole geophysical and geological fields. For the benefit of fundamental geophysical and geological geodynamic research, with straightforward applications to practice like the construction, operation and safety of nuclear power plants, high level radioactive waste repository, deep mines, great water dams, etc.,the proposed investigation should include the study of further geophysical fields and processes. In such a case the fundamental research of the regional and global areas due to our ideas (Nedoma (1991)) have to include:

(a) Long-term tectonophysical research of the present motion of the lithosphere by means of the stress-strain field observations in accessible places of the lithosphere, i.e. in deep boreholes at various depths, in galleries, etc., where the overcoring method and the stress-strain monitoring in boreholes are applied to study the present stress-strain states in the rock massif. The hydrofracturing method in our opinion is not so good, as the neigbourhood of the borehole is destroyed and therefore the stress-strain monitoring cannot be applied. From this point of view the overcoring method is much better. Additional methods include observation of the recent motion of the Earth's crust, i.e. of both vertical and horizontal components, and of the seismicity in the area (detectors, i.e. wide-band geophones, are placed in a probe to measure the stress-strain fields in the borehole during long time) and detailed determination of the geological structure of the area by methods of seismic and electromagnetic tomography in boreholes, seismic microzoning, etc..

(b) Theoretical analyses of the stress-strain fields in the area complete with the maps of the stress-strain fields and their components, etc. and with a database for various depths; comparison of theoretical results for various depths based on mathematical modelling and simulation of tectonophysical processes in the regional and global areas with the results of measurements in situ; extrapolation of the results of measurements in situ to longer distances and greater depths from the place of measurements in situ; choice of a most probable geodynamical model of the area based on expert systems and further improvement of this model facilitated by continuing research.

(c) Measurements of the seismoacoustic activities in the long-term observation points, where detectors — geophones — are placed in the probes for the long-term measurements of the stress-strain fields in the boreholes.

(d) Long-term observations of the stress-strain fields caused by tidal forces in the area of interest, and the subsequent comparison with analogous measurements in tidal research observatories and stations in other areas.

(e) Observation of the distribution and orientation of macro- and micro-cracks by pattern recognition methods and determination of the stress in the rock massif by means of the study of the deformicity of rock crystalline structure and of the acquisition of rock cores from places where the overcoring method is applied to study the present stress-strain states in rock massif.

(f) Long-term measurements of temperature and heat flow in observations of the stress-strain fields, where detectors — thermometers — are placed in the probes at long-term stress-strain field observation points in accessible places of the lithosphere.

(g) All the currently performed measurements of physical quantities, such as density, Young modulus of elasticity, Poisson's constant, porosity, etc. of the rock-cores obtained in the overcoring method.

(h) Paleomagnetic measurements of the rock core samples from the places of application of the overcoring method and of long-term observation points. They enable the paleomagnetic reconstruction of the orientation of the magnetic field and the determination of motion and historical rotation of the individual geological blocks and plates that form the area of interest.

(i) Magnetic measurements at long-term observation points.

(j) Magnetotelluric measurements carried out at long-term observation points in item (a) and, moreover, on exactly defined profiles that characterize the tectonics of the regional and global areas.

(k) Magnetometric measurements for determining fractures and their motion.

(l) Deep seismic sounding on the profiles in item (j) and points in item (a) for determination of the geological structure of the area of interest.

(m) Analyses of aerial and satellite photographs of the considered areas followed by evaluation.

(n) Gradual creation of databases for the area of interest containing the results of the proposed complex research of the whole area investigated.

Further improvement of geodynamical models of the regional and global areas based on the actual level of knowledge compiled by the special types of expert systems. This will allow the elimination of the influence of civilization on the areas considered and will facilitate finding criteria for the prediction of earthquakes in tectonically active regions as well as for prediction of minor shocks for requirements of practice.

We admit that the task is extremely difficult as it is connected with a great deal of problems not yet solved which are outside the scope of geology and geophysics. This book aims to give several methods for solving some problems of the class of main geophysical field problems.

The inexorable logic of this note presenting the importance of organizing the complex geological and geophysical research lies in the fact that this project improves the fundamental geological and geophysical research, as well as improving the safety of the operation of nuclear power plants, high level radioactive waste repositories and activity in other branches of the economy, as in mining, etc.. The contribution of fundamental research will apparently be much greater than the investment put in. But we cannot proceed without changes of the present legal regulations, financial resources, development of new mathematical as well as numerical methods and algorithms, development of new instruments and measurement devices, and also a new, more flexible approach to urgent economic problems represented by the energy balance of the present, but above all the future, generations and a new way of economic as well as political thinking on both a local and an international scale. An example that hampers the development of complex geological and geophysical research is the problem of the residual value of the borehole after its completion for geological prospecting. The price of a borehole depends on its depth, and ranges from several ten or hundred thousands for shallow boreholes to several ten or hundred millions of dollars for deep boreholes. After completing a borehole, the geological survey is not interested in these structural boreholes, they are closed up and destroyed. But such boreholes could be a valuable laboratory for complex geological and geophysical research, in particular for the study of temperature, heat flux, stress-strain fields, seismicity, monitoring the seismoacoustics, as well as of the useful values of the physical quantities in the lithosphere.

4.2. GEODYNAMIC PROCESSES IN THE UPPER PARTS OF THE EARTH: AN INTRODUCTION

The main geodynamic processes passing through the upper parts of the Earth depend on the density differentiation at the core/mantle boundary and the geodynamic processes in the lower mantle. Light materials, which, after accumulation below the core/mantle boundary, form an unstable layer, display a tendency to produce a stable state by escaping through the GTLV-channels into the upper parts of the Earth. The light material rises to the lower boundary of the lithosphere which it heats and melts and strains mechanically in the vertical direction; after effusing below the lithosphere it also strains it horizontally. The mechanical strain in the vertical and horizontal directions, together with its continual melting, results in the opening of a rift. In the first stage of rifting, the vertical components are predominant (e.g. the Baykal rift zone, the East African rift, etc.), in the second stage, horizontal components prevail (Rhine graben, Red Sea and the Bay of Aden, etc.). Due to viscous and Coulombian friction, the opening of the rift represents the beginning of the action of oblique and horizontal forces on a large scale. The immediate consequence of these horizontal motions are transverse fault zones, along which parts of the ridge in the rift region are displaced with respect to another (e.g. the mid-ocean ridges). The forming of these transverse fault zones is also related to the inhomogeneous influence of friction. The lithosphere behaves like a constrained plate on which non-uniformly distributed horizontal forces act in the horizontal direction. The absolute magnitude of the individual components of the horizontal forces depends on the effusion of light masses and their spreading below the lithosphere.

So, as a result of geodynamic processes in the lower parts of the Earth, rift zones open at various locations on the Earth's surface. Consequently, the lithosphere is strained by horizontal forces acting from various sides. The lithosphere behaves like a stress guide; it is strained, and stresses are generated, thus accumulating in various parts of the lithosphere. Hence in the plate, as a stress guide, compressive and tensile stresses are generated. If the magnitude of the stress exceeds the marginal value in straining, the lithosphere is divided by depth fractures and faults into several megaplates and blocks. During the subsequent stage of development, the lithospheric plates collide with the blocks into which the original lithosphere has broken up (see for example Uyeda (1982)). The evolution of the lithosphere can be characterized by several stages, which will be discussed next in some detail.

4.2.1. Rheology of the Lithosphere and of the Upper Mantle

The state of deformation of the lithosphere and of the upper mantle may depend not only on the level of stresses, but also on the variation of stress in time. As a result of an operation of stresses, the rocks of the lithosphere and of the upper mantle are subject to elastic and plastic deformation, and under the great stresses, the material of the lithosphere and the upper mantle can be destroyed. Manifestations of these last effects are smaller or higher earthquakes. In this chapter we will be interested in the behaviour of stages of evolution of the lithosphere in the process of deformation and of describing mathematically the mechanism of evolution of geodynamic processes taking place in the upper parts of the Earth. We will draw attention to the methods of description of the complex state of deformation of the lithosphere and the upper mantle in the framework of continuum theory. The geodynamic processes in the lithosphere which take place beyond the elasticity limit can be described by the rheology in the domain of elasticity, mainly characterized by our knowledge of distribution of seismic P and S waves as well as of density ρ in the investigated region. Other descriptions, through constitutive relations between the deformations and stresses, are of phenomenological nature. Along with material constants, we must sometimes include in such a description additional parameters characterizing the changes in the medium structure and the increase in the density of defects resulting from the deformation process. Another method in the phenomenological approach is to write constitutive relations in different ranges of stress. Beyond the elasticity limit we have visco-elastic, up to visco-plastic domains. The individual ranges are divided by more or less distinct threshold conditions, characterized for example by the plastic flow threshold limit. To obtain more precise information about the behaviour of lithospheric rocks we can make an analogy between the rock deformation processes on a small, laboratory scale and the natural conditions in the large scale of the Earth. It is mainly advisable for such laboratory experiments to be carried out on the samples from the boreholes, where complete investigations were made. To obtain precise data for rheological studies about lithospheric and upper mantle rocks the complete investigations (see the discussions above in Section 4.1) must be made. The rheology of rocks can be described by means of some additional parameters in the stress-strain relation (See Ranalli (1987)), which we will study next.

Ranalli and Murphy (1987) introduced several definitions of the lithosphere. In a seismological sense, the lithosphere represents the rocks above the upper mantle low-velocity zone, the so-called asthenosphere. In a tectonic sense, it may be defined as the outermost shell that translates coherently during plate movements. In a rheological sense, the lithosphere consists of the rocks above the thermally controlled depth where ductile behaviour becomes predominant, and is characterized by effective viscosity and creep strength. Melosh (1977) defined the lithosphere as being like an elastic solid over a geological period of time; moreover, it can support applied shear stress by deforming elastically. Due to Melosh (1977) the asthenosphere and the mantle below it do not have this property, and the materials of which they are composed creep steadily under applied shear stress. The shear stress τ_{xz} acting on the base of the lithosphere must be maintained by velocity gradients in the asthenosphere and mantle. The asthenosphere is a zone of low viscosity in the upper mantle, and velocity gradients are concentrated within it. This situation is potentially unstable to thermal runaways which can cause drastic modifications of the thermal structure of the asthenosphere. This is a direct result of the stress-strain relation for hot crystalline solids (Weertman and Weertman (1975)). The velocity gradients in the asthenosphere generate shear heat of the amount $\tau_{xz}\dot{e}_{xz}$ ergcm^{-3}.

As was shown by Oliver, Cook and Brown (1983) the asthenosphere, the low-velocity layer below the lithosphere, is absent in many areas of the upper parts of the Earth, and the thickness and material properties vary with age and with tectonic areas. Moreover, the rheological properties are a function of geodynamic processes taking place in the investigated region and depend on many influences, namely on magnitude and rate of stress and strain, pressure, temperature, density, intensity of magnetic field (Maxwell's stresses), porosity, etc.. Then we speak about the rheological stratification of the lithosphere and the upper mantle.

Rock rheology in general can be described by means of some additional parameters in the stress-strain relation. Let us denote by U the strain potential function and by W the stress potential function. Between them the following relation

$$W + U = e\tau$$

holds, where e and τ are the strain and stress tensors, respectively.

The stress potential function W is a function of some material constants $\alpha, \beta, \ldots$ which are functions of a rheological parameter φ, i.e. $W(\tau, \alpha(\varphi), \beta(\varphi))$, where the parameter φ reflects the irreversible changes in the rock.

The elasto-plastic behaviour of materials can now be expressed by the incremental law

$$de = \frac{\partial^2 W}{\partial\tau\partial\tau}\mathrm{d}\tau + \frac{\partial^2 W}{\partial\varphi\partial\tau} + \mathrm{d}e^p = \mathrm{d}e^e + \mathrm{d}e^r + \mathrm{d}e^p,$$

where the first term represents the elastic range, the second represents rheological changes, and the last represents the plastic yielding.

The theory of the hysteresis behaviour of materials in general is given for example in Krasnosel'skij and Pokrovskij (1983). The theory of hysteresis behaviour of geological rocks and the functional dependence of material parameters on the rheological parameter φ are very important studies in geodynamics.

In the linear theory of elasticity

$$W = \tfrac{1}{2}(\alpha(\varphi)I_1^2 + \beta(\varphi)I_2),$$

where $I_1 = \tau_{ii}, I_2 = \tau_{ij}\tau_{ij}$ and

$$\mathrm{d}e_{ij}^e = \frac{\partial^2 W}{\partial \tau_{ij}\partial\varphi}\mathrm{d}\varphi = \left(\frac{\mathrm{d}\alpha}{\mathrm{d}\varphi}\tau_{kk}\delta_{ij} + \frac{\mathrm{d}\beta}{\mathrm{d}\varphi}\tau_{ij}^D\right)\mathrm{d}\varphi,$$

where τ_{ij}^D is a deviator of the stress tensor τ_{ij} and α and β are functions defined by

$$\alpha = \frac{1}{3}\left(\frac{1}{K(\varphi)}\right) = \frac{1}{3}\left(\frac{1}{3K_o\,(1 - h\varphi + h\varphi^2)}\right), \quad \beta = \frac{1}{2\mu(\varphi)} = \frac{1}{2\mu_o\,(1 - b\varphi^n)},$$

where K, K_o are incompressibility moduli, h is a material parameter, μ, μ_o are shear moduli, and b, n are additional material parameters. Hence

$$\mathrm{d}e_{kk}^e = 3\tau_{kk}\frac{\mathrm{d}\alpha}{\mathrm{d}\varphi}\mathrm{d}\varphi, \quad \mathrm{d}e_{ij}^{eD} = \tau_{ij}^D\frac{\mathrm{d}\beta}{\mathrm{d}\varphi}\mathrm{d}\varphi,$$

and thus

$$\mathrm{d}e_{ij}^D = \mu_o^{-1}\mathrm{d}\tau_{ij}^D + \tau_{ij}^D\frac{nb\varphi^{n-1}}{2\mu_o}\mathrm{d}\varphi + \mathrm{d}e_{ij}^{pD},$$

$$\mathrm{d}e_{kk} = \frac{1}{3\mu_o}\mathrm{d}\tau_{kk} + \tau_{kk}\frac{h(1-2\varphi)}{3K_o}\mathrm{d}\varphi + \mathrm{d}e_{kk}^p,$$

which represent the deviatoric and volumetric parts of the incremental strain.

Strains in the Earth's crust have their origin in tidal loadings and temperature variations, in the effect of the variation of atmospheric pressure and mainly in the effect of tectonic and geodynamic processes. The last two effects are the main interest in geophysics, namely in the seismic and tectonic areas. Therefore, the influences of the other effects are filtered out from the instrumental measurements. For such analyses the strain gauge measurements are applied; these are very important but also very difficult due to the ground topography, rock inhomogeneities, anisotropy and mainly the perturbed rock massif by the micro- and macro-cracks. For such measurements the strain instruments are then installed in a cave or in the boreholes (see for examples Kovári (1985),(1988), Kovári and Peter (1983), Kovári and Amstad (1983), Nedoma (1990b), Mantič and Nedoma (1989), etc.).

To determine the rheological changes in the rock massif several methods can be used. One method resulting from the formation of cracks was developed by O'Connell and Budiansky (1974) and analyses the changes of elastic moduli of the solid with isotropic crack distribution. The effect of the number of cracks on the value of elastic parameters was studied in numerous experiments with dry, saturated and partially saturated cracked rocks. It was shown that the velocities of the P- and S-waves decrease with increasing crack density, so that the ratio v_p/v_s decreases for dry cracks and increases for saturated cracked rocks. The visco-elastic moduli for rocks containing fluid-filled cracks was investigated by O'Connell and Budiansky (1977). In the early concept of plate tectonics the analyses started from the Bussinesq approximation of moving fluid and assumed that the Rayleigh number Ra is large and constant. The Rayleigh number Ra is defined as

$$Ra = \alpha g d^3 \Delta T \kappa^{-1} \nu^{-1}, \tag{4.1}$$

where d is the thickness of the fluid layer (e.g. the lithosphere and asthenosphere, etc. or the depth z), α is the coefficient of volume expansion, g is the gravitational acceleration, $\kappa = k\rho^{-1}c_p^{-1}$ is the coefficient of thermometric conductivity, ρ is the density, c_p is the specific heat at constant pressure, ν is the kinematic viscosity, and ΔT denotes the difference in temperature at the lower and upper boundary. In the case where the lithosphere or the mantle is heated by sources located within them, then

$$Ra = \rho \alpha g \theta d^5 \kappa^{-1} \nu^{-1} k^{-1}, \tag{4.2}$$

where θ are heat sources per unit mass, and k is the coefficient of thermal conductivity. The Rayleigh number varies between 10^6-10^9 in the mantle. Besides the Rayleigh number, which is a very important quantity for convection, the Reynolds number Re, the dissipative number $\mathscr{D}$, the Prandtl number $\mathscr{P}$ and the Nusselt number $\mathscr{N}$ are used.

The Prandtl number is defined as

$$\mathscr{P} = \nu \kappa^{-1}, \tag{4.3}$$

where $\nu = \eta\rho^{-1}$ denotes kinematic viscosity, and η is the coefficient of viscosity and is of the order 10^{22} poise. The dissipative number $\mathscr{D}$ is defined as

$$\mathscr{D} = \alpha g d c_p^{-1}, \tag{4.4}$$

and expresses the relation of the work of the gravitational forces carried out on the volume element to the change in the thermal energy of the volume element. The Reynolds number Re is defined as

$$Re = U d \nu^{-1}, \tag{4.5}$$

where U denotes a characteristic velocity — in practise practically the average velocity — and in the mantle is about 10^{-10}. The Nusselt number is defined as a ratio of the total heat transfer through a fluid layer to the heat transfer which would be transfered through the same layer by conduction alone, $\mathscr{N} = a\,Ra^b = 0.2697\,Ra^{0.3185}$. Moreover, $\mathscr{N} = q/q_c$ and $\mathscr{W} = \Delta T_c/\Delta T$, where q_c and ΔT_c represent heat flux and temperature drop across the layer in the absence of convection, while their unsubscripted equivalents represent total heat flow and temperature drop in the presence of convection ($q > q_c$, $\Delta T_c > \Delta T$) (see Raualli (1987), p. 173-174).

But new investigations show that the state of deformation of upper parts of the Earth, the lithosphere and the upper mantle, may depend not only on the level of stresses, but also on the variation of stresses in time. Due to stresses acting in time, the rocks of upper parts of the Earth are subject to elastic and plastic deformation as well as, in special cases, to the destruction of the rocks taking place. The viscosity η, in terms of the deviatoric stress field and the strain rate tensor $\dot{\varepsilon}$, is defined by

$$\tau(\boldsymbol{x}, t) = 2\eta(\boldsymbol{x}, t)\dot{\varepsilon}(\boldsymbol{x}, t).$$

Moreover, present analyses show that the effective viscosity of mantle minerals such as olivine is an extremely strong function of temperature. The lithospheric upper mantle is assumed to have olivine rheology. The rheology of the lithospheric materials, on the whole, is characterized by silicate polycrystals. But silicate polycrystals exhibit rheological properties that vary as a function of temperature and pressure, and therefore also of depth. Another long-standing observation concerning the Earth is that the sublithospheric mantle is assumed to have a relatively constant viscosity, of the order of 10^{22} poise. Just as surface plate velocity found an incorrect explanation in the case of constant viscosity convection, so constant sub-lithospheric viscosity has an incorrect explanation in terms of variable temperature-dependent viscosity.

The experimentally determined flow law for practically all crustal and upper mantle rocks was given by Kirby (1983) and is of the form

$$D \equiv \dot{e} = A\tau^n \exp(-Q/RT), \tag{4.6}$$

where $D \equiv \dot{e}$ is the strain rate, τ is the stress difference $\tau = \tau_1 - \tau_3$, where τ_1, τ_3 are maximum and minimum compressive stresses, R is the gas constant, T is the temperature and A, n, Q are material parameters which are approximately stress- and temperature-dependent in the range of interest. The parameter Q represents the creep activation energy. The parameter A weakly depends on temperature T, but this dependence is negligible when compared with the exponential term.

From Table 4.1 we see that for quartz-rich rocks, i.e. for granite and quartzite, the activation energy Q is in the range 100–150 kJmol^{-1}, while for basic rocks, i.e. diabase and quartz diorite, is in the 200–250 kJmol^{-1} range. Ultrabasic, olivine-rich rocks are the hardest.

The variation of creep strength can be calculated as

$$\tau = \dot{e}^{1/n} A^{-1/n} \exp(Q/nRT) \equiv e^{1/n}(\dot{u}) A^{-1/n} \exp(Q/nRT). \tag{4.7}$$

The curves of creep strength/depth can be found in the paper of Ranalli and Murphy (1987).

To estimate the viscosity η in the upper parts of the Earth, i.e. in the lithosphere and in the asthenosphere, we shall start from the equation of moving incompressible viscous fluid. The equation of state of the upper part of the Earth for long-term deformation in many cases is described by the Bussinesq approximation for the density ρ as

$$\rho = \rho_o \left[1 - \alpha\left(\Theta - \Theta_o\right)\right],$$

Table 4.1. Creep Parameters for Lithospheric Rocks

Material	$A[GPa^{-n}\text{s}^{-1}]$	n	Q(kJmol^{-1}]
Granite	5.0	3.2	123
Wet Granite	100.0	1.9	137
Quartzite	100.0	2.4	156
Wet Quartzite	2000.0	2.3	154
Quartz diorite	2.10^4	2.4	219
Diabase	3.210^6	3.4	260
Olivine	4.10^{15}	3.0	540

where ρ_o is the reference density and $\rho_o \sim 3000$ kgm^{-3}, α is the coefficient of voluntaric thermal expansion and $\alpha \sim 3.10^{-5}$K^{-1}, Θ_o is the reference temperature associated with ρ_o and Θ is the temperature in absolute degrees. In our investigations we shall start from the equation of moving incompressible viscous fluid in the Navier-Stokes form:

$$\text{div } \boldsymbol{v} = 0, \tag{4.8}$$

$$\rho \partial \boldsymbol{v} / \partial t + \rho \boldsymbol{v} \text{ grad } \boldsymbol{v} = -\text{grad } p + \eta \Delta \boldsymbol{v} + \rho \boldsymbol{g}, \tag{4.9}$$

where ρ is the density of a fluid, $\boldsymbol{v}$ its velocity, p pressure, η viscosity, t time and $\boldsymbol{g}$ acceleration of gravity. The left hand side has the meaning of inertia forces, $-$grad p represents a force due to overfall of pressure, the term $\eta\Delta\boldsymbol{v}$ represents forces of viscous friction and $\rho\boldsymbol{g}$ are gravity forces. Since

$$\rho\, |\partial \boldsymbol{v} / \partial t| \sim \rho v t^{-1} \sim 3 \text{ gcm}^{-3} \text{ 10 cm/yr/yr} \sim 3.02\ 10^{-14} \text{ dyncm}^{-3},$$

$$\rho\, |\boldsymbol{v} \text{grad } \boldsymbol{v}\,| \sim \rho v^2 r^{-1} \sim 3 \text{ gcm}^{-3} (10 \text{ cm/yr})^2 / 1000 \text{ km} \sim 3.016\ 10^{-21} \text{ dyncm}^{-3}$$

and therefore can be neglected. Then Equation 4.9 yields

$$-\text{grad } p + \eta \Delta \boldsymbol{v} + \rho \boldsymbol{g} = 0. \tag{4.10}$$

Moreover, we can find that for the pressure, which is of lithostatic character, $p \sim \rho g z$ and its gradient grad $p \sim \rho g z / l$, where l is the measure of the lithospheric block investigated ($l \sim 1000$ km). In the case where l is less than or equal to the thickness d of the lithospheric block investigated (or the layer) then the term of viscous friction $\eta\Delta\boldsymbol{v} \sim \eta v l^{-2}$. Then from Equation 4.10 we obtain

$$-\rho g z l^{-1} + \eta v l^{-2} + \rho g = 0.$$

Hence

$$\eta = \rho g z l^{-1} v^{-1} l^2 - \rho g l^2 v^{-1} = \rho g l v^{-1} (z - l), \tag{4.11}$$

and

$$v \sim \rho g l \eta^{-1} (z - l). \tag{4.12}$$

In the case where $l \gg$ d, then the flow in the layer takes place in the horizontal direction. Then the term $\eta\Delta\boldsymbol{v} \sim \eta v_x \text{d}^{-2}$ and terms $-$grad $p + \rho\boldsymbol{g}$ are of the same order $\rho g z l^{-1}$. Then

$$\eta = \rho g z \text{d}^2 v_x^{-1} l^{-1}, \tag{4.13}$$

and

$$v_x \sim \rho g z \text{d}^2 \eta^{-1} l^{-1}. \tag{4.14}$$

Denoting $\tau \sim z v^{-1}$ the characteristic time of decay of perturbation, then for $z \gg l$

$$\tau \sim \eta (\rho g l)^{-1}. \tag{4.15}$$

Since the fluid is incompressible then for the case where $l \gg \mathrm{d}$ we can find $v_z \sim v_x \mathrm{d} l^{-1}$ * and then

$$v = v_z \sim Pgz\mathrm{d}\eta^{-1}\mathrm{d}^2 l^{-2} = \rho gz\mathrm{d}^3\eta^{-1}l^{-2} \tag{4.16}$$

and

$$\tau = \eta(\rho g\mathrm{d})^{-1}l^2\mathrm{d}^{-2} = \eta l^2\rho^{-1}g^{-1}\mathrm{d}^{-3}. \tag{4.17}$$

To derive the rheological relation between the strain $e(t)$ and stress $\tau(t)$, several ways, according to the physical model problem, can be used (see Kirby (1983), Cathles (1975), Lliboutry (1971), Rektorys (1982), Fowler (1983), (1985), (1993), Quareni *et al.* (1985), Weertman (1978), etc.). Two ideas for derivation of the rheological relations will now be given.

In classical one-dimensional elasticity the stress-strain relation, the so-called Hooke's law, is of the form

$$\tau = Ee \text{ or } e = E^{-1}\tau. \tag{4.18}$$

But in rheology this relation becomes

$$\tau = Fe \text{ or } e = F^{-1}\tau, \tag{4.19}$$

where F represents an operator containing time t, and F^{-1} is its inverse operator. In general we can write

$$e(t) = F^{-1}\tau = E^{-1}\tau(t) + E^{-1}\int_o^t K(t, s)\tau(s)\mathrm{d}s, \tag{4.20}$$

where the function $K(t, s)$, the so-called kernel, is a given function which in general contains singularities. A simple law of creep is of the form

$$e(t) = F^{-1}\tau = E^{-1}\tau(t) + E^{-1}\int_o^t \exp(-(t - s))\tau(s)\mathrm{d}s, \tag{4.21}$$

where the kernel $K(t, s) = \exp(-(t - s))$.

In the theory of rheology another method can be used. We shall start from the rheological analogue to the differential equation for the deflection of a sample bar, loaded by periodic or unperiodic forces, in classical elasticity, which can be written as the following problem:

Rheology problem $\mathcal{P}_{r}1$:

$$\begin{aligned} IFw^{(4)} + Pw'' &= f(x) \qquad \text{in } (a, b),\\ w(a) &= w(b) = 0,\\ w''(a) &= w''(b) = 0, \end{aligned} \tag{4.22}$$

* Let us assume a layer of length l and thickness d. Let the velocity of fluid, which streams into the layer, be denoted by v_x. Then the volume of the matter in the unit of time is $\sim v_x l\mathrm{d}$. The motion of the surface of the above volume in the vertical direction is $\sim v_z l^2$ per unit time. Since a fluid is assumed to be incompressible then $v_x l\mathrm{d} \sim v_z l^2$. Hence $v_z \sim v_x \mathrm{d} l^{-1}$.

where I represents the moment of inertia of the cross-section of the bar with respect to the bending axis, F is the operator defined by Equations 4.19-4.21, $P(x)$ and $f(x)$ are horizontal force and vertical loading respectively, and $w(x)$ represents the deflection of the bar.

To solve Equation 4.22 the method of discretization in time or the Laplace transform, FEM,FDM, etc., can be used. Here the method of discretization in time will be applied. This method consists in the following:

Let $h > 0$ be a step in time and let

$$t_i = ih, \quad i = 0, 1, 2, \ldots.$$

Then we construct the discrete approximation F_i^{-1} of the operator F^{-1} at the point t_i, and on this base we construct the discrete approximation F_i of F. Then, for $i = 0, 1, 2, \ldots$ we replace F in Equation 4.22 by this discrete approximation. According to Equation 4.21, the integral occuring here will be replaced by a sum. Hence, putting $\tau_i = \tau(t_i)$, we find

$$\int_o^{t_i} \exp(-(t_i - s))\tau(s)\mathrm{d}s = \exp(-t_i) \int_o^{t_i} \exp(s)\tau(s)\mathrm{d}s$$
$$\sim (1 - \exp(-h)) \sum_{j=0}^{i-1} \exp(j - (i-1))h\tau_j.$$

Then

$$\varepsilon_i = E^{-1}\tau_i + E^{-1}(1 - \exp(-h)) \sum\nolimits_{j=0}^{i-1} \exp((j - (i-1))h)\tau_j$$
$$= E^{-1}\tau_i + E^{-1}\alpha \sum\nolimits_{j=0}^{i-1} \exp((j - (i-1))h)\tau_j, \tag{4.23}$$

which is equivalent to

$$\varepsilon_i = F_i^{-1}\tau, \quad i = 0, 1, 2, \ldots \tag{4.24}$$

and where $\alpha = 1 - \exp(-h)$. From Equation 4.23 for $i = 0$ we find

$$\varepsilon_o = E^{-1}\tau_o, \tag{4.25}$$

which shows that for time $t = 0$ the law of creep is identical with the classical Hooke's law, and from Equation 4.22 we obtain

$$IF_o w_o^{(4)} + Pw_o'' = f(x), \tag{4.26}$$

which according to Equation 4.25 is equivalent to

$$IEw_o^{(4)} + Pw_o'' = f(x). \tag{4.27}$$

Solving this equation we obtain the same results as in the classical theory of elasticity. Then we shall seek solutions for $i = 1, 2, \ldots$. In general we solve for $t = t_i$

$$IF_i w_i^{(4)} + Pw_i'' = f(x). \tag{4.28}$$

The special case $f(x) = c \sin x$, $x \in (0, \pi)$ is analysed in detail in Rektorys (1982). For this case the solution of Equation 4.28 is of the form

$$w_i = A_i \sin x, \tag{4.29}$$

where

$$A_i = A_o[1 + \alpha IE(IE - P)^{-1} \sum_{j=0}^{i-1} (\alpha IE(IE - P)^{-1} + (\exp(-h) - \alpha))^j], \quad i = 0, 1, 2, \ldots . \tag{4.30}$$

and $A_o = c(IE - P)^{-1}$, which can be proved by the induction technique.

Let $t = t_i = ih$ be fixed. Then (see (Rektorys (1982))

$$\begin{aligned}
\lim_{h\to 0} A_i/A_o &= \lim_{h\to 0} \left[1 + \alpha IE(IE - P)^{-1} \sum_{j=0}^{i-1} (\alpha IE(IE - P)^{-1} + (\exp(-h) - \alpha))^j\right], \\
&\qquad i = 0, 1, 2, \ldots . \\
&= \lim_{h\to 0} \{2[2 - IE(IE - P)^{-1}]^{-1} - IE(IE - P)^{-1}[2 - IE(IE - P)^{-1}]^{-1} \\
&\quad \times [\alpha IE(IE - P)^{-1} + (\exp(-h) - \alpha)]^i\} \\
&= \lim_{h\to 0} \{2[2 - IE(IE - P)^{-1}]^{-1} - IE(IE - P)^{-1}[2 - IE(IE - P)^{-1}]^{-1} \\
&\quad \times [\exp(-h)[2 - IE(IE - P)^{-1} + (IE(IE - P)^{-1} - 1)]^{t/h}\} \\
&= 2[2 - IE(IE - P)^{-1}]^{-1} - IE(IE - P)^{-1}[2 - IE(IE - P)^{-1}]^{-1} \\
&\quad \times \exp[-(2 - IE(IE - P)^{-1})]^{-1} t \\
&= [2 - IE(IE - P)^{-1}]^{-1}\{2 - IE(IE - P)^{-1} \\
&\quad \times \exp[-(2 - IE(IE - P)^{-1})]^{-1} t\},
\end{aligned} \tag{4.31}$$

where the l'Hospital rule was used. Finally, we obtain

$$w(x, t) = A \sin x, \tag{4.32}$$

where

$$A = 2c(IE - 2P)^{-1} - cIE[(IE - P)(IE - 2P)]^{-1} \exp\{-(IE - 2P(IE)^{-1} - P)t\}.$$

Remark: The same result can be found by applying the Laplace transform.

For the determination of the viscosity, the Maxwell liquid is also used. The rheological equation of a Maxwell liquid is given by

$$\dot{e}_{xz} = \mu^{-1} \dot{\tau}_{xz} + \eta^{-1} \tau_{xz}, \tag{4.33}$$

where $\dot{e}_{xz}$ is the shearing strain rate, $\dot{\tau}_{xz}$ is the shearing stress rate, η is the viscosity and μ is the rigidity. For the case of a constant strain rate $\dot{e}_{xz} = c$=constant., then

$$e_{xz} = ct. \tag{4.34}$$

Moreover,

$$(\tau_{xz})_{t=0} = 0 \tag{4.35}$$

represents the initial condition. Hence, and using Equation 4.33, we find

$$\tau_{xz} = \eta c\{1 - \exp(-\eta^{-1}\mu t\}. \tag{4.36}$$

The pair of equations, Equations 4.34 and 4.36, represent a relation between stress τ_{xz} and strain e_{xz}, where t is a parameter. Let us denote as $t_r = \mu^{-1}\eta$ a relaxation time of the Maxwell liquid. If t is smaller than the relaxation time t_r, then Equations 4.34 and 4.36 become

$$\tau_{xz} = \mu ct = \mu e_{xz} \qquad \text{for } t \ll t_r, \tag{4.37}$$

which is the Hooke equation of solids. If t is equal to the relaxation time t_r, then

$$\tau_{xz} = (1 - e^{-1})\mu e_{xz} \qquad \text{for } t = t_r, \tag{4.38}$$

where e represents a strain. If t is much larger than the relaxation time t_r, then

$$\tau_{xz} = \eta c = \eta \dot{e}_{xz} \qquad \text{for } t \gg t_r, \tag{4.39}$$

which represents the equation for the Newtonian liquid.

To obtain some information about rheology of the upper parts of the Earth the analyses based on thermodynamics can be applied. As it was shown earlier (and also for example see Stacey (1977)) standard thermodynamic potentials were used for finding equilibrium conditions and for the derivation of relations, which are important for better understanding of the physical processes taking place in the Earth's interior. From the first and second law of thermodynamics it follows

$$\mathrm{d}U + \mathrm{d}W \le T\mathrm{d}S - p\mathrm{d}V, \quad \mathrm{d}W = \xi \mathrm{d}x, \tag{4.40}$$

where $\mathrm{d}W$ is the increment of work, $\mathrm{d}U$ the increment of internal energy, T, S, p, V are the temperature, entropy, presure and volume, respectively, ξ is the generalized force and x the generalized coordinate. For the reversible processes in Equation 4.40 the equality holds, while for an isolated system $\mathrm{d}U = 0$, $\mathrm{d}V = 0$, $\mathrm{d}x = 0$ and therefore, $\mathrm{d}S \ge 0$, i.e. to reach a state of equilibrium the entropy attains its maximum. Moreover, for constant S and V the state of equilibrium yields

$$\mathrm{d}U + \mathrm{d}W = 0, \tag{4.41}$$

which for $\mathrm{d}W = 0$ is related to a minimum of internal energy. Since the enthalpy H is defined by $H = U + pV$ then

$$\mathrm{d}H + \mathrm{d}W \le T\mathrm{d}S + V\mathrm{d}p. \tag{4.42}$$

For constant S and p a state of equilibrium is reached when

$$\mathrm{d}H + \mathrm{d}W = 0.$$

Similarly for the Helmholtz free energy $F = U - TS$ and the Gibbs free energy $G = H - TS$ we find

$$\mathrm{d}F + \mathrm{d}W \leq -S\mathrm{d}T - p\mathrm{d}V, \tag{4.43}$$

$$\mathrm{d}G + \mathrm{d}W \leq -S\mathrm{d}T + V\mathrm{d}p. \tag{4.44}$$

For constant T and V a state of equilibrium is reached when

$$\mathrm{d}F + \mathrm{d}W = 0$$

and for constant T and p when

$$\mathrm{d}G + \mathrm{d}W = 0.$$

The additional work $\mathrm{d}W$ can often be neglected. The minimum of the Gibbs free energy represents the condition of equilibrium under constant T and p. The minima of solid and liquid phases should meet at the melting point. Stacey (1977) shows that this condition leads to the Claudius-Clapeyron equation in the form

$$\frac{\mathrm{d}T_m}{\mathrm{d}_p} = \frac{V_L - V_S}{S_L - S_S}, \tag{4.45}$$

where T_m is the melting temperature, p the pressure, and indices L and S denote the liquid and solid phases. Moreover, we have

$$\frac{\mathrm{d}U}{\mathrm{d}m} = \frac{\mathrm{d}H}{\mathrm{d}m} = \frac{\mathrm{d}F}{\mathrm{d}m} = \frac{\mathrm{d}G}{\mathrm{d}m} = g, \tag{4.46}$$

where m represents the mass, and g is the specific Gibbs free energy potential or the so-called chemical potential, and is equal to the derivative of any potential related to a change in the mass of a substance. Since the equation of state $p = f(T, S)$ then

$$\left(\frac{\partial p}{\partial T}\right)_S \left(\frac{\partial T}{\partial S}\right)_p \left(\frac{\partial S}{\partial p}\right)_T = 1$$

holds.

Let us assume the reversible processes and let $\mathrm{d}W = 0$, then the so-called Maxwell relations are

$$\left(\frac{\partial T}{\partial V}\right)_S = -\left(\frac{\partial p}{\partial S}\right)_V = -\gamma T V^{-1}, \tag{4.47}$$

$$\left(\frac{\partial S}{\partial V}\right)_V = \left(\frac{\partial p}{\partial T}\right)_V = \gamma \rho c_v = \alpha K_T, \tag{4.48}$$

$$\left(\frac{\partial T}{\partial p}\right)_S = \left(\frac{\partial V}{\partial S}\right)_p = \gamma T K_s^{-1}, \tag{4.49}$$

$$-\left(\frac{\partial S}{\partial p}\right)_T = \left(\frac{\partial V}{\partial T}\right)_p = \alpha V, \tag{4.50}$$

where γ is the Grüneisen ratio, $c_p = T(\partial S/\partial T)_p$, $c_v = T(\partial S/\partial T)_V$ are the specific heat at constant pressure and constant volume (per unit mass), respectively, and K_s, K_T are the adiabatic and isothermic bulk modulus, respectively. Furthermore,

$$c_p/c_v = K_s/K_T = 1 + \gamma\alpha T, \tag{4.51}$$

where α is the volume expansion coefficient or the coefficient of thermal expansion. The Grüneisen ratio follows from the Maxwell relations

$$\gamma = \alpha K_T \rho^{-1} c_v^{-1} = \alpha K_s \rho c_p^{-1}. \tag{4.52}$$

The coefficient of thermal expansion can be found as

$$\alpha = \gamma c_p (v_p^2 - 4/3 v_s^2)^{-1}, \tag{4.53}$$

where v_p, v_s are velocities of P and S waves, respectively. It is evident that the coefficient α depends on the depth, as the velocities v_p and v_s in the Earth's interior depend also on the depth. For example $\alpha = 5.5\ 10^{-5}\ {}^o\mathrm{C}^{-1}$ at 33 km whilst $\alpha = 0.5\ 10^{-5}\ {}^o\mathrm{C}^{-1}$ at the core/mantle boundary at ~2900 km. The bulk modulus K_s is known from the velocity distribution in the Earth's interior. Due to Anderson and Sumino (1980) γ and α vary between $\gamma = 1.172, \alpha = 1.912\ 10^{-5}$ at the depth $z = 670$ km and $\gamma = 1.053, \alpha = 7.77\ 10^{-6}$ at $z = 2885.3$ km.

The rheology of the lithosphere is also characterized by porosity. The porosity of the lithospheric rocks has its origin in the microcracks, microcavities, granular porosity, mineral composition of the matrix and the binding of the individual grains of the minerals, etc.. The term porous medium is usually applied to a multiphase medium. Theories describing such media are mainly based on the theory of deformation of a continuous medium. But it is very difficult to define final results concerning the effective constants, which must be compared with experiment. The theory of a porous elastic rheology containing a fluid was firstly presented by Biot (1956 a,b and later papers), where the description of such a solid starts from Lagrange's equations of motion

$$\frac{\partial}{\partial t}\left(\frac{\partial E_k}{\partial v_i}\right) + \frac{\partial E_d}{\partial v_i} = f_i, \quad \frac{\partial}{\partial t}\left(\frac{\partial E_k}{\partial w_i}\right) + \frac{\partial E_d}{\partial w_i} = F_i, \tag{4.54}$$

where E_k, E_d are the kinetic energy and the dissipation functions, respectively; $\boldsymbol{v}$, $\boldsymbol{w}$ are the velocities of the solid and the fluid, respectively; and f_i, F_i are the concentrated and fluid parts. Hence Biot (1956a,b) found

$$\tau_{ij,j} = \rho_{11}\frac{\partial^2 u_i}{\partial t^2} + b\left(\frac{\partial u_i}{\partial t} - \frac{\partial U_i}{\partial t}\right) + \rho_{12}\frac{\partial^2 U_i}{\partial t^2}, \tag{4.55}$$

$$\tau_{ij,j} = \rho_{22}\frac{\partial^2 U_i}{\partial t^2} + b\left(\frac{\partial U_i}{\partial t} - \frac{\partial u_i}{\partial t}\right) + \rho_{12}\frac{\partial^2 u_i}{\partial t^2}, \tag{4.56}$$

where $v_i = \partial u_i/\partial t$, $w_i = \partial U_i/\partial t$, $\tau = -\Phi p$, Φ is the porosity, τ_{ij} is the stress tensor in the solid, $\tau_{ij,j}$ its derivative, $\boldsymbol{u}$ and $\boldsymbol{U}$ are the displacement vectors of the solid matrix and the fluid, respectively; $\rho_{11} = \rho_1 + \rho_a$, $\rho_{22} = \rho_2 + \rho_a$, $\rho_{12} = -\rho_a < 0$, where $\rho_1 > 0$ is the mass of solid per unit volume of the rock created by the solid matrix and the fluid, $\rho_2 > 0$ is the mass of fluid per unit volume of the rock, ρ_{12} is a mass coupling between the fluid and the solid. The displacement vector of the fluid relative to the solid matrix is given by $\boldsymbol{s} = \Phi(\boldsymbol{U} - \boldsymbol{u})$. Instead of Equation 4.55 and 4.56, the Darcy law of fluid transport

$$\text{grad } p = -b(\partial u_i/\partial t - \partial U_i/\partial t) \tag{4.57}$$

can be given, where p is the fluid pressure, $b = \nu\Phi^2 k^{-1}$, where ν is the viscosity of the fluid, k is the Darcy coefficient of permeability, and Φ is the porosity. The system of Equations 4.55–4.57 represents the constitutive equations of the porous media.

For the strains of the solid phase and fluid mass content under isothermal conditions deduced from Biot equations one can find

$$e_{ij} = \mu^{-1}\{(\tau_{ij} + p\delta_{ij}) - \sigma(1+\sigma)^{-1}(\tau_{kk} + 3p)\delta_{ij} + 2/3\mu(H^{-1} - K^{-1})p\delta_{ij}\}, \tag{4.58}$$

where σ is the Poisson's ratio, and $H^{-1} - K^{-1} = K_s^{-1}$, where K_s is related to the bulk modulus of the solid phase.

In the case of undrained conditions we assume that the change of mass $\Delta m = m - m_o = 0$, where $m = \rho_L v$, ρ_L is the fluid density, v is the apparent fluid volume. It means that under stress loading the pore fluid pressure does not change due to diffusive transport to or from the neighbourhood elements. In the case of drained conditions we must add the constitutive law governing the pore fluid diffusion $\mathbf{d} = -\rho_o k\nu^{-1}\text{grad } p$, where ρ_o is the fluid density relating to the reference state (unstressed state), k is permeability, ν is the fluid viscosity, p is the porous pressure, $\mathbf{d}$ is the pore fluid diffusion, representing the Darcy law equivalent to Equation 4.57, and the mass conservation equation for an infiltrating pore fluid is

$$\text{div } \mathbf{d} + \partial m/\partial t = 0, \tag{4.59}$$

which, under the equilibrium conditions, leads to the homogeneous diffusion equations for the fluid mass m.

From the studies of seismic wave propagation, the velocity of seismic waves depends on the porosity of the rocks. The velocity of seismic waves depends on the mineral structure of the rock, on the texture of the rock, on the mineral composition of the matrix of the porous rock, on the geometry and size of the pores, on the outer and inner pressure in the pores, so that the velocity, for example of P waves, is a function of porosity Φ, $v_p = f(\Phi)$, where porosity changes only within a small range. The relation between the velocity of seismic waves and porosity was found as

$$v_p^{-1} = v_m^{-1}(1-\Phi) + v_f^{-1}\Phi, \tag{4.60}$$

where v_p is the longitudinal wave velocity in the medium, v_m, v_f are the longitudinal wave velocities in the matrix or in the pore-saturating medium, respectively, and Φ is the fractional porosity. The relation between the bulk modulus k and porosity Φ is found as

$$K^{-1} = K_m^{-1}(1-\Phi) + K_f^{-1}\Phi, \tag{4.61}$$

where K_m, K_f are bulk moduli of the matrix (solid) and the liquid, and Φ is the fractional porosity. The other relations can be found, for example in O'Connell and Budiansky (1977).

It can be remarked that considerations of the mechanism of porous media deformation are of importance in seismic wave propagation, in tectonics and geodynamics as well as of the earthquake mechanism.

The other technique for determination of the viscosity, in the case of temperature-dependent viscosity, will be based on the analysis of the Bussinesq equations (see, for example, Fowler (1985), (1993), Quareni *et al.* (1985)). The Bussinesq equations of thermal convection at infinite Prandtl number and for an incompressible fluid can be written in the following form (see Fowler (1993))

$$\operatorname{div} \boldsymbol{v} = 0, \tag{4.62}$$

$$-\operatorname{grad} p + \operatorname{grad} \tau + Ra T \boldsymbol{k} = 0, \tag{4.63}$$

$$\partial T/\partial t + \boldsymbol{v} \operatorname{grad} T = \Delta T, \tag{4.64}$$

where $\boldsymbol{v}$ is the velocity vector, p is the pressure minus the lithostatic pressure, τ is the deviatoric stress tensor, $\boldsymbol{k}$ is a unit vertical vector, Ra is the Rayleigh number, T is the temperature and

$$T(x, z, t) = \overline{T}(z, t) + \Theta(z, t) \cos kx \tag{4.65}$$

in the mean $2D$ field approach. Let us assume the $2D$ velocity field $\boldsymbol{v}$ in the form

$$\boldsymbol{v} = (-k^{-1} w_z \sin kx, w \cos kx), \tag{4.66}$$

where the Cartesian system (x, z) is oriented in such a direction that z points upwards. The components τ_1, τ_2 of the stress tensor $\tau = \begin{pmatrix} \tau_1 & \tau_2 \\ \tau_2 & -\tau_1 \end{pmatrix}$ are defined as

$$\tau_1 = -2\eta w_z \cos kx, \; \tau_2 = -k^{-1}\eta(\partial^2 w/\partial z^2 + k^2 w) \sin kx. \tag{4.67}$$

Then from Equation 4.62-4.64 we find

$$(\partial^2/\partial z^2 + k^2)[\eta(\partial^2 w/\partial z^2 + k^2 w)] - 4k^2 \partial/\partial z(\eta w_z) - Ra\, k^2 \Theta = 0, \tag{4.68}$$

where Θ is a measure of buoyancy, $\eta = \eta(\overline{T}, z)$ is the viscosity, and $\overline{T}$ is the horizontally averaged temperature.

From Equation 4.64 using Equation 4.65 then for $\overline{T}$ and Θ we find

$$\partial \overline{T}/\partial t + \tfrac{1}{2}\partial/\partial z(\Theta w) = \partial^2 \overline{T}/\partial z^2, \tag{4.69}$$

$$\partial \Theta/\partial t + w \partial \overline{T}/\partial z = \partial^2 \Theta/\partial z^2 - k^2 \Theta. \tag{4.70}$$

Let us assume that the axis z points downwards (i.e. then $\partial/\partial z \rightarrow -\partial/\partial z$) and that the layer of convection flow is bounded by $z = 0$ and $z = 1$. Moreover, firstly let us assume that in Equation 4.68 $k \gg 0$, hence $w = Ra\, k^{-2}\eta^{-1}\Theta$, and secondly let us assume that horizontal conduction in Equation 4.69 and 4.70 is small, then the term

$-k^2\Theta$ can be neglected. Then Equation 4.69 and 4.70 yield

$$\partial T/\partial t - \partial/\partial z(\Theta w) = \partial^2 T/\partial z^2, \quad (4.71)$$

$$\partial\Theta/\partial t - w\partial T/\partial z = \partial^2\Theta/\partial z^2, \quad (4.72)$$

where the overbar on the average temperature was omitted. Moreover, $w = Ra\ k^{-2}\eta^{-1}\Theta$ yields that for $\Theta > 0$ then $w > 0$, i.e. that a positive buoyancy Θ in a convecting flow is associated with upwelling.

The efficiency of convection in transfering heat is characterized by the Nusselt number $\mathcal{N}$. We define the Nusselt number $\mathcal{N}$ to be the value of $\partial T/\partial z$ on $z = 0$. In a steady state, Equation 4.71 can be integrated. Integrating Equation 4.71 and rescaling Θ with $(\mathcal{N}/R)^{1/2}$ then Equation 4.71 and 4.72 yield

$$\partial T/\partial z = \mathcal{N}(1 - \Theta^2 f), \quad (4.73)$$

$$\partial\Theta/\partial t + R\Theta f\partial T/\partial z = 0, \quad (4.74)$$

where $f = \eta^{-1}$ is the so-called fluidity and is equal to the inverse value of the viscosity η, and $R = Ra/k^2$. Boundary conditions prescribed on $z = 0$ and $z = 1$ are the following

$$T = -1, \quad \Theta = 0 \text{ on } z = 0, \quad (4.75)$$

$$T = 0, \quad \Theta = 0 \text{ on } z = 1. \quad (4.76)$$

The rheology in Equation 4.73 and 4.74 will be defined by $f = \eta^{-1} = 1$(constant viscosity) or $h = \eta^{-1} = \exp(T/\varepsilon)$, where $\varepsilon \ll 1$, respectively. Detailed analysis of Equation 4.73-4.76 is made in Fowler (1993), where it is shown that

$$\mathcal{N} \sim R, \quad (4.77)$$

which confirms that $\mathcal{N} \gg 1$ when $R \gg 1$, the case of constant viscosity, and

$$\mathcal{N} \sim 1.44\varepsilon^2 R[1 + O(\varepsilon ln(\varepsilon^{-1})] \quad (4.78)$$

in the case of temperature-dependent viscosity, so that $N \gg 1$ when $R \gg \varepsilon^{-2} \gg 1$.

4.2.2. Tectonic Evolution of Rift Zones

To discuss the geodynamic processes in the upper parts of the Earth, i.e. in the lithosphere, asthenosphere and mesosphere, we recall processes in the mantle. The lower mantle is characterized by the thermo-visco-plastic materials heated from the bottom and cooled from overhead. Resultant effects are convective currents, convective cells, and channels of higher temperature and low viscosity (the GTLV-channel). Due to the convective currents the light hot viscous materials from the mantle'core boundary flow through the GTLV-channel below the lithosphere, which is then heated and melted, and also mechanically stressed vertically as well as horizontally, as the light hot materials penetrate under the lithosphere, where they create the asthenosphere. Hence, the asthenosphere is characterized by a low-velocity channel of seismic waves, which differs for *P*-waves, where its thickness is down to depths of about 70 to 120 km, with a maximum of 150 km, and for *S*-waves, where its thickness varies from 70 km to 250–350 km. At depths exceeding 250 km the density and velocity of seismic waves increase again. The lithosphere, after the formulation of the asthenosphere, is strained by vertical as well as horizontal forces in the places where the GTLV-channels reach the bottom of the lithosphere, due to the participation of friction (Coulombian and viscous).

Due to convective currents, rift zones are opening on different places of the Earth's surface. Owing to mechanical strains and continual melting the rifts are opening there. As a consequence of the rift opening, cross disturbance zones are created. Absolute values of horizontal and vertical forces operating in the rift zones depend on the outbreak of the light hot viscous materials from the GTLV-channels and their penetration under the lithosphere (see Figure 4.1). The intensity of the effusion of light materials under the lithosphere changed, so that the brief opening of the rift was probably followed by its closure and again by reopening. This process is repeated several times and is in close relation with geological epochs.

Therefore, the evolution of the rift has several stages. In the first stage of rifting, the vertical components of the acting forces are predominant (e.g. the Baykal rift zone, the East African rift, etc.); in the second stage, horizontal force components start to prevail (e.g. Rhone Graben, Red Sea and Bay of Aden, etc.). In the third stage the horizontal force components prevail (the mid-oceanic rift zones). New rift zones appear in the vicinity of old rifts, which is the result of the generation of convective cells that have not changed their location too much in the course of geological epochs. Their changes of location depend on the rotation of the Earth's body and geodynamic processes inside it. Since opening of the rift zones depends on the effusion of the light materials in time, older and younger sediments accumulated in the open rift. As

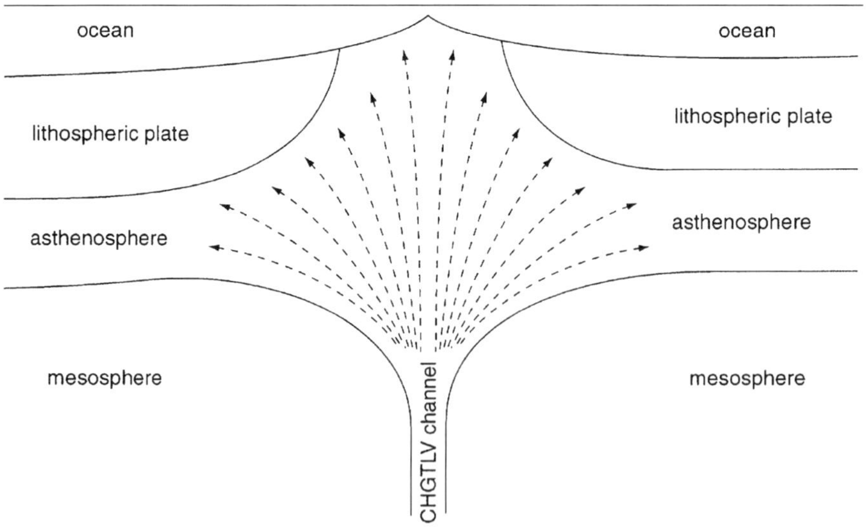

Figure 4.1. The Rift Zone: Schematic Cross-Section

a result of a substantial effusion of light materials taking place over a geologically relatively long interval, both lithospheric plates become separated and move away from each other. We refer to divergent motions. Thus older sediments can be found along the edge of one of the lithospheric plates, e.g. at the eastern edge of the American plate the sediments are about 155 million years old, whereas the sediments near the other plate are considerably younger, e.g. close to the north-western edge of the African plate the sediments as of about 110 million years.

Resultant effects of convective currents are the hexagonal cells on the Earth's surface characterized by the rift zones, with their bifurcation to the three branches containing angles about 120° between them and the channel of higher temperature and lower viscosity (the GTLV-channel) in the lower mantle. Due to acting frictions, the opening of the rift represents the beginning of the action of oblique and horizontal forces on a large scale. The immediate consequence of these horizontal motions are transverse fault zones along which parts of the ridge in the rift region are displaced with respect to one another (e.g. the mid-ocean ridges in the Atlantic, Pacific and Indian Oceans). The forming of these transverse fault zones is also related to the inhomogeneous influence of friction. Compressive and tensile stresses are thus generated, and these are responsible (due to shear) for forming transverse fault zones perpendicular or oblique to the rift zone, i.e. the mid-ocean ridge.

The originally compact lithosphere (of the oceanic and continental types) separates in the neighbourhood of the rift into sub-blocks, which rotate due to the action of horizontal and vertical forces, as well as to the collisions between the sub-blocks them selves. The special case of the rift zone is the region of San Andreas, where the rift and subduction zones are very close together. The rift zones are accompanied by volcanism, the composition of volcanic rocks are the other as in the subduction zones.

The evolution of the rift zone can be described by several main stages. In the first stage, i.e. in the stage of the beginning of the convective currents in the primary mantle, due to the originated convective currents the hot viscous light materials from the mantle-core boundary flow through the primary GTLV-channels below the primary upper mantle. The bottom of the lower part of the upper mantle (i.e. the part which can be denoted as the primary mesosphere) is heated, melted and strained. During the second stage, the primary mesosphere is further heated, melted and strained for a long time so that the light materials through the channel originated by acting forces and due to the further melting of the primary upper mantle, effuse on to the primary Earth's surface, which initiated the primary continent Pangea. A similar phenomenon was modelled and observed by the author if a viscous gelatinous liquid exhibiting a strong inhomogeneity was heated on one side and cooled on the opossite. Then the more heated and, therefore, lighter and less viscous liquid effuses to the surface, at first in one place located close to the inhomogeneity, this mass effusion being large, and later at other places but with smaller mass effusion as the main part of the light materials was drained earlier. During this geological period the density differentiation took place and the structures of the lithosphere, asthenosphere and mesosphere stage by stage were created. After finishing the effusion the solidification of the materials of a new super continent Pangaea took place.

After formation of the lithosphere, asthenosphere and mesosphere and after accumulation of new light hot materials at the mantle-core boundary, the light hot materials effuse below the lithosphere, which is strained vertically as well as horizontally and also melted. In this stage of rifting, the vertical components of the acting forces are predominant. The next stage of evolution is then characterized by the acting forces, where horizontal components prevail, and by continual melting of the lithosphere. The mechanical strains together with continual melting result in the opening of the rift. These new rift zones appear in the immediate vicinity of old rifts. As a result of a continuous effusion of light materials taking place over a geologically relatively long interval, the lithospheric plates become separated and move away from each other. We speak about divergent motions and about spreading of the sea floor.

To derive the mathematical model of the rift zone we shall assume that the upper part of the Earth is created by the visco-plastic materials under great temperatures, gravity field and magnetic field. The thermo-visco-plastic materials can be approximated by the thermo-visco-plastic Bingham fluid in the first case and by the thermo-elasto-visco-plasticity in the second. Moreover, in the stages when the material of the upper parts of the Earth is melted, then in both cases we have the coupled Stefan-visco-plastic approximations of the Bingham fluid type or of the elasto-visco-plastic type, respectively, under the gravity and magnetic fields.

The conservation of energy, also known as the first principle of thermodynamics, yields (see Equation 3.145)

$$\rho \frac{\mathrm{d}e}{\mathrm{d}t} = \tau_{ij} D_{ij} + \rho w - q_{i,i},$$

where the term $\tau_{ij} D_{ij}$, where $D_{ij} = \frac{1}{2}(\partial v_i/\partial x_j + \partial v_j/\partial x_i)$, is the so-called dissipation function. Under the assumption that the dissipation function depends only on the strain rate D_{ij} (see for example Prager (1961)), then putting $\mathscr{D} = \tau_{ij} D_{ij}$, we find

$$\tau_{ij} D_{ij} = \frac{\partial \mathscr{D}_1}{\partial D_{ij}} D_{ij} + \frac{1}{2} \frac{\partial \mathscr{D}_2}{\partial D_{ij}} D_{ij},$$

where $\mathscr{D}_i$, $i = 1, 2$, are positive homogeneous functions of order 1 and 2, respectively, in the components of the strain rate tensor. This identity holds under constraints $D_{kk} \equiv \operatorname{div} \boldsymbol{v} = 0$. For the Bingham visco-plastic fluid the dissipative functions $\mathscr{D}_i$, $i = 1, 2$, are defined by (see for example Prager (1961))

$$\mathscr{D}_1 = 2\hat{g}(D_{\mathrm{II}})^{1/2}, \ \mathscr{D}_2 = 4\hat{\mu} D_{\mathrm{II}},$$

where $D_{\mathrm{II}} = \frac{1}{2} D_{ij} D_{ij}$ and where $\hat{g}$ and $\hat{\mu}$ are the threshold of plasticity (or the yield limit) and of viscosity, respectively. Then the constituent stress-strain rate relation for the thermo-visco-plastic Bingham fluid under magnetic and gravity fields is defined by

$$\begin{aligned} \tau_{ij} &= -p\delta_{ij} + \hat{g} D_{ij} D_{II}^{(-1/2)} + 2\hat{\mu} D_{ij} + \beta_{ij}(T - T_o) \\ &\quad + (E_i D_j + E_j D_i + H_i B_j + H_j B_i - \delta_{ij}(\boldsymbol{ED}) - \delta_{ij}(\boldsymbol{HB})) \\ &= \tau_{ij}^B + \tau_{ij}^T + \tau_{ij}^M, \end{aligned} \tag{4.79}$$

which has a sense when $D_{\mathrm{II}} \neq 0$ and where p is the spherical part of the stress tensor and has a meaning as the pressure, $\tau_{ij}^B = \beta_{ij}(T - T_o)$ is the thermal stress and

$\tau_{ij}^M = (E_iD_j + E_jD_i + H_iB_j + H_jB_i - ((\boldsymbol{ED}) + (\boldsymbol{HB}))\delta_{ij})$ is the Maxwell's stress. The detailed discussion of the general problem will be given in the section on the global geodynamic model of the Earth.

Since we shall study the rift zone in the relatively small region of the main part of the rift, where the materials are melted, the contact conditions in these regions have not occured yet. Moreover, since the light materials have their origin at the mantle-core boundary, the volume of gases is very small so that its influence can be neglected. Also the influence of the magnetic and gravity fields can be neglected, but the influence of the geothermal field is substantial. Moreover, due to the fact that the light material had the origin at the mantle-core boundary, where it was under high pressure, we can assume that it is incompressible or very weakly compressible. In our further analyses we shall assume that it is incompressible.

Next we shall study the incompressible thermo-visco-plastic Bingham fluid without Maxwell's stresses and without the gravity effect only. Then Equation 4.79 leads to

$$\tau_{ij} = -p\delta_{ij} + \hat{g}D_{ij}D_{II}^{(-1/2)} + 2\hat{\mu}D_{ij} + \beta_{ij}(T - T_o) = \tau_{ij}^B + \tau_{ij}^T. \tag{4.80}$$

For the inverse of τ_{ij}^B in Equation 4.80, we set

$$\tau_{\mathrm{II}} = \tfrac{1}{2}\tau_{ij}^D\tau_{ij}^D,$$

where τ_{ij}^D is the deviation of the stress τ_{ij}^B. Hence and using Equation 4.80 we find

$$\tau_{\mathrm{II}} = (\hat{g} + 2\hat{\mu}D_{\mathrm{II}}^{1/2})^2 \tag{4.81}$$

and

$$D_{ij} = (2\hat{\mu})^{-1}(1 - \hat{g}\tau_{\mathrm{II}}^{-1/2})\tau_{ij}^D + \gamma_{ij}(T - T_o). \tag{4.82}$$

Moreover, Equation 4.81 implies $\tau_{\mathrm{II}}^{1/2} \geq \hat{g}$. Then the inverse constituent law for the thermo-Bingham fluid can be written

$$\begin{aligned} &\tau_{\mathrm{II}}^{1/2} < \hat{g} \Leftrightarrow D_{ij} = 0, \\ &\tau_{\mathrm{II}}^{1/2} \geq \hat{g} \Leftrightarrow D_{ij} = (2\hat{\mu})^{-1}(1 - \hat{g}\tau_{\mathrm{II}}^{-1/2})\tau_{ij}^D + \gamma_{ij}(T - T_o). \end{aligned} \tag{4.83}$$

We see that this constituent law describes the flow type problems where the displacements of material points are large. Such type of rheology is useful for the study of effusion of light hot mantle-core materials during the whole geological periods, i.e. during very long periods. Here $\boldsymbol{u}(\boldsymbol{x}, t)$ is the velocity at the instant time t of the rock particle at the point $\boldsymbol{x} \in \Omega$ at this instant time t, $D_{ij}(\boldsymbol{v})$ represents the strain rate tensor, where $\boldsymbol{v} = \boldsymbol{v}(\boldsymbol{x}, t)$ is the velocity of the particle at the point $\boldsymbol{x} \in \Omega$ at the instant time t. We see that the flow of the Bingham fluid is described in the Euler co-ordinates.

In the second case, the thermo-elasto-visco-plastic approximation of the rocks of the lithosphere, asthenosphere and mesosphere, the constituent law is expressed by

$$D_{ij} \equiv \dot{e}_{ij}(\boldsymbol{u}) = s_{ijkl}\dot{\tau}_{kl} + \lambda_{ij} - \gamma_{ij}(T - T_o), \tag{4.84}$$

where s_{ijkl} are the elasticity coefficients and satisfy the conditions of symmetry and of positive definiteness

$$s_{ijkl} = s_{jikl} = s_{ijlk} = s_{klij}, \tag{4.85}$$

$$s_{ijkl}\tau_{ij}\tau_{kl} \geq s_o\tau_{ij}\tau_{ij}, \quad s_o = \text{constant} > 0, \tag{4.86}$$

and where $s_{ijkl}\dot{\tau}_{kl}$ represents the elastic part of the strain rate tensor, $\tau_{ij} = \tau_{ji}$, and λ_{ij} represents the velocity of plastic deformation. The velocity of plastic deformation is defined as follows:

Let us define the convex, and continuous with respect to τ_{ij}, function $\mathscr{F}(\tau_{ij})$, so that the domain of R^6, where $\mathscr{F}(\tau_{ij}) \leq a_o$, $a_o > 0$, is a closed convex set of the six scalar variables τ_{ij} and where $\mathscr{F}(\tau_{ij})$ depends on the deviator τ_{ij}^D only. Further, we define the closed convex set K by

$$K = \{\tau \mid \tau = (\tau_{ij}) \in R^6, \tau_{ij} = \tau_{ji}, \mathscr{F}(\tau_{ij}) \leq a_o\}$$

and the mapping $\tau \to P_k(\tau)$ representing the orthogonal projection of $\tau \in R^6$ on the convex set K. Then the velocity of plastic deformation is defined by

$$\lambda_{ij} = 0 \quad \text{if } \mathscr{F}(\tau_{ij}) < a_o,$$

$$\lambda_{ij} = (2\mu)^{-1}[\tau_{ij} - (P_k\tau)_{ij}] \quad \text{if } \mathscr{F}(\tau_{ij}) \geq a_o. \tag{4.87}$$

Let us introduce the mapping $\mathscr{J}_\mu(\tau) : R^6 \to R$ by

$$\mathscr{J}_\mu(\tau) = (4\mu)^{-1}(\tau_{ij} - (P_k\tau)_{ij})(\tau_{ij} - (P_k\tau)_{ij}), \quad \mu > 0. \tag{4.88}$$

Then $\lambda_{ij} \in \mathscr{J}_\mu(\tau)$, where $\mathscr{J}_\mu(\tau)$ denotes the subgradient of $\mathscr{J}_\mu(\tau)$ in τ and can be written

$$\dot{e}_{ij}(\boldsymbol{u}) = s_{ijkl}\dot{\tau}_{kl} + (\mathscr{J}_\mu(\tau))_{ij} + \gamma_{ij}(T - T_o).$$

We can rewrite Equation 4.84-4.87 as

$$\dot{e}_{ij}(\boldsymbol{u}) = s_{ijkl}\dot{\tau}_{kl} + \lambda_{ij}, \tag{4.89}$$

$$\mathscr{J}_\mu(\sigma) - \mathscr{J}_\mu(\tau) \geq \lambda_{ij}(\sigma_{ij} - \tau_{ij}) \quad \forall\sigma_{ij} \in R. \tag{4.90}$$

Remark: The convex function $\mathscr{F}(\tau_{ij})$ can be defined by the Von Mises relation

$$\mathscr{F}(\tau_{ij}) = \tfrac{1}{2}\tau_{ij}^D\tau_{ij}^D - k^2, \tag{4.91}$$

and then we speak about Von Mises' model of plasticity which is described by

$$\dot{e}_{ij}(\boldsymbol{u}) = s_{ijkl}\dot{\tau}_{kl} + \lambda_{ij}, \tag{4.92}$$

$$\lambda_{ij} = 0 \quad \text{for } \mathscr{F}(\tau_{ij}) < a_o, \tag{4.93}$$

$$\lambda_{ij} = (2\mu)^{-1}\tau_{\mathrm{II}}^{-1/2}(\tau_{\mathrm{II}}^{1/2} - k)\tau_{ij}^D \quad \text{for } \mathscr{F}(\tau_{ij}) \geq a_o, \tag{4.94}$$

where $\tau_{\mathrm{II}} = \frac{1}{2}\tau_{ij}^D\tau_{ij}^D$, and τ_{ij}^D is the deviator of the stress tensor. If the set $K = \{\tau \mid \tau = (\tau_{ij})\}$ is a spherical tensor, then $(P_k\tau)_{ij} = -p\delta_{ij}$ and we have the rheology of the Newtonian liquid, and then the constituent law is defined as $\tau_{ij} = -p\delta_{ij} + 2\mu\dot{e}_{ij}(\boldsymbol{u})$. If $K = R^6$ then $\dot{e}_{ij}(\boldsymbol{u}) = s_{ijkl}\dot{\tau}_{kl}$ and if we assume that at an instant time t_o the rocks were in a stress free state and that the displacements are measured from this time, then after integration over $\langle t_o, t\rangle$ we find $e_{ij}(\boldsymbol{u}) = s_{ijkl}\tau_{kl}$, which represents the inverse Hooke's law in elastic rheology.

In this case the deformations of rocks, in its non-deformed state, occupies a domain Ω of R^6 and the displacements of the mass particle at points $\boldsymbol{x} \in \Omega$ are small. Here $\boldsymbol{u}(\boldsymbol{x}, t)$ represents the displacement vector at the point $\boldsymbol{x} \in \Omega$ at this instant time t. Since $\boldsymbol{u}(\boldsymbol{x}, t)$ is small, we can assume $\dot{\boldsymbol{u}}(\boldsymbol{x}, t) = \partial\boldsymbol{u}(\boldsymbol{x}, t)/\partial t$ to be the velocity and $\partial^2\boldsymbol{u}(\boldsymbol{x}, t)/\partial t^2$ to be the acceleration vector, for the same particle at the point $\boldsymbol{x}$ at the instant time t.

Moreover, the conservation of mass

$$\partial\rho/\partial t + \mathrm{div}(\rho\boldsymbol{v}) = 0, \tag{4.95}$$

or

$$\mathrm{div}\ \boldsymbol{v} = 0 \tag{4.96}$$

for the incompressible rocks, and the conservation of momentum

$$\tau_{ij,j} + f_i = \rho(\partial v_i/\partial t + v_k\partial v_i/\partial x_k), \tag{4.97}$$

where $\rho, \boldsymbol{v}$ are density and velocity vector, and $\boldsymbol{f}$ body forces, are satisfied. The Einstein's summation convention is used.

The geophysical observations prove that the rocks in the region of the rift zone are heated. Thus the thermal regime in these regions will be described by the generalized expanded equation of heat conduction in the form

$$\rho\beta_{ij}T_o\dot{e}_{ij}(\boldsymbol{u}) + \rho c_e(\partial T/\partial t + v_k\partial T/\partial x_k) = W - \partial/\partial x_i(\kappa_{ij}\partial T/\partial x_j), \tag{4.98}$$

where the first term on the left hand side represents the energy dissipated in the form of heat in the moving medium in the upper parts of the Earth in the region of the rift zones and generated by their deformation. The second term on the left hand side and both terms on the right hand side represent the classical expanded equation of heat conduction, and describe heat conduction in this region. Further, W represents heat sources in this region evoked by the radioactive sources, etc., β_{ij} is the coefficient of thermal expansion, T_o the input temperature, characterizing the initial stress-strain state of rocks in this regions, T is the temperature, $\dot{e}_{ij}$ is the strain rate tensor, κ_{ij} is the thermal conductivity, and c_e the specific heat under constant strain.

Moreover, the geological observations show that rocks in the rift zones are also melted and near these regions are also recrystallized, so that the continuum representing these regions may have either a liquid or solid phase in the rift zones or more solid phases in the case of recrystallization. The governing equation follows from an integral formulation of the principle of heat conservation, which is a simplified version of the principle of energy conservation under the assumption that both phases are incompressible. The phase change is assumed to take place at some specified temperature, which assumes the existence of some moving surface (we denote it by $R(t)$) such that the rocks are in their solid phase on one side of this surface (the boundary of melting) and in their liquid phase on the other side, in the case of melting; or that the rocks are in their first solid phase on one side of this moving surface (the boundary of recrystallization) and in their second solid phase on the other side, in the case of recrystallization. Energy balance arguments then show that the temperature is continuous everywhere, i.e. also across the boundary of melting or recrystallization. The heat flux has a jump discontinuity across this boundary. These surfaces $R^s(t)$ divide Ω into regions Ω^s_S, Ω^s_L representing, respectively, the solid and the liquid phases as well as recrystallized phases in "s"rocks. Thus on $R^s(t)$ at time t

$$T^s_S = T^s_L = T^s_R, \tag{4.99}$$

$$\rho^s L^s v^s_\nu + (\kappa_{ij}\partial T/\partial x_j \nu_i)^s_S - (\kappa_{ij}\partial T/\partial x_j \nu_i)^s_L = 0, \tag{4.100}$$

where ρ^s is density, T^s_R is the temperature of the phase transition, T^s_S, T^s_L are the temperatures at the phase change boundary $R^s(t)$ in the solid and liquid phases, or in both solid phases, respectively, ν^s_i are components of the unit normal to $R^s(t)$ pointing towards Ω^s_S, v^s_ν is the speed of $R^s(t)$ along ν^s, and L^s is the latent heat of phase transition in the "s" rock. This approximation, known as the Stefan problem, is only a simplification of the real situation. Moreover, the heat equation (Equation 4.98) describes the distribution of the geothermal field in every Ω^s_S and Ω^s_L, i.e. in $\Omega \backslash \cup R^s(t)$.

4.2.3. Tectonic Evolution of Collision Zones

The plate tectonic hypothesis assumes that the lithosphere is divided into a small number of rigid plates interacting only near their common boundaries. There exist large regions (of continental or oceanic nature), where the lithosphere undergoes large-scale deformation, and a diffuse seismicity prevails. We speak of collision or subduction zones. The subduction zones are thus localities of fundamental geodynamic, geophysical and geological processes. They are probably the source of major plate-driving forces; they seem to be the sites at which a new continental lithosphere is differentiated, and they give rise to most of the tectonic imprint in continents. This evokes the questions, which have not been answered without knowledge of:

(a) the stress distribution in these regions — in the subducted (descended) and obducted (overridden) lithosphere,

(b) the behaviour of the subducted lithosphere in the adjacent upper mantle,

(c) the rheology and geometry of these regions.

A number of authors show that the rheology of the lithosphere, based on theoretical considerations and experimental data, can be considered elasto-visco-plastic up to thermo-elasto-visco-plastic with thermo-diffusion processes in the lithosphere (see, for example, Nedoma (1986a-c)) (i.e. gases in rocks or a fluid in porous visco-elastic media).

Though the kinetic description of the Earth's surface motion has been generally accepted, there is much less agreement about how the motions are maintained. Observed subduction zones are those expressed topographically by long linear oceanic trenches flanked by low outer rises on the seaward sides and island arcs on the landward sides. These topographic structures are in close connection with gravity anomalies. So Melosh and Raefsky (1980) show that the similarity of both topographic and free air gravity profiles for nearly all subduction zones strongly suggests that a single process acts to produce these features. The existence of large free air anomalies further suggests that this process is dynamic and that the forces generated during subduction act on the Earth's surface to produce the observed topography. The dynamic problem consists of two separate but related questions: how does the mantle convert heat into mechanical work to maintain the observed plate motions, and why is the large-scale flow, of which the plate motions form part, stable to smaller scale motions? The answer to these questions is the result of much research during the last twenty years. One of the complete geodynamic models of the Earth based on the plate tectonic hypothesis is given in the author's papers. By definition in the literature, the lithosphere behaves like an elastic solid and stress guide over geological periods. On the other hand, the lithosphere's creep is observed. Thus a realistic description of the rheology of the lithosphere must allow its viscosity to decrease with depth as was discussed in the previous section. Moreover, the greater part of the geological and geodynamical processes take place over extremely long time periods, during which the mechanical and rheological properties of the rocks change. Therefore, for a better understanding of the tectonic evolution of the subduction zones, the mechanical and rheological properties of rocks are substantial. Hence, the lithosphere in the subduction zones can be assumed to be thermo-elastic (for short time periods from the geological point of view) up to thermo-elasto-visco-plastic, and displays magnetic fields and thermo-diffusion processes (gases in rocks, fluid in porous rocks). The asthenosphere and mesosphere below it do not have such ambiguous properties, but have relatively simple properties like lithospheric materials. The materials of the asthenosphere are characterized by a steady creep under the applied shear stress τ_{13} (Melosh (1977)). The shear stress acting on the base of the lithospheric plate must be maintained by velocity gradients. Since the asthenosphere is a zone of especially low viscosity, velocity gradients will concentrate within it. This situation is potentially unstable due to the thermal structure of the asthenosphere. Its thermal instability is the direct result of the stress-strain rate relation for hot crystalline solids. As was proved in the literature, the lithosphere is an excellent insulator, having a thermal line constant of the order 100 myr^{-1}. The velocity gradients in the asthenosphere generate shear heat in the amount of $\tau_{13}\dot{e}_{13}$ (erg cm^{-3} sec). Unless this heat is conducted away through the lithosphere, the temperature of the asthenosphere must also rise due to the conversion of friction to heat.

Further, some authors presume that if the lithospheric plate is descending faster than the surrounding asthenosphere and mesosphere, a shear stress acting on the surfaces of the lithospheric plate will resist its descent and the plate will descend slower than the surrounding mantle. They also assume the resistance to the leading edge of the plate to be blunt. For greater depth they assume viscous resistance and positive buoyancy due either to a depressed phase boundary within the plate or a change in the surrounding asthenospheric and mesospheric compositions.

Nedoma (1984) and in later papers draws attention to the fact that, due to the friction in the contact zone, overriding and subducting (descending) plates are heated and melted, and in the case of more deeply subducted plates the masses of the asthenosphere and mesosphere are also heated, where they create the anomalous zone. In the contact zone between both obducting and subducting plates, the molten materials form magmatic deposits (volcanic chambers), from which the magma, together with gases, ascends to the Earth's surface, where it forms volcanoes. The magma ascends to the Earth's surface along faults, which are the result of enormous stresses in the obducting plate (Nedoma (1984) and later papers). In some regions of the subduction zones such as the Peri-Pieninean geophysical lineament (zone) of the Carpathian, recrystallization of melted rocks occurs in the collision zones after retardation of tectonic activity. Their electric conductivity and density are changed. Manifestation of these local inhomogeneities are thus gravity and magnetic (Wiese's vector) anomalies.

4.2.3.1. Mechanics of subducted lithosphere

The mechanics of the subducted lithosphere is closely connected with the geodynamic processes at the outer core/mantle boundary and in the mantle of the Earth. To investigate the geodynamic processes in the upper parts of the Earth, above all in the lithosphere, the rheological and specific mechanical properties of rocks are of decisive importance. As we saw above, the thermo-visco-elastic to thermo-elasto-visco-plastic rheology of the materials of the upper parts of the Earth, displayed by the gravity and magnetic fields as well as the thermo-diffusive processes, can be considered. If our investigations are restricted to the mechanical manifestation of the geodynamic processes only, the effects of an induced magnetic field can be neglected. But if besides the mechanical manifestations of the geodynamic processes we are also studying magnetic and gravity anomalies in the subduction zones, the magnetic field cannot be neglected.

Due to the convective currents and effusion of light materials from the mantle/core boundary, rift zones open at different places on the Earth's surface and thus the lithosphere is strained by the horizontal forces from different sides. The lithosphere behaves like a stress guide; it is strained and stresses are generated, thus accumulating in various parts of the lithosphere. If the magnitude of the stress exceeds the marginal value in straining, the lithosphere is divided by deep fractures and faults into several megaplates and blocks.

The equations describing mathematically this discussed reality are based on the law of conversation of mass, the law of momentum conservation, the general conservation law of energy, Maxwell's equations, the thermo-diffusion equation and the equations of gravity potential and gravity acceleration $\boldsymbol{g}$.

The law of conservation of mass which gave a precise characterization of the idea that a specified portion of Earth's material (which is assumed to be incompressible and in the general case to be compressible) occupies the Earth's body at time t, is of the form

$$\frac{\partial \rho(\boldsymbol{x}, t, \ldots)}{\partial t} + \operatorname{div}(\boldsymbol{v}(\boldsymbol{x}, t, \ldots)\rho(\boldsymbol{x}, t, \ldots)) = 0. \tag{4.101}$$

Since the law of momentum conservation is independent of the relation between the strain and stress, we have

$$\tau_{ij,j}(\boldsymbol{x}, t, \ldots) + F_i(\boldsymbol{x}, t, \ldots) = \rho(\boldsymbol{x}, t, \ldots)(v_{i,t} + v_j v_{i,j}), \quad i = 1, 2, 3, \tag{4.102}$$

where $F_i(\boldsymbol{x}, t, \ldots) = f_{ai} + f_{bi} + f_{ci} + f_{vi} + f_{npi}, \boldsymbol{f_a} = (f_{ai}) = \rho D\boldsymbol{g}$ represents the Archimedean force, $D \ll 1$ is a function representing the density inhomogeneities in the Earth's interior, $\boldsymbol{f_b} = (f_{bi}) = [\boldsymbol{j}, \boldsymbol{B}]$ represents the volume density of the electromagnetic force, $\boldsymbol{f_v} = (f_{vi})$ is the viscous force evoked by viscous stress $\simeq \frac{1}{2}\eta_{ijkl} v_{i,j} v_{k,l}, \boldsymbol{f_c} = (f_{ci}) = 2\rho[\vec{\omega}, \boldsymbol{v}]$ is the Coriolis force, and $\boldsymbol{f_{np}} = (f_{npi})$ represents the force due to precession, mutation, etc.. But for the investigation of the geodynamic processes in the lithosphere $\boldsymbol{f_c}$ and $\boldsymbol{f_{np}}$ can be neglected with sufficient accuracy.

The constitutional stress-strain relations are defined by

$$D_{ij} = D_{ij}^e + D_{ij}^{p+v} + D_{ij}^T + D_{ij}^M, \tag{4.103}$$

where $D_{ij}^e = \frac{1}{2}(v_{i,j} + v_{j,i}) + \frac{1}{2}v_{k,j}v_{k,j}$ $\left(\text{or } D_{ij}^e = \frac{1}{2}(v_{i,j} + v_{j,i})\right)$ is the finite (or small) strain rate tensor, $D_{ij}^e = (s_{ijkl}\tau_{kl})_{,t}$, where the coefficients s_{ijkl} are the so-called elastic moduli fulfilling the symmetry conditions $s_{ijkl} = s_{jikl} = s_{ijlk} = s_{klij}$. Further D_{ij}^T, D_{ij}^M are the strain rate tensors due to the thermal, $\tau_{ij}^T = \beta_{ij}(T - T_o)$, and the magnetic (Maxwell's stress τ_{ij}^M), $\tau_{ij}^M = E_i D_j + E_j D_i + H_i B_j + H_j B_i + ((\boldsymbol{ED}) + (\boldsymbol{HB}))\delta_{ij}$. Further, D_{ij}^{p+v} represents the viscoplastic rate tensor. The domain of elasticity and visco-plasticity is defined by the function of plasticity $\mathscr{F}(\tau_{ij})$, where $\tau_{ij} = \tau_{ji}$. The function $\mathscr{F}$ is convex and continuous in τ_{ij} and $\mathscr{F}(0) = 0, \mathscr{F}(\tau_{ij}) < a_o$ for $\tau_{ij} \neq 0$. Indeed, $\mathscr{F}(\tau_{ij})$ includes the deviator τ_{ij}^D of the stress tensor τ_{ij} only. To define the relation between stress and strain we define the set K by

$$K = \left\{\tau \mid \tau = (\tau_{ij}) \in R^6, \tau_{ij} = \tau_{ji}, \mathscr{F}(\tau_{ij}) \leq a_o\right\}, \tag{4.104}$$

where $\mathscr{F}$ is a convex continuous function of six scalar variables, defined by

$$\mathscr{F}(\tau_{ij}) < a_o, \quad \tau_{ij} \neq 0, \quad a_o = \text{constant} > 0 \text{ for the domain of elasticity,} \tag{4.105}$$

$$\mathscr{F}(\tau_{ij}) \geq a_o, \quad \tau_{ij} \neq 0, \quad a_o = \text{constant} > 0 \text{ for the domain of visco-plasticity.}$$

Further, the mapping $\tau \to P_k(\tau)$ is the orthogonal projection which maps $\tau \in R^6$ onto the convex set K. Then the visco-plastic strain rate tensor is defined by

$$D_{ij}^{v+p} = \begin{cases} 0 \text{ in the case of elasticity, i.e. } \mathscr{F}(\tau_{ij}) < a_o, \\ (2\eta)^{-1}[\tau_{ij} - (P_k\tau)_{ij}] \text{ in the case of visco-plasticity, i.e. } \mathscr{F}(\tau_{ij}) \geq a_o, \end{cases} \tag{4.106}$$

where η is the coefficient of viscosity.

The conservation law of energy gives the generalized expanded equation of heat conduction

$$\rho\beta_{ij}T_o(D_{ij} + e_{ij,k}^{e+v+p}) + \rho c_e(T_{,t} + v_k T_{,k}) = W - (\kappa_{ij}T_{,j})_{,i} - \sigma_{ij}^{-1} j_i j_j + T j_i(\alpha_{ij}^o)_{,j}, \tag{4.107}$$

where the first term on the left hand side represents the energy dissipated in the form of heat in the moving masses of the upper parts of the Earth and generated by their deformation, and the second term on the left hand side and the first two terms on the right hand side represent the well-known classical expanded equation of heat conduction. Here $W = \Sigma_i W_i$ represents heat sources such as the radioactive sources W_1, sources W_2 evoked by the viscous stresses $\simeq \frac{1}{2}\eta_{ijkl}v_{i,j}v_{k,l}$, where η_{ijkl} is the tensor of the 4-th order, the so-called coefficient of viscosity, sources W_3 and W_4 evoked by the viscous and Coulombian frictions, etc.. The third term on the right hand side represents Joule's heat, the fourth term represents the specific thermoelectric effects.

Due to some geodynamic processes the materials of the descending lithospheric plates in the subduction zones are melted, the gases contained in them are released and one part of them is again dissolved in these materials whilst the other escapes. Moreover, the upper part of the lithosphere is also built of porous rocks. The fluid then pervades these materials. Therefore, diffusion processes exist in the lithosphere and thus must be considered in the model problem. Since the distribution of concentration of gases and fluids in the Earth's rocks changed during the geological period, the diffusion processes in the lithosphere and the upper mantle are one of the sources of dissipative energy. The point of departure for our consideration is the free energy F for thermo-diffusion in a solid, where the chemical potential M is defined as $M = \partial F/\partial \mathrm{d}$. The equation for the flux of mass $\boldsymbol{i}$ has the form

$$\rho(\mathrm{d}_{,t} + \boldsymbol{v}\ \mathrm{grad}\ \mathrm{d}) + \mathrm{div}\ \boldsymbol{i} = h, \tag{4.108}$$

where d is the concentration, and h is the intensity of the source of diffusion mass.

The equation for the concentration d is derived on the basis of the relation between the vector mass flux $\boldsymbol{i}$ and the chemical potential M by the relation $\boldsymbol{i} = L_{\mathrm{dd}}T_o^{-1}$ grad M, where L_{dd} is the so-called Onsager constant. Putting $D_{\mathrm{d}} \simeq K_{\mathrm{d}}L_{pp}T_o^{-1}$, where K_{d} is a material constant, we obtain

$$\boldsymbol{i} = \rho D_{\mathrm{d}}\mathrm{grad}\ \mathrm{d} + K_T(T)\mathrm{grad}\ T + P_p(p)\mathrm{grad}\ \mathrm{div}\ \boldsymbol{u}, \tag{4.109}$$

where D_{d} is the coefficient of diffusion, $K_T(T)$ is the coefficient of thermo-diffusion, $P_p(p)$ is the coefficient of baro-diffusion, and p is the pressure. From Equation 4.109 it is evident that the mass flux vector depends on concentration, temperature and dilatation gradients.

The lithosphere and upper mantle are under a persistent effect of the inner and outer magnetic field. Geophysical measurements show that the quantities characterizing the magnetic field of the Earth are measurable on the Earth's surface as well as inside and outside the Earth. The analyses of those geophysical measurements then show that they provide further information about the structure and conductivity of the lithosphere and the upper mantle. Maxwell's equation or the well-known magnetodynamic equation, i.e. the equation

$$\boldsymbol{B}_{,t} - \mathrm{rot}[\boldsymbol{v}, \boldsymbol{b}] = (\mu\sigma)^{-1}\Delta\boldsymbol{B} - [\nabla(\mu\sigma)^{-1}, \mathrm{rot}\ \boldsymbol{B}] + \mathrm{rot}\ \boldsymbol{D}_{,t} \tag{4.110}$$

describes the behaviour of the magnetic field in the upper parts of the Earth's body.

All particles of the Earth are in its gravity field. The same also holds for the particles of the lithosphere and the upper mantle. The gravity field of the Earth is characterized by the potential of gravity Φ and the gravity acceleration vector $\boldsymbol{g}$ by

$$\Delta\Phi(\boldsymbol{x}, t, \ldots) = -4\pi G_c p(\boldsymbol{x}, t, \ldots)K^{-1}(\boldsymbol{x}, t, \ldots)\rho(\boldsymbol{x}, t, \ldots) + 2\omega^2(t), \quad (4.111)$$

$$\boldsymbol{g} = \boldsymbol{g}(\boldsymbol{x}, t, \ldots) = \text{grad } \Phi(\boldsymbol{x}, t, \ldots), \quad (4.112)$$

where G_c is the gravitational constant, p the pressure, K the bulk modulus, and ω the angular velocity of the rotating Earth.

The evolution of the subducted lithosphere can be characterized by several stages, which will now be discussed in detail.

STAGE 1: Fracture Zones as the First Stage of the Evolution

The opening of rifts causes the lithospheric plates to move, the megaplates and blocks are deformed, and stresses are generated and accumulated within them. Microcracks are created as the result of straining of the lithospheric rocks. Due to further loading and straining of the lithospheric rocks, the microcracks grow and propagate, and cracks are generated. Boundary conditions are created on the faces of the cracks. In the first approximation this means that the surface traction vanishes, $\tau_{ij}n_j = 0$. As the cracks, due to further loading and straining, propagate and grow, the boundary condition on the faces of the macrocracks and fractures will change. Thus the boundary conditions follow from the following deliberation (see Figure 4.2).

The traction vector on the boundary of the macrocrack $\tau_i = \tau_{ij}n_j$ can be decomposed into the normal and tangential components

$$\tau_n = \tau_i n_i = \tau_{ij} n_j n_i, \quad \vec{\tau}_t = \tau_i t_i = \tau_{ij} n_j t_i,$$

where $\boldsymbol{n} = (n_i)$, $\boldsymbol{t} = (t_i)(i = 1, 2, 3)$ are the outward normal and tangential unit vectors. The displacement vector $\boldsymbol{u}$ can be decomposed into the normal and tangential components $u_n = u_i n_i$, $\boldsymbol{u}_t = u_i t_i$.

It is well known that tectonic microcracks are elongated, configured into some systems and intersect the mineral grains or the boundaries between different rocks.

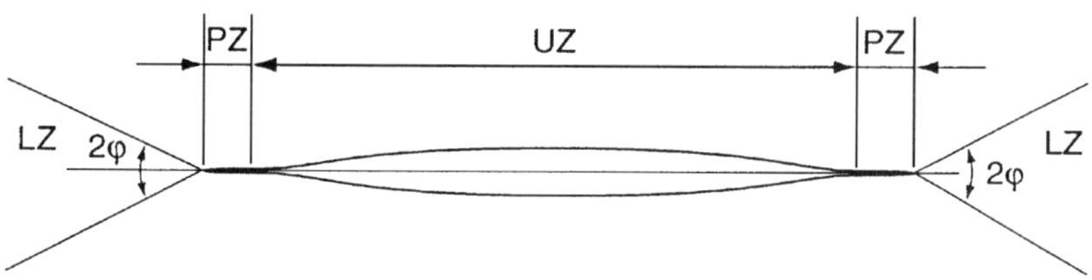

LZ-the elasto-plastic loading zone, *PZ*-the elasto-plastic zone, *UZ*-the elastic unloading zone. The elastic unloading zone and the elasto-plastic (in the literature known also as process zone, plastic zone, damage zone, failure zone) and the elastic unloading zone are characterized by the Signorini type conditions with Coulombian friction.

Figure 4.2. Crack-tip Geometry and Zones of Loading and Unloading

After further loading the microcracks rise and therefore macrocracks are created, and after even further loading the fracture and deep faults are originated. Both the opposite parts of the fractured lithospheric rocks cannot penetrate, thus $u_n^k - u_n^l \leq 0$ at the fracture boundary. If $u_n^k - u_n^l < 0$ at some point of the fracture boundary, the support does not produce a reaction and thus $\tau_n^k = -\tau_n^l \equiv \tau_n = 0$. If $u_n^k - u_n^l = 0$, i.e. if the opposite parts of the fracture boundary are in mutual contact, it exerts a normal force $\tau_n^k = -\tau_n^l \equiv \tau_n \leq 0$. Hence

$$u_n^k - u_n^l \leq 0,\ \tau_n \leq 0,\ \tau_n(u_n^k - u_n^l) = 0 \text{ at the fracture boundary.} \tag{4.113}$$

As the friction acts on the fracture boundary, the Coulombian law of friction can be used. Thus,

$$\begin{aligned}
&\text{if } u_n^k - u_n^l = 0\\
&\vec{\tau}_t^k = -\vec{\tau}_t^l \equiv \vec{\tau}_t,\ |\vec{\tau}_t| \leq \mathscr{F}_c\,|\tau_n|\,,\\
&\text{and then}\\
&\text{if } |\vec{\tau}_t| < \mathscr{F}_c\,|\tau_n| \text{ then } \partial/\partial t(u_n^k - u_n^l) = 0,\\
&\text{if } |\vec{\tau}_t| = \mathscr{F}_c\,|\tau_n| \text{ then there exists a constant } \theta \geq 0 \text{ such}\\
&\quad \text{that } \partial/\partial t(u_n^k - u_n^l) = -\theta\vec{\tau}_t \text{ at the fracture boundary.}
\end{aligned} \tag{4.114}$$

The analysis of this problem requires the solution of variational inequalities in linear elasticity, thermo-elasticity or thermo-visco-plasticity. This problem will be discussed later in connection with the analysis of the global plate tectonic model.

Cracks and faults in a lithospheric plate give rise to a surface discontinuity of the displacement vector $\boldsymbol{u}$. In the classical investigation of this problem fracture mechanics can be used.

From classical fracture mechanics it is known that the investigations are concerned with the analysis of the stability of cracks. Although fracture involves processes on the atomic and molecular level, it is a matter of wide experience that useful quantitative information can be obtained by basing computations on a continuum model for the elastic medium. A continuum model, based on the presumption that the medium is homogeneous, isotropic and linearly elastic, makes it feasible to carry out a mathematical analysis of the displacement and stress-strain fields in the neighbourhood of a crack. A fracture criterion can be employed to determine the conditions for crack propagation. Most results of fracture mechanics are based on the analytical methods and especially on quasi-static elasticity theory. But in our geodynamic studies the dynamic effects are very important. There are two main classes of fracture mechanics problems that have to be treated as dynamic problems of the propagation of deep fractures in the lithosphere:

(i) cracked lithosphere subjected to rapidly varying loads,
(ii) lithospheric blocks containing rapidly propagating cracks.

In both cases the crack tip is an environment disturbed by wave motions. We now briefly review some topics in the main area of elastodynamic fracture mechanics which are relevant to the study of the earthquake mechanism.

For the governing equation for such studies, the equation of motion

$$\tau_{ij,j} + f_i = \rho u_{i,tt}, \tag{4.115}$$

and the Hooke's law

$$\tau_{ij} = \lambda e_{kk}(\boldsymbol{u})\delta_{ij} + 2\mu e_{ij}(\boldsymbol{u}), \tag{4.116}$$

where the small strain tensor is defined by

$$e_{ij}(\boldsymbol{u}) = \tfrac{1}{2}(u_{i,j} + u_{j,i}), \tag{4.117}$$

will be given. The boundary conditions on the faces of the cracks and faults are such that the surface traction vanishes, i.e. $\tau_{ij}n_j = 0$. When the field variables are independent of one of the Cartesian co-ordinates, say x_3, the elastic wave motions uncouple into anti-plane and in-plane motions. The anti-plane motion is described by the displacement component $u_3(x_1, x_2, t)$, which is governed by equations $u_{3,ii}(x_1, x_2, t) = c_s^{-2}u_{3,tt}$, $i = 1, 2$. Motions in plane strain are described by displacement components $u_i(x_1, x_2, t)$, $i = 1, 2$. A crack in a lithospheric block gives rise to a surface of discontinuity of the displacement vector. In general all three displacement vector components suffer discontinuities. For simplicity, we shall explain crack propagation in the 2*D* case only. In this case one can find that in the neighbourhood of the crack tip, the stress components τ_{ij} in a local co-ordinate system (r, φ) at the crack tip (see Figure 4.3) are of the form

$$\tau_{ij}(r, \varphi, t) = r^{-1/2} f_{ij}(\varphi, t), \tag{4.118}$$

where r, φ are polar co-ordinates centred at the crack tip and the function $f_{ij}(\varphi, t)$ is a smooth function. Separate expressions for $f_{ij}(\varphi, t)$ for in-plane displacements can be found in symmetric and anti-symmetric parts, with respect to the plane of the crack, and similarly for anti-plane displacements. These symmetric and anti-symmetric parts of the displacements correspond to Mode I and Mode II fracture, respectively. Anti-plane displacements correspond to Mode III fracture. The function $f_{ij}(\varphi, t)$ can be written by using the stress intensity factors and the angular distribution. Overall geometry and loading is included in the stress intensity factor. It is well-known (see foe example Aschenbach (1978), Nedoma (1990b)) that for the elasto-static case the

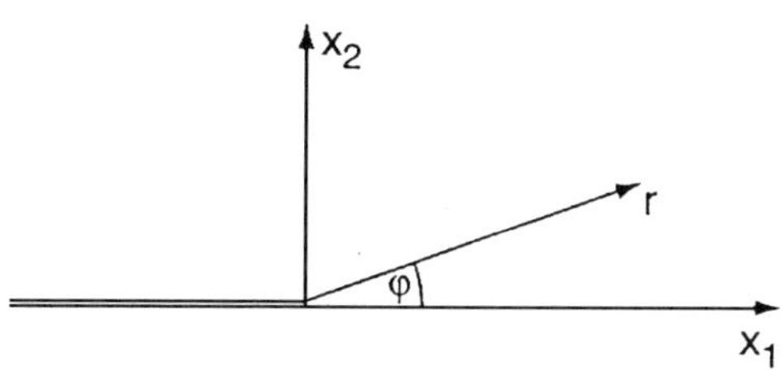

Figure 4.3. Propagating Crack Tip

behaviour of the stress is given by its components τ_φ (Mode I) and $\tau_{\varphi 3}$ (Mode III), which can be written as follows:

$$\tau_\varphi = (2\pi r)^{-1} k_I(t) F^I_\varphi(\varphi),\ \tau_{\varphi 3} = (2\pi r)^{-1} k_{III}(t) F^{III}_\varphi(\varphi) \tag{4.119}$$

with $F^I_\varphi(0) = F^{III}_\varphi(0) = 1$, where for the elasto-static case the time t is a parameter connected with the loading or with the length of the crack for the case when the crack is propagating. For a stationary crack tip, Equation 4.119 holds and only the stress intensity factors are different. For the elasto-dynamic propagating crack tip the angular distribution changes and can be denoted by $F^I_\varphi(\varphi, v)$ and $F^{III}_\varphi(\varphi, v)$ with $F^I_\varphi(0, v) = F^{III}_\varphi(0, v) = 1$, where $v = v(t)$ is the speed of propagation of the crack and $F^I_\varphi(\varphi, v)$ is independent of the geometry and the loading. The stress intensity factors are functions of time and the speed of propagation of the crack.

The crack opening displacement u_φ on the upper crack surface just behind the moving crack tip is defined by

$$\begin{aligned} u_\varphi = {} & (2r)^{1/2} \pi^{(-1/2)} v^2(t) (\mu c_s^2)^{-1} (1 - v^2(t)/c_p^2)^{1/2} k_I(t, v) [(v(t)/c_s^2 - 2)^2 \\ & - 4(1 - v^2(t)/c_s^2)^{1/2} (1 - v^2(t)/c_p^2)^{1/2}]^{-1}, \end{aligned} \tag{4.120}$$

where $c_p = [(\lambda + 2\mu)/\rho]^{1/2}$, $c_s = (\mu/\rho)^{1/2}$ are the velocities of elastic P and S waves, λ and μ are Lamé's coefficients, and ρ is the density.

The expressions of this section also apply to $3D$ problems for plane cracks with a smoothly curved edge, where the Cartesian co-ordinate system is as follows: let 0, a point on the edge, serve as origin of this co-ordinate system; the x_3-axis is directed along the tangent to the edge of the crack, the x_2-axis is normal to the plane of the closed crack, and the x_1-axis is directed into the solid. Hence, in the $x_3 = 0$ plane and in the neighbourhood of 0, the stress components are also given by Equations 4.118 and 4.119. The geometry of the crack edge is characterized by the magnitude of the stress intensity factor.

Criteria for crack propagation are in close connection with the stress intensity factor $k_I(t, v)$, which reaches an experimentally determined critical value. Hence the initiation of crack growth $k_I(t, 0) = k_{in}(\dot{\tau})$ and for a propagating crack $k_I(t, v) = k_p(v)$, where $\dot{\tau}$ is a loading rate, and k_p is the so-called dynamic fracture toughness.

On the other hand, a propagating crack tip behaves like an energy sink. Mechanical energy of the lithospheric plate is dissipated during the fracture process (for perfectly elastic materials; for a thermo-elastic and for an elasto-plastic crack growth the analysis is much more complicated). We limit ourselves to perfectly elastic material only.

To describe the process of crack propagation and to give a condition for the crack tip motion, we assume that the materials have the property of resistance of fracture, characterized by the function $r(v)$, which is a function of the speed of propagation of the crack. It represents the energy dissipated in the fracture processes accompanying crack propagation. The energy release rate per unit thickness E_r can be written in the following form:

$$E_r = v^{-1}(t)(\mathrm{d}P/\mathrm{d}t - \mathrm{d}W/\mathrm{d}t - \mathrm{d}U/\mathrm{d}t), \tag{4.121}$$

where P is the work of external loads, W is the strain energy and U is the kinetic energy. The energy release rate E_r reaches its critical value at the value corresponding to the rate of the material's resistance to fracture, defined by the function $r(v)$. Thus $E_r = r(v)$. The flux of energy into the crack tip is defined as a product of the energy release rate and the speed of crack propagation, i.e.

$$Q = E_r v(t) = r(v)v(t). \tag{4.122}$$

The condition in the above equation represents the necessary condition for fracture. As has been shown (see Aschenbach (1978), Nedoma (1990b)), the flux of energy into the crack tip (in a $2D$ geometry) can be expressed in terms of the stress intensity factor $k_I(t, v)$, $k_{\mathrm{II}}(t, v)$ and $k_{III}(t, v)$, as follows:

$$\begin{aligned} Q = & -v^3(t)\left\{2\mu c_s^2\left[(v^2(t)/c_s^2 - 2)^2 - 4(1 - v^2(t)/c_p^2)^{1/2}(1 - v^2(t)/c_s^2)^{1/2}\right]\right\}^{-1} \\ & \times\left\{(1 - v^2(t)/c_p^2)^{1/2}k_I^2(t, v) + (1 - v^2(t)/c_s^2)^{1/2}k_{\mathrm{II}}^2(t, v)\right\} \\ & + v(t)\left[2\mu(1 - v^2(t)/c_s^2)^{1/2}\right]^{-1} k_{III}^2(t, v). \end{aligned} \tag{4.123}$$

For a slowly propagating crack tip, for which $v(t) \rightarrow 0$, from Equation 4.123 we obtain

$$Q = c_p^2 k_I^2(t)v(t)/[4\mu(c_p^2 - c_s^2)]^{-1}. \tag{4.124}$$

Hence and using Equation 4.122 it follows that k_I is equal to the toughness for fracture initiation k_c, i.e.

$$k_I = k_c = [r(0)E/(1 - \nu^2)]^{1/2}, \tag{4.125}$$

where E is Young's modulus and ν is the Poisson constant. Hence we see that for a slowly propagating crack the usual fracture toughness enters as the material property, and the stress intensity factor equals the fracture toughness k_c while the crack propagates.

In the case when a rapidly running crack departs from a straight trajectory and continues to move at an angle to the plane of the initial crack, we have so-called kinking. The causes of this are the asymmetry of the loading, and inhomogeneities in the rocks. A rapidly propagating crack consists of two stages: firstly, in arresting of the initial crack at time t_{in} and, secondly, in synchronous emanation of the crack kink. If the crack kink propagates with speed $v(t) < c_s$, it propagates into a field that has already received signals from the arrest of the initial crack, since the latter propagates with velocity c_s. Thus a necessary condition of directional instability of a rapidly propagating crack tip is reached when the flux of energy into the crack tip is higher if the crack tip deviates from its initial straight trajectory. This is connected with critical speed of crack propagation. It is evident that the analysis of the stress singularities at the tip of a kinked crack is much more complicated and does not lead to the required goal. The numerical approach leads to the required goal and therefore will be one of the main topics of our investigations of the mathematical simulation of geodynamic processes in the upper parts of the Earth and of the plate tectonics.

The detailed discussion will be given in the next studies about FEM analyses of geodynamic processes in the lithosphere and the upper mantle.

STAGE 2: Enlarging Contact Zones as the Second Stage of the Evolution

The second stage of the evolution of the lithosphere is characterized by an enlarging contact with a relatively, from the geological point of view, short time period. In the regions of the fundamental deep faults into which the original lithosphere has broken up, the neighbouring lithospheric plates are in contact. They are pressed against each other. It may also occur that the closing of rifts which are in the relatively great distances of the enlarging contact zone, relieves both plates; the contact surfaces will become "smoothed" due to relative mutual displacements. Due to the further loading of the megaplates, the near-surface parts of the faults thus begin to open up, and at the same time, but not so markedly, also the bottom parts of the faults (see Figure 4.4).

To illustrate this situation, we shall assume that the lithospheric plates are big and have a layered structure. Let us denote the neighbouring broken up lithospheric plates before their further deformation by Ω^k and Ω^l, and their common contact boundary by Γ_c^k and Γ_c^l at time $t = t_o$. Let $t \in I = (t_o, t_1)$. Let (x, y, z) be a local co-ordinate system, where the x-axis coincides with the direction of the unit normal $\boldsymbol{n}_k$, the z-axis with the common tangent to $\overline{\Gamma}_c^k \cap \overline{\Gamma}_c^l \equiv \Gamma_c^{kl}$ at a central point of the common contact boundary and the y-axis is perpendicular to the (x, z)-plane. Let f_k and f_l be parametric functions of the contact boundaries Γ_c^k and Γ_c^l, continuous in the enlarging interval $\langle z_o, z_1 \rangle$. The contact boundaries of colliding plates can then be defined as

$$\Gamma_c^k = \{(x, y, z, t), z_o \le z(t) \le z_1, x(t) = f_k(y, z, t) \text{ for every } y \text{ at } t \in I\},$$

$$\Gamma_c^l = \{(x, y, z, t), z_o \le z(t) \le z_1, x(t) = f_l(y, z, t) \text{ for every } y \text{ at } t \in I\}.$$

Then for every $z \in \langle z_o, z_1 \rangle$ and every y at time $t \in I$ we have

$$u_x^k(x(t), y(t), z(t), t) - u_x^l(x(t), y(t), z(t), t) \le f_l(y(t), z(t), t) - f_k(y(t), z(t), t) \quad \text{for every } y \text{ at time } t \in I, \tag{4.126}$$

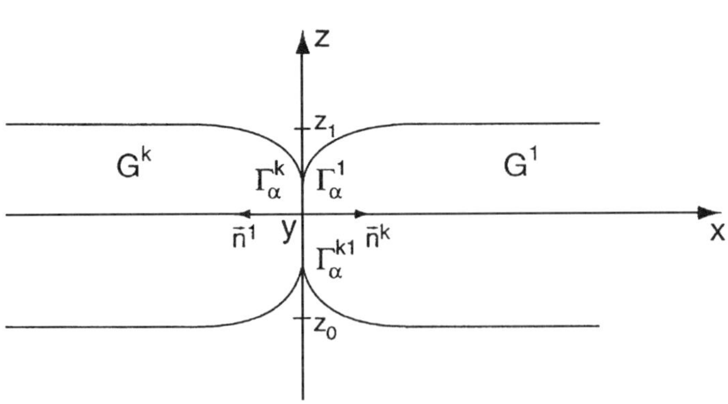

Figure 4.4. Cross-Section through the Enlarging Contact Zone in the Lithosphere

where $u_x^k(x(t), y(t), z(t), t)$, $u_x^l(x(t), y(t), z(t), t)$ are the normal ($x-$) components of the thermo-elasto-visco-plastic displacement vector $\boldsymbol{u}$ in the (x, y, z) co-ordinate system and

$$f_l(y(t), z(t), t) - f_k(y(t), z(t), t) = h(y(t), z(t), t) \text{ for every } y \text{ at time } t \in I$$

is the distance of the opposite points of the contact boundaries Γ_c^k and Γ_c^l of the enlarging contact zone at time $t \in I$.

The contact forces follow from the law of action and reaction

$$\tau_n^k = -\tau_n^l \equiv \tau_n^{kl}, \quad \vec{\tau}_t^k = -\vec{\tau}_t^l \equiv \vec{\tau}_t^{kl} \text{ at time } t \in I.$$

Then for normal components, which are not tensile forces, we have

$$\tau_x^k(x(t), y(t), z(t), t)(\cos\beta_k(t))^{-1} = -\tau_x^l(x(t), y(t), z(t), t)(\cos\beta_l(t))^{-1} \leq 0 \quad (4.127)$$

for all $z(t) \in \langle z_o, z_1\rangle$ and $t \in I$, and

$$\cos\beta_k(t) = \{1 + [\mathrm{d}f_k(y(t), z(t), t)/\mathrm{d}z(t)]^2\}^{1/2},$$
$$\cos\beta_l(t) = \{1 + [\mathrm{d}f_l(y(t), z(t), t)/\mathrm{d}z(t)]^2\}^{1/2},$$

for every $t \in I$, where $\beta_k(t)$, $\beta_l(t)$ are the angles between the z-axis and the tangent to Γ_c^k and Γ_c^l at time $t \in I$.

If there is no contact, then $u_x^k - u_x^l - h < 0$ at $t \in I$ and the contact forces are zero, i.e. $\tau_x^k(x(t), y(t), z(t), t) = -\tau_x^l(x(t), y(t), z(t), t) = 0$. If the contact exists then $u_x^k - u_x^l - h = 0$ and $\tau_x^k(x(t), y(t), z(t), t) = -\tau_x^l(x(t), y(t), z(t), t) \leq 0$. Such situations are described by the following condition

$$(u_x^k(x(t), y(t), z(t), t) - u_x^l(x(t), y(t), z(t), t) - h(y(t), z(t), t))\tau_x^k(x(t), y(t), z(t), t) = 0 \quad (4.128)$$

for all $z(t) \in \langle z_o, z_1\rangle$, and for all $y(t)$ at time t from the geological period $I = (t_o, t_1)$. If there is a contact then the friction force $g_c^{kl} = \mathscr{F}_c^{kl}\left|\tau_n^{kl}\right|$, which is assumed to be proportional to the normal component of the stress vector τ_n and the coefficient of Coulombian friction $\mathscr{F}_c^{kl}$, acts at the contact zone, where the normal component of stress is defined by $\tau_n^{kl} = \tau_x^k(x(t), y(t), z(t), t)$. If the tangential stress component $\tau_t^{kl} = \tau_z^k(x(t), y(t), z(t), t)$ is less than the friction force g_c^{kl}, then the tangential force precludes the mutual motion of the contact points. If the absolute value of the tangential stress is equal to the friction force g_c^{kl}, the contact points change their position. Then we have

$$\left|\vec{\tau}_t^{kl}\right| \equiv \left|\tau_z^k(x(t), y(t), z(t), t)\right| = \left|-\tau_z^k(x(t), y(t), z(t), t)\right|$$
$$\leq g_c^{kl}(t) \equiv \mathscr{F}_c^{kl}\left|\tau_x^k(x(t), y(t), z(t), t)\right|. \quad (4.129)$$

If

$$u_x^k(x(t), y(t), z(t), t) - u_x^l(x(t), y(t), z(t), t) - h(y(t), z(t), t) = 0, \quad (4.130)$$

then

$$\text{if } \left|\vec{\tau}_t^{kl}\right| \equiv \left|\tau_z^k(x(t), y(t), z(t), t)\right| < g_c^{kl}(t) \equiv \mathscr{F}_c^{kl} \left|\tau_x^k(x(t), y(t), z(t), t)\right|$$

then

$$\partial/\partial t(u_x^k(x(t), y(t), z(t), t) - u_x^l(x(t), y(t), z(t), t)) = 0, \tag{4.131}$$

i.e. the friction forces are sufficient to preclude the mutual motion of both lithospheric plates.

$$\text{If } \left|\vec{\tau}_t^{kl}\right| \equiv \left|\tau_z^k(x(t), y(t), z(t), t)\right| = g_c^{kl}(t) \equiv \mathscr{F}_c^{kl} \left|\tau_x^k(x(t), y(t), z(t), t)\right|$$

then there exists $\vartheta \geq 0$ such that

$$\partial/\partial t(u_x^k(x(t), y(t), z(t), t) - u_x^l(x(t), y(t), z(t), t)) = -\vartheta\tau_z^k(x(t), y(t), z(t), t) \tag{4.132}$$

for all $z(t) \in \langle z_o, z_1 \rangle$ and all y and all t from the geological period $I = (t_o, t_1)$, i.e. the frictional forces are not sufficient to preclude the mutual motion of both lithospheric plates and then the plates slip one along the other, where $\mathscr{F}_c^{kl}(x, y, z, t) \geq \mathscr{F}_{co}^{kl} > 0$ for all $(x, y, z) \in \overline{\Gamma}_c^k \cap \overline{\Gamma}_c^l \equiv \Gamma_c^{kl}$ and all $t \in I$.

At the same time, due to the deformation of both plates and to the friction, the contact zone and its surroundings are heated. But at the contact boundary $\overline{\Gamma}_c^k \cap \overline{\Gamma}_c^l \equiv \Gamma_c^{kl}$, the temperature and heat flux are continuous. Thus

$$T^k = T^l, \tag{4.133}$$

$$\kappa_{ij}T_{,j}n_i\big|_{(k)} = \kappa_{ij}T_{,j}n_i\big|_{(l)}, \tag{4.134}$$

where κ_{ij} is the coefficient of thermal conductivity and T is the temperature.

Due to horizontal forces acting on the bottom of the lithosphere, the lower part of the lithosphere is pressed more strongly. The near-surface part of the fault thus begins to open up, and the extent of the contact zone in the deep fault increases even more due to the deformation of lithospheric plates, with the bottom part of the contact zone between the lithospheric plates being deformed most. The contact surface will become "smoothed" due to relatively small mutual displacements. While the opened upper part of the contact zone is filled with sediments, the opened lower part of the contact zone near the asthenosphere is filled with asthenospheric materials. Further geodynamic processes in the upper parts of the Earth cause the lithospheric plates to move. The collision zone begins to diminish and, as a result of the horizontal forces acting at the lower part of the lithospheric plates, the asthenospheric material from the opened lower part of the colliding plates is pressed out. The upper opened part of the contact zone at the surface of the lithosphere contines to open up to the detriment of one of the plates. When one of the plates begins to descend below the other, the second one begins to override the downgoing plate.

STAGE 3: The Transient Stage of Evolution

The third stage is a transient stage of evolution of collision zones. One plate begins to descend below the other, while the second one overrides the downgoing plate.

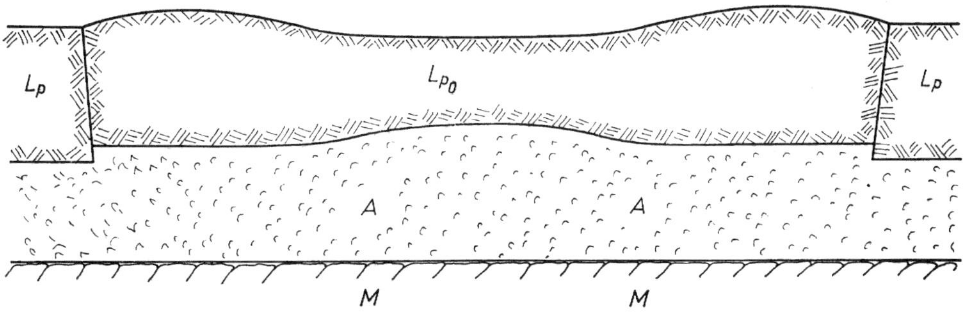

Figure 4.5. Cross-Section through the Deformed Lithospheric Plate Strained by Two Lithospheric Plates of the Same Thickness (e.g. the African plate or the Pannonian Basin)

Further evolution of the collision zone depends on the speeds of both plates as well as on their thickness. If the plates are of the same thickness, the forces acting on both plates and the actual deformation of the lower part of the contact zone of the deep fault determine which of the plates will descend below and which of them will override the other. The contact zone is strained and broken up and the thinner plate begins to subduct under the thick one. Convergent motion of lithospheric plates thus occurs and we speak of subduction and obduction of lithospheric plates.

If the speed of both colliding plates is low and both plates are of approximately the same thickness, the enlarged contact is compressed, both plates will be deformed and their thickness will increase at their head, folded sediments will be thrust over them and the centre of the deformed lithospheric plate will bulge out at the bottom and grow thinner (e.g. the African plate or the Pannonian Basin, see Figure 4.5 for the cross-section). The space, where the lithosphere has bulged out is immediatelly filled with melted asthenospheric materials. If the bulge is larger, the centre of the plate or geological block will display an anomalous heat flow (as for example in the Pannonian Basin and the African plate), the latter being closely related to tectonic processes. The boundary conditions at the contact boundaries will be analogous to the next stage, where they will be derived and discussed in detail.

STAGE 4: The Subducted Lithosphere as the Fourth Stage of Evolution

The fourth stage of the evolution of collision zones (if it exists, i.e. if conditions for the origin of a subduction process materialize) is characterized by subduction processes during which the thinner plates descend below the others. In the opposite case, i.e. in the case when both plates are of approximately the same thickness, the colliding plates are further compressed and strained. The mechanism of this case was discussed in the previous stage and we will not discuss it here.

During this fourth period, the heads of both colliding plates are strained and broken up into several plates and blocks, which rotate with respect to one another, and the heads of the lithospheric continental blocks and the sedimentary fillings of the contact zone and of the basins between continental blocks or plates are folded. Bigger and smaller plates and blocks of the broken-up lithosphere of continental, quasi-oceanic or purely oceanic types collide. However, for subduction processes to occur, the invasion plates must move with a sufficiently large convergent speed, of about 4–8 cm yr^{-1} or more. The analyses of the convergent speeds of lithospheric plates (e.g. from the results of the JOIDES project, geodynamic projects, etc.) show the following:

(i) If the convergent speed of the collisional lithospheric plates is 4–8 cm yr^{-1} and if at least one of them is of the oceanic type, deep-sea trenches are generated in places where the oceanic lithospheric plate descends below the continental or oceanic plate. At the head of the plate which is colliding with a continental or oceanic lithospheric plate, sedimentary fillings of basins are folded and even whole mountain ranges of the Cordillerian type (the Andes, Cordilleras) or of the collision type (the Himalayas, Carpathians) are generated. We then speak of the Cordillerian and collision models of the collision zones. According to the characterized parts of the colliding lithospheric plates and blocks we speak about continent/continent or oceanic/oceanic or oceanic/continent collision models.

More precisely by the Cordillerian type we understand a model based on the convergent contact of two plates, where one of them is of an oceanic type. The head of the invasion lithospheric plate is crushed and broken up, and the sedimentary fillings of the periphery of the plates are folded. Along the faults at the head of the plate, magmas of acid volcanism, originated from melted light sediments, ascend. On the other hand, along the faults far away from the head of the colliding plates, originated owing to deep subduction, magmas of basaltic and andesitic origin ascend. Owing to convergent motions, the inner transport of lithospheric masses and the motions along the system of faults in the obducting plate, together with thermal and phase change of rocks, whole mountain ranges are elevated, e.g. the Andes, Cordilleras, etc..

The essence of the collision model is also the convergent motion of both colliding plates. It originates by folding of quasioceanic or oceanic type plates with sediments between two blocks of the continental lithosphere. At the subduction, the basins of the quasi-oceanic or oceanic lithosphere are firstly compressed and then the sediments are folded. Due to friction, rocks are heated and then melted and also recrystallized; the rocks are metamorphosed and the oceanic crust ascends along the fault zones into the folding layers (ophiolites zones). The inner transport of lithospheric masses into regions of elevated mountain ranges also occurs. The Himalayas and Carpathians are examples of this.

But there also exists the collision-subduction type of model of collison zones. The essence of this model is also the convergent motion of both colliding plates, which are of the continental and oceanic types. The model is characterized by properties of both the previous models. Such a situation exists in the case of the East Himalayas, where the collision model of the Central Himalayas changes into the subduction of the oceanic plate, here of the lithosphere of the Bengal basin descending under the Burmese plate of the continental type.

(ii) For convergent speeds of plates less than 4 cm yr^{-1} subduction processes are not observed. This type of collision is characterized by the deformation of plates with greater thickness at the head of the plates near the collision zones and by their

building out and weakening in the central parts of the blocks or plates as well as by observed anomalous heat flow. This collision model represents for example the collision models of the African plate or the Pannonian basin. Some details of this model were discussed in the transient stage of evolution.

(iii) For convergent speeds much larger than 8 cm yr^{-1} subduction processes are observed but the downgoing plate descends below the overriding plate without greater folding of materials at the head of the obducting plates.

The retardation of the shifting movements of the subducted plates are observed in the course of subduction processes. Some authors explain this effect by the uneven surface of the subducted lithosphere. Nedoma (1984) presented the idea that besides the uneven surface of the subducted lithospheric plates, the main role is played by the Coulombian type of friction acting at the contact boundaries.

The contact zone in the domain of the subducted lithosphere is also characterized by a zone of deep earthquake foci, the Wadati-Benioff zone which, according to several authors, is divided by aseismic faults into two or more seismically active zones. The subducted (descended) lithosphere reaches different depths of 100 km to 700 km, i.e. it reaches down to the mesosphere. As observed, the subducted lithosphere is broken up. Furthermore, in the mesosphere the emerged and broken-up parts of the subducted oceanic lithosphere are observed in the close neighbourhood of the subduction zones. If the convective currents reach the lower boundary of the lithosphere, then these parts of the broken-up and emerged subducted oceanic lithosphere must be pulled to greater distances from the subduction zones. Moreover, such parts of the emerged oceanic lithosphere would be further strained and also broken up. Earthquakes, located in these places far from the subduction zones, would originate. But such phenomena are not observed and thus the assumption of convective currents under the lithosphere is not likely and, therefore, the idea that the mesosphere (about which we assume that it is of the visco-plastic state similar to the state of asphalt) is without streams seems to be more probable. But the anomalous zone would originate only in the surrounding of the subducted lithosphere (see Figure 4.6).

To study these processes we shall find conditions describing the behaviour of the colliding plates and blocks in the collision zones. In the following we shall assume that the colliding plates are sufficiently large and geometrically situated as in the case of enlarging contact zones, i.e. in the second period of evolution. Suppose that before the deformation both colliding plates, denoted by Ω^k and Ω^l, were in contact along the contact boundary $\Gamma_c^{kl} = \overline{\Gamma}_c^k \cap \overline{\Gamma}_c^l$.

Assume that $\boldsymbol{n} = (n_1, n_2, n_3), \boldsymbol{t} = (t_1, t_2, t_3)$ are the outward normal and tangential unit vectors, then the traction vector $\tau_i = \tau_{ij} n_j$ at the boundary can be resolved into its normal and tangential components

$$\tau_n = \tau_i n_i = \tau_{ij} n_j n_i, \quad \vec{\tau}_t = \vec{\tau} - \vec{\tau}_n \boldsymbol{n} = \tau_i t_i = \tau_{ij} n_j t_i$$

and the displacement vector can be resolved into its normal and tangential components

$$u_n = u_i n_i, \quad \boldsymbol{u}_t = u_i t_i \equiv \boldsymbol{u} - u_n \boldsymbol{n}.$$

Further, let $\boldsymbol{u}^k, \boldsymbol{u}^l, T^k, T^l$ (the indices k, l denote neighbouring plates) be the displacements and temperatures in the neighbouring plates. Let the (x, y, z)-co-ordinate system

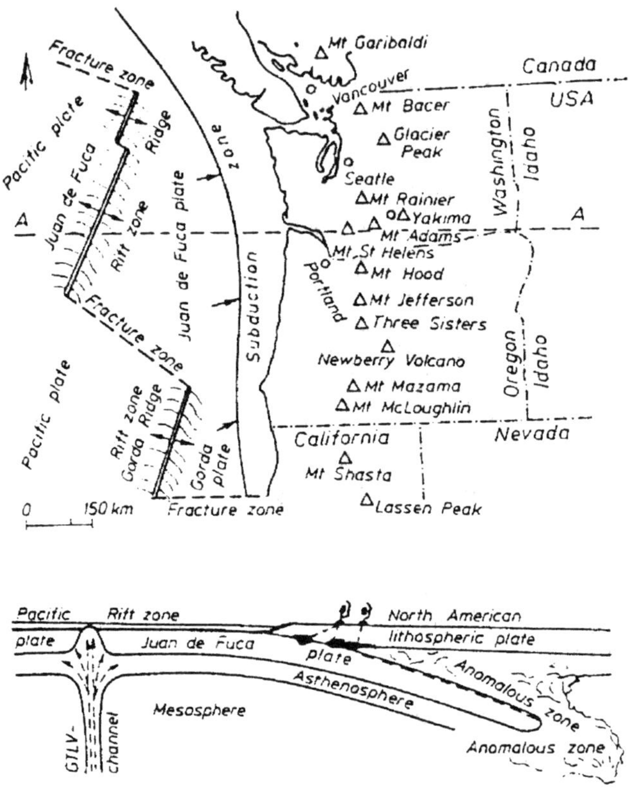

(a)

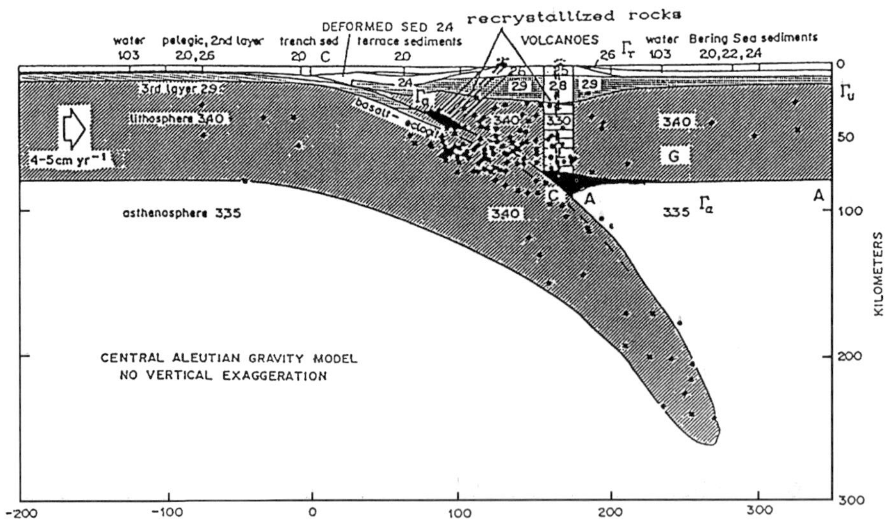

(b)

Figure 4.6. Subducted Lithosphere — Schematic Cross-Section: (a) model of the Juan de Fuca plate, (b) model of the Central Aleutian arc system

at the point $O \in \Gamma_c^{kl}$ ($O^k \equiv O^l \equiv O$, $O^k \in \Gamma_c^k$, $O^l \in \Gamma_c^l$, $\Gamma_c^{kl} = \overline{\Gamma}_c^k \cap \overline{\Gamma}_c^l$) and time $t \in I \equiv (t_o, t_1)$ be placed in such a way that the x-axis coincides with the direction of the unit normal $\boldsymbol{n}_k$, the z-axis with the common unit tangent of Γ_c^{kl} and the y-axis is perpendicular to the (x, z)-plane (see Figure 4.7). Let Γ_c^{kl} be described by the function $h(y, z, t)$. During the deformation, points $O^k \in \Gamma_c^k$ and $O^l \in \Gamma_c^l$ are transformed in a different way. Since the lithospheric materials are assumed to thermo-elasto-visco-plastic, both colliding plates cannot penetrate. This condition means that

$$-u_1^l(0, y, 0, t) \geq u_1^k(\underline{x}, y, \underline{z}, t) - h(y, \underline{z}, t) \text{ at time } t \text{ for every } y, \qquad (4.135)$$

where point $(\underline{x}, y, \underline{z}, t)$ is unknown, but fulfiles the conditions $u_3^l(\underline{x}, y, \underline{z}, t) = u_3^l(0, y, 0, t) - \underline{z}$ at time t and for every y. To simplify the condition of Equation 4.135 we shall assume that, in a small neighbourhood of point $O \in \Gamma_c^{kl}$, the curvature is small, so that $h(y, \underline{z}, t) \doteq 0$. Applying Taylor's expansion to $u_1^l(\underline{x}, y, \underline{z}, t)$ in the neighbourhood of point $O \in \Gamma_c^{kl}$, and limiting ourselves to linear terms only, the mutual shifts $|u_3^k - u_3^l|$ as well as the derivatives $|\partial u_1^k / \partial z|$ are so small that they can be neglected. Hence from Equation 4.135 we obtain

$$u_n^k(0, y, 0, t) - u_n^l(0, y, 0, t) \leq 0 \text{ for every } y \text{ at time } t \in I,$$

which is equivalent to the condition in the original co-ordinate system (x_1, x_2, x_3)

$$u_n^k(\boldsymbol{x}, t) - u_n^l(\boldsymbol{x}, t) \leq 0 \quad \text{at time } t \in I. \qquad (4.136)$$

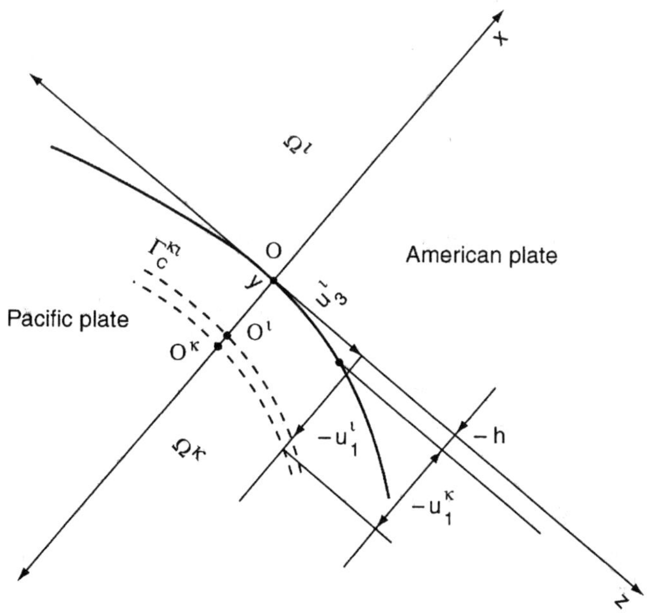

Figure 4.7. Derivation of the Contact Conditions on the Common Boundary of Colliding Lithospheric Plates

For the contact forces, due to the law of action and reaction, we have

$$\tau_n^k(\boldsymbol{x},t) = -\tau_n^l(\boldsymbol{x},t) \equiv \tau_n^{kl}(\boldsymbol{x},t), \quad \tau_t^k(\boldsymbol{x},t) = -\tau_t^l(\boldsymbol{x},t) \equiv \tau_t^{kl}(\boldsymbol{x},t). \tag{4.137}$$

Since the normal components are not tensile forces then

$$\tau_n^k(\boldsymbol{x},t) = -\tau_n^l(\boldsymbol{x},t) \equiv \tau_n^{kl}(\boldsymbol{x},t) \le 0 \text{ at time } t \in I. \tag{4.138}$$

During the deformation, the colliding plates are in or not in contact. If the colliding plates are not in contact then $u_n^k(\boldsymbol{x},t) - u_n^l(\boldsymbol{x},t) < 0$ at time $t \in I$ and the contact forces are equal to zero, i.e. $\tau_n^k(\boldsymbol{x},t) = -\tau_n^l(\boldsymbol{x},t) \equiv \tau_n^{kl}(\boldsymbol{x},t) = 0$ at every time $t \in I$. If the plates are in contact, then $u_n^k(\boldsymbol{x},t) - u_n^l(\boldsymbol{x},t) = 0$ at time $t \in I$ and the contact forces $\tau_n^k(\boldsymbol{x},t) = -\tau_n^l(\boldsymbol{x},t) \equiv \tau_n^{kl}(\boldsymbol{x},t) < 0$ at every time $t \in I$. Both cases are included in the following condition

$$(u_n^k(\boldsymbol{x},t) - u_n^l(\boldsymbol{x},t))\tau_n^{kl}(\boldsymbol{x},t) = 0 \text{ at every time } t \in I. \tag{4.139}$$

If both plates are in contact, Coulombian friction is generated at the contact boundary Γ_c^{kl}. The friction forces g_c^{kl} acting on the contact boundary Γ_c^{kl} are, in their absolute value, proportional to the normal stress component, where the coefficient of proportionality is the coefficient of friction $\mathscr{F}_c^{kl}(\boldsymbol{x},t)$, i.e.

$$g_c^{kl}(\boldsymbol{x},t) = \mathscr{F}_c^{kl}\left|\tau_n^{kl}(\boldsymbol{x},t)\right|.$$

Moreover, $\mathscr{F}_c^{kl}(\boldsymbol{x},t)$ can be estimated by

$$\min_{\Omega}(c_s(\boldsymbol{x},t)/c_p(\boldsymbol{x},t))^{1/2} \tag{4.140}$$

or

$$\min_{\Omega}(\mu(\boldsymbol{x},t)/(\lambda(\boldsymbol{x},t) + 2\mu(\boldsymbol{x},t))^{1/4}, \tag{4.141}$$

and/or

$$\min_{\Omega}(\mu(\boldsymbol{x},t)/(2\lambda(\boldsymbol{x},t) + 3\mu(\boldsymbol{x},t))^{1/2}, \tag{4.142}$$

with t-fixed, where c_s, c_p, λ, μ are velocities of S and P waves and Lamé's coefficients (see Nedoma (1986a-c), (1990a,b), Hlaváček, Haslinger, Nečas and Lovíšek (1982), Jarušek (1983)). Due to the forces acting on both plates and to the friction forces affecting the contact boundary Γ_c^{kl}, the following two situations can ocuur. Firstly, the absolute value of the tangential stress component $\tau_t^{kl}(\boldsymbol{x},t)$ is less than the friction forces $g_c^{kl}(\boldsymbol{x},t)$, so that the tangential force precludes the mutual motion of both colliding plates. Secondly, the absolute value of the tangential stress component $\tau_t^{kl}(\boldsymbol{x},t)$ is equal to the friction force $g_c^{kl}(\boldsymbol{x},t)$, so that there are no forces which can preclude the mutual (bilateral) motion of both plates. Thus the contact points change their position in the direction opposite to that in which the tangential stress component acts. Thus we have the following conditions describing the above discussed situation:

$$\begin{aligned} &\text{If } u_n^k(\boldsymbol{x},t) - u_n^l(\boldsymbol{x},t) = 0 \text{ then} \\ &\left|\tau_t^{kl}(\boldsymbol{x},t)\right| \le g_c^{kl}(\boldsymbol{x},t) = \mathscr{F}_c^{kl}\left|\tau_n^{kl}(\boldsymbol{x},t)\right| \text{ at time } t \in I; \\ &\text{if } \left|\tau_t^{kl}(\boldsymbol{x},t)\right| < g_c^{kl}(\boldsymbol{x},t) = \mathscr{F}_c^{kl}\left|\tau_n^{kl}(\boldsymbol{x},t)\right| \text{ then} \\ &\partial/\partial t(u_n^k(\boldsymbol{x},t) - u_n^l(\boldsymbol{x},t)) = 0 \text{ at time } t \in I, \end{aligned} \tag{4.143}$$

which means that the friction forces are sufficient to preclude the mutual shifting of both lithospheric plates, if $\left|\tau_t^{kl}(\boldsymbol{x},t)\right| = g_c^{kl}(\boldsymbol{x},t) = \mathscr{F}_c^{kl}\left|\tau_n^{kl}(\boldsymbol{x},t)\right|$ then there exists a function $\vartheta \ge 0$ such that

$$\partial/\partial t(u_n^k(\boldsymbol{x},t) - u_n^l(\boldsymbol{x},t)) = -\vartheta\tau_c^{kl}(\boldsymbol{x},t) \text{ at time } t \in I,$$

which means that the friction forces are sufficient to preclude the mutual shifting of both plates. This shift as the opposite direction to the acting tangential forces.

The temperature and the heat flux are continuous at the contact boundary of both colliding plates Γ_c^{kl}, so that

$$T^k = T^l, \tag{4.144}$$

$$\kappa_{ij}T_{,j}n_i\big|_{(k)} = \kappa_{ij}T_{,j}n_i\big|_{(l)}. \tag{4.145}$$

Moreover, at the same time due to the deformation of both plates and to the function of Coulombian friction acting on the contact boundary Γ_c^{kl}, which is assumed to be uneven, the materials in the contact zones are heated and melted, as well as recrystallized or later solidified after losing tectonic motion. Across any phase change boundary the conservation law of energy momentum and mass must be fulfilled and the material density is required to be continuous. Generally the continuum has two (solid-liquid) or more (solid-liquid-gases) phases in which some material properties are markedly different owing to non-uniform heating and straining. Most of the interest in the investigation of the problem lies in the properties of the boundary surface between the phases (see Chapter 3). If T represents temperature, then latent heat is added in the case of melting, or taken off in the case of recrystallization or later solidification at the phase change boundary. Energy balance arguments then show that although the temperature is continuous everywhere, the heat flux has a jump discontinuity across the phase change boundary surface, which we denote $\Gamma_R(t)$. But the real situation is much more complicated as the phase changes are accompanied by transition layers characterized by mushy zones where both phases co-exist (see Nedoma (1986a-c),(1997)). The investigation of mushy zones is important for detailed studies but not in our global studies of collision zones.

Let $\Gamma_R(t)$ denote the phase change boundary surface (the head of the recrystallization wave), let $G_i(t)$, $i = 1, \dots, n$ denote the unrecrystallized (unmelted, unsolidified) and recrystallized (melted, solidified) parts of the lithosphere and upper mantle at time $t \in (t_o, t_1)$, and let Θ_i denote the temperature in $G_i(t)$. Further, let

$$\kappa_k = \lim_{\Theta \to T_R^-} \kappa(\Theta, p, \dots), \tag{4.146}$$

$$\kappa_l = \lim_{\Theta \to T_R^+} \kappa(\Theta, p, \ldots). \tag{4.147}$$

Generally, at the phase change boundary surface $R(\boldsymbol{x}, t) = 0$ and we have

$$\Theta_k(\boldsymbol{x}, t) = \Theta_l(\boldsymbol{x}, t) = T_R, \quad q_k\boldsymbol{n} - q_l\boldsymbol{n} = \rho L v_n, \tag{4.148}$$

where $v_n = \mathrm{d}R/\mathrm{d}t$ is the speed of the phase change boundary surface $\Gamma_R(t)$ along the normal $\boldsymbol{n}$, $R(t)$ is the position vector of a point on $\Gamma_R(t)$, $L(T, p, \ldots)$ is the latent heat, and $q_i = -\kappa_i \operatorname{grad} \Theta$ is the heat flux in $G_i(t)$. Thus Equation 4.148 leads to

$$\Theta_k(\boldsymbol{x}, t) = \Theta_l(\boldsymbol{x}, t) = T_R, \quad \kappa_k \nabla\Theta_k \nabla R - \kappa_l \nabla\Theta_k \nabla R = \rho L v_n \tag{4.149}$$

and, specially for the case of melting, to

$$\Theta_k(\boldsymbol{x}, t) = \Theta_l(\boldsymbol{x}, t) = T_R, \quad \kappa_k \nabla\Theta_k \nabla R = -Q_M + \rho L V_n, \tag{4.150}$$

where Q_M is the heat flux input from the liquid to the solid at the phase boundary surface $\Gamma_R(t)$.

At the part of the contact boundaries Γ_c^{kl} in the lithosphere at which materials are not melted, the temperature and heat flux are continuous and given by Equation 4.149. For the case of two incompressible phases of different densities ρ_k and ρ_l, where k refers to solid and l to melted rocks (liquid up to strongly visco-plastic), the liquid phase must have a velocity v_l satisfying the mass continuity condition

$$\rho_k v_n \equiv \rho_k \mathrm{d}R/\mathrm{d}t = \rho_l(v_k - v_l) \equiv \rho_l(\mathrm{d}R/\mathrm{d}t - v_l), \tag{4.151}$$

which is important in the domains containing the anomalous zones.

In the next stage, in which the geometry of the contact boundary of both colliding plates or blocks is characterized by their bulging out, the domains, originated in this way, are filled by the melted materials. Thus the situation is described at every time $t \in I$ by the conditions

$$u_n^k(\boldsymbol{x}, t) - u_n^l(\boldsymbol{x}, t) \le 0, \tag{4.152}$$

$$\tau_n^k(\boldsymbol{x}, t) = -\tau_n^l(\boldsymbol{x}, t) \equiv \tau_n^{kl}(\boldsymbol{x}, t) \le 0, \tag{4.153}$$

$$(u_n^k(\boldsymbol{x}, t) - u_n^l(\boldsymbol{x}, t))\tau_n^{kl}(\boldsymbol{x}, t) = 0, \tag{4.154}$$

$$\tau_t^k(\boldsymbol{x}, t) = -\tau_t^l(\boldsymbol{x}, t) \equiv \tau_t^{kl}(\boldsymbol{x}, t) \tag{4.155}$$

and

$$\left|\tau_t^{kl}(\boldsymbol{x}, t)\right| \le g_c^{kl}(\boldsymbol{x}, t) = \mathscr{F}_c^{kl}(\boldsymbol{x}, t)\left|\tau_n^{kl}(\boldsymbol{x}, t)\right| \text{ at time } t \in I. \tag{4.156}$$

If $u_n^k(\boldsymbol{x}, t) - u_n^l(\boldsymbol{x}, t) = 0$ then $\tau_n^k(\boldsymbol{x}, t) = -\tau_n^l(\boldsymbol{x}, t) \equiv \tau_n^{kl}(\boldsymbol{x}, t) \le 0$ and moreover

$$\text{if } \left|\tau_t^{kl}(\boldsymbol{x}, t)\right| < g_c^{kl}(\boldsymbol{x}, t) = \mathscr{F}_c^{kl}\left|\tau_n^{kl}(\boldsymbol{x}, t)\right| \text{ then}$$

$$\partial/\partial t(u_n^k(\boldsymbol{x}, t) - u_n^l(\boldsymbol{x}, t)) = 0 \text{ and } T^k = T^l, \tag{4.157}$$

and

$$\kappa_{ij} T_{,j} n_i + \rho T_o \beta_{ij} e_{ij,t}(\boldsymbol{u})\big|_{(k)} = \kappa_{ij} T_{,j} n_i + \rho T_o \beta_{ij} e_{ij,t}(\boldsymbol{u})\big|_{(l)}.$$

if $\left|\tau_t^{kl}(\boldsymbol{x}, t)\right| = g_c^{kl}(\boldsymbol{x}, t) = \mathscr{F}_c^{kl}\left|\tau_n^{kl}(\boldsymbol{x}, t)\right|$ then there exists a function $\vartheta \geq 0$ such that

$$\partial/\partial t(u_n^k(\boldsymbol{x}, t) - u_n^l(\boldsymbol{x}, t)) = -\vartheta\tau_c^{kl}(\boldsymbol{x}, t)$$

and

$$\kappa_{ij}T,_j n_i + \rho T_o \beta_{ij} e_{ij,t}(\boldsymbol{u})\big|_{(k)} = q_1,$$

$$\kappa_{ij}T,_j n_i + \rho T_o \beta_{ij} e_{ij,t}(\boldsymbol{u})\big|_{(l)} = q_2$$

at time $t \in I$, where q_p, $p = 1, 2$, represent a certain heat flow in such a bulging out domain.

On the surface of the lithospheric plates and blocks the conditions

$$\tau_{ij}n_j = P_i, \tag{4.158}$$

$$\kappa_{ij}T_{,j}n_i + \rho T_o \beta_{ij} e_{ij,t}(\boldsymbol{u}) = 0 \tag{4.159}$$

are fulfilled.

During the next period, which is short from the geological point of view, these domains are filled by the melted materials from their surroundings.

Moreover, the lithosphere is heated from the asthenosphere-ward side and cooled from overhead. Since the tectonically active periods depend on the geodynamic processes in the Earth's interior, above all at the mantle/core boundary, and display different intensities in the course of geological epochs, the geodynamic processes also display different intensities in the lithosphere. Therefore, the thermal processes change from time to time during geological epochs. At the same time, in the course of melting, gases are released and again dissolved in rocks. The appropriate discontinuities and governing diffusion equations can also be derived from the integral formulation of the law of conservation of energy. Across any phase change boundary surface, the conservation of energy, momentum and mass must be continuous and thus the required material density must be continuous. Hence the concentration $\mathrm{d}(\boldsymbol{x}, t)$ and flux of mass $\boldsymbol{i}(\boldsymbol{x}, t)$ must also be continuous, so that

$$\mathrm{d}^k = \mathrm{d}^l, \tag{4.160}$$

$$\boldsymbol{i}^k = \boldsymbol{i}^l \tag{4.161}$$

or

$$\mathrm{d}^k = \mathrm{d}^l, \tag{4.162}$$

$$\rho_k D_\mathrm{d}^k(\nabla \mathrm{d}^k + K_T^k(T)\nabla T^k + P^k(p)\nabla p^k) = \rho_l D_\mathrm{d}^l(\nabla \mathrm{d}^l + K_T^l(T)\nabla T^l + P^l(p)\nabla p^l). \tag{4.163}$$

In our next investigations the simplified model problems will be studied.

4.3 MODEL PROBLEMS OF PLATE TECTONICS

4.3.1. Introduction

In this Section, we consider the geodynamic problems whose simulations will be based on elasticity, thermo-elasticity, thermo-visco-elasticity as well as on thermo-visco-plasticity. Therefore, our models will be investigated as:

(i) classical boundary value problems of elasticity, thermo-elasticity, thermo-visco-elasticity and thermo-visco-plasticity,

(ii) problems with conditions of friction at the contact boundaries between colliding lithospheric plates and blocks, which lead to new types of variational problems involving inequalities.

From the first part of this chapter we recall that thermo-visco-elasticity and thermo-visco-plasticity differ from thermo-elasticity, which will be discussed in more detail in the first part of the next section. It will be shown that the state of stress at present depends on all deformations undergone in the past in the case of thermo-visco-elasticity, or it depends on the threshold of plasticity. Therefore, thermo-visco-elastic rocks are characterized by materials endowed with memory. In the case of plastic rocks a threshold of plasticity can be of two types. Either the threshold is fixed, or it depends on the history of the deformations, and in this case we speak about rocks with strain hardening. In thermo-visco-plasticity the stress-strain relations are represented by non-linear constituent laws.

4.3.2. Elastic, Thermo-Elastic and Thermo-Visco-Elastic Stress-Strain Analysis of the Geodynamic Mechanism

4.3.2.1. Elastic and thermo-elastic case

In attempting to explain the origin of the stress field and the thermal field associated with it in the lithosphere and asthenosphere to a first approximation, we shall draw on the deliberations of elasticity and thermo-elasticity and the ideas of the new global tectonics. We shall derive the governing equations in thermo-elasticity only and shall presuppose the elastic case to be a special case of the thermo-elastic one.

In order to derive the equation of motion of a lithospheric plate moving along the asthenosphere and to expand the equation of heat conduction, the so-called conduction-convection problem describing the thermal regime of the lithospheric plate and of the asthenosphere, we shall assume, to a first approximation, that they are formed by elastically isotropic, or anisotropic, homogeneous or non-homogeneous, in general compressible, materials of an absolute temperature T_o and that they are in a strainless and stressless state. Due to the external load, caused by the motion along the asthenosphere, mutual collisions with another lithospheric plate or block (which occurs from the very beginning of the development of a lithospheric plate) and by heating of the lithosphere, the plate becomes deformed. This generates a displacement in the lithospheric plate, $\boldsymbol{u} = (u_i)(i = 1, 2, 3)$, a strain e_{ij} and stress $\tau_{ij}(i, j = 1, 2, 3)$ (in our considerations we shall limit ourselves to small strains only), and its temperature will increase by $\Theta = T - T_o$, where u_i, e_{ij}, τ_{ij} and the absolute temperature T are functions of the co-ordinates $\mathbf{x} = (x_i)(i = 1, 2, 3)$ and of time t. To simplify our considerations we shall assume that the temperature change Θ is small compared to the temperature T_o, i.e. $|\Theta/T_o| \ll 1$, and also that the increase in temperature by Θ has no effect on the material, elastic and thermal parameters of the rocks forming the lithosphere and, therefore, that we are able to consider them independent of temperature to a first approximation. We are thus restricting ourselves to the case of linear thermo-elasticity.

We shall now introduce the notations $u_{i,j} = \partial u_i/\partial x_j$, $u_{i,tt} = \partial^2 u_i/\partial t^2$, $e_{ij,t} = \partial e_{ij}/\partial t = \frac{1}{2}(u_{i,jt} + u_{j,it})$, $S_{,e_{ij}} = (\partial S/\partial e_{ij})_T$ (the latter representing the derivative of entropy with respect to the strain at constant temperature), $e_{ij,kl} = \partial^2 e_{ij}/\partial x_k \partial x_l$, and $\mathrm{d}f/\mathrm{d}t = \partial f/\partial t + \partial f/\partial x_j v_j$ (-the first term in the last expression being the local derivative and the second one being the convective derivative). Their physical meaning is evident. Also in this section the Einstein's summation rule will be used.

The small strain will be defined by the relation

$$e_{ij}(\boldsymbol{u}) = \tfrac{1}{2}(u_{i,j} + u_{j,i}) \quad i, j = 1, 2, 3. \tag{4.164}$$

This represents a system of six equations, which have to be used to determine the three required functions u_i, the component of the displacement vector $\boldsymbol{u}$. For the system in Equation 4.164 to be solvable, the following conditions of strain compatibility have to be satisfied:

$$e_{ij,kl} + e_{kl,ij} + e_{jl,ik} - e_{ik,jl} = 0 \quad i, j, k, l = 1, 2, 3. \tag{4.165}$$

To derive the relations between the components of stress and of strain we shall draw on the thermodynamics of strain of an elastic body. In these deliberations we shall start with thermodynamic identity, the so-called Gibbs equation, for a strained body. Determining the relation between the changes of internal energy U, the strain work $\tau_{ij}\,\mathrm{d}e_{ij}$ and the entropy S, leads to

$$\mathrm{d}U = \tau_{ij}\mathrm{d}e_{ij} + T\mathrm{d}S, \tag{4.166}$$

with the independent variables e_{ij} and T. Equation 4.166 is in fact the Second law of thermodynamics (for details see chapter 3). If we introduce Helmholtz's free energy $F = F(e_{ij}, T) = U - ST$, then instead of Equation 4.166 we have

$$\mathrm{d}F = \tau_{ij}\mathrm{d}e_{ij} - S\mathrm{d}T \tag{4.167}$$

with the independent variables e_{ij} and T. Equations 4.166 and 4.167 yield

$$\tau_{ij} = (U_{,e_{ij}})_S = (F_{,e_{ij}})_T, \tag{4.168}$$

which is similar to the relation $\tau_{ij} = \partial W/\partial e_{ij} = W_{,e_{ij}}$, where W is the density of the strain energy from the classical linear theory of elasticity, which allows the stress τ_{ij} to be expressed in terms of the derivative of the internal energy with respect to e_{ij} at constant entropy S, and in terms of the derivative of the free energy F with respect to e_{ij} at constant temperature T. If we develop the free elastic energy in the neighbourhood of point $(0, T_o)$ into the Taylor series, assuming the linear theory and isothermal state, we arrive at

$$F(e_{ij}, T) = \tfrac{1}{2}c_{ijkl}e_{ij}e_{kl} - \beta_{ij}e_{ij}(T - T_o) + \tfrac{1}{2}m(T - T_o)^2 \tag{4.169}$$

and according to Equation 4.168 we have the so-called Duhamel-Neumann's law

$$\tau_{ij} = c_{ijkl}e_{kl} - \beta_{ij}(T - T_o), \tag{4.170}$$

where $c_{ijkl} = F(0, T_o)_{,e_{ij}e_{kl}}$, $\beta_{ij} = -F(0, T_o)_{,e_{ij}T}$, and $m = F(0, T_o)_{,TT}$. Then $F(0, T_o)_{,e_{ij}} = 0$, $F(0, T_o) = 0$ and, as follows from Equation 4.167, $F(0, T_o)_{,T} = -S(0, T_o) = 0$, because the natural state of the lithospheric plate is determined if $\tau_{ij} = 0$ for $T = T_o$ and $e_{ij} = 0$. Equation 4.170 yields

$$c_{ijkl} = (\tau_{ij,e_{kl}})_T, \quad \beta_{ij} = -(\tau_{ij,T})_e. \tag{4.171}$$

For the coefficients of elasticity c_{ijkl}, the well-known conditions of symmetry hold. Particularly, for homogeneous isotropic materials

$$F(e_{ij}, T) = \mu e_{ij}^2 + \tfrac{1}{2}\lambda e_{kk}^2 - \beta e_{kk}(T - T_o) - \tfrac{1}{2}m(T - T_o)^2. \tag{4.172}$$

Therefore,

$$\tau_{ij} = 2\mu e_{ij} + [\lambda e_{kk} - \beta(T - T_o)]\delta_{ij}, \tag{4.173}$$

where λ and μ are Lamé's elastic constants, $\beta = (3\lambda + 2\mu)\alpha_t$, and α_t is the coefficient of linear thermal expansion. Equation 4.173 yields

$$(\tau_{ij,T})_e = -\beta\delta_{ij}.$$

From Equation 4.169 and 4.172 it follows that the righthand side is also a scalar, because the free energy is a scalar. Equation 4.172 also means that the square of the volume strain and the sum of the squares $(e_{ij})^2$ of all strain components e_{ij} are scalars. The third term is a linear scalar relative to the volume strain. The last term is also a scalar, because m is a material constant and Θ^2 is the square of the temperature increment, and β is a material constant. A similar deliberation proves the statement with regard to Equation 4.169.

The law of momentum conservation can be expressed in integral form as follows

$$\frac{\mathrm{d}}{\mathrm{d}t}\int_\Omega \rho v_i \mathrm{d}\boldsymbol{x} = \int_{\partial\Omega} \tau_{ij}n_j \mathrm{d}s + \int_\Omega \rho f_i \mathrm{d}\boldsymbol{x}, \tag{4.174}$$

where Ω is the volume of the body in the initial undeformed state, $\partial\Omega$ is its surface, n_j are the directional cosines of the normal $\boldsymbol{n}$ to surface $\partial\Omega$, and $\mathrm{d}s$ is the surface element, or in an equivalent differential form

$$\tau_{ij,j} + \rho f_i = \rho(u_{i,tt} + u_{j,t}u_{i,tj}), \quad \tau_{ij} = \tau_{ji}, \tag{4.175}$$

where the term $u_{j,t}u_{i,tj}$ expresses convection, which, in special cases, may be neglected in Equation 4.175-4.177 and 4.193-4.195.

Since the law of momentum conservation is independent of the relations between the stress and strain of the lithospheric plate, the dynamic equations (Equation 4.175) are identical to those of the classical theory of elasticity. Equations 4.175 yields the equations of motion of the moving lithospheric plate in the form

$$(c_{ijkl}e_{kl})_{,j} - [\beta_{ij}(T - T_o)]_{,j} + \rho f_i = \rho(u_{i,tt} + u_{j,t}u_{i,tj}), \tag{4.176}$$

for the case of an anisotropic lithospheric plate, and

$$\mu u_{i,jj} + (\lambda + \mu)u_{j,ji} + \rho f_i - \beta(T - T_o)_{,i} = \rho(u_{i,tt} + u_{j,t}u_{i,tj}), \tag{4.177}$$

for the case of a homogeneous isotropic lithospheric plate.

Let us now derive the equation describing the thermal regime in the moving lithospheric plate. The heat transfer there takes place by heat conduction on the one hand, which represents heat transfer from places of higher to places of lower temperature, and by convection on the other. From thermodynamics we know that processes of this kind are spontaneous and irreversible, and that they are associated with the increase in entropy. Gibbs' equation (Equation 4.166), which gives the relation between the changes of the internal energy, strain work and strain, yields

$$\begin{aligned} dU = \tau_{ij}\mathrm{d}e_{ij} + T\mathrm{d}S &= \tau_{ij}\mathrm{d}e_{ij} + T[(S_{,e_{ij}})_T\mathrm{d}e_{ij} + (S_{,T})_e\mathrm{d}T] \\ &= [T(S_{,e_{ij}})_T + \tau_{ij}]\mathrm{d}e_{ij} + T(S_{,T})_e\mathrm{d}T \\ &= (U_{,T})_e\mathrm{d}T + (U_{,e_{ij}})_T de_{ij}, \end{aligned}$$

because $\mathrm{d}S$ is a total differential, and also the increment of the internal energy, $\mathrm{d}U$, is a total differential of the variables T and $\mathrm{d}e_{ij}$. By comparison

$$(U_{,T})_e = T(S_{,T})_e,\ (U_{,e_{ij}})_T = T(S_{,e_{ij}})_T + \tau_{ij}. \tag{4.178}$$

The condition that $\mathrm{d}U$ is a total differential with respect to T, as well as e_{ij}, yields

$$[T(S_{,T})_e]_{,e_{ij}} = [T(S_{,e_{ij}})_T + \tau_{ij}]_{,T}, \tag{4.179}$$

and by differentiating we arrive at

$$0 = (S_{,e_{ij}})_T + (\tau_{ij,T})_e. \tag{4.180}$$

However, from the Duhamel-Neumann's relation (Equation 4.170), it follows that

$$(\tau_{ij,e_{kl}})_T = c_{ijkl}, \quad (\tau_{ij,T})_e = -\beta_{ij}. \tag{4.181}$$

Substituting the second relation of Equation 4.181 into Equation 4.180 yields

$$(S_{,e_{ij}})_T = \beta_{ij}. \tag{4.182}$$

Denote the specific heat under constant strain in Equation 4.178 as $(U_{,T})_e = T(S_{,T})_e = c_e$. Then

$$\mathrm{d}U = c_e\mathrm{d}T + c_{ijkl}e_{kl}\mathrm{d}e_{ij} + \beta_{ij}T_o\mathrm{d}e_{ij}.$$

Assuming that the displacement $\boldsymbol{u} = 0$ if $e_{ij} = 0$ and $T = T_o$ under natural conditions in the lithospheric plate, for the internal energy, integration will yield

$$U = \int \mathrm{d}U = \tfrac{1}{2}c_{ijkl}e_{ij}e_{kl} + T_o\beta_{ij}e_{ij} + c_e(T - T_o) \tag{4.183}$$

For the increment of entropy we have

$$\mathrm{d}S = (S_{,e_{ij}})_T\mathrm{d}e_{ij} + (S_{,T})_e\mathrm{d}T = \beta_{ij}\mathrm{d}e_{ij} + c_e\mathrm{d}T/T,$$

and, assuming $S = 0$ for $e_{ij} = 0$ and $T = T_o$, integration of the above relation yields

$$S = \int \mathrm{d}S = \beta_{ij}e_{ij} + c_e \ln[1 + (T - T_o)/T_o] \cong \beta_{ij}e_{ij} + c_e(T - T_o)/T_o, \quad (4.184)$$

because $|(T - T_o)/T_o| \ll 1$. In Equation 4.183 the expression $\frac{1}{2}c_{ijkl}e_{ij}e_{kl}$ is strain work, $c_e(T - T_o)$ is the amount of heat contained in a unit volume of the lithospheric plate, and the term $T_o\beta_{ij}e_{ij}$ is the result of the combined effects of strain and heat conduction. In Equation 4.184 the term $\beta_{ij}e_{ij}$ expresses the combining of the thermal field with the field of strain, and the term $c_e \ln[1 + (T - T_o)/T_o]$ symbolizes the increase of entropy caused by heat conduction in the lithospheric plate. The term characterizing the purely elastic effect is lacking, since the elastic process is reversible and does not result in a change of entropy. As regards the quantity m in Equation 4.169 and 4.171, we obtain $m = -\frac{1}{2}c_e/T_o$.

In deriving the equation describing the thermal regime of the lithospheric plate, we shall begin with the law of entropy balance. Thus,

$$\frac{\mathrm{d}}{\mathrm{d}t}\int_{\Omega_1} \rho S \mathrm{d}\boldsymbol{x} = \int_{\Omega_1} \rho\sigma \mathrm{d}\boldsymbol{x} + \int_{\partial\Omega_1} S_k \mathrm{d}s_k = \int_{\Omega_1} \rho\sigma \mathrm{d}\boldsymbol{x} + \int_{\Omega_1} S_{k,k} \mathrm{d}\boldsymbol{x}, \quad (4.185)$$

where ρ is density, S the entropy density relative to unit mass, σ the rate at which entropy is generated, S_k is the vector of entropy flow entering the elementary volume of the lithospheric plate through its surface, Ω_1 is the elementary volume delimited from the lithospheric plate Ω, and $\partial\Omega_1$ is its boundary. Since

$$\frac{\mathrm{d}}{\mathrm{d}t}\int_{\Omega_1} \rho S \mathrm{d}\boldsymbol{x} = \int_{\Omega_1} \rho \mathrm{d}S/\mathrm{d}t \mathrm{d}\boldsymbol{x} = \int_{\Omega_1} \rho(S_{,t} + S_{,j}v_j)\mathrm{d}\boldsymbol{x},$$

Equation 4.185 and the above relation yield

$$\rho(S_{,t} + S_{,j}v_j) = \rho\sigma + S_{j,j} = T^{-1}(W - q_{j,j}), \quad (4.186)$$

where $W = W(\tau_{ij}, \dot{e}_{ij}, \mathscr{F}, \ldots)$ is the dissipation function, $S = S(e_{ij}, T, \ldots)$ the entropy density, $\mathscr{F}$ is the friction force and $\boldsymbol{q} = (q_i)$ is the heat flow, whose components q_i will be defined below. By calculation we find that

$$\begin{aligned}\int_{\Omega_1} T^{-1}(W - q_{j,j})\mathrm{d}\boldsymbol{x} &= \int_{\Omega_1} [T^{-1}W - (T^{-1}q_j)_{,j} - q_jT^{-2}T_{,j}]\mathrm{d}\boldsymbol{x} \\ &= \int_{\Omega_1} WT^{-1}\mathrm{d}\boldsymbol{x} - \int_{\partial\Omega_1} T^{-1}q_n \mathrm{d}s - \int_{\Omega_1} q_jT^{-2}T_{,j}\mathrm{d}\boldsymbol{x}. \quad (4.187)\end{aligned}$$

Here, the first integral characterizes the distribution of the heat sources in the lithospheric plate, the second term represents the entropy flow, i.e. the exchange of entropy with the environmemt, and the third term represents the increment of entropy as a result of the irreversible process of heat conduction. The term $T^{-1}W - q_jT_{,j}T^{-2} = \overline{\sigma}$ represents sources of entropy, $\boldsymbol{q}$ is the heat flow, and κ_{ij} is the coefficient of heat conductivity of anisotropic materials of the lithospheric plate. Note $\kappa_{ij} > 0$, because the heat flow proceeds from a place with higher to a place with lower temperature, and $\boldsymbol{q}$ and $T_{,j}$ are of the opposite sense. Then

$$\int_{\Omega_1} T^{-1} q_{j,j} \mathrm{d}\boldsymbol{x} = -\int_{\Omega_1} T^{-1} (\kappa_{ij} T_{,j})_{,i} \mathrm{d}\boldsymbol{x},$$

and we thus arrive at

$$\rho(S_{,t} + v_i S_{,i}) = T^{-1}(W + (\kappa_{ij} T_{,j})_{,i}). \tag{4.188}$$

But Equation 4.184 yields

$$T_o S_{,t} = T_o \beta_{ij} e_{ij,t} + c_e T_{,t}, \tag{4.189}$$

$$T_o S_{,k} = T_o \beta_{ij} e_{ij,k} + c_e T_{,k}. \tag{4.190}$$

The righthand side of Equation 4.189 represents the amount of heat accumulated in a unit volume per unit time, and is equal to the sum of the heat, $c_e T_{,t} = c_e \Theta_{,t}$, which was transferred into the investigated volume element of the lithosphere Ω_1 by means of heat conduction, and the term $T_o \beta_{ij} e_{ij,t}$ represents the energy dissipated in the form of heat in the lithosphere and generated by its deformation. Equation 4.188 then yields the expanded equation of heat conduction:

$$\rho \beta_{ij} T_o (e_{ij,t} + v_k e_{ij,k}) + \rho c_e (T_{,t} + v_k T_{,k}) = W + (\kappa_{ij} T_{,j})_{,i} \tag{4.191}$$

or, in the special case,

$$\rho \beta_{ij} T_o e_{ij,t} + \rho c_e (T_{,t} + v_k T_{,k}) = W + (\kappa_{ij} T_{,j})_{,i}, \tag{4.192}$$

where $W = W_1 + W_2 + \cdots$ is the dissipation function which represents heat sources, e.g. from radioactive sources, friction, viscous stresses $\overline{\sigma}_{ij}$, i.e. $W_3 = \overline{\sigma}_{ij} v_{,j}$, where $\overline{\sigma}_{ij} = \hat{\mu}(v_{i,j} + v_{j,i})$ is the tensor of viscous stresses, v_i is the velocity component, and $\hat{\mu}$ the coefficient of viscosity. The term $\rho \beta_{ij} T_o e_{ij,t}$ relates the temperature increment to the rate of relative change of volume of the deformed lithosphere.

Discussion of the Equations and the Limits of their Validity in Geodynamics

In order to obtain the complete set of equations, as well as the dynamic equations and the expanded equation of heat conduction for a moving lithospheric plate, we must also give the defining relations, i.e. the relations between the stress and strain tensor, the tensor of the strain rate, temperature, etc., and accurately define the heat sources in the lithosphere.

The fundamental dynamic equations for a moving lithospheric plate can be expressed as

$$(c_{ijkl} e_{kl})_{,j} - [\beta_{ij}(T - T_o)]_{,j} + \rho f_i = \rho(u_{i,tt} + u_{i,tj} u_{j,t}) \tag{4.193}$$

if the lithospheric plate is anisotropic,

and
$$\mu u_{i,jj} + (\lambda + \mu) u_{j,ji} + \rho f_i - \beta(T - T_o)_{,i} = \rho(u_{i,tt} + u_{i,tj} u_{j,t}) \tag{4.194}$$

if the lithospheric plate is isotropic. The expanded equation of heat conduction for the lithospheric plate can be written in the following form:

$$\rho\beta_{ij}T_o(e_{ij,t} + v_k e_{ij,k}) + \rho c_e(T_{,t} + v_k T_{,k}) = W + (\kappa_{ij}T_{,j})_{,i}, \tag{4.195}$$

if we assume the lithosphere to be formed by an anisotropic medium. If the medium is homogeneous and isotropic, $\kappa_{ij} = \kappa\delta_{ij}(\kappa = \text{constant})$. In Equation 4.194 the quantities μ and λ are assumed for an isothermal state, i.e. $T = \text{constant}$, under the assumption that we are investigating the processes over relatively short periods of time on the geological scale.

The causes of the strain and the associated thermal field are:

(i) material forces and sources distributed within the lithosphere,

(ii) external forces such as the forces responsible for the motion of the lithospheric plate and the thermal effect of the medium surrounding the plate under investigation, in this case mainly the thermal effect of the asthenosphere.

The strain and thermal fields are generated by each of the effects mentioned, which are considered independently here. If an irreversible process of heat conduction is occurring in the lithospheric plate as a result of a temperature gradient, this automatic process is connected with an elastic deformation of the plate. The process of deformation is a secondary phenomenon here, forced by and taking place to the detriment of the heat conduction energy. The process of heat conduction is associated with an increase in entropy. The derivative of these entropies is positive at each point of the lithospheric plate and is responsible for the generation of uncompensated heat, of the so-called dissipation energy.

On the other hand, the lithospheric plate is subject to the process of deformation due to its external loading, e.g. by sliding in the sense of the new global tectonics, as well as by the effect of material forces (e.g. tidal forces, etc.). This deformation process results in a forced secondary process of heat conduction. Both processes are closely associated and connect processes which take place in the lithosphere under the collision of lithospheric plates or blocks.

For the sake of simplifying some of our subsequent considerations we assume that the lithosphere is formed by isotropic materials with a constant heat conductivity κ. Let us put $\lambda_o = \kappa/c_e$, $Q = W/c_e$, and $\eta = \beta T_o/\kappa$. The thermal regime in the lithospheric plate is then described by the equation

$$\frac{\rho}{\lambda_o}(T_{,t} + v_j T_{,j}) + \rho\eta e_{ij,t} = \frac{Q}{\lambda_o} + T_{,ji}. \tag{4.196}$$

The Quasi-Stationary Case

If the variability in time of the forces in the lithosphere and asthenosphere, and of the body and surface forces is slow, and if we assume that the motion of the lithospheric plate is uniform, e.g. over the last 10 or 12 million years, the problem can be studied as a quasi-stationary one, and the forces of inertia can be omitted from the equations of motion. If, moreover, the motion, in addition to being uniform, is slow, we can neglect the term $\rho c_e(T_{,t} + v_k T_{,k})$ in Equation 4.195 and the term $\rho\lambda_o^{-1}T_{,t}$ in Equation 4.196 describing the thermal regime of the plate. Equations 4.193-4.196 then yield

$$(c_{ijkl}e_{kl})_{,j} - [\beta_{ij}(T - T_o)]_{,j} + \rho f_i = 0 \tag{4.197}$$

$$\rho\beta_{ij}T_o e_{ij,t} = W + (\kappa_{ij}T_{,j})_{,i}, \tag{4.198}$$

if the lithospheric plate is anisotropic, and

$$\mu u_{i,jj} + (\lambda + \mu)u_{j,ji} + \rho f_i - \beta(T - T_o)_{,i} = 0 \tag{4.199}$$

$$\rho\beta\delta_{ij}T_o e_{ij,t} = W + (\mathscr{H}\delta_{ij}T_{,j})_{,i}, \tag{4.200}$$

if the lithospheric plate is homogeneous and isotropic.

The Case of a Stationary Lithospheric Plate

Assume that the thermal field and the field of strain are independent of time. This corresponds to the steady state in a stationary lithospheric plate. The field of stresses and strains is then described by the classical equation of elasticity:

$$(c_{ijkl}e_{kl})_{,j} - (\beta_{ij}T)_{,j} + \rho f_i = 0, \tag{4.201}$$

if the lithospheric plate is anisotropic,

and $$\mu u_{i,jj} + (\lambda + \mu)u_{j,ji} + \rho f_i - \beta T_{,i} = 0, \tag{4.202}$$

if the lithospheric plate is isotropic.

Let us discuss the thermal field in this special case. In an irreversible process the time change of the entropy is composed of the exchange of entropy with the environment and of the generation of entropy (entropy source $\overline{\sigma} > 0$). However, in a steady state the change of entropy with time is equal to zero, $S_{,t} = 0$. Therefore, Equation 4.185 yields

$$\rho S_{,t} = T^{-1}(W - q_{i,i}) = T^{-1}W - (T^{-1}q_i)_{,i} - T^{-2}q_i T_{,i}.$$

In an unsteady state the constant generation of entropy is compensated for by the exchange of entropy with the surrounding medium. This entropy exchange is negative and equal to the absolute increment of entropy in the lithospheric plate. Thus,

$$\rho S_{,t} = T^{-1}(W - q_{i,i}) = T^{-1}[W - (\kappa_{ij}T_{,j})_{,i}].$$

Since $S_{,t} = 0$, it follows that the temperature satisfies Poisson's equation:

$$(\kappa_{ij}T_{,j})_{,i} = W \tag{4.203}$$

in the anistropic case

and $$(\kappa\delta_{ij}T_{,j})_{,i} = W \tag{4.204}$$

in the isotropic case.

Equations 4.201 and 4.202 and Equations 4.203 and 4.204 are independent of each other. We therefore determine the temperature T from Equations 4.203 and 4.204 and then substitute into Equation 4.201 and 4.202 for the displacement. We speak about the quasi-coupled thermo-elastic problem.

The Case of Slow Heat Exchange between the Lithospheric Plates and the Asthenosphere

Assume that the heat exchange between two lithospheric plates which are in collision with the asthenosphere is very slow. The estimate of the distance of thermal penetration is several metres to tens of metres per 1000 years. In this case we may consider both the lithospheric plates and the asthenosphere to be nearly thermally insulated and the whole process to be adiabatic. The Lamé elastic parameters λ and μ, and the coefficients of elasticity for anisotropic materials, c_{ijkl}, should then be considered adiabatic. From the point of view of the geological time scale this case is most probable; therefore we shall investigate it.

The material constant in the relations between the stresses and strains should in this case be taken to be adiabatic. The appropriate equations for the thermo-elastic displacements and stresses can be derived from the theory of thermo-elasticity provided the process is adiabatic. For the entropy in this process $S_{,t} = 0$ and, therefore, Equation 4.189 yields the relation

$$T_o\beta_{ij}e_{ij,t} + c_eT_{,t} = T_o\beta_{ij}e_{ij,t} + c_e(T - T_o)_{,t} = 0.$$

By integrating the latter in time, assuming that $e_{ij} = 0$ and $T - T_o = 0$ for $t = 0$, we arrive at

$$T - T_o = -T_oc_e^{-1}\beta_{ij}e_{ij}.$$

Then from the latter relation and the equation of motion (Equation 4.176), or the analogous relation for isotropic materials (Equations 4.177), for the adiabatic anisotropic material constants we obtain the relations

$$(c_{ijkl})_S = (c_{ijkl})_T + (\beta_{ij})_Tc_e^{-1}T_o,$$

and for the adiabatic isotropic material constants, λ_S and μ_S, we have the relations

$$(\mu)_S = (\mu)_T, \quad (\lambda)_S + 2(\mu)_S = (\lambda)_T + 2(\mu)_T + \lambda_o\eta(\beta)_T,$$

where κ, η and $(\beta)_T$ have been defined above.

Limits of Validity of the Coefficient $\rho\beta_{ij}T_o$ in the Equation for the Thermal Regime of the Lithospheric Plate

For the sake of simplicity we shall consider an isotropic lithospheric plate. The Duhamel-Neumann equation

$$\tau_{ij} = 2\mu e_{ij} + [\lambda e_{kk} - \beta(T - T_o)]\delta_{ij}.$$

applies to isotropic materials. Put

$$e_{ij} = (e_{ij} - \tfrac{1}{3}e_{kk}\delta_{ij}) + \tfrac{1}{3}e_{kk}\delta_{ij} = e^D_{ij} + \tfrac{1}{3}\vartheta\delta_{ij}, \quad e^D_{ij} = e_{ij} - \tfrac{1}{3}\vartheta\delta_{ij}, \quad \vartheta = e_{kk},$$

where the second-order tensor e^D_{ij}, the so-called strain deviator, describes the deformation in shape, and the invariant ϑ characterizes the volume deformation, i.e. the volume dilatation. Denote the modulus of compressibility by $k = \lambda + 2/3\mu$. Then

$$\tau_{ij} = 2\mu e^D_{ij} + [k\vartheta - \beta(T - T_o)]\delta_{ij}. \tag{4.205}$$

By reducing the stress tensor we obtain

$$\tau_{kk} = 3[k\vartheta - \beta(T - T_o)],\ e^D_{kk} = 0. \tag{4.206}$$

We shall now attempt to determine the relations characterizing the free expansion of the lithospheric plate under which no internal stress is generated, i.e. $\tau_{ij} = 0$. Equation 4.206 then yields

$$\vartheta^o = k^{-1}\beta(T - T_o) = \alpha(T - T_o),$$

where ϑ^o represents the change of the volume dilatation and $\alpha = k^{-1}\beta$ the coefficient of volume expansion. The second relation of Equation 4.206 plus Equation 4.205 yield

$$e^o_{ij} = (3k)^{-1}\beta(T - T_o)\delta_{ij} = \alpha_t(T - T_o)\delta_{ij}.$$

This equation shows that no changes in shape take place under a change of temperature in the lithospheric plate or block, but only a change in volume. Here, $\alpha_t = (3k)^{-1}\beta$ is the coefficient of linear expansion. Note also $\beta = (3\lambda + 2\mu)\alpha_t$.

If we put $\eta = \kappa^{-1}\beta T_o$ for the case of an isotropic plate, then $\eta = (3\lambda + 2\mu)\alpha_t\kappa^{-1}T_o = 3\alpha_t k\kappa^{-1}T_o$.

Two alternatives may occur:

(i) the case in which $\eta \neq 0$, because all quantities differ from zero,

(ii) the case in which $\eta = 0$.

Since we are assuming a real medium, the modulus of compressibility $k \neq 0$, the initial temperature $T_o \neq 0$, and the coefficient of thermal conductivity $\kappa \neq 0$. Therefore, only $\alpha_t = 0$. But this means that the change of the volume dilatation ϑ^o due to the temperature change is zero, since $\vartheta^o = e^o_{ii} = \alpha_t(T - T_o)$. This would mean that the coefficient of thermal conductivity would be equal to zero. We can thus see that we must consider the case in which $\eta \neq 0$. An analogous deliberation applies to the case of an anisotropic medium.

4.3.2.2. *The Thermo-Visco-Elastic Case*

In this section we shall briefly discuss several types of constituent laws for thermo-visco-elastic and thermo-elasto-visco-plastic materials. The visco-elastic rocks are rocks characterized by a memory of short or long type, in the sense that the state of stress at the investigated time depends on all the deformations which have occured in the investigated rock in previous time. Similarly as in linear elasticity, the rocks will be linear if the constituent laws establish linear relations between the stress tensor and the strain tensor. And the rocks will be non-linear, if the constituent laws establish non-linear relations between the stress tensor and the strain tensor.

An important class of visco-elastic rocks is represented by rocks which are characterized by the linear visco-elastic constituent laws in which there exist relationships between the stress tensor and strain rate tensor. We speak about the materials of strain rate type. For such type of rocks the constituent law is as follows:

$$\tau_{ij}(t) + C^{(1)}_{ijkl}\partial\tau_{kl}(t)/\partial t + \cdots + C^{(m)}_{ijkl}\partial^m_{\tau_{kl}}(t)/\partial t^m$$
$$= c^{(0)}_{ijkl}e_{kl}(t) + c^{(1)}_{ijkl}\partial e_{kl}(t)/\partial t + \cdots + c^{n}_{ijkl}\partial^n e_{kl}(t)/\partial t^n - \beta_{ij}(T - T_o), \quad (4.207)$$

where the derivatives with respect to t are to be understood in the sense of distributions $\tau_{ij}(t)$, $e_{ij}(t) = e_{ij}(\boldsymbol{u}(t))$ are the stress and strain tensors and are equal to zero for $t < 0$; $C^{(r)}_{ijkl}$ and $c^{(s)}_{ijkl}$ are the material constants, β_{ij} is the coefficient of thermal expansion; and T, T_o are the absolute temperature and the initial temperature, respectively.

For the case when $n \geq m$ it follows that

$$\tau_{ij}(t) = c^{(0)}_{ijkl}e_{kl}(t) + c^{(1)}_{ijkl}\partial e_{kl}(t)/\partial t + \cdots + c^{(p)}_{ijkl}\partial^p e_{kl}(t)/\partial t^p$$
$$+ \int_o^t b_{ijkl}(t-s)e_{kl}(s)\mathrm{d}s - \beta_{ij}(T - T_o), \quad (4.208)$$

where $p = n - m$ and $c^{(k)}_{ijkl}$, $k = 0, \ldots, p$ are material constants, and b_{ijkl} is a regular function for $t > 0$ and continuous for $t \geq 0$.

For the case when $m = 0, n = 1$ we have

$$\tau_{ij}(t) = c^{(0)}_{ijkl}e_{kl}(t) + c^{(1)}_{ijkl}\partial e_{kl}(t)/\partial t - \beta_{ij}(T - T_o), \quad (4.209)$$

which represents the constituent law for the rocks with a short memory. In this type of material the state of stress at time t depends on the deformation at the instant time t and on the deformation at the immediately preceding time moment. The coefficients $c^{(0)}_{ijkl}$ and $c^{(1)}_{ijkl}$ represent the coefficients of elasticity and of viscosity, as they involve the strain tensor or the strain rate tensor, respectively. The coefficients $c^{(0)}_{ijkl}$, $c^{(1)}_{ijkl}$ and β_{ij} satisfy the conditions of symmetry

$$c^{(0)}_{ijkl} = c^{(0)}_{jilk} = c^{(0)}_{ijlk} = c^{(0)}_{klij}, \quad (4.210)$$

$$c^{(1)}_{ijkl} = c^{(1)}_{jilk} = c^{(1)}_{ijlk} = c^{(1)}_{klij}, \quad (4.211)$$

$$\beta_{ij} = \beta_{ji}. \quad (4.212)$$

Moreover, they satisfy the conditions

$$c^{(0)}_{ijkl}e_{ij}(\boldsymbol{u}(t))e_{kl}(\boldsymbol{u}(t)) \geq c_o e_{ij}(\boldsymbol{u}(t))e_{ij}(\boldsymbol{u}(t)), \quad (4.213)$$

$$c^{(1)}_{ijkl}e_{ij}(\dot{\boldsymbol{u}}(t))e_{kl}(\dot{\boldsymbol{u}}(t)) \geq c_1 e_{ij}(\dot{\boldsymbol{u}}(t))e_{ij}(\dot{\boldsymbol{u}}(t)), \quad (4.214)$$

and have the meaning of inner (strain) energy and dissipative energy, respectively.

For the case when $m = n = 1$ we find

$$\tau_{ij}(t) = c^{(0)}_{ijkl} e_{kl}(t) + \int_o^t b_{ijkl}(t-s) e_{kl}(s) \mathrm{d}s - \beta_{ij}(T - T_o), \tag{4.215}$$

and we speak about the visco-elastic rocks with a long memory, because the state of stresses at the instant time t depends on the deformation at the instant time t and on all deformations at the time period $\langle 0, t \rangle$. The coefficients $c^{(0)}_{ijkl}$, the so-called coefficients of instantaneous elasticity, depend on $\boldsymbol{x}$, are bounded in $\boldsymbol{x}$ and satisfy the conditions of symmetry (Equations 4.210-4.212); and the coefficients $b_{ijkl} = b_{ijkl}(\boldsymbol{x}, t)$, which depend on t and $\boldsymbol{x}$, are bounded in $\boldsymbol{x}, t$ and satisfy the symmetry condition

$$b_{ijkl}(\boldsymbol{x}, t) = b_{jikl}(\boldsymbol{x}, t) \tag{4.216}$$

and the regularity assumptions

$$b_{ijkl}(\boldsymbol{x}, t), \partial b_{ijkl}(\boldsymbol{x}, t)/\partial t, \partial^2 b_{ijkl}(\boldsymbol{x}, t)/\partial t^2 \in L^\infty(\Omega \times I), \quad I = (0, T). \tag{4.217}$$

If $\int_o^\infty b_{ijkl}(t)dt = b^\infty_{ijkl}$ is finite we speak about visco-elastic rocks of solid type, and the coefficients $c^\infty_{ijkl} = c_{ijkl} + b^\infty_{ijkl}$ are the so-called coefficients of retarded elasticity which satisfy the conditions of symmetry similar to Equation 4.210-4.212 and the conditions similar to Equation 4.213 and 4.214. The coefficients β_{ij} satisfy the conditions of symmetry in Equation 4.212 as above.

For the case when $n < m$ it follows that

$$\tau_{ij}(t) = \int_o^t b_{ijkl}(t-s) e_{kl}(s) \mathrm{d}s - \beta_{ij}(T - T_o), \tag{4.218}$$

where b_{ijkl} has the same properties as b_{ijkl} in Equation 4.208. In generall this law cannot be used as it has not agreed with experiences.

4.3.3. Thermo-Elasto-Visco-Plastic Stress-Strain Analysis of the Geodynamic Mechanism

The lithospheric rocks can be interpreted (simulated) by a very extensive class of elasto-visco-plastic materials under the influence of a geothermal field, which can be described by the deformation theory of thermo-plasticity, by the perfectly plastic media, up to flow theory of thermo-plasticity. In this section we shall limit ourselves to several plastic types of media and the corresponding constituent laws only. These types of plastic rocks are characterized by the threshold, which can be fixed, or it can depend on the history of the deformations. In this last case we speak about rocks undergoing work hardening. Plastic rocks with the threshold being fixed are for example the rigid visco-plastic, elastic perfectly plastic, or rigid perfectly plastic materials, and in general the elasto-visco-plastic materials.

4.3.3.1. Elasto-plastic rocks: deformation theory of thermo-plasticity

The materials of the lithosphere and the upper mantle are characterized by properties of plasticity. In the first approximation the lithospheric materials can be modelled by using the constituent law of the deformation theory of thermo-plasticity.

In the deformation theory of thermo-plasticity the Lamé coefficients λ and μ depend on the first two invariants ϑ_1 and ϑ_2 of the small strain tensor $e_{ij}(\boldsymbol{u})$, and on the assumption that the relation between the first invariants of stress and strain is linear. Let $k(\boldsymbol{x}) = \lambda(\boldsymbol{x}) + 2/3\mu(\boldsymbol{x})$ be the bulk modulus and let

$$\begin{aligned}\gamma^2 &= \tfrac{1}{3}[(e_{11}-e_{22})^2 + (e_{22}-e_{33})^2 + (e_{11}-e_{33})^2 + 6(e_{12}^2 + e_{23}^2 + e_{13}^2)] \\ &= -\tfrac{1}{3}\vartheta_1^2 + e_{ij}e_{ij} = 2/9(\vartheta_1^2 - 3\vartheta_2),\end{aligned}$$

where $\vartheta_1 = e_{11} + e_{22} + e_{33} \equiv e_{kk}$, and $\vartheta_2 = e_{11}e_{22} + e_{11}e_{33} + e_{22}e_{33} - (e_{12}^2 + e_{13}^2 + e_{23}^2)$. Then the constituent law is of the form

$$\begin{aligned}\tau_{ij}(\boldsymbol{x}) = [k(\boldsymbol{x}) - 2/3\mu(\boldsymbol{x},\gamma^2)]\vartheta_1(\boldsymbol{u})\delta_{ij} + 2\mu(\boldsymbol{x},\gamma^2)e_{ij}(\boldsymbol{u}) \\ - \beta_{ij}(T - T_o), \quad i, j = 1, 2, 3.\end{aligned} \tag{4.219}$$

Let us introduce the strain and stress deviators by

$$e_{ij}^D = e_{ij} - \tfrac{1}{3}e_{kk}\delta_{ij},\ \tau_{ij}^D = \tau_{ij} - \tfrac{1}{3}\tau_{kk}\delta_{ij} = 2\mu e_{ij}^D,$$

then

$$\gamma = \{e_{ij}^D e_{ij}^D\}^{1/2}, \tag{4.220}$$

$$S = 2\mu(\boldsymbol{x},\gamma)\gamma. \tag{4.221}$$

Equation 4.221 is the so-called Hencky relation, where

$$0 < \mu_o \le \mu(\boldsymbol{x},\gamma) \le \mu_1 < \infty. \tag{4.222}$$

Let us suppose that $\mathrm{d}S/\mathrm{d}\gamma > 0$, hence $\mathrm{d}S/\mathrm{d}\gamma = 2(\mu + \mu'\gamma) > 0$.

Let us suppose that

$$\mu + \mu'\gamma \ge M > 0. \tag{4.223}$$

Similarly as in Section 4.3.2.1, we can find

$$\begin{aligned}F(e_{ij}, T) &\equiv \tfrac{1}{2}k(\boldsymbol{x})\vartheta_1^2 + 2\int_0^{\gamma}\mu(t)t\,\mathrm{d}t - \beta_{ij}(\boldsymbol{x})e_{ij}(\boldsymbol{u})(T - T_o) + \tfrac{1}{2}m(T - T_o)^2 \\ &= \tfrac{1}{2}k(\boldsymbol{x})\vartheta_1^2 + \int_0^{\gamma}S(t)\,\mathrm{d}t - \beta_{ij}(\boldsymbol{x})e_{ij}(\boldsymbol{u})(T - T_o) + \tfrac{1}{2}m(T - T_o)^2.\end{aligned} \tag{4.224}$$

Hence, similarly as above (compare with Equation 4.168, Section 4.3.2.1)

$$\tau_{ij} = \left(\frac{\partial F}{\partial e_{ij}}\right)_T, \tag{4.225}$$

which represents a general non-linear Hooke's (Duhamel-Neumann's) law in the small displacement theory and, together with Equation 4.224, it represents a special case of a non-linear Hooke's law for thermo-elasto-plastic rocks.

4.3.3.2. Perfectly plastic and rigid perfectly plastic rocks under geothermal field

Such types of rocks are characterized by the fixed threshold. Let us denote by $\hat{g}$ the threshold of plasticity or the yield limit. To derive the constituent law for example in the 1Dcase as a function $e = e(\tau)$ or $\tau = \tau(e)$, we shall start from the state (e, τ), $\tau \leq \hat{g}$ and we increase e by a small amount de. Then τ changes by $d\tau$ so that

$$\mathrm{d}e = s\mathrm{d}\tau + \lambda, \tag{4.226}$$

where

$$\lambda = 0 \text{ when } \tau < \hat{g} \text{ or when } \tau = \hat{g}, \text{ and } \mathrm{d}\tau < 0,$$

$$\lambda \geq 0 \text{ when } \tau = \hat{g} \text{ and } \mathrm{d}\tau = 0.$$

But in the real lithospheric as well as the upper mantle rocks the plastic deformation takes place in time. Thus the increase de_{ij} takes place during time dt. Then Equation 4.226 yields

$$e_{ij,t} = s_{ijkl}\tau_{kl,t} + \lambda_{ij} - (\gamma_{ij}(T - T_o))_{,t}, \tag{4.227}$$

where

$$\lambda_{ij} = 0 \text{ when } \tau_{ij} < \hat{g} \text{ or when } \tau_{ij} = \hat{g} \text{ and } \tau_{ij,t} < 0,$$

$$\lambda_{ij} \geq 0 \text{ when } \tau_{ij} = \hat{g} \text{ and } \tau_{ij,t} = 0, \tag{4.228}$$

and where τ_{ij}, e_{ij} are the stress and (small) strain tensors satisfying the condition of symmetry $\tau_{ij} = \tau_{ji}, e_{ij} = e_{ji}$; γ_{ij} are the inverse to β_{ij}; and the coefficients s_{ijkl}, the so-called inverse elasticity coefficients, have the usual properties of symmetry and of Lipschitz continuity, i.e.

$$s_{ijkl} = s_{jikl} = s_{ijlk} = s_{klij}, \tag{4.229}$$

$$s_{ijkl}\tau_{ij}\tau_{kl} \geq s_o\tau_{ij}\tau_{ij}, \quad s_o = \text{ constant } > 0. \tag{4.230}$$

The quantity $s_{ijkl}\tau_{kl,t}$ represents the elastic part of the strain rate tensor $e_{ij,t}(u)$, and λ_{ij} is the velocity of plastic deformation.

The equivalent closed form of Equation 4.227 and 4.228 is represented by

$$e_{ij,t} = s_{ijkl}\tau_{kl} + \lambda_{ij} - (\gamma_{ij}(T - T_o))_{,t}, \tag{4.231}$$

$$\lambda_{ij}(\sigma_{ij} - \tau_{ij}) \leq 0 \quad \forall\tau_{ij} \leq \hat{g}, \tag{4.232}$$

$$\lambda_{ij}\tau_{ij,t} = 0. \tag{4.233}$$

For the case when the coefficients s_{ijkl} tend to zero, then Equations 4.231-4.233 lead to

$$e_{ij,t} = \lambda_{ij} - (\gamma_{ij}(T - T_o))_{,t}, \tag{4.234}$$

$$\lambda_{ij}(\sigma_{ij} - \tau_{ij}) \leq 0 \quad \forall \tau_{ij} \leq \hat{g}, \tag{4.235}$$

$$\lambda_{ij}\tau_{ij,t} = 0, \tag{4.236}$$

representing the constituent law for rigid perfectly plastic rocks. It is evident that the rocks undergo no deformation if $\tau_{ij} < \hat{g}$, and they deform if $\tau_{ij} = \hat{g}$.

The more generalized formulation, which will be used in our investigations, is based on the definition of the domain of elasticity and plasticity. Let $\mathscr{F}$ be a convex and continuous function with respect to τ_{ij}. Then the domain of $\boldsymbol{R}^6$, where $\mathscr{F}(\tau_{ij}) \leq a_o$, is a closed convex set. Then the elastic zone is defined by $\mathscr{F}(\tau_{ij}) < a_o$ and the plastic zone by $\mathscr{F}(\tau_{ij}) = a_o$. Actually, $\mathscr{F}(\tau_{ij})$ involves the deviator τ_{ij}^D of the stress tensor only.

In practice the function $\mathscr{F}(\tau_{ij})$ is frequently of the form

$$\mathscr{F}(\tau_{ij}) = \tfrac{1}{2}\tau_{ij}^D\tau_{ij}^D - k^2, \quad k = \text{ constant }, \tag{4.237}$$

known as the von Mises' model of plasticity,

and

$$\mathscr{F}(\tau_{ij}) = \sup_{I,J,I\neq J} |\tau_I - \tau_J| - \hat{g}, \tag{4.238}$$

known as the Tresca's model of plasticity, where τ_I are the eigenvalues of stress tensor τ_{ij} and $\hat{g} = \text{ constant } > 0$.

We see that in the von Mises' and Tresca's models of plasticity, the function $\mathscr{F}(\tau_{ij})$ does not depend on the deformation and, therefore, both models describe perfectly thermo-plastic rocks.

Moreover, Equation 4.231-4.233 can be written as

$$e_{ij,t} = s_{ijkl}\tau_{kl,t} + \lambda_{ij} - (\gamma_{ij}(T - T_o))_{,t}, \tag{4.239}$$

$$\lambda_{ij}(\sigma_{ij} - \tau_{ij}) \leq 0 \quad \forall \tau \text{ with } \mathscr{F}(\tau_{ij}) \leq a_o, \tag{4.240}$$

$$\lambda_{ij}\tau_{ij,t} = 0. \tag{4.241}$$

Since we assume that the functions $t \rightarrow \tau_{ij}(t)$ are differentiable in t, then Equation 4.241 follows from Equation 4.240. Really, from Equation 4.240 for $\mathrm{d}t > 0$ we find $\sigma_{ij} = \tau_{ij}(t \pm \mathrm{d}t)$. Hence, after dividing by $\mathrm{d}t$ in the limit for $\mathrm{d}t \rightarrow 0$ we have $\lambda_{ij}\tau_{ij,t}(t) \leq 0$ for $t + dt$, or $\lambda_{ij}\tau_{ij,t}(t) \geq 0$ for $t - \mathrm{d}t$. Hence Equation 4.241 follows immediately.

4.3.3.3. Elasto-visco-plastic rocks under geothermal field

The above formulations of thermo-plastic models in the Prandtl-Reuss sense show that the speed of plastic deformation at constant stress can reach large absolute values. Such models represent models of the lithosphere and the upper mantle in idealized cases, i.e. consisting of perfectly plastic or thermo-plastic rocks. But in the real lithosphere and the upper mantle, viscosity effect and creep are observed. Therefore, we must find a better approximation of rocks of the lithosphere and the upper mantle. For this reason, we shall give a more generalized model, characterized by a law of elasto-visco-plastic behaviour, in which the rocks will be elastic below the threshold of plasticity and visco-plastic beyond the threshold of plasticity.

Let us introduce the closed convex set

$$K = \{\tau \mid \tau = (\tau_{ij}) \in R^6, \quad \tau_{ij} = \tau_{ji}, \quad \mathscr{F}(\tau_{ij}) \leq a_o\}, \tag{4.242}$$

where $\mathscr{F}$ is a convex continuous function of six components of tensor functions, everyone of which is a scalar function. Let the mapping $\tau \to P_K(\tau)$ be an orthogonal projection in the usual Eucleidean structure of $\tau \in R^6$ on the closed convex set K. Then the constituent law can be written as follows:

$$e_{ij,t}(\boldsymbol{u}) = s_{ijkl}\tau_{kl,t} + \lambda_{ij} - (\gamma_{ij}(T - T_o))_{,t}, \tag{4.243}$$

$$\lambda_{ij} = 0 \text{ for the case when } \mathscr{F}(\tau_{ij}) < a_o, \text{ i.e. for the elastic case,} \tag{4.244}$$

$$\lambda_{ij} = (2\hat{\mu})^{-1}[\tau_{ij} - (P_K\tau)_{ij}] \text{ for the case when } \mathscr{F}(\tau_{ij}) \geq a_o, \text{ i.e. for the visco-plastic case,} \tag{4.245}$$

where $\hat{\mu}$ is a positive scalar, known as a coefficient of viscosity. If $\tau_{ij,t} = 0$, i.e. if τ_{ij} is constant, then

$$\tau_{ij} = (P_K\tau)_{ij} + 2\hat{\mu}e_{ij,t}(\boldsymbol{u}) - \gamma_{ij}(T - T_o). \tag{4.246}$$

If $K = \{\tau \mid \tau = (\tau_{ij}), \tau$ is a spherical tensor$\}$ then $(P_K\tau)_{ij} = -p\delta_{ij}$, and then

$$\tau_{ij} = -p\delta_{ij} + 2\hat{\mu}e_{ij,t}(\boldsymbol{u}) - \gamma_{ij}(T - T_o), \tag{4.247}$$

which represents the Stokes-Navier constituent law for fluids.

Let $\tau_{ij}^D = \tau_{ij} - \frac{1}{3}\tau_{kk}\delta_{ij} = 2\hat{\mu}e_{ij}^D$ be the stress deviator, where e_{ij}^D is the strain deviator, $e_{ij}^D = e_{ij} - \frac{1}{3}e_{kk}\delta_{ij}$. Let the function $\mathscr{F}$ be defined by the von Mises' condition (Equation 4.237), i.e. $\mathscr{F}(\tau_{ij}) = \frac{1}{2}\tau_{ij}^D\tau_{ij}^D - k^2, k =$ constant. Then the constituent law will be of the form

$$e_{ij,t}(\boldsymbol{u}) = s_{ijkl}\tau_{kl,t} + \lambda_{ij} - (\gamma_{ij}(T - T_o))_{,t}, \tag{4.248}$$

$$\lambda_{ij} = 0 \text{ for the case when } \mathscr{F}(\tau_{ij}) < a_o, \text{ i.e. for the elastic case,} \tag{4.249}$$

$$\lambda_{ij} = (2\hat{\mu})^{-1}[\tau_{\mathrm{II}}^{1/2} - k]\tau_{\mathrm{II}}^{-1/2}\tau_{ij}^D \text{ for the case when } \mathscr{F}(\tau_{ij}) \geq a_o, \text{ i.e. for the visco-plastic case,} \tag{4.250}$$

where $\tau_{\mathrm{II}} = \frac{1}{2}\tau_{ij}^D\tau_{ij}^D$.

Let us introduce the mapping $\mathscr{J}_\mu(\tau) : R^6 \to R$ by

$$\mathscr{J}_\mu(\tau) = (4\hat{\mu})^{-1}[\tau_{ij} - (P_K\tau)_{ij}][\tau_{ij} - (P_K\tau)_{ij}]. \tag{4.251}$$

Then the constituent law can be rewritten in the following form:

$$e_{ij,t}(\boldsymbol{u}) = s_{ijkl}\tau_{kl,t} + \lambda_{ij} - (\gamma_{ij}(T - T_o))_{,t}, \tag{4.252}$$

$$\mathscr{J}_\mu(\sigma) - \mathscr{J}(\tau) \geq \lambda_{ij}(\sigma_{ij} - \tau_{ij}), \quad \forall\sigma \in R^6, \tag{4.253}$$

or in the equivalent formulation

$$e_{ij,t}(\boldsymbol{u}) = s_{ijkl}\tau_{kl,t} + (\partial \mathcal{J}_\mu(\tau))_{ij} - (\gamma_{ij}(T - T_o))_{,t}, \tag{4.254}$$

where $\lambda_{ij} \in \partial \mathcal{J}_\mu(\tau)$, and $\partial \mathcal{J}_\mu(\tau)$ is a subgradient of $\mathcal{J}_\mu(\tau)$ in τ.

Let the viscosity coefficient tend to zero. Then the function $\mathcal{J}_\mu(\tau)$ tends toward the indicator function of the convex set $K \in R^6$, i.e. it is equal to zero on K and equal to $+\infty$ outside K. Then

$$e_{ij,t}(\boldsymbol{u}) = s_{ijkl}\tau_{kl,t} + \lambda_{ij} - (\gamma_{ij}(T - T_o))_{,t}, \tag{4.256}$$

$$\lambda_{ij}(\sigma_{ij} - \tau_{ij}) \le 0, \quad \forall \sigma \in K, \quad \tau \in K. \tag{4.257}$$

The last constituent law (Equations 4.256 and 4.256) represents the Prandtl-Reuss law of plasticity discussed above.

For the case when all s_{ijkl} in Equation 4.243-4.245 are equal to zero, then we have the constituent law for rigid-visco-plastic rock models. If all s_{ijkl} in Equations 4.256 and 4.257 are equal to zero, then we have the constituent law for rigid-perfectly plastic rock models.

For the case when $K = R^6$, then Equations 4.243-4.245 lead to

$$e_{ij,t}(\boldsymbol{u}) = s_{ijkl}\tau_{kl,t} - (\gamma_{ij}(T - T_o))_{,t}. \tag{4.258}$$

At an initial time t_o, let the lithospheric and upper mantle rocks be in a stress-strain free state. Then integrating Equation 4.258 from t_o to t we obtain the constituent law for elastic rocks

$$e_{ij}(\boldsymbol{u}) = s_{ijkl}\tau_{kl} - \gamma_{ij}(T - T_o). \tag{4.259}$$

Remark: For the strain rate tensor we introduced the notation $D_{ij}(\boldsymbol{u}) \equiv e_{ij,t}(\boldsymbol{u}) \equiv e_{ij}(\boldsymbol{v})$, where $\boldsymbol{v} = \boldsymbol{u}_{,t}$. Then the constituent law Equation 4.243-4.245 can be rewritten as

$$D_{ij}(\boldsymbol{u}) = s_{ijkl}\tau_{kl,t} + \lambda_{ij} - (\gamma_{ij}(T - T_o))_{,t}, \tag{4.260}$$

$$\lambda_{ij} = 0 \text{ for the case when } \mathcal{F}(\tau_{ij}) < a_o, \text{ i.e.for the elastic case,} \tag{4.261}$$

$$\lambda_{ij} = (2\hat{\mu})^{-1}[\tau_{ij} - (P_K\tau)_{ij}] \text{ for the case when } \mathcal{F}(\tau_{ij}) \ge a_o,$$
$$\text{i.e. for the visco-plastic case.} \tag{4.262}$$

Furthermore, the constituent law with function $\mathcal{F}$ defined by the von Mises' condition (Equation 4.237), i.e. by $\mathcal{F}(\tau_{ij}) = \frac{1}{2}\tau_{ij}^D\tau_{ij}^D - k^2$, $k =$ constant, will be of the form

$$D_{ij}(\boldsymbol{u}) = s_{ijkl}\tau_{kl,t} + \lambda_{ij} - (\gamma_{ij}(T - T_o))_{,t}, \tag{4.263}$$

$$\lambda_{ij} = 0 \text{ for the case when } \mathcal{F}(\tau_{ij}) < a_o, \text{ i.e. for the elastic case,} \tag{4.264}$$

$$\lambda_{ij} = (2\hat{\mu})^{-1}[\tau_{\text{II}}^{1/2} - k]\tau_{\text{II}}^{-1/2}\tau_{ij}^D \text{ for the case when } \mathcal{F}(\tau_{ij}) \ge a_o,$$
$$\text{i.e. for the visco-plastic case,} \tag{4.265}$$

where $\tau_{\text{II}} = \frac{1}{2}\tau_{ij}^D\tau_{ij}^D$.

4.3.3.4. Lithospheric and Upper Mantle Rocks as a Rigid Visco-Plastic Fluid under the Geothermal Field

As was said above, the upper mantle as well as parts of the lithosphere, namely in regions of rift zones, are characterized by the rocks, which in very short time behave as elastic, while in very long time they behave as a fluid. We saw above that for such types of rocks the constituent laws in Equation 4.243–4.247 are good approximations. But in this section we shall give another way for investigating one class of problems taking place in the upper parts of the Earth, namely the analysis of rift zone regions.

In Chapter 3 the conservation laws were given. We saw that the term $\tau_{ij}D_{ij}$ in the equation of energy represents the dissipation function. Let us assume that the dissipation function $\tau_{ij}D_{ij}$ depends on the strain rate tensor only. Then we can write

$$\tau_{ij}D_{ij} = \mathscr{D}_1(D) + \mathscr{D}_2(D), \tag{4.266}$$

where $\mathscr{D}_1$ and $\mathscr{D}_2$ are positive homogeneous functions in the components of the strain rate tensor $D = (D_{ij})$, i.e.

$$\mathscr{D}_1(D) = (\partial\mathscr{D}_1/\partial D_{ij})D_{ij}, \quad \mathscr{D}_2(D) = \tfrac{1}{2}(\partial\mathscr{D}_2/\partial D_{ij})D_{ij}. \tag{4.267}$$

The relation in Equation 4.266 must be satisfied for all components of the strain rate tensor $D = (D_{ij})$, and if subject to the constraint

$$D_{ii}(\boldsymbol{u}) = 0, \tag{4.268}$$

then the stress component τ_{ij} satisfies

$$\tau_{ij} = -p\delta_{ij} + \partial\mathscr{D}_1/\partial D_{ij} + \tfrac{1}{2}\partial\mathscr{D}_2/\partial D_{ij} - \beta_{ij}(T - T_o), \tag{4.269}$$

where p is a scalar, which is independent of the strain rate tensor $D = (D_{ij})$ and, therefore, has the meaning of pressure. Let us assume that the materials of the lithosphere and the upper mantle are isotropic. Then the functions $\mathscr{D}_1$ and $\mathscr{D}_2$ are scalar functions of the invariants of the strain rate tensor D only, and are of the form

$$\mathscr{D}_1(D) = (\partial\mathscr{D}_1/\partial D_{ij})D_{ij} = 2\hat{g}D_{\mathrm{II}}^{1/2} = 2\hat{g}(\tfrac{1}{2}D_{ij}D_{ij})^{1/2}, \tag{4.270}$$

$$\mathscr{D}_2(D) = \tfrac{1}{2}(\partial\mathscr{D}_2/\partial D_{ij})D_{ij} = 4\hat{\mu}D_{\mathrm{II}} = 2\hat{\mu}D_{ij}D_{ij}, \tag{4.271}$$

where $D_{\mathrm{II}} = \frac{1}{2}D_{ij}D_{ij}$ is the invariant of the strain rate tensor $D = (D_{ij})$, $\hat{g}$ is the threshold of plasticity or the yield point or the yield stress, and $\hat{\mu}$ is the viscosity of the visco-plastic Bingham fluid.

Then the constituent law (Equation 4.269) can be rewritten as follows

$$\tau_{ij} = -p\delta_{ij} + \hat{g}D_{ij}D_{\mathrm{II}}^{-1/2} + 2\hat{\mu}D_{ij} - \beta_{ij}(T - T_o), \tag{4.272}$$

which is valid when $D_{\mathrm{II}} \neq 0$. The inverse constituent law can be found as follows.

Let $\tau_{\mathrm{II}} = \frac{1}{2}\tau_{ij}^D\tau_{ij}^D$, where τ_{ij}^D are the components of the stress deviator. Then together with Equation 4.272 we find

$$\tau_{\mathrm{II}} = (\hat{g} + 2\hat{\mu}D_{\mathrm{II}}^{1/2})^2. \tag{4.273}$$

Hence, $\tau_{\mathrm{II}}^{1/2} \geq \hat{g}$ and then

$$D_{ij} = (2\hat{\mu})^{-1}(1 - \hat{g}\tau_{\mathrm{II}}^{-1/2})\tau_{ij}^D. \tag{4.274}$$

For the case $D_{\rm II} = 0$, i.e. the case for which the stress tensor is undetermined, it follows that $\tau_{\rm II}^{1/2} \le \hat{g}$. Hence the inverse constituent law can be written as follows

$$\tau_{\rm II}^{1/2} < \hat{g} \Leftrightarrow D_{ij} = 0, \tag{4.275}$$

$$\tau_{\rm II}^{1/2} \ge \hat{g} \Leftrightarrow D_{ij} = (2\hat{\mu})^{-1}(1 - \hat{g}\tau_{\rm II}^{-1/2})\tau_{ij}^D - \gamma_{ij}(T - T_o). \tag{4.276}$$

If the threshold of plasticity $\hat{g} = 0$, then we have the classical constituent law for a viscous incompressible Newtonian fluid. For $\hat{g}$ being small, then this constituent law describes the medium behaving nearly as a viscous incompressible fluid. If $\hat{g}$ is strictly positive, then the rigid zones can be observed; if $\hat{g}$ increases, these rigid zones become larger and block the flow when $\hat{g}$ is sufficiently large. For $\hat{g} \to +\infty$ the medium behaves although absolutely rigid. For $\hat{\mu} = 0$ we obtain the constituent law for rigid perfectly plastic materials:

$$\tau_{\rm II} < \hat{g} \Leftrightarrow D_{ij} = 0, \tag{4.277}$$

$$\tau_{\rm II} = \hat{g} \Rightarrow \exists \lambda \ge 0, \quad \text{such that } D_{ij} = \lambda \tau_{ij}^D. \tag{4.278}$$

Hence, we see that when the representative point in R^6 of the stress deviator remains in the interior of the sphere defined by $\tau_{\rm II} \le \hat{g}^2$, then the material of rock massif is in a rigid state. When it is on the boundary of this sphere, then the material of the rock massif behaves as plastic.

From these short analyses we see that it will be a very good approximation, for example for models of the rift zones or for investigations of stability of slopes with glaciers and slopes of mountains containing below the radioactive waste repository.

REFERENCES

Anderson, O.L. and Sumino, Y. (1980), "Elastic Constants of Minerals", in *Handbook of Physical Properties of Rocks, Vol. III,* (Carmichael, R.S.(Ed.)) CRS Press, Inc..

Andrews, D.J. and Sleep, N.P. (1974), "Numerical Modelling of Tectonic Flow Behind Island Arc", *Geophys. J. Roy. Astr. Soc.*, **38**, 237-251.

Aschenbach, J.D. (1972), "Dynamic Effects in the Brittle Fracture", in *Mechanics Today, Vol. 1, (Nemant-Nasser, S. (Ed.))*, Pergamon Press, New York, 1-57.

Aschenbach, J.D. (1978), "Elastodynamic Fracture Mechanics", in *Fracture Mechanics, SIAM-AMS Proceedings, Vol. XII (Burridge, R.(Ed.))*, AMS, New York, 3-20.

Ascione, L. and Grimaldi, A. (1984), "Unilateral Contact Problem Between a Plate and an Elastic Foundation", *Meccanica*, **19**, 223-233.

Biot, (1956a,b), "Theory of Propagation of Elastic Waves in a Fluid Saturated Porous Solid: (I) Low-frequency Range", *J. Acoust. Soc. Am.*, **28**, 168-178; "(II) High Frequency Range", *J. Acoust. Soc. Am.*, **28**, 179-191.

Bird, P. (1978), "Finite Element Modelling of Lithosphere Deformation: The Zagros Collision Orogeny", *Tectonophysics*, **50**, 307-336.

Bird, P. and Baumgardner, J. (1984), "Fault Friction, Regional Stress, and Crust-Mantle Coupling in Southern California from Finite Element Models", *J. Geophys. Res.*, **89**, **B3**, 1932-1944.

Cathles, L.M. (1975), *The Viscosity of the Earth's Mantle*, Princeton University Press, Princeton, N.J.

Coppetti, M. and Elliott, C. (1993), "A One Dimensional Quasisteady Contact Problem in Linear Thermoelasticity", *Eur. J. Appl. Math.*, **4**, 151-174.

Desai, C.S. and Gioda, G. (Eds.) (1990), *Numerical Methods and Constitutive Modelling in Geomechanics*, Wien, N.Y., IUTAM, Springer Vlg..

Duvaut, G. and Lions, J.L. (1972). *Les Inéquations en mécanique et en Physique*, Dunod, Paris.

Duvaut, G. and Lions, J.L. (1976), *Inequalities in Mechanics and Physics*, Springer Verlag, Berlin, Heidelberg, New York.

Elliott, C.M. and Tang Qi (1994), "A Dynamic Contact Problem in Thermo-elasticity", *Nonlinear Analysis, Theory and Applications* **Vol. 23**, **No. 7**, 883-898.

Fowler, A.C. (1983), "On the Thermal State of the Earth's Mantle", *J. Geophys.*, **53**, 42-51.

Fowler, A.C. (1985), "Fast Thermo-viscous Convection", *Stud. Appl. Math.*, **72**, 189-219.

Fowler, A.C. (1993), "Toward a Description of Convection with Temperature-and-Pressure-Dependent Viscosity", *Stud. Appl. Math.*, **88**, 113-139.

Gilbert, R.P., Shi, P. and Shillor, M. (1990), "A Quasi-static Contact Problem in Linear Thermoelasticity", *Rc. Mat.*, **Serie VII.**, 785-880.

Gioda, G. (1990), "Some Applications of Mathematical Programming in Geomechanics", in *Numerical Methods and Constitutive Modelling in Geomechanics* (Desai and Gioda (Eds.)), Wien. N. Y., IUTAM, Springer Vlg., 319-350.

Haslinger, J. and Hlaváček, I. (1980),(1981), "Contact Between Elastic Bodies: I. Continuous Problems", *Apl. Mat.*, **25(1980)**, 324-347; "II. Finite Element Analysis", *Apl. Mat.*, **26(1981)**, 263-290; "III. Dual Finite Element Analysis", *Apl. Mat.*, **26(1981)**, 321-344.

Hlaváček, I., Haslinger, J., Nečas, J. and Lovíšek, J. (1982), "Solution of Variational Inequalities in Mechanics", ALFA, Bratislava; SNTL, Prague (in Slovak).

Hlaváček, I. and Lovíšek, J. (1977), "A Finite Element Analysis for the Signorini Problem in Plane Elasto-statics", *Apl. Mat.*, **22**, 215-228.

Jarušek, J. (1983), "Contact Problems with Bounded Friction: Coercive Case", *Czech. Math. J.*, **33, 2**, 237-261.

Kikuchi, N. and Oden, J.T. (1988), *Contact Problems in Elasticity: A Study of Variational Inequalities and Finite Element Methods*, SIAM, Philadelphia, PA.

Kirby, S.H. (1983), "Rheology of the Lithosphere", *Rev. Geophys. Space Phys.*, **21**, 1458-1487.

Kovári, K. (1985), "Detection and Monitoring of Structural Deficiencies in the Rock Foundation of Large Dams", *Comm. Int. des Grandes Barrages, Quinziéme Congres des Grands Barrages*, Lausane.

Kovári, K. (1988), "Methods of Monitoring Landslides", *V-th International Symposium of Landslides*, Lausanne.

Kovári, K. and Amstad, Ch. (1983), "Fundamentals of Deformation Measurements", *Proc. Int. Symp. on Field Measurements in Geomech.*, Zürich, Balkema Publisher, Rotterdam.

Kovári, K. and Peter, G. (1983), "Continuous Strain Monitoring in the Rock Foundation of a Large Gravity Dam", *Rock Mechanics and Rock Engineering*, **16**, 157-171.

Krasnosel'skij, M.A. and Pokrovskij, A.V. (1983), *Systems with Hysteresis*, Nauka, Moscow (in Russian).

Kuczma, M.S. and Rakowski, J. (1994), "The Interaction Problem of a Plate on Unilateral Foundation", in *Proc. of the 3-rd Inter. Workshop on the Design and Evaluation of Concrete Pavements*, Knembach, Austria, Sept. 29-31, 1994.

Kuczma, M.S. and Whiteman, J.R. (1995), "Variational Inequality Formulation for Flow Theory of Plasticity", *Internat. J. Engng. Sci.*, **33(8)**, 1153-1169.

Lliboutry, L.A. (1971), "Rheological Properties of the Asthenosphere from Fennoscandian Data", *J. Geophys. Res.*, **76**, 1433-1446.

Mantič, V. and Nedoma, J. (1990), "Optimal Design of Tensometric Probes for In-Situ Stress Measurements by Finite Element Analysis", in *Proc. of the 4-th Int. Symposium on the Analysis of Seismicity and Seismic Risk (Schenková, Z.(Ed.))*, Geophysical Institute Czech. Acad. Sci, Prague.

McAdoo, D.C., Caldwell, J.G. and Turcotte, D.L. (1978), "On the Elastic-Perfectly-Plastic Bending of the Lithosphere Under Generalized Loading with Application to the Kuril Trench", *Geophys. J. R. Astr. Soc.*, **54**, 11-26.

Melosh, H.J. (1977), "Shear Stress on the Base of a Lithospheric Plate" *Pure Appl. Geophys.*, **115**, 429-439.

Melosh, H.J. and Raefsky, A. (1980), "The Dynamical Origins of Subduction Zone Topography", *Geophys. J. R. Astr. Soc.*, **60**, 333-354.

Nečas, J. and Hlaváček, I. (1981), *Mathematical Theory of Elastic and Elasto-Plastic Bodies: An Introduction*, Elsevier.

Nedoma, J. (1983), "On One Type of Signorini Problem Without Friction in Linear Thermoelasicity", *Apl. Mat.*, **28, 6**, 393-407.

Nedoma, J. (1984), "On the Global Geodynamic Model of the Earth", *Gerlands Beitr. Geophysik*, **93,4**, 241-260.

Nedoma, J. (1986a-c), "On the Geodynamic Model of the Earth and the Plate Tectonic Hypothesis. Part I-III", *Gerlands Beitr. Geophysik*, **95, 1**, 36-62; **95, 2**, 89-105.

Nedoma, J. (1987), "On the Signorini Problem with Friction in Linear Thermoelasticity: The Quasi-Coupled Case", *Apl. Mat.*, **32, 3**, 186-199.

Nedoma, J. (1990a), "Tectonic Evolution of Collision Zones. I. Mechanics of Tectonic Processes", *Gerlands Beitr. Geophysik,* (I) **99, 2**, 97-108; (II) **99, 2**, 109-121.

Nedoma, J. (1990b). "Plate Tectonics and the Analysis of Fracture Mechanics in a Collision Zone. I. Formulation of the Problem; *Contr. Geophys. Inst. Slov. Acad. Sci.*, **(I) 20**, 53-69; II Crack Mechanics, **(II) 21**, 38-57.

Nedoma, J. (1990c), "On the stress-strain Monitoring methods In-Situ", *Technical Report V-477*, ICS AS CR, Prague (in Czech).

Nedoma, J. (1991), "On the Protection of Nuclear Power Plants and Great Structures in Central Europe with Respect to the Tectonics of European Alpids: A Conception Study", *Acta Geod. Geoph. Mont. Hung.*, **26(1-4)** 287-300.

Nedoma, J. (1993), "Finite Element Analysis in Nuclear Safety", MAFELAP 93, Brunel University, London April 1993, in *Technical Report V-550*, Institute of Computer Science AS CR, Prague; Abstract in *The Mathematics of Finite Elements and Applications VIII* (Whiteman J.(Ed.)), J. Wiley & Sons, New York.

Nedoma, J. (1994a), "Finite Element Analysis of Contact Problems in Thermo-elasticity:. The Semi-Coercive Case", *J. Comput. Appl. Math.*, **50**, 411-423.

Nedoma, J. (1994b), "Numerical Modelling of Tectonic Evolution of Collision Zones", *20-th Inter. Conference on Mathematical Geophysics: Complex Space-Time Geophysical Structures, La Citadelle, Villefranche-sur-mer, France*, June 19-24, 1994.

Nedoma, J. (1997), "On a Coupled Stefan-like Problem in the Thermo-Visco-Plastic Rheology", *J. Comput. Appl. Math.*, (in print).

Nedoma, J., Haslinger, J. and Hlaváček, I. (1989), "The Problem of an Obducting Lithospheric Plate in the Aleutian Arc System: A Finite Element Analysis in the Frictionless Case", *Math. Comput. Modelling*, **12, 1**, 61-75.

Nedoma, J. and Dvořák, J. (1995), "On the FEM Solution of a Coupled Contact Two-Phase Stefan Problem in Thermo-elasticity: Coercive Case". *J. Comp. Appl. Math.*, **63**, 411-420.

O'Connell, R.J. and Budiansky, B. (1974), "Seismic Velocities in Dry and Saturated Cracked Solids", *J. Geophys. Res.*, **79**, 5412-5426.

O'Connell, R.J. and Budiansky, B. (1977), "Visco-elastic Properties of Fluid-Saturated Cracked Solids", *J. Geophys. Res.*, **82**, 5719-5735.

Oliver, J., Cook, F. and Brown, L. (1983), "COCORP and the Continental Crust", *J. Geophys. Res.*, **88**, 3329-3347.

Panagiotopoulos, P.D. (1985), *Inequality Problems in Mechanics and Applications: Convex and Non-convex Energy Functions*. Birkhäuser, Boston.

Panagiotopoulos, P.D. (1993), *Hemi-variational Inequalities*, Springer Verlag, Berlin, Heidelberg, New York, London, Paris, Tokyo, Hong Kong, Barcelona, Budapest.

Park, M.J. and Westbrook, G.K. (1983), "Visco-elastic Modelling of Lithospheric Stress Arising from Density Contract", *Geophys. J. R. Astr. Soc.*, **74**, 905-914.

Peltier, W.R. (1989), *Mantle Convection, Plate Tectonics and Global Dynamics*, Gordon & Breach Science Publishers, New York, London, Paris, Montreux, Tokyo, Melbourne.

Prager, W. (1961), *Introduction to Mechanics of Continua*, Ginn and Company.

Quareni, D.A. *et al.* (1985). "High Rayleigh Number Convection with Strongly Variable Viscosity", *J. Geophys. Res.*, **90**, 12633-12644.

Ranalli, G. (1987), *Rheology of the Earth*, Allen & Unwin, Boston, London, Sydney, Wellington.

Ranalli, G. and Murphy, D.C. (1987), "Rheological Stratification of the Lithosphere", *Tectonophysics*, **132**, 281-295.

Rektorys, K. (1982), *The Method of Discretization in Time and Partial Differential Equations*, D. Reidel Publishing Company, Dordrecht, Boston, London.

Sato, K., Bhatia, S.C. and Gupta, H.K. (1996), "Three-Dimensional Numerical Modelling of Deformation and Stress in the Himalaya and Tibetan Plateau with a Simple Geometry", *J. Phys. Earth*, **44**, 227-254.

Scheidegger, A.E. (1982), *Principles of Geodynamics*, Springer Verlag, Berlin.

Schmeling, H. and Jacoby, W.R. (1981), "On Modelling the Lithosphere in Mantle Convection", *J. Geophys.*, **50**, 89-100.

Shi, P. and Shillor, M. (1990), "Uniqueness and Stability of the Solution to a Thermo-elastic Contact Problem", *Eur. J. Appl. Math.*, **1**, 371-387.

Shi, P., Shillor, M. and Zou, X. (1991). "Numerical Solution to One-Dimensional Problems of Thermo-elastic Contact", *Comput. Math. Appl.*, **22**, 65-78.

Stacey, F.D. (1977), *Physics of the Earth*, J. Wiley & Sons, New York.

Toksösz, M.N. and Hsui, A.T. (1978), "Numerical Studies of Back-arc Convection and the Formation of Marginal Basins", *Tectonophysics*, **50**, 177-196.

Turcotte, D.L., McAdoo, D.C. and Caldwell, J.G. (1978), "An Elastic-Perfectly-Plastic Analysis of the Bending of the Lithosphere at a Trench", *Tectonophysics*, **47**, 193-205.

Turcotte, D.L. and Schubert, G. (1982), *Geodynamics Applications of Continuum Physics to Geological Problems*. J. Wiley & Sons, New York.

Uyeda, S. (1982), "Subduction Zones: An Introduction to Comparative Subductology", *Tectonophysics*, **81**, 133-159.

Vanpé, J.M. (1984), "Thermo-mechanical Convection in a Subduction Zone, and Initiation of Back-arc Spreading", *Anales Geophysicae*, **2, 3**, 343-352.

Weertman, J. (1978), "Creep Laws for the Mantle of the Earth", *Philosophical Transactions of the Royal Society, London*, **A288**, 9-26.

Weertman, J. and Weertman, J.R. (1975), "High Temperature Creep of Rock and Mantle Viscosity", *Ann. Rev. Earth. Planet. Sci.*, **3**, 293-315.

Zhao, W.L. and Morgan, W.J. (1987), "Injection of Indian Crust Into Tibetan Lower Crust: A Two-Dimensional Finite Element Model Study", *Tectonics*, **6**, 489-504.

5

Mathematical Models of the Rift Zones

5.1. INTRODUCTION

In Chapter 1 we saw that the evolution of the rift zones depends on the light materials, which, after accumulation below the core-mantle boundary, form an unstable layer which displays a tendency to produce a stable situation by escaping through the GTLV-channels into the upper parts of the Earth. The light material penetrate below the lithosphere, where it forms a layer of low viscosity, the so-called asthenosphere. After the formation of the asthenosphere, the lithosphere is strained by horizontal forces. The mechanical strain in the vertical and horizontal directions, together with the continual melting, results in the opening of a rift. The mechanism of the evolution of the rift zones is described in detail in Chapter 1 and the rift zone in its schematic cross-section is given in Figure 5.1.

5.2. MODELS BASED ON THERMO-BINGHAM RHEOLOGY

5.2.1. Mathematical Model of the Rift Zone based on the Thermo-Elasto-Visco-Plastic Bingham Fluid Without Melting and Recrystallization

Let us denote by Ω the region occupied at time t_o by the part of the lithosphere and of the upper mantle where the rift zone is situated, and by Ω_t the region occupied by this designated portion at time t, $\Omega_t = \Omega \times I$, $I = (t_o, t_1)$. In our simple model we shall assume the region Ω to be two dimensional only, with Lipschitz boundary $\partial\Omega = \Gamma_u \cup \Gamma_\tau \cup R$ occupying the part of the lithosphere and the upper mantle including the rift, where $\Gamma_u = {}^1\Gamma_u \cup {}^2\Gamma_u$, $\Gamma_\tau = {}^1\Gamma_\tau \cup {}^2\Gamma_\tau$ and R is the boundary of the zero measure. The fundamental equations for the moving thermo-visco-plastic media follow from the general conservation law of energy and the law of momentum conservation, which

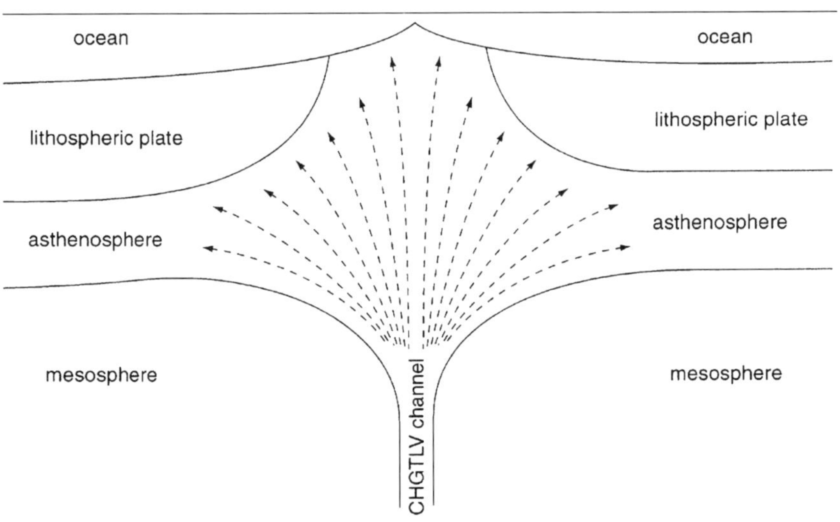

Figure 5.1. The Rift Zone: Schematic Cross-Section

in differential form are as follows

$$\rho c_e \frac{\mathrm{d}T}{\mathrm{d}t} = Q - \frac{\partial}{\partial \boldsymbol{x}_i}\left(\kappa_{ij}\frac{\partial T}{\partial \boldsymbol{x}_j}\right), \quad i, j = 1, 2, \tag{5.1}$$

$$\rho \frac{\mathrm{d}u_i}{\mathrm{d}t} = \frac{\partial \tau_{ij}}{\partial \boldsymbol{x}_j} + F_i \quad i, j = 1, 2, \tag{5.2}$$

where ρ is density, T is temperature, c_e is specific heat, Q represents thermal sources, κ_{ij} are coefficients of thermal conductivity, t represents time and $\boldsymbol{x} = (x_i)$ are the Cartesian co-ordinates. Furthermore, $\boldsymbol{u} = (u_i)_{i=1,2}$ is a flow velocity, τ_{ij} is the stress tensor of the Bingham fluid defined by the constitutional stress-strain rate relation, and F_i represents body forces. Moreover, we denote $\mathrm{d}/\mathrm{d}t = \partial/\partial t + u_j \partial/\partial x_j$.

The constitutional stress-strain rate and strain rate-displacement relations are defined by

$$\tau_{ij} = -p\delta_{ij} + \hat{g}D_{ij}D_{\mathrm{II}}^{-1/2} + 2\hat{\mu}D_{ij} - \beta_{ij}(T - T_o), \tag{5.3}$$

$$D_{ij} = \tfrac{1}{2}(\partial u_i/\partial \boldsymbol{x}_j + \partial u_j/\partial \boldsymbol{x}_i), \quad D_{\mathrm{II}} = \tfrac{1}{2}D_{ij}D_{ij}, \tag{5.4}$$

where $\boldsymbol{u}$ represents a flow velocity vector, p is pressure, $\hat{g}$ and $\hat{\mu}$ represent thresholds of plasticity (defined by the Mises type relation) and of viscosity of a Bingham medium respectively, β_{ij} is a coefficient of thermal expansion, and T_o is an initial temperature. The last term in Equation 5.3 represents thermal stresses. If the threshold of plasticity $\hat{g}$ is equal to zero then the medium is identical with the Newtonian liquid, and if the threshold of plasticity is equal to infinity then the medium behaves as though absolutely rigid.

We consider the following boundary conditions. On the boundaries ${}^1\Gamma_u$ and ${}^2\Gamma_u$, which are parts of the rock massif and which make it possible to describe the influences of the other part of the rock massif on the region of interest, the velocity fields

$\boldsymbol{u}_1(\boldsymbol{x}, t)$ and $\boldsymbol{u}_2(\boldsymbol{x}, t)$ as well as the heat flux $q_1(\boldsymbol{x}, t)$ and $q_2(\boldsymbol{x}, t)$ or the temperatures $T_1(\boldsymbol{x}, t)$ and $T_2(\boldsymbol{x}, t)$ are known, so that we have

$$\boldsymbol{u}(\boldsymbol{x}, t) = \boldsymbol{u}_1(\boldsymbol{x}, t),\ \kappa_{ij}\partial T/\partial x_j n_i = q_1(\boldsymbol{x}, t) \text{ or } T(\boldsymbol{x}, t) = T_1(\boldsymbol{x}, t), \tag{5.5}$$

on the boundary ${}^1\Gamma_u$

and $$\boldsymbol{u}(\boldsymbol{x}, t) = \boldsymbol{u}_2(\boldsymbol{x}, t),\ \kappa_{ij}\partial T/\partial x_j n_i = q_2(\boldsymbol{x}, t) \text{ or } T(\boldsymbol{x}, t) = T_2(\boldsymbol{x}, t), \tag{5.6}$$

on the boundary ${}^2\Gamma_u$.

Since the rock massif undergoes loading $\boldsymbol{P}_1(\boldsymbol{x}, t)$ on the boundary ${}^1\Gamma_\tau$ under temperature $T_3(\boldsymbol{x}, t)$ then we have

$$\tau_{ij}n_j = P_{1i}(\boldsymbol{x}, t),\ T(x, t) = T_3(\boldsymbol{x}, t), \tag{5.7}$$

on the boundary ${}^1\Gamma_\tau$. On the surface of the high level radioactive waste repository structure, the spreading of the loading $\boldsymbol{P}_2(\boldsymbol{x}, t)$ and the temperature $T_4(\boldsymbol{x}, t)$ or the heat flux $q_3(\boldsymbol{x}, t)$ are measured so that

$$\tau_{ij}n_j = P_{2i}(\boldsymbol{x}, t),\ T(\boldsymbol{x}, t) = T_4(\boldsymbol{x}, t) \text{ or } \kappa_{ij}\partial T/\partial x_j n_i = q_3(\boldsymbol{x}, t). \tag{5.8}$$

For simplicity we shall assume that the rock massif is created by an incompressible medium so that

$$\operatorname{div} \boldsymbol{u} = 0 \quad \text{in } \Omega_t. \tag{5.9}$$

At the initial time $t = t_o$ the initial velocity and temperature fields are prescribed, so that

$$T(\boldsymbol{x}, t_o) = T_o(\boldsymbol{x}), \tag{5.10}$$

$$\boldsymbol{u}(\boldsymbol{x}, t_o) = \boldsymbol{u}_o(\boldsymbol{x}). \tag{5.11}$$

5.2.1.1. *Mathematical formulation and variational solution of the problem*

Let Ω be a bounded domain in R^2 with a smooth boundary $\partial\Omega = \Gamma_u \cup \Gamma_\tau$, where $\Gamma_u = {}^1\Gamma_u \cup^2 \Gamma_u$, and $\Gamma_\tau = {}^1\Gamma_\tau \cup^2 \Gamma_\tau$. Let $I = (t_o, t_1)$, $0 < t_1 < +\infty$, $\Omega_t = \Omega \times I$, $\partial\Omega_t = \partial\Omega \times I$.

Let us investigate for simplicity the following non-stationary incompressible thermo-Bingham model problem:

Problem ($\mathscr{P}$):

$$\rho c_e \frac{\mathrm{d}T}{\mathrm{d}t} = Q - \frac{\partial}{\partial \boldsymbol{x}_i}\left(\kappa_{ij}\frac{\partial T}{\partial x_j}\right) \quad \text{in } \Omega_t, \quad i, j = 1, 2, \tag{5.12}$$

$$\rho\frac{\mathrm{d}u_i}{\mathrm{d}t} = \frac{\partial \tau_{ij}}{\partial x_j} + F_i \quad \text{in } \Omega_t \quad i, j = 1, 2, \tag{5.13}$$

$$\operatorname{div} \boldsymbol{u} = 0 \quad \text{in } \Omega_t, \tag{5.14}$$

$$T(\boldsymbol{x}, t) = T_1(\boldsymbol{x}, t), \quad \tau_{ij}n_j = P_i(\boldsymbol{x}, t) \quad \text{on } \Gamma_\tau \times I, \quad i, j = 1, 2, \tag{5.15}$$

$$\kappa_{ij}\partial T/\partial x_j n_i = q(\boldsymbol{x}, t), \quad \boldsymbol{u}(x, t) = \boldsymbol{u}_2(\boldsymbol{x}, t) \quad \text{on } \Gamma_u \times I, \quad i, j = 1, 2, \tag{5.16}$$

$$T(\boldsymbol{x}, t_o) = T_o(\boldsymbol{x}), \quad \boldsymbol{u}(\boldsymbol{x}, t_o) = \boldsymbol{u}_o(\boldsymbol{x}) \quad \text{a.e. in } \Omega, \tag{5.17}$$

where the stress-strain relation and the strain rate tensor are defined by

$$\tau_{ij}(\boldsymbol{u}) = -p\delta_{ij} + \hat{g}D_{ij}(\boldsymbol{u})D_{\rm II}^{-1/2}(\boldsymbol{u}) + 2\hat{\mu}D_{ij}(\boldsymbol{u}) - \beta_{ij}(T - T_o), \tag{5.17}$$

$$D_{ij}(\boldsymbol{u}) = \tfrac{1}{2}(\partial u_i/\partial x_j + \partial u_j/\partial x_i) \quad i, j = 1, 2, \tag{5.18}$$

and where $D_{\rm II}(\boldsymbol{u}) = \frac{1}{2}D_{ij}(\boldsymbol{u})D_{ij}(\boldsymbol{u})$, $\boldsymbol{u} = (u_i)_{i=1,2}$ is the flow velocity, τ_{ij} is the stress tensor of the Bingham fluid, p is the pressure, $\hat{g}$ and $\hat{\mu}$ are the threshold of plasticity and viscosity of the Bingham fluid respectively, $\boldsymbol{F}$ represents the body forces and $\boldsymbol{P}$ the surface forces, δ_{ij} is the Kronecker symbol, and $\boldsymbol{u}_o$,$\boldsymbol{u}_2$,T_o,T_1 represent initial and prescribed flow velocity and temperature, respectively. Furthermore, ρ is the density, c_e is the specific heat, Q are the thermal sources, κ_{ij} are coefficients of thermal conductivity, β_{ij} is a coefficient of thermal expansion, and q is the heat flow. Moreover, we denote $\mathrm{d}/\mathrm{d}t = \partial/\partial t + u_j\partial/\partial x_j$, $\boldsymbol{u}' = \partial\boldsymbol{u}/\partial t$.

Since in the variational formulation the term $\hat{g}D_{ij}(\boldsymbol{u})D_{\rm II}^{-1/2}(\boldsymbol{u})$ creates the non-differential functional $j(\boldsymbol{u})$, we introduce the multipliers $m = (m_{ij})$, $ij = 1, 2$ by

$$m_{ij} \in L^\infty(\Omega_t), \quad m_{ij} = m_{ji}, \quad \forall i, j, \quad m_{\rm ii} = 0, \tag{5.19}$$

$$m_{ij}m_{ij} \le 1 \quad \text{a.e. in } \Omega_t, \tag{5.20}$$

$$m_{ij}D_{ij}(\boldsymbol{u}) = (D_{ij}(\boldsymbol{u})D_{ij}(\boldsymbol{u}))^{1/2} = (2)^{1/2}D_{\rm II}^{1/2}(\boldsymbol{u}) \quad \text{a.e. in } \Omega_t. \tag{5.21}$$

Then Equation 5.17 can be written in the form

$$\tau_{ij}^m = -p\delta_{ij} + 2\hat{\mu}D_{ij}(\boldsymbol{u}) + 2^{1/2}\hat{g}m_{ij} - \beta_{ij}(T - T_o) \tag{5.22}$$

and then Equation 5.13 yields

$$\rho\frac{\partial u_i}{\partial t} - \hat{\mu}\Delta u_i + \rho u_j\frac{\partial u_i}{\partial x_j} - 2^{1/2}\hat{g}\frac{\partial m_{ij}}{\partial x_j} = F_i - \frac{\partial p}{\partial x_i} - \frac{\partial}{\partial x_j}\left(\beta_{ij}(T - T_o)\right). \tag{5.23}$$

VARIATIONAL FORMULATION

We shall employ the Sobolev spaces of functions having generalized derivatives of the order s of the type $[H^s(\Omega)]^k$. The norm will be denoted by $\|.\|$, the scalar product by (.,.) for each k. We set $H^o(\Omega) \equiv L^2(\Omega)$ and denote by (.,.) the scalar product in $L^2(\Omega)$. We denote by $L^\infty(\Omega)$ the set of all measurable functions f defined almost everywhere on Ω and such that $\|f\|_\infty = \text{ess sup}_\Omega |f(\boldsymbol{x})|$ is finite. We shall denote $|\boldsymbol{u}|^2 = \int_\Omega e_{ij}(\boldsymbol{u})e_{ij}(\boldsymbol{u})\mathrm{d}\boldsymbol{x} + \|\boldsymbol{u}\|^2$.

Let us define the spaces

$${}^1V = \{z \mid z \in H^1(\Omega), \quad z = 0 \text{ on } \Gamma_\tau \text{ in the sense of traces}\},$$

$$V = \{\boldsymbol{v} \mid \boldsymbol{v} \in [H^1(\Omega)]^2, \quad \operatorname{div} \boldsymbol{v} = 0, \quad \boldsymbol{v} = 0 \text{ on } \Gamma_u \text{ in the sense of traces}\},$$

$$H = \{\boldsymbol{v} \mid \boldsymbol{v} \in [L^2(\Omega)]^2, \quad \operatorname{div} \boldsymbol{v} = 0, \quad \boldsymbol{v} = 0 \text{ on } \Gamma_u \text{ in the sense of traces}\},$$

${}^1V'$, V' be the dual spaces of ${}^1V$ or V respectively, ${}^1V \subset L^2(\Omega) \subset {}^1V'$, $V \subset H \subset V'$.

Throughout, we suppose that $W(\boldsymbol{x}, t) \in L^2(I; L^2(\Omega))$, $\boldsymbol{F}(\boldsymbol{x}, t) \in L^2(I; [L^2(\Omega)]^2)$, $\boldsymbol{P}(\boldsymbol{x}, t) \in L^2(I; [L^2(\Gamma_\tau)]^2)$, and we shall suppose that $\partial\beta_{ij}/\partial x_j \in L^\infty(\Omega)\,\forall i, j$.

Moreover, we suppose that $\kappa_{ij} \in L^{\infty}(\Omega)$ are Lipschitz on Ω and fulfil the symmetry condition $\kappa_{ij} = \kappa_{ji}$ on Ω for every $i, j = 1, 2$, and

$$\kappa_{ij}\xi_i\xi_j \geq c_o \|\xi\|^2, \boldsymbol{x} \in \Omega, \quad \xi \in R^4, \quad c_o = \text{constant} > 0,$$

and that all of the other physical parameters are positive constants, and q, $\boldsymbol{u}_2$, T_1 are for simplicity equal to zero.

We multiply Equation 5.12 by $z - T(t)$ and Equation 5.13 by $\boldsymbol{v} - \boldsymbol{u}(t)$, then integrating over Ω, using the Green theorem with respect to the boundary conditions and after some modifications, we obtain the following variational formulation of our problem:

Problem ($\mathscr{P}_V$): Find a pair of functions $(T, \boldsymbol{u})$, $T \in {}^1V$, $\boldsymbol{u} \in V$ such that

$$(\rho c_e T'(t), z - T(t)) + b_o(\boldsymbol{u}(t), T(t), z - T(t)) + a_T(T(t), z - T(t)) - (Q(t), z - T(t)) \geq 0 \ \forall z \in {}^1V, \tag{5.24}$$

$$(\rho \boldsymbol{u}'(t), \boldsymbol{v} - \boldsymbol{u}(t)) + \hat{\mu} a(\boldsymbol{u}(t), \boldsymbol{v} - \boldsymbol{u}(t)) + b(\boldsymbol{u}(t), \boldsymbol{u}(t), \boldsymbol{v} - \boldsymbol{u}(t)) + b_s(T(t) - T_o, \boldsymbol{v} - \boldsymbol{u}(t))$$
$$+ \hat{g} j(\boldsymbol{v}) - \hat{g} j(\boldsymbol{u}(t)) \geq (\boldsymbol{f}(t), \boldsymbol{v} - \boldsymbol{u}(t)) \quad \forall \boldsymbol{v} \in V, \tag{5.25}$$

$$T(\boldsymbol{x}, t_o) = T_o(\boldsymbol{x}), \tag{5.26}$$

$$\boldsymbol{u}(\boldsymbol{x}, t_o) = \boldsymbol{u}_o(\boldsymbol{x}), \tag{5.27}$$

where for $T, z \in {}^1V$, $\boldsymbol{u}, \boldsymbol{v} \in V$ we put

$$(\rho c_e T'(t), z) = \int_\Omega \rho c_e T'(t) z \, d\boldsymbol{x}, \quad a_T(T(t), z) = \int_\Omega \kappa_{ij} \partial T/\partial x_i \partial z/\partial x_j \, d\boldsymbol{x},$$

$$b_o(\boldsymbol{u}(t), g, z) = \int_\Omega \rho c_e u_k(t) \partial g/\partial x_k z \, d\boldsymbol{x}, \quad (W(t), z) = \int_\Omega Q(t) z \, d\boldsymbol{x} - \int_{\Gamma_1} q z \, ds,$$

$$(\rho \boldsymbol{u}'(t), \boldsymbol{v}) = \int_\Omega \rho u_i'(t) v_i \, d\boldsymbol{x}, \quad a(\boldsymbol{u}, \boldsymbol{v}) = 2 \int_\Omega D_{ij}(\boldsymbol{u}) D_{ij}(\boldsymbol{v}) \, d\boldsymbol{x},$$

$$j(\boldsymbol{v}) = 2 \int_\Omega (D_{\text{II}}(\boldsymbol{v}))^{1/2} d\boldsymbol{x}, \quad b(\boldsymbol{u}(t), \boldsymbol{v}, \boldsymbol{w}) = \int_\Omega \rho u_i(t) \partial v_i/\partial x_j w_j \, d\boldsymbol{x},$$

$$(\boldsymbol{f}(t), \boldsymbol{v}) = \int_\Omega F_i(t) v_i \, d\boldsymbol{x} - \int_\Omega P_i v_i \, ds,$$

$$b_s(T(t), \boldsymbol{v}) = \int_\Omega \partial/\partial x_j (\beta_{ij}(T(t) - T_o)) v_i \, d\boldsymbol{x}.$$

The bilinear forms $T, z \to a_T(T, z)$ and $\boldsymbol{u}, \boldsymbol{v} \to a(\boldsymbol{u}, \boldsymbol{v})$ are symmetric and continuous on ${}^1V$ or V respectively, and coercive, i.e. from the above assumption the symmetry $a_T(T, z) = a_T(z, T)$, $a(\boldsymbol{u}, \boldsymbol{v}) = a(\boldsymbol{v}, \boldsymbol{u})$ and, moreover, there exists $c_o > 0$ such that $a_T(T, T) \geq c_o \|T\|^2 \ \forall T \in {}^1V$, and constant $c_1 > 0$ such that $a(\boldsymbol{u}, \boldsymbol{u}) \geq c_1 \|\boldsymbol{u}\|^2 \ \forall \boldsymbol{u} \in V$. Furthermore, we have

$$D_{ij}(\boldsymbol{u}) D_{ij}(\boldsymbol{v}) \leq 2 D_{\text{II}}^{1/2}(\boldsymbol{u}) D_{\text{II}}^{1/2}(\boldsymbol{v}), \ D_{ij} = \tfrac{1}{2} \hat{\mu}^{-1} (1 - \hat{g} \tau_{\text{II}}^{-1/2}) \tau_{ij}^D, \quad \boldsymbol{u}, \boldsymbol{v} \in V,$$

where τ_{ij}^D are components of the stress tensor deviator and where $\tau_{\text{II}} = (\hat{g} + 2\hat{\mu} D_{\text{II}}^{1/2})^2$. Moreover, $b(\boldsymbol{u}, \boldsymbol{u}, \boldsymbol{v}) = -b(\boldsymbol{u}, \boldsymbol{v}, \boldsymbol{u})$ for $\boldsymbol{u}, \boldsymbol{v} \in V$, and $b(\boldsymbol{v}, \boldsymbol{v}, \boldsymbol{v}) = 0$. Similarly $b_o(\boldsymbol{u}, y, z) + b_o(\boldsymbol{u}, z, y) = 0 \ y, z \in {}^1V$, $\boldsymbol{u} \in V$. Due to the Hölder inequality applying to $b(\boldsymbol{u}, \boldsymbol{v}, \boldsymbol{w})$, we have (see for example Temam (1976))

$$|b(\boldsymbol{u}, \boldsymbol{v}, \boldsymbol{w})| \leq c \, \|\boldsymbol{u}\|_{[L^p(\Omega)]^2} \, \|\boldsymbol{w}\|_{[L^p(\Omega)]^2} \sum_{i,j} \left\| D_i v_j \right\|_{[L^2(\Omega)]^2}$$

$$\underset{2p^{-1}+1/2=1}{\leq} c \, \|\boldsymbol{u}\|_{[L^p(\Omega)]^2} \, \|\boldsymbol{w}\|_{[L^p(\Omega)]^2} \, \|\boldsymbol{v}\|, \qquad (5.28)$$

Theorem 5.1: Let the assumptions about $a_T(T, z)$ and $a(\boldsymbol{u}, \boldsymbol{v})$ be fulfilled, let $Q \in L^2(I, {}^1V')$, $F \in L^2(I, V')$, $T_o \in L^2(\Omega)$, and $\boldsymbol{u}_o \in H$. Then there exists a pair of functions $(T, \boldsymbol{u})$ such that $T \in L^2(I, {}^1V)$, $T' \in L^2(I, {}^1V')$, $\boldsymbol{u} \in L^2(I, V)$, $\boldsymbol{u}' \in L^2(I, V')$; $(T, \boldsymbol{u})$ satisfies Equations 5.24 and 5.25 and $T(\boldsymbol{x}, t_o) = T_o(\boldsymbol{x})$, $\boldsymbol{u}(\boldsymbol{x}, t_o) = \boldsymbol{u}_o(\boldsymbol{x})$.

Proof: To prove the theorem, the regularization of $j(\boldsymbol{v})$ by the regularized functional $j_\varepsilon(\boldsymbol{v})$ will be used. The regularized functional $j_\varepsilon(\boldsymbol{v})$ is defined by

$$j_\varepsilon(\boldsymbol{v}) = 2(1+\varepsilon)^{-1} \int_\Omega (D_{\mathrm{II}}(\boldsymbol{v}))^{1/2(1+\varepsilon)} \mathrm{d}\boldsymbol{x}, \qquad \varepsilon > 0.$$

Then

$$(j'_\varepsilon(\boldsymbol{u}), \boldsymbol{v}) = \int_\Omega D_{\mathrm{II}}(\boldsymbol{u})^{1/2(\varepsilon-1)} D_{ij}(\boldsymbol{u}) D_{ij}(\boldsymbol{v}) \mathrm{d}\boldsymbol{x}.$$

Hence Equations 5.24 and 5.25 yield

$$(\rho c_e T'_\varepsilon(t), z - T_\varepsilon(t)) + b_o(\boldsymbol{u}_\varepsilon(t), T_\varepsilon(t), z - T_\varepsilon(t)) + a_T(T_\varepsilon(t), z - T_\varepsilon(t))$$

$$- (W(t), z - T_\varepsilon(t)) \geq 0 \quad \forall z \in {}^1V, \qquad (5.29)$$

$$(\rho \boldsymbol{u}'_\varepsilon(t), \boldsymbol{v} - \boldsymbol{u}_\varepsilon(t)) + \hat{\mu} a(\boldsymbol{u}_\varepsilon(t), \boldsymbol{v} - \boldsymbol{u}_\varepsilon(t)) + b(\boldsymbol{u}_\varepsilon(t), \boldsymbol{u}_\varepsilon(t), \boldsymbol{v} - \boldsymbol{u}_\varepsilon(t)) + b_s(T_\varepsilon(t) - T_o, \boldsymbol{v} - \boldsymbol{u}_\varepsilon(t))$$

$$+ \hat{g} j_\varepsilon(\boldsymbol{v}) - \hat{g} j_\varepsilon(\boldsymbol{u}_\varepsilon(t)) \geq (\boldsymbol{f}(t), \boldsymbol{v} - \boldsymbol{u}_\varepsilon(t) \quad \forall \boldsymbol{v} \in V. \qquad (5.30)$$

The technique of the proof is parallel to that of Theorem 3.2, Chapter VI of Duvaut and Lions (1972) (See also Nedoma (1988), (1995)) and it consists of the following:

a) the proof of the existence of the solution of Equations 5.29 and 5.30 is based on the Galerkin approximation,
b) a priori estimates I and II independent of ε will be given,
c) limitation processes over m and ε will be performed,
d) the uniqueness of the solution will be proved.

The existence of $(T, \boldsymbol{u}_\varepsilon)$ will be proved by means of the finite-dimensional approximation. We construct separable bases of spaces ${}^1V$ and V, and since elements of these bases are linearly independent their linear combinations are dense in ${}^1V$ and V, respectively. Then the approximate solution of order m satisfies the regular system of ordinary differential equations of the first order

$$(\rho c_e T'_m(t), z_j - T_m(t)) + b_o(\boldsymbol{u}_m(t), T_m(t), z_j - T_m(t)) + a_T(T_m(t), z_j - T_m(t))$$

$$- (W(t), z_j - T_m(t)) \geq 0 \quad \forall z_j \in {}^1V^m, \qquad (5.31)$$

$$(\rho \boldsymbol{u}'_m(t), \boldsymbol{v}_j - \boldsymbol{u}_m(t)) + \hat{\mu} a(\boldsymbol{u}_m(t), \boldsymbol{v}_j - \boldsymbol{u}_m(t)) + b(\boldsymbol{u}_m(t), \boldsymbol{u}_m(t), \boldsymbol{v}_j - \boldsymbol{u}_m(t))$$

$$+ b_s(T_m(t) - T_o, \boldsymbol{v}_j - \boldsymbol{u}_m(t)) + \hat{g} j(\boldsymbol{v}_j) - \hat{g} j(\boldsymbol{u}_m(t)) \geq (\boldsymbol{f}(t), \boldsymbol{v}_j - \boldsymbol{u}_m(t)) \quad \forall \boldsymbol{v}_j \in V^m, \qquad (5.32)$$

with initial conditions

$$T_m(\boldsymbol{x}, t_o) = T_o(\boldsymbol{x}), \qquad (5.33)$$

$$\boldsymbol{u}_m(\boldsymbol{x}, t_o) = \boldsymbol{u}_o(\boldsymbol{x}), \qquad (5.34)$$

where ${}^1V^m$, V^m are spaces created by z_j, v_j, $1 \le j \le m$. Since $\{z_j\}$ and $\{\boldsymbol{v}_j\}$ are linearly independent, then $(T_m, \boldsymbol{u}_m)$ is uniquely defined by Equations 5.31-5.34 on the interval $I_m = (t_o, t_m)$.

A priori estimates show that $t_m = t_1$.

Adding Equations 5.31 and 5.32, and after some modification, we obtain

$$\frac{1}{2}\frac{\mathrm{d}}{\mathrm{d}t}\{|\boldsymbol{u}_m(t)|^2 + |T_m(t)|^2\} + \hat{\mu}a(\boldsymbol{u}_m(t), \boldsymbol{u}_m(t)) + a_T(T_m(t), T_m(t)) + \hat{g}(j'_\varepsilon(\boldsymbol{u}_m(t)), \boldsymbol{u}_m(t))$$
$$+ b_s(T_m(t), \boldsymbol{u}_m(t)) = (Q(T), T_m(t)) + (\boldsymbol{f}(t), \boldsymbol{u}_m(t)).$$

Integrating over t from t_o to t and using the Gronwall lemma, after some modification we obtain estimates

$$|\boldsymbol{u}_m(t)| \le c,\ |T_m(t)| \le c,\ \int_I \|\boldsymbol{u}_m(\tau)\|^2\, d\tau \le c, \quad \int_I \|T_m(\tau)\|^2\, \mathrm{d}\tau \le c, \tag{5.35}$$

which represent the first type of a priori estimates.

For $\boldsymbol{v} \in V$ it holds $b(\boldsymbol{u}_m(t), \boldsymbol{u}_m(t), \boldsymbol{v}) = -b(\boldsymbol{u}_m(t), \boldsymbol{v}, \boldsymbol{u}_m(t))$ and due to Equations 5.28 and 5.35 we have

$$|b(\boldsymbol{u}_m(t), \boldsymbol{u}_m(t), \boldsymbol{v})| \le c\,\|\boldsymbol{u}_m\|\,\|\boldsymbol{v}\|\,. \tag{5.36}$$

Hence

$$b(\boldsymbol{u}_m(t), \boldsymbol{u}_m(t), \boldsymbol{v}) + b_s(T_m(t), \boldsymbol{v}) = (\boldsymbol{h}_m(t), \boldsymbol{v}) \quad \forall \boldsymbol{v} \in V,$$

where $\boldsymbol{h}_m \in a$ bounded subset of $L^2(I, V')$, $\|\boldsymbol{h}_m\|_{L^2(I,V')} \le c > 0$,

$$b_o(\boldsymbol{u}_m(t), T_m(t), z) = (g_m(t), z),$$

where $g_m \in a$ bounded subset of $L^2(I, {}^1V')$, $\|g_m\|_{L^2(I,{}^1V')} \le c > 0$.

Bilinear forms $a_T(T, z)$ and $a(\boldsymbol{u}, \boldsymbol{v})$ are continuous on ${}^1V$ or V, respectively so that

$$a_T(T, z) = (A_T T, z), \quad a(\boldsymbol{u}, \boldsymbol{v}) = (A\boldsymbol{u}, \boldsymbol{v}), \quad A_T \in \mathscr{L}({}^1V, {}^1V'), A \in \mathscr{L}(V, V').$$

Then Equations 5.31 and 5.32 can be written in the following form

$$(\rho c_e T'_m(t) + A_T T_m(t) + g_m(t) - Q(t), z_j) = 0 \quad 1 \le j \le m, \tag{5.37}$$
$$(\rho \boldsymbol{u}'_m(t) + \hat{\mu}A\boldsymbol{u}_m(t) + \hat{g}j'(\boldsymbol{u}_m(t)) + \boldsymbol{h}_m(t) - \boldsymbol{f}(t), \boldsymbol{v}_j) = 0, \quad 1 \le j \le m. \tag{5.38}$$

Let P_m,R_m be orthogonal projections $L^2(\Omega) \to {}^1V^m$ or $H \to V^m$, respectively. Then

$$P_m T = \sum\nolimits_{i=1}^m (T, T_j)T_j, \quad R_m\boldsymbol{u} = \sum\nolimits_{i=1}^m (\boldsymbol{u}, \boldsymbol{v}_j)\boldsymbol{v}_j, \quad 1 \le j \le m,$$

and since $P_m T' = T'$, $R_m\boldsymbol{u}' = \boldsymbol{u}'$ then

$$T'_m = P_m(Q - A_T T_m - g_m), \tag{5.39}$$
$$\boldsymbol{u}'_m = R_m(\boldsymbol{F} - \hat{\mu}A\boldsymbol{u}_m - \hat{g}j'_\varepsilon(\boldsymbol{u}_m) - \boldsymbol{h}_m). \tag{5.40}$$

Due to the definition of $(j'_\varepsilon(\boldsymbol{u}), \boldsymbol{v})$ we obtain

$$\left\|j'_\varepsilon(u)\right\|_* \le c\left(\int_\Omega (D_{\mathrm{II}}(\boldsymbol{u}))^\varepsilon \mathrm{d}\boldsymbol{x}\right)^{1/2}$$

and hence $j'_\varepsilon(\boldsymbol{u}) \in a$ bounded subset of $L^2(I, V')$. Thus Equations 5.37 and 5.38 indicate that $T'_m = P_m p_m$, $\boldsymbol{u}'_m = R_m\boldsymbol{r}_m$, $p_m \in \mathscr{P}_m \subset L^2(I, {}^1V')$, $\boldsymbol{r}_m \in \mathscr{R}_m \subset L^2(I, V')$. Since $\|P_m z\|_{{}^1V'} \le c\,\|z\|_{{}^1V'}$, $\|R_m\boldsymbol{v}\|_{V'} \le c\,\|\boldsymbol{v}\|_{V'}$ then $(T'_{m'}, \boldsymbol{u}'_m) \in a$ bounded subset of $L^2(I, {}^1V') \times L^2(I, V')$, which completes this part of the proof.

The convergence of the finite-dimensional approximation will be proved under the assumption that ε is fixed. From the a priori estimates the subsequence $(T_\mu, \boldsymbol{u}_\mu)$ of the sequence $(T_m, \boldsymbol{u}_m)$ can be taken such that

$$\begin{aligned}
(T_\mu, \boldsymbol{u}_\mu) &\to (T_{\varepsilon\mu}, \boldsymbol{u}_{\varepsilon\mu}) \quad \text{in } L^2(I, L^2(\Omega)) \times L^2(I, H) * \text{weakly},\\
(T_\mu, \boldsymbol{u}_\mu) &\to (T_{\varepsilon\mu}, \boldsymbol{u}_{\varepsilon\mu}) \quad \text{in } L^2(I, {}^1V) \times L^2(I, V) \text{ weakly},\\
(T'_m, \boldsymbol{u}'_m) &\to (T'_{\varepsilon\mu}, \boldsymbol{u}'_{\varepsilon\mu}) \quad \text{in } L^2(I, {}^1V') \times L^2(I, V') \text{ weakly},\\
(T_\mu, \boldsymbol{u}_\mu) &\to (T_{\varepsilon\mu}, \boldsymbol{u}_{\varepsilon\mu}) \quad \text{in } L^2(I, {}^1V) \times L^2(I, V) \text{ strongly}.
\end{aligned}$$

Thus for the appropriate components $u_{i\mu}$ of $\boldsymbol{u}_\mu$ we have

$$\boldsymbol{u}_{i\mu} \to \boldsymbol{u}_{i\varepsilon\mu} \text{ weakly a.e. in } \Omega \times I.$$

Furthermore, $\{j'_\varepsilon(\boldsymbol{u}_m)\}$, $\{u_{i\mu}u_{j\mu}\}$ are bounded subsets of $L^2(I, V')$ or $L^2(I, L^{1/2p}(\Omega))$, respectively. Then we can assume that

$$\begin{aligned}
j'_\varepsilon(\boldsymbol{u}_\mu) &\to \chi \quad \text{weakly in } L^2(I, V'),\\
u_{i\mu}u_{j\mu} &\to \Theta_{ij} \quad \text{weakly in } L^2(I, L^{p/2}(\Omega)).
\end{aligned}$$

Hence and as $u_{i\mu}u_{j\mu} \to u_{i\varepsilon\mu}u_{j\varepsilon\mu}$ in the sense of distributions in $\Omega \times I$ we obtain $\Theta_{ij} = u_{i\varepsilon\mu}u_{j\varepsilon\mu}$. But

$$b(\boldsymbol{u}_\mu, \boldsymbol{u}_\mu, \boldsymbol{v}_j) = -b(\boldsymbol{u}_\mu, \boldsymbol{v}_j, \boldsymbol{u}_\mu) \to -b(\boldsymbol{u}_{\varepsilon\mu}, \boldsymbol{v}_j, \boldsymbol{u}_{\varepsilon\mu}) \text{ weakly in } L^2(I) \quad \forall \boldsymbol{v}_j$$

(see Duvaut and Lions (1972), Nedoma (1988)). For $m = \mu$, then from Equation 5.32 we obtain

$$(\rho\boldsymbol{u}'_{\varepsilon\mu}(t), \boldsymbol{v}_j) + \hat{\mu}a(\boldsymbol{u}_{\varepsilon\mu}(t), \boldsymbol{v}_j) + b(\boldsymbol{u}_{\varepsilon\mu}(t), \boldsymbol{u}_{\varepsilon\mu}(t), \boldsymbol{v}_j) + b_s(T_{\varepsilon\mu}(t) - T_o, \boldsymbol{v}_j) + \hat{g}(\chi, \boldsymbol{v}_j) \geq (\boldsymbol{f}(t), \boldsymbol{v}_j) \quad \forall j,$$

where we denote $\chi = j'_\varepsilon(\boldsymbol{u}_{\varepsilon\mu})$. Since the system $\{\boldsymbol{v}_j\}$ is complete in V then

$$(\rho\boldsymbol{u}'_{\varepsilon\mu}(t), \boldsymbol{v}) + \hat{\mu}a(\boldsymbol{u}_{\varepsilon\mu}(t), \boldsymbol{v}) + b(\boldsymbol{u}_{\varepsilon\mu}(t), \boldsymbol{u}_{\varepsilon\mu}(t), \boldsymbol{v}) + b_s(T_{\varepsilon\mu}(t) - T_o, \boldsymbol{v}) + \hat{g}(\chi, \boldsymbol{v}) \geq (\boldsymbol{f}(t), \boldsymbol{v}) \quad \forall \boldsymbol{v} \in V.$$

By using a similar technique as in Duvaut and Lions (1972), Nedoma (1988), where we put

$$\begin{aligned}
X_\mu = {} & \hat{g}\int_I (j'_\varepsilon(\boldsymbol{u}_\mu) - j'_\varepsilon(\boldsymbol{v}), \boldsymbol{u}_\mu - \boldsymbol{v})\mathrm{d}t + \hat{\mu}\int_I a(\boldsymbol{u}_\mu - \boldsymbol{v}, \boldsymbol{u}_\mu - \boldsymbol{v})\mathrm{d}t\\
& + \int_I (\boldsymbol{u}'_\mu - \boldsymbol{v}', \boldsymbol{u}_\mu - \boldsymbol{v})\mathrm{d}t + \int_I a_T(T_\mu - z, T_\mu - z)\mathrm{d}t + \int_I (T'_\mu - z, T_\mu - z)\mathrm{d}t,
\end{aligned}$$

then we find that

$$\chi(t) = j'_\varepsilon(\boldsymbol{u}_{\varepsilon\mu}(t)),$$

which completes this part of the proof.

Then we proved that

$$\left.\begin{aligned}
&\boldsymbol{u}_\varepsilon \in a \text{ bounded set of } L^2(I, V) \cap L^\infty(I, H),\\
&\boldsymbol{u}'_\varepsilon \in a \text{ bounded set of } L^2(I, V'),\\
&T_\varepsilon \in a \text{ bounded set of } L^2(I, {}^1V) \cap L^\infty(I, L^2(\Omega)),\\
&T'_\varepsilon \in a \text{ bounded set of } L^2(I, {}^1V').
\end{aligned}\right\} \tag{5.41}$$

To prove the limit $\varepsilon \to 0$ we put

$$\begin{aligned}
Y_\varepsilon = \int_I \{ & (\boldsymbol{v}', \boldsymbol{v} - \boldsymbol{u}_\varepsilon) + \hat{\mu}a(\boldsymbol{u}_\varepsilon, \boldsymbol{v} - \boldsymbol{u}_\varepsilon) + b(\boldsymbol{u}_\varepsilon, \boldsymbol{u}_\varepsilon, \boldsymbol{v} - \boldsymbol{u}_\varepsilon) + \hat{g}j_\varepsilon(\boldsymbol{v}) - \hat{g}j_\varepsilon(\boldsymbol{u}_\varepsilon) + (z', z - T_\varepsilon)\\
& + a_T(T_\varepsilon, z - T_\varepsilon) + b_o(\boldsymbol{u}_\varepsilon, T_\varepsilon, z - T_\varepsilon) + b_s(T_\varepsilon - T_o, \boldsymbol{v} - \boldsymbol{u}_\varepsilon) - (Q, z - T_\varepsilon) - (\boldsymbol{F}, \boldsymbol{v} - \boldsymbol{u}_\varepsilon)\}\mathrm{d}t.
\end{aligned}$$

Due to Equations 5.29 and 5.30

$$Y_\varepsilon = \int_I \{(v' - u'_\varepsilon, v - u_\varepsilon) + (z' - T'_\varepsilon, z - T_\varepsilon)\}dt + \hat{g}\int_I \{j_\varepsilon(v) - j_\varepsilon(u_\varepsilon) - (j'_\varepsilon(u_\mu), v - u_\mu)\}dt.$$

The first term gives

$$\tfrac{1}{2}|v(t_1) - u_\varepsilon(t_1)|^2 + \tfrac{1}{2}|z(t_1) - T_\varepsilon(t_1)|^2 - \tfrac{1}{2}|v(t_0) - u_\varepsilon(t_0)|^2 - \tfrac{1}{2}|z(t_0) - T_\varepsilon(t_0)|^2$$

and since the functional $v \to j_\varepsilon(v)$ is convex then the second term is ≥ 0 and therefore $Y_\varepsilon \geq 0$. Hence and due to the a priori estimates there exist subsequences; we denote them $\{T_\varepsilon\}$, $\{u_\varepsilon\}$ such that $\{(T_\varepsilon, u_\varepsilon)\} \to (T, u)$ in topology corresponding to Equation 5.41. Hence

$$\int_I \{(v', v - u) + \hat{\mu}a(u, v - u) + b(u, u, v) + \hat{g}j(v) + (z', z - T) + a_T(T, z - T) + b_o(u, T, z)$$

$$+ b_s(T - T_o, v) - (Q, z - T) - (f, v - u)\}dt \geq \liminf \int_I \{a_T(T_\varepsilon, T_\varepsilon) + \hat{\mu}a(u_\varepsilon, u_\varepsilon) + \hat{g}j_\varepsilon(u_\varepsilon)\}dt$$

$$\geq \int_I \{a_T(T, T) + \hat{\mu}a(u, u) + \hat{g}j(u)\}dt$$

as $\liminf \int_I j_\varepsilon(u_\varepsilon)dt \geq \int_I j(u)dt$. Therefore, we proved that (T, u) is a pair of functions solving our problem.

To prove the uniqueness of the solution, we shall assume that (T_1, u_1) and (T_2, u_2) are two solutions of Equations 5.24 and 5.25. Then $(\underline{T}, \underline{u}) = (T_1 - T_2, u_1 - u_2)$. Then making some modifications we find

$$\frac{d}{dt}\{|\underline{T}(t)|^2 + |\underline{u}(t)|^2\} \leq ch(t)\{|\underline{T}(t)|^2 + |\underline{u}(t)|^2\},$$

where $h(t) = \|\underline{T}(t)\|^2 + \|\underline{u}(t)\|^2$, $h(t) \in L^1(I)$ and where estimates based on the properties of bilinear and trilinear forms were used. Then

$$|\underline{T}|^2 + |\underline{u}|^2 \leq c\int_I h(\tau)(|\underline{T}(\tau)|^2 + |\underline{u}(\tau)|^2)d\tau,$$

Hence $\underline{T} = 0$, $\underline{u} = 0$, which completes the proof.

The functional $j(u)$ is a non-differential functional and it will be approximated by the use of multipliers m_{ij} defined by Equations 5.19-5.21. Therefore $j(u) = 2^{1/2}(m_{ij}, D_{ij}(u))$. Then we will solve the equivalent variational problem:

Problem $(\mathscr{P}_m)_V$: Find a pair of functions (T, u), $T \in {}^1V$, $u \in V$ such that

$$(\rho c_e T'(t), z) + b_o(u(t), T(t), z) + a_T(T(t), z) - (Q(t), z) = 0 \quad \forall z \in {}^1V, \tag{5.42}$$

$$(\rho u'(t), v) + \hat{\mu}a(u(t), v) + b(u(t), u(t), v) + b_s(T(t) - T_o, v) + 2^{1/2}\hat{g}(m_{ij}, D_{ij}(v)) - (f(t), v) = 0 \quad \forall v \in V, \tag{5.43}$$

$$T(x, t_o) = T_o(x), \tag{5.44}$$

$$u(x, t_o) = u_o(x). \tag{5.45}$$

As a result we have the following theorem:

Theorem 5.2: Let $n = 2$, let $Q \in L^2(I; {}^1V')$, $f \in L^2(I; V')$, $T_o \in L^2(\Omega)$ and $u_o \in H$. Let a pair of functions (T, u) be a solution of the problem $(\mathscr{P})_v$ defined in Theorem 5.1. Then there exist functions $m = \{m_{ij}\}_{i,j=1,2}$ and p defined on Ω_t and satisfying Equations 5.19-5.21 and 5.23, i.e.

$$m_{ij} \in L^\infty(\Omega_t), \quad m_{ij} = m_{ji}, \quad \forall i, j, \quad m_{\text{ii}} = 0,$$

$$m_{ij} m_{ij} \le 1 \quad \text{a.e. in } \Omega_t,$$

$$m_{ij} D_{ij}(\boldsymbol{u}) = (D_{ij}(\boldsymbol{u}) D_{ij}(\boldsymbol{u}))^{1/2} = 2^{1/2} D_{\text{II}}^{1/2}(\boldsymbol{u}) \quad \text{a.e. in } \Omega_t,$$

$$\rho \frac{\partial u_i}{\partial t} - \hat{\mu} \Delta u_i + \rho u_j \frac{\partial u_i}{\partial x_j} - 2^{1/2} \hat{g} \frac{\partial m_{ij}}{\partial x_j} = F_i - \frac{\partial p}{\partial x_i} - \frac{\partial}{\partial x_j}(\beta_{ij}(T - T_0)), \quad i = 1, 2 \text{ a.e. in } \Omega_t.$$

Vice versa if the given T, $\boldsymbol{u}$, m, p are such that

$$T \in L^2(I; {}^1V), T' \in L^2(I; {}^1V'), \quad \boldsymbol{u} \in L^2(I; V), \quad \boldsymbol{u}' \in L^2(I; V'),$$
$$T(\boldsymbol{x}, t_o) = T_o(\boldsymbol{x}), \boldsymbol{u}(\boldsymbol{x}, t_o) = \boldsymbol{u}_o(\boldsymbol{x}), \tag{5.46}$$

and Equations 5.19-5.21 and 5.23 are satisfied then a pair of functions $(T, \boldsymbol{u})$ is a solution of the problem $(\mathscr{P})_v$ proved in Theorem 5.1.
To prove this theorem we need the following lemma

Lemma 5.1: Let $\mathscr{D}(\Omega)$ be the space of all infinitely continuously differentiable functions with compact support in $\Omega \subset R^2$. Let $\boldsymbol{u} = (u_i)$, $u_i \in \mathscr{D}(\Omega)'$, let $W = \{\boldsymbol{v} \mid \boldsymbol{v} \in [\mathscr{D}(\Omega)]^2, \operatorname{div} \boldsymbol{v} = 0\}$. Then there exists $p \in \mathscr{D}(\Omega)$ such that $\boldsymbol{u} = \operatorname{grad} p \Leftrightarrow (\boldsymbol{u}, \boldsymbol{v}) = 0$.

Proof of Theorem 5.2: It is sufficient to prove that the solution of the problem $(\mathscr{P}_m)_v$ also solves the problem $(\mathscr{P})_v$. Since the problem is quasi-coupled we shall analyse the second part of the problem only, where by f_i we denote i-th component of the body forces. Multiplying Equation 5.23 by $\boldsymbol{v} - \boldsymbol{u}(t)$, integrating over Ω, applying the Green theorem and the boundary conditions we obtain

$$(\rho \boldsymbol{u}'(t), \boldsymbol{v} - \boldsymbol{u}(t)) + \hat{\mu} a(\boldsymbol{u}(t), \boldsymbol{v} - \boldsymbol{u}(t)) + b(\boldsymbol{u}(t), \boldsymbol{u}(t), \boldsymbol{v} - \boldsymbol{u}(t)) + \hat{g} j(\boldsymbol{v}) - \hat{g} j(\boldsymbol{u}(t)) - (\boldsymbol{f}(t), \boldsymbol{v} - \boldsymbol{u}(t))$$
$$= \hat{g} j(\boldsymbol{v}) - \hat{g} j(\boldsymbol{u}(t)) - 2^{1/2}(m_{ij}, D_{ij}(\boldsymbol{v})) + 2^{1/2}(m_{ij}, D_{ij}(\boldsymbol{u}(t)) \equiv X. \tag{5.47}$$

According to Equation 5.21 $j(\boldsymbol{u}) = 2^{1/2}(m_{ij}, D_{ij}(\boldsymbol{u}))$, and due to Equations 5.19 and 5.20 we have

$$2^{1/2}(m_{ij}, D_{ij}(\boldsymbol{v})) \le 2^{1/2} \int_\Omega (m_{ij} m_{ij})^{1/2} (D_{ij}(\boldsymbol{v}) D_{ij}(\boldsymbol{v}))^{1/2} \mathrm{d}\boldsymbol{x}$$
$$\le 2^{1/2} \int_\Omega (D_{ij}(\boldsymbol{v}) D_{ij}(\boldsymbol{v}))^{1/2} \mathrm{d}\boldsymbol{x} = j(\boldsymbol{v}).$$

Hence and using Equation 5.47 it follows that $X \ge 0$, which proves that $\boldsymbol{u}$ is a solution of the problem $(\mathscr{P})_v$.

Now we shall prove the existence of m and p. The existence of p follows from Lemma 5.1. To prove the existence of m we define $B(\boldsymbol{u}, \boldsymbol{v}) \in V$ by $b(\boldsymbol{u}, \boldsymbol{v}, \boldsymbol{w}) = (B(\boldsymbol{u}, \boldsymbol{v}), \boldsymbol{w})$, $\boldsymbol{u}, \boldsymbol{v}, \boldsymbol{w} \in V$. Let us put

$$\mathscr{F} = \rho \boldsymbol{u}' - \hat{\mu} \Delta \boldsymbol{u} + B(\boldsymbol{u}, \boldsymbol{u}) - \boldsymbol{f} \in L^2(I, V'). \tag{5.48}$$

Then the problem $(\mathscr{P})_v$ can be rewritten as follows

$$(\mathscr{F}(t), \boldsymbol{v}) + \hat{g} j(\boldsymbol{v}) - ((\mathscr{F}(t), \boldsymbol{u}(t)) + \hat{g} j(\boldsymbol{u}(t))) \ge 0 \quad \forall \boldsymbol{v} \in V. \tag{5.49}$$

Let $\lambda \ge 0$ and let us put $\boldsymbol{v} = \pm \lambda \boldsymbol{v}$ in Equation 5.49. Then

$$\lambda(\pm(\mathscr{F}(t), \boldsymbol{v}) + \hat{g} j(\boldsymbol{v})) - ((\mathscr{F}(t), \boldsymbol{u}(t)) + \hat{g} j(\boldsymbol{u}(t))) \ge 0. \tag{5.50}$$

We shall put $v = v(t) \in L^2(I, V)$ in Equation 5.50 and integrating over I we obtain

$$\lambda\left\{\pm\int_I(\mathscr{F}(t), v(t))\mathrm{d}t + \int_I\hat{g}(j(v(t))\mathrm{d}t\right\} - \left\{\int_I((\mathscr{F}(t), u(t)) + \hat{g}j(u(t))\mathrm{d}t\right\} \geq 0$$

$$\forall v \in L^2(I, V), \quad \forall\lambda \geq 0. \tag{5.51}$$

Hence

$$\left|\int_I(\mathscr{F}(t), v(t))\mathrm{d}t\right| \leq \int_I\hat{g}j(v(t))\mathrm{d}t \quad \forall v \in L^2(I, V). \tag{5.52}$$

Then

$$\int_I((\mathscr{F}(t), u(t)) + \hat{g}j(u(t)))\mathrm{d}t = 0 \tag{5.53}$$

as from Equation 5.51 $\int_I((\mathscr{F}(t), u(t)) + \hat{g}j(u(t)))\mathrm{d}t \leq 0$, and from Equation 5.52 we have $\int_I((\mathscr{F}(t), u(t)) + \hat{g}j(u(t)))\mathrm{d}t \geq 0$ and therefore Equation 5.53 holds.

Now the corollary of the Hahn-Banach theorem (see Ekeland and Temam (1976)) will be used. Let us define the space

$$W = \{w \mid w = (w_{ij}), i, j = 1, 2, w_{ij} = w_{ji}, w_{ij} \in L^1(\Omega_t)\}, \|w\| = \int_{\Omega t}(w_{ij}w_{ij})^{1/2}\mathrm{d}x\,\mathrm{d}t.$$

Let the mapping $\Pi : v \to \{D_{ij}(v)\} = \{\frac{1}{2}\partial v_i/\partial x_j + \partial v_j/\partial x_i)\}$, $i, j = 1, 2$, functions from $L^2(I, V)$ into W. Then Equation 5.52 can be rewritten into the following form

$$\left|\int_I(\mathscr{F}(t), v(t))\mathrm{d}t\right| \leq \hat{g}2^{1/2}\,\|\Pi v\|\,.$$

Then due to the corollary of the Hahn-Banach theorem there exists $m \in W' = \{w \mid w = (w_{ij}), w_{ij} = w_{ji}, w_{ij} \in L^\infty(\Omega_t)\}$ such that

$$\int_I(\mathscr{F}(t), v(t))\mathrm{d}t = -\hat{g}2^{1/2}\int_{\Omega_t} m_{ij}D_{ij}(v(t))\mathrm{d}x\,\mathrm{d}t, \tag{5.54}$$

$$\|m\|_W \leq 1, \tag{5.55}$$

where we took $\{m_{ij}\} \in W'$ such that $D_{ii}(v) = 0$ and $m_{ii} = 0$. The last inequality (Equation 5.55) is equivalent to $m_{ij}m_{ij} \leq 1$, i.e. to Equation 5.20.
Due to Equation 5.48, then Equation 5.23 follows from Equation 5.54 immediately. Let us put $v = u$ in Equation 5.54. Hence and using Equation 5.53 we obtain

$$0 = \int_I(\mathscr{F}(t), u(t)) + \hat{g}j(u(t)))\mathrm{d}t = -\hat{g}2^{1/2}\int_{\Omega t} m_{ij}D_{ij}(u)\mathrm{d}x\,\mathrm{d}t + \hat{g}2^{1/2}\int_{\Omega_t} D_{ij}(u(t))D_{ij}(u)^{1/2}\mathrm{d}x\,\mathrm{d}t$$

$$= \hat{g}2^{1/2}\int_{\Omega t}((D_{ij}(u(t))D_{ij}(u(t)))^{1/2} - m_{ij}D_{ij}(u(t)))\mathrm{d}x\,\mathrm{d}t,$$

which proves Equation 5.21.

5.2.1.2. Finite element approximation

Let us suppose that Ω be approximated by a bounded polygonal domain Ω_h of R^2. Let $\{\mathscr{T}_h\}$ be a family of triangulations $\mathscr{T}_h$ of Ω such that $\overline{\Omega} = \underline{T}_h \cup_{\in\mathscr{T}_h} \underline{T}_h$. We set $h = h(\underline{T}_h)$ equal to the length of the greatest side of a triangle $\underline{T}_h : h = \max h(\underline{T}_h)$. Furthermore, we suppose that the points of Ω_h are vertices of the triangulation $\mathscr{T}_h$. For the given triangulation $\mathscr{T}_h$ we denote by B_i, $i = 1, \ldots, m$, all centres of sides of all triangles $\underline{T}_h \in \mathscr{T}_h$; by A_i, $i = 1, \ldots, m$, all vertices of all triangles $\underline{T}_h \in \mathscr{T}_h$; by $\mathscr{U}_h^B$

the set of centres B_i, $i = 1, \ldots, m$, laying inside Ω_h; and by $\mathscr{U}_h^K$ the set of all vertices laying inside Ω_h. Let us define the following finite element spaces

$$^1V_h = \{z \,|\, z \in C(\Omega), z|_{\underline{T}_h} \in P_1, z = 0 \text{ on } \Gamma_1, \quad \forall \underline{T}_h \in \mathscr{T}_h\},$$

where P_1 denotes the space of all confirming linear polynomials, and $W_h = \{\boldsymbol{v}_h \mid \boldsymbol{v}_h = (v_{h1}, v_{h2}), v_{hi} \in \underline{P}_1 \, \forall \underline{T}_h \in \mathscr{T}_h$, continuous in points B_j, $j = 1, \ldots, m$, equal to zero in B_i laying on $\Gamma_{2h}, \boldsymbol{v}_h(x) = 0$ for $x \in \Omega \backslash \overline{\Omega}_h\}$, where $\underline{P}_1$ denotes the space of all non-conforming linear polynomials, i.e. of all $\boldsymbol{w}_h : \boldsymbol{w}_h(B) = 1$, $\boldsymbol{w}_h(B_i) = 0$, $B_i \neq B$, $i = 1, \ldots, m$, $B \in \mathscr{U}_h^B$. Moreover, $\boldsymbol{w}_h \boldsymbol{e}_i (i = 1, 2, \boldsymbol{e}_1 = (1, 0), \boldsymbol{e}_2 = (0, 1), B \in \mathscr{U}_h^B)$ creates the basis of W_h. Let

$$V_h = \{\boldsymbol{v}_h \mid \boldsymbol{v}_h \in W_h, \sum_{i=1,2} D_i u_{\text{hi}} = 0\},$$

where we denote by div $\boldsymbol{u}_h = \sum_{i=1,2} D_i u_{\text{hi}}$ and $V_h \subset W_h \subset [L^2(\Omega)]^2 \, \forall h$. We denote by

$$(T_h, z_h)_h = \sum_{i=1}^{2} (D_i T_h, D_i z_h)_o \quad \forall T_h, z_h \in {}^1V_h, \tag{5.56}$$

$$(\boldsymbol{u}_h, \boldsymbol{v}_h)_h = \sum_{i=1}^{2} (D_i \boldsymbol{u}_h, D_i \boldsymbol{v}_h)_o \quad \forall \boldsymbol{u}_h, \boldsymbol{v}_h \in W_h, \tag{5.57}$$

$$(u_h, v_h)_h = (u_h, v_h)_o + \sum_{i=1}^{2} (D_i u_h, v_h)_o \quad \forall u_h, v_h \in W_h \tag{5.58}$$

scalar products in ${}^1V_h$ or W_h, respectively. These scalar products induce norms $|.|_h = |.|$, $\|.\|_h = \|.\|$ in $L^2(\Omega)$, ${}^1V_h$, V_h, W_h, respectively. For $\boldsymbol{u}_h \in W_h$ partial derivative $D_i \boldsymbol{u}_h = (D_i u_{h1}, D_i u_{h2})^T$ is defined inside every triangle $\underline{T}_h \in \mathscr{T}_h$, $D_i u_{hj} =$ constant, $\boldsymbol{x} \in \underline{T}_h$, $i, j = 1, 2$.

For $T_h, z_h \in {}^1V_h$, $\boldsymbol{u}_h, \boldsymbol{v}_h, \boldsymbol{w}_h \in W_h$

$$a_{Th}(T_h, z_h) = \sum_{\underline{T}_h \in \mathscr{T}_h} \int_{\underline{T}_h} \kappa_{ij} \partial T_h / \partial x_i \partial z_h / \partial x_j \mathrm{d}\boldsymbol{x}, \tag{5.59}$$

$$b_{oh}(\boldsymbol{u}_h, g_h, z_h) = \sum_{\underline{T}_h \in \mathscr{T}_h} \int_{\underline{T}_h} \rho c_e u_{hj} \partial g_h / \partial x_j z_h \mathrm{d}\boldsymbol{x}, \tag{5.60}$$

$$a_h(\boldsymbol{u}_h, \boldsymbol{v}_h) = \sum_{\underline{T}_h \in \mathscr{T}_h} \int_{\underline{T}_h} D_{ij}(\boldsymbol{u}_h) D_{ij}(\boldsymbol{v}_h) \mathrm{d}\boldsymbol{x}, \tag{5.61}$$

$$b_h(\boldsymbol{u}_h, \boldsymbol{v}_h, \boldsymbol{w}_h) = \sum_{\underline{T}_h \in \mathscr{T}_h} \int_{\underline{T}_h} \rho u_{hi} \partial v_{hj} / \partial x_i w_{hj} \mathrm{d}\boldsymbol{x}, \tag{5.62}$$

$$b_{sh}(T_h, \boldsymbol{v}_h) = \sum_{\underline{T} \in \mathscr{T}_h} \int_{\underline{T}_h} \partial / \partial x_j (\beta_{ij} (T_h - T_o)) v_{hi} \mathrm{d}\boldsymbol{x}, \tag{5.63}$$

$$j_h(\boldsymbol{u}_h) = 2 \sum_{\underline{T}_h \in \mathscr{T}_h} \int_{\underline{T}_h} (D_{II}(\boldsymbol{u}_h))^{1/2} \mathrm{d}\boldsymbol{x}. \tag{5.64}$$

Let the interval I be divided into n sub-intervals $I_i = \langle t_{i-1}, t_i \rangle$, $i = 1, \ldots, n$, of lengths $k = t_1/n$, $t_i = ik$. Let (T_h^i, u_h^i) be an approximation of $(T_h(x, t), \boldsymbol{u}_h(\boldsymbol{x}, t))$ for $t = t_i$. Furthermore, we define

$$\boldsymbol{f}_k^{i+\Theta} = k^{-1} \int_{ik}^{(i+1)k} \boldsymbol{f}(t)\mathrm{d}t, \boldsymbol{f}_k^{i+\Theta} \in [L^2(\Omega)]^2 \text{ for } O \le \Theta \le 1, \quad i = 0, 1, \ldots, n-1, \tag{5.65}$$

$$W_k^{i+\Theta} = k^{-1} \int_{ik}^{(i+1)k} W(t)\mathrm{d}t, W_k^{i+\Theta} \in L^2(\Omega) \text{ for } O \le \Theta \le 1, \quad i = 0, 1, \ldots, n-1. \tag{5.66}$$

Multipliers $m = \{m_{ij}\}_{i,j=1,2}$ are approximated by functions $m_h = \{m_{hij}\}_{i,j=1,2}$, which are piecewise constant on every $\underline{T}_h \in \mathscr{T}_h$.

If $(T_h^i, \boldsymbol{u}_h^i)$, $(T_h^{i+1}, \boldsymbol{u}_h^{i+1})$ are known then

$$T_h^{i+\Theta} = \Theta T_h^{i+1} + (1 - \Theta)T_h^i, \quad i = 0, 1, \ldots, n-1, \quad 0 \le \Theta \le 1, \tag{5.67}$$

$$\boldsymbol{u}_h^{i+\Theta} = \Theta \boldsymbol{u}_h^{i+1} + (1 - \Theta)\boldsymbol{u}_h^i, \quad i = 0, 1, \ldots, n-1, \quad 0 \le \Theta \le 1, \tag{5.68}$$

$$T_n(t) = T_h^i + (t - t_i)/(\Theta k)^{-1}(T_h^{i+\Theta} - T_h^i)$$
$$\text{for } t_i < t \le t_{i+1}, \quad 0 \le \Theta \le 1, \quad i = 0, 1, \ldots, n-1, \tag{5.69}$$

$$\boldsymbol{u}_n(t) = \boldsymbol{u}_h^i + (t - t_i)/(\Theta k)^{-1}(\boldsymbol{u}_h^{i+\Theta} - \boldsymbol{u}_h^i)$$
$$\text{for } t_i < t \le t_{i+1}, \quad 0 \le \Theta \le 1, \quad i = 0, 1, \ldots, n-1, \tag{5.70}$$

where $T_h(t)$, $\boldsymbol{u}_h(t)$ are linear on $\langle ik, (i+1)k \rangle$, $i = 0, 1, \ldots, n-1$, and $T_n(t) : \langle t_o, t_1 \rangle \to {}^1V_h$, $\boldsymbol{u}_n(t) : \langle t_o, t_1 \rangle \to W_h$. The corresponding step-functions

$$\overline{T}_n(t) = T_h^{i+\Theta}, ik < t \le (i+1)k, \quad i = 0, 1, \ldots, n-1, \ 0 \le \Theta \le 1, \ \overline{T}_n(0) = T_o, \tag{5.71}$$

$$\overline{\boldsymbol{u}}_n(t) = \boldsymbol{u}_h^{i+\Theta}, ik < t \le (i+1)k, \quad i = 0, 1, \ldots, n-1, \ 0 \le \Theta \le 1, \ \overline{\boldsymbol{u}}_n(0) = \boldsymbol{u}_o, \tag{5.72}$$

are piecewise constant on $\langle t_o, t_1 \rangle$. Furthermore,

$$T'(t) \simeq T_h'(t) = (T_h^{i+1} - T_h^i)/k = (T_h^{i+\Theta} - T_h^i)/(\Theta k), \quad ik < t \le (i+1)k,$$
$$0 \le \Theta \le 1, \quad i = 0, 1, \ldots, n-1, \tag{5.73}$$

$$\boldsymbol{u}'(t) \simeq \boldsymbol{u}_h'(t) = (\boldsymbol{u}_h^{i+1} - \boldsymbol{u}_h^i)/k = (\boldsymbol{u}_h^{i+\Theta} - \boldsymbol{u}_h^i)/(\Theta k), \quad ik < t \le (i+1)k,$$
$$0 \le \Theta \le 1, \quad i = 0, 1, \ldots, n-1. \tag{5.74}$$

Then we have the following scheme

Problem ($\mathscr{P}_h$): Let $(T_h^o, \boldsymbol{u}_h^o)$, $(T_h^1 \boldsymbol{u}_h^1)$, ..., $(T_h^i, \boldsymbol{u}_h^i)$ be known, then $(T_h^{i+\Theta}, \boldsymbol{u}_h^{i+\Theta})$ and $m_h^{i+\Theta}$ can be determined by

$$(\rho c_e T_h^{i+\Theta}, z_h) + k\Theta a_{Th}(T_h^{i+\Theta}, z_h) + k\Theta b_{\text{oh}}(\boldsymbol{u}_h^i, T_h^i, z_h) = k\Theta(W_k^{i+\Theta}, z_h) + (\rho c_e T_h^i, z_h) \quad \forall z_h \in {}^1V_h, \tag{5.75}$$

$$m_{hjl}^{i+\Theta} m_{hjl}^{i+\Theta} \leq 1 \text{ on every } \underline{T}_h \in \mathscr{T}_h, m_{h12}^{i+\Theta} = m_{h21}^{i+\Theta}, \tag{5.76}$$

$$m_{hjl}^{i+\Theta} D_{jl}(\boldsymbol{u}_h^{i+\Theta}) = (D_{jl}(\boldsymbol{u}_h^{i+\Theta}) D_{jl}(\boldsymbol{u}_h^{i+\Theta}))^{1/2} \text{ on every } \underline{T}_h \in \mathscr{T}_h, \tag{5.77}$$

$$(\rho \boldsymbol{u}_h^{i+\Theta}, \boldsymbol{v}_h) + k\Theta\hat{\mu} a_h(\boldsymbol{u}_h^{i+\Theta}, \boldsymbol{v}_h) + 2^{1/2} k\Theta\hat{g}(m_{hjl}^{i+\Theta}, D_{jl}(\boldsymbol{v}_h)) + k\Theta b_{sh}(T_h^i, \boldsymbol{v}_h) + k\Theta b_h(\boldsymbol{u}_h^i, \boldsymbol{u}_h^i, \boldsymbol{v}_h)$$

$$= k\Theta(\boldsymbol{f}_k^{i+\Theta}, \boldsymbol{v}_h) + (\rho \boldsymbol{u}_h^i, \boldsymbol{v}_h), \boldsymbol{v}_h \in V_h \quad \text{for } i = 0, \ldots, n-1. \tag{5.78}$$

Then $(T_h^{i+1}, \boldsymbol{u}_h^{i+1})$ can be determined from Equations 5.67 and 5.68.

First, we shall prove the stability conditions of the scheme given by Equations 5.75-5.78; second, we shall show that the approximate solution of Equations 5.75-5.78 converges to the solution of the problem $(\mathscr{P}_m)_v$ (i.e. of Equations 5.42-5.45. According to Temam (1976), Nedoma (1988) it holds

$$|\boldsymbol{u}_h| \leq d_o \, \|\boldsymbol{u}_h\| \quad \forall \boldsymbol{u}_h \in W_h, d_o \text{ does not depend on } h, \tag{5.79}$$

$$\|\boldsymbol{u}_h\| \leq S(h) \, |\boldsymbol{u}_h| \quad \forall \boldsymbol{u}_h \in W_h, S(h) \text{ depends on } h, \tag{5.80}$$

$$|b_{oh}(\boldsymbol{u}_h, T_h, z_h)| \leq c_o \, |\boldsymbol{u}_h| \, |T_h| \, \|z_h\| \quad \forall \boldsymbol{u}_h \in V_h, T_h, z_h \in {}^1V_h, \tag{5.81}$$

$$|b_h(\boldsymbol{u}_h, \boldsymbol{v}_h, \boldsymbol{w}_h)| \leq d_1 \, \|\boldsymbol{u}_h\| \, \|\boldsymbol{v}_h\| \, \|\boldsymbol{w}_h\| \quad \forall \boldsymbol{u}_h, \boldsymbol{v}_h, \boldsymbol{w}_h \in V_h, \tag{5.82}$$

where d_1 does not depend on h. Moreover,

$$|b_h(\boldsymbol{u}_h, \boldsymbol{u}_h, \boldsymbol{v}_h)| \leq d_1 \, \|\boldsymbol{u}_h\|^2 \, \|\boldsymbol{v}_h\| \leq d_1 S^2(h) \, |\boldsymbol{u}_h| \, \|\boldsymbol{u}_h\| \, |\boldsymbol{v}_h| \leq S_1(h) \, |\boldsymbol{u}_h| \, \|\boldsymbol{u}_h\| \, |\boldsymbol{v}_h| \quad \forall \boldsymbol{u}_h, \boldsymbol{v}_h \in V_h, \tag{5.83}$$

where $S_1(h) \leq d_1 S^2(h)$ and

$$b_h(\boldsymbol{u}_h, \boldsymbol{u}_h, \boldsymbol{u}_h) = 0 \quad \forall \boldsymbol{u}_h \in V_h. \tag{5.84}$$

Furthermore,

$$b_{\text{oh}}(\boldsymbol{u}_h, T_h, z_h) + b_{\text{oh}}(\boldsymbol{u}_h, z_h, T_h) = 0 \quad \forall T_h, z_h \in {}^1V_h, \boldsymbol{u}_h \in V_h, \tag{5.85}$$

$$|b_{sh}(T_h, \boldsymbol{u}_h)| \leq c(1 + \|T_h\| \, |\boldsymbol{u}_h|) \quad \forall T_h \in {}^1V_h, \boldsymbol{u}_h \in V_h. \tag{5.86}$$

Theorem 5.3: Let k, h fulfill the following conditions

$$kS_1(h) \leq d_2, \tag{5.87}$$

$$kS^2(h) \leq d_3, \tag{5.88}$$

where constants d_2, d_3 do not depend on k, h. Then $(T_h^i, \boldsymbol{u}_h^i)$ are defined by the scheme given by Equations 5.75-5.78, and it holds

$$\|T_h^i\|^2 \leq c_T, \tag{5.89}$$

$$k \sum_{i=0}^{n-1} \|T_h^{i+\Theta}\|^2 \leq c_T, \tag{5.90}$$

$$|\boldsymbol{u}_h^i|^2 \leq c_u, \tag{5.91}$$

$$k \sum_{i=0}^{n-1} \left\| \boldsymbol{u}_h^{i+\Theta} \right\|^2 \le c_u, \tag{5.92}$$

$$\sum_{i=0}^{n=1} \left| \boldsymbol{u}_h^{i+1} - \boldsymbol{u}_h^{i} \right|^2 \le c_u, \tag{5.93}$$

where c = constant is independent of k and h and $\frac{1}{2} < \Theta \le 1$. Moreover, the scheme is stable and

$$\left\| T_h - \overline{T}_n(t) \right\| \to 0 \text{ strongly in } L^2(I; L^2(\Omega)) \text{ for } k, h \to 0, \tag{5.94}$$

$$\left\| \boldsymbol{u}_h - \overline{\boldsymbol{u}}_n(t) \right\| \to 0 \text{ strongly in } L^2(I; [L^2(\Omega)]^2) \text{ for } k, h \to 0. \tag{5.95}$$

Let a pair of functions $(T, \boldsymbol{u})$ be the solution of the problem $(\mathscr{P}_m)_v$, and $(T_h, \boldsymbol{u}_h)$ be an approximate solution of $(\mathscr{P}_m)_v$, $\frac{1}{2} < \Theta \le 1$. Let k, h satisfy Equations 5.87 and 5.88; then for $(k, h) \to 0$ a pair of functions $(T_h, \boldsymbol{u}_h) \to (T, \boldsymbol{u})$ weakly in $L^2(I; {}^1V_h) \times L^2(I; V_h)$.

Sketch of the proof : according to Equations 5.73 and 5.74 we put $z_h = T_h^{i+\Theta}$, $v_h = \boldsymbol{u}_h^{i+\Theta}$ in Equations 5.75-5.78. Since $2ab \le a^2 + b^2$, $\mu \ge 0$ and due to Equations 5.79-5.82

$$\left| 2k(Q_k^{i+\Theta}, T_h^{i+\Theta}) \right| \le 2kd_o \left| Q_k^{i+\Theta} \right| \left\| T_h^{i+\Theta} \right\|,$$

$$\left| 2k(\boldsymbol{f}_k^{i+\Theta}, \boldsymbol{u}_h^{i+\Theta}) \right| \le 2kd_o \left| \boldsymbol{f}_k^{i+\Theta} \right| \left\| \boldsymbol{u}_h^{i+\Theta} \right\| \le k\mu \left\| \boldsymbol{u}_h^{i+\Theta} \right\|^2 + \mu k d_o^2 \left| \boldsymbol{f}_k^{i+\Theta} \right|^2,$$

$$\left| 2kb_h(\boldsymbol{u}_h^i, \boldsymbol{u}_h^i, \boldsymbol{u}_h^{i+\Theta}) \right| = \left| 2kb_h(\boldsymbol{u}_h^i, \boldsymbol{u}_h^i, \boldsymbol{u}_h^{i+\Theta} - \boldsymbol{u}_h^i) \right| \le c_1 k^2 \Theta^2 S_1^2(h) \left| \boldsymbol{u}_h^i \right|^2 \left\| \boldsymbol{u}_h^i \right\|^2 + c_1^{-1} \left| \boldsymbol{u}_h^{i+1} - \boldsymbol{u}_h^i \right|^2,$$

where $c_1 > 0$ and such that $(2\Theta - 1)\rho - c_1^{-1} > 0$,

$$\left| 2kb_{\rm sh}(T_h^i, \boldsymbol{u}_h^i) \right| \le ck(1 + \left\| T_h^i \right\| \left| \boldsymbol{u}_h^i \right|).$$

Furthermore, for $T_h \in {}^1V_h$, $\boldsymbol{u}_h \in W_h$ we have

$$a_{Th}(T_h^i, T_h^i) \ge c \left\| T_h^i \right\|^2, a_h(\boldsymbol{u}_h^i, \boldsymbol{u}_h^i) \ge \sum_{\underline{T}_h \in \mathscr{T}_h} \int_{\underline{T}_h} (D_{ij}(\boldsymbol{u}_h^i))^2 \mathrm{d}\boldsymbol{x} = \left\| \boldsymbol{u}_h^i \right\|^2,$$

$$2k2^{1/2} \hat{g}(m_{hjl}^{i+\Theta}, D_{jl}(\boldsymbol{u}_h^{i+\Theta})) \ge 0.$$

Summing up Equations 5.75-5.78 ($\boldsymbol{v}_h = \boldsymbol{u}_h^{n+\Theta}$) for $i = 0, \dots, r-1$, we then obtain

$$\left| T_h^r \right|^2 + ck \sum_{j=1}^{r-1} \left\| T_h^j \right\|^2 \le |T_{\rm oh}|^2 + ck \sum_{j=1}^{r-1} \left(\left\| Q_k^{j+\Theta} \right\|_*^2 + \left| T_h^i \right|^2 \right), \tag{5.96}$$

$$\rho \left| \boldsymbol{u}_h^r \right|^2 + ((2\Theta - 1) - c_1^{-1}) \sum_{j=0}^{r-1} \left| \boldsymbol{u}_h^{j+1} - \boldsymbol{u}_h^j \right|^2 + k\hat{\mu} \sum_{j=0}^{r-1} \left\| \boldsymbol{u}_h^{j+\Theta} \right\|^2$$

$$- c_1 k^2 \Theta S_1^2(h) \sum_{j=1}^{r-1} \left| \boldsymbol{u}_h^j \right|^2 \left\| \boldsymbol{u}_h^i \right\|^2 + ck \sum_{j=1}^{r-1} \left(1 - \left\| T_h^j \right\| \left| \boldsymbol{u}_h^j \right| \right) \le \rho \left| \boldsymbol{u}_{\rm oh} \right|^2 + \mu k d_o^2 \sum_{j=0}^{r-1} \left| \boldsymbol{f}_k^{j+\Theta} \right|^2$$

$$+ c_1 k^2 \Theta^2 S_1^2(h) \left| \boldsymbol{u}_{oh} \right|^2 \left\| \boldsymbol{u}_{\rm oh} \right\|^2 + ck\,(1 - \left\| T_{\rm oh} \right\| \left| \boldsymbol{u}_{\rm oh} \right|) \equiv \lambda_k, \tag{5.97}$$

where $r = 1, \dots, n$ and $\rho^{-1} c_1 \Theta S_1^2(h) \lambda_n \le \mu - \delta$ for an arbitrary δ, $0 < \delta < \mu$. The value λ_n depends on the data of the problem only. Then Equations 5.96 and 5.97 give Equations 5.89 to 5.93. In particular, from Equations 5.89 and 5.91 we have

$$\sup_{t \in I} |T_n(t)| \le c_T^{1/2}, \sup_{t \in I} |\boldsymbol{u}_n(t)| \le c_U^{1/2},$$

$$\int_0^{t_1} \|p_h T_n(t)\|^2 \, \mathrm{d}t \le \int_0^{t_1} c_t \|T_n(t)\|^2 \, \mathrm{d}t \le c_t c_T, \quad (\|p_h\| = 1, \|p_h T_h\| \le c_t \|T_h\| \; \forall T_h \in {}^1V_h, c_t \text{ does not depend on } h),$$

$$\int_0^{t_1} \|p_h \boldsymbol{u}_n(t)\|^2 \, \mathrm{d}t \le \int_0^{t_1} c_u \|\boldsymbol{u}_n(t)\|^2 \, \mathrm{d}t \le c_u c_U, \quad (\|p_h\| = 1, \|p_h u_h\| \le c_U \|\boldsymbol{u}_h\| \; \forall \boldsymbol{u}_h \in V_h, c_U \text{ does not depend on } h),$$

and then the scheme is conditionally stable if Equations 5.87 and 5.88 are valid (where p_h is the so-called operator of extension).

By means of $(T_h^i, \boldsymbol{u}_h^i)$ $(i = 0, 1, \ldots, n-1)$ we construct $(\overline{T}_n(t), \overline{\boldsymbol{u}}_n(t))$ from Equations 5.71,5.72 and $(T_n(t), \boldsymbol{u}_n(t))$ from Equations 5.69, 5.70. The estimates in Equations 5.89-5.93 can be rewritten in the form

$$|\mathrm{d}T_n(t)/\mathrm{d}t| \le c \text{ for a.e. } t \in I, \left\|\overline{T}_n(t)\right\| \le c \quad \forall n \ge n_o, t \in I, \tag{5.98}$$

$$|\mathrm{d}\boldsymbol{u}_n(t)/\mathrm{d}t| \le c \text{ for a.e. } t \in I, \|\overline{\boldsymbol{u}}_n(t)\| \le c \quad \forall n \ge n_o, t \in I. \tag{5.99}$$

We rewrite Equations 5.75 and 5.78 in the form

$$\left(\rho c_e \frac{\mathrm{d}\overline{T}_n(t)}{\mathrm{d}t}, z_h\right) + a_{Th}^n(\overline{T}_n(t), z_h) + b_{\mathrm{oh}}(\overline{\boldsymbol{u}}_n(t), \overline{T}_n(t), z_h) = (\overline{Q}_n(t), z_h) \quad \forall z_h \in {}^1V_h \text{ and a.e. } t \in I, \tag{5.100}$$

$$\left(\rho \frac{\mathrm{d}\overline{\boldsymbol{u}}_n(t)}{\mathrm{d}t}, \boldsymbol{v}_h\right) + \hat{\mu} a_h^n(\overline{\boldsymbol{u}}_n(t), \boldsymbol{v}_h) + 2^{1/2}\hat{g}(m_{\mathrm{hjl}}, D_{\mathrm{jl}}(\boldsymbol{v}_h)) + b_h(\overline{\boldsymbol{u}}_n(t), \overline{\boldsymbol{u}}_n(t), \boldsymbol{v}_h) + b_{\mathrm{sh}}(\overline{T}_n(t), \overline{\boldsymbol{u}}_n(t))$$
$$= (\overline{\boldsymbol{f}}_n(t), \boldsymbol{v}_h) \quad \forall \boldsymbol{v}_h \in V_h \text{ and a.e. } t \in I, \tag{5.101}$$

where $a_{Th}^n(T, z) = a_{Th}(T, z)$, $a_h^n(\boldsymbol{u}, \boldsymbol{v}) = a_h(\boldsymbol{u}, \boldsymbol{v})$ for $t_i < t \le t_{i+1}$, $i = 0, 1, \ldots, n-1$. Subtracting Equations 5.100 and 5.101 for $n = r$ and $n = s$, where $z = \overline{T}_r(t) - \overline{T}_s(t)$, $\boldsymbol{v} = \overline{\boldsymbol{u}}_r(t) - \overline{\boldsymbol{u}}_s(t)$, integrating over (t_o, t), owing to the estimates

$$\left|T_n(t) - \overline{T}_n(t)\right| \le c/n \text{ for } t_i \le t \le t_{i+1}, \quad i = 0, 1, \ldots, n-1,$$

$$\left|a_{Th}(\overline{T}_r(t), \overline{T}_r(t) - \overline{T}_s(t)) - a_{Th}^r(\overline{T}_r(t), \overline{T}_r(t) - \overline{T}_s(t))\right| \le c/r,$$

$$\left\|\overline{W}_r(t) - W(t)\right\|_* \left\|\overline{T}_r(t) - \overline{T}_s(t)\right\| \le (2\varepsilon)^{-1} \left\|\overline{Q}_r(t) - Q(t)\right\|_*^2 + \tfrac{1}{2}\varepsilon \left\|\overline{T}_r(t) - \overline{T}_s(t)\right\|^2$$

$$|\boldsymbol{u}_n(t) - \overline{\boldsymbol{u}}_n(t)| \le c/n \text{ for } t_i \le t \le t_{i+1}, \quad i = 0, 1, \ldots, n-1,$$

$$\left|a_h(\overline{\boldsymbol{u}}_r(t), \overline{\boldsymbol{u}}_r(t) - \overline{\boldsymbol{u}}_s(t)) - a_h^r(\overline{\boldsymbol{u}}_r(t), \overline{\boldsymbol{u}}_r(t) - \overline{\boldsymbol{u}}_s(t))\right| \le c/r,$$

$$\left\|\overline{\boldsymbol{f}}_r(t) - \boldsymbol{f}(t)\right\|_* \|\overline{\boldsymbol{u}}_r(t) - \overline{\boldsymbol{u}}_s(t)\| \le (2\varepsilon)^{-1} \left\|\overline{\boldsymbol{f}}_r(t) - \boldsymbol{f}(t)\right\|_*^2 + \tfrac{1}{2}\varepsilon \|\overline{\boldsymbol{u}}_r(t) - \overline{\boldsymbol{u}}_s(t)\|^2$$

and using the Gronwall Lemma, and Lemmas 1.3.18 and 1.3.15 of Kačur (1985), we obtain the weak and the strong convergences of $T_n \to T$ in $C(I; L^2(\Omega))$, $\overline{T}_n \to T$ in $L^2(I, {}^1V)$, $\mathrm{d}T_n/\mathrm{d}t \to \mathrm{d}T/\mathrm{d}t$ weakly in $L^2(I; L^2(\Omega))$, $\overline{T}_n \to T$ weakly in ${}^1V \; \forall t \in I$. The second part of this assertion is parallel to that of Fortin (1972). The uniqueness of the solution can be proved by the standard technique.

5.2.1.3. *Algorithm*

Since the problem is quasi-coupled, the algorithm can be constructed separately for the thermal and the visco-plastic parts of the problem.

The algorithm of the thermal part of the problem is based on the method of discretization in time and finite element approximation (Kolář, Kratochvíl, Leitner and Ženíšek (1979), Zlámal (1974), Oden and Reddy (1976)). The obtained scheme is the stable scheme. In every time level the obtained system of algebraic equations is solved by the conjugate gradient method.

The algorithm of the visco-plastic part of the problem is based on the scheme given in Equations 5.76–5.78, which is also the stable scheme, but a certain problem represents finding the basis of V_h with div $\boldsymbol{v}_h = 0$. In our contribution the Uzawa algorithm for such a problem will be used. Moreover, we shall show that the problem in Equations 5.76–5.78 is equivalent to finding a saddle point (see for example Céa (1971), Ekeland and Temam (1976)). As the functional $j_h(\boldsymbol{v}_h)$ is a non-differentiable functional, the Uzawa algorithm will be used. Both problems then lead us to solve the systems of linear algebraic equations with sparse symmetric positive definite matrices. For their solution the conjugate gradient method will be used. Since the problem is very similar to that investigated in Chapter 12, the algorithm will be discussed in detail there.

5.2.2. Models based on the Stefan-like Problem and Visco-Plastic Bingham Rheology: the Case with Melting and Solidification

To discuss the geodynamic processes in the lithosphere we must study processes passing through the mantle. We saw that the lower mantle is characterized by the thermo-visco-plastic materials heated from the bottom and cooled from overhead. Resultant effects which ensued are convective currents, the convective cells and the channels of higher temperature and low viscosity (the GTLV-channels). The light hot viscous materials from the mantle-core boundary then flow through the GTLV-channel, where they are heated up to 1300°C and reach the lithosphere. The lithosphere is then heated and melted as well as mechanically stressed vertically and horizontally. The light hot viscous materials penetrate under the lithosphere, and one part of these materials also reach the Earth's surface (in the middle ocean's regions), where they then solidify. In the neighbourhood of the GTLV-channels the neighbouring rocks are melted, and at greater distances are recrystallized. Due to the convective currents the rift zones are opening on the different places of the Earth's surface.

To describe this problem mathematically we shall assume the visco-plastic Bingham rheology, as this is an optimal rheology from the physical and the approximation of the reality point of view, as well as from the computation point of view. It is evident that due to the extremely complicated shape of the front for melting, recrystallization and solidification, respectively, the description of the reality in the rift zones cannot be described in full. As a result practically, instead of the streams of light materials flowing through the GTLV-channels, we obtain the phase change zones, as the forecast of the fine geometry would need a very expensive computation as well as the necessity of regularity of the front, where both phases are parallel at the same time — the so-called mushy zones. In these mushy zones both parallel phases are microscopically present.

The geophysical model of the rift zones is derived from the conservation principles. The equation of momentum in its differential form gives the equation of motion, and the mass conservation specifies the rocks as incompressible or weakly compressible, and also gives the solute equation. The energy conservation then gives the heat equation. Moreover, the stress-strain rate relation for the visco-plastic Bingham rheology will be given. Since the problem represents a part of the problem describing the global model of the Earth, we give here the main idea of the problem only and the detailed analysis will be given in Chapter 12 and Nedoma (1997).

In the case of thermo-visco-plastic Bingham's rheology, from the law of conservation of momentum the equations of motion

$$\rho\left(\frac{\partial u_i}{\partial t}+u_k\frac{\partial u_i}{\partial x_k}\right)=\frac{\partial \tau_{ij}}{\partial x_j}+\rho f_i \text{ in } \Omega\times I,\quad i,j=1,\dots,N,\ N=2,3,\ I=(t_o,t_1), \tag{5.102}$$

where u_i are components of the velocity vector, ρf_i components of the body forces, and τ_{ij} the stress tensor defined by

$$\tau_{ij}=-p\delta_{ij}+\hat{g}D_{ij}D_{\mathrm{II}}^{-1/2}+2\hat{\mu}D_{ij}-\beta_{ij}(T-T_o), \tag{5.103}$$

$$D_{ij}=\tfrac{1}{2}(\partial u_i/\partial x_j+\partial u_j/\partial x_i),\ D_{\mathrm{II}}=\tfrac{1}{2}D_{ij}D_{ij}\neq 0, \tag{5.104}$$

where the first three terms on the right hand side of Equation 5.103 represent the stress-strain rate relation in the Bingham rheology and the last term represents the thermal stresses.

We shall assume that the investigated rocks in the rift zone are incompressible so that

$$\operatorname{div}\boldsymbol{u}=0 \quad \text{in } \Omega\times I. \tag{5.105}$$

The diffusion equation for rock mixtures follows from the mass conservation law and therefore, we have

$$\rho\left(\frac{\partial c}{\partial t}+\boldsymbol{u}\operatorname{grad} c\right)=-\operatorname{div}\boldsymbol{i} \quad \text{in } \Omega\times I, \tag{5.106}$$

where c denotes the concentration and $\boldsymbol{i}$ is the density of the diffusion flow defined as

$$\boldsymbol{i}=-\rho D(\operatorname{grad} c+k_T T^{-1}\operatorname{grad} T+k_p p^{-1}\operatorname{grad} p), \tag{5.107}$$

and where D is the coefficient of diffusion, $k_T D$ and $k_p D$ are the coefficients of thermodiffusion and barodiffusion, T is the temperature and p is the pressure.

The energy conservation law gives the generalized heat equation in the form

$$\rho c_e\left(\frac{\partial T}{\partial t}+\boldsymbol{u}\operatorname{grad} T\right)+\gamma\beta_{ij}T_o e_{ij}(\boldsymbol{u})=\operatorname{div}(\kappa(T)\operatorname{grad} T)+Q \quad \text{in } \Omega\times I. \tag{5.108}$$

We assume that the investigated region $\Omega=\cup_{\iota=1}^{S}\Omega^{\iota}$ occupied by the rift zone has the lipschitzian boundary $\partial\Omega=\Gamma_u\cup\Gamma_\tau\cup\mathscr{R}$. Most of the interest of the investigated problem lies in the properties of the boundaries between the phases. In such cases the latent heat is released or absorbed at the phase temperature and then we distinguish firstly the case of melting or recrystallization and secondly the case of solidification.

The appropriate discontinuities and the governing diffusion equation can be derived from an integral formulation of the principle of heat conservation. This principle is a simplified version of the principle of energy conservation in which both phases are assumed to be incompressible and at rest. Across any phase change boundary, denoted as $R^{\iota,s}(t)$, there must be fulfilled the conservation of energy, momentum and mass, and the latter then requires that the material density is continuous. This phase change is assumed to take place at some specified temperature-the temperature of melting, recrystallization and/or solidification. These problems are known as Stefan-like problems. The usual treatment of a Stefan-like problem assumes the existence of some moving surfaces such that the material is in its solid, more visco-plastic phase on one side of the phase change surface and in its liquid, more strongly visco-plastic phase near liquid phase on the other side. Energy balance arguments then show that although the temperature is continuous everywhere, the heat flux has a jump discontinuity across the phase change surface, known as the Stefan conditions, i.e.

$$T_S^{\iota,s} = T_L^{\iota,s} = T_R^{\iota,s}, \tag{5.109}$$

$$\rho^{\iota,s} L^{\iota,s} v_n^{\iota,s} + \kappa_{ij}^{\iota} \partial T^{\iota}/\partial x_j \nu_i^{\iota}\Big|_S^s - \kappa_{ij}^{\iota} \partial T^{\iota}/\partial x_j \nu_i^{\iota}\Big|_L^s = 0 \text{ on } R^{\iota,s}(t), \tag{5.110}$$

where $R^{\iota,s}(t)$ are phase change boundaries between two different phases, $L^{\iota,s}$ are latent heats, $T_S^{\iota,s}$, $T_L^{\iota,s}$, $T_R^{\iota,s}$, $\rho^{\iota,s}$ are temperatures of solid-visco-plastic state, of strongly visco-plastic with low threshold of plasticity $\hat{g}$ or temperature of phase changes and densities associated with $\Omega^{\iota,s}$, and $\vec{\nu}^{\iota,s}$ are normals to phase change boundaries $R^{\iota,s}(t)$.

Detailed analyses of the model problem of the evolution of rift zones are similar to that of the global geodynamic model of the Earth (see Nedoma (1997)), which will be studied in detail in Chapter 12. Therefore, we do not analyse the model problem here; analyses of the problem are given in the Chapter 12.

REFERENCES

Céa, J. (1971), *Optimisation: Théorie et Algorithms*, Dunod, Paris.

Duvaut, G. and Lions, J.L. (1972), *Les Inéquations en Mécanique et en Physique*, Dunod, Paris.

Ekeland, I. and Temam, R. (1976), *Convex Analysis and Variational Problems*, North-Holland, Amsterdam.

Fortin, M. (1972), *Calcul numérique des Ecoulements des Fluides de Bingham et des Fluides Newtonians Incompressible par la Méthode de Elements Finis*, Thése d'Etat Université Pierre et Marie Curie, Paris.

Kačur, J. (1985), "Method of Rothe in Evolution Equations", *Teubner-Texte zur Mathematik*, B.G. Teubner Vlg., Leipzig.

Kolář, V., Kratochvíl, J., Leitner, F. and Ženíšek, A.(1979), *The Computation of Two- and Three-Dimensional Structures by the Finite Element Method*, SNTL, Prague (in Czech).

Nedoma, J. (1988), "On the New Problem in the Potential Theory", in *Potential Theory* (Král, J.*et al.* (eds)), Pergamon Press, New York.

Nedoma, J. (1995), "Equations of Magnetodynamics of Incompressible Thermo-Bingham's Fluid Under the Gravity Effect", *J. Comput. Appl. Math.*, **59**,109-128.

Nedoma, J. (1997), "On a Coupled Stefan-like Problem in the Thermo-Visco-Plastic Rheology", J. Comput. Appl. Math., 84, 45-80

Oden, J.T. and Reddy, J.N. (1976), *An Introduction to the Mathematical Theory of Finite Elements*, John Wiley and Sons, New York, London, Sydney, Toronto.

Temam, R. (1976), *Navier-Stokes Equations, Theory and Numerical Analysis*, North-Holland Publ. Co., Amsterdam, New York, Oxford.

Zlámal, M. (1974), "Finite Element Methods for Parabolic Equations", *Mathematics of Computation*, **28**, 393-404.

6

Simple Models of Collision and Subduction Zones

6.1. INTRODUCTION

Let us define the mathematical models of the collision and subduction zones, which in the first approximation will be based on the basic boundary value problems of elasticity, visco-elasticity, thermo-elasticity and plasticity. In this section we shall discuss several model problems. The first type of model will be based on linear elasticity the isotropic and anisotropic cases. The second one, due to the composite structure of the rocks, will be based on the homogenization method, discussed in detail in Chapter 3 with connection to the overcoring method and the stress-strain monitoring in situ. These types of model lead to the first type of anisotropic model, and will be regarded as a special case there of. The third, fourth and fifth types of model will be based on boundary value problems of thermo-elasticity and plasticity. The final model will be based on the so-called contact problems of Signorini type and the penalty technique.

6.2. MATHEMATICAL MODELS BASED ON LINEAR ELASTICITY

Let Ω be the domain in R^N, $N = 2, 3$, with a Lipschitz boundary $\partial\Omega$. Let the boundary $\partial\Omega$ be divided into disjoint parts Γ_τ, Γ_u, Γ_c such that $\partial\Omega = \Gamma_\tau \cup \Gamma_u \cup \Gamma_c \cup \mathscr{R}$, where the surface measure of $\mathscr{R}$ is zero and each of the boundaries Γ_τ, Γ_u and Γ_c are open sets in $\partial\Omega$. Let us suppose that the domain Ω is occupied by the lithosphere, about which we shall assume that it is isotropic or anisotropic and that the Lamé coefficients λ and μ as well as the elastic coefficients c_{ijkl} are bounded measurable functions, i.e.

$$\lambda(\boldsymbol{x}),\ \mu(\boldsymbol{x}),\ c_{ijkl}(\boldsymbol{x}) \in L^\infty(\Omega).$$

The relation between the displacement vector $\boldsymbol{u} = (u_i)$ and small strain tensor $e_{ij}(\boldsymbol{u})$ is defined by

$$e_{ij}(\boldsymbol{u}) = \tfrac{1}{2}(u_{i,j} + u_{j,i}) \equiv \tfrac{1}{2}(\partial u_i/\partial x_j + \partial u_j/\partial x_i), \quad i, j = 1, \dots, N,$$

where we use the notation $\partial u_i/\partial x_j \equiv u_{i,j}$.

The relation between the stress and strain tensors is defined by the linear Hooke's law as

$$\tau_{ij}(\boldsymbol{x}) = c_{ijkl}(\boldsymbol{x})e_{kl}(\boldsymbol{u}), \quad i, j, = 1, \dots, N, \tag{6.1}$$

where the coefficients c_{ijkl} form a matrix (e.g. of the type $(3 \times 3 \times 3 \times 3)$ for $N = 3$) and satisfy

$$a_o \le c_{ijkl}\xi_{ij}\xi_{kl}\,|\xi|^{-2} \le A_o < +\infty \quad \forall \boldsymbol{x} \in \Omega, \quad \xi = (\xi_{ij}) \in R^9, \tag{6.2}$$

$$c_{ijkl} = c_{jikl} = c_{klij} = c_{ijlk}, \quad i, j, k, l = 1, \dots, N, \tag{6.3}$$

where a_o, A_o are constants independent of $\boldsymbol{x} \in \Omega$ and $\xi \in R^9$, and where $a_o = 2\min\{\mu(\boldsymbol{x}); \boldsymbol{x} \in \Omega\}$, $A_o = \max\{2\mu(\boldsymbol{x}) + 3\lambda(\boldsymbol{x}); \boldsymbol{x} \in \Omega\}$.
In the dual formulation we use the inverse Hooke's law defined by

$$e_{ij}(\boldsymbol{u}) = s_{ijkl}(\boldsymbol{x})\tau_{kl}(\boldsymbol{x}), \tag{6.4}$$

where coefficients $s_{ijkl}(\boldsymbol{x})$ are of the same type as $c_{ijkl}(\boldsymbol{x})$, i.e. a tensor of the 4th order, and satisfy

$$A_o^{-1} \le s_{ijkl}\zeta_{ij}\zeta_{kl}\,|\zeta|^{-2} \le a_o^{-1} < +\infty \quad \forall \boldsymbol{x} \in \Omega, \zeta = (\zeta_{ij}) \in R^9, \tag{6.5}$$

$$s_{ijkl} = s_{jikl} = s_{klij} = s_{ijlk}, \quad i, j, k, l, = 1, \dots, N, \tag{6.6}$$

and

$$c_{ijkl}s_{ijkl} = \tfrac{1}{2}(\delta_{ik}\delta_{jl} + \delta_{il}\delta_{jk}), \quad i, j, k, l, = 1, \dots, N, \tag{6.7}$$

where δ_{ij} is the Kronecker's symbol.
Particularly, for an isotropic material with Lamé's constants $\lambda(\boldsymbol{x})$ and $\mu(\boldsymbol{x})$ we have

$$c_{ijkl} = \lambda\delta_{ij}\delta_{kl} + \mu(\delta_{ik}\delta_{jl} + \delta_{il}\delta_{jk}), \quad i, j, k, l, = 1, \dots, N, \tag{6.8}$$

$$s_{ijkl} = (4\mu)^{-1}(\delta_{ik}\delta_{jl} + \delta_{il}\delta_{jk}) - \lambda[2\mu(3\lambda + 2\mu)]^{-1}\delta_{ij}\delta_{kl'} \quad i, j, k, l, = 1, \dots, N. \tag{6.9}$$

The stress tensor satisfies the equilibrium conditions

$$\tau_{ij,j}(\boldsymbol{x}) + f_i(\boldsymbol{x}) = 0 \quad \forall \boldsymbol{x} \in \Omega. \tag{6.10}$$

On the boundary $\partial\Omega$ we define the stress vector $\vec{\tau}$ and the displacement vector $\boldsymbol{u}$, and the normal and tangential components by

$$\tau_i(\boldsymbol{x}) = \tau_{ij}(\boldsymbol{x})n_j, \; \tau_n(\boldsymbol{x}) = \tau_i(\boldsymbol{x})n_i = \tau_{ij}(\boldsymbol{x})n_in_j,$$

$$\vec{\tau}_t(\boldsymbol{x}) = \vec{\tau}(\boldsymbol{x}) - \tau_n(\boldsymbol{x})\boldsymbol{n} = (\tau_{ti}), \; \tau_{ti}(\boldsymbol{x}) = \tau_{ij}(\boldsymbol{x})n_j - \tau_n(\boldsymbol{x})n_i,$$

$$u_n(\boldsymbol{x}) = u_i(\boldsymbol{x})n_i, \; \boldsymbol{u}_t(\boldsymbol{x}) = \boldsymbol{u} - u_n\boldsymbol{n} = (u_{ti}), \; u_{ti} = u_i - u_nn_i, \quad i = 1, \dots, N, \quad N = 2, 3.$$

In our further investigations we shall consider the following three types of boundary conditions:

$$\tau_{ij}(\boldsymbol{x})n_j = P_i(\boldsymbol{x}) \quad \text{on } \Gamma_\tau, \tag{6.11}$$

which represents the loading of one part of the lithospheric plate, in many cases the Earth's surface of the lithospheric plate, and where $\boldsymbol{P}$ represents the loading forces acting on sites Γ_τ of the geological block;

$$\boldsymbol{u}(\boldsymbol{x}) = \boldsymbol{u}_o(\boldsymbol{x}) \quad \text{on } \Gamma_u, \tag{6.12}$$

by which we prescribe the movements of the lithospheric plates or blocks, or their components; and

$$u_n(\boldsymbol{x}) = 0, \vec{\tau}_t(\boldsymbol{x}) = 0 \quad \text{on } \Gamma_c, \tag{6.13}$$

which describes the contact of the lithospheric plates or blocks with their perfectly (absolutely) rigid and perfectly smooth support, and without friction. These conditions are also valid in the plane of symmetry of the investigated region, that is both loaded and supported symmetrically with respect to this plane.

It is evident that in special model problems some part of the boundary $\partial\Omega$ may be absent. In many present geophysical papers the boundary $\Gamma_c \equiv 0$. In these papers a part of the boundary is replaced by a boundary, denoted by Γ'_u, and the condition in Equation 6.13 is replaced by the so-called roller condition, given by Equation 6.12, in which the vertical component $u_{o3}(\boldsymbol{x})$ is equal to zero and the components $u_{o1}(\boldsymbol{x})$ and $u_{o2}(\boldsymbol{x})$ describe the horizontal components of the movement of the lithospheric plates or blocks. That the so-called roller condition (Equation 6.12) is of a type of $\boldsymbol{u}(\boldsymbol{x}) = (u_{o1}(\boldsymbol{x}), u_{o2}(\boldsymbol{x}), 0)$. Since the lithospheric plates and blocks during the investigated (from the geological point of view very short) time period are deformed, the vertical component $u_3(\boldsymbol{x})$ on Γ'_u cannot be equal to zero, but, on the contrary, it is a result of the deformation of the lithospheric plate at its bottom parts. Then, for better approximation of the situation at the bottom of the lithospheric plate, we will prescribe in Equation 6.12 one or both horizontal components of the displacement vector $\boldsymbol{u}_o(\boldsymbol{x})$; whilst the vertical (and/or one horizontal) displacement component on Γ'_u is unknown (free), and it (they) will be determined by the numerical computation of the model problem.

In addition to the cases discussed above, on this part of the boundary Γ_c we can consider the conditions of elastic support in the normal direction, i.e.

$$u_n = -g_n(\boldsymbol{x})\tau_n(\boldsymbol{x}), g_n(\boldsymbol{x}) \geq g_o > 0, \tag{6.14}$$

along with the conditions in the tangent directions (for example in special cases in the horizontal directions) such that

$$u_{t_j} = 0, \quad j = 1, 2, (3), \tag{6.15}$$

or

$$\tau_{t_j} = 0, \quad j = 1, 2, (3), \tag{6.16}$$

or

$$u_{tj} = -g_t(\boldsymbol{x})\tau_{tj}(\boldsymbol{x}), \quad g_t(\boldsymbol{x}) \geq g_o > 0. \tag{6.17}$$

Remark: In Equation 6.13, only three conditions are independent since the components τ_{tj} are constrained by the relation $\tau_{tj}n_j = 0$.

Since it is very difficult to determine the horizontal components of the velocity of the lithosphere at its bottom boundary with the asthenosphere, i.e. to determine the roller conditions, then the conditions of Equation 6.13 or of Equations 6.14 and 6.15, or 6.14 and 6.16, or 6.14 and 6.17 must be given. But it is well known that the results obtained from geodynamic models with roller conditions in a certain domain of the lithosphere, namely near the contact boundary with the asthenosphere, cannot be used for the geophysical analysis of plate tectonics, as the numerical results obtained are not an adequate representation of the physical reality. Similarly, the conditions in Equation 6.14 and Equation 6.15 or 6.16 or 6.17 cannot adequately describe the physical reality in this domain, i.e. in the region near the lithosphere/asthenosphere contact boundary, as it is very difficult to precisely determine the functions $g_n(\boldsymbol{x})$ and $g_t(\boldsymbol{x})$.

From all these arguments it follows that we must find a type of boundary conditions which better describe physical conditions on the bottom of the lithosphere, namely on the contact of the lithosphere with the asthenosphere. Such types of conditions will be discussed in Section 6.6.

As in Chapter 3 the elasticity problem can be formulated in terms of displacements as well as in terms of stresses, but in the next section we shall limit ourselves namely to primary types of problems (i.e. problems formulated in the displacements) and to dual problems, which will be discussed for the contact problem only.

Since these types of problems are not our main interest, we shall only give an idea of how to solve these problems. We will pay greater attention to the realistic models of the collision as well as subduction zones, and these will be studied in more detail in Chapters 7 and 8.

6.3. SIMPLE MODELS OF COLLISION AND SUBDUCTION ZONES BASED ON THE BASIC BOUNDARY VALUE PROBLEMS OF ELASTICITY

6.3.1. Formulation of the Model Problem

Let us assume that the lithosphere and the asthenosphere occupy the domain $\Omega \in R^3$ with a Lipschitz boundary $\partial\Omega$. Let the boundary $\partial\Omega$ be divided into disjoint parts Γ_τ, Γ_u, Γ_c and $\mathscr{R}$ such that $\partial\Omega = \Gamma_\tau \cup \Gamma_u \cup \Gamma_c \cup \mathscr{R}$, as defined in the previous section. We shall assume that the lithosphere and the asthenosphere are isotropic or anisotropic, and that the Lamé coefficients λ and μ as well as the elastic coefficients c_{ijkl} are bounded measurable functions, i.e.

$$\lambda(\boldsymbol{x}),\ \mu(\boldsymbol{x}),\ c_{ijkl}(\boldsymbol{x}) \in L^\infty(\Omega).$$

The stress tensor τ_{ij} and the small strain tensor $e_{ij}(\boldsymbol{u})$ are defined by

$$\tau_{ij}(\boldsymbol{x}) = c_{ijkl}(\boldsymbol{x})e_{kl}(\boldsymbol{u}), \quad e_{ij}(\boldsymbol{u}) = \tfrac{1}{2}(u_{i,j} + u_{j,i}), \tag{6.18}$$

where the coefficients c_{ijkl} satisfy

$$a_o \leq c_{ijkl}\xi_{ij}\xi_{kl}\,|\xi|^{-2} \leq A_o < +\infty \quad \forall \boldsymbol{x} \in \Omega, \quad \xi = (\xi_{ij}) \in R^9, \tag{6.19}$$

where a_o, A_o are constants independent of $\boldsymbol{x} \in \Omega$ and $\xi \in R^9$, and

$$c_{ijkl} = c_{jikl} = c_{klij} = c_{ijlk}. \tag{6.20}$$

The inverse Hooke's law is defined by

$$e_{ij}(\boldsymbol{u}) = s_{ijkl}(\boldsymbol{x})\tau_{kl}(\boldsymbol{x}), \tag{6.21}$$

where the coefficients $s_{ijkl}(\boldsymbol{x})$ satisfy

$$A_o^{-1} \leq s_{ijkl}\zeta_{ij}\zeta_{kl}\,|\zeta|^{-2} \leq a_o^{-1} < +\infty \quad \forall \boldsymbol{x} \in \Omega, \quad \zeta = (\zeta_{ij}) \in R^9, \tag{6.22}$$

where $a_o = 2\min\{\mu(\boldsymbol{x}); \boldsymbol{x} \in \Omega\}$, $A_o = \max\{2\mu(\boldsymbol{x}) + 3\lambda(\boldsymbol{x}); \boldsymbol{x} \in \Omega\}$, and

$$s_{ijkl} = s_{jikl} = s_{klij} = s_{ijlk}, \tag{6.23}$$

$$c_{ijkl}s_{ijkl} = \tfrac{1}{2}(\delta_{ik}\delta_{jl} + \delta_{il}\delta_{jk}), \tag{6.24}$$

where δ_{ij} is the Kronecker's symbol. For an isotropic material

$$c_{ijkl} = \lambda\delta_{ij}\delta_{kl} + \mu(\delta_{ik}\delta_{jl} + \delta_{il}\delta_{jk}), \tag{6.25}$$

$$s_{ijkl} = (4\mu)^{-1}(\delta_{ik}\delta_{jl} + \delta_{il}\delta_{jk}) - \lambda[2\mu(3\lambda + 2\mu)]^{-1}\delta_{ij}\delta_{kl}. \tag{6.26}$$

The stress tensor satisfies the equilibrium conditions

$$\tau_{ij,j}(\boldsymbol{x}) + f_i(\boldsymbol{x}) = 0 \quad \forall \boldsymbol{x} \in \Omega. \tag{6.27}$$

On the boundary $\partial\Omega$ with stress vector $\vec{\tau}$ and displacement vector $\boldsymbol{u}$ the normal and tangential components are defined by

$$\tau_i(\boldsymbol{x}) = \tau_{ij}(\boldsymbol{x})n_j,\ \tau_n(\boldsymbol{x}) = \tau_i(\boldsymbol{x})n_i = \tau_{ij}(\boldsymbol{x})n_in_j,$$

$$\vec{\tau}_t(\boldsymbol{x}) = \vec{\tau}(\boldsymbol{x}) - \tau_n(\boldsymbol{x})\boldsymbol{n} = (\tau_{ti}),\ \tau_{ti}(\boldsymbol{x}) = \tau_{ij}(\boldsymbol{x})n_j - \tau_n(\boldsymbol{x})n_i, \quad i = 1, 2, 3,$$

$$u_n(\boldsymbol{x}) = u_i(\boldsymbol{x})n_i,\ \boldsymbol{u}_t(\boldsymbol{x}) = \boldsymbol{u} - u_n\boldsymbol{n} = (u_{ti}),\ u_{ti} = u_i - u_nn_i, \quad i = 1, 2, 3.$$

On the boundary Γ_τ the condition

$$\tau_{ij}(\boldsymbol{x})n_j = P_i(\boldsymbol{x}) \tag{6.28}$$

is given, where $\boldsymbol{P}$ represents the loading forces acting on sites Γ_τ of the geological block; on the boundary Γ_u the displacement

$$\boldsymbol{u}(\boldsymbol{x}) = \boldsymbol{u}_o(\boldsymbol{x}) \tag{6.29}$$

is prescribed, relating to the movements of the lithospheric plates or blocks,or their components; and on the boundary Γ_c the contact boundary

$$u_n(\boldsymbol{x}) = 0, \quad \vec{\tau}_t(\boldsymbol{x}) = 0 \tag{6.30}$$

is given.

As we said above, in many present geophysical papers the boundary $\Gamma_c \equiv 0$. Such problems lead to simple model problems, which will be denoted in the next section as the problem $\mathscr{P}1$.

Next we shall denote by Γ_k the part of the boundary $\partial\Omega$ where the so-called unilateral contact condition

$$u_n(\boldsymbol{x}) = 0, \quad \vec{\tau}_t(\boldsymbol{x}) = 0, \quad \text{on } \Gamma_k, \tag{6.31}$$

which describes the contact of the lithospheric plates or blocks with their perfectly (absolutely) rigid and perfectly smooth support, is given. In addition, on the boundary Γ_c we shall investigate the conditions of elastic support in the normal direction, i.e.

$$u_n = -g_n(\boldsymbol{x})\tau_n(\boldsymbol{x}), \quad g_n(\boldsymbol{x}) \geq g_o > 0, \tag{6.32}$$

along with the conditions in the tangent directions, such that

$$u_{tj} = 0 \quad j = 1, 2, (3), \tag{6.33}$$

or

$$\tau_{tj} = 0, \quad j = 1, 2, (3), \tag{6.34}$$

or

$$u_{tj} = -g_t(\boldsymbol{x})\tau_{tj}(\boldsymbol{x}), \quad g_t(\boldsymbol{x}) \geq g_o > 0, \tag{6.35}$$

are considered. This type of problem will be investigated as the problem $\mathscr{P}2$.

6.3.2. Variational Formulations and Weak Solutions of Primary and Dual Problems in Linear Elasticity

Let Ω be a domain occupied by the part of the lithosphere and/or together with a part of the upper mantle, as defined above, with a Lipschitz boundary $\partial\Omega = \Gamma_\tau \cup \Gamma_u \cup \mathscr{R}$. Let us assume that $\boldsymbol{f} \in [L^2(\Omega)]^3$, and $\boldsymbol{P} \in [L^2(\Gamma_\tau)]^3$. Next, we shall deal with the following problem:

Problem ($\mathscr{P}1$): Find a vector function $\boldsymbol{u}$ satisfying

$$(c_{ijkl}(\boldsymbol{x})e_{kl}(\boldsymbol{u})),_j + f_i(\boldsymbol{x}) = 0 \quad i = 1, 2, 3, \text{ in } \Omega, \tag{6.36}$$

$$\boldsymbol{u}(\boldsymbol{x}) = \boldsymbol{u}_o(\boldsymbol{x}) \quad \text{on } \Gamma_u, \tag{6.37}$$

$$\tau_{ij}(\boldsymbol{x})n_j = P_i(\boldsymbol{x}) \quad \text{on } \Gamma_\tau. \tag{6.38}$$

Let us define the following: the space of virtual displacements

$$V_o = \{\boldsymbol{v} \mid \boldsymbol{v} \in W = [H^1(\Omega)]^3, \quad \boldsymbol{v} = 0 \text{ in the sense of traces}\}, \tag{6.39}$$

a geometrically admissible displacement field

$$V = \{\boldsymbol{v} \mid \boldsymbol{v} \in W, \boldsymbol{u} - \boldsymbol{u}_o \in V_o\}, \tag{6.40}$$

further spaces of symmetric tensors

$$H = \{\tau \in [L^2(\Omega)]^9, \tau_{ij} = \tau_{ji}\}, \tag{6.41}$$

$$\Sigma = \left\{\tau \mid \tau \in H, \int_\Omega \tau_{ij}e_{ij}(\boldsymbol{v})\mathrm{d}\boldsymbol{x} = 0 \quad \forall \boldsymbol{v} \in V\right\}, \tag{6.42}$$

and a statically admissible stress field

$$\Sigma_o = \left\{ \tau \mid \tau \in H, \int_\Omega \tau_{ij} e_{ij}(\boldsymbol{v}) \mathrm{d}\boldsymbol{x} = \int_\Omega f_i v_i d\boldsymbol{x} + \int_{\Gamma_\tau} P_i v_i \mathrm{d}s \quad \forall \boldsymbol{v} \in V \right\}. \tag{6.43}$$

Definition 6.1: We say that a function $\boldsymbol{u} \in V$ is the so-called weak solution of the problem in Equation 6.36-6.38, if

$$\boldsymbol{u} - \boldsymbol{u}_o \in V,$$

and

$$\int_\Omega c_{ijkl}(\boldsymbol{x}) e_{ij}(\boldsymbol{u}) e_{kl}(\boldsymbol{v}) \mathrm{d}\boldsymbol{x} = \int_\Omega f_i(\boldsymbol{x}) v_i \mathrm{d}\boldsymbol{x} + \int_{\Gamma_\tau} P_i(\boldsymbol{x}) v_i \mathrm{d}s \quad \forall \boldsymbol{v} \in V. \tag{6.44}$$

From the theory of elasticity it is known that the formulation of the weak (generalized) solution is equivalent to the principle of virtual displacements. We shall formulate the problem in the form of basic principles of elasticity.

Formulation in terms of displacements:

Problem 1 (Principle of virtual displacements): Find $\boldsymbol{u} \in \boldsymbol{u}_o + V$ such that

$$\int_\Omega c_{ijkl}(\boldsymbol{x}) e_{ij}(\boldsymbol{u}) e_{kl}(\boldsymbol{v}) \mathrm{d}\boldsymbol{x} = \int_\Omega f_i(\boldsymbol{x}) v_i \mathrm{d}\boldsymbol{x} + \int_{\Gamma_\tau} P_i(\boldsymbol{x}) v_i \mathrm{d}s \quad \forall \boldsymbol{v} \in V. \tag{6.45}$$

Problem 2 (Principle of minimum potential energy): Find $\boldsymbol{u} \in \boldsymbol{u}_o + V$ minimizing the functional

$$L_o(\boldsymbol{u}) = \tfrac{1}{2} a(\boldsymbol{u}, \boldsymbol{u}) - \int_\Omega f_i(\boldsymbol{x}) u_i \mathrm{d}\boldsymbol{x} - \int_{\Gamma_\tau} P_i(\boldsymbol{x}) u_i \mathrm{d}s \tag{6.46}$$

on the set V, where $a(\boldsymbol{u}, \boldsymbol{v})$ is the bilinear form defined by

$$a(\boldsymbol{u}, \boldsymbol{v}) = \int_\Omega c_{ijkl}(\boldsymbol{x}) e_{ij}(\boldsymbol{u}) e_{kl}(\boldsymbol{v}) \mathrm{d}\boldsymbol{x}. \tag{6.47}$$

Remark: In the case when $\Gamma_u \equiv 0$ then the condition of total equilibrium

$$\int_\Omega f_i(\boldsymbol{x}) v_i \mathrm{d}\boldsymbol{x} + \int_{\Gamma_\tau} P_i(\boldsymbol{x}) v_i \mathrm{d}s = 0 \tag{6.48}$$

holds for all $\boldsymbol{v} \in \mathscr{P} = \{\boldsymbol{w} \mid \boldsymbol{w} \in W = [H^1(\Omega)]^3, \boldsymbol{w} = \boldsymbol{a} + \boldsymbol{b} \times \boldsymbol{x}\}$.

Formulation in terms of stresses:

Problem 3 (Modification of the principle of virtual stresses): Find $\tau \in \Sigma$ such that

$$\int_\Omega s_{ijkl}(\boldsymbol{x}) \tau_{ij}(\boldsymbol{x}) \sigma_{kl}(\boldsymbol{x}) \mathrm{d}\boldsymbol{x} = \int_\Omega e_{ij}(\boldsymbol{u}_o) \sigma_{ij}(\boldsymbol{x}) \mathrm{d}\boldsymbol{x} \quad \forall \tau \in \Sigma. \tag{6.49}$$

Problem 4 (Principle of minimum complementary energy): Find $\tau \in \Sigma_o$ minimizing the functional

$$\mathscr{S}\tau = \tfrac{1}{2} b(\tau, \tau) - \int_\Omega e_{ij}(\boldsymbol{u}_o) \tau_{ij}(\boldsymbol{x}) \mathrm{d}\boldsymbol{x} \tag{6.50}$$

on the set Σ_o of statically admissible stress fields, where $b(\tau, \tau)$ is the bilinear form defined by

$$b(\tau, \sigma) = \int_\Omega s_{ijkl}(\boldsymbol{x})\tau_{ij}(\boldsymbol{x})\sigma_{kl}(\boldsymbol{x})\mathrm{d}\boldsymbol{x}. \tag{6.51}$$

Both the formulations in Problem 1 and Problem 2, and both formulations in Problem 3 and Problem 4 are equivalent. Under the above assumptions the bilinear forms $a(\boldsymbol{u}, \boldsymbol{v})$ and $b(\tau, \sigma)$ and the functionals $L_o(\boldsymbol{v})$ and $\mathscr{S}(\sigma)$ are coercive. Then the solutions $\boldsymbol{u}$ of Problems 1 and 2 and τ of Problems 3 and 4 exist, are unique and are connected by Equation 6.18 or Equation 6.21.

Let us prove these assertions. It is known that if the functional $L_o : V \to R^1$ is coercive and weakly lower semi-continuous, then it attains its minimum in V. The functional L_o possesses a Gâteaux differential at every point $\boldsymbol{u} \in W = [H^1(\Omega)]^3$ and

$$\begin{aligned} DL_o(\boldsymbol{u}, \boldsymbol{v}) &= \mathrm{d}/\mathrm{d}t\{\tfrac{1}{2}\int_\Omega c_{ijkl}(\boldsymbol{x})e_{ij}(\boldsymbol{u}+t\boldsymbol{v})e_{kl}(\boldsymbol{u}+t\boldsymbol{v})\mathrm{d}\boldsymbol{x} \\ &\quad - \int_\Omega f_i(\boldsymbol{x})(u_i + tv_i)\mathrm{d}\boldsymbol{x} - \int_{\Gamma_\tau} P_i(\boldsymbol{x})(u_i + tv_i)\mathrm{d}s\}\Big|_{t=0} \\ &= \int_\Omega c_{ijkl}(\boldsymbol{x})e_{ij}(\boldsymbol{u})e_{kl}(\boldsymbol{v})\mathrm{d}\boldsymbol{x} - \int_\Omega f_i(\boldsymbol{x})v_i\mathrm{d}\boldsymbol{x} - \int_{\Gamma_\tau} P_i(\boldsymbol{x})v_i\mathrm{d}s. \end{aligned} \tag{6.52}$$

Let us put $\boldsymbol{u} = \boldsymbol{u}_o + \boldsymbol{w}$ for $\boldsymbol{w} \in V$, then $L_o(\boldsymbol{u}) = L_o(\boldsymbol{u}_o + \boldsymbol{w}) = \Psi(\boldsymbol{w})$. Hence

$$D\Psi(\boldsymbol{w}, \boldsymbol{v}) = DL_o(\boldsymbol{u}_o + \boldsymbol{w}, \boldsymbol{v}) = DL_o(\boldsymbol{u}, \boldsymbol{v}) \quad \forall \boldsymbol{v}, \boldsymbol{w} \in V. \tag{6.53}$$

The functional L_o attains its minimum at the point $\boldsymbol{u}^o = \boldsymbol{u}_o + \boldsymbol{w}^o$ if, and only if, the functional Ψ attains its minimum at the point $\boldsymbol{w}^o$; i.e. $D\Psi(\boldsymbol{w}^o, \boldsymbol{v}) = 0\, \forall \boldsymbol{v} \in V$, which implies that $DL_o(\boldsymbol{u}^o, \boldsymbol{v}) = 0\, \forall \boldsymbol{v} \in$V. From Equations 6.52 and 6.44 we see that $\boldsymbol{u}^o$ is a weak solution of our investigated problem.

It is known that if the functional is coercive and weak lower semi-continuous then it attains its minimum in V (see Chapter 2). Then the existence of a weak solution of our problem follows from the existence of the minimum of $\Psi(\boldsymbol{w})$, for which we must verify its coerciveness and weak lower semi-continuity.

To prove its coerciveness we use Equation 6.19, then

$$c_{ijkl}(\boldsymbol{x})e_{ij}(\boldsymbol{u})e_{kl}(\boldsymbol{u}) \geq \mu_o(\boldsymbol{x})e_{ij}(\boldsymbol{u})e_{ij}(\boldsymbol{u}) \quad \text{a.e. in } \Omega,$$

and then

$$\begin{aligned} L_o(\boldsymbol{u}_o + \boldsymbol{w}) &= \tfrac{1}{2}\int_\Omega c_{ijkl}(\boldsymbol{x})e_{ij}(\boldsymbol{u}_o + \boldsymbol{w})e_{kl}(\boldsymbol{u}_o + \boldsymbol{w})\mathrm{d}\boldsymbol{x} \\ &\quad - \int_\Omega f_i(\boldsymbol{x})(u_{oi} + w_i)\mathrm{d}\boldsymbol{x} - \int_{\Gamma_\tau} P_i(\boldsymbol{x})(u_{oi} + w_i)\mathrm{d}s \\ &\geq \mu_o(\boldsymbol{x})\int_\Omega e_{ij}(\boldsymbol{u}_o + \boldsymbol{w})e_{ij}(\boldsymbol{u}_o + \boldsymbol{w})\mathrm{d}\boldsymbol{x} - \int_\Omega f_i(\boldsymbol{x})(u_{oi} + w_i)\mathrm{d}\boldsymbol{x} \\ &\quad - \int_{\Gamma_\tau} P_i(\boldsymbol{x})(u_{oi} + w_i)\mathrm{d}s \end{aligned}$$

$$\geq c_1 \|\boldsymbol{w}\|_W^2 - c_2 \|\boldsymbol{u}_o\|_W \|\boldsymbol{w}\|_W - c_3 \|\boldsymbol{f}\|_{[L^2(\Omega)]^3} \|\boldsymbol{w}\|_W - c_4 \|\boldsymbol{P}\|_{[L^2(\Gamma_\tau)]^3} \|\boldsymbol{w}\|_W - c_5, \tag{6.54}$$

where all constants c_1-c_5 are positive and independent of $\boldsymbol{w}$. Hence

$$\Psi(\boldsymbol{w}) = L_o(\boldsymbol{u}_o + \boldsymbol{w}) \to +\infty \quad \text{as } \|\boldsymbol{w}\|_W \to +\infty.$$

Thus we have proved that the functional $\Psi(\boldsymbol{w})$ is coercive.

To prove its lower semi-continuity then for points $\boldsymbol{w} + \boldsymbol{v}$ and $\boldsymbol{v}$ we have

$$D\Psi(\boldsymbol{w} + \boldsymbol{v}, \boldsymbol{v}) - D\Psi(\boldsymbol{w}, \boldsymbol{v}) = \int_\Omega c_{ijkl}(\boldsymbol{x}) e_{ij}(\boldsymbol{v}) e_{kl}(\boldsymbol{v}) \mathrm{d}\boldsymbol{x} \geq c \|\boldsymbol{v}\|_W^2 \quad \forall \boldsymbol{v} \in V,$$

where Equations 6.52 and 6.53, the positivity of elastic coefficients and the Korn inequality were used. From Chapter 2, the differential $D\Psi(\boldsymbol{w}, \boldsymbol{v})$ is continuous on any line segment in the variable $\boldsymbol{w}$, and hence the functional $\Psi(\boldsymbol{w})$ is weakly lower semi-continuous.

To prove the uniqueness of the weak solution, the standard technique can be used. Let $\boldsymbol{u}_1$ and $\boldsymbol{u}_2$ be two weak solutions of our problem. Let $\boldsymbol{w} = \boldsymbol{u}_1 - \boldsymbol{u}_2$, then $\boldsymbol{w} \in V$. Then from the definition of the weak solution given in Equation 6.44 we have

$$\int_\Omega c_{ijkl}(\boldsymbol{x}) e_{ij}(\boldsymbol{w}) e_{kl}(\boldsymbol{v}) \mathrm{d}\boldsymbol{x} = 0 \quad \forall \boldsymbol{v} \in V.$$

Hence, putting $\boldsymbol{w} = \boldsymbol{v}$ and since $c_{ijkl}(\boldsymbol{x}) > 0$, then using the Korn inequality, we find $\|\boldsymbol{w}\|_V = 0$, which is the same as $\boldsymbol{u}_1 = \boldsymbol{u}_2$. The uniqueness of the weak solution is then proved.

To prove the dependence of the solution on the boundary conditions, and body and surface forces we use the following idea:

From Equation 6.44, substituting $\boldsymbol{u} = \boldsymbol{u}_o + \boldsymbol{w}$, we obtain

$$\int_\Omega c_{ijkl}(\boldsymbol{x}) e_{ij}(\boldsymbol{w}) e_{kl}(\boldsymbol{v}) \mathrm{d}\boldsymbol{x} = -\int_\Omega c_{ijkl}(\boldsymbol{x}) e_{ij}(\boldsymbol{u}_o) e_{kl}(\boldsymbol{v}) \mathrm{d}\boldsymbol{x} + \int_\Omega f_i(\boldsymbol{x}) v_i \mathrm{d}\boldsymbol{x} + \int_{\Gamma_\tau} P_i(\boldsymbol{x}) v_i \mathrm{d}s \quad \forall \boldsymbol{v} \in V.$$

Hence, putting $\boldsymbol{v} = \boldsymbol{w}$ and using the Korn inequality, we have

$$c \|\boldsymbol{w}\|_W^2 \leq \int_\Omega c_{ijkl}(\boldsymbol{x}) e_{ij}(\boldsymbol{w}) e_{kl}(\boldsymbol{w}) \mathrm{d}\boldsymbol{x} \leq c \|\boldsymbol{w}\|_W (\|\boldsymbol{u}_o\|_W + \|\boldsymbol{f}\|_{[L^2(\Omega)]^3} + \|\boldsymbol{P}\|_{[L^2(\Gamma_\tau)]^3}).$$

Hence

$$\|\boldsymbol{u}\|_W \leq c(\|\boldsymbol{u}_o\|_W + \|\boldsymbol{f}\|_{[L^2(\Omega)]^3} + \|\boldsymbol{P}\|_{[L^2(\Gamma_\tau)]^3}) \leq C. \tag{6.55}$$

Then we have proved the following theorem.

Theorem 6.1: Let $\boldsymbol{u}_o \in W$, $\boldsymbol{f} \in [L^2(\Omega)]^3$, and $\boldsymbol{P} \in [L^2(\Gamma_\tau)]^3$ be given functions. Let $\Gamma_u \neq 0$ and let $c_{ijkl}(\boldsymbol{x}) \geq c_o > 0$ a.e. $\boldsymbol{x} \in \Omega$. Then there exists one and only one weak solution of the

Problem $\mathscr{P}1$ and

$$\|\boldsymbol{u}\|_W \leq c(\|\boldsymbol{u}_o\|_W + \|\boldsymbol{f}\|_{[L^2(\Omega)]^3} + \|\boldsymbol{P}\|_{[L^2(\Gamma_\tau)]^3}). \tag{6.56}$$

Problem 4 has a unique solution as the bilinear form $b(\tau, \sigma)$ is symmetric and H-elliptic, and Σ_o is non-empty, closed in H and convex. Moreover, following a similar method as above, $\|\tau\|_H \leq C$.

Theorem 6.2: Let the assumptions of the previous theorem, which guarantee the existence of the weak solution $\boldsymbol{u}$ of our investigated problem, be satisfied. Then the functional of complementary energy (Equation 6.50),

$$\mathscr{S}(\tau) = \tfrac{1}{2}b(\tau, \tau) - \int_\Omega e_{ij}(\boldsymbol{u}_o)\tau_{ij}(\boldsymbol{x})\mathrm{d}\boldsymbol{x},$$

where (from Equation 6.51)

$$b(\tau, \sigma) = \int_\Omega s_{ijkl}(\boldsymbol{x})\tau_{ij}(\boldsymbol{x})\sigma_{kl}(\boldsymbol{x})\mathrm{d}\boldsymbol{x},$$

achieves its minimum on the set Σ_o of statically admissible stress fields if and only if $\tau(\boldsymbol{u})$ is the stress tensor with components $\tau_{ij}(\boldsymbol{u}) = c_{ijkl}(\boldsymbol{x})e_{kl}(\boldsymbol{u})$.

Proof: Let us put $\boldsymbol{u} = \boldsymbol{u}_o + \boldsymbol{w}$, $\boldsymbol{w} \in V$. Then $\tau(\boldsymbol{u}) = \tau(\boldsymbol{u}_o) + \tau(\boldsymbol{w})\forall\tau(\boldsymbol{w}) \in H_1$, where $H_1 = \{\tau \mid \tau \in H, \exists \boldsymbol{v} \in V \text{ such that } \tau_{ij} = c_{ijkl}(\boldsymbol{x})e_{kl}(\boldsymbol{v})\forall i, j\}$ is the so-called set of compatible stress tensors. From the definition of the weak solution $\boldsymbol{u}$ there follows the existence of $\tau(\boldsymbol{u}) \in \Sigma_o$. For an arbitrary $\sigma \in \Sigma_o$ we have $\sigma - \tau(\boldsymbol{u}) \in \Sigma$ (see Lemma 5.2 of Nečas and Hlaváček (1981), p. 105), and then the functional

$$\mathscr{J}(\sigma) = \|\sigma - \tau(\boldsymbol{u}_o)\|_H^2 = \|\sigma - \tau(\boldsymbol{u}) + \tau(\boldsymbol{u}) - \tau(\boldsymbol{u}_o)\|_H^2 = \|\sigma - \sigma(\boldsymbol{u})\|_H^2 + \|\tau(\boldsymbol{w})\|_H^2$$

achieves its minimum on Σ_o if and only if $\sigma = \tau(\boldsymbol{u})$.

Due to Equation 6.51

$$\begin{aligned}\mathscr{J}(\sigma) - \|\tau(\boldsymbol{u}_o)\|_H^2 &= \|\tau\|_H^2 - 2b(\sigma, \tau(\boldsymbol{u}_o)) \\ &= \int_\Omega s_{ijkl}(\boldsymbol{x})\sigma_{ij}\sigma_{kl}\mathrm{d}\boldsymbol{x} - 2\int_\Omega \sigma_{ij}e_{ij}(\boldsymbol{u}_o)\mathrm{d}\boldsymbol{x} = 2\mathscr{S}(\sigma).\end{aligned}$$

Hence Equation 6.50 follows immediately.

Then from this theorem the weak solution of the dual problem can be defined as follows:

Definition 6.2: We say that a tensor τ is the so-called weak solution of the dual problem if

$$\tau \in \Sigma_o \text{ and } \int_\Omega s_{ijkl}(\boldsymbol{x})\tau_{ij}\sigma_{kl}\mathrm{d}\boldsymbol{x} = \int_\Omega e_{ij}(\boldsymbol{u}_o)\sigma_{ij}\mathrm{d}\boldsymbol{x}, \quad \forall\sigma \in \Sigma. \tag{6.57}$$

Remark: The present analyses of 2D and 3D models simulating deformed lithospheric plates and geological blocks presented by many authors are based on 2D and 3D finite element methods and classical linear elasticity, and were discussed in Chapter 3 (for details see also for example Kolář *et al.* (1972), Nečas and Hlaváček (1981), Oden and Reddy (1976)).

Next we shall investigate a mixed contact problem in elasticity. Let Ω be a domain occupied by a part of the lithosphere and/or together with a part of the upper mantle, with a Lipschitz boundary $\partial\Omega = \Gamma_\tau \cup \Gamma_u \cup \Gamma_k \cup \Gamma_c \cup \mathscr{R}$ as defined above. Let us assume that $\boldsymbol{f} \in [L^2(\Omega)]^3$, and $\boldsymbol{P} \in [L^2(\Gamma_\tau)]^3$. We shall now consider the following problem.

Problem ($\mathscr{P}2$): Find a vector function $\boldsymbol{u}$ satisfying

$$(c_{ijkl}(\boldsymbol{x})e_{kl}(\boldsymbol{u}))_{,j} + f_i(\boldsymbol{x}) = 0 \quad i = 1, \ldots, N, \quad N = 2, 3 \quad \text{in } \Omega, \tag{6.58}$$

$$\boldsymbol{u}(\boldsymbol{x}) = \boldsymbol{u}_o(\boldsymbol{x}) \quad \text{on } \Gamma_u, \tag{6.59}$$

$$\tau_{ij}(\boldsymbol{x})n_j = P_i(\boldsymbol{x}), \quad j = 1, \ldots, N, \quad N = 2, 3 \quad \text{on } \Gamma_\tau, \tag{6.60}$$

$$u_n = 0, \tau_{tj} = 0 \quad j = 1, \ldots, N, \quad N = 2, 3 \quad \text{on } \Gamma_k, \tag{6.61}$$

and one of the conditions of Equation 6.32 and 6.33, or 6.34 or 6.35 i.e.

$$u_n = -g_n(\boldsymbol{x})\tau_n(\boldsymbol{x}) \text{ and } u_{tj} = 0, j = 1, \ldots N, \quad N = 2, 3 \quad \text{on } \Gamma_c \tag{6.62}$$

or

$$\tau_{tj} = 0 \text{ or } u_{tj} = -g_t(\boldsymbol{x})\tau_{tj}(\boldsymbol{x}), j = 1, \ldots N, \quad N = 2, 3 \quad \text{on } \Gamma_c, \tag{6.63}$$

where $g_n(\boldsymbol{x}) \geq g_{on} > 0$, $g_t(\boldsymbol{x}) \geq g_{to} > 0$.

Let us define the space

$$\mathscr{V} = \{\boldsymbol{v} \mid \boldsymbol{v} \in [C^1(\overline{\Omega})]^3, \ \text{supp } v_n \cap (\Gamma_u \cup \Gamma_k) = 0, \ \text{supp } u_{tj} \cap (\Gamma_u \cup \Gamma_c) = 0\}$$

and let V be the closure of the space $\mathscr{V}$ in $W = [H^1(\Omega)]^3$. Let us introduce the bilinear form $a(\boldsymbol{u}, \boldsymbol{v})$ as above and the bilinear form $a_o(\boldsymbol{u}, \boldsymbol{v})$ as

$$a_o(\boldsymbol{u}, \boldsymbol{v}) = \int_{\Gamma_c} g_n^{-1} u_n v_n \mathrm{d}s \left(+ \int_{\Gamma_c} g_t^{-1} u_{tj} v_{tj} \mathrm{d}s \right),$$

where the last term in $a_o(\boldsymbol{u}, \boldsymbol{v})$ holds for Equation 6.63.

Since in the Problem $\mathscr{P}2$ the boundary Γ_u is assumed to be a non-empty set in $\partial\Omega$, then the condition $\boldsymbol{v} = 0$ on Γ_u eliminates arbitrary rigid-body translations and rotations. In the case when $\Gamma_u = 0$ then we define the space of virtual displacements of a rigid body on stiffened supports as

$$\mathscr{P} = \{\boldsymbol{v} \mid \boldsymbol{v} \in V, \boldsymbol{v} = \boldsymbol{a} + \boldsymbol{b} \times \boldsymbol{x}, a_o(\boldsymbol{v}, \boldsymbol{v}) = 0\}. \tag{6.64}$$

It is evident that $\mathscr{P} = 0$ for the case $\Gamma_u \neq 0$.

Definition 6.3: We say that a function $\boldsymbol{u} \in W$ is the so-called weak solution of the problem given in Equations 6.58-6.63, where $\boldsymbol{u}_o \in W, \boldsymbol{f} \in [L^2(\Omega)]^N$, and $\boldsymbol{P} \in [L^2(\Gamma_\tau)]^N$ are given functions, if

$$\boldsymbol{u} - \boldsymbol{u}_o \in V, \quad \text{and}$$

$$\int_\Omega c_{ijkl}(\boldsymbol{x})e_{ij}(\boldsymbol{u})e_{kl}(\boldsymbol{v})\mathrm{d}\boldsymbol{x} + a_o(\boldsymbol{u}, \boldsymbol{v}) = \int_\Omega f_i(\boldsymbol{x})v_i \mathrm{d}\boldsymbol{x} + \int_{\Gamma_\tau} P_i(\boldsymbol{x})v_i \mathrm{d}s \quad \forall \boldsymbol{v} \in V, \tag{6.65}$$

where $a_o(\boldsymbol{u}, \boldsymbol{v})$ is defined by

$$a_o(\boldsymbol{u}, \boldsymbol{v}) = \int_{\Gamma_c} g_n^{-1} u_n v_n \mathrm{d}s \left(+ \int_{\Gamma_c} g_t^{-1} u_{tj} v_{tj} \mathrm{d}s \right), \tag{6.66}$$

and where the last term in Equation 6.66 holds for the last part of Equation 6.63 only.

As a result we have the following theorem.

Theorem 6.3: Let $\Gamma_u \neq 0$, then $\mathscr{P} = 0$. Let $\boldsymbol{u}_o \in W, \boldsymbol{f} \in [L^2(\Omega)]^N$ and $\boldsymbol{P} \in [L^2(\Gamma_\tau)]^N$ be given functions. Then there exists one and only one weak solution of the Problem $\mathscr{P}2$ and

$$\|\boldsymbol{u}\|_W \leq c(\|\boldsymbol{u}_o\|_W + \|\mathbf{f}\|_{[L^2(\Omega)]^3} + \|\mathbf{P}\|_{[L^2(\Gamma_\tau)]^3}). \tag{6.67}$$

Proof: The proof is parallel to that of Equation 6.54 for the functional of potential energy

$$L_o(\boldsymbol{u}) = \tfrac{1}{2}a(\boldsymbol{u}, \boldsymbol{u}) + \tfrac{1}{2}a_o(\boldsymbol{u}, \boldsymbol{u}) - \int_\Omega f_i(\boldsymbol{x})u_i \mathrm{d}\boldsymbol{x} - \int_{\Gamma_\tau} P_i(\boldsymbol{x})u_i \mathrm{d}s,$$

where the Korn inequality in the form

$$\int_\Omega e_{ij}(\boldsymbol{v})e_{kl}(\boldsymbol{v})\mathrm{d}x + a_o(\boldsymbol{v}, \boldsymbol{v}) \geq c\,\|\boldsymbol{v}\|_{1,2}^2 \quad \forall \boldsymbol{v} \in V \tag{6.68}$$

was used, and where $a(\boldsymbol{u}, \boldsymbol{v}) = \int_\Omega c_{ijkl}(\boldsymbol{x})e_{ij}(\boldsymbol{u})e_{kl}(\boldsymbol{v})\mathrm{d}\boldsymbol{x}$ and $a_o(\boldsymbol{v}, \boldsymbol{v})$ is defined by Equation 6.66.

Numerical solutions and algorithms are parallel to those derived in Chapter 3 for the deformation of geological bodies, and therefore they will not be discussed in this section.

6.4 MODELS BASED ON THE ANALYSIS OF PERIODIC GEOLOGICAL STRUCTURES: METHOD OF HOMOGENIZATION

6.4.1 Formulation of the Problem and the Main Results

In the previous section we approximated the real lithospheric rocks by isotropic and anisotropic materials. But in the case of complicated anisotropic geological structures such models are very trivial and do not correspond to the reality. To conduct a theoretical stress analysis a mathematical model of the lithosphere which describes its mechanical behaviour with an adequate degree of accuracy has to be developed. The real lithospheric plate is characterized by its complicated structure. We can assume that real rocks are created by materials with periodically heterogeneous structures which correspond to composite materials. In these cases such models will be non-trivial and practically unsolved, and here homogenization is an important method to solve such problems. From the physical point of view, homogenization means replacing the periodically heterogeneous material by an equivalent homogeneous one. From the mathematical point of view it means that the equation with highly oscillating periodic coefficients is approximated by an equation with constant coefficients, which describes the anisotropic behaviour of the rocks. The presented results are based on results obtained by myself with my student Z. Milka, and on the results of Chapter 3. We shall give the ideas of the primary and dual problems only.

Let Ω be a domain occupied by the rocks with periodic structure with a Lipschitz boundary $\partial\Omega = \Gamma_\tau \cup \Gamma_u \cup \mathscr{R}$. Let us assume that $\boldsymbol{f} \in [L^2(\Omega)]^N$ and $\boldsymbol{P} \in$

$[L^2(\Gamma_\tau)]^N$, $N = 2, 3$. The relation between the stress and strain tensors is defined by the linear Hooke's law for the periodic structure of the material defined by

$$\tau_{ij}^\varepsilon(\boldsymbol{x}) = c_{ijkl}(\boldsymbol{x}/\varepsilon)e_{kl}(\boldsymbol{u}^\varepsilon), \tag{6.69}$$

$$e_{ij}^\varepsilon(\boldsymbol{x}) = s_{ijkl}(\boldsymbol{x}/\varepsilon)\tau_{kl}^\varepsilon, \tag{6.70}$$

where $c_{ijkl}(\boldsymbol{x}/\varepsilon) \equiv c_{ijkl}^\varepsilon(\boldsymbol{x})$ and $s_{ijkl}(\boldsymbol{x}/\varepsilon) \equiv s_{ijkl}^\varepsilon(\boldsymbol{x})$ are Y-periodic in all variables and satisfy

$$a_o^\varepsilon \leq c_{ijkl}^\varepsilon(\boldsymbol{x})\xi_{ij}\xi_{kl}\,|\xi|^{-2} \leq A_o^\varepsilon < +\infty \quad \forall \boldsymbol{x} \in \Omega, \quad \xi = (\xi_{ij}) \in R^{N\times N}, \tag{6.71}$$

and

$$(A_o^\varepsilon)^{-1} \leq s_{ijkl}^\varepsilon(\boldsymbol{x})\zeta_{ij}\zeta_{kl}\,|\zeta|^{-2} \leq (a_o^\varepsilon)^{-1} < +\infty \quad \forall \boldsymbol{x} \in \Omega, \zeta = (\zeta_{ij}) \in R^{N\times N}, \tag{6.72}$$

where a_o^ε, A_o^ε are constants independent of $\boldsymbol{x} \in \Omega$ and $\xi \in R^{N\times N}$, and

$$c_{ijkl}^\varepsilon(\boldsymbol{x}) = c_{jikl}^\varepsilon(\boldsymbol{x}) = c_{klij}^\varepsilon(\boldsymbol{x}) = c_{ijlk}^\varepsilon(\boldsymbol{x}), \tag{6.73}$$

$$s_{ijkl}^\varepsilon(\boldsymbol{x}) = s_{jikl}^\varepsilon(\boldsymbol{x}) = s_{klij}^\varepsilon(\boldsymbol{x}) = s_{ijlk}^\varepsilon(\boldsymbol{x}), \tag{6.74}$$

and

$$c_{ijkl}^\varepsilon(\boldsymbol{x})s_{ijkl}^\varepsilon(\boldsymbol{x}) = \tfrac{1}{2}(\delta_{ik}\delta_{jl} + \delta_{il}\delta_{jk}), \tag{6.75}$$

where $a_o^\varepsilon = 2\min\{\mu^\varepsilon(\boldsymbol{x}); \boldsymbol{x} \in \Omega\}$, $A_o^\varepsilon = \max\{2\mu^\varepsilon(\boldsymbol{x}) + 3\lambda^\varepsilon(\boldsymbol{x}); \boldsymbol{x} \in \Omega\}$ and δ_{ij} is the Kronecker symbol.

Particularly, for an isotropic material with Lamé's constants $\lambda^\varepsilon = \lambda(\boldsymbol{x}/\varepsilon)$ and $\mu^\varepsilon = \mu(\boldsymbol{x}/\varepsilon)$ we have

$$c_{ijkl}^\varepsilon = \lambda^\varepsilon\delta_{ij}\delta_{kl} + \mu^\varepsilon(\delta_{ik}\delta_{jl} + \delta_{il}\delta_{jk}), \tag{6.76}$$

$$s_{ijkl}^\varepsilon = (4\mu^\varepsilon)^{-1}(\delta_{ik}\delta_{jl} + \delta_{il}\delta_{jk}) - \lambda^\varepsilon(2\mu^\varepsilon)^{-1}(3\lambda^\varepsilon + 2\mu^\varepsilon)^{-1}\delta_{ij}\delta_{kl}. \tag{6.77}$$

The stress tensor satisfies the equilibrium conditions

$$\tau_{ij,j}^\varepsilon(\boldsymbol{x}) + f_i(\boldsymbol{x}) = 0 \quad \forall \boldsymbol{x} \in \Omega. \tag{6.78}$$

For our model problem we use the following boundary conditions:

$$\tau_{ij}^\varepsilon(\boldsymbol{x})n_j = P_i(\boldsymbol{x}) \quad \text{on } \Gamma_\tau, \tag{6.79}$$

$$u^\varepsilon(\boldsymbol{x}) = u_o(\boldsymbol{x}) \quad \text{on } \Gamma_u. \tag{6.80}$$

Let us investigate the case when $\varepsilon \to 0_+$. Coefficients $c_{ijkl}^\varepsilon(\boldsymbol{x})$ oscillate strongly in the interval Y, therefore the function $\boldsymbol{u}^\varepsilon(\boldsymbol{x})$ also strongly changes, but it is not a periodic function. We look for the asymptotic expansion of $\boldsymbol{u}^\varepsilon$ and τ^ε in the form

$$\boldsymbol{u}^\varepsilon(\boldsymbol{x}) = \boldsymbol{u}^o(\boldsymbol{x}, \boldsymbol{x}/\varepsilon) + \varepsilon\boldsymbol{u}^1(\boldsymbol{x}, \boldsymbol{x}/\varepsilon) + \varepsilon^2\boldsymbol{u}^2(\boldsymbol{x}, \boldsymbol{x}/\varepsilon) + \cdots, \tag{6.81}$$

$$\tau^\varepsilon(\boldsymbol{x}) = \tau^o(\boldsymbol{x}, \boldsymbol{x}/\varepsilon) + \varepsilon\tau^1(\boldsymbol{x}, \boldsymbol{x}/\varepsilon) = \varepsilon^2\tau^2(\boldsymbol{x}, \boldsymbol{x}/\varepsilon) + \cdots, \tag{6.82}$$

where the functions $\boldsymbol{u}^i(x, y)$ are independent of ε and periodic in the variable y. Now

$$\frac{\mathrm{d}\boldsymbol{v}(\boldsymbol{x},\boldsymbol{x}/\varepsilon)}{\mathrm{d}x_i}=\frac{\partial\boldsymbol{v}(\boldsymbol{x},\boldsymbol{x}/\varepsilon)}{\partial x_i}+\varepsilon^{-1}\frac{\partial\boldsymbol{v}(\boldsymbol{x},\boldsymbol{x}/\varepsilon)}{\partial y_i}\bigg|_{\boldsymbol{y}=\boldsymbol{x}/\varepsilon}.$$

Let us put $\boldsymbol{u}^\varepsilon(\boldsymbol{x})=\boldsymbol{u}^\varepsilon(\boldsymbol{x},\boldsymbol{x}/\varepsilon)$ into Equation 6.78, and with Equation 6.69 we obtain

$$\begin{aligned}&\varepsilon^{-2}(c_{ijkl}u^\varepsilon_{k,y_l})_{,y_j}-\varepsilon^{-1}[(c_{ijkl}u^\varepsilon_{k,l})_{,y_j}+(c_{ijkl}u^\varepsilon_{k,l})_{,j}]\\&\quad-(c_{ijkl}u^\varepsilon_{k,l})_{,j}=f_i\quad(i,j=1,2,3)\quad\text{in }\Omega.\end{aligned}\tag{6.83}$$

Hence

$$-(c^\varepsilon_{ijkl}u^\varepsilon_{k,l})_{,j}=(A^\varepsilon\boldsymbol{u}^\varepsilon)_i=((\varepsilon^{-2}A^o+\varepsilon^{-1}A^1+A^o)\boldsymbol{u}^\varepsilon)_i=f_i,\tag{6.84}$$

where we have denoted

$$\begin{aligned}&(A^\varepsilon\boldsymbol{u}^\varepsilon)_i=-(c_{ijkl}u^\varepsilon_{k,l})_{,j},\qquad(A^o\boldsymbol{u}^\varepsilon)_i=-(c_{ijkl}u^\varepsilon_{k,y_l})_{,y_j},\\&(A^1\boldsymbol{u}^\varepsilon)_i=-(c_{ijkl}u^\varepsilon_{k,l})_{,y_j}-(c_{ijkl}u^\varepsilon_{k,y_l})_{,j},\\&(A^2\boldsymbol{u}^\varepsilon)_i=-(c_{ijkl}u^\varepsilon_{k,l})_{,j}.\end{aligned}$$

Inserting Equations 6.81 and 6.82 into Equations 6.79, 6.80 and 6.84, and comparing the coefficients at the corresponding powers of ε (for all integers $k\geq-2$), we have

$$A^o\boldsymbol{u}^o=0\quad\text{in }\Omega\times Y,\tag{6.85}$$

$$A^o\boldsymbol{u}^1+A^1\boldsymbol{u}^o=0\quad\text{in }\Omega\times Y,\tag{6.86}$$

$$c_{ijkl}(\boldsymbol{y})u^o_{k,y_l}(\boldsymbol{x},\boldsymbol{y})n_j(\boldsymbol{x})=0\quad\text{on }\Gamma_\tau\times Y,\tag{6.87}$$

$$A^o\boldsymbol{u}^2+A^1\boldsymbol{u}^1+A^2\boldsymbol{u}^o=f\quad\text{in }\Omega\times Y,\tag{6.88}$$

$$c_{ijkl}(\boldsymbol{y})(u^1_{k,y_l}+u^o_{k,l})n_j(\boldsymbol{x})=P_i(\boldsymbol{x})\quad\text{on }\Gamma_\tau\times Y,\tag{6.89}$$

$$u^o_i(\boldsymbol{x})=u_{oi}(\boldsymbol{x})\quad\text{on }\Gamma_u\times Y,\tag{6.90}$$

$$A^o\boldsymbol{u}^3+A^1\boldsymbol{u}^2+A^2\boldsymbol{u}^1=0\quad\text{in }\Omega\times Y,\tag{6.91}$$

$$c_{ijkl}(\boldsymbol{y})(u^2_{k,y_l}+u^1_{k,l})n_j(\boldsymbol{x})=0\quad\text{on }\Gamma_\tau\times Y,\tag{6.92}$$

$$u^1_i(\boldsymbol{x})=0\quad\text{on }\Gamma_u\times Y.\tag{6.93}$$

Remark: As in Chapter 3, subscript "per" denotes periodic functions.

Lemma 6.1: Let $\boldsymbol{f}\in[L^2_{\text{per}}(Y)]^3$. $A^o\boldsymbol{u}=\boldsymbol{f}$ has a solution $\boldsymbol{u}\in[H^1_{\text{per}}(Y)]^3$ if, and only if, $\mathcal{M}(\boldsymbol{f})=0$. This solution is unique up to an additive constant.

For the proof see Chapter 3.

Using the same technique as in Chapter 3 we obtain the following result:

Problem ($\mathscr{P}_h$): We have to solve the so-called homogenized problem

$$\left.\begin{aligned} &\tau^o_{ij,j} = f_i \quad \text{in } \Omega,\\ &\tau^o_{ij} = c^o_{ijkl} e_{kl}(\boldsymbol{u}^o) \quad \text{in } \Omega,\\ &\tau^o_{ij} n_j = P_i \quad \text{on } \Gamma_\tau,\\ &u^o_i = u_{oi} \quad \text{on } \Gamma_u \end{aligned}\right\} \quad i, j = 1, \ldots, N, \tag{6.94}$$

with c^o_{ijkl} defined by

$$c^o_{ijkl} = \mathscr{M}(c_{ijgh}(\boldsymbol{y})(\delta_{gh}\delta_{hl} - \chi^{kl}_{g,y_h}(\boldsymbol{y}))), \tag{6.95}$$

where $\chi \in [H^1_{\text{per}}(Y)]^N$ satisfies

$$(c_{ijgh}(\boldsymbol{y})(\chi^{kl}_{g,y_h}(\boldsymbol{y}) - \delta_{gk}\delta_{hl}))_{,y_j} = 0. \tag{6.96}$$

Constants c^o_{ijkl} are the so-called homogenization coefficients with regard to the initial one. Similarly, if s^o_{ijkl} denotes the inverse matrix to the matrix c^o_{ijkl}, then the equation

$$e_{ij}(\boldsymbol{u}^o) = s^o_{ijkl}\tau^o_{kl} \tag{6.97}$$

represents the inverse Hooke's law for the homogenized problem. Moreover, the coefficients c^o_{ijkl} and s^o_{ijkl} have the same symmetry conditions and ellipticity conditions as the initial problem; i.e. for homogenization coefficients it holds that

$$a^o_o \le c^o_{ijkl}\xi_{ij}\xi_{kl}\,|\xi|^{-2} \le A^o_o < +\infty \quad \forall \boldsymbol{x} \in \Omega, \xi = (\xi_{ij}) \in R^{N\times N}, \tag{6.98}$$

$$(A^o_o)^{-1} \le s^o_{ijkl}(\boldsymbol{x})\zeta_{ij}\zeta_{kl}\,|\zeta|^{-2} \le (a^o_o)^{-1} < +\infty \quad \forall \boldsymbol{x} \in \Omega, \zeta = (\zeta_{ij}) \in R^{N\times N}, \tag{6.99}$$

where a^o_o and A^o_o are constants independent of $\boldsymbol{x} \in \Omega$ and $\xi \in R^{N\times N}$,

$$c^o_{ijkl} = c^o_{jikl} = c^o_{klij} = c^o_{ijlk}, \tag{6.100}$$

$$s^o_{ijkl} = s^o_{jikl} = s^o_{klij} = s^o_{ijlk}, \tag{6.101}$$

and $a^o_o = 2\min\{\mu^o(\boldsymbol{x}); \boldsymbol{x} \in \Omega\}$, $A^o_o = \max\{2\mu^o(\boldsymbol{x}) + 3\lambda^o(\boldsymbol{x}); \boldsymbol{x} \in \Omega\}$.

Theorem 6.4: Let us assume that c_{ijkl}, c^o_{ijkl}, s_{ijkl} and s^o_{ijkl} satisfy the symmetry conditions and the conditions of ellipticity. Let $\boldsymbol{u}^\varepsilon$ be the solution of the problem

$$\begin{aligned} &\boldsymbol{u}_o - \boldsymbol{u}^\varepsilon \in V = \{\boldsymbol{v} \mid \boldsymbol{v} \in [H^1(\Omega)]^N, \quad \boldsymbol{v} = 0 \text{ on } \Gamma_u\},\\ &a^\varepsilon(\boldsymbol{u}^\varepsilon, \boldsymbol{v}) = (\boldsymbol{F}, \boldsymbol{v}) \quad \forall \boldsymbol{v} \in V, \end{aligned} \tag{6.102}$$

where

$$a^\varepsilon(\boldsymbol{u}, \boldsymbol{v}) = \int_\Omega c^\varepsilon_{ijkl}(\boldsymbol{x}) e_{ij}(\boldsymbol{u}) e_{kl}(\boldsymbol{v})\mathrm{d}\boldsymbol{x}, \quad (\boldsymbol{F}, \boldsymbol{v}) = \int_\Omega f_i v_i \mathrm{d}\boldsymbol{x} - \int_{\Gamma_\tau} P_i v_i \mathrm{d}s.$$

Let $\boldsymbol{u}^o$ be the solution of the homogenized problem

$$\begin{aligned} &\boldsymbol{u}_o - \boldsymbol{u}^o \in V = \{\boldsymbol{v} \mid \boldsymbol{v} \in [H^1(\Omega)]^N, \quad \boldsymbol{v} = 0 \text{ on } \Gamma_u\},\\ &a^o(\boldsymbol{u}^o, \boldsymbol{v}) = (\boldsymbol{F}, \boldsymbol{v}) \quad \forall \boldsymbol{v} \in V, \end{aligned} \tag{6.103}$$

where

$$a^o(\boldsymbol{u},\boldsymbol{v}) = \int_\Omega c^o_{ijkl}(\boldsymbol{x})e_{ij}(\boldsymbol{u})e_{kl}(\boldsymbol{v})\mathrm{d}\boldsymbol{x}, \quad (\boldsymbol{F},\boldsymbol{v}) = \int_\Omega f_i v_i \mathrm{d}\boldsymbol{x} - \int_{\Gamma_\tau} P_i v_i \mathrm{d}s.$$

Then for $\varepsilon \to 0$

$$\boldsymbol{u}^\varepsilon \to \boldsymbol{u} \quad \text{in } V \text{ weakly,} \tag{6.104}$$

and

$$\tau^\varepsilon \to \tau^o \quad \text{in } H \text{ weakly.} \tag{6.105}$$

The proof is similar to that of Birolli (1977), and Bensoussan, Lions and Papanicolaou (1978), pp. 24-28.

To solve the problem numerically the finite element technique will be used.

Let a domain occupied by the lithosphere and the upper mantle be approximated by an N-dimensional domain Ω, the boundary $\partial\Omega$ of which is a polygonal in the $2D$ case and polyhedral in the $3D$ case. Let a domain Ω be divided into a finite number of triangles T_h in the $2D$ case and tetrahedrons T_h in the $3D$ case, such that two triangles have either no common point or have a common vertex or a common side in the $2D$ case and such that two arbitrary tetrahedrons have either no common point or have a common vertex or a common edge or a common side. We restrict our considerations to the space V_h of all continuous functions which are linear on each element of the given division. Let h be the largest diameter of all the elements of the given division. We shall assume that the used division is regular. Let V_h be the finite element approximation of V consisting of all functions from V which on the division T_h are equal to interpolation polynomials of the first degree. Then the problem leads to the following problem:

Problem $(\mathscr{P}_h)_h$: Find $\boldsymbol{u}^o_h \in \boldsymbol{u}_o + V_h$ minimizing the functional

$$J(\boldsymbol{u}) = \tfrac{1}{2}a^o(\boldsymbol{u},\boldsymbol{u}) - \int_\Omega f_i u_i \mathrm{d}\boldsymbol{x} - \int_{\Gamma_\tau} P_i u_i \mathrm{d}s \tag{6.106}$$

on the set V_h, where $a^o(\boldsymbol{u},\boldsymbol{v})$ is the bilinear form defined by

$$a^o(\boldsymbol{u},\boldsymbol{v}) = \int_\Omega c^o_{ijkl} e_{ij}(\boldsymbol{u})e_{kl}(\boldsymbol{v})\mathrm{d}\boldsymbol{x}. \tag{6.107}$$

Let us denote by $\varepsilon(h) = \|\boldsymbol{u}^o - \boldsymbol{u}^o_h\|$ the error between the exact and the approximate solutions, where $\boldsymbol{u}^o_h \in V_h$. We shall prove that $\varepsilon(h) \to 0$ for $h \to 0_+$.

Theorem 6.5: Let $a^o : V \times V \to R_1$ be a symmetric, bounded and V-elliptic bilinear form on V. Let V_h be the finite element approximation of V defined above with the following properties:

$$\forall \boldsymbol{v} \in \boldsymbol{u}_o + V \quad \exists \boldsymbol{v}_h \in \boldsymbol{u}_o + V_h : \boldsymbol{v}_h \to \boldsymbol{v}, h \to 0_+; \tag{6.108}$$

$$\text{from } \boldsymbol{v}_h \to \boldsymbol{v} \text{ weakly,} \quad \boldsymbol{v}_h \in \boldsymbol{u}_o + V_h \Rightarrow \boldsymbol{v} \in \boldsymbol{u}_o + V. \tag{6.109}$$

Then

$$\|\boldsymbol{u}^o - \boldsymbol{u}^o_h\| \to 0 \text{ for } h \to 0_+. \tag{6.110}$$

The proof is similar of that of Chapter 3.

Since S^o_{ijkl} represents the inverse matrix to the matrix c^o_{ijkl} then, using Equation 6.97 and the principle of virtual stresses or the principle of minimum complementary energy, we can solve the dual problem.

6.5. MODELS OF COLLISION ZONES BASED ON THE BASIC INITIAL BOUNDARY VALUE PROBLEMS OF THERMO-VISCO-ELASTICITY

6.5.1. Introduction

It is well known that subduction and collision zones are the sites of fundamental geological processes, which are probably the source of major plate driving forces they seem to be the sites where the lithosphere is differentiated and they give rise to most of the tectonic imprint in continents. But these geological processes are the result of geodynamic processes in the upper parts of the Earth. In addition, there are still fundamental questions which have not been answered concerning the stress distribution in the subducted lithosphere, its mechanical coupling to the surface lithosphere and its interaction with flow and temperature in the mantle, with observed ascending magma as well as with back-arc spreading. In modelling these geological processes in the upper parts of the Earth, the main role is played by the rheology of upper parts of the Earth. The stresses in the lithosphere arise from three sources: tectonic stresses applied to its boundaries, thermally induced stresses, and stresses resulting from the action of gravitational body forces on the density distribution in the lithosphere.

As was shown in Nedoma, Haslinger and Hlaváček (1989), for time periods shorter than $t_c = \nu/\mu$ (known also as a Maxwell's time) the upper parts of the Earth behave thermo-elastically, while for longer times they behave thermo-visco-elastically and thermo-visco-plastically. Here ν is the effective viscosity at a given stress level,

$$\nu = \frac{\tau_2[(\frac{2}{3})(e_{ij}(\dot{\boldsymbol{u}})e_{ij}(\dot{\boldsymbol{u}})]^{1/2}\underline{\gamma}^{-1})^{1/n}}{3^{1/2}2^{1/4}(e_{ij}(\dot{\boldsymbol{u}})e_{ij}(\dot{\boldsymbol{u}}))^{1/2}},$$

where
$$e_{ij}(\dot{u}) = \frac{1}{2}\left(\frac{\partial \dot{u}_i}{\partial x_j} + \frac{\partial \dot{u}_j}{\partial x_i}\right),$$

τ_2 is the yield stress of the continental lithosphere, and

$$\underline{\gamma} = L^{-1}\int_L \underline{\gamma}_o \exp(-Q/RT)\mathrm{d}x_3,$$

where Q is an activation energy, R is the universal gas constant and L is the thickness of the lithosphere, and μ is the rigidity. For $\underline{\gamma} \to \infty$ we have rigid perfectly plastic flow. In the next section we shall analyze the model problems based on the thermo-visco-elastic rheology with short and long memory.

6.5.2. Models Based on Linear Thermo-Visco-Elasticity with Short Memory

6.5.2.1. Dynamic Case

FORMULATION OF THE PROBLEM

Let the investigated upper part of the Earth occupy a region $\Omega = \cup_{\iota=1}^{S}\Omega^{\iota} \subset R^N$, $N = 2, 3$, with Lipschitz boundary $\partial\Omega = \Gamma_u \cup \Gamma_\tau \cup \mathscr{R}$, where $\mathscr{R}$ is the set of zero measure, and let $\bar{I} = \langle t_o, t_1\rangle$.

Then we have to solve the following problem.

Problem ($\mathscr{P}$): Find a pair of functions $(T, \boldsymbol{u})$ satisfying

$$\rho c_e \frac{\partial T}{\partial t} + \rho\beta_{ij} T_o(\mathbf{x}) e_{ij}(\dot{\boldsymbol{u}}) = \frac{\partial}{\partial x_i}\left(\kappa_{ij}\frac{\partial T}{\partial x_j}\right) + Q_o \quad \text{in } \Omega, t \in I, \tag{6.111}$$

$$\rho \frac{\partial^2 u_i}{\partial t^2} = \frac{\partial \tau_{ij}}{\partial x_j} + f_{oi} \quad \text{in } \Omega, t \in I, \tag{6.112}$$

boundary conditions

$$T(\boldsymbol{x}, t) = T_1(\boldsymbol{x}, t); \quad \text{on } \Gamma_u, t > 0, \tag{6.113}$$

$$\boldsymbol{u}(\boldsymbol{x}, t) = \boldsymbol{u}_1(\boldsymbol{x}, t) \quad \text{on } \Gamma_u, t > 0, \tag{6.114}$$

$$\kappa_{ij}(\boldsymbol{x})\frac{\partial T}{\partial x_j} n_i = q(\boldsymbol{x}, t) \quad \text{on } \Gamma_\tau, t > 0 \tag{6.115}$$

$$\tau_{ij}(\boldsymbol{u}(\boldsymbol{x}, t)) n_j = P_{oi}(\boldsymbol{x}, t) \quad \text{on } \Gamma_\tau, t > 0 \tag{6.116}$$

and initial conditions

$$T(\boldsymbol{x}, t_o) = T_o(\boldsymbol{x}), \quad \boldsymbol{x} \in \Omega, \tag{6.117}$$

$$\boldsymbol{u}(\boldsymbol{x}, t_o) = \boldsymbol{u}_o(\boldsymbol{x}), \quad \boldsymbol{x} \in \Omega, \tag{6.118}$$

$$\frac{\partial \boldsymbol{u}}{\partial t}(\boldsymbol{x}, t_o) = \boldsymbol{v}_o(\boldsymbol{x}), \quad \boldsymbol{x} \in \Omega, \tag{6.119}$$

where $Q_o(\boldsymbol{x}, t)$ denotes thermal sources, $c_e(\boldsymbol{x})$ is a specific heat, $\rho(\boldsymbol{x})$ is a density, $\kappa_{ij}(\boldsymbol{x})$ is a thermal conductivity, satisfying

$$\kappa_{ij}(\boldsymbol{x}) = \kappa_{ji}(\boldsymbol{x}), \tag{6.120}$$

$$\kappa_{ij}(\boldsymbol{x})\xi_i\xi_j \geq \kappa_o |\xi|^2, \quad \xi \in R^N, N = 2, 3, \tag{6.121}$$

q is a heat flow, $T_o(\boldsymbol{x})$ is an initial temperature, $\boldsymbol{f}_o(\boldsymbol{x}, t) = (f_{oi}(\boldsymbol{x}, t))$ are body forces, $\boldsymbol{P}(\boldsymbol{x}, t) = (P_i(\boldsymbol{x}, t))$ are surface forces, $\beta_{ij}(\boldsymbol{x})$ is a coefficient of linear thermal expansion, and τ_{ij} is the stress tensor defined as

$$\tau_{ij}(\boldsymbol{u}(\boldsymbol{x}, t)) = c_{ijkl}(\boldsymbol{x}) e_{kl}(\boldsymbol{u}(\boldsymbol{x}, t)) + C_{ijkl}(\boldsymbol{x})\frac{\partial e_{kl}(\boldsymbol{u}(\boldsymbol{x}, t))}{\partial t} - \beta_{ij}(\boldsymbol{x})(T(\boldsymbol{x}, t) - T_o(\boldsymbol{x})), \tag{6.122}$$

representing the stress-strain relation for rocks with short memory. Here $c_{ijkl}(\boldsymbol{x})$ are the elastic coefficients and $C_{ijkl}(\boldsymbol{x})$ are coefficients of viscosity satisfying conditions of symmetry

$$c_{ijkl}(\boldsymbol{x}) = c_{jikl}(\boldsymbol{x}) = c_{ijlk}(\boldsymbol{x}) = c_{klij}(\boldsymbol{x}), \tag{6.123}$$

$$C_{ijkl}(\boldsymbol{x}) = C_{jikl}(\boldsymbol{x}) = C_{ijlk}(\boldsymbol{x}) = C_{klij}(\boldsymbol{x}), \tag{6.124}$$

and conditions of ellipticity

$$c_{ijkl}(\boldsymbol{x}) e_{ij}(\boldsymbol{u}) e_{kl}(\boldsymbol{u}) \geq c e_{ij}(\boldsymbol{u}) e_{ij}(\boldsymbol{u}), \quad c > 0, \tag{6.125}$$

$$C_{ijkl}(\boldsymbol{x}) e_{ij}(\boldsymbol{u}') e_{kl}(\boldsymbol{u}') \geq C e_{ij}(\boldsymbol{u}') e_{ij}(\boldsymbol{u}'), \quad C > 0. \tag{6.126}$$

The term (scalar) $c_{ijkl}(\boldsymbol{x}) e_{ij}(\boldsymbol{u}) e_{kl}(\boldsymbol{u})$ has the meaning of deformation energy, while the term (scalar) $C_{ijkl}(\boldsymbol{x}) e_{ij}(\boldsymbol{u}') e_{kl}(\boldsymbol{u}')$ then has the meaning of dissipation of energy; both of which are positive.

VARIATIONAL FORMULATION OF THE PROBLEM

Let us introduce the space of virtual temperatures and of virtual velocities

$$
\begin{aligned}
{}^1V_o &= \{z \mid z \in H^1(\Omega), \quad z = T_1(t) \text{ on } \Gamma_u \text{ in the sense of traces}\},\\
\mathscr{U}_{\text{ad}} &= \{\boldsymbol{v} \mid \boldsymbol{v} \in [H^1(\Omega)]^N, \quad N = 2, 3, \boldsymbol{v} = \boldsymbol{u}'(t) \text{ on } \Gamma_u \text{ in the sense of traces}\}.
\end{aligned}
$$

Multiplying Equation 6.111 by $z - T(t)$, and Equation 6.112 by $\boldsymbol{v} - \boldsymbol{u}'(t)$, integrating over Ω, and using the Green's theorem, then after some algebra, we obtain the following variational problem:

Problem $(\mathscr{P}_O)_V$**:** Find a pair of functions $(T(t), \mathbf{u}'(t))$ such that

$$
T(t) \in {}^1V_o, \boldsymbol{u}'(t) \in \mathscr{U}_{\text{ad}}(t) \quad \forall t, \text{ and}
$$

$$
(T'(t), z - T(t)) + a_T(T(t), z - T(t)) + b_p(\boldsymbol{u}'(t), z - T(t)) \geq s(z - T(t)) \quad \forall z \in {}^1V_o, \tag{6.127}
$$

$$
(\boldsymbol{u}''(t), \boldsymbol{v} - \boldsymbol{u}'(t)) + A(\boldsymbol{u}'(t), \boldsymbol{v} - \boldsymbol{u}'(t)) + a(\boldsymbol{u}(t), \boldsymbol{v} - \boldsymbol{u}'(t)) + b_s(T(t), \boldsymbol{v} - \boldsymbol{u}'(t)) \geq S(\boldsymbol{v} - \boldsymbol{u}'(t)) \quad \forall \boldsymbol{v} \in \mathscr{U}_{\text{ad}}(t), \tag{6.128}
$$

and with the initial conditions

$$
T(\boldsymbol{x}, t_o) = T_o(\boldsymbol{x}), \quad \boldsymbol{x} \in \Omega, \tag{6.129}
$$

$$
\boldsymbol{u}(\boldsymbol{x}, t) = \boldsymbol{u}_o(\boldsymbol{x}), \quad \boldsymbol{x} \in \Omega, \tag{6.130}
$$

$$
\boldsymbol{u}'(\boldsymbol{x}, t_o) = \boldsymbol{v}_o(\boldsymbol{x}), \quad \boldsymbol{x} \in \Omega, \tag{6.131}
$$

where

$$
\begin{aligned}
a_T(T, z) &= \int_\Omega \kappa_{ij}(\boldsymbol{x}) \frac{\partial T}{\partial x_j} \frac{\partial z}{\partial x_i} \mathrm{d}\boldsymbol{x},\\
b_p(\boldsymbol{u}', z) &= \int_\Omega \rho \beta_{ij} T_o(\boldsymbol{x}) e_{ij}(\dot{\boldsymbol{u}}) z \mathrm{d}\boldsymbol{x},\\
a(\boldsymbol{u}, \boldsymbol{v}) &= \int_\Omega c_{ijkl}(\boldsymbol{x}) e_{ij}(\boldsymbol{u}) e_{kl}(\boldsymbol{v}) \mathrm{d}\boldsymbol{x},\\
A(\boldsymbol{u}, \boldsymbol{v}) &= \int_\Omega C_{ijkl}(\boldsymbol{x}) e_{ij}(\boldsymbol{u}) e_{kl}(\boldsymbol{v}) \mathrm{d}\boldsymbol{x},\\
b_s(T, \boldsymbol{v}) &= \int_\Omega \frac{\partial}{\partial x_i} (\beta_{ij}(\boldsymbol{x})(T - T_o) z \mathrm{d}\boldsymbol{x},\\
s(z) &= \int_\Omega Q z \mathrm{d}\boldsymbol{x} - \int_{\Gamma_\tau} q z \mathrm{d}s,\\
S(\boldsymbol{v}) &= \int_\Omega f_i v_i \mathrm{d}\boldsymbol{x} - \int_{\Gamma_\tau} P_i v_i \mathrm{d}s.
\end{aligned}
$$

It is evident that from $\boldsymbol{u}'(t) \in \mathscr{U}_{\text{ad}}(t)$ and Equation 6.130 and 6.131, it follows that $\boldsymbol{u}(\boldsymbol{x}, t) = \boldsymbol{u}_1(\boldsymbol{x}, t) + (\boldsymbol{u}_o(\boldsymbol{x}) - \boldsymbol{u}_1(\boldsymbol{x}, t_o))$ on Γ_u. Hence, the boundary condition $\boldsymbol{u}(\boldsymbol{x}, t) = \boldsymbol{u}_1(\boldsymbol{x}, t)$ follows if $\boldsymbol{u}_o(\boldsymbol{x}) = \boldsymbol{u}_1(\boldsymbol{x}, t_o)$ on Γ_u.

In the next section we shall give the equivalent formulation of the problem, which will be useful for further investigation.

Let us introduce functions $\bar{z}(t) \equiv \bar{z}(\boldsymbol{x}, t)$ and $\boldsymbol{w}(t) \equiv \boldsymbol{w}(\boldsymbol{x}, t)$ satisfying

$$\bar{z}(t) \in H^1(\Omega), \boldsymbol{w}(t) \in [H^1(\Omega)]^N, \quad N = 2, 3,$$

$$\bar{z}(t) = T_1(\boldsymbol{x}, t), \boldsymbol{w}(t) = \boldsymbol{u}_1(\boldsymbol{x}, t) \text{ on } \Gamma_u,$$

and let us put $T(t) \cong T(t) - \bar{z}(t)$, $\boldsymbol{u}(t) \cong \boldsymbol{u}(t) - \boldsymbol{w}(t)$.

Then the boundary conditions (Equation 6.113 and 6.114) are transformed to

$$T(\boldsymbol{x}, t) = 0, \quad \text{on } \Gamma_u, t > 0, \tag{6.132}$$

$$u(\boldsymbol{x}, t) = 0 \quad \text{on } \Gamma_u, t > 0, \tag{6.133}$$

and $Q(\boldsymbol{x}, t), \boldsymbol{f}(\boldsymbol{x}, t), \boldsymbol{P}(\boldsymbol{x}, t), T_o(\boldsymbol{x}), \boldsymbol{u}_o(\boldsymbol{x}), \boldsymbol{v}_o(\boldsymbol{x}), q(\boldsymbol{x}, t)$ are changed in the usual way.

Let us introduce the spaces of virtual temperatures ${}^1V$ and displacements V as

$${}^1V = \{z \mid z \in H^1(\Omega), \quad z = 0 \text{ on } \Gamma_u \text{ in the sense of traces}\},$$

$$V = \{\boldsymbol{v} \mid \boldsymbol{v} \in [H^1(\Omega)]^N, \quad N = 2, 3, \boldsymbol{v} = 0 \text{ on } \Gamma_u \text{ in the sense of traces}\}.$$

We set ${}^1V \subset L^2(\Omega) \subset {}^1V'$, $V \subset [L^2(\Omega)]^N \subset V'$.

Multiplying Equation 6.111 by $z - T(t)$, and Equation 6.112 by $\boldsymbol{v} - \boldsymbol{u}'(t)$, integrating over Ω, and using the Green's theorem, then after some algebra we obtain the following variational problem:

Problem $(\mathscr{P})_V$: Find a pair of functions $(T(t), \mathbf{u}'(t))$ such that

$$T(t) \in {}^1V, \boldsymbol{u}'(t) \in V,$$

$$(T'(t), z - T(t)) + a_T(T(t), z - T(t)) + b_p(\boldsymbol{u}'(t), z - T(t)) \geq s(z - T(t)) \quad \forall z \in {}^1V, \tag{6.134}$$

$$(\boldsymbol{u}''(t), \boldsymbol{v} - \boldsymbol{u}'(t)) + A(\boldsymbol{u}'(t), \boldsymbol{v} - \boldsymbol{u}'(t)) + a(\boldsymbol{u}(t), \boldsymbol{v} - \boldsymbol{u}'(t))$$
$$+ b_s(T(t), \boldsymbol{v} - \boldsymbol{u}'(t)) \geq S(\boldsymbol{v} - \boldsymbol{u}'(t)) \quad \forall \boldsymbol{v} \in V, \tag{6.135}$$

and with the initial conditions

$$T(\boldsymbol{x}, t_o) = T_o(\boldsymbol{x}), \tag{6.136}$$

$$\boldsymbol{u}(\boldsymbol{x}, t) = \boldsymbol{u}_o(\boldsymbol{x}), \tag{6.137}$$

$$\boldsymbol{u}'(\boldsymbol{x}, t_o) = \boldsymbol{v}_o(\boldsymbol{x}), \tag{6.138}$$

where

$$s(z) = (W(t), z) = \int_\Omega Qz\mathrm{d}\boldsymbol{x} - \int_{\Gamma_\tau} qz\mathrm{d}s - (\bar{z}'(t), z) - a_T(\bar{z}(t), z) - b_p(\boldsymbol{w}'(t), z)$$

$$S(\boldsymbol{v}) = (\boldsymbol{F}(t), \boldsymbol{v}) = \int_\Omega f_i v_i \mathrm{d}\boldsymbol{x} - \int_{\Gamma_\tau} P_i v_i \mathrm{d}s - (\boldsymbol{w}''(t), \boldsymbol{v}) - A(\boldsymbol{w}'(t), \boldsymbol{v}) - a(\boldsymbol{w}(t), \boldsymbol{v}) - b_s(\bar{z}(t), \boldsymbol{v})$$

Then as a main result we have the following theorem:

Theorem 6.6: Let $Q, Q' \in L^2(I; L^2(\Omega))$, $\boldsymbol{f}, \boldsymbol{f}', \boldsymbol{f}'' \in L^2(I; [L^2(\Omega)]^N)$, $\bar{z}, \bar{z}' \in L^2(I; {}^1V)$, $\boldsymbol{w}, \boldsymbol{w}', \boldsymbol{w}'' \in L^2(I; V), \boldsymbol{w}''', \boldsymbol{w}'''' \in L^2(I; [L^2(\Omega)]^N)$, $T_o \in {}^1V, \boldsymbol{u}_o \in V, \boldsymbol{v}_o \in [L^2(\Omega)]^N$. Then there exists one and only one pair of functions $(T, \boldsymbol{u})$ such that

$$T, T' \in L^\infty(I; {}^1V) \cap L^2(I; {}^1V),$$

$$u \in L^\infty(I; V), \boldsymbol{u}' \in L^2(I; V) \cap L^\infty(I; [L^2(\Omega)]^N), \boldsymbol{u}'' \in L^2(I; V')$$

and satisfying Equations 6.134 and 6.135.

Proof: To prove the uniqueness we shall assume that $(T^1, \boldsymbol{u}^1)$, and $(T^2, \boldsymbol{u}^2)$ are two possible solutions. Taking $z = T^1(t), \boldsymbol{v} = \boldsymbol{u}^{1'}(t)$ (and then $z = T^2(t)$, $\boldsymbol{v} = \boldsymbol{u}^{2'}(t)$) in Equation 6.134 and 6.135, and adding, setting $\Theta = T^1 - T^2$, $\boldsymbol{U} = \boldsymbol{u}^1 - \boldsymbol{u}^2$, we obtain

$$-(\Theta', \Theta) - a_T(\Theta, \Theta) - b_p(\boldsymbol{U}', \Theta) - (\boldsymbol{U}'', \boldsymbol{U}') - A(\boldsymbol{U}', \boldsymbol{U}') - a(\boldsymbol{U}, \boldsymbol{U}') - b_s(\Theta, \boldsymbol{U}') \geq 0,$$

i.e.

$$\frac{\mathrm{d}}{\mathrm{d}t}(\|\Theta(t)\|_o^2) + a_T(\Theta(t), \Theta(t)) + \frac{\mathrm{d}}{\mathrm{d}t}[\|\boldsymbol{U}'(t)\|_o^2 + a(\boldsymbol{U}(t), \boldsymbol{U}(t))]$$
$$+A(\boldsymbol{U}'(t), \boldsymbol{U}'(t)) \leq c\,\|\Theta(t)\|_o \,\|\boldsymbol{U}'(t)\|_o .$$

Since $\Theta(t_o) = 0$, $\boldsymbol{U}(t_o) = 0$, $a_T(z, z) \geq 0$, $a(\boldsymbol{v}, \boldsymbol{v}) \geq 0$ and $A(\boldsymbol{v}, \boldsymbol{v}) \geq 0$, then we have $\Theta(t) = 0, \boldsymbol{U}'(t) = 0$ and therefore $\boldsymbol{U}(t) = 0$, which proves the uniquenes of the solution.

To prove the existence of the solution the Galerkin approximation will be used (the index m will be omitted). It is sufficient to find a priori estimates.

(i) First a priori estimates:

We note that, $\forall\lambda > 0$, there exist c, c_o, C_o with

$$a_T(z, z) + \lambda \|z\|_o^2 \geq c\,\|z\|^2, c > 0, \quad \forall z \in {}^1V,$$

$$a(\boldsymbol{v}, \boldsymbol{v}) + \lambda \|\boldsymbol{v}\|_o^2 \geq c_o \|\boldsymbol{v}\|_1^2, c_o > 0, \quad \forall \boldsymbol{v} \in V,$$

$$A(\boldsymbol{v}, \boldsymbol{v}) + \lambda \|\boldsymbol{v}\|_o^2 \geq C_o \|\boldsymbol{v}\|_1^2, C_o > 0, \quad \forall \boldsymbol{v} \in V,$$

and we can take $\lambda = 0$ if Γ_u has positive measure (i.e. if $\Gamma_u \neq 0$). We take $z = T(t) + \bar{z}(t)$, and $\boldsymbol{v} = \boldsymbol{u}'(t) + \boldsymbol{w}'(t)$, then after some modification we obtain

$$\tfrac{1}{2}\frac{\mathrm{d}}{\mathrm{d}t}[\|T(t)\|_o^2 + \|\boldsymbol{u}'(t)\|_o^2 + a(\boldsymbol{u}(t), \boldsymbol{u}(t))] + a_T(T(t), T(t)) + A(\boldsymbol{u}'(t), \boldsymbol{u}'(t))$$
$$+b_p(\boldsymbol{u}'(t), T(t)) + b_s(T(t), \boldsymbol{u}'(t)) \leq (W(t), T(t)) + (\boldsymbol{F}(t), \boldsymbol{u}'(t)).$$

After integrating in t, we obtain

$$\|T(t)\|_o^2 + \|\boldsymbol{u}'(t)\|_o^2 + a(\boldsymbol{u}(t), \boldsymbol{u}(t)) + \int_{t_o}^{t} a_T(T(\tau), T(\tau))\mathrm{d}\tau + \int_{t_o}^{t} A(\boldsymbol{u}'(\tau), \boldsymbol{u}'(\tau))\mathrm{d}\tau$$

$$= \|T_o\|_o^2 + \|\boldsymbol{v}_o\|_o^2 + 2\int_{t_o}^{t} (W(\tau), T(\tau))\mathrm{d}\tau - \int_{t_o}^{t} [b_p(\boldsymbol{u}'(\tau), T(\tau)) + b_s(T(\tau), \boldsymbol{u}'(\tau))]\mathrm{d}\tau$$

$$\leq \|T_o\|_o^2 + \|\boldsymbol{v}_o\|_o^2 + 2(W(t), T(t)) - 2(W(t_o), T_o) - 2\int_{t_o}^{t} (W'(\tau), T(\tau))\mathrm{d}\tau + 2(\boldsymbol{F}(t), \boldsymbol{u}(t))$$

$$- 2(\boldsymbol{F}(t_o), \boldsymbol{u}_o) - 2\int_{t_o}^{t} (\boldsymbol{F}'(\tau), \boldsymbol{u}(\tau))\mathrm{d}\tau - \int_{t_o}^{t} [b_p(\boldsymbol{u}'(\tau), T(\tau)) + b_s(T(\tau), \boldsymbol{u}'(\tau))]\mathrm{d}\tau,$$

$$\boldsymbol{u}(t) = \boldsymbol{u}_o + \textstyle\int_{t_o}^{t} \boldsymbol{u}(\tau)\mathrm{d}\tau.$$

Hence, with the 'c's denoting various constants > 0,

$$a_T(z,z) \geq c\,\|z\|_1^2\,,\ a(\boldsymbol{v},\boldsymbol{v}) \geq c_o\,\|\boldsymbol{v}\|_1^2 - c\,\|\boldsymbol{v}\|_o^2\,,\ A(\boldsymbol{v},\boldsymbol{v}) \geq C_o\,\|\boldsymbol{v}\|_1^2 - c\,\|\boldsymbol{v}\|_o^2\,,$$

$$\|\boldsymbol{v}_{om}\|_o \leq c\,\|\boldsymbol{v}\|_o\,,\ 2\,|(W(t),T(t))| \leq \tfrac{1}{2}\kappa_o\,\|T(t)\|_1^2 + c\,\|W(t)\|_*^2\,,$$

$$2\,|(\boldsymbol{F}(t),\boldsymbol{u}(t))| \leq \tfrac{1}{2}c_o\,\|\boldsymbol{u}(t)\|_1^2 + c\,\|\boldsymbol{F}(t)\|_*^2\,,\ \|\boldsymbol{u}(t)\|_o^2 \leq 2\,\|\boldsymbol{u}_o\| + c\int_{t_o}^{t}\boldsymbol{u}(\tau)\mathrm{d}\tau,$$

$$\left|b_s(T-T_o,\boldsymbol{v}) + b_p(\boldsymbol{v},T)\right| \leq c(1 + \|T(t)\|_1\,\|\boldsymbol{v}(t)\|_0 + \|T(t)\|_0\,\|\boldsymbol{v}(t)\|_1).$$

Hence, after some modification and using the Gronwall's lemma, it follows that

$$\|T(t)\|_o \leq c, \tag{6.139}$$

$$\int_{t_o}^{t}\|T(\tau)\|_1^2\,\mathrm{d}\tau \leq c, \tag{6.140}$$

$$\|\boldsymbol{u}(t)\|_1 + \left\|\boldsymbol{u}'(t)\right\|_o \leq c, \tag{6.141}$$

$$\int_{t_o}^{t}\left\|\boldsymbol{u}'(\tau)\right\|_1^2\mathrm{d}\tau \leq c. \tag{6.142}$$

(ii) Second a priori estimates:

We need an estimate for $T'(t_o)$ and $\boldsymbol{u}'(t_o)$. We set $t = t_o$ in Equations 6.134 and 6.135 and then by the usual method we find

$$\left\|T'(t_o)\right\|_o + \left\|\boldsymbol{u}''(t_o)\right\|_o \leq c. \tag{6.143}$$

Now we differentiate Equations 6.134 and 6.135 with respect to t, and take $z = T'(t) + \overline{z}'(t)$ and $\boldsymbol{v} = \boldsymbol{u}''(t) + \boldsymbol{w}''(t)$, integrating in t from t_o to t. Then after some modifications, as in the first a priori estimates, and using the estimate

$$\left|b_p(\boldsymbol{u}''(t),T(t)) + b_s(T'(t),\boldsymbol{u}''(t))\right| \leq c\left\|T'(t)\right\|_0\left\|\boldsymbol{u}''(t)\right\|_0$$

and then applying the Gronwall's lemma, we find

$$\left\|T'(t)\right\|_0 \leq c,\ \int_{t_o}^{t_1}\left\|T'(t)\right\|_1^2\mathrm{d}t,$$

$$\left\|\boldsymbol{u}''(t)\right\|_0 + \left\|\boldsymbol{u}'(t)\right\|_1 \leq c,\ \int_{t_o}^{t_1}\left\|\boldsymbol{u}''(t)\right\|_1^2\mathrm{d}t.$$

Then we pass to the limit process over m by the usual way, which completes the proof.

Remark: If at least one part of the boundary $\Gamma_u = \cup_{\iota=1}^{s}\Gamma_u^{\iota}$ is of zero measure, then a global motion of the visco-elastic body can occur, and this can be decomposed into a solid rigid motion and a rotation. Such a solid motion will generally destroy the imposed boundary conditions and, therefore, we will consider the solution which satisfies the conditions of total equilibrium $\forall t > 0$.

6.5.2.2 *Quasi-static case*

THE MODEL

In this case the constituent law depends on time because of the viscosity term, i.e.

$$\tau_{ij}(\boldsymbol{u}(\boldsymbol{x},t)) = c_{ijkl}(\boldsymbol{x})e_{kl}(\boldsymbol{u}(\boldsymbol{x},t)) + C_{ijkl}(\boldsymbol{x})\frac{\partial}{\partial t}e_{kl}(\boldsymbol{u}(\boldsymbol{x},t)) - \beta_{ij}(\boldsymbol{x})(T(\boldsymbol{x},t) - T_o(\boldsymbol{x})). \tag{6.145}$$

We assume that the acting forces and displacements vary little for $t \geq t_o > 0$, and as our interest is in the solution for $t \geq t_o$ it is possible to neglect the linearized acceleration term $u''(t)$ in the equation of motion, representing the inertia forces. We shall call such a problem the quasi-static problem.

Let $\Omega = \cup_{\iota=1}^{s} \Omega^{\iota} \subset R^N$, $N = 2, 3$, with Lipschitz boundary $\partial\Omega = \Gamma_u \cup \Gamma_\tau \cup \mathscr{R}$, where $\mathscr{R}$ is the set of zero measure, be defined as above in the dynamic case, and let $\bar{I} = \langle t_o, t_1 \rangle$.

Then we have to solve the following problem:

Problem ($\mathscr{P}$): Find a pair of functions $(T, \boldsymbol{u})$ satisfying

$$\rho c_e \frac{\partial T}{\partial t} + \rho \beta_{ij} T_o(\boldsymbol{x}) \frac{\partial}{\partial t} e_{ij}(\boldsymbol{u}) = \frac{\partial}{\partial x_i}\left(\kappa_{ij} \frac{\partial T}{\partial x_j}\right) + Q_o \quad \text{in } \Omega, t \in I, \tag{6.146}$$

$$\frac{\partial \tau_{ij}}{\partial x_j} + f_{oi} = 0 \quad \text{in } \Omega, t \in I, \tag{6.147}$$

boundary conditions

$$T(\boldsymbol{x}, t) = T_1(\boldsymbol{x}, t) \quad \text{on } \Gamma_u, t > 0, \tag{6.148}$$

$$\boldsymbol{u}(\boldsymbol{x}, t) = \boldsymbol{u}_1(\boldsymbol{x}, t) \quad \text{on } \Gamma_u, t > 0, \tag{6.149}$$

$$\kappa_{ij}(\boldsymbol{x}) \frac{\partial T}{\partial x_j} n_i = q(\boldsymbol{x}, t), \quad \text{on } \Gamma_\tau, t > 0, \tag{6.150}$$

$$\tau_{ij}(\boldsymbol{u}(\boldsymbol{x}, t)) n_j = P_{oi}(\boldsymbol{x}, t) \quad \text{on } \Gamma_\tau, t > 0, \tag{6.151}$$

and initial conditions

$$T(\boldsymbol{x}, t_o) = T_o(\boldsymbol{x}), \tag{6.152}$$

$$\boldsymbol{u}(\boldsymbol{x}, t_o) = \boldsymbol{u}_o(\boldsymbol{x}), \tag{6.153}$$

where $Q_o(\boldsymbol{x}, t)$ denotes thermal sources, $c_e(\boldsymbol{x})$ is a specific heat, $\rho(\boldsymbol{x})$ is a density, $\kappa_{ij}(\boldsymbol{x})$ is a thermal conductivity, satisfying

$$\kappa_{ij}(\boldsymbol{x}) = \kappa_{ji}(\boldsymbol{x}), \tag{6.154}$$

$$\kappa_{ij}(\boldsymbol{x}) \xi_i \xi_j \geq \kappa_o |\xi|^2, \quad \xi \in R^N, \quad N = 2, 3, \tag{6.155}$$

q is a heat flow, $T_o(\boldsymbol{x})$ is an initial temperature, $\boldsymbol{f}_o(\boldsymbol{x}, t) = (f_{oi}(\boldsymbol{x}, t))$ are body forces, $\boldsymbol{P}(\boldsymbol{x}, t) = (P_i(\boldsymbol{x}, t))$ are surface forces and $\beta_{ij}(\boldsymbol{x})$ is a coefficient of linear thermal expansion. Furthermore, $c_{ijkl}(\boldsymbol{x})$ are the elastic coefficients and $C_{ijkl}(\boldsymbol{x})$ are coefficients of viscosity satisfying conditions of symmetry

$$c_{ijkl}(\boldsymbol{x}) = c_{jikl}(\boldsymbol{x}) = c_{ijlk}(\boldsymbol{x}) = c_{klij}(\boldsymbol{x}), \tag{6.156}$$

$$C_{ijkl}(\boldsymbol{x}) = C_{jikl}(\boldsymbol{x}) = C_{ijlk}(\boldsymbol{x}) = C_{klij}(\boldsymbol{x}) \tag{6.157}$$

and conditions of ellipticity

$$c_{ijkl}(\boldsymbol{x}) e_{ij}(\boldsymbol{u}) e_{kl}(\boldsymbol{u}) \geq c e_{ij}(\boldsymbol{u}) e_{ij}(\boldsymbol{u}), \; c > 0, \tag{6.158}$$

$$C_{ijkl}(\boldsymbol{x}) e_{ij}(\boldsymbol{u}') e_{kl}(\boldsymbol{u}') \geq C e_{ij}(\boldsymbol{u}') e_{ij}(\boldsymbol{u}'), \; C > 0. \tag{6.159}$$

VARIATIONAL FORMULATION OF THE PROBLEM

In this section we shall start from the formulation of the Problem $(\mathscr{P})_v$ for the case when the inertia forces are neglected. Then under some assumptions and using the same notation as above,

i.e.
$$\bar{z}(t) \in H^1(\Omega), \quad w(t) \in [H^1(\Omega)]^N, \quad N = 2, 3,$$
$$\bar{z}(t) = T_1(\boldsymbol{x}, t), \quad \boldsymbol{w}(t) = \boldsymbol{u}_1(\boldsymbol{x}, t) \quad \text{on } \Gamma_u,$$

and with
$$T(t) \cong T(t) - \bar{z}(t), \quad \boldsymbol{u}(t) \cong \boldsymbol{u}(t) - w(t),$$

the boundary conditions (Equations 6.148 and 6.149) are transformed to

$$T(\boldsymbol{x}, t) = 0 \quad \text{on } \Gamma_u, t > 0, \tag{6.160}$$

$$u(\boldsymbol{x}, t) = 0 \quad \text{on } \Gamma_u, t > 0, \tag{6.161}$$

and $Q(\boldsymbol{x}, t)$, $\boldsymbol{f}(\boldsymbol{x}, t)$, $\boldsymbol{P}(\boldsymbol{x}, t)$, $T_o(\boldsymbol{x})$, $\boldsymbol{u}_o(\boldsymbol{x})$ and $q(\boldsymbol{x}, t)$ change similarly as in the previous case.

We introduce the spaces of virtual temperatures ${}^1V$ and velocities V as

$${}^1V = \{z \mid z \in H^1(\Omega), \quad z = 0 \text{ on } \Gamma_u \text{ in the sense of traces }\},$$
$$V = \{\boldsymbol{v} \mid \boldsymbol{v} \in [H^1(\Omega)]^N, \quad N = 2, 3, \quad \boldsymbol{v} = 0 \text{ on } \Gamma_u \text{ in the sense of traces }\}.$$

Hence, ${}^1V \subset L^2(\Omega) \subset {}^1V'$, $V \subset [L^2(\Omega)]^N \subset V'$.

Multiplying Equation 6.146 by $z - T(t)$, and Equation 6.147 by $\boldsymbol{v} - \boldsymbol{u}'(t)$, integrating over Ω, and using the Green's theorem, then after some algebra we obtain the following variational problem:

Problem $(\mathscr{P})_v$: Find a pair of functions (T(t),$\boldsymbol{u}$(t)) such that

$$T(t) \in {}^1V, \boldsymbol{u}'(t) \in V,$$

$$(T'(t), z - T(t)) + a_T(T(t), z - T(t)) + b_p(\boldsymbol{u}'(t), z - T(t)) \geq s(z - T(t)) \quad \forall z \in {}^1V, \tag{6.162}$$

$$A(\boldsymbol{u}'(t), \boldsymbol{v} - \boldsymbol{u}'(t)) + a(\boldsymbol{u}(t), \boldsymbol{v} - \boldsymbol{u}'(t)) + b_s(T(t), \boldsymbol{v} - \boldsymbol{u}'(t)) \geq S(\boldsymbol{v} - \boldsymbol{u}'(t)) \quad \forall \boldsymbol{v} \in V, \tag{6.163}$$

and the initial conditions

$$T(\boldsymbol{x}, t_o) = T_o(\boldsymbol{x}), \tag{6.164}$$

$$\boldsymbol{u}(\boldsymbol{x}, t) = \boldsymbol{u}_o(\boldsymbol{x}), \tag{6.165}$$

where

$$T_o(\boldsymbol{x}) \cong T_o(\boldsymbol{x}) - \bar{z}(\boldsymbol{x}, t_o), \quad \boldsymbol{u}_o(\boldsymbol{x}) \cong \boldsymbol{u}_o(\boldsymbol{x}) - \boldsymbol{w}(\boldsymbol{x}, t_o),$$

$$s(z) = (W(t), z) = \int_\Omega Qz\mathrm{d}\boldsymbol{x} - \int_{\Gamma_\tau} qz\mathrm{d}s - (\bar{z}'(t), z) - a_T(\bar{z}(t), z) - b_p(\boldsymbol{w}'(t), \boldsymbol{z}),$$

$$S(\boldsymbol{v}) = (\boldsymbol{F}(t), \boldsymbol{v}) = \int_\Omega f_i v_i \mathrm{d}\boldsymbol{x} - \int_{\Gamma_\tau} P_i v_i \mathrm{d}s - A(\boldsymbol{w}'(t), \boldsymbol{v}) - a(\boldsymbol{w}(t), \boldsymbol{v}) - b_s(\bar{z}(t), \boldsymbol{v}).$$

The main result gives the next theorem:

Theorem 6.7: Let us assume that $Q \in L^2(I; L^2(\Omega))$, $\boldsymbol{f} \in L^2(I; [L^2(\Omega)]^N)$, $\bar{z} \in L^2(I; {}^1V)$, $\boldsymbol{w}, \boldsymbol{w}' \in L^2(I; V)$, $T_o \in {}^1V$, $\boldsymbol{u}_o \in V$. Then there exists one and only one pair of functions $(T, \boldsymbol{u})$ such that

$$T \in L^\infty(I; {}^1V), \quad T' \in L^2(I; {}^1V),$$
$$\boldsymbol{u} \in L^\infty(I; V), \quad \boldsymbol{u}' \in L^2(I; V),$$

and that Equations 6.162-6.165 hold.

Proof: To prove the uniqueness of the solution we shall assume that $(T^1, \boldsymbol{u}^1)$ and $(T^2, \boldsymbol{u}^2)$ are two possible solutions of Equations 6.162-6.165. Taking $z = T^1(t)$, $\boldsymbol{v} = \boldsymbol{u}^{1'}(t)$ (and then $z = T^2(t)$, $\boldsymbol{v} = \boldsymbol{u}^{2'}(t)$) in Equations 6.162 and 6.163 and adding both inequalities, then after some algebra we find, if $\Theta = T^1 - T^2$, $\boldsymbol{U} = \boldsymbol{u}^1 - \boldsymbol{u}^2$, and $\Theta(t_o) = 0$, $\boldsymbol{U}(t_o) = 0$, that $a_T(z, z) \geq c\,\|z\|_1^2$, $c > 0$, $\forall z \in {}^1V$, $a(\boldsymbol{v}, \boldsymbol{v}) \geq c_o\,\|\boldsymbol{v}\|_1^2$, $c_o > 0$, $A(\boldsymbol{v}, \boldsymbol{v}) \geq C_o\,\|\boldsymbol{v}\|_1^2$, $C_o > 0$, $\forall \boldsymbol{v} \in V$ and therefore $\Theta(t) = 0$, $\boldsymbol{U}(t) = 0$, which proves the uniqueness of the solution.

To prove the existence of the solution the usual technique will be used. Then taking $z = T(t) + \bar{z}(t)$ and $\boldsymbol{v} = \boldsymbol{u}'(t) + \boldsymbol{w}'(t)$ in Equations 6.162 and 6.163, integrating in t from t_o to t, then after some modifications and using the properties of bilinear forms, the estimate of $|b_p(.,.) + b_s(.,.)|$ and the Gronwall's lemma, we find

$$\|T(t)\|_o \leq c, \int_{t_o}^{t_1} \|T(t)\|_1^2\,dt \leq c, \|\boldsymbol{u}(t)\|_o \leq c, \int_{t_o}^{t_1} \|\boldsymbol{u}'(t)\|_1^2\,dt \leq c.$$

To finish the proof the usual technique will be used.

Remark: Numerically both problems, the dynamic and the quasi-static, can be approximated by the finite element method, and are not of further interest here. The main goal of plate tectonic problems will be the realistic model problems discussed in more details in the next chapter.

REFERENCES

Bensoussan, A., Lions, J.L. and Papanicolaou, G. (1978), *Asymptotic Analysis for Periodic Structures*, North-Holland.

Birolli, M. (1977), "G-Convergence for Elliptic Equations, Variational Inequalities and Quasi-variational Inequalities", *Red. Sem. Mat. Fiz.*, Milano, **47**, 269-328.

Kolář, V., Kratochvíl, J., Leitner, F. and Ženíšek, A. (1972), The Computation of Two- and Three-Dimensional Structures by the Finite Element Method, SNTL, Prague (in Czech).

Nečas, J. and Hlaváček, I. (1981). *Mathematical Theory of Elastic and Elasto-Plastic Bodies: An Introduction*, Elsevier, Amsterdam, Oxford, New York.

Nedoma, J., Haslinger, J. and Hlaváček, I. (1989), "The Problem of an Obducting Lithospheric Plate in the Aleutian Arc System: A Finite Element Analysis in the Frictionless Case", Math. Comput. Modelling **12**, **1**, 61-75.

Oden, J.T. and Reddy, J.N. (1976), *An Introduction to the Mathematical Theory of Finite Elements*, J.Wiley & Sons, New York, London, Sydney, Toronto.

7

Realistic Mathematical Models of Collision Zones Based on Variation Inequalities

7.1. MODELS OF COLLISION ZONES BASED ON CONTACT PROBLEMS WITH OR WITHOUT FRICTION IN LINEAR ELASTICITY

7.1.1. Introduction

In order to review the general characteristics of the subduction and collision zones, the N-th dimensional ($N = 2, 3$) model problem will be investigated.

The stresses in the lithospheric plate arise from several fundamental sources: tectonic stresses, thermally induced stresses, stresses resulting from the action of gravitational and magnetic body forces as well as of tidal body forces acting on the density distribution in the lithospheric plate. In the study, the influence of stresses resulting from the action of magnetic and tidal bodies will be neglected.

Many authors have analysed the deformation of the lithosphere in the subduction and collision zones using thin elastic plate theory, based on the classical boundary value problems. Generally the model considered can be elastic or thermo-elastic. It is well-known that, for time periods shorter than a characteristic time $t_c = \nu/\mu$, the lithosphere and the upper mantle behave elastically or thermo-elastically, while for longer times they behave thermo-visco-elastically and plastically. Here ν represents the effective viscosity at a given stress level defined by

$$\nu = \frac{\tau_2 + [(\frac{2}{3}e_{ij}(\dot{\boldsymbol{u}})e_{ij}(\dot{\boldsymbol{u}}))^{1/2}\underline{\gamma}^{-1}]^{1/n}}{3^{1/2}2^{1/4}(e_{ij}(\dot{\boldsymbol{u}})e_{ij}(\dot{\boldsymbol{u}}))^{1/2}},$$

where $e_{ij}(\dot{\boldsymbol{u}})$ is the strain rate tensor, τ_2 is the yield stress or the so-called mean flow limit of the continental lithosphere,

$$\underline{\gamma} = L^{-1} \int_L \underline{\gamma}_o \exp(-Q/RT)\mathrm{d}x_3,$$

($\underline{\gamma} \to \infty$ for the rigid perfectly plastic flow), where Q is an activation energy, R is the universal gas constant, L the thickness of the lithosphere and μ is the elastic rigidity. As it is known that $t_c \sim 10^0 - 10^5$ years the rheology can be taken as elastic. Then the stress-strain relations of such materials are given by the Hooke linear law.

7.1.2. The Models

Let $0x_1 \ldots x_N$ be the orthogonal Cartesian coordinate system. Let the subducted and obducted zones and part of the upper mantle, which are approximated by an elastic anisotropic or isotropic plate, occupy the region Ω with boundary $\partial\Omega$ at a time $t \in \langle t_o, t_1 \rangle$. The total boundary of this colliding zone consists of three parts: the Earth's surface Γ_τ, the contact boundary Γ_c and the boundary Γ_u. The boundary Γ_c consists of the contact boundaries Γ_c^{kl} between the obducting and subducting plates as well as the asthenosphere. For simplicity we assume collision of subducting and obducting lithospheric plates which rest on the asthenosphere. Hence $\Gamma_c = {}_{k,} \cup_l \Gamma_c^{kl}$. The boundary Γ_u is a fictitious boundary and is represented by the boundary defined at a sufficiently large distance from the collision zone. It is convenient to choose the origin as the point nearest to the trench axis (see Figure 7.1).

As discussed in the previous chapter, the stress field in the lithospheric plate satisfies the equilibrium equations

$$\tau_{ij,j} + f_i = 0, \quad i = 1, \ldots, N, \tag{7.1}$$

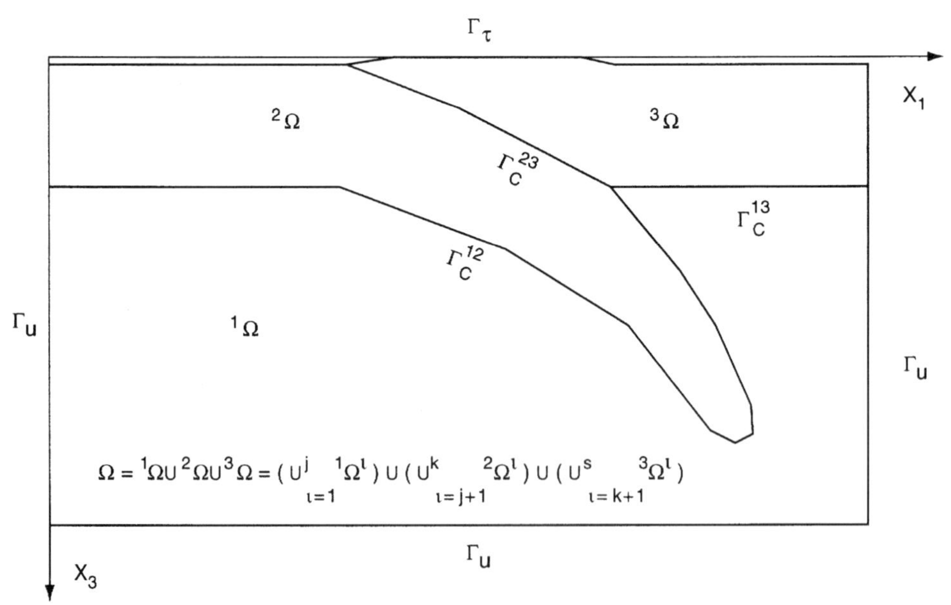

Figure 7.1. The model

where $\boldsymbol{f} = (f_i)$ denotes the vector of body forces and τ_{ij} represents the elastic stress tensor. Henceforth we use the summation Einstein's convention.

The traction vector $\vec{\tau} = (\tau_i)$, where

$$\tau_i = \tau_{ij} n_j, \tag{7.2}$$

on the boundary can be decomposed into the normal and tangential components

$$\tau_n = \tau_i n_i = \tau_{ij} n_i n_j, \quad \vec{\tau}_t = (\tau_i t_i) = \vec{\tau} - \tau_n \boldsymbol{n}, \tag{7.3}$$

where $\boldsymbol{n} = (n_i)$ and $\boldsymbol{t} = (t_i)$ are the outward normal and tangential unit vectors. The displacement vector can be decomposed into the normal and tangential components

$$u_n = u_i n_i, \boldsymbol{u}_t = (u_i t_i) = \boldsymbol{u} - u_n \boldsymbol{n}. \tag{7.4}$$

Due to our presumption the collision model can be investigated within the range of linear elasticity. Therefore the stress-strain relation is given by the linear Hooke's law

$$\tau_{ij} = c_{ijkl} e_{kl}(\boldsymbol{u}) \tag{7.5}$$

for anisotropic rocks, or

$$\tau_{ij} = \lambda e_{kk}(\boldsymbol{u}) \delta_{ij} + 2\mu e_{ij}(\boldsymbol{u}) \tag{7.6}$$

for isotropic rocks, where c_{ijkl}, λ, and $\mu \in L^\infty(\overline{\Omega})$ (or $\in C^1(\overline{\Omega})$) are the elastic or the so-called Lamé coefficients satisfying

$$c_{ijkl} = c_{jikl} = c_{klij} = c_{ijlk} \quad \text{a.e. in } \Omega, \tag{7.7}$$

and $\exists c_o = \text{constant} > 0$ such that

$$c_{ijkl} e_{ij}(\boldsymbol{u}) e_{kl}(\boldsymbol{u}) \geq c_o e_{ij}(\boldsymbol{u}) e_{ij}(\boldsymbol{u}) \quad \forall e_{ij}(\boldsymbol{u}) = e_{ji}(\boldsymbol{u}) \quad \text{a.e. in } \Omega \tag{7.8}$$

in the case of anisotropic rocks,
or $\exists c_o = \text{constant} > 0$ such that

$$\tfrac{1}{2}(\lambda \vartheta^2(\boldsymbol{u}) + 2\mu e_{ij}(\boldsymbol{u}) e_{ij}(\boldsymbol{u})) \geq \mu_o e_{ij}(\boldsymbol{u}) e_{ij}(\boldsymbol{u}) \quad \forall e_{ij}(\boldsymbol{u}) = e_{ji}(\boldsymbol{u}) \quad \text{a.e. in } \Omega \tag{7.9}$$

in the case of isotropic rocks,
where $\vartheta(\boldsymbol{u}) = e_{kk}(\boldsymbol{u})$ and

$$e_{ij}(\boldsymbol{u}) = \tfrac{1}{2}(u_{i,j} + u_{j,i}), \quad i, j = 1, \ldots, N \tag{7.10}$$

is a small strain tensor.

7.1.3. Boundary Conditions

To investigate the stress-strain field in the obducting and subducting plates and in the asthenosphere in a neighbourhood of the collision zone, conditions should usually be given characterizing the effect of the environment of the plate on the processes taking place within a limited region containing the collision parts of the lithospheric plates under study and the upper mantle. The interaction between the colliding plates and their environment is modelled by the boundary conditions for the displacement vector $\boldsymbol{u}$. In the model problem the boundary $\partial\Omega$ of the investigated region Ω occupied by the obducting and subducting plates and the asthenosphere with the mesosphere consists of three parts Γ_u, Γ_τ and Γ_c for the $2D$ model problem and of four parts Γ_u, Γ_τ, Γ_o and Γ_c for the $3D$ model problem, where Γ_o has a meaning as a boundary of symmetry or a contact boundary with perfectly rigid and perfectly smooth support without friction. Then on the Earth's surface Γ_τ the surface forces $\boldsymbol{P}$ are prescribed as caused by the weight of the oceans, of the atmosphere, and of the mountains above some approximate surface. Thus one has

$$\tau_{ij}n_j = P_i \quad \text{on } \Gamma_\tau. \tag{7.11}$$

In the model problem the weight of the oceans, atmosphere, etc. can be neglected so that $P_i = 0$.

On the contact boundary $\Gamma_c = {}_{k,} \cup_l \Gamma_c^{kl}$ the boundary conditions are deduced from the following consideration (see also Chapter 4, regarding the evolution of collision zones, where the contact conditions were derived): at time $t \in \langle t_o, t_1 \rangle$ we assume that before the deformation the colliding plates Ω^ι, $\iota = 1, \ldots, s$, and the part of the upper mantle were in contact along all contact boundaries $\cup\Gamma_c^{kl}$. During the deformation process the contact points will be displaced in a different way, but colliding plates can not penetrated,

$$u_n^k - u_n^l \leq 0, \tag{7.12}$$

where the direction of the normal $\boldsymbol{n}$ to the boundary of the plate denoted as Ω^l is taken in relation to the plate Ω^k.

For the contact forces one obtains from the law of action and reaction $\tau_n^k = -\tau_n^l$ and $\vec{\tau}_t^k = -\vec{\tau}_t^l$. Since normal components cannot be positive (traction) then

$$\tau_n^k = -\tau_n^l \leq 0. \tag{7.13}$$

During the deformation of both colliding plates, the colliding plates are in contact (i.e. $u_n^k - u_n^l = 0$) or are not in contact (i.e. $u_n^k - u_n^l < 0$); therefore the normal components of stresses are nonzero in the first case and are equal to zero in the second case. Hence

$$\tau_n^k(u_n^k - u_n^l) = 0. \tag{7.14}$$

Equation 7.12-7.14 are known as the Signorini conditions.
If $u_n^k - u_n^l = 0$ then $\left|\vec{\tau}_t^k\right| \leq g_c^{kl}$ on the contact boundary $\Gamma_c = {}_{k,} \cup_l \Gamma_c^{kl}$ and

$$\left|\vec{\tau}_t^k\right| < g_c^{kl} \Rightarrow \boldsymbol{u}_t^k - \boldsymbol{u}_t^l = 0, \tag{7.15}$$

$$\left|\vec{\tau}_t^k\right| = g_c^{kl} \Rightarrow \exists\lambda \geq 0 \text{ such that } \boldsymbol{u}_t^k - \boldsymbol{u}_t^l = -\lambda\vec{\tau}_t^k, \tag{7.16}$$

where it is assumed that $g_c^{kl} \in L^\infty(\cup\Gamma_c^{kl})$, $g_c^{kl} \geq 0$ a.e. on $\cup\Gamma_c^{kl}$.
If $g_c^{kl} = 0$ then

$$\vec{\tau}_t^k = -\vec{\tau}_t^l = 0. \tag{7.17}$$

This condition means that on the contact boundary there is no friction.

In the case of the lithospheric plate being in contact with absolutely rigid lithospheric plates and the absolutely rigid asthenosphere, then Equations 7.12 to 7.14 reduce to

$$u_n \leq 0, \tag{7.18}$$

$$\tau_n \leq 0, \tag{7.19}$$

$$u_n \tau_n = 0. \tag{7.20}$$

To model the shifting of the obducted and subducted plates, the value of the displacement vector is determined at a distance sufficiently far from the collision zone. This boundary is denoted by Γ_u. The values of the displacement vector components can be determined from the knowledge (obtained for example, from geological or satellite observations) of the speed of the lithospheric plates and of speeds in the upper mantle, if they exist. Thus

$$\boldsymbol{u} = \boldsymbol{u}_o \text{ on } \Gamma_u. \tag{7.21}$$

On the boundary Γ_o we have the conditions

$$u_n = 0, \quad \tau_{tj} = 0, \quad j = 1, \ldots, N. \tag{7.22}$$

These conditions (Equation 7.22) are called the one side contact conditions because they describe the contact of the body Ω with its perfectly rigid and perfectly smooth support for the case without friction. In addition, the same conditions hold in the plane of symmetry of a symmetric body that is both loaded and supported symmetrically with respect to the plane of symmetry. Therefore, we may solve the problem on one half of the region only.

The problem is coercive for $\Gamma_u \neq 0$ and semi-coercive for $\Gamma_u = 0$ or if one part of $\Gamma_u = \cup_{\iota=1}\Gamma_u^\iota$, say Γ_u^s, is equal to zero. We shall consider the case $\Gamma_u \neq 0$, i.e. the coercive case. The semi-coercive case will be investigated later.

7.1.4. Classical Formulation of the Problem

Let $\Omega = \cup_{\iota=1}^{S}\Omega^\iota \subset R^N, N = 2, 3$, be a domain with Lipschitz boundary $\partial\Omega$, occupied by a part of the lithosphere and the upper mantle (the asthenosphere and the mesosphere), and which we assume to be elastic. Let the boundary $\partial\Omega$ consist of three parts Γ_τ, Γ_u and Γ_c, $\partial\Omega = \overline{\Gamma}_\tau \cup \overline{\Gamma}_u \cup \overline{\Gamma}_c \cup R$, where $\overline{\Gamma}_c = {}_k\cup_l \overline{\Gamma}_c^{kl}$ represents the boundary between bodies in contact, and R is a set of zero measure. Moreover, we denote the displacement vector by $\boldsymbol{u} = (u_1, \ldots, u_N)$. Let $\boldsymbol{f} \in [L^2(\Omega)]^N, \boldsymbol{P} \in [L^2(\Gamma_\tau)]^N$, and $\boldsymbol{u}_o \in [C(\Gamma_u)]^N$. Next we shall deal with the following problem.

Problem ($\mathscr{P}'$): Find a vector function $\boldsymbol{u}$ satisfying

$$(c_{ijkl}e_{kl}(\boldsymbol{u}))_{,j} + f_i = 0, \quad i, j, k, l = 1, \ldots, N, \text{ in } \Omega, \tag{7.23}$$

$$\boldsymbol{u} = \boldsymbol{u}_o \text{ on } \Gamma_u, \tag{7.24}$$

$$\tau_{ij} n_j = P_i, \quad i, j = 1, \ldots, N, \text{ on } \Gamma_\tau, \tag{7.25}$$

$$u_n^k - u_n^l \leq 0, \quad \tau_n^k(\boldsymbol{u}) = -\tau_n^l(\boldsymbol{u}) \leq 0, \quad (u_n^k - u_n^l)\tau_n^k = 0, \text{ on } \cup_{k,l} \Gamma_c^{kl} \tag{7.26}$$

$$\text{if } u_n^k - u_n^l = 0 \text{ then } |\vec{\tau}_t^k| \leq g_c^{kl} \text{ on } \cup_{k,l} \Gamma_c^{kl} \tag{7.27}$$

$$|\vec{\tau}_t^k| < g_c^{kl} \Rightarrow \boldsymbol{u}_t^k - \boldsymbol{u}_t^l = 0, \tag{7.28}$$

$$|\vec{\tau}_t^k| = g_c^{kl} \Rightarrow \exists \lambda \geq 0 \text{ such that } \boldsymbol{u}_t^k - \boldsymbol{u}_t^l = -\lambda \vec{\tau}_t^k, \text{ on } \cup_{k,l} \Gamma_c^{kl} \tag{7.29}$$

$$u_n = 0, \quad \vec{\tau}_t = 0 \text{ on } \Gamma_o, \tag{7.30}$$

where the last condition (Equation 7.30) is very suitable for $3D$ problems, as it describes the fact that the geodynamic problem can be studied in a finite closed region. In some problems, namely $2D$ problems, this condition can be omitted.

Next we give the following lemma.

Lemma 7.1: Let $\boldsymbol{u} = (u_i)$ and $\boldsymbol{v} = (v_i)$, $i = 1, \ldots, N$, be vector functions and let $\boldsymbol{u}, \boldsymbol{v} \in [C^2(\overline{\Omega})]^N$. Then

$$\int_\Omega \tau_{ij}(\boldsymbol{u}) e_{ij}(\boldsymbol{v}) \mathrm{d}\boldsymbol{x} = \int_\Omega c_{ijkl} e_{kl}(\boldsymbol{u}) e_{ij}(\boldsymbol{v}) \mathrm{d}\boldsymbol{x} = -\int_\Omega \partial\tau_{ij}(\boldsymbol{u})/\partial x_j v_i \mathrm{d}\boldsymbol{x} + \int_{\partial\Omega} \tau_{ij}(\boldsymbol{u}) v_i n_j \mathrm{d}s. \tag{7.31}$$

Proof: By using the Green theorem we obtain

$$\int_\Omega \tau_{ij}(\boldsymbol{u}) e_{ij}(\boldsymbol{v}) \mathrm{d}\boldsymbol{x} = \int_\Omega \sum_{i,j=1}^N \tau_{ij}(\boldsymbol{u}) e_{ij}(\boldsymbol{v}) \mathrm{d}\boldsymbol{x} = \frac{1}{2}\left(\int_\Omega \sum_{i,j=1}^N \tau_{ij}(\boldsymbol{u})(\partial v_i/\partial x_j + \partial v_j/\partial x_i) \mathrm{d}\boldsymbol{x}\right)$$

$$= \frac{1}{2}\left(\int_\Omega \sum_{i,j=1}^N \tau_{ij}(\boldsymbol{u}) \partial v_i/\partial x_j \mathrm{d}\boldsymbol{x} + \int_\Omega \sum_{i,j=1}^N \tau_{ij}(\boldsymbol{u}) \partial v_j/\partial x_i \, \mathrm{d}\boldsymbol{x}\right)$$

$$= \frac{1}{2}\left(-\int_\Omega \sum_{i,j=1}^N \partial\tau_{ij}(\boldsymbol{u})/\partial x_j v_i \mathrm{d}\boldsymbol{x} + \int_{\partial\Omega} \sum_{i,j=1}^N \tau_{ij}(\boldsymbol{u}) v_i n_j \mathrm{d}s\right.$$

$$\left. - \int_\Omega \sum_{i,j=1}^N \partial\tau_{ij}(\boldsymbol{u})/\partial x_i v_j \mathrm{d}\boldsymbol{x} + \int_{\partial\Omega} \sum_{i,j=1}^N \tau_{ij}(\boldsymbol{u}) v_j n_i \mathrm{d}s\right)$$

$$= -\int_\Omega \sum_{i,j=1}^N \partial\tau_{ij}(\boldsymbol{u})/\partial x_j v_i \mathrm{d}\boldsymbol{x} + \int_{\partial\Omega} \sum_{i,j=1}^N \tau_{ij}(\boldsymbol{u}) v_i n_j \mathrm{d}s$$

$$= -\int_\Omega \partial\tau_{ij}(\boldsymbol{u})/\partial x_j v_i \, \mathrm{d}\boldsymbol{x} + \int_{\partial\Omega} \tau_{ij}(\boldsymbol{u}) v_i n_j \, \mathrm{d}s,$$

where the Einstein's convention was assumed.

Remark: Lemma 7.1 is valid also for $\boldsymbol{u}, \boldsymbol{v} \in W^2 = [H^2(\Omega)]^N$.

Lemma 7.2 (Korn inequality): Let $\Omega \in R^N$, $N = 2, 3$, be a domain with Lipshitz boundary $\partial\Omega$ and let $\Gamma_u \in \partial\Omega$ be a non-empty open subset. Let a function $\boldsymbol{v} \in W^1 = [H^1(\Omega)]^N$ and $\boldsymbol{v} = 0$

on Γ_u. Let $e_{ij}(\boldsymbol{u})$ be defined by Equation 7.10. Then there exists c = constant > 0 such that

$$\int_{\Omega} e_{ij}(\boldsymbol{v})e_{ij}(\boldsymbol{v})\mathrm{d}\boldsymbol{x} \geq c\,|\boldsymbol{v}|_1^2 . \tag{7.32}$$

For the proof see Nečas and Hlaváček (1981).

7.1.5. Variational Formulation and Weak Solution

Let us denote $W^k = [H^k(\Omega)]^N, N = 2, 3$. Let us transform the problem to one with homogeneous boundary conditions. In the equations let us replace $\boldsymbol{u}$ by $\boldsymbol{u} + \boldsymbol{w}$, where $\boldsymbol{w} \in W^2$ is a sufficiently smooth vector function in $\overline{\Omega} = \Omega \cup \partial\Omega$ satisfying Equations 7.24 and 7.26 (in the case if the contact boundary Γ_c is perpendicular to Γ_u then it is sufficient to put $\boldsymbol{w} = 0$ on Γ_c). According to this transformation the boundary conditions, and the surface ($\boldsymbol{P}$) and body ($\boldsymbol{f}$) forces will be changed. Thus we obtain the following equivalent formulation of our problem (we will use the same symbols $\boldsymbol{u}$, $\vec{\tau} = \vec{\tau}(\boldsymbol{u})$ and $\boldsymbol{f}$ as above).

Problem ($\mathscr{P}$): Find a vector function $\boldsymbol{u}$ satisfying

$$(c_{ijkl}e_{kl}(\boldsymbol{u}))_{,j} + f_i = 0 \quad i, j, k, l = 1, \ldots, N, \quad N = 2, 3, \text{ in } \Omega, \tag{7.33}$$

$$\boldsymbol{u} = 0 \text{ on } \Gamma_u, \tag{7.34}$$

$$\tau_{ij}n_j = P_i, \quad i, j = 1, \ldots, N, \quad N = 2, 3 \text{ on } \Gamma_\tau, \tag{7.35}$$

$$u_n^k - u_n^l \leq 0, \quad \tau_n^k(\boldsymbol{u}) = -\tau_n^l(\boldsymbol{u}) \leq 0, \quad (u_n^k - u_n^l)\tau_n^k = 0, \text{ on } \Gamma_c, \tag{7.36}$$

$$\text{if } u_n^k - u_n^l = 0 \text{ then } \left|\vec{\tau}_t^k\right| \leq g_c^{kl} \text{ on } \cup_{k,l}\Gamma_c^{kl} \tag{7.37}$$

$$\left|\vec{\tau}_t^k\right| < g_c^{kl} \Rightarrow \boldsymbol{u}_t^k - \boldsymbol{u}_t^l = 0, \text{ on } \cup_{k,l}\Gamma_c^{kl}, \tag{7.38}$$

$$\left|\vec{\tau}_t^k\right| = g_c^{kl} \Rightarrow \exists\lambda \geq 0 \text{ such that } \boldsymbol{u}_t^k - \boldsymbol{u}_t^l = -\lambda\vec{\tau}_t^k, \text{ on } \cup_{k,l}\Gamma_c^{kl} \tag{7.39}$$

$$u_n = 0, \vec{\tau}_t = 0 \text{ on } \Gamma_o, \tag{7.40}$$

where $f_i \equiv f_i + (c_{ijkl}e_{kl}(\boldsymbol{w}))_{,j} \in L^2(\Omega)$, $P_i \equiv P_i - c_{ijkl}e_{kl}(\boldsymbol{w})n_j \in L^2(\Gamma_\tau)$.

Let $\mathscr{E}(\overline{\Omega})$ be the space of all real infinitely diferentiable functions on Ω such that all of their derivatives are continuously extended up to the boundary $\partial\Omega$. By $\mathscr{D}(\Omega)$ we denote the subspace of $\mathscr{E}(\overline{\Omega})$ containing all functions with compact support in Ω.

Let us define the space of virtual displacements as

$$V = \{\boldsymbol{v} \mid \boldsymbol{v} \in W^1 = [H^1(\Omega)]^N, \quad \boldsymbol{v} = 0, v_n = 0 \text{ on } \Gamma_o \text{ in the sense of traces}\}$$

and the set of admissible displacements

$$K = \{\boldsymbol{v} \mid \boldsymbol{v} \in V, v_n^k - v_n^l \leq 0 \text{ in the sense of traces}\}.$$

Lemma 7.3: The set K is non-empty, closed and convex in V.

Proof: K is evidently non-empty, as a zero function f is also a member of K. The set K is closed, which follows from the fact that the space $[H^1(\Omega)]^N$ is complete and from the trace theorem. Let

$\boldsymbol{u}, \boldsymbol{v} \in K$, $\alpha \in \langle 0, 1 \rangle$, then the function $\boldsymbol{y} = \alpha \boldsymbol{v} + (1-\alpha)\boldsymbol{w} \in V$ and for its normal component y_n it holds

$$y_n = y_i n_i = \alpha v_i n_i + (1-\alpha) w_i n_i \le 0 \text{ on } \cup \Gamma_c^{kl}. \text{ Hence } K \text{ is convex.}$$

Multiplying Equation 7.33 by $v_i - u_i$, integrating over Ω, using the Green theorem and the boundary conditions we obtain the variational formulation of the problem $(\mathscr{P})$ as follows.

Problem $(\mathscr{P})_v$: Find $\boldsymbol{u} \in K$ such that

$$a(\boldsymbol{u}, \boldsymbol{v} - \boldsymbol{u}) + j(\boldsymbol{v}) - j(\boldsymbol{u}) \ge \int_\Omega f_i(v_i - u_i)\mathrm{d}\boldsymbol{x} + \int_{\Gamma_\tau} P_i(v_i - u_i)\mathrm{d}s \quad \forall \boldsymbol{v} \in K, \tag{7.41}$$

where $a(\boldsymbol{u}, \boldsymbol{v})$ is the bilinear form defined by

$$a(\boldsymbol{u}, \boldsymbol{v}) = \int_\Omega c_{ijkl} e_{ij}(\boldsymbol{u}) e_{kl}(\boldsymbol{v})\mathrm{d}\boldsymbol{x} \equiv (c_{ijkl} e_{kl}(\boldsymbol{u}), e_{ij}(\boldsymbol{v})), \tag{7.42}$$

$$j(\boldsymbol{v}) = \int_{\cup_{k,l}\Gamma_c^{k1}} g_c^{k1} \, |\boldsymbol{v}_t| \, \mathrm{d}s \tag{7.43}$$

where $g_c^{kl} \in L^\infty(\cup \Gamma_c^{kl})$, $g_c^{kl} \ge 0$ a.e. on $\cup \Gamma_c^{kl}$.

This variational formulation represents variational inequality and is equivalent to finding a minimum of the functional of potential energy on the set of admissible displacement K.

Further, let us put

$$L(\boldsymbol{v}) = L_o(\boldsymbol{v}) + j(\boldsymbol{v}), \tag{7.44}$$

$$L_o(\boldsymbol{v}) = \tfrac{1}{2} a(\boldsymbol{v}, \boldsymbol{v}) - S(\boldsymbol{v}), \tag{7.45}$$

$$S(\boldsymbol{v}) = \int_\Omega f_i v_i \mathrm{d}\boldsymbol{x} + \int_{\Gamma_\tau} P_i v_i \mathrm{d}s = (\boldsymbol{f}, \boldsymbol{v}) + \int_{\Gamma_\tau} P_i v_i \mathrm{d}s. \tag{7.46}$$

Then we have the following equivalent variational formulation of our problem $(\mathscr{P})_v$.

Find $\boldsymbol{u} \in K$ such that

$$L(\boldsymbol{u}) \le L(\boldsymbol{v}) \quad \forall \boldsymbol{v} \in K. \tag{7.47}$$

Lemma 7.4: The bilinear form $a(\boldsymbol{u}, \boldsymbol{v})$ defined by Equation 7.42 is

(i) bounded on V, i.e. $\exists M > 0$ such that

$$|a(\boldsymbol{u}, \boldsymbol{v})| \le M \, \|\boldsymbol{u}\|_1 \, \|\boldsymbol{v}\|_1 \quad \forall \boldsymbol{u}, \boldsymbol{v} \in V;$$

(ii) V-elliptic on V, i.e. $\exists \alpha > 0$ such that

$$a(\boldsymbol{v}, \boldsymbol{v}) \ge \alpha \, \|\boldsymbol{v}\|_1^2 \quad \forall \boldsymbol{v} \in V;$$

(iii) symmetric on V, i.e.

$$a(\boldsymbol{u}, \boldsymbol{v}) = a(\boldsymbol{v}, \boldsymbol{u}) \quad \forall \boldsymbol{u}, \boldsymbol{v} \in V;$$

where $\|.\|_1$ is the norm in $[H^1(\Omega)]^N$.

Proof: (i) From the definition of e_{ij}, $\|.\|_o$, $\|.\|_1$ and the Schwarz inequality it follows

$$\sum_{i,j=1}^N \|e_{ij}(\boldsymbol{u})\|_o^2 = \frac{1}{4} \sum_{i,j=1}^N \int_\Omega [(\partial u_i/\partial x_j)^2 + 2(\partial u_i/\partial x_j \partial u_j/\partial x_i) + (\partial u_j/\partial x_i)^2]\mathrm{d}\boldsymbol{x}$$

$$\leq \frac{1}{2}\sum_{i,j=1}^{N}\int_{\Omega}(\partial u_i/\partial x_j)^2\mathrm{d}\boldsymbol{x} + \frac{1}{2}\sum_{i,j=1}^{N}\left[\int_{\Omega}(\partial u_i/\partial x_j)^2\mathrm{d}\boldsymbol{x}\right]^{1/2}\left[\int_{\Omega}(\partial u_j/\partial x_i)^2\mathrm{d}\boldsymbol{x}\right]^{1/2}$$

$$\leq \tfrac{1}{2}(\|\boldsymbol{u}\|_1^2 + 4\,\|\boldsymbol{u}\|_1^2) = \tfrac{5}{2}\,\|\boldsymbol{u}\|_1^2 .$$

Let us put $c = \sup_{\substack{\boldsymbol{x}\in\Omega \\ i,j,k,l=1,\ldots,N}} |c_{ijkl}(\boldsymbol{x})|$. Then applying the Schwarz inequality we have

$$|a(\boldsymbol{u},\boldsymbol{v})| = |(c_{ijkl}e_{kl}(\boldsymbol{u}), e_{ij}(\boldsymbol{v}))| \leq c\sum_{i,j=1}^{N}(e_{kl}(\boldsymbol{u}), e_{ij}(\boldsymbol{v})) \leq c\sum_{i,j=1}^{N}\|e_{ij}(\boldsymbol{u})\|_o \sum_{i,j=1}^{N}\|e_{ij}(\boldsymbol{v})\|_o .$$

Since

$$\left(\sum_{i,j=1}^{N}\|e_{ij}(\boldsymbol{u})\|_o\right)^2 \leq 4\sum_{i,j=1}^{N}\|e_{ij}(\boldsymbol{u})\|_o^2 \leq 10\,\|\boldsymbol{u}\|_1^2 ,$$

we find

$$|a(\boldsymbol{u},\boldsymbol{v})| \leq 10c\,\|\boldsymbol{u}\|_1\,\|\boldsymbol{v}\|_1 .$$

(ii) Due to Equation 7.8 there exists $c_o > 0$, and due to the Korn inequality a constant $c > 0$ such that

$$a(\boldsymbol{u},\boldsymbol{u}) = (c_{ijkl}e_{kl}(\boldsymbol{u}), e_{ij}(\boldsymbol{u})) \geq c_o(e_{kl}(\boldsymbol{u}), e_{kl}(\boldsymbol{u})) \geq c_o c\,\|\boldsymbol{u}\|_1^2 \quad \forall\boldsymbol{u}\in V.$$

Then $\alpha = c_o c$, which proves the assertion.

(iii) Since Equation 7.7 holds then

$$a(\boldsymbol{u},\boldsymbol{v}) = (c_{ijkl}e_{kl}(\boldsymbol{u}), e_{ij}(\boldsymbol{v})) = (c_{ijkl}e_{ij}(v), e_{kl}(\boldsymbol{u})) = a(\boldsymbol{v},\boldsymbol{u}),\ \boldsymbol{u},\boldsymbol{v}\in V.$$

Lemma 7.5: The functional L_o defined by Equation 7.44 on the space V is

(i) G-differentiable (i.e. Gâteaux differentiable), if

$$DL_o(\boldsymbol{u},\boldsymbol{v}) = a(\boldsymbol{u},\boldsymbol{v}) - S(\boldsymbol{v}),$$

$$D^2L_o(\boldsymbol{u};\boldsymbol{v},\boldsymbol{v}) = a(\boldsymbol{v},\boldsymbol{v}),$$

$$L_o(\boldsymbol{u}) = L_o(\boldsymbol{v}) + DL_o(\boldsymbol{u},\boldsymbol{u}-\boldsymbol{v}) + \tfrac{1}{2}D^2L_o(\boldsymbol{u};\boldsymbol{u}-\boldsymbol{v},\boldsymbol{u}-\boldsymbol{v}) \quad \forall\boldsymbol{u},\boldsymbol{v}\in V;$$

(ii) purely convex, i.e.

$$L_o(\alpha\boldsymbol{u} + (1-\alpha)\boldsymbol{v}) < \alpha L_o(\boldsymbol{u}) + (1-\alpha)L_o(\boldsymbol{v}) \text{ for all } \alpha\in(0,1), \quad \boldsymbol{u},\boldsymbol{v}\in V;$$

(iii) coercive, i.e.

$$\lim_{\substack{\|\boldsymbol{u}\|\to\infty \\ \boldsymbol{u}\in V}} L_o(\boldsymbol{u}) = +\infty;$$

(iv) weakly lower semi-continuous, i.e.

$$\lim_{\boldsymbol{v}\in V} L_o(\boldsymbol{v}) \geq L_o(\boldsymbol{u}),$$

$$\boldsymbol{v}\to\boldsymbol{u} \text{ weakly}.$$

Proof:

(i) For $\boldsymbol{u},\boldsymbol{v}\in V$ and $t\in R^1$, from L_o and the symmetry condition it follows

$$\lim_{t\to 0} t^{-1}(L_o(\boldsymbol{u}+t\boldsymbol{v}) - L_o(\boldsymbol{u})) = \lim_{t\to 0} t^{-1}(a(\boldsymbol{u}+t\boldsymbol{v},\boldsymbol{u}+t\boldsymbol{v}) - S(\boldsymbol{u}+t\boldsymbol{v}) - \tfrac{1}{2}a(\boldsymbol{u},\boldsymbol{u}) + S(\boldsymbol{u}))$$

$$= a(\boldsymbol{u},\boldsymbol{v}) - S(\boldsymbol{v}).$$

Similarly for $D^2L_o(\boldsymbol{u};\boldsymbol{v},\boldsymbol{v})$. The third assertion is the Taylor's expansion of L_o at $\boldsymbol{u}$ and follows from the definition of $L_o(\boldsymbol{v})$ and definitions of $DL_o(\boldsymbol{v};\boldsymbol{v})$ and $D^2L_o(\boldsymbol{u};\boldsymbol{v},\boldsymbol{v})$.

(ii) The functional L_o is G-differentiable. From $D^2L_o(\boldsymbol{u};\boldsymbol{v},\boldsymbol{v}) = a(\boldsymbol{v},\boldsymbol{v})$ and the V-ellipticity of the form $a(\boldsymbol{u},\boldsymbol{v})$ on V the purely convexity of L_o follows immediately (see e.g. Céa (1971), Ekeland and Temam (1976)).

(iii) The functional S is linear and continuous on $[H^1(\Omega)]^N$. Then there exists a constant $c > 0$ such that $S(\boldsymbol{u}) \leq c\,\|\boldsymbol{u}\|_1$ for every $\boldsymbol{u} \in [H^1(\Omega)]^N$. From the definition of L_o and the V-ellipticity of the bilinear form $a(\boldsymbol{u},\boldsymbol{v})$ for $\boldsymbol{u} \in V$ we obtain

$$L_o(\boldsymbol{u}) = \tfrac{1}{2}a(\boldsymbol{u},\boldsymbol{u}) - S(\boldsymbol{u}) \geq \tfrac{1}{2}\alpha\,\|\boldsymbol{u}\|_1 - c\,\|\boldsymbol{u}\|_1 .$$

In the limit $\|\boldsymbol{u}\| \to \infty$ the coerciveness follows immediately.

(iv) The functional L_o is convex and continuous and, therefore, weakly lower semi-continuous (see e.g. Céa (1971), Ekeland and Temam (1976)).

Theorem 7.1: A vector function $\boldsymbol{u} \in K$ solves the problem $(\mathscr{P})_v$, i.e. it satisfies Equation 7.47, if and only if

$$a(\boldsymbol{u},\boldsymbol{v}-\boldsymbol{u}) + j(\boldsymbol{v}) - j(\boldsymbol{u}) \geq S(\boldsymbol{v}-\boldsymbol{u}) \quad \forall \boldsymbol{v} \in K. \tag{7.48}$$

Proof: The set K is non-empty, closed and convex in V. The functional $L_o(\boldsymbol{v})$ is convex and differentiable. The functional $j(.)$ is convex and non-differentiable. We put $L(\boldsymbol{v}) = L_o(\boldsymbol{v}) + j(\boldsymbol{v})$.

(a) The case without friction: In this case $g_c^{kl} = 0$ and then $j(\boldsymbol{v}) = 0$.

(i) Let $\boldsymbol{u} \in K$ satisfy Equation 7.48 and let $\boldsymbol{v}$ be an arbitrary element of K. Then from Lemma 7.5, Equation 7.48 and the V-ellipticity of the bilinear form $a(\boldsymbol{u},\boldsymbol{v})$, we obtain

$$L_o(\boldsymbol{v}) - L_o(\boldsymbol{u}) = a(\boldsymbol{u},\boldsymbol{v}-\boldsymbol{u}) - S(\boldsymbol{v}-\boldsymbol{u}) + \tfrac{1}{2}a(\boldsymbol{v}-\boldsymbol{u},\boldsymbol{v}-\boldsymbol{u}) \geq \tfrac{1}{2}\alpha\,\|\boldsymbol{v}-\boldsymbol{u}\|_1^2 \geq 0.$$

Therefore, $\boldsymbol{u}$ solves the problem $(\mathscr{P})_v$.

(ii) On the other hand let $\boldsymbol{u} \in K$ be a solution of the problem $(\mathscr{P})_v$.
Let us define for an arbitrary function $\boldsymbol{v} \in K$ the function

$$\varphi(t) = L_o(\boldsymbol{u} + t(\boldsymbol{v}-\boldsymbol{u})),\, t \in \langle 0, 1\rangle.$$

Since K is convex, then $\boldsymbol{u} + t(\boldsymbol{v}-\boldsymbol{u}) \in K$. The function is differentiable, and due to Equation 7.47 it gets its minimum in the point $t = 0$. Thus $\varphi'(0) = 0$ as $\boldsymbol{u} \in \operatorname{int} K$ or $\varphi'(0) \geq 0$ as $\boldsymbol{u} \in \partial K$. Similarly as in Lemma 7.5 we find that

$$\varphi'(0) = \lim_{t\to 0^+} t^{-1}(L_o(\boldsymbol{u} + t(\boldsymbol{v}-\boldsymbol{u})) - L_o(\boldsymbol{u})) = a(\boldsymbol{u},\boldsymbol{v}-\boldsymbol{u}) - S(\boldsymbol{v}-\boldsymbol{u})$$

and therefore $\boldsymbol{u}$ satisfies Equation 7.48.

(b) The case with friction: For the case with friction $g_c^{kl} \in L^\infty(\cup\Gamma_c^{kl})$, $g_c^{kl} \geq 0$ a.e. on $\cup\Gamma_c^{kl}$ and the functional $j(\boldsymbol{v})$ is defined by Equation 7.43. Then Equation 7.47 is equivalent to

$$DL_o(\boldsymbol{u},\boldsymbol{v}-\boldsymbol{u}) + j(\boldsymbol{v}) - j(\boldsymbol{u}) \geq 0 \text{ for all } \boldsymbol{v} \in K.$$

But since the functional L_o is G-differentiable (see Lemma 7.5) then the last relation is equivalent to Equation 7.48

Definition 7.1: A function $\boldsymbol{u} \in K$ satisfying Equation 7.47 is called a weak solution of Equation 7.33-7.40.

It can be proved that any classical solution of the problem without as well as with friction represented by the solution of Equation 7.33-7.40 is a weak solution and, on the other hand, if the weak solution is smooth enough then it represents a classical solution of our problem.

Theorem 7.2: Under the assumption that the weak solution $\boldsymbol{u}$ is smooth enough, then every variational solution is a classical solution of the problem without or with friction, and every classical solution of the problem $(\mathscr{P})$ is a solution of the variational problem $(\mathscr{P})_v$.

Proof: Firstly, we shall study the case without friction and then the case with friction.

(a) The case without friction: In this case $j(\boldsymbol{v}) \equiv 0$ as $g_c^{kl} = 0$ and, therefore, $\vec{\tau}_t^k = 0$ at Γ_c^{kl}.

(i) Let $\boldsymbol{u} \in [H^2(\Omega)]^N \cap K$ be a solution of the variational problem $(\mathscr{P})_v$ for $j(\boldsymbol{v}) = 0$, i.e. it satisfies Equation 7.47, and let $\boldsymbol{v} \in K$ be an arbitrary function. Then due to Theorem 7.1 $\boldsymbol{u}$ satisfies the variational inequality of Equation 7.48, i.e.

$$a(\boldsymbol{u}, \boldsymbol{v} - \boldsymbol{u}) \geq S(\boldsymbol{v} - \boldsymbol{u}) \quad \forall \boldsymbol{v} \in K.$$

Let $\boldsymbol{w} \in [\mathscr{D}(\Omega)]^N$ and let us put $\boldsymbol{v} = \boldsymbol{u} \pm \boldsymbol{w}$. Evidently, $\boldsymbol{w} \in K$. Then

$$a(\boldsymbol{u}, \pm\boldsymbol{w}) \geq \int_\Omega f_i(\pm w_i)\mathrm{d}\boldsymbol{x} + \int_{\Gamma_\tau} P_i(\pm w_i)\mathrm{d}s.$$

Hence

$$\int_\Omega \tau_{ij}(\boldsymbol{u})e_{ij}(\boldsymbol{w})\mathrm{d}\boldsymbol{x} = \int_\Omega f_i w_i \mathrm{d}\boldsymbol{x}.$$

Applying Lemma 7.1 we have

$$-\int_\Omega \partial\tau_{ij}(\boldsymbol{u})/\partial x_j w_i \mathrm{d}\boldsymbol{x} + \int_{\partial\Omega} \tau_{ij}(\boldsymbol{u}) w_i n_j \mathrm{d}s = \int_\Omega f_i w_i \mathrm{d}\boldsymbol{x}.$$

Hence

$$\partial\tau_{ij}(\boldsymbol{u})/\partial x_j + f_i = 0 \quad \text{a.e. in } \Omega,\ i = 1, \ldots, N.$$

The condition given in Equation 7.34 is satisfied due to the definition of the set K. Let us put $\boldsymbol{v} = \boldsymbol{u} \pm \boldsymbol{w}, \boldsymbol{w} \in [\mathscr{E}(\overline{\Omega})]^N$, where the traces of w_i on the boundaries have the support in Γ_τ. Evidently, $\boldsymbol{v} \in K$. Then we have

$$-\int_\Omega \partial\tau_{ij}(\boldsymbol{u})/\partial x_j w_i \,\mathrm{d}\boldsymbol{x} + \int_{\partial\Omega} \tau_{ij}(\boldsymbol{u}) w_i n_j \mathrm{d}s = \int_\Omega f_i w_i \mathrm{d}\boldsymbol{x} + \int_{\Gamma_\tau} P_i w_i \mathrm{d}s.$$

According to the condition of equilibrium we have

$$\int_{\Gamma_\tau} (\tau_{ij} n_j - P_i) w_i \mathrm{d}s = 0$$

and hence Equation 7.35 follows. Now let the condition of equilibrium in Equation 7.41 be used. Then

$$\int_{\partial\Omega} \tau_{ij}(\boldsymbol{u})(v_i - u_i) n_j \mathrm{d}s \geq \int_{\Gamma_\tau} P_i(v_i - u_i)\mathrm{d}s. \tag{7.49}$$

Since Equation 7.34 and 7.35 are satisfied, then it holds

$$\begin{aligned}
&\int_{\Gamma_o} \tau_{ij}(\boldsymbol{u})(v_i - u_i) n_j \mathrm{d}s + \int_{\cup\Gamma_c^{kl}} \tau_{ij}(\boldsymbol{u})(v_i^k - v_i^l - (u_i^k - u_i^l)) n_j \mathrm{d}s \\
&= \int_{\Gamma_o} \vec{\tau}_t(\boldsymbol{u})(\boldsymbol{v}_t - \boldsymbol{u}_t)\mathrm{d}s + \int_{\Gamma_o} \vec{\tau}_t(\boldsymbol{u})(\boldsymbol{v}_t - \boldsymbol{u}_t)\mathrm{d}s + \int_{\cup\Gamma_c^{kl}} \vec{\tau}_t^k(\boldsymbol{u})(\boldsymbol{v}_t^k - \boldsymbol{v}_t^l - (\boldsymbol{u}_t^k - \boldsymbol{u}_t^l))\mathrm{d}s \\
&+ \int_{\cup\Gamma_c^{kl}} \tau_n^k(\boldsymbol{u})(v_n^k - v_n^l - (u_n^k - u_n^l)\mathrm{d}s \geq 0.
\end{aligned}$$

Let us put $\boldsymbol{v} = \boldsymbol{u} \pm \boldsymbol{w}$, $\boldsymbol{w} \in [\mathscr{E}(\overline{\Omega})]^N$, where traces of w_i have the support in Γ_o, and $w_n = 0$ on Γ_o. Evidently, $\boldsymbol{v} \in K$. From the previous inequality we have

$$\int_{\Gamma_o} \vec{\tau}_t(\boldsymbol{u})\boldsymbol{w}_t \mathrm{d}s = 0 \Rightarrow \vec{\tau}_t(\boldsymbol{u}) = 0 \quad \text{a.e. on } \Gamma_o.$$

Hence, and from the definition of the set K, Equation 7.40 follows immediately. By using Equation 7.34, 7.35 and 7.40 with Equation 7.49 we find

$$\int_{\cup\Gamma_c^{kl}} \vec{\tau}_t^k(\boldsymbol{u})(\boldsymbol{v}_t^k - \boldsymbol{v}_t^l - (\boldsymbol{u}_t^k - \boldsymbol{u}_t^l))\mathrm{d}s + \int_{\cup\Gamma_c^{kl}} \tau_n^k(\boldsymbol{u})(v_n^k - v_n^l - (u_n^k - u_n^l))\mathrm{d}s \geq 0. \qquad (7.50)$$

Let $\varphi_i \in \mathscr{E}(\overline{\Omega})$ and let the trace of $\vec{\varphi}$ on $\partial\Omega$ have the support in $\cup\Gamma_c^{kl}$. Let us put $\boldsymbol{w} \in V$ such that $w_n^k - w_n^l = 0$ and $\boldsymbol{w}_t^k - \boldsymbol{w}_t^l = \vec{\varphi}^k - \vec{\varphi}^l$ on $\cup\Gamma_c^{kl}$, and let us put $\boldsymbol{v} = \boldsymbol{u} \pm \boldsymbol{w}$. It is evident that $\boldsymbol{v} \in K$. Then from Equation 7.50 we find

$$\int_{\cup\Gamma_c^{kl}} \vec{\tau}_t^k(\boldsymbol{u})(\vec{\varphi}_t^k - \vec{\varphi}_t^l)\mathrm{d}s = 0 \Rightarrow \vec{\tau}_t^k(\boldsymbol{u}) = 0 \text{ a.e. on } \cup\Gamma_c^{kl}.$$

Hence, we have

$$\int_{\cup\Gamma_c^{kl}} \tau_n^k(v_n^k - v_n^l - (u_n^k - u_n^l))\mathrm{d}s \geq 0 \Rightarrow \int_{\cup\Gamma_c^{kl}} \tau_n^k(v_n^k - v_n^l)\mathrm{d}s \geq \int_{\cup\Gamma_c^{kl}} \tau_n^k(u_n^k - u_n^l)\mathrm{d}s.$$

Let $\boldsymbol{w} \in V$ be such that $w_n^k - w_n^l = u_n^k - u_n^l$ and let us put $\boldsymbol{v} = \boldsymbol{u} \pm \boldsymbol{w}$. It is evident that $\boldsymbol{v} \in K$. Then from the above relation we find

$$\int_{\cup\Gamma_c^{kl}} \tau_n^k(w_n^k - w_n^l)\mathrm{d}s = \int_{\cup\Gamma_c^{kl}} \tau_n^k(u_n^k - u_n^l)\mathrm{d}s = 0 \qquad (7.51)$$

and therefore we find

$$\int_{\cup\Gamma_c^{kl}} \tau_n^k(v_n^k - v_n^l)\mathrm{d}s \geq 0 \quad \forall \boldsymbol{v} \in K.$$

Due to the definition of the set K, $v_n^k - v_n^l \leq 0$ a.e. on $\cup\Gamma_c^{kl}$, therefore $\tau_n^k \leq 0$ a.e. on $\cup\Gamma_c^{kl}$. From the definition of K and from Equation 7.51 we obtain

$$\tau_n^k(u_n^k - u_n^l) = 0 \quad \text{a.e. on } \Gamma_c^{kl}.$$

Since the first part of Equation 7.36, i.e. $u_n^k - u_n^l \leq 0$, holds due to the definition of K, then the assertion of the theorem is proved.

(ii) Let $\boldsymbol{u}$ be a classical solution of the problem $(\mathscr{P})$, then $\tau_{ij} = c_{ijkl}e_{kl}(\boldsymbol{u})$ satisfies the equilibrium condition. Let $\boldsymbol{v} \in V$ be an arbitrary function. Let us multiply the equilibrium condition (Equation 7.33) by v_i, sum over $i = 1, \ldots, N$ and integrate over Ω; then we obtain

$$\begin{aligned} 0 &= \int_\Omega (-c_{ijkl}e_{kl}(\boldsymbol{u})\partial v_i/\partial x_j + f_i v_i)\mathrm{d}\boldsymbol{x} + \int_{\cup(\partial\Omega^k \cup \partial\Omega^l)} \tau_{ij}(\boldsymbol{u})n_j v_i \, \mathrm{d}s \\ &= -a(\boldsymbol{u}, \boldsymbol{v}) + S(\boldsymbol{v}) + \int_{\Gamma_o} (\tau_n(\boldsymbol{u})v_n + \vec{\tau}_t(\boldsymbol{u})\boldsymbol{v}_t)\mathrm{d}s + \int_{\cup\Gamma_c^{kl}} (\tau_n^k(\boldsymbol{u})v_n^k + \vec{\tau}_t^k(\boldsymbol{u})\boldsymbol{v}_t^k - \tau_n^l(\boldsymbol{u})v_n^l - \vec{\tau}_t^l(\boldsymbol{u})\boldsymbol{v}_t^l)\mathrm{d}s. \end{aligned}$$

The integral $\int_{\Gamma_o}(\tau_n(\boldsymbol{u})v_n + \vec{\tau}_t(\boldsymbol{u})\boldsymbol{v}_t)\mathrm{d}s = 0$ as $v_n = 0$, $\vec{\tau}_t = 0$ on Γ_o. Due to Equations 7.36 and 7.37, we find

$$0 = -a(\boldsymbol{u}, \boldsymbol{v}) + S(\boldsymbol{v}) + \int_{\cup\Gamma_c^{kl}} \tau_i(\boldsymbol{u})v_i \mathrm{d}s.$$

Due to Theorem 7.1 and the fact that $\tau_i u_i = \tau_n u_n + \vec{\tau}_t \boldsymbol{u}_t$, it is sufficient to show that

$$\int_{\cup\Gamma_c^{kl}} (\tau_n^k(\boldsymbol{u})(v_n^k - v_n^l) + \vec{\tau}_t^k(\boldsymbol{u})(\boldsymbol{v}_t^k - \boldsymbol{v}_t^l))\mathrm{d}s = 0. \tag{7.52}$$

Let $\boldsymbol{v} \in K$ be an arbitrary function and let us put $\boldsymbol{v} = \boldsymbol{w} - \boldsymbol{u}$. Due to the third part of Equation 7.36 for points where $u_n^k - u_n^l < 0$, we have $\tau_n^k(\boldsymbol{u}) = 0$. For points where $u_n^k - u_n^l = 0$, it holds that $v_n^k - v_n^l = w_n^k - w_n^l \leq 0$, $\tau_n^k(\boldsymbol{u}) = 0$. Hence

$$\int_{\cup\Gamma_c^{kl}} \tau_n^k(\boldsymbol{u})(v_n^k - v_n^l)\mathrm{d}s \geq 0 \quad \forall \boldsymbol{v} \in K, \quad \boldsymbol{v} = \boldsymbol{w} - \boldsymbol{u}$$

and from $\vec{\tau}_t(\boldsymbol{u}) = 0$ Equation 7.52 follows.

(b) The case with friction:

(i) Let $\boldsymbol{u}$ be a variational solution of the variational problem $(\mathscr{P})_v$, i.e. it satisfies Equation 7.47, and let us assume that $\boldsymbol{u} \in [H^2(\Omega)]^N \cap K$. Let $\boldsymbol{v} \in K$ be an arbitrary function. Then due to Theorem 7.1, $\boldsymbol{u}$ satisfies the variational inequality in Equation 7.48, i.e.

$$a(\boldsymbol{u}, \boldsymbol{v} - \boldsymbol{u}) + j(\boldsymbol{v}) - j(\boldsymbol{u}) \geq S(\boldsymbol{v} - \boldsymbol{u}) \quad \forall \boldsymbol{v} \in K.$$

The first part of the proof, i.e. to prove the conditions of Equations 7.34, 7.35 and 7.40, is similar to that of the case without friction. It only remains to prove the conditions of Equations 7.36 and 7.37.

Let $\boldsymbol{v} \in K$ be an arbitrary function. From the equilibrium equation it follows that

$$-\int_\Omega \partial\tau_{ij}/\partial x_j (v_i - u_i)\,\mathrm{d}\boldsymbol{x} = \int_\Omega f_i(v_i - u_i)\mathrm{d}\boldsymbol{x}.$$

Applying Lemma 7.1 we find

$$\int_\Omega \tau_{ij}(\boldsymbol{u})e_{ij}(\boldsymbol{v} - \boldsymbol{u})\,\mathrm{d}\boldsymbol{x} - \int_{\partial\Omega} \tau_{ij}(\boldsymbol{u})(v_i - u_i)n_j\mathrm{d}s = \int_\Omega f_i(v_i - u_i)\mathrm{d}\boldsymbol{x}.$$

Since $\partial\Omega = \Gamma_u \cup \Gamma_\tau \cup \Gamma_o \cup R$, $\Gamma_c = \cup_{k,l}\Gamma_c^{kl}$, then due to Equations 7.34, 7.35 and 7.40 we find

$$\begin{aligned} &a(\boldsymbol{u}, \boldsymbol{v} - \boldsymbol{u}) - S(\boldsymbol{v} - \boldsymbol{u}) + j(\boldsymbol{v}) - j(\boldsymbol{u}) \\ &= \int_{\cup\Gamma_c^{kl}} (\tau_n^k(\boldsymbol{u})(v_n^k - v_n^l - (u_n^k - u_n^l)) + \vec{\tau}_t^k(\boldsymbol{u})(\boldsymbol{v}_t^k - \boldsymbol{v}_t^l - (\boldsymbol{u}_t^k - \boldsymbol{u}_t^l)))\mathrm{d}s + j(\boldsymbol{v}) - j(\boldsymbol{u}). \end{aligned}$$

According to Equation 7.48 we have

$$\int_{\cup\Gamma_c^{kl}} (\tau_n^k(\boldsymbol{u})(v_n^k - v_n^l - (u_n^k - u_n^l)) + \vec{\tau}_t^k(\boldsymbol{u})(\boldsymbol{v}_t^k - \boldsymbol{v}_t^l - (\boldsymbol{u}_t^k - \boldsymbol{u}_t^l)))\mathrm{d}s + j(\boldsymbol{v}) - j(\boldsymbol{u}) \geq 0. \tag{7.53}$$

Let us denote $\mathscr{D}(\cup\Gamma_c^{kl}) = \{\varphi \mid \varphi \in C^\infty(\partial\Omega),\ \text{support}\ \varphi_i \subset \cup\Gamma_c^{kl}\}$. Let $\varphi \in \mathscr{D}(\cup\Gamma_c^{kl})$. Let us take function $\boldsymbol{v} \in V$ such that $v_n = u_n$, $\boldsymbol{v}_t = \pm\lambda_o\vec{\varphi}$, $\lambda_o \geq 0$, on $\cup\Gamma_c^{kl}$. It is evident that $\boldsymbol{v} \in$K. Then Equation 7.53 yields

$$\begin{aligned} &\int_{\cup\Gamma_c^{kl}} \vec{\tau}_t^k(\pm\lambda_o(\vec{\varphi}^k - \vec{\varphi}^l) - (\boldsymbol{u}_t^k - \boldsymbol{u}_t^l))\mathrm{d}s + \int_{\cup\Gamma_c^{kl}} g_c^{kl}\left(|\lambda_o(\vec{\varphi}^k - \vec{\varphi}^l)| - |\boldsymbol{u}_t^k - \boldsymbol{u}_t^l|\right)\mathrm{d}s \\ &= \lambda_o \int_{\cup\Gamma_c^{kl}} \left(\pm\vec{\tau}_t^k(\vec{\varphi}^k - \vec{\varphi}^l) + g_c^{kl}\left(|\vec{\varphi}^k - \vec{\varphi}^l|\right)\mathrm{d}s - \int_{\cup\Gamma_c^{kl}} (\vec{\tau}_t^k(\boldsymbol{u}_t^k - \boldsymbol{u}_t^l) + g_c^{kl}\left|\boldsymbol{u}_t^k - \boldsymbol{u}_t^l\right|\right)\mathrm{d}s \geq 0 \\ &\hspace{30em} \forall\lambda_o \geq 0. \end{aligned} \tag{7.54}$$

Hence

$$\int_{\cup\Gamma_c^{kl}} (\pm\vec{\tau}_t^k(\vec{\varphi}^k - \vec{\varphi}^l) + g_c^{kl}\,|\vec{\varphi}^k - \vec{\varphi}^l|)\mathrm{d}s \geq 0,$$

i.e. $\int_{\cup\Gamma_c^{kl}} g_c^{kl}\,|\vec{\varphi}^k - \vec{\varphi}^l|\,\mathrm{d}s \geq \pm\int_{\cup\Gamma_c^{kl}} \vec{\tau}_t^k(\vec{\varphi}^k - \vec{\varphi}^l)\mathrm{d}s$ and hence

$$\int_{\cup\Gamma_c^{kl}} g_c^{kl}\,|\vec{\varphi}^k - \vec{\varphi}^l|\,\mathrm{d}s \geq \left|\int_{\cup\Gamma_c^{kl}} \vec{\tau}_t^k(\vec{\varphi}^k - \vec{\varphi}^l)\mathrm{d}s\right|. \tag{7.55}$$

Since $g_c^{kl} \in L^\infty(\Gamma_c^{kl})$, $g_c^{kl} \geq 0$ a.e. on Γ_c^{kl}, then

$$\int_{\cup\Gamma_c^{kl}} g_c^{kl}\,|\vec{\varphi}^k - \vec{\varphi}^l|\,\mathrm{d}s \leq \|g_c^{kl}\|_{L^\infty(\cup\Gamma_c^{kl})} \int_{\cup\Gamma_c^{kl}} |\vec{\varphi}^k - \vec{\varphi}^l|\,\mathrm{d}s. \tag{7.56}$$

On $\mathscr{D}(\cup\Gamma_c^{kl})$ let the linear functional $F(\boldsymbol{v})$ be defined by

$$F(\vec{\varphi}) = \int_{\cup\Gamma_c^{kl}} \vec{\tau}_t^k(\boldsymbol{u})(\vec{\varphi}^k - \vec{\varphi}^l)\mathrm{d}s.$$

Since $\mathscr{D}(\cup\Gamma_c^{kl})$ is dense in $L^1(\cup\Gamma_c^{kl})$, then due to Equations 7.55 and 7.56 the functional F is continuous on $\mathscr{D}(\cup\Gamma_c^{kl})$ in the topology induced by $L^1(\cup\Gamma_c^{kl})$. From the theorem about the generalized form of a linear continuous functional on the space $L^1(\cup\Gamma_c^{kl})$ we find

$$\vec{\tau}_t^k \in L^\infty(\cup\Gamma_c^{kl}) \text{ and } \left\|\vec{\tau}_t^k(\boldsymbol{u})\right\|_{L^\infty(\cup\Gamma_c^{kl})} = \|F\|\,.$$

Hence, and from Equations 7.55 and 7.56, it follows that

$$\left\|\vec{\tau}_t^k(\boldsymbol{u})\right\|_{L^\infty(\cup\Gamma_c^{kl})} \leq \|g_c^{kl}\|_{L^\infty(\cup\Gamma_c^{kl})}$$

and hence

$$\left|\vec{\tau}_t^k(\boldsymbol{u})\right| \leq g_c^{kl} \text{ a.e. on } \cup\Gamma_c^{kl}.$$

Hence, and from Equation 7.54 where we put $\lambda_o = 0$, we find

$$g_c^{kl}\left|\boldsymbol{u}_t^k - \boldsymbol{u}_t^l\right| + \vec{\tau}_t^k(\boldsymbol{u})(\boldsymbol{u}_t^k - \boldsymbol{u}_t^l) = 0 \text{ a.e. on } \cup\Gamma_c^{kl}. \tag{7.57}$$

At some point $\boldsymbol{x} \in \cup\Gamma_c^{kl}$ let $g_c^{kl} > \left|\vec{\tau}_t^k(\boldsymbol{u}(\boldsymbol{x}))\right|$. Then from Equation 7.57 it follows that $\boldsymbol{u}_t^k - \boldsymbol{u}_t^l = 0$, i.e Equation 7.38. Furthermore, at some point $\boldsymbol{x} \in \cup\Gamma_c^{kl}$ let $g_c^{kl} = \left|\vec{\tau}_t^k(\boldsymbol{u}(\boldsymbol{x}))\right|$. Then from Equation 7.57 we find

$$\left|\vec{\tau}_t^k(\boldsymbol{u}(x))\right|\left|\boldsymbol{u}_t^k - \boldsymbol{u}_t^l\right| + \vec{\tau}_t^k(\boldsymbol{u})(\boldsymbol{u}_t^k - \boldsymbol{u}_t^l) = 0.$$

Hence it follows that there exists $\lambda_o \geq 0$ such that

$$\boldsymbol{u}_t^k - \boldsymbol{u}_t^l = -\lambda_o\vec{\tau}_t^k(\boldsymbol{u}(\boldsymbol{x})),$$

which is Equation 7.39. Therefore, Equation 7.37 is proved and then $\boldsymbol{u}$ is the solution of the problem $(\mathscr{P})$.

(ii) Let $\boldsymbol{u}$ be a classical solution of $(\mathscr{P})$, i.e. $\boldsymbol{u} \in [H^2(\Omega)]^N \cap K$, and let $\boldsymbol{v} \in K$ be an arbitrary function. Let us multiply the equilibrium condition (Equation 7.33) by v_i, sum over $i = 1, \ldots, N$ and integrate over Ω, then similarly as in the case without friction, we find

$$0 = -a(\boldsymbol{u}, \boldsymbol{v}) + S(\boldsymbol{v}) + \int_{\cup\Gamma_c^{kl}} \tau_i(\boldsymbol{u})v_i\mathrm{d}s. \tag{7.58}$$

In Equation 7.58 let us put $\boldsymbol{v} = \boldsymbol{w} - \boldsymbol{u}$, where $\boldsymbol{v} \in K$ is an arbitrary function. Let us add to both sides of Equation 7.58 the term $\int_{\cup\Gamma_c^{kl}} g_c^{kl}(\left|\boldsymbol{w}_t^k - \boldsymbol{w}_t^l\right| - \left|\boldsymbol{u}_t^k - \boldsymbol{u}_t^l\right|)\mathrm{d}s$.

Then

$$a(\boldsymbol{u}, \boldsymbol{w} - \boldsymbol{u}) - S(\boldsymbol{w} - \boldsymbol{u}) + \int_{\cup\Gamma_c^{kl}} g_c^{kl} \left(\left| \boldsymbol{w}_t^k - \boldsymbol{w}_t^l \right| - \left| \boldsymbol{u}_t^k - \boldsymbol{u}_t^l \right| \right) \mathrm{d}s$$
$$= \int_{\cup\Gamma_c^{kl}} g_c^{kl} \left(\left| \boldsymbol{w}_t^k - \boldsymbol{w}_t^l \right| - \left| \boldsymbol{u}_t^k - \boldsymbol{u}_t^l \right| \right) \mathrm{d}s + \int_{\cup\Gamma_c^{kl}} \tau_i(\boldsymbol{u})(w_i - u_i)\mathrm{d}s.$$

According to the equivalent variational formulation it is sufficient to show that the righthand side is non-negative. It holds that $\tau_i v_i = \tau_n v_n + \vec{\tau}_t \boldsymbol{v}_t$ and $\int_{\cup\Gamma_c^{kl}} \tau_n^k(\boldsymbol{u})(\left|w_n^k - w_n^l\right| - \left|u_n^k - u_n^l\right|)\mathrm{d}s \geq 0$.
At some point $\boldsymbol{x} \in \cup\Gamma_c^{kl}$ let $\left|\vec{\tau}_t^k(\boldsymbol{u}(\boldsymbol{x}))\right| < g_c^{kl}$. Then due to Equation 7.37 $\boldsymbol{u}_t^k - \boldsymbol{u}_t^l = 0$ and $g_c^{kl} \left|\boldsymbol{w}_t^k - \boldsymbol{w}_t^l\right| + \vec{\tau}_t^k(\boldsymbol{u})(\boldsymbol{w}_t^k - \boldsymbol{w}_t^l) = 0$. Furthermore, at some point $\boldsymbol{x} \in \cup\Gamma_c^{kl}$ let $\left|\vec{\tau}_t^k(\boldsymbol{u}(\boldsymbol{x}))\right| = g_c^{kl}$, then due to Equation 7.37 we find

$$g_c^{kl} \left(\left| \boldsymbol{w}_t^k - \boldsymbol{w}_t^l \right| - \left| \boldsymbol{u}_t^k - \boldsymbol{u}_t^l \right| \right) + \vec{\tau}_t^k(\boldsymbol{w})(\boldsymbol{w}_t^k - \boldsymbol{w}_t^l - (\boldsymbol{u}_t^k - \boldsymbol{u}_t^l))$$
$$= \left|\vec{\tau}_t^k\right| \left(\left| \boldsymbol{w}_t^k - \boldsymbol{w}_t^l \right| - \lambda_o \left|\vec{\tau}_t^k\right| \right) + \vec{\tau}_t^k(\boldsymbol{w}_t^k - \boldsymbol{w}_t^l - \lambda_o \vec{\tau}_t^k)$$
$$= \left|\vec{\tau}_t^k(\boldsymbol{w}_t^k - \boldsymbol{w}_t^l)\right| + \vec{\tau}_t^k(\boldsymbol{w}_t^k - \boldsymbol{w}_t^l) \geq 0.$$

Next we shall prove the existence and uniqueness of the weak solution.

Theorem 7.3: Let $\boldsymbol{f} \in [L^2(\Omega)]^N$ and $\boldsymbol{P} \in [L^2(\Gamma_\tau)]^N$. Then there exists only one weak solution of the problem with or without friction.

Proof:

(a) The case without friction: The set K is closed and convex (see Lemma 7.3) in V. We shall show that the functional L_o is purely convex and weakly lower semi-continuous and that

$$\lim_{\|\boldsymbol{u}\|_{1,2} \to \infty} L_o(\boldsymbol{u}) = +\infty \tag{7.59}$$

The proof of the existence and uniqueness of the minimum of the functional L_o under these assumptions is similar of that of Céa (1971). At first we shall prove that the bilinear form $a(\boldsymbol{u}, \boldsymbol{v})$ is V-elliptic. We have

$$a(\boldsymbol{v}, \boldsymbol{v}) = \int_\Omega c_{ijkl} e_{kl}(\boldsymbol{v}) e_{ij}(\boldsymbol{v}) \mathrm{d}\boldsymbol{x} \geq c_o \int_\Omega e_{ij}(\boldsymbol{v}) e_{ij}(\boldsymbol{v}) \mathrm{d}\boldsymbol{x} \quad \forall \boldsymbol{v} \in V, \tag{7.60}$$

where Equation 7.8 was used.

Due to the Korn inequality

$$a(\boldsymbol{v}, \boldsymbol{v}) \geq c_1 \|\boldsymbol{v}\|_{1,2}^2 \quad \forall \boldsymbol{v} \in V$$

and

$$L_o(\boldsymbol{v}) = \tfrac{1}{2} a(\boldsymbol{v}, \boldsymbol{v}) - S(\boldsymbol{v}) \geq c_1 \|\boldsymbol{v}\|_{1,2}^2 - c_2 \|\boldsymbol{v}\|_{1,2}.$$

The last expression converges to $+\infty$ for $\|\boldsymbol{v}\|_{1,2} \to +\infty$.

Since $\quad L_o(\boldsymbol{u}) = \tfrac{1}{2} a(\boldsymbol{u}, \boldsymbol{u}) - S(\boldsymbol{u})$ then

$$L_0'(\boldsymbol{u}, \boldsymbol{v}) = a(\boldsymbol{u}, \boldsymbol{v}) - S(\boldsymbol{v}) \quad \forall \boldsymbol{u}, \boldsymbol{v} \in [H^1(\Omega)]^N,$$
$$L_0''(\boldsymbol{u}; \boldsymbol{v}, \boldsymbol{w}) = a(\boldsymbol{v}, \boldsymbol{w}) \quad \forall \boldsymbol{u}, \boldsymbol{v}, \boldsymbol{w} \in [H^1(\Omega)]^N.$$

Hence

$$L_0''(\boldsymbol{u}; \boldsymbol{v}, \boldsymbol{w}) = a(\boldsymbol{v}, \boldsymbol{w}) \geq \alpha \|\boldsymbol{v}\|_{1,2}^2 > 0 \quad \forall \boldsymbol{u}, \boldsymbol{v}, \boldsymbol{w} \in [H^1(\Omega)]^N, \boldsymbol{v} \neq 0.$$

Therefore L_o is purely convex in V (see, for example, Céa (1971)). Since L_o' is linear and continuous in $\vec{v}$ then L_o is weakly lower semi-continuous (see, for example Céa (1971)) and then the theorem is proved.

(b) The case with friction: The set K is non-empty, closed and convex (see Lemma 7.3) in V. The functional L_o is purely convex, weakly lower semi-continuous and coercive, as was proved above. The functional $j(\boldsymbol{v})$ is convex and continuous and, therefore, it is weakly lower semi-continuous. Hence the functional L is purely convex and weakly lower semi-continuous. It is sufficient to prove that

$$\lim_{\|\boldsymbol{u}\|_{1,2}\to\infty} L(\boldsymbol{u}) = +\infty.$$

But

$$L(\boldsymbol{u}) = L_o(\boldsymbol{u}) + j(\boldsymbol{u}) \geq c_o \|\boldsymbol{u}\|_{1,2}^2 - c_1 \|\boldsymbol{u}\|_{1,2} + j(\boldsymbol{u}).$$

Since $j(\boldsymbol{u}) = \int_{\cup\Gamma_c^{kl}} g_c^{kl} \left|\boldsymbol{u}_t^k - \boldsymbol{u}_t^l\right| \mathrm{d}\boldsymbol{x} \geq 0$, as $g_c^{kl} \geq 0$ a.e. on Γ_c^{kl}, then

$$L(\boldsymbol{u}) \to +\infty \text{ for } \|\boldsymbol{u}\|_{1,2} \to \infty,$$

which completes the proof.

7.1.6. Numerical Solution

As we saw in the previous chapter, in geophysical problems the so-called roolling conditions are used in many cases. But the contact conditions are a much better approximation of real situations. Therefore, we shall also discuss the contact problems from the numerical point of view. Numerical solution of the problem will be based on the finite element approximation and the Ritz method.

Let the domain $\Omega \subset R^N$ be a bounded domain and let it be "triangulated". We divide $\overline{\Omega} = \Omega \cup \partial\Omega$ into a system of m triangles T_{hi} in the $2D$ case and into a system of m tetrahedrons T_{hi} in the $3D$ case, generating a triangulation $\mathscr{T}_h$ such that $\overline{\Omega} = \cup_{i=1}^m T_{hi}$ and such that two neighbouring triangles have only a vertex or an entire side common in the $2D$ case, and that two neighbouring tetrahedra have only a vertex or an entire edge or an entire face common in the $3D$ case. Denote by h the maximal side in the $2D$ case or the maximal edge in the $3D$ case in $\mathscr{T}_h$, i.e. $h = \max_{1\leq i\leq m}(\operatorname{diam} T_{hi})$, and by ϑ_h an arbitrary interior angle in an arbitrary triangle in the $2D$ case and an arbitrary tetrahedron in the $3D$ case. We say that a family of triangulations $\{\mathscr{T}_h\}$, $0 < h \leq h_o < \infty$, is regular if there exists a constant $\vartheta_o > 0$ such that

$$\min_{\vartheta_h\in\mathscr{T}_h} \vartheta_h \geq \vartheta_o \text{ for all } h \in (0, h_o\rangle.$$

We further assume that the end points $\overline{\Gamma}_u \cap \overline{\Gamma}_\tau$, $\overline{\Gamma}_u \cap \overline{\Gamma}_c$, $\overline{\Gamma}_\tau \cap \overline{\Gamma}_c$, $\overline{\Gamma}_u \cap \overline{\Gamma}_o$, $\overline{\Gamma}_o \cap \overline{\Gamma}_\tau$ coincide with the vertices of T_{hi}.

Let $P_i \in \Omega$ be an arbitrary interior vertex of the triangulation $\mathscr{T}_h$. The basis function $\boldsymbol{v}_h^i$ is defined to be a function linear (quadratic, etc.) on each element $T_{hi} \in \mathscr{T}_h$ and taking the values $\boldsymbol{v}_h^i(P_j) = \delta_{ij}$ at the vertices of the triangulation, where δ_{ij} is the Kronecker symbol. The function $\mathbf{v}_h^i$ represents a pyramid of height 1 with its vertex above the point P_i and with its support (supp $\boldsymbol{v}_h^i$) consisting of those triangles or tetrahedra which have the vertex P_i in common. It is evident that the basis function has small support since diam(supp v_h^i) $\leq 2h$ and the parameter h is taken to be very small, i.e. $h \to 0$.

Let V_h be the space of all continuous piecewise linear (quadratic, etc.) functions on the triangulation $\mathscr{T}_h$, vanishing on the boundary Γ_u, where the displacement vector is prescribed, i.e.

$$V_h = \{v \mid v \in \prod_{\iota=1}^{S}[C(\Omega^\iota)]^N, v \mid_{T_h} \in [P_1]^N, v = 0 \text{ on } \Gamma_u, v_n = 0 \text{ on } \Gamma_o \quad \forall T_h \in \mathscr{T}_h\},$$

where P_1 is the space of all linear polynomials (similarly for polynomials of the higher order). For any h we define

$$K_h = \{v \mid v \in V_h, v_n^k - v_n^l \leq 0 \text{ on } {}_k\cup_{k,l}\Gamma_c^{kl}\} = V_h \cap \mathrm{K}.$$

It is evident that K_h is a convex and closed subset of $V_h \forall$h.

Lemma 7.6: Let V, K, V_h, and K_h be spaces and sets as defined above. Then $K_h \subset K$ for all h.

Proof: We shall prove this assertion for $N = 2$. Then $V_h \subset V$ holds. Let $v \in K_h \Rightarrow v \in V$. It is sufficient to prove that $v_n^k - v_n^l \leq 0$ on $\cup\Gamma_c^{kl}$, where $\cup\Gamma_c^{kl} = \cup_{i=1}^{n(h)}\overline{a_i a_{i+1}}$, where a_i, a_{i+1} are two neighbouring points of the triangulation $\mathscr{T}_h$.

Let us show that $v_n^k - v_n^l \leq 0$ on $\overline{a_i a_{i+1}}$ for an arbitrary i. Due to the definition of K_h, $(v_n^k - v_n^l)(a_{i+j}) \leq 0$, $j = 0, 1$ holds. Let $\boldsymbol{n}$ be the normal unit vector to $\overline{a_i a_{i+1}}$. Hence $\boldsymbol{n}$ is constant on the whole length of $\overline{a_i a_{i+1}}$. Let $v = (v_i)_{i=1}^N \in [P_1]^N \Rightarrow v_i \in P_1$ and therefore $v_n^k - v_n^l$ is linear on the whole length of $\overline{a_i a_{i+1}}$. Hence $v_n^k - v_n^l \leq 0$ on $\overline{a_i a_{i+1}}$. Since $\cup\Gamma_c^{kl} = \cup_{i=1}^{n(h)}\overline{a_i a_{i+1}}$, then $v_n^k - v_n^l \leq 0$ on $\cup\Gamma_c^{kl}$ and therefore $v \in K$, which finishes the proof.

Definition 7.2: We say that $\boldsymbol{u}_h \in K_h$ is a finite element approximation of the **problem** $(\mathscr{P})_{vh}$, if

$$L(\boldsymbol{u}_h) \leq L(v) \quad \forall v \in K_h, \tag{7.61}$$

where $L(v) = \frac{1}{2}a(v, v) - S(v) + j(v)$.

It is evident that the problem in Equation 7.61 has a unique solution.

Theorem 7.4: The problem $(\mathscr{P})_{vh}$ has a unique solution.

Proof: This follows from the properties of the functional L and from the fact that the set K_h is closed and convex in the space V_h, and is similar of that of Theorem 7.3.

To find the relation between the solutions of problems $(\mathscr{P})_v$ and $(\mathscr{P})_{vh}$ we prove the following theorem.

Theorem 7.5: Let $\{K_h\}$ be the system of closed convex subspaces of V satisfying the following conditions:

(i) $\forall v \in K\ \exists v_h \in K_h$ such that $\|v - v_h\|_{1,2} \to 0$ for $h \to 0_+$,
(ii) let $v_h \in K_h$, then $v_h \to v$ weakly for $h \to 0_+$ in the space $V \Rightarrow v \in K$.

Then

$$\|\boldsymbol{u} - \boldsymbol{u}_h\|_{1,2} \to 0 \text{ for } h \to 0_+. \tag{7.62}$$

Proof: Let $\boldsymbol{u}$ be a solution of the problem $(\mathscr{P})_v$. According to (i), $\exists \boldsymbol{v}_h \in K_h$ such that

$$\|\boldsymbol{u} - \boldsymbol{u}_h\|_{1,2} \to 0 \text{ for } h \to 0_+.$$

Let $\boldsymbol{u}_h$ be a solution of the problem $(\mathscr{P})_{vh}$, then due to the continuity of the functional L and the fact that $\boldsymbol{u}_h$ is a solution of the problem $(\mathscr{P})_{vh}$, we have

$$L(\boldsymbol{u}_h) \le L(\boldsymbol{v}_h) \to L(\boldsymbol{v})\ h \to 0_+. \tag{7.63}$$

Since $\lim_{\|\boldsymbol{v}\|^{1,2} \to \infty} L(\boldsymbol{v}) = +\infty$, i.e. the functional L is coercive, then the sequence $\{\boldsymbol{u}_h\}$ is bounded, i.e. $\exists c > 0$ such that $\|\boldsymbol{u}_h\| \le c\ \forall h,\ 0 \le h \le h_h < +\infty$. Then the subsequence $\{\boldsymbol{u}_{\hbar}\}$ can be taken out such that $\boldsymbol{u}_{\hbar} \to \boldsymbol{u}'$ weakly in V for $\hbar \to 0_+$, $\boldsymbol{u}' \in V$. According to (ii), $\boldsymbol{u}' \in K$. Moreover, L is weakly lower semi-continuous, hence

$$\lim_{\hbar \to 0_+} \inf L(\boldsymbol{u}_{\hbar}) \ge L(\boldsymbol{u}).$$

Due to Equation 7.63, we have $L(\boldsymbol{u}') \le L(\boldsymbol{u})$. Since the problem $(\mathscr{P})_v$ has only one solution, then $\boldsymbol{u}' = \boldsymbol{u}$. Hence, it is evident that the limit $\boldsymbol{u}'$ does not depend on the choice of the subsequence $\{\boldsymbol{u}_{\hbar}\}$ and therefore the whole subsequence $\{\boldsymbol{u}_h\}$ converges weakly to $\boldsymbol{u}$.

Furthermore, the functional $L_o(\boldsymbol{v})$ is twice differentiable. Hence, $L_o(\boldsymbol{u}_h) = L_o(\boldsymbol{u}) + L_o'(\boldsymbol{u}, \boldsymbol{u}_h - \boldsymbol{u}) + \frac{1}{2}a(\boldsymbol{u}_h - \boldsymbol{u}, \boldsymbol{u}_h - \boldsymbol{u})$. Thus $\forall \boldsymbol{v}_h \in K_h$ we have

$$\begin{aligned} L(\boldsymbol{v}_h) \ge L(\boldsymbol{u}_h) &= L_o(\boldsymbol{u}_h) + j(\boldsymbol{u}_h) \\ &= L_o(\boldsymbol{u}) + a(\boldsymbol{u}, \boldsymbol{u}_h - \boldsymbol{u}) - S(\boldsymbol{u}_h - \boldsymbol{u}) + \tfrac{1}{2}a(\boldsymbol{u}_h - \boldsymbol{u}, \boldsymbol{u}_h - \boldsymbol{u}) + j(\boldsymbol{u}_h) \\ &\ge L_o(\boldsymbol{u}) + a(\boldsymbol{u}, \boldsymbol{u}_h - \boldsymbol{u}) - S(\boldsymbol{u}_h - \boldsymbol{u}) + j(\boldsymbol{u}_h) + \tfrac{1}{2}\alpha\, \|\boldsymbol{u}_h - \boldsymbol{u}\|_{1,2}^2, \end{aligned}$$

where α is a coefficient of ellipticity of the bilinear form $a(.,.)$. Hence

$$\begin{aligned} \tfrac{1}{2}\alpha\, \|\boldsymbol{u}_h - \boldsymbol{u}\|_{1,2}^2 &\le L(\boldsymbol{u}_h) - L_o(\boldsymbol{u}) - a(\boldsymbol{u}, \boldsymbol{u}_h - \boldsymbol{u}) + S(\boldsymbol{u}_h - \boldsymbol{u}) - j(\boldsymbol{u}_h) \\ &\le L_o(\boldsymbol{v}_h) + j(\boldsymbol{v}_h) - L_o(\boldsymbol{u}) - a(\boldsymbol{u}, \boldsymbol{u}_h - \boldsymbol{u}) + S(\boldsymbol{u}_h - \boldsymbol{u}) - j(\boldsymbol{u}_h). \end{aligned}$$

Since the last inequality holds for all $\boldsymbol{v}_h \in K$, we can take $\boldsymbol{v}_h \in K_h$ such that $\boldsymbol{v}_h \to \boldsymbol{u}$. Then

$$L_o(\boldsymbol{v}_h) \to L_o(\boldsymbol{u}),\ j(\boldsymbol{v}_h) \to j(\boldsymbol{u}) \text{ for } h \to 0_+.$$

Since the sequence $\{\boldsymbol{u}_h\} \to \boldsymbol{u}$ weakly, then

$$\begin{aligned} \lim_{h\to 0_+} \sup \tfrac{1}{2}\alpha\, \|\boldsymbol{u} - \boldsymbol{u}_h\|^2 &\le \lim_{h\to 0_+} \{L_o(\boldsymbol{v}_h) - L_o(\boldsymbol{u})\} + \lim_{h\to 0_+} j(\boldsymbol{v}_h) \\ &- \lim_{h\to 0_+} a(\boldsymbol{u}, \boldsymbol{u}_h - \boldsymbol{u}) + \lim_{h\to 0_+} S(\boldsymbol{u}_h - \boldsymbol{u}) + \lim_{h\to 0_+} \sup(-j(\boldsymbol{u}_h)) \\ &= j(\boldsymbol{u}) - \lim_{h\to 0_+} \inf(j(\boldsymbol{u}_h)). \end{aligned}$$

Since $\lim_{h\to 0_+} \inf(j(\boldsymbol{u}_h)) \ge j(\boldsymbol{u}) \Rightarrow -\lim_{h\to 0_+} \inf(j(\boldsymbol{u}_h)) + j(\boldsymbol{u}) \le 0$, then $\lim_{h\to 0_+} \sup \|\boldsymbol{u} - \boldsymbol{u}_h\|^2 \le 0$ and thus $\|\boldsymbol{u} - \boldsymbol{u}_h\| \to 0$ for $h \to 0_+$, which completes the proof.

Definition 7.3: Let $\boldsymbol{v} \in [C(\overline{\Omega})]^N \cap K, 0 \le h \le h_o < +\infty$. Let us denote $a_i, i = 1, \dots, n(h)$, the nodes of the triangulation $\mathscr{T}_h$. We say that $r_h\boldsymbol{v} \equiv \boldsymbol{v}_{LI}$ is a piecewise linear interpolation of $\boldsymbol{v}$, when $r_h\boldsymbol{v} \in V_h$, $r_h\boldsymbol{v}(a_i) = \boldsymbol{v}(a_i), i = 1, \dots, n(h)$.

Lemma 7.7: For every function $\boldsymbol{v} \in [C(\overline{\Omega})]^N \cap K$ then $r_h\boldsymbol{v} \in K_h$ holds. As $\{\mathscr{T}_h\}$ is a regular system of triangulations of $\overline{\Omega}$, then

$$\|\boldsymbol{v} - r_h\boldsymbol{v}\|_{1,2} \le \mathrm{ch} \mid \boldsymbol{v} \mid_{2,2} \quad \forall \boldsymbol{v} \in [H^2(\Omega)]^N \tag{7.64}$$

holds, where $|\cdot|_{2,2}$ is a semi-norm in $H^2(\Omega)$ and $c > 0$ is a positive constant independent of h and $\boldsymbol{v}$.

The proof follows from Definition 7.3 and properties of the Lagrange interpolation.

Lemma 7.8: Let $\cup\overline{\Gamma}_c^{kl} \cap \overline{\Gamma}_u = 0$ and let there exist a finite number of boundary points $\cup\overline{\Gamma}_c^{kl} \cap \overline{\Gamma}_\tau$, $\cup\overline{\Gamma}_c^{kl} \cap \overline{\Gamma}_o$ and $\overline{\Gamma}_\tau \cap \overline{\Gamma}_u$. Denote $\mathscr{M} = K \cap [C^\infty(\overline{\Omega})]^N$. Then $\mathscr{M}$ is dense in K.
For the proof see Hlaváček and Lovíšek (1977).

Theorem 7.6: Let the presumption of Lemma 7.8 be satisfied. Let K, and K_h be defined as above. Then for all $\boldsymbol{v} \in K$ there exists $\boldsymbol{v}_h \in K_h$ such that

$$\|\boldsymbol{v} - \boldsymbol{v}_h\|_{1,2} \to 0 \text{ for } h \to 0_+. \tag{7.65}$$

Proof: Let $\boldsymbol{v} \in K$. Then according to Lemma 7.8 there exists a sequence $\{\boldsymbol{v}_m\} \in \mathscr{M} = K \cap [C^\infty(\overline{\Omega})]^N$ such that

$$\|\boldsymbol{v} - \boldsymbol{v}_m\|_{1,2} \to 0 \text{ for } m \to \infty.$$

Then towards the element $\boldsymbol{v}_m \in \mathscr{M} = K \cap [C^\infty(\overline{\Omega})]^N$ the element $r_h\boldsymbol{v}_m \in K_h$ can be found, and we denote it by $\boldsymbol{v}_h$ so that

$$\|\boldsymbol{v}_h - \boldsymbol{v}_m\|_{1,2} \le \text{ch} \mid \boldsymbol{v}_m \mid_{2,2}.$$

Thus

$$\|\boldsymbol{v} - \boldsymbol{v}_h\|_{1,2} \le \|\boldsymbol{v} - \boldsymbol{v}_m\|_{1,2} + \|\boldsymbol{v}_m - \boldsymbol{v}_h\|_{1,2} \to 0 \text{ for } h \to 0_+, m \to \infty.$$

Lemma 7.9: Let $\boldsymbol{u} \in K$ be a solution of the problem $(\mathscr{P})_v$, i.e. $L(\boldsymbol{u}) \le L(\boldsymbol{v})$ for all $\boldsymbol{v} \in K$, and $\boldsymbol{u}_h \in K_h$ be a solution of the problem $(\mathscr{P})_{vh}$, i.e. $L(\boldsymbol{u}_h) \le L(\boldsymbol{v}_h)$ for all $\boldsymbol{v}_h \in K_h$. Then there exists a constant $c > 0$ such that

$$\begin{aligned}\|\boldsymbol{u} - \boldsymbol{u}_h\|_{1,2}^2 \le c\{&a(\boldsymbol{u}_h - \boldsymbol{u}, \boldsymbol{v}_h - \boldsymbol{u}) + a(\boldsymbol{u}, \boldsymbol{v} - \boldsymbol{u}_h) + a(\boldsymbol{u}, \boldsymbol{v}_h - \boldsymbol{u}) - S(\boldsymbol{v} - \boldsymbol{u}_h)\\ &- S(\boldsymbol{v}_h - \boldsymbol{u}) + j(\boldsymbol{v}_h) - j(\boldsymbol{u}) + j(\boldsymbol{v}) - j(\boldsymbol{u}_h)\} \quad \forall \boldsymbol{v} \in K, \boldsymbol{v}_h \in K_h.\end{aligned} \tag{7.66}$$

Proof: For arbitrary $\boldsymbol{v} \in K$ and $\boldsymbol{v}_h \in K_h$ the following inequalities are satisfied:

$$a(\boldsymbol{u}, \boldsymbol{v} - \boldsymbol{u}) - S(\boldsymbol{v} - \boldsymbol{u}) - j(\boldsymbol{v}) + j(\boldsymbol{u}) \ge 0, \tag{7.67}$$

$$a(\boldsymbol{u}_h, \boldsymbol{v}_h - \boldsymbol{u}_h) - S(\boldsymbol{v}_h - \boldsymbol{u}_h) - j(\boldsymbol{v}_h) + j(\boldsymbol{u}_h) \ge 0. \tag{7.68}$$

Adding both inequalities and then to the resulting inequality adding and subtracting terms $a(\boldsymbol{u}, \boldsymbol{v} - \boldsymbol{u}_h) - a(\boldsymbol{u}_h, \boldsymbol{u})$; and after some modifications and using the ellipticity of the bilinear form $a(.,.)$, we find

$$\begin{aligned}\alpha \|\boldsymbol{u} - \boldsymbol{u}_h\|_{1,2}^2 &\le a(\boldsymbol{u} - \boldsymbol{u}_h, \boldsymbol{u} - \boldsymbol{u}_h)\\ &\le a(\boldsymbol{u}_h - \boldsymbol{u}, \boldsymbol{v}_h - \boldsymbol{u}) + a(\boldsymbol{u}, \boldsymbol{v} - \boldsymbol{u}_h) + a(\boldsymbol{u}, \boldsymbol{v}_h - \boldsymbol{u}) + S(\boldsymbol{u}_h - \boldsymbol{v})\\ &\quad + S(\boldsymbol{u} - \boldsymbol{v}_h) + j(\boldsymbol{v}_h) - j(\boldsymbol{u}) + j(\boldsymbol{v}) - j(\boldsymbol{u}_h).\end{aligned}$$

Hence Equation 7.66 follows immediately.

Lemma 7.10: Let the above assumptions be satisfied. Let $K_h \subset K\ \forall h$. Then

(i) the assumption (ii) of Theorem 7.5 is satisfied,

(ii) the estimate

$$\|\boldsymbol{u}-\boldsymbol{u}_h\|_{1,2}^2 \leq c\{a(\boldsymbol{u}-\boldsymbol{u}_h, \boldsymbol{u}-\boldsymbol{v}_h)+a(\boldsymbol{u}, \boldsymbol{v}-\boldsymbol{u}_h)+a(\boldsymbol{u}, \boldsymbol{v}_h-\boldsymbol{u})-S(\boldsymbol{u}-\boldsymbol{v}_h) + j(\boldsymbol{v}_h)-j(\boldsymbol{u})\} \quad \forall \boldsymbol{v}_h \in K_h. \tag{7.69}$$

holds.

Proof:

(i) Let $\boldsymbol{v}_h \in K_h$, then $\boldsymbol{v}_h \in K$ and $\boldsymbol{v}_h \to \boldsymbol{v}$ weakly in V. Since K is closed, then it is weakly closed and thus $\boldsymbol{v} \in K$.

(ii) $K_h \subset K \Rightarrow \boldsymbol{u}_h \in K$. Let us put $\boldsymbol{v} = \boldsymbol{u}_h$ in Equation 7.66. Then

$$\|\boldsymbol{u}-\boldsymbol{u}_h\|_{1,2}^2 \leq c\{a(\boldsymbol{u}_h-\boldsymbol{u}, \boldsymbol{v}_h-\boldsymbol{u})+0+a(\boldsymbol{u}, \boldsymbol{v}_h-\boldsymbol{u})-S(\boldsymbol{u}-\boldsymbol{v}_h)+0+j(\boldsymbol{v}_h)-j(\boldsymbol{u})+0\} \quad \forall \boldsymbol{v}_h \in K_h,$$

which completes the proof.

Theorem 7.7: Let the presumption of Lemma 7.8 be satisfied. Let the space V_h and the set K_h be defined as above, and let the functional L be defined by $L(\boldsymbol{v}) = \frac{1}{2}a(\boldsymbol{v}, \boldsymbol{v}) - S(\boldsymbol{v}) + j(\boldsymbol{v})$. Then the approximation of the problem $(\mathscr{P})_{vh}$ based on the finite element method $\boldsymbol{u}_h$ converges to the weak solution of the problem $(\mathscr{P})_v$ in the norm of the space $[H^1(\Omega)]^N$ and it holds

$$\|\boldsymbol{u}-\boldsymbol{u}_h\|_{1,2}^2 \leq \alpha^{-1}\{a(\boldsymbol{u}-\boldsymbol{u}_h, \boldsymbol{u}-\boldsymbol{v}_h)+S(\boldsymbol{u}-\boldsymbol{v}_h)+a(\boldsymbol{u}, \boldsymbol{v}_h-\boldsymbol{u})+j(\boldsymbol{v}_h)-j(\boldsymbol{u})\} \quad \forall \boldsymbol{v}_h \in K_h, \tag{7.70}$$

where α is a coefficient of ellipticity of the bilinear form $a(\boldsymbol{u}, \boldsymbol{v})$.

Proof: Due to Theorem 7.6, presumption (i) of Theorem 7.5 is satisfied. Since $K_h \subset K$ and due to Lemma 7.10, presumption (ii) of Theorem 7.5 is satisfied. Therefore, the assumptions of Theorem 7.5 are satisfied and, therefore, $\|\boldsymbol{u}-\boldsymbol{u}_h\|_{1,2} \to 0$ for $h \to 0_+$. The estimate in Equation 7.70 follows immediately from Lemma 7.10.

Theorem 7.8: Let Γ_c be polygonal. Let $\boldsymbol{u} \in [H^2(\Omega)]^N \cap K$, $\tau_n(\boldsymbol{u}) \in L^2(\cup\Gamma_c^{kl})$, $g_c^{kl} \in L^\infty(\cup\Gamma_c^{kl})$ and let $K_h \subset K$. Let the changes $u_n^k - u_n^l < 0 \to u_n^k - u_n^l = 0$ and $\boldsymbol{u}_t^k - \boldsymbol{u}_t^l = 0 \to \boldsymbol{u}_t^k - \boldsymbol{u}_t^l \neq 0$ occur at finitely many points of $\cup\Gamma_c^{kl}$ only. Then

$$\|\boldsymbol{u}-\boldsymbol{u}_h\|_{1,2} = O(h^{3/4}). \tag{7.71}$$

Proof: We have $K_h \subset K\, \forall h$. According to Equation 7.70 given is Theorem 7.7 it holds that

$$\|\boldsymbol{u}-\boldsymbol{u}_h\|_{1,2}^2 \leq \alpha^{-1}\{a(\boldsymbol{u}-\boldsymbol{u}_h, \boldsymbol{u}-\boldsymbol{v}_h)+S(\boldsymbol{u}-\boldsymbol{v}_h)+a(\boldsymbol{u}, \boldsymbol{v}_h-\boldsymbol{u})+j(\boldsymbol{v}_h)-j(\boldsymbol{u})\} \quad \forall \boldsymbol{v}_h \in K_h.$$

The bilinear form a(.,.) is bounded, i.e.

$$a(\boldsymbol{u}-\boldsymbol{u}_h, \boldsymbol{u}-\boldsymbol{v}_h) \leq C\,\|\boldsymbol{u}-\boldsymbol{u}_h\|_{1,2}\,\|\boldsymbol{u}-\boldsymbol{v}_h\|_{1,2} \text{ for arbitrary } \boldsymbol{v}_h \in K_h.$$

Since the last inequality holds for arbitrary $\boldsymbol{v}_h \in K_h$, then due to Lemma 7.7 it also holds for $\boldsymbol{v}_h = r_h\boldsymbol{u}$. Since $ab \leq \frac{1}{2}\varepsilon a^2 + (2\varepsilon)^{-1}b^2$, $a, b \in R$, $\varepsilon > 0$, then from this fact and Lemma 7.7 we obtain

$$a(\boldsymbol{u}-\boldsymbol{u}_h, \boldsymbol{u}-\boldsymbol{v}_h) \leq \tfrac{1}{2}\varepsilon C\,\|\boldsymbol{u}-\boldsymbol{u}_h\|_{1,2}^2 + \tfrac{1}{2}C\varepsilon^{-1}\,\|\boldsymbol{u}-\boldsymbol{v}_h\|_{1,2}^2 \leq \tfrac{1}{2}\varepsilon C\,\|\boldsymbol{u}-\boldsymbol{u}_h\|_{1,2}^2 + ch^2. \tag{7.72}$$

From Equation 7.70, where the second and third terms are modified, and due to Lemma 7.1 we obtain

$$a(\boldsymbol{u}, \boldsymbol{v}_h-\boldsymbol{u}) - S(\boldsymbol{v}_h-\boldsymbol{u}) = \int_\Omega \tau_{ij}(\boldsymbol{u})e_{ij}(\boldsymbol{v}_h-\boldsymbol{u})\mathrm{d}\boldsymbol{x} - \int_\Omega f_i(v_{ih}-u_i)\mathrm{d}\boldsymbol{x} - \int_{\Gamma_\tau} P_i(v_{ih}-u_i)\mathrm{d}s$$

$$= -\int_{\Omega} \partial\tau_{ij}(\boldsymbol{u})/\partial x_j(\boldsymbol{v}_h - \boldsymbol{u})\mathrm{d}\boldsymbol{x} + \int_{\partial\Omega} \tau_{ij}(\boldsymbol{u})(\boldsymbol{v}_h - \boldsymbol{u})n_j\mathrm{d}s$$

$$-\int_{\Omega} f_i(v_{ih} - u_i)\mathrm{d}\boldsymbol{x} - \int_{\Gamma_\tau} P_i(v_{ih} - u_i)\mathrm{d}s$$

$$= \int_{\Omega} (\partial\tau_{ij}(\boldsymbol{u})/\partial x_j + f_i)(u_i - v_{ih})\mathrm{d}\boldsymbol{x} + \int_{\Gamma_\tau} (-\tau_{ij}(\boldsymbol{u})n_j + P_i)(u_i - v_{ih})\mathrm{d}s$$

$$-\int_{\Gamma_u} \tau_{ij}(\boldsymbol{u})(u_i - v_{ih})\mathrm{d}s - \int_{\Gamma_o} (\tau_n^k(\boldsymbol{u})(u_n^k - u_n^l - (v_{nh}^k - v_{nh}^l))$$

$$+ \vec{\tau}_t^k(\boldsymbol{u})(\boldsymbol{u}_t^k - \boldsymbol{u}_t^l - (\boldsymbol{v}_{th}^k - \boldsymbol{v}_{th}^l))\mathrm{d}s - \int_{\cup\Gamma_c^{kl}} \tau_{ij}(\boldsymbol{u})(u_i - v_{ih})n_j\mathrm{d}s.$$

According to boundary conditions the first four integrals are equal to zero and then we obtain

$$a(\boldsymbol{u}, \boldsymbol{v}_h - \boldsymbol{u}) - S(\boldsymbol{v}_h - \boldsymbol{u}) = \int_{\cup\Gamma_c^{kl}} \tau_{ij}(\boldsymbol{u})(v_{ih} - u_i)n_j\mathrm{d}s$$

$$= \int_{\cup\Gamma_c^{kl}} (\tau_n^k(\boldsymbol{u})(v_{nh}^k - v_{nh}^l - (u_n^k - u_n^l)))\mathrm{d}s + \int_{\cup\Gamma_c^{kl}} \vec{\tau}_t^k(\boldsymbol{u})(\boldsymbol{v}_{th}^k - \boldsymbol{v}_{th}^l - (\boldsymbol{u}_t^k - \boldsymbol{u}_t^l))\mathrm{d}s. \quad (7.74)$$

To estimate both last integrals, we define the following sets

$$\Gamma_c^+ = \{\boldsymbol{x} \mid \boldsymbol{x} \in \cup\Gamma_c^{kl}, u_n^k - u_n^l = 0\}, \Gamma_c^- = \{\boldsymbol{x} \mid \boldsymbol{x} \in \cup\Gamma_c^{kl}, u_n^k - u_n^l < 0\}.$$

Let $\cup_{k,l}\Gamma_c^{kl}$ be constituted by piecewise linear segments, i.e. $\Gamma_c^{kl} = \cup_{i=1}^r \overline{a_i a_{i+1}}$.

(i) To estimate the first integral on the right hand side of Equation 7.74 along $\overline{a_i a_{i+1}}$, we denote it by I_1, we have three cases:

— the case $\overline{a_i a_{i+1}} \subset \Gamma_c^-$: since $\tau_n^k(u_n^k - u_n^l) = 0$ then $\tau_n^k = 0$ on $\overline{a_i a_{i+1}}$, hence $I_1 = 0$.

— the case $\overline{a_i a_{i+1}} \subset \Gamma_c^+$: Let $\boldsymbol{v}_h = r_h\boldsymbol{u}$. Since the normal vector $\boldsymbol{n}$ along $\overline{a_i a_{i+1}}$ is constant and the operator r_h is linear then

$$(r_h\boldsymbol{u})_n = (r_h u)_i n_i = r_h(u_i n_i) = r_h u_n \text{ (where } i = 1, .., r).$$

Therefore, $(r_h\boldsymbol{u})_n(a_{i+j}) = u_n(a_{i+j})$, where $i = 1, .., r$, $j = 0, 1$ and thus $r_h u_n = 0$ along $\overline{a_i a_{i+1}}$. Hence

$$\int_{\overline{a_i a_{i+1}}} (\tau_n^k(\boldsymbol{u})(r_h u_n^k - r_h u_n^l - (u_n^k - u_n^l))\mathrm{d}s = 0.$$

— the case $u_n^k - u_n^l < 0 \rightarrow u_n^k - u_n^l = 0$ along $\overline{a_i a_{i+1}}$: Let $\boldsymbol{v}_h = r_h\boldsymbol{u}$. Applying the Schwarz's inequality and the following assertion (the proof of which is a modification of that of Aziz (1972))

$$\boldsymbol{u} \in [H^2(\Omega)]^N \cap K \Rightarrow \boldsymbol{u} \in [H^{3/2}(\overline{a_i a_{i+1}})]^N \text{ and}$$
$$\|r_h u_n^k - r_h u_n^l - (u_n^k - u_n^l)\|_{L^2(\overline{a_i a_{i+1}})} = O(h^{3/2}), \quad (7.75)$$
$$\|r_h \boldsymbol{u}_t^k - r_h \boldsymbol{u}_t^l - (\boldsymbol{u}_t^k - \boldsymbol{u}_t^l)\|_{L^2(\overline{a_i a_{i+1}})} = O(h^{3/2}).$$

Hence

$$\left|\int_{\overline{a_i a_{i+1}}} (\tau_n^k(\boldsymbol{u})(r_h u_n^k - r_h u_n^l - (u_n^k - u_n^l)))\mathrm{d}s\right|$$

$$\leq \int_{\overline{a_i a_{i+1}}} (\tau_n^k(\boldsymbol{u}))^2\mathrm{d}s)^{1/2} \left(\int_{\overline{a_i a_{i+1}}} (r_h u_n^k - r_h u_n^l - (u_n^k - u_n^l))^2\mathrm{d}s\right)^{1/2} \leq c_r h^{3/2}.$$

Thus $\int_{\overline{a_i a_{i+1}}}(\tau_n^k(\boldsymbol{u})(r_h u_n^k - r_h u_n^l - (u_n^k - u_n^l)))\mathrm{d}s \le c_r h^{3/2}$.
Hence $|I_1| \le c_r h^{3/2}$.

(ii) To estimate the second integral on the right hand side of Equation 7.74 along $\overline{a_i a_{i+1}}$ and $j(\boldsymbol{v}_h) - j(\boldsymbol{u})$, we denote it by

$$I_2 = \int_{\cup\Gamma_c^{kl}} \vec{\tau}_t^k(\boldsymbol{u})(\boldsymbol{v}_{th}^k - \boldsymbol{v}_{th}^l - (\boldsymbol{u}_t^k - \boldsymbol{u}_t^l))\mathrm{d}s + j(\boldsymbol{v}_h) - j(\boldsymbol{u}),$$

we have four cases. Let us denote by

$$\Gamma_{tc}^{+} = \{\boldsymbol{x} \mid \boldsymbol{x} \in \cup\Gamma_{tc}^{kl}, \boldsymbol{u}_t^k - \boldsymbol{u}_t^l > 0\}, \quad \Gamma_{tc}^{-} = \{\boldsymbol{x} \mid \boldsymbol{x} \in \cup\Gamma_c^{kl}, \boldsymbol{u}_t^k - \boldsymbol{u}_t^l < 0\},$$
$$\Gamma_{tc}^{o} = \{\boldsymbol{x} \mid \boldsymbol{x} \in \cup\Gamma_c^{kl}, \boldsymbol{u}_t^k - \boldsymbol{u}_t^l = 0\}.$$

Then we investigate the following cases:

— the case $\overline{a_i a_{i+1}} \in \Gamma_{tc}^{+}$: Equation 7.57 gives

$$g_c^{kl} \mid \boldsymbol{u}_t^k - \boldsymbol{u}_t^l \mid + \vec{\tau}_t^k(\boldsymbol{u})(\boldsymbol{u}_t^k - \boldsymbol{u}_t^l) = 0 \text{ a.e. on } \cup\Gamma_c^{kl}.$$

Thus $g_c^{kl} = -\tau_t^k(\boldsymbol{u})$ a.e. on $\overline{a_i a_{i+1}}$. Let us put $\boldsymbol{v}_h = r_h \boldsymbol{u}$. The normal $\boldsymbol{n}$ is constant along $\overline{a_i a_{i+1}}$, the operator r_h is linear and, therefore, $(r_h \boldsymbol{u})_t = (r_h \boldsymbol{u})_i t_i = r_h(u_i t_i) = r_h \boldsymbol{u}_t$, where $\boldsymbol{t} = (t_i)$, $i = 1, \ldots, N$. Hence,

$$(r_h \boldsymbol{u})_t(a_{i+j}) = \boldsymbol{u}_t(a_{i+j}), \quad i = 1, \ldots, r, \quad j = 0, 1.$$

Therefore, $r_h \boldsymbol{u}_t > 0$ on $\overline{a_i a_{i+1}}$. Thus

$$\int_{\overline{a_i a_{i+1}}} (-g_c^{kl}(r_h \boldsymbol{u}_t^k - r_h \boldsymbol{u}_t^l - (\boldsymbol{u}_t^k - \boldsymbol{u}_t^l)) + g_c^{kl}(r_h \boldsymbol{u}_t^k - r_h \boldsymbol{u}_t^l - (\boldsymbol{u}_t^k - \boldsymbol{u}_t^l))\mathrm{d}s = 0.$$

— the case $\overline{a_i a_{i+1}} \subset \Gamma_{tc}^{o}$: Let $\boldsymbol{v}_h = r_h \boldsymbol{u}$. Since $(r_h \boldsymbol{u})_t = r_h \boldsymbol{u}_t$ then $r_h \boldsymbol{u}_t = 0$ on $\overline{a_i a_{i+1}}$, and thus

$$\int_{\overline{a_i a_{i+1}}} (\vec{\tau}_t^k(\boldsymbol{u})(r_h \boldsymbol{u}_t^k - r_h \boldsymbol{u}_t^l - (\boldsymbol{u}_t^k - \boldsymbol{u}_t^l)) + g_c^{kl}\left(\left|r_h \boldsymbol{u}_t^k - r_h \boldsymbol{u}_t^l\right| - \left|\boldsymbol{u}_t^k - \boldsymbol{u}_t^l\right|\right)\right) \mathrm{d}s = 0.$$

— the case $\overline{a_i a_{i+1}} \subset \Gamma_{tc}^{-}$: Since $\boldsymbol{u}_t < 0$, then $\left|\boldsymbol{u}_t^k - \boldsymbol{u}_t^l\right| = -(\boldsymbol{u}_t^k - \boldsymbol{u}_t^l)$ and thus $r_h \boldsymbol{u}_t^k - r_h \boldsymbol{u}_t^l < 0$. Hence and from Equation 7.57 we find $g_c^{kl} = \vec{\tau}_t^k(\boldsymbol{u})$ a.e. on $\overline{a_i a_{i+1}}$. Let us put $\boldsymbol{v}_h = r_h \boldsymbol{u}$. Then

$$\int_{\overline{a_i a_{i+1}}} (g_c^{kl}(r_h \boldsymbol{u}_t^k - r_h \boldsymbol{u}_t^l - (\boldsymbol{u}_t^k - \boldsymbol{u}_t^l)) + g_c^{kl}(-(r_h \boldsymbol{u}_t^k - r_h \boldsymbol{u}_t^l) + (\boldsymbol{u}_t^k - \boldsymbol{u}_t^l)))\mathrm{d}s = 0.$$

— the case $\boldsymbol{u}_t^k - \boldsymbol{u}_t^l = 0 \to \boldsymbol{u}_t^k - \boldsymbol{u}_t^l \ne 0$ on $\overline{a_i a_{i+1}}$: Let us put $\boldsymbol{v}_h = r_h \boldsymbol{u}$. Applying Schwarz's inequality and Equation 7.75, then

$$\left|\int_{\overline{a_i a_{i+1}}} (\vec{\tau}_t^k(\boldsymbol{u})(r_h \boldsymbol{u}_t^k - r_h \boldsymbol{u}_t^l - (\boldsymbol{u}_t^k - \boldsymbol{u}_t^l)))\mathrm{d}s + \int_{\overline{a_i a_{i+1}}} g_c^{kl}\left(\left|r_h \boldsymbol{u}_t^k - r_h \boldsymbol{u}_t^l\right| - \left|\boldsymbol{u}_t^k - \boldsymbol{u}_t^l\right|\right)\mathrm{d}s\right|$$
$$\le \int_{\overline{a_i a_{i+1}}} \left(\left|\vec{\tau}_t^k(\boldsymbol{u})\right| + g_c^{kl}\right)\left|r_h \boldsymbol{u}_t^k - r_h \boldsymbol{u}_t^l - (\boldsymbol{u}_t^k - \boldsymbol{u}_t^l)\right|\mathrm{d}s$$
$$\le \left(\int_{\overline{a_i a_{i+1}}} \left(\left|\vec{\tau}_t^k(\boldsymbol{u})\right| + g_c^{kl}\right)^2 \mathrm{d}s\right)^{1/2} \left(\int_{\overline{a_i a_{i+1}}} \left|r_h \boldsymbol{u}_t^k - r_h \boldsymbol{u}_t^l - (\boldsymbol{u}_t^k - \boldsymbol{u}_t^l)\right|^2 \mathrm{d}s\right)^{1/2}$$
$$\le c_s' h^{3/2},$$

as $g_c^{kl} \in L^\infty(\overline{a_i a_{i+1}})$ and using the proof of Theorem 7.2, thus $\vec{\tau}_t^k(\boldsymbol{u}) \in L^\infty(\overline{a_i a_{i+1}})$. Hence $|I_2| \le c_s h^{3/2}$.

Since $a(\boldsymbol{u} - \boldsymbol{u}_h, \boldsymbol{u} - \boldsymbol{v}_h) \le \frac{1}{2}\varepsilon C \|\boldsymbol{u} - \boldsymbol{u}_h\|_{1,2}^2 + ch^2$, $|I_1| \le c_r h^{3/2}$ and $|I_2| \le c_s h^{3/2}$, then from Equation 7.70 we finally obtain $\|\boldsymbol{u} - \boldsymbol{u}_h\|_{1,2}^2 \le ch^{3/2}$, from which the assertion of the theorem follows immediately.

A better idea for the estimate of $\|\boldsymbol{u} - \boldsymbol{u}_h\|_{1,2}$, based on the results of Mosco and Strang (1974), is given, for example, in Hlaváček and Lovíšek (1977) for the coercive $2D$ case without friction, $\Gamma_o \equiv 0$ and for $\iota = 1$ only, when the elastic body is in contact with absolutely rigid bodies. Under the assumptions that $\boldsymbol{u} \in [H^2(\Omega)]^2$ and $u_n \in H^2(\Gamma_c \cap \Gamma_m)$, where Γ_m, $m = 1, \ldots, k$, represent the sides of the polygonal boundary of the region Ω, we find the estimate $\|\boldsymbol{u} - \boldsymbol{u}_h\|_{1,2} = O(h)$ for the case without friction. We need the following lemmas and theorems.

Lemma 7.11: Let $K \subset B$, where B is a Banach reflexive space, be a closed and convex set, and let L_o be a twice differentiable functional in B such that there exist constants $\alpha > 0$ and $c > 0$ and the second differential of L satisfies the inequalities

$$\alpha \|\boldsymbol{w}\|^2 \le D^2 L_o(\boldsymbol{u}; \boldsymbol{w}, \boldsymbol{w}) \le c \|\boldsymbol{w}\|^2 \quad \forall \boldsymbol{u} \in K,\ \boldsymbol{w} \in B. \tag{7.76}$$

Moreover, let $K_h \subset K$ be a closed convex set. Let $\boldsymbol{u} \in K$ and $\boldsymbol{u}_h \in K_h$ be the minimizer of L_o over K or K_h, respectively, i.e. elements $\boldsymbol{u} \in K$ and $\boldsymbol{u}_h \in K_h$ satisfy

$$L_o(\boldsymbol{u}) \le L_o(\boldsymbol{v}) \quad \forall \boldsymbol{v} \in K \tag{7.77}$$

or

$$L_o(\boldsymbol{u}_h) \le L_o(\boldsymbol{v}) \quad \forall \boldsymbol{v} \in K_h. \tag{7.78}$$

Assume that there exists $\boldsymbol{w}_h \in K_h$ such that $2\boldsymbol{u} - \boldsymbol{w}_h \in K$. Then

$$\|\boldsymbol{u} - \boldsymbol{u}_h\| \le c^{1/2}\alpha^{-1/2} \|\boldsymbol{u} - \boldsymbol{w}_h\|. \tag{7.79}$$

Proof: The Taylor's expansion of L for arbitrary $\boldsymbol{v} \in K$ with a centre in $\boldsymbol{u}$ gives

$$L_o(\boldsymbol{v}) = L_o(\boldsymbol{u}) + DL_o(\boldsymbol{u}; \boldsymbol{v} - \boldsymbol{u}) + \tfrac{1}{2} D^2 L_o(\boldsymbol{u} + \vartheta(\boldsymbol{v} - \boldsymbol{u}); \boldsymbol{v} - \boldsymbol{u}, \boldsymbol{v} - \boldsymbol{u}), \quad \vartheta \in \langle 0, 1\rangle. \tag{7.80}$$

Since $\boldsymbol{u}$ satisfies Equation 7.77 then $DL_o(\boldsymbol{u}, \boldsymbol{u}_h - \boldsymbol{u}) \ge 0$. Let us put $\boldsymbol{v} = \boldsymbol{u}_h$, then using Equations 7.76, 7.77 and 7.78 we obtain

$$L_o(\boldsymbol{u}_h) \ge L_o(\boldsymbol{u}) + \tfrac{1}{2}\alpha \|\boldsymbol{u}_h - \boldsymbol{u}\|^2. \tag{7.81}$$

Since $\boldsymbol{w}_h \in K_h \subset K$ such that $2\boldsymbol{u} - \boldsymbol{w}_h \in K$ and $\boldsymbol{u}$ satisfies Equation 7.77 we find that $DL_o(\boldsymbol{u}, \boldsymbol{w}_h - \boldsymbol{u}) \ge 0$, $DL_o(\boldsymbol{u}, 2\boldsymbol{u} - \boldsymbol{w}_h - \boldsymbol{u}) \ge 0$ and thus $DL_o(\boldsymbol{u}, \boldsymbol{w}_h - \boldsymbol{u}) = 0$.

Let us put $\boldsymbol{v} = \boldsymbol{w}_h$ into Equation 7.80; then from the last equality and from Equations 7.76 and 7.77 we obtain

$$L_o(\boldsymbol{u}_h) \le L_o(\boldsymbol{u}) + \tfrac{1}{2} c \|\boldsymbol{w}_h - \boldsymbol{u}\|^2.$$

From the last inequality and Equation 7.81 we have

$$\|\boldsymbol{u}_h - \boldsymbol{u}\|^2 \le 2\alpha^{-1}(L_o(\boldsymbol{u}_h) - L_o(\boldsymbol{u})) \le c\alpha^{-1} \|\boldsymbol{w}_h - \boldsymbol{u}\|^2,$$

from which the assertion of the lemma follows immediately.

To obtain the estimate of $\|\boldsymbol{u} - \boldsymbol{u}_h\|_{1,2}$ we need the next two lemmas, which we shall introduce without proofs.

Lemma 7.12: (One side approximation of u_n on the boundary): Let $\boldsymbol{u} \in [H^2(\Omega)]^2$ and $u_n \in H^2(\Gamma_c \cap \Gamma_m) \forall m$, where Γ_m denote parts of the polygonal boundary of the domain Ω. Let $\{\mathcal{T}_h\}$ be a system of ϑ-regular triangulations of the domain Ω.

Then there exist linear spline functions $\varphi_h^{(m)} \in C(\Gamma_m)$ with mesh points in the vertices of the triangulation $\mathcal{T}_h$ such that

$$0 \geq \varphi_h^m \geq u_n \text{ on } \Gamma_c \cap \Gamma_m, \quad \left\|u_{nLI} - \varphi_h^m\right\|_m^2 \leq h^3 \int_{\Gamma_m \cap \Gamma_c} [d^2 u_n / ds^2]^2 ds$$

holds $\forall m$, where $r_h u_n \equiv u_{nLI}$ is the linear Lagrange interpolate of u_n on $\overline{\Gamma}_m$, and where $\|\varphi\|_m = \sup_{s \in \Gamma_m} |\varphi(s)|$.
For the proof see Hlaváček and Lovíšek (1977).

Lemma 7.13: Let $\varphi_m \in C(\overline{\Gamma}_m) \forall m$ be linear spline-functions with mesh points in the vertices of the triangulation $\mathcal{T}_h$ such that $\varphi_m = 0$ on $\overline{\Gamma}_m$. Then there exists $\boldsymbol{v}_h \in V_h$ such that $v_{hn} = \varphi_m$ on $\Gamma_m \forall m$ and

$$\|\boldsymbol{v}_h\|_{1,2} \leq c h^{1/2} \max_m \|\varphi_m\|_{\Gamma_m} .$$

For the proof see Hlaváček and Lovíšek (1977).

As a result for the case without friction we then obtain the following theorem.

Theorem 7.9: Let $\boldsymbol{u} \in [H^2(\Omega)]^2$ and $u_n \in H^2(\Gamma_c \cap \Gamma_m)$, where Γ_m denote parts of the polygonal boundary of the domain Ω. Let $\{\mathcal{T}_h\}$ be a system of ϑ-regular triangulations of the domain Ω. Then there exists $\boldsymbol{w}_h \in V_h$ such that

$$0 \geq w_{hn} \geq u_n \text{ on } \Gamma_c, \tag{7.82}$$

$$\|\boldsymbol{u} - \boldsymbol{w}_h\|_{1,2} \leq c h \|\boldsymbol{u}\|_{2,2} + \max_m \|u_n\|_{H^2(\Gamma_c \cap \Gamma_m)} , \tag{7.83}$$

where c is a constant independent of h and $\boldsymbol{u}$.

Proof: Let φ_h^m be functions as defined in Lemma 7.12. Let $\Psi_m = u_{nLI} - \varphi_h^m \forall m$, where Ψ_m satisfies the assumptions of Lemma 7.13. Let $\boldsymbol{v}_h \in V_h$ be the vector function defined by Lemma 7.13. Then the function $\boldsymbol{w}_h = \boldsymbol{u}_{LI} - \boldsymbol{v}_h$, where $\boldsymbol{u}_{LI}$ is the linear Lagrange interpolate of $\boldsymbol{u}$ over the triangulation $\mathcal{T}_h$ and $\boldsymbol{v}_h$ is the function from Lemma 7.13, satisfies Equations 7.82 and 7.83 as $\boldsymbol{w}_h \in V_h$ and on every Γ_m it holds that $w_{hn} = u_{LIn} - v_{hn} = u_{LIn} - \psi_m = \varphi_h^m$. Then the assertion of Lemma 7.12, i.e. $0 \geq \varphi_h^m \geq u_n$ on $\Gamma_c \cap \Gamma_m$, implies Equation 7.82 which proves the first part of the proof.

To prove the second part of the proof, it is known that

$$\|\boldsymbol{u} - \boldsymbol{u}_h\|_{1,2} \leq c h \|\boldsymbol{u}\|_{2,2} . \tag{7.84}$$

Obviously

$$\|\boldsymbol{u} - \boldsymbol{w}_h\|_{1,2} \leq \|\boldsymbol{u} - \boldsymbol{u}_{LI}\|_{1,2} + \|\boldsymbol{u}_{LI} - \boldsymbol{w}_h\|_{1,2} \leq \|\boldsymbol{u} - \boldsymbol{u}_{LI}\|_{1,2} + \|\boldsymbol{v}_h\|_{1,2} \tag{7.85}$$

where as above we put $\boldsymbol{w}_h = \boldsymbol{u}_{LI} - \boldsymbol{v}_h$. Due to Lemma 7.13 and since $\Psi_m = u_{nLI} - \varphi_h^m$ we find

$$\|\boldsymbol{v}_h\|_{1,2} \leq c h^{-1/2} \max_m \|\Psi_m\|_{\Gamma_m} .$$

From Lemma 7.12 and $\psi_m = u_{LIn} - \varphi_h^m$

$$\|\Psi_m\|_{\Gamma_m}^2 = \left\|u_{nLI} - \varphi_h^m\right\|_{\Gamma_m}^2 \leq h^3 \int_{\Gamma_m \cap \Gamma_c} [d^2 u_n / ds^2]^2 ds \leq h^3 \|u_n\|_{H^2(\Gamma_m \cap \Gamma_c)}^2 . \tag{7.86}$$

Hence, using Equations 7.84 and 7.85 we find

$$\|\boldsymbol{u} - \boldsymbol{w}_h\|_{1,2} \le ch\,\|\boldsymbol{u}\|_{2,2} + ch^{-1/2}h^{3/2}\max_m \|u_n\|_{H^2(\Gamma_c\cap\Gamma_m)},$$

which completes the proof.

Theorem 7.10: Let $\boldsymbol{u}$ and $\boldsymbol{u}_h$ be the solution of the problem $(\mathscr{P})_v$ and the problem $(\mathscr{P})_{vh}$ without friction under the above assumptions, respectively. Then

$$\|\boldsymbol{u} - \boldsymbol{u}_h\|_{1,2} = O(h). \tag{7.87}$$

Proof: The proof is based on Lemma 7.11 and Theorem 7.9. According to Theorem 7.9 there exists $\boldsymbol{w}_h \in V_h$ such that Equation 7.83 holds. Equation 7.82 yields that $\boldsymbol{w}_h \in K_h$, $2\boldsymbol{u} - \boldsymbol{w}_h \in K$, which proves the assumption of Lemma 7.11.

The functional $L_o(\boldsymbol{v}) = \frac{1}{2}a(\boldsymbol{v},\boldsymbol{v}) - S(\boldsymbol{v})$ is twice differentiable. The inequalities (Equation 7.76) of Lemma 7.11 follow from the properties of L_o given in Lemmas 7.4 and 7.5. Evidently, $K_h \subset K$.

Therefore, the assumptions of Lemma 7.11 are satisfied and then from the assertions of Lemma 7.11 (i.e. Equation 7.79) and Theorem 7.9 (i.e. Equation 7.83) we obtain

$$\|\boldsymbol{u} - \boldsymbol{u}_h\|_{1,2} \le c^{1/2}\alpha^{-1/2}\,\|\boldsymbol{u} - \boldsymbol{w}_h\|_{1,2} \le c^{1/2}\alpha^{-1/2}c'h\{\|\boldsymbol{u}\|_{2,2} + \max_m \|\boldsymbol{u}_n\|_{H^2(\Gamma_c\cap\Gamma_m)}\},$$

which completes the proof.

7.1.7. Construction of the Stiffness Matrix and of the Right Hand Side Vector

Numerical solution of the problem investigated requires the solution of the contact problem with friction by using the numerical approximation of a saddle point. By an approximate saddle point is meant a point $(\boldsymbol{u}_{sh}, \lambda_h) \subset K_h \times \Lambda \subset K \times \Lambda$ of a functional $\mathscr{L}$, the so-called Lagrangian, on $K_h \times \Lambda$, if a saddle point $(\boldsymbol{u}_s, \lambda) \in K \times \Lambda$ exists and if

$$\mathscr{L}(\boldsymbol{u}_{sh}, \mu) \le \mathscr{L}(\boldsymbol{u}_{sh}, \lambda_h) \le \mathscr{L}(\boldsymbol{v}, \lambda_h) \tag{7.88}$$

for all pairs of functions $(\boldsymbol{v}, \mu) \in K_h \times \Lambda$.

The problem is equivalent to the following problem: find $\boldsymbol{u}_h \in K_h$ such that

$$L(\boldsymbol{u}_h) = \min_{\boldsymbol{v}\in K_h}\sup_{\lambda\in\Lambda}\mathscr{L}(\boldsymbol{v},\mu) = \min_{\boldsymbol{v}\in K_h}\sup_{\mu\in\Lambda}\{L_o(\boldsymbol{v}) + \int_{\cup\Gamma_c^{kl}} \mu^{kl} g_c^{kl}(\boldsymbol{v}_t^k - \boldsymbol{v}_t^l)\mathrm{d}s\}, \tag{7.89}$$

where $\Lambda = \cup\{\mu^{kl} \mid \mu^{kl} \in L^2(\Gamma_c^{kl}), |\mu^{kl}| \le 1 \text{ a.e. on } \Gamma_c^{kl}\}$, $\mathscr{L}$ is the Lagrangian, and the non-differentiable functional $j(\boldsymbol{v})$ is replaced by

$$j(\boldsymbol{v}) = \int_{\cup\Gamma_c^{kl}} g_c^{kl}\left|\boldsymbol{v}_t^k - \boldsymbol{v}_t^l\right|\mathrm{d}s = \int_{\cup\Gamma_c^{kl}} \mu^{kl} g_c^{kl}(\boldsymbol{v}_t^k - \boldsymbol{v}_t^l)\mathrm{d}s,$$

where μ^{kl} is the Lagrange's multiplier.

It can be shown that the approximate saddle point exists and is unique (see Chapter 2, as well as, for example, Céa (1971), Ekeland and Temam (1976)). If

$(\boldsymbol{u}_{sh}, \lambda_h)$ is a saddle point of $\mathscr{L}$ on $K_h \times \Lambda$, then we have

$$L_o(\boldsymbol{u}_{sh}) + \int_{\cup\Gamma_c^{kl}} \mu^{kl} g_c^{kl}((\boldsymbol{u}_{sh}^k)_t - (\boldsymbol{u}_{sh}^l)_t)\mathrm{d}s$$

$$\leq L_o(\boldsymbol{u}_{sh}) + \int_{\cup\Gamma_c^{kl}} \lambda_h^{kl} g_c^{kl}((\boldsymbol{u}_{sh}^k)_t - (\boldsymbol{u}_{sh}^l)_t)\mathrm{d}s \text{ for all } \mu = (\mu^{kl}) \in \Lambda.$$

Hence

$$\int_{\cup\Gamma_c^{kl}} (\mu^{kl} - \lambda_h^{kl}) g_c^{kl}((\boldsymbol{u}_{sh}^k)_t - (\boldsymbol{u}_{sh}^l)_t)\mathrm{d}s \leq 0 \text{ for all } \mu \in \Lambda$$

and

$$\int_{\cup\Gamma_c^{kl}} (\lambda_h^{kl} + \rho g_c^{kl}((\boldsymbol{u}_{sh}^k)_t - (\boldsymbol{u}_{sh}^l)_t - \lambda_h^{kl})(\mu^{kl} - \lambda_h^{kl})\mathrm{d}s \text{ for all } \rho > 0 \text{ and all } \lambda \in \Lambda.$$

Thus

$$\lambda_h^{kl} = P(\lambda_h^{kl} + \rho g_c^{kl}((\boldsymbol{u}_h^k)_t - (\boldsymbol{u}_h^l)_t), \quad \rho > 0, \tag{7.90}$$

where P is a projection of $L^2(\Gamma_c^{kl})$ onto Λ. Due to Proposition 1.7 of Ekeland and Temam (1976)

$$L_o(\boldsymbol{u}_{sh}) + \int_{\cup\Gamma_c^{kl}} \lambda_h^{kl} g_c^{kl}((\boldsymbol{u}_{sh}^k)_t - (\boldsymbol{u}_{sh}^l)_t)\mathrm{d}s$$

$$\leq L_o(\boldsymbol{v}) + \int_{\cup\Gamma_c^{kl}} \lambda^{kl} g_c^{kl}((\boldsymbol{v}_h^k)_t - (\boldsymbol{v}_h^l)_t)\mathrm{d}s \text{ for all } \boldsymbol{v} \in K_h. \tag{7.91}$$

The inequalities of Equations 7.91 and 7.90 give an idea for the numerical solution of the problem, which can be realized by the Uzawa's or the Arrow-Hurwicz's algorithms, respectively. For our numerical realization the Uzawa's algorithm was used, i.e. we used the following algorithm.

Let us choose $\lambda_h^o \in \Lambda$, e.g. $\lambda_h^o = 0$. Let $\lambda_h^i \in \Lambda$ be known, then we find $u_{hi} \in K_h$ by solving the following problem

$$\min_{\boldsymbol{v}\in K_h} \left\{ L_o(\boldsymbol{v}) + \int_{\cup\Gamma_c^{kl}} \lambda_h^{i,kl} g_c^{kl}(\boldsymbol{v}_t^k - \boldsymbol{v}_t^l)\mathrm{d}s \right\} \tag{7.92}$$

and $\lambda_h^{i+1,kl}$ we find from the relation

$$\lambda_h^{i+1,kl} = P(\lambda_h^{i,kl} + \rho_i g_c^{kl}((u_{hi}^k)_t - (u_{hi}^l)_t)), \tag{7.93}$$

where P is a projection $P: L^2(\Gamma_c^{kl}) \to \Lambda$ defined as follows:

$$\begin{aligned} &\mu \in L^2(\Gamma_c^{kl}),\ y \in \Gamma_c^{kl} \\ &P\mu(y) = \mu(y) \quad \text{for } |\mu(y)| \leq 1, \\ &P\mu(y) = 1 \quad \text{for } \mu(y) > 1, \\ &P\mu(y) = -1 \quad \text{for } \mu(y) < -1, \end{aligned}$$

and where $0 < \rho_1 \leq \rho_i \leq \rho_2$, where ρ_1, and ρ_2 are sufficiently small numbers following from the convergence theorem of Uzawa's algorithm (Ekeland and Temam (1976)).

The finite element approximation leads to minimizing the functional $L(\boldsymbol{v}_h)$ on the set K_h. From the theory it is known (see Ekeland and Temam (1976)) that this problem is equivalent to the problem of finding the minimum of the quadratic functional

$$f(\boldsymbol{w}) = \tfrac{1}{2}\boldsymbol{w}^T B_h \boldsymbol{w} - \boldsymbol{b}^T \boldsymbol{w} \tag{7.94}$$

with linear constraints

$$A\boldsymbol{w} \leq \boldsymbol{d}. \tag{7.95}$$

Next we shall assume two model problems — the $2D$ and $3D$ model problems.

CASE 1: TWO DIMENSIONAL PROBLEMS — PROBLEMS OF PLANE ELASTICITY

We shall consider the model of lithospheric plates which are in collision and which occupy a general domain in R^3. Further, we shall assume that lithospheric plates are elastic and that the material coefficients, body forces, and boundary conditions on the surface of the investigated domain do not depend on the coordinate x_2, provided that the x_2 axis is perpendicular to the vertical cross section of the lithospheric plates in collision. Then it is sufficient to investigate only the vertical cross section of the lithospheric plates in collision. Therefore, our model reduces to the $2D$ problem. Let us denote the domain occupied by the lithospheric plates in their vertical cross section by $\Omega \subset R^2$, and let all physical coefficients, body forces, and boundary conditions on the surface of the investigated domain Ω be functions of two variables x_i, $i = 1, 2$. Principally, we have three cases of plane elasticity problems, i.e. problems of plane strain, plane stress and generalized plane stress, but we shall study here the first two cases only.

A) Plane Strain

We shall consider the $2D$ model of the elastic lithospheric plates in collision, about which we shall assume that they occupy the domain $\Omega \in R^2$. Let all physical parameters and boundary conditions satisfy the conditions given above. Then $u_i = u_i(x_1, x_2)$. Finally, let us suppose that $f_3(\boldsymbol{x}) = P_3(\boldsymbol{x}) = u_{o3}(\boldsymbol{x}) = 0$. Under these assumptions $u_3 = 0$ and $\partial u_1/\partial x_3 = \partial u_2/\partial x_3 = 0$. Hence, $e_{33}(\boldsymbol{u}) = e_{13}(\boldsymbol{u}) = e_{23}(\boldsymbol{u}) = 0$ and then

$$e_{ij} = \tfrac{1}{2}(\partial u_i/\partial x_j + \partial u_j/\partial x_i), \quad i, j = 1, 2 \tag{7.96}$$

and

$$\begin{aligned} \tau_{ij}(\boldsymbol{u}) &= \lambda e_{kk}\delta_{ij} + 2\mu e_{ij}(\boldsymbol{u}), \quad i, j = 1, 2, \\ \tau_{i3}(\boldsymbol{u}) &= \tau_{3i}(\boldsymbol{u}) = 0, \quad i = 1, 2, \ \tau_{33} = \lambda(e_{11} + e_{22}). \end{aligned} \tag{7.97}$$

From the inverse Hooke's law we find

$$2\mu e_{33}(\boldsymbol{u}) = \tau_{33}(\boldsymbol{u}) - \lambda/(3\lambda + 2\mu)(\tau_{11} + \tau_{22} + \tau_{33}) = 0,$$

hence

$$\tau_{33}(\boldsymbol{u}) = \lambda(2(\lambda + \mu))^{-1}(\tau_{11} + \tau_{22}) = \sigma(\tau_{11} + \tau_{22}),$$

where σ is Poisson's ratio. From the inverse Hooke's law we find

$$e_{ij}(\boldsymbol{u}) = (2\mu)^{-1}\tau_{ij}(\boldsymbol{u}) - \lambda(4\mu(\lambda + \mu))^{-1}(\tau_{11}(\boldsymbol{u}) + \tau_{22}(\boldsymbol{u}))\delta_{ij}, \quad i, j = 1, 2. \tag{7.98}$$

or

$$\begin{aligned} e_{11}(\boldsymbol{u}) &= (1 + \sigma)E^{-1}[\tau_{11}(\boldsymbol{u}) - \sigma(\tau_{11}(\boldsymbol{u}) + \tau_{22}(\boldsymbol{u}))], \\ e_{12}(\boldsymbol{u}) &= (1 + \sigma)E^{-1}\tau_{12}(\boldsymbol{u}), \\ e_{22}(\boldsymbol{u}) &= (1 + \sigma)E^{-1}[\tau_{22}(\boldsymbol{u}) - \sigma(\tau_{11}(\boldsymbol{u}) + \tau_{22}(\boldsymbol{u}))], \end{aligned} \tag{7.99}$$

where E is Young's modulus.

The equations of equilibrium and the boundary conditions are functions of $x_i, i = 1, 2$.

B) Plane Stress

Consider the $2D$ model of the elastic lithospheric plates as described above occupying the domain $\Omega \in R^2$. The lithosphere is assumed very thin, approximately $\pm h$, i.e. $x_3 = \pm h$, and subject to no load on the ends of $x_3 = \pm h$, i.e. $\tau_{3i}(\boldsymbol{u}) = 0, x_3 = \pm h, i = 1, 2, 3$. Then $u_i = u_i(x_1, x_2)$, and let us assume that $f_3(\boldsymbol{x}) = P_3(\boldsymbol{x}) = 0$ and $\tau_{3i}(\boldsymbol{u}) = 0$, $i = 1, 2, 3$ holds throughout the body. Then from $\tau_{33}(\boldsymbol{u}) = 0$ for isotropic rocks we find $e_{33}(\boldsymbol{u}) = -\lambda(\lambda + 2\mu)^{-1}(e_{11}(\boldsymbol{u}) + e_{22}(\boldsymbol{u}))$. Hence and using the generalized Hooke's law yields

$$\tau_{ij}(\boldsymbol{u}) = \lambda^*\vartheta'\delta_{ij} + 2\mu e_{ij}(\boldsymbol{u}), \quad i, j = 1, 2, \tag{7.100}$$

where $\lambda^* = 2\lambda\mu(\lambda + 2\mu)^{-1} = E\sigma(1 - \sigma^2)^{-1}$, $\vartheta' = (\tau_{11} + \tau_{22})(2(\lambda^* + \mu))^{-1}$.
The inverse Hooke's law yields

$$e_{ij}(\boldsymbol{u}) = (2\mu)^{-1}\tau_{ij}(\boldsymbol{u}) - \lambda^*(4\mu(\lambda^* + \mu))^{-1}(\tau_{11}(\boldsymbol{u}) + \tau_{22}(\boldsymbol{u}))\delta_{ij}, \quad i, j = 1, 2 \tag{7.101}$$

or

$$\begin{aligned} & e_{11}(\boldsymbol{u}) = E^{-1}[\tau_{11}(\boldsymbol{u}) - \sigma\tau_{22}(\boldsymbol{u})], \quad e_{12}(\boldsymbol{u}) = (1 + \sigma)E^{-1}\tau_{12}(\boldsymbol{u}), \\ & e_{22}(\boldsymbol{u}) = E^{-1}[\tau_{22}(\boldsymbol{u}) - \sigma\tau_{11}(\boldsymbol{u})], \quad e_{13}(\boldsymbol{u}) = e_{23}(\boldsymbol{u}) = 0, \\ & e_{33}(\boldsymbol{u}) = -\sigma E^{-1}[\tau_{11}(\boldsymbol{u}) + \tau_{22}(\boldsymbol{u})], \end{aligned} \tag{7.102}$$

where E is Young's modulus.

The equations of equilibrium and the boundary conditions are as in the case of plane strain, but the components of stress, displacement and strain are functions of x_3.

In our case, $f(\boldsymbol{w})$ is the functional $L(\boldsymbol{v}_h)$; B_h is the stiffness matrix ($\dim n \times n$), which is positive definite and generated by the term $\frac{1}{2}\int_\Omega c_{ijkl}(\boldsymbol{x})e_{ij}(\boldsymbol{v})e_{kl}(\boldsymbol{v})\mathrm{d}\boldsymbol{x}$; $\boldsymbol{b}(\dim n)$ is the vector generated by the body and surface forces, i.e. by the term $\int_\Omega F_i v_i \mathrm{d}\boldsymbol{x} + \int_{\Gamma_\tau} P_i v_i \mathrm{d}s$, as well as the friction forces $\int_{\cup\Gamma_c^{kl}} g_c^{kl}\left(\left|\boldsymbol{v}_t^k - \boldsymbol{v}_t^l\right|\right)\mathrm{d}s$ and it changes due to the friction forces in every step of the Uzawa's algorithm; A is the matrix of constraints (dim $m \times n$) generated by the condition $u_n^k - u_n^l \leq 0$ on Γ_c^{kl} in the definition of the admissible set of displacements K_h; and $\boldsymbol{d} \equiv \boldsymbol{0}$. Here n represents double number of points of triangulation in which the displacement is not prescribed and m is the number of points of triangulation on the contact boundary Γ_c^{kl}. The method of minimization of the quadratic functional solving our problem is discussed in this chapter, where $\boldsymbol{d} = \boldsymbol{0}$ due to $u_n^k - u_n^l \leq 0$ on Γ_c^{kl}. Denote $s_1, \ldots, s_M$ as the mesh points on $\cup\Gamma_c^{kl}$, and $n_1, \ldots, n_m$ as the outward unit normal onto $\cup\Gamma_c^{kl}$ in mesh points $s_1, \ldots, s_m$. The contact boundary $\cup\Gamma_c^{kl}$ is approximated by the final number of segments. Let the mesh points s_j, s_k lie on two neighbouring segments and let the mesh point s_l be their common outward unit normal to Γ_c^{kl}. Then we define $n_i^{(l)} = \frac{1}{2}(n_i^{(j)} + n_i^{(k)})$ for $i = 1, 2$ (for the 2D case). The body forces are generated by the gravitational forces, which are approximately of the order of $\rho \boldsymbol{g} x_3$.

To derive the matrix B_h the well-known FEM technique (see, for example, Aziz (1972), Ciarlet (1978)) was used. The stiffness matrix B_h is generated by the term

$$\frac{1}{2}\int_G c_{ijkl} e_{ij}(\boldsymbol{v}) e_{kl}(\boldsymbol{v}) \mathrm{d}\boldsymbol{x} = \sum_{T_{hi}} \frac{1}{2}\int_{T_{hi}} c_{ijkl} e_{ij}(\boldsymbol{v}) e_{kl}(\boldsymbol{v}) \mathrm{d}\boldsymbol{x} = \boldsymbol{w}^T B_h \boldsymbol{w}.$$

It can be shown that the matrix B_{hi} is created by the elementary 6×6 matrices $\overline{B}_{hi}$. We have (and if we put $x \equiv x_1$, $y \equiv x_3$)

$$\int_{T_{hi}} c_{ijkl} e_{ij} e_{kl} \mathrm{d}\boldsymbol{x} = J \int_{\underline{T}_o} J^{-2} \boldsymbol{w}^T A \boldsymbol{w} \, \mathrm{d}\xi \mathrm{d}\zeta = \frac{1}{2} J^{-1} \boldsymbol{w}^T A \boldsymbol{w} = \boldsymbol{w}^T \overline{B}_{hi} \boldsymbol{w},$$

where $\underline{T}_o$ is the unit triangle defined by the vertices (0,0), (0,1), (1,0). The transformation between T_{hi} and $\underline{T}_o$ is defined by $x = x_1 + \overline{x}_2\xi + \overline{x}_3\zeta$, $y = y_1 + \overline{y}_2\xi + \overline{y}_3\zeta$, where $\overline{x}_2 = x_2 - x_1$, $\overline{y}_2 = y_2 - y_1$, $\overline{x}_3 = x_3 - x_1$, $\overline{y}_3 = y_3 - y_1$ and where (x_i, y_i), $i = 1, 2, 3$ are the coordinates of the vertices of the triangle T_{hi}. Further,

$$J = \left|\frac{D(x, y)}{D(\xi, \zeta)}\right| = \left|\begin{matrix} \overline{x}_2, \overline{x}_3 \\ \overline{y}_2, \overline{y}_3 \end{matrix}\right| = \overline{x}_2\overline{y}_3 - \overline{x}_3\overline{y}_2 \text{ and } \overline{B}_{hi} = \frac{1}{2} J^{-1} \frac{E}{1+\sigma} C,$$

where $C = (c_{ij})$, E is Young's modulus and σ is Poisson's ratio. Let us denote $s = (1-\sigma)/(1-2\sigma)$, $r = \frac{1}{2} + \sigma$, $x_{13} = x_1 - x_3$, $x_{23} = x_2 - x_3$, $x_{12} = x_1 - x_2$, $y_{13} = y_1 - y_3$, $y_{23} = y_2 - y_3$, $y_{12} = y_1 - y_2$, then

$$c_{11} = s y_{13}^2 + \tfrac{1}{2} x_{13}^2,\ c_{12} = 1 - r x_{23} y_{23},\ c_{13} = -s y_{13} y_{23} - \tfrac{1}{2} x_{13} x_{23},$$

$$c_{14} = \sigma x_{13} y_{23} + \tfrac{1}{2} x_{23} y_{23},\ c_{15} = s y_{12} y_{23} + \tfrac{1}{2} x_{12} x_{23},\ c_{16} = -\sigma x_{12} y_{23} - \tfrac{1}{2} x_{23} y_{12},$$

$$c_{22} = s x_{23}^2 + \tfrac{1}{2} y_{23}^2,\ c_{23} = \sigma x_{23} y_{13} + \tfrac{1}{2} x_{13} y_{23},\ c_{24} = -s x_{13} x_{23} - \tfrac{1}{2} y_{13} y_{23},$$

$$c_{25} = -\sigma x_{23} y_{12} - \tfrac{1}{2} x_{12} y_{13},\ c_{26} = s x_{12} x_{23} + \tfrac{1}{2} y_{12} y_{23},$$

$$c_{33} = s y_{13}^2 + \tfrac{1}{2} x_{13}^2,\ c_{34} = -r x_{13} y_{13},\ c_{35} = s y_{12} y_{13} - \tfrac{1}{2} x_{12} x_{13},$$

$$c_{36} = \sigma x_{12} y_{13} - \tfrac{1}{2} x_{13} y_{12},$$

$$c_{44} = sy_{13}^2 + \tfrac{1}{2}x_{13}^2,\ c_{45} = \sigma x_{13}y_{12} + \tfrac{1}{2}x_{12}y_{13},\ c_{46} = -sx_{12}x_{13} - \tfrac{1}{2}y_{12}y_{13},$$
$$c_{55} = sy_{12}^2 + \tfrac{1}{2}x_{12}^2,\ c_{56} = rx_{12}y_{12},\ c_{66} = sx_{12}^2 + \tfrac{1}{2}y_{12}^2,\ c_{ij} = 0 \text{ for } i > j.$$

To calculate $\int_\Omega (f_1u_1 + f_2u_2)\mathrm{d}\boldsymbol{x}$ we use the standard technique. We have

$$\int_\Omega (f_1u_1 + f_2u_2)\mathrm{d}\boldsymbol{x} = \sum_{T_{hi}} \int_{T_{hi}} (f_1u_1 + f_2u_2)\mathrm{d}\boldsymbol{x} = \sum_{T_{hi}} \boldsymbol{w}_h^T \boldsymbol{F}_h,$$

where the transformation between T_{hi} and $\underline{T}_o$ is made in a similar way as in the previous case.

To calculate $\int_{\Gamma_\tau} P_iu_i\mathrm{d}s$, we approximate the boundary Γ_τ by the piecewise linear segments l_h and then the segment can be written in its parametric form $x = x_k + \overline{x}_k t$, $y = y_k + \overline{y}_k t$, $\overline{x}_k = x_j - x_k$, $\overline{y}_k = y_j - y_k$, $t \in \langle 0, 1\rangle$. Then $\int_{\Gamma_\tau} P_iu_i\mathrm{d}s = \sum_{l_h} \int_{l_h} P_iu_i\mathrm{d}s = \sum_{l_h} \boldsymbol{w}_h^T \boldsymbol{P}_h$.

By a similar technique we approximate the contribution of the frictional forces.

It can be noted that if the coefficients of elasticity are constant on every element, and the components f_i of body forces are constant or can be approximated by polynomials, then for the integration over $\underline{T}_o$ the formula

$$\int_{\underline{T}_o} \xi^a \eta^b \mathrm{d}\xi \mathrm{d}\eta = \frac{a!b!}{(a+b+2)!},$$

where ξ, η represent the coordinate system of the unit element $\underline{T}_o$, will be used. In the opposite case these integrals must be computed numerically. Similarly in the case of the loading forces, if these can be approximated by polynomials then for the integration over the unit segment l_{ho} the formula

$$\int_0^1 t^n \mathrm{d}t = (n+1)^{-1}$$

will be used. In the opposite case these integrals must be computed numerically. The same is valid in the case of the frictional forces.

CASE 2: THE THREE DIMENSIONAL PROBLEM

The three dimensional case was studied in detail in connection with the homogenization method and the stress-strain post-processing in Chapter 3 for the cases of linear polynomials and of polynomials of the third order. The constructions of the stiff matrix B_h, generated by the term $\frac{1}{2}\int_\Omega c_{ijkl}(\boldsymbol{x})e_{ij}(\boldsymbol{v})e_{kl}(\boldsymbol{v})\mathrm{d}\boldsymbol{x}$, and the vector $\boldsymbol{b}$, generated by the body and surface forces, i.e. by the terms $\int_\Omega F_iv_i\mathrm{d}\boldsymbol{x} + \int_{\Gamma_\tau} P_iv_i\mathrm{d}s$ and $\int_{\cup\Gamma_c^{kl}} g_c^{kl}\left(\left|\boldsymbol{v}_t^k - \boldsymbol{v}_t^l\right|\right)\mathrm{d}s$, are similar to the algorithm discussed in Chapter 3. The matrix A, the matrix of constraints (dim $m \times n$), is generated by the condition $u_n^k - u_n^l \le 0$ on Γ_c^{kl} in the definition of the admissible set of displacements K_h, and $\boldsymbol{d} \equiv 0$. The contact boundary $\cup\Gamma_c^{kl}$ is approximated by the final number of plane segments. Similarly as for the $2D$ case the common outward unit normal $\boldsymbol{n}^{(l)}$ to the contact boundary Γ_c^{kl} in points of common edges (case (1)) and of common corners (case (2)), we then define for case (1) $n_i^{(l)} = \frac{1}{2}(n_i^{(j)} + n_i^{(k)})$, $i = 1, 2, 3$, where $n_i^{(j)}$, and $n_i^{(k)}$ are normals to two planes (j) and (k), and for case (2) $n_i^{(l)} = p^{-1}\Sigma_{j=1}^{p} n_i^{(j)}$, $i = 1, 2, 3$, where p represents the number of planes which cut in the common corner.

7.2. MATHEMATICAL MODELS OF PLATE TECTONICS BASED ON CONTACT PROBLEMS WITH FRICTION IN THERMO-ELASTICITY

7.2.1. Quasi-Coupled Plate Tectonic Models based on Contact Problems with Friction: The Coercive case

7.2.1.1. Introduction

In this section the problem of geomechanical processes, which take place during quasi-static loading in the contact area between several colliding lithospheric plates, will be discussed.

The mathematical problem discussed in this section represents a simulation of geodynamic processes in the subduction zones and/or in the collision zones between several colliding lithospheric plates or geological blocks, respectively, during relatively short geological time, and is based on the hypothesis of the new global tectonics (plate tectonics) and the theory of thermo-elasticity. The aim of this section is to give a new mathematical theory, which better describes the situation in the subducting and obducting lithospheric plates or in the regions where several lithospheric plates collide. The theory is able to describe the stress-strain situations on contact boundaries between colliding lithospheric plates.

Since the thermal regime in the lithosphere is characterized during this short geological period by a relatively low and constant temperature, the problem can be investigated from the point of view of an N-dimensional ($N = 2, 3$) quasi-coupled thermo-elastic analysis (see, for example, Nedoma (1982b),(1983),(1987)). But in our studies we shall limit ourselves to the three dimensional problems in which elastic parameters and all other physical parameters depend on one horizontal and the vertical coordinates only and are independent of the second horizontal coordinate. Such type of problems require the solution of the two dimensional ($N = 2$) problems only.

Similarly as in the case of simple models based on contact problems with or without friction in linear elasticity, in thermo-elasticity for time periods shorter than a characteristic time t_c (see the discussion in previous sections) the lithosphere and the upper parts of the mantle behave elastically. Therefore, our model can be studied as the quasi-coupled thermo-elastic model. In this section we shall discuss the 2*D* coercive case only, i.e. the case in which the displacement is prescribed on the part of the boundary of colliding lithospheric plates, donated by Γ_u. Moreover, we shall assume that the lithospheric plates move by uniform speeds so that the inertia term can be neglected, and also that the heat is transfered from the lower parts of the Earth to the upper parts of the Earth as well as from the more heated lithospheric plates to the less warm lithospheric plate.

7.2.1.2. *Formulation of the problem*

THE EQUILIBRIUM EQUATION AND THE HEAT CONDUCTION EQUATION

Let us denote by Ω the $2D$ region, with a Lipschitz boundary $\partial\Omega$, occupied by a part of the loading lithosphere of our interest (see Figure 7.1, which is repeated here for convenience). Moreover, we shall assume that $\Omega = \cup_{\iota=1}^{S}\Omega^{\iota}$, where Ω^{ι} denote separate parts of the lithosphere and the upper mantle studied. Let the boundary $\partial\Omega$ consist of three parts Γ_τ, Γ_u, Γ_c such that $\partial\Omega = \overline{\Gamma}_\tau \cup \overline{\Gamma}_u \cup \overline{\Gamma}_c \cup \mathscr{R}$, where $\mathscr{R}$ is a set of zero measure. The equilibrium equations in the differential form for every subdomain Ω^{ι} of Ω is as follows

$$\tau^{\iota}_{ij,j} + f^{\iota}_i = 0, \quad i, j = 1, \dots, N, \quad N = 2, \quad \iota = 1, \dots, s. \tag{7.103}$$

In the lithosphere and the upper mantle, heat transmission is by conduction. From the second thermodynamic theorem for a quasi-static case (see Nedoma (1982a),(1990), Nowacki (1970), and also Chapter 3) we have

$$\operatorname{div}(\kappa^{\iota}\operatorname{grad}T^{\iota}) + \rho c_e v_j T_{,j} + Q^{\iota} = 0, \quad \iota = 1, \dots, s, \text{ in } \Omega^{\iota}, \tag{7.104}$$

where κ^{ι} represents the thermal conductivity, T^{ι} is the temperature, T^{ι}_o is the initial temperature in which rocks are in an initial strain and stress state, Q^{ι} and W^{ι} are the thermal sources, $Q^{\iota} = W^{\iota} - \rho^{\iota}c^{\iota}_e T^{\iota}_o e_{kk}(\dot{u}^{\iota})$, where the second term represents deformation energy which changed into heat, ρ^{ι} is the density, c_e is the specific heat, $e_{ij}(\boldsymbol{u})$ and $e_{ij}(\dot{\boldsymbol{u}})$ are, respectively the small strain tensor defined later and the

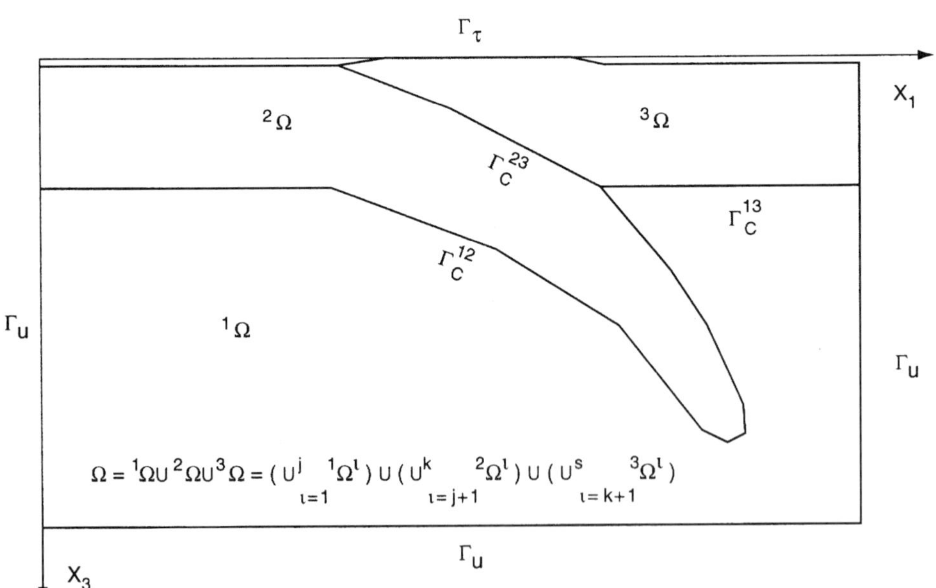

Figure 7.1. The model

strain rate tensor (in this section we shall assume that the dissipative thermal energy $\rho^\iota c_e^\iota T_o^\iota e_{kk}(\dot{u}^\iota)$ is one part of the thermal sources). The term $\rho c_e v_j T_{,j}$ represents the convective term. The stress-strain relation is given by the well-known Duhamel-Neumann's (Hooke's) relation

$$\tau_{ij} = 2\mu e_{ij} + (\lambda e_{kk} - \beta T)\delta_{ij} \equiv 2\mu e_{ij} + ((k - \tfrac{2}{3}\mu)e - \beta T)\delta_{ij} \qquad (7.105)$$

in the isotropic case and

$$\tau_{ij} = c_{ijkl}e_{kl} - \beta_{ij}T, \qquad (7.106)$$

in the anisotropic case, where $k = \lambda + \frac{2}{3}\mu$ is the bulk modulus, λ and μ are Lamé coefficients, β and β_{ij} are coefficients of thermal expansion, and e_{ij} is the small strain tensor defined by

$$e_{ij} = \tfrac{1}{2}(u_{i,j} + u_{j,i}), \quad e = e_{kk}. \qquad (7.107)$$

Since processes and materials in the lithosphere and the upper mantle are assumed to be elastic, the Lamé coefficients are taken in the isothermic state.

Moreover, the notations $\partial f/\partial x_i \equiv f_{,i}$, $\partial u_i/\partial x_j \equiv u_{i,j}$ are used.

BOUNDARY CONDITIONS

Similarly as in the previous sections the boundary $\partial\Omega$ of the investigated region Ω, occupied by the broken up obducting and subducting lithospheric plates and the asthenosphere with the mesosphere, consists of three parts $\Gamma_u, \Gamma_\tau, \Gamma_c$, i.e $\partial\Omega = \Gamma_u \cup \Gamma_\tau \cup \Gamma_c \cup \mathscr{R}$, in the $2D$ model and of four parts $\Gamma_u, \Gamma_\tau, \Gamma_c, \Gamma_o$ in the $3D$ model, where Γ_o has a meaning as a boundary of symmetry or a contact boundary with perfectly rigid and perfectly smooth support without friction (for details see the previous section). The Earth surface Γ_τ is loaded by the weight of the oceans, of the atmosphere, of the mountains above some approximate surface, etc.. Then on the Earth surface we have the condition

$$\tau_{ij}n_j = P_i \text{ on } \Gamma_\tau. \qquad (7.108)$$

We can also assume that the loading forces $\boldsymbol{P} = 0$.

From paleomagnetic studies, satellite measurements and analyses as well as from geological observations, the movements of lithospheric plates can be deduced. Then by modelling these movements the displacement vectors at sufficiently great distances from the collision zones can be determined. Since the speeds of lithospheric plates are known from the observations discussed above, we can determine the required displacement vectors, denoted it by $\boldsymbol{u}_o = (u_{oi})_{i=1}^N$. Let us describe this part of the boundary by Γ_u. Thus we have the condition

$$u_i = u_{oi} \text{ on } \Gamma_u. \qquad (7.109)$$

From Equations 7.103, 7.106 and 7.108 it follows that the effect of a change of temperature due to the deformation is equivalent to the replacement of mass forces by forces $F_i - \beta_{ij}\partial T/\partial x_j$ and of surface loading by surface forces $P_i + \beta_{ij}Tn_j$

under boundary conditions. Thereby the term $\beta\delta_{ij}Tn_j$ has the character of an omnidirectional pressure such as hydrostatic pressure in hydrodynamics. We shall assume that on the boundary $\Gamma_u \cup \Gamma_\tau$ the temperature or the heat flow are known, so that

$$T = T_1 \text{ or } \kappa_{ij}T_{,j}n_i = q_o \text{ on } \Gamma_\tau, \tag{7.110}$$

$$T = T_1 \quad \text{on } \Gamma_u. \tag{7.111}$$

Let us assume that the part of the colliding lithospheric plates of our interest consists of s subdomains Ω^ι, $\iota = 1, \ldots, s$, so that $\Omega = \cup_{\iota=1}^{S}\Omega^\iota$, with the boundary $\partial\Omega = \cup_{\iota=1}^{S}\partial\Omega^\iota$. Let us denote by Γ_c^{kl} the common contact boundary before deformation between neighbouring lithospheric plates and/or the asthenosphere, denoted by Ω^k and Ω^l. Let $\boldsymbol{n} = (n_i) = (n_1, n_2)$, $\boldsymbol{t} = (t_i) = (n_2, -n_1)$ be the outward unit normal and unit tangential vector, respectively, to the contact boundary Γ_c^{kl}. Further, let us denote the stress vector by $\tau_i = \tau_{ij}n_j$, its normal and tangential components by $\tau_n = \tau_i n_i = \tau_{ij}n_i n_j$, $\vec{\tau}_t = \vec{\tau} - \tau_n\boldsymbol{n}$ and the normal and tangential components of the displacement vector by $u_n = u_i n_i$ and $\boldsymbol{u}_t = \boldsymbol{u} - u_n\boldsymbol{n}$. Let us denote by $\boldsymbol{u}^k, \boldsymbol{u}^l, T^k, T^l$ (where indices k, l denote the neighbouring lithospheric plates or geological blocks) the displacements and temperatures in the neighbouring lithospheric plates or geological blocks. All these quantities are functions of spatial coordinates. Let (x, z) be the local Cartesian coordinate system. The x-axis is identical with the positive direction of the outward normal $\boldsymbol{n}_k$ (associated with Ω^k) to the contact boundary Γ_c^{kl}, and the z-axis is identical with the unit tangential vector $\boldsymbol{t}_k$ to the contact boundary Γ_c^{kl} at the point $O \in \Gamma_c^{kl}$, where $\Gamma_c^{kl} = \Gamma_c^k \cap \Gamma_c^l$ is the common contact boundary between the lithospheric plates and/or between the lithospheric plate and the asthenosphere in the non-deformable state, denoted as Ω^k and Ω^l (see Figure 7.2).

Let the contact boundary of both lithospheric plates Γ_c^{kl} be mathematically described by the function $h(z)$. The origin of the local coordinate system shall be denoted by $O \in \Gamma_c^{kl}$, and let $O \equiv O^k \equiv O^l$, $O^k \in \Gamma_c^k$, $O^l \in \Gamma_c^l$. During the deformation of the lithosphere the points O^k and O^l are generally transferred in a different way, but such that the lithospheric plate Ω^l cannot penetrate into the lithospheric plate Ω^k (or the asthenosphere), as both lithospheric plates are assumed to be elastic. From this condition we see that

$$u_1^l(0, 0) \le u_l^k(\bar{x}, \bar{z}) + h(z), \tag{7.112}$$

where the point $(\bar{x}, \bar{z}) \in \Gamma_c^{kl}$ is an unknown point, but satisfies $u_3^k(\bar{x}, \bar{z}) = u_3^l(O, O) - \bar{z}$. To obtain a simpler condition, we shall assume that in a small neighbourhood of the point $O \in \Gamma_c^{kl}$ the curvature of the contact boundary Γ_c^{kl} is small, so that $h(\bar{z}) \doteq 0$. Applying the Taylor's expansion to $u_1^l(\bar{x}, \bar{z})$ in the neighbourhood of $O \in \Gamma_c^{kl}$ and restricting ourselves to the linear terms only, then the mutual shifts $\left|u_3^k - u_3^l\right|$ and the derivatives $\left|\partial u_1^k/\partial z\right|$ are so small that they can be neglected. Therefore, $u_1^k(\bar{x}, \bar{z}) \doteq u_1^k(0, 0)$. Since $u_1^k \equiv u_n^k$, $u_1^l \equiv u_n^l$, then (Equation 7.112) can be written in the form $u_n^k(0, 0) - u_n^l(0, 0) \le 0$, which generally gives

$$u_n^k(\boldsymbol{x}) - u_n^l(\boldsymbol{x}) \le 0. \tag{7.113}$$

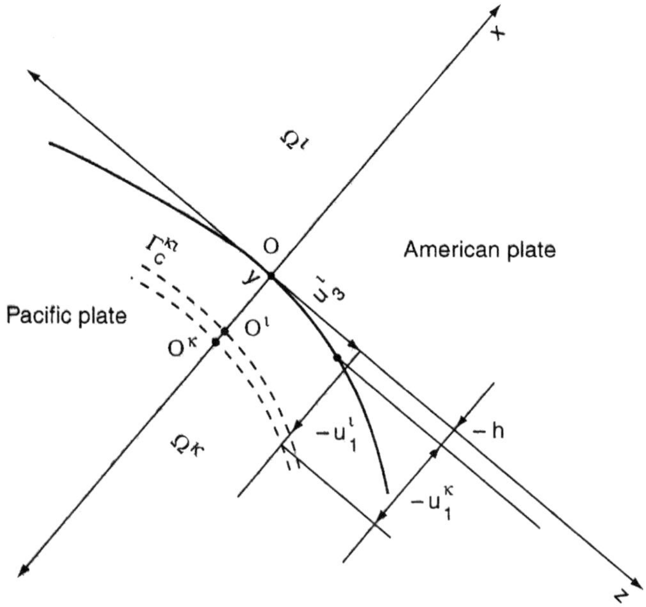

Figure 7.2. Derivation of the conditions on the contact boundary

From the principle of action and reaction for the contact forces we have

$$\tau_n^k(\boldsymbol{x}) = -\tau_n^l(\boldsymbol{x}) \equiv \tau_n^{kl}(\boldsymbol{x}), \quad \tau_t^k(\boldsymbol{x}) = -\tau_t^l(\boldsymbol{x}) \equiv \tau_t^{kl}(\boldsymbol{x}). \tag{7.114}$$

Since the normal components of contact forces cannot be positive, i.e. cannot be tensile, then

$$\tau_n^k(\boldsymbol{x}) = -\tau_n^l(\boldsymbol{x}) \equiv \tau_n^{kl}(\boldsymbol{x}) \leq 0. \tag{7.115}$$

During the deformation of both lithospheric plates they are in contact or are not in contact. If they are not in contact, then $u_n^k - u_n^l < 0$ and the contact forces are equal to zero, i.e. $\tau_n^k(\boldsymbol{x}) = -\tau_n^l(\boldsymbol{x}) = 0$. If the lithospheric plates are in contact, i.e. $u_n^k - u_n^l = 0$, then there exist non zero contact forces $\tau_n^k = -\tau_n^l \equiv \tau_n^{kl} < 0$. These conditions can be written in the following form

$$(u_n^k(\boldsymbol{x}) - u_n^l(\boldsymbol{x}))\tau_n^{kl}(\boldsymbol{x}) = 0. \tag{7.116}$$

Further, if both lithospheric plates are in contact, then on the contact boundary acts the friction in the Coulombian sense and then on the contact boundary act friction forces $g_c = \mathscr{F}_c^{kl} |\tau_n^{kl}|$, which in the absolute value are proportional to the absolute value of acting normal forces τ_n. The coefficient of proportionality $\mathscr{F}_c^{kl}$ is the coefficient of the Coulombian friction. Its magnitude is unknown but can be estimated by its upper estimation. Due to the acting friction forces we have the following cases.

If the absolute value of the tangential forces $\vec{\tau}_t^{kl}(\boldsymbol{x})$ is less than the frictional forces $g_c^{kl} = \mathscr{F}_c^{kl} |\tau_n^{kl}|$, then the frictional forces preclude the mutual shifts of both colliding lithospheric plates. If the tangential forces $\vec{\tau}_t^{kl}$ are equal in their absolute value to the frictional forces, then both lithospheric plates mutually shift. At the same time points of the contact boundary Γ_c^{kl} change their position in the direction opposite to the tangential forces. These situations are described by the following conditions:

$$\left|\vec{\tau}_t^{kl}(\boldsymbol{x})\right| \leq g_c^{kl}(\boldsymbol{x}), \tag{7.117}$$

if $u_n^k(\boldsymbol{x}) - u_n^l(\boldsymbol{x}) = 0$ then if

$$\left|\vec{\tau}_t^{kl}(\boldsymbol{x})\right| < g_c^{kl} \quad \text{then } \boldsymbol{u}_t^k(\boldsymbol{x}) - \boldsymbol{u}_t^l(\boldsymbol{x}) = 0, \tag{7.118}$$

i.e. the frictional forces preclude the mutual shifts of colliding lithospheric plates or geological blocks. If

$$\left|\vec{\tau}_t^{kl}(\boldsymbol{x})\right| = g_c^{kl}(\boldsymbol{x}) \tag{7.119}$$

then there exists function $\Theta \geq 0$ such that $u_t^k(\boldsymbol{x}) - u_t^l(\boldsymbol{x}) = -\Theta\tau_t^{kl}(\boldsymbol{x})$,
i.e. the friction forces cannot preclude the mutual shifts of lithospheric plates or geological blocks. On the contact boundary between the colliding lithospheric plates the temperatures and heat flow are continuous, i.e.

$$T^k = T^l, \tag{7.119}$$

$$\kappa_{ij}\partial T/\partial x_j n_i\big|_{(k)} = \kappa_{ij}\partial T/\partial x_j n_i\big|_{(l)}. \tag{7.120}$$

But in the real situation the contact boundaries between lithospheric plates, geological blocks and the asthenosphere are partially melted, the points on the contact boundary change their position in horizontal and vertical directions (contact conditions given in Equations 7.113 to 7.119), and then the conditions in Equations 7.119 and 7.120 are not satisfied. We shall assume that the difference of the temperatures T^k and T^l changes at about the value T_2^{kl}, and similarly that the heat flux q change is about some value q_2^{kl}. Hence, instead of the conditions in Equations 7.119 and 7.120 we find the conditions

$$T^k - T^l \leq T_2^{kl}, \tag{7.121}$$

$$q \leq q_2^{kl}, \tag{7.122}$$

$$(T^k - T^l - T_2^{kl})(q - q_2^{kl}) = 0. \tag{7.123}$$

As was shown in Hlaváček *et al*(1982) for the elastic case and in Nedoma (1987) for the thermo-elastic case (in which $s = 1$) the coefficient of friction can be estimated by

$$\left\|\mathscr{F}_c^{kl}\right\| < (2\mu/(\lambda + 3\mu))^{1/2} \sim \left(\tfrac{1}{2}(c_p^2 + c_s^2)c_s^{-2}\right)^{-\frac{1}{2}} \tag{7.124}$$

$$\text{or} \quad \left\|\mathscr{F}_c^{kl}\right\| < (\mu/(\lambda + 2\mu))^{1/4} \sim (c_s/c_p)^{1/2}, \tag{7.125}$$

where λ and μ are the Lamé coefficients and c_p and c_s are velocities of P and S waves. From here we see that the coefficient of friction depends on the material only.

Mathematically, the problem was discussed in detail in Nedoma (1987) and will also be discussed in detail in the next part of this section for the thermo-elastic case.

7.2.1.3. Variational solution of the problem

Let $\Omega \subset R^2$ be the plane with a Lipschitz boundary $\partial\Omega$, occupied by either obducting or subducting plates at the moment t-fixed. The boundary $\partial\Omega$ consists of the parts $\Gamma_\tau, \Gamma_u, \Gamma_c$, such that $\partial\Omega = \overline{\Gamma}_\tau \cup \overline{\Gamma}_u \cup \overline{\Gamma}_c \cup \mathscr{R}$. Let $\boldsymbol{x} = (x_1, x_2)$ be Cartesian coordinates. Let $\boldsymbol{n} = (n_1, n_2)$ and $\boldsymbol{t} = (-n_2, n_1)$ denote the unit outward normal and tangential vectors to the boundary $\partial\Omega$. Let us look for the temperature $T \in H^1(\Omega)$ and the displacement vector $\boldsymbol{u} = (u_1, u_2) \in W^1 = [H^1(\Omega)]^2$, where $H^k(\Omega)$ denotes the Sobolev space in the usual sense. We let $e_{ij}(\boldsymbol{u})$ be the strain tensor, $\tau_{ij}(\boldsymbol{u})$ the stress tensor, $T_o = T_o(\boldsymbol{x})$ the input temperature at which the rocks are in an initial strain and stress state, $\beta_{ij}(\boldsymbol{x}) \in C^1(\overline{\Omega})$ be the coefficient of thermal expansion and $f_i \in L^2(\Omega)$ be components of the body forces. The elastic coefficients $c_{ijkl}(\boldsymbol{x}) \in C^1(\overline{\Omega})$ satisfy the usual symmetry condition and the conditions of Lipschitz continuity and ellipticity, i.e.

$$c_{ijkl} = c_{jikl} = c_{klij}, \tag{7.126}$$

$$0 < a_o \leq (c_{ijkl}(\boldsymbol{x})\xi_{kl}\xi_{ij})\,|\xi|^{-2} \leq A_o < \infty \quad \forall \boldsymbol{x} \in \Omega,$$

$$\xi = (\xi_{ij})_{i,j=1}^2 \in R^4, \tag{7.127}$$

where a_o, A_o are constants independent of $\boldsymbol{x} \in \Omega$ and $\xi \in R^4$.

Let $\kappa_{ij} \in C^1(\overline{\Omega})$ be the thermal conductivity satisfying

$$\kappa_{ij} = \kappa_{ji}, \tag{7.128}$$

$$\kappa_{ij}\zeta_i\zeta_j \geq c\,|\zeta|^2, \quad \forall \boldsymbol{x} \in \Omega, \quad \forall \zeta \in R^2, c = \text{constant} > 0. \tag{7.129}$$

Let $W = W(\boldsymbol{x}) \in L^2(\Omega)$ be the heat sources in the lithospheric plate, and $\rho(\boldsymbol{x}) \in C(\overline{\Omega})$ and $c_e(\boldsymbol{x}) \in C(\overline{\Omega})$ be the density and the specific heat.

The stress tensor satisfies the equilibrium equation (Equation 7.103). On $\partial\Omega$ we define the stress vector $\vec{\tau}$, its normal and tangential components by $\tau_i = \tau_{ij}(\boldsymbol{x})n_j$, $\tau_n = \tau_i n_i = \tau_{ij}n_i n_j$, $\vec{\tau}_t = \vec{\tau} - \tau_n \boldsymbol{n} = \tau_{ij}n_j t_i$ and the normal and tangential components of the displacement vector by $u_n = u_i n_i$ and $\boldsymbol{u}_t = u_i t_i$.

Further, denote by (.,.) the inner (scalar) product in $L^2(\Omega)$ and $[L^2(\Omega)]^2$, by $\langle .,. \rangle$ the inner (scalar) product in $[L^2(\Gamma_c)]^2$ and by [.,.] the inner (scalar) product in $[L^2(\partial\Omega)]^2$. Denote by $H^{-1/2}(\Gamma_c)$ the dual space of $\mathring{H}^{1/2}(\Gamma_c) = \{\boldsymbol{v} \mid \boldsymbol{v} \in H^{1/2}(\partial\Omega), \boldsymbol{v}_{|cl(\partial\Omega\setminus\Gamma_c)} = 0$, with the norm of $H^{1/2}(\partial\Omega)\}$.

We shall look for T and $\boldsymbol{u}$ such that T is replaced by $T + z$ and, $\boldsymbol{u}$ is replaced by $\boldsymbol{u} + \boldsymbol{w}$, where z is a sufficiently smooth scalar function in $\overline{\Omega} = \Omega \cup \partial\Omega$ satisfying Equations 7.110 and 7.111 and $z = 0$ on Γ_c, and $\boldsymbol{w}$ is a sufficiently smooth vector function in $\overline{\Omega} = \Omega \cup \partial\Omega$ satisfying the Dirichlet type condition on Γ_u, where it is prescribed, and the Signorini type conditions on Γ_c (or $\boldsymbol{w} = 0$ on Γ_c in the case when the boundary is perpendicular to the boundary $\partial\Omega$). Then we have to solve the following problem:

Problem ($\mathscr{P}_f$): Find a scalar function T and a vector function $\boldsymbol{u}$ satisfying

$$-(\kappa_{ij}(\boldsymbol{x})T_{,j})_{,i} + \rho c_e v_j T_{,j} = Q, \tag{7.130}$$

$$(c_{ijkl}(\boldsymbol{x})e_{kl}(\boldsymbol{u}))_{,j} + F_i = 0, i = 1, 2 \quad \text{in } \Omega, \tag{7.131}$$

$$T = 0(\text{or } \kappa_{ij}T_{,j}n_i = q_o) \text{ on } \Gamma_\tau, \tag{7.132}$$

$$\tau_{ij} = P_i \quad \text{on } \Gamma_\tau, \tag{7.133}$$

$$T = 0 \text{ on } \Gamma_u, \tag{7.134}$$

$$u_i = 0, i = 1, 2 \quad \text{on } \Gamma_u, \tag{7.135}$$

$$T^k - T^l \leq T_2^{kl}, \tag{7.136}$$

$$q \leq q^{kl}, \tag{7.137}$$

$$(T^k - T^l - T_2^{kl})(q - q^{kl}) = 0 \quad \text{on } \Gamma_c, \tag{7.138}$$

$$u_n^k(\boldsymbol{x}) - u_n^l(\boldsymbol{x}) \leq 0, \tag{7.139}$$

$$\tau_n^{kl}(\boldsymbol{x}) \leq 0 \tag{7.140}$$

$$(u_n^k(\boldsymbol{x}) - u_n^l(\boldsymbol{x}))\tau_n^{kl}(\boldsymbol{x}) = 0 \quad \text{on } \Gamma_c, \tag{7.141}$$

and the Coulombian law

$$\left|\vec{\tau}_t^{kl}(\boldsymbol{x})\right| \leq g_c^{kl}(\boldsymbol{x}), \tag{7.142}$$

if $u_n^k(\boldsymbol{x}) - u_n^l(\boldsymbol{x}) = 0$ then if $\left|\vec{\tau}_t^{kl}(\boldsymbol{x})\right| < g_c^{kl}$ then $\boldsymbol{u}_t^k(\boldsymbol{x}) - \boldsymbol{u}_t^l(\boldsymbol{x}) = 0$,
if $\left|\vec{\tau}_t^{kl}(\boldsymbol{x})\right| = g_c^{kl}(\boldsymbol{x})$ then there exists function $\Theta \geq 0$ such that $\boldsymbol{u}_t^k(\boldsymbol{x}) - \boldsymbol{u}_t^l(\boldsymbol{x}) = -\Theta\vec{\tau}_t^{kl}(\boldsymbol{x})$,
which can also be written in the following equivalent form

$$\left|\vec{\tau}_t^{kl}(\boldsymbol{u})\right| \leq \mathscr{F}_c^{kl}\left|\tau_n^{kl}(\boldsymbol{u})\right|, \tag{7.143}$$

$$\left|\boldsymbol{u}_t^k - \boldsymbol{u}_t^l\right|\left(\left|\vec{\tau}_t^{kl}(\boldsymbol{u})\right| - \mathscr{F}_c^{kl}\left|\tau_n^{kl}\right|\right) = 0,$$

$$(\mathscr{F}_c^{kl}(\boldsymbol{u}))(\boldsymbol{x}) < 0 \Rightarrow \exists\Theta \geq 0,$$

$$\boldsymbol{u}_t^k(\boldsymbol{x}) - \boldsymbol{u}_t^l(\boldsymbol{x}) = -(\Theta\vec{\tau}_t)(\boldsymbol{x}),$$

where $F_i = f_i - (\beta_{ij}(T - T_p))_{,j} + (c_{ijkl}(\boldsymbol{w}))_{,j} \in L^2(\Omega)$, $Q = W + (\kappa_{ij}z_{,j})_{,i} - \rho c_e v_j z_{,j} \in L^2(\overline{\Omega})$ and where $\mathscr{F}_c^{kl}$ is the coefficient of Coulombian friction, Θ is a non-negative function on Γ_c^{kl}, $q \in L^2(\Gamma_\tau)$ is the heat flow, $\boldsymbol{F} \in [L^2(\overline{\Omega})]^2$, $\boldsymbol{P} \in [H^{-1/2}(\Gamma_\tau)]^2$, $\boldsymbol{u}_o \in [H^{1/2}(\Gamma_u)]^2$, $T_o \in H^{-1/2}(\Omega)$, $T_1 \in H^{-1/2}(\Gamma_u)$, $T_2^{kl} \in H^{-1/2}(\Gamma_c^{kl})$, $Q \in L^2(\overline{\Omega})$, $q_o \in L^2(\Gamma_\tau)$. Let us suppose that $\Gamma_c = \cup_{k,l}\Gamma_c^{kl} \in C^{2,1}(\Gamma_c)$, $q_2^{kl} \in L^2(\Gamma_c^{kl})$ and let $\mathscr{F}_c^{kl} \in C^{o,1}$ have a compact support and, further more, let dist(supp $\mathscr{F}_c^{kl}$, $\partial\Omega\backslash \cup \Gamma_c^{kl}) > 0$. Let us denote by

$$^1V = \{w \mid w \in H^1(\Omega), \quad w = 0 \quad \text{on} \quad \Gamma_u \cup \Gamma_\tau \quad \text{in the sense of traces}\},$$

$$V = \{\boldsymbol{v} \mid \boldsymbol{v} \in W^1, \quad \boldsymbol{v} = 0 \quad \text{on} \quad \Gamma_u \quad \text{in the sense of traces}\}$$

the spaces of virtual temperatures and displacements, respectively, and by

$$^1K = \{w \mid w \in {}^1V, \quad w^k - w^l \leq T_2^{kl} \quad \text{on} \quad \Gamma_c^{kl} \quad \text{in the sense of traces}\},$$

$$K = \{\boldsymbol{v} \mid \boldsymbol{v} \in V, \quad v_n^k - v_n^l \leq 0 \quad \text{on} \quad \Gamma_c^{kl} \quad \text{in the sense of traces}\}$$

the sets of admissible virtual temperatures and admissible virtual displacements, respectively. Furthermore, denote

$$\mathscr{H} = \cup\mathscr{H}^{kl} = \{g_n^{kl} \in H^{-1/2}(\cup\Gamma_c^{kl}), g_n^{kl} \leq 0 \text{ in the dual sense to the ordering on}$$
$$\{w \in H^{1/2}(\partial\Omega), w\,|_{cl(\partial\Omega\backslash\cup\Gamma_c^{kl})} = 0, \text{provided with the norm of } H^{1/2}(\partial\Omega)\}$$
$$\text{given by the restriction of the canonical ordering on } L^2(\cup\Gamma_c^{kl})\}.$$

As our quasi-coupled problem is indeed not coupled, both the problems in thermics and elasticity can be solved separately and the coupling term $(\beta_{ij}(T - T_o))_{,j}$ have a meaning of body forces.

We shall introduce an auxiliary problem, in which $\tau_n^{kl}(\mathbf{u})$ in Equation 7.143 is replaced by g_n^{kl}. Let $g_n^{kl} \in \mathscr{H}^{kl}$ be arbitrary. For $T, w \in H^1(\Omega)$, $\boldsymbol{u}, \boldsymbol{v} \in W^1$ we put

$$b(T, w) = \sum_{\iota=1}^{s} b^{\iota}(T^{\iota}, w^{\iota}) = b_1(T, w) + b_2(T, w) = \int_{\Omega} (\kappa_{ij}(\boldsymbol{x})T_{,j}w_{,i} + \rho c_e v_j T_{,j} w)\mathrm{d}\boldsymbol{x}, \tag{7.144}$$

$$s(w) = \sum_{\iota=1}^{s} s^{\iota}(w^{\iota}) = \int_{\Omega} Qw\mathrm{d}\boldsymbol{x}, \left(\text{or } s(w) = \sum_{\iota=1}^{s} s^{\iota}(w^{\iota}) = \int_{\Omega} Qw\mathrm{d}\boldsymbol{x} + \int_{\Gamma_{\tau}} q_o w \mathrm{d}s\right), \tag{7.145}$$

$$l(w) = \tfrac{1}{2} b(w, w) - s(w), \tag{7.146}$$

$$a(\boldsymbol{u}, \boldsymbol{v}) = \sum_{\iota=1}^{s} a^{\iota}(\boldsymbol{u}^{\iota}, \boldsymbol{v}^{\iota}) = \int_{\Omega} c_{ijkl} e_{ij}(\boldsymbol{u}) e_{kl}(\boldsymbol{v})\mathrm{d}\boldsymbol{x}, \tag{7.147}$$

$$S(\boldsymbol{v}) = \int_{\Omega} F_i v_i \mathrm{d}\boldsymbol{x} + \int_{\Gamma_{\tau}} P_i v_i \mathrm{d}s, \tag{7.148}$$

$$L(\boldsymbol{v}) = L_o(\boldsymbol{v}) + j_{gn}(\boldsymbol{v}), \tag{7.149}$$

$$L_o(\boldsymbol{v}) = \tfrac{1}{2} a(\boldsymbol{v}, \boldsymbol{v}) - S(\boldsymbol{v}), \tag{7.150}$$

$$j_{gn}(\boldsymbol{v}) = \int_{\cup_{k,l}\Gamma_c} \mathscr{F}_c^{kl} \left|g_n^{kl}\right| \left|\boldsymbol{v}_t^k - \boldsymbol{v}_t^l\right| \mathrm{d}s = \left\langle \mathscr{F}_c^{kl} \left|g_n^{kl}\right|, \left|\boldsymbol{v}_t^k - \boldsymbol{v}_t^l\right| \right\rangle. \tag{7.151}$$

Definition 7.4: By a variational (weak) solution of the auxiliary problem $(\mathscr{P}_f)_v$ we mean a pair of functions $(T, \boldsymbol{u})$, $T \in^1 K$, $\boldsymbol{u} \in K$, such that

$$b(T, w - T) \geq s(w - T), \quad \forall w \in {}^1K, \tag{7.152}$$

$$a(\boldsymbol{u}, \boldsymbol{v} - \boldsymbol{u}) + \langle \mathscr{F}_c^{kl} \left|g_n^{kl}\right|, \left|\boldsymbol{v}_t^k - \boldsymbol{v}_t^l\right| \rangle \geq S(\boldsymbol{v} - \boldsymbol{u}), \quad \forall \boldsymbol{v} \in K, \tag{7.153}$$

where $\left|g_n^{kl}\right|$ means $-g_n^{kl}$ for $g_n^{kl} \in \mathscr{H}^{kl}$. For $g_n^{kl} = 0$ we have the case without friction.

For $\boldsymbol{u} \in W^1$ we define $\tau_n^{kl}(\boldsymbol{u})$ by

$$\langle \tau_n^{kl}(\boldsymbol{u}), v_n^k - v_n^l \rangle = a(\boldsymbol{u}, \boldsymbol{v}) - (\boldsymbol{F}, \boldsymbol{v}) \quad \forall \boldsymbol{v} \in W^1, \boldsymbol{v}_t^k - \boldsymbol{v}_t^l \,|_{cl(\partial\Omega \setminus \cup \Gamma_c^{kl}} = 0. \tag{7.154}$$

Definition 7.5: A solution $(T, \boldsymbol{u})$, $T \in {}^1K$, $\boldsymbol{u} \in K$ of Equations 7.152 and 7.153 such that $\mathscr{F}_c^{kl} \tau_n^{kl}(\boldsymbol{u}) = \mathscr{F}_c^{kl} g_n^{kl}$ with $\tau_n^{kl}(\boldsymbol{u})$ from Equation 7.154 is called a weak solution of the problem with friction in linear quasi-coupled thermo-elasticity.

As the problem is quasi-coupled, it is equivalent to the following variational formulation.

Find $(T, \boldsymbol{u})$, $T \in {}^1K$, $\boldsymbol{u} \in K$ such that

$$l(T) \leq l(w), \quad \forall w \in {}^1K, \tag{7.155}$$

$$L(\boldsymbol{u}) \leq L(\boldsymbol{v}) \quad \forall \boldsymbol{v} \in K. \tag{7.156}$$

Theorem 7.11: A pair of functions $(T, \boldsymbol{u})$, $T \in {}^1K$, $\boldsymbol{u} \in K$ satisfy Equations 7.155 and 7.156 if and only if

$$b(T, w - T) \geq s(w - T). \quad \forall w \in {}^1K, \tag{7.157}$$

$$a(\boldsymbol{u}, \boldsymbol{v} - \boldsymbol{u}) + \langle g_c^{kl}, |\boldsymbol{v}_t^k - \boldsymbol{v}_t^l| \rangle \geq S(\boldsymbol{v} - \boldsymbol{u}). \quad \forall \boldsymbol{v} \in K, \tag{7.158}$$

Proof: The sets ${}^1K$ and K are non-empty, closed and convex subsets of ${}^1V$ and V, respectively, and the functionals $l(w)$ and $L_o(\boldsymbol{v})$ are convex and differentiable. The functional $j_{gn}(\boldsymbol{v}) = \langle g_c^{kl}, |\boldsymbol{v}_t^k - \boldsymbol{v}_t^l| \rangle = \langle \mathscr{F}_c^{kl} |g_n^{kl}|, |\boldsymbol{v}_t^k - \boldsymbol{v}_t^l| \rangle$, $g_c^{kl} = \mathscr{F}_c^{kl} |g_n^{kl}|$, is convex and non-differentiable. Hence and due to Céa (1971), Equations 7.157 and 7.158 are equivalent to Equations 7.155 and 7.156, which completes the proof.

Similarly as in the previous sections it can be shown that any classical solution of the problem discussed is a weak solution, and conversely if the weak solution of the discussed problem is smooth enough, then it represents a classical solution of the problem discussed.

The next theorem gives the existence and uniqueness of the solution of the problem $(\mathscr{P}_f)_v$ and an estimate of it. The theorem states that there is a unique solution and that the problem is well-posed. By the well-posedness we shall mean the existence, uniqueness, and continuous dependence of the solution on the given data $(T_o, T_1, Q, \boldsymbol{u}_o, \boldsymbol{F}, \boldsymbol{P})$. From the physical point of view, it reflects the fact that the solution only changes a little for little changes in the displacements, surface and body forces, thermal sources and temperatures.

Theorem 7.12: Let Equations 7.126 to 7.129 hold. Then for every $Q \in L^2(\Omega)$, $q_o \in L^2(\Gamma_\tau)$, $g_c^{kl} \in L^\infty(\Gamma_c^{kl})$, $\boldsymbol{P} \in [L^2(\Gamma_\tau)]^2$, $\boldsymbol{F} \in [L^2(\Omega)]^2$ there exists a unique solution of Problem $(\mathscr{P}_f)_v$. Furthermore, there exist constants c_o and c_1 independent of g_n^{kl} such that

$$\|T\|_1 \leq c_o(\|q_o\|_{L^2(\Gamma_\tau)} + \|Q\|_{L^2(\Omega)} + \|T_1\|_1 + \|T_o\|_1),$$

$$\|\boldsymbol{u}\|_{W^1} \leq c_1(\|\boldsymbol{P}\|_{[L^2(\Gamma_\tau)]^2} + \|\boldsymbol{F}\|_{[L^2\Omega)]^2} + \|\boldsymbol{u}_o\|_{W^1}).$$

Proof: The bilinear form $b : {}^1V \times {}^1V \to R^1$ is ${}^1V$-elliptic and bounded on ${}^1V$ and $Q \in ({}^1V)'$. To complete the first part of the proof for the thermics, the analogy of the Lax-Milgram theorem for variational inequalities can be used. For the second part of the proof for elasticity, the set K is closed and convex in ${}^1W$, hence it is weakly closed. The functional L_o is strictly convex and weakly lower semi-continuous (see Nedoma (1983)). The functional $j_{gn} = \langle g_c^{kl}, |\boldsymbol{v}_t^k - \boldsymbol{v}_t^l| \rangle$ is convex and continuous, and thus it is weakly lower semi-continuous (see Ekeland and Temam (1976)). According to the Korn inequality $a(\boldsymbol{u}, \boldsymbol{u}) \geq c_1 \|\boldsymbol{u}\|_{W^1}$, which is satisfied on V and $j_{gn}(\boldsymbol{v}) = \langle g_c^{kl}, |\boldsymbol{v}_t^k - \boldsymbol{v}_t^l| \rangle \geq 0$ (for an arbitrary $g_c^{kl} \in L^\infty(\Gamma_c^{kl})$, we have $\langle g_c^{kl}, |\boldsymbol{v}_t^k - \boldsymbol{v}_t^l| \rangle \geq 0$ as $g_c^{kl} \geq 0$ a.e. on Γ_c^{kl}), we find $L(\boldsymbol{u}) \to 0$ as $\|\boldsymbol{u}\|_{W^1} \to \infty$. Further, substituting $\boldsymbol{v} = \boldsymbol{u}_o$, $\boldsymbol{u}_o \in W^1$, $\boldsymbol{u}_{o|\Gamma_u} = 0$, into Equation 7.158 we get $a(\boldsymbol{u}, \boldsymbol{v} - \boldsymbol{u}_o) \geq S(\boldsymbol{u}_o - \boldsymbol{u})$. Applying the Korn inequality, we conclude that

$$\begin{aligned} c_1 \|\boldsymbol{u}\|_{W^1}^2 &\leq a(\boldsymbol{u}, \boldsymbol{u}) \leq a(\boldsymbol{u}, \boldsymbol{u} - \boldsymbol{u}_o) + S(\boldsymbol{u} - \boldsymbol{u}_o) \\ &\leq \|\boldsymbol{u}\|_{W^1} \|\boldsymbol{u}_o\|_{W^1} + \|\boldsymbol{F}\|_{[L^2(\Omega)]^2} \|\boldsymbol{u}\|_{W^1} + \|\boldsymbol{P}\|_{[L^2(\Gamma_\tau)]^2} \|\boldsymbol{u}\|_{W^1}, \end{aligned}$$

which completes the proof.

To prove the existence theorem of the contact problem with bounded friction, estimates of the admissible coefficient of friction are necessary (see Theorem 7.13). Further, we need the next lemma.

Lemma 7.14: A mapping $\Phi : \mathscr{F}_c^{kl} g_n^{kl} \to \mathscr{F}_c^{kl} \tau_n^{kl}(\boldsymbol{u})$ is continuous on $H^{-1/2}(\cup\Gamma_c^{kl})$ and $\tau_n^{kl}(\boldsymbol{u}) \in \mathscr{H}$ for every $g_n^{kl} \in \mathscr{H}$.
The proof is a modification of that of Jarušek (1983).

The next lemma gives the regularity of Φ.

Lemma 7.15: A mapping Φ acts from $H^{-1/2+\gamma}(\cup\Gamma_c^{kl})$ to $H^{-1/2+\gamma}(\cup\Gamma_c^{kl})$, $\gamma \in (0, \frac{1}{2})$ such that

$$\|\Phi(g_n^{kl})\|_{-1/2+\gamma} \leq c(\|\mathscr{F}_c^{kl}\|_\infty) \|g_n^{kl}\|_{-1/2+\gamma} + c_1(\boldsymbol{F}, \boldsymbol{P}, \boldsymbol{u}_o), \tag{7.159}$$

where $\|.\|_{-1/2+\gamma}$ is the norm in $H^{-1/2+\gamma}(\cup\Gamma_c^{kl})$, $\|.\|_\infty$ is the norm in $L^\infty(\cup\Gamma_c^{kl})$.
The proof is based on the use of certain renormation techniques by means of shifts in arguments in Equation 7.153, the Fourier transformation and the method of local coordinates.
For the proof see Jarušek (1983).

Finally, we have the following result.

Theorem 7.13: Let Ω be the domain with the boundary $\partial\Omega = \Gamma_\tau \cup \Gamma_u \cup (\cup\Gamma_c^{kl})$ defined above. Let $\kappa_{ij}(\boldsymbol{x})$, $c_{ijkl}(\boldsymbol{x})$ satisfy Equations 7.126 and 7.129. Let $T_o \in H^{-1/2}(\Gamma_\tau)$, $T_1 \in H^{-1/2}(\Gamma_u)$, $T_2^{kl} \in H^{-1/2}(\Gamma_c^{kl})$ and $\boldsymbol{u}_o \in W^1$ such that $\boldsymbol{u}_{o|cl(\partial\Omega\setminus\Gamma_u)} = 0$. Let $\boldsymbol{F} \in [L^2(\overline{\Omega})]^2$, $Q \in L^2(\overline{\Omega})$, $q_o \in L^2(\Gamma_\tau)$, $q_2^{kl} \in L^2(\Gamma_c^{kl})$, $\boldsymbol{P} \in [H^{-1/2}(\Gamma_\tau)]^2$ such that $[\boldsymbol{P}, \boldsymbol{v}] = 0$ for every $\boldsymbol{v} \in [H^{1/2}(\partial\Omega)]^2$ such that $\boldsymbol{v}_{|cl(\partial\Omega\setminus\Gamma_\tau)} = 0$. Let $\mathscr{F}_c^{kl} \in C^{0,1}(\cup\Gamma_c^{kl})$ have a compact support in $\cup\Gamma_c^{kl}$. Let

$$\|\mathscr{F}_c^{kl}\|_\infty (\min(\mu^k/(\lambda^k + 2\mu^k), \mu^l/(\lambda^l + 2\mu^l))^{1/4} \sim \min(c_s^k/c_p^k, c_s^l/c_p^l)^{1/2} \tag{7.160}$$

for the non-homogeneous isotropic case, where $c_s^2 = \mu\rho^{-1}$, $c_p^2 = (\lambda + 2\mu)\rho^{-1}$, and where λ and μ are Lamé coefficients, ρ the density, c_s and c_p the velocities of S and P waves, respectively, or

$$\|\mathscr{F}_c^{kl}\| < \min(a_o^k/(2A_o^k), a_o^l/(2A_o^l))^{1/2} \tag{7.161}$$

for the non-homogeneous anisotropic case. Then there exists at least one solution of the Signorini problem with friction for the quasi-coupled case in thermo-elasticity, describing the geodynamic problem $(\mathscr{P}_f)_v$.

Proof: The proof of the thermic part of this theorem is similar to that of Theorem 7.12. The proof of the elastic part is a modification of that of Nedoma (1987). Further, the coupling term $(\beta_{ij}(T - T_o))_{,j} \in L^2(\Omega)$ acts as a body force. If $\beta_{ij} \in C^1(\overline{\Omega})$ and $T, T_o \in H^1(\Omega)$, then $\beta_{ij}(T - T_o) \in H^1(\Omega)$ and then $(\beta_{ij}(T - T_o))_{,j} \in L^2(\Omega)$. (In Nedoma (1983) it is shown that if $c_{ijkl} \in L^\infty(\overline{\Omega})$, $\kappa_{ij} \in L^\infty(\overline{\Omega})$, $\beta_{ij} \in L^\infty(\overline{\Omega})$ then $(\beta_{ij}(T - T_o))_{,j} \in H^{-1}(\Omega)$). As the estimate of $\mathscr{F}_c^{kl}$ does not directly depend on the body and surface forces, it does not depend on the term $(\beta_{ij}(T - T_o))_{,j}$, either. Then similarly as in Nedoma (1987), using the estimation of Lemma 7.15, the above mentioned properties of Φ on $H^{-1/2}(\cup\Gamma_c^{kl})$, the Sobolev imbedding theorem and the reflexivity of the spaces involved, the weak continuity of Φ on $\mathscr{H} \cap H^{-1/2+\gamma}(\cup\Gamma_c^{kl})$ is proved. Further, a convex subset $M \subset \mathscr{H} \cap H^{-1/2+\gamma}(\cup\Gamma_c^{kl})$ which is mapped by Φ into itself is found, provided $c(\|\mathscr{F}_c^{kl}\|_\infty) < 1$, for which the estimate in Equation 7.160 or 7.161 must be fulfilled. Applying Tichonov's fixed point theorem we prove the existence of a fixed point of Φ, for which the appropriate $\boldsymbol{u}$ is a solution of the contact problem with friction considered.

Remark: Theorem 7.13 and especially the estimates of Equations 7.160 and 7.161 have practical significance in geophysics, above all in the analysis of the geodynamic processes taking place in the subduction regions as well as in the regions of collision of lithospheric plates and blocks. The meaning of these estimates is the following.

The expression $\mathscr{F}_c^{kl}\left|g_n^{kl}\right| = \mathscr{F}_c^{kl}\left|\tau_n^{kl}\right|$ represents the friction forces acting on the contact boundaries Γ_c^{kl}, which must be overridden in order to shift the lithospheric plates or blocks. The above estimates and the estimates based on the numerical analysis give us some information about the distribution of the stress field in the collision zone, and allow us interpret some observed anomalous geophysical fields discussed by the author for the Carpathians (see Nedoma (1980), (1982b)).

7.2.1.4. *Numerical solution*

For the numerical solution the finite element method, based on the Galerkin technique, will be used. Let us assume the case with the given friction $g_c^{kl} = \mathscr{F}_c^{kl}\left|g_n^{kl}\right| \geq 0$, $g_c^{kl} \in L^\infty(\Gamma_c^{kl})$ a.e. on $\cup\Gamma_c^{kl}$, where g_c^{kl} is the given frictional force. Then we define

$$j_{gn}(\boldsymbol{v}) = \left\langle g_c^{kl}, \left|\boldsymbol{v}_t^k - \boldsymbol{v}_t^l\right|\right\rangle = \int_{\cup\Gamma_c^{kl}} g_c^{kl}\left|\boldsymbol{v}_t^k - \boldsymbol{v}_t^l\right| \mathrm{d}s. \tag{7.162}$$

Let the given bounded domain $\Omega \subset R^2$ with the polygonal boundary $\partial\Omega$ be "triangulated", as in Section 7.1.6. Let $\{\mathscr{T}_h\}$ be a system of regular triangulations defined by a system of triangles $\underline{T}_h$ as in Section 7.1.6 with the end points $\overline{\Gamma}_u \cap \overline{\Gamma}_\tau$, $\overline{\Gamma}_u \cap \overline{\Gamma}_c^{kl}$, $\overline{\Gamma}_c^{kl} \cap \overline{\Gamma}_\tau$ coinciding with the vertices of the triangles $\underline{T}_h$.

Let ${}^1V_h$, V_h be the spaces of linear finite elements (or elements of higher order, but in our study we shall assume linear elements only), where

$${}^1V_h = \{w \mid w \in C(\overline{\Omega}), w_{|\underline{T}_h} \in P_1, w = 0 \text{ on } \Gamma_u \cup \Gamma_\tau, \quad \forall \underline{T}_h \in \mathscr{T}_h\},$$

$$V_h = \{\boldsymbol{v} \mid \boldsymbol{v} \in [C(\overline{\Omega})]^2, \boldsymbol{v}_{|\underline{T}_h} \in [P_1]^2, \boldsymbol{v} = 0 \text{ on } \Gamma_u, \quad \forall \underline{T}_h \in \mathscr{T}_h\},$$

are the spaces of all linear FEM approximations of virtual temperatures and displacements, respectively, and where P_1 is the space of linear polynomials. Further, define

$${}^1K_h = \{w \mid w \in {}^1V_h, w^k - w^l \leq T_2^{kl} \text{ on } \cup\Gamma_c^{kl}\} = {}^1V_h \cap {}^1K, \text{ hence } {}^1K_h \subset {}^1K \quad \forall h,$$

$$K_h = \{\boldsymbol{v} \mid \boldsymbol{v} \in V_h, \boldsymbol{v}_n^k - \boldsymbol{v}_n^l \leq 0 \text{ on } \cup\Gamma_c^{kl}\} = V_h \cap K, \text{ hence } K_h \subset K \quad \forall h,$$

as the sets of all linear FEM approximations of admissible temperatures and displacements, respectively.

Definition 7.6: We say that a pair of functions $(T_h, \boldsymbol{u}_h)$, $T_h \in {}^1K_h$, $\boldsymbol{u}_h \in K_h$ is a finite element approximation of the problem $(\mathscr{P}_f)_v$, if

$$b(T, w - T_h) \geq s(w - T_h) \quad \forall w \in {}^1K_h, \tag{7.163}$$

$$a(\boldsymbol{u}, \boldsymbol{v} - \boldsymbol{u}_h) + \left\langle g_c^{kl}, \left|\boldsymbol{v}_t^k - \boldsymbol{v}_t^l\right| - \left|(u_h^k)_t - (u_h^l)_t\right|\right\rangle \geq S(\boldsymbol{v} - \boldsymbol{u}_h) \quad \forall \boldsymbol{v} \in K_h. \tag{7.164}$$

Remark: As the problem is quasi-coupled, a finite element approximation as given in Equation 7.164 of the primary problem in Equation 7.153 is equivalent to the following formulation.

Find a pair of functions $(T_h, \boldsymbol{u}_h)$, $T_h \in {}^1K_h$, $\boldsymbol{u}_h \in K_k$ such that

$$l(T_h) \leq l(w), \quad \forall w \in {}^1K_h, \tag{7.165}$$

$$L(\boldsymbol{u}_h) \leq L(\boldsymbol{v}) \quad \forall \boldsymbol{v} \in K_h. \tag{7.166}$$

Theorem 7.14: There exists a unique solution of the finite element approximation in Equations 7.163 and 7.164 $\forall h, h \in (0, 1)$.

Proof: Since ${}^1K_h$ and K_h are non-empty, closed and convex subsets of $H^1(\Omega)$ and W^1, respectively, they are weakly closed. Thus the proof is similar to that of Theorem 7.13, in which the functional $j_{gn}(\boldsymbol{v}) = \langle g_c^{kl}, |\boldsymbol{v}_t^k - \boldsymbol{v}_t^l| \rangle$ is convex.

Lemma 7.16: For $T \in {}^1K$, $\boldsymbol{u} \in K$, $T_h \in {}^1K_h$, $\boldsymbol{u}_h \in K_h$ we have

$$\|T - T_h\|_1 \le c\,\{b(T_h - T, w_h - T) + b(T, w - T_h) + b(T, w_h - T) - (Q, w - T_h) - (Q, w_h - T)\}^{1/2} \quad \forall w \in {}^1K,\ w_h \in {}^1K_h,\ c = \text{ constant } > 0, \tag{7.167}$$

$$\|\boldsymbol{u} - \boldsymbol{u}_h\|_{W^1} \le c_o\{a(\boldsymbol{u}_h - \boldsymbol{u}, \boldsymbol{v}_h - \boldsymbol{u}) + a(\boldsymbol{u}, \boldsymbol{v} - \boldsymbol{u}_h) + a(\boldsymbol{u}, \boldsymbol{v}_h - \boldsymbol{u}) - (\boldsymbol{F}, \boldsymbol{v} - \boldsymbol{u}_h) - (\boldsymbol{F}, \boldsymbol{v}_h - \boldsymbol{u}) + j_{gn}(\boldsymbol{v}_h) - j_{gn}(\boldsymbol{u}) + j_{gn}(\boldsymbol{v}) - j_{gn}(\boldsymbol{u}_h)\}^{1/2} \quad \forall \boldsymbol{v} \in K,\ \boldsymbol{v}_h \in K_h,\ c_o = \text{ constant } > 0. \tag{7.168}$$

The proof is similar of that of Falk (1974), Nedoma (1983) and of Lemma 7.9 and Lemma 7.10 of Section 7.1.6.

Theorem 7.15: Let $\Gamma_c = \cup\Gamma_c^{kl}$ be polygonal. Let $T_2^{kl} \in H^2(\cup\Gamma_c^{kl}) \cap H^1(\cup\Gamma_c^{kl})$, $\forall k, l, T \in {}^1K \cap H^2(\Omega)$, $T_{|\cup\Gamma_c^{kl}} \in H^2(\cup\Gamma_c^{kl})$, $\boldsymbol{u} \in K \cap W^2$, $\vec{\tau}(\boldsymbol{u}) \in [L^\infty(\cup\Gamma_c^{kl})]^2$, $\boldsymbol{u} \in [W^{1,\infty}(\cup\Gamma_c^{kl})]^2$, $g_c^{kl} \in L^\infty(\cup\Gamma_c^{kl})$. Let ${}^1K_h \subset {}^1K$, $K_h \subset$K. Let the changes $u_n^k - u_n^l < 0 \to u_n^k - u_n^l = 0$ and $\boldsymbol{u}_t^k - \boldsymbol{u}_t^l = 0 \to \boldsymbol{u}_t^k - \boldsymbol{u}_t^l \neq 0$ occur at finitely many points of $\cup\Gamma_c^{kl}$ only. Then

$$\|T - T_h\|_1 = O(h), \tag{7.169}$$

$$\|\boldsymbol{u} - \boldsymbol{u}_h\|_1 = O(h). \tag{7.170}$$

Proof: To prove this theorem, the technique of the proofs of Theorems 2.6 and 2.10 of Nedoma (1983) and of Theorem 3.2 of Nedoma (1987) will be used. We have ${}^1K_h \subset {}^1K$, $K_h \subset K$, $\forall h \in (0, 1)$. Using Lemma 7.16, Green's theorem and the assumption that T and $\boldsymbol{u}$ are sufficiently regular we obtain

$$\|T - T_h\|_1 \le c\left\{\tfrac{1}{2}\varepsilon\,\|T - T_h\|_1^2 + \tfrac{1}{2}\varepsilon^{-1}\,\|T - w_h\|_1^2 + c_3\,\|T_h - T_1\|_1\,\|w_h - T\|_{L^2(\Omega)} + \int_{\cup\Gamma_c^{kl}} T_{,n}\,(w - T_h)\mathrm{d}s + c_2\,\|w_h - T\|_{L^2(\cup\Gamma_c^{kl})}\right\}^{1/2},\ \varepsilon > 0 \text{ arbitrary}, \tag{7.171}$$

$$\|\boldsymbol{u} - \boldsymbol{u}_h\|_{W^1} \le c_o\left\{\tfrac{1}{2}\varepsilon\,\|\boldsymbol{u}_h - \boldsymbol{u}\|_{W^1}^2 + \tfrac{1}{2}\varepsilon^{-1}\,\|\boldsymbol{v}_h - \boldsymbol{u}\|_{W^1}^2 + \int_{\partial\Omega} \tau_{ij}(\boldsymbol{u})(\boldsymbol{v}_h - \boldsymbol{u})_i\mathrm{d}s\right\}^{1/2},\quad \varepsilon > 0 \text{ arbitrary}. \tag{7.172}$$

To estimate the right hand side of Equation 7.171 we put $w_h = T_{LI}$, where $T_{LI} \in {}^1V_h$ is the Lagrange interpolation of T on the triangulation $\mathcal{T}_h$. As $(T_{LI}^k)_n - (T_{LI}^l)_n \le T_2^{kl}$ on $\cup\Gamma_c^{kl}$, we have $T_{LI} \in {}^1K$ and since $T_{LI} \in {}^1V_h$, then $T_{LI} \in {}^1K_h$. Thus

$$\|T_{LI} - T\|_1 \le c_r h\,\|T\|_2, \tag{7.173}$$

$$\left\|(T_{LI}^k)_n - (T_{LI}^l)_n - (T_n^k - T_n^l)\right\|_{L^2(\cup\Gamma_c^{kl})} \le c_s h^2 \sum \left\|T_n^k - T_n^l\right\|_{H^2(\cup\Gamma_c^{kl})}, \tag{7.174}$$

$$\|T_{LI} - T\|_{L^2(\Omega)} \le c_p h^2\,\|T\|_2. \tag{7.175}$$

As in Nedoma (1983) we find

$$\int_{\cup\Gamma_c^{kl}} T,_n (w - (T_h^k - T_h^l))\mathrm{d}s = O(h^4). \tag{7.176}$$

Similarly as in (Nedoma (1983)) we define

$$w = \sup(T_h^k - T_h^l, T_2^{kl}) \text{ on } \cup\Gamma_c^{kl}, w = 0 \text{ on } \partial\Omega\backslash \cup\Gamma_c^{kl}.$$

Then $w \in H^1(\partial\Omega)$, $w^k - w^l \le T_2^{kl}$ on $\cup\Gamma_c^{kl}$ and there exists a function $z \in H^1(\Omega)$ such that $z = w$ on $\partial\Omega$. Then $z \in {}^1K$, $T_h^k - T_h^l - w = 0$ for $T_h^k - T_h^l \le T_2^{kl}$ and $T_h^k - T_h^l - w = T_h^k - T_h^l - T_2^{kl}$ for $T_h^k - T_h^l > T_2^{kl}$. Let s_i be the vertices of the triangulation $\mathcal{T}_h$ on $\cup\Gamma_c^{kl}$. Then, as $T_h^k(s_i) - T_h^l(s_i) \le T_2^{kl}(s_i)$, $i = 1, \ldots, n$, we have $T_h^k - T_h^l \le (T_2^{kl})_{LI}$ on $\cup\Gamma_c^{kl}$, where $(T_2^{kl})_{LI}$ is the linear Lagrange interpolation of T_2^{kl} on $\cup\Gamma_c^{kl}$. Let

$${}^1\Gamma_c = \cup^1\Gamma_c^{kl} = \{\boldsymbol{x} \in \cup\Gamma_c^{kl} \mid T_h^k(\boldsymbol{x}) - T_h^l(\boldsymbol{x}) > T_2^{kl}\}.$$

Then

$$\int_{\cup\Gamma_c^{kl}} (w - (T_h^k - T_h^l))^2\mathrm{d}s = \int_{{}^1\Gamma_c} (T_2^{kl} - (T_h^k - T_h^l))^2\mathrm{d}s \le \int_{{}^1\Gamma_c} ((T_2^{kl})_{LI} - T_2^{kl})^2\mathrm{d}s = O(h^4),$$

as on ${}^1\Gamma_c$, $0 < T_h^k(\boldsymbol{x}) - T_h^l(\boldsymbol{x}) - T_2^{kl}(\boldsymbol{x}) \le ((T_2^{kl}(\boldsymbol{x}))_{LI} - T_2^{kl}(\boldsymbol{x}))$ holds, which proves Equation 7.176. Hence

$$\begin{aligned}\|T - T_h\|_1 &\le c\{\tfrac{1}{2}\varepsilon c_{r1} h \|T\|_2 + \tfrac{1}{2}\varepsilon^{-1} c_{r2} h \|T\|_2 + c_3 c_r h \|T\|_2 + O(h^4) \\ &\quad + c_2 c_s h^2 \sum \left\|T_n^k - T_n^l\right\|_{H^2(\cup\Gamma_c^{kl})}\} = O(h),\end{aligned} \tag{7.177}$$

which proves the first part of the theorem.

To estimate the right hand side of Equation 7.172 we set $\boldsymbol{v}_h = \boldsymbol{u}_{LI}$, where $\boldsymbol{u}_{LI} \in V_h$ is the Lagrange interpolation of $\boldsymbol{u}$ on the triangulation $\mathcal{T}_h$. As $(u_{LI}^k)_n - (u_{LI}^l)_n \le 0$ on $\cup\Gamma_c^{kl}$, $\boldsymbol{u}_{LI} \in K$. Since $\boldsymbol{u}_{LI} \in V_h$, we have $\boldsymbol{u}_{LI} \in K_h$. Thus

$$\|\boldsymbol{u}_{LI} - \boldsymbol{u}\|_{W^1} \le c_r h \|\boldsymbol{u}\|_{W^2},$$

$$\left\|(u_{LI}^k)_n - (u_{LI}^l)_n - (u_n^k - u_n^l)\right\|_{[L^2(\cup\Gamma_c^{kl})]^2} \le c_s h^2 \sum \left\|u_n^k - u_n^l\right\|_{[H^2(\cup\Gamma_c^{kl})]^2}.$$

In Equation 7.168 the terms $a(\boldsymbol{u}, \boldsymbol{v} - \boldsymbol{u}_h) - (\boldsymbol{F}, \boldsymbol{v} - \boldsymbol{u}_h)$ and $a(\boldsymbol{u}, \boldsymbol{v}_h - \boldsymbol{u}) - (\boldsymbol{F}, \boldsymbol{v}_h - \boldsymbol{u})$ are estimated by using Green's theorem and later by using a suitable choice of $\boldsymbol{v}_h \in K_h$, $\boldsymbol{v} \in K$. Thus applying Green's theorem we obtain

$$a(\boldsymbol{u}, \boldsymbol{v}_h - \boldsymbol{u}) - (\boldsymbol{F}, \boldsymbol{v}_h - \boldsymbol{u}) = \int_{\cup\Gamma_c^{kl}} \tau_n^{kl}(v_h - u)_n \mathrm{d}s + \int_{\cup\Gamma_c^{kl}} \vec{\tau}_t^{kl}(\boldsymbol{v}_h - \boldsymbol{u})_t \mathrm{d}s \le 0.$$

To estimate the integrals

$$J_1(\cup\Gamma_c^{kl}) = \int_{\cup\Gamma_c^{kl}} \tau_n^{kl}(v_h - u)_n \mathrm{d}s \text{ and } J_2(\cup\Gamma_c^{kl}) = \int_{\cup\Gamma_c^{kl}} \vec{\tau}_t^{kl}(\boldsymbol{v}_h - \boldsymbol{u})_t \mathrm{d}s$$

we assume that $\Gamma_c = \cup_i\Gamma_{ci} = \cup_i, \cup_{k,l}\Gamma_{ci}^{kl}$, where $\cup_i\Gamma_{ci}^{kl}$ approximate piecewise linearly the boundary $\Gamma_c = \cup\Gamma_c^{kl}$. To estimate the first integral we distinguish several cases:

— the case $u_n^k(\boldsymbol{x}) - u_n^l(\boldsymbol{x}) < 0$, $\boldsymbol{x} \in \overline{\Gamma}_{ci}^{kl}$. Then due to $(u_n^k - u_n^l)\tau_n^{kl} = 0$, the integral $J_1(\overline{\Gamma}_{ci}^{kl}) = 0$.

— the case $u_n^k(\boldsymbol{x}) - u_n^l(\boldsymbol{x}) = 0$ and $u_n^k(\boldsymbol{x}) - u_n^l(\boldsymbol{x}) \to u_n^k(\boldsymbol{x}) - u_n^l(\boldsymbol{x}) = 0$, $\boldsymbol{x} \in \overline{\Gamma}_{ci}^{kl}$. Let us put $\boldsymbol{v}_h = \boldsymbol{u}_{LI}$. Due to the properties of $\boldsymbol{u}_{LI}$ discussed above, either $(u_{LI}^k)_n - (u_{LI}^l)_n < 0$ and then $J_1(\overline{\Gamma}_{ci}^{kl}) = 0$

or $u_n^k(\boldsymbol{x}) - u_n^l(\boldsymbol{x}) < 0$ and then $\left|J_1(\overline{\Gamma}_{ci}^{kl})\right| \le \left\|(u_{LI}^k(\boldsymbol{x}))_n - (u_{LI}^l(\boldsymbol{x}))_n - (u_n^k(\boldsymbol{x}) - u_n^l(\boldsymbol{x}))\right\|_{L^\infty(\overline{\Gamma}_{ci}^{kl})}$ $\int_{\overline{\Gamma}_{ci}^{kl}} \left|\tau_n^{kl}\right| \mathrm{d}s \le c_{si}h^2$. The last inequalities hold because for $\boldsymbol{u} \in [W^{1,\infty}(\cup\Gamma_c^{kl})]^2$, and $\vec{\tau} \in [L^\infty(\cup\Gamma_c^{kl})]^2$, $u_n^k, u_n^l \in W^{1,\infty}(\Gamma_{ci}^{kl})$, $\tau_n^{kl} \in L^\infty(\overline{\Gamma}_{ci}^{kl})$, and because $(u_{LI}^k)_n - (u_{LI}^l)_n$ is the Lagrange interpolation of $u_n^k - u_n^l$. Hence $\left|J_1(\overline{\Gamma}_{ci}^{kl})\right| \le c_s h^2$.

Now we estimate $\left|\int_{\cup\Gamma_c^{kl}} \vec{\tau}_t^{kl}(\boldsymbol{u})(\boldsymbol{v}_h - \boldsymbol{u})_t \mathrm{d}s + j_{gn}(\boldsymbol{v}_h) - j_{gn}(\boldsymbol{u})\right|$. We have several cases:

— the case $\boldsymbol{u}_t^k(\boldsymbol{x}) - \boldsymbol{u}_t^l(\boldsymbol{x}) > 0$ for $\boldsymbol{x} \in \cup\Gamma_c^{kl}$. Due to Equation 7.164 we have

$$J_1(\cup\Gamma_c^{kl}) + J_2(\cup\Gamma_c^{kl}) + j_{gn}(\boldsymbol{v}) - j_{gn}(\boldsymbol{u}) \ge 0.$$

Further, $g_c^{kl}\left|\boldsymbol{u}_t^k - \boldsymbol{u}_t^l\right| + \vec{\tau}_t^{kl}(\boldsymbol{u}_t^k - \boldsymbol{u}_t^l) = 0$ a.e. on $\cup\Gamma_c^{kl}$. Hence $g_c^{kl} = -\tau_t^{kl}$ a.e. on $\cup\overline{\Gamma}_{ci}^{kl}$. Let us put $\boldsymbol{v}_h = \boldsymbol{u}_{LI}$. Thus $(\boldsymbol{u}_{LI}^k - \boldsymbol{u}_{LI}^l)_t = (\boldsymbol{u}_t^k - \boldsymbol{u}_t^l)_{LI}$ and $(\boldsymbol{u}_{LI}^k - \boldsymbol{u}_{LI}^l)_t(s_i)$, $s_i \in \cup\Gamma_{ci}^{kl}$, s_i are the points of the triangulation on $\cup\Gamma_c^{kl}$. Then $(\boldsymbol{u}_t^k - \boldsymbol{u}_t^l)_{LI} > 0$ on $\overline{\Gamma}_{ci}^{kl}$ and

$$\int_{\cup\Gamma_{ci}^{kl}} \{-g_c^{kl}[(\boldsymbol{u}_t^k - \boldsymbol{u}_t^l)_{LI} - (\boldsymbol{u}_t^k - \boldsymbol{u}_t^l)] - g_c^{kl}[(\boldsymbol{u}_t^k - \boldsymbol{u}_t^l)_{LI} - (\boldsymbol{u}_t^k - \boldsymbol{u}_t^l)]\}\mathrm{d}s = 0.$$

— the case $\boldsymbol{u}_t^k(\boldsymbol{x}) - \boldsymbol{u}_t^l(\boldsymbol{x}) = 0$ for $\boldsymbol{x} \in \cup\Gamma_c^{kl}$. As $(\boldsymbol{u}_t^k - \boldsymbol{u}_t^l)_{LI} = 0$ on $\cup\overline{\Gamma}_{ci}^{kl}$, we have

$$\int_{\cup\Gamma_{ci}^{kl}} \{\vec{\tau}_t^{kl}(\boldsymbol{u})[(\boldsymbol{u}_t^k - \boldsymbol{u}_t^l)_{LI} - (\boldsymbol{u}_t^k - \boldsymbol{u}_t^l)] + g_c^{kl}[\left|(\boldsymbol{u}_t^k - \boldsymbol{u}_t^l)_{LI}\right| + \left|\boldsymbol{u}_t^k - \boldsymbol{u}_t^l\right|]\}\mathrm{d}s = 0.$$

— the case $\boldsymbol{u}_t^k(\boldsymbol{x}) - \boldsymbol{u}_t^l(\boldsymbol{x}) < 0$ for $\boldsymbol{x} \in \cup\Gamma_c^{kl}$. Since $\boldsymbol{u}_t^k - \boldsymbol{u}_t^l < 0$ on $\cup\Gamma_c^{kl}$, we have $\left|\boldsymbol{u}_t^k - \boldsymbol{u}_t^l\right| = -(\boldsymbol{u}_t^k - \boldsymbol{u}_t^l)$ and $(\boldsymbol{u}_t^k - \boldsymbol{u}_t^l)_{LI} < 0$. Thus $g_c^{kl} = \tau_t^{kl}$ a.e. on $\cup\overline{\Gamma}_{ci}^{kl}$. Let us set $\boldsymbol{v}_h = \boldsymbol{u}_{LI}$. Then

$$\int_{\cup\overline{\Gamma}_{ci}^{kl}} \{g_c^{kl}[(\boldsymbol{u}_t^k - \boldsymbol{u}_t^l)_{LI} - (\boldsymbol{u}_t^k - \boldsymbol{u}_t^l)] + g_c^{kl}[-(\boldsymbol{u}_t^k - \boldsymbol{u}_t^l)_{LI} + (\boldsymbol{u}_t^k - \boldsymbol{u}_t^l)]\}\mathrm{d}s = 0.$$

— the case when $\boldsymbol{u}_t^k(\boldsymbol{x}) - \boldsymbol{u}_t^l(\boldsymbol{x}) = 0$ changes to $\boldsymbol{u}_t^k(\boldsymbol{x}) - \boldsymbol{u}_t^l(\boldsymbol{x}) \ne 0$ for $\boldsymbol{x} \in \cup\Gamma_c^{kl}$. Let $\boldsymbol{v}_h = \boldsymbol{u}_{LI}$. Since $\boldsymbol{u} \in [W^{1,\infty}(\cup\Gamma_c^{kl})]^2$, $\vec{\tau} \in [L^\infty(\cup\Gamma_c^{kl})]^2$, $g_c^{kl} \in L^\infty(\cup\Gamma_c^{kl})$, we have $\boldsymbol{u}_t^k - \boldsymbol{u}_t^l \in [W^{1,\infty}(\cup\overline{\Gamma}_{ci}^{kl})]^2$ and $\vec{\tau}_t^{kl} \in [L^\infty(\cup\overline{\Gamma}_{ci}^{kl})]^2$.
Then

$$\int_{\cup\Gamma_{ci}^{kl}} \left\{\vec{\tau}_t^{kl}(\boldsymbol{u})[(\boldsymbol{u}_t^k - \boldsymbol{u}_t^l)_{LI} - (\boldsymbol{u}_t^k - \boldsymbol{u}_t^l)] + g_c^{kl}\left[\left|(\boldsymbol{u}_t^k - \boldsymbol{u}_t^l)_{LI}\right| - \left|\boldsymbol{u}_t^k - \boldsymbol{u}_t^l\right|\right]\right\}\mathrm{d}s$$

$$\le \left\|(\boldsymbol{u}_t^k - \boldsymbol{u}_t^l)_{LI} - (\boldsymbol{u}_t^k - \boldsymbol{u}_t^l)\right\|_{L^\infty(\cup\overline{\Gamma}_{ci})} \int_{\cup\Gamma_{ci}} \left(\left|\vec{\tau}_t^{kl}(\boldsymbol{u})\right| + g_c^{kl}\right)\mathrm{d}s \le c_{ri}h^2.$$

Thus

$$\left|\int_{\cup\Gamma_c^{kl}} \vec{\tau}_t^{kl}(\boldsymbol{u})[(\boldsymbol{u}_t^k - \boldsymbol{u}_t^l)_{LI} - (\boldsymbol{u}_t^k - \boldsymbol{u}_t^l)]\mathrm{d}s + j_{gn}(\boldsymbol{u}_{LI}) - j_{gn}(\boldsymbol{u})\right| \le c_r h^2.$$

Finally, we obtain

$$\|\boldsymbol{u} - \boldsymbol{u}_h\|_{W^1} \le c_o\{\varepsilon h^2\|\boldsymbol{u}\|_{W^1}^2 + \varepsilon^{-1}h^2\|\boldsymbol{u}\|_{W^1}^2 + c_s h^2 + c_r h^2\}^{1/2} = O(h), \tag{7.178}$$

which completes the proof.

7.2.1.5. Algorithm

Since the problem is quasi-coupled the algorithm is divided into two parts. The first, for the thermal part, is based on the technique of quadratic programming (i.e. minimization of a quadratic functional with constraints) for a scalar variable. The second, for the elastic contact problem with friction, is based on the numerical approximation of a saddle point and on the technique of quadratic programming for a vector variable.

The thermal problem leads to finding a minimum of the functional $l(w)$ on the set ${}^1K_h$, i.e.

$$l(T_h) = \min_{w \in {}^1K_h} l(w). \tag{7.179}$$

The problem is equivalent to that of finding a minimizer of quadratic functional $g(\boldsymbol{x})$ of n variables with linear constraints of the form

$$g(\boldsymbol{x}) = \tfrac{1}{2}\boldsymbol{x}^T\boldsymbol{K}\boldsymbol{x} + \boldsymbol{Q}^T\boldsymbol{x}, \tag{7.180}$$

$$\mathscr{A}\boldsymbol{x} \le \boldsymbol{c}. \tag{7.181}$$

From the finite element analysis it is known that

$$l(T_h) = \sum_{k=1}^{n} \tfrac{1}{2}\vec{\Delta}_k^T \underline{\boldsymbol{K}}_{hk} \vec{\Delta}_k - \vec{\Delta}_k^T \boldsymbol{Q}_k = \tfrac{1}{2}\vec{\Delta}^T \underline{\boldsymbol{K}}_h \vec{\Delta} - \vec{\Delta}^T \boldsymbol{Q}_h, \tag{7.182}$$

where $\underline{\boldsymbol{K}}_h$ and $\boldsymbol{Q}_h$ are the conduction matrix and the vector generated by thermal sources. The matrix $\underline{\boldsymbol{K}}_h$ is positive definite. The matrix of constraints $\mathscr{A}(\dim m \times n)$ and the vector $\boldsymbol{c}$ are generated by the condition $T^k - T^l \le T_2^{kl}$ on Γ_c^{kl} in the definition of the set of admissible temperatures ${}^1K_h$. Here n represents the number of points of triangulation in which the temperature is not prescribed and m is the number of points of triangulation on the contact boundary $\cup\Gamma_c^{kl}$. The method of minimization of the quadratic functional solving our problem is discussed in Daniel (1971), Luenberger (1973), (1984), Pschenichny and Danilin (1978), Fletcher (1981) and will also be discussed in the last section of this chapter.

The usual technique for the 2D isotropic case (for which $\kappa_{ij} = \kappa\delta_{ij}$) gives

$$\int_{T_{hk}} \kappa[(\partial T/\partial x_1)^2 + (\partial T/\partial x_2)^2]\mathrm{d}\boldsymbol{x} = \vec{\Delta}_k^T \underline{\boldsymbol{K}}_k \vec{\Delta}_k, \tag{7.183}$$

where $\vec{\Delta}_k = (T_i, T_j, T_m)^T$, $T_l (l = i, j, m)$ represents the value of T in the vertices i, j, m of the triangle $\underline{T}_{hk}$ with coordinates (x_l, y_l), $\underline{\boldsymbol{K}}_{hk}$ is a conduction matrix due to $\underline{T}_{hk}$ which is given by

$$\underline{\boldsymbol{K}}_{hk} = \frac{1}{2}|J|^{-2}\kappa \begin{pmatrix} b_i^2 + c_i^2, & b_ib_j + c_ic_j, & b_ib_m + c_ic_m \\ b_ib_j + c_ic_j, & b_j^2 + c_j^2, & b_jb_m + c_jc_m \\ b_ib_m + c_ic_m, & b_jb_m + c_jc_m, & b_m^2 + c_m^2 \end{pmatrix} \tag{7.184}$$

where $|J| = c_mb_j - c_jb_m = \pm 2\underline{T}_{ijm}^h$, where $\underline{T}_{ijm}^h$ is the area of the triangle $\underline{T}_{hk}$, and

$$a_i = x_jy_m - x_my_i, \quad b_i = y_j - y_m, \quad c_i = x_m - x_j,$$

$$a_j = x_m y_i - x_i y_m, \quad b_j = y_m - y_i, \quad c_j = x_i - x_m,$$
$$a_m = x_i y_j - x_j y_i, \quad b_m = y_i - y_j, \quad c_m = x_j - x_i.$$

Further,

$$\int_{\underline{T}_{hk}} QT\mathrm{d}\boldsymbol{x} = \Delta_k^T \boldsymbol{Q}_{hk}, \tag{7.185}$$

where

$$\boldsymbol{Q}_{hk} = J^{-1}\begin{pmatrix} a_i I_1 + b_i I_2 + c_i I_3 \\ a_j I_1 + b_j I_2 + c_j I_3 \\ a_m I_1 + b_m I_2 + c_m I_3 \end{pmatrix} \tag{7.186}$$

where

$$I_1 = \int_{\underline{T}_{hk}} Q_h(x_1, x_2)\mathrm{d}\boldsymbol{x} \doteq \frac{1}{2}\,|J|\,\overline{Q}^h,$$
$$I_2 = \int_{\underline{T}_{hk}} x_1 Q_h(x_1, x_2)\mathrm{d}\boldsymbol{x} \doteq \frac{1}{6}\,|J|\,(x_i + x_j + x_m)\overline{Q}^h,$$
$$I_3 = \int_{\underline{T}_{hk}} x_2 Q_h(x_1, x_2)\mathrm{d}\boldsymbol{x} \doteq \frac{1}{6}\,|J|\,(y_i + y_j + y_m)\overline{Q}^h.$$

The second part of the problem, the elastic contact problem with friction, is based on the numerical approximation of a saddle point (see, for example, Ekeland and Temam (1976), Nedoma (1987) as well as Chapter 2). By an approximate saddle point is meant a point $(\boldsymbol{u}_{sh}, \lambda_h) \in K_h \times \Lambda \subset K \times \Lambda$ of a functional $\mathscr{L}$, the so-called Lagrangian, on $K_h \times \Lambda$, if a saddle point $(\boldsymbol{u}_s, \lambda) \in K \times \Lambda$ exists and if $\mathscr{L}(\boldsymbol{u}_{sh}, \mu) \leq \mathscr{L}(\boldsymbol{u}_{sh}, \lambda_h) \leq \mathscr{L}(\boldsymbol{v}, \lambda_h)$ for all pairs of functions $(\boldsymbol{v}, \mu) \in K_h \times \Lambda$. The problem in Equation 7.158 is equivalent to the following problem.

Find $\boldsymbol{u}_h \in K_h$ such that

$$L(\boldsymbol{u}_h) = \min_{\boldsymbol{v}\in K_h}\sup_{\lambda\in\Lambda} \mathscr{L}(\boldsymbol{v}, \mu) = \min_{\boldsymbol{v}\in K_h}\sup_{\lambda\in\Lambda}\left\{L_o(v) + \int_{\cup\Gamma_c^{kl}} \mu^{kl} g_c^{kl}(\boldsymbol{v}_t^k - \boldsymbol{v}_t^l)\mathrm{d}s\right\}, \tag{7.187}$$

where $\Lambda = \cup\left\{\mu^{kl} \mid \mu^{kl} \in L^2(\Gamma_c^{kl}), \left|\mu^{kl}\right| \leq 1 \text{ a.e. on } \Gamma_c^{kl}\right\}$, $\mathscr{L}$ is the Lagrangian and where the non-differentiable functional $j(\boldsymbol{v})$ is replaced by

$$j(\boldsymbol{v}) = \int_{\cup\Gamma_c^{kl}} g_c^{kl}\left|\boldsymbol{v}_t^k - \boldsymbol{v}_t^l\right|\mathrm{d}s = \int_{\cup\Gamma_c^{kl}} \mu^{kl} g_c^{kl}(\boldsymbol{v}_t^k - \boldsymbol{v}_t^l)\mathrm{d}s, \tag{7.188}$$

where μ^{kl} is the Lagrange's multiplier.

It can be shown that the approximate saddle point exists and is unique (Hlaváček et al. (1982), Nedoma (1987)). As $(\boldsymbol{u}_{sh}, \lambda_h)$ is a saddle point of $\mathscr{L}$ on $K_h \times \Lambda$, then we have

$$L_o(\boldsymbol{u}_{sh}) + \int_{\cup\Gamma_c^{kl}} \mu^{kl} g_c^{kl}((\boldsymbol{u}_{sh}^k)_t - (\boldsymbol{u}_{sh}^l)_t)\mathrm{d}s$$
$$\leq L_o(\boldsymbol{u}_{sh}) + \int_{\cup\Gamma_c^{kl}} \lambda_h^{kl} g_c^{kl}((\boldsymbol{u}_{sh}^k)_t - (\boldsymbol{u}_{sh}^l)_t)\mathrm{d}s \text{ for all } \mu = (\mu^{kl}) \in \Lambda.$$

Hence

$$\int_{\cup\Gamma_c^{kl}} (\mu^{kl} - \lambda_h^{kl}) g_c^{kl}((\boldsymbol{u}_{sh}^k)_t - (\boldsymbol{u}_{sh}^l)_t)\mathrm{d}s \leq 0 \text{ for all } \mu \in \Lambda$$

and

$$\int_{\cup\Gamma_c^{kl}} ((\lambda_h^{kl} + \rho g_c^{kl}((\boldsymbol{u}_{sh}^k)_t - (\boldsymbol{u}_{sh}^l)_t)) - \lambda_h^{kl})(\mu^{kl} - \lambda_h^{kl})\mathrm{d}s \leq 0$$

$$\text{for all } \rho > 0 \text{ and all } \lambda \in \Lambda.$$

Thus

$$\lambda_h^{kl} = P(\lambda_h^{kl} + \rho g_c^{kl}((\boldsymbol{u}_h^k)_t - (\boldsymbol{u}_h^l)_t)), \quad \rho > 0, \tag{7.189}$$

where P is a projection of $L^2(\Gamma_c^{kl})$ onto Λ. Due to Proposition 1.7 of Ekeland and Temam (1976)

$$L_o(\boldsymbol{u}_{sh}) + \int_{\cup\Gamma_c^{kl}} \lambda_h^{kl} g_c^{kl}((\boldsymbol{u}_{sh}^k)_t - (\boldsymbol{u}_{sh}^l)_t)\mathrm{d}s$$
$$\leq L_o(\boldsymbol{v}) + \int_{\cup\Gamma_c^{kl}} \lambda^{kl} g_c^{kl}((\boldsymbol{v}_h^k)_t - (\boldsymbol{v}_h^l)_t)\mathrm{d}s \text{ for all } \boldsymbol{v} \in K_h. \tag{7.190}$$

The inequalities of Equations 7.190 and 7.189 give an idea for the numerical solution of the problem, which can be realized by the Uzawa's or the Arrow-Hurwicz's algorithms, respectively. For our numerical realization the Uzawa's algorithm was used, i.e. the following algorithm was used.

Let us choose $\lambda_h^o \in \Lambda$, e.g. $\lambda_h^o = 0$. Let $\lambda_h^i \in \Lambda$ be known, then we find $\boldsymbol{u}_{hi} \in K_h$ solving the following problem

$$\min_{\boldsymbol{v} \in K_h} \left\{ L_o(\boldsymbol{v}) + \int_{\cup\Gamma_c^{kl}} \lambda_h^{i,kl} g_c^{kl}(\boldsymbol{v}_t^k - \boldsymbol{v}_t^l)\mathrm{d}s \right\} \tag{7.191}$$

and then we find $\lambda_h^{i+1,kl}$ from the relation

$$\lambda_h^{i+1,kl} = P(\lambda_h^{i,kl} + \rho_i g_c^{kl}((\boldsymbol{u}_{hi}^k)_t - (\boldsymbol{u}_{hi}^l)_t)), \tag{7.192}$$

where P is a projection $P : L^2(\Gamma_c^{kl}) \to \Lambda$ defined as follows: For $\mu \in L^2(\Gamma_c^{kl})$, $y \in \Gamma_c^{kl}$,

$$P\mu(y) = \mu(y) \text{ for } |\mu(y)| \leq 1,$$
$$P\mu(y) = 1 \text{ for } \mu(y) > 1,$$
$$P\mu(y) = -1 \text{ for } \mu(y) < -1,$$

$\rho > 0$, $0 < \rho_1 \leq \rho_i \leq \rho_2$, where ρ_1 and ρ_2 are sufficiently small numbers which follow from the convergence theorem of Uzawa's algorithm (Ekeland and Temam (1976)).

The finite element approximation of Equation 7.158 leads to minimizing the functional $L(\boldsymbol{v}_h)$ on the set K_h. From the theory (Ekeland and Temam (1976)) it is known that this problem is equivalent to the problem of finding the minimum of the quadratic functional

$$f(\boldsymbol{w}) = \tfrac{1}{2}\boldsymbol{w}^T B_h \boldsymbol{w} + \boldsymbol{b}^T \boldsymbol{w} \tag{7.193}$$

with linear constraints

$$A\boldsymbol{w} \le \boldsymbol{d}. \tag{7.194}$$

In our case $f(\boldsymbol{w})$ is the functional $L(\boldsymbol{v}_h)$, B_h is the stiff matrix (dim $n \times n$), which is positive definite and generated by the term $\frac{1}{2}\int_\Omega c_{ijkl}(\boldsymbol{x})e_{ij}(\boldsymbol{v})e_{kl}(\boldsymbol{v})\mathrm{d}x$, $\boldsymbol{b}$ (dim n) is the vector generated by the body and surface forces, i.e. by $\int_\Omega F_i v_i \mathrm{d}\boldsymbol{x} + \int_{\Gamma_\tau} P_i v_i \mathrm{d}s$, as well as the friction forces $\int_{\cup\Gamma_c^{kl}} g_c^{kl}\left(\left|\boldsymbol{v}_t^k\right| - \left|\boldsymbol{v}_t^l\right|\right)\mathrm{d}s$ and it changes due to the friction forces in every step of the Uzawa's algorithm, and A is the matrix of constraints (dim $m \times n$) generated by the condition $u_n^k - u_n^l \le 0$ on Γ_c^{kl} in the definition of the admissible set of displacements K_h. Here n represents double the number of points of triangulation in which the displacement is not prescribed and m is the number of points of triangulation on the contact boundary Γ_c^{kl}. The method of minimization of the quadratic functional solving our problem is discussed in Pschenichny and Danilin (1978)*, where $\boldsymbol{d} = 0$ due to $u_n^k - u_n^l \le 0$ on Γ_c^{kl} and will also be discussed in the last section of this chapter. Denote by $s_1, \ldots, s_m$ the mesh points on $\cup\Gamma_c^{kl}$, and by $\boldsymbol{n}_1, \ldots, \boldsymbol{n}_m$ the outward unit normal to $\cup\Gamma_c^{kl}$ at mesh points $s_1, \ldots, s_m$. The contact boundary $\cup\Gamma_c^{kl}$ is approximated by the final number of segments. Let the mesh points s_j and s_k lie on two neighbouring segments and let the mesh point s_l be their common outward unit normal to Γ_c^{kl}, then we define $n_i^{(i)} = \frac{1}{2}\left(n_i^{(j)} + n_i^{(k)}\right)$ for $i = 1, 2$ (for the $2D$ case). The body forces are generated by the gravity as well as by the thermal stresses.

To derive the matrix B_h the usual technique was used. The stiff matrix B_h is generated by the term

$$\frac{1}{2}\int_\Omega c_{ijkl}e_{ij}(\boldsymbol{v})e_{kl}(\boldsymbol{v})\mathrm{d}\boldsymbol{x} = \sum_{T_{hi}} \frac{1}{2}\int_{T_{hi}} c_{ijkl}e_{ij}(\boldsymbol{v})e_{kl}(\boldsymbol{v})\mathrm{d}\boldsymbol{x} = \boldsymbol{w}^T B_h \boldsymbol{w}.$$

It can be shown that the matrix B_{hi} is created by the elementary 6×6 matrices B_{hi}. Now (and if we write $x = x_1$, $y \equiv x_2$)

$$\int_{T_{hi}} c_{ijkl}e_{ij}e_{kl}\mathrm{d}x\mathrm{d}y = J\int_{\underline{T}_o} J^{-2}\boldsymbol{w}^T C\boldsymbol{w}\,\mathrm{d}\xi\mathrm{d}\zeta = \frac{1}{2}J^{-1}\boldsymbol{w}^T C\boldsymbol{w} = \boldsymbol{w}^T \underline{B}_{hi}\boldsymbol{w},$$

where $\underline{T}_o$ is the unit triangle defined by the vertices (0,0), (0,1), (1,0). The transformation between T_{hi} and $\underline{T}_o$ is defined as $x = x_1 + \overline{x}_2\xi + \overline{x}_3\zeta$, $y = y_1 + \overline{y}_2\xi + \overline{y}_3\zeta$, where $\overline{x}_2 = x_2 - x_1$, $\overline{y}_2 = y_2 - y_1$, $\overline{x}_3 = x_3 - x_1$, $\overline{y}_3 = y_3 - y_1$ and where (x_i, y_i), $i = 1, 2, 3$, are the coordinates of the vertices of the triangle T_{hi}. Further,

$$J = \left|\frac{D(x, y)}{D(\xi, \zeta)}\right| = \begin{vmatrix} \overline{x}_2, \overline{x}_3 \\ \overline{y}_2, \overline{y}_3 \end{vmatrix} = \overline{x}_2\overline{y}_3 - \overline{x}_3\overline{y}_2 \text{ and } \underline{B}_{hi} = \frac{1}{2}J^{-1}E(1+\sigma)^{-1}\underline{C},$$

where $\underline{C} = (\underline{c}_{ij})$, E is Young's modulus and σ is Poisson's ratio.

* The method of Axelsson and Kaporin (1995) can be also used.

Let us denote $s = (1-\sigma)/(1-2\sigma)$, $r = \frac{1}{2} + \sigma$, $x_{13} = x_1 - x_3$, $x_{23} = x_2 - x_3$, $x_{12} = x_1 - x_2$, $y_{13} = y_1 - y_3$, $y_{23} = y_2 - y_3$, $y_{12} = y_1 - y_2$, then

$$\begin{aligned}
&\underline{c}_{11} = sy_{13}^2 + \tfrac{1}{2}x_{13}^2, \quad \underline{c}_{12} = 1 - rx_{23}y_{23}, \quad \underline{c}_{13} = -sy_{13}y_{23} - \tfrac{1}{2}x_{13}x_{23},\\
&\underline{c}_{14} = \sigma x_{13}y_{23} + \tfrac{1}{2}x_{23}y_{23}, \quad \underline{c}_{15} = sy_{12}y_{23} + \tfrac{1}{2}x_{12}x_{23},\\
&\underline{c}_{16} = -\sigma x_{12}y_{23} - \tfrac{1}{2}x_{23}y_{12},\\
&\underline{c}_{22} = sx_{23}^2 + \tfrac{1}{2}y_{23}^2, \quad \underline{c}_{23} = \sigma x_{23}y_{13} + \tfrac{1}{2}x_{13}y_{23}, \quad \underline{c}_{24} = -sx_{13}x_{23} - \tfrac{1}{2}y_{13}y_{23},\\
&\underline{c}_{25} = -\sigma x_{23}y_{12} - \tfrac{1}{2}x_{12}y_{13}, \quad \underline{c}_{26} = sx_{12}x_{23} + \tfrac{1}{2}y_{12}y_{23},\\
&\underline{c}_{33} = sy_{13}^2 + \tfrac{1}{2}x_{13}^2, \quad \underline{c}_{34} = -rx_{13}y_{13}, \quad \underline{c}_{35} = sy_{12}y_{13} - \tfrac{1}{2}x_{12}x_{13},\\
&\underline{c}_{36} = \sigma x_{12}y_{13} - \tfrac{1}{2}x_{13}y_{12},\\
&\underline{c}_{44} = sy_{13}^2 + \tfrac{1}{2}x_{13}^2, \quad \underline{c}_{45} = \sigma x_{13}y_{12} + \tfrac{1}{2}x_{12}y_{13}, \quad \underline{c}_{46} = -sx_{12}x_{13} - \tfrac{1}{2}y_{12}y_{13},\\
&\underline{c}_{55} = sy_{12}^2 + \tfrac{1}{2}x_{12}^2, \quad \underline{c}_{56} = rx_{12}y_{12},\\
&\underline{c}_{66} = sx_{12}^2 + \tfrac{1}{2}y_{12}^2,\\
&\underline{c}_{ij} = 0 \text{ for } i > j.
\end{aligned}$$

To calculate $\int_{T_{hi}}(F_1u_1 + F_2u_2)\mathrm{d}\boldsymbol{x}$ and the effect of the frictional and loading forces we use the standard technique discussed in the previous section.

7.2.2. Quasi-Coupled Plate Tectonic Models based on Contact Problems with Friction: the Semi-Coercive Case

7.2.2.1. Introduction

The problem investigated in this section represents simulation of geodynamic processes in collision and subduction zones as well as in regions where great structures are situated, simulation of influences evoked by the geodynamic processes on a great structure, like the radioactive waste disposal system (RWDS) as well as influences of high temperatures generated in a RWDS at the catastrophic situation onto the neighbouring rock massif. The model problem can be investigated from the point of view of $2D$ quasi-coupled thermo-elasticity as the quasi-dynamic case.

We shall deal with the quasi-dynamic problem solved in Nedoma (1994) consisting of the heat conduction equation

$$\frac{\partial}{\partial x_i}\left(\kappa_{ij}^{\iota}(\boldsymbol{x})\frac{\partial T^{\iota}}{\partial x_i}\right) + W^{\iota}(\boldsymbol{x}) = 0 \quad \text{in} \quad \Omega^{\iota}, \quad i, j = 1, 2, \iota = 1, \ldots, s \tag{7.195}$$

and the equilibrium equation

$$\frac{\partial}{\partial x_j}(c_{ijkl}(\boldsymbol{x})e_{kl}(\boldsymbol{u}) - \beta_{ij}(\boldsymbol{x})(T - T_o)) + F_i(\boldsymbol{x}) = 0 \text{ in } \Omega^{\iota}, \quad i, j = 1, 2, \ \iota = 1, \ldots, s, \tag{7.196}$$

where $\Omega = \cup_{\iota=1}^{S} \Omega^{\iota}$ is a plane domain with Lipschitz boundary $\partial\Omega = \cup_{\iota=1}^{S} \partial\Omega^{\iota}$ occupied by an elastic body (i.e. by subduction and/or collision zones, by a rock massif including the high level radioactive waste repository or including the deep mine, etc.). The boundary consists of three parts Γ_τ, Γ_u and Γ_c, $\partial\Omega = \overline{\Gamma}_\tau \cup \overline{\Gamma}_u \cup \overline{\Gamma}_c \cup \mathscr{R}$, $\Gamma_\tau = {}^1\overline{\Gamma}_\tau \cup {}^2\overline{\Gamma}_\tau$, $\overline{\Gamma}_u = \cup_\iota \overline{\Gamma}_u^\iota$, $\overline{\Gamma}_c = \cup_{k,l} \overline{\Gamma}_c^{kl}$. Moreover, we denote by T a temperature and by $\boldsymbol{u} = (u_1, u_2)$ a displacement vector.

We consider the following boundary conditions:

$$\left.\begin{aligned} &\kappa_{ij}^{\iota} \frac{\partial T^{\iota}}{\partial x_j} n_i^{\iota} = 0 \qquad (7.197)\\ &\tau_{ij}^{\iota} n_j^{\iota} = P_i^{\iota} \qquad (7.198)\end{aligned}\right\} \text{ on } \cup_\iota {}^1\Gamma_\tau^\iota, \quad \iota = 1, \ldots, s, \quad i, j = 1, 2,$$

$$\left.\begin{aligned} &T^{\iota} = T_1^{\iota} \qquad (7.199)\\ &\tau_{ij}^{\iota} n_j^{\iota} = 0 \qquad (7.200)\end{aligned}\right\} \text{ on } \cup_\iota {}^2\Gamma_\tau^\iota, \quad \iota = 1, \ldots, s, \quad i, j = 1, 2,$$

$$\left.\begin{aligned} &T^k = T^l \qquad (7.201)\\ &\kappa_{ij} \frac{\partial T}{\partial x_j} n_i \bigg|_{(k)} = \kappa_{ij} \frac{\partial T}{\partial x_j} n_i \bigg|_{(l)} \qquad (7.202)\end{aligned}\right\} \text{ on } \cup_{k,l} \Gamma_c^{kl}$$

$$\left.\begin{aligned} &u_n^k - u_n^l \le 0 \qquad (7.203)\\ &\tau_n^k = -\tau_n^l \le 0 \qquad (7.204)\\ &\tau_n^k (u_n^k - u_n^l) = 0 \qquad (7.205)\end{aligned}\right\} \text{ on } \cup_{k,l} \Gamma_c^{kl}$$

$$\text{if } u_n^k - u_n^l = 0, \qquad (7.206)$$

$$\text{then } \left|\vec{\tau}_t^k\right| \le \mathscr{F}_c^{kl} \left|\tau_n^k\right| \equiv g_c^{kl} \quad \text{on} \quad \cup_{k,l} \Gamma_c^{kl} \qquad (7.207)$$

and if

$$\left|\vec{\tau}_t^k\right| < g_c^{kl} \Rightarrow \boldsymbol{u}_t^k - \boldsymbol{u}_t^l = 0, \qquad (7.208)$$

if

$$\left|\vec{\tau}_t^k\right| = g_c^{kl} \Rightarrow \exists \quad \lambda \ge 0 \quad \text{such that} \quad \boldsymbol{u}_t^k - \boldsymbol{u}_t^l = -\lambda \vec{\tau}_t^k, \qquad (7.209)$$

$$\left.\begin{aligned} &\kappa_{ij}^{\iota} \frac{\partial T^{\iota}}{\partial x_j} n_i^{\iota} = 0 \qquad (7.210)\\ &u_i^{\iota} = u_{oi}^{\iota} \qquad (7.211)\end{aligned}\right\} \text{ on } \cup_\iota \Gamma_u^\iota, \iota \in \langle 1, r-1\rangle \cup \langle r+1, s\rangle, \ i, j = 1, 2,$$

where u_n^ι, $\boldsymbol{u}_t^\iota$ are the normal and tangential components of the displacement vector $\boldsymbol{u}$ and $\tau_n^\iota(\boldsymbol{u}^\iota) = \tau_{ij}^\iota(\boldsymbol{u}^\iota) n_i^\iota n_j^\iota$, $\vec{\tau}_t^\iota = \vec{\tau}^\iota - \tau_n^\iota \boldsymbol{n}^\iota$ are the normal and tangential components of the stress vector $\tau_i = \tau_{ij} n_i$, τ_{ij} is the stress tensor and e_{ij} is the small strain tensor defined by

$$\tau_{ij}^\iota(\boldsymbol{u}^\iota) = c_{ijkl}^\iota e_{kl}(\boldsymbol{u}^\iota) - \beta_{ij}^\iota (T^\iota - T_o^\iota), \quad i, j, k, l = 1, 2. \qquad (7.212)$$

$$e_{ij}(\boldsymbol{u}^\iota) = \frac{1}{2}\left(\frac{\partial u_i^\iota}{\partial x_j} + \frac{\partial u_j^\iota}{\partial x_i}\right), \quad i, j = 1, 2. \qquad (7.213)$$

Furthermore, $\boldsymbol{n}^\iota = (n_i^\iota)$ and $\boldsymbol{t}^\iota = (t_i^\iota)$ denote the unit outward normal and the tangential vector to the boundary $\partial\Omega^\iota$, $\boldsymbol{P}^\iota = (P_i^\iota)$ and $\boldsymbol{F}^\iota = (F_i^\iota)$ denote the surface loads on ${}^1\Gamma_\tau^\iota$ and the body forces in Ω^ι respectively, W^ι denotes the thermal sources in Ω^ι, $\boldsymbol{u}_o^\iota$ is a displacement vector prescribed on Γ_u, and T_o^ι, and T_1^ι are the given temperatures in Ω^ι or on ${}^2\Gamma_\tau^\iota$, respectively. Furthermore, κ_{ij}^ι are the thermal conductivity coefficients in Ω^ι, c_{ijkl}^ι are the elastic coefficients in Ω^ι, β_{ij}^ι are the coefficients of thermal expansion in Ω^ι, $g_c^{kl} \equiv \mathscr{F}_c^{kl} \left|\tau_n^k\right|$ represents the given friction forces acting on the contact boundary Γ_c^{kl}, and $\mathscr{F}_c^{kl}$ is a Coulombian friction coefficient.

Remark: The conditions in Equations 7.203 to 7.205 represent the so-called Signorini conditions, where Equation 7.203 represents the condition of non-penetration of colliding bodies, Equation 7.204 represents the condition for the contact forces following from the law of action and reaction. The last condition (Equation 7.205) together with Equations 7.203 and 7.204 describes the situation during the deformation of both colliding bodies being or not being in contact. Moreover, if both colliding bodies are in contact (see Equation 7.206), the Coulombian friction, defined by the Coulombian law of friction (Equations 7.206 to 7.209), acts at the contact boundary Γ_c^{kl}. The friction forces g_c^{kl} acting on the contact boundary Γ_c^{kl} are, in their absolute value, proportional to the normal stress component (see Equation 7.207), where the proportionality coefficient is the coefficient of friction $\mathscr{F}_c^{kl}$. If the absolute value of the tangential forces is less then the friction forces g_c^{kl} then the tangential forces preclude the mutual motion of both colliding bodies (see Equation 7.208). If the absolute value of the tangential forces reaches the value of the friction forces then there are no forces, which can preclude the mutual (bilateral) motion of both colliding plates.

Similarly as in Theorem 2.2 of Nedoma (1987) (see also in the previous section), the Coulombian coefficient of friction $\mathscr{F}_c^{kl}$ can be estimated by

$$\left\|\mathscr{F}_c^{kl}\right\|_\infty < \min((c_o^k/(2c_1^k))^{1/2}, (c_o^l/(2c_1^l))^{1/2}),$$

or

$$\left\|\mathscr{F}_c^{kl}\right\|_\infty < \min((\mu^k/(\lambda^k + 2\mu^k))^{1/4}, (\mu^l/(\lambda^l + 2\mu^l))^{1/4}) \sim \min(c_s^k/c_p^k, c_s^l/c_p^l)^{1/2}$$

for the non-homogeneous anisotropic or isotropic cases, respectively, where c_o^r and $c_1^r (r = k, l)$ will be defined in the next section and c_s^r and $c_p^r (r = k, l)$ are the velocities of S and P waves in the regions Ω^k and Ω^l.

The problem is semi-coercive for $\Gamma_u = 0$ or if at least one of Γ_u^ι is of zero measure. For simplicity we shall assume that $\boldsymbol{u}_o^\iota = 0$ on ${}^1\Gamma_u$, $\Gamma_u^\iota = 0$ for $\iota = 2, \dots, r-1, r+1, \dots, s$ and $T_1^\iota = 0$ on ${}^2\Gamma_\tau^\iota$. Furthermore, we denote ${}^1\Gamma_u \equiv \Gamma_u$.

7.2.2.2 *Variational formulation and weak solution*

Let $\Omega \subset R^2$ be the region with a Lipschitz boundary $\partial\Omega$ consisting of three parts $\Gamma_\tau, \Gamma_u, \Gamma_c$, $\partial\Omega = \overline{\Gamma}_\tau \cup \overline{\Gamma}_u \cup \overline{\Gamma}_c \cup \mathscr{R}$. Let $\boldsymbol{x} = (x_1, x_2)$ be Cartesian coordinates and let $\boldsymbol{n} = (n_1, n_2)$, $\boldsymbol{t} = (t_1, t_2) = (-n_2, n_1)$ be the outward normal and tangential vectors to $\partial\Omega$.

Let us denote by (.,.) the scalar product in $[L^2(\Omega)]^2$, by $\langle .,. \rangle$ the scalar product in $[L^2(\Gamma_c)]^2$, by $\|.\|_k$ the norm in $[H^k(\Omega)]^2$ (k being integer) and by $\|.\| \equiv \|.\|_1$ the norm in $H^1(\Omega)$, where $H^k(\Omega^\iota)$, $k \in R^1$, $\iota = 1, \ldots, s$, denotes the Sobolev space in the usual sense. Let $|\boldsymbol{v}| = \left(\int_\Omega e_{ij}(\boldsymbol{v})e_{ij}(\boldsymbol{v})\mathrm{d}\boldsymbol{x}\right)^{1/2}$ denote the semi-norm on Q_p defined later. Let us denote by

$$\begin{aligned} {}^1V = \{z \mid z \in {}^1W(\Omega) \equiv H^1(\Omega^1) \times \ldots \times H^1(\Omega^k) \\ \times H^1(\Omega^l) \times \ldots \times H^1(\Omega^s), z = 0 \quad \text{on} \quad \cup_\iota {}^2\Gamma_\tau^\iota\}, \\ V = \{\boldsymbol{v} \mid \boldsymbol{v} \in W(\Omega) \equiv [H^1(\Omega^1)]^2 \times \ldots \times [H^1(\Omega^k)]^2 \\ \times [H^1(\Omega^l)]^2 \times \ldots \times [H^1(\Omega^s)]^2, \boldsymbol{v} = 0 \quad \text{on} \quad \Gamma_u\} \end{aligned}$$

the spaces of virtual temperatures and virtual displacements respectively, and by

$$K = \{\boldsymbol{v} \mid \boldsymbol{v} \in V, \quad v_n^k - v_n^l \leq 0 \quad \text{on} \quad \cup_{k,l} \Gamma_c^{kl}\}$$

the set of all admissible displacements (which is the so-called convex cone with vertex at the origin. The set K is called the convex cone with vertex at the origin, if it is closed and moreover, together with elements $\boldsymbol{u}, \boldsymbol{v} \in K$ also element $t\boldsymbol{u} + s\boldsymbol{v} \in K, t \geq 0, s \geq 0$).

We suppose that all κ_{ij}^ι and c_{ijkl}^ι are Lipschitz on Ω^ι, $\iota = 1, \ldots, s$, and fulfil the usual symmetry conditions

$$\kappa_{ij}^\iota = \kappa_{ji}^\iota, \quad \text{on} \quad \Omega^\iota \tag{7.214}$$

$$c_{ijkl}^\iota = c_{jikl}^\iota = c_{klij}^\iota, \text{ on } \Omega^\iota \tag{7.215}$$

for every $i, j, k, l = 1, 2$, $\iota = 1, \ldots, s$ and

$$0 < \kappa_o^\iota \leq (\kappa_{ij}^\iota(\boldsymbol{x})\zeta_i^\iota\zeta_j^\iota)\left|\zeta^\iota\right|^{-2} \leq \kappa_1^\iota < +\infty \quad \forall \boldsymbol{x} \in \Omega^\iota,\ \zeta^\iota = (\zeta_i^\iota) \in R^2, \tag{7.216}$$

$$0 < c_0^\iota \leq (c_{ijkl}^\iota(\boldsymbol{x})\xi_{ij}^\iota\xi_{kl}^\iota)\,|\xi|^{-2} \leq c_1^\iota < +\infty \quad \forall \boldsymbol{x} \in \Omega^\iota,\ \xi^\iota = (\xi_{ij}^\iota) \in R^4, \tag{7.217}$$

where κ_o^ι, κ_1^ι, c_o^ι, c_1^ι are constants independent of $\boldsymbol{x} \in \Omega^\iota$ and $\zeta^\iota \in R^2$, $\xi^\iota \in R^4$.

Let us look for the pair of functions $(T^\iota, \boldsymbol{u}^\iota)$, $T^\iota(\boldsymbol{x}) \in H^1(\Omega^\iota)$, $\boldsymbol{u}^\iota(\boldsymbol{x}) \in [H^1(\Omega^\iota)]^2$. Let $\beta_{ij}^\iota(\boldsymbol{x}) \in C^1(\overline{\Omega}^\iota)$, $c_{ijkl}^\iota(\boldsymbol{x}) \in C^1(\overline{\Omega}^\iota)$, $T_1^\iota(\boldsymbol{x}) \in L^2(\Gamma_\tau^\iota)$, $T_p^\iota(\boldsymbol{x}) \in H^1(\Omega^\iota)$, $g_c^{kl}(\boldsymbol{x}) \in L^\infty(\Gamma_c^{kl})$, $\kappa_{ij}^\iota(\boldsymbol{x}) \in C^1(\Omega^\iota)$. The coupling term $\partial/\partial x_j(\beta_{ij}(T^\iota - T_p^\iota)) \in L^2(\Omega^\iota)$. If $\beta_{ij}^\iota \in C^1(\overline{\Omega}^\iota)$ and $T^\iota, T_p^\iota \in H^1(\Omega^\iota)$, then $\beta_{ij}^\iota(T^\iota - T_p^\iota) \in H^1(\Omega^\iota)$ and then $\partial/\partial x_j(\beta_{ij}^\iota(T^\iota - T_p^\iota)) \in L^2(\Omega^\iota)$. Then $f_i^\iota = F_i^\iota - \partial/\partial x_j(\beta_{ij}^\iota(T^\iota - T_p^\iota)) \in L^2(\Omega^\iota)$.

Definition 7.7: By a variational (weak) solution of the problem we mean a pair of functions $(T, \boldsymbol{u})$, $T \in {}^1V$, $\boldsymbol{u} \in K$ such that

$$b(T, z - T) \geq s(z - T) \quad \forall z \in {}^1V, \tag{7.218}$$

$$a(\boldsymbol{u}, \boldsymbol{v} - \boldsymbol{u}) + \left\langle g_c^{kl}, \left|\boldsymbol{v}_t^k - \boldsymbol{v}_t^l\right| - \left|\boldsymbol{u}_t^k - \boldsymbol{u}_t^l\right|\right\rangle \geq S(\boldsymbol{v} - \boldsymbol{u}) \quad \forall \boldsymbol{v} \in K, \tag{7.219}$$

where

$$b(T, z) = \int_\Omega \kappa_{ij} \frac{\partial T}{\partial x_j} \frac{\partial z}{\partial x_i} \mathrm{d}\boldsymbol{x} = \sum_{\iota=1}^{s} b^\iota(T^\iota, z^\iota),$$

$$s(z) = \int_\Omega Qz\mathrm{d}\boldsymbol{x} = \sum_{\iota=1}^{s} s^\iota(z^\iota),$$

$$a(\boldsymbol{u}, \boldsymbol{v}) = \int_\Omega c_{ijkl}(\boldsymbol{x}) e_{ij}(\boldsymbol{u}) e_{kl}(\boldsymbol{v})\mathrm{d}\boldsymbol{x} = \sum_{\iota=1}^{s} a^\iota(\boldsymbol{u}^\iota, \boldsymbol{v}^\iota),$$

$$S(\boldsymbol{v}) = \int_\Omega f_i v_i \mathrm{d}\boldsymbol{x} + \int_{{}^1\Gamma_\tau} P_i v_i \mathrm{d}s = \sum_{\iota=1}^{s} S^\iota(\boldsymbol{v}^\iota),$$

$$j_{gn}(\boldsymbol{v}) \equiv \langle g_c^{kl}, |\boldsymbol{v}_t^k - \boldsymbol{v}_t^l| \rangle = \int_{\cup_{k,l}\Gamma_c^{kl}} g_c^{kl} \left|\boldsymbol{v}_t^k - \boldsymbol{v}_t^l\right| \mathrm{d}s.$$

The problem in Equations 7.218 and 7.219 is equivalent to the following variational formulation.

Find a pair of functions $(T, \boldsymbol{u})$, $T \in {}^1V$, $\boldsymbol{u} \in K$ such that

$$l(T) \le l(z) \quad \forall z \in {}^1V, \tag{7.220}$$

$$L(\boldsymbol{u}) \le L(\boldsymbol{v}) \quad \forall \boldsymbol{v} \in K, \tag{7.221}$$

where

$$l(z) = \tfrac{1}{2} b(z, z) - s(z), \tag{7.222}$$

$$L(\boldsymbol{v}) = \tfrac{1}{2} a(\boldsymbol{v}, \boldsymbol{v}) - S(\boldsymbol{v}) + \left\langle g_c^{kl}, \left|\boldsymbol{v}_t^k - \boldsymbol{v}_t^l\right| \right\rangle = L_o(\boldsymbol{v}) + j_{gn}(\boldsymbol{v}), \tag{7.223}$$

where $L_o(\boldsymbol{v}) = \frac{1}{2} a(\boldsymbol{v}, \boldsymbol{v}) - S(\boldsymbol{v})$ and $j_{gn}(\boldsymbol{v}) = \langle g_c^{kl}, |\boldsymbol{v}_t^k - \boldsymbol{v}_t^l| \rangle$.

It can be proved that any classical solution of the problem in Equations 7.195 to 7.213 is a weak solution and conversely, if the weak solution is smooth enough it represents a classical solution.

To prove the existence and uniqueness of the variational (weak) solution we introduce the set of all displacements and rotations (the set of rigid displacements)

$$P = \{\boldsymbol{v} \mid \boldsymbol{v} = \boldsymbol{a} + (b \times \boldsymbol{x})\} = \{\boldsymbol{v} \mid \boldsymbol{v} = (v_1, v_2), v_1 = a_1 - bx_2, v_2 = a_2 + bx_1\}, \tag{7.224}$$

where a_1, a_2 and b are arbitrary real constants. Let Q_p be the orthogonal decomposition of P in W, i.e. $Q_p \oplus P = W$.

Lemma 7.17: Let $\Omega \subset R^2$, $(T, \boldsymbol{u}) \in {}^1W \times W$. Then

$$e_{ij}(\boldsymbol{u}) = 0 \quad \forall i, j, = 1, 2 \Leftrightarrow \boldsymbol{u} = \boldsymbol{a} + (b \times \boldsymbol{x}). \tag{7.225}$$

Proof: Since the problem is quasi-coupled it is sufficient to investigate the elastic case only. Then the proof is parallel to that of Theorem 6.3.2 of Nečas and Hlaváček (1981).

Lemma 7.18: Necessary conditions for the existence of the solution are the conditions of total equilibrium

$$\int_\Omega f_i w_i \mathrm{d}\boldsymbol{x} + \int_{{}^1\Gamma_\tau} P_i w_i \mathrm{d}s + \int_{\cup_{k,l}\Gamma_c^{kl}} g_c^{kl} \left|\boldsymbol{w}_t^k - \boldsymbol{w}_t^l\right| \mathrm{d}s < 0 \quad \forall \boldsymbol{w} \in K \cap P. \tag{7.226}$$

Proof: Since the problem is quasi-coupled we can investigate the elastic part of the problem only.

Putting $\boldsymbol{v} \in P$ for the trial functions into Equation 7.219, then according to Lemma 7.17 $e_{ij}(\boldsymbol{v}) = 0$ and therefore $a(\boldsymbol{v}, \boldsymbol{v}) = 0$. Since $j_{gn}(\boldsymbol{v}) \equiv \langle g_c^{kl}, |\boldsymbol{v}_t^k - \boldsymbol{v}_t^l| \rangle \geq 0$ (as $g_c^{kl} \geq 0$ a.e. on Γ_c^{kl}) then for $\boldsymbol{v} = \boldsymbol{u} + \boldsymbol{w}$

$$\int_\Omega f_i w_i \mathrm{d}\boldsymbol{x} + \int_{^1\Gamma_\tau} P_i w_i \mathrm{d}s + \int_{\cup\Gamma_c^{kl}} g_c^{kl} \left|\boldsymbol{w}_t^k - \boldsymbol{w}_t^l\right| \mathrm{d}s \leq 0 \quad \forall \boldsymbol{w} \in K \cap P \tag{7.227}$$

as for $\boldsymbol{u} \in K$ also $\boldsymbol{v} \in K$.

Let $P_o = \{\boldsymbol{v} \in P \mid v_n^k - v_n^l = 0 \text{ on } \Gamma_c^{kl}\}$, then $\boldsymbol{v} \in P_o \Leftrightarrow \pm\boldsymbol{v} \in K \cap P$. Conversely, if for $\boldsymbol{w} \in K \cap P$ Equation 7.227 holds and for $\boldsymbol{w} \in K \cap (P \backslash P_o)$

$$\int_\Omega f_i w_i \mathrm{d}\boldsymbol{x} + \int_{^1\Gamma_\tau} P_i w_i \mathrm{d}s + \int_{\cup\Gamma_c^{kl}} g_c^{kl} \left|\boldsymbol{v}_t^k - \boldsymbol{v}_t^l\right| \mathrm{d}s < 0 \tag{7.228}$$

holds, then the functional $L(\boldsymbol{v})$ is coercive on K and achieves its minimum on K. Furthermore, we shall assume that $P_o = \{0\}$.

Moreover, we shall introduce the penalty functional by

$$\mathscr{P}(\boldsymbol{u}) \equiv \mathscr{P}(\boldsymbol{u}, \boldsymbol{u}) = \frac{1}{2} \int_{\cup_{k,l}\Gamma_c^{kl}} \left([u_n^k - u_n^l]^+\right)^2 \mathrm{d}s. \tag{7.229}$$

The penalty functional is Lipschitz continuous and monotone and for $t > 0$ is $\mathscr{P}(t\boldsymbol{u}) = t^2 \mathscr{P}(\boldsymbol{u})$. Then we have the equivalent variational problem.

Find $(T, \boldsymbol{u}) \in {}^1V \times V$ such that

$$b(T, z - T) \geq s(z - T) \quad \forall z \in {}^1V, \tag{7.230}$$

$$a(\boldsymbol{u}, \boldsymbol{v} - \boldsymbol{u}) + (\varepsilon_m)^{-1} \int_{\cup\Gamma_c^{kl}} [u_n^k - u_n^l]^+ ((v_n^k - v_n^l) - (u_n^k - u_n^l)) \mathrm{d}s$$
$$+ \langle g_c^{kl}, |\boldsymbol{v}_t^k - \boldsymbol{v}_t^l| - |\boldsymbol{u}_t^k - \boldsymbol{u}_t^l| \rangle \geq S(\boldsymbol{v} - \boldsymbol{u}) \quad \forall \boldsymbol{v} \in V,\ \varepsilon_m > 0,\ \varepsilon_m \to 0, \tag{7.231}$$

or in the equivalent form

find $(T, u) \in {}^1V \times V$ such that

$$b(T, z) = s(z) \quad \forall z \in {}^1V, \tag{7.232}$$

$$a(\boldsymbol{u}, \boldsymbol{v}) + (\varepsilon_m)^{-1} \int_{\cup\Gamma_c^{kl}} [u_n^k - u_n^l]^+ (v_n^k - v_n^l) \mathrm{d}s$$
$$+ \langle g_c^k, |\boldsymbol{v}_t^k - \boldsymbol{v}_t^l| \rangle = S(\boldsymbol{v}) \quad \forall \boldsymbol{v} \in V, \quad \varepsilon_m > 0, \quad \varepsilon_m \to 0. \tag{7.233}$$

Then our problem leads to minimizations of functionals $l(z)$ and $L(\boldsymbol{v})$ over ${}^1V$ or V, respectively. The solution $(T, \boldsymbol{u})$ can be found as

$$(T, \boldsymbol{u}) = \lim_{m \to \infty} (T, \boldsymbol{u}_m), \tag{7.234}$$

where $(T, \boldsymbol{u}_m)$ are solutions of the variational equations above (Equations 7.232 and 7.233) which can be written as

$$Dl(T, z) = 0 \quad \forall z \in {}^1V, \tag{7.235}$$

$$DL_m(\boldsymbol{u}_m, \boldsymbol{v}) = 0 \quad \forall \boldsymbol{v} \in V, \tag{7.236}$$

where $L_m(\boldsymbol{v}) = L_m(\boldsymbol{v}, \boldsymbol{v}) = L(\boldsymbol{v}) + \varepsilon_m^{-1} \mathscr{P}(\boldsymbol{v})$, $L(\boldsymbol{v}) = L_o(\boldsymbol{v}) + \langle g_c^{kl}, |\boldsymbol{v}_t^k - \boldsymbol{v}_t^l| \rangle$. Since the problem is quasi-coupled it is sufficient to investigate both problems separately. Therefore we will investigate only the functional $L_m(\boldsymbol{v})$ in detail as it is known that the functional $l(z)$ is convex and differentiable and has a minimum over ${}^1V$ at the point T. The functional $L_m(\boldsymbol{v})$ is coercive and weakly lower semi-continuous as the functional $L_o(\boldsymbol{v})$ is strictly convex and weakly lower semi-continuous on Q_p, the functional $\langle g_c^{kl}, |\boldsymbol{v}_t^k - \boldsymbol{v}_t^l|\rangle$ is convex and continuous and thus weakly lower semi-continuous and, moreover, $\langle g_c^{kl}, |\boldsymbol{v}_t^k - \boldsymbol{v}_t^l|\rangle \geq 0$ for an arbitrary $g_c^{kl} \in L^\infty(\Gamma_c^{kl})$ (as $g_c^{kl} \geq 0$ a.e. on Γ_c^{kl}), the penalty functional $\mathscr{P}(\boldsymbol{v})$ is monotone and has a Lipschitz continuous differential in V, and moreover, the Korn inequality $|\boldsymbol{v}| \geq c\,\|\boldsymbol{v}\|_1$ is valid on the orthogonal complement Q_p of P in W. The functional $L_m(\boldsymbol{v})$ has at a point $\boldsymbol{u}_m$ the Gâteaux differential $DL_m(\boldsymbol{u}, \boldsymbol{v})$. Furthermore, we have

$$DL_m(\boldsymbol{u}, \boldsymbol{u}) \geq c_0 \int_\Omega e_{ij}(\boldsymbol{u})e_{ij}(\boldsymbol{u})\mathrm{d}\boldsymbol{x} + \mathscr{P}(\boldsymbol{u}) - S(\boldsymbol{u}) \geq c_0\,|\boldsymbol{u}|^2 + \mathscr{P}\boldsymbol{u} - S(\boldsymbol{u})$$
$$\geq c_1\,\|\boldsymbol{u}\|_1^2 - c_2\,\|\boldsymbol{u}\|_1 \geq c_1\,\|\boldsymbol{u}\|_1 - c_2, \quad c_1 > 0, \quad c_2 \geq 0 \quad \forall \boldsymbol{u} \in V. \tag{7.237}$$

Thus the functional $L_m(\boldsymbol{v})$ has a minimum over Q_p at the point $\boldsymbol{u}_m$, so that $\boldsymbol{u}_m$ solve variational equations

$$DL(\boldsymbol{u}_m, \boldsymbol{v}) + \varepsilon_m^{-1} D\mathscr{P}(\boldsymbol{u}_m, \boldsymbol{v}) = 0 \quad \forall \boldsymbol{v} \in V. \tag{7.238}$$

Let $\boldsymbol{u}_m$ be the sequence of solutions of Equation 7.238, and let $\boldsymbol{v} \in K$. Then

$$DL(\boldsymbol{u}_m, \boldsymbol{u}_m - \boldsymbol{v}) - DL(\boldsymbol{u}, \boldsymbol{u}_m - \boldsymbol{v}) + \varepsilon_m^{-1} D\mathscr{P}(\boldsymbol{u}_m, \boldsymbol{u}_m - \boldsymbol{v}) - \varepsilon_m^{-1} D\mathscr{P}(\boldsymbol{u}, \boldsymbol{u}_m - \boldsymbol{v}) = -DL(\boldsymbol{u}, \boldsymbol{u}_m - \boldsymbol{v}). \tag{7.239}$$

Since the functional L is Lipschitz continuous and strictly monotone, the functional $\mathscr{P}$ has a Lipschitz continuous differential, $D\mathscr{P}$ is monotone, and then Equation 7.239 yields

$$c_1\,\|\boldsymbol{u}_m - \boldsymbol{v}\|_1^2 \leq c_2\,\|\boldsymbol{u}_m - \boldsymbol{v}\|_1\,(1 + \|\boldsymbol{v}\|_1),$$

hence $\|\boldsymbol{u}_m\|_1 \leq c_s < +\infty$. Then we can choose the extracted subsequence such that $\boldsymbol{u}_m \to \boldsymbol{u}$ weakly. To prove that $\boldsymbol{u} \in K$, the Minty trick was used (see, for example, Nečas and Hlaváček (1981)). Let $\boldsymbol{v} \in K$, then from Equation 7.239 for $\boldsymbol{v} = \boldsymbol{u}$ we obtain

$$c\,\|\boldsymbol{u}_m - \boldsymbol{u}\|_1^2 \leq -DL_m(\boldsymbol{u}, \boldsymbol{u}_m - \boldsymbol{u})$$

from which $\boldsymbol{u}_m \to \boldsymbol{u}$. For $\boldsymbol{w} \in K$

$$0 = DL(\boldsymbol{u}_m, \boldsymbol{u}_m - \boldsymbol{w}) + \varepsilon^{-1}(D\mathscr{P}(\boldsymbol{u}_m, \boldsymbol{u}_m - \boldsymbol{w}) - D\mathscr{P}(\boldsymbol{w}, \boldsymbol{u}_m - \boldsymbol{w})).$$

Hence $DL(\boldsymbol{u}_m, \boldsymbol{u}_m - \boldsymbol{w}) \leq 0$, which for $m \to \infty$ gives $DL(\boldsymbol{u}, \boldsymbol{w} - \boldsymbol{u}) \geq 0\,\forall \boldsymbol{w} \in K$. Since the last variational inequality has one and only one solution, we have $\boldsymbol{u}_m \to \boldsymbol{u}$ for the whole sequence $\{\boldsymbol{u}_m\}$.

The uniqueness of the solution can be proved as follows. If $\boldsymbol{u}_1$ and $\boldsymbol{u}_2$ are two solutions of $DL(\boldsymbol{u}, \boldsymbol{w} - \boldsymbol{u}) \geq 0\,\forall \boldsymbol{w} \in K$ then $DL(\boldsymbol{u}_2, \boldsymbol{u}_1 - \boldsymbol{u}_2) \geq 0$, $DL(\boldsymbol{u}_1, \boldsymbol{u}_2 - \boldsymbol{u}_1) \geq 0$, from which $DL(\boldsymbol{u}_2, \boldsymbol{u}_2 - \boldsymbol{u}_1) - DL(\boldsymbol{u}_1, \boldsymbol{u}_2 - \boldsymbol{u}_1) \leq 0$. Since the functional L is assumed to be twice differentiable in the Gâteaux sense and continuous in the variable $\boldsymbol{u}$ for fixed $\boldsymbol{w}$ then $D^2L(\boldsymbol{u}, \boldsymbol{w}, \boldsymbol{w}) \geq c\,\|\boldsymbol{w}\|^2$ and since

$$\int_0^1 D^2L(\boldsymbol{u}_1 + t(\boldsymbol{u}_2 - \boldsymbol{u}_1); \boldsymbol{u}_2 - \boldsymbol{u}_1, \boldsymbol{u}_2 - \boldsymbol{u}_1)\mathrm{d}t = DL(\boldsymbol{u}_2, \boldsymbol{u}_2 - \boldsymbol{u}_1) - DL(\boldsymbol{u}_1, \boldsymbol{u}_2 - \boldsymbol{u}_1) \leq 0$$

then the uniqueness of the solution follows immediately.

Then we have proved the following theorem.

Theorem 7.16: Let $\partial\Omega = \overline{\Gamma}_\tau \cup \overline{\Gamma}_u u \overline{\Gamma}_c u \mathscr{R}$, $\Gamma_\tau = {}^1\overline{\Gamma}_\tau \cup^2 \overline{\Gamma}_\tau$, $\Gamma_c = \cup_{k,l}\overline{\Gamma}_c^{kl}$. Let us assume that Equations 7.216 and 7.217 hold. Let us suppose that $f_i^\iota(\boldsymbol{x}) \in L^2(\overline{\Omega}^\iota)$, $P_i^\iota(\boldsymbol{x}) \in L^2(\Gamma_\tau)$, $T_p^\iota(\boldsymbol{x}) \in$

$H^1(\overline{\Omega}^\iota)$, $g_c^{kl}(\boldsymbol{x}) \in L^\infty(\Gamma_c^{kl})$, $c_{ijkl}^\iota(\boldsymbol{x}) \in C^1(\overline{\Omega}^\iota)$, $\kappa_{ij}^\iota(\boldsymbol{x}) \in C^1(\overline{\Omega}^\iota)$, $\beta_{ij}^\iota(\boldsymbol{x}) \in C^1(\overline{\Omega}^\iota)$, $W^\iota(\boldsymbol{x}) \in C^1(\overline{\Omega}^\iota)$.
Let us define

$$^1V = \{z \mid z \in {}^1W(\Omega), z = 0 \quad \text{on} \quad \cup_\iota {}^2\Gamma_\tau^\iota\}, \quad V = \{\boldsymbol{v} \mid \boldsymbol{v} \in W(\Omega), \boldsymbol{v} = 0 \quad \text{on} \quad \Gamma_u\},$$

$$K = \{\boldsymbol{v} \mid \boldsymbol{v} \in V, v_n^k - v_n^l \leq 0 \quad \text{on} \quad \Gamma_c^{kl}\}, \quad P = \{\boldsymbol{v} \mid \boldsymbol{v} = \boldsymbol{a} + (b \times \boldsymbol{x})\},$$

and let us suppose that

$$\int_\Omega f_i v_i \mathrm{d}\boldsymbol{x} + \int_{^1\Gamma_\tau} P_i v_i \mathrm{d}s + \int_{\cup\Gamma_c^{kl}} g_c^{kl} \left|\boldsymbol{v}_t^k - \boldsymbol{v}_t^l\right| \mathrm{d}s < 0 \quad \forall \boldsymbol{v} \in K \cap P, \boldsymbol{v} \neq 0. \tag{7.240}$$

Then there exists a pair of functions $(T, \boldsymbol{u})$, which is a solution of the variational equality and inequality in Equations 7.218 and 7.219 or 7.220 and 7.221 respectively, and every solution of Equations 7.218 and 7.219 is of the form $(T, \boldsymbol{u} + \boldsymbol{v})$, $\boldsymbol{v} \in P$ and

$$\int_\Omega f_i v_i \mathrm{d}\boldsymbol{x} + \int_{^1\Gamma_\tau} P_i v_i \mathrm{d}s + \int_{\cup\Gamma_c^{kl}} g_c^{kl} \left|\boldsymbol{v}_t^k - \boldsymbol{v}_t^l\right| \mathrm{d}s = 0.$$

Moreover, the solution $(T, \boldsymbol{u})$ can be found as $(T, \boldsymbol{u}) = \lim_{m\to\infty}(T, \boldsymbol{u}_m)$, where $(T, \boldsymbol{u}_m)$ are solutions of the variational equations

$$Dl(T, z) = 0 \quad \forall z \in {}^1V, \tag{7.241}$$

$$DL_m(\boldsymbol{u}_m, \boldsymbol{v}) = 0 \quad \forall \boldsymbol{v} \in V, \tag{7.242}$$

where $L_m(\boldsymbol{v}) = L(\boldsymbol{v}) + \varepsilon^{-1}\mathscr{P}(\boldsymbol{v})$, $L(\boldsymbol{v}) = L_o(\boldsymbol{v}) + \left\langle g_c^{kl}, \left|\boldsymbol{v}_t^k - \boldsymbol{v}_t^l\right|\right\rangle$. $(T, \boldsymbol{u}_m)$ can be found as a minimum of functionals $l(z)$ and $L_m(\boldsymbol{v})$ over ^1V and V respectively, and furthermore

$$DL_m(\boldsymbol{u}, \boldsymbol{u}) \geq c_1 \|\boldsymbol{u}\|_1 - c_2, \quad c_1 > 0, \quad c_2 \geq 0 \quad \forall \boldsymbol{u} \in V. \tag{7.243}$$

From every sequence of solutions $(T, \boldsymbol{u}_m)$ of the variational equations (Equations 7.241 and 7.242) an extracted subsequence converges strongly to $(T, \boldsymbol{u})$, representing the solution of Equations 7.218 and 7.219.

7.2.2.3. *Finite element approximation*

For the numerical solution the finite element method will be used. Let the given bounded domain $\Omega \subset R^2$ with the polygonal boundary $\partial\Omega$ be triangulated, i.e. $\overline{\Omega} = \Omega \cup \partial\Omega$ is covered by a finite number of triangles T_h, forming a triangulation $\mathscr{T}_h$. We further assume that the end points $\overline{\Gamma}_u \cap \overline{\Gamma}_\tau$, $\overline{\Gamma}_u \cap \overline{\Gamma}_c$, $\overline{\Gamma}_c \cap \overline{\Gamma}_\tau$ coincide with vertices of $\mathscr{T}_h$. The family $\{\mathscr{T}_h\}$ of triangulations, $0 < h \leq h_o$, is assumed to be regular. Let ^1V_h and V_h be the sets of linear finite elements, i.e.

$$^1V_h = \{z \,|\, z \in C(\Omega), z|_{T_h} \in P_1, z = 0 \quad \text{on} \quad {}^2\Gamma_\tau, \quad \forall T_h \in \mathscr{T}_h\},$$

$$V_h = \{\boldsymbol{v} \left| \boldsymbol{v} \in [C(\Omega)]^2, \boldsymbol{v} \right|_{T_h} \in [P_1]^2, \boldsymbol{v} = 0 \quad \text{on} \quad \Gamma_u, \quad \forall T_h \in \mathscr{T}_h\},$$

where P_1 is the space of all linear polynomials. Let

$$K_h = \{\boldsymbol{v} \mid \boldsymbol{v} \in V_h, v_n^k - v_n^l \leq 0 \quad \text{on} \quad \cup_{k,l} \Gamma_c^{kl}\} = V_h \cap K, \text{ hence } K_h \subset K \quad \forall h.$$

Definition 7.8: We say that a pair of functions $(T_h, \boldsymbol{u}_h)$, $T_h \in {}^1V_h$, $\boldsymbol{u}_h \in K_h$ is a finite element approximation of our problem, if

$$b_h(T_h, z - T_h) \geq s(z - T_h) \quad \forall z \in {}^1V_h, \tag{7.244}$$

$$a(\boldsymbol{u}_h, \boldsymbol{v} - \boldsymbol{u}_h) + \langle g_c^{kl}, |\boldsymbol{v}_t^k - \boldsymbol{v}_t^l| - |(\boldsymbol{u}_h^k)_t - (\boldsymbol{u}_h^l)_t|\rangle \geq S(\boldsymbol{v} - \boldsymbol{u}_h) \quad \forall \boldsymbol{v} \in K_h, \tag{7.245}$$

or in the equivalent form

$$l(T_h) \leq l(z) \quad \forall z \in {}^1V_h, \tag{7.246}$$

$$L(\boldsymbol{u}_h) \leq L(\boldsymbol{v}) \quad \forall \boldsymbol{v} \in K_h. \tag{7.247}$$

Theorem 7.17: There exists a unique solution of the finite element approximation of Equations 7.244 and 7.245 (or 7.246 and 7.247) ∀ h.

Proof: As the problem is quasi-coupled it is sufficient to investigate the problem in Equation 7.247, as the problem in Equation 7.246 is the standard FEM problem. The set K_h is a non-empty, closed and convex subset of K and of Q_p. Theorem 7.16 implies that functionals $l(z)$ and $L(\boldsymbol{v})$ are coercive over ${}^1V_h$ and K_h respectively, and the functional $j_{gn}(\boldsymbol{v}) = \langle g_c^{kl}, |\boldsymbol{v}_t^k - \boldsymbol{v}_t^l|\rangle$ is convex. Hence the existence of $(T_h, \boldsymbol{u}_h)$ follows.

The uniqueness can be proved by the following technique. Let $(T_h^1, \boldsymbol{u}_h^1) \in {}^1V_h \times K_h$ and $(T_h^2, \boldsymbol{u}_h^2) \in {}^1V_h \times K_h$ be two approximate solutions of Equations 7.246 and 7.247. Then we have

$$c_o \left\|T_h^1 - T_h^2\right\|^2 \leq b(T_h^1 - T_h^2, T_h^1 - T_h^2) \leq 0, \quad c_1 \left|\boldsymbol{u}_h^1 - \boldsymbol{u}_h^2\right|^2 \leq a(\boldsymbol{u}_h^1 - \boldsymbol{u}_h^2, \boldsymbol{u}_h^1 - \boldsymbol{u}_h^2) \leq 0.$$

Hence the uniqueness of the solution follows immediately.

To estimate the rates of convergence of $\|T - T_h\|$ and $|\boldsymbol{u} - \boldsymbol{u}_h|$ the method proposed by Falk (1974) will be applied. Since the problem is quasi-coupled we prove the following lemma.

Lemma 7.19: Let $|.|$ be the semi-norm defined above. Then it holds

$$|\boldsymbol{u} - \boldsymbol{u}_h| \leq c_o\{a(\boldsymbol{u}_h - \boldsymbol{u}, \boldsymbol{v}_h - \boldsymbol{u}) + a(\boldsymbol{u}, \boldsymbol{v} - \boldsymbol{u}_h) + a(\boldsymbol{u}, \boldsymbol{v}_h - \boldsymbol{u}) - S(\boldsymbol{v} - \boldsymbol{u}_h)$$
$$- S(\boldsymbol{v}_h - \boldsymbol{u}) + j_{gn}(\boldsymbol{v}_h) - j_{gn}(\boldsymbol{u}) + j_{gn}(\boldsymbol{v}) - j_{gn}(\boldsymbol{u}_h)\}^{1/2} \quad \forall \boldsymbol{v} \in K, \quad \forall \boldsymbol{v}_h \in K_h, \tag{7.248}$$

where $c_o = \text{constant} > 0$, $j_{gn}(\boldsymbol{v}) = \langle g_c^{kl}, |\boldsymbol{v}_t^k - \boldsymbol{v}_t^l|\rangle$.

Proof: The proof follows from the conditions

$$a(\boldsymbol{u}, \boldsymbol{v} - \boldsymbol{u}) + \langle g_c^{kl}, |\boldsymbol{v}_t^k - \boldsymbol{v}_t^l| - |\boldsymbol{u}_t^k - \boldsymbol{u}_t^l|\rangle - S(\boldsymbol{v} - \boldsymbol{u}) \geq 0 \quad \forall \boldsymbol{v} \in K,$$

$$a(\boldsymbol{u}_h, \boldsymbol{v}_h - \boldsymbol{u}_h) + \langle g_c^{kl}, |(\boldsymbol{v}_h^k)_t - (\boldsymbol{v}_h^l)_t| - |(\boldsymbol{u}_h^k)_t - (\boldsymbol{u}_h^l)_t|\rangle - S(\boldsymbol{v}_h - \boldsymbol{u}_h) \geq 0 \quad \forall \boldsymbol{v}_h \in K_h.$$

Adding these inequalities, adding and subtracting the term $a(\boldsymbol{u}, \boldsymbol{u}_h) - a(\boldsymbol{u}_h, \boldsymbol{u})$ to the resulting inequality and performing some modifications, we obtain

$$a(\boldsymbol{u} - \boldsymbol{u}_h, \boldsymbol{u} - \boldsymbol{u}_h) \leq a(\boldsymbol{u}_h - \boldsymbol{u}, \boldsymbol{v}_h - \boldsymbol{u}) + a(\boldsymbol{u}, \boldsymbol{v} - \boldsymbol{u}_h) + a(\boldsymbol{u}, \boldsymbol{v}_h - \boldsymbol{u}) - S(\boldsymbol{v} - \boldsymbol{u}_h) - S(\boldsymbol{v}_h - \boldsymbol{u})$$
$$+ j_{gn}(\boldsymbol{v}_h) - j_{gn}(\boldsymbol{u}) + j_{gn}(\boldsymbol{v}) - j_{gn}(\boldsymbol{u}_h).$$

This immediately implies the assertion in Equation 7.248.

Corollary: Let $K_h \subset K$. Then substituting $\boldsymbol{v} = \boldsymbol{u}_h$ in Equation 7.248 we obtain

$$|\boldsymbol{u} - \boldsymbol{u}_h| \leq c_o\{a(\boldsymbol{u}_h - \boldsymbol{u}, \boldsymbol{v}_h - \boldsymbol{u}) + a(\boldsymbol{u}, \boldsymbol{v}_h - \boldsymbol{u}) - S(\boldsymbol{v}_h - \boldsymbol{u}) + j_{gn}(\boldsymbol{v}_h) - j_{gn}(\boldsymbol{u})\}^{1/2} \quad \forall \boldsymbol{v}_h \in K_h. \tag{7.249}$$

As a result we shall prove the following theorem.

Theorem 7.18: Let $\partial\Omega$ and its parts Γ_u, Γ_τ, Γ_c be piecewise polygonal. Let $T \in {}^1V$, $\boldsymbol{u} \in K \cap [H^2(\Omega)]^2$, $\vec{\tau}(\boldsymbol{u}) \in [L^\infty(\cup_{k,l}\Gamma_c^{kl})]^2$, $\boldsymbol{u} \in [W^{1,\infty}(\cup_{k,l}\Gamma_c^{kl})]^2$, $g_c^{kl} \in L^\infty(\Gamma_c^{kl})$. Let ${}^1V_h \subset {}^1V$, $K_h \subset K$. Let the changes $u_n^k - u_n^l < 0 \to u_n^k - u_n^l = 0$ and $\boldsymbol{u}_t^k - \boldsymbol{u}_t^l = 0 \to \boldsymbol{u}_t^k - \boldsymbol{u}_t^l \neq 0$ occur at only finitely many points of $\cup_{k,l}\Gamma_c^{kl}$. Then

$$\|T - T_h\| = O(h), \tag{7.250}$$

$$|\boldsymbol{u} - \boldsymbol{u}_h| = O(h). \tag{7.251}$$

Proof: As the problem is quasi-coupled we will analyse both problems separately.

The first part of the proof represents classical finite element analysis in the heat conduction problem. The approximate solution T_h is the function from ${}^1V_h$ minimizing $l(z)$ over ${}^1V_h$, i.e. $l(T_h) = \min_{z\in {}^1V_h} l(z)$. As $b(T, z) = s(z)$ for any $z \in {}^1V$, then $b(T, T) = s(T)$ and we have $l(z) - l(T) = \min_{z\in {}^1V_h} \frac{1}{2}b(T - z, T - z)$. Furthermore,

$$\min_{z\in {}^1V_h} l(z) - l(T) = \min_{z\in {}^1V_h} \tfrac{1}{2}b(T - z, T - z)$$

so that

$$b(T - T_h, T - T_h) \leq b(T - z, T - z) \quad \forall z \in {}^1V_h.$$

Since the bilinear form $b(T, z)$ is continuous on $H^1(\Omega) \times H^1(\Omega)$ and bounded (i.e. $|b(T, z)| \leq M\,\|T\|\,\|z\|\ \forall T, z \in H^1(\Omega)$ and M = constant independent of T, z) and V-elliptic (i.e. $b(z, z) \geq c\,\|z\|^2\ \forall z \in {}^1V$ and c = constant > 0), we easily obtain

$$\|T - T_h\| \leq (M/c)^{1/2}\,\|T - z\| \quad \forall z \in {}^1V_h. \tag{7.252}$$

From the well-known interpolating theorem (see, for example, Bramble and Zlámal (1970)) it follows that

$$\left\|z - z_h^n\right\| \leq c_1 M_{n+1} h^{n+1}, \tag{7.253}$$

where z_h^n is the function which on every triangle of the given triangulation $\mathcal{T}_h$ is equal to the polynomial interpolating the function z, i.e. it is a piecewise polynomial of n-th degree. In our case we assume that $n = 1$.

We put $z = z_h$, where $z_h \in {}^1V_h$ is the function which on every triangle of the given triangulation is equal to the polynomial interpolating the exact solution T. Then from Equation 7.252 with $z = z_h$, and from Equation 7.253 with $z = T$ we get

$$\|T - T_h\| \leq (M/c)^{1/2} c_1\, h. \tag{7.254}$$

The second part of the proof is based on the technique of Hlaváček and Lovíšek (1980) and Nedoma (1987).

Using Lemma 7.19 we estimate Equation 7.248. This estimate can be applied provided the solution u is sufficiently regular. Green's theorem implies

$$\begin{aligned}
&a(\boldsymbol{u}, \boldsymbol{v}_h - \boldsymbol{u}) - S(\boldsymbol{v}_h - \boldsymbol{u}) + j_{gn}(\boldsymbol{v}_h) - j_{gn}(\boldsymbol{u}) \\
&= \int_\Omega -\frac{\partial}{\partial x_j}(c_{ijkl}e_{kl}(\boldsymbol{u}))(\boldsymbol{v}_h - \boldsymbol{u})_i \mathrm{d}\boldsymbol{x} + \int_{\partial\Omega} \tau_{ij}n_j(\boldsymbol{v}_h - \boldsymbol{u})_i \mathrm{d}s - \int_{\Gamma_\tau} P_i(\boldsymbol{v}_h - \boldsymbol{u})_i \mathrm{d}s \\
&\quad - \int_\Omega f_i(\boldsymbol{v}_h - \boldsymbol{u})_i \mathrm{d}\boldsymbol{x} + j_{gn}(\boldsymbol{v}_h) - j_{gn}(\boldsymbol{u}) \\
&= \int_\Omega -\frac{\partial}{\partial x_j}(c_{ijkl}e_{kl}(\boldsymbol{u}))(\boldsymbol{v}_h - \boldsymbol{u})_i \mathrm{d}\boldsymbol{x} + \int_{\Gamma_\tau} \tau_{ij}n_j(\boldsymbol{v}_h - \boldsymbol{u})_i \mathrm{d}s - \int_{\Gamma_\tau} P_i(\boldsymbol{v}_h - \boldsymbol{u})_i \mathrm{d}s \\
&\quad + \int_{\Gamma_u} \tau_{ij}n_j(\boldsymbol{v}_h - \boldsymbol{u})_i \mathrm{d}s + \int_{\cup\Gamma_c^{kl}} \tau_{ij}n_j(\boldsymbol{v}_h - \boldsymbol{u})_i \mathrm{d}s - \int_\Omega f_i(\boldsymbol{v}_h - \boldsymbol{u})_i \mathrm{d}\boldsymbol{x} \\
&\quad + j_{gn}(\boldsymbol{v}_h) - j_{gn}(\boldsymbol{u})
\end{aligned}$$

$$= \int_{\cup\Gamma_c^{kl}} \tau_{ij}(\boldsymbol{u})n_j(\boldsymbol{v}_h - \boldsymbol{u})_i \mathrm{d}s + j_{gn}(\boldsymbol{v}_h) - j_{gn}(\boldsymbol{u})$$

$$= \int_{\cup\Gamma_c^{kl}} \tau_n(\boldsymbol{u})(\boldsymbol{v}_h - \boldsymbol{u})_n \mathrm{d}s + \int_{\cup\Gamma_c^{kl}} \vec{\tau}_t(\boldsymbol{u})(\boldsymbol{v}_h - \boldsymbol{u})_t \mathrm{d}s + j_{gn}(\boldsymbol{v}_h) - j_{gn}(\boldsymbol{u}) \geq 0 \quad \forall \boldsymbol{v}_h \in K_h, \tag{7.255}$$

since

$$-\frac{\partial}{\partial x_j}(c_{ijkl}e_{kl}(\boldsymbol{u})) = f_i \quad i = 1, 2, \quad \text{a.e. in} \quad \Omega.$$

And in virtue of the corollary of Lemma 7.19 we have

$$|\boldsymbol{u} - \boldsymbol{u}_h| \leq c_o\{a(\boldsymbol{u}_h - \boldsymbol{u}, \boldsymbol{v}_h - \boldsymbol{u}) + a(\boldsymbol{u}, \boldsymbol{v}_h - \boldsymbol{u}) - S(\boldsymbol{v}_h - \boldsymbol{u}) + j_{gn}(\boldsymbol{v}_h) - j_{gn}(\boldsymbol{u})\}^{1/2} \quad \forall \boldsymbol{v}_h \in K_h. \tag{7.256}$$

Let us denote by $\boldsymbol{u}_I$ the Lagrange linear interpolation of $\boldsymbol{u}$, and by $\boldsymbol{u}_{LI}$ the orthogonal projection of the Lagrange linear interpolation of $\boldsymbol{u}$ on the triangulation $\{\mathcal{T}_h\}$.

The bilinear form a(.,.) is bounded, so that

$$a(\boldsymbol{u}_h - \boldsymbol{u}, \boldsymbol{v}_h - \boldsymbol{u}) \leq M\,|\boldsymbol{u}_h - \boldsymbol{u}|\,|\boldsymbol{v}_h - \boldsymbol{u}|\,.$$

Since for every $\boldsymbol{v} \in [C(\overline{\Omega})]^2 \cap K$ we have $\boldsymbol{v}_{LI} \in K_h$ and since the last inequality is valid for an arbitrary $\boldsymbol{v}_h \in K_h$, it is valid also for $\boldsymbol{v}_h = \boldsymbol{u}_{LI}$. As $ab \leq \frac{1}{2}\varepsilon a^2 + (2\varepsilon)^{-1}b^2$, $a, b \in R^1$, $\varepsilon > 0$, then applying the interpolation theorem we obtain

$$a(\boldsymbol{u}_h - \boldsymbol{u}, \boldsymbol{v}_h - \boldsymbol{u}) \leq \tfrac{1}{2}M\varepsilon\,|\boldsymbol{u}_h - \boldsymbol{u}|^2 + M(2\varepsilon)^{-1}\,|\boldsymbol{v}_h - \boldsymbol{u}| \leq \tfrac{1}{2}M\varepsilon\,|\boldsymbol{u}_h - \boldsymbol{u}|^2 + ch^2. \tag{7.257}$$

Choosing $\boldsymbol{v}_h = \boldsymbol{u}_{LI}$ and since $\boldsymbol{u}_{LI} \in V_h$ then $\boldsymbol{v}_h = \boldsymbol{u}_{LI} \in K_h = Q_p \cap K \cap V_h$. Moreover,

$$(\boldsymbol{u}_{LI}^k - \boldsymbol{u}_{LI}^l)_n = (\boldsymbol{u}_I^k - \boldsymbol{u}_I^l)_n \quad \text{on } \cup_{k,l}\,\Gamma_c^{kl}, \tag{7.258}$$

$$(\boldsymbol{u}_{LI}^k - \boldsymbol{u}_{LI}^l)_t = (\boldsymbol{u}_I^k - \boldsymbol{u}_I^l)_t \quad \text{on } \cup_{k,l}\,\Gamma_c^{kl}. \tag{7.259}$$

Thus

$$(\boldsymbol{u}_{LI}^k - \boldsymbol{u}_{LI}^l)_n \leq 0 \quad \text{on } \cup_{k,l}\,\Gamma_c^{kl}, \qquad \text{so that } \boldsymbol{u}_{LI} \in K.$$

Since $\boldsymbol{u}_{LI} \in V_h$, then $\boldsymbol{u}_{LI} \in K_h$. Furthermore, we have $|\boldsymbol{u}_{LI} - \boldsymbol{u}| \leq c_r h\,|\boldsymbol{u}|_2$,

$$\left\|(\boldsymbol{u}_{LI}^k - \boldsymbol{u}_{LI}^l)_n - (\boldsymbol{u}_n^k - \boldsymbol{u}_n^l)\right\|_{L^2(\cup\Gamma_c^{kl})} = \left\|(\boldsymbol{u}_I^k - \boldsymbol{u}_I^l)_n - (\boldsymbol{u}_n^k - \boldsymbol{u}_n^l)\right\|_{L^2(\cup\Gamma_c^{kl})}$$

$$\leq ch^2 \sum \left\|\boldsymbol{u}_n^k - \boldsymbol{u}_n^l\right\|_{H^2(\cup\Gamma_c^{kl})},$$

where Equations 7.258 and 7.259 were used.

To estimate the integrals $J_1(\cup\Gamma_c^{kl}) = \int_{\cup\Gamma_c^{kl}} \tau_n(\boldsymbol{u})(\boldsymbol{v}_h - \boldsymbol{u})_n \mathrm{d}s$ and $J_2(\cup\Gamma_c^{kl}) = \int_{\cup\Gamma_c^{kl}} \vec{\tau}_t(\boldsymbol{u})(\boldsymbol{v}_h - \boldsymbol{u})_t \mathrm{d}s$, the same technique as in the previous section will be used. Then $\left|J_1(\cup\Gamma_c^{kl})\right| \leq c_s h^2$ and $\left|\int_{\cup\Gamma_c^{kl}} \vec{\tau}_t^{kl}(\boldsymbol{u})[(\boldsymbol{u}_t^k - \boldsymbol{u}_t^l)_{LI} - (\boldsymbol{u}_t^k - \boldsymbol{u}_t^l)]\mathrm{d}s + j_{gn}(\boldsymbol{u}_{LI}) - j_{gn}(\boldsymbol{u})\right| \leq c_r h^2$.

From all these estimates we obtain

$$|\boldsymbol{u} - \boldsymbol{u}_h| \leq c_N\{\varepsilon h^2|\boldsymbol{u}|_2^2 + \varepsilon^{-1} h^2\,|\boldsymbol{u}|^2 + c_s h^2 + c_r h^2\}^{1/2}$$

from which we obtain the estimate in Equation 7.251, choosing ε sufficiently small.

7.3. MATHEMATICAL MODELS OF PLATE TECTONICS BASED ON COUPLED CONTACT-TWO-PHASE STEFAN-LIKE PROBLEMS IN LINEAR THERMO-ELASTICITY: PROBLEMS OF RECRYSTALLIZATION OF ROCKS IN SUBDUCTION AND COLLISION ZONES

7.3.1. Introduction

In this section we shall investigate the model of subduction and collision zones, which covers recrystallization and partial melting of rocks in colliding zones and their close and distant neighbourhoods. From the literature it is known that mathematical investigations of phase change transitions lead to the study of the so-called Stefan-like problems. The problem investigated represents a simulation of geodynamic processes in collision zones, based on the hypothesis of the plate tectonics and on contact problems with or without friction in thermo-elasticity as well as on the two-phase Stefan-like problem. In this section we shall assume the case of uniformly moving colliding lithospheric plates and geological bodies with or without friction between them, where time is taken to be a parameter. Shortly we shall mention the case of non-uniformly moving colliding lithospheric plates and blocks. The weak formulation of this contact part of the problem thus results in an elliptic variational inequality, while an implicit discretization in time of the weak formulation of the Stefan problem using the enthalpy approximation results in a non-linear elliptic boundary value problem, and therefore can be formulated as an elliptic variational inequality. Numerically then the problem yields the finite element approximation of a coupled contact-Stefan-like problem in linear thermo-elasticity. The linear finite elements will be used and the convergence of the FEM approximation of the solution to the weak solution of the coupled contact-two-phase Stefan-like problem will be proved. At the end we also give an idea of the algorithm solving these types of problems.

As we saw in previous sections, the upper parts of the Earth for geological time shorter than a characteristic time $t_c = \nu/\mu$, where ν is the effective viscosity at a given stress level and μ is the elastic rigidity, both defined in previous sections, behave elastically, while for geological times longer than the characteristic time t_c they behave thermo-visco-elastically or thermo-visco-plastically. In this section we shall assume that $t_c \le 2.10^5$ years and, therefore, we can use the thermo-elastic rheology only. The rocks are melted in the bulged-out regions as the normal stresses lead to zero, and then due to the phase (P/T) diagrams the rocks are quickly melted. But the process of melting must be described by the other model problems, e.g. based on the Stefan-like problem in a visco-plastic Bingham's rheology (see Nedoma (1997a) and Chapter 12).

The model problem solved in this section represents model problems of plate tectonics as well as model problems which model the stress-strain situation in regions where the great structures are or will be situated as the high-level radioactive waste disposal systems, deep mines, etc. The investigated model allows us to study the recrystallization of rocks and their partial melting for the coercive case only.

7.3.2. Numerical Solutions of Coupled Two-Phase Stefan-Like-Contact Problems with Friction Arising from an Implicit Time Discretization: The Coercive Case

7.3.2.1. The model problem

We shall consider a model plate tectonic problem, based on the plate tectonic hypothesis and on theories of contact and two-phase Stefan problems. Without generality we shall limit ourselves to the $2D$ model case only.

Let the stress-strain relation, the so-called Duhamel-Neumann's law, be defined as

$$\tau_{ij} = c_{ijkl}e_{kl}(\boldsymbol{u}) - \beta_{ij}(\Theta - \Theta_o), \tag{7.260}$$

where e_{ij} is a small strain tensor defined by

$$e_{ij} = \frac{1}{2}\left(\frac{\partial u_i}{\partial x_j} + \frac{\partial u_j}{\partial x_i}\right), \quad i, j = 1, 2, \tag{7.261}$$

$\beta_{ij} \in C^1(\overline{\Omega})$ is a coefficient of linear thermal expansion, $\Theta_o = \Theta_o(\boldsymbol{x}) \in H^1(\Omega)$ is an initial temperature at $t = t_o$, and the coefficients of elasticity $c_{ijkl}(\boldsymbol{x}) \in C^1(\overline{\Omega})$ satisfy

$$c_{ijkl}(\boldsymbol{x}) = c_{jikl}(\boldsymbol{x}) = c_{klij}(\boldsymbol{x}) \tag{7.262}$$

and

$$c_{ijkl}(\boldsymbol{x})e_{ij}(\boldsymbol{u})e_{kl}(\boldsymbol{u}) \geq c_o e_{ij}(\boldsymbol{u})e_{ij}(\boldsymbol{u}), \quad c_o = \text{ constant} > 0, \quad \forall e_{ij} = e_{ji}. \tag{7.263}$$

Let the coefficients of thermal conductivity $\kappa_{ij}(\boldsymbol{x}) \in C^1(\overline{\Omega})$ satisfy

$$\kappa_{ij}(\boldsymbol{x}) = \kappa_{ji}(\boldsymbol{x}) \tag{7.264}$$

and

$$\kappa_{ij}(\boldsymbol{x})\zeta_i\zeta_j \geq c_1 |\zeta|^2 \quad \forall \zeta \in R^2, \text{ a.e. in } \Omega. \tag{7.265}$$

Moreover, assume the density ρ and specific heat c are bounded, i.e.

$$0 < \rho_1 \leq \rho(\boldsymbol{x}) \leq \rho_2, \tag{7.266}$$

$$0 < c_1 \leq c(\boldsymbol{x}) \leq c_2, \tag{7.267}$$

and a function $W(\Theta, \boldsymbol{x}, t)$ satisfies Lipschitzian conditions according to Θ and t, i.e.

$$|W(\Theta_1, \boldsymbol{x}, t) - W(\Theta_2, \boldsymbol{x}, t)| \leq C_1 |\Theta_1 - \Theta_2|, \quad C_1 = \text{constant}, \boldsymbol{x} \in \Omega, t \in I = (t_o, t_1), \tag{7.268}$$

$$|W(\Theta, \boldsymbol{x}, t_1) - W(\Theta, \boldsymbol{x}, t_2)| \leq C_2 |t_1 - t_2|, \quad C_2 = \text{constant}, \boldsymbol{x} \in \Omega, t \in I = (t_o, t_1). \tag{7.269}$$

Let $\overline{I} = \langle t_o, t_1 \rangle$ and let $\boldsymbol{n}$ be the unit outward normal to $\Gamma_c = \cup \Gamma_c^{kl}$ related to Ω^k, $u_n = u_i n_i$ and $\boldsymbol{u}_t = \boldsymbol{u} - u_n \boldsymbol{n}$ be the normal and tangential components of the displacement vector $\boldsymbol{u}$, $\tau_n(\boldsymbol{u})$ and $\vec{\tau}_t(\boldsymbol{u})$ be the normal and tangential components of the stress vector $\vec{\tau}(\boldsymbol{u}) = (\tau_i(\boldsymbol{u}))$, $\tau_i(\boldsymbol{u}) = \tau_{ij}(\boldsymbol{u})n_j$ and $\tau_n = \tau_{ij}(\boldsymbol{u})n_i n_j$, $\vec{\tau}_t = \vec{\tau} - \tau_n \boldsymbol{n}$ and τ_{ij} be the stress tensor, and $\mathscr{F}_c^{kl}$ be the coefficient of Coulombian friction. If $\mathscr{F}_c^{kl} \equiv 0$ we have a case without friction and then $\vec{\tau}_t = 0$. Furthermore, Einstein's summational convention will also be used. Moreover, we denote $\dot{\boldsymbol{u}} = \partial \boldsymbol{u}/\partial t$.

Given a smoothly bounded convex domain $\Omega = \cup_{\iota=1}^S \Omega^\iota \subset R^2$ with boundary $\partial\Omega$, $\partial\Omega = \overline{\Gamma}_u \cup \overline{\Gamma}_\tau$, $\overline{\Gamma}_u = \cup_{\iota=1}^S \overline{\Gamma}_u^\iota$, $\overline{\Gamma}_\tau = \cup_{\iota=1}^S \overline{\Gamma}_\tau^\iota$, occupied by colliding bodies Ω^ι with boundaries $\partial\Omega^\iota = \overline{\Gamma}_u^\iota \cup \overline{\Gamma}_\tau^\iota \cup \overline{\Gamma}_c^\iota$, $\overline{\Gamma}_c^\iota = \cup_{\substack{k,l\\k\neq l}} \overline{\Gamma}_c^{kl}$, $\Gamma_c^{kl} = \overline{\Omega}^k \cap \overline{\Omega}^l$, we consider the following problem.

Problem ($\mathscr{P}_O$): Find a pair of functions $(\Theta, \boldsymbol{u})$ satisfying the heat conduction equation

$$\rho(\boldsymbol{x})c(\boldsymbol{x})\frac{\partial\Theta(\boldsymbol{x},t)}{\partial t} + \rho(\boldsymbol{x})\beta_{ij}(\boldsymbol{x})\Theta_o(\boldsymbol{x})e_{ij}(\dot{\boldsymbol{u}}(\boldsymbol{x},t)) - \frac{\partial}{\partial x_i}\left(\kappa_{ij}(\boldsymbol{x})\frac{\partial\Theta(\boldsymbol{x},t)}{\partial x_j}\right) = W(\boldsymbol{x},t)$$

$$\text{for a.e. } (\boldsymbol{x},t) \in \Omega \times I, \quad i,j = 1,2, \tag{7.270}$$

and the equation of motion for the case of uniformly moving lithospheric plates and blocks in linear thermo-elasticity

$$\frac{\partial}{\partial x_j}[c_{ijkl}(\boldsymbol{x})e_{kl}(\boldsymbol{u}(\boldsymbol{x},t)) - \beta_{ij}(\boldsymbol{x})(\Theta(\boldsymbol{x},t) - \Theta_o(\boldsymbol{x}))] - f_i(\boldsymbol{x},t) = 0$$

$$\text{for a.e. } (\boldsymbol{x},t) \in \Omega \times I, \quad i,j,k,l = 1,2, \tag{7.271}$$

and the following boundary conditions

$$\left.\begin{aligned} &\Theta = 0 && (7.272)\\ &\tau_{ij}n_j = P_i, && (7.273)\end{aligned}\right\} (\boldsymbol{x},t) \in \Gamma_\tau, \quad \forall t,$$

$$\left.\begin{aligned} &\Theta = \Theta_1 \left(\text{or } \kappa_{ij}\frac{\partial\Theta}{\partial x_j}n_i = q\right), && (2.274)\\ &\boldsymbol{u} = \boldsymbol{u}_o, && (2.275)\end{aligned}\right\} (\boldsymbol{x},t) \in \Gamma_u, \quad \forall t,$$

$$\left.\begin{aligned} &\Theta^k = \Theta^l, && (2.276)\\ &\kappa_{ij}\left.\frac{\partial\Theta}{\partial x_j}\right|_{(k)} = \kappa_{ij}\left.\frac{\partial\Theta}{\partial x_j}\right|_{(l)}, && (7.277)\end{aligned}\right\} (\boldsymbol{x},t) \in \cup_{k,l}\Gamma_c^{k,l}, \quad \forall t,$$

$$\left.\begin{aligned} &u_n^k - u_n^l \le 0, && (7.278)\\ &\tau_n^k = -\tau_n^l \le 0, && (7.279)\\ &\tau_n^k(u_n^k - u_n^l) = 0, && (7.280)\end{aligned}\right\} (\boldsymbol{x},t) \in \cup_{k,l}\Gamma_c^{k,l}, \quad \forall t,$$

(i.e. the so-called Signorini condition) and the Coulombian law of friction in the form

$$\left.\begin{aligned} &u_n^k - u_n^l = 0 \text{ then } |\vec{\tau}_t| \le g_c^{kl} = \mathscr{F}_c^{kl}\left|\tau_n^k\right|, && (7.281)\\ &\text{if } |\vec{\tau}_t| < g_c^{kl} = \mathscr{F}_c^{kl}\left|\tau_n^k\right| \text{ then } \frac{\partial}{\partial t}\left(\boldsymbol{u}_t^k - \boldsymbol{u}_t^l\right) = 0, && (7.282)\\ &\text{if } |\vec{\tau}_t| \le g_c^{kl} = \mathscr{F}_c^{kl}\left|\tau_n^k\right| \text{ then there exists } \lambda \ge 0 \text{ such that } \frac{\partial}{\partial t}\left(\boldsymbol{u}_t^k - \boldsymbol{u}_t^l\right) = -\lambda\vec{\tau}_t && (7.283)\end{aligned}\right\} \text{on } \cup\Gamma_c^{kl}, \quad \forall t.$$

Let R^m be the phase change boundaries between two different phases, recrystallized phases (we denote them by indices S and L), at time t. Then these boundaries divide the domain Ω into domains Ω_S^m and Ω_L^m. On the boundary $R^m(t)$ the following conditions are given:

$$\Theta_S^m = \Theta_L^m = \Theta_R^m, \tag{7.284}$$

$$\kappa_{ij}^m \frac{\partial \Theta^m}{\partial x_j} \nu_i \bigg|_{(S)}^m - \kappa_{ij}^m \frac{\partial \Theta^m}{\partial x_j} \nu_i \bigg|_{(L)}^m = -\rho^m L^m v_\nu^m, \tag{7.285}$$

where ρ^m is the density, Θ_R^m the temperature of the phase transition, Θ_S^m and Θ_L^m the temperatures at the phase change boundary $R^m(t)$ in the recrystallized (solid and liquid) phases, respectively, ν^m the unit normal to $R^m(t)$ pointing towards Ω_S^m, v_ν^m is the speed of $R^m(t)$ along ν^m, and L^m is the latent heat of phase transition or recrystallization. Furthermore, we shall assume that c_S^m, c_L^m, ρ_S^m and ρ_L^m are specific heats and densities in both recrystallized phases, and $\rho_S^m = \rho_L^m = \rho^m$ on $R^m(t)$.

Moreover, the initial condition

$$\Theta(\boldsymbol{x}, t_o) = \Theta_o(\boldsymbol{x}) \tag{7.286}$$

is given.

Let us transform the coupled contact-Stefan-like problem to the problem with homogeneous boundary conditions. Let us replace Θ by $\Theta + z$, where z is a sufficiently smooth scalar function in $\overline{\Omega}$ satisfying Equation 7.274, and $\boldsymbol{u}$ by $\boldsymbol{u} + \boldsymbol{w}$, where $\boldsymbol{w}$ is a sufficiently smooth vector function in $\overline{\Omega}$ satisfying Equations 7.275 and 7.278 to 7.280, or in a special case $\boldsymbol{w} = 0$ on $\cup \Gamma_c^{kl}$ as in the previous sections (in the following we will use the same symbols as above). According to these transformations the boundary conditions, thermal sources $W(\boldsymbol{x})$, surface force $\boldsymbol{P}(\boldsymbol{x})$ and body forces $\boldsymbol{f}(\boldsymbol{x}, t)$ will be changed. Then Equations 7.274 and 7.275 lead to

$$\Theta(\boldsymbol{x}, t) = 0 \tag{7.287}$$

and

$$\boldsymbol{u}(\boldsymbol{x}, t) = 0 \quad \text{on } \Gamma_u, \quad \forall t. \tag{7.288}$$

Moreover, we obtain $W(\boldsymbol{x}, t) \equiv W(\boldsymbol{x}, t) + \partial/\partial x_i \left(\kappa_{ij}(\boldsymbol{x})(\partial z(\boldsymbol{x}, t)/\partial x_j)\right)$, $f_i(\boldsymbol{x}) \equiv f_i(\boldsymbol{x}) + \partial/\partial x_j (c_{ijkl}(\boldsymbol{x}) e_{kl}(\boldsymbol{w}))$, $P_i(\boldsymbol{x}) \equiv P_i(\boldsymbol{x}) - c_{ijkl}(\boldsymbol{x}) e_{kl}(\boldsymbol{w}) n_j$.

To solve the thermal part of the problem, the technique of the enthalpy method will be used. We define a generalized enthalpy $H(\Theta)$ as the subdifferential of the functional $\Phi(\Theta)$ defined by

$$\Phi(\Theta) = \tfrac{1}{2}\rho_L^m c_L^m (\Theta - \Theta_R^m)_+^2 + \tfrac{1}{2}\rho_S^m c_S^m (\Theta - \Theta_R^m)_-^2 + \rho^m L^m (\Theta - \Theta_R^m)_+, \tag{7.289}$$

i.e., $\partial\Phi(.) : R \to R$ is represented by a monotonically increasing multi-valued function of temperature with a jump discontinuity at the phase change temperature Θ_R. Hence

$$\partial\Phi(\Theta) \equiv H(\Theta) = \begin{cases} \rho_L^m c_L^m (\Theta - \Theta_R^m) + \rho^m L^m & \text{for } \Theta > \Theta_R^m, \\ \rho_S^m c_S^m (\Theta - \Theta_R^m) & \text{for } \Theta < \Theta_R^m, \\ \langle 0, \rho^m L^m \rangle & \text{for } \Theta = \Theta_R^m, \end{cases} \tag{7.290}$$

where $\langle 0, \rho^m L^m \rangle$ is the enthalpy interval and $\Phi : R \to R$ is a piecewise quadratic convex function and locally Lipschitz function, defined by Equation 7.289, satisfying

$$|\Phi(\Theta_1) - \Phi(\Theta_2)| \leq \max\{|\partial\Phi(\Theta_1)|, |\partial\Phi(\Theta_2)|\} |\Theta_1 - \Theta_2| \leq c |\Theta_1 - \Theta_2|, \quad \text{when } |\Theta_1|, |\Theta_2| \leq \mathrm{K}. \tag{7.291}$$

Since $\dfrac{\partial H}{\partial t} = \rho c \dfrac{\partial \Theta}{\partial t}$,then Equation 7.270 leads to

$$\frac{\partial H(\Theta(\boldsymbol{x}, \boldsymbol{t}))}{\partial \boldsymbol{t}} + \rho(\boldsymbol{x})\beta_{ij}(\boldsymbol{x})\Theta_o(\boldsymbol{x})e_{ij}(\dot{\boldsymbol{u}}(\boldsymbol{x}, t)) - \frac{\partial}{\partial x_i}\left(\kappa_{ij}(\boldsymbol{x})\frac{\partial\Theta(\boldsymbol{x}, t)}{\partial x_j}\right) = W(\boldsymbol{x}, t)$$
$$\text{a.e. } (\boldsymbol{x}, t) \in \Omega \times I, \quad i, j = 1, 2. \tag{7.292}$$

Moreover, Equations 7.272, 7.274 and 7.286 become

$$H(\boldsymbol{x}, t) = 0, \quad (\boldsymbol{x}, t) \in \partial\Omega \times I, \tag{7.293}$$

$$H(\boldsymbol{x}, t_o) = H_o(\boldsymbol{x}). \tag{7.294}$$

In the following we shall assume that the solution of the problem ($\mathscr{P}$) exists and is unique. In the case of the two-phase Stefan-like problem described in Equations 7.292, 7.293, 7.276, 7.277 and 7.294, an implicit discretization in time will be used. Then the problem leads, at every time t, to a boundary value problem for a non-linear partial differential equation of elliptic type, while the problem in Equations 7.271, 7,273, 7.275 and 7.278 to 7.283 represents the contact problem with friction. The dissipative term $\rho(\boldsymbol{x})\beta_{ij}(\boldsymbol{x})\Theta_o(\boldsymbol{x})e_{ij}(\dot{\boldsymbol{u}}(\boldsymbol{x}, t))$ and the term $(\partial/\partial x_i)(\beta_{ij}(\boldsymbol{x})(\Theta(\boldsymbol{x}, t) - \Theta_o(\boldsymbol{x}))$ represent the coupling terms, where the first term represents the dissipative energy evoked by the deformation of rocks with dispersion of heat, and the second term, $\beta_{ij}(x)(\Theta(\boldsymbol{x}, t) - \Theta_o(\boldsymbol{x}))$, represents thermal stresses. Equations 7.292 and 7.271 and the boundary conditions in Equations 7.293, 7,273 to 7.281, 7.284, 7.285 and 7.294 represent the boundary value and contact problem of elliptic type for a coupled contact-two-phase Stefan problem in an enthalpy formulation (see, for example, Nedoma and Dvořák (1995)). Hence, after discretizing

$$\frac{\partial H(\Theta)}{\partial t} = \frac{H(\Theta^{n+1}) - H(\Theta^n)}{\Delta t}, \ \Theta^n = \Theta(\boldsymbol{x}, n\Delta t),$$

then Equation 7.292 yields

$$H(\Theta^{n+1}) - \Delta t \frac{\partial}{\partial x_i}\left(\kappa_{ij}\frac{\partial\Theta^{n+1}}{\partial x_j}\right) + \Delta t \rho \Theta_o \beta_{ij} e_{ij}\left(\dot{\boldsymbol{u}}^{n+1}\right) = H(\Theta^n) + \Delta t W^{n+1}, \tag{7.295}$$

where

$$W^{n+1} = \Delta t^{-1} \int_{n\Delta t}^{(n+1)\Delta t} W(\underline{\tau}) \mathrm{d}\underline{\tau}, \quad \dot{\boldsymbol{u}}^{n+1} = \Delta t^{-1}\left(\boldsymbol{u}^{n+1} - \boldsymbol{n}^n\right).$$

To solve the two-phase Stefan problem numerically, the formulation in the form of a generalized non-linear boundary value problem (see Barbu (1976) and Glowinski

(1979)) of the form

$$-\mu\Delta w + \alpha w + \partial\Phi(w) \in g(\boldsymbol{x}), \quad \text{a.e. } \boldsymbol{x} \in \Omega, \tag{7.296}$$

$$w = 0, \quad \text{a.e. } \boldsymbol{x} \in \partial\Omega \tag{7.297}$$

will be used. Since $w \in H^2(\Omega)$, $g(\boldsymbol{x}) \in L^2(\Omega)$, then the Equations 7.296 and 7.297 have a unique solution (see Barbu (1976)). This differential equation is discontinuous (as Θ passes through Θ_R^m) and hence, together with the above boundary conditions, it represents the free boundary problem. The corresponding free boundaries R^m represent the phase change boundaries between two different phases, recrystallized phases, at time t.

For simplicity let us put $\Theta \equiv \Theta^{n+1}$, $\boldsymbol{u} \equiv \boldsymbol{u}^{n+1}$, $Q = H(\Theta^n) + \Delta t W^{n+1} + \alpha\Theta_R$, $\mu \equiv \Delta t$, $\alpha\Theta + \partial\Phi(\Theta) \equiv H(\Theta) + \alpha\Theta_R$, where $0 < \alpha \leq \min\{\rho_S^m c_S^m, \rho_L^m c_L^m\}$. Then Equations 7.295 and 7.296 yield

$$\alpha\Theta - \mu\frac{\partial}{\partial x_i}\left(\kappa_{ij}\frac{\partial\Theta}{\partial x_j}\right) + \partial\Phi(\Theta) + \mu\rho\Theta_o\beta_{ij}e_{ij}(\dot{\boldsymbol{u}}) \ni Q(\boldsymbol{x}), \text{ a.e. } \boldsymbol{x} \in \Omega, \tag{7.298}$$

$$\Theta = 0, \quad \text{a.e. } \boldsymbol{x} \in \partial\Omega, \tag{7.299}$$

which are equivalent to Equations 7.296 and 7.297. Therefore, the theories of Glowinski (1979), for the thermal part of the problem, and of Nedoma (1983),(1987),(1994), for the elastic part of the problem, can be applied.

7.3.2.2. Variational formulation of the elliptic problem arising from an implicit time discretization

Let $\Omega \subset R^2$ be the domain with Lipschitz boundary at the moment t, t being fixed, defined above. Let $\boldsymbol{x} = (x_1, x_2)$ be Cartesian coordinates and $\boldsymbol{n} = (n_1, n_2)$ denote the unit outward normal defined above.

Let us denote by $(.,.)$ and $\langle .,. \rangle$ scalar products in $L^2(\Omega)$ or $[L^2(\Omega)]^2$ and in $[L^2(\cup\Gamma_c^{kl})]^2$, by $\|.\|_{k,2}$ the norm in $[H^k(\Omega)]^2$, k is an integer, $k \in R^1$, $H^k(\Omega)$ is the Sobolev space in the usual sense, i.e. $\|.\|_{0,1}$ is the norm in $H^o(\Omega)$, $\|.\|_{1,1}$ the norm in $H^1(\Omega)$ and $\|.\|_{1,2}$ the norm in $[H^1(\Omega]^2$.

Let $\Theta(\boldsymbol{x}) \in H^1(\Omega)$ be the temperature, $\boldsymbol{u}(\boldsymbol{x}) = (u_i(\boldsymbol{x})) \in [H^1(\Omega)]^2$ be the displacement vector and $e_{ij}(\boldsymbol{u})$ and $\tau_{ij}(\boldsymbol{u})$ be the small strain and stress tensors defined above. Let $\beta_{ij} \in C^1(\overline{\Omega})$ be a coefficient of linear thermal expansion, $\Theta_o = \Theta_o(\boldsymbol{x}) \in H^1(\Omega)$ be an initial temperature at $t = t_o$, and let coefficients of thermal conductivity $\kappa_{ij}(\boldsymbol{x}) \in C^1(\overline{\Omega})$ and elastic coefficients $c_{ijkl}(\boldsymbol{x}) \in C^1(\overline{\Omega})$ be Lipschitz and satisfy the symmetry conditions, i.e. satisfy conditions

$$\kappa_{ij}(\boldsymbol{x}) = \kappa_{ji}(\boldsymbol{x}) \quad \text{on } \Omega, \tag{7.300}$$

$$c_{ijkl}(\boldsymbol{x}) = c_{jikl}(\boldsymbol{x}) = c_{klij}(\boldsymbol{x}) \quad \text{on } \Omega, \tag{7.301}$$

$$\kappa_{ij}(\boldsymbol{x})\zeta_i\zeta_j \geq c_1 |\zeta|^2 \quad c_1 = \text{constant} > 0, \quad \forall\zeta \in R^2, \quad \forall\boldsymbol{x} \in \Omega, \tag{7.302}$$

$$0 < c_o \leq (c_{ijkl}(\boldsymbol{x})\xi_{ij}\xi_{kl})\,|\xi|^{-2} \leq C_o < +\infty \quad c_o, C_o = \text{constant} > 0,$$
$$\forall \xi \in R^4, \quad \forall \boldsymbol{x} \in \Omega. \tag{7.303}$$

Let heat sources $Q(\boldsymbol{x}) \in L^2(\overline{\Omega})$, body forces $\boldsymbol{f}(\boldsymbol{x}) \in [L^2(\Omega)]^2$, surface forces $\boldsymbol{P}(\boldsymbol{x}) \in [L^2(\Gamma_\tau)]^2$ and let a density $\rho(\boldsymbol{x}) \in C(\overline{\Omega})$, a specific heat $c(\boldsymbol{x}) \in C(\overline{\Omega})$ and $g_c^{kl} \in L^\infty(\Gamma_c^{kl})$. Moreover, since $\beta_{ij} \in C^1(\overline{\Omega})$, $\Theta, \Theta_o \in H^1(\Omega)$, $\boldsymbol{u} \in [H^1(\Omega)]^2$, therefore $\partial/\partial x_j \beta_{ij}(\boldsymbol{x})(\Theta(\boldsymbol{x}) - \Theta_o(\boldsymbol{x})) \in L^2(\Omega)$, and $\mu\rho(\boldsymbol{x})\Theta_o(\boldsymbol{x})\beta_{ij}(\boldsymbol{x})e_{ij}(\dot{\boldsymbol{u}}) \in L^2(\Omega)$.

Thus we have to solve the following problem.

Problem ($\mathscr{P}$): Find $(\Theta, \boldsymbol{u})$ satisfying

$$\alpha\Theta - \mu\frac{\partial}{\partial x_i}\left(\kappa_{ij}\frac{\partial\Theta}{\partial x_j}\right) + \partial\Phi(\Theta) + \mu\rho\Theta_o\beta_{ij}e_{ij}(\dot{\boldsymbol{u}}) \ni Q(\boldsymbol{x}), \qquad \text{for a.e. } \boldsymbol{x} \in \Omega, \tag{7.304}$$

$$\frac{\partial}{\partial x_i}[c_{ijkl}(\boldsymbol{x})e_{kl}(\boldsymbol{u}(\boldsymbol{x},t)) - \beta_{ij}(\boldsymbol{x})(\Theta(\boldsymbol{x},t) - \Theta_o(\boldsymbol{x}))] - f_i(\boldsymbol{x},t) = 0$$
$$\text{for a.e. } \boldsymbol{x} \in \Omega, \quad i,j,k,l = 1,2,\ t \text{ being fixed}, \tag{7.305}$$

and the following boundary conditions

$$\left.\begin{aligned} &H = 0 && (7.306)\\ &\tau_{ij}n_j = P_i, && (7.307)\end{aligned}\right\} (\boldsymbol{x},t) \in \Gamma_\tau,\ t \text{ being fixed}$$

$$\left.\begin{aligned} &H = 0 && (7.308)\\ &\boldsymbol{u} = 0, && (7.309)\end{aligned}\right\} (\boldsymbol{x},t) \in \Gamma_u,\ t \text{ being fixed}$$

$$\left.\begin{aligned} &\Theta^k = \Theta^l, && (7.310)\\ &\kappa_{ij}\frac{\partial\Theta}{\partial x}n_i\Big|_{(k)} = \kappa_{ij}\frac{\partial\Theta}{\partial x_j}n_i\Big|_{(l)}, && (7.311)\end{aligned}\right\} \boldsymbol{x} \in \cup_{k,l}\Gamma_c^{k,l},\ t \text{ being fixed}$$

$$\left.\begin{aligned} &u_n^k - u_n^l \leq 0, && (7.312)\\ &\tau_n^k = -\tau_n^l \leq 0 && (7.313)\\ &\tau_n^k(u_n^k - u_n^l) = 0, && (7.314)\end{aligned}\right\} \boldsymbol{x} \in \cup_{k,l}\Gamma_c^{k,l},\ t \text{ being fixed}$$

(i.e. the so-called Signorini conditions)

$$u_n^k - u_n^l = 0 \text{ then } |\vec{\tau}_t| \leq g_c^{kl} = \mathscr{F}_c^{kl}\left|\tau_n^k\right|, \tag{7.315}$$
$$\text{if } |\vec{\tau}_t| < g_c^{kl} = \mathscr{F}_c^{kl}\left|\tau_n^k\right| \text{ then } \boldsymbol{u}_t^k - \boldsymbol{u}_t^l = 0, \tag{7.316}$$
$$\text{if } |\vec{\tau}_t| \leq g_c^{kl} = \mathscr{F}_c^{kl}\left|\tau_n^k\right| \text{ then there exists } \lambda \geq 0 \text{ such that } \boldsymbol{u}_t^k - \boldsymbol{u}_t^l = -\lambda\vec{\tau}_t \quad \text{on } \cup\Gamma_c^{kl},\ t \text{ being fixed}, \tag{7.317}$$

(i.e. the Coulombian law of friction), where we denoted $\alpha\Theta + \partial\Phi(\Theta) \equiv H(\Theta^{n+1}) + \alpha\Theta_R^m$, $Q = H(\Theta^n) + \Delta t W^{n+1} + \alpha\Theta_R^m$ and $\mu \equiv \Delta t$.

Let us define the space of virtual temperatures as

$$^1V = \{z \mid z \in H^1(\Omega),\ z = 0 \text{ on } \partial\Omega \text{ in the sense of traces}\},$$

the space of virtual displacements as

$$V = \{\boldsymbol{v} \mid \boldsymbol{v} \in W = [H^1(\Omega)]^2, \boldsymbol{v} = 0 \text{ on } \Gamma_u \text{ in the sense of traces } \}$$

and the set of admissible displacements as

$$K = \{\boldsymbol{v} \mid \boldsymbol{v} \in V, \quad v_n^k - v_n^l \leq 0 \text{ on } \cup \Gamma_c^{kl}\}.$$

On ${}^1V$ we define the bilinear form $a_\Theta(.,.)$, scalar product $(.,.)$ and functionals $s(.)$, $b_p(.,.)$ and $j(.)$ by

$$a_\Theta(\Theta, z) = \int_\Omega \kappa_{ij} \frac{\partial \Theta}{\partial x_i} \frac{\partial z}{\partial x_j} \mathrm{d}\boldsymbol{x}, \tag{7.318}$$

$$(\Theta, z) = \mu \int_\Omega \Theta z \mathrm{d}\boldsymbol{x}, \tag{7.319}$$

$$s(z) = \mu \int_\Omega Q z \mathrm{d}\boldsymbol{x} \left(-\mu \int_{\Gamma_u} \kappa_{ij} \frac{\partial \Theta}{\partial x_j} n_i z \mathrm{d}s \right), \tag{7.320}$$

$$b_p(\boldsymbol{v}, z) = \mu \int_\Omega \rho \Theta_o \beta_{ij} \partial v_i / \partial x_j z \mathrm{d}\boldsymbol{x}, \tag{7.321}$$

$$j(z) = \mu \int_\Omega \Phi(z(\boldsymbol{x})) \mathrm{d}\boldsymbol{x}, \tag{7.322}$$

and on V the bilinear form $a(.,.)$ and functionals $b_s(\Theta, .)$, $j_{gn}(.)$ and $S(.)$ by

$$a(\boldsymbol{u}, \boldsymbol{v}) = \int_\Omega c_{ijkl} e_{ij}(\boldsymbol{u}) e_{kl}(\boldsymbol{v}) \mathrm{d}\boldsymbol{x}, \tag{7.323}$$

$$b_s(\Theta, \boldsymbol{v}) = \int_\Omega \frac{\partial}{\partial x_j} (\beta_{ij} \Theta) v_i \mathrm{d}\boldsymbol{x}, \tag{7.324}$$

$$j_{gn} = \int_{\cup \Gamma_c^{kl}} g_c^{kl} \left| \boldsymbol{v}_t^k - \boldsymbol{v}_t^l \right| \mathrm{d}s = \left\langle g_c^{kl}, \left| \boldsymbol{v}_t^k - \boldsymbol{v}_t^l \right| \right\rangle, \tag{7.325}$$

$$S(\boldsymbol{v}) = \int_\Omega f_i v_i \mathrm{d}\boldsymbol{x} + \int_{\Gamma_\tau} P_i v_i \mathrm{d}s, \tag{7.326}$$

where $\langle .,. \rangle$ denotes a scalar product in $[L^2(\Gamma_c)]^2$.

Then the variational or weak formulation of the problem given is as follows.

Problem $(\mathscr{P})_V$: Find a pair of functions $(\Theta, \boldsymbol{u})$, $\Theta \in {}^1V, \boldsymbol{u} \in K$, such that

$$\alpha(\Theta, z - \Theta) + \mu a_\Theta(\Theta, z - \Theta) + b_p(\boldsymbol{u}, z - \Theta) + j(z) - j(\Theta) \geq s(z - \Theta) \quad \forall z \in {}^1V, \tag{7.327}$$

$$a(\boldsymbol{u}, \boldsymbol{v} - \boldsymbol{u}) + b_s(\Theta, \boldsymbol{v} - \boldsymbol{u}) + j_{gn}(\boldsymbol{v}) - j_{gn}(\boldsymbol{u}) \geq S(\boldsymbol{v} - \boldsymbol{u}) \quad \forall \boldsymbol{v} \in K. \tag{7.328}$$

Definition 7.9: A pair of functions $(\Theta, \boldsymbol{u})$, $\Theta \in {}^1V$, $\boldsymbol{u} \in K$ satisfying Equations 7.327 and 7.328 is said to be a weak solution of the Problem $(\mathscr{P})$, i.e. of Equations 7.304 to 7.317.

Lemma 7.20: The set K is non-empty, closed and convex in V.

Proof: K is evidently non-empty as a zero function f is also a member of K. The set K is closed, which follows from the fact that the space $[H^1(\Omega)]^2$ is complete and from the trace theorem. Let $\boldsymbol{v}, \boldsymbol{w} \in K$, $\alpha \in \langle 0, 1\rangle$, then the function $\boldsymbol{y} = \alpha\boldsymbol{v} + (1-\alpha)\boldsymbol{w} \in V$ and for its normal component y_n it holds that $y_n = y_i n_i = \alpha v_i n_i + (1-\alpha) w_i n_i \leq 0$ on $\cup\Gamma_c^{kl}$. Hence K is convex.

Theorem 7.19: Let Equations 7.306 to 7.326 hold, then there exists a unique solution of the problem $(\mathscr{P})_v$, i.e. of Equations 7.327 and 7.328.

Proof: Since Equation 7.327 is a non-linear elliptic variational inequality of the second kind and since $\Phi(z)$ is a continuous convex function, and $j(z)$ is a convex continuous functional on $H_o^1(\Omega)$, then it is equivalent to the generalized formulation of the non-linear elliptic variational inequality of the second kind of Glowinski (1979). Moreover, as mentioned above, $\partial/\partial x_j \beta_{ij}(\boldsymbol{x})(\Theta(\boldsymbol{x}) - \Theta_o(\boldsymbol{x})) \in L^2(\Omega)$ and $\mu\rho(\boldsymbol{x})\Theta_o(\boldsymbol{x})\beta_{ij}(\boldsymbol{x})e_{ij}(\dot{\boldsymbol{u}}) \in L^2(\Omega)$. Due to Lemma 7.20 the set K is closed and convex in $[H^1(\Omega)]^2$, hence it is weakly closed, as a closed and convex ball in a Hilbert space is weakly closed. Let us denote $L(\boldsymbol{v}) = L_o(\boldsymbol{v}) + j_{gn}(\boldsymbol{v})$, where $L_o(\boldsymbol{v}) = \frac{1}{2}a(\boldsymbol{v}, \boldsymbol{v}) + b_s(\Theta, \boldsymbol{v}) - S(\boldsymbol{v})$. The functional $L_o(\boldsymbol{v})$ is coercive and weakly lower semi-continuous, and the functional $j_{gn}(\boldsymbol{v})$ is convex and continuous, and thus it is weakly lower semi-continuous, and $j_{gn}(\boldsymbol{v}) = \langle g_c^{kl}, |\boldsymbol{v}_t^k - \boldsymbol{v}_t^l|\rangle \geq 0$ (as for arbitrary $g_c^{kl} \in L^\infty(\Gamma_c^{kl})$, we have $\langle g_c^{kl}, |\boldsymbol{v}_t^k - \boldsymbol{v}_t^l|\rangle \geq 0$ as $g_c^{kl} \geq 0$ a.e. on Γ_c^{kl}). The coerciveness follows from Equations 7.308 and 7.309 and Korn's inequality (see Nedoma (1987)), i.e.

$$L(\boldsymbol{v}) \geq c\,\|\boldsymbol{v}\|_{1,2}^2 - c_1\,\|\boldsymbol{v}\|_{1,2} \quad \forall \boldsymbol{v} \in \mathrm{V}.$$

For two points $\boldsymbol{v} + \boldsymbol{u}$ and $\boldsymbol{u}$, using Equations 7.308 and 7.309 and Korn's inequality we find

$$DL(\boldsymbol{v} + \boldsymbol{u}, \boldsymbol{v}) - DL(\boldsymbol{u}, \boldsymbol{v}) \geq c_o \int_\Omega e_{ij}(\boldsymbol{v})e_{ij}(\boldsymbol{v})\mathrm{d}\boldsymbol{x} \geq \|v\|_{1,2} \quad \forall \boldsymbol{v} \in V,$$

i.e. we have proved the weakly lower semi-continuity.

Let us assume that $\partial\beta_{ij}/\partial x_j \in L^\infty(\Omega)\,\forall i, j$. Then, moreover,

$$\begin{aligned} |b_s(\Theta - \Theta_o, \boldsymbol{v}) + b_p(\boldsymbol{v}, \Theta)| &= \left| \int_\Omega (\partial/\partial x_j(\beta_{ij}(\Theta - \Theta_o)v_i + \mu\rho\Theta_o\beta_{ij}\partial v_i/\partial x_j\Theta)\mathrm{d}\boldsymbol{x} \right| \\ &\leq c((1 + \|\Theta(t)\|_{1,1})\,\|\boldsymbol{v}(t)\|_{0,2} + \|\Theta(t)\|_{0,1}\,\|\boldsymbol{v}(t)\|_{1,2}). \end{aligned} \tag{7.329}$$

As it is known that the generalized formulation of the non-linear elliptic variational inequality of the second kind (see Glowinski (1979)) has a unique solution, and since the functional $L(\boldsymbol{v})$ is coercive and weakly lower semi-continuous, it has a minimum in V. Hence, and from all the above facts, there also exists a unique solution of Equations 7.327 and 7.328.

Remark: In the case of non-uniformly moving colliding lithospheric plates and blocks, the inertia force term $\rho\partial^2\boldsymbol{u}/\partial t^2 \equiv \rho\partial\boldsymbol{v}/\partial t$, where $\boldsymbol{v} = \partial\boldsymbol{u}/\partial t$, is added in Equation 7.271 and the boundary and initial conditions (Equations 7.272 to 7.286) are modified. But an implicit discretization, similar as in the Stefan-like problem, leads to the variational formulation of the type given by Equations 7.327 and 7.328, where the bilinear form is denoted by $A(\boldsymbol{u}, \boldsymbol{v})$, where $A(\boldsymbol{u}, \boldsymbol{v}) \equiv a(\boldsymbol{u}, \boldsymbol{v}) + (\rho\boldsymbol{u}, \boldsymbol{v})$, $\boldsymbol{u} \equiv \boldsymbol{u}^{n+1}$ and $S(\boldsymbol{v}) \equiv S(\boldsymbol{v}) + (\rho\boldsymbol{u}^n, \boldsymbol{v})$ and $a(\boldsymbol{u}, \boldsymbol{v})$ and $S(\boldsymbol{v})$ are defined above. It can be shown that the bilinear form $A(\boldsymbol{u}, \boldsymbol{v})$ and the functional $\underline{S}(\boldsymbol{v})$ have the same properties as the bilinear form $a(\boldsymbol{u}, \boldsymbol{v})$ and the functional $S(\boldsymbol{v})$.

7.3.2.3. The finite element approximation

Let us assume that $\Omega \subset R^2$ is polygonal. Let the domain Ω be triangulated, i.e. $\overline{\Omega} = \Omega \cup \partial\Omega$ is covered by a finite number of triangles T_h, forming a triangulation

$\mathcal{T}_h$. Further, we assume that the end points $\overline{\Gamma}_u \cap \overline{\Gamma}_\tau$, $\overline{\Gamma}_u \cap \overline{\Gamma}^k_c$, $\overline{\Gamma}^k_c \cap \overline{\Gamma}_\tau$ coincide with the vertices of T_h. Let $h_o > 0$. The family $\{\mathcal{T}_h\}$, $0 < h \leq h_o$, of triangulations is assumed to be regular. Let us define the spaces of finite elements

$$^1V_h = \{z \mid z \in C(\overline{\Omega}), \quad z_{|T_h} \in P_1, z = 0 \text{ on } \partial\Omega, \quad \forall T_h \in \mathcal{T}_h\},$$

$$V_h = \{\boldsymbol{v} \mid \boldsymbol{v} \in [C(\overline{\Omega})]^2, \quad \boldsymbol{v}_{|T_h} \in [P_1]^2, \boldsymbol{v} = 0 \text{ on } \Gamma_u, \quad \forall T_h \in \mathcal{T}_h\},$$

where P_1 is the space of all linear polynomials. Let

$$K_h = \{\boldsymbol{v} \mid \boldsymbol{v} \in V_h, v^k_n - v^l_n \leq 0 \text{ on } \cup \Gamma^{kl}_c\}$$

be the set of finite element approximations of the set of admissible displacements.

Definition 7.10: A pair of functions $(\Theta_h, \boldsymbol{u}_h)$ is said to be a finite element approximation of the problem $(\mathcal{P})_v$, we denote it as the problem $(\mathcal{P})_{vh}$, if

$$\alpha(\Theta_h, z_h - \Theta_h) + \mu a_\Theta(\Theta_h, z_h - \Theta_h) + b_p(\boldsymbol{u}_h, z_h - \Theta_h) + j(z_h) - j(\Theta_h) \geq s(z_h - \Theta_h) \quad \forall z_h \in {}^1V_h, \tag{7.330}$$

$$a(\boldsymbol{u}_h, \boldsymbol{v}_h - \boldsymbol{u}_h) + b_s(\Theta_h, \boldsymbol{v}_h - \boldsymbol{u}_h) + j_{gn}(\boldsymbol{v}_h) - j_{gn}(\boldsymbol{u}_h) \geq S(\boldsymbol{v}_h - \boldsymbol{u}_h) \quad \forall \boldsymbol{v}_h \in K_h. \tag{7.331}$$

Let us define

$$\Omega^m_+ = \{\boldsymbol{x} \mid \Theta(\boldsymbol{x}) > \Theta^m_R\}, \tag{7.332}$$

$$\Omega^m_- = \{\boldsymbol{x} \mid \Theta(\boldsymbol{x}) < \Theta^m_R\}, \tag{7.333}$$

$$\Omega^m_o = \{\boldsymbol{x} \mid \Theta(\boldsymbol{x}) = \Theta^m_R\}, \tag{7.334}$$

$$\Omega^m_{+h} = \{T_h \in \mathcal{T}_h \mid \Theta > \Theta^m_R \text{ on } T_h\}, \tag{7.335}$$

$$\Omega^m_{-h} = \{T_h \in \mathcal{T}_h \mid \Theta < \Theta^m_R \text{ on } T_h\}, \tag{7.336}$$

$$\Omega^m_{oh} = \{T_h \in \mathcal{T}_h \mid \Theta = \Theta^m_R \text{ on } T_h\}. \tag{7.337}$$

Let $\Omega^m_{*h} = \Omega \backslash \cup (\Omega^m_{+h} \cup \Omega^m_{-h} \cup \Omega^m_{oh})$ be the set of all triangles which are closely situated on the phase change boundary. It is evident that Ω^m_{*h} is not empty as the phase change boundary is contained in some triangles, i.e. $\Omega \neq \cup(\Omega^m_{+h} \cup \Omega^m_{-h} \cup \Omega^m_{oh}) \neq 0$. The length of the phase change boundary satisfies

$$\begin{aligned}\text{meas}\{\cup R^m(t)\} &= \text{meas}\{\cup(\partial\Omega^m_+ \cap \partial\Omega^m_-)\} + \text{meas}\{\cup(\partial\Omega^m_o \cap \partial\Omega^m_-)\}\\ &\quad + \text{meas}\{\cup(\partial\Omega^m_o \cap \partial\Omega^m_+)\} < M\end{aligned}$$

i.e. it is finite, and, moreover, there exists a constant c, independent of h and such that the number of triangles N^m_{*h} in Ω^m_{*h} is bounded by $N^m_{*h} < ch^{-1}$.

Theorem 7.20: Let Equations 7.306 to 7.326 hold. Then there exists a unique finite element approximation of the problem $(\mathcal{P})_v$ for all h.

Proof: Since $\Phi(z)$ is a continuous convex function, $j(z)$ is a convex continuous functional on $H^1_o(\Omega)$, then for the discrete approximate problem given by Equation 7.330 there exists a unique

Θ_h such that it also uniquely solves the optimization problem

$$l_h(\Theta_h) = \min_{z \in {}^1V_h} l_h(z) \equiv \tfrac{1}{2}\alpha(z,z)_h + \tfrac{1}{2}\mu a_\Theta(z,z) + b_p(\boldsymbol{v}, z) + j_h(z) - (Q,z)_h, \tag{7.338}$$

where $(.,.)_h$ is the discrete approximation of the scalar product in L^2 and

$$(\Theta, z)_h = \sum_{T_h \in \mathscr{T}_h} \frac{1}{3} \text{meas } (T_h) \sum_{i=1}^{3} \Theta_i^T z_i^T, \tag{7.339}$$

$$j_h(z) = \mu \int_\Omega \Phi(z(\boldsymbol{x})_{LI}) \mathrm{d}\boldsymbol{x} = \mu \sum_{T_h \in \mathscr{T}_h} \frac{1}{3} \text{meas } (T_h) \sum_{i=1}^{3} \Phi(z_i^T), \tag{7.340}$$

where meas(T_h) is the area of the triangle T_h, and z_i^T is the value of z at the node i of the triangle T_h; the same stands for $b_{ph}(\boldsymbol{v}, z)$, when we put $g = \mu \rho \Theta_o \beta_{ij} \partial v_i / \partial x_j$, which will be taken as a part of the thermal sources.

The set K_h is closed and convex, therefore it is weakly closed. Let $L(\boldsymbol{u}_h)$ be a functional defined by $L(\boldsymbol{v}) = \frac{1}{2}a(\boldsymbol{v}, \boldsymbol{v}) - S(\boldsymbol{v}) + j_{gn}(\boldsymbol{v})$, where the coupled term is taken as a part of the body forces. Theorem 7.19 implies that the functional $L(\boldsymbol{v})$ is coercive over K_h and $j_{gn}(\boldsymbol{v})$ is convex and that Equation 7.329 holds. Then the proof is similar to that of Theorem 7.19.

The uniqueness can be proved by the usual techniques. Let $(\Theta_h^1, \boldsymbol{u}_h^1) \in {}^1V_h \times K_h$ and $(\Theta_h^2, \boldsymbol{u}_h^2) \in {}^1V_h \times K_h$ be two approximate solutions of the problem $(\mathscr{P})_v$. Then $c_1 \left\|\Theta_h^1 - \Theta_h^2\right\|^2 \leq 0$, $c_2 \left\|\boldsymbol{u}_h^1 - \boldsymbol{u}_h^2\right\|^2 \leq 0$, from which the uniqueness follows immediately.

To find an estimate of the rates of convergence of $\|\Theta - \Theta_h\|$ and $\|\boldsymbol{u} - \boldsymbol{u}_h\|$ the technique proposed by Falk (1974) and modified by Nedoma (1983),(1987),(1994) will be used.

For any pair of functions $(z, \boldsymbol{v})$ on Ω the Lagrange interpolate $(z_{LI}, \boldsymbol{v}_{LI})$, where $z_{LI} \in {}^1V_h$, $\boldsymbol{v}_{LI} \in V_h$ and $(z_{LI}, \boldsymbol{v}_{LI}) = (z, \boldsymbol{v})$ at each vertex of triangulation $\mathscr{T}_h$, is defined in the usual way. Then we have the well-known interpolate estimate (see Ciarlet (1978))

$$\|z - z_{LI}\|_{0,1} + h\,\|z - z_{LI}\|_{1,1} \leq ch^2, \quad \forall z \in H_o^1(\Omega) \cap H^2(\Omega), \tag{7.341}$$

$$\|\boldsymbol{v} - \boldsymbol{v}_{LI}\|_{0,2} + h\,\|\boldsymbol{v} - \boldsymbol{v}_{LI}\|_{1,2} \leq ch^2, \quad \forall \boldsymbol{v} \in [H_o^1(\Omega)]^2 \cap [H^2(\Omega)]^2. \tag{7.342}$$

For $\Theta, z \in {}^1V_h$ we have Equations 7.339 and 7.340. Further, we need the following lemma.

Lemma 7.21 (Ciavaldini,Raviart): The following estimates

$$|(\Theta, z)_h - (\Theta, z)| \leq ch^2 \|\Theta\|_{1,1} \|z\|_{1,1} \quad \forall \Theta, z \in^1 V_h, \tag{7.343}$$

$$\|z\|_{0,1} \leq (z,z)_h^{1/2} \leq c\,\|z\|_{0,1} \quad \forall z \in^1 V_h, \tag{7.344}$$

hold.
For the proof see Ciavaldini (1975), Raviart (1972).

Lemma 7.22: It holds that

$$j(z) \leq j_h(z) \quad \forall z \in {}^1V_h, \tag{7.345}$$

$$\|j(\Theta_{LI}) - j(\Theta)\| \leq c\,\|\Theta - \Theta_{LI}\|_{0,1}\,, \tag{7.346}$$

$$\|j_h(\Theta_{LI}) - j(\Theta_{LI})\| \leq ch^2. \tag{7.347}$$

Proof: The proof follows from the results of Elliott (1981). To prove Euqation 7.345 for any element $T_h \in \mathcal{T}_h$ we denote by $w_i(\boldsymbol{x})$, $0 \leq w_i(\boldsymbol{x}) \leq 1$, the basis functions for each node i. Then $\sum_{i=1}^{3} w_i(\boldsymbol{x}) = 1$. Since the functional $\Phi(z)$ is a continuous convex function, then

$$\Phi(z(\boldsymbol{x})) = \Phi\left(\sum_{i=1}^{3} z_i w_i(\boldsymbol{x})\right) \leq \sum_{i=1}^{3} w_i(\boldsymbol{x})\Phi(z_i).$$

Since $j(z)$ is a convex continuous functional on $H_o^1(\Omega)$, then

$$j(z) = \mu\int_\Omega \Phi(z(\boldsymbol{x}))\mathrm{d}\boldsymbol{x} = \mu\sum_{T_h\in\mathcal{T}_h}\int_{T_h}\Phi(z(\boldsymbol{x}))\mathrm{d}\boldsymbol{x} \leq \mu\sum_{T_h\in\mathcal{T}_h}\int_{T_h} w_i(\boldsymbol{x})\Phi(z_i)\mathrm{d}\boldsymbol{x}$$

$$= \mu\int_\Omega (\Phi(z(\boldsymbol{x})))_{LI}\mathrm{d}\boldsymbol{x} = j_h(z).$$

To prove Equation 7.346 then, from the fact that $\Phi : R \to R$ is a piecewise quadratic convex function defined by Equation 7.289 and locally Lipschitz continuous, i.e. that Equation 7.291,

$$|\Phi(\Theta_1) - \Phi(\Theta_2)| \leq \max\{|\partial\Phi(\Theta_1)|\,,|\partial\Phi(\Theta_2)|\}\,|\Theta_1 - \Theta_2| \leq c\,|\Theta_1 - \Theta_2|\,, \text{ when } |\Theta_1|\,,|\Theta_2| \leq K,$$

holds and that Θ is continuous, we obtain

$$|j(\Theta_{LI}) - j(\Theta)| = \left|\mu\int_\Omega(\Phi(\Theta_{LI}) - \Phi(\Theta))\mathrm{d}\boldsymbol{x}\right| \leq c(\|\Theta\|_{L^\infty(\Omega)})\int_\Omega |\Theta - \Theta_{LI}|\,\mathrm{d}\boldsymbol{x} \leq c\,\|\Theta - \Theta_{LI}\|_{0,1}\,.$$

To prove Equation 7.347 we firstly have

$$|j(\Theta_{LI}) - j_h(\Theta_{LI})| = \left|\mu\sum_{T_h\in\mathcal{T}_h}\int_{T_h}[\Phi(\Theta_{LI}) - \Phi(\Theta_{LI}(\boldsymbol{x}))_{LI})\mathrm{d}\boldsymbol{x}\right|. \tag{7.348}$$

Now we shall investigate the behaviour of the last integral of Equation 7.348 in the neighbourhood of the free boundary $R^m(t)$. Therefore, we shall investigate the last integral of Equation 7.348 for $T_h \in \Omega^m_{+h}, \Omega^m_{-h}, \Omega^m_{oh}$ and for $T_h \in \Omega^m_{*h} = \Omega\backslash \cup (\Omega^m_{+h} \cup \Omega^m_{-h} \cup \Omega^m_{oh})$, which are defined by Equations 7.335 to 7.337 separately.

Let $T_h \in \Omega^m_{oh}$, then $\Theta_{LI} \equiv \Theta^m_R$ and $\Phi(\Theta_{LI}) \equiv 0$. Hence,

$$\int_{T_h\in\Omega^m_{oh}}[\Phi(\Theta_{LI}) - \Phi(\Theta_{LI}(\boldsymbol{x}))_{LI}]\mathrm{d}\boldsymbol{x} = 0. \tag{7.349}$$

Let $T_h \in \Omega^m_{+h}$ and $T_h \in \Omega^m_{-h}$. Due to Equation 7.289 $\Phi(.)$ is a quadratic function of Θ_{LI} and therefore, similarly as for Lemma 7.20, we have

$$\left|\mu\int_{T_h\in\Omega^m_{+h}}[\Phi(\Theta_{LI}) - \Phi(\Theta_{LI}(\boldsymbol{x}))_{LI}]\mathrm{d}\boldsymbol{x}\right| \leq ch^2, \tag{7.350}$$

$$\left|\mu\int_{T_h\in\Omega^m_{-h}}[\Phi(\Theta_{LI}) - \Phi(\Theta_{LI}(\boldsymbol{x}))_{LI}]\mathrm{d}\boldsymbol{x}\right| \leq ch^2, \tag{7.351}$$

Now let $T_h \in \Omega^m_{*h}$. Then due to the Lipschitz continuity of $\Phi(.)$, defined by Equation 7.291, and since we assumed that the number of triangles N^m_{*h} in Ω_{*h} is bounded by $N^m_{*h} < ch^{-1}$, we find

$$\left|\mu\int_{T_h\in\Omega^m_{*h}}(\Phi(\Theta_{LI}) - \Phi(\Theta_{LI}(\boldsymbol{x}))_{LI})\mathrm{d}\boldsymbol{x}\right| = \left|\mu\left(\int_{T_h\in\Omega^m_{*h}}\Phi(\Theta_{LI})\mathrm{d}\boldsymbol{x} - \frac{1}{3}\mathrm{meas}(T_h)\sum_{i=1}^{3}\Phi(\Theta_i^T)\right)\right|$$

$$= \frac{1}{3}\left|\sum_{i=1}^{3}\mu\int_{T_h\in\Omega^m_{*h}}(\Phi(\Theta_{LI})\mathrm{d}\boldsymbol{x} - \Phi(\Theta_i^T)\right|,$$

where as above we denote by Θ_i^T the value of Θ at the node i of T_h. Since due to Equation 7.291 $\Phi(.)$ is locally Lipschitz continuous then

$$\begin{aligned}\frac{1}{3}\left|\sum_{i=1}^{3}\mu\int_{T_h\in\Omega_{*h}^m}\Phi(\Theta_{LI})\mathrm{d}\boldsymbol{x}-\Phi(\Theta_i^T)\right| &\leq \frac{1}{3}\sum_{i=1}^{3}c\int_{T_h\in\Omega_{*h}^m}\left|\Theta_{LI}(\boldsymbol{x})-\Theta_i^T\right|\mathrm{d}\boldsymbol{x}\\ &= \frac{1}{3}\sum_{i=1}^{3}c\int_{T_h\in\Omega_{*h}^m}|(\boldsymbol{x}-\boldsymbol{x}_i).\text{grad }\Theta_{LI}|\,\mathrm{d}\boldsymbol{x}\\ &\leq \frac{1}{3}\sum_{i=1}^{3}c\left(\int_{T_h\in\Omega_{*h}^m}|\boldsymbol{x}-\boldsymbol{x}_i|^2\right)^{1/2}\left(\int_{T_h\in\Omega_{*h}^m}|\text{grad }\Theta_{LI}|^2\right)^{1/2}\mathrm{d}\boldsymbol{x}\\ &\leq ch^2\left(\int_{T_h\in\Omega_{*h}^m}|\text{grad }\theta|^2\right)^{1/2}\mathrm{d}x+\left(\int_{T_h\in\Omega_{*h}^m}|\text{grad }(\boldsymbol{x}-\boldsymbol{x}_i)|^2\right)^{1/2}\\ &\leq ch^2(h\,\|\Theta\|_{W^{1,\infty}}+ch).\end{aligned}$$

Hence

$$\left|\mu\int_{T_h\in\Omega_{*h}^m}(\Phi(\Theta_{LI}(\boldsymbol{x}))-\Phi(\Theta_{LI}(\boldsymbol{x}))_{LI})\mathrm{d}\boldsymbol{x}\right| \leq \sum_{T_h\in\Omega_{*h}^m} ch^3 \leq ch^2. \tag{7.352}$$

From Equations 7.348 to 7.352 we obtain

$$|j_h(\Theta_{LI})-j(\Theta_{LI})|\leq ch^2,$$

which completes the proof.

Lemma 7.23: For $\Theta\in{}^1V$, $\boldsymbol{u}\in K$, $\Theta_h\in{}^1V_h$,$\boldsymbol{u}_h\in K_h$ we have

$$\begin{aligned}\alpha\,\|\Theta-\Theta_h\|_{0,1}^2+\mu\,\|\Theta-\Theta_h\|_{1,1}^2 \leq\; & c[\alpha(\Theta_h-\Theta,z_h-\Theta)+\mu a_\Theta(\Theta_h-\Theta,z_h-\Theta)+\alpha(\Theta,z-\Theta_h)\\ &+\mu a_\Theta(\Theta,z-\Theta_h)+\alpha(\Theta,z_h-\Theta)+\mu a_\Theta(\Theta,z_h-\Theta)\\ &+b_p(\boldsymbol{u},z_h-\Theta)+b_p(\boldsymbol{u},z-\Theta_h)+j(z)-j(\Theta_h)+j(z_h)-j(\Theta)\\ &-s(z-\Theta_h)-s(z_h-\Theta)]^{1/2}\qquad \forall z\in{}^1V,\ z_h\in{}^1V_h,\end{aligned}\tag{7.353}$$

$$\begin{aligned}\|\boldsymbol{u}-\boldsymbol{u}_h\|_{1,2}\leq\; & c_o[a(\boldsymbol{u}_h-\boldsymbol{u},\boldsymbol{v}_h-\boldsymbol{u})+a(\boldsymbol{u},\boldsymbol{v}-\boldsymbol{u}_h)+a(\boldsymbol{u},\boldsymbol{v}_h-\boldsymbol{u})-S(\boldsymbol{v}-\boldsymbol{u}_h)\\ &-S(\boldsymbol{v}_h-\boldsymbol{u})+j_{gn}(\boldsymbol{v}_h)-j_{gn}(\boldsymbol{u})+j_{gn}(\boldsymbol{v})-j_{gn}(\boldsymbol{u}_h)+b_s(\Theta,\boldsymbol{v}-\boldsymbol{u}_h)\\ &+b_s(\Theta,\boldsymbol{v}_h-\boldsymbol{u})]^{1/2}\quad \forall\boldsymbol{v}\in K,\boldsymbol{v}_h\in K_h,\ c_o=\text{constant}>0.\end{aligned}\tag{7.354}$$

Proof: Let $\Theta\in{}^1V$, $\Theta_h\in{}^1V_h$, $\boldsymbol{u}\in K$, $\boldsymbol{u}_h\in K_h$. Using Equations 7.330, 7.331, 7,327 and 7.328, and then adding $\alpha(\Theta,\Theta-\Theta_h)-\alpha(\Theta_h,\Theta)+\mu a_\Theta(\Theta,\Theta-\Theta_h)-\mu a_\Theta(\Theta_h,\Theta)$ and $a(\boldsymbol{u},\boldsymbol{u}-\boldsymbol{u}_h)-a(\boldsymbol{u}_h-\boldsymbol{u})$ to both parts in Θ and $\boldsymbol{u}$, then we obtain directly

$$\begin{aligned}&\alpha(\Theta,\Theta-\Theta_h)-\alpha(\Theta_h,\Theta)-\alpha(\Theta,\Theta-\Theta_h)+\alpha(\Theta_h,\Theta)+\alpha(\Theta,z-\Theta)+\alpha(\Theta_h,z_h-\Theta_h)\\ &\quad+\mu a_\Theta(\Theta,z-\Theta)+\mu a_\Theta(\Theta_h,z_h-\Theta_h)+\mu a_\Theta(\Theta,\Theta-\Theta_h)-\mu a_\Theta(\Theta_h,\Theta)-\mu a_\Theta(\Theta,\Theta-\Theta_h)\\ &\quad+\mu a_\Theta(\Theta_h,\Theta)+b_p(\boldsymbol{u},z-\Theta)+b_p(\boldsymbol{u}_h,z_h-\Theta_h)+j(z)-j(\Theta)+j(z_h)-j(\Theta_h)\\ &\quad-s(z-\Theta)-s(z_h-\Theta_h)\geq 0.\end{aligned}$$

$$\begin{aligned}&a(\boldsymbol{u},\boldsymbol{v}-\boldsymbol{u})+a(\boldsymbol{u}_h,\boldsymbol{v}_h-\boldsymbol{u}_h)+a(\boldsymbol{u},\boldsymbol{u}-\boldsymbol{u}_h)-a(\boldsymbol{u}_h,\boldsymbol{u})-a(\boldsymbol{u},\boldsymbol{u}-\boldsymbol{u}_h)+a(\boldsymbol{u}_h,\boldsymbol{u})+b_s(\Theta,\boldsymbol{v}-\boldsymbol{u})\\ &\quad+b_s(\Theta_h,\boldsymbol{v}_h-\boldsymbol{u}_h)-S(\boldsymbol{v}-\boldsymbol{u})-S(\boldsymbol{v}_h-\boldsymbol{u}_h)+j_{gn}(\boldsymbol{v}_h)-j_{gn}(\boldsymbol{u})+j_{gn}(\boldsymbol{v})-j_{gn}(\boldsymbol{u}_h)\geq 0.\end{aligned}$$

From here it follows that

$$\begin{aligned}
&\alpha(\Theta, \Theta - \Theta_h) - \alpha(\Theta_h, \Theta - \Theta_h) + \mu a_\Theta(\Theta, \Theta - \Theta_h) - \mu a_\Theta(\Theta_h, \Theta - \Theta_h)\\
&\quad = \alpha(\Theta - \Theta_h, \Theta - \Theta_h) + \mu a_\Theta(\Theta - \Theta_h, \Theta - \Theta_h)\\
&\quad \leq \alpha(\Theta_h - \Theta, z_h - \Theta) + \mu a_\Theta(\Theta_h - \Theta, z_h - \Theta) + \alpha(\Theta, z - \Theta_h) + \mu a_\Theta(\Theta, z - \Theta_h)\\
&\qquad + \alpha(\Theta, z_h - \Theta) + \mu a_\Theta(\Theta, z_h - \Theta) + b_p(\boldsymbol{u}, z_h - \Theta) + b_p(\boldsymbol{u}_h, z - \Theta_h)\\
&\qquad + j(z) - j(\Theta_h) + j(z_h) - j(\Theta) + s(\Theta_h - z) + s(\Theta - z_h).\\
&a(\boldsymbol{u}, \boldsymbol{u} - \boldsymbol{u}_h) - a(\boldsymbol{u}_h, \boldsymbol{u} - \boldsymbol{u}_h) = a(\boldsymbol{u} - \boldsymbol{u}_h, \boldsymbol{u} - \boldsymbol{u}_h)\\
&\quad \leq a(\boldsymbol{u}_h - \boldsymbol{u}, \boldsymbol{v}_h - \boldsymbol{u}) + a(\boldsymbol{u}, \boldsymbol{v} - \boldsymbol{u}_h) + a(\boldsymbol{u}, \boldsymbol{v}_h - \boldsymbol{u}) + b_s(\Theta, \boldsymbol{v} - \boldsymbol{u}_h) + b_s(\Theta_h, \boldsymbol{v}_h - \boldsymbol{u})\\
&\qquad + S(\boldsymbol{u}_h - \boldsymbol{v}) + S(\boldsymbol{u} - \boldsymbol{v}_h) + j_{gn}(\boldsymbol{v}_h) - j_{gn}(\boldsymbol{u}) + j_{gn}(\boldsymbol{v}) - j_{gn}(\boldsymbol{u}_h).
\end{aligned}$$

Hence after some algebra the assertions in Equations 7.353 and 7.354 of the lemma follow immediately.

Corollary: Let ${}^1V_h \subset {}^1V$, $K_h \subset$K. Then

$$\begin{aligned}
&\alpha \|\Theta - \Theta_h\|_{0,1}^2 + \mu \|\Theta - \Theta_h\|_{1,1}^2 + \|\boldsymbol{u} - \boldsymbol{u}_h\|_{1,2}^2 \leq c[\alpha(\Theta_h - \Theta, z_h - \Theta) + \mu a_\Theta(\Theta_h - \Theta, z_h - \Theta)\\
&\quad + a(\boldsymbol{u}_h - \boldsymbol{u}, \boldsymbol{v}_h - \boldsymbol{u}) + \alpha(\Theta, z_h - \Theta) + \mu a_\Theta(\Theta, z_h - \Theta) + a(\boldsymbol{u}, \boldsymbol{v}_h - \boldsymbol{u}) + b_p(\boldsymbol{u}, \Theta_h - \Theta)\\
&\quad + b_p(\boldsymbol{u}, z_h - \Theta_h) + b_s(\Theta, \boldsymbol{u}_h - \boldsymbol{u}) + b_s(\Theta, \boldsymbol{v}_h - \boldsymbol{u}_h) + j(z_h) - j(\Theta) + j_{gn}(\boldsymbol{v}_h) - j_{gn}(\boldsymbol{u})\\
&\quad + s(\Theta - z_h) + S(\boldsymbol{u} - \boldsymbol{v}_h)] \quad \forall z_h \in {}^1V_h, \boldsymbol{v}_h \in K_h.
\end{aligned} \tag{7.355}$$

Proof: Let ${}^1V_h \subset {}^1V$, $K_h \subset K$. Let us put $z = \Theta_h$ in Equation 7.353 and $\boldsymbol{v} = \boldsymbol{u}_h$ in Equation 7.354. When the resulting inequalities are added, it leads immediately to Equation 7.355.

The main result of this section is the next theorem.

Theorem 7.21: Let $\Gamma_c = \cup_{k,l}\Gamma_c^{kl}$ be polygonal, let free boundary $\cup_m R^m(t)$ be finite. Let $\Theta \in W^{1,\infty}(\Omega) \cap H^2(\Omega)$, $\boldsymbol{u} \in K \cap [H^2(\Theta)]^2$, $\boldsymbol{u}\,|_{\Gamma_c} \in [H^2(\Gamma_c)]^2$. Let Ω_+^m, Ω_-^m, Ω_o^m, Ω_{+h}^m, Ω_{-h}^m, Ω_{oh}^m be defined by Equations 7.332 to 7.337 and let $N_{*h}^m < c/h$. Then

$$\alpha \|\Theta - \Theta_h\|_{0,1}^2 + \mu \|\Theta - \Theta_h\|_{1,1}^2 + \|\boldsymbol{u} - \boldsymbol{u}_h\|_{1,2}^2 \leq ch^2. \tag{7.356}$$

Proof: A technique similar to that of Nedoma (1983),(1987), Nedoma and Dvořák (1995) can be used. The proof follows from the Corollary of Lemma 7.23. We estimate Equation 7.355 and we shall assume that Θ and $\boldsymbol{u}$ are sufficiently regular. Terms $\alpha(\Theta, z - \Theta_h) + \mu a_\Theta(\Theta, z - \Theta_h) - s(z - \Theta_h)$, $\alpha(\Theta, z_h - \Theta) + \mu a_\Theta(\Theta, z_h - \Theta) - s(z_h - \Theta)$ and $a(\boldsymbol{u}, \boldsymbol{v} - \boldsymbol{u}_h) - (\boldsymbol{f}, \boldsymbol{v} - \boldsymbol{u}_h)$, $a(\boldsymbol{u}, \boldsymbol{v}_h - \boldsymbol{u}) - (\boldsymbol{f}, \boldsymbol{v}_h - \boldsymbol{u})$ are estimated by applying the Green lemma and later by using a suitable choice of $z \in {}^1V$, $z_h \in {}^1V_h$, $\boldsymbol{v} \in K$, $\boldsymbol{v}_h \in K_h$. Then

$$\begin{aligned}
&\alpha \|\Theta - \Theta_h\|_{0,1}^2 + \mu \|\Theta - \Theta_h\|_{1,1}^2 + \|\boldsymbol{u} - \boldsymbol{u}_h\|_{1,2}^2\\
&\quad \leq c[\alpha(\Theta_h - \Theta, z_h - \Theta) + \mu a_\Theta(\Theta_h - \Theta, z_h - \Theta) + a(\boldsymbol{u}_h - \boldsymbol{u}, \boldsymbol{v}_h - \boldsymbol{u})\left(+\mu \int_{\Gamma_u} q(z_h - \Theta)\mathrm{d}s\right)\\
&\qquad + \int_{\partial\Omega} \tau_{ij}(\boldsymbol{u})n_j(\boldsymbol{v}_h - \boldsymbol{u})_i \mathrm{d}s - \int_{\Gamma_\tau} P_i(\boldsymbol{v}_h - \boldsymbol{u})_i \mathrm{d}s + j(z_h) - j(\Theta) + j(z) - j(\Theta_h)\\
&\qquad + b_p(\boldsymbol{u}, z_h - \Theta) + b_s(\Theta - \Theta_o, \boldsymbol{v}_h - \boldsymbol{u}) + b_p(\boldsymbol{u}, z - \Theta_h) + b_s(\Theta - \Theta_o, \boldsymbol{v} - \boldsymbol{u}_h)]
\end{aligned}$$

$$\leq c[\alpha \|\Theta_h - \Theta\|_{0,1} \|z_h - \Theta\|_{0,1} + \mu \|\Theta_h - \Theta\|_{1,1} \|z_h - \Theta\|_{1,1} + \|\boldsymbol{u}_h - \boldsymbol{u}\|_{1,2} \|\boldsymbol{v}_h - \boldsymbol{u}\|_{1,2}$$

$$+ \int_{\cup\Gamma_c^{kl}} ((\boldsymbol{v}_h^k - \boldsymbol{v}_h^l)_n - (\boldsymbol{u}_h^k - \boldsymbol{u}_h^l)_n)\mathrm{d}s + \int_{\Gamma_u} \Theta,_n (z - \Theta_h)\mathrm{d}s + j(z_h) - j(\Theta) + j(z) - j(\Theta_h)$$

$$+ b_p(\boldsymbol{u}, z_h - \Theta) + b_s(\Theta - \Theta_o, \boldsymbol{v}_h - \boldsymbol{u}) + b_p(\boldsymbol{u}, z - \Theta_h) + b_s(\Theta - \Theta_o, \boldsymbol{v} - \boldsymbol{u}_h)]$$

$$\leq c\left[\frac{1}{2}\varepsilon \|\Theta - \Theta_h\|_{0,1}^2 + \frac{1}{2}\varepsilon^{-1} \|\Theta - z_h\|_{0,1}^2 + \frac{1}{2}\varepsilon \|\Theta - \Theta_h\|_{1,1}^2 + \frac{1}{2}\varepsilon^{-1} \|\Theta - z_h\|_{1,1}^2\right.$$

$$+ \frac{1}{2}\varepsilon \|u_h - u\|_{1,2}^2 + \frac{1}{2}\varepsilon^{-1} \|\boldsymbol{v}_h - \boldsymbol{u}\|_{1,2}^2 + \int_{\cup\Gamma_c^{kl}} \tau_{ij}(\boldsymbol{u}) n_j ((\boldsymbol{v}_h^k - \boldsymbol{v}_h^l)_n - (u_n^k - u_n^l))\mathrm{d}s$$

$$+ \int_{\Gamma_u} \Theta,_n (z - \Theta_h)\mathrm{d}s + j(z_h) - j(\Theta) + j(z) - j(\Theta_h) + b_p(\boldsymbol{u}, z_h - \Theta) + b_s(\Theta - \Theta_o, \boldsymbol{v}_h - \boldsymbol{u})$$

$$\left. + b_p(\boldsymbol{u}, z - \Theta_h) + b_s(\Theta - \Theta_o, \boldsymbol{v} - \boldsymbol{u}_h)\right].$$

To estimate the last inequality, we put $z_h = \Theta_{LI}$, and $\boldsymbol{v}_h = \boldsymbol{u}_{LI}$, where $\Theta_{LI} \in {}^1V_h$ is the Lagrange interpolation of Θ on triangulation $\mathscr{T}_h$ and $\boldsymbol{u}_{LI} \in V_h$ is the Lagrange interpolation of $\boldsymbol{u}$ on triangulation $\mathscr{T}_h$. The bilinear form $a(.,.)$ is bounded so that $a(\boldsymbol{u}_h - \boldsymbol{u}, \boldsymbol{v}_h - \boldsymbol{u}) \leq M \|\boldsymbol{u}_h - \boldsymbol{u}\|_{1,2} \|\boldsymbol{v}_h - \boldsymbol{u}\|_{1,2}$. Since for $\boldsymbol{v} \in [C(\overline{\Omega})]^2 \cap K$ we have $\boldsymbol{v}_{LI} \in K_h$ and since the last inequality is valid for an arbitrary $\boldsymbol{v}_h \in K_h$, it is also valid for $\boldsymbol{v}_h = \boldsymbol{u}_{LI}$. The interpolation theorem yields (see Nedoma (1987), Nedoma and Dvořák (1995)) that

$$a(\boldsymbol{u}_h - \boldsymbol{u}, \boldsymbol{v}_h - \boldsymbol{u}) \leq \tfrac{1}{2}M\varepsilon \|\boldsymbol{u}_h - \boldsymbol{u}\|^2 + ch^2.$$

And as $(\boldsymbol{u}_{LI}^k - \boldsymbol{u}_{LI}^l)_n \leq 0$ on $\cup\Gamma_c^{kl}$, thus $\boldsymbol{u}_{LI} \in K$. Since $\boldsymbol{u}_{LI} \in V_h$, then $\boldsymbol{u}_{LI} \in K_h$. Thus

$$\|\boldsymbol{u}_{LI} - \boldsymbol{u}\|_{1,2} \leq c_r h \|\boldsymbol{u}\|_{1,2},$$

$$\left\|(\boldsymbol{u}_{LI}^k - \boldsymbol{u}_{LI}^l)_n - (u_n^k - u_n^l)\right\| \leq c_s h^2 \sum_{\cup\Gamma_c^{kl}} \left\|\boldsymbol{u}_n^k - \boldsymbol{u}_n^l\right\|_{[H^2(\Gamma_\tau)]^2}.$$

According to Lemma 7.21 and Lemma 7.22 we find

$$j(\Theta_h) - j(\Theta) + j_h(\Theta_{LI}) - j_h(\Theta_h) \leq j_h(\Theta_{LI}) - j(\Theta)$$

$$= j_h(\Theta_{LI}) - j(\Theta_{LI}) + j(\Theta_{LI}) - j(\Theta) \leq ch^2 + c\|\Theta - \Theta_{LI}\|_{0,1},$$

$$\alpha[(\Theta_h, \Theta_h - \Theta_{LI}) - (\Theta_h, \Theta_h - \Theta_{LI})_h] \leq ch^2,$$

$$\|\Theta_{LI} - \Theta\|_{1,1} \leq c_r h^2 \|\Theta\|_{2,1}, \quad \|\Theta_{LI} - \Theta\|_{0,1} \leq c_p h^2 \|\Theta\|_{2,1},$$

$$\alpha(\Theta_h, \Theta_{LI} - \Theta) + \mu a_\Theta(\Theta_h - \Theta, \Theta_{LI} - \Theta) - s(\Theta_{LI} - \Theta)$$

$$\leq \alpha \|\Theta_h\|_{0,1} \|\Theta_{LI} - \Theta\|_{0,1} + \mu \|\Theta_h - \Theta\|_{1,1} \|\Theta_{LI} - \Theta\|_{1,1} + \|Q\|_{0,1} \|\Theta_{LI} - \Theta\|_{0,1}$$

$$\leq c\|\Theta_{LI} - \Theta\|_{0,1} + \tfrac{1}{2}\varepsilon\mu \|\Theta_h - \Theta\|_{1,1}^2 + \tfrac{1}{2}\varepsilon^{-1}\mu \|\Theta_{LI} - \Theta\|_{1,1}^2.$$

Due to Equation 7.329 putting $z_h = \Theta_{LI}$ and $\boldsymbol{v}_h = \boldsymbol{u}_{LI}$,

$$\left|b_s(\Theta - \Theta_o, \boldsymbol{v}_h - \boldsymbol{u}) + b_p(\boldsymbol{u}, z_h - \Theta)\right| \leq c((1 + \|\Theta\|_{1,1}) \|\boldsymbol{u}_{LI} - \boldsymbol{u}\|_{0,2} + \|\Theta_{LI} - \Theta\|_{0,1} \|\boldsymbol{u}\|_{1,2})$$

and putting $z = \Theta_{LI} \in {}^1V_h$ and $\boldsymbol{v} = \boldsymbol{u}_{LI} \in V_h$

$$\left|b_s(\Theta - \Theta_o, \boldsymbol{u}_{LI} - \boldsymbol{u}_h) + b_p(\boldsymbol{u}, \Theta_{LI} - \Theta_h)\right| \leq c((1 + \|\Theta\|_{1,1}) \|\boldsymbol{u}_{LI} - \boldsymbol{u}_h\|_{0,2} + \|\Theta_{LI} - \Theta_h\|_{0,1} \|\boldsymbol{u}\|_{1,2}).$$

To estimate the integrals

$$J_1\left(\cup\Gamma_c^{kl}\right) = \int_{\cup\Gamma_c^{kl}} \tau_n(\boldsymbol{u})(\boldsymbol{v}_h - \boldsymbol{u})_n \mathrm{d}s \text{ and } J_2\left(\cup\Gamma_c^{kl}\right) = \int_{\cup\Gamma_c^{kl}} \vec{\tau}_t(\boldsymbol{u})(\boldsymbol{v}_h - \boldsymbol{u})_t \mathrm{d}s$$

we shall assume that $\Gamma_c = \cup\Gamma_c^{kl}$, where $\cup\overline{\Gamma}_c^{kl}$ approximates piece-wise linearly the boundary Γ_c. The results of Nedoma (1987),(1994) as well as of previous sections will be used. Then $|J_1(\Gamma_c^{kl})| \leq c_s h^2$ and

$$\left| \int_{\cup\Gamma_c^{kl}} \{\vec{\tau}_t^{kl}(\boldsymbol{u})[(\boldsymbol{u}_t^k - \boldsymbol{u}_t^l)_{LI} - (\boldsymbol{u}_t^k - \boldsymbol{u}_t^l)]\}\, \mathrm{d}s + j_{gn}(\boldsymbol{u}_{LI}) - j_{gn}(\boldsymbol{u}) \right| \leq c_r h^2.$$

As a result of all the above estimates, we find

$$|\boldsymbol{u} - \boldsymbol{u}_h| \leq c_N \left\{\varepsilon h^2 \,|\boldsymbol{u}|_2^2 + \varepsilon^{-1} h^2 \,|\boldsymbol{u}|^2 + c_s h^2 + c_r h^2\right\}^{1/2}, \tag{7.357}$$

ε being sufficiently small, which completes the proof.

7.3.2.4. Algorithm

Since we shall assume the case of uniformly moving colliding geological bodies with a friction between them, then the contact part of the problem leads to the Signorini problem with friction. The numerical solution of the Stefan problem in enthalpy approximation thus results in an implicit discretization in time and the FEM approximation in spatial coordinates, while the numerical solution of the contact part of the problem thus results in an FEM approximation of elliptic variational inequality. Since $\partial/\partial x_j(\beta_{ij}(\boldsymbol{x})(\Theta(\boldsymbol{x}) - \Theta_o(\boldsymbol{x}))) \in L^2(\Omega)$ and $\mu\rho(\boldsymbol{x})\Theta_o(\boldsymbol{x})\beta_{ij}(\boldsymbol{x})e_{ij}(\dot{\boldsymbol{u}}) \in L^2(\Omega)$ hold for the coupled terms and have the meaning of body forces and thermal sources, the algorithm used will be based on an implicit scheme and thus in every time level we minimize the functional

$$l_\Theta(z) = \tfrac{1}{2} z^T B z - d^T z + \phi(z) \text{ without constraints,} \tag{7.358}$$

where $B = \alpha I + \mu B_o$, and B_o represents the finite element approximation of $a_\Theta(.,.)$ in the case of the Stefan-like part of the problem and on the numerical approximation of a saddle point (Céa (1971),Ekeland and Temam (1976), Nedoma (1987)) and on methods of quadratic programming, i.e. minimization of the functional

$$L(\boldsymbol{v}) = \tfrac{1}{2}\boldsymbol{v}^T C \boldsymbol{v} + \boldsymbol{b}^T \boldsymbol{v} \text{ with constraints } A\boldsymbol{v} \leq 0, \tag{7.359}$$

for the elastic part of the problem. These problems represent non-linear minimization problems in the first case and the Uzawa or Arrow-Hurwicz algorithms and optimization problems with constraints in the second. The algorithm in the first case will be based, for example, on the non-linear (preconditioned) conjugate gradient or S.O.R. methods and in the second one the algorithm will be based on the active set method and the preconditioning conjugate gradient method.

It is evident that both problems are of non-linear type. Numerical experiments, studying both problems separately (Dvořák (1996), Kestřánek and Nedoma (1996)), show that the conjugate gradient method is one of the best methods which can be applied to both parts of the investigated problem. From this point of view in this chapter we shall investigate the following coupled type of algorithms.

Firstly, at every time we shall minimize the functional $l_\Theta(z)$, defined by Equation 7.358 in the case of the Stefan-like problem, based on the non-linear conjugate gradient method for the non-linear system of equations

$$Bz - d + \partial\Phi(z) = 0,$$

where $\partial\Phi(z)$ is the subdifferential of the functional $\Phi(z) = \sum_{i=1}^{n} \phi(z_i)$, where $\partial\Phi(z) = (\partial\phi(z_1), \ldots, \partial\phi(z_n))$, which is equivalent to

$$l_\Theta(\Theta) = \inf_{z \in R^n} l_\Theta(z),$$

and recendly, since the contact part of the investigated coupled problem leads to the numerical approximation of a saddle point, we shall find it by the Uzawa algorithm and the constrained conjugate gradient method.

Next we shall discuss the basic idea of the algorithm.

Let $(z_i)_{i=1}^n$ be a basis of the space ${}^1V_h$. Then every element of ${}^1V_h$ can be found as a linear combination of the basis $\{z_i\}$; similarly for Θ and Q. Then by the usual way we obtain a non-linear system of algebraic equations

$$F(z) \equiv Bz - d + \partial\Phi(z) = 0, \tag{7.360}$$

where $B = \alpha I + \mu B_o$, $\mu = \Delta t$, and where elements of B_o are generated by $\int_{\Omega_h} \kappa_{ij} \nabla z_i \nabla z_j d\boldsymbol{x}$ and the matrix B is sparse and symmetric positive definite and $\partial\Phi(z) = MH(z)$, where $M = (M_{ij}) = \{\int_{\Omega_h} z_i z_j \mathrm{d}\boldsymbol{x}\}_{i,j=1}^n$ and, moreover, where F denotes the non-linear differentiable mapping $F : R^n \to R^n$, where F is differentiable in R^n and has a meaning of the gradient of the functional $l_\Theta(z)$. Equation 7.360 is a necessary condition for the existence of a minimum of the functional $l_\Theta(z)$. Then, if the matrices M and B are positive definite, the functional l_Θ has a Găteaux derivative and is monotone on R^n. Then the functional l_Θ is uniformly convex on R^n and, therefore, there exists a minimizer z and hence Θ^{n+1}, H^{n+1} represent a unique solution of the investigated problem. To finish the proof of this assertion it is sufficient to prove the next lemma (Dvořák (1993)).

Lemma 7.24: Matrices M and B_o are sparse regular matrices.

Proof: We shall prove that M and B_o are the Gramm's matrices. Let $(\Theta, z)_h = \int_{\Omega_h} \Theta z \mathrm{d}\boldsymbol{x}$, $a_{\Theta h}(\Theta, z) = \int_{\Omega_h} \mathrm{grad}\, \Theta \, \mathrm{grad}\, z \mathrm{d}\boldsymbol{x}$ $\forall \Theta, z \in {}^1V_h$. If we prove that $(\Theta, z)_h$ and $a_{\Theta h}(\Theta, z)$ have on ${}^1V_h$ properties of inner products, then matrices M and B_o are Gramm's matrices and as the bases functions $\{\varphi_i\}$, $i = 1, \ldots, n$ of ${}^1V_h$ are linearly independent, then the matrices M and B_o are regular. We have

$$(\Theta, z)_h = \int_{\Omega_h} \Theta z \mathrm{d}\boldsymbol{x} = \int_{\Omega_h} z \Theta \mathrm{d}\boldsymbol{x} = (z, \Theta)_h, \tag{7.361}$$

$$(r\Theta + sz, w)_h = \int_{\Omega_h} (r\Theta + sz) w \mathrm{d}\boldsymbol{x} = r \int_{\Omega_h} \Theta w \mathrm{d}\boldsymbol{x} + s \int_{\Omega_h} zw \mathrm{d}\boldsymbol{x} = r(\Theta, w)_h + s(z, w)_h, \tag{7.362}$$

$$(\Theta, \Theta)_h = \int_{\Omega_h} \Theta^2 \mathrm{d}\boldsymbol{x} \geq 0. \tag{7.363}$$

Hence and using the property of the space ${}^1V_h$ it follows that

$$(\Theta, \Theta)_h = 0 \Leftrightarrow \Theta = 0 \text{ a.e.} \quad on \quad \Omega_h.$$

Similarly,

$$a_{\Theta h}(\Theta, z) = \int_{\Omega_h} \mathrm{grad}\, \Theta \, \mathrm{grad}\, z \mathrm{d}\boldsymbol{x} = \int_{\Omega_h} \mathrm{grad}\, z \, \mathrm{grad}\, \Theta \mathrm{d}\boldsymbol{x} = a_{\Theta h}(z, \Theta)_h, \tag{7.364}$$

$$a_{\Theta h}(r\Theta + sz, w) = \int_{\Omega_h} \text{grad } (r\Theta + sz) w \mathrm{d}\boldsymbol{x} = r \int_{\Omega_h} \text{grad } \Theta \text{ grad } w \mathrm{d}\boldsymbol{x} + s \int_{\Omega_h} \text{grad } z \text{ grad } w \mathrm{d}\boldsymbol{x}$$

$$= r a_{\Theta h}(\Theta, w) + s a_{\Theta h}(z, w), \tag{7.365}$$

$$a_{\Theta h}(\Theta, \Theta) = \int_{\Omega_h} (\text{ grad } \Theta)^2 \mathrm{d}\boldsymbol{x}. \tag{7.366}$$

Hence grad Θ grad $\Theta = 0$ if and only if all partial derivatives Θ are equal to zero a.e. on Ω_h, i.e. $\Theta =$ constant a.e. on Ω_h. It is valid for the function $\Theta = 0$ a.e. on Ω_h only as $\Theta \in {}^1V_h$ and, therefore, $\Theta = 0$ on Γ_u and Θ is approximated by linear polynomials and constant on triangles close to the boundary $\Gamma_u \cup \Gamma_\tau$. From here and the continuity of the function Θ it follows that $\Theta = 0$ a.e. in Ω_h. Therefore, we have proved that

$$a_{\Theta h}(\Theta, \Theta) = 0 \Leftrightarrow \Theta = 0 \text{ a.e. } \quad \text{in} \quad \Omega_h.$$

Then we have proved that $(\Theta, z)_h$ and $a_{\Theta h}(\Theta, z)$ have the properties of inner products and, therefore, the matrices M and B_o are regular. From the properties of $(\Theta, z)_h$ and $a_{\Theta h}(\Theta, z)$ it follows that M and B_o are positive semi-definite. Since they are regular then $\det(M) \neq 0$ and $\det(B_o) \neq 0$, and M and B_o are positive semi-definite. Hence, using lemma 7.25 the matrices M and B_o are symmetric, positive definite and, moreover, due to their construction they are also sparse.

Lemma 7.25: Let $A \in R^n$ be positive semi-definite, then A is positive definite $\Leftrightarrow \det(A) \neq 0$. For the proof see Fiedler (1981).

The second part of the algorithm of our coupled-contact two-phase Stefan-like problem is based on the numerical approximation of a saddle point and on the quadratic programming with constraints. Sufficient conditions for the existence of a saddle point can be found in Céa (1971), Ekeland and Temam (1976) and in the previous sections.

Let us define the set Λ as

$$\Lambda = \cup\{\mu^{kl} \mid \mu^{kl} \in L^2(\Gamma_c^{kl}), \quad \left|\mu^{kl}\right| \leq 1 \text{ a.e. } \quad \text{on} \quad \Gamma_c^{kl}\},$$

and let us denote by $\mathscr{L}(\boldsymbol{v}, \mu)$ the Lagrangian defined by

$$\mathscr{L}(\boldsymbol{v}, \mu) = L_o(\boldsymbol{v}) + b_s(\Theta, \boldsymbol{v}) + \int_{\cup\Gamma_c^{kl}} \mu^{kl} g_c^{kl}(\boldsymbol{v}_t^k - \boldsymbol{v}_t^l)\mathrm{d}s, \tag{7.367}$$

as $\int_{\cup\Gamma_c^{kl}} g_c^{kl} \left|\boldsymbol{v}_t^k - \boldsymbol{v}_t^l\right| \mathrm{d}s = \sup_{\mu\in\Lambda} \int_{\cup\Gamma_c^{kl}} \mu^{kl} g_c^{kl}(\boldsymbol{v}_t^k - \boldsymbol{v}_t^l)\mathrm{d}s$, where $g_c^{kl}(\boldsymbol{v}_t^k - \boldsymbol{v}_t^l)$ is a Lipschitz operator mapping $V \to L^2(\cup\Gamma_c^{kl})$. Moreover, $K_h \subset K \subset V$.

Definition 7.11: A point $(\boldsymbol{u}_{sh}, \lambda_h) \in K_h \times \Lambda \subset K \times \Lambda$ is said to be an approximate saddle point of a functional $\mathscr{L}$ on $K_h \times \Lambda$ if a saddle point $(\boldsymbol{u}_s, \Lambda) \in K \times \Lambda$ exists and if

$$\mathscr{L}(\boldsymbol{u}_{sh}, \mu) \leq \mathscr{L}(\boldsymbol{u}_{sh}, \lambda_h) \leq \mathscr{L}(\boldsymbol{v}, \lambda_h) \quad \forall(\boldsymbol{v}, \mu) \in K_h \times \Lambda. \tag{7.368}$$

Then the problem is equivalent to the following problem.

Problem $(\mathscr{P}_S)_h$: Find $(\Theta_h, \boldsymbol{u}_h)$, $\Theta_h \in {}^1V_h$, $\boldsymbol{u}_h \in K_h$ such that

$$l_h(\Theta_h) = \min_{z\in {}^1V}\{l_h(z) \equiv \tfrac{1}{2}\alpha(z, z)_h + \tfrac{1}{2}\mu a_\Theta(z, z) + b_p(\boldsymbol{u}_h, z) + j_h(z) - (Q, z)_h\}, \tag{7.369}$$

$$L_h(\boldsymbol{u}_h) = \min_{\boldsymbol{v}\in K_h} \sup_{\lambda\in\Lambda}\{\mathscr{L}(\boldsymbol{v},\mu) = L_o(\boldsymbol{v}) + b_s(\Theta_h,\boldsymbol{v}) + \int_{\cup\Gamma_c^{kl}} \mu^{kl} g_c^{kl}(\boldsymbol{v}_t^k - \boldsymbol{v}_t^l)\mathrm{d}s\}, \qquad (7.370)$$

where $L_o(\boldsymbol{v}) = \frac{1}{2}a(\boldsymbol{v},\boldsymbol{v}) - S(\boldsymbol{v})$ and the bilinear form $a(\boldsymbol{u}, v)$ and the functional $S(\boldsymbol{v})$ were defined above.

The first part of the problem was briefly discussed above. Now we shall discuss the second part, which is parallel to that of the algorithm from the previous section. We shall assume that the coupled terms have a meaning of body forces and thermal sources. We put $(Q, z)_h \equiv (Q, z)_h + b_p(\boldsymbol{u}_h, z)$ and $S(\boldsymbol{v}) \equiv \int_{\Omega_h} F_i v_i \mathrm{d}\boldsymbol{x} + \int_{\Gamma_\tau} P_i v_i \mathrm{d}s + b_s(\Theta_h, \boldsymbol{v})$.

The Propositions 1.2 and 2.2 of Ekeland and Temam (1976) and the Korn inequality prove the existence of a saddle point and, moreover, Proposition 2.2 of Ekeland and Temam (1976) proves the uniqueness of its approximation $(\boldsymbol{u}_{sh,}\lambda_h) \in K_h \times \Lambda$.

Let $\boldsymbol{v} \in K_h$. Then there exists $\boldsymbol{v}_h \in K_h$ such that $\|\boldsymbol{v}_h - \boldsymbol{v}\| \xrightarrow[h\to 0]{} 0$. Therefore, a sequence $\{\boldsymbol{v}_h\}$, $\boldsymbol{v}_h \in K_h$, can be chosen such that $\boldsymbol{v}_h \to \boldsymbol{u}_s$ in V. Propositions 1.2 and 2.2 of Ekeland and Temam (1976) and this assumption yield

$$\mathscr{L}(\boldsymbol{u}_{sh},\lambda_h) = \min_{\boldsymbol{v}\in K_h}\sup_{\mu\in\Lambda}\mathscr{L}(\boldsymbol{v},\mu) = \min_{\boldsymbol{v}\in K_h} L(\boldsymbol{v}) = L(\boldsymbol{u}_{sh}) \le L(\boldsymbol{v}_h) \xrightarrow[h\to 0]{} L(\boldsymbol{u}_s), \qquad (7.371)$$

as the functional L is continuous. But the set Λ is closed and since $\lim_{\|\boldsymbol{v}\|\to\infty} L(\boldsymbol{v}) \to \infty$ the subsequences $\{\boldsymbol{u}_{shi}\}$, $\{\lambda_{hi}\}$ of $\{\boldsymbol{u}_{sh}\}$, $\{\lambda_h\}$ can be taken such that $\boldsymbol{u}_{shi} \to \boldsymbol{u}_o$ weakly in V, $\lambda_{hi} \to \lambda_o$ weakly in Λ. But $\boldsymbol{v}_h \in K_h$, $\boldsymbol{v}_h \to \boldsymbol{v}$ weakly in V and we have $\boldsymbol{v} \in K$ and thus $\boldsymbol{u}_o \in$K. As Λ is closed, it is weakly closed and therefore $\lambda_o \in \Lambda$.

It holds that $L(\boldsymbol{u}_{sh}) \le L(\boldsymbol{v}_h) \xrightarrow[h\to 0]{} L(\boldsymbol{u}_s)$ and since L is weakly lower semi-continuous, then

$$L(\boldsymbol{u}_o) \le \lim_{h_i\to 0}\inf L(\boldsymbol{u}_{shi}) \le L(\boldsymbol{u}_s). \qquad (7.372)$$

Furthermore, since $\boldsymbol{u}_s$ is a solution of the problem in Equation 7.370, we find that $\boldsymbol{u}_s = \boldsymbol{u}_o$ and the sequence $\{\boldsymbol{u}_{shi}\}$ can be taken arbitrarily. Hence $\{\boldsymbol{u}_{sh}\} \to \boldsymbol{u}_s$. As

$$L(\boldsymbol{v}_h) \ge L(\boldsymbol{u}_{sh}) = L_o(\boldsymbol{u}_{sh}) + j_{gn}(\boldsymbol{u}_{sh}) \ge L_o(\boldsymbol{u}_{sh}) + a(\boldsymbol{u}_s, \boldsymbol{u}_{sh} - \boldsymbol{u}_s) - S(\boldsymbol{u}_{sh} - \boldsymbol{u}_s)$$
$$+ j_{gn}(\boldsymbol{u}_{sh}) + \tfrac{1}{2}c_o\|\boldsymbol{u}_{sh} - \boldsymbol{u}_s\|^2, \quad \forall \boldsymbol{v}_h \in K_h,$$

then

$$\|\boldsymbol{u}_s - \boldsymbol{u}_{sh}\| \le c\{L_o(\boldsymbol{v}_h) + j_{gn}(\boldsymbol{v}_h) - L_o(\boldsymbol{u}_{sh}) + a(\boldsymbol{u}_s, \boldsymbol{u}_s - \boldsymbol{u}_{sh})$$
$$- S(\boldsymbol{u}_s - \boldsymbol{u}_{sh}) - j_{gn}(\boldsymbol{u}_{sh})\}^{1/2}, \quad \forall \boldsymbol{v}_h \in K_h.$$

Putting $\boldsymbol{v}_h \in K_h$ such that $\boldsymbol{v}_h \to \boldsymbol{u}_s$, then $L_o(\boldsymbol{v}_h) \to L_o(\boldsymbol{u}_s)$, $j_{gn}(\boldsymbol{v}_h) \to j_{gn}(\boldsymbol{u}_s)$ for $h \to 0_+$. Then we find

$$\lim_{h\to 0_+}\sup\|\boldsymbol{u}_s - \boldsymbol{u}_{sh}\| \le c\{j_{gn}(\boldsymbol{u}_s) - \lim_{h\to 0_+}\inf j_{gn}(\boldsymbol{u}_{sh})\}^{1/2} \le 0$$

as the functional j_{gn} is weakly lower semi-continuous. Therefore, $\|\boldsymbol{u}_s - \boldsymbol{u}_{sh}\| \to 0$ for $h \to 0_+$.

Since $(\boldsymbol{u}_{sh}, \lambda_h)$ is a saddle point of $\mathscr{L}$ on $K_h \times \Lambda$, then

$$L_o(\boldsymbol{u}_{sh})+\int_{\cup\Gamma_c^{kl}} \mu^{kl} g_c^{kl}(\boldsymbol{u}_{sh}^k-\boldsymbol{u}_{sh}^l)_t \mathrm{d}s \le L_o(\boldsymbol{u}_{sh})+\int_{\cup\Gamma_c^{kl}} \lambda_h^{kl} g_c^{kl}(\boldsymbol{u}_{sh}^k-\boldsymbol{u}_{sh}^l)_t \mathrm{d}s, \quad \forall\mu^{kl} \in \Lambda.$$

Thus

$$\int_{\cup\Gamma_c^{kl}} (\mu^{kl} - \lambda_h^{kl}) g_c^{kl}(\boldsymbol{u}_{sh}^k - \boldsymbol{u}_{sh}^l)_t \mathrm{d}s \le 0 \quad \forall\mu^{kl} \in \Lambda$$

and,

$$\int_{\cup\Gamma_c^{kl}} (\lambda_h^{kl} \rho g_c^{kl}(\boldsymbol{u}_{sh}^k - \boldsymbol{u}_{sh}^l)_t - \lambda_h^{kl})(\mu^{kl} - \lambda_h^{kl}) \mathrm{d}s \le 0 \quad \forall\rho > 0, \quad \forall\mu^{kl} \in \Lambda.$$

Hence

$$\lambda_h^{kl} = P(\lambda_h^{kl} + \rho g_c^{kl}(\boldsymbol{u}_h^k - \boldsymbol{u}_h^l)_t), \quad \rho > 0, \tag{7.373}$$

where P is a projection of $L^2(\cup\Gamma_c^{kl})$ onto Λ.

Due to Proposition 1.7 of Ekeland and Temam (1976), we have

$$L_o(\boldsymbol{u}_{sh}) + \int_{\cup\Gamma_c^{kl}} \lambda_h^{kl} g_c^{kl}(\boldsymbol{u}_{sh}^k - \boldsymbol{u}_{sh}^l)_t \mathrm{d}s \le L_o(\boldsymbol{v}) + \int_{\cup\Gamma_c^{kl}} \lambda^{kl} g_c^{kl}(\boldsymbol{v}_t^k - \boldsymbol{v}_t^l) \mathrm{d}s \quad \forall\boldsymbol{v} \in K_h. \tag{7.374}$$

This and Equation 7.373 give an idea for a numerical solution. Such type of problems can be solved by the Uzawa or/and Arrow-Hurwicz algorithms.

The Uzawa algorithm is based on the construction of two sequences of elements $\boldsymbol{u}_h^n \in K_h$, $\lambda_h^{n,kl} \in \Lambda$, defined in the following way.

Starting with any $\lambda_h^{0,kl} \in \Lambda$, we calculate $\boldsymbol{u}_h^0$, and then $\lambda_h^{1,kl}$, $\boldsymbol{u}_h^1$, etc.. Let $\lambda_h^{n,kl}$ be known. We determine $\boldsymbol{u}_h^n$ as the element of K_h which minimizes the functional

$$L_o(\boldsymbol{v}) + \int_{\cup\Gamma_c^{kl}} \lambda_h^{n,kl} g_c^{kl}(\boldsymbol{v}_t^k - \boldsymbol{v}_t^l) \mathrm{d}s. \tag{7.375}$$

Then we define

$$\lambda_h^{n+1,kl} = P(\lambda_h^{n,kl} + \rho^n g_c^{kl}(\boldsymbol{u}_h^{n,k} - \boldsymbol{u}_h^{n,l})_t) \quad \rho^n > 0, \tag{7.376}$$

where P is a projection $P : L^2(\cup\Gamma_c^{kl}) \to \Lambda$ defined as

$$\begin{cases} \mu^{kl} \in L^2(\Gamma_c^{kl}),\ y \in \Gamma_c^{kl} & \\ P\mu^{kl}(y) = \mu^{kl}(y) & \text{for } \left|\mu^{kl}(y)\right| \le 1, \\ P\mu^{kl}(y) = 1 & \text{for } \mu^{kl}(y) > 1, \\ P\mu^{kl}(y) = -1 & \text{for } \mu^{kl}(y) < -1, \end{cases} \tag{7.377}$$

$\rho^n > 0$, $0 < \rho_1 \le \rho^n \le \rho_2$, ρ_1, ρ_2 are sufficiently small numbers which follow from the convergence theorem of the Uzawa algorithm (see Ekeland and Temam (1976).

The FEM approximation leads to minimizing the functional of Equation 7.375 on the set K_h, which is equivalent to the minimization of the quadratic functional

$$L(\boldsymbol{v}) = \tfrac{1}{2}\boldsymbol{v}^T C \boldsymbol{v} - \boldsymbol{b}^T \boldsymbol{v} \text{ with constraints } A\boldsymbol{v} \le 0, \tag{7.378}$$

where C is the stiff matrix ($\dim n \times n$), which is positive definite and generated by the term $\frac{1}{2}\int_\Omega c_{ijkl}(\boldsymbol{x}) e_{ij}(\boldsymbol{v}) e_{kl}(\boldsymbol{v}) \mathrm{d}\boldsymbol{x}$, $\boldsymbol{b}$ is the vector generated by the body and surface forces and the coupled term ($\dim n$), i.e. by the term $\int_\Omega F_i v_i \mathrm{d}\boldsymbol{x} + \int_{\Gamma_\tau} P_i v_i \mathrm{d}s +$

$b_s(\Theta, \boldsymbol{v})$ as well as the friction forces $\int_{\cup\Gamma_c^{kl}} g_c^{kl}(|\boldsymbol{v}_t^k| - |\boldsymbol{v}_t^l|)\mathrm{d}s$ and it changes due to the friction forces in every step of the Uzawa algorithm, A is the matrix of constraints ($\dim m \times n$) and is generated by the condition of non-penetration $u_n^k - u_n^l \leq 0$ on $\cup\Gamma_c^{kl}$ in the definition of the admissible set of displacements K_h. The dimension n represents a double number of points of triangulation in which the displacement is not prescribed and m is the number of points of triangulation on the contact boundary Γ_c^{kl}.

Next we shall investigate solutions for both parts of the problem.

7.4. MODELS OF COLLISION ZONES BASED ON A COUPLED CONTACT-STEFAN-LIKE PROBLEM ARISING FROM THE TIME DISCRETIZATION IN DEFORMATION THEORY OF THERMO-PLASTICITY

7.4.1. Introduction

In this section we shall investigate the plate tectonic model problem based on the contact problem in the deformation theory of thermo-plasticity and the two-phase Stefan-like problem. The model problem admits to study recrystallization of rocks in the collision zones as well as in their close neighbourhoods due to their heating resulting from the effect of friction acting on the common contact boundaries between colliding lithospheric plates and blocks and deformation energy changing to heat. We shall assume that the investigated collision region is occupied by lithospheric rocks simulated by plastic bodies. The model problem also allows us to study recrystallization of rocks in regions where high-level radioactive waste repositories are situated, as well as recrystallization of rocks during the catastrophic situation in the high-level radioactive waste repository, deep mines, tunnels, etc.. In this section we shall assume the case of uniformly moving colliding lithospheric plates and geological bodies with or without friction between them. Then the contact part of the problem leads to the generalized Signorini problem with or without friction in the deformation theory of plasticity, and the thermal part leads to the two-phase Stefan-like problem. The weak formulation of this contact part of the problem is equivalent to finding a minimum of the potential energy. The secant modules iterative method is used to solve this problem numerically, each step of which corresponds to a contact problem in elasticity. An implicit discretization in time of the weak formulation of the Stefan-like problem in the enthalpy formulation results in an non-linear elliptic boundary value problem and, therefore, leads to an elliptic variational inequality. The linear finite elements will be used and the convergence of the FEM approximation of the solution to the weak solution of the coupled contact-two-phase Stefan problem will be proved. This section is based on the results of Nedoma (1997b).

7.4.2. Formulation of the Coupled Field Problem

We shall consider the N-dimensional ($N = 2, 3$) model problem based on the contact problem in the deformation theory of thermo-plasticity and the two-phase Stefan-like problem.

Let plastic bodies, which are in contact, occupy a bounded domain $\Omega \subset R^N$, $N = 2, 3$, with a Lipschitz boundary $\partial\Omega = \overline{\Gamma}_u \cup \overline{\Gamma}_\tau \cup \overline{\Gamma}_c$.

Let the stress-strain relation, the so-called Hencky-Mises stress-strain relation, be defined by

$$\tau_{ij} = (k(\boldsymbol{x}) - \tfrac{2}{3}\mu(\boldsymbol{x}, \gamma^2))e_{kk}(\mathbf{u})\delta_{ij} + 2\mu(\boldsymbol{x}, \gamma^2)e_{ij}(\boldsymbol{u}) - \beta_{ij}(\boldsymbol{x})(\Theta - \Theta_o), \quad i, j = 1, \ldots, N, \tag{7.379}$$

where e_{ij} is the small strain tensor defined by

$$e_{ij} = \frac{1}{2}\left(\frac{\partial u_i}{\partial x_j} + \frac{\partial u_j}{\partial x_i}\right), \quad i, j = 1, 2, 3, \tag{7.380}$$

$\vartheta(\boldsymbol{u}) = e_{kk}(\boldsymbol{u}) \equiv \operatorname{div} \boldsymbol{u}$, $k(\boldsymbol{x}) = \lambda(\boldsymbol{x}) + \frac{2}{3}\mu(\boldsymbol{x})$ is a bulk modulus,

$$\gamma(\boldsymbol{u}, \boldsymbol{v}) = -\tfrac{2}{3}\vartheta(\boldsymbol{u})\vartheta(\boldsymbol{v}) + 2e_{ij}(\boldsymbol{u})e_{ij}(\boldsymbol{v}), \tag{7.381}$$

$$\gamma(\boldsymbol{u}, \boldsymbol{u}) \equiv \gamma^2(\boldsymbol{u}) \equiv \gamma^2,$$

$$\gamma^2(\boldsymbol{u}) = e_{ij}^D e_{ij}^D, \quad e_{ij}^D = e_{ij}^{-1/3} e_{kk}\delta_{ij}, \quad s = \tau_{ij}^D\tau_{ij}^D, \quad \tau_{ij}^D = 2\mu e_{ij}^D$$

and

$$0 < k_o \leq k(\boldsymbol{x}) \leq k_1 < +\infty \text{ for all } \boldsymbol{x} \in \overline{\Omega}, \tag{7.382}$$

$$0 < \mu_o \leq \mu(\boldsymbol{x}, s) \leq \tfrac{3}{2}k(\boldsymbol{x}) \text{ for all } \boldsymbol{x} \in \overline{\Omega}, \quad s \geq 0, \tag{7.383}$$

$$0 < a_o \leq \mu(\boldsymbol{x}, s) + 2s\frac{\partial\mu(\boldsymbol{x}, s)}{\partial s} \leq A_o < +\infty \text{ for all } \boldsymbol{x} \in \overline{\Omega},\ s \geq 0, \tag{7.384}$$

where the functions $k(\boldsymbol{x})$ and $\mu(\boldsymbol{x}, s)$ are supposed to be continuous and μ to be continuously differentiable with respect to s, and where $\beta_{ij} \in C^1(\overline{\Omega})$ is a coefficient of linear thermal expansion, and $\Theta_o = \Theta_o(\boldsymbol{x}) \in H^1(\Omega)$ is an initial temperature at $t = t_o$.

Let a coefficient of thermal conductivity $\kappa_{ij}(\boldsymbol{x}) \in C^1(\overline{\Omega})$ satisfy

$$\kappa_{ij}(\boldsymbol{x}) = \kappa_{ji}(\boldsymbol{x}), \quad i, j = 1, \ldots, N \tag{7.385}$$

and

$$\kappa_{ij}(\boldsymbol{x})\xi_i\xi_j \geq c_1 |\xi|^2 \quad \forall \xi \in R^N, \text{ a.e. in } \Omega. \tag{7.386}$$

Let $I = (t_o, t_1)$, let $\boldsymbol{n}^\iota$ be the unit outward normal to $\partial\Omega^\iota = \overline{\Gamma}_u^\iota \cup \overline{\Gamma}_\tau^\iota \cup \overline{\Gamma}_c^\iota$ (in the case of $\Gamma_c = \cup\overline{\Gamma}_c^{kl}$ related to Ω^k), $u_n = u_i n_i$, and $\boldsymbol{u}_t = \boldsymbol{u} - u_n\boldsymbol{n}$ be the normal and tangential components of the displacement vector $\boldsymbol{u}$, $\tau_n(\boldsymbol{u})$ and $\vec{\tau}_t(\boldsymbol{u})$ be the normal and tangential components of the stress vector $\vec{\tau}(\boldsymbol{u}) = (\tau_i(\boldsymbol{u}))$, $\tau_i(\boldsymbol{u}) = \tau_{ij}(\boldsymbol{u})n_j$, $\tau_n = \tau_{ij}(\boldsymbol{u})n_i n_j$, $\vec{\tau}_t = \vec{\tau} - \tau_n\boldsymbol{n}$ and τ_{ij} is the stress tensor. Let $\mathcal{F}_c^{kl}$ be the coefficient of Coulombian friction. If $\mathcal{F}_c^{kl} \equiv 0$ we have the case without friction and then $\vec{\tau}_t = 0$. Furthermore, the Einstein's summational repeated index convention is also used.

Given a smoothly bounded convex domain $\Omega = \cup_{\iota=1}^S \Omega^\iota \subset R^N$ with boundary $\partial\Omega$, $\partial\Omega = \overline{\Gamma}_u \cup \overline{\Gamma}_\tau(\cup\overline{\Gamma}_o) \cup \mathscr{R}$ (meas $\mathscr{R} = 0$), $\Gamma_u = \cup_{\iota=1}^S \overline{\Gamma}_u^\iota$, $\Gamma_\tau = \cup_{\iota=1}^S \overline{\Gamma}_\tau^\iota$ $(\Gamma_o = \cup_{\iota=1}^S \overline{\Gamma}_o^\iota)$

occupied by colliding lithospheric plates and blocks Ω^ι with boundaries $\partial\Omega^\iota = \overline{\Gamma}_u^\iota \cup \overline{\Gamma}_\tau^\iota \cup \overline{\Gamma}_c^\iota (\cup \overline{\Gamma}_o^\iota)$, $\Gamma_c = \cup_{\substack{k,l \\ k \neq 1}} \overline{\Gamma}_c^{kl}$, $\Gamma_c^{kl} = \overline{\Omega}^k \cap \overline{\Omega}^l$, $k \neq l$, we consider the following problem.

Problem ($\mathscr{P}_O$): Find a pair of functions $(\Theta, \boldsymbol{u})$ satisfying the heat conduction equation

$$\rho(\boldsymbol{x})c(\boldsymbol{x})\frac{\partial\Theta(\boldsymbol{x},t)}{\partial t} + \rho\beta_{ij}\Theta_o e_{ij}(\dot{\boldsymbol{u}}) - \frac{\partial}{\partial x_i}\left(\kappa_{ij}(\boldsymbol{x})\frac{\partial\Theta(\boldsymbol{x},t)}{\partial x_j}\right) = W(\boldsymbol{x},t)$$

$$\text{for a.e. } (\boldsymbol{x},t) \in \Omega \times I, \quad i,j = 1,\ldots,N, \tag{7.387}$$

and the equation of motion for the case of uniformly moving lithospheric plates and blocks in linear thermo-elasticity

$$\frac{\partial}{\partial x_i}\left[(k - \frac{2}{3}\mu(\gamma^2))e_{kk}(\boldsymbol{u})\delta_{ij} + 2\mu(\gamma^2)e_{ij}(\boldsymbol{u}) - \beta_{ij}(\boldsymbol{x})(\Theta(\boldsymbol{x},t) - \Theta_o(\boldsymbol{x}))\right] - f_i(\boldsymbol{x},t) = 0$$

$$\text{for a.e. } (\boldsymbol{x},t) \in \Omega \times I, \quad i,j,k,l = 1,\ldots,N \tag{7.388}$$

and the following boundary conditions

$$\left.\begin{aligned} &\Theta = 0, && (7.389)\\ &\tau_{ij}n_j = P_i, && (7.390)\end{aligned}\right\} (\boldsymbol{x},t) \in \Gamma_\tau \times I,$$

$$\left.\begin{aligned} &\Theta = \Theta_1 \left(\text{or } \kappa_{ij}\frac{\partial\Theta}{\partial x_j}n_i = q\right), && (7.391)\\ &\boldsymbol{u} = \boldsymbol{u}_o, && (7.392)\end{aligned}\right\} (\boldsymbol{x},t) \in \Gamma_u \times I,$$

$$\left.\begin{aligned} &\Theta^k = \Theta^l, && (7.393)\\ &\kappa_{ij}\frac{\partial\Theta}{\partial x_j}n_i\Big|_{(k)} = \kappa_{ij}\frac{\partial\Theta}{\partial x_j}n_i\Big|_{(l)}, && (7.394)\end{aligned}\right\} (\boldsymbol{x},t) \in \cup_{k,l}\Gamma_c^{k,l} \times I, \quad i,j = 1,\ldots,N,$$

$$\left.\begin{aligned} &u_n^k - u_n^l \le 0, && (7.395)\\ &\tau_n^{kl} \equiv \tau_n^k = -\tau_n^l \le 0, && (7.396)\\ &\tau_n^{kl}(u_n^k - u_n^l) = 0, && (7.397)\end{aligned}\right\} (\boldsymbol{x},t) \in \cup_{k,l}\Gamma_c^{k,l} \times I$$

(i.e. the so-called Signorini condition) and the Coulombian law of friction in the form

$$\left.\begin{aligned} &u_n^k - u_n^l = 0 \text{ then } \left|\vec{\tau}_t^{kl}\right| \le g_c^{kl} = \mathscr{F}_c^{kl}\left|\tau_n^{kl}\right|, && (7.398)\\ &\text{if } \left|\vec{\tau}_t^{kl}\right| < g_c^{kl} = \mathscr{F}_c^{kl}\left|\tau_n^{kl}\right| \text{ then } \boldsymbol{u}_t^k - \boldsymbol{u}_t^l = 0, && (7.399)\\ &\text{if } \left|\vec{\tau}_t^{kl}\right| \le g_c^{kl} = \mathscr{F}_c^{kl}\left|\tau_n^{kl}\right| \text{ then there exists } \lambda^{kl} \ge 0 \text{ such that } \boldsymbol{u}_t^k - \boldsymbol{u}_t^l = -\lambda^{kl}\vec{\tau}_t^{kl} && (7.400)\end{aligned}\right\} \text{on } \cup_{k,l}\Gamma_c^{kl}.$$

Moreover, let R^m be the phase change boundaries between two different recrystallized phases at time t. Then these boundaries divide the domain Ω into domains Ω_S^m and Ω_L^m, $m = 1,\ldots,r$. On the boundary $R^m(t)$ the following conditions are given:

$$\Theta_S^m = \Theta_L^m = \Theta_R^m, \tag{7.401}$$

$$\kappa_{ij}^m\frac{\partial\Theta^m}{\partial x_j}\nu_i^m\Big|_{(S)} - \kappa_{ij}^m\frac{\partial\Theta^m}{\partial x_j}\nu_i^m\Big|_{(L)} = -\rho^m L^m v_\nu^m, \tag{7.402}$$

where ρ^m is the density, Θ_R^m the temperature of the phase transition, Θ_S^m and Θ_L^m the temperatures at the phase change boundary $R^m(t)$ in the recrystallized (solid and liquid) phases respectively, v^m the unit normal to $R^m(t)$ pointing towards Ω_S^m, v_v^m is the speed of $R^m(t)$ along v^m, and L^m is the latent heat of phase transition (recrystallization). Furthermore, we shall assume that c_S^m, c_L^m, ρ_S^m, ρ_L^m are specific heats and densities in recrystallized phases, and $\rho_S^m = \rho_L^m = \rho^m$ on $R^m(t)$.

According to Equation 7.379 the above conditions are non-linear. In the 3D case, as in the previous models, it is pertinent for the definition of the region investigated to use the bilateral contact conditions and the condition of thermal isolation of the form

$$\kappa_{ij}\frac{\partial\Theta}{\partial x_j}n_i = 0, \quad u_n = 0, \quad \vec{\tau}_t = 0 \text{ on } \Gamma_o \times I, \tag{7.403}$$

where the second and third conditions describe the contact of the body with its perfectly rigid and perfectly smooth support without friction, and the first boundary condition says that the region investigated from the side of Γ_o is thermally isolated.

Moreover, we assume the initial condition

$$\Theta(\boldsymbol{x}, t_o) = \Theta_o(\boldsymbol{x}). \tag{7.404}$$

We shall assume the coupled contact-two-phase Stefan-like problem with homogeneous boundary value conditions. Then let us replace Θ by $\Theta + z$, where z is a sufficiently smooth scalar function in $\overline{\Omega}$ satisfying Equation 7.391, and we shall use the same symbols as above. According to these transformations the boundary conditions and thermal sources $W(\boldsymbol{x})$ will be changed. Then Equations 7.391 and 7.392 lead to

$$\Theta(\boldsymbol{x}, t) = 0 \text{ on } \Gamma_u \times I, \tag{7.405}$$

$$\boldsymbol{u}(\boldsymbol{x}, t) = \boldsymbol{u}_o(\boldsymbol{x}). \tag{7.406}$$

For the thermal part of the problem the technique of the enthalpy method will be used. The generalized enthalpy $H(\Theta)$ is defined as the subdifferential of the functional $\Phi(\Theta)$, where

$$\Phi(\Theta) = \tfrac{1}{2}\rho_L^m c_L^m(\Theta - \Theta_R^m)_+^2 + \tfrac{1}{2}\rho_S^m c_S^m(\Theta - \Theta_R^m)_-^2 + \rho^m L^m(\Theta - \Theta_R^m)_+, \tag{7.407}$$

i.e. $\partial\Phi(.) : R \to R$ is represented by a monotonically increasing multi-valued function of temperature with a jump discontinuity at the phase change temperature Θ_R. Hence

$$\partial\Phi(\Theta) \equiv H(\Theta) \begin{cases} \rho_L^m c_L^m\left(\Theta - \Theta_R^m\right) + \rho^m L^m & \text{for } \Theta > \Theta_R^m, \\ \rho_S^m c_S^m\left(\Theta - \Theta_R^m\right) & \text{for } \Theta < \Theta_R^m, \\ \langle 0, \rho^m L^m\rangle & \text{for } \Theta = \Theta_R^m, \end{cases} \tag{7.408}$$

where $\langle 0, \rho^m L^m\rangle$ is the enthalpy interval and $\Phi : R^1 \to R^1$ is a piecewise quadratic convex function given by Equation 7.407 as well as a locally Lipschitz continuous function, i.e.

$$|\Phi(\Theta_1) - \Phi(\Theta_2)| \le \max\{|\partial\Phi(\Theta_1)|, |\partial\Phi(\Theta_2)|\}\,|\Theta_1 - \Theta_2| \le c\,|\Theta_1 - \Theta_2|, \quad \text{when } |\Theta_1|, |\Theta_2| \le K. \tag{7.409}$$

Since $(\partial H/\partial t) = \rho c(\partial\Theta/\partial t)$, then Equation 7.387 leads to

$$\frac{\partial H(\Theta(\boldsymbol{x}, t))}{\partial t} + \rho\beta_{ij}\Theta_o e_{ij}(\dot{\boldsymbol{u}}) - \frac{\partial}{\partial x_i}\left(\kappa_{ij}(\boldsymbol{x})\frac{\partial\Theta(\boldsymbol{x}, t)}{\partial x_j}\right) = W(\boldsymbol{x}, t) \quad \text{for a.e. } (\boldsymbol{x}, t) \in \Omega \times I, \quad i, j = 1, \ldots, N. \tag{7.410}$$

Moreover, Equations 7.389, 7.391 and 7.404 become

$$H(\boldsymbol{x},t)=0, \quad (\boldsymbol{x},t)\in\partial\Omega\times I, \tag{7.411}$$

$$H(\boldsymbol{x},t_o)=H_o(\boldsymbol{x}). \tag{7.412}$$

Further, we shall assume that the solution of the problem ($\mathscr{P}$) exists and is unique. In the case of the two-phase Stefan problem (Equations 7.410, 7.411, 7.393 and 7.412), an implicit discretization in time will be used. Then the problem leads at every time t to the boundary value problem for non-linear partial differential equations of elliptic type, while the problem in Equations 7.388, 7.390, 7.392 and 7.395 to 7.400 represents the contact problem with friction at the given time level. The term $\partial/\partial x_i(\beta_{ij}(\boldsymbol{x})(\Theta(\boldsymbol{x},t)-\Theta_o(\boldsymbol{x})))$ represents the deformation energy due to the thermal stresses, where $\beta_{ij}(x)(\Theta(\boldsymbol{x},t)-\Theta_o(\boldsymbol{x}))$ is the thermal stress and $\rho\beta_{ij}\Theta_o e_{ij}(\dot{\boldsymbol{u}})$ is the dissipative heat due to the deformation. Equations 7.410 and 7.388, and the boundary conditions given in Equations 7.411, 7.390, 7.392 to 7.400 and 7.412 represent the coupled contact-two-phase Stefan problem in enthalpy formulation.

Thus, on discretizing

$$\frac{\partial H(\Theta)}{\partial t}=\frac{H(\Theta^{n+1})-H(\Theta^n)}{\Delta t}, \quad \Theta^n=\Theta(\boldsymbol{x},n\Delta t),$$

Equation 7.410 yields

$$H(\Theta^{n+1})-\Delta t\frac{\partial}{\partial x_i}\left(\kappa_{ij}\frac{\partial\Theta^{n+1}}{\partial x_j}\right)+\Delta t\rho\Theta_o\beta_{ij}e_{ij}(\dot{\boldsymbol{u}}^{n+1})=H(\Theta^n)+\Delta t W^{n+1}, \tag{7.413}$$

where we denote

$$W^{n+1}=\int_{n\Delta t}^{(n+1)\Delta t}W(s)\mathrm{d}s, \quad \dot{\boldsymbol{u}}^{n+1}=\Delta t^{-1}(\boldsymbol{u}^{n+1}-\boldsymbol{u}^n).$$

For the numerical solution of the two-phase Stefan problem, the formulation in the form of the generalized non-linear boundary value problem (see Barbu (1976), Glowinski (1979)) of the form

$$-\Delta t\Delta w+\alpha w+\partial\phi(w)\ni g(\boldsymbol{x}), \quad \text{a.e. } \boldsymbol{x}\in\Omega, \tag{7.414}$$

$$w=0, \quad \text{a.e. } \boldsymbol{x}\in\partial\Omega, \tag{7.415}$$

will be used. Since $w\in H^2(\Omega)$, $g(\boldsymbol{x})\in L^2(\Omega)$, then Equations 7.414 and 7.415 have a unique solution (see Barbu (1976), Chapter II,p.87). This differential equation is discontinuous (as Θ passes through Θ_R^m) and hence, together with the above boundary conditions, it represents the free boundary problem. The corresponding free boundaries R^m represent the phase change boundaries between two different phases (recrystallized phases) at time t.

We put $\Theta \equiv \Theta^{n+1}$, $\boldsymbol{u} \equiv \boldsymbol{u}^{n+1}$, $Q = H(\Theta^n) + \Delta t W^{n+1} + \alpha\Theta_R$, $\alpha\Theta + \partial\Phi(\Theta) \equiv H(\Theta) + \alpha\Theta_R$, where $0 < \alpha \leq \min\{\rho_S^m c_S^m, \rho_L^m c_L^m\}$. Then Equation 7.413 yields

$$\alpha\Theta - \Delta t \frac{\partial}{\partial x_i}\left(\kappa_{ij}\frac{\partial\Theta}{\partial x_j}\right) + \Delta t \rho\beta_{ij}\Theta_o e_{ij}(\dot{\boldsymbol{u}}) + \partial\Phi(\Theta) \ni Q(\boldsymbol{x}), \qquad \text{a.e. } \boldsymbol{x} \in \Omega, \tag{7.416}$$

$$\Theta = 0, \qquad \text{a.e. } \boldsymbol{x} \in \partial\Omega, \tag{7.417}$$

which are equivalent to Equations 7.414 and 7.415. Hence, the theory of Glowinski (1979) for the thermal part of the problem and of Fučík, Kratochvíl and Nečas (1975), Nečas and Hlaváček (1981), (1983), Nedoma (1983), (1987), (1997b) for the plastic part of the problem will be used.

The problem is coercive if $\Gamma_u \neq 0$ and is semi-coercive if $\Gamma_u = 0$ or if at least one part of $\Gamma_u = \cup_{\iota=1}^{S}\Gamma_u^\iota$ is equal to zero. In our model problem we shall assume that, for example, $\Gamma_u^k \cup \Gamma_u^m \cup \Gamma_u^n$ are equal to zero. Such model problems represent realistic models of collision zones, but we are aware that the more realistic models are much more complicated.

7.4.3. Variational Formulation of the Problem

In this section we shall analyze the problem arising from an implicit time discretization only.

Let $\Omega = \cup_{\iota=1}^{S}\Omega^\iota \subset R^N$ be the domain with Lipschitz boundary at time moment t defined above. Let $\boldsymbol{x} = (x_i)_{i=1}^N$ be Cartesian coordinates and $\boldsymbol{n} = (n_i)_{i=1}^N$ denote the unit outward normal defined above.

We denote by $(.,.)$ and $\langle .,. \rangle$ scalar products in $L^2(\Omega)$ or $[L^2(\Omega)]^N$ respectively, and in $[L^2(\cup\Gamma_c^{kl})]^N$, by $\|.\|_{k,N}$ the norm in $[H^k(\Omega)]^N$, k an integer, $k \in R^1$, $H^k(\Omega)$ is the Sobolev space in the usual sense, i.e. $\|.\|_{0,1}$ is the norm in $H^o(\Omega) = L^2(\Omega)$, $\|.\|_{1,1}$ the norm in $H^1(\Omega)$ and $\|.\|_{1,N}$ the norm in $[H^1(\Omega)]^N$.

Let $\Theta(\boldsymbol{x}) \in H^1(\Omega)$ be the temperature, $\boldsymbol{u}(\boldsymbol{x}) = (u_i(\boldsymbol{x})) \in [H^1(\Omega)]^N$ be the displacement vector and $e_{ij}(\boldsymbol{u})$ and $\tau_{ij}(\boldsymbol{u})$ be the small strain and stress tensors defined above. Let $\beta_{ij} \in C^1(\overline{\Omega})$ be a coefficient of linear thermal expansion, $\Theta_o = \Theta_o(\boldsymbol{x}) \in H^1(\Omega)$ be an initial temperature at $t = t_o$, and let coefficients of thermal conductivity $\kappa_{ij}(\boldsymbol{x}) \in C^1(\overline{\Omega})$ be Lipschitz and satisfy the symmetry conditions

$$\kappa_{ij}(\boldsymbol{x}) = \kappa_{ji}(\boldsymbol{x}), \quad \kappa_{ij}(\boldsymbol{x})\xi_i\xi_j \geq c_1 |\xi|^2 \quad \forall\xi \in R^N, \text{ a.e. in } \Omega.$$

Moreover, the bulk modulus $k(\boldsymbol{x})$ satisfies $0 < k_o \leq k(\boldsymbol{x}) \leq k_1 < +\infty$ and $0 < \mu_o \leq \mu(\boldsymbol{x}, s) \leq \frac{3}{2}k(\boldsymbol{x})$, $0 < a_o \leq \mu(\boldsymbol{x}, s) + 2s(\partial\mu(\boldsymbol{x}, s)/\partial s) \leq A_o < +\infty$ for all $\boldsymbol{x} \in \overline{\Omega}$, $s \geq 0$, where the functions $k(\boldsymbol{x})$ and $\mu(\boldsymbol{x}, s)$ are assumed to be continuous and μ to be continuously differentiable with respect to s. Let heat sources $Q(\boldsymbol{x}) \in L^2(\overline{\Omega})$, body forces $\boldsymbol{f}(\boldsymbol{x}) \in [L^2(\overline{\Omega})]^N$, surface forces $\boldsymbol{P}(\boldsymbol{x}) \in [L^2(\Gamma_\tau)]^N$, and let a density $\rho(\boldsymbol{x}) \in C(\overline{\Omega})$, a specific heat $c(\boldsymbol{x}) \in C(\overline{\Omega})$ and frictional forces $g_c^{kl} \in L^\infty(\Gamma_c^{kl})$. For our considerations the coupling terms satisfy the condition $(\beta_{ij}(\Theta - \Theta_o))_{,j} \in L^2(\Omega)$ as $\beta_{ij} \in C^1(\overline{\Omega})$, $\Theta, \Theta_o \in H^1(\Omega)$, so that $\beta_{ij}(\Theta - \Theta_o) \in H^1(\Omega)$ and therefore $(\beta_{ij}(\Theta -$

$\Theta_o)),_j \in L^2(\Omega)$. Moreover, $\rho \in C(\overline{\Omega})$, $\Theta_o \in H^1(\Omega)$ and $\Delta t e_{ij}(\dot{\boldsymbol{u}}) \in L^2(\Omega)$, so that $\Delta t \rho \Theta_o \beta_{ij} e_{ij}(\dot{\boldsymbol{u}}) \in L^2(\Omega)$.

Thus we have to solve the following problem arising from an implicit discretization.

Problem ($\mathscr{P}$): Find $(\Theta, \boldsymbol{u})$ satisfying

$$\alpha\Theta - \Delta t \frac{\partial}{\partial x_i}\left(\kappa_{ij}\frac{\partial \Theta}{\partial x_j}\right) + \partial\Phi(\Theta) + \Delta t \rho \Theta_o \beta_{ij} e_{ij}(\dot{\boldsymbol{u}}) \ni Q(\boldsymbol{x}), \qquad \text{for a.e. } \boldsymbol{x} \in \Omega, t \text{ being fixed}, \tag{7.418}$$

$$\frac{\partial}{\partial x_i}\left[(k(\boldsymbol{x}) - \frac{2}{3}\mu(\gamma^2))e_{kk}(\boldsymbol{u})\delta_{ij} + 2\mu(\gamma^2)e_{ij}(\boldsymbol{u}) - \beta_{ij}(\boldsymbol{x})(\Theta(\boldsymbol{x}, t) - \Theta_o(\boldsymbol{x}))\right] - f_i(\boldsymbol{x}, t) = 0$$

$$\text{for a.e. } \boldsymbol{x} \in \Omega, \quad i, j, k, l = 1, \ldots, N, t \text{ being fixed}, \tag{7.419}$$

and with the following boundary conditions

$$\left.\begin{aligned} H(\Theta) &= 0, \\ \tau_{ij} n_j &= P_i, \end{aligned}\right\} \ \boldsymbol{x} \in \Gamma_\tau, t \text{ being fixed}, \tag{7.420, 7.421}$$

$$\left.\begin{aligned} H(\Theta) &= 0, \\ \boldsymbol{u} &= \boldsymbol{u}_o, \end{aligned}\right\} \ \boldsymbol{x} \in \Gamma_u, t \text{ being fixed}, \tag{7.422, 7.423}$$

$$\left.\begin{aligned} &\Theta^k = \Theta^l, && (7.424)\\ &\kappa_{ij}\frac{\partial\Theta}{\partial x} n_i\Big|_{(k)} = \kappa_{ij}\frac{\partial\Theta}{\partial x_j} n_i\Big|_{(l)}, && (7.425)\\ &u_n^k - u_n^l \leq 0, && (7.426)\\ &\tau_n^{kl} \equiv \tau_n^k = -\tau_n^l \leq 0, && (7.427)\\ &\tau_n^{kl}(u_n^k - u_n^l) = 0, && (7.428) \end{aligned}\right\} \ \boldsymbol{x} \in \cup_{k,l}\Gamma_c^{k,l}, t \text{ being fixed},$$

(i.e. the so-called Signorini conditions) and the Coulombian law of friction in the form

$$\left.\begin{aligned} &u_n^k - u_n^l = 0 \text{ then } \left|\vec{\tau}_t^{kl}\right| \leq g_c^{kl} = \mathscr{F}_c^{kl}\left|\tau_n^{kl}\right|, && (7.429)\\ &\text{if } \left|\vec{\tau}_t^{kl}\right| < g_c^{kl} \text{ then } \boldsymbol{u}_t^k - \boldsymbol{u}_t^l = 0, && (7.430)\\ &\text{if } \left|\vec{\tau}_t^{kl}\right| \leq g_c^{kl} \text{ then there exists } \lambda \geq 0 \text{ such that } \boldsymbol{u}_t^k - \boldsymbol{u}_t^l = -\lambda^{kl}\vec{\tau}_t^{kl} && (7.431) \end{aligned}\right\} \text{ on } \cup\Gamma_c^{kl},$$

where we denote $\alpha\Theta + \partial\Phi(\Theta) \equiv H(\Theta^{n+1}) + \alpha\Theta_R^m$, $Q = H(\Theta^n) + \Delta t W^{n+1} + \alpha\Theta_R^m$ and $\vec{\tau}_t^{kl} \equiv \vec{\tau}_t^k \equiv -\vec{\tau}_t^l$.

We define the space of virtual temperatures as

$${}^1V = \{z \mid z \in H^1(\Omega), z = 0 \text{ on } \partial\Omega \text{ in the sense of traces }\},$$

the space of virtual displacements as

$$V = \{\boldsymbol{v} \mid v \in W = [H^1(\Omega)]^N, \quad \boldsymbol{v} = \boldsymbol{u}_o \text{ on } \Gamma_u \text{ in the sense of traces}\}$$

and the set of admissible displacements as

$$K = \{\boldsymbol{v} \mid \boldsymbol{v} \in V, \quad v_n^k - v_n^l \leq 0 \text{ on } \cup\Gamma_c^{kl}\}.$$

Remark: The set K is a closed convex subset of the Hilbert space V and represents a closed convex cone with vertex at the origin.

On ${}^1V$ we define the bilinear form $a_\Theta(.,.)$, scalar product $(.,.)$ and functionals $s(.)$, $b_p(.,.)$, and $j(.)$ by

$$a_\Theta(\Theta, z) = \int_\Omega \kappa_{ij} \frac{\partial \Theta}{\partial x_i} \frac{\partial z}{\partial x_j} \mathrm{d}\boldsymbol{x}, \tag{7.432}$$

$$(\Theta, z) = \int_\Omega \Theta z \mathrm{d}\boldsymbol{x}, \tag{7.433}$$

$$s(z) = \Delta t \int_\Omega Q z \mathrm{d}\boldsymbol{x} \left(-\Delta t \int_{\Gamma_u} q z \mathrm{d}s \right), \tag{7.434}$$

$$b_p(\boldsymbol{v}, z) = \Delta t \int_\Omega \rho \Theta_o \beta_{ij} \partial v_i / \partial x_j z \mathrm{d}\boldsymbol{x}, \tag{7.435}$$

$$j(z) = \Delta t \int_\Omega \Phi(z(\boldsymbol{x})) \mathrm{d}\boldsymbol{x}, \tag{7.436}$$

and on V the bilinear form a(.;.,.) and functionals $b_s(\Theta, .)$, $j_{gu}(.)$ and $S(.)$ by

$$a(\boldsymbol{w}; \boldsymbol{u}, \boldsymbol{v}) = \int_\Omega \left[(k - \frac{2}{3}\mu(\gamma^2(\boldsymbol{w})))\vartheta(\boldsymbol{u})\vartheta(\boldsymbol{v}) + 2\mu(\gamma^2(\boldsymbol{w}))e_{ij}(\boldsymbol{u})e_{ij}(\boldsymbol{v}) \right] \mathrm{d}\boldsymbol{x}, \tag{7.437}$$

$$b_s(\Theta, \boldsymbol{v}) = \int_\Omega \frac{\partial}{\partial x_j} (\beta_{ij}(\Theta - \Theta_o)) v_i \mathrm{d}\boldsymbol{x}, \tag{7.438}$$

$$S(\boldsymbol{v}) = \int_\Omega f_i v_i \mathrm{d}\boldsymbol{x} + \int_{\Gamma_\tau} P_i v_i \mathrm{d}s, \tag{7.439}$$

$$j_{gn} = \int_{\cup \Gamma_c^{kl}} g_c^{kl} \left| \boldsymbol{v}_t^k - \boldsymbol{v}_t^l \right| \mathrm{d}s = \left\langle g_c^{kl}, \left| \boldsymbol{v}_t^k - \boldsymbol{v}_t^l \right| \right\rangle, \tag{7.440}$$

where $\langle .,. \rangle$ denotes scalar product in $[L^2(\cup \Gamma_c^{kl})]^N$.

Then the variational or weak formulation of the given problem is as follows.

Problem ($\mathscr{P}_V$): Find a pair of functions $(\Theta, \boldsymbol{u})$, $\Theta \in {}^1V$, $\boldsymbol{u} \in K$, such that

$$\alpha(\Theta, z - \Theta) + \Delta t a_\Theta(\Theta, z - \Theta) + b_p(\boldsymbol{v}, \Theta) + j(z) - j(\Theta) \geq s(z - \Theta) \quad \forall z \in {}^1V, \tag{7.441}$$

$$a(\boldsymbol{u}; \boldsymbol{u}, \boldsymbol{v} - \boldsymbol{u}) + b_s(\Theta, \boldsymbol{v} - \boldsymbol{u}) + j_{gn}(\boldsymbol{v}) - j_{gn}(\boldsymbol{u}) \geq S(\boldsymbol{v} - \boldsymbol{u}) \quad \forall \boldsymbol{v} \in K. \tag{7.442}$$

Definition 7.12: A pair of functions $(\Theta, \boldsymbol{u})$, $\Theta \in {}^1V$, $\boldsymbol{u} \in K$ satisfying Equations 7.441 and 7.442 is said to be a weak solution of the Problem ($\mathscr{P}$), i.e. of Equations 7.418 to 7.431.

Since $\partial\Phi(.) : R^1 \to R^1$ is represented by a monotonically increasing multi-valued function of temperature with a jump discontinuity at the phase change temperature

Θ_R, i.e.

$$\partial\Phi(\Theta) \equiv H(\Theta) = \begin{cases} \rho_L^m c_L^m(\Theta - \Theta_R^m) + \rho^m L^m & \text{for } \Theta > \Theta_R^m, \\ \rho_S^m c_S^m(\Theta - \Theta_R^m) & \text{for } \Theta < \Theta_R^m, \\ \langle 0, \rho^m L^m \rangle & \text{for } \Theta = \Theta_R^m, \end{cases}$$

where $\langle 0, \rho^m L^m\rangle$ is the enthalpy interval and $\Phi(\Theta)$ is a locally Lipschitz continuous function

$$|\Phi(\Theta_1) - \Phi(\Theta_2)| \leq \max\{|\partial\Phi(\Theta_1)|, |\partial\Phi(\Theta_2)|\} |\Theta_1 - \Theta_2| \leq c |\Theta_1 - \Theta_2|,$$

when $|\Theta_1|, |\Theta_2| \leq K$, and the function $\Phi(\Theta)$ is piecewise quadratic convex function, then the functional $j(z) = \Delta t \int_\Omega \Phi(z(\boldsymbol{x}))\mathrm{d}\boldsymbol{x}$ is a convex continuous functional on ${}^1V$.

As the function Φ is the piecewise quadratic convex function, it is lower semi-continuous, and $\Phi(z(\boldsymbol{x}))$ is measurable in $\boldsymbol{x}$ for every measurable function z on Ω. Moreover, since Φ is bounded from below by an affine function (see Ekeland and Temam (1976)), it follows that $j(z) > -\infty$ for any $z \in L^2(\Omega)$; also $j(z) \neq +\infty$. The convexity of $j(z)$ is immediately obvious from the convexity of Φ. To prove that $j(z)$ is lower semi-continuous on $L^2(\Omega)$ it suffices to show that, for every λ, the set $C = \left\{z \in L^2(\Omega) \mid \int_\Omega \Phi(z(\boldsymbol{x}))\mathrm{d}\boldsymbol{x} \leq \lambda\right\}$ is a closed subset of $L^2(\Omega)$. Let $\{z_n\}$ be such that $z_n \to z$ in $L^2(\Omega)$ as $n \to \infty$. Then there exists $\{z_{n_k}\} \subset \{z_n\}$ such that $\lim_{n_k\to\infty} z_{n_k}(\boldsymbol{x}) = z(\boldsymbol{x})$ a.e. $\boldsymbol{x} \in \Omega$.

Let $\underline{\alpha}, \underline{\beta} \in R^1$ be such that $\Phi(z) \geq \underline{\alpha} z + \underline{\beta}$ for every $z \in R^1$. Then $\Phi(z_{n_k}(\boldsymbol{x})) - \underline{\alpha} z_{n_k}(\boldsymbol{x}) - \underline{\beta} \geq 0$ a.e. on Ω and $\lim_{n_k\to\infty} \inf(\Phi(z_{n_k}(\boldsymbol{x})) - \underline{\alpha} z_{n_k} - \underline{\beta}) \geq \Phi(z(\boldsymbol{x})) - \underline{\alpha} z - \underline{\beta}$ a.e. $\boldsymbol{x} \in \Omega$. Then, the Fatou lemma gives

$$\int_\Omega \Phi(z(\boldsymbol{x}))\mathrm{d}\boldsymbol{x} \leq \lim_{n_k\to\infty} \inf \int_\Omega \Phi(z_{n_k}))\mathrm{d}\boldsymbol{x}.$$

Hence $z \in C$, as is required. Moreover, $\left|b_p(\boldsymbol{v}, \Theta)\right| \leq c \|\Theta\|_{0,1} \|\boldsymbol{v}\|_{1,N}$. Then Propositions 3.7. and 3.8. of Barbu (1976) imply that for any $Q \in L^p(\Omega)$, $1 < p < \infty$, the non-linear boundary value problem in Equations 7.416 and 7.417 has a unique solution $\Theta \in H^p(\Omega)$, $1 < p < \infty$.

To analyze the plastic part of the problem at a given time level, we firstly analyze the property of the functional of potential energy $L(\boldsymbol{u})$.

To derive the functional of the potential energy $L(\boldsymbol{u})$ the same technique as in Nečas and Hlaváček (1981) will be used. Then we find that

$$\begin{aligned} L(\boldsymbol{u}) &= \frac{1}{2}\int_\Omega \left[k(\boldsymbol{x})\vartheta^2(\boldsymbol{u}) + \int_0^{\gamma(\boldsymbol{u},\boldsymbol{u})} \mu(\sigma)\mathrm{d}\sigma\right] \mathrm{d}\boldsymbol{x} + b_s(\Theta, \boldsymbol{u}) + j_{gn}(\boldsymbol{u}) \\ &\quad - \int_\Omega F_i u_i \mathrm{d}\boldsymbol{x} - \int_{\Gamma_\tau} P_i u_i \mathrm{d}s \\ &= L_o(\boldsymbol{u}) + b_s(\Theta, \boldsymbol{u}) + j_{gn}(\boldsymbol{u}), \end{aligned} \tag{7.443}$$

where $L_o(\boldsymbol{u}) = L_1(\boldsymbol{u}) - S(\boldsymbol{u})$.

The functional L_o is coercive and weakly lower semi-continuous, i.e. $\lim L_o(\boldsymbol{u}) = +\infty$ as $\|\boldsymbol{u}\| \to \infty$ and $\boldsymbol{u}_n \to \boldsymbol{u} \Rightarrow \liminf L_o(\boldsymbol{u}_n) \geq L_o(\boldsymbol{u})$, where $\boldsymbol{u}_n \to \boldsymbol{u}$ weakly.

Now we shall prove the coercivness and lower semi-continuity of the functional $L(\boldsymbol{u})$.

At first, since $\mathrm{d}/\mathrm{d}t\gamma^2(\boldsymbol{u}+t\boldsymbol{v})\big|_{t=0} = 2\gamma(\boldsymbol{u},\boldsymbol{v})$, $\boldsymbol{u}\in W=[H^1(\Omega)]^N$, $\boldsymbol{v}\in V$, then by virtue of the definition of the G-differential of the functional L_o, denoted by $DL_o(\boldsymbol{u},\boldsymbol{v})$, we have

$$\frac{\mathrm{d}}{\mathrm{d}t}L_o(\boldsymbol{u}+t\boldsymbol{v})\bigg|_{t=0} = DL_o(\boldsymbol{u},\boldsymbol{v})$$
$$= \int_\Omega [k(\boldsymbol{x})\vartheta(\boldsymbol{u})\vartheta(\boldsymbol{v}) + \mu(\gamma^2(\boldsymbol{u}))\gamma(\boldsymbol{u},\boldsymbol{v})]\mathrm{d}\boldsymbol{x} - \int_\Omega F_i v_i \mathrm{d}\boldsymbol{x} - \int_\Gamma P_i v_i \mathrm{d}s. \tag{7.444}$$

We define the second Gâteaux differential of L_o as

$$D^2L_o(\boldsymbol{u};\boldsymbol{v},\boldsymbol{w}) = \frac{\mathrm{d}}{\mathrm{d}t}DL_o(\boldsymbol{u}+t\boldsymbol{w},\boldsymbol{v})\bigg|_{t=0}$$
$$= \int_\Omega [k(\boldsymbol{x})\vartheta(\boldsymbol{w})\vartheta(\boldsymbol{v}) + 2\mu'(\gamma^2(\boldsymbol{u}))\gamma(\boldsymbol{u},\boldsymbol{w})\gamma(\boldsymbol{u},\boldsymbol{v}) + \mu(\gamma^2(\boldsymbol{u}))\gamma(\boldsymbol{w},\boldsymbol{v})]\mathrm{d}\boldsymbol{x}, \tag{7.445}$$

where we use the theorem about differentiating under the integral sign.

Denote $\Omega = \Omega_1\cup\Omega_2 = \{\boldsymbol{x}\in\Omega \mid \mu'(\gamma^2(\boldsymbol{u}))(\boldsymbol{x})\geq 0\}\cup\{\boldsymbol{x}\in\Omega \mid \mu'(\gamma^2(\boldsymbol{u}))(\boldsymbol{x})<0\}$. Then we have

$$DL_o(\boldsymbol{u}+\boldsymbol{v},\boldsymbol{v}) - DL_o(\boldsymbol{u},\boldsymbol{v}) = \int_0^1 DL_o(\boldsymbol{u}+t\boldsymbol{v};\boldsymbol{v},\boldsymbol{v})\mathrm{d}t$$
$$= \int_{\Omega_1\cup\Omega_2} [k(\boldsymbol{x})\vartheta(\boldsymbol{u}+t\boldsymbol{v})\vartheta(\boldsymbol{v}) + 2\mu'(\gamma^2(\boldsymbol{u}+t\boldsymbol{v}))\gamma(\boldsymbol{u}+t\boldsymbol{v},\boldsymbol{v})\gamma(\boldsymbol{u}+t\boldsymbol{v},\boldsymbol{v})$$
$$+ \mu(\gamma^2(\boldsymbol{u}+t\boldsymbol{v}))\gamma(\boldsymbol{v},\boldsymbol{v})]\mathrm{d}\boldsymbol{x}\,\mathrm{d}t$$
$$\geq \mu_o\int_0^1\int_{\Omega_1}\gamma(\boldsymbol{v},\boldsymbol{v})\mathrm{d}\boldsymbol{x}\,\mathrm{d}t + \frac{2}{3}\mu_o\int_0^1\int_{\Omega_1}\vartheta^2(\boldsymbol{v})\mathrm{d}\boldsymbol{x}\,\mathrm{d}t + \frac{2}{3}\mu_o\int_0^1\int_{\Omega_2}\vartheta^2(\boldsymbol{v})\mathrm{d}\boldsymbol{x}\,\mathrm{d}t$$
$$+ \int_0^1\int_{\Omega_2}[\mu(\gamma^2(\boldsymbol{u}))\gamma(\boldsymbol{v},\boldsymbol{v}) - 2\left|\mu'(\gamma^2(\boldsymbol{u}))\right|(\gamma(\boldsymbol{u},\boldsymbol{v}))^2]\mathrm{d}\boldsymbol{x}\,\mathrm{d}t$$
$$\geq 2\mu_o\int_0^1\int_{\Omega_1} e_{ij}(\boldsymbol{v})e_{ij}(\boldsymbol{v})\mathrm{d}\boldsymbol{x}\,\mathrm{d}t + \frac{2}{3}\mu_o\int_0^1\int_{\Omega_2}\vartheta^2(\boldsymbol{v})\mathrm{d}\boldsymbol{x}\,\mathrm{d}t$$
$$+ \int_0^1\int_{\Omega_2}[\mu(\gamma^2(\boldsymbol{u}))\gamma(\boldsymbol{v},\boldsymbol{v}) + 2\mu'(\gamma^2(\boldsymbol{u}))\gamma(\boldsymbol{u},\boldsymbol{u})\gamma(\boldsymbol{v},\boldsymbol{v})]\mathrm{d}\boldsymbol{x}\,\mathrm{d}t$$
$$\geq 2\mu_o\int_0^1\int_{\Omega_1} e_{ij}(\boldsymbol{v})e_{ij}(\boldsymbol{v})\mathrm{d}\boldsymbol{x}\,\mathrm{d}t + \frac{2}{3}\mu_o\int_0^1\int_{\Omega_2}\vartheta^2(\boldsymbol{v})\mathrm{d}\boldsymbol{x}\,\mathrm{d}t + \kappa\int_0^1\int_{\Omega_2}\gamma(\boldsymbol{v},\boldsymbol{v})\mathrm{d}\boldsymbol{x}\,\mathrm{d}t$$
$$\geq 2\mu_o\int_0^1\int_{\Omega_1} e_{ij}(\boldsymbol{v})e_{ij}(\boldsymbol{v})\mathrm{d}\boldsymbol{x}\,\mathrm{d}t + \frac{2}{3}(\mu_o-\kappa)\int_0^1\int_{\Omega_2}(e_{kk}(\boldsymbol{v}))^2\mathrm{d}\boldsymbol{x}\,\mathrm{d}t$$
$$+ 2\kappa\int_0^1\int_{\Omega_2} e_{ij}(\boldsymbol{v})e_{ij}(\boldsymbol{v})\mathrm{d}\boldsymbol{x}\,\mathrm{d}t$$

$$\geq c \int_0^1 \int_\Omega e_{ij}(\boldsymbol{v}) e_{ij}(\boldsymbol{v}) \mathrm{d}\boldsymbol{x}\, \mathrm{d}t, \tag{7.446}$$

where $(\gamma(\boldsymbol{u}, \boldsymbol{v}))^2 \leq \gamma(\boldsymbol{u}, \boldsymbol{u})\gamma(\boldsymbol{v}, \boldsymbol{v})$ (due to the Schwarz inequality) and $\mu_o \geq \kappa$ were used. Applying the Korn inequality

$$\int_\Omega e_{ij}(\boldsymbol{v}) e_{ij}(\boldsymbol{v}) \mathrm{d}\boldsymbol{x} \geq c \|\boldsymbol{v}\|^2_{[H^1(\Omega)]^N}, \tag{7.447}$$

then we find

$$DL_o(\boldsymbol{u} + \boldsymbol{v}, \boldsymbol{v}) - DL_o(\boldsymbol{u}, \boldsymbol{v}) \geq c \int_\Omega e_{ij}(\boldsymbol{v}) e_{ij}(\boldsymbol{v}) \mathrm{d}\boldsymbol{x} \geq c \|\boldsymbol{v}\|^2_{[H^1(\Omega)]^N}. \tag{7.448}$$

The coerciveness of $L_o(\boldsymbol{u})$, $\boldsymbol{u} \in V$ follows from the following idea. We see that

$$L_o(\boldsymbol{u}) = \int_0^1 DL_o(t\boldsymbol{u}, \boldsymbol{u}) \mathrm{d}t = \int_0^1 t^{-1} DL_o(t\boldsymbol{u}, t\boldsymbol{u}) \mathrm{d}t. \tag{7.449}$$

From Equation 7.448, putting $\boldsymbol{u} = 0$ and $\boldsymbol{v} = t\boldsymbol{u}$, we have

$$DL_o(t\boldsymbol{u}, t\boldsymbol{u}) - DL_o(0, t\boldsymbol{u}) \geq c \int_\Omega e_{ij}(t\boldsymbol{u}) e_{ij}(t\boldsymbol{u}) \mathrm{d}\boldsymbol{x}.$$

Thus

$$DL_o(t\boldsymbol{u}, t\boldsymbol{u}) \geq ct^2 \int_\Omega e_{ij}(\boldsymbol{u}) e_{ij}(\boldsymbol{u}) \mathrm{d}\boldsymbol{x} + tDL_o(0, \boldsymbol{u}).$$

Hence and using Equation 7.449 we obtain

$$L_o(\boldsymbol{u}) = \int_o^1 t^{-1} DL_o(t\boldsymbol{u}, t\boldsymbol{u}) \mathrm{d}t \geq \int_0^1 ct \int_\Omega e_{ij}(\boldsymbol{u}) e_{ij}(\boldsymbol{u}) \mathrm{d}\boldsymbol{x}\, \mathrm{d}t + \int_0^1 DL_o(0, \boldsymbol{u}) \mathrm{d}t.$$

Using the Korn inequality, then for any $\boldsymbol{u} \in K$ we find

$$\begin{aligned} L_o(\boldsymbol{u}) &\geq c \|\boldsymbol{u}\|_{[H^1(\Omega)]^N} - c_1 \|\boldsymbol{F}\|_{[L^2(\Omega)]^N} \|\boldsymbol{u}\|_{[H^1(\Omega)]^{N-}} - c_2 \|\boldsymbol{P}\|_{[L^2(\Gamma_\tau)]^N} \|\boldsymbol{u}\|_{[H^1(\Omega)]^N} \\ &\geq c \|\boldsymbol{u}\|^2_{1,2} - c_o \|\boldsymbol{u}\|_{1,2} \quad \forall \boldsymbol{u} \in V \end{aligned} \tag{7.450}$$

and

$$\lim L_o(\boldsymbol{u}) \to \infty \text{ as } \|\boldsymbol{u}\| \to \infty,$$

which proves the coerciveness of the functional $L_o(\boldsymbol{u})$.

Now we shall prove the lower semi-continuity of the functional $L_o(\boldsymbol{u})$.

The functional L_o possesses a Gâteaux differential at any point. Let $\Psi(t) = L_o(\boldsymbol{u} + t(\boldsymbol{u}_n - \boldsymbol{u}))$, $t \in \langle 0, 1 \rangle$. The functional $\Psi(t)$ is continuous on the line segment in the variable $\boldsymbol{u}$. To prove it, we integrate $\Psi'(t)$ for $t \in \langle 0, 1 \rangle$. Then

$$\Psi(1) - \Psi(0) = L_o(\boldsymbol{u}_n) - L_o(\boldsymbol{u}) = \int_0^1 DL_o(\boldsymbol{u} + t(\boldsymbol{u}_n - \boldsymbol{u}), \boldsymbol{u}_n - \boldsymbol{u}) \mathrm{d}t.$$

Now

$$L_o(\boldsymbol{u}_n) - L_o(\boldsymbol{u}) - DL_o(\boldsymbol{u}, \boldsymbol{u}_n - \boldsymbol{u})$$
$$= \int_0^1 DL_o(\boldsymbol{u} + t(\boldsymbol{u}_n - \boldsymbol{u}), \boldsymbol{u}_n - \boldsymbol{u})\mathrm{d}t - DL_o(\boldsymbol{u}, u_n - \boldsymbol{u})$$
$$= \int_0^1 t^{-1}[DL_o(\boldsymbol{u} + t(\boldsymbol{u}_n - \boldsymbol{u}), t(\boldsymbol{u}_n - \boldsymbol{u}))\mathrm{d}t - DL_o(\boldsymbol{u}, t(u_n - \boldsymbol{u}))]\mathrm{d}t \geq 0.$$

Hence $\liminf[L_o(\boldsymbol{u}_n) - L_o(\boldsymbol{u}) - DL_o(\boldsymbol{u}, \boldsymbol{u}_n - \boldsymbol{u})] \geq 0$, and since $DL_o(\boldsymbol{u}, \boldsymbol{u}_n - \boldsymbol{u}) \to 0$ as $\boldsymbol{u}_n \to \boldsymbol{u}$ weakly, then

$$\liminf L_o(\boldsymbol{u}_n) - L_o(\boldsymbol{u}) \geq 0 \text{ as } \boldsymbol{u}_n \to \boldsymbol{u} \text{ weakly.} \tag{7.451}$$

Thus $DL_o(\boldsymbol{u}, \boldsymbol{v})$ is continuous on any line segment in the space V.

According to Equation 7.448, $DL_o(\boldsymbol{u} + \boldsymbol{v}, \boldsymbol{v}) - DL_o(\boldsymbol{u}, \boldsymbol{v}) \geq 0$ for $\boldsymbol{u}, \boldsymbol{v} \in V$. Hence and since $DL_o(\boldsymbol{u}, \boldsymbol{v})$ is continuous on any line segment in V then L_o is weakly lower semi-continuous.

The set K is closed and convex in $W = [H^1(\Omega)]^N$, hence it is weakly closed, as a closed and convex ball in a Hilbert space is weakly closed. The functional L_o is then convex and differentiable, the functional $j_{gn}(\boldsymbol{v}) = \langle g_c^{kl}, |\boldsymbol{v}_t^k - \boldsymbol{v}_t^l|\rangle$, $g_c^{kl} = \mathscr{F}_c^{kl}\, |\tau_n^{kl}|$, is convex and non-differentiable. Let us assume that $\partial\beta_{ij}/\partial x_j \in L^\infty(\Omega)\, \forall i, j$, then $|b_s(\Theta - \Theta_o, \boldsymbol{v})| \leq c(1 + \|\Theta\|_{1,1}) \|\boldsymbol{v}\|_{0,N} \leq c\, \|\boldsymbol{v}\|_{1,N}$, since $\|\boldsymbol{v}\|_{0,N} \leq c\, \|\boldsymbol{v}\|_{1,N}$. Then the functional $L(\boldsymbol{v})$ defined by Equation 7.442 is coercive and weakly lower semi-continuous. From the convex analysis (see, for example, Céa (1971), Ekeland and Temam (1976)) it is known that if the functional $L(\boldsymbol{v})$ is coercive and weakly lower semi-continuous then it has a minimum in a Hilbert space. Since the set of admissible displacements K is a subset of the space of virtual displacements V, then it is a closed convex subset of the Hilbert space V, and therefore the functional $L(\boldsymbol{v})$ attains its minimum. Let us put $\boldsymbol{v} = \boldsymbol{u}_o + \boldsymbol{w}$ and let $\varphi(\boldsymbol{w}) = L(\boldsymbol{u}_o + \boldsymbol{w})$. Then the functional $\varphi(\boldsymbol{w})$ is coercive on K and lower semi-continuous. Therefore, the functional $\varphi(\boldsymbol{w})$ attains its minimum in K. Therefore, there exists an element $\boldsymbol{w} \in K$, which is a minimum of φ in K. For this $\boldsymbol{w}$ and any $\boldsymbol{v} \in K$ we have $DL(\boldsymbol{u}_o + \boldsymbol{w}, \boldsymbol{v}) \geq 0$, representing the definition of the weak solution of Equation 7.441, so that $\boldsymbol{u}_o + \boldsymbol{w} = \boldsymbol{u}$ is a weak solution.

To prove the uniqueness of the solution we shall assume that $\boldsymbol{u}_1$ and $\boldsymbol{u}_2$ are two weak solutions. Then

$$0 \geq DL(\boldsymbol{u}_2, \boldsymbol{u}_2 - \boldsymbol{u}_1) - DL(\boldsymbol{u}_1, \boldsymbol{u}_2 - \boldsymbol{u}_1) \geq c\, \|\boldsymbol{u}_2 - \boldsymbol{u}_1\|_{1,2}^2,$$

since $\boldsymbol{u}_2 - \boldsymbol{u}_1 \in K$ as $\boldsymbol{u} - \boldsymbol{u}_o \in K$. Hence $\boldsymbol{u}_1 = \boldsymbol{u}_2$.

Then we have proved the following theorem.

Theorem 7.22: Let $\Gamma_u \neq 0$ and let $Q(\boldsymbol{x}) \in L^2(\overline{\Omega})$, $\boldsymbol{f}(\boldsymbol{x}) \in [L^2(\overline{\Omega})]^N$, $\boldsymbol{P}(\boldsymbol{x}) \in [L^2(\Gamma_\tau)]^N$, $\beta_{ij} \in C^1(\overline{\Omega})$, $\rho(\boldsymbol{x}) \in C(\overline{\Omega})$, $c(\boldsymbol{x}) \in C(\overline{\Omega})$ and $g_c^{kl} \in L^\infty(\Gamma_c^{kl})$. Let coefficients of thermal conductivity $\kappa_{ij}(\boldsymbol{x}) \in C^1(\overline{\Omega})$ be symmetric and satisfy Lipschitz conditions

$$\kappa_{ij}(\boldsymbol{x}) = \kappa_{ji}(\boldsymbol{x}), \quad \kappa_{ij}(\boldsymbol{x})\xi_i\xi_j \geq c_1\, |\xi|^2 \quad \forall \xi \in R^N, \text{ a.e. in } \Omega.$$

Further, let

$$\gamma(\boldsymbol{u}, \boldsymbol{v}) = -\tfrac{2}{3}\vartheta(\boldsymbol{u})\vartheta(\boldsymbol{v}) + 2e_{ij}(\boldsymbol{u})e_{ij}(\boldsymbol{v}), \quad \gamma(\boldsymbol{u}, \boldsymbol{u}) \equiv \gamma^2(\boldsymbol{u}) \equiv \gamma^2, \quad \text{and}$$

$$0 < k_o \leq k(\boldsymbol{x}) \leq k_1 < +\infty, \quad 0 < \mu_o \leq \mu(\boldsymbol{x}, s) \leq \tfrac{3}{2}k(\boldsymbol{x}),$$

$$0 < a_o \leq \mu(\boldsymbol{x}, s) + 2s\frac{\partial \mu(\boldsymbol{x}, s)}{\partial s} \leq A_o < +\infty \quad \text{for all } \boldsymbol{x} \in \overline{\Omega}, \gamma \geq 0,$$

where the functions $k(\boldsymbol{x})$ and $\mu(\boldsymbol{x}, s)$ are supposed to be continuous and μ to be continuously differentiable with respect to s.

Then there exists a unique weak solution $(\Theta, \boldsymbol{u})$ of the problem $(\mathscr{P})$. This solution minimizes the functionals $l(z)$ and $L(\boldsymbol{v})$ over the space ${}^1V$ and the set K, where

$$l(z) = \tfrac{1}{2}\alpha(z, z) + \tfrac{1}{2}\Delta t a_\Theta(z, z) + b_p(\boldsymbol{v}, z) + j(z) - s(z),$$

and

$$L(\boldsymbol{v}) = \frac{1}{2}\int_\Omega [k(\boldsymbol{x})\vartheta^2(\boldsymbol{v}) + \int_0^{\gamma(\boldsymbol{v},\boldsymbol{v})} \mu(\sigma)\mathrm{d}\sigma]\mathrm{d}\boldsymbol{x} - \int_\Omega F_i v_i \mathrm{d}\boldsymbol{x} - \int_{\Gamma_\tau} P_i v_i \mathrm{d}s + b_s(\Theta, \boldsymbol{v}) + j_{gn}(\boldsymbol{v})$$

$$= L_o(\boldsymbol{v}) + b_s(\Theta, \boldsymbol{v}) + j_{gn}(\boldsymbol{v})$$

and where $b_p(\boldsymbol{v}, z)$, $S(z)b_s(\Theta, \boldsymbol{v})$ and $j_{gn}(\boldsymbol{v})$ are defined by Equations 7.435, 7.434, 7.438 and 7.440.

7.4.4. Numerical Solution

To solve the problem $(\mathscr{P})$ numerically the technique of Nedoma (1997b) in the thermal part of the problem, and the method of secant modules in the plastic part of the problem, which consists of solving a sequence of variational inequalities corresponding to the linear Hooke's law, will be used.

Let us assume that $\Omega \subset R^N$, $N = 2, 3$, is polygonal for $N = 2$ or polyhedral for $N = 3$. Let the domain Ω be triangulated, i.e. $\overline{\Omega} = \Omega \cup \partial\Omega$ is covered by a finite number of triangles T_h for $N = 2$ or tetrahedra for $N = 3$, forming a triangulation $\mathscr{T}_h$. Further, we assume that the end points $\overline{\Gamma}_u \cap \overline{\Gamma}_\tau$, $\overline{\Gamma}_u \cap \overline{\Gamma}_c^k$, $\overline{\Gamma}_c^k \cap \overline{\Gamma}_\tau$ coincide with the vertices of T_h. Let $h_o > 0$. The family of triangulations $\{\mathscr{T}_h\}$, $0 < h \leq h_o$, is assumed to be regular. Let us define the spaces of finite elements

$$ {}^1V_h = \{z \mid z \in C(\overline{\Omega}), \quad z\,|_{T_h} \in P_1, \quad z = 0 \text{ on } \partial\Omega, \quad \forall T_h \in \mathscr{T}_h\},$$

$$V_h = \{\boldsymbol{v} \mid \boldsymbol{v} \in [C(\overline{\Omega})]^N, \quad \boldsymbol{v}_{|T_h} \in [P_1]^N, \quad \boldsymbol{v} = \boldsymbol{u}_o \text{ on } \Gamma_u, \quad \forall T_h \in \mathscr{T}_h\},$$

where P_1 is the space of all linear polynomials. Let

$$K_h = \{\boldsymbol{v} \mid \boldsymbol{v} \in V_h, v_n^k - v_n^l \leq 0 \text{ on } \cup \Gamma_c^{kl}\}$$

be the set of finite element approximations of the set of admissible displacements.

Definition 7.13: A pair of functions $(\Theta_h, \boldsymbol{u}_h)$ is said to be a finite element approximation of the problem $(\mathscr{P})_v$, denoted as the problem $(\mathscr{P})_{vh}$, if

$$\alpha(\Theta_h, z_h - \Theta_h) + \Delta t a_\Theta(\Theta_h, z_h - \Theta_h) + b_p(\boldsymbol{u}_h, z_h - \Theta_h) + j(z_h) - j(\Theta_h) \geq s(z_h - \Theta_h) \quad \forall z_h \in {}^1V_h, \tag{7.452}$$

$$a(\boldsymbol{u}_h; \boldsymbol{u}_h, \boldsymbol{v}_h - \boldsymbol{u}_h) + b_s(\Theta_h, \boldsymbol{v}_h - \boldsymbol{u}_h) + j_{gn}(\boldsymbol{v}_h) - j_{gn}(u_h) \geq S(\boldsymbol{v}_h - \boldsymbol{u}_h) \quad \forall \boldsymbol{v}_h \in K_h, \tag{7.453}$$

where $a_\Theta(.,.)$, $b_p(\boldsymbol{u},.)$, $j(.)$, $s(.)$, $a(\boldsymbol{u};.,.)$, $b_s(\Theta,.)$, $j_{gn}(.)$, and $S(.)$ are defined by Equations 7.432 to 7.440.

Let us define

$$\Omega_+^m = \{\boldsymbol{x} \mid \Theta(\boldsymbol{x}) > \Theta_R^m\}, \tag{7.454}$$

$$\Omega_-^m = \{\boldsymbol{x} \mid \Theta(\boldsymbol{x}) < \Theta_R^m\}, \tag{7.455}$$

$$\Omega_o^m = \{\boldsymbol{x} \mid \Theta(\boldsymbol{x}) = \Theta_R^m\}, \tag{7.456}$$

$$\Omega_{+h}^m = \{T_h \in \mathcal{T}_h \mid \Theta > \Theta_R^m \text{ on } T_h\}, \tag{7.457}$$

$$\Omega_{-h}^m = \{T_h \in \mathcal{T}_h \mid \Theta < \Theta_R^m \text{ on } T_h\}, \tag{7.458}$$

$$\Omega_{oh}^m = \{T_h \in \mathcal{T}_h \mid \Theta = \Theta_R^m \text{ on } T_h\}. \tag{7.459}$$

Further, let $\Omega_{*h} = \Omega \setminus \cup (\Omega_{+h}^m \cup \Omega_{-h}^m \cup \Omega_{oh}^m)$ be the set of all triangles in the $2D$ case and of all tetrahedra in the $3D$ case, which are situated close to the phase change boundary. It is evident that Ω_{*h} is not empty as the phase change boundary is contained in some triangles or tetrahedra, i.e. $\Omega \neq \cup(\Omega_{+h}^m \cup \Omega_{-h}^m \cup \Omega_{oh}^m) \neq 0$. The length of the phase change boundary satisfies

$$\begin{aligned}\text{meas}\{\cup R^m(t)\} &= \text{meas}\{\cup(\partial\Omega_+^m \cap \partial\Omega_-^m)\} + \text{meas}\{\cup(\partial\Omega_o^m \cap \partial\Omega_-^m)\} \\ &\quad + \text{meas}\{\cup(\partial\Omega_o^m \cap \partial\Omega_+^m)\} < M,\end{aligned}$$

i.e. it is finite, and there exists a constant c independent of h and such that the number of triangles or tetrahedra N_{*h}^m in Ω_{*h}^m is bounded by $N_{*h}^m < ch^{-1}$.

The thermal as well as the plastic parts of the problem are non-linear. For such type of problems it is convenient to transform the problem into a sequence of linear problems. The secant modules method represents a very useful method, and will be used for the plastic part of the problem.

The idea of the method consists of the following procedure.

Let $\boldsymbol{u}^n$ be the n-th approximation of the solution to the above problem $(\mathcal{P})_v$. We seek $\boldsymbol{u}^{n+1}$ as the solution of the problem for the linear Hooke's law

$$\tau_{ij} = (k - \tfrac{2}{3}\mu(\gamma^2(\boldsymbol{u}^n))e_{kk}\delta_{ij} + 2\mu(\gamma^2(\boldsymbol{u}^n))e_{ij}.$$

This leads to finding the weak solution of the linear problem

$$\begin{aligned}&\int_\Omega \left[\left(k - \frac{2}{3}\mu(\gamma^2(\boldsymbol{u}^n))\right) e_{kk}(\boldsymbol{u}^{n+1})e_{kk}(\boldsymbol{v} - \boldsymbol{u}^{n+1})\right. \\ &\quad \left. + 2\mu(\gamma^2(\boldsymbol{u}^n))e_{ij}(\boldsymbol{u}^{n+1})e_{ij}(\boldsymbol{v} - \boldsymbol{u}^{n+1})\right] \mathrm{d}\boldsymbol{x} - S(\boldsymbol{v} - \boldsymbol{u}^{n+1}) \geq 0 \quad \forall \boldsymbol{v} \in K\end{aligned}$$

and such that

$$\boldsymbol{u}^n \to \boldsymbol{u} \text{ in } K.$$

As the main result we shall prove the following theorem.

Theorem 7.23: Let the assumptions and conditions of Theorem 7.22 mentioned above be fulfilled. Let $\partial\Phi(.) : R^1 \to R^1$ be the subdifferential of the piecewise quadratic convex function

$\Phi : R^1 \to R^1$ defined by Equation 7.407 and let the functional $j(z) = \Delta t \int_\Omega \Phi(z(\boldsymbol{x}))\mathrm{d}\boldsymbol{x}$ be the convex continuous functional on ${}^1V$. Let the functional L be given on K and let L have the second Gâteaux differential $D^2L(\boldsymbol{u};\boldsymbol{v},\boldsymbol{w})$. Let the mapping $\boldsymbol{u} \to D^2L(\boldsymbol{u};\boldsymbol{v},\boldsymbol{w})$ be continuous on every line segment. Let

$$D^2L(\boldsymbol{u};\boldsymbol{v},\boldsymbol{v}) \geq c\,\|\boldsymbol{v}\|^2, \quad c = \text{constant} > 0, \tag{7.460}$$

$$a(\boldsymbol{u};\boldsymbol{v},\boldsymbol{v}) \geq c_1\,\|\boldsymbol{v}\|^2, \quad c_1 = \text{constant} > 0, \tag{7.461}$$

$$|a(\boldsymbol{w};\boldsymbol{u},\boldsymbol{v})| \leq c_2\,\|\boldsymbol{u}\|\,\|\boldsymbol{v}\|, \tag{7.462}$$

$$a(\boldsymbol{u};\boldsymbol{u},\boldsymbol{v}) = DL(\boldsymbol{u},\boldsymbol{v}), \tag{7.463}$$

$$\tfrac{1}{2}a(\boldsymbol{u};\boldsymbol{v},\boldsymbol{v}) - \tfrac{1}{2}a(\boldsymbol{u};\boldsymbol{u},\boldsymbol{u}) - L(\boldsymbol{v}) - L(\boldsymbol{u}) \geq 0 \quad \forall \boldsymbol{u},\boldsymbol{v} \in V \tag{7.464}$$

hold. Let $\boldsymbol{f} \in V'$ be given.

Then the problem leads to solving the following two steps problem. Find a pair of functions $(\Theta_h, \boldsymbol{u}_h)$, $\Theta_h \in {}^1V_h$, $\boldsymbol{u}_h \in K_h$ such that Θ_h satisfies

$$\alpha(\Theta_h, z_h-\Theta_h)+\Delta t a_\Theta(\Theta_h, z_h-\Theta_h)+b_p(u_h, z_h - \Theta_h)+j(z_h) - j(\Theta_h)\geq s(z_h - \Theta_h) \quad \forall z_h \in {}^1V_h, \tag{7.465}$$

and $\boldsymbol{u}_h$ satisfies

$$a(\boldsymbol{u}_h;\boldsymbol{u}_h, \boldsymbol{v}_h - \boldsymbol{u}_h) + b_s(\Theta_h, \boldsymbol{v}_h - \boldsymbol{u}_h) + j_{gn}(\boldsymbol{v}_h) - j_{gn}(\boldsymbol{u}_h) \geq S(\boldsymbol{v}_h - \boldsymbol{u}_h) \quad \forall \boldsymbol{v}_h \in K_h, \tag{7.466}$$

and such that there exists a sequence $\{\boldsymbol{u}_h^n\} \in K_h$, where $\boldsymbol{u}_h^n$ is the n-th iterate $(n = 1, 2, \ldots .)$ of the secant modules method and the $(n+1)$-th iterate $\boldsymbol{u}^{n+1}$ is the solution of the linear problem

$$a(\boldsymbol{u}_h^n;\boldsymbol{u}_h^{n+1}, \boldsymbol{v}_h - \boldsymbol{u}_h^{n+1}) + b_s(\Theta_h, \boldsymbol{v}_h - \boldsymbol{u}_h^{n+1}) + j_{gn}(\boldsymbol{v}_h) - j_{gn}(u_h^{n+1}) \geq S(\boldsymbol{v}_h - \boldsymbol{u}_h^{n+1}) \quad \forall \boldsymbol{v}_h \in K, \tag{7.467}$$

and

$$\lim_{n\to\infty} \boldsymbol{u}_h^n = \boldsymbol{u}_h \text{ in } K_h. \tag{7.468}$$

Proof: Let the basis functions for each node i of any element T_h be denoted as $w_i(\boldsymbol{x})$, $0 \leq w_i(\boldsymbol{x}) \leq 1$ and $\Sigma_{i=1}^p w_i(\boldsymbol{x}) = 1$, where $p = 3$ for the $2D$ case and $p = 4$ for the $3D$ case. Due to Equation 7.407 $\Phi(z)$ is a continuous convex function, and then

$$\Phi(z(\boldsymbol{x})) = \Phi\left(\sum_{i=1}^p z_i w_i(\boldsymbol{x})\right) \leq \sum_{i=1}^p w_i(\boldsymbol{x})\Phi(z_i). \tag{7.469}$$

Since $j(z)$ is a convex continuous functional on $H_o^1(\Omega)$, then

$$\begin{aligned} j(z) &= \int_\Omega \Phi(z(\boldsymbol{x}))\mathrm{d}\boldsymbol{x} = \sum_{T_h\in\mathcal{T}_h} \int_{T_h} \Phi(z(\boldsymbol{x}))\mathrm{d}\boldsymbol{x} \leq \sum_{T_h\in\mathcal{T}_h} \int_{T_h} \sum_{i=1}^p w_i(\boldsymbol{x})\Phi(z_i)\mathrm{d}\boldsymbol{x} \\ &= \int_\Omega \Phi(z(\boldsymbol{x}))_{LI}\mathrm{d}\boldsymbol{x} = j_h(z), \end{aligned} \tag{7.470}$$

where $z_i(T_h)$ is the value of z at node i of the element T_h and meas(T_h) is the area of element T_h. Hence

$$j(z) \leq j(z_h) \quad \forall z \in {}^1V_h, \tag{7.471}$$

and since

$$\left|b_p(\boldsymbol{u}_h, \Theta_h)\right| \leq c\,\|\Theta_h\|_{0,1}\,\|\boldsymbol{u}_h\|_{1,2}, \tag{7.472}$$

then the existence and uniqueness of the thermal part of the problem is similar to that of Theorem 7.20. Moreover, for the discrete problem in Equation 7.452 there exists a unique Θ_h that also uniquely solves the optimization problem

$$l_h(\Theta_h) = \min_{z \in {}^1V_h} l_h(z) \equiv \tfrac{1}{2}\alpha(z, z)_h + \tfrac{1}{2}\Delta t a_{\Theta h}(z, z) + b_{ph}(\boldsymbol{u}_h, z) + j_h(z) - s_h(z), \tag{7.473}$$

where $(.,.)_h$, $a_{\Theta h}(.,.)$, and $j_h(.)$ are the discrete approximations of the scalar product in L^2, of the bilinear form $a_\Theta(.,.)$ and of the functional $j(.)$, respectively.

To prove the second part of the proof, we shall firstly prove that the secant modules method converges to the exact solution $\boldsymbol{u} \in K$.

We see that the conditions in Equations 7.461 and 7.462 are evident, and Equation 7.459 was proved above. Since $0 < \mu_o \leq \mu(\boldsymbol{x}, s) \leq 3/2k(\boldsymbol{x}) \equiv \mu_1 < +\infty$, $s \geq 0$, $\boldsymbol{x} \in \Omega$ and

$$\int_\Omega e_{ij}(\boldsymbol{v})e_{ij}(\boldsymbol{v})\mathrm{d}\boldsymbol{x} \geq \frac{1}{2}\int_\Omega (\partial v_i/\partial x_j)^2 \mathrm{d}\boldsymbol{x}, \tag{7.474}$$

then Equation 7.462 holds. The condition in Equation 7.464 is satisfied as follows.

Let us put (for $\gamma \equiv \gamma^2(\boldsymbol{u})$)

$$\mathscr{M}(\gamma) = \int_0^\gamma \mu(t)\mathrm{d}t.$$

Then Equation 7.464 yields

$$\int_\Omega [\mu(\gamma(\boldsymbol{u}))(\gamma(\boldsymbol{v}) - \gamma(\boldsymbol{u})) - (\mathscr{M}(\gamma(\boldsymbol{v})) - \mathscr{M}(\gamma(\boldsymbol{u})))]\mathrm{d}\boldsymbol{x} \geq 0. \tag{7.475}$$

The function $\mathscr{M}(\gamma)$ is concave if $\mathscr{M}'' = \mu'(\gamma) \leq 0$. Hence Equation 7.464 is satisfied.

Now we shall prove that $\boldsymbol{u}_h^n \to \boldsymbol{u}_h$ in K. Let us put

$$\mathscr{L}_n(\boldsymbol{v}) = L_1(\boldsymbol{u}_h^n) + b_s(\Theta, \boldsymbol{v}) + j_{gn}(\boldsymbol{v}) - S(\boldsymbol{v}) + \tfrac{1}{2}a(\boldsymbol{u}_h^n; \boldsymbol{v}, \boldsymbol{v}) - \tfrac{1}{2}a(\boldsymbol{u}_h^n; \boldsymbol{u}_h^n, \boldsymbol{u}_h^n). \tag{7.476}$$

Hence for $\boldsymbol{v} = \boldsymbol{u}_h^{n+1}$ and Equation 7.464 (with $\boldsymbol{u}_h = \boldsymbol{u}_h^n$, $\boldsymbol{v} = \boldsymbol{u}_h^{n+1}$ and $\Theta \equiv \Theta_h$) we have

$$\begin{aligned}\mathscr{L}_n(\boldsymbol{u}_h^{n+1}) &= L_1(\boldsymbol{u}_h^n) + b_s(\Theta, \boldsymbol{u}_h^{n+1}) + j_{gn}(\boldsymbol{u}_h^{n+1}) - S(\boldsymbol{u}_h^{n+1}) + \tfrac{1}{2}a(\boldsymbol{u}_h^n; \boldsymbol{u}_h^{n+1}, \boldsymbol{u}_h^{n+1}) \\ &\quad - \tfrac{1}{2}a(\boldsymbol{u}_h^n; \boldsymbol{u}_h^n, \boldsymbol{u}_h^n) - L_1(\boldsymbol{u}_h^{n+1}) + L_1(\boldsymbol{u}_h^{n+1}) \\ &\geq L_1(\boldsymbol{u}_h^{n+1}) + b_s(\Theta, \boldsymbol{u}_h^{n+1}) + j_{gn}(\boldsymbol{u}_h^{n+1}) - S(\boldsymbol{u}_h^{n+1}) \equiv \varphi(\boldsymbol{u}_h^{n+1}),\end{aligned} \tag{7.477}$$

where

$$\varphi(\boldsymbol{v}) = L_1(\boldsymbol{v}) + b_s(\Theta, \boldsymbol{v}) + j_{gn}(\boldsymbol{v}) - S(\boldsymbol{v}). \tag{7.478}$$

Due to Equation 7.466, we find

$$\tfrac{1}{2}a(\boldsymbol{u}_h^n; \boldsymbol{u}_h^{n+1}, \boldsymbol{u}_h^{n+1}) + b_s(\Theta, \boldsymbol{u}_h^{n+1}) + j_{gn}(\boldsymbol{u}_h^{n+1}) - S(\boldsymbol{u}_h^{n+1}) \leq \tfrac{1}{2}a(\boldsymbol{u}_h^n; \boldsymbol{v}, \boldsymbol{v}) + b_s(\Theta, \boldsymbol{v}) + j_{gn}(\boldsymbol{v}) - S(\boldsymbol{v}) \quad \forall \boldsymbol{v} \in K. \tag{7.479}$$

Hence and using Equation 7.477 we obtain

$$\mathscr{L}_n(\boldsymbol{u}_h^n) = \varphi(\boldsymbol{u}_h^n) \geq \mathscr{L}_n(\boldsymbol{u}_h^{n+1}) \geq \varphi(\boldsymbol{u}_h^{n+1}). \tag{7.480}$$

According to Equation 7.460 the functional $\varphi(\boldsymbol{v})$ is coercive, so that

$$\varphi(\boldsymbol{v}) \geq c\,\|\boldsymbol{v}\|^2 - c_1\,\|\boldsymbol{v}\| - c_2 \quad \forall \boldsymbol{v} \in V. \tag{7.481}$$

Hence and using Equation 7.480 we find

$$\lim_{n\to\infty} \varphi(\boldsymbol{u}_h^n) = c < -\infty. \tag{7.482}$$

Since

$$c\left\|\boldsymbol{u}_h^{n+1} - \boldsymbol{u}_h^n\right\|^2 \le a(\boldsymbol{u}_h^n; \boldsymbol{u}_h^{n+1} - \boldsymbol{u}_h^n, \boldsymbol{u}_h^{n+1} - \boldsymbol{u}_h^n) = a(\boldsymbol{u}_h^n; \boldsymbol{u}_h^n, \boldsymbol{u}_h^n) + a(\boldsymbol{u}_h^n; \boldsymbol{u}_h^{n+1}, \boldsymbol{u}_h^{n+1} - 2\boldsymbol{u}_h^n), \tag{7.483}$$

and since because of Equation 7.466

$$2\varphi(\boldsymbol{u}_h^n) - 2\mathscr{L}_n(\boldsymbol{u}_h^{n+1}) \ge a(\boldsymbol{u}_h^n; \boldsymbol{u}_h^n, \boldsymbol{u}_h^n) - a(\boldsymbol{u}_h^n; \boldsymbol{u}_h^{n+1}, \boldsymbol{u}_h^{n+1}) + 2a(\boldsymbol{u}_h^n; \boldsymbol{u}_h^{n+1} - \boldsymbol{u}_h^n, \boldsymbol{u}_h^{n+1} - \boldsymbol{u}_h^n), \tag{7.484}$$

then we obtain

$$2\varphi(\boldsymbol{u}_h^n) - 2\mathscr{L}_n(\boldsymbol{u}_h^{n+1}) \ge a(\boldsymbol{u}_h^n; \boldsymbol{u}_h^n, \boldsymbol{u}_h^n) + a(\boldsymbol{u}_h^n; \boldsymbol{u}_h^{n+1}, \boldsymbol{u}_h^{n+1} - 2\boldsymbol{u}_h^n) \ge c\left\|\boldsymbol{u}_h^{n+1} - \boldsymbol{u}_h^n\right\|^2. \tag{7.485}$$

Using Equation 7.480 and 7.477, we find

$$\lim_{n\to\infty} \left\|\boldsymbol{u}_h^{n+1} - \boldsymbol{u}_h^n\right\| = 0. \tag{7.486}$$

Because of Equation 7.466 we obtain

$$\begin{aligned}\tfrac{1}{2}c\left\|\boldsymbol{u}_h^n - \boldsymbol{u}\right\|^2 &\le DL_1(\boldsymbol{u}_h^n, \boldsymbol{u}_h^n - \boldsymbol{u}_h) - DL_1(\boldsymbol{u}_h, \boldsymbol{u}_h^n - \boldsymbol{u}_h) = a(\boldsymbol{u}_h^n; \boldsymbol{u}_h^n, \boldsymbol{u}_h^n - \boldsymbol{u}_h) - DL_1(\boldsymbol{u}_h, \boldsymbol{u}_h^n - \boldsymbol{u}_h)\\ &\le a(\boldsymbol{u}_h^n; \boldsymbol{u}_h^n, \boldsymbol{u}_h^n - \boldsymbol{u}_h) + b_s(\Theta, \boldsymbol{u}_h^n - \boldsymbol{u}_h) + j_{gn}(\boldsymbol{u}_h^n) - j(\boldsymbol{u}_h) - S(\boldsymbol{u}_h^n - \boldsymbol{u}_h). \end{aligned}\tag{7.487}$$

According to Equation 7.466 we find

$$\begin{aligned} &a(\boldsymbol{u}_h^n; \boldsymbol{u}_h^n, \boldsymbol{u}_h^n - \boldsymbol{u}_h) + b_s(\Theta, \boldsymbol{u}_h^n - \boldsymbol{u}_h) + j_{gn}(\boldsymbol{u}_h^n) - j(\boldsymbol{u}_h) - S(\boldsymbol{u}_h^n - \boldsymbol{u}_h)\\ &= a(\boldsymbol{u}_h^n; \boldsymbol{u}_h^n - \boldsymbol{u}_h^{n+1}, \boldsymbol{u}_h^n - \boldsymbol{u}_h) + a(\boldsymbol{u}_h^n; \boldsymbol{u}_h^{n+1}, \boldsymbol{u}_h^n - \boldsymbol{u}_h) + b_s(\Theta, \boldsymbol{u}_h^n - \boldsymbol{u}_h)\\ &\quad + j_{gn}(\boldsymbol{u}_h^n) - j(\boldsymbol{u}_h) - S(\boldsymbol{u}_h^n - \boldsymbol{u}_h)\\ &= a(\boldsymbol{u}_h^n; \boldsymbol{u}_h^n - \boldsymbol{u}_h^{n+1}, \boldsymbol{u}_h^n - \boldsymbol{u}_h) + a(\boldsymbol{u}_h^n; \boldsymbol{u}_h^{n+1}, \boldsymbol{u}_h^n - \boldsymbol{u}_h^{n+1}) + b_s(\Theta, \boldsymbol{u}_h^n - \boldsymbol{u}_h^{n+1})\\ &\quad + j_{gn}(\boldsymbol{u}_h^n) - j(\boldsymbol{u}_h^{n+1}) - S(\boldsymbol{u}_h^n - \boldsymbol{u}_h^{n+1}) + a(\boldsymbol{u}_h^n; \boldsymbol{u}_h^{n+1}, \boldsymbol{u}_h^{n+1} - \boldsymbol{u}_h)\\ &\quad + b_s(\Theta, \boldsymbol{u}_h^{n+1} - \boldsymbol{u}_h) + j_{gn}(\boldsymbol{u}_h^{n+1}) - j(\boldsymbol{u}_h) - S(\boldsymbol{u}_h^{n+1} - \boldsymbol{u}_h)\\ &\le a(\boldsymbol{u}_h^n; \boldsymbol{u}_h^n - \boldsymbol{u}_h^{n+1}, \boldsymbol{u}_h^n - \boldsymbol{u}_h) + a(\boldsymbol{u}_h^n; \boldsymbol{u}_h^{n+1}, \boldsymbol{u}_h^n - \boldsymbol{u}_h^{n+1}) + b_s(\Theta, \boldsymbol{u}_h^{n+1} - \boldsymbol{u}_h^n)\\ &\quad + j_{gn}(\boldsymbol{u}_h^{n+1}) - j(\boldsymbol{u}_h^n) + S(\boldsymbol{u}_h^{n+1} - \boldsymbol{u}_h^n). \end{aligned}$$

Hence and using Equations 7.487, 7.462 and 7.486, and the boundedness of $\boldsymbol{u}^n$, we find successively

$$\tfrac{1}{2}c\left\|\boldsymbol{u}_h^n - \boldsymbol{u}\right\|^2 \le c_o\left\|\boldsymbol{u}_h^n - \boldsymbol{u}_h^{n+1}\right\|\left\|\boldsymbol{u}_h^n - \boldsymbol{u}_h\right\|$$

and $\lim_{n\to\infty} \boldsymbol{u}_h^n = \boldsymbol{u}_h$, which completes this part of the proof.

The set K_h is non-empty, closed and convex in V_h, therefore it is weakly closed. Let $L(\boldsymbol{u}_h^n)$ be the linearized functional defined by $L(\boldsymbol{v}) = \frac{1}{2}a(\boldsymbol{v}, \boldsymbol{v}) + b_s(\Theta, \boldsymbol{v}) + j_{gn}(\boldsymbol{v}) - S(\boldsymbol{v})$. Theorem 7.19 implies that the functional $L(\boldsymbol{v})$ is coercive over K_h, and $j_{gnh}(\boldsymbol{v})$ is convex, which completes the proof of the existence of the solution.

The uniqueness can be proved by the usual techniques. Let $(\Theta_h^1, \boldsymbol{u}_h^1) \in {}^1V_h \times K_h$ and $(\Theta_h^2, \boldsymbol{u}_h^2) \in {}^1V_h \times K_h$ be two approximate solutions of the problem $(\mathscr{P})_v$. Then $c_1\left\|\Theta_h^1 - \Theta_h^2\right\|^2 \le 0$, $c_2\left\|\boldsymbol{u}_h^1 - \boldsymbol{u}_h^2\right\|^2 \le 0$, from which the uniqueness follows immediately.

To find an estimate of the rates of convergence of $\|\Theta - \Theta_h\|$ and $\|\boldsymbol{u} - \boldsymbol{u}_h\|$ the technique proposed by Falk (1974) and modified by Nedoma (1983), (1987), (1994), (1997b) will be used.

For any pair of functions $(z, \boldsymbol{v})$ on Ω we define the Lagrange linear interpolate $(z_I, \boldsymbol{v}_I)$ and the orthogonal projection of the Lagrange linear interpolate $(z_{LI}, \boldsymbol{v}_{LI})$ on the triangulation $\mathcal{T}_h$ such that for $z_{LI} \in {}^1V_h$, $\boldsymbol{v}_{LI} \in V_h$ $(z_{LI}, \boldsymbol{v}_{LI}) = (z, \boldsymbol{v})$ at each vertex of T_h. Then the well-known interpolate estimates (see Ciarlet (1978)) hold, i.e.

$$\|z - z_{LI}\|_{0,1} + h\,\|z - z_{LI}\|_{1,1} \le ch^2, \quad \forall z \in H_o^1(\Omega) \cap H^2(\Omega), \tag{7.488}$$

$$\|\boldsymbol{v} - \boldsymbol{v}_{LI}\|_{0,2} + h\,\|\boldsymbol{v} - \boldsymbol{v}_{LI}\|_{1,2} \le ch^2, \quad \forall \boldsymbol{v} \in [H_o^1(\Omega)]^2 \cap [H^2(\Omega)]^2. \tag{7.489}$$

For $\Theta, z \in {}^1V_h$ let us put

$$(\Theta, z)_h = \sum_{T_h \in \mathcal{T}_h} p^{-1}\mathrm{meas}(T_h)\Theta_i^T z_i^T,$$

$$j_h(z) = \int_\Omega (\Phi(z))_{LI}\mathrm{d}x = \sum_{T_h \in \mathcal{T}_h} p^{-1}\mathrm{meas}(T_h)\sum_{i=1}^{p}\Phi(z_i^T),$$

where z_i^T denotes the value of z at node i of T_h, $p = 3$ in the $2D$ case and $p = 4$ in the $3D$ case, and $\mathrm{meas}(T_h)$ is the area of T_h. Further, we need the following lemma.

Lemma 7.26 (Ciavaldini,Raviart): The following estimates

$$|(\Theta, z)_h - (\Theta, z)| \le ch^2\,\|\Theta\|_{1,1}\,\|z\|_{1,1} \quad \forall \Theta, z \in {}^1V_h, \tag{7.490}$$

$$\|z\|_{0,1} \le (z, z)_h^{1/2} \le c\,\|z\|_{0,1} \quad \forall z \in {}^1V_h, \tag{7.491}$$

hold.

For the proof see Raviart (1972), Ciavaldini (1975).

Lemma 7.27: It holds that

$$|j(\Theta_{LI}) - j(\Theta)| \le c\,\|\Theta - \Theta_{LI}\|_{0,1}, \tag{7.492}$$

$$|j_h(\Theta_{LI}) - j(\Theta_{LI})| \le ch^2. \tag{7.493}$$

Proof: The proof is a modification of the results of Elliott (1981) and previous results for the thermo-elastic cases. To prove Equation 7.492 for any element $T_h \in \mathcal{T}_h$ we denote by $w_i(\boldsymbol{x})$, $0 \le w_i(\boldsymbol{x}) \le 1$, the basis functions for each node i. Then $\Sigma_{i=1}^p w_i(\boldsymbol{x}) = 1$, where $p = 3$ in the $2D$ and $p = 4$ in the $3D$ cases. Since the functional $\Phi(z)$ is a continuous convex function, then

$$\Phi(z(\boldsymbol{x})) = \Phi\left(\sum_{i=1}^{p} z_i w_i(\boldsymbol{x})\right) \le \sum_{i=1}^{p} w_i(\boldsymbol{x})\Phi(z_i).$$

Since $j(z)$ is a convex continuous functional on $H_o^1(\Omega)$, then

$$j(z) = \int_\Omega \Phi(z(\boldsymbol{x}))\mathrm{d}\boldsymbol{x} = \sum_{T_h \in \mathcal{T}_h}\int_{T_h}\Phi(z(\boldsymbol{x}))\mathrm{d}\boldsymbol{x} \le \sum_{T_h \in \mathcal{T}_h}\int_{T_h} w_i(\boldsymbol{x})\Phi(z_i)\mathrm{d}\boldsymbol{x} = \int_\Omega (\Phi(z(\boldsymbol{x})))_{LI}\mathrm{d}x = j_h(z).$$

To prove Equation 7.492 then, from the fact that $\Phi : R^1 \to R^1$ is a piecewise quadratic convex function defined by Equation 7.407 and locally Lipschitz continuous, i.e. that we have Equation 7.409

$$|\Phi(\Theta_1) - \Phi(\Theta_2)| \leq \max\{|\partial\Phi(\Theta_1)|, |\partial\Phi(\Theta_2)|\} |\Theta_1 - \Theta_2| \leq c |\Theta_1 - \Theta_2|, \text{ when } |\Theta_1|, |\Theta_2| \leq K,$$

and that Θ is continuous, we obtain

$$|j(\Theta_{LI}) - j(\Theta)| = \left| \int_\Omega (\Phi(\Theta_{LI}) - \Phi(\Theta))\mathrm{d}\boldsymbol{x} \right| \leq c \left(\|\Theta\|_{L^\infty(\Omega)} \right) \int_\Omega |\Theta - \Theta_{LI}| \, \mathrm{d}\boldsymbol{x} \leq c \, \|\Theta - \Theta_{LI}\|_{0,1} .$$

To prove Equation (7.493) we firstly have

$$|j(\Theta_{LI}) - j_h(\Theta_{LI})| = \left| \sum_{T_h \in \mathcal{T}_h} \int_{T_h} [\Phi(\Theta_{LI}) - \Phi(\Theta_{LI}(\boldsymbol{x}))_{LI})\mathrm{d}\boldsymbol{x} \right| . \tag{7.494}$$

Now we shall investigate the behaviour of the last integral of Equation 7.494 in the neighbourhood of the free boundary $R^m(t)$. Therefore, we shall investigate this last integral for $T_h \in \Omega^m_{+h}, \Omega^m_{-h}, \Omega^m_{oh}$ and for $T_h \in \Omega^m_{*h} = \Omega \backslash \cup (\Omega^m_{+h} \cup \Omega^m_{-h} \cup \Omega^m_{oh})$, which are defined by Equations 7.454 to 7.459.

Let $T_h \in \Omega^m_{oh}$, then $\Theta_{LI} \equiv \Theta^m_R$ and $\Phi(\Theta_{LI}) \equiv 0$. Hence,

$$\int_{T_h \in \Omega^m_{oh}} [\Phi(\Theta_{LI}) - \Phi(\Theta_{LI}(\boldsymbol{x}))_{LI}]\mathrm{d}\boldsymbol{x} = 0. \tag{7.495}$$

Let $T_h \in \Omega^m_{+h}$ and $T_h \in \Omega^m_{-h}$. Due to Equation 7.404 $\Phi(.)$ is a quadratic function of Θ_{LI} and therefore, similarly to Lemma 7.26, we have

$$\left| \int_{T_h \in \Omega^m_{+h}} [\Phi(\Theta_{LI}) - \Phi(\Theta_{LI}(\boldsymbol{x}))_{LI}]\mathrm{d}\boldsymbol{x} \right| \leq ch^2, \tag{7.496}$$

$$\left| \int_{T_h \in \Omega^m_{-h}} [\Phi(\Theta_{LI}) - \Phi(\Theta_{LI}(\boldsymbol{x}))_{LI}]\mathrm{d}\boldsymbol{x} \right| \leq ch^2. \tag{7.497}$$

Now let $T_h \in \Omega^m_{*h}$. Then due to the Lipschitz continuity of $\Phi(.)$, defined by Equation 7.408, and since we assumed that the number of triangles N^m_{*h} in Ω_{*h} is bounded by $N^m_{*h} < ch^{-1}$, we find

$$\left| \int_{T_h \in \Omega^m_{*h}} (\Phi(\Theta_{LI}) - \Phi(\Theta_{LI}(\boldsymbol{x}))_{LI})\mathrm{d}\boldsymbol{x} \right| = \left| \int_{T_h \in \Omega^m_{*h}} \Phi(\Theta_{LI})\mathrm{d}\boldsymbol{x} - p^{-1}\mathrm{meas}(T_h) \sum_{i=1}^{p} \Phi(\Theta^T_i) \right|$$
$$= p^{-1} \left| \sum_{i=1}^{p} \int_{T_h \in \Omega^m_h} (\Phi(\Theta_{LI})\mathrm{d}\boldsymbol{x} - \Phi(\Theta^T_i) \right| ,$$

where we denoted as above by Θ^T_i the value of Θ at the node i of T_h, and $p = 3$ or $p = 4$ for the $2D$ or $3D$ cases, respectively. Since, due to Equation 7.409, $\Phi(.)$ is locally Lipschitz continuous then

$$p^{-1} \left| \sum_{i=1}^{p} \int_{T_h \in \Omega^m_{*h}} (\Phi(\Theta_{LI}))\mathrm{d}\boldsymbol{x} - \Phi(\Theta^T_i) \right| \leq p^{-1} \sum_{i=1}^{p} c \int_{T_h \in \Omega^m_{*h}} |\Theta_{LI}(\boldsymbol{x}) - \Theta^T_i| \, \mathrm{d}\boldsymbol{x}$$
$$= p^{-1} \sum_{i=1}^{p} c \int_{T_h \in \Omega^m_{*h}} |(\boldsymbol{x} - \boldsymbol{x}_i).\mathrm{grad}\ \Theta_{LI}| \, \mathrm{d}\boldsymbol{x}$$
$$\leq p^{-1} \sum_{i=1}^{p} c \left(\int_{T_h \in \Omega^m_{*h}} |\boldsymbol{x} - \boldsymbol{x}_i|^2 \right)^{1/2} \left(\int_{T_h \in \Omega^m_{*h}} |\mathrm{grad}\ \Theta_{LI}|^2 \right)^{1/2} \mathrm{d}\boldsymbol{x}$$

$$\leq ch^2 \left(\int_{T_h \in \Omega^m_{*h}} |\text{grad } \Theta|^2 \right)^{1/2} \mathrm{d}\boldsymbol{x} + \left(\int_{T_h \in \Omega^m_{*h}} |\text{grad } (\boldsymbol{x} - \boldsymbol{x}_i)|^2 \right)^{1/2} \leq ch^2 (h \, \|\Theta\|_{W^{1,\infty}} + ch).$$

Hence

$$\left| \int_{T_h \in \Omega^m_{*h}} (\Phi(\Theta_{LI}(\boldsymbol{x})) - \Phi(\Theta_{LI}(\boldsymbol{x}))_{LI}) \mathrm{d}\boldsymbol{x} \right| \leq \sum_{T_h \in \Omega^m_{*h}} ch^3 \leq ch^2. \tag{7.498}$$

From Equation 7.492 to 7.497 we obtain

$$|j_h(\Theta_{LI}) - j(\Theta_{LI})| \leq ch^2,$$

which completes the proof.

Lemma 7.28: For $\Theta \in {}^1V$, $\boldsymbol{u} \in K$, $\Theta_h \in {}^1V_h$, and $\boldsymbol{u}_h^n \in K_h$ we have

$$\begin{aligned}
&\alpha \, \|\Theta - \Theta_h\|^2_{0,1} + \Delta t \, \|\Theta - \Theta_h\|^2_{1,1} \\
&\quad \leq c[\alpha(\Theta_h - \Theta, z_h - \Theta) + \Delta t a_\Theta(\Theta_h - \Theta, z_h - \Theta) + \alpha(\Theta, z - \Theta_h) + \Delta t a_\Theta(\Theta, z - \Theta_h) \\
&\qquad + \alpha(\Theta, z_h - \Theta) + \Delta t a_\Theta(\Theta, z_h - \Theta) + b_p(\boldsymbol{u}, z_h - \Theta) + b_p(\boldsymbol{u}, z - \Theta_h) + j(z) - j(\Theta_h) \\
&\qquad + j(z_h) - j(\Theta) - s(z - \Theta_h) - s(z_h - \Theta)] \quad \forall z \in {}^1V, z_h \in {}^1V_h,
\end{aligned} \tag{7.499}$$

$$\begin{aligned}
\left\|\boldsymbol{u} - \boldsymbol{u}_h^n\right\|^2_{1,2} &\leq c_o[a(\boldsymbol{u}^n; \boldsymbol{u}_h^n - \boldsymbol{u}, \boldsymbol{v}_h - \boldsymbol{u}) + a(\boldsymbol{u}^n; \boldsymbol{u}, \boldsymbol{v} - \boldsymbol{u}_h) + a(\boldsymbol{u}^n; \boldsymbol{u}, \boldsymbol{v}_h - \boldsymbol{u}) - S(\boldsymbol{v} - \boldsymbol{u}_h^n) \\
&\quad - S(\boldsymbol{v}_h - \boldsymbol{u}) + j_{gn}(\boldsymbol{v}_h) - j_{gn}(\boldsymbol{u}) + j_{gn}(\boldsymbol{v}) - j_{gn}(\boldsymbol{u}_h^n) + b_s(\Theta, \boldsymbol{v} - \boldsymbol{u}_h^n) \\
&\quad + b_s(\Theta, \boldsymbol{v}_h - \boldsymbol{u})] \quad \forall \boldsymbol{v} \in K, \boldsymbol{v}_h \in K_h, c_o = \text{constant} > 0.
\end{aligned} \tag{7.500}$$

Proof: Let $\Theta \in {}^1V$, $\Theta_h \in {}^1V_h$, $\boldsymbol{u} \in K$, and $\boldsymbol{u}_h^n \in K_h$. Using Equation 7.452, 7.453, 7.441 and 7.442, adding and subtracting $\alpha(\Theta, \Theta - \Theta_h) - \alpha(\Theta_h, \Theta) + \Delta t a_\Theta(\Theta, \Theta - \Theta_h) - \Delta t a_\Theta(\Theta_h, \Theta)$ and $a(\boldsymbol{u}^n; \boldsymbol{u}, \boldsymbol{u} - \boldsymbol{u}_h) - a(\boldsymbol{u}^n; \boldsymbol{u}, \boldsymbol{u}_h^n - \boldsymbol{u})$ to both parts in Θ and $\boldsymbol{u}$, then we obtain

$$\begin{aligned}
&\alpha(\Theta, \Theta - \Theta_h) - \alpha(\Theta_h, \Theta) - \alpha(\Theta, \Theta - \Theta_h) + \alpha(\Theta, z - \Theta) + \alpha(\Theta_h, z_h - \Theta_h) + \Delta t a_\Theta(\Theta, z - \Theta) \\
&\quad + \Delta t a_\Theta(\Theta_h, z - \Theta_h) + \Delta t a_\Theta(\Theta, \Theta - \Theta_h) - \Delta t a_\Theta(\Theta_h, \Theta) - \Delta t a_\Theta(\Theta, \Theta - \Theta_h) \\
&\quad + \Delta t a_\Theta(\Theta_h, \Theta) + b_p(\boldsymbol{u}, z - \Theta) + b_p(\boldsymbol{u}_h, z_h - \Theta_h) + j(z) - j(\Theta) + j(z_h) \\
&\quad - j(\Theta_h) - s(z - \Theta) - s(z_h - \Theta_h) \geq 0.
\end{aligned}$$

$$\begin{aligned}
&a(\boldsymbol{u}; \boldsymbol{u}, \boldsymbol{v} - \boldsymbol{u}) + a(\boldsymbol{u}; \boldsymbol{u}_h, \boldsymbol{v}_h - \boldsymbol{u}_h) + a(\boldsymbol{u}; \boldsymbol{u}, \boldsymbol{u} - \boldsymbol{u}_h) - a(\boldsymbol{u}; \boldsymbol{u}_h, \boldsymbol{u}) - a(\boldsymbol{u}; \boldsymbol{u}, \boldsymbol{u} - \boldsymbol{u}_h) \\
&\quad + a(\boldsymbol{u}; \boldsymbol{u}_h, \boldsymbol{u}) + b_s(\Theta, \boldsymbol{v} - \boldsymbol{u}) + b_s(\Theta_h, \boldsymbol{v}_h - \boldsymbol{u}_h) - S(\boldsymbol{v} - \boldsymbol{u}) - S(\boldsymbol{v}_h - \boldsymbol{u}_h) + j_{gn}(\boldsymbol{v}_h) \\
&\quad - j_{gn}(\boldsymbol{u}) + j_{gn}(\boldsymbol{v}) - j_{gn}(\boldsymbol{u}_h) \geq 0.
\end{aligned}$$

From here it follows that

$$\begin{aligned}
&\alpha(\Theta, \Theta - \Theta_h) - \alpha(\Theta_h, \Theta - \Theta_h) + \Delta t a_\Theta(\Theta, \Theta - \Theta_h) - \Delta t a_\Theta(\Theta_h, \Theta - \Theta_h) \\
&\quad = \alpha(\Theta - \Theta_h, \Theta - \Theta_h) + \Delta t a_\Theta(\Theta - \Theta_h, \Theta - \Theta_h) \\
&\quad \leq \alpha(\Theta_h - \Theta, z_h - \Theta) + \Delta t a_\Theta(\Theta_h - \Theta, z_h - \Theta) + \alpha(\Theta, z - \Theta_h) + \Delta t a_\Theta(\Theta, z - \Theta_h) \\
&\qquad + \alpha(\Theta, z_h - \Theta) + \Delta t a_\Theta(\Theta, z_h - \Theta) + b_p(\boldsymbol{u}, z_h - \Theta) + b_p(\boldsymbol{u}_h, z - \Theta_h) + j(z) \\
&\qquad - j(\Theta) + j(z_h) - j(\Theta_h) + s(\Theta_h - z) + s(\Theta - z_h).
\end{aligned}$$

$$a(\boldsymbol{u};\boldsymbol{u},\boldsymbol{u}-\boldsymbol{u}_h)-a(\boldsymbol{u};\boldsymbol{u}_h,\boldsymbol{u}-\boldsymbol{u}_h)=a(\boldsymbol{u};\boldsymbol{u}-\boldsymbol{u}_h,\boldsymbol{u}-\boldsymbol{u}_h)$$
$$\leq a(\boldsymbol{u};\boldsymbol{u}_h-\boldsymbol{u},\boldsymbol{v}_h-\boldsymbol{u})+a(\boldsymbol{u};\boldsymbol{u},\boldsymbol{v}-\boldsymbol{u}_h)+a(\boldsymbol{u};\boldsymbol{u},\boldsymbol{v}_h-\boldsymbol{u})+b_s(\Theta,\boldsymbol{v}-\boldsymbol{u})+b_s(\Theta_h,\boldsymbol{v}_h-\boldsymbol{u}_h)$$
$$+S(\boldsymbol{u}_h-\boldsymbol{v})+S(\boldsymbol{u}-\boldsymbol{v}_h)+j_{gn}(\boldsymbol{v}_h)-j_{gn}(\boldsymbol{u})+j_{gn}(\boldsymbol{v})-j_{gn}(\boldsymbol{u}_h).$$

Hence, and after some algebra, the assertions in Equation 7.449 and 7.500 of the lemma follow immediately.

Corollary: Let ${}^1V_h\subset{}^1V$, $K_h\subset K$. Then

$$\alpha\,\|\Theta-\Theta_h\|_{0,1}^2+\Delta t\,\|\Theta-\Theta_h\|_{1,1}^2+\|\boldsymbol{u}-\boldsymbol{u}_h\|_{1,2}^2\leq c[\alpha(\Theta_h-\Theta,z_h-\Theta)$$
$$+\Delta t a_\Theta(\Theta_h-\Theta,z_h-\Theta)+a(\boldsymbol{u};\boldsymbol{u}_h-\boldsymbol{u},\boldsymbol{v}_h-\boldsymbol{u})+\alpha(\Theta,z_h-\Theta)+\Delta t a_\Theta(\Theta,z_h-\Theta)$$
$$+a(\boldsymbol{u};\boldsymbol{u},\boldsymbol{v}_h-\boldsymbol{u})+b_p(\boldsymbol{u},\Theta_h-\Theta)+b_p(\boldsymbol{u},z_h-\Theta_h)+b_s(\Theta,\boldsymbol{u}_h-\boldsymbol{u})$$
$$+b_s(\Theta,\boldsymbol{v}_h-\boldsymbol{u}_h)+j(z_h)-j(\Theta)+j_{gn}(\boldsymbol{v}_h)-j_{gn}(\boldsymbol{u})+s(\Theta-z_h)+S(\boldsymbol{u}-\boldsymbol{v}_h)]$$
$$\forall z_h\in{}^1V_h,\ \boldsymbol{v}_h\in K_h. \quad (7.501)$$

Proof: Let ${}^1V_h\subset{}^1V$, $K_h\subset K$. Let us put $z=\Theta_h$ in Equation 7.499 and $\boldsymbol{v}=\boldsymbol{u}_h$ in Equation 7.500, and then add the resulting inequalities. Then Equation 7.501 follows immediately.

Lemma 7.29: Let $\partial\beta_{ij}/\partial x_j\in L^\infty(\Omega)\,\forall i,j$. Then

$$\left|b_s(\Theta-\Theta_o,\boldsymbol{v})+b_p(\boldsymbol{v},\Theta)\right|\leq c\left((1+\|\Theta\|_{1,1})\,\|\boldsymbol{v}\|_{0,N}+\|\Theta\|_{0,1}\,\|\boldsymbol{v}\|_{1,N}\right).$$

For the proof see Nedoma (1995), as well as Chapter 12.

The main result of this section is the following theorem.

Theorem 7.24: Let $\Gamma_c=\cup_{k,l}\Gamma_c^{kl}$ be polygonal in the $2D$ case and polyhedral in the $3D$ case, and let free boundary $\cup_m R^m(t)$ be finite. Let $\Theta\in W^{1,\infty}(\Omega)\cap H^2(\Omega)$, $\boldsymbol{u}\in K\cap[H^2(\Theta)]^N$ and $\boldsymbol{u}\,|_{\Gamma_c}\in[H^2(\Gamma_c)]^N$. Let Ω_+^m, Ω_-^m, Ω_o^m, Ω_{+h}^m, Ω_{-h}^m, Ω_{oh}^m be defined by Equations 7.454 to 7.459 and let $N_{*h}^m<c/h$. Then

$$\alpha\,\|\Theta-\Theta_h\|_{0,1}^2+\Delta t\,\|\Theta-\Theta_h\|_{1,1}^2+\|\boldsymbol{u}-\boldsymbol{u}_h\|_{1,N}^2\leq ch^2. \quad (7.502)$$

Proof: The technique similar to that of Nedoma (1983),(1987), (1994),(1997b), Nedoma and Dvořák (1995) as well as of the previous sections of this chapter can be used. The proof follows from the Corollary of Lemma 7.28. We estimate Equation 7.501 and we shall assume that Θ and $\boldsymbol{u}$ are sufficiently regular. Terms $\alpha(\Theta,z-\Theta_h)+\Delta t a_\Theta(\Theta,z-\Theta_h)-s(z-\Theta_h)$, $\alpha(\Theta,z_h-\Theta)+\Delta t a_\Theta(\Theta,z_h-\Theta)-s(z_h-\Theta)$ and $a(\boldsymbol{u};\boldsymbol{u},\boldsymbol{v}-\boldsymbol{u}_h)+j_{gn}(\boldsymbol{v})-j_{gn}(\boldsymbol{u}_h)-S(\boldsymbol{v}-\boldsymbol{u}_h)$, $a(\boldsymbol{u};\boldsymbol{u},\boldsymbol{v}_h-\boldsymbol{u})+j_{gn}(\boldsymbol{v}_h)-j_{gn}(\boldsymbol{u})-S(\boldsymbol{v}_h-\boldsymbol{u})$ are estimated by applying the Green lemma and later by using a suitable choice of $z\in{}^1V$, $z_h\in{}^1V_h$, $\boldsymbol{v}\in K$, and $\boldsymbol{v}_h\in K_h$. Then

$$a(\boldsymbol{u};\boldsymbol{u},\boldsymbol{v}_h-\boldsymbol{u})+j_{gn}(\boldsymbol{v}_h)-j_{gn}(\boldsymbol{u})-S(\boldsymbol{v}_h-\boldsymbol{u})$$
$$=-\int_\Omega\partial\tau_{ij}(\boldsymbol{u})/\partial x_j(\boldsymbol{v}_h-\boldsymbol{u})_i\mathrm{d}\boldsymbol{x}+\int_{\partial\Omega}\tau_{ij}n_j(\boldsymbol{v}_h-\boldsymbol{u})_i\mathrm{d}s+j_{gn}(\boldsymbol{v}_h)-j_{gn}(\boldsymbol{u})$$
$$-\int_{\Gamma_\tau}P_i(\boldsymbol{v}_h-\boldsymbol{u})_i\mathrm{d}s-\int_\Omega f_i(\boldsymbol{v}_h-\boldsymbol{u})_i\mathrm{d}\boldsymbol{x}$$
$$=\int_{\cup\Gamma_c^{kl}}\tau_n(\boldsymbol{u})\left[(v_n^k-v_n^l)-(u_n^k-u_n^l)\right]\mathrm{d}s+\int_{\cup\Gamma_c^{kl}}\vec{\tau}_t(\boldsymbol{u})\left[(\boldsymbol{v}_t^k-\boldsymbol{v}_t^l)-(\boldsymbol{u}_t^k-\boldsymbol{u}_t^l)\right]\mathrm{d}s$$
$$+j_{gn}(\boldsymbol{v}_h)-j_{gn}(\boldsymbol{u})\geq 0\quad\forall\boldsymbol{v}_h\in K_h,$$

since

$$-\partial\tau_{ij}/\partial x_j = f_i, \quad i = 1, \ldots, N, \text{ a.e. in } \Omega.$$

The bilinear forms $a_\Theta(.,.)$ and $a(.;.,.)$ are bounded, so that

$$a_\Theta(\Theta_h - \Theta, z_h - \Theta) \le c_o \|\Theta_h - \Theta\| \|z_h - \Theta\|,$$

$$a(\boldsymbol{u};\boldsymbol{u}_h - \boldsymbol{u}, \boldsymbol{u}_h - \boldsymbol{u}) \le c \|\boldsymbol{u}_h - \boldsymbol{u}\| \|v_h - \boldsymbol{u}\|.$$

Since for every $\boldsymbol{v} \in [C(\overline{\Omega})]^N \cap K$ we have $\boldsymbol{v}_{LI} \in K_h$ and, since the last inequality is valid for an arbitrary $\boldsymbol{v}_h \in K_h$, it is also valid for $\boldsymbol{v}_h = \boldsymbol{u}_{LI}$, and for every $\Theta \in C(\overline{\Omega}) \cap {}^1V$, $\Theta_h \in {}^1V_h$ the first inequality is valid. As $ab \le \frac{1}{2}\varepsilon a^2 + (2\varepsilon)^{-1}b^2$, $a, b \in R^1$, $\varepsilon > 0$, then applying the interpolation theorem we obtain

$$a_\Theta(\Theta_h - \Theta, z_h - \Theta) \le \tfrac{1}{2}c\varepsilon \|\Theta - \Theta_h\|_{1,1}^2 + \tfrac{1}{2}c\varepsilon^{-1} \|\Theta - z_h\|_{1,1}^2 \le \tfrac{1}{2}c\varepsilon \|\Theta - \Theta_h\|_{1,1}^2 + C_o h^2,$$

$$a(\boldsymbol{u};\boldsymbol{u}_h - \boldsymbol{u}, \boldsymbol{v}_h - \boldsymbol{u}) \le \tfrac{1}{2}c\varepsilon \|\boldsymbol{u} - \boldsymbol{u}_h\|_{1,N}^2 + \tfrac{1}{2}c\varepsilon^{-1} \|\boldsymbol{v}_h - \boldsymbol{u}\|_{1,N}^2 \le \tfrac{1}{2}c\varepsilon \|\boldsymbol{u}_h - \boldsymbol{u}\|_{1,N}^2 + C_1 h^2.$$

Due to Lemmas 7.26 and 7.27 it holds

$$j(\Theta_h) - j(\Theta) + j_h(\Theta_{LI}) - j_h(\Theta_h)$$
$$\le j_h(\Theta_{LI}) - j(\Theta) = j(\Theta_{LI}) - j(\Theta_{LI}) + j(\Theta_{LI}) - j(\Theta) \le ch^2 + c\|\Theta - \Theta_{LI}\|_{0,1},$$

$$\alpha[(\Theta_h, \Theta_h - \Theta_{LI}) - (\Theta_h, \Theta_h - \Theta_{LI})_h] \le ch^2,$$

$$\|\Theta_{LI} - \Theta\|_{1,1} \le c_{ro}h^2 \|\Theta\|_{2,1}, \quad \|\Theta_{LI} - \Theta\|_{0,1} \le c_{po}h^2 \|\Theta\|_{2,1},$$

$$\alpha(\Theta_h, \Theta_{LI} - \Theta) + \Delta t a_\Theta(\Theta_h - \Theta, \Theta_{LI} - \Theta) - s(\Theta_{LI} - \Theta)$$
$$\le \alpha \|\Theta_h\|_{0,1} \|\Theta_{LI} - \Theta\|_{0,1} + \Delta t \|\Theta_h - \Theta\|_{1,1} \|\Theta_{LI} - \Theta\|_{1,1} + \|\Theta\|_{0,1} \|\Theta_{LI} - \Theta\|_{0,1}$$
$$\le c\{\|\Theta_{LI} - \Theta\|_{0,1} + \tfrac{1}{2}\varepsilon\Delta t \|\Theta_h - \Theta\|_{1,1}^2 + \tfrac{1}{2}\varepsilon^{-1}\Delta t \|\Theta_{LI} - \Theta\|_{1,1}^2\}$$
$$\le c\{\|\Theta_{LI} - \Theta\|_{0,1} + \tfrac{1}{2}\varepsilon\Delta t \|\Theta_h - \Theta\|_{1,1}^2 + c_o h^2\}.$$

Choosing $\boldsymbol{v}_h = \boldsymbol{u}_{LI}$ and since $\boldsymbol{u}_{LI} \in V_h$, then $\boldsymbol{v}_h = \boldsymbol{u}_{LI} \in K_h$. Moreover, $(\boldsymbol{u}_{LI}^k - \boldsymbol{u}_{LI}^l)_n = (\boldsymbol{u}_L^k - \boldsymbol{u}_L^l)_n$, $(\boldsymbol{u}_{LI}^k - \boldsymbol{u}_{LI}^l)_t = (\boldsymbol{u}_L^k - \boldsymbol{u}_L^l)_t$ on $\cup\Gamma_c^{kl}$. Thus

$$(\boldsymbol{u}_{LI}^k - \boldsymbol{u}_{LI}^l)_n \le 0 \text{ on } \cup\Gamma_c^{kl},$$

so that $\boldsymbol{u}_{LI} \in K$ and since $\boldsymbol{u}_{LI} \in V_h$, then $\boldsymbol{u}_{LI} \in K_h$. Furthermore, we have

$$\|\boldsymbol{u}_{LI} - \boldsymbol{u}\|_{1,N} \le c_r h \|\boldsymbol{u}\|_{2,N},$$

$$\left\|(\boldsymbol{u}_{LI}^k - \boldsymbol{u}_{LI}^l)_n - (u_n^k - u_n^l)\right\|_{[L^2(\cup\Gamma_c^{kl})]^N} = \left\|(\boldsymbol{u}_L^k - \boldsymbol{u}_L^l)_n - (u_n^k - u_n^l)\right\|_{[L^2(\cup\Gamma_c^{kl})]^N}$$
$$\le c_s h^2 \sum_{\Gamma_c^{kl}} \left\|u_n^k - u_n^l\right\|_{[H^2(\cup\Gamma_c^{kl})]^N}.$$

To investigate the integrals

$$J_1(\cup\Gamma_c^{kl}) = \int_{\cup\Gamma_c^{kl}} \tau_n(\boldsymbol{u})\left[(\boldsymbol{v}_h^k - \boldsymbol{v}_h^l)_n - (u_n^k - u_n^l)\right] \mathrm{d}s,$$

$$J_2(\cup\Gamma_c^{kl}) = \int_{\cup\Gamma_c^{kl}} \vec{\tau}_t(\boldsymbol{u})\left[(\boldsymbol{v}_h^k - \boldsymbol{v}_h^l)_t - (\boldsymbol{u}_t^k - \boldsymbol{u}_t^l)\right] \mathrm{d}s$$

we shall assume that the boundaries Γ_c^{kl} are approximated piecewise linearly and then the results of Nedoma (1987),(1994) as well as of previous sections can be used. Thus we obtain

$$\left|J_1(\cup\Gamma_c^{kl})\right| \le c_s h^2$$

and

$$\left| \int_{\Gamma_c^{kl}} \vec{\tau}_t^{kl}(\boldsymbol{u})[(\boldsymbol{u}_t^k - \boldsymbol{u}_t^l)_{LI} - (\boldsymbol{u}_t^k - \boldsymbol{u}_t^l)]\mathrm{d}s + j_{gn}(\boldsymbol{u}_{LI}) - j_{gn}(\boldsymbol{u}) \right| \leq c_r h^2.$$

From the Corollary of Lemma 7.28, Lemma 7.29 and these estimates, we obtain

$$\begin{aligned}
&\alpha \|\Theta - \Theta_h\|_{0,1}^2 + \Delta t \|\Theta - \Theta_h\|_{1,1}^2 + \|\boldsymbol{u} - \boldsymbol{u}_h\|_{1,N}^2 \\
&\quad \leq c[\alpha(\Theta_h - \Theta, z_h - \Theta) + \Delta t a_\Theta(\Theta_h - \Theta, z_h - \Theta) + a(\boldsymbol{u}; \boldsymbol{u}_h - \boldsymbol{u}, \boldsymbol{v}_h - \boldsymbol{u}) \\
&\quad \left(+ \int_{\Gamma_u} q(z_h - \Theta)\mathrm{d}s \right) + \int_{\partial\Omega} \tau_{ij}(\boldsymbol{u}) n_j (\boldsymbol{v}_h - \boldsymbol{u})_i \mathrm{d}s + j(z_h) - j(\Theta) + j(z) - j(\Theta_h) \\
&\quad + b_p(\boldsymbol{u}, z_h - \Theta) + b_s(\Theta - \Theta_o, \boldsymbol{v}_h - \boldsymbol{u}) + b_p(\boldsymbol{u}, z - \Theta_h) + b_s(\Theta - \Theta_o, \boldsymbol{v} - \boldsymbol{u}_h)] \\
&\quad \leq c[\alpha \|\Theta_h - \Theta\|_{0,1} \|z_h - \Theta\|_{0,1} + \Delta t \|\Theta_h - \Theta\|_{1,1} \|z_h - \Theta\|_{1,1} + \|\boldsymbol{u}_h - \boldsymbol{u}\|_{1,N} \|\boldsymbol{v}_h - \boldsymbol{u}\|_{1,N} \\
&\quad + \int_{\cup\Gamma_c^{kl}} ((\boldsymbol{v}_h^k - \boldsymbol{v}_h^l)_n - (\boldsymbol{u}_h^k - \boldsymbol{u}_h^l)_n)\mathrm{d}s + \int_{\Gamma_u} \Theta_{,n} (z - \Theta_h)\mathrm{d}s + j(z_h) - j(\Theta) + j(z) - j(\Theta_h) \\
&\quad + b_p(\boldsymbol{u}, z_n - \Theta) + b_s(\boldsymbol{u}, z_h - \Theta) + b_s(\Theta - \Theta_o, \boldsymbol{v}_h - \boldsymbol{u}) + b_p(\boldsymbol{u}, z - \Theta_h) + b_s(\Theta - \Theta_o, \boldsymbol{v} - \boldsymbol{u}_h)] \\
&\quad \leq c[\tfrac{1}{2}\varepsilon \|\Theta - \Theta_h\|_{0,1}^2 + \tfrac{1}{2}\varepsilon^{-1} \|\Theta - z_h\|_{0,1}^2 \\
&\quad + \tfrac{1}{2}\varepsilon \|\Theta - \Theta_h\|_{1,1}^2 + \tfrac{1}{2}\varepsilon^{-1} \|\Theta - z_h\|_{1,1}^2 + \tfrac{1}{2}\varepsilon \|\boldsymbol{u}_h - \boldsymbol{u}\|_{1,N}^2 + \tfrac{1}{2}\varepsilon^{-1} \|\boldsymbol{v}_h - \boldsymbol{u}\|_{1,N}^2 \\
&\quad + J_1(\cup\Gamma_c^{kl}) + J_2(\cup\Gamma_c^{kl}) + j(z_h) - j(\Theta) + j(z) - j(\Theta_h) + b_p(\boldsymbol{u}, z_h - \Theta) \\
&\quad + b_s(\boldsymbol{u}, z_h - \Theta) + b_s(\Theta - \Theta_o, \boldsymbol{v}_h - \boldsymbol{u}) + b_p(\boldsymbol{u}, z - \Theta_h) + b_s(\Theta - \Theta_h, \boldsymbol{v} - \boldsymbol{u}_h)] \leq ch^2.
\end{aligned}$$

Hence we proved Equation 7.502.

We saw above when solving the problem $(\mathscr{P})$ numerically that for the thermal part the technique, as well as algorithms from previous sections concerning the contact-two-phase Stefan-like problem can also be applied in this case. For the plastic part of the problem the method of secant modules was used. It was shown that this consists of solving a sequence of variational inequalities corresponding to the linear Hooke's law. Therefore, the algorithm parallel to that of the thermo-elastic case of the contact-two-phase Stefan-like problem can be applied for the plastic part of the problem investigated. Hence solvers which will be discussed in the next sections can also be useful in this case.

7.5. COMPUTER IMPLEMENTATION OF THE ALGORITHMS FOR CONTACT PROBLEMS

7.5.1. Introduction

Discretization of the Lagrangian $\mathscr{L}(v, \lambda)$ on $K_h \times \Lambda$ leads to minimizing the quadratic functional with constraints, which can be rewritten in the matrix form as

$$\begin{aligned}
f(\boldsymbol{x}) \equiv \mathscr{L}_h(\boldsymbol{x}, \lambda_h) &= L_o(\boldsymbol{x}) + (\boldsymbol{S}(\lambda_h), \boldsymbol{x}) \\
&= \tfrac{1}{2}(\boldsymbol{x}, C\boldsymbol{x}) - (\boldsymbol{F}, \boldsymbol{x}) + (\boldsymbol{S}(\lambda_h), \boldsymbol{x}) = \tfrac{1}{2}(\boldsymbol{x}, C\boldsymbol{x}) + (\boldsymbol{b}, \boldsymbol{x}), \qquad (7.503)
\end{aligned}$$

$$(\boldsymbol{a}_i, \boldsymbol{x}) \leq 0 \quad i \in J^-,$$
$$(\boldsymbol{a}_i, \boldsymbol{x}) = 0 \quad i \in J^o, \tag{7.504}$$

or in the matrix form

$$A\boldsymbol{x} \leq 0, \tag{7.505}$$

where C is the stiffness matrix, which is symmetric and positive (semi-)definite, $\boldsymbol{b} = \boldsymbol{S}(\lambda_h) - \boldsymbol{F}$, $\boldsymbol{F}$ is the vector originated by integration of the body and surface forces, $\boldsymbol{S}(\lambda_h)$ is the vector originated by integration of the friction forces on the boundary $\cup\Gamma_c^{kl}$, $A = (a_{ij})$ is the matrix of constraints ($\dim m \times n$) generated by the conditions $u_n^k - u_n^l \leq 0$ on $\cup\Gamma_c^{kl}$ in the definition of the admissible set of displacements K_h (here n represents double the number of points of triangulation in which the displacement is not prescribed and m is the number of points of triangulation on the contact boundary Γ_c^{kl}) and further $\boldsymbol{x}, \boldsymbol{a}_i, \boldsymbol{b} \in R^n$, J^-, J^o are finite sets of indices $J^- \cup J^o = \{1,..,m\}$ and $(.,.)$ denotes the inner product in R^n.

The non-differentiable functional

$$\int_{\cup\Gamma_c^{kl}} g_c^{kl} \left|\boldsymbol{v}_t^k - \boldsymbol{v}_t^l\right| \mathrm{d}s = \int_{\cup\Gamma_c^{kl}} \lambda^{kl} g_c^{kl} (\boldsymbol{v}_t^k - \boldsymbol{v}_t^l)\mathrm{d}s$$

can be approximated with sufficient accuracy by the trapezoidal rule. In Section 7.2 we saw that the thermal part of the geodynamic problem leads to minimization of the quadratic functional $g(\boldsymbol{x})$ with constraints, i.e. to the following problem

$$g(\boldsymbol{x}) = \tfrac{1}{2}(\boldsymbol{x}, \underline{\boldsymbol{K}}x) + (\boldsymbol{Q}, \boldsymbol{x}), \tag{7.506}$$

$$\mathscr{A}\boldsymbol{x} \leq c, \tag{7.507}$$

where the $n \times n$ matrix $\boldsymbol{K}$ is a positive definite matrix, $\mathscr{A}$ the $m \times n$ matrix of constraints, $\boldsymbol{c} \in R^m$ the vector generated by the condition $T^k - T^l \leq T_2^{kl}$ in the definition of the set of admissible temperatures ${}^1K_h$ (here n represents the number of points of triangulation in which the temperature is not prescribed and m is the number of points of triangulation on the contact boundary Γ_c^{kl}, where the condition $T^k - T^l \leq T_2^{kl}$ is prescribed), and $\boldsymbol{Q} \in R^n$ is the vector generated by thermal sources.

Computational methods for such problems were developed very intensively during recent years. The list of literature on these subjects contains hundreds of papers and books. As already emphasized, problems with constraints are the most frequent in geodynamics as these problems lead to variational inequalities.

However, constrained optimization methods often derive from unconstrained ones. The gradient projection method is generally efficient when used to solve problems with linear constraints. This method can be regarded as the steepest descent method adapted to constrained optimization problems. In some way some other unconstrained optimization methods can be applied to constrained problems by adequate projections. From these types of method we shall note the Newton projection method, quasi-Newton projection method and the conjugate gradient projection methods. As well as these methods the class of penalty methods, which transform the constrained minimization problem into sequences of unconstrained minimization problems, are also useful. Next we shall give some useful methods and algorithms.

7.5.2. Conjugate Gradient Method with Constraints

The problem discussed represents the problem of the minimization of a quadratic function with linear constraints (the so-called problem of quadratic programming), which will be written in the form

$$f(\boldsymbol{x}) = \tfrac{1}{2}(\boldsymbol{x}, C\boldsymbol{x}) + (\boldsymbol{b}, \boldsymbol{x}), \tag{7.508}$$

$$\begin{aligned}(\boldsymbol{a}_i, \boldsymbol{x}) &\leq d_i, \quad i \in J^-,\\ (\boldsymbol{a}_i, \boldsymbol{x}) &= d_i, \quad i \in J^o,\end{aligned} \tag{7.509}$$

or in the matrix form

$$A\boldsymbol{x} \leq d, \tag{7.510}$$

where $\boldsymbol{x} \in R^n, \boldsymbol{a}_i \in R^n, \boldsymbol{b} \in R^n, \boldsymbol{d} = (d_i) \in R^m$, d_i are numbers, C is $n \times n$ symmetric, positive definite, i.e. $(\boldsymbol{x}, C\boldsymbol{x}) \geq 0 \, \forall \boldsymbol{x}, A = (a_{ij})$ is the matrix of constraints ($\dim m \times n$) $\underline{J} = J^- \cup J^o$, J^-, J^o are finite sets of indices.

The main idea of the method is as follows. We minimize Equation 7.508 on certain facets (faces of the polyhedral set defined by the linear inequalities of Equation 7.509) created by the constraints in Equation 7.509 among which the equalities are satisfied. We find a minimum of Equation 7.508 on each such facet by using the conjugate gradient method, which converges after a finite number of iterate steps. Hence, and since the number of facets is limited, then the algorithm for Equations 7.508 and 7.509 also converges after a finite number of steps. For more details see Daniel (1971), Luenberger (1973),(1984), Pschenichny and Danilin (1978), Brousse (1988).

Let $J(\boldsymbol{x})$ be a subset of $\underline{J}$ defined by

$$J(\boldsymbol{x}) = \{i \mid (\boldsymbol{a}_i, \boldsymbol{x}) = d_i, \quad i \in J^- \cup J^o\}, \tag{7.511}$$

where $\boldsymbol{a}_i$ are linearly independent vectors. Let $J^o \neq 0$ then $J(\boldsymbol{x})$ is non-empty and $\boldsymbol{x}$ satisfies Equation 7.509.

Take a set of indices $J = J(\boldsymbol{x})$ and construct the matrix A_J in such a way that its rows are vectors $\boldsymbol{a}_i$, $i \in J$. The matrix A_J is $m \times n$-dimensional, where m is the number of elements of the set J, which corresponds to the number of all nodal points on $\cup \Gamma_c^{kl}$, and n is the dimension of $\boldsymbol{x}$. Since vectors $\boldsymbol{a}_i$, $i \in J$ are linearly independent, then the matrix $A_J A_J^T$ is regular. To prove this assertion we shall assume that there exists a non-zero vector $\boldsymbol{y} = (y_i) \in R^m$ such that

$$A_J A_J^T \boldsymbol{y} = 0. \tag{7.512}$$

Then

$$\boldsymbol{y}^T A_J A_J^T \boldsymbol{y} = (A_J^T \boldsymbol{y})^T A_J^T \boldsymbol{y} = (A_J^T \boldsymbol{y}, A_J^T \boldsymbol{y}) = \left\|A_J^T \boldsymbol{y}\right\|^2 = 0, \quad A_J^T \boldsymbol{y} = 0. \tag{7.513}$$

Due to our assumption, $A_j^T \boldsymbol{y}$ is a linear combination of vectors $\boldsymbol{a}_i$, $i \in J$ with coefficients y_i, $i = 1, 2, \ldots, m$, where y_i are components of $\boldsymbol{y} \in R^m$. Since $\boldsymbol{a}_i$, $i \in J$, are linearly independent, this combination cannot be zero, so that Equation 7.512 and 7.513 are valid only for a zero vector $\boldsymbol{y}$ and therefore the matrix $A_J A_J^T$ is regular.

We construct the operator P_J by

$$P_J = A_J^T (A_J A_J^T)^{-1} A_J. \tag{7.514}$$

The operator P_J has the following properties

$$\begin{aligned}
&P_J P_J = A_J^T (A_J A_J^T)^{-1} A_J A_J^T (A_J A_J^T)^{-1} A_J = P_J, \\
&P_J^T = (A_J^T (A_J A_J^T)^{-1} A_J)^T = A_J^T (A_J A_J^T)^{-1} A_J = P_J, \\
&P_J (I - P_J) = (I - P_J) P_J = P_J - P_J P_J = 0.
\end{aligned} \tag{7.515}$$

The operator P_J is an orthogonal projection into a subspace spanned by vectors $\boldsymbol{a}_i$, $i \in J$. Therefore, for any vector $\boldsymbol{x} \in R^n$

$$\boldsymbol{x} = P_J \boldsymbol{x} + (I - P_J)\boldsymbol{x}.$$

Thus, due to Equation 7.515,

$$(P_J \boldsymbol{x}, (I - P_J)\boldsymbol{x}) = (\boldsymbol{x}, P_J^T (I - P_J)\boldsymbol{x}) = 0.$$

Let us define the vector $\boldsymbol{u}$ by

$$\boldsymbol{u} = (A_J A_J^T)^{-1} A_J \boldsymbol{x}, \tag{7.516}$$

then

$$P_J \boldsymbol{x} = A_J^T \boldsymbol{u} = \sum_{i \in J} a_i u_i, \tag{7.517}$$

so that the vector $P_J \boldsymbol{x}$ is wholly in the subspace spanned by vectors $\boldsymbol{a}_i$, $i \in J$. Furthermore,

$$A_J (I - P_J) = A_J - (A_J A_J^T)(A_J A_J^T)^{-1} A_J = 0, \tag{7.518}$$

and thus for any $\boldsymbol{x} \in R^n$, vector $\boldsymbol{y} = (I - P_J)\boldsymbol{x}$ satisfies the system of equations $A_J \boldsymbol{y} = 0$.

Let us minimize a quadratic functional $f(\boldsymbol{x})$ from Equation 7.508 with constraints given by Equation 7.509, where $i \in J$. Let $\boldsymbol{x}_o \in R^n$ be an arbitrary point satisfying the second part of Equation 7.509, i.e. $A_J \boldsymbol{x}_o = 0$. Let us introduce the transformation $\boldsymbol{x} = \boldsymbol{x}_o + (I - P_J)\boldsymbol{y}$. Then the quadratic functional $f(\boldsymbol{x})$ can be written as $\varphi(\boldsymbol{y}) = f(\boldsymbol{x}_o + (I - P_J)\boldsymbol{y})$.

We shall show that the above transformation reduces the initial problem with constraints to the minimization of the quadratic functional $\varphi(\boldsymbol{y})$ without constraints and that the conjugate gradient method, which can be applied to the minimization of $\varphi(\boldsymbol{y})$, converges.

Let us denote by $f'(\boldsymbol{x}) = (C, \boldsymbol{x}) + \boldsymbol{b}$ the gradient of $f(\boldsymbol{x})$. Then for the gradient of $\varphi(\boldsymbol{y})$, denoted by $\varphi'(\boldsymbol{y})$, the following relation, $\varphi'(\boldsymbol{y}) = (I - P_J) f'(\boldsymbol{x})$, is valid.

We define the following process: let $\boldsymbol{x}_o$ be given, then

$$\begin{aligned}
&\text{(i)} \quad \boldsymbol{p}_1 = -(I - P_J) f'(\boldsymbol{x}_o), \\
&\text{(ii)} \quad \boldsymbol{x}_{k+1} = \boldsymbol{x}_k + \alpha_{k+1} \boldsymbol{p}_{k+1}, \\
&\text{(iii)} \quad \boldsymbol{p}_{k+1} = -(I - P_J) f'(\boldsymbol{x}_k) + \frac{\left\|(I - P_J) f'(\boldsymbol{x}_k)\right\|^2}{\|(I - P_J) f'(\boldsymbol{x}_{k-1})\|^2} \boldsymbol{p}_k \\
&\text{(iv)} \quad \alpha_{k+1} = \frac{(f'(\boldsymbol{x}_k), \boldsymbol{p}_{k+1})}{(\boldsymbol{p}_{k+1}, C\boldsymbol{p}_{k+1})}, \qquad k = 0, 1, 2, \ldots
\end{aligned} \tag{7.519}$$

Theorem 7.25: Let $\{\boldsymbol{x}_k\}$, $k = 0, 1, 2, \ldots$, be a sequence given by Equation 7.519, and let $\boldsymbol{x}_o$ be an initial approximation. Then the problem of the minimization of the quadratic function $f(\boldsymbol{x})$ with constraints given by Equation 7.509, $i \in J$, is solved after a finite number of steps $n \geq 1$.

Proof: Due to $u_n^k - u_n^l \leq 0$ (or $T^k - T^l \leq T_2^{kl}$), the vectors $\boldsymbol{a}_i$, $i \in J$, are linearly independent. Let $\boldsymbol{x}_o$ be an initial approximation satisfying Equation 7.509, $i \in J$, i.e. $A_J \boldsymbol{x}_o = 0$, $i \in J$. Let us introduce a new variable $\boldsymbol{y}$ by

$$\boldsymbol{x} = \boldsymbol{x}_o + (I - P_J)\boldsymbol{y}, \tag{7.520}$$

then

$$\varphi(\boldsymbol{y}) = f(\boldsymbol{x}_o + (I - P_J)\boldsymbol{y}). \tag{7.521}$$

Hence

$$\varphi'(\boldsymbol{y}) = (I - P_J) f'(\boldsymbol{x}). \tag{7.522}$$

Let $\boldsymbol{y}^o$ be the point of absolute minimum of $\varphi(\boldsymbol{y})$, then at point $\boldsymbol{y}^o$ the gradient $\varphi'(\boldsymbol{y})$ becomes zero, i.e. $\varphi'(\boldsymbol{y}^o) = 0$. Hence and using Equation 7.522 we have

$$(I - P_J) f'(\boldsymbol{x}^o) = 0 \tag{7.523}$$

or

$$f'(\boldsymbol{x}^o) - A_J^T (A_J A_J^T)^{-1} A_J f'(\boldsymbol{x}^o) = f'(\boldsymbol{x}^o) + A_J^T \boldsymbol{u}(\boldsymbol{x}^o) = 0, \tag{7.524}$$

where $\boldsymbol{u}(\boldsymbol{x}) = -(A_J A_J^T)^{-1} A_J f'(\boldsymbol{x})$. Since, due to Equation 7.518, $A_J(I - P_J) = 0$ then

$$A_J \boldsymbol{x}^o = A_J \boldsymbol{x}_o + A_J (I - P_J)\boldsymbol{y}^o = A_J \boldsymbol{x}_o = \boldsymbol{d}_J \tag{7.525}$$

and therefore $\boldsymbol{x}^o$ satisfies Equation 7.509, $i \in J$. Thus $\boldsymbol{x}^o$ is the feasible point, and the last relation in Equation 7.524 at this point is satisfied. These conditions represent necessary and sufficient conditions for $\boldsymbol{x}^o$ being the minimum point of $f(\boldsymbol{x})$ with constraints given by Equation 7.509, $i \in J$.

Then the problem leads to the minimization of the quadratic functional $\varphi(\boldsymbol{y})$ without constraints for which the method of conjugate gradients can be used. To minimize $\varphi(\boldsymbol{y})$ we apply the conjugate gradient method:

$$\begin{aligned}
&\boldsymbol{y}_o = 0, \quad \boldsymbol{p}_1 = -\varphi'(0), \\
&\boldsymbol{y}_{k+1} = \boldsymbol{y}_k + \alpha_{k+1} \boldsymbol{p}_{k+1}, \\
&\boldsymbol{p}_{k+1} = -\varphi'(\boldsymbol{y}_k) + \frac{\|\varphi'(\boldsymbol{y}_k)\|^2}{\|\varphi'(\boldsymbol{y}_{k-1})\|^2} \boldsymbol{p}_k,
\end{aligned} \tag{7.526}$$

where

$$\alpha_{k+1} = -\frac{(\varphi'(\boldsymbol{y}_k), \boldsymbol{p}_{k+1})}{(\boldsymbol{p}_{k+1}, (I - P_J) C (I - P_J) \boldsymbol{p}_{k+1})},$$

since the matrix, which determines the quadratic term of $\varphi(\boldsymbol{y})$, has the form $(I - P_J)C(I - P_J)$. We proved that the process involving the new variable is convergent, as the CGM is convergent.

Now we will prove that the algorithm in the primary variable $\boldsymbol{x}$ is also convergent. Firstly we prove that

$$(I - P_J)\boldsymbol{p}_k = \boldsymbol{p}_k, \quad k = 1, 2, \ldots \tag{7.527}$$

Now for $k = 1$ since $\boldsymbol{p}_1 = -\varphi'(0)$ (see Equation 7.526) it holds

$$(I - P_J)\boldsymbol{p}_1 = -(I - P_J)\varphi'(0). \tag{7.528}$$

Hence and using Equation 7.522

$$(I - P_J)\boldsymbol{p}_1 = -(I - P_J)(I - P_J)f'(\boldsymbol{x}_o).$$

Since

$$(I - P_J)(I - P_J) = I - P_J - (I - P_J)P_J = I - P_J,$$

then

$$(I - P_J)\boldsymbol{p}_1 = -(I - P_J)f'(\boldsymbol{x}_o) = -\varphi'(0) = \boldsymbol{p}_1.$$

Now we shall assume that Equation 7.527 is valid for k, and we prove that it also holds for $k + 1$. Then due to Equation 7.526, 7.527 and 7.522 we have

$$\begin{aligned}(I - P_J)\boldsymbol{p}_{k+1} &= -(I - P_J)\left[-\varphi'(\boldsymbol{y}_k) + \frac{\|\varphi'(\boldsymbol{y}_k)\|^2}{\|\varphi'(\boldsymbol{y}_{k-1})\|^2}\boldsymbol{p}_k\right] \\ &= -(I - P_J)(I - P_J)f'(\boldsymbol{x}_k) + \frac{\|\varphi'(\boldsymbol{y}_k)\|^2}{\|\varphi'(\boldsymbol{y}_{k-1})\|^2}\boldsymbol{p}_k \\ &= -(I - P_J)f'(\boldsymbol{x}_k) + \frac{\|\varphi'(\boldsymbol{y}_k)\|^2}{\|\varphi'(\boldsymbol{y}_{k-1})\|^2}\boldsymbol{p}_k = \boldsymbol{p}_{k+1}\end{aligned} \tag{7.529}$$

Then from Equation 7.520 we have

$$\boldsymbol{x}_{k+1} = \boldsymbol{x}_o + (I - P_J)\boldsymbol{y}_{k+1}, \boldsymbol{x}_k = \boldsymbol{x}_o + (I - P_J)\boldsymbol{y}_k.$$

Hence and using Equation 7.526

$$\boldsymbol{x}_{k+1} = \boldsymbol{x}_k + (I - P_J)(\boldsymbol{y}_{k+1} - \boldsymbol{y}_k) = \boldsymbol{x}_k + (I - P_J)\alpha_{k+1}\boldsymbol{p}_{k+1}$$

and due to Equation 7.529 we have

$$\boldsymbol{x}_{k+1} = \boldsymbol{x}_k + \alpha_{k+1}\boldsymbol{p}_{k+1}. \tag{7.530}$$

Since

$$\boldsymbol{p}_{k+1} = -\varphi'(\boldsymbol{y}_k) + \frac{\|\varphi'(\boldsymbol{y}_k)\|^2}{\|\varphi'(\boldsymbol{y}_{k-1})\|^2}\boldsymbol{p}_k, \quad \alpha_{k+1} = -\frac{(\varphi'(\boldsymbol{y}_k), \boldsymbol{p}_{k+1})}{(\boldsymbol{p}_{k+1}, (I - P_J)C(I - P_J)\boldsymbol{p}_{k+1})},$$

then after applying Equation 7.522 and some modifications we obtain

$$\boldsymbol{p}_{k+1} = -(I - P_J)f'(\boldsymbol{x}_k) + \frac{\|(I - P_J)f'(\boldsymbol{x}_k)\|^2}{\|(I - P_J)f'(\boldsymbol{x}_{k-1})\|^2}\boldsymbol{p}_k$$

and

$$\alpha_{k+1} = -\frac{((I - P_J)f'(\boldsymbol{x}_k), \boldsymbol{p}_{k+1})}{((I - P_J)\boldsymbol{p}_{k+1}, C(I - P_J)\boldsymbol{p}_{k+1})} = -\frac{(f'(\boldsymbol{x}_k), \boldsymbol{p}_{k+1})}{(\boldsymbol{p}_{k+1}, C\boldsymbol{p}_{k+1})}$$

Since the algorithm in Equation 7.526 is convergent then we have proved that the algorithm in Equation 7.579 is also convergent, which completes the proof.

Now we shall define the algorithm minimizing $f(\boldsymbol{x})$ under the constraints of Equation 7.509.

Algorithm 7.1:

(i) let us choose $\boldsymbol{x}_o \in R^n$ satisfying Equation 7.509,

(ii) we determine a set $J = J(\boldsymbol{x}_o) = \{i \mid (\boldsymbol{a}_i, \boldsymbol{x}_o) - d_i = 0, \quad i \in J^- \cup J^o\}$, and form a matrix $A_{J(\boldsymbol{x}_o)}$,

(iii) we calculate a vector $(I - P_J)f'(\boldsymbol{x}_o)$, where

$$f'(\boldsymbol{x}_o) = C\boldsymbol{x}_o + \boldsymbol{b}, \quad P_J = A^T_{J(\boldsymbol{x}_o)}(A_{J(\boldsymbol{x}_o)}A^T_{J(\boldsymbol{x}_o)})^{-1}A_{J(\boldsymbol{x}_o)},$$

(iv) there are two cases:

a) $(I - P_J)f'(\boldsymbol{x}_o) \neq 0$, here we apply the algorithm in (i) to (iv) of Equation 7.519. Then for every $k = 0, 1, 2, \ldots$ we calculate (i.e. we check whether the points obtained are feasible or not)

$$\alpha_{k+1} = -\frac{(f'(\boldsymbol{x}_k), \boldsymbol{p}_{k+1})}{(\boldsymbol{p}_{k+1}, C\boldsymbol{p}_{k+1})} \text{ and } \underline{\alpha}_{k+1} = \min_{\substack{i \\ (\boldsymbol{a}_i, \boldsymbol{p}_{k+1})>0}} \frac{d_i - (\boldsymbol{a}_i, \boldsymbol{x}_k)}{(\boldsymbol{a}_i, \boldsymbol{p}_{k+1})}$$

Then we put $\boldsymbol{x}_{k+1} = \boldsymbol{x}_k + \alpha_{k+1}\boldsymbol{p}_{k+1}$ provided $\alpha_{k+1} \leq \underline{\alpha}_{k+1}$ and then continue in the algorithm given by Equation 7.519. In the case if $\boldsymbol{x}_{k+1}$ is the minimum of $f(\boldsymbol{x})$ with constraints as in Equation 7.509, $i \in J$, we put $\boldsymbol{x}_o = \boldsymbol{x}_{k+1}$ and go to (v) below.
In the case if $\alpha_{k+1} > \underline{\alpha}_{k+1}$ we put $\boldsymbol{x}_{k+1} = \boldsymbol{x}_k + \underline{\alpha}_{k+1}\boldsymbol{p}_{k+1}$ and the algorithm continues onto (ii).

b) $(I - P_J)f'(\boldsymbol{x}_o) = 0$, then we continue onto (v).

(v) We calculate $\boldsymbol{u}(\boldsymbol{x}_o) = -(A_J A_J^T)^{-1} A_J f'(\boldsymbol{x}_o)$.
If $(\boldsymbol{u}(\boldsymbol{x}_o))_i \geq 0 \, \forall i \in J(\boldsymbol{x}_o) \cap J^-$ we find the minimum point of $f(\boldsymbol{x})$ and the process stops.
If there exists an index $j \in J(\boldsymbol{x}_o) \cap J^-$ such that

$$(\boldsymbol{u}(\boldsymbol{x}_o))_j < 0 \text{ we put } J(\boldsymbol{x}_o) = J(\boldsymbol{x}_o) - \{j\} \text{ and go to (iii).}$$

The sequence $\{f(\boldsymbol{x}_k)\}$, $k = 0, 1, 2, \ldots$, generated by Algorithm 7.1 from the sequence $\{\boldsymbol{x}_k\}$ is strictly decreasing. Then there exists a number n_o such that $f(\boldsymbol{x}_{n_o})$ minimizes $f(\boldsymbol{x})$ with constraints given by Equation 7.509.

Theorem 7.26: The necessary and sufficient conditions for the convex function $f(\boldsymbol{x})$ with constraints as in Equation 7.509 to attain its minimum at the point $\boldsymbol{x} \in R^n$ are as follows. There exist $(\boldsymbol{u}(\boldsymbol{x}))_i$, $i \in J^- \cup J^o$ such that

$$f'(\boldsymbol{x}) + \sum_{i \in J^- \cup J^o} \boldsymbol{a}_i(\boldsymbol{u}(\boldsymbol{x}))_i = 0, \tag{7.531}$$

$$\begin{aligned} &(\boldsymbol{u}(\boldsymbol{x}))_i = 0 \text{ if } (\boldsymbol{a}_i, \boldsymbol{x}) - d_i < 0 \quad i \in J^-, \\ &(\boldsymbol{u}(\boldsymbol{x}))_i \geq 0, \quad i \in J^-. \end{aligned} \tag{7.532}$$

Proof: The proof is parallel to that of Theorem 3.8, Chapter I of Pschenichny and Danilin (1978). Thus, due to Algorithm 7.1, the point $\boldsymbol{x} \in R^n$ is the minimum of the function $f(\boldsymbol{x})$ provided $(I -$

$P_J)f'(\boldsymbol{x}) = 0$ and $(\boldsymbol{u}(\boldsymbol{x}))_i \geq 0$ for $i \in J \cap J^-$, where $\boldsymbol{u}(\boldsymbol{x})$ is defined as $\boldsymbol{u}(\boldsymbol{x}) = -(A_J A_J^T)^{-1} A_J f'(\boldsymbol{x})$. Now, if $(\boldsymbol{u}(\boldsymbol{x}))_i = 0$ provided $i \in (J^- \cup J^o)\backslash J$, then

$$f'(\boldsymbol{x}) + \sum_{i \in J^- \cup J^o} \boldsymbol{a}_i(\boldsymbol{u}(\boldsymbol{x}))_i = f'(\boldsymbol{x}) + \sum_{i \in J} \boldsymbol{a}_i(\boldsymbol{u}(\boldsymbol{x}))_i$$

$$= f'(\boldsymbol{x}) + A_J^T \boldsymbol{u}(\boldsymbol{x}) = (I - P_J)f'(\boldsymbol{x}) = 0,$$

and $(\boldsymbol{u}(\boldsymbol{x}))_i$ satisfies Equation 7.532. Therefore, the conditions of algorithm 7.1 are sufficient to attain the minimum of the quadratic form with constraints given by Equation 7.509.

7.5.3. Elimination Method

In this section we shall discuss one type of algorithm which is useful for contact problems, and which is very useful from the economical point of view (Kikuchi and Oden (1988), Haslinger and Tvrdý (1983), Tvrdý (1980)). The author applied this method for the preconditioning technique onto Equations 7.508 and 7.509 (Nedoma (1993)). The basic idea of the preconditioning technique is to reduce the initial minimizing problem to that of the related problem for which the rate of convergence of the conjugate gradient method is better than for the initial one. For details about the preconditioning technique see Chapter 2. Then we have the problem of minimizing

$$\underline{f}(z) = \tfrac{1}{2}(z, \overline{C}z) - (\overline{\boldsymbol{b}}, z), \quad z \in R^n, \tag{7.533}$$

$$\overline{A}z \leq 0, \tag{7.534}$$

where $z = D^T\boldsymbol{x}$, $\overline{C} = (\overline{C}_{ij}) = D^{-1}CD^{-T}$, $\overline{\boldsymbol{b}} = D^{-1}\boldsymbol{b}$, $D^{-T} = (D^T)^{-1}$ and E is a positive definite matrix such that $E = DD^T$.

In the case of the Uzawa algorithm, if we calculate the projection $\overline{P}$, only variables, denoted by $z_1, \ldots, z_m$, $m = m(h)$, corresponding to values u_{hi} in points of the triangulation $\mathcal{T}_h$ laying on $\cup\overline{\Gamma}_c^{kl}\backslash\Gamma_u$, are taken into account. Similarly, the right hand side vector $\boldsymbol{S}(\lambda_h)$ depends on λ_h only in components with indices $i_1, \ldots, i_m$. We shall assume that unknown variables, corresponding to u_{hi} in nodes of the triangulation $\mathcal{T}_h$ laying on $\cup\overline{\Gamma}_c^{kl}\backslash\Gamma_u$, are situated on the last $m = m(h)$ places of a vector z. Then we may split a vector $z \in R^n$ in such a way that $z = (z_1, z_2) \in R^{n-m} \times R^m$, where z_1 and z_2 correspond to the free and constrained variables, respectively. Then we define the set

$$K_{ho} = \{z \in R^n, \quad z = (z_1, z_2), \quad z_1 \in R^{n-m}, \quad z_2 \in R^m, \quad \overline{A}z_2 \leq 0\}.$$

Due to the splitting of the vector z, we can split the vector $\overline{\boldsymbol{d}} = \overline{\boldsymbol{S}}(\lambda_h) - \overline{\boldsymbol{F}}$ and decompose the matrix $\overline{C}$ by

$$\overline{C} = \begin{pmatrix} \overline{C}_{11} & \overline{C}_{12} \\ \overline{C}_{21} & \overline{C}_{22} \end{pmatrix}, \tag{7.535}$$

where $\overline{C}_{11}$ and $\overline{C}_{22}$, are $(n-m) \times (n-m)$ or $m \times m$ matrices, respectively and $\overline{C}_{12}$ and $\overline{C}_{21}$ are rectangular $(n-m) \times m$ or $m \times (n-m)$ matrices, respectively. Since $\overline{C}$ is symmetric, then $\overline{C}_{21} = \overline{C}_{12}^T$.

The problem in Equations 7.533 and 7.534 is equivalent to the following problem. Find $\boldsymbol{z} = (z_1, z_2)^T \in K_{ho}$, $z_1 \in R^{n-m}$, $z_2 \in R^m$ and $\overline{A}z_2 \leq 0$ such that

$$(\overline{C}\boldsymbol{z}, \boldsymbol{v} - \boldsymbol{z}) \geq (\overline{\boldsymbol{b}}, \boldsymbol{v} - \boldsymbol{z}) \quad \forall \boldsymbol{v} \in K_{ho}. \tag{7.536}$$

Let $\boldsymbol{v} = (\boldsymbol{v}_1, \boldsymbol{v}_2)^T, \overline{\boldsymbol{b}} = (\boldsymbol{b}_1, \boldsymbol{b}_2)^T$ such that $\boldsymbol{v}_1, \boldsymbol{b}_1 \in R^{n-m}$, $\boldsymbol{v}_2, \boldsymbol{b}_2 \in R^m, \overline{A}\boldsymbol{v}_2 \leq 0$. Let us put in Equation 7.536 $\boldsymbol{v} = (\boldsymbol{v}_1, \boldsymbol{v}_2)$ such that $\boldsymbol{v}_1 = z_1 \pm \boldsymbol{y}_1$, $\boldsymbol{v}_2 = z_2$, where $\boldsymbol{y}_1 \in R^{n-m}$ is arbitrary. Then from Equation 7.536 we obtain

$$\begin{pmatrix} z_1 \pm \boldsymbol{y}_1 - z_1 \\ z_2 - z_2 \end{pmatrix}^T \begin{pmatrix} \overline{C}_{11} & \overline{C}_{12} \\ \overline{C}_{21} & \overline{C}_{22} \end{pmatrix} \begin{pmatrix} z_1 \\ z_2 \end{pmatrix} \geq \begin{pmatrix} z_1 \pm \boldsymbol{y}_1 - z_1 \\ z_2 - z_2 \end{pmatrix}^T \begin{pmatrix} \overline{\boldsymbol{b}}_1 \\ \overline{\boldsymbol{b}}_2 \end{pmatrix}.$$

Hence

$$\boldsymbol{y}_1^T(\overline{C}_{11}z_1 + C_{12}z_2) = \boldsymbol{y}_1^T\overline{\boldsymbol{b}}_1 \quad \forall \boldsymbol{y}_1 \in R^{n-m}$$

and thus

$$\overline{C}_{11}z_1 + \overline{C}_{12}z_2 = \overline{\boldsymbol{b}}_1. \tag{7.537}$$

Moreover, let us put in Equation 7.536 $\boldsymbol{v}$ such that $\boldsymbol{v}_1 = z_1$, $\boldsymbol{v}_2 = \boldsymbol{y}_2$, where $\boldsymbol{y}_2 \in R^m$, $\overline{A}\boldsymbol{y}_2 \leq 0$ and $\boldsymbol{y}_2$ is arbitrary. Then

$$(\boldsymbol{y}_2 - z_2)^T(\overline{C}_{21}z_1 + \overline{C}_{22}z_2) \geq (\boldsymbol{y}_2 - z_2)^T\overline{\boldsymbol{b}}_2. \tag{7.538}$$

Since from Equation 7.537

$$z_1 = \overline{C}_{11}^{-1}(\overline{\boldsymbol{b}}_1 - \overline{C}_{12}z_2)$$

then from Equation 7.538 we obtain

$$(\boldsymbol{y}_2 - z_2)^T(\overline{C}_{21}\overline{C}_{11}^{-1}(\overline{\boldsymbol{b}}_1 - \overline{C}_{12}z_2) + \overline{C}_{22}z_2) \geq (\boldsymbol{y}_2 - z_2)^T\overline{\boldsymbol{b}}_2$$

and hence

$$(\boldsymbol{y}_2 - z_2)^T((\overline{C}_{22} - \overline{C}_{21}\overline{C}_{11}^{-1}\overline{C}_{12})z_2) \geq (\boldsymbol{y}_2 - z_2)^T(\overline{\boldsymbol{b}}_2 - \overline{C}_{21}\overline{C}_{11}^{-1}\overline{\boldsymbol{b}}_1).$$

Let us denote $C_o = \overline{C}_{22} - \overline{C}_{12}^T\overline{C}_{11}^{-1}\overline{C}_{12}$, $\overline{\boldsymbol{b}}_{20} = \overline{\boldsymbol{b}}_2 - \overline{C}_{21}\overline{C}_{11}^{-1}\overline{\boldsymbol{b}}_1$, $K_{hY} = \{z \mid z \in R^m, \ \overline{A}z \leq 0\}$ then

$$(\boldsymbol{y}_2 - z_2)^T C_o z_2 \geq (\boldsymbol{y}_2 - z_2)^T\overline{\boldsymbol{b}}_{20}. \tag{7.539}$$

Theorem 7.27: Let $\overline{C}$ be a symmetric, positive semi-definite matrix of the order n. Then the matrix $C_o = \overline{C}_{22} - \overline{C}_{12}^T\overline{C}_{11}^{-1}\overline{C}_{12}$ is a symmetric and positive semi-definite matrix of the order m.

Proof: Since $\overline{C}$ is symmetric, then $\overline{C}_{11}$ is also symmetric and therefore $\overline{C}_{11}^{-1}$ is symmetric. Hence C_o is also symmetric.

The matrix $\overline{C}$ is positive semi-definite, i.e. $z^T\overline{C}z \geq 0 \, \forall z \in R^n$. Since

$$\overline{C} = \begin{pmatrix} \overline{C}_{11} & \overline{C}_{12} \\ \overline{C}_{21} & \overline{C}_{22} \end{pmatrix}, \quad z = (z_1, z_2)^T, \quad z_1 \in R^{n-m}, \quad z_2 \in R^m$$

then

$$z_1^T\overline{C}_{11}z_1 + z_1^T\overline{C}_{12}z_2 + z_2^T\overline{C}_{21}z_1 + z_2^T\overline{C}_{22}z_2 = z_1^T\overline{C}_{11}z_1 + z_1^T\overline{C}_{12}z_2 + z_2^T\overline{C}_{12}^Tz_1 + z_2^T\overline{C}_{22}z_2 \geq 0$$
$$\forall z_1 \in R^{n-m}, z_2 \in R^m.$$

Let $v \in R^m$ be arbitrary. Let us put

$$z_1 = \overline{C}_{11}^{-1}\overline{C}_{12}v, \quad z_2 = (1 + 2^{1/2})v$$

then

$$v^T\overline{C}_{12}^T\overline{C}_{11}^{-1}\overline{C}_{12}v + (1+2^{1/2})v^T\overline{C}_{12}^T\overline{C}_{11}^{-1}\overline{C}_{12}v + (1+2^{1/2})v^T\overline{C}_{12}^T\overline{C}_{11}^{-1}\overline{C}_{12}v + (1+2^{1/2})^2v^T\overline{C}_{22}v \geq 0.$$

Thus

$$(1 + 2^{1/2})^2(v^T\overline{C}_{12}^T\overline{C}_{11}^{-1}\overline{C}_{12}v + v^T\overline{C}_{22}v) \geq 0.$$

Since v is an arbitrary vector, then the matrix C_o is positive semi-definite.

Then from Equations 7.539 and 7.537 our original problem is equivalent to the following problem.

Algorithm 7.2: Find $z = (z_1, z_2)^T$ such that

$$\tfrac{1}{2}(z_2, C_o z_2) - (\overline{b}_{20}, z_2) = \min_{z \in K_{hy}} \{\tfrac{1}{2}(z, C_o z) - (\overline{b}_{20}, z)\}, \tag{7.540}$$

and z_1 is a solution of Equation 7.537.

From Equation 7.539 we see that z_2 is a solution of the variational inequality

$$(y_2 - z_2, C_o z_2) \geq (\overline{b}_{20}, y_2 - z_2) \quad \forall y_2 \in R^m,$$

and from Equation 7.537

$$z_1 = C_{11}^{-1}(\overline{b}_1 - \overline{C}_{12} z_2).$$

It is evident that the matrix C_o and the vector $\overline{b}_{20}$ can be obtained by Gaussian elimination from the initial matrix $\overline{C}$ and the vector $\overline{b}$.

Theorem 7.28: Let $\overline{C}$ be the $n \times n$ matrix defined by Equation 7.535, and let $\overline{b} = (\overline{b}_1, \overline{b}_2)^T$ be an n-dimensional vector. Let C_1 be the $n \times (n+1)$ dimensional matrix

$$C_1 = \begin{pmatrix} \overline{C}_{11} & \overline{C}_{12} & \overline{b}_1 \\ \overline{C}_{21} & \overline{C}_{22} & \overline{b}_2 \end{pmatrix}$$

Let B be a matrix generated by the elimination of the first n columns of the matrix C_1, i.e.

$$B = \begin{pmatrix} B_{11} & B_{12} & \underline{b}_1 \\ 0 & B_{22} & \underline{b}_2 \end{pmatrix}.$$

Then $B_{22} = \overline{C}_{22} - \overline{C}_{12}^T\overline{C}_{11}^{-1}\overline{C}_{12} = C_o$, $\underline{b}_2 = b_2 - \overline{C}_{12}^T\overline{C}_{11}^{-1}\overline{b}_1 = \overline{b}_{20}$.

Proof: The Gaussian elimination gives

$$B = B_n B_{n-1} B_{n-2} \ldots B_1 C_1 = \underline{B} C_1$$

where

$$
B_{i+1} = \begin{pmatrix}
1 & & & & & & \\
& \ddots & & & 0 & & \\
& & 1 & & & & \\
& & -\dfrac{\overline{b}^{(i)}_{i+2,\,i+1}}{\overline{b}^{(i)}_{i+1,\,i+1}} & & & & \\
& & \vdots & 1 & & & \\
& & & & \ddots & & \\
0 & & -\dfrac{\overline{b}^{(i)}_{i+1,\,i+1}}{\overline{b}^{(i)}_{i+1,\,i+1}} & 0 & & & 1
\end{pmatrix},
$$

where $(\overline{b}^{(i)}_{jk})$, $j, k = 1, 2, \ldots, n$ are elements of the matrix $\overline{B}_i = B_i B_{i-1} \ldots B_1 C_1$, $\overline{B}_o = C_1$. Then the matrix $\underline{B}$ can be written in the form

$$
\underline{B} = \left(\begin{array}{c:c}
\underline{B}_{11} & 0 \\
\hdashline
\underline{B}_{21} & I_m
\end{array}\right),
$$

where $\underline{\boldsymbol{B}}_{11}$ is the triangular matrix of order $n-m$, $\underline{B}_{21}$ is an $m \times n-m$ matrix and I_m is the unit matrix of order m. Let us denote

$$
\boldsymbol{B} = \underline{B}^{-1} = \begin{pmatrix} \boldsymbol{B}_{11} & \boldsymbol{B}_{12} \\ \boldsymbol{B}_{21} & \boldsymbol{B}_{22} \end{pmatrix}.
$$

Since $\boldsymbol{B}\underline{B} = \underline{B}^{-1}\underline{B} = I_n$ then

$$
\boldsymbol{B}_{11} = \underline{B}_{11}^{-1}, \quad \underline{B}_{11}\boldsymbol{B}_{11} = 0 \text{ then } \boldsymbol{B}_{12} = 0, \quad \underline{B}_{21}\boldsymbol{B}_{11} + \boldsymbol{B}_{21} = 0, \quad \boldsymbol{B}_{22} = I_m.
$$

Since $C_1 = \boldsymbol{B}B$ then

$$
\begin{aligned}
\overline{C}_{11} &= \boldsymbol{B}_{11}B_{11} \text{ then } \overline{C}_{11}^{-1} = B_{11}^{-1}\underline{B}_{11}, \\
\overline{C}_{21} &= \boldsymbol{B}_{21}B_{11}, \\
\overline{C}_{12} &= \boldsymbol{B}_{11}B_{12}, \\
\overline{C}_{22} &= B_{22} + \boldsymbol{B}_{21}B_{12}, \\
\overline{\boldsymbol{b}}_1 &= \boldsymbol{B}_{11}\underline{\boldsymbol{b}}_1, \quad \overline{\boldsymbol{b}}_2 = \boldsymbol{B}_{21}\underline{\boldsymbol{b}}_1 + \underline{\boldsymbol{b}}_2.
\end{aligned}
$$

Therefore,

$$
C_o = \overline{C}_{22} - \overline{C}_{12}^T \overline{C}_{11}^{-1} \overline{C}_{12} = B_{22} + \boldsymbol{B}_{21}B_{12} - \boldsymbol{B}_{21}B_{11}B_{11}^{-1}\underline{B}_{11}\boldsymbol{B}_{11}B_{12} = B_{22}
$$

and

$$
\overline{\boldsymbol{b}}_{20} = \overline{\boldsymbol{b}}_2 - \overline{C}_{12}^T \overline{C}_{11}^{-1} \overline{\boldsymbol{b}}_1 = \underline{\boldsymbol{b}}_2 + \boldsymbol{B}_{21}\underline{\boldsymbol{b}}_1 - \boldsymbol{B}_{21}B_{11}B_{11}^{-1}\underline{B}_{11}\boldsymbol{B}_{11}\underline{\boldsymbol{b}}_1 = \underline{\boldsymbol{b}}_2,
$$

which completes the proof.

Remark: From the last theorem and Equation 7.540 we see that it is sufficient to eliminate the matrix $\overline{C}$ onto C_o, then by applying the minimizing algorithm onto the form $\frac{1}{2}(z, C_o z) + (\overline{\boldsymbol{b}}_{20}, z)$

we find the last m unknown variables, and by back elimination we find the remaining $(n-m)$ variables.

7.6. COMPUTER IMPLEMENTATION OF THE ALGORITHMS FOR COUPLED CONTACT-STEFAN-LIKE PROBLEMS

7.6.1. Introduction

Numerical solution of the coupled Stefan-like-contact problems at every time level leads to solving:

(i) The Stefan-like problem with coupled term

$$\Delta t \rho(\boldsymbol{x})\Theta_o(\boldsymbol{x})\beta_{ij}(\boldsymbol{x})e_{ij}(\dot{\boldsymbol{u}}(\boldsymbol{x},t)) \in L^2(\Omega),\ t \text{ being fixed},$$

having a meaning of thermal sources which, in the enthalpy approximation, then leads to minimizing the functional

$$L_\Theta(z) = \tfrac{1}{2}z^T Bz - \boldsymbol{d}^T z + \Phi(z) \tag{7.541}$$

without constraints, where $B = \alpha I + \mu B_o \equiv M + \mu B_o$, where $\mu = \Delta t$, B_o represents the finite element approximation of $a_\Theta(.,.)$, and I is the unit matrix;

(ii) the contact problem with friction with coupled term

$$\frac{\partial}{\partial x_j}(\beta_{ij}(\boldsymbol{x}))(\Theta(\boldsymbol{x},t) - \Theta_o(\boldsymbol{x})) \in L^2(\Omega),\ t \text{ being fixed},$$

having a meaning of body forces, which leads to finding an approximate saddle point and minimization of quadratic functional with linear constraints

$$L(\boldsymbol{v}) = \tfrac{1}{2}\boldsymbol{v}^T C\boldsymbol{v} + \boldsymbol{b}^T\boldsymbol{v} \text{ with constraints } A\boldsymbol{v} \le 0. \tag{7.542}$$

These coupled problems lead at every time level to a non-linear minimization problem in the first case (i), and the Uzawa or Arrow-Hurwicz algorithms, where in every step of the Uzawa algorithm we solve an optimization problem with constraints in case (ii). Hence, there is a question of how to solve such types of coupled problems. Practically we have three posibilities:

(P1) Since coupled terms have a meaning of thermal sources and of body forces, then at every time level we can solve both problems, i.e. the Stefan-like problem and the contact problem with friction, separately and by different methods which are optimal for each problem. Due to our studies the non-linear Gauss-Seidel or non-linear SOR methods are optimal in the case of the Stefan-like problem (Dvořák (1993),(1996)), and the Uzawa algorithm with the preconditioned conjugate gradient method with constraints is optimal in the second case (Kestřánek and Nedoma (1996)), i.e. in the case of the contact part of the investigated coupled problem.

(P2) Since the problem is coupled, then at every time level we will solve the problem as a coupled one, i.e. both parts of this coupled problem are solved by the same type of method, optimal from the coupled problem point of view.

(P3) Since the problem investigated is coupled, then at every time level we will formulate the problem as fully coupled, i.e. we will formulate the problem as one optimization non-linear problem, involving both problems (i) and (ii), i.e. we will find a saddle point and then we will find

$$\min \mathscr{L}(\boldsymbol{w}) \text{ with constraints } \mathscr{A}\boldsymbol{w} \leq 0, \quad \boldsymbol{w} = (\boldsymbol{z}, \boldsymbol{v}),$$

where $\mathscr{L}(\boldsymbol{w})$ and $\mathscr{A}$ are the corresponding quadratic functional and the matrix of constraints of the fully coupled problem, respectively.

From these three posibilities, we shall investigate the problem denoted as $(P2)$ in more detail. The results of Daniel (1967a, b), Ortega and Rheinboldt (1970), Pschenichny and Danilin (1978), Fletcher (1981), Axelsson (1982),(1994), Axelsson and Vassilevski (1992), Chronopoulos (1992), Axelsson and Chronopoulos (1994), Dvořák (1996), Kestřánek and Nedoma (1996) Kestřánek, Dvořák and Nedoma (1997) will be used.

7.6.2. Algorithm

7.6.2.1. Basic idea of the algorithm

It is evident that both problems are of non-linear type. Our numerical results of both problems, being solved separately (Dvořák (1996), Kestřánek and Nedoma (1996)), show that conjugate gradient methods are one of the best methods which can be applied to both parts of the investigated problem. From this point of view, in this section we shall investigate the following coupled type algorithm at every time:

(A) we shall minimize the functional $L_\Theta(\boldsymbol{z})$, defined by Equation 7.541 in the case of the Stefan-like problem, based on the non-linear conjugate gradient method for the non-linear system of equations

$$B\boldsymbol{z} - \boldsymbol{d} + \partial\Phi(\boldsymbol{z}) = 0,$$

where $\partial\Phi(\boldsymbol{z})$ is the sub-differential of the functional $\Phi(\boldsymbol{z}) = \Sigma_{i=1}^{n}\phi(z_i)$, where $\partial\Phi(\boldsymbol{z}) = (\partial\phi(z_1), \ldots, \partial\phi(z_n))$;

(B) since the contact part of the investigated coupled problem leads to the numerical approximation of a saddle point, we shall find it by the Uzawa algorithm and the constrained conjugate gradient method.

The basic idea of the algorithm was discussed in more detail in Section 7.5.2; here we repeat the basic ideas only.

Since $\{z_i\}_{i=1}^{n}$ is assumed to be a basis of the space ${}^1V_h$, then every element of ${}^1V_h$ can be found as a linear combination of the basis $\{z_i\}$, and similarly for Θ and Q.

Then in the usual way we obtain a non-linear system of algebraic equations

$$F(z) \equiv Bz - \boldsymbol{d} + \partial\Phi(z) = 0, \quad B = \alpha I + \mu B_o, \tag{7.543}$$

where elements of B_o are generated by $\int_{\Omega_h} \kappa_{ij} \nabla z_i \nabla z_j \mathrm{d}\boldsymbol{x}$ and the matrix B is sparse and symmetric positive definite, and $\partial\Phi(z) = MH(z)$, where $M = \{M_{ij}\} = \{\int_{\Omega_h} z_i z_j \mathrm{d}\boldsymbol{x}\}_{i,j=1}^n$ is positive definite and where F denotes the non-linear differentiable mapping $F : R^n \to R^n$, where F is differentiable in R^n and has a meaning of the gradient of the functional $L_\Theta(z)$. Equation 7.543 is a necessary condition for the existence of a minimum of functional $L_\Theta(z)$ given by Equation 7.541. Due to the above assumptions, the G-derivative of L_Θ is a monotone mapping on R^n. Therefore L_Θ is convex and has a global minimizer, representing the solution of Equation 7.543.

The second part of the algorithm of our coupled Stefan-like-contact problem is based on the numerical approximation of a saddle point and on quadratic programming.

To find a saddle point the Uzawa or/and Arrow-Hurwicz algorithms can be used. We shall describe the Uzawa algorithm only. The Uzawa's algorithm is based on the construction of two sequences of elements $\boldsymbol{u}_h^n \in K_h$, $\lambda_h^{n,kl} \in \Lambda$, defined in the following way.

Start with any $\lambda_h^{o,kl} \in \Lambda$ and calculate $\boldsymbol{u}_h^o$, then $\lambda_h^{1,kl}$, $\boldsymbol{u}_h^1$, etc.. Let $\lambda_h^{n,kl}$ be known. We determine $\boldsymbol{u}_h^n$ as the element of K_h which minimizes

$$L_o(\boldsymbol{v}) + \int_{\cup\Gamma_c^{kl}} \lambda_h^{n,kl} g_c^{kl} (\boldsymbol{v}_t^k - \boldsymbol{v}_t^l) \mathrm{d}s. \tag{7.544}$$

Then we define

$$\lambda_h^{n+1,kl} = P(\lambda_h^{n,kl} + \rho^n g_c^{kl} (\boldsymbol{u}_h^{n,k} - \boldsymbol{u}_h^{n,l})_t), \quad \rho^n > 0, \tag{7.545}$$

where P is a projection $P : L^2(\cup\Gamma_c^{kl}) \to \Lambda$ defined by Equation 7.377.

Minimization of the functional given by Equation 7.544 on the set K_h is equivalent to minimization of the quadratic functional

$$L(\boldsymbol{v}) = \tfrac{1}{2}\boldsymbol{v}^T C\boldsymbol{v} + \boldsymbol{b}^T\boldsymbol{v} \text{ with constraints } A\boldsymbol{v} \le 0, \tag{7.546}$$

where C is the stiff matrix ($\dim n \times n$), which is positive definite and generated by the term $\frac{1}{2}\int_\Omega c_{ijkl}(\boldsymbol{x})e_{ij}(\boldsymbol{v})e_{kl}(\boldsymbol{v})\mathrm{d}\boldsymbol{x}$, $\boldsymbol{b}$ is the vector generated by the body and surface forces and the coupled term ($\dim n$), i.e. by the term $\int_\Omega F_i v_i \mathrm{d}\boldsymbol{x} + \int_{\Gamma_\tau} P_i v_i \mathrm{d}s + b_s(\Theta, \boldsymbol{v})$ as well as the friction forces $\int_{\cup\Gamma_c^{kl}} g_c^{kl} \left(\left|\boldsymbol{v}_t^k\right| - \left|\boldsymbol{v}_t^l\right|\right) \mathrm{d}s$, and A is the matrix of constraints ($\dim m \times n$) and is generated by the condition of nonpenetration $u_n^k - u_n^l \le 0$ on $\cup\Gamma_c^{kl}$ in the definition of the admissible set of displacements K_h, discussed in Section 7.2.1.4.

7.6.2.2. Non-linear solution for the Stefan part of the problem based on non-linear conjugate gradient methods

We saw above that the Stefan part of the problem leads to solving the non-linear system of algebraic equations

$$F(z) = 0. \tag{7.547}$$

We shall describe the non-linear conjugate gradient method (Chronopoulos (1992), Axelsson and Chronopoulos (1994)). Consider the non-linear system of equations in Equation 7.547, where F is a non-linear operator from a real n-dimensional Euclidean space into itself. Let $F'(z)$ be the Jacobian and $F''(z)$ the Hessian such that there exist positive constants c_1, c_2, c_3, $c_1 \leq c_2$ and

$$c_1 \|w\|_1 \leq (F'(z)w, w) \quad \forall w \in R^n, \tag{7.548}$$

$$\left\|F'(z)w\right\|_1 \leq c_2 \|w\|_1 \quad \forall w \in R^n, \tag{7.549}$$

$$\|F''(z)\|_1 \leq c_3, \tag{7.550}$$

where Equation 7.548 means that the Jacobian is uniformly positive definite and non-singular. From Equation 7.548 then follows.

Lemma 7.30: Let $F : \Omega \subset R^n \rightarrow R^n$, where Ω is convex, and let F be differentiable on Ω. Let c_1 be a non-negative constant. Then the following three relations are equivalent.

$$(F'(z)w, w) > c_1 \|w\|_1^2 \quad \forall z \in \Omega, \quad w \in R^n \tag{7.551}$$

$$(F(x) - F(y), x - y) \geq c_1(x - y, x - y) \quad \forall x, y \in \Omega, \tag{7.552}$$

the so-called condition of monotonicity.

Let $x(t), y(t)$ be two solutions of the ordinary differential equation

$$x'(t) = -F(x(t)), t > 0. \tag{7.553}$$

Then

$$\|x(t) - y(t)\| \leq \exp(-c_1 t) \|x(0) - y(0)\|. \tag{7.554}$$

Proof: It holds that

$$(F(x) - F(y), x - y) = \left(\int_o^1 F'(z(t))(x - y)\mathrm{d}t, x - y\right), \quad z(t) = y + t(x - y).$$

To prove Equation 7.552, the mean value theorem will be applied. Hence Equation 7.552 holds for some z on the line segment between x and y. we have

$$\frac{\mathrm{d}(\|x(t) - y(t)\|^2)}{\mathrm{d}t} = 2(x' - y', x - y) = -2(F(x(t)) - F(y(t)), x(t) - y(t)) \leq -2c_1 \|x(t) - y(t)\|^2.$$

Hence, by integrating, the last relation, Equation 7.553, follows immediately. Equation 7.554 yields

$$\frac{\mathrm{d}(\|x(t) - y(t)\|^2)}{\mathrm{d}t} = -2(F(x(t)) - F(y(t)), x(t) - y(t)) \leq -2c_1 \exp(-2c_1 t) \|x(0) - y(0)\|^2.$$

Hence

$$(F(x(t)) - F(y(t)), x(t) - y(t)) \leq c_1 \exp(-2c_1 t) \|x(0) - y(0)\|^2,$$

which for $t \to 0$ proves Equation 7.552. Let us set $\boldsymbol{x} = z$, $\boldsymbol{x} - \boldsymbol{y} = t\boldsymbol{w}$ in Equation 7.552 and let $t \to 0_+$. Then Equation 7.551 is also proved, which completes the proof.

Lemma 7.31: The equation $F(z) = 0$, i.e. Equation 7.547, has a unique solution.

Proof: The existence follows from the condition of monotonicity (Equation 7.552) of $F(z)$ for $c_1 > 0$. Then F is strongly monotone and therefore Equation 7.547 has a unique solution.

By analogy with the linear case of the conjugate gradient method we shall consider the following iteration process, the so-called non-linear conjugate gradient (NCG) algorithm.

Algorithm 7.3(NCG): Let z_o be an initial approximation and let $\boldsymbol{r}_o = \boldsymbol{p}_o = -F(z_o)$. Then, for $k \in N = \{0, 1, \dots, \}$, the k-th iteration is made by the scheme:

begin

$z_{k+1} = z_k + \alpha_k \boldsymbol{p}_k$,

where α_k is the smallest positive root of $(F(z_k + \alpha_k \boldsymbol{p}_k), \boldsymbol{p}_k) = 0$;

$\boldsymbol{r}_{k+1} = -F(z_{k+1})$;

$\beta_{k+1} = -(\boldsymbol{r}_{k+1}, F'(z_{k+1})\boldsymbol{p}_k)/(\boldsymbol{p}_k, F'(z_{k+1})\boldsymbol{p}_k)$;

$\boldsymbol{p}_{k+1} = \boldsymbol{r}_{k+1} + \beta_{k+1}\boldsymbol{p}_k$;

end.

Since the assumptions on $F(z)$ and $F'(z)$ guarantee that $F(z)$ is the gradient of the functional $L_\Theta(z)$ then α, satisfying $(F(z + \alpha \boldsymbol{p}), \boldsymbol{p}) = 0$, must yield the minimum of $L_\Theta(z + \alpha \boldsymbol{p})$ as a function of α.

Lemma 7.32: The following relations

$$(\boldsymbol{p}_k, F'(z_k)\boldsymbol{p}_{k-1}) = 0, \tag{7.555}$$

$$(\boldsymbol{r}_{k+1}, \boldsymbol{p}_k) = 0, \tag{7.556}$$

$$(\boldsymbol{r}_k, \boldsymbol{p}_k) = \|\boldsymbol{r}_k\|^2, \tag{7.557}$$

$$(\boldsymbol{p}_k, F'(z_k)\boldsymbol{r}_k) = (\boldsymbol{p}_k, F'(z_k)\boldsymbol{p}_k), \tag{7.558}$$

$$\|\boldsymbol{p}_k\|^2 = \|\boldsymbol{r}_k\|^2 + \beta_{k-1}^2 \|\boldsymbol{p}_{k-1}\|^2, \tag{7.559}$$

$$(\boldsymbol{r}_k, F'(z_k)\boldsymbol{r}_k) = (\boldsymbol{p}_k, F'(z_k)\boldsymbol{p}_k) + \beta_{k-1}^2(\boldsymbol{p}_{k-1}, F'(z_k)\boldsymbol{p}_{k-1}), \tag{7.560}$$

hold.

The proof follows from the properties of $F'(z)$ and Algorithm 7.3.

Lemma 7.33: Let us assume that $F'(z)$ is self-adjoint and let $0 < aI \le F'(z) \le AI$, where a and A are scalars. Then it holds that

$$\|\boldsymbol{r}_k\|^2 \le \|\boldsymbol{p}_k\|^2 \le a^{-1}A \|\boldsymbol{r}_k\|^2 . \tag{7.561}$$

Proof: The first inequality follows from Lemma 7.32. The second one then follows from the fact that $\boldsymbol{p}_k$ and $\boldsymbol{p}_{k-1}$ are conjugate.

The next theorem gives the global convergence of the NCG algorithm (Algorithm 7.3).

Theorem 7.29: Let z_o be an initial approximation. Under the above assumptions, the sequence $\{z_k\}$, generated by Algorithm 7.3 converges to the unique solution $\underline{z}$ of the non-linear system of equations $F(z) = 0$,

$$L_\Theta(\underline{z}) = \min\{L_\Theta(z), \quad z \in R^n\}$$

and

$$\left\|z_k - \underline{z}\right\|_1 \le a^{-1} \|\boldsymbol{r}_k\|_1 = a^{-1} \|F(z_k)\|_1 . \tag{7.562}$$

Proof: Let ${}^kL_\Theta(\alpha) = L_\Theta(z_k + \alpha \boldsymbol{p}_k)$. Then

$$ {}^kL'_{\Theta'}(\alpha) = (F(z_k + \alpha\boldsymbol{p}_k), \boldsymbol{p}_k),$$

$$a\|\boldsymbol{p}_k\|^2 \le {}^kL''_\Theta(\alpha) = (F'(z_k + \alpha\boldsymbol{p}_k), \boldsymbol{p}_k) \le A\|\boldsymbol{p}_k\|^2 .$$

Since $L'_\Theta(0) = -(\boldsymbol{r}_k, \boldsymbol{p}_k) = -\|\boldsymbol{r}_k\|^2 < 0$, then α_k exists and

$$\alpha_k \ge A^{-1}\|\boldsymbol{r}_k\|^2 \|\boldsymbol{p}_k\|^{-2} > aA^{-2}.$$

Let $0 \le t \le 1$, then for some t

$$\begin{aligned} L_\Theta(z_k + \alpha\boldsymbol{p}_k) &= L_\Theta(z_k) + \alpha(F(z_k), \boldsymbol{p}_k) + \tfrac{1}{2}\alpha^2(F'(z_k + t\alpha\boldsymbol{p}_k), \boldsymbol{p}_k) \\ &\le L_\Theta(z_k) - \alpha\|\boldsymbol{r}_k\|^2 + \tfrac{1}{2}\alpha^2 a^{-1}A^2\|\boldsymbol{r}_k\|^2 \quad \forall\alpha \le \alpha_k. \end{aligned}$$

Hence

$$\begin{aligned} L_\Theta(z_{k+1}) &\le L_\Theta(z_k + aA^{-2}\boldsymbol{p}_k) \le L_\Theta(z_k) - aA^{-2}\|\boldsymbol{r}_k\|^2 + \tfrac{1}{2}aA^{-2}\|\boldsymbol{r}_k\|^2 \\ &\le L_\Theta(z_k) - \tfrac{1}{2}aA^{-2}\|\boldsymbol{r}_k\|^2 . \end{aligned}$$

Since the functional $L_\Theta(z)$ is bounded below, then $F(z_k) \to 0$. But

$$L_\Theta(z) \ge L_\Theta(z_o) - \|z - z_o\| \, \|F(z_o)\| + \tfrac{1}{2}a\|z - z_o\|^2 .$$

The set of z for which $L_\Theta(z) \le L_\Theta(z_o)$ is bounded, then $\left\|z_{k+j} - z_k\right\|$ is bounded. Since

$$a\left\|z_{k+j} - z_k\right\|^2 \le (F(z_{k+j}) - F(z_k), z_{k+j} - z_k) \to 0,$$

then there exists $\underline{z}$ such that $z_k \to \underline{z}$ and $F(\underline{z}) = 0$, $L_\Theta(\underline{z}) = \min\{L_\Theta(z), z \in R^n\}$, where $\underline{z}$ is the minimizer of $L_\Theta(z)$.

To prove that $\underline{z}$ is unique, then due to Lemma 7.30, we obtain

$$\|F(\boldsymbol{x}) - F(\boldsymbol{y})\| \, \|\boldsymbol{x} - \boldsymbol{y}\| \ge (F(\boldsymbol{x}) - F(\boldsymbol{y}), \boldsymbol{x} - \boldsymbol{y}) \ge a(\boldsymbol{x} - \boldsymbol{y}, \boldsymbol{x} - \boldsymbol{y}) \ge a\|\boldsymbol{x} - \boldsymbol{y}\|^2 .$$

Putting $\boldsymbol{x} = \underline{z}$, $\boldsymbol{y} = z_k$ the uniqueness follows immediately.

Lemma 7.34: Given any initial vector $z_o \in R^n$ all iterations are well defined and $\|\boldsymbol{r}_{k+1}\|_1 \le \gamma\|\boldsymbol{r}_k\|_1$.

Proof: It holds that

$$\|\boldsymbol{r}_{k+1}\| \le \|\boldsymbol{r}_{k+1} - \boldsymbol{r}_k\| + \|\boldsymbol{r}_k\| \le A\|z_{k+1} - z_k\| + \|\boldsymbol{r}_k\| \le A\|\alpha_k\boldsymbol{p}_k\| + \|\boldsymbol{r}_k\| \le (1 + a^{-1}A)\|\boldsymbol{r}_k\| = \gamma\|\boldsymbol{r}_k\| ,$$

where we used Lemma 7.33 and the fact that $\|\boldsymbol{p}_k\|^2 \le a^{-1}A\|\boldsymbol{r}_k\|^2$.

For the method the preconditioned technique can also be used.

7.6.2.3. *Solutions for the contact part of the problem based on constrained conjugate gradient methods*

The elastic part of the problem leads to minimizing the functional

$$L(\boldsymbol{v}) = \tfrac{1}{2}(\boldsymbol{v}, C\boldsymbol{v}) + (\boldsymbol{b}, \boldsymbol{v}), \tag{7.563}$$

$$A\boldsymbol{v} \le \boldsymbol{d}, \quad \text{i.e.} \quad (\boldsymbol{a}_i, \boldsymbol{x}) - d_i \le 0, \quad i \in J^-, \tag{7.564}$$

$$(\boldsymbol{a}_i, \boldsymbol{x}) - d_i = 0, \quad i \in J^o, \tag{7.565}$$

where $\boldsymbol{x}, \boldsymbol{a}_i, \boldsymbol{d} \in R^n$, C is the $n \times n$ matrix, symmetric and positive (semi-) definite, $d_i \in R^1$, $J = J^- \cup J^o$.

The effective numerical method solving constrained function minimization problems is the method of conjugate gradients. The main idea of the application of this method is based on the active set method and was discussed in Chapter 2 and in Section 7.5. In this section we shall limit ourselves to the positive definite case only. The derivation of the algorithm used is similar to that in Section 7.5 and therefore we give the result only.

Similarly as in the previous section we shall use the following notation: $J \equiv J(\boldsymbol{x}) = \{i \mid (\boldsymbol{a}_i, \boldsymbol{x}) - d_i = 0,\ i \in J^- \cup J^o\}$, where $\boldsymbol{a}_i$ are linearly independent vectors. If $J^o \neq 0$ then $J(\boldsymbol{x})$ is non-empty and $\boldsymbol{x}$ satisfies Equations 7.564 and 7.565. We construct the matrix A_J such that the rows are vectors $\boldsymbol{a}_i$, $i \in J$. We denote by $f'(\boldsymbol{x})$ the gradient of $f(\boldsymbol{x})$, i.e. $f'(\boldsymbol{x}) = C\boldsymbol{x}+\boldsymbol{b}$. By P we denote the operator defined as $P = A_J^T(A_J A_J^T)^{-1}A_J$ and by $\boldsymbol{u}(\boldsymbol{x})$ the vector $\boldsymbol{u} = (u_i) \in R^m$, defined as $\boldsymbol{u} = -(A_J A_J^T)^{-1}A_J f'(\boldsymbol{x})$. Then we have the following theorem.

Theorem 7.30: The problem of minimization of the quadratic functional $f(\boldsymbol{x})$ of Equation 7.563 with constraints $(\boldsymbol{a}_i, \boldsymbol{x}) - d_i = 0$, $i \in J$, is solved after a finite number of iteration steps by the following process.

Let the initial guess $\boldsymbol{x}_o$ satisfying (Equation 7.565) be given. Let $J(\boldsymbol{x}_o) = J_o$. Then

$$\boldsymbol{p}_1 = -(I - P_{J_o})f'(\boldsymbol{x}_o), \tag{7.566}$$

$$\boldsymbol{x}_{k+1} = \boldsymbol{x}_k + \alpha_{k+1}\boldsymbol{p}_{k+1}, \tag{7.567}$$

$$\beta_k = \frac{\left\|(I - P_{J_o})f'(\boldsymbol{x}_k)\right\|^2}{\left\|(I - P_{J_o})f'(\boldsymbol{x}_{k-1})\right\|}, \tag{7.568}$$

$$\boldsymbol{p}_{k+1} = -(I - P_{J_o})f'(\boldsymbol{x}_k) + \beta_k\boldsymbol{p}_k, \tag{7.569}$$

$$\alpha_{k+1} = -(f'(\boldsymbol{x}_k), \boldsymbol{p}_{k+1})/(\boldsymbol{p}_{k+1}, C\boldsymbol{p}_{k+1}), \quad k = 0, 1, 2, \ldots\ldots. \tag{7.570}$$

The additional part of the problem of quadratic programming represents the problem of minimization of Equation 7.563 with linear constraints

$$A\boldsymbol{x} - \boldsymbol{d} \le 0,\ (\boldsymbol{a}_i, \boldsymbol{x}) - d_i = \boldsymbol{a}_i^T\boldsymbol{x} - d_i \le 0, \quad i \in J^-,$$

$$\text{i.e. } (\boldsymbol{a}_i, \boldsymbol{x}) - d_i = \boldsymbol{a}_i^T\boldsymbol{x} - d_i = 0, \quad i \in J^o.$$

The algorithm is the following.

Algorithm 7.4: This algorithm consist of:

(i) let us take $\boldsymbol{x}_o \in R^n$ such that $\boldsymbol{x}_o$ satisfies Equations 7.564 and 7.565;

(ii) we determine the set $J(\boldsymbol{x}_o) \equiv J_o$ and form a matrix A_{J_o} and the operator P_{J_o};

(iii) we compute the vector $(I - P_{J_o})f'(\boldsymbol{x}_o)$;

(iv) we have two cases:

(a) If $(I - P_{J_o})f'(\boldsymbol{x}_o) \neq 0$, then the algorithm in Equations 7.566 to 7.570 is applied, and we calculate

$$\underline{\alpha}_{k+1} = \min_i \frac{d_i - (\boldsymbol{a}_i, \boldsymbol{x}_k)}{(\boldsymbol{a}_i, \boldsymbol{p}_{k+1})}, \quad \forall i \text{ for which } (\boldsymbol{a}_i, \boldsymbol{p}_{k+1}) > 0,$$

where $\boldsymbol{x}_k$ is the point constructed by the CG-algorithm and $\boldsymbol{p}_{k+1}$ is the conjugate direction at $\boldsymbol{x}_{k+1}$.
Let α_{k+1} represent the value of α in the k-th iterate of the CG-method. Then we shall study two cases:

(1) the case $\alpha_{k+1} \leq \underline{\alpha}_{k+1}$, then $\boldsymbol{x}_{k+1} = \boldsymbol{x}_k + \alpha_{k+1}\boldsymbol{p}_{k+1}$ and the iteration CG-process continues. If $\boldsymbol{x}_{k+1}$ is a point of minimum of Equation 7.563 with constraints given by Equation 7.565 $i \in J_o$, then $\boldsymbol{x}_o = \boldsymbol{x}_{k+1}$ and go to (v);

(2) the case $\alpha_k > \underline{\alpha}_k$, then $\boldsymbol{x}_{k+1} = \boldsymbol{x}_k + \underline{\alpha}_{k+1}\boldsymbol{p}_{k+1}$ and $\boldsymbol{x}_o = \boldsymbol{x}_{k+1}$ and go to (ii);

(b) If $(I - P_{J_o})f'(\boldsymbol{x}_o) = 0$ go to (v);

(v) compute $\boldsymbol{u}(\boldsymbol{x}_o) = -(A_{J_o}A_{J_o}^T)^{-1}A_{J_o}f'(\boldsymbol{x}_o)$. If $(\boldsymbol{u}(\boldsymbol{x}_o))_i \geq 0\,\forall i \in J_o \cap J^-$ then we find the minimum and the process finishes.
If there exists $i \in J_o \cap J^-$ such that $(\boldsymbol{u}(\boldsymbol{x}_o))_i < 0$ then we set $J_o = J_o - \{i\}$ and go to (iii).

It is evident that this step is a limiting factor for efficiency of the Uzawa algorithm.

Lemma 7.35: The sequence $\{f(\boldsymbol{x}_k)\}$, $k = 0, 1, 2, \ldots$, obtained from the sequence $\{\boldsymbol{x}_k\}$ by Algorithm 7.4 is strongly decreasing.

The algorithm can be improved by using the preconditioned technique (Kestřánek (1995)).

REFERENCES

Axelsson, O. (1982), "On Global Convergence of Iterative Methods", in *Iterative Solution of Nonlinear Systems of Equations* (Eds. Ansorge, R., Meis, T. and Törning, E., Springer Verlag, 1-13.

Axelsson, O. (1994), *Iterative Solution Methods*, Cambridge, New York, Melbourne.

Axelsson, O. and Chronopoulos, A.T. (1994), "On Nonlinear Generalized Conjugate Gradient Methods", *Numerische Mathematik*, **69**, 1-15.

Axelsson, O. and Koporin, I.E. (1995), "On the Solution of Non-Linear Equations for Nondifferentiable Mappings." In: Hacktush, W. and Wittum G. (Eds), Fast Solvers for Flow Problems, *NNFM* **49**, Vieweg Vlq, Braunschweig Wiesbaden 1-14.

Axelsson, O. and Vassilevski, P.S. (1992), "Construction of Variable-step Preconditioners for Inner-outer Iteration Methods", in *Iterative Methods in Linear Algebra* (Eds Beauwens, R. and deGroen, P.), North-Holland, 1-14.

Aziz, A.K. (Ed.) (1972), *The Mathematical Foundations of the Finite Element Method with Applications to Partial Differential Equations*, Academic Press, New York.

Barbu, V. (1976), *Nonlinear Semigroup and Differential Equations in Banach Spaces*, Noordhoff International Publishing, Leyden.

Bramble, J.H. and Zlámal, M. (1970), "Triangular Elements in the Finite Element Method", *Math. Comput.*, **24, 112**, 809-820.

Brousse, P. (1988), *Optimization in Mechanics: Problems and Methods*, North-Holland, Amsterdam, New York, Oxford, Tokyo.

Céa, J. (1971), *Optimisation: Théorie et Algorithmes*, Dunod, Paris.

Chronopoulos, A.T. (1992), "A Non-linear CG-like Iterative Method," *J. Comp. Appl. Math.*, **40**.

Ciarlet, P. (1978), *The Finite Element Method for Elliptic Problems*, North-Holland, Amsterdam.

Ciavaldini, J.F. (1975), "Analyse Numérique d'un Probléme de Stefan á Deux Phases par un Méthod d'éléments Finis," *SIAM J. Numer. Anal.*, **12**, 464-489.

Daniel, J.W. (1967a), "Convergence of the Conjugate Gradient Method with Computationally Convenient Modifications", *Numer. Math.*, **10**, 125-131.

Daniel, J.W. (1967b), "The Conjugate Gradient Method for Linear and Non-linear Operator Equations." *SIAM J. Numer. Anal.*, **4, 1**, 10-26.

Daniel, J.W. (1971), *The Approximate Minimization of Functionals*, Publishers Englewood Cliff, NJ.

Dvořák, J. (1993), *Numerical Solution of Stefan Problem in Inhomogeneous Media*, Diploma Thesis, Faculty of Mathematics and Physics of the Charles University, Prague (in Czech).

Dvořák, J. (1996), "Numerical Modelling of the Olivine-Spinel Phase Change Based on the Stefan-like Problem." *Geomechanics 96*, Balkema 181-184.

Ekeland, I. and Temam, R. (1976), *Convex Analysis and Variational Problems*, North-Holland, Amsterdam.

Elliott, Ch.M. (1981), "On the Finite Element Approximation of an Elliptic Variational Inequality Arising from an Implicit Time Discretization of the Stefan Problem", *IMA J. Numer. Anal.*, **1**, 115-125.

Falk, R.S. (1974), "Error Estimates for Approximation of a Class of Variational Inequalities", *Math. of Comp.*, **28**, 963-971.

Fiedler, M. (1981), *Special Matrices and Their Application in Numerical Mathematics.*, SNTL, Prague (in Czech).

Fletcher, R. (1981), *Practical Methods of Optimization Vol. 2: Constrained Optimization*, J. Wiley & Sons, Chichester, New York, Brisbane, Toronto.

Fučík, S., Kratochvíl, A. and Nečas, J. (1975), "Kačanov's Method and Its Applications", *Revue Roumaine de Mathém. Pures et Appliquées*, **20, 8**, 909-915, Bucharest.

Glowinski, R. (1979), "Finite Elements and Variational Inequalities", in, *The Mathematics of Finite Elements and Applications III* (Ed: Whiteman J.R.), Academic Press, New York, 135-171.

Haslinger, J. and Tvrdý, M. (1983), "Approximation and Numerical Solution of Contact Problems with Friction", *Apl. Mat.*, **28, 1**, 55-71.

Hlaváček, I. and Lovíšek, J. (1977), "A Finite Element Analysis for the Signorini Problem in Plane Elastostatics", *Apl.Mat.*, **22**, 215-228.

Hlaváček, I., Haslinger, J., Nečas, J. and Lovíšek, J. (1982), *Solution of Variational Inequalities in Mechanics*, ALFA, Bratislava, SNTL, Prague.

Jarušek, J. (1983), "Contact Problems with Bounded Friction: Coercive Case", *Czech. Math. J.*, **33, 2**, 237-261.

Kestřánek, Z. (1995), "Numerical Analysis of the Contact Problem: Comparison of Methods for Finding the Approximate Solution", *Technical Report No V-648*, Inst. of Computer Science AS CR, Prague.

Kestřánek, Z. and Nedoma, J. (1996), "Numerical Simulation of a Bridge on a Nonstable Slope: Comparison of Results Based on the Contact Problem Without and With Friction", *Geodynamics 96*, Balkema.

Kestřánek, Z., Dvořák, J. and Nedoma, J. (1997). Iterative Solvers for Coupled Contact-Stefan-like Problems. In: Sydow, A. Seliofex, R.P., Rufeges, W., and Lekman, H. Computational Physics, Chemistry and Biology, Proc. 15th IMACS World Compress on Scientific Computation, Modelling and Applied Mathematics, Berlin, August 1997, Vol. 3, Wissenschaft and Technik Verlag, Berlin, 467-471.

Kikuchi, N. and Oden, J.T. (1988). *Contact Problems in Elasticity: A Study of Variational Inequalities and Finite Element Methods*, SIAM, Philadelphia.

Luenberger, D.G. (1973), *Introduction to Linear and Nonlinear Programming*, Addison-Wesley Publishing Company, Reading, Massachusetts, Menlo Park, California, London, Amsterdam, Ontario, Sydney.

Luenberger, D.G. (1984), *Linear and Nonlinear Programing*, Addison-Wesley Publishing Company, Reading, Massachusetts, Menlo Park, California, London, Amsterdam, Ontario, Sydney.

Mosco, U. and Strang, G. (1974), "One-Sided Approximations and Variational Inequalities." *Bull. Amer. Math. Soc.* **80**, 308-312.

Nečas, J. and Hlaváček, I. (1981), *Mathematical Theory of Elastic and Elasto-Plastic Bodies: An Introduction*, Elsevier, Amsterdam, Oxford, New York.

Nečas, J. and Hlaváček, I. (1983), "Solution of Signorini's Contact Problem in the Deformation Theory of Plasticity by Secant Modules Method," *Apl. Mat.*, **28, 3**, 169-214.

Nedoma, J. (1980), "Model of the Carpathian Continent/Continent Collision." *Proc. of the 17th Assembly of the ESC*, Budapest.

Nedoma, J. (1982a), "Thermo-elastic Stress-Strain Analysis of the Geodynamic Mechanism", Gerlands Beitr. Geophysik, **91, 1**, 75-89.

Nedoma, J. (1982b), "Contribution to the Study of the Geotectonic Stress Field in the Region of the West Carpathians", in: *Proc. of the 2nd Int. Symposium on the Analysis of Seismicity and Seismic Risk* (Liblice Castle May 18-23, 1981) (Ed. Schenkova, Z.), Geophysical Institute Czech. Acad. Sci., Prague, 328-338.

Nedoma, J. (1983), "On One Type of Signorini Problem Without Friction in Linear Thermoelasticity", *Apl. mat.*, **28, 6**, 393-407.

Nedoma, J. (1987), "On the Signorini Problem with Friction in Linear Thermoelasticity: The Quasi-Coupled 2D-case", *Apl. Mat.*, **32, 3**, 186-199.

Nedoma, J. (1990), "Tectonic Evolution of Collision Zones, I. Mechanics of Tectonic Processes, II Subducted Lithosphere, *Gerlands Beitr. Geophysics*, (I) **99, 2**, 97-108, (II) **99, 2**, 109-121.

Nedoma, J. (1993), "Finite Element Analysis in Nuclear Safety", *Technical Report V-550*, Inst. of Computer Science AS CR, Prague.

Nedoma, J. (1994), "Finite Element Analysis of Contact Problems in Thermoelasticity: The Semi-Coercive Case", *J. Comput. Appl. Math.*, **50**, 411-423.

Nedoma, J. (1995), "Equations of Magnetodynamics of Incompressible Thermo-Bingham's Fluid Under the Gravity Effect", *J. Comput. Appl. Math.*, **59**, 109-128.

Nedoma, J. (1997a), "On a Coupled Stefan-Like Problem in the Thermo-Visco-Plastic Rheology", *J. Comput. Appl. Math.*, **84**, 45-80.

Nedoma, J. (1997b). "Finite Element Approximation of a Coupled Contact-Stefan-like Problem Arising from the Time Discretization in Deformation Theory of Thermo-Plasticity", 7th Int. Congress on Computational and Applied Mathematics, Leuven July 1996, *J. Comput. Appl. Math.*, **82**, 313-334.

Nedoma, J. and Dvořák, J. (1995), "On the FEM Solution of a Coupled Contact-Two-Phase Stefan Problem in Thermo-Elasticity: Coercive Case", *J. Comput. Appl. Math.*, **63**, 411-420.

Nowacki, W. (1970), *Dynamic Problems of Thermoelasticity*, Mir, Moscow (in Russian).

Ortega, J.M. and Rheibolt, W.C. (1970), *Iterative Solution of Nonlinear Equations in Several Variables*, Academic Press, New York.

Pschenichny, B.N. and Danilin, Yu. M. (1978), *Numerical Methods in Extremal Problems*. Mir, Moscow.

Raviart, P.A. (1972), "The Use of Numerical Integration in Finite Element Methods for Solving Parabolic Equations", Conference on Numerical Analysis, Royal Irish Academy, Dublin.

Tvrdý, M. (1980), *The Signorini Problem with Friction*, Thesis, Fac. Math. Phys., Charles Univeristy, Prague (in Czech).

8

Numerical Simulation of Tectonic Evolution of Collision and Subduction Zones

8.1. SIMULATION OF TECTONIC PROCESSES IN THE TIME DOMAIN: ORIGIN OF THE EARTHQUAKE

8.1.1. Introduction

The aim of this chapter is the discussion of the mechanism of geodynamic processes in collision and subduction zones, their tectonic evolution and their mathematical simulation. It will be shown that these types of problems lead to solving contact problems of Signorini type with friction in thermo-elasticity and thermo-plasticity discussed in the previous chapter, as well as their generalization in the time domain. The problems discussed lead to the variational inequalities in the linear coupled thermo-elasticity and thermo-plasticity. The numerical solution based on the finite element method will be briefly considered. On the basis of the algorithm discussed, the theory of earthquake and mine shock mechanisms will be covered.

The present plate tectonic hypothesis assumes that the lithosphere is divided into a small number of rigid plates and blocks which are in a collision. There exist large regions of continental and oceanic nature where the lithosphere undergoes large-scale deformations and a diffuse seismicity prevails. Nedoma (1984a), (1990a) draws attention to the fact that, due to friction in the collision zone, the overriding and subducting plates are heated and melted, and then create the volcanic chambers. The mechanics of the subducted lithosphere is closely connected with the geodynamic processes at the boundary between the core and the mantle (Nedoma, 1986a-c). As it was shown in Chapter 4, tectonic evolution of collision zones has four stages. In this chapter we shall analyse the fourth stage of tectonic evolution of collision zones, which is represented by the models of subduction and collision.

Generally, these problems lead to solving problems of Signorini type in thermo-elasto-visco-plasticity. The lithospheric rocks are characterized by the creep, so that the strain rate $D_{ij} = e_{ij}(\dot{\boldsymbol{u}})$ is a non-linear function of stress and temperature. But for the case when time periods are shorter than a characteristic time t_c,

$$t_c = \mu^{-1}(\tau_\varepsilon + ((\tfrac{2}{3}D_{ij}D_{ij})^{1/2}\gamma^{-1})^{1/n}(3\sqrt{2D}_{ij}D_{ij})^{-1/2},$$

where τ_ε is the yield stress and $\gamma = L^{-1}\int_L \gamma_o \exp(-Q/RT)\mathrm{d}x$ (where $\gamma \to \infty$ for the rigid perfectly plastic flow), Q is an activation energy, R is the universal gas constant, L is the thickness of the lithosphere and μ is the rigidity, the lithosphere behaves elastically and the geodynamic problem can be studied in the same way as the thermo-elastic one. On the other hand the model problem investigated also involves, as its consequence, the theory of the earthquake. For a numerical solution of the problem the method of discretization in time and the finite element method will be used.

8.1.2. The Model and Constituting Equations

In our considerations we shall use a model of subducted lithosphere of Cordillerian type (e.g. models of the Andes or Cordilleras) and of collision type (e.g. models of the Himalayas, Carpathians). In both types of model, as well as the uneven surface of the subducted or collided lithospheric plates the main role is played by the Coulombian type friction acting at the contact boundaries between colliding lithospheric plates and blocks. Seismological analyses show that the contact zone in the domain of the subducted lithosphere is also characterized by a zone of deep earthquake foci, the so-called Wadati-Benioff zone, which, according to several authors, is divided by aseismic faults into two or more seismically active zones. The subducted lithosphere reaches different depths of 100 km to 700 km, so that it reaches down to the mesosphere. As observed, the subducted lithosphere is broken up and in the mesosphere the emerged and broken-up parts of the subducted oceanic lithosphere are observed in the close neighbourhood of the subduction zones.

To study geodynamic processes in the lithosphere, namely the evolution of the collision zones, we shall find governing equations and conditions describing the behaviour of the colliding plates and blocks in the collision zones.

Let us consider lithospheric plates and blocks, consisting of several deformable bodies Ω^ι, $\iota = 1, \ldots, s$, which are in a mutual contact. Let us assume that on common surfaces a friction obeying the Coulombian law acts. Then we shall use the following notation: Ω and $\partial\Omega$ are an open domain in R^N and its boundary, $G = \Omega \times I$, and $\partial G = \partial\Omega \times I$ denote a time-space domain and its boundary, N is the dimension of the space, $\boldsymbol{x} = (x_i)$ is a position vector, $\boldsymbol{n}$ is an outward normal to the boundary, Γ_τ, Γ_u, and Γ_α are parts of the boundary $\partial\Omega$ with, $\partial\Omega = \overline{\Gamma}_\tau \cup \overline{\Gamma}_u \cup \overline{\Gamma}_c$, $\Gamma_\tau \neq 0$, $\Gamma_u \neq 0$, $\Gamma_c \neq 0$, t_o and t_1 denote initial and final values of a time interval $\bar{I} = \langle t_o, t_1\rangle$, T and T_o denote a temperature and initial temperature, W and Q are thermal sources, c is a specific heat, κ_{ij} a thermal conductivity tensor, q denotes a heat flow, ρ a density, β_{ij} a coefficient of thermal expansion, $\boldsymbol{f}$ and $\boldsymbol{F}$ denote vectors of body forces, c_{ijkl} denote elastic coefficients, $\boldsymbol{u}$ and $\boldsymbol{u}_o$ a displacement vector and an initial displacement vector, $\boldsymbol{u}_1$ an initial velocity vector, u_n and $\boldsymbol{u}_t$

denote normal and tangential components of the displacement vector, τ_{ij} and τ denote a stress tensor and a stress vector, τ_n and $\vec{\tau}_t$ are normal and tangential components of the stress vector, e_{ij} is a small strain tensor, $D_{ij} \equiv e_{ij,t} \equiv e_{ij}(\dot{\boldsymbol{u}})$ is a strain rate tensor, $\boldsymbol{u}_{,t} \equiv \dot{\boldsymbol{u}} \equiv \boldsymbol{u}' \equiv v$ and $\boldsymbol{u}'' \equiv \boldsymbol{u}_{,tt}$ are velocity and acceleration vectors, $\boldsymbol{P}$ denotes a surface force vector, $\mathscr{F}_c^{kl}$ a coefficient of the Coulombian friction, z and $\underline{\boldsymbol{w}}$ are sufficiently smooth scalar and vector functions in the investigated domain, w and $\boldsymbol{v}$ are test scalar and vector functions, ${}^1V$ denotes a space of virtual temperatures, V a space of virtual displacements, K a set of admissible displacements, and ${}^1V'$ and V' are their dual spaces.

Suppose that elastic bodies occupy domains $\Omega^\iota \subset R^N$, $\iota = 1, \ldots, s$, $N = 2, 3$, with Lipschitz-continuous boundaries. Let $\Omega = \cup_{\iota=1}\Omega^\iota$ and let its boundary $\partial\Omega$ be divided into disjoint parts Γ_τ, Γ_u and Γ_c such that $\partial\Omega = \overline{\Gamma}_\tau \cup \overline{\Gamma}_u \cup \overline{\Gamma}_c$, $\Gamma_\tau \neq 0$, $\Gamma_u \neq 0$, $\Gamma_c \neq 0$ (then problems will be coercive iff $\Gamma_u \neq 0$ and are semi-coercive iff at least one of part of $\Gamma_u = 0$). Let $G = \Omega \times (t_o, t_1)$ and $\partial G = \partial\Omega \times (t_o, t_1)$. Let $\boldsymbol{x} = (x_i)$ be the Cartesian coordinates.

Let us look for the displacement vector $u^\iota = u^\iota(\boldsymbol{x}, t) = (u_i^\iota(\boldsymbol{x}, t))$ on Ω^ι, $i = 1, \ldots, N$, $\iota = 1, \ldots, s$, and the temperature $T^\iota = T^\iota(\boldsymbol{x}, t)$ on Ω^ι, $\iota = 1, \ldots, s$. Let $e_{ij}(\boldsymbol{u}^\iota)$ be the small strain tensor defined by

$$e_{ij}(\boldsymbol{u}^\iota) = \tfrac{1}{2}(u_{i,j}^\iota + u_{j,i}^\iota), \quad \iota = 1, \ldots, s, \tag{8.1}$$

where $u_{i,j} = \partial u_i/\partial x_j$.

On every Ω^ι, $\iota = 1, \ldots, s$, let the Duhamel-Neumann's law

$$\tau_{ij}^\iota = c_{ijkl}^\iota(\boldsymbol{x}, t)e_{kl}(\boldsymbol{u}^\iota) - \beta_{ij}^\iota(\boldsymbol{x}, t)(T^\iota(\boldsymbol{x}, t) - T_o^\iota(\boldsymbol{x})) \quad i, j = 1, \ldots, N, \quad \iota = 1, \ldots, s \tag{8.2}$$

(in the sense of the obvious Einstein's summation convention over repeated indices) be fulfilled, where $\tau_{ij}^\iota = \tau_{ij}^\iota(\boldsymbol{x}, t)$ is the stress tensor defined on every Ω^ι, $T_o^\iota(\boldsymbol{x})$ is an input temperature at which the materials are in initial strain and stress state, and $\beta_{ij}^\iota(\boldsymbol{x}, t)$ is the coefficient of a thermal expansion. Elastic coefficients $c_{ijkl}^\iota(\boldsymbol{x}, t)$ satisfy the obvious symmetric conditions on Ω^ι, $\iota = 1, \ldots, s$, and the conditions of Lipschitz continuity and ellipticity

$$c_{ijkl}^\iota(\boldsymbol{x}, t) = c_{jikl}^\iota(\boldsymbol{x}, t) = c_{ijlk}^\iota(\boldsymbol{x}, t) = c_{klij}^\iota(\boldsymbol{x}, t), \tag{8.3}$$

$$0 < \mu_o \leq c_{ijkl}^\iota(\boldsymbol{x}, t)\xi_{ij}\xi_{kl}\,|\xi|^{-2} \leq M_o < +\infty \tag{8.4}$$

for every $\boldsymbol{x} \in \Omega^\iota$ and all t (e.g. $\forall t$) and $\xi = (\xi_{ij})_{i,j=1}^N \in R^{N^2}$, $\mu_o = 2\min\{\mu(\boldsymbol{x}, t); \boldsymbol{x} \in \Omega, \forall t\}$ and $M_o = \max\{2\mu(\boldsymbol{x}, t) + 3\lambda(\boldsymbol{x}, t); \boldsymbol{x} \in W, \forall t\}$ are constants independent of $\boldsymbol{x} \in \Omega^\iota$. For each $\iota = 1, \ldots, s$ let the stress tensor τ_{ij} satisfy the equation of motion (see e.g. Nowacki (1970))

$$\tau_{ij,j}^\iota + f_i^\iota = \rho^\iota \mathrm{d}v_i^\iota/\mathrm{d}t, \quad i, j = 1, \ldots, N, \quad \iota = 1, \ldots, s \text{ in } G^\iota, \tag{8.5}$$

and the temperature T satisfies the expanded equation of heat conduction (see e.g. Nowacki (1970), Nedoma (1982))

$$\rho^\iota \beta_{ij}^\iota T_o^\iota e_{ij,t}(u^\iota) + \rho^\iota c^\iota \mathrm{d}T^\iota/\mathrm{d}t = W^\iota + (\kappa_{ij}^\iota T^\iota{}_{,j})_{,i} \text{ in } G^\iota, \quad i, j = 1, \ldots, N, \ \iota = 1, \ldots, s, \tag{8.6}$$

where the first term on the left hand side represents the energy dissipated in the form of heat in the moving masses and generated by their deformation, the other terms represent the classical expanded equation of heat conduction, $W^{\iota}(\boldsymbol{x}, t)$ are the heat sources, $\kappa^{\iota}_{ij}(\boldsymbol{x}, t)$ is a thermal conductivity tensor satisfying

$$\kappa^{\iota}_{ij} = \kappa^{\iota}_{ji}, \tag{8.7}$$

$$\kappa^{\iota}_{ij}(\boldsymbol{x}, t)\zeta^{\iota}_i\zeta^{\iota}_j \geq \kappa^{\iota}_o \left|\zeta^{\iota}\right|^2, \kappa^{\iota}_o = \text{constant.} > 0, \tag{8.8}$$

for every $\boldsymbol{x} \in \Omega^{\iota}$ and $\forall t \in I$ and $\zeta^{\iota} = (\zeta^{\iota}_i)^N_{i=1} \in R^N$, $\kappa^{\iota}_o > 0$ is a constant independent of $\boldsymbol{x} \in \Omega^{\iota}$ and $\zeta^{\iota} \in R^N$, and, further, $\rho^{\iota}(\boldsymbol{x}, t)$ is a density, $c^{\iota}(\boldsymbol{x}, t)$ is a specific heat in Ω^{ι} and $f^{\iota}(\boldsymbol{x}, t)$ is a body force vector in Ω^{ι}, and we have $\mathrm{d}/\mathrm{d}t = \partial/\partial t + v_j\partial/\partial x_j$ and $\partial \boldsymbol{u}/\partial t = \boldsymbol{u}' \equiv \boldsymbol{u}_{,t}$. On the boundary $\partial\Omega$ the stress vector $\tau(\boldsymbol{x}, t)$ is already defined by

$$\tau_i(\boldsymbol{x}, t) \equiv \tau_i(\boldsymbol{u}) \equiv \tau_i(\boldsymbol{u}(\boldsymbol{x}, t)) = \tau_{ij}(\boldsymbol{u}(\boldsymbol{x}, t))n_j$$

and its normal and tangential components by

$$\tau_n(\boldsymbol{u}) = \tau_i(\boldsymbol{u})n_i = \tau_{ij}(\boldsymbol{u})n_j n_i, \vec{\tau}_t(\boldsymbol{u}) = \vec{\tau}(\boldsymbol{u}) - \tau_n(\boldsymbol{u})\boldsymbol{n}.$$

The normal and tangential displacement components are defined by

$$u_n = \boldsymbol{un} = (u_i n_i), \quad \boldsymbol{u}_t = \boldsymbol{u} - u_n\boldsymbol{n},$$

where $\boldsymbol{n} = (n_i)^N_{i=1}$ is the unit outward normal vector, and the heat flow q is defined by

$$q = \kappa_{ij}T_{,j}\, n_i.$$

The boundary $\partial\Omega$ of Ω consists of three parts $\partial\Omega = \overline{\Gamma}_u \cup \overline{\Gamma}_\tau \cup \overline{\Gamma}_c$. On the boundary $\Gamma_u \times I$, the displacement and temperature are prescribed (or the heat flow is equal to zero), i.e.

$$\boldsymbol{u}(\boldsymbol{x}, t) = \boldsymbol{u}_2(\boldsymbol{x}, t) \tag{8.9}$$

and

$$T(\boldsymbol{u}, t) = T_2(\boldsymbol{x}, t) \text{ or} \tag{8.10}$$

$$\kappa_{ij}(\boldsymbol{x}, t)T_{,j}\, n_i = 0. \tag{8.11}$$

On the boundary $\Gamma_\tau \times I$, the surface loading and the temperature or the heat flow are prescribed, i.e.

$$\tau_{ij}(\boldsymbol{x}, t)n_j = P_{oi}(\boldsymbol{x}, t) \tag{8.12}$$

and

$$T(\boldsymbol{x}, t) = T_1(\boldsymbol{x}, t) \text{ or} \tag{8.13}$$

$$\kappa_{ij}(\boldsymbol{x}, t)T_{,j}(\boldsymbol{x}, t)n_i = q(\boldsymbol{x}, t). \tag{8.14}$$

To study geodynamic processes in the collision zone we must find conditions describing the behaviour of the colliding plates and blocks and their mutual motions, as well as deformation in the collision zones. To derive the conditions on the contact boundary between two colliding plates Ω^k and Ω^l we shall suppose that before the deformation both plates were in contact along the contact boundary Γ^{kl}_c.

We shall denote temperatures and displacement vectors in both colliding plates by $T^k, T^l, \boldsymbol{u}^k, \boldsymbol{u}^l$. Let (x, y, z), a local coordinate system at the point $O \in \Gamma_c^{kl}$ $(O \equiv O^k \equiv O^l$, $O \in \Gamma_c^{kl}$, $O^k \in \Gamma_c^k$, $O^l \in \Gamma_c^l$, $\Gamma_c^{kl} = \overline{\Gamma}_c^k \cap \overline{\Gamma}_c^l)$ and time $t \in \langle t_o, t_1 \rangle$, be placed in such a way that the x-axis coincides with the direction of the unit normal n_k, the z-axis with the common unit tangent of Γ_c^{kl} and the y-axis be perpendicular to the (x, z)-plane. Let Γ_c^{kl} be described by the function $h(y, z, t)$. During the deformation points O^k and O^l are transferred in a different way. Since the lithospheric plates are created from the thermo-elasto-visco-plastic rocks, both colliding plates cannot penetrate, i.e.

$$u_n^k(\boldsymbol{x}, t) - u_n^l(\boldsymbol{x}, t) \leq 0 \quad \text{at time } t \in (t_o, t_1).$$

Due to the law of action and reaction for the contact forces we have

$$\tau_n^k(\boldsymbol{x}, t) = -\tau_n^l(\boldsymbol{x}, t) = \tau_n(\boldsymbol{x}, t), \quad \vec{\tau}_t^k(\boldsymbol{x}, t) = -\vec{\tau}_t^l(\boldsymbol{x}, t) = \vec{\tau}_t(\boldsymbol{x}, t).$$

Since the normal forces are not tensile forces then

$$\tau_n^k(\boldsymbol{x}, t) = -\tau_n^l(\boldsymbol{x}, t) = \tau_n(\boldsymbol{x}, t) \leq 0 \quad \text{at time } t \in (t_o, t_1).$$

During the deformation, the colliding plates and blocks are in or not in some contact. If the colliding plates are not in contact then $u_n^k(\boldsymbol{x}, t) - u_n^l(\boldsymbol{x}, t) < 0$ and the contact forces are equal to zero, i.e. $\tau_n^k(\boldsymbol{x}, t) = -\tau_n^l(\boldsymbol{x}, t) = \tau_n(\boldsymbol{x}, t) = 0$ for every $t \in (t_o, t_1)$. If the plates and blocks are in contact then $u_n^k(\boldsymbol{x}, t) - u_n^l(\boldsymbol{x}, t) = 0$ and $\tau_n^k(\boldsymbol{x}, t) = -\tau_n^l(\boldsymbol{x}, t) = \tau_n^{kl}(\boldsymbol{x}, t) < 0$ for every $t \in (t_o, t_1)$. Both cases can be included in the condition

$$(u_n^k(\boldsymbol{x}, t) - u_n^l(\boldsymbol{x}, t))\tau_n^{kl}(\boldsymbol{x}, t) = 0 \quad \text{at every moment } t \in (t_o, t_1).$$

In the case when both plates are in contact, then Coulombian friction forces g_c act at the contact boundary Γ_c^{kl}. These friction forces g_c are in their absolute value proportional to the normal stress component, where the coefficient of proportionality is the coefficient of Coulombian friction $\mathscr{F}_c(\boldsymbol{x}, t)$, so that $g_c(\boldsymbol{x}, t) = \mathscr{F}_c(\boldsymbol{x}, t)\,|\tau_n(\boldsymbol{x}, t)|$, where the coefficient of the Coulombian friction $\mathscr{F}_c(\boldsymbol{x}, t)$ can be estimated by $(c_s/c_p)^{1/2}$, t-fixed, where c_s, and c_p are velocities of S and P waves (Nedoma(1987)). During the deformation of the lithospheric plates two situations can occur. In the first case, the absolute value of the tangential stress component $\vec{\tau}_t(\boldsymbol{x}, t)$ is less than the friction force $g_c(\boldsymbol{x}, t)$, and then the tangential force precludes the mutual motion of both colliding plates and in the lithosphere the deformation energy accumulates. In the second case, the absolute value of the tangential stress component $\vec{\tau}_t(\boldsymbol{x}, t)$ is equal to the friction force $g_c(\boldsymbol{x}, t)$, and then there are no forces which can preclude the mutual motion of both plates. Thus the contact points change their position in the direction opposite to that in which the tangential stress component acts. Then on the contact boundaries Γ_c^{kl} we have the following conditions

$$u_n^k - u_n^l \leq 0, \quad \tau_n^k = -\tau_n^l \leq 0, \quad \tau_n^k(u_n^k - u_n^l) = 0 \tag{8.15}$$

and

$$\vec{\tau}_t^k = \vec{\tau}_t^l \equiv \vec{\tau}_t^{kl}, \quad \left|\vec{\tau}_t^k\right| \leq \mathscr{F}_c^{kl}\left|\tau_n^k\right| \equiv g_c^{kl} \tag{8.16}$$

$$\left|\vec{\tau}_t^k\right| < \mathscr{F}_c^{kl}\left|\tau_n^k\right| \Rightarrow \partial/\partial t(\boldsymbol{u}_t^k(\boldsymbol{x},t) - \boldsymbol{u}_t^l(\boldsymbol{x},t)) = 0 \tag{8.17}$$

$$\left|\vec{\tau}_t^k\right| = \mathscr{F}_c^{kl}\left|\tau_n^k\right| \Rightarrow \exists\Theta^k \geq 0 \text{ such that } \partial/\partial t(\boldsymbol{u}_t^k(x,t) - \boldsymbol{u}_t^l(x,t)) = -\Theta^k\vec{\tau}_t^k(\boldsymbol{x},t) \tag{8.18}$$

(the so-called Coulombian law of friction)

$$T^k = T^l, \tag{8.19}$$

$$\kappa_{ij}T_{,j}\,n_{i|(k)} = \kappa_{ij}T_{,j}\,n_{i|(l)}, \quad i,j=1,\ldots,N, l\neq k, \quad \iota=1,\ldots,s, \iota=k, \quad \forall t, \tag{8.20}$$

where for the couple $\boldsymbol{u}^\iota = (u_i^\iota)$ $(i = 1,\ldots,N, \iota = 1,\ldots,s)$ the stress vector $\tau^\iota(\boldsymbol{x},t) = \tau^\iota(u^\iota(\boldsymbol{x},t))$ and

$$\tau_i^\iota(\boldsymbol{u}^\iota) = (-1)^r(\tau_{ij}^\iota(\boldsymbol{u}^\iota)n_j), \quad \iota = 1,\ldots,s \text{ with } r = 0 \text{ for } \iota = k \text{ and } r = 1 \text{ for } \iota = l,$$

and where $\boldsymbol{n}$ is the unit outward normal to Γ_c^{kl} with respect to Ω^k, the normal components of $\boldsymbol{u}^\iota$ and $\vec{\tau}^\iota$ are considered with respect to this vector, $\Theta^\iota \geq 0$ is a non-negative function on Γ_c and $\mathscr{F}_c^{kl}$ is the coefficient of Coulombian friction.

At the same time, due to the deformation of both colliding plates and blocks and to the acting Coulombian friction, the rocks in the contact zones are heated and melted, as well as recrystallized and, later, after losing tectonic motion and relaxation of the situation in the colliding zone, the rocks are solidified. Across any phase change boundary the conservation law of energy momentum and mass must be satisfied and the rock density is required to be continuous. In rocks there are two or more phases in which rock properties are markedly different owing to non-uniform heating and straining. The great interest is to find the boundary between two phases (e.g. liquid-solid, or between two recrystallized phases). If T represents temperature, then the latent heat is added in the case of melting and released in the case of recrystallization and solidification. Energy balance arguments then show that, although the temperature is continuous everywhere, the heat flux has a jump discontinuity across the phase change boundary $\Gamma_R(t)$— the head of the recrystallization wave in the case of recrystallization or the phase change boundary between solid and liquid phases.

Let $G_i(t)$, $i = 1,\ldots,n$, denote the unrecrystallized (unmelted) and recrystallized (melted) parts of the lithosphere, asthenosphere and mesosphere at time $t \in I$ and let $\Theta_i(t)$ denote the temperature in $G_i(t)$. Furthermore, let

$$\kappa_i = \lim_{\Theta\to T_R^-} \kappa(\Theta, p, \ldots) \text{ and } \kappa_j = \lim_{\Theta\to T_R^+} \kappa(\Theta, p, \ldots). \tag{8.21}$$

Parametrically the phase change boundary is described by $R(\boldsymbol{x},t) = 0$. At the phase change boundary it holds that

$$\Theta_i(\boldsymbol{x},t) = \Theta_j(\boldsymbol{x},t) = T_R, \quad q_i\boldsymbol{n} - q_j\boldsymbol{n} = \rho L v_n,$$

where T_R is a temperature on the phase change boundary, $v_n = \mathrm{d}R/\mathrm{d}t$ is the speed of the phase change boundary along the normal $\boldsymbol{n}$, $R(t)$ represents the position vector of a point on $\Gamma_R(t)$, $L(\Theta, p, \ldots)$ is the latent heat, and $q_i = -\kappa_i$ grad Θ is the heat flux in $G_i(t)$. Thus the last relations lead to

$$\Theta_i(\boldsymbol{x}, t) = \Theta_j(\boldsymbol{x}, t) = T_R, \tag{8.22}$$

$$\kappa_i \text{ grad } \Theta_i \text{ grad } R - \kappa_j \text{ grad } \Theta_j \text{ grad } R = \rho L v_n. \tag{8.23}$$

Further, the initial conditions at $t = t_o$

$$T(\boldsymbol{x}, t_o) = T_o(\boldsymbol{x}), \quad \boldsymbol{u}(\boldsymbol{x}, t_o) = \boldsymbol{u}_o(\boldsymbol{x}), \quad \boldsymbol{u}_{,t}(\boldsymbol{x}, t_o) = \boldsymbol{u}_1(\boldsymbol{x}), \tag{8.24}$$

are given.

Remark: If $\mathscr{F}_c^{kl} \left|\tau_n^k(\boldsymbol{x}, t)\right| = g_c^{kl}(\boldsymbol{x}, t) \geq 0$ is the known contact force on Γ_c^{kl} then we have the contact problem with the given friction for $g_c^{kl}(\boldsymbol{x}, t) > 0$ and the contact problem without friction for $g^{kl}(\boldsymbol{x}, t) = 0$. If the contact force g^{kl} depends on the space coordinates only, then $\mathscr{F}_c^{kl} \left|\tau_n^k(\boldsymbol{x})\right| = g^{kl}(\boldsymbol{x})$ and we have the contact problem with the given friction independent of time t. If $\iota = 2$ and if the corresponding body is rigid, then $\boldsymbol{u}^2 = 0$; for $\iota > 2$ and we have the model contact problem without or with the given friction, in which the thermo-elastic body (e.g. the lithospheric plate) is, along Γ_c, unilaterally supported by a rigid foundation. In this dynamic Signorini problem the estimate of the coefficient of the Coulombian friction is not known and mathematically represents an open problem.

The problem formulated above (Equations 8.1 to 8.24) is in its general form up to date unsolved. Here only the problem with the given friction will be investigated. The terms $\rho^\iota c^\iota u^\iota_{j,t} T^\iota{}_{,j}$ and $\rho^\iota u^\iota_{j,t} u^\iota_{i,tj}$ are the convective terms. If the first term in Equation 8.6 is neglected, then we have the problem studied in the quasi-coupled thermo-elasticity. The case of recrystallization, corresponding to the problem studied, can be investigated next.

8.1.3. Variational Solution of the Problem with the Given Friction

For a numerical solution we shall formulate the above problem with homogeneous boundary conditions and by using the enthalpy. The enthalpy will be defined by

$$\partial\phi(T) \equiv H(T) = \begin{cases} c_L^s(T - T_R) & \text{for } T < T_m, \\ c_s^s(T - T_R) + \rho L & \text{for } T > T_m, \\ [0, \rho L] & \text{for } T = T_m, \end{cases} \tag{8.25}$$

where $T_m < T_R$ is a given temperature less than a temperature of a phase change. Then $\mathrm{d}H/\mathrm{d}t = \rho c \mathrm{d}T/\mathrm{d}t$ and Equation 8.6 is transformed to the equation of heat in an enthalpy formulation.

Let us denote by $C^k(\Omega^\iota)$ the space of all functions defined on Ω^ι and k-times continuous differentiable in $\Omega^\iota \forall t$, $k \geq 0$ integer or $k = \infty$, by $L^\infty(\Omega^\iota)$ the space of all bounded measured functions defined a.e. in Ω^ι with the norm $\|u^\iota\|_{L^\infty(\Omega^\iota)} =$ vrai $_{\boldsymbol{x}\in\Omega^\iota} \max |\boldsymbol{u}^\iota(\boldsymbol{x})|$, and by $H^k(\Omega^\iota) = \{v \mid v \in L^2(\Omega^\iota), D^\alpha v^\iota \in L^2(\Omega^\iota) \forall \alpha, |\alpha| \leq k\}$ the Sobolev space with scalar product $(u, v)_k = \sum_{0\leq|\alpha|\leq k} (D^\alpha u, D^\alpha v)$.

Let

$$V = \left\{ \boldsymbol{v} \mid \boldsymbol{v} \in [H^1(\Omega)]^N = \prod_{\iota=1}^{S} [H^1(\Omega^\iota)]^N, \quad \boldsymbol{v} = 0 \text{ on } \Gamma_u \right\},$$

$$^1V = \left\{ w \mid w \in H^1(\Omega) = \prod_{\iota=1} H^1(\Omega^\iota), w = 0 \text{ on } \Gamma_\tau \quad (\text{or on } \Gamma_u) \right\},$$

be the space of virtual displacements or the space of virtual temperatures, respectively, and let

$$K = \{\boldsymbol{v} \mid \boldsymbol{v} \in V, v_n^k - v_n^l \leq 0 \text{ on }_k, \cup_l \Gamma_\alpha^{kl}\}$$

be the set of the admissible displacements. Let

$$H = [L^2(\Omega)]^N = \prod_{\iota=1}^{S} [L^2(\Omega^\iota)]^N, \quad \mathscr{H} = L^2(\Omega) = \prod_{\iota=1}^{S} L^2(\Omega^\iota)$$

and V' and $^1V'$ be dual spaces of V and ^1V, so that $V \subset H \subset V'$, $^1V \subset \mathscr{H} \subset {}^1\mathrm{V}'$. Let us denote by $(.,.)$ the scalar product in $L^2(\Omega)$, $|.|_{\mathscr{H}}$ the norm in $\mathscr{H}$, $|.|_H$ the norm in H. The norm in V is defined by $\|v\|_V^2 = \sum_{\iota=1}^N \|v_i\|_1^2$, $\|.\|_{1_V}$ is the norm in ^1V, $\|.\|_1$ is the norm in $H^1(\Omega)$. Further, we shall use the notation $u_{i,j} = \partial u_i/\partial x_j$, $u_{i,t} = u_i' \equiv \dot{u}_i$, $u_{i,tt} = u_i''$.

In our model problem we shall assume that on $\Gamma_u \times I$ the condition in Equation 8.9 is prescribed. On the boundary $\Gamma_u \times I$ let the temperature $T_2(\boldsymbol{x}, t) \equiv 0$. We shall assume that $\beta_{ij}^\iota \in C^1(\overline{\Omega}^\iota)$, $c_{ijkl}^\iota(\boldsymbol{x}, t) \in C^1(\overline{\Omega}^\iota)$, $\kappa_{ij}^\iota(\boldsymbol{x}, t) \in C^1(\overline{\Omega}^\iota)$, $c^\iota(\boldsymbol{x}, t) \in C(\overline{\Omega}^\iota)$, $\rho^\iota(\boldsymbol{x}, t) \in C(\overline{\Omega}^\iota)$, $f^\iota(x, t) \in [L^2(\overline{\Omega}^\iota)]^N$, $W^\iota(x, t) \in L^2(\overline{\Omega}^\iota)$ for all $t \in I$. Let $\boldsymbol{w}^\iota(x, t)$ be a sufficiently smooth vector function in $G = \cup_{\iota=1}^S G \cup \partial G$ satisfying Equations 8.9 and 8.13 and $\underline{\boldsymbol{w}}^\iota = 0$ on Γ_c^{kl} (as we shall assume that the end of the contact boundary is perpendicular to the boundary Γ_u).

Let us denote

$$(\boldsymbol{u}''(t), v) = \int_\Omega \rho u_i' v_i \mathrm{d}\boldsymbol{x} = \sum_{\iota=1}^{S} \int_{\Omega^\iota} \rho^\iota u_{i,tt}^\iota v_i^\iota \mathrm{d}\boldsymbol{x},$$

$$(H'(t), w) = \int_\Omega H' w \mathrm{d}\boldsymbol{x} = \sum_{\iota=1}^{S} (H^\iota{}_{,t}, w^\iota),$$

$$b(t; T(t), w) = \int_\Omega \kappa_{ij}(\boldsymbol{x}, t) T_{,j} w_{,i} \, \mathrm{d}\boldsymbol{x} = \sum_{\iota=1}^{S} b^\iota(t; T^\iota(t), w^\iota),$$

$$B(t; \boldsymbol{u}(t), \boldsymbol{v}) = \int_\Omega c_{ijkl}(\boldsymbol{x}, t) e_{ij}(\boldsymbol{u}) e_{kl}(\boldsymbol{v}) \mathrm{d}\boldsymbol{x} = \sum_{\iota=1}^{S} B^\iota(t; \boldsymbol{u}^\iota(t), \boldsymbol{v}^\iota),$$

$$(\Phi(t), w) = (Q(t), w) - \int_{\Gamma_u \cup \Gamma_\tau} q w \mathrm{d}s = \sum_{\iota=1}^{S} (\Phi^\iota(t), w^\iota),$$

$$(\vec{\psi}(t), \boldsymbol{v}) = (\boldsymbol{F}(t), \boldsymbol{v}) - \int_{\Gamma_\tau} P_i v_i \mathrm{d}s = \sum_{\iota=1}^{S} (\vec{\psi}^\iota(t), \boldsymbol{v}^\iota),$$

$$j_{gn}(\boldsymbol{v}) = \int_{\Gamma_c} \mathscr{F}_c |g_n| \left| \boldsymbol{v}_t^k - \boldsymbol{v}_t^l \right| \mathrm{d}s \equiv \int_{\Gamma_c} g \left| \boldsymbol{v}_t^k - \boldsymbol{v}_t^l \right| \mathrm{d}s = \sum_{\iota=1}^{S} j_{gn}^\iota(\boldsymbol{v}^\iota),$$

$$D(T(t), \boldsymbol{v}) = \int_\Omega (\beta_{ij}(\boldsymbol{x}, t)(T(t) - T_o)),_j v_i \mathrm{d}\boldsymbol{x} = \sum_{\iota=1}^{S} D^\iota(T^\iota(t), \boldsymbol{v}^\iota),$$

$$\mathrm{d}(\boldsymbol{u}, w) = \int_\Omega \rho \beta_{ij} T_o e_{ij}(\boldsymbol{u}) w \mathrm{d}\boldsymbol{x} = \sum_{\iota=1}^{S} d^\iota(\boldsymbol{u}^\iota, w^\iota).$$

Replacing $\boldsymbol{u}^\iota$ by $\boldsymbol{u}^\iota + \underline{\boldsymbol{w}}^\iota$, multiplying Equation 8.6 by $w^\iota - T^\iota(t)$ and Equation 8.5 by $\boldsymbol{v}^\iota - \boldsymbol{u}'^\iota(t)$, integrating over Ω^ι, $\forall\iota$, $\forall t$, applying the divergence theorems for scalar and vector functions and the boundary conditions, then we have the following variational problem.

Problem $(\mathscr{P}_c)_{\boldsymbol{v}}$: Find a pair of functions $(T, \boldsymbol{u})$ satisfying

$$(H'(t), w - T(t)) + b(t; T(t), w - T(t)) + \mathrm{d}(\boldsymbol{u}'(t), w - T(t)) \geq (\Phi(t), w - T(t)) \quad \forall w \in {}^1V, \quad (8.26)$$

$$(\boldsymbol{u}''(t), \boldsymbol{v} - \boldsymbol{u}'(t)) + B(t; \boldsymbol{u}(t), \boldsymbol{v} - \boldsymbol{u}'(t)) + D(T(t), \boldsymbol{v} - \boldsymbol{u}'(t)) + j_{gn}(\boldsymbol{v} + \underline{\boldsymbol{w}}'(t)) - j_{gn}(\boldsymbol{u}'(t) + \underline{\boldsymbol{w}}'(t))$$
$$\geq (\vec{\psi}(\boldsymbol{x}, t), \boldsymbol{v} - \boldsymbol{u}'(t)) \quad \forall \boldsymbol{v} \in K, \quad (8.27)$$

and initial conditions

$$H(\boldsymbol{x}, t_o) = H_o(\boldsymbol{x}), \quad (8.28)$$

$$\boldsymbol{u}(\boldsymbol{x}, t_o) = \boldsymbol{u}_o(\boldsymbol{x}), \quad (8.29)$$

$$\boldsymbol{u}'(\boldsymbol{x}, t_o) = \boldsymbol{u}_1(\boldsymbol{x}), \quad (8.30)$$

where

$$F_i^\iota = f_i^\iota - (\beta_{ij}^\iota(T^\iota - T_o^\iota)),_j + (c_{ijkl}^\iota e_{kl}(\underline{\boldsymbol{w}}^\iota)),_j - \underline{w}_{i,tt}^\iota \in L^2(I; L^2(\overline{\Omega}^\iota)),$$

$$Q^\iota = W^\iota \in L^2(I; L^2(\overline{\Omega}^\iota)), \quad P_i^\iota = P_{oi}^\iota - c_{ijkl}^\iota e_{kl}(\underline{\boldsymbol{w}}^\iota) n_j,$$

$$\boldsymbol{u}_o^\iota(\boldsymbol{x}) \cong \boldsymbol{u}_o^\iota(\boldsymbol{x}) + \underline{\boldsymbol{w}}^\iota(\boldsymbol{x}, t_o), \quad \boldsymbol{u}_1^\iota(\boldsymbol{x}, t) \cong \boldsymbol{u}_1^\iota(\boldsymbol{x}) + \underline{\boldsymbol{w}},_t(\boldsymbol{x}, t_o).$$

If the term $\rho^\iota \beta_{ij}^\iota T_o^\iota e_{ij,t}(\boldsymbol{u}^\iota)$ is neglected then we have the quasi-coupled Signorini problem $(\mathscr{P}_{qc})$ with the given Coulombian friction, while if this term is not neglected then we have the Signorini type problem $(\mathscr{P}_c)$ with the given Coulombian friction in the coupled thermo-elasticity. The analysis of the problem represents an open mathematical problem.

8.1.4. Numerical Model: Numerical Solution of Discretized Coupled Equations

The functional j_{gn} is a non-differentiable functional. Then the first method can be based on the regularization of the functional j_{gn} by its convex regularization $j_{gn\varepsilon}$

and the penalty technique. Then we solve the problem given by Equations 8.26 to 8.30, which is represented by the parabolic and hyperbolic equalities. The problem analysed in such a way was discussed in Nedoma (1990c).

The second method will be based on the method of discretization in time for non-linear parabolic variational equalities and hyperbolic variational inequalities.

For the numerical solution we shall use the method of discretization in time and the finite element method (FEM).

Let the given bounded domain $\Omega \subset R^2$ with the polygonal boundary $\partial\Omega$ be "triangulated", i.e. the domain $\overline{\Omega} = \Omega \cup \partial\Omega$ be covered by a finite number of triangles $\underline{T}_{hi}$, $i = 1, \ldots, n(h)$, $h \in \langle 0, h_o \rangle$, forming a triangulation with h being a step of triangulation. Let $\{\mathcal{T}_h\}$ be a system of regular triangulation with end points $\overline{\Gamma}_u \cap \overline{\Gamma}_\tau$, $\overline{\Gamma}_u \cap \overline{\Gamma}_c$, $\overline{\Gamma}_c \cap \overline{\Gamma}_\tau$ coinciding with the vertices of the triangles $\underline{T}_h$. We say that the triangulation is ϑ-regular if there exists $\vartheta > 0$ such that the inner angles of all triangles $\underline{T}_{hi}$, $i = 1, \ldots, n(h)$, are greater than ϑ. The system of triangulation $\{\mathcal{T}_h\}$, $0 \le h \le h_o$ is ϑ-regular if there exists ϑ independent of h such that every triangulation $\mathcal{T}_h$ is ϑ-regular where h is a step of triangulation.

Let ${}^1V_h$, V_h be the spaces of linear finite elements

$$ {}^1V_h = \left\{ w \,\middle|\, w \in C(\overline{\Omega}), w\big|_{\underline{T}_h} \in P_1, w = 0 \quad \text{on} \quad \Gamma_u(\cup\Gamma_\tau), \quad \forall \underline{T}_h \in \mathcal{T}_h \right\}, $$

$$ V_h = \left\{ \boldsymbol{v} \,\middle|\, \boldsymbol{v} \in [C(\overline{\Omega})]^2, \boldsymbol{v}\Big|_{\underline{T}_h} \in [P_1]^2, \boldsymbol{v} = 0 \quad \text{on} \quad \Gamma_u, \quad \forall \underline{T}_h \in \mathcal{T}_h \right\}, $$

$$ K_h = \{\boldsymbol{v} \mid \boldsymbol{v} \in V_h, v_n^k - v_n^l \le 0 \quad \text{on} \quad \Gamma_\alpha\}, $$

where P_1 is the space of linear polynomials. Then ${}^1V_h \subset {}^1V$, $V_h \subset V$, $K_h \subset K$.

We shall start from the variational formulation given by Equations 8.26 to 8.30 i.e. from the problem $(\mathcal{P}_c)_v$.

Further, we shall apply the technique of discretization in time and the finite element method. Let $N = 2, 3, \ldots$ and let p' be a positive integer. Let the interval $\overline{I} = \langle t_o, t_1 \rangle$ be divided into $p = 2^{n-1}p'$ sub-intervals $\overline{I}_i = \langle t_{i-1}, t_i \rangle$, $i = 1, \ldots, p$, and let $k \equiv k_n = (t_1 - t_o)/p$, $t_i = t_o + ik$. Let H^i, u^i be an approximation of $H(\boldsymbol{x}, t)$ and $\boldsymbol{u}(\boldsymbol{x}, t)$ for $t = t_i$, $\delta_k H_i = {}^{(1)}H^i = (H^i - H^{i-1})k^{-1}$, $\delta_k^{(o)} H^i \equiv {}^{(o)}H^i = H^i$, $\delta_k \boldsymbol{u}^i = {}^{(1)}\boldsymbol{u}^i = k^{-1}(\boldsymbol{u}^i - \boldsymbol{u}^{i-1})$, $\delta_k^2 \boldsymbol{u}^i = {}^{(2)}\boldsymbol{u}^i = k^{-2}(\boldsymbol{u}^i - 2\boldsymbol{u}^{i-1} + \boldsymbol{u}^{i-2})$, $\delta_k^{(o)}\boldsymbol{u}^i = {}^{(o)}\boldsymbol{u}^i = \boldsymbol{u}^i$.

Further, let $T_h^i \in {}^1V_h$ and $\boldsymbol{u}_h^i \in V_h$ be finite element approximations of T and u at $t = t_i$, then $T_h^i = T_h(t_i)$, $\boldsymbol{u}_h^i = \boldsymbol{u}_h(t_i)$. Then the completely discrete approximation of Equations 8.26 to 8.30 leads to the following discrete problem.

Problem $(\mathcal{P}_{pv})_d$: Let $T_h^o \in {}^1V_h$, $\boldsymbol{u}_h^o \in V_h$ be given, let $n = 2, 3, \ldots, k = (t_1 - t_o)/p$, where $p = 2^{n-1}p'$ is an integer. Then we shall find a pair of discrete functions $(T_h^i, \boldsymbol{u}_h^i)$, $T_h^i \in {}^1V_h$, $\boldsymbol{u}_h^i \in V_h$, $i = 1, \ldots, p$, such that

$$ (\delta_k H_h^i, w - T_h^i) + b(t_i; T_h^i, w - T_h^i) + d(\delta_k u_h^i, w - T_h^i) \ge (\Phi_h^i, w - T_h^i) \quad \forall w \in {}^1V_h, \tag{8.31} $$

$$ \begin{aligned} &(\delta_k^2 \boldsymbol{u}_h^i, \boldsymbol{v} - \delta_k \boldsymbol{u}_h^i) + B(t_i, \boldsymbol{u}_h^i, \boldsymbol{v} - \delta_k \boldsymbol{u}_h^i) + D(T_h^i - T_o, \boldsymbol{v} - \delta_k \boldsymbol{u}_h^i) \\ &\quad + j_{gn}(\boldsymbol{v}) - j_{gn}(\delta_k \boldsymbol{u}_h^i) \ge (\vec{\Psi}_h^i, \boldsymbol{v} - \delta_k \boldsymbol{u}_h^i) \quad \forall \boldsymbol{v} \in K_h, \end{aligned} \tag{8.32} $$

where $T_h^o = T_o$, $\boldsymbol{u}_h^o = \boldsymbol{u}_o$, and with discrete approximation of the initial conditions given by Equations 8.28 to 8.30.

8.1.5. Algorithm and Mechanisms of Earthquakes

The algorithm can be based on the implicit or semi-implicit schemes, in the linear coupled thermo-elasticity. Here the algorithm will be based on the semi-implicit scheme.

Since $\beta_{ij}^\iota \in C^1(\overline{\Omega}^\iota)$, $\rho^\iota \in C(\overline{\Omega}^\iota)$ and $T_o \in H^1(\overline{\Omega})$, then $\beta_{ij}^\iota(T^\iota - T_o^\iota) \in H^1(\overline{\Omega}^\iota)$, $(\beta_{ij}^\iota(T^\iota - T_o^\iota))_{,j} \in L^2(\overline{\Omega}^\iota)$, and $\rho^\iota \beta_{ij} T_o^\iota e_{ij,t}(\boldsymbol{u}^\iota) \in L^2(\overline{\Omega}^\iota)$. Therefore, the coupled terms can be taken as thermal sources or body forces, respectively.

Our problem in every time level leads to finding $T_h^i \in {}^1V_h$, $\boldsymbol{u}_h^i \in K_h$ in every time level such that

$$l(T_h^i) \leq l(w) \quad \forall w \in {}^1V_h, \tag{8.33}$$

$$L(\boldsymbol{u}_h^i) \leq L(\boldsymbol{v}) \quad \forall \boldsymbol{v} \in K_h, \tag{8.34}$$

where

$$l(w) = \tfrac{1}{2}(A_{Th}w, w) - (\Phi^i, w) - (H^{i-1}, w)k^{-1} - d(\delta_k u^i, w) \quad \forall w \in {}^1V, \tag{8.35}$$

$$L(\boldsymbol{v}) = L_o(\boldsymbol{v}) + j_{gn}(\boldsymbol{v}), \tag{8.36}$$

with

$$L_o(\boldsymbol{v}) = \tfrac{1}{2}(A_{uh}\boldsymbol{v}, \boldsymbol{v}) - (\vec{\Psi}^i, \boldsymbol{v}) - (2\boldsymbol{u}^{i-1} - \boldsymbol{u}^{i-2}, \boldsymbol{v})k^{-2} - D(T^i - T_o, \boldsymbol{v}) \quad \forall \boldsymbol{v} \in V,$$

where we define operators A_{Th} and A_{uh} by

$$(A_{Th}z, w) = k^{-1}(H, w) + b(t; z, w) \quad \forall z, w \in {}^1V,$$

$$(A_{uh}\boldsymbol{u}, \boldsymbol{v}) = k^{-2}(\boldsymbol{u}, \boldsymbol{v}) + B(t; \boldsymbol{u}, \boldsymbol{v}) \quad \forall \boldsymbol{u}, \boldsymbol{v} \in \mathrm{V}.$$

The problem in Equation 8.33 leads to solving the system of non-linear algebraic equations

$$MH(\omega) + \Delta tKT(\omega) + \Delta tN\Psi(\omega) = W, \tag{8.37}$$

where M, K and N are constant band matrices, which at every time level are positive definite matrices and the matrix N is positive semi-definite. Matrices M and K are Gramm's matrices as (.,.) and $b(.,.)$ have properties of scalar products and the basis functions are linearly independent and therefore they are also regular. This non-linear system of equations can be solved by the Newtonian method and then the resulting linear system by the conjugate gradient method or by the SSOR or Gauss-Seidel methods, respectively.

Since the functional $L_o(\boldsymbol{v})$ is convex and differentiable (since it is coercive and weakly lower semi-continuous) and the functional j_{gn} is convex and non-differentiable then by introducing the Lagrangian $\mathscr{L}$ the problem in Equation 8.34 is shown to be related to the min – max problem (Céa (1971), Ekeland and Temam (1976)), i.e. to the problem of finding $u_h^i \in K_h$ such that

$$L(u) = \min_{v \in K_h} \sup_{\mu \in \Lambda} \mathscr{L}(v, \mu), \tag{8.38}$$

where $\Lambda = \cup_{k,l}\{\mu^{kl} \mid \mu^{kl} \in L^2(\Gamma_c^{kl}), |\mu^{kl}| \leq 1 \text{ a.e. on } \Gamma_c^{kl}\}$,

$$\mathscr{L}(v, \mu) = L_o(v) + \int_{\cup_{k,l}\Gamma_c^{kl}} \mu^{kl} g_n^{kl} (v_t^k - v_t^l) \mathrm{d}s \tag{8.39}$$

is the Lagrangian, and

$$j_{gn}(v) = \int_{\cup_{k,l}\Gamma_c^{kl}} g_n^{kl} \left| v_t^k - v_t^l \right| \mathrm{d}s = \sup_{\mu \in \Lambda} \int_{\cup_{k,l}\Gamma_c^{kl}} \mu^{kl} g_n^{kl} (v_t^k - v_t^l) \mathrm{d}s.$$

Then similarly as in Chapter 7 the problem leads to finding a saddle point. We see that the Uzawa algorithm can also be used, as follows.

We construct two sequences $u_h^{ni} \in K_h$ and $\lambda_h^n \in \Lambda$, defined in the following way. Start with $\lambda_h^o \in \Lambda$ and calculate u_h^{oi}, and then λ_h^n, u_h^{ni}, $n = 1, 2, \ldots$. Let λ_h^n be known, then we determinate $u_h^{ni} \in K_h$ as the element of K_h which satisfies

$$\min_{v \in K_h} \left\{ L_o(v) + \int_{\cup_{k,l}\Gamma_c^{kl}} \lambda_h^{kl} g_n^{kl} (v_t^k - v_t^l)^i \mathrm{d}s \right\},$$

i.e. by using the conjugate gradient, SSOR, or Gauss-Seidel methods with linear constraints (i.e. by methods of quadratic programming). Then we calculate

$$\lambda_h^{(n+1)kl} = P(\lambda_h^{nkl} + \rho_n g_n^{kl} (u_h^k - u_h^l)_t^i),$$

where $0 < \rho_o \leq \rho_n \leq \rho_1$, ρ_o, ρ_1 being sufficiently small, and the projection P is defined by

$$P\mu(z) = \mu(z) \quad \text{for } |\mu(z)| \leq 1,$$
$$P\mu(z) = 1 \quad \text{for } \mu(z) > 1,$$
$$P\mu(z) = -1 \quad \text{for } \mu(z) < -1.$$

The detailed numerical analysis of the investigated dynamic problem is an open problem.

We now consider the mechanisms of earthquakes. At present the accepted model of an earthquake origin is the dynamic spreading of a shear rupture on a fault plane and radiation from crack models (see Das and Aki (1977), Madariaga (1979), Aki (1979), etc.). But there exists another model of the earthquake source mechanism. (see Nedoma (1990c)) However from the practical point of view the problem is complicated due to the intrinsic 3D nature of seismic sources, the heterogeneity of elastic properties, and the non-uniformity of the stress-strain distribution on the uneven contact boundary Γ_c^{kl} between two colliding geological blocks or plates. The radiation from the collision zones between lithospheric plates or blocks, in the case of earthquake origin or smaller geological blocks in the case of mine shock origin, may be conveniently studied assuming firstly that the slip history on the fault has already been obtained from the solution of a crack problem by using the adaptive finite element methods, i.e. the h-version, p-version, $h-p$-version or m-version, etc. of FEM (see e.g. Nedoma (1990*b*)), and secondly that the frictional forces cannot yet preserve the acting tangential forces. We shall discuss each case separately.

8.1.5.1. *The case of a crack propagation*

Let $(T^i, \boldsymbol{u}^i)$ be the exact solution of the problem discussed and $(T_h^i, \boldsymbol{u}_h^i)$ its finite element approximation in the time $t = t_i$. Then

$$\tfrac{1}{2}\left\|T^i - T_h^i\right\|^2 = \left|l(T^i) - l(T_h^i)\right| = \left|\mathscr{W}(T^i) - \mathscr{W}(T_h^i)\right|, \tag{8.40}$$

$$\tfrac{1}{2}\left\|\boldsymbol{u}^i - \boldsymbol{u}_h^i\right\|^2 \leq \left|L(\boldsymbol{u}^i) - L(\boldsymbol{u}_h^i)\right| \leq \left|\boldsymbol{W}(\boldsymbol{u}^i) - \boldsymbol{W}(\boldsymbol{u}_h^i)\right|. \tag{8.41}$$

The finite element solution minimizes the error in the energy norm (see Babuška and Szabó (1981)). Therefore, the magnitude of the error in energy norm is a good measure of the overall quality of the approximation. Asymptotic error estimates are useful for the methods discussed. Further, we shall assume that the points of singularity — the crack tips — are located at the nodal points, as we shall discuss the h-version of the FEM only.

In the case of the h-version of the FEM, when a sequence of meshes is obtained through uniform refinement, then the asymptotic error estimate is of the form

$$\left\|\boldsymbol{u}^i - \boldsymbol{u}_h^i\right\|_E \leq C(\Omega, h, \mathscr{T}_h, p)N^{-1/2}\min(p, \alpha), \tag{8.42}$$

where C is a constant independent of $\boldsymbol{u}^i$ and N, the absolute value of the exponent of N is called the rate of convergence, N being the number of freedoms, $\mathscr{T}_h$ is the family of regular triangulations, p is the polynomial degree of elements and α measures the smoothness of the exact solution. In the special case of equilibrated (optimal) meshes, in which the sequence of meshes is constructed in such a way that the error associated with each element is approximately the same, the estimate leads to

$$\left\|\boldsymbol{u}^i - \boldsymbol{u}_h^i\right\|_E < CN^{-p/2}, \tag{8.43}$$

in which the rate of convergence is independent of α and can be made arbitrarily large due to the choice of p.

The error in strain energy can also be used for the construction of meshes. The error in strain energy can be written in the form

$$\left|\boldsymbol{W}(\boldsymbol{u}^i) - \boldsymbol{W}(\boldsymbol{u}_h^i)\right| \leq CN_{W(h)}^{-\beta}, \quad c = \text{constant} > 0, \tag{8.44}$$

where $\boldsymbol{W}(\boldsymbol{u}^i)$ and $\boldsymbol{W}(\boldsymbol{u}_h^i)$ are the exact and the computed strain energy, $N_{W(h)}$ is the corresponding number of degrees of freedom due to the polynomial degree p and the mesh with step h, $\beta = \min(p, \alpha)$ is the rate of convergence and α is defined above. Since $\boldsymbol{W}(\boldsymbol{u}^i) > \boldsymbol{W}(\boldsymbol{u}_k^i)$, then $\boldsymbol{W}(\boldsymbol{u}^i) - \boldsymbol{W}(\boldsymbol{u}_h^i) \leq CN_{W(h)}^{-\beta}$. Let us assume that the finite element solution is in the asymptotic range and β is known. Then

$$C = \frac{\boldsymbol{W}(\boldsymbol{u}_h^i) - \boldsymbol{W}(\boldsymbol{u}_{h-1}^i)}{N_{W(h-1)}^{-\beta} - N_{W(h)}^{-\beta}},$$

where $\boldsymbol{W}(\boldsymbol{u}_{k-1}^i)$ and $\boldsymbol{W}(\boldsymbol{u}_k^i)$ are the strain energies corresponding to the coarse and softer meshes. For the exact strain energy we obtain

$$\boldsymbol{W}(\boldsymbol{u}^i) = \frac{\boldsymbol{W}(u_h^i)N_{W(h)}^{\beta} - \boldsymbol{W}(u_{h-1}^i)N_{W(h-1)}^{\beta}}{N_{W(h)}^{\beta} - N_{W(h-1)}^{\beta}} \quad (8.45)$$

From Equation 8.45 one can see that the strain energy can be estimated by the numerical processes (i.e. by h, p, $h - p$, etc. versions of the FEM), which is a useful idea for the earthquake mechanism if its origin is a crack propagation.

8.1.5.2. *The case of Coulombian friction forces acting on the contact boundary*

The analysis of the earthquake and mine shock mechanisms is based on numerical analysis of the contact conditions given in Equations 8.15 to 8.20 at every time level t_i. On the contact boundaries Γ_c^{kl} during the deformation the points of the boundaries of both colliding geological blocks and plates are displaced in different ways, but block Ω^k never penetrates into block Ω^l, i.e. $u_n^k - u_n^l \leq 0$. For the contact forces, from the law of action and reaction it follows that $\tau_n^k = -\tau_n^l (\equiv \tau_n^{kl})$, $\vec{\tau}_t^k = -\vec{\tau}_t^l (\equiv \vec{\tau}_t^{kl})$. The tangential component vanishes in the case of zero friction and the normal contact force cannot be tensile, i.e. $\tau_n^{kl} \leq 0$. Therefore, if the colliding blocks are not in contact, then $u_n^k - u_n^l < 0$ and the contact forces are zero, i.e. $\tau_n^{kl} = 0$. If there is a contact, then $u_n^k - u_n^l = 0$ and $\tau_n^{kl} < 0$. Both cases are included in the condition $(u_n^k - u_n^l)\tau_n^{kl} = 0$ at every time t. If both colliding geological blocks or plates are in a contact, i.e. if $u_n^k - u_n^l = 0$ and $\tau_n^{kl} < 0$, Coulombian friction acts at the contact boundary Γ_c^{kl}. The frictional forces g_c^{kl} acting on the contact boundaries Γ_c^{kl}, are in their absolute value, proportional to the normal stress component, where the coefficient of proportionality is the coefficient of Coulombian friction $\mathscr{F}_c^{kl}(\boldsymbol{x}, t)$, i.e. $g_c^{kl} = \mathscr{F}_c^{kl}(\boldsymbol{x}, t) \left|\tau_n^{kl}(\boldsymbol{x}, t)\right|$. Due to Equations 8.17 and 8.18, the two following situations can occurs. Firstly, if the absolute value of the tangential stress component $\vec{\tau}_t$ is less than the frictional force g_c^{kl}, then the tangential force precludes the mutual motion of both colliding blocks or plates. Secondly, if the absolute value of the tangential stress component $\vec{\tau}_t$ is equal to the frictional force g_c^{kl}, then there are no forces which can preclude the mutual shifting of both colliding geological blocks or plates. Thus the contact points change their position in the direction opposite to that in which the tangential stress component acts. At the same time one part of the deformation energy, accumulated in the colliding region, is released in the form of an earthquake. Since this phenomenon acts from the geological point of view in a very short time, the problem can be solved as a problem where all physical quantities during this time are practically time independent. In this case the problem of seismic radiation or of mine shock radiation is also described by Equations 8.26 to 8.30 and their discrete approximations discussed above in the previous chapter, but with physical quantities which are functions of spatial coordinates only and very small time steps (the time step in the geodynamic problem is of about tens of thousand years, while in the seismic radiation the time step is split second only). By the algorithm discussed above, we calculate stresses on every triangle of the triangulation $\mathcal{T}_h$ used as well as stresses, strains and displacements on the contact boundary Γ_c^{kl}. Further, we validate the conditions in Equations 8.17 and 8.18 on the contact boundary Γ_c^{kl} (the critical problem of the algorithm discussed). As was shown in Nedoma (1984b), the step of triangulation is estimated by the wavelength of S waves in the investigated region, i.e. by estimation $h < \lambda_s/2\pi$, where λ_s is the smallest wavelength of S waves in the investigated region Ω. Such condition must also be fulfilled in this case. This condition leads to a substantial thickering of the mesh of triangulation used. Whilst in the case of the geodynamic problem we must recalculate the triangulation used in every time step t_i as the mesh changes in time (due to the displacement of the materials which build the geological block or plates in time), in the case of seismic and mine shock radiations the mesh is fixed, but it must be constructed very softly at the moment when the block or plates shift and part of the deformation energy, accumulated in the collision zone, is released.

The algorithm used for numerical realization involves both cases discussed above at the same time, as the cracks in geological blocks and plates can be propagated and on the other hand geological blocks and plates can also be shifted.

8.2. NUMERICAL MODELLING OF INTERPLATE DEFORMATION IN THE CENTRAL ALEUTIAN ARC SYSTEM: TECTONIC STRESSES IN THE LITHOSPHERE

8.2.1. Introduction

The contact zone between the subducted Pacific lithospheric plate and the obducted North American lithospheric plate is an important tectonic area of the world. A small part of this zone is represented by the Aleutian arc system situated in the north of this region, and is characterized by a collision between the oceanic and continental lithospheric plates. The American seismological studies showed that the present-day earthquake activity at the Aleutian arc is controlled by the geometry of the northern boundary between the North American and the Pacific lithospheric plates and that the earthquakes are located near the intersection of the Aleutian arc and the Bower Ridge (see, for example, Stauder (1968), (1972), Grow (1973), Grow and Atwater (1973)). The Aleutian volcanoes are situated at Buldir Island eastward in the Aleutian arc and have a connection with the earthquake activity beneath a depth of about 110 km (Engdhal (1973)). The maximum depth of the earthquakes is observed as approximately 250–300 km, in the Central Aleutian area at 200–300 km. As was shown by Spence (1977) the dominant tectonic process occurring at the central Aleutian arc is the episodic occurrence of large thrust earthquakes, and the subsequent subduction process is characterized by episodic downslab pressure overload pulses.

Recent seismic evidence indicates that the Wadati-Benioff zone is segmented along its length by breaks (see Stauder (1972), Van Wormers, Davies and Gedney (1974), etc.). The present relative motion of the Pacific plate in the region of the Aleutian arc and Alaska, based on the analysis of spreading in the Pacific and the Atlantic ocean, is about 4-5 cm/yr (Atwater (1970), Grow (1973), Grow and Atwater (1973)).

In our studies of the Central Aleutian region we are going to carry out an approximation of the behaviour of the lithosphere over long periods of time. But as we saw above, it is well established that for geological periods shorter than the characteristic time t_c, the lithosphere behaves as an elastic medium. Practically, $t_c \leq 2.10^5$ years (Cathles (1975)). Laboratory experiments on lithospheric rocks (Kirby (1983), Brace and Kohlstedt (1980)), investigating the pressure and temperature conditions, show that they are functions of depth. For the condition on the contact boundaries in the collision zones and on the contact boundary between the lithosphere and the asthenosphere, Nedoma (1987) gives an upper estimate of the coefficient of the Coulombian friction $\mathscr{F}_c^{kl} \sim (c_s/c_p)^{1/2}$, where c_s and c_p are velocities of S and P waves. The lithosphere presents a layer of strength to low deformation that extends between 20 km and 70 km for dry olivine and between 20 km and 55 km for wet olivine. The main constituent of the upper mantle is assumed to be olivine (Liebermann (1992), Goetze (1978)), dunite and lherzolite. In the lower parts of the upper mantle the dominant mechanism is non-linear creep. For simplicity we shall assume in our further investigations that the mechanical behaviour of the upper parts of the upper mantle over geological periods less than a characteristic time t_c can also be characterized by elastic rocks.

In order to solve the model problem of the Central Aleutian, based on the author's theory from the previous chapter and on the Grow's model (Grow (1973), Grow and Atwater (1973)) for a finite section of the American continental lithospheric plate and the Pacific lithospheric plate, we have to improve boundary conditions on the plate limits. These conditions replace the relative motion of both lithospheric plates at greater distances from the collision zones. Furthermore, we assume that the Earth's surface is approximately load free in one case, and that in the second case it is loaded by the weight of the ocean and the atmosphere. At greater depths under the lithosphere we shall assume that on these surfaces the displacements are very small, practically equal to zero.

In our considerations we shall use a simpler model. This model assumes an American lithospheric plate 80 km thick, of an average density 3400 kgm^{-3} in the deeper parts of the lithosphere and of 2800 kgm^{-3} in the crustal part of the lithosphere. Note also the implied accretion of about 15 km of deformed sediments of average density 2400 kgm^{-3} at the leading edge of the overriding plate (see Figure 8.1 and Table 8.1). The geometrical surface configuration of the lithosphere is derived from the distribution of intermediate and deep earthquakes.

Table 8.1. Data for the Aleutian Island Arc Model

ρ[kg/m^3]	2400	2900	3400	2600	2500	2800	3300	3400	2900	2300
c_p[m/sec]	3800	6410	7600	6200	6100	6300	7780	7900	6400	5950
c_s[m/sec]	2100	3700	4400	3600	3500	3680	4500	4560	3700	3400

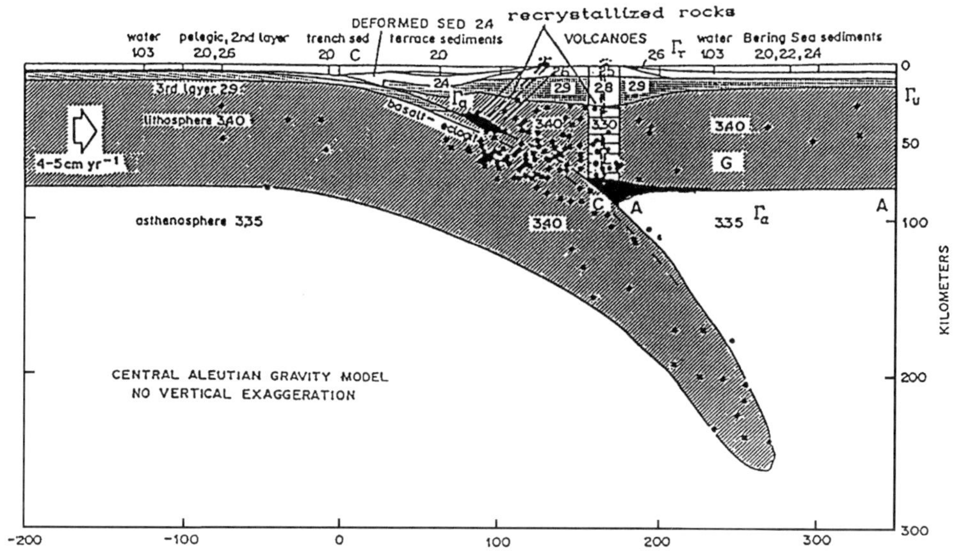

Figure 8.1. Central Aleutian Model (modified after Grow (1973))

In order to give the equations of motion of the obducting plate moving along the asthenosphere, we shall assume that the obducting plate is formed by inhomogeneous isotropic materials. Following our considerations about the rheology in the lithosphere (see Nedoma (1982) and previous chapters) we shall assume that the variability in time of sources in the lithospheric plate and the asthenosphere and of the body and surface forces is slow, and moreover that the motion of the lithospheric plate is uniform over the last several million years. Then the problem can be studied as a quasi-stationary and elastic one and the inertia forces can be neglected. We shall limit ourselves to the 2-*D* case only. Further for simplicity we assume that the obducting elastic plate (the American lithospheric plate) occupying the domain Ω rests on the rigid support — the asthenosphere and the subducting oceanic plate. It is convenient to choose the origin (0,0) as the point nearest to the trench axis.

It is well known that in such a case the stress field in the plate satisfies the equilibrium equation. To investigate the stress-strain field in the obducting plate in a neighbourhood of the collision zone, conditions should usually be given characterizing the effect of the environment of the plate on the processes taking place within a limited region containing the collision parts of the lithospheric plates under study and the asthenosphere. The interaction between the colliding plates and their environment is modelled by the boundary conditions for the displacement vector $\boldsymbol{u}$. In the model problem the boundary $\partial\Omega$ of the investigated region occupied by the obducting American plate Ω consists of three parts Γ_u, Γ_τ and Γ_c. Then on the Earth's surface Γ_τ the surface forces $\boldsymbol{P}$ are prescribed as caused by the weight of the ocean, seas, etc., i.e. $\tau_{ij}n_j = P_i$. On the contact boundary Γ_c the contact conditions of Signorini type are given. We shall assume the case without friction. To model the relative shifting of the obducted North American plate the value of the displacement vector is determined at a distance sufficiently far from the collision zone (it is the boundary Γ_u).

8.2.2. Numerical Results

The initiation of subduction is a key element in plate tectonic schemes for the evolution of the whole Earth's lithosphere. But nevertheless, up to the present, the underlying mechanism has not been well understood (Kanamori (1980), Kirby (1983), Grow and Sheridan (1981)). The numerical results based on the author's theory of

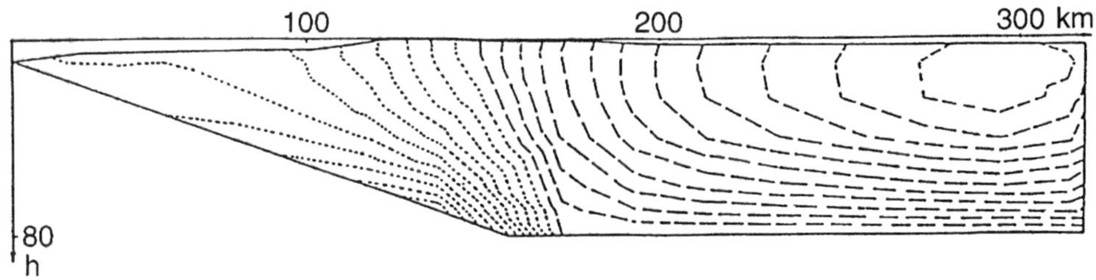

Short-dashed lines correspond to the elevation, long-dashed lines to the depression. The greatest value of elevation is about 385m at a distance of about 140km from the Aleutian trench and is about 33m at the head of ther plate. The greatest value of depression is about 55m at a distance 280km from the Aleutian trench (t 10 y

Figure 8.2. Vertical Component of Displacement **u**

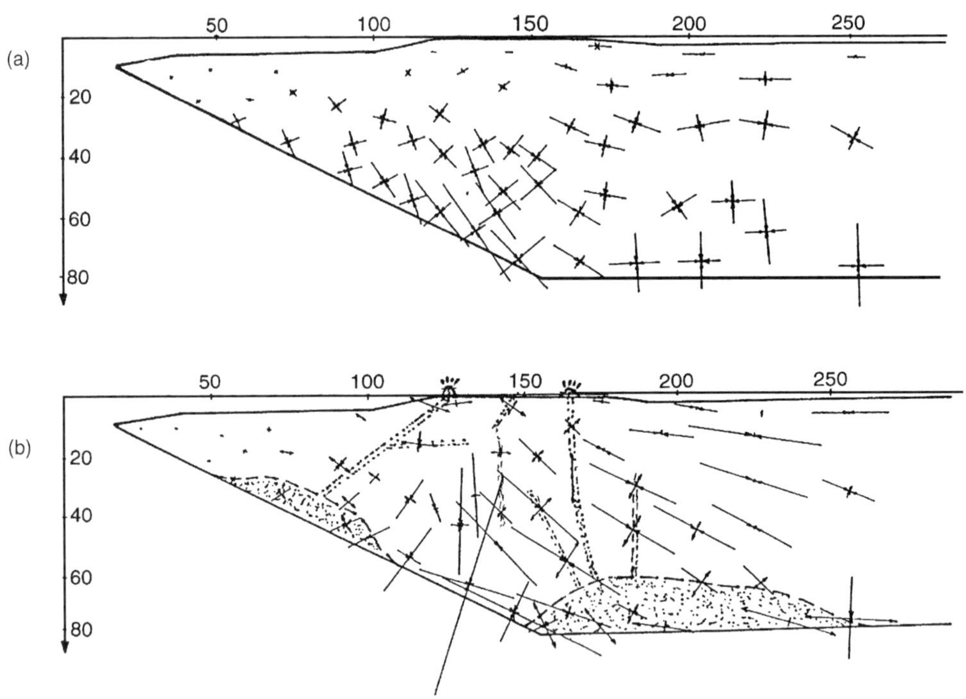

Figure 8.3. Principal Stresses τ_1, τ_2 in the obducting American plate at time: (a) $t = 10^4$ yrs in $10^8\,\mathrm{Nm}^{-2}$, (b) $t = 10^5$ yrs in $10^9\,\mathrm{Nm}^{-2} \mapsto 10^8\,\mathrm{Nm}^{-2}$

the evolution of collision zones (see Nedoma (1990a)) and on analyses of collision of several elastic (visco-elastic and plastic) bodies indicate a new view of the mechanism of evolution of subduction zones. The analyses will be made on the subduction zone of the Central Aleutian area (see Figure 8.1). For numerical analyses the FEM technique will be used. The obducted American lithospheric plate will be analyzed in more detail (see Figures 8.2 to 8.5).

In papers based on classical boundary value problems in elasticity the interaction between the lithosphere and the asthenosphere is assumed, although the associated shear stresses are probably lower than those created by collision. Such model problems are a first approximation only; a better approximation which is close to the geological reality is based on contact problems in elasticity and visco-plasticity.

In this section we shall investigate the Central Aleutian model within a plate-tectonic concept. We briefly review some observations relevant to the interplate deformation and the geodynamic history of the region of the Central Aleutian arc. The contact between the American and Pacific plates was completely established about 30 million years ago. Since then, the Pacific plate has continued to underthrust below the

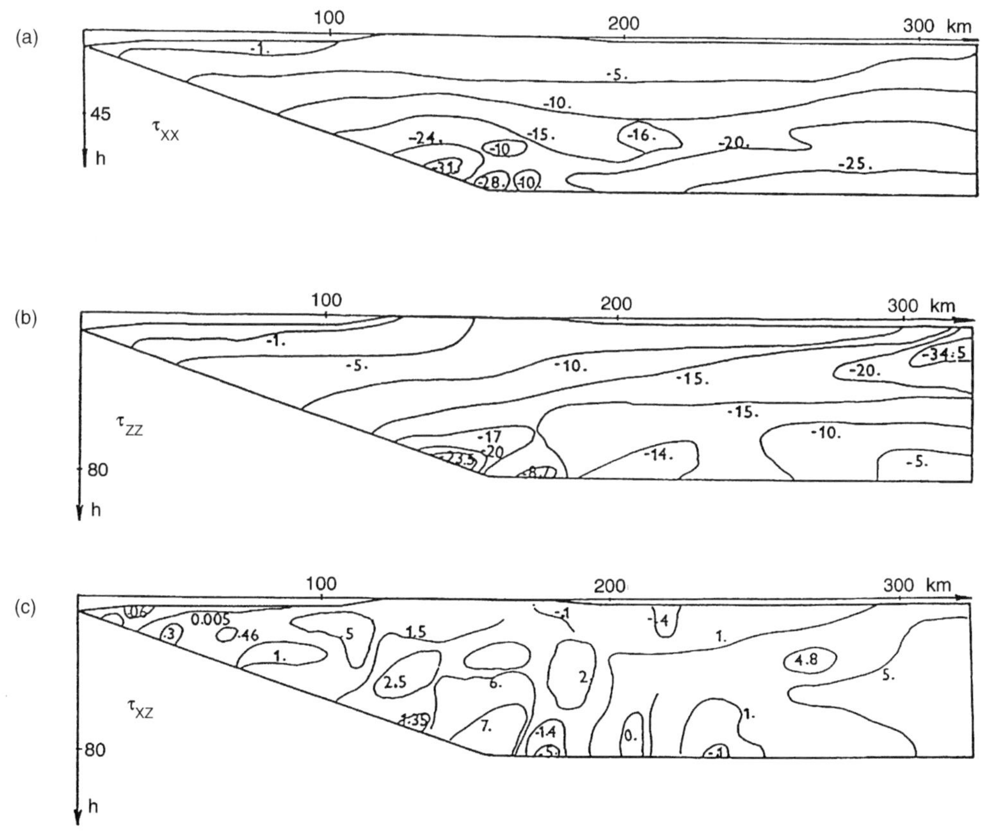

Figure 8.4. Stress Components: (a) $\tau_{xx} \equiv \tau_{11}$, (b) $\tau_{zz} \equiv \tau_{33}$, (c) $\tau_{xz} \equiv \tau_{13}$ in $10^9\,\mathrm{Nm}^{-2}$ at time $t = 10^4$ yrs

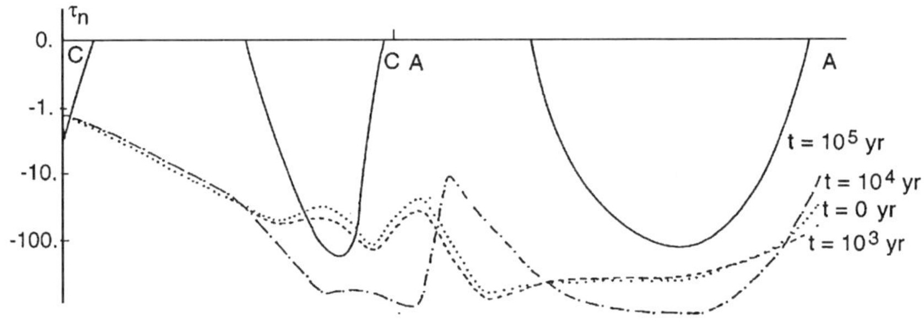

Figure 8.5. Normal Component of the Stress Vector τ_n in $10^9\,\mathrm{Nm}^{-2}$

American plate at a mean rate of 4-5 cm/yr, while the eastern part of the American plate moves at a mean rate of 4-5 cm/yr and the moving western part at a mean rate of 1 cm/yr. The Central Aleutian arc model was studied in two versions. The first model is represented by the obducting American lithospheric plate only while the subducting Pacific plate and the asthenosphere are assumed to be absolutely rigid, and the second model is represented by the collision of the obducting American plate with the subducting Pacific lithospheric plate, which submerges into the asthenosphere and the mesosphere. Due to a short geological period ($t < t_c \leq 10^5$ years) the lithosphere as well as its floor are assumed to be elastic. The lithosphere is modelled as a non-homogeneous rigid medium resting without friction over the asthenosphere. The curvature of both plates is considered due to the possibilities of the finite element technique (FEM) and because of the possibility of obtaining much more accurate results than the classic FEM, namely on the contact zones of the colliding plates. We shall discuss the first model.

8.2.3. Vertical Motions

We have calculated the vertical displacement component u_3 for the plane elastic model. In Figure 8.2 we show the two-dimensional plot of the vertical component of displacement u_3. The most important features of the vertical displacement component are firstly the uplift (elevation) in the area of the head of the obducting American plate, representing the elevation $\sim$33 m/10^4 yrs of the Aleutian arc system, and secondly the downward tendency in the area behind the Aleutian arc creating the great depression $\sim$55 m/10^4 yrs in the Bering Sea (the Bering and Bowers basins). Due to further motions of colliding plates, their evolution continues in the present time, which also proves our numerical experiments.

8.2.4. Interplate Stresses

Figure 8.3 illustrates the display of principal stresses at time $t = 10^4$ yrs and $t = 10^5$ yrs, and Figure 8.4 shows the stress components τ_{11}, τ_{13} and τ_{33} in the obducted American lithospheric plate at $t = 10^4$ yrs. At time $\sim 10^4$ yrs the principal stresses are characterized by compression only, while at time $\sim 10^5$ yrs there are observed regions with great compression and regions where the tensile stresses occur. It is interesting that these tensile stresses occur in the regions of deep faults, characterized by places where volcanoes occur. In Figure 8.3b the normal component of displacement u_n (u_n is drawn in its enlargement to show the shapes of the volcanic chambers as u_n bounds the shape of the cavity) is also entered at the contact boundaries between the obducted American plate and the subducted Pacific plate and the asthenosphere. The normal component of displacement u_n defines the bulging out regions at the contact boundary in which the magma deposits, the volcanic chambers, are created. These numerical results represent a new hypothesis about the evolution of volcanoes. This hypothesis is based on the idea that the rocks in these regions bulge out after the downward tendency of stresses (see Figure 8.5 representing the normal component of stresses τ_n) and these spaces are immediately filled by melted materials from subducted zones mainly from places near the contact surfaces. From the thermodynamics it is known that the temperature of melting of rocks depends on the temperature and pressure. If the pressure increases then the temperature of melting also increases. If the pressure decreases then the temperature of melting also decreases. Since the pressure falls strongly the rocks are rather quickly melted, gases are quickly released and crystallizations of materials in volcanic chambers are also influenced. Moreover, the phase transitions also depend on the pressure, namely on its decrease. In this case, at the subduction zone and in its greater neighbourhoods, by phase transitions is meant the melting and the recrystallization of the lithospheric and upper mantle rocks. In Figure 8.3a, where the principal stresses corresponding to $t \sim 10^4$ yrs are displaced, compressive stresses only occur in the whole American lithospheric plate, with greatest values in the lower part of the contact in the subduction zone. In Figure 8.3b, the principal stresses corresponding to $t \sim 10^5$ yrs in the obducted American lithospheric plate are characterized by great compressive stresses in places near the lower part of the contact of the obducting and subducting plates in the subduction zone, and on both sides of them are areas with tensile stresses. These tensile stresses are observed in two zones. The first one rises from the bulging out area and is directed to the Earth's surface in the head of the Aleutian arc system, and the second one rises from the bulging out area of the lower lithosphere and the upper asthenosphere and is directed practically vertically to the Earth's surface on the back part of the Aleutian arc system. Due to further loading, faults appear as a result of the compressive and mainly of the tensile stresses. These deep faults occur in the regions of tensile stress zones. The normal component of displacement indicates the size of volcanic chambers. The melted rocks (the so-called magma) in the subducted zone create volcanic chambers by the mechanism discussed above. Rocks absorb gases, which after their release from the rocks, compress the melted rocks, and then transport the melted materials (the magma) along these deep faults up to the Earth's surface, where volcanoes are created.

Generally, considering the origin of both volcanoes, which on the Earth's surface are at a distance of several tens of kilometers (50–120 km) only, the chemical compositions of both volcanoes are diverse. This effect is observed for example in the Japanese arc system, in the African volcanic system, etc.. Although the distance of both volcanoes is small, one of them can be of an eruptive type, while the second one is of an effusion type. During the effusion of magma materials from the volcanic chambers, the neighbouring rocks are then heated and recrystallized. Moreover, due to the strained rocks as well as to the acting friction on the contact surfaces between colliding lithospheric plates and blocks themselves and between lithospheric plates and the asthenosphere, the lithosphere is also heated and as a result lithospheric materials recrystallize. In regions in the lithosphere where the lithosphere is strongly strained, one part of the deformation energy changes into heat and then, if the temperature of the strained rocks is high, the strained lithospheric rocks are also recrystallized. Since the lithospheric rocks are viscous, only one part of energy is released as earthquakes by the mechanisms discussed above.

Theoretical gravity profiles immediately follow directly from the best-fitting deflection profiles $u_3(\boldsymbol{x}, t)$ by using the well-known Bouguer relation.

8.3. MODEL OF THE CARPATHIAN CONTINENT/CONTINENT COLLISION

The theory presented in Nedoma (1980), (1984a), (1986a-c), (1990a,b) and in previous chapters based on contact problems (see Nedoma (1983), (1987), (1994), (1997) and Chapter 7) was firstly verified on a model problem of the Central Aleutian arc system. The results have been published and discussed in detail in Nedoma, Haslinger and Hlaváček (1989). The numerical results obtained for the area of the Central Aleutian Islands imply that the methodology can also be used without any change in the case of a geodynamic model of the Carpathians and all regions of collision and subduction zones.

8.3.1. Geodynamic Model of Tectonics of the Western Carpathians

8.3.1.1. Introduction

The Carpathian system is a young orogenic system surrounded by a stable ancient platform. The geological structure of the Carpathians is now viewed as a result of the plate tectonics (Bleahu, Bocaletti, Manetti and Peltz (1973), Radulescu and Sandulescu (1973), Hertz and Savu (1974), Channell and Horváth (1976), Krs and Roth (1979), Horváth and Stegena (1977), Krs, Krsová and Pruner (1996), Morley (1996), etc.).

The geodynamic model of the Carpathians is a collision model by nature. The collision model is based on a convergent motion of both colliding geological blocks where the velocity of motion of one of the blocks may even be zero. During this process, the quasi-oceanic or oceanic lithosphere with sedimentary filling between both colliding blocks of the continental lithosphere is folded. In the initial phase, basins of the quasi-oceanic or oceanic lithosphere are compressed and only then are the sediments folded. The rocks are heated up due to the friction between the blocks and to the change of deformation energy into heat. The temperature increases, and as a consequence the rocks firstly metamorphose and then are partially melted or melted down in places where normal forces decrease on the contact between the rock blocks in agreement with the phase diagrams of rocks. At greater distances of these melted regions the rocks are also metamorphosed. The oceanic (quasi-oceanic) crust is uplifted along fault zones to folded layers. The internal displacement of mass inside the thrusted and moving blocks into the areas of elevating mountain ridges is confirmed by numerical results of the author (Nedoma, Haslinger and Hlaváček (1989), Nedoma (1990a)).

The area of the Carpathian Arc including its western part, i.e. the Western Carpathians, is the result of the collision of the African and Eurasian lithospheric plates that has been proceeding with varying intensity since the Late Mesozoic up to recent time (Figure 8.6). The folding of the West Carpathians can be described with sufficient accuracy (as far as the ideas of tectonics of lithospheric blocks are concerned) by the collision model of the invasive Carpathian Block with the platform lithosphere of continental type connected with a foreland of quasi-oceanic type. It therefore represents a dynamic collision model. As the invasive Carpathian Block was approaching the platform, the quasi-oceanic lithosphere was absorbed and compressed between the continental lithospheric blocks. The discussed region of the Western Carpathians is located in the central part of Europe with its centre in the area of contact of the Carpathian Arc with the North European Platform, and it includes areas of the Tertiary Carpathian Block, areas of the Alpine-Carpathian Foredeep, the Bavarian Block and the Block of Moravian Foredeeps, the Bohemian Massif, and the southern part of the Polish Paleozoic Platform. The West Carpathians consists of two different zones separated by the Peri-Pieninean Lineament — the Inner and Outer Carpathians (Roth (1980)). Their development is described in detail in Roth (1980) and Jiříček (1979).

The Outer Western Carpathians represent a part of the European Alpides. The European Alpides are formed by mountain ridges originated in the Alpine orogenesis that are in an acute tectonic contact with the platform part of the European plate and, moreover, separated by the Alpine tectonic nappes. From the orographic point of view, the European Alpides are a pronounced European phenomenon characterized by their dominant continuity over older parts of the Variscides or possibly even older mountain belts. Geologically, the Eastern Alps and Western Carpathians are connected by a Flysch zone* The inner zone of both these systems is rather discontinuous on the surface. The Flysch zone of the Western Carpathians forms their principal part from the geologic as well as orographic point of view. In the Alps, on the contrary, the Flysch zone is subordinate to the central mountain ridge both tectonically and orographically. Even though the outer zone can geologically be followed continuously from the Alps to the Carpathians, it is impossible to determine an exact boundary between the Eastern Alps and Western Carpathians, or an exact boundary between the Outer and Inner Carpathians, either. A better knowledge or a more detailed determination of these boundaries will be one of the important future

* Flysch denotes a rather thick sequence of rhythmically repeating layers of sediments of the terrigenous type (psammites, pelites, sandstones, silistones, sandy slates, sideritic limestones, etc.). A rhythmic alternation of rock species in the Flysch sequence is the primary feature of the Flysch sedimentation where the determining role was played by sedimentation from muddy flows which resulted in the gradational stratification of sediments that were compacted by pressure during time (Roth (1980)).

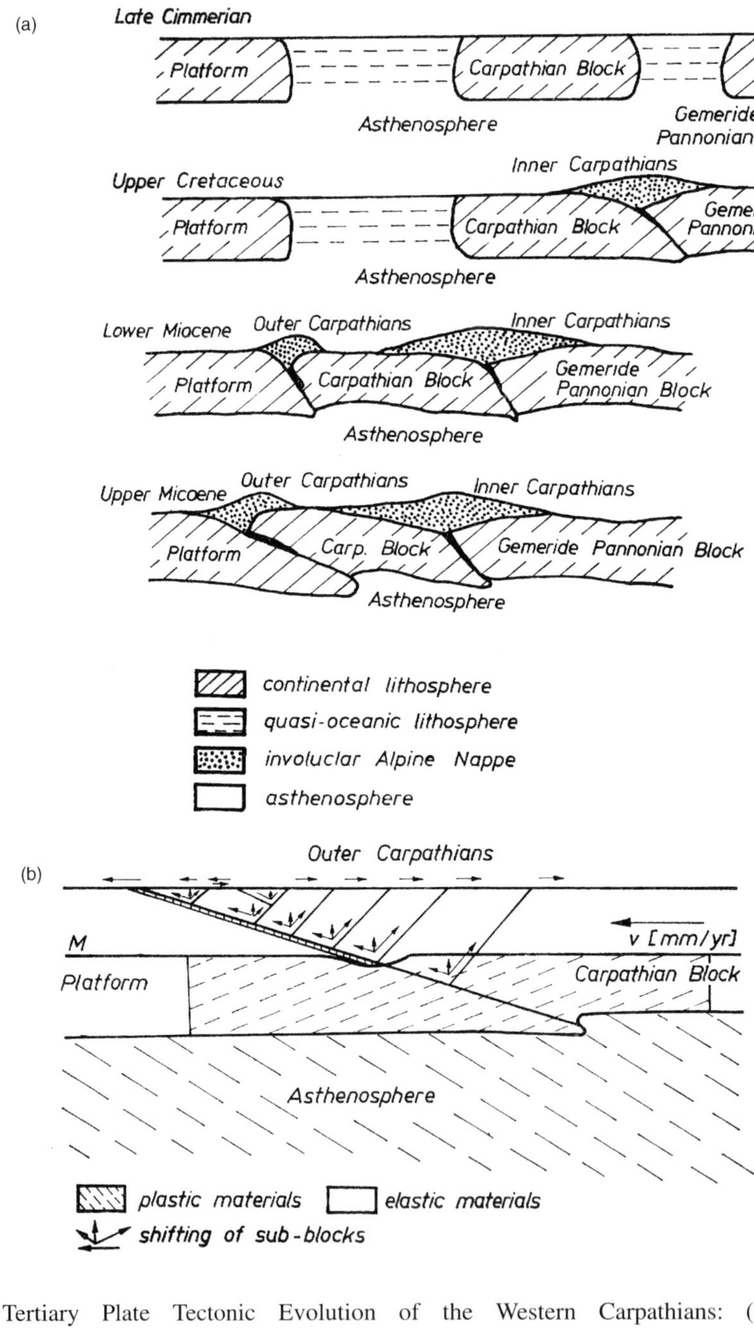

Figure 8.6. Tertiary Plate Tectonic Evolution of the Western Carpathians: (a) Scheme, (b) Schematic Cross-Section

contributions of the 3-D modelling of geodynamic processes in the lithosphere in this area. Main results can be expected only from 3-D models of the European Alpides of this area together with the results of the monitoring of the stress field in situ. At present, geologists assume that this boundary coincides with the flow of the Danube River. The determination of the boundary between the Western and Eastern Carpathians, if it exists, is more difficult, and a discussion on this subject is beyond our scope here.

The European Alpides are formed by two longitudinal zones, of which the outer one (whose folding is of Alpine type) is characteristic of the Alps as well as the Carpathians. It consists of Peninicum, Austroalpinicum, Helveticum and a Flysch zone in the Alps, and a Flysch zone and Klippen Belts in the Carpathians. The inner zone is formed by the Insubrian-Carpathian system of lithospheric blocks. In the Alps it consists of the doubleblock of the Southern Alps split by the Guidar fault into the western and eastern blocks, while in the Carpathians it consists of the Tatride, Pannonian and Apuseni-Transylvanian Blocks.

The folded zone was folded by an intensive Neoalpine (Tertiary) folding* (called the Paleoalpine orogenesis) both in the Alps and Carpathians in the Cretaceous period (Argyriadis (1974)). This ridge of Tertiary folding in both the formations, i.e. in the Alps as well as the Carpathians, diagonally crosses the zone folded in the Paleoalpine time. In the western part of the Alps this area of the Paleoalpine folding is situated in the foreland of the Alps, in the Eastern Alps the areas overlap, and in the Carpathians this area is located south of the Tertiary tectonics. The boundary that separates both the major zones of the Tertiary tectonics is formed by the Kanaves, Tonali, Guidicar, and Pustertal-Gailtal fault zones. In the Eastern Alps, fault zones are covered by the Neogene of the Styrian and Vienna Basins. The boundary thus continues from Graz to Vienna and further to Myjava, where the boundary of the inner and outer Neoalpine zone forms the tectonic contact between the Outer and Inner Western Carpathians. The tectonization is considerable in this zone. In the Alps as well as the Carpathians, the tectonic boundary of the inner and outer Neoalpine zone is one of the most apparent discontinuities from the geological point of view. It is formed by Late Tertiary longitudinal faults that are steeply inclined to the north near the surface. Along this steep fault system the Insubrian-Carpathian blocks were underthrusted to the north, below the Alpine-type folded Tertiary zone (Trümpy (1975), Roth (1980)) in the Tertiary. According to Roth (1980), the tectonic position of the southern boundary of the common Flysch zone principally changes on the boundary between the Eastern Alps and the Western Carpathians. In the Eastern Alps, the boundary between the Flysch zone and the Northern Alps lies inside the Tertiary zone of the Alpine type. In

* The Neoalpine folding started in the Upper Eocene by the horizontal movement of the Juvanids and Tyrolids together and continued until the end of the Pliocene, while the Paleoalpine folding in the Alps started in the Barremian. In the Western Carpathians it started by the Laramian movements, i.e. about 65 million years ago, and by the Paleoalpine remobilization caused by the Austrian movement about 100 million years ago. The culmination of the Paleoalpine movements in the Carpathians represented the sub-Hercynian phase about 88 million years ago that gave rise to a large nappe structure. The Epipaleoalpine Quasiplatform of the Inner Carpathians, that became the basis of the Tertiary microcontinents, is formed inside the Carpathians by the reconsolidation of this phase. The Paleoalpine tectonic phases show most significantly in the Inner Carpathians and in the Eastern Carpathians. On the contrary, the Neoalpine tectonic phases gave rise to the Outer Carpathians.

the Western Carpathians, the boundary between the Outer and Inner West Carpathians is placed into Carpathian nappes.

The oldest parts of the tectonically delimited West Carpathians are Proterozoic in age (Cambel *et al.* (1977)). Most of the crystalline of the West Carpathians, however, originated in the Paleozoic when the rocks were metamorphosed. Granitoids are Paleozoic (Bagdasarjan *et al.* (1977)). The continental basement of the West Carpathians was an original part of the Epi-variscan platform. The relics of quasi-oceanic lithosphere, originally distributed among the individual partial blocks of the continental lithosphere, form fragments of the ophiolithitic association of rocks inside the continental lithosphere at present. Their origin is placed in the Jurassic and Early Cretaceous period or the Lower Triassic at most. The Outer West Carpathians are formed by a folded cover that is separated from its original basement and called the Outer Carpathian Allochton. For example, the fragments of the platform basement in basal parts of the Silesian unit are fragments that were torn away from the basement by the moving nappe and shifted some distance further. They are separated from the nappe by a thrust zone and, borehole prospection has shown, their displacement is substantially smaller than the movement of nappes (Roth (1980)). The Outer Carpathians are thus not likely to form an independent lithospheric block. The neoid sedimentary cover separated tectonically from the basement was, in the frontal part of the Carpathian-Transylvanian Block, gradually rolled on the platform foreland of the Carpathians as a relatively narrow zone to the distance of 20 to 150 km.

The development of the Outer West Carpathians began by the folding of an area in the frontal part of the quasi-platform, consolidated in the Paleoalpine time, of the Inner Carpathians during the Laramian phase (Roth (1980)). The extent of the Outer West Carpathians increased during the next phases and the Outer Carpathians were developing new forms. These forms are denoted by tectonic phases or names of tectonic units formed by them. These phases are as follows: Laramian-Pieniny, Early Pyrenean-Magurian, Late Pyrenean-Silesian, Savian, Early Styrian-Ždánice, and Late Styrian-Moldavian. The most important phases for the origin of the cover structure of the West Carpathians was the Early Pyrenean phase about 40 million years ago, the Late Pyrenean phase between the Lower Oligocene and Egerian, and the Early Styrian between the Carpathian and Lower Badenian. The Laramian phase (Danian-Paleocene) and Late Styrian phase (Moldavian, following the Lower Badenian and mostly in the end of the Sarmatian) had a limited importance in forming the structure. Late Styrian (Moldavian) movements formed a new unit in the Outer Carpathians from older Outer Carpathian covers, i.e. the inner Late Styrian (Moldavian) unit that turned slightly, about 6°, to the left with respect to the Alps as well as to the platform basement. It can be expected that this unit was part of a large entirety that included the Carpathian-Transylvanian Block together with the Dinarides and a part of the Adriatic Block. In addition to these rotational movements, a slight left-hand rotation of the block with respect to the Outer Carpathians was also likely to occur, the angle being about 11°, when the Carpathian-Pannonian Block was being pushed under the Outer Carpathians. The total rotation of the Carpathian-Pannonian Block with respect to the platform basement thus amounts to about 17°. Roth (1980), however, shows that this relative movement of the Carpathian-Pannonian Block with respect to the inner unit of the Outer Carpathians was also accompanied by a right-hand rotation and thus the actual turn of the block with respect to the platform is less, about 11° to 17°. The rotation of the front of the Carpathians with respect to the platform took place in the Lower Badenian, i.e. about 12 to 15 million years ago, while the relative movement of the Carpathian-Pannonian Block with respect to the inner unit of the Outer Carpathians probably occurred in the Upper Badenian. It is assumed that the Carpathian-Pannonian Block started to carry away the Moldavian inner unit of the Outer Carpathians only during its Late Styrian turn. This rotation of the Carpathian-Pannonian Block enables us to explain some of the Tertiary fault zones as well as the origin of Late Styrian volcanic rocks in the Carpathian-Pannonian Block (volcanic rocks between Zázrivá and Nové Mesto-on-Váh, volcanic rocks in the Prešov Range, etc.).

The Early Styrian displacement gave rise to compression and translation movements in the Carpathians and also in the Alps, but no rotation movements as for the Late Styrian (Moldavia) displacement. The Early Styrian front of the Carpathians that was formed before the Badenian is preserved only partly, while it is mostly covered by the Late Styrian outer margin of the Carpathians.

From a detailed study of the West Carpathians it is apparent that in younger tectonic deformations, tectonic deformations of previous geological periods are also concealed. Tectonic forms develop all the time and the older formations, therefore, cannot vanish completely. This is, in a rough approximation, also shown by the numerical results of the analysis of geodynamic processes in the area of the Central Aleutian Islands. The case of the European Alpides is similar.

8.3.1.2. Model of the Western Carpathians

As has been shown above, the model of the Western Carpathians is a collision model. The model of the Western Carpathians will be designed as a collision model. The section of the Western Carpathians will arise from the geologic-seismic profile VI (Ilavský *et al.* (1979); see also Čech (1982)) whose construction is based on results of the deep seismic sounding. The section of length 600 km will include the Bohemian massif, the platform foreland of the West Carpathians, the southern part of the Outer West Carpathians, and a part of the Pannonian Block, i.e. the profile VI in the segment between the Krušné hory Mountains and Budapest (Figure 8.7) and in the future between the Krušné hory Mountains and the city of Bucharest. (The second profile, after some modification, can be constructed on the basis of the kinematic model of the Western Carpathians (Grecula and Roth (1978))). This complex section is only a rough simplification of the reality from the geological point of view. In its upper part, the model is taken from the deep seismic sounding profiles according to Ilavský *et al.* (1979); however, the lower part of the lithosphere, the asthenosphere, and mesosphere are created in addition by the author. In the model the lower boundary of the lithosphere was taken from the map of the lower boundary of the lithosphere in Central Europe. This map of the lower boundary of the lithosphere of the greater part of

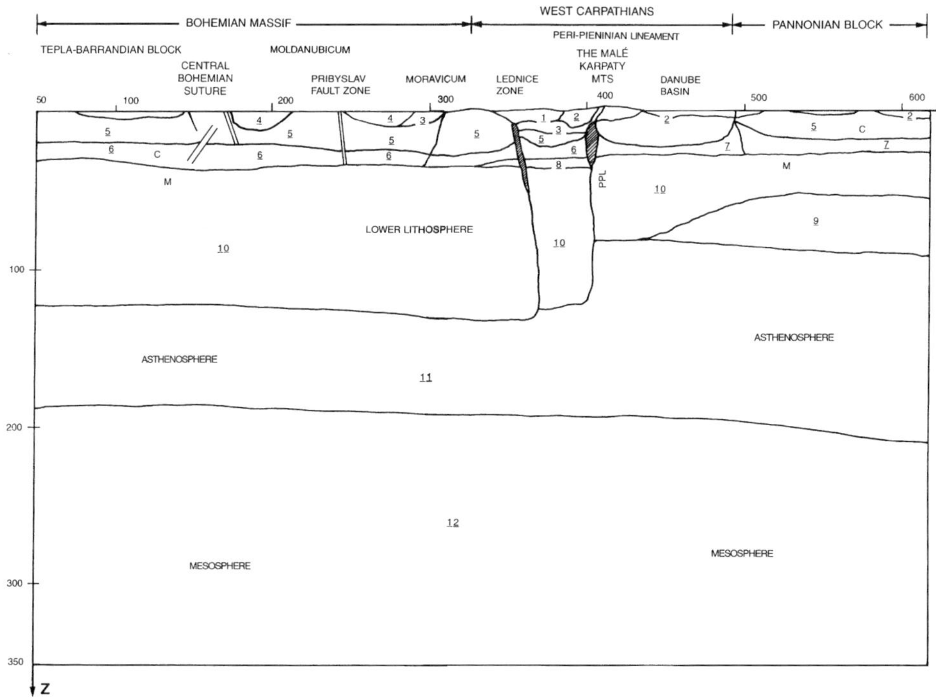

Figure 8.7. Model of the West Carpathians, Profile VI

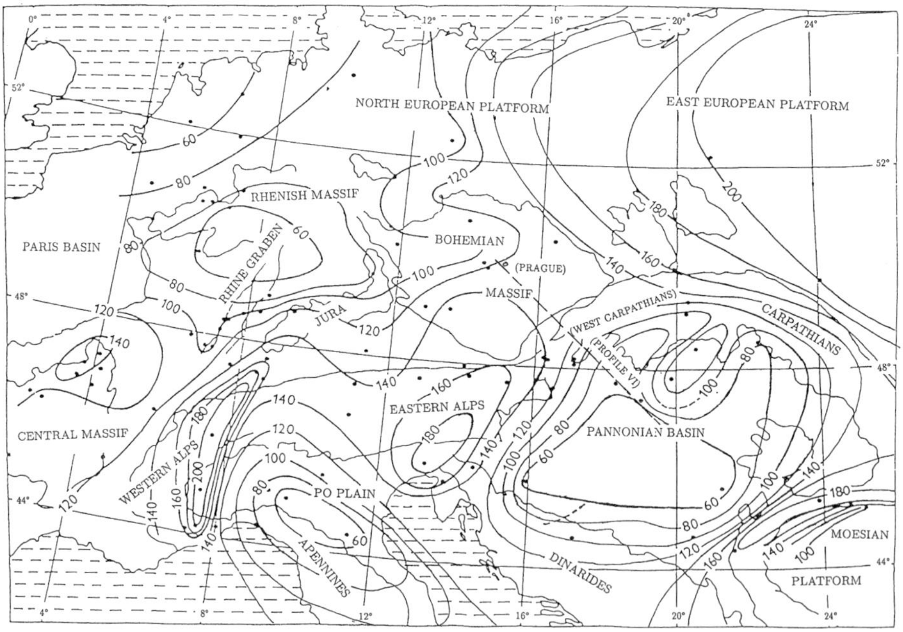

Figure 8.8. Map of the Lower Boundary of the European Lithosphere

Europe (Figure 8.8) was constructed by the author on the basis of his knowledge about this topic, and on results of contemporary geophysics (see, for example, A'dám (1976), Babuška, Plomerová, Šílený and Janatková (1986), Babuška and Plomerová (1995), Wielandt, Plešinger, Sigg and Horálek (1987)) that are modified by the other information, and information from the mathematical modelling of the Central Aleutian Islands and the Himalayas. We should realize that the extent of geological and geophysical information decreases quickly with depth and thus the lower parts of the sections bear only an informative character in order that the dynamics of the area, i.e. the possibility of displacements in time on the contact surfaces of deep faults and those of the lithosphere on its basement, could be described in the section. The section is a collection of qualitative results of geology and geophysics, i.e. results of all the geophysical fields and branches.

8.3.2. Numerical Verification of Geodynamic Processes in the Western Carpathians

In order to solve numerically the model of the West Carpathians, based on the plate tectonics hypothesis and the author's theory from the previous chapter, the elastic rheology will be assumed. The model assumes, due to the map of the lower boundary of the lithosphere, the thickness of the platform lithosphere ~120 km, while in the West Carpathians 80-110 km and the Pannonian block 60 km thick. We shall assume that the average density in the deeper parts is ~3400 kg/m^3, while in the upper parts ~2850 kg/m^3 and the upper sedimentary rocks ~1400-1670 kg/m^3 (see Table 8.2, where the velocities of P and S waves are in m/sec and the density is in kg/m^3). Velocities of seismic waves P and S were taken from the interpretations of the deep seismic sounding results (Beránek and Zátopek (1981)). The area of the West Carpathians is relatively quiet. The present relative motion is assumed to be ~3-6 mm/yr.

On the contact boundaries the Coulombian friction is also assumed. An upper estimate of the coefficient of the Coulombian friction was found from the estimate $\mathscr{F}_c^{kl} \sim (c_s/c_p)^{1/2}$, where c_s and c_p are velocities of seismic S and P waves (see above and Nedoma (1987)).

We have calculated the interplate stresses. The principal stresses are illustrated in Figure 8.9, and the components of the stress tensor in Figure 8.10 The distribution of principal stresses indicates a cumulation of compression in practically the whole investigated area of the upper parts of the Pannonian block, the Carpathian block and the Bohemian massif with one outweighing direction of the principal stress directed aslant downwards. Numerical results indicate greater cumulation of

Table 8.2. The model-physical data

ρ	1300	1480	1670	1780	2100	3200	3280	3320	3330	3400	3450	3550
c_p	3400	4500	5000	5500	6000	6500	7000	7500	8000	8200	8500	8800
c_s	2040	2650	2950	3250	3520	3850	4120	4340	4780	5020	5100	5150

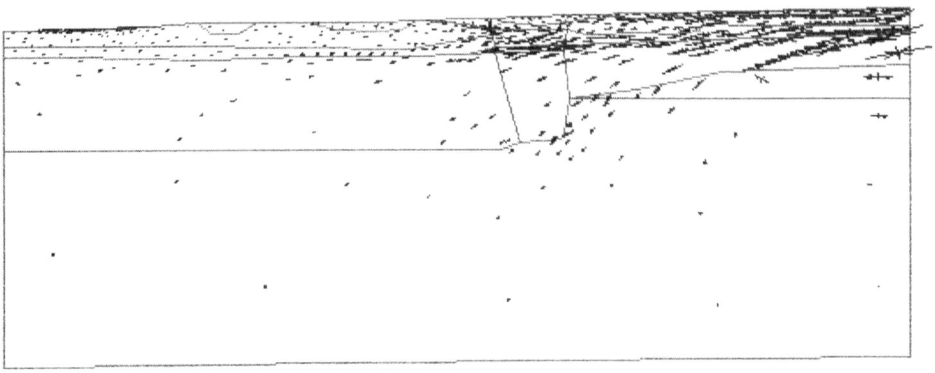

Figure 8.9. Principal Stresses on the Profile VI

compression in the area of the Pannonian block and the Carpathian block, while in the Bohemian massif the distribution of stresses decreases with the distance from the West Carpathians. It is also the reason for the existence of earthquakes in the Malé Karpaty Mountains and the Danube Basin. Moreover, the numerical results indicate tensile stresses on the lower lithosphere and upper asthenosphere in the area of the Pannonian Basin, which indicates the bulged out region with partially molten rocks which are the result of bulged out lower lithosphere and upper asthenosphere accompanied by decreasing normal stresses. Since the temperature of melting depends on the pressure, and increases with increasing pressure, then with decreasing pressure in this area the temperature of melting also decreases and rocks below the lithosphere of the Pannonian Basin start partially melting. Since the thickness of the lithosphere here is smaller then the anomalous thermal field is observed.

8.4. GEODYNAMIC EVOLUTION OF THE HIMALAYAN COLLISION REGION AND THE INTRAPLATE TECTONICS IN THE NORTHERN INDIAN OCEAN, AND GEODYNAMIC MODELS OF THE HIMALAYA AND THE INDIAN PLATE MARGIN

8.4.1. Introduction

The geology and tectonics as well as the geodynamics of the Himalaya, the peninsular Indian Shield, the Pakistan-Kashmir-Karakorum region, the Indus-Ganga-Brahmaputra basin and the Andaman-Nicobar arc system have been studied by a large number of people. From the resulting large number of papers we shall mention the following: Wadia (1931), (1975), Verma (1991), Harrison, Copeland, Kidd and Yin (1992), Powell and Canaghan (1973), Scotese (1991), Gupta and Delany (1981), Valdiya (1976), Verma, Roonwal and Gupta (1993), Yu and McNutt (1996), Tapponnier and Molnar (1977), Wu and Wang (1988), Gansser (1964), (1991), Molnar and Lyon-Caen (1989), Molnar and Tapponnier (1975), (1978), Molnar, Fitch and

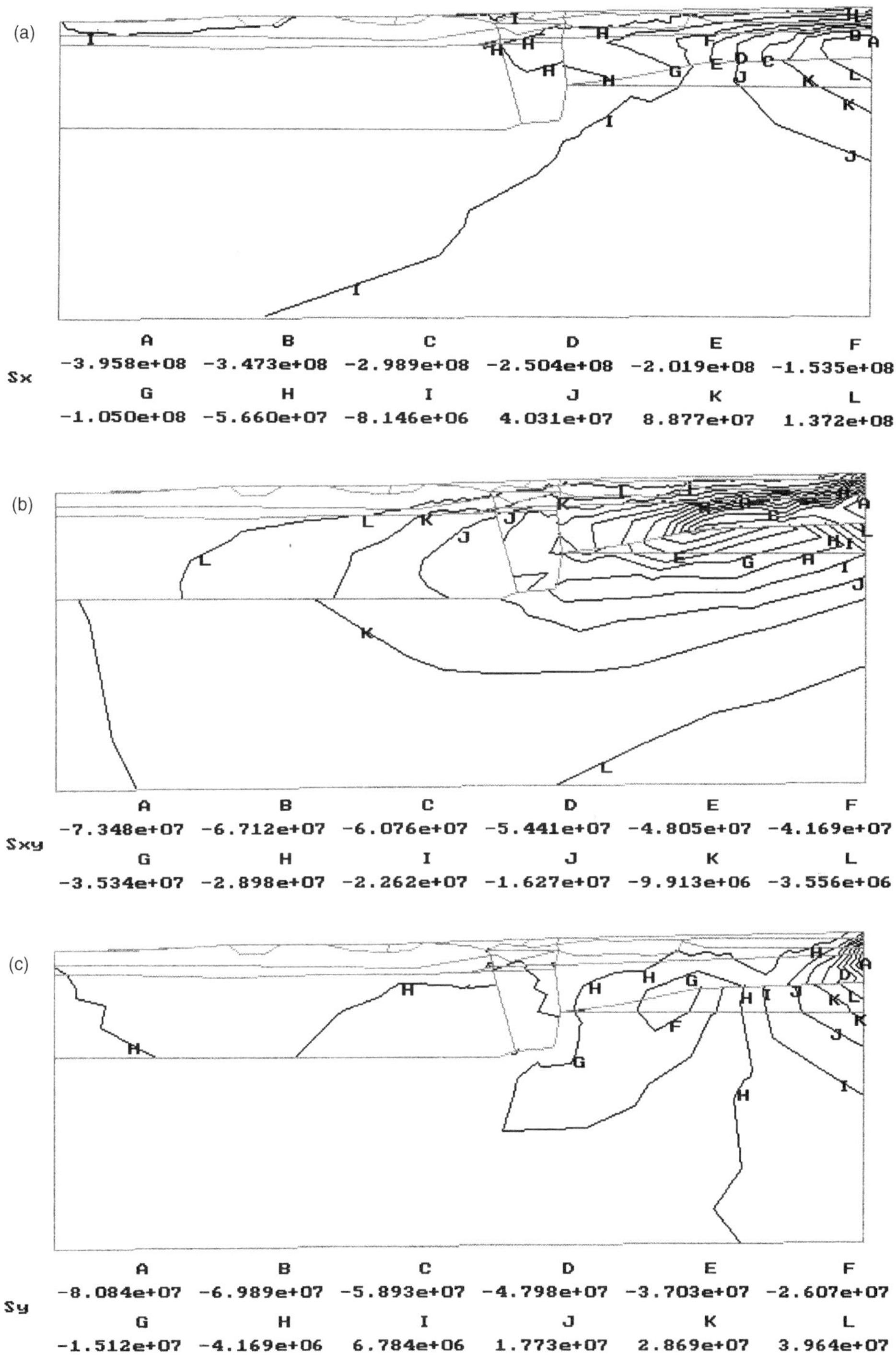

Figure 8.10. Stress Tensor Components along the Profile VI (a) horizontal stresses, (b) shear stresses, (c) vertical stresses

Wu (1973), Ni and Barazangi (1984), (1985), Ni *et al.* (1989), Tandon and Srivastava (1975), Verma and Prasad (1988), Verma and Reddy (1988), Verma, Mukhopadhyay and Roy (1977), Valdiya (1984a,b), (1985), Sato, Bhatia and Gupta (1996), Sharma (1983), Holt, Ni, Wallace and Haines (1991), Rastogi (1976), Westaway (1995), Thakur (1987), (1992), etc. referring to the region of Western, Central and Eastern Himalayas, and Verma (1991), Verma and Mukhopadhyay (1977), Verma and Kumar (1987), Verma, Mukhopadhyay and Ahluwalia (1976), Verma, Mukhopadhyay and Bhuin (1978), Verma, Mukhopadhyay and Nag (1980), Mukhopadhyay (1984), (1988), Mukhopadhyay and Dasgupta (1988), Evans (1964), Dasgupta and Mukhopadhyay (1993), Bergman and Solomon (1985), Curray, Emmel, Moore and Raitt (1982), Eguchi, Uyeda and Maki (1979), Hamilton (1979), Mukhopadhyay and Krishna (1991), etc. dealing with the Burma and Andaman-Nicobar regions.

The origin of the Himalaya, the highest mountain chain in the world, has been the most interesting problem from the geodynamic point of view. Plate tectonics, which considers the lithosphere as a number of rigid plates which are in a contact along their common boundaries, can explain many of the features of the Indian subcontinent. Molnar and Tapponnier (1975) consider most of the large scale tectonics of Asia to be a result of the India-Eurasia continental collision, which apparently not only created the Himalayas but also rejuvenated an old orogenic belt (Tien Shan) approximately 1000 km north of the Indus-Tsangpo Thrust. The collision of the Indian plate and Eurasia involves the largest scale of active continental deformation, and therefore a rather diffuse seismicity prevails in this region. At present authors consider the collision of the Indian and Eurasian plates, with the Indus-Tsangpo Thrust marking their contact, to give rise to the Himalaya. According to these authors, the seismicity pattern in the Tibet Himalayan region is quite diffuse. The evolution of the Himalayas is characterized by several stages of evolution, starting from the Aptian up to the present. During the first two stages of evolution of the Himalayan region, the subduction of the oceanic lithosphere under the continental one of Eurasia was characterized by different speed of movement of the Indian plate relative to the Eurasia one.

The Aptian (Middle Cretaceous ~110 Ma) is the time when India, together with Madagascar, splits apart from East Africa, and as a result the Somalian basin within the Indian Ocean arises. At the period of Cretaceous-Paleocene (~65 Ma), India moves to the north at a very high velocity of about 18.4 cmyr^{-1} with the India-Eurasia pole of relative motion in North Africa. In the Late Eocene and Oligocene between 65 and 54 Ma, it moves in the north direction at a velocity of 12 cmyr^{-1}, and between 54 to 35 Ma by the same northward motion at a speed of 8 cmyr^{-1}. During Late Eocene-Oligocene, oceanic areas in the Neo-Tethys are shortened by 300 km to the west, 600 to 1000 km off Iran and by 3000 km between India and Eurasia. Then, between 35 and 20 Ma (Early Miocene ~20 Ma), the speed of motion of the Indian plate began to slow down to 5 cmyr^{-1} due to the collision of India with Eurasia. The India-Eurasia pole of relative motion is affected in the same way as in the preceding period. However, it abruptly moves east of the Horn of Africa, which results in a dextral northwestward motion of India, due to the collision of the two continental plates — the Indian and Eurasian plates. During this period, the whole Alpine-Himalayan system along approximately the same latitudes was created.

This is the time when continental collision operates along all the Eurasian margins. As a result, a complex system of intra-continental strike-slip faults begins to appear. During Early Miocene ($\sim$20 Ma) India moves northwestward. The Flysch sedimentation was observed in the basin thrust over the Indian plate where it forms the present Waziristan, Zhob and Quetta arcs. During the Middle Miocene ($\sim$10 Ma), the India-Eurasia pole of rotation moved back from the eastern Indian Ocean into North Africa, into the region of northern Egypt. The speed of motion of the India plate to the north is about 4 $cmyr^{-1}$ and it became north-northeastward at about 5 $cmyr^{-1}$ after the Upper Miocene. The continent/continent collision operates along the Neo-Tethys margin with the rate of convergence increasing from west to east. Folded mountain ranges develop, for example the Pamir-Hindu Kush and Himalayas, in the region of collision of the Indian plate with the Eurasian plate. These mountain ranges continue via the Zagros-Lesser Caucasus-Anatolia, North-east and North Apulia to the Dinarides and the Alps. Formation of the Hindu Kush and Elbrus mountain ranges is achieved. These mountain belts continue westward as the high reliefs of the Pamir, Karakorum and Himalayas. To the north of them, molasses are deposited in a marine and in a continental environment.

The relative position of India and Eurasia at four different times 71, 55, 38 and 10 Ma (million years ago) is shown in Figure 13.2a,b of Verma (1991), p. 280.

The tectonic evolution of the region has continued from the last 10 Ma up to the present. To understand the stress-strain state in the northern part of the Indian plate — the Indian subcontinent, we can divide it into four different regions: the Himalaya, the Indus-Gangetic basin, the Peninsular Indian Shield, and the Andaman-Nicobar arc region.

The Himalayas are geographically defined as the 2500 km long mountain range 200–250 km wide, from Nanga Parbat (8125 m) in the west-northwest to Namche Barwa (7755 m) in the east, and convex towards the south. South of the Himalayas lies the Indus-Ganga-Brahmaputra basin, and south of the Ganga basin lies the Peninsular Indian Shield. The western part of the Himalayas is bounded by the Sulaiman and Kirthar ranges; by the Hindu Kush ranges on the north-west and by the Tibetan Plateau on the north. To the east, the Himalayas are bordered by the Arakan-Burma ranges. The Andaman-Nicobar ridge system forms the southward continuation of the Arakan-Burma arc and it represents the boundary between the Indian and Eurasian plate in the east.

The Himalayas consist of sedimentary rocks, deposited over millions of years in shallow seas. To understand the evolution of the Himalayas, we must firstly understand the relationship between the Indian and Tibetan landmass, the evolution and the history of Gondwana land and the Asian plate.

The border between the Indian lithospheric plate and the Eurasian plate, according to Ghosh (1988), Kaila and Narain (1976), Verma (1991) and others, is situated in the Indus-Tsangpo Thrust (ITT). They believe that the ITT (they denote it as the Indus Suture Zone (ISZ) or the Indus-Tsampo Suture (ITS)), with the presence of ophiolitic suite, is possibly the relict of old Tethys and the junction of two continental lithospheric plates. It might represent an earlier subduction zone and a collision boundary. In the light of the plate tectonic hypothesis, the general mechanism of plate movement appears to involve a push and pull mechanism (Ghosh (1988)). Details of the collision process between the Indian and Eurasian plates have been suggested by Valdiya (1984b). The stage of the subducted Indian plate below the Eurasian plate during the period between the Late Cretaceous and the Palaeocene is presented in Verma (1991), p. 282. The process of subduction led to the formation of granitic plutons along the southern margin of the Eurasian plate (Tibetan Plateau). The granitic plutons form a magmatic arc along the north margin of the Indus-Tsangpo Thrust (ITT). In addition, a volcanic arc developed north of the trench axis, and it extends from the Kohistan region, situated west of Nanga Parbat, to Shigatse, situated northwest of Lhasa, in the east part of this volcanic arc. In the author's geodynamic model, which will be discussed later, the border between the Indian lithospheric plate and the Eurasian plate is situated in the Main Boundary Thrust (MBT), which is explained by the numerical results and by their good correlation with geological observations.

Valdiya (1984a,b) supposes that actual collision between India and Eurasia took place during the Eocene to Oligocene, and it leads to the obduction of deep oceanic crustal material in the form of ophiolites, which in this area consist of a suite of rocks including harzburgite, pillow lavas, dunite, sheets of dolerites, layered gabbro, etc.. These rocks are in close neigbourhood with trench sediments such as pelagic clays of the Upper Cretaceous age, radiolarites, turbidites, etc.. South of the Indus-Tsangpo Thrust, plutonic bodies of quartz diorite-granodiorite-granite association were also formed. The rise of the crystalline basement occurred south of the Indus-Tsangpo Thrust. Ophiolite nappes of the Malla Johar phase formed as a result of the geotectonic movements in this area. During Middle Eocene, a back-thrust developed north of ITT.

8.4.2. Geological and Tectonic Characterization of the Himalayas and the Tibetan Plateau

Based on geological and tectonic studies the Himalayas (see Gansser (1964), Das and Filson (1975), Rastogi (1974), Chandra (1978), Ni and Barazangi (1984), Baranowski, Armbruster, Seeber and Molnar (1984), Molnar and Lyon-Caen (1989), Molnar (1990), Verma, Mukhopadhyay and Roy (1977), Verma (1991), Khattri (1992), Chandra (1992)) can be subdivided into two main areas: the Siwalik Belt, and the Lesser and High Himalayas.

8.4.2.1. The Siwalik belt

The Siwalik Belt (see Valdiya (1976)) is characterized by rugged topography, being built by Tertiary and Early Pleistocene molassic sediments. The belt is bounded by the Main Boundary Thrust (MBT) in the north and separated by the Himalayan Frontal Fault (HFF) from the planes in the south. The MBT consists of a series of thrusts that separate the predominantly pre-tertiary Lesser Himalayan sediments from the Tertiary Siwalik sedimentary belt (the Sub-Himalayan belt). The MBT has developed during Pliocene and the thrust zone was active through the Pleistocene (Valdiya (1981)). However, modelling results presented by the author firstly in the Indian School of Mines, Dhanbad and in the Wadia Institute of Himalayan Geology, Dehra Dun 1996 and now in this book indicate that the MBT represents the main contact boundary between the Indian lithospheric plate and the Eurasian plate as will be discussed later with analysis of the obtained results.

According to Gansser (1964) the Siwalik belt is built up of structures with a normal and not a reversed sedimentary succession. More or less wide synclines alternate with steep and often faulted asymmetric anticlines. The strike-slip faults in this area are steep at the surface, but probably dip towards the mountains at depth (i.e. northwards). In the upper Siwalik belt, the younger layers outcrop in the north along the Main Boundary Thrust (MBT), whereas monoclinally folded older Siwalik members outcrop in the south. Their southern limit is mostly erosional and their course is interrupted by the large fans of the Ganges alluvial deposits. Folding and faulting in this area are of the early to middle Pleistocene age, and probably have preceded the overthrusting of the Lesser Himalayas along the MBT. The strike of the Siwalik does not always conform to that of the fault. North of the Main Boundary Thrust, the Siwalik does not overlay the Lesser Himalayas. They are overthrust by the pre-Siwalik rocks of the Lesser Himalayas and, therefore, they do not form a normal cover for these ranges. Moreover, the northern hinterland of the Himalayas is a highly elevated mass, which is drained by southward-flowing rivers.

The Main Boundary Thrust forms the southern limit of the Lesser Himalayas. Secondary thrust or faults branch off the Main Boundary Thrust (e.g. in Punjab, where they separate the Siwalik and Murrees). These secondary thrust zones diverge westwards and then join with the main fault towards the east. One of them is the Krol thrust sheet. The Krol thrust sheet coincides, for over 400 km, with the Main Boundary Thrust. On the other hand in some places, as in south Nepal, the higher, internal crystalline thrust sheets lie in contact with Siwalik. Gondwana formations are situated between the MBT and the next higher, internal thrusts, as was observed in the Damudas. Along the frontal outcrops, the MBT and the internal thrusts show a similar dip and run parallel to each other.

8.4.2.2. The Lesser and High Himalayas

The topography of the Himalayas (Ni and Barazangi (1984), Verma (1991)) is characterized by:

(i) an abrupt rise of the Sub-Himalayas and the Lesser Himalayas above the close-to-sea-level Ganga basin;

(ii) a gradual increase in elevation across the Lesser Himalayas;

(iii) a steep slope between the northern Lesser Himalayas and the High Himalayan range;

(iv) a uniform elevation of about 5 km north of the High Himalayan range.

The uplift of the Himalayas is a consequence of the horizontal convergence between India and Eurasia. The Lesser Himalayas and Sub-Himalayas represent ranges with elevation from about 1500 to 3000 m with dissected valleys. The Lesser Himalayas consist of various types of sediments and metamorphics which have been strongly folded and thrusted along several North-South thrusts.

The Lesser Himalayas are separated from the High Himalayas by the Main Central Thrust (MCT), which has a northward dip of about 30–45°. The MCT is known under different names in different parts of the Himalayas, i.e. as the Panjal thrust in the Kashmir Himalayas, the Vaikrita thrust in the Kumaun Himalayas and the Darjeeling thrust in the Eastern Himalayas. The thrust is characterized by a zone of an abrupt change in the structure, intense shearing, and in the grade of metamorphism representing the main manifestation of the presence of the MCT. The MCT has existed since mid-Tertiary time, and there are some geological indications of minor recent movements along the MCT (Valdiya (1980), (1981), Gansser (1980), Verma (1991)). According to the author's results, it originated during the subduction of the oceanic (Tethyan) lithosphere under the Eurasian plate, and therefore it is younger than MBT and older than the Indus-Tsangpo Thrust.

The Lesser Himalayas on their southern side are limited by the Main Boundary Thrust. As was observed in situ, a border facies of the Peninsular Shield makes up a relatively large volume of the formations of the Lesser Himalayas, which can have their origin in the young Himalayan phase during the mid-Tertiary age. Gansser (1964) noted that the primary differences in metamorphism, reflecting the history of the shield rocks, were overprinted by the metamorphism of this young Himalayan phase. Moreover, the striking changes in metamorphism are mostly restricted to the older rocks. In the Himalayas no higher grade metamorphism has been observed in formations younger than Ordovician. Only along the ophiolite belts, such as the Tibetan zone and in the Karakorum, were younger beds of the Jurassic age (e.g. in southern Tibet) and Permo-Carboniferous age (e.g. in the Karakorum) affected by higher-grade metamorphism. The Himalayan metamorphism which produced migmatization and finally young mobilization (resulting in tectonically undisturbed tourmaline granites) were restricted to the previously metamorphosed sections of the crystalline thrust sheets. The youngest granites which intrude even Mesozoic sediments have not produced any marked contact metamorphism.

Along the MBT, a strong tectonization was observed, but this structural zone is not a sharply delimited feature. The strong tectonization is manifested in the foothills of Bhutan. Within this tectonized zone along the Himalayan south border, independent thrust sheets, such as the Krol thrust, are developed. Where the Krol thrusts are absent, a segment of steeply thrust Gondwana rocks occur. They extend along practically the whole length of the Nepalese foothills, and continue through Sikkim and Bhutan up to the frontal thrust region in the NEFA Himalayas.

The structures of the older, inner thrust masses of the Lesser Himalayas (Gansser (1964)) depend on the lithology of the involved formations which varies from place to place.

Between the Main Central Thrust and the Indus-Tsangpo Thrust is situated the 2500 km long mountain range of the High Himalayas, which is about 200–250 km wide. It starts in the area of the Nanga Parbat on the west-northwest and ends in the area of Namche Barwa on the east.

The High Himalayas are built by monoclinal in the centre, and/or gently folded in the western and eastern part, Precambrian metamorphics intruded by middle Tertiary granites. Valdiya (1979), (1981) shows that the High Himalayan rocks represent basement rocks. A previous study of elevation in the High Himalayan-Tibetan region (Bird (1978)) indicates that the mean elevation of the High Himalayas is about 5000 m, similarly as in Tibet, but with maximal elevation more than 8000 m. In the northern half of the Himalayas there are fewer earthquakes. In contrast, there is a maximal concentration of seismicity in the southern half.

North of the High Himalayas lies a vast stretch of the Tethys Himalayas, made up of fossiliferous sediments deposited in the basin of the Tethys Sea during the Late Precambrian to the Upper Cretaceous period. The Tethys sediments occur in several basins that extend along almost the entire length of the northern margin of the Himalayas, and which are suddenly cut off by the Indus-Tsangpo Thrust (ITT). The Tethys sediments are preserved in several basins all along the length of the Himalayas, i.e. the Kashmir and Spiti basins in the *NW* Himalayas, the Kuman basin in the central part, the West Nepal basin in the Nepal Himalayas and the Kampa Dzong basin in the eastern part of the Himalayas.

The Indus-Tsangpo Zone (according to our results we speak about thrust and denote it as the Indian-Tsangpo Thrust, ITT) according to Gansser (1964), (1980) denotes the northern limit of the Indian subcontinent following the Late Cretaceous-early Tertiary closure of the Neo-Tethys. The ITT comprises pillow lavas, volcanics, and basic and ultrabasic rocks as well as radiolarite (see Sharma and Gupta (1983)). The author, due to his own numerical results, shows that the Indus-Tsangpo Thrust is the secondary deep fault of the third period of evolution that originates in areas with shear stresses.

North of the Indus-Tsangpo Thrust, a long belt of Trans-Himalayan plutons is situated. The belt extends from Pakistan, i.e. from Swat 36°N, 65°E, situated west of the Nanga Parbat-Hamosh massif, to the eastern margin of the ITT $\sim 30^0$N, 90°E, where it forms an Andean-type magmatic arc. The plutons located along the belt are known as Swat, Ladakh, Kailas, Tsangpo, Kyichu (Lhasa), and Eastern Tsangpo. The Swat and Ladakh plutons are separated by a transverse structure, the Nanga Parbat high, which forms part of the western Himalayas. The Trans-Himalayan Plutons have a genetic link to the deep fault zone, which suggests that magma production is related to the processes that occurred during the subduction of the Indian plate. The Trans-Himalayan plutons are similar to those of the Andean subduction zone (the Andean batholiths from the zone of subduction of the Nazca plate below the South American plate).

The High Himalayan pluton belt extends over a large part of the Himalayas and it runs close to Kedarnable, Badrinath, Mustang, Mugu, Manaslu, Makalu, Kanchanjangha, Thirupu and Chekha in Bhutan on the east. The plutons are of two types. The older: 580 ± 10 Ma — the Manali-Rohtang area, 495 ± 16 Ma — Lahsel-Spiti area, 467 ± 46 Ma — biotite granite from Manikaran, 486 ± 10 Ma — Nepal Himalayas; and the younger (Upper Tertiary): 28 Ma-18 Ma — Manaslu granite, 18 Ma — Badrinath granite, 19 Ma — biotite from Everest granite. These younger plutons were formed by remelting of the pre-existing older rocks. Mehta (1980) shows that the younger ages fall into three groups related to different phases of the Himalayan orogeny. The first group 75-50 Ma is related to folding and metamorphism; the second one 40-25 Ma appears to be related to the uplift of the Himalaya; the third one 25-10 Ma is related to the major uplift, formation of nappe structures and regional metamorphism. The Himalayan crustal thrusting of hot slab over a rather cold volcano-sedimentary pile provides the necessary fluids, which pass along the MCT and produce a leucogranite mass.

8.4.2.3. The Tibetan plateau

The Tibetan Plateau is situated north from the High Himalayas. The southern and main Kun-Lun borders the High Tibetan Plateau to the north. The geology of the region of Central Tibet is not so complicated. It is characterized by elongated ranges striking mainly east-west. Marine Jurassic is succeeded by Cretaceous and widespread Tertiary formation of mostly continental type. The distribution of young volcanic rocks in the Tibetan Plateau and along the Kun-Lun Range is quaint (Gansser (1964)). As often suggested, the youngest basaltic lavas together with relic volcanic cones are probably of Pleistocene age. In the southern Kun-Lun, there are indications of post-Pleistocene volcanism (Leuchs (1937)).

The highland to the west is situated between the Karakorum in the Southwest and the Kun-Lun in the north. Tethys-type marine formations are widespread in Inner Tibet. The Tibetan high plateau is limited to the east by the Eastern Himalayan syntaxis. Its north-east border is formed by the Nan-Shan ranges. The Nan-Shan ranges end in the Kuhn-Nor depression. The E-W striking element is the Tsingling-Shan Range. A major structural element is formed by the Tsinking-Shang Range of metamorphic and basic rocks.

Northeast and east Tibet are built by older formations. In the Tang La Range and *NW* Yunnan, the Devonian formations and Hercynian core of the range are observed. From the Lhasa-Bomi folded belt, thick Devonian and Carboniferous deposits with intercalated submarine volcanics representing the deeper elements of the Tibetan platform are described. In eastern Tibet (Chando region), deposits of the Lower Palaeozoic are situated. Many authors discuss the question of whether or not the Tibetan Plateau was really consolidated in the Hercynian period. The influence of the Himalayas is already observed in southern Tibet, where the Trans-Himalayas (Kailas Range) as a continuation of the Karakorum and Ladakh ranges to the SE represent a structural element lying south of the main Pamir's region. The author shows that the Tibetan Plateau was mainly formed during the first two evolution periods, namely when the speed of the Indian plate with respect to the Eurasian plate was less than 8 $cmyr^{-1}$.

Structures of the Tibetan/Tethys Himalayas are of several types. The crystalline thrust sheet of the Higher Himalaya is normally covered by a great thickness of mostly pelitic sediments. These sediments are welded to the schists and form the highest part of the main crystalline thrust mass in several parts of this area. It was in the Ordovician that more differentiated sediments, as quartzites, shales, limestones, are observed. South-vergent folds and south-directed thrusts are mainly observed, but in northern Nepal the north-vergent folding plays an important role. Towards the western Himalayas, where the range widens and the elevations of the main range decrease, the folding of the Tibetan Himalayas is gentler and more of a Jura-type. The Mesozoic section of the Tibetan Himalayas is characterized by differentiation into plastic shales, thick limestones and dolomites. Disharmonic folding is observed in these areas.

8.4.2.4 Seismic activity

The Himalayan region is one of the most seismically active regions in the world (see Figures 13.10, 13.13 and 13.15 of Verma (1991). The seismo-tectonic activity in the Himalayan region is primarily confined to the region between the MBT and MCT. A notable feature of the seismicity of the Himalayan arc is that earthquakes of large magnitude (exceeding 8.0) have occurred close to the MBT of the Himalayas. These include the Kangra earthquake of 1905 (of magnitude ~8.4, see Gutenberg and Richter (1954)), the Bihar earthquake of 1834 (of magnitude ~8.3), the Assam earthquake of 1950 (of magnitude 8.6), and probably the Shillong Plateau earthquake of 1897 (of magnitude ~8.7), as well as earthquakes of the magnitude ~7.0, such as the Kashmir earthquake of 1985, the Kuman Himalaya earthquake of 1803, the Nepal Himalayan earthquake of 1833, as well as two earthquakes that occurred between the 1897 Shillong earthquake and the 1950 Assam earthquake of magnitude 7.2 (1943) and 7.7, and which are also located in this region. The seismic activity is quite intense and it is situated in shallow depths in the eastern and western Himalayas. The seismicity in the central Himalayas, as observed by seismologists, is proved to be rather low and irregular in time and space. The linear alignments of the earthquakes, situated transverse to the Himalayan arc system, are also evident. As was shown by Chandra (1992), some transverse tectonic structures, like grabens but not so well documented, can be seen. The earthquakes in the neighbourhood of these transverse structures are of a strike-slip character, with one of the planes transverse to the Himalayan arc system, as is documented by the earthquakes of Bihar-Nepal (1934, magnitude ~8.4, see Singh and Gupta (1980)), Bihar-Nepal (1988, magnitude ~6.6), and Uttarkashi (1991, magnitude ~6.6). North of the Indus-Tsangpo Thrust, i.e. in the region of the Tibetan Plateau, the seismic activity, with shallow earthquake depths, is irregular in time and space. In this Tibetan region, however, intermediate-depth earthquakes were also observed. As was shown in Molnar, Fitch and Wu (1973), Biswas (1989), Molnar and Lyon-Caen (1989), Molnar (1992), the fault plane solutions point to thrust, strike-slip and normal types. However, the source mechanisms of most of the earthquakes in the Himalayan arc system are predominantly of the thrust type, as was shown in Fitch (1970), Rastogi (1974), Chandra (1978), Baranowski, Armbruster, Seebar and Molnar (1984), Ni and Barazangi (1984), Molnar and Lyon-Caen (1989), Molnar (1990), Khattri (1992). At the end we can introduce another result of observation: during the last century, the location of four great earthquakes were situated in the western and eastern Himalayas. The central Himalayas have been relatively quiet.

8.4.3. Geodynamic Model of the Himalayas, Profile — Nanda Devi

The Himalayas, the highest mountains in the world, form a well defined arc to the north of the Indian Peninsula and extend over a distance of about 2500 km from the west-northwest (Nanga Parbat 8125 m/Indus River) to the east-northeast (Namcha Barwa 7755 m/Brahmaputra River). They comprise more than 30 peaks, rising to the height of 7300 m above the mean sea level. The formation of the Himalayas is the result of the collision between the Indian and Eurasian plates. The Himalayas can be divided physiographically into four parallel, longitudinal belts: the Outer or Sub-Himalayas, the Lesser or Lower Himalayas, the Higher or Great Himalayas and the Tibetan or Tethys Himalayas. Farther north, as a part of the Himalayan orogenic belt, the Trans Himalayas are situated.

The investigated area comprises five main lithotectonic zones of the Himalayas and its foreground: the Indus-Ganga-Brahmaputra plain, the Siwalik belt of the Outer

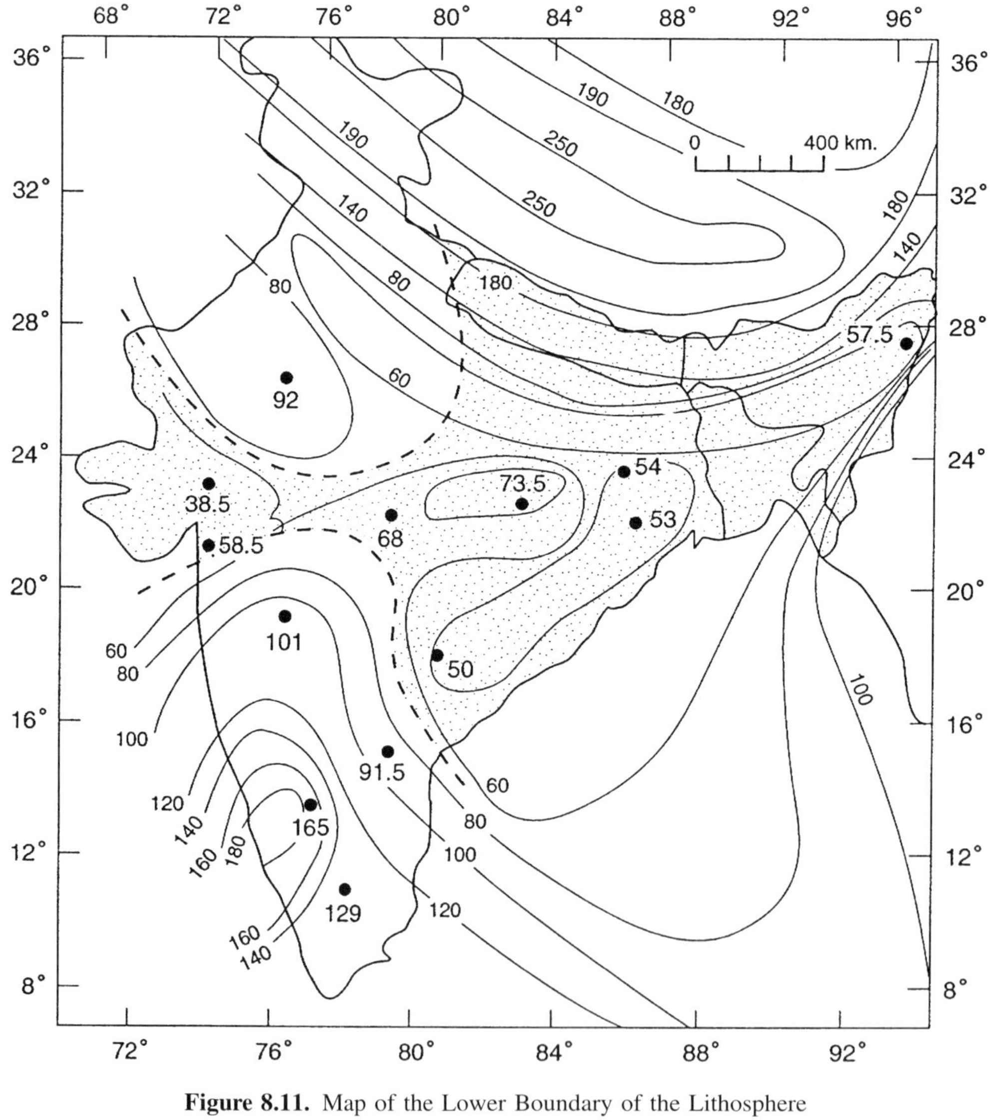

Figure 8.11. Map of the Lower Boundary of the Lithosphere

or Sub-Himalayas, the sedimentary belt of the Lesser Himalayas with subordinate volcanics, the Central Crystalline zone of the Higher Himalayas and the Tibetan Himalayas.

The Ganga basin appears to have had a very long and complex history of evolution dating back to Precambrian times. At present, the Ganga basin essentially consists of the Upper Tertiary sediments, mainly of Siwalik formation, overlying the Vindhyan formation and Aravalli-Delhi meta-sediments. Analyses of geophysical data show that the Vindhyan basin extends northward, far beyond its surface exposures in the northern shield, and that a considerable thickness of these sediments underlies the Ganga basin, the deepest part being close to the Himalayan foothills. The Ganga basin is characterized by a small thickness of the lithosphere of about 60 km, and therefore by an anomalous geothermal field. The thickness of the lithosphere was determined from the author's map of the lower boundary of the lithosphere (Figure 8.11), based on the analysis of the geothermal field made by Negi, Pandey and Agraqual (1986) and on the author's geodynamical analysis of the Himalayan arc system. This analysis shows that the Indian plate, which is in a collision with the Eurasian plate, starts to thicken in the area of Peninsular India, whereas in the area of the Ganga plain the bottom of the lithosphere bulged out and the upper parts of the lithosphere subsided, which resulted in weakening of the lithosphere (Figure 8.12). Moreover, from the thermodynamics it is known that the temperature of melting also depends on the pressure. If the pressure increases then the temperature of melting also increases. Since the bottom of the lithosphere bulged out, the pressure in this area dropped and therefore the temperature of melting rapidly decreased. Then the rocks of the lower

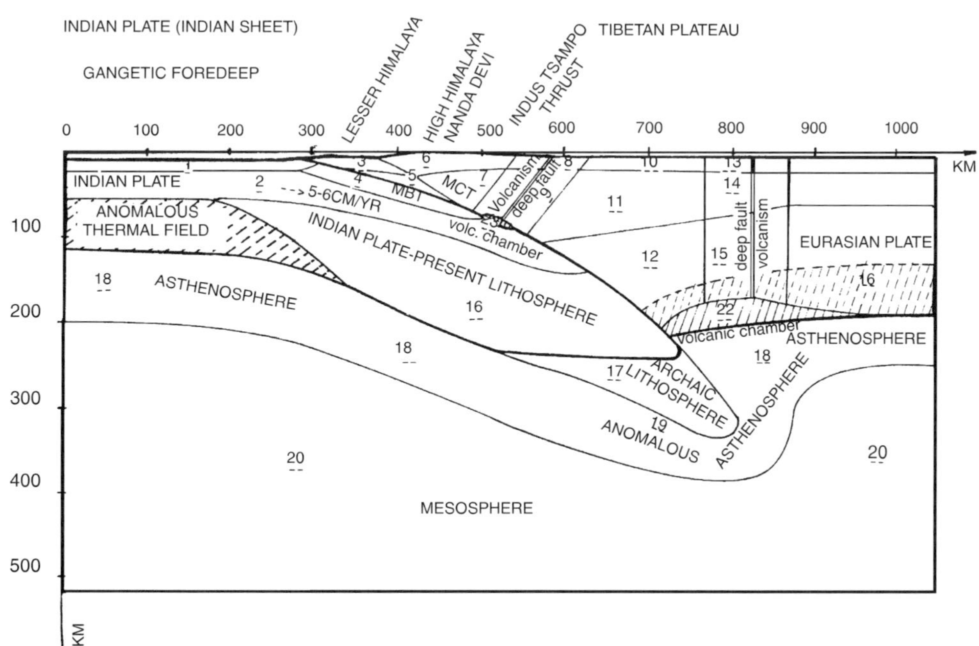

Figure 8.12. Geodynamic Model of Himalayas, Profile Nanda Devi

lithosphere and asthenosphere are melted or partially melted, which together with the weakened lithosphere produces an anomalous geothermal field.

Analyses of geodynamic processes, and geological and geophysical data of the area of the Himalayas, as well as results of previous similar analyses from the Central Aleutian arc system indicate that the major contact between the Indian lithospheric plate and the Eurasian lithospheric plate is not the Indus-Tsangpo Thrust (ITT) but the Main Boundary Thrust (MBT). The evolution of the Himalayas has three major periods.

The first and second stages are characterized by the position of India with respect to the rest of Eurasia in the period of 110 Ma to 40 Ma, i.e. by the period where the Indian continental lithosphere was situated between the African and Eurasian plates. These periods are represented by the subduction of the Tethyan lithosphere of oceanic type below the Eurasian plate and by the development of the island arc system (which is observed practically in all subduction zones, where the speed of the subducted plate is $\sim$ 8 cm/yr)-the Tibet. This configuration is similar to that of the present subduction of the Pacific plate below the American plate. By analogy with Grow's model of the Central Aleutian arc system we can deduce that the maximal depth of the subducted Tethyan lithosphere is about 300 km. A similar estimate follows from our observation of the thickness of the lithosphere in the area of the collision of the Adriatic microplate with the Eurasian plate. From the analysis of the thickness of the European lithosphere it follows that the ratio between its thickness below the Alps and Dinarides is $\sim$3/2. Since there is a certain similarity between mechanisms of present evolutionary stages in Europe and in India, we can assume that the ratio between the thickness below the Himalayas and below Burma is also $\sim$3/2. The present thickness of the lithosphere below the Burmese plate is about 160 km–220 km. We can assume that the mechanism of subduction below the Eurasian plate in the direction *S-N* and below the Burmese plate during the first period of evolution was at least slightly similar as at present in the Aleutian arc system. Then for the maximal depth of the subduction of the Tethyan plate of oceanic type below the Eurasian plate we find the value of about 240 km to 330 km. Since this archaic Tethyan lithosphere is partially melted, then at present it cannot be manifested by earthquakes, as the deformation energy accumulated in this Tethyan lithosphere is released in other ways, for example creep, chemical processes, heat, solidification and recrystallization of rocks, the second phase change, etc.. During this period deep faults probably originated in similar positions as in the case of the subduction of the Pacific plate below the American plate, as well as in the arc system in the Tethyan Sea. The fault near the archaic trench is the Main Central Thrust. In this geological period, the MCT was directed obliquely across the Eurasian plate. The second deep fault situated north of the MCT could or did not have to reach the Earth surface. During these first geological periods the Tibet originated as an island arc system and in the next phase the Tibetan Himalayas and the Trans Himalayas farther north were folded; according to our estimates, the uplifts of the Tibetan and Trans Himalayas during the first period of the evolution, when the speed of the subducted Tethyan lithospheric plate below the Eurasian plate was more than 8 cm/yr, was very small (only a few mm/yr), while during the second period of evolution, when the speed of the subducted lithosphere was less than 8 cm/yr, the uplift was more than 10 mm/yr. During the last 10 Ma the Himalayan arc system above all was developed.

The third geological period of the evolution starts at the moment of the first contact of the Indian continental lithosphere with the Eurasian plate 40 Ma. This contact resulted in the folding in Hindu Kush and Pamir. During the last ten million years the Indian plate has been rotating with respect to the Eurasian plate, which leads to the concave shape form of the Himalayan arc system. At every time step of evolution, two different deep faults originated. One of them is the Indus-Tsangpo Thrust, whilst the second one is situated north of the ITT in the Tibetan Plateau. The rock formations of the Lesser and High Himalayas are folded. It is evident that the MCT during the second and mainly the third period of the evolution process changes its position, from the initial oblique direction across the Eurasian plate with an angle of about 115° to the present oblique direction with an angle of about 20°. The rocks between the MBT and MCT as well as between the MCT and ITT are deformed, which reflects the uplifts of the Lesser and High Himalayas. However, since the rocks between the MBT and MCT are strained by the colliding Indian plate then, due to a small angle between the MBT and MCT direction and also to the acting Coulombian friction, the uplift of the Lesser Himalayas is less than in the case of the High Himalayas. In the latter region, the angle between the MCT and ITT direction is greater than 90°. These uplifts are of a different origin than earlier thought based on the isostasy theory. Our numerical investigations indicate two types of volcanism — the calc-alkaline type in the ITT area and the basaltic type to the north.

The Sub Himalayas consist of the Siwalik Hills, with height varying from 250 m to 800 m and with width of about 25 km to 100 km, situated to the south of the Main Boundary Thrust (MBT). The Siwaliks comprise molassic sediments of Neogene and younger rocks, such as red shale, red sandstones, grey clay in the Lower Siwaliks, micaceous, friable grey sandstone in the Middle Siwaliks and boulder conglomerate, pebble horizons and sands in the Upper Siwaliks, with its thickness varying from 4500 m to 6000 m. The Siwalik Hills might represent a relic of an island arc system from the first two periods of evolution discussed above. Since the Siwalik sediments are of Neogene and younger age and since the island arc system can be of the same age, it is probable that the Siwaliks are the relic of this arc system. Similarly, the Main Frontal Thrust (MFT), situated to the south of the MBT, is the relic of a deep fault from the first and mainly the second period of evolution of the Himalayas. Folding and faulting in the Siwaliks are of early Middle Pleistocene age, and must have preceded the overthrusting of the Lesser Himalayas along the Main Boundary Thrust (Gansser (1964)), which is a result of the following geological period. The mechanism of evolution of the Siwalik belt can be based on the idea discussed later (see p. 558 of this monography).

The Lesser Himalayas are bounded by the Main Boundary Thrust (MBT) in the south and the Main Central Thrust (MCT) in the north and is built by south-directed nappes involving rocks of Precambrian to Mesozoic age, mainly devoid of fossils. The Lesser Himalayas consist of thick sedimentary sequences, low grade metamorphics and volcanics which are mainly of Early Proterozoic to Lower Paleozoic ages. All along the Lesser Himalayas, thick argillaceous formations with some quartzites and limestones are observed.

In the Lesser Himalayas of the Garhwal and Kumaun sections, four principal lithotectonic units are distinguished: Autochthonous-Paraautochthonous sedimentary belt, Berinag Unit, Ramgarth Unit and Almora Nappe Unit.

The autochthonous unit of Precambrian sedimentary rocks is observed in the northern (inner) belt of the Lesser Himalayas. The paraautochthonous nappe of the southern (outer Krol) belt of the Lesser Himalayas comprises the Jaunsar and Mussoorie groups of sediments of the Late Precambrian to Lower Cambrian ages. The Berinag Formation is formed of a huge succession of massive coarse-grained to pebbly and often sericitic quartz arenite with metamorphosed basalts and tuffites. The Berinag Quartzite and Chamoli Quarzite and the volcanics of the Karanprayag-Chamoli area are observed in the Alaknanda valley and Uttarkashi area. The Ramgarh Nappe is built by the crystalline rocks, and it consists of two units: the Ramgarh porphyroids with sericite quartzite and phyllite, and the Amritpur granite. These units are separated by an intervening unit of the Bhowali-Bhimtal volcanics and the Nagthat quartzite. The porphyric and mylonitised granitic gneiss of the Ramgarh Nappe is about 1765 ± 60 Ma old (Trivedi, Gopalan and Valdiya (1984)). The Almora Nappe is built by a variety of schists, micaceous quartzite and amphibolite-facies gneisses together with intrusions of granite and granitodiorites (Valdiya (1980)). The Almora granite is of 560 ± 20 Ma in age.

The High Himalayas consisting of a 10–15 km thick sequence of metamorphic rocks and granite represent the region of greatest uplift; our numerical results indicate

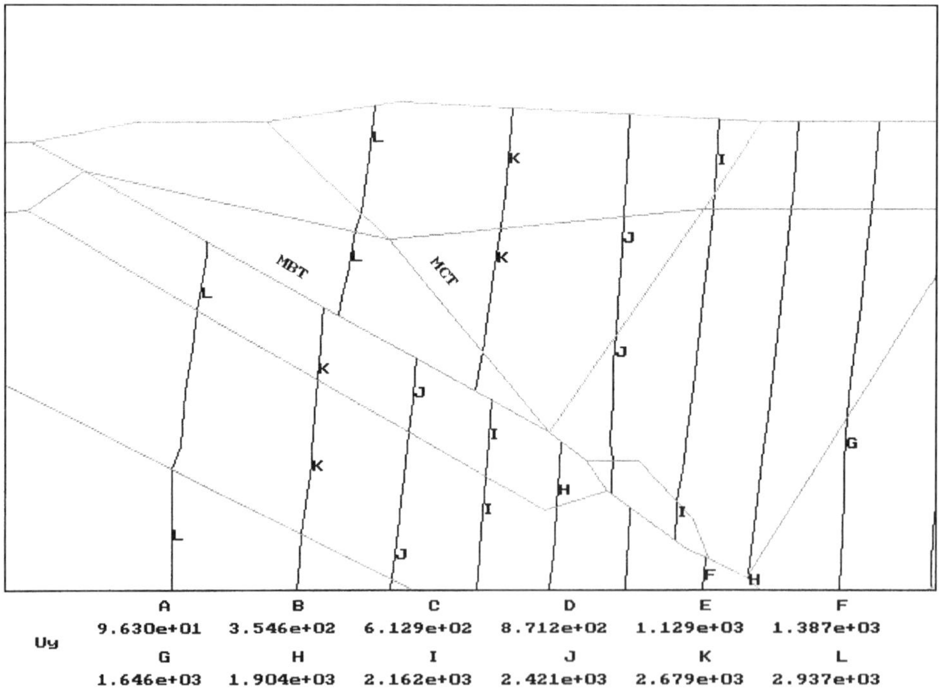

Figure 8.13. Vertical Component of the Displacement Vector in the Area of the High Himalayas — Profile Nanda Devi

Table 8.3. Data — Profile Nanda Devi

	1	2	3	4	5	6	7	8	9	10
ρ[kg/m^3]	1300	2280	1400	2100	2000	1780	2500	1770	2100	1680
c_p[m/sec]	3400	7000	3500	6900	6800	5500	7100	5500	6000	5000
c_s[m/sec]	2050	4120	2100	4100	4050	3280	4150	3250	3520	2950
	11	12	13	14	15	16	17	18	19	20
ρ[kg/m^3]	2850	3400	1690	2800	3380	3400	3320	3420	3450	3550
c_p[m/sec]	7500	8500	5000	7300	8450	8200	7500	8400	8500	8800
c_s[m/sec]	4350	5100	2950	4280	5080	5020	4340	5050	5100	5160
	21	22	23	24	25	26	27	28	29	30
ρ[kg/m^3]	2350	3420	2250							
c_p[m/sec]	7200	8300	6300							
c_s[m/sec]	4300	5100	3700							

an uplift of about 19.58-20.73 mm/yr at maximum (obtained from numerical data computed for times $\sim 1.5E+5$ and/or $1.0E+5$ yrs, see, for example, Figure 8.13 and Table 8.3). The High Himalayas are bounded by the Main Central Thrust (MCT) in the south and by the Indus-Tsangpo Thrust (ITT) in the north. As indicated by the author's model discussed, this boundary (MCT) between rocks of the Lesser and Higher Himalayas is very sharp and has a dip of about 30° to 40° to the north. The contact between the crystalline rocks and the underlying sediments is exposed. The striking difference between the sediments north and south of this main thrust is the most important observation in the Himalayan geology and, as we shall see from obtained numerical results, this fact confirms the author's idea that the MCT is a deep fault formed during the first and second evolution periods, i.e. the period of subduction of the Tethys oceanic lithosphere below the Eurasian plate with relative speeds of the Indian plate to the Eurasian plate being more than 8 cmyr^{-1} or less than 8 cmyr^{-1}. The basalt part of the MCT is formed by gneisses, migmatites, crystalline schists, thick quartzites and some tectonized granite intrusions. In the upper part of the crystalline thrust mass follows distinct marble and lime silicate horizons with amphibolites and psammite gneisses. The High Himalayan crystalline rocks show progressive regional metamorphism ranging from greenschist to upper amphibolite facies. The grade of metamorphism appears to increase northward from chlorite to garnet grade. In the area near the Nanda Devi, the coarse-grained kyanite, garnet and biotite-bearing gneiss of Vaikrita Formation dipping to the north is observed.

North of the Indus-Tsangpo Thrust lies a vast stretch of the Tibetan or Tethys Himalayas. The Tibetan Himalayas are made up of a pile of richly fossiliferous sediments in the basin of the Tethys Sea of the Late Precambrian to Upper Cretaceous ages, folded into a synclinorium during the first two evolution periods of the Himalayas, i.e during the period when the Tethyan lithosphere of oceanic type subducted below the Eurasian plate with two different types of speed of subducting Tethyan lithosphere. The Tethyan sediments occur in several basins that lie along almost the entire length of the northern margin of the Himalayas, and that are abruptly cut by the Indus-Tsangpo Thrust. According to our numerical modelling, the ITT represents a deep fault which originated during the third period of evolution of the Himalayas. It brought up ophiolites and melanges. The ITT is accompanied by the calc-alkaline volcanism from the boundary between the colliding Indian and Eurasian plates, represented by pillow lavas, volcanics, and basic and ultrabasic rocks, containing also radiolarite. The geology of Central Tibet is not so complicated. It is characterized by elongated ranges striking mainly *E-W*. The young volcanic rocks are observed in the Tibetan Plateau and along the Kun-Lun Range. These very young basaltic lavas of the third period of the evolution of the Himalayas are a product of basaltic volcanism with a source at the lithosphere/asthenosphere boundary. With these volcanics are associated older volcanic structures, which represent a relic of basaltic volcanism of the previous evolutionary period of the Himalayas.

Analyses of the numerical results show that the seismicity in the investigated Himalayan region will be in the upper parts of the lithosphere (Figure 8.14), where

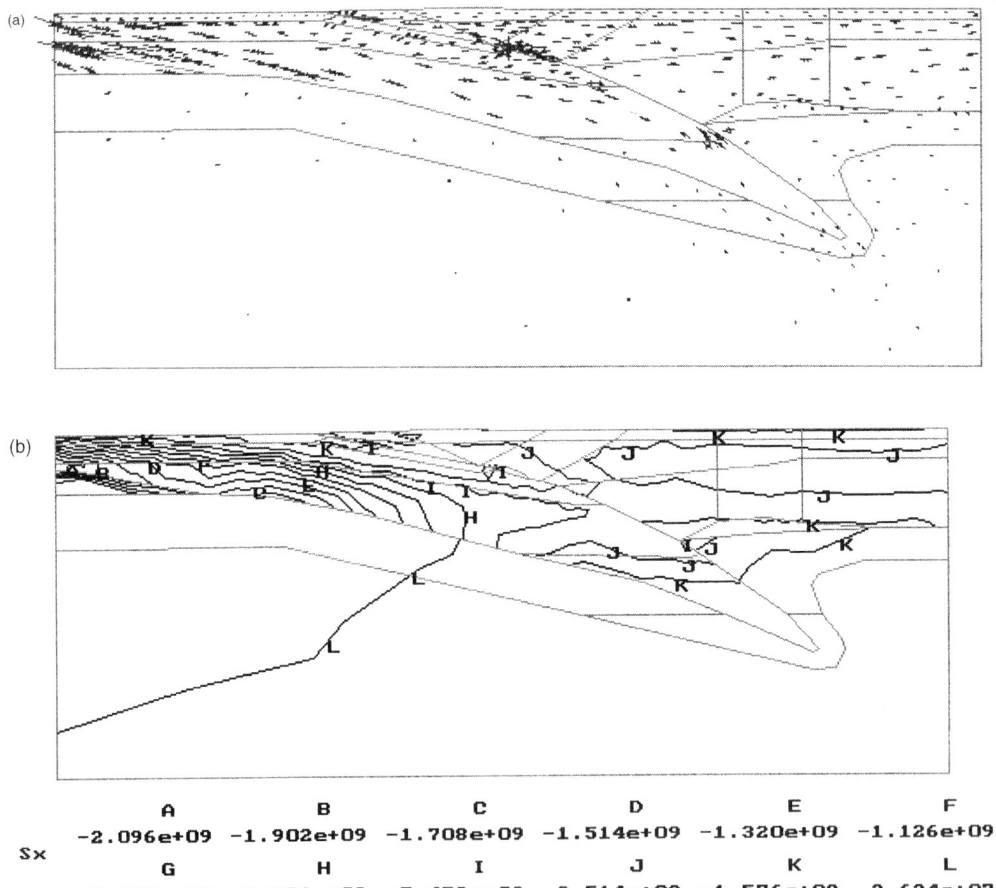

Figure 8.14. Principle Stresses and Horizontal, Shear and Vertical Components of Stresses — Profile Nanda Devi

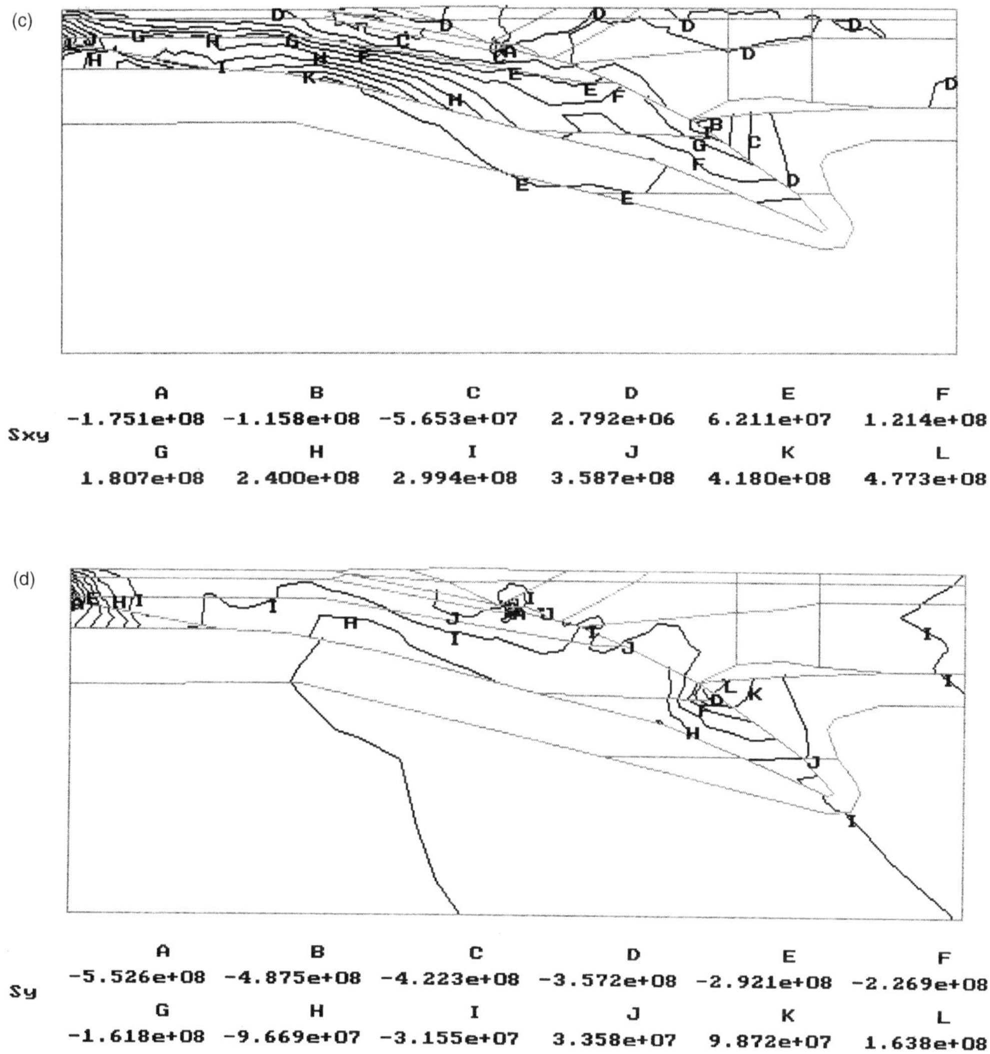

Figure 8.14. (*Continued*)

the lithospheric rocks behave elastically, while in deeper parts the lithospheric rocks are partially melted due to the change of the deformation energy into a heat, and to the change of the effect of Coulombian friction into a heat, as well as to chemical processes taking part in this part of the investigated region. Since the deformation energy accumulated in this region is of very high value, the dissipative heat will also be of very high value and similarly for the effect of Coulombian friction, and therefore the rocks of the lower parts of the lithosphere are partially melted. Hence the deformation energy accumulated in the lower part of the lithosphere will be released by the creep, by visco-plastic effects, by chemical processes, etc.

The geodynamic model of the Himalayas in its cross-section, profile Nanda Devi, is presented in Figure 8.12. Figures 8.13 and 8.14 present the vertical component of the displacement vector in the detailed area of the High Himalayas, and the distribution of the principal stresses as well as the horizontal, vertical and shear components of the stresses in the collision zone of the Himalayas for the time period $\sim 1.5E + 5$ yrs. The above-given analyses of our numerical results show that the previous models of the Himalayas must be revised. We suggest that the main contact between the Indian plate and the Eurasian plate is not the Indus-Tsangpo Thrust (ITT) but the Main Boundary Thrust (MBT). In comparison with previous models of the Himalayas, the presented geodynamic model elucidates the evolution process from the initial state when the Indian subcontinent was a part of the African plate up to the present state. It is evident that geodynamic processes taking place during the evolution of the Himalayas are the same, or practically similar, as in the case of the evolution of the Central Aleutian arc system, the Mediterranean region, and the Japanese arc system, as well as the other collision and subduction regions on the Earth's surface. We see that the mechanism is the same; only the physical parameters and the geometry of the regions are different. We also see that our numerical results are in good agreement with geological observations and/or enable us to interpret correctly some geological phenomena that were misinterpreted or unclear. From the presented geodynamic model of the Himalayas we see that all geological phenomena observed on the Earth's surface, such as MBT, MCT, ITT, volcanisms, etc., are situated in places where they are observed in situ.

8.5. EASTERN HIMALAYA AND NORTHERN BURMA, AND SURROUNDING REGIONS

8.5.1. Introduction

The northeastern part of the Indian plate and the northern part of the Burma plate is limited by the area between the latitude 22° to 30°N and longitude 88° to 100°E. This area is one of the most interesting parts of the world, as the collision of the central Himalaya in this region changes to the subduction of the Indian plate underneath the Burma plate. The large convergent movements directed from northwest and southeast resulted in the formation of the Eastern Himalayas with the highest peak Namche Barwa (7755 m) in the north, the Naga Hills in the south and the Assam-Brahmaputra valley in the middle. Large-scale vertical movements resulted in the uplift of the Shillon Plateau and Mikir Hills in the central part, and the deposition of a great thickness of sediments, of approximately 12-13 km in the Surma Valley, of 12-15 km in the Bengal basin, and of 4-6 km in the Assam-Brahmaputra valley. This area is built by the following tectonic units (see Verma (1991)):

(i) Eastern Himalayas,
(ii) Mishmi Block,
(iii) Naga Hills,
(iv) Assam-Brahmaputra Valley,

(v) Shillong Plateau,
(vi) Mikir Hills,
(vii) Arakan-Yoma fold belt,
(viii) Bengal basin,
(ix) Surma Valley.

The area of the Eastern Himalayas lies approximately 27°-19°N and 88°-96°E and is characterized by Paleozoic, Mezozoic and Tertiary rocks. The tectonics of the region is dominated by an extensive thrust sheet, which has moved from the north towards the southeast or south. The outer ranges of the Eastern Himalayas consist of several overthrust units. A major part of the ranges is formed by Tertiary rocks, which are overlaid by a thrust sheet composed of Gondwana strata. This sheet has been overthrust by a sequence of older crystalline rocks.

The Himalayan fold belt lies between 94° to 98°E and has a strike of *NE-SW* with a sharp turn near 96°E, where the Abor-Mishmi Hills were formed. Near the southern part of the Mishmi Hills, the Himalayas bifurcate into two parts — the Naga-Luskai Hills trending *NE-SW* and the Arakan-Yoma fold belt trending NNE-SSW in the northern part and *N-S* in the southern part. The Naga Hills are formed by a complete Tertiary succession of Eocene and Paleocene rocks.

The Assam-Brahmaputra Valley lies between the Eastern Himalayas and the Naga Hills in the eastern part and between the Himalayas and the Shillong Plateau in its western part. This valley has been formed as a result of the transfer of sediments that have their source area in the Himalaya, Naga Hills and Shillong Plateau. The main direction of this valley is *NE-SW* in its eastern part.

The Shillong Plateau and Mikir Hills, the so-called Meghalaya Plateau, lying between 25° to 26°N and 90° to 93°E form a part of the northern Indian plate. The height of this plateau is about 1000 m and it is built by an Archean gneiss complex with a Proterozoic intracratonic Shillong Series.

The Arakan-Yoma fold belt extends from 16°N to 27°N and consists of a series of thrusts and thrust slices. Rocks in this region represent Paleogene flysch in its western part, whereas the Central Burma basin is formed by molasse of the Quaternary age. In the Arakan-Yoma suture zone, highly deformed sediments, fossiliferous limestone from Cretacerous-Eocene, deeply exposed Paleozonic metamorphics with ophiolites and also Paleozoic sediments occur.

The Bengal basin is situated south of the Shillong Plateau and is bounded by the Tripera fold belt on the east and by the Indian Peninsular Shield on the west. It was formed during the Eocene to Pleistocene period and comprises sediments of 12-15 km thickness. In the western part, Tertiary rocks are found to rest over older Gondwana formations, whereas in the eastern part the rocks are folded due to thrust movements directed from east to west. This part is known as the Tripera Fold Belt.

8.5.2. Seismo-tectonic Activity of the NE Indian Plate

The Indian oceanic plate is of interest to geophysicists and geologists as it can contribute to our understanding of the history of the last period of existence of

Gondwana land and the evolution of India, Australia, Antarctica as well as Africa with the island of Madagascar and the small islands in the Indian Ocean. From the topographic point of view the Indian plate can be divided into the following topographic elements:

(i) Central Indian Ocean Ridge,
(ii) South Eastern Indian Ocean Ridge,
(iii) South Western Indian Ocean Ridge,
(iv) Ninetyeast Ridge,
(v) Chagos-Laccadine Ridge,
(vi) Broken Ridge,
(vii) Kerguelen Plateau in the southern part,
(viii) Crozet Plateau in the southwestern part,
(ix) Mascarene Plateau with Seychelles Islands in the central part,
(x) Chagos-Laccadine Islands on the western side of the Indian Peninsula,
(xi) Andaman and Nicobar Islands in the eastern side of the Indian peninsula.

The concept of sea-floor spreading was given by Hess (1960), (1962), Dietz (1961), (1962), Green and Ringwood (1967). The last postulated that the upper mantle material of pyrolite composition moves from a low velocity zone to the upper crustal layers along the ridge axis and that magma is generated by phase transformation from eclogite to basalt. Vine and Mathews (1963) interpreted magnetic anomalies observed over the Carlsberg Ridge in the Indian Ocean. They showed that the observed magnetic anomalies could not be explained in terms of uniform magnetisation of the ocean floor, and they suggested that different blocks of the oceanic crust are normally and reversely magnetised. They stated that if the oceanic rocks are formed over a connecting up-current in the mantle at the centre of an oceanic ridge, the magma would be magnetised in the direction of the magnetic field prevailing at that time. The oceanic rocks create plates (blocks, zones), which will move away from the ridge axis and form a new oceanic lithosphere. Now, if the magnetic field changes its polarity with time, the nature of the magnetisation of the oceanic rocks that are formed will also change.

Transform faults represent strike-slip faults along which motion takes place. In a trans-current fault, shear motion takes place on both sides of the fault indefinitely. However, in a transform fault, the lateral expansion of a ridge ends abruptly at one end of the transform fault and the same continues at the other end. A fault which terminates abruptly with mid-oceanic ridge at both ends is known as a ridge-ridge type of transform fault. Transform faults have played a major role in the evolution of the Indian Ocean. The Chagos-Laccadive-Maldives ridge and Mascarene plateau formed a part of the transform fault connecting the southeast Indian Ridge with the Carlsberg Ridge. Several fracture zones in the Indian Ocean are related to transform faults. These include the Ninetyeast, the Amsterdam fracture zone in the south, the Rodriguez fracture zone in the central part and the Owen fracture zone with its continuation known as the Chaman fault in the north. All transform faults are seismically active.

The Ninetyeast Ridge is a transform fault and represents a major topographic feature situated in the eastern part of the Indian Ocean. It extends from 30°S to 9°N and is of several hundred kilometres long. It stopped being active during the Paleocene-Eocene period. Large magnitude earthquakes of 1928 (magnitude 7.7) and 1939 (magnitude 7.2) have occurred along the northern part of the Ninetyeast Ridge. The analyses show strike-slip faulting. The sense of motion is left lateral.

The Chagos-Laccadive Ridge in its southern part, known as the Chagos Bank, is a region with high seismicity (magnitude 5.1 to 6.2). Seismicity along the Chagos-Laccadive area indicates a complex pattern of deformation, as focal mechanisms show thrust, normal as well as strike-slip faulting.

The Andaman-Nicobar arc in the northeastern part of the Indian plate together with the Burmese arc define nearly a 2100 km long margin of the subducting Indian plate below the Burmese plate. The two arc systems represent an important link between the Himalayan collision zone in the north and a major island arc trench system of southern Asia, the Indonesian arc. Major tectonic elements of the Andaman-Nicobar arc system include the Andaman Trench, Andaman-Nicobar Ridge, Nicobar Deep, West Andaman fault, Andaman Sea and the Mergui terrace. The volcanic arc forms a part of the Andaman Sea. The Andaman-Nicobar islands are built by sediments of the Bay of Bengal.

The seismicity of this region was presented by Mukhopadhyay (1984), who showed that it is seismically very active. This active region extends from 4°N to 20°N and 93°E to 97°E. Seismicity is particularly associated with the Andaman-Nicobar Ridge and the Andaman Spreading Ridge. The Andaman-Nicobar Ridge seismic zone is about 90 km to 160 km wide and it extends eastwards up to the volcanic islands, containing shocks up to 150 km depth. The gravity field is highly anomalous, the free-air values reach values of as much as −190 mgal. The back-arc seismic zone related to the Andaman-Nicobar Ridge is relatively narrow and it shows less intense seismic activity. The activity is well defined along the ridge axis.

8.5.3. Seismo-tectonic Activity of the Eastern Himalayan Area and Surrounding Areas

The Eastern Himalayan Syntaxis joins the Main Boundary Thrust (MBT) in the Himalaya with the Burma Arc. In the Himalayan region, the Indian plate is underthrust from the beginning at small angles, while at the Burma Arc system the Indian subducted plate dips at angle 30°-60° and is seismically active up to depths of 150-200 km (Ni *et al.* (1989), Holt, Ni, Wallace and Haines (1991)). The seismicity of this investigated region is a consequence of collision tectonics in the Himalaya and the subduction tectonics below the Burmese plate.

Most of the seismic activity is associated with a few major geological units in this region, such as the Ganges foredeep, the Shillong-Mikir Precambrian plateau, the Kopili lineament, the Bomdila lineament, the Naga Hills, the Assam Valley, the Tripura fold belt. Seismic activity in the Ganga foredeep corresponds to the Tista and to parallel lineaments. The seismic activity then extends northwest to the Sikkim Himalayas, where the Gangtok lineament represents the eastern activity region. The Shillong-Mikir Precambrian plateau is seismically active as a whole. During a period of 86 years, three big earthquakes, the first in 1897 ($m = 8.7$), the second in 1930 ($m = 7.1$) and the last in 1943 ($m = 7.2$), occurred here. In addition there occurred earthquakes of smaller magnitudes. The activity of the plateau is attributed also to its northern lineaments, the most extensive of which is the Brahmaputra lineament. The Brahmaputra lineament crosses the Himalayan foredeep. However, no earthquakes are observed beyond the east edge of the Mikir massif. The Dauki fault and Yamuna lineament, representing the south and west margins of the plateau, are seismically inactive. The Kopili lineament of the Kopili Valley is located between the Shillong and Mikir massifs. It is very active and the focal depths of earthquakes are about 70 km. Its activity extends from the Himalayas to the Naga Hills of the Burmese arc system in a NNW-SSE direction. The Bombila lineament trends across the eastern Himalayas to the east of the Kopili lineament. It is possible that the lineament extends to the Naga Hills across the Assam Valley through the east edge of the Mikir massif (Nandy (1980), Rastogi, Singh and Verma (1973)). The lineament is active within the Himalayan domain, and on the last part is inactive. East of the Bombila lineament several thrusts of the eastern Himalayas are highly active. Similarly, the eastern syntaxis of the Himalayas is highly seismically active. The northwestern fault system of the eastern syntaxis produced the big Assam earthquake of 1950 ($m = 8.7$) with several of its major aftershocks (Ben-Menahem, Aboodi and Schild (1974), Chandra (1978), Chhibber (1934)).

The Mishmi thrust, which borders the Assam Valley, is also active, but with epicentral distribution different from those of the neighbouring Himalayan and Burmese arcs. Seismic activity in the Naga Hills is located in the area of their southern thrusts. The Assam Valley, which is situated between the Himalayas and the Burmese arc, is aseismic. The western part of the Burmese arc, the Tripura fold belt, is seismically active and correlates with north-striking structures of the arc, reflecting the westward convexity of the arc. Seismicity of the Bengal basin is located to the Padma, Madhupur, Sylhet and their parallel lineaments. The Madhupur lineament produced the 1918 Srimangal earthquake ($m = 7.6$).

8.5.4. Seismo-tectonic Activity in Burma

East of the Eastern Himalayas and their foredeep is situated Burma, and north-east is southwest-south China. This area is located between 19°N and 35°N and between 91°E and 108°E, and it includes several geotectonic units defined by several prominent thrusts and faults.

It was suggested above that the Indian plate actively subducts below the Burmese plate, as shown by an east-dipping Wadati-Benioff zone that extends to about 180 km depth below the central lowlands east of the Arakan-Yoma mountain. Burma is divided into three major parts trending north-south: the Shan plateau in the east, the Central Belt including the basins of the Irrawaddy, Chindwin and Sittang rivers, and the Burmese Fold Mountain Belt in the west. The Fold Mountain Belt includes the Arakan-Yoma mountains and the Chin, Naga, Manipur, Luskai and Patkai Hills. Westwards is situated the Bengal Basin and the Bengal Fan. The Shan plateau, as the easternmost tectonic unit, forms the eastern highlands, which are elevated ~700 m above the Burmese plains and built up of Tertiary deposits. West of the Shan plateau, Burma is probably formed by the old Gondwanaland. The Mogok metamorphic belt, which is built up of metamorphic rocks with acid intrusives, runs more than 1000 km north-south along the western edge of the Shan plateau. Its width is about 24-40 km. The Mogok Series represent migmatites, garnet-biotite gneiss, biotite gneiss, garnet-graphite-sillimanite gneiss, calcareous rocks such as marble, mafic rocks of the Late Eocene-Early Oligocene age, pegmatites of Middle Miocene age, and Kabaing granites (largest and youngest intrusive) of the Late Miocene-Pliocene age.

The Central Belt and the Burmese Fold (Mountain) Belt form a 1000 km long and 250-700 km wide arc convex towards the foreland. This belt is a product of the Tertiary orogeny, and it can be divided into seven zones, from east to west: the Pegu-Sagaing Rise (a molasse basin resting on a Paleozoic basement), the Central Belt molasse basin (post-Eocene molasse rests on flysch and folded older metamorphics), the Inner Thrust zone (sediments with pre-Alpine metamorphics including Cretaceous-Miocene flysch, resting on abyssal Cretaceous-Mesozoic strata with ophiolites), the coastal Ramri-Andaman Ridge (Cretaceous-Eocene strata, strongly folded), an outer molasse basin (Tertiary sediments about 20 km thick — Tripura fold belt), the Bengal-Surma foredeep (beyond the Indian border) and the Bengal-Assam foreland (Pleistocene deposits). In the Central Belt, young andesitic and basaltic volcanics occur along the active Mt. Popa-Chindwin-Wuntho volcanic system located in south Burma. It continues through the Jude area in north Burma with Cretaceous ophiolites, and basic and ultrabasic submarine flows (Chhibber (1934)). The Upper Cretaceous-Eocene flysch of the Burmese Fold (Mountain) Belt is bounded by faults and thrusts on the east and partly on the west side. The Eastern Boundary Thrust is a major tectonic element of the area of ophiolitic rocks, and it defines the contact between the Burmese Fold Mountain Belt and the Central Belt throughout Burma.

The Bengal basin is formed by at least 13 km thick of Cretaceous-Tertiary deposits. The Hinge zone represents the Eocene limestone reflector. The Hinge zone is an ~500 km long area between Calcutta on the south and the Dauki fault on the north, and it continues into the Bay of Bengal. Its width is between 25 km and 110 km.

We now consider the seismicity of Burma. The Naga Hills area, the Chin Hills area, and the central portion of the Burmese arc are seismically very active, whereas coastal Burma, where the arc orientation changes to NW-SE, is far less seismic. The earthquake foci range in depth from near the surface to around 180 km. The foci distribution in the inclined seismic zone is used here to define the boundaries of the subducting Indian lithospheric plate below the Burmese plate. The lower boundary of the lithosphere is below the lowest foci at a depth of 80 km. This boundary dips to the east in a region below the Burmese Fold (Mountain) Belt and reaches a depth of about 180 km; in the area of the Naga Hills, the Chin Hills and in the central portion of the Burmese arc it reaches a depth of about 160 km; while in coastal Burma only about 100 km. The thickness of the Wadati-Benioff zone below the Central Belt is approximately 60 km and it has an average dip of about 45°

according to several authors. The Burmese volcanic arc is located above the deepest part of the Wadati-Benioff zone. East of the Wadati-Benioff zone within the overriding Burma plate in the Naga Hills, the Chin Hills areas and in the central portion of the Burmese arc, two shallower seismic zones are situated. The relatively shallow seismic zone occurs below the Chinswin forearc basin. We therefore speak about the forearc seismicity, whereas in the case of the crustal seismic zone we speak about backarc activity. The backarc seismicity is most intense in the Jade Mines area in north Burma. The forearc and backarc seismic zones are separated by an aseismic zone between them. The island seismic slab underlying the Central Belt is about 50-80 km thick.

The distribution of density in the lower lithosphere and asthenosphere is assumed to be primarily due to temperature perturbations in the upper mantle. Similarly as in Grow (1973) it is assumed that the subcrustal lithosphere has a higher density ($\sim$3.40 gcm^{-3}) than the upper parts of the asthenosphere ($\sim$3.35 gcm^{-3}). At destructive plate margins at depths below 150 km, the asthenosphere is assumed to be denser than 3.35 gcm^{-3}. In the case of the descending lithosphere, the density is probably higher than 3.40 gcm^{-3}(similarly as in Grow (1973)). These considerations concerning densities in these regions follow from the gravity, as the positive gravity contributions must be compensated by a lower density zone above the descending slab, in order to fit the observed gravity anomalies over the Central Belt and the andesitic volcanic zones. The density in the upper and lower part of the andesitic volcanic zone is from 2.60 gcm^{-3} to 2.80 gcm^{-3}. The density of 2.60 gcm^{-3} for the upper portion of the volcanic arc corresponds to the rock density of the volcanic rocks of the andesitic volcanic zone being measured.

The Indian shield crust underlying the western part of the Bengal basin is almost 30 km thick, and it is built of two layers of densities of 2.70 gcm^{-3} and 2.90 gcm^{-3} in its upper and lower parts, respectively. The observed anomalies of density that change in the oceanic crust from 2.90 to 3.40 gcm^{-3} at about 30 km depth can, according to Grow (1973), be explained by the phase change in the oceanic crust (i.e. basalt to eclogite) at pressures between 10 to 20 kb (corresponding to 30-60 km depth) in a Wadati-Benioff zone environment.

8.5.5. Seismo-tectonic Activity in the Area of South China

The area is characterized by several geotectonic units defined by the prominent thrusts and faults, (see Figure 3, Verma, Mukhopadhyay and Nag (1980)). The major faults are the Athyn-Tagh, Kansu, Kun Lun, Kang Ting and Red River faults, which run for a few hundered kilometres. Many of these deep faults run in an east-west direction over the central part of Asia and Tibet. Near eastern Tibet they change their trend to the south-east and to the south in south China. The mechanism of movement is predominantly strike-slip faulting.

A major change in the structural trend is observed at the Assam wedge, where the NNE-SSW trending Arakan-Yoma fold belt of west Burma joints the ENE-WSW trending Himalayan ranges. To the east of the Arakan-Yoma ranges is situated the

Burmese plain (the so-called Irrawaddy basin), with thick sequence of sediments, and the Shan plateau of Burma. To the north-east and east of the Shan plateau, the Hengtnan mountain arc is located, characterized by a north-south strike in regions of Yunnan, Sikan and south-west Szechwan. The tectonics and seismicity of south China are more complex in regions where the north-south structures are cut at right angles by east-west trending mountain ranges. The seismicity is widespread in the whole region discussed. Intense seismicity is observed along the whole Burmese arc, including the Arakan-Yoma mountains and the Burmese plain, the western part of Shan plateau, the Assam wedge, Hengtnan ranges, Kang Ting fault, the Red River fault, Kang Tien ranges and western Szechwan region.

In the Hengtnan ranges and in the east Yunnan-Sikang belt, a combination of thrusting, normal and strike-slip faulting were observed. Over the Yunnan-Sikang belt and the Shan plateau, normal, strike-slip, and thrusting are observed. The majority of the earthquakes in the Yunnan-Sikang region is characterized by shallow foci and therefore that deformation mostly involves the upper part of the lithosphere. The NW-SE trending Red River fault and other parallel faults in south China are seismically active. Ben-Menahem, Aboodi and Schild (1974) presented the analysis of the Assam earthquake of 1950, and they showed that it was of strike-slip character. They have chosen the NNW-SSE trending model plane as the fault plane on the basis of aftershock distribution and, moreover, they have suggested that this earthquake was caused by a motion of the Asian plate relative to the eastern flank of the Indian plate, where the northeast Assam geological block has a tendency to rotate.

8.5.6. Geodynamic Model of the Andaman Arc System: Profile across North Andaman Island, Barren-Narcondam Volcanic Arc System and Burmese Plate

The Andaman-Nicobar island arc system is situated in the northeastern Indian Ocean. This island arc system is 2200 km long and is formed by sedimentary rocks of very young age. Its origin is related to the subduction of the Indian plate below the Burmese plate. The active subduction is documented by the active Barren-Narcondam volcanism (the last eruption of the Barren volcano was in June 1991), by the Wadati-Benioff zone with earthquakes with focal depth up to 200 km, and by anomalous geothermal and gravity fields (Mukhopadhyay (1988), Mukhopadhyay and Krishna (1991), Verma (1991), Dasgupta and Mukhopadhyay (1993)). The geology and tectonics of this region is characterized by active faults such as the Semanko fault near Sumatra, the West Andaman fault in the Andaman arc system, the Sagaing transform in the Burmese plate and the Neogene Andaman-Nicobar back-arc spreading ridge, where the convergence of the Indian plate against the Eurasian plate is realized. The greater part of the Andaman Sea is occupied by the Andaman basin.

In this section we present a geodynamic model in cross-section along a regional traverse extending from the Bengal Fan across the North Andaman Island and Barren-Narcondam volcanic arc system to the Burmese plate (see Figure 8.15). We use

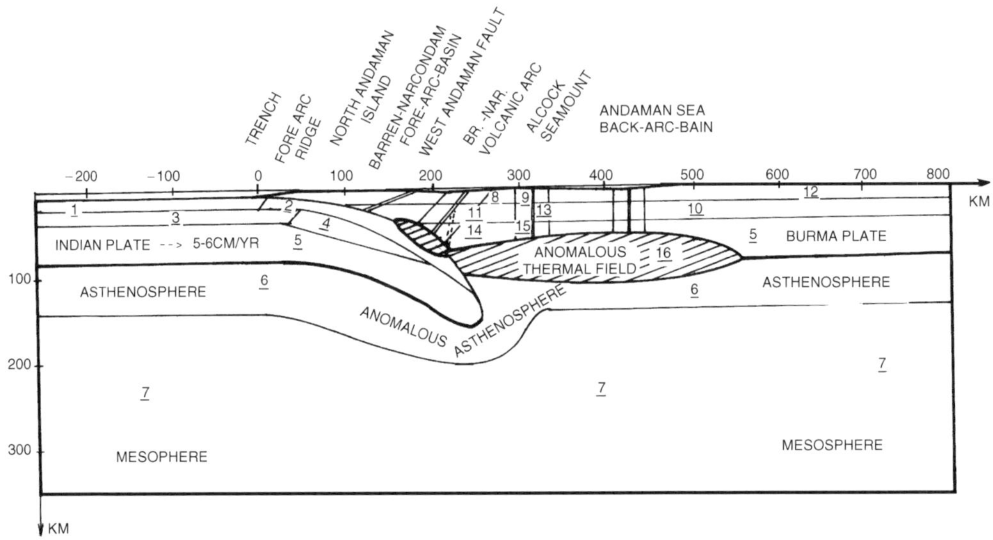

Figure 8.15. Geodynamic Model of the North Andaman Island Arc System

the available geological and seismic data of this region (Verma and Mukhopadhyay (1977), Curray *et al.* (1979), Eguchi, Uyeda and Maki (1979), Mukhopadhyay (1984), (1988), Mukhopadhyay and Dasgupta (1988), Mukhopadhyay and Krishna (1991), Verma (1991), Dasgupta and Mukhopadhyay (1993)). The model corresponds to the NNE-trending Andaman arc system where the Barren and Nancondam volcanic islands represent an intra-oceanic magmatic arc and where the Alcock Seamount divides the Andaman Sea into fore- and back-arc basins. The thickness of the Indian subducting plate is assumed to be about 80 km, and the Burmese plate is then 85 km thick. The Indian plate is subducted below the Burmese plate to a depth of about 160 km in the studied region, of about 180 km in the region of the Nicobar island arc system and to 200 km in the region of Sumatra. The dip of the Wadati-Benioff zone is about 40° in the studied region and changes to 40°-45° in the Nicobar island and Sumatra region. The seismic zone in the studied region which displays the back-arc seismicity related to spreading of the Burma plate (Dasgupta and Mukhopadhyay (1993)) is situated relatively shallowly below the Andaman Sea.

The geodynamic model of the North Andaman Island was constructed on the basis of results of Verma (1991), Dasgupta and Mukhopadhyay (1993) and modified by the author's results in a cross-section along regional transverse extending from the Bengal Fan across the North Andaman Island and Narcondam-Barren volcanic arc system to the Burmese plate. The profile of the present geodynamic model is taken in the little modified than in Dasgupta and Mukhopadhyay (1993). The Andaman Island arc system was originated as a result of the direct motion of the Indian plate and its rotation. Since we simulate the geodynamic process in this region, we must take into the consideration the main direction of the motion from the rift zone. This was the reason for the modification of the investigated profile. Nevertheless, the geodynamic model in its general case must take in an account both these motions.

The volcanism in this investigated area is a part of the magmatic arc of Burma, which extends from the Jade Mines in the North Burma and continues through Taungthorlon and Wuntho, Lower Chindwin, Mt Popa and Pegu Yoma in central and south Burma to the Narcondam volcanic is geologically of the andesite-dacite suite, i.e. it is a typical calc-alkaline type of volcanism. Presence of amphibole in calc-alkaline suites suggests for the hydrous nature of the parental magma, which corresponds with the fact that the melted materials in a previous time before subduction were situated on the bottom of the ocean. On the other hand the Barren volcano is of about 300m high above sea-level and geologically is of the basalt-andesite suite. The Labradorite-augite basalt was also observed on the Barren volcanic island. The older lava of the Barren volcano is vesicular olivine basalt with plagioclase, augite, magmatite and apatite set in a microcrystalline groundmass, while younger lavas are olivine basalt and porphyritic olivine andesite. One can assume that the basaltic lavas are from a greater depth than in the previous Narcondam volcano case. We see that both volcanoes are of the different compositions (Jafri, Balaram and Govil (1993), Jafri, Charan and Govil (1995), Dasgupta, Mukhopadhyay and Nandy (1990)). The Narcondam and Barren volcanoes are dissected by the West Andaman fault. The volcanism was originated by the Miocene with the opening of the Andaman Sea. The area is characterized by an anomalous geothermal field. The present volcanic shapes are of Pleistocene to Sub-Recent age. Eastward of the Barren-Narcondam volcanic arc the Alcock Seamount and the "back-arc rift" (Dasgupta and Mukhopadhyay (1997)) are situated and the anamolous geothermal field is observed (Curray et al. (1979)). The origin of the first one is older than the second one, but the mechanism of both has the same mechanism, which was discussed in the previous chapter and is based on the bulged out bottom of the Burmese plate and the fact that the temperature of melting decreases with decreasing of the pressure as well as on the fact that the Burmese plate is deformed by the effect of the subduction of the Philippine plate below the China plate (therefore, the second one is not a back-arc rift. Hence we see that the anomalous area eastward from Alcock Seamount is not a rift zone as it was described in (Dasgupta and Mukhopadhyay (1997)), but hot spots forming by the above mechanism (see also the section about hot spots in this monography).

Seismic observations of the subducting Indian plate indicate an intermediate seismic gap between 90–110 km of the Benioff-Wadati zone. The mechanism of the origin of this aseismic zone can be cleared up by the existence of two deep faults in the subducting lithosphere. It is clear that at the moment when the length of the subducted lithosphere is sufficiently descended below the Burmese plate then, according to a great value of the excerting pressure of the Burmese plate the descending lithosphere is broken off. The subduction process continues up to a moment when the descending part of the lithosphere is newly sufficiently large, then the descending lithosphere is broken off again. The zone between both deep faults in the descending lithosphere represents the aseismic zone as the deformation energy is accumulated outside this zone.

The density distribution in the lower lithosphere and asthenosphere is assumed to be primarily due to temperature perturbations in an upper mantle and, similarly as in Grow (1973), we assume for the lower lithosphere a higher density (3.40 gcm^{-3}) than for the upper asthenosphere (3.35 gcm^{-3}) which results from the anomalous gravity field. At depths greater than 150 km, the density is probably higher than 3.40 gcm^{-3}. Even though the density contrast of about 0.05 gcm^{-3} between the dipping lithosphere below the Andaman Sea and its surroundings is relatively small, the positive gravity contribution of the descending lithosphere reaches a maximum of about 20 mgal at the trench axis and 35-40 mgal over the volcanic arc. The observed anomalous gravity field is supposedly a combination of this deep gravity effect and the effect of the overlying layers. According to Grow (1973), this positive gravity anomaly must be compensated by a lower density zone above the descending slab (below 75 km) in order to fit the observed anomalous gravity field. For the Bengal basin we assume the crustal layer to be 30 km thick with densities of about 2.70 gcm^{-3} to 2.90 gcm^{-3} in its upper and lower parts, respectively. Moreover, it has been argued that heating due to friction between the subducting and obducting plates may produce a high-temperature, low-density zone generally below the volcanic arc. For oceanic crust we assume densities between 2.50 gcm^{-3} to 2.90 gcm^{-3}. These densities were estimated on the basis of observed seismic velocities under the Ganges Cone and Andaman arc-trench areas (Curray *et al.* (1979)). The sedimentary layer of the Campanian to Holocene age which underlies this part of the Bengal Fan is assumed to be 4.5 km-5 km thick; the thickness is quite variable. The velocity of P waves is assumed to be 6.22 kms^{-1}. The thickness of the Cretaceous-Tertiary sediments below the Andaman-Nicobar sedimentary islands (also known as as the Andaman-Nicobar Ridge) is about 5-6 km to 13 km. Similarly as in Grow (1973), the effect of the gravity anomalous field on the density gives the density of the oceanic lithosphere between 2.90 gcm^{-3} and 3.4 gcm^{-3}. For the northern part of the Indian oceanic crust the seismic velocities of compressional waves were estimated as 5.04 ± 0.69 kms^{-1} to 6.73 ± 0.19 kms^{-1} on the top and bottom of the crust, respectively.

Figure 8.15 represents the geodynamic model of the subduction of the Indian oceanic plate of Bengal Bay below the Burmese plate, with a presumption that the lithosphere and its support behave elastically. We assume that the time of our investigations is less than a characterized time $t_c = \nu/\mu$, where ν is the effective

Table 8.4. The model-physical data

	1	2	3	4	5	6	7	8	9
ρ[kgm^{-3}]	2500	2700	2900	2900	3400	3350	3400	2600	2700
c_p[ms^{-1}]	3900	6200	6410	6400	7900	7800	7900	6200	6300
c_s[ms^{-1}]	2140	3600	3700	3700	4560	4520	4560	3600	3700

	10	11	12	13	14	15	16
ρ[kgmm^{-3}]	2900	2900	2600	2800	3400	3300	3350
c_p[ms^{-1}]	6500	6450	6200	5950	7900	7780	7850
c_s[ms^{-1}]	3720	3710	3600	3400	4560	4500	4600

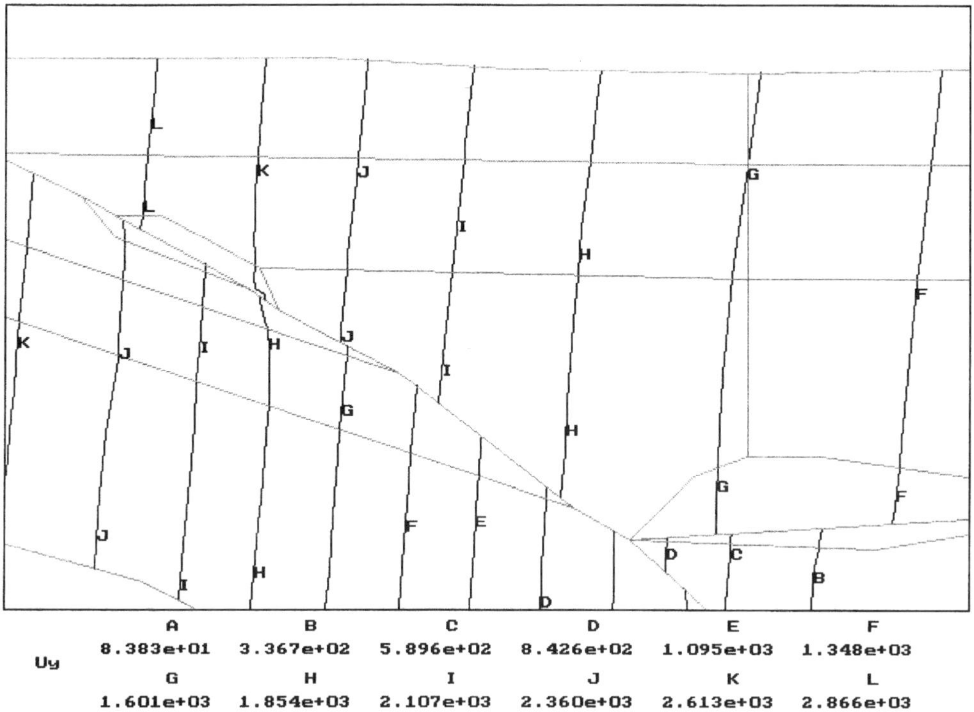

Figure 8.16. Geodynamic Model of the North Andaman Island Arc System: Vertical Component of Displacement **u** in the Detailed Area of the Andaman Island Arc System

viscosity at a given stress level and μ is the elastic rigidity. For a mathematical model the author's theory presented in Chapter 7 was used. The main results are illustrated in Figures 8.16 to 8.18 for time period 1.5E + 5 yrs. Physical parameters were derived from results of authors referred to above and they are given in Table 8.4. The geometry of the region investigated is based on the gravity and seismic data of this region and uses knowledge derived from studied geodynamic problems. Numerical results (here the SI system was used) indicate that the horizontal component of displacement has a value of about 20.506 mm/yr to 17.946 mm/yr in the Bengal Bay and the vertical component has a maximum value 19.106 mm/yr to 14.047 mm/yr in the region of the Andaman islands, which corresponds to the observed values. Similar results were obtained for time period 1.0E + 4 yrs, where the values of the horizontal and vertical components of displacement were about several millimeters per year smaller. The resulting shift in the area of Bengal Bay is about 18 mm/yr, and the resulting uplift of the Andaman islands is about 20 mm/yr. In Figure 8.16 the vertical component of the displacement vector (time period 1.5E + 5 yrs) in the detailed area of the Andaman island arc is presented. Figure 8.17 demonstrates horizontal, vertical and shear components of the stress tensor, while Figure 8.18 demonstrates the distribution of the principal stresses in the investigated region. The distribution of principal stresses indicates the bulged-out regions where the materials

(a)

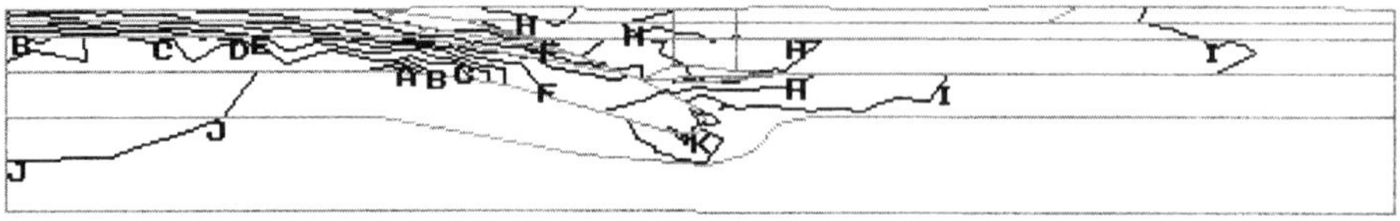

	A	B	C	D	E	F
Sx	-1.700e+09	-1.504e+09	-1.308e+09	-1.112e+09	-9.167e+08	-7.208e+08
	G	H	I	J	K	L
	-5.250e+08	-3.291e+08	-1.333e+08	6.253e+07	2.583e+08	4.542e+08

(b)

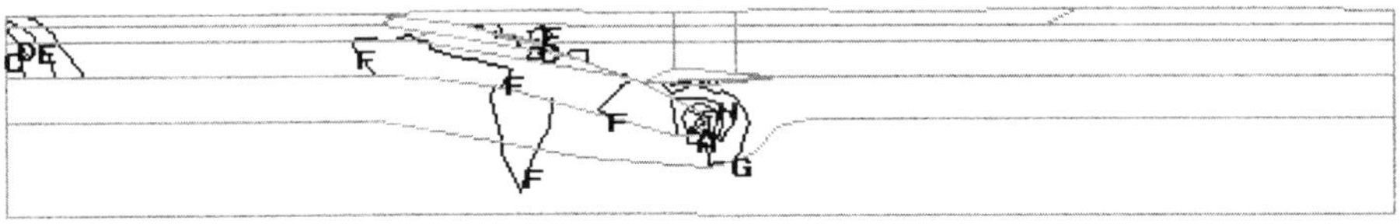

	A	B	C	D	E	F
Sy	-7.444e+08	-6.058e+08	-4.671e+08	-3.285e+08	-1.899e+08	-5.126e+07
	G	H	I	J	K	L
	8.736e+07	2.260e+08	3.646e+08	5.032e+08	6.419e+08	7.805e+08

(c)

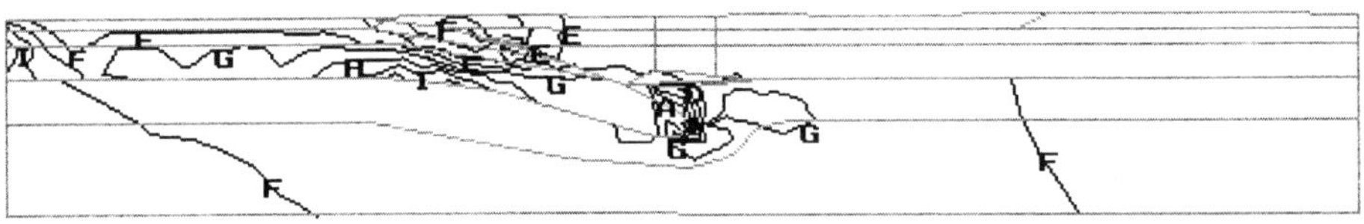

	A	B	C	D	E	F
Sxy	-2.893e+08	-2.213e+08	-1.534e+08	-8.545e+07	-1.749e+07	5.046e+07
	G	H	I	J	K	L
	1.184e+08	1.863e+08	2.543e+08	3.223e+08	3.902e+08	4.582e+08

Figure 8.17. Geodynamic Model of the North Andaman Island Arc System: (a) Horizontal, (b) vertical, (c) Shear Components of Stresses

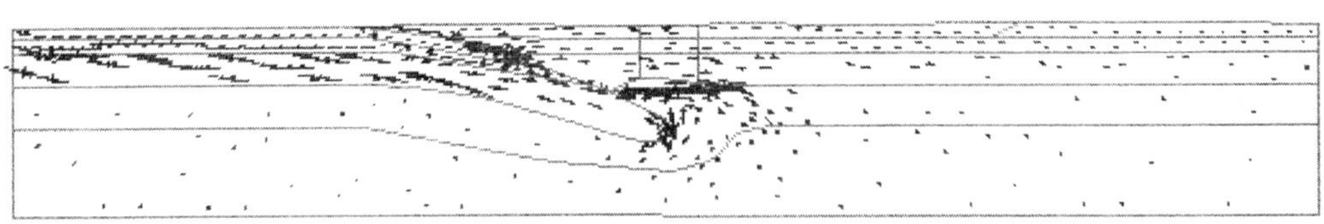

Figure 8.18. Geodynamic Model of the North Andaman Island Arc System: Distribution of Principal Stresses

are quickly melted. These regions indicate the volcano chambers (the black areas in Figure 8.18), namely of the Barren-Narcondam active volcanic arc (the Barren last erupted in June 1991). Places of great compression are indicated below the Andaman island arc system at a depth ~25 km and below the volcanic arc system at a depth ~150–160 km.

8.6. GEODYNAMIC PROCESSES IN SOME VOLCANICALLY ACTIVE AREAS

8.6.1. Introduction

The study of geodynamic processes taking place within the Earth in relationship with the origin of orogenic arc systems as well as with the origin of volcanoes in such areas, which has been carried out over the last seventy years, is founded on various, sometimes even contradictory, assumptions and hypotheses. The hypothesis which is most probable at present is the plate tectonic hypothesis. It can be used as a basis for developing a theory which will enable us to describe the intra-terrestrial geodynamic processes in the whole Earth, including the origin of volcanoes. We shall next describe several typical types of such regions, for example the Mediterranean region, the Japanese arc system, the hot spots regions, etc..

The volcanically active zones are located in rift zones as well as in collision and subduction zones. Rift zones were discussed in Chapter 5. Therefore, the tectonic evolution of volcanically active areas in some collision, subduction and hot-spot zones will be the main topic of this section. The existence of volcanoes in the oceanic plate, i.e. the hot spots zones such as the Hawaii islands and the East African system, evoke questions which have not been accountably answered. These cannot be answered without knowledge of the stress-strain distribution in these areas — in the subducted, obducted and bulged-out lithosphere; the behaviour of the subducted lithosphere in the adjacent upper mantle; and the rheology and geometry of these regions.

If the plates are approximately of the same thickness, the forces acting on both plates and the actual deformation of the lower part of the contact zone of the deep fault determine which of the plates will descend below the other and which of them will override. The contact zone is deformed and broken up and the thinner plate begins to subduct under the thicker one. If the speed of both colliding plates is low and both plates are practically of the same thickness, then the enlarged contact is compressed and both plates will be deformed. Their thickness will increase at their head, folded sediments will be thrust over them and the centre of the deformed lithospheric plate will bulge out at the bottom and will decline at the top, so that it will be thinner, resulting in the small thickness of the lithosphere as well as the anomalous geothermal field. This is observed, for example, in the case of the central part of the African plate and/or in the Pannonian block as well as in the Indus-Ganga-Brahmaputra plain. In the case of the African plate in its east part, the Red Sea, the Aden Gulf rift zone has a great influence on the evolution of this region.

Similar effects can also be observed in the case of very great oceanic megaplates, such as the Pacific plate. The megaplates shift from one side, from the riftward side, and subduct under the lithosphere of continental or oceanic origins and/or collide with the lithosphere of continental type on the other side, resulting in their deformation.

Furthermore, there exist regions where the rift and subducted zones are very near one another. Such regions are observed in regions of subduction of the Juan de Fuca plate under the American plate (see Figure 8.19) and in the San Andreas region.

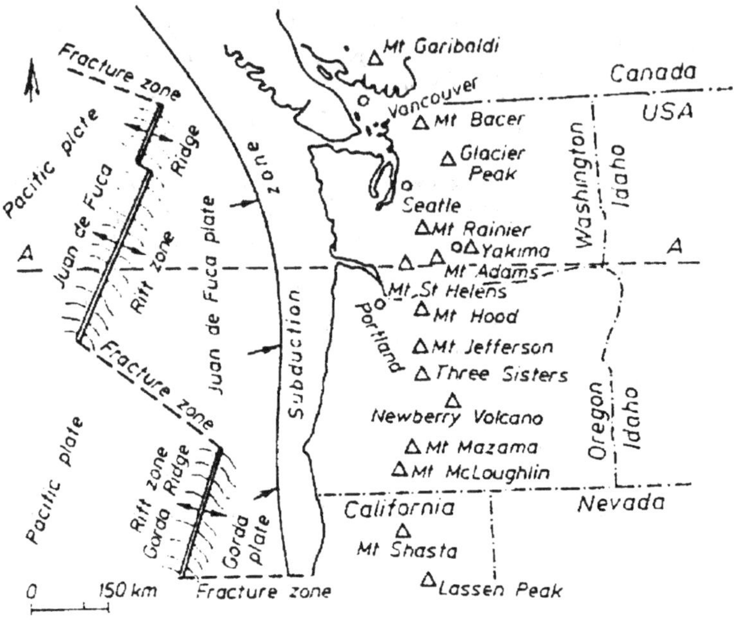

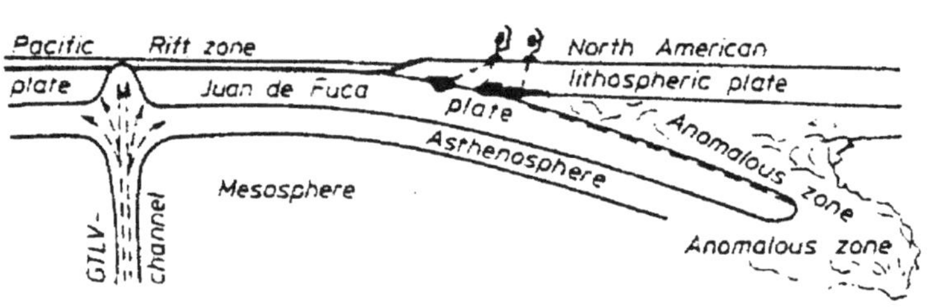

Figure 8.19. The Region of the Juan de Fuca and Gorda Plates (modified after Atwater (1970), Nedoma (1984a))

Next we shall investigate the Mediterranean area, the Japanese arc system, the hot spots area in the Pacific Ocean and anomalous thermal fields in the Pannonian Basin and in Central Africa.

8.6.2. Mechanics of the Subducted Lithosphere in the Hot Spot Areas, Mediterranean Area and Japanese Area

Numerical results obtained in previous sections of this chapter can also be applied in the first approximation to other volcanic regions.

The first of these will be the Mediterranean area. In this area previous and recent geological, geophysical and volcanological research has provided many new diagnostic facts (Ritsema (1970a,b), Barberi *et al.* (1983), Royden and Patecca (1987), Barberi, Gasparini, Innocenti and Villari (1974), D'Ingeo, Calcaguile and Panza

(1980), Gaudiosi, Luongo and Ricciardi (1984), Barberi *et al.* (1984), Gasparini, Iaunaccone, Scandone and Scarpa (1983), Gresta, Longo and Viavattene (1990), Neri (1990)). It was suggested that the African lithospheric plate thrusts under the European plate, causing orogenic, volcanic, seismic, and other geonomic phenomena in the Mediterranean mobile belt. The diameter of the African plate is about 5000 km and its northern dimension is about 3500 km so that the African plate is a mega-tectonic plate. The contact front with the Eurasian plate is over a width of thousands of kilometres. The uniform compression evoked by the African plate is held responsible for geodynamic processes, which characterize the Cenozoic evolution of the Mediterranean belt system. It formed the Alpine mountain system, the Mediterranean sea basins, the intrusive and extrusive volcanic processes, the seismicity and the gravimetric and geomagnetic anomalies.

The actual geodynamic movements of the Mediterranean system are very complicated and are characterized by the compressive, tensional and shear effects in various directions. In the circum-Tyrrhenian sector of the Alpine mountain system, earthquake foci occur at intermediate depth and are interpreted in the sense of a Wadati-Benioff zone (see Barogangi and Issaks (1971)). The other sectors of the African front are characterized by compressive stresses in diverging directions and the absence of intermediate earthquake foci N-NE in Arabia, NW in Libya and NW in Morocco. The system of movements in this area is very complicated and is the result of the movement of the African plate and movement due to the Red Sea rift zone, so that there are several smaller units acting with different forces and with different rates in different directions. Present seismic analysis shows that geodynamic processes of a more restricted extent are active in the mobile belt. The Mediterranean seismicity is partly related to the formation of orogenic arcs and collapse basins at their concave inner side, and partly belongs to great shear zones extending E-W over thousands of kilometres. According to classical plate tectonics many authors thought this area to be an arc-trench system which developed from the subduction of the Ionian plate in a NW-SE direction beneath the Calabrian arc, taking as evidence the presence of intermediate and deep earthquakes in the southern Tyrrhenian sea (Gaudiosi, Luongo and Ricciardi (1984)). Due to the opinion of those authors the present structure and dynamics of the arc is due to the strain of an elastic plate in shear bending under the stress produced by the African-European collision, which started 10 million years ago when the Tyrrhenian sea began to open. The Apennine chain forms a wide-arc convex in the direction of the Ionian sea. The tectonic analysis of this area allows the identification of a series of faults longitudinal and transverse to the axis of the chain. The longitudinal structures show a direction which changes from NW-SE in Campania, to N-S and NE-SW in Calabria and E-W in Sicily. The azimutal changes of direction occur at the places where the longitudinal structures cross the transverse ones. The volcanic situation of this region is characterized by present and recent active volcanoes (Barberi, Gasparini, Innocenti and Villari (1974)). The chemical composition of the magma changes from place to place. So abyssal tholeites occur in the centre of the Tyrrhenian sea, K and high-K volcanic magma in the Napolitan area, calc-alcaline magma in the Aeolian islands, and intraplate-type destensive volcanism in eastern Sicily in the area of Etna and Iblei (Marinelli (1975)). The main shallow earthquake activity is concentrated along the chain, while the biggest part of the seismic activity is situated along the NE-SW line going from Catanzoro through to the Etnean volcanic area. The seismicity of the southern Tyrrhenian sea is characterized by intermediate and deep earthquake foci, which define the subducted lithosphere under the dip of about 55° in the NW direction and at a depth of about 500 km (Royden and Patecca (1987), Gasparini, Iaunaccone, Scandone and Scarpa (1983)).

With all these objections and numerical results obtained in the case of the Central Aleutian island arc and other regions, the model of the global plate tectonics of the Mediterranean arc can be constructed. The model of this area is illustrated in Figure 8.20. The lithosphere with diapiric rise of the asthenosphere is discussed in Berry and Knopoff (1969) (see also van Bemmelen (1972)). The bulged-out lithosphere represents the magmatic chamber filled by the molten rocks at the bottom of the lithosphere and the asthenospheric material originated by the mechanism described above in the case of the Central Aleutian island arc. Along the deep fault the molten material ascends through the system of deep faults, together with gases, to the Earth's surface, where it forms volcano Stromboli. The eruption of magma depends on the magma further pressed in the bulged-out domain and on the concentration of gases in the volcanic chamber. The bulged-out domain below Calabria and filled by the molten lithospheric materials, though it is not manifested as the volcano, can be proved by the magneto-telluric and gravimetric measurements. The effect of such type of bulged-out domain is manifested by magnetic and gravimetric anomalies (for the case of the Carpathians see for example Wiese (1963), Nedoma (1980), (1986a-c)). Furthermore, the fact is that in this region aseismic zones are observed, which is in agreement with observations of the Wadati-Benioff zone in other areas.

The origin and evolution of the West Pacific marginal seas and islands arcs are discussed by many authors. The geodynamic processes in this region, which have been going on for some 100 million years, evidently have their mechanical cause at the concave inner side of the orogenic arcs. So far the model proposed conforms to our concept, illustrating the case of tectonic histories and evolution of the Aleutian island arc system and of the other areas discussed. The high heat flow under the Japan Sea cannot be explained by mere conduction of this frictional heat. The correct explanation of this fact can be provided by the mechanism discussed above. In Figure 8.21 the model of the West Pacific island arc system and the Japan island arc system is illustrated. The distance between separate volcanoes in the Japan island arc system is about 95 km, while in the Aleutian arc island system it is about 75 km. The distances between the neighbouring volcanoes also characterize the slope of the subducted lithospheric plate. The rock composition of the magma depends on the depth of the bulged-out domains. The sodium (Na_2O) and the potassium (K_2O) change their content in magmatic rocks in a dependence on depth, as is observed for example in the Japan island arc system.

On the Earth's surface we even observe cases in which the rift zone and the subduction zone are located closer to, or further away from, each other (for example the San Andreas area, the Juan de Fuca subduction zone). For illustration of such a situation, the Juan de Fuca subduction zone in the cross-section Juan de Fuca plate-Mt.St.Helens-Mt.Adams is given (see Figure 8.19), whose modification corresponds

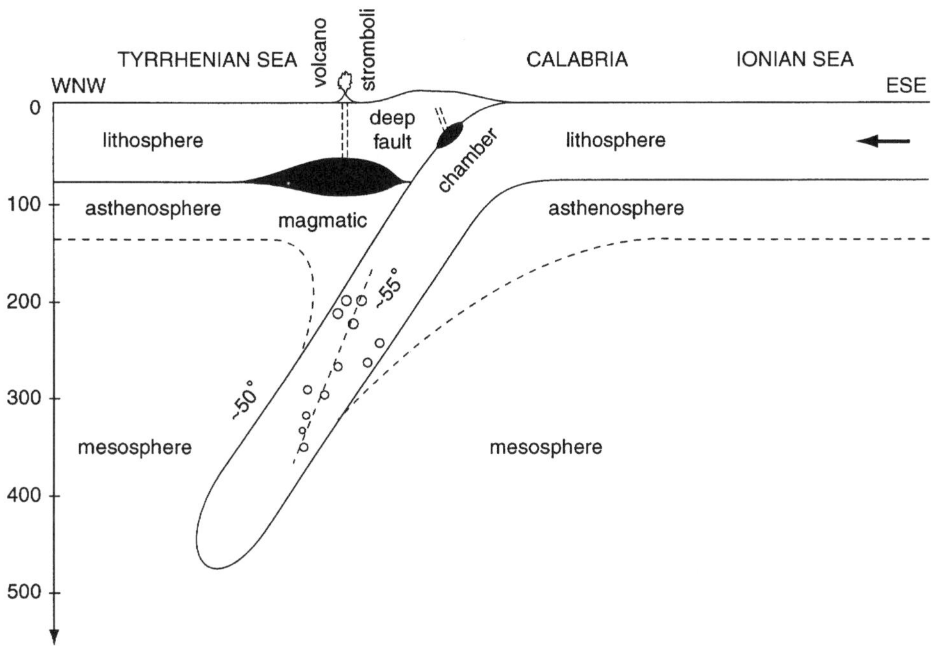

Figure 8.20. The Geodynamic Model of the Mediterranean Arc

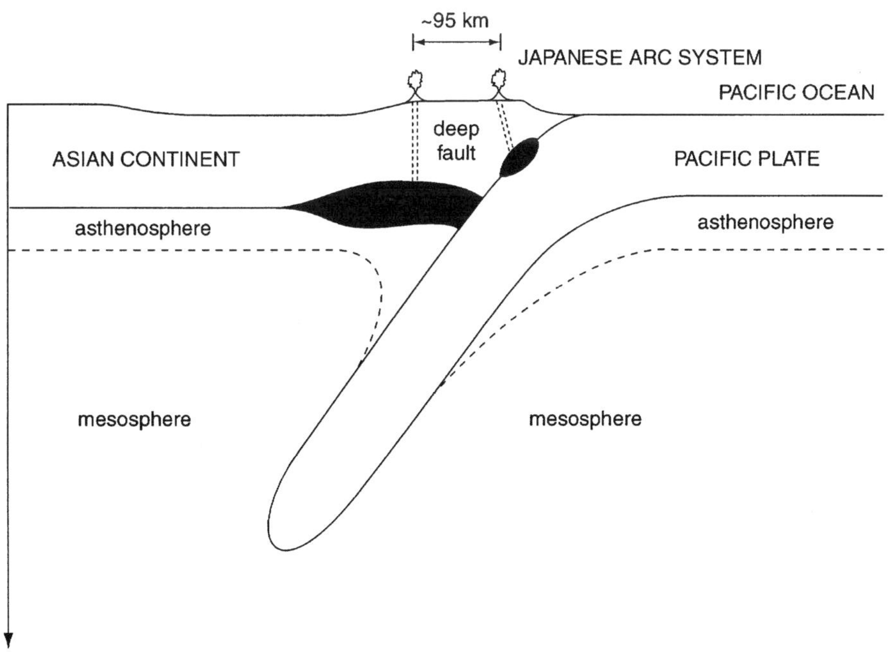

Figure 8.21. Geodynamic Model of the West Pacific Area

to the concept of the presented geodynamic model. The fully seismic underthrusting of the plate is observed in the Alaska area while in the Japan area it is of about 1/4 seismic and 3/4 aseismic. Practically it depends on the viscosity of the materials of the plate, on the magnitude of the Coulombian friction, and on the quality of the downgoing plate, as well as on the strongly heated and partially melted contact boundaries of subducting plates. The size and depth of the bulged-out areas of the lithosphere and the asthenosphere can only be determined from detailed numerical analysis.

If the speed of both colliding plates or geological blocks is low and the thickness of both plates is approximately the same, the collision will not be reflected in the subduction, but both plates are deformed and their thickness increased. At the same time, the centre of the deformed lithospheric block or megaplate will bulge out at the bottom and will decline from above and thus grow thinner. The space where the lithosphere has bulged out is immediately filled with melted materials from the lithosphere-asthenosphere interface. If the lithospheric bulge is large, the centre of the megaplate will display an anomalous heat flow, the latter being closely related to tectonic processes. This heat flow has anomalous values in several areas of the Earth‘s surface, such as in the Pannonian Basin and the African plate (see Figure 8.22), where in the second case the influence of the Red Sea rift zone on the geodynamic processes for the evolution there of hot spots is suitable. Since the area is strained, there occur deep faults which, due to the concept discusssed above, represent the origin of the volcanism in this area. The above mechanism could also describe the mechanism of the evolution of the Siwalik belt in the Sub-Himalayan region.

The evolution of the hot spots in the Pacific plate, for the most part, has a similar origin, since the Pacific plate is large and is strained from the rift-ward side, and on the second side it is subducted. Due to the Coulombian friction, and especially due to the condition that if both plates are in a contact, i.e. if

$$u_n^k(\boldsymbol{x}, t) - u_n^l(\boldsymbol{x}, t) = 0 \text{ then } \left|\vec{\tau}_t^k(\boldsymbol{x}, t)\right| \leq g_c^{kl}(\boldsymbol{x}, t) \text{ at every time } t,$$

and moreover, if $|\vec{\tau}_t(\boldsymbol{x}, t)| < g_c^{kl}(x, t)$ then $\partial/\partial t(\boldsymbol{u}_t^k(\boldsymbol{x}, t) - \boldsymbol{u}_t^l(\boldsymbol{x}, t)) = 0$ at every time t, and if $|\vec{\tau}_t(\boldsymbol{x}, t)| = g_c^{kl}(\boldsymbol{x}, t)$ then there exists a function $\vartheta \geq 0$ such that $\partial/\partial t(\boldsymbol{u}_t^k(\boldsymbol{x}, t) - \boldsymbol{u}_t^l(\boldsymbol{x}, t)) = -\vartheta\vec{\tau}_t^k(\boldsymbol{x}, t)$ at every time t, then this shift has the opposite direction than the acting tangential forces. These conditions have the following

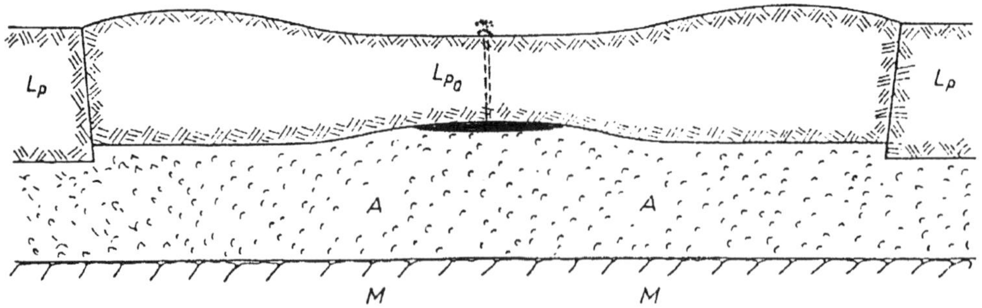

Figure 8.22. Geodynamic Model of Africa and Pannonian Basin $L_{p1}L_{p0}$ - the lithosphere M - the mesosphere A - the asthenosphere

physical meaning: if the absolute value of the tangential stress component $\vec{\tau}_t$ is less than the frictional force g_c^{kl}, then the tangential force precludes the mutual motion of both plates and the deformation energy a cummulates in the Pacific and continental plates. If the absolute value of the tangential stress component $\vec{\tau}_t$ reaches the value of the frictional force g_c^{kl}, then there are no forces, which can preclude the mutual shifting of both plates. Thus the contact points change their position in the direction opposite to that in which the tangential stress component acts. At the same time one part of the deformation energy accumulated in the collision region is released in the form of an earthquake. This is one of the earthquake's mechanisms; the second one corresponds to the fracture mechanics and represents the evolution of originating deep faults (the algorithm solving such problems was discussed in Chapter 7). Contemporary with geodynamic processes taking place in the rift and the subduction zones, the Pacific plate is strained, and the centre of the Pacific plate bulges out at the bottom and declines from above, so that it grows thinner and is broken by a system of deep faults. At the same time the lithospheric and asthenospheric rocks of the bulged-out area are melted by a similar mechanism as discussed above and then fill these domains. The molten material from this bulged-out domain ascends to the Earth's surface through deep faults situated in places with shear stresses, and therefore at places on the Earth's surface where volcanoes are observed. This is the mechanism of the origin and evolution of the hot spots. From the presented mechanism of the hot spots it is evident that hot spots can change their position in time. Hence it follows that after they have changed to their new position, the primary place of the hot spot starts to cool then solidifies and recrystallizes, resulting in extinct volcanoes on the bottom of the Pacific Ocean, which is observed by geologists. However, in general their motion in time depends on the configuration of the plate tectonic system. The schematic cross-section through the Pacific Ocean is given in Figure 8.23.

The magma in the volcanic chambers is densely differentiated, heavy minerals such as augit and olivine crystallize like the first, while gases such as chlorides, oxides of sulphur and carbon, nitrogen, chlorine, hydrogen, hydrogen chloride, etc. are accumulated below the ceiling of the bulged-out area (the volcanic chamber). The pressure in this bulged-out area grows, up to the moment when the gases are released into the atmosphere and the volcano begins to erupt. Such types of explosive volcanoes are situated in subduction zones near the trench, i.e. typically in the periphery of

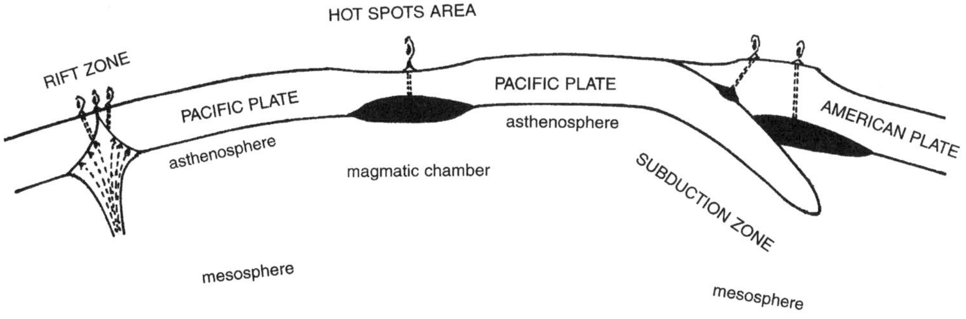

Figure 8.23. Geodynamic Model of the Pacific Ocean: Model of the Hot Spots

great oceans like the Indian, Pacific, and Atlantic Oceans. In the areas of rift zones, of subduction zones at greater distances from the trench and of hot spots the effusive type of volcanoes occur, as the magma is composed from materials melted on the contact boundary between the lithosphere and the asthenosphere, i.e. from materials with a small volume of gaseous components. Observations in volcanic areas prove our theoretical conclusions.

We saw above that the distance between two neighbouring volcanoes which are of different types, i.e. of effusive and eruptive types, depends on the angle of the contact boundary (deep fault) between two colliding lithospheric plates or biger geological bodies. On the Earth's surface there exist places where the directions of these deep faults are near the vertical direction, for example the Great Rift Valley and, west of it, the area of Great Lakes and the Virunga Mountain. In such cases it was observed that the composition of gases in lavas, defined by the ratio between gases and lavas, is about 1% in the first case and about 30% in the second, and the distances between two different volcanoes is about 8 to 15 km. The mechanism of volcanism in such regions can be clarified by the same mechanism as above, where the angle of the contact boundary between two colliding lithospheric blocks is near to the vertical direction only. That fact can have an influence on the composition of gases in lavas. In the case of the volcano of effusion type, the composition of gases in lavas can also change due to the possibility of melted rocks in the upper part of the lithosphere situated in volcano chambers near the Earth's surface where the lavas obtain more gases than the lavas at the bottom of the lithosphere and the asthenosphere.

8.7. PROGRAMME ON THE ASSESSMENT OF EARTHQUAKE HAZARD FOR AREAS OF CATASTROPIC EARTHQUAKES

8.7.1. Some Rudimentary Definitions and Present Studies of Seismic Hazard

By critical structure is meant a structure of high national economy and, moreover, which in the case of its destruction by earthquake represents a great hazard for great areas of the Earth. As an example we note power plants, nuclear power plants, great waterdams, factories, deep mines, bridges, etc..

Risk is defined (see Fournier d'Albe (1977)) as the possibility of a loss. Then the financial risk is defined as the possibility of a financial loss, and the economic risk, in the earthquake context, as the possibility of the loss of property or loss of function of structures, utilities, buildings, etc..

Seismic hazard is defined as the probability of occurrence, at a given or within a given area and within a given period of time, of ground motion due to an earthquake capable of causing significant loss or damage. It may be represented by sets of figures indicating the probability of occurrence of certain ground accelerations, velocities and displacements, of ground movements of various duration, or of any other physical parameters which are of significance for the vulnerability of buildings or structures.

Vulnerability is defined as a measure of the proportion of the capital value or of production capacity in the case of factories, power plants, waterdams, etc. which may be expected to be lost as a result of a given earthquake.

The present studies of seismic hazard analysis involve:

(i) definition of earthquake source regions in terms of boundaries, recurrence intervals and upper threshold magnitudes;

(ii) knowledge of attenuation of intensity, acceleration and other quantities defining severe ground motion with distance, magnitude and focal depth;

(iii) development of sound statistical and deterministic procedures for data processsing.

8.7.2. Objectives Qualifying the Programme

The concentration of population in earthquake prone areas in the world, for example the San Andreas region, Japan, Mexico, India, etc., is enormous. The number of critical structures with extraordinary hazard for great regions of the world, for example power plants, nuclear power plants, high-level radioactive waste repositories, deep mines, great waterdams, factories, utilities, etc. and facilities exposed to earthquake effects, is growing. Consequently, seismic hazard assessments for the epi-central areas of great earthquakes as well as for areas with active earthquakes, where the nuclear power plants (e.g. NPPs in Armenia situated on the seismically active Ararat's deep fault, Bulgaria, and NPP Jaslovské Bohunice in Slovakia), great waterdams (e.g. in the Himalayas), deep mines (e.g. in the West Carpathians, South Africa, the Caucasus, etc.) are built and operated, must be performed with the highest responsibility.

At present several methods and procedures are applied for hazard assessment; however these procedures and methods are not always well defined and do not find the correct causes of earthquakes. Catastrophic earthquakes in areas with great concentration of population (e.g. in regions where catastrophic earthquakes are expected such as Japan, San Andreas region, California in USA, India, Mexico, etc.) and in areas with construction and operation of nuclear power plants (for example the NPP in Armenia situated on the Ararat deepfault, taking into account the consequences of the catastrophic situation in the NPP Chernobyl) showed an immediate need to elaborate the new research programme based on the complex geophysical research including all geophysical fields, but above all based on the monitoring and theoretical analyses of the stress-strain fields.

8.7.3. Assessment of Seismic Hazard and Earthquake Prediction-Proposed Programme

The present methods are based on calculations of seismological data and on evaluation of geological-geophysical data from the area concerned. In contrast to the present methods used, the proposed programme requires complex tectono-physical research of the present-day motion of the lithosphere based on the analyses of the stress-strain state in global regions of interest. Then from the results of such analyses the assessments of hazard and earthquake predictions can be given with greater accuracy because of the analysis of the stress-strain fields as well as of other geophysical fields.

The proposed programme is based on the project representing the fundamental geophysical and geological research project (proposed in my conception studies and my earlier papers and contributions, See Nedoma (1986a-c), (1990a,b), (1991), Mantič and Nedoma (1990)) consisting of the following recommendations:

(1) Long-term tectono-physical research of the present-day motion of the lithosphere by means of the stress-strain fields observed, and monitoring in accessible places of the lithosphere, i.e. in deep boreholes at various geological depths, in adits, etc.. In addition, observation of the recent motion of the Earth's crust, (i.e. of both components-vertical and horizontal) and of the seismicity in the area (i.e. detectors — wide-band geophones — will be placed in a probe to monitor the stress-strain field in the borehole during the long term), and detailed determination of the geological structure of the area by deep seismic sounding, by methods of seismic tomography, electromagnetic tomography, seismic microzoning and magneto-telluric measurements.

(2) Theoretical analysis of the stress-strain fields in the area, based on plate tectonic hypothesis and thermo-elastic as well as thermo-plastic contact problems, complete with the map of the stress-strain fields and the database system for various geological depths. Compression of theoretical results based on the mathematical modelling and simulation of tectono-physical processes in the global area with the results of measurements in situ (see point (1)). Extrapolation of obtained results to longer distances and greater depths from the place of measurements in situ in this way. Choice of a most probable geodynamical model of the area based on expert systems and further improvement of this model facilitated by continuing research.

(3) Measurement of seismo-acoustic activity in the places of long-term observation of the stress-strain fields, where detectors, i.e. geophones, are placed in the probe for long-term measurement of the stress-strain fields in the boreholes.

(4) Determination of all physical parameters of rocks, for example density, Young's modulus, Poisson's constant, porosity, etc., and determination of their age from the samples from the places where the overcoring method was applied.

(5) Observation of the distribution and orientation of micro-and macro-cracks by pattern recognition methods, and determination of stress in the rock massif by means of the study of the determination of the rock crystallic structure and its evolution in time, and the acquisition of rock cores from the places where the overcoring method was applied to study the present-day stress-strain state in the rock massif. Determination of paleo-stresses in the rocks.

(6) Long-term observation of stress-strain fields caused by tidal forces in the area of interest, and the following comparison with similar measurements in research tidal stations and stations in other areas.

(7) Long-term measurement of temperature and heat flow in the places of observation of the stress-strain fields, where detectors, i.e. thermometers, geophones, etc., are placed in the probes for long-term observation of stress-strain fields in accessible places of the lithosphere.

(8) Paleomagnetic measurements of the rock-core samples from the places of application of the overcoring method and of long-term observation of components of the stress-strain tensors. They will sure for paleomagnetic reconstruction of the orientation of the magnetic field and for the determination of motion and historical rotation of the individual blocks that form the area of interest.

(9) Magnetometric measurements and long-distance extrapolation of the Earth, i.e. the pattern recognition analysis of aerial and satellite photography for determination of fractures, faults and their motions.

(10) Gradual creation of databases for areas of interest containing the results of the proposed complex research of the global area. Further improvement of geodynamical models of the global area, based on the actual level of knowledge, will allow the elimination of the influence of civilization processes on the area considered to facilitate finding criteria for the prediction of earthquakes in the global area.

8.7.4. Assessment of Hazard of Critical Structures-Proposed Programme

One of the principal factors that determine the safety of critical structures, such as nuclear power plants, high-level radioactive waste repositories, great waterdams, etc., is the maximal possible earthquake which still cannot endanger their operation. Several methods are used to estimate this. They are based on calculations from simple given data or on an evaluation of geological-geophysical data from the area concerned. So in the case of the nuclear power plants (NPPs) the methods involve estimation of the maximal magnitude of earthquakes in the region of the NPP by experts from the field of seismology, statistical data analysis with the help of the Gumbel distribution, and analyses of seismic, gravimetric, and geodetic data from the investigated area. All these methods evaluate the results of observations on the Earth's surface. None of them takes into account the accumulation of stress-strain fields in geological blocks defined by tectonic lines in the area considered and their broader vicinity and, therefore, none of them studies the causes of activity in these areas. The project proposed is based only on stress-strain data and fully enters into relation with a programme for assessment of seismic hazard and earthquake prediction, discussed above.

The proposed programme consists of the following recommendations for the construction and operation of critical structures:

(1) Long-term tectono-physical research of the present-day motion of the lithosphere in the area of critical structures, based on monitoring of geomechanic quantities, i.e. by means of the stress-strain fields in accessible places of the lithospheric plate (in deep boreholes at various geological depths, in adits, etc.). In addition, observation of the recent motion of the Earth's crust and of the seismicity in the area (i.e. detectors — geophones — are placed in probes monitoring the stress-strain fields), monitoring the temperature at the same places as the probes monitor the stress-strain field, and detailed determination of the geological structure of the area by using the methods of deep seismic sounding, magneto-telluric sounding, methods of seismic and electro-magnetic tomography, seismic microzoning, etc..

(2) Monitoring of seismo-acoustic activity in the same places as the monitoring of stress-strain fields, i.e. detectors — geophones — will be placed in the probe for the monitoring of the stress-strain fields in the boreholes.

(3) Observation of the area tectonic effects on the structure of the critical structure (i.e of NPPs, high-level radioactive waste repositories, deep mines, great waterdams, etc.) by means of placing detectors, i.e. tensometers, tensometric rosettes and space tensometric probes, in foundations and walls of the critical structure.

(4) Mathematical modelling and simulation of tectono-physical processes in the area concerned and its broader vicinity, and comparing numerical results obtained with the in-situ monitoring data from points (1) and (2).

(5) Measurement of the inclination of the foundation plate of the critical structure, especially of the nuclear power plants and high-level radioactive waste repositories, great waterdams, etc..

(6) Observation of the temperature in the masonry of the NPP and of the high-level radioactive waste repository in relation to the changes of stress-strain state in the masonry and the processes in nuclear piles and/or in local waste deposits.

8.7.5. Mitigation of Risk

At present various procedures are under taken to reduce earthquake risk. Leaving aside the possibility of reducing seismic hazard by action directed at the natural phenomenon itself (e.g. earthquake control by the injection of fluids into the Earth's crust, etc.), there remain various means of reducing the exposed value and of reducing vulnerability. But the mitigation of risk in reality cannot be effectively ensured without detailed knowledge of the stress-strain state in time in the colliding lithospheric blocks. From this point of view complex information about the stress-strain state of the lithosphere is desirable. The reduction of vulnerability of the regions with critical structures such as nuclear power plants and great waterdams must be ensured (in addition to the mitigation of earthquake risk in regions with great concentration of population) by the consequential monitoring of the stress-strain state of lithospheric rocks and, on this basis, then formulating earthquake prediction conditions for the region of interest.

At the end I should like to emphasise that for the mitigation of risk the complex geological and geodynamical research discussed above will be necessary. Such complex research of the region of interest involves complex information on every rock sample (such as all physical parameters of the rock, its age, its accurate location in the region of interest, i.e. all its spatial coordinates, the distribution of microcracks, paleostresses, porosity, all information of geophysical fields in the place where the rock sample was taken etc.) from the investigated region. Only in such a way we can give accurate information for a tectonic evolution of the region of interest and therefore more accurate information for the prediction of earthquakes.

REFERENCES

A'dám, A. (Ed). (1976), Geoelectric and Geothermal Studies. *KAPG Geophys.* Monograph, Budapest, 752pp.

Aki, K. (1979), "Evolution of Quantitative Models of Earthquakes", in *Fracture Mechanics* (Burridge, R. (Ed.)), *SIAM-AMS Proceedings*, **Vol. XII**, New York, AMS, 43-58.

Argyriadis, I. (1974), "Sur L'Orogenése Mésagéenne des Temps crétacés", *Rew. Géorg. Phys. Geol. Dynam.*, **16, 1**, 23-59, Paris.

Atwater, T. (1970), "Implications of Plate Tectonics for the Cenozoic Evolution of Western North America", *Geol. Soc. Am. Bull.*, **81**, 3513-3536.

Babuška, I. and Szabó, B. (1981), "On the Rate of Convergence of the FEM", *Int. J. Numer. Methods Eng.*, **28**, 323-341.

Babuška, V., Plomerová, J., Šílený, J. and Janatková, Z. (1986), *Structural Model of the Deep Lithospheric Structure: Geophysical Model of the Lithosphere*, Geophysical Institute Czech. Acad. Sci. Prague; Geofysika, Brno; Geophysical Institute Slov. Acad. Sci., Bratislava, 245-252.

Babuška, V. and Plomerová, J. (1995), "Regional Fabric of the Deep Lithosphere in Central Europe from Seismic Anisotropy", *Studia Geoph. et Geod*, **39**, 219-226.

Bagdasarjan, G. P. *et al.* (1977), "Determination of the Age of Crystalline Systems of the West Carpathians and Preliminary Interpretation of Results", *Geol. Sbor. Slov. Acad. Sci.*, **28, 2**, 219-242, Bratislava (in Russian).

Baranowski, J., Armbruster, J., Seeber, L. and Molnar, P. (1984), "Focal Depths and Fault Plane Solutions of Earthquakes and Active Tectonics of the Himalaya", *J. Geophys. Res.*, **89**, 6918-6928.

Barazangi, M. and Issacks, B. (1971), "Lateral Variations of Seismic Wave Attenuation in the Upper Mantle Above the Inclined Earthquake Zone at the Tonga Island Arc", *J. Geophys. Res.*, **76**, 8493-8516.

Barberi, F., Gasparini, P., Innocenti, F. and Villari, L. (1974), "Volcanism of the Southern Tyrrhenian Sea and its Geodynamic Implications", *J. Geophys. Res.*, **78(23)**, 5221-5223.

Barberi, F., Bigi, G., Gandiosi, G., Luongo, G., Panizza, A. and Santacroce, R. (1983), "Dinamica del Mediterraneo: Guida alla Escursione nei Campi Flegrei ed al Somme-Vesuvio con note Sulla Dinamica del Mediterraneo", *Osservatio Vesuviano*, Napoli.

Barberi, F., Hill, D. P., Innocenti, F. I., Luongo, G. and Treuil, M. (Eds) (1984), "The 1982-1984 Bradyseismic Crisis at Phlegranean Fields (Italy)", *Bulletin Volcanologigue*, **47, 2**, (Special Issue).

Bemmelen, van B. W. (1972), *Geodynamic Models*, Elsevier, Amsterdam, London, New York.

Ben-Menahem, A., Aboodi, J. and Schild, R. (1974), "The Source of the Great Assam Earthquake, an Interplate Wedge Motion", *Phy. Earth and Plan. Inter.*, **9**, 265-289.

Beránek, B. and Zátopek, A. (1981), "Preliminary Results of Geophysical Synthesis in Czechoslovakia and Central Europe Based on Explosion Seismology Until 1980", *Geophysical Synthesis in Czechoslovakia*, VEDA, Bratislava, 469-497.

Bergman, E.A. and Solomon, S.C. (1985), "Earthquake Source Mechanisms From Body-Waveform Inversion and Interplate Tectonics in the Northern Indian Ocean", *Phy. Earth and Planet. Inter.*, **40**, 1-23.

Berry, M.J. and Knopoff, L. (1969), "Structure of the Upper Mantle under the Western Mediterranean", *J. Geophys. Res.*, **72(14)**, 3613-3626.

Bird, P. (1978), "Finite Element Modelling of Lithospheric Deformation: the Zagros Collision Orogeny", *Tectonophysics*, **50**, 307-336.

Biswas, S. (1989), "State of Stress and Its Bearing on Tectonics Under the Tibetan Plateau", *Phys. Earth and Planet. Inter.*, **54**, 4-9.

Bleahu, M., Bocaletti, M., Manetti, P. and Peltz, S. (1973), "Neogene Carpathian Arc: A Continental Arc Displaying the Fractures of an "Island Arc"", *J. Geophys. Res.*, **78**, 5025-5032.

Brace, W.F. and Kohlstedt, D.L. (1980), "Limits on Lithospheric Stress Imposed by Laboratory Experiments", *J. Geophys. Res.*, **85**, 6248-6252.

Cambel, B. *et al.* (1977), "Some Informations about Geochronology of the Crystalline Westerm Carpathians Based on U-Th-Pb Method", *Geol. Sbor. Slov. Acad. Sci*, **28, 2**, 243-259 (in Russian).

Cathles, L.M. (1975), *The Viscosity of the Earth's Mantle*, Princeton University Press, Princeton, NJ.

Céa, J. (1971), *Optimisation, Théorie et Algorithmes*, Dunod, Paris.

Čech, F. (1982), *Deposits of Fuels: Relation to the Deep Structure of the Panonnian Basin and the Carpathian Arc*, D. Štúr Geological Institute, Bratislava.

Chandra, U. (1978), "Seismicity, Earthquake Mechanisms and Tectonics Along the Himalayan Mountain Range and Vicinity", *Phys. Earth and Planet. Inter.*, **16**, 109-131.

Chandra, U. (1992), "Seismo-tectonics of Himalaya", *Curr. Sci*, **62**, 40-71.

Channell, J. E. T. and Horváth, F. (1976), "The African/Adriatic Promontory as a Paleogeographical Premise for Alpine Orogeny and Plate Movements in the Carpatho-Balkan Region", *Tectonophysics*, **35**, 71-101.

Chhibber, H. L. (1934), *Geology of Burma*, McMillan, London. Curray, J. R. *et al.* (1979), "Tectonics of the Andaman Sea and Burma", *Mem. Am. Assoc. Pet. Geol.*, **29**, 189-198.

Christiansen, N. (1996). Poisson's Ratio and Crustal Seismology, *J. Geophys. Res.* **102**, 3139-3156.

Curray .J. R., Emmel, F. J., Moore, D. G. and Raitt, R.W. (1982), "Structure, Tectonics and Geological History of the Northeastern Indian Ocean", in *The Ocean Basins and Margins* (Naim, E.M and Stehli, F. (Eds.)) Plenum Publishing Corporation, 399-450.

Das, S. and Aki, K. (1977), "A Numerical Study of Two-Dimensional Spontaneous Rupture Propagation", *Geophys. J. R. Astr. Soc.*, **50**, 643-668.

Das, K. K. and Filson, J. R. (1975), "On the Tectonics of Asia", *Earth and Planetary Science Letters*, **28**, 241-253.

Dasgupta, S., M. Mukhopadhyay and D.R. Nandy (1990). Magmatism and Tectonic Setting in the Burmese-Andaman Arc. Indian J.Geol. v. 62, 117-141.

Dasgupta, S. and Mukhopadhyay, M. (1993), "Seismicity and Plate Deformation Below the Andaman Arc, Northeastern Indian Ocean", *Tectonophysics*, **225**, 529-542.

Dasgupta, S. and Mukhopadhyay, M. (1997). Aseismicity of the Andaman Subduction Zone and Recent Volcanism. J. Geophysical Society of India, vol. 49, 513-521.

Dietz, R. S. (1961), "Continent and Ocean Basin Evolution by Spreading of the Sea Floor", *Nature*, **190**, 854-857.

Dietz, R. S. (1962), "Ocean Basin Evolution by Sea Floor Spreading", in *The Crust of the Pacific Basin*, (Mac Divakara, Rao and Kemo, H. (Eds.)), Geophys. Mono, **6**, Am. Geophys. Union, Washington.

D'Ingeo, F., Calcaguile, F. and Panza, G. F. (1980), "On the Fault Plane Solutions in the Central-Eastern Mediterranean Region", *Boll. Geof. Teor. e Appl.*, **85**, 13-22.

Eguchi, T., Uyeda, S. and Maki, T. (1979), "Seismo-tectonics and Tectonic History of the Andaman Sea", *Tectonophysics*, **57**, 35-51.

Ekeland, I. and Temam, R. (1976), *Convex Analysis and Variational Problems*, North-Holland, Amsterdam.

Engdahl, E. R. (1973), "Relocation of Intermediate Depth in the Earthquakes in the Central Aleutians by Seismic Ray Tracing", *Nature*, **245**, 23-25.

Evans, P. (1964), "The Tectonic Framework of Assam", *J. Geopl. Soc. Ind.*, **5**, 80-86.

Fitch, T. J. (1970), "Earthquake Mechanisms in the Himalayan, Burmese, and Andaman Regions, and Continental Tectonics in Central Asia", *J. Geophys. Res.*, **75**, 2699-2709.

Fournier d'Albe, E. M. (1978), "The UNESCO Programme on the Assessment and Mitigation of Earthquake Risk", in *Proc. of the Symposium on the Analysis of Seismicity and on Seismic Risk* Schenková, Z. (Ed.), Geophys. Inst. Czech. Acad. Sci, Prague, 15-23.

Gansser, A. (1964), *Geology of the Himalaya*, Interscience Publishers, New York.

Gansser, A. (1980), "The Significance of the Himalayan Suture Zone", *Tectonophysics*, **62**, 37-52.

Gansser, A. (1991), "Facts and Theories on the Himalayas", *Eclogae Geol. Helv.*, **84**, 33-59.

Gasparini, C., Iaunaccone, G., Scandone, P. and Scarpa, R. (1983), "Seismo-tectonics of the Calabrian Arc", *Tectonophysics*, **84**, 267-286.

Gaudiosi, G., Luongo, G. and Ricciardi, G. P. (1984), "A Bending Model for the Calabrian Arc", in *Problems and Methods for Lithospheric Exploration* (Cassinis, R. (Ed.)), Plenum Publ. Corp..

Ghosh, D. K. (1988), "On the Collision Controlled Structures and Consequent Seismic Environment in the Indian Block — An Overview", *Journal of Engineering Geology*, **XII, 3-4**, 39-45.

Goetze, C. (1978), "The Mechanism of Creep in Olivine", *Philos. Trans. R. Soc. London*, **Ser.A., 288**, 99-119.

Grecula, P. and Roth, Z. (1978), "Kinetic Model of the Western Carpathians in the Collected Profile", *Sborník geologických věd-Geologie*, **32**, 49-73 (in Czech).

Green, D. H. and Ringwood, A. E. (1967), "Genesis of Basaltic Magmas", *Beitr. Mineral. Petrog.*, **15**, 90-103.

Gresta, S., Longo, V. and Viavattene, A. (1990), "Geodynamic Behaviour of the Eastern and Western Sides of Mt. Etna", *Tectonophysics*, **179**, 81-92.

Grow, J. A. (1973), "Crustal and Upper Mantle Structure of the Aleutian Arc", *Geol. Soc. Am. Bull.*, **84**, 2169-2192.

Grow, J. A. and Atwater, T. (1973), "Mid-Tertiary Tectonic Transition in the Aleutian Arc", *Geol. Soc. Am. Bull.*, **81**, 3715-3721.

Grow, J. A. and Sheridan, R. E. (1981), "Deep Structure and Evolution of the Continental Margin of the Eastern United States", *Ocealogica Acta, Colloque C3, Geology of Continental Margins*, 26th Int. Geol. Congr., Paris, 11-19.

Gupta, H. K. and Delany, F. M. (Eds) (1981), "Zagros, Hindu Kush and Himalaya Geodynamic Evolution", *Geodyn.*, **Ser.3.**, Am.Geophys. Union.

Gutenberg, B. and Richter, C. F. (1954), *Seismicity of the Earth and Associated Phenomena*, Princeton Univ. Press, Princeton, USA.

Hamilton, W. (1979), "Tectonics of the Indonesian Region", *USGS. Prof. Pap.*, **1078**, 345.

Harrisson, T., Copeland, M. P., Kidd, W. S. F. and Yin, An (1992), "Raising Tibet", *Science*, **255**, 1663-1670.

Hess, H. H. (1960), "Evolution of Ocean Basins", *Report to Office of Naval Research*, ONR Contract No.1858.

Hess, H. H. (1962), "History of Ocean Basins", in *Petrologic Studies* (Engel, A. E. J., James, H. L. and Leonard B. F. (Eds.)), Geol. Soc.Am., 599-620.

Hentz, N. and Savu, H. (1974), "Plate Tectonic History of Romania", *Am. Geol. Soc. Bull.*, **85**, 1429-1440, Denver.

Holt, W. E., Ni, J. F., Wallace, T. C. and Haines, A. J. (1991), "The Active Tectonics of the Eastern Himalayan Syntaxis and Surrounding Regions", *J. Geophys. Res.*, **96**, 14595-14632.

Horváth, F. and Stegena, L. (1977), "The Pannonian Basin: A Mediterranean Inter-arc Basin", in *Structural History of the Mediterranean Basins* (Biju-Duval, B. and Montadert, L. (Eds.)), Editions Techniq, Paris, 333-340.

Ilavský, J. *et al.* (1979), *Metallogese del'Europe Alpine Centrale et du Sud-Est*, D. Štúr Geological Institute, Bratislava.

Jafri, S.H., V. Balaram and P.K. Govil (1993). Depositional Environments of Cretaceous Radiolarian Cherts from Andaman-Nicobar Islands, Northeastern Indian Ocean. Marine Geology, 112, 291-301.

Jafri, S.H., S.N. Charan and P.K. Govil (1995). Plagiogranite from the Andaman Ophiolite Belt, Bay of Bengal, India. J. of the Geological Society, London, vol. 152, 681-687.

Jiříček, R. (1979), "Tectono-genetic Development of the Carpathian Arc during Oligocene and Neogene", in *Tectonic Profiles of the West Carpathians* (Mahel M. (Ed.)), D. Štúr Geological Institute, Bratislava (in Czech).

Kaila, K. L. and Narain, H. (1976), "Evolution of the Himalaya Based on Seismo-tectonics and Deep Seismic Soundings", *Himalayan Geology Seminar*, **Sec. II**, NGRI, Hyderabad, New Delhi, India.

Kanamori, H. (1980), "Stress in the Lithosphere", in *Physics of the Earth's Interior* Dziewonski A. M. and Boschi, E. (Eds.), North Holland, Amsterdam, 712-713.

Khattri, K. N. (1992), "Seismic Hazard in Indian Region", *Curr. Sci.*, **62**, 109-116.

Kirby, S. H. (1983), "Rheology of the Lithosphere", *Recs. Geophys. Space Phys.*, **21**, 1458-1487.

Krs, M. and Roth, Z. (1979), "The Insubric-Carpathian Tertiary Block System: Its Origin and Disintegration", *Geologický sborník-Geologica Carpathica*, **30, 1**, 1-15.

Krs, M., Krsová, M. and Pruner, P. (1996), "Paleomagnetism and Paleogeography of the Western Carpathians from the Permian to the Neogene", in *Paleomagnetism and Tectonics of the Mediterranean Region* (Morris, A. and Tarling, D. H. (Eds.)), **Spec. Publ. No 105**, Geological Society, London, 175-184.

Leuchs, K. (1937), *Geologie von Asien*, Borntraeger, Berlin.

Liebermann, R. C. (1992), "Mineral Physics and Seismology", *Proc. of the XXVIII General Assembly of the European Seismological Commision*, (Prague, Czechoslovakia, 7-12 September 1992), Geophys. Institute Czech. Acad. Sci. 295-302.

Madariaga, R. (1979), "Seismic Radiation From Earthquake Models Based on Fracture Mechanics", in *Fracture Mechanics* (Burridge, R. (Ed.)), *SIAM-AMS Proceedings*, **Vol. XII**, New York, AMS, 59-77.

Makovsky, Y. *et al.* (1996). Structural Elements of the Southern Thethyan Himalaya Crust from Wide-Angle Seismic Date. **Tectonics 15**, 997-1005.

Mantič, V. and Nedoma, J. (1990), "Optimal Design of Tensometric Probes for In-Situ Stress Measurements by Finite Element Analysis and Their Optimal Dislocation in the Collision Region", in *Proc.of the 4th Int. Symposium on the Analysis of Seismicity and Seismic Risk* (Schenková, z. (Ed.)), Geophys. Inst. Czech. Acad. Sci, Prague.

Marinelli, G. (1975), "Magma Evolution in Italy", in *Geology in Italy* (Squyres, H. (Ed.)).

Mehta, P. K. (1980), "Tectonic Significance of the Young Mineral Dates and the Rates of Cooling and Uplift in the Himalaya", *Tectonophysics*, **60**, 205-217.

Molnar, P. (1990), "A Review of Seismicity and the Rates of Active Underthrusting and Deformation at the Himalaya", *J. Himalayan Geol.*, **1**, 131-154.

Molnar, P. (1992), "A Review of Seismicity, Recent Faulting and Active Deformation of the Tibetan Plateau", *J. Himalayan Geol.*, **3**, 43-78.

Molnar, P., Fitch, W. P. and Wu, F. T. (1973), "Fault Plane Solutions of Shallow Earthquakes: Contemporary Tectonics of Asia", *Earth and Planet. Sci. Lett.*, **19**, 101-112.

Molnar, P. and Lyon-Caen, H. (1989), "Fault Plane Solutions of Earthquakes and Active Tectonics of the Tibetan Plateau and its Margins", *Geophys. J. Int.*, **99**, 123-153.

Molnar, P. and Tapponnier, P. (1975), "Cenozoic Tectonics of Asia: Effect of Continental Collision", *Science*, **189**, 419-426.

Molnar, P. and Tapponnier, P. (1978), "Active Tectonics of Tibet", *J. Geophys. Res.*, **83**, 5377-5387.

Morley, C. K. (1996), "Models for Relative Motion of Crustal Blocks Within the Carpathian Region, Based on Restorations of the Outer Carpathian Thrust Sheets", *Tectonics*, **15, 4**, 885-904.

Mukhopadhyay, M. (1984), "Seismo-tectonics of Subduction and Back-Arc Rifting under the Andaman Sea", *Tectonophysics*, **108**, 229-239.

Mukhopadhyay, M. (1988), "Gravity Anomalies and Deep Structure of the Andaman Arc", *Marine Geophys. Res.*, **9**, 197-210.

Mukhopadhyay, M. and Dasgupta, S. (1988), "Deep Structure and Tectonics of Burmese Arc: Constraints From Earthquake and Gravity Data", *Tectonophysics*, **149**, 299-322.

Mukhopadhyay, M. and Krishna, M. R. (1991), "Gravity Field and Deep Structure of the Bengal Fan and its Surrounding Continental Margins, Northeast Indian Ocean", *Tectonophysics*, **186**, 365-386.

Nandy, D. R. (1980), "Tectonic Patterns in Northeastern India", *Ind. J. Earth. Sci.*, **7**, 103-107.

Nedoma, J. (1980), "Model of the Carpathian Continent/Continent Collision", *Proc. of the 17th Assembly of the ESC*, Budapest, 629-637.

Nedoma, J. (1982), "Thermo-Elastic Stress-Strain Analysis of the Geodynamic Mechanism", *Gerlands Beitr.Geophysik*, **91, 1**, 75-89.

Nedoma, J. (1983), "On One Type of Signorini Problem Without Friction in Linear Thermo-elasticity", *Apl. Mat.*, **28, 6**, 393-407.

Nedoma, J. (1984a), "On the Global Geodynamic Model of the Earth", *Gerlands Beitr. Geophysik*, **93, 3**, 241-260.

Nedoma, J. (1984b), "On One Type of Finite Element Simulation in Dynamic Elasticity", *Mathematics and Computers in Simulation*, **XXVI**, 189-209.

Nedoma, J. (1986a-c), "On the Geodynamic Model of the Earth and the Plate Tectonic Hypothesis: I-III", *Gerlands Beitr. Geophysik*, **95, 1**, 36-62; **95, 2**, 89-105.

Nedoma, J. (1987), "On the Signorini Problem With Friction in Linear Thermo-Elasticity: The Quasi-Coupled 2D-Case", *Apl. Mat.*, **32, 3**, 186-199.

Nedoma, J. (1990a), "Tectonic Evolution of Collision Zones: I-II", *Gerlands Beitr. Geophysik*, **99, 2**, 97-108; **99, 2**, 109-121.

Nedoma, J. (1990b), "Plate Tectonics and the Analysis of Fracture Mechanics in a Collision Zone: I-II", *Contr. Geophys. Inst. Slov. Acad. Sci.*, **20**, 53-69, **21**, 38-58.

Nedoma, J. (1990c), "On a New Theory of Earthquake Mechanism", in *Proc. 4th Intl Symp. on Analysis of Seismicity and Seismic Risk*, (Schenkova, Z. (Ed.)), Geophys. Inot. Czech. Acad. Sci., Prague, 339-346.

Nedoma, J. (1991), "On the Protection and Operation of Nuclear Power Plants and Great Structures in Central Europe With Respect to the Tectonics of European Alpids: A Conception Study", *Acta Geod. Geoph. Mont. Hung.*, **26 (1-4)**, 287-300.

Nedoma, J. (1994), "Finite Element Analysis of Contact Problems in Thermo-Elasticity: The Semi-Coercive Case", *J. Comput. Appl. Math.*, **50**, 411-423.

Nedoma, J. (1997), "Finite Element Approximation of a Coupled Contact-Stefan-Like Problem Arising from the Time Discretization in Deformation Theory of Thermo-Plasticity", *J. Comp. Appl. Math.*, 82, 313-331.

Nedoma, J., Haslinger, J. and Hlaváček, I. (1989), "The Problem of an Obducting Lithospheric Plate in the Aleutian Arc System: A Finite Element Analysis in the Frictionless Case", *Math. Comput. Modelling*, **12, 1**, 61-75.

Negi, J. G., Pandey, O. P. and Agrawal, P. K. (1986), "Super-Mobility of Hot Indian Lithosphere", *Tectonophysics*, **131**, 147-156.

Nelson, D. *et al.* (1996). Partially Molten Middle Crust Beneath Southern Tibet: Synthesis of Project INDEPTH Results. *Science 274*, 1684-1687.

Neri, G. (1990), "Earthquake Clustering on Volcanic and Tectonic Structures", *Tectonophysics*, **175**, 35-45.

Ni, J. and Barazangi, M. (1984), "Seismo-tectonics of the Himalayan Collision Zone: Geometry of the Underthrusting Indian Plate Beneath the Himalaya", *J. Geophys. Res.*, **89, B2**, 1147-1163.

Ni, J. and Barazangi, M. (1985), "Active Tectonics of Western Thethyan Himalaya above the Underthrusting Indian Plate: The Upper Sutlej River as a Pull-Apart Structure", *Tectonophysics*, **12**, 277-295.

Ni, J. *et al.* (1989), "Accretionary Tectonics of Burma Subduction", *Geology*, **17**, 68-71.

Nowacki, W. (1970), *Dynamic Problems of Thermo-Elasticity*, Mir, Moscow (in Russian).

Owens, T.J. and Zandt, G. (1997). Implication of Crustal Property Variations for Models of Tibetian Plateau Evolution. Nature, **387/1**, May, 37-43.

Powell, C. M. and Canagham, P. Z. (1973), "Plate Tectonics and the Himalayas", *Earth and Planet. Sci lett.*, **29**, 1-12.

Radulescu, D. P. and Sandulescu, M. (1973), "The Plate-Tectonics Concept and the Geological Structure of the Carpathians", *Tectonophysics*, **16**, 155-161.

Rastogi, B. K. (1974), "Earthquake Mechanisms and Plate Tectonics in the Himalayan Region", *Tectonophysics*, **21**, 47-56.

Rastogi, B. K. (1976), "Source Mechanism Studies and Contemporary Tectonics in Himalaya and Nearby Regions", *Bull. Int. Inst. Seis.*, **14**, 99-134.

Rastogi, B. K., Singh, J. and Verma, R. K. (1973), "Earthquake Mechanisms and Tectonics in Assam-Burma Region", *Tectonophysics*, **18**, 355-366.

Ritsema, A. R. (1970a), "On Plate Tectonics in the Mediterranean", *Proc. Eur. Seism. Comm. Gen. Ass.*, Luxenbourg.

Ritsema, A. R. (1970b), "On the Origin of the Western Mediterranean Sea Basins", *Tectonophysics*, **10(5/6)**, 609-624.

Roth, Z. (1980), *Western Carpathians — Tertiary Structure of Central Europe*, Academia, Prague.

Royden, L. and Patecca, E. (1987), "Segmentation and Configuration of Subducted Lithosphere in Italy: An Important Control on Thrust-Belt and Foredeep-Basin Evolution", *Geology*, **15**, 714-717.

Sato, K., Bhatia, S. C. and Gupta, H. K. (1996), "Three-Dimensional Numerical Modelling of Deformation and Stress in the Himalaya and Tibetan Plateau with a Simple Geometry", *J. Phys. Earth.*, **44**, 227-254.

Scotese, C. R. C. (1991), "Jurassic and Cretaceous Plate Tectonic Reconstructions", *Paleogeogr., Palaeoclimatol., Palaeoecol.*, **87**, 493-501.

Sharma, K. K. (1983), "Granitoid Belts of Himalayas", in *Granites of Himalaya, Karakorum and Hindukush* (Shams, F. A. (Ed.)), Panjub University Publ., Lahore, Pakistan, 11-13.

Sharma, K. K. and Gupta, K. R. (1983), "Calc-Alkaline Island Arc Volcanism in Indus-Tsangpo Suture Zone", in *Geology of Indus Suture Zone of Ledakh* (Thakur, V. C. and Sharma, K. K. (Eds.)) Wadia Institute of Himalayan Geology, Dehra Dun, 71-78.

Singh, D. D. and Gupta, H. K. (1980), "Source Dynamics of Two Great Earthquakes of the Indian Subcontinent", *Bull. Seism. Soc. Am.*, **66**, 887-901.

Spence, W. (1977), "The Aleutian Arc: Tectonic Blocks, Episodic Subduction, Strain Diffusion and Magma Generation", *J. Geophys. Res.*, **82**, 213-230.

Stauder, W. (1968), "Tensional Character of Earthquake Foci Beneath the Aleutian Trench with Relation to Sea Floor Spreading", *J. Geophys. Res.*, **73**, 7693-7701.

Stauder, W. (1972), "Fault Motion and Spatially Bounded Character of Earthquakes in the Amchitka Pass and Delaraf Islands", *J. Geophys. Res.*, **77**, 2072-2080.

Tandon, A. N. and Srivastava, H. N. (1975), "Focal Mechanisms of Some Recent Himalayan Earthquakes and Regional Plate Tectonics", *Bull. Seis. Soc. Am.*, **65**, 963-969.

Tapponnier, P. and Molnar, P. (1977), "Active Faulting and Tectonics in China", *J. Geophys. Res.*, **82**, 2905-2930.

Thakur, V. C. (1981), "Regional Framework and Geodynamic Evolution of the Indus-Tsangpo Suture Zone in Ladakh Himalaya", *Trans. Roy. Soc. Edingburgh, Earth Sci.*, **72**, 89-97.

Thakur, V. C. (1987), "Plate Tectonic Interpretation of the Western Himalayas", *Tectonophysics*, **134**, 91-102.

Thakur, V. C. (1992), "Geology of the Western Himalaya", *Physics & Chemistry of the Earth*, **19**, 363-374.

Trivedi, J. R., Gopalan, K. and Valdiya, K. S. (1984), "Rb-Sr Ages of Granitic Rocks within the Lesser Himalayan Nappes, Kumaun, India", *J. Geol. Soc. India*, **25**, 641-654.

Trümpy, R. (1975), "On Crustal Subductions in the Alps", in *Tectonic Problems of the Alpine System (Mahel, M. (Ed.))*, VEDA, Bratislava, 121-130.

Valdiya, K. S. (1976), "Himalayan Transverse Faults and Folds: their Parallelism with Subsurface Structures of North Indian Plains", *Tectonophysics*, **32**, 353-386.

Valdiya, K. S. (1979), "An Outline of Structural Set-up of the Kuman Himalaya", *J. Geol. Soc. Ind.*, **20**, 145-157.

Valdiya, K. S. (1980), *Geology of Kuman Lesser Himalaya*, Publ Wadia Inst. Himalayan Geology, Dehra Dun, India.

Valdiya, K. S. (1981), "Tectonics of Central Sector of the Himalaya", in *Zagros-Hindukush-Himalaya-Geodynamic Evolution* (Gupta, H. K. and Delaney, F. M. (Eds.)), Geodynamic, **Ser. 3**, Pub. Am. Geophys. Un., Washington DC, 87-110.

Valdiya, K. S. (1984a), "Evolution of the Himalaya", *Tectonophysics*, **105**, 229-248.

Valdiya, K. S. (1984b), *Aspects of Tectonics-Focus on South Central Asia*, Tata McGraw Hill Pub. Co, New Delhi, pp. 261, 137-150.

Valdiya, K. S. (1985), "Tectonic Setting of Himalayan Granites", in *Granites of Himalaya, Karakorum and Hindukush* (Shams, F. A. (Ed.)), Pub. Punjab University, Lahore.

Van Wormers, J. D., Davies, J. and Gedney, L. (1974), "Seismicity and Plate Tectonics in South-central Alaska", *Bull. Seism. Soc. Am.*, **64**, 1467-1475.

Verma, R. K. (1991), *Geodynamics of the Indian Peninsula and the Indian Plate Margin*, Oxford & IBH Publishing Co. PVT. LTD., New Delhi, Bombay, Calcutta.

Verma, R. K. and Kumar, G. V. K. (1987), "Seismicity and Nature of Plate Movement Along the Himalayan Arc, NE India and Arakan Yoma", *Tectonophysics*, **134**, 153-175.

Verma, R. K. and Mukhopadhyay, M. (1977), "An Analysis of the Gravity Field in Northeastern India", *Tectonophysics*, **42**, 283-317.

Verma, R. K. and Prasad, K. A. V. K. (1988), "Analysis of the Gravity Field in Nepalese Himalayas-Tibetan Region in the Light of Plate Tectonics", *Tectonics*, **147**, 59-70.

Verma, R. K. and Reddy, Y. S. K. (1988), "Seismicity, Focal Mechanism and Their Correlation with the Geological/Tectonic History of Tibetan Plateau", *Tectonophysics*, **156**, 107-131.

Verma, R. K., Mukhopadhyay, M. and Ahluwalia, M. S. (1976), "Seismicity, Gravity and Tectonics of Northeast India and Northern Burma", *Bull. Seis. Soc. Am.*, **66**, 1683-1694.

Verma, R. K., Mukhopadhyay, M. and Bhuin, N. C. (1978), "Seismicity, Gravity and Tectonics in Andaman Sea", *J. Phy. Earth*, **26**, 233-248.

Verma, R. K., Mukhopadhyay, M. and Nag, A. K. (1980), "Seismicity and Tectonics in South China and Burma", *Tectonophysics*, **64**, 85-96.

Verma, R. K., Mukhopadhyay, M. and Roy, B. N. (1977), "Seismo-tectonics in Himalaya and Continental Plate Convergence", *Tectonophysics*, **42**, 319-335.

Verma, R. K., Roonwal, G. S. and Gupta, Y. (1993), "Statistical Analysis of Seismicity of NE India and Northern Burma during the Period 1979-1990", *J. of Himalayan Geology*, **4(1)**, 71-79.

Vine, F. J. and Mathews, D. H. (1963), "Magnetic Anomalies Over Oceanic Ridges", *Nature*, **199**, 947.

Wadia, D. N. (1931), "The Syntaxis of NW Himalaya, its Rocks, Tectonic and Orogeny", *Rec. Geol. Surv. Ind.*, **65**, 189-220.

Wadia, D. N. (1975), *Geology of India*, Tata McGraw Hill Pub. Co. New Delhi, India.

Westaway, R. (1995), "Crustal Volume Balance during the India-Eurasia Collision and Altitude of the Tibetan Plateau: A Working Hypothesis", *J. Geophys. Res.*, **100, B8**, 15173-15192.

Wielandt, E., Plešinger, A., Sigg, A. and Horálek, J. (1987), "Deep Structure of the Bohemian Massif for Phase Velocities of Rayleigh and Love Waves", *Studia Geoph. et Geod.*, **31**, 1-7.

Wiese, R. (1963), "Geomagnetische Teifentellurik 3. Die Geomagnetischen Variationen in Mittel- und Südost-Europe als Indicator der Streichtung Grossräumiger Elektrischer Untergrundstrukturen", *Geofis. Pura Appl.*, **56**, 101-114.

Wu, F. and Wang, P. (1988), "Tectonics of Western Yunnan Province, China", *Geology*, **16**, 153-157.

Yu, Jin and McNutt, M. K. (1996), "Mapping the Descent of India and Eurasia Plates Beneath the Tibetan Plateau from Gravity Anomalies", *J. Geophys. Res.*, **101, B5**, 11275-11290.

PART V

Mathematical Problems of Geophysical Fields

9

Mathematical Models of the Geothermal Field of the Earth

9.1. INTRODUCTION

Geothermics is the study of the geothermal field inside the Earth representing the distribution of the temperature in the interior of the Earth.

Temperature plays a central role in geodynamics. It strongly controls the rheology of the inner parts of the Earth. It is related to the thickness of the lithosphere, and to viscous flow in the mantle; and, through various kinds of energy flow, it effects all geotectonic processes. The main intrinsic rheological parameters affecting the mechanical behaviour of materials are time, temperature and pressure, as well as chemical environment (diffusion).

The geophysical and geological investigations show that the geothermal field in the Earth changes with time. In the Earth's body there exist sources of heat as well as places with lower temperature, and then heat flows from places with higher temperature into places with lower temperature. The surface of the Earth radiates thermal energy into the outer space of the Earth. It is probable that geothermal sources play a fundamental role in the evolution of the Earth. Moreover, it is evident that the distribution of energy sources and their magnitudes are fundamental for all geophysical processes in the deeper as well as upper parts of the Earth, as all geophysical and geological processes occurring in the Earth need heat as their main source (for example, creation of magma, metamorphism of rocks, etc. need heat for these material processes). Temperature and pressure are the main physical parameters characterizing the situation in all parts of the Earth's interior.

9.2. THERMAL PROPERTIES OF ROCKS AND THERMAL REGIME INSIDE THE EARTH

The heat propagates through the Earth by conduction, radiation, convection and excitation transfer. In the upper parts of the Earth heat propagates mainly by conduction, which is characterized by a coefficient of thermal conduction κ_c and by a heat flow q. Under high temperatures in deeper parts of the Earth's mantle the thermal energy propagates by radiation, i.e. by electromagnetic waves, which can be characterized by a heat flow q and by a coefficient of radiative conduction κ_r. Since both problems can be described by similar equations, we combine both problems into one with a coefficient of conduction $\kappa = \kappa_c + \kappa_r$. As shown by investigations in nuclear physics during the last 25-30 years, the thermal energy also propagates by excitation transfer, i.e. by a mediation of excitons in solid rocks. The theory of excitons is similar to the theory of photons in quantum mechanics. The atom, which absorbs a certain quantum of energy, starts to be in an excited state. At the same time the exciton is radiated and then absorbed by atoms in its neighbourhood in the crystalline lattice.

The conduction of heat is physically studied by the quantum theory of solid physics. In this theory every atom in the crystalline lattice oscillates around its equilibrium position. This oscillation of atoms depends on the temperature. Atoms in a hot part of the rock oscillate more quickly than in a cooler one and thus transfer energy to atoms in cooler parts of rocks. These quantum of transferring energy, as in quantum theory, are known as phonons. Hence the coefficient of phonon conductivity (lattice conductivity) κ_{ph} is determined from the results of the theory of solid physics, and was determined as

$$\kappa_{ph} = \kappa_o \rho^{3/2} v^{7/2} T^{-5/4}, \tag{9.1}$$

where ρ is a density of rock, κ_o represents an empirical constant characterizing the rock under normal temperature and pressure, v is an average velocity of seismic waves defined by

$$v = [3v_p^3 v_s^3 (2v_p^3 + v_s^3)]^{1/3} = 3^{1/3} v_p v_s (2v_p^3 + v_s^3)^{1/3} \doteq 0.76969 v_p^3, \tag{9.2}$$

where v_p and v_s are velocities of P and S waves, and T is a temperature. From Equation 9.1 it is evident that κ_{ph} decreases with increasing temperature. Moreover, since ρ and v depend on pressure, κ_{ph} increases with increasing pressure. Thus the effect of temperature and pressure is mutually reciprocal. For depth greater than 100-150 km the effect of pressure is greater than the effect of temperature (see Figure 9.1). Under higher temperatures a deviation of κ_{ph} was observed. This deviation is due to heat transfer induced by radiation. The coefficient of radiative thermal conductivity is defined by

$$\kappa_r = 16/3 n^2 \delta \varepsilon^{-1} T^3, \tag{9.3}$$

where n is an average index of refraction, δ is the Stefan-Boltzmann constant, ε is an excitation coefficient characterizing a permeability of rock due to thermal radiation (see, for example, Landau and Lifshitz (1984)) and T is a temperature. It is known that for depths less than 100 km the coefficient κ_r is very small and for depths greater than 100 km then κ_r rapidly increases, but in the lower mantle its value increases more slowly due to an increase of ε.

In the lower parts of the mantle the excitation propagation of energy plays an important role. Then we also introduce the coefficient of excitation κ_{ex}. Figure 9.1

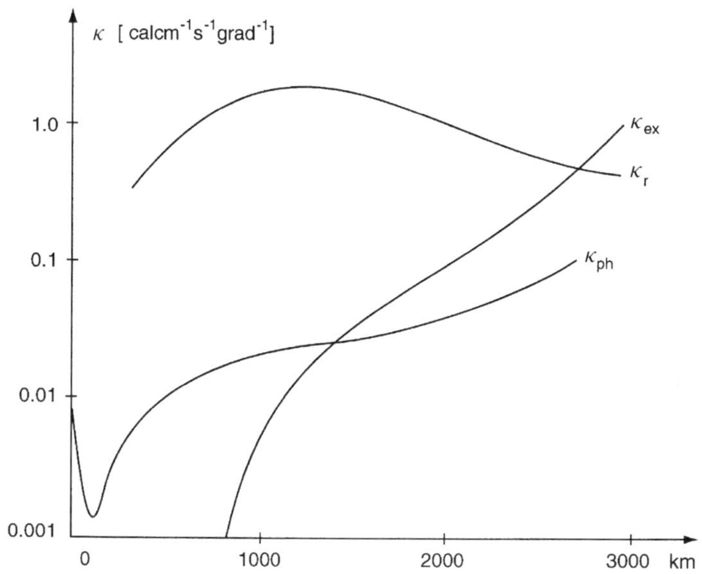

Figure 9.1. Distribution of coefficients of conductivity κ_{ph}, κ_r and κ_{ex} in the lithosphere and the mantle

shows the distribution of coefficients of conductivity κ_{ph}, κ_r and κ_{ex} in the lithosphere and the mantle.

In the mantle and in the outer core the heat also propagates due to thermal convection. Moreover, in the outer core a thermo-magneto-hydrodynamic turbulence is assumed.

In the outer core thermal conductivity is defined by the Wiedamann-Franz's law, which defines the mutual dependence of electrical and thermal conductivity in metals or in metallized rocks as follows

$$\kappa = 3K^2 e^{-2} \sigma T, \tag{9.4}$$

where K is the Boltzman's constant, e is a charge of electron, σ is electrical conductivity and T temperature.

Thermal sources in the Earth are of several types. The outer source represents sun shine. Surface changes of temperature penetrate only into small depths and the energy of the Sun's origin is radiated again into the outer space of the Earth. The average density of the heat flow from the Sun on the Earth's surface is of about 10^{-2} cal cm^{-2}sec^{-1}, while from the inner Earth it is approximately 10^{-6} cal cm^{-2}sec^{-1}. Therefore, the effect of the heat from the inner parts of the Earth is, from the global point of view, very small. For the geothermal balance accounts the inner sources are fundamental only.

In the lithosphere a great role as a source of heat is played by the radioactive rocks, i.e. all rocks containing isotopes with atomic numbers greater than 83, and also some other isotopes with atomic numbers less than 83. The natural radioactivity consists in the fact that atoms of some chemical elements spontaneously radiate α and

β elements as well as electromagnetic radiation and other elements. For generation of heat in the Earth's interior the α and β radiations are fundamental. From nuclear physics it is known that the emissions of α- and β-elements are characterized by the probability that during some time interval a certain number of atoms will be split. This probability is proportional to the number of atoms which are not split. Let N be the number of atoms which at time t are not split. Then for the probability investigated we have

$$dN/dt = -\lambda N, \tag{9.5}$$

hence

$$N = N_o \exp(-\lambda t),$$

where N_o is the number of atoms which are not split at time $t = 0$, and λ is the so-called splitting constant for the investigated chemical element. For the half-time of splitting, τ, we have

$$\tfrac{1}{2}N_o = N_o \exp(-\lambda\tau), \quad \tau = \lambda^{-1}\ln 2. \tag{9.6}$$

Elements α and β are absorbed in the neighbouring rocks, where their kinetic energy changes into heat. Table 9.1 gives the production of the radioactive isotopes, representing isotopes which occur in the rocks in sufficient quantity. Table 9.2 gives the concentration of radioactive isotopes in rocks.
Table 9.3 gives the production of heat in the rocks in 10^{-14} cal g^{-1} $\sec^{-1}$.

One part of the heat in the upper parts of the Earth is due to the tidal friction and is assumed to be characterized by about 20–30% of the thermal energy of radioactive origin. Further sources of heat are represented by absorption of seismic energy in rocks, by changing of deformation energy into heat, by changing of kinetic energy of meteorites into heat (this probably represents ~1000–2000°C during the evolution of

Table 9.1. Production of the radioactive isotopes

isotope	production of heat in cal $g^{-1}yr^{-1}$	half-time of splitting
U_{238}	0.71	4.5 10^9 yrs
U_{235}	4.3	7.1 10^8 yrs
Th_{232}	0.2	1.4 10^{10} yrs
K_{40}	0.22	1.3 10^9 yrs

Table 9.2. Concentration of radioactive isotopes in rocks

isotopes / rocks	U_{238}	U_{235}	Th_{232}	K_{40}
granite	2.9 10^{-6}	2.1 10^{-8}	1.0 10^{-7}	3.1 10^{-6}
transient	1.4 10^{-6}	1.0 10^{-8}	4.3 10^{-6}	1.3 10^{-6}
basalt	0.9 10^{-6}	0.65 10^{-8}	3.2 10^{-8}	1.1 10^{-6}
dunite	1.3 10^{-8}	0.96 10^{-10}	4.6 10^{-8}	1.2 10^{-9}

Table 9.3. Production of heat in rocks

isotopes / rocks	U	Th	K	$\sum$
granite	9.2	8.2	3.0	20.4
basalt	2.3	1.9	1.1	5.30
gabbro	0.05	0.04	0.03	0.12

the Earth), by changing of potential energy into heat ($\sim$1.0 10^{39} ergs) as well as of energy due to a gravity differentiation into heat ($\sim$1.5 10^{38} ergs), which can also be assumed in our mathematical models of collision zones.

To derive the generalized heat equation we shall start from the conservation of energy law in the form of the equation of entropy transfer, representing the second law of thermodynamics, in the form

$$T(\mathrm{d}S/\mathrm{d}t) = -\mathrm{div}\ \boldsymbol{q} + W \equiv Q + W, \tag{9.7}$$

where T is the temperature, S is the entropy, $\boldsymbol{q} = (q_i)$ represents the heat flow, W is the amount of heat per unit time and unit volume and $Q = -\mathrm{div}\boldsymbol{q}$. Equation 9.7 can be rewritten as

$$\mathrm{d}S/\mathrm{d}t = -T^{-1}\mathrm{div}\ \boldsymbol{q} + T^{-1}W = -T^{-1}\ \mathrm{div}\ \boldsymbol{q} - q_i T^{-2}\partial T/\partial x_i + T^{-1}W = -T^{-1}\mathrm{div}\ \boldsymbol{q} + \sigma, \tag{9.8}$$

where

$$\sigma = -q_i T^{-2}\partial T/\partial x_i + T^{-1}W = q_i F_i + T^{-1}W \tag{9.9}$$

is the so-called source of entropy and

$$F_i = -T^{-2}\partial T/\partial x_i \tag{9.10}$$

is the thermodynamic force per unit mass, which is proportional to the gradient of temperature.

In the thermodynamics of irreversible processes it is assumed that for laminar flows the following linear relation

$$q_i = L_{ij}\rho F_j \tag{9.11}$$

holds,

$$L_{ij} = L_{ji}. \tag{9.12}$$

From Equations 9.10 and 9.11 we find

$$q_i = -T^{-2}\rho L_{ij}\partial T/\partial x_j = -\kappa_{ij}\partial T/\partial x_j, \tag{9.13}$$

where we put $\kappa_{ij} = T^{-2}\rho L_{ij} > 0$. Equation 9.13 is known as Fourier's law. Hence and using Equation 9.7 we find

$$T\mathrm{d}S/\mathrm{d}t = \partial/\partial x_i(\kappa_{ij}\partial T/\partial x_j) + W. \tag{9.14}$$

We have

$$\mathrm{d}S = (\partial S/\partial e_{ij})_T \mathrm{d}e_{ij} + (\partial S/\partial T)_e \mathrm{d}T = \beta_{ij}\mathrm{d}e_{ij} + c_e T^{-1}\mathrm{d}T, \qquad (9.15)$$

where $(\partial S/\partial e_{ij})_T = \beta_{ij}$ and $(\partial S/\partial T)_e = c_e$ represent the linear expansion coefficient (where β_{kk} is the volume expansion coefficient) and the specific heat at constant deformation per unit mass, respectively. Then, integrating Equation 9.15 over t, and if $e_{ij} = 0$, $T = T_o$ at $t = t_o$ then $S = 0$, and putting $\Theta = T - T_o$, we find

$$S = \beta_{ij}e_{ij} + c_e \ln(1 + T_o^{-1}\Theta), \qquad (9.16)$$

and assuming that $\left|T_o^{-1}\Theta\right| \ll 1$, then

$$S = \beta_{ij}e_{ij} + T_o^{-1}c_e\Theta. \qquad (9.17)$$

Hence,

$$T\mathrm{d}S/\mathrm{d}t = T\beta_{ij}\dot{e}_{ij} + T_o^{-1}c_e T\partial\Theta/\partial t. \qquad (9.18)$$

Thus, from Equations 9.14 and 9.18 with linearization of the resulting equation, we find

$$\partial/\partial x_i(\kappa_{ij}\partial\Theta/\partial x_j) - \rho c_e\partial\Theta/\partial t - \rho T_o\beta_{ij}\dot{e}_{ij} = -\rho W, \qquad (9.19)$$

representing the generalized (and linearized) heat equation, where the term $\rho T_o\beta_{ij}\dot{e}_{ij}$ represents the dissipation of deformation energy into heat.

But there are other heat sources inside the Earth. The rocks deeper inside the Earth are less or more metallized. Let us now consider non-uniformly heated rocks, which cannot be in thermodynamic equilibrium. Then the current density $\boldsymbol{j}$ can flow in non-uniformly heated rocks even if the field $\boldsymbol{E}$ is zero, i.e.

$$\boldsymbol{j} = \sigma(\boldsymbol{E} - \alpha \text{ grad } T), \text{ i.e. } j_i = \sigma_{ij}(E_j - \alpha_{jk}\partial T/\partial x_k), \qquad (9.20)$$

where σ is the ordinary conductivity of rocks, α is another quantity characterizing the electrical property of rocks, and T is a temperature. As well as the electric density $\boldsymbol{j}$, we can consider the density flux energy $\boldsymbol{q}$. This quantity contains an amount $\varphi\boldsymbol{j}$, which results from the fact that each charged particle "e" carries with it an energy $e\varphi$. Then the difference $\boldsymbol{q} - \varphi\boldsymbol{j}$ does not depend on the potential, and thus can be written as a linear function of the gradients grad $\varphi = -\boldsymbol{E}$ and grad T, i.e. $\varphi = \partial F/\partial r$, where F is the free energy of the metallized rocks and r is the charge on the conduction electrons per unit volume $\mathrm{d}\boldsymbol{x}$. Therefore,

$$\boldsymbol{q} - \varphi\boldsymbol{j} = -\beta \text{ grad } \varphi - \gamma \text{ grad } T = \beta\boldsymbol{E} - \gamma \text{ grad } T. \qquad (9.21)$$

Let us calculate the rate of change of the total entropy of the metallized rocks, which thus behave as a conductor. The evolution of heat results in an increase in the entropy. The amount of heat per unit time and volume $\mathrm{d}\boldsymbol{x}$ is equal to $-\text{div } \boldsymbol{q}$, so that

$$\mathrm{d}S/\mathrm{d}t = -\int T^{-1} \text{ div } \boldsymbol{q}\,\mathrm{d}\boldsymbol{x}. \qquad (9.22)$$

Since div $\boldsymbol{j} = 0$, then

$$T^{-1}\text{div } \boldsymbol{q} = T^{-1}(\text{div}(\boldsymbol{q} - \varphi\boldsymbol{j}) + \text{div } \varphi\boldsymbol{j}) = T^{-1}[\text{div}(\boldsymbol{q} - \varphi\boldsymbol{j}) - \boldsymbol{E}\boldsymbol{j}].$$

Hence, integrating by parts,

$$dS/dt = \int T^{-1}\boldsymbol{Ej}\,d\boldsymbol{x} - \int T^{-2}(\boldsymbol{q} - \varphi\boldsymbol{j})\ \text{grad}\ T d\boldsymbol{x}. \tag{9.23}$$

Since the rate of change of the total entropy of the body can be written (see Landau and Lifshitz (1984) p. 88, or Chapter 3) as

$$dS/dt = \int_{\Omega} \sum_{i} X_i \partial x_i/\partial t d\boldsymbol{x}, \tag{9.24}$$

then the components $\partial x_i/\partial t$ are components of $\boldsymbol{j}$ and $\boldsymbol{q} - \varphi\boldsymbol{j}$, and the corresponding quantities X_i are components of vectors $-T\boldsymbol{E}^{-1}$ and T^{-2} grad T. Hence,

$$\boldsymbol{j} = \sigma T\boldsymbol{E}/T - \sigma\alpha T^2\ \text{grad}\ T/T^2, \tag{9.25}$$

$$\boldsymbol{q} - \varphi\boldsymbol{j} = \beta T\boldsymbol{E}/t - \gamma T^2\ \text{grad}\ T/T^2, \tag{9.26}$$

and therefore coefficients $\sigma\alpha T^2$ and βT must be equal, so that $\beta = \sigma\alpha T$. Hence, $\boldsymbol{q} - \varphi\boldsymbol{j} = \sigma\alpha T\boldsymbol{E} - \gamma$ grad T and since due to Equation 9.7 $\boldsymbol{E} = \sigma^{-1}\boldsymbol{j} + \alpha$ grad T, i.e. $E_i = \sigma_{ij}^{-1} j_j + \alpha_{ij}\partial T/\partial x_j$, we find

$$\boldsymbol{q} = (\varphi + \alpha T)\boldsymbol{j} - \kappa\ \text{grad}\ T,\ \text{i.e.}\ q_i - \varphi j_i = T\alpha_{ji} j_j - \kappa_{ij}\partial T/\partial x_j, \tag{9.27}$$

where $\kappa = \gamma - T\alpha^2\sigma$, with $\kappa_{ij} = \kappa_{ji}$ and $\sigma_{ij} = \sigma_{ji}$, is the ordinary thermal conductivity giving the heat flux in the absence of an electric current. Then Equations 9.23 and 9.27 and $\boldsymbol{E} = \sigma^{-1}\boldsymbol{j} + \alpha$ grad T, i.e. $E_i = \sigma_{ij}^{-1} j_j + \alpha_{ij}\partial T/\partial x_j$ give

$$dS/dt = \int_{\Omega} ((\sigma T)^{-1} j^2 + T^{-2}\kappa(\text{grad}\ T)^2)d\boldsymbol{x} > 0, \tag{9.28}$$

from which we see that the coefficients of thermal and electrical conductivity are positive. The divergence of Equation 9.15 yields

$$Q = -\text{div}\ \boldsymbol{q} = \text{div}(\kappa\ \text{grad}\ T) + \boldsymbol{Ej} + \boldsymbol{j}\ \text{grad}\ (\alpha T) = \text{div}(\kappa\ \text{grad}\ T) + \sigma^{-1} j^2 - T\boldsymbol{j}\,\text{grad}\,\alpha, \tag{9.29}$$

where the first term represents thermal conduction, the second term is the Joule heat and the third term represents the thermo-electric effects.

To discuss the thermo-electric effects we shall assume that the metallized rocks are homogeneous. Then the change in α is due to the temperature gradient only, and

$$\text{grad}\ \alpha = (d\alpha/dt)\ \text{grad}\ T,$$

if the pressure is constant through the investigated part of the Earth, $d\alpha/dT \sim (\partial\alpha/\partial T)_p$. Hence,

$$-T\boldsymbol{j}\ \text{grad}\ \alpha = -T\boldsymbol{j}\,d\alpha/dT\ \text{grad}\ T = \rho\boldsymbol{j}\ \text{grad}\ T, \tag{9.30}$$

where $\rho = -T d\alpha/dT$ is the so-called Thompson coefficient and Equation 9.30 represents the so-called Thompson effect, which is proportional to the first power of the current $\boldsymbol{j}$. If $\rho > 0$, the Thompson heat is emitted when the current flows in the direction of increasing temperature, and is absorbed when it flows in the opposite direction; if $\rho < 0$ the reverse is valid.

Let us assume two metallized rocks, which we denote by indices k and l. Then at the contact surface between them the Peltier effect is observed. At this surface the temperature, the potential and the normal components of the current density and energy flux are continuous. For normal components of $\boldsymbol{q}$ on the two sides we have

$$-\kappa_{ij}\partial T/\partial x_j n_i \,|_k^l = -j_n T(\alpha_l - \alpha_k). \tag{9.31}$$

Let us assume that the positive direction of the normal be from the rock k to the rock l; then the left hand side of Equation 9.19 represents the amount of heat taken from the surface per unit time and area by thermal conduction. This heat loss is balanced by the evolution at the junction of an amount of heat which is represented by the right hand side of Equation 9.31. Thus the amount of heat generated per unit time and area is given by

$$j\Pi_{kl} = -jT(\alpha_l - \alpha_k), \tag{9.32}$$

representing the Peltier effect, which is also proportional to the first power of the current and changes sign when the direction of the current is reversed. The coefficient Π_{kl} is the so-called Peltier coefficient which is additive.

Now let us assume the effect of an electric field, a magnetic field, and a thermal field representing an effect of temperature gradient. To derive this effect we use a technique similar of that of an electric field. We shall assume that

$$j_i = a_{ij}T^{-1}E_j + b_{ij}\partial/\partial x_j(T^{-1}), \tag{9.33}$$

$$q_i - \varphi j_i = c_{ij}T^{-1}E_j + \mathrm{d}_{ij}\partial/\partial x_j(T^{-1}), \tag{9.34}$$

and that all the coefficients are functions of the magnetic field. From the symmetry of the kinetic coefficients we find

$$a_{ij}(\boldsymbol{H}) = a_{ji}(-\boldsymbol{H}), \tag{9.35}$$

$$\mathrm{d}_{ij}(\boldsymbol{H}) = \mathrm{d}_{ji}(-\boldsymbol{H}), \tag{9.36}$$

$$b_{ij}(\boldsymbol{H}) = c_{ji}(-\boldsymbol{H}). \tag{9.37}$$

Then expressing $\boldsymbol{E}$ and $\boldsymbol{q} - \varphi\boldsymbol{j}$ in terms of $\boldsymbol{j}$ and grad T from Equations 9.33 and 9.34 we find

$$E_i = \sigma_{ij}^{-1} j_j + \alpha_{ij}\partial T/\partial x_j, \tag{9.38}$$

$$q_i - \varphi j_i = \beta_{ij} j_j - \kappa_{ij}\partial T/\partial x_j, \tag{9.39}$$

where tensors $\sigma^{-1}, \alpha, \beta, \kappa$ are certain functions of tensors a, b, c, d, and satisfy the following symmetry conditions

$$\sigma_{ij}^{-1}(\boldsymbol{H}) = \sigma_{ji}^{-1}(-\boldsymbol{H}), \tag{9.40}$$

$$\kappa_{ij}(\boldsymbol{H}) = \kappa_{ji}(-\boldsymbol{H}), \tag{9.41}$$

$$\beta_{ij}(\boldsymbol{H}) = T\alpha_{ji}(-\boldsymbol{H}), \tag{9.42}$$

which follow from Equations 9.35 to 9.37. In anisotropic metallized rocks the thermo-electric tensors α_{ij} and β_{ij} are not, in general, symmetric even when there is no magnetic field.

Like any tensors of rank two, the tensors σ^{-1}, κ and $B + T\alpha$ can be divided into symmetric and antisymmetric parts. In a weak magnetic field, the symmetrical parts may be regarded as constants independent of $\boldsymbol{H}$, while the antisymmetric parts are linear in $\boldsymbol{H}$. Then for isotropic rocks

$$\boldsymbol{E} = \sigma^{-1}\boldsymbol{j} + \alpha \text{ grad } T + R[\boldsymbol{H},\boldsymbol{j}] + N[\boldsymbol{H}, \text{ grad } T], \tag{9.43}$$

$$\boldsymbol{q} - \varphi\boldsymbol{j} = \alpha T\boldsymbol{j} - \kappa \text{ grad } T + NT[\boldsymbol{H},\boldsymbol{j}] + L[\boldsymbol{H}, \text{ grad } T], \tag{9.44}$$

where σ, κ are coefficients of electrical and thermal conductivity, α is the thermo-electric coefficient (as $\boldsymbol{E} = \sigma^{-1}\boldsymbol{j} + \alpha$ grad T), R is the Hall coefficient, N and L are new material coefficients, and [.,.] denotes the vector product. The term $N[\boldsymbol{H}, \text{ grad } T]$ represents the effect of the magnetic field on the thermo-electro-magnetic field, known as the Nernst effect, and the term $L[\boldsymbol{H}, \text{ grad } T]$ represents the effect of this field on the thermal conduction, known as the Leduc-Righi effect. The term $NT[\boldsymbol{H},\boldsymbol{j}]$ represents the influence of the magnetic field on the Peltier effect, known as the Ettingshausen effect. Then the amount of heat generated in the rocks per unit time and volume is $Q = -\text{div}\boldsymbol{q}$, where $\boldsymbol{q}$ is given by Equation 9.44 and $\boldsymbol{E} = -$ grad φ in accordance with Equation 9.43. As a result we obtain

$$Q = -\text{div}\,\boldsymbol{q} = \sigma^{-1} j^2 + \text{div}(\kappa \text{ grad } T) - T\boldsymbol{j} \text{ grad } \alpha + (\sigma T)^{-1} \text{d}/\text{d}T(\sigma N T^2)[\boldsymbol{j},\boldsymbol{H}] \text{ grad } T,$$

where we assumed that α, N, L, etc. are functions of temperature and, therefore, their gradients are proportional to grad T. Moreover, we put rot $(\sigma^{-1}\boldsymbol{j}) \sim$ rot $\boldsymbol{E} = 0$, rot $\boldsymbol{H} = 0$, div $\boldsymbol{j} = 0$. We see that the third term represents the Thompson effect and the last term represents the change in the Thompson effect under the influence of the magnetic field.

The volume expansion coefficient $\alpha \equiv \beta_{kk}$ is defined as

$$\alpha = V^{-1}(\partial V/\partial T)_p = -\rho^{-1}(\partial \rho/\partial T)_p \tag{9.45}$$

in terms of the volume or density changes. Since $\alpha > 0$, a temperature increase ($\Delta T > 0$) generates a volume increase or density decrease. The expansion coefficient has dimension $[T^{-1}]$ and for silicates is of the order of about $10^{-5}K^{-1}$, where K denotes the Kelvin.

The bulk modulus or incompressibility is defined as

$$\beta_{T,S} = -V^{-1}(\partial V/\partial p)_{T,S} = \rho^{-1}(\partial \rho/\partial p)_{T,S}. \tag{9.46}$$

Hence and using equation 9.45 yields

$$(\partial \alpha/\partial p)_T = -(\partial \beta_T/\partial T)_p, \tag{9.47}$$

from which we see that increasing pressure reduces α in a way related to the change in β_T with temperature. But in the seismology we introduce the adiabatic incompress-ibility K_s, as we assume that during seismic wave propagation the conditions can be taken as adiabatic. The relation between isothermal and adiabatic values of β_T and β_s is given as

$$\beta_T = \beta_s + (\rho c_p)^{-1}\alpha^2 T, \tag{9.48}$$

where ρ is a density, T a temperature and c_p is a specific heat (per unit mass) at constant pressure, where c_p for silicates is of the order 1 kJkg^{-1}K^{-1} (see Birch (1952), Stacey (1977)). In geodynamics the Grüneisen parameter is defined as

$$\gamma = (\rho c_p)^{-1}\alpha K_s = (\rho c_v)^{-1}\alpha K_T, \tag{9.49}$$

where c_p, c_v and K_s, K_T are specific heats and incompressibilities (bulk modules) under the appropriate conditions. This parameter γ is a dimensionless parameter and can be defined as the ratio between thermal pressure and thermal energy per unit volume. From Equations 9.48 and 9.49 we find

$$\beta_T/\beta_s = K_s/K_T = 1 + \gamma\alpha T. \tag{9.50}$$

The heat flow is defined as a flux of energy through a unit area per unit time ([mWm^{-2}] or [μcal cm^{-2}s^{-1}] or HFU = heat flow unit, 1 HFU = 42 mWm^{-2}), and then the thermal conductivity κ in the Fourier's law of heat conduction has dimension [mlt^{-3}T^{-1}]. In the SI system it is measured in Wm^{-1}K^{-1} and for silicates it varies in the interval 1-5 Wm^{-1}K^{-1} for isotropic rocks, i.e. for the polycrystallic rocks. In reality the coefficient of thermal conductivity (or shortly the thermal conductivity) κ is a tensor of second order and is denoted as κ_{ij}, which is symmetric because heat flow from an isolated point occurs along straight lines. In the geodynamics the thermal diffusivity $\underline{\kappa} = \rho^{-1}c_v^{-1}\kappa$ is often used. Its dimensions are $[l^2t^{-1}]$ and for silicates it is of the order 1.10^{-6}m^{2}s^{-1}. Although the scatter of measurements is very large, heat flow in general decreases with increasing age. The averages of heat flow observations in the oceans and in the continents are very similar, 66 and 57 mWm^{-2}, respectively. The total estimated heat loss for the Earth is about 42.10^{12} W; for continents $\sim\frac{1}{3}$ and for oceans $\sim\frac{2}{3}$ of this value. The estimated present heat generation is ~31.3 10^{12}W representing 11.33 10^{12}W in the case of U_{238}, 0.49 10^{12}W for U^{235}, 11.18 10^{12}W for Th and 8.41 10^{12}W in the case of K_{40} isotope. Heat generation in various rocks in 10^{-12} Wkg^{-1} is given in Table 9.4, which is a compilation of several papers (Clark (1966), Stacey (1977), etc.), and for rocks from different parts of the world in 10^{-13}cal cm^{-3} sec and μ Wm^{-3} are shown in Table 9.5.

Table 9.4. Heat generation in rocks

rocks	heat generation 10^{-12}Wkg^{-1}
granites	1050
alkali basalts	180
tholeitic basalts	27
eclogites	9.2
periodites	1.5
chondrites	5.2-5.9

Table 9.5. Heat generation in rocks

rocks	heat generation	
location	10^3 cal cm^{-3} sec	μWm^{-3}
Granites, world average	6.7	2.81
Granites, Australia	17.13	7.17
Granites, Switzerland	7.6	3.2
Magmatic granites, India	2.69	1.13
Diorite	2.94	1.23
Granodiorites	3.78	1.58
Greenstone area Australia	2.0	0.84
Hornblende schist	2.23	0.93
Hornblende schist, India	0.53	0.22
Almandine amphibolite facies, Australia	5.0	2.1
Granodiorite gneiss of amphibolite facies, Canada	3.0	1.26
Pyroxene granulite facies, Australia	1.25	0.66
Hornblende granulite facies, Canada	2.20	0.92
Archaean shield:		
a) Canada	3.9	1.63
b) Australia	6.0	2.51
Central Stable region USA	5.37	2.24
Gabbro, USA	2.49	1.04
Basalt, world average	1.1	0.46
Eclogite, low - average -	0.084	0.033
Eclogite, high - average -	0.36	0.14
Ultrabasic rocks - average -	0.007	0.003
Chondrite, - average -	0.044	0.018
Migmatite gneisses, India	8.85	3.71
Gneisses, India	6.70	2.8
Schists, India	5.24	2.19
phyllites, India	6.22	2.60
Anorthosite, India	0.5	0.21
Amphibolites, India	0.44	0.18
Amphibolites, average	0.45	0.19

The coefficient of thermal conductivity for several types of rock is given in Table 9.6 (for more data see for example Clark (1966)).

The coefficient of thermal conductivity depends mainly on the density, pressure, temperature and porosity. Its dependence on density is practically linear and it increases with increasing density. The dependence on the pressure is similar as for elastic coefficients and can be taken as

$$\kappa \equiv \kappa_p = \kappa_o(1 + c_o p)^{-1}, \tag{9.51}$$

Table 9.6. Thermal conductivity of rocks

rocks location	thermal conductivity 10^{-3} cal cm^{-1} sec°C average	 range	age formation
granite, Colorado	7.89	6.7-8.6	
granite, Switzerland	7.77	6.2-9.0	
granite, India	7.13	6.61-8.31	
granodiorite, USA	6.64	6.2-6.9	
gneiss, USA	7.74	4.0-11.0	
crystalline schist	7.74	4.0-11.0	
rhyolite, Canada	8.23	7.4-8.8	
syenite, Canada	7.66	6.3-9.5	
syenite porphyry	7.66	6.3-9.5	
quartz porphyry	8.0	7.6-8.6	
quartzite, USA	16.05	14.2-17.6	
quartzite, India	10.47	8.26-13.63	
quartzite, phyllite, graphitic phyllite, biotite quartzite, India	8.36	6.9-11.5	Upper Proterozoic
amphibolite, USA	6.92	6.1-9.1	
amphibolite, India	7.36	6.47-9.03	
dolomite, Switzerland	11.93	8.9-13.9	
anhydrite	11.93	8.9-13.9	
limestone, Toronto	6.12	4.7-7.1	
sandstone, USA	6.68	5.1-10.2	
Permian marlite ,USA	5.22	4.2-6.6	
Carboniferous sandstone, UK	6.62	6.0-7.7	
Triassic marlite	3.52	2.2-5.3	
biotite schist, India	7.26	6.2-8.77	Upper Proterozoic
felspathic schist chlorite schist sericite schist basalt, India	5.98	5.0-7.6	
Coarse- grained sandstones, India	9.35	6.79-10.66	Gondwana
Medium-grained sandstones, India	7.69	6.96-8.66	Gondwana
Siltstone, India	5.60	5.0-6.5	Tertiary
Limestone, India	5.15	4.9-5.34	Tertiary
Claystones and clay shale, India	3.0	2.8-3.2	Tertiary
Horneblende schists India	7.26	6.2-8.77	Archean
Shale, India	3.60	2.47-4.36	Gondwana
Black shale, India	4.48	4.21-4.81	Gondwana

where κ_o represents a coefficient of thermal conductivity under normal pressure, c_o is a constant, and p is a pressure. For small pressures, $p \sim 0 \div 1$ kbar, the coefficient of thermal conductivity can increase due to a closing of microcracks, pores, etc. as well as to a better contact between crystals.

The dependence on the temperature in the crust and the upper mantle, represented by the temperature interval $0 \div 1600°$C, is determined by the phonon's conductivity and radiation conductivity; while the phonon's conductivity κ_{ph} decreases with increasing temperature

$$\kappa_{ph} \sim T^{-1},$$

the radiation conductivity increases with increasing temperature as

$$\kappa_r \sim T^3.$$

Practically, for temperatures $T < 600°$C the linear dependence on the temperature is used, i.e.

$$\kappa \equiv \kappa_T = \kappa_o(1 + cT)^{-1}, \tag{9.52}$$

where κ_o is a thermal conductivity under 0°C which, for rocks of the upper crust, was found to be ~3.0 $\mathrm{Wm^{-1}K^{-1}}$, for rocks of the lower crust ~2.0 $\mathrm{Wm^{-1}K^{-1}}$, for rocks of the subcrustal lithosphere ~2.5 $\mathrm{Wm^{-1}K^{-1}}$, and for rocks of the asthenosphere ~3.0 $\mathrm{Wm^{-1}K^{-1}}$, and the constant c for rocks of the upper crust was found to be ~ -0.001 $\mathrm{K^{-1}}$, for rocks of the lower crust ~0 $\mathrm{K^{-1}}$, for rocks of the subcrustal lithosphere ~ -0.00025 $\mathrm{K^{-1}}$, for rocks of the asthenosphere ~ -0.0004 $\mathrm{K^{-1}}$.

For some rocks this dependence leads to

granite : $\kappa_T = 3.45 + 600T^{-1}$, $[\kappa_T] = 10^{-3}$ cal cm sec° C, $[T] = K$(Kelvin),

gabbro : $\kappa_T = 5.2 + 140T^{-1}$,

Olivine (30%) + gabbro : $\kappa_T = 7.1 + 110T^{-1}$,

forsterite : $\kappa_T = 11.5$,

feldspar : $\kappa_T \doteq$ constant.

For the dependence on temperature and pressure we find

$$\kappa \equiv \kappa_{Tp} = \kappa_o(1 + cT + c_o p)^{-1} \text{ or } \kappa \equiv \kappa_{Tp} = \kappa_o(1 + c(T - T_o) + c_o(p - p_o))^{-1}. \tag{9.53}$$

We said above that, owing to an anisotropy of rocks, the coefficient of thermal conductivity is represented by a symmetric tensor of order 2; we denote it by κ_{ij}. It holds

$$\kappa_{ij} = \kappa_{ji}, \quad i, j = 1, 2, 3. \tag{9.54}$$

For minerals of the cubic system $\kappa_{ij} = \kappa\delta_{ij}$, in the square and six-fold system $\kappa_{ij} = tr(\kappa_{11}, \kappa_{22}, \kappa_{33})$, etc., as in the case of elastic parameters. To model the anisotropy thermal conductivity the homogenization technique (the idea of which is the same as in the elastic case discussed earlier) can be used. Putting

$$\underline{\kappa} = (\rho c_e)^{-1}\kappa, \tag{9.55}$$

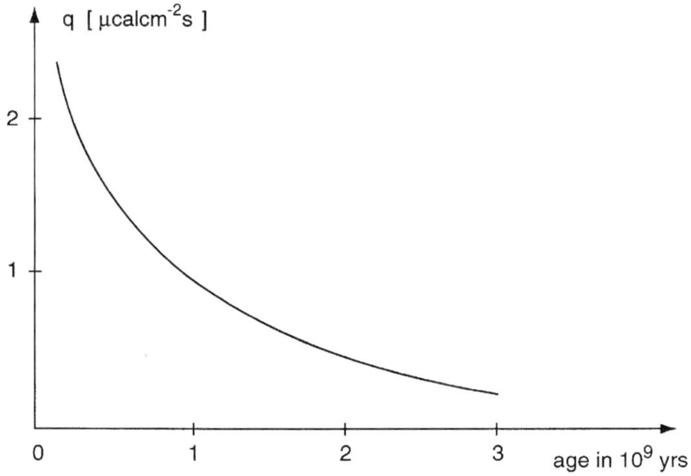

Figure 9.2. The dependence of geothermal heat on age

we obtain the so-called coefficient of thermal diffusivity (cm^{2}sec^{-1}), where c_e is a specific heat, κ is a coefficient of thermal conductivity and ρ is a density. Since the specific heat for practically all rocks is in the interval 0.18–0.23 cal g^{-1} then, if the density and coefficient of thermal conductivity are known, we can estimate the temperature conductivity. Its value, for practically all rocks is in the interval 0.005–0.015 cm^{2}sec^{-1}, with $\underline{\kappa} < 0.005$ for unhardened sediments and $\underline{\kappa} > 0.015$ for crystallic rocks. The thermal diffusivity depends on the saturation of the rocks by water and will increase if the capacity of water increases.

The heat flow is analysed on the basis of spherical harmonic analysis and the obtained results are presented in the maps of heat flow. The dependence of geothermal heat on age is given in Figure 9.2, completing the results of our present knowledge (Lee and Uyeda (1965), Verma (1991), etc.).

Between thermophysical quantities in SI and cgs systems the following relations hold:

heat flow q:	
Wm^{-2},	1 Wm^{-2} = 2.388 10^{-5} cal cm^{-2} sec,
cal cm^{-2},	1 μ cal cm^{-2} sec = 41.87 mWm^{-2},
thermal conductivity κ:	
Wm^{-1}K^{-1},	1 Wm^{-1}K^{-1} = 2.388 10^3 cal cm^{-1} sec°C,
cal cm^{-1} sec°C,	1 mcal cm^{-1} sec°C = 0.4187 W^{-1}K^{-1},
thermal diffusivity $\underline{\kappa}$:	
m^{2}s^{-1},	1m^{2}s^{-1} = 1.0 10^4 cm^2 sec^{-1},
cm^2 sec^{-1},	1cm^2 sec^{-1} = 1.0 10^{-4} m^2 s^{-1},
specific heat c_e:	
J kg^{-1}K^{-1},	1 Jkg^{-1}K^{-1} = 2.388 10^{-4} cal g^{-1}°C,
cal g^{-1}°C,	1 cal g^{-1}°C = 4.187 kJ kg^{-1}K^{-1},

production of heat P:

Wm^{-3},	$1Wm^{-3} = 2.388\ 10^{-7}$ cal cm^{-3} sec,
cal cm^{-3} sec,	$1.0\ 10^{-13}$ cal cm^{-3} sec $= 0.4187\ \mu\ Wm^{-3}$.

The geothermal gradient must decrease below the lithosphere, otherwise the bulk of the mantle would be molten. Conduction alone is unable to account for the temperature distribution in the Earth's interior. Thermal conductivity generally increases with depth in the mantle, perhaps by one order of magnitude. Thermal diffusivity is approximately constant at $\sim 10^{-6}m^2s^{-1}$ as density increases with depth. That such a thermal diffusivity is insufficient to remove heat from the deeper parts of the Earth can be seen from the solution to the generalized heat (conduction) equation. But the heat is also transported by convection, i.e. heat transfer by mass transfer, which will be discussed in Chapter 12.

The temperature gradient in a convecting system is close to the adiabatic gradient except in the thermal boundary layers. By the adiabatic gradient is meant the rate of increase of temperature with depth resulting from the increase in pressure. For the case of incompressible rocks there are no adiabatic temperature increases. But silicates are suffcently compressible for significant adiabatic gradients to be formed (for more details see Shimatzu (1954), Ranalli (1987).

To derive the adiabatic gradient, the thermodynamic relation between entropy per unit mass S, temperature T and pressure p will be used. Then we have

$$\mathrm{d}S = T^{-1}c_p\mathrm{d}T - \alpha\rho^{-1}\mathrm{d}p.$$

For a reversible adiabatic process ($\mathrm{d}S = 0$) we find

$$(\mathrm{d}T/\mathrm{d}p)_S = c_p^{-1}\rho^{-1}\alpha T = K_s^{-1}\gamma T, \tag{9.56}$$

where γ is the Grüneisen parameter. Since for homogeneous regions $\mathrm{d}p/\mathrm{d}x_3 = \rho g$, then the adiabatic increase of temperature with depth is

$$(\mathrm{d}T/\mathrm{d}x_3)_S = (\mathrm{d}T/\mathrm{d}p)_S\mathrm{d}p/\mathrm{d}z = \alpha g c_p^{-1}T = \gamma g K^{-1}\rho T,$$

where K is a bulk modulus, ρ is a density, and $K\rho^{-1} = v_p^2 - 4/3v_s^2 \equiv \varphi$ is the seismic parameter. Integrating the last relation and using the definition of incompressibility, $(\partial\rho/\partial p)_S = \rho K_s^{-1}$, we find

$$T_2/T_1 \sim (\rho_2/\rho_1)^{\gamma}, \tag{9.57}$$

which gives the adiabatic temperature change across a finite depth interval in terms of the ratio of densities. The adiabatic gradient is the approximate temperature gradient in a convecting system. For this system, a necessary but not sufficient condition for convective instability is $(\mathrm{d}T/\mathrm{d}x_3) > (\mathrm{d}T/\mathrm{d}x_3)_S$. This condition means that material displaced adiabatically upwards (downwards) is hotter (cooler) then the surrounding material and therefore unstable in the gravity field.

Materials at different places inside the Earth are melted. In geothermics an upper limit to the mantle temperature is mentioned as the so-called solidus temperature T_m, which can be estimated (by several empirical equations) under high pressure by the Lindemann's law of melting (see for example Stacey (1977)) as

$$T_m^{-1}(\mathrm{d}T_m/\mathrm{d}p) = 2(\gamma - \tfrac{1}{3})K_m^{-1}, \tag{9.58}$$

where γ is the Grüneisen parameter and K_m is the incompressibility along the melting curve.

But adiabatic gradient and melting temperature provide constraints on the temperature variation with depth. Since the properties of physical parameters change with increasing temperature, we may obtain other useful information from the study of their physical properties. One of these is the electrical conductivity σ and another the phase equilibrium of minerals. Electrical conductivity depends on the temperature as

$$\sigma = \sigma_o \exp(-k^{-1}T^{-1}E), \tag{9.59}$$

where σ is the electrical conductivity, k is the Boltzman constant, and E is an activation energy of the process. The electrical conductivity changes from $\sigma = 1.0\ 10^{-2}\ \mathrm{Sm}^{-1}$ (Siemens per metre) in the upper mantle to $\sigma = 1\text{-}10\ \mathrm{Sm}^{-1}$ in the lower mantle, with the change concentrated in the transition zone. In the lithosphere and asthenosphere it has lateral variations, with changes of $10^{-1} - 1.0\ \mathrm{Sm}^{-1}$ corresponding to the asthenosphere in oceanic and continental orogenic regions and absent elsewhere under continents, which is probably due to partial melting.

Investigations of mineral phase equilibria estimate the temperatures of equilibrations of certain mineral assemblages in upper mantle rocks tectonically transported to the Earth's surface (temperatures down to about 200 km), and the so-called (p, T)-relations pertaining to the olivine → spinel and spinel → post-spinel transitions (depths of about 400 km and 670 km and temperatures $T = 1880°\mathrm{K}$ and $2300°\mathrm{K}$). Partial melting is also an attractive explanation in view of the high conductivity mainly in the low velocity regions. But superadiabatic thermal boundary zones are also situated in the mantle. The region called "the 'D' layer" (~2695 km) can be determined not only from the seismological observations but also on the basis of thermal investigations. At depth $x_3 = 670$ km asume the temperature $T \sim 2300°\mathrm{K}$, and a lower mantle temperature gradient $\mathrm{d}T/\mathrm{d}x_3 \sim 0.3°\ \mathrm{K\ km}^{-1}$, and the temperature at the mantle/core boundary would be below the melting point of core materials. Since the heat flow is from the core into the mantle, then a thermal boundary layer occurs at the base of the lower mantle, with temperature probably increasing by 300-800°K over a depth interval 100-200 km. Hence, $T > T_m$ in the upper parts of the core. It is estimated that $T_m \sim 2800\text{-}3000°\mathrm{K}$ for the outermost core under the assumption that the light material present in the core is sulphur, and $T_m \sim 3200\text{-}3500°\mathrm{K}$ for oxygen.

The olivine (Mg_2SiO_4, Fe_2SiO_4), the principal component of the mantle rocks, undergoes a phase change to a spinel structure at a depth of about 400 km. The spinel is about 10% denser than the olivine. This phase change represents recrystallization of the rocks. In the mantle the 400 km olivine → spinel transition is probably exothermic, with a latent heat $L \sim 160\ \mathrm{kJkg}^{-1}$, with temperature increase $\Delta T \sim 100\text{-}150°\mathrm{K}$ for flow across it. For the next 670 km phase change the latent heat L is not known; and phase changes in the lower mantle, if it exists there, are probably very small. The occurrence of thermal boundary zones in the mantle would have some implications for the thermal, chemical, and geodynamic evolution of the Earth.

Another phase change in the mantle is melting. Since the molten rocks are less dense there is an upward buoyancy force which enhances the instability. The melting phase change absorbs heat which tends to stabilize the flow. Again the relative importance of the density change and the heat release must be considered. In a solid-liquid phase change, segregation of the liquid occurs. When sufficient quantities of liquid are formed they form liquid inclusions, which are observed for example in volcanic regions.

As we saw in Chapter 3, by a phase change we meant the change of one state of rock to another, known as recrystallization, solidification or melting. At this phase change boundary a latent heat is absorbed or released. On the boundary between two different phases the thermodynamic potential φ is continuous, but its first derivatives $\partial\varphi/\partial p = v$ and $\partial\varphi/\partial T = -s$, where v is specific volume and s is a specific entropy. A jump of entropy is equal to the latent heat divided by an absolute temperature. At the change of mass $\mathrm{d}m_2$ from a first phase into a second one a heat $\mathrm{d}Q = q_{12}\mathrm{d}m_2$ is absorbed, where q_{12} is a latent heat of unit mass. But for every reversible process $\mathrm{d}s = (s_2 - s_1)\mathrm{d}m_2$, so that $q_{12} = T(s_2 - s_1)$. At the same time the temperature across the boundary between two different phases is continuous and equal to the temperature of recrystallization or solidification or melting, respectively, i.e.

$$T_1 = T_2 = T_m, \tag{9.60}$$

and for the heat flow

$$\kappa_{ij}\partial T/\partial x_j n_i \big|_{(1)} - \kappa_{ij}\partial T/\partial x_j n_i \big|_{(2)} = -\rho_1 L\mathrm{d}r(t)/\mathrm{d}t, \tag{9.61}$$

where $x = r(t)$ is a parametric representation of the phase change boundary and L is a latent heat.

To derive the governing diffusion equation and appropriate discontinuities, an integral formulation of the principle of conservation of heat is a simplified version of conservation of energy in which both phases are assumed to be incompressible and at rest. Across any phase change boundary there must be conservation of energy, momentum and mass, and the latter then requires that the rock density is continuous.

For simplicity let us assume the one dimensional case only. In this case it holds

$$\int_R \{H(T)\mathrm{d}x + \kappa\partial T/\partial x\mathrm{d}t\} = 0 \tag{9.62}$$

for any closed contour R in the x, t plane, where

$$H(T) = T \quad \text{for} \quad T > T_m, \tag{9.63}$$

$$H(T) = T - L \quad \text{for} \quad T \leq T_m. \tag{9.64}$$

Here $H(T)$ and $L > 0$ represent the heat content and latent heat, and T_m the phase change temperature. In any region where $T \neq T_m$ we rewrite Equation 9.62 as

$$\int\int \{\partial/\partial t(H(T)) - \partial/\partial x(\kappa\partial T/\partial x)\}\mathrm{d}x\mathrm{d}t = 0$$

over an arbitrary area and hence satisfy

$$\partial T/\partial t = \partial/\partial x(\kappa\partial T/\partial x), \tag{9.65}$$

where κ is the thermal diffusion coefficient and where source densities were neglected and $\rho \equiv 1$, $c_e \equiv 1$, and $\kappa > 0$ is a piecewise differentiable function. If $T = T_m$ on $x = r(t)$, by considering a contour R which is tangent to a section of $x = r(t)$ in the limit $x \to r(t)^+$ and $x \to r(t)^-$, then across such a boundary

$$[H(T)]_1^2\dot{r}(t) + [\kappa\partial T/\partial x]_1^2 = 0,$$

where suffices 1 and 2 denote limiting values in regions where $T < T_m$ (solid) and $T > T_m$ (liquid), respectively. Hence

$$x = r(t), \quad T = T_m, \quad [\kappa\partial T/\partial x]_2^1 = -L\dot{r} \tag{9.66}$$

or

$$x = r(t), \quad T = T_m, \quad \kappa\partial T/\partial x\,|_1 - \kappa\partial T/\partial x|_2 = -L\dot{r} \tag{9.67}$$

is the so-called Stefan condition.

In the case if we investigated two incompressible phases of different densities ρ_1 and ρ_2, the first of which is a solid and the second one is liquid, then the liquid phase must have a velocity v_n to satisfy the mass continuity condition, where

$$\rho_1\dot{r} = \rho_2(\dot{r} - v_n) \quad \text{on } x = r(t), \tag{9.68}$$

where v_n is a velocity in the normal direction to $x = r(t)$. Thus in the liquid it is necessary to add a convection term $v\partial T/\partial x$ to the left hand side of Equation 9.65 so that

$$\partial T/\partial t + v\partial T/\partial x = \partial/\partial x(\kappa\partial T/\partial x). \tag{9.69}$$

In a one-dimensional case the continuity and momentum equations for a liquid require that $\partial v/\partial x = 0$. In a three-dimensional case the convective term will be equal to $\boldsymbol{v}$ grad T and the phase change boundary will be of the form $R(\boldsymbol{x}, t) = 0$ and the Stefan conditions generalize to

$$T_1 = T_2 = T_m, \tag{9.70}$$

$$\kappa \text{ grad } T \text{ grad } R\,\Big|_{(1)} - \kappa \text{ grad } T \text{ grad } R\Big|_{(2)} = \rho_1 L\partial R/\partial t. \tag{9.71}$$

Solving these problems we seek the position of the phase change boundary in time with determination of the thermal field in the whole region investigated. But we shall study another problem. We shall determine a temperature T and a heat flow across

Table 9.7.

depth	p	ρ	range ρ	K	μ	ν	g
0	0.0	2.840	1.0-3.0	0.650	0.358	10^{28}	982.20
33	0.009	2.840	2.7-3.32	0.650	0.358	|	984.50
33	0.009	3.320	3.32	1.155	0.629	10^{21}	984.50
182	0.059	3.434	3.35-3.46	1.359	0.722	10^{20}	989.25
200	0.065	3.492	3.46-3.5	1.393	0.739	10^{17}	989.75
400	0.138	3.843	3.75-3.875	1.815	0.937	10^{23}	993.50
670	0.225	4.145	3.95-4.25	2.575	1.320	10^{23}	994.30
800	0.302	4.341	4.30-4.385	3.085	1.632	10^{23}	993.40
984	0.382	4.476	4.45-4.50	3.431	1.806	10^{23}	992.00
984	0.382	4.650	4.65-4.68	3.510	1.806	10^{17}	992.00
1200	0.479	4.690	4.68-4.70	3.715	1.944	|	991.00
1400	0.572	4.711	4.705-4.75	4.021	2.066	|	991.10
1600	0.666	4.819	4.80-4.84	4.335	2.184	|	992.70
1800	0.763	4.924	4.90-4.96	4.669	2.297	|	996.25
2000	0.863	5.026	5.00-5.10	5.014	2.410	|	1002.20
2200	0.965	5.125	5.10-5.175	5.342	2.525	|	1011.40
2400	1.070	5.225	5.20-5.285	5.673	2.646	|	1024.60
2600	1.179	5.324	5.30-5.45	6.002	2.771	|	1043.00
2695	1.232	5.371	5.365-5.385	6.162	2.829	|	1053.90
2900	1.340	5.654	5.650-5.70	6.497	3.016	10^{26}	1078.20
2900	1.340	3.00	3.00-3.105	6.497	0	10^{20}	1078.20
3000	1.508	3.40	3.375-3.45	1.34	0	10^{20}	1050.00
3000	1.508	9.834	9.785-9.88	7.460	0	10^{9}	1007.80
3600	2.086	10.855	10.80-10.90	8.925	0	|	911.10
4000	2.466	11.200	11.15-11.25	10.277	0	|	807.10
4400	2.812	11.325	11.32-11.33	11.440	0	|	699.90
4561	2.940	11.400	11.38-11.42	11.928	0	|	657.30
4703	3.054	11.500	11.485-11.52	12.363	0	10^{-2}	616.50
4703	3.054	12.000	11.95-12.10	12.363	0.523	10^{10}	616.50
5154	3.370	12.500	12.48-12.85	13.603	0.576	10^{14}	476.80
5154	3.370	13.000	12.98-13.40	13.603	2.580	10^{14}	476.80
5600	3.615	13.400	13.39-13.93	14.593	2.767	|	317.10
6000	3.755	13.950	13.90-14.75	15.174	2.877	|	165.20
6378	3.799	14.000	14.00-15.18	15.357	2.912	10^{18}	0

the phase change boundary $R(\boldsymbol{x}, t) = 0$ such that the rocks are melted, recrystallized or solidified along the known phase change boundary $R(\boldsymbol{x}, t) = 0$.

Table 9.7 gives the variation of the pressure p, density ρ, incompressibility K, rigidity μ, viscosity ν, and gravitational intensity g; and Table 9.8 gives velocities v_p and v_s of the P and S waves, Young's modulus E and Poisson's ratio σ. The units in the vertical scale are 10^{12}dyn cm^{-2} for p, K and μ, g cm^{-3} for ρ, 100 cm sec^{-2} for g, and poise for ν, and km sec^{-2} for v_p and v_s representing one of the possible models of the Earth. The values represent the average values of the above physical parameters. For example, the density in the lithosphere changes from 1.2-3.4 gcm^{-3}, in the asthenosphere from 3.33-3.35 g cm^{-3}, in the mesosphere from

Table 9.8.

z	v_p	v_p-interval	v_s	v_s-interval	E	σ
0	6.300	3.400-6.500	3.550	1.965-3.757	0.907	0.267
33	6.300	6.300-6.500	3.550	3.550-3.757	0.907	0.267
33	7.750	7.700-7.800	4.353	4.345-4.418	1.597	0.270
182	8.198	8.100-8.250	4.573	4.525-4.590	1.841	0.274
200	8.256	8.200-8.300	4.601	4.590-4.658	1.884	0.275
400	8.923	8.855-8.950	4.938	4.912-4.980	2.398	0.279
670	10.550	10.500-10.650	6.098	6.069-6.120	4.220	0.275
800	11.009	10.950-11.050	6.131	6.125-6.186	4.161	0.275
984	11.420	11.400-11.450	6.351	6.330-6.354	4.608	0.276
984	11.450	11.450-11.500	6.355	6.345-6.370	4.640	0.276
1200	11.706	11.695-11.705	6.500	6.460-6.560	4.965	0.277
1400	11.992	11.950-12.050	6.622	6.598-6.634	5.291	0.281
1600	12.262	12.240-12.400	6.730	6.712-6.742	5.607	0.284
1800	12.530	12.500-12.600	6.829	6.812-6.835	5.919	0.298
2000	12.794	12.750-12.850	6.924	6.912-6.934	6.231	0.293
2200	13.034	12.980-13.110	7.018	6.995-7.056	6.543	0.296
2400	13.270	13.200-13.300	7.117	7.098-7.122	6.871	0.298
2600	13.495	13.450-13.350	7.214	7.197-7.234	7.204	0.300
2695	13.600	13.580-13.650	7.258	7.268-7.287	7.361	0.301
2900	13.640	13.600-13.640	7.304	7.296-7.318	7.836	0.299
2900	8.128	8.100-8.135	0	0	0	0.5
3000	8.250	8.200-8.280	0	0	0	0.5
3000	8.300	8.290-8.350	0	0	0	0.5
3600	9.068	8.995-9.100	0	0	0	0.5
4000	9.539	8.520-8.550	0	0	0	0.5
4400	9.905	9.895-9.950	0	0	0	0.5
4561	10.058	10.000-10.100	0	0	0	0.5
4703	10.058	10.000-10.100	0	0	0	0.5
4703	10.336	10.330-10.337	2.060	1.995-2.087	1.547	0.479
5154	10.338	10.337-10.339	2.070	2.046-2.084	1.704	0.479
5154	11.258	11.250-11.260	4.380	4.355-4.392	7.279	0.411
5600	11.258	11.250-11.260	4.380	4.355-4.392	7.279	0.411
6000	11.258	11.250-11.260	4.380	4.355-4.392	7.279	0.411
6378	11.258	11.250-11.260	4.380	4.355-4.392	7.279	0.411

3.42-3.8 g cm^{-3}, in the lower mantle from 4.65-5.68 g cm^{-3}, in the non-stable layer from 3.0-3.4 g cm^{-3}, in the outer core from 9.8-11.5 g cm^{-3}, in the transition layer from 12-12.5 g cm^{-2} and in the inner core from 13.0-14.0 g cm^{-3}. The viscosity changes in the lithosphere from 10^{21}-10^{28} poise, in the asthenosphere from 10^{17}-10^{20} poise, in the lower mantle from 10^{14}-10^{17}-10^{26} poise, in the non-stable layer $\geq 10^{20}$ poise, in the outer core from 10^{-2}-10^{9} poise, in the transition layer $\geq 10^{10}$ poise and in the inner core $\geq 10^{14}$ poise. The dependence v_p, v_s, v_p/v_s and Poisson's ratio on pressure can be found in Christensen (1996).

The heat production values in μWm^{-3} for Precambrin (Pc) and Phanerozoic (Ph) rocks determined from seismic velocities v_p are given in Table 9.9.

The heat production Q in μ Wm^{-3} can be found for crystalline and metamorphosed rocks from our knowledge of the distribution of the heat flow on the Earth's surface as a function of depth as

$$Q(x_3) = Q_o \exp(-x_3/z_s) \doteq 0.4qz_s^{-1} \exp(-x_3/z_s), \quad \text{(minimal version)} \quad (9.72)$$

$$Q(x_3) \doteq 0.4q(1 - e^{-1})z_s^{-1} \exp(-x_3/z_s), \quad \text{(maximal version)} \quad (9.73)$$

where z_s represents the thickness of sediments which are situated above the crystalline or metamorphosed rock layer, x_3 is the depth, and terms $0.4qz_s^{-1}$ and/or $0.4q(1 - e^{-1})z_s^{-1}$ represent the minimal and maximal surface value of heat production (Roy, Blackwell and Birch (1968), Lachenbruch (1968)). But it seems that the last relations are universally valid.

The realistic models of the Earth's crust were presented by Clark (1962), and their modifications were then used and applied by many authors for local or global regions of the Earth and on these obtained results maps of heat flows were constructed.

Table 9.10 gives parameter estimates on the phase boundaries at 400 km and 670 km depths

In Figures 9.3 to 9.5, the temperature dependence of thermal conductivity $\kappa(T)$, thermal diffusivity $\underline{\kappa}(T)$, and specific heat $c_p(T)$ will be given.

To find the value of the coefficient of thermal expansion (thermal dilatability), $\gamma = l^{-1}(dl/dT)[K^{-}n^{1}]$, see Clark (1966).

Table 9.9.

v_p	heat production		v_p	heat production	
	Pc	Ph		Pc	Ph
sediments	1.20	1.00			
≤ 6	0.82	2.60	6.0-6.1	0.65	2.00
6.1-6.2	0.56	1.55	6.2-6.3	0.48	1.20
6.3-6.4	0.39	0.92	6.4-6.5	0.32	0.72
6.5-6.6	0.27	0.53	6.6-6.7	0.20	0.41
6.7-6.8	0.16	0.31	6.8-6.9	0.12	0.23
6.9-7.0	0.10	0.18	7.0-7.1	0.08	0.13
7.1-7.2	0.06	0.01	7.2-7.3	0.05	0.08
7.3-7.4	0.035	0.06	7.4-7.5	0.035	0.06
7.5-7.6	0.02	0.035	7.6-7.7	0.02	0.035
7.7-7.8	0.015	0.02	7.8-7.9	0.015	0.02
7.9-8.0	0.01	0.015	8.0-8.1	0.01	0.015
8.1-8.2	0.01	0.015	8.2-8.3	0.01	0.015
8.3-8.4	0.005	0.005	8.4-8.5	0.005	0.005
8.5-8.6	0.005	0.005	8.6-8.7	0.005	0.005
≥ 8.5	0.004	0.004			
asthenosphere rocks	0.05	0.05			

Table 9.10.

physical parameter	400 km phase boundary	670 km phase boundary
bulk modulus, K	50 GPa	50 GPa
Density of high-pressure phase, ρ_2	3.80 gcm^{-3}	4.40 gcm^{-3}
Density of low-pressure phase, ρ_1	3.60 gcm^{-3}	4.10 gcm^{-3}
Characteristic dimension of grains, R	1.50 cm	1.50 cm
Velocity of phase boundary propagation, v_n	2.6 10^{-7} cm s^{-1}	2.6 10^{-7} cm s^{-1}
Pressure, p	13 GPa	24 GPa
Temperature, T	1700-1880 K	2000-2300-2500 K
High-temperature diffusion, D_o	1.2 10^6 cm^2 s^{-1}	1.2 10^6 cm^2 s^{-1}
Activation energy, E^*	644 kJ mol^{-1}	644 kJ mol^{-1}
Superplastic activation energy, $\frac{1}{2}E^*$	322 KJ mol^{-1}	322 KJ mol^{-1}
Activation volume, μ	11 cm^3 mol^{-1}	11 cm^3 mol^{-1}

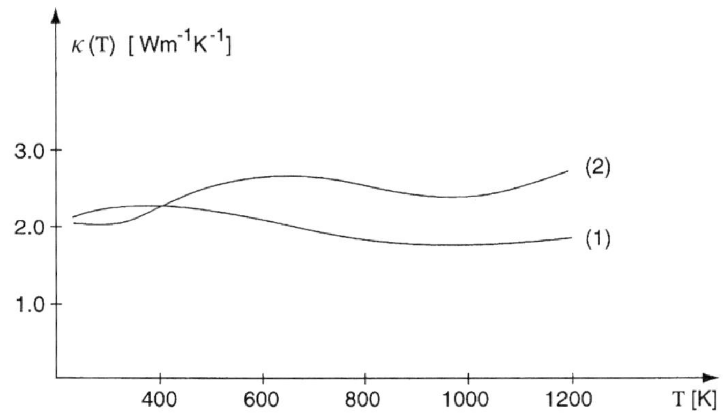

Figure 9.3.

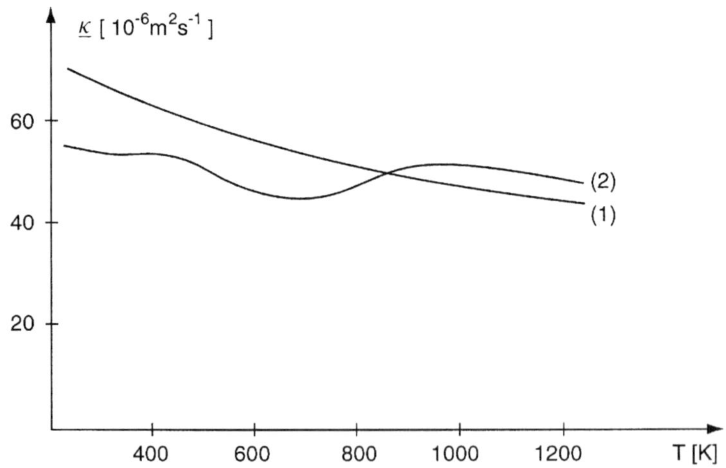

Figure 9.4.

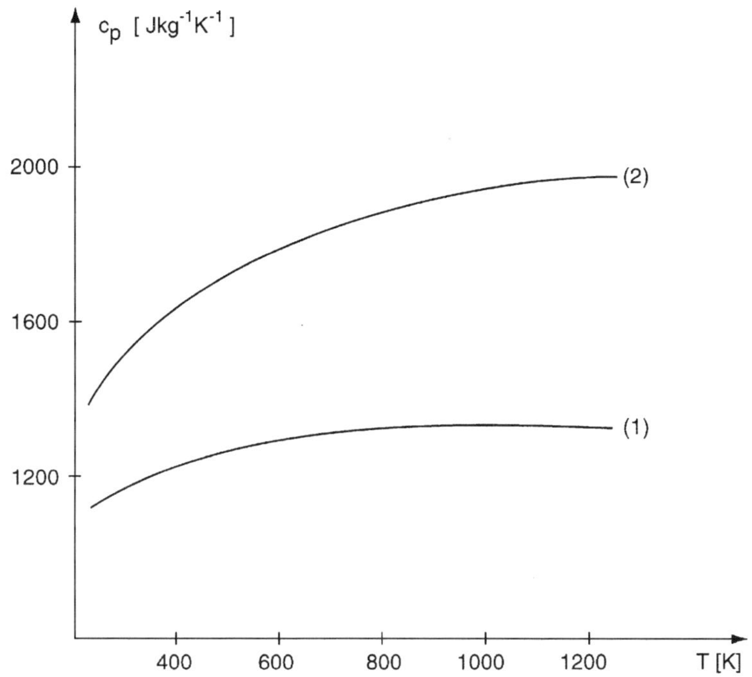

Figure 9.5.

9.3. STEADY-STATE THERMAL FIELD OF THE EARTH

9.3.1. Introduction

The heat in the lithosphere is transferred by conduction, except in anomalous regions where volcanic or hydrothermal processes play a significant role. Thus the temperature distribution can be determined by solving the relevant form of the heat conduction equation. It is evident that the conditions of heat transfer are different for the continental and oceanic lithosphere.

Let us assume the case of the continental lithosphere. The first geothermal maps of the continental lithosphere were constructed on one-dimensional models of the lithosphere and the solution was given in an analytical form (see Carslaw and Jaeger (1950), Ranalli (1987)). The solutions were found for the steady-state cases as a function of the depth x_3, and the geological structure was assumed to be layered. The heat sources were taken in the form

$$Q = Q_o \exp(-x_3/h), \tag{9.74}$$

where Q_o is the radioactive heat generation at the surface, and h is the depth at which $Q = Q_o \exp(-1)$. An exponential decrease of heat production with depth characterizes the radioactive sources in the lithosphere. Then using the boundary condition $T = T_o$

at the Earth's surface, i.e. for $x_3 = 0$, we find the solution in the form

$$T(x_3) = T_o + (\rho Q_o h^2 \kappa^{-1})(1 - \exp(-h^{-1}x_3)) + \kappa^{-1} q_m x_3$$
$$= T_o + [(q_o - q_m)\kappa^{-1} h](1 - \exp(-h^{-1}x_3)) + \kappa^{-1} q_m x_3, \qquad (9.75)$$

where q_o is the observed heat flow at the Earth's surface and q_m is the so-called reduced heat flow for which $q_o = q_m + \rho Q_o h$ holds, i.e. the observed heat flow in the continental lithosphere is a linear function of the heat generated by the radioactive elements in the surface rock layer. Theoretically, q_m represents the heat flow at large depth ($x_3 \to \infty$). Practically, the equation $q_o = q_m + \rho Q_o h$ is satisfied with $q_m \sim$ 20–30 mWm^{-2} and $h \sim 10$ km.

In the case of the oceanic lithosphere the steady-state temperature field is assumed i.e. $\partial T/\partial t = 0$, but the convective term $v\partial T/\partial x_3$ cannot be neglected. The boundary conditions for the oceanic lithosphere are $T = T_o$ at $x_3 = 0$, and $T = T_m$ at the bottom of the lithosphere. Then we find

$$T(x_1, x_3) = T_o + \operatorname{erf}\left(2^{-1}(\kappa x_1) - \tfrac{1}{2}v^{1/2}\right)(T_m - T_o), \qquad (9.76)$$

where $\operatorname{erf}(y) = 2\pi^{-1/2}\int_0^y \exp(-\xi^2)\mathrm{d}\xi$ is the error function and the transformation $x_1 = vt$ was used. In other words, the problem of finding $T(x_1, x_3)$ in a moving plate by the above transformation is equivalent to that of finding $T(x_3, t)$ in a stationary plate.

If we assume that the thermal conductivity κ and heat production Q are constant in every layer and change with a jump on every boundary between two neighbouring layers then for the temperature $T(x_3)$ in the i-th layer we find for $x_3 > x_{3i}$

$$T(x_3) = T_i + \kappa_i^{-1} q x_3 - \tfrac{1}{2}\kappa_i^{-1} Q_o\left(x_3^2 - \tfrac{1}{3}c_o x_3^3\right), \qquad (9.77)$$

where x_{3i} denotes the depth x_3 in the i-th layer and where the linear production of heat $Q = Q_o(1 - c_o x_3)$, Q_o represents production of heat at $x_3 = 0$ and $c_o =$ constant, T_i is temperature at the upper surface fo the layer, κ_i is the thermal conductivity in the i-th layer, $\kappa = \kappa_o(1 + cT)^{-1}$, $\kappa_o = 0.007$ cal cm^{-1} sec°C, $\kappa_{200} =$ 0.0055 – 0.006 cal cm^{-1} sec°C, $\kappa_{400} = 0.005$ cal cm^{-1} sec°C.

For the case of the evolution type of geothermal field it was assumed that the thermal sources are generated by a periodically dependent source, e.g. generated by sunshine with period one day or one year or with an eleven-year cycle period, etc.. Then the solution was sought in an exponential form. Since we assume that the heat production is of an exponential form $Q(x_3, t) = A(x_3)\exp(i\omega t)$, then the solution can be written in symbolic form as

$$T(x_3, t) = A(x_3)\exp(\alpha x_3 + \beta t). \qquad (9.78)$$

In the case of the evolution type of geothermal field in the continental lithosphere, the solution of the homogeneous equation can be found as

$$T(x_3, t) = X(x_3)\Theta(t).$$

Then we will solve the system

$$X'' + \lambda X = 0, \qquad \Theta' + \kappa\lambda\Theta = 0,$$

where $\kappa^{-1}\Theta'/\Theta = X''/X = -\lambda$, λ = constant for all $x_3 \in \langle 0, h \rangle$. The solution is the following (see for example Tikhonov and Samarskij (1955))

$$T(x_3, t) = \int_{-\infty}^{\infty} G(x_3, \xi, t)\varphi(\xi)\mathrm{d}\xi$$

$$= \frac{1}{2}\pi^{-1/2}\int_{-\infty}^{\infty}(\kappa t)^{-1/2}\exp(-(4\kappa t)^{-1}(x_3 - \xi)^2)\varphi(\xi)\mathrm{d}\xi. \qquad (9.79)$$

Then we shall seek the solution of the inhomogeneous heat equation. The resulting solution is then the sum of the solutions of homogeneous and inhomogeneous problems (Tikhonov and Samarskij (1955)). But in the next section we shall investigate the modern and very useful methods based on the numerical analyses of linear and nonlinear steady-state and evolution conduction problems, which can also be applied in geophysical and geological practice.

9.3.2. The Linear Steady-State Case

9.3.2.1 Introduction

The heat equation describes the propagation of heat in the interior of the Earth. If the variation of temperature with time is very slow then the term $\rho c_e \partial T/\partial t$ can be neglected, and if the dissipative term $\rho T_o \beta_{ij} e_{ij}(\dot{\boldsymbol{u}})$ representing the dissipation of deformation energy into heat is taken as the thermal sources, then the equation of heat conduction can be written in the form

$$-\frac{\partial}{\partial x_i}\left(\kappa_{ij}(\boldsymbol{x})\frac{\partial T(\boldsymbol{x})}{\partial x_j}\right) = Q(\boldsymbol{x}), \quad i, j = 1, \ldots, N, \quad N = 2, 3, \qquad (9.80)$$

where the coefficient of thermal conductivity $\kappa = (\kappa_{ij})$ does not depend on the temperature T and will be assumed to be a function of spatial coordinates $\boldsymbol{x} = (x_i)$ only. We shall assume that the geological body (geological block, lithospheric plate, etc.) occupies a bounded domain $\Omega \subset R^N$, $N = 2, 3$, with boundary $\partial\Omega$. We shall assume that the coefficient of thermal conductivity κ_{ij} is symmetric and uniformly positive definite, i.e. that for $\boldsymbol{x} \in \Omega$

$$\kappa_{ij}(\boldsymbol{x}) = \kappa_{ji}(\boldsymbol{x}), \qquad (9.81)$$

$$\kappa_{ij}(\boldsymbol{x})\xi_i\xi_j \geq \kappa_o |\xi|^2, \quad \kappa_o = \text{constant} > 0. \qquad (9.82)$$

For the heat production we shall assume the thermal sources to be sufficiently smooth functions, so that $Q(\boldsymbol{x}) \in L_2(\Omega)$. Equation 9.80 describes a steady-state thermal field or a geothermal field for a very short, from the geological point of view, time period.

Let us briefly discuss the boundary value conditions. On the Earth's surface Γ_1, the temperature or the heat flow can be prescribed, i.e. they are known, for example from the measurements

$$T(\boldsymbol{x}) = T_1(\boldsymbol{x}) \quad \text{or} \quad -\kappa_{ij}(\boldsymbol{x})\frac{\partial T(\boldsymbol{x})}{\partial x_j}n_i = q(\boldsymbol{x}).$$

On the other part of the boundary of the investigated domain Ω occupied by the studied part of the lithosphere denoted by Γ_2, we have the following boundary conditions

$$-\kappa_{ij}(\boldsymbol{x})\frac{\partial T(\boldsymbol{x})}{\partial x_j}n_i = q_o(\boldsymbol{x}) \quad \text{or} \quad -\kappa_{ij}(\boldsymbol{x})\frac{\partial T(\boldsymbol{x})}{\partial x_j}n_i = 0,$$

where $q(\boldsymbol{x})$ and $q_o(\boldsymbol{x})$ represent heat flows, and the last condition represents the condition of thermal isolation.

For simplicity we shall analyse the following problem.

Problem ($\mathscr{P}$)**:** Let $\Omega \subset R^N$, $N = 2, 3$, be the domain occupied by a lithospheric plate or geological block with boundary $\partial\Omega = \Gamma_1 \cup \Gamma_2 \cup \Gamma_3$. Let the steady-state thermal field satisfy the equation of heat conduction (Equation 9.80)

$$-\frac{\partial}{\partial x_i}\left(\kappa_{ij}(\boldsymbol{x})\frac{\partial T(\boldsymbol{x})}{\partial x_j}\right) = Q(\boldsymbol{x}), \quad i, j = 1, \ldots, N, \quad N = 2, 3, \text{ in } \Omega,$$

where the coefficients of thermal conduction satisfy Equations 9.81 and 9.82, ie

$$\kappa_{ij}(\boldsymbol{x}) = \kappa_{ji}(\boldsymbol{x}), \quad \kappa_{ij}(\boldsymbol{x})\xi_i\xi_j \geq \kappa_o\,|\xi|^2, \quad \kappa_o = \text{constant} > 0$$

and also the boundary value conditions

$$T(\boldsymbol{x}) = T_1(\boldsymbol{x})(\equiv 0) \quad \text{on the Earth's surface } \Gamma_1, \tag{9.83}$$

$$-\kappa_{ij}(\boldsymbol{x})\frac{\partial T(\boldsymbol{x})}{\partial x_j}n_i = 0 \quad \text{on } \Gamma_2, \tag{9.84}$$

$$-\kappa_{ij}(\boldsymbol{x})\frac{\partial T(\boldsymbol{x})}{\partial x_j}n_i = q(\boldsymbol{x}) \quad \text{on } \Gamma_3. \tag{9.85}$$

Remark: Since the average temperature on the Earth's surface on the studied part of the lithosphere is constant, we may put $T_1(\boldsymbol{x}) \equiv 0$. In the opposite case if the temperature on Γ_1 changes from place to place, then we transform the problem with the nonhomogeneous condition (Equation 9.83) into the problem with homogeneous condition $T(\boldsymbol{x}) = 0$ on Γ_1 by the transformation $T(\boldsymbol{x}) \sim T(\boldsymbol{x}) + w(\boldsymbol{x})$, where $w(\boldsymbol{x})$ is a sufficiently smooth function satisfying $w(\boldsymbol{x}) = T_1(\boldsymbol{x})$ on Γ_1. It is evident that $Q(\boldsymbol{x})$ and $q(\boldsymbol{x})$ change.

Consider Problem ($\mathscr{P}$) with the solution $T(\boldsymbol{x}) \in C^2(\overline{\Omega}) \subset H^1(\Omega)$ and choose $z \in H^1(\Omega)$ arbitrarily, where $H^1(\Omega)$ is the Sobolev space in the usual sense. Then applying Green's theorem we find

$$\int_\Omega \sum_{ij} \kappa_{ij}\frac{\partial T}{\partial x_j}\frac{\partial z}{\partial x_i}\mathrm{d}\boldsymbol{x} + \int_{\Gamma_3} qz\mathrm{d}s = \int_\Omega Qz\mathrm{d}\boldsymbol{x}.$$

The function $T(\boldsymbol{x})$ is the classical solution of Problem ($\mathscr{P}$). It is known that the classical solution $T(\boldsymbol{x}) \in C^2(\overline{\Omega})$ need not exist. To generalize the classical problems, we keep the weaker regularity assumptions.

Let us assume that $\kappa_{ij} \in L^\infty(\Omega)$, $\Theta(\boldsymbol{x}) \in L^2(\Omega)$ and $q(\boldsymbol{x}) \in L^2(\Gamma_3)$. Let us define the space of virtual temperatures V by

$$V = \{z \mid z \in H^1(\Omega), \quad z = 0 \text{ on } \Gamma_1 \text{ in the sense of traces}\}.$$

Multiplying Equation 9.80 by $z \in V$, and integrating over Ω we obtain

$$-\int_\Omega \frac{\partial}{\partial x_i}(\kappa_{ij}(\boldsymbol{x})\frac{\partial T(\boldsymbol{x})}{\partial x_j}z\mathrm{d}\boldsymbol{x} = \int_\Omega Qz\mathrm{d}\boldsymbol{x}, \tag{9.86}$$

then using the Green's theorem and after some algebra we obtain

$$\int_\Omega \sum_{ij} \kappa_{ij}\frac{\partial T}{\partial x_j}\frac{\partial z}{\partial x_i}\mathrm{d}\boldsymbol{x} + \int_{\Gamma_3} qz\mathrm{d}s = \int_\Omega Qz\mathrm{d}\boldsymbol{x} \quad \forall z \in V. \tag{9.87}$$

Functions z are called the test or trial functions.

Definition 9.1: We say that T is a weak solution of the studied problem $(\mathscr{P})$, we call it problem $(\mathscr{P})_v$, if

$$a(T, z) = (f, z) \equiv F(z) \quad \forall z \in V, \tag{9.88}$$

where

$$a(T, z) = \int_\Omega \sum_{ij} \kappa_{ij}\frac{\partial T}{\partial x_j}\frac{\partial z}{\partial x_i}\mathrm{d}x, \tag{9.89}$$

$$F(z) \equiv (f, z) = \int_\Omega Qz\mathrm{d}\boldsymbol{x} - \int_{\Gamma_3} qz\mathrm{d}s. \tag{9.90}$$

A mapping $a(.,.) : V \times V \to R^1$ is the bilinear form which is symmetric, i.e. $a(T, z) = a(z, T)\,\forall T, z \in V$, and let $F : V \to R^1$ be the linear form (functional), where F is defined by Equation 9.90.

Lemma 9.1: (i) Let the problem $(\mathscr{P})$ have a classical solution T. Then T is a solution of a weak problem.

(ii) On the contrary let the problem $(\mathscr{P})$ have a solution T. If the data of the problem $(\mathscr{P})_v$ as well as the solution T are sufficiently smooth, then T is a classical solution of the problem $(\mathscr{P})$.

Proof: (i) From Equations 9.86, 9.87, 9.89 and 9.90 it follows that the classical solution T satisfies Equation 9.88.

(ii) Let $a(T, z)$ be written in the following form

$$a(T, z) = \int_\Omega \frac{\partial}{\partial x_i}\left(\kappa_{ij}\frac{\partial T}{\partial x_j}z\right)\mathrm{d}\boldsymbol{x} - \int_\Omega \frac{\partial}{\partial x_i}\left(\kappa_{ij}\frac{\partial T}{\partial x_j}\right)z\mathrm{d}\boldsymbol{x} = \int_{\Gamma_3}\kappa_{ij}\frac{\partial T}{\partial x_j}zn_i\mathrm{d}s - \int_\Omega \frac{\partial}{\partial x_i}\left(\kappa_{ij}\frac{\partial T}{\partial x_j}\right)z\mathrm{d}\boldsymbol{x},$$

where the Green's theorem and the definition of the space V were used. Hence

$$\int_\Omega \left[\frac{\partial}{\partial x_i}\left(\kappa_{ij}\frac{\partial T}{\partial x_j}\right) + Q\right]z\mathrm{d}\boldsymbol{x} + \int_{\partial\Omega}\left[\kappa_{ij}\frac{\partial T}{\partial x_j}n_i - q\right]z\mathrm{d}s = 0.$$

As $H_o^1(\Omega) \subset V$ we can consider the last relation for $z \in H_o^1(\Omega)$ only. Thus

$$\int_\Omega \left[\frac{\partial}{\partial x_i}\left(\kappa_{ij}\frac{\partial T}{\partial x_j}\right) + Q\right]z\mathrm{d}\boldsymbol{x} = 0 \quad \forall z \in H_o^1(\Omega).$$

Since the space $H_o^1(\Omega)$ is dense in $L_2(\Omega)$ then

$$\frac{\partial}{\partial x_i}\left(\kappa_{ij}\frac{\partial T}{\partial x_j}\right) + Q = 0 \quad \text{in } \Omega.$$

As Equation 9.80 holds, then from Equation 9.89 follows

$$\int_{\partial\Omega} \left[\kappa_{ij} \frac{\partial T}{\partial x_j} n_i - q \right] z \mathrm{d}s = 0 \quad \forall z \in V,$$

and from here Equation 9.84 follows for $q = 0$ and Equation 9.85 follows immediately. Equation 9.83 is satisfied due to the definition of the space V, which completes the proof.

The existence and uniqueness of the weak solution gives the next theorem.

Theorem 9.1: Let V be a Hilbert space with norm $\|.\|_V$. Then the problem $(\mathscr{P})_v$, find $T \in V$ such that

$$a(T, z) = F(z) \quad \forall z \in V, \tag{9.92}$$

has one and only one solution.

Proof: The space of virtual temperatures satisfies all conditions for the Hilbert space, so that V is the Hilbert space. It is evident that the bilinear form is symmetric and we shall show that it is continuous. We have (using Equation 9.89 and the Cauchy-Schwarz inequality)

$$|a(T, z)| \leq \sum_{i,j=1}^{N} \left\|\kappa_{ij}\right\|_{L^\infty(\Omega)} \left\| \frac{\partial T}{\partial x_j} \right\|_{0,\Omega} \left\| \frac{\partial z}{\partial x_i} \right\|_{0,\Omega} \leq c \|T\|_{1,\Omega} \|z\|_{1,\Omega} .$$

The linearity of $F(.)$ on V is obvious and the continuity follows from the trace theorem in $L^2(\Gamma_3)$, i.e.

$$\begin{aligned} |F(z)| &= \left| \int_\Omega Q z \mathrm{d}\boldsymbol{x} - \int_{\Gamma_3} q z \mathrm{d}s \right| \\ &\leq \|Q\|_{0,\Omega} \|z\|_{0,\Omega} + \|q\|_{L^2(\Gamma_3)} \|z\|_{L^2(\Gamma_3)} \leq \|Q\|_{0,\Omega} \|z\|_{1,\Omega} + \|q\|_{L^2(\Gamma_3)} \|z\|_{L^2(\Gamma_3)} , \end{aligned}$$

since $\|.\|_{0,\Omega} \leq \|.\|_{1,\Omega}$.

Moreover, the bilinear form $a(T, z)$ is V-elliptic. Hence due to Equation 9.82 and the Freidrich's inequality we see that

$$a(T, T) \geq c \int_\Omega \sum_{i,j=1}^{N} \left(\frac{\partial T}{\partial x_i} \right)^2 \mathrm{d}\boldsymbol{x} \geq c \|T\|_{1,\Omega}^2 .$$

Then due to the Lax-Milgram theorem (see for example Chapter 2 on (Nečas (1967)) there exists one and only one solution $T(\boldsymbol{x}) \in V$.

Remark: The relation in Equation 9.92 is called the variational equality.

Theorem 9.2: Let the assumptions of Theorem 9.1 be satisfied. Then the problem $(\mathscr{P})_v$, i.e. Equation 9.92, is equivalent to the following problem $(\mathscr{P})_f$: find $T \in V$ such that

$$L(T) = \inf_{z \in V} L(z), \tag{9.93}$$

where

$$L(z) = \tfrac{1}{2} a(z, z) - F(z). \tag{9.94}$$

Proof: Let $T \in V$ be the solution of Equation 9.92. Since $a(.,.)$ is symmetric and bilinear, then

$$L(z) = \tfrac{1}{2} a(z, z) - a(T, z) = \tfrac{1}{2} a(z - T, z - T) - \tfrac{1}{2} a(T, T) \tag{9.95}$$

and $L(z)$ attains its minimum just for $z - T$ as the term is constant.

On the other hand, since $a(.,.)$ is V-elliptic, then from Equation 9.95 we see that the element T minimizes the functional L, where T is the solution of Equation 9.92 and therefore the problems in Equation 9.92 and in Equations 9.93 and 9.94 are equivalent.

9.3.2.2. Numerical solution

Let us approximate the space of virtual temperatures V by its finite element approximation V_h.

Let the domain Ω be approximated by the polygonal domain Ω_h. We establish a triangulation $\mathscr{T}_h$ over the domain $\overline{\Omega}_h$ so that the domain $\overline{\Omega}_h$ is subdivided into a finite number of subsets $\underline{T}_h$, i.e. triangles $\underline{T}_h$ in the $2D$ case and tetrahedra $\underline{T}_h$ in the $3D$ case, etc., in such a way that the following properties are satisfied:

(i) for each $\underline{T}_h \in \mathscr{T}_h$, the set $\underline{T}_h$ is closed and its interior $\mathring{\underline{T}}_h$ is non-empty and connected;

(ii) for each $\underline{T}_h \in \mathscr{T}_h$ the boundary $\partial \underline{T}_h$ is Lipschitz-continuous;

(iii) $\overline{\Omega}_h = \cup_{\underline{T}_h \in \mathscr{T}_h} \underline{T}_h$;

(iv) for two sets $\underline{T}_h^1, \underline{T}_h^2 \in \mathscr{T}_h$ then $\mathring{\underline{T}}_h^1 \cap \mathring{\underline{T}}_h^2 = 0$.

Definition 9.2: We say that a family of triangulations (divisions) $\mathscr{T}_h$ is regular if there exists a constant $\alpha > 0$ such that for any triangulation $\mathscr{T}_h \in \{\mathscr{T}_h\}$ and for any element $\underline{T}_h \in \mathscr{T}_h$ there exists a ball $\mathscr{B}$ of radius ρ such that $\mathscr{B} \subset T_h$ and $h \leq \rho$, where $h \leq \operatorname{diam} T_h$ is diameter of T_h.

Let the FEM approximation of the space of virtual temperatures be defined as

$$V_h = \{z_h \,|z_h \in C(\overline{\Omega}), z_h\big|_{\underline{T}_h} \in P_1, z_h = 0 \text{ on } \Gamma_1 \quad \forall \underline{T}_h \in \mathscr{T}_h\},$$

where P_1 is the space of all polynomials of the first order.

Remark: In a similar way we can define the finite element approximation of the space of virtual temperatures by using the polynomials of higher order n.

Definition 9.3: We say that $T_h \in V_h$ is the finite element approximation of the problem $(\mathscr{P})_v$ if

$$a(T_h, z_h) = F(z_h) \quad \forall z_h \in V_h. \tag{9.96}$$

Theorem 9.3: There exists one and only one solution T_h of the problem in Equation 9.96. We shall call such a solution the discrete or FEM solution.

Proof: The proof immediately follows from the Lax-Milgram lemma.

Since $a(.,.)$ is symmetric, the FEM solution is also characterized by the property

$$L(T_h) = \inf_{z_h \in V_h} L(z_h), \tag{9.97}$$

where L is defined by Equation 9.94, known as the Ritz approximation. Note that

$$L(T) \leq L(T_h)$$

since $V_h \subset V$. From Equations 9.92 and 9.96 we obtain

$$a(T - T_h, z_h) = 0 \quad \forall z_h \in V_h,$$

which represents the orthogonality relation for the error $T - T_h$. We see that the error $T - T_h$ is orthogonal to V_h with respect to the inner product $a(.,.)$.

Let us put $z_h = T_h$, then

$$0 \leq a(T - T_h, T - T_h) = a(T, T) - 2a(T, T_h) + 2a(T_h, T_h) - a(T_h, T_h) = a(T, T) - a(T_h, T_h).$$

Hence,

$$a(T, T) \geq a(T_h, T_h) \quad \text{and} \quad F(T) \geq F(T_h).$$

From the orthogonality relation it follows that T_h is the projection of T on V_h with respect to $a(.,.)$ and that

$$a(T - T_h, T - T_h) = \inf_{z_h \in V_h} a(T - z_h, T - z_h).$$

This last relation says that the Ritz approximation is the best approximation with respect to the energy norm.

Let us construct the basis $\{z_h^i\}_{i=1}^m$ of V_h, where $m = \dim V_h$. Let the discrete solution T_h be expressed as a linear combination of the basis functions

$$T_h = \sum_{i=1}^{m} c_i z_h^i. \tag{9.98}$$

Then putting Equation 9.98 into Equation 9.96 we obtain

$$a\left(\sum_{i=1}^{m} c_i z_h^i, z_h^j\right) = F(z_h^j) \quad \forall j = 1, \ldots, m.$$

Hence,

$$\sum_{i=1}^{m} a(z_h^i, z_h^j) c_i = F(z_h^j) \quad \forall j = 1, \ldots, m, \tag{9.99}$$

representing the system of linear algebraic equations for unknown c_i, $i = 1, \ldots, m$. Then Equation 9.99 leads to

$$Aw = f, \tag{9.100}$$

where $A = a(z_h^i, z_h^j)_{i,j=1}^m$, $f = (F(z_h^j))_{j=1}^m$. It is evident that the matrix A is nonsingular, which follows from the linearity and V-ellipticity of the linear form $a(.,.)$. Hence

$$(Ay, y) = \sum_{ij} a(z_h^i, z_h^j) y_i y_j = a\left(\sum_i y_i z_h^i, \sum_j y_j z_h^j\right)$$
$$= a(v, v) \geq c \|v\|_V^2 > 0 \quad \forall y = ((y_i)_{i=1}^m)^T \in R^m, y \neq 0,$$

where $v = \sum_i y_i z_h^i$, and $w \neq 0$ as $\{z_h^i\}$ is a basis of V_h and $y \neq 0$. Hence, the homogeneous system of equations $Ay = 0$ cannot have a non-trivial solution $y \neq 0$.

Since $a(.,.)$ is symmetric then A is the Gramm matrix and, moreover, also symmetric, and as $(Ay, y) > 0 \, \forall y \neq 0$, it is also positive definite.

The following generalized lemma holds (Ciarlet (1980)).

Lemma 9.2: Let $\mathcal{T}_h$ be the given triangulation of Ω and let $r_h^n T \in V_h$ be the interpolation of $T \in H^{n+1}(\Omega)$, i.e. the function from V_h which is, on each triangle $\underline{T}_h \in \mathcal{T}_h$, the interpolation polynomial of degree n of the function T, so that $(r_h^n T)(R_i) = T(R_i)$, $i = 1, \ldots, n$. Then

$$\left\|T - r_h^n T\right\|_k \leq ch^{n+1-k} \|T\|_{n+1}, \quad k = 0, 1,$$

where $c =$ constant is independent of h and T, and where the index n denotes the polynomial degree.

To prove the convergence of the FEM approximation T_h to the exact solution T we shall start from the next theorem, which is a modification of that of Strang and Fix (1973).

Theorem 9.4: Let $T \in H^1(\Omega)$ be the solution of the problem $(\mathcal{P})_v$, where $a(T, z)$ is V-elliptic. Then

$$\|T - T_h\|_1 \leq c \inf_{z \in V_h} \|T - z\|_1, \tag{9.101}$$

where c is a constant independent of h, T and z.
The proof is parallel of that of Strang and Fix (1973), p. 200, where the Dirichlet type boundary condition is prescribed on the whole boundary $\partial\Omega$.

Theorem 9.5: Let $\mathcal{T}_h$ be the given triangulation of Ω. Let the form $a(T, z)$ be V-elliptic and let $T \in H^2(\Omega)$, where T is the solution of problem $(\mathcal{P})_v$. Then

$$\|T - T_h\|_1 \leq ch \|T\|_2, \tag{9.102}$$

where $c =$ constant does not depend on h and T.

Proof: Let $r_h T$ denote the interpolation of $T \in C(\overline{\Omega})$ in V_h, i.e. the function from V_h which is, on each triangle $\overline{T}_h \in \mathcal{T}_h$, the interpolation polynomial of degree one of the function T, then $(r_h T)(R_i) = T(R_i)$, $i = 1, \ldots, n$. Since $r_h T \in V_h$, then due to Theorem 9.4

$$\|T - T_h\|_1 \leq c \|T - r_h T\|_1 .$$

Applying Lemma 9.2 for $n = 1$, $k = 1$, then Equation 9.102 follows immediately.

This theorem represents the maximum rate of convergence.

Theorem 9.6: Let the bilinear form be V-elliptic. Then

$$\lim_{h \to 0} \|T - T_h\|_1 = 0, \tag{9.103}$$

where T and T_h are exact and approximate solutions of problems $(\mathcal{P})_v$ and $(\mathcal{P})_{vh}$.

Proof: We proved above that the problem $(\mathcal{P})_v$ has a unique solution $T \in H^1(\Omega)$ and that the bilinear form is V-elliptic. Let $\delta > 0$ be given. Since the space $C^\infty(\overline{\Omega}) \cap V$ is dense in V with respect to the norm $\|.\|_V = \|.\|_1$ there exists $z \in C^\infty(\overline{\Omega}) \cap V$ such that

$$\|T - z\|_1 \leq \tfrac{1}{2}\delta. \tag{9.104}$$

Since $\|z - r_h z\|_1 \leq ch \|z\|_2 \; \forall z \in H^2(T_h)$, then for a sufficiently small h we find

$$\|z - r_h z\|_1 \leq \tfrac{1}{2}\delta. \tag{9.105}$$

Using Equations 9.101, 9.104 and 9.105 we have

$$\|T - T_h\|_1 \le c\,\|T - r_h z\|_1 \le c(\|T - z\|_1 + \|z - r_h z\|_1) \le c\delta,$$

where c = constant

The next theorem gives the error estimate in the $L^2(\Omega)$-norm, which gives the estimate of the order $O(h^2)$ under some H^2-regularity assumptions.

Theorem 9.7: Let $\mathcal{T}_h$ be a regular triangulation of Ω, and let $T \in V$ belong to $H^2(\Omega)$ and $V \subset H^1(\Omega)$ be equipped with the norm $\|.\|_V = \|.\|_1$. Then there exist $h_o, c > 0$ such that, for any $\underline{T}_h \in \mathcal{T}_h$ with $h \in (0, h_o)$, the following estimate

$$\|T - T_h\|_o \le ch^2\,|T|_2 = O(h^2) \tag{9.106}$$

holds.

Proof: Let $e = T - T_h \in V \subset L^2(\Omega)$. We define $z \in V$, $z_h \in V_h$ by

$$a(z, v) = (e, v) \quad \forall v \in V, \tag{9.107}$$

$$a(z_h, v_h) = (e, z_h) \quad \forall v_h \in V_h. \tag{9.108}$$

As $z \in H^2(\Omega)$ then

$$\|z - z_h\|_1 \le ch\,\|z\|_2 \le cc_1 h\,\|e\|_o\,. \tag{9.109}$$

Since the bilinear form $a(.,.)$ is symmetric, then

$$a(z_h, e) = a(T - T_h, z_h) = 0. \tag{9.110}$$

Then Equations 9.107, 9.108, 9.110, 9.109 and 9.102 yield

$$\|T - T_h\|_o^2 = \|e\|_o^2 = (e, e)_o = a(z, e) = a(z - z_h, e) \le c\,\|z - z_h\|_1\,\|e\|_1 \le ch\,\|e\|_o\,c_o h\,|T|_2 = O(h^2),$$

which completes the proof.

9.3.2.3. *Algorithm*

Theorem 9.2 shows that the solution of the problem $(\mathcal{P})_v$, i.e. finding the distribution of the steady-state geothermal field, leads to minimizing the functional $L(z) = \frac{1}{2}a(z, z) - F(z)$ over the FEM approximation of the space of virtual temperatures V_h. Then this problem leads to the following steps. Firstly the domain Ω will be covered by a system of tetrahedra $\underline{T}_h \equiv \underline{T}_h^{ijkl}$, and we let m be the total number of nodes of a given triangulation $\mathcal{T}_h$ of Ω and s be the number of nodes not on the boundary $\overline{\Gamma}_1$ where the temperature is prescribed (i.e. on the Earth's surface). We denote by R_i, $i = 1, \ldots, s$ nodes not lying on $\overline{\Gamma}_1$ and by R_j, $j = s + 1, \ldots, m$ the nodes lying on $\overline{\Gamma}_1$. Secondly, to construct the FEM approximation of the functional L we define the function $w_i(\boldsymbol{x}) \in V_h$ as $w_i(R_j) = \delta_{ij}$, where δ_{ij} is the Kronecker delta. Then due to the definition of the space V_h the function $w_i(\boldsymbol{x})$ is different from zero only on the tetrahedra having the node R_i as a common vertex, whilst at the other places it is equal to zero. Then we see that the graph of $w_i(\boldsymbol{x})$ is a pyramid with light equal to unity at R_i, $\neq 0$ at the support of w_i and equal to zero outside the support of w_i.

Every function $z(\boldsymbol{x}) \in V_h$ can be written as

$$z(\boldsymbol{x}) = \sum_{i=1}^{m} z(R_i) w_i(\boldsymbol{x}),$$

where the set $\{w_1(\boldsymbol{x}), \ldots, w_m(\boldsymbol{x})\}$ is a base of $V_h (\dim V_h = m)$ and function values $z(R_1), \ldots, z(R_m)$ are coordinates of $z(\boldsymbol{x})$ in this base.

Now $T_h(R_j) = 0$ for $R_j \in \overline{\Gamma}_1$ and as we do not know the function values of $T_h(\boldsymbol{x})$ at the nodes R_i, $i = 1, \ldots, s$, we denote them by $v_{h1}, \ldots, v_{hs}$. Then

$$T_h = \sum_{i=1}^{s} v_{hj} w_j(\boldsymbol{x}). \tag{9.111}$$

Since, as we saw above,

$$a(T_h, z) = F(z) \quad \forall z \in V_h \tag{9.112}$$

and

$$L(T_h) = \inf_{z_h \in V_h} L(z_h), \quad L(z) = \tfrac{1}{2} a(z, z) - F(z) \tag{9.113}$$

are equivalent, then inserting Equation 9.111 into Equation 9.112, setting $z = w_i$, $i = 1, \ldots, s$ and using the symmetry of the bilinear form $a(.,.)$, we obtain a system of linear algebraic equations

$$\sum_{i=1}^{s} a(w_i, w_j) v_{hj} = F(w_i), \quad i = 1, \ldots, s, \tag{9.114}$$

i.e.

$$K_h \boldsymbol{v}_h = \boldsymbol{b}_h, \tag{9.115}$$

where $K_h = (K_{hij}) = (a(w_i, w_j))$, $\boldsymbol{b}(b_i) = (F(w_i))$.

To construct the conductivity matrix A we shall first construct elementary matrices A_{ijkl} for every tetrahedron $\underline{T}_h^{ijkl}$ and then construct the whole conductivity matrix A for the whole domain Ω.

Let us assume the tetrahedra $\underline{T}_h^{ijkl} \in \mathscr{T}_h$ and let R_s^k, $s = i, j, k, l$ be the vertices of $\underline{T}_h^{ijkl}$ (where indices i, j, k, l denote vertices of the k-th tetrahedra). We shall assume that $\underline{T}_h$ is approximated on the tetrahedra by a polynomial of the first order, denoted by

$$\varphi^{ijkl}(\boldsymbol{x}) \equiv \varphi(\boldsymbol{x}). \tag{9.116}$$

We put $x_1 \equiv x$, $x_2 \equiv y$, $x_3 \equiv z$ and

$$\varphi(x, y, z) = \boldsymbol{G}^T \boldsymbol{a},$$

where $\boldsymbol{G} = (x, y, z, 1)^T$ and $\boldsymbol{a} = (a_1, a_2, a_3, a_4)^T$. The polynomial of first order $\varphi(x, y, z)$ on the tetrahedra $\underline{T}_h^{ijkl}$ is explicitly given by the vector $\boldsymbol{v}_h$ of functional values in vertices R_i, R_j, R_k, R_l, i.e.

$$\boldsymbol{v}_h = (\varphi(R_i), \varphi(R_j), \varphi(R_k), \varphi(R_l))^T, \tag{9.117}$$

where the superscript T denotes the transposition. It holds that

$$\boldsymbol{v}_h = S\boldsymbol{a}, \tag{9.118}$$

where

$$S = (\boldsymbol{G}(R_i), \boldsymbol{G}(R_j), \boldsymbol{G}(R_k), \boldsymbol{G}(R_l))^T = \begin{pmatrix} x_i & y_i & z_i & 1 \\ x_j & y_j & z_j & 1 \\ x_k & y_k & z_k & 1 \\ x_l & y_l & z_l & 1 \end{pmatrix} \tag{9.119}$$

Let us put

$$\underline{X}_i = x - x_i, \quad \underline{Y}_i = y - y_i, \quad \underline{Z}_i = z - z_i, \tag{9.120}$$

then the matrix S is transformed into the matrix S_o such that

$$\boldsymbol{a} = S_o^{-1}\boldsymbol{v}_h, \tag{9.121}$$

where

$$S_o = \begin{pmatrix} 0 & 0 & 0 & 1 \\ \underline{X}_j & \underline{Y}_j & \underline{Z}_j & 1 \\ \underline{X}_k & \underline{Y}_k & \underline{Z}_k & 1 \\ \underline{X}_l & \underline{Y}_l & \underline{Z}_l & 1 \end{pmatrix} \tag{9.122}$$

$$S_o^{-1} = \begin{pmatrix} s_1 & s_2 & s_3 & s_4 \\ s_5 & s_6 & s_7 & s_8 \\ s_9 & s_{10} & s_{11} & s_{12} \\ s_{13} & s_{14} & s_{15} & s_{16} \end{pmatrix}. \tag{9.123}$$

Let us assume that $(\kappa_{ij}) = (\kappa_{kk}\delta_{ij})$. Then for the functional L approximated on the element $\underline{T}_h^{ijkl}$ we have

$$L_{ijkl}^o = \frac{1}{2}\int_{\underline{T}_h^{ijkl}} \left(\frac{\partial\varphi}{\partial x}, \frac{\partial\varphi}{\partial y}, \frac{\partial\varphi}{\partial z}\right) \begin{pmatrix} \kappa_{11} & & 0 \\ & \kappa_{22} & \\ 0 & & \kappa_{33} \end{pmatrix} \begin{pmatrix} \partial\varphi/\partial x \\ \partial\varphi/\partial y \\ \partial\varphi/\partial z \end{pmatrix} \mathrm{d}x\mathrm{d}y\mathrm{d}z. \tag{9.124}$$

Hence and applying Equations 9.116 to 9.118 and 9.120 we obtain

$$L_{ijkl}^o = \tfrac{1}{2}\boldsymbol{v}_h^T(S_o)^{-1}I_o^T\underline{K}_h I_o S_o^{-1}\boldsymbol{v}_h.\mathrm{meas}(\underline{T}_h^{ijkl}) = \tfrac{1}{2}\boldsymbol{v}_h^T K_h^{ijkl}\boldsymbol{v}_h, \tag{9.125}$$

where

$$I_o = \begin{pmatrix} 1 & 0 & 0 & 0 \\ 0 & 1 & 0 & 0 \\ 0 & 0 & 1 & 0 \end{pmatrix}, \tag{9.126}$$

$$K_h = \begin{pmatrix} \kappa_{11} & & 0 \\ & \kappa_{22} & \\ 0 & & \kappa_{33} \end{pmatrix} \tag{9.127}$$

and

$$K_h^{ijkl} = (S_o^{-1})^T I_o^T \underline{K}_k I_o S_o^{-1} \boldsymbol{v}_h.\text{meas}(\underline{T}_h^{ijkl})$$

$$= \begin{pmatrix} s_1 & s_5 & s_9 \\ s_2 & s_6 & s_{10} \\ s_3 & s_7 & s_{11} \\ s_4 & s_8 & s_{12} \end{pmatrix} \underline{K}_h \begin{pmatrix} s_1 & s_2 & s_3 & s_4 \\ s_5 & s_6 & s_7 & s_8 \\ s_9 & s_{10} & s_{11} & s_{12} \end{pmatrix} \cdot \text{meas}(T_h^{ijkl}), \quad (9.128)$$

where meas $(\underline{T}_h^{ijkl}) = \frac{1}{6}\begin{pmatrix} 0 & 0 & 0 & 1 \\ \underline{X}_j & \underline{Y}_j & \underline{Z}_j & 1 \\ \underline{X}_k & \underline{Y}_k & \underline{Z}_k & 1 \\ \underline{X}_l & \underline{Y}_l & \underline{Z}_l & 1 \end{pmatrix}$ and s_i, $i = 1, \ldots, 12$ are determined from s_o and Equation 9.120.

In a similar way we find the approximation of $F(w_i)$, $i = 1, \ldots, s$. Then after some algebra

$$L_{ijkl} = \tfrac{1}{2}\Delta_{ijkl}^T K_h^{ijkl} \Delta_{ijkl} - \Delta_{ijkl}^T F_{ijkl}. \quad (9.129)$$

Summing for the whole domain Ω_h we obtain

$$L = \tfrac{1}{2}\boldsymbol{v}^T K_h \boldsymbol{v} - \boldsymbol{v}^T \boldsymbol{b}, \quad (9.130)$$

where K_h is the conductivity matrix.

To find a minimum of the functional L, we have $\delta L = 0$, which is the same as

$$\frac{\partial L}{\partial \boldsymbol{v}_h} = 0.$$

Since $(\partial L/\partial \boldsymbol{v}_h) = K_h \boldsymbol{v}_h - \boldsymbol{b}$, we obtain the system of linear algebraic equations with positive definite matrix K in the form

$$K_h \boldsymbol{v}_h = \boldsymbol{b}. \quad (9.131)$$

The system of linear algebraic equations is then solved by direct or iterative methods such as the Gauss method, Choleski method, Gauss-Seidel and SOR methods or preconditioned conjugate gradient methods, etc.

9.3.3. The Non-Linear Steady-State Geothermal Models of the Earth

9.3.3.1. Introduction

In this section we consider the model problem similar to that of the linear case in which the coefficients of thermal conductivity κ_i depend on temperature and gradient of temperature. If the investigated upper part of the Earth consists of several different rocks, then the heat conduction coefficients κ_i and the heat sources density Q are, in general, discontinuous. Some results concerning the properties and numerical solution of these problems can be found for example in Lions (1969), Gajewski, Gröger and

Zacharis (1974), Nečas (1983), Glowinski (1984), Boccardo and Buttazzo (1987), Ženíšek (1990), Feistauer and Sobotíková (1990), etc..

Then we shall solve the following non-linear problem. Let $\Omega \in R^N$, $\overline{\Omega} = \cup_{\iota=1}^{S} \overline{\Omega}^{\iota}$, and $\Omega^{\iota} \subset R^N$ be bounded domains with Lipschitz continuous boundaries $\partial\Omega$, $\partial\Omega^{\iota}$, $\iota = 1, \ldots, s$ and let $\Omega^k \cap \Omega^l = 0$ for $k, l = 1, \ldots, s, k \neq l$ and, moreover, let $\partial\Omega = \overline{\Gamma}_u \cup \overline{\Gamma}_\tau$, $\Gamma_u \cap \Gamma_\tau = 0$, meas$(\Gamma_u) \neq 0$. Furthermore, $\overline{\Omega}$, $\overline{\Gamma}_u$, $\overline{\Gamma}_\tau$, etc. denote the closures of Ω, Γ_u, Γ_τ, etc. Further, we set $\overline{\Gamma}^{kl} = \overline{\Gamma}^{lk} \equiv \partial\Omega^k \cap \partial\Omega^l$, $k, l = 1, \ldots, s$, $k \neq l$, $\overline{\Gamma}_u^{\iota} = \overline{\Gamma}_u \cap \partial\Omega^{\iota}$, $\overline{\Gamma}_\tau^{\iota} = \overline{\Gamma}_\tau \cap \partial\Omega^{\iota}$, $\iota = 1, \ldots, s$. We shall assume that Γ_u, Γ_τ, Γ^{kl} are formed by a finite number of open arcs (i.e. arcs without their end-points) or in a simple version by lines. Then $\partial\Omega^{\iota} = \overline{\Gamma}_u^{\iota} \cup \overline{\Gamma}_\tau^{\iota} \cup (\cup_{k=1,k\neq l}^{s} \overline{\Gamma}^{kl})$, $\overline{\Gamma}_u = \cup_{\iota=1}^{s} \overline{\Gamma}_u^{\iota}$, $\overline{\Gamma}_\tau = \cup_{\iota=1}^{s} \overline{\Gamma}_\tau^{\iota}$, where some of the sets $\overline{\Gamma}_u^{\iota}$, $\overline{\Gamma}_\tau^{\iota}$, $\overline{\Gamma}^{kl}$ can also be empty. For simplicity let $N = 2$.

The spaces used will be denoted by the symbols $C^k(\overline{\Omega})$, $C^k(\overline{\Omega}^{\iota})$, $C(\overline{\Omega}) \equiv C^o(\overline{\Omega})$, $L^p(\Omega)$, $L^p(\partial\Omega)$, $L^p(\Gamma_\tau)$, $W^{k,p}(\Omega)$, $H^k(\Omega) \equiv W^{k,2}(\Omega)$, $W^{k,\infty}(\Omega)$, etc., which represent spaces of continuously-differentiable functions, and Lebesgue and Sobolev spaces of measurable functions equipped with their usual norms. We denote by $\|.\|_{o,\Omega}$, $\|\cdot\|_{o,\partial\Omega}$, $\|\cdot\|_{o,p,\Omega}$, $\|\cdot\|_{o,p,\partial\Omega}$, $\|\cdot\|_{o,\Omega}$, $\|\cdot\|_{o,p,\Omega}$ the norms in the spaces $L^2(\Omega)$, $L^2(\partial\Omega)$, $L^p(\Omega)$, $L^p(\partial\Omega)$, $H^k(\Omega)$, $W^{k,p}(\Omega)$, respectively. Moreover, we define in $H^1(\Omega)$ the norm $\|u\|_{1,\Omega}$ and the semi-norm $|u|_{1,\Omega}$ by

$$\|u\|_{1,\Omega} = \left(\int_\Omega (u^2 + |\nabla u|^2) \mathrm{d}\boldsymbol{x} \right)^{1/2}, \quad |u|_{1,\Omega} = \left(\int_\Omega |\nabla u|^2 \, \mathrm{d}\boldsymbol{x} \right)^{1/2},$$

where the norm $\|.\|_{1,\Omega}$ in $H^1(\Omega)$ is induced by the scalar product $(u, u,)_{1,\Omega}$, defined on $H^1(\Omega) \times H^1(\Omega)$, $(u, v)_{1,\Omega} = \int_\Omega (uv + \nabla u \nabla v) \mathrm{d}\boldsymbol{x}$.

Further, we shall denote by $\underline{\Omega}^{\iota}$ a bounded domain such that $\underline{\Omega}^{\iota} \supset \Omega^{\iota} \cup \Omega_h^{\iota}$, $\forall h \in (0, h_o)$, $h_o > 0$, $\iota = 1, \ldots, s$, where Ω_h, Ω_h^{ι} are polygonal approximations of Ω and Ω^{ι} for $h \in (0, h_o)$, $h_o > 0$.

Let us assume that $Q^{\iota} \in W^{1,\infty}(\Omega^{\iota})$, $q \in L^\infty(\Gamma_\tau)$, $T_1 = z\,|_{\Gamma_u}$, $z \in W^{1,p}(\Omega)$ with $p > 2$, which means that T_1 can be extended on $\overline{\Gamma}_u$ such that $T_1 \in C(\overline{\Gamma}_u)$. Moreover, the coefficients of thermal conductivity $\kappa_i^{\iota} = \kappa_i^{\iota}(\boldsymbol{x}, \xi)$, $\boldsymbol{x} = (x_1, x_2) \in \underline{\Omega}^{\iota}$, $\xi = (\xi_i) \in R^2$, $i = 1, 2$, $\iota = 1, \ldots, s$ have the following properties:

(i) $\kappa_i^{\iota}(\boldsymbol{x}, \xi)$, $i = 1, 2$, are continuous in $\underline{\Omega}^{\iota} \times R^2$; there exists a constant $c > 0$ such that

$$\left|\kappa_i^{\iota}(\boldsymbol{x}, \xi)\right| \leq c \left(1 + \sum_{j=1}^{2} |\xi_j| \right) \quad \forall \boldsymbol{x} \in \underline{\Omega}^{\iota}, \forall \xi = (\xi_i) \in R^2,$$

$$i = 1, 2, \quad \iota = 1, \ldots, s; \tag{9.132}$$

(ii) the derivatives $\partial\kappa_i^{\iota}(\boldsymbol{x}, \xi)/\partial\xi_j$, $i, j = 1, 2$, are continuous and bounded in $\underline{\Omega}^{\iota} \times R^2$,

$$\left|\partial\kappa_i^{\iota}(\boldsymbol{x}, \xi)/\partial\xi_j\right| \leq c_o \quad \forall (\boldsymbol{x}, \xi) \in \underline{\Omega}^{\iota} \times R^2, \quad \iota = 1, \ldots, s; \tag{9.133}$$

(iii) there exist constants $c_1 > 0$, $c_2 \geq 0$ such that

$$\sum_{i=1}^{2} \kappa_i(\boldsymbol{x}, \xi)\xi_i \geq c_1(\xi_1^2 + \xi_2^2) - c_2\left(1 + \sum_{i=1}^{2} |\xi_i|\right) \quad \forall \boldsymbol{x} \in \underline{\Omega}^\iota, \forall \xi \in R^2, \quad \iota = 1, \ldots, s; \tag{9.134}$$

(iv) there exists a constant $\alpha > 0$ such that

$$\sum_{i,j=1}^{2} \partial\kappa_i^\iota(\boldsymbol{x}, \xi)/\partial\xi_j \zeta_i \xi_j \geq \alpha(\zeta_1^2 + \zeta_2^2) \quad \forall \boldsymbol{x} \in \underline{\Omega}^\iota, \forall \zeta, \xi \in R^2, \quad \iota = 1, \ldots, s; \tag{9.135}$$

(v) the derivatives $\partial\kappa_i^\iota/\partial x_j$ are continuous in $\Omega^\iota \times R^2$ and

$$\left|\partial\kappa_i^\iota(\boldsymbol{x}, \xi)/\partial x_j\right| \leq c_o\left(1 + \sum_{k=1}^{2} |\xi_k|\right) \quad \forall (\boldsymbol{x}, \xi) \in \underline{\Omega}^\iota \times R^2, \quad i, j = 1, 2, \quad \iota = 1, \ldots, s; \tag{9.136}$$

(vi) we assume that

$$\kappa_i(\boldsymbol{x}, \xi) = \kappa_i^\iota(\boldsymbol{x}, \xi) \quad \forall \boldsymbol{x} \in \Omega^\iota, \forall \xi \in R^2, \quad i = 1, 2, \quad \iota = 1, \ldots, s. \tag{9.137}$$

Thus, the coefficients of thermal conductivity κ_i and thermal sources can have discontinuities across Γ^{kl}.

Problem ($\mathscr{P}_N$): Find $T : \overline{\Omega} \to R^1$ satisfying

$$-\sum_{i=1}^{2} \frac{\partial}{\partial x_i} \kappa_i(\boldsymbol{x}, T(\boldsymbol{x}), \nabla T(\boldsymbol{x})) = Q(\boldsymbol{x}), \quad \boldsymbol{x} \in \Omega \tag{9.138}$$

and boundary conditions

$$T(\boldsymbol{x}) = T_1(\boldsymbol{x}), \quad \boldsymbol{x} \in \Gamma_u, \tag{9.139}$$

$$\sum_{i=1}^{2} \kappa_i^\iota(\boldsymbol{x}, T^\iota(\boldsymbol{x}), \nabla T^\iota(\boldsymbol{x})) n_i^\iota(\boldsymbol{x}) = q(\boldsymbol{x}), \quad \boldsymbol{x} \in \Gamma_\tau, \quad \iota = 1, \ldots, s, \tag{9.140}$$

and the transient conditions

$$\left.\sum_{i=1}^{2} \kappa_i^\iota(\boldsymbol{x}, T^\iota(\boldsymbol{x}), \nabla T^\iota(\boldsymbol{x})) n_i^\iota(\boldsymbol{x})\right|_{(k)} = \left.\sum_{i=1}^{2} \kappa_i^\iota(\boldsymbol{x}, T^\iota(\boldsymbol{x}), \nabla T^\iota(\boldsymbol{x})) n_i^\iota(\boldsymbol{x})\right|_{(l)}, \quad \boldsymbol{x} \in \Gamma^{kl}, \quad k, l = 1, \ldots, s, \quad k \neq l, \tag{9.141}$$

where $\boldsymbol{n}^\iota(\boldsymbol{x}) = (n_1^\iota(\boldsymbol{x}), n_2^\iota(\boldsymbol{x}))$ denote the unit outer normal to $\partial\Omega^\iota$. Obviously, $\boldsymbol{n}^k(\boldsymbol{x}) = -\boldsymbol{n}^l(\boldsymbol{x})$ on Γ^{kl}.

Definition 9.4: We say that $T : \overline{\Omega} \to R^1$ is a classical solution of the model problem ($\mathscr{P}_N$) defined by Equations 9.132 to 9.141 if $T \in C(\overline{\Omega})$, $T^\iota \in C^2(\overline{\Omega}^\iota)$, $T^\iota = T|_{\Omega^\iota}$, $\iota = 1, \ldots, s$ and Equations 9.132 to 9.141 are satisfied.

To define the weak formulation, we define the spaces

$$\mathscr{V} = \{z \mid z \in C^{\infty}(\overline{\Omega}),\ \text{supp } z \subset \Omega \cup \Gamma_\tau\}, \quad V = \{z \mid z \in H^1(\Omega), z = 0 \text{ on } \Gamma_u\},$$

where supp z represents the support of the function z, and the space V is the closure of $\mathscr{V}$ in $H^1(\Omega)$.

Since meas$(\Gamma_u) \neq 0$, the semi-norm $|.|_{1,\Omega}$ is a norm in V and is equivalent to $\|.\|_{1,\Omega}$, i.e.

$$\|z\|_{1,\Omega} \leq c\,|z|_{1,\Omega} \quad \forall z \in V, \quad c = \text{constant} > 0 \text{ independent of } z.$$

Let us multiply Equation 9.138 by an arbitrary $z \in \mathscr{V}$, integrate over each Ω^ι, apply the Green's theorem for each Ω^ι, $\iota = 1, \ldots, s$, then we obtain

$$\int_{\cup\Omega^\iota} \sum_{\iota=1}^{2} \kappa_i(\boldsymbol{x}, T, \nabla T)\partial z/\partial x_i d\boldsymbol{x} = \int_{\cup\Omega^\iota} Qz\mathrm{d}\boldsymbol{x} + \int_{\Gamma_\tau} qz\mathrm{d}s \quad \forall z \in \mathscr{V}. \tag{9.142}$$

Let us denote

$$a(T, z) = \int_{\cup\Omega^\iota} \sum_{\iota=1}^{2} \kappa_i(\boldsymbol{x}, T, \nabla T)\partial z/\partial x_i \mathrm{d}\boldsymbol{x} \text{ for } T, z \in H^1(\Omega), \tag{9.143}$$

$$L(z) = \int_{\cup\Omega^\iota} Qz\mathrm{d}\boldsymbol{x} + \int_{\Gamma_\tau} qz\mathrm{d}s \text{ for } z \in H^1(\Omega). \tag{9.144}$$

Definition 9.5: We say that the form $a : H^1(\Omega) \times H^1(\Omega) \to R^1$ is $H^1(\Omega)$-Lipschitz continuous if there exists a constant $c > 0$ such that for all $T, z, w \in H^1(\Omega)$

$$|a(T, z) - a(w, z)| \leq c\,\|T - w\|_{1,\Omega}\,\|z\|_{1,\Omega} \tag{9.145}$$

holds, is $H^1(\Omega)$-strongly monotone with respect to the semi-norm $|.|_{1,\Omega}$ if there exists a constant $c_1 > 0$ such that for all $T, z \in H^1(\Omega)$

$$a(T, T - z) - a(z, T - z) \geq c_1\,|T - z|^2_{1,\Omega} \tag{9.146}$$

holds, and is $H^1(\Omega)$-bounded if there exists a constant $c_2 > 0$ such that for all $T, z \in H^1(\Omega)$

$$|a(T, z)| \leq c_2(1 + \|T\|_{1,\Omega})\,\|z\|_{1,\Omega} \tag{9.147}$$

holds. Moreover,

$$|L(z)| \leq c\,\|z\|_{1,\Omega} \quad \forall z \in H^1(\Omega). \tag{9.148}$$

Hence, for each $T \in H^1(\Omega)$ the functionals $a(T, .)$ and $L(.)$ are continuous and linear on $H^1(\Omega)$. Moreover, we will assume that $Q \in L^2(\Omega)$ and $q \in L^2(\Gamma_\tau)$, and there exists $\overline{T}_1 \in H^1(\Omega)$ such that $T_1 = \overline{T}_1\,|_{\Gamma_u}$. For simplicity instead of $\overline{T}_1$ on Γ_u we shall write T_1 again.

Definition 9.6: We say that $T : \overline{\Omega} \to R^1$ is a weak (variational) solution of the problem $(\mathscr{P}_N)$ defined by Equations 9.138 to 9.141 if $T \in H^1(\Omega)$, $T - T_1 \in V$, and

$$a(T, z) = L(z) \quad \forall z \in V; \tag{9.149}$$

we denote this as the problem $(\mathscr{P}_N)_v$.

It is possible to show that any classical solution is a weak solution and, on the contrary, if T is a weak solution and $T^\iota \in C^2(\overline{\Omega}^\iota)$, $\iota = 1, \ldots, s$, then T is a classical solution.

Theorem 9.8: Problem $(\mathscr{P}_N)_v$, defined by Equation 9.149, has a unique solution.

Proof: The proof is a modification of that of Theorem 3.3.23 of Nečas (1983). Since the proof of the existence of the discrete approximation of the investigated problem is a modification of the proof of this theorem, we give the proof for the discrete approximation problem only.

9.3.3.2. *Numerical solution*

To solve the problem $(\mathscr{P}_N)$ numerically, we shall start with its weak formulation (Equation 9.149).

Let domains Ω and Ω^ι be triangulated. Then Ω and Ω^ι are approximated by Ω_h and Ω_h^ι respectively, where $h \in (0, h_o)$, $h_o > 0$, $\iota = 1, \ldots, s$, with properties:

(i) $\overline{\Omega}_h = \cup_{\iota=1}^s \overline{\Omega}_h^\iota$, $\quad \Omega_h^k \cap \Omega_h^l = 0$ for $k \neq l$, $\quad k, l = 1, \ldots, s$;

(ii) $\partial\Omega_h$ and $\partial\Omega_h^\iota$ are formed by finite numbers of simple closed piecewise linear curves, the vertices of which are lying on $\partial\Omega$ and $\partial\Omega^\iota$, respectively.

Let us define triangulations $\mathscr{T}_h$ and $\mathscr{T}_h^{\ \iota}$ of Ω_h and Ω_h^ι, respectively, formed by finite numbers of closed triangles, as

(i) $\mathscr{T}_h = \cup_{\iota=1}^s \mathscr{T}_h^{\ \iota}$, $\quad \overline{\Omega}_h = \cup_{\underline{T}_h \in \mathscr{T}_h} \underline{T}_h$, $\quad \overline{\Omega}_h^\iota = \cup_{\underline{T}_h \in \mathscr{T}_h^{\ \iota}} \underline{T}_h^\iota$;

(ii) if $T_{hi}, T_{h,i+1} \in \mathscr{T}_h$, $\quad \underline{T}_{hi} \neq \underline{T}_{h,i+1}$, then either $\underline{T}_{hi} \cap \underline{T}_{h,i+1} = 0$ or $\underline{T}_{hi} \cap \underline{T}_{h,i+1}$ is a common vertex or a common side of $\underline{T}_{hi}$ and $\underline{T}_{h,i+1}$;

(iii) if $\underline{T}_h \in \mathscr{T}_h^{\ \iota}$, $\quad \iota = 1, \ldots, s$, then at most two vertices of $\underline{T}_h$ are lying on $\partial\Omega^\iota$.

We say that the triangulation $\mathscr{T}_h$ is regular, if there exist $h_o > 0$, $\vartheta_o > 0$ such that

$$\vartheta_h = \min_{\underline{T}_h \in \mathscr{T}_h} \vartheta_{\underline{T}_h} \geq \vartheta_o > 0 \quad \forall h, h = \max_{\underline{T}_h \in \mathscr{T}_h} h_{\underline{T}_h} \geq h_o > 0,$$

where $h_{\underline{T}_h}$ and $\vartheta_{\underline{T}_h}$ are the length of the maximal side and the magnitude of the minimal angle of $\underline{T}_h \in \mathscr{T}_h$. Furthermore, we denote by $\sigma_h = \{R_i\}_{i=1}^m$ and by σ_h^ι the set of all vertices of $\mathscr{T}_h$ and $\mathscr{T}_h^{\ \iota}$, respectively, such that

(i) $\sigma_h \subset \overline{\Omega}$, $\quad \sigma_h^\iota \subset \overline{\Omega}^\iota$, $\quad \sigma_h \cap \partial\Omega_h \subset \partial\Omega$, $\quad \sigma_h^\iota \cap \partial\Omega_h^\iota \subset \partial\Omega^\iota$, $\quad \iota = 1, \ldots, s$;

(ii) $\overline{\Gamma}_u \cap \overline{\Gamma}_\tau \subset \sigma_h$;

(iii) the points from $\cup_{\iota=1}^s \partial\Omega^\iota$ are elements of σ_h.

Hence, it follows that

a) for each triangle $\underline{T}_h \in \mathscr{T}_h$ there exists exactly one $\iota \in \{1, \ldots, s\}$ such that $\underline{T}_h \subset \overline{\Omega}_h^\iota$, i.e. $\underline{T}_h \in \mathscr{T}_h^{\ \iota}$;

b) $\sigma_h = \cup_{\iota=1}^{s} \sigma_h^{\iota}$;

c) $\partial\Omega \cap \overline{\Gamma}^{kl} \subset \sigma_h$ for $k \neq l$ and $\overline{\Gamma}^{k'l'} \cap \overline{\Gamma}^{k''l''} \subset \sigma_h$ for $(k', l') \neq (k'', l''), k' \neq l', k'' \neq l''$.

Let us define the space of conforming piecewise linear elements $X_h \subset H^1(\Omega_h)$, i.e.

$$X_h = \{z_h \mid z_h \in C(\overline{\Omega}_h), z_h \mid_{\underline{T}_h} \in P_1, \quad \underline{T}_h \in \mathcal{T}_h\},$$

where P_1 is the space of all linear polynomials. Then the space $\mathcal{V}$ can be approximated by

$$V_h = \{z_h \mid z_h \in X_h, z_h = 0 \text{ on } \Gamma_{uh}\},$$

and we let

$$W_h = \{z_h \mid z_h \in X_h, z_h(R_i) = T_1(R_i) \quad \forall R_i \in \Gamma_{uh}\}.$$

Then the semi-norm $|.|_{1,\Omega_h}$ is a norm on V_h, uniformly equivalent to $\|.\|_{1,\Omega_h}$, i.e. there exists a constant c_o independent of $z_h \in V_h$ and h, such that

$$\|.\|_{1,\Omega_h} \leq c_o |.|_{1,\Omega_h} \quad \forall z_h \in V_h, \forall h \in (0, h_o), \quad h_o > 0.$$

The function $q : \overline{\Gamma}_\tau \to R^1$ will be approximated by $q_h : \overline{\Gamma}_{\tau h} \to R^1$. Moreover, we define the Lagrange interpolate operator $r_h : H^1(\Omega_h) \cap C(\overline{\Omega}_h) \to X_h$ by

$$r_h z \in X_h \text{ for } z \in H^1(\Omega_h) \cap C(\overline{\Omega}_h), \quad r_h z(R_i) = z(R_i) \quad \forall R_i \in \sigma_h.$$

Then the bilinear form $a(T, z)$ and the linear form $L(z)$ will be approximated as follows:

$$\underline{a}_h(T, z) = \int_{\cup_{\iota=1}^{s} \Omega_h^{\iota}} \sum_{i=1}^{2} \kappa_i^{\iota}(., T, \nabla T) \partial z^{\iota} / \partial x_i \mathrm{d}\boldsymbol{x}, \quad T, z \in H^1(\Omega_h), \tag{9.150}$$

$$\underline{L}_h(z) = \int_{\cup_{\iota=1}^{s} \Omega_h^{\iota}} Q^{\iota} z^{\iota} \mathrm{d}\boldsymbol{x} + \int_{\Gamma_{\tau h}} q_h z \mathrm{d}s, \quad z \in H^1(\Omega_h) \tag{9.151}$$

and then we have the following problem $(\mathcal{P}_N)_{vh}$: find $T_h^* \in V_h$, $T_h^* - T_{1h} \in V_h$ such that

$$\underline{a}(T_h^*, z_h) = \underline{L}_h(z_h) \quad \forall z_h \in V_h. \tag{9.152}$$

To obtain the fully discrete approximation of the bilinear and linear forms we shall introduce numerical integration in Equations 9.150 and 9.151. We shall use the following numerical quadratures:

$$\int_{\Omega_h} f \mathrm{d}\boldsymbol{x} = \sum_{\underline{T}_h \in \mathcal{T}_h} \int_{\mathcal{T}_h} f \mathrm{d}\boldsymbol{x}, \quad \int_{\underline{T}_h} f \mathrm{d}\boldsymbol{x} \doteq \operatorname{meas}(\underline{T}_h) \sum_{k=1}^{k_T} w_{T,k} f(\boldsymbol{x}_{T,k}), \text{ if } f \in C(T_h),$$

where $\boldsymbol{x}_{T,k} \in T_h$, $w_{T,k} \in R^1$ and $w_{T,k} > 0$, $\sum_{k=1}^{k_T} w_{T,k} = 1$;

$$\int_{\Gamma_{\tau h}} f \mathrm{d}s = \sum_{S \subset \overline{\Gamma}_{\tau h}} \int_S f \mathrm{d}s, \quad \int_S f \mathrm{d}s \doteq s(S) \sum_{k=1}^{k_s} v_{s,k} f(\boldsymbol{x}_{s,k}), \text{ if } f \in C(S),$$

where $s(S) = \text{meas}(S)$ represents the length of the side $S \subset \overline{\Gamma}_{\tau h}$ of the triangle T_h, $\boldsymbol{x}_{S,h} \in S$, $v_{s,k} \in R^1$. Hence due to Equations 9.150 to 9.152 we have the following fully discrete problem.

Problem $(\mathscr{P}_N)_{vh}$: Find $T_h : \overline{\Omega}_h \to R^1$, $T_h \in X_h$, and $T_h - T_{1h} \in V_h$ such that

$$a_h(T_h, z_h) = L_h(z_h) \quad \forall z_h \in V_h, \tag{9.153}$$

where

$$a_h(T_h, z_h) = \sum_{\iota=1}^{s} \sum_{\underline{T}_h \in \mathscr{T}_h} \text{meas}\,(\underline{T}_h) \left[\sum_{i=1}^{2} \frac{\partial z_h}{\partial x_i}\bigg|_{\underline{T}_h} \sum_{j=1}^{k_T} w_{T,j} \kappa_i^{\iota}(\boldsymbol{x}_{T,j}, T_h(\boldsymbol{x}_{T,j}), \nabla T_h \,|_{\underline{T}_h}) \right], \tag{9.154}$$

$$L_h(z_h) = \sum_{\iota=1}^{s} \sum_{\underline{T}_h \in \mathscr{T}_h} \text{meas}\,(T_h) \sum_{j=1}^{k_T} w_{T,j} Q^{\iota}(\boldsymbol{x}_{T,j}) z_h(\boldsymbol{x}_{Tj}) + \sum_{S \subset \overline{\Gamma}_{\tau h}} \text{meas}\,(S) \sum_{j=1}^{k_s} v_{s,j} q_h(\boldsymbol{x}_{s,j}) z_h(\boldsymbol{x}_{s,j}). \tag{9.155}$$

Definition 9.7: We say that the form $\underline{a}_h : H^1(\Omega_h) \times H^1(\Omega_h) \to R^1$, $h \in (0, h_o)$, is uniformly $H^1(\Omega_h)$-Lipschitz continuous if there exists a constant $c > 0$ such that

$$|a_h(T, z) - \underline{a}(w, z)| \le c \, \|T - w\|_{1,\Omega_h} \|z\|_{1,\Omega_h} \quad \forall T, w, z \in H^1(\Omega_h), \forall h \in (0, h_o),$$

is uniformly $H^1(\Omega_h)$-strongly monotone with respect to the semi-norm $|.|_{1,\Omega_h}$ if there exists a constant $c_o > 0$ such that

$$\underline{a}_h(T, T - z) - \underline{a}_h(z, T - z) \ge c_o \, |T - z|^2_{1,\Omega_h} \quad \forall T, z \in H^1(\Omega_h), \forall h \in (0, h_o),$$

and is $H^1(\Omega_h)$-bounded if there exists a constant $c_1 > 0$ such that

$$\left|\underline{a}_h(T, z)\right| \le c_1(1 + \|T\|_{1,\Omega_h}) \|z\|_{1,\Omega_h} \quad \forall T, z \in H^1(\Omega_h) \forall h \in (0, h_o).$$

We say that the form $a_h : X_h \times X_h \to R^1$, $h \in (0, h_o)$ is uniformly X_h-Lipschitz continuous if there exists $c_o > 0$ such that

$$|a_h(T, z) - a_h(w, z)| \le c_o \, \|T - w\|_{1,\Omega_h} \|z\|_{1,\Omega_h} \quad \forall T, w, z \in X_h, \forall h \in (0, h_o),$$

and is uniformly X_h-strongly monotone with respect to the semi-norm $|.|_{1,\Omega_h}$ if there exists constant $c_1 > 0$ such that

$$a_h(T, T - z) - a_h(z, T - z) \ge c_1 \, |T - z|^2_{1,\Omega_h} \quad \forall T, z \in X_h, \forall h \in (0, h_o),$$

and is uniformly X_h-bounded if there exists $c_2 > 0$ such that

$$|a_h(T, z)| \le c_2(1 + \|T\|_{1,\Omega_h}) \|z\|_{1,\Omega_h} \quad \forall T, z \in X_h, \forall h \in (0, h_o).$$

Lemma 9.3: There exists constant $c > 0$ such that

$$\left|a_h(T, z) - \underline{a}_h(T, z)\right| \le ch(1 + \|T\|_{1,\Omega_h}) \|z\|_{1,\Omega_h} \quad \forall T, z \in X_h, \forall h \in (0, h_o), \tag{9.156}$$

$$\left|L_h(z) - \underline{L}_h(z)\right| \le ch \|z\|_{1,\Omega_h} \quad \forall z \in X_h, \forall h \in (0, h_o), \tag{9.157}$$

$$\left|\underline{a}_h(T, z)\right| \le ch(1 + \|T\|_{1,\Omega_h}) \|z\|_{1,\Omega_h} \quad \forall T, z \in H^1(\Omega_h), \forall h \in (0, h_o), \tag{9.158}$$

$$\left|\underline{L}_h(z)\right| \le c\,\|z\|_{1,\Omega_h} \qquad \forall z \in H^1(\Omega_h),\ \forall h \in (0, h_o), \tag{9.159}$$

$$|a_h(T, z)| \le c(1 + \|T\|_{1,\Omega_h})\,\|z\|_{1,\Omega_h} \qquad \forall T, z \in X_h,\ \forall h \in (0, h_o), \tag{9.160}$$

$$|L_h(z)| \le c\,\|z\|_{1,\Omega_h} \qquad \forall z \in X_h,\ \forall h \in (0, h_o). \tag{9.161}$$

The proof is a modification of that of Feistauer and Ženíšek (1987).

The solvability of problem $(\mathscr{P}_N)_{vh}$ will be based on the next theorem.

Theorem 9.9: There exist constants $c_o, c_1 > 0$ and c_2, c_3 such that

$$\underline{a}_h(T_{1h} + z, z) - \underline{L}_h(z) \ge c_2 c_3^{-2}\,\|z\|^2_{1,\Omega_h} - c_o(1 + \|z\|_{1,\Omega_h} + \|T_{1h}\|_{1,\Omega_h})(1 + \|T_{1h}\|_{1,\Omega_h})$$
$$\forall z \in V_h,\ \forall h \in (0, h_o), \tag{9.162}$$

$$a_h(T_{1h} + z, z) - L_h(z) \ge c_1 c_3^{-2}\,\|z\|^2_{1,\Omega_h} - c(1 + \|z\|_{1,\Omega_h} + \|T_{1h}\|_{1,\Omega_h})(1 + \|T_{1h}\|_{1,\Omega_h})$$
$$\forall z \in V_h,\ \forall h \in (0, h_o), \tag{9.163}$$

where $T_{1h} = r_h T_1$, $T_{1h} \in X_h$.

Proof: Due to the properties given in Equations 9.132 and 9.134, $\Omega_h^\iota \subset \Omega^\iota$ and as $\eta = (\vartheta + \eta) - \vartheta$ then

$$\sum_{i=1}^{2} \kappa_i^\iota(\boldsymbol{x}, \vartheta + \eta)\eta_i \ge c(\eta_1^2 + \eta_2^2) - c\left(1 + \sum_{i=1}^{2}(|\eta_i| + |\vartheta_i|)\left(1 + \sum_{i=1}^{2}|\vartheta_i|\right)\right)$$
$$\forall \iota = 1, \ldots, s,\ \forall \boldsymbol{x} \in \Omega_h^\iota,\ \forall \vartheta = (\vartheta_i),\quad i = 1, 2,\quad \forall \eta = (\eta_i) \in R^2,\quad i = 1, 2, \tag{9.164}$$

where the constant c depends on constants from Equations 9.132 and 9.134.

Let $z \in V_h$, then Equations 9.150 and 9.164 yield

$$\underline{a}_h(T_{1h} + z, z) = \int_{\cup_{\iota=1}^s \Omega_h^\iota} \kappa_i^\iota(., T_{1h} + z, \nabla(T_{1h} + z))\frac{\partial z}{\partial z_i}\mathrm{d}x \ge c_1 \int_{\cup_{\iota=1}^s \Omega_h^\iota} |\nabla z|^2\,\mathrm{d}\boldsymbol{x} - c\sum_{\iota=1}^{s} J^\iota, \tag{9.165}$$

where

$$\sum_{i=1}^{s} J^\iota = \int_{\Omega_h} (1 + |z| + |\nabla z| + |T_{1h}| + |\nabla T_{1h}|)(1 + |T_{1h}| + |\nabla T_{1h}|)\mathrm{d}\boldsymbol{x}$$
$$\le c(1 + \|z\|_{1,\Omega_h} + \|T_{1h}\|_{1,\Omega_h})(1 + \|T_{1h}\|_{1,\Omega_h}),$$

and where the Cauchy inequality was applied. Then, $\|.\|_{1,\Omega_h} \le c\,|.|_{1,\Omega_h}\ \forall z_h \in V_h, \forall h$ and using Equations 9.159 and 9.165 the estimate in Equation 9.162 follows immediately.

To prove Equation 9.163 the technique used is similar to that above. Let $z \in V_h$, then

$$a_h(T_{1h} + z, z) = \sum_{\iota=1}^{s} \sum_{T_h^\iota \in \mathscr{T}_h^\iota} \text{meas}\,(T_h) \sum_{j=1}^{k_T} w_{T,j} \sum_{i=1}^{2} \kappa_i^\iota\Big(\boldsymbol{x}_{T,j}, (T_{1h} + z)(\boldsymbol{x}_{T,j}), \nabla(T_{1h} + z)\big|_{\underline{T}_h}\Big) \left.\frac{\partial z}{\partial x_i}\right|_{\underline{T}_h}.$$

Using

$$\sum_{i=1}^{2} \kappa_i^\iota\left(\boldsymbol{x}_{T,j}, (T_{1h} + z)(\boldsymbol{x}_{T,j}), \nabla(T_{1h} + z)\,|_{\underline{T}_h}\right) \left.\frac{\partial z}{\partial x_i}\right|_{\underline{T}_h}$$
$$\ge c_1\,|\nabla z|^2_{1,\underline{T}_h} - c\left(1 + \left|z(\boldsymbol{x}_{T,j})\right| + \left|(\nabla z\,|_{\underline{T}_h})\right| + \left|T_{1h}(\boldsymbol{x}_{T,j})\right| + \left|(\nabla T_{1h}\,|_{\underline{T}_h})\right|\right)$$
$$\times\left(1 + \left|T_{1h}(\boldsymbol{x}_{T,j})\right| + \left|(\nabla T_{1h}\,|_{\underline{T}_h})\right|\right)$$

$$\geq c_1 |\nabla z|^2_{1,\underline{T}_h} - c\left(1 + \max_{\underline{T}_h} |z| + \left|(\nabla z\, |_{\underline{T}_h})\right| + \max_{\underline{T}_h} |T_{1h}| + \left|(\nabla T_{1h}\, |_{\underline{T}_h})\right|\right)$$

$$\times \left(1 + \max_{\underline{T}_h} |T_{1h}| + \left|(\nabla T_{1h}\, |_{\underline{T}_h})\right|\right)$$

$$\equiv \int_{\Omega_h^\iota} \sum_{i=1}^{2} \kappa_i(., T_{1h} + z, \nabla(T_{1h} + z)) \frac{\partial z}{\partial x_i} \mathrm{d}\boldsymbol{x},$$

and

$$\max_{\underline{T}_h} |z| \leq c(\text{ meas } (T_h))^{-1/2} \|z\|_{o,\underline{T}_h} \quad \forall z \in X_h, \forall \underline{T}_h \in \mathscr{T}_h, \forall h \in (0, h_o),$$

where c is independent of z, $\underline{T}_h$ and h, and

$$\text{meas } (\underline{T}_h) \left|(\nabla z\, |_{\underline{T}_h})\right|^2 = |z|^2_{1,\underline{T}_h}$$

$$\text{meas } (\underline{T}_h) \left|(\nabla z\, |_{\underline{T}_h})\right| = \int_{\underline{T}_h} |\nabla z|\, \mathrm{d}\boldsymbol{x}, \quad z \in X_h$$

and applying the Cauchy inequality we have proved Equation 9.163.

Lemma 9.4: The following estimates

$$\left|\underline{a}_h(T^1, z) - \underline{a}_h(T^2, z)\right| \leq c \left\|T^1 - T^2\right\|_{1,\Omega_h} \|z\|_{1,\Omega_h} \quad \forall T^1, T^2, z \in H^1(\Omega_h), h \in (0, h_o),$$

$$\left|a_h(T^1, z) - a_h(T^2, z)\right| \leq c \left\|T^1 - T^2\right\|_{1,\Omega_h} \|z\|_{1,\Omega_h} \quad \forall T^1, T^2, z \in H^1(\Omega_h), h \in (0, h_o),$$

hold, where $c =$ constant is independent of T^1, T^2, z and h.
The proof is a modification of that of Lemma 2.3.2 of Feistauer and Sobotíková (1990).

The main result, the solvability theorem for the discrete approximation of the non-linear heat problem, will be given in the next theorem.

Theorem 9.10: There exists at least one solution T_h^* of the problem in Equation 9.152, and at least one solution T_h of the problem in Equations 9.153 to 9.155. Moreover, if

$$\|T_{1h}\|_{1,\Omega_h} \leq c_s \quad \forall h \in (0, h_o), \tag{9.166}$$

where c_s is a constant independent of h, then there exists a constant $c > 0$ such that

$$\left\|T_h^*\right\|_{1,\Omega_h} \leq c, \quad \forall h \in (0, h_o), \tag{9.167}$$

$$\|T_h\|_{1,\Omega_h} \leq c \quad \forall h \in (0, h_o). \tag{9.168}$$

Proof: We shall seek T_h as $T_h = T_{1h} + w_h$, where $w_h \in V_h$. Due to Equations 9.160 and 9.161 the mapping

$$z_h \in V_h \to a_h(T_{1h} + w_h, z_h) = L_h(z_h) \in R^1, \quad \forall w_h \in V_h,$$

is a continuous linear functional. Then, applying the Riesz theorem, which establishes a close relationship between Hilbert spaces and their duals (see Chapter 2), we have

$$a_h(T_{1h} + w_h, z_h) - L_h(z_h) = (\Pi_h(w_h), z_h)_{1,\Omega_h},$$

where $(.,.)_{1,\Omega_h}$ is the scalar product in $H^1(\Omega_h)$, $\|.\|_{1,\Omega_h}$ the induced norm in $H^1(\Omega_h)$, so that

$$\|\Pi_h(w_h)\|_{1,\Omega_h} = \sup_{z_h \in V_h, z_h \neq 0} |a_h(T_{1h} + w_h, z_h) - L_h(z_h)| \, \|z_h\|^{-1}_{1,\Omega_h},$$

and therefore $\Pi_h : V_h \to V_h$. Thus, the problem $(\mathscr{P}_N)_{vh}$ is equivalent to $\Pi_h(w_h) = 0$, $w_h \in V_h$ and, moreover, due to Lemma 9.4, the operator Π_h is continuous.

Due to Theorem 9.9 the following estimate

$$(\Pi_h(z), z)_{1,\Omega_h} \geq c_1 c_3^{-1} \|v\|_{1,\Omega_h} - c(1 + \|T_{1h}\|_{1,\Omega_h}) \|v\|_{1,\Omega_h} - c(1 + \|T_{1h}\|_{1,\Omega_h})^2 \tag{9.169}$$

holds, where c = constant independent of h and $v \in V_h$.

Hence, there exists a constant $c_k > 0$ such that $(\Pi_h(z), z) \geq 0 \, \forall z \in V_h$ with $\|z\|_{1,\Omega_h} = c_k$. According to Lemma 4.3 of Chapter I of Lions (1969) there exists at least one solution $w_h \in V_h$ of the equation $\Pi_h(w_h) = 0$ and therefore a solution $T_h = T_{1h} + w_h$ of the problem $(\mathscr{P}_N)_{vh}$.

Let Equation 9.166 hold. Then from $\Pi_h(w_h) = 0$ and Equation 9.169 it follows that

$$\begin{aligned} 0 &= (\Pi_h(w_h), w_h)_{1,\Omega_h} \\ &\geq c_1 c_3^{-2} \|w_h\|_{1,\Omega_h} - c(1 + \|T_{1h}\|_{1,\Omega_h}) \|w_h\|_{1,\Omega_h} - c(1 + \|T_{1h}\|_{1,\Omega_h})^2 \quad \forall h \in (0, h_o). \end{aligned}$$

Since $c_1 c_3^{-2} > 0$, $c(1 + \|T_{1h}\|_{1,\Omega_h}) > 0$ and $c(1 + \|T_{1h}\|_{1,\Omega_h})^2 > 0$ are constants independent of h, then there exists a constant $c_r > 0$ such that $\|w_h\|_{1,\Omega_h} \leq c_r \, \forall h \in (0, h_o)$. Hence, T_h satisfies Equation 9.168 with $c = c_r + \|T_{1h}\|_{1,\Omega_h}$.

To prove the existence of a solution of the problem in Equations 9.152 and 9.167, a similar method can be used.

The convergence theorem and the error estimate will be based on the results of Feistauer (1987), Feistauer and Ženíšek (1987), (1988), Feistauer and Sobotíková (1990), Ženíšek (1990).

Let $z \in H^k(\Omega)$, $k = 1, 2$. Then we define z_c as $z_c = E(z)$ if $E : H^1(\Omega) \to H^1(R^2)$ and $z_{cc} = E(z)$ if $E : H^2(\Omega) \to H^2(R^2)$. Let us set

$$\tau_h^\iota = \Omega_h^\iota - \overline{\Omega}^\iota, \quad \omega_h^\iota = \Omega^\iota - \overline{\Omega}_h^\iota, \quad \tau_h = \Omega_h - \overline{\Omega}, \quad \omega_h = \Omega - \overline{\Omega}_h, \quad \iota = 1, \ldots, s,$$

and (see Feistauer (1987))

$$\operatorname{meas}(\tau_h \cup \omega_h), \ \operatorname{meas}(\tau_h^\iota \cup \omega_h^\iota) \leq ch^2 \quad \forall h \in (0, h_o), \quad \iota = 1, \ldots, s,$$

where c = constant independent of h.

Let us assume that $T^\iota := T \,|_{\Omega}^{\iota} \in H^2(\Omega^\iota)$, $\iota = 1, \ldots, s$. Further, we denote $\tau^\iota = \cup_{h \in (0,h_o)} (\tau_h^\iota - \overline{\Omega})$, $\iota = 1, \ldots, s$ and let $\tilde{\Omega} = \cup_{h \in (0,h_o)} (\Omega_h \cup \Omega)$. If $T^\iota \in H^2(\Omega^\iota)$, $\iota = 1, \ldots, s$, then we define the function $\tilde{T} \in H^1(\tilde{\Omega})$ by $\tilde{T} := T$ in $\overline{\Omega}$, $\tilde{T} := T^\iota_{cc}$ in τ^ι, $\iota = 1, \ldots, s$, where $T^\iota_{cc} \in H^2(R^2)$ is the Calderon extension of $T^\iota \in H^2(\Omega^\iota)$.

Let us introduce theorems which will be used for proving the convergence theorem and for the error estimate. These theorems will be introduced without proofs (see Ženíšek (1990)).

Theorem 9.11: Let $w \in W^{1,p}(\Omega)$, $p > 2$. Then for every $z \in w + V$ and $\delta > 0$, we can find $z_\delta \in W^{1,p}(\Omega)$ and $h_\delta > 0$ such that

$$\|z_c - r_h z_\delta\|_{1,\Omega_h} \leq \delta \quad \forall h \in (0, h_o).$$

Theorem 9.12 (Calderon extension theorem): Let $\Omega \in \mathscr{C}^{0,1}$, $N \geq 2$, $k \in \{1, \ldots\ldots\}$, $p \in \langle 0, \infty)$. Then there exists a bounded linear operator $E : W^{k,p}(\Omega) \to W^{k,p}(R^N)$ such that $E(u)\,|_{\Omega} = u$ on $\Omega\, \forall u \in W^{k,p}(\Omega)$.

Lemma 9.5: Let the boundary $\partial\Omega$ of the given domain Ω, $\Omega \in \mathscr{C}^{o,1}$, be piecewise of class C^2. Then for $h \in (0, h_o)$

$$\|v\|_{o,\varepsilon_h} \leq ch\, \|v\|_{1,\tilde{\Omega}} \quad \forall v \in H^1(\tilde{\Omega}), \quad \varepsilon_h = \tau_h, \omega_h, \quad \tau_h = \Omega_h - \overline{\Omega}, \quad \omega_h = \Omega - \overline{\Omega}_h,$$

where c = constant which does not depend on h and v, and $\tilde{\Omega}$ is a bounded domain satisfying $\tilde{\Omega} \supset \overline{\Omega} \cup \overline{\Omega}_h\, \forall h \in (0, h_o)$.

Theorem 9.13 (interpolation theorem): Let $\underline{T}_h \in \mathscr{T}_h$, let $T_I \in p(n)$, for which $T_I(R_i) = T(R_i)$, $i = 1, \ldots, m$, where $n = \frac{1}{2}(m+1)(m+2)$, be the interpolation polynomial of T and let $T \in H^k(\underline{T}_h)$, $k \in \langle 2, n+1\rangle$. Then

$$\|T - T_I\|_{s,\underline{T}_h} \leq ch_{\underline{T}_h}^{k-s}\, |T|_{k,\underline{T}_h}, \quad 0 \leq s \leq k,$$

where c = constant independent of $\underline{T}_h$ and T.

Theorem 9.14: Under the assumptions in Equations 9.132 to 9.136 and $T^\iota \in H^2(\Omega^\iota)$, $\iota = 1, \ldots, s$, we have

$$\lim_{h\to 0_+} \inf_{z\in W_h} \|T_c - z\|_{1,\Omega_h} = 0, \tag{9.170}$$

$$\lim_{z\in W_h} \left\|\tilde{T} - T_h\right\|_{1,\Omega_h} \leq ch \quad \forall h \in (0, h_o), \tag{9.171}$$

where c = constant independent of h, and where $\tilde{T} \in H^1(\tilde{\Omega})$, $\tilde{T} := T$ in $\overline{\Omega}$, $\tilde{T} := T^\iota_{cc}$ in τ^ι, $\iota = 1, \ldots, s$.

Proof: Since the proof is a modification of that of Theorem 38.6 of Ženíšek (1990) we give only the idea of the proof.

Due to Theorem 9.8 and due to the assumptions in Equations 9.132 to 9.136 there exists a solution of the discussed problem such that $T \in w + V$, $w \in W^{1,p}(\Omega)$, $p > 2$. Then due to Theorem 9.11 for a given $\delta > 0$ there exist $T_\delta \in W^{1,p}(\Omega)$, $h_\delta > 0$ such that

$$\|T_c - r_h T_\delta\|_{1,\Omega_h} < \delta \quad \forall h \in (0, h_\delta).$$

Since $r_h T_\delta \in W_h$ then

$$\inf_{z\in W_h} \|T_c - z\|_{1,\Omega_h} \leq \|T_c - r_h T_\delta\|_{1,\Omega} < \delta \quad \forall h \in (0, h_\delta).$$

In the limit $h \to 0_+$ Equation 9.170 follows immediately.

To prove Equation 9.171 we see that due to Ženíšek (1990) the implication

$$T \in H^1(\Omega), \quad T^\iota \in H^2(\Omega^\iota), \quad \iota = 1, \ldots, s \Rightarrow T \in C(\overline{\Omega}) \tag{9.172}$$

holds. Then $r_h T$ exists on every triangulation $\mathscr{T}_h$ and as $r_h T \in W_h$ then

$$\inf_{z\in W_h} \left\|\tilde{T} - z\right\|_{1,\Omega_h} \leq \left\|\tilde{T} - r_h T\right\|_{1,\Omega_h}. \tag{9.173}$$

Since $r_h T = r_h^\iota T^\iota$ on $\overline{\Omega}_h^\iota$, $\iota = 1, \ldots, s$, then

$$\left\|\tilde{T} - r_h T\right\|_{1,\Omega_h} \leq \sum_{\iota=1}^{s} \left\|\tilde{T} - r_h^\iota T^\iota\right\|_{1,\Omega_h^\iota} \leq \sum_{\iota=1}^{s} \left(\left\|\tilde{T} - T_{cc}^\iota\right\|_{1,\Omega_h^\iota} + \left\|T_{cc}^\iota - r_h^\iota T^\iota\right\|_{1,\Omega_h}\right). \tag{9.174}$$

Due to the Calderon extension theorem (Theorem 9.12) and the interpolation theorem (Theorem 9.13) we obtain

$$\left\|T_{cc}^\iota - r_h^\iota T^\iota\right\|_{1,\Omega_h^\iota} \leq ch \left\|T^\iota\right\|_{2,\Omega^\iota}. \tag{9.175}$$

Then due to the definition of τ^ι, τ_h^ι, τ_h, ω_h and $\tilde{T}$ we have

$$\left\|\tilde{T} - T_{cc}^\iota\right\|_{1,\Omega_h^\iota} = \left\|T - T_{cc}^\iota\right\|_{1,\tau_h^\iota - \tau_h} \leq \left\|T\right\|_{1,\tau_h^\iota - \tau_h} + \left\|T_{cc}^\iota\right\|_{1,\tau_h}. \tag{9.176}$$

Due to Lemma 9.5 and the Calderon extension theorem (Theorem 9.12) we have

$$\left\|T_{cc}^\iota\right\|_{1,\tau_h^\iota} \leq ch \left\|T_{cc}^\iota\right\|_{2,\tilde{\Omega}^\iota} \leq ch \left\|T^\iota\right\|_{2,\Omega^\iota}. \tag{9.177}$$

Since $\tau_h^\iota - \tau_h \subset \cup_{\iota=1}^{s} \omega_h^\iota$, $\mathrm{meas}(\partial\Omega^k \cap \partial\Omega^l) > 0$, $k \neq l$, then applying Lemma 9.5 we obtain

$$\left\|T\right\|_{1,\tau_h^\iota - \tau_h} \leq \sum_{\iota=1}^{s} \left\|T^\iota\right\|_{1,\omega_h^\iota} \leq ch \sum_{\iota=1}^{s} \left\|T^\iota\right\|_{2,\Omega^\iota}.$$

Hence from Equations 9.173 to 9.177 the estimate in Equation 9.171 follows.

Theorem 9.15: Under the same assumptions as in Theorem 9.14

$$\lim_{h \to 0_+} \left\|T - T_h\right\|_{1,\Omega \cap \Omega_h} = 0, \tag{9.178}$$

$$\left\|T - T_h\right\|_{1,\Omega \cap \Omega_h} \leq ch, \tag{9.179}$$

where $\Omega = \cup_{\iota=1}^{s} \Omega^\iota$, $\Omega_h = \cup_{\iota=1}^{s} \Omega_h^\iota$.

Proof: The proof is the corollary of the existence theorems for continuous and discrete problems, the results of Theorem 9.14, the fact that

$$\inf_{z \in W_h} \sup_{v \in V_h} \left|\underline{a}_h(z, v) - a_h(z, v)\right| \left\|v\right\|_{1,\Omega_h}^{-1} \leq ch(1 + \left\|T\right\|_1),$$

c = constant independent of h and T, where $T \in H^1(\Omega)$ is the solution of the problem $(\mathscr{P}_N)_v$, as well as Equation 9.172, and since $\lim_{h \to 0_+} \left\|T_c - T_h\right\|_{1,\Omega_h} = 0$, $\left\|\tilde{T} - T_h\right\|_{1,\Omega_h} \leq ch$.

9.4. EVOLUTION MODELS OF THE GEOTHERMAL FIELD OF THE EARTH

9.4.1. Introduction

One of the significant geophysical manifestations from the energy point of view is the energy which is transferred by the Earth's heat flow. Its value for the whole Earth per year is estimated at about 2.0 10^{17}kcal yr^{-1} and it is greater than the energy released by earthquakes, tidal friction, volcanic manifestations, etc.. One part of this energy is consumed into chemical processes in the Earth's interior, and another part is released into the world-wide space. Instead of a heat flow the temperature is a further fundamental geophysical parameter, which determines the state of the Earth's interior mass. Therefore, the investigations of thermal flow as well as a distribution of temperature in the Earth's interior have the principal significance for our knowledge of the state of the Earth.

9.4.2. The Model Governing Equations and Initial and Boundary Value Conditions

Certain concepts of classical geothermics are utilized to derive the governing equations of heat conduction. The intent of this section is to provide the formulation of a unified analysis for adequate partial differential equations of evolution type, such as the thermal diffusion equation.

The general form of a partial differential equation describing diffusion of heat in the upper parts of the Earth, as was derived in Section 9.2, is given by

$$\rho(\boldsymbol{x})c_e(\boldsymbol{x})\frac{\partial T(\mathbf{x},t)}{\partial t} = \frac{\partial}{\partial x_i}\left(\kappa_{ij}(\mathbf{x})\frac{\partial T(\boldsymbol{x},t)}{\partial x_j}\right) + Q(\boldsymbol{x},t),$$

$$i, j = 1, \ldots, N, \quad N = 2, 3, \text{ on } \Omega \times I, \quad \bar{I} = \langle t_o, t_1 \rangle, \qquad (9.180)$$

where T is the temperature, $\boldsymbol{x} = (x_i)$, x_i is the spatial coordinate, t is time, $\kappa_{ij}(\boldsymbol{x})$ is the tensor of diffusion coefficients of heat or coefficient of thermal conduction and $Q(\boldsymbol{x}, t)$ is the generalized thermal source term.

In order to describe the kinematics of a medium, reference is made to two configurations: the reference or initial configuration at time $t = t_o$ and the present configuration at time t, with T_o and T as values of the absolute temperatures, respectively. A dimensionless temperature is defined in terms of T and T_o if one considers the ratio of the change ΔT of the absolute temperatures from the reference to the present state over the absolute value of that variable at the reference state

$$\Theta(\boldsymbol{x}, t) = T_o^{-1}(\boldsymbol{x})(T(\boldsymbol{x}, t) - T_o(\boldsymbol{x})) \equiv T_o^{-1}(\boldsymbol{x})\Delta T(\boldsymbol{x}, t). \qquad (9.181)$$

Equations 9.180 and 9.181 in terms of the dimensionless variable Θ yield

$$\rho(\boldsymbol{x})c_e(\boldsymbol{x})\frac{\partial \Theta(\boldsymbol{x},t)}{\partial t} = \frac{\partial}{\partial x_i}\left(\kappa_{ij}(\boldsymbol{x})\frac{\partial \Theta(\boldsymbol{x},t)}{\partial x_j}\right) + T_o^{-1}(\boldsymbol{x})Q(\boldsymbol{x},t),$$

$$i, j = 1, \ldots, N, \quad N = 2, 3, \text{ on } \Omega \times I. \qquad (9.182)$$

This equation also represents the transport of heat by natural convection or its changes in time from the reference state.

To determine the distribution of the temperature in the lithosphere, similarly as in the steady-state case, we prescribe the boundary value conditions and in addition the initial condition, describing the initial or reference state of a distribution of the temperature in the upper parts of the Earth. So that on the Earth's surface Γ_1 the temperature or heat flow are prescribed, i.e.

$$T(\boldsymbol{x}, t) = T_1(\boldsymbol{x}, t) \quad \text{on } \Gamma_1 \times I \qquad (9.183)$$

or

$$\kappa_{ij}(\boldsymbol{x})\frac{\partial T(\boldsymbol{x},t)}{\partial x_j}n_i = q_1(\boldsymbol{x},t) \quad \text{on } \Gamma_1 \times I, \tag{9.184}$$

where $\boldsymbol{n} = (n_i)_{i=1}^N$, $N = 2, 3$, is the normal to Γ_1 and $q_1(\boldsymbol{x}, t)$ represents the heat flow across the Earth's surface. Since the heat of the Earth is also radiated into the world-wide space, the condition describing radiation of heat across the Earth's surface can also be used, i.e.

$$\kappa(\boldsymbol{x})\frac{\partial T(\boldsymbol{x},t)}{\partial n} + \sigma(\boldsymbol{x})(T^4(\boldsymbol{x},t) - T_m^4(\boldsymbol{x},t)) \quad \text{on } \Gamma_1 \times I, \tag{9.185}$$

where $\kappa(\boldsymbol{x})$ is the coefficient of heat conductivity, $\boldsymbol{n}$ is the normal to the Earth's surface, $\sigma(\boldsymbol{x})$ is an emissivity constant and $T_m(\boldsymbol{x}, t)$ is the temperature of the ambient medium, in this case of the atmosphere. The condition in Equation 9.183 also includes the condition of a periodical warming of the Earth's surface by sunshine.

On the lower boundary Γ_3 of the investigated region (i.e. on the bottom boundary of the lithosphere) we often prescribe the heat flow or the temperature, so that

$$\kappa_{ij}(\boldsymbol{x})\frac{\partial T(\boldsymbol{x},t)}{\partial x_j}n_i = q_2(\boldsymbol{x},t) \quad \text{on } \Gamma_3 \times I \tag{9.186}$$

or

$$T(\boldsymbol{x},t) = T_2(\boldsymbol{x},t) \quad \text{on } \Gamma_3 \times I, \tag{9.187}$$

where $\boldsymbol{n} = (n_i)$ is the normal to Γ_3, $q(\boldsymbol{x}, t)$ is the heat flow across this lower boundary from the Earth's interior and $T_2(\boldsymbol{x}, t)$ is the known temperature on this boundary.

Often we assume that the investigated region is thermally isolated from the outside parts of the lithosphere or of the investigated domain of the upper parts of the Earth. For such boundary the heat flow is equal to zero, so that on this boundary, we denoted by Γ_2, we prescribe the condition

$$\kappa_{ij}(\boldsymbol{x})\frac{\partial T(\boldsymbol{x},t)}{\partial x_j}n_i = 0 \quad \text{on } \Gamma_2 \times \text{I}. \tag{9.188}$$

The initial state of temperature distribution in the investigated region (or the reference temperature in this region at time $t = t_o$) is prescribed by the initial condition at $t = t_o$

$$T(\boldsymbol{x},t_o) = T_o(\boldsymbol{x}). \tag{9.189}$$

But upper parts of the Earth are created by different rocks, which are or are not in close contact. Let us investigate the case if the neigbouring rocks are not in close contact (see Figure 9.6).

Let us assume an elementary volume ABCD≡V, created by four planes $\overline{AB}$, $\overline{CD}$, $\overline{AC}$, $\overline{BD}$, from which $\overline{AB}$ and $\overline{CD}$ are parallel with an ideal interface boundary Γ_c^{kl} and $\overline{AC}$ and $\overline{BD}$ are perpendicular to it. The thermal balance for the volume V gives

$$-\int_{AB}\kappa^k\frac{\partial T^k}{\partial n}\mathrm{d}s + \int_{CD}\kappa^l\frac{\partial T^1}{\partial n}\mathrm{d}s = \overline{\zeta C_e}\frac{\partial \overline{T}}{\partial t}l\int_{AB}\mathrm{d}s, \tag{9.190}$$

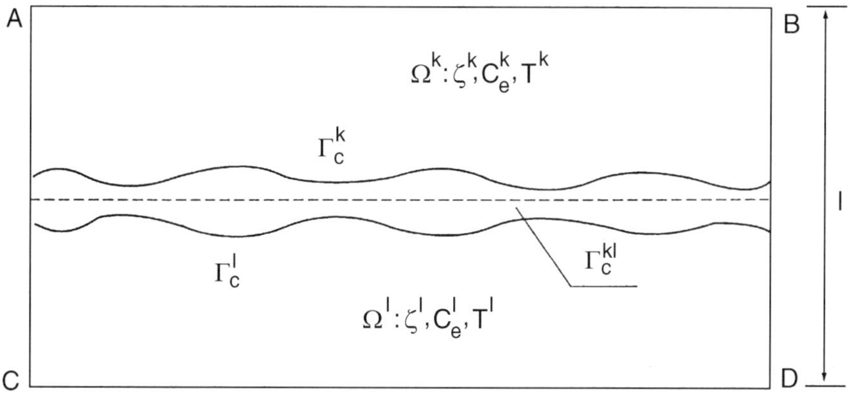

Figure 9.6.

where integrals in Equation 9.190 are the surface integrals, l is the distance of both planes $\overline{AB}$ and $\overline{CD}$, and $\overline{\rho}$, $\overline{c_e}$, and $\overline{T}$ represent mean values of density, specific heat and temperature according to the assumed volume. In the limit $l \to 0$ and in the case when the temperature T changes very slowly then Equation 9.190 leads to

$$T^k(\boldsymbol{x}, t) = T^l(\boldsymbol{x}, t), \tag{9.191}$$

$$\kappa(\boldsymbol{x}, t)\frac{\partial T(\boldsymbol{x},t)}{\partial n}\bigg|_{(k)} = \kappa(\boldsymbol{x}, t)\frac{\partial T(\boldsymbol{x},t)}{\partial n}\bigg|_{(1)} \quad \text{on } \Gamma_c^{kl} \times I. \tag{9.192}$$

All physical parameters can be functions of spatial coordinates; spatial coordinates and time; spatial coordinates, time and temperature; as well as of spatial coordinates, time, temperature and pressure; and eventually also of gradients of temperature and pressure. Then we speak about linear or non-linear geothermal evolution models.

As we saw in the introductory part of this chapter, the equation of heat also describes the geothermal regime in the deep parts of the Earth and, therefore, the theory, which will be investigated in the next section, can also be used for investigation of the global geothermal field of the Earth as well as in modification for global geodynamic models of the Earth.

9.4.3. Geothermal Models: Linear Case

9.4.3.1. Mathematical analysis of the continuous part of the problem

Let us assume that the investigated part of the Earth (the lithosphere, upper mantle, etc.) occupies the domain Ω. Let us assume that the domain Ω is a bounded domain in R^N, $N = 2, 3$, with Lipschitz boundary $\partial\Omega = \overline{\Gamma}_1 \cup \overline{\Gamma}_2 \cup \overline{\Gamma}_3$. Let $\overline{I} = \langle t_o, t_1\rangle$ be the time interval. Then we have to solve the following problem.

Problem $(\mathscr{P})$: Find a temperature $T(\boldsymbol{x},t)$ satisfying

$$\rho(\boldsymbol{x})c_e(\boldsymbol{x})\frac{\partial T(\boldsymbol{x},t)}{\partial t} = \frac{\partial}{\partial x_i}\left(\kappa_{ij}(\boldsymbol{x})\frac{\partial T(\boldsymbol{x},t)}{\partial x_i}\right) + Q(\boldsymbol{x},t) \text{ in } \Omega_T = \Omega \times (t_o, t_1), \tag{9.193}$$

boundary conditions

$$T(\boldsymbol{x}, t) = T_1(\boldsymbol{x}, t) \quad \text{on } \Gamma_1 \times (t_o, t_1), \tag{9.194}$$

$$\kappa_{ij}(\boldsymbol{x}) \frac{\partial T(\boldsymbol{x}, t)}{\partial x_j} n_i = 0 \quad \text{on } \Gamma_2 \times (t_o, t_1), \tag{9.195}$$

$$\kappa_{ij}(\boldsymbol{x}) \frac{\partial T(\boldsymbol{x}, t)}{\partial x_j} n_i = q(\boldsymbol{x}, t) \quad \text{on } \Gamma_3 \times (t_o, t_1) \tag{9.196}$$

and initial condition

$$T(\boldsymbol{x}, t_o) = T_o(\boldsymbol{x}). \tag{9.197}$$

Definition 9.8: We say that the operator $\rho c_e \partial/\partial t + L$ is of parabolic type if L, defined by $LT = -\partial/\partial x_i(\kappa_{ij}(\boldsymbol{x})\partial T(\boldsymbol{x}, t)/\partial x_i)$, is a second order differential operator of elliptic type, i.e. if it satisfies

$$\sum_{i,j=1}^{N} \kappa_{ij}\xi_i\xi_j \geq c_o \, |\xi|^2 , \quad \xi \in R^N, \text{a.e. } \boldsymbol{x} \in \Omega, \, c_o = \text{constant.} > 0. \tag{9.198}$$

Remark: For simplicity we shall assume that on the Earth's surface the temperature $T_1(\boldsymbol{x})$ can be put equal to zero. This is possible for the case when we do not study the influence of the periodic effect of sunshine, i.e. if we study the global geothermal field of the Earth. Then the mean value of the temperature on the Earth's surface can be taken to be equal to zero.

Let us denote by $H^1(\Omega)$, $H_o^1(\Omega)$ the Sobolev spaces in the usual sense. Let $D(I)$ denote the space of infinitely differentiable functions having compact support I, and let

$$V = \{z \mid z \in H^1(\Omega), z = 0 \text{ on } \Gamma_1 \text{ in the sense of traces}\},$$

then $H_o^1(\Omega) \subset V \subset H^1(\Omega)$. Let us denote by $L^2(I; V)$ the space

$$L^2(I; V) = \{z \,|z : I \rightarrow V| \, z \text{ is measurable and } \int_I \|z(t)\|_1^2 \, \mathrm{d}t < \infty\},$$

and similarly other functional spaces for time-space functions.

To formulate problem $(\mathscr{P})$ in its weak (variational) formulation, we multiply Equation 9.193 by $z \in V$, integrate over Ω, apply the Green's theorem and the boundary conditions, then we obtain the following problem.

Problem $(\mathscr{P})_V$: Given $Q(\boldsymbol{x}, t) \in L^2(\Omega_T)$, $T_o(\boldsymbol{x}) \in L^2(\Omega)$, $\kappa_{ij}(\boldsymbol{x}) \in L^\infty(\overline{\Omega})$, ρ, $c_e \in C(\overline{\Omega})$; find $T \in L^2(I; V) \cap C^\circ(I; L^2(\Omega))$ such that

$$\left(\frac{\partial T(t)}{\partial t}, z\right) + a(T(t), z) = (Q(t), z) \quad \forall z \in V, \tag{9.199}$$

$$T(\boldsymbol{x}, t_o) = T_o(\boldsymbol{x}), \tag{9.200}$$

where (.,.) denotes the scalar product in $L^2(\Omega)$ and a (.,.) is the bilinear form defined by

$$a(T(t), z) = \int_\Omega \kappa_{ij}(\boldsymbol{x}) \frac{\partial T(\boldsymbol{x}, t)}{\partial x_i} \frac{\partial z}{\partial x_i} \mathrm{d}\boldsymbol{x}, \tag{9.201}$$

$$(\partial T(t)/\partial t, z) = \int_\Omega \rho c_e \partial T(t)/\partial t z \mathrm{d}\boldsymbol{x}, \tag{9.202}$$

$$(Q(t), z) = \int_\Omega Q(t) z \mathrm{d}\boldsymbol{x}. \tag{9.203}$$

Moreover, Equation 9.199 is intended to be in the sense of distributions in $I = (t_o, t_1)$.

The existence and uniqueness of the solution gives the next theorem

Theorem 9.16: Let the bilinear form a (.,.) be continuous on $V \times V$ and

$$a(z, z) \geq c\,\|z\|^2, \quad \forall z \in V. \tag{9.204}$$

Let $Q(\boldsymbol{x}, t) \in L^2(\Omega_T)$, $T_o(\boldsymbol{x}) \in L^2(\Omega)$, $\kappa_{ij}(\boldsymbol{x}) \in L^\infty(\overline{\Omega})$, $\rho(\boldsymbol{x})$, $c_e(\boldsymbol{x}) \in C(\overline{\Omega})$.
Then there exists a unique solution $T \in L^2(I; V) \cap C^\circ(I; L^2(\Omega))$ of Equations 9.199 and 9.200. Moreover, $\partial T/\partial t \in L^2(I; V')$ and

$$\max_{t \in I} \|T(t)\|_o^2 + c \int_I \|T(t)\|_1^2 \,\mathrm{d}t \leq \|T_o\|_o^2 + c^{-1} \int_I \|Q(t)\|_o^2 \,\mathrm{d}t, \tag{9.205}$$

the so-called energy estimate, holds.

Proof:The proof will be based on the Faedo-Galerkin method. Let us construct an approximate sequence solving useful finite dimensional problems.

The space V is a closed subspace of the Sobolev space $H^1(\Omega)$ and it is a separable Hilbert space. Let us denote by $\{v_j\}$, $j = 1, \ldots, n$ a complete orthonormal basis of V and let $V_n = \mathrm{span}\{w_1, .., w_n\}$.
Then let us consider the following approximate problem: find $T_n(t) \in V_n$, $\forall t \in I$, such that

$$(T_n'(t), w_j) + a(T_n(t), w_j) = (Q(t), w_j) \quad \forall j = 1, \ldots, n, t \in I, \tag{9.206}$$

$$T_n(t_o) = T_{\mathrm{on}}(\boldsymbol{x}) \equiv \Pi_n(T_o) = \sum_{i=1}^n (T_o, w_i) w_i, \tag{9.207}$$

where Π_n represents the orthogonal projection in $L^2(\Omega)$ on V_n. Since $\{w_j\}$, $j = 1, \ldots, n$, is a basis of V_n, then Equation 9.206 is valid for each $w_n \in V_n$.

Let

$$T_n(t) = \sum_{i=1}^n \alpha_i^n(t) w_i, \tag{9.208}$$

then Equations 9.206 and 9.207 yield to solving the system of linear differential equations

$$C\alpha_n'(t) + A\alpha_n(t) = B(t), \tag{9.209}$$

$$\alpha_n(t_o) = \alpha_o(\boldsymbol{x}), \tag{9.210}$$

where $C = (C_{ij})$, $A = (A_{ij})$, $B = (B_i)$, $\alpha_o = (\alpha_{oi})$, $C_{ij} = (w_j, w_i)$, $A_{ij} = a(w_j, w_i)$, $B_i(t) = (Q(t), w_i)$, and $\alpha_{oi} = (T_o, w_i)$. The matrix C is positive definite. Then there exists a unique solution $\alpha_n(t)$ of Equations 9.209 and 9.210. Since $B(t) \in L^2(I)$, then $\alpha_n(t) \in H^1(I)$, i.e. $T_n \in H^1(I; V)$. Let us put $w_j = T_n(t)$ in Equation 9.206, then

$$(T_n'(t), T_n(t)) + a(T_n(t), T_n(t)) = (Q(t), T_n(t)),$$

and hence

$$\frac{1}{2}\frac{\mathrm{d}}{\mathrm{d}t} \|T_n(t)\|_o^2 + c\,\|T_n(t)\|_1^2 \leq \|Q(t)\|_o \,\|T_n(t)\|_o\,.$$

Then integrating over (t_o, t), $t \in (t_o, t_1\rangle$, we obtain

$$\|T_n(t)\|_o^2 + c\int_{t_o}^{t} \|T_n(\tau)\|_1^2\,\mathrm{d}\tau \le \|T_o\|_o^2 + c^{-1}\int_{t_o}^{t} \|Q(\tau)\|_o^2\,\mathrm{d}\tau. \tag{9.211}$$

Since the sequence $\{T_n(t)\}$ is bounded in $L^\infty(I; L^2(\Omega)) \cap L^2(I; V)$ then we can select a subsequence, we denote it by the same symbol $\{T_n(t)\}$, converging *-weakly in $L^\infty(I; L^2(\Omega))$ and weakly in $L^2(I; V)$. Then there exists $T \in L^\infty(I; L^2(\Omega)) \cap L^2(I; V)$ such that

$$\int_I (T_n(t), v(t))\mathrm{d}t \to \int_I (T(t), v(t))\mathrm{d}t \quad \text{as } n \to \infty \quad \forall v \in L^1(I; L^2(\Omega))$$

and

$$\int_I (\nabla T_n(t), v(t))\mathrm{d}t \to \int_I (\nabla T(t), v(t))\mathrm{d}t \quad \text{as } n \to \infty \quad \forall v \in L^2(I; L^2(\Omega)).$$

Moreover, due to the weak convergence

$$\sup_{t\in I} \|T(t)\|_o^2 + c\int_I \|T(\tau)\|_1^2\,\mathrm{d}\tau \le \lim_{n\to\infty} \inf(\sup_{t\in I} \|T_n(t)\|_o^2 + c\int_I \|T_n(\tau)\|_1^2\,\mathrm{d}\tau.$$

Hence, in the limit we obtain the assertion (Equation 9.205) of the theorem.

Let us investigate the limit of Equations 9.206 and 9.207. Let $v \in C^1(I)$, $v(t_1) = 0$. We multiply Equation 9.206 by v, integrate over $I = (t_o, t_1)$, use the Green's theorem, then we find

$$\int_I (T_n'(t), w_j)v(t)\mathrm{d}t = -\int_I (T_n(t), w_j)\frac{\mathrm{d}v(t)}{\mathrm{d}t}\mathrm{d}t - (T_{\mathrm{on}}, w_j)v(t_o).$$

Since $T_{\mathrm{on}} \to T_o$ in $L^2(\Omega)$, $n \ge n_o$, n_o being arbitrary, then in the limit in Equations 9.206 and 9.207 we obtain

$$\int_I (T(t), w_j)\frac{\mathrm{d}v(t)}{\mathrm{d}t}\mathrm{d}t - (T_o, w_j)v(t_o) + \int_I a(T(t), w_j)v(t)\mathrm{d}t = \int_I (Q(t), w_j)v(t)\mathrm{d}t \quad \forall j = 1, \ldots, n_o. \tag{9.212}$$

But the linear combinations of w_j are dense in V, and then Equation 9.212 is valid for each $w \in V$. For $v \in D(I)$ then Equation 9.212 means nothing other than Equations 9.199 and 9.200, and thus we will construct a solution $T(t) \in L^\infty(I; L^2(\Omega)) \cap L^2(I; V)$ of Equations 9.199 and 9.200

It is sufficient to show that $T(t_o) = T_o$ and $T'(t) \in L^2(I; V')$. Since $T \in L^2(I; V) \cap H^1(I; V')$ then due to Lions and Magenes (1968) p. 23, $T \in C^\circ(I; L^2(\Omega))$. From Equations 9.199 and 9.200 it follows that $(T(t), w) \in H^1(I)$ and, therefore, it is a continuous function on $\bar{I} = \langle t_o, t_1\rangle$. Multiplying Equation 9.199 by $v \in C^1(I)$, with $v(t_1) = 0$ and integrating over I then

$$-\int_I (T(t), w)\frac{\mathrm{d}v(t)}{\mathrm{d}t}\mathrm{d}t - (T(t_o), w)v(t_o) + \int_I a(T(t), w)v(t)\mathrm{d}t = \int_I (Q(t), w)v(t)\mathrm{d}t \quad \forall w \in V.$$

Setting $v(t_o) \equiv 1$, then $(T(t_o) - T_o, w) = 0\,\forall w \in V$, which proves $T(t_o) = T_o$. Moreover, from Equations 9.199 and 9.200 it follows that

$$\rho c_e \frac{\partial T(t)}{\partial (t)} + LT(t) = \rho(\boldsymbol{x})c_e(\boldsymbol{x})\frac{\partial T(\boldsymbol{x},t)}{\partial t} + \frac{\partial}{\partial x_i}\left(\kappa_{ij}(\boldsymbol{x})\frac{\partial T(\boldsymbol{x},t)}{\partial x_j}\right) = Q(\boldsymbol{x}, t)$$

in the sense of distributions on Ω_T. And since $LT(t) \in L^2(I; V')$ then $T' \in L^2(I; V')$.

To prove the uniqueness of the solution we assume that $\tilde{T}_1(t)$ and $\tilde{T}_2(t)$ are two solutions of the problem in Equations 9.199 and 9.200. Then $\overline{T}(t) = \tilde{T}_1(t) - \tilde{T}_2(t)$ also belongs to $L^2(I; V) \cap H^1(I; V')$.

Putting $z = \overline{T}(t)$ in Equation 9.199 we find

$$\frac{1}{2}\frac{\mathrm{d}}{\mathrm{d}t}\left\|\overline{T}(t)\right\|_o^2 + a(\overline{T}(t), \overline{T}(t)) = (Q(t), \overline{T}(t)).$$

Hence, by a similar way as above, we find the a-priori estimate of the type in Equation 9.205, from which the uniqueness follows immediately.

Moreover, the following additional a-priori estimates hold.

Theorem 9.17: Let the bilinear form $a(T, z)$ be coercive, let $Q \in L^2(\Omega_T)$, $T_o \in V$, $\kappa_{ij} \in C^1(\overline{\Omega})$, $\rho, c_e \in C(\overline{\Omega})$, and $\overline{I} = \langle t_o, t_1\rangle$. Then the solution T of Equations 9.199 and 9.200, $T \in L^\infty(I; V) \cap H^1(I; L^2(\Omega))$, satisfies the energy estimate

$$\sup_{t\in I}\|T(t)\|_1^2 + \int_I \left\|\frac{\partial T(t)}{\partial t}\right\|_o^2 \mathrm{d}t \le c\left(\|T_o\|_1^2 + \int_I \|Q(t)\|_o^2\,\mathrm{d}t\right), \qquad (9.213)$$

where $c > 0$ is a constant independent of $\underline{T}_h$.

Moreover, if the solution $T(t)$ satisfies

$$\|T(t)\|_2^2 \le c(\|LT(t)\|_o^2 + \|T(t)\|_1^2) \text{ a.e. in } \overline{I}, \qquad (9.214)$$

where $LT(t) = -\partial/\partial x_i(\kappa_{ij}\partial T(t)/\partial x_j)$, then $T \in L^2(I; H^2(\Omega)) \cap H^1(I; L^2(\Omega)) \cap C^\circ(I; V)$ and satisfies the estimate

$$\max_{t\in I}\|T(t)\|_1^2 + \int_I \left(\left\|\frac{\partial T(t)}{\partial t}\right\|_o^2 + \|T(t)\|_2^2\right)\mathrm{d}t \le c\left(\|T(t)\|_1^2 + \int_I \|Q(t)\|_o^2\,\mathrm{d}t\right). \qquad (9.215)$$

Proof: When $\kappa_{ij} \in C^1(\overline{\Omega})$, it is not restrictive assuming that $\kappa_{ij} = \kappa_{ji}$ for each $i, j = 1, .., N$, as we can write

$$\frac{\partial}{\partial x_i}\left(\kappa_{ij}\frac{\partial T(t)}{\partial x_j}\right) = \frac{\partial}{\partial x_i}\left(\frac{1}{2}(\kappa_{ij} + \kappa_{ji})\frac{\partial T(t)}{\partial x_j}\right) + \frac{\partial}{\partial x_i}\left(\frac{1}{2}(\kappa_{ij} - \kappa_{ji})\frac{\partial T(t)}{\partial x_j}\right)$$

and the coefficients $\kappa_{ij} = \frac{1}{2}(\kappa_{ij} + \kappa_{ji})$ satisfy the ellipticity assumption (Equation 9.198) with the same constant $c_o > 0$ and therefore we can assume that $\kappa_{ij} = \kappa_{ji}$.

Now, we multiply Equation 9.199 by $T'(t) \in L^2(I, V)$ and integrate over Ω. Since $(\partial T/\partial t) = 0$ on $\partial\Omega \times I$, then for almost each $t \in I$ we have

$$\int_\Omega [(T')^2 + \mathrm{LTT'}]\mathrm{d}\boldsymbol{x} = \int_\Omega \left[(T')^2 + \kappa_{ij}(\boldsymbol{x})\frac{\partial T}{\partial x_j}\frac{\partial T'}{\partial x_i}\right]\mathrm{d}\boldsymbol{x} = \int_\Omega QT'\mathrm{d}\boldsymbol{x}.$$

Applying the Green's theorem and integrating over (t_o, t), then from the ellipticity assumption we obtain

$$\|T(t)\|_1^2 + \int_{t_o}^t \left\|T'(\tau)\right\|_o^2 \mathrm{d}\tau \le c(\|T_o\|_1^2 + \int_{t_o}^t (\|T(\tau)\|_1^2 + \|Q(\tau)\|_o \left\|T'(\tau)\right\|_o)\mathrm{d}\tau$$

and hence

$$\|T(t)\|_1^2 + \int_{t_o}^t \left\|T'(\tau)\right\|_o^2 \mathrm{d}\tau \le c(\|T_o\|_1^2 + \int_{t_o}^t (\|T(\tau)\|_1^2 + \|Q(\tau)\|_o^2)\mathrm{d}\tau.$$

Using Equation 9.205 then Equation 9.213 follows.

To prove the second part of the assertion of this theorem then the estimate in Equation 9.215 follows from Equations 9.214, 9.213 and 9.205 and $LT(t) = Q(t) - \rho c_e \partial T(t)/\partial t$. Due to Theorem 1.3.8 of Lions and Magenes (1968), we find that $T \in L^2(I; H^2(\Omega)) \cap H^1(I; L^2(\Omega)) \cap C^\circ(I; H^1(\Omega))$.

Remark: In the mathematical literature the proof of the existence and uniqueness of the solution to the problem $(\mathscr{P})$ can also be based on the semigroup theory (see for example Pazy (1983)) setting

$$T(t) = \exp(-tA)T_o + \int_{t_o}^{t} \exp(-(t-\tau)A)Q(\tau)\mathrm{d}\tau,$$

where A represents a useful realization of the operator L with associated boundary conditions on $\partial\Omega$.

9.4.3.2. Numerical solution

Numerical solution of the linear geothermal problem discussed can be based on several types of methods, for example finite difference, finite element and finite volume methods, wavelet method, Galerkin (Faedo-Galerkin) method, operator-splitting methods, decomposition methods, spectral collocation method and their modification for semi-implicit and fully implicit discrete approximations as well as the method of discretization in time (known also as the Rothe method). The weak formulation of the geothermal problem is then one of the bases for numerical solution of the problem discussed. Firstly, if we discretize the space variables only then this leads to solving a system of ordinary differential equations, where the solution is an approximation of the exact solution of the geothermal problem for each $t \in \bar{I} \equiv \langle t_o, t_1 \rangle$. Secondly, if we discretize the space variables as well as time then we obtain a relatively great number of fully discretized schemes, which are very poor or very good stable and quickly converge. But since the volume of this section is limited, we shall discuss only few methods. For more details of these methods the reader should consult texts dealing with these methods from the mathematical point of view (See for example Quarteroni, Valli (1991), Ženíšek (1990)).

SEMI-DISCRETE APPROXIMATION AND THE FEM METHOD

The variational formulation of the geothermal problem $(\mathscr{P})_v$ defined by Equations 9.199 and 9.200 leads in the first approximation to a semi-discrete problem by approximating the space V by a finite dimensional space V_h and discretizing the time derivative by its difference. As we saw in the proof of Theorem 9.16 the Faedo-Galerkin method, consisted in choosing the approximate space $V_h \equiv V_n = \text{span}\{w_1, \ldots, w_n\}$, where $\{w_i\}$ represents a complete orthonormal basis of the space of virtual temperatures V. But the space V_h can be constructed in another way, based on the finite element method.

The problem $(\mathscr{P})_v$ in its semi-discrete version leads to the following problem

Problem $(\mathscr{P}_{sd})_{vh}$: Find $T_h(t) \in V_h, t \in I$ such that

$$\frac{\mathrm{d}}{\mathrm{d}t}(T_h(t), z_h) + a(T_h(t), z_h) = (Q(t), z_h) \quad \forall z_h \in V_h, t \in I, \tag{9.216}$$

$$T_h(\boldsymbol{x}, t_o) = T_{oh}(\boldsymbol{x}), \tag{9.217}$$

where $Q \in L^2(\Omega_T)$, and $T_{oh}(\boldsymbol{x}) \in V_h$ represents a suitable approximation of the initial value of temperature $T_o(\boldsymbol{x}) \in L^2(\Omega)$.

Moreover, we shall assume that the domain Ω is a polygonal domain and its boundary $\partial\Omega$ is Lipschitzian. Let us put

$$T_h(t) = \sum_{i=1}^{n(h)} \alpha_i(t) w_i(\boldsymbol{x}),$$

where $\{w_i\}$, $i = 1, \ldots, n(h)$, is a basis of V_h, and $T_{oh} = \sum_{i=1}^{n(h)} \alpha_{oi} w_i$. Then Equations 9.216 and 9.217 lead to a system of ordinary differential equations

$$C\alpha'(t) + A\alpha(t) = B(t), \tag{9.218}$$

$$\alpha(t_o) = \alpha_o, \tag{9.219}$$

where $C = (C_{ij}) \equiv (C_{\text{FEM}})_{ij} = (w_i, w_j)$, $A = (A_{ij})$, $A_{ij} = (A_{\text{FEM}})_{ij} = a(w_j, w_i)$, $B = (B_i)$, $B_i = (B_{\text{FEM}})_i(t) = (Q(t), w_i)$, $i, j = 1, \ldots, n(h)$. The matrix C is positive definite. Then there exists a unique solution $\alpha(t)$ of Equations 9.218 and 9.219. By a technique similar to that of the proof of Theorem 9.16, we find that the solution $T_h(t)$ satisfies an estimate similar to that of Equation 9.205 with the property that T_{oh} converges to T_o in $L^2(\Omega)$, which proves the stability of the method.

The convergence of $T_h(t)$ to $T(t)$ gives the next theorem.

Theorem 9.18: Let $\{\mathcal{T}_h\}$ be a regular family of piecewise linear triangulations. Let the bilinear form be weakly coercive in V and let $Q \in L^2(\Omega_T)$, $T_o \in V$, $\kappa_{ij} \in C^1(\overline{\Omega})$, and $\rho, c_e \in C(\overline{\Omega})$. Then the solutions T and T_h of problem $(\mathscr{P})_v$ and $(\mathscr{P}_{sd})_{vh}$, respectively, satisfy

$$\max_{t\in I} \|T(t) - T_h(t)\|_o^2 + c\int_I \|T(t) - T_h(t)\|_1^2$$
$$\leq \|T_o - T_{\text{oh}}\|_o^2 + c_1(c,\gamma)h^2\left(\|T_{\text{oh}}\|_1^2 + \|T_o\|_1^2 + \int_I \|Q(t)\|_o^2\,\mathrm{d}t\right), \tag{9.220}$$

where c is the coerciveness constant, γ is a constant of continuity of bilinear form a (.,.), and $c_1(c,\gamma) =$ constant. > 0 independent of h. Moreover, T_{oh} can be taken such that $\|T_{\text{oh}}\|_1 \leq \|T_o\|_1$.

Proof:Let us denote $e_h(t) = T(t) - T_h(t)\,\forall t \in$I. Then from Equations 9.216, 9.217, and 9.199 and 9.200 we obtain

$$\frac{\mathrm{d}}{\mathrm{d}t}(e_h(t), z_h) + a(e_h(t), z_h) = \left(\frac{\partial e_h(t)}{\partial t}, z_h\right) + a(e_h(t), z_h) = 0 \quad \forall z_h \in V_h. \tag{9.221}$$

Hence, putting $z_h = T_h(t) - v_h$, $v_h \in V_h$, t being fixed, then for each $\varepsilon > 0$ and for almost any $t \in I$ we have

$$\frac{1}{2}\frac{\mathrm{d}}{\mathrm{d}t}(e_h(t), e_h(t)) + a(e_h(t), e_h(t)) = \left(\frac{\partial e_h(t)}{\partial t}, T(t) - v_h\right) + a(e_h(t), T(t) - v_h)$$
$$\leq \left\|\frac{\partial e_h(t)}{\partial t}\right\|_o \|T(t) - v_h\|_o + \gamma\,\|e_h(t)\|_1\,\|T(t) - v_h\|_1$$
$$\leq \left\|\frac{\partial e_h(t)}{\partial t}\right\|_o \|T(t) - v_h\|_o + (4\varepsilon)^{-1}\gamma^2\,\|T(t) - v_h\|_1^2 + \varepsilon\,\|e_h(t)\|_1^2. \tag{9.222}$$

From Theorem 9.17 it follows that $T \in L^2(I; H^2(\Omega))$. Let us set $v_h = r_h T(t)$ for almost any $t \in I$, where r_h is an interpolation operator, then

$$\|T(t) - r_h T(t)\|_o^2 + h^2 \|T(t) - r_h T(t)\|_1^2 \leq ch^4 \|T(t)\|_2^2 .$$

Integrating Equation 9.222 over (t_o, t) and putting $\varepsilon = \frac{1}{2}c_o$ we obtain

$$\|e_h(t)\|_o^2 + c_o \int_{t_o}^{t} \|e_h(\tau)\|_1^2 \, d\tau \leq \|T_o - T_{oh}\|_o^2 + c_1(c, \gamma) h^2 \int_{t_o}^{t} \left(\|T'(\tau)\|_o^2 + \|T_h'(\tau)\|_o^2 + \|T(\tau)\|_2^2 \right) d\tau.$$

Due to Theorem 9.17, the estimate in Equation 9.213 gives

$$\int_{t_o}^{t} \|T'(\tau)\|_o^2 \, d\tau \leq C(c) \left(\|T_{oh}\|_1^2 + \int_{t_o}^{t} \|Q(\tau)\|_o^2 \, d\tau \right)$$

and due to Equation 9.215 the estimate in Equation 9.220 follows.

Remark: If $T_{oh} \in V_h$ is taken in such a way that

$$\|T_o - T_{oh}\|_o + h \|T_o - T_{oh}\|_1 \leq ch \|T_o\|_1 , \tag{9.223}$$

then the error

$$\|T - T_h\| = O(h) \tag{9.224}$$

in the space $C^o(I; L^2(\Omega)) \cap L^2(I; H^1(\Omega))$.

Remark: The L^2-error estimate was given by Wheeler (1973). We assume that the bilinear form is symmetric and weakly coercive, $T_o \in H^{k+1}(\Omega)$, $k \geq 1$ and the solution of Equations 9.199 and 9.200 is such that $T' \in L^1(I; H^{k+1}(\Omega))$. Then, if the finite elements of V_h are defined piecewise-polynomials of degree less than or equal to k, $\forall t \in I$ the solution T_h satisfies

$$\|T(t) - T_h(t)\|_o \leq \|T_o - T_{oh}\|_o + ch^{k+1} \left(\|T_o\|_{k+1} + \int_{t_o}^{t} \|T'(t)\|_{k+1} \, d\tau \right), \tag{9.225}$$

where $c =$ constant > 0 and is independent of h.

The L^∞-error estimate was studied in Schatz, Thomeé and Wahlbin (1980). Here we also assume that the bilinear form is symmetric and weakly coercive. Then for each $t \in I$ we have the error estimate

$$\|T(t) - T_n(t)\|_{L^\infty(\Omega)} \leq ch^2 |\log h|^2 \left(\|T_o\|_{W^{2,\infty}(\Omega)} + \int_{t_o}^{t} \|T(\tau)\|_{W^{2,\infty}(\Omega)} \, d\tau \right). \tag{9.226}$$

FULLY-DISCRETE APPROXIMATION AND THE FEM METHOD

The weak formulation of the problem discussed needs to be discretized with respect to both the time and space variables. As we saw in the previous section one way is the Faedo-Galerkin method (and also the generalized Galerkin and collocation methods (see Quarteroni and Valli (1991))), where we approximated $T_h(\boldsymbol{x}, t)$ by $T_h(\boldsymbol{x}, t) = \sum_{i=1}^{n(h)} \alpha_i(t) w_i(\boldsymbol{x})$ and we spoke on a semi-discrete (or continuous-in-time) approximation. But the discretization in time of the weak formulation can be accomplished in several other ways. One of them is a fully-discrete approximation of Equations 9.199

and 9.200. Let a uniform mesh for the time variable t with $\Delta t = (t_1 - t_o)/n > 0$ as the time step be defined. Then

$$t_i = i\Delta t, \quad i = 0, 1, 2, \dots, n$$

(where the notation for $i = 1$, $t_1 = \Delta t$, does not represent the end of the interval $\bar{I}$).

Let us apply the so-called Θ-scheme to Equations 9.216 and 9.217, then we have the following problem

Problem $(\mathscr{P}_{fd})_{vh}$**:** Find $T_h^n \in V_h$ such that

$$\Delta t^{-1}(T_h^{n+1} - T_h^n, z_h) + a(\Theta T_h^{n+1} + (1-\Theta)T_h^n, z_h) = (\Theta Q(t_{n+1}) + (1-\Theta)Q(t_n), z_h) \quad \forall z_h \in V_h, \tag{9.227}$$

$$T_h^o = T_{oh}, \tag{9.228}$$

for each $n = 0, 1, .., n-1$; we also assume that Q is everywhere defined on $\bar{I} = \langle t_o, t_1 \rangle$, $\Theta \in \langle 0, 1 \rangle$.

Let us put

$$T_h^{n+1} = \sum_{i=1}^{n(h)} \alpha_i^{n+1} w_i, \tag{9.229}$$

where w_j are the basis functions of V_h, then at each time step we solve the linear system

$$(C + \Theta\Delta t A)\vec{\alpha}^{n+1} = \boldsymbol{b}^n, \tag{9.230}$$

where $\boldsymbol{b}^n$ is determined from the previous iterate steps, the matrix $C = (C_{ij})$, $C_{ij} = (w_i, w_j)$, $A = (A_{ij})$, $A_{ij} = a(w_j, w_i)$. Then the matrix $(C + \Theta\Delta t A)$ is positive definite, so that the linear system (Equation 9.230) has a unique solution. Moreover, $(C + \Theta\Delta t A)$ is symmetric as the bilinear form a(.,.) is symmetric, i.e. $a(T, z) = a(z, T)\,\forall T, z \in V$. For $\Theta = 0$ or $\Theta = 1$ the scheme in Equations 9.227 and 9.228 is known as the forward Euler method or backward Euler method, respectively. For $\Theta = \frac{1}{2}$ it is known as the Cranck-Nichelson method and it may suffer from unexpected instabilities especially when rough perturbations are introduced in the data. For $\Theta \geq \frac{1}{2}$ the scheme is A-stable * (see for example Lambert (1991), p. 244).

Now we shall prove that the Θ-scheme in Equations 9.227 and 9.228 $\frac{1}{2} \leq \Theta \leq 1$, is unconditionally stable with respect to the $L^2(\Omega)$-norm and convergent. Let us introduce the following nottaion: for any function $f \in L^2(\Omega)$, we define

$$\|f\|_{-1,h} = \sup_{z_h \in V_h} \frac{(f, z_h)}{\|z_h\|_1} \tag{9.231}$$

which is a norm on V_h (as $\|f\|_{-1,h} \leq \|f\|_o$ for each $f \in L^2(\Omega)$).

* A scheme is said to be absolutely stable for an assigned value $\mu\Delta t$ if the approximate solution y^n of $dy(t)/dt = \mu y(t)$, $\mu \in C$, generated by a discrete scheme vanishes as $n \to \infty$. The region of absolute stability is the set of all $\mu\Delta t$ for which the method is absolutely stable. A scheme is called A-stable if its region of absolute stability includes the region $\{z \in C \mid Re\, z < 0\}$.

Theorem 9.19: Let us assume that the bilinear form a (.,.) is weakly coercive in V (i.e. $a(T,T) \geq \alpha \|T\|_1^2 \ \forall T \in V, \alpha > 0$), and that the mapping $t \to \|Q(t)\|_o$ is bounded in $\bar{I} = \langle t_o, t_1 \rangle$. When $0 \leq \Theta < \frac{1}{2}$, assume that $\mathcal{T}_h$ is a quasi-uniform family of triangulations ** and that the restriction on the time-step

$$\Delta t(1 + ch^{-2}) \leq 2c_o(1 - 2\Theta)^{-1}c_1^{-2} \tag{9.232}$$

holds, where $c > 0$ is a constant for which $\|\nabla T_h\|_o^2 \leq ch^{-2}\|T_h\|_o^2$, c_o and c_1 are the coerciveness and continuity constants of bilinear form a (.,.). Then T_h^n defined by Equations 9.227 and 9.228 satisfies

$$\left\|T_h^n\right\|_o \leq c_\Theta(\|T_{oh}\|_o + \max_{t \in I}\|Q(t)\|_o),\ n = 0, 1, \ldots, n, \tag{9.233}$$

where $c_\Theta =$ constant > 0 is a non-decreasing function of c_o^{-1}, c_1 and $\underline{T}_h$ and independent of n, Δt and h.

Proof: Let us put $z_h = \Theta T_h^{n+1} + (1-\Theta)T_h^n$ in Equations 9.227 and 9.228. Then

$$\tfrac{1}{2}\left\|T_h^{n+1}\right\|_o^2 - \tfrac{1}{2}\left\|T_h^n\right\|_o^2 + \left(\Theta - \tfrac{1}{2}\right)\left\|T_h^{n+1} - T_h^n\right\|_o^2 + \Delta t a(\Theta T_h^{n+1} + (1-\Theta)T_h^n, \Theta T_h^{n+1} + (1-\Theta)T_h^n)$$
$$= \Delta t(\Theta Q(t_{n+1}) + (1-\Theta)Q(t_n), \Theta T_h^{n+1} + (1-\Theta)T_h^n).$$

Since the bilinear form a (.,.) is coercive, then

$$\left\|T_h^{n+1}\right\|_o^2 - \left\|T_h^n\right\|_o^2 + (2\Theta - 1)\left\|T_h^{n+1} - T_h^n\right\|_o^2 + 2(1-\varepsilon)c_o\Delta t\left\|\Theta T_h^{n+1} + (1-\Theta)T_h^n\right\|_1^2$$
$$\leq \Delta t(2\varepsilon c_o)^{-1}\|\Theta Q(t_{n+1}) + (1-\Theta)Q(t_n)\|_{-1,h}^2.$$

Now let us assume that $\frac{1}{2} \leq \Theta \leq 1$. Then the left hand side $> \left\|T_h^{n+1}\right\|_o^2 - \left\|T_h^n\right\|_o^2$, where we put $\varepsilon = 1$. If $0 \leq \Theta < \frac{1}{2}$, then putting $z_h = T_h^{n+1} - T_h^n$ in Equations 9.227 and 9.228 gives

$$\left\|T_h^{n+1} - T_h^n\right\|_o^2$$
$$= -\Delta t a(\Theta T_h^{n+1} + (1-\Theta)T_h^n, T_h^{n+1} - T_h^n) + \Delta t(\Theta Q(t_{n+1}) + (1-\Theta)Q(t_n), T_h^{n+1} - T_h^n)$$
$$\leq c_1\Delta t\left\|\Theta T_h^{n+1} + (1-\Theta)T_h^n\right\|_1\left\|T_h^{n+1} - T_h^n\right\|_1 + \Delta t\|\Theta Q(t_{n+1}) + (1-\Theta)Q(t_n)\|_{-1,h}\left\|T_h^{n+1} - T_h^n\right\|_1.$$

Since $\left\|\nabla T_h^n\right\|_o^2 \leq ch^{-2}\left\|T_h^n\right\|_o^2$, then we find

$$\left\|T_h^{n+1} - T_h^n\right\|_o^2 \leq \Delta t(1 + ch^{-2})^{\frac{1}{2}}[c_1\left\|\Theta T_h^{n+1} + (1-\Theta)T_h^n\right\|_1 + \|\Theta Q(t_{n+1}) + (1-\Theta)Q(t_n)\|_{-1,h}].$$

For each $\beta > 0$ let

$$c_\beta = [2(1-\varepsilon)c_o - (1-2\Theta)c_1(c_1 + \beta)\Delta t(1 + ch^{-2})],$$

then

$$\left\|T_h^{n+1}\right\|_o^2 - \left\|T_h^n\right\|_o^2 + \Delta t c_\beta\left\|\Theta T_h^{n+1} + (1-\Theta)T_h^n\right\|_1^2 \leq c_{\varepsilon\beta}\Delta t(1 + \Delta t h^{-2})\|\Theta Q(t_{n+1}) + (1-\Theta)Q(t_n)\|_{-1h}^2.$$

Now let us choose ε, β small enough, then $c_\beta > 0$ for $\Delta t(1 + ch^{-2}) < 2c_o(1-2\Theta)^{-1}c_1^{-2}$, and moreover, $1 + \Delta t h^{-2} \leq c_2, c_2 > 0$.

** A family of triangulations $\mathcal{T}_h$, $h > 0$, is said to be quasi-uniform if it is regular and moreover, there exists a constant $c > 0$ such that

$$\min_{\underline{T}_h \in \mathcal{T}_h} h_{\underline{T}_h} \geq ch \quad \forall h > 0.$$

Hence,

$$\left\|T_h^{n+1}\right\|_o^2 - \left\|T_h^n\right\|_o^2 \le c_{\varepsilon\beta}\Delta t \left\|\Theta Q(t_{n+1}) + (1-\Theta)Q(t_n)\right\|_{-1,h}^2 .$$

Let $1 \le k \le n$, k being fixed. Then for $0 \le \Theta < \frac{1}{2}$, $\frac{1}{2} \le \Theta \le 1$, summing up from $n = 0$ to $n = k - 1$, we obtain

$$\left\|T_h^n\right\|_o^2 \le \|T_{oh}\|_o^2 + c\Delta t \sum_{n=0}^{k-1} \|\Theta Q(t_{n+1}) + (1-\Theta)Q(t_n)\|_{-1,h}^2 .$$

Hence Equation 9.233 follows immediately.

Theorem 9.20: Let us assume that the bilinear form a (.,.) is weakly coercive and that $\partial T_h(t_o)/\partial t \in L^2(\Omega)$, $\quad Q \in L^2(\Omega_T)$, $(\partial Q/\partial t) \in L^2(\Omega_T)$.
When $0 \le \Theta < \frac{1}{2}$, assume that $\mathscr{T}_h$ is a quasi-uniform family of triangulations and that for the time-step

$$\Delta t(1 + ch^{-2}) < 2c_o(1 - 2\Theta)^{-1}c_1^{-2}$$

holds. Then

$$\left\|T_h^n - T_h(t_n)\right\|_o \le c_\Theta \Delta t \left(\left\| \frac{\partial T_h}{\partial t}(t_o) \right\|_o^2 + \int_I \left\| \frac{\partial Q(\tau)}{\partial t} \right\|_o^2 \mathrm{d}\tau \right)^{1/2}, \quad n = 0, \ldots, n(h), \tag{9.234}$$

where T_h^n is defined by Equations 9.227 and 9.228, $T_h(t_n)$ by Equations 9.216 and 9.217, $c_\Theta > 0$ is a non-decreasing function of $c_o^{-1}, c_1, \underline{T}_h$ and independent of $n(h)$, Δt, h.

Let $\Theta = \frac{1}{2}$, $\partial^2 Q/\partial t^2 \in L^2(\Omega_T)$, $\partial^2 T_h/\partial t^2(t_o) \in L^2(\Omega)$, then

$$\left\|T_h^n - T_h(t_n)\right\|_o \le c(\Delta t)^2 \left(\left\|\partial^2 T_h(t_o)/\partial t^2\right\|_o^2 + \int_I \left\|\partial^2 Q(\tau)/\partial t^2\right\|_o^2 \right)^{1/2}, \quad n = 0, 1, \ldots, n(h), \tag{9.235}$$

where $c > 0$ is a non-decreasing function of $c_o^{-1}, c_1, \underline{T}_h$ and independent of $n(h)$, Δt and h.

Proof: From Equations 9.216 and 9.217 it follows

$$\begin{aligned}
&\Delta t^{-1}(T_h(t_{n+1}) - T_h(t_n), z_h) + a(\Theta T_h(t_{n+1}) + (1-\Theta)T_h(t_n), z_h)\\
&= (\Delta t^{-1}(T_h(t_{n+1}) - T_h(t_n)) - \Theta\partial T_h(t_{n+1})/\partial t - (1-\Theta)\partial T_h(t_n)/\partial t, z_h) + (\Theta Q(t_{n+1}) + (1-\Theta)Q(t_n), z_h)\\
&= -\Delta t^{-1} \left(\int_{t_n}^{t_{n+1}} [\tau - (1-\Theta)t_{n+1} - \Theta t_n]\partial^2 T_h(\tau)/\partial t^2 \mathrm{d}\tau, z_h) + (\Theta Q(t_{n+1}) + (1-\Theta)Q(t_n), z_h \right).
\end{aligned} \tag{9.236}$$

Putting $e_h^n \equiv T_h^n - T_h(t_n)$, then from Equations 9.227, 9.228 and 9.236 we find

$$\begin{aligned}
&\Delta t^{-1}(e_h^{n+1} - e_h^n, z_h) + a(\Theta e_h^{n+1} + (1-\Theta)e_h^n, z_h)\\
&= \Delta t^{-1} \left(\int_{t_n}^{t_{n+1}} [\tau - (1-\Theta)t_{n+1} - \Theta t_n]\partial^2 T_h(\tau)/\partial t^2, z_h \right) \quad \forall z_h \in V_h.
\end{aligned} \tag{9.237}$$

Hence, putting $z_h = \Theta e_h^{n+1} + (1-\Theta)e_h^n$ and similarly to the proof of Theorem 9.19 we obtain

$$\begin{aligned}
&\left\|e_h^{n+1}\right\|_o^2 - \left\|e_h^n\right\|_o^2 + (2\Theta - 1)\left\|e_h^{n+1} - e_h^n\right\|_o^2 + 2(1-\varepsilon)c_o\Delta t \left\|\Theta e_h^{n+1} + (1-\Theta)e_h^n\right\|_1^2\\
&\le (\Delta t)^2(2\varepsilon c_o)^{-1} \int_{t_n}^{t_{n+1}} \left\|\partial^2 T_h(\tau)/\partial t^2\right\|_{-1,h}^2 .
\end{aligned}$$

We then distinguish the cases $0 \le \Theta < \frac{1}{2}$ and $\frac{1}{2} \le \Theta \le 1$. Let $\frac{1}{2} \le \Theta \le 1$, then the left hand side $> \left\|e_h^{n+1}\right\|_o^2 - \left\|e_h^n\right\|_o^2$. Let $0 \le \Theta < \frac{1}{2}$. To evaluate $\left\|e_h^{n+1} - e_h^n\right\|_o^2$ we use the technique of

Theorem 9.19. Since $\Delta t(1+ch^{-2}) < 2c_\circ[(1-2\Theta)c_1^2]^{-1}$, then we finally obtain

$$\left\|e_h^{n+1}\right\|_o^2 - \left\|e_h^n\right\|_o^2 \le C(\Delta t)^2 \int_{t_n}^{t_{n+1}} \left\|\partial^2 T_h(\tau)/\partial t^2\right\|_{-1,h}^2 \mathrm{d}\tau.$$

Let $1 \le k \le n(h)$, k being fixed. Summing up from $n=0$ to $k-1$ and as $e_h^o = 0$, we obtain

$$\left\|e_h^k\right\|_o^2 \le c(\Delta t)^2 \int_{t_o}^{t_k} \left\|\partial^2 T_h(\tau)/\partial t^2\right\|_{-1,h}^2 \mathrm{d}\tau.$$

Now, differentiating Equation 9.216 with respect to t, we obtain

$$(\partial^2 T_h(t)/\partial t^2, z_h) + a(\partial T_h(t)/\partial t, z_h) = (\partial Q(t)/\partial t, z_h) \quad \forall z_h \in V_h. \tag{9.238}$$

Hence

$$\left\|\partial^2 T_h(t)/\partial t^2\right\|_{-1,h} \le c(\|\partial T_h(t)/\partial t\|_1 + \|\partial Q(t)/\partial t\|_{-1,h}).$$

Then for each $t \in I$, t being fixed, we put $z_h = \partial T_h(t)/\partial t$ in Equation 9.238. Then applying the technique of the proof of Theorem 9.16, we find

$$\int_{t_o}^{t_k} \|\partial T_h(\tau)/\partial t\|_1^2 \,\mathrm{d}\tau \le c(\|\partial T_h(t_o)/\partial t\|_o^2 + \int_{t_o}^{t_k} \|\partial Q(\tau)/\partial t\|_{-1,h}^2 \,\mathrm{d}\tau),$$

and hence Equation 9.234 follows immediately.

Let $\Theta = \frac{1}{2}$. Then the first term on the right hand side of Equation 9.236 becames

$$\begin{aligned}(\Delta t^{-1}&(T_h(t_{n+1}) - T_h(t_n)) - \Theta\partial T_h(t_{n+1})/\partial t - (1-\Theta)\partial T_h(t_n)/\partial t, z_h)\\ &= (2\Delta t)^{-1}\left(\int_{t_n}^{t_{n+1}} (t_{n+1}-\tau)(t_n-\tau)\partial^3 T_h(\tau)/\partial t^3 \mathrm{d}\tau, z_h\right).\end{aligned}$$

Similarly as above we find

$$\left\|e_h^k\right\|_o^2 \le c(\Delta t)^4 \int_{t_o}^{t_k} \left\|\partial^3 T_h(\tau)/\partial t^3\right\|_{-1,h}^2 \mathrm{d}\tau.$$

Differentiating Equation 9.238 with respect to t we obtain

$$(\partial^3 T_h(t)/\partial t^3, z_h) + a(\partial^2 T_h(t)/\partial t^2, z_h) = (\partial^2 Q(t)/\partial t^2, z_h) \quad \forall z_h \in V_h$$

and proceeding as above, the estimate in Equation 9.235 follows immediately.

If the bilinear form is symmetric we can prove the following results.

Theorem 9.21: Let the bilinear form a (.,.) be symmetric and coercive. Let $T_o \in V$ and let $T' \in L^1(I;V)$, $\partial^2 T/\partial t^2 \in L^1(I;L^2(\Omega))$.

Assume that the map $t \to \|Q(t)\|_o$ is bounded in $\bar{I} = \langle t_o, t_1\rangle$. Moreover, when $0 \le \Theta < \frac{1}{2}$ the time-step Δt satisfies

$$\Delta t \mu_{n(h),h} \le 2(1-2\Theta)^{-1},$$

where $\mu_{n(h),h} \ge \ldots \ge \mu_{1,h} > c_o$ is a sequence of eigenvalues corresponding to an $L^2(\Omega)$-orthogonal basis of eigenvectors $w_{i,h} \in V_h$, $i = 1, \ldots, n(h)$ (i.e. any $z_h = \sum_{i=1}^{n(h)} (z_h, w_{i,h})w_{i,h}$); then the solution T_h^n of the fully discrete problem $(\mathscr{P}_{fd})_{vh}$ satisfies

$$\left\|T_h^n\right\|_o \le \|T_{oh}\|_o + \mathrm{meas}(\underline{T}_h) \max_{t \in I} \|Q(t)\|_o, \; n = 0, 1, \ldots, n(h),$$

i.e the stability condition.

Moreover, let us assume that the approximation property

$$\left\|z - \Pi_{1h}^k(z)\right\|_o + h\left\|z - \Pi_{1h}^k(z)\right\|_1 \le ch^{k+1}|z|_{k+1} \quad \forall z \in H^{k+1}(\Omega)$$

is satisfied and let $T_o \in H^{k+1}(\Omega)$, $\partial T/\partial t \in L^1(I; H^{k+1}(\Omega))$. Then

$$\left\|T_h^n - T(t_n)\right\|_o \leq \|T_{oh} - T_o\|_o + ch^{k+1}(|T_o|_{k+1} + \int_{t_o}^{t_n} |\partial T(\tau)/\partial t|_{k+1} \, d\tau$$

$$+ c_\Theta \Delta t \int_{t_o}^{t_n} \left\|\partial^2 T(\tau)/\partial t^2\right\|_o d\tau, \quad n = 0, 1, \ldots, n(h).$$

For $\Theta = \frac{1}{2}$, under the further assumption that $\partial^3 T/\partial t^3 \in L^1(I; L^2(\Omega))$, then

$$\left\|T_h^n - T(t_n)\right\|_o \leq \|T_{oh} - T_o\|_o + ch^{k+1}(|T_o|_{k+1} + \int_{t_o}^{t_n} |\partial T(\tau)/\partial t|_{k+1} \, d\tau$$

$$+ \frac{1}{8}(\Delta t)^2 \int_{t_o}^{t_n} \left\|\partial^3 T(\tau)/\partial t^3\right\| d\tau,$$

where $c > 0$ is independent of $n(h)$, Δt and h.

The proof is a modification of that of Quarteroni and Valli (1991).

9.4.4. Evolution Geothermal Models: Non-Linear Case

9.4.4.1. Mathematical analysis of the continuous part of the problem

We shall assume that the investigated part of the Earth occupies the domain Ω, about which we assume that it is bounded in R^N, $N = 2, 3$. About its boundary $\partial\Omega$ we shall assume that it is Lipschitzian and that $\partial\Omega = \overline{\Gamma}_1 \cup \overline{\Gamma}_2 \cup \overline{\Gamma}_3$. Let $\overline{I} = \langle t_o, t_1 \rangle$ be the time interval. Moreover, we shall assume that the physical parameters, the thermal conductivity $\kappa = \kappa(T(\boldsymbol{x}, t))$ and thermal capacity $c = c(T(\boldsymbol{x}, t)) = \rho(T(\boldsymbol{x}, t))c_e(T(\boldsymbol{x}, t))$, depend on the temperature $T = T(\boldsymbol{x}, t)$. Then we have to solve the following problem.

Problem $(\mathscr{P})$**:** Find a temperature $T(\boldsymbol{x}, t)$ satisfying

$$c(T(\boldsymbol{x}, t))\frac{\partial T(\boldsymbol{x}, t)}{\partial t} = \text{div } (\kappa(T(\boldsymbol{x}, t))\text{grad } T(\boldsymbol{x}, t)) + Q(\boldsymbol{x}, t, T(\boldsymbol{x}, t)) \text{ in } \Omega, \quad t \in I, \tag{9.239}$$

the boundary conditions

$$T(\boldsymbol{x}, t) = T_b(\boldsymbol{x}, t) \text{ for } \boldsymbol{x} \in \Gamma_1 \cup \Gamma_2, \quad t \in I, \tag{9.240}$$

$$\frac{\partial T(\boldsymbol{x}, t)}{\partial n} = q(\boldsymbol{x}, t, T(\boldsymbol{x}, t) \text{ for } \boldsymbol{x} \in \Gamma_3, \quad t \in I \tag{9.241}$$

and initial condition

$$T(\boldsymbol{x}, t_o) = T_o(\boldsymbol{x}) \text{ for } \boldsymbol{x} \in \Omega. \tag{9.242}$$

We shall assume that $c(T(\boldsymbol{x}, t))$ and $\kappa(T(\boldsymbol{x}, t))$ are piecewise continuously differentiable functions bounded from below as well as above by positive constants, i.e.

$$0 < c_1 \leq c(T(\boldsymbol{x}, t)) \leq c_2, \quad \forall T \in (-\infty, \infty), \tag{9.243}$$

$$0 < \kappa_1 \leq \kappa(T(\boldsymbol{x}, t)) \leq \kappa_2, \quad \forall T \in (-\infty, \infty) \tag{9.244}$$

and thermal sources $Q(\boldsymbol{x}, t, T(\boldsymbol{x}, t))$ satisfy

$$|Q(\boldsymbol{x}, t, T_1) - Q(\boldsymbol{x}, t, T_2)| \leq c\,|T_1 - T_2|\,, \tag{9.245}$$

$$|Q(\boldsymbol{x}, t_1, T) - Q(\boldsymbol{x}, t_2, T)| \leq c\,|t_1 - t_2|\,, \quad \boldsymbol{x} \in \overline{\Omega},\ t \in I,\ T_1, T_2 \in (-\infty, \infty). \tag{9.246}$$

Moreover, we shall assume that $T_o(\boldsymbol{x}) \in H^2(\Omega)$; $T_b(\boldsymbol{x}, t)$ is a continuous function on $\overline{\Omega} \times I$, $\boldsymbol{n}$ is the outward normal to $\Gamma_1 \cup \Gamma_2$, $q(\boldsymbol{x}, t, T(\boldsymbol{x}, t))$ is continuous for $T \in (-\infty, \infty)$, $\boldsymbol{x} \in \overline{\Omega}$, $t \in I$ and

$$q(\boldsymbol{x}, t, T_2) - q(\boldsymbol{x}, t, T_1) \geq 0 \quad \forall T_1, T_2, \quad T_1 \geq T_2. \tag{9.247}$$

To analyse the problem $(\mathscr{P})$, we shall introduce the following spaces and notations:

$$H^k(\Omega) = \{z \mid z \in L^2(\Omega), D^\alpha z \in L^2(\Omega), \quad \forall\,|\alpha| \leq k\},$$

$$W^{k,\infty}(\Omega) = \{z \mid z \in L^\infty(\Omega), D^\alpha z \in L^\infty(\Omega), \quad \forall\,|\alpha| \leq k\},$$

$$V = \{z \,\big|\, z \in H^1(\Omega), z\big|_{\Gamma_1 \cup \Gamma_2} = 0 \text{ in the sense of traces}\}.$$

Let X be a Banach space, $L^\infty(I; X) = L^\infty(X)$ denotes the space of functions $z(t)$: $t \to z(t) \in X$, $t \in I$, which are measurable and such that

$$\operatorname*{ess\,sup}_{t \in I} \|z(t)\|_X = \|z\|_{L^\infty(X)} < \infty.$$

We introduce the enthalpy by

$$H(T) = \int_0^T c(\tau)\mathrm{d}\tau, \tag{9.248}$$

and the Kirchhoff transformation by

$$G(T) = \int_0^T \kappa(\tau)\mathrm{d}\tau. \tag{9.249}$$

Let us assume $T(\boldsymbol{x}, t)$ to be sufficiently smooth, i.e. that $G(T) \in L^\infty(W^{2,\infty}(\Omega))$, $\partial G(T)/\partial t \in L^\infty(W^{2,\infty}(\Omega))$, $\partial^2 H(T)/\partial t^2 \in L^\infty(L^\infty(\Omega))$, then multiplying Equation 9.239 by a function $z \in V$ and applying the Green's theorem we obtain a weak formulation in the enthapy formulation, denoted it shown below.

Problem $(\mathscr{P})_v$: Find functions $H(\boldsymbol{x}, t)$ and $T(\boldsymbol{x}, t)$ satisfying

$$\left(\frac{\partial H(T)}{\partial t}, z\right) + a(G(T), z) + \int_{\Gamma_3} q(\boldsymbol{x}, t, T(\boldsymbol{x}, t))z\mathrm{d}s = \int_\Omega Q(\boldsymbol{x}, t, T(\boldsymbol{x}, t))z\mathrm{d}\boldsymbol{x}, \quad \forall z \in V,\ t \in I, \tag{9.250}$$

or

$$(H'(T), z) + a(G(T), z) + \langle q, z\rangle = (Q, z) \quad \forall z \in V, \quad t \in I, \tag{9.251}$$

where we denoted $(H', z) = \int_\Omega H' z\mathrm{d}\boldsymbol{x}$, $a(G, z) = \int_\Omega \nabla G \nabla z \mathrm{d}\boldsymbol{x}$, $(Q, z) = \int_\Omega Q z \mathrm{d}\boldsymbol{x}$, $\langle q, z\rangle = \int_{\Gamma_3} qz\mathrm{d}s$.

Then we can prove the following theorem.

Theorem 9.22: The problem $(\mathscr{P})$ has a unique solution.

9.4.4.2. *Numerical solution*

Numerical solution of the problem can be based on several techniques, for example the predictor-corrector method, the Crank-Nicholson method, the Rothe method, the discrete approximation of a formulation in variational inequalities (which was used in the coupled contact-two-phase Stefan problem in Chapter 7), etc.. In the following we shall use the results of Zlámal (1980), as that method gives good results for rapid variation of heat capacity and when Δt is not very small.

Therefore, solving the problem $(\mathscr{P})$ numerically, we shall introduce a family $\mathscr{T}_h$ of triangulations consisting of triangles $\underline{T}_h$ in the $2D$ case and tetrahedra $\underline{T}_h$ in the $3D$ case, with vertices lying in $\overline{\Omega}$, where we denote by $\underline{T}_h$ a triangle in the $2D$ case or a tetrahedron in the $3D$ case, respectively, i.e. an element of $\mathscr{T}_h$ for which $h = \text{diam}(T_h)$, $\rho_{\underline{T}_h} = \sup\{\text{diam}(S);\ S$ being a ball contained in $\underline{T}_h\}$, $\overline{\Omega}_h = \cup_{\underline{T}_h \in \mathscr{T}_h} \underline{T}_h$ (but in general $\Omega_h \neq \Omega$), $\partial\Omega_h = \cup_{i=1}^{3} \Gamma_{ih}$, $i = 1, 2, 3$. Finally, we assume that the family $\mathscr{T}_h$ is regular, i.e.

(i) there exists a constant $\beta > 0$ such that $\rho^{-1} h \mid_{\underline{T}_h} \leq \beta\, \forall \underline{T}_h \in \cup_h \mathscr{T}_h$;

(ii) the quantity $h = \max_{\underline{T}_h \in \mathscr{T}_h} h \mid_{\underline{T}_h}$, $h \to 0$,
and is of acute type if

(iii) all angles of the triangles and all angles made by adjacent faces and edges of the tetrahedra are not greater than $\frac{1}{2}\pi$.

Then due to each triangulation $\mathscr{T}_h$ we associate the finite dimensional space

$$W_h = \{z_h \mid z_h \in C^o(\overline{\Omega}_h),\ z_h \text{ is piecewise linear on } \mathscr{T}_h\}$$

and

$$V_h = \{z_h \mid z_h \in W_h,\ z_h|_{\Gamma_{1h} \cup \Gamma_{2h}} = 0\}.$$

Let $\{x_k\}_{k=1}^{s}$ be the set of all nodes of $\mathscr{T}_h$ and $\{x_k\}_{k=1}^{r}$ the set of nodes from $\Omega_h \cup \Gamma_3$. Let us denote by $z_k(\boldsymbol{x})$ the basis function associated with the node x_k ($z_k(\boldsymbol{x}) \in W_h$, $z_k(\boldsymbol{x}^i)$ for $k \neq i$, $z_k(\boldsymbol{x}^k) = 1$). Let

$$t_i = i\Delta t, \quad i = 1, \ldots, n; n < \Delta t^{-1}(t_1 - t_o), \quad 0 < t_1 < \ldots . < t_n.$$

Let the value $T^i = T(\boldsymbol{x}, t_i)$ of the exact solution $T(\boldsymbol{x}, t)$ at $t = t_i$ be approximated by

$$\Theta^i = \sum_{j=1}^{s} T^i_j z_j(\boldsymbol{x}), \quad \Theta^i_j = T_b(\boldsymbol{x}^j, t_i), \quad j = r+1, \ldots, s. \tag{9.252}$$

Then for approximations of $H^i = H(T^i)$, $G^i = G(T^i)$, $q^i = q(\boldsymbol{x}, t_i, T^i)$, $Q^i = Q(\boldsymbol{x}, t_i, T^i)$ we take

$$\mathscr{H}^i = \sum_{j=1}^{s} H(\Theta^i_j) z_j(\boldsymbol{x}), \tag{9.253}$$

$$Y^i = \sum_{j=1}^{s} G(\Theta^i_j) z_j(\boldsymbol{x}), \tag{9.254}$$

$$\Psi^i = \sum_{j=1}^{s} q(\boldsymbol{x}^j, t_i, \Theta^i_j) z_j(\boldsymbol{x}), \tag{9.255}$$

$$W^i = \sum_{j=1}^{s} Q(\boldsymbol{x}^j, t_i, \Theta^i_j) z_j(\boldsymbol{x}), \tag{9.256}$$

representing interpolates of the functions H^i, G^i, q^i and Q^i. For approximations of scalar products, bilinear and linear forms we shall use the following quadrature formulas

$$\int_\Omega f(\boldsymbol{x}) \mathrm{d}\boldsymbol{x} = (m+1)^{-1} \mathrm{meas}(\Omega) \sum_{j=1}^{m+1} f(x^j), \tag{9.257}$$

$$\int_\Gamma g(\boldsymbol{x}) \mathrm{d}s = (m+1)^{-1} \mathrm{meas}(\Gamma) \sum_{j=1}^{m+1} g(x^j). \tag{9.258}$$

Then for the discrete analog of Equation 9.251 we put $t = t_{i+1}$ and replace $\partial H^{i+1}/\partial t \sim (\Delta t)^{-1}(\mathscr{X}^{i+1} - \mathscr{X}^i)$, $G^{i+1} \sim Y^{i+1}$, $q^{i+1} \sim \Psi^{i+1}$, $Q^i \sim W^i$, $(T, z) \sim (T, z)_h$, $a(T, z) \sim a_h(T, z)$, $T_o \sim T_{oI}$.

Then Equation 9.251 yields

$$(\mathscr{X}^{i+1} - \mathscr{X}^i, z)_h + \Delta t a_h(Y^{i+1}, z) + \Delta t \langle \Psi^{i+1}, z\rangle_h = \Delta t (W^i, z)_h$$
$$\forall z \in V_h, \quad i = 0, 1, \ldots, n-1, \tag{9.259}$$
$$\Theta^o = T_{oI}. \tag{9.260}$$

Let us denote by $\{x^i\}_{i=1}^{r'}$ nodes from Ω_h and by

$$\boldsymbol{X} = \{X_j\}_{j=1}^{r}, \quad X_j = \Theta_j^{i+1},$$

$$\boldsymbol{H}(\boldsymbol{X}) = (H(X_1), \ldots, H(X_r))^T, \quad \boldsymbol{G}(\boldsymbol{X}) = (G(X_1, \ldots, G(X_r))^T,$$

$$\boldsymbol{q}(\boldsymbol{X}) = (0, \ldots, 0, q(x^{r'+1}, t_{i+1}, X_{r'+1}), \ldots, q(x^r, t_{i+1}, X_r))^T,$$

$$C = (C_{ij}) = \{(z_i, z_j)_h\}_{i,j=1}^{r}, \; A = (A_{ij}) = \{a_h(z_i, z_j)\}_{i,j=1}^{r}, \; B = (B_{ij}) = \{\langle z_i, z_j\rangle_h\}_{i,j=1}^{r}.$$

The matrices C, A and B are $r \times r$ band matrices, C and A are positive definite and B is positive semi-definite. It can be shown that due to the special choice of the quadrature formula (Equations 9.257 and 9.258), matrices C and B are diagonal. Then Equations 9.259 and 9.260 lead to the non-linear system of equations

$$C\boldsymbol{H}(\boldsymbol{X}) + \Delta t A \boldsymbol{G}(\boldsymbol{X}) + \Delta t B \boldsymbol{q}(\boldsymbol{X}) = \boldsymbol{f}. \tag{9.261}$$

Let us introduce new variables

$$y_j = G(X_j), \quad j = 1, \ldots, r \tag{9.262}$$

and let the diagonal entries of matrices C and B be denoted by

$$C_j \equiv C_{jj} > 0, \quad B_j \equiv B_{jj} \geq 0, \quad j = 1, \ldots, r.$$

Due to the assumptions in Equations 9.243 and 9.244 the mapping in Equation 9.262 maps R^r one-to-one to R^r. Putting

$$\sigma_j(\xi) = C_j H(\xi) + \Delta t B_j q(x^j, t_{i+1}, \xi), \ \boldsymbol{F}(\boldsymbol{y}) = (F_1(y_1), \ldots, F_r(y_r))^T, \ F_j(\boldsymbol{y}) = \sigma_j G^{-1}. \tag{9.263}$$

Then from Equations 9.261 and 9.263 we find

$$\boldsymbol{F}(\boldsymbol{y}) + \Delta t A \boldsymbol{y} = \boldsymbol{f}. \tag{9.264}$$

From the theory (see Ortega and Rheiboldt (1970)) it is known that Equation 9.264 represents a necessary condition for the minimum of the functional

$$J(\boldsymbol{y}) = \frac{1}{2}\Delta t \boldsymbol{y}^T A \boldsymbol{y} - \boldsymbol{f}^T \boldsymbol{y} + \sum_{j=1}^{r} \int_o^{y_j} F_j(\xi) \mathrm{d}\xi. \tag{9.265}$$

Theorem 9.23: The functional $J(.)$ is uniformly convex on R^r and thus it has a unique global minimizer.

Proof: Due to Ortega and Rheiboldt (1970), p. 141-145 and Equations 9.243, 9.244 and 9.248, and as the matrices C and A are positive definite, B is positive semi-definite and since due to the quadrature formula used the matrices C and B are diagonal matrices, then the G-derivative of J is a uniformly monotone mapping on R^r. Then J is convex on R^r if and only if $(J'(\boldsymbol{y}) - J'(\boldsymbol{x}))(\boldsymbol{y} - \boldsymbol{x}) \geq 0 \, \forall \boldsymbol{x}, \boldsymbol{y} \in R^r$. Moreover, it is uniformly convex on R^r as $(J'(\boldsymbol{y}) - J'(\boldsymbol{x}))(\boldsymbol{y} - \boldsymbol{x}) \geq 2c \, \|\boldsymbol{y} - \boldsymbol{x}\|^2 \; \forall \boldsymbol{x}, \boldsymbol{y} \in R^r$, $c > 0$. As it is uniformly convex on R^r, then it has a unique local and global minimizer (see Ortega and Rheiboldt (1970), p. 106). Hence, if J has a minimum at the point $\boldsymbol{y} = \boldsymbol{y}^o$, then $X_j^o = G^{-1}(y_j^o)$ is the solution of Equation 9.261.

Theorem 9.17 proved the existence of the unique local and global minimum. Therefore, if $z \in W_h$ attains a local minimum (maximum) at a point from $\overline{\Omega}_h$ then it also attains this minimum (maximum) at a node. Then z attains a local minimum (maximum) at a node x^j iff the values of z at the neighbouring nodes are not smaller (greater) than $z(x^j)$.

Lemma 9.6: Let us assume that the triangulation $\mathcal{T}_h$ is of acute type. If $z \in W_h$ attains a local minimum (maximum) at the node x^j then

$$a_h(z, z_j) \leq 0 (\geq 0),$$

where z_j is the basis function associated with the node x^j.
For the proof see Zlámal (1980).

Remark: From this lemma the discrete maximum-minimum principle can be deduced. We have, $\{\Theta_i\}_{i=1}^n$ are functions from W_h which satisfy

$$(\mathscr{X}^{i+1} - \mathscr{X}^{i}, z)_h + \Delta t a_h(Y^{i+1}, z) + \Delta t \langle \Psi^{i+1}, z \rangle_h = \Delta t (W^i, z)_h$$
$$\forall z \in V_h, i = 0, 1, \ldots, n-1.$$

If $Q \geq 0 (Q \leq 0)$ then

$$\Theta_j^i \geq \min_{(k,l)\in S} \Theta_k^l \left(> \max_{(k,l)\in S} \Theta_k^l \right), \quad j = 1, \dots, s, \quad i = 1, \dots, n, \tag{9.266}$$

where S is the set of ordered couples (k, l), $k = 1, \dots, s$; $l = 0, 1, \dots, n-1$ (where s denotes the number of all nodes of the triangulation $\mathcal{T}_h$) such that either $\boldsymbol{x}^k \in \Gamma_{1h} \cup \Gamma_{2h}$ and $l = 0, 1, \dots, n-1$ or $x^k \in \Omega_h$ and $l = 0$.

Remark: In the maximum norm we have the following error estimate. Let $w \in W^{2,\infty}(\Omega)$, then

$$\|w - w_h\|_{L^\infty(\Omega\cap\Omega_h)} \leq c\,\|w\|_{W^{2,\infty}(\Omega)}\,\vartheta(h). \tag{9.267}$$

Theorem 9.24: Let us assume the case with the triangulations $\mathcal{T}_h$ of acute type. Let us assume that the solution $T(\boldsymbol{x}, t)$ of the problem $(\mathcal{P})$ is sufficiently smooth, i.e. we assume that

$$G(T), \frac{\partial G(T)}{\partial t} \in L^\infty(W^{2,\infty}(\Omega)), \quad \partial^2 H(T)/\partial t^2 \in L^\infty(L^\infty(\Omega)).$$

Then

$$\left\|T^i - \Theta^i\right\|_{L^\infty(\Omega\cap\Omega_h)} \leq c(\vartheta(h) + \Delta t), \quad i = 1, 2, \dots, n, \tag{9.268}$$

where $c > 0$ does not depend on $\vartheta(h)$, Δt and i.

Proof: From Equations 9.251 and 9.241 it follows that $G^i = G(T^i) = G(T(\boldsymbol{x}, t_i))$ satisfies

$$a(G^i, z) + \langle q^i, z\rangle = (Q^i - (H')^i, z) \quad \forall z \in V, \tag{9.269}$$

$$G^i\,|_{\Gamma_1\cup\Gamma_2} = G(T_b^i(\boldsymbol{x}^j))\,|_{\Gamma_1\cup\Gamma_2} \quad \forall \boldsymbol{x}^j \in \Gamma_{1h} \cup \Gamma_{2h}. \tag{9.270}$$

Let $g^i \in W_h$ be the approximate solution of Equation 9.269. Then it satisfies

$$a_h(g^i, z) + \langle q^i, z\rangle_h = (Q^i - (H')^i, z)_h \quad \forall z \in V_h, \tag{9.271}$$

$$g^i(\boldsymbol{x}^j) = G(T_b^i(\boldsymbol{x}^j)) \quad \forall \boldsymbol{x}^j \in \Gamma_{1h} \cup \Gamma_{2h}. \tag{9.272}$$

Since we assume that $G(T) \in L^\infty(W^{2,\infty}(\Omega))$, then from Equation 9.267 we obtain

$$\left\|G^i - g^i\right\|_{L^\infty(\Omega\cap\Omega_h)} \leq c\vartheta(h). \tag{9.273}$$

Hence

$$\left\|G^{i+1} - G^i - (g^{i+1} - g^i)\right\|_{L^\infty(\Omega\cap\Omega_h)} \leq c\left\|G^{i+1} - G^i\right\|_{W^{2,\infty}(\Omega)} \vartheta(h) \leq c\Delta t\vartheta(h). \tag{9.274}$$

Since $(w, z)_h = (w_I, z)_h$, $\langle w, z\rangle_h = \langle w_I, z\rangle_h$, $z \in W_h$, w defined on $\overline{\Omega}_h$, then Equation 9.271 yields

$$(H_I^{i+1} - H_I^i, z)_h + \Delta t a_h(g^{i+1}, z)_h + \Delta t\langle q_I^{i+1}, z\rangle_h = \Delta t(\Delta t^{-1}[H^{i+1} - H^i] - (H')^{i+1}, z)_h + \Delta t(Q^{i+1}, z)_h$$
$$= \Delta t(r^i, z)_h + \Delta t(Q^{i+1}, z)_h \quad \forall z \in V_h, \tag{9.275}$$

and

$$\left|r_j^i\right| \leq c\Delta t, \quad j = 1, \dots, s; i = 0, \dots, n-1; \quad r_j^i = r(\boldsymbol{x}^j, t_i), \tag{9.276}$$

as all nodes of $\mathcal{T}_h$ lie in $\overline{\Omega}$, $\partial^2 H(T)/\partial t^2 \in L^\infty(L^\infty(\Omega))$ and the implicit Euler method is used.

Let us subtract Equation 9.275 from Equation 9.259 and put

$$\boldsymbol{x}^i = H_I^i - \boldsymbol{x}^i, \quad \gamma^i = g^i - Y^i, \quad \eta^i = q_I^i - \Psi^i, \quad e^i = T_I^i - \Theta^i, \tag{9.277}$$

then we obtain

$$(\boldsymbol{x}^{i+1}-\boldsymbol{x}^{i},z)_h+\Delta t a_h(\gamma^{i+1},z)+\Delta t\langle\eta^{i+1},z\rangle_h=\Delta t(W^i-Q^i,z)_h-\Delta t(r^i,z)_h \quad \forall z\in V_h. \tag{9.278}$$

By means of Equations 9.245, 9.246 and 9.256 we estimate $W^i - Q^{i+1}$, i.e.

$$\begin{aligned}Q(\boldsymbol{x}^j,t_i,\Theta^i_j)-Q(\boldsymbol{x}^j,t_{i+1},T^{i+1}_j)&=Q(\boldsymbol{x}^j,t_i,\Theta^i_j)-Q(\boldsymbol{x}^j,t_i,T^i_j)+Q(\boldsymbol{x}^j,t_i,T^i_j)-Q(\boldsymbol{x}^j,t_{i+1},T^i_j)\\&\quad+Q(\boldsymbol{x}^j,t_{i+1},T^i_j)-Q(\boldsymbol{x}^j,t_{i+1},T^{i+1}_j)\\&=O\left(\left|e^i_j\right|+\Delta t\right).\end{aligned} \tag{9.279}$$

Since due to the mean-value theorem

$$\gamma^i_j=g^i_j-G(T^i_j)+G(T^i_j)-G(\Theta^i_j)=O(\vartheta(h))+\kappa(\tau^i_j)e^i_j,$$

then Equation 9.273 yields

$$\left|e^i_j\right|\le c\left(\vartheta(h)+\left|\gamma^i_j\right|\right) \tag{9.280}$$

and therefore

$$\left|W^i_j-Q\left(\boldsymbol{x}^j,t_{i+1},T^{i+1}_j\right)\right|\le c\left(\vartheta(h)+\Delta t+\left|\gamma^i_j\right|\right). \tag{9.281}$$

Then Equation 9.278 yields

$$(\boldsymbol{x}^{i+1}-\boldsymbol{x}^{i},z)_h+\Delta t a_h(\gamma^{i+1},z)+\Delta t\langle\eta^{i+1},z\rangle_h=\Delta t(r^i,z)_h \quad \forall z\in V_h, \tag{9.282}$$

$$\left|r^i_j\right|\le c\left(\delta+\left|\gamma^i_j\right|\right),\qquad \delta=\vartheta(h)+\Delta t.$$

We shall denote $\Delta\boldsymbol{x}^i_j=\boldsymbol{x}^{i+1}_j-\boldsymbol{x}^i_j$, $\Delta\gamma^i_j=\gamma^{i+1}_j-\gamma^i_j$, $\Delta e^i_j=e^{i+1}_j-e^i_j$, etc.. Then

$$\Delta\gamma^i_j=\Delta g^i_j-\Delta G(T^i_j)+\Delta G(T^i_j)-\Delta G(\Theta^i_j)=O(\Delta t\vartheta(h))+\Delta G(T^i_j)-\Delta G(\Theta^i_j). \tag{9.283}$$

The mean-value theorem gives

$$\begin{aligned}\Delta G(T^i_j)-\Delta G(\Theta^i_j)&=\kappa(\xi^i_j)\Delta T^i_j-\kappa(\zeta^i_j)\Delta\Theta^i_j=\kappa(\zeta^i_j)\Delta e^i_j+(\kappa(\xi^i_j)-\kappa(\zeta^i_j))\Delta T^i_j\\&=\kappa^i_j\Delta e^i_j+(\kappa(\xi^i_j)-\kappa(\zeta^i_j))O(\Delta t),\end{aligned}$$

where $\xi^i_j\in\langle T^i_j,T^{i+1}_j\rangle$, $\zeta^i_j\in\langle\Theta^i_j,\Theta^{i+1}_j\rangle$, $\kappa^i_j=\kappa(\zeta^i_j)$, and where numbers ξ^i_j,ζ^i_j are defined as

$$\xi^i_j=(1-\alpha)T^i_j+\alpha T^{i+1}_j,\quad \zeta^i_j=(1-\beta)\Theta^i_j+\beta\Theta^{i+1}_j,\quad 0<\alpha<1,\quad 0<\beta<1.$$

Hence $\xi^i_j-\zeta^i_j=(1-\beta)e^i_j+\beta e^{i+1}_j+(\alpha-\beta)\Delta T^i_j$ and $\Delta\gamma^i_j=\kappa^i_j\Delta e^i_j+\Delta t O\left(\delta+\left|e^i_j\right|+\left|e^{i+1}_j\right|\right)$.
In similar way we find

$$\Delta\boldsymbol{x}^i_j=c^i_j\Delta e^i_j+\Delta t O\left(\Delta t+\left|e^i_j\right|+\left|e^{i+1}_j\right|\right). \tag{9.284}$$

Then Equations 9.284, 9.283, 9.280, 9.243 and 9.244 yield

$$\Delta\boldsymbol{x}^i_j=c^i_j\Delta\gamma^i_j+\Delta t O\left(\delta+\left|\gamma^i_j\right|+\left|\gamma^{i+1}_j\right|\right),\quad 0<\kappa_2^{-1}c_1\le d^i_j\le\kappa_1^{-1}c_2. \tag{9.285}$$

Since γ^i is piecewise linear, instead of $\|\vec{\gamma}^i\|_{L^\infty(\Omega_h)}$ we can estimate the term $\max_j\left|\gamma^i_j\right|$ only. We denote $\vec{\gamma}^{i+1}=(\gamma^i_1,\ldots,\gamma^i_s)^T$, $\|\vec{\gamma}^{i+1}\|_\infty=\max_j\left|\gamma^i_j\right|$. Let $\|\vec{\gamma}^{i+1}\|_\infty=\left|\gamma^{i+1}_j\right|$ and firstly let $\gamma^{i+1}_j e^{i+1}_j\ge$

0. If $\gamma_j^{i+1} > 0$ then $e_j^{i+1} \geq 0$ and due to Equation 9.247 we find that $\eta_j^{i+1} \geq 0$. Putting $z = z_j$ in Equation 9.282, applying Lemma 9.6 and Equation 9.285, then we find

$$C_j d_j^i (\gamma_j^{i+1} - \gamma_j^i) = C_j d_j^i \Delta\gamma_j^i \leq c C_j \Delta t \left(\delta + \left\|\vec{\gamma}^i\right\|_\infty + \gamma_j^{i+1}\right)$$

and then

$$\left\|\vec{\gamma}^{i+1}\right\|_\infty \leq (1 + c\Delta t) \left\|\vec{\gamma}^i\right\|_\infty + c\Delta t\delta. \tag{9.286}$$

Similarly, Equation 9.286 can be proved for $\gamma_j^{i+1} < 0$; if $\gamma_j^{i+1} = 0$ then $\left\|\vec{\gamma}^{i+1}\right\|_\infty = 0$.

Secondly, let $\gamma_j^{i+1} e_j^{i+1} < 0$. If $\gamma_j^{i+1} > 0$ then $e_j^{i+1} < 0$ and, as

$$\gamma_j^{i+1} = O(\vartheta(h)) + \kappa(\tau_j^{i+1}) e_j^{i+1},$$

then $e_j^{i+1} > -c\vartheta(h)$ and since e_j^{i+1} is negative (see above), then $e_j^{i+1} = O(\vartheta(h))$ and thus

$$\left\|\vec{\gamma}^{i+1}\right\|_\infty \leq c\vartheta(h). \tag{9.287}$$

Similarly, Equation 9.287 can be proved for $\gamma_j^{i+1} < 0$, $e_j^{i+1} > 0$. Since $\|\vec{\varepsilon}^o\|_\infty \leq c\vartheta(h)$, then Equations 9.286 and 9.287 yield that

$$\|\vec{\gamma}^o\|_\infty \leq c\delta, \quad \left\|\vec{\gamma}^{i+1}\right\|_\infty \leq \max\{(1 + c\Delta t) \left\|\vec{\gamma}^i\right\|_\infty + c\Delta t\delta, c\delta\}, \quad i = 0, 1, \ldots, n-1. \tag{9.288}$$

Putting $\alpha^o = c\delta$, $\alpha^{i+1} = \beta_o\alpha^i + c\Delta t\delta$ and $\beta_o = 1 + c\Delta t$, then $\alpha^i \geq c\delta$ and the induction technique proves that $\left\|\vec{\gamma}^i\right\|_\infty \leq \alpha^i$. But

$$\alpha^i \leq c\exp(ct_1)\delta, \quad i = 1, \ldots, n,$$

then $\left\|\vec{\gamma}^i\right\|_\infty \leq c\delta$ and Equation 9.280 yields $\left\|\boldsymbol{e}^i\right\|_\infty \leq c\delta$, $i = 1, \ldots, n$. As $\boldsymbol{e}^i$ is piecewise linear, then $\left\|\boldsymbol{e}^i\right\|_{L^\infty(\Omega_h)} \leq c\delta$.

To finish the proof we shall write

$$T^i - \Theta^i = T^i - T_I^i + T_I^i - \Theta^i.$$

Due to Ciarlet (1980), $\left\|T^i - T_I^i\right\|_{L^\infty(\Omega\cap\Omega_h)} \leq ch^2$ and $\left\|T_I^i - \Theta^i\right\|_{L^\infty(\Omega\cap\Omega_h)} \leq c\delta$, from which Equation 9.268 follows immediately.

9.5. THERMAL MODELS OF THE EARTH'S INTERIOR WITH PHASE TRANSITION ZONES

9.5.1. Introduction

In Section 9.1 the model problems of the thermal distribution in the Earth's interior were discussed in detail. In this section we shall discuss from the mathematical point of view the so-called Stefan problem describing problems of melting, recrystallization and solidification. Here we shall discuss the problem of melting and recrystallization only as the solidification of a rock in a liquid state is governed by the heat and solute equations coupled with Navier-Stokes equations. The study of such processes is of great importance in geophysics. It is known that these processes are described by the so-called two- (three-) phase Stefan problem, involving quasi-linear parabolic equations and non-linear boundary conditions with coefficients explicitly dependent upon space and time variables.

The problem, formulated by Stefan (1889), was studied in most other investigated papers, e.g. Kamenomotskaya (1958), (1961), Oleinik (1960), Friedman (1968), (1982), Rubenstein (1971), Ciavaldini (1975), Meyer (1973), Damlamian (1977), Jerome (1977), (1978), (1983), Ockendon and Hodgkins (1975), Duvaut (1975), Cannon and Di Benedetto (1980), Jerome and Rose (1982), White (1982a,b), Visintin (1981), Niezgódka and Pawlow (1983), Cannon, Di Benedetto and Knightly (1980a,b), Elliott (1981), Roubíček (1989), Nochetto, Paolini and Verdi (1991a,b), Nedoma and Dvořák (1995), Picasso (1995), Nedoma (1997a,b).

In this section we use the concept of a weak solution of the Stefan problem as was introduced by Kamenomotskaya (1961), Friedman (1968), (1982) and Niezgódka and Pawlow (1983). The problem investigated in this section is motivated by a concrete geophysical task connected with recrystallization and melting in the different depths inside the Earth such as the olivine-spinel phase change or in the neighbourhood of volcanoes. It is evident that the melting temperature also depends on the pressure. Moreover, the heat equation is non-linear, mainly due to a phase transition (at different temperatures about 900 K up to 2500 K) arising from recrystallization and melting of rocks. The phase transition will be described as a two-phase Stefan problem with boundary conditions which can also be non-linear, namely of the Stefan-Boltzman type, if that describes the investigated real situation.

Mathematically, the Stefan problems represent free boundary type problems for parabolic equations with the initial-boundary conditions, where a part of the boundary is free and where additional conditions are given at this free boundary. A two-phase Stefan problem is such that on both sides of the free boundary there are given parabolic equations and initial boundary conditions, and neither of the solutions is identically constant.

9.5.2. Classical Formulation of the Non-Linear Two-Phase Stefan Problem

Let $\bar{I} = \langle t_o, t_1 \rangle$, $\Omega = \cup_{\iota=1}^m \Omega^\iota \subset R^N$, $N = 2(3)$ be a convex polygonal (polyhedral) (and therefore bounded) region occupied by elastic or visco-elastic or visco-plastic geological bodies Ω^ι with boundaries $\partial\Omega^\iota = \Gamma_u^\iota \cup \Gamma_\tau^\iota$. Geometrically boundaries $\partial\Omega^\iota$ represent hypersurfaces. Let $\Omega_T = \Omega \times \langle t_o, t_1 \rangle$, $\Omega_t = \Omega \times (t_o, t_1)$, $\Omega(t) = \Omega \times \{t\}$, and $\partial\Omega_T = \partial\Omega \times \langle t_o, t_1 \rangle$, $\partial\Omega_t = \partial\Omega \times (t_o, t_1)$, $\partial\Omega(t) = \partial\Omega \times \{t\}$, $t_o < t < t_1$, their boundaries. Let us denote $T^\iota = T^\iota(\boldsymbol{x}, t)$ the temperature in Ω_t, let $\rho^\iota = \rho^\iota(\boldsymbol{x}, T^\iota)$ be the density in Ω^ι, $c_e^\iota = c_e^\iota(\boldsymbol{x}, T^\iota)$ be the specific heat in Ω^ι, and $\kappa^\iota = \kappa^\iota(\boldsymbol{x}, T^\iota)$ be the thermal conductivity in Ω_t^ι.

Then we seek the distribution of the temperature $T(\boldsymbol{x}, t)$ in Ω_t and surfaces of phase transitions $R^{\iota,s}(t) = \{(\boldsymbol{x}, t) \in \Omega_t^\iota, \ \Phi^\iota(\boldsymbol{x}, t) = 0\}$ satisfying

$$\rho^\iota(\boldsymbol{x}, T^\iota)c_e^\iota(\boldsymbol{x}, T^\iota)\frac{\partial T^\iota(\boldsymbol{x}, T)}{\partial t} - \operatorname{div}(\kappa^\iota(\boldsymbol{x}, T^\iota)\nabla T^\iota) = Q_o(\boldsymbol{x}, t, T^\iota)$$

$$\text{for a.e. } (\boldsymbol{x}, t) \in \Omega_t^\iota, \quad \iota = 1, \ldots, m, \quad (9.289)$$

$$T^\iota(\boldsymbol{x},t) = T_1^\iota(\boldsymbol{x},t) \equiv 0 \text{ for } (\boldsymbol{x},t) \in \Gamma_{ut} = \Gamma_u \times (t_o, t_1), \quad \iota = 1,\dots,m, \tag{9.290}$$

$$\kappa^\iota(\boldsymbol{x},T^\iota)\frac{\partial T^\iota(\boldsymbol{x},t)}{\partial n} = q_o^\iota(\boldsymbol{x},t,T^\iota) \text{ for } (\boldsymbol{x},t) \in \Gamma_{\tau t} = \Gamma_\tau \times (t_o, t_1), \quad \iota = 1,\dots,m, \tag{9.291}$$

$$T^\iota(\boldsymbol{x},t) = T_o^\iota(\boldsymbol{x}) \text{ on } \Omega^\iota(t_o), \quad \iota = 1,\dots,m, \tag{9.292}$$

$$T_S^{\iota,s} = T_L^{\iota,s} = T_R^{\iota,s} \text{ for } (\boldsymbol{x},t) \in R^{\iota,s}(t), \tag{9.293}$$

$$\left(\kappa^\iota(\boldsymbol{x},T^\iota)\frac{\partial T^\iota(\boldsymbol{x},t)}{\partial n}\right)_S^s - \left(\kappa^\iota(\boldsymbol{x},T^\iota)\frac{\partial T^\iota(\boldsymbol{x},t)}{\partial n}\right)_L^s = -\rho^{\iota,s}(\boldsymbol{x},T^\iota)L^{\iota,s}(\boldsymbol{x},t)v_\nu^{\iota,s} \quad \text{for } (\boldsymbol{x},t) \in R^{\iota,s}(t), \tag{9.294}$$

where $\rho^{\iota,s}$ is density, $T_R^{\iota,s}$ is the temperature at the phase change boundary $R^{\iota,s}(t)$, $T_S^{\iota,s}$ and $T_L^{\iota,s}$ are temperatures in the solid and liquid phases, respectively, $\vec{\nu}^{\iota,s}$ is the unit normal vector to $R^{\iota,s}(t)$ pointing towards $\Omega_S^{\iota,s}$ (solid), $R^{\iota,s}(t)$ represents the hypersurface laying in Ω_t^ι, $\Omega_{St}^{\tau,s}$ is the domain lying in Ω_t^ι and bounded by $R^{\iota,s}(t)$ and $\partial\Omega_t^\iota$, $v_\nu^{\iota,s} = \partial R^{\iota,s}(t)/\partial t$ is the speed of $R^{\iota,s}(t)$ along $\vec{\nu}^{\iota,s}$, $L^{\iota,s}$ is the latent heat of the phase change, $T_o^\iota(\boldsymbol{x})$ is the initial temperature, $q_o^\iota(\boldsymbol{x},t,T^\iota)$ is the heat flow, $T_1^\iota(\boldsymbol{x},t)$ the given temperature on Γ_{ut}, and $Q_o^\iota(\boldsymbol{x},t,T^\iota)$ are the thermal sources in $\Omega^\iota(t)$. Further, we denote the heat capacity by $c(\boldsymbol{x},t,T) = \rho(\boldsymbol{x},T)c_e(\boldsymbol{x},T)$.

Let $\Phi^{\iota,s}(\boldsymbol{x},t)$ be a C^1 function in $\overline{\Omega}_T$ such that

$$R^{\iota,s}(t) = \{(\boldsymbol{x},t) \in \overline{\Omega}_T, \quad \Phi^{\iota,s}(\boldsymbol{x},t) = 0\}, \quad \nabla_x\Phi^{\iota,s}(\boldsymbol{x},t) \neq 0 \quad \text{on} \quad R^{\iota,s}(t),$$
$$\Phi^{\iota,s}(\boldsymbol{x},t) < 0 \text{ in } \Omega_S^{\iota,s} \text{ and } \Phi^{\iota,s}(\boldsymbol{x},t) > 0 \text{ in } \Omega_L^{\iota,s}.$$

Then Equations 9.293 and 9.294 can be written in the equivalent form

$$T_S^{\iota,s} = T_L^{\iota,s} = T_R^{\iota,s} \quad \text{for} \quad (\boldsymbol{x},t) \in R^{\iota,s}(t), \tag{9.295}$$

$$(\kappa^\iota(\boldsymbol{x},t,T^\iota)\nabla T^\iota\nabla\Phi^\iota)_S^s - (\kappa^\iota(\boldsymbol{x},t,T^\iota)\nabla T^\iota\nabla\Phi^\iota)_L^s = \alpha^{\iota,s}\frac{\partial\Phi^{\iota,s}}{\partial t}, \tag{9.296}$$

where $\alpha^{\iota,s}$ are given positive constants. We shall refer to the Stefan problem in Equations 9.289 to 9.294 or 9.296 as Problem ($\mathscr{P}$).

Further, for simplicity whilst proving the existence and uniqueners theorem we reduce the studied problem to one phase change boundary only.

Furthermore, we shall assume that:

(P1) $\Omega \subset R^N$, $N = 2(3)$, is an open and bounded domain satisfying the Sobolev embedding and trace theorem, which can be approximated by a family of domains $\{\Omega^M\}_M \subset R^N$, $M \in \mathscr{N}$, with boundaries of the Hölder class $\mathscr{H}^{2+\beta}$, $\beta \in (0,1)$ such that

(i) $\Omega^M \supset \Omega^{M+1} \supset \Omega \quad \forall M \in \mathscr{N}$,

(ii) its Lebesgue's measure $\mu_N(\Omega^M) \xrightarrow[M]{} \mu_N(\Omega)$,

(iii) Lebesgue's measure $\mu_{N-1}(\partial\Omega^M \cap \partial\Omega) \xrightarrow[M]{} \mu_{N-1}(\partial\Omega)$ and $\exists c =$ constant such that $\mu_{N-1}(\partial\Omega^M) \leq c\mu_{N-1}(\cap\partial\Omega)\, \forall M \in \mathcal{N}$;

(P2) $c(\boldsymbol{x}, t, \xi) \in C^{o,1}(\overline{\Omega}_t)$ for each $\xi \in \mathcal{R}$ and $\exists c_m =$ constant > 0, $c_M =$ constant > 0 such that $c_m \leq c(\boldsymbol{x}, t, \xi) \leq c_M$ for $(\boldsymbol{x}, t, \xi) \in \overline{\Omega}_t \times \mathcal{R}$;

(P3) $\kappa(\boldsymbol{x}, t, \xi) = \kappa_1(\boldsymbol{x}, t)\kappa_2(t, \xi)$

(P4) (i) $\kappa(\boldsymbol{x}, t, \xi) \in C^{2,1}(\Omega_t)$ for each $\xi \in \mathcal{R}$, $\kappa(\boldsymbol{x}, t, .) \in C(\mathcal{R}\backslash\{T_R(\boldsymbol{x}, t)\})$,

(ii) $\exists \underline{\kappa}_o =$ constant > 0, $\underline{\kappa}_1 =$ constant > 0 such that $\underline{\kappa}_o \leq \kappa(\boldsymbol{x}, t, \xi) \leq \underline{\kappa}_1$ for $(\boldsymbol{x}, t, \xi) \in \overline{\Omega}_t \times \mathcal{R}$,

(P5) (i) $q(\boldsymbol{x}, t, \xi)$ is continuous in $\xi \in \mathcal{R}$ for almost all $(\boldsymbol{x}, t) \in \Gamma_{\tau t}$ and measurable in $(\boldsymbol{x}, t) \in \Gamma_{\tau t}\, \forall \xi \in \mathcal{R}$ (the so-called Carathédory conditions),

(ii) $\exists c_3 =$ constant > 0, $c_4 =$ constant > 0 such that $\xi q(\boldsymbol{x}, t, \xi) \leq c_3\xi^2 + c_4$, $(\boldsymbol{x}, t, \xi) \in \Gamma_{\tau t} \times \mathcal{R}$;

(iii) q is Lipschitz continuous with respect to t, i.e. $\exists c_5 =$ constant > 0, $c_6 =$ constant > 0 such that, for almost all $(\boldsymbol{x}, t) \in \Gamma_{\tau t}$ and all $\xi \in \mathcal{R}_o \subset \mathcal{R}$ $|q(\boldsymbol{x}, t + k, \xi) - q(\boldsymbol{x}, t, \xi)| \leq (c_5 + c_6\, |\xi|)\, |k|$, for almost all k such that $t + k \in \overline{I} \equiv \langle t_o, t_1\rangle$, where $\mathcal{R}_o$ is a compact subset of $\mathcal{R}$,

(iv) for each bounded $\mathcal{R}_o \subset \mathcal{R}$, function $q(\boldsymbol{x}, t, \xi)$ is Lipschitz continuous in $\xi \in \mathcal{R}_o$, uniformly for almost all $(\boldsymbol{x}, t) \in \Gamma_{\tau t}$,

(v) function $q(\boldsymbol{x}, t, \xi)$ is dissipative function, i.e. $(q(\boldsymbol{x}, t, \xi) - q(\boldsymbol{x}, t, \zeta))$ $(\xi - \zeta) \leq 0\, \forall \xi, \zeta \in \mathcal{R}$, a.e. on $\Gamma_{\tau t}$;

(P6) (i) $Q_o(\boldsymbol{x}, t, \xi)$ is continuous in $\xi \in \mathcal{R}$ for almost all $(\boldsymbol{x}, t) \in \Omega_t$ and measurable in $(\boldsymbol{x}, t) \in \Omega_t\, \forall \xi \in \mathcal{R}$ (the so-called Carathédory conditions),

(ii) $\exists c_1 =$ constant > 0, $c_2 =$ constant > 0 such that $\xi Q_o(\boldsymbol{x}, t, \xi) \leq c_1\xi^2 + c_2$,

(iii) for each bounded $\mathcal{R}_o \subset \mathcal{R}$ function $Q_o(\boldsymbol{x}, t, \xi)$ is Lipschitz continuous in $\xi \in \mathcal{R}_o$, uniformly for almost all $(\boldsymbol{x}, t) \in \Omega_t$, where $\mathcal{R}_o$ is a compact subset of $\mathcal{R}$;

(P7) $T_o(\boldsymbol{x}) \in H^1(\Omega) \cap L^\infty(\Omega)$;

(P8) $T_R(\boldsymbol{x}, t) \in C^{2,1}(\Omega_t)$ is a given function, $\alpha > 0$ is a given constant;

(P9) Concerning the velocity field $\boldsymbol{v}$ we shall assume $\boldsymbol{v} \in L^2(I; [H^1(\Omega)]^2) \cap L^\infty(\Omega_t)$, div $\boldsymbol{v} = 0$ in Ω_t and the prescribed boundary conditions on $\partial\Omega$.

Definition 9.9: We say that a pair $(T(\boldsymbol{x}, t), R(\boldsymbol{x}, t))$, being sufficiently regular and satisfying Equations 9.289 to 9.294, is a classical solution of the problem $(\mathcal{P})$.

Proving the existence of the classical solution of the Stefan problem in its generalized formulation is an open problem at present.

9.5.3. Variational Formulation

We shall now formulate the problem ($\mathscr{P}$) in its weak (variational) formulation. We shall use the standard notation regarding function spaces.

Further, we introduce a new variable y by the Kirchhoff transformation

$$y = G(\boldsymbol{x}, t, T) = \int_o^{T(\boldsymbol{x},t)} \kappa(\boldsymbol{x}, t, \xi) d\xi, \quad (\boldsymbol{x}, t) \in \overline{\Omega}_t. \tag{9.297}$$

From (P4)(ii), the mapping $T \to y = G(T)$ is one-to-one. We denote by $G^{-1}(y)$ its inverse, so that $T = G^{-1}(y)$. If T_R is the temperature of a phase change, then $y_R(\boldsymbol{x}, t) = G(\boldsymbol{x}, t, T_R)$. We introduce the enthalpy by

$$H(\boldsymbol{x}, t, T) = \int_o^T c((G^{-1}(\zeta))\kappa^{-1}(G^{-1}(\zeta))d\zeta + \begin{cases} 0 & \text{if } T < y_R(\boldsymbol{x}, t), \\ \langle 0, \alpha \rangle & \text{if } T = y_R(\boldsymbol{x}, t), \\ \alpha & \text{if } T > y_R(\boldsymbol{x}, t). \end{cases} \tag{9.298}$$

Definition 9.10: We define the above mapping, by w i.e. a measurable function such that

$$w(\boldsymbol{x}, t) \in H(\boldsymbol{x}, t, y(\boldsymbol{x}, t)) \text{ for almost all } (\boldsymbol{x}, t) \in \Omega_t. \tag{9.299}$$

The following estimates hold.

Lemma 9.7: It holds that

$$c_m \tilde{\kappa}_1^{-1} \|y\|_{L^p(\Omega_t)} \le \|w\|_{L^p(\Omega_t)} \le c_M \tilde{\kappa}_o^{-1} \|y\|_{L^p(\Omega_t)} + \alpha[C\mu_N(\Omega)]^{1/p}, \quad 1 \le p < \infty, \tag{9.300}$$

$$c_m \tilde{\kappa}_1^{-1} \|y\|_{L^\infty(\Omega_t)} \le \|w\|_{L^\infty(\Omega_t)} \le c_M \tilde{\kappa}_o^{-1} \|y\|_{L^\infty(\Omega_t)} + \alpha. \tag{9.301}$$

Proof: The estimates in Equations 9.300 and 9.301 follow from the definition of G and the Minkowski inequality.

Definition 9.11: By the Nemytskii's operator $\overline{H} : L^2(\Omega_t) \to L^2(\Omega_t)$ we shall understand the mapping (multi-valued) corresponding to $H(\boldsymbol{x}, t, T)$ defined by

$$\overline{H}(\boldsymbol{x}, t, y) = H(\boldsymbol{x}, t, y(\boldsymbol{x}, t)) \text{ for almost all } (\boldsymbol{x}, t) \in \Omega_t. \tag{9.302}$$

Regarding its properties and Carathéodory conditions see, for example, Fučík and Kufner (1980).

We shall assume that $w_o \in L^\infty(\Omega)$ is given and such that $w_o(\boldsymbol{x}) \in H(\boldsymbol{x}, t_o, G(\boldsymbol{x}, t_o, T(\boldsymbol{x}, t_o))$ a.e. in Ω. Then we can define

$$y_o(\boldsymbol{x}) = G(\boldsymbol{x}, t_o, T_o) = H^{-1}(\boldsymbol{x}, t_o, w_o(\boldsymbol{x})) \text{ a.e. in } \Omega.$$

Since from (P7) $T_o \in H^1(\Omega) \cap L^\infty(\Omega)$ then $y_o \in L^\infty(\Omega) \cap H^1(\Omega)$. By the above transformation, the problem ($\mathscr{P}$) can be written in the new variables as follows.

Problem ($\mathscr{P}_{\text{enth}}$): Find the distribution of (w, y) in Ω_t and surface of phase transition $R(\boldsymbol{x}, t)$ satisfying

$$\frac{\partial w}{\partial t} = \Delta y + W(\boldsymbol{x}, t, y, \nabla y), \quad (\boldsymbol{x}, t) \in \Omega_t \backslash R(\boldsymbol{x}, t), \quad t_o < t < t_1, \tag{9.303}$$

$$w(\boldsymbol{x}, t) = w_1(\boldsymbol{x}, t) \equiv 0, \quad (\boldsymbol{x}, t) \in \Gamma_{ut}, \quad t_o < t < t_1, \tag{9.304}$$

$$\frac{\partial y}{\partial n} = q(\boldsymbol{x}, t, y), \quad (\boldsymbol{x}, t) \in \Gamma_{\tau t}, \quad t_o < t < t_1, \tag{9.305}$$

$$w(\boldsymbol{x}, 0) = w_o(\boldsymbol{x}), \tag{9.306}$$

$$y(\boldsymbol{x}, t) = y_R(\boldsymbol{x}, t), \quad (\boldsymbol{x}, t) \in R(\boldsymbol{x}, t), \tag{9.307}$$

$$(\text{grad } y \text{ grad } \Phi)_S - (\text{grad } y \text{ grad } \Phi)_L = \alpha \frac{\partial \Phi}{\partial t}, \ (\boldsymbol{x}, t) \in R(\boldsymbol{x}, t), \tag{9.308}$$

where

$$\begin{aligned} W(\boldsymbol{x}, t, y, \nabla y) &= (\kappa(G^{-1}(y)))^{-1} \nabla y(G^{-1}(y)) - \int_o^{G^{-1}(y)} \Delta\kappa(\boldsymbol{x}, t, \xi) \mathrm{d}\xi + (\kappa(G^{-1}(y)))^{-1} \\ &\times \int_o^{G^{-1}(y)} \nabla\kappa(G^{-1}(y)) \nabla\kappa(\boldsymbol{x}, t, \xi) \mathrm{d}\xi + \int_o^{G^{-1}(y)} \frac{\partial c(\boldsymbol{x}, t, \xi)}{\partial t} \mathrm{d}\xi + Q(\boldsymbol{x}, t, y) \\ &= \sum_{k=1}^{N} W_k(\boldsymbol{x}, t) \frac{\partial y}{\partial x_j} + W_o(\boldsymbol{x}, t, y) (\text{ due to assumption (P3)}), \end{aligned} \tag{9.309}$$

$$Q(\boldsymbol{x}, t, y) = Q_o(\boldsymbol{x}, t, G^{-1}(y)), \tag{9.310}$$

$$q(\boldsymbol{x}, t, y) = q_o(\boldsymbol{x}, t, G^{-1}(y)) + \int_o^{G^{-1}(y)} \frac{\partial \kappa(\boldsymbol{x}, t, \xi)}{\partial n} \mathrm{d}\xi, \tag{9.311}$$

$$y_R(\boldsymbol{x}, t) = G(\boldsymbol{x}, t, T_R), \tag{9.312}$$

and where

$$W_k(\boldsymbol{x}, t) = -\kappa_1^{-1}(\boldsymbol{x}, t) \frac{\partial \kappa_1(\boldsymbol{x}, t)}{\partial x_k}, \quad k = 1, \dots, N, \quad N = 2, 3, \tag{9.313}$$

$$\begin{aligned} W_o(\boldsymbol{x}, t, y) &= \int_o^{G^{-1}(y)} \frac{\partial c(\boldsymbol{x}, t, \xi)}{\partial t} \mathrm{d}\xi - \int_o^{G^{-1}(y)} \Delta\kappa(\boldsymbol{x}, t, \xi) \mathrm{d}\xi \\ &+ \sum_{k=1}^{N} W_k(\boldsymbol{x}, t) \int_o^{G^{-1}(y)} \frac{\partial \kappa}{\partial x_k}(\boldsymbol{x}, t, \xi) \mathrm{d}\xi + Q(\boldsymbol{x}, t, y). \end{aligned} \tag{9.314}$$

Let us define the space

$$V(\Omega_t) = \{v \mid v \in C^{2,1}(\overline{\Omega}_t), v(\boldsymbol{x}, t_1) = 0, \boldsymbol{x} \in \Omega\}.$$

Multiplying Equation 9.303 by $v \in V(\Omega_t)$, integrating over $\Omega \backslash R(\boldsymbol{x}, t)$ and taking into account Equations 9.304 to 9.308, we obtain

$$\begin{aligned} &\int_{\Omega_t} \left[w \frac{\partial v}{\partial t} + y \Delta v - \sum_{k=1}^{N} W_k(\boldsymbol{x}, t) y \frac{\partial v}{\partial x_k} + (W(\boldsymbol{x}, t, y, \nabla y) - y \sum_{k=1}^{N} \frac{\partial W_k(\boldsymbol{x}, t)}{\partial x_k} v \right] \mathrm{d}\boldsymbol{x} \, \mathrm{d}t \\ &+ \int_{\Gamma_{\tau t}} \left[q_o(\boldsymbol{x}, t, y) v - \sum_{k=1}^{N} W_k(\boldsymbol{x}, t) y \cos(\boldsymbol{n}, x_k) v - y \frac{\partial v}{\partial n} \right] \mathrm{d}s \mathrm{d}t + \int_{\Omega} w_o(\boldsymbol{x}) v(\boldsymbol{x}, 0) \mathrm{d}\boldsymbol{x} = 0. \end{aligned} \tag{9.315}$$

Definition 9.12: We say that a pair or couple function (y, w) is a weak solution of Problem $(\mathscr{P}_{\text{enth}})$ if $y \in L^{\infty}(\Omega_t)$, $w \in L^{\infty}(\Omega_t)$, w is defined by Equation 9.299 and the pair (y, w) satisfies the integral identity (Equation 9.315) for all $v \in V(\Omega_t)$.

To prove the existence of the weak solution of Problem $(\mathscr{P}_{\text{enth}})$ the technique of Kamenomotskaya (1961), Oleinik (1960), Friedman (1968), Ladyženskaya, Solonikov and Ural'ceva (1968), Niezgódka and Pawlow (1983) will be used. This technique consists of the construction of a family of regularized non-linear parabolic boundary value problems, the classical solutions of which converge to the solution of Problem $(\mathscr{P}_{\text{enth}})$. It means that we construct the auxiliary parabolic problems which will be regularized in the sense of smoothing of all data, i.e. all coefficients as well as the boundary and initial conditions.

To regularize the auxiliary problems we introduce a family of domains (Ω^n), $n \in \mathscr{N}$, with boundaries $\partial\Omega^n$ of the Hölder class $\mathscr{H}^{2+\beta}$, $\beta \in (0, 1)$, approximating Ω in the sense of the assumption (P1) and such that $\cup_{n\in\mathscr{N}}\Omega^n = \Omega^1$, $\cup_{n\in\mathscr{N}}\Omega_t^n = \Omega_t^1$, where $\Omega_t^n = \Omega^n \times (t_o, t_1)$, $\partial\Omega_t^n = \partial\Omega^n \times (t_o, t_1)$. Such a choice of the external approximations Ω^n, $n \in \mathscr{N}$ is given for technical reasons only for our further investigations. It is evident that approximations in the form $\Omega^n \subset \Omega$ can also be used.

For the case of regular approximations we shall assume that:

(R1) $\{H^n(\boldsymbol{x}, t, \xi)\} \subset C^\infty(\overline{\Omega}_t^n \times \mathscr{R})$ satisfy

(i) $H^n(\xi) \xrightarrow[n]{\text{Weakly}} w$ in $L^2(\Omega_t)$, uniformly with respect to ξ from the bounded subsets of $C(\overline{\Omega}_t) \cap H^1(\Omega_t)$,

(ii) there exists a measurable function

$$H_m(\boldsymbol{x}, t, \xi) : \cup_{n\in\mathscr{N}}\Omega_t^n \times \mathscr{R} \to (0, \infty)$$

bounded over bounded sets and such that for all $n \in \mathscr{N}$

$$\left|H^n(\boldsymbol{x}, t, \xi)\right| \le H_m(\boldsymbol{x}, t, \xi) \text{ a.e. in } (\Omega_t^n \backslash \Omega_t) \times \mathscr{R},$$

(iii)
$$0 < c_m\tilde{\kappa}_1^{-1} \le \frac{\partial H^n(x, t, \xi)}{\partial \xi}, \quad (\boldsymbol{x}, t, \xi) \in \overline{\Omega}_t^n \times \mathscr{R},$$

$$\frac{\partial H^n(x, t, \xi)}{\partial \xi} \le 1, \quad (\boldsymbol{x}, t, \xi) \in \overline{\partial\Omega}_t^n \times \mathscr{R};$$

(R2) $\{W_k^n(\boldsymbol{x}, t)\} \subset C^\infty(\overline{\Omega}_t^n)$, $k = 1, \ldots, N$ such that

(i) $W_k^n \xrightarrow[n]{\text{Weakly}} W_k$ in $L^2(\Omega_t)$, $\dfrac{\partial W_k^n}{\partial x_j} \xrightarrow[n]{\text{Weakly}} \dfrac{\partial W_k}{\partial x_j}$ in $L^1(\Omega_t)$, $k = 1, \ldots, N$,

(ii) W_k^n, $\dfrac{\partial W_k^n}{\partial x_j}$, $k = 1, \ldots, N$ are uniformly bounded in $L^1(\Omega_t^n \backslash \Omega_t)$ for all $n \in \mathscr{N}$;

(R3) $\{W^n(\boldsymbol{x}, t, \xi)\} \subset C^\infty(\overline{\Omega}_t^n \times \mathscr{R})$, such that

(i) $W^n(\xi) \xrightarrow[n]{\text{Weakly}} W(\xi)$ in $L^1(\Omega_t)$, uniformly with respect to ξ from bounded subsets of $C(\overline{\Omega}_t) \cap H^1(\Omega_t)$,

(ii) there exists a measurable function $W_m(\boldsymbol{x}, t, \xi) : \Omega_t^n \times \mathscr{R} \to (0, \infty)$ bounded over bounded sets and such that for all $n \in \mathscr{N}$ $|W^n(\boldsymbol{x}, t, \xi)| \leq W_m(\boldsymbol{x}, t, \xi)$ a.e. in $(\Omega_t^n \backslash \Omega_t) \times \mathscr{R}$;

(R4) $\{q^n(\boldsymbol{x}, t, \xi)\} \subset C^\infty(\Gamma_{\tau t}^n \times \mathscr{R})$ such that

(i) $q^n \xrightarrow[n]{\text{Weakly}} q(\xi)$ in $L^1(\Gamma_{\tau t})$, uniformly with respect to ξ from bounded subsets of $C(\overline{\Gamma}_{\tau t})$,

(ii) there exists a measurable function

$$q_m(\boldsymbol{x}, t, \xi) : \cup_{\in \mathscr{N}} \overline{(\overline{\Omega}_t^n \backslash \Omega_t)} \to (0, \infty)$$

bounded over bounded sets and such that for all $n \in \mathscr{N}$

$$\left|q^n(\boldsymbol{x}, t, \xi)\right| \leq q_m(\boldsymbol{x}, t, \xi) \text{ a.e. in } (\Gamma_{\tau t}^n \backslash \Gamma_{\tau t}) \cup (\Gamma_{\tau t} \backslash \Gamma_{\tau t}^n) \times \mathscr{R},$$

(iii) $q^n(\boldsymbol{x}, t_o, \xi) = 0$, $(\boldsymbol{x}, t) \in \Gamma_\tau^n \times \mathscr{R}$;

(R5) $\{w_o^n(\boldsymbol{x})\} \subset C^\infty(\overline{\Omega}^n)$, $y_o^n(\boldsymbol{x}) = G^{-1}(\boldsymbol{x}, t_o, w_o^n(\boldsymbol{x}))$, $\boldsymbol{x} \in \overline{\Omega}^n$, such that

(i) $w_o^n \xrightarrow[n]{\text{Weakly}} w_o$ in $L^1(\Omega_t)$,

(ii) w_o^n, $n \in \mathscr{N}$ are uniformly bounded in $C(\overline{\Omega}^n)$,

(iii) $\dfrac{\partial y_o^n(\boldsymbol{x})}{\partial n} = 0$, $\boldsymbol{x} \in \partial\Omega^n$.

Then the following lemma holds.

Lemma 9.8: There exist constants $c_o > 0$, $c_1 > 0$ such that for all $n \in \mathscr{N}$

$$q^n(\boldsymbol{x}, t, \xi) \leq c_o |\xi| + c_1, \; (\boldsymbol{x}, t, \xi) \in \overline{\Omega}_t^n \times \mathscr{R}, \tag{9.316}$$

$$\left|\frac{\partial q^n(\boldsymbol{x}, t, \xi)}{\partial t}\right| \leq c_o |\xi| + c_1, \; (\boldsymbol{x}, t, \xi) \in \overline{\Omega}_t^n \times \mathscr{R}, \tag{9.317}$$

and constants $c_2 > 0$, $c_3 > 0$, $c_4 > 0$ such that for all $n \in \mathscr{N}$

$$\sum_{k=1}^N W_k^n(\boldsymbol{x}, t) q_k + W^n(\boldsymbol{x}, t, \xi) \leq c_2 \sum_{k=1}^N |q_k| + c_3 |\xi| + c_4, \quad N = 2(3), \tag{9.318}$$

$$(\boldsymbol{x}, t, \xi, q_1, \ldots, q_N) \in \overline{\Omega}_t^n \times \mathscr{R}^{N+1}, \quad N = 2(3). \tag{9.319}$$

Proof: The proof follows from the definition of q and assumptions (P5)(i),(ii), (R4) and (P2)-(P4), (R2) and (R3).

Then we have to solve the following regularized problem.

Problem ($\mathscr{P}_{\text{Rn}}$): Let $n \in \mathscr{N}$, then

$$\frac{\partial y}{\partial t} = \left(\frac{\partial H^n(\boldsymbol{x}, t, y)}{\partial y}\right)^{-1} \left(\Delta y + \sum_{k=1}^N W_k^n(\boldsymbol{x}, t) y_k + W^n(\boldsymbol{x}, t, y) - \frac{\partial H^n(\boldsymbol{x}, t, y)}{\partial t}\right), \quad (\boldsymbol{x}, t) \in \Omega_t^n, \tag{9.320}$$

$$y = 0, \quad (\boldsymbol{x}, t) \in \Gamma_{ut}, \tag{9.321}$$

$$\frac{\partial y}{\partial n} = q^n(\boldsymbol{x}, t, y), \quad (\boldsymbol{x}, t) \in \Gamma_{\tau t}, \tag{9.322}$$

$$y(\boldsymbol{x}, 0) = y_o(\boldsymbol{x}), \quad \boldsymbol{x} \in \Omega^n. \tag{9.323}$$

Theorem 9.25: The regularized Problem $(\mathscr{P}_{Rn})$, $n \in \mathcal{N}$, has a unique classical solution $y^n \in \mathscr{H}^{2+\beta, 1+1/2\beta}(\Omega_t^n)$.

Proof: The proof of the existence and uniqueness of the solution follows immediately from Theorem 7.4 of Ladyženskaya, Solonikov and Ural'ceva (1968), Chapter V.

We shall next give the a priori estimates of the classical solutions of problem $(\mathscr{P}_{Rn})$ based on the results of Ladyženskaya, Solonikov and Ural'ceva (1968) (Chapter V, Section 7) and Niezgódka and Pawlow (1983). Therefore, we give the next results without proofs.

Lemma 9.9: Let assumptions (P1)-(P4), (P5)(ii), (P6)(i) and (P7) be satisfied. Then there exists constant C_o, independent of $n \in \mathcal{N}$, such that all solutions y^n of the regularized problems $(\mathscr{P}_{Rn})$, $n \in \mathcal{N}$, can be a priori estimated by:

$$\|y^n\|_{C(\overline{\Omega}_t^n)} \leq C_o, \quad n \in \mathcal{N}. \tag{9.324}$$

If assumption (P5)(iii) is satisfied, then there exists constant C_1, independent of n, such that

$$\|y^n\|_{H^1(\Omega_t^n)} \leq C_1, \quad n \in \mathcal{N}, \tag{9.325}$$

so that the set of functions $\{y^n \mid n \in \mathcal{N}\}$ is relatively weakly compact in the space $H^1(\Omega_t)$.

Moreover, for every y, being an accumulation point $\{y^n\}$ in the weak topology of the space $H^1(\Omega_t)$,

$$\|y\|_{L^\infty(\Omega_t)} \leq C_o, \tag{9.326}$$

$$\|y\|_{H^1(\Omega_t)} \leq C_1. \tag{9.327}$$

The next theorem gives the main results concerning the existence and uniqueness of the weak solution of Problem $(\mathscr{P})$. Firstly we give the following lemma, which is needed for the proof of the existence of the solution.

Lemma 9.10: It holds that

$$H^n(y^n) \xrightarrow[n]{} w \quad \text{in} \quad L^2(\Omega_t) \text{ weakly}, \tag{9.328}$$

where $w = \overline{H}(y)$.

For the proof see Niezgódka and Pawlow (1983).

Theorem 9.26: Let assumptions (P1)-(P8) be satisfied. Then the Stefan problem $(\mathscr{P}_{\text{enth}})$ has at least one weak solution (y, w). This solution can be constructed as a limit of the unique classical solutions y^n of the regularized Problems $(\mathscr{P}_{Rn})$, $n \in \mathcal{N}$ as

$$y^n \xrightarrow[n]{} y \text{ in } H^1(\Omega_t) \text{ weakly}, \quad H^n(y^n) \xrightarrow[n]{} w \text{ in } L^2(\Omega_t) \text{ weakly}. \tag{9.329}$$

Moreover, in addition, under the assumption that the whole boundary $\partial\Omega$ of the domain Ω is of the Hölder class $\mathscr{H}^{2+\beta}$, then the weak solution (y, w) of Problem $(\mathscr{P}_{\text{enth}})$ is unique.

Proof: The first part of the proof, proving the existence of the weak solution of the Problem ($\mathscr{P}_{\text{enth}}$), is based on the construction of such a solution, which will be a limit of solutions of regularized Problems ($\mathscr{P}_{Rn}$), $n \in \mathcal{N}$.

Let y^n be the classical solutions of regularized Problem ($\mathscr{P}_{Rn}$), $n \in \mathcal{N}$. Then y^n satisfy the following integral identity

$$\int_{\Omega_t^n} (H^n(y^n)\frac{\partial v}{\partial t} + y^n \Delta v - \sum_{k=1}^{N} W_k^n(\boldsymbol{x},t) y^n \frac{\partial v}{\partial x_k} + W^n(\boldsymbol{x},t,y^n) - y^n \sum_{k=1}^{N} \frac{\partial W_k^n(\boldsymbol{x},t)}{\partial x_j} \mathrm{d}\boldsymbol{x}\,\mathrm{d}t$$

$$+ \int_{\Gamma_{\tau t}} \left(q^n(y^n)v - y^n \frac{\partial v}{\partial n}\right) \mathrm{d}s\mathrm{d}t + \int_{\Gamma_{\tau t}^n} \left(q^n(y^n)v - y^n \frac{\partial v}{\partial n}\right) \mathrm{d}s\mathrm{d}t + \int_{\Omega^n} w_o^n(\boldsymbol{x}) v(\boldsymbol{x},t_o)\mathrm{d}\boldsymbol{x} = 0$$

$$\forall v \in V(\cup_{n\in\mathcal{N}} \Omega_t^n). \quad (9.330)$$

Due to Lemma 9.9 (i.e. Equation 9.325), we can take a subsequence $\{y^n\}$, using the same indices as in the whole sequence, then

$$y^n \xrightarrow[n]{} y. \quad (9.331)$$

Now we shall prove that y satisfies the integral identity in Equation 9.330. We shall show that

$$\int_{\Omega_t^n} H^n(y^n)\frac{\partial v}{\partial x}\mathrm{d}\boldsymbol{x}\,\mathrm{d}t = \int_{\Omega_t} H^n(y^n)\frac{\partial v}{\partial x}\mathrm{d}\boldsymbol{x}\,\mathrm{d}t + \int_{\Omega_t^n \setminus \Omega_t} H^n(y^n)\frac{\partial v}{\partial t}\mathrm{d}\boldsymbol{x}\,\mathrm{d}t \xrightarrow[n]{\text{Weakly}} \int_{\Omega_t} w\frac{\partial v}{\partial t}\mathrm{d}\boldsymbol{x}\,\mathrm{d}t,$$

where $v \in V(\cup_{n\in\mathcal{N}} \Omega_t^n)$, and the assumption (R1)(ii) and Lemma 9.10 were used.

Moreover, due to assumptions (P1),(R2),(R3) and Lemma 9.9 (i.e. Equations 9.324 and 9.325) we have

$$\int_{\Omega_t^n} \left(y^n \Delta v - \sum_{k=1}^{N} W_k^n(\boldsymbol{x},t) y^n \frac{\partial v}{\partial x_k} + \left(W^n(\boldsymbol{x},t,y^n) - y^n \sum_{k=1}^{N} \frac{\partial W_k^n(\boldsymbol{x},t)}{\partial x_k} \right) v \right) \mathrm{d}\boldsymbol{x}\,\mathrm{d}t \xrightarrow[n]{}$$

$$\int_{\Omega_t} \left(y\Delta v - \sum_{k=1}^{N} W_k(\boldsymbol{x},t) y \frac{\partial v}{\partial x_k} + \left(W(\boldsymbol{x},t,y) - y \sum_{k=1}^{N} \frac{\partial W_k(\boldsymbol{x},t)}{\partial x_k} \right) v \right) \mathrm{d}\boldsymbol{x}\,\mathrm{d}t. \quad (9.332)$$

as we have:

(a) $$- \int_{\Omega_t} W_k^n(\boldsymbol{x},t) y^n \frac{\partial v}{\partial x_k}\mathrm{d}\boldsymbol{x}\,\mathrm{d}t - \int_{\Omega_t^n \setminus \Omega_t} W_k^n(\boldsymbol{x},t) y^n \frac{\partial v}{\partial x_k}\mathrm{d}\boldsymbol{x}\,\mathrm{d}t$$

$$= - \int_{\Omega_t} W_k^n(\boldsymbol{x},t)(y^n - y)\frac{\partial v}{\partial x_k}\mathrm{d}\boldsymbol{x}\,\mathrm{d}t - \int_{\Omega_t} W_k^n(\boldsymbol{x},t)\frac{\partial v}{\partial x_k}\mathrm{d}\boldsymbol{x}\,\mathrm{d}t \xrightarrow[n]{} - \int_{\Omega_t} W_k(\boldsymbol{x},t) y \frac{\partial v}{\partial x_k}\mathrm{d}\boldsymbol{x}\,\mathrm{d}t,$$

$$k = 1, \ldots, N,$$

as $\int_{\Omega_t^n \setminus \Omega_t} W_k^n(\boldsymbol{x},t) y^n (\partial v/\partial x_t)\mathrm{d}\boldsymbol{x}\,\mathrm{d}t \xrightarrow[n]{} 0$ due to assumptions (P1) and (R2)(ii), and where the assumption (R2)(i) and the uniform boundedness of $W_k^n(\boldsymbol{x},t)$, $n \in \mathcal{N}$, in $L^2(\Omega_t)$ were used;

(b) $$- \sum_{k=1}^{N} \int_{\Omega_t^n} \frac{\partial W_k^n(\boldsymbol{x},t)}{\partial x_k} y^n v\mathrm{d}\boldsymbol{x}\,\mathrm{d}t + \int_{\Omega_t^n} W^n(\boldsymbol{x},t,y^n) v\mathrm{d}\boldsymbol{x}\,\mathrm{d}t$$

$$= - \sum_{k=1}^{N} \int_{\Omega_t} \left(\frac{\partial W_k^n(\boldsymbol{x},t)}{\partial x_k} y^n - \frac{\partial W_k(\boldsymbol{x},t)}{\partial x_k} y \right) v\mathrm{d}\boldsymbol{x}\,\mathrm{d}t - \sum_{k=1}^{N} \int_{\Omega_t^n \setminus \Omega_t} \frac{\partial W_k^n(\boldsymbol{x},t)}{\partial x_k} y^n v\mathrm{d}\boldsymbol{x}\,\mathrm{d}t$$

$$- \sum_{k=1}^{N} \int_{\Omega_t} \frac{\partial W_k(\boldsymbol{x},t)}{\partial x_k} yv\mathrm{d}\boldsymbol{x}\,\mathrm{d}t \xrightarrow[n]{} - \sum_{k=1}^{N} \int_{\Omega_t} \frac{\partial W_k(\boldsymbol{x},t)}{\partial x_k} yv\mathrm{d}\boldsymbol{x}\,\mathrm{d}t,$$

due to assumptions (P1) and (R2) and as

$$\int_{\Omega_t}\left(\frac{\partial W_k^n(\boldsymbol{x},t)}{\partial x_k}y^n-\frac{\partial W_k(\boldsymbol{x},t)}{\partial x_k}y\right)v\mathrm{d}\boldsymbol{x}\,\mathrm{d}t$$
$$=\int_{\Omega_t}\frac{\partial W_k^n(\boldsymbol{x},t)}{\partial x_k}(y^n-y)v\mathrm{d}\boldsymbol{x}\,\mathrm{d}t+\int_{\Omega_t}\left(\frac{\partial W_k^n(\boldsymbol{x},t)}{\partial x_k}-\frac{\partial W_k(\boldsymbol{x},t)}{\partial x_k}\right)yv\mathrm{d}\boldsymbol{x}\,\mathrm{d}t\xrightarrow[n]{}0.$$

Moreover, according to (P6)(i) we see that $\int_{\Omega_t}(Q(\boldsymbol{x},t,y^n)-Q(\boldsymbol{x},t,y))v\mathrm{d}\boldsymbol{x}\,\mathrm{d}t\xrightarrow[n]{}0$, and hence, using assumptions (P2)-(P4), we obtain

$$\int_{\Omega_t}(w(\boldsymbol{x},t,y^n)-W(\boldsymbol{x},t,y)v\mathrm{d}\boldsymbol{x}\,\mathrm{d}t\xrightarrow[n]{}0. \tag{9.333}$$

Using the assumption (R3)(i) and the diagonalization procedure, then

$$\int_{\Omega_t}(W^n(\boldsymbol{x},t,y^n)-W(\boldsymbol{x},t,y^n))v\mathrm{d}\boldsymbol{x}\,\mathrm{d}t\xrightarrow[n]{}0.$$

Hence and using Equation 9.333 with assumptions (P1) and (R3)(ii) we find that

$$\int_{\Omega_t^n}W^n(\boldsymbol{x},t,y^n)\mathrm{d}\boldsymbol{x}\,\mathrm{d}t-\int_{\Omega_t}W(\boldsymbol{x},t,y)v\mathrm{d}\boldsymbol{x}\,\mathrm{d}t\xrightarrow[n]{}0$$

and therefore Equation 9.332 is fully proved.

According to Lemma 9.10 and assumption ($R5$) we find that

$$\int_{\Omega^n}w_o^n(\boldsymbol{x})v(\boldsymbol{x},t_o)\mathrm{d}\boldsymbol{x}\xrightarrow[n]{}\int_{\Omega}w_o(\boldsymbol{x})v(\boldsymbol{x},t_o)\mathrm{d}\boldsymbol{x}.$$

According to Equation 9.331, $y^n\xrightarrow[n]{}y$ in $H^1(\Omega_t)$ weakly, and assumption (P1) we have

$$\int_{\Gamma_{\tau t}^n}y^n\frac{\partial v}{\partial n}\mathrm{d}s\mathrm{d}t\xrightarrow[n]{}\int_{\Gamma_{\tau t}^n}y\frac{\partial v}{\partial n}\mathrm{d}s\mathrm{d}t. \tag{9.334}$$

To complete the proof we must prove that

$$\int_{\Gamma_{\tau t}^n}q^n(\boldsymbol{x},t,y^n)v\mathrm{d}s\mathrm{d}t\xrightarrow[n]{}\int_{\Gamma_{\tau t}}q(\boldsymbol{x},t,y)v\mathrm{d}\boldsymbol{x}\,\mathrm{d}t.$$

Since $\Gamma_{\tau t}^n=\Gamma_{\tau t}\cup(\Gamma_{\tau t}^n\backslash\Gamma_{\tau t})\backslash(\Gamma_{\tau t}\backslash\Gamma_{\tau t}^n)$ and due to assumptions (P1), (R4)(ii) and Lemma 9.9 (i.e. Equations 9.324 and 9.325 we can find

$$\int_{\Gamma_{\tau t}^n\backslash\Gamma_{\tau t}}q^n(\boldsymbol{x},t,y^n)v\mathrm{d}s\mathrm{d}t\xrightarrow[n]{}0,\quad\int_{\Gamma_{\tau t}\backslash\Gamma_{\tau t}^n}q^n(\boldsymbol{x},t,y^n)v\mathrm{d}s\mathrm{d}t\xrightarrow[n]{}0. \tag{9.335}$$

By the assumpton (R4)(i) and the diagonalization procedure we obtain

$$g^n(\boldsymbol{x},t,y^n)-q(\boldsymbol{x},t,y^n)\xrightarrow[n]{}0\quad\text{in}\quad L^1(\Gamma_{\tau t}). \tag{9.336}$$

The Nemytskii's operator $\overline{q}(\boldsymbol{x},t,y):L^2(\Gamma_{\tau t})\to L^2(\Gamma_{\tau t})$, corresponding to $q(\boldsymbol{x},t,\xi)$, is, due to assumption (P5)(i), continuous. Due to Equation 9.331

$$\overline{q}(\boldsymbol{x},t,y^n)-\overline{q}(\boldsymbol{x},t,y)\xrightarrow[n]{}0\quad\text{in}\quad L^2(\Gamma_{\tau t})\text{ weakly.}$$

Hence, due to Equations 9.335 and 9.336 the convergence of Equation 9.334 follows immediately, which completes the first part of the proof.

To prove the uniqueness of the weak solution, we shall assume that there exist two weak solutions (y_1, w_1) and (y_2, w_2) of our problem $(\mathscr{P})$. Then Equation 9.315 yields

$$\int_{\Omega_t}\left[(w_1 - w_2)\frac{\partial v}{\partial t} + (y_1 - y_2)\Delta v + \sum_{k=1}^{N}(W_k(\boldsymbol{x}, t)y_1 - W_k(\boldsymbol{x}, t)y_2)\frac{\partial v}{\partial x_k} + (W(\boldsymbol{x}, t, y_1)\right.$$

$$\left. - W(\boldsymbol{x}, t, y_2))v\right] \mathrm{d}\boldsymbol{x}\,\mathrm{d}t - \int_{\Gamma_{\tau t}}\left[(y_1 - y_2)\frac{\partial v}{\partial n} - (q(\boldsymbol{x}, t, y_1) - q(\boldsymbol{x}, t, y_2))v\right] \mathrm{d}s\mathrm{d}t = 0$$

$$\forall v \in V(\Omega_t). \quad (9.337)$$

We define the test function v as a solution of a certain auxiliary boundary value problem for a parabolic equation. Hence and using Equation 9.337 it follows that $w_1 = w_2$ a.e. in Ω_t, and thus also $y_1 = y_2$ a.e. in Ω_t, which completes the proof.

9.5.4. Numerical Solution

Let the domain Ω be covered by a uniform system of triangles $\underline{T}_h$, where h denotes the length of the largest triangle edge, with disjoint interiors and such that each triangle edge is either the edge of the neighbouring triangle or has its vertices on $\partial\Omega$, and which create a triangulation $\mathscr{T}_h$. Further, we shall assume that the smallest angle of the triangulation for any h is bounded below, independently of h. Let $\{\mathscr{T}_h\}$ be a system of regular triangulations.

In the previous section we proved that a weak solution of the problem $(\mathscr{P}_{\text{enth}})$ exists and is unique. Next we shall use the semi-implicit discretization of the problem $(\mathscr{P}_{\text{enth}})$, which leads at every time to the boundary value problem for non-linear partial differential equations of elliptic type. Since

$$\frac{\partial w}{\partial t} \sim \frac{w(y^{m+1}) - w(y^m)}{\Delta t}, \qquad y^m = y(\boldsymbol{x}, m\Delta t),$$

then Equation 9.303 can be expressed as

$$w(y^{m+1}) - \Delta t \text{ div grad } y^{m+1} = w(y^m) + \Delta t W^{m+1}, \tag{9.338}$$

where

$$W^{m+1} = \int_{m\Delta t}^{(m+1)\Delta t} W(\tau)\mathrm{d}\tau.$$

If $\Delta t \to 0$, W depends monotonically on y and then the solution of Equation 9.338 converges to the solution of the problem $(\mathscr{P}_{\text{enth}})$ (Jerome (1977)). For simplicity we put $y = y^{m+1}$, $\mu = \Delta t$, $W = w(y) + \Delta t W^{m+1} + \alpha y_R$, $\alpha y + \partial\Phi(y) \equiv w(y) + \alpha y_R$, where $0 < \alpha \leq \min_s\{c_S^s, c_L^s\}$, div grad $\equiv \Delta$. Then Equation 9.338 yields

$$\alpha y - \mu \text{ div grad } y + \partial\Phi(y) = W, \tag{9.339}$$

which corresponds to the generalized form of Glowinski (1979), Barbu (1976), i.e. to the problem: find $u \in H^2(\Omega)$ satisfying

$$\alpha u - \mu\Delta u + \partial\Phi(u) \ni f(\boldsymbol{x}) \quad \text{a.e.} \quad \boldsymbol{x} \in \Omega, \tag{9.340}$$

$$u = 0, \quad \text{a.e.} \quad \boldsymbol{x} \in \Gamma_u, \tag{9.341}$$

$$\frac{\partial u}{\partial n} = q, \quad \text{a.e.} \quad \boldsymbol{x} \in \partial\Omega\backslash\Gamma_u \equiv \Gamma_\tau, \tag{9.342}$$

where $\partial\Phi(.) : R \to R$ is the multi-valued monotone mapping defined as

$$\partial\Phi(\lambda) = \begin{cases} \alpha_2(\lambda - u_o) + L, & \lambda > u_o, \\ \langle 0, L\rangle, & \lambda = u_o, \\ \alpha_1(\lambda - u_o), & \lambda < u_o. \end{cases} \quad (9.343)$$

Now $\Phi(.)$ is locally Lipschitz continuous since

$$|\Phi(\lambda) - \Phi(\mu)| \leq \max\{|\partial\Phi(\lambda)|, |\partial\Phi(\mu)|\} |\lambda - \mu| \leq c |\lambda - \mu| \text{ when } |\lambda|, |\mu| \leq \sigma. \quad (9.344)$$

The functional $\Phi(.)$ is a continuous convex function and can be taken to be a piecewise quadratic convex function. Hence, α_i, $i = 1, 2$, from Equation 9.343 are, in our case, $\alpha_1 = c_S - \alpha$, $\alpha_2 = c_L - \alpha$, $\lambda \sim y$, $u_o \sim y_R$ and L is the latent heat.

We shall define by

$$V = \{z \mid z \in H^1(\Omega), z = 0 \quad \text{on} \quad \Gamma_u \text{ in the sense of traces}\}$$

the space of virtual temperatures, and by

$$V_h = \{z \,|z \in C(\overline{\Omega})|\, z_{|\underline{T}_h} \in P_1, \quad z = 0 \text{ on } \Gamma_u, \quad \forall \underline{T}_h \in \mathscr{T}_h\},$$

its finite element approximation, where P_1 represents the space of linear polynomials.

For numerical solution of our Stefan problem, we shall use the definition of weak (variational) formulation based on Equation 9.339.

Definition 9.13: By a variational (weak) solution of the discretized problem $(\mathscr{P}_{\text{enth}})_D$ we understand a function $y \in V$ such that

$$\alpha(y, z - y) + \Delta t a(y, z - y) + j(z) - j(y) \geq s(z - y) \quad \forall z \in V, \quad (9.345)$$

where

$$a(y, z) = \int_\Omega \frac{\partial y}{\partial x_i}\frac{\partial z}{\partial x_j} \mathrm{d}\boldsymbol{x}, \quad (y, z) = \int_\Omega yz\mathrm{d}\boldsymbol{x},$$

$$s(z) = -\Delta t \int_{\Gamma_\tau} \frac{\partial y}{\partial n} z\mathrm{d}s + \Delta t \int_\Omega \underline{W} z\mathrm{d}\boldsymbol{x}, \quad j(z) = \int_\Omega \Phi(z(\boldsymbol{x}))\mathrm{d}\boldsymbol{x}.$$

Definition 9.14: The function y_h is said to be a finite element approximation of the discretized problem $(\mathscr{P}_{\text{enth}})_D$ if

$$\alpha(y_h, z_h - y_h) + \Delta t a(y_h, z_h - y_h) + j(z_h) - j(y_h) \geq s(z_h - y_h) \quad \forall z_h \in V_h. \quad (9.346)$$

Put

$$\begin{aligned} \Omega_+^s &= \{\boldsymbol{x} \mid y(\boldsymbol{x}) > y_R^s\}, & \Omega_{+h}^s &= \{\underline{T}_h \in \mathscr{T}_h \mid y > y_R^s \text{ on } \underline{T}_h\}, \\ \Omega_-^s &= \{\boldsymbol{x} \mid y(\boldsymbol{x}) < y_R^s\}, & \Omega_{-h}^s &= \{\underline{T}_h \in \mathscr{T}_h \mid y < y_R^s \text{ on } \underline{T}_h\}, \\ \Omega_o^s &= \{\boldsymbol{x} \mid y(\boldsymbol{x}) = y_R^s\}, & \Omega_{oh}^s &= \{\underline{T}_h \in \mathscr{T}_h \mid y = y_R^s \text{ on } \underline{T}_h\}, \end{aligned} \quad (9.347)$$

where s corresponds to phase change boundaries.

Let $\Omega_{*h} = \Omega \backslash \cup (\Omega^s_{+h} \cup \Omega^s_{-h} \cup \Omega^s_{oh})$. Since the free boundary is contained in some triangles, Ω_{*h} is not empty. The length of the free boundary $\cup_s R^s(\boldsymbol{x}, t)$ satisfies

$$\begin{aligned} \text{meas}\{\cup_s R^s(\boldsymbol{x}, t)\} &= \text{meas}\{\cup(\partial\Omega^s_+ \cap \partial\Omega^s_-)\} + \text{meas}\{\cup(\partial\Omega^s_- \cap \partial\Omega^s_o)\} \\ &\quad + \text{meas}\{\cup(\partial\Omega^s_o \cap \partial\Omega^s_+)\} \end{aligned}$$

i.e., it is finite. Then there exists a constant c independent of h and such that the number of triangles N^s_{*h} in Ω^s_{*h} is bounded by $N^s_{*h} < ch^{-1}$.

Let $z_{LI} \in V_h$ be the Lagrange interpolate of z, and $z_{LI} = z$ at each vertex of triangulation $\mathscr{T}_h$. Then the following estimate

$$\|z - z_{LI}\|_{o,1} + h\,\|z - z_{LI}\|_{1,1} \le ch^2 \quad \forall z \in H^1_o(\Omega) \cap H^2(\Omega) \tag{9.348}$$

holds (Ciarlet (1980)). For $y, z \in V_h$ we put

$$(y, z)_h = \sum_{\underline{T}_h \in \mathscr{T}_h} \frac{1}{3}\text{meas}(\underline{T}_h) \sum_{i=1}^{3} y_i^T z_i^T, \tag{9.349}$$

$$j_h(z) = \int_\Omega (\Phi(z))_{LI} \mathrm{d}\boldsymbol{x} = \sum_{\underline{T}_h \in \mathscr{T}_h} \frac{1}{3}\text{meas}(\underline{T}_h) \sum_{i=1}^{3} \Phi(z_i^T), \tag{9.350}$$

where meas $(\underline{T}_h)$ represents the area of triangle $\underline{T}_h$, and z_i represents the value of z at node i of triangle $\underline{T}_h$.

Lemma 9.11 (Ciavaldini, Raviart): The following estimates hold:

$$|(y, z)_h - (y, z)| \le ch^2\,\|y\|_{1,1}\,\|z\|_{1,1} \quad \forall y, z \in V_h, \tag{9.351}$$

$$\|z\|_{o,1} \le (z, z)_h^{1/2} \le c\,\|z\|_{o,1} \quad \forall z \in V_h. \tag{9.352}$$

For the proof see Ciavaldini (1975).

Next we shall prove the lemma of Falk's type.

Lemma 9.12: Let $y \in V$, $y_h \in V_h$, then

$$\begin{aligned} \alpha\,\|y - y_h\|^2_{o,1} + \Delta t\,\|y - y_h\|^2_{1,1} &\le c[\alpha(y_h - y, z_h - y) + \Delta t a(y_h - y, z_h - y) + \alpha(y, z - y_h) \\ &\quad + \Delta t a(y, z - y_h) + \alpha(y, z_h - y) + \Delta t a(y, z_h - y) + j(z) - j(y_h) \\ &\quad + j(z_h) - j(y) - s(z - y_h) - s(z_h - y)]^{1/2} \\ &\qquad \forall z \in V, z_h \in V_h, \end{aligned} \tag{9.353}$$

where c = constant > 0.

Proof: Let $y \in V$, $y_h \in V_h$. Using Equations 9.344 and 9.345 and adding and subtracting terms $\alpha(y, y - y_h) - \alpha(y_h, y) + \Delta t a(y, y - y_h) - \Delta t a(y_h, y)$ then we obtain

$$\begin{aligned} &\alpha(y, y - y_h) - \alpha(y_h, y) - \alpha(y, y - y_h) + \alpha(y_h, y) + \alpha(y, z - y) + \alpha(y_h, z_h - y_h) \\ &\quad + \Delta t a(y, z - y) + \Delta t a(y_h, z_h - y_h) + \Delta t a(y, y - y_h) - \Delta t a(y_h, y) - \Delta t a(y, y - y_h) \\ &\quad + \Delta t a(y_h, y) + j(z) - j(y) + j(z_h) - j(y_h) - s(z - y) - s(z_h - y_h) \ge 0. \end{aligned}$$

Hence,

$$\begin{aligned}
&\alpha(y, y - y_h) - \alpha(y_h, y - y_h) + \Delta t a(y, y - y_h) - \Delta t a(y_h, y - y_h) \\
&\quad = \alpha(y - y_h, y - y_h) + \Delta t a(y - y_h, y - y_h) \\
&\quad \leq \alpha(y_h - y, z_h - y) + \Delta t a(y_h - y, z_h - y) + \alpha(y, z - y_h) + \Delta t a(y, z - y_h) + \alpha(y, z_h - y) \\
&\qquad + \Delta t a(y, z_h - y) + j(z) - j(y) + j(z_h) - j(y_h) + s(y_h - z) + s(y - z_h).
\end{aligned}$$

From here Equation 9.353 follows immediately.

Corollary: Let $V_h \subset V$. Then by substituting $z = y_h$ in Equations 9.351 and 9.352 we have

$$\begin{aligned}
\alpha \|y - y_h\|_{o,1}^2 + \Delta t \|y - y_h\|_{1,1}^2 &\leq c[\alpha(y_h - y, z_h - y) + \Delta t a(y_h - y, z_h - y) \\
&\quad + \alpha(y, z_h - y) + \Delta t a(y, z_h - y) + j(z_h) - j(y) \\
&\quad + s(y - z_h)] \quad \forall z_h \in V_h. \qquad (9.354)
\end{aligned}$$

Lemma 9.13: It holds that

$$j(y) \leq j_h(y) \quad \forall y \in V_h, \qquad (9.355)$$

$$|j(y_{LI}) - j(y)| \leq c \|y - y_{LI}\|_{o,1}, \qquad (9.356)$$

$$|j_h(y_{LI}) - j(y_{LI})| \leq ch^2. \qquad (9.357)$$

Proof: The proof is a modification of that of Elliott (1981). To prove Equation 9.355 for any element $\underline{T}_h \in \mathcal{T}_h$ we denote by $w_i(\boldsymbol{x})$, $0 \leq w_i(\boldsymbol{x}) \leq 1$, the basis functions for each node i. Hence, $\sum_{i=1}^3 w_i(\boldsymbol{x}) = 1$. The functional $\Phi(y(\boldsymbol{x}))$ is a continuous and convex function, so that

$$\Phi(y(\boldsymbol{x})) = \Phi\left(\sum_{=1}^{3} y_i w_i(\boldsymbol{x})\right) \leq \sum_{i=1}^{3} w_i(\boldsymbol{x})\Phi(y_i).$$

The functional $j(y)$ is a continuous convex functional on $H_o^1(\Omega)$, so that

$$j(y) = \int_\Omega \Phi(y(\boldsymbol{x}))\mathrm{d}\boldsymbol{x} = \sum_{\underline{T}_h \in \mathcal{T}_h} \int_{\underline{T}_h} \Phi(y(\boldsymbol{x}))\mathrm{d}\boldsymbol{x} \leq \sum_{\underline{T}_h \in \mathcal{T}_h} \int_{\underline{T}_h} w_i(\boldsymbol{x})\Phi(y_i)\mathrm{d}\boldsymbol{x} = \int_\Omega \Phi(y(\boldsymbol{x}))_{LI}\mathrm{d}\boldsymbol{x} = j_h(\boldsymbol{y}).$$

To prove Equation 9.356 then, due to Equation 9.344 and the fact that y is continuous, it holds that

$$|j(y)_{LI} - j(y)| = \int_\Omega (\Phi(y_{LI}) - \Phi(y))\mathrm{d}\boldsymbol{x} \; |\leq c[\|y\|_{L^\infty(\Omega)} \int_\Omega |y - y_{LI}| \,\mathrm{d}\boldsymbol{x} \leq c \|y - y_{LI}\|_{o,1}.$$

To prove Equation 9.357, we have

$$|j(y_{LI}) - j_h(y_{LI})| = \left| \sum_{\underline{T}_h \in \mathcal{T}_h} \int_{\underline{T}_h} [\Phi(y_{LI}) - \Phi(y_{LI}(\boldsymbol{x}))_{LI}]\mathrm{d}\boldsymbol{x} \right| \equiv \left| \sum_{\underline{T}_h \in \mathcal{T}_h} J(T_h) \right|. \qquad (9.358)$$

Now we shall investigate the integral in Equation 9.358 in the neighbourhood of the free boundary $R^s(\boldsymbol{x}, t)$, i.e. for $\underline{T}_h \in \Omega^s_{+h}, \Omega^s_{-h}, \Omega^s_{oh}$ and for $\underline{T}_h \in \Omega^s_{*h} = \Omega \backslash \cup_s (\Omega^s_{+h} \cup \Omega^s_{-h} \cup \Omega^s_{oh})$.

Firstly, let $\underline{T}_h \in \Omega^s_{oh}$. Then $y_{LI} = y^s_R$ and $\Phi(y_{LI}) = 0$. Hence,

$$J(\underline{T}_h) = 0. \qquad (9.359)$$

Now, let $\underline{T}_h \in \Omega^s_{+h}$ and $\underline{T}_h \in \Omega^s_{-h}$. Since $\Phi(.)$ is a quadratic function of y_{LI}, then due to Lemma 9.11 we have

$$\left| \int_{\underline{T}_h \in \Omega^s_{+h}} (\Phi(y_{LI}) - \Phi(y_{LI}(\boldsymbol{x}))_{LI}) \mathrm{d}\boldsymbol{x} \right| \leq ch^2, \tag{9.360}$$

$$\left| \int_{\underline{T}_h \in \Omega^s_{-h}} (\Phi(y_{LI}) - \Phi(y_{LI}(\boldsymbol{x}))_{LI}) \mathrm{d}\boldsymbol{x} \right| \leq ch^2. \tag{9.361}$$

Now, let $\underline{T}_h \in \Omega^s_{*h}$. Since $\Phi(.)$ is Lipschitz continuous and since $N^s_{*h} < ch^{-1}$, then

$$\begin{aligned} \left| \int_{\underline{T}_h \in \Omega^s_{*h}} (\Phi(y_{LI}) - \Phi(y_{LI}(\boldsymbol{x}))_{LI}) \mathrm{d}\boldsymbol{x} \right| &\leq \left| \int_{\underline{T}_h \in \Omega^s_{+h}} \Phi(y_{LI}) \mathrm{d}\boldsymbol{x} - \frac{1}{3} \mathrm{meas}(\underline{T}_h) \sum_{i=1}^{3} \Phi(y_i^T) \right| \\ &= \frac{1}{3} \left| \sum_{i=1}^{3} \int_{\underline{T}_h \in \Omega^s_{*h}} [\Phi(y_{LI}) - \Phi(y_i^T)] \mathrm{d}\boldsymbol{x} \right|, \end{aligned}$$

where y_i^T denotes the value of y at the node i of the triangle $\underline{T}_h$. Since $\Phi(.)$ is locally Lipschitz continuous then

$$\begin{aligned} \frac{1}{3} \left| \sum_{i=1}^{3} \int_{\underline{T}_h \in \Omega^s_{*h}} [\Phi(y_{LI}) - \Phi(y_i^T)] \mathrm{d}\boldsymbol{x} \right| &\leq \frac{1}{3} \left| \sum_{i=1}^{3} c \int_{\underline{T}_h \in \Omega^s_{*h}} |y_{LI}(\boldsymbol{x}) - y_i^T| \, \mathrm{d}\boldsymbol{x} \right| \\ &= \frac{1}{3} \left| \sum_{i=1}^{3} c \int_{\underline{T}_h \in \Omega^s_{*h}} |\boldsymbol{x} - \boldsymbol{x}_i| \, \mathrm{grad} \, y_{LI} \mathrm{d}\boldsymbol{x} \right| \\ &\leq \frac{1}{3} \left| \sum_{i=1}^{3} c \left(\int_{\underline{T}_h \in \Omega^s_{*h}} |\boldsymbol{x} - \boldsymbol{x}_i|^2 \mathrm{d}\boldsymbol{x} \right)^{1/2} \left(\int_{\underline{T}_h \in \Omega^s_{*h}} |\mathrm{grad} \, y_{LI}|^2 \mathrm{d}\boldsymbol{x} \right)^{1/2} \right| \\ &\leq ch^2 \left[\left(\int_{\underline{T}_h \in \Omega^s_{*h}} |\mathrm{grad} \, y|^2 \mathrm{d}\boldsymbol{x} \right)^{1/2} + \left(\int_{\underline{T}_h \in \Omega^s_{*h}} |\mathrm{grad}(\boldsymbol{x} - \boldsymbol{x}_i)|^2 \mathrm{d}\boldsymbol{x} \right)^{1/2} \right] \\ &\leq ch^2 (h \, \|y\|_{W^{1,\infty}} + ch). \end{aligned}$$

Thus

$$\left| \int_{\underline{T}_h \in \Omega^s_{*h}} (\Phi(y_{LI}(\boldsymbol{x})) - \Phi(y_{LI}(\boldsymbol{x}))_{LI}) \mathrm{d}\boldsymbol{x} \right| \leq \sum_{\underline{T}_h \in \Omega^s_{*h}} ch^3 \leq ch^2. \tag{9.362}$$

Hence and using Equations 9.358 to 9.361 yields

$$|j_h(y_{LI}) - j(y_{LI})| \leq ch^2,$$

which completes the proof.

The main result gives the next theorem.

Theorem 9.27: Let the free boundary $\cup_s R^s(\boldsymbol{x}, t)$ be finite. Let $y \in W^{1,\infty}(\Omega) \cap H^2(\Omega)$. Let Ω_+, Ω_-, Ω_o, Ω_{+h}, Ω_{-h}, Ω_{oh} be defined as in Equation 9.347 and let $N_{*h} < ch^{-1}$. Then

$$\alpha \, \|y - y_h\|_{o,1} + \Delta t \, \|y - y_h\|_{1,1} \leq ch. \tag{9.363}$$

Proof: To prove this theorem the usual technique will be used. To achieve this we shall assume that y is sufficiently regular. Let us estimate Equation 9.363 by using Lemma 9.12. Terms $\alpha(y, z - y_h) + \Delta t a(y, z - y_h) - s(z - y_h)$, $\alpha(y, z_h - y) + \Delta t a(y, z_h - y) - s(z_h - y)$ are estimated by applying the Green's theorem and later by a suitable choice of $y_h \in V_h$, $y \in$V.

From the Corollary of Lemma 9.12 we obtain

$$\begin{aligned}
\alpha \|y - y_h\|_{o,1}^2 + \Delta t \|y - y_h\|_{1,1}^2 &\leq c \Big[\alpha(y_h - y, z_h - y) + \Delta t a(y_h - y, z_h - y) \\
&\quad + \int_{\Gamma_\tau} q(z_z - y) \mathrm{d}s + j(z_h) - j(y) + j(z) - j(y_h)\Big] \\
&\leq c \Big[\alpha \|y_h - y\|_{o,1} \|z_h - y\|_{o,1} + \Delta t \|y_h - y\|_{1,1} \|z_h - y\|_{1,1} \\
&\quad + \int_{\Gamma_\tau} \frac{\partial y}{\partial n}(z - y_h) \mathrm{d}s + j(z_h) - j(y) + j(z) - j(y_h)\Big] \qquad (9.364) \\
&\leq c \Big[\tfrac{1}{2}\varepsilon \|y - y_h\|_{o,1}^2 + \tfrac{1}{2}\varepsilon^{-1} \|y - z_h\|_{o,1}^2 + \tfrac{1}{2}\varepsilon^{-1} \|y - z_h\|_{1,1}^2 \\
&\quad + \int_{\Gamma_\tau} \frac{\partial y}{\partial n}(z - y_h) \mathrm{d}s + j(z_h) - j(y) + j(z) - j(y_h)\Big].
\end{aligned}$$

In the last inequality, let us put $z_h = y_{LI}$, where $y_{LI} \in V_h$ is the Lagrange interpolation of y on the triangulation $\mathcal{T}_h$. To estimate $j(z_h) - j(y) + j_h(z) - j_h(y_h)$, Lemmas 9.11 and 9.13 will be used. Then

$$j(z_h) - j(y) + j(z) - j(y_h) \leq j_h(y_{LI}) - j(y) = j_h(y_{LI}) - j(y_{LI}) + j(y_{LI}) - j(y) \leq ch^2 + c \|y - y_{LI}\|_{o,1}.$$

Moreover, it holds

$$\begin{aligned}
&\alpha[(y_h, y_h - y_{LI})(y_h, y_h - y_{LI})_h] \leq ch^2, \\
&\|y_{LI} - y\|_{1,1} \leq ch^2 \|y\|_{2,1}, \\
&\|y_{LI} - y\|_{o,1} \leq c_p h^2 \|y\|_{2,1}, \\
&\alpha(y_h, y_{LI} - y) + \Delta t a(y_h - y, y_{LI} - y) - s(y_{LI} - y) \\
&\quad \leq \alpha \|y_h\|_{o,1} \|y_{LI} - y\|_{o,1} + \Delta t \|y_h - y\|_{1,1} \|y_{LI} - y\|_{1,1} + \|\underline{W}\|_{o,1} \|y_{LI} - y\|_{o,1} \\
&\quad \leq c \|y_{LI} - y\|_{o,1} + \tfrac{1}{2}\varepsilon \Delta t \|y_h - y\|_{1,1}^2 + \tfrac{1}{2}\varepsilon^{-1} \Delta t \|y_{LI} - y\|_{1,1}^2.
\end{aligned}$$

From Equation 9.364 and these estimates, then Equation 9.363 follows immediately.

Remark: The method was tested for the case of the phase transition olivine-spinel at depth 400 km. The phase change boundary was found numerically to be near the 400 km depth.

REFERENCES

Barbu, V. (1976), *Non-linear Semi-groups and Differential Equations in Banach Spaces*, Noordhoff International Publishing, Leyden.

Birch, F. (1952), "Elasticity and Constitution of the Earth's Interior", *J. Geophys. Res.*, **57**, 227-286.

Boccardo, L. and Buttazzo, G. (1987), "Quasi-linear Elliptic Equations with Discontinuous Coefficients", *Universita Degli Studi Di Piza*, Dipartmento Di Matematica, Luglio, **Numero 206**.

Cannon, J.R. and Di Benedetto, E. (1980), "On the Existence of Weak Solutions to an n-Dimensional Stefan Problem with Non-linear Boundary Conditions", *SIAM J. Math. Anal.*, **11**, 632-645.

Cannon, J.R., Di Benedetto, E. and Knightly, G.H. (1980a), "The Steady State Stefan Problem with Convection", *Archive for Rational Mechanics and Analysis*, **73**, 79-97.

Cannon, J.R., Di Benedetto E. and Knightly, G.H. (1980b), "The Bi-dimensional Stefan Problem with Convection: The Time Dependent Case", *Partial Differential Equations*, **14**, 1549-1604.

Carslaw, H.S., Jaeger, J.C. (1950), *Conduction of Heat in Solid*, Clarendon Press, Oxford.

Christensen, N.I. (1996). Poisson's Ratio and Crustal Seismology. *J. Geophys. Res.* **101**, No. B2, 3139-3156.

Ciarlet, P. (1980), *The Finite Element Method for Elliptic Problems*, North-Holland, Amsterdam.

Ciavaldini, J.F. (1975), "Analyse Numérique d'un Probléme de Stefan á Deux Phases par un Méthod d'Éléments Finis", *SIAM J. Numer. Anal.*, **12**, 464-489.

Clark, S.P. (1962), "Temperature in the Continental Crust", in *Temperature, Its Measurements and Control in Science and Technology* Herzfeld, Ch.M. (Ed.), **Vol. 3, Part I**, Reihold Publ. Corp., New York, 779-790.

Clark, S.P. (1966), "Handbook of Physical Constants", *Geol. Soc. Am.*, **Memoir 97**, New York.

Damlamian, A. (1977), "Some Results on the Multi-Phase Stefan Problem", *Commun.*, **PDE 2**, 1017-1044.

Duvaut, G. (1975), "The Solution of a Two-Phase Stefan Problem by Variational Inequalities", in *Moving Boundary Problems in Heat Flow and Diffusion* Ockendon, J.R. and Hodkins, W.R. (Eds.), Clarendon Press, Oxford, 173-182.

Elliott, C.M. (1981), "On the Finite Element Approximation of an Elliptic Variational Inequality arising from an Implicit Time Discretization of the Stefan Problem", *IMA J. Numer. Anal.*, **1**, 115-125.

Feistauer, M. (1987), "On the Finite Element Approximation of a Cascade Flow Problem", *Numer. Math.*, **50**, 655-684.

Feistauer, M. and Sobotíková, V. (1990), "Finite Element Approximation of Non-linear Elliptic Problems with Discontinuous Coefficients", *M^2AN Mathematical Modelling and Numerical Analysis*, **24, 4**, 457-500.

Feistauer, M. and Ženíšek, A. (1987), "Finite Element Solution of Non-linear Elliptic Problems", *Numer. Math.*, **50**, 451-475.

Feistauer, M. and Ženíšek, A. (1988), "Compactness Method in the Finite Element Theory of Non-linear Elliptic Problems", *Numer. Math.*, **52**, 147-163.

Friedman, A. (1968), "The Stefan Problem in Several Space Variables", *Trans. Amer. Math. Soc.*, **133**, 51-87.

Friedman, A. (1982), *Variational Principles and Free Boundary Problem*, Wiley-Interscience, New York.

Fučík, S. and Kufner, A. (1980). *Non-linear Differential Equations*, Elsevier, Amsterdam, Oxford, New York.

Gajewski, H., Gröger, K. and Zacharias, K. (1974), *Nichtlineare Operator-gleichungen und Operatordifferentialgleichungen*,. Akademie Verlag, Berlin.

Glowinski, R.(1979), "Finite Element and Variational Inequalities", in: *The Mathematics of Finite Elements and Applications III* Whiteman, J.R.(Ed.), Academic Press, New York, 135-171.

Glowinski, R. (1984), *Numerical Methods for Non-linear Variational Problems. Series in Computational Physics*, Springer, Berlin, Heidelberg, New York, Tokyo.

Jerome, J.W. (1977), "Non-linear Equations of Evolution and a Generalized Stefan Problem", *J. Diff. Equations*, **26**, 240-261.

Jerome, J.W. (1978), "Existence and Approximation of Weak Solutions of Non-linear Dirichlet Problems with Discontinuous Coefficients", *SIAM J. Math. Anal.*, **9**, 730-742.

Jerome, J.W. (1983), *Approximation of Non-linear Evolution Systems*, Academic Press, New York, London.

Jerome, J.W. and Rose, M.E. (1982), "Error Estimates for Multi-Dimensional Two-Phase Stefan Problem", *Math. Comp.*, **39**, 377-414.

Kamenomotskaya, S. (1958), "On the Stefan Problem", *Naučnye Dokl. Vysšei Školy*, **1**, 60-62. (in Russian).

Kamenomotskaya, S. (1961), "On the Stefan Problem", *Matem. Sbornik*, **53, 4**, 489-514 (in Russian).

Lachenbruch, A. (1968), "Rapid Estimation of Topographic Correction to Superficial Geothermal Gradients", *Rev. Geophys.*, **6**, 365-400.

Ladyženskaya, O.A., Solonikov, V.A. and Ural'ceva, N. (1968), *Linear and Quasi-linear Equations of Parabolic Type*, American Mathematical Society, RI.

Lambert, J.D. (1991), *Numerical Methods for Ordinary Differential Systems: The Initial Value Problem*, J.Wiley&Sons, Chichester.

Landau, L.D. and Lifshitz, E.M. (1984), *Electrodynamics of Continuous Media*, Pergamon Press, Oxford, New York, Beijing, Frankfurt, Sao Paulo, Sydney, Tokyo, Toronto.

Lee, W.H.K. and Uyeda, S. (1965), "Review of Heat Flow Data", in "Terrestrial Heat Flow" (Lee, W.H.K. (Ed.)). *Am. Geophys. Union Monograph*, **8**, 87-190.

Lions, J.L. (1969), *Quelques Méthodes de Résolution des Problémes aux Limites Non Linéaires*", Dunod, Paris.

Lions, J.L. and Magenes, E. (1968), *Problémes aux Limites Non Homogénes et Applications I*", Dunod, Paris.

Meyer, G.H. (1973), "Multi-dimensional Stefan Problems", *SIAM J. Numer. Anal.*, **10**, 522-539.

Nečas, J. (1967), *Les Méthodes Directes en Théorie des Équations Elliptique*, Academia, Prague.

Nečas, J. (1983), "Introduction to the Theory of Non-linear Elliptic Equations", *Teubner-Texte zur Mathematik*, **Band 52**, BSB B.G. Teubner Verlagsgesgesellschaft, Leipzig.

Nedoma, J. (1997a), "Finite Element Approximation of a Coupled Contact-Stefan-Like Problem Arising from the Time Discretization in Deformation Theory of Thermo-Plasticity", 7th *Int. Congress on Computational and Applied Mathematics*, Leuven, July 1996; *J. Comput. Appl. Math.* **82**, 313-331.

Nedoma, J. (1997b), "On a Coupled Stefan-Like Problem in Thermo-Visco-Plastic Rheology", 7th *Int. Congress on Computational and Applied Mathematics*, Leuven, July 1996; *J. Comput. Math.* **84**, 45-80.

Nedoma, J. and Dvořák, J. (1995), "On the FEM solution of a Coupled Contact-Two-Phase Stefan Problem in Thermo-Elasticity: Coercive Case", *J. Comput. Appl. Math.*, **63**, 411-420.

Niezgódka, M. and Pawlow, I. (1983), "A Generalized Stefan Problem in Several Space Variables", *Appl. Math. Optim.*, **9**, 193-224.

Nochetto, R.H., Paolini, M. and Verdi, C. (1991a), "An Adaptive Finite Element Method for Two-Phase Stefan Problems in Two Space Variables", *Math. Comp.*, **57**, 73-108.

Nochetto, R.H., Paolini, M. and Verdi, C. (1991b), "Local Mesh Refinements for Two-Phase Stefan Problem in Two Space Variables", *SIAM J. Sc. Statistical Comput.*, **12**, 1207-1244.

Ockendon, J.R. and Hodgkins, W.R. (Eds.) (1975), *Moving Boundary Problems in Heat Flow and Diffusion*, Clarendon Press, Oxford, 173-182.

Oleinik, O.A. (1960), "A Method of Solution of the General Stefan Problem", *Dokl. AN SSSR*, **135**, 1054-1057.

Ortega, J.M. and Rheiboldt, W.C. (1970), *Iterative Solution of Non-Linear Equations in Several Variables*, Academic Press, New York, London.

Pazy, A. (1983), *Semi-groups of Linear Operators and Applications to Partial Differential Equations*, Springer Verlag, New York.

Picasso, M. (1995), "An Adaptive Finite Element Algorithm for a Two-Dimensional Stationary Stefan-Like Problem", *Comput. Methods Appl. Mech. Engrg.*, **124**, 213-230.

Quarteroni, A. and Valli, A. (1991), *Numerical Approximation of Partial Differential Equations*, Springer Verlag, Berlin, Heidelberg, New York, London, Paris, Tokyo, Hong Kong, Barcelona, Budapest.

Ranalli, G. (1987), *Rheology of the Earth*, Allen & Unwin, Boston, London, Sydney, Wellington.

Roubíček, T. (1989), "The Stefan Problem in Heterogeneous Media", *Ann. Inst. Henri Poincaré, Analyse Non Linéaire*, **6(6)**, 481-501.

Roy, R.F., Blackwell, D.D. and Birch, F. (1968), "Heat Generation of Plutonic Rocks and Continental Heat Flow Provinces", *Earth and Planet. Sci. Lett.*, **5**, 1-12.

Rubinstein, L.I. (1971), "The Stefan Problem", *Trans. Math. Monographs*, **27**, American Mathematical Society, Providence, RI.

Schatz, A.H., Thomée, V. and Wahlbin, L.B. (1980), "Maximum Norm Stability and Error Estimates in Parabolic Finite Element Equations", *Comm. Pure Appl. Math.*, **33**, 265-304.

Shimatzu, Y. (1954), "Equation of State of Materials Composing the Earth's Interior", *J. Earth Sci, Nagoya Univ.*, **2**, 15-172.

Stacey, F.D. (1977), *Physics of the Earth*, J.Wiley & Sons, New York.

Stefan, J. (1889), "Über Einige Probleme der Theorie der Wärmeleitung", *Akad. Mat. Natur.*, **98**, S.-B.Wien.

Strang, G. and Fix, G. (1973), *An Analysis of the Finite Element Method*, Prentice Hall, New York.

Tikhonov, A.N. and Samarskij, A.A. (1955), *Equations of Mathematical Physics*, NČSAV, Prague (in Czech),

Verma, R.K. (1991), *Geodynamics of the Indian Peninsula and the Indian Plate Margin*, Oxford & IBH Publishing Co. PVT. LTD, New Delhi, Bombay, Calcutta.

Visintin, A. (1981), "Sur le Probléme de Stefan avec Flux Non Linéaire", *Boll UMI*, **C18**, 63-86.

Wheeler, M.F. (1973), "A Priori L_2 Error Estimates for Galerkin Approximations to Parabolic Partial Differential Equations", *SIAM J. Numer. Anal.*, **10**, 723-759.

White, R.E. (1982a), "An Enthalpy Formulation of the Stefan Problem", *SIAM J. Numer. Anal.*, **19**, 1129-1157.

White, R.E. (1982b), "A Numerical Solution of the Enthalpy Formulation of the Stefan Problem", *SIAM J. Numer. Anal.*, **19**, 1158-1172.

Zlámal, M. (1980), "A Finite Element Solution of the Non-Linear Heat Equation", *RAIRO Anal. Numer.*, **14**, 203-216.

Ženíšek, A. (1990), *Non-linear Elliptic and Evolution Problems and their Finite Element Approximations*, Academic Press, London, San Diego, New York, Boston, Sydney, Tokyo, Toronto

10

Mathematical Problems of the Earth's Electromagnetic Field. Numerical Analyses of the Magneto-Telluric Models of the Earth

10.1. INTRODUCTION

The study of natural and artificial electromagnetic fields provides us with information on the distribution of inhomogeneities in the lithosphere and outer mantle as well as on the general structure of the Earth's body. While natural fields characterized by frequencies in the interval from 10^{-5} to 10^3 Hz allow us to study the Earth's lithosphere and outer mantle to a depth up to 300 or 350 km, artificial fields induced by artificial electric sources and characterized by frequencies in the interval from 100 to 5000 Hz make study of the lithosphere possible only in a very thin layer to a depth of about 300 m. The penetration depth of variations of the natural electromagnetic field that generate these fields grows towards the Earth's interior proportionally with the decrease of frequency. The natural electromagnetic field is caused by variations of the electromagnetic field of outer sources. Since the Earth's lithosphere, mantle, and core are electric conductors (some parts of the lithosphere can also be dielectric), variable electromagnetic fields of electric currents (geomagnetic induced field) originate inside the Earth's body by the influence of the outer electromagnetic field, i.e. by the induction. The physical quantity that characterizes the electric properties of rocks that form the Earth's body is the electric conductivity σ or its reciprocal value $r = \sigma^{-1}$, called the electric resistance. Electric conductivity thus provides information on the structure of the Earth's body, on the state of rocks the Earth's body is composed of, on the location of the individual inhomogeneities, faults, sedimentary basins and zones of increased temperature, and on the determination of horizontal interfaces of electric conductivity that are connected with changes in stratigraphy or with changes of temperature. The comparison of the study of the electromagnetic field with the results of study of other geophysical fields and with the results of laboratory measurements of electric conductivity at different temperatures and pressures, or possibly also with some other laboratory measurements, allows us to better assess the most likely composition of rocks that form the Earth's body in the domain studied. Zones of electric conductivity in the Earth's crust and outer mantle that agree with the results of geothermal observations and which are being explained by domains of probable local partial melting of rocks have been found in this way (Rikitake (1966)). It is known from long-term observations of the Earth's electromagnetic field that the Earth's currents possess long-term variations of period of 1 month and short-term ones of period of several days. A relation has been found between the Earth's currents and geomagnetic fluctuations of the field with oscillation periods from 1 second to 1 day. These variations of the Earth's currents together with geomagnetic fluctuations form the initial information for study of local and near-surface inhomogeneities (disturbation mass) and for study of conductivity of the upper part of the lithosphere. They are, however, completely improper for study of the Earth's global field. Detailed analyses of the materials measured, laboratory measurements, and theoretical results show that the Earth is not a homogeneous body, but an inhomogeneous body of conductivity σ that varies from place to place in the interval from about 10^{-12} to 10^{-15} emu. Moreover, a certain anisotropy of the Earth's currents is observed that cannot be explained only by geology and topography of surface layers, but primarily by likely inhomogeneities in the mantle (Rikitake (1966)).

Processes of mutual influence of the impact of corpuscular radiation from the Sun and the Earth's own magnetic field give rise to a time variable system of electric currents in the magnetosphere. This time variable system of electric currents induces an electromagnetic undulation that is incident with the Earth's surface. Matsushita and Cambell (1967) showed that in middle latitudes this field can be considered as plane electromagnetic waves. The incident wave reflects and refracts on the Earth's surface. The reflected wave propagates through the lithosphere and the Earth's mantle where it induces secondary, time variable electric currents that cause a secondary undulation. The secondary undulation interferes with the primary field and the field of the reflected wave. Its character is, however, influenced by the character of the incident wave as well as the character of the medium in the Earth's interior through which it propagates; primarily by the conductivity of the rock medium. The analysis of the electromagnetic field allows us to study electric conductivities of the rock medium investigated as well as further physical parameters of the rock medium, and to make conclusions about these phenomena, i.e. also on the structure of the Earth's body.

Electromagnetic sounding is carried out by two methods — geometric sounding and parametric sounding. These deep sounding methods are performed in several ways.

Geomagnetic deep sounding is used to determine lateral inhomogeneities of conductivity in the Earth's crust and outer mantle (Price (1949), (1950), Rikitake (1966), Weaver (1970), Jones and Price (1970), (1971), Nedoma (1973), Nedoma and Praus (1972), Reddy and Rankin (1975)). The method consists of the measurement of three orthogonal components of variations of the geomagnetic field. Based on these measurements, the behaviour of the individual geomagnetic components or, possibly, their ratio are then studied along profiles obtained on the Earth's surface for various periods of the outer field. Their evaluation leads to conclusions on the structure of the outer parts of the Earth's body (Rikitake (1966)). Anomalous behaviour of the vertical component of the magnetic field H_z is observed, in particular, in the vicinity of inhomogeneities like a vertical or oblique contact, or inhomogeneities formed by disturbance bodies in the lithosphere. The magnitude of this component attains a distinct maxima there while it almost vanishes in relatively homogeneous geological structures. In contrast, the horizontal components H_x and H_y mutually complement and amplify each other.

The magneto-telluric method allows us to study and determine the individual layers of the Earth's lithosphere and outer mantle. It employs variable parts of the Earth's natural electromagnetic field called the magneto-telluric field. Orthogonal horizontal components of the electric and magnetic field are measured as time series on the Earth's surface. Based on the study of the Earth's surface impedance, i.e. of the ratio of horizontal components of the electric and magnetic natural field on the Earth's surface, we obtain certain ideas on electric properties of the lithosphere, asthenosphere, and possibly also mesosphere. Moreover, based on synthetic analyses of all geophysical fields, we arrive at an idea of the structure of the outer parts of the Earth's body.

Analysis of the boundary value problem for the electromagnetic field of a homogeneous Earth's body shows that the resistivity ρ, which is a function of the frequency ω, is given by the relation

$$\rho(\omega) = (\mu\omega)^{-1} |J(\omega)|^2 = (\mu\omega)^{-1} \left|E_x/H_y\right|^2 = (\mu\omega)^{-1} \left|E_y/H_x\right|^2 \tag{10.1}$$

where $J(\omega)$ is the inner impedance of the Earth, μ is the magnetic permeability of the Earth, E_x, E_y, H_x, H_y are components of the intensity of electric and magnetic fields of the Earth and ω is the circular frequency. The resistivity $\rho(\omega)$ and the inner impedance of the Earth $J(\omega)$ are independent of the frequency if we consider a really homogeneous medium. If the Earth's body is approximated by n isotropic homogeneous horizontal layers parallel to the plane surface then the resistivity is a function of the frequency, electric conductivity, and thickness of layers. Then

$$\rho_a(\omega) = (\mu\omega)^{-1} |Z(\omega)|^2 , \tag{10.2}$$

where $\rho_a(\omega)$ is an average weighted value of the resistivity over all the layers and $Z(\omega)$ is the input impedance on the Earth's surface. If we consider an anisotropic or laterally inhomogeneous Earth's body then the impedance is a tensor and the relation among electric and magnetic components of the field can be expressed with the help of the impedance tensor as

$$\begin{pmatrix} E_x \\ E_y \end{pmatrix} = \begin{pmatrix} Z_{11} & Z_{12} \\ Z_{21} & Z_{22} \end{pmatrix} \begin{pmatrix} H_x \\ H_y \end{pmatrix}, \tag{10.3}$$

where E_x, E_y, H_x, H_y and Z_{ij}, $i, j = 1, 2$, in the frequency domain are functions of the frequency ω. In matrix notation Equation 10.3 can be written as follows

$$\boldsymbol{E}(\omega) = \boldsymbol{Z}(\omega)\boldsymbol{H}(\omega). \tag{10.4}$$

The tensor Z_{ij} in Equation 10.3 may be quite general in dependence on the character of the medium.

In the time-domain approach the impedance tensor $\boldsymbol{Z}(t)$ depends on the noise produced by spikes. Then

$$\begin{pmatrix} E_x(t) \\ E_y(t) \end{pmatrix} = \begin{pmatrix} Z_{11}(t) & Z_{12}(t) \\ Z_{21}(t) & Z_{22}(t) \end{pmatrix} \begin{pmatrix} H_x(t) \\ H_y(t) \end{pmatrix} + \begin{pmatrix} N_x(t) \\ N_y(t) \end{pmatrix} \tag{10.5}$$

or in the matrix notation

$$\boldsymbol{E}(t) = \boldsymbol{Z}(t)\boldsymbol{H}(t) + \boldsymbol{N}(t), \tag{10.6}$$

where $\boldsymbol{N}(t)$ denotes the additive noise. In the time-domain approach the pulse or impulse response matrix $\boldsymbol{Z}(t)$, the impedance tensor, is determined directly from the time measured data, while in the frequency domain approach it is determined from the Fourier transform of the elements of the pulse response matrix $\boldsymbol{Z}(t)$.

An important group of real medium models is formed by $2D$ (two-dimensional) or $3D$ (three-dimensional) inhomogeneous media. In $2D$ media, the distribution of electric conductivity changes substantially on certain parallel planes perpendicular to the Earth's surface and does not change in the direction perpendicular to these planes (the case of the $2D$ model). This distinguished direction is then called the direction of homogeneity of the medium. This property characterizes for example media with vertical or oblique contacts of geological blocks with different electric conductivities, corrugated interfaces, mildly corrugated interfaces of a synclinal or anticlinal type, and their combinations. In what follows we will be concerned with $2D$ models only.

The existence of the direction of homogeneity of the medium means that, for the corresponding tensor Z_{ij} in Equation 10.3, two mutually perpendicular directions can be found in which direct proportionality among components of the electric field and the component of the magnetic field in the perpendicular direction again holds. These proportionality functions are then called the principal directions of impedance, and the relation in Equation 10.3 is transformed into the form

$$\begin{pmatrix} \hat{E}_x \\ \hat{E}_y \end{pmatrix} = \begin{pmatrix} 0 & \hat{Z}_{12} \\ \hat{Z}_{21} & 0 \end{pmatrix} \begin{pmatrix} H_x \\ H_y \end{pmatrix}. \tag{10.7}$$

In interpreting magneto-telluric data, the measured values of impedance are compared, as functions of frequency, with the values obtained from theoretical studies of model problems. It means that, for the true interpretation of magneto-telluric measurements according to Equation 10.7, the curves of impedances obtained in a general measurement system have first to be transformed into the principal directions of impedance (if, however, they do exist) and only then they can be interpreted. In this case, the principal directions of impedance Z_{12} and Z_{21} contain only information on the distribution of the electric conductivity with depth measured from the Earth's surface without directional influences. Analysis of Maxwell equations shows that the principal directions of impedance correspond to the two basic modes of electromagnetic field, i.e. the TM and TE modes. The TM mode describes the H-polarization of field under which the distribution of the field can be uniquely determined from the distribution of the magnetic component of field in the direction of homogeneity. The TE mode describes the E-polarization of field under which the distribution of the field can uniquely be determined from the distribution of the electric component of field in the direction of homogeneity.

10.2. VARIATIONAL PROBLEMS IN THE THEORY OF MAGNETO-TELLURIC FIELD AND DEEP GEOMAGNETIC SOUNDING. MODEL PROBLEMS

10.2.1. Introduction

The study of induced geomagnetic field provides us with very valuable information on the conductivity of rocks that form the lithosphere and outer mantle. By means of conductivity, together with synthetic studies based on the deep investigation of other geophysical fields (Zátopek (1979), Rikitake (1966)), this information is extended to the structure of the Earth's body. Methods discussed in the introductory part are used to determine the electric conductivity in the lithosphere and the Earth's mantle. One of the methods discussed is the magneto-telluric method which employs variations of the natural electromagnetic field of the Earth. These variations include frequencies in the range 10^{-5} to 10^{3} Hz. Penetration depth of these variations in the direction of the Earth's core increases with the decrease of frequency according to the electromagnetic skin effect. This fact allows us to study electric conductivity with depth. In the case of a horizontally homogeneous medium, it is possible, from the knowledge of the analytic solution, to determine the dependence of the surface impedance on frequency. Moreover, comparing this behaviour with the values measured on the Earth's surface, we can determine the structure of the lithosphere in dependence on the depth, i.e. the number of layers, their profile, electric conductivities, and their location in the lithosphere. In addition to the frequency dependence of surface impedancies, it proves necessary to study and analyse the behaviour of the vertical component of the magnetic field on the Earth's surface.

10.2.2. Physical Formulation. Mathematical Models in Frequency and Time Domains

Spectral analyses of the magneto-telluric field from magneto-telluric data for impedance tensor estimation demands the stationarity of both H and E measured components of the magneto-telluric field. The time-domain technique for estimation of the magneto-telluric impedance tensor is less sensitive to noise which is produced by spikes (McMechan and Barrodale (1985), Yee, Kosteniuk and Paulson (1988), Larsen (1989), Spagnolini (1994)) and without the noise mathematically formulated and analysed by Nedoma (1974), (1975), (1977), (1984/1991).

Let us consider the Earth as an infinite, in general inhomogeneous medium characterized by properties of an isotropic or possibly anisotropic inhomogeneous conductor. In our considerations, we will now confine ourselves to an inhomogeneous isotropic medium, the extension for an anisotropic inhomogeneous case is evident.

In our model problems, we will assume that all the quantities that characterize the electromagnetic field depend only on the $x_2 = y$ and $x_3 = z$ coordinates. Let us first derive the corresponding equations and relations for general $3D$ models for which we suppose that all the parameters of the medium are functions of all three variables x_i, $i = 1, 2, 3$.

We will start from the general equations of an electromagnetic field in inhomogeneous isotropic conductors that are of the form

$$\frac{\partial \boldsymbol{D}}{\partial t} + \boldsymbol{J} - \text{rot } \boldsymbol{H} = 0, \tag{10.8}$$

$$\text{div } \boldsymbol{D} = q, \tag{10.9}$$

$$\frac{\partial \boldsymbol{B}}{\partial t} + \text{rot } \boldsymbol{E} = 0, \tag{10.10}$$

$$\text{div } \boldsymbol{B} = 0, \tag{10.11}$$

with the definition relations

$$\boldsymbol{D} = \varepsilon \boldsymbol{E}, \tag{10.12}$$

$$\boldsymbol{B} = \mu \boldsymbol{H}, \tag{10.13}$$

$$\boldsymbol{J} = \sigma(\boldsymbol{E} + \underline{\boldsymbol{E}}), \tag{10.14}$$

where $\boldsymbol{H}$ is the vector of intensity of the magnetic field, $\boldsymbol{E}$ the vector of intensity of electric field, $\boldsymbol{D}$ the vector of electric induction, $\boldsymbol{B}$ the vector of magnetic induction, $\boldsymbol{J}$ the vector of density of electric current (Equation 10.14 is the Ohm law), q the density of electric charge, ε dielectric constant, μ permeability, σ conductivity, $r = \sigma^{-1}$ resistivity ($[\Omega m]$), t time, and $\underline{\boldsymbol{E}}$ the vector of impressed force.

Let us consider a model of an inhomogeneous infinite conductor that approximates the Earth with the parameters ε_i, μ_i and σ_i, $i = 1, 2, \ldots, n$, surrounded by dielectric with the parameters ε', μ' and $\sigma' = 0$. Let us further suppose that the conductor (conductive Earth) is piecewise homogeneous and, therefore, $\underline{\boldsymbol{E}} = 0$. Then

$$\boldsymbol{J} = \sigma \boldsymbol{E} = \sigma \varepsilon^{-1} \boldsymbol{D}. \tag{10.15}$$

We apply the divergence operation to Equation 10.15 and, using Equation 10.9, we arrive at

$$\text{div } \boldsymbol{J} = \sigma\varepsilon^{-1} \text{div } \boldsymbol{D} = \sigma\varepsilon^{-1} q.$$

The continuity equation, i.e. the equation

$$\text{div } \boldsymbol{J} + \frac{\partial \rho}{\partial t} = 0 \tag{10.16}$$

implies

$$-\frac{\partial q}{\partial t} = \sigma\varepsilon^{-1} q.$$

Separating the variables and integrating, we obtain $\ln q = -\sigma\varepsilon^{-1} t$, i.e.,

$$q(\boldsymbol{r}, t) = q(\boldsymbol{r}, 0) \exp(-\sigma\varepsilon^{-1} t) = q(\boldsymbol{r}, 0) \exp(-t/\hat{T}), \tag{10.17}$$

where

$$\hat{T} = \varepsilon\sigma^{-1}([\hat{T}] = \text{sec}) \tag{10.18}$$

is the relaxation time, which is extremely short for conductors since $\sigma \sim 10^{17}\text{s}^{-1}$. The quantity $q_o = q(\boldsymbol{r}, 0)$ is the initial volume density of the free charge. We see from this that the initial distribution of the charge density q_o in a homogeneous conductor, if it is left without outer influences, cannot be held and decreases exponentially to zero during time q. In this process, the charge is dispersed to the surface of the conductor and leaks to the surrounding space. The situation is analogical in a piecewise homogeneous infinite conductor. Let us thus consider the Earth as a conductor in the field of outer sources. Then we can use the general equations of field in an inhomogeneous isotropic conductor. Apply the operation rot(curl) to Equations 10.8 and 10.10. Since

$$\text{rot rot } \boldsymbol{H} = \text{grad div } \boldsymbol{H} - \Delta\boldsymbol{H},$$

we have

$$\text{grad div } \boldsymbol{H} - \Delta\boldsymbol{H} - \frac{\partial}{\partial t}(\varepsilon \text{ rot } \boldsymbol{E}) = \text{ rot } \boldsymbol{J} = \sigma \text{ rot } \boldsymbol{E}.$$

But div $\boldsymbol{B} = 0$, therefore div $\boldsymbol{H} = 0$ and the first term vanishes. Using Equation 10.10, we obtain $-\Delta\boldsymbol{H} + (\partial/\partial t)(\varepsilon(\partial\boldsymbol{B}/\partial t)) = -\sigma(\partial\boldsymbol{B}/\partial t)$, i.e.

$$\Delta\boldsymbol{H} - \varepsilon\mu\frac{\partial^2\boldsymbol{H}}{\partial t^2} - \sigma\mu\frac{\partial\boldsymbol{H}}{\partial t} = 0. \tag{10.19}$$

Analogically

$$\text{rot rot } \boldsymbol{E} + \text{ rot } \frac{\partial\boldsymbol{B}}{\partial t} = \text{grad div } \boldsymbol{E} - \Delta\boldsymbol{E} + \mu \text{ rot } \frac{\partial\boldsymbol{H}}{\partial t} = 0.$$

Since the relaxation time for conductors is very short we can assume that the charge density q inside the conductive Earth is zero, i.e. that

$$q = q_o \exp(-\sigma\varepsilon^{-1}t) = 0.$$

Then

$$\text{div } \boldsymbol{D} = \varepsilon\text{div } \boldsymbol{E} = 0, \quad \text{i.e.} \quad \text{div } \boldsymbol{E} = 0.$$

This implies grad div $\boldsymbol{E} = 0$ and we finally obtain

$$\Delta\boldsymbol{E} - \mu \text{ rot } \frac{\partial\boldsymbol{H}}{\partial t} = 0.$$

We have from the first series of Maxwell equations, i.e. from Equation 10.8, that

$$\text{rot } \boldsymbol{H} = \frac{\partial\boldsymbol{D}}{\partial t} + \boldsymbol{J}.$$

Then

$$\Delta\boldsymbol{E} - \mu\frac{\partial}{\partial t}\left(\frac{\partial\boldsymbol{D}}{\partial t} + \boldsymbol{J}\right) = 0$$

and from this,

$$\Delta \boldsymbol{E} - \varepsilon\mu \frac{\partial^2 \boldsymbol{E}}{\partial t^2} - \sigma\mu \frac{\partial \boldsymbol{E}}{\partial t} = 0. \tag{10.20}$$

follows. Symmetry of Equations 10.19 and 10.20 shows that it is possible to neglect the term $(\partial \boldsymbol{D}/\partial t)$ that represents the density of displacement currents in the conductor but, at the same time, to retain the term $(\partial \boldsymbol{B}/\partial t)$ that represents the density of induction flux despite their order being the same. The physical explanation is based on the fact that the value of conductivity σ in the conductor is large in the conductive Earth and thus the quantities $\boldsymbol{J} = \sigma \boldsymbol{E}$ and rot $\boldsymbol{H}$ that appear in Equation 10.8 of the first series of Maxwell equations are large while the term $(\partial \boldsymbol{D}/\partial t)$ is small in magnitude, i.e.

$$\left| \frac{\partial \boldsymbol{D}}{\partial t} \right| \ll |\boldsymbol{J}|. \tag{10.21}$$

In fact, it has been observed that the displacement currents can be neglected for the periods considered. The quantities $\boldsymbol{E}$, rot $\boldsymbol{E}$, $(\partial \boldsymbol{D}/\partial t)$, and $(\partial \boldsymbol{B}/\partial t)$ are small and comparable as far as their magnitude is concerned. This implies that the term $(\partial \boldsymbol{B}/\partial t)$ cannot be neglected since it appears in an equation whose terms are of the same order of magnitude. Then Equations 10.8 to 10.11 and the corresponding equations, Equations 10.19 and 10.20, can be written in the form

$$\text{rot}\, \boldsymbol{H} = \boldsymbol{J}, \tag{10.22}$$

$$\text{div}\, \boldsymbol{D} = 0, \tag{10.23}$$

$$\text{rot}\, \boldsymbol{E} + \frac{\partial \boldsymbol{B}}{\partial t} = 0, \tag{10.24}$$

$$\text{div}\, \boldsymbol{B} = 0, \tag{10.25}$$

$$\Delta \boldsymbol{H} - \sigma\mu \frac{\partial \boldsymbol{H}}{\partial t} = 0, \tag{10.26}$$

$$\Delta \boldsymbol{E} - \sigma\mu \frac{\partial \boldsymbol{E}}{\partial t} = 0. \tag{10.27}$$

If displacement currents are neglected the decisive parameter of medium is its conductivity σ that is considered as a variable $\sigma = \sigma(x_i)$ while the remaining two parameters of medium, i.e. the dielectric constant ε and magnetic permeability μ, are considered constant.

Remark: The dielectric constant ε and magnetic permeability μ are tensors in the case of anisotropic medium. Moreover, they need not be Hermitian and their components are functions of all three variables x_i, $i = 1, 2, 3$.

If the inducing field is periodically variable with the factor $\exp(i\omega t)$ then Equations 10.26 and 10.27 transform into the form

$$\Delta \boldsymbol{H} - i\omega\sigma\mu \boldsymbol{H} = 0, \tag{10.28}$$

$$\Delta \boldsymbol{E} - i\omega\sigma\mu \boldsymbol{E} = 0, \tag{10.29}$$

and Equations 10.22 to 10.25 transform to the form

$$\text{rot}\, \boldsymbol{H} - \sigma \boldsymbol{E} = 0, \tag{10.30}$$

$$\text{div}\, \boldsymbol{D} = 0, \tag{10.31}$$

$$\text{rot}\, \boldsymbol{E} + i\omega\mu \boldsymbol{H} = 0, \tag{10.32}$$

$$\text{div}\, \boldsymbol{B} = 0. \tag{10.33}$$

In our considerations we took the Earth for an infinite plane conductor filling the half-space $z > 0$ of the Cartesian coordinate system. These considerations can be assumed since the plane conductive Earth can be understood as a limit case of a spheric symmetric approximation of the conductive Earth if its radius $R \to \infty$. This, nevertheless, implies that investigating the Earth's body as an infinite plane conductor we have to take into account physical properties of the spheric symmetric conductive Earth. The induced electric currents in the Earth are the consequence of a time variable outer magnetic field. Moreover, these currents influence again the outer magnetic field in the sense of the above stated considerations. Srivastava (1966) showed that the curvature can be neglected in certain cases. According to his results, it can be neglected for conductivities $\sigma \sim 10^{-12}$ to $10^{-3}\Omega^{-1}\text{m}^{-1}$ and periods less than one day.

In our further considerations, we will confine ourselves to a simpler case with symmetry in the direction of the x_1 axis. We will assume in the model problem that the isotropic conductive Earth is approximated by an infinite, in general inhomogeneous, isotropic medium that fills the half-space $z > 0$ of the Cartesian coordinate system. In this case, we again take the direction tending to the isotropic medium given for the positive direction of the x_3 axis. Let the inducing current system be in the half-space $z < 0$.

Consider thus in the symmetry given that all the quantities characterizing the electromagnetic field depend only on the x_2 and x_3 coordinates and are completely independent of the x_1 coordinate. In this symmetry, Equations 10.28 and 10.29 take the form

$$\Delta H_x - \sigma\mu \frac{\partial H_x}{\partial t} = 0 \tag{10.34}$$

and

$$\Delta E_x - \sigma\mu \frac{\partial E_x}{\partial t} = 0. \tag{10.35}$$

Equations 10.34 and 10.35 can also be derived in the following way (see also Nedoma (1973), (1975)). In this symmetry, Equations 10.8 to 10.11 imply

$$\frac{\partial H_z}{\partial x_2} - \frac{\partial H_y}{\partial x_3} = \sigma E_x, \tag{10.36}$$

$$\frac{\partial H_x}{\partial x_3} = \sigma E_y, \tag{10.37}$$

$$-\frac{\partial H_x}{\partial x_2} = \sigma E_z, \tag{10.38}$$

as far as the first series of Maxwell equations is concerned, and the equations

$$\frac{\partial E_z}{\partial x_2} - \frac{\partial E_y}{\partial x_3} = -\mu \frac{\partial H_x}{\partial t}, \tag{10.39}$$

$$\frac{\partial E_x}{\partial x_3} = -\mu \frac{\partial H_y}{\partial t}, \tag{10.40}$$

$$-\frac{\partial E_x}{\partial x_2} = -\mu \frac{\partial H_z}{\partial t}, \tag{10.41}$$

follow from the second series of Maxwell equations.

Equations 10.36 to 10.38 and 10.39 and 10.41 can be rewritten in such a way that they contain either only E_y, E_z, and H_x, or only E_x, H_z, and H_y. We will see that this combination of the components of the vectors $\boldsymbol{E}$ and $\boldsymbol{H}$ possesses a special symmetry. The first combination of components of the vectors $\boldsymbol{E}$ and $\boldsymbol{H}$ is then called the H-polarization, the other combination the E-polarization. Then we can rewrite Equations 10.36 to 10.38 and 10.39 to 10.41 into the form

$$\frac{\partial E_z}{\partial x_2} - \frac{\partial E_y}{\partial x_3} = -\mu \frac{\partial H_x}{\partial t}, \tag{10.42}$$

$$\frac{\partial H_x}{\partial x_3} = \sigma E_y, \tag{10.43}$$

$$-\frac{\partial H_x}{\partial x_2} = \sigma E_z, \tag{10.44}$$

for the H-polarization, and

$$\frac{\partial H_z}{\partial x_2} - \frac{\partial H_y}{\partial x_3} = \sigma E_x, \tag{10.45}$$

$$\frac{\partial E_x}{\partial x_3} = -\mu \frac{\partial H_y}{\partial t}, \tag{10.46}$$

$$-\frac{\partial E_x}{\partial x_2} = -\mu \frac{\partial H_z}{\partial t}, \tag{10.47}$$

for the E-polarization.

We differentiate Equations 10.37 and 10.38 with respect to x_3 and x_2 to obtain

$$\frac{\partial^2 H_x}{\partial x_3^2} = \sigma \frac{\partial E_y}{\partial x_3}, \quad -\frac{\partial^2 H_x}{\partial x_2^2} = \sigma \frac{\partial E_z}{\partial x_2}, \quad \text{or, generally}$$

$$\frac{\partial}{\partial x_3}\left(\sigma^{-1} \frac{\partial H_x}{\partial x_3}\right) = \frac{\partial E_y}{\partial x_3}, \quad -\frac{\partial}{\partial x_2}\left(\sigma^{-1} \frac{\partial H_x}{\partial x_2}\right) = \frac{\partial E_z}{\partial x_2},$$

and from Equation 10.42 for the H-polarization we obtain

$$\frac{\partial^2 H_x}{\partial x_2^2} + \frac{\partial^2 H_x}{\partial x_3^2} = \sigma\mu \frac{\partial H_x}{\partial t}, \quad \text{or, generally} \quad \frac{\partial}{\partial x_2}\left(\sigma^{-1} \frac{\partial H_x}{\partial x_2}\right) + \frac{\partial}{\partial x_3}\left(\sigma^{-1} \frac{\partial H_x}{\partial x_3}\right) = \mu \frac{\partial H_x}{\partial t}. \tag{10.48}$$

Analogically we obtain for the E-polarization

$$\frac{\partial^2 E_x}{\partial x_2^2}+\frac{\partial^2 E_x}{\partial x_3^2}=\sigma\mu\frac{\partial E_x}{\partial t}, \quad \text{or, generally} \quad \frac{\partial}{\partial x_2}\left(\sigma^{-1}\frac{\partial E_x}{\partial x_2}\right)+\frac{\partial}{\partial x_3}\left(\sigma^{-1}\frac{\partial E_x}{\partial x_3}\right)=\mu\frac{\partial E_x}{\partial t}. \tag{10.49}$$

In the case of the H-polarization, the primary magnetic field is oriented along the homogeneity axis x_1 and the direction of the induced electric field and electric currents is perpendicular to this axis across the geological structure. In the case of the E-polarization, the primary electric field is oriented along the homogeneity axis and the corresponding magnetic field is perpendicular to it. The induced electric current flows along the homogeneity axis and, therefore, along the geological structure.

10.2.2.1. *The frequency domain approach*

Suppose that the outer inducing magnetic field varies periodically with the factor $\exp(i\omega t)$. The equations describing the electromagnetic field in this case can be derived in the same way as above. Writing the Maxwell equations in components in the symmetry considered we arrive at

$$\frac{\partial H_z}{\partial x_2}-\frac{\partial H_y}{\partial x_3}=\sigma E_x, \qquad \frac{\partial H_x}{\partial x_3}=\sigma E_y, \qquad -\frac{\partial H_x}{\partial x_2}=\sigma E_z$$

and

$$\frac{\partial E_z}{\partial x_2}-\frac{\partial E_y}{\partial x_3}=-i\omega\mu H_x, \qquad \frac{\partial E_x}{\partial x_3}=-i\omega\mu H_y, \qquad -\frac{\partial E_x}{\partial x_2}=-i\omega\mu H_z.$$

Let us order these equations again in such a way that they contain only certain components, i.e. in such a way that they contain either only E_y, E_z, and H_x, or only E_x, H_z, and H_y. Then we have for the H-polarization (TM-mode)

$$\frac{\partial E_z}{\partial x_2}-\frac{\partial E_y}{\partial x_3}=-i\omega\mu H_x, \qquad \frac{\partial H_x}{\partial x_3}=\sigma E_y, \qquad -\frac{\partial H_x}{\partial x_2}=\sigma E_z,$$

and for the E-polarization (TE-mode) we have

$$\frac{\partial H_z}{\partial x_2}-\frac{\partial H_y}{\partial x_3}=\sigma E_x, \qquad \frac{\partial E_x}{\partial x_3}=-i\omega\mu H_y, \qquad -\frac{\partial E_x}{\partial x_2}=-i\omega\mu H_z.$$

A procedure analogous to the above leads, in the case of the H-polarization, to

$$\frac{\partial^2 H_x}{\partial x_2^2}+\frac{\partial^2 H_x}{\partial x_3^2}-i\omega\sigma\mu H_x=0, \tag{10.50}$$

while for the E-polarization we obtain

$$\frac{\partial^2 E_x}{\partial x_2^2}+\frac{\partial^2 E_x}{\partial x_3^2}-i\omega\sigma\mu E_x=0, \tag{10.51}$$

which are identical to Equations 10.48 and 10.49 for the time dependence $\exp(\pm i\omega t)$.

In an analogical way we derive also the equation

$$\frac{\partial}{\partial x_2}\left(\sigma^{-1}\frac{\partial H_x}{\partial x_2}\right)+\frac{\partial}{\partial x_3}\left(\sigma^{-1}\frac{\partial H_x}{\partial x_3}\right)-i\omega\mu H_x=0$$

for the H-polarization and

$$\frac{\partial}{\partial x_2}\left(\sigma^{-1}\frac{\partial E_x}{\partial x_2}\right)+\frac{\partial}{\partial x_3}\left(\sigma^{-1}\frac{\partial E_x}{\partial x_3}\right)-i\omega\mu E_x=0$$

for the E-polarization.

Remark: We introduce the electric and magnetic operators L_E and L_H. Since rot rot $=$ grad div $-\ \Delta$ the operators L_E and L_H can be written in the form

$$L_E=\mu^{-1}\text{grad div }-\mu^{-1}\Delta,\quad L_H=\varepsilon^{-1}\text{grad div}-\varepsilon^{-1}\Delta,$$

or in a general form

$$L_E=\text{grad}(\mu^{-1}\text{div})-\nabla(\mu^{-1}\nabla),\quad L_H=\text{grad}(\varepsilon^{-1}\text{div})-\nabla(\varepsilon^{-1}\nabla),$$

where $\mu=$ constant and $\varepsilon=$ constant . In the case of the H- and E-polarization considered we have $L_E=\Delta$ and $L_H=\Delta$ since the terms with grad div vanish with respect to Equations 10.23 and 10.25.

BOUNDARY CONDITIONS

We saw that the electromagnetic field considered is quasi-stationary. Therefore, the condition

$$\text{Div}\,\boldsymbol{J}={}^1J_n-J_n=0$$

will be satisfied on the interface of conductor and dielectric. Since $\sigma_1=0$ in the dielectric, then also ${}^1J_n=0$ there, and thus

$$J_n=0. \tag{10.52}$$

This implies that the current density, and also the intensity of electric field, can have only the tangential component different from zero on the inner side of the interface in the conductor (Earth's body). We have, however, ${}^1J_n=0$ on the outer side, i.e. in the dielectric (in the Earth's outer space), but $\boldsymbol{E}\neq 0$ (see Votruba and Muzikář (1955), Landau Lifshitz (1984)). The normal component is determined by the total charge on the surface of the conductor (the Earth's surface) as in electrostatics. It holds that

$$\text{Div}\,\boldsymbol{D}=q={}^1D_n-D_n=\varepsilon_1{}^1E_n-\varepsilon E_n. \tag{10.53}$$

But the change of surface charge q induces the current

$$J_n=\sigma E_n=\frac{\mathrm{d}q}{\mathrm{d}t}$$

and thus

$$\text{Div}\,\boldsymbol{D}=q=\varepsilon_1{}^1E_n-\varepsilon\sigma^{-1}\frac{\mathrm{d}q}{\mathrm{d}t}.$$

Hence

$$\frac{\mathrm{d}q}{\mathrm{d}t}+\sigma\varepsilon^{-1}q=\sigma\varepsilon^{-1}\varepsilon_1{}^1E_n.$$

As we neglect displacement currents we have

$$q\sim\varepsilon_1{}^1E_n$$

and, moreover,

$$E_n = \sigma^{-1} J_n = \sigma^{-1} \frac{\mathrm{d}q}{\mathrm{d}t} \ll \varepsilon^{-1} q \doteq \varepsilon^{-1} \varepsilon_1 {}^1E_n. \tag{10.54}$$

This, therefore, implies that the normal component E_n is negligible compared with the component in dielectric. The normal component of current density caused by the change of surface charge is

$$\frac{\mathrm{d}q}{\mathrm{d}t} = \varepsilon_1 \frac{\mathrm{d}^1 E_n}{\mathrm{d}t}$$

as follows from Equation 10.53 by neglecting the term with E_n according to Equation 10.54. The order of its magnitude is $c^{-2}E/T$. An estimate of the density of conducting current caused by the magnetically induced field follows from the definition of current density $\boldsymbol{J} = \sigma \boldsymbol{E}$ and from the condition showing that the displacement currents can be neglected, i.e.

$$\boldsymbol{J} = \sigma \boldsymbol{E} \gg \varepsilon E T^{-1} = c^{-2} E T^{-1}.$$

Vertical currents caused by the change of surface charge are therefore negligible compared with conducting currents caused by the magnetic induction of outer sources.

It is known from the theory of electromagnetic field that on the interface between conductors as well as on the Earth's surface, i.e. on the interface conductor-dielectric ($z = 0$, where for simplicity we put $x_3 \equiv z$), we have

(i) all components of the magnetic intensity $\boldsymbol{H}$ are continuous;

(ii) tangential components of the electric intensity $\boldsymbol{E}$ are continuous.

Let us derive the corresponding conditions. We know from the theory of electromagnetic field (quasi-stationary field) that

$$\operatorname{Div} \boldsymbol{J} = {}^1J_n - J_n = \sigma_1 {}^1E_n - \sigma E_n = \operatorname{Div} \sigma \boldsymbol{E} = 0, \tag{10.55}$$

on the interface between two media of conductivities σ and σ_1, since otherwise the surface electric charge would accumulate or disappear on the interface and the current would not be stationary. Further,

$$\operatorname{Rot} \boldsymbol{E} = 0, \text{ i.e. } {}^1E_t - E_t = 0. \tag{10.56}$$

Moreover, the boundary conditions div $\boldsymbol{B} = 0$ hold on the interface between two different media. With respect to the relation $\boldsymbol{B} = \mu \boldsymbol{H}$, we further have

$$\operatorname{Div} \boldsymbol{H} = {}^1H_n - H_n = 0 \tag{10.57}$$

and

$$\operatorname{Rot} \boldsymbol{H} = 0, \text{ i.e. } {}^1H_t - H_t = 0. \tag{10.58}$$

The conditions in Equations 10.57 and 10.58 imply the continuity of all components of the magnetic intensity vector $\boldsymbol{H}$.

For the sake of simplicity, let us consider the case of a periodic inducing outer magnetic field with factor $\exp(-i\omega t)$.

Let us first study the boundary conditions for the H-polarization. For the outer space where $\sigma = 0$, Equations 10.37 and 10.38 imply that

$$\frac{\partial H_x}{\partial x_3} = \frac{\partial H_x}{\partial x_2} = 0,$$

and H_x thus does not depend on the x_2 and x_3 coordinates. It does not depend on the x_1 coordinate with regard to the symmetry considered, either. Then

$$H_x = \text{constant} = H_o \tag{10.59}$$

on the Earth's surface (and thus $\boldsymbol{H}$ is homogeneous) just with respect to the condition of continuity of $\boldsymbol{H}$ for $z = 0$ (see the conditions in Equations 10.57 and 10.58. This condition, i.e. the condition $\partial H_x/\partial x_2 = 0$ together with Equation 10.38, gives $\sigma E_z = 0$.
This condition agrees with the condition in Equation 10.52, i.e. $J_n = 0$, which means that the normal component of current density vanishes on the surface. This condition shows that no current can flow from the Earth to the half-space with conductivity $\sigma = 0$.

Let us now investigate the interface $x_2 = 0$ or $x_3 = z_s$. The requirement on the continuity of $\boldsymbol{H}$, and Equations 10.57 and 10.58 imply

$$H_x|_{(l)} = H_x|_{(m)} \,. \tag{10.60}$$

At the same time,

$$\left.\frac{\partial H_x}{\partial x_3}\right|_{(l)} = \left.\frac{\partial H_x}{\partial x_3}\right|_{(m)}$$

also holds for $x_2 = 0$ or $x_3 = z_s$. This requirement is in accord with Equation 10.55 which represents the condition of continuity of the current density, i.e. the condition

$${\sigma_1}^1 E_n = \sigma E_n$$

or, according to Equation 10.37,

$$\left.\frac{\partial H_x}{\partial x_3}\right|_{(l)} = \left.\frac{\partial H_x}{\partial x_3}\right|_{(m)} .$$

Equation 10.56 which requires continuity of the tangential component E_t, i.e. the component E_2 or E_y in the simplified case studied, implies that

$$E_z|_{(l)} = E_z|_{(m)} \text{ or } E_y|_{(l)} = E_y|_{(m)}$$

holds on the interfaces $x_2 = y_r$ or $z = z_s$, $r = 1, 2, \ldots, m$, $s = 1, 2, \ldots, n$. This gives, with respect to Equation 10.38, the conditions

$$\sigma^{-1}\left.\frac{\partial H_x}{\partial x_2}\right|_{(l)} = \sigma^{-1}\left.\frac{\partial H_x}{\partial x_2}\right|_{(m)} \quad \text{for } x_2 = y_r, \tag{10.61}$$

or

$$\sigma^{-1}\left.\frac{\partial H_x}{\partial x_3}\right|_{(l)} = \sigma^{-1}\left.\frac{\partial H_x}{\partial x_3}\right|_{(m)} \quad \text{for } x_3 = z_s. \tag{10.62}$$

The behaviour of the field at large distances from the interface is similar to that for a homogeneous conductor and the field is independent of x_2 or x_3. Let us investigate the former case. Here we have

$$\frac{\partial^2 H_x}{\partial x_3^2} = i\omega\mu\sigma_k H_x,$$

where $\sigma_k = \sigma_l$ or σ_m. The solution of this ordinary differential equation has the form (we write z in the direction x_3)

$$H_x = H_o \exp(-(i\omega\mu\sigma_k)^{1/2} z).$$

We have

$$H_x = H_o \exp\left(-\tfrac{1}{2}2^{1/2}(1+i)(\omega\mu\sigma_k)^{1/2} z\right) \tag{10.63}$$

for $x_2 \to \pm\infty$, where $\sigma_k = \sigma_l$ or σ_m. Moreover,

$$H_x = 0 \tag{10.64}$$

for $x_3 \to \infty$. Consider the interface $z = z_s$ and suppose, for the sake of simplicity, that it is horizontal and located at a distance $z = z_s$ from the Earth's surface. Then we have

$$\frac{\partial^2 H_x}{\partial x_3^2} = i\eta^2 H_x \tag{10.65}$$

for large values of $|x_2|$ in the region of conductivity σ, where $\eta^2 = \sigma\omega\mu$, and

$$\frac{\partial^2 H_x}{\partial x_3^2} = i\eta_s^2 H_x \tag{10.66}$$

in the region of conductivity σ_1, where $\eta_s^2 = \sigma_1\omega\mu$. The solution is of the form

$$\begin{aligned} H_x &= H_x^{(o)} \exp(-(i\omega\mu\sigma_1)^{1/2} z) + H_x^{(1)} \exp((i\omega\mu\sigma_1)^{1/2} z) \\ &= H_x^{(o)} \exp\left(-\eta_s \tfrac{1}{2}2^{1/2}(1+i)z\right) + H_x^{(1)} \exp\left(\eta_s \tfrac{1}{2}2^{1/2}(1+i)z\right), \end{aligned} \tag{10.67}$$

where $H_x^{(o)}$ and $H_x^{(1)}$ are yet undetermined constants. The first term in Equation 10.67 represents the passing wave, the other the reflected wave. Now $H_x = 0$ for $z \to \infty$. We have $H_x^{(1)} = 0$ from this and thus

$$H_x = H_x^{(o)} \exp(-(i)^{1/2}\eta_s z) = H_x^{(o)} \exp\left(-\tfrac{1}{2}2^{1/2}\eta_s(1+i)z\right). \tag{10.68}$$

Now $H_x = H_x^{(z_s)}$ for $z = z_s$, which implies

$$H^{(o)} = H_x^{(z_s)} \exp\left(\tfrac{1}{2}2^{1/2}(1+i)\eta_s z_s\right)$$

and thus

$$H_x = H_x^{(z_s)} \exp\left(-\tfrac{1}{2}2^{1/2}\eta_s(z - z_s)\right). \tag{10.69}$$

The solution of Equation 10.65 is analogical to that of Equation 10.66 but a bit more complex. We have to take into account that the unknown constants $H_x^{(2)}$ and $H_x^{(3)}$ in the solution

$$\begin{aligned} H_x &= H_x^{(2)} \exp(-\eta_1 (i)^{1/2} z) + H_x^{(3)} \exp((i)^{1/2} \eta_1 z) \\ &= H_x^{(2)} \exp\left(-\eta_1 \tfrac{1}{2} 2^{1/2}(1+i)z\right) + H_x^{(3)} \exp\left(\eta_1 \tfrac{1}{2} 2^{1/2}(1+i)z\right) \end{aligned} \tag{10.70}$$

depend on conductivities σ and σ_1, the depth of interface z_s, and the intensity H_o on the Earth's surface. The first term in Equation 10.70 again represents the reflected (backward) wave. It is now necessary to determine the unknown constants $H_x^{(2)}$ and $H_x^{(3)}$, and the yet unknown value of $H_x^{(z_s)}$ on the interface $z = z_s$. We find them from a linear algebraic system we are now going to derive. With respect to Equation 10.59,

$$H_o = H_x^{(2)} + H_x^{(3)} = \text{ constant.} \tag{10.71}$$

holds on the Earth's surface $z = 0$. We further have

$$H_x^{(z_s)} = H_x^{(2)} \exp(-(i)^{1/2} \eta z_s) + H_x^{(3)} \exp((i)^{1/2} \eta z_s) \tag{10.72}$$

for $z = z_s$. The conditions in Equations 10.60 and 10.62 hold on the interface $z = z_s$. This implies for $z = z_s$ that

$$\begin{aligned} &\sigma^{-1} \left[-(i)^{1/2} \eta H_x^{(2)} \exp(-(i)^{1/2} \eta z_s) + (i)^{1/2} \eta H_x^{(3)} \exp((i)^{1/2} \eta z_s)\right] \\ &\quad = \sigma_1^{-1} \left[-(i)^{1/2} \eta_s H_x^{(2)} \exp(-(i)^{1/2} \eta_s z_s) + (i)^{1/2} \eta_s H_x^{(3)} \exp((i)^{1/2} \eta_s z_s)\right], \end{aligned}$$

i.e.

$$\begin{aligned} &\eta \sigma_1 (\eta_s \sigma)^{-1} \left[-H_x^{(2)} \exp(-(i)^{1/2} \eta z_s) + H_x^{(3)} \exp((i)^{1/2} \eta z_s)\right] \\ &\quad = -H_x^{(2)} \exp(-(i)^{1/2} \eta_s z_s) + H_x^{(3)} \exp((i)^{1/2} \eta_s z_s). \end{aligned}$$

The relation

$$\begin{aligned} H_x^{(z_s)} = \eta \sigma_1 (\eta_s \sigma)^{-1} &\left[-H_x^{(2)} \exp(-(i)^{1/2} \eta z_s) + H_x^{(3)} \exp((i)^{1/2} \eta z_s)\right. \\ &\left. + 2(\eta \sigma_1)^{-1} \eta_s \sigma H_x^{(2)} \exp(-(i)^{1/2} \eta_s z_s)\right] \end{aligned}$$

follows from the above formula, Equation 10.72 and the condition of continuity of the vector $\boldsymbol{H}$ across the interface $z = z_s$. Put

$$\begin{aligned} D_1 &= \exp\left(-\tfrac{1}{2} 2^{1/2}(1+i)\eta z_s\right), \\ D_2 &= \exp\left(\tfrac{1}{2} 2^{1/2}(1+i)\eta z_s\right), \\ D_3 &= \eta \sigma_1 (\eta_s \sigma)^{-1} \left[-\exp\left(\tfrac{1}{2} 2^{1/2}(1+i)\eta z_s\right) + 2(\eta \sigma_1)^{-1} \eta_s \sigma \exp\left(-\tfrac{1}{2} 2^{1/2}(1+i)\eta_s z_s\right)\right], \\ D_4 &= \eta \sigma_1 (\eta_s \sigma)^{-1} D_2. \end{aligned}$$

Then

$$H_x^{(2)} + H_x^{(3)} = H_o, \quad H_x^{(z_s)} - D_1 H_x^{(2)} - D_2 H_x^{(3)} = 0, \quad H_x^{(z_s)} - D_3 H_x^{(2)} - D_4 H_x^{(3)} = 0.$$

From this,

$$
\begin{aligned}
H_x^{(2)} &= H_o(D_2 - D_4)(D_3 + D_4 - D_1 - D_2)^{-1}, \\
H_x^{(3)} &= H_o - H_x^{(2)} = H_o(1 - (D_2 - D_1)(D_3 + D_4 - D_1 - D_2)^{-1}), \\
H_x^{(z_s)} &= H_o(D_2 - D_4)(D_3 + D_4 - D_1 - D_2)^{-1} D_1 \\
&\quad + H_o[1 - (D_2 - D_1)(D_3 + D_4 - D_1 - D_2)^{-1}] D_2 \\
&= H_o(D_3 + D_4 - D_1 - D_2)^{-1}\{(D_2 - D_4)D_1 + [D_3 - D_1 + 2(D_4 - D_2)]D_2\}.
\end{aligned}
$$

Substituting now for $H_x^{(z_s)}$ into Equations 10.68 and 10.69, we finally have the boundary conditions for this case.

We will now determine boundary conditions for the E-polarization. Analogically to the H-polarization, it now also holds that

$$\frac{\partial^2 E_x}{\partial x_3^2} = i\eta^2 E_x \tag{10.73}$$

for $z > 0$, $|x_2| \to \infty$. This implies

$$E_x = E_o \exp\left(-\tfrac{1}{2}2^{1/2}(1+i)\eta z\right). \tag{10.74}$$

We have from the Maxwell equation

$$\frac{\partial E_x}{\partial x_3} = -i\omega\mu H_y$$

and from Equation 10.74 that

$$-i\omega\mu H_y = -\eta\tfrac{1}{2}2^{1/2}(1+i)E_o \exp\left(-\tfrac{1}{2}2^{1/2}(1+i)\eta z\right). \tag{10.75}$$

Note that $H_y = H_o =$ constant outside the conductive medium, i.e. for $\sigma = 0$, as follows from the Maxwell equation

$$\frac{\partial H_z}{\partial x_2} - \frac{\partial H_y}{\partial x_3} = \sigma E_x.$$

This and Equation 10.75 imply for $z = 0$ that

$$i\omega\mu H_o = \eta\tfrac{1}{2}2^{1/2}(1+i)E_o$$

and thus

$$H_o = \tfrac{1}{2}2^{1/2}(1+i)(i\omega\mu)^{-1}\eta E_o.$$

As

$$\frac{\partial E_x}{\partial x_3} = -i\omega\mu H_y = -i\omega\mu H_o = -\tfrac{1}{2}2^{1/2}(1+i)\eta E_o,$$

according to the Maxwell equation, we obtain from this by integration for $z < 0$ that

$$E_x = -\tfrac{1}{2}2^{1/2}(1+i)\eta E_o z + E_o = E_o\left(1 - \tfrac{1}{2}2^{1/2}(1+i)\eta z\right), \tag{10.76}$$

since E_x is a continuous function for $z = 0$. The conditions are thus determined by these formulae for large $|x_2|$, i.e. for sufficiently large distances from the inhomogeneity.

As for the H-polarization, $E_x = 0$ for large $z > 0$.

Let us investigate conditions for large $z < 0$, i.e. for such $z = -z_2$ that are sufficiently far from the Earth's surface $z = 0$. The thickness of this air layer (dielectric) has to be sufficiently large in order that the damping of secondary fields from the half-space $z > 0$, occupied by the Earth's lithosphere and outer mantle, can take place. In this case,

$$\frac{\mathrm{d}^2 E_x}{\mathrm{d}x_3^2} = 0 \tag{10.77}$$

and thus

$$E_x = C_1 + C_2 z \qquad \text{for } z < 0, \text{ i.e. } E_x = E_o(1 - (i)^{1/2}\eta)z. \tag{10.78}$$

If $E_x = E_{01}$ or E_{02} on $x_2 = -R_1$ or $x_2 = R_2$ then the condition $-\partial E_x/\partial x_2 = -i\omega H_z$ implies

$$E_x = -(R_1 + R_2)^{-1}(E_{02} - E_{01})x_2 + \tfrac{1}{2}(E_{02} + E_{01}) \text{ on } x_3 = -z_2. \tag{10.79}$$

10.2.2.2. Time variable inducing magnetic field and the time domain approach

We saw above that for spectral analyses of the magneto-telluric field from magneto-telluric data for impedance tensor estimations, the stationarity of measured H and E components of the magneto-telluric field is demanded. Since the time-domain technique for estimation of the magneto-telluric impedance tensor is less sensitive to noise, which is produced by spikes, then the impedance tensor relation in the matrix notation can be written in the form

$$\boldsymbol{E}(t) = \boldsymbol{Z}(t)\boldsymbol{H}(t) + \boldsymbol{N}(t), \qquad t \in \langle 0, T\rangle,$$

where $\boldsymbol{E}(t)$ and $\boldsymbol{H}(t)$ are noise free and $\boldsymbol{N}(t)$ represents the additive noise. The time-domain approach estimates the pulse and impulse response matrix $\boldsymbol{Z}(t)$ directly from the measured time data. The full problem leads to analysing the stochastic description of this magneto-telluric problem as a stochastic system. For an analysis of such type of stochastic systems see for example Bendat (1958), Davenport and Root (1958), Otnes and Enochson (1972), Adomian (1983), Kleiber and Hien (1992). Here we shall limit ourselves to the noise free field only.

In the conclusion we will sketch the principal ideas of the physical derivation of a mathematical model for a time variable inducing magnetic field. As we showed above the electromagnetic field is described by Equations 10.48 or 10.49. The conditions for the vectors of intensity of magnetic and electric field on the boundary between geological structures (geological regions, geological layers, geological bodies and blocks) of different conductivity as well as those on the Earth's surface ($z = 0$) follow from the fact that, firstly, all components of the vector $\boldsymbol{H}$ are continuous on this boundary, and secondly tangential components of the vector $\boldsymbol{E}$ are also continuous there (Equations 10.57 and 10.56). Further, the normal component of current density has to be continuous across these interfaces and has to vanish on the Earth's surface, i.e. for $z = 0$. Maxwell equations for the Earth's surface (i.e. Equations 10.43 and 10.44) imply that $\partial H_x/\partial x_2 = \partial H_x/\partial x_3 = 0$ for the space outside the Earth, where $\sigma = 0$, and H_x thus does not depend on x_3 and x_2 coordinates, i.e.

$$H_x = H_o \tag{10.80}$$

on the Earth's surface. From Div $\boldsymbol{B} = 0$ and Rot $\boldsymbol{H} = 0$, and Div $\boldsymbol{J} = 0$ and Rot $\boldsymbol{E} = 0$, it follows, with respect to $\partial H_x/\partial x_3 = \sigma E_y$, that

$$H_x|_{(l)} = H_x|_{(m)}, \quad \sigma^{-1}\left.\frac{\partial H_x}{\partial \boldsymbol{n}}\right|_{(l)} = \sigma^{-1}\left.\frac{\partial H_x}{\partial \boldsymbol{n}}\right|_{(m)} \quad \text{on } \gamma_l^m, \tag{10.81}$$

which are transition conditions on interfaces between two geological structures of different conductivity for the H_x component. The other boundary conditions show that E_x, H_y, and H_z are continuous on the Earth's surface. Maxwell equations and the relations Rot $\boldsymbol{E} = 0$, Div $\boldsymbol{B} = 0$, and Rot $\boldsymbol{H} = 0$ yield conditions on the interface between geological structures of different conductivity in the form

$$E_x|_{(l)} = E_x|_{(m)}, \quad \left.\frac{\partial E_x}{\partial \boldsymbol{n}}\right|_{(l)} = \left.\frac{\partial E_x}{\partial \boldsymbol{n}}\right|_{(m)} \quad \text{on } \gamma_l^m. \tag{10.82}$$

In the case of E-polarization, the field in the domain $x_3 < 0$, where $\sigma_o = 0$, is not constant as in the case of H-polarization and, therefore, it is necessary to determine it at a sufficient distance $x_3 = -Z_o$,

$$E_x = E_o, \quad x_3 = -Z_o, \quad t \geq 0. \tag{10.83}$$

The field behaves as an inhomogeneous conductor at a sufficient distance from the inhomogeneities and their interfaces, and it thus depends on the x coordinate only. Then

$$\frac{\partial^2 H_x}{\partial x_3^2} = \sigma_k \mu \frac{\partial H_x}{\partial t}, \quad \text{where } k = 1, 2, \ldots, p, \tag{10.84}$$

$$\frac{\partial^2 E_x}{\partial x_3^2} = \sigma_k \mu \frac{\partial E_x}{\partial t}, \quad \text{where } k = 1, 2, \ldots, p, \tag{10.85}$$

with given initial conditions $H_x = H_o$, $E_x = E_o$. We can find components of the field from this, either analytically or numerically, assuming that steps in the variables x_3 and t are the same as for the discretization of Equations 10.48 and 10.49, i.e.

$$H_x = H_1(x_2, x_3, t) \quad \text{on } \partial\Omega\backslash\Gamma_1, \tag{10.86}$$

$$E_x = E_1(x_2, x_3, t) \quad \text{on } \partial\Omega\backslash\Gamma_1. \tag{10.87}$$

The components H_x and $E_x \to 0$ as $x_3 \to +\infty$. We further assume that some initial field is distributed in the inhomogeneous conductor, i.e.

$$H_x = H_{01} \quad \text{in } \overline{\Omega}, \quad t = 0, \tag{10.88}$$

$$E_x = E_{01} \quad \text{in } \overline{\Omega}, \quad t = 0. \tag{10.89}$$

10.2.3. Magneto-Telluric Field. Mathematical Models and Variational Solutions

10.2.3.1. Mathematical formulation of the problem in the frequency and time domains

Let a Cartesian coordinate system (x_1, x_2, x_3) be given such that the plane (x_1, x_3) is identical with the Earth's surface. Further let the x_3 axis point to inhomogeneous, possibly layered, medium characterized by electric conductivity σ_l (in emu). We suppose that the conductivity does not depend on the x_1 coordinate and that $\sigma_l \neq 0$ for $l = 1, 2, \ldots, p$, $x_3 \geq 0$, and $\sigma_l = 0$ for $l = 0$, $z < 0$, i.e. for the Earth's outer space. The index l is assumed to denote the respective geological structure or, in layered medium, the respective layer.

Let us first investigate magneto-telluric field that is periodic with the time factor $\exp(\pm i\omega t)$. We will include both cases of H- and E-polarization (TM- and TE-mode) into one study.

The field in each geological structure (geological region, layer, body, block) is described by the partial differential equation with complex coefficients

$$LH \equiv \frac{\partial^2 H_l}{\partial x_2^2} + \frac{\partial^2 H_l}{\partial x_3^2} - i\omega\mu_l\sigma_l H_l = 0, \quad l = 1, 2, \ldots, p, \tag{10.90}$$

where $H \equiv H_x$ in the case of H-polarization, or $H \equiv E_x$ in the case of E-polarization. Denote by Ω_l and Ω_m two adjacent geological structures (geological regions, bodies, layers, blocks) with conductivities σ_l and σ_m. Denote further their common interface by $\gamma_l^m = \overline{\Omega}_l \cap \overline{\Omega}_m$, where $\sigma_l \geq 0$ for $l = 1, 2, \ldots, p$, $z \geq 0$, and $\sigma_l = 0$ for $l = 0$, $z < 0$, i.e. for the space outside the Earth's surface. The conditions

$$H_l = H_m, \quad \alpha \left.\frac{\partial H}{\partial \boldsymbol{n}}\right|_{(l)} = \alpha \left.\frac{\partial H}{\partial \boldsymbol{n}}\right|_{(m)}, \tag{10.91}$$

where

$\alpha_j = \sigma_j^{-1}$ in the case of H-polarization, $j = l$ or m

$\alpha_j = 1$ in the case of E-polarization, $j = l$ or m

and where $\boldsymbol{n}$ is the direction of the outer normal to γ_l^m, hold on the interfaces γ_l^m between the individual geological structures Ω_l and Ω_m.

We showed in the previous section that we can confine ourselves to a finite domain in the vicinity of the inhomogeneity in our considerations. Denote this finite domain by $\Omega = \cup_{i=1}^{p} \Omega_i$ in the case of H-polarization and $\Omega = \cup_{i=0}^{p} \Omega_i$ in the case of E-polarization, where $\Omega_l \cap \Omega_m = 0$, $l \neq m$. Let Ω be a connected domain with a piecewise smooth boundary $\partial\Omega = \cup_j \Gamma_j$ and the boundaries between geological structures $\gamma_l^m = \overline{\Omega}_l \cap \overline{\Omega}_m$, $\overline{\Omega}_l = \Omega_l \cup \partial\Omega_l$, $\overline{\Omega}_l \cap \overline{\Omega}_m \neq 0$ for $|l - m| > 1$. On this domain we solve Equation 10.90 with the interface conditions of Equation 10.91 on interfaces γ_l^m and with the conditions on the boundary $\partial\Omega$ of the form

$$H|_{\partial\Omega} = \varphi(x_2, x_3), \tag{10.92}$$

where $\varphi(x_2, x_3)$ has been determined from the known properties of electromagnetic field discussed in the previous section. It is assumed that the presented properties of the field hold at distances of several wavelengths from the last interface γ_l^m.

In the case of more general geological structures, the magneto-telluric field is described by the boundary value problem (for both H- and E-polarization)

$$-\frac{\partial}{\partial x_2}\left(\alpha\frac{\partial H}{\partial x_2}\right) - \frac{\partial}{\partial x_3}\left(\alpha\frac{\partial H}{\partial x_3}\right) + i\omega\mu H = 0 \quad \text{in } \Omega, \tag{10.93}$$

$$H|_{\partial\Omega} = \varphi(x_2, x_3), \tag{10.94}$$

where $\sigma = \sigma(x_2, x_3)$ and $\varphi(x_2, x_3)$ has again been determined from the known properties of electromagnetic field discussed above.

The other case discussed is the case of a magneto-telluric field caused by time variable inducing magnetic field. We will again include both possibilities, i.e. the H- as well as the E-polarization (TM- and TE-mode) into one study. The electromagnetic field in each structure is described by the partial differential equation of evolutionary type

$$LH \equiv \frac{\partial}{\partial x_2}\left(\alpha\frac{\partial H}{\partial x_2}\right) + \frac{\partial}{\partial x_3}\left(\alpha\frac{\partial H}{\partial x_3}\right) = \mu\frac{\partial H}{\partial t} \quad \text{in } \Omega_t = \Omega \times (0, T), \quad 0 \le T < \infty, \tag{10.95}$$

where Ω is defined in the same way as above, $H = H_x$ in the case of H-polarization and $H = E_x$ in the case of E-polarization.

We then have, on the boundary ∂G of the domain chosen from infinite inhomogeneous medium, the condition of Dirichlet type

$$H(x_2, x_3, t) = H_1(x_2, x_3, t) \quad \text{on } \partial\Omega, \quad t > 0, \tag{10.96}$$

where the function $H_1(x_2, x_3, t)$ has been determined from the behaviour of the time variable field either analytically or numerically, by solving a one-dimensional equation of the type

$$\frac{\partial}{\partial x_3}\left(\alpha\frac{\partial H}{\partial x_3}\right) = \mu\frac{\partial H}{\partial t} \quad \text{for } x_3 \ge 0, \quad t > 0,$$

with the initial condition

$$H(x_3, 0) = \overline{H}_o(x_3),$$

where $\overline{H}_o(x_3)$ is derived from $H_o(x_2, x_3)$, which is a given function describing the initial state of the field at time $t = 0$. In addition to Equation 10.95 and the boundary condition in Equation 10.96, we also have the initial condition

$$H(x_2, x_3, 0) = H_o(x_2, x_3) \quad (x_2, x_3) \in \Omega, \quad t = 0, \tag{10.97}$$

that describes the magneto-telluric field at the moment $t = 0$.

10.2.3.2. Variational formulation, existence and uniqueness of the solution of a model magneto-telluric problem. The frequency and time domains

In this section, we will present a few model problems that will be discussed from various points of view.

First we will start from the problem in Equations 10.93 and 10.94, i.e. from the problem

$$LH \equiv -\left(\frac{\partial}{\partial x_2}\left(\alpha\frac{\partial H}{\partial x_2}\right)+\frac{\partial}{\partial x_3}\left(\alpha\frac{\partial H}{\partial x_3}\right)\right)+i\omega\mu H = 0 \text{ in } \Omega, \tag{10.98}$$

$$H|_{\partial\Omega} = \varphi(x_2, x_3), \tag{10.99}$$

where we assume that $\alpha = \alpha(x_2, x_3)$ and $\mu = \mu(x_2, x_3)$ are functions bounded and measurable in the domain Ω. As $\alpha(x_2, x_3)$ is a real function we have

$$\exists c > 0 \quad \forall \boldsymbol{x} \in \overline{\Omega} \qquad \sum_{i,j=2,3} \alpha\xi_i\bar{\xi}_j \geq c \sum_{i=2,3} |\xi_i|^2, \tag{10.100}$$

where $\xi = (\xi_2, \xi_3)$ is an arbitrary non-zero vector. Further let $\mu(x_2, x_3) \in C^1(\Omega)$ be a real function.

We will first reformulate the problem in Equations 10.98 and 10.99 to a problem with homogeneous boundary conditions.

Let w be an arbitrary sufficiently smooth function satisfying the boundary condition in Equation 10.99 and such that $H = H - w$ (where we retain the same notation for the function sought as in the original problem). The original problem in Equations 10.98 and 10.99 is then equivalent to the problem: find a function H that satisfies the equation

$$LH \equiv -\left(\frac{\partial}{\partial x_2}\left(\alpha\frac{\partial H}{\partial x_2}\right)+\frac{\partial}{\partial x_3}\left(\alpha\frac{\partial H}{\partial x_3}\right)\right)+i\omega\mu H = f \text{ in } \Omega, \tag{10.101}$$

$$H|_{\partial\Omega} = 0 \tag{10.102}$$

where

$$f = -\left(\frac{\partial}{\partial x_2}\left(\alpha\frac{\partial w}{\partial x_2}\right)+\frac{\partial}{\partial x_3}\left(\alpha\frac{\partial w}{\partial x_3}\right)\right)+i\omega\mu w. \tag{10.103}$$

Let

$$V = \{h \mid h \in H^1(\Omega), \quad h = 0 \text{ on } \partial\Omega \text{ in the sense of traces }\} \equiv H_o^1(\Omega) \tag{10.104}$$

be the space of test functions. For a function $H \in H^1(\Omega)$ we put

$$a(H, h) = \int_\Omega \left(\alpha\frac{\partial H}{\partial x_2}\frac{\partial \bar{h}}{\partial x_2} + \alpha\frac{\partial H}{\partial x_3}\frac{\partial \bar{h}}{\partial x_3} + i\omega\mu H\bar{h}\right) \mathrm{d}\boldsymbol{x} \quad \forall h \in V = H_o^1(\Omega). \tag{10.105}$$

The function $\hat{H} \in H^1(\Omega)$ is called a weak or generalized solution of the problem in Equations 10.101 and 10.102 if

$$\hat{H} \in V, \quad a(\hat{H}, h) = (f, h) \quad \text{for all } h \in H^1(\Omega),$$

where $a(H, h)$ is defined by Equation 10.105.

Lemma 10.1: The operator L from Equation 10.101 is V-elliptic if the inequality

$$\alpha\xi_i\bar{\xi}_j \geq c\,|\xi_i|^2, \quad i, j = 2, 3, \tag{10.106}$$

holds a.e. for any n-tuple ($n = 2$) of complex numbers and $Re\ b \geq 0$, $b = i\omega\mu$.

Proof: We have

$$|a(H,H)| \geq Re(a(H,H)) = Re \int_\Omega \left(\alpha \frac{\partial H}{\partial x_2}\frac{\partial \overline{H}}{\partial x_2} + \alpha \frac{\partial H}{\partial x_3}\frac{\partial \overline{H}}{\partial x_3} + i\omega\mu H\overline{H} \right) \mathrm{d}\boldsymbol{x}$$

$$= \int_\Omega \left(\alpha \frac{\partial H}{\partial x_2}\frac{\partial \overline{H}}{\partial x_2} + \alpha \frac{\partial H}{\partial x_3}\frac{\partial \overline{H}}{\partial x_3} \right) \mathrm{d}\boldsymbol{x} \geq c \int_\Omega \sum_{i=2,3} \left| \frac{\partial H}{\partial x_i} \right|^2 \mathrm{d}\boldsymbol{x} = c\,\|H\|^2_{H^1_o(\Omega)}.$$

Definition 10.1: A sesquilinear form $\boldsymbol{a}(\boldsymbol{u},\boldsymbol{v})$ is $\boldsymbol{V}$-elliptic if

$$|a(u,v)| \geq c_o\,\|v\|^2_{H^1_o(\Omega)}, \quad c_o = \text{constant} > 0 \quad \forall v \in V,$$

where $V = \left\{ v \,\middle|\, v \in H^1(\Omega), v\big|_{\partial\Omega} = 0 \right\} \equiv H^1_o(\Omega)$.

Lemma 10.1 immediately implies the V-ellipticity of the sesquilinear form in Equation 10.105 since

$$|a(H,h)| \geq |a(H,H)| \geq Re(a(H,H)) \geq c\,\|H\|^2_{H^1_o(\Omega)}.$$

The following statement follows from the Lax-Milgram theorem (Nečas (1967)).

Theorem 10.1: There exists one and only one generalized (weak) solution $\hat{H} \in H^1(\Omega)$ of the problem in Equations 10.101 and 10.102.

Corollary: There exists one and only one solution of the original problem in Equations 10.98 and 10.99.

Proof: It is sufficient to show that $\hat{H}$ does not depend on the choice of w we employ to construct $\hat{H} = H - w$. Therefore, let H_1 and H_2 be two different solutions of the problem in Equations 10.98 and 10.99. Put $H_3 = H_1 - H_2$. Then Equation 10.101 with $f = 0$ and Equation 10.102 hold for H_3. In other words, H_3 is equal to the generalized solution of the problem in Equations 10.101 and 10.102 with $f = 0$. This solution, however, equals zero according to Theorem 10.1 and thus $H_3 = 0$, i.e. $H_1 = H_2$, which was the proof required.

Let us now study the problem in Equations 10.90 to 10.92, i.e. the problem

$$LH \equiv \frac{\partial^2 H_l}{\partial x_2^2} + \frac{\partial^2 H_l}{\partial x_3^2} - i\omega\mu_l\sigma_l H_l = f \text{ in } \Omega_l, \quad l = 1, 2, \ldots, p, \tag{10.107}$$

$$H_l = H_m, \quad \alpha \frac{\partial H}{\partial \boldsymbol{n}}\bigg|_{(l)} = \alpha \frac{\partial H}{\partial \boldsymbol{n}}\bigg|_{(m)} \quad \text{on } \gamma_l^m, \tag{10.108}$$

$$H|_{\partial\Omega} = \varphi(x_2, x_3). \tag{10.109}$$

We will proceed analogically to the solution of the previous problem.

Let $\alpha_l > 0$ for $l = 1, 2, \ldots, p$ and $\alpha_l = 0$ for $l = 0$ be real, bounded, and measurable functions in Ω_l. Let $\varphi(x_2, x_3)$, $(x_2, x_3) \in \partial\Omega$, be a trace of a function $w \in H^1(\Omega)$.

Let $\tilde{V}$ be a space of test functions defined by

$$\tilde{V} = \{v \mid v \in H^1(\Omega)\} \equiv H^1(\Omega).$$

We look for a function $H \in H^1(\Omega)$ such that

$$H - w \in \tilde{V}, \tag{10.110}$$

$$a(H, v) = \sum_l \left(\int_{\Omega_l} \alpha_l \frac{\partial H_l}{\partial x_2} \frac{\partial \bar{v}}{\partial x_2} + \alpha_l \frac{\partial H_l}{\partial x_3} \frac{\partial \bar{v}}{\partial x_3} \right) \mathrm{d}\boldsymbol{x} + \int_{\Omega_l} i\omega\mu H_l \bar{v} \mathrm{d}\boldsymbol{x} = 0 \quad \forall v \in \tilde{V}, \tag{10.111}$$

where $l = 1, 2, \ldots, p$ in the case of H-polarization and $l = 0, 1, 2, \ldots, p$ in the case of E-polarization. Put $H = w + h$, where w is a sufficiently smooth function with the trace $w = \varphi(x_2, x_3)$ on $\partial\Omega$. Then the function h satisfies

$$Lh \equiv \frac{\partial^2 h_l}{\partial x_2^2} + \frac{\partial^2 h_l}{\partial x_3^2} - i\omega\mu_l\sigma_l h_l = -g \text{ in } \Omega_l, \quad l = 1, 2, \ldots, p, \tag{10.112}$$

$$h_l = h_k, \quad \alpha \frac{\partial h}{\partial \boldsymbol{n}}\bigg|_{(l)} = \alpha \frac{\partial h}{\partial \boldsymbol{n}}\bigg|_{(m)} \quad \text{on } \gamma_l^m, \tag{10.113}$$

$$h|_{\partial\Omega} = 0, \tag{10.114}$$

where $g = (\partial^2 w_l / \partial x_2^2) + (\partial^2 w_l / \partial x_3^2) - i\omega\mu_l\sigma_l w_l$.
Then the variational formulation of the problem in Equations 10.112 to 10.114 assumes the form: find

$$h \in V \equiv H_o^1(\Omega) \quad (V \text{ is defined in Equation 10.104}), \tag{10.115}$$

$$\begin{aligned} a(h, v) &= \sum_l \left[\int_{\Omega_l} \left(\alpha_l \frac{\partial h_l}{\partial x_2} \frac{\partial \bar{v}}{\partial x_2} + \alpha_l \frac{\partial h_l}{\partial x_3} \frac{\partial \bar{v}}{\partial x_3} \right) \mathrm{d}\boldsymbol{x} + \int_{\Omega_l} i\omega\mu_l\sigma_l h_l \bar{v} \mathrm{d}\boldsymbol{x} - \int_{\partial\Omega_l} \alpha_l h_l \bar{v} \mathrm{d}s \right] \\ &= -\sum_l \left[\int_{\Omega_l} \left(\alpha_l \frac{\partial w_l}{\partial x_2} \frac{\partial \bar{v}}{\partial x_2} + \alpha_l \frac{\partial w_l}{\partial x_3} \frac{\partial \bar{v}}{\partial x_3} \right) \mathrm{d}x - \int_{\Omega_l} i\omega\mu_l\sigma_l w_l \bar{v} \mathrm{d}\boldsymbol{x} \right] \\ &\quad - \int_{\partial\Omega_l} \alpha_l w_l \bar{v} \mathrm{d}s = (-g, v) \quad \forall v \in V, \end{aligned} \tag{10.116}$$

where $g = (\partial^2 w_l / \partial x_2^2) + (\partial^2 w_l / \partial x_3^2) - i\omega\mu_l\sigma_l w_l$.
We will prove the existence and uniqueness of this model problem from the theory of magneto-telluric field. The proof will be based on the paper by Rojtberg and Scheftel' (1962).

Let us consider the direct sum of Sobolev spaces

$$H^k(\Omega) = \tilde{H}^k(\Omega_1) \oplus \tilde{H}^k(\Omega_2) \oplus \ldots \oplus \tilde{H}^k(\Omega_p).$$

Denote by $H^2(S)$ a set of functions $h \in H^2$ that satisfy the conditions

$$h|_{\partial\Omega} = 0,$$

$$h_l = h_m, \quad \alpha \frac{\partial h}{\partial \boldsymbol{n}}\bigg|_{(l)} = \alpha \frac{\partial h}{\partial \boldsymbol{n}}\bigg|_{(m)} \quad \text{on } \gamma_l^m.$$

Then $H^2(S)^+$ is a subspace of H^2, where $H^2(S)^+$ is the set of such functions $v \in H^2$ for which

$$(Lh, \alpha v) = (h, L^+(\alpha v)), \quad h \in H^2(S),$$

where

$$L^+h \equiv \frac{\partial^2 h_l}{\partial x_2^2} + \frac{\partial^2 h_l}{\partial x_3^2} - i\omega\mu_l\sigma_l h_l \text{ in } \Omega_l.$$

The space $H^2(S)^+$ is then a subspace of H^2 and $H^2(S)^{++} = H^2(S)$. Denote by H'^1 the closure of $H^2(S)^+$, i.e. $H'^1 = \overline{H^2(\Omega)^+}$ in the metric of the space H^1, and let H''^1 denote the subspace of elements αv, where $v \in H'^1$. The spaces H'^1 and H''^1 are also subspaces of the space H^1.

Definition 10.2: The function $h \in H'^1$ is called a generalized (weak) solution of the problem in Equations 10.112 and 10.114 if Equation 10.116 holds for any $v \in H'^1$.

Denote the positive Hilbert space by $H^+ = H''^1(\Omega)$, the zero Hilbert space by $H^o = L^2(\Omega)$, and the negative Hilbert space, i.e. the space with negative norm (see Berezanskii (1965)), by $H^- = H^{-1}(\Omega)$.

Then we can state the following theorem.

Theorem 10.2: For any $g \in H''^{-1}$, there exists one and only one generalized (weak) solution $h \in H'^1$ of the problem in Equations 10.112 and 10.114.

Corollary: There exists one and only one generalized solution of the original problem in Equations 10.107 to 10.109.
The proof is analogical to that of the Corollary of Theorem 10.1.

Let us now solve the problem in Equations 10.95 to 10.97, i.e. the problem

$$LH \equiv \frac{\partial}{\partial x_2}\left(\alpha\frac{\partial H}{\partial x_2}\right) + \frac{\partial}{\partial x_3}\left(\alpha\frac{\partial H}{\partial x_3}\right) = \mu\frac{\partial H}{\partial t}, \quad (x_2, x_3, t) \in \Omega \times (0, T), 0 < T < \infty, \tag{10.117}$$

$$H(x_2, x_3, t) = H_1(x_2, x_3, t), \quad (x_2, x_3, t) \in \partial\Omega \times (0, T), 0 < T < \infty, \tag{10.118}$$

$$H(x_2, x_3, 0) = H_o(x_2, x_3), \quad (x_2, x_3) \in \Omega, \tag{10.119}$$

where $H_o \in L^2(\Omega), H_1 \in L^2(\partial\Omega \times (0, T))$, and where we can denote $\Omega_t = \Omega \times (0, T)$, $\partial\Omega_t = \partial\Omega \times (0, T)$.

A function $H(x_2, x_3, t)$ of the class $C^2(\Omega \times (0, T)) \cap C(\Omega \times (0, T))$ is called the classical solution of the problem above if it satisfies Equation 10.117 in the time cylinder $\Omega_t = \Omega \times (0, T)$, the boundary condition in Equation 10.118, and the initial condition in Equation 10.119. Necessary conditions for the existence of the classical solution of Equations 10.117 to 10.119 are the smoothness conditions $H_o \in C(\overline{\Omega})$ and $H_1 \in C(\partial\Omega \times (0, T))$, and the compatibility condition

$$H_o(x_2, x_3)|_{\partial\Omega} = H_1(x_2, x_3, 0).$$

Let us introduce the notion of a generalized (weak) solution of Equations 10.117 to 10.119. Let there exist a sequence of functions $H_{1j} \in C(\partial\Omega_t)$, and $H_{oj} \in C(\overline{\Omega})$, $j = 1, 2, \ldots$, such that $H_{1j} \to H_1$ in $C(\partial\Omega_t)$ as $j \to \infty$, where $T > 0$ is arbitrary,

$H_{oj} \to H_o$ in $C(\overline{\Omega})$, and let there exist a classical solution of the problem

$$LH_j \equiv \frac{\partial}{\partial x_2}\left(\alpha\frac{\partial H_j}{\partial x_2}\right) + \frac{\partial}{\partial x_3}\left(\alpha\frac{\partial H_j}{\partial x_3}\right) = \mu\frac{\partial H_j}{\partial t}, \qquad \text{in } \Omega \times (0,T), 0 < T < \infty,$$

$$H_j(x_2, x_3, t) = H_{1j}(x_2, x_3, t), \qquad \text{on } \partial\Omega \times (0,T), 0 < T < \infty,$$

$$H_j(x_2, x_3, 0) = H_{oj}(x_2, x_3), \qquad \text{in } \Omega,$$

for any $j = 1, 2, \ldots$. Then there exists a function $H(x_2, x_3, t)$ continuous in $\overline{\Omega}_t$ and such that $H_j \to H$ as $j \to \infty$ in $C(\overline{\Omega}_t)$ for an arbitrary $T > 0$. The function $H(x_2, x_3, t)$ is called a generalized (weak) solution of Equations 10.117 to 10.119.

Conversely, it is possible to prove that if the generalized (weak) solution is sufficiently smooth then it is classical as well. The proof of this statement can be found in any beginner's textbook on partial differential equations of parabolic type.

The problem in Equations 10.117 to 10.119 will be transformed into a problem with homogeneous boundary conditions. As usual we put $h(x_2, x_3, t) = H(x_2, x_3, t) - w(x_2, x_3, t)$, where H is the sought solution of Equations 10.117 to 10.119 and w is a sufficiently smooth function that satisfies the condition $w|_{\partial\Omega_t} = H_1$. Then we are to solve the problem

$$\frac{\partial h}{\partial t} - Lh \equiv \frac{\partial h}{\partial t} - \frac{\partial}{\partial x_2}\left(\mu^{-1}\alpha\frac{\partial h}{\partial x_2}\right) - \frac{\partial}{\partial x_3}\left(\mu^{-1}\alpha\frac{\partial h}{\partial x_3}\right) = f \qquad \text{in } \Omega_t, \tag{10.120}$$

$$h|_{\partial\Omega_t} = 0, \tag{10.121}$$

$$h(x_2, x_3, 0) = h_o(x_2, x_3), \tag{10.122}$$

where $h_o \in L^2(\Omega)$, the operator L satisfies the condition of uniform parabolicity

$$\beta_o\xi^2 \le \mu^{-1}\alpha\xi_i\xi_j \le \beta_1\xi^2, \qquad \beta_o, \beta_1 = \text{constant} > 0,$$

and where

$$f = \frac{\partial w}{\partial t} - \frac{\partial}{\partial x_2}\left(\mu^{-1}\alpha\frac{\partial w}{\partial x_2}\right) - \frac{\partial}{\partial x_3}\left(\mu^{-1}\alpha\frac{\partial w}{\partial x_3}\right).$$

We will now prove the existence and uniqueness of the generalized solution of the problem considered. However, we will prove the continuous dependence of the solution h on H_o and f first. We use the Galerkin method for the proof, which can be carried out in various ways (see for example Vladimirov (1967), Friedman (1964), Ladyzenskaya, Solonikov and Uralćeva (1968), Landis (1971)). The technique of Blanchard and Brüning (1982) and Renardy and Rogers (1993) can also be used. We are to solve the following problem.

Find a function $h \in V$ such that

$$\left(\frac{\partial h}{\partial t}(t) - Lh(t) - f(t), v\right) = 0 \quad \forall v \in V, t \in (0,T),$$

$$(h(0), v) = (h_o, v).$$

We will carry out the proof in more detail. Calculate the inner product of the function $h(\boldsymbol{x}, t)$ with Equation 10.120 and integrate in the limits from 0 to t (we write $(\boldsymbol{x}, t)$

instead of (x_2, x_3, t)) to obtain

$$\int_0^t \left[\left(\frac{\partial h(t')}{\partial t'}, h(t')\right) - (Lh(t'), h(t'))\right] \mathrm{d}t' = \int_o^t (f(t'), h(t'))\mathrm{d}t'. \tag{10.123}$$

The integration yields

$$\frac{1}{2}\|h(\boldsymbol{x}, t)\|_{2,\Omega}^2 + \int_0^t (Lh(t'), h(t'))\mathrm{d}t' = \frac{1}{2}\|h(\boldsymbol{x}, 0)\|_{2,\Omega}^2 + \int_0^t (f(t'), h(t'))\mathrm{d}t'.$$

From this, we have

$$\frac{1}{2}\|h(\boldsymbol{x}, t)\|_{2,\Omega}^2 + \beta_o \left\|\frac{\partial h}{\partial x_i}\right\|_{2,\Omega_t}^2 \leq \frac{1}{2}\|h(\boldsymbol{x}, 0)\|_{2,\Omega}^2 + \|f\|_{2,1,\Omega_t} \max_{\tau} \|h(\boldsymbol{x}, t)\|_{2,\Omega}.$$

Put

$$\|h(\boldsymbol{x}, t)\|_{2,\Omega}^2 = u(t)\,\|h(\boldsymbol{x}, 0)\|_{2,\Omega},$$

where

$$u(t) = \max_{0\leq\tau\leq t} \|h(\boldsymbol{x}, \tau)\|.$$

Then

$$\|h(\boldsymbol{x}, t)\|_{2,\Omega}^2 + 2\beta_o \left\|\frac{\partial h}{\partial x_i}\right\|_{2,\Omega_t}^2 \leq u(t)\,\|h(\boldsymbol{x}, 0)\|_{2,\Omega} + 2u(t)\,\|f\|_{2,1,\Omega_t}.$$

Two inequalities follow from the above:

$$\max_{0\leq\tau\leq t} \|h(\boldsymbol{x}, \tau)\|_{2,\Omega}^2 = u^2(t) \leq F(t),$$

$$\left\|\frac{\partial h}{\partial x_i}\right\|_{2,\Omega_t}^2 \leq \frac{1}{2}\beta_o^{-1} F(t).$$

Taking the square root and summing, we arrive at

$$\begin{aligned}
|h|_{\Omega_t} \equiv u(t) + \left\|\frac{\partial h}{\partial x_i}\right\|_{2,\Omega_t} &\leq \left(1 + \frac{1}{2}2^{1/2}\beta_o^{-1/2}\right) F^{1/2}(t) \\
&\leq \left(1 + \tfrac{1}{2}2^{1/2}\beta_o^{-1/2}\right)\left[u(t)\,\|h(\boldsymbol{x}, 0)\|_{2,\Omega} + 2u(t)\,\|f\|_{2,1,\Omega_t}\right]^{1/2} \\
&= \left(1 + \tfrac{1}{2}2^{1/2}\beta_o^{-1/2}\right) u^{1/2}(t)\left[\|h(\boldsymbol{x}, 0)\|_{2,\Omega} + 2\,\|f\|_{2,1,\Omega_o}\right]^{1/2} \\
&\leq \left(1 + \tfrac{1}{2}2^{1/2}\beta_o^{-1/2}\right) |h|_{\Omega_t}^{1/2}\left[\|h(\boldsymbol{x}, 0)\|_{2,\Omega} + 2\,\|f\|_{2,1,\Omega_t}\right]^{1/2}.
\end{aligned}$$

This implies

$$|h|_{\Omega_t}^{1/2} \leq \left(1 + \tfrac{1}{2}2^{1/2}\beta_o^{-1/2}\right)\left[\|h(\boldsymbol{x}, 0)\|_{2,\Omega} + 2\,\|f\|_{2,1,\Omega_t}\right]^{1/2}$$

i.e.

$$|h|_{\Omega_t}^{1/2} \leq \left(1+\tfrac{1}{2}2^{1/2}\beta_o^{-1/2}\right)\left(\|h(\boldsymbol{x}, 0)\|_{2,\Omega}+2\,\|f\|_{2,1,\Omega_t}\right) = c_o\left(\|h(\boldsymbol{x}, 0)\|_{2,\Omega}+2\,\|f\|_{2,1,\Omega_t}\right). \tag{10.124}$$

The estimate in Equation 10.124 holds in the interval $\langle 0, T)$ for any t since

$$\|h(\boldsymbol{x}, t)\|_{2,\Omega} \leq |h|_{\Omega_t} .$$

We have thus proved the following theorem.

Theorem 10.3: Let h be a solution of the problem in Equations 10.120 to 10.122. Then $h(\boldsymbol{x}, t)$ given by Equation 10.124 depends continuously on f and H_o, where we write $h(\boldsymbol{x}, t) \equiv h(x_2, x_3, t)$.

Definition 10.3: The function $h \in H_o^{1,0}(\Omega_t)$ is called the generalized (weak) solution of the problem in Equations 10.120 and 10.122 if it satisfies the relation

$$\int_{\Omega_t}\left(-\frac{\partial h}{\partial t}u+\mu^{-1}\alpha\frac{\partial h}{\partial x_2}\frac{\partial u}{\partial x_2}+\mu^{-1}\alpha\frac{\partial h}{\partial x_3}\frac{\partial u}{\partial x_3}\right)\mathrm{d}\boldsymbol{x}\,\mathrm{d}t=\int_{\Omega}u(\boldsymbol{x},0)h_o(\boldsymbol{x})\mathrm{d}\boldsymbol{x}+\int_{\Omega_t}(f,u)\mathrm{d}\boldsymbol{x}\,\mathrm{d}t \quad \forall u \in H^1(\Omega_t) \tag{10.125}$$

and vanishes for $t = T$.

Now we can prove the following theorem.

Theorem 10.4: The problem in Equations 10.120 to 10.122 has one and only one generalized solution from the space $H_o^1(\Omega_t)$.

Proof: Apparently, the set of generalized solutions form a linear space. Let $\{\varphi_k(\boldsymbol{x})\}_{k=1}^{n}$ be the system of basis functions in the space $H_o^1(\Omega)$. Let $h^n(\boldsymbol{x}, t)$, $\boldsymbol{x} = (x_2, x_3)$, be an approximate solution and let

$$h^n(\boldsymbol{x}, t) = \sum_{i=1}^{n} a_i^n(t)\varphi_i(\boldsymbol{x}).$$

Further let

$$\left(\frac{\partial h^n(\boldsymbol{x}, t)}{\partial t}, \varphi_l(\boldsymbol{x})\right) - (Lh^n(t), \varphi_l(\boldsymbol{x})) = (f, \varphi_l) \quad l = 1, 2, \ldots, n, \tag{10.126}$$

$$a_l^n(0) = (h_o, \varphi_l). \tag{10.127}$$

Equations 10.126 and 10.127 represent a system of ordinary differential equations with non-negative coefficients $a_l(t)$, $l = 1, 2, \ldots, n$, with principal terms $\mathrm{d}a_l(t)/\mathrm{d}t$, with coefficients at $a_l(t)$ that are bounded functions of the variable t, and with terms (f, φ_l) that are functions integrable on the interval $(0, T)$. It is known from the theory of ordinary differential equations that such a system has a unique solution and defines absolutely continuous functions $a_l^n(t)$ on $\langle 0, T\rangle$.

To obtain an estimate for $h^n(t)$, multiply each of Equations 10.126 by the corresponding a_l and integrate over $t \in \langle 0, T\rangle$. Then we have

$$\frac{1}{2}\|h^n(\boldsymbol{x}, t)\|_{2,\Omega}^2 + \int_o^t (Lh^n(t'), h^n(t'))\mathrm{d}t' = \frac{1}{2}\|h^n(\boldsymbol{x}, 0)\|_{2,\Omega}^2 + \int_o^t (f(t'), h^n(t'))\mathrm{d}t'.$$

An estimate analogical to the estimate in Equation 10.124 follows from this. As $\|h^n(\boldsymbol{x}, 0)\|_{2,\Omega} \leq \|h_o(\boldsymbol{x})\|_{2,\Omega}$, we obtain

$$|h^n|_{\Omega_t} \leq c_1, \tag{10.128}$$

where the constant c_1 does not depend on n. With respect to the above inequality (Equation 10.128), we can extract, from the sequence $\{h^n\}$, $n = 1, 2, \ldots$, a sequence $\{h^{n_i}\}$, $i = 1, 2, \ldots$, that weakly converges in $L^2(\Omega_t)$ together with its derivatives $\partial h^{n_i}/\partial x_j$ to an element $h \in H_o^{0,1}(\Omega_t)$. It can be shown that this element is a generalized solution of the problem in Equations 10.120 to 10.122. The rest of the proof is similar to that in Ladyzenskaja (1973), p. 169.

To prove the existence theorem, the technique of Renardy and Rogers (1993) can also be used.

10.2.3.3. Numerical solutions of magneto-telluric model problems. The frequency and time domain

10.2.3.3.1. The frequency domain

Numerical solution of the H- and E-polarization will be based on the finite element method. In the usual manner for the finite element method, we cover the domain Ω by a regular triangulation, i.e. by a grid of disjoint triangles (squares, rectangles, etc.), in such a way that the interface γ_l^m contains either the whole side or the vertex of a triangle (or square, rectangle, etc.). The condition just stated is substantial for approximation of interface conditions. It is the essence of the finite element method that natural interface conditions are approximated for the constructed approximate solutions with a higher and higher accuracy when the triangulation is refined while the basis functions do not satisfy these conditions. However, these interface conditions can be approximated in such a way that we use basis functions that satisfy interface conditions either exactly or approximately. Then the corresponding approximate solution will also have this property (Marek and Nedoma (1973), Najzar (1975)). In what follows, we consider several cases of the approximation of interface conditions on the interface γ_l^m (in the sense of the work by Marek and Nedoma (1973)):

(i) We use a triangulation refinement in the vicinity of the interface γ_l^m and, at the same time, the interface conditions are not fulfilled by the basis functions generated by linear hill functions. This basis will be denoted by LH.

(ii) The normal triangulation is used in the vicinity of the interface γ_l^m. The basis functions are generated by functions piecewise linear on triangles that have no common point with any of the interfaces γ_l^m and by functions piecewise cubic on triangles that have a non-empty intersection with some of the interfaces. Proceeding in this way, we require that the basis functions are continuous or that they satisfy the interface conditions. Both these requirements can be met due to our choice of the basis functions. Moreover, we assume that the basis functions whose support has a non-empty intersection with the boundary of the domain Ω satisfy the Dirichlet boundary conditions where they are prescribed. We use the notation LCI (linear-cubic hill functions with the interface condition) for the basis constructed in this way.

(iii) Choosing an $h > 0$, we construct a square grid on the domain Ω (or a general trapezoidal grid on the domain Ω) in which all squares have straight sides parallel to the coordinate axes x_2 and x_3. Instead of the original problem, we investigate the problem where the originally discontinuous coefficients are replaced by continuous coefficients that coincide with the original ones outside the vicinity of interfaces and that are linearly linked in the vicinity of interfaces γ_l^m in such a way that they attain a value equal to the average of the values in both the adjacent subdomains. Moreover, we construct a further square grid with the same nodes as in the given square grid in such a way that the sides of the squares of the new grid are diagonals of the squares of the original grid. We choose piecewise linear basis functions on each of the squares of the new grid in such a way that the basis function attains the value 1 at the centre of the square (of the new grid), vanishes on its sides, and is identically equal to zero outside the given square. We again assume that the basis functions fulfil the Dirichlet boundary conditions where they are prescribed. We introduce the notation SLH (square linear hill functions) for the discretization constructed in this way. A generalization to an irregular trapezoidal grid (TLH — trapezoidal linear hill functions) is possible.

We first consider the case of a magneto-telluric field caused by the inducing field which is periodically variable with the factor $\exp(i\omega t)$.

LCI CASE

Denote by V_h the space of functions piecewise linear or cubic constructed by means of LCI for each vertex of the triangular grid given.

Definition 10.4: Let $\{V_h\}$, $h \in (0, 1)$, be a class of finite dimensional subspaces V. We say that $V_h \to V$ as $h \to 0$ if

$$\lim_{h\to 0_+} (\text{dist}\,[V_h, v]) = 0$$

for any $v \in V$.

Definition 10.5: A function $h_{ho} \in V_h$ is called the approximate solution of the problem of H- and E-polarization if

$$a(h_{ho}, v) = (\alpha^{-1}g, v) \quad \forall v \in V_h, \tag{10.129}$$

where

$$a(u, v) = a_1(u, v) + a_2(u, v) \tag{10.130}$$

and

$$a_1(u, v) = -\int_\Omega \left(\alpha \frac{\partial u}{\partial x_2}\frac{\partial \overline{v}}{\partial x_2} + \alpha \frac{\partial u}{\partial x_3}\frac{\partial \overline{v}}{\partial x_3}\right) \mathrm{d}\boldsymbol{x}, \tag{10.131}$$

$$a_2(u, v) = \int_\omega g u \overline{v} \mathrm{d}\boldsymbol{x}. \tag{10.132}$$

Write the approximate solution h_h in the form

$$h_h = \sum_{k=1}^{n} c_k^{(h)} \varphi_k^{(h)}, \tag{10.133}$$

where $\{\varphi_k^{(h)}\}_{k=1}^n$ is the basis of the space V_h. Substituting Equation 10.133 into Equation 10.129, we then obtain

$$\sum_{k=1}^{n} c_k^{(h)} a\left(\varphi_k^{(h)}, \varphi_j^{(h)}\right) = \left(\alpha^{-1} g, \varphi_j^{(h)}\right), \quad j = 1, 2, \ldots, n, \tag{10.134}$$

which is a linear algebraic system for the determination of unknown coefficients $c_k^{(h)}$, $k = 1, 2, \ldots, n$. The matrix $(a_1(\varphi_k^{(h)}, \varphi_j^{(h)}))$, $j, k = 1, \ldots, n$, is negative definite. We further have

$$\frac{a_2\left(\varphi_l^{(h)}, \varphi_m^{(h)}\right)}{a_1\left(\varphi_l^{(h)}, \varphi_m^{(h)}\right)} \to 0 \text{ as } h \to 0 \tag{10.135}$$

for those subscripts l and m for which the denominator does not vanish. If the denominator vanishes identically then the numerator possesses the same property with respect to the construction of the LCI discretization. Put

$$\begin{aligned} a_1\left(\varphi_k^{(h)}, \varphi_j^{(h)}\right) &= e_{jk} \quad \text{for } k < j, \\ a_1\left(\varphi_j^{(h)}, \varphi_j^{(h)}\right) &= d_j, \\ a_2\left(\varphi_j^{(h)}, \varphi_j^{(h)}\right) &= \hat{d}_j, \\ a_2\left(\varphi_k^{(h)}, \varphi_j^{(h)}\right) &= f_{jk} \quad \text{for } k \neq j. \end{aligned} \tag{10.136}$$

Denote the corresponding matrices by symbols E_h, D_h, $\hat{D}_h$, F_h. The relation in Equation 10.135 and the positive definiteness of the matrix $-(a_1(\varphi_k^{(h)}, \varphi_j^{(h)}))$ imply that

$$(D_h - E_h)^{-1}(\hat{D}_h + F_h) \to 0 \text{ as } h \to 0. \tag{10.137}$$

This condition, as we will see in what follows, guarantees the convergence of the Gauss-Seidel iterative method for solving the problem in Equation 10.134 for every $h > 0$.

The convergence of the LCI method is a straightforward consequence of the Babuška theorem (Babuška (1971), see also Theorem GLM in Chapter 11 of this book.) since $V_1 = V_2 = H^1(\Omega)$ and $M_1 = M_2 = V_h$ in our case. The conditions (4.2) to (4.4) of the Babuška theorem are satisfied; for example, the condition (4.3) is fulfilled as the inequality

$$|a(h, h)| \geq c\,\|h\|^2$$

holds. The conditions (4.5) and (4.6) are satisfied for the same reason (see the paper quoted). The condition (4.8) is fulfilled (see the paper quoted) with $\delta = \delta(h)$ such that $\lim_{h\to 0} \delta(h) = 0$, since the solutions h_o sought can be approximated by functions constructed with the help of the LCI method with an arbitrary accuracy. This is obvious outside the vicinity of the interface as the rate of approximation is $\delta(h) \sim h$ and the vicinity of the interface can be made sufficiently small in dependence on h. It thus suffices to guarantee the existence of $\hat{u}_o \equiv h_o$ from the closure of V_h. In this case, it is sufficient to apply the Lax-Milgram theorem in the closure of V_h in the $H^1(\Omega)$ norm since its assumptions are satisfied as they are satisfied in the whole of $H^1(\Omega)$. Therefore, the following theorem holds.

Theorem 10.5: There exists one and only one LCI approximate solution $h_{ho} \in V_h$ of the problem of H- and E-polarization, with the bilinear form $a : V_h \times V_h$ given by Equations 10.131 and 10.132, for sufficiently small $h > 0$. The relation

$$\lim_{h \to 0} \|h_o - h_{ho}\|_1 = 0 \tag{10.138}$$

holds for the solution of the problem of H- and E-polarization.

Remark: The solution h_o is not, in general, smooth in the vicinity of vertices and interfaces of the domain Ω. A solution h_o corresponding to a particular right-hand part g is smooth in the whole domain Ω and we can then show, in addition to Equation 10.138, the rate of convergence in a way analogical to the papers by Babuška (1972).

SLH CASE

We denote by the symbol S_h a space of functions constructed by the SLH method. In other words, S_h is a class of functions whose support is always a square (a trapezoid in the general case). Denote the basis in the space S_h by $\{\psi_i^h\}$, $i = 1, \ldots, n$. Suppose further that the basis functions fulfil Dirichlet boundary conditions on all parts of the boundary where they are prescribed for the problem.

Definition 10.6: A function $h_{ho} \in S_h$ is called the SLH approximate solution of the problem of H- and E-polarization if

$$a(h_{ho}, v) = (\alpha^{-1} g, v) \quad \forall v \in S_h. \tag{10.139}$$

Look for $h_h \in S_h$ in the form

$$h_h = \sum_{k=1}^{n} b_k^{(h)} \psi_k^{(h)}. \tag{10.140}$$

Substituting into Equation 10.139, we arrive at

$$\sum_{k=1}^{n} b_k^{(h)} a\left(\psi_k^{(h)}, \psi_j^{(h)}\right) = \left(\alpha^{-1} g, \psi_j^{(h)}\right), \quad j = 1, \ldots, n, \tag{10.141}$$

which is a linear algebraic system for the determination of unknown coefficients $b_k^{(h)}$, $k = 1, \ldots, n$, of Equation 10.140.

Analogically to the previous case, we have

$$\frac{\left|a_2\left(\psi_l^{(h)}, \psi_m^{(h)}\right)\right|}{\left|a_1\left(\psi_l^{(h)}, \psi_m^{(h)}\right)\right|} \to 0 \text{ as } h \to 0, \tag{10.142}$$

for those subscripts l and m for which the denominator does not vanish. As with the LCI method above, the fact that the denominator in Equation 10.142 is zero implies here that the numerator is also zero. Putting then

$$\begin{aligned} a_1\left(\psi_k^{(h)}, \psi_j^{(h)}\right) &= e_{jk} \quad \text{for } k < j, \\ a_1\left(\psi_j^{(h)}, \psi_j^{(h)}\right) &= d_j, \\ a_2\left(\psi_j^{(h)}, \psi_j^{(h)}\right) &= \hat{d}_j, \\ a_2\left(\psi_k^{(h)}, \psi_j^{(h)}\right) &= f_{jk} \quad \text{for } k \neq j. \end{aligned} \tag{10.143}$$

and denoting the corresponding matrices by E_h, D_h, $\hat{D}_h$, F_h, we obtain $e_{jk} = f_{jk} = 0$ for those subscripts j and k that correspond to vertices of those squares which are not immediate neighbours in the original square grid. This follows from the construction of the basis $\{\psi_k\}$, $k = 1, \ldots, n$, of the space S_h. This further implies that the matrices $E_h + E_h^*$ (E_h^* is the Hermitian adjoint) and F_h contain only a non-zero first superdiagonal and a non-zero first subdiagonal. The matrix $a_1(\psi_k^{(h)}, \psi_j^{(h)})$, $j, k = 1, \ldots, n$, is irreducible and diagonally dominant, and the corresponding Jacobi matrix $D_h^{-1}(E_h + E_h^*)$ is convergent. This and Equation 10.142 imply the following theorem.

Theorem 10.6: For any $h > 0$, there exists one and only one SLH approximate solution h_{ho} of the problem of H- and E-polarization with the bilinear form given by Equations 10.131 and 10.132. Moreover, the Jacobi matrix $(D_h + \hat{D}_h)^{-1}(E_h + E_h^* + F_h)$ is convergent as well as the corresponding matrix of moduli

$$\left|(D_h + \hat{D}_h)^{-1}\right| \left(\left|E_h + E_h^*\right| + \mid F_h \mid\right). \tag{10.144}$$

Both the LCI and SLH methods are based on the work by Marek and Nedoma (1973).

LH CASE

We denote by the symbol S^h a space of functions constructed by the LH method, i.e. piecewise linear functions which do not fulfil the interface conditions on interfaces γ_l^m. Let their support be a hexagon. These basis functions attain the value 1 at the centre of gravity of the hexagon, vanish on the opposite sides of the individual triangles that form the support, and their restrictions to triangles are linear functions (linear hills).

Definition 10.7: A function $h_{ho} \in S^h$ is called an approximate solution of the problem of H- and E-polarization if

$$a(h_{ho}, v) = (\alpha^{-1}g, v) \quad \forall v \in S^h, \tag{10.145}$$

where

$$a(u_h, v) = -\sum_{l=1}^{s} \alpha_l \int_{\Omega_l} \left(\frac{\partial u_h}{\partial x_2}\frac{\partial \overline{v}}{\partial x_2} + \frac{\partial u_h}{\partial x_3}\frac{\partial \overline{v}}{\partial x_3}\right) \mathrm{d}\boldsymbol{x} + \int_{\Omega_l} \omega\mu i u_h \overline{v} \mathrm{d}\boldsymbol{x}, \tag{10.146}$$

$$(\alpha^{-1}g, v) = -\sum_{l=1}^{s} \left(\alpha_l \int_{\Omega_l} \left(\frac{\partial w}{\partial x_2}\frac{\partial \overline{v}}{\partial x_2} + \frac{\partial w}{\partial x_3}\frac{\partial \overline{v}}{\partial x_3}\right) \mathrm{d}\boldsymbol{x} + \int_{\Omega_l} \omega\mu i w \overline{v} \mathrm{d}\boldsymbol{x}\right), \tag{10.147}$$

and the function w was defined above.

Construct the finite dimensional space

$$S^h = \left\{ u^h \left| u^h = \sum_{j=1}^{m} \sum_{k=1}^{n} s_{jk}^h u_{jk}^h, \, u_{jk}^h \right|_{\partial\Omega} = 0 \right\}, \qquad h = \max(h_i, h_j).$$

The basis of the space S^h is $\{u_{jk}^h\}_{j,k=1}^{m,n}$ and we have $\lim_{h\to 0} S^h = V$. According to the Galerkin method, there exists one and only one $h_h \in S^h$ such that Equation 10.145 holds for all $v \in S^h$. Looking for $h_h \in S^h$ in the form

$$h_h = \sum_{j=1}^{m} \sum_{k=1}^{n} s_{jk}^{(h)} v_{jk}^h, \tag{10.148}$$

and substituting successively the individual basis functions v_{rs}^h for functions $v \in S^h$ into Equation 10.145, we obtain a system of equations for unknown coefficients s_{jk}^h, i.e. the system

$$\sum_{j=1}^{m} \sum_{k=1}^{n} s_{jk}^{(h)} a(v_{jk}^h, v_{rs}^h) = -\left(\alpha^{-1} g, v_{rs}^h\right), \quad r = 1, \ldots, m, \; s = 1, \ldots, n, \tag{10.149}$$

which is equivalent to the five point scheme of the finite difference approximation. The corresponding integrals in inner products in Equation 10.149 are computed analytically or numerically. The function w that appears on the right-hand part of Equation 10.149 in g is chosen such that it is equal to 0 in Ω, is non-zero in the vicinity of the boundary $\partial\Omega$, and satisfies boundary conditions of Dirichlet type on all parts of the boundary $\partial\Omega$ where they are prescribed by the problem.

Note that Equation 10.141 can be written in the form of an operator equation

$$Lh - f = L_o h + Kh - f = 0, \tag{10.150}$$

where L_o is a positive definite and symmetric operator, and K is a bounded operator (i.e. there exists a constant M such that $\|Kh\| \leq M \|h\|$ for an arbitrary $h \in H^2(S)$). The system in Equation 10.149 can then be rewritten in the form

$$\sum_{j=1}^{m} \sum_{k=1}^{n} s_{jk}^{(h)} \left\{ \left(L_o v_{jk}^h, v_{rs}^h\right) + \left(K v_{jk}^h, v_{rs}^h\right) \right\} = -\left(\alpha^{-1} g, v_{rs}^h\right), \quad r = 1, \ldots, m, \; s = 1, \ldots, n. \tag{10.151}$$

The operator equation (Equation 10.150) is equivalent to the operator equation

$$h + Th = F, \tag{10.152}$$

where $T = L_o^{-1} K$, $F = L_o^{-1} f$, T is totally continuous in H'^1, and L_o^{-1} is also a totally continuous operator (see Mihlin (1970)) since it is represented by an operator of discrete spectrum. This implies the following theorem (see Mihlin (1970)).

Theorem 10.7: The approximate solution of Equation 10.150 converges in the sense of convergence in the space H'^1 to the exact solution of the problem in Equations 10.141 to 10.143 if the following conditions are fulfilled:

(i) Equation 10.150 has at most one solution in H'^1;

(ii) The operator $T = L_o^{-1}K$ is totally continuous in H'^1.

Let us now prove the convergence of the finite element method. Let $\{S^h\}$ be a sequence of subspaces S^h of dimension $n_h = \dim S^h < \infty$ and such that $S^h \subset H'^1$. Recall that the sequence of subspaces $\{S^h\}$ is dense in H'^1 if and only if, for any $\varepsilon > 0$ and $h \in H'^1$, there exists an element $v \in S^h$ such that

$$\|h - v\|_{H'^1} \leq \varepsilon.$$

We introduce a system $\{u_{jk}^h\}_{j,k=1}^{m,n(h)}$ of basis functions generated by the LH hill functions in the space S^h. An approximate solution of Equation 10.150 will be of the form of Equation 10.148 where m and n are limited by the dimension $n(h)$ of the subspace S^h. This leads to the linear algebraic system

$$P^h\left(L_o h_h^h + K h_h^h - f\right) = 0,$$

where P^h are orthogonal projectors that map H'^1 onto subspaces $S^h \subset H'^1$. Then the following theorem (analogical to Theorem 16.2 of Krasnosel'skii *et al.* (1969)) holds.

Theorem 10.8: The approximate solution of Equation 10.150 converges in the sense of convergence in the space H'^1 to the exact (generalized) solution of the problem in Equations 10.141 to 10.143 if the following conditions are fulfilled:

(i) The operator $L_o^{-1}K$ is totally continuous in the space H'^1;

(ii) The operator equation (Equation 10.150) has at most one solution in H'^1;

(iii) The sequence of subspaces S^h is dense in H'^1.

The system in Equation 10.149 is the linear algebraic system

$$A\boldsymbol{h} = \vec{\psi}. \tag{10.153}$$

As was shown in the paper by Nedoma (1975) the matrix A is a block, irreducible and diagonally dominant, and thus nonsingular, matrix. The corresponding Jacobi matrix $B = -D^{-1}(L + U)$, where $A = L + U + D$, $D =$ blockdiag (A_{kk}), L is a lower block triangular matrix with zero blocks on the diagonal, and U is an upper block triangular matrix with zero blocks on the diagonal, is convergent and, moreover, 2-cyclic. It is also possible to show that the Gauss-Seidel matrix $\mathscr{L}_1$ and over-relaxation matrix $\mathscr{L}_\omega$ are convergent. Then the Jacobi, Gauss-Seidel, and SOR methods converge and

$$0 < \rho(\mathscr{L}_\omega) < \rho(\mathscr{L}_1) < \rho(B) < 1, \tag{10.154}$$

where $\rho(.)$ denotes, in the sequel, the spectral radius of the over-relaxation, Gauss-Seidel, and Jacobi matrix, and where ω is a relaxation parameter, $0 < \omega < 2$. The optimum relaxation parameter ω satisfies the inequality

$$0 < \omega < 2(1 + D)^{-1},$$

where $D = \rho(B) > 0$ (Axelsson (1996)).

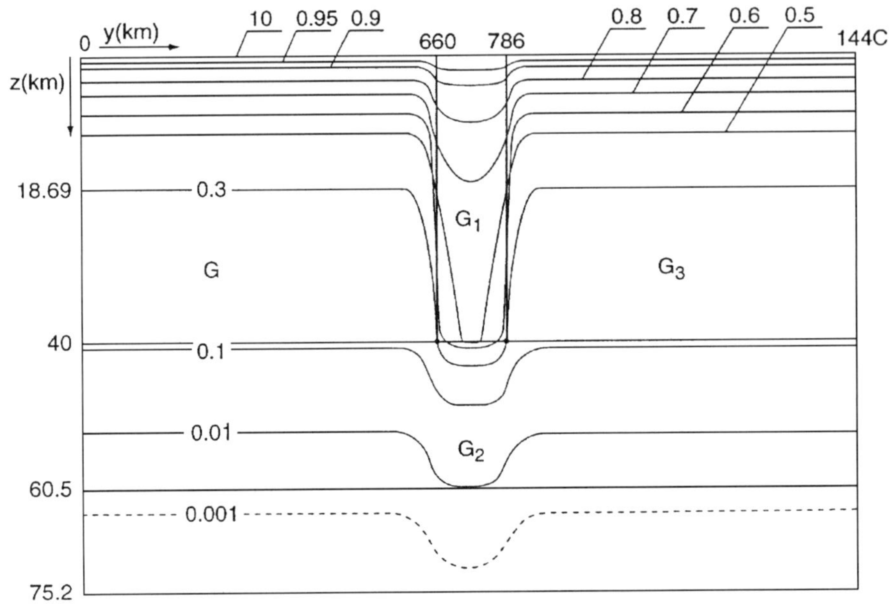

Electrical conductivities in emu are $\sigma_1 = 10^{-13}$, $\sigma_2 = 10^{-14}$, $\sigma_3 = 10^{-13}$ and $\sigma_4 = 10^{-12}$

Figure 10.1. Model with horizontal layer underlying the vertical contact — the frequency domain case

Moreover, the bi-conjugate gradient method can also be used (see Axelsson (1996), Chapter II of this book, Křížek and Neittaanmäki (1990)).

Estimates of error of the solution can also be found in the papers of Babuška and Kellogg (1971), and Babuška (1971).

For illustration isolines of the magnetic modulus in a model with a horizontal layer underlying the vertical contacts for $T = 10\,\text{sec}$ is given (see Figure 10.1).

10.2.3.3.2. Aperiodic inducing magneto-telluric field — the time domain case

Let us moreover study the case of an aperiodic inducing field. We will start our considerations from the problem

$$Lh + \frac{\partial h}{\partial t} \equiv -\alpha\mu^{-1}\Delta h + \frac{\partial h}{\partial t} = -f \quad \text{for } (x_2, x_3) \in \Omega, t > 0, t \in (0, T), \tag{10.155}$$

with the initial and boundary conditions

$$h(x_2, x_3, 0) = h_o(x_2, x_2) \quad \text{for } (x_2, x_3) \in \overline{\Omega}, t = 0, \tag{10.156}$$

$$h(x_2, x_3, t) = 0 \quad \text{for } (x_2, x_3) \in \partial\Omega, t > 0, \tag{10.157}$$

where $f = -\alpha\mu^{-1}\Delta w + (\partial w/\partial t)$, w is a sufficiently smooth function satisfying boundary conditions of Dirichlet type on $\partial\Omega$, $t > 0$, and where

$$h_o(x_2, x_3) \in L^2(\Omega), \quad f \in H_o^{-1}(\Omega_t).$$

If we use the Faedo-Galerkin method the problem in Equations 10.155 to 10.157 is equivalent to the problem: find a function $h \in H_o^1(\Omega) \times C^1(0,T)$ that satisfies

$$\left(\frac{\partial h}{\partial t}, v\right) + (Lh, v) = -(f, v) \quad \forall v \in H_o^1(\Omega), t \in (0,T), \tag{10.158}$$

$$(h(x_2, x_3, 0), v) = (h_o, v) \quad (x_2, x_3) \in \overline{\Omega}, \tag{10.159}$$

where $h(x_2, x_3, t) \in H_o^1(\Omega_t)$, $\Omega_t = \Omega \times (0,T)$, i.e. $h = H - w \in H_o^1(\Omega)$ for any $t \in (0,T)$, and where

$$(Lh, v) = -\int_\Omega \left(\alpha\mu^{-1}\frac{\partial h}{\partial x_2}\frac{\partial v}{\partial x_2} + \alpha\mu^{-1}\frac{\partial h}{\partial x_3}\frac{\partial v}{\partial x_3}\right)\mathrm{d}\mathbf{x}, \tag{10.160}$$

$$(f, v) = \int_\Omega f v \mathrm{d}\mathbf{x} \tag{10.161}$$

with the inner product defined by the relation

$$(u, v) = \int_\Omega u v \mathrm{d}\mathbf{x}.$$

We look for the approximate solution in the form

$$h^h(x_2, x_3, t) = \sum_{k=1}^{n} \hat{\alpha}_k(t) v_k(x_2, x_3) \tag{10.162}$$

with unknown coefficients $\hat{\alpha}_k(t)$ that satisfy the system

$$\left(\frac{\partial h^h(x_2, x_3, t)}{\partial t}, v_k\right) + (Lh^h(t), v_k) = -(f, v_k) \quad t \in (0,T), \quad \forall v_k \in V_h, k = 1, 2, \ldots, n, \tag{10.163}$$

and the initial condition

$$(h^h(x_2, x_3, 0), v_k) = (h_o, v_k) \quad (x_2, x_3) \in \overline{\Omega}, t = 0, \quad \forall v_k \in V_h. \tag{10.164}$$

This is a system of ordinary differential equations of a very high order that can be rewritten in the form

$$C(\boldsymbol{x})\frac{\mathrm{d}\hat{\alpha}(t)}{\mathrm{d}t} + A(\boldsymbol{x})\hat{\alpha}(t) = F(t), \quad t > 0, \tag{10.165}$$

$$C(\boldsymbol{x})\hat{\alpha}(0) = \hat{\alpha}_o(\boldsymbol{x}), \tag{10.166}$$

where C and A are positive definite matrices given as

$$C = (C_{kl}), \quad C_{kl} = (v_k, v_l),$$

$$A = (A_{kl}), \quad A_{kl} = \int_\Omega \left(\alpha\mu^{-1}\frac{\partial v_k}{\partial x_l}\frac{\partial v_l}{\partial x_k}\right)\mathrm{d}\boldsymbol{x},$$

$$\hat{\alpha}(t) = (\hat{\alpha}_1(t), \ldots, \hat{\alpha}_n(t))^T, \quad \hat{\alpha}_{ok} = (h_o, v_k),$$

$$F(t) = (F_1(t), \ldots, F_n(t)), \quad F_k(t) = (f(t), v_k),$$

and where $V_h \subset H_o^1$ is defined above. The system in Equations 10.165 and 10.166 can be rewritten as

$$\frac{\mathrm{d}\hat{\alpha}}{\mathrm{d}t} + C^{-1}A\hat{\alpha} = C^{-1}F, \quad \hat{\alpha}(0) = C^{-1}\hat{\alpha}_o. \tag{10.167}$$

Equation 10.167 has, in accord with the theory of ordinary differential equations, a unique solution of the form

$$\hat{\alpha} = C^{-1}\hat{\alpha}_o \exp(C^{-1}At) + \int_o^t \exp\{-C^{-1}A(t-t')\}C^{-1}F(t')\mathrm{d}t', \tag{10.168}$$

where

$$e^{Dt} = I + (1!)^{-1}Dt + (2!)^{-1}D^2t^2 + \ldots + (k!)^{-1}D^kt^k + \ldots$$

Our above discussion thus implies that the approximate solution of Equation 10.162 does exist and is unique.

We now find estimates of the rate of convergence of the approximate solution to the exact one. We first prove a discrete analog of Theorem 10.7. The following theorem thus holds.

Theorem 10.9: Let h^h be an approximate solution of the problem in Equations 10.155 to 10.157. Then $h^h(x_2, x_3, t)$ depends continuously on f and h_o.

Proof: Multiply each part of Equation 10.155 by the corresponding function $\hat{\alpha}_k(t)$, sum up over k, and integrate with respect to $t' \in (0, t)$. We arrive at

$$\int_o^t \left[\left(\frac{\partial h^h(t')}{\partial t}, h^h(t')\right) + (Lh^h(t'), h^h(t'))\right]\mathrm{d}t' = \int_o^t (f(t'), h^h)\mathrm{d}t'.$$

From this

$$\frac{1}{2}\left\|h^h(t)\right\|^2 + \int_o^t (Lh^h(t'), h^h(t'))\mathrm{d}t' = \frac{1}{2}\left\|h^h(\boldsymbol{x}, 0)\right\| + \int_o^t (f(t'), h^h(t'))\mathrm{d}t'$$

follows. But since

$$(h^h(0), v_k) = (h_o, v_k), \quad \left(h^h(0), \sum_{k=1}^n \hat{\alpha}_k(0)v_k\right) = \left(h_o, \sum_{k=1}^n \hat{\alpha}_k(0)v_k\right),$$

$$\left\|h^h(0)\right\|^2 = (h_o, h^h(0)), \quad \left\|h^h(0)\right\| \le \|h_o\|,$$

we then have

$$\frac{1}{2}\left\|h^h(t)\right\| + \int_o^t (Lh^h(t'), h^h(t'))\mathrm{d}t'$$

$$\le \left(\int_o^t \left\|F(t')\right\|^2 \mathrm{d}t'\right)^{1/2} \left(\int_o^t \left\|h^h(t')\right\|^2 \mathrm{d}t'\right)^{1/2} + \frac{1}{2}\|h_o\|^2$$

$$\le \left(\int_o^t \left\|F(t')\right\|^2 \mathrm{d}t'\right)^{1/2} C\left(\frac{1}{2}\left\|h^h(t)\right\|^2 + \int_o^t (Lh^h(t'), h^h(t'))\mathrm{d}t'\right)^{1/2} + \frac{1}{2}\|h_o\|^2,$$

where $c > 0$. We obtain from this, analogically as in the proof of Theorem 10.3, that

$$\left\|h^h(t)\right\|^2 + \int_o^t (Lh^h(t'), h^h(t'))\mathrm{d}t' \le c\left(\int_o^t \left\|F(t')\right\|^2 \mathrm{d}t' + \|h_o\|^2\right)$$

and

$$\max_{t\in(0,T)} \left\|h^h(t)\right\|^2 + \int_o^T (Lh^h(t'), h^h(t'))\mathrm{d}t' \leq c\left(\int_o^T \left\|F(t')\right\|^2 \mathrm{d}t' + \|h_o\|^2\right), \quad c > 0,$$

which was to be proved.

Let us now prove the following theorem stating the rate of convergence of the approximate solution to the exact solution.

Theorem 10.10: We have

$$\max \left\|h(t) - h^h(t)\right\| + \left(\int_o^T (L(h - h^h), h - h^h)\mathrm{d}t\right)^{1/2} \leq ch, \quad c > 0, \tag{10.169}$$

where h on the right hand side denotes the step of the mesh.

Proof: Put $u_h = h - h^h$. Then

$$\left(\frac{\partial u_h}{\partial t}, w_h\right) + (Lu_h, w_h) = 0, \quad (u_h(0), w_h) = 0 \quad \forall w_h,$$

where

$$w_h = \sum_{k=1}^{n} a_k(t)v_k.$$

This implies

$$\left(\frac{\partial u_h}{\partial t}, u_h\right) + (Lu_h, u_h) = \left(\frac{\partial u_h}{\partial t}, h - w_h\right) + (L(h - h^h), h - w_h).$$

Hence,

$$\frac{1}{2}\|u_h(t)\|^2 + \int_o^t (Lu_h, u_h)\mathrm{d}t' = \int_o^t (Lu_h, h - w_h)\mathrm{d}t' + \int_o^t \left(\frac{\partial u_h}{\partial t}, h - w_h\right)\mathrm{d}t' + \frac{1}{2}\|u_h(0)\|^2.$$

Since

$$\int_o^t \left(\frac{\partial u_h}{\partial t}, h - w_h\right)\mathrm{d}t' = (u_h(t), h(t) - w_h(t)) - (u_h(t), h(0) - w_h(0)) - \int_o^t \left(u_h, \frac{\partial h}{\partial t} - \frac{\partial w_h}{\partial t}\right)\mathrm{d}t'$$
$$= (u_h, h(t) - w_h) - (u_h, h_o - w_h(0)) - \int_o^t \left(u_h, \frac{\partial h}{\partial t} - \frac{\partial w_h}{\partial t}\right)\mathrm{d}t',$$

we then arrive at

$$\frac{1}{2}\|u_h(t)\|^2 + \int_o^t (Lu_h, u_h)\mathrm{d}t' \leq \left(\int_o^t (Lu_h, u_h)\mathrm{d}t'\right)^{1/2} \left(\int_o^t (L(h - w_h, h - w_h)\mathrm{d}t'\right)^{1/2}$$
$$+ \frac{1}{2}\|u_h(0)\|^2 + \|u_h(t)\|\,\|h(t) - w_h(t)\| + \|u_h(0)\|\,\|h_o - w_h(0)\|$$
$$+ \left(\int_o^t \|u_h\|^2 \mathrm{d}t'\right)^{1/2} \left(\int_o^t \left\|\frac{\partial h}{\partial t} - \frac{\partial w_h}{\partial t}\right\| \mathrm{d}t'\right)^{1/2}.$$

But $\|u_h(0)\| \leq \|h_o - w_h(0)\|$. Using, in addition, the inequality $|ab| \leq (4\varepsilon)^{-1}a^2 + \varepsilon b^2$, we obtain

$$\frac{1}{2}\|u_h(t)\|^2 + \int_o^t (Lu_h, u_h)\mathrm{d}t' \leq \|h(t) - w_h(t)\| + c\left(\int_o^t (L(h - w_h, h - w_h)\mathrm{d}t' + \|h_o - w_h(0)\|^2\right.$$
$$\left. + \int_o^t \left\|\frac{\partial h}{\partial t} - \frac{\partial w_h}{\partial t}\right\|^2 \mathrm{d}t'\right), \quad c > 0, \tag{10.170}$$

and this yields

$$\max_t \|u_h(t)\|^2 + \int_o^T (Lu_h, u_h)\mathrm{d}t \le \max_t \|h(t) - w_h(t)\|^2 + c\left(\int_o^T (L(h - w_h, h - w_h)\mathrm{d}t\right.$$

$$\left. \le \|h_o - w(0)\|^2 + \int_o^T \left\|\frac{\partial h}{\partial t} - \frac{\partial w_h}{\partial t}\right\|^2 \mathrm{d}t\right).$$

Suppose that w_h has coefficients $\alpha_k = h(x_k, t)$. Then Equation 10.170 implies the convergence $h^h \to h$ as $k \to 0$ holds for the triangulation step, i.e.

$$\max_t \left\|h(t) - h^h(t)\right\|^2 + \int_o^T (L(h - h^h, h - h^h)\mathrm{d}t \to 0 \quad \text{as } k \to 0.$$

Supposing $h_o \in L^2(\Omega)$, $\partial h/\partial x_i \in L^2((0, T); \Omega)$, $\partial h/\partial t \in L^2((0, T); \Omega)$ and, moreover, $\partial^2 h/\partial x_i^2 \in L^2((0, T); \Omega)$, $\partial^2 h/\partial x_i \partial t \in L^2((0, T); \Omega)$, we have

$$\max_t \left\|h(t) - h^h(t)\right\|^2 + \left(\int_o^T (L(h - h^h, h - h^h)\mathrm{d}t\right)^{1/2} \le ch, \quad c > 0,$$

which was to be proved.

Corollary: The inequality in Equation 10.169 implies the convergence of the approximate solution $H^h(x_2, x_3, t)$ to the exact solution $H(x_2, x_3, t)$ of the problem in Equations 10.117 to 10.119, as $h \to 0$ holds for the triangulation step.

Remark: The time-dependent model problem was solved together with my diploma student Mgr.Lukeš by using the multigrid technique (see Lukeš (1996)), where the additive noise was neglected.

10.3. VARIATIONAL PROBLEMS IN THE THEORY OF ELECTROMAGNETIC FIELD. ELECTROMAGNETIC INDUCTION

10.3.1. Introduction

Geophysical measurements show that the quantities that characterize the Earth's magnetic field can be measured both on the Earth's surface and outside the Earth. Analyses of geophysical measurements then prove that the Earth's body is located in a magnetic field. The Earth's magnetic field is characterized by a quantity that is called the total intensity $\boldsymbol{H}_T$ of the magnetic field. A detailed study of the world maps of isolines as well as maps of individual partial regions of the Earth's surface, together with the mathematical analysis of the measured Earth's magnetic field lead to the conclusion that the magnetic field measured on the Earth's surface is a sum of several fields that originate from different sources. Denoting the intensity of the magnetic field by $\boldsymbol{H}_T$, we have

$$\boldsymbol{H}_T = \boldsymbol{H}_o + \boldsymbol{H}_m + \boldsymbol{H}_a + \boldsymbol{H}_e + \delta\boldsymbol{H},$$

where $\boldsymbol{H}_o$ represents the intensity of the homogeneously magnetized spheric Earth, $\boldsymbol{H}_m$ the intensity of the magnetic field that originates by inner reasons and is connected with inhomogeneous structures of the Earth's inner layers, $\boldsymbol{H}_a$ corresponds to the magnetized upper layer of the Earth's crust, $\boldsymbol{H}_e$ is connected with outer reasons and $\delta\boldsymbol{H}$ represents a field of variations that has its reasons outside the Earth's body. The field described by the vector $\boldsymbol{H}_a$ is called the anomal field, the field $\boldsymbol{H}_m$ is called the continental or residual field. The anomal field $\boldsymbol{H}_a$ is composed of two fields — ${}^1\boldsymbol{H}_a$, the field caused by deep sources and having a regional character (the anomalies are thus called regional), and ${}^2\boldsymbol{H}_a$, the field caused by less deep sources and having a local character (we thus speak about local anomalies of magnetic field). The sum of the field of the homogeneously magnetized Earth, continental and outer field is often called the normal field and for its intensity $\boldsymbol{H}_n$ we have that

$$\boldsymbol{H}_n = \boldsymbol{H}_o + \boldsymbol{H}_m + \boldsymbol{H}_e.$$

Then the total field, if variations of magnetic field are excluded, is given as the sum of the normal and total field,

$$\boldsymbol{H}_T = \boldsymbol{H}_n + \boldsymbol{H}_a.$$

If we know the distribution of the normal field on the Earth's surface we can determine the anomal part of the magnetic field in this way and try to answer the question of the geological structure of the lithosphere or its upper parts. Depending on the decision of which anomaly is to be studied, either $\boldsymbol{H}_o + \boldsymbol{H}_m + \boldsymbol{H}_e$ or $\boldsymbol{H}_o + \boldsymbol{H}_m + \boldsymbol{H}_e + \boldsymbol{H}_a$ is taken for the intensity of the normal field. If we are to study the continental magnetic field we take the field of a homogeneously magnetized sphere for the normal field.

The Earth's magnetic field (geomagnetic field) is either permanent, that is also subject to changes but only of a long time character, or the field of short time changes that are either of a periodic character (variations) or of an aperiodic character (disturbances). The permanent field is either interior, i.e. normal, anomal, and secular variation field, or exterior, i.e. non-potential field. Secular variations show a significant relation to changes of the activity of the Sun. The variable field interferes with the permanent Earth's magnetic field. The magnitude of this field is substantially less when compared with the constant field. Its connection with geophysical processes that take place in the ionosphere provides a chance to judge the influence of these processes on the processes in the Earth's lithosphere and to study reasons for the causes of these processes.

The variable Earth's magnetic field represents the difference between the magnitude of intensity of the magnetic field and its mean value over a certain time interval (e.g. a month, year). This difference defines the vector of magnetic variations. Magnetic variations consist of a large number of periodic variations (e.g. the solar daily variation with period of one day), and aperiodic and irregular fluctuations. In addition to the solar variations, there exist variations of other periods that include monthly, daily, and annual variations.

The theory of electromagnetic induction in a conductive body shows that the relation between the exterior and interior parts of the variable field depends on electric conductivity and magnetic permeability. Several types of geomagnetic variations can be split into the effects of the exterior and interior field if a proper analysis is applied.

This analysis, based on the idea of electromagnetic induction, provides us with a way to study electric and magnetic properties inside the Earth. In practice, however, we consider a suitable model and investigate if it corresponds to the surveyed relation of interior and exterior parts of geomagnetic variations. If the model yields an image identical with the variations of the considered types then it is one of the possible approximations to the conductivity distribution inside the Earth.

In the beginning of this century, Schuster (1908) and Chapman (1919) showed that part of the solar daily variations originate in the Earth's body and cause electric currents induced in the Earth's body as a consequence of changes in the exterior magnetic field. Therefore, the theory of electromagnetic induction inside the Earth's body caused by a variable magnetic field is important for finding the relation between exterior and interior parts of short-period geomagnetic variations.

We start from Maxwell equations, i.e. from equations

$$\frac{\partial \boldsymbol{D}}{\partial t} + \boldsymbol{J} - \operatorname{rot} \boldsymbol{H} = 0, \qquad \text{the 1st series of Maxwell equations,} \tag{10.171}$$

$$\operatorname{div} \boldsymbol{D} = q, \tag{10.172}$$

$$\frac{\partial \boldsymbol{B}}{\partial t} + \operatorname{rot} \boldsymbol{E} = 0, \qquad \text{the 2nd series of Maxwell equations,} \tag{10.173}$$

$$\operatorname{div} \boldsymbol{B} = 0, \tag{10.174}$$

where $\boldsymbol{H}$ is the intensity of magnetic field, $\boldsymbol{J}$ current density, $\boldsymbol{D}$ vector of electric induction, $\boldsymbol{E}$ intensity of electric field, $\boldsymbol{B}$ vector of magnetic induction, and q volume density of electric charge. Assume that the medium is isotropic and that

$$\boldsymbol{D} = \varepsilon \boldsymbol{E}, \quad \boldsymbol{B} = \mu \boldsymbol{H}, \quad \boldsymbol{J} = \sigma \boldsymbol{E},$$

where ε is the dielectric constant, μ magnetic permeability, and σ electric conductivity. Apply the operation of divergence to Equation 10.171. Then

$$\operatorname{div} \boldsymbol{J} = -\operatorname{div} \frac{\partial \boldsymbol{D}}{\partial t} = -\frac{\partial q}{\partial t}, \quad \text{i.e. } \operatorname{div} \boldsymbol{J} = -\frac{\partial q}{\partial t},$$

as $\operatorname{div} \operatorname{rot} \boldsymbol{u} = 0$. This implies that

$$\operatorname{div} \boldsymbol{D} = \operatorname{div} \varepsilon \boldsymbol{E} = \operatorname{div} \varepsilon\sigma^{-1}\boldsymbol{J} = \varepsilon\sigma^{-1}\operatorname{div} \boldsymbol{J} + \boldsymbol{J}\operatorname{grad}(\varepsilon\sigma^{-1}) = -\varepsilon\sigma^{-1}\frac{\partial q}{\partial t} + \boldsymbol{J}\operatorname{grad}(\varepsilon\sigma^{-1}).$$

Further, $(\partial q/\partial t) + \sigma\varepsilon^{-1}\operatorname{div} \boldsymbol{D} = \sigma\varepsilon^{-1}\boldsymbol{J}\operatorname{grad}(\varepsilon\sigma^{-1})$, then

$$\frac{\partial q}{\partial t} + q\sigma\varepsilon^{-1} = \sigma\varepsilon^{-1}\boldsymbol{J}\operatorname{grad}(\varepsilon\sigma^{-1}). \tag{10.175}$$

We see from the last equation that q does not depend on the field if it is constant or if $\boldsymbol{J}$ is parallel to grad (ε/σ). The density of electric charge satisfies the exponential damping law in these cases. Suppose that $\varepsilon, \mu, \sigma =$ constant and put $q = 0$. Then

$$\operatorname{div} \boldsymbol{D} = \operatorname{div} \varepsilon \boldsymbol{E} = 0, \quad \text{i.e. } \operatorname{div} \boldsymbol{E} = 0. \tag{10.176}$$

Apply now the operation of rotation to Equation 10.171. We obtain successively

$$\text{rot rot } \boldsymbol{H} = \text{rot } \boldsymbol{J} + \text{rot } \frac{\partial \boldsymbol{D}}{\partial t},$$

$$\text{grad div } \boldsymbol{H} - \Delta \boldsymbol{H} = \sigma \text{ rot } \boldsymbol{E} + \varepsilon \text{rot } \frac{\partial \boldsymbol{E}}{\partial t} = -\sigma \frac{\partial \boldsymbol{B}}{\partial t} + \varepsilon \frac{\partial}{\partial t} \text{rot } \boldsymbol{E}$$

$$= -\sigma\mu \frac{\partial \boldsymbol{H}}{\partial t} + \varepsilon \frac{\partial}{\partial t}\left(-\frac{\partial \mu \boldsymbol{H}}{\partial t}\right) = -\sigma\mu \frac{\partial \boldsymbol{H}}{\partial t} - \varepsilon\mu \frac{\partial^2 \boldsymbol{H}}{\partial t^2},$$

since div $\boldsymbol{B}$ = div $\mu \boldsymbol{H} = 0$, i.e.

$$\Delta \boldsymbol{H} = \mu \left(\sigma \frac{\partial \boldsymbol{H}}{\partial t} + \varepsilon \frac{\partial^2 \boldsymbol{H}}{\partial t^2}\right), \qquad (10.177)$$

Analogically, for the component $\boldsymbol{E}$ we obtain

$$\text{rot rot } \boldsymbol{E} = -\text{rot } \frac{\partial \boldsymbol{B}}{\partial t} = -\frac{\partial}{\partial t} \text{rot } \boldsymbol{B} = -\mu \frac{\partial}{\partial t} \text{rot } \boldsymbol{H},$$

$$\text{grad div } \boldsymbol{E} - \Delta \boldsymbol{E} = -\mu \frac{\partial}{\partial t}\left(\boldsymbol{J} + \frac{\partial \boldsymbol{D}}{\partial t}\right) = -\mu \left(\sigma \frac{\partial \boldsymbol{E}}{\partial t} + \varepsilon \frac{\partial^2 \boldsymbol{E}}{\partial t^2}\right).$$

With respect to the relation div $\boldsymbol{E} = 0$ we arrive at

$$\Delta \boldsymbol{E} = \mu \left(\sigma \frac{\partial \boldsymbol{E}}{\partial t} + \varepsilon \frac{\partial^2 \boldsymbol{E}}{\partial t^2}\right) \qquad (10.178)$$

If the vector of magnetic induction $\boldsymbol{H}$ does not change too quickly then $\varepsilon(\partial \boldsymbol{H}/\partial t) \ll \sigma \boldsymbol{H}$ holds in the conductor, which is equivalent to the condition $|\partial \boldsymbol{D}/\partial t| \ll |\boldsymbol{J}|$, so that

$$\Delta \boldsymbol{H} = \sigma\mu \frac{\partial \boldsymbol{H}}{\partial t}. \qquad (10.179)$$

On the other hand, the conductivity in the dielectric is so small that the first term on the right-hand part of Equation 10.177 can be neglected. Then

$$\Delta \boldsymbol{H} = \mu\varepsilon \frac{\partial^2 \boldsymbol{H}}{\partial t^2}. \qquad (10.180)$$

The last equation, which is the wave equation, implies that electromagnetic effects propagate with velocity $v = (\mu\varepsilon)^{-1/2}$. We can neglect the right-hand part of Equation 10.180 for sufficiently slow variations of magnetic field with respect to the fact that $v^2 = \mu\varepsilon$ and thus

$$\Delta \boldsymbol{H} = 0. \qquad (10.181)$$

It is known from the theory of electromagnetic field that the intensity of magnetic field $\boldsymbol{H}$ can be derived in dielectric from a scalar potential W. Put

$$\boldsymbol{H} = -\text{grad } W. \qquad (10.182)$$

We can define a vector potential $\boldsymbol{A}$ in conductor by the relation

$$\operatorname{rot} \boldsymbol{A} = \boldsymbol{B}. \tag{10.183}$$

The relations

$$\operatorname{rot} \boldsymbol{E} = -\frac{\partial \boldsymbol{B}}{\partial t} = -\frac{\partial}{\partial t}\operatorname{rot} \boldsymbol{A}, \quad \text{i.e. } \operatorname{rot} \boldsymbol{E} = -\operatorname{rot} \frac{\partial \boldsymbol{A}}{\partial t},$$

follow from Equation 10.173 and thus

$$\boldsymbol{E} = -\frac{\partial \boldsymbol{A}}{\partial t} - \operatorname{grad} \varphi, \tag{10.184}$$

where φ is an arbitrary function.

Equation 10.179 is equivalent to Equation 10.171. Let us show that. Multiply Equation 10.171 by the permeability μ. Then

$$\operatorname{rot} \boldsymbol{B} = \mu \boldsymbol{J}, \text{ i.e. } \operatorname{rot} \operatorname{rot} \boldsymbol{A} = \mu\sigma \boldsymbol{E} = \mu\sigma\left(-\frac{\partial \boldsymbol{A}}{\partial t} - \operatorname{grad} \varphi\right)$$

and

$$\operatorname{grad} \operatorname{div} \boldsymbol{A} - \Delta \boldsymbol{A} = \mu\sigma\left(-\frac{\partial \boldsymbol{A}}{\partial t} - \operatorname{grad} \varphi\right).$$

From this, we obtain

$$\Delta \boldsymbol{A} - \mu\sigma\frac{\partial \boldsymbol{A}}{\partial t} = \operatorname{grad} (\operatorname{div} \boldsymbol{A} + \mu\sigma\varphi).$$

Equation 10.179 can be satisfied if

$$\operatorname{div} \boldsymbol{A} + \sigma\varphi = 0, \tag{10.185}$$

which is called the Lorentz condition. Choose $\varphi = 0$. Then

$$\operatorname{div} \boldsymbol{A} = 0, \tag{10.186}$$

as also follows from Equation 10.176 since

$$\operatorname{div}\left(-\frac{\partial \boldsymbol{A}}{\partial t}\right) = -\frac{\partial}{\partial t}\operatorname{div} \boldsymbol{A} = 0.$$

Then the relation in Equation 10.184 attains the form

$$\boldsymbol{E} = -\frac{\partial \boldsymbol{A}}{\partial t} \tag{10.187}$$

if and only if $\boldsymbol{A}$ satisfies the relation

$$\Delta \boldsymbol{A} = \mu\sigma\frac{\partial \boldsymbol{A}}{\partial t} \tag{10.188}$$

or, in the general form,

$$\frac{\partial}{\partial x_i}\left(\sigma^{-1}\frac{\partial \boldsymbol{A}}{\partial x_j}\right) = \mu\frac{\partial \boldsymbol{A}}{\partial t}, \quad \text{i.e.} \quad \operatorname{div}(\sigma^{-1}\operatorname{grad} \boldsymbol{A}) = \mu\frac{\partial \boldsymbol{A}}{\partial t}. \tag{10.189}$$

If we investigate geomagnetic variations on a bounded surface we can then can study the Earth as an infinite conductive half-space with a plane (wavy) boundary. Consider a homogeneous conductive medium with electric conductivity σ = constant and magnetic permeability μ. The field inside the conductive medium is then described by an equation for some vector potential $\boldsymbol{F}$ of the form

$$\Delta \boldsymbol{F} = \mu\sigma\frac{\partial \boldsymbol{F}}{\partial t}, \tag{10.190}$$

where the displacement currents are neglected on the assumption that variations are slow; and the field outside the conductive medium is described by the equation

$$\Delta \boldsymbol{F} = 0. \tag{10.191}$$

If we solve a problem that stems from the scalar and vector potential we can define a magnetic potential W outside the conductive medium and a vector potential inside the conductor. We obtain for them

$$\Delta W = 0, \tag{10.192}$$

$$\Delta \boldsymbol{A} = \mu\sigma\frac{\partial \boldsymbol{A}}{\partial t}. \tag{10.193}$$

Supposing that σ and μ are functions of coordinates, we have

$$\operatorname{div}\,(\sigma^{-1}\operatorname{grad} \boldsymbol{A}) = \mu\frac{\partial \boldsymbol{A}}{\partial t}. \tag{10.194}$$

These potentials are connected with the vectors of field through the equations

$$\boldsymbol{H} = -\operatorname{grad} W, \tag{10.195}$$

$$\boldsymbol{E} = -\frac{\partial \boldsymbol{A}}{\partial t}. \tag{10.196}$$

10.3.2. Electromagnetic Induction in a Conductive Layer

Let us investigate the case of electromagnetic induction in a thin conductive layer starting from Price's theory (Price (1949), (1950)). Let an inhomogeneous conductive layer be given in free space (in insulator). We suppose that the magnetic permeability $\mu = 1$, and that the rate of time variations of magnetic and electric fields is so small that we can neglect displacement currents. Then

$$\operatorname{rot} \boldsymbol{H} = \boldsymbol{J}, \tag{10.197}$$

$$\operatorname{rot} \boldsymbol{E} = -\frac{\partial \boldsymbol{H}}{\partial t}, \tag{10.198}$$

$$\boldsymbol{J} = \sigma\boldsymbol{E}, \quad \sigma = \sigma(x_1, x_2, x_3). \tag{10.199}$$

Integrating Equation 10.199 over the thickness of the layer (denoted by h) we arrive at

$$\boldsymbol{J}_s = \int_o^h \boldsymbol{J}\,\mathrm{d}\boldsymbol{x}$$

and, analogically,

$$\sigma_s = r^{-1} = \int_o^h \sigma \mathrm{d}\boldsymbol{x}\,,$$

where r is the electric resistance. Further,

$$\boldsymbol{E}_s = r\boldsymbol{J}_s. \tag{10.200}$$

For small parts on the surface of the layer we obtain from Equation 10.198 that

$$\text{rot}\ \boldsymbol{E}_s = -\boldsymbol{n}\frac{\partial \boldsymbol{H}_n}{\partial t}, \tag{10.201}$$

where $\boldsymbol{n}$ is a unit vector pointing from the negative (i) to the positive (e) side of the layer. From Equations 10.201 and 10.200 we obtian

$$r\ \text{rot}\ \boldsymbol{J}_s + [\text{grad}\ r, \boldsymbol{J}_s] = -\boldsymbol{n}\frac{\partial \boldsymbol{H}_n}{\partial t}. \tag{10.202}$$

Let two sides of a small rectangular loop parallel to the layer be of length ds and let the other two sides be situated on the inner and outer boundary of the loop. Then Equation 10.197 yields

$$((H_s)_+ - (H_s)_-)\mathrm{d}s = J_s \mathrm{d}s \sin\Theta,$$

where Θ is the angle between ds and $\boldsymbol{J}_s$. From this,

$$\boldsymbol{J}_s = \{\boldsymbol{n}, \boldsymbol{H}_+ - \boldsymbol{H}_-]. \tag{10.203}$$

Let $\psi = W_+ - W_-$ be functions of fluxes, where W is magnetic potential in the free space (insulator). Then from Equation 10.203 we obtain

$$\boldsymbol{J}_s = [\boldsymbol{n}, \text{grad}\ \psi].$$

Substituting for $\boldsymbol{J}_s$ from Equation 10.203 into Equation 10.202 we have, after some manipulation, that

$$r\ \text{div}\ (\boldsymbol{H}_+ - \boldsymbol{H}_-) + \text{grad}\ r(\boldsymbol{H}_+ - \boldsymbol{H}_-) = -\frac{\partial \boldsymbol{H}_n}{\partial t}. \tag{10.204}$$

Denote by W_e the magnetic potential of the variable magnetic field that originates from exterior sources, and by W_i the potential of the magnetic field caused by induced fluxes. Further, let $W = W_e + W_i$. Then Equation 10.204 implies

$$r\ \text{div grad}\ (W_{i+} - W_{i-}) + \text{grad}\ r\ \text{grad}\ (W_{i+} - W_{i-}) = -\frac{\partial}{\partial t}\left(\frac{\partial W_e}{\partial \boldsymbol{n}} + \frac{\partial W_i}{\partial \boldsymbol{n}}\right), \tag{10.205}$$

since the induction field, and grad W_e as well, are continuous at the passage through a conductive layer. Equation 10.205 can be rewritten in the form

$$\operatorname{div}\,(r \text{ grad } \psi) = -\frac{\partial}{\partial t}\left(\frac{\partial W_e}{\partial \boldsymbol{n}} + \frac{\partial W_i}{\partial \boldsymbol{n}}\right), \tag{10.206}$$

and, considering the induction field with the factor $\exp(-i\omega t)$, we arrive at

$$\operatorname{div}\,(\text{grad } \psi) = -i\omega(B_n^e + B_n^i), \tag{10.207}$$

which are Price's equations, where B_n^e and B_n^i are normal components (with respect to the layer) of the corresponding given exterior magnetic field and induced magnetic field.

If the conductivity is very high or if the exterior magnetic field changes very rapidly then the problem leads to the equation

$$\Delta W = 0, \tag{10.208}$$

with the boundary condition on the surface of the conductor

$$\frac{\partial W}{\partial \boldsymbol{n}} = 0, \tag{10.209}$$

where $\boldsymbol{n}$ is the normal to the surface (outer normal). Equations 10.206 and 10.207 can be rewritten in the curvilinear coordinates as

$$\frac{\partial}{\partial \xi}\left[\alpha(\xi,\eta)\frac{\partial \psi}{\partial \xi}\right] + \frac{\partial}{\partial \eta}\left[\beta(\xi,\eta)\frac{\partial \psi}{\partial \eta}\right] = -\frac{\partial}{\partial t}\left(\frac{\partial W_e(\xi,\eta)}{\partial \boldsymbol{n}} + \frac{\partial W_i(\xi,\eta)}{\partial \boldsymbol{n}}\right), \tag{10.210}$$

$$\frac{\partial}{\partial \xi}\left[\alpha(\xi,\eta)\frac{\partial \psi}{\partial \xi}\right] + \frac{\partial}{\partial \eta}\left[\beta(\xi,\eta)\frac{\partial \psi}{\partial \eta}\right] = F(\xi,\eta), \tag{10.211}$$

where $\xi = x$, $\eta = y$, $\alpha = \beta = r$, $F = i\omega(B_n^e - B_n^i)$, in the planar case, and $\xi = \varphi$, $\eta = \Theta$, $\alpha = R\sin^{-1}\Theta$, $F = \sin\Theta i\omega(B_n^e + B_n^i)$, $B_n^e = B_n^e(\varphi,\Theta)$, $B_n^i = B_n^i(\varphi,\Theta)$ in the case of the surface of the sphere.

The relation between the electric field and the flux function has the form

$$E_x = r\frac{\partial \psi}{\partial x_2} = \sigma^{-1}\frac{\partial \psi}{\partial x_2}, \quad E_y = -r\frac{\partial \psi}{\partial x_1} = \sigma^{-1}\frac{\partial \psi}{\partial x_1}. \tag{10.212}$$

Inhomogeneities of conductivity lead to a change of polarization of the electric field (the circular polarization becomes elliptic).

In particular, let us study the following problem. Investigate the influence of inhomogeneities in some region of the Earth's body. A problem of this type leads to the solution of a two-dimensional (plane) problem. Suppose that a homogeneous side flux flows through a conductive plane that is homogeneous everywhere except for a bounded domain Ω_o with an arbitrarily distributed surface conductivity σ. In this case, $B_n^e = 0$. The inhomogeneity of conductivity σ in the domain Ω_o causes a disturbance of this flux which is damped at large distances from the domain Ω_o.

Consider a rectangular domain Ω that contains the domain Ω_o. We suppose that the boundary $\partial\Omega$ of the domain Ω is sufficiently distant from the domain Ω_o so that the disturbance of flux can be neglected in these places. Then we are to solve the problem

$$\frac{\partial}{\partial x_1}\left(r\frac{\partial \psi}{\partial x_1}\right) + \frac{\partial}{\partial x_2}\left(r\frac{\partial \psi}{\partial x_2}\right) = 0 \quad \text{in } \Omega, \tag{10.213}$$

$$\psi = \psi_o \quad \text{on } \partial\Omega, \tag{10.214}$$

where ψ_o is the value of the side flux given (e.g. $\psi_o = -\sigma_o x_1$ or $\psi_o = -\sigma_o x_2$). The relation between the electric field and the flux function is given by the formulae

$$E_x = \sigma^{-1}\frac{\partial \psi}{\partial x_2} = r\frac{\partial \psi}{\partial x_2}, \quad E_y = -\sigma^{-1}\frac{\partial \psi}{\partial x_1} = -r\frac{\partial \psi}{\partial x_1}. \tag{10.215}$$

The problem in Equations 10.213 and 10.214 is transformed into a problem with homogeneous boundary conditions in the usual way. Let w be a sufficiently smooth function that satisfies the boundary condition in Equation 10.214 on $\partial\Omega$. The problem in Equations 10.213 and 10.214 is then equivalent to the problem

$$Au \equiv \frac{\partial}{\partial x_1}\left(r\frac{\partial u}{\partial x_1}\right) + \frac{\partial}{\partial x_2}\left(r\frac{\partial u}{\partial x_2}\right) = f \quad \text{in } \Omega, \tag{10.216}$$

$$u = 0 \quad \text{on } \partial\Omega, \tag{10.217}$$

where $u = \psi + w$, $\quad f = \partial/\partial x_1(r(\partial w/\partial x_1)) + \partial/\partial x_2(r(\partial w/\partial x_2)) \in L^2(\Omega)$, and r is a continuous function of variables x_1 and x_2. Further let

$$r(\xi_1^2 + \xi_2^2) \geq r_o \|\xi\|^2, \quad r_o = \text{constant}, \quad \forall(\xi_1, \xi_2) \in R^2, \forall \mathbf{x} = (x_1, x_2) \in \overline{\Omega}.$$

The operator A is symmetric as

$$(Au, v) = \int_\Omega \left(r\frac{\partial u}{\partial x_1}\frac{\partial v}{\partial x_1} + r\frac{\partial u}{\partial x_2}\frac{\partial v}{\partial x_2}\right) \mathrm{d}\mathbf{x} = (Av, u) \quad \forall u, v \in V,$$

where $V = \{u \mid u \in H^1(\Omega), u = 0 \text{ on } \partial\Omega \text{ in the sense of traces}\}$.

Let us prove the existence of the unique generalized (weak) solution of the problem in Equations 10.216 and 10.217. Form the inner product of Equation 10.216 with a test function $v \in V$ and use the Green theorem. Then

$$a(u, v) = f(v),$$

where

$$a(u, v) = \int_\Omega \left(r\frac{\partial u}{\partial x_1}\frac{\partial v}{\partial x_1} + r\frac{\partial u}{\partial x_2}\frac{\partial v}{\partial x_2}\right) \mathrm{d}\mathbf{x}, \quad f(v) = \int_\Omega f v \mathrm{d}\mathbf{x}.$$

We see that the bilinear form $a(u, v)$ is symmetric. The problem in Equations 10.216 and 10.217 is thus transformed into the problem of finding a function $u \in V$ such that it satisfies the relation

$$a(u, v) = f(v)$$

for any $f \in V'$ and any $v \in V$. The bilinear form $a : V \times V \to R_1$ is bounded and also V-elliptic on V according to the Friedrichs inequality. The existence of the unique solution u follows from the Lax-Milgram theorem.

Let us now introduce the quadratic functional

$$J(u) = \frac{1}{2}a(u, u) - f(u) = \frac{1}{2}\int_{\Omega}\left(r\left(\frac{\partial u}{\partial x_1}\right)^2 + r\left(\frac{\partial u}{\partial x_2}\right)^2\right)\mathrm{d}\boldsymbol{x} - 2\int f u \mathrm{d}\boldsymbol{x}. \tag{10.218}$$

Our task is to find the minimum of the functional $J(u)$. Construct a triangulation $\mathcal{T}_h$. We assume that the triangulation $\mathcal{T}_h$ is regular. Let V_h be the space of functions $v(x_1, x_2)$ such that they are piecewise polynomials of the first degree on the individual triangles of the triangulation given and such that they fulfil the boundary condition in Equation 10.217. Every function $v(x_1, x_2)$ is given uniquely by its values at vertices P_i of the triangulation considered, i.e. $v_i = v(P_i)$. The boundary condition (Equation 10.217) is satisfied in such a way that $v_i = 0$ at vertices lying on $\partial\Omega$. The number of parameters that determine the function v is thus equal to the number of vertices of the triangulation given that do not coincide with $\partial\Omega$. Therefore, the space V_h is a finite dimensional subspace of the space V.

Thus there is one and only one function $u_h \in V_h$ such that

$$J(u_h) = \min_{v \in V_h} J(v),$$

where the function u_h is the approximate solution of our problem. Moreover, it is true for the space V_h, that the system of such subspaces, i.e. $\{V_h\}$, $\forall h$, is complete in V and thus

$$\overline{\cup_{h>o} V_h} = V, \tag{10.219}$$

and that $V_h \subset V\, \forall h$. Practically this means that functions from V_h fulfil the boundary condition in Equation 10.217. The condition in Equation 10.219 is equivalent to the condition: Let $\vartheta \subseteq V$ be dense in V and let there be operators $r_h : \vartheta \to V_h$ such that

$$\|v - r_h v\| \to 0, \qquad \text{for } h \to 0. \tag{10.220}$$

Since $a : V \times V \to R_1$ is a bounded and V-elliptic bilinear form and since Equation 10.220 holds we arrive at

$$\|u - u_h\| \to 0, \qquad \text{for } h \to 0_+ \quad \text{for } \forall v \in \vartheta.$$

To find an approximate solution u_h requires to express the energy functional $J(u)$ as a function of parameters u_i and then find the minimum of this function. As a result, we obtain a linear algebraic system with a positive definite matrix where the unknowns are the parameters u_i, which are just the values of the approximate solution at vertices of the triangulation. In accord with the relation $\psi = u - w$ and Equation 10.215, we directly obtain the values of the components of the electric field at vertices of the triangulation. A detailed analysis of the problem can be found in any elementary textbook on the finite element method. This is the reason for not presenting the analysis here.

Remark: The problem of determination of the field of an infinitely long straight cable on a homogeneous ground was solved by Bezvoda and Segeth (1985), under the assumption of the harmonically time varying field. Putting $\boldsymbol{H} = \operatorname{rot} \boldsymbol{A}$, where $\boldsymbol{A}$ is the vector potential, then the problem leads to solving $\Delta \boldsymbol{A} - k^2 \boldsymbol{A} = 0$, where $k^2 = 4\pi j\omega\mu\sigma c^{-2}$, with certain boundary conditions. Then the analysis of such a problem is similar to above.

REFERENCES

Adomian, G. (1983), *Stochastic Systems*, Academic Press, New York, London.

Axelsson, O. (1996), *Iterative Solution Methods*, Cambridge University Press.

Babuška, I. (1971), "Error-bounds for the Finite Element Method", *Numer Math.*, **16**, 322-333.

Babuška, I. (1972), "Lectures on Mathematical Foundations of Finite Element Method", *Technical Note*, **BN-748**, University of Maryland, College Park.

Babuška, I. and Kellogg, B. (1971), "Numerical Solution of the Neutron Differential Equation in the Presence of Corners and Interfaces", *Technical Note*, **BN-720**, University of Maryland, College Park.

Bendat, J.S. (1958), *Principles and Applications of Random Noise Theory*, J. Wiley & Sons, Chichester, New York.

Berezanskii, Ju.M. (1965), "Expansions in Eigenfunctions of Self-adjoint Operators", Transl.of Math. Monograph, *AMS*, **17**, Providence, RI.

Bezvoda, V. and Segeth, K. (1985), "The Electromagnetic Response of an Inhomogeneous Layered Earth — A General One-Dimensional Approach", *Geophysics*, **50**, 434-442.

Blanchard, P. and Brüning, E. (1982), *Direkte Methoden der Variationsrechnung*, Springer Verlag, Wien, New York.

Chapman, S. (1919), "The Solar and Lunar Diurnal Variation of the Earth's Magnetism", *Phil. Trans. Roy. Soc. London*, **Ser. A, 218**, 1-118.

Davenport, W.B. and Root, R.L. (1958), *An Introduction to the Theory of Random Signals and Noise*, McGraw-Hill, London.

Friedman, A. (1964), *Partial Differential Equations of Parabolic Type*, Prentice-Hall, Englewood Cliffs, New Jersey.

Jones, F.W. and Price, A.T. (1970), "The Perturbations of Alternating Geomagnetic Fields by Conductivity Anomalies", *Geophys. J. Roy. Astr. Soc.*, **20**, 317-334.

Jones, F.W. and Price, A.T. (1971), "The Geomagnetic Effects of Two-Dimensional Conductivity Inhomogeneities at Different Depths", *Geophys. J. R. Astr. Soc.*, **22**, 333-345.

Kleiber, M. and Hien, T.D. (1992), *The Stochastic Finite Element Method*, J.Wiley & Sons, Chichester, New York, Brisbane, Toronto, Singapore.

Krasnosel'skii, M.A., Vainiko, G.M., Zabreiko, P.P., Rushickii, Ja.B. and Stecenko, V.Ja. (1969), *Approximate Solution of Operator Equations*, Nauka, Moscow (in Russian),

Křížek, M. and Neittaanmäki, P. (1990), *Finite Element Approximation of Variational Problems and Applications*, Longman Scientific & Technical, Harlow, England.

Ladyzenskaya, O.A. (1973), *Boundary Value Problems of Mathematical Physics*, Nauka, Moscow (in Russian),

Ladyzenskaya, O.A., Solonikov, V.A. and Ural'ceva, N. (1968), *Linear and Quasi-linear Equations of Parabolic Type*, American Mathematical Society, RI.

Landau, L.D. and Lifshitz, E.M. (1984), *Electrodynamics of Continous Medium*, Pergamon Press, Oxford, New York, Beijing, Frankfurt, Sao Paulo, Sydney, Tokyo, Toronto.

Landis, E.M. (1971), *Equations of the 2nd Order of Elliptic and Parabolic Types*, Nauka, Moscow (in Russian),

Larsen, J.C. (1989), "Transfer Functions: Smooth Robust Estimates by Least-Squares and Remote Reference Methods", *J. Geophys. Int.*, **99**, 645-663.

Lukeš, D. (1996), *Solution of Parabolic Equation for Electromagnetic Field in Quasi-Stationary Approximation for Nonhomogeneous Model of the Earth* (Thesis), Faculty of Mathematics and Physics of the Charles University, Prague.

Marek, I. and Nedoma, J. (1973), "Finite Element Method in the Theory of SH-Wave Propagation", in *Numerische Methoden in der Geophysik*, Nedoma, J. and Pěč, K. (Eds.), Geophysical Institute Czech. Acad. Sci., Prague.

Matsushita, S. and Cambell, W.H. (1967), *Physics of Geomagnetic Phenomena*, Academic Press, New York, London.

McMechan, G.A. and Barrodale, I. (1985), "Processing Electromagnetic Data in Time", *Geophys. J. Roy. Astr. Soc.*, **81**, 277-293.

Mihlin, S.G. (1970), *Variational Methods in Mathematical Physics*, Nauka, Moscow. (in Russian).

Najzar, K. (1975), "Algorithms of Some Geophysical Problems", *Tech. Report*, **KNM 010514/75**, Faculty of Mathematics and Physics of the Charles University, Prague.

Nečas, J. (1967), *Les Méthodes Directes en Théorie des Équations Elliptiques*, Academia, Prague.

Nedoma, J. (1973), "Mathematical Analysis of the Problem of Investigation of Electromagnetic Field in Inhomogeneous Medium, H- and E- Polarization", in *Proc. of the Symposium Numerische Methoden in der Geophysik*, Nedoma, J. and Pěč, K. (Eds.), Liblice Castle, 1972, Geophysical Institute Czech. Acad. Sci., Prague.

Nedoma, J. (1974), "Electromagnetic Induction in the Conducting Layered Media", *Proc. Symposium on Interpretation Methods in Geophysics*, Bratislava 1974, Geophysical Institute Slovak Acad. Sci., Bratislava.

Nedoma, J. (1975), "The General Theory of the Electromagnetic Induction (H- and E-Polarization) in Inhomogeneous Models of the Earth", in *Numerische Methoden in der Geophysik*, Nedoma, J. and Pěč, K. (Eds.), Geophysical Institute Czech. Acad. Sci., Prague.

Nedoma, J. (1977), "Contribution to the Solution of a Magneto-telluric Problem for the Time Dependent Induced Source Field", *Acta Geodetica, Geophysica et Montanica Acad. Sci. Hung.* **12**, 223-231.

Nedoma, J. (1984/91), *Mathematical Problems of Contemporary Geodynamics* (PhD Thesis), Institute of Computer Science Czech. Acad. Sci., Prague.

Nedoma, J. and Praus, O. (1972), "Contribution to the Numerical Treatment of the Electromagnetic Field (H-Polarization) in Horizontally Nonhomogeneous Models of the Earth", *Studia Geophys. et Geod.*, **16**, 69-76.

Otnes, R.K. and Enochson, L. (1972), *Digital Time Series Analysis*, J. Wiley & Sons, Chichester, New York.

Price, A.T. (1949), "The Induction of Electric Currents in Non-uniform Thin Sheets and Shells", *Quart. J. Mech. Appl. Math.*, **2**, 283-310.

Price, A.T. (1950), "Electromagnetic Induction in a Semi-infinite Conductor with a Plane Boundary", *Quart. J. Mech. Appl. Math.*, **3**, 385-410.

Reddy, I.K. and Rankin, D. (1975), "Magneto-telluric Response of Laterally Inhomogeneous and Anisotropic Structure", *Geophysics*, **40**, 1035-1045.

Renardy, M. and Rogers, R.C. (1993), *An Introduction to Partial Differential Equations*, Springer Verlag, New York.

Rikitake, T. (1966), *Electromagnetism and the Earth's Interior*, Elsevier, Amsterdam, London, New York.

Rojtberg, Ja.A. and Scheftel', Z.G. (1962), "On the Equations of Elliptical Type with Noncontinuous Coefficients", *DAN SSSR*, **146, 6**, 1275-1278 (in Russian),

Schuster, A. (1908), "The Diurnal Variation of Terrestrial Magnetism", *Phil. Trans. Roy. Soc. London, Ser A, 208*, 163-204.

Spagnolini, U. (1994), "Time-Domain Estimation of MT Impedance Tensor", *Geophysics*, **59, 5**, 712-721.

Srivastava, S.P. (1966), "Theory of the Magneto-telluric Method for a Spherical Conductor", *Geophys. J. R. Astr. Soc.*, **11**, 373--387.

Vladimirov, B.C. (1967), *Equations of Mathematical Physics*, Nauka, Moscow.

Votruba, V. and Muzikář, C. (1955), *Theory of Electromagnetic Field*, NČSAV, Prague (in Czech),

Weaver, J.T. (1970), "The General Theory of Electromagnetic Induction in a Conducting Halfspace", *Geophys. J. Roy. Astr. Soc.*, **22**, 83-100.

Yee, E., Kosteniuk, P.R. and Paulson, K.V. (1988), "The Reconstruction of the Magnetotelluric Impedance Tensor: An Adaptive Parametric Time-Domain Approach", *Geophysics*, **53**, 1080-1087.

Zátopek, A. (1979), "On Geodynamical Aspects of Geophysical Synthesis in Central Europe", *Geodynamic Investigations in Czechoslovakia*, Bratislava, 91-111.

11

Some Mathematical Problems of Seismic Wave Propagation

11.1. INTRODUCTION

Problems of seismic wave propagation play an important role in geodynamic investigations. Modern geophysical studies have done remarkably little to change the main features of the Earth models suggested more than fifty years ago by Gutenberg and Jeffreys and their co-workers. Recent modelling of the Earth's compressional-wave velocity-profile emphasises details in the velocity gradient, and does little to change the average velocity over scales of hundred kilometres in the radial direction. But it is just these details which are very important for the development of compositional and geodynamic Earth models. The first evidence for anomalies in the velocity gradient is the observation of body waves which appear to be reflected from some depths within the mantle itself. Further seismic wave investigations find the strongest evidence for reflectors at depths 50, 130, 280, 410, 520, 630, 940 and 1250 km. Furthermore they have associated a particular compositional feature with each reflector. The furthest reflection boundaries in the Earth's interior were investigated.

The propagation of seismic waves through the Earth, characterized by an inhomogeneous medium with elastic parameters which are functions of place, was studied theoretically by means of simplified models in several papers and books (Alekseyev, Babich and Gel'chinskyi (1961), Burridge and Knopoff (1964), Bullen (1965), Richtmayer and Morton (1967), Bath (1968), Alterman and Karal (1968), Hron (1968), Boore (1970), (1972), Pěč and Nedoma (1970), Aboudi (1971), Alterman and Aboudi (1970), Boore, Larner and Aki (1971), Červený and Ravindra (1971), Fuchs and Müller (1971), Richards (1971), Babich and Buldyrev (1972), Clearbout and Johnson (1971), Bolt (1972), Nedoma (1972), Aschenbach (1973), Marek and Nedoma (1973), Officer (1974), Janovský, Marek and Nedoma (1974), Aschenbach and Harris (1978), Stephen (1983), Mikhailenko (1984), Nedoma (1984), Olson, Orcutt and Frazier (1984), Morton (1985), Aboudi (1986), Halpern and Rauch (1987), Chen

(1988), Despres (1991), Cohen, Halpern and Joly (1991), Nedoma (1992), Douglas, Hensley and Roberts (1993), Douglas, Santos, Sheen and Bennethum (1993), Kim (1994), Mark and Meister (1995), etc.).

Theoretical seismographs, constructed on the basis of results obtained by various methods, do not yield an exhausting answer to the question of the structure of the Earth's interior — the Earth's crust, upper and lower mantle as well as the core. Their physical structure is so complicated that the present-day methods are incapable of describing their nature in detail. The actual structure of the upper parts of the Earth contains a number of three-dimensional, spatially very complicated discontinuities, ranging from first to very high orders, corrugated interfaces, discontinuous boundaries, transition zones, inhomogeneities of various shape and nature, low-velocity and high-velocity channels of seismic waves, phase change zones of first and second orders, etc.. As a result of this, one uses simplified physical concepts in studying them and the studied phenomena are investigated by means of relatively simple models. One has to bear in mind that essentially all geophysical problems are very difficult and therefore also incorrect. The study of model problems is only one way of studying the geophysical reality. The "full answers" for the above discussed type of questions give model problems which are discussed in Chapters 8 and 12 based on dynamic contact-two-phase Stefan problems in thermo-visco-elasticity or thermo-visco-plasticity. Unfortunately, at present these types of problems are, from the point of view of present computer techniques, unsolved. They need larger computer techniques and will be very time consuming. It is evident that they represent the future problems of geophysical investigations. Unfortunately, they will only be a better approximation of reality. One of the goals of this monography is to point out to readers these unsolved problems representing the future problems of geophysical investigations.

Of the methods being used today there is, firstly, the wave method, which allows one to study the fields of S and P waves in analytical form. It is based on formal integral solutions of equations of motion, satisfying boundary conditions relevant to the studied problem of the propagation of seismic waves. However its use is strongly limited to models with homogeneous horizontal layers. If the model of the Earth is more complicated, one encounters difficulties with this method as a result of singular points, i.e. poles and saddle points, which necessarily occur in the solutions. The study of seismic waves using this method or methods modified for inhomogeneous media leads to a solution in closed form only in a few types of media, in which the velocity of the seismic waves was a function of depth, and in relatively simple models. The solution is sought in integral form, of the Sommerfeld integral type, as a superposition of plane waves. These methods cannot so far be used to treat oblique interfaces and interfaces perpendicular to the surface, because the integrals obtained are too complicated and the discussion of their physical meaning and interpretation are extremely difficult.

Other frequently used methods are the ray, Gaussian beam, matrix and propagator methods. These methods can be used to solve most problems approximately under certain assumptions. For example the ray method is limited to the high-frequency part of the spectrum and to distances sufficiently remote from the source, which does not quite correspond to physical facts. Its large drawback is that it does not yield an estimate of the errors incurred in the solution. It allows one to study inhomogeneous media under the assumption that the velocity variations at distances proportional to the wavelength are small. The numerical solution leads to a system of ordinary differential equations which can be solved by standard methods of the Runge-Kutta, predictor-corrector, etc. types. Certain modifications of the ray method can be used. One of these methods is based on the method of parabolic equations (see, for example, Babich and Buldyrev (1972)), another on the geometrical theory of diffraction as well as the exact ray method.

With the development of computers, methods of numerical analysis of solving partial differential equations also evolved and these methods were also applied to the solution of the problems of propagation of seismic waves in the stratified homogeneous and non-homogeneous medium in the time domain (hyperbolic equations) and the frequency domain (elliptic equations). Such models allow the simulation of propagations of seismic waves inside the Earth closed to the real rheology of the investigated part of the Earth.

The method of propagators, the asymptotic methods, the finite difference method and the finite element method have been elaborated. The first papers on the finite difference method appeared at the end of the Sixties and the beginning of the Seventies. Alterman and Karal (1968), Boore (1970), (1972) Alterman and Aboudi (1971), Nedoma (1972), Stephen (1983), Kim (1994) and others have studied the field of seismic waves in the time realm (hyperbolic type of differential equations) and in the frequency realm (elliptical type of differential equations). Both approaches to the study of seismic waves allow one to study the field more generally. The disadvantage of studying the field in the time realm is the instability of the differential patterns which may also affect the study of a number of problems.

The SH wave as well as the P and SV wave field in a nonhomogeneous stratified and generally spatially inhomogeneous media have been studied using the finite element method by Nedoma (1972), Marek and Nedoma (1973), Janovský, Marek and Nedoma (1974), Handlová, Nedoma and Schmid (1976), Nedoma (1979), (1984), (1990), (1992), Despres (1991), Cohen, Halpern and Joly (1991), Douglas, Hensley and Roberts (1993), Douglas, Santos, Sheen and Bennethum (1993), Kim (1994), etc. in the frequency realm and in the time realm. These methods may be used with great success to study the propagation of waves in inhomogeneous media with local inhomogeneities and in anisotropic as well as composite media (Nedoma (1979), Aboudi (1986), etc.). Some of the problems mentioned above will also be discussed in this monography.

The method of numerical holography is based on numerical methods. In numerical holography one constructs in a computer a map of the locations of reflecting interfaces. Any position in space where the material impedance undergoes a jump is called a reflector. Reflectors are defined as the loci of points in space where an impulse of incident radiation is time coincident with the reflected radiation. Such type of definition may be recast into the frequency domain, may be extended to multiple reflections, and may be elaborated in a number of other ways (Clearbout (1970), (1976), Metherell (1970), etc.).

11.2. THEORY AND PHYSICAL FORMULATION OF THE PROBLEM

In the general theory of the mechanics of the continuum (Hooke's continuum) the equations of motion are derived from the static condition of equilibrium, using the d'Alembert's principle. In dynamic problems the displacement $\boldsymbol{u} = (u_i)$, the stress τ_{ij} and the body forces $\boldsymbol{F} = (F_i)$, together with the force due to damping, are, in general, functions of the coordinates x_i and of time t, and they have continuous derivatives of the required order. Using d'Alembert's principle, the condition of equilibrium yields the dynamic equations. According to this, the equations are obtained by adding the inertial force to the acting body and surface forces (Landau and Lifshitz (1954), Brdička (1959), Nowacki (1970)). The equations of motion then read

$$\frac{\partial \tau_{ij}(\boldsymbol{x}, t)}{\partial x_j} + \rho(\boldsymbol{x}, t)F_i(\boldsymbol{x}, t) = \rho(\boldsymbol{x}, t)\frac{\partial^2 u_i(\boldsymbol{x}, t)}{\partial t^2} + \alpha(\boldsymbol{x})\frac{\partial u_i(\boldsymbol{x}, t)}{\partial t}, \quad i = 1, \dots, N, \tag{11.1}$$

where $N = 2, 3$, $\rho = \rho(\boldsymbol{x}, t)$ is the density of the medium, F_i is the i-th component of the body forces per unit volume, $\partial^2 u_i(\boldsymbol{x}, t)/\partial t^2$ is the i-th component of the acceleration, $\rho\partial^2 u_i(\boldsymbol{x}, t)/\partial t^2$ represents the i-th component of the inertia force and $\partial u_i(\boldsymbol{x}, t)/\partial t$ the i-th component of the velocity of the considered volume element, and $\alpha = \alpha(\boldsymbol{x})$ represents the coefficient of elastic wave damping.

The generalized Hooke's law,

$$\tau_{ij} = c_{ijkl}(\boldsymbol{x}, t)e_{kl}(\boldsymbol{u}(\boldsymbol{x}, t)), \tag{11.2}$$

where c_{ijkl} is a fourth order tensor, the so-called coefficients of elasticity, and $e_{ij}(\boldsymbol{x}, t)$ is the small strain tensor defined by

$$e_{ij}(\boldsymbol{u}(\boldsymbol{x}, t)) = \frac{1}{2}\left(\frac{\partial u_i(\boldsymbol{x}, t)}{\partial x_j} + \frac{\partial u_j(\boldsymbol{x}, t)}{\partial x_i}\right), \quad e_{ij}(\boldsymbol{u}(\boldsymbol{x}, t)) = e_{ji}(\boldsymbol{u}(\boldsymbol{x}, t)), \tag{11.3}$$

and Equation 11.1 yield the general equation for Hooke's continuum:

$$\frac{\partial}{\partial x_j}\left(c_{ijkl}(\boldsymbol{x}, t)\frac{\partial u_k(\boldsymbol{x}, t)}{\partial x_l}\right) + \rho(\boldsymbol{x}, t)F_i(\boldsymbol{x}, t) = \rho(\boldsymbol{x}, t)\frac{\partial^2 u_i(\boldsymbol{x}, t)}{\partial t^2} + \alpha\frac{\partial u_i(\boldsymbol{x}, t)}{\partial t}, \quad i = 1, \dots, N. \tag{11.4}$$

We now consider an isotropic inhomogeneous medium and we shall assume that the elastic coefficients and the density are functions of spatial coordinates only. Then the Hooke's law is described by the well-known relation

$$\tau_{ij}(\boldsymbol{x}, t) = \lambda(\boldsymbol{x})\delta_{ij}\vartheta(\boldsymbol{u}(\boldsymbol{x}, t)) + 2\mu e_{ij}(\boldsymbol{u}(\boldsymbol{x}, t)), \tag{11.5}$$

where the symmetric tensor e_{ij} is defined by Equation 11.3, $\vartheta(\boldsymbol{u}(\boldsymbol{x}, t))$ is the so-called cubic dilatation of the body defined as $\vartheta(\boldsymbol{u}(\boldsymbol{x}, t)) = \operatorname{div} \boldsymbol{u}(\boldsymbol{x}, t) = \partial u_i/\partial x_i$, δ_{ij} is the Kronecker's delta, and λ and μ are Lamé's elastic coefficients. By substituting

Equation 11.5 into Equation 11.1, and after some necessary adjustments, we arrive at the equation of motion:

$$(\lambda + \mu)\ \text{grad div}\ \boldsymbol{u} + \mu\Delta\boldsymbol{u} + \ \text{grad}\ \lambda\ \text{div}\ \boldsymbol{u} + (\text{grad}\ \mu \times\ \text{rot}\ \boldsymbol{u})$$
$$+ 2(\text{grad}\ \mu\ \text{grad})\boldsymbol{u} + \rho\boldsymbol{F} = \rho\partial^2\boldsymbol{u}/\partial t^2 + \alpha\partial\boldsymbol{u}/\partial t. \tag{11.6}$$

In this section we shall limit ourselves to a special case of the propagation of seismic waves in a stratified or/and in a generally inhomogeneous medium, distributed in an infinite half-space and excited by a linear source. We shall assume that inhomogeneous absorption (damping), defined by the coefficient of absorption (damping) α, is also present in the medium. Moreover, in this special case we assume that the elastic coefficients λ and μ and the density ρ do not depend on one co-ordinate and time. Assume that the linear source of elastic waves is parallel with the axis $x_2 \equiv y$ and that all the physical parameters do not depend on this co-ordinate. Equation 11.1 then yields

$$\frac{\partial\tau_{11}(\boldsymbol{x},t)}{\partial x_1}+\frac{\partial\tau_{13}(\boldsymbol{x},t)}{\partial x_3}+\rho(\boldsymbol{x})F_1(\boldsymbol{x},t)=\rho(\boldsymbol{x})\frac{\partial^2 u_1(\boldsymbol{x},t)}{\partial t^2}+\alpha(\boldsymbol{x})\frac{\partial u_1(\boldsymbol{x},t)}{\partial t}, \tag{11.7}$$

$$\frac{\partial\tau_{21}(\boldsymbol{x},t)}{\partial x_1}+\frac{\partial\tau_{23}(\boldsymbol{x},t)}{\partial x_3}+\rho(\boldsymbol{x})F_2(\boldsymbol{x},t)=\rho(\boldsymbol{x})\frac{\partial^2 u_2(\boldsymbol{x},t)}{\partial t^2}+\alpha(\boldsymbol{x})\frac{\partial u_2(\boldsymbol{x},t)}{\partial t}, \tag{11.8}$$

$$\frac{\partial\tau_{31}(\boldsymbol{x},t)}{\partial x_1}+\frac{\partial\tau_{33}(\boldsymbol{x},t)}{\partial x_3}+\rho(\boldsymbol{x})F_3(\boldsymbol{x},t)=\rho(\boldsymbol{x})\frac{\partial^2 u_3(\boldsymbol{x},t)}{\partial t^2}+\alpha(\boldsymbol{x})\frac{\partial u_3(\boldsymbol{x},t)}{\partial t}. \tag{11.9}$$

It then follows from Hooke's law (Equation 11.5) that

$$\tau_{11}(\boldsymbol{x},t) = (\lambda(\boldsymbol{x}) + 2\mu(\boldsymbol{x}))\frac{\partial u_1(\boldsymbol{x},t)}{\partial x_1} + \lambda(\boldsymbol{x})\frac{\partial u_3(\boldsymbol{x},t)}{\partial x_3},$$

$$\tau_{33}(\boldsymbol{x},t) = (\lambda(\boldsymbol{x}) + 2\mu(\boldsymbol{x}))\frac{\partial u_3(\boldsymbol{x},t)}{\partial x_3} + \lambda(\boldsymbol{x})\frac{\partial u_1(\boldsymbol{x},t)}{\partial x_1},$$

$$\tau_{13}(\boldsymbol{x},t) = \tau_{31}(\boldsymbol{x},t) = \mu(\boldsymbol{x})\left(\frac{\partial u_1(\boldsymbol{x},t)}{\partial x_3} + \frac{\partial u_3(\boldsymbol{x},t)}{\partial x_1}\right),$$

$$\tau_{12}(\boldsymbol{x},t) = \tau_{21}(\boldsymbol{x},t) = \mu(\boldsymbol{x})\frac{\partial u_2(\boldsymbol{x},t)}{\partial x_1},$$

$$\tau_{23}(\boldsymbol{x},t) = \tau_{32}(\boldsymbol{x},t) = \mu(\boldsymbol{x})\frac{\partial u_2(\boldsymbol{x},t)}{\partial x_3}.$$

Then Equation 11.8 yields

$$\frac{\partial}{\partial x_1}\left(\mu(\boldsymbol{x})\frac{\partial u_2(\boldsymbol{x},t)}{\partial x_1}\right) + \frac{\partial}{\partial x_3}\left(\mu(\boldsymbol{x})\frac{\partial u_2(\boldsymbol{x},t)}{\partial x_3}\right) + \rho(\boldsymbol{x})F_2(\boldsymbol{x},t)$$
$$= \rho(\boldsymbol{x})\frac{\partial^2 u_2(\boldsymbol{x},t)}{\partial t^2} + \alpha(\boldsymbol{x})\frac{\partial u_2(\boldsymbol{x},t)}{\partial t}. \tag{11.10}$$

This equation is separated for the u_2 component and describes the field of the SH waves. Equations 11.7 and 11.9 become

$$\frac{\partial}{\partial x_1}\left((\lambda(\boldsymbol{x})+2\mu(\boldsymbol{x}))\frac{\partial u_1(\boldsymbol{x},t)}{\partial x_1}+\lambda(\boldsymbol{x})\frac{\partial u_3(\boldsymbol{x},t)}{\partial x_3}\right)+\frac{\partial}{\partial x_3}\left(\mu(\boldsymbol{x})\left(\frac{\partial u_1(\boldsymbol{x},t)}{\partial x_3}+\frac{\partial u_3(\boldsymbol{x},t)}{\partial x_1}\right)\right)$$
$$+\rho(\boldsymbol{x})F_1(\boldsymbol{x},t)=\rho(\boldsymbol{x})\frac{\partial^2 u_1(\boldsymbol{x},t)}{\partial t^2}+\alpha(\boldsymbol{x})\frac{\partial u_1(\boldsymbol{x},t)}{\partial t}, \tag{11.11}$$

$$\frac{\partial}{\partial x_1}\left(\mu(\boldsymbol{x})\left(\frac{\partial u_1(\boldsymbol{x},t)}{\partial x_3}+\frac{\partial u_3(\boldsymbol{x},t)}{\partial x_1}\right)\right)+\frac{\partial}{\partial x_3}\left((\lambda(\boldsymbol{x})+2\mu(\boldsymbol{x}))\frac{\partial u_3(\boldsymbol{x},t)}{\partial x_3}+\lambda(\boldsymbol{x})\frac{\partial u_1(\boldsymbol{x},t)}{\partial x_1}\right)$$
$$+\rho(\boldsymbol{x})F_3(\boldsymbol{x},t)=\rho(\boldsymbol{x})\frac{\partial^2 u_3(\boldsymbol{x},t)}{\partial t^2}+\alpha(\boldsymbol{x})\frac{\partial u_3(\boldsymbol{x},t)}{\partial t}, \tag{11.12}$$

and represent the system of equations of motion for P and SV waves.

Now we shall define the boundary conditions. At the surface of the medium, the surface of the Earth, which is identified with the x_1 axis, there are only three boundary conditions, $\tau_{33}=\tau_{31}=\tau_{32}=0$, i.e.

$$(\lambda(\boldsymbol{x})+2\mu(\boldsymbol{x}))\frac{\partial u_3(\boldsymbol{x},t)}{\partial x_3}+\lambda(\boldsymbol{x})\frac{\partial u_1(\boldsymbol{x},t)}{\partial x_1}=0 \tag{11.13}$$

$$\mu(\boldsymbol{x})\left(\frac{\partial u_1(\boldsymbol{x},t)}{\partial x_3}+\frac{\partial u_3(\boldsymbol{x},t)}{\partial x_1}\right)=0, \tag{11.14}$$

$$\mu(\boldsymbol{x})\frac{\partial u_2(\boldsymbol{x},t)}{\partial x_3}=0 \tag{11.15}$$

The last equation is the boundary condition for the SH waves, and the other two equations boundary conditions for the P and SV waves. In the general case of a curvilinear boundary the condition leads to $\tau_{ij}n_j=0$.

At the boundary between the two media, the normal and tangential components of the displacement and the stress vectors must be continuous, i.e. for the layered medium we have

$$u_1(\boldsymbol{x},t)|_{(l)}=u_1(\boldsymbol{x},t)|_{(m)}, \tag{11.16}$$

$$u_2(\boldsymbol{x},t)|_{(l)}=u_2(\boldsymbol{x},t)|_{(m)}, \tag{11.17}$$

$$u_3(\boldsymbol{x},t)|_{(l)}=u_3(\boldsymbol{x},t)|_{(m)} \tag{11.18}$$

and

$$\tau_{33}|_{(l)}=\tau_{33}|_{(m)}, \tag{11.19}$$

$$\tau_{31}|_{(l)}=\tau_{31}|_{(m)}, \tag{11.20}$$

$$\tau_{32}|_{(l)}=\tau_{32}|_{(m)}, \tag{11.21}$$

or Equations 11.16 to 11.18 and

$$\left.(\lambda(\boldsymbol{x})+2\mu(\boldsymbol{x}))\frac{\partial u_1(\boldsymbol{x},t)}{\partial x_1}+\lambda(\boldsymbol{x})\frac{\partial u_3(\boldsymbol{x},t)}{\partial x_3}\right|_{(l)}$$

$$=\left.(\lambda(\boldsymbol{x})+2\mu(\boldsymbol{x}))\frac{\partial u_1(\boldsymbol{x},t)}{\partial x_1}+\lambda(\boldsymbol{x})\frac{\partial u_3(\boldsymbol{x},t)}{\partial x_3}\right|_{(m)}, \tag{11.22}$$

$$\left.\mu(\boldsymbol{x})\left(\frac{\partial u_1(\boldsymbol{x},t)}{\partial x_3}+\frac{\partial u_3(\boldsymbol{x},t)}{\partial x_1}\right)\right|_{(l)}=\left.\mu(\boldsymbol{x})\left(\frac{\partial u_1(\boldsymbol{x},t)}{\partial x_3}+\frac{\partial u_3(\boldsymbol{x},t)}{\partial x_1}\right)\right|_{(m)}, \tag{11.23}$$

$$\left.\mu(\boldsymbol{x})\frac{\partial u_2(\boldsymbol{x},t)}{\partial x_3}\right|_{(l)}=\left.\mu(\boldsymbol{x})\frac{\partial u_2(\boldsymbol{x},t)}{\partial x_3}\right|_{(m)}. \tag{11.24}$$

Here, Equations 11.17 and 11.24 represent the transition conditions at the boundary between the layers or inhomogeneous areas with elasticity parameters μ_l, λ_l, ρ_l and μ_m, λ_m, ρ_m for the SH waves, and the remaining equations are for the P and SV waves. For the generalized curvilinear boundaries between neighbouring rocks, the conditions will be as follows

$$\boldsymbol{u}(\boldsymbol{x},t)|_{(k)}=\boldsymbol{u}(\boldsymbol{x},t)|_{(l)}, \tag{11.25}$$

$$\tau_{ij}(\boldsymbol{u})n_j\big|_{(k)}=\tau_{ij}(\boldsymbol{u})n_j\big|_{(l)}. \tag{11.26}$$

If the source of the seismic waves is within the medium, Sommerfeld's radiation conditions $\boldsymbol{u}=O(\boldsymbol{r}^{-1})$ and $(\partial\boldsymbol{u}/\partial\boldsymbol{r})+ik\boldsymbol{u}=O(\boldsymbol{r}^{-1})$, where $\boldsymbol{r}$ is the radius vector, must be satisfied at the largest distances, i.e. in infinity.

In our next studies we shall consider a linear time-dependent and harmonic sources (i.e. with dependence $\boldsymbol{u}(\boldsymbol{x},t)=\exp(\pm i\omega t)\boldsymbol{v}(\boldsymbol{x}))$ of seismic waves, located at depth z_o beneath the Earth's surface. The problem, defined above, can be studied by finite element (finite difference) methods in the time and/or frequency domains. For this purpose it is necessary to define finite regions, as these methods for such studies can be applied with advantage on the bounded finite domain. We shall introduce an inhomogeneous absorption (damping) coefficient α, such that it corresponds to physical reality in the "working region" Ω_p of propagated seismic waves, and such that outside the working region Ω_p it becomes completely absorbed at a certain distance, which we shall denote by D, i.e.

$$\alpha(\boldsymbol{x})=\begin{cases}\alpha_o=\text{constant for points } \boldsymbol{x}=(x_i) \text{ for which } |x_i|<X_{oi},\\ \alpha(\boldsymbol{x},\overline{\lambda}) \text{ for points } \boldsymbol{x}=(x_i) \text{ for which } |x_i|\geq X_{oi}, \text{ i.e.}\\ \quad\alpha \text{ is a function of the wave length } \overline{\lambda},\end{cases} \tag{11.27}$$

where X_{oi}, $i=1,\ldots,N$, $N=2,3$, are some suitable numbers defining the dimensions of the "working volume", i.e. the volume in which the physical properties of the medium are preserved. To determine this we shall assume that the waves propagated in this absorbed region are plane waves. Then the absorption coefficient α is a function of place and the physical parameters of the medium, and/or of the wave length $\overline{\lambda}$, and is constant and equal to the real absorption of the medium in the working region Ω_p

and a function of velocity of seismic waves propagated through the absorbed region, e.g. for the cylindrical type of wave

$$|\boldsymbol{u}(X_{oi}) - \boldsymbol{u}(x_i)| \sim \left|\exp(-i\alpha X_{oi})X_{oi}^{-1/2} - \exp(-i\alpha(X_{oi}+D)(X_{oi}+D)^{-1/2}\right| \sim 0.001,$$

After some algebra we obtain

$$\alpha(\boldsymbol{x}) = \begin{cases} \alpha_o = \text{constant for points } \boldsymbol{x} = (x_i) \text{ for which } |x_i| < X_{oi}, \\ \alpha(\boldsymbol{x}, \bar{\lambda}) = 8\pi^{-1}D^{-2}c^{-1}(4\bar{\lambda}^2 + \pi^2 D^2)^{1/2} \text{ for points } \boldsymbol{x} = (x_i) \\ \quad \text{for which } |x_i| \geq X_{oi}, \end{cases} \tag{11.28}$$

where X_{oi} is defined above and c represents the velocity of seismic wave propagated in $\Omega\backslash\Omega_o$. Hence, due to the damping of the seismic waves, at distance $|\boldsymbol{x} - \boldsymbol{X_o}| = D$ the seismic waves are fully absorbed, i.e. the displacements there are equal to zero, so that the boundary condition can be found as

$$\boldsymbol{u} = 0, \tag{11.29}$$

which bounds the investigated domain.

The sources of seismic waves can be of point, line, etc. types, symmetric (e.g. an explosion) or non-symmetric (e.g. the majority of earthquakes), and are characterized by their directional characteristics (represented by the function $g(\alpha, \eta)$, where α is an angle between the x_3 axis and a direction of the ray, and η represents the location of the plane; if $g(\alpha, \eta) = \text{constant}$ then we have the symmetric point source). In the close neighbourhood of the source the linearized Equation 11.6 is not valid as the processes taking place in the region close to the source behave non-linearly. Then the sources can be generated in two ways: firstly by using the body forces (see for example Aboudi (1971), or secondly by using the boundary condition in the vicinity of the source of the seismic waves (see for example Nedoma (1972)), which eliminates the non-linear effects near the source.

In the first case a numerical treatment of body forces is used. This treatment allows the formulation of sources of several different orders and with different spatial and temporal dependence. The body force techniques (Burridge and Knopoff (1964), Alterman and Aboudi (1970), Aboudi (1971), etc.), based on a certain type of elastic dislocation, are useful for numerical modelling of sources of seismic waves. The body force approximations depend only upon the source mechanism and the elastic properties of the medium in the immediate vicinity of the source and not upon any reflecting surface or other inhomogenities in the Earth's interior. The numerical treatment of the forces can be applied to more complicated geometries as well as to nonhomogeneous anisotropic rocks. This technique can also be used for numerical approximation of seismic sources which have different extensions in the co-ordinate axes. We shall consider three type of seismic sources. The spatial and time dependence of seismic sources is determined by the function $F_i(x_j, t)$, $i, j = 1, \ldots, N$, $N = 2, 3$, characterizing the type and spatial extension of the source. For the definition of the source function $F(\boldsymbol{x})$ let the following function $f(x, \Delta x)$ (we write x, y, z instead of x_1, x_2, x_3) be defined as

$$f(x, \Delta x) = 1 - \delta_{2x}[\tfrac{1}{2}x^2 H(x)], \ \text{for } x \geq 0, \ \text{with } f(-x, \Delta x) = f(x, \Delta x), \tag{11.30}$$

where $\delta_{2x}(g(x)) = (\Delta x)^{-2}(g(x) - 2g(x - \Delta x) + g(x - 2\Delta x))$ is a second finite difference operator and $H(x)$ is the Heaviside step function. The derivative of the function $f(x, \Delta x)$ is given by

$$f'(x, \Delta x) = -\delta_{2x}[xH(x)], \quad \text{for } x \geq 0, \quad \text{with } f'(-x, \Delta x) = -f'(x, \Delta x). \tag{11.31}$$

The source time dependence is defined by the function $\tau(t)$ as

$$\tau(t) = -f(t - 2\Delta t, \Delta t), \qquad \text{for } t \geq 0. \tag{11.32}$$

The functions $f(x, \Delta x)$ and its derivative and $\tau(t)$ are given in Figure 1 of Aboudi (1971), and the displacement components for several types of seismic sources are presented in the further figures of Aboudi (1971).

Let us define the following types of numerical approximation of seismic sources.

Seismic sources of the first order which are of several types:

(i) a vertical finite-line source of first order located at point (0,0,Z) which radiates in the direction x is defined as

$$\boldsymbol{F}(x, t) = \begin{pmatrix} 0 \\ f(x, \Delta x)f(z - Z, \Delta z)\tau(t) \end{pmatrix}, \tag{11.33}$$

where the functions f and τ are defined above.

(ii) a horizontal finite-line source of first order located at point $(0, 0, Z)$ which radiates in the direction x is defined as

$$\boldsymbol{F}(x, t) = \begin{pmatrix} f(x, \Delta x)f(z - Z, \Delta z)\tau(t) \\ 0 \end{pmatrix}, \tag{11.34}$$

where the functions f and τ are defined above.

Seismic sources of the second order which are of several types:

(i) a couple without moment in the z direction of second order located at point $(0, 0, Z)$ which radiates in the direction x is defined as

$$\boldsymbol{F}(x, t) = \begin{pmatrix} 0 \\ f(x, \Delta x)f'(z - Z, \Delta z)\tau(t) \end{pmatrix}; \tag{11.35}$$

(ii) a couple without moment in the x direction of second order located at point $(0, 0, Z)$ which radiates in the direction x is defined as

$$\boldsymbol{F}(x, t) = \begin{pmatrix} f'(x, \Delta x)f(z - Z, \Delta z)\tau(t) \\ 0 \end{pmatrix}; \tag{11.36}$$

(iii) a double force with moment in the vertical direction located at point $(0, 0, Z)$ is defined as

$$\boldsymbol{F}(x, t) = \begin{pmatrix} 0 \\ f'(x, \Delta x)f(z - Z, \Delta z)\tau(t) \end{pmatrix}; \tag{11.37}$$

(iv) a double force with moment in the horizontal direction located at point $(0, 0, Z)$ is defined as

$$\boldsymbol{F}(x, t) = \begin{pmatrix} f(x, \Delta x) f'(z - Z, \Delta z)\tau(t) \\ 0 \end{pmatrix}; \qquad (11.38)$$

(v) a combination of three mutually perpendicular double forces without moment located at point $(0, 0, Z)$ is defined as

$$\boldsymbol{F}(x, t) = \begin{pmatrix} f'(x, \Delta x) f(z - Z, \Delta z)\tau(t) \\ f(x, \Delta x) f'(z - Z, \Delta z)\tau(t) \end{pmatrix}, \qquad (11.39)$$

where the force is a gradient of a scalar function and therefore represents the source of P waves (the explosive source) only;

(vi) a centre of rotation given by a superposition of two double forces with moment located at point $(0, 0, Z)$ is defined as

$$\boldsymbol{F}(x, t) = \begin{pmatrix} f(x, \Delta x) f'(z - Z, \Delta z)\tau(t) \\ -f'(x, \Delta x) f(z - Z, \Delta z)\tau(t) \end{pmatrix}; \qquad (11.40)$$

(vii) a double couple without moment is defined as

$$\boldsymbol{F}(x, t) = \begin{pmatrix} f(x, \Delta x) f'(z - Z, \Delta z)\tau(t) \\ f'(x, \Delta x) f(z - Z, \Delta z)\tau(t) \end{pmatrix}; \qquad (11.41)$$

(viii) a body force caused by a dip-slip dislocation is defined as

$$\boldsymbol{F}(x, t) = \begin{pmatrix} -f(x, \Delta x) f'(z - Z, \Delta z)\tau(t) \\ -f'(x, \Delta x) f(z - Z, \Delta z)\tau(t) \end{pmatrix} \cos 2\Theta + \begin{pmatrix} f'(x, \Delta x) f(z - Z, \Delta z)\tau(t) \\ -f(x, \Delta x) f'(z - Z, \Delta z)\tau(t) \end{pmatrix} \sin 2\Theta. \qquad (11.42)$$

Seismic sources of the third order which are formed by a superposition of second order sources at the limit are defined as

$$\boldsymbol{F}(x, t) = \begin{pmatrix} f_1(x, \Delta x) f_1''(z - Z, \Delta z)\tau(t) \\ 0 \end{pmatrix}, \qquad (11.43)$$

where

$$f_1(x, \Delta x) = 1 - \delta_{3x}[x^3 H(x)/6], \quad x \geq 0 \text{ with } f_1(-x, \Delta x) = f_1(x, \Delta x), \qquad (11.44)$$

as the second derivative of $f(x, \Delta x)$ does not exist, and where δ_{3x} is a third finite difference operator.

In the second case described earlier we take full advantage of the properties of seismic waves in the neighbourhood of the source. There is a singularity of $\boldsymbol{r}^{-1/2}$-type. Since only a singular component of the field has a meaning in the immediate vicinity of the source, we shall describe a cylindrical surface in its neighbourhood with a sufficiently small radius r_p. On this cylindrical surface we prescribe the effect of non-damped direct waves only.

For the linear harmonic sources Equations 11.10 to 11.12 will then become

$$\frac{\partial}{\partial x_1}\left(\mu(\boldsymbol{x})\frac{\partial u_2(\boldsymbol{x},t)}{\partial x_1}\right)+\frac{\partial}{\partial x_3}\left(\mu(\boldsymbol{x})\frac{\partial u_2(\boldsymbol{x},t)}{\partial x_3}\right)+[\rho(\boldsymbol{x})\omega^2-i\omega\alpha(\boldsymbol{x},\overline{\lambda})]u_2+\rho(\boldsymbol{x})F_2(\boldsymbol{x},t)=0, \tag{11.45}$$

$$\frac{\partial}{\partial x_1}\left((\lambda(\boldsymbol{x})+2\mu(\boldsymbol{x}))\frac{\partial u_1(\boldsymbol{x},t)}{\partial x_1}+\lambda(\boldsymbol{x})\frac{\partial u_3(\boldsymbol{x},t)}{\partial x_3}\right)+\frac{\partial}{\partial x_3}\left(\mu(\boldsymbol{x})\left(\frac{\partial u_1(\boldsymbol{x},t)}{\partial x_3}+\frac{\partial u_3(\boldsymbol{x},t)}{\partial x_1}\right)\right)$$
$$+[\rho(\boldsymbol{x})\omega^2-i\omega\alpha(\boldsymbol{x},\overline{\lambda})]u_1+\rho(\boldsymbol{x})F_1(\boldsymbol{x},t)=0, \tag{11.46}$$

$$\frac{\partial}{\partial x_1}\left(\mu(\boldsymbol{x})\left(\frac{\partial u_1(\boldsymbol{x},t)}{\partial x_3}+\frac{\partial u_3(\boldsymbol{x},t)}{\partial x_1}\right)\right)+\frac{\partial}{\partial x^3}\left((\lambda(\boldsymbol{x})+2\mu(\boldsymbol{x}))\frac{\partial u_3(\boldsymbol{x},t)}{\partial x_3}+\lambda(\boldsymbol{x})\frac{\partial u_1(\boldsymbol{x},t)}{\partial x_1}\right)$$
$$+[\rho(\boldsymbol{x})\omega^2-i\omega\alpha(\boldsymbol{x},\overline{\lambda})]u_3+\rho(\boldsymbol{x})F_3(\boldsymbol{x},t)=0. \tag{11.47}$$

Similarly as above we shall find the boundary conditions.

At the surface of the medium, the surface of the Earth, which is identified with the x_1 axis, there are only three boundary conditions, $\tau_{33}=\tau_{31}=\tau_{32}=0$, i.e.

$$(\lambda(\boldsymbol{x})+2\mu(\boldsymbol{x}))\frac{\partial u_3(\boldsymbol{x})}{\partial x_3}+\lambda(\boldsymbol{x})\frac{\partial u_1(\boldsymbol{x})}{\partial x_1}=0, \tag{11.48}$$

$$\mu(\boldsymbol{x})\left(\frac{\partial u_1(\boldsymbol{x})}{\partial x_3}+\frac{\partial u_3(\boldsymbol{x})}{\partial x_1}\right)=0, \tag{11.49}$$

$$\mu(\boldsymbol{x})\frac{\partial u_2(\boldsymbol{x})}{\partial x_3}=0. \tag{11.50}$$

At the boundary between the two media, the normal and tangential components of the displacement and the stress vectors must be continuous, i.e. for the planar interface boundary we require

$$u_1|_{(l)}=u_1|_{(m)}, \tag{11.51}$$

$$u_2|_{(l)}=u_2|_{(m)}, \tag{11.52}$$

$$u_3|_{(l)}=u_3|_{(m)}, \tag{11.53}$$

and

$$\tau_{33}|_{(l)}=\tau_{33}|_{(m)}, \tag{11.54}$$

$$\tau_{31}|_{(l)}=\tau_{31}|_{(m)}, \tag{11.55}$$

$$\tau_{32}|_{(l)}=\tau_{32}|_{(m)}, \tag{11.56}$$

or Equations 11.51 and 11.53 and

$$\left.(\lambda(\boldsymbol{x})+2\mu(\boldsymbol{x}))\frac{\partial u_1(\boldsymbol{x})}{\partial x_1}+\lambda(\boldsymbol{x})\frac{\partial u_3(\boldsymbol{x})}{\partial x_3}\right|_{(l)}=\left.(\lambda(\boldsymbol{x})+2\mu(\boldsymbol{x}))\frac{\partial u_1(\boldsymbol{x})}{\partial x_1}+\lambda(\boldsymbol{x})\frac{\partial u_3(\boldsymbol{x})}{\partial x_3}\right|_{(m)}, \tag{11.57}$$

$$\mu(\boldsymbol{x})\left(\frac{\partial u_1(\boldsymbol{x})}{\partial x_3}+\frac{\partial u_3(\boldsymbol{x})}{\partial x_1}\right)\Bigg|_{(l)}=\mu(\boldsymbol{x})\left(\frac{\partial u_1(\boldsymbol{x})}{\partial x_3}+\frac{\partial u_3(\boldsymbol{x})}{\partial x_1}\right)\Bigg|_{(m)}, \tag{11.58}$$

$$\mu(\boldsymbol{x})\frac{\partial u_2(\boldsymbol{x})}{\partial x_3}\Bigg|_{(l)}=\mu(\boldsymbol{x})\frac{\partial u_2(\boldsymbol{x})}{\partial x_3}\Bigg|_{(m)}. \tag{11.59}$$

For curvilinear interface boundaries the equivalent conditions of Equations 11.25 and 11.26 are also valid for this case.

Due to the damping of the seismic waves we find

$$\boldsymbol{u}(x)=0, \tag{11.60}$$

which bounds the investigated domain.

We see that the equation and conditions for the component u_2 describe the boundary value problem for the SH waves.

The sources of seismic waves, depending only on spatial co-ordinates, can be defined in a similar way as above.

Remark: For ideal homogeneous isotropic rocks $\lambda =$ constant > 0 and $\mu =$ constant > 0, and for ideal homogeneous fluid $\mu = 0$.

Remark: Equations 11.10 to 11.12 can be rewritten in vector form as

$$(\lambda+\mu)\text{ grad div }\boldsymbol{u}+\mu\Delta\boldsymbol{u}+\text{grad }\lambda\text{ div }\boldsymbol{u}+(\text{grad }\mu\times\text{ rot }\boldsymbol{u})+2(\text{grad }\mu\text{ grad })\boldsymbol{u}+\rho\boldsymbol{f}=\rho\partial^2\boldsymbol{u}/\partial t^2. \tag{11.61}$$

Since $\Delta\boldsymbol{u} =$ grad div $\boldsymbol{u}$ − rot rot $\boldsymbol{u}$, then we obtain

$$(\lambda+2\mu)\text{grad div }\boldsymbol{u}-\mu\text{ rot rot}\boldsymbol{u}+\text{grad }\lambda\text{ div}\boldsymbol{u}+(\text{grad }\mu\times\text{rot }\boldsymbol{u})+2(\text{grad }\mu\text{ grad})\boldsymbol{u}+\rho\boldsymbol{f}=\rho\partial^2\boldsymbol{u}/\partial t^2. \tag{11.62}$$

Let us assume the case of ideal homogeneous isotropic rocks and let the body forces $\boldsymbol{f} = 0$, then Equation 11.62 yields

$$(\lambda+2\mu)\text{grad div }\boldsymbol{u}-\mu\text{rot rot }\boldsymbol{u}=\rho\partial^2\boldsymbol{u}/\partial t^2. \tag{11.63}$$

Hence, applying the operation "divergence", and since

$$\text{div rot }\boldsymbol{u}=0,\ \text{ div grad }\vartheta=\Delta\vartheta\text{ and }\vartheta=\text{div }\boldsymbol{u},$$

we have

$$(\lambda+2\mu)\Delta\vartheta=\rho\partial^2\vartheta/\partial t^2\text{ or }v_p^2\Delta\vartheta=\rho\partial^2\vartheta/\partial t^2, \tag{11.64}$$

where $v_p = [(\lambda + 2\mu)/\rho]^{1/2}$ is the velocity of the P waves.

From Equation 11.63, applying the operation "rot", and since

$$\text{rot }\boldsymbol{u}=\vec{\Omega},\ \text{ rot grad }\vec{\Omega}=0\text{ and }\Delta\vec{\Omega}=\text{ grad div }\overline{\Omega}-\text{ rot rot }\vec{\Omega},$$

then

$$-\mu\text{ rot rot }\vec{\Omega}=\rho\text{ rot }\partial^2\boldsymbol{u}/\partial t^2=\rho\partial^2/\partial t^2\text{ rot }\boldsymbol{u}=\rho\partial^2\vec{\Omega}/\partial t^2.$$

Hence

$$\Delta\vec{\Omega} - \text{grad div } \vec{\Omega} = v_s^{-2}\partial^2\vec{\Omega}/\partial t^2$$

and then

$$\Delta\vec{\Omega} = v_s^{-2}\partial^2\vec{\Omega}/\partial t^2, \tag{11.65}$$

as div rot $\boldsymbol{u}$ = div $\vec{\Omega} = 0$, where $v_s = (\mu/\rho)^{1/2}$ is the velocity of the S waves.

Equation 11.64 characterizes seismic waves of dilatation (P) type, whilst Equation 11.65 characterizes seismic waves of rotational or of transversal (S) types.

Another technique of separation is based on the Helmholtz theorem from vector analysis.

Theorem 11.1 (Helmholtz): An arbitrary vector field $\boldsymbol{u}$, which is continuous, unambiguous, finite and in the infinity equal to zero, can be unambiguously written in the form

$$\boldsymbol{u} = \boldsymbol{u}_1 + \boldsymbol{u}_2, \text{ where } \boldsymbol{u}_1 = \text{rot } \boldsymbol{u}_1 = 0 \text{ and div } \boldsymbol{u}_2 = 0, \tag{11.66}$$

where $\boldsymbol{u}_1$ is of unvorticity type as rot $\boldsymbol{u}_1 = 0$, and $\boldsymbol{u}_2$ is of unsourced type as div $\boldsymbol{u}_2 = 0$.

Using this theorem we put

$$\boldsymbol{u} = \text{ grad } \varphi + \text{rot } \vec{\psi}, \text{ where div } \vec{\psi} = 0.$$

Since div grad $\varphi = \Delta\varphi$, rot grad $\varphi = 0$, and div rot $\vec{\psi} = 0$ then from Equation 11.63 we obtain

$$\text{grad}[(\lambda + 2\mu) \text{ div grad } \varphi - \rho\partial^2\varphi/\partial t^2] + \text{rot } [-\mu \text{rot rot}\vec{\psi} - \rho\partial^2\vec{\psi}/\partial t^2] = 0$$

and hence

$$\Delta\varphi = v_p^{-2}\partial^2\varphi/\partial t^2, \tag{11.67}$$

$$\Delta\vec{\psi} = v_s^{-2}\partial^2\vec{\psi}/\partial t^2, \tag{11.68}$$

where $v_p = [(\lambda + 2\mu)/\rho]^{1/2}$, $v_s = (\mu/\rho)^{1/2}$, φ is a scalar potential or a potential of longitudinal (compressional or P) waves and $\vec{\psi}$ is a vector potential or a potential of transversal (S) waves.

Since every impulsive seismic wave can be understood as a superposition of harmonic waves, we firstly solve boundary value problems for harmonic sources for all angular frequencies ω. By harmonic waves we mean waves with harmonic time dependence of the form $\exp(\pm i\omega t)$.

Let us assume the wave equation in the generalized form

$$L\boldsymbol{u}(\boldsymbol{x}, t) \equiv c^2\Delta\boldsymbol{u}(\boldsymbol{x}, t) = \partial^2\boldsymbol{u}(\boldsymbol{x}, t)/\partial t^2, \tag{11.69}$$

where L is a linear differential operator. Let us put

$$\boldsymbol{u}(\boldsymbol{x}, t) = \boldsymbol{W}(\boldsymbol{x})T(t). \tag{11.70}$$

Then the Fourier method of separation of variables can be used. Since $L\boldsymbol{u}(\boldsymbol{x}, t) = T(t)L\boldsymbol{W}(\boldsymbol{x})$ and $\partial^2\boldsymbol{u}(\boldsymbol{x}, t)/\partial t^2 = \boldsymbol{W}(\boldsymbol{x})\partial^2 T/\partial t^2 = \boldsymbol{W}(\boldsymbol{x})\mathrm{d}^2T/\mathrm{d}t^2$, then from Equation 11.69 we obtain

$$LW(x)/W(x) = T^{-1}(t)\mathrm{d}^2T(t)/\mathrm{d}t^2 = \gamma,$$

where γ is arbitrary. Putting $\gamma = -\omega^2$ we then find

$$\mathrm{d}^2T(t)/\mathrm{d}t^2 + \omega^2 T(t) = 0, \quad LW(x) + \omega^2 W(x) = 0.$$

Hence

$$T(t) = A\exp(i\omega t) + B\exp(-i\omega t),$$

then

$$u(x, t) = T(t)W(x) = W(x)\exp(\pm i\omega t), \tag{11.71}$$

where $W(x)$ is a solution of the equation $LW(x) + \omega^2 W(x) = 0$.

For the special scalar case of the wave equation,

$$L\varphi \equiv c^2\Delta\varphi = \partial^2\varphi/\partial t^2, \tag{11.72}$$

we find

$$\varphi(x, t) = \tilde{\varphi}(x)\exp(\pm i\omega t), \tag{11.73}$$

where $\tilde{\varphi}(x)$ is a solution of the equation $L\varphi(x) \equiv c^2\Delta\varphi + \omega^2\varphi \equiv \Delta\varphi + k^2\varphi = 0$, where we put $k = \omega/c$ and k is called the wave-number.

The generalized solution of Equation 11.69 is of the form

$$u(x, t) = f(\omega)W(x, \omega)\exp(\pm i\omega t), \tag{11.74}$$

which represents the harmonic wave. By a superposition of all harmonic waves we obtain

$$u(x, t) = \int_{-\infty}^{\infty} f(\omega)W(x, \omega)\exp(i\omega t)\mathrm{d}\omega, \tag{11.75}$$

representing the solution of Equation 11.69. Denoting

$$f(\omega) = S(\omega)/2\pi,$$

then

$$u(x, t) = (2\pi)^{-1}\int_{-\infty}^{\infty} S(\omega)W(x, \omega)\exp(i\omega t)\mathrm{d}\omega, \tag{11.76}$$

which is called the Fourier integral, also representing the solution of the initial equation, Equation 11.69. We see that instead of the initial-boundary value problem, represented by Equation 11.69 with the given boundary and initial conditions, we can solve the system of the other problems for harmonic waves, described by $LW(x) + \omega^2 W(x) = 0$ with corresponding boundary conditions, and then the solution of the initial-boundary value problem can be written in the form of the Fourier integral, Equation 11.76. The function $S(\omega)$ allows the introduction of the boundary and initial conditions.

From the Fourier analysis it is known that if the function $f(t)$ together with its first derivative is continuous in parts in the interval $(-\infty, \infty)$ and if it is absolutely integrable, i.e. if $\int_{-\infty}^{\infty} |f(t)|\,dt < 0$ and if the integral $\int_{-\infty}^{\infty} |f(t)|\,dt$ exists, then

$$f(t) = (2\pi)^{-1} \int_{-\infty}^{\infty} S(\omega) \exp(i\omega t)\mathrm{d}\omega, \tag{11.77}$$

$$S(\omega) = \int_{-\infty}^{\infty} f(\tau) \exp(-i\omega\tau)\mathrm{d}\tau, \tag{11.78}$$

where $f(t)$ is a pulse function.

The Fourier integral can also be written in the following form

$$f(t) = (\pi)^{-1} \int_{0}^{\infty} |S(\omega)| \cos(\omega t + \varphi(\omega))\mathrm{d}\omega, \tag{11.79}$$

$$S(\omega) = |S(\omega)| \exp(i\varphi(\omega)). \tag{11.80}$$

The function $S(\omega)$ is called the complex spectrum of the function $f(t)$.

The seismic signals are always absolutely integrable and always continuous. In the case when the function $f(t)$ is discontinuous (the number of them is finite) then in the point of discontinuity we put

$$f(t) = \tfrac{1}{2}(f(t+0) + f(t-0)). \tag{11.81}$$

Remark: If $f(t)$ is defined analytically, then Equation 11.78 is integrated numerically. Many seismologists use the Fillon numerical quadrature for numerical integration of Equation 11.78. Unfortunately, the Fillon numerical quadrature is numerically unstable (see Babuška, Práger and Vitásek (1966)) and, therefore, the obtained results are not correct.

To define the initial boundary condition we assume that the medium at time $t = 0$ is in a quiescent state, and it starts to oscillate when the wave is coming. The boundary condition follows from the quality of the boundary. The boundary can be of free type (e.g. the Earth's surface), and of n-th order type (i.e. the $(n-1)$th derivative of density and velocities of seismic wave change), and the normal and tangential components of the displacement and stress vectors are continuous. In the case of the boundary of the first order the density and velocities of seismic waves have finite discontinuities.

The planar wave propagated in the direction of the x-axis is of the form $\boldsymbol{u} = f(t - x/c)$ for the wave propagated in the positive direction of the x-axis and $\boldsymbol{u} = f(t + x/c)$ for the wave in the negative direction of the x-axis. In the infinity we have the so-called Sommerfeld's conditions

$$\boldsymbol{u} = O(r^{-1}), \quad \partial \boldsymbol{u}/\partial r + ik\boldsymbol{u} = o(r^{-1}), \quad \text{where } k = \omega/c. \tag{11.82}$$

For spherical waves in infinity we have the Sommerfeld's type conditions

$$\partial \boldsymbol{u}/\partial x + c^{-1}\partial \boldsymbol{u}/\partial t = o(r^{-1}) \quad \text{for the divergent pulse wave,}$$

$$\text{or } \partial \boldsymbol{u}/\partial x + ik\boldsymbol{u} = o(r^{-1}) \quad \text{for the divergent harmonic wave,}$$

$$\partial \boldsymbol{u}/\partial x - c^{-1}\partial \boldsymbol{u}/\partial t = o(r^{-1}) \quad \text{for the convergent pulse wave,}$$

$$\text{or } \partial \boldsymbol{u}/\partial x - ik\boldsymbol{u} = o(r^{-1}) \quad \text{for the convergent harmonic wave.}$$

For the spherical wave from the scalar wave equation $\partial^2 u/\partial t^2 = c^2 \Delta u$ putting $u = F(r,t)r^{-1}$, where $F(r,t) = A_1 G_1(r - ct) + A_2 G_2(r + ct)$, we find

$$u = A_1 r^{-1} G_1(r - ct) + A_2 r^{-1} G_2(r + ct).$$

Since the convergent wave does not have a physical meaning then we obtain

$$u = Ar^{-1}G(r - ct) = Ar^{-1}G(\tau),$$

where $\tau = r - ct$. For the constant phase surface $\tau = \tau_o = \text{constant} > 0$ we have $r = \text{constant} > 0$, which represents the sphere surface. For the harmonic spherical wave we find

$$u = Ar^{-1}\exp(-ikr)\exp(i\omega t) = Ar^{-1}\exp(ikr)\exp(-i\omega t),$$

where $k = \omega/c$. The spherical wave can be written as a superposition of the planar waves by using the Fourier integral.

The planar wave can be found in the form

$$u = AF(r - at),$$

and for the harmonic planar wave

$$u = A\exp[-ik(r - ct)] = A\exp(-ikr)\exp(i\omega t) = A\exp(ikr)\exp(-i\omega t).$$

In the case of the cylindrical wave the source is linear, i.e. it is independent of one co-ordinate. Then the wave equation in the cylindrical co-ordinate system will be of the form

$$c^2 r^{-1}\partial/\partial r(r\partial u/\partial r) = \partial^2 u/\partial t^2.$$

Its solution can be found in the form

$$u = Ar^{-1/2}f(r - ct).$$

For the harmonic cylindrical wave we assume that

$$u = f(r)\exp(i\omega t).$$

Then the wave equation is transferred to the well-known Helmholtz equation

$$r^{-1}\mathrm{d}/\mathrm{d}r(r\mathrm{d}f/\mathrm{d}r) + k^2 f = 0, \quad k = \omega/c.$$

Its solution can be found by using the Hankel functions of the first and second orders in the form

$$f = AH_o^{(1)}(kr) + BH_o^{(2)}(kr).$$

If $kr \to 0$ then $H_o^{(1)}(kr) \to 0$ as lg r. If $kr \gg 1$, known as the wave zone, then the Hankel functions can be expressed as

$$H_o^{(1)}(kr) \sim 2^{1/2}(\pi kr)^{-1/2}\exp[i(kr - \tfrac{1}{4}\pi)],$$

$$H_o^{(2)}(kr) \sim 2^{1/2}(\pi kr)^{-1/2}\exp[-i(kr - \tfrac{1}{4}\pi)]$$

and then

$$u = (AH_o^{(1)}(kr) + BH_o^{(2)}(kr))\exp(i\omega t) \sim (kr)^{-1/2}(C\exp(ikr) + D\exp(-ikr))\exp(i\omega t).$$

For the divergent wave we have

$$u \sim (kr)^{-1/2}D\exp(-ikr)\exp(i\omega t) = (kr)^{-1/2}D\exp[i(\omega t - kr)].$$

We see that the cylindrical wave decreases as $r^{-1/2}$, while the spherical wave decreases as r^{-1}.

For the vector wave equation the wave number is a vector and then $\boldsymbol{k} = (k_j)$, $k_j = \alpha_j k$, $j = 1, 2, 3$, where α_j are the directional cosines. For example for the plane harmonic wave we find

$$\boldsymbol{u} = (u_j) = A\exp[-i(\boldsymbol{kr} - \omega t)] = A\exp(-i\boldsymbol{kr})\exp(i\omega t) = A\exp(i\boldsymbol{kr})\exp(-i\omega t),$$

where $u_j = A_j\exp(ikx_j\cos\alpha_j)\exp(-i\omega t)$, $j = 1, 2, 3$, $k = \omega/c$.

In the wave equations, Equations 11.69 and 11.72, we assume that the seismic waves in the medium propagate only by the velocities of their phases, by the so-called phase velocities c. But the waves which are displayed by the dispersion effect are created by a group of waves and, therefore, they propagate by a velocity v of this group of waves, i.e. by the so-called group velocity. It is evident that the wave created by a system of harmonic waves is not strictly a harmonic wave as its amplitude will change in time.

Let us denote by k_o the middle value of the wave number k and let the amplitude A depend on k. Then

$$u = \int_{k_o-\Delta k}^{k_o+\Delta k} a(k)\exp[i(kr - \omega(k)t]\mathrm{d}k,$$

where $a(k)$ denotes the amplitude of individual waves. Let us put $kr - \omega t = k_o r - \omega_o t + (k - k_o)r - (\omega - \omega_o)t$, then $u = A\exp[i(k_o r - \omega_o t)]$, where the amplitude A can be written in the form

$$A = \int_{k_o-\Delta k}^{k_o+\Delta k} a(k)\exp[i(k - k_o)r - (\omega - \omega_o)t]\mathrm{d}k.$$

Hence for the velocity of the phase propagation we obtain

$$v = \lim_{k\to k_o}(\omega - \omega_o)(k - k_o)^{-1} = \frac{\mathrm{d}\omega}{\mathrm{d}k}.$$

Since $\omega = ck$, $\omega = 2\pi/T$ and $k = \omega/c = 2\pi/cT = 2\pi\lambda^{-1}$, where T is a period and λ a wavelength of a seismic wave, then for the group velocity of the seismic wave v we obtain

$$v = c - \lambda\frac{\mathrm{d}c}{\mathrm{d}\lambda}. \qquad (11.83)$$

11.3. PROPAGATION OF PSEUDO *P* AND *SV* WAVES IN AN INHOMOGENEOUS, GENERALLY ANISOTROPIC MEDIA GENERATED BY HARMONIC LINE SOURCES

11.3.1. Introduction

This section is concerned with the Galerkin solution to $3D$ mixed boundary value problems of the equation of motion in the dynamic theory of elasticity. As a special case, the problem solved in this section describes the propagation of elastic pseudo P and S waves in an inhomogeneous, generally anisotropic, media generated by a harmonic line source. Generally such type of problems tends to the solution of the equation of motion

$$\frac{\partial \tau_{ij}}{\partial x_j} - \rho f_{oi} = \rho \frac{\partial^2 v_i}{\partial t^2} + R \frac{\partial v_i}{\partial t}, \quad i = 1, 2, 3, \text{ in } \Omega \times (t_o, t_1), \tag{11.84}$$

where the stress tensor τ_{ij} is defined by the generalized Hooke's law, $\boldsymbol{v} = (v_i)$ is the displacement vector, $\rho = \rho(\boldsymbol{x})$ is the density, $F_{oi} = \rho f_{oi}$ are components of body forces, $R = R(\boldsymbol{x})$ is a Rayleigh's damping coefficient of elastic waves, $\boldsymbol{x}$ is a point in a Cartesian co-ordinate system (x_1, x_2, x_3) and t is time. The form of damping can easily be taken account of in the numerical analysis, because $R \rightarrow R_h = a_1 M + a_2 K$, where R_h is the damping matrix, M is the mass matrix, K is the stiffness matrix, and a_1 and a_2 = constants, so that no storage and no multiplications for a damping matrix are required.

Further, we shall assume that all elastic coefficients c_{ijkl}, and all the other physical parameters of the medium as well as acting body and surface forces are independent of the x_2 co-ordinate. We shall also assume that the volume forces are given by $\boldsymbol{F}_o = \boldsymbol{F} \exp(\pm i\omega t)$, and the displacement vector by $\boldsymbol{v} = \boldsymbol{u} \exp(\pm i\omega t)$. Thus the displacement components v_2 or u_2 are separated, and the equations for them describe the field of the SH waves generated by the pulse or harmonic source, respectively, discussed in previous sections. According to these assumptions the equation of motion, Equation 11.84, is changed by

$$\frac{\partial \tau_{ij}}{\partial x_j} - \rho f_{oi} = -g u_i, \quad i = 1, 3 \text{ in } \Omega, \tag{11.85}$$

where g is a given complex function.

11.3.2. Formulation of the Problem

Let $\Omega \subset R^2$ be a simply connected bounded domain with polygonal boundary $\partial\Omega = \cup_{i=1}^{4} \Gamma_i$. Note that a normal to the boundary exists almost everywhere, the outward unit normal always being denoted by $\boldsymbol{n}$. Let $\boldsymbol{x} = (x_1, x_3)$ be a point of a Cartesian co-ordinate system.

Let $\boldsymbol{u}(\boldsymbol{x}) = (u_i(\boldsymbol{x}))$, $i = 1, 3$, be the displacement vector, and e_{ij} be the small strain tensor defined by

$$e_{ij} = \tfrac{1}{2}(\partial u_i/\partial x_j + \partial u_j/\partial x_i), \quad i = 1, 3. \tag{11.86}$$

The stress-strain relation is defined by the generalized Hooke's law for anisotropic media

$$\tau_{ij} = c_{ijkl} e_{kl}(\boldsymbol{u}), \quad i = 1, 3, \tag{11.87}$$

and particularly for isotropic media by

$$\tau_{ij} = \lambda e_{kk}(\boldsymbol{u})\delta_{ij} + 2\mu e_{ij}(\boldsymbol{u}), \quad i = 1, 3, \tag{11.88}$$

and we let c_{ijkl} satisfy the conditions of symmetry and positive definitness

$$c_{ijkl} = c_{jikl} = c_{klij}, \tag{11.89}$$

$$c_{ijkl} e_{ij} e_{kl} \geq c_o e_{ij} e_{ij}, \quad c_o = \text{constant} > 0, \quad \forall e_{ij} = e_{ji} \text{ a.e. in } \Omega. \tag{11.90}$$

A summation convention over a repeated subscript is adopted. The subscripts pass the integer 1 and 3. The notation $f_{i,j} = \partial f_i/\partial x_i$ can also be used.

The stress tensor satisfies Equation 11.85, i.e.

$$\frac{\partial \tau_{ij}}{\partial x_j} + g u_i = F_i, \quad i = 1, 3 \text{ in } \Omega, \tag{11.91}$$

where $\boldsymbol{F} = \boldsymbol{F}(x_1, x_3)$ is the vector of the body forces, $\rho = \rho(x_1, x_3)$ is the density and $g = g(x_1, x_3) = g_1(x_1, x_3) - i g_2(x_1, x_3)$ is the given complex function, $g \geq g_o > 0$.

Further, on the boundary $\partial\Omega$, let the stress vector $\vec{\tau} = (\tau_i)$ given by

$$\tau_i = \tau_{ij}(\boldsymbol{x}) n_j \tag{11.92}$$

be defined, and its normal component τ_n be given by

$$\tau_n = \tau_i n_i = \tau_{ij}(\boldsymbol{x}) n_j n_i. \tag{11.93}$$

Let the boundary $\partial\Omega$ consist of two disjoint parts, $\partial\Omega = \partial\Omega_1 \cup \partial\Omega_2$, with $\partial\Omega_1 = \Gamma_1$, $\partial\Omega_2 = \Gamma_2 \cup \Gamma_3 \cup \Gamma_4$. Then we shall solve the following problems.

Problem ($\mathscr{P}_1$): Find a displacement vector $\boldsymbol{u} = \boldsymbol{u}(\boldsymbol{x})$ satisfying

$$\frac{\partial}{\partial x_j}(c_{ijkl}(\boldsymbol{x}) e_{kl}(u)) + g(\boldsymbol{x}) u_i = F_i, \quad i = 1, 3 \text{ in } \Omega, \tag{11.94}$$

and the boundary conditions

$$\tau_{ij} n_j = P_i (= 0) \quad \text{on } \partial\Omega_1, \tag{11.95}$$

$$u_i = 0 \quad \text{on } \partial\Omega_2. \tag{11.96}$$

Some of the problems can lead to the following problem.

Problem ($\mathscr{P}_2$): Find a displacement vector $\boldsymbol{u} = \boldsymbol{u}(\boldsymbol{x})$ satisfying

$$\frac{\partial}{\partial x_j}(c_{ijkl}(\boldsymbol{x}) e_{kl}(\boldsymbol{u})) + g(\boldsymbol{x}) u_i = F_i, \quad i = 1, 3 \text{ in } \Omega, \tag{11.97}$$

and the boundary condition

$$u_i = 0 \text{ on } \partial\Omega. \tag{11.98}$$

Remark: Problem ($\mathscr{P}_2$) is a special case of Problem ($\mathscr{P}_1$) if $\partial\Omega_1 \equiv 0$. Moreover, the elastic coefficients c_{ijkl} in both Problems ($\mathscr{P}_1$) and ($\mathscr{P}_2$) can be determined by using the homogenization method described in Chapter 3 and in the special case in the previous section.

Remark: In the theory of elasticity the following notation is often used for simplicity: $\partial u_i/\partial x_j \equiv u_{i,j}$, $\partial u_i/\partial t \equiv u_{i,t}$, $\partial^2 u_i/\partial t^2 \equiv u_{i,tt}$, $\partial\tau_{ij}/\partial x_j \equiv \tau_{ij,j}$, $\partial f/\partial x_i \equiv f_{,i}$, etc.. This notation will be used in the next section.

11.3.3. Variational Formulation and Weak Solution

Let $H^1(\Omega)$ be the Sobolev space, and let $W = [H^1(\Omega)]^2$. Let us define the space of virtual displacements as $V = \{\boldsymbol{v} \mid v \in W, \boldsymbol{v} = 0 \text{ on } \partial\Omega_2\}$.

Let us assume that $c_{ijkl}(\boldsymbol{x})$, $\lambda(\boldsymbol{x})$, $\mu(\boldsymbol{x}) \in C^1(\Omega)(\text{or } \in L^\infty(\Omega))$, $\boldsymbol{F} \in [L^2(\overline{\Omega})]^2$, $g(\boldsymbol{x}) \in C^o(\overline{\Omega})$, and $\rho(\boldsymbol{x}) \in C^o(\overline{\Omega})$.

Definition 11.1: The vector function $\boldsymbol{u}_o$ is called the generalized or weak solution of Problem ($\mathscr{P}_1$) or ($\mathscr{P}_2$) if

$$\boldsymbol{u}_o \in V, \quad a(\boldsymbol{u}, \boldsymbol{v}) = (\boldsymbol{F}, \boldsymbol{v}) \quad \forall \boldsymbol{v} \in W, \tag{11.99}$$

where

$$\begin{aligned} a(\boldsymbol{u}, \boldsymbol{v}) &= \int_\Omega c_{ijkl}(\boldsymbol{x}) e_{kl}(\boldsymbol{u}) \overline{e_{ij}(\boldsymbol{v})} \mathrm{d}\boldsymbol{x} + \int_\Omega g(\boldsymbol{x}) u_i \overline{v_i} \mathrm{d}\boldsymbol{x} \\ &= -\int_\Omega \tau_{ij,j} \overline{v_i} \mathrm{d}\boldsymbol{x} + \int_\Omega g(\boldsymbol{x}) u_i \overline{v_i} \mathrm{d}\boldsymbol{x} = a_1(\boldsymbol{u}, \boldsymbol{v}) + a_2(\boldsymbol{u}, \boldsymbol{v}) \end{aligned} \tag{11.100}$$

and

$$(\boldsymbol{u}, \boldsymbol{v}) = \int_\Omega u_i \overline{v_i} \mathrm{d}\boldsymbol{x}, \quad \boldsymbol{u}, \boldsymbol{v} \in [L^2\Omega)]^2. \tag{11.101}$$

Theorem 11.2: Let $a : V \times V \to R$ be bounded and of V-elliptic bilinear form on V. The problem

$$a(\boldsymbol{u}, \boldsymbol{v} - \boldsymbol{u}) \geq (\tilde{\boldsymbol{F}}, \boldsymbol{v} - \boldsymbol{u}) \quad \forall \boldsymbol{v} \in V, \tag{11.102}$$

for an arbitrary right side $\boldsymbol{F} \in V'$, has only one solution $\boldsymbol{u}$. Moreover,

$$\|\boldsymbol{u}\| \leq \alpha^{-1} \left\|\tilde{\boldsymbol{F}}\right\|_* . \tag{11.103}$$

The proof is parallel to that of Yosida (1965), Nečas (1967).

Remark: It can be shown that any "classical" solution of Problem ($\mathscr{P}_1$) or ($\mathscr{P}_2$) is also a weak solution. On the other hand, if the solution is smooth enough, then it is also a classical solution of our problem.

Theorem 11.3: Let $F \in [L^2(\Omega)]^2$, $g \in C^o(\overline{\Omega})$. Then there exists just one solution of the problem ($\mathscr{P}_1$) or ($\mathscr{P}_2$) and, moreover,

$$\|\boldsymbol{u}\|_W \leq c \|\boldsymbol{F}\|_{[L^2(\Omega)]^2} , \tag{11.104}$$

where $c > 0$ is independent of $\boldsymbol{u}$ and $\boldsymbol{F}$.

Proof: As $c_{ijkl}(\boldsymbol{x}) \in C^1(\overline{\Omega})$, then according to Equations 11.101 and 11.102 the bilinear form $a(\boldsymbol{u}, \boldsymbol{v})$ is bounded and symmetric on V and

$$a(\boldsymbol{v}, \boldsymbol{v}) = \tfrac{1}{2}(c_{ijkl}e_{kl}(\boldsymbol{v}), e_{ij}(\boldsymbol{v})) + \int_{\Omega} g(\boldsymbol{x})\boldsymbol{v}\overline{\boldsymbol{v}}\mathrm{d}\boldsymbol{x} \geq \tfrac{1}{2}\mu_o(e_{ij}(\boldsymbol{v}), e_{ij}(\boldsymbol{v})) + g_o(\boldsymbol{v}, \boldsymbol{v}) \geq \tfrac{1}{2}\mu_o(e_{ij}(\boldsymbol{v}), e_{ij}(\boldsymbol{v})).$$

By using the Korn's inequality (Hlaváček and Nečas (1970), Duvaut and Lions (1972)) we have

$$(e_{ij}(\boldsymbol{v}), e_{ij}(\boldsymbol{v})) \geq c\,\|\boldsymbol{v}\|_W^2, \quad \forall \boldsymbol{v} \in V.$$

Then

$$a(\boldsymbol{v}, \boldsymbol{v}) \geq c\,\|\boldsymbol{v}\|_W^2,$$

so that the bilinear form is V-elliptic. Then, according to Theorem 11.2, there exists only one solution. Furthermore, for $\left\|\tilde{\boldsymbol{F}}\right\|_*$ from Equation 11.103 we have

$$\left\|\tilde{\boldsymbol{F}}\right\|_* = c\,\|\boldsymbol{F}\|_{[L^2(\Omega)]^2},$$

and thus

$$\|\boldsymbol{u}\|_W \leq c\,\|\boldsymbol{F}\|_{[L^2(\Omega)]^2},$$

which completes the proof.

11.3.4. Numerical Solution

Numerically Problems $(\mathscr{P}_1)$ and $(\mathscr{P}_2)$ will be solved by the finite element method (FEM). We will assume that the domain Ω is a bounded domain with a polygonal boundary $\partial\Omega$. Let the domain Ω be "triangulated", i.e. $\overline{\Omega} = \Omega \cup \partial\Omega$ is covered by a finite number of triangles T_h, forming a triangulation $\mathscr{T}_h$. We further assume that the points of $\partial\Omega$ coincide with the vertices of T_h. The family $\{\mathscr{T}_h\}$, $0 < h \leq h_o < \infty$, of the triangulations $\mathscr{T}_h$ is assumed to be regular. A family $\{\mathscr{T}_h\}$, $0 < h \leq h_o < \infty$, of triangulations $\mathscr{T}_h$ will be called regular, if a positive constant α exists independent of h and such that no interior angle ϑ_h is less than ϑ_o. Let V_h be the set of linear finite elements (or cubic finite elements), i.e. the space of all continuous vector functions in $\overline{\Omega}$ which are piecewise linear over $\mathscr{T}_h$ and fulfill the Dirichlet's boundary condition, where it is prescribed.

Definition 11.2: The vector function $\boldsymbol{u}_{oh} \in V_h$ will be called an approximate solution of Problem $(\mathscr{P}_1)$ or $(\mathscr{P}_2)$, and we denote it as Problem $(\mathscr{P}_1)_h$ or $(\mathscr{P}_2)_h$, if

$$\boldsymbol{u}_{oh} \in V_h, \quad a(\boldsymbol{u}_{oh}, \boldsymbol{v}) = (\boldsymbol{F}, \boldsymbol{v}) \quad \forall \boldsymbol{v} \in V_h, \tag{11.105}$$

where

$$a(\boldsymbol{u}, \boldsymbol{v}) = a_1(\boldsymbol{u}, \boldsymbol{v}) + a_2(\boldsymbol{u}, \boldsymbol{v}), \tag{11.106}$$

$$a_1(\boldsymbol{u}, \boldsymbol{v}) = -\int_{\Omega} \tau_{ij,j}\overline{v}_i\mathrm{d}\boldsymbol{x} = \int_{\Omega} c_{ijkl}e_{kl}(\boldsymbol{u})\overline{e_{ij}(\boldsymbol{v})}\mathrm{d}\boldsymbol{x}, \tag{11.107}$$

$$a_2(\boldsymbol{u}, \boldsymbol{v}) = \int_{\Omega} g u_i \overline{v}_i \mathrm{d}\boldsymbol{x}. \tag{11.108}$$

The proof of the convergence of the finite element approximation $\boldsymbol{u}_h$ to the exact solution is the consequence of the generalized Lax-Milgram theorem (Oden and Reddy (1976), Handlová, Nedoma and Schmid (1975), Nedoma (1984)).

Theorem 11.4 (GLM — Generalized Lax-Milgram theorem): Let H_1 and H_2 be two Hilbert (complex and complete) spaces with scalar product $(.,.)_{H_1}((.,.)_{H_2})$. Further, let $a(\boldsymbol{u},\boldsymbol{v})$ be a bilinear form on $H_1 \times H_2$, $\boldsymbol{u} \in H_1$, $\boldsymbol{v} \in H_2$ such that

$$|a(\boldsymbol{u},\boldsymbol{v})| \leq c_1 \|\boldsymbol{u}\|_{H_1} \|\boldsymbol{v}\|_{H_2}, \tag{11.109}$$

$$\inf_{\substack{\boldsymbol{u}\in H_1 \\ \|\boldsymbol{u}\|_{H_1}=1}} \sup_{\substack{\boldsymbol{v}\in H_2 \\ \|\boldsymbol{v}\|_{H_2}=1}} |a(\boldsymbol{u},\boldsymbol{v})| \geq c_2 > 0, \tag{11.110}$$

$$\sup_{\boldsymbol{u}\in H_1} |a(\boldsymbol{u},\boldsymbol{v})| > 0, \quad 0 \neq \boldsymbol{v} \in H_2, \tag{11.111}$$

where c_1 and c_2 are positive constants independent of $\boldsymbol{u}$ and $\boldsymbol{v}$.

Further, let f be a linear functional on H_2, and let there be given the subspaces $M_1 \subset H_1$ and $M_2 \subset H_2$, i.e. the linear manifolds such that

$$\inf_{\substack{\boldsymbol{u}\in M_1 \\ \|\boldsymbol{u}\|_{H_1}=1}} \sup_{\substack{\boldsymbol{v}\in M_2 \\ \|\boldsymbol{v}\|_{H_2}=1}} |a(\boldsymbol{u},\boldsymbol{v})| \geq d_2(M_1, M_2) \|\boldsymbol{u}\|_{H_1} \tag{11.112}$$

with $d_2(M_1, M_2) > 0$ which is independent of $\boldsymbol{u}$ and $\boldsymbol{v}$. For every vector $\boldsymbol{v} \in M_2$ let

$$\sup_{\boldsymbol{u}\in M_1} |a(\boldsymbol{u},\boldsymbol{v})| > 0. \tag{11.113}$$

Let f be given on H_2, and let us denote by $\boldsymbol{U}_o$ an element of H_1 (its existence and uniqueness follow from Equations 11.109-11.111) so that

$$a(\boldsymbol{U}_o, \boldsymbol{v}) = f(\boldsymbol{v}) \quad \forall \boldsymbol{v} \in H_2. \tag{11.114}$$

and $\|\boldsymbol{U}_o\|_{H_1} \leq C^{-1} \|f\|_{H_2'}$, where H_2' denotes the dual space.

Let there exist $\boldsymbol{w} \in M_1$ such that

$$\|\boldsymbol{U}_o - \boldsymbol{w}\|_{H_1} \leq \delta. \tag{11.115}$$

Further, let $\hat{\boldsymbol{u}}_o \in M_1$ be such that

$$a(\hat{\boldsymbol{u}}_o, \boldsymbol{v}) = f(\boldsymbol{v}) \quad \forall \boldsymbol{v} \in M_2. \tag{11.116}$$

Then

$$\left\|\boldsymbol{U}_o - \hat{\boldsymbol{u}}_o\right\|_{H_1} = (\boldsymbol{U}_o - \hat{\boldsymbol{u}}_o, \boldsymbol{U}_o - \hat{\boldsymbol{u}}_o)_{H_1}^{1/2} \leq (1 + c_1 \mathrm{d}_2^{-1}(M_1, M_2))\delta. \tag{11.117}$$

The proof is parallel to that of Oden and Reddy (1976).

Theorem 11.5: There exists just one approximate solution $\boldsymbol{u}_{oh} \in V_h$ of the problem in Equation 11.105 with the bilinear form $a(\boldsymbol{u},\boldsymbol{v})$, defined by Equations 11.106-11.108 for sufficiently small h. Then for Problems $(\mathscr{P}_1)$ and $(\mathscr{P}_2)$ it holds that

$$\|\boldsymbol{u}_o - \boldsymbol{u}_{oh}\| \to 0 \quad \text{for } h \to 0.$$

Proof: The proof is a direct consequence of the Generalized Lax-Milgram theorem, where we set $H_1 = H_2 = W$ and $M_1 = M_2 = V_h$. Let us verify the presumptions and assertions of this theorem (i.e. Equations 11.109-11.117 of Theorem 11.4). Equation 11.109 follows from the definition of $a(\boldsymbol{u}, \boldsymbol{v})$ in Equations 11.106 to 11.108 and the Hölder inequality. Equation 11.110 follows immediately from

$$|a(\boldsymbol{u}, \boldsymbol{u})| \geq c\,\|\boldsymbol{u}\|_W^2\,,$$

and thus Equation 11.111 is also satisfied. Similarly Equations 11.112 and 11.113 are also satisfied. Equation 11.115 is satisfied for $\delta = \delta(h)$, $\delta(h) \to 0$ for $h \to 0$, as the finite element approximation $\boldsymbol{u}_o$ (according to $\boldsymbol{u}_h = \Sigma_{i=1}^{n(h)} \alpha_{hi}\varphi_i^h$, where $\{\varphi_i^h\}$ is the basis of the space V_h) can be ensured with sufficiently accuracy. For the proof of the assertions in Equations 11.116 and 11.117 it is enough to ensure the existence of $\boldsymbol{u}_o$ from the closure of $V_h \subset W$. For this it is sufficient to apply the Lax-Milgram theorem on the closure of V_h in the norm of the space W. The presumptions of that are satisfied, as they are even satisfied on the whole space W. Thus Equation 11.117 is also satisfied, which completes the proof.

11.3.5. Algorithm

The approximate solutions of Problems $(\mathscr{P}_1)_h$ and $(\mathscr{P}_2)_h$, will be sought in the form

$$\boldsymbol{u} \sim \boldsymbol{u}_{oh} = \sum_{i=1}^{n(h)} \alpha_{hi}\varphi_i^h(x_1, x_3), \tag{11.118}$$

where $\{\varphi_i^h(x_1, x_3)\}$, $i = 1, \ldots, n(h)$, is the basis of the space V_h, generated by the polynomials of s-order and satisfying boundary value conditions of Dirichlet type everywhere, where they are prescribed by the problems $(\mathscr{P}_1)$ and $(\mathscr{P}_2)$

By introducing Equation 11.118 into Equations 11.105-11.108 we obtain

$$\sum_{i=1}^{n(h)} \alpha_{hi} a(\varphi_i^h, \varphi_j^h) = (\boldsymbol{F}, \varphi_j^h), \quad j = 1, \ldots, n(h), \tag{11.119}$$

which represents a system of linear algebraic equations for unknown coefficients α_{hi}. This system will be solved by the Gauss-Seidel method and the SOR method. To prove the convergence of both methods we shall firstly prove the following lemma.

Lemma 11.1: It holds that

$$\frac{a_2(\varphi_i^h, \varphi_j^h)}{a_1(\varphi_i^h, \varphi_j^h)} \to 0 \quad \text{as } h \to 0 \tag{11.120}$$

for such indices i, j for which the denominator is non-zero.

Proof: In the case when the denominator is equal to zero, the numerator, due to the construction of Equation 11.118, is also equal to zero, so we shall only study the case where the denominator is non-zero. Firstly, let us estimate $\left|a_1(\varphi_i^h, \varphi_j^h)\right|$ and $\left|a_2(\varphi_i^h, \varphi_j^h)\right|$ separately. We have

$$\left|a_1(\varphi_k^h, \varphi_l^h)\right| = \left|\int_\Omega c_{ijmn}\varphi_{km,i}^h \varphi_{ln,j}^h \mathrm{d}\boldsymbol{x}\right| \geq \left|\int_{\bigcup_{t=1}^{M} T_h^t} c_{ijmn}\varphi_{km,i}^h \varphi_{ln,j}^h \mathrm{d}\boldsymbol{x}\right| \geq \min_t \left|\mu_{to}^h\right| \left|\int_{\bigcup_{t=1}^{M} T_h^t} \varphi_{km,i}^h \varphi_{ln,j}^h \mathrm{d}\boldsymbol{x}\right|$$

$$\geq \min_t \left|\mu_{to}^h\right| c_o \left|\int_{\text{supp}} \varphi_{km,i}^h \varphi_{ln,j}^h \mathrm{d}\boldsymbol{x}\right| \geq \min_t \left|\mu_{to}^h\right| c_o h_{i+s} k_{j+r} \left|\int_{\text{supp}^\text{o}} \varphi_{km,i}^{oh} \varphi_{ln,j}^{oh} \mathrm{d}x' \mathrm{d}z'\right|$$

$$\geq \min_t |\mu_{to}^h| c_1 \left\{ \frac{k_{j+r}}{h_{i+s}} A_{kl}^{ij} + \frac{h_{i+s}}{k_{j+r}} B_{kl}^{ij} + C_{kl}^{ij} + D_{kl}^{ij} \right\}, \ |a_2(\varphi_k^h, \varphi_l^h)| = \left| \int_\Omega g \varphi_k^h \varphi_l^h \mathrm{d}\mathbf{x} \right|$$

$$\leq \max_t |g_t^h| \left| \int_{\bigcup_{i=1}^M T_h^t} \varphi_k^h \varphi_l^h \mathrm{d}\mathbf{x} \right| \leq \max_t |g_t^h| \mid h_{i+s} k_{j+r} \left| \int_{\mathrm{supp}^o} \varphi_k^o \varphi_l^o \mathrm{d}x' \mathrm{d}z' \right| = \max_t |g_t^h| h_{i+s} k_{j+r} I_{kl},$$

where the Fubini theorem, the substitution theorem and the theorem for derivation of a function of multiple variables were used, and where c_o, c_1 are some constants, h_{i+s}, $k_{j+r} (r, s = -1, 0)$ denote the steps of the useful uneven division, φ_i^o are the basis functions on the fundamental support in the local coordinate system (x', z'), $\mathrm{supp}^o \equiv \{(x', z') \mid -1 \leq x' \leq 1, -1 \leq z' \leq 1, f(x', z') \neq 0$; and $f(x', z') = 0$ everywhere else$\}$ and which are generated by linear (or cubic) polynomials with value equal to 1 at point (0,0) and at the other six vertices of the fundamental support $\mathrm{supp}^o = \cup_{i=1}^6 T_i^o$ are equal to zero. Further, let us denote by

$$I_{kl} = \left| \int_{\mathrm{supp}^o} \varphi_k^o \varphi_l^o \mathrm{d}x' \mathrm{d}z' \right|,$$

$$A_{kl}^{ij} = \left| \int_{\mathrm{supp}^o} \varphi_{ki,1}^o \varphi_{lj,1}^o \mathrm{d}x' \mathrm{d}z' \right|, \quad i, j = 1, 3,$$

$$B_{kl}^{ij} = \left| \int_{\mathrm{supp}^o} \varphi_{ki,3}^o \varphi_{lj,3}^o \mathrm{d}x' \mathrm{d}z' \right|, \quad i, j = 1, 3,$$

$$C_{kl}^{ij} = \left| \int_{\mathrm{supp}^o} \varphi_{ki,1}^o \varphi_{lj,3}^o \mathrm{d}x' \mathrm{d}z' \right|, \quad i, j = 1, 3,$$

$$D_{kl}^{ij} = \left| \int_{\mathrm{supp}^o} \varphi_{ki,3}^o \varphi_{lj,1}^o \mathrm{d}x' \mathrm{d}z' \right|, \quad i, j = 1, 3.$$

Then

$$\frac{|a_2(\varphi_k^h, \varphi_l^h)|}{|a_1(\varphi_k^h, \varphi_l^h)|} \leq \frac{\max_t |g_t^h| h_{i+s} k_{j+r} I_{kl}}{\min_t |\mu_{to}^h| c_1 \left\{ k_{j+r} h_{i+s}^{-1} A_{kl}^{ij} + h_{i+s} k_{j+r}^{-1} B_{kl}^{ij} + C_{kl}^{ij} + D_{kl}^{ij} \right\}}$$

$$\leq \frac{h^2 I_{kl}}{A_{kl}^{ij} + B_{kl}^{ij} + C_{kl}^{ij} + D_{kl}^{ij}} \to 0 \quad \text{for } h \to 0,$$

as $h_{i+s}, k_{j+r} \leq \max_{\mathcal{T}_h} (h_{i+s}, k_{j+r}) \leq h$, and as $h \to 0$ also $h_{i+s} \to 0$ and $k_{j+r} \to 0$. Thus

$$\frac{|a_2(\varphi_k^h, \varphi_l^h)|}{|a_1(\varphi_k^h, \varphi_l^h)|} \to 0 \text{ and also } \frac{a_2(\varphi_k^h, \varphi_l^h)}{a_1(\varphi_k^h, \varphi_l^h)} \to 0,$$

which completes the proof.

Let us denote

$$d_{ii} = a_1(\varphi_i^h, \varphi_i^h) \quad \text{for } i = j,$$

$$e_{ij} = a_1(\varphi_i^h, \varphi_j^h) \quad \text{for } i > j,$$

$$u_{ij} = a_1(\varphi_i^h, \varphi_j^h) \quad \text{for } i < j,$$

$$\hat{d}_{ii} = a_2(\varphi_i^h, \varphi_i^h) \quad \text{for } i = j,$$

$$\hat{e}_{ij} = a_2(\varphi_i^h, \varphi_j^h) \quad \text{for } i > j,$$

$$\hat{u}_{ij} = a_2(\varphi_i^h, \varphi_j^h) \quad \text{for } i < j,$$

and the corresponding matrices by the symbols D_h, E_h, U_h, $\hat{D}_h$, $\hat{E}_h$, $\hat{U}_h$, where $E_h = U_h^T$, $\hat{E}_h = \hat{U}_h^T$ and where the symbol A^T denotes the transpose matrix of the matrix A.

According to the substance of the FEM, the matrix $(a_1(\varphi_i^h, \varphi_j^h))$ is positive definite. Let

$$(a_1(\varphi_i^h, \varphi_j^h)) = (D_h - U_h - E_h), \quad (a_2(\varphi_i^h, \varphi_j^h)) = (\hat{D}_h + \hat{U}_h + \hat{E}_h) \tag{11.121}$$

be the decomposition of the matrices $(a_1(\varphi_i^h, \varphi_j^h))$ and $(a_2(\varphi_i^h, \varphi_j^h))$ into the diagonal, upper and lower triangular matrices.

Let us prove the following lemma.

Lemma 11.2: It holds that

$$(D_h - E_h)^{-1}(\hat{D}_h + \hat{U}_h + \hat{E}_h) \to 0 \quad \text{as } h \to 0. \tag{11.122}$$

Proof: According to Lemma 11.1 and Equation 11.121 we have

$$(D_h - U_h - E_h)^{-1}(\hat{D}_h + \hat{U}_h + \hat{E}_h) \to 0 \quad \text{as } h \to 0.$$

From there

$$(I - (D_h - E_h)^{-1} U_h)^{-1}(D_h - E_h)^{-1}(\hat{D}_h + \hat{U}_h + \hat{E}_h) \to 0 \quad \text{as } h \to 0. \tag{11.123}$$

As the matrix $(a_1(\varphi_i^h, \varphi_j^h))$ is positive definite, then

$$\left\|(D_h - E_h)^{-1} U_h\right\| < 1 \text{ and thus } \left\|I - (D_h - E_h)^{-1} U_h\right\| \le 2,$$

where $\|A\| = \max_k \Sigma_{k=1}^{n(h)} \left\|A_{ij}\right\|$.
According to Equation 11.123

$$\left\|(I - (D_h - E_h)^{-1} U_h)^{-1}(D_h - E_h)^{-1}(\hat{D}_h + \hat{U}_h + \hat{E}_h)\right\| \to 0 \quad \text{as } h \to 0.$$

Then

$$\begin{aligned}
&\left\|(D_h - E_h)^{-1}(\hat{D}_h + \hat{U}_h + \hat{E}_h)\right\| \\
&= \left\|(I - (D_h - E_h)^{-1} U_h)(I - (D_h - E_h)^{-1} U_h)^{-1}(D_h - E_h)^{-1}(\hat{D}_h + \hat{U}_h + \hat{E}_h)\right\| \\
&\le \left\|I - (D_h - E_h)^{-1} U_h\right\| \left\|(I - (D_h - E_h)^{-1} U_h)^{-1}(D_h - E_h)^{-1}(\hat{D}_h + \hat{U}_h + \hat{E}_h)\right\| \\
&\le 2 \left\|(I - (D_h - E_h)^{-1} U_h)^{-1}(D_h - E_h)^{-1}(\hat{D}_h + \hat{U}_h + \hat{E}_h)\right\| \to 0 \text{ as } h \to 0.
\end{aligned}$$

Moreover,

$$(D_h - E_h)^{-1}(\hat{D}_h + \hat{U}_h + \hat{E}_h) \to 0 \text{ as } h \to 0,$$

which completes the proof.

This lemma guarantees the convergence of the Gauss-Seidel method. Problems $(\mathscr{P}_1)_h$ and $(\mathscr{P}_2)_h$ can be written in the operator form as follows

$$L_h \boldsymbol{u}_h + K_h \boldsymbol{u}_h + \boldsymbol{F}_h = 0, \tag{11.124}$$

where L_h is a symmetric positive definite matrix, $L_h = D_h - U_h - E_h$ and $K_h = \hat{D}_h + \hat{U}_h + \hat{E}_h$, where D_h, U_h, E_h, $\hat{D}_h$, $\hat{U}_h$, $\hat{E}_h$ were defined above. Then

$$(D_h - E_h)^{-1} K_h \to 0 \quad \text{as } h \to 0. \tag{11.125}$$

Now we shall prove the convergence of the Jacobi, Gauss-Seidel and SOR methods. We have the following results.

Theorem 11.6: There exist $h_o > 0$, $\delta > 0$ such that the system of algebraic equations

$$\sum_{j=1}^{n(h)} u_j^h a(\varphi_j^h(\boldsymbol{x}), \varphi_i^h(\boldsymbol{x})) = (\boldsymbol{F}, \varphi_i^h(\boldsymbol{x})), \quad i = 1, \ldots, n(h) \tag{11.126}$$

arising from the FEM approximation of Problems $(\mathscr{P}_1)$ and $(\mathscr{P}_2)$ is solvable for $0 < h \le h_o$ and the Jacobi, Gauss-Seidel and SOR methods are convergent and $\omega \in \langle 1, 1 + \delta \rangle$.

Proof: The first part of the assertion is the consequence of Theorem 11.5 and of the fact that the matrix of the system in Equation 11.126 is regular. From the construction of the basis $\{\varphi_j^h\}_{j=1}^{n(h)}$ it follows that the matrix $A = (a(\varphi_i^h, \varphi_j^h))$, $i, j = 1, \ldots, n(h)$ is a band block matrix with non-zero elements, corresponding to non-zero intersections of the neighbourhood basis functions. Moreover, it is a block irreducible diagonal dominant and then regular matrix. Let us prove this assertion in detail.

Firstly we shall prove that the submatrix $(a(\varphi_i^h, \varphi_j^h))$ is regular. We have $\det(a(\varphi_i^h, \varphi_j^h)) \neq 0$ for all inner points of the triangulation used, inclusive of points in the neighbourhood of the boundary and points of the part of boundary where the Neumann's boundary condition is prescribed. Further, it holds that

$$\left(\left\|a(\varphi_i^h, \varphi_i^h)^{-1}\right\|\right)^{-1} \ge \left(\left\|a_o(\varphi_i^h, \varphi_i^h)^{-1}\right\|\right)^{-1},$$

where $(a_o(\varphi_i^h, \varphi_i^h))$ is a diagonal element of the FEM approximation of the real operator $(L_o u)_i = (c_{ijkl}(\boldsymbol{x}) e_{kl}(\boldsymbol{u}))_{,j}$, for which the diagonal dominant condition holds, i.e.

$$\left(\left\|a_o(\varphi_i^h, \varphi_i^h)^{-1}\right\|\right)^{-1} \ge \sum_{j=1}^{n(h)} \left\|a_o\left(\varphi_i^h, \varphi_j^h\right)^{-1}\right\|, \quad k = 1, \ldots, n(h)$$

and where $a_o(.,.)$ is the corresponding bilinear form of the operator L_o.

For non-diagonal submatrices $(a(\varphi_i^h, \varphi_j^h))$, $i \neq j$, we obtain $\det(a(\varphi_i^h, \varphi_j^h)) \neq 0$, $i \neq j$ and

$$\sum_{\substack{j=1 \\ j \neq i}}^{n(h)} \left\|a_1(\varphi_i^h, \varphi_j^h)\right\| = \sum_{\substack{j=1 \\ i \neq j}}^{n(h)} \left\|a_o(\varphi_i^h, \varphi_j^h)\right\|, \quad i = 1, \ldots, n(h).$$

Then

$$\left(\left\|a(\varphi_i^h, \varphi_i^h)^{-1}\right\|\right)^{-1} \ge \left(\left\|a_o(\varphi_i^h, \varphi_i^h)^{-1}\right\|\right)^{-1} \ge \sum_{\substack{j=1 \\ j \neq i}}^{n(h)} \left\|a_o(\varphi_i^h, \varphi_j^h)\right\| - \sum_{\substack{j=1 \\ i \neq j}}^{n(h)} \left\|a_1(\varphi_i^h, \varphi_j^h)\right\| \ge \sum_{\substack{j=1 \\ j \neq i}}^{n(h)} \left\|a(\varphi_i^h, \varphi_j^h)\right\|.$$

Thus the matrix $A = (a(\varphi_i^h, \varphi_j^h))$, $i, j = 1, \ldots, n(h)$ is a block diagonally dominant matrix.

To prove that the matrix $A = (a(\varphi_i^h, \varphi_j^h))$, $i, j = 1, \dots, n(h)$, is irreducible the technique of Varga (1962) and Axelsson (1996) can be used. Let us construct the directed graph of the matrix $\tilde{A} = (\|A_{ij}\|)$, $1 \le i, j \le n(h)$. The directed graph of the matrix $\tilde{A}$ is strongly connected and thus the matrix $\tilde{A}$ is irreducible. Then the matrix A is also irreducible.

As the matrix A is block diagonally dominant and irreducible, then it is also regular. Moreover, as A is regular, the Jacobi matrix $B = (D_h + \hat{D}_h)^{-1}(U_h + E_h + \hat{U}_h + \hat{E}_h)$ is convergent.

To prove the convergence of the Gauss-Seidel and SOR methods, we must prove that the corresponding spectral radius of the Gauss-Seidel matrix G_h or the SOR matrix $\mathscr{L}_h(\omega)$ is less than 1, i.e. $\rho(G_h) < 1$ or $\rho(\mathscr{L}_h(\omega)) < 1$ for $0 < h \le h_o$.
Let us prove the first assertion. It holds that

$$((D_h - \hat{D}_h) - (E_h + \hat{E}_h))^{-1} = ((D_h - E_h) - (\hat{D}_h + \hat{E}_h))^{-1} = [I_h - (D_h - E_h)^{-1}(\hat{D}_h + \hat{E}_h)]^{-1}(D_h - E_h)^{-1}.$$

Hence

$$G_h = (D_h - E_h)^{-1}U_h + W_h,$$

where

$$W_h = (D_h - E_h)^{-1}U_h + \sum_{k=1}^{\infty}(-1)^k[(D_h - E_h)^{-1}(\hat{D}_h + \hat{E}_h)]^k(D_h - E_h)^{-1}(U_h + \hat{U}_h),$$

and where the relation

$$[I_h - (D_h - E_h)^{-1}(\hat{D}_h + \hat{E}_h)]^{-1} = I_h - \sum_{k=1}^{\infty}(-1)^k[(D_h - E_h)^{-1}(\hat{D}_h + \hat{E}_h)]^k$$

was used. For $h \to 0$ then $W_h \to 0$. Thus there exists $h_o > 0$ such that W_h for $0 < h \le h_o$ will be so small that

$$\rho(G_h) = \rho((D_h - E_h)^{-1}U_h + W_h) < 1,$$

which completes the proof of the Gauss-Seidel method.

In an analogous way we shall prove the convergence of the SOR method. It holds that

$$\begin{aligned}[(D_h - \hat{D}_h) - \omega(E_h + \hat{E}_h)]^{-1} &= [(D_h - \omega E_h) - (\hat{D}_h + \omega\hat{E}_h)]^{-1} \\ &= [I_h - (D_h - \omega E_h)^{-1}(\hat{D}_h + \omega\hat{E}_h)]^{-1}(D_h - \omega E_h)^{-1} \\ &= [I_h - \sum_{k=1}^{\infty}(-1)^k[(D_h - \omega E_h)^{-1}(\hat{D}_h + \omega\hat{E}_h)]^k(D_h - \omega E_h)^{-1},\end{aligned}$$

as

$$I_h - (D_h - \omega E_h)^{-1}(\hat{D}_h + \omega\hat{E}_h)]^{-1} = I_h - \sum_{k=1}^{\infty}(-1)^k[(D_h - \omega E_h)^{-1}(\hat{D}_h + \hat{E}_h)]^k.$$

Hence it follows that

$$\mathscr{L}_h(\omega) = [(D_h - \hat{D}_h) - \omega(E_h + \hat{E}_h)]^{-1}[(1-\omega)(D_h - \hat{D}_h) + \omega(U_h + \hat{U}_h)] + W_h(\omega),$$

where

$$W_h(\omega) = \sum_{k=1}^{\infty}(-1)^k[(D_h - \omega E_h)^{-1}(\hat{D}_h + \omega\hat{E}_h)]^k(D_h - \omega E_h)^{-1}(U_h + \hat{U}_h).$$

For $h \to 0$ it follows that $W_h(\omega) \to 0$. Then there exists $h_o > 0$ and $\delta > 0$ such that $W_h(\omega)$ for $0 < h \le h_o$, $1 \le \omega < 1 + \delta$ is so small that

$$\rho(\mathscr{L}_h(\omega)) = \rho([(D_h - \hat{D}_h) - \omega(E_h + \hat{E}_h)]^{-1}[(1-\omega)(D_h - \hat{D}_h) + \omega(U_h + \hat{U}_h)] + W_h(\omega)) < 1,$$

which completes the proof of the convergence of the SOR method.

To estimate the dependence of the magnitude of the step of the triangulation used on the elastic coefficients c_{ijkl}, we shall start from Theorem 11.6. As the matrix $A = (A_{ij}) = (a(\varphi_i^h, \varphi_j^h))$ is regular, i.e. it is irreducible and block diagonally dominant, then

$$\max_i \left(\sum_{\substack{j=1\\ j\neq i}}^{n(h)} \|A_{ij}\| \left\|A_{ii}^{-1}\right\|^{-1} \right) < 1, \quad i = 1, \dots, n(h)$$

or

$$\max_i \left(\sum_{\substack{j=1\\ j\neq i}}^{n(h)} \left\|(a(\varphi_i^h, \varphi_j^h))\right\| \left\|(a(\varphi_i^h, \varphi_i^h))^{-1}\right\|^{-1} \right) < 1, \quad i = 1, \dots, n(h).$$

Let us denote $\min_\Omega(c_{ijkl}) = \min_\Omega(\mu) = \mu_o$.
Then

$$\max_i \left(\sum_{\substack{j=1\\ j\neq i}}^{n(h)} \|A_{ij}\| \left\|A_{ii}^{-1}\right\|^{-1} \right) \geq \max_i \left(\mu_o \sum_{\substack{j=1\\ j\neq i}}^{n(h)} \left\|(a(\varphi_i^h, \varphi_j^h))\right\| \left\|(a(\varphi_i^h, \varphi_i^h))^{-1}\right\|^{-1} \right)$$

$$\geq \max_i \left(\mu_o \sum_{\substack{j=1\\ j\neq i}}^{n(h)} \left\|(\hat{a}(\varphi_i^h, \varphi_j^h))\right\| \left\|(\hat{a}(\varphi_i^h, \varphi_i^h))^{-1}\right\|^{-1} \right),$$

where â is the bilinear form of the operator

$$\hat{L}\boldsymbol{u} \equiv \partial/\partial x_1(\mu\partial \boldsymbol{u}/\partial x_1) + \partial/\partial x_2(\mu\partial \boldsymbol{u}/\partial x_3) + g_o(x_1, x_3)\boldsymbol{u},$$

where $g_o(x_1, x_3)$ is a given complex function. For simplicity let us assume that h and k are steps of the useful division in the direction x_1 and x_3, respectively. Then

$$\frac{\mu_o(hk^{-1} + kh^{-1} + kh^{-1})}{\mu_0(kh^{-1} + hk^{-1} + kh^{-1} + hk^{-1}) - |g|\, hk} < 1 \tag{11.127}$$

and thus

$$h > (\mu_o/\,|g|)^{1/2}.$$

With respect to the construction of the Rayleigh's damping, the imaginary part of g is less than its real part and than $|g| \sim \omega^2\rho(\boldsymbol{x})$, and we have the estimation

$$h < (\mu_o(\omega^2\rho)^{-1})^{1/2} \sim (2\pi)^{-1}\lambda_s, \tag{11.128}$$

where λ_s is the smallest wavelength of S waves in the investigated region Ω.

It is possible to assert that $a_1(\varphi_i^h, \varphi_j^h)$ and $a_2(\varphi_i^h, \varphi_j^h)$ are approximated by the integrals $c_m^h \int_{\text{supp}} \varphi_{jm,k}^h \varphi_{in,l}^h \mathrm{d}\boldsymbol{x}$ and $g_m^h \int_{\text{supp}} \varphi_j^h \varphi_i^h \mathrm{d}\boldsymbol{x}$, respectively, and integrals $\int_{\text{supp}} \varphi_{jm,k}^h \varphi_{in,l}^h \mathrm{d}\boldsymbol{x}$ and $\int_{\text{supp}} \varphi_j^h \varphi_i^h \mathrm{d}\boldsymbol{x}$ are computed directly and exactly, where the symbol $\int_{\text{supp}}$ means the integration over the support of the basis functions φ_i^h. The values c_m^h and g_m^h are approximate values of $c_{ijkl}(\boldsymbol{x})$ and $g(\boldsymbol{x})$ on the non-zero intersection of supports of the basis functions. Generally $(\boldsymbol{F}, \varphi_i^h) \neq 0$, i.e. in the case where the source of elastic waves is modelled by the body forces, they can be computed numerically, e.g. by using the Gauss formula or by the trapezoidal or the Simpson's formulas, etc.

Remark: The problem in Equation 11.126 can also be solved by the conjugate gradient and bi-conjugate gradient methods, as discussed in Chapter 2 (see also Jacobs (1986), Křížek and Neitteanmäki (1990), Freund (1992), Křížek and Mlýnek (1994)).

11.3.6. Models and their Classification

All models are assumed to be homogeneous isotropic layered media with a channel of low velocity of seismic (elastic) waves (LVCH — the low velocity channel) between the system of layers. Physically the models are characterized by the velocities of the *P* and *S* waves and densities ρ which are functions of the depth. For velocities of *S* waves the relation $c_s \doteq c_p/1.732$, where c_p are velocities of *P* waves, is assumed. In our symmetry, the 3*D* models lead to 2*D* models only. In our LVCH models we approximate the interface conditions between the low velocity layer and the roof (overburden) as well as the bed (floor) by the layers of relatively small thickness, where the physical parameters change linearly from the higher values to the lower values.

The model of the asthenosphere was constructed as two similar types of model, which differed in the thickness of the LVCH only.

11.3.6.1. Model M1 — The asthenosphere

The model is constructed as a system of five layers with the LVCH in the fourth layer at depths between

(i) 120 km to 400 km (see Table 11.1) and
(ii) 70 km to 120 km (see Table 11.2).

In both models, in the first two layers the velocities of *P* and *S* waves and the densities are linear functions of depth, whilst in the other layers the velocities of *P* and *S* waves and densities are constant, with a very small transitient layer between

Table 11.1. Model M1(i)-data

depth [km]	0	15	45	120	121	400	401	540
c_p [ms^{-1}]	5600	6300	8200	8200	7400	7400	8300	8300
ρ [kgm^{-3}]	2670	2900	3400	3300	3300	3300	3400	3400

Table 11.2. Model M1(ii)-data

depth [km]	0	15	45	70	71	120	121	200
c_p [ms^{-1}]	5600	6300	8200	8200	7400	7400	8200	8200
ρ [kgm^{-3}]	2670	2900	3400	3300	3300	3300	3400	3400

LVCH and the roof and bed, and with the free boundary simulating the Earth's surface. Velocities and densities are represented by values characterizing the crust, the lithosphere and the upper mantle. Velocities in the LVCH are lower than in the roof and/or in the bed, and the thickness of the LVCH is taken due to S waves in one case and due to P waves in the other.

The model is also very interesting due to the fact that during the seventies the author proved the existence of the transverse channel wave and explained its genesis firstly in the asthenosphere, and later in all low velocity channels including exploration seimology in coal mining. Channels of low velocity are observed in different places of the crust and mantle of the Earth. For example channels of low velocities of seismic waves were also observed closely below the Conrad discontinuity in the depth between 10-15 km and closely above the Moho discontinuity at a depth of about 25-27 km. In the neighbourhood of the Conrad discontinuity such channels were observed in the Bohemian massif. Moreover, these channels are observed in the sedimentary layers near the Earth's boundary (see below). Velocities of seismic waves as well as densities in the upper parts of the lithosphere are very different. The conspicuous boundaries are the Conrad discontinuity at depth around 10-15 km, the Moho discontinuity at depth around 15-25 km and more, the boundaries at depths around 70-120 km, and at 150 km, 400 km, 670 km. At the depth around 70-120 km the velocities of seismic waves rapidly decrease and the densities of rocks decrease by about 0.05-0.1 gcm^{-3}. In the areas of rift zones the thickness of channels of low velocities is about 200-300 km. The geophysical studies show that the majority of all observed discontinuities are not sharp discontinuities, but rather the transitient zones (hence it follows the technique of approximation of transition conditions). The velocities of P and S waves change with depth as well as with block geological structure, as they are different under the platform lithosphere, under the island arcs, under the ocean lithosphere and under the rift zones. The structure of the Earth was studied in Chapter 9, where the velocities of P and S waves and densities of rocks were discussed.

To verify the genesis of propagated waves in the LVCH-asthenosphere as well as for mine practice, the three layers model with the same physical properties in the roof and the bed was taken.

11.3.6.2. Model M2 — The asthenosphere between two halfspaces

The model represents the asthenosphere between two rocks of the same physical parameters (see Table 11.3). The model is derived from Model M1(ii), where the velocities of seismic waves in the roof and the bed are greater than in the asthenosphere, and are characterized by their values for the lithosphere and the upper mantle.

Table 11.3. Model M2-data

depth [km]	$-\infty$	69.5	70	120	120.5	$+\infty$
c_p [ms^{-1}]	8200	8200	7400	7400	8200	8200
ρ [kgm^{-3}]	3400	3400	3300	3300	3400	3400

The thickness of the asthenosphere is taken as in the previous model, M1. According to the symmetry of the geometry of the LVCH, one half of the model is given with the symmetry axis situated into the axis of the asthenosphere. Our axis of symmetry in $3D$ reality represents the plane (x_1, x_2) in the depth of the centre of the asthenosphere. It is evident that the model is symmetrical about the axis of the channel as well as about the plane (x_2, x_3), representing in our $2D$ model the axis x_3. The limitation of our model depends on the properties of Rayleigh's waves. For numerical computations the symmetry due to the middle plane of the LVCH was used.

11.3.6.3. Model M3 — The exploration of seismological practice

The model is constructed as a three layer model with the LVCH between two layers, the overburden and the bed having the same physical parameters which are greater than in the LVCH (see Table 11.4, representing one special case of the coal layer of $1m$ thickness at depth ~500 m). The material data for the model represent the real data of the cool mine in the region of Ostrava-Karviná in the Czech Republic (see Staš (1978), Nedoma (1981), (1984)). The model problem was studied for thickness of coal layers from 10 cm to 20 m. For numerical computations the symmetry due to the middle plane of the LVCH was used. The technique presented is useful for verification of continuity of the coal layers, as well as for investigations of coal layers with inhomogeneities and for investigations of bifurcated coal layers.

11.3.6.4. Model M4 — The layer of sediments between two halfspaces

The upper part of the Earth's crust is strongly disintegrated and is characterized by low velocities of seismic waves. Moreover, in areas near the lower part of the zone with low velocities of seismic waves, channels of low velocities are also observed. Velocities of seismic waves in these channels are about 0.6–1.7 km/sec. Changes in velocities of elastic waves in horizontal directions are generated by changes in geological structure of rocks in these channels, and also depend on the degree of water saturation as well as on changes of the Earth's relief. The thickness of this layer is about 10–20–120 m and more. This layer has unfavourable influence on our measurements practised on the Earth's surface. From this point of view investigations of channel waves in the low velocity channels are very important for our understanding of their genesis, their characters and their application in geological and geophysical practice. The boundary between the low velocity layer and the bed, and also the roof, is very expressive. Inside the layer of lower velocities elastic waves are strongly absorbed; therefore they change their own spectrum and lose a great deal of energy.

Table 11.4. Model M3-data

depth [m]	$-\infty$	499.95	500	501	501.05	$+\infty$
c_p [ms^{-1}]	3850	3850	1650	1650	3850	3850
ρ [kgm^{-3}]	2650	2650	1400	1400	2650	2650

Table 11.5. Model M4-data

depth [m]	0	39.9	40	50	50.1	250
c_p [ms^{-1}]	3600	3600	1500	1500	3600	3600
ρ [kgm^{-3}]	2350	2350	1450	1450	2350	2350

The model, which is similar to Model M1, is constructed as a three layer medium with free boundary, simulating the Earth's surface (see Table 11.5). The physical parameters in the roof and the bed of the LVCH are greater than in the LVCH.

RAYLEIGH'S AND STONELEY'S WAVES

From the theory of continuum mechanics it is known that in the unbounded space P and S waves are propagated. But, besides P and S waves, Love's, Rayleigh's and Stoneley's waves are also propagated. The Love's waves are of transversal type and oscillate in the plane parallel to the Earth's surface and perpendicular to the direction of their transfer. The Rayleigh's and Stoneley's waves are characterized by the fact that they oscillate in the plane defined by the direction of the wave transfer and the normal to the Earth's surface in the case of the Rayleigh's waves and to the interface between two different rocks. Both waves propagate in a very small layer close to the Earth's surface in the case of Rayleigh's waves and in a small layer near the boundary between two different rocks in the case of Stoneley's waves.

Since in our models the Stoneley's waves play an important role, we shall study the Stoneley's waves only. Let us assume that the rock denoted by "*I*" is characterized by physical parameters ρ_1 — the density, μ_1, λ_1 — the Lamé parameters and c_{p1}, c_{s1} — the velocities of P and S waves, and that the rock denoted by "*II*" is characterized by analogous parameters ρ_2, μ_2, λ_2, c_{p2} and c_{s2}.

Let φ and ψ be two elastic potentials. Then components of the displacement $\boldsymbol{u}$ in our symmetry can be expressed as

$$u_1 = \partial\varphi/\partial x_1 + \partial\psi/\partial x_3, \quad u_3 = \partial\varphi/\partial x_3 - \partial\psi/\partial x_1. \tag{11.129}$$

Then from the equation of motion for the harmonic source we obtain

$$\left.\begin{aligned} &\Delta\varphi_1 + (\omega/c_{p1})^2\varphi_1 = 0, && (11.130)\\ &\Delta\psi_1 + (\omega/c_{s1})^2\psi_1 = 0, && (11.131)\end{aligned}\right\} \text{ in the rock "I"},$$

$$\left.\begin{aligned} &\Delta\varphi_2 + (\omega/c_{p2})^2\varphi_2 = 0, && (11.132)\\ &\Delta\psi_2 + (\omega/c_{s2})^2\psi_2 = 0, && (11.133)\end{aligned}\right\} \text{ in the rock "II"},$$

Let us denote by c_{SR} the phase velocity of Stoneley's waves. Then the solution can be found in the form

$$\left.\begin{aligned} &\varphi_i = f_i(x_3)\exp(i\omega(t - x_1/c_{RS})), && (11.134)\\ &\psi_i = g_i(x_3)\exp(i\omega(t - x_1/c_{RS})), && (11.135)\end{aligned}\right\} \quad i = 1, 2.$$

Hence and (using Equations 11.130-11.133) we find

$$(i\omega/c_{RS})^2 f_i(x_3) + f_i''(x_3) + \omega^2/c_{pi}^2 f(x_3) = 0,$$

$$(i\omega/c_{RS})^2 g_i(x_3) + g_i''(x_3) + \omega^2/c_{si}^2 g(x_3) = 0,$$

or

$$f_i''(x_3) + \omega^2/c_{pi}^2(1 - c_{pi}^2/c_{RS}^2) f_i(x_3) = 0,$$

$$g_i''(x_3) + \omega^2/c_{si}^2(1 - c_{si}^2/c_{RS}^2) g_i(x_3) = 0.$$

Their solution can be found in the form

$$f_i(x_3) = {}^1A_i \exp(i\omega/c_{pi}(1 - c_{pi}^2/c_{RS}^2)^{1/2} x_3) + {}^2A_i \exp(-i\omega/c_{pi}(1 - c_{pi}^2/c_{RS}^2)^{1/2} x_3), \tag{11.136}$$

$$g_i(x_3) = {}^1B_i \exp(i\omega/c_{si}(1 - c_{si}^2/c_{RS}^2)^{1/2} x_3) + {}^2B_i \exp(-i\omega/c_{si}(1 - c_{si}^2/c_{RS}^2)^{1/2} x_3), \tag{11.137}$$

where ${}^kA_i$, ${}^kB_i$, $k = 1, 2$ are some constants. Let us assume that the solution decreases with the depth and therefore $c_{RS} < c_{pi}$ and $c_{RS} < c_{si}$. Then ${}^2A_i = {}^2B_i = 0$. For simplicity let us put ${}^1A_i \equiv A_i$ and ${}^1B_i \equiv B_i$, $i = 1, 2$. Then

$$f_i(x_3) = A_i \exp(i\omega/c_{pi}(1 - c_{pi}^2/c_{RS}^2)^{1/2} x_3) = A_i \exp(-r_i x_3), \tag{11.138}$$

$$g_i(x_3) = B_i \exp(i\omega/c_{si}(1 - c_{si}^2/c_{RS}^2)^{1/2} x_3) = B_i \exp(-s_i x_3), \tag{11.139}$$

where

$$r_i = \omega/c_{pi}(c_{pi}^2/c_{RS}^2 - 1)^{1/2},\ s_i = \omega/c_{si}(c_{si}^2/c_{RS}^2 - 1)^{1/2} \tag{11.140}$$

Then for stress components we find

$$\tau_{33} = \lambda\vartheta + 2\mu\partial u_3/\partial x_3 = \lambda\Delta\varphi + 2\mu(\partial^2\varphi/\partial x_3 + \partial^2\psi/\partial x_1\partial x_3), \tag{11.141}$$

$$\tau_{31} = \mu(\partial u_3/\partial x_1 + \partial u_1/\partial x_3) = \mu(2\partial^2\varphi/\partial x_1\partial x_3 - \partial^2\psi/\partial x_1^2 + \partial^2\psi/\partial x_3^2), \tag{11.142}$$

where λ and μ are Lamé parameters. On the interface boundary the stress components τ_{33}, τ_{32} and τ_{31} must be continuous and similarly components of the displacement vector u_1, u_2, u_3 must be continuous. Conditions $u_2|_{(1)} = u_2|_{(2)}$ and $\tau_{32}|_{(1)} = \tau_{32}|_{(2)}$ are automatically satisfied due to the symmetry used. We find

$$\varphi_i(x_3) = A_i \exp(-r_i x_3)\exp(i\omega(t - x_1/c_{RS}) = A_i \exp(-r_i x_3 + i\omega t - ipx_1), \tag{11.143}$$

$$\psi_i(x_3) = B_i \exp(-s_i x_3)\exp(i\omega(t - x_1/c_{RS}) = B_i \exp(-s_i x_3 + i\omega t - ipx_1). \tag{11.144}$$

From $u_1|_{(1)} = u_1|_{(2)}$ it follows that

$$u_1|_{(1)} = \partial\varphi_1/\partial x_1 + \partial\psi_1/\partial x_3 = \partial\varphi_2/\partial x_1 + \partial\psi_2/\partial x_3 = u_1|_{(2)},$$

and since

$$\partial\varphi_i/\partial x_1 = A_i(-ip)\exp(-r_i x_3 + i\omega t - ipx_1) = (-ip)\varphi_i(x_3),$$

$$\partial\psi_i/\partial x_3 = B_i(-s_i)\exp(-s_i x_3 + i\omega t - ipx_1) = (-s_i)\psi_i(x_3),$$

then we find

$$-ipA_1 \exp(-r_1 x_3) - s_1 B_1 \exp(-s_1 x_3) + ipA_2 \exp(-r_2 x_3) + s_2 B_2 \exp(-s_2 x_3) = 0. \tag{11.145}$$

From $u_3|_{(1)} = u_3|_{(2)}$ it follows

$$u_3|_{(1)} = \partial\varphi_1/\partial x_3 - \partial\psi_1/\partial x_1 = \partial\varphi_2/\partial x_3 - \partial\psi_2/\partial x_1 = u_3|_{(2)}\,.$$

To find the unknown coefficients A_i, B_i, $i = 1, 2$, under some modifications as in the case of Rayleigh waves, we find the characteristic equation for the phase velocity of Stoneley's waves (Stoneley (1924), Yamaguchi and Satô (1955), Ginzbarg and Strick (1958)) in the form

$$\begin{aligned}&[2 - \tau Q - sQ(2-\tau)]^2 + 4(1-sQ)^2\alpha_1\beta_1\beta_2 - A_1\beta_1[2 - \tau Q - 2sQ]^2 \\ &\quad - \alpha_2\beta_2[2 - sQ(2-\tau)]^2 - sQ^2T^2(a_1\beta_2 + \beta_1\alpha_2) = 0,\end{aligned} \tag{11.146}$$

where

$$\begin{aligned}&s = \rho_1/\rho_2,\ \tau^2 = c_{RS}/c_{s1},\ Q^2 = c_{s1}/c_{s2},\\ &\alpha_1 = [1 - (c_{s1}/c_{p1})^2(c_{RS}/c_{s1})^2]^{1/2} = [1 - (c_{RS}/c_{p1})^2]^{1/2},\\ &\alpha_2 = [1 - (c_{s2}/c_{p2})^2(c_{s1}/c_{s2})^2(c_{RS}/c_{s1})^2]^{1/2} = [1 - (c_{RS}/c_{p2})^2]^{1/2},\\ &\beta_1 = (1-\tau),\ \beta_2 = (1 - Q\tau)^{1/2} = [1 - (c_{RS}/c_{s2})^2]^{1/2}.\end{aligned}$$

Equation 11.146 represents a quadratic equation in $s = \rho_1/\rho_2$ so that

$$s = \rho_1/\rho_2 = [B \pm (B^2 - AC)]^{1/2}(QA)^{-1}, \tag{11.147}$$

where

$$\begin{aligned}&A = (1 - \alpha_2\beta_2)[(2-\tau)^2 - 4\alpha_1\beta_1], \quad C = (1 - \alpha_1\beta_1)[(2 - Q\tau)^2 - 4\alpha_2\beta_2],\\ &B = (2 - Q\tau - 2\alpha_2\beta_2)(2 - \tau - 2\alpha_1\beta_1) + \tfrac{1}{2}Q\tau^2(\alpha_1\beta_2 + \beta_1\alpha_2).\end{aligned}$$

Analyses show that the Stoneley's waves exist only for $0 \le s \le \infty$ (see Ginzbarg and Strick (1958)), more precisely for the cases (i) $0 \le s \le 1$ and (ii) $1 \le s \le \infty$, and satisfy

(i) $0.8742 \le c_{RS}/c_{s2} \le c_{s1}/c_{s2} \le 1$ for $\rho_1 < \rho_2$,
or
$0.8742c_{s2} \le c_{RS} \le c_{s1} \le c_{s2}$ for $\rho_1 < \rho_2$,

(ii) $0.8742 \le c_{RS}/c_{s1} \le 1$ for $\rho_1 > \rho_2$,
or
$0.8742c_{s1} \le c_{RS} \le c_{s1}$ for $\rho_1 > \rho_2$.

Similarly as in the case of Rayleigh's waves, the amplitudes of Stoneley's waves decrease with depth H as $\exp(-\gamma H)$. From here the depth of maximal effect of Stoneley's waves $\sim \Lambda/2\pi$, where Λ is the wave length of Stoneley's waves and, therefore, the depth is less than their wavelength.

NUMERICAL RESULTS AND THEIR ANALYSES

Computer-generated printer-plot contours of displacement are shown in vertical section for one value of frequency, one displacement component and one model problem. The spreading of black type areas with a different intensity of growing black and black areas in the figure represent the growth of the instability of the displacement field of propagated elastic waves. Certain special information about the propagated waves can be extracted from the profiles of the displacement vector components as well as of the strain and stress tensor components and profiles of deformation energy, characterized by the spatial behaviour of the propagated waves. Theoretical seismograms, which may also be constructed from the obtained components of the displacement vector by using the Fourier's transform as well as phase and group velocities, are based on the theory presented above.

Analyses of Model M1 show one very complicated situation of the wave field, namely that for higher frequencies. In Figure 11.1 the computer-generated printer-plot contours of the displacement components u_1 and u_3 for Model M1(i) (see

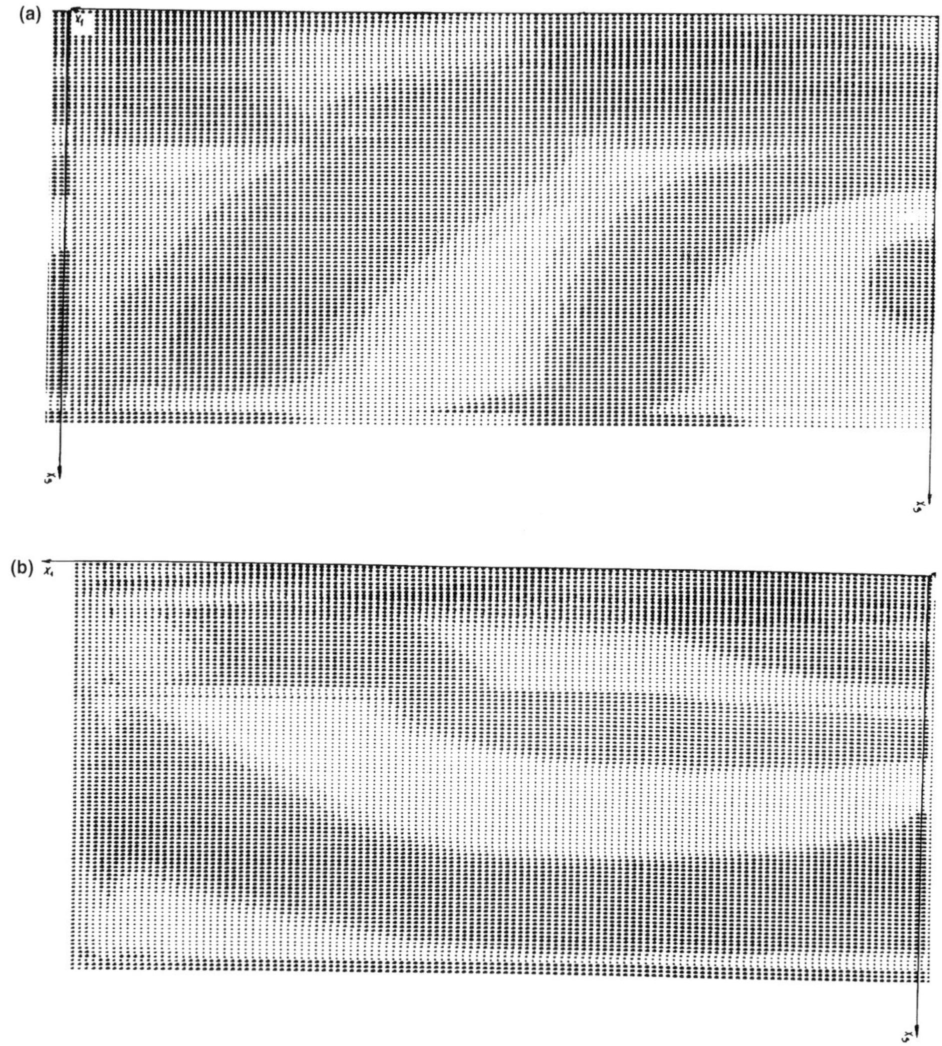

Figure 11.1. Model M(i),(ii)

(c)

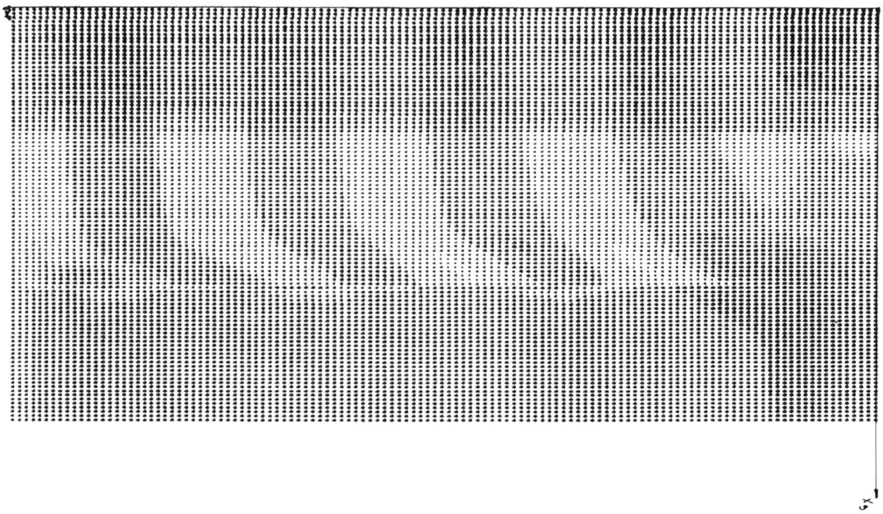

Figure 11.1. (*Continued*)

Figure 11.1a, b) and u_1 for Model M1(ii) (see Figure 11.1c) are presented. Analyses indicate the Rayleigh wave close to the Earth's surface and Stoneley's waves close to the interfaces channel/overburden and channel/floor (details will be given for Model M3). In the asthenosphere (i.e. the LVCH), the wave having the character of the transversal wave is generated. Since the thickness of the floor, characterized by higher velocities of seismic waves, is limited, the energy transmitted by this wave radiates more into the mesosphere than into the lithosphere. We denote this type of wave as the generalized transverse channel wave.

(a)

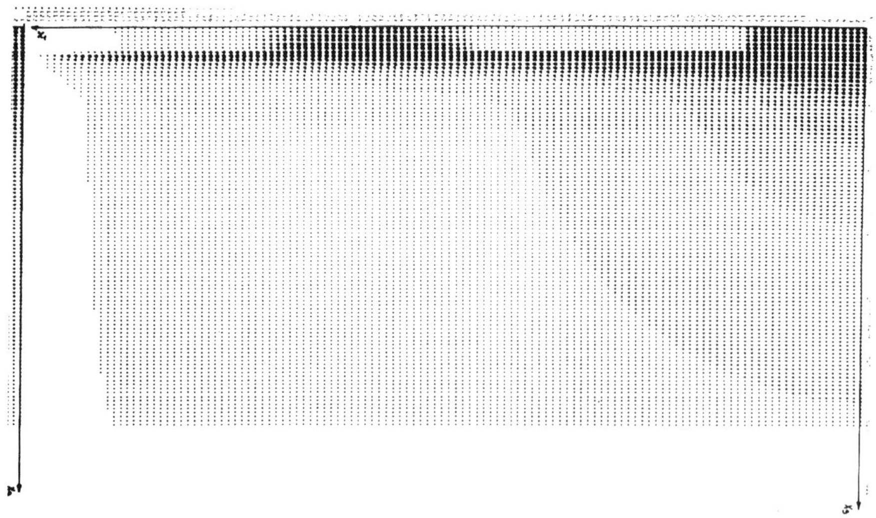

Figure 11.2. Model M2

(b)

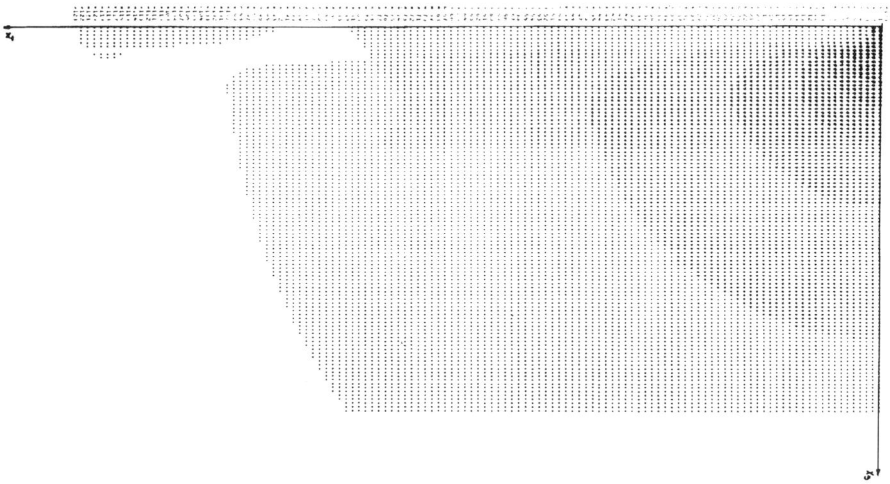

Figure 11.2. (*Continued*)

In Figure 11.2 the wave field of Model M2, verifying the genesis of transverse channel waves, is presented. The wave field is characterized by expressive Stoneley's waves on the interfaces asthenosphere/lithosphere and asthenosphere/mesosphere. Inside the LVCH — the asthenosphere — the wave of transverse character is generated. This wave propagates in the x_1 direction and oscillates in the x_1x_2-plane. We denote this type of wave as the transverse channel wave. This wave was also verified in the cases of prospecting seismology and coal mines, discussed later.

Figures 11.3 and 11.4 illustrate the origin of the channel waves in the model of type M3, propagated by the velocity of the Stoneley's type waves c_{RS} and their behaviour. Model M2 is derived from one type of Model M1 and is characterized by the low velocity channel laying between two infinite halfspaces with the same physical parameters. The observed transverse and longitudinal channel waves are generated under the conditions that

(i) the velocity c_p of the longitudinal wave, which also propagates through the channel, is less than the velocity of the symmetric Stoneley's type waves, rising on the interfaces channel/overburden and channel/floor (i.e. the interfaces channel/halfspaces) and propagating in the overburden and in the floor, and exponentially decreasing with the distance from the interfaces;

(ii) the physical parameters of the overburden and of the floor, i.e. the velocities c_p and c_s of P and S waves and the density ρ, are greater than in the low velocity channel.

The transverse and longitudinal channel waves are generated by the effect of the symmetrical Stoneley's type waves. Their existence and character are confirmed by the courses of the displacement components and of the stress tensor components. Figures 11.4a, b show horizontal and vertical displacement components u_1 and u_3 in

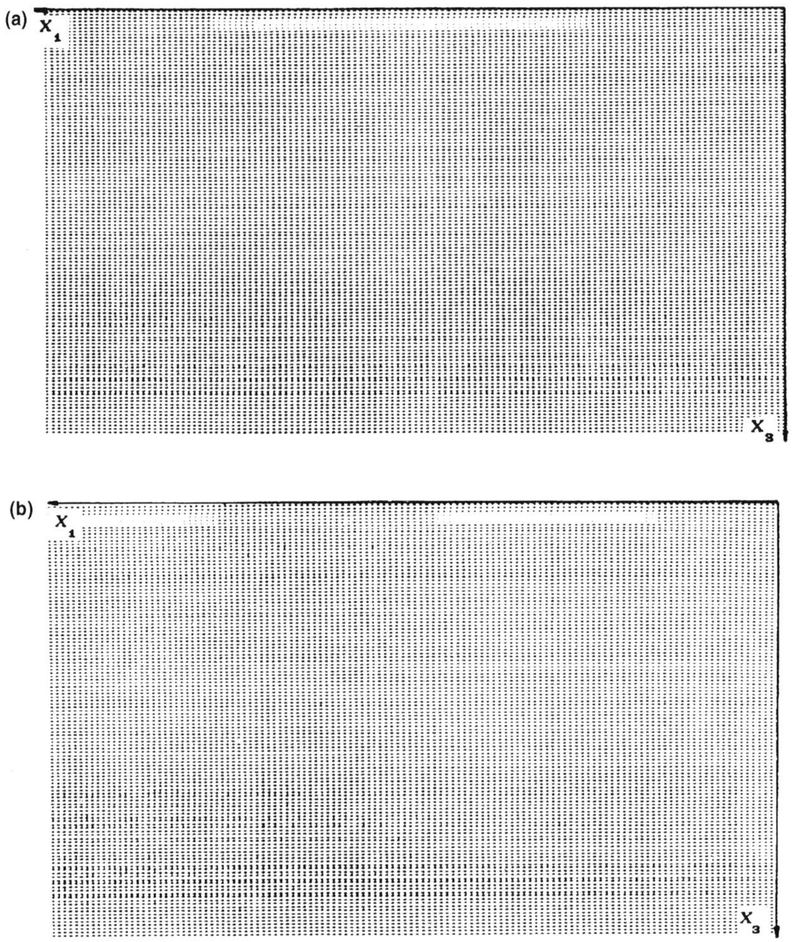

Figure 11.3. Model M3 — the model of Seismological prospection in coal mines

dependence on the x_1 and x_3 co-ordinates, respectively. The damping of the channel waves and their spatial profiles at several distances from the line source are shown. Figures 11.4 c–f illustrate the stress tensor components as a function of x_1 and x_3 co-ordinates. The damping of the stress tensor components with epicentral distances from the line sources is presented. The non-zero stress component τ_{22} is due to the compression of the channel material by the Stoneley's waves. The genesis of the transverse channel wave (TCHW) is illustrated in Figure 11.5 (Nedoma (1981) for the model case of the coal layer between the elastic roof and elastic floor.

Figure 11.3 represents the model problem of seismological prospection in coal mines. The layer of low velocity elastic waves represents the coal layer. The model problem was verified in the real conditions of the Ostrava-Karviná coal mines in the dependence of the continuity of coal layers in these mines.

The variability of the geological conditions has a meaningful effect in the genesis of useful elastic waves acceptable for analyses of localities, which are of special

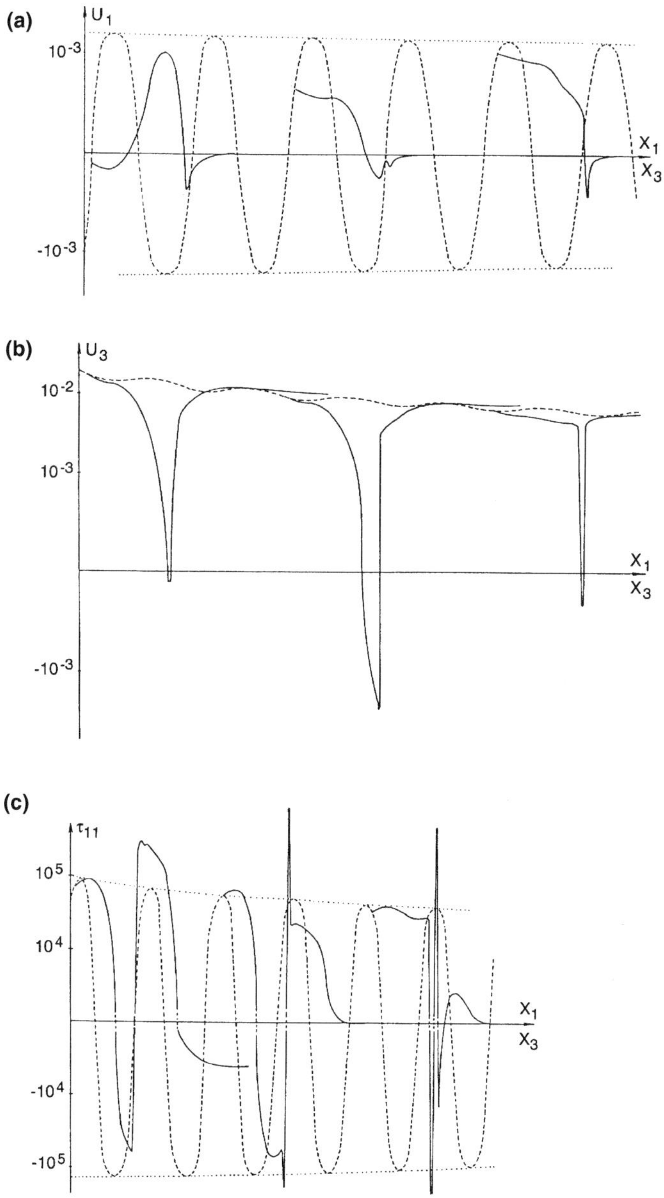

Figure 11.4. Model M3 — horizontal and vertical displacement components and the Stress tension components

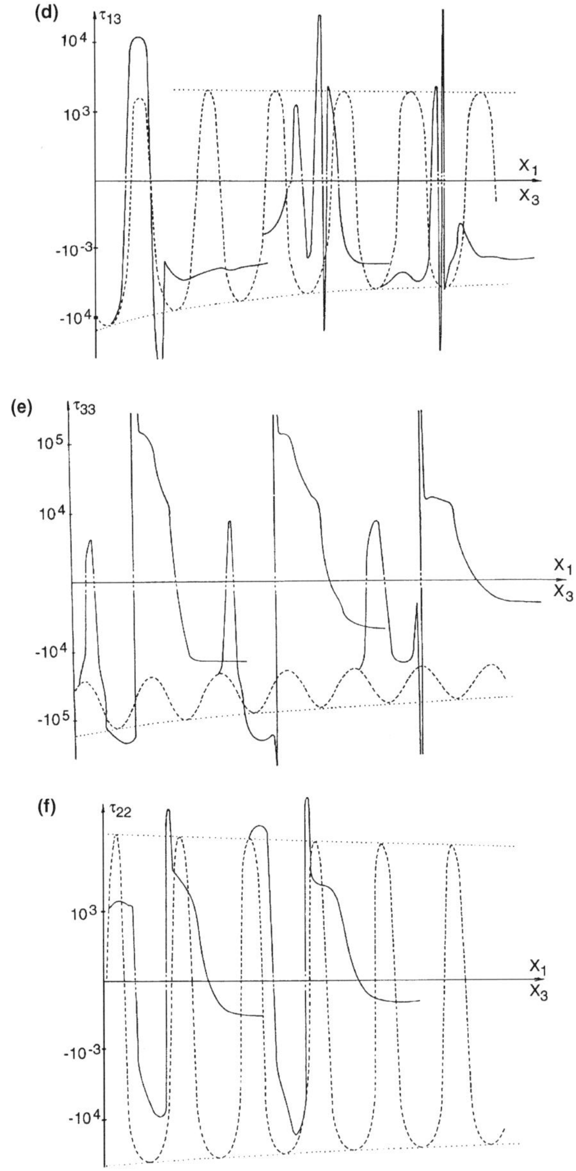

Figure 11.4. (*Continued*)

interest. From the point of view of the geological structure there were found two types of rocks, characterized by

(i) the same average velocity of overburden's and floor's Carboniferous rocks, between them the coal layer is situated;

(ii) the same average velocity of overburden's and floor's Carbonifereous rocks — below the floor's Carboniferous rocks the layer of proterozoic's rocks or

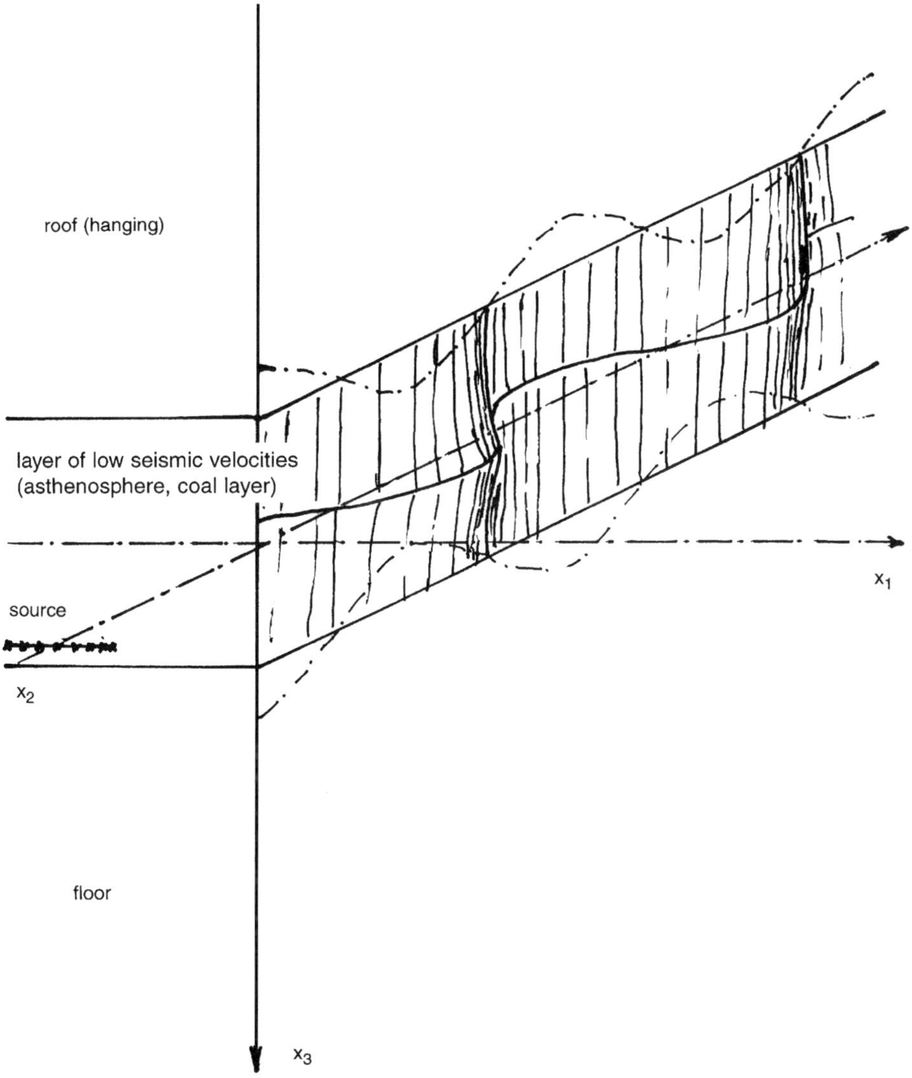

Figure 11.5. Genesis of the TCHW Wave

systems of conglomerate rocks at a certain depth is situated, between them the coal layer is situated.

In the coal layer, the transverse channel type wave (TCHW) is generated in a similar way to the asthenosphere, i.e. by symmetric Stoneley's waves, which are generated on the interfaces overburden/channel and channel/floor. In Figure 11.3 the computer-generated printer-plot contours of the u_1 and u_3 displacement components are presented. Figure 11.6 illustrates the dependence of the wavelength of the transverse channel wave on the thickness of the low velocity channel for one given frequency of the excited wave. The results obtained are in good agreement with observations (Staš

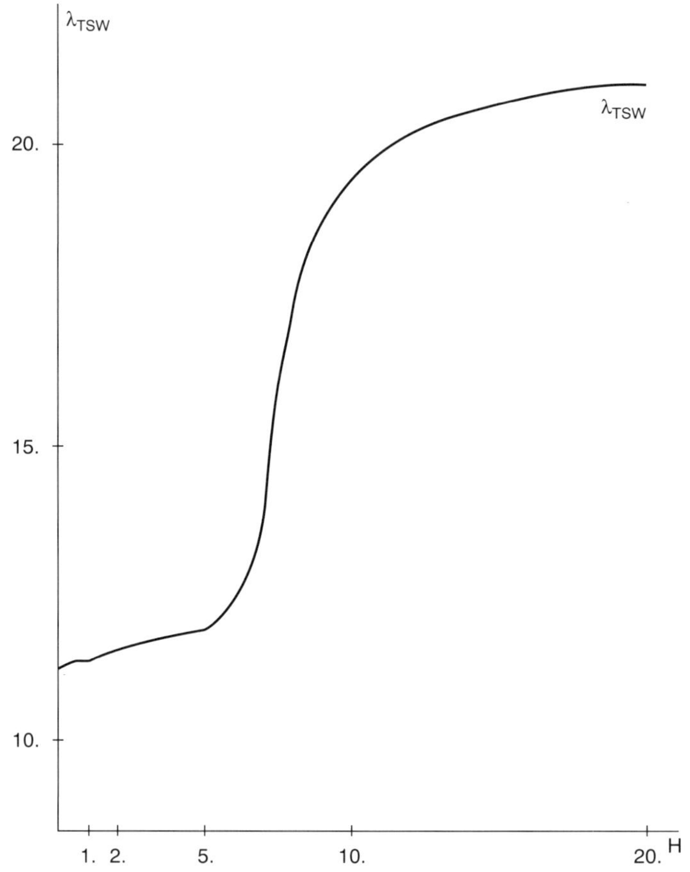

Figure 11.6. Dependence of the wavelength of the TCHW on the thickness of the low velocity channel for the given frequency of the excited wave

(1978)). The measurements in situ observed the existence of the TCHW in the coal layers minimally of thickness 45 cm and maximally of 6.36 m. But our numerical computations proved the genesis of the TCHW in coal layers of thickness minimally of 10, 20, 40 cm and maximally to the thickness of 10 m and of 20 m. Knowledge about the genesis of the TCHWs used in the seismological prospection and about their wavelength are the main conditions for an application of these waves in situ practice in coal mines. The in-situ observations showed that decreasing thicknesses of coal layers lead to a decrease of the wavelengths, which in the course of in-situ measurements lead to higher frequencies of measurements.

In Figure 11.7 the model of the layer of sediments between two halfspaces, situated at small depths in the Earth's crust, is simulated. On the interfaces channel/overburden and channel/floor Stoneley's waves of wavelengths $\Lambda_{RS} \sim 11.5$ m and $\Lambda_{RS} \sim 13.5$ m were observed. These Stoneley's waves generate inside the channel the TCHW wave oscillating in the $\{x_1 x_2\}$-plane with maximal deviation in the x_2-direction. The figure shows that the Stoneley's waves are damped with distance from the line sources of generation of elastic waves inside the channel.

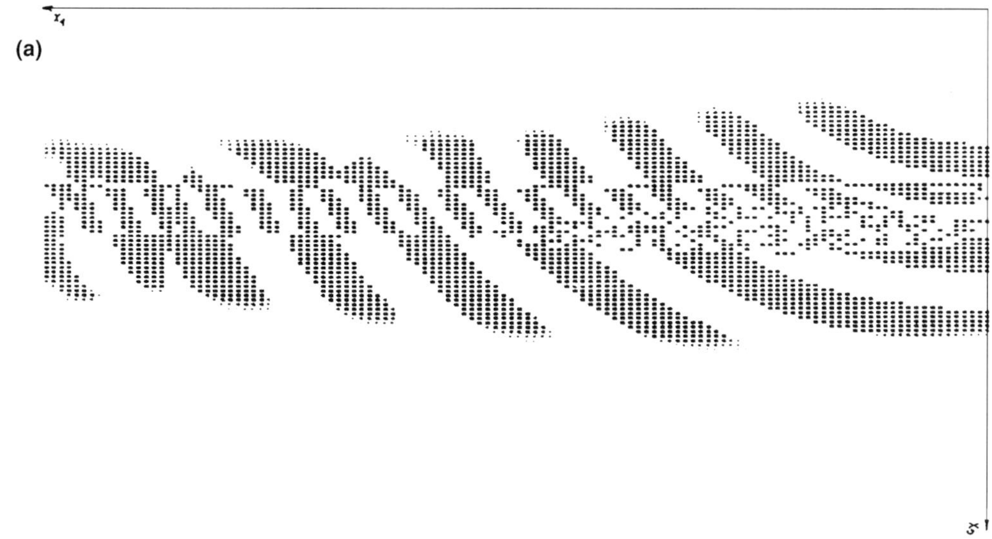

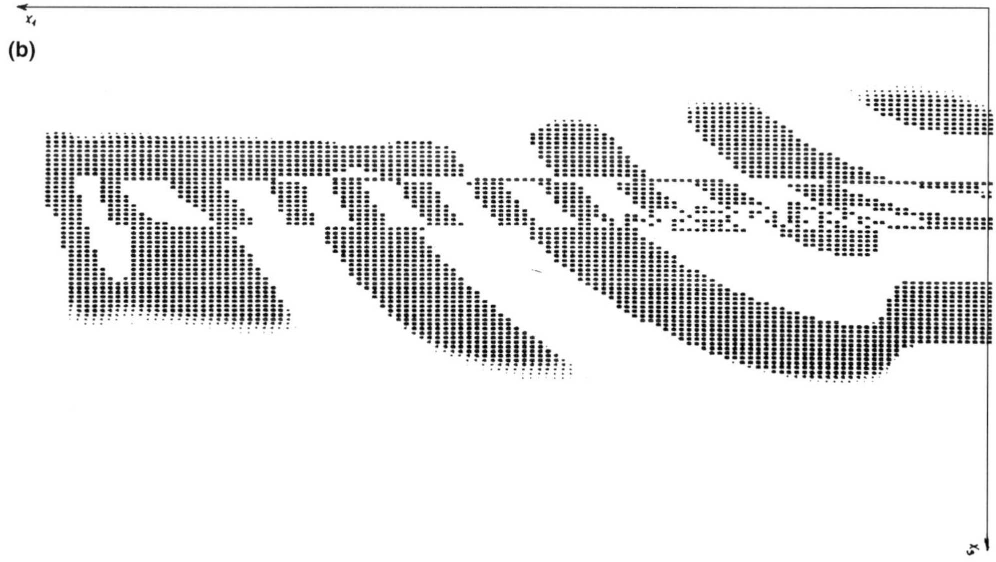

Figure 11.7. The model of the layer of sediments between two halfspaces

Table 11.6. The Bohemian Massif Model — data

depth [km]	0	15	45	70	71	120	121	200
c_p [ms^{-1}]	4500	6500	8200	8200	7500	7500	8200	8200
ρ [kgm^{-3}]	1500	2140	3400	3400	3350	3350	3400	3400

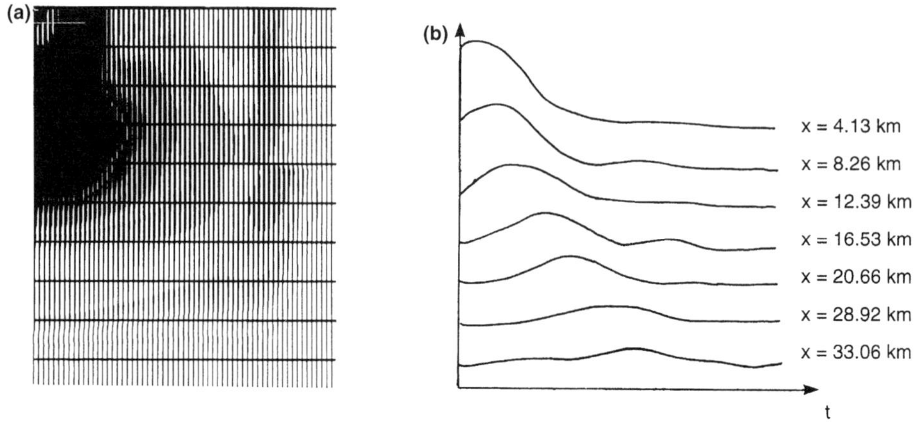

Figure 11.8. The Bohemian Massif Model — the computer-generated plot contours of the horizontal displacement component and the corresponding theoretical seismograms

The model for the Bohemian Massif is of the type of M1 model discussed above (see Table 11.6). For illustration the computer-generated plot contours of the displacement components u_1 and the corresponding theoretical seismograms are given (see Figure 11.8).

11.4. PROPAGATION OF SEISMIC SH WAVES IN UNBOUNDED DOMAINS

We saw above that the Helmholtz equation with a suitable radiation at infinity serves as a model for a variety of wave propagation problems. During the last few years several papers have presented mathematical and numerical analyses of such types of problems (Chan *et al.* (1990), Cohen, Halpern and Joly (1991), Douglas, Hensley and Roberts (1993), Douglas *et al.* (1993), Kim (1994)). Such types of problems can be solved numerically by first truncating the given unbounded domain, imposing a suitable outgoing radiation condition on the artificial boundary of the truncated bounded domain and then solving the resulting problem by use of the finite difference and finite element methods as well as the mixed methods in time and/or frequency domains.

Let Ω be the N-dimensional domain ($N = 2, 3$) occupying the investigated part of the Earth, with Lipschitz boundary $\partial\Omega$. We shall assume that the rocks of the investigated part of the Earth are isotropic. Further, we shall assume that the seismic SH waves are generated by the line harmonic source, as in the above investigated problems.

Let us consider the following problem.

Problem ($\mathscr{P}$): Find a displacement component $u_2(\boldsymbol{x}) \equiv u(\boldsymbol{x})$ satisfying the complex-valued Helmholtz equation

$$\Delta u + (\omega^2/c^2(\boldsymbol{x}) - i\omega\alpha(\boldsymbol{x}, \overline{\lambda}))u = f(\boldsymbol{x}), \quad \boldsymbol{x} \in \Omega, \tag{11.148}$$

$$\partial u/\partial n + i\omega c^{-1}(\boldsymbol{x})u = 0, \quad \boldsymbol{x} \in \partial\Omega, \tag{11.149}$$

where $\Omega \in R^N$, $N = 2, 3$, with the boundary $\partial\Omega$, $c(\boldsymbol{x})$ is the SH wave velocity and is bounded below and above by positive constants c_o and C_o, respectively, n is the outer unit normal to $\partial\Omega$ and the angular frequency $\Omega > 0$, f represents body forces and $\alpha =$ constant ≥ 0 in the whole domain Ω characterizing the natural damping of rocks only.

Equation 11.149 represents a first-order absorbing boundary condition that allows normally incident waves to pass out of Ω transparently.

We shall solve the problem by the decomposition technique and the finite difference method.

Let the investigated domain Ω be partitioned as $\overline{\Omega} = \cup_{\iota=1}^{s}\overline{\Omega}^{\iota}$, $\Omega^k \cap \Omega^l = 0$, $k \neq l$, where Ω^{ι}, $\iota = 1, \ldots, s$ represent a partition of the domain Ω, which we assume to be convex. Let us denote $\Gamma^{\iota} = \partial\Omega \cap \partial\Omega^{\iota}$, $\iota = 1, \ldots, s$, $\Gamma^{kl} = \partial\Omega^k \cap \partial\Omega^l$, $S = \cup_{k,l}\Gamma^{kl}$.

Instead of the problem in Equations 11.148 and 11.149 we shall consider the decomposition problem over $\{\Omega^{\iota}\}$.

Problem $(\mathscr{P})_d$**:** Let u^{ι}, $\iota = 1, \ldots, s$, satisfy

$$\Delta u^{\iota} + (\omega^2/(c^{\iota})^2(\boldsymbol{x}) - i\omega\alpha^{\iota}(\boldsymbol{x}, \overline{\lambda}^{\iota}))u^{\iota} = f^{\iota}(\boldsymbol{x}), \quad \boldsymbol{x} \in \Omega^{\iota}, \tag{11.150}$$

$$\partial u^{\iota}/\partial n + i\omega(c^{\iota})^{-1}(\boldsymbol{x})u^{\iota} = 0, \quad \boldsymbol{x} \in \Gamma^{\iota}, \tag{11.151}$$

and the consistency conditions

$$u^k = u^l, \partial u^k/\partial n^k + \partial u^l/\partial n^l = 0, \quad \boldsymbol{x} \in \Gamma^{kl}, \tag{11.152}$$

where n^{ι} is the outward normal to Ω^{ι}. The last condition in Equation 11.152 can be replaced by the so-called Robin type of interface boundary condition

$$\partial u^k/\partial n^k + i\Theta u^k = -\partial u^l/\partial n^l + i\Theta u^l, \quad \boldsymbol{x} \in \Gamma^{kl}, \tag{11.153}$$

where Θ is a complex function on S with $Re(\Theta) > 0$. For such values of Θ the subproblems in Equations 11.150, 11.151 and 11.153 are well-posed (see, For example, Lions (1990), Chan *et al.* (1990), Despres (1991)).

As is well known the idea of the domain decomposition method leads to computing the smaller subdomain problems on smaller subdomains and then the resulting solution of the initial problem will be obtained by the iterative algorithm. The iterative algorithm is as follows.

Let u_o^{ι}, $\iota = 1, \ldots, s$, be given arbitrarily, then u_j^{ι}, $j \geq 1$, can be given recursively by solving

$$\Delta u_j^{\iota} + (\omega^2/(c^{\iota})^2(\boldsymbol{x}) - i\omega\alpha^{\iota}(\boldsymbol{x}, \overline{\lambda}^{\iota}))u_j^{\iota} = f^{\iota}(\boldsymbol{x}), \quad \boldsymbol{x} \in \Omega^{\iota}, \tag{11.154}$$

$$\partial u_j^{\iota}/\partial n + i\omega(c^{\iota})^{-1}(\boldsymbol{x})u_j^{\iota} = 0, \quad \boldsymbol{x} \in \Gamma^{\iota}, \tag{11.155}$$

$$\partial u_j^k/\partial n^k + i\Theta u_j^k = -\partial u_{j-1}^l/\partial n^l + i\Theta u_{j-1}^l, \quad \boldsymbol{x} \in \Gamma^{kl}. \tag{11.156}$$

This method converges for $\Theta = \omega$. For the proof see Despres (1991). However for the discretized problems generated by the finite difference and finite element methods and/or by mixed methods, the convergence analyses are not know in detail. The method can also be applied for propagations of P and SV waves in inhomogeneous media.

Remark: Initial-boundary value problems in linear visco-elasticity and in linear elasticity on the halfspace were solved in Meister *et al.* (1991), Mark and Meister (1995).

11.5. PROPAGATION OF PULSE SEISMIC WAVES IN INHOMOGENEOUS MEDIA

For simplicity we shall investigate the SH waves only. Let us consider the propagation of SH waves in an inhomogeneous medium generated by a line pulse source. The orthogonal co-ordinate system $0x_1x_2x_3$ is chosen in such a way that the x_3-axis is directed toward the medium and is perpendicular to the line source, which is located at depth z_o from the origin of the co-ordinate system. We shall assume the medium to have non-zero absorption, described by an absorption coefficient $\alpha = \text{constant} > 0$. The model problem is valid for such times t_1 which are less than time t during which the SH wave reaches the boundary $\partial\Omega$ of the investigated domain. Further, we shall put $u_2(\boldsymbol{x}) \equiv u(\boldsymbol{x})$.

Let the investigated domain Ω occupy the part of the Earth characterized by the isotropic rocks. Let the boundary $\partial\Omega$ of Ω be Lipschitzian and let $\partial\Omega = \cup_{j=1}^{4}\Gamma_j$. Let us denote the time interval by $I = (t_o, t_1)$.

Then we have to solve the following problem.

Problem ($\mathscr{P}$): Find $u(\boldsymbol{x}, t) \in C^2(\Omega \times I) \cap C^1(\overline{\Omega} \times I)$ satisfying

$$c^{-2}\partial^2 u(\boldsymbol{x}, t)/\partial t^2 + \alpha\partial u(\boldsymbol{x}, t)/\partial t - \Delta u(\boldsymbol{x}, t) = f(\boldsymbol{x}, t) \quad \text{in } \Omega \times I, \tag{11.157}$$

and boundary and initial conditions

$$\partial u/\partial n = 0 \quad \text{on } \Gamma_1 \times I, \tag{11.158}$$

$$u = 0 \quad \text{on } (\Gamma_2 \cup \Gamma_3 \cup \Gamma_4) \times I, \tag{11.159}$$

$$u(\boldsymbol{x}, t_o) = u_o(\boldsymbol{x}), \quad \boldsymbol{x} \in \Omega, \tag{11.160}$$

$$\partial u(\boldsymbol{x}, t_o)\partial t = u_1(\boldsymbol{x}), \tag{11.161}$$

where $u_o(\boldsymbol{x})$ and $u_1(\boldsymbol{x})$ are the given function.

Remark: In this case we shall assume that the SH waves are generated by the body forces, where the pulse sources are defined above. The second formulation can be the modification of the formulation of the frequency domain.

The variational formulation of Problem ($\mathscr{P}$) is as follows.

Problem $(\mathscr{P})_V$: Find $u \in C(\bar{I}, V) \cap C^1(\bar{I}; L^2(\Omega))$ satisfying

$$c^{-2}\mathrm{d}^2/\mathrm{d}t^2(u(\boldsymbol{x}, t), v) + \alpha\mathrm{d}/\mathrm{d}t(u(\boldsymbol{x}, t), v) + a(u(\boldsymbol{x}, t), v) = (f(\boldsymbol{x}, t), v) \quad \forall v \in V, \tag{11.162}$$

$$u(\boldsymbol{x}, t_o) = u_o(\boldsymbol{x}), \quad \boldsymbol{x} \in \Omega \tag{11.163}$$

$$\mathrm{d}u(\boldsymbol{x}, t_o)\mathrm{d}t = u_1(\boldsymbol{x}), \tag{11.164}$$

in the sense of distribution in I, where

$$V = \{v \mid v \in H^1(\Omega), v = 0 \text{ on } \Gamma_2 \cup \Gamma_3 \cup \Gamma_4 \text{ in the sense of traces}\},$$

$$a(u, v) = \int_\Omega (\partial u/\partial x_1 \partial v/\partial x_1 + \partial u/\partial x_3 \partial v/\partial x_3)\mathrm{d}\boldsymbol{x}, \; (u, v) = \int_\Omega uv\mathrm{d}\boldsymbol{x}.$$

Assuming that $u_o \in V$, $u_1 \in L^2(\Omega)$ and $f \in L^2(\Omega \times I)$, then the problem has a unique solution. The proof is similar of that of Lions and Magenes (1968) and/or of Raviart and Thomas (1983).

For the numerical solution the semi-discrete scheme will be used. Let V_h be a finite dimensional subspace of V, generated by the finite elements of the first order, and let us assume that $u_{o,h}$, $u_{1,h} \in V_h$ are suitable approximations of u_o and u_1. Then we have the following discrete problem.

Problem $(\mathscr{P})_h$**:** For each $t \in \bar{I}$ we look for a function $u_h(\boldsymbol{x}, t) \in V_h$ such that it satisfies

$$c^{-2}\mathrm{d}^2/\mathrm{d}t^2(u_h(\boldsymbol{x}, t), v_h) + \alpha\mathrm{d}/\mathrm{d}t(u_h(\boldsymbol{x}, t), v) + a(u_h(\boldsymbol{x}, t), v_h) = (f(\boldsymbol{x}, t), v_h) \quad \forall v_h \in V_h, t \in I, \tag{11.165}$$

$$u_h(\boldsymbol{x}, t_o) = u_{o,h}(\boldsymbol{x}), \tag{11.166}$$

$$\mathrm{d}u_h(\boldsymbol{x}, t_o)/\mathrm{d}t = u_{1,h}(\boldsymbol{x}). \tag{11.167}$$

The proof of the existence of the unique solution of Equations 11.165–11.167 is parallel to the proof of the unique solution of Equations 11.162–11.164 as the bilinear form $a(u, v)$ is coercive in V and symmetric.

Let us denote $\{w_j(\boldsymbol{x})\}$, $j = 1, \ldots, n(h)$ a basis of V_h. Let us set

$$u_h(\boldsymbol{x}, t) = \sum_{j=1}^{n(h)} \xi_j(t) w_j(\boldsymbol{x}), \; u_{o,h}(\boldsymbol{x}, t_o) = \sum_{j=1}^{n(h)} \xi_{o,j} w_j(\boldsymbol{x}),$$

$$u_{1,h}(\boldsymbol{x}, t_o) = \sum_{j=1}^{n(h)} \xi_{1,j} w_j(\boldsymbol{x}).$$

Then Problem $(\mathscr{P})_h$ is equivalent to the problem

$$M\frac{\mathrm{d}^2\vec{\xi}}{\mathrm{d}t^2}(t) + N\frac{\mathrm{d}\vec{\xi}}{\mathrm{d}t}(t) + A\vec{\xi}(t) = \boldsymbol{F}(t), \quad t \in I, \tag{11.168}$$

$$\vec{\xi}(t_o) = \vec{\xi}_o, \tag{11.169}$$

$$\frac{\mathrm{d}\vec{\xi}}{\mathrm{d}t}(t_o) = \vec{\xi}_1, \tag{11.170}$$

where $M_{ij} = c^{-2}(w_i, w_j)$, $N_{ij} = \alpha(w_i, w_j)$, $A_{ij} = a(w_i, w_j)$, $F_i(t) = (f(t), w_i)$ and $\vec{\xi}_o$, $\vec{\xi}_1$ correspond to the approximation of the initial conditions.

The time discretizaton leads to simple ordinary differential equations of second order in time of the form

$$y''(t) = \varphi(t, y(t), y'(t)), \quad t \in I, \tag{11.171}$$

$$y(t_o) = y_o, \tag{11.172}$$

$$y'(t_o) = y_1, \tag{11.173}$$

where $\varphi: I \times R^2 \to R$ is a continuous function. The problem in Equations 11.171-11.173 can be solved numerically by several methods, e.g. the Runge-Kutta method, Crank-Nichelson method, Newmark method, etc..

Due to Raviart and Thomas (1983) and Quarteroni and Valli (1991), the Newmark method can be used. Let the interval $\overline{I} = \langle t_o, t_1 \rangle$ be divided into sub-intervals $\langle t^k, t^{k+1} \rangle$ with $t^o = t_o$, $t^n = t_1$, $t^{k+1} = t^k + \Delta t = t_o + k\Delta t$, $k = 0, \ldots, n-1$. Denote $y^k = y(t^k)$ and set $y^o = y_o$, $z^o = y_1$. Then the solution of Equations 11.171-11.173 can be found in the form

$$y^{k+1} = y^k + \Delta t z^k + (\Delta t)^2 \left[\beta \varphi^{k+1} + \left(\tfrac{1}{2} - \beta\right) \varphi^k\right], \tag{11.174}$$

$$z^{k+1} = z^k + \Delta t[\gamma \varphi^{k+1} + (1-\gamma)\varphi^k], \quad k = 0, 1, \ldots, n-1, \tag{11.175}$$

where β, γ are some non-negative parameters, $\varphi^k = \varphi(t_k, y^k, z^k)$, and z^k is an approximation of $y'(t_k)$. The scheme is explicit for $\beta = \gamma = 0$. If $\gamma \geq \frac{1}{2}$ then the scheme is stable. If $\gamma = \frac{1}{2}$ and $\beta = \frac{1}{4}$ the scheme is unconditionally stable. But for long time integration for $\gamma > \frac{1}{2}$ we must take $\beta \geq \frac{1}{4}(\gamma + \frac{1}{2})^2$.

When $\alpha = 0$, then Problem $(\mathscr{P})$ and its finite differential approximation lead to the following.

Problem $(\mathscr{P}1)_h$**:** For each $t \in \overline{I}$ we look for a function $u_h(\boldsymbol{x}, t) \in V_h$ that satisfies

$$c^{-2}\mathrm{d}^2/\mathrm{d}t^2(u_h(\boldsymbol{x}, t), v_h) + a(u_h(\boldsymbol{x}, t), v_h) = (f(\boldsymbol{x}, t), v_h) \quad \forall v_h \in V_h, t \in I, \tag{11.176}$$

$$u_h(\boldsymbol{x}, t_o) = u_{o,h}(\boldsymbol{x}), \tag{11.177}$$

$$\mathrm{d}u_h(\boldsymbol{x}, t_o)/\mathrm{d}t = u_{1,h}(\boldsymbol{x}), \tag{11.178}$$

which is equivalent to the problem

$$M \frac{\mathrm{d}^2\vec{\xi}}{\mathrm{d}t^2}(t) + A\vec{\xi}(t) = \boldsymbol{F}(t), \quad t \in I, \tag{11.179}$$

$$\vec{\xi}(t_o) = \vec{\xi}_o, \tag{11.180}$$

$$\frac{\mathrm{d}\vec{\xi}}{\mathrm{d}t}(t_o) = \vec{\xi}_1, \tag{11.181}$$

where $M_{ij} = c^{-2}(w_i, w_j)$, $A_{ij} = a(w_i, w_j)$, and $F_i(t) = (f(t), w_i)$.

This method is valid for times t less than time t_1 during which the SH wave reaches the boundary $\partial\Omega$ of the investigated domain.

Then the Newmark scheme is as follows.

$$(\Delta t)^{-2}M(\vec{\xi}^{k+1} - \vec{\xi}^{k} - \Delta t\vec{\zeta}) + A[\beta\vec{\xi}^{k+1} + (\tfrac{1}{2} - \beta)\vec{\xi}^{k}] = \beta \boldsymbol{F}(t_{k+1}) + (\tfrac{1}{2} - \beta)\boldsymbol{F}(t_k), \tag{11.182}$$

$$(\Delta t)^{-1}M(\vec{\zeta}^{k+1} - \vec{\zeta}^{k}) + A[\gamma\vec{\xi}^{k+1} + (1 - \gamma)\vec{\xi}^{k}] = \gamma \boldsymbol{F}(t_{k+1}) + (1 - \gamma)\boldsymbol{F}(t_k),$$
$$k = 0, 1, \ldots, n-1, \tag{11.183}$$

where $\vec{\xi}^{o} = \vec{\xi}_{o}$ and $\vec{\zeta}^{o} = \vec{\zeta}_{1}$.

At each step of Equations 11.182 and 11.183 we solve a linear system

$$[M + \beta(\Delta t)^2 A]\vec{\xi}^{k+1} = \boldsymbol{b}^k, \tag{11.184}$$

with symmetric and positive definite matrix. If $\beta = 0$ and M is diagonal then Equation 11.184 provides $\vec{\xi}^{k+1}$ explicitly.

The above technique can also be applied for P and S waves. Moreover, the finite difference method is also used (see, For example, Alterman and Rotenberg (1968), Boore (1970), (1972), Strikwerda (1989)).

REFERENCES

Aboudi, J. (1971), "Numerical Simulation of Seismic Sources", *Geophysics*, **36, 5**, 810-821.

Aboudi, J. (1986), "Harmonic Waves in Composite Materials", *Wave Motion*, **8**, 289-303.

Alekseyev, A.S., Babich, V.M and Gel'chinskyi, B.Ja (1961), "Ray Method for the Computation of the Intensity Wave Fronts", in: *Problems of the Dynamic Theory of Propagation of Seismic Waves, Vol. 5* (Petrashen, G.I. (Ed.)), Leningrad University Press, Leningrad, 3-24, (in Russian).

Alterman, Z.S. and Aboudi, J. (1970), "Source of Finite Extent, Applied Force and Couple in an Elastic Half-Space", *Geophys. J. Roy. Astr. Soc.*, **21**, 47-64.

Alterman, Z.S. and Karal, F.C. (1968), "Propagation of Elastic Waves in Layered Media by Finite-Difference Methods", *Bull. Seism. Soc. Am.*, **50**, 367.

Alterman, Z.S. and Rotenberg, I., (1968), "Seismic Waves in a Quarter Plane", *Bull. Seism. Soc. Am.*, **59**, 347.

Aschenbach, J.D. (1973), *Waves Propagation in Elastic Solids, Series in Applied Mathematics and Mechanics*, North-Holland, Amsterdam.

Aschenbach, J.D. and Harris, J.G. (1978), "Ray Method for Elastodynamic Radiation from a Slip Zone of Arbitrary Shape", *J. Geophys. Res.*, **83, B5**, 2283-2291.

Axelsson, O. (1996), *Iterative Solution Methods*, Cambridge University Press, Cambridge.

Babich, V.M. and Buldyrev, V.S. (1972), *Asymptotic Methods in Problems of Diffraction of Short Waves*, Nauka, Moscow (in Russian).

Babuška, I., Práger, M. and Vitásek, E. (1966), *Numerical Processes is Differential Equations*, J. Wiley & Sons, London, New York, Sydney.

Bath, M. (1968), *Mathematical Aspects of Seismology*, Elsevier, Amsterdam.

Bolt, B.A. (1972), *Methods in Computational Physics, Vol. 11*, Academic Press, New York.

Boore, D.M. (1970), "Love Waves in Non-uniform Wave Guides: Finite Difference Calculations", *J. Geopys. Res.*, **75**, 1512-1528.

Boore, D.M. (1972), "Finite Difference Methods for Seismic Wave Propagation in Heterogeneous Materials", *Methods in Computational Physics*, **11**, 1-37.

Boore, D.M., Larnet, K.L. and Aki, K. (1971), "Comparison of Two Independent Methods for the solution of Wave-Scattering Problems: Response of a Sedimentary Basin to Vertically Incident SH Waves", *J. Geopys. Res.*, **10**, 558-569.

Brdička, M. (1959), *Continuum Mechanics*, NČSAV, Prague.

Burridge, R. and Knopoff, L. (1964), "Body Force Equivalents for Seismic Dislocation", *Bull. Seism. Soc. Am.*, **54**, 1875-1888.

Bullen, K.E. (1965), *An Introduction to the Theory of Seismology*, Cambridge University Press, Cambridge.

Červený, V. and Ravindra, R. (1971), *Theory of Seismic Head Waves*, University of Toronto Press, Toronto.

Chan, R., Glowinski, J., Périaux, J. and Widlund, O.B. (Eds.) (1990), *Third International Symposium on Domain Decomposition Methods for Partial Differential Equations*, SIAM, Philadelphia, PA.

Chen, Y.M. (1988), "Parallelism by Hierarchy of GPST Inversion Algorithm for Elastic Wave Equation", *Applied Numerical Mathematics*, **4**, 83-95.

Clearbout, J.F. (1970), "Numerical Holography", in *Acoustical Holography, Vol. 3* Metherell, A.F. (Ed.), Plenum Publishing Corp., New York.

Clearbout, J.F. (1976), *Fundamentals of Geophysical Data Processing*, McGraw-Hill, New York.

Clearbout, J.F. and Johnson, A.G. (1971), "Extrapolation of Time-Dependent Waveforms along their Path of Propagation", *Geophys. J. R. Astr.Soc.*, **26**, 285-293.

Cohen, G., Halpern, L. and Joly, P. (Eds.) (1991), *Mathematical and Numerical Aspects of Wave Propagation Phenomena*, SIAM, Philadelphia, PA.

Despres, B. (1991), "Domain Decomposition Method and the Helmholtz Problem", in *Mathematical and Numerical Aspects of Wave Propagation Phenomena*, Cohen, G., Halpern, L. and Joly, P. (Eds.), SIAM, Philadelphia, PA.

Douglas, Jr J., Hensley, J.L. and Roberts, J.E. (1993), "An Alternating-Direction Iteration Method for Helmholtz Problems", *Tech. Report No 214*, Mathematics Department, Purdue University, West Lafayette, IN.

Douglas, Jr. J., Santos, J.E., Sheen, D. and Bennethum, L.S. (1993), "Frequency Domain Treatment of One-Dimensional Scalar Waves", *Math. Models Method Appl. Sci.*, **13**, 171-194.

Duvaut, G. and Lions, J.L. (1972), *Lés Inequalities en Mécanique et en Physique*, Dunod, Paris.

Freund, R.W. (1992), "Conjugate Gradient Type Methods for Linear Systems with Complex Symmetric Coefficient Matrices", *SIAM J. Sci. Statist. Comput.*, **13**, 1-23.

Fuchs, K. and Müller, G. (1971), "Computation of Synthetic Seismograms with Reflectivity Method and Comparison with Observations", *Geophys. J.*, **23**, 417-433.

Ginzbarg, A.S. and Strick, E. (1958), "Stoneley-Wave Velocities for a Solid-Solid Interface", *Bull. Seism. Sos. Am.*, **48**, 51-63.

Halpern, L. and Rauch, J. (1987), "Error Analysis for Absorbing Boundary Conditions", *Numer. Math.*, **51**, 459-467.

Handlová, E., Nedoma, J. and Schmid, J. (1976), "Investigation of Propagation of Elastic SH-Waves in Layered Media with Local Inhomogeneities by Methods of Numerical Analysis, *Numerische Methoden in der Geophysik*, Geophys. Inst. Czech. Acad. Sci, Prague, 177-195 (in Russian).

Hlaváček, I., and Nečas, J. (1970), "On Inequalities of Korn's Type", *Arch. Rational Mech. Anal.*, **36**, 305-311.

Hron, F. (1968), *Introduction to the Ray Theory in a Broader Sense. Applications to Seismology* (Lecture Notes), Laboratoire de Physique de l'École Normale Supérieure, Université de Paris, Paris.

Jacobs, D.A.H. (1986), "A Generalization of the Conjugate Gradient method to Solve Complex Systems", *IMA J. Numer. Anal.*, **6**, 447-452.

Kim, S. (1994), "A Parallelizable Iterate Procedure for the Helmholtz Problem", *Applied Numerical Mathematics*, **14**, 435-449.

Křížek, M. and Mlýnek, J. (1994), "On the Preconditioned Bi-conjugate Gradients for Solving Linear Complex Equations Arising from Finite Elements", *Numerical Analysis and Mathematical Modelling*, **29**, Banach Centre Publ., Inst. of Mathematics, Polish Acad. Sci, Waszawa, 195-205.

Křížek, M. and Neittaanmäki, P. (1990), *Finite Element Approximation of Variational Problems and Applications*, Longman, Harlow.

Landau, L.D. and Lifshitz, E.M. (1954), *Mechanics of Continuous Media*, GITTL, Moscow (in Russian).

Lions, J.L. and Magenes, E. (1968), *Problémes aux Linites non Homogéneous et Applications, 2*, Dunod, Paris.

Lions, P.L. (1990), "On the Schwarz Alternating Method III: A Variant for Non-overlapping Subdomains", in *Third International Symposium on Domain Decomposition Methods for Partial Differential Equations* Chan *et al.* (Eds.), SIAM, Philadelphia, PA, 202-223.

Marek, I. and Nedoma, J. (1973), "Finite Element Method in the Theory of SH-Wave Propagation", in *Numerische Methoden in der Geophysik* Pěč, K. and Nedoma, J. (Eds.), Geophysical Institute of the Czech. Acad. Sci, Prague, 31-37.

Marek, I., Janovský, V., and Nedoma, J. (1974), "Investigation of SH-Waves in the Layered Media with Local Inhomogeneities by the Finite Element Method", *Proc. Internal Symposium on "The Method of Geophysical Interpretation"*, Geophysical Institute of the Slovak Acad. Sci., Bratislava, 117-123.

Mark, J. and Meister, E. (1995), "Initial-Boundary Value Problems in Linear Viscoelasticity on the Halfspace", *Mathematical Methods in Applied Sciences*, **18**, 1181-1214.

Metherell, A.F. (Ed.) (1970), *Acoustical Holography, Vol. 3*, Plenum Publishing Corp., New York.

Mikhailenko, B.G. (1984), "Numerical Experiment in Seismic Investigations", *J. Geophys.*, **58**, 101-124.

Morton, K.W. (1985), "Generalized Galerkin Methods for Hyperbolic Problems", *Computer Methods in Applied Mechanics and Engineering*, **52**, 847-871.

Nečas, J. (1967), *Les Méthodes Directes en Théorie des Equations Elliptiques*, Academia, Prague.

Nedoma, J. (1972), "Investigation of Linear Harmonic Source Field of SH-Waves in a Stratified Inhomogeneous Medium Using the Finite Difference Method," *Zeitschrift für Geophysik*, **38**, 529-542.

Nedoma, J. (1979), "SH-Wave Propagation in Quasi-anisotropic Rocks: Problem of Partial Differential Equation with Rapidly Oscillating Coefficients," in *Inverse and Improperly Posed Problems in Differential Equations. Band 1* Anger, G. (Ed.), Akademie Verlab, Berlin, 181-188.

Nedoma, J. (1981), "Propagation of Elastic Waves Through the Low Velocity Channels", *Tech. Rep.*, Geophys. Inst. CSAS, Prague (in Czech).

Nedoma, J. (1984), "On One Type of Finite Element Simulation in Dynamic Elasticity," *Mathematics and Computers in Simulation*, **XXVII**, 189-209.

Nedoma, J. (1990), "On a New Theory of Earthquake Mechanism", in *Proc. of the Symposium on Analysis of Seismicity and Seismic Risk*, Schenková, Z. (Ed.), Geophys. Inst. Czech. Acad. Sci., Prague, 339-346.

Nedoma, J. (1992), "On the Genesis of the Transverse Channel Wave", in *Geomechanics 91* Rakowski, Z. (Ed.), Balkema, Rotterdam, 159-166.

Nowacki, W. (1970), *Dynamical Problems of Thermoelasticity*, Mir, Moscow (in Russian).

Oden, J.T. and Reddy, J.N. (1976), *An Introduction to the Mathematical Theory of Finite Elements*, J. Wiley & Sons, New York, London, Sydney, Toronto.

Officer, C.H.B. (1974), *Introduction to Theoretical Geophysics*, Springer Verlag, New York.

Olson, A.H., Orcutt, J.A. and Frazier, G.A. (1984), "The Discrete Wavenumber/Finite Element Method for Synthetic Seismograms", *Geophys. J. R. Astr. Soc.*, **77**, 421-460.

Pěč, K. and Nedoma, J. (1970), "Analysis of a Field of SH Waves in a Layer Indicated by a Point Harmonic Source", *Geofyzikální Sborník XVIII* Travaux De l'Institute Géophysique de l'Academie Tchécoslovaque des Sciences, 81-189.

Quarteroni, A. and Valli, A. (1991), *Numerical Approximation of Partial Differential Equations*, Springer Verlag, Berlin, Heidelberg, New York, London, Paris, Tokyo, Hong Kong, Barcelona, Budapest.

Raviart, P.A. and Thomas, J.M. (1983), *Introduction á l'Analyse Numérique des Équations aux Dérivées Partielles*, Masson, Paris.

Richards, P.G. (1971), "Elastic Wave Solutions in Stratified Media", *Geophysics*, **36**, 798-809.

Richtmayer, R.D. and Morton, K.W. (1967), *Difference Methods for Initial-Value Problems*, J. Wiley & Sons, New York.

Staš, B. (1978), "Seismic Methods for Exploration of the Geological Structure of Rock Massifs", *VOKD/EZ-30*, Polytechna Praha.

Stephen, R.A. (1983), "A Comparison of Finite Difference and Reflectivity Seismograms for Marine Models", *Geophys. J. R. Astr. Soc.*, **72**, 39-57.

Stoneley, R. (1924), "Elastic Waves at the Surface of Separation of Two Solids", *Proc. Roy. Soc. London*, **106**, 416.

Strikwerda, J.C. (1989), *Finite Difference Schemes and Partial Differential Equations*, Wadsworth & Brooks-Cole, Pacific Grove.

Varga, R. (1962), *Matrix Iterative Analysis*, Prentice-Hall, Englewood Cliffs, New Jersey.

Yamaguchi, R. and Sato, Y. (1955), "Stoneley Wave — Its Velocity, Orbit, and Distribution of Amplitude", *Bull. Earthq. Res. Inst.*, **33**, 549.

Yosida, K. (1965), *Functional Analysis*, Springer Verlag, Berlin.

12

Mathematical Problems of Global Geodynamic, Geomagnetic, Geothermal and Gravity Models of the Earth and Planets and a Global Plate Tectonic Hypothesis

12.1. INTRODUCTION

The results of studies and analyses by geologists, geochemicists, theoretical and experimental geophysicists as well as physicists working above all in the domain of physics of the solid phase and mathematics show that the observed tectonic motions of the Earth's lithosphere are the result of the geodynamic processes passing in the greater depths of the Earth's body. These processes contribute to the change of the inner structures, the physical-chemical properties of the Earth and their compositions. The geodynamic processes in the Earth's interior are known only a little. On the Earth's surface or in relatively shallow depths we observe only their demonstrations and their consequences. Their profound understanding is based on an analysis of:

(i) the origin of the deep processes and their sources in space and time from fundamental geological and geophysical observations;

(ii) the spreading and geological shape of inhomogeneities in the upper parts of the Earth, rising during geological development by these geodynamic processes;

(iii) the manifestation of deep geodynamic processes on the Earth's surface at the present time and at ancient (foregoing) geological epochs;

(iv) models, both mathematical and experimental, and comparison of their results with observed reality.

Thus an assignment of the theory of the Earth is to find a mutual relation between surface and deep geodynamic processes. Geodynamic processes passing in the Earth's body were interpreted from different points of view as well as on different presumptions, e.g. the plutonic, expansional, contractional, isostatic (however effects of isostasy can be explained by results of mechanical processes mathematically simulated by contact problems in thermo-elasticity and thermo-plasticity), pulsational hypotheses, oscillational rotationally dynamic hypothesis, hypotheses of extra-terrestrial influences, convective currents, and continental drift and, more recently, on the hypothesis of plate tectonics.

The idea of plate tectonics, verified with geographical and geological data, was first given by Wegener (1912),(1915), (1929). Before Wegener's fundamental works, the idea of plate tectonics was darkly presented by Taylor (1910) and earlier by Sir Francis Bacon (1561-1626), Francis Placet (1866, in the book *La Corruption du Grand et Petit Monde*), Theodor Christoph Lilienthal (in the book *Die gute Sache der Göttlichen Offenbarung*), A. von Humboldt (1769-1859), and later in the book and papers of Baker (1914), Du Toit (1937) and Holmes (1965).

Convection, as a thermodynamic engine, was discussed by Holmes (1931). The hypothesis of Holmes was later modified by Hess (1962) and Dietz (1961). Their hypothesis about sea floor spreading assumes that a new oceanic bottom is created in the region of oceanic ridges and spread in the directions perpendicular to them. On the opposite side, in the regions near the border of continents, is submerged into the mantle, i.e. in the regions of the subduction zones. Vening-Meinesz applied the hypothesis to explain the down-coming of the Earth's crust seen in the oceanic trenches. Runcorn (1962) attempted to interpret the newly developing paleomagnetic data for continental drift in the conceptions of convection.

In the motions of Hess and Dietz, the sea floor spreading resumes the plate tectonic hypothesis, which was proposed by McKenzie and Parker (1967) and Morgan (1968). During the last 35-40 years the new conception of global tectonics, plate tectonics, is generally accepted and the principles have been worked up. This period is characterized by the works of Knopoff, Ritsema, Isacks, Molnar, Uyeda, Elsasser, Turcotte, and LePichon, and presently by for example, O'Connell, Peltier, Glatzmaier, Bercovici, Yuen, Fleitout, Ranalli, Hide, Braginski, Busse, and Moffat, and the further works of many specialists of the older and present periods (some appear in the list of the references).

12.2. GLOBAL GEODYNAMIC, GEOMAGNETIC, GEOTHERMAL AND GRAVITY MODELS OF THE EARTH

12.2.1. Introduction

The discussed geodynamic model of the Earth (see Figure 12.1) will be built on several fundamental assumptions (see Nedoma (1984), (1986)) such that:

(i) the Earth was probably created by an accumulation of interstellar matter on material condensation nuclei;

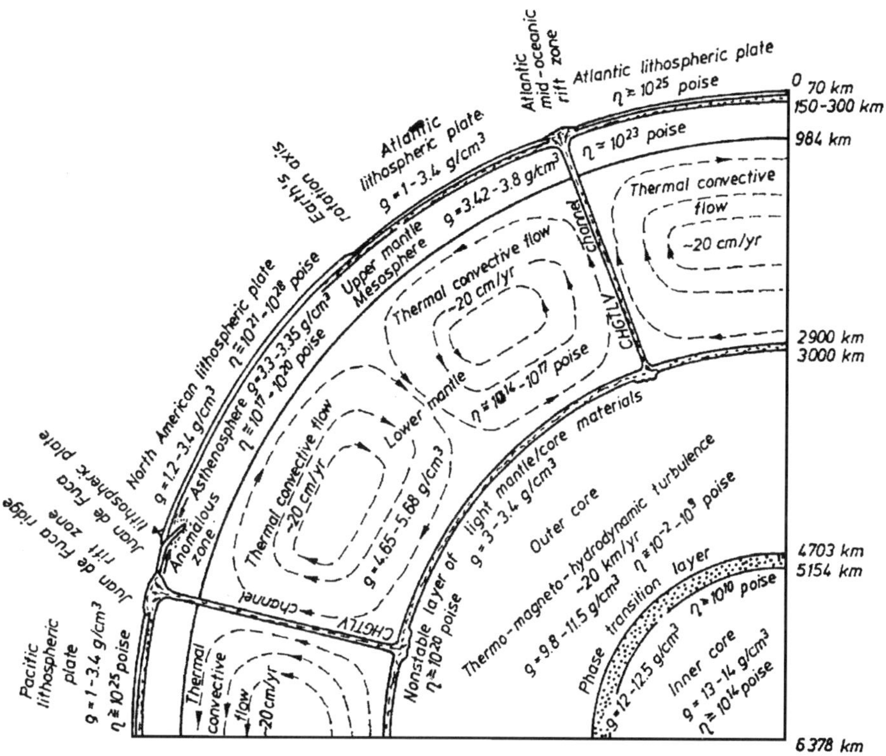

Figure 12.1. Global geodynamic model of the Earth, schematic cross-section

(ii) in the course of the development of the Earth's body the pronounced factors were gravity, radioactivity, temperature, density differentiation associated with phase transitions of the Earth's material under high temperatures, great pressures and chemical reactions under the same conditions;

(iii) the mass element in the Earth's body can be characterized by a thermo-elasto-visco-plastic state in a relatively weak magnetic field and, consequently for the derivation of an equation describing geodynamic processes in an irregular rotating Earth's body, the thermo-elasto-visco-plastic theory as well as electromagnetic theory, mainly the theory of thermo-magneto-hydrodynamics, can be applied.

According to Nedoma (1984),(1986) the presented geodynamic model of the Earth is based on the following ideas.

1. The Earth was probably created by an accumulation of interstellar matter on material condensation nuclei. After a sufficient amount of interstellar matter had been accumulated, the temperature within the original Earth increased due to the effect of the gravity force and the inherent radioactivity. The rocks of the interstellar matter in the neighbourhood of the centre of gravity were melted and the first density differentiation occurred. The heavier matter remained in the vicinity of the centre of gravity, whilst the lighter substances and the substances of the more easily melted rocks had a tendency to rise. The Earth's core began to form.

2. The matter of the core was thus differentiated, and the inner and outer cores were created.

3. With the increasing size of the core its temperature also began to increase and similarly the pressure, and first phase transition occurred within the core. The temperature in the neighbourhood of the core also increased. The matter of the initial Earth outside its core was thus subjected to considerable temperatures and pressures, and the Earth's mantle began to form.

4. The materials of the mantle at the core/mantle boundary were still being melted. The heavier substances, attracted by the matter of the inner core, descended and increased the size of the inner core, whereas the lighter substances and the more easily melted rocks gradually combined with the materials of the outer core. These were formed by melted rocks in liquid state and also contained various gaseous components which were, moreover, strongly electrically and heat conducting.

5. Since the surface of the outer liquid core with the strongly heated materials was cooled from the mantleward side and heated from the inner coreward side, heat convection was generated in the outer liquid core. This is also contributed to by the transport of the heavier substances towards the inner core, the effect of the different dynamic ellipticity of the core and mantle, the effect of the precision of the Earth's axis of rotation and the effects of the Coriolis' force $\sim 2\rho \boldsymbol{v} \times \vec{\omega}$, the Archimedean force $\sim \rho \boldsymbol{g} C$ (where $C > 1$ is a function representing density inhomogeneities in the Earth's core), the electromagnetic forces $\sim \boldsymbol{j} \times \boldsymbol{B}$ (due to Maxwell's stresses) and viscous forces $\sim \rho\nu\nabla^2\boldsymbol{v}$ (ν is the kinematic viscosity coefficient). In the deeper part of the outer liquid core, the molten liquid matter has a tendency to rotate faster on average than its upper parts which rotate more slowly due to the viscosity of the materials and in order to preserve angular momentum. These effects result in a kind of stratification of the outer core. The velocity of the convective flow is about $\sim$ 20 km/yr, the density $\rho \sim 9.8 - 12.5$ gcm^{-3}, and the viscosity coefficient $\eta \sim 10^{-2} - 10^9$ poise. Since the viscous and Coulombian friction at the mantle/outer-core boundary is small due to the assumption of the thermo-visco-plastic model of the light materials there (the Coulombian friction is assumed to be small), the mantle is not coupled with the core by large mechanical forces and thus the mantle rotates faster than the core. This is responsible for the observed westward drift of the magnetic field, which is also a consequence of the different velocities of rotation of the layers in the Earth's core, located at various distances from the axis of rotation under conditions of developed convection. This leads to transfer of mass and moments of rotation between the higher and lower layers. It is also a consequence of an effect of tidal forces.

6. The mantle/outer core boundary, i.e. the boundary at which the density differentiation of the melted material takes place, is itself uneven due to the non-uniform melting of the mantle rocks, or to possible recrystallization processes in accumulated light substances and the possible consequences of the shape non-uniformity of the initial core/mantle boundary. This in turn may be responsible for temporal changes in the rotation of the mantle and of the whole Earth and may have a feedback effect on the primary geomagnetic field (time variations of the primary geomagnetic field) and on the geomagnetic processes in the Earth's core and mantle.

7. The deflection of the axis of the magnetic field from the axis of rotation ($\sim 10^\circ$) is probably due to the asymmetrically distributed convective cells and the inhomogeneous mass distribution in the outer core and mantle.

8. The inhomogeneities of the mass distribution in the outer core and mantle and the non-uniformity in the rotation of the Earth are responsible for the non-uniformity in the action of Coriolis' and Poincaré's forces. This in turn is reflected in the effect on the thermo-magneto-hydrodynamic turbulence in the outer core. The changes in the motion of masses of the outer core and their irregularities may cause inhomogeneities in the regeneration of the magnetic field and, consequently, also an inversion of the Earth's magnetic field.

9. The inner core is probably formed by a stratified polycrystalline structure in the thermo-visco-plastic state, characterized by a high density $\rho \sim 13$-$14\ \mathrm{gcm}^{-3}$, with the phase transition layer at depths between 4700-5160 km (or 4703-5154 km).

10. The electrical conductivity of matter depends on temperature, pressure and its chemical physical compositions. As the compression of silicates under a constant pressure of about 0.9 Mbar leads to their metallization, the core will probably consist of oxides and other compounds in a metallized state.

11. The density differentiation at the core/mantle boundary is continuous and has continued until the present time. The bottom of the mantle is characterized by a thermo-visco-plastic medium strongly heated from below (from the side of the outer core) and cooled from above. This generates convective heat flows and the so-called convective cells and channels of increased temperature and decreased viscosity (GTLV channels) between them. In the literature these are also known as plumes. The light strongly heated viscous material from the core/mantle boundary, which accumulated here in a layer during the first density differentiation in the primary Earth, represents a very unstable layer with bumps (see Hide (1970), Hide and Malin (1971)). The first density differentiation resulted in spatially non-uniform accumulation of light materials. Also the creation of convective cells was spatially non-uniform, because the masses of the primary Earth's mantle were still considerably inhomogeneous compared with the present. This caused the effusion of light masses at those places where the GTLV channels were greatest and the mantle of the primary Earth was thinner towards the Earth's surface. After spreading over the surface and solidifying, this created the primary continent of Pangaea. This model was verified experimentally by the model of a strongly viscous fluid, which was firstly cooled and which therefore simulated a visco-plastic medium. In this visco-plastic medium was located an inhomogeneous medium, which modelled the inhomogeneity of the primary mantle. Such a model was then heated from below and cooled from above.

The GTLV channel was observed and the light material moved upwards, effused on the surface and then stiffened there.

12. The effusion of light material from the core/mantle boundary, the lower boundary of the lithosphere which it heats, melts and strains mechanically, vertically and horizontally, and which also penetrates below the lithosphere, causes the creation of a low-viscosity-layer — the asthenosphere — as well as geodynamic processes in the lithosphere.

Thus the geodynamic processes in the Earth's interior are closely connected to the thermo-magneto-hydrodynamic ("thermo-magneto-strongly-plasto-dynamic") processes and are mutually influenced. For the geodynamic processes in the Earth's interior, as well as the generation of the Earth's geomagnetic field and the geothermal field, it is evident that the released gravity energy (e.g. $\sim 2.10^8 \mathrm{Js}^{-1}$) is the basic source of all observed phenomena. Therefore, it is necessary to synchronously investigate all considered physical influences on the geodynamic processes in the Earth's interior. Moreover, due to enormous pressures and temperatures the material is metallized and ionized. The theory given in this chapter can be used for all such phenomena, but, unfortunately, many mathematical models are at present unsolved. However, we give an idea of how to solve such problems, i.e. the algorithms will be presented. The mathematical analyses will be given for relatively simple model problems only.

12.2.2. The Fundamental Equations of the Geodynamic Mechanism

In order to obtain the complete system of equations describing the geodynamic processes in the interior of the Earth, we must give the equation of continuity and the equations of motion, describing the motions of the Earth's masses, the generalized equation of heat, describing the thermal regime in the interior of the Earth, the diffusion equation, describing the diffusion processes in the Earth's interior and the concentration of strongly thermo-plastic (liquid) masses in the Earth's interior. We also require equations for gravitational potential and gravity acceleration, and the system of Maxwell's equations, describing the Earth's magnetic field, and the defining relations, the so-called constitutional relations, i.e. the relations between stress and strain tensors, the tensor of the strain rate, the temperature, the intensity of the electric and magnetic fields, etc. We must accurately define the heat sources, the magnetic sources and the other type of sources in the Earth, having an influence on the geodynamic processes. Then the Earth can be described as a thermo-elasto-visco-plastic medium in an electromagnetic field.

As a continuum moves, it is natural to try to specify the position of each point as a function of time t. To identify a given point, we specify its position $\boldsymbol{x}^o$ at a certain initial time t_o. If $\boldsymbol{x}$ is the position vector of this point at the moment t, the motion of the medium can be described by $\boldsymbol{x} = \boldsymbol{x}(\boldsymbol{x}^o, t)$, $\boldsymbol{x}(\boldsymbol{x}^o, t_o) = \boldsymbol{x}^o$. We shall denote by $\Omega(0)$ the region occupied at the moment $t = t_o$ by the designated portion of the Earth's material. Then the region $\Omega(t)$ occupied by the designated Earth's material at time t is the image $\Omega(0)$ at time t under the mapping $\boldsymbol{x} = \boldsymbol{x}(\boldsymbol{x}^o, t)$. Let $V(0)$ denote the volume of $\Omega(t)$, then

$$V(t) = \int_{\Omega(0)} \mathrm{d}x_1^o \mathrm{d}x_2^o \mathrm{d}x_3^o, \tag{12.1}$$

$$V(t) = \int_{\Omega(t)} \mathrm{d}x_1 \mathrm{d}x_2 \mathrm{d}x_3. \tag{12.2}$$

If in Equation 12.2 we change the variables of integration x_1^o, x_2^o, x_3^o, where $\boldsymbol{x}^o = \boldsymbol{x}^o(\boldsymbol{x}, t)$, then

$$V(t) = \int_{\Omega(0)} J(x_1^o, x_2^o, x_3^o, t) \mathrm{d}x_1^o \mathrm{d}x_2^o \mathrm{d}x_3^o, \tag{12.3}$$

where $J(x_1^o, x_2^o, x_3^o) = \partial(x_1, x_2, x_3)/\partial(x_1^o, x_2^o, x_3^o)$ is the Jacobian. From Equation 12.3 it follows that if $V(0)$ is the volume of a very small region centred about $\boldsymbol{x}^o$ then $V(t)/V(0) \sim J$ and therefore the Jacobian at point $\boldsymbol{x}^o$ and time t represents the dilatation of an infinitesimal volume initially at the point $\boldsymbol{x}^o$, where the dilatation is the ratio of a volume occupied by the infinitesimal material region at time t to its initial volume. These considerations give a precise characterization of the idea that a specified portion of Earth's material occupies the region $\Omega(t)$ at time t. That the mass of a portion of Earth's material does not vary as time increases is one way to state what is meant by conservation of mass, i.e.

$$\frac{\mathrm{d}}{\mathrm{d}t} \int_{\Omega(t)} \rho(\boldsymbol{x}, t) \mathrm{d}\boldsymbol{x} = \frac{\mathrm{d}}{\mathrm{d}t} \int_{\Omega(0)} \rho[\boldsymbol{x}(\boldsymbol{x}^o, t)] J(\boldsymbol{x}^o, t) \mathrm{d}\boldsymbol{x}^o = 0, \quad \Omega(t) \text{ arbitrary}, \tag{12.4}$$

where $J(\boldsymbol{x}^o, t)$ is the Jacobian and $\mathrm{d}\boldsymbol{x} = \mathrm{d}x_1 \mathrm{d}x_2 dx_3$, $\mathrm{d}\boldsymbol{x}^o = \mathrm{d}x_1^o \mathrm{d}x_2^o \mathrm{d}x_3^o$. Equation 12.4 is the conservation of mass in its integral form (see Chapter 3). The differential equation form of mass conservation, the so-called continuity equation, is an alternative form of Equation 12.4, i.e.

$$\frac{\partial \rho \boldsymbol{x}, t}{\partial t} + \mathrm{div}(\boldsymbol{v}(\boldsymbol{x}, t) \rho(\boldsymbol{x}, t)) = 0 \tag{12.5}$$

or

$$\frac{\partial \rho(\boldsymbol{x}, t)}{\partial t} + (\boldsymbol{v}(\boldsymbol{x}, t) \nabla) \rho(\boldsymbol{x}, t) + \rho(\boldsymbol{x}, t) \mathrm{div}(\boldsymbol{v}(\boldsymbol{x}, t)) = 0. \tag{12.6}$$

The fundamental dynamic equation for the thermo-elasto-visco-plastic media follows from the law of momentum conservation (the rate of change of linear momentum)

$$\frac{\mathrm{d}}{\mathrm{d}t} \int_{\Omega(t)} \rho(\boldsymbol{x}, t) v_i(\boldsymbol{x}, t) \mathrm{d}\boldsymbol{x} = \int_{\partial\Omega(t)} \tau_{ij}(\boldsymbol{x}, t) n_j \mathrm{d}\boldsymbol{x} + \int_{\Omega(t)} \rho(\boldsymbol{x}, t) f_i(\boldsymbol{x}, t) \mathrm{d}\boldsymbol{x} + \int_{\Omega(t)} \tilde{f}_i(\boldsymbol{x}, t) \mathrm{d}\boldsymbol{x}, \tag{12.7}$$

i.e. the rate of change of linear momentum in a material region $\Omega(t)$ equals the contribution of the body forces plus the contribution of the surface forces.

Let $(0x_1x_2x_3)$ be the given Cartesian co-ordinate system in R^3, t being time the $\boldsymbol{x} = (x_1, x_2, x_3)$ a point in $G(t)$. In our considerations we shall assume that the matter of the Earth is formed by anisotropic, inhomogeneous, generally compressible thermo-elasto-visco-plastic substances of an absolute temperature T_o, and that at the moment $t = t_o$ they are in an initial strain and stress state. In view of the extremal and internal loads as well as the heating of the Earth's interior, and due to the thermo-magneto-hydrodynamic processes passing through the Earth's interior, the Earth's body becomes deformed. This generates a displacement in the Earth's body $\boldsymbol{u}(\boldsymbol{x}, t) = (u_i(\boldsymbol{x}, t))$ $i = 1, 2, 3$, a strain $e_{ij}(\boldsymbol{x}, t)$, a rate of elasto-visco-plastic strain D_{ij}^{e+p+v}, a stress τ_{ij} and its temperature will increase by $\Theta = T - T_o$.

Generally u_i, e_{ij}, D_{ij}^{e+p+v}, and τ_{ij} are functions of the co-ordinates $\boldsymbol{x} = (x_i)$, $i = 1, 2, 3$ and time as well as usually also of the temperature T and the electric and magnetic intensities $\boldsymbol{E}$, and $\boldsymbol{H}$. Moreover, $\boldsymbol{E}$ and $\boldsymbol{H}$ are also functions of co-ordinates $\boldsymbol{x}$, time t, temperature T and stress τ_{ij} and the concentration of a component of the materials forming the Earth's interior d.

Now we shall introduce the notations:
$u_{i,j} = \partial u_i/\partial x_j$, $u_{i,t} = \partial u_i/\partial t$, $u_{i,tt} = \partial^2 u_i/\partial t^2$, $\mathrm{d}f/\mathrm{d}t = \partial f/\partial t + v_j \partial f/\partial x_j = f_{,t} + v_j f_{,j}$, where the first term in the last expression is the local derivative and the second is the convective derivative; their physical meaning is evident. Further, the expression $[\boldsymbol{ab}] = \boldsymbol{a} \times \boldsymbol{b}$ means the vector product of the vectors $\boldsymbol{a}$ and $\boldsymbol{b}$. The Einstein's summation rule will be used.

As the law of momentum conservation is independent of the relation between the strain and stress, then in our geodynamic consideration Equation 12.7 can be written in the equivalent differential form

$$\tau_{ij,j} + \rho f_i + f_{Bi} + f_{vi} + f_{ci} + f_{pi} = \rho(u_{i,tt} + u_{j,t}u_{j,tj}) \quad i = 1, 2, 3, \tag{12.8}$$

where $\boldsymbol{F} = (\rho f_i)$, $i = 1, 2, 3$, represents the body forces of the type of the gravity forces, e.g. first of all the Archimedes force $\boldsymbol{F}_a = \rho D \boldsymbol{g}$, $D \ll 1$, is a function representing density inhomogeneities in the interior of the Earth, $\boldsymbol{F}_B = (f_{Bi}) = [\boldsymbol{j}, \boldsymbol{B}]$, $i = 1, 2, 3$, represents the volume density of the electromagnetic force, the Lorentz's force, $\boldsymbol{F}_\nu = \rho\nu[\nabla^2 \boldsymbol{v} + \frac{1}{3}\nabla(\nabla, \boldsymbol{v})]$ is the viscous force, ν is the coefficient of kinematic viscosity, $\boldsymbol{F}_c = 2\rho[\vec{\omega}, \boldsymbol{v}]$ is the Coriolis' force; and where the term on the right hand side represents the inertia forces. Also τ_{ij} is the stress tensor, ρ is the density, $\boldsymbol{j}$ the current density, $\boldsymbol{B}$ the magnetic induction vector, $\boldsymbol{B} = \mu \boldsymbol{H}$, $\boldsymbol{H}$ the intensity of the magnetic field, μ the permeability, $\vec{\omega}$ the angular velocity of the Earth, $\boldsymbol{v} = (u_{i,t})$, $i = 1, 2, 3$, the velocity of the moving hot material of the Earth, $\boldsymbol{g}$ the Earth's gravity acceleration vector, and $\boldsymbol{F}_p$ is the force due to precession and diffusion, etc.

The constitutional relations are defined by

$$D_{ij}^{e+p+v} = D_{ij}^e + D_{ij}^{p+v} + \cdots, \tag{12.9}$$

where

$$D_{ij}^e = \tfrac{1}{2}(u_{i,jt} + u_{j,it}) = e_{ij}^e(\dot{\boldsymbol{u}}) = (s_{ijkl}\tau_{kl})_{,t} \tag{12.10}$$

are components of the elastic part of the strain rate tensor,

$$e_{ij}^e = \tfrac{1}{2}(u_{i,j} + u_{j,i}), \quad e_{ij}^e = e_{ji}^e, \tag{12.11}$$

is the small strain tensor,

$$e_{ij} = \tfrac{1}{2}(u_{i,j} + u_{j,i}) + \tfrac{1}{2}u_{k,i}u_{k,j} \tag{12.12}$$

is the finite strain tensor, the coefficients s_{ijkl} are the so-called elastic modulus and

$$s_{ijkl} = s_{jikl} = s_{ijlk} = s_{klij}. \tag{12.13}$$

Further, D_{ij}^{v+p} represents the viscous and plastic parts of the strain rate tensor. The domain of elasticity and visco-plasticity is defined by the function of plasticity $\mathscr{F}(\tau_{ij})$, where $\tau_{ij} = \tau_{ji}$. The function $\mathscr{F}$ is a convex and continuous function in τ_{ij} and $\mathscr{F}(0) = 0$, $\mathscr{F}(\tau_{ij}) < a_o$ for $\tau_{ij} \neq 0$. In fact, $\mathscr{F}(\tau_{ij})$ includes the deviator τ_{ij}^D of the stress tensor τ_{ij} only. To define the relation between stress and strain we define the set K by

$$K = \{\tau \mid \tau = (\tau_{ij}) \in R^6, \quad \tau_{ij} = \tau_{ji}, \mathscr{F}(\tau_{ij}) \leq a_o\}, \tag{12.14}$$

where $\mathscr{F}$ is a convex continuous function of six scalar variables, defined by

$$\begin{aligned} &\mathscr{F}(\tau_{ij}) < a_o, \quad \tau_{ij} \neq 0, a_o = \text{constant} > 0 \text{ for the domain of elasticity,} \\ &\mathscr{F}(\tau_{ij}) \geq a_o, \quad \tau_{ij} \neq 0, a_o = \text{constant} > 0 \text{ for the domain of visco-plasticity.} \end{aligned} \tag{12.15}$$

Let the mapping $\tau \to P_k(\tau)$ be an orthogonal projection that maps $\tau \in R^6$ onto the convex set K. Then the visco-plastic strain rate tensor is defined by

$$D_{ij}^{v+p} = \begin{cases} 0 & \text{for } \mathscr{F}(\tau_{ij}) < a_o, \\ \frac{1}{2}\eta^{-1}[\tau_{ij} - (P_k\tau)_{ij}] & \text{for } \mathscr{F}(\tau_{ij}) \geq a_o, \end{cases} \tag{12.16}$$

where η is a coefficient of viscosity. But in the real Earth (planet) all geophysical fields, i.e. the gravity, stress-strain fields, the geothermal field and the electromagnetic field are coupled so that the geodynamic processes in the Earth must be described as a problem of thermo-elasto-visco-plasticity and thermo-magneto-hydrodynamics. Thus the stress-strain relation for such media can be expressed in the form

$$\begin{aligned} D_{ij}^{e+p+v} = {} & (s_{ijkl}\tau_{kl})_{,t} + D_{ij}^{p+v}(\boldsymbol{u}) - (\alpha_{ij}(T - T_o))_{,t} + \tfrac{1}{8}\pi^{-1}(E_iD_j \\ & + E_jD_i + H_iB_j + H_jB_i - \delta_{ij}((\boldsymbol{ED}) + (\boldsymbol{HB})))_{,t}^{D}, ^{(*)} \end{aligned} \tag{12.17}$$

where the first term on the right hand side represents an elastic strain rate tensor, D_{ij}^{v+p} is a visco-plastic strain rate tensor, $(\alpha_{ij}(T - T_o))_{,t}$ represents a strain rate tensor evoked by a heat process and the last term represents the strain rate tensor evoked by the Maxwell's stress tensor.

The thermal energy propagates in the Earth's interior by convection, conduction, and by means of electromagnetic waves, by radiation and excitation. The first three

---

(*) To derive the rate of a Maxwell's stress tensor we have $\tau_{ij,t} = (\tilde{F}\delta_{ij} + (\partial\tilde{F}/\partial e_{ij})_{T,E} + \frac{1}{8}\pi^{-1}(E_iD_j + E_jD_i + H_iB_j + H_jB_i))_{,t}$, $D_i = \varepsilon_{ij}E_j$, $B_i = \mu_{ij}H_j$, where $\tilde{F} = F_o - \frac{1}{8}\pi^{-1}(\varepsilon_{ij}E_iE_j + \mu_{ij}H_iH_j)$ is the free energy, F_o is the free energy in the sense of the thermo-elasto-visco-plasticity, $\varepsilon_{ij} = \varepsilon_{ij}^o + a_{ijkl}e_{kl}$, $\mu_{ij} = \mu_{ij}^o + b_{ijkl}e_{kl}$ are the dielectric constant and magnetic permeability, respectively, $\boldsymbol{E}$ and $\boldsymbol{H}$ are vectors of electric and magnetic intensity, and a_{ijkl}, and b_{ijkl} are tensors of the fourth order, symmetric in i, j and k, l.

types of propagation of thermal energy can be described by the uniform type of equations. The fourth type of propagation of thermal energy is connected with modern quantum mechanics and nuclear physics. For details see Chapter 9.

Thus the thermal regime in the Earth is described by the generalized expanded equation of heat conduction. Such type of equation follows from the general conservation law of energy, which generally can be written in its integral form as follows

$$\int_{\Omega(t)} \frac{\mathrm{d}e}{\mathrm{d}t}\mathrm{d}\boldsymbol{x} = \int_{\Omega(t)} \tau_{ij}D_{ij}\mathrm{d}\boldsymbol{x} + \int_{\Omega(t)} (\rho w - q_{i,i})\mathrm{d}\boldsymbol{x}, \tag{12.18}$$

where e is a scalar, characterizing the intrinsic energy, w is an influx of energy, and $\boldsymbol{q}$ is a vector of energy transport. Then the generalized expanded equation can be found in the following form

$$\begin{aligned}\rho\beta_{ij}T_o(D_{ij}^{e+p+v} &+ u_{k,t}e_{ij,k}^{e+p+v} + \rho c_e(T_{,t} + u_{k,t}T_{,k})\\ &= W - (\kappa_{ij}T_{,j})_{,i} - \sigma_{ij}^{-1}j_ij_j + Tj_i(\overline{\alpha}_{ij})_{,i},\end{aligned} \tag{12.19}$$

where the first term on the left hand side represents the energy dissipated in the form of heat in the moving masses of the Earth and generated by their deformation, whilst the second term on the left hand side and the first two terms on the right hand side represent the well-known classical expanded equation of heat conduction and describe the heat conduction. Further $W = \Sigma_i W_i$ is the dissipation function, representing heat sources as the radioactive sources W_1; sources W_2 evoked by the viscous stresses $\hat{\sigma}_{ij}$ and defined by

$$W_2 \sim \tfrac{1}{2}\eta_{ijkl}v_{i,j}\ v_{k,l} \tag{12.20}$$

for anisotropic materials, where η_{ijkl} is a tensor of fourth order, called the coefficient of viscosity, and

$$W_2 \sim \hat{\sigma}_{ij}v_{i,j} = [\eta(v_{i,j} + v_{j,i}2\tfrac{1}{3}v_{k,k}\delta_{ij}) + \xi v_{k,k}\delta_{ij}]v_{i,j} \tag{12.21}$$

for isotropic materials, where η, ξ are the coefficients of viscosity, and δ_{ij} is the Kronecker's delta; as well as sources W_3 and W_4 evoked by the viscous and Coulombian frictions, etc.. The other terms on the right hand side represent the following: the third term represents the Joule's heat, the fourth term the specific thermoelastic effects, discussed later. Further β_{ij} is the coefficient of thermal expansion, T_o the input temperature, characterizing the initial strain and stress state of the Earth matter, e_{ij} is the finite strain tensor, D_{ij}^{e+p+v} the elasto-visco-plastic strain rate tensor, c_e the specific heat under constant strain, T the temperature, and $\sigma = (\sigma_{ij})$ and $\kappa = (\kappa_{ij})$ the electrical and the thermal conductivities. Generally, the thermal conductivity $\kappa = (\kappa_{ij})$ consists of three parts with different origins, i.e.

$$\kappa = \kappa_{ph} + \kappa_{rad} + \kappa_{exc},$$

where κ_{ph} is the phonon conductivity, κ_{rad} the radioactive conductivity and κ_{exc} the excited conductivity. Note that κ_{ph} decreases with the temperature as $T^{-1o}K$, κ_{rad} is expected to increase with temperature but not faster than $(T^oK)^3$ and κ_{exc} becomes effective only at temperatures higher than $\sim 1000^\circ$C.

Besides the thermal sources discussed there also arise further more or less meaningful sources. One part of the Earth's rotation energy is converted into thermal energy at the tidal friction (in our model problem regarded as body sources). Ljubimova (1967) estimates that the heat quantity of this origin created from the beginning of the Earth is approximately 20–30% of the thermal energy, disengaged by radioactive decay. A quantity of thermal energy, disengaged by absorption of seismic waves, energy released at chemical reactions and recrystallizational processes, amounts to merely a fraction of the thermal energy disengaged at radioactive decay. When meteors and meteoric dust fall onto the Earth surface their kinetic energy is changed into heat, one part of it is radiated away whilst the second is absorbed by the growing planet. Benfield and Safronov estimate the value of the Earth's temperature to be about 1000–1500°C during its development (see for example Benfield (1950)). But the heat released due to gravitational differentiation and deformation processes of the Earth's masses, which are of tectonic origin and act during the whole terrestial development, contributes to a greater part of the Earth's whole geothermal balance. With the enlargement of the Earth's body its potential energy increases. Still increasing upper parts of the Earth effect the lower and deform it. Thus a part of the potential energy was disengaged and is transformed into energy of deformation as well as thermal energy. For this Ljubimova (1967) mentiones a value about 10^{39} ergs and for gravitational differentiation a value of about 1.5×10^{38} ergs. New analyses of the thermal history of the Earth are made in Shape and Peltier (1979), (see also Peltier (1989)).

In a deforming thermo-visco-plastic Earth's body, processes of thermo-diffusion are also observed. Let c be the function characterizing the concentration of a component of the materials forming the Earth's interior. Since the distribution of a component of the Earth's matter changes, during the geological period the diffusive processes in the Earth's interior are one of the sources of dissipative energy.

Due to Landau and Lifshitz (1984) the thermo-diffusion process can be described by the equation of diffusion

$$\rho(c_{,t} + \boldsymbol{v}\,\text{grad}\ c) + \text{div}\ \boldsymbol{i} = h, \tag{12.22}$$

where $\boldsymbol{i}$ is the density of the diffusion flow and

$$\boldsymbol{i} = \rho D_c(\text{grad}\ c + K_c \text{grad}\ T + P_c \text{grad}\ p), \tag{12.23}$$

where ρ is the density, D_c the coefficient of diffusion, $K_c = K_c(T)$ is the coefficient of thermo-diffusion, p the pressure and $P_c = P_c(p)$ the coefficient of baro-diffusion.

The behaviour of the magnetic field in the Earth is described by the Maxwell's equations

$$\text{rot}\ \boldsymbol{H} = \boldsymbol{j} + \boldsymbol{D}_{,t}, \tag{12.24}$$

$$\text{rot}\ \boldsymbol{E} = -\boldsymbol{H}_{,t}, \tag{12.25}$$

$$\text{div}\ \boldsymbol{D} = \rho_e, \tag{12.26}$$

$$\text{div}\ \boldsymbol{B} = 0, \tag{12.27}$$

where $\boldsymbol{B} = \mu\boldsymbol{H}$, $\boldsymbol{j} = \sigma(\boldsymbol{E} + [\boldsymbol{v}, \boldsymbol{B}])$, and $\boldsymbol{D} = \varepsilon(\boldsymbol{E} + [\boldsymbol{v}, \boldsymbol{B}])$. The dielectric tensor ε_{ij} and magnetic permeability tensor μ_{ij} are defined by

$$\varepsilon_{ij} = \varepsilon^o_{ij} + a^o_{ijkl} e_{kl} + \varepsilon^1_{ij}(T, p), \quad \mu_{ij} = \mu^o_{ij} + b^o_{ijkl} e_{kl} + \mu^o_{ij}(T, p), \tag{12.28}$$

and ε^o_{ij} and μ^o_{ij} are the dielectric constant and magnetic permeability of the undeformed material at normal temperature, a^o_{ijkl} and b^o_{ijkl} are symmetric tensors of fourth order in the indices i, j and k, l, and $\varepsilon^1_{ij}(T, p)$ and $\mu^1_{ij}(T, p)$ are the parts of the dieletric constant and magnetic permeability which depend on the temperature and the pressure, etc..

The electric conductivity tensor σ_{ij} is defined by

$$\sigma_{ij} = \sigma^o_{ij} + d^o_{ijkl} H_k H_l + a_{ij} + \sigma^1_{ij}(T, p), \tag{12.29}$$

where σ^o_{ij} is the electric conductivity tensor of the non-deformed material at normal temperature and without exposure to a magnetic field, d^o_{ijkl} is a symmetric tensor of the fourth order in the indices i, j and k, l, a_{ij} is an anti-symmetric tensor of the second order and $\sigma^1_{ij}(T, p)$ represents that part of the electric conductivity which depends on the temperature and pressure.

The materials of the Earth's interior are created by conductors as well as by dielectrics, in a rigid state as well as in visco-plastic and liquid states. Moreover, the materials of the interior of the Earth move with velocity $\boldsymbol{v} = \boldsymbol{u}_{,t}$ so that the arising electrical field is described by the intensity vector of the electric field $\boldsymbol{E}_v = \mu[\boldsymbol{v}, \boldsymbol{H}] = [\boldsymbol{v}, \boldsymbol{B}]$, the total intensity vector of the electric field $\boldsymbol{E}$ by

$$\boldsymbol{E} = \boldsymbol{E}_o + \mu[\boldsymbol{v}, \boldsymbol{H}], \tag{12.30}$$

and the current density vector $\boldsymbol{j}$ by

$$\boldsymbol{j} = \sigma(\boldsymbol{E}_o + \mu[\boldsymbol{v}, \boldsymbol{H}]). \tag{12.31}$$

Generally the Earth and its separate parts such as the core, the mantle, etc. are characterized by a conductivity. Then the Earth as a conductor occurs in an outer magnetic field. Thus (for an isotropic conductor) there exists a current perpendicular to the electric field and with a magnitude, which is proportional to the intensity of the magnetic field. Conversely there exists an electric field which is perpendicular to the current $\boldsymbol{j}$ whose magnitude is proportional to the intensity of the magnetic field and the current $\boldsymbol{j}$. Then for an isotropic conductor

$$\boldsymbol{E} = \sigma_o \boldsymbol{j} + a_1[\boldsymbol{j}, \boldsymbol{H}] + a_2 H^2 \boldsymbol{j} + a_3 \boldsymbol{H}(\boldsymbol{H}\boldsymbol{j}), \tag{12.32}$$

where σ_o is the conductivity of materials with the absence of a magnetic field, a_1 is a real (positive or negative) constant, and a_2 and a_3 are constants. For general terrestrial materials with anisotropic conductivity, such current does not need to be unique and perpendicular to $\boldsymbol{E}$. It can be shown that Joule's heat due to this current is equal to

$$\boldsymbol{j}\boldsymbol{E} = \sigma_{ij} E_i E_j = (s_{ij} + c_{ij}) E_i E_j = s_{ij} E_i E_j, \tag{12.33}$$

where s_{ij} is a symmetric and c_{ij} an anti-symmetric part of the conductivity tensor σ_{ij}. Moreover, $c_{ij} E_j = [\boldsymbol{E} c]_i$, as every anti-symmetric tensor of second order is equivalent to some axial vector, and as vector $[\boldsymbol{E} c]$ is perpendicular to $\boldsymbol{E}$, then $c_{ij} E_i E_j = 0$. We see that the influence of the outer magnetic field can also be one of the heat sources in the Earth.

From geophysical measurements it is evident that the Earth's body is non-uniformly heated, and likewise the core. The core as well as the mantle can be investigated as a heated conductor in thermodynamic unbalance. Then the intensity of the electric field $\boldsymbol{E}$ is different from zero in the absence of a current $\boldsymbol{j}$. Thus

$$E_i = \sigma_{ij}^{-1} j_j + \overline{\alpha}_{ij} T_{,j}, \tag{12.34}$$

where σ_{ij} is the conductivity tensor and the tensor $\overline{\alpha}_{ij}$ characterizes electric properties of a conductor, and $\overline{\alpha}_{ij}$ is a symmetric, but generally non-symmetric, tensor. From Equation 12.34 we obtain

$$j_i = \sigma_{ij}(E_j - \overline{\alpha}_{ij} T_{,j}) \tag{12.35}$$

from which we can obtain that in a non-uniformly heated Earth's core and mantle an electric current can also flow if the intensity of the electric field is equal to zero. All coefficients are also functions of the magnetic field (i.e. of $\boldsymbol{H}$) (*). In addition to these effects the effect of the influence of a magnetic field onto the thermo-electro-motive force can be considered. For an isotropic conductor it is proportional to $\hat{\beta}[\boldsymbol{H}, \text{grad } T]$, where $\hat{\beta}$ is a function of temperature T.

From the theory of thermodynamics it is known that a quantity of heat, selected at a unit of time in a unit volume of a conductor, is equal to $-\text{div } \boldsymbol{q}$, where $\boldsymbol{q}$ is the density of energy flux. Then from Landau and Lifshitz (1984)

$$Q = \text{div}(\kappa \text{ grad } T) + \sigma^{-1} j^2 - T\boldsymbol{j} \,\text{grad } \overline{\alpha}, \tag{12.36}$$

where the first term is connected with heat conductivity, the second one represents the Joule's heat and the third thermo-electrical effects. Due to Landau and Lifshitz (1984) the Thompson's and Peltier's effects can play a role in the Earth's interior. The Thompson's effect $\sim -T\overline{\alpha}_{,T}\boldsymbol{j}$ grad T is proportional to the first power of the current $\boldsymbol{j}$ and the gradient of temperature. The Peltier's effect arises if the current penetrates through a contact boundary of two different conductors. This effect is also proportional to the first power of the current $\boldsymbol{j}$ and is characterized by the term $-T(\overline{\alpha}_k - \overline{\alpha}_l) j_n$, where the indices k, l represent the neighbouring conductors.

Besides the thermo-electric effects, the thermo-galvano-magnetic effects, i.e. the effects originating from flowing currents at simultaneous existence of electric and magnetic fields and a gradient of temperature, also arise in the interior of the Earth. In this case

$$j_i = a_{ij} E_j T^{-1} + b_{ij}(T^{-1})_{,j}, \quad q_i - \varphi j_i = c_{ij} E_j T^{-1} + d_{ij}(T^{-1})_{,j},$$

where all coefficients are a function of the magnetic field and $a_{ij}(\boldsymbol{H}) = a_{ji}(-\boldsymbol{H})$, $d_{ij}(\boldsymbol{H}) = d_{ji}(-\boldsymbol{H})$, and $b_{ij}(\boldsymbol{H}) = c_{ji}(-\boldsymbol{H})$. Then from there it follows that

$$E_i = \sigma_{ij}^{-1} j_j + \overline{\alpha} T_{,j}, \tag{12.37}$$

$$q_i - \varphi j_i = \beta_{ij} j_j - \kappa_{ij} T_{,j}.^{\dagger} \tag{12.38}$$

(*) $\sigma_{ij}^{-1}(\boldsymbol{H}) = \sigma_{ij}^{-1}(-\boldsymbol{H})$, where σ_{ij} is a symmetric, but generally non-symmetric, tensor.

† where $\sigma_{ij}(\boldsymbol{H}) = \sigma_{ji}^{-1}(-\boldsymbol{H})$, $\kappa_{ij}(\boldsymbol{H}) = \kappa_{ji}(-\boldsymbol{H})$, and $\overline{\beta}_{ij}(\boldsymbol{H}) = T\alpha_{ji}(-\boldsymbol{H})$.

If we approximate the Earth's interior by isotropic conductive layers then

$$E_i = \sigma_{ij}^{-1} j_j + \overline{\alpha}_{ij} T_{,j} + R[\boldsymbol{H}, \boldsymbol{j}]_i + N[\boldsymbol{H}, \text{grad } T]_i, \tag{12.39}$$

$$q_i - \varphi j_i = \overline{\alpha}_{ij} T j_j - \kappa_{ij} T_{,j} + NT[\boldsymbol{H}, \boldsymbol{j}]_i + L[\boldsymbol{H}, \text{grad } T]_i, \tag{12.40}$$

where the term $N[\boldsymbol{H}, \text{ grad } T]$ represents the influence of the magnetic field onto the thermo-electro-motive force, and the term $L[\boldsymbol{H}, \text{grad } T]$ characterizes the influence of the magnetic field on the thermo-conductivity. Further, the term $NT[\boldsymbol{H}, \boldsymbol{j}]$ designates the influence of the magnetic field onto the Peltier's effect (i.e. the effect originating from the current penetrating through the contact boundary between two different conductors) as on the boundary between two conductors besides the vectors $\boldsymbol{j}$ and $\boldsymbol{q}$ also a vector $-\kappa$ grad $T + \alpha T \boldsymbol{j} + NT[\boldsymbol{H}, \boldsymbol{j}] + L[\boldsymbol{H}, \text{grad } T]$ is continuous. At the end the quantity of heat is given by $-\text{div } \boldsymbol{q}$ from which we obtain

$$Q = -\text{div } \boldsymbol{q} = \sigma_{ij}^{-1} j_i j_j + (\kappa_{ij} T_{,j})_{,i} - T j_i \overline{\alpha}_{,i} + T^{-1} \sigma_{ij}^{-1} (\sigma_{ij} N T^2)_{,T} [\boldsymbol{j}, \boldsymbol{H}]_i T_{,i}, \tag{12.41}$$

where the last term represents a variation of the Thompson's effect (i.e. a thermal effect which is proportional to the first order of the current $\boldsymbol{j}$ and the gradient of temperature $\sim c\boldsymbol{j}$ grad T) due to the existence of a magnetic field.

Finally, there is also the equation of state in the form

$$p = f(\rho, T, \boldsymbol{E}, \boldsymbol{H}, \ldots.). \tag{12.42}$$

As the effect of gravity cannot be negligible then $\boldsymbol{g} \neq 0$ and thus it is hardly a restriction to assume a gravitational potential Φ and

$$\Delta\Phi(\boldsymbol{x}, t, T, p, \boldsymbol{E}, \boldsymbol{H}, c, \ldots) = -4\pi G_c K^{-1} p\rho + 2\overline{\omega}^2, \quad \boldsymbol{g} = \text{grad } \Phi, \tag{12.43}$$

where Φ is the gravitational potential, G_c is the gravitational constant, ρ the density, p the pressure, K the bulk modulus and $\vec{\omega}$ the angular velocity of the Earth.

12.2.3. Discussion of Equations, Limits of their Validity and their Physical Consequences

12.2.3.1. The case of geodynamic processes in the core

The Earth's core is comprised of two large parts: the outer core — from about 2900 km to 5154 km, with the viscosities η within 10^{-2} to 10^9 poise and densitities of about 9.8 – 10.3 gcm^{-3} $\leq \rho \leq$ 12.1 – 12.5 gcm^{-3} and with great thermal and electrical conductivities; and the inner core — from about 5154 km to 6371 km, with the viscosities $\eta \sim 10^{10}$ to 10^{14} poise, the densities range within 13-14 gcm^{-3} and also with great thermal and electrical conductivities. The inner core is created by the thermo-visco-plastic matter in metallized and ionized states, while the outer core is composed of thermo-visco-strongly plastic up to thermo-visco-liquid molten materials also in metallized and ionized states. Hence the geodynamic processes can be described by the system (A) of Equations 12.4-12.8, 12.18, 12.19, 12.22 and 12.24-12.27 with the constitutional and defined relations of Equation 12.9-12.17,

12.20, 12.21 and 12.28–12.31. Moreover the materials of the Earth's core move with a velocity $\boldsymbol{v}$ (for the outer core $\sim$ 20 km yr^{-1}). Since all geodynamic processes in the Earth's core have a close relation to the thermal processes as well as to the thermo-magneto-hydrodynamic processes, Equations 12.4–12.8, 12.18, 12.19, 12.22, and 12.24–12.27, which are non-linear, are fully coupled. Moreover, all physical quantities such as ρ, β_{ij}, κ_{ij}, σ_{ij}, etc. are also functions of the temperature, pressure, electric and magnetic intensities, the spatial co-ordinates and time.

The outer core is characterized by thermo-visco-liquid up to thermo-strongly-visco-plastic matter. In this case either $\tau_{ij,t}$, i.e. the stress components τ_{ij} are constants, or the set $K = \{\tau \mid \tau = (\tau_{ij})$ is the spherical tensor$\}$, then $(P_K\tau)_{ij} = -p\delta_{ij}$ represents hydrostatic pressure. Therefore,

$$\tau_{ij} = -p\delta_{ij} + 2\mu e_{ij}(\boldsymbol{u}) - \beta_{ij}(T - T_o) + \tfrac{1}{8}\pi^{-1}(E_iD_j + E_jD_i + H_iB_j + H_jB_i - \delta_{ij}((\boldsymbol{ED}) + (\boldsymbol{HB}))). \tag{12.44}$$

We see that this system of equations (we denote it as system (B)), describing the thermo-visco-liquid materials of the outer core, is the limiting case of the general system of equations (A), defined by Equation 12.16 and 12.17. (In this case Equation 12.8 represents a generalized Navier-Stokes' equation; if we neglect the coupled terms then we obtain the classical Navier-Stokes' equation, which is in the Boussinesq's approximation used in the present geodynamics.) It can be said that the system of equations (A) generally describes the geodynamic processes in the whole core. Since all geodynamic processes in the outer core also have a close relation with the thermal and the thermo-magneto-hydrodynamic processes, the system of equations (B) is also fully coupled.

The phase transition layer and the inner core are characterized by thermo-rigid-visco-plastic materials. For such materials the coefficients s_{ijkl} in Equation 12.17 are equal to zero, i.e. $s_{ijkl} = 0$. On the other hand the term $\tau_{ij}D_{ij}$ in a conservation law of energy can be written in the form

$$\tau_{ij}D^e_{ij} = 2\hat{g}(D^e_{\mathrm{II}})^{1/2} + 4\hat{\mu}D^e_{\mathrm{II}}, \quad D_{\mathrm{II}} = \tfrac{1}{2}D^e_{ij}D^e_{ij}, \tag{12.45}$$

where D^e_{II} is an invariant of the tensor $D^e = (D^e_{ij})$, $\hat{g}$ and $\hat{\mu}$ are a yield point or a threshold of plasticity and of viscosity of a Bingham's fluid, respectively. Then for simplicity, due to the assumption that the materials of the inner core and of the phase transition zone are in a strongly metallized state and, moreover, are approximately incompressible, we assume that $D_{kk} = 0$. Then the stress tensor components τ_{ij} for our model problem can be written in the form

$$\tau_{ij} = -p\delta_{ij} + \hat{g}D^e_{ij}(D^e_{\mathrm{II}})^{1/2} + 2\hat{\mu}D^e_{ij} - \beta_{ij}(T - T_o) + \tfrac{1}{8}\pi^{-1}(E_iD_j + E_jD_i + H_iB_j + H_jB_i - \delta_{ij}((\boldsymbol{ED}) + (\boldsymbol{HB}))), \quad D^e_{\mathrm{II}} \neq 0, \tag{12.46}$$

where p is the spherical part of the stress tensor τ_{ij} and has the meaning of a pressure. For $\hat{g}$ strictly positive, which is the case for the inner core and the phase transition zone, there exist zones where the matter (liquid or strongly visco-plastic) behaves as a rigid body (i.e. the phase transition zone), and for $\hat{g}$ much more positive the materials behave as rigid in the whole region, i.e. the case of the inner core. It is evident that if the value of $\hat{g}$ is near zero then the materials behave like thermo-strongly(near liquid)-visco-plastic materials and this type of constitutive law be can also used for the description of the outer core.

From the theory of magneto-hydrodynamics it is known that in the outer core thermal convection can result if the Rayleigh's number $R = \overline{\alpha} g \overline{\beta} l^4 \kappa^{-1} \nu^{-1}$, (where $\overline{\alpha}$ is the coefficient of thermal expansion, g the gravity acceleration, $\overline{\beta}$ represents the thermal gradient, l is the thickness of the core, κ the thermal conductivity and $\nu = \overline{\eta}\rho^{-1}$ the kinematic viscosity, $\overline{\eta}$ is the dynamic viscosity, and ρ the density) is greater than the critical value $27/4\pi^4$. Moreover, beside the thermal convection, due to gravitational energy (about which we often spoke as a primary source of energy for convection in the Earth), there is also a chemical-density convection in the outer core. For this it follows that $R_d = \delta\rho g l^3 D^{-3}$, where D is the diffusion coefficient, and R_d the Rayleigh's number. The influence of the rotation of the Earth onto the convection in the outer core is substantial as the Taylor's number $T_n = 4\omega^2 \mathrm{d}^4 \nu^{-2}$ is much greater then 1. Here ω is the angular velocity of the Earth (in rad^{-1}), d the thickness of the outer core, ν its kinematic viscosity, $\nu = \overline{\eta}\rho^{-1}$, $\overline{\eta}$ is the dynamic viscosity (in poise), and ρ the density. All these estimations represent Rayleigh's, Jeffrey's, Knopoff's, Thompson's and Chandrasekhar's results based on the Newtonian liquid. As the Newtonian liquid is a special case of the assumed thermo-elasto-visco-plastic medium, discussed above, then all these estimations can be derived directly from the model equations discussed.

For large-scale irregularities, the magnetic Reynold's number, defined by $R_m = l v D_m^{-1}$, is very large ($\sim 10^{-1}$), where l is the characteristic horizontal scale of the irregularity, v is the velocity of the fluid in the rotating co-ordinate system, and D_m the coefficient of magnetic diffusion. Anufriev and Braginsky (1975) discussed the flow over small bumps, considered by Hide (1970) and Hide and Malin (1971). Their size is large enough so that $R_m \gg 1$ holds true, but, on the other hand, it is small enough to render the magnetic force in the core's interior larger than Coriolis' force. The magnetic field $\boldsymbol{B}$, as they showed, vanishes at the boundary and increases with distance from it. Therefore, the magnetic force near the boundary is less than Coriolis' force, and it increases with distance from the boundary. The papers of Moffatt and Dillon (1976), Hide (1977), Moffatt (1977), Parker (1979), Hide (1979), Moffatt (1978), Cupal (1983), Soward (1983), Cupal and Hejda (1989), Hejda (1982), Hejda and Anufriev (1993), Mauersberger (1956), Nedoma (1993) and many others are devoted to this problem and the gobal geomagnetic field of the Earth.

From the magneto-hydrodynamics (Landau and Lifshitz (1984)) it is known that the turbulent movement of the conducting liquid reduces to a spontaneous origin comparatively great magnetic field. The analysis tends to the theory of the turbulent dynamo discussed in Roberts and Stix (1971), Roberts and Soward (1972),(1975) and in many other papers. The turbulent dynamo models of the Earth are characterized by very drastic simplifications, in particular with respect to the electro-motive force due to the fluctuating motions. In spite of this, in all cases extensive numerical calculations had to be carried out.

The westward drift as well as the possible mechanisms of generating secular variations are based on the existence (presumptions) of Alfvén waves. It is known that in an electrically conducting medium under the presence of the magnetic field, transverse oscillations propagating along the magnetic field may exist; these are the so-called Alfvén waves. Hide (1966) interprets the westward drift in terms of magneto-hydrodynamic waves propagating along the field lines of the toroidal field in the Earth's core. The deviations of the magnetic field from axial symmetry can be interpreted as magneto-hydrodynamic waves (Alfvén waves), the velocity of which, having the value c in a non-rotating conductor, is reduced to about $(1/2)c^2\omega^{-1}r_s^{-1}$ by the Coriolis' forces. If it is assumed that the Lorentz's forces and Coriolis' forces are approximately equal, then the velocity c is approximately equal to the observed value of the westward drift. Due to the fact that the Alfvén's waves are described by a relatively simple system of equations without speculations about further influences, for example the asymmetry of the geoid, the asymmetry of the magnetic processes in the core, the electromagnetic damping, temperature, gravity, etc., then we may deduce that the nature will be more complicated and thus better approximated by the presented global geodynamic model, comprehending all the discussed problems as a special case.

Further it must be commented that the westward drift was interpreted also in another way, e.g. as the subject of the complications of magnetic (Lorentz's), Archimedean and Coriolis' forces (MAC forces), and as the subject of the terrestrial conductions, etc. (see for example. Braginsky (1967) and many other papers).

Models of α^2-type, models of $\alpha\omega$-type, models of Z-type for planetary dynamos and models of the nearly symmetric dynamo were studied by Steenbeck and Krause (1966), Tough (1967), Tough and Roberts (1968), Roberts and Stix (1972), Braginsky (1975), Rochester *et al.* (1975), Hide (1979),(1982), Hejda (1982), Cupal (1983), Braginsky and Roberts (1987), Cupal and Hejda (1989), Hejda and Anufriev (1993), Anufriev, Cupal and Hejda (1993), etc.. For the citation see for example Bucha (1983).

The thermal regime in the core is characterized by greater temperature and larger pressures, so that one can assume that the materials of the core are in a metallic state under a pressure about 1.4 Mbar and temperature of about 5000°C. The laboratory experiments of recent time have proved this theory that for example the compression of silicates under a constant pressure of 0.9 Mbar leads to their metallization. Special attention will be paid to the problem of boundary inhomogeneities in the core due to metallization and recrystallization of the core materials. All these problems will be studied in detail later.

12.2.3.2. The case of geodynamic processes in the mantle

The thermo-visco-plasticity up to thermo-visco-liquidity (i.e. the thermo-Newtonian viscosity) is assumed for the mantle materials. The Rayleigh's number R is given by $R = \overline{\alpha} g \overline{\beta} l^4 \kappa^{-1} \nu^{-1}$, where $\overline{\alpha}(\sim 2.10^{-5}/°\mathrm{C})$ is the coefficient of thermal expansion, g $(\sim 10^3\ \mathrm{cms}^{-2})$ the acceleration of gravity, $\beta(\sim 10^{-6}°\mathrm{Ccm}^{-1} = 0.1°\ \mathrm{Ckm}^{-1})$ the vertical temperature gradient at the onset of convection, $l(\sim 3.10^8\ \mathrm{cm})$ the thickness of the layer, κ the thermal conductivity and ν $(\sim 10^{20}$ to $10^{22}\ \mathrm{cm}^2 s^{-1})$ the kinematic viscosity (see Knopoff (1964), Cathles (1975), McKenzie (1983), Ranalli (1987), Peltier (1989) and the discussion and citation in Chapter 9).

This value is far in excess of the value of the critical Rayleigh number ($R \sim 10^3$) for thermal convection in a (Newtonian) liquid mantle. This general situation is unchanged when the effects of sphericity and rotational motion of the Earth are taken into account. When l is taken as the thickness of the upper mantle, the value of R becomes comparable with the critical value. In the actual Earth, physical properties such as heat flow and thermal conductivity are expected to have a strong layered structure and the heating must be internal and also from below. Chemical segregation and reactions of many kinds will accompany the convective motions (see Chandrasekhar (1961), Knopoff (1964), Peltier (1989), etc.). For the chemical-gravity convection the Rayleigh's number $R_d = \delta\rho g l^3 D^{-4}$, where D is the diffusion coefficient, is in the interval $\langle 10^{17}, 10^{34}\rangle$ which is much greater than the critical value. The influence of the rotation of the Earth for the Earth's mantle is insubstantial as the Taylor's number is given by $T_a = 4\omega^2 l^4 \nu^{-2} \sim 10^{-16}$ ($\omega = 7.27 \times 10^{-5}$ rad/s, $l = 2.9 \times 10^8$ cm, $\nu = 10^{21}$ cm^{2}s^{-1}).

It can be remarked that two different types of mantle convection models are discussed in the world literature (McKenzie (1983), Peltier (1989) and the citations in this book). One is the flow caused by vertical temperature differences, the second is the flow driven by horizontal temperature differences.

In the present model of the modified mantle the convection current hypothesis is supported. The lower mantle is characterized by the powerful thermo-visco-plastic media up to thermo-rigid-visco-plastic liquid (so that due to the above discussion the simple models are described by the thermo-Newtonian-viscous liquid), which are heated from beneath and cooled from above. Resultant effects of this process are the ensuing convective currents — the hexagonal convective cells — on the Earth's surface, characterized by the rift zones with their typical bifurcation to three branches containing angles about 120° between themselves, and the channel of higher temperature and lower viscosity (GTLV channels, in the literaure known also as plumes) in the lower mantle (see Figure 12.1). The light hot viscous material from the core/mantle boundary, which was accumulated here in large amounts, represents a very non-stable thermo-visco-plastic (liquid) layer. Light materials display a tendency to create a stable situation by escaping through GTLV channels into the upper parts of the Earth. By passing through the GTLV channel, the light materials are heated up to 1000-1300°C and rise to the lower boundary of the lithosphere. The lithosphere is then heated, melted and strained mechanically in the vertical direction, and after these materials effuse below the lithosphere they also strain it horizontally. The light hot viscous materials penetrate below the lithosphere, where they create the layer of lower viscosity — the asthenosphere — characterized by the channel of lower seismic velocities and lower viscosities, in the depth between 70-120 km for P waves and a maximum 250-350 km for S waves. The channel from 70 km to 120 km is more plastic and more viscous' whilst from 120 km to 250-350 km it is less plastic and less viscous than in 70 km to 120 km. But nevertheless it is still considerably plastic and viscous, so that for P waves a downward tendency of plasticity and viscosity displays an increase of velocities of P waves, while for S waves the influence of this tendency of plasticity and viscosity is not yet displayed. It is also probable that in the asthenosphere, in the domains near the rift zones, "turbulent" movements can occur.

The geodynamic processes in the asthenosphere can be described by thermo-elasto-visco-plastic materials or a Bingham's rigid-visco-plastic liquid, as in the deeper parts the matter can be regarded as rigid.

The zone between 350 km and 850 km-984 km is referred to as the transitional zone, characterized by a layered structure with a relatively high viscosity. It is probable that the differentiation and the phase transition of the matter of the asthenosphere, and also of the materials of lithospheric plates, which have penetrated it in the subduction zones, proceeded and continue to proceed gradually. We speak about this as of the mesosphere. In this zone the convection currents do not occur, which can be deduced from the fact that the oldest subducted lithosphere is observed near the present subducted lithosphere. In the opposite case, if we assume convection currents under the lithosphere, then the broken-up lithosphere must be shifted away from the subducting lithosphere and in these places we would observe the foci of deep earthquakes. But that is not observed in practice.

Matter of the upper and lower mantle can generally be described by the non-linear compressible and incompressible thermo-ela-sto-visco-plastic rheology (see, for example, Equations 12.14-12.17 and 12.46), which in a special case leads to a thermo-rigid-visco-plastic rheology, the so-called thermo-Bingham's rheology (i.e. Equation 12.44) and its limiting case, the thermo-Newtonian liquid (for which $\tau_{ij,j} = 0$ in Equation 12.16 and 12.17, or $\hat{g} = 0$ in Equation 12.46), which occur in the magnetic field. Let us analyse the geodynamic processes and their description in more detail.

As was said above, the upper and lower mantle can be characterized in general by a thermo-elasto-visco-plastic material. The mantle is characterized by a layered structure. The behaviour of the separate layers of the mantle is different in the separate layers, as in the asthenosphere, the mesosphere or in the layers of the lower mantle, and is the result and consequence of convective currents in the mantle (Ritsema (1972), Turcotte and Schubert (1982), Ranalli (1987), Dziewouski, Hager and O'Connell (1977), Hide (1985), Le Pichon, Fraucheteau and Bouniz (1973) and the previous chapters of this book).

According to Prager (1961) the dissipative term $\tau_{ij}D_{ij}$ can be written as (putting $\tau^{e+v+p} \equiv \tau$)

$$\tau_{ij}D_{ij} = \mathscr{D}_1(\boldsymbol{D}) + \mathscr{D}_2(\boldsymbol{D}), \tag{12.47}$$

where $\boldsymbol{D} = (D_{ij})$ is the strain rate tensor with components $D_{ij} = (1/2)(v_{i,j} + v_{j,i})$, $v_i = u_{i,t}$, $\mathscr{D}_1 = \mathscr{D}_{1,D_{ij}}D_{ij} \sim 2\hat{g}(D_{\mathrm{II}})^{1/2} = 2\hat{g}((1/2)D_{ij}D_{ij})^{1/2}$, $D_{\mathrm{II}} = (1/2)D_{ij}D_{ij}$ the invariant of the tensor $\boldsymbol{D}$, $\hat{g}$ is the threshold of plasticity (the yield point, the yield stress) and $\mathscr{D}_2 = (1/2)\mathscr{D}_{2,D_{ij}}D_{ij} \sim 4\hat{\mu}D_{\mathrm{II}} = 2\hat{\mu}D_{ij}D_{ij}$, where $\hat{\mu}$ is the threshold of viscosity. For simplicity if we assume an incompressible material then

$$\begin{aligned}\tau_{ij} &= -p\delta_{ij} + \mathscr{D}_{1,D_{ij}} + \tfrac{1}{2}\mathscr{D}_{2,D_{ij}} - \beta_{ij}(T - T_o) + \tfrac{1}{8}\pi^{-1}(E_iD_j + E_jD_i + H_iB_j \\ &\quad + (\boldsymbol{HB}))\delta_{ij} + H_jB_i - ((\boldsymbol{ED}) \\ &= -p\delta_{ij} + \hat{g}D_{ij}D_{\mathrm{II}}^{-1/2} + 2\hat{\mu}D_{ij} - \beta_{ij}(T - T_o) + \tfrac{1}{8}\pi^{-1}(E_iD_j + E_jD_i \\ &\quad + H_iB_j + H_jB_i - ((\boldsymbol{ED}) + (\boldsymbol{HB}))\delta_{ij},\end{aligned} \tag{12.48}$$

where $D_{\mathrm{II}} \neq 0$ (as for $D_{\mathrm{II}} = 0$ the stress tensor τ_{ij} is undefined). Here p is the symmetric part of the stress tensor τ_{ij} and is equivalent to the pressure. For the compressible materials of the mantle the density ρ is a function of the pressure p, the temperature T, the intensity of magnetic field $\boldsymbol{H}$, spatial co-ordinates $\boldsymbol{x}$ and time t, etc.. For slightly compressible materials the relation between density, temperature and pressure is given by

$$\rho = \rho_o[1 + c_o(T - T_o) + c_1(p - p_o)],$$

where c_o, and c_1 are the thermal and compressibility coefficients, scalar and small compared to unity, in the equation for conservation of mass, and, moreover, where $\rho\boldsymbol{v}$ differs little from $\rho_o\boldsymbol{v}$ (see Chapter 3).

For the invariant of the stress tensor we have

$$\tau_{\mathrm{II}}^{1/2} = (\tfrac{1}{2}\tau_{ij}^D\tau_{ij}^D)^{1/2} = \hat{g} + 2\hat{\mu}(D_{\mathrm{II}})^{1/2},$$

where τ_{ij}^D are components of the deviator of the stress tensor τ. From there we see that $\tau_{\mathrm{II}}^{1/2} \geq \hat{g}$. Hence from Equation 12.48 we obtain

$$D_{ij} = (2\hat{\mu})^{-1}(1 - \hat{g}\tau_{\mathrm{II}}^{-1/2})\tau_{ij}^D - (\alpha_{ij}(T - T_o))_{,t} + \{\tfrac{1}{8}\pi^{-1}(E_iD_j + E_jD_i + H_iB_j + H_jB_i - ((\boldsymbol{ED}) + (\boldsymbol{HB}))\delta_{ij})\}_{,t}^D, \tag{12.49}$$

where $\{\ldots\}^D$ represents the inverse relation to $\{\ldots\}$. From there we notice that the parameters $\hat{g}$ and $\hat{\mu}$, the threshold of plasticity and viscosity, thus characterize the structure of the mantle. If a square root of an invariant of the stress tensor $\tau_{\mathrm{II}}^{1/2}$ is less than the threshold of plasticity $\hat{g}$, then the materials of the mantle behave as thermo-visco strongly visco-plastic (up to liquid). Using such materials we can approximate the layer of low velocity of seismic waves and low viscosity, the asthenosphere and the GTLV channels (plumes) as well as all anomalous heated zones and places in the mantle (such as the subduction areas). If $\tau_{\mathrm{II}}^{1/2} \geq \hat{g}$ then the materials of the mantle behave as thermo-visco- plastic matter. If $\hat{g}$ is strictly positive then in the stream there exist zones in which the liquid-plastic material of the mantle behaves like a thermo-visco-rigid solid. If $\hat{g}$ grows then these zones enlarge and for $\hat{g}$ sufficiently large the streams are blocked and create the layered structure of the mantle in general (in special cases they create inhomogeneous areas). The convenction currents in the mantle were and are studied (e.g. Runcorn (1972)) as currents in the Newtonian liquid. Then the Navier-Stokes equation for convection in the mantle simplifies to the equation of "creeping flow" (see Runcorn (1972)). In Equation 12.48 for $\hat{g} = 0$ the materials of the mantle behave like a thermo-Newtonian viscous liquid. Such types of materials often also characterize the GTLV channel's materials, the asthenosphere substances and the downward matter of the asthenosphere and mesosphere under the collision and subduction zones (i.e. materials in the anomalous zones), and in the first approximation they characterize the materials of the whole mantle in agreement with Knopoff's results (Knopoff (1964), (1967)) and the results of specialists discussed in the beginning of this section. From the above discussion the equations of state of the mantle materials can be written as

$$\begin{aligned}
&\tau_{\mathrm{II}}^{1/2} < \hat{g} \Leftrightarrow D_{ij} = 0,\\
&\tau_{\mathrm{II}}^{1/2} \geq \hat{g} \Leftrightarrow D_{ij} = (2\hat{\mu})^{-1}(1 - \hat{g}\tau_{\mathrm{II}}^{1/2})\tau_{ij}^{D} - (\alpha_{ij}(T - T_o))_{,t}\\
&\qquad + \{\tfrac{1}{8}\pi^{-1}(E_iD_j + E_jD_i + H_iB_j + H_jB_i - ((\boldsymbol{ED}) + (\boldsymbol{HB}))\delta_{ij}\}_{,t}^{D},
\end{aligned} \tag{12.50}$$

which remain the case if $\hat{\mu} = 0$. In this case the first relation in Equation 12.50 holds, while the second does not make sense. Then the inverse relation must be rewritten. It holds that

$$\tau_{ij}^{D} = \hat{g}D_{ij}(D_{\mathrm{II}})^{-1/2},$$

where τ_{ij}^{D} is a deviator of the stress tensor. Moreover, $\tau_{\mathrm{II}}^{1/2} = \hat{g}$ and thus D_{ij} is proportional to τ_{ij}^{D}. Hence

$$\begin{aligned}
&\tau_{\mathrm{II}} < \hat{g} \Leftrightarrow D_{ij} = 0,\\
&\tau_{\mathrm{II}} < \hat{g} \Leftrightarrow \text{ there exists } \lambda \geq 0 \text{ such that}\\
&\quad D_{ij} = \lambda\tau_{ij}^{D} - (\alpha_{ij}(T - T_o))_{,t} + \{\tfrac{1}{8}\pi^{-1}(E_iD_j + E_jD_i\\
&\qquad + H_iB_j + H_jB_i - ((\boldsymbol{ED}) + (\boldsymbol{HB}))\delta_{ij})\}_{,t}^{D}.
\end{aligned} \tag{12.51}$$

Such materials are called ideally thermo-rigid-plastic or thermo-Mises' materials and are also used in geological practice.

The structure of the mantle materials is represented by a rotating unevenly heated non-homogeneous conductor exposed to an electro-magnetic field. Then in the conductive materials of the mantle thermo-electric and thermo-magnetic effects arise. As above there also appear in the mantle a current perpendicular to the electric field and with a magnitude, which is proportional to the intensity of the magnetic field, and an electric field which is perpendicular to the current $\boldsymbol{j}$ and with a magnitude proportional to the magnetic field and the current $\boldsymbol{j}$. Since the mantle is composed of anisotropic conductive materials such currents need not be unique and perpendicular to $\boldsymbol{E}$. A detailed analysis similar to that in the case of the core discussed above shows that the outer magnetic field contribute to the heat in the mantle.

If the Rayleigh's number is higher than its critical value, then convective currents occur and these may be characterized by a turbulent motion of the mantle masses. Such a state is designated by changes in the temperature, in the density and in the velocities in the horizontal and vertical planes, and depends strongly on the rheological properties of the mantle materials. As the Reynold's number of the mantle convection is very small ($\sim 10^{-20}$) ordinary turbulence is not expected, but a much higher than critical Rayleigh number is reasonably associated with a much higher harmonic pattern than would be expected from the case of marginal stability (Runcorn (1972), Peltier (1989)). Such reasons are correct if we assume that the mantle is a Newtonian liquid. But the nature is much more complicated since the mantle is created by thermo-elasto-visco-plastic materials. Runcorn (1967) first derived the relation between parameters of the gravity field and the stress in the upper mantle under the assumption that the mantle is a viscous Newtonian liquid. Further studies deduce that convection currents reach the lithosphere, and that the field of convection currents under the lithosphere is divided into many cells and connected with strongly inhomogeneous spreading of horizontal stress components and of radial stress components in the asthenospheric materials. Due to their conception, oceanic ridges are "spontaneous fountains" of an asthenospheric magma displaced onto the divergent boundary from the asthenosphere. They admit also a possibility of increased tectonic convection due to changes in the number of all-mantle convective cells. Thus their conception consists of the fact that the analysis of the gravity satellite data is related to the bottom boundary of the lithosphere. Namely since these gravity data are due to the effects of all inhomogeneities in the Earth's interior, then their results describe a projection of all such effects onto the bottom lithospheric boundary. On the other hand if we accept the author's idea discussed above and presented firstly in Nedoma (1984), then all geodynamic processes can be cleared up from the mechanical point of view in connection with the manifestation of all geophysical fields. For example, the "spontaneous fountains" of asthenospheric magma and an observed system of small convective cells related to one greater lithospheric plate are a mixture of projecting effects of deeper Earth's inhomogeneities and of the effects of turbulence on the bottom lithospheric boundary with gravity effects of the uppermost inhomogeities. Furthermore, in the mesosphere the emerged and broken parts of the subducted oceanic lithosphere are observed in the close neighbourhood of the subduction zones. In the case when the convective currents reach the lower boundary of the lithosphere, then these parts of broken and emerged subducted oceanic lithosphere must be pulled to greater distances of the subduction zones. Moreover, such parts of the emerged oceanic lithosphere would be further strained and broken. Then an earthquake located in such places of the mesosphere would originate. But such phenomena are not observed and thus the assumption of the convective currents under the lithosphere is shown as unlikely.

12.2.3.3. Thermal regime in the core and in the mantle of the earth and their structure

The geophysical observations prove the layered structure of the Earth, created of the hot materials under high pressure. Then the hot materials of the Earth's interior under various pressures may have more phases in which some material properties of rocks are markedly different, e.g. the continuum may have

(i) either a strongly plastic up to liquid or solid phase, i.e. the case of the mantle core boundary or of the outer core/inner core boundary, the case of the GTLV channels (plumes) in the mantle or of the magmatic intrusion in the lithosphere;

(ii) more solid phases, i.e. the case of the layered structure in the mantle or the fundamental layers in the lithosphere;

(iii) more strongly plastic up to liquid phases, i.e the case of the layered structure in the outer core or the case of cloud liquid due to turbulent transport in the outer core;

(iv) liquid-solid "mushy" zones (layered regions) of a solidifying and melting Earth's interior material.

Most of the interest in the solution of the problem lies in the properties of the boundary or boundary surface between the phases. In such cases latent heat is released or absorbed at the phase temperature and we then distinguish the case of melting or solidification. The appropriate discontinuities and the governing diffusion equation can be derived from an integral formulation of the heat conservation. This principle is a simplified version of the principle of energy conservation in which both phases are assumed to be incompressible and at rest. Generally for compressible and unrest phases the principle of conversation of energy must be used. Across any phase transition boundary the conservation of energy, momentum and mass must be fulfilled, and the latter then requires that the material density is continuous. This phase change is assumed to take place at some specified temperature. Such problems are frequently referred to as Stefan-like problems (the historic paper by Joseph Stefan (1889) gave the initial impulse to the study of this family of problems). The usual treatment of a Stefan-like problem assumes the existence of some moving surface such that the material is in its solid phase on one side of the surface and its liquid phase on the other side or that the material is in its first solid phase (liquid phase) on one side of the surface and in its second solid phase (liquid phase) on the other side. Similar properties are satisfied in the case of recrystallization. Solid and liquid phases are delimited by a melting (solidification, recrystallization) front. The shape of the front is dendritic for quick melting (solidification); dendrites result from the unstable growth of the front. The geometry at the front is relatively small. However we must state that the mushy zone is not a third phase. Microscopically each phase (solid and liquid) is clearly delimited by the front of melting (solidification) which has null measure. The mushy zones refer to a region where both phases are microscopically present. Hence phases will be associated to microscopic thermo-dynamical states, and zones to macroscopical averaged variables. Since our geodynamic studies are investigations of global processes in the Earth's interior, then investigations of mushy zones can be omitted (these problems are substantial in mineralogy and petrology). Nevertheless, the phase diagrams, giving the dependence between temperature and concentration and/or temperature and pressure and/or concentration and pressure, are very important for better understanding of the behaviour of rocks under high temperatures and pressures. Moreover, they make possible the formulation of model problems of the geodynamic processes and thermal behaviour of rocks under extremal conditions (e.g. we can also formulate these problems as variational inequalities). Energy balance arguments then show that although the temperature is continuous everywhere, the heat flux has a jump discontinuity across the phase change surface Γ_R at time t, i.e.

$$T^k = T^l = T_R, \tag{12.52}$$

$$\kappa_{ij}\frac{\partial T}{\partial x_j}n_i\bigg|_k - \kappa_{ij}\frac{\partial T}{\partial x_j}n_i\bigg|_l = \rho L \partial_t v_n, \tag{12.53}$$

where Γ_R is a front of the wave of melting or recrystallization or solidification — a moving phase change boundary, T_R is the melting (recrystallization, solidification) temperature, indices k and l characterize domains of both phases at time t, v_n is the normal component of the motion of the phase change boundary, and

$$\kappa_{ij}^{k} = \lim_{T \to T_R^-} \kappa_{ij}(T), \quad \kappa_{ij}^{l} = \lim_{T \to T_R^+} \kappa_{ij}(T) \tag{12.54}$$

are the thermal conductivities in both sides of the phase change boundary. The thermal regime in this region is also described by Equation 12.19.

The Stefan problem is only a simplification of the real situation. Solving the problem discussed above we find only a front of the recrystallizational (melting, solidificational) wave, a head of the moving phase change boundary. But the real situation is much more complicated. The phase changes in the Earth's interior are accompanied by the transition layers characterized also by jumps in velocities of seismic waves.

The aim of our study will be to give an idea of how to obtain estimations of the occurrence and position of phase transitions in the Earth's interior from a global point of view, i.e. at different temperatures and pressures at the various depths. According to these different situations in the Earth's interior the phase change was studied from a different point of view for special local regions. Hence Schubert and Turcotte (1971), Turcotte (1975), Hodkins (1975), Turcotte and Schubert (1982) show that in thermodynamic equilibrium the pressure p as the phase change occurs is related to the temperature by

$$\frac{\mathrm{d}p_{kl}}{\mathrm{d}T} = \frac{L_{kl}}{T(\rho_k^{-1} - \rho_l^{-1})}, \tag{12.55}$$

where L_{kl} is the heat evolved per unit mass when the light phase k of density ρ_k transforms into a heavy phase l of density ρ_l, and that the phase change instability due to convection is characterized by three dimensionless parameters

$$R_\beta = \alpha\beta g d^4 \kappa^{-1} \nu^{-1}, \; R_Q = \alpha L_k g d^3 c_p^{-1} \kappa^{-1} \nu^{-1}, \quad S = \Delta\rho(\rho\alpha d)^{-1} \left(\frac{\rho g}{\mathrm{d}p_{kl}/\mathrm{d}T} - \beta \right)^{-1}. \tag{12.56}$$

The parameter R_β is the usual Rayleigh number based on the coefficient of thermal expansion α. The parameter S is a measure of the importance of the phase change instability to Rayleigh instability. The parameter R_Q is a measure of the stabilizing influence of the latent heat associated with the phase change. These show that for small values of R_Q the release of latent heat is unimportant and the stability of the fluid layer is determined by R_β and S. Increasing values of S decrease the critical Rayleigh number for the onset of thermal convection. For large values of R_Q the influence of the latent heat in stabilizing the flow dominates the influence of S. For a given value of S there exists a value of R_Q which inhibits the phase change instability.

For slow flows the problem is isothermal, essentially identical to the case with no heat conduction at large distances from the phase change boundary. For upward flow the phase change is endothermic and the temperature falls as the phase change occurs. For fast flow through the phase boundary the process is adiabatic (i.e. no transport of heat by conduction).

Turcotte and Schubert (1971) show that the transition from the discontinuity between the phases to a two-phase region occurs at a critical velocity

$$v_{\text{crit}} \sim \kappa_k g \left(\frac{\mathrm{d}p_{kl}}{\mathrm{d}T} T_k (S_k - S_l) \right)^{-1}, \tag{12.57}$$

where κ_k is the thermal conductivity of phase k, and $S_k - S_l$ is the entropy change associated with the phase change. At flow velocities less than the critical velocity a phase discontinuity exists, whilst for flow velocities greater than the critical value a two-phase region exists. For the olivine-spinel change the critical value of velocity is $v_{\text{crit}} \sim 0.2 \text{ cmyr}^{-1}$ at about 8 cmyr^{-1} of mantle convection.

Later Christensen and Yuen (1984) analysed a finite element model of time-dependent convection to determine the conditions for penetration of the subducted plate into the lower mantle. In this paper and in Christensen (1984) the authors show that the 650 km discontinuity is taken as either a chemical or phase boundary or as a combination of both. They also show that non-Newtonian convection might be essential for facilitating flows in the Earth's lower mantle, for making the upper thermal boundary layer behave like a quasi-rigid but subductable plate, or for stabilizing a hot low-viscosity boundary layer.

The survey of new results in this field of problems can be found in Peltier (1989). In this book our investigations will be based on the modern analysis of Stefan-like problems. In this chapter the author's new results of melting, solidification and of recrystallization will be investigated and discussed.

12.3. GEODYNAMIC PROCESSES IN THE MANTLE

12.3.1. Introduction

The investigation of a driving mechanism of plate motions is one of the most fundamental problems that have been studied since the beginning of the sixties in connection with the origin of global plate tectonics. Over time scales of several ten thousands or millions of years and longer, the rocks of the Earth's mantle behave like a thermo-visco-plastic medium, while over time scales of several seconds they behave like a thermo-elastic or brittle medium. The recent geophysical literature shows that over time scales of millions of years and longer the mantle's rocks behave like a fluid. As the Earth is heated from below and cooled from above, mass transport by thermal convection is realized. But the key role in the mantle convection is played by density heterogeneities which generate body forces driving the convection. The recent studies of mass heterogeneities in the Earth's mantle, based on the studies of seismic velocity anomalies, show that these heterogeneities induce the flow in the mantle. Nevertheless, the velocities of P and S waves are influenced by effects such as anisotropy, partial melting, recrystallization, solidification and chemical heterogeneities, etc.. These effects can in some cases be more important than the temperature and pressure variations. Mantle convection is generally accepted as the engine of plate tectonics in that the rigid plates on the surface of the Earth, whose relative motions create mountains arcs, volcanoes and earthquakes, are assumed to be an integral part of the convective system and the surface expression of mantle convection. Thus the major plate tectonic results such as the subduction zones (i.e. regions of convergence overlaying colder downwelling currents), the oceanic trenches, mid-oceanic ridges (i.e. the spreading regions delineating upwelling, known also as rift zones), and the hotspots (i.e. the regions with anomalously high volcanism and heat flow indicating upwelling) are thought to clear up the nature of geodynamic processes in the mantle. Although the idea that plate motions are the results of mantle convection has been accepted since the sixties and seventies, a realistic model of geodynamic processes in the mantle has not yet been given. Since the transport of mantle rocks is associated with heat transfer, the convection is described by the laws of mass, momentum and energy conservation, by the equation of state as well and by the second thermodynamic law. This then leads to a coupled system of evolution partial differential equations. To solve them it is necessary to prescribe the appropriate boundary and initial conditions. In addition to this information the rheology of the mantle is also necessary. Unfortunately, for contemporary numerical calculations a number of simplifications above all in the rheology, in the determination of the behaviour of mantle rocks under various conditions (namely temperatures and pressures, and also intensity of electromagnetic fields, etc.), must be made. The problem remains, however, of which type of rheology should be used for approximating the properties of the mantle rocks. Although the power-law creep can be dominant in the mantle, a linear Newtonian rheology is often used for approximation of the whole mantle. This leads to solving the Stokes problem. Using a non-linear Newtonian rheology for the mantle then the Navier-Stokes (steady-state or evolution) problem in the classical or in the Boussinesq's approximations will be solved. Later we shall show that both approximations are special cases of the Bingham's visco-plastic rheology. Therefore, the main results of this part of the book will be based on the Bingham's visco-plastic rheology.

The philosophy of convection modelling in the mantle has no unique principle and idea. Different models of the mantle convection incorporating different physical processes may each satisfy the observed reality. The basic philosophy of the modelling approach is to isolate the various processes in idealized models and assess their relative importance. More complex models are developed to study the interactions of the individual effects. One may say that each new modelling study adds to our cumulative understanding of the physics of convection in the mantle. Therefore, the main goal of this chapter is to give new ideas of how to solve the above problems and the methods for their solution. Since the author does not believe that the present description of mantle convection is realistic he seeks new methods which better approximate the above problems.

12.3.2. Mathematical Models Based on the Linear and Nonlinear Newtonian Fluid Rheology

12.3.2.1. Introduction

By a thermal convection is meant the motions induced by thermal buoyancy in a layer of strongly visco-plastic fluid which is heated from below and cooled from above.

In the mathematical formulation of problems of convection, it is useful to consider the energy balance which characterizes the instability of a static layer heated from below and cooled from above. The potential energy can be estimated from above by $\rho_o \alpha \Delta T \boldsymbol{g v}$, where $\boldsymbol{g}$ is the gravity acceleration, ρ is the density defined as $\rho = \rho_o(1 - \alpha(T - T_o))$, Δt is the temperature difference across the layer, α is the coefficient of thermal expansion and $\boldsymbol{v}$ represents velocities. The characteristic velocity of isotherms returning to their equilibrium position in the absence of fluid motion is given by kd^{-1}, where k is the thermometric conductivity and d is the thickness of the layer. For small velocities $\boldsymbol{v}$, the fraction vdk^{-1} of the density difference across the layer enters a realistic expression for the potential energy release per unit time,

$$L_p = c_o \rho_o \alpha \Delta T g v^2 \mathrm{d} k^{-1},$$

where c_o is a constant depending on the motion and boundary conditions. The condition for convection occurring is given by $L_p \geq L_D$, i.e. the convection occurs if the potential energy L_p exceeds the energy L_D dissipated per unit time by viscous friction. The viscous friction in the Navier-Stokes equations can be estimated by $\rho_o \nu \boldsymbol{v} d^{-2}$, where $\nu = \mu \rho^{-1}$is the kinematic viscosity, and μ the dynamic viscosity. Then

$$L_D = C_o \rho_o \nu v^2 d^{-2},$$

where C_o is a constant depending as above on the motion and boundary conditions. The condition of convection $L_p \geq L_D$ yields

$$R \equiv \alpha \Delta T g d^3 (k\nu)^{-1} \geq C_o c_o^{-1} (\equiv R_c), \tag{12.58}$$

where R is the so-called Rayleigh number. The lowest value of $c_o^{-1} C_o$ for given physical conditions of the layer is the critical Rayleigh number R_c, and represents the fundamental parameters in the theory of convection. The ratio $\nu k^{-1} \equiv P$ is called the Prandlt number.

Then the governing equations describing motions in a mantle approximated by a chemically homogeneous fluid follow from the principle of conservation of mass, momentum and energy (see Chapter 3) as

$$\partial_t \rho + \partial_j (\rho v_j) = 0, \tag{12.59}$$

$$\partial_t \rho v_i + \partial_j \rho v_j v_i = \partial_j \tau_{ij} + \rho f_i, \tag{12.60}$$

$$\partial_t \rho e + \partial_j \rho v_j e = \tau_{ij} \partial_j v_i + \rho Q - \partial_j q_j \equiv \tau_{ij} D_{ij}(\boldsymbol{v}) + \rho Q - \partial_j q_j, \tag{12.61}$$

where $\boldsymbol{v} = (v_i)$ is the velocity vector, $D_{ij} \equiv \frac{1}{2}(\partial_j v_i + \partial_i v_j)$ is the strain rate tensor, τ_{ij} is the stress tensor, e is the internal energy, $\boldsymbol{q} = (q_i)$ is the heat flux vector, Q is the rate of internal heating per unit volume and $\boldsymbol{f} = (f_i)$ represents the body forces per unit mass. The symbol $\partial_j f \equiv f_{,j} \equiv \partial f / \partial x_j$ denotes the differentiation with respect to the co-ordinate x_j of a Cartesian co-ordinate system. The stress tensor τ_{ij} is defined by

$$\tau_{ij} = -p\delta_{ij} + \mu(\partial_i v_j + \partial_j v_i - 2\tfrac{1}{3}\partial_k v_k \delta_{ij}) = -p\delta_{ij} + \mu(\partial_i v_j + \partial_j v_i) + \lambda \partial_k v_k \delta_{ij}, \tag{12.62}$$

where p is the pressure, μ is the dynamic viscosity, and $\lambda = -2(1/3)\mu$. The dissipation function $\mathscr{D}$ is defined as

$$\mathscr{D}(\boldsymbol{v}) = \rho^{-1}\mu(\partial_i v_j + \partial_j v_i - 2\tfrac{1}{3}\partial_k v_k \delta_{ij})\partial_j v_i \tag{12.63}$$

and

$$p = p_e - \underline{\mu}\partial_k v_k, \tag{12.64}$$

where p_e is the thermodynamic equilibrium pressure and $\underline{\mu}$ is the bulk viscosity. In addition to Equations 12.59-12.61, the state equation

$$\rho = \rho(p_e, T) \tag{12.65}$$

must also be given. In general Equation 12.65 can be written as

$$\rho = \rho_o(p_s, T_s)[1 - \alpha(T - T_s) + K_T^{-1}(p - p_s)], \tag{12.66}$$

where p_s, and T_s are the so-called isentropic hydrostatic reference state, and K_T denotes the isothermal modulus of bulk compresibility. If in Equation 12.59 $\rho = \rho_o =$ constant then

$$\partial_j v_j \equiv \text{div } \boldsymbol{v} = 0, \tag{12.67}$$

which represents the condition for non-compressible mantle rocks, while Equation 12.59 describes the compressible mantle rocks. As was shown in Chapter 3, the principle of conservation of mass in its linearized case can be written in the form

$$\partial_t \rho(\boldsymbol{x}, t) + \rho(\boldsymbol{x}, t)\partial_j v_j(\boldsymbol{x}, t) = 0,$$

which for $\boldsymbol{x} =$ constant then gives

$$\mathrm{d}\rho/\rho + \partial_j v_j \mathrm{d}t = 0$$

and after integration over $\langle t_o, t\rangle$ we have

$$\ln(\rho(\boldsymbol{x}, t)\rho^{-1}(\boldsymbol{x}, t_o)) + \int_{t_o}^{t} \partial_j v_j(\boldsymbol{x}, s)\mathrm{d}s = 0.$$

Since $u_j(\boldsymbol{x}, t)$ (where $\boldsymbol{u}$ is the displacement vector) is small, then we can assume that $\partial_j v_j = \partial^2 u_j/\partial t \partial x_j \equiv \partial_{tj} u_j$ is also small, so that

$$\rho(\boldsymbol{x}, t) = \rho_o(\boldsymbol{x})\left(1 - \int_{t_o}^{t} \partial_j v_j(\boldsymbol{x}, s)\mathrm{d}s\right), \tag{12.68}$$

where we put $\rho_o(\boldsymbol{x}) \equiv \rho_o(\boldsymbol{x}, t_o)$.

From Equations 12.60 and 12.62 we obtain the well-known Navier-Stokes equation

$$\rho(\partial_t v_i + \partial_j v_j v_i) = -\partial_j p + \partial_j[\mu(\partial_i v_j + \partial_j v_i - 2\tfrac{1}{3}\partial_k v_k \delta_{ij})] + \rho f_i \tag{12.69}$$

or in vector form

$$\rho \mathrm{d}\boldsymbol{v}/\mathrm{d}t = -\text{grad } p + \mu\Delta\boldsymbol{v} + \rho\boldsymbol{f}, \tag{12.70}$$

where $\mathrm{d}v_i/\mathrm{d}t \equiv \partial_t v_i + v_j \partial_j v_i \equiv v_{i,t} + v_j v_{i,j}$.

Since the thickness of the Earth's mantle is certainly much less than the density scale height, then the variation of density with depth has only a minor influence on convection in the Earth's mantle. Hence, the Boussinesq approximation can be used for theoretical investigations. In general the density depends on temperature and pressure (see, for example, Equation 12.66). The Boussinesq approximation follows from the following assumptions:

(i) density is independent of pressure and depends only weakly on temperature;

(ii) the thermodynamic parameters such as thermal conductivity $\mathscr{K}$, coefficient of thermal expansion α, specific heat at constant pressure c_p, gravitational acceleration $\boldsymbol{g}$, and dynamic viscosity μ are independent of temperature and pressure, so that they are constants.

The condition of negligible dependence of the density on $p - p_s$ is of the form

$$\rho g \mathrm{d} K_T^{-1} \ll 1. \tag{12.71}$$

This condition shows that the change of density throughout the layer of thickness d owing to pressure variation is small. Moreover, a weak dependence of density upon temperature requires that the maximum value of $\alpha \Delta T$ is small compared to one, i.e.

$$\alpha(\max \Delta T) \ll 1. \tag{12.72}$$

This represents the condition that the density is nearly constant. Furthermore, the Boussinesq approximation requires that all material properties other than density are constant.

Under these conditions the principle of conservation of mass, of momentum and of energy can be given as follows:

$$\rho_o(\partial_t \boldsymbol{v} + \boldsymbol{v}\,\mathrm{grad}\ \boldsymbol{v}) = -\mathrm{grad}\ p + \rho \boldsymbol{f}(T) + \mu \Delta \boldsymbol{v}, \tag{12.73}$$

$$\mathrm{div}\ \boldsymbol{v} = 0, \tag{12.74}$$

$$\rho_o c_p(\partial_t T + \boldsymbol{v}\,\mathrm{grad}\ T) = \partial_i(\kappa \partial_j T) + Q. \tag{12.75}$$

These equations are frequently expressed in terms of dimensionless variables by scaling length d, time $t = d/v$, mass $\rho_o d^3$ and temperature $\max(\Delta T)$ by values which are characteristic of the studied problem, where v is the characteristic velocity, and $\max(\Delta T)$ was defined above. Assuming a balance between buoyancy and viscous forces then $v = g\alpha\nu^{-1}\max(\Delta T)d^2$, where $\nu = \mu\rho^{-1}$ is the kinematic viscosity. Thus all the variables in Equations 12.73-12.75 can be expressed in terms of these characteristic values and dimensionless variables in the form

$$\boldsymbol{x} = \mathrm{d}\boldsymbol{x}', \quad \boldsymbol{x} = (x_i), \quad t = (tv^{-1})t', \quad \boldsymbol{v} = v\boldsymbol{v}', \quad \rho = \rho_o\rho' = \rho_o(1 - \alpha\max(\Delta T)T'),$$

$p = \rho_o g \mathrm{d} p'$, $T = \max(\Delta T)T' + T_o$. Substituting these into Equations 12.73-12.75 we obtain

$$RP^{-1}(\partial_{t'}\boldsymbol{v}' + \boldsymbol{v}'\mathrm{grad}\ \boldsymbol{v}') = (\alpha\max(\Delta T))^{-1}\mathrm{grad}\ p' - \rho'(\alpha\max(\Delta T))^{-1}\boldsymbol{f}(T') + \Delta\boldsymbol{v}', \tag{12.76}$$

$$\mathrm{div}\ \boldsymbol{v}' = 0, \tag{12.77}$$

$$(\partial_{t'}T' + \boldsymbol{v}'\ \mathrm{grad}\ T') = R^{-1}\Delta T' + Q', \tag{12.78}$$

where $R = \alpha(\max(\Delta T))gd^3K^{-1}\nu^{-1}$ is the Rayleigh number, $k = \kappa\rho_o^{-1}c_p^{-1}$ is the thermal diffusivity, $P = \nu k^{-1}$ is the Prandtl number, $Q' = Qd(\rho_o c_p \max(\Delta T)\nu)^{-1} = Q\nu(\rho_o c_p g\alpha \max(\Delta T)^2 d)^{-1}$ represents the internal heating rate, and R^{-1} in Equation 12.78 has a meaning of a dimensionless thermal diffusivity. It is evident that the Rayleigh number strongly depends upon d, the thickness of the mantle, and upon the thermodynamic parameters. From this point of view there are discussions about convection in the mantle. There are two hypotheses: the first assumes that the convection is confined to the layer between the bottom of the lithosphere and the lower mantle (~670 km) and the second convection is confined to the lower mantle from the depth ~670 km to ~2903 km, i.e. to the core/mantle interface; the second assumes convection in the whole mantle, i.e. from the bottom of the lithosphere to depth ~2903 km. The discussion about the plausibility of the mantle convection is open.

Physical parameters for both models are modified from Peltier (1974), Stacey (1977), McKenzie *et al.*(1974), Dziewonski, Hales and Lapwood (1975), Sclater, Jaupart and Galson (1980), Jarvis and Peltier (1982), Wu and Peltier (1983), Peltier (1989).

Upper mantle model	*Whole mantle model*
$R = 1.38\ 10^6$	$R = 6.04\ 10^7$
$d = 570$ km	$d = 2803$ km
$g = 9.8\ \mathrm{ms}^{-2}$	$g = 9.8\ \mathrm{ms}^{-2}$
$\alpha = 2.0\ 10^{-5}\ \mathrm{K}^{-1}$	$\alpha = 1.4\ 10^{-5}\ \mathrm{K}^{-1}$
$k = 1.5\ 10^{-6}\ \mathrm{m^2s^{-1}}$	$k = 2.5\ 10^{-6}\ \mathrm{m^2s^{-1}}$
$\nu = 2.7\ 10^{17}\ \mathrm{m^2s^{-1}}$	$\nu = 4.3\ 10^{17}\ \mathrm{m^2s^{-1}}$
$\rho_o = 3700\ \mathrm{kgm^{-3}}$	$\rho_o = 4700\ \mathrm{kgm^{-3}}$
$\mu = 1.0\ 10^{21}\ \mathrm{kgm^{-1}s^{-1}}$	$\mu = 2.0\ 10^{21}\ \mathrm{kgm^{-1}s^{-1}}$
$= 10^{22}$ poise	$= 2.0\ 10^{22}$ poise
$c_p = 1.26\ 10^3\ \mathrm{Jkg^{-1}}K^{-1}$	$c_p = 1.26\ 10^3\ \mathrm{Jkg^{-1}}K^{-1}$
$E = 99\ \mathrm{mWm^{-2}}$	$E = 99\ \mathrm{mWm^{-2}}$
$\kappa = 6.66\ \mathrm{Wm^{-1}K^{-1}}$	$\kappa = 14.8\ \mathrm{Wm^{-1}K^{-1}}$.

The author presents his global geodynamic model of the Earth as well as of the mantle in Nedoma (1984),(1986).

12.3.2.2. Models of mantle convection based on numerical analyses of the Navier-Stokes problem in the Boussinesq approximation

A. LINEAR CASE

Let Ω be a bounded domain in R^N, $N = 2, 3$, with Lipschitzian boundary $\partial\Omega$ and let $I = (t_o, t_1)$, $\Omega_T = \Omega \times I$. Then the model problem of mantle convection leads to our solving the following quasi-coupled problem.

Problem ($\mathscr{P}_L$): Find a triple of scalar and vector functions $(T, \boldsymbol{v}, p)$, $T : \Omega \times I \to R^N$, $\boldsymbol{v} : \Omega \times I \to R^N$, $p : \Omega \times I \to R$ such that

$$\rho(\boldsymbol{x})c_e(\boldsymbol{x})\partial_t T(\boldsymbol{x}, t) = Q(\boldsymbol{x}, t) - \partial_i(\kappa_{ij}(\boldsymbol{x}, t)\partial_j T(\boldsymbol{x}, t)) \quad \text{in } \Omega \times I, \qquad (12.79)$$

$$\rho(\boldsymbol{x})\partial_t \boldsymbol{v}(\boldsymbol{x}, t) - \nu(\boldsymbol{x})\Delta\boldsymbol{v}(\boldsymbol{x}, t) + \text{grad } p(\boldsymbol{x}, t) = \boldsymbol{f}(\boldsymbol{x}, t, T(\boldsymbol{x}, t)) \quad \text{in } \Omega \times I, \tag{12.80}$$

$$\text{div } \boldsymbol{v}(\boldsymbol{x}, t) = 0 \quad \text{in } \Omega \times I, \tag{12.81}$$

$$\left.\begin{aligned} T(\boldsymbol{x}, t) &= T_1(\boldsymbol{x}, t), \quad (12.82)\\ \boldsymbol{v}(\boldsymbol{x}, t) &= \boldsymbol{v}_1(\boldsymbol{x}, t) \quad (12.83)\end{aligned}\right\} \text{ on } \partial\Omega \times I,$$

$$\left.\begin{aligned} T(\boldsymbol{x}, t_o) &= T_o(\boldsymbol{x}), \quad (12.84)\\ \boldsymbol{v}(\boldsymbol{x}, t_o) &= \boldsymbol{v}_o(\boldsymbol{x}) \quad (12.85)\end{aligned}\right\} \text{ in } \Omega,$$

where T is the temperature, $\boldsymbol{v}$ is the velocity and p is the pressure, and where the functions $Q(\boldsymbol{x}, t)$, $\rho(\boldsymbol{x})$, $c_e(\boldsymbol{x})$, $\nu(\boldsymbol{x})$, $T_1(\boldsymbol{x}, t)$, $\boldsymbol{v}_1(\boldsymbol{x}, t)$, $T_o(\boldsymbol{x})$, and $\boldsymbol{v}_o(\boldsymbol{x})$ are given functions and $\boldsymbol{f}(\boldsymbol{x}, t, T) = f_o(\boldsymbol{x}, t) - \partial_j(\beta_{ij}(\boldsymbol{x})(T(\boldsymbol{x}, t) - T_o(\boldsymbol{x})))$, where $\boldsymbol{f}_o(\boldsymbol{x}, t)$, $\beta_{ij}(\boldsymbol{x})$ are given functions.

Since Problem ($\mathscr{P}_L$) is linear and non-homogeneous, then using the transformation $T \sim T + z$, $\boldsymbol{v} \sim \boldsymbol{v} + \boldsymbol{w}$, where z and $\boldsymbol{w}$ are sufficiently smooth scalar and vector, respectively, functions satisfying Equations 12.82 and 12.83, then we have to solve the problem ($\mathscr{P}$) with the homogeneous boundary conditions of Equations 12.82 and 12.83 and where $Q(\boldsymbol{x}, t)$, $\boldsymbol{f}(\boldsymbol{x}, t)$, $T_o(\boldsymbol{x})$ and $\boldsymbol{v}_o(\boldsymbol{x})$ are changed in the usual way.

Remark: If the term $\rho\partial_t\boldsymbol{v}$ is neglected, i.e. if the inertia forces are equal to zero, then the steady-state problem will be investigated.

Next for simplicity we shall assume the quasi-coupled steady-state model problem of incompressible Newtonian fluid in the Boussinesq approximation only. Then we will solve the following problem.

Problem ($\mathscr{P}_S$): Find a triple of functions $(T, \boldsymbol{v}, p)$ such that

$$\partial_i(\kappa_{ij}(\boldsymbol{x})\partial_j T(\boldsymbol{x})) - Q(\boldsymbol{x}) = 0 \quad \text{in } \Omega \times I, \tag{12.86}$$

$$-\nu(\boldsymbol{x})\Delta\boldsymbol{v}(\boldsymbol{x}) + \text{grad } p(\boldsymbol{x}) = \boldsymbol{f}(\boldsymbol{x}, T(\boldsymbol{x})) \quad \text{in } \Omega \times I, \tag{12.87}$$

$$\text{div } \boldsymbol{v} = 0, \quad \text{in } \Omega \times I, \tag{12.88}$$

$$T(\boldsymbol{x}) = 0, \quad \boldsymbol{v}(\boldsymbol{x}) = 0 \quad \text{on } \partial\Omega \times I, \tag{12.89}$$

where $\nu > 0$ is a given constant, $\boldsymbol{f} : \Omega \to R^N$ is a given function depending on temperature T, and $\boldsymbol{v} : \Omega \to R^N$, and $p : \Omega \to R$ are unknown functions — the velocity and pressure.

Since the thermal part of the problem is the well known BVP for elliptic partial differential equations (see Chapter 9, Section 9.3), we shall analyse the second part of the problem only.

Let us assume that $\boldsymbol{f} \in [L^2(\Omega)]^N$ and let us introduce the spaces

$$\mathscr{V}_d = \{\boldsymbol{w} \mid w_i \in \mathscr{D}(\Omega), \quad \text{div } \boldsymbol{w} = 0\}, \quad V = [H_o^1(\Omega)]^N,$$

where $\mathscr{D}(\Omega) = C_o^\infty(\Omega)$ denotes the space of infinitely differentiable functions having compact support (i.e. vanishing outside a bounded open set $\Omega' \subset \Omega$, which has a positive distance from the boundary $\partial\Omega$ of Ω). Then multiplying Equation 12.87 by $\boldsymbol{w} \in [\mathscr{D}(\Omega)]^N$, and using Green's lemma, we obtain

$$\nu(\text{grad } \boldsymbol{v}, \text{ grad } \boldsymbol{w}) - (p, \text{div } \boldsymbol{w}) = (\boldsymbol{f}, \boldsymbol{w}) \quad \forall \boldsymbol{w} \in [\mathscr{D}(\Omega)]^N,$$

and therefore

$$\nu(\text{grad } \boldsymbol{v}, \text{grad } \boldsymbol{w}) = (\boldsymbol{f}, \boldsymbol{w}) \quad \forall \boldsymbol{w} \in \mathscr{V}_d.$$

Let us define the space

$$V_d = \{\boldsymbol{w} \mid \boldsymbol{w} \in V, \quad \text{div } \boldsymbol{w} = 0\}, \quad V_d \subset V,$$

$$W = L_o^2(\Omega) = \left\{q \mid q \in L^2(\Omega), \quad \int_\Omega q \text{d}\boldsymbol{x} = 0\right\}.$$

By virtue of the Poincaré inequality the space of admissible velocities V_d is a Hilbert space with norm $\|\boldsymbol{w}\| = \|\text{grad } \boldsymbol{w}\|_o$. The space W, the space of admissible pressure, is also a Hilbert space.

Let

$$a(\boldsymbol{v}, \boldsymbol{w}) = \nu(\text{grad } \boldsymbol{v}, \text{grad } \boldsymbol{w}) \quad \forall \boldsymbol{v}, \boldsymbol{w} \in V$$

be the bilinear form on $V_d \times V_d$. Then we have the following variational (weak) problem.

Problem $(\mathscr{P}_S)_V$: Find $\boldsymbol{v} \in V_d$ such that

$$a(\boldsymbol{v}, \boldsymbol{w}) = (\boldsymbol{f}, \boldsymbol{w}) \equiv F(\boldsymbol{w}) \quad \forall \boldsymbol{w} \in V_d, \tag{12.90}$$

where F is a linear operator and $F(\boldsymbol{w}) = \int_\Omega \boldsymbol{f}\boldsymbol{w} \, \text{d}\boldsymbol{x}$.

Theorem 12.1: The problem in Equation 12.90 has a unique solution.

Proof: The bilinear form $a(\boldsymbol{v}, \boldsymbol{w})$ is coercive. The functional $\boldsymbol{w} \to (\boldsymbol{f}, \boldsymbol{w})$ is linear and continuous over V_d. Then applying the Lax-Milgram lemma the assertion follows immediately.

Lemma 12.1: Let $\Omega \subset R^N$ be a bounded domain with a Lipschitz continuous boundary. Let L be an element of V' (i.e. a linear continuous functional on V). Then L vanishes identically on V_d if and only if there exists a function $p \in L^2(\Omega)$ such that

$$L(\boldsymbol{w}) = (p, \text{div } \boldsymbol{w}) \quad \forall \boldsymbol{w} \in V. \tag{12.91}$$

Furthermore, Equation 12.91 defines a unique function p up to an additive constant.
For the proof see Girault and Raviart (1986), p 22.

Theorem 12.2: Let $\Omega \subset R^N$ be a bounded domain with a Lipschitz continuous boundary, let $\boldsymbol{f} \in [L^2(\Omega)]^N$ and let $\boldsymbol{v}$ be the solution of Equation 12.90. Then there exists a function $p \in L^2(\Omega)$, which is unique up to an additive constant, such that

$$a(\boldsymbol{v}, \boldsymbol{w}) - (p, \text{div } \boldsymbol{w}) = (\boldsymbol{f}, \boldsymbol{w}) \quad \forall \boldsymbol{w} \in V. \tag{12.92}$$

Proof: Let $\boldsymbol{v}$ be the solution of Equation 12.90. Let us put

$$L(\boldsymbol{w}) = a(\boldsymbol{v}, \boldsymbol{w}) - (\boldsymbol{f}, \boldsymbol{w}) \quad \forall \boldsymbol{w} \in V.$$

Then due to Equation 12.90 L satisfies Equation 12.91 and due to Lemma 12.1 there exists $p \in L^2(\Omega)$ such that $(\boldsymbol{v}, p)$ is a solution of Equation 12.92, which completes the proof.

Let us define the bilinear form $b(\boldsymbol{w}, q)$ by

$$b(\boldsymbol{w}, q) = -(q, \operatorname{div} \boldsymbol{w}) \quad \forall \boldsymbol{w} \in V, q \in W. \tag{12.93}$$

Then the weak (variational) formulation can also be written in the following form.

Problem $(\mathscr{P}_S)_{bv}$: Find $\boldsymbol{v} \in V$, $p \in W$ such that

$$a(\boldsymbol{v}, \boldsymbol{w}) + b(\boldsymbol{w}, p) = (\boldsymbol{f}, \boldsymbol{w}) \quad \forall \boldsymbol{w} \in V, \tag{12.94}$$

$$b(\boldsymbol{v}, q) = 0 \quad \forall q \in W. \tag{12.95}$$

Theorem 12.3: The problem $(\mathscr{P}_S)_{bv}$ defined by Equations 12.94 and 12.95 has a unique solution, where the pressure is still defined up to an additive constant.
The proof is the consequence of Theorem 12.2.

Remark: The solution $(\boldsymbol{v}, p)$ of Equations 12.94 and 12.95 can be found as a saddle point of the Lagrangian functional

$$\mathscr{L}(\boldsymbol{w}, q) = \tfrac{1}{2}a(\boldsymbol{w}, \boldsymbol{w}) + b(\boldsymbol{w}, q) - (\boldsymbol{f}, \boldsymbol{w}) \quad \forall \boldsymbol{w} \in V, q \in W, \tag{12.96}$$

i.e.

$$\mathscr{L}(\boldsymbol{v}, p) = \min_{\boldsymbol{w} \in V} \max_{q \in W} \mathscr{L}(\boldsymbol{w}, q). \tag{12.97}$$

For the proof see Temam (1977).

Remark: The weak formulation of Equations 12.94 and 12.95 can be reformulated as follows. Let $W_d = V_D = V \times W$. Then we find $\boldsymbol{u} = (\boldsymbol{v}, p) \in W_d$ such that

$$A(\boldsymbol{u}, z) = F(z) \quad \forall z = (\boldsymbol{w}, q) \in V_D, \tag{12.98}$$

where

$$A(\boldsymbol{u}, z) = \nu(\operatorname{grad} \boldsymbol{u}, \operatorname{grad} z) - \int_\Omega p \operatorname{div} z \,\mathrm{d}\boldsymbol{x} + \int_\Omega q \operatorname{div} \boldsymbol{u} \,\mathrm{d}\boldsymbol{x},$$

$$F(z) = \int_\Omega \boldsymbol{f} z \,\mathrm{d}\boldsymbol{x}.$$

NUMERICAL SOLUTION

For the numerical solution several techniques can be used, for example the Galerkin method, FEM, FVM, FDM, spectral method, spectral Galerkin method, spectral collocation method, spectral generalized Galerkin method, the pressure-matrix method, penalty method, etc. (see, for example, Girault and Raviart (1986), Peltier (1989), Thomaset (1981), Temam (1977), (1983), Quarteroni and Valli (1994)). In geodynamics the spherical harmonic decomposition method with the FEM is also used (see for example, Martinec *et al.* (1993)).

To numerically analyse the above problem, we solve Equation 12.90 find the velocity field $\boldsymbol{v}$ and then the pressure p can be determined from Equation 12.92. Alternately we can solve Equations 12.94, 12.95. 12.97 and 12.98.

We shall use the weak formulation of Equation 12.90 and the Galerkin method.

Let the domain Ω be covered by Ω_h, where $h > 0$ is the mesh spacing of the finite elements. Let $\{V_{dh}, h > 0\}$ be a family of finite dimensional subspaces of V_d. We assume that the consistency condition

$$\forall \boldsymbol{w} \in V_d \quad \inf_{w_h \in V_{dh}} \|\boldsymbol{w} - \boldsymbol{w}_h\| \to 0 \text{ as } h \to 0 \tag{12.99}$$

holds. Then we have the following Galerkin approximation problem.

Problem $(\mathscr{P}_S)_g$: Find $\boldsymbol{v}_h \in V_{dh}$ such that

$$a(\boldsymbol{v}_h, \boldsymbol{w}_h) = (\boldsymbol{f}, \boldsymbol{w}_h) \quad \forall \boldsymbol{w}_h \in V_{dh}. \tag{12.100}$$

The existence and uniqueness of the solution of Equation 12.100 gives the next theorem.

Theorem 12.4 (Céa): Under the assumptions of the Lax-Milgram lemma there exists a unique solution $\boldsymbol{v}_h$ of Equation 12.99 which, furthermore, is stable since

$$\|\boldsymbol{v}_h\| \leq \alpha^{-1} \|\boldsymbol{F}\|_{V_d'}. \tag{12.101}$$

Moreover, if $\boldsymbol{v}$ is the solution of Equation 12.90, then there exists constant $c > 0$ such that

$$\|\boldsymbol{v} - \boldsymbol{v}_h\| \leq c \inf_{\boldsymbol{v}_h \in V_{dh}} \|\boldsymbol{v} - \boldsymbol{v}_h\|. \tag{12.102}$$

Hence $\boldsymbol{v}_h \to \boldsymbol{v}$, owing to Equation 12.99.

Proof: Since $V_h \subset V$, then the assumptions of the Lax-Milgram lemma are satisfied also for Equation 12.100. Hence the existence and uniqueness of $\boldsymbol{v}_h$ is proved. Putting $\boldsymbol{w}_h = \boldsymbol{v}_h$ and using the fact that $a(\boldsymbol{v}_h, \boldsymbol{w}_h)$ is coercive, then

$$c_o \|\boldsymbol{v}_h\|^2 \leq a(\boldsymbol{v}_h, \boldsymbol{v}_h) = F(\boldsymbol{v}_h) \leq \|\boldsymbol{F}\|_{V'} \|\boldsymbol{v}_h\|$$

(where V' denotes the dual space of V), which proves Equation 12.101.

Subtracting Equation 12.100 from Equation 12.90 we have

$$a(\boldsymbol{v} - \boldsymbol{v}_h, \boldsymbol{w}_h) = 0 \quad \forall \boldsymbol{w}_h \in V_{dh}.$$

Hence, putting $\boldsymbol{w}_h = \boldsymbol{u}_h - \boldsymbol{v}_h$, $\boldsymbol{u}_h \in V_{dh}$ then

$$c_o \|\boldsymbol{v} - \boldsymbol{v}_h\|^2 \leq a(\boldsymbol{v} - \boldsymbol{v}_h, \boldsymbol{v} - \boldsymbol{v}_h) = a(\boldsymbol{v} - \boldsymbol{v}_h, \boldsymbol{v} - \boldsymbol{u}_h) \leq c_1 \|\boldsymbol{v} - \boldsymbol{v}_h\| \|\boldsymbol{v} - \boldsymbol{u}_h\| \quad \forall \boldsymbol{u}_h \in V_{dh},$$

which proves Equation 12.102.

This theorem proves the existence and uniqueness of the approximation of the velocity field $\boldsymbol{v}_h$, but it is not possible to use it for numerical computation. Firstly, it is very difficult to construct the approximation of the space V_d, i.e. a finite dimensional space of divergence-free vector-valued functions with a good convergence behaviour with respect to h.

Secondly, the construction of a basis of V_{dh} is very difficult for almost all problems in practice. Some modification of the finite dimensional space V_{dh} in the case of finite elements is possible. The idea consists in considering the following problem instead of the problem in Equation 12.100.

Find $\boldsymbol{v}_h \in \tilde{V}_h$ such that

$$a(\boldsymbol{v}_h, \boldsymbol{w}_h) = (\boldsymbol{f}, \boldsymbol{w}_h) \quad \forall \boldsymbol{w}_h \in \tilde{V}_h, \tag{12.103}$$

where

$$\tilde{V}_h = \{\boldsymbol{w}_h \mid \boldsymbol{w}_h \in V_h, (q_h, \text{div } \boldsymbol{w}_h) = 0 \quad \forall q_h \in W_h\}, \tag{12.104}$$

where W_h is an approximation of the space W. The space $\tilde{V}_{dh}$ represents the space of discretely divergence-free functions associated with the finite dimensional spaces, and generally $\tilde{V}_{dh} \not\subset V_{dh}$. Then Equation 12.100 would provide a non-conforming approximation to Equation 12.90. Hence $\tilde{V}_{dh}$ is a non-conforming finite dimensional space, whose merit is simpler possibility of constructing the basis functions of the space $\tilde{V}_{dh}$.

We shall now discuss the numerical solution of Equations 12.94 and 12.95, which can be a better approximation of the initial Stokes problem. We introduce finite dimensional subspaces $V_h \subset V$, $W_h \subset W$ depending on h. Then we have to solve the following problem.

Problem $(\mathscr{P}_S)_{bh}$: Find $\boldsymbol{v}_h \in V_h$, $p_h \in W_h$ such that

$$a(\boldsymbol{v}_h, \boldsymbol{w}_h) + b(\boldsymbol{w}_h, p_h) = (\boldsymbol{f}, \boldsymbol{w}_h) \quad \forall \boldsymbol{w}_h \in V_h, \tag{12.105}$$

$$b(\boldsymbol{v}_h, q_h) = 0 \quad \forall q_h \in W_h. \tag{12.106}$$

Let $\tilde{V}_h$ be the space of discretely divergence-free functions associated with the finite dimensional spaces. Then the bilinear form $a(.,.)$ is coercive in $\tilde{V}_h$, as $\tilde{V}_h \subset V$ and it is coercive in V. Thus, there exists $c > 0$ such that

$$\nu \left\|\text{grad } \boldsymbol{w}_h\right\|_o^2 > c \left\|\boldsymbol{w}_h\right\|^2 \quad \forall \boldsymbol{w}_h \in \tilde{V}_h. \tag{12.107}$$

Applying the Poincaré inequality, then there exist constants $c_1 > 0$, $c_2 > 0$ such that

$$|\nu(\text{grad } \boldsymbol{u}, \text{grad } \boldsymbol{w})| \leq c_1 \|\boldsymbol{u}\| \, \|\boldsymbol{w}\| \quad \forall \boldsymbol{u}, \boldsymbol{w} \in V, \tag{12.108}$$

i.e. the bilinear form $a(.,.)$ is continuous over $V \times V$, and

$$|(q, \text{div } \boldsymbol{w})| \leq c_2 \|\boldsymbol{w}\| \, \|q\|_o \quad \forall \boldsymbol{w} \in V, q \in W, \tag{12.109}$$

i.e. the bilinear form $b(.,.)$ is continuous over $V \times W$.

We shall assume that spaces V_h, W_h have the following property: $\exists c_3 > 0$ such that

$$\forall q_h \in W_h, \quad \exists \boldsymbol{w}_h \in V_h, \quad \boldsymbol{w}_h \neq 0: \quad (q_h, \text{div } \boldsymbol{w}_h) \geq c_3 \|\boldsymbol{w}_h\| \, \|q_h\|_o,$$

the so-called condition of compatibility.

Theorem 12.5: The problem in Equations 12.105 and 12.106 satisfying the above assumptions has a unique solution. This solution satisfies

$$\|\boldsymbol{v}_h\| \leq c^{-1} \|\boldsymbol{f}\|_{V'}, \quad \|p_h\|_o \leq c_3^{-1}(1 + c_1 c^{-1}) \|\boldsymbol{f}\|_{V'} \tag{12.110}$$

and the solution is stable if c_3 is independent of h.

The proof is the consequence of Theorems 7.4.1 and 7.4.2 of Quarteroni and Valli (1991), where the term generated by the thermal stresses is connected with the body forces term.

The convergence theorem will be given in the next theorem.

Theorem 12.6: Let Equations 12.108 and 12.109 and the compatibility conditions be satisfied. Let the bilinear form $a(.,.)$ be coercive. Then

$$\|v - v_h\| \leq (1 + c_1 c^{-1}) \inf_{w_h \in \tilde{V}_h} \|v - w_h\| + c_2 c^{-1} \inf_{q_h \in W_h} \|p - p_h\|_o, \tag{12.111}$$

$$\|p - p_h\|_o \leq c_1 c_3^{-1}(1 + c_1 c^{-1}) \inf_{w_h \in \tilde{V}_h} \|v - w_h\| + (1 + c_2 c_3^{-1} + c_1 c_2 c^{-1} c_3^{-1}) \inf_{q_h \in W_h} \|p - q_h\|_o. \tag{12.112}$$

Moreover, if the estimate

$$\inf_{w_h \in W_h} \|v - w_h\| \leq (1 + c_2 c_3^{-1}) \inf_{w_h \in V_h} \|v - w_h\|, \tag{12.113}$$

holds, then the convergence estimate is optimal provided c_3 is independent of h.
The proof is a modification of that of Theorem 7.4.3 of Quarteroni and Valli (1991), where the term generated by the thermal stresses is connected with body forces.

To solve Equation 12.100, let $\{z_j \mid j = 1, \ldots, n(h)\}$ be a basis of the subspace V_{dh}, and $n(h)$ be the dimension of the subspace V_{dh}, h being fixed. Putting

$$v_h(x) = \sum_{i=1}^{n(h)} \xi_i z_i(x),$$

then Equation 12.100 yields the system

$$A\vec{\xi} = b, \tag{12.114}$$

where $A_{ij} = a(z_j, z_i)$ and $b_i = (f, z_i)$, where the matrix A is a symmetric and positive definite matrix and therefore the system in Equation 12.114 is solvable.

To solve Equations 12.105 and 12.106, let us denote by $n(h)$ and $k(h)$ the dimensions of V_h and W_h, respectively, and let $\{z_j \mid j = 1, \ldots, n(h)\}$, $\{\omega_i \mid i = 1, \ldots, k(h)\}$ be the bases of V_h and W_h. Putting

$$v_h(x) = \sum_{i=1}^{n(h)} v_i z_i, \quad p_h(x) = \sum_{i=1}^{k(h)} p_i \omega_i(x),$$

then Equations 12.105 and 12.106 yield the system

$$\begin{pmatrix} A & B^T \\ B & 0 \end{pmatrix} \begin{pmatrix} v \\ p \end{pmatrix} = \begin{pmatrix} f \\ 0 \end{pmatrix}, \tag{12.115}$$

where $A_{ij} = a(z_j, z_i)$, $B_{ki} = b(z_i, \omega_k)$, $f_i = (f, z_i)$. The $n(h) \times n(h)$ matrix A is symmetric and positive definite, the $k(h) \times n(h)$ matrix B is a rectangular matrix. The matrix $\begin{pmatrix} A & B^T \\ B & 0 \end{pmatrix}$ of dimension $n(h) \times k(h)$ is non-singular. Solving Equation 12.115 we obtain

$$\boldsymbol{v} = A^{-1}(\boldsymbol{f} - B^T\boldsymbol{p}), \tag{12.116}$$

$$BA^{-1}B^T\boldsymbol{p} \equiv R\boldsymbol{p} = BA^{-1}\boldsymbol{f}, \tag{12.117}$$

where $R = BA^{-1}B^T$, $\dim k(h) \times k(h)$, is the pressure matrix, which is symmetric and is positive definite if $\ker B^T = 0$. Then $\boldsymbol{v}$ and $\boldsymbol{p}$ are uniquely defined by Equations 12.116 and 12.117.

The system in Equation 12.115 can be solved by several methods such as the Uzawa, Arrow-Hurwitz, penalty methods, the pressure-matrix method, methods based on pressure solvers, preconditioning techniques of different types, multi-grid techniques, etc.. For more details see Thomaset (1981), Temam (1983), Girault and Raviart (1986), Quarteroni and Valli (1991).

B. NON-LINEAR CASE

Let Ω be a bounded domain in R^N, $N = 2, 3$, with Lipschitzian boundary $\partial\Omega$ and let $I = (t_o, t_1)$. Then the convection model leads to our solving the following non-linear problem.

Problem ($\mathscr{P}_N$): Find a triple of scalar and vector functions $(T, \boldsymbol{v}, p)$ satisfying

$$\rho(\boldsymbol{x})c_e(\boldsymbol{x})(\partial_t T(\boldsymbol{x},t) + \boldsymbol{v}(\boldsymbol{x},t)\operatorname{grad} T(\boldsymbol{x},t)) = Q(\boldsymbol{x},t) - \partial_i(\kappa_{ij}(\boldsymbol{x},t)\partial_j T(\boldsymbol{x},t)) \quad \text{in } \Omega \times I, \tag{12.118}$$

$$\rho(\boldsymbol{x})(\partial_t \boldsymbol{v}(\boldsymbol{x},t) + \boldsymbol{v}(\boldsymbol{x},t)\operatorname{grad} \boldsymbol{v}(\boldsymbol{x},t)) - \nu(\boldsymbol{x})\Delta\boldsymbol{v}(\boldsymbol{x},t)$$
$$+\operatorname{grad} p(\boldsymbol{x},t) = \boldsymbol{f}(\boldsymbol{x},t,T(\boldsymbol{x},t)) \quad \text{in } \Omega \times I, \tag{12.119}$$

$$\operatorname{div} \boldsymbol{v}(\boldsymbol{x},t) = 0 \quad \text{in } \Omega \times I, \tag{12.120}$$

$$\left.\begin{aligned} T(\boldsymbol{x},t) &= T_1(\boldsymbol{x},t), \quad (12.121) \\ \boldsymbol{v}(\boldsymbol{x},t) &= \boldsymbol{v}_1(\boldsymbol{x},t) \quad (12.122) \end{aligned}\right\} \text{ on } \partial\Omega \times I,$$

$$\left.\begin{aligned} T(\boldsymbol{x},t_o) &= T_o(\boldsymbol{x}), \quad (12.123) \\ \boldsymbol{v}(\boldsymbol{x},t_o) &= \boldsymbol{v}_o(\boldsymbol{x}) \quad (12.124) \end{aligned}\right\} \text{ in } \Omega,$$

where T is the temperature, $\boldsymbol{v}$ the velocity vector and p is the pressure, and where the functions $Q(\boldsymbol{x},t)$, $\rho(\boldsymbol{x})$, $c_e(\boldsymbol{x})$, $\nu(\boldsymbol{x})$, $T_1(\boldsymbol{x},t)$, $\boldsymbol{v}_1(\boldsymbol{x},t)$, $T_o(\boldsymbol{x})$ and $\boldsymbol{v}_o(\boldsymbol{x})$ are given functions and $\boldsymbol{f}(\boldsymbol{x},t,T) = \boldsymbol{f}_o(\boldsymbol{x},t) - \partial_j(\beta_{ij}(\boldsymbol{x})(T(\boldsymbol{x},t) - T_o(\boldsymbol{x})))$, where $\boldsymbol{f}_o(\boldsymbol{x},t)$ and $\beta_{ij}(\boldsymbol{x})$ are given functions.

Remark: For the boundary conditions we can take into account one of the following possibilities: an imposed temperature field, an imposed heat flux and/or an imposed velocity field, an imposed stress field, a classical free slip boundary condition, or if we assume that the boundary $\partial\Omega = \Gamma_u \cup \Gamma_\tau$, where Γ_u, Γ_τ denote the parts of boundary $\partial\Omega$ where the Dirichlet's type or Neumann's type of boundary conditions are prescribed; then on $\partial\Omega = \Gamma_u \cup \Gamma_\tau$ mixed versions of the above boundary conditions can be prescribed.

Remark: In geodynamics the free slip boundary conditions, i.e.

$$\boldsymbol{v}.\boldsymbol{n} = 0 \quad \text{on } \partial\Omega,$$

$$\nu(\partial_i v_j + \partial_j v_i) n_i t_i^k = g^k \quad \text{on } \partial\Omega, \quad k = 1, \ldots, N-1,$$

where $t^k = t^k(\boldsymbol{x})$ are linearly independent tangent vectors at $\boldsymbol{x} \in \partial\Omega$, and g^k are the given data, are often used.

Since in the next sections the non-linear case, based on the thermo-Bingham rheology, will be discussed in more detail, and as this model problem is a special case of the model problem in thermo-Bingham's rheology, then in this section we shall analyze the quasi-coupled non-linear steady-state case only. Therefore, it is sufficient to investigate in more detail the BVP for the non-linear Stokes equation only. The thermal part of the problem was studied in Chapter 9, Section 9.3, with sufficient accuracy.

B1. STEADY-STATE CASE

For simplicity we shall limit ourselves, as in the linear case, to the model problem of incompressible Newtonian fluid in the Boussinesq approximation in the following form.

Problem ($\mathscr{P}_{SN}$): Find a pair of functions $(\boldsymbol{v}, p)$ satisfying

$$-\nu\Delta\boldsymbol{v} + \boldsymbol{v}\,\text{grad}\boldsymbol{v} + \text{grad } p = f(T) \quad \text{in } \Omega, \tag{12.125}$$

$$\text{div } \boldsymbol{v} = 0 \quad \text{in } \Omega, \tag{12.126}$$

$$\boldsymbol{v} = 0 \quad \text{on } \partial\Omega. \tag{12.127}$$

Let us define the spaces

$$V = [H_o^1(\Omega)]^N,\ W = L_o^2(\Omega) = \{q \mid q \in L^2(\Omega),\ \int_\Omega q\text{d}\boldsymbol{x} = 0\},$$

$$V_d = \{\boldsymbol{w} \mid \boldsymbol{w} \in V,\ \text{div } \boldsymbol{w} = 0\},\ \tilde{V}_d = \text{ the closure of } V_d \text{ in } V \cap L^N(\Omega),$$

and the bilinear and trilinear forms by

$$a(\boldsymbol{w}, \boldsymbol{u}) = \nu(\text{ grad } \boldsymbol{w}, \text{grad } \boldsymbol{u}),\ b_o(\boldsymbol{u}, q) = -(q,\ \text{div } \boldsymbol{u}),$$

$$b(\boldsymbol{w}, \boldsymbol{z}, \boldsymbol{u}) = \int_\Omega w_j \frac{\partial z_i}{\partial x_j} u_i \text{d}\boldsymbol{x},$$

where the trilinear form is associated with the non-linear convective term. Let $\boldsymbol{f} \in [L^2(\Omega)]^N$. Multiplying Equation 12.125 by $\boldsymbol{w} \in V$, integrating over Ω, and using the boundary condition, then the weak formulation is as follows.

Problem ($\mathscr{P}_{SN})_V$: Find $\boldsymbol{v} \in V,\ p \in W$ such that

$$a(\boldsymbol{v}, \boldsymbol{w}) + b(\boldsymbol{v}, \boldsymbol{v}, \boldsymbol{w}) + b_o(\boldsymbol{w}, \boldsymbol{p}) = (\boldsymbol{f}, \boldsymbol{w}) \quad \forall \boldsymbol{w} \in V, \tag{12.128}$$

$$b_o(\boldsymbol{v}, q) = 0 \quad \forall q \in W. \tag{12.129}$$

Lemma 12.2: Let $s = (1/2)N$, then $\partial_j w_i \in L^N(\Omega)$ for each $\boldsymbol{w} \in V_d$, $N = 2, 3$. For the proof see Duvaut and Lions (1972a).

The trilinear form $b(., ., .)$ is continuous on $[H^1(\Omega)]^N$, $N = 2, 3$. From the Hölder inequality applied to $b(\boldsymbol{w}, \boldsymbol{z}, \boldsymbol{u})$ we have

$$| b(\boldsymbol{w}, \boldsymbol{z}, \boldsymbol{u}) | \leq c \, \|\boldsymbol{w}\|_{L_p^N(\Omega)} \|\boldsymbol{u}\|_{L_p^N(\Omega)} \Sigma_{i,j} \left\|D_i z_j\right\|_{L_N^N(\Omega)}, \, 2p^{-1} + N^{-1} = 1. \quad (12.130)$$

Applying Lemma 12.2 we have

$$| b(\boldsymbol{w}, \boldsymbol{z}, \boldsymbol{u}) | \leq c_1 \, \|\boldsymbol{w}\|_{L_p^N(\Omega)} \|\boldsymbol{u}\|_{L_p^N(\Omega)} \|\boldsymbol{z}\|_{s,N}, \, s = \tfrac{1}{2}N \quad (12.131)$$

and

$$| b(\boldsymbol{w}, \boldsymbol{z}, \boldsymbol{u}) | \leq c(n) \, \|\boldsymbol{w}\|_V \|\boldsymbol{z}\|_V \|\boldsymbol{u}\|_{V \cap L^N(\Omega)} . \quad (12.132)$$

Moreover, for any open set Ω (see Temam (1984))

$$b(\boldsymbol{u}, \boldsymbol{u}, \boldsymbol{u}) = 0 \quad \forall \boldsymbol{u} \in V \cap L^N(\Omega), \quad (12.133)$$

$$b(\boldsymbol{w}, \boldsymbol{u}, \boldsymbol{u}) = 0 \quad \forall \boldsymbol{w} \in V_d, \boldsymbol{u} \in V \cap L^N(\Omega), \quad (12.134)$$

$$b(\boldsymbol{w}, \boldsymbol{u}, \boldsymbol{z}) = -b(\boldsymbol{w}, \boldsymbol{z}, \boldsymbol{u}) \quad \forall \boldsymbol{w} \in V_d, \boldsymbol{u}, \boldsymbol{z} \in V \cap L^N(\Omega). \quad (12.135)$$

For any fixed $\boldsymbol{w} \in V$, the mapping $\boldsymbol{u} \to b(\boldsymbol{w}, \boldsymbol{w}, \boldsymbol{u})$ is linear and continuous on V, so that it is an element of V'.

Now we shall formulate the weak formulation of the studied problem as follows.

Problem $(\mathscr{P}_{SN})_{dv}$: Find $\boldsymbol{v} \in V_d$ such that

$$a(\boldsymbol{v}, \boldsymbol{w}) + b(\boldsymbol{v}, \boldsymbol{v}, \boldsymbol{w}) = (\boldsymbol{f}, \boldsymbol{w}) \forall \boldsymbol{w} \in \tilde{V}_d, \quad (12.136)$$

where $\boldsymbol{f} \in L^2(\Omega)$.

The next lemma shows that solutions of both problems $(\mathscr{P}_{SN})_v$ and $(\mathscr{P}_{SN})_{dv}$ are equivalent. For this we need the following lemma.

Lemma 12.3: Let $\Omega \subset R^N$ be a bounded domain with a Lipschitz continuous boundary $\partial\Omega$. Let $L(\boldsymbol{w}) \in V'$. Then L vanishes identically on V_d if and only if there exists a function $p \in L^2(\Omega)$ such that

$$L(\boldsymbol{w}) = (p, \operatorname{div} \boldsymbol{w}) \quad \forall \boldsymbol{w} \in V. \quad (12.137)$$

Moreover, Equation 12.93 defines a unique function p up to an additive constant.
For the proof see Girault and Raviart (1986), p 22.

If $(\boldsymbol{v}, p)$ is a solution of Equations 12.128 and 12.129, then $\boldsymbol{v}$ is a solution of Equation 12.136. Then it is sufficient to prove the next lemma.

Lemma 12.4: Let $\boldsymbol{v}$ be a solution of Equation 12.136. Then there exists a unique $p \in W$ such that $(\boldsymbol{v}, p)$ is a solution of the problem in Equations 12.128 and 12.129.

Proof: The mapping $w \to a(v, w) + b(v, v, w) - (f, w) \in V'$ and vanishes on V_d. Applying Lemma 12.3 the assertion follows immediately.

Theorem 12.5: Let $\Omega \subset R^N$, $N = 2, 3$, be a bounded domain, and let $f \in [H^{-1}(\Omega)]^N$.

Then Problem $(\mathscr{P}_{SN})_{\mathrm{dv}}$ has at least one solution $v \in V_d$ and there exists $p \in L^1_{\mathrm{loc}}(\Omega)$ such that Equations 12.125-12.127 are satisfied.

Proof: We have only to prove the existence of w; the existence of p has already been shown. For the proof of the existence of v the Galerkin technique will be used. Then the a priori estimate will be found and thus pass to the limit.

The space $\tilde{V}_d$ is separable as a subspace of V. Let $\{z_i\}$ be a countable basis of $\tilde{V}_d$, i.e. each finite subset $\{z_i\}$ is linearly independent and span $\{z_i \mid i = 1, 2, \ldots\}$ is dense in $\tilde{V}_d$, as $\tilde{V}_d$ is separable. Let W^m be the space spanned by $\{z_j \mid 1 \le j, k \le m\}$. Then we define an approximate solution

$$v_m = \sum_{j=1}^{m} \alpha_{i,m} z_i, \qquad \alpha_{i,m} \in R, \tag{12.138}$$

$$a(v_m, w_k) + b(v_m, v_m, w_k) = (f, w_k), \quad k = 1, \ldots, m. \tag{12.139}$$

Then Equations 12.138, and 12.139 represent a system of non-linear equations for $\alpha_{1,m}, \ldots, \alpha_{m,m}$. It is known that such a system has a unique solution.

Let us multiply Equation 12.139 by $\alpha_{k,m}$ and add corresponding equalities for $k = 1, \ldots, m$. Hence

$$\nu \|v_m\|^2_{1,N} + b(v_m, v_m, v_m) = (f, v_m)$$

and thus

$$\nu \|v_m\|^2_{1,N} = (f, v_m) \le \|f\|_{V'} \|v_m\|_{1,N}$$

as $b(v_m, v_m, v_m) = 0$. Hence

$$\|v_m\|_{1,N} \le \nu^{-1} \|f\|_{V'} \tag{12.140}$$

representing the a priori estimate. From this estimate we obtain

$$\{v_{m'}, m' \in N\} \text{ is a bounded subset in } V_d.$$

Due to the compactness theorem(*).

$$v_{m'} \to v \text{ in the norm of } L^2(\Omega).$$

From the estimate in Equation 12.140 it follows that there exists a subsequence $\{v_\mu, \mu \in N\}$ of $\{v_m, m \in N\}$ such that

$$v_\mu \to v \quad \text{in } V_d \text{ weakly,} \tag{12.141}$$

$$v_\mu \to v \quad \text{in } V_d \text{ strongly.} \tag{12.142}$$

Then $b(v_\mu, v_\mu, u) \to b(v, v, u)\, \forall u \in V_d$.
Actually,

$$b(v_\mu, v_\mu, u) = -b(v_\mu, u, v_\mu) = -\int_\Omega v_{\mu i} v_{\mu j} D_i u_j \mathrm{d}x.$$

(*) Compactness theorem: Let $\Omega \in R^N$ be a Lipschitz open set. Then the set of vector functions belonging to $\mathscr{D}(\overline{\Omega})$ is dense in $E(\Omega)$, where $E(\Omega) = \{u \mid u \in L^2(\Omega);\ \mathrm{div}\ u \in L^2(\Omega)\}$ is a Hilbert space with the scalar product $(u, v)_{E(\Omega)} = (u, v) + (\mathrm{div}\ u, \mathrm{div}\ v)$

But due to the a priori estimate $\boldsymbol{u}_{\mu i} \to u_i$ in $L^2(\Omega)$ strongly (see Equation 12.142) and since $D_i u_j \in L^\infty(\Omega)$ then

$$\int_\Omega v_{\mu i} v_{\mu j} D_i u_j \mathrm{d}\boldsymbol{x} \to \int_\Omega v_i v_j D_i u_j \mathrm{d}\boldsymbol{x}.$$

Therefore,

$$b(\boldsymbol{v}_\mu, \boldsymbol{u}, \boldsymbol{v}_\mu) \to b(\boldsymbol{v}, \boldsymbol{u}, \boldsymbol{v}) = -b(\boldsymbol{v}, \boldsymbol{v}, \boldsymbol{u}).$$

To prove the uniqueness of the solution for $N = 2$ then we put $\boldsymbol{w} = \boldsymbol{v}$ in Equation 12.136 since $\tilde{V}_d = V_d$, and since Equation 12.134 holds then

$$\nu \|\boldsymbol{v}\|_{1,2}^2 = (\boldsymbol{f}, \boldsymbol{v}) \leq \|\boldsymbol{f}\|_{V'} \|\boldsymbol{v}\|_{1,2}$$

and thus

$$\|\boldsymbol{v}\|_{1,2} \leq \nu^{-1} \|\boldsymbol{f}\|_{V'}. \tag{12.143}$$

Now let $\boldsymbol{v}_1$ and $\boldsymbol{v}_2$ be two solutions of Equation 12.136 and let us put $\boldsymbol{v} = \boldsymbol{v}_1 - \boldsymbol{v}_2$. We subtract Equation 12.136 corresponding to $\boldsymbol{v}_1$ and $\boldsymbol{v}_2$, and then we obtain

$$a(\boldsymbol{v}, \boldsymbol{w}) + b(\boldsymbol{v}_1, \boldsymbol{v}, \boldsymbol{w}) + b(\boldsymbol{v}, \boldsymbol{v}_2, \boldsymbol{w}) = 0 \quad \forall \boldsymbol{w} \in V.$$

Hence, putting $\boldsymbol{w} = \boldsymbol{v}$ and using Equation 12.134, we find that

$$\nu \|\boldsymbol{v}\|_{1,2}^2 = -b(\boldsymbol{v}, \boldsymbol{v}_2, \boldsymbol{v}).$$

Using Equations 12.132 and 12.143 for $\boldsymbol{v} = \boldsymbol{v}_1$ then

$$\nu \|\boldsymbol{v}\|_{1,2}^2 \leq c(n) \|\boldsymbol{v}\|^2 \|\boldsymbol{v}_2\| \leq \nu^{-1} c(n) \|\boldsymbol{f}\|_{V'} \|\boldsymbol{v}\|^2,$$

thus

$$(\nu - \nu^{-1} c(n) \|\boldsymbol{f}\|_{V'}) \|\boldsymbol{v}\|_{1,2}^2 \leq 0 \Rightarrow \|\boldsymbol{v}\|_{1,2} = 0,$$

which completes the proof.

12.3.3. Mathematical Models Based on the Thermo-Bingham Rheology

12.3.3.1. Introduction

During a long time, over approximately millions of years, the earth's mantle behaves like a highly viscous fluid, while the upper part of the Earth, the lithosphere, behaves like a visco-elastic and visco-plastic medium. Since the Earth's mantle is heated from below and cooled from above, the mantle rocks remove heat from the deeper parts of the Earth by the transport of mass, i.e. by thermal convection. Mantle convection is manifested in the motions of lithospheric plates, in the seismically and gravitationally inferred thermal heterogeneity of the upper and lower mantle, and in the topographic undulation of the core-mantle boundary (the unstable layer of light mantle materials resulting in the density differentiation of the here melted mantle materials). The outer core is of relatively small visco-plastic fluid but is strongly metallized and ionized. The topography of the core-mantle boundary suggests the coupling of mantle and core dynamics, and the generation of the geomagnetic field of the Earth. The highly non-linear nature of the core and the mantle dynamics, and the geodynamic processes in the lithosphere require that investigations be done through laboratory and numerical experiments. Since laboratory experiments are very difficult, only numerical experimentation will be suitable for our understanding of the geodynamic processes in the Earth's interior. Since present methods, based on the Boussinesq's approximation of fluid, cannot give satisfactory results in the mantle-lithosphere coupling as well as being unable to facilitate the study of the outer core, the mantle and the lithosphere as a whole (as one body), it is necessary to seek new methods. In this chapter we shall give several methods which make possible such studies. In this section we shall present the first of these based on the thermo-rigid-visco-plastic Bingham's rheology. The physical discussion was given in the introductory part of this chapter. Here we shall discuss the mathematical model, the mathematical analysis of this model problem, and the numerical methods and algorithms solving the model problem. We shall give better approximations of the mantle rheology and therefore better methods for investigation of geodynamic processes in the mantle. The connection with the above method based on the Boussinesq's approximation will be given.

We shall assume that the mantle is created by thermo-elasto-plastic materials. Such type of materials can be described by the thermo-plastic rheology. For such rheologies there exists a threshold of plasticity separating two types of behaviour. In the first case this threshold of plasticity can be fixed, whilst in the second it can depend on the history of deformations. In this latter case the rocks undergo work hardening. Since the $2D$ case will not be investigated in this book, we limit ourselves to the case with a fixed threshold of plasticity and to the case of thermo-rigid-visco-plastic, the so-called thermo-Bingham, rheology. For both cases we will state the elasto-visco-plastic constituent law and we will show that in the limit we obtain the above discussed Newtonian fluid rheologies.

Let us investigate the case of a fixed threshold of plasticity. We shall assume that the thermo-elastic zone is defined by $\mathscr{F}(\tau_{ij}) < a_o$ and the thermo-plastic zone by $\mathscr{F}(\tau_{ij}) = a_o$, where a_o is a constant, τ_{ij} represents stress and $\mathscr{F}$ is a convex and continuous with respect to τ_{ij} function, characterizing the domains of properties of the rocks. It is evident that the domain of R^6, where $\mathscr{F}(\tau_{ij}) \leq a_o$, is a closed convex set. Moreover, $\mathscr{F}(\tau_{ij})$ involves the deviator $\{\tau_{ij}^D\}$ of the tensor $\{\tau_{ij}\}$. The plastic deformations of the mantle rocks, which are of large amplitude, show a great viscosity effect. From this point of view we shall give a model of the mantle rheology, characterized by a law of thermo-elasto-visco-plastic behaviour, where the mantle rocks will be thermo-elastic below the threshold of plasticity and thermo-visco-plastic beyond the threshold of plasticity.

Let us introduce the closed convex set

$$K = \{\tau \mid \tau = \{\tau_{ij}\} \in R^6, \tau_{ij} = \tau_{ji}, \mathscr{F}(\tau_{ij}) \leq a_o\}, \tag{12.144}$$

where $\mathscr{F}$ is a convex continuous function of the six scalar variables τ_{ij} (e.g. $\mathscr{F}(\tau_{ij}) = (1/2)\tau_{ij}^D\tau_{ij}^D - k^2$, where k is a given constant and τ_{ij}^D is the deviator of the stress tensor, characterizing the Von Mises model of plasticity). Furthermore, let us introduce the mapping $\tau \to \mathscr{P}_K(\tau)$, the orthogonal projection, in the usual Euclidean structure of $\tau \in R^6$ on the convex set K. Let us define the small strain tensor $e_{ij}(\boldsymbol{u})$ and the strain rate tensor $D_{ij}(\boldsymbol{v})$, where $\boldsymbol{u}$ and $\boldsymbol{v}$ are the displacement and velocity vectors, by

$$e_{ij}(\boldsymbol{u}) = \tfrac{1}{2}(\partial_j u_i + \partial_i u_j),\ D_{ij}(\boldsymbol{v}) = e_{ij}(\dot{\boldsymbol{u}}) = \tfrac{1}{2}(\partial_j v_i + \partial_i v_j),\ \boldsymbol{v} = \dot{\boldsymbol{u}} \equiv \partial_t \boldsymbol{u}. \quad (12.145)$$

Then the constituent law of thermo-elasto-visco-plastic rheology can be written in the following form

$$e_{ij}(\dot{\boldsymbol{u}}) \equiv D_{ij}(\boldsymbol{v}) = s_{ijkl}\dot{\tau}_{kl} + \lambda_{ij} - \alpha_{ij}(T - T_o), \quad (12.146)$$

where
$$\lambda_{ij} = 0 \quad \text{for } \mathscr{F}(\tau_{ij}) < a_o,$$
$$\lambda_{ij} = (2\mu)^{-1}[\tau_{ij} - (\mathscr{P}_K(\tau))_{ij}] \quad \text{for } \mathscr{F}(\tau_{ij}) \geq a_o,$$

where $\mu > 0$ is a viscosity coefficient.

If $\dot{\tau}_{ij} = 0$ (i.e. if $\{\tau_{ij}\}$ is constant), then Equation 12.146 yields

$$\tau_{ij} = (\mathscr{P}_K(\tau))_{ij} + 2\mu e_{ij}(\dot{\boldsymbol{u}}) - \beta_{ij}(T - T_o) = (\mathscr{P}_K(\tau))_{ij} + 2\mu D_{ij}(\boldsymbol{v}) - \beta_{ij}(T - T_o). \quad (12.147)$$

In the case when $K = \{\tau \mid \{\tau_{ij}\}$ be a spherical tensor$\}$ then $(\mathscr{P}_k(\tau))_{ij} = -p\delta_{ij}$ and Equation 12.146 yields

$$\tau_{ij} = -p\delta_{ij} + 2\mu e_{ij}(\dot{\boldsymbol{u}}) - \beta_{ij}(T - T_o) = -p\delta_{ij} + 2\mu D_{ij}(\boldsymbol{v}) - \beta_{ij}(T - T_o), \quad (12.148)$$

which is identical with the Newtonian viscous rheology. Then

$$\partial_t \rho + \partial_j(\rho v_j) = 0, \quad (12.149)$$

$$\partial_t \rho v_i + \partial_j \rho v_j v_i = \partial_j \tau_{ij} + \rho f_i, \quad (12.150)$$

$$\partial_t \rho e + \partial_j \rho v_j e = \tau_{ij}\partial_j v_i + \rho Q - \partial_j q_j \equiv \tau_{ij} D_{ij}(\boldsymbol{v}) + \rho Q - \partial_j q_j, \quad (12.151)$$

and Equation 12.146 or 12.147 or 12.148, respectively, give the governing equations for global thermo-elasto-visco-plastic models of the mantle with a fixed threshold of plasticity.

The further rheology studied in this book will be the thermo-rigid-plastic rheology. The dissipation function $\mathscr{D} = \tau_{ij} D_{ij}$ in Equation 12.151 will be assumed to depend on the strain rate tensor only as

$$\tau_{ij} D_{ij} = \mathscr{D}_1(\boldsymbol{D}) + \mathscr{D}_2(\boldsymbol{D}), \quad (12.152)$$

where $\mathscr{D}_1$ and $\mathscr{D}_2$ are positive homogeneous functions of order one and two, respectively, in components of the tensor $\boldsymbol{D} = \{D_{ij}\}$ and such that

$$\mathscr{D}_1 = \frac{\partial \mathscr{D}_1}{\partial D_{kl}} D_{kl},\ \mathscr{D}_2 = \frac{1}{2}\frac{\partial \mathscr{D}_2}{\partial D_{kl}} D_{kl}. \quad (12.153)$$

Since we shall assume the case of incompressible mantle rock the identity in Equation 12.152 must satisfy the condition

$$D_{kk} = 0, \text{ i.e. } \operatorname{div} \boldsymbol{v} = 0, \quad (12.154)$$

and then the components of the stress tensor will be given by

$$\tau_{ij} = -p\delta_{ij} + \frac{\partial \mathscr{D}_1}{\partial D_{ij}} + \frac{1}{2}\frac{\partial \mathscr{D}_2}{\partial D_{ij}}, \quad (12.155)$$

where p is a scalar, independent of the D_{ij}, and as we shall assume the isotropic case of rocks, the scalars $\mathscr{D}_1$ and $\mathscr{D}_2$ are functions of the tensor $\boldsymbol{D} = \{D_{ij}\}$ only.

Let us put

$$\mathscr{D}_1 = 2\hat{g}D_{\text{II}}^{1/2}, \mathscr{D}_2 = 4\hat{\mu}D_{\text{II}}, \tag{12.156}$$

where

$$D_{\text{II}} = \tfrac{1}{2}D_{ij}D_{ij} \tag{12.157}$$

is the invariant of the tensor $\boldsymbol{D}$, and $\hat{g}$ and $\hat{\mu}$ are thresholds of plasticity and viscosity of the Bingham rigid visco-plastic fluid. Putting Equation 12.156 into Equation 12.155 we obtain

$$\tau_{ij} = -p\delta_{ij} + \hat{g}D_{ij}(D_{\text{II}})^{-1/2} + 2\hat{\mu}D_{ij}, \tag{12.158}$$

the constituent law for the Bingham fluid. Equation 12.158 is valid only when $D_{\text{II}} \neq 0$. In Equation 12.158, p represents the spherical part of the stress tensor and has a meaning of pressure.

For our further investigations and analyses the inverse constituent law for the Bingham fluid must be given. To invert Equation 12.158, we put

$$\tau_{\text{II}} = \tfrac{1}{2}\tau_{ij}^D\tau_{ij}^D, \tag{12.159}$$

where τ_{ij}^D is the stress deviator. Then Equation 12.158 yields

$$\tau_{\text{II}} = (\hat{g} + 2\hat{\mu}D_{\text{II}}^{1/2})^2. \tag{12.160}$$

Hence $\tau_{\text{II}}^{1/2} \geq \hat{g}$, and then Equation 12.158 can be inverted to

$$D_{ij} = (2\hat{\mu})^{-1}(1 - \hat{g}\tau_{\text{II}}^{1/2})\tau_{ij}^D. \tag{12.161}$$

In the case when $D_{\text{II}} = 0$ the stress tensor is undefined, but it is still true that $\tau_{\text{II}}^{1/2} \leq \hat{g}$, as in the opposite case Equation 12.161 gives $D_{\text{II}} = 1/2D_{ij}D_{ij} > 0$.

Then the constituent law for the Bingham fluid is of the form

$$\begin{aligned} D_{ij} &= 0 \quad \text{for } \tau_{\text{II}}^{1/2} < \hat{g}, \\ D_{ij} &= (2\hat{\mu})^{-1}(1 - \hat{g}\tau_{\text{II}}^{1/2})\tau_{ij}^D \quad \text{for } \tau_{\text{II}}^{1/2} \geq \hat{g}. \end{aligned} \tag{12.162}$$

For $\hat{g} = 0$ the constituent laws in Equations 12.158 and 12.162 are identical for the viscous Newtonian fluid. If $\hat{g} \to +\infty$ the mantle rocks will be characterized as absolutely rigid rocks. For $\hat{g}$ small then the mantle rocks behave as a medium close to a Newtonian fluid. If $\hat{g}$ is strictly positive, the rigid zones in the interior of the flow will be observed. If $\hat{g}$ increases, these rigid zones become larger and larger and they correspond to visco-plastic rocks. From the condition $\tau_{\text{II}}^{1/2} \geq \hat{g}$, i.e. $(1/2)\tau_{ij}^D\tau_{ij}^D \geq \hat{g}^2$, when the velocities of P and S seismic waves and the densities are known, the threshold of plasticity $\hat{g}$ can be estimated.

Then Equations 12.149-12.151 together with Equation 12.158 or 12.162 give the governing equations for our geodynamic models of the mantle.

Finally it can be shown that Equations 12.149–12.151 and 12.158 for $\hat{g} = 0$, give the classical convection model of the mantle. On the other hand, both models presented by the author in this book render possible a study of models of the whole Earth, i.e. together with the lithosphere, outer and inner cores (as was noted in connection with the core study). Such a global geodynamic model of the Earth will be studied later. Based on this global geodynamic model, the global gravity and geomagnetic model of the Earth and planets will be discussed.

12.3.3.2. Analysis of geodynamic processes in the mantle based on thermo-Bingham approximation

INTRODUCTION

Let Ω denote a bounded domain of R^N, $N = 2, 3$, with a Lipschitz boundary $\partial\Omega = \Gamma_u \cup \Gamma_\tau \cup R$, occupied at time t_o by the Earth or its parts, and let Ω_t denote the region occupied by their designated portion at time t, $\Omega_t = \Omega \times (t_o, t_1)$, $\partial\Omega_t = \partial\Omega \times (t_o, t_1)$, and $\Omega_T = \Omega \times \bar{I}$, $\partial\Omega_T = \partial\Omega \times \bar{I}$, $\bar{I} = \langle t_o, t_1\rangle$, where $\Gamma_u = {}^1\Gamma_u \cup {}^2\Gamma_u$, $\Gamma_\tau = {}^1\Gamma_\tau \cup {}^2\Gamma_\tau$ and R be the boundary of the zero measure. The fundamental equations for the moving thermo-visco-plastic media follow from the general conservation law of energy and the law of momentum conservation, which in differential form are as follows

$$\rho(\boldsymbol{x})\beta_{ij}(\boldsymbol{x})T_o(\boldsymbol{x})D_{ij}(\boldsymbol{u})+\rho(\boldsymbol{x})c_e(\boldsymbol{x})\frac{\mathrm{d}T(\boldsymbol{x},t)}{\mathrm{d}t} = W(\boldsymbol{x},t) - \frac{\partial}{\partial x_i}\left(\kappa_{ij}(\boldsymbol{x})\frac{\partial T(\boldsymbol{x},t)}{\partial x_j}\right), \quad \text{in } \Omega_t,$$
$$i, j = 1, \dots, N, \qquad (12.163)$$

$$\rho(\boldsymbol{x})\frac{\mathrm{d}u_i(\boldsymbol{x},t)}{\mathrm{d}t} = \frac{\partial\tau_{ij}(\boldsymbol{u})}{\partial x_j} + F_i(\boldsymbol{x},t), \quad \text{in } \Omega_t, \quad i, j = 1, \dots, N, \qquad (12.164)$$

where ρ is a density, T is a temperature, c_e is a specific heat, W represents thermal sources, κ_{ij} are coefficients of thermal conductivity, β_{ij} is a coefficient of linear thermal expansion, t represents time and $\boldsymbol{x} = (x_i)$ are the Cartesian co-ordinates. Furthermore, $\boldsymbol{u} = (u_i)_{i=1,2}$ is a flow velocity, τ_{ij} is the stress tensor of the Bingham fluid defined by the constitutional stress-strain rate relation, and F_i represents body forces. Moreover, we denote $\mathrm{d}/\mathrm{d}t = \partial/\partial t + u_j\partial/\partial x_j$, $\partial_t \equiv \partial/\partial t$, $\partial_j \equiv \partial/\partial x_j$, $\boldsymbol{u}' \equiv \dot{\boldsymbol{u}} = \partial\boldsymbol{u}/\partial t \equiv \partial_t\boldsymbol{u}$.

The constitutional stress-strain rate and strain rate-displacement relations are defined by

$$\tau_{ij} = -p\delta_{ij} + \hat{g}D_{ij}D_{\mathrm{II}}^{-1/2} + 2\hat{\mu}D_{ij} - \beta_{ij}(T - T_o), \qquad (12.165)$$
$$D_{ij} = \tfrac{1}{2}(\partial_j u_i + \partial_i u_j) \equiv \tfrac{1}{2}(\partial u_i/\partial x_j + \partial u_j/\partial x_i), \quad D_{\mathrm{II}} = \tfrac{1}{2}D_{ij}D_{ij},$$

where $\boldsymbol{u}$ represents a flow velocity vector, p is a pressure, $\hat{g}$ and $\hat{\mu}$ represent thresholds of plasticity (defined by the Mises type relation) and of viscosity of a Bingham medium, respectively, β_{ij} is a coefficient of linear thermal expansion, and T_o is an initial temperature. The last term in Equation 12.165 represents thermal stresses. If the threshold of plasticity $\hat{g}$ is equal to zero then the medium is identical with the Newtonian liquid, and if the threshold of plasticity is equal to infinity then the medium behaves as though absolutely rigid.

We consider the following boundary conditions. On the boundaries ${}^1\Gamma_u$ and ${}^2\Gamma_u$, which are parts of the Earth's surface and which make it possible to describe the influences of the other part of the rock massif on the region of interest, the velocity fields $\boldsymbol{u}_1(\boldsymbol{x}, t)$ and $\boldsymbol{u}_2(\boldsymbol{x}, t)$ as well as the heat flux $q_1(\boldsymbol{x}, t)$ and $q_2(\boldsymbol{x}, t)$ or the temperatures $T_1(\boldsymbol{x}, t)$ and $T_2(\boldsymbol{x}, t)$ are known, so that we have

$$\boldsymbol{u}(\boldsymbol{x}, t) = \boldsymbol{u}_1(\boldsymbol{x}, t), \quad \kappa_{ij}\partial T/\partial x_j n_i = q_1(\boldsymbol{x}, t) \text{ or } T(\boldsymbol{x}, t) = T_1(\boldsymbol{x}, t), \tag{12.166}$$

on the boundary ${}^1\Gamma_u$ and

$$\boldsymbol{u}(\boldsymbol{x}, t) = \boldsymbol{u}_2(\boldsymbol{x}, t), \quad \kappa_{ij}\partial T/\partial x_j n_i = q_2(\boldsymbol{x}, t) \text{ or } T(\boldsymbol{x}, t) = T_2(\boldsymbol{x}, t), \tag{12.167}$$

on the boundary ${}^2\Gamma_u$. Since the rock massif is loaded by the loading $\boldsymbol{P}_1(\boldsymbol{x}, t)$ on the boundary ${}^1\Gamma_\tau$ under the temperature $T_3(\boldsymbol{x}, t)$ then we have

$$\tau_{ij}n_j = P_{1i}(\boldsymbol{x}, t), \quad T(\boldsymbol{x}, t) = T_3(\boldsymbol{x}, t) \tag{12.168}$$

on the boundary ${}^1\Gamma_\tau$. On the surface of the Earth (e.g. in the locality of the high-level radioactive waste repository structure), the spread of the loading $\boldsymbol{P}_2(\boldsymbol{x}, t)$ and the temperature $T_4(\boldsymbol{x}, t)$ or heat flux $q_3(\boldsymbol{x}, t)$ are measured so that

$$\tau_{ij}n_j = P_{2i}(\boldsymbol{x}, t), \quad T(\boldsymbol{x}, t) = T_4(\boldsymbol{x}, t) \text{ or } \kappa_{ij}\partial T/\partial x_j n_i = q_3(\boldsymbol{x}, t), \tag{12.169}$$

For simplicity we shall assume that the rock massif is created by an incompressible medium so that

$$\operatorname{div} \boldsymbol{u} = 0 \quad \text{in } \Omega_t. \tag{12.170}$$

At the initial time $t = t_o$ the initial velocity and temperature fields are prescribed, so that

$$T(\boldsymbol{x}, t_o) = T_o(\boldsymbol{x}), \tag{12.171}$$

$$\boldsymbol{u}(\boldsymbol{x}, t_o) = \boldsymbol{u}_o(\boldsymbol{x}). \tag{12.172}$$

MATHEMATICAL FORMULATION OF THE PROBLEM

Without any detriment of the universality of the problem studied we can solve the following problem.

If the data of the geothermal part of the problem are sufficiently smooth, then the problem, due to the well known substitution $T \to T + T_1 + T_o$, leads to the problem with homogeneous initial-boundary value conditions.

Let Ω be a bounded domain in R^N, $N = 2, 3$, with a smooth boundary $\partial\Omega = \Gamma_u \cup \Gamma_\tau$, where $\Gamma_u = {}^1\Gamma_u \cup {}^2\Gamma_u$, $\Gamma_\tau = {}^1\Gamma_\tau \cup {}^2\Gamma_\tau$. Let $\bar{I} = \langle t_o, t_1\rangle$, $0 < t_1 < +\infty$, $\Omega_t = \Omega \times (t_o, t_1)$, $\partial\Omega_t = \partial\Omega \times (t_o, t_1)$.

For simplicity let us investigate the following non-stationary incompressible thermo-Bingham model problem.

Problem ($\mathscr{P}$):

$$\rho(\boldsymbol{x})\beta_{ij}(\boldsymbol{x})T_o(\boldsymbol{x})D_{ij}(\boldsymbol{u}) + \rho(\boldsymbol{x})c_e(\boldsymbol{x})\frac{\mathrm{d}T(\boldsymbol{x}, t)}{\mathrm{d}t} = W(\boldsymbol{x}, t) - \frac{\partial}{\partial x_i}\left(\kappa_{ij}(\boldsymbol{x})\frac{\partial T(\boldsymbol{x}, t)^-}{\partial x_j}\right), \quad \text{in } \Omega_t,$$

$$i, j = 1, \ldots, N, \tag{12.173}$$

$$\rho(\boldsymbol{x})\frac{\mathrm{d}u_i(\boldsymbol{x},t)}{\mathrm{d}t} = \frac{\partial\tau_{ij}(\boldsymbol{u})}{\partial x_j} + F_i(\boldsymbol{x},t), \quad \text{in } \Omega_t, \quad i,j = 1,\dots,N, \tag{12.174}$$

$$\operatorname{div} \boldsymbol{u}(\boldsymbol{x},t) = 0 \quad \text{in } \Omega_t, \tag{12.175}$$

$$\left.\begin{aligned} &T(\boldsymbol{x},t) = 0, \\ &\tau_{ij}n_j = P_i(\boldsymbol{x},t) \quad i,j = 1,\dots,N, \end{aligned}\right\} \text{ on } \Gamma_\tau \times (t_o,t_1), \tag{12.176, 12.177}$$

$$\left.\begin{aligned} &\kappa_{ij}\partial T/\partial x_j n_i = q(\boldsymbol{x},t), \\ &\boldsymbol{u}(\boldsymbol{x},t) = 0 \end{aligned}\right\} \text{ on } \Gamma_u \times (t_o,t_1), \quad i,j = 1,\dots,N, \tag{12.178, 12.179}$$

$$\left.\begin{aligned} &T(\boldsymbol{x},t_o) = 0, \\ &\boldsymbol{u}(\boldsymbol{x},t_o) = 0 \end{aligned}\right\} \text{ a.e. in } \Omega, \tag{12.180, 12.181}$$

where the stress-strain relation and the strain rate tensor are defined by

$$\tau_{ij}(\boldsymbol{u}) = -p\delta_{ij} + \hat{g}D_{ij}(\boldsymbol{u})D_{\mathrm{II}}^{-1/2}(\boldsymbol{u}) + 2\hat{\mu}D_{ij}(\boldsymbol{u}) - \beta_{ij}(T - T_o), \tag{12.182}$$

$$D_{ij}(\boldsymbol{u}) = \tfrac{1}{2}(\partial_j u_i + \partial_i u_j) \equiv \tfrac{1}{2}(\partial u_i/\partial x_j + \partial u_j/\partial x_i) \quad i,j = 1,\dots,N \tag{12.183}$$

and where $D_{\mathrm{II}}(\boldsymbol{u}) = (1/2)D_{ij}(\boldsymbol{u})D_{ij}(\boldsymbol{u})$, $\boldsymbol{u} = (u_i)_{i=1,2}$ is the flow velocity, τ_{ij} is the stress tensor of the Bingham fluid, p is the pressure, $\hat{g}$ and $\hat{\mu}$ are the thresholds of plasticity and of viscosity of the Bingham fluid, $\boldsymbol{F}$ represents the body forces, $\boldsymbol{P}$ the surface forces, and δ_{ij} is the Kronecker symbol. Furthermore, ρ is the density, c_e is the specific heat, W are the thermal sources, κ_{ij} are coefficients of thermal conductivity, β_{ij} are coefficients of thermal expansion.

VARIATIONAL FORMULATION

We shall employ the Sobolev spaces of functions having generalized derivatives of order s of the type $[H^s(\Omega)]^k$. The norm will be denoted by $\|\cdot\|_{s,k}$, and the scalar product by (.,.) for each k. We set $H^o(\Omega) \equiv L^2(\Omega)$ and denote by (.,.) the scalar product in $L^2(\Omega)$. By $L^\infty(\Omega)$ we denote the set of all measurable functions f defined almost everywhere on Ω and such that $\|f\|_\infty = \text{ ess sup}_\Omega |f(x)|$ is finite. We shall denote by $|u|^2 = \int_\Omega e_{ij}(u)e_{ij}(u)\mathrm{d}x + \|u\|^2_{0,N}$.

Let us define the spaces

$$\begin{aligned} {}^1V &= \{z \mid z \in H^1(\Omega), \quad z = 0 \text{ on } \Gamma_\tau \text{ in the sense of traces }\}, \\ V &= \{\boldsymbol{v} \mid v \in [H^1(\Omega)]^N, \quad \operatorname{div} \boldsymbol{v} = 0, \boldsymbol{v} = 0 \text{ on } \Gamma_u \text{ in the sense of traces }\}, \\ H &= \{\boldsymbol{v} \mid v \in [L^2(\Omega)]^N, \quad \operatorname{div} \boldsymbol{v} = 0, \boldsymbol{v} = 0 \text{ on } \Gamma_u \text{ in the sense of traces }\}, \end{aligned}$$

and let ${}^1V'$, and V' be the dual spaces of ${}^1V$ and V respectively, and ${}^1V \subset L^2(\Omega) \subset {}^1V'$, $V \subset H \subset V'$. Let $D(\Omega)$ be the space of all functions from C^∞ with compact support in Ω. Then

$$\mathscr{V} = \{\boldsymbol{v} \mid \boldsymbol{v} \in [D(\Omega)]^N, \quad \operatorname{div} \boldsymbol{v} = 0\},$$

V_s is the closure of $\mathscr{V}$ in $[H^s(\Omega)]^N$, $s = (1/2)N$, where $H^s(\Omega)$ is a space of restriction on Ω of all elements from $H^s(R^N)$ (see Duvaut and Lions (1972a)), V_s is a Hilbert space with norm $\|\boldsymbol{v}\|^2_{s,N} \equiv \|\boldsymbol{v}\|^2_s = ((\boldsymbol{v},\boldsymbol{v}))_s$ if $((\boldsymbol{v},\boldsymbol{w}))_s = (v_i, w_i)_{H^s(\Omega)}$. Further, let us put $V_1 = V$, $\|\boldsymbol{v}\|_1 = \|\boldsymbol{v}\|_{1,N}$, $V_o = H$, $\|\boldsymbol{v}\|_o = \|\boldsymbol{v}\|_{0,N}$, $((\boldsymbol{v},\boldsymbol{w}))_o = (\boldsymbol{v},\boldsymbol{w})$ and for $s > 1$ $V_s \subset V \subset H \subset V' \subset V'_s$, where V' and V'_s, are dual spaces of V and V_s and $H \equiv H'$.

We shall denote by $\boldsymbol{w}_j$ the eigen-functions of a canonical isomorphism Λ_s (see Duvaut and Lions (1972)), $V_s \to V'_s$, i.e.

$$((\boldsymbol{w}_j,\boldsymbol{v}))_s = \lambda_i(\boldsymbol{w}_j,\boldsymbol{v}) \quad \forall \boldsymbol{v} \in V_s, \qquad |\boldsymbol{w}_j| = 1.$$

Furthermore, ${}^1\mathscr{V} = \{z \mid z \in D(\Omega), \quad z = 0 \text{ on } \partial\Omega\}$, ${}^1V_s$ is the closure of ${}^1\mathscr{V}$ in $H^s(\Omega)$ and ${}^1V_s$ is a Hilbert space with scalar product $((z,y))_s$ and with norm $\|z\|_s = ((z,z))_s$ if $((z,y))_s = (z,y)_{H^s(\Omega)}$. Moreover, for $s > 1$ ${}^1V_s \subset {}^1V \subset {}^1H \subset {}^1V' \subset {}^1V'_s$, where ${}^1V'$ and ${}^1V'_s$ are dual spaces of ${}^1V$ and ${}^1V_s$. Similarly as above we define by $z_j \in {}^1V_s$ the eigen-functions of canonical isomorphisms ${}^1\Lambda_s$, ${}^1V_s \to {}^1V'_s$.

Lemma 12.5: Let $s = (1/2)N$, then $v_{i,j} \in L^N(\Omega)$, $\forall \boldsymbol{v} \in V_s$.

For the proof see Duvaut and Lions (1972a).

Further, we define the space

$$W = \{\boldsymbol{v} \mid \boldsymbol{v} \in L^2(I; V_s), \boldsymbol{v}' \in L^2(I; H), \boldsymbol{v}(t) = 0\},$$
$${}^1W = \{z \mid z \in L^2(I; {}^1V_s), z' \in L^2(I; {}^1H), z(t_o) = 0\}.$$

Lemma 12.6: There exist constants $c_T > 0$ and $c_B > 0$ such that

$$a(\boldsymbol{v},\boldsymbol{v}) \geq c_B \|\boldsymbol{v}\|^2_{1,N}, \quad A(T,T) \geq c_T \|T\|^2_{1,1} \quad \forall \boldsymbol{v} \in V, \forall T \in {}^1V.$$

For the proof see Duvaut and Lions (1972a) and Nečas (1967).

Further we have

$$D_{ij}(\boldsymbol{u})D_{ij}(\boldsymbol{v}) \leq 2D_{\mathrm{II}}^{1/2}(\boldsymbol{u})D_{\mathrm{II}}^{1/2}(\boldsymbol{v}), \quad D_{ij} = (2\hat{\mu})^{-1}(1 - \hat{g}\tau_{\mathrm{II}}^{-1/2})\tau_{ij}^D,$$

where τ_{ij}^D are components of the deviator of the stress tensor.

For $T, z \in {}^1V$, $\boldsymbol{u}, \boldsymbol{v} \in V$ let us define

$$(T'(t), z) = \int_\Omega \rho c_e T'(t) z \,\mathrm{d}\boldsymbol{x},$$

$$a_T(T(t), z) = \int_\Omega \kappa_{ij} \partial T(t)/\partial x_i \partial z/\partial x_j \,\mathrm{d}\boldsymbol{x},$$

$$b_p(\boldsymbol{v}, z) = \int_\Omega \rho T_o \beta_{ij} \partial v_i/\partial x_j z \,\mathrm{d}\boldsymbol{x},$$

$$(Q, z) = \int_\Omega W(t) z \,\mathrm{d}\boldsymbol{x} - \int_{\Gamma_u} qz \,\mathrm{d}s$$

$$(\boldsymbol{u}', \boldsymbol{v}) = \int_\Omega \rho u'_i v_i \,\mathrm{d}\boldsymbol{x},$$

$$a(\boldsymbol{u},\boldsymbol{v}) = 2\int_\Omega D_{ij}(\boldsymbol{u})D_{ij}(\boldsymbol{v})\mathrm{d}\boldsymbol{x},$$

$$j(\boldsymbol{v}) = 2\int_\Omega (D_{\mathrm{II}}(\boldsymbol{v}))^{1/2}\mathrm{d}\boldsymbol{x},$$

$$b(\boldsymbol{u},\boldsymbol{v},\boldsymbol{w}) = \int_\Omega \rho u_i \partial v_i/\partial x_j w_j \mathrm{d}\boldsymbol{x},$$

$$b_o(\boldsymbol{u},g,z) = \int_\Omega \rho c_e u_k \partial g/\partial x_k z \mathrm{d}\boldsymbol{x},$$

$$b_s(T,\boldsymbol{v}) = \int_\Omega \partial/\partial x_j(\beta_{ij}T)v_i \mathrm{d}\boldsymbol{x},$$

$$(\boldsymbol{f},v) = \int_\Omega F_i v_i \mathrm{d}\boldsymbol{x} - \int_{\Gamma_\tau} P_i v_i \mathrm{d}s.$$

Throughout the section we suppose that $W(\boldsymbol{x},t) \in L^2(I; L^2(\Omega))$, $\boldsymbol{F}(\boldsymbol{x},t) \in L^2(I; [L^2(\Omega)]^N)$, and $\boldsymbol{P}(\boldsymbol{x},t) \in L^2(I; [L^2(\Gamma_\tau)]^N)$, and about β_{ij} we shall suppose that $\partial\beta_{ij}/\partial x_j \in L^\infty(\Omega)\, \forall i, j$. Moreover, we suppose that $\kappa_{ij} \in L^\infty(\Omega)$ are Lipschitz on Ω and fulfil the symmetry condition $\kappa_{ij} = \kappa_{ji}$ on Ω for every $i, j = 1, \ldots, N$ and

$$\kappa_{ij}\xi_i\xi_j \geq c_o\,|\xi|^2, \quad x \in \Omega, \quad \xi \in R^{N^2}, \quad c_o = \text{ constant} > 0,$$

and that all of the other physical parameters are positive constants.

It holds (see, for example, Temam (1977)) that

(i) $b(\boldsymbol{v},\boldsymbol{v},\boldsymbol{v}) = 0.$

(ii) From the Hölder inequality applied to $b(\boldsymbol{u},\boldsymbol{v},\boldsymbol{w})$, we have

$$|b(\boldsymbol{u},\boldsymbol{v},\boldsymbol{w})| \leq c_1\,\|\boldsymbol{u}\|_{[L^p(\Omega)]^N}\|\boldsymbol{w}\|_{[L^p(\Omega)]^N}\sum_{i,j}\left\|\partial_i v_j\right\|_{[L^N(\Omega)]^N},\; 2p^{-1}+N^{-1}=1.$$

Applying Lemma 12.6 we have

$$|b(\boldsymbol{u},\boldsymbol{v},\boldsymbol{w})| \leq c_2\,\|\boldsymbol{u}\|_{[L^p(\Omega)]^N}\,\|\boldsymbol{w}\|_{[L^p(\Omega)]^N}\,\|\boldsymbol{v}\|_s, \quad s = \tfrac{1}{2}N. \tag{12.184}$$

Since

$$\|\boldsymbol{v}\|_{[H^{1/2}(\Omega)]^N} \leq c\|\boldsymbol{v}\|^{1/2}_{[L^2(\Omega)]^N}\|\boldsymbol{v}\|^{1/2}_{[H^1(\Omega)]^N} \text{ and } H^{1/2}(\Omega)\subset L^p(\Omega),\, 2p^{-1}+N^{-1}=1 \tag{12.185}$$

then

$$\|\boldsymbol{v}\|_{[L^p(\Omega)]^N} \leq c_3\,\|\boldsymbol{v}\|^{1/2}_{[H^1(\Omega)]^N}\,\|\boldsymbol{v}\|^{1/2}_{[L^2(\Omega)]^N} \quad \forall \boldsymbol{v} \in [H^1_o(\Omega)]^N,\, 2p^{-1}+N^{-1}=1.$$

From Equations 12.184 and 12.185 we obtain

$$|b(\boldsymbol{u},\boldsymbol{v},\boldsymbol{w})| \leq c_4\,\|\boldsymbol{u}\|^{1/2}_{1,N}\,\|\boldsymbol{u}\|^{1/2}_{0,N}\,\|\boldsymbol{w}\|^{1/2}_{1,N}\,\|\boldsymbol{w}\|^{1/2}_{0,N}\,\|\boldsymbol{v}\|_s, \quad s = \tfrac{1}{2}N.$$

(iii) If $\boldsymbol{u},\boldsymbol{v},\boldsymbol{w} \in \mathscr{V}$ then $b(\boldsymbol{u},\boldsymbol{v},\boldsymbol{w}) + b(\boldsymbol{u},\boldsymbol{w},\boldsymbol{v}) = 0$. The last relation is valid also for $\boldsymbol{u} \in [L^p(\Omega)]^N$, $\boldsymbol{v},\boldsymbol{w} \in V_s$.

(iv) $b(\boldsymbol{u},\boldsymbol{u},\boldsymbol{v}) = -b(\boldsymbol{u},\boldsymbol{v},\boldsymbol{u}).$

Multiplying Equation 12.184 by $z - T(t)$ and Equation 12.185 by $\boldsymbol{v} - \boldsymbol{u}(t)$, then integrating over Ω, using the Green theorem, and respecting the boundary conditions, then after some modifications we obtain the following variational formulation of our problem.

Problem $(\mathscr{P})_V$**:** Find a pair of functions (T, u), $T \in {}^1V$, $\boldsymbol{u} \in V$ such that

$$(T'(t), z - T(t)) + b_o(\boldsymbol{u}(t), T(t), z - T(t)) + a_T(T(t), z - T(t)) + b_p(\boldsymbol{u}(t), z - T(t))$$
$$-(W(t), z - T(t)) \geq 0 \quad \forall z \in {}^1V, \tag{12.186}$$

$$(\boldsymbol{u}'(t), \boldsymbol{v} - \boldsymbol{u}(t)) + \hat{\mu}a(\boldsymbol{u}(t), \boldsymbol{v} - \boldsymbol{u}(t)) + b(\boldsymbol{u}(t), \boldsymbol{u}(t), \boldsymbol{v} - \boldsymbol{u}(t)) + b_s(T(t) - T_o, \boldsymbol{v} - \boldsymbol{u}(t))$$
$$+\hat{g}j(\boldsymbol{v}) - \hat{g}j(\boldsymbol{u}(t)) \geq (\boldsymbol{f}(t), \boldsymbol{v} - \boldsymbol{u}(t)) \quad \forall \boldsymbol{v} \in V, \tag{12.187}$$

$$T(\boldsymbol{x}, t_o) = 0, \tag{12.188}$$

$$\boldsymbol{u}(x, t_o) = 0. \tag{12.189}$$

The bilinear forms $T, z \to a_T(T, z)$ and $\boldsymbol{u}, \boldsymbol{v} \to a(\boldsymbol{u}, \boldsymbol{v})$ are symmetric and continuous on ${}^1V$ or V respectively, and coercive (i.e. from the above assumption the symmetry $a_T(T, z) = a_T(z, T)$, $a(\boldsymbol{u}, \boldsymbol{v}) = a(\boldsymbol{v}, \boldsymbol{u})$ hold and, moreover,there exists $c_o > 0$ such that $a_T(T, T) \geq c_o \|T\|^2$ $\forall T \in V$, and constant $c_1 > 0$ such that $a(\boldsymbol{u}, \boldsymbol{u}) \geq c_1 \|\boldsymbol{u}\|_{1,N}^2$ $\forall \boldsymbol{u} \in V$).

From Equations 12.186 and 12.187 it follows that

$$(T'(t), z - T(t)) + (\boldsymbol{u}'(t), \boldsymbol{v} - \boldsymbol{u}(t)) + a_T(T(t), z - T(t)) + \hat{\mu}a(\boldsymbol{u}(t), \boldsymbol{v} - \boldsymbol{u}(t))$$
$$+ b_o(\boldsymbol{u}(t), T(t), z - T(t)) + b(\boldsymbol{u}(t), \boldsymbol{u}(t), \boldsymbol{v} - \boldsymbol{u}(t)) + b_p(\boldsymbol{u}(t), z - T(t))$$
$$+ b_s(T(t) - T_o, \boldsymbol{v} - \boldsymbol{u}(t)) + \hat{g}j(\boldsymbol{v}) - \hat{g}j(\boldsymbol{u}(t))$$
$$\geq (Q(t), z - T(t)) + (\boldsymbol{f}(t), \boldsymbol{v} - \boldsymbol{u}(t)) \quad \forall z \in {}^1V, \forall \boldsymbol{v} \in V, \tag{12.190}$$

and the initial conditions in Equations 12.188 and 12.189 hold.

The main result of this section is contained in the following theorem.

Theorem 12.6: Let $N \geq 2$, $s = (1/2)N$. Let the assumptions about $a_T(T, z)$ and $a(\boldsymbol{u}, \boldsymbol{v})$ be fulfilled, let $Q \in L^2(I; {}^1V_s')$, $\boldsymbol{f} \in L^2(I; V_s')$, $T_o \in L^2(\Omega)$, $\boldsymbol{u}_o \in H$. Then there exists a pair of functions $(T, \boldsymbol{u})$ such that $T \in L^2(I; {}^1V_s) \cap L^\infty(I; {}^1H)$, $T' \in L^2(I; {}^1V_s')$, $\boldsymbol{u} \in L^2(I; V_s) \cap L^\infty(I; H)$, $\boldsymbol{u}' \in L^2(I; V_s')$ and $T(\boldsymbol{x}, t_o) = T_o(\boldsymbol{x})$, $\boldsymbol{u}(\boldsymbol{x}, t_o) = \boldsymbol{u}_o(\boldsymbol{x})$, and $(T, \boldsymbol{u})$ satisfies the relation

$$\int_I \{(T'(t), z - T(t)) + (\boldsymbol{u}'(t), \boldsymbol{v} - \boldsymbol{u}(t)) + a_T(T(t), z - T(t)) + \hat{\mu}a(\boldsymbol{u}(t), \boldsymbol{v} - \boldsymbol{u}(t))$$
$$+ b_o(\boldsymbol{u}(t), T(t), z - T(t)) + b(\boldsymbol{u}(t), \boldsymbol{u}(t), \boldsymbol{v} - \boldsymbol{u}(t)) + b_p(\boldsymbol{u}(t), z - T(t)) + b_s(T(t) - T_o, \boldsymbol{v} - \boldsymbol{u}(t))$$
$$+ \hat{g}j(\boldsymbol{v}) - \hat{g}j(\boldsymbol{u}(t)) - (Q(t), z - T(t)) - (\boldsymbol{f}(t), \boldsymbol{v} - \boldsymbol{u}(t))\}\mathrm{d}t \geq 0 \quad \forall z \in {}^1V, \forall \boldsymbol{v} \in V, \tag{12.191}$$

and such that

$$z(t) \in L^2(I; {}^1V_s), z'(t) \in L^2(I; {}^1H), \quad T(t_o) = 0,$$
$$\boldsymbol{u}(t) \in L^2(I; V_s), \boldsymbol{u}'(t) \in L^2(I; H), \quad \boldsymbol{u}(t_o) = 0.$$

Proof: To prove the theorem the double regularization will be used. Firstly, the regularization of $j(\boldsymbol{v})$ by the regularized functional $j_\varepsilon(\boldsymbol{v})$ will be used. The regularized functional $j_\varepsilon(\boldsymbol{v})$ is defined by

$$j_\varepsilon(\boldsymbol{v}) = 2(1+\varepsilon)^{-1} \int_\Omega (D_{\mathrm{II}}(\boldsymbol{v}))^{(1+\varepsilon)/2} \mathrm{d}\boldsymbol{x}, \qquad \varepsilon > 0,$$

for which

$$(j'_\varepsilon(\boldsymbol{u}), \boldsymbol{v}) = \int_\Omega D_{\mathrm{II}}(\boldsymbol{u})^{(1/2)(1-\varepsilon)} D_{ij}(\boldsymbol{u}) D_{ij}(\boldsymbol{v}) \mathrm{d}\boldsymbol{x}.$$

Further the viscosity regularized terms, i.e. terms $\eta((T_{\varepsilon\eta}, z))_s$ and $\eta((\boldsymbol{u}_{\varepsilon\eta}, \boldsymbol{v}))_s$, $s = (1/2)N$, $\eta > 0$, will be added. For $N = 2$ then $s = 1$ and thus ${}^iV_s = {}^iV$, $i = 0, 1$, and the viscosity regularized terms are of the same order as the bilinear forms and therefore can be omitted.

Therefore, we shall solve the double regularized problem as follows

Problem $(\mathscr{P}_r)_v$:

$$\begin{aligned} &(T'_{\varepsilon\eta}(t), z) + b_o(\boldsymbol{u}_{\varepsilon\eta}(t), T_{\varepsilon\eta}(t), z) + a_T(T_{\varepsilon\eta}(t), z) \\ &\quad + b_p(\boldsymbol{u}_{\varepsilon\eta}(t), z) - (Q(t), z) = 0 \quad \forall z \in {}^1V_s, \end{aligned} \tag{12.192}$$

$$\begin{aligned} &(\boldsymbol{u}'_{\varepsilon\eta}(t), \boldsymbol{v}) + \hat{\mu} a(\boldsymbol{u}_{\varepsilon\eta}(t), \boldsymbol{v}) + b(\boldsymbol{u}_{\varepsilon\eta}(t), \boldsymbol{u}_{\varepsilon\eta}(t), \boldsymbol{v}) + b_s(T_\varepsilon(t) - T_o, \boldsymbol{v}) \\ &\quad + \hat{g}(j'_\varepsilon(\boldsymbol{u}_{\varepsilon\eta}(t), \boldsymbol{v}) = (\boldsymbol{f}(t), \boldsymbol{v}) \quad \forall \boldsymbol{v} \in V_s, \end{aligned} \tag{12.193}$$

with the initial conditions

$$T_{\varepsilon\eta}(\boldsymbol{x}, t_o) = 0, \tag{12.194}$$

$$\boldsymbol{u}_{\varepsilon\eta}(\boldsymbol{x}, t_o) = 0. \tag{12.195}$$

The idea of the proof is:

(i) proving the existence of the solution to Equations 12.192-12.195, based on the Galerkin approximation;

(ii) deriving a priori estimates I and II independent of ε, η;

(iii) performing the limitation procedures over m and ε, η;

(iv) proving the uniqueness of the solution.

The existence of $(T_{\varepsilon\eta}, \boldsymbol{u}_{\varepsilon\eta})$ will be proved by means of the finite-dimensional approximation (Galerkin). We construct separable bases $\{T_i\}$, $\{\boldsymbol{v}_i\}$ of spaces ${}^1V_s$ and V_s, and since elements of these bases are linearly independent then their linear combinations are dense in ${}^1V_s$ and V_s respectively, as ${}^1V_s$ and V_s are separable spaces. Let ${}^1\mathscr{W}^m$ and $\mathscr{W}^m$ be spaces created by $z_1, \ldots, z_m$ and $\boldsymbol{v}_1, \ldots, \boldsymbol{v}_m$, respectively. Then the approximate solution of order m satisfies the regular system of ordinary differential equations of first order

$$\begin{aligned} &(T'_m(t), z_j) + b_o(\boldsymbol{u}_m(t), T_m(t), z_j) + a_T(T_m(t), z_j) \\ &\quad + b_p(\boldsymbol{u}_m(t), z_j) - (Q(t), z_j) \geq 0 \quad \forall z_j \in {}^1V^m, \end{aligned} \tag{12.196}$$

$$\begin{aligned} &(\boldsymbol{u}'_m(t), \boldsymbol{v}_j) + \hat{\mu} a(\boldsymbol{u}_m(t), \boldsymbol{v}_j) + b(\boldsymbol{u}_m(t), \boldsymbol{u}_m(t), \boldsymbol{v}_j) + b_s(T_m(t) - T_o, \boldsymbol{v}_j) + \hat{g} j(\boldsymbol{v}_j) - \hat{g} j(\boldsymbol{u}_m(t)) \\ &\quad \geq (\boldsymbol{f}(t), \boldsymbol{v}_j - \boldsymbol{u}_m(t)) \quad \forall \boldsymbol{v}_j \in V^m, \end{aligned} \tag{12.197}$$

with initial conditions

$$T_m(\boldsymbol{x}, t_o) = 0, \tag{12.198}$$

$$\boldsymbol{u}_m(\boldsymbol{x}, t_o) = 0, \tag{12.199}$$

where ${}^1V^m$, V^m are spaces created by z_j, $\boldsymbol{v}_j$, $1 \le j \le m$. Since $\{z_j\}$, $\{\boldsymbol{v}_j\}$ are linearly independent, then $(T_m, \boldsymbol{u}_m)$ is uniquely defined by Equations 12.196-12.199 on the interval $\overline{I}_m = \langle t_o, t_m \rangle$.

A priori estimates show that $t_m = t_1$.

We have a priori estimate I. It holds that

$$b(\boldsymbol{v}_m(t), \boldsymbol{v}_m(t), \boldsymbol{v}_m(t)) = 0, \tag{12.200}$$

$$b_o(\boldsymbol{v}_m(t), T_m(t), T_m(t)) = 0, \tag{12.201}$$

$$\begin{aligned} |X_m| &= \left| b_s(T_m(t) - T_o, \boldsymbol{v}_m(t)) + b_p(\boldsymbol{v}_m(t), T_m(t)) \right| \\ &\le c(1 + \|T_m(t)\|_{1,1} \|\boldsymbol{v}_m(t)\|_{o,N} + \|T_m(t)\|_{o,1} \|\boldsymbol{v}_m(t)\|_{1,N}), \end{aligned} \tag{12.202}$$

where c is a positive constant independent of m and ε, as $\boldsymbol{v}_m(t) = 0$ on $\Gamma_u \times (t_o, t_1)$, $T_m(t) = 0$ on $\Gamma_\tau \times (t_o, t_1)$ and $\partial_j \beta_{ij} \in L^\infty(\Omega)$, $\forall i, j$. Further,

$$a_T(T, z) = a_T(z, T), \quad a_T(z, z) \ge \alpha_o \|z\|^2 \quad \forall z \in {}^1V,$$

$$(j'(\boldsymbol{v}), \boldsymbol{v}) \ge 0, \quad a(\boldsymbol{u}, \boldsymbol{v}) = a(\boldsymbol{v}, \boldsymbol{u}), \quad a(\boldsymbol{v}, \boldsymbol{v}) \le \alpha \|\boldsymbol{v}\|^2 \quad \forall \boldsymbol{v} \in V.$$

Adding Equations 12.196-12.199 and after some modification we obtain

$$\begin{aligned} &\frac{1}{2}\frac{\mathrm{d}}{\mathrm{d}t}\{\|\boldsymbol{u}_m(t)\|_{0,N}^2 + \|T_m(t)\|_{0,1}^2\} + \hat{\mu} a(\boldsymbol{u}_m(t), \boldsymbol{u}_m(t)) + a_T(T_m(t), T_m(t)) + \hat{g}(j'_\varepsilon(\boldsymbol{u}_m(t)), \boldsymbol{u}_m(t)) \\ &\quad + \eta \|\boldsymbol{u}_m(t)\|_s^2 + \eta \|T_m(t)\|_s^2 + b_p(\boldsymbol{u}'_m(t), T_m(t)) + b_s(T_m(t), \boldsymbol{u}_m(t)) \\ &= (Q(T), T_m(t)) + (\boldsymbol{f}(t), \boldsymbol{u}_m(t)). \end{aligned}$$

Integrating over t from t_o to t, we obtain

$$\begin{aligned} &\|\boldsymbol{u}_m(t)\|_{0,N}^2 + \|T_m(t)\|_{0,1}^2 + 2\alpha\hat{\mu} \int_{t_o}^t \|\boldsymbol{u}_m(\tau)\|_{1,N}^2 \,\mathrm{d}\tau + 2\alpha_o \int_{t_o}^t \|T_m(\tau)\|_{1,1}^2 \,\mathrm{d}\tau \\ &\quad + 2\eta \int_{t_o}^t \|\boldsymbol{u}_m(\tau)\|_s^2 \,\mathrm{d}\tau + 2\eta \int_{t_o}^t \|T_m(\tau)\|_s^2 \,\mathrm{d}\tau \\ &\le 2c \int_{t_o}^t ((1 + \|T_m(\tau)\|_{1,1} + \|\boldsymbol{f}(\tau)\|_*) \|\boldsymbol{v}_m(\tau)\|_{1,N} + \|T_m(\tau)\|_{0,1} \|\boldsymbol{v}_m(\tau)\|_{1,N})\mathrm{d}\tau \\ &\quad + c \int_{t_o}^t \|Q(\tau)\|_* \|T_m(t)\|_{1,1} \,\mathrm{d}\tau. \end{aligned}$$

Using the Gronwall's lemma after some modification we obtain estimates

$$\|\boldsymbol{u}_m(t)\|_{0,N} \le c, \tag{12.203}$$

$$\|T_m(t)\|_{0,1} \le c, \tag{12.204}$$

$$\int_I \|\boldsymbol{u}_m(\tau)\|_{1,N}^2 \,\mathrm{d}\tau \le c, \tag{12.205}$$

$$\int_I \|T_m(\tau)\|_{1,1}^2 \,\mathrm{d}\tau \le c, \tag{12.206}$$

$$\eta \int_I \|\boldsymbol{u}_m(\tau)\|_s^2 \,\mathrm{d}\tau \le c, \tag{12.207}$$

$$\eta \int_I \|T_m(\tau)\|_s^2 \,\mathrm{d}\tau \le c, \tag{12.208}$$

where $c =$ constant > 0 independent of m.

From these estimates we obtain

$$T_m(t) \in \text{a bounded set of the space } L^2(I;{}^1V) \cap L^\infty(I;{}^1H), \tag{12.209}$$

$$\boldsymbol{u}_m(t) \in \text{a bounded set of the space } L^2(I;V) \cap L^\infty(I;H), \tag{12.210}$$

$$\eta^{1/2}T_m(t) \in \text{a bounded subset of the space } L^2(I;{}^1V_s), \tag{12.211}$$

$$\eta^{1/2}\boldsymbol{u}_m(t) \in \text{a bounded subset of the space } L^2(I;{}^1V_s). \tag{12.212}$$

Now consider a priori estimate II. We shall show that

$$T'_m(t) \in \text{a bounded subset of the space } L^2(I;{}^1V'_s), \tag{12.213}$$

$$\boldsymbol{u}'_m(t) \in \text{a bounded subset of the space } L^2(I;V'_s), \tag{12.214}$$

For $\boldsymbol{v} \in V_s$ it holds that $b(\boldsymbol{u}_m(t), \boldsymbol{u}_m(t), \boldsymbol{v}) = -b(\boldsymbol{u}_m(t), \boldsymbol{v}, \boldsymbol{u}_m(t))$, and due to Equations 12.196, 12.197 and 12.203-12.208 we have

$$|b(\boldsymbol{u}_m(t), \boldsymbol{u}_m(t), \boldsymbol{v})| \leq c\,\|\boldsymbol{u}_m(t)\|_{1,N}\,\|\boldsymbol{u}_m(t)\|_{0,N}\,\|v\|_s \leq c\,\|\boldsymbol{v}_m(t)\|_{1,N}\,\|\boldsymbol{v}\|_s\,, \tag{12.215}$$

so that due to Equations 12.203-12.208

$$b(\boldsymbol{u}_m(t), \boldsymbol{u}_m(t), \boldsymbol{v}) + b_s(T_m(t), \boldsymbol{v}) = (\boldsymbol{h}_m(t), \boldsymbol{v}) \quad \forall \boldsymbol{v} \in V,$$

where $\boldsymbol{h}_m \in$ a bounded subset of $L^2(I, V'_s)$, $\|\boldsymbol{h}_m\|_{L^2(I,V_s)} \leq c > 0$, $c =$ constant < 0 independent of m, ε, η, and

$$b_o(\boldsymbol{u}_m(t), T_m(t), z) + b_p(\boldsymbol{v}'_m(t), z) = (g_m(t), z),$$

where $g_m \in$ a bounded subset of $L^2(I, {}^1V'_s)$, $\|g_m\|_{L^2(I,{}^1V'_s)} \leq c > 0$.
Bilinear forms $T \to a_T(T, z)$ and $\boldsymbol{u} \to a(\boldsymbol{u}, \boldsymbol{v})$ are continuous on ${}^1V$ and V respectively, so that

$$a_T(T, z) = (A_T T, z), a(\boldsymbol{u}, \boldsymbol{v}) = (A\boldsymbol{u}, \boldsymbol{v}), A_T \in \mathscr{L}({}^1V, {}^1V'), A \in \mathscr{L}(V, V'),$$

$$((T, z))_s = (\mathscr{T}_s T, z), \mathscr{T}_s \in \mathscr{L}({}^1V_s, {}^1V'_s),$$

$$((\boldsymbol{u}, \boldsymbol{v}))_s = (\Lambda_s \boldsymbol{u}, \boldsymbol{v}), \Lambda_s \in \mathscr{L}(V_s, V'_s).$$

Then Equations 12.192 and 12.193 can be written in the following form

$$(T'_m(t) + A_T T_m(t) + \eta \mathscr{T}_s T_m + g_m(t) - Q(t), z_j) = 0 \quad 1 \leq j \leq m, \tag{12.216}$$

$$(u'_m(t) + \hat{\mu} A\boldsymbol{u}_m(t) + \hat{g} j'_\varepsilon(\boldsymbol{u}_m(t)) + \boldsymbol{h}_m(t) - \boldsymbol{f}(t), \boldsymbol{v}_j) = 0, \quad 1 \leq j \leq m. \tag{12.217}$$

Let P_m and R_m be orthogonal projections $L^2(\Omega) \to {}^1\underline{W}^m = \{z_i\}_{i=1}^m$ or $H \to \underline{W}^m = \{\boldsymbol{v}_i\}_{i=1}^m$, respectively. Then

$$P_m g = \sum_{j=1}^m (g, z_j) z_j, \quad R_m \boldsymbol{h} = \sum_{i=1}^m (\boldsymbol{h}, \boldsymbol{v}_j)\boldsymbol{v}_j \quad 1 \leq j \leq m,$$

and since $P_m T'_m = T'_m$, $R_m \boldsymbol{u}'_m = \boldsymbol{u}'_m$ then

$$T'_m = P_m(Q - A_T T_m - \eta \mathscr{T}_s T_m - g_m), \quad \boldsymbol{u}'_m = R_m(\boldsymbol{f} - \hat{\mu} A\boldsymbol{u}_m - \eta \Lambda_s \boldsymbol{u}_m - \hat{g} j'_\varepsilon(\boldsymbol{u}_m) - \boldsymbol{h}_m). \tag{12.218}$$

Then due to Equation 12.209-12.212 we find

$$A_T T_m \in \text{ a bounded subset } P_T \text{ of the space } L^2(I;{}^1V') \cap L^2(I;{}^1V'_s),$$

$$A\boldsymbol{u}_m \in \text{ a bounded subset } P_B \text{ of the space } L^2(I;V') \cap L^2(I;V'_s),$$

and therefore due to Equation 12.209-12.212

$$\|\eta \mathscr{T}_s T_m\|_{L^2(I;{}^1V')} = O(\eta^{1/2}), \quad \|\eta \Lambda_s \boldsymbol{u}_m\|_{L^2(I;V')} = O(\eta^{1/2}).$$

Due to the definition of $(j'_\varepsilon(\boldsymbol{u}), \boldsymbol{v})$ we obtain

$$\left\|j'_\varepsilon(\boldsymbol{u})\right\|_* \leq c\left(\int_\Omega (D_{\mathrm{II}}(\boldsymbol{u}))^\varepsilon \mathrm{d}\boldsymbol{x}\right)^{1/2}$$

and hence $j'_\varepsilon(\boldsymbol{u}_m) \in$ a bounded subset of $L^2(I, V')$.
Thus Equations 12.216 and 12.217 indicate that $T'_m = P_m p_m$, $\boldsymbol{u}'_m = R_m \boldsymbol{r}_m$, $p_m \in \mathscr{P}_m \subset L^2(I, {}^1V'_s)$, $\boldsymbol{r}_m \in \mathscr{R}_m \subset L^2(I, V'_s)$, where $\mathscr{P}_m$, and $\mathscr{R}_m$ are bounded subsets of $L^2(I; {}^1V'_s)$ and $L^2(I; V'_s)$, respectively. We have

$$\|P_m p\|_{{}^1V'_s} \leq c\,\|p\|_{{}^1V'_s} = c\left\|\mathscr{T}_s^{-1} p\right\|_{{}^1V_s} \text{ and } \|R_m \boldsymbol{r}\|_{V'_s} \leq c\,\|\boldsymbol{r}\|_{V'_s} = c\left\|\Lambda_s^{-1}\boldsymbol{r}\right\|_{V_s}. \tag{12.219}$$

This holds because $\lambda^{1/2}\boldsymbol{v}_j$ is a total orthogonal system in V'_s (according to the norm $\|\boldsymbol{r}\|^2_{V_s} = \left\|\Lambda_s^{-1}\boldsymbol{r}\right\|_{V_s}$) and thus $\|\boldsymbol{r}\|^2_{V'_s} = \sum_{j=1}^m (\boldsymbol{r}, \boldsymbol{v}_j\lambda_j^{1/2})$, $\|R_m\boldsymbol{r}\|^2_{V'_s} = \sum_{j=1}^m (\boldsymbol{r}, \boldsymbol{v}_j\lambda^{1/2})$. Similarly for P_m. From there the estimates in Equation 12.219 follow immediately, where $c = 1$. Then $(T'_m, \boldsymbol{u}'_m) \in$ a bounded subset of $L^2(I, {}^1V') \times L^2(I, V')$, which completes this part of the proof.

The convergence of the finite-dimensional approximation will be proved under the assumption that ε and η are fixed. From the a priori estimates the subsequence (T_μ, u_μ) of the sequence (T_m, u_m) can be taken such that

$$\begin{aligned}
(T_\mu, \boldsymbol{u}_\mu) &\to (T_{\varepsilon\eta}, \boldsymbol{u}_{\varepsilon\eta}) && \text{in } L^2(I, L^2(\Omega)) \times L^2(I, H) * - \text{ weakly,}\\
(T_\mu, \boldsymbol{u}_\mu) &\to (T_{\varepsilon\eta}, \boldsymbol{u}_{\varepsilon\eta}) && \text{in } L^2(I, {}^1V) \times L^2(I, V) \text{ weakly,}\\
(T'_\mu, \boldsymbol{u}'_\mu) &\to (T'_{\varepsilon\eta}, \boldsymbol{u}'_{\varepsilon\eta}) && \text{in } L^2(I, {}^1V') \times L^2(I, V') \text{ weakly,}\\
(T_\mu, \boldsymbol{u}_\mu) &\to (T_{\varepsilon\eta}, \boldsymbol{u}_{\varepsilon\eta}) && \text{in } L^2(I, L^2(\Omega)) \times L^2(I, L^2(\Omega)) \text{ strongly.}
\end{aligned}$$

Thus for the appropriate components $u_{i\mu}$ of $\boldsymbol{u}_\mu$ we have

$$v_{i\mu} \to v_{i\varepsilon\eta} \text{ weakly a.e. in } \Omega \times I.$$

Furthermore, $\{j'_\varepsilon(\boldsymbol{u}_m)\}$ and $\{u_{i\mu}u_{j\mu}\}$ are from bounded subsets of $L^2(I, V')$ or $L^2(I, L^{p/2}(\Omega))$, respectively. Then we can assume that

$$\begin{aligned}
j'_\varepsilon(\boldsymbol{u}_\mu) &\to \chi && \text{weakly in } L^2(I, V'),\\
u_{i\mu}u_{j\mu} &\to \Theta_{ij} && \text{weakly in } L^2(I, L^{p/2}(\Omega)).
\end{aligned}$$

Hence and as $u_{i\mu}u_{j\mu} \to u_{i\varepsilon\eta}u_{j\varepsilon\eta}$ in the sense of distributions in $\Omega \times I$ we obtain that $\Theta_{ij} = u_{i\varepsilon\eta}u_{j\varepsilon\eta}$. But

$$b(\boldsymbol{u}_\mu, \boldsymbol{u}_\mu, \boldsymbol{v}_j) = -b(\boldsymbol{u}_\mu, \boldsymbol{v}_j, \boldsymbol{u}_\mu) \to -b(\boldsymbol{u}_{\varepsilon\eta}, \boldsymbol{v}_j, \boldsymbol{u}_{\varepsilon\eta}) \text{ weakly in } L^2(I) \quad \forall \boldsymbol{v}_j$$

(see Duvaut and Lions (1972a)). For $m = \mu$ then from Equation 12.197 we obtain

$$\begin{aligned}
&(\boldsymbol{u}'_{\varepsilon\eta}(t), \boldsymbol{v}_j) + \hat{\mu}a(\boldsymbol{u}_{\varepsilon\eta}(t), \boldsymbol{v}_j) + b(\boldsymbol{u}_{\varepsilon\eta}(t), \boldsymbol{u}_{\varepsilon\eta}(t), \boldsymbol{v}_j) + b_s(T_{\varepsilon\eta}(t) - T_o, \boldsymbol{v}_j) + \hat{g}(\chi, \boldsymbol{v}_j)\\
&\quad + \eta((\boldsymbol{u}_{\varepsilon\eta}, \boldsymbol{v}_j))_s \geq (\boldsymbol{f}(t), \boldsymbol{v}_j) \quad \forall j,
\end{aligned} \tag{12.220}$$

where we denote $\chi = j'_\varepsilon(\boldsymbol{u}_{\varepsilon\mu})$. Since the system $\{\boldsymbol{v}_j\}$ is complete in V_s, then from Equation 12.220 it follows that

$$\begin{aligned}
&(\boldsymbol{u}'_{\varepsilon\eta}(t), \boldsymbol{v}) + \hat{\mu}a(\boldsymbol{u}_{\varepsilon\eta}(t), \boldsymbol{v}) + b(\boldsymbol{u}_{\varepsilon\eta}(t), \boldsymbol{u}_{\varepsilon\eta}(t), \boldsymbol{v}) + b_s(T_{\varepsilon\eta}(t) - T_o, \boldsymbol{v})\\
&\quad + \eta((\boldsymbol{u}_{\varepsilon\eta}, \boldsymbol{v}))_s + \hat{g}(\chi, \boldsymbol{v}) \geq (\boldsymbol{f}(t), \boldsymbol{v}) \quad \forall \boldsymbol{v} \in V.
\end{aligned}$$

Since the systems of basis functions $\{T_j\}$ are also complete in ${}^1V_s$, then Equation 12.192 is satisfied. Hence it is enough to prove that

$$X = j'_\varepsilon(\boldsymbol{v}_{\varepsilon\eta}). \tag{12.221}$$

To prove Equation 12.221 the property of monotonicity will be used. Let

$$T \in L^2(I; {}^1V_s) \text{ such that } T' \in L^2(I; {}^1V'_s), \quad T(t_o) = 0,$$
$$\boldsymbol{v} \in L^2(I; V_s) \text{ such that } \boldsymbol{v}' \in L^2(I; V'_s), \quad \boldsymbol{v}(t_o) = 0.$$

By using a similar technique as in Duvaut and Lions (1972a,b), where we put

$$X_\mu = \hat{g}\int_I (j'_\varepsilon(\boldsymbol{u}_\mu) - j'_\varepsilon(\boldsymbol{v}), \boldsymbol{u}_\mu - \boldsymbol{v})\mathrm{d}t + \hat{\mu}\int_I a(\boldsymbol{u}_\mu - \boldsymbol{v}, \boldsymbol{u}_\mu - \boldsymbol{v})\mathrm{d}t + \eta\int_I \|\boldsymbol{u}_\mu - \boldsymbol{v}\|_s^2\,\mathrm{d}t$$
$$+ \int_I (\boldsymbol{u}'_\mu - \boldsymbol{v}', \boldsymbol{u}_\mu - \boldsymbol{v})\mathrm{d}t + \int_I a_T(T_\mu - z, T_\mu - z)\mathrm{d}t + \int_I (T'_\mu - z, T_\mu - z)\mathrm{d}t + \eta\int_I \|\boldsymbol{u}_\mu - \boldsymbol{v}\|_s^2\,\mathrm{d}t.$$

From Equation 12.197 it follows

$$X_\mu = \int_I (\boldsymbol{f}, \boldsymbol{u}_\mu)\mathrm{d}t - \hat{g}\int_I \{(j'_\varepsilon(\boldsymbol{u}_\mu), \boldsymbol{v}) + (j'_\varepsilon(\boldsymbol{v}), \boldsymbol{u}_\mu - \boldsymbol{v})\}\mathrm{d}t - \hat{\mu}\int_I \{a(\boldsymbol{u}_\mu, \boldsymbol{v}) + a(\boldsymbol{v}, \boldsymbol{u}_\mu - \boldsymbol{v})\}\mathrm{d}t$$
$$- \eta\int_I \{((\boldsymbol{u}_\mu, \boldsymbol{v}))_s + ((\boldsymbol{u}_\mu, \boldsymbol{u}_\mu - \boldsymbol{v}))_s\}\mathrm{d}t + \int_I (Q, T_\mu)\mathrm{d}t - \int_I \{a_T(T_\mu, z) + a_T(z, T_\mu - z)\}\mathrm{d}t$$
$$- \eta\int_I \{((T_\mu, z))_s - ((z, T_\mu - z))_s\}\mathrm{d}t - \int_I \{(\boldsymbol{u}'_\mu, \boldsymbol{v}) + (\boldsymbol{u}'_\mu, \boldsymbol{u}_\mu - \boldsymbol{v})\}\mathrm{d}t - \int_I \{(T'_\mu, z) + (z', T_\mu - z)\}\mathrm{d}t.$$

Hence $X_\mu \to X$, where

$$X = \int_I \{(\boldsymbol{f}, \boldsymbol{u}_{\varepsilon\eta})\mathrm{d}t - \hat{g}[(X, \boldsymbol{v}) - (j'_\varepsilon(\boldsymbol{v}), \boldsymbol{u}_{\varepsilon\eta} - \boldsymbol{v})] - \hat{\mu}\{a(\boldsymbol{u}_{\varepsilon\eta}, \boldsymbol{v}) + a(\boldsymbol{v}, \boldsymbol{u}_{\varepsilon\eta} - \boldsymbol{v})\}$$
$$- \eta\{((\boldsymbol{u}_{\varepsilon\eta}, \boldsymbol{v}))_s + ((\boldsymbol{u}_{\varepsilon\eta}, \boldsymbol{u}_{\varepsilon\eta} - \boldsymbol{v}))_s\} - (Q, T_{\varepsilon\eta}) - \{a_T(T_{\varepsilon\eta}, z) + a_T(z, T_{\varepsilon\eta} - z)\}$$
$$- \eta\{((T_{\varepsilon\eta}, z))_s - ((z, T_{\varepsilon\eta} - z))_s\} - \{(\boldsymbol{u}'_{\varepsilon\eta}, \boldsymbol{v}) + (\boldsymbol{u}'_{\varepsilon\eta}, \boldsymbol{u}_{\varepsilon\eta} - \boldsymbol{v})\} - \{(T'_{\varepsilon\eta}, z) + (z', T_{\varepsilon\eta} - z)\}\mathrm{d}t.$$

Since $X_\mu \geq 0\,\forall\mu$ then $X \geq 0$.

Let us put $z = T_{\varepsilon\eta} - \lambda s$, $\boldsymbol{v} = \boldsymbol{u}_{\varepsilon\eta} - \lambda\boldsymbol{w}$, $\lambda > 0$, where

$$s \in L^2(I; {}^1V_s) \text{ such that } s' \in L^2(I; {}^1V'_s), \quad s(t_o) = 0,$$
$$\boldsymbol{u} \in L^2(I; V_s) \text{ such that } \boldsymbol{u}' \in L^2(I; V'_s), \quad \boldsymbol{u}(t_o) = 0.$$

Substituting for z and $\boldsymbol{v}$ and dividing by λ, then in the limit $\lambda \to 0$ we successively have

$$\hat{g}\int_I (X - j'_\varepsilon(\boldsymbol{u}_{\varepsilon\eta} - \lambda\boldsymbol{w}), \boldsymbol{w})\mathrm{d}t + \int_I [\hat{\mu}a(\boldsymbol{w}, \boldsymbol{w}) + \eta\|\boldsymbol{w}\|_s^2 + (\boldsymbol{w}', \boldsymbol{w}) + a_T(s, s) + \eta\|s\|_s^2 + (s', s)]\mathrm{d}t \geq 0$$

and hence in the limit $\lambda \to 0$ we obtain

$$\hat{g}\int_I (X - j'_\varepsilon(\boldsymbol{u}_{\varepsilon\eta}), \boldsymbol{w})\mathrm{d}t \geq 0 \quad \forall\boldsymbol{w},$$

from which

$$X = j'_\varepsilon(\boldsymbol{u}_{\varepsilon\eta}), \boldsymbol{w}),$$

which proves the fact that there exist functions $T_{\varepsilon\eta}$ and $\boldsymbol{v}_{\varepsilon\eta}$ satisfying Equations 12.196-12.199. Then we proved that

$$\begin{aligned}
&T_{\varepsilon\eta} \quad \text{is bounded in the space } L^2(I;{}^1V)\cap L^\infty(I;{}^1H),\\
&T'_{\varepsilon\eta} \quad \text{is bounded in the space } L^2(I;{}^1V'_s),\\
&\eta^{1/2}T_{\varepsilon\eta} \quad \text{is bounded in the space } L^2(I;{}^1V'_s),\\
&\boldsymbol{u}_{\varepsilon\eta} \quad \text{is bounded in the space } L^2(I;V)\cap L^\infty(I;H),\\
&\boldsymbol{u}'_{\varepsilon\eta} \quad \text{is bounded in the space } L^2(I;V'_s),\\
&\eta^{1/2}\boldsymbol{u}_{\varepsilon\eta} \quad \text{is bounded in the space } L^2(I;V'_s).
\end{aligned}$$

Limitation process ε, $\eta \to 0$.

To prove the limit $\varepsilon \to 0$ we put

$$\begin{aligned}
Y_{\varepsilon\eta} = \int_I \{&\boldsymbol{v}', \boldsymbol{v}-\boldsymbol{u}_{\varepsilon\eta})+\hat{\mu}a(\boldsymbol{u}_{\varepsilon\eta}, \boldsymbol{v}-\boldsymbol{u}_{\varepsilon\eta})+b(\boldsymbol{u}_{\varepsilon\eta}, \boldsymbol{u}_{\varepsilon\eta}, \boldsymbol{v}-\boldsymbol{u}_{\varepsilon\eta})+\eta((\boldsymbol{u}_{\varepsilon\eta}, \boldsymbol{v}-\boldsymbol{u}_{\varepsilon\eta}))_s+\hat{g}j_\varepsilon(\boldsymbol{v})-\hat{g}j_\varepsilon(\boldsymbol{u}_{\varepsilon\eta})\\
&+(z', z-T_{\varepsilon\eta})+a_T(T_{\varepsilon\eta}, z-T_{\varepsilon\eta})+b_o(\boldsymbol{u}_{\varepsilon\eta}, T_{\varepsilon\eta}, z-T_{\varepsilon\eta})+b_s(T_{\varepsilon\eta}-T_o, \boldsymbol{v}-\boldsymbol{u}_{\varepsilon\eta})\\
&+b_p(\boldsymbol{u}'_{\varepsilon\eta}, z-T_{\varepsilon\eta})+\eta((T_{\varepsilon\eta}, z-T_{\varepsilon\eta}))_s-(Q, z-T_{\varepsilon\eta})-(\boldsymbol{f}, \boldsymbol{v}-\boldsymbol{u}_{\varepsilon\eta})\}\mathrm{d}t.
\end{aligned}$$

Due to Equation 12.196 and 12.197

$$Y_{\varepsilon\eta} = \int_I \{\boldsymbol{v}'-\boldsymbol{u}'_{\varepsilon\eta}, \boldsymbol{v}-\boldsymbol{u}_{\varepsilon\eta})+(z'-T'_{\varepsilon\eta}, z-T_{\varepsilon\eta})\}\mathrm{d}t+\hat{g}\int_I \{j_\varepsilon(\boldsymbol{v})-j_\varepsilon(\boldsymbol{u}_{\varepsilon\eta})-(j'_\varepsilon(\boldsymbol{u}_{\varepsilon\eta}), \boldsymbol{v}-\boldsymbol{u}_{\varepsilon\eta})\}\mathrm{d}t.$$

The first term gives

$$\tfrac{1}{2}\left\|\boldsymbol{v}(t_1)-\boldsymbol{u}_{\varepsilon\eta}(t_1)\right\|^2_{0,N}+\frac{1}{2}\left\|z(t_1)-T_{\varepsilon\eta}(t_1)\right\|^2_{0,1},$$

and since the functional $\boldsymbol{v} \to j_\varepsilon(\boldsymbol{v})$ is convex then the second term is ≥ 0 and therefore $Y_{\varepsilon\eta} \geq 0$, i.e.

$$\begin{aligned}
\int_I \{&(\boldsymbol{v}', \boldsymbol{v}-\boldsymbol{u}_{\varepsilon\eta})+\hat{\mu}a(\boldsymbol{u}_{\varepsilon\eta}, \boldsymbol{v})+b(\boldsymbol{u}_{\varepsilon\eta}, \boldsymbol{u}_{\varepsilon\eta}, \boldsymbol{v})+\eta((\boldsymbol{u}_{\varepsilon\eta}, \boldsymbol{v}))_s+\hat{g}j_\varepsilon(\boldsymbol{v})+(z', z-T_{\varepsilon\eta})+a_T(T_{\varepsilon\eta}, z)\\
&+b_o(\boldsymbol{u}_{\varepsilon\eta}, T_{\varepsilon\eta}, z)+b_s(T_{\varepsilon\eta}-T_o, \boldsymbol{v})+b_p(\boldsymbol{u}'_{\varepsilon\eta}, z)+\eta((T_{\varepsilon\eta}, z))_s-(Q, z-T_{\varepsilon\eta})-(\boldsymbol{f}, \boldsymbol{v}-\boldsymbol{u}_{\varepsilon\eta})\}\mathrm{d}t\\
\geq \int_I \{&\hat{\mu}a(\boldsymbol{u}_{\varepsilon\eta}, \boldsymbol{u}_{\varepsilon\eta})+\hat{g}j_\varepsilon(\boldsymbol{u}_{\varepsilon\eta})+a_T(T_{\varepsilon\eta}, T_{\varepsilon\eta})\}\mathrm{d}t. \qquad (12.222)
\end{aligned}$$

Due to Equation 12.209-12.214 there exist subsequences, denoted by $T_{\varepsilon\eta}$ and $\boldsymbol{u}_{\varepsilon\eta}$, such that

$$\begin{aligned}
&T_{\varepsilon\eta} \to T \quad \text{in } L^\infty(I;{}^1H) *-\text{ weakly and weakly in } L^2(I;{}^1V),\\
&T'_{\varepsilon\eta} \to T' \quad \text{in } L^2(I;{}^1V'_S),\\
&\boldsymbol{u}_{\varepsilon\eta} \to \boldsymbol{u} \quad \text{in } L^\infty(I;H) *-\text{ weakly and weakly in } L^2(I;V),\\
&\boldsymbol{u}_{\varepsilon\eta} \to \boldsymbol{u}' \quad \text{in } L^2(I;V).
\end{aligned} \qquad (12.223)$$

From Equation 12.222, 12.223, 12.213 and 12.214 follows that

$$\begin{aligned}
\int_I \{&\boldsymbol{v}', \boldsymbol{v}-\boldsymbol{u})+\hat{\mu}a(\boldsymbol{u}, \boldsymbol{v}-\boldsymbol{u})+b(\boldsymbol{u}, \boldsymbol{u}, \boldsymbol{v})+\hat{g}j(\boldsymbol{v})+(z', z-T)+a_T(T, z-T)+b_o(\boldsymbol{u}, T, z)\\
&+b_s(T-T_o, \boldsymbol{v})+b_p(\boldsymbol{u}', z)-(Q, z-T)-(\boldsymbol{f}, \boldsymbol{v}-\boldsymbol{u})\}\mathrm{d}t\\
\geq \liminf &\int_1 \{a_T(T_{\varepsilon\eta}, T_{\varepsilon\eta})+\hat{\mu}a(\boldsymbol{u}_{\varepsilon\eta}, \boldsymbol{u}_{\varepsilon\eta})+\hat{g}j_\varepsilon(\boldsymbol{u}_{\varepsilon\eta})\}\mathrm{d}t. \qquad (12.224)
\end{aligned}$$

Since functions $\boldsymbol{v} \rightarrow \int_{\mathrm{I}} a(\boldsymbol{v}, \boldsymbol{v})\mathrm{d}t$, $z \rightarrow \int_{\mathrm{I}} a_T(z, z)\mathrm{d}t$ are l.sc. in $L^2(I; V)$, $L^2(I; {}^1V)$ respectively, then

$$\begin{aligned} \liminf \int_{\mathrm{I}} a(\boldsymbol{u}_{\varepsilon\eta}, \boldsymbol{u}_{\varepsilon\eta})\mathrm{d}t &\geq \int_{\mathrm{I}} a(\boldsymbol{u}, \boldsymbol{u})\mathrm{d}t, \\ \liminf \int_{\mathrm{I}} a_T(T_{\varepsilon\eta}, T_{\varepsilon\eta})\mathrm{d}t &\geq \int_{\mathrm{I}} a_T(T, T)\mathrm{d}t \end{aligned} \tag{12.225}$$

and since

$$\int_{\mathrm{I}} j(\boldsymbol{u})\mathrm{d}t \leq \left[\int_{\Omega_t} D_{\mathrm{II}}(\boldsymbol{u})^{1/2(1+\varepsilon)}\mathrm{d}\boldsymbol{x}\,\mathrm{d}t\right]^{1/(1+\varepsilon)} \left[\int_{\Omega_t} \mathrm{d}\boldsymbol{x}\,\mathrm{d}t\right]^{\varepsilon/(1+\varepsilon)},$$

hence

$$\int_{\mathrm{I}} j_\varepsilon(\boldsymbol{u}_{\varepsilon\eta})\mathrm{d}t \geq c\left(\int_{\mathrm{I}} j(\boldsymbol{u}_{\varepsilon\eta})\mathrm{d}t\right)^{(1+\varepsilon)}, \tag{12.226}$$

$c = c(\varepsilon)$, depending on Ω_t,

$$\liminf \int_{\mathrm{I}} j_\varepsilon(\boldsymbol{u}_{\varepsilon\eta})\mathrm{d}t \geq \liminf \int_{\mathrm{I}} j(\boldsymbol{u}_{\varepsilon\eta})\mathrm{d}t,$$

and since the function $\boldsymbol{v} \rightarrow \int_{\mathrm{I}} j(\boldsymbol{v})\mathrm{d}t$ is convex and continuous on $L^2(I; V)$, then it is l.sc. in the weak topology of the space $L^2(I; V)$ and thus

$$\liminf \int_{\mathrm{I}} j(\boldsymbol{u}_{\varepsilon\eta})\mathrm{d}t \geq \int_{\mathrm{I}} j(\boldsymbol{u})\mathrm{d}t,$$

which together with Equation 12.225 proves

$$\liminf \int_{\mathrm{I}} j_\varepsilon(\boldsymbol{u}_{\varepsilon\eta})\mathrm{d}t \geq \int_{\mathrm{I}} j(\boldsymbol{u})\mathrm{d}t. \tag{12.227}$$

Then Equation 12.224-12.226 prove that $(T, \boldsymbol{u})$ satisfy

$$\begin{aligned} &\int_{\mathrm{I}} \{\boldsymbol{v}', \boldsymbol{v} - \boldsymbol{u}) + \hat{\mu} a(\boldsymbol{u}, \boldsymbol{v} - \boldsymbol{u}) - b(\boldsymbol{u}, \boldsymbol{v}, \boldsymbol{u}) + \hat{g} j(\boldsymbol{v}) - \hat{g} j(\boldsymbol{u}) + (z', z - T) + a_T(T, z - T) \\ &+ b_o(\boldsymbol{u}, T, z - T) + b_s(T - T_o, \boldsymbol{v} - \boldsymbol{u}) + b_p(\boldsymbol{u}', z - T) - (Q, z - T) - (\boldsymbol{f}, \boldsymbol{v} - \boldsymbol{u})\}dt \geq 0 \\ &\qquad \forall z \in \{y \mid y \in L^2(I; {}^1V_s),\ y' \in L^2(I; {}^1H),\ y(t_o) = 0\}, \\ &\qquad \forall \boldsymbol{v} \in \{\boldsymbol{w} \mid \boldsymbol{w} \in L^2(I; V_s), \boldsymbol{w}' \in L^2(I; H), \boldsymbol{w}(t_o) = 0\}, \end{aligned}$$

i.e. $(T, \boldsymbol{u})$ is the solution of the problem given. Therefore, we have proved that $(T, \boldsymbol{u})$ is a pair of functions solving our problem.

To prove the uniqueness of the solution, we shall assume that $(T, \boldsymbol{u})$ and $(T_1, \boldsymbol{u}_1)$ are two solutions of our given problem and we let $N = 2$. Then $(\underline{T}, \boldsymbol{U}) = (T - T_1, \boldsymbol{u} - \boldsymbol{u}_1)$. Then $s = 1$ and thus $V_s = V$, ${}^1V_s = {}^1V$ and the terms of viscosity regularization $\eta((\boldsymbol{u}, \boldsymbol{v}), \eta((T, z))$ are of the same order as $\hat{\mu} a(\boldsymbol{u}, \boldsymbol{v})$ and $a_T(T, z)$, respectively, and therefore can be neglected. Let us put $\boldsymbol{v} = \boldsymbol{u}$, $z = T$ ($\boldsymbol{v} = \boldsymbol{u}_1$, $z = T_1$, respectively) into Equation 12.186-12.189. then

$$\begin{aligned} &-\frac{1}{2}\frac{\mathrm{d}}{\mathrm{d}t}\{|\,\boldsymbol{U}(t)\,|^2 + |\,\underline{T}(t)\,|^2\} - \hat{\mu} a(\boldsymbol{U}(t), \boldsymbol{U}(t)) - a_T(\underline{T}(t), \underline{T}(t)) - b(\boldsymbol{u}(t), \boldsymbol{u}(t), \boldsymbol{U}) + b(\boldsymbol{u}_1(t), \boldsymbol{u}_1(t), \boldsymbol{U}) \\ &\quad - [b_s(\underline{T}, \boldsymbol{U}) + b_p(\boldsymbol{U}, \underline{T})] + b_o(\boldsymbol{u}(t), T(t), \underline{T}(t)) - b_o(\boldsymbol{u}_1(t), T_1(t), \underline{T}(t)) \geq 0. \end{aligned}$$

Hence

$$\begin{aligned} &\frac{1}{2}\frac{\mathrm{d}}{\mathrm{d}t}\{|\,\boldsymbol{U}(t)\,|^2 + |\,\underline{T}(t)\,|^2\} + \hat{\mu} a(\boldsymbol{U}(t), \boldsymbol{U}(t)) + a_T(\underline{T}(t), \underline{T}(t)) \\ &\qquad \leq -b(\boldsymbol{u}(t) - \boldsymbol{U}, \boldsymbol{u}(t) - \boldsymbol{U}, \boldsymbol{U}) - b(\boldsymbol{u}(t), \boldsymbol{u}(t), \boldsymbol{U}) + G, \end{aligned} \tag{12.228}$$

where $G = b_o(\boldsymbol{U}, \underline{T}, \underline{T}) + c(\|\underline{T}\|_{1,1}\,\|\boldsymbol{U}\|_{0,N} + \|\underline{T}\|_{0,1}\,\|\boldsymbol{U}\|_{1,N})$ and where

$$|\,b_s(T, \boldsymbol{U}) + b_p(\boldsymbol{U}, \underline{T})\,| \leq c(\|\underline{T}\|_{1,1}\,\|\boldsymbol{U}\|_{0,N} + \|T\|_{0,1}\,\|\boldsymbol{U}\|_{1,N}).$$

Since as above

$$|b_s(z, \boldsymbol{v}) + b_p(\boldsymbol{v}, z)| \leq c(\|z\|_{1,1} \|\boldsymbol{v}\|_{0,N} + \|z\|_{0,1} \|\boldsymbol{v}\|_{1,N}),$$
$$|b(\boldsymbol{v}, \boldsymbol{u}, \boldsymbol{v})| \leq c_3 \|\boldsymbol{v}\|_{1,N} \|\boldsymbol{u}\|_{0,N} \|\boldsymbol{v}\|_{s,N} \leq c_1 \|\boldsymbol{v}\|_{1,N}^2 + c_2 \|\boldsymbol{u}\|_{0,N}^2 \|\boldsymbol{v}\|_{s,N}^2,$$

then

$$|b(\boldsymbol{U}, \boldsymbol{v}, \boldsymbol{U})| \leq c_p \|\boldsymbol{U}\|_{1,N}^2 + K(c_p) \|\boldsymbol{v}\|_{1,N}^2 \|\boldsymbol{U}\|_{S,N}^2,$$
$$\hat{\mu} a(\boldsymbol{U}, \boldsymbol{U}) + a_T(\underline{T}, \underline{T}) \geq c(\|\boldsymbol{U}\|_{1,N}^2 + \|\underline{T}\|_{1,1}^2).$$

Then Equation 12.228 yields

$$\frac{\mathrm{d}}{\mathrm{d}t}\{|\ \boldsymbol{U}(t)\ |^2 + |\ \underline{T}(t)\ |^2\} \leq c(\|\boldsymbol{U}(t)\|_{1,N}^2 + \|\underline{T}(t)\|_{1,1}^2)(\|\boldsymbol{U}(t)\|_{0,N}^2 + \|\underline{T}(t)\|_{1,1}^2)$$
$$= cg(t)(|\ \boldsymbol{U}(t)\ |^2 + |\ \underline{T}(t)\ |^2),$$

where $g(t)$ is an integrable function in variable t, i.e. $g(t) \in L^1(I)$ and thus

$$\|\boldsymbol{U}(t)\|_{0,N}^2 + \|\underline{T}(t)\|_{0,1}^2 \leq \int_I cg(\tau)(\|\boldsymbol{U}(\tau)\|_{0,N}^2 + \|\underline{T}(\tau)\|_{0,1}^2)\mathrm{d}\tau.$$

Hence $\boldsymbol{U}(t) = 0$, $\underline{T}(t) = 0$, which completes the proof.

In the following we shall assume that $N = 2$.

Since in the variational formulation the term $\hat{g} D_{ij}(\boldsymbol{u}) D_{\mathrm{II}}^{-1/2}(\boldsymbol{u})$ creates the non-differential functional $j(\boldsymbol{u})$, we introduce the multipliers $m = (m_{ij})_{ij=1,2}$ defined by

$$m_{ij} \in L^\infty(\Omega_t),\ m_{ij} = m_{ji}, \quad \forall i, j, m_{\mathrm{ii}} = 0, \tag{12.229}$$
$$m_{ij} m_{ij} \leq 1 \text{ a.e. in } \Omega_t, \tag{12.230}$$
$$m_{ij} D_{ij}(\boldsymbol{u}) = (D_{ij}(\boldsymbol{u}) D_{ij}(\boldsymbol{u}))^{1/2} = (2)^{1/2} D_{\mathrm{II}}^{1/2}(\boldsymbol{u}) \text{ a.e. in } \Omega_t. \tag{12.231}$$

Then Equations 12.182 and 12.183 can be re-written in the form

$$\tau_{ij}^m = -p\delta_{ij} + 2\hat{\mu} D_{ij}(\boldsymbol{u}) + (2)^{1/2} \hat{g} m_{ij} - \beta_{ij}(T - T_o),$$

and then Equation 12.174 yields

$$\rho \frac{\partial u_i}{\partial t} - \hat{\mu} \Delta u_i + \rho u_j \frac{\partial u_i}{\partial x_j} - (2)^{1/2} \hat{g} \frac{\partial m_{ij}}{\partial x_j} = F_i - \frac{\partial p}{\partial x_i} - \frac{\partial}{\partial x_j}(\beta_{ij}(T - T_o)). \tag{12.232}$$

The functional $j(\boldsymbol{u})$ is a non-differential functional and it can be approximated by the use of multipliers m_{ij} defined by Equations 12.229 and 12.230. Therefore $j(\boldsymbol{u}) = 2^{1/2}(m_{ij}, D_{ij}(\boldsymbol{u}))$.

Then we will solve the equivalent variational problem.

Problem $(\mathscr{P}_m)_v$: Find a pair of functions $(T, \boldsymbol{u})$, $T \in {}^1V$, $\boldsymbol{u} \in V$ such that

$$(\rho c_e T'(t), z - T(t)) + b_o(\boldsymbol{u}(t), T(t), z - T(t)) + a_T(T(t), z - T(t))$$
$$+ b_p(\boldsymbol{u}(t), z - T(t)) - (Q(t), z - T(t)) = 0 \qquad \forall z \in {}^1V, \tag{12.233}$$

$$(\rho \boldsymbol{u}'(t), \boldsymbol{v} - \boldsymbol{u}(t)) + \hat{\mu} a(\boldsymbol{u}(t), \boldsymbol{v} - \boldsymbol{u}(t)) + b(\boldsymbol{u}(t), \boldsymbol{u}(t), \boldsymbol{v} - \boldsymbol{u}(t)) + b_s(T(t), \boldsymbol{v} - \boldsymbol{u}(t))$$
$$+2^{1/2}\hat{g}(m_{ij}, D_{ij}(\boldsymbol{v}) - D_{ij}(\boldsymbol{u}(t)) - (\boldsymbol{f}(t), \boldsymbol{v} - \boldsymbol{u}(t)) = 0 \quad \forall \boldsymbol{v} \in V, \tag{12.234}$$
$$T(\boldsymbol{x}, t_o) = T_o(\boldsymbol{x}), \tag{12.235}$$
$$\boldsymbol{u}(\boldsymbol{x}, t_o) = \boldsymbol{u}_o(\boldsymbol{x}). \tag{12.236}$$

As a result we have the following theorem.

Theorem 12.7: Let $n = 2$, and let $Q \in L^2(I; {}^1V')$, $\boldsymbol{f} \in L^2(I; V')$, $T_o \in L^2(\Omega)$, $\boldsymbol{u}_o \in H$. Let a pair of functions $(T, \boldsymbol{u})$ be a solution of Problem $(\mathscr{P})_v$ (defined in Theorem 12.6). Then there exist functions $m = \{m_{ij}\}_{i,j=1,2}$ and p defined on Ω_t and satisfying Equation 12.229, 12,230 and 12.232.

Conversely, if the given $T, \boldsymbol{u}, m, p$ are such that $T \in L^2(I; {}^1V)$, $T' \in L^2(I; {}^1V')$, $\boldsymbol{u} \in L^2(I; V)$, $\boldsymbol{u}' \in L^2(I; V')$, and satisfy Equations 12.235, 12.236, and Equations 12.229, 12.231 and 12.232 are satisfied, then a pair of functions (T, u) will be a solution of Problem $(\mathscr{P})_v$ as proved in Theorem 12.6.

To prove Theorem 12.7 we need the following lemma.

Lemma 12.7: Let $\mathscr{D}(\Omega)$ be the space of all infinitely continuously differentiable functions with compact support in $\Omega \subset R^2$. Let $\boldsymbol{u} = (u_i)$, $u_i \in \mathscr{D}(\Omega)'$, $i = 1, \ldots, N$ and let

$$W = \{\boldsymbol{v} \mid \boldsymbol{v} \in [\mathscr{D}(\Omega)]^2, \ \operatorname{div} \boldsymbol{v} = 0\}.$$

Then there exists $p \in \mathscr{D}(\Omega)'$ such that $\boldsymbol{u}$=grad $p \Leftrightarrow (\boldsymbol{u}, \boldsymbol{v}) = 0$, $\forall \boldsymbol{v} \in W$.

Proof of Theorem 12.7: It is sufficient to prove that the solution of Problem $(\mathscr{P}_m)_v$ also solves Problem $(\mathscr{P})_v$. The problem is coupled and therefore the proof is similar to that above. Hence we shall show the difference of the above, so that we shall analyse the second part of the problem only. Multiplying Equation 12.232 by $\boldsymbol{v} - \boldsymbol{u}(t)$, integrating over Ω, applying the Green theorem and the boundary conditions we obtain

$$(\rho \boldsymbol{u}'(t), \boldsymbol{v} - \boldsymbol{u}(t)) + \hat{\mu} a(\boldsymbol{u}(t), \boldsymbol{v} - \boldsymbol{u}(t)) + b(\boldsymbol{u}(t), \boldsymbol{u}(t), \boldsymbol{v} - \boldsymbol{u}(t)) + \hat{g} j(\boldsymbol{v}) - \hat{g} j(\boldsymbol{u}(t))$$
$$-(\boldsymbol{F}(t), \boldsymbol{v} - \boldsymbol{u}(t)) = \hat{g} j(\boldsymbol{v}) - \hat{g} j(\boldsymbol{u}(t)) - 2^{1/2}(m_{ij}, D_{ij}(\boldsymbol{v})) + 2^{1/2}(m_{ij}, D_{ij}(\boldsymbol{u}(t))) \equiv X. \tag{12.237}$$

According to Equation 12.231 $j(\boldsymbol{u}) = 2^{1/2}(m_{ij}, D_{ij}(\boldsymbol{u}))$ and due to Equation 12.230 we have

$$2^{1/2}(m_{ij}, D_{ij}(\boldsymbol{v})) \leq 2^{1/2} \int_\Omega (m_{ij} m_{ij})^{1/2} (D_{ij}(\boldsymbol{v}) D_{ij}(\boldsymbol{v}))^{1/2} \mathrm{d}x \leq 2^{1/2} \int_\Omega (D_{ij}(\boldsymbol{v}) D_{ij}(\boldsymbol{v}))^{1/2} \mathrm{d}x = j(\boldsymbol{v}).$$

Hence and using Equation 12.237 it follows that $X \geq 0$, which proves that $\boldsymbol{u}$ is a solution of the problem $(\mathscr{P})_v$.

Now we shall prove the existence of m and p. The existence of p follows from Lemma 12.5. To prove the existence of m we define $B(\boldsymbol{u}, \boldsymbol{v}) \in V$ by $b(\boldsymbol{u}, \boldsymbol{v}, \boldsymbol{w}) = (B(\boldsymbol{u}, \boldsymbol{v}), \boldsymbol{w})$, $\boldsymbol{u}, \boldsymbol{v}, \boldsymbol{w} \in V$. Let us put

$$\vec{\mathscr{F}} = \rho \boldsymbol{u}' - \hat{\mu} \Delta \boldsymbol{u} + B(\boldsymbol{u}, \boldsymbol{u}) - \boldsymbol{F} \in L^2(I, V). \tag{12.238}$$

Then Problem $(\mathscr{P})_v$ can be rewritten as

$$(\vec{\mathscr{F}}(t), \boldsymbol{v}) + \hat{g} j(\boldsymbol{v}) - ((\vec{\mathscr{F}}(t), \boldsymbol{u}(t)) + \hat{g} j(\boldsymbol{u}(t)) \geq 0 \quad \forall \boldsymbol{v} \in V. \tag{12.239}$$

Let $\lambda > 0$ and let us put $\boldsymbol{v} = \pm \lambda \boldsymbol{v}$ in Equation 12.239. Then

$$\lambda(\pm(\vec{\mathscr{F}}(t), \boldsymbol{v}) + \hat{g} j(\boldsymbol{v})) - ((\vec{\mathscr{F}}(t), \boldsymbol{u}(t)) + \hat{g} j(\boldsymbol{u}(t))) \geq 0. \tag{12.240}$$

Putting $\boldsymbol{v} = \boldsymbol{v}(t) \in L^2(I, V)$ in Equation 12.240 and integrating over I we obtain

$$\lambda\left\{\pm\int_{\mathrm{I}}(\vec{\mathcal{F}}(t), \boldsymbol{v}(t))\mathrm{d}t + \int_{\mathrm{I}}\hat{g}j(\boldsymbol{v}(t))\mathrm{d}t\right\} - \left\{\int_{\mathrm{I}}(\vec{\mathcal{F}}(t), \boldsymbol{u}(t)) + \hat{g}j(\boldsymbol{u}(t))\mathrm{d}t\right\} \geq 0$$
$$\forall \boldsymbol{v} \in L^2(I, V), \quad \forall\lambda \geq 0. \quad (12.241)$$

Hence

$$\left|\int_{\mathrm{I}}(\vec{\mathcal{F}}(t), \boldsymbol{v}(t))\mathrm{d}t\right| \leq \int_{\mathrm{I}}\hat{g}j(\boldsymbol{v}(t))\mathrm{d}t \quad \forall \boldsymbol{v} \in L^2(I, V). \quad (12.242)$$

$$\int_{\mathrm{I}}((\vec{\mathcal{F}}(t), \boldsymbol{u}(t)) + \hat{g}j(\boldsymbol{u}(t)))\mathrm{d}t = 0 \quad (12.243)$$

as from Equation 12.241 $\int_{\mathrm{I}}((\vec{\mathcal{F}}(t), \boldsymbol{u}(t)) + \hat{g}j(\boldsymbol{u}(t)))\mathrm{d}t \leq 0$, and from Equation 12.242 we have $\int_{\mathrm{I}}((\vec{\mathcal{F}}(t), \boldsymbol{u}(t)) + \hat{g}j(\boldsymbol{u}(t)))\mathrm{d}t \geq 0$ and therefore Equation 12.243 holds.

The corollary of the Hahn-Banach theorem* (see Ekeland and Temam (1976)) will now be used. Let us define the space

$$W = \left\{\boldsymbol{w} \mid \boldsymbol{w} = (w_{ij}), i, j = 1, 2, w_{ij} = w_{ji}, w_{ij} \in L^1(\Omega_t), \|\boldsymbol{w}\| = \int_{\Omega_t}(w_{ij}w_{ij})^{1/2}\mathrm{d}x\mathrm{d}t\right\}.$$

Let the mapping $\Pi : v \to \{D_{ij}(\boldsymbol{v})\} = \{1/2(\partial v_i/\partial x_j + \partial v_j/\partial x_i)\}$, $i, j = 1, 2$, functions from $L^2(I, V)$ into W. Then Equation 12.242 can be rewritten in the following form

$$\left|\int_{\mathrm{I}}(\vec{\mathcal{F}}(t), \boldsymbol{v}(t))\mathrm{d}t\right| \leq \hat{g}2^{1/2}\,\|\Pi\boldsymbol{v}\|.$$

Then due to the corollary of the Hahn-Banach theorem there exists $m \in W' = \{\boldsymbol{w} \mid \boldsymbol{w} = (w_{ij}), w_{ij} = w_{ji}, w_{ij} \in L^\infty(\Omega_t)\}$ such that

$$\int_{\mathrm{I}}(\vec{\mathcal{F}}(t), \boldsymbol{v}(t))\mathrm{d}t = -\hat{g}2^{1/2}\int_{\Omega_t} m_{ij}D_{ij}(\boldsymbol{v}(t))\mathrm{d}\boldsymbol{x}\,\mathrm{d}t, \quad (12.244)$$

$$\|m\|_{W'} \leq 1, \quad (12.245)$$

where we took $\{m_{ij}\} \in W'$ such that $D_{ii}(\boldsymbol{v}) = 0$ and $m_{ii} = 0$. The last inequality (Equation 12.245) is equivalent to $m_{ij}m_{ij} \leq 1$, i.e. to Equation 12.230.

Due to Equation 12.238 and 12.244, Equation 12.232 follows immediately. Let us put $\boldsymbol{v} = \boldsymbol{u}$ in Equation 12.244. Hence and using Equation 12.243 we obtain

$$0 = \int_I((\vec{\mathcal{F}}(t), \boldsymbol{u}(t)) + \hat{g}j(\boldsymbol{u}(t)))\mathrm{d}t = -\hat{g}2^{1/2}\int_{\Omega_t} m_{ij}D_{ij}(\boldsymbol{u})\mathrm{d}\boldsymbol{x}\,\mathrm{d}t + \hat{g}2^{1/2}\int_{\Omega_t} D_{ij}(\boldsymbol{u}(t))D_{ij}(\boldsymbol{u})^{1/2}\mathrm{d}\boldsymbol{x}\,\mathrm{d}t$$

$$= \hat{g}2^{1/2}\int_{\Omega_t}((D_{ij}(\boldsymbol{u}(t))D_{ij}(\boldsymbol{u}(t)))^{1/2} - m_{ij}D_{ij}(\boldsymbol{u}(t)))\mathrm{d}\boldsymbol{x}\,\mathrm{d}t,$$

which proves Equation 12.231.

---

* **Hahn-Banach theorem**: Let V be a real vector space, let j be a sub-linear function over V, i.e. a mapping j of $V \to \overline{R}$ satisfying

$$j(\lambda\boldsymbol{u}) = \lambda j(\boldsymbol{u}) \quad \forall\boldsymbol{u} \in V, \quad \forall\lambda > 0,$$

$$j(\boldsymbol{u} + \boldsymbol{v}) \leq j(\boldsymbol{u}) + j(\boldsymbol{v}) \quad \forall(\boldsymbol{u}, \boldsymbol{v}) \in V \times V,$$

let M be a vector subspace of V, and L a linear functional over M, which is everywhere less than j. Then there exists a linear functional $\tilde{L}$ over V, which extends L and is everywhere less than j.

Corollary: Let V be a normed space, let M be a topological vector subspace, L a continuous linear functional over M. Then L can be extended into a continuous functional over V with the same norm.

FINITE ELEMENT APPROXIMATION

For simplicity we shall assume the 2D model problem only. Then we can investigate the above studied model problem by applying the non-conforming and conforming finite elements.

Firstly, we shall investigate the approximation by the polynomial of first order. Because of the condition div $\boldsymbol{u} = 0$, the space cannot be approximated by the piecewise linear continuous functions. Fortin (1972) shows that the condition div $\boldsymbol{u} = 0$ can be approximated by piecewise linear but discontinuous functions, i.e. by non-conforming elements of first order only. To approximate the space W we will firstly approximate the space $[H_o^1(\Omega)]^2$ by non-conforming piecewise linear finite elements in the case of the visco-plastic part of the problem, whilst the thermal part of the problem will be approximated by the conforming linear finite elements.

Let us suppose that Ω is approximated by a bounded polygonal domain Ω_h of R^2. Let $\{\mathcal{T}_h\}$ be a family of triangulations $\mathcal{T}_h$ of Ω such that $\overline{\Omega} = \cup_{\underline{T}_h \in \mathcal{T}_h} \underline{T}_h$. We set $h = h(\underline{T}_h)$ equal to the length of the greatest side of a triangle $\underline{T}_h$, $h = \max h(\underline{T}_h)$. Furthermore, we suppose that the points of Ω_h are vertices of the triangulation $\mathcal{T}_h$. Let $\mathcal{T}_h$ be a triangulation of Ω created by a system of triangles $\underline{T}_h$. We shall assume that the triangulation $\mathcal{T}_h$ is regular. If $\underline{T}_h \in \mathcal{T}_h$, we denote by A_i, $i = 1, \ldots, m$, its vertices, and by B_i, $i = 1, \ldots, m$, the barycentre of the face of the triangle, i.e. the centres of sides of all triangles $\underline{T}_h \in \mathcal{T}_h$. Let us denote by $\mathcal{U}_h^B$ the set of barycentres of $\underline{T}_h$, i.e. centres B_i, $i = 1, \ldots, m$, laying in the interior of Ω_h, and by $\mathcal{U}_h^K$ the set of all vertices laying in the interior of Ω_h. We shall define the following finite element spaces ${}^1V_h$ and W_h by

$$ {}^1V_h = \{z \,|z \in C(\Omega), z|_{\underline{T}_h} \in P_1, \quad z = 0 \text{ on } \Gamma_\tau, \quad \forall \underline{T}_h \in \mathcal{T}_h\}, $$

where P_1 denotes the space of all conforming linear polynomials,

$$ W_h = \{\boldsymbol{v}_h \mid \boldsymbol{v}_h = (v_{h1}, v_{h2}), \quad v_{hi} \in \underline{P}_1 \quad \forall \underline{T}_h \in \mathcal{T}_h, \text{ continuous in points } B_j, $$
$$ j = 1, \ldots, m, \text{ equal to zero in } B_i \text{ laying on } \Gamma_{uh}, \boldsymbol{v}_h(x) = 0 \text{ for } \boldsymbol{x} \in \Omega \backslash \overline{\Omega}_h\}, $$

where $\underline{P}_1$ denotes the space of all non-conforming linear polynomials. A function $\boldsymbol{v}_h \in W_h$ is characterized by its values at the points $B_i \in \mathcal{U}_h^B$. Let us denote by w_h a scalar function such that w_h: $w_h(B) = 1$, $w_h(B_i) = 0$, $B_i \neq B$, $i = 1, \ldots, m$, $B \in \mathcal{U}_h^B$ and satisfying the boundary and matching conditions. Such a function w_h has a support equal to the two simplices (triangles) which are neighbours to the point B. Moreover, $w_h e_i$ ($i = 1, 2$, $e_1 = (1, 0)$, $e_2 = (0, 1)$, $B \in \mathcal{U}_h^B$) creates the basis of the space W_h (for the proof of this assertion see Lemma 4.15, Temam (1977)). Then any function $\boldsymbol{v}_h \in W_h$ can be written as

$$ \boldsymbol{v}_h = \sum_{B \in \mathcal{U}_h^B} \boldsymbol{v}_h(B) w_h. $$

Let

$$ V_h = \left\{ \boldsymbol{v}_h \mid \boldsymbol{v}_h \in W_h, \sum_{i=1,2} \partial_i v_{hi} = 0 \right\}, $$

where we denote div $\boldsymbol{v}_h = \Sigma_{i=1,2}\partial_i v_{hi}$ and $V_h \subset W_h \subset [L^2(\Omega)]^2\, \forall h$. The space W_h is not included in $H_o^1(\Omega)$. Actually, since the derivative $\partial_i \boldsymbol{v}_h$ is the sum of Dirac distributions located on the faces of the simplices-triangles $\underline{T}_h$ and of a step function $\partial_{ih}\boldsymbol{v}_h$ defined a.e. by $\partial_{ih}\boldsymbol{v}_h(\boldsymbol{x}) = \partial_i \boldsymbol{v}_h(\boldsymbol{x})\, \forall \boldsymbol{x} \in \underline{T}_h$, $\forall \underline{T}_h \in \mathscr{T}_h$, and since $\boldsymbol{v}_h$ is linear on $\underline{T}_h$, then $\partial_{ih}\boldsymbol{v}_h$ is constant on each triangle $\underline{T}_h$.

Let $F = [L^2(\Omega)]^2$ be equipped with the natural Hilbertian scalar product. Let us define the mapping $\underline{w}$ (isomorphism) by

$$\boldsymbol{v} \in H_o^1(\Omega) \to \underline{w}\boldsymbol{v} = (\boldsymbol{v}, \partial_1\boldsymbol{v}, \partial_2\boldsymbol{v}) \in F \tag{12.246}$$

and similarly the prolongation operator p_h by

$$\boldsymbol{v}_h \in W_h \to p_h\boldsymbol{v}_h = (\boldsymbol{v}_h, \partial_1\boldsymbol{v}_h, \partial_2\boldsymbol{v}_h) \in F, \tag{12.247}$$

where $\|p_h\| = 1$ and, therefore, the operators of prolongation are stable. Furthermore, we define the restriction operators $r_h\boldsymbol{v} = \boldsymbol{v}_h$, $\boldsymbol{v} \in \mathscr{D}(\Omega)$, by

$$\boldsymbol{v}_h(B) = \boldsymbol{v}(B) \quad \forall B \in \mathscr{U}_h^B. \tag{12.248}$$

We denote

$$(T_h, z_h)_h = \sum_{i=1}^{2}(\rho c_e D_i T_h, D_i z_h)_o \quad \forall T_h, z_h \in {}^1V_h, \tag{12.249}$$

$$(\boldsymbol{u}_h, \boldsymbol{v}_h)_h = \sum_{i=1}^{2}(\rho D_i\boldsymbol{u}_h, D_i\boldsymbol{v}_h)_o \quad \forall \boldsymbol{u}_h, \boldsymbol{v}_h \in W_h, \tag{12.250}$$

$$(\boldsymbol{u}_h, \boldsymbol{v}_h)_h = (\boldsymbol{u}_h, \boldsymbol{v}_h)_o + \sum_{i=1}^{2}(D_i\boldsymbol{u}_h, \boldsymbol{v}_h)_o \quad \forall \boldsymbol{u}_h, \boldsymbol{v}_h \in W_h \tag{12.251}$$

the scalar products in ${}^1V_h$ or W_h, respectively, where $(.,.)_o$ denotes the scalar product in $L^2(\Omega)$. These scalar products induce norms, which for simplicity we denote as $|\cdot|_h = |\cdot|$, $\|.\|_h = \|.\|$ in $L^2(\Omega)$, or ${}^1V_h$, V_h, W_h, respectively. For $\boldsymbol{v}_h \in W_h$, the partial derivative $\partial_i\boldsymbol{v}_h = (\partial_i u_{h1}, \partial_i u_{h2})^T$ is defined inside every triangle $\underline{T}_h \in \mathscr{T}_h$, $\partial_i u_{hj} =$ constant, $\boldsymbol{x} \in \underline{T}_h$, $i, j = 1, 2$.

For $T_h, z_h \in {}^1V_h$, and $\boldsymbol{u}_h, \boldsymbol{v}_h, \boldsymbol{w}_h \in W_h$ we define

$$(T'_h, z_h) = \sum_{\underline{T}_h \in \mathscr{T}_h} \int_{\underline{T}_h} \rho c_e T'_h(t) z \mathrm{d}\boldsymbol{x}, \tag{12.252}$$

$$a_{Th}(T_h, z_h) = \sum_{\underline{T}_h \in \mathscr{T}_h} \int_{\underline{T}_h} \kappa_{ij} \partial T_h/\partial x_i \partial z_h/\partial x_j \mathrm{d}\boldsymbol{x}, \tag{12.253}$$

$$b_{oh}(\boldsymbol{u}_h, g_h, z_h) = \sum_{\underline{T}_h \in \mathscr{T}_h} \int_{\underline{T}_h} \rho c_e u_{hj} \partial g_h/\partial x_j z_h \mathrm{d}\boldsymbol{x}, \tag{12.254}$$

$$b_{ph}(\boldsymbol{u}_h, z_h) = \sum_{I_h \in \mathscr{T}_n} \int_{\underline{T}_h} \rho T_o \rho_{ij} \partial v_t/\partial x_j z \mathrm{d}*$$

$$(\boldsymbol{u}', \boldsymbol{v}) = \sum_{\underline{T}_h \in \mathscr{T}_h} \int_{\underline{T}_h} \rho u_{hi} v_{hi} \mathrm{d}\boldsymbol{x}, \tag{12.255}$$

$$a_h(\boldsymbol{u}_h, \boldsymbol{v}_h) = \sum_{\underline{T}_h \in \mathscr{T}_h} \int_{\underline{T}_h} D_{ij}(\boldsymbol{u}_h) D_{ij}(\boldsymbol{v}_h) \mathrm{d}\boldsymbol{x}, \tag{12.256}$$

$$b_h(\boldsymbol{u}_h, \boldsymbol{v}_h, \boldsymbol{w}_h) = \sum_{\underline{T}_h \in \mathscr{T}_h} \int_{\underline{T}_h} \rho u_{hi} \partial v_{hj} / \partial x_i w_{hj} \mathrm{d}\boldsymbol{x}, \tag{12.257}$$

$$b_{sh}(T_h, \boldsymbol{v}_h) = \sum_{\underline{T}_h \in \mathscr{T}_h} \int_{\underline{T}_h} \partial / \partial x_j (\beta_{ij}(T_h)) v_{hi} \mathrm{d}\boldsymbol{x}, \tag{12.258}$$

$$j_h(\boldsymbol{u}_h) = 2 \sum_{\underline{T}_h \in \mathscr{T}_h} \int_{\underline{T}_h} (D_{\mathrm{II}}(\boldsymbol{u}_h))^{1/2} \mathrm{d}\boldsymbol{x}. \tag{12.259}$$

Let the interval $\bar{I}$ be divided into n sub-intervals $\bar{I}_i = \langle t_{i-1}, t_i \rangle$, $i = 1, \ldots, n$, of lengths $k = t_1/n$, $t_i = ik$. Let $(T_h^i, \boldsymbol{u}_h^i)$ be an approximation of $(T_h(\boldsymbol{x}, t), \boldsymbol{u}_h(\boldsymbol{x}, t))$ for $t = t_i$. Furthermore, we define

$$\boldsymbol{f}_k^{i+\Theta} = k^{-1} \int_{ik}^{(i+1)k} \boldsymbol{f}(t) \mathrm{d}t, \quad \boldsymbol{f}_k^{i+\Theta} \in [L^2(\Omega)]^2 \text{ for } 0 \le \Theta \le 1, \quad i = 0, 1, \ldots, n-1, \tag{12.260}$$

$$Q_k^{i+\Theta} = k^{-1} \int_{ik}^{(i+1)k} Q(t) \mathrm{d}t, \quad Q_k^{i+\Theta} \in L^2(\Omega) \text{ for } 0 \le \Theta \le 1, \quad i = 0, 1, \ldots, n-1, \tag{12.261}$$

Let us define

$$T_h^{i+\Theta} = \Theta T_h^{i+1} + (1 - \Theta) T_h^i, \quad 0 \le \Theta \le 1, \tag{12.262}$$

$$\boldsymbol{u}_h^{i+\Theta} = \Theta \boldsymbol{u}_h^{i+1} + (1 - \Theta) \boldsymbol{u}_h^i, \quad 0 \le \Theta \le 1. \tag{12.263}$$

Suppose that $(T_h^o, \boldsymbol{u}_h^o), \ldots, (T_h^i, \boldsymbol{u}_h^i)$ are known, then we determine $(T_h^{i+1}, \boldsymbol{u}_h^{i+1})$ as a solution in ${}^1V_h, V_h$ by

$$\begin{aligned} &(T_h^{i+1} - T_h^i, z_h - T_h^{i+\Theta}) + k a_{Th}(T_h^{i+\Theta}, z_h - T_h^{i+\Theta}) + k b_{oh}(\boldsymbol{u}_h^i, T_h^{i+\Theta}, z_h) \\ &\quad + k b_{ph}(\boldsymbol{u}_h^i, z_h - T_h^{i+\Theta}) \ge k(Q_k^{i+\Theta}, z_h - T_h^{i+\Theta}) \quad \forall z_h \in {}^1V_h, \end{aligned} \tag{12.264}$$

$$\begin{aligned} &(\boldsymbol{u}_h^{i+1} - \boldsymbol{u}_h^i, \boldsymbol{v}_h - \boldsymbol{u}_h^{i+\Theta}) + k\hat{\mu} a_h(\boldsymbol{u}_h^{i+\Theta}, \boldsymbol{v}_h - \boldsymbol{u}_h^{i+\Theta}) + k b_h(\boldsymbol{u}_h^{i+\Theta}, \boldsymbol{u}_h^{i+\Theta}, \boldsymbol{v}_h) \\ &\quad + k b_{sh}(T_h^i - T_o, \boldsymbol{v}_h - \boldsymbol{u}_h^{i+\Theta}) + k\hat{g} j_h(\boldsymbol{v}_h) - k\hat{g} j_h(\boldsymbol{u}_h^{i+\Theta}) \ge k(\boldsymbol{f}_k^{i+\Theta}, \boldsymbol{v}_h - \boldsymbol{u}_h^{i+\Theta}) \\ &\quad \forall \boldsymbol{v}_h \in V_h. \end{aligned} \tag{12.265}$$

The scheme is valid for $\Theta \ge 1/2$. For $\Theta = 1$ we obtain the implicit scheme, for $\Theta = 1/2$ the Cranck-Nicholson scheme. The scheme for $\Theta = 0$ has no justification. Let $\Theta > 0$, then

$$T_h^{i+1} = \Theta^{-1} T_h^{i+\Theta} + \Theta^{-1}(1 - \Theta) T_h^{i+\Theta}, \tag{12.266}$$

$$k^{-1}(T_h^{i+1} - T_h^i) = (\Theta k)^{-1}(T_h^{i+\Theta} - T_h^i), \tag{12.267}$$

$$\boldsymbol{u}_h^{i+1} = \Theta^{-1}\boldsymbol{u}_h^{i+\Theta} + \Theta^{-1}(1-\Theta)\boldsymbol{u}_h^{i+\Theta}, \tag{12.268}$$

$$k^{-1}(\boldsymbol{u}_h^{i+1} - \boldsymbol{u}_h^i) = (\Theta k)^{-1}(\boldsymbol{u}_h^{i+\Theta} - \boldsymbol{u}_h^i). \tag{12.269}$$

Then after transforming we obtain

$$\begin{aligned}&(T_h^{i+\Theta}, z_h - T_h^{i+\Theta}) + k\Theta a_{\mathrm{Th}}(T_h^{i+\Theta}, z_h - T_h^{i+\Theta}) + k\Theta b_{oh}(\boldsymbol{u}_h^i, T_h^{i+\Theta}, z_h)\\ &\quad + k\Theta b_{ph}(\boldsymbol{u}_h^i, z_h - T_h^{i+\Theta}) \geq (T_h^i, z_h - T_h^{i+\Theta}) + k\Theta(Q_k^{i+\Theta}, z_h - T_h^{i+\Theta}),\end{aligned} \tag{12.270}$$

$$\begin{aligned}&(\boldsymbol{u}_h^{i+\Theta}, \boldsymbol{v}_h - \boldsymbol{u}_h^{i+\Theta}) + k\Theta\hat{\mu}a_h(\boldsymbol{u}_h^{i+\Theta}, \boldsymbol{v}_h - \boldsymbol{u}_h^{i+\Theta}) + k\Theta b_h(\boldsymbol{u}_h^{i+\Theta}, \boldsymbol{u}_h^{i+\Theta}, \boldsymbol{v}_h)\\ &\quad + k\Theta b_{sh}(T_h^i - T_o, \boldsymbol{v}_h - \boldsymbol{u}_h^{i+\Theta}) + k\Theta\hat{g}j_h(\boldsymbol{v}_h) - k\Theta\hat{g}j_h(\boldsymbol{u}_h^{i+\Theta}) \geq (\boldsymbol{u}_h^i, \boldsymbol{v} - \boldsymbol{u}_h^{i+\Theta})\\ &\quad + k\Theta(\boldsymbol{f}_k^{i+\Theta}, \boldsymbol{v}_h - \boldsymbol{u}_h^{i+\Theta}).\end{aligned} \tag{12.271}$$

Remark: The schemes in Equations 12.264 and 12.265, and Equations 12.270 and 12.271 can be modified if we put

$$b_{ph}(\boldsymbol{u}_h^{i+\Theta}, T_h^{i+\Theta}) \text{ and } b_{sh}(T_h^{i+\Theta} - T_o, \boldsymbol{u}_h^{i+\Theta}),$$

or

$$b_{ph}(\boldsymbol{u}_h^i, T_h^{i+\Theta}) \text{ and } b_{sh}(T_h^i - T_o, \boldsymbol{u}_h^{i+\Theta}),$$

and similarly for the trilinear forms $b_{oh}(.,.,.)$ and $b_h(.,.,.)$ and their further combinations.

The semi-implicit scheme is as follows:

$$\begin{aligned}&k^{-1}(T_h^{i+1} - T_h^i, z_h - T_h^{i+1}) + a_{Th}(T_h^{i+1}, z_h - T_h^{i+1}) + b_{oh}(\boldsymbol{u}_h^i, T_h^i, z_h - T_h^{i+1})\\ &\quad + b_{ph}(\boldsymbol{u}_h^i, z_h - T_h^{i+1}) \geq (Q_k^{i+1}, z_h - T_h^{i+1}) \quad \forall z_h \in {}^1V_h,\end{aligned} \tag{12.272}$$

$$\begin{aligned}&k^{-1}(\boldsymbol{u}_h^{i+1} - \boldsymbol{u}_h^i, \boldsymbol{v}_h - \boldsymbol{u}_h^{i+1}) + \hat{\mu}a_h(\boldsymbol{u}_h^{i+1}, \boldsymbol{v}_h - \boldsymbol{u}_h^{i+1}) + b_h(\boldsymbol{u}_h^i, \boldsymbol{u}_h^i, \boldsymbol{v}_h - \boldsymbol{u}_h^{i+1})\\ &\quad + \hat{g}j(\boldsymbol{v}_h) - \hat{g}j(\boldsymbol{u}_h^{i+1}) + b_{sh}(T_h^i, \boldsymbol{v}_h - \boldsymbol{u}_h^{i+1}) \geq (\boldsymbol{f}_k^{i+1}, \boldsymbol{v}_h - \boldsymbol{u}_h^{i+1})\\ &\qquad \forall \boldsymbol{v}_h \in V_h.\end{aligned} \tag{12.273}$$

Multipliers $m = \{m_{ij}\}_{i,j=1,2}$ are approximated by functions $m_h = \{m_{hij}\}_{i,j=1,2}$, which are piecewise constant on every $\underline{T}_h \in \mathscr{T}_h$.

Then if $(T_h^i, \boldsymbol{u}_h^i)$, $(T_h^{i+1}, \boldsymbol{u}_h^{i+1})$ are known then

$$T_h^{i+\Theta} = \Theta T_h^{i+1} + (1-\Theta)T_h^i, \quad i = 0, 1, \ldots, n-1, \quad 0 \leq \Theta \leq 1, \tag{12.274}$$

$$\boldsymbol{u}_h^{i+\Theta} = \Theta \boldsymbol{u}_h^{i+1} + (1-\Theta)\boldsymbol{u}_h^i, \quad i = 0, 1, \ldots, n-1, \quad 0 \leq \Theta \leq 1, \tag{12.275}$$

$$T_n(t) = T_h^i + (t - t_i)/(\Theta k)^{-1}(T_h^{i+\Theta} - T_h^i) \quad \text{for } t_i < t \leq t_{i+1}, 0 \leq \Theta \leq 1, \quad i = 0, 1, \ldots, n-1, \tag{12.276}$$

$$\boldsymbol{u}_n(t) = \boldsymbol{u}_h^i + (t - t_i)/(\Theta k)^{-1}(\boldsymbol{u}_h^{i+\Theta} - \boldsymbol{u}_h^i) \quad \text{for } t_i < t \leq t_{i+1}, 0 \leq \Theta \leq 1, \quad i = 0, 1, \ldots, n-1, \tag{12.277}$$

where $T_h(t)$, $\boldsymbol{u}_h(t)$ are linear on $\langle ik, (i+1)k\rangle$, $i = 0, 1, \ldots, n-1$ and $T_n(t)$: $\langle t_o, t_1\rangle \to {}^1V_h$, $\boldsymbol{u}_n(t)$: $\langle t_o, t_1\rangle \to W_h$. The corresponding step-functions

$$\overline{T}_n(t) = T_h^{i+\Theta}, \quad ik < t \leq (i+1)k,$$
$$i = 0, 1, \ldots, n-1, 0 \leq \Theta \leq 1, \overline{T}_n(0) = T_o, \tag{12.278}$$
$$\overline{\boldsymbol{u}}_n(t) = \boldsymbol{u}_h^{i+\Theta}, \quad ik < t \leq (i+1)k,$$
$$i = 0, 1, \ldots, n-1, 0 \leq \Theta \leq 1, \overline{\boldsymbol{u}}_n(0) = \boldsymbol{u}_o, \tag{12.279}$$

are piecewise constant on $\langle t_o, t_1 \rangle$. Furthermore,

$$T'(t) \simeq T'_h(t) = (T_h^{i+1} - T_h^i)/k = (T_h^{i+\Theta} - T_h^i)/(\Theta k), \quad ik < t \leq (i+1)k,$$
$$0 \leq \Theta \leq 1, \quad i = 0, 1, \ldots, n-1, \tag{12.280}$$
$$\boldsymbol{u}'(t) \simeq \boldsymbol{u}'_h(t) = (\boldsymbol{u}_h^{i+1} - \boldsymbol{u}_h^i)/k = (\boldsymbol{u}_h^{i+\Theta} - \boldsymbol{u}_h^i)/(\Theta k), \quad ik < t \leq (i+1)k,$$
$$0 \leq \Theta \leq 1, \quad i = 0, 1, \ldots, n-1. \tag{12.281}$$

Then we have the following scheme.

Problem $(\mathscr{P}_h)$: Let $(T_h^o, \boldsymbol{u}_h^o), (T_h^1, \boldsymbol{u}_h^1), \ldots, (T_h^i, \boldsymbol{u}_h^i)$ be known. Then $(T_h^{i+1}, \boldsymbol{u}_h^{i+1}), (T_h^{i+\Theta}, \boldsymbol{u}_h^{i+\Theta})$ and $m_h^{i+\Theta}$ can be determined by

$$(T_h^{i+\Theta}, z_h) + k\Theta a_{Th}(T_h^{i+\Theta}, z_h) + k\Theta b_{oh}(\boldsymbol{u}_h^i, T_h^{i+\Theta}, z_h) + k\Theta b_{ph}(\boldsymbol{u}_h^i, z_h) = k\Theta(Q_k^{i+\Theta}, z_h) + (T_h^i, z_h)$$
$$\forall z_h \in {}^1V_h, \tag{12.282}$$

$$m_{hjl}^{i+\Theta} m_{hjl}^{i+\Theta} \leq 1 \text{ on every } \underline{T}_h \in \mathscr{T}_h, \quad m_{hjl}^{i+\Theta} = m_{hlj}^{i+\Theta}, m_{hjl}^{i+\Theta} \in L^\infty(\Omega), \quad l \leq j, l \leq 2, \tag{12.283}$$

$$m_{hjl}^{i+\Theta} D_{jl}(\boldsymbol{u}_h^{i+\Theta}) = (D_{jl}(\boldsymbol{u}_h^{i+\Theta}) D_{jl}(\boldsymbol{u}_h^{i+\Theta}))^{1/2} \text{ on every } \underline{T}_h \in \mathscr{T}_h, \tag{12.284}$$

$$(\boldsymbol{u}_h^{i+\Theta}, \boldsymbol{v}_h) + k\Theta\hat{\mu} a_h(\boldsymbol{u}_h^{i+\Theta}, \boldsymbol{v}_h) + 2^{1/2} k\Theta\hat{g}(m_{hjl}^{i+\Theta}, D_{jl}(\boldsymbol{v}_h)) + k\Theta b_{sh}(T_h^i - T_o, \boldsymbol{v}_h) + k\Theta b_h(\boldsymbol{u}_h^{i+\Theta}, \boldsymbol{u}_h^{i+\Theta}, \boldsymbol{v}_h)$$
$$= k\Theta(\boldsymbol{f}_k^{i+\Theta}, \boldsymbol{v}_h) + \boldsymbol{u}_h^i \boldsymbol{v}_h) \quad \forall \boldsymbol{v}_h \in \boldsymbol{V}_h, \quad \text{for } i = 0, 1, \ldots, n-1, \tag{12.285}$$

and $(T_h^{i+1}, \boldsymbol{u}_h^{i+1})$ will be determined from Equations 12.266 and 12.268, where Einstein's summation convention is also used.

First, we shall prove the stability conditions of the scheme in Equations 12.282-12.285, then second, we shall show that the approximate solution of these equations converges to the solution of Problem $(\mathscr{P}_m)_v$. According to Fortin (1972), Temam (1977), Nedoma (1988),(1995) it holds that

$$|T_h| \leq d_o \|T_h\| \quad \forall T_h \in {}^1V_h, \quad d_o \text{ does not depend on } h, \tag{12.286}$$
$$\|T_h\| \leq S_1(h)\,|T_h| \quad \forall T_h \in {}^1V_h, \quad S_1(h) \text{ depends on } h, \tag{12.287}$$
$$|\boldsymbol{u}_h| \leq d_2 \|\boldsymbol{u}_h\| \quad \forall \boldsymbol{u}_h \in W_h, \quad d_2 \text{ does not depend on } h, \tag{12.288}$$
$$\|\boldsymbol{u}_h\| \leq S(h)\,|\boldsymbol{u}_h| \quad \forall \boldsymbol{u}_h \in W_h, \quad S(h) \text{ depends on } h, \tag{12.289}$$
$$|b_{oh}(\boldsymbol{u}_h, T_h, z_h)| \leq c_o\,|\boldsymbol{u}_h|\,|T_h|\,\|z_h\| \quad \forall \boldsymbol{u}_h \in V_h, T_h, z_h \in {}^1V_h, \tag{12.290}$$
$$|b_h(\boldsymbol{u}_h, \boldsymbol{v}_h, \boldsymbol{w}_h)| \leq d_1 \|\boldsymbol{u}_h\|\,\|\boldsymbol{v}_h\|\,\|\boldsymbol{w}_h\| \quad \forall \boldsymbol{u}_h, \boldsymbol{v}_h, \boldsymbol{w}_h \in V_h, \tag{12.291}$$

where d_1 does not depend on h, and the stability constant $S(h)$ was determined by the following technique.

Let the space V_h be a space of polynomials of the first order. Then the constant $S(h)$ is a bound of the square root of the supremum

$$\sup_{\boldsymbol{u}_h \in V_h} \left\{ \sum_{i=1}^{N} \int_{\Omega} |\text{grad } u_{hi}(\boldsymbol{x})|^2 \, \mathrm{d}\boldsymbol{x} \left(\sum_{i=1}^{N} \int_{\Omega} |u_{ih}(\boldsymbol{x})|^2 \, \mathrm{d}\boldsymbol{x} \right)^{-1} \right\}, \quad N = 2.$$

This supremum is related to

$$\sup_{\Phi \in \boldsymbol{V}_h} \left\{ \sum_{i=1}^{N} \int_{\Omega} |\text{grad } \Phi(\boldsymbol{x})|^2 \, \mathrm{d}\boldsymbol{x} \left(\sum_{i=1}^{N} \int_{\Omega} |\Phi(\boldsymbol{x})|^2 \, \mathrm{d}\boldsymbol{x} \right)^{-1} \right\}$$

among all functions Φ which are polynomials of first degree on all $\underline{T}_h \in \mathscr{T}_h$. Let $\underline{\mu}_1$ be a bound for this supremum. Then for each $i = 1, 2$ and such a $\underline{T}_h \in \mathscr{T}_h$

$$\int_{\underline{T}_h} |\text{grad } u_{hi}(\boldsymbol{x})|^2 \, \mathrm{d}\boldsymbol{x} \leq \underline{\mu}_1 \int_{\underline{T}_h} |\text{grad } u_{hi}(\boldsymbol{x})|^2 \, \mathrm{d}\boldsymbol{x}, \quad \boldsymbol{u}_h(\boldsymbol{x}) \in V_h. \tag{12.292}$$

Hence

$$\|\boldsymbol{u}_h\|_h^2 = \sum_{i=1}^{N} \int_{\Omega} |\text{grad } u_{ih}(\boldsymbol{x})|^2 \, \mathrm{d}\boldsymbol{x} \leq \underline{\mu}_1 \sum_{i=1}^{N} \int_{\Omega} |u_{ih}(\boldsymbol{x})|^2 \, \mathrm{d}\boldsymbol{x} = \underline{\mu}_1 \, |\boldsymbol{u}_h|^2, N = 2,$$

so that

$$S(h) = \underline{\mu}_1^{1/2}.$$

To estimate Equation 12.292 we use a linear mapping Λ, $\boldsymbol{x} = \Lambda \boldsymbol{x}$, which maps $\underline{T}_h$ on a reference $\underline{T}_{oh}$, such that

$$0 \leq \Theta_i \leq 1, \quad \sum_{i=1}^{N} \Theta_i = 1.$$

Similarly as in Temam (1977)

$$\underline{\mu}_1 = \underline{\tilde{\mu}}_1 \rho^2(\underline{T}_{ho})(\rho'(h))^{-2},$$

where $\underline{\tilde{\mu}}_1$ is the supremum among all linear functions ψ of the ratio $\left\{ \int_{\underline{T}_h} |\text{grad } \Phi(\boldsymbol{x})|^2 \, \mathrm{d}\underline{x} / (\int_{\underline{T}_{ho}} |\psi(\boldsymbol{x})|^2 \, \mathrm{d}\boldsymbol{x} \right\}$ and where ρ is the diameter of the smallest ball containing $\underline{T}_h$, and ρ' is the diameter of the largest ball containing $\underline{T}_h$. Thus

$$S(h) = c_o(N)(\rho'(h))^{-1}, \quad c_o(N) = \underline{\tilde{\mu}}_1 \rho(\underline{T}_{ho}), \quad N = 2.$$

Moreover,

$$\begin{aligned} |b_h(\boldsymbol{u}_h, \boldsymbol{u}_h, \boldsymbol{v}_h)| &\leq d_1 \, \|\boldsymbol{u}_h\|^2 \, \|\boldsymbol{v}_h\| \leq d_1 S^2(h) \, |\boldsymbol{u}_h| \, \|\boldsymbol{u}_h\| \, |\boldsymbol{v}_h| \\ &\leq S_1(h) \, |\boldsymbol{u}_h| \, \|\boldsymbol{u}_h\| \, |\boldsymbol{v}_h| \quad \forall \boldsymbol{u}_h, \boldsymbol{v}_h \in V_h, \end{aligned} \tag{12.293}$$

where $S_1(h) \le d_1(\Omega, \alpha, \varepsilon) S^{(1+\varepsilon)}(h) \le d_1 S^2(h)$, $0 < \varepsilon < 1$, $\rho(\underline{T}_h)/\rho'(\underline{T}_h) \le \alpha < +\infty$ (see Temam (1977)) and

$$b_h(\boldsymbol{u}_h, \boldsymbol{u}_h, \boldsymbol{u}_h) = 0 \quad \forall \boldsymbol{u}_h \in V_h. \tag{12.294}$$

Furthermore,

$$b_{oh}(\boldsymbol{u}_h, T_h, z_h) + b_{oh}(\boldsymbol{u}_h, z_h, T_h) = 0 \quad \forall T_h, z_h \in {}^1V_h, \boldsymbol{u}_h \in V_h, \tag{12.295}$$

$$\left|b_{sh}(T_h - T_o, \boldsymbol{u}_h) + b_{ph}(\boldsymbol{u}_h, T_h)\right| \le c(1 + \|T_h\| \, |\boldsymbol{u}_h| + |T_h| \, \|\boldsymbol{u}_h\|) \quad \forall T_h \in {}^1V_h, \boldsymbol{u}_h \in V_h. \tag{12.296}$$

Theorem 12.8 (Stability): Let $k \equiv \Delta t$, h fulfil the following conditions

$$kS_o(h) < d_o, \tag{12.297}$$

$$kS_1(h) \le d_2, \tag{12.298}$$

$$kS^2(h) \le d_3, \tag{12.299}$$

where constants d_o, d_2, d_3 do not depend on k, h. Let $\Theta \ge (1/2)$. Then $(T_h^i, \boldsymbol{u}_h^i)$ are defined by the scheme in Equations 12.264 and 12.265 and it holds that

$$\left|T_h^i\right|^2 \le c_T, \quad 0 \le i \le n, \tag{12.300}$$

$$k \sum_{i=0}^{n-1} \left\|T_h^{i+\Theta}\right\|^2 \le c_T, \tag{12.301}$$

$$k \sum_{i=0}^{n-1} \left|T_h^{i+1} - T_h^i\right|^2 \le c_T, \tag{12.302}$$

$$\left|\boldsymbol{u}_h^i\right|^2 \le c_u, \quad 0 \le i \le n, \tag{12.303}$$

$$k \sum_{i=0}^{n-1} \left\|\boldsymbol{u}_h^{i+\Theta}\right\|^2 \le c_u, \tag{12.304}$$

$$k \sum_{i=0}^{n-1} \left|\boldsymbol{u}_h^{i+1} - \boldsymbol{u}_h^i\right|^2 \le c_u, \tag{12.305}$$

where c_T, c_u = constants are independent of k and h.

Proof: Take $z_h = 0$, $\boldsymbol{v}_h = 0$ in Equation 12.264 and 12.265, then

$$\left|T_h^{i+1}\right|^2 - \left|T_h^i\right|^2 + (2\Theta - 1)\left|T_h^{i+1} - T_h^i\right|^2 + 2k\left\|T_h^{i+\Theta}\right\|^2 - ck\left|T_h^{i+\Theta}\right| \left\|\boldsymbol{u}_h^i\right\| \le 2k(Q_k^{i+\Theta}, T_h^{i+\Theta}), \tag{12.306}$$

$$\begin{aligned}&\left|\boldsymbol{u}_h^{i+1}\right|^2 - \left|\boldsymbol{u}_h^i\right|^2 + (2\Theta - 1)\left|\boldsymbol{u}_h^{i+1} - \boldsymbol{u}_h^i\right|^2 + 2k\hat{\mu}\left\|\boldsymbol{u}_h^{i+\Theta}\right\|^2 - ck\hat{g}j_h(\boldsymbol{u}_h^{i+\Theta}) - ck\left(1 + \left\|T_h^i\right\| \left|\boldsymbol{u}_h^{i+\Theta}\right|\right)\\ &\le 2k(\boldsymbol{f}_k^{i+\Theta}, \boldsymbol{u}_h^{i+\Theta}).\end{aligned} \tag{12.307}$$

Due to Equations 12.286 and 12.287 and since $2ab \le \varepsilon a^2 + \varepsilon^{-1} b^2$, $0 < \varepsilon \le 1$, and as $\hat{\mu} \ge o$, then

$$\left|2k(Q_k^{i+\Theta}, T_h^{i+\Theta})\right| \le 2kd_o\left|Q_k^{i+\Theta}\right|_* \left\|T_h^{i+\Theta}\right\| \le k\varepsilon\left\|T_h^{i+\Theta}\right\|^2 + k\varepsilon^{-1}\left|Q_h^{i+\Theta}\right|_*^2,$$

$$\left|2k(\boldsymbol{f}_k^{i+\Theta}, u_h^{i+\Theta})\right| \le 2kd_2\left|\boldsymbol{f}_k^{i+\Theta}\right| \left\|\boldsymbol{u}_h^{i+\Theta}\right\| \le k\hat{\mu}d_2^2\left\|\boldsymbol{u}_h^{i+\Theta}\right\|^2 + \hat{\mu}^{-1}kd_2^2\left|\boldsymbol{f}_k^{i+\Theta}\right|^2.$$

Thus for $\Theta \geq (1/2)$, adding Equations 12.306 and 12.307 (with $\varepsilon = 1$)

$$\left|T_h^{i+1}\right|^2 - \left|T_h^i\right|^2 + (2\Theta - 1)\left|T_h^{i+1} - T_h^i\right|^2 + 2k\left\|T_h^{i+\Theta}\right\|^2 + \left|\boldsymbol{u}_h^{i+1}\right|^2 - \left|\boldsymbol{u}_h^i\right|^2 + (2\Theta - 1)\left|\boldsymbol{u}_h^{i+1} - \boldsymbol{u}_h^i\right|^2$$
$$+ 2k\hat{\mu}\left\|\boldsymbol{u}_h^{i+\Theta}\right\|^2 - 2k\hat{g}\,j(\boldsymbol{u}_h^{i+\Theta}) - ck\left(1 + \left\|T_h^i\right\|\left|\boldsymbol{u}_h^{i+\Theta}\right| + \left|T_h^{i+\Theta}\right|\left\|\boldsymbol{u}_h^i\right\|\right) \leq k\left|Q_k^{i+\Theta}\right|_*^2 + k\hat{\mu}^{-1}\left|\boldsymbol{f}_k^{i+\Theta}\right|_*^2.$$

Adding over i from 0 to n, using Equations 12.260 and 12.261 and since $\hat{g} \geq 0$, then $k j_h(\boldsymbol{u}_h^{i+\Theta}) \geq 0$, and we obtain

$$\left|T_h^{n+1}\right|^2 + k\sum_{i=0}^{n}\left\|T_h^{i+\Theta}\right\|^2 + \left|\boldsymbol{u}_h^{n+1}\right|^2 + k\hat{\mu}\sum_{i=0}^{n}\left\|\boldsymbol{u}_h^{i+\Theta}\right\|^2 \leq \left|T_h^o\right|^2 + \left|\boldsymbol{u}_h^o\right|^2 + k\sum_{i=0}^{n}\left|Q_k^{i+\Theta}\right|_*^2$$
$$+ k\hat{\mu}^{-1}\sum_{i=0}^{n}\left|\boldsymbol{f}_k^{i+\Theta}\right|_*^2 \leq c\left[|T_o|^2 + |\boldsymbol{u}_o|^2 + \int_I |Q(t)|_*^2\,\mathrm{d}t + \hat{\mu}^{-1}\int_I |\boldsymbol{f}(t)|_*^2\,\mathrm{d}t\right],$$

and the first part of these estimates follows immediately.

Now let us put $z_h = T_h^i$, $\boldsymbol{v}_h = \boldsymbol{u}_h^i$, $\Theta \geq (1/2)$ in Equations 12.264 and 12.265. Then

$$\left|T_h^{i+1} - T_h^i\right|^2 + ka_T(T_h^{i+\Theta}, T_h^{i+1} - T_h^i) \leq -kb_{oh}(\boldsymbol{u}_h^{i+\Theta}, T_h^{i+\Theta}, T_h^{i+1} - T_h^i) - kb_{ph}(\boldsymbol{u}_h^i, T_h^{i+1} - T_h^i)$$
$$+ k(Q_k^{i+\Theta}, T_h^{i+1} - T_h^i), \tag{12.308}$$

$$\left|\boldsymbol{u}_h^{i+1} - \boldsymbol{u}_h^i\right|^2 + k\hat{\mu}a_h(\boldsymbol{u}_h^{i+\Theta}, \boldsymbol{u}_h^{i+1} - \boldsymbol{u}_h^i) \leq -kb_h(\boldsymbol{u}_h^{i+\Theta}, \boldsymbol{u}_h^{i+\Theta}, \boldsymbol{u}_h^{i+1} - \boldsymbol{u}_h^i) - kb_{sh}(T_h^i, \boldsymbol{u}_h^{i+1} - \boldsymbol{u}_h^i)$$
$$+ k\Theta^{-1}\hat{g}\,j_h(\boldsymbol{u}_h^i) - k\Theta^{-1}\hat{g}\,j_h(\boldsymbol{u}_h^{i+\Theta}) + k(\boldsymbol{f}_k^{i+\Theta}, \boldsymbol{u}_h^{i+1} - \boldsymbol{u}_h^i). \tag{12.309}$$

Since

$$\left|kb_{oh}(\boldsymbol{u}_h^i, T_h^{i+\Theta}, T_h^{i+1} - T_h^i)\right| \leq ck\left|\boldsymbol{u}_h^{i+\Theta}\right|\left\|T_h^{i+\Theta}\right\|\left\|T_h^{i+1} - T_h^i\right\| \leq ck\left\|T_h^{i+\Theta}\right\|\left\|T_h^{i+1} - T_h^i\right\|$$
$$\leq ck\left\|T_h^{i+\Theta}\right\|^2 + kS_o^2(h)d_o^{-1}\left|T_h^{i+1} - T_h^i\right|^2, \tag{12.310}$$

$$\left|k(Q_k^{i+\Theta}, T_h^{i+1} - T_h^i)\right| \leq ck\left|Q_k^{i+\Theta}\right|_*^2 + kS_o^2(h)d_o^{-1}\left|T_h^{i+1} - T_h^i\right|^2, \tag{12.311}$$

$$\left|kb_h(\boldsymbol{u}_h^{i+\Theta}, \boldsymbol{u}_h^{i+\Theta}, \boldsymbol{u}_h^{i+1} - \boldsymbol{u}_h^i)\right| \leq ck\left|\boldsymbol{u}_h^{i+\Theta}\right|\left\|\boldsymbol{u}_h^{i+\Theta}\right\|\left\|\boldsymbol{u}_h^{i+1} - \boldsymbol{u}_h^i\right\| \leq ck\left\|\boldsymbol{u}_h^{i+\Theta}\right\|\left\|\boldsymbol{u}_h^{i+1} - \boldsymbol{u}_h^i\right\|$$
$$\leq ck\left\|\boldsymbol{u}_h^{i+\Theta}\right\|^2 + kS^2(h)d_3^{-1}\left|\boldsymbol{u}_h^{i+1} - \boldsymbol{u}_h^i\right|^2, \tag{12.312}$$

$$\left|k(\boldsymbol{f}_k^{i+\Theta}, \boldsymbol{u}_k^{i+1} - \boldsymbol{u}_h^i)\right| \leq ck\left|\boldsymbol{f}_k^{i+\Theta}\right|_* + kS^2(h)d_3^{-1}\left|\boldsymbol{u}_h^{i+1} - \boldsymbol{u}_h^i\right|^2, \tag{12.313}$$

$$k\Theta^{-1}\hat{g}\,j_h(\boldsymbol{u}_h^i) - k\Theta^{-1}\hat{g}\,j_h(\boldsymbol{u}_h^{i+\Theta}) \leq k\sum_{j,l=1}^{2}\int_\Omega m_{hjl}D_{jl}(\boldsymbol{u}_h^{i+1} - \boldsymbol{u}_h^i)\mathrm{d}\boldsymbol{x} \leq kc + kS^2(h)d_3^{-1}\left|\boldsymbol{u}_h^{i+1} - \boldsymbol{u}_h^i\right|^2, \tag{12.314}$$

then Equations 12.308 and 12.309 yield

$$(1 - 3kS_o^2(h)d_o^{-1})\left|T_h^{i+1} - T_h^i\right|^2 + \tfrac{1}{2}k\left(\left\|T_h^{i+1}\right\|^2 - \left\|T_h^i\right\|^2\right) + (\Theta - \tfrac{1}{2})k\left\|T_h^{i+1} - T_h^i\right\|^2$$
$$+ (1 - 3kS^2(h)d_3^{-1})\left|\boldsymbol{u}_h^{i+1} - \boldsymbol{u}_h^i\right|^2 + \tfrac{1}{2}k\hat{\mu}\left(\left\|\boldsymbol{u}_h^{i+1}\right\|^2 - \left\|\boldsymbol{u}_h^i\right\|^2\right) + (\Theta - \tfrac{1}{2})k\hat{\mu}\left\|\boldsymbol{u}_h^{i+1} - \boldsymbol{u}_h^i\right\|^2$$
$$\leq ck\left(1 + \left|Q_h^{i+\Theta}\right|_*^2 + \left|\boldsymbol{f}_k^{i+\Theta}\right|_*^2 + \left\|T_h^{i+\Theta}\right\|^2 + \left\|\boldsymbol{u}_h^{i+\Theta}\right\|^2\right).$$

Summing Equations 12.306 and 12.307 over i from 0 to n, majorization as above, then we obtain

$$k\sum_{i=0}^{n-1}\left|T_h^{i+1} - T_h^i\right|^2 \leq c_T, \quad k\sum_{i=0}^{n-1}\left|\boldsymbol{u}_h^{i+1} - \boldsymbol{u}_h^i\right|^2 \leq c_u.$$

To obtain estimates

$$k \sum_{i=0}^{n-1} \left\| T_h^{i+1} - T_h^i \right\|^2 \le c_T, \tag{12.315}$$

$$k \sum_{i=0}^{n-1} \left\| \boldsymbol{u}_h^{i+1} - \boldsymbol{u}_h^i \right\|^2 \le c_u, \tag{12.316}$$

it is sufficient to majorize instead of Equations 12.310-12.314 the corresponding differences, i.e.

$$\begin{aligned} \left| k b_{oh}(\boldsymbol{u}_h^i, T_h^{i+\Theta}, T_h^{i+1} - T_h^i) \right| &\le ck \left| \boldsymbol{u}_h^i \right| \left\| T_h^{i+\Theta} \right\| \left\| T_h^{i+1} - T_h^i \right\| \le ck \left\| T_h^{i+\Theta} \right\| \left\| T_h^{i+1} - T_h^i \right\| \\ &\le \tfrac{1}{4}\varepsilon k(\Theta - \tfrac{1}{2}) \left\| T_h^{i+1} - T_h^i \right\|^2 + kc^2(\Theta - \tfrac{1}{2})^{-1} \left\| T_h^{i+\Theta} \right\|^2, \quad 0 < \varepsilon < 1, \end{aligned} \tag{12.317}$$

$$\left| k(Q_k^{i+\Theta}, T_h^{i+1} - T_h^i) \right| \le \tfrac{1}{4} k\varepsilon(\Theta - \tfrac{1}{2}) \left\| T_h^{i+1} - T_h^i \right\|^2 + k(\Theta - \tfrac{1}{2})^{-1} \left| Q_k^{i+\Theta} \right|_*^2, \tag{12.318}$$

$$\begin{aligned} \left| k b_h(\boldsymbol{u}_h^{i+\Theta}, \boldsymbol{u}_h^{i+\Theta}, \boldsymbol{u}_h^{i+1} - \boldsymbol{u}_h^i) \right| &\le ck \left| \boldsymbol{u}_h^{i+\Theta} \right| \left\| \boldsymbol{u}_h^{i+\Theta} \right\| \left\| \boldsymbol{u}_h^{i+1} - \boldsymbol{u}_h^i \right\| \le ck \left\| \boldsymbol{u}_h^{i+\Theta} \right\| \left\| \boldsymbol{u}_h^{i+1} - \boldsymbol{u}_h^i \right\| \\ &\le \tfrac{1}{4}\hat{\mu}(\Theta - \tfrac{1}{2}) \left\| \boldsymbol{u}_h^{i+1} - \boldsymbol{u}_h \right\|^2 + kc^2(\Theta - \tfrac{1}{2})^{-1}\hat{\mu}^{-1} \left\| \boldsymbol{u}_h^{i+\Theta} \right\|^2, \end{aligned} \tag{12.319}$$

$$\left| k(\boldsymbol{f}_k^{i+\Theta}, \boldsymbol{u}_h^{i+1} - \boldsymbol{u}_h^i) \right| \le \tfrac{1}{4} k\hat{\mu}(\Theta - \tfrac{1}{2}) \left\| \boldsymbol{u}_h^{i+1} - \boldsymbol{u}_h^i \right\|^2 + k(\Theta - \tfrac{1}{2})^{-1}\hat{\mu}^{-1} \left| \boldsymbol{f}_k^{i+\Theta} \right|_*^2, \tag{12.320}$$

$$k\Theta^{-1}\hat{g} j_h(\boldsymbol{u}_h^i) - k\Theta^{-1}\hat{g} j_h(\boldsymbol{u}_h^{i+\Theta}) \le kc + \tfrac{1}{4} k\hat{\mu}(\Theta - \tfrac{1}{2}) \left\| \boldsymbol{u}_h^{i+1} - \boldsymbol{u}_h^i \right\|^2 . \tag{12.321}$$

Thus

$$\begin{aligned} &\left| T_h^{i+1} - T_h^i \right|^2 + \tfrac{1}{2}k \left\| T_h^{i+1} \right\|^2 - \tfrac{1}{4}k \left\| T_h^i \right\|^2 + \tfrac{1}{2}k(\Theta - \tfrac{1}{2}) \left\| T_h^{i+1} - T_h^i \right\|^2 + \left| \boldsymbol{u}_h^{i+1} - \boldsymbol{u}_h^i \right|^2 \\ &\quad + \tfrac{1}{2}k\hat{\mu} \left\| \boldsymbol{u}_h^{i+1} \right\|^2 - \tfrac{1}{4}k\hat{\mu} \left\| \boldsymbol{u}_h^i \right\|^2 + \tfrac{1}{2}k(\Theta - \tfrac{1}{2})\hat{\mu} \left\| \boldsymbol{u}_h^{i+1} - \boldsymbol{u}_h^i \right\|^2 \le ck \Big(1 + \left| Q_k^{i+\Theta} \right|_*^2 \\ &\quad + \left| \boldsymbol{f}_k^{i+\Theta} \right|_*^2 + \left\| T_h^{i+\Theta} \right\|^2 + \left\| \boldsymbol{u}_h^{i+\Theta} \right\|^2 \Big), \end{aligned}$$

and similarly as above, by using the summational technique, we obtain the estimates in Equations 12.315 and 12.316.

For a pair of step functions $(\overline{T}_h(t), \overline{\boldsymbol{u}}_h(t))$ we have the following result.

Theorem 12.9: The pair of functions $(T_h(t), \boldsymbol{u}_h(t))$ and $(\overline{T}_h(t), \overline{\boldsymbol{u}}_h(t))$ are bounded on $L^\infty(I; L^2(\Omega)) \times L^\infty(I, H)$ and $(\overline{T}_h(t), \overline{\boldsymbol{u}}_h(t))$ on $L^2(I; {}^1V) \times L^2(I, V)$ and are independent of k and h. For $\Theta > (1/2)$, $(T_h(t), \boldsymbol{u}_h(t))$ is bounded on $L^2(I; {}^1V) \times L^2(I; V)$. In the case when $\Theta = 1/2$, then $kS_o^2(h) \le d_o$, $kS^2(h) \le d_3$ must be satisfied.

Proof: $(T_h(t), \boldsymbol{u}_h(t))$ is bounded on $L^2(I; {}^1V) \times L^2(I, V)$. Thus

$$\int_I \|T_h(t)\|^2 \, \mathrm{d}t \le c \left(k \sum_{i=0}^{n-1} \left\| T_h^{i+1} \right\|^2 + k \left\| T_h^o \right\|^2 \right),$$

$$\int_I \|\boldsymbol{u}_h(t)\|^2 \, \mathrm{d}t \le c \left(k \sum_{i=0}^{n-1} \left\| \boldsymbol{u}_h^{i+1} \right\|^2 + k \left\| \boldsymbol{u}_h^o \right\|^2 \right).$$

But

$$\left\| T_h^{i+1} \right\|^2 \le 2 \left\| T_h^{i+\Theta} \right\|^2 + 2 \left\| T_h^{i+1} - T_h^i \right\|^2, \; \left\| \boldsymbol{u}_h^{i+1} \right\|^2 \le 2 \left\| \boldsymbol{u}_h^{i+\Theta} \right\|^2 + 2 \left\| \boldsymbol{u}_h^{i+1} - \boldsymbol{u}_h^i \right\|^2,$$

therefore,

$$k\sum_{i=0}^{n-1}\|T_h^{i+1}\|^2 \le ck\sum_{i=0}^{n-1}\|T_h^{i+\Theta}\|^2 + ck\sum_{i=0}^{n-1}\|T_h^{i+1}-T_h^i\|^2,$$

$$k\sum_{i=0}^{n-1}\|\boldsymbol{u}_h^{i+1}\|^2 \le ck\sum_{i=0}^{n-1}\|\boldsymbol{u}_h^{i+\Theta}\|^2 + ck\sum_{i=0}^{n-1}\|\boldsymbol{u}_h^{i+1}-\boldsymbol{u}_h^i\|^2,$$

and since, according to Theorem 12.8, $\|T_h\| \le S_o(h)\,|T_h|$, $\|\boldsymbol{u}_h\| \le S(h)\,|\boldsymbol{u}_h|$, then the theorem is completed.

To prove the convergence theorem we need the following theorem and lemmas.

Theorem 12.10: The approximation $(T_h(t), \boldsymbol{u}_h(t))$ satisfies

$$\int_{t_o}^{t_1-\tau} |T_h(t+\tau) - T_h(t)|\,\mathrm{d}t \le c\tau, \tag{12.322}$$

$$\int_{t_o}^{t_1-\tau} |\boldsymbol{u}_h(t+\tau) - \boldsymbol{u}_h(t)|\,\mathrm{d}t \le c\tau \tag{12.323}$$

and $kS_o^2(h) < d_o$, $kS^2(h) < d_3$. Then for every $\varepsilon > 0$

$$(T_h(t), \boldsymbol{u}_h(t)) \text{ is bounded in } H^{(1/2)-\varepsilon}(I; L^2(\Omega)) \times H^{(1/2)-\varepsilon}(I; H). \tag{12.324}$$

Proof: From Equations 12.264 and 12.265 we obtain

$$k^{-1}(T_h^{i+1} - T_h^i, T_h^{i+\Theta} - z_h(t)) \le a_{Th}(T_h^{i+\Theta}, T_h^{i+\Theta} - z_h(t)) - b_{oh}(\boldsymbol{u}_h^i, T_h^{i+\Theta}, T_h^{i+\Theta} - z_h(t))$$
$$- b_{ph}(\boldsymbol{u}_h^i, T_h^{i+\Theta} - z_h(t)) + (Q_k^{i+\Theta}, T_h^{i+\Theta} - z_h(t)) \quad \forall z_h \in L^2(I; {}^1V_h), \quad \forall t \in (ik, (i+1)k\rangle, \tag{12.325}$$

$$k^{-1}(\boldsymbol{u}_h^{i+1} - \boldsymbol{u}_h^i, \boldsymbol{u}_h^{i+\Theta} - \boldsymbol{v}_h(t)) \le \hat{\mu} a_h(\boldsymbol{u}_h^{i+\Theta}, \boldsymbol{u}_h^{i+\Theta} - \boldsymbol{v}_h(t)) - b_h(\boldsymbol{u}_h^{i+\Theta}, \boldsymbol{u}_h^{i+\Theta}, \boldsymbol{u}_h^{i+\Theta} - \boldsymbol{v}_h(t))$$
$$- b_{sh}(T_h^i, \boldsymbol{u}_h^{i+\Theta} - \boldsymbol{v}_h(t)) + \hat{g} j_h(\boldsymbol{v}_h(t)) - \hat{g} j_h(\boldsymbol{u}_h^{i+\Theta}) + (\boldsymbol{f}_k^{i+\Theta}, \boldsymbol{u}_h^{i+\Theta} - \boldsymbol{v}_h(t))$$
$$\forall \boldsymbol{v}_h \in L^2(I; V_h), \quad \forall t \in (ik, (i+1)k\rangle. \tag{12.326}$$

Due to the definition of $(T_h(t), \boldsymbol{u}_h(t))$ we have

$$k^{-1}(T_h^{i+1} - T_h^i, T_h^{i+\Theta} - z_h(t)) = k^{-1}(T_h^{i+1} - T_h^i, T_h^{i+\Theta}) - (T_h'(t), z_h(t)), \tag{12.327}$$

$$k^{-1}(\boldsymbol{u}_h^{i+1} - \boldsymbol{u}_h^i, \boldsymbol{u}_h^{i+\Theta} - \boldsymbol{v}_h(t)) = k^{-1}(\boldsymbol{u}_h^{i+1} - \boldsymbol{u}_h^i, \boldsymbol{u}_h^{i+\Theta}) - (\boldsymbol{u}_h'(t), \boldsymbol{v}_h(t)). \tag{12.328}$$

Let us put $z_h(t) = \tau^{-1}\int_{t-\tau}^{t} T_h(\sigma)\mathrm{d}\sigma$, $\boldsymbol{v}_h(t) = \tau^{-1}\int_{t-\tau}^{t} \boldsymbol{u}_h(\sigma)\mathrm{d}\sigma$, where we assume that $T_h(\sigma) = T_h^o$, $\boldsymbol{u}_h(\sigma) = \boldsymbol{u}_h^o$ for $\sigma \le 0$. Integrating the last terms in Equations 12.327 and 12.328 from t_o to t_1, we obtain

$$-\int_I (T_h'(t), z_h(t))\mathrm{d}t = \int_I (T_h(t), z_h'(t))\mathrm{d}t + (T_h(t_o), z_h(t_o)) - (T_h(t_1), z_h(t_1))$$
$$= \int_I \tau^{-1}(T_h(t), T_h(t) - T_h(t-\tau))\mathrm{d}t + \left|T_h^o\right|^2 - (T_h(t_1), z_h(t_1)),$$

$$-\int_I (\boldsymbol{u}_h'(t), \boldsymbol{v}_h(t))\mathrm{d}t = \int_I (\boldsymbol{u}_h(t), \boldsymbol{v}_h'(t))\mathrm{d}t + (\boldsymbol{u}_h(t_o), \boldsymbol{v}_h(t_o)) - (\boldsymbol{u}_h(t_1), \boldsymbol{v}_h(t_1))$$
$$= \int_I \tau^{-1}(\boldsymbol{u}_h(t), \boldsymbol{u}_h(t) - \boldsymbol{u}_h(t-\tau))\mathrm{d}t + \left|\boldsymbol{u}_h^o\right|^2 - (\boldsymbol{u}_h(t_1), \boldsymbol{v}_h(t_1)).$$

Moreover,

$$\int_I k^{-1}(T_h^{i+1} - T_h^i, T_h^{i+\Theta})\mathrm{d}t = \sum_{i=0}^{n-1}(T_h^{i+1} - T_h^i, T_h^{i+\Theta}) = \left(\Theta - \frac{1}{2}\right)\sum_{i=0}^{n-1}\left|T_h^{i+1} - T_h^i\right|^2 + \frac{1}{2}\left|T_h^n\right|^2 - \frac{1}{2}\left|T_h^o\right|^2,$$

$$\int_I k^{-1}(\boldsymbol{u}_h^{i+1} - \boldsymbol{u}_h^i, \boldsymbol{u}_h^{i+\Theta})\mathrm{d}t = \sum_{i=0}^{n-1}(\boldsymbol{u}_h^{i+1} - \boldsymbol{u}_h^i, \boldsymbol{u}_h^{i+\Theta}) = \left(\Theta - \frac{1}{2}\right)\sum_{i=0}^{n-1}\left|\boldsymbol{u}_h^{i+1} - \boldsymbol{u}_h^i\right|^2 + \frac{1}{2}\left|\boldsymbol{u}_h^n\right|^2 - \frac{1}{2}\left|\boldsymbol{u}_h^o\right|^2,$$

and

$$\begin{aligned}
&\int_I \tau^{-1}(T_h(t), T_h(t) - T_h(t-\tau))\mathrm{d}t \\
&= \int_I \tau^{-1}\left|T_h(t)\right|^2 \mathrm{d}t - \int_\tau^{t_1} \tau^{-1}(T_h(t), T_h(t-\tau))\mathrm{d}t - \int_{t_o}^{\tau} \tau^{-1}(T_h(t), T_h^o)\mathrm{d}t \\
&= \frac{1}{2}\int_\tau^{t_1} \tau^{-1}\left|T_h(t) - T_h(t-\tau)\right|^2\mathrm{d}t + \frac{1}{2}\int_{t_o}^{\tau} \tau^{-1}\left|T_h(t)\right|^2\mathrm{d}t - \int_{t_o}^{\tau}\tau^{-1}(T_h(t), T_h^o)\mathrm{d}t + \frac{1}{2}\int_{t_1-\tau}^{t_1}\tau^{-1}\left|T_h(t)\right|^2\mathrm{d}t,
\end{aligned}$$

$$\begin{aligned}
&\int_I \tau^{-1}(\boldsymbol{u}_h(t), \boldsymbol{u}_h(t) - \boldsymbol{u}_h(t-\tau))\mathrm{d}t \\
&= \int_I \tau^{-1}\left|\boldsymbol{u}_h(t)\right|^2 \mathrm{d}t - \int_\tau^{t_1} \tau^{-1}(\boldsymbol{u}_h(t), \boldsymbol{u}_h(t-\tau))\mathrm{d}t - \int_{t_o}^{\tau} \tau^{-1}(\boldsymbol{u}_h(t), \boldsymbol{u}_h^o)\mathrm{d}t \\
&= \frac{1}{2}\int_\tau^{t_1} \tau^{-1}\left|\boldsymbol{u}_h(t) - \boldsymbol{u}_h(t-\tau)\right|^2\mathrm{d}t + \frac{1}{2}\int_{t_o}^{\tau} \tau^{-1}\left|\boldsymbol{u}_h(t)\right|^2\mathrm{d}t - \int_{t_o}^{\tau}\tau^{-1}(\boldsymbol{u}_h(t), \boldsymbol{u}_h^o)\mathrm{d}t + \frac{1}{2}\int_{t_1-\tau}^{t_1}\tau^{-1}\left|\boldsymbol{u}_h(t)\right|^2\mathrm{d}t,
\end{aligned}$$

Then integrating Equations 12.325 and 12.326 over t we have

$$\begin{aligned}
&\frac{1}{2}\int_\tau^{t_1} \tau^{-1}\left|T_h(t) - T_h(t-\tau)\right|^2 \mathrm{d}t + \left(\Theta - \frac{1}{2}\right)\sum_{i=0}^{n-1}\left|T_h^{i+1} - T_h^i\right|^2 + \frac{1}{2}\left|T_h^n\right|^2 \\
&\quad + \frac{1}{2}\int_{t_1-\tau}^{t_1} \tau^{-1}\left|T_h(t)\right|^2 \mathrm{d}t - (T_h^n, z_h(t_1)) + \frac{1}{2}\left|T_h^o\right|^2 + \frac{1}{2}\int_{t_o}^{\tau}\tau^{-1}\left|T_h(t)\right|^2\mathrm{d}t - \left(T_h^o, \tau^{-1}\int_{t_o}^{\tau} T_h(t)\mathrm{d}t\right) \\
&\quad \le \int_I (Q_h(t), \overline{T}_h(t) - z_h(t))\mathrm{d}t + \int_I b_{ph}(\boldsymbol{u}_h(t), \overline{T}_h(t) - z_h(t))\mathrm{d}t - \int_I b_{oh}(\boldsymbol{u}_h(t), \overline{T}_h(t), \overline{T}_h(t) - z_h(t))\mathrm{d}t,
\end{aligned} \tag{12.329}$$

$$\begin{aligned}
&\frac{1}{2}\int_\tau^{t_1} \tau^{-1}\left|\boldsymbol{u}_h(t) - \boldsymbol{u}_h(t-\tau)\right|^2 \mathrm{d}t + \left(\Theta - \frac{1}{2}\right)\sum_{i=0}^{n-1}\left|\boldsymbol{u}_h^{i+1} - \boldsymbol{u}_h^i\right|^2 + \frac{1}{2}\left|\boldsymbol{u}_h^n\right|^2 + \frac{1}{2}\int_{t_1-\tau}^{t_1}\tau^{-1}\left|\boldsymbol{u}_h(t)\right|^2\mathrm{d}t \\
&\quad - (\boldsymbol{u}_h^n, \boldsymbol{v}_h(t_1)) + \frac{1}{2}\left|\boldsymbol{u}_h^o\right|^2 + \frac{1}{2}\int_{t_o}^{\tau}\tau^{-1}\left|\boldsymbol{u}_h(t)\right|^2\mathrm{d}t - \left(\boldsymbol{u}_h^o, \tau^{-1}\int_{t_o}^{\tau}\boldsymbol{u}_h(t)\mathrm{d}t\right) \le \int_I (\boldsymbol{f}_h(t), \overline{\boldsymbol{u}}_h(t) - \boldsymbol{v}_h(t))\mathrm{d}t \\
&\quad + \int_I b_{sh}(T_h(t), \overline{\boldsymbol{u}}_h(t) - \boldsymbol{v}_h(t))\mathrm{d}t + \hat{g}\int_I j_h(\boldsymbol{v}_h(t)\mathrm{d}t - \int_I b_h(\boldsymbol{u}_h(t), \overline{\boldsymbol{u}}_h(t), \overline{\boldsymbol{u}}_h(t), -\boldsymbol{v}_h(t))\mathrm{d}t.
\end{aligned} \tag{12.330}$$

Furthermore,

$$\left|(T_h^n, z_h(t_1))\right| \le \tfrac{1}{2}\left|T_h^n\right|^2 + \tfrac{1}{2}\left|z_h(t_1)\right|^2, \tag{12.331}$$

$$\left|z_h(t_1)\right|^2 \le \tau^{-2}\left|\int_{t_1-\tau}^{t_1} T_h(t)\mathrm{d}t\right|^2 \le \tau^{-1}\int_{t_1-\tau}^{t_1}\left|T_h(t)\right|^2\mathrm{d}t, \tag{12.332}$$

$$\left|(\boldsymbol{u}_h^n(t), \boldsymbol{v}_h(t_1))\right| \le \tfrac{1}{2}\left|\boldsymbol{u}_h^n\right|^2 + \tfrac{1}{2}\left|\boldsymbol{v}_h(t_1)\right|^2, \tag{12.333}$$

$$|\boldsymbol{v}_h(t_1)|^2 \leq \tau^{-2}\left|\int_{t_1-\tau}^{t_1} \boldsymbol{u}_h(t)\mathrm{d}t\right|^2 \leq \tau^{-1}\int_{t_1-\tau}^{t_1} |\boldsymbol{u}_h(t)|^2\,\mathrm{d}t, \tag{12.334}$$

$$\left|\left(T_h^o, \tau^{-1}\int_{t_o}^{\tau} T_h(t)\mathrm{d}t\right)\right| \leq \tfrac{1}{2}\left|T_h^o\right|^2 + \tfrac{1}{2}\int_{t_o}^{\tau} \tau^{-1}\,|T_h(t)|^2\,\mathrm{d}t, \tag{12.335}$$

$$\left|\left(\boldsymbol{u}_h^o, \tau^{-1}\int_{t_o}^{\tau} \boldsymbol{u}_h(t)\mathrm{d}t\right)\right| \leq \tfrac{1}{2}\left|\boldsymbol{u}_h^o\right|^2 + \frac{1}{2}\int_{t_o}^{\tau} \tau^{-1}\,|\boldsymbol{u}_h(t)|^2\,\mathrm{d}t. \tag{12.336}$$

Thus

$$\frac{1}{2}\int_{\tau}^{t_1} \tau^{-1}\,|T_h(t) - T_h(t-\tau)|^2\,\mathrm{d}t \leq \int_I (Q_h(t), \overline{T}_h(t) - z_h(t))\mathrm{d}t, \tag{12.337}$$

$$\frac{1}{2}\int_{\tau}^{t_1} \tau^{-1}\,|\boldsymbol{u}_h(t) - \boldsymbol{u}_h(t-\tau)|^2\,\mathrm{d}t \leq \int_I (\boldsymbol{f}_h(t), \overline{\boldsymbol{u}}_h(t) - \boldsymbol{v}_h(t))\mathrm{d}t + \hat{g}\int_I j_h(\boldsymbol{v}_h)\mathrm{d}t. \tag{12.338}$$

Now we shall estimate the other members of Equations 12.325 and 12.326. We have

$$\begin{aligned}\int_I a_{Th}(\overline{T}_h(t), \overline{T}_h(t) - z_h(t))\mathrm{d}t &\leq \frac{1}{2}\int_I \left\|\overline{T}_h(t)\right\|^2\mathrm{d}t + \frac{1}{2}\int_I \left\|\overline{T}_h(t) - z_h(t)\right\|^2\mathrm{d}t \\ &\leq \frac{1}{2}\int_I \left\|\overline{T}_h(t)\right\|^2\mathrm{d}t + \int_I \left\|\overline{T}_h(t)\right\|^2\mathrm{d}t + \int_I \|z_h(t)\|^2\,\mathrm{d}t,\end{aligned}$$

and due to Theorem 12.9

$$\int_I \|z_h(t)\|^2\,\mathrm{d}t \leq c\int_I \|T_h(t)\|^2\,\mathrm{d}t \leq \text{constant}$$

if $kS_o^2(h) \leq d_o$. Furthermore,

$$\begin{aligned}\hat{\mu}\int_I a_h(\overline{\boldsymbol{u}}_h(t), \overline{\boldsymbol{u}}_h(t) - \boldsymbol{v}_h(t))\mathrm{d}t &\leq \frac{1}{2}\hat{\mu}\int_I \|\boldsymbol{u}_h(t)\|^2\,\mathrm{d}t + \frac{1}{2}\hat{\mu}\int_I \|\overline{\boldsymbol{u}}_h(t) - \boldsymbol{v}_h(t)\|^2\,\mathrm{d}t \\ &\leq \frac{1}{2}\hat{\mu}\int_I \|\overline{\boldsymbol{u}}_h(t)\|^2\,\mathrm{d}t + \hat{\mu}\int_I \|\overline{\boldsymbol{u}}_h(t)\|^2\,\mathrm{d}t + \hat{\mu}\int_I \|\boldsymbol{v}_h(t)\|^2\,\mathrm{d}t,\end{aligned}$$

and due to Theorem 12.9

$$\int_I \|\boldsymbol{v}_h(t)\|^2\,\mathrm{d}t \leq c\int_I \|\boldsymbol{u}_h(t)\|^2\,\mathrm{d}t \leq \text{constant}$$

if $kS^2(h) \leq d_3$. Moreover,

$$\int_I (Q_k^{i+\Theta}, \overline{T}_h(t) - z_h(t))\mathrm{d}t \leq c\int_I |Q(t)|_*^2\,\mathrm{d}t + \text{constant},$$

$$\int_I (\boldsymbol{f}_k^{i+\Theta}, \overline{\boldsymbol{u}}_h(t) - \boldsymbol{v}_h(t))\mathrm{d}t \leq c\int_I |\boldsymbol{f}(t)|_*^2\,\mathrm{d}t + \text{constant},$$

and

$$\int_I b_{oh}(\overline{\boldsymbol{u}}_h(t), \overline{T}_h(t), \overline{T}_h(t) - z_h(t))\mathrm{d}t = \int_I b_{oh}(\overline{\boldsymbol{u}}_h(t), z_h(t), \overline{T}_h(t))\mathrm{d}t$$

as $b_{oh}(\boldsymbol{u}_h, z_h, z_h) = 0$. Due to the Cagliardo's inequality$^{(*)}$ then

$$\left|\int_I b_{oh}(\overline{\boldsymbol{u}}_h(t), z_h(t), \overline{T}_h(t))\mathrm{d}t\right| \le c\int_I |\overline{\boldsymbol{u}}_h(t)|\,\left\|\overline{T}_h(t)\right\|\,\|z_h(T)\|\,\mathrm{d}t$$

$$\le c\int_I \left\|\overline{T}_h(t)\right\|^2 \mathrm{d}t + c\int_I \|z_h(t)\|^2\,\mathrm{d}t \le \text{constant}, \quad (12.339)$$

and also

$$\int_I b_h(\overline{\boldsymbol{u}}_h(t), \overline{\boldsymbol{u}}_h(t), \overline{\boldsymbol{u}}_h(t) - \boldsymbol{v}_h(t))\mathrm{d}t = \int_I b_h(\overline{\boldsymbol{u}}_h(t), \boldsymbol{v}_h(t), \overline{\boldsymbol{u}}_h(t))\mathrm{d}t$$

as $b_h(\overline{\boldsymbol{u}}_h, \overline{\boldsymbol{u}}_h, \overline{\boldsymbol{u}}_h) = 0$. Due to the Cagliardo's inequality then

$$\left|\int_I b_h(\overline{\boldsymbol{u}}_h(t), \boldsymbol{v}_h(t), \overline{\boldsymbol{u}}_h(t))\mathrm{d}t\right| \le c\int_I |\overline{\boldsymbol{u}}_h(t)|\,\|\overline{\boldsymbol{u}}_h(t)\|\,\|\boldsymbol{v}_h(t)\|\,\mathrm{d}t$$

$$\le c\int_I \|\overline{\boldsymbol{u}}_h(t)\|^2\,\mathrm{d}t + c\int_I \|\boldsymbol{v}_h(t)\|^2\,\mathrm{d}t \le \text{constant}. \quad (12.340)$$

At the end we estimate

$$\hat{g}\int_I j_h(\boldsymbol{v}_h(t))\mathrm{d}t \le c + \hat{g}\int_I \|\boldsymbol{v}_h(t)\|^2\,\mathrm{d}t \le \text{constant}. \quad (12.341)$$

Using all these estimates, then the proof is complete.

To prove the convergence theorem we also need the following lemmas.

Lemma 12.8: The subsequences $(T_h, \boldsymbol{u}_h)$ and $(\overline{T}_h, \overline{\boldsymbol{u}}_h)$ are weakly star convergent on $L^\infty(I \times L^2(\Omega)) \times L^\infty(I \times H)$, and are weakly convergent on $L^2(I; {}^1V) \times L^2(I; V)$ and converge to the element $(Z, \boldsymbol{w}) \in (L^\infty(I; L^2(\Omega)) \cap L^2(I; {}^1V)) \times (L^\infty(I; H) \cap L^2(I; V))$.

Proof: According to Theorem 12.9 we consider $(T_h, \boldsymbol{u}_h)$ and $(\overline{T}_h, \overline{\boldsymbol{u}}_h)$ which are bounded on $(L^\infty(I; L^2(\Omega)) \cap L^2(I; {}^1V)) \times (L^\infty(I; H) \cap L^2(I; V))$. It is sufficient to show that the subsequences converge to the same limit. From Theorem 12.8 it follows that

$$\int_I \left|T_h(t) - \overline{T}_h(t)\right|^2 \mathrm{d}t \le ck\sum_{i=0}^{n-1}\left|T_h^{i+1} - T_h^i\right|^2 \le c_o k, \quad (12.342)$$

$$\int_I |\boldsymbol{u}_h(t) - \overline{\boldsymbol{u}}_h(t)|^2\,\mathrm{d}t \le ck\sum_{i=0}^{n-1}\left|\boldsymbol{u}_h^{i+1} - \boldsymbol{u}_h^i\right|^2 \le c_1 k. \quad (12.343)$$

Hence, in the limit the assertion follows immediately.

Lemma 12.9: The extracted subsequences $(T_h, \boldsymbol{u}_h)$ and $(\overline{T}_h, \overline{\boldsymbol{u}}_h)$ converge strongly to $(Z, \boldsymbol{w})$ in $L^2(I; L^2(\Omega)) \times L^2(I; H)$.

Proof: According to Theorem 12.10, $(T_h, \boldsymbol{u}_h)$ is independent of h and k and bounded on $H^{(1/2)-\varepsilon}(I; L^2(\Omega)) \times H^{(1/2)-\varepsilon}(I; H)$. Let us define the spaces

$$\boldsymbol{X} \equiv \{\varphi \mid \varphi \in L^2(I; {}^1V),\quad \varphi \in H^{(1/2)-\varepsilon}(I; L^2(\Omega))\} \subset L^2(I; L^2(\Omega)),$$

$$\boldsymbol{Y} \equiv \{\vec{\psi} \mid \vec{\psi} \in L^2(I; V),\quad \vec{\psi} \in H^{(1/2)-\varepsilon}(I; H)\} \subset L^2(I; H).$$

$^{(*)}$ Cagliardo's inequality: Let $N = 2$. then

$$|\boldsymbol{u}|_{[L^4(\Omega)]^2} \le c\,|\boldsymbol{u}|^{1/2}\,\|\boldsymbol{u}\|^{1/2}.$$

Imbedding $V \subset H$ is compact, so that $\boldsymbol{Y} \subset L^2(I;H)$ is also compact, and similarly $\boldsymbol{X} \subset L^2(I, L^2(\Omega))$ is also compact. Then a subsequence $(T_h, \boldsymbol{u}_h)$ converges weakly to $(Z, \boldsymbol{w})$ in $\boldsymbol{X} \times \boldsymbol{Y}$ and, due to the compactness, strongly in $L^2(I; {}^1V) \times L^2(I;H)$. The result for $(\overline{T}_h, \overline{\boldsymbol{u}}_h)$ follows immediately from Equations 12.342 and 12.343.

In the following we shall verify that $(Z(t), \boldsymbol{w}(t)) = (T(t), \boldsymbol{u}(t))$ and that $(T(t), \boldsymbol{u}(t))$ is a unique solution of Problem $(\mathscr{P})_v$.

Let $z(t) \in C^1(I; {}^1V)$, $\quad z(t_1) = 0$, where $z(t) \in {}^1\mathscr{V} = {}^1V \cap \mathscr{D}(\Omega)$,

$$\boldsymbol{v}(t) \in C^1(I;V), \quad \boldsymbol{v}(t_1) = 0, \text{ where } \boldsymbol{v}(t) \in \mathscr{V} = V \cap [\mathscr{D}(\Omega)]^2,$$

for every $t_o \le t \le t_1$. We define a pair of functions $(z_h(t), \boldsymbol{v}_h(t))$ as follows:

$(z_h(t), \boldsymbol{v}_h(t))$ is linear in t in the interval $(ik, (i+1)k\rangle$,

$(z_h(ik), \boldsymbol{v}_h(ik)) = (r_h(z(ik), r_h\boldsymbol{v}(ik))$, where r_h is a restriction operator.

Lemma 12.10: Let $(z_h, \boldsymbol{v}_h)$ and $(z'_h, \boldsymbol{v}'_h)$ converge strongly in $L^2(I; {}^1V) \times L^2(I;V)$ to $(z, \boldsymbol{v})$ and $(z', \boldsymbol{v}')$ respectively, if k and h tend to zero.

Proof: The scheme in Equations 12.264 and 12.265 can be rewritten in the following form:

$$\begin{aligned}
&\int_{ik}^{(i+1)k} (T'_h, z_h(t))\mathrm{d}t + \int_{ik}^{(i+1)k} a_{Th}(\overline{T}_h(t), z_h(t))\mathrm{d}t + \int_{ik}^{(i+1)k} b_{oh}(\overline{\boldsymbol{u}}_h(t), \overline{T}_h(t), z_h(t))\mathrm{d}t \\
&\quad + \int_{ik}^{(i+1)k} b_{ph}(\overline{\boldsymbol{u}}_h(t), z_h(t))\mathrm{d}t \ge (T_h^{i+1} - T_h^i, T_h^{i+\Theta}) + \int_{ik}^{(i+1)} a_{Th}(\overline{T}_h(t), \overline{T}_h(t))\mathrm{d}t \\
&\quad + \int_{ik}^{(i+1)k} b_{ph}(\overline{\boldsymbol{u}}_h(t), \overline{T}_h(t))\mathrm{d}t + \int_{ik}^{(i+1)k} (Q_k(t), z_h(t) - \overline{T}_h(t))\mathrm{d}t,
\end{aligned} \tag{12.344}$$

$$\begin{aligned}
&\int_{ik}^{(i+1)k} (\boldsymbol{u}'_h(t), \boldsymbol{v}_h(t))\mathrm{d}t + \hat{\mu} \int_{ik}^{(i+1)k} a_h(\overline{\boldsymbol{u}}_h(t), \boldsymbol{v}_h(t))\mathrm{d}t + \int_{ik}^{(i+1)k} b_h(\overline{\boldsymbol{u}}_h(t), \overline{\boldsymbol{u}}_h(t), \boldsymbol{v}_h(t))\mathrm{d}t \\
&\quad + \int_{ik}^{(i+1)k} b_{sh}(\overline{T}_h(t), \boldsymbol{v}_h(t))\mathrm{d}t + \hat{g} \int_{ik}^{(i+1)k} j_h(\boldsymbol{v}_h(t))\mathrm{d}t \ge (\boldsymbol{u}_h^{i+1} - \boldsymbol{u}_h^i, \boldsymbol{u}_h^{i+\Theta}) + \hat{\mu} \int_{ik}^{(i+1)} a_h(\overline{\boldsymbol{u}}_h(t), \overline{\boldsymbol{u}}_h(t))\mathrm{d}t \\
&\quad + \int_{ik}^{(i+1)k} b_{sh}(\overline{T}_h(t), \overline{\boldsymbol{u}}_h(t))\mathrm{d}t + \hat{g} \int_{ik}^{(i+1)k} j_h(\overline{\boldsymbol{u}}_h(t))\mathrm{d}t + \int_{ik}^{(i+1)k} (\boldsymbol{f}_k(t), \boldsymbol{v}_h(t) - \overline{\boldsymbol{u}}_h(t))\mathrm{d}t,
\end{aligned} \tag{12.345}$$

where we put $Q_k(t) = Q_k^{i+\Theta}$ and $\boldsymbol{f}_k(t) = \boldsymbol{f}_k^{i+\Theta}$ for $t \in (ik, (i+1)k\rangle$. But we have

$$\int_{ik}^{(i+1)k} (T'_h(t), z_h(t))\mathrm{d}t = -\int_{ik}^{(i+1)k} (T_h(t), z'(t))\mathrm{d}t + (T_h((i+1)k), z_h((i+1)k)) - (T_h(ik), z_h(ik)),$$

$$\int_{ik}^{(i+1)k} (\boldsymbol{u}'_h(t), \boldsymbol{v}_h(t))\mathrm{d}t = -\int_{ik}^{(i+1)k} (\boldsymbol{u}_h(t), \boldsymbol{v}'(t))\mathrm{d}t + (\boldsymbol{u}_h((i+1)k), \boldsymbol{v}_h((i+1)k)) - (\boldsymbol{u}_h(ik), \boldsymbol{v}_h(ik)),$$

$$(T_h^{i+1} - T_h^i, T_h^{i+\Theta}) = \tfrac{1}{2}\left|T_h^{i+1}\right|^2 - \tfrac{1}{2}\left|T_h\right|^2 + (\Theta - \tfrac{1}{2})\left|T_h^{i+1} - T_h^i\right|^2,$$

$$(\boldsymbol{u}_h^{i+1} - \boldsymbol{u}_h^i, \boldsymbol{u}_h^{i+\Theta}) = \tfrac{1}{2}\left|\boldsymbol{u}_h^{i+1}\right|^2 - \tfrac{1}{2}\left|\boldsymbol{u}_h\right|^2 + (\Theta - \tfrac{1}{2})\left|\boldsymbol{u}_h^{i+1} - \boldsymbol{u}_h^i\right|^2.$$

Summing Equations 12.344 and 12.345 over i, $0 \le i \le n-1$, we obtain

$$- \int_I (T_h(t), z_h'(t))\mathrm{d}t + \int_I a_{Th}(\overline{T}_h(t), z_h(t))\mathrm{d}t + \int_I b_{oh}(\overline{\boldsymbol{u}}_h(t), \overline{T}_h(t), z_h(t))\mathrm{d}t$$

$$+ \int_I b_{ph}(\overline{\boldsymbol{u}}_h(t), z_h(t))\mathrm{d}t - (T_h^o, z_h(t_o)) \geq |T_h(t_1)|^2 - |T_h(t_o)|^2 + \int_I a_{Th}(\overline{T}_h(t), \overline{T}_h(t))\mathrm{d}t$$

$$+ \int_I b_{ph}(\overline{\boldsymbol{u}}_h(t), \overline{T}_h(t))\mathrm{d}t + \int_I (Q_k(t), z_h(t) - \overline{T}_h(t))\mathrm{d}t,$$

$$- \int_I (\boldsymbol{u}_h(t), \boldsymbol{v}_h'(t))\mathrm{d}t + \hat{\mu} \int_I a_h(\overline{\boldsymbol{u}}_h(t), \boldsymbol{v}_h(t))\mathrm{d}t + \int_I b_h(\overline{\boldsymbol{u}}_h(t), \overline{\boldsymbol{u}}_h(t), \boldsymbol{v}_h(t))\mathrm{d}t$$

$$+ \int_I b_{sh}(\overline{T}_h(t), \boldsymbol{v}_h(t))\mathrm{d}t + \hat{g} \int_I j_h(\boldsymbol{v}_h(t))\mathrm{d}t - (\boldsymbol{u}_h^o, \boldsymbol{v}_h(t_o)) \geq |\boldsymbol{u}_h(t_1)|^2 - |\boldsymbol{u}_h(t_o)|^2$$

$$+ \hat{\mu} \int_I a_h(\overline{\boldsymbol{u}}_h(t), \overline{\boldsymbol{u}}_h(t))\mathrm{d}t + \int_I b_{sh}(\overline{T}_h(t), \overline{\boldsymbol{u}}_h(t))\mathrm{d}t + \hat{g} \int_I j_h(\overline{\boldsymbol{u}}_h(t))\mathrm{d}t + \int_I (\boldsymbol{f}_k(t), \boldsymbol{v}_h(t) - \overline{\boldsymbol{u}}_h(t))\mathrm{d}t.$$

Then, similarly as in Duvaut and Lions (1972a), Nedoma (1995) there exist convergent subsequences, and taking the above lemmas and applying the limit inferior we have

$$- \int_I (Z(t), z'(t))\mathrm{d}t + \int_I a_T(Z(t), z(t)\mathrm{d}t + \int_I b_o(\boldsymbol{w}(t), Z(t), z(t))\mathrm{d}t + \int_I b_p(\boldsymbol{w}(t), z(t))\mathrm{d}t - (T_o, z(t_o))$$

$$\geq \frac{1}{2}|Z(t_1)|^2 - \frac{1}{2}|Z(t_o)|^2 + \int_I a_T(Z(t), Z(t))\mathrm{d}t + \int_I b_p(\boldsymbol{w}(t), Z(t))\mathrm{d}t + \int_I (Q(t), z(t) - Z(t))\mathrm{d}t, \tag{12.346}$$

$$- \int_I (\boldsymbol{w}(t), \boldsymbol{v}'(t))\mathrm{d}t + \hat{\mu} \int_I a(\boldsymbol{w}(t), \boldsymbol{v}(t))\mathrm{d}t + \int_I b(\boldsymbol{w}(t), \boldsymbol{w}(t), \boldsymbol{v}(t))\mathrm{d}t + \int_I b_s(Z(t), \boldsymbol{v}(t))\mathrm{d}t$$

$$+ \hat{g} \int_I j(\boldsymbol{v}(t))\mathrm{d}t - (\boldsymbol{u}_o, \boldsymbol{v}(t_o)) \geq \frac{1}{2}|\boldsymbol{w}(t_1)|^2 - \frac{1}{2}|\boldsymbol{w}(t_o)|^2 + \hat{\mu} \int_I a(\boldsymbol{w}(t), \boldsymbol{w}(t))\mathrm{d}t$$

$$+ \int_I b_s(Z(t), \boldsymbol{w}(t))\mathrm{d}t + \hat{g} \int_I j(\boldsymbol{w}(t))\mathrm{d}t + \int_I (\boldsymbol{f}(t), \boldsymbol{v}(t) - \boldsymbol{w}(t))\mathrm{d}t. \tag{12.347}$$

Now we shall verify that $(z', \boldsymbol{w}') \in L^2(I; {}^1V') \times L^2(I; V')$. We can write

$$\frac{1}{2}|Z(t_1)|^2 - \frac{1}{2}|Z(t_o)|^2 = \int_I (Z', Z)\mathrm{d}t,$$

$$\frac{1}{2}|\boldsymbol{w}(t_1)|^2 - \frac{1}{2}|\boldsymbol{w}(t_o)|^2 = \int_I (\boldsymbol{w}', \boldsymbol{w})\mathrm{d}t,$$

Then

$$\int_I (Z'(t), z(t) - Z(t))\mathrm{d}t + \int_I a_T(Z(t), z(t) - Z(t))\mathrm{d}t + \int_I b_o(\boldsymbol{w}(t), Z(t), z(t))\mathrm{d}t$$

$$+ \int_I b_p(\boldsymbol{w}(t), z(t) - Z(t))\mathrm{d}t \geq \int_I (Q(t), z(t) - Z(t))\mathrm{d}t + (T_o, z(t_o)) - (Z(t_o), z(t_o)),$$

$$\int_I (\boldsymbol{w}'(t), \boldsymbol{v}(t) - \boldsymbol{w}(t))\mathrm{d}t + \hat{\mu} \int_I a(\boldsymbol{w}(t), \boldsymbol{v}(t) - \boldsymbol{w}(t))\mathrm{d}t + \int_I b(\boldsymbol{w}(t), \boldsymbol{w}(t), \boldsymbol{v}(t))\mathrm{d}t$$

$$+ \int_I b_s(Z(t), \boldsymbol{v}(t) - \boldsymbol{w}(t))\mathrm{d}t + \hat{g} \int_I j(\boldsymbol{v}(t))\mathrm{d}t - \hat{g} \int_I j(\boldsymbol{w}(t))\mathrm{d}t$$

$$\geq \int_I (\boldsymbol{f}(t), \boldsymbol{v}(t) - \boldsymbol{w}(t))\mathrm{d}t + (\boldsymbol{u}_o, \boldsymbol{v}(t_o)) - (\boldsymbol{w}(t_o), \boldsymbol{v}(t_o)).$$

Since $Z(t_o) = T_o$, $\boldsymbol{w}(t_o) = \boldsymbol{u}_o$ then $(z(t), \boldsymbol{w}(t))$ is a solution of the problem studied.

Now we shall prove that $(Z', \boldsymbol{w}') \in L^2(I; {}^1V') \times L^2(I; V')$. Then Equations 12.346 and 12.347 yield

$$\int_I \{(Z(t), z'(t)) - a_T(Z(t), z(t)) - b_o(\boldsymbol{w}(t), Z(t), z(t)) - b_p(\boldsymbol{w}(t), Z(t)) + (Q(t), z(t))\}\mathrm{d}t - (T_o z(t_o)) \leq c, \tag{12.348}$$

$$\int_I \{(\boldsymbol{w}(t), \boldsymbol{v}'(t)) - \hat{\mu} a(\boldsymbol{w}(t), \boldsymbol{v}(t)) - b(\boldsymbol{w}(t), \boldsymbol{w}(t), \boldsymbol{v}(t)) - b_s(Z(t), \boldsymbol{v}(t)) + \int_I (\boldsymbol{f}(t), \boldsymbol{v}(t))\}\mathrm{d}t - (\boldsymbol{u}_o, \boldsymbol{v}(t_o)) \leq \hat{g} \int_I j(\boldsymbol{w}(t))\mathrm{d}t + c, \tag{12.349}$$

where c are constants dependent on Z or $\boldsymbol{w}$, respectively. Let us put

$$l(z) = \int_I \{(Z(t), z'(t)) - a_T(Z(t), Z(t)) - b_o(\boldsymbol{w}(t), Z(t), z(t)) - b_p(\boldsymbol{w}(t), Z(t)) + (Q(t), z(t))\}\mathrm{d}t,$$

$$L(\boldsymbol{w}) = \int_I \{(\boldsymbol{w}(t), \boldsymbol{v}'(t)) - \hat{\mu} a(\boldsymbol{w}(t), \boldsymbol{w}(t)) - b(\boldsymbol{w}(t), \boldsymbol{w}(t), \boldsymbol{v}(t)) - b_s(Z(t), \boldsymbol{v}(t)) + \int_I (\boldsymbol{f}(t), \boldsymbol{v}(t))\}\mathrm{d}t.$$

Taking $z \in \mathscr{D}(I; {}^1V)$, $\boldsymbol{v} \in \mathscr{D}(I; V)$ and replacing z by $\pm\lambda z$ and $\boldsymbol{v}$ by $\pm\nu\boldsymbol{v}$ in Equations 12.348 and 12.349, then for $\lambda > 0$, $\nu > 0$, we obtain

$$\pm\lambda l(z) \leq c \quad \forall\lambda > 0, \quad \pm\nu L(\boldsymbol{v}) \leq \nu\hat{g} \int_I j(\boldsymbol{v})\mathrm{d}t + c \quad \forall\nu > 0,$$

and $|l(z)| \leq 0$, $|L(\boldsymbol{v})| \leq \hat{g} \int_I j(\boldsymbol{v}(t))\mathrm{d}t$.
Thus $l(z)$ and $L(\boldsymbol{v})$ are continuous on $\mathscr{D}(I; {}^1V)$ and $\mathscr{D}(I; V)$, respectively. Thus there exist $X \in L^\infty(I; {}^1V')$, $\boldsymbol{Y} \in L^\infty(I; V')$ such that

$$l(z) = \int_I (X(t), z(t))\mathrm{d}t, \quad L(\boldsymbol{v}) = \int_I (\boldsymbol{Y}(t), \boldsymbol{v}(t))\mathrm{d}t.$$

According to the definition of $l(z)$ and $L(\boldsymbol{v})$ then

$$z' \in L^2(I; {}^1V'), \quad \boldsymbol{w}' \in L^2(I; V').$$

To prove that $(T_h(t), \boldsymbol{u}_h(t))$ and $(\overline{T}_h(t), \overline{\boldsymbol{u}}(t))$ converge weakly to $(T(t), \boldsymbol{u}(t))\, \forall t$, we take the subsequence of a pair of functions $(T_{h'}(t), \boldsymbol{u}_{h'}(t))$ of $(T_h(t), \boldsymbol{u}_h(t))$ converging weakly to $(X(t), \boldsymbol{Y}(t)) \in L^2(\Omega) \times H$; then

$$(T_{h'}(t), z) \to (X(t), z) \quad \forall z \in L^2(\Omega),$$

$$(\boldsymbol{u}_{h'}(t), \boldsymbol{v}) \to (\boldsymbol{Y}(t), \boldsymbol{v}) \quad \forall \boldsymbol{v} \in H.$$

Let $\varphi_h(t) = (T_{h'}(t), z)$, $\vec{\psi}_h(t) = (\boldsymbol{u}_{h'}(t), \boldsymbol{v})$, then

$$|\varphi_{h'}(t)| \leq c\,|z|, \quad \left|\vec{\psi}_{h'}(t)\right| \leq c\,|\boldsymbol{v}|.$$

Thus, the Lebesgue theorem gives

$$\int_I (T_{h'}(t), z)\alpha(t)\mathrm{d}t \to \int_I (X(t), z)\alpha(t)\mathrm{d}t,$$

$$\int_I (\boldsymbol{u}_{h'}(t), \boldsymbol{v})\beta(t)\mathrm{d}t \to \int_I (\boldsymbol{Y}(t), \boldsymbol{v})\beta(t)\mathrm{d}t,$$

for every $\alpha(t) \in C(I; {}^1V)$, $\beta(t) \in C(I; V)$. But $(T_h, \boldsymbol{u}_h)$ converges weakly star to $(T(t), \boldsymbol{u}(t))$ in $L^\infty(I; L^2(\Omega)) \times L^\infty(I; H)$, then

$$\int_I (T_{h'}(t), z)\alpha(t)\mathrm{d}t \to \int_I (T(t), z)\alpha(t)\mathrm{d}t,$$

$$\int_I (\boldsymbol{u}_{h'}(t), \boldsymbol{v})\beta(t)\mathrm{d}t \to \int_I (\boldsymbol{u}(t), \boldsymbol{v})\beta(t)\mathrm{d}t.$$

Thus

$$X(t) = T(t) \text{ and } \boldsymbol{Y}(t) = \boldsymbol{u}(t),$$

which completes the proof.

Then summing all the obtained results we have proved the following convergence theorem.

Theorem 12.11: Let $\Theta \geq (1/2)$ and let the assumptions of Theorem 12.8 be satisfied. Then

$$(T_h, \boldsymbol{u}_h) \text{ and } (\overline{T}_h, \overline{\boldsymbol{u}}_h) \text{ converge weakly star to } (T, \boldsymbol{u}) \text{ in the space } L^\infty(I; L^2(\Omega)) \times L^\infty(I; H), \tag{12.350}$$

$$(T_h, \boldsymbol{u}_h) \text{ and } (\overline{T}_h, \overline{\boldsymbol{u}}_h) \text{ converge weakly to } (T, \boldsymbol{u}) \text{ in the space } L^2(I; {}^1V) \times L^2(I; V), \tag{12.351}$$

$$(T_h(t), \boldsymbol{u}_h(t)) \text{ and } (\overline{T}_h(t), \overline{\boldsymbol{u}}_h(t)) \text{ converge weakly to } (T(t), \boldsymbol{u}(t)) \text{ in } L^2(\Omega) \times H \quad \forall t. \tag{12.352}$$

ALGORITHM

The algorithm based on the scheme in Equations 12.264 and 12.265 represents the coupled non-linear problem, as terms $b_{oh}(\boldsymbol{u}_h, T_h, z_h)$, $b_h(\boldsymbol{u}_h, \boldsymbol{u}_h, \boldsymbol{v}_h)$ and $j_h(\boldsymbol{v}_h)$ are non-linear. Moreover, the functional $j_h(\boldsymbol{v}_h)$ is a non-differentiable functional. The difficulty in practical computations is connected with the approximation of the constraint "div $\boldsymbol{v}_h = 0$" (i.e. the incompressibility of Bingham's fluid). There is a question of how to build this constraint condition into the definition of the space V_h. The further numerical problem represents the discrete approximation of the pressure. In the converse to the solution of the equation of motion in Bingham's rheology, the solution of the thermal part of the problem is relatively easy and represents the approximation of the non-linear problem, with the numerical difficulties which prevail in non-linear problems evoked by the convective term $(\boldsymbol{v} \cdot \nabla)T$ only. The scheme in Equations 12.264 and 12.265 is the stable scheme, but as we said above there exists a certain problem of finding the basis of V_h with div $\boldsymbol{v}_h = 0$. In our contribution the Uzawa's and the Arrow-Hurwicz's algorithms for such type of problems will be used. Moreover, we shall show that in the problem in Equations 12.264 and 12.265, $j_h(\boldsymbol{v}_h)$ is a non-differential functional, and then the problem leads to finding a saddle point and, therefore, the Uzawa's algorithm can also be used (see Temam (1977)). Then both problems lead to solving the systems of non-linear algebraic equations with sparse symmetric (or in a special case, not solved in this paper, with sparse non-symmetric) positive definite matrices. For their solution the conjugate gradient method will be used.

Spaces ${}^1V_h$, W_h and V_h were defined above. Let us consider $m = \{m_{ij}\}$, $1 \le i, j \le 2$, $m \in [L^2(\Omega)]^4$, $m_{ij} = m_{ji}$ and let us put

$$m_{ij}D_{ij}(\boldsymbol{v}) = \sum_{i,j=1}^{2} m_{ij}D_{ij}(\boldsymbol{v}),\ (m, D\boldsymbol{v}) = \int_\Omega m_{ij}D_{ij}(\boldsymbol{v})\mathrm{d}\boldsymbol{x}.$$

Let $K = \{m \mid m_{ij} \in L^\infty(\Omega), \|m\|_\infty \le 1\}$. For the given triangulation $\mathscr{T}_h$ we define

$$K_h = \{m_h \mid \|m_h\|_\infty \le 1, m_{hjl} = \text{ constant on every } \underline{T}_h, m_{hjl} = 0 \text{ on } G\backslash\overline{G}_h, \quad j,l = 1,2, \quad m_{hjl} = m_{hjl}\}. \tag{12.353}$$

Furthermore, let us define the bilinear forms A_{Th} and A_h by

$$A_{Th}(T_h, z_h) = (T_h, z_h) + k\Theta a_{Th}(T_h, z_h) \quad \forall T_h, z_h \in {}^1V_h, \tag{12.354}$$

$$A_h(\boldsymbol{u}_h, \boldsymbol{v}_h) = (\boldsymbol{u}_h, \boldsymbol{v}_h) + k\Theta\hat{\mu}a_h(\boldsymbol{u}_h, \boldsymbol{v}_h) \quad \forall \boldsymbol{u}_h, \boldsymbol{v}_h \in V_h, \tag{12.355}$$

and operators $\underline{A}_{Th}$ and $\underline{A}_h$ on ${}^1V_h$ and V_h respectively, by

$$(\underline{A}_{Th}T_h, z_h) = (T_h, z_h) + k\Theta a_{Th}(T_h, z_h) + k\Theta b_{oh}(\boldsymbol{u}_h, T_h, z_h), \tag{12.356}$$

$$(\underline{A}_h\boldsymbol{u}_h, \boldsymbol{v}_h) = (\boldsymbol{u}_h, \boldsymbol{v}_h) + k\Theta\hat{\mu}a_h(\boldsymbol{u}_h, \boldsymbol{v}_h) + k\Theta b_h(\boldsymbol{u}_h, \boldsymbol{u}_h, \boldsymbol{v}_h). \tag{12.357}$$

We denote $D\boldsymbol{v}_h = \{D_{ij}(\boldsymbol{v}_h)\}_{i,j=1,2}$. Then Problem $\mathscr{P}_h$ can be re-written in the following form.

Problem ($\mathscr{P}_{hA}$): Let $(T_h^o, \boldsymbol{u}_h^o), (T_h^1, \boldsymbol{u}_h^1), (T_h^2, \boldsymbol{u}_h^2), \ldots, (T_h^i, \boldsymbol{u}_h^i)$ be known. Then $(T_h^{i+\Theta}, \boldsymbol{u}_h^{i+\Theta})$ will be determined from the following conditions:

$$A_{Th}(T_h^{i+\Theta}, z_h) + k\Theta b_{oh}(\boldsymbol{u}_h^i, T_h^{i+\Theta}, z_h) = k\Theta(Q_k^{i+\Theta}, z_h) + (T_h^i, z_h) - k\Theta b_{ph}(\boldsymbol{u}_h^i, z_h) \quad \forall z_h \in {}^1V_h, \tag{12.358}$$

$$\begin{aligned}&A_h(\boldsymbol{u}_h^{i+\Theta}, \boldsymbol{v}_h) + k\Theta b_h(\boldsymbol{u}_h^{i+\Theta}, \boldsymbol{u}_h^{i+\Theta}, \boldsymbol{v}_h) + k\Theta 2^{1/2}\hat{g}(m_h^{i+\Theta}, D\boldsymbol{v}_h)\\ &\quad = (\boldsymbol{u}_h^i, \boldsymbol{v}_h) + k\Theta(\boldsymbol{f}_k^{i+\Theta}, \boldsymbol{v}_h) - k\Theta b_{sh}(T_h^i - T_o, \boldsymbol{v}_h) \quad \forall \boldsymbol{v}_h \in V_h,\end{aligned} \tag{12.359}$$

$$(D\boldsymbol{u}_h^{i+\Theta}, m - m_h^{i+\Theta}) \le 0 \quad \forall m \in K, \tag{12.360}$$

$$K = \{m \mid m_{jl} = m_{lj} \in L^\infty(\Omega), \|m\|_\infty \le 1\}, \quad j,l = 1,2, \quad i = 0,1,\ldots,n-1, \quad \tfrac{1}{2} \le \Theta \le 1, \tag{12.361}$$

and where the condition $\|m\|_\infty \le 1$ is equivalent to Equation 12.283, whilst Equation 12.360 is equivalent to Equation 12.284, as for every $m \in K$

$$2^{1/2}(m, D\boldsymbol{v}_h) \le 2^{1/2}\int_\Omega (m_{ij}m_{ij})^{1/2}(D_{ij}(\boldsymbol{v}_h)D_{ij}(\boldsymbol{v}_h))^{1/2}\mathrm{d}\boldsymbol{x} \le 2^{1/2}\int_\Omega (D_{ij}(\boldsymbol{v}_h)D_{ij}(\boldsymbol{v}_h))^{1/2}\mathrm{d}x = j_h(\boldsymbol{v}_h),$$

so that Equation 12.360 is equivalent to $2^{1/2}(m_h^{i+\Theta}, D\boldsymbol{u}_h^{i+\Theta}) = j_h(\boldsymbol{u}_h^{i+\Theta})$.

We introduce the functional $l_h(z_h)$ and the Lagrangian $\mathscr{L}_h(\boldsymbol{v}_h, m_h)$ by

$$l_h(z_h) = \tfrac{1}{2}(\underline{A}_{Th}z_h, z_h) - \langle s_h^i, z_h\rangle, \tag{12.362}$$

$$\mathscr{L}_h(\boldsymbol{v}_h, m_h) = \tfrac{1}{2}(\underline{A}_h\boldsymbol{v}_h, \boldsymbol{v}_h) + k\Theta 2^{1/2}\hat{g}(m_h, D\boldsymbol{v}_h) - \langle S_h^i, \boldsymbol{v}_h\rangle = \mathscr{J}(\boldsymbol{v}_h) + (m_h, \Psi(\boldsymbol{v}_h)), \tag{12.363}$$

where

$$\langle s_h^i, z_h\rangle = (T_h^i, z_h) + k\Theta(Q_k^{i+\Theta}, z_h) - k\Theta b_{ph}(\boldsymbol{u}_h^i, z_h), \qquad i = 0, 1, \ldots, n-1,$$
$$\langle S_h^i, \boldsymbol{v}_h\rangle = (\boldsymbol{u}_h^i, \boldsymbol{v}_h) + k\Theta(\boldsymbol{f}_k^{i+\Theta}, \boldsymbol{v}_h) - k\Theta b_{sh}(T_h^i - T_o, \boldsymbol{v}_h), \quad i = 0, 1, \ldots, n-1,$$
$$\Psi(\boldsymbol{v}_k) = k\Theta\hat{g}2^{1/2}D_h\boldsymbol{v}_h.$$

Then the problem is equivalent to finding a minimum of the functional $l_h(z_h) : {}^1V_h \to R$ and a saddle point of the Lagrangian $\mathscr{L}_h : V_h \times K_h \to R$ such that

$$l_h(T_h^{i+\Theta}) \le l_h(z_h) \quad \forall z_h \in {}^1V_h, \tag{12.364}$$
$$\mathscr{L}_h(\boldsymbol{u}_h^{i+\Theta}, m_h^{i+\Theta}) \le \mathscr{L}_h(\boldsymbol{v}_h, m_h^{i+\Theta}) \quad \forall \boldsymbol{v}_h \in V_h, \tag{12.365}$$
$$\mathscr{L}_h(\boldsymbol{u}_h^{i+\Theta}, m_h) \le \mathscr{L}_h(\boldsymbol{u}_h^{i+\Theta}, m_h^{i+\Theta}) \quad \forall m_h \in K_h, \tag{12.366}$$
$$i = 0, 1, \ldots, n-1, \ \tfrac{1}{2} < \Theta \le 1.$$

Then due to Equations 12.286 and 12.287, the stability of the scheme under the condition $kS^2(h) \le d_3$ and the presumptions about the thermal sources Q and the body forces $\boldsymbol{f}$, we find

$$\left|\langle s_h^i, z_h\rangle\right| \le c\,\|z_h\|, \tag{12.367}$$
$$\left|\langle S_h^i, \boldsymbol{v}_h\rangle\right| \le c\,\|\boldsymbol{v}_h\|, \tag{12.368}$$

where c are constants independent of k and h.
Therefore, $s_h^i \in {}^1V_h'$, $S_h^i \in V_h'$. Furthermore, bilinear forms $A_{Th}(T_h, z_h)$ and $A_h(\boldsymbol{u}_h, \boldsymbol{v}_h)$ are, on ${}^1V_h \times {}^1V_h$ and $W_h \times W_h$ respectively, symmetric, continuous and positive definite and

$$a_{Th}(T_h, T_h) \ge c\,\|T_h\|^2 \quad \forall T_h \in {}^1V_h,$$
$$a_h(\boldsymbol{u}_h, \boldsymbol{u}_h) \ge \sum_{\underline{T}_h \in \mathscr{T}_h} \int_{\underline{T}_h} \sum_{j,l=1,2} (D_j u_{hl})^2 \mathrm{d}\boldsymbol{x} = \|\boldsymbol{u}_h\|^2 \quad \forall \boldsymbol{u}_h \in W_h,$$

where the norm $\|\cdot\|$ in V_h is induced by the scalar product $(\boldsymbol{u}_h, \boldsymbol{v}_h)_h = \Sigma_{i=1,2}(D_i\boldsymbol{u}_h, D_i\boldsymbol{v}_h)$ and operators $\underline{A}_{Th}$ and $\underline{A}_h$ are strongly monotone (see the next lemma).

Lemma 12.11: Operators $\underline{A}_{Th}$ and $\underline{A}_h$ are strongly monotone, i.e.

$$(\underline{A}_{Th}T_h - \underline{A}_{Th}z_h, T_h - z_h) \ge \alpha_o\,\|T_h - z_h\|^2 \quad \forall T_h, z_h \in {}^1V_h,$$
$$(\underline{A}_h\boldsymbol{u}_h - \underline{A}_h\boldsymbol{v}_h, \boldsymbol{u}_h - \boldsymbol{v}_h) \ge \alpha\,\|\boldsymbol{u}_h - \boldsymbol{v}_h\|^2 \quad \forall \boldsymbol{u}_h, \boldsymbol{v}_h \in V_h,$$

and

$$k\Theta c_o S_1(h)\,\|\boldsymbol{u}_h\| \le 1, \; k\Theta c_1 S(h)\,\|\boldsymbol{u}_h\| \le 1,$$

where c_o, c_1 are constants independent of h.

Proof: The proof is a modification of that of Fučík and Kufner (1980), p. 240-241. We have

$$\begin{aligned}&(A_{Th}T_h - \underline{A}_{Th}z_h, T_h - z_h)\\&= |T_h - z_h|^2 + k\Theta a_{Th}(T_h - z_h, T_h - z_h) + k\Theta b_{oh}(\boldsymbol{u}_h, T_h - z_h, T_h) - k\Theta b_{oh}(\boldsymbol{u}_h, T_h - z_h, z_h)\\&= |T_h - z_h|^2 + k\Theta a_{Th}(T_h - z_h, T_h - z_h) + k\Theta b_{oh}(\boldsymbol{u}_h, T_h - z_h, T_h - z_h).\end{aligned}$$

Since

$$|b_{oh}(\boldsymbol{u}_h, T_h - z_h, T_h - z_h)| \le c_o \|\boldsymbol{u}_h\| \, |T_h - z_h| \, \|\boldsymbol{T}_h - z_h\| \le c_o S_1(h) \|\boldsymbol{u}_h\| \, |T_h - z_h|^2,$$

then

$$(\underline{A}_{Th}T_h - \underline{A}_{Th}z_h, T_h - z_h) \ge (1 - k\Theta c_o S_1(h) \|\boldsymbol{u}_h\|) \, |T_h - z_h|^2 + k\Theta \|T_h - z_h\|^2,$$

where Equation 12.287 was also used. Hence $k\Theta c_o S_1(h) \|\boldsymbol{u}_h\| \le 1$, $\alpha_o = k\Theta$.

Furthermore, we have

$$\begin{aligned}&(\underline{A}_h\boldsymbol{u}_h - \underline{A}_h\boldsymbol{v}_h, \boldsymbol{u}_h - \boldsymbol{v}_h)\\&= |\boldsymbol{u}_h - \boldsymbol{v}_h|^2 + k\Theta\hat{\mu} a_h(\boldsymbol{u}_h - \boldsymbol{v}_h, \boldsymbol{u}_h - \boldsymbol{v}_h) + k\Theta b_h(\boldsymbol{u}_h, \boldsymbol{u}_h, \boldsymbol{u}_h - \boldsymbol{v}_h) - k\Theta b_h(\boldsymbol{v}_h, \boldsymbol{v}_h, \boldsymbol{u}_h - \boldsymbol{v}_h)\\&= |\boldsymbol{u}_h - \boldsymbol{v}_h|^2 + k\Theta\hat{\mu} a_h(\boldsymbol{u}_h - \boldsymbol{v}_h, \boldsymbol{u}_h - \boldsymbol{v}_h) + k\Theta b_{oh}(\boldsymbol{u}_h - \boldsymbol{v}_h, \boldsymbol{u}_h, \boldsymbol{u}_h - \boldsymbol{v}_h).\end{aligned}$$

Since

$$|b_{oh}(\mathbf{u}_h - \boldsymbol{v}_h, \boldsymbol{u}_h, \boldsymbol{u}_h - \boldsymbol{v}_h)| \le c_o \, |\boldsymbol{u}_h - \mathbf{v}_h| \, \|\boldsymbol{u}_h\| \, \|\boldsymbol{u}_h - \boldsymbol{v}_h\| \le c_1 S(h) \|\boldsymbol{u}_h\| \, |\boldsymbol{u}_h - \boldsymbol{v}_h|^2,$$

then

$$(\underline{A}_h\boldsymbol{u}_h - \underline{A}_h\boldsymbol{v}_h, \boldsymbol{u}_h - \boldsymbol{v}_h) \ge (1 - k\Theta c_1 S(h) \|\boldsymbol{u}_h\|) \, |\boldsymbol{u}_h - \boldsymbol{v}_h|^2 + k\Theta\hat{\mu} \, \|\boldsymbol{u}_h - \boldsymbol{v}_h\|^2,$$

where Equation 12.289 was also used. Hence $k\Theta c_1 S(h) \|\boldsymbol{u}_h\| \le 1$, $\alpha_o = k\Theta\hat{\mu}$.

Lemma 12.12: Linear functionals $\langle s_h^i, z_h\rangle \in {}^1V_h'$ and $\langle S_h^i, \boldsymbol{v}_h\rangle - k\Theta 2^{1/2}\hat{g}(m, D\boldsymbol{v}_h) \in W_h'$ and, therefore, the second is also an element of V_h' for every $m \in K$, $i = 0, 1, \dots, n-1$.

Proof: The first part of the proof is evident. To prove the second part we shall assume that $\boldsymbol{v}_h \in W_h$. Then

$$|(m, D\boldsymbol{v}_h)| \le |m| \, |D\boldsymbol{v}_h| = \left(\int_\Omega \sum_{j,l=1,2} (m_{jl})^2 \mathrm{d}\boldsymbol{x}\right)^{1/2} \left(\int_\Omega \sum_{j,l=1,2} (D_{jl}(\boldsymbol{v}_h))^2 \mathrm{d}\boldsymbol{x}\right)^{1/2}. \tag{12.369}$$

Since

$$\int_\Omega \sum_{j,l=1,2} (D_{jl}(\boldsymbol{v}_h))^2 \mathrm{d}\boldsymbol{x} \le \|\boldsymbol{v}_h\|^2$$

and $\int_\Omega \Sigma_{j,l=1,2}(m_{jl})^2 \mathrm{d}\boldsymbol{x} \le \int_\Omega 1 \mathrm{d}\boldsymbol{x} = \text{meas }(\Omega)\, \forall m \in K$, and Equation 12.369 is valid then we obtain

$$\left|\langle S_h^i, \boldsymbol{v}_h\rangle - k\Theta 2^{1/2}\hat{g}(m, D\boldsymbol{v}_h)\right| \le (c + k\Theta 2^{1/2}\hat{g}(\text{meas }(\Omega))^{1/2}) \|\boldsymbol{v}_h\|$$

$$\forall \boldsymbol{v}_h \in W_h, m \in K, \quad i = 0, 1, \dots, n-1,$$

from which the assertion of the lemma follows immediately.

In order to state the next result it is necessary to introduce the notation of a saddle point and to recall the theorems giving an existence criteria for it. Our starting point is a function $\mathscr{L} \equiv \mathscr{L}(\boldsymbol{v}, \mu)$, defined on a set $V \times \Lambda$ with real values, i.e. $\mathscr{L}(\boldsymbol{v}, \mu) : V \times \Lambda \to R$.

Definition 12.1: A point $(\boldsymbol{u}_s, \Lambda) \in V \times \Lambda$ is said to be a saddle point of a functional $\mathscr{L}(\boldsymbol{v}, \mu)$: $V \to R^1$ if

$$\mathscr{L}(\boldsymbol{u}_s, \mu) \leq \mathscr{L}(\boldsymbol{u}_s, \lambda) \leq \mathscr{L}(\boldsymbol{v}, \lambda) \quad \forall(\boldsymbol{v}, \mu) \in V \times \Lambda.$$

Proposition 12.1: If $\mathscr{L}$ is a functional defined on $V \times \Lambda$ with real values, then

$$\sup_{\mu \in \Lambda} \inf_{\boldsymbol{v} \in V} \mathscr{L}(\boldsymbol{v}, \mu) \leq \inf_{\boldsymbol{v} \in V} \sup_{\mu \in \Lambda} \mathscr{L}(\boldsymbol{v}, \mu).$$

Proposition 12.2: The functional $\mathscr{L}$ defined on $V \times \Lambda$ with real values possesses a saddle point $(\boldsymbol{u}_s, \lambda)$ on $V \times \Lambda$ if and only if

$$\max_{\mu \in \Lambda} \inf_{\boldsymbol{v} \in V} \mathscr{L}(\boldsymbol{v}, \mu) = \min_{\boldsymbol{v} \in V} \sup_{\mu \in \Lambda} \mathscr{L}(\boldsymbol{v}, \mu)$$

and this number is equal to $\mathscr{L}(\boldsymbol{u}_s, \lambda)$.

For the proof see Ekeland and Temam (1976), p. 167.

Generally, the problem can be written as follows.

Problem ($\mathscr{P}_{\mathscr{L}}$): Find $\boldsymbol{u} \in V$ such that

$$L(\boldsymbol{u}) = \min_{\boldsymbol{v} \in V} \sup_{\mu \in \Lambda} \mathscr{L}(\boldsymbol{v}, \mu).$$

According to Proposition 12.1 instead of studying Problem ($\mathscr{P}_{\mathscr{L}}$) it is enough to investigate the existence and uniqueness of a saddle point. For this we recall a useful characterization of a saddle point.

Proposition 12.3: If there exist $\boldsymbol{u}_s \in V$, $\lambda \in \Lambda$ and $\alpha \in R$ such that

(i) $\mathscr{L}(\boldsymbol{u}_s, \lambda) \leq \alpha, \quad \forall \lambda \in \Lambda,$

(ii) $\mathscr{L}(\boldsymbol{u}, \lambda) \geq \alpha, \quad \forall \boldsymbol{u} \in V,$

then $(\boldsymbol{u}_s, \lambda)$ is a saddle point of $\mathscr{L}$ and

$$\alpha = \inf_{\boldsymbol{u} \in V} \sup_{\lambda \in \Lambda} \mathscr{L}(\boldsymbol{u}, \lambda) = \sup_{\lambda \in \Lambda} \inf_{\boldsymbol{u} \in V} \mathscr{L}(\boldsymbol{u}, \lambda).$$

For the proof see Ekeland and Temam (1976), p. 168.

Proposition 12.4: Let W and Z be two reflexive Banach spaces. Let

(i) $V \subset W$ be convex, closed and non-empty,
(ii) $\Lambda \subset Z$ be convex, closed and non-empty,
(iii) $\forall \boldsymbol{v} \in V$, $\mu \to \mathscr{L}(\boldsymbol{v}, \mu)$ is concave and upper semi-continuous (u.s.c),
(iv) $\mu \in \Lambda$, $\boldsymbol{v} \to \mathscr{L}(\boldsymbol{v}, \mu)$ is convex and lower semi-continuous (l.s.c.).

The set of saddle points of $\mathscr{L}$ is of the form $V_o \times \Lambda_o$, where $V_o \subset V$, $\Lambda_o \subset \Lambda$. Then under the hypotheses (i)-(iv), the set $V_o \times \Lambda_o$ of the saddle points of $\mathscr{L}$ is convex. Moreover, if $\{\lambda \to$

$\mathscr{L}(\boldsymbol{v}, \lambda)\}$ is strictly concave $\forall \boldsymbol{v} \in V$, then Λ_o contains at most one point. If $\{\boldsymbol{v} \to \mathscr{L}(\boldsymbol{v}, \lambda)\}$ is strictly convex $\forall \lambda \in \Lambda$, then V contains at most one point.

For the proof see Ekeland and Temam (1976), pp 168-169.

Proposition 12.5: Let W and Z be two reflexive Banach spaces and let conditions (i)-(iv), of Proposition 12.4 be satisfied. Let

(i) $\exists \mu_o \in \Lambda$ such that

$$\lim_{\boldsymbol{v} \in V, \|\boldsymbol{v}\| \to \infty} \mathscr{L}(\boldsymbol{v}, \mu_o) = +\infty,$$

(ii) $\exists\ \boldsymbol{u}_o \in V$ such that

$$\lim_{\mu \in \Lambda, \|\mu\| \to \infty} \mathscr{L}(\boldsymbol{u}_o, \mu) = -\infty.$$

Then $\mathscr{L}$ possesses at least one saddle point on $V \times \Lambda$ and

$$\mathscr{L}(\boldsymbol{u}_s, \lambda) = \min_{\boldsymbol{v} \in V} \sup_{\lambda \in \Lambda} \mathscr{L}(\boldsymbol{v}, \mu) = \max_{\mu \in \Lambda} \inf_{\boldsymbol{v} \in V} \mathscr{L}(\boldsymbol{v}, \mu).$$

For the proof see Ekeland and Temam (1976), p. 173.

Proposition 12.6: We assume, in addition to conditions (i)-(iv) of Proposition 12.4, that

(i) $\forall \boldsymbol{v} \in V$, $\mu \to \mathscr{L}(\boldsymbol{v}, \mu)$ is Gâteaux-differentiable,

(ii) $\forall \mu \in \Lambda$, $\boldsymbol{v} \to \mathscr{L}(\boldsymbol{v}, \mu)$ is Gâteaux-differentiable.

Then $(\boldsymbol{u}_s, \lambda) \in V \times \Lambda$ is the saddle point of $\mathscr{L}$ if and only if

$$\langle \partial \mathscr{L}(\boldsymbol{u}_s, \lambda)/\partial \boldsymbol{v}, \boldsymbol{v} - \boldsymbol{u}_s \rangle \geq 0 \quad \forall \boldsymbol{v} \in V,$$

$$\langle \partial \mathscr{L}(\boldsymbol{u}_s, \lambda)/\partial \mu, \mu - \lambda \rangle \geq 0 \quad \forall \mu \in \Lambda,$$

where $\langle .,. \rangle$ denotes a bilinear pairing in the usual sense.

For the proof see Ekeland and Temam (1976), p. 170.

For a numerical approximation of a saddle point we recall the following definition and propositions.

Definition 12.2: A point $(\boldsymbol{u}_{sh}, \Lambda_h) \in V_h \times \Lambda \subset V \times \Lambda$ is said to be an approximate saddle point of a functional $\mathscr{L}$ on $V_h \times \Lambda$ if an approximate saddle point $(\boldsymbol{u}_{sh}, \lambda_h) \in V_h \times \Lambda \subset V \times \Lambda$ exists and if

$$\mathscr{L}(\boldsymbol{u}_{sh}, \mu) \leq \mathscr{L}(\boldsymbol{u}_{sh}, \lambda_h) \leq \mathscr{L}(\boldsymbol{v}, \lambda_h) \quad \forall (\boldsymbol{v}, \mu) \in V_h \times \Lambda.$$

The assertion of Propositions 12.4 and 12.5 may easily be extended for the case where V_h are compact subsets of W and Z. Then the problem $(\mathscr{P}_v)_h$ can be written in the following form.

Problem $(\mathscr{P}_{\mathscr{L}})_h$: Find $\boldsymbol{u}_h \in V_h$ such that

$$L(\boldsymbol{u}_h) = \min_{\boldsymbol{v} \in V_h} \sup_{\mu \in \Lambda} \mathscr{L}(\boldsymbol{v}, \mu).$$

As $V_h \subset V \subset W$, $\forall h$ and letting $\mathscr{L}_h$ be an approximation of the Lagrangian $\mathscr{L}$, then we have the following proposition.

Proposition 12.7: Let $V_h \subset V \subset W$, $\forall h$, and let

(i) $V_h \subset W$ be convex, closed and non-empty,

(ii) $\Lambda \subset Z$ be convex, closed and non-empty,

(iii) $\forall \boldsymbol{v} \in V_h$, $\mu \to \mathscr{L}_h(\boldsymbol{v}, \mu)$ be concave and u.s.c.,

(iv) $\mu \in \Lambda$, $\boldsymbol{v} \to \mathscr{L}_h(\boldsymbol{v}, \mu)$ is convex and l.s.c.,

(v) $\exists \mu_o \in \Lambda$ such that

$$\lim_{\boldsymbol{v} \in V_h, \|\boldsymbol{v}\| \to \infty} \mathscr{L}_h(\boldsymbol{v}, \mu_o) = +\infty,$$

(vi) $\exists\ \boldsymbol{u}_o \in V_h$ such that

$$\lim_{\mu \in \Lambda, \|\mu\| \to \infty} \mathscr{L}_h(\boldsymbol{u}_o, \mu) = -\infty.$$

Then $\mathscr{L}_h$ possesses at least one saddle point on $\mathscr{L}_h$ and

$$\mathscr{L}_h(\boldsymbol{u}_s, \lambda) = \min_{\boldsymbol{v} \in V_h} \sup_{\mu \in \Lambda} \mathscr{L}_h(\boldsymbol{v}, \mu) = \max_{\mu \in \Lambda} \inf_{\boldsymbol{v} \in V_h} \mathscr{L}_h(\boldsymbol{v}, \mu).$$

For the proof see Ekeland and Temam (1976).

The problems investigated can be solved by the Uzawa's or Arrow-Hurwicz's algorithms, respectively. Both algorithms and their convergences are discussed, in for example, Fortin (1972), Temam (1977), pp. 138–147, Ekeland and Temam (1976), pp. 186–193. Next we shall introduce the discrete version of the Uzawa and Arrow-Hurwicz's algorithms.

THE DISCRETE VERSION OF THE UZAWA ALGORITHM

Let us solve the following problem. There exists a pair of functions $(\boldsymbol{u}, \lambda)$ such that

$$(A\boldsymbol{u} - \boldsymbol{f}, \boldsymbol{v} - \boldsymbol{u}) + (\lambda, G\boldsymbol{v} - G\boldsymbol{u}) \geq 0 \quad \forall \boldsymbol{v} \in V, \tag{12.370}$$

$$(\mu - \lambda, G\boldsymbol{u}) \leq 0 \quad \forall \mu \in \Lambda, \Lambda \text{ being a convex set.} \tag{12.371}$$

Let us assume that A is strictly monotone and coercive (but not necessarily linear). Let us assume that $\boldsymbol{u}$ is unique.

Let $\boldsymbol{v}^o$, μ^o be uniquely given on $V \times \Lambda$. Then let us assume that a pair $(\boldsymbol{v}^i, \mu^i)$ be known, then $(\boldsymbol{v}^{i+1}, \mu^{i+1})$ will be find by solving

$$A(\boldsymbol{v}^{i+1} - \boldsymbol{f}, \boldsymbol{v} - \boldsymbol{v}^{i+1}) + (\mu^i, G\boldsymbol{v} - G\boldsymbol{v}^{i+1}) \geq 0 \quad \forall \boldsymbol{v} \in V, \tag{12.372}$$

$$(\mu^{i+1} - \mu^i, \mu - \mu^{i+1}) - \rho(G\boldsymbol{v}^{i+1}, \mu - \mu^{i+1}) \geq 0 \quad \forall \mu \in \Lambda. \tag{12.373}$$

Let us denote by $\|\boldsymbol{v}\|$ the norm of $\boldsymbol{v}$ on V and by $|\boldsymbol{v}|$ the norm of $\boldsymbol{v}$ on the dual space of V.

Lemma 12.13: Let the operator A be strongly monotone, i.e.

$$(A\boldsymbol{u} - A\boldsymbol{v}, \boldsymbol{u} - \boldsymbol{v}) \geq \alpha \|\boldsymbol{u} - \boldsymbol{v}\|^2 \quad \forall \boldsymbol{u}, \boldsymbol{v} \in V. \tag{12.374}$$

Let G satisfy

$$|G(\boldsymbol{u}) - G(\boldsymbol{v})| \leq M \|\boldsymbol{u} - \boldsymbol{v}\| \quad \forall \boldsymbol{u}, \boldsymbol{v} \in V, \tag{12.375}$$

and let ρ be chosen such that

$$\rho \leq 2\alpha\delta M^{-2} - \varepsilon, \varepsilon > 0, \quad 0 < \delta < 1. \tag{12.376}$$

Thus

(i) $\boldsymbol{v}^i$ converges strongly to $\boldsymbol{u} \in V$,

(ii) μ^i is bounded on the dual space V,

(iii) $|\mu^{i+1} - \mu^i| \to 0$,

(iv) for every λ^* of μ^i, $(\boldsymbol{u}, \lambda^*)$ is a saddle point solving Equations 12.370 and 12.371.

Proof: Let us put $\boldsymbol{v} = \boldsymbol{v}^{i+1}$ in Equation 12.370 and $\boldsymbol{v} = \boldsymbol{u}$ in Equation 12.372, then adding both inequalities, we find

$$(A\boldsymbol{v}^{i+1} - A\boldsymbol{u}, \boldsymbol{v}^{i+1} - \boldsymbol{u}) + (\mu^i - \mu, G\boldsymbol{v}^{i+1} - G\boldsymbol{u}) \leq 0. \tag{12.377}$$

On the other hand, let us multiply Equation 12.371 by ρ, putting $\mu = \mu^{i+1}$, and put $\mu = \lambda$ in Equation 12.373, denoting $\lambda^i = \mu^i - \lambda$, then

$$(\lambda^{i+1} - \lambda^i, \lambda^{i+1}) - \rho(G\boldsymbol{v}^{i+1} - G\boldsymbol{u}, \lambda^{i+1}) \leq 0.$$

Hence

$$|\lambda^{i+1}|^2 - |\lambda^i|^2 + |\lambda^{i+1} - \lambda^i|^2 \leq 2\rho(G\boldsymbol{v}^{i+1} - G\boldsymbol{u}, \lambda^i) + 2\rho(G\boldsymbol{v}^{i+1} - G\boldsymbol{u}, \lambda^{i+1} - \lambda^i).$$

Let us put $\boldsymbol{u}^i = \boldsymbol{v}^i - \boldsymbol{u}$, and summing with Equation 12.377, we obtain

$$\begin{aligned} |\lambda^{i+1}|^2 - |\lambda^i|^2 + |\lambda^{i+1} - \lambda^i|^2 + 2\rho\alpha \|\boldsymbol{u}^{i+1}\|^2 &\leq 2\rho(G\boldsymbol{v}^{i+1} - G\boldsymbol{u}, \lambda^{i+1} - \lambda^i) \\ &\leq 2\rho |G\boldsymbol{v}^{i+1} - G\boldsymbol{u}| \, |\lambda^{i+1} - \lambda^i| \end{aligned} \tag{12.378}$$

and thus

$$|\lambda^{i+1}|^2 - |\lambda^i|^2 + |\lambda^{i+1} - \lambda^i|^2 + 2\rho\alpha \|\boldsymbol{u}^{i+1}\|^2 \leq \delta |\lambda^{i+1} - \lambda^i|^2 + \rho^2 M^2 \delta^{-1} \|\boldsymbol{u}^{i+1}\|^2 \tag{12.379}$$

and

$$|\lambda^{i+1}|^2 - |\lambda^i|^2 + (1 - \delta) |\lambda^{i+1} - \lambda^i|^2 + \rho(2\alpha - \rho M^2 \delta^{-1} \|\boldsymbol{u}^{i+1}\|^2) \leq 0. \tag{12.380}$$

Hence Equation 12.376 is satisfied. Summing Equation 12.380 over i from 0 to n, we obtain

$$\rho\varepsilon \sum_{i=0}^{n} \|\boldsymbol{u}^{i+1}\|^2 + (1 - \delta) \sum_{i=0}^{n} |\lambda^{i+1} - \lambda^i|^2 + |\lambda^{n+1}|^2 \leq |\lambda^o|^2 \quad \forall n.$$

Since $\|\boldsymbol{u}^{i+1}\|^2 \to 0$ as $|\lambda^{i+1} - \lambda^i|^2 = |\mu^{i+1} - \mu^i|^2 \to 0$, and as μ^i remains bounded, then conditions (i), (ii) and (iii) are proved.

Note that $\mu^i \to \lambda^*$ weakly. If $\boldsymbol{v}^{i+1} \to \boldsymbol{u}$ strongly, then taking the limit in Equation 12.372, we obtain Equation 12.370. On the other hand, if $\mu^{i+1} - \mu^i$ tends strongly to zero, then the limit of Equation 12.373 leads to Equation 12.371.

THE DISCRETE VERSION OF THE ARROW-HURWICZ'S ALGORITHM

Let V and $\boldsymbol{v}$ be finite dimensional spaces. Let us denote the scalar products in V and $\boldsymbol{v}$ by (.,.), and the associated norms satisfy

$$c_o\,|\boldsymbol{u}| \leq \|\boldsymbol{u}\| \leq S(n)\,|\boldsymbol{u}| \quad \forall u \in V,$$

where $S(n) \to \infty$, $n = \dim V$.

Let $\boldsymbol{v}^o$, μ^o be initial approximations on $V \times \Lambda$ and let $\boldsymbol{v}^i$, μ^i be known. Then $\boldsymbol{v}^{i+1}$, μ^{i+1} can be determined by

$$(\boldsymbol{v}^{i+1} - \boldsymbol{v}^i, \boldsymbol{v} - \boldsymbol{v}^{i+1}) + \rho(A\boldsymbol{v}^i - \boldsymbol{f}, \boldsymbol{v} - \boldsymbol{v}^{i+1}) + \rho(\mu^i, G\boldsymbol{v} - G\boldsymbol{v}^{i+1}) \geq 0 \quad \forall \boldsymbol{v} \in V, \tag{12.381}$$

$$\beta(\mu^{i+1} - \mu^i, \mu - \mu^{i+1}) - \rho(G\boldsymbol{v}^{i+1}, \mu - \mu^{i+1}) \geq 0 \quad \forall \mu \in \Lambda. \tag{12.382}$$

Lemma 12.14: Let the operator A be strongly monotone, i.e

$$(A\boldsymbol{u} - A\boldsymbol{v}, \boldsymbol{u} - \boldsymbol{v}) \geq \alpha\,\|\boldsymbol{u} - \boldsymbol{v}\|^2 \quad \forall \boldsymbol{u}, \boldsymbol{v} \in V. \tag{12.383}$$

Let

$$\|A\boldsymbol{u} - A\boldsymbol{v}\|_{V'} \leq c\,\|\boldsymbol{u} - \boldsymbol{v}\| \quad \forall \boldsymbol{u}, \boldsymbol{v} \in V, \tag{12.384}$$

$$|G\boldsymbol{u} - G\boldsymbol{v}| \leq M\,\|\boldsymbol{u} - \boldsymbol{v}\| \quad \forall \boldsymbol{u}, \boldsymbol{v} \in V. \tag{12.385}$$

Let α and c satisfy

$$2\alpha - c = \alpha_1 > 0. \tag{12.386}$$

Let ρ and γ be taken such that

$$\rho \leq 2\alpha\delta\gamma M^{-2} - \varepsilon,\ \varepsilon > 0, \quad 0 < \delta < 1, \tag{12.387}$$

$$\rho c S^2(n) \leq 1. \tag{12.388}$$

Thus $\boldsymbol{v}^{i+1}$ converges strongly to $\boldsymbol{u}$ in V, the sequence μ^i being bounded and for every λ^* of μ^i a pair $(\boldsymbol{u}, \lambda^*)$ is a saddle point of equations 12.370 and 12.371.

Proof: Let us put $\boldsymbol{v} = \boldsymbol{v}^{i+1}$ in Equation 12.370 and $\boldsymbol{v} = \boldsymbol{u}$ in Equation 12.381 and sum. Putting $\boldsymbol{u}^i = \boldsymbol{v}^i - \boldsymbol{u}$, $\lambda^i = \mu^i - \lambda$ then

$$\left|\boldsymbol{u}^{i+1}\right|^2 - \left|\boldsymbol{u}^i\right|^2 + \left|\boldsymbol{u}^{i+1} - \boldsymbol{u}^i\right|^2 + 2\rho(A\boldsymbol{v}^i - A\boldsymbol{u}, \boldsymbol{v}^{i+1} - \boldsymbol{u}) + 2\rho(\lambda^i, G\boldsymbol{v}^{i+1} - G\boldsymbol{u}) \leq 0.$$

But

$$2\rho(A\boldsymbol{v}^i - A\boldsymbol{u}, \boldsymbol{v}^{i+1} - \boldsymbol{u}) = 2\rho(A\boldsymbol{v}^{i+1} - A\boldsymbol{u}, \boldsymbol{v}^{i+1} - \boldsymbol{u}) - 2\rho(A\boldsymbol{v}^{i+1} - A\boldsymbol{v}^i, \boldsymbol{v}^{i+1} - \boldsymbol{u}),$$

hence

$$\begin{aligned}\left|\boldsymbol{u}^{i+1}\right|^2 - \left|\boldsymbol{u}^i\right|^2 + \left|\boldsymbol{u}^{i+1} - \boldsymbol{u}^i\right|^2 + 2\rho\alpha\left\|\boldsymbol{u}^{i+1}\right\|^2 + 2\rho(\lambda^i, G\boldsymbol{v}^{i+1} - G\boldsymbol{u}) &\leq 2\rho c\left\|\boldsymbol{v}^{i+1} - \boldsymbol{v}^i\right\|\left\|\boldsymbol{u}^{i+1}\right\| \\ &\leq \rho c S^2(n)\left|\boldsymbol{u}^{i+1} - \boldsymbol{u}^i\right|^2 + \rho c\left\|\boldsymbol{u}^{i+1}\right\|^2.\end{aligned} \tag{12.389}$$

On the other hand similarly as in Lemma 12.13 we find

$$\gamma\left|\lambda^{i+1}\right|^2 - \gamma\left|\lambda^i\right|^2 + \gamma\left|\lambda^{i+1} - \lambda^i\right|^2 \leq 2\rho(G\boldsymbol{v}^{i+1} - G\boldsymbol{u}, \lambda^i) + 2\rho(G\boldsymbol{v}^{i+1} - G\boldsymbol{u}, \lambda^{i+1} - \lambda^i). \tag{12.390}$$

But

$$2\rho(G\boldsymbol{v}^{i+1} - G\boldsymbol{u}, \lambda^{i+1} - \lambda^i) \le \gamma\delta\left|\lambda^{i+1} - \lambda^i\right|^2 + \rho^2 M^2\gamma^{-1}\delta^{-1}\left\|\boldsymbol{u}^{i+1}\right\|^2.$$

Hence, using Equation 12.390 and summing together with Equation 12.389, we obtain

$$\left|\boldsymbol{u}^{i+1}\right|^2 - \left|\boldsymbol{u}^i\right|^2 + (1 - \rho c S^2(n))\left|\boldsymbol{u}^{i+1} - \boldsymbol{u}^i\right|^2 + \gamma\left|\lambda^{i+1}\right|^2 - \gamma\left|\lambda^i\right|^2 + (1-\delta)\gamma\left|\lambda^{i+1} - \lambda^i\right|^2$$
$$+ \rho(2\alpha - c - \rho M^2\gamma^{-1}\delta^{-1})\left\|\boldsymbol{u}^{i+1}\right\| \le 0.$$

The proposition follows as in Lemma 12.13.

To find a saddle point of equations 12.365 and 12.366, the Uzawa algorithm will be used (see Ekeland and Temam (1976)). First, we verify conditions which are necessary for its convergence. In our investigated case we set $V = V_h$, $\Lambda = K_h$. We see that V_h is a non-empty, closed and convex set and $K_h \subset [L^2(\Omega)]^4$ is a non-empty, closed, convex and bounded set. The second assertion follows from the fact that for m_{h1}, $m_{h2} \in K_h$, $\underline{\lambda} \in \langle 0, 1\rangle$, the convex combination is piecewise constant and it holds that

$$\operatorname*{ess\,sup}_{\boldsymbol{x}\in\Omega} \sum_{j,l=1,2} (\underline{\lambda} m_{h1jl} + (1-\underline{\lambda})m_{h2jl})^2$$
$$= \operatorname*{ess\,sup}_{\boldsymbol{x}\in\Omega}\left(\sum_{j,l=1,2} \underline{\lambda}^2 m_{h1jl}^2 + \sum_{j,l=1,2} 2\underline{\lambda}(1-\underline{\lambda})m_{h1jl}m_{h2jl} + \sum_{j,l=1,2}(1-\underline{\lambda})^2(m_{h2jl})^2\right)$$
$$\le \underline{\lambda}^2 + 2\underline{\lambda}(1-\underline{\lambda}) + (1-\underline{\lambda})^2 = 1$$

and

$$\sum_{j,l=1,2}\int_\Omega m_{hjl}^2 \mathrm{d}\boldsymbol{x} \le \mathrm{meas}(\Omega) \quad \forall m_h \in K_h.$$

We denote by $\mathscr{L}_h(\boldsymbol{v}_h, m_h) = \mathscr{J}(\boldsymbol{v}_h) + (m_h, \Psi(\boldsymbol{v}_h))$ the Lagrangian, where $\mathscr{J}(\boldsymbol{v}_h) = \frac{1}{2}(\underline{A}_h\boldsymbol{v}_h, \boldsymbol{v}_h) - \langle S_h^i, \boldsymbol{v}_h\rangle$ and $\Psi(\boldsymbol{v}_h) = k\Theta 2^{1/2}D\boldsymbol{v}_h$ are mappings such that $\mathscr{J} : V_h \to R$, $\Psi : V_h \to [L^2(\Omega)]^4$. The functional $\mathscr{J}$ is G-differentiable. Furthermore, there exists $\alpha > 0$ such that

$$(\mathscr{J}'(\boldsymbol{u}_h) - \mathscr{J}'(\boldsymbol{v}_h), \boldsymbol{u}_h - \boldsymbol{v}_h)$$
$$= (\underline{A}_h\boldsymbol{u}_h, \boldsymbol{u}_h - \boldsymbol{v}_h) - \langle S_h^i, \boldsymbol{u}_h - \boldsymbol{v}_h\rangle - (\underline{A}_h\boldsymbol{v}_h, \boldsymbol{u}_h - \boldsymbol{v}_h) + \langle S_h^i, \boldsymbol{u}_h - \boldsymbol{v}_h\rangle$$
$$= (\underline{A}_h\boldsymbol{u}_h - \underline{A}_h\boldsymbol{v}_h, \boldsymbol{u}_h - \boldsymbol{v}_h) \ge \alpha\left\|\boldsymbol{u}_h - \boldsymbol{v}_h\right\|^2 \quad \forall \boldsymbol{u}_h, \boldsymbol{v}_h \in V_h, \tag{12.391}$$

where $\alpha = k\Theta\hat{\mu}$ in the case of the norm induced by the scalar product defined by Equation 12.250, $\alpha = \min(\rho, k\Theta\hat{\mu})$ in the case of the norm induced by the scalar product defined by Equation 12.251, and $(\mathscr{J}'(\boldsymbol{u}_h), \boldsymbol{u}_h - \boldsymbol{v}_h)$ represents the G-derivative of $\mathscr{J}$ in the direction $\boldsymbol{u}_h - \boldsymbol{v}_h$ at the point $\boldsymbol{u}_h$. For all $m_h \in K_h$ the mapping $M : \boldsymbol{v}_h \to (m_h, \Psi(\boldsymbol{v}_h))$ is convex and l.s.c. in V_h, as the mapping $M : \boldsymbol{v}_h \to (m_h, \Psi(\boldsymbol{v}_h))$ is linear and therefore convex and G-differentiable in V_h. For a convex G-differentiable function M it holds that

$$M(\boldsymbol{v}_h) \ge M(\boldsymbol{u}_h) + (M'(\boldsymbol{u}_h), \boldsymbol{v}_h - \boldsymbol{u}_h) \quad \forall \boldsymbol{u}_h, \boldsymbol{v}_h \in V_h.$$

Hence for $\boldsymbol{v}_h \to \boldsymbol{u}_h$ in V_h we have $\lim\ \inf_{\boldsymbol{v}_h \to \boldsymbol{u}_h} M(\boldsymbol{v}_h) \geq M(\boldsymbol{u}_h)$, which proves the l.s.c. The mapping Ψ is Lipschitzian as

$$|\Psi(\boldsymbol{u}_h) - \Psi(\boldsymbol{v}_h)| = k\Theta 2^{1/2}\hat{g}\left(\sum_{\underline{T}_h \in \mathscr{T}_h} \int_{\underline{T}_h} \sum_{j,l=1}^{2} (D_{hjl}(\boldsymbol{u}_h - \boldsymbol{v}_h))^2 \mathrm{d}\boldsymbol{x}\right)^{1/2}$$

$$\leq k\Theta 2^{1/2}\hat{g}\left(\sum_{\underline{T}_h \in \mathscr{T}_h} \int_{\underline{T}_h} \sum_{j,l=1,2} (D_{hl}(u_{hj} - v_{hj}))^2 \mathrm{d}\boldsymbol{x}\right)^{1/2} \leq c\,\|\boldsymbol{u}_h - \boldsymbol{v}_h\|$$

$$\forall \boldsymbol{u}_h, \boldsymbol{v}_h \in V_h, \qquad (12.392)$$

where $c = k\Theta 2^{1/2}\hat{g}$ in both cases of norms induced by scalar products defined by equations 12.250 and 12.251. Moreover, in Lemma 12.11 we proved that the operator $\underline{A}_h$ is strongly monotone. Hence the existence conditions for a saddle point and then for the Uzawa algorithm are verified.

Theorem 12.12: Let l_h and $\mathscr{L}_h$ be the nonlinear functional and the Lagrangian defined by equations 12.362 and 12.363. Then the functional l_h has on ${}^1V_h$ a minimum at the point T_h, and the Lagrangian $\mathscr{L}_h$ possesses at least one saddle point $(\boldsymbol{u}_h^o, m_h^o)$ on $V_h \times K_h$. The first component $\boldsymbol{u}_h$ is uniquely determined and is the solution of the optimization problem

$$\inf_{\boldsymbol{u}_h \in V_h} \left\{ \sup_{m_h \in K_h} \mathscr{L}_h(\boldsymbol{u}_h, m_h) \right\}.$$

Proof: The problem is coupled and, as was shown above, the coupled terms are from L^2-spaces.

The first part of the proof follows from the fact that the bilinear form $A_{Th}(T_h, z_h)$ on ${}^1V_h \times {}^1V_h$ is symmetric, continuous and positive definite, and the functional $l_h(z_h)$ is continuous, strictly convex and coercive on ${}^1V_h$. Then the functional $l_h(z_h) = \frac{1}{2}(\underline{A}_{Th} z_h, z_h) - \langle s_h^i, z_h \rangle$ has a minimum on ${}^1V_h$ at a point T_h.

In the second part of the proof the existence of a saddle point $(\boldsymbol{u}_h, m_h)$ on $V_h \times K_h$ follows from Proposition 12.7.

Let $m_{ho} \in K_h$, then

$$\mathscr{L}_h(\boldsymbol{u}_h, m_h) \geq \mathscr{J}(\boldsymbol{u}_h) + (m_{ho}, \Psi(\boldsymbol{u}_h)). \qquad (12.393)$$

Since Ψ is Lipschitzian then due to Equation 12.392

$$|\Psi(\boldsymbol{u}_h) - \Psi(\boldsymbol{u}_{ho})| \leq c\,\|\boldsymbol{u}_h - \boldsymbol{u}_{ho}\|, \qquad (12.394)$$

$\boldsymbol{u}_{ho}$ being a fixed element of V_h, and

$$|\Psi(\boldsymbol{u}_h)| \leq |\Psi(\boldsymbol{u}_{ho})| + c\,\|\boldsymbol{u}_h - \boldsymbol{u}_{ho}\|. \qquad (12.395)$$

Then

$$\mathscr{L}_h(\boldsymbol{u}_h, m_h) \geq \mathscr{J}(\boldsymbol{u}_h) - \|m_{ho}\|\,(|\Psi(\boldsymbol{u}_{ho})| - c\,\|\boldsymbol{u}_h - \boldsymbol{u}_{ho}\|). \qquad (12.396)$$

Hence, according to Proposition 12.3 condition (i) and Proposition 12.7 condition (i) it follows that $\mathscr{L}_h$ possesses at least one saddle point. Since the mapping $\boldsymbol{u} \to \mathscr{L}_h(\boldsymbol{u}_h, m_h)$ is strictly convex $\forall m_h \in K_h$, then due to Proposition 12.4 the uniqueness of $\boldsymbol{u}_{ho}$ follows immediately. Moreover, since $(\boldsymbol{u}_{ho}, m_{ho})$ is a saddle point of $\mathscr{L}_h$ on $V_h \times K_h$, $\boldsymbol{u}_{ho}$ is the solution of the optimization problem $\inf_{\boldsymbol{v}_h \in V_h} \sup_{m_h \in K_h} \mathscr{L}_h(\boldsymbol{v}_h, m_h)$.

Theorem 12.13: Let a triplet $\{T_h^{i+\Theta}, \boldsymbol{u}_h^{i+\Theta}, m_h^{i+\Theta}\}$, $i = 0, 1, \ldots, n-1$ be a solution of Problem $\mathscr{P}_{hA}$, then a triplet $\{T_h^{i+\Theta}, \boldsymbol{u}_h^{i+\Theta}, m_h^{i+\Theta}\}$, $i = 0, 1, \ldots, n-1$ also solves the problem in equations 12.358-12.360, and on the contrary if the triplet $\{T_h^{i+\Theta}, \boldsymbol{u}_h^{i+\Theta}, m_h^{i+\Theta}\}$ is a solution of equations 12.358-12.360 then it is also a solution of Problem $\mathscr{P}_{hA}$.

Proof: Since the problem is coupled the proof can be made for coupled terms from L^2-spaces. The first part of the proof follows from the fact that the bilinear form $A_{Th}(T_h, z_h)$ on ${}^1V_h \times {}^1V_h$ is symmetric, continuous and positive definite.

Let $(\boldsymbol{u}_h^{i+\Theta}, m_h^{i+\Theta})$ be a solution of Problem $\mathscr{P}_{hA}$. Then Equation 12.366 follows immediately from Equation 12.360. The bilinear form $A_h(\boldsymbol{u}_h, \boldsymbol{v}_h)$ on $V_h \times V_h$ is symmetric, continuous and positive definite, and then the functional $\frac{1}{2}(\underline{A}_h \boldsymbol{v}_h, \boldsymbol{v}_h) - \langle S_h^i, \boldsymbol{v}_h\rangle + k\Theta 2^{1/2}\hat{g}(m, D_h\boldsymbol{v}_h) \equiv \mathscr{L}_h(\boldsymbol{v}_h, m_h^{i+\Theta})$ has a minimum at the point $\boldsymbol{u}_h^{i+\Theta}$, which is equivalent to Equation 12.365.

On the contrary let a triplet $\{T_h^{i+\Theta}, \boldsymbol{u}_h^{i+\Theta}, m_h^{i+\Theta}\}$ be a solution of the problem in equations 12.364-12.366. Then from Equation 12.366 we obtain

$$\mathscr{L}_h(\boldsymbol{u}_h^{i+\Theta}, m_h) - \mathscr{L}_h(\boldsymbol{u}_h^{i+\Theta}, m_h^{i+\Theta}) = k\Theta 2^{1/2}(D_h\boldsymbol{u}_h^{i+\Theta}, m_h - m_h^{i+\Theta}) \leq 0$$

and therefore equations 12.360 and 12.361 are valid.

Let $\boldsymbol{u}_h, \boldsymbol{v}_h \in V_h$ be arbitrary. The G-derivative of $\mathscr{L}_h(., m_h^{i+\Theta})$ at $\boldsymbol{u}_h$ in the direction of $\boldsymbol{v}_h$ is

$$\frac{\partial \mathscr{L}_h(\boldsymbol{u}_h, m_h^{i+\Theta})}{\partial \boldsymbol{v}_h} = \underline{A}_h(\boldsymbol{u}_h, \boldsymbol{v}_h) - \langle S_h^i, \boldsymbol{v}_h\rangle + k\Theta 2^{1/2}\hat{g}(m^{i+\Theta}, D\boldsymbol{v}_h), \tag{12.397}$$

and the second G-derivative is

$$\frac{\partial \mathscr{L}_h^2(\boldsymbol{u}_h, m_h^{i+\Theta})}{\partial \boldsymbol{v}_h^2} = \underline{A}_h(\boldsymbol{v}_h, \boldsymbol{v}_h) \leq 0. \tag{12.398}$$

Then the Lagrangian $\mathscr{L}_h(\boldsymbol{v}_h, m_h^{i+\Theta})$ is convex, and for its minimum $\boldsymbol{u}_h^{i+\Theta}$ on $V_h \times K_h$ we have that $\partial\mathscr{L}_h(\boldsymbol{u}_h^{i+\Theta}, m_h^{i+\Theta})/\partial \boldsymbol{v}_h = 0$ holds. Putting $\boldsymbol{u}_h = \boldsymbol{u}_h^{i+\Theta}$ in Equation 12.397 we obtain

$$\underline{A}_h(\boldsymbol{u}_h^{i+\Theta}, \boldsymbol{v}_h) - \langle S_h^i, \boldsymbol{v}_h\rangle + k\Theta 2^{1/2}\hat{g}(m^{i+\Theta}, D_h\boldsymbol{v}_h) = 0 \quad \forall \boldsymbol{v}_h \in V_h, \tag{12.399}$$

which is equivalent to Equation 12.359.

Remark: It holds that

$$(m_h, D_h\boldsymbol{u}_h^{i+\Theta}) = \int_\Omega \left(\sum_{j,l=1,2} (D_{hjl}(\boldsymbol{u}_h^{i+\Theta}))^2\right)^{1/2} \mathrm{d}\boldsymbol{x}, \leq \left\|m_h^{i+\Theta}\right\|_\infty \int_\Omega \left(\sum_{j,l=1,2} (D_{hjl}(\boldsymbol{u}_h^{i+\Theta}))^2\right)^{1/2} \mathrm{d}\boldsymbol{x}.$$

Therefore, $\left\|m_h^{i+\Theta}\right\|_\infty = 1$ provided $D\boldsymbol{u}_h^{i+\Theta}$ is not equal to zero on the whole Ω.

Due to Céa (1971) and Ekeland and Temam (1976) it is convenient to use a saddle point solver like the following algorithm of Uzawa's type.

Algorithm 12.1: Let $T_h^o \in {}^1V_h$, $\boldsymbol{u}_h^o m_h^o \in K_h$ be arbitrary, then we calculate $T_h^1, \boldsymbol{u}_h^1, m_h^1, \ldots$ etc., where $T_h^i \in {}^1V_h$, $\boldsymbol{u}_h^i \in V_h$, $m_h^i \in K_h$. If m_h^i be is known, we determine $(T_h^{i+1}, \boldsymbol{u}_h^{i+1})$ such that

$$l_h(T_h^{i+1}) = \min_{z_h \in {}^1V_h} l_h(z_h), \tag{12.400}$$

$$\mathscr{L}_h(\boldsymbol{u}_h^{i+1}, m_h^i) = \min_{\boldsymbol{v}_h \in V_h} \mathscr{L}_h(\boldsymbol{v}_h, m_h^i), \tag{12.401}$$

$$m_h^{i+1} = P_{K_h}(m_h^i + \rho_i \Psi(\boldsymbol{u}_h^{i+1})) = P_{K_h}(\boldsymbol{u}_h^i + \rho_i k\Theta 2_i^{1/2}\hat{g} D_h \boldsymbol{u}_h^{i+1}), \quad i = 0, 1, \ldots, n-1, \tag{12.402}$$

where P_{K_h} is the mapping from $[L^2(\Omega)]^4$ into K_h and defined by

$$P_{K_h}(m_h) = m_h \quad \text{for } m_h \in K_h, \tag{12.403}$$

$$P_{K_h}(m_h) = m_h \,\|m_h\|_\infty^{-1} \text{ for } m_h \text{ being piecewise constant on every } \underline{T}_h \in \mathscr{T}_h,\ \|m_h\|_\infty > 1, \tag{12.404}$$

$$P_{K_h}(m_h) = e = \{e_{jl}\}_{j,l=1,2} \text{ for the other } m_h \in [L^2(\Omega)]^4 \text{ and where } e_{jl} = 0 \text{ in } \Omega. \tag{12.405}$$

Since m_h^i are piecewise constant and equal to zero outside Ω_h, then the case in Equation 12.405 cannot occur.

Minimizations of $l_h(z_h) \forall z_h \in {}^1V_h$ and $\mathscr{L}_h(\boldsymbol{v}_h, m_h^i) \forall \boldsymbol{v}_h \in V_h$ are equivalent to solving non-linear systems

$$(\underline{A}_{Th} T_h^{i+1}, z_h) = \langle s_h^i, z_h \rangle \quad \forall z_h \in {}^1V_h, \tag{12.406}$$

$$(\underline{A}_h \boldsymbol{u}_h^{i+1}, \boldsymbol{v}_h) = \langle S_h^i, \boldsymbol{v}_h \rangle - k\Theta 2^{1/2} \hat{g}(m_h^i, D_h \boldsymbol{v}_h) \quad \forall \boldsymbol{v}_h \in V_h. \tag{12.407}$$

Since $A_{Th}(.,.)$, $A_h(.,.)$ are symmetric, continuous and positive definite bilinear forms on ${}^1V_h \times {}^1V_h$ or $V_h \times V_h$ respectively, and the operators $\underline{A}_{Th}$ and $\underline{A}_h$ are monotone, then in every step of the algorithm for the given m_h^i there exists a unique pair of functions $(T_h^{i+1}, \boldsymbol{u}_h^{i+1})$, $T_h^{i+1} \in {}^1V_h$, $\boldsymbol{u}_h^{i+1} \in V_h$.

Theorem 12.14: Since ${}^1V_h \subset {}^1V$, $V_h \subset V$, $K_h \subset K$ are non-empty closed convex sets, K_h is a bounded set, and under the hypothesis that

$$\mathscr{L}_h(\boldsymbol{u}_h, m_h) = \mathscr{J}(\boldsymbol{u}_h) + (m_h, \Psi(\boldsymbol{u}_h)), \tag{12.408}$$

$\mathscr{J}$ is G-differentiable from V_h into R,

$$(\mathscr{J}'(\boldsymbol{u}_h) - \mathscr{J}'(\boldsymbol{v}_h), \boldsymbol{u}_h - \boldsymbol{v}_h) \geq \alpha \|\boldsymbol{u}_h - \boldsymbol{v}_h\|^2, \alpha > 0, \quad \forall \boldsymbol{u}_h, \boldsymbol{v}_h \in V_h, \tag{12.409}$$

$$\forall m_h \in K_h, \text{ the function } \boldsymbol{v}_h \to (m_h, \Psi(\boldsymbol{v}_h)) \text{ is convex and l.s.c. on } V_h, \tag{12.409a}$$

Ψ is Lipschitzian from V_h into K, i.e.

$$|\Psi(\boldsymbol{u}_h) - \Psi(\boldsymbol{v}_h)| \leq c \|\boldsymbol{u}_h - \boldsymbol{v}_h\| \quad \forall \boldsymbol{u}_h, \boldsymbol{v}_h \in V_h \quad \text{(see above)}, \tag{12.410}$$

then Algorithm 12.1 is convergent in the following sense

$$(T_h^i, \boldsymbol{u}_h^i) \to (T_{ho}, \boldsymbol{u}_{ho}) \quad \text{in} {}^1V_h \times V_h, \tag{12.411}$$

where $(T_{ho}, \boldsymbol{u}_{ho})$ is the solution of the problem

$$\min_{z_h \in {}^1V_h} l_h(z_h), \ \inf_{\boldsymbol{v}_h \in V_h} \sup_{m_h \in K_h} \mathscr{L}_h(\boldsymbol{v}_h, m_h)$$

provided the ρ_i satisfy

$$0 < \rho_o \leq \rho_i \leq \rho_1 = 2\alpha c^{-2}, \tag{12.412}$$

where $\alpha = k\Theta\hat{\mu}$ for the case where the norm is induced by the scalar product defined by Equation 12.250, $\alpha = \min(\rho, k\Theta\hat{\mu})$ for the case where the norm is induced by the scalar product defined by Equation 12.251, and $c = k\Theta 2^{1/2}\hat{g}$, will ρ_1 being sufficiently small.

The proof in its first part is parallel to minimizing the functional $l_h(z_h)$ and in its second part is parallel to that of Lemma 12.13.

Lemma 12.15: Let k and ρ satisfy the conditions

$$k\Theta c_1 S(h)\left\|\boldsymbol{u}_h^{i+\Theta}\right\| \leq 1, \tag{12.413}$$

$$\rho \leq \hat{\mu}(k\Theta\hat{g}^2)^{-1} - \varepsilon, \quad \varepsilon > 0, \tag{12.414}$$

Then the Uzawa algorithm solving equations 12.358 and 12.359 converges in the sense of Lemma 12.13.

Proof: It is sufficient to verify that the condition of strong monotonicity can be replaced by the condition of weakness:

$$(\underline{A}\boldsymbol{u}_h - \underline{A}\boldsymbol{u}_h^{i+\Theta}, \boldsymbol{u}_h - \boldsymbol{u}_h^{i+\Theta}) \geq \alpha\left\|\boldsymbol{u}_h - \boldsymbol{u}_h^{i+\Theta}\right\|^2 \quad \forall \boldsymbol{u}_h \in V_h, \tag{12.415}$$

and equations 12.413 and 12.414 are verified thanks to Lemma 12.13.

The algorithm of Uzawa can be written as follows. Let us define

$$(\underline{A}_h\boldsymbol{u}_h, \boldsymbol{v}_h) = (\boldsymbol{u}_h, \boldsymbol{v}_h) + k\Theta\hat{\mu}a_h(\boldsymbol{u}_h, \boldsymbol{v}_h) + k\Theta b_h(\boldsymbol{u}_h, \boldsymbol{u}_h, \boldsymbol{v}_h)$$

and let m_h^o be given. Then

$$(\underline{A}_h\boldsymbol{v}_h^{s+1}, \boldsymbol{v}_h) + k\Theta 2^{1/2}\hat{g}(m_h^s, D_h v_h) + k\Theta b_{sh}(T_h^s, \boldsymbol{v}_h) = (\boldsymbol{u}_h^i, \boldsymbol{v}_h) + k\Theta(\boldsymbol{f}_k^{i+\Theta}, \boldsymbol{v}_k) \ \ \forall \boldsymbol{v}_h \in V_h,$$

$$m_{hjl}^{s+1/2} = m_{hjl}^s + \rho\Theta k 2^{1/2} D_{hjl}\boldsymbol{v}_h^{s+1},$$

$$m^{s+1} = P_k(m_h^{s+1/2}),$$

where P_k is a mapping from K_h to $[L^2(\Omega)]^4$. We define P_k as follows. Let $\boldsymbol{x} \in \Omega$ then

$$(P_k m)(\boldsymbol{x}) = \begin{cases} m(\boldsymbol{x}) & \text{as } |m(\boldsymbol{x})| \leq 1, \\ m(\boldsymbol{x})\,|m(\boldsymbol{x})|^{-1} & \text{as } |m(\boldsymbol{x})| > 1. \end{cases}$$

Similarly for the Arrow-Hurwicz algorithm we have the following. Let λ be a parameter, then

$$(\underline{A}_h\boldsymbol{u}_h, \boldsymbol{v}_h) = (\boldsymbol{u}_h, \boldsymbol{v}_h) + k\Theta\hat{\mu}a_h(\boldsymbol{u}_h, \boldsymbol{v}_h) + \lambda k\Theta b_h(\boldsymbol{u}_h, \boldsymbol{u}_h, \boldsymbol{v}_h).$$

Then

$$(\boldsymbol{v}_h^{s+1} - \boldsymbol{v}_h^s, \boldsymbol{v}_h) + \rho(\boldsymbol{v}^s, \boldsymbol{v}_h) + \rho k\Theta\hat{\mu}a_h(\boldsymbol{v}_h^s, \boldsymbol{v}_h) + \lambda\rho k\Theta b_h(\boldsymbol{v}_h^s, \boldsymbol{v}_h^s, \boldsymbol{v}_h) + k\Theta b_{sh}(T_h^s, \boldsymbol{v}_h)$$

$$= \rho(\boldsymbol{u}_h^i, \boldsymbol{v}_h) + \rho k\Theta(\boldsymbol{f}_k^{i+\Theta}, \boldsymbol{v}_h) \quad \forall \boldsymbol{v}_h \in V_h, \tag{12.416}$$

$$\gamma m_{hjl}^{s+1/2} = \gamma m_{hjl}^s + \rho k\Theta 2^{1/2} D_{hjl}\boldsymbol{v}_h^{s+1}, \tag{12.417}$$

$$m^{s+1} = P_k(m^{s+1/2}), \tag{12.418}$$

where P_k is defined as above.

The case $\lambda = 0$ represents the case without the convective term, the case $\lambda = 1$ represents the case with non-linear convective term.

Lemma 12.16: Let $\lambda = 0$, then the algorithm in equations 12.416-12.418 converges and the conditions

$$\rho(1 - k\Theta\hat{\mu}S^2(h)) \leq 1, \tag{12.419}$$

$$\rho\gamma^{-1} \leq \tfrac{1}{2}\hat{\mu}k^{-1}\Theta^{-1}\hat{g}^{-1} - \varepsilon, \quad \varepsilon > 0, \tag{12.420}$$

$$2\rho\gamma^{-1}k^2\Theta^2S^2(h)\hat{g}^2 \leq 1, \tag{12.421}$$

hold.
The proof is similar of that of Lemma 12.15.

The construction of the space V_h brings certain difficulties. The practical computation of the element $\boldsymbol{u}_h^i \in V_h$, defined by the above discussed schemes, is not easy. The difficulty is connected with an approximation of the constraint div $\boldsymbol{u} = 0$ built into the definition of the space V_h. Practically, the problem is the same as in the cases of the Navier-Stokes and Stokes equations. Therefore, the same technique discussed in Temam (1984)) can also be used in the Bingham case. At the same time we shall introduce the discrete approximation of the pressure, which is implicitly contained in Equation 12.359. We shall approximate functions $\boldsymbol{u}_h \in V_h$ by exactly the same type of functions, but no divergence condition is imposed on them. We denote the corresponding space as $\underline{W}_h$. Moreover, the element $\boldsymbol{u}_h^i \in V_h$ in the Uzawa algorithm will be approximated by a sequence of elements of $\underline{W}_h$. Let the space $\underline{W}_h$ be the space of step functions (see Temam (1977), Chapter I, Sections 3-4)

$$\boldsymbol{u}_h = \sum_{\underline{T}_h} \boldsymbol{u}_h(\underline{T}_h)\chi_{\underline{T}_h}(\boldsymbol{x}), \quad \boldsymbol{u}_h(\underline{T}_h) \equiv \xi_{\underline{T}_h} \in R^2,$$

which are discretely divergence free in the following sense:

$$\sum_{i=1} \nabla_{hi}u_{hi}(\underline{T}_h) = 0 \quad \forall \underline{T}_h,$$

which amounts to saying

$$\sum_{i=1} \nabla_{hi}u_{hi}(\boldsymbol{x}) = 0 \quad \forall \boldsymbol{x} \in \Omega_h.$$

Then we define the discrete divergence $D_h\boldsymbol{u}_h \equiv \text{div}_h\boldsymbol{u}_h$, $\boldsymbol{u}_h \in \underline{W}_h$, as the step function

$$D_h\boldsymbol{u}_h = \sum_{\underline{T}_h} (D_h\boldsymbol{u}_h(\underline{T}_h))\chi_{\underline{T}_h},$$

where $D_h\boldsymbol{u}_h(\underline{T}_h) = \Sigma_{j=1,2}\nabla_{hj}u_{hj}(\underline{T}_h)\forall \underline{T}_h$, i.e. $D_h\boldsymbol{u}_h = \Sigma_{j=1,2}D_{hj}u_{hj} = \Sigma_{\underline{T}_h \in \mathscr{T}_h}\eta_{\underline{T}_h}\chi_{\underline{T}_h}$, $\eta_{\underline{T}_h} = (\text{meas}\underline{T}_h)^{-1}\int_{\underline{T}_h} \text{div } \boldsymbol{u}_h \text{d}x$, where $\chi_{\underline{T}_h}$ are characteristic functions of triangles $\underline{T}_h \in \mathscr{T}_h$. Then an element $\boldsymbol{u}_h \in \underline{W}_h$ belongs to V_h if and only if $D_h\boldsymbol{u}_h = 0$ inside every triangle $\underline{T}_h \in \mathscr{T}_h$.

The discrete approximation of pressure is an element of the space of step functions of type

$$p_h = \sum_{\underline{T}_h} p_h(\underline{T}_h)\chi_{\underline{T}_h}, \quad p_h(\underline{T}_h) \equiv \eta_{\underline{T}_h} \in R,$$

where $\chi_{\underline{T}_h}$ is the characteristic function of the triangle $\underline{T}_h$.

Then for the scheme in Equation 12.265 (and similarly for the above mentioned schemes with some modifications as in Temam (1977)) there exists some step function p_h^i of type given by Equation 12.422 such that

$$A_h(\boldsymbol{u}_h^{i+1}, \boldsymbol{v}_h) + k\Theta b_h(\boldsymbol{u}_h^i, \boldsymbol{u}_h^{i+1}, \boldsymbol{v}_h) + k\Theta b_{sh}(T_h^i - T_o, \boldsymbol{v}_h) + k\Theta 2^{1/2}\hat{g}(m_h^i, D_h\boldsymbol{v}_h)$$
$$- k\Theta(p_h^{i+1}, D_h\boldsymbol{v}_h) = (\boldsymbol{u}_h^i, \boldsymbol{v}_h) + k\Theta(\boldsymbol{f}^{i+1}, \boldsymbol{v}_h) \quad \forall \boldsymbol{v}_h \in W_h. \tag{12.423}$$

Let $\boldsymbol{u}_h^{i+1} \in V_h$ be the solution of Equation 12.423. Since the bilinear form $A_h(\boldsymbol{u}_h, \boldsymbol{v}_h)$ is symmetric, continuous and positive definite on $W_h \times W_h$, the operator $(\underline{A}_h\boldsymbol{u}_h, \boldsymbol{v}_h)$ is strictly monotone and the functional $\langle S_h^i, \boldsymbol{v}_h\rangle - k\Theta 2^{1/2}\hat{g}(m_h, D\boldsymbol{v}_h) \in W_h'$ and to V_h' for every $m_h \in K_h$, $i = 0, 1, \ldots, n-1$, then the mapping $\boldsymbol{v}_h \to \underline{A}_h(\boldsymbol{u}_h^{i+1}, \boldsymbol{v}_h) - \langle S_h^i, \boldsymbol{v}_h\rangle + k\Theta 2^{1/2}\hat{g}(m_h, D\boldsymbol{v}_h)$ is also nonlinear and continuous on W_h. Then there exists a set of real numbers $\{\eta_{\underline{T}_h}\}$, $\underline{T}_h \in \mathcal{T}_h$ such that

$$(\underline{A}_h\boldsymbol{u}_h^{i+1}, \boldsymbol{v}_h) - \langle S_h^i, \boldsymbol{v}_h\rangle + k\Theta 2^{1/2}\hat{g}(m_h, D\boldsymbol{v}_h) = k\Theta \sum_{\underline{T}_h \in \mathcal{T}_h} \eta_{\underline{T}_h} \int_{\underline{T}_h} D_h\boldsymbol{v}_h \mathrm{d}\boldsymbol{x} \quad \forall \boldsymbol{v}_h \in W_h. \tag{12.424}$$

Since functions of Equation 12.422 are elements of $L^2(\Omega)$ then we can put $p_h(T_h) = \eta_{\underline{T}_h}$, $\underline{T}_h \in \mathcal{T}_h$. Hence and from Equation 12.424 we obtain

$$(\underline{A}_h\boldsymbol{u}_h^{i+1}, \boldsymbol{v}_h) - \langle S_h^i, \boldsymbol{v}_h\rangle + k\Theta 2^{1/2}\hat{g}(m_h, D_h\boldsymbol{v}_h) = k\Theta(p_h, D_h\boldsymbol{v}_h) \quad \forall \boldsymbol{v}_h \in W_h. \tag{12.425}$$

Then we have the following theorem

Theorem 12.15: Let $\boldsymbol{u}_h^{i+1} \in V_h$ be a solution of Equation 12.359 (or Equation 12.423), then there exists a piecewise constant function p_h defined by Equation 12.422 such that Equation 12.425 holds. Conversely, let $\boldsymbol{u}_h^{i+1} \in V_h$ and let p_h defined by Equation 12.422 fulfils Equation 12.425, then $\boldsymbol{u}_h^{i+1}$ fulfils Equation 12.359 (or Equation 12.423).

To solve Equation 12.425 the following Uzawa's type of algorithm will be used.

Algorithm 12.2: Let p_h^o be an arbitrary piecewise constant function. Then for $j \geq 0$ we obtain $\boldsymbol{w}_h^j$, p_h^{j+1} ($\boldsymbol{w}_h^j \in W_h$, p_h^j is a piecewise constant function) from p_h^j by the following algorithm.

Compute $\boldsymbol{w}_h^j \in V_h$ such that

$$(\underline{A}_h\boldsymbol{w}_h^j, \boldsymbol{v}_h) - \langle S_h^{j-1}, \boldsymbol{v}_h\rangle + k\Theta 2^{1/2}\hat{g}(m_h^j, D_h\boldsymbol{v}_h) = k\Theta(p_h^j, D_h\boldsymbol{v}_h) \quad \forall \boldsymbol{v}_h \in W_h, \tag{12.426}$$

$$p_h^{j+1} = p_h^j - \rho D_h\boldsymbol{w}_h^j \quad \forall \underline{T}_h \in \mathcal{T}_h, \tag{12.427}$$

where $D_h\boldsymbol{w}_h^j \equiv \mathrm{div}_h\boldsymbol{w}_h^j$ represents a value of the divergence of $\boldsymbol{w}_h^j \in V_h$, and $\rho = \text{constant} > 0$.

Next, we shall prove the convergence theorem of Agorithm 12.2.

Theorem 12.16: Let $0 < \rho < \rho_1$. Then $\boldsymbol{w}_h^j$, p_h^j determined by the algorithm defined by equations 12.426 and 12.427 exists and is unique for every j, and converges to the solution of Equation 12.425 as follows

$$\boldsymbol{w}_h^j \to \boldsymbol{u}_h \text{ strongly in the norm of the space } W_h, \boldsymbol{u}_h \in V_h, j \to \infty,$$

$$p_h^j \to p_h \text{ strongly in the norm of the space } L^2(\Omega), j \to \infty.$$

Proof: It holds that $|D_h v_h| \leq \|v_h\| \; \forall v_h \in W_h$, where $\|\cdot\|$ is the norm induced by Equation 12.250 or 12.251. Then

$$(p_h^j, D_h v_h) \leq \left|p_h^j\right| |D_h v_h| \leq \left|p_h^j\right| \|v_h\| .$$

Moreover, the linear functional $\langle S_h^j, v_h\rangle - k\Theta 2^{1/2}\hat{g}(m_h, D_h v_h) \in W_h'$, therefore $\langle S_h^j, v_h\rangle - k\Theta 2^{1/2}\hat{g}(m_h, D_h v_h) + (p_h^j, D_h v_h) \in W_h'$ The existence and uniqueness of the solution $w_h^j \in W_h$ for every j of equations 12.426 and 12.427 then follow from the Lax-Milgram theorem. The convergence of the Uzawa algorithm is parallel to that of Lemma 12.13.

Both Uzawa's algorithms can be compiled into one algorithm only. Then the algorithm used can be written as follows.

Algorithm 12.3: Let m_h^o be arbitrary. Let u_h^k, m_h^k be known. Then u_h^{k+1}, m_h^{k+1} can be determined as follows.

Let p_h^o be arbitrary. Let w_h^j, p_h^j be known. Then w_h^{j+1}, p_h^{j+1} will be determined as follows. Find $w_h^{j+1} \in W_h$ such that

$$\min\{\tfrac{1}{2}(\underline{A}_h w_h, w_h) - \langle S_h^i, w_h\rangle + k\Theta 2^{1/2}\hat{g}(m_h^k, D_h w_h) - (p_h^j, D_h w_h)\}, \tag{12.428}$$

$$p_h^{j+1}(\underline{T}_h) = p_h^j(T_h) - \rho D_h w_h^{j+1} \quad \forall T_h \in \mathcal{T}_h,$$

$$u_h^{k+1} = w_h^{j(k)}, \text{ where } w_h^{j(k)} \text{ depends on the requested accuracy of the approximation,}$$

$$m_h^{k+1} = P_k(m_h^k + k\Theta 2^{1/2} D_h u_h^{k+1}),$$

$$u_h^{i+\Theta} = u_h^k, u_h^{i+1} = \Theta^{-1}(u_h^{i+\Theta} - (1-\Theta)u_h), \quad i = 0, 1, \ldots, n-1, \Theta \in \langle \tfrac{1}{2}, 1).$$

In the above, the algorithm in Equation 12.428 leads to finding a minimum of a non-linear functional, for which the non-linear SOR, conjugate gradient, etc. methods can be used. But since the SOR method depends on the size of the mesh, then it will not be very useful for large systems of non-linear equations.

12.3.3.3. Analysis of geodynamic processes in the earth based on coupled Stefan-like problems in a thermo-visco-plastic Bingham rheology

INTRODUCTION

Many physical processes connected with heat flow and diffusion involving phase-change phenomena give rise to free boundary problems for parabolic partial differential equations of the Stefan type. Historically the first paper was given by Joseph Stefan (1835–1893) (Stefan (1889)) and was concerned with the melting of ice at 0°C. A concept of a weak solution of the Stefan problem was introduced by Kamenomotskaya (1958), (1961) and was next analysed by means of smoothing techniques as developed in Friedman (1968), Ladyzenskaya (1969), Oleinik (1960) and of monotonicity methods developed by Brezis (1971) and Lions (1969). A very rich literature exists for numerical solutions of Stefan problems. The problem is non-linear so that numerically the problem was solved by finite difference and finite element methods, See, for example, Jerome and Rose (1977), Fox (1975), Ockendon and Hodgkins (1975), Ciavaldini (1975), Jerome (1977), (1983), Ladyzenskaya, Solonnikov and Ural'cova (1968) Elliott (1981), Nochetto, Paollini and Verdi

(1991a,b), Picasso (1995), etc., and by the variational inequality approach in Duvaut (1975), Pawlow (1982), etc. The algorithms are then based on the non-linear SOR method, developed for example in White (1982a,b), Elliott (1981), or on the non-linear conjugate gradient method. The coupled contact-two-phase Stefan-like problems are solved in Nedoma (1996a,b), (1997b), Nedoma and Dvořák (1995).

The interest of geophysical disciplines for numerical modelling has encouraged the development of corresponding numerical methods. Likewise in geodynamics, where instead of processes connected with moving masses, processes connected with heating (melting and recrystallization) and freezing (solidification) also play an important role. As a first step, simulations related to this topic consist of studying the macroscopic heat transfer mechanism, modelled by the so-called two-phase Stefan problem, which was investigated in more detail in Chapter 9. Since problems connected with melting, recrystallization and solidification play an important role in geodynamic processes in the core, the mantle and the lithosphere of the Earth, investigations of the corresponding high performance numerical methods have a great importance for simulation and further understanding of geodynamic processes inside the Earth, and at present have application to the modelling and simulation geomechanic and geodynamic processes in regions where the high-level radioactive waste repositories will be built. In "classical" problems of technology the two-phase Stefan problems with convection in the fluid phase are investigated (Blanc and Gasser (1996), Cannon, DiBenedetto and Knightly (1983)). Similar problems are also investigated in geodynamics (Peltier (1989)). In both problems the rheology is taken as Newtonian rheology, often in the Boussinesq's approximation (Blanc and Gasser (1996), Gasser (1995), Peltier (1989)). As we saw in Section 12.3.1 this approximation depends on the Rayleigh's, Prandtl's, etc. numbers, which change in relatively large value intervals. Therefore, the main goal of this section is to give an optimal method (See Nedoma (1997a)), i.e. optimal from the better physical and rheological approximation point of view as well as from the high performance computation possibility point of view. We saw in Chapter 5 and in Section 12.3.3 that the Bingham rheology is an optimal rheology for present computations, as for the case when the threshold of plasticity $\hat{g}$ is equal to zero then we have the usual present case of Newtonian rheology, and if the threshold of plasticity $\hat{g} \to \infty$ then the medium is absolutely rigid and between these cases can model all types of visco-plastic materials. It is evident that both results can be compared, and as the threshold of plasticity is determined by the Mises relation and hence by the velocities of seismic P and S waves and the density (Lamé coefficients and density), then the approximation of the rheology inside the Earth is close to reality. Moreover, together with the two-phase Stefan problem the approximation of reality is much more realistic. Similar considerations are also valid for the case of rift zones.

It is evident that due to the extremely complicated shape of the front for melting, recrystallization or solidification, these cannot be described in full. As a result, practically we obtain the phase change zones only as the forecast of the fine geometry would need a very expensive computation as well as the necessity of regularity of the front, where both phases are parallel at the same time (the so-called mushy zone). To study these problems we propose to average variables over the phases

(Whitaker (1969)) so that instead of the fine geometry a smoother phase change zone is investigated. The principle of this average is based on the integration of the variables over an elementary domain, where each variable is in its own phase. The effect of these averaging processes is to define zones of intermediate state (mushy zones), in between both phases. Within these mushy zones, the relative proportion of each phase is given by the volumic fractions f_S, f_L, where $f_S + f_L = 1$. In the mushy zones both phases are microscopically parallel. From the mathematical point of view, by the mushy zones is meant the regions where the operator of the heat equation is degenerated. In the phase diagram, this corresponds to the phase changes either at null concentration and temperature of melting (fusion), or at concentration c from the interval, corresponding to the concentration at which the solid rock begins to melt and the concentration at which the liquid rock begins to solidify, and eutectic temperature, i.e. the temperature under which no liquid remains whatever the concentration. The latent heat represents the energy necessary for a complete change of phase of a unit volume at temperature of melting (fusion) and null concentration.

PHYSICAL MODEL

The physical model is derived from the conservation principle (see Chapter 3). The conservation of momentum in its differential form gives the equation of motion, whilst the mass conservation specifies the rocks as incompressible rocks and also gives the solute equation. The energy conservation then gives the heat equation. The "classical" approach in the phase change problems of first order (melting, recrystallization, solidification) is based upon Boussinesq's hypothesis assuming that the density variation is neglected except in the force term. Then the obtained equation is valid in the liquid phase only as the solid phase is assumed to be static. These assumptions lead to the Navier-Stokes equations. But in this section for better approximation of the rheology we shall assume that the rheology is of the visco-plastic Bingham's type.

In the case of thermo-visco-plastic Bingham's rheology from the conservation of momentum we obtain the equations of motion

$$\rho\left(\frac{\partial u_i}{\partial t} + u_k\frac{\partial u_i}{\partial x_k}\right) = \frac{\partial \tau_{ij}}{\partial x_j} + \rho f_i \quad \text{in } \Omega \times I, \quad i, j = 1, 2(3), \quad I = (t_o, t_1), \tag{12.429}$$

where u_i are components of the flow velocity vector, and ρf_i are components of body forces.

The stress-strain rate relation, the constituent law, can be derived from the dissipation function $\tau_{ij}D_{ij}$, where τ_{ij} is the stress tensor and D_{ij} is the strain-rate tensor defined as

$$D_{ij} = \frac{1}{2}\left(\frac{\partial u_i}{\partial x_j} + \frac{\partial u_j}{\partial x_i}\right), \quad i, j = 1, 2(3), \tag{12.430}$$

assuming that the dissipation function depends only on the strain rate tensor, and where $\boldsymbol{u} = (u_i)$ is the flow velocity vector. Then we can write

$$\tau_{ij}D_{ij} = \mathscr{D}_1(\boldsymbol{D}) + \mathscr{D}_2(\boldsymbol{D}),$$

where $\mathscr{D}_1$ and $\mathscr{D}_2$ are positive homogeneous functions of order 1 and 2 respectively in the components of the strain rate tensor $\boldsymbol{D}$. Let us put

$$\mathscr{D}_1 = \frac{\partial \mathscr{D}_1}{\partial D_{kl}} D_{kl}, \tag{12.431}$$

$$\mathscr{D}_2 = \frac{1}{2}\frac{\partial \mathscr{D}_2}{\partial D_{kl}} D_{kl}. \tag{12.432}$$

Since we shall assume that the rocks are incompressible, then

$$\operatorname{div} \boldsymbol{v} = 0, \text{ i.e. } D_{kk} = 0, \tag{12.433}$$

which follows from the mass conservation law. Hence, using Equations 12.431, 12.432 and 12.430 we find

$$\tau_{ij} = -p\delta_{ij} + \frac{\partial \mathscr{D}_1}{\partial D_{ij}} + \frac{1}{2}\frac{\partial \mathscr{D}_2}{\partial D_{ij}}, \tag{12.434}$$

where p is a scalar, independent of the strain rate tensor D_{ij}, which represents the spherical part of the stress tensor and has a meaning of the pressure, and δ_{ij} is the Kronecker symbol.

Next we shall assume that the rocks are isotropic, and then the scalars $\mathscr{D}_1$ and $\mathscr{D}_2$ are functions of the invariants of the strain rate tensor D_{ij} only. For the case of the Bingham's rheology

$$\mathscr{D}_1 = 2\hat{g}D_{\text{II}}^{1/2}, \tag{12.435}$$

$$\mathscr{D}_2 = 4\hat{\mu}D_{\text{II}}, \tag{12.436}$$

$$D_{II} = \tfrac{1}{2}D_{ij}D_{ij}, \tag{12.437}$$

where D_{II} is the invariant of the strain rate tensor D_{ij}, and $\hat{g}$, μ represent the thresholds of plasticity and viscosity. Due to Nowacki (1970) the thermal stress τ_{ij}^T satisfies

$$\tau_{ij}^T = \beta_{ij}(T - T_o),$$

where β_{ij} is a coefficient of thermal expansion, and $T(\boldsymbol{x}, t)$, and $T_o(\boldsymbol{x})$, $T_o(\boldsymbol{x}) > 0$, are a temperature and an initial temperature in which the medium is in the initial stress-strain state. Thus the constituent law is as

$$\tau_{ij} = -p\delta_{ij} + \hat{g}D_{ij}D_{\text{II}}^{-1/2} + 2\hat{\mu}D_{ij} - \beta_{ij}(T - T_o), \tag{12.438}$$

having sense for $D_{\text{II}} \neq 0$ only. To derive the inversion stress-strain rate relations for the Bingham's part of the stress tensor we firstly define the invariant of the stress tensor as

$$\tau_{\text{II}} = \tfrac{1}{2}\tau_{ij}^D\tau_{ij}^D, \tag{12.439}$$

where $\tau_{ij}^D = \tau_{ij} - (1/3)\tau_{kk}\delta_{ij}$ is the deviator of the stress tensor. Then with Equation 12.438 we find

$$\tau_{\text{II}} = (\hat{g} + 2\hat{\mu}D_{\text{II}}^{1/2})^2. \tag{12.440}$$

Hence,

$$\tau_{\mathrm{II}}^{1/2} \geq \hat{g}, \tag{12.441}$$

and thus

$$D_{ij} = (2\hat{\mu})^{-1}(1 - \hat{g}\tau_{\mathrm{II}}^{1/2})\tau_{ij}^{D}. \tag{12.442}$$

Hence and using Equation 12.437

$$D_{\mathrm{II}} = \tfrac{1}{2}D_{ij}D_{ij} = (2\hat{\mu})^{-2}(\tau_{\mathrm{II}}^{1/2} - \hat{g})^2.$$

Then

$$D_{\mathrm{II}} = 0 \text{ for } \tau_{\mathrm{II}}^{1/2} - \hat{g} \leq 0, \text{ i.e. for } \tau_{\mathrm{II}}^{1/2} \leq \hat{g}$$

and

$$D_{\mathrm{II}} \neq 0 \text{ for } \tau_{\mathrm{II}}^{1/2} - \hat{g} > 0, \text{ i.e. for } \tau_{\mathrm{II}}^{1/2} > \hat{g}.$$

Then the inverse constituent law in the thermo-Bingham rheology can be written in the form

$$\begin{aligned} &\tau_{\mathrm{II}}^{1/2} \leq \hat{g} \Leftrightarrow D_{ij} = 0, \\ &\tau_{II}^{1/2} > \hat{g} \Leftrightarrow D_{ij} = (2\hat{\mu})^{-1}(1 - \hat{g}\tau_{II}^{-1/2})\tau_{ij}^{D} - \alpha_{ij}(T - T_o) \end{aligned} \tag{12.443}$$

If $\hat{g} = 0$ then Equation 12.438 represents the constituent law for the Newtonian fluid, i.e. for the well-known viscous incompressible fluid. For $\hat{g} \to \infty$ we have the absolute rigid types of materials, and for small $\hat{g}$ we have the Bingham rheology close to a Newtonian fluid. The last types of materials are close to real rocks in the melted (strongly visco-plastic) parts of the Earth. Moreover, we see that the threshold of plasticity (yield point or yield limit) $\hat{g}$ is defined by the Mises' type relation

$$\tfrac{1}{2}\tau_{ij}^{D}\tau_{ij}^{D} \leq \hat{g}^2. \tag{12.444}$$

The momentum equation in Boussinesq's approximation, representing the classical approach for melting and solidification, is valid in the liquid zone since the solid is assumed to be static. The additional term $M(f_s(c, T))\boldsymbol{v}$ in Equation 12.429, based on empirical observations, describes the density variation in the mushy zone and has a meaning of body forces. The density and the dynamic viscosity $\hat{\mu}$ are assumed to be piecewise constant. For the mushy zones the additional term $M(f_S)\boldsymbol{v}$ was found (Blanc and Gasser (1996), Gasser (1995)) as

$$M(y) = C_o y^2 (1 - y)^{-3}, \; C_o \text{ an empirical constant}, \tag{12.445}$$

and where the mushy zones are empirically modelled as porous media. In Bingham's approximation both phases, liquid and solid, are represented by a strongly visco-plastic fluid, close to a viscous liquid and characterized by the lower value of the threshold of plasticity $\hat{g}$ in the first case, and by the viscous fluid with sufficiently high threshold of plasticity $\hat{g}$ in the second one, for which the volumic fractions are satisfied.

The diffusion equation for the rock's liquid mixture (alloy) follows from the mass conservation law

$$\frac{\partial \rho}{\partial t} + \operatorname{div}\,(\rho \boldsymbol{u}) = 0. \tag{12.446}$$

For rock mixtures (metallic alloys) the definition of velocity must be defined in a new way.

In the case when the diffusion is absent, then the composition of every given element of fluid remains unknown during the movement.
This means that $\mathrm{d}c/\mathrm{d}t = 0$, i.e.

$$\frac{\partial c}{\partial t} + \boldsymbol{u}\ \operatorname{grad}\, c = 0. \tag{12.447}$$

Hence and with Equation 12.446 we have

$$\frac{\partial(\rho c)}{\partial t} + \operatorname{div}\,(\boldsymbol{u}\,\rho c) = 0. \tag{12.448}$$

In integral form Equation 12.448 is of the form

$$\frac{\partial}{\partial t}\int \rho c \mathrm{d}\boldsymbol{x} = -\oint \rho c \boldsymbol{u}\,\mathrm{d}s. \tag{12.449}$$

In the case when diffusion is assumed, then

$$\frac{\partial}{\partial t}\int \rho c \mathrm{d}\boldsymbol{x} = -\oint \rho c \boldsymbol{u}\,\mathrm{d}s - \oint \boldsymbol{i}\,\mathrm{d}s, \tag{12.450}$$

where $\boldsymbol{i}$ denotes the density of diffusion flow. In differential form Equation 12.450 is of the form

$$\frac{\partial(\rho c)}{\partial t} = -\operatorname{div}\,(\rho c \boldsymbol{u}) - \operatorname{div}\,\boldsymbol{i}. \tag{12.451}$$

Hence and using Equation 12.446 we find

$$\rho\left(\frac{\partial c}{\partial t} + \boldsymbol{u}\ \operatorname{grad}\, c\right) = -\operatorname{div}\,\boldsymbol{i} \quad \text{in } \Omega \times I, \tag{12.452}$$

representing the equation of diffusion, and where (see Landau and Lifshitz (1959))

$$\boldsymbol{i} = -\rho D(\operatorname{grad}\, c + k_T T^{-1} \operatorname{grad}\, T + k_p p^{-1}\ \operatorname{grad}\, p), \tag{12.453}$$

where D is the coefficient of diffusion, $k_T D$ represents the coefficient of thermo-diffusion and $k_p D$ represents the coefficient of baro-diffusion, T is the temperature, and p is the pressure.

Finally, the energy conservation law gives the generalized heat equation in the form

$$\rho c_e\left(\frac{\partial T(\boldsymbol{x}\,t)}{\partial t} + \boldsymbol{u}\ \operatorname{grad}\, T(\boldsymbol{x}, t)\right) + \rho \beta_{ij} T_o D_{ij}(\boldsymbol{u}) = \operatorname{div}\,(\kappa(T)\operatorname{grad} T) + Q_o(\boldsymbol{x}, t, T) \text{ in } \Omega_t. \tag{12.454}$$

The coupled system of Equations 12.429, 12.438, and 12.452-12.454 gives governing equations for melted and solidified materials. The boundary and initial conditions must then describe the real situation of the investigated problem.

Remark: By a "liquid" phase we shall mean the strongly visco-plastic material with low threshold of plasticity $\hat{g}(\hat{g} \to 0)$, and by a "solid" phase we shall mean the visco-plastic material with threshold of plasticity $\hat{g}$ sufficiently large. Moreover, we can put $d = \rho D$, where e.g. $d \sim 10^{-9}$ for Al-Si (Al_2SiO_4) alloy. Furthermore, the liquid concentration c_L is the concentration c in the liquid zone ($c = f_L c_L + f_S c_S$, where f_L, f_S are the relative proportion of both phases solid (visco-plastic) (S) and liquid (L), and $f_L + f_S = 1$), and c_S is the concentration c in the solid visco-plastic phase. The diffusion factor d does not account for the different diffusivity of the solute in the liquid and solid (visco-plastic) phases. But both are very small, and the convective effect easily overrides the diffusive one (e.g. $d \sim 10^{-9}$ for Al-Si alloy). Moreover, compared to the heat diffusivity, the effect of the diffusion is much slower for the concentration. The body force term in Equation 12.429 can be found as

$$\boldsymbol{F}(\boldsymbol{x}, t, c, T) = \rho g x_3 + \operatorname{div}(\beta(T - T_o)) + \rho \boldsymbol{g}(c_o c_L(\boldsymbol{x}, t, c, T) + c_1 T + c_2), \tag{12.455}$$

where the first term is body forces due to the gravity effect, the second one represents thermal stresses and the third one represents the effect due to the concentration of the "liquid" phase (or gas) in the rocks. The effects of friction and Joule's heat can also be assumed.

We shall assume that the investigated domain $\overline{\Omega} = \cup_{\iota=1}^{m} \cup_{s=1}^{r} \overline{\Omega}^{\iota,s}$, where Ω^{ι} denotes subdomains of Ω characterized by the material properties and indices "s" characterize the domains with the phase changes of rocks (melting, recrystallization, solidification).

MATHEMATICAL FORMULATION OF THE PROBLEM

Let $\Omega = \cup_{\iota=1}^{m} \Omega^{\iota} \subset R^N$, $N = 2(3)$ be a bounded domain occupied by the interior of the Earth or its parts, or by geological bodies in a geological practice, with a smooth boundary $\partial\Omega = \Gamma_u \cup \Gamma_\tau \cup \mathscr{R}$, where $\mathscr{R}$ is the set of zero measure. Let $I = (t_o, t_1)$ and let $\Omega_T = \Omega \times \overline{I}$, $\partial\Omega_T = \partial\Omega \times \overline{I}$, $\Omega_t = \Omega \times (t_o, t)$, $\partial\Omega_t = \partial\Omega \times (t_o, t)$, $t \in I$. We suppose that the components of $\partial\Omega$ are smooth enough to admit a normal $\boldsymbol{n}$ almost everywhere in the sense of the surface measure on $\partial\Omega$. Moreover, we shall assume incompressible materials. Furthermore, we shall assume that $\Gamma^{\iota,s}$, $\iota = 1, \ldots, m$, $s = 1, \ldots, r$, denote $n-1$ dimensional open sets in the relative topology of Ω, i.e. surfaces which divide Ω into r open sets $\Omega^{\iota,s}$. We denote the respective portions of the boundaries of $\Omega^{\iota,s}$ and $\Omega^{\iota,s+1}$ by $\partial\Omega^{\iota,s}$ and $\partial\Omega^{\iota,s+1}$. We shall assume that the components $\Gamma^{\iota,s} \equiv R^{\iota,s}(t)$ of $\Gamma =_\iota \cup_s \Gamma^{\iota,s}$ are smooth enough to also admit normals $\vec{\nu}^{\iota,s}$ almost everywhere in the sense of the surface measures on $\Gamma^{\iota,s} \equiv R^{\iota,s}(t)$. We shall assume that $\Omega^{\iota,s}$, $\Omega^{\iota,s+1}$ constitute other states of materials, i.e. "solid", "liquid" or other type of recrystallized rocks. Thus $\Gamma^{\iota,s} \equiv R^{\iota,s}(t)$ denote the interfaces between two different phase states of rocks and we speak about the phase change boundaries or phase transitient boundaries (zones). Here $\Omega^{\iota,s}$, $\Omega^{\iota,s+1}$, $\Gamma^{\iota,s}$ are unknowns that must be determined as part of the solution. We denote by $T^{\iota,s}(\boldsymbol{x}, t)$ the temperature of rocks in $\Omega^{\iota,s}$, by $\boldsymbol{u}^{\iota,s}(\boldsymbol{x}, t)$ and $p^{\iota,s}(\boldsymbol{x}, t)$ the velocity and pressure in $\Omega^{\iota,s}$, and by $c^{\iota,s}(\boldsymbol{x}, t)$ and $\varphi^{\iota,s}(\boldsymbol{x}, t)$ the concentrations and sources of concentrations in $\Omega^{\iota,s}$.

Then we shall investigate the non-stationary incompressible coupled two-phase Stefan-like-Bingham problem.

Problem ($\mathscr{P}_O$): Consider the problem of finding functions $T(\boldsymbol{x}, t)$, $\boldsymbol{u}(\boldsymbol{x}, t)$, $p(\boldsymbol{x}, t)$ and $c(\boldsymbol{x}, t)$ defined on the closure $\overline{\Omega}^{\iota,s}$ of open sets $\Omega^{\iota,s}$ such that $\Omega = \cup_{\iota,s}\overline{\Omega}^{\iota,s}$ and satisfying

$$\rho^\iota c_e^\iota\left(\frac{\partial T^\iota}{\partial t}+u\frac{\iota\partial T^\iota}{k\partial x_k}\right)+\rho^\iota\beta_{ij}^\iota T_o^\iota D_{ij}(\boldsymbol{u}^\iota)=Q_o^\iota(\boldsymbol{x},t)+\frac{\partial}{\partial x_i}\left(\kappa_{ij}^\iota\frac{\partial T^\iota}{\partial x_j}\right) \quad \text{in } \Omega^\iota\times I,\ i,j=1,2(3), \tag{12.456}$$

$$\rho^\iota\left(\frac{\partial u_i^\iota}{\partial t}+u_k^\iota\frac{\partial u_i^\iota}{\partial x_k}\right)=\frac{\partial \tau_{ij}^\iota}{\partial x_j}+F_i^\iota \quad \text{in } \Omega^\iota\times I, \quad i,j=1,2(3), \tag{12.457}$$

$$\operatorname{div}\boldsymbol{u}^\iota=0 \quad \text{in } \Omega^\iota\times I, \tag{12.458}$$

$$\frac{\partial c^\iota}{\partial t}+\boldsymbol{u}^\iota \operatorname{grad} c^\iota=\lambda^\iota\Delta c^\iota+\varphi^\iota \quad \text{in } \Omega^\iota\times I, \tag{12.459}$$

and boundary and initial conditions

$$\left.\begin{aligned} &T(\boldsymbol{x},t)=T_1(\boldsymbol{x},t)(=0), && (12.460)\\ &\tau_{ij}(\boldsymbol{x},t)n_j=P_{oi}(\boldsymbol{x},t), && (12.461)\\ &\frac{\partial c(\boldsymbol{x},t)}{\partial \boldsymbol{n}}=0 && (12.462)\end{aligned}\right\}\ \text{on } \Gamma_\tau\times I, \quad i,j=1,2(3),$$

$$\left.\begin{aligned} &\kappa_{ij}\frac{\partial T(\boldsymbol{x}t)}{\partial x_j}n_i=q(\boldsymbol{x},t), && (12.463)\\ &\boldsymbol{u}(\boldsymbol{x},t)=\boldsymbol{u}_1(\boldsymbol{x},t), && (12.464)\\ &\frac{\partial c(\boldsymbol{x},t)}{\partial \boldsymbol{n}}=0 && (12.465)\end{aligned}\right\}\ \text{on } \Gamma_u\times I, \quad i,j=1,2(3),$$

$$\left.\begin{aligned} &T_S^{\iota,s}=T_L^{\iota,s}=T_R^{\iota,s}, && (12.466)\\ &\left(\kappa_{ij}^\iota\frac{\partial T^\iota}{\partial x_j}\nu_i^\iota\right)_S^s-\left(\kappa_{ij}^\iota\frac{\partial T}{\partial x_j}\nu^\iota i\right)_L^s=-\rho^{\iota,s}L^{\iota,s}v_\nu^{\iota,s} && (12.467)\end{aligned}\right\}\ \text{on } R^{\iota,s}(t),$$

$$\left.\begin{aligned} &T^\iota(\boldsymbol{x},t_o)=T_o^\iota(\boldsymbol{x}), && (12.468)\\ &\boldsymbol{u}^\iota(\boldsymbol{x},t_o)=\boldsymbol{u}_o^\iota(\boldsymbol{x}), && (12.469)\\ &c^\iota(\boldsymbol{x},t_o)=c_o^\iota(\boldsymbol{x}) && (12.470)\end{aligned}\right\}\ \text{in } \Omega^\iota,$$

where τ_{ij} is defined by the constituent law of Equation 12.438; and where $\rho^{\iota,s}$ is a density associated with $\Omega^{\iota,s}$; $T_S^{\iota,s}$, $T_L^{\iota,s}$, $T_R^{\iota,s}$ are temperatures in $\Omega^{\iota,s}$, where rocks are in a solid-visco-plastic state ($T_S^{\iota,s}$) or in a liquid (strongly visco-plastic with low threshold of plasticity $\hat{g}$) state ($T_L^{\iota,s}$), or temperatures of phase changes; $\vec{\nu}^{\iota,s}$ is a normal to the phase change boundary $R^{\iota,s}(t)\equiv\Gamma^{\iota,s}$ pointing towards $\Omega_S^{\iota,s}$; $v_\nu^{\iota,s}=\partial\Phi^{\iota,S}/\partial t\equiv\partial_t\Phi^{\iota,s}$ are the speeds of $R^{\iota,s}(t)$ along $\vec{\nu}^{\iota,s}$; $L^{\iota,s}$ are the latent heats of the phase changes — melting, recrystallization, solidification (the latent heat represents the energy necessary for a complete phase change of a unit volume at melting temperature and null concentration; it is positive in the case of melting and recrystallization and negative in the case of solidification); $R^{\iota,s}(t)$ represent hyper-surfaces lying in $\Omega_t^{\iota,s}$; and $\Omega_{St}^{\iota,s}$ are domains lying in $\Omega_t^{\iota,s}$ and bounded by $R^{\iota,s}(t)$ and $\partial\Omega_t^{\iota,s}$. Furthermore, let $\Phi^{\iota,s}(\boldsymbol{x},t)$ be a C^1 function in $\overline{\Omega}_T$ such that

$$R^{\iota,s}(t)=\{(\boldsymbol{x},t)\in\overline{\Omega}_T,\ \Phi^{\iota,s}(\boldsymbol{x},t)=0\},\ \nabla_x\Phi^{\iota,s}(\boldsymbol{x},t)\neq 0 \quad \text{on } R^{\iota,s}(t),$$

$$\Phi^{\iota,s}(\boldsymbol{x},t)<0 \quad \text{in } \Omega_S^{\iota,s}\times I,\ \Phi^{\iota,s}(\boldsymbol{x},t)>0 \quad \text{in } \Omega_L^{\iota,s}\times I,$$

and $R(t) = \cup_{\iota,s} R^{\iota,s}(t)$, $t_o \le t < t_1$, is the phase change or free boundary. The functions $T_o^\iota(\boldsymbol{x})$, $T_1^\iota(\boldsymbol{x}, t)$, $\boldsymbol{u}_o(\boldsymbol{x})$, $\boldsymbol{u}_1(\boldsymbol{x}, t)$, $c_o(\boldsymbol{x})$ are the initial and boundary data on $\Gamma_\tau \times I$ or $\Gamma_u \times I$ respectively; $c_e^\iota(\boldsymbol{x})$ are specific heat in Ω^ι; $\kappa_{ij}^\iota(\boldsymbol{x})$ are the thermal conductivity in Ω^ι; and $Q^\iota(\boldsymbol{x}, t)$, $\varphi^\iota(\boldsymbol{x}, t)$ are thermal sources and sources of concentration respectively, in $\Omega^\iota \times I$, $\iota = 1, \ldots, m$. All the functions used here are real valued. Further, we shall investigate the $2D$ model problem ($N = 2$) only.

To analyse the Stefan problem we shall introduce a new variable Θ by the Kirchhoff transformation

$$\Theta = G(\boldsymbol{x}, t, T) = \int_0^{T(\boldsymbol{x},t)} \kappa(\boldsymbol{x}, t, \xi)\mathrm{d}\xi,\ (\boldsymbol{x}, t) \in \overline{\Omega}_t.$$

Since $\kappa_o \le \kappa(\boldsymbol{x}, t, \xi) \le \kappa_1$ for $(\boldsymbol{x}, t, \xi) \in \overline{\Omega} \times I \times R$, $\kappa_o, \kappa_1 =$ constant > 0, then the mapping $T \to \Theta(\boldsymbol{x}, t, T)$ is one-to-one and $T = G^{-1}(\boldsymbol{x}, t, \Theta)$, where $G^{-1}(\boldsymbol{x}, t, \Theta)$ is its inverse. We introduce the enthalpy by

$$H(\boldsymbol{x}, t, T) = \int_0^{T(\boldsymbol{x},t)} \underline{c}((G^{-1}(\zeta))\kappa^{-1}(G^{-1}(\zeta)))\mathrm{d}\zeta + \begin{cases} 0 & \text{if } T < \Theta_R(\boldsymbol{x}, t), \\ \langle 0, \alpha\rangle & \text{if } T = \Theta_R(\boldsymbol{x}, t), \\ \alpha & \text{if } T < \Theta_R(\boldsymbol{x}, t), \end{cases}$$

where we denoted $\underline{c}(\boldsymbol{x}, t) = \rho(\boldsymbol{x}, t)c_e(\boldsymbol{x}, t)$, and $\Theta_R(\boldsymbol{x}, t) = G(\boldsymbol{x}, t, T_R)$, T_R is the temperature of a phase change, and $\alpha = \rho L$.

Then we shall investigate the non-stationary incompressible coupled two-phase Stefan problem in the Bingham rheology and in the enthaply formulation.

Problem ($\mathscr{P}$): Consider the problem of finding functions $H^{\iota,s}(\boldsymbol{x}, t)$, $\Theta^{\iota,s}(\boldsymbol{x}, t)$, $c^{\iota,s}(\boldsymbol{x}, t)$, $\boldsymbol{u}^{\iota,s}(\boldsymbol{x}, t)$ (and $p^{\iota,s}(\boldsymbol{x}, t)$) defined on the closure $\overline{\Omega}^{\iota,s}$ of open sets $\Omega^{\iota,s}$ such that $\Omega = \cup_{\iota,s}\overline{\Omega}^{\iota,s}$ and satisfying

$$\left(\frac{\partial H^\iota}{\partial t} + u_k^\iota \frac{\partial \Theta^\iota}{\partial x_k}\right) + \rho^\iota \beta_{ij}^\iota \Theta_o^\iota D_{ij}(\boldsymbol{u}^\iota) = Q^\iota(\boldsymbol{x}, t) + \Delta\Theta^\iota(\boldsymbol{x}, t) \text{ in } \Omega^\iota \times I, \tag{12.471}$$

$$\rho^\iota\left(\frac{\partial u_i^\iota(\boldsymbol{x}, t)}{\partial t} + u_k^\iota \frac{\partial u_i^\iota(\boldsymbol{x}, t)}{\partial x_k}\right) = \frac{\partial \tau_{ij}^\iota(\boldsymbol{x}, t)}{\partial x_j} + F_i^\iota(\boldsymbol{x}, t) \quad \text{in } \Omega^\iota \times I,\ i, j = 1, 2, \tag{12.472}$$

$$\operatorname{div} \boldsymbol{u}^\iota = 0 \quad \text{in } \Omega^\iota \times I, \tag{12.473}$$

$$\frac{\partial c^\iota(\boldsymbol{x}, t)}{\partial t} + \boldsymbol{u}^\iota(\boldsymbol{x}, t) \operatorname{grad} c_L^\iota(c^\iota(\boldsymbol{x}, t), \Theta^\iota(\boldsymbol{x}, t)) = \lambda^\iota(\boldsymbol{x})\Delta c^\iota(\boldsymbol{x}, t) + \varphi^\iota(\boldsymbol{x}, t) \quad \text{in } \Omega^\iota \times I, \tag{12.474}$$

$$\Theta^\iota(\boldsymbol{x}, t) = T^\iota(c^\iota(\boldsymbol{x}, t), H^\iota(\boldsymbol{x}, t)) \quad \text{in } \Omega^\iota \times I, \tag{12.475}$$

and boundary and initial conditions

$$\left.\begin{aligned} H(\boldsymbol{x}, t) &= 0, && (12.476) \\ \tau_{ij}(\boldsymbol{x}, t)n_j &= P_{oi}(\boldsymbol{x}, t), && (12.477) \\ \frac{\partial c(\boldsymbol{x}, t)}{\partial \boldsymbol{n}} &= 0 && (12.478) \end{aligned}\right\} \text{ on } \Gamma_\tau \times I,$$

$$\left.\begin{aligned} \frac{\partial \Theta(\boldsymbol{x}, t)}{\partial n} &= q(\boldsymbol{x}, t) && (12.479) \\ \boldsymbol{u}(\boldsymbol{x}, t) &= \boldsymbol{u}_1(\boldsymbol{x}, t), && (12.480) \\ \frac{\partial c(\boldsymbol{x}, t)}{\partial \boldsymbol{n}} &= 0 && (12.481) \end{aligned}\right\} \text{ on } \Gamma_u \times I,$$

$$\left.\begin{aligned} H^{\iota}(\boldsymbol{x}, t)_o) &= H^{\iota}_o(\boldsymbol{x}), \\ \boldsymbol{u}^{\iota}(\boldsymbol{x}, t)_o) &= \boldsymbol{u}^{\iota}_o(\boldsymbol{x}), \\ c^{\iota}(\boldsymbol{x}, t_o) &= c^{\iota}_o(\boldsymbol{x}) \end{aligned}\right\} \text{ in } \Omega^{\iota}, \quad \begin{aligned} &(12.482)\\ &(12.483)\\ &(12.484)\end{aligned}$$

where

$$\tau_{ij} = -p\delta_{ij} + \hat{g}D_{ij}D_{\mathrm{II}}^{-1/2} + 2\hat{\mu}D_{ij} - \beta_{ij}(T - T_o), \tag{12.485}$$

and where $D_{ij} = (1/2)(\partial_j u_i + \partial_i u_j)$, $D_{\mathrm{II}} = (1/2)D_{ij}D_{ij}$, and $H^{\iota}_o(\boldsymbol{x})$, $c^{\iota}_o(\boldsymbol{x})$, $\lambda^{\iota}(\boldsymbol{x})$, $\boldsymbol{u}^{\iota}_o(\boldsymbol{x})$ are the given positive functions. For simplicity we shall assume that $\boldsymbol{u}_1(\boldsymbol{x}, t) = 0$.

We shall also use the following notation: $\partial_t r = \partial r/\partial t$, where r is a scalar or vector function, and $\partial_s r(y_1, y_2) = \partial r(y_1, y_2)/\partial y_s$, $s = 1, 2$, and similarly, $\partial_j v_i = v_{i,j} = \partial v_i/\partial x_j$. Further, for simplicity indices ι and s will be omitted.

We shall further assume that

(A1) $\Omega = \cup_{\iota=1}^{m} \Omega^{\iota} \subset R^2$ is an open bounded connected domain with Lipschitzian boundary $\partial\Omega = \Gamma_u \cup \Gamma_\tau \cup \mathscr{R}$.

On the domain Ω for functions T, c_L, multipliers m and the initial and boundary data we shall assume:

(A2) For the function $T^{\iota} : R^2_+ \to R$, where R_+ denotes the set of non-negative reals, we assume that

(i) T^{ι} is uniformly Lipschitzian,

(ii) $T^{\iota}(x, 0) = 0 \quad \forall x \in R_+$,

(iii) there exist (possibly empty) open sets $O_i \subset R^2_+$, $i \in \mathcal{N}$, such that T^{ι} is constant on each O_i and such that there is a function $\alpha^{\iota} \in C^o(\langle 0, \infty)) \cap C^2((0, \infty))$ verifying: $(\alpha'')^{\iota}$ is decreasing, $\lim_{x\to\infty}(\alpha''(x))^{\iota} > 0$, and $\partial_2 T^{\iota}(x, y) \geq ((\alpha''(x))^{\iota})^{-1}$ for a.e. $(x, y) \in R'_+ \times R_+ \backslash \cup_{i\in\mathcal{N}} O_i$.

(A3) Since the term $\hat{g}^{\iota}D^{\iota}_{ij}(\boldsymbol{u}^{\iota})D^{1/2}_{II}(\boldsymbol{u}^{\iota})$ creates in the variational formulation the non-differential functional, then we introduce the multipliers $m^{\iota} = (m^{\iota}_{ij})$, $i, j = 1, 2$, by

(i) $m^{\iota}_{ij} \in L^{\infty}(\Omega^{\iota} \times I)$, $m^{\iota}_{ij} = m^{\iota}_{ji}, \quad \forall i, j = 1, 2, m^{\iota}_{\mathrm{ii}} = 0$,

(ii) $m^{\iota}_{ij}m^{\iota}_{ij} \leq 1$ a.e. in $\Omega^{\iota} \times I$,

(iii) $m^{\iota}_{ij}D_{ij}(\boldsymbol{u}^{\iota}) = (D_{ij}(\boldsymbol{u}^{\iota})D_{ij}(\boldsymbol{u}^{\iota}))^{1/2} = 2^{1/2}D^{1/2}_{\mathrm{II}}(\boldsymbol{u}^{\iota})$ a.e. in $\Omega^{\iota} \times I$,

(iv) the constituent law in Equation 12.438 can be rewritten as $\tau^{\iota}_{ij} = -p^{\iota}\delta_{ij} + 2\hat{g}^{\iota}D_{ij}(\boldsymbol{u}^{\iota}) + 2^{1/2}\hat{g}^{\iota}m^{\iota}_{ij} - \beta^{\iota}_{ij}(T^{\iota} - T^{\iota}_o)$.

(A4) The functions $c^{\iota}_L : R^2_+ \to R$, representing a concentration in a liquid (strongly visco-plastic with low $\hat{g}$) part of the medium and stemming from the constitutive law (phase diagram), satisfy

(i) $c_L^\iota \in C^o(R_+^2) \cap L^\infty(R_+^2)$,

(ii) $c_L^\iota(0, y) = 0 \quad \forall y \in R_+$,

(iii) $\exists c_1 =$ constant. > 0 such that $\left|c_L^\iota(x, y)\right| \le c_1[(\alpha''(x))^\iota]^{-1/2} \quad \forall (x, y) \in R'_+ \times R_+$, where α^ι satisfies (A2)(iii).

(A5) The initial data satisfy

(i) $H_o(\boldsymbol{x}) \in L^2(\Omega), H_o(\boldsymbol{x}) \ge 0$ a.e in Ω,

(ii) $\boldsymbol{u}_o(\boldsymbol{x}) \in H(\Omega) = \{\boldsymbol{v} \mid \boldsymbol{v} \in [L^2(\Omega)]^2,\ \text{div}\,\boldsymbol{v} = 0, \boldsymbol{v} = 0 \text{ on } \Gamma_u\}, \boldsymbol{u}_o(\boldsymbol{x}) \ge 0$ a.e. in Ω,

(iii) $c_o(\boldsymbol{x}) \in L^2(\Omega), c_o(\boldsymbol{x}) \ge 0$ a.e. in Ω.

(A6) The physical data are sufficiently smooth, i.e.

$$
\begin{aligned}
&Q^\iota(\boldsymbol{x}, t) \in L^2(I; L^2(\Omega^\iota)), \quad \boldsymbol{F}^\iota(\boldsymbol{x}, t) \in L^2(I; [L^2(\Omega^\iota)]^2),\\
&\varphi^\iota(\boldsymbol{x}, t) \in L^2(I; L^2(\Omega^\iota)), \quad q(\boldsymbol{x}, t) \in L^2(I; L^2(\Gamma_u)),\\
&\boldsymbol{P}(\boldsymbol{x}, t) \in L^2(I; [L^2(\Gamma_\tau)]^2), \quad \kappa^\iota(\boldsymbol{x}, t) \in L^2(I; C^1(\overline{\Omega}^\iota)),\\
&\hat{\mu}^\iota(\boldsymbol{x}), \hat{g}^\iota(\boldsymbol{x}) = \text{constant} > 0, \quad \partial\beta_{ij}^\iota(\boldsymbol{x})/\partial x_j \in L^\infty(\Omega^\iota) \quad \forall i, j,\\
&\lambda^\iota(\boldsymbol{x}) \in C^1(\overline{\Omega}^\iota).
\end{aligned}
$$

Remark: The conditions (iii) in assumptions (A2) and (A4) are satisfied if there exist positive constants c_1, c_2 and a real $\delta \in \langle 0, 2)$ such that (the symbol "ι" is omitted)

(i) $\partial_2 T(x, y) \ge c_1 \min(x^\delta, 1)$ for almost every $(x, y) \in R_+^2 \setminus \cup_{i \in \mathcal{N}'} O_i$, and

(ii) $|c_L(x, y)| \le c_2 \min(x^{(1/2)\delta}, 1) \quad \forall (x, y) \in R_+^2$.

In that case, the function α is such that

(iii) $$\alpha''(x) = \begin{cases} c^{-1} x^{-\delta}, & x < 1,\\ c_1^{-1}, & x \ge 1. \end{cases}$$

Because of assumption (A2)(iii), the system of equations for determining $T(\boldsymbol{x}, t)$ and $c(\boldsymbol{x}, t)$ is generally degenerated and Equation 12.471 is similar to the classical Stefan's equation.

Remark: The function c in the mushy zone can be written as (see the phase diagram, Figure 12.2) $c = c_L f_L + c_S f_S = c_L(1 - f_S) + c_S f_S$, where f_L, f_S represent the relative portion of visco-plastic (solid) and strongly visco-plastic with low $\hat{g}$ (liquid) phases, respectively and will be determined as a result of computations, where the portion of "solid" and "liquid" parts are controlled means of the threshold of plasticity $\hat{g}$.

In the phase diagram (Figure 12.2), T_M is the temperature of melting; T_E is the eutectic temperature, under which no "liquid" remains; $c_S(T)$ is the concentration of the solute for which the "solid" rock begins to melt; $c_L(T)$ is the concentration of the

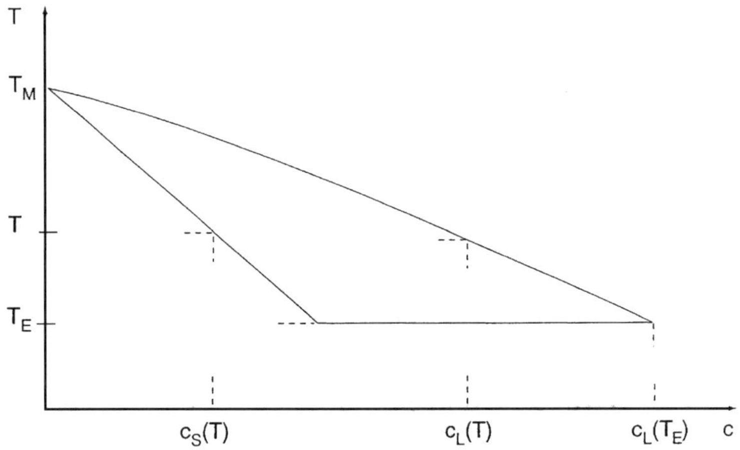

Figure 12.2. Phase Diagram (c, T) (simple version)

solute for which the "liquid" rock begins to solidify. The mushy zone corresponds to points (c, T), where $c \in \langle c_S(T), c_L(T)\rangle$ and $T \in \langle T_E, T_M\rangle$.

VARIATIONAL FORMULATION

Let us denote by $C^k(\Omega)$, $0 \le k < \infty$, the space of all functions f defined on Ω which have continuous derivatives up to order k on Ω; for $k = 0$ we put $C^o(\Omega) = C(\Omega)$. We denote by $H^1(\Omega)$ the Sobolev space of scalar (or vector) functions such that $f \in L^2(\Omega)$, $\partial_i f \in L^2(\Omega)$, $i = 1, \dots, N$, where $L^2(\Omega)$ is the space of square summable scalar (or vector) functions on Ω and the derivatives $\partial_i f$ are taken in the sense of distributions on Ω. By $L^\infty(\Omega)$ we denote the space of all measurable functions f defined a.e. on Ω and such that there exists a constant $c > 0$ such that

$$|f(\boldsymbol{x})| \le c \quad \forall \boldsymbol{x} \in \Omega \backslash E, E \subset \Omega, \text{ meas } (E) = 0.$$

By $L^p(I; X)$, where X is a functional space, $1 \le p < \infty$, we denote the space of functions $f : I \to X$ such that $\|f(.)\|_X \in L^p(I)$ and by $H^1(I; X)$ the space of functions $f : I \to X$ for which $\|f(.)\|_X \in H^1(I)$.

Let us define the spaces

$$
\begin{aligned}
&{}^1V = \{z \mid z \in H^1(\Omega), z = 0 \quad \text{on } \Gamma_\tau\},\ {}^2V = \{w \mid w \in H^1(\Omega)\},\\
&V = \{\boldsymbol{v} \mid \boldsymbol{v} \in [H^1(\Omega)]^2, \quad \text{div } \boldsymbol{v} = 0, \boldsymbol{v} = 0 \text{ on } \Gamma_u\},\\
&H(\Omega) = \{\boldsymbol{v} \mid \boldsymbol{v} \in [L^2(\Omega)]^2, \quad \text{div } \boldsymbol{v} = 0, \boldsymbol{v} = 0 \text{ on } \Gamma_u\},\\
&{}^1V', {}^2V', V' \text{ are dual spaces of } {}^1V, {}^2V \text{ and } V, \text{ respectively},\\
&{}^1V \subset L^2(\Omega) \subset {}^1V', {}^2V \subset L^2(\Omega) \subset {}^2V', V \subset H \subset V', {}^1V' \equiv H^{-1}(\Omega), {}^2V' \equiv (H^1(\Omega))',\\
&{}^1W = \{z \mid z \in L^2(I; {}^1V), z(\boldsymbol{x}, t_o) = 0\}, {}^2W = \{w \mid w \in L^2(I; {}^2V), w(\boldsymbol{x}, t_o) = 0\},\\
&W = \{\boldsymbol{v} \mid \boldsymbol{v} \in L^2(I; V), \boldsymbol{v}' \in L^2(I; H), \boldsymbol{v}(\boldsymbol{x}, t_o) = 0\}.
\end{aligned}
$$

Then we have the following variational (weak) formulation of the coupled problem defined in enthalpy.

Problem $(\mathscr{P}_{\text{centh}})_v$**:** Find a tetrad $\{H(\boldsymbol{x},t), \Theta(\boldsymbol{x},t), \boldsymbol{u}(\boldsymbol{x},t), c(\boldsymbol{x},t)\}$, i.e. enthalpy H, generalized temperature Θ, velocity $\boldsymbol{u}$ and concentration c, satisfying

$$\int_I \{(\partial_t H(t), z - \Theta(t)) + (\boldsymbol{u}(t) \text{ grad } \Theta, z - \Theta(t)) + a_\Theta(\Theta(t), z - \Theta(t))$$

$$+ b_p(\boldsymbol{u}(t), z - \Theta(t)) - (Q(t), z - \Theta(t))\}\mathrm{d}t \geq 0 \quad \forall z \in {}^1W, \tag{12.486}$$

$$\int_I \{(\partial_t \boldsymbol{u}(t), \boldsymbol{v} - \boldsymbol{u}(t)) + \hat{\mu} a(\boldsymbol{u}(t), \boldsymbol{v} - \boldsymbol{u}(t)) + b(\boldsymbol{u}(t), \boldsymbol{u}(t), \boldsymbol{v} - \boldsymbol{u}(t)) + \hat{g} j(\boldsymbol{v}) - \hat{g} j(\boldsymbol{u}(t))$$

$$+ b_s(\Theta(t) - \Theta_o, \boldsymbol{v} - \boldsymbol{u}(t)) - (\boldsymbol{F}(t), \boldsymbol{v} - \boldsymbol{u}(t))\}\mathrm{d}t \geq 0 \quad \forall \boldsymbol{v} \in W, \tag{12.487}$$

$$\int_I \{(\partial_t c(t), d - c(t)) + (\boldsymbol{u}(t) \text{ grad } c_L(c, \Theta), d - c(t)) + a_c(c(t), d - c(t)) - (\varphi(t),$$

$$d - c(t))\}\mathrm{d}t \geq 0, \quad \forall \mathrm{d} \in {}^2W, \tag{12.488}$$

$$\Theta(t) = T(c(t), H(t)), H \geq 0, c \geq 0 \quad \text{a.e. in } \Omega \times I, \tag{12.489}$$

$$H(\boldsymbol{x}, t_o) = H_o(\boldsymbol{x}), \tag{12.490}$$

$$\boldsymbol{u}(\boldsymbol{x}, t_o) = \boldsymbol{u}_o(\boldsymbol{x}), \tag{12.491}$$

$$c(\boldsymbol{x}, t_o) = c_o(\boldsymbol{x}) \quad \text{a.e. in } \Omega, \tag{12.492}$$

where $H^{-1}(\Omega)$ stands for $(H_o^1(\Omega))'$ and where $\boldsymbol{u}, \boldsymbol{v} \in [H^1(\Omega)]^2$, $\Theta, z, c, d \in H^1(\Omega)$, $\varphi, Q \in L^2(\Omega)$, $\boldsymbol{F} \in [L^2(\Omega)]^2$ and

$$a_\Theta(\Theta, z) = \int_\Omega \text{grad } \Theta \text{ grad } z\mathrm{d}\boldsymbol{x},\ a(\boldsymbol{u}, \boldsymbol{v}) = 2\int_\Omega D_{ij}(\boldsymbol{u})D_{ij}(\boldsymbol{v})\mathrm{d}\boldsymbol{x},$$

$$a_c(c, d) = \int_\Omega \lambda \text{ grad } c \text{ grad } d\mathrm{d}\boldsymbol{x},\ b_\Theta(\boldsymbol{u}, \Theta, z) = (\boldsymbol{u} \text{ grad } \Theta, z),$$

$$b(\boldsymbol{u}, \boldsymbol{v}, \boldsymbol{w}) = \int_\Omega \rho u_i v_{j,i} w_j \mathrm{d}\boldsymbol{x} = (\rho \boldsymbol{u} \text{ grad } \boldsymbol{v}, \boldsymbol{w}),$$

$$b_s(\Theta, \boldsymbol{v}) = \int_\Omega \frac{\partial}{\partial x_j}(\beta_{ij}\Theta)v_i\mathrm{d}\boldsymbol{x},\ b_p(\boldsymbol{v}, g) = \int_\Omega \rho\Theta_o\beta_{ij}v_{i,j}g\mathrm{d}\boldsymbol{x},$$

$$b_c(\boldsymbol{u}, c, d) = (\boldsymbol{u} \text{ grad } c, d),\ (\boldsymbol{u}, \boldsymbol{v}) = \int_\Omega u_i v_i \mathrm{d}\boldsymbol{x},$$

$$(\Theta, z) = \int_\Omega \Theta z\mathrm{d}\boldsymbol{x},\ (Q, z) = \int_\Omega Qz\mathrm{d}\boldsymbol{x} - \int_{\Gamma_u} qz\mathrm{d}s,$$

$$(\boldsymbol{F}, \boldsymbol{v}) = \int_\Omega F_i v_i \mathrm{d}\boldsymbol{x} - \int_{\Gamma_\tau} P_{oi}v_i\mathrm{d}\boldsymbol{x},\ j(\boldsymbol{v}) = 2\int_\Omega (D_{\mathrm{II}}(\boldsymbol{v}))^{1/2}\mathrm{d}\boldsymbol{x}.$$

Definition 12.3: We say that the tetrad $\{H, \Theta, \boldsymbol{u}, c\}$ is a weak solution of the Problem $(\mathscr{P}_{\text{centh}})$ if $H \in L^\infty(I; L^2(\Omega)) \cap H^1(I; H^{-1}(\Omega))$, $\Theta \in L^2(I; {}^1V)$, $\boldsymbol{u} \in L^\infty(I; L^2(\Omega)) \cap L^2(I; V)$, $c \in L^2(I; {}^2V) \cap H^1(I; (H^1(\Omega))')$, $H \geq 0$, $c \geq 0$ and $\Theta = T(c, H)$ a.e. in $\Omega \times I$ and Equations 12.486-12.488, and 12.490-12.492 hold.

We now have preliminary results and the main theorem. In the sequel the following lemmas will be used.

Lemma 12.17: It holds that

$$v_{i,j} \in L^2(\Omega) \quad \forall \boldsymbol{v} \in V, \quad i, j = 1, 2.$$

Proof: If $\boldsymbol{v} \in V$, then $v_{i,j} \in L^2(\Omega)$. The proof is a consequence of Lemma 3.1 of Duvaut and Lions (1976).

Further, we have the following estimates. Firstly, from the above given assumptions, the symmetry conditions

$$a_\Theta(\Theta, z) = a_\Theta(z, \Theta), a(\boldsymbol{u}, \boldsymbol{v}) = a(\boldsymbol{v}, \boldsymbol{u}), a_c(c, d) = a_c(d, c)$$

hold. Moreover, they yield that for $\Theta \in {}^1V, \boldsymbol{u} \in V, c \in {}^2V$ there exist constant $c_\Theta > 0$, $c_u > 0$, $c_c > 0$ such that

$$a_\Theta(\Theta, \Theta) \geq c_\Theta \|\Theta\|_{1,1}^2 \quad \forall \Theta \in {}^1V, \quad a(\boldsymbol{u}, \boldsymbol{u}) \geq c_u \|\boldsymbol{u}\|_{1,2}^2 \quad \forall \boldsymbol{u} \in V,$$
$$a_c(c, c) \geq c_c \|c\|_{1,1}^2 \quad \forall c \in {}^2V.$$

For $\boldsymbol{u}$, y, z it holds that $b_\Theta(\boldsymbol{u}, y, z) + b_\Theta(\boldsymbol{u}, z, y) = 0$, $\boldsymbol{u} \in H$, $y, z \in H_o^1(\Omega)$ and there exist positive constants c_5, c_6, c_7 independent of $\boldsymbol{u}$, y, z such that

$$\begin{aligned} |b_\Theta(\boldsymbol{u}, y, z)| &\leq c_5 \|\boldsymbol{u}\|_{[L^p(\Omega)]^2} \|y\|_{L^p(\Omega)} \|\{D_i z\}\|_{0,1} \\ &\leq c_6 \|\boldsymbol{u}\|_{[L^p(\Omega)]^2} \|y\|_{L^p(\Omega)} \|z\|_{1,1} \\ &\leq c_7 \|\boldsymbol{u}\|_{1,2}^{1/2} \|\boldsymbol{u}\|_{0,2}^{1/2} \|y\|_{1,1}^{1/2} \|y\|_{0,1}^{1/2} \|z\|_{1,1} \end{aligned}$$

and similarly for $b_c(\boldsymbol{u}, c, d)$.

For a vector field $\boldsymbol{u}, \boldsymbol{v}, \boldsymbol{w}$ on Ω we put

$$b(\boldsymbol{u}, \boldsymbol{v}, \boldsymbol{w}) = \int_\Omega u_i v_{i,j} w_i \mathrm{d}\boldsymbol{x}$$

and $$|b(\boldsymbol{u}, \boldsymbol{v}, \boldsymbol{w})| \leq c \|\boldsymbol{u}\|_{[L^p(\Omega)]^2} \|\boldsymbol{w}\|_{[L^p(\Omega)]^2} \Sigma_{i,j} \left\|D_i v_j\right\|_{0,2}, \quad \frac{2}{p} + \frac{1}{2} = 1.$$

Using Lemma 12.17 we find that

$$|b(\boldsymbol{u}, \boldsymbol{v}, \boldsymbol{w})| \leq c \|\boldsymbol{u}\|_{[L^p(\Omega)]^2} \|\boldsymbol{w}\|_{[L^p(\Omega)]^2} \|\boldsymbol{v}\|_{1,2} \,.$$

Moreover, the inequality of convexity

$$\|\boldsymbol{v}\|_{[L^p(\Omega)]^2} \leq c \|\boldsymbol{v}\|_{1,2}^{1/2} \|\boldsymbol{v}\|_{0,2}^{1/2} \quad \forall \boldsymbol{v} \in H_o^1(\Omega), \; \frac{1}{p} = \frac{1}{2} - \frac{1}{2} N^{-1} = \frac{1}{4}$$

holds and

$$|b(\boldsymbol{u}, \boldsymbol{v}, \boldsymbol{w})| \leq c \|\boldsymbol{u}\|_{1,2}^{1/2} \|\boldsymbol{u}\|_{0,2}^{1/2} \|\boldsymbol{w}\|_{1,2}^{1/2} \|\boldsymbol{w}\|_{0,2}^{1/2} \|\boldsymbol{v}\|_{1,2} \,.$$

For $\boldsymbol{u}, \boldsymbol{v}, \boldsymbol{w}$ it holds that

$$b(\boldsymbol{u}, \boldsymbol{v}, \boldsymbol{w}) + b(\boldsymbol{u}, \boldsymbol{w}, \boldsymbol{v}) = 0, b(\boldsymbol{v}, \boldsymbol{v}, \boldsymbol{v}) = 0.$$

The main result gives the following theorem.

Theorem 12.17: Let assumptions $(A1)$-$(A6)$ be satisfied for every $t \in (0, 1\rangle$. Then there exist scalar functions H, Θ and c and a vector function $\boldsymbol{u}$ such that

$$H \in L^2(I; H^1_o(\Omega)) \cap L^\infty(I; L^2(\Omega)), \partial_t H \in L^2(I; H^{-1}(\Omega)),$$

$$\boldsymbol{u} \in L^2(I; V) \cap L^\infty(I; H), \partial_t \boldsymbol{u} \in L^2(I; V'),$$

$$c \in L^2(I; H^1(\Omega)) \cap L^\infty(I; L^2(\Omega)), \partial_t c \in L^2(I; (H^1(\Omega))')$$

and satisfying Equation 12.486-12.492.

PROOF OF THEOREM 12.17

To prove Theorem 12.17 the regularization of Equations 12.486-12.488 will be used. The technique used will be similar to that of Nedoma (1995), and results of Blanc and Gasser (1996) will also be used. Firstly, we define the extension of functions T and c_L on R^2 by setting $T(x, y) = T(x^+, y^+)$ and $c_L(x, y) = c_L(x^+, y^+)$, where $r^+ = \max(r, 0)$. Then the regularization of T and c_L is as follows.

Let $\mathscr{D}(R^2)$ be the space of functions C^∞ with compact support in R^2, and let $\{R_\varepsilon\} \subset \mathscr{D}(R^2)$ be a family of mollifiers such that

$$\text{supp } (R_\varepsilon) \subset B((0, 0), \varepsilon) = \{(x, y) \in R^2 \mid x^2 + y^2 < \varepsilon^2\}, \varepsilon \in (0, 1\rangle.$$

The regularizations of T and c_L are defined by

$$T^\varepsilon(x, y) = \varepsilon y + \int_{R^2} R_\varepsilon(x + 2\varepsilon - r, y - \varepsilon - s) T(r^+, s^+) \mathrm{d}r\mathrm{d}s,$$

$$c_L^\varepsilon(x, y) = \int_{R^2} R_\varepsilon(x - \varepsilon - r, y - s) c_L(r^+, s^+) \mathrm{d}r\mathrm{d}s,$$

where $r^+ = \max(r, 0)$, $s^+ = \max(s, 0)$. These regularized functions Θ^ε, c_L^ε have the properties given in the next lemma (see Blanc and Gasser (1996), Gasser (1995)).

Lemma 12.18: Let assumptions (A1), (A2), (A4), (A5) be satisfied. Let α be the function satisfying (A2)(iii), (A4)(iii) and let $\varepsilon \in (0, 1\rangle$. Then the regularized functions T^ε and c_L^ε satisfy

(i) $T^\varepsilon(x, 0) = 0 \quad \forall x \in R,$

(ii) $c_L^\varepsilon(0, y) = 0 \quad \forall y \in R,$

(iii) there exists a function $C = C(\varepsilon)$ such that

$$\left|c_L^\varepsilon(x, y)\right| \le C\,|x| \quad \forall (x, y) \in R \times R,$$

(iv) there exists c_1 = constant, satisfying $(A4)(iii)$ such that

$$\left|c_L^\varepsilon(x, y)\right| \le c_1(\alpha''(x + \varepsilon))^{-1} \quad \forall (x, y) \in R_+ \times R,$$

where R_+ is a set of non-negative reals,

(v) there exists $c_2 =$ constant, independent of ε, such that

$$\frac{|\partial_1 T^\varepsilon(x, y)|}{\partial_2 T^\varepsilon(x, y)} \le c_2 \alpha''(x+\varepsilon) \quad \forall (x, y) \in R_+ \times R.$$

Proof: The proof of conditions (i)-(iv) follows from the definitions of T^ε and c_L^ε and from assumptions $(A2)$, $(A4)$.

To prove (i), then from the definition of T_m^ε we find

$$T^\varepsilon(x, 0) = \int_{R^2} R_\varepsilon(x + 2\varepsilon - r, -\varepsilon - s) T(r^+, s^+) \mathrm{d}r \mathrm{d}s = 0$$

as $T(r^+, s^+) > 0$ only for $s > 0$ and $R_\varepsilon(x - 2\varepsilon - r, -\varepsilon - s) = 0$.

To prove (ii) the same technique for $c_L^\varepsilon(0, y)$ can be used.

To prove (iii), We see that regularization c_L^ε is of class $C^1(R^2)$. Since $c_L^\varepsilon(0, y) = 0$ we can write

$$c_L^\varepsilon(x, y) = c_L^\varepsilon(x, y) - c_L^\varepsilon(0, y) = \int_0^x \partial_1 c_L^\varepsilon(r, y) \mathrm{d}r.$$

Hence

$$\left|c_L^\varepsilon(x, y)\right| = \left|\int_0^x \partial_1 c_L^\varepsilon(r, y)\mathrm{d}r\right| \le \int_0^x c \left\|\partial_1 c_L^\varepsilon(r, y)\right\|_{L^\infty(R^2)} \mathrm{d}r \le C\,|x|,$$

where $\left|\partial_1 c_L^\varepsilon(r, y)\right| \le c \left\|\partial_1 c_L^\varepsilon\right\|_{L^\infty(R^2)}$ depending on ε.

To prove (iv) we start from the definition of c_L^ε. The assumption (A4)(iii), applying to the integrand, was also used.

To prove (v), we assume that $\Omega_\varepsilon = B((x + 2\varepsilon, y - \varepsilon), \varepsilon) \cap (R_+^2 \setminus \cup_{i \in \mathcal{N}} O_i)$ and $(x, y) \in R_+ \times R$, where $B((x + 2\varepsilon, y - \varepsilon), \varepsilon)$ is the open ball with centre at $(x + 2\varepsilon, y - \varepsilon)$ and radius ε.

In the case when $\mathrm{meas}(\Omega_\varepsilon) = 0$ then since $\partial_1 T_\varepsilon(x, y) = 0$ the assertion trivially follows. In the case when $\mathrm{meas}(\Omega_\varepsilon) > 0$ then due to the definition of T^ε we have

$$\frac{|\partial_1 T^\varepsilon(x, y)|}{|\partial_2 T^\varepsilon(x, y)|} \le \frac{\int_{\Omega_\varepsilon} R_\varepsilon(x + 2\varepsilon - r, y - \varepsilon - s)\,|\partial_1 T(r, s)|\,\mathrm{d}r\mathrm{d}s}{\varepsilon + \int_{\Omega_\varepsilon} R_\varepsilon(x + 2\varepsilon - r, y - \varepsilon - s)\partial_2 T(r, s)\mathrm{d}r\mathrm{d}s} \equiv \frac{I_{1\varepsilon}}{\varepsilon + I_{2\varepsilon}}.$$

Since, due to the definition of the mollifier,

$$I_\varepsilon \equiv \int_\Omega R_\varepsilon(x + 2\varepsilon - r, y - \varepsilon - s)\mathrm{d}r\mathrm{d}s \le 1,$$

then

$$\frac{I_{1\varepsilon}}{\varepsilon + I_{2\varepsilon}} \le \frac{I_\varepsilon \ \mathrm{vrai} \max_{(x,y)\in R_+^2} |\partial_1 T(x, y)|}{\varepsilon + I_\varepsilon \ \mathrm{vrai} \max_{(x,y)\in\Omega_\varepsilon} \partial_2 T(x, y)} \le \frac{\mathrm{vrai} \max_{(x,y)\in R_+^2} |\partial_1 T(x, y)|}{\varepsilon + (\alpha''(x+\varepsilon))^{-1}}$$

$$\le \alpha''(x+\varepsilon) \ \mathrm{vrai} \max_{(x,y)\in R_+^2} |\partial_1 T(x, y)| \le c_2 \alpha''(x+\varepsilon),$$

which completes the proof.

Since $j(\boldsymbol{v}(t)) = 2\int_\Omega (D_{\mathrm{II}}(\boldsymbol{v}))^{1/2} \mathrm{d}\boldsymbol{x}$ is the non-differentiable functional and since $J(\boldsymbol{v}) = \int_\Omega \hat{g} j(\boldsymbol{v}(t))\mathrm{d}t$ then $J_\varepsilon(\boldsymbol{v})$ denotes a regularized functional $J(\boldsymbol{v})$, defined by

$$J_\varepsilon(\boldsymbol{v}) = \int_I \hat{g} j_\varepsilon(\boldsymbol{v}(t))\mathrm{d}t,$$

where

$$j_\varepsilon(\boldsymbol{v}(t)) = \frac{2}{2+\varepsilon}\int_\Omega D_{\mathrm{II}}(\boldsymbol{v}(t))^{1/2(1+\varepsilon)} \mathrm{d}\boldsymbol{x}, \ \varepsilon > 0,$$

for which

$$(J'_\varepsilon(\boldsymbol{v}), \boldsymbol{w}) = \int_{\Omega\times I} \hat{g}(D_{\mathrm{II}}(\boldsymbol{v}))^{1/2(\varepsilon-1)} D_{ij}(\boldsymbol{v}) D_{ij}(\boldsymbol{w}) \mathrm{d}\boldsymbol{x}\, \mathrm{d}t.$$

Then the weak regularized solution to Problem $(\mathscr{P}_{\mathrm{centh}})_v$ can be defined. The idea is to approximate the variational inequality by the variational equality.

Let assumptions (A1)-(A5) be satisfied for every $\varepsilon \in (0, 1\rangle$. Let $H^\varepsilon \in L^2(I; {}^1V) \cap H^1(I; H^{-1}(\Omega))$, $\boldsymbol{u}^\varepsilon \in L^\infty(I; H) \cap L^2(I; V)$, $c^\varepsilon \in L^2(I; {}^2V) \cap H^1(I; (H^1(\Omega))')$. Then the tetrad $\{H^\varepsilon, \Theta^\varepsilon, \boldsymbol{u}^\varepsilon, c^\varepsilon\}$ is a weak solution of the regularized problem to Problem $(\mathscr{P}_{\mathrm{centh}})_v$ if the following hold:

$$\int_I \{(\partial_t H^\varepsilon(t), z - \Theta^\varepsilon(t)) + b_\Theta(\boldsymbol{u}^\varepsilon(t), \Theta^\varepsilon(t), z - \Theta^\varepsilon(t)) + a_\Theta(\Theta^\varepsilon(t), z - \Theta^\varepsilon(t))$$
$$+ b_p(\boldsymbol{u}^\varepsilon(t), z - \Theta^\varepsilon(t)) - (Q(t), z - \Theta^\varepsilon(t))\} \mathrm{d}t \geq 0 \quad \forall z \in {}^1W, \tag{12.493}$$

$$\int_I \{(\partial_t \boldsymbol{u}^\varepsilon(t), \boldsymbol{v} - \boldsymbol{u}^\varepsilon(t)) + \hat{g}a(\boldsymbol{u}^\varepsilon(t), \boldsymbol{v} - \boldsymbol{u}^\varepsilon(t)) + b(\boldsymbol{u}^\varepsilon(t), \boldsymbol{u}^\varepsilon(t), \boldsymbol{v} - \boldsymbol{u}^\varepsilon(t)) + \hat{g} j_\varepsilon(\boldsymbol{v})$$
$$- \hat{g} j_\varepsilon(\boldsymbol{u}^\varepsilon(t)) + b_s(\Theta^\varepsilon(t) - \Theta_o, \boldsymbol{v} - \boldsymbol{u}^\varepsilon(t)) - (\boldsymbol{F}(t), \boldsymbol{v} - \boldsymbol{u}^\varepsilon(t))\} \mathrm{d}t \geq 0 \quad \forall \boldsymbol{v} \in W, \tag{12.494}$$

$$\int_I \{(\partial_t c^\varepsilon(t), d - c^\varepsilon(t)) + b_c(\boldsymbol{u}^\varepsilon(t), c_L^\varepsilon(t), d - c^\varepsilon(t)) + a_c(c^\varepsilon(t), d - c^\varepsilon(t))$$
$$- (\varphi(t), d - c^\varepsilon(t))\} \mathrm{d}t \geq 0, \quad \forall \mathrm{d} \in {}^2W, \tag{12.495}$$

$$H^\varepsilon(\boldsymbol{x}, t_o) = H_o(\boldsymbol{x}), \tag{12.496}$$
$$\boldsymbol{u}^\varepsilon(\boldsymbol{x}, t_o) = \boldsymbol{u}_o(\boldsymbol{x}), \tag{12.497}$$
$$c^\varepsilon(\boldsymbol{x}, t_o) = c_o(\boldsymbol{x}) \quad \text{a.e. in } \Omega, \tag{12.498}$$

where $\Theta^\varepsilon = T^\varepsilon(c^\varepsilon, H^\varepsilon)$.

The method of the proof is as follows:

(a) the existence of the solution of Equations 12.493-12.498 based on the Galerkin approximation will be proved,

(b) a priori estimates I and II independent of ε will be derived,

(c) limitation processes over m and ε will be given,

(d) limitation process $\hat{g} \to 0$ and the existence of multipliers.

The existence of $\{H^\varepsilon, \Theta^\varepsilon, \boldsymbol{u}^\varepsilon, c^\varepsilon\}$ will be proved by means of the finite-dimensional approximation. Let $\{\xi_j\}$, $\{\zeta_j\}$ be two orthogonal bases of the space $L^2(\Omega)$, composed of eigen-functions of the operator -Δ over the domain Ω, where the first is relative to the Dirichlet homogeneous boundary conditions and the second to the Neumann homogeneous boundary conditions. Let $\{\boldsymbol{w}_j\}$ be a countable basis of the space V composed of eigen-functions of the canonical isomorphism Λ of $V \to V'$, i.e. $((\boldsymbol{w}_j, \boldsymbol{v})) = \lambda_j(\boldsymbol{w}_j, \boldsymbol{v}) \forall \boldsymbol{v} \in V$, $|\boldsymbol{w}_j| = 1$, i.e. each finite subset of bases $\{\xi_j\}$, $\{\zeta_j\}$ and $\{\boldsymbol{w}_j\}$ are linearly independent and span $\{\xi_j \mid j = 1, 2, \ldots\}$, span $\{\boldsymbol{w}_j \mid j = 1, 2, \ldots\}$ and span $\{\zeta_j \mid j = 1, 2, \ldots.\}$ are dense in $H_o^1(\Omega)$, or in V, or in $H^1(\Omega)$, respectively, as $H_o^1(\Omega), V, H^1(\Omega)$ are separable spaces. Let ${}^1V^m$, V^m, ${}^2V^m$ be spaces spanned by $\{\xi_j \mid$

$1 \le j \le m\}$, $\{\boldsymbol{w}_j \mid 1 \le j \le m\}$ and $\{\zeta_j \mid 1 \le j \le m\}$, respectively. Then the approximations $H_m^\varepsilon \in {}^1V^m$, $\boldsymbol{u}_m^\varepsilon \in V^m$, $c_m^\varepsilon \in {}^2V^m$ of order m satisfy

$$(\partial_t H_m^\varepsilon(t), z_j) + b_\Theta(\boldsymbol{u}_m^\varepsilon(t), \Theta_m^\varepsilon(t), z_j) + a_\Theta(\Theta_m^\varepsilon(t), z_j) + b_p(\boldsymbol{u}_m^\varepsilon(t), z_j) - (Q(t), z_j) = 0, \quad 1 \le j \le m, \tag{12.499}$$

$$(\partial_t \boldsymbol{u}_m^\varepsilon(t), \boldsymbol{v}_j) + \hat{\mu} a(\boldsymbol{u}_m^\varepsilon(t), \boldsymbol{v}_j) + b(\boldsymbol{u}_m^\varepsilon(t), \boldsymbol{u}_m^\varepsilon(t), \boldsymbol{v}_j) + \hat{g}(j_\varepsilon(\boldsymbol{u}_m^\varepsilon(t)), \boldsymbol{v}_j) + b_s(\Theta_m^\varepsilon(t) - \Theta_o, \boldsymbol{v}_j) - (\boldsymbol{F}(t), \boldsymbol{v}_j) = 0, \quad 1 \le j \le m, \tag{12.500}$$

$$(\partial_t c_m^\varepsilon(t), d_j) + b_c(\boldsymbol{u}_m^\varepsilon(t), c_m^\varepsilon(t), d_j) + a_c(c_m^\varepsilon(t), d_j) - (\varphi(t), d_j) = 0, \quad 1 \le j \le m, \tag{12.501}$$

verified for almost every $t \in I$, for every $z \in {}^1V^m$, $\boldsymbol{v} \in V^m$, $\mathrm{d} \in {}^2V^m$ and the initial conditions

$$H_m^\varepsilon(\boldsymbol{x}, t_o) = \sum_{i=1}^m (H_o, \xi_i)\xi_i(\boldsymbol{x}), \tag{12.502}$$

$$\boldsymbol{u}_m^\varepsilon(\boldsymbol{x}, t_o) = \sum_{i=1}^m (\boldsymbol{u}_o, \boldsymbol{w}_i)\boldsymbol{w}_i(\boldsymbol{x}), \tag{12.503}$$

$$c_m^\varepsilon(\boldsymbol{x}, t_o) = \sum_{i=1}^m (c_o, \zeta_i)\zeta_i(\boldsymbol{x}), \tag{12.504}$$

verified for a.e. $\boldsymbol{x} \in \Omega$. Since $\{\xi_j\}_{j=1}^m$, $\{\boldsymbol{w}_j\}_{j=1}^m$, $\{\zeta_j\}_{j=1}^m$ are linearly independent, the system in Equations 12.499-12.504 is a regular system of ordinary differential equations of first order, and therefore Equations 12.499-12.504 uniquely define $\{H_m^\varepsilon, \Theta_m^\varepsilon, \boldsymbol{u}_m^\varepsilon, c_m^\varepsilon\}$ on the interval $I_m = \langle t_o, t_m \rangle$. Therefore, Equations 12.499-12.504 valid for every test functions

$$z(t) = \sum_{i=1}^m a_i(t)\xi_i, \quad \boldsymbol{v}(t) = \sum_{i=1}^m b_i(t)\boldsymbol{w}_i, \quad \mathrm{d}(t) = \sum_{i=1}^m \gamma_i(t)\mathrm{d}_i, \quad t \in I_m,$$

where a_i, b_i, γ_i are continuously differentiable functions on I_m, $i = 1, \ldots, m$. Particularly, it holds for $z(t) = H_m^\varepsilon(t)$, $v(t) = \boldsymbol{u}_m^\varepsilon(t)$, $\mathrm{d}(t) = c_m^\varepsilon(t)$, $t \in I_m$.

We now consider the a priori estimates I and II: firstly a priori estimates I.

Let us introduce the notation

$$X_m = b_s(\Theta_m^\varepsilon - \Theta_o, \boldsymbol{u}_m^\varepsilon) + b_p(\boldsymbol{u}_m^\varepsilon, \Theta_m^\varepsilon).$$

Let $\partial\beta_{ij}/\partial x_j \equiv \partial_j\beta_{ij} \in L^\infty(\Omega)\, \forall i, j$. Then there exists a positive constant c, independent of m and ε, such that

$$|X_m| = \left|b_s(\Theta_m^\varepsilon - \Theta_o, \boldsymbol{u}_m^\varepsilon) + b_p(\boldsymbol{u}_m^\varepsilon, \Theta_m^\varepsilon)\right| \le c\left(1 + \left\|\Theta_m^\varepsilon(t)\right\|_{1,1} \left\|\boldsymbol{u}_m^\varepsilon(t)\right\|_{0,2} + \left\|\Theta_m^\varepsilon(t)\right\|_{0,1} \left\|\boldsymbol{u}_m^\varepsilon(t)\right\|_{1,2}\right). \tag{12.505}$$

We shall denote all used constants by type c without any indices (if possible). Via the integration of Equations 12.502-12.504 with $z = H_m^\varepsilon(t)$, $\boldsymbol{v}(t) = \boldsymbol{u}_m^\varepsilon(t)$, $d(t) = c_m^\varepsilon(t)$ in time t over $I_m = (t_o, t_m)$, we obtain

$$\int_{I_m} \{(\partial_t H_m^\varepsilon(t), H_m^\varepsilon(t)) + (\partial_t \boldsymbol{u}_m^\varepsilon(t), \boldsymbol{u}_m^\varepsilon(t)) + (\partial_t c_m^\varepsilon(t), c_m^\varepsilon(t))$$

$$+ a_{\Theta}(\Theta_m^{\varepsilon}(t), H_m^{\varepsilon}(t)) + \hat{\mu} a(\boldsymbol{u}_m^{\varepsilon}(t), \boldsymbol{u}_m^{\varepsilon}(t)) + a_c(c_m^{\varepsilon}(t), c_m^{\varepsilon}(t))$$
$$+ b_{\Theta}(\boldsymbol{u}_m^{\varepsilon}(t), \Theta_m^{\varepsilon}(t), H_m^{\varepsilon}(t)) + b(\boldsymbol{u}_m^{\varepsilon}(t), \boldsymbol{u}_m^{\varepsilon}(t), \boldsymbol{u}_m^{\varepsilon}(t)) + b_c(\boldsymbol{u}_m^{\varepsilon}(t), c_m^{\varepsilon}(t), c_m^{\varepsilon}(t))$$
$$+ b_s(\Theta_m^{\varepsilon}(t) - \Theta_o, \boldsymbol{u}_m^{\varepsilon}(t)) + b_p(\boldsymbol{u}_m^{\varepsilon}(t), \Theta_m^{\varepsilon}(t)) + \hat{g}(j'_{\varepsilon}(\boldsymbol{u}_m^{\varepsilon}(t)), \boldsymbol{u}_m^{\varepsilon}(t))$$
$$- (Q(t), H_m^{\varepsilon}(t)) - (\boldsymbol{F}(t), \boldsymbol{u}_m^{\varepsilon}(t)) - (\varphi(t), c_m^{\varepsilon}(t))\}\mathrm{d}t = 0.$$

We find

$$b(\boldsymbol{u}_m^{\varepsilon}(t), \boldsymbol{u}_m^{\varepsilon}(t), \boldsymbol{u}_m^{\varepsilon}(t)) = 0,$$
$$(J'_{\varepsilon}(\boldsymbol{u}_m^{\varepsilon}(t)), \boldsymbol{u}_m^{\varepsilon}(t)) = \int_{I_m} \hat{g}(j'_{\varepsilon}(\boldsymbol{u}_m^{\varepsilon}(t)), \boldsymbol{u}_m^{\varepsilon}(t))\mathrm{d}t \geq 0, \quad \text{as } (j'_{\varepsilon}(\boldsymbol{v}), \boldsymbol{v}) \geq 0.$$
$$\int_{I_m} (\partial_t H_m^{\varepsilon}(t), H_m^{\varepsilon}(t))\mathrm{d}t = \frac{1}{2}\int_{\Omega} (H_m^{\varepsilon}(t))^2 \mathrm{d}\boldsymbol{x} - \frac{1}{2}\int_{\Omega} (H_m^{\varepsilon}(t_o))^2 \mathrm{d}\boldsymbol{x}$$

and similarly for the remaining terms.

Since $\Theta^{\varepsilon} = T^{\varepsilon}(c_m^{\varepsilon}, H_m^{\varepsilon})$, then using Green's theorem, and after modification of grad T^{ε} by the chain rule, using (A2)(i)(ii), $\partial_2 T^{\varepsilon} \geq \varepsilon$ and the resulting estimate $|T^{\varepsilon}(x, y)| \leq \|\partial_2 T^{\varepsilon}\|_{L^{\infty}(R^2)} |y|$, we obtain

$$\int_{I_m} (a_{\Theta}(\Theta_m^{\varepsilon}(t), H_m^{\varepsilon}(t)) + b_{\Theta}(\boldsymbol{u}_m^{\varepsilon}(t), \Theta_m^{\varepsilon}(t), H_m^{\varepsilon}(t))\mathrm{d}t$$
$$= \int_{I_m}\int_{\Omega} \text{grad } T^{\varepsilon}(c_m^{\varepsilon}(t), H_m^{\varepsilon}(t)) \text{grad } H_m^{\varepsilon}(t) \mathrm{d}\boldsymbol{x}\,\mathrm{d}t$$
$$+ \int_{I_m}\int_{\Omega} \boldsymbol{u}_m^{\varepsilon}(t) \text{grad } T^{\varepsilon}(c_m^{\varepsilon}(t), H_m^{\varepsilon}(t)) H_m^{\varepsilon}(t) \mathrm{d}\boldsymbol{x}\,\mathrm{d}t$$
$$= \int_{I_m}\int_{\Omega} \text{grad } T^{\varepsilon}(c_m^{\varepsilon}(t), H_m^{\varepsilon}(t)) \text{grad } H_m^{\varepsilon}(t) \mathrm{d}\boldsymbol{x}\,\mathrm{d}t$$
$$- \int_{I_m}\int_{\Omega} T^{\varepsilon}(c_m^{\varepsilon}(t), H_m^{\varepsilon}(t)) \boldsymbol{u}_m^{\varepsilon}(t) \text{grad } H_m^{\varepsilon}(t) \mathrm{d}\boldsymbol{x}\,\mathrm{d}t$$
$$\leq \varepsilon \int_{I_m}\int_{\Omega} \left\|\text{grad } H_m^{\varepsilon}(t)\right\|_{1,1}^2 \mathrm{d}\boldsymbol{x}\,\mathrm{d}t - \left\|\partial_2 T^{\varepsilon}(t)\right\|_{L^{\infty}(R^2)} \int_{I_m}\int_{\Omega} \left|H_m^{\varepsilon}(t)\right|$$
$$\times \left|\boldsymbol{u}_m^{\varepsilon}(t) \text{grad } H_m^{\varepsilon}(t)\right| \mathrm{d}\boldsymbol{x}\,\mathrm{d}t - \int_{I_m}\int_{\Omega} \left|\partial_1 T^{\varepsilon}(c_m^{\varepsilon}(t), H_m^{\varepsilon}(t))\right|$$
$$\times \left|\text{grad } c_m^{\varepsilon}(t) \text{grad } H_m^{\varepsilon}(t)\right| \mathrm{d}\boldsymbol{x}\,\mathrm{d}t.$$

Likewise we estimate $\int_{I_m} (a_c(c_m^{\varepsilon}(t), c_m^{\varepsilon}(t)) + b_c(\boldsymbol{u}_m^{\varepsilon}(t), c_m^{\varepsilon}(t), c_m^{\varepsilon}(t)))\mathrm{d}t$. We have the estimate

$$\left\|\partial_2 T^{\varepsilon}(t)\right\|_{L^{\infty}(R^2)} \int_{I_m}\int_{\Omega} \left|H_m^{\varepsilon}(t)\right| \left|\boldsymbol{u}_m^{\varepsilon}(t) \text{grad } H_m^{\varepsilon}(t)\right| \mathrm{d}\boldsymbol{x}\,\mathrm{d}t$$
$$\leq \left\|\partial_2 T^{\varepsilon}(t)\right\|_{L^{\infty}(R^2)} \int_{I_m}\int_{\Omega} \left|H_m^{\varepsilon}(t)\right| \left\|\boldsymbol{u}_m^{\varepsilon}(t)\right\| \left\|\text{grad } H_m^{\varepsilon}(t)\right\| \mathrm{d}\boldsymbol{x}\,\mathrm{d}t$$

$$\leq \gamma_o \left\| \boldsymbol{u}_m^\varepsilon(t) \right\|_{L^\infty(\Omega\times I)} \int_{I_m} \int_\Omega \left| H_m^\varepsilon(t) \right| \left\| \text{grad } H_m^\varepsilon(t) \right\| \mathrm{d}\boldsymbol{x}\,\mathrm{d}t$$

$$\leq \gamma_1^2 \varepsilon^{-1} \int_{I_m} \int_\Omega (H_m^\varepsilon(t))^2 \mathrm{d}\boldsymbol{x}\,\mathrm{d}t + \frac{1}{4}\varepsilon \int_{I_m} \int_\Omega \left\| \text{grad } H_m^\varepsilon(t) \right\|^2 \mathrm{d}\boldsymbol{x}\,\mathrm{d}t,$$

where $\gamma_1 = \gamma_o \left\| \boldsymbol{u}_m^\varepsilon(t) \right\|_{L^\infty(\Omega\times I)}$, γ_o is the Lipschitz constant and the Young's inequality was applied. Similarly,

$$\int_{I_m} \int_\Omega \left| \partial_1 T^\varepsilon(c_m^\varepsilon(t), H_m^\varepsilon(t)) \right| \left| \text{grad } c_m^\varepsilon(t) \text{grad } H_m^\varepsilon(t) \right| \mathrm{d}\boldsymbol{x}\,\mathrm{d}t$$

$$\leq \left\| \partial_1 T^\varepsilon(c_m^\varepsilon(t), H_m^\varepsilon(t)) \right\|_{L^\infty(R^2)} \int_{I_m} \int_\Omega \left\| \text{grad } c_m^\varepsilon(t) \right\| \left\| \text{grad } H_m^\varepsilon(t) \right\| \mathrm{d}\boldsymbol{x}\,\mathrm{d}t$$

$$\leq \gamma_1^2 \varepsilon^{-1} \int_{I_m} \int_\Omega \left\| \text{grad } c_m^\varepsilon(t) \right\| \mathrm{d}\boldsymbol{x}\,\mathrm{d}t + \frac{1}{4}\varepsilon \int_{I_m} \int_\Omega \left\| \text{grad } H_m^\varepsilon(t) \right\|^2 \mathrm{d}\boldsymbol{x}\,\mathrm{d}t.$$

Moreover,

$$\int_{I_m} (\boldsymbol{F}(t), \boldsymbol{u}_m^\varepsilon(t))\mathrm{d}t \leq \int_{I_m} [\|\boldsymbol{f}(t)\|_* \left\| \boldsymbol{u}_m^\varepsilon(t) \right\|_{1,2} + \|\boldsymbol{P}_o(t)\|_* \left\| \boldsymbol{u}_m^\varepsilon(t) \right\|_{1,2}]\mathrm{d}t,$$

where $\|\boldsymbol{f}\|_* = \sup_{\boldsymbol{v}\in V} |(\boldsymbol{f}, \boldsymbol{v})| \, \|\boldsymbol{v}\|^{-1}$ denotes the norm in V', and furthermore,

$$2\int_{I_m} \|\boldsymbol{F}(t)\|_* \left\| \boldsymbol{u}_m^\varepsilon(t) \right\|_{1,2} \mathrm{d}t \leq c\hat{\mu} \int_{I_m} \left\| \boldsymbol{u}_m^\varepsilon(t) \right\|_{1,2}^2 \mathrm{d}t + (c\hat{\mu})^{-1} \int_{I_m} \|\boldsymbol{F}(t)\|_*^2 \mathrm{d}t.$$

Similarly, we estimate $\int_{I_m} (Q(t), H_m^\varepsilon(t))\mathrm{d}t$ and $\int_{I_m} (\varphi(t), c_m^\varepsilon(t))\mathrm{d}t$. Summing all the above obtained estimates, using Equation 12.505 and the estimates of bilinear and trilinear forms, and applying the Gronwall's lemma, then after some modifications we find the estimates

$$\left.\begin{aligned} &\left\| H_m^\varepsilon(t) \right\|_{0,1} \leq c_o(\varepsilon), && t \in I_m, && \int_{I_m} \left\| H_m^\varepsilon(t) \right\|_{1,1} \mathrm{d}t \leq c_o(\varepsilon), \\ &\left\| \boldsymbol{u}_m^\varepsilon(t) \right\|_{0,2} \leq c, && t \in I_m, && \int_{I_m} \left\| \boldsymbol{u}_m^\varepsilon(t) \right\|_{1,2}^2 \mathrm{d}t \leq c, \\ &\left\| c_m^\varepsilon(t) \right\|_{0,1} \leq c, && t \in I_m, && \int_{I_m} \left\| c_m^\varepsilon(t) \right\|_{1,1} \mathrm{d}t \leq c, \ \varepsilon \in (0,1\rangle, \end{aligned}\right\} \tag{12.506}$$

where constants c are independent of m and ε and c_o is independent of m.

From these estimates we obtain

$$\{H_m^\varepsilon(t), m \in \mathcal{N}, \varepsilon \in (0,1\rangle\} \text{ is a bounded subset in } L^2(I; {}^1V) \cap L^\infty(I; L^2(\Omega)), \tag{12.507}$$

$$\{\boldsymbol{u}_m^\varepsilon(t), m \in \mathcal{N}, \varepsilon \in (0,1\rangle\} \text{ is a bounded subset in } L^2(I; V) \cap L^\infty(I; H(\Omega)), \tag{12.508}$$

$$\{c_m^\varepsilon(t), m \in \mathcal{N}, \varepsilon \in (0,1\rangle\} \text{ is a bounded subset in } L^2(I; {}^2V) \cap L^\infty(I; L^2(\Omega)). \tag{12.509}$$

Now consider the a priori estimate II. We shall show that

$$\{\partial_t H_m^\varepsilon(t),\, m \in \mathcal{N},\, \varepsilon \in (0,1\rangle\} \text{ is a bounded subset in } L^2(I; H^{-1}(\Omega)), \quad (12.510)$$

$$\{\partial_t \boldsymbol{u}_m^\varepsilon(t),\, m \in \mathcal{N},\, \varepsilon \in (0,1\rangle\} \text{ is a bounded subset in } L^2(I; V'), \quad (12.511)$$

$$\{\partial_t c_m^\varepsilon(t),\, m \in \mathcal{N},\, \varepsilon \in (0,1\rangle\} \text{ is a bounded subset in } L^2(I; (H^1(\Omega))'), \quad (12.512)$$

independent of m and ε.

By virtue of the estimates given above

$$\left|b(\boldsymbol{u}_m^\varepsilon(t), \boldsymbol{u}_m^\varepsilon(t), \boldsymbol{v})\right| \leq c \left\|\boldsymbol{u}_m^\varepsilon(t)\right\|_{1,2} \left\|\boldsymbol{u}_m^\varepsilon(t)\right\|_{0,2} \|\boldsymbol{v}\|_{1,2}$$

and due to Equation 12.506

$$\left|b(\boldsymbol{u}_m^\varepsilon(t), \boldsymbol{u}_m^\varepsilon(t), \boldsymbol{v})\right| \leq c \left\|\boldsymbol{u}_m^\varepsilon(t)\right\|_{1,2} \|\boldsymbol{v}\|_{1,2}\,.$$

Then

$$b(\boldsymbol{u}_m^\varepsilon(t), \boldsymbol{u}_m^\varepsilon(t), \boldsymbol{v}) = (h_{um}(t), \boldsymbol{v}), \quad \forall \boldsymbol{v} \in V,$$

and similarly,

$$b_\Theta(\boldsymbol{u}_m^\varepsilon(t), \Theta_m^\varepsilon(t), z) = (h_{\Theta m}(t), \boldsymbol{v}), \quad \forall z \in {}^1V,$$

$$b_c(\boldsymbol{u}_m^\varepsilon(t), c_m^\varepsilon(t), d) = (h_{cm}(t), d), \quad \forall d \in {}^2V,$$

$h_{um}, h_{\Theta m}, h_{cm}$ remaining in bounded sets of $L^2(I; V')$ or $L^2(I; {}^1V')$ or $L^2(I; {}^2V')$, respectively. The coupled terms are from $L^2(\Omega)$ or $[L^2(\Omega)]^2$, and therefore are included in Q and $\boldsymbol{F}$.

Linear form $\Theta \to a_\Theta(\Theta, z)$ for a fixed $z \in {}^1V$ is continuous on ${}^1V$ so that

$$a_\Theta(\Theta(t), z) = (A_\Theta \Theta(t), z), \quad A_\Theta \in \mathcal{L}({}^1V, {}^1V'). \quad (12.513)$$

Similarly, forms $\boldsymbol{u} \to a(\boldsymbol{u}, \boldsymbol{v})$ and $c \to a_c(c, w)$ for fixed $\boldsymbol{u} \in V$ or $c \in {}^2V$ are continuous on V or ${}^2V$ respectively, so that

$$a(\boldsymbol{u}(t), \boldsymbol{v}) = (A_u \boldsymbol{u}(t), \boldsymbol{v}), \quad A_u \in \mathcal{L}(V, V'). \quad (12.514)$$

$$a_c(c(t), \mathrm{d}) = (A_c c(t), \mathrm{d}), \quad A_c \in \mathcal{L}({}^2V, {}^2V'). \quad (12.515)$$

Then Equations 12.499–12.501 are equivalent to

$$(\partial_t H_m^\varepsilon + A_\Theta \Theta_m^\varepsilon + h_{\Theta m} - Q, z_j) = 0, \quad 1 \leq j \leq m, \quad (12.516)$$

$$(\partial_t \boldsymbol{u}_m^\varepsilon + \hat{\mu} A_u \boldsymbol{u}_m^\varepsilon + h_{um} + \hat{g} j'_\varepsilon(\boldsymbol{u}_m^\varepsilon) - \boldsymbol{F}, \boldsymbol{v}_j) = 0, \quad 1 \leq j \leq m, \quad (12.517)$$

$$(\partial_t c_m^\varepsilon + A_c c_m^\varepsilon + h_{cm} - \varphi, \mathrm{d}_j) = 0, \quad 1 \leq j \leq m. \quad (12.518)$$

Let $S_{\Theta m}$ be the orthogonal projection $L^2(\Omega) \to W_\Theta^m = \mathrm{span}\{z_j \mid 1 \leq j \leq m\}$,

S_{um} be the orthogonal projection $H \to W^m = \mathrm{span}\{\boldsymbol{v}_j \mid 1 \leq j \leq m\}$,

S_{cm} be the orthogonal projection $L^2(\Omega) \to W_c^m = \mathrm{span}\{\mathrm{d}_j \mid 1 \leq j \leq m\}$,

then

$$S_{\Theta m} h_\Theta = \sum_{j=1}^m (h_\Theta, z_j) z_j,\; S_{um} \boldsymbol{h}_u = \sum_{j=1}^m (\boldsymbol{h}_u, \boldsymbol{v}_j) \boldsymbol{v}_j,\; S_{cm} h_c = \sum_{j=1}^m (h_c, \mathrm{d}_j) \mathrm{d}_j,$$

where $\{z_j\}$, $\{\boldsymbol{v}_j\}$, $\{\mathrm{d}_j\}$ are orthogonal bases of W^m_Θ, W^m, W^m_c, respectively. Then from Equations 12.516-12.518 and from the fact that $S_{\Theta m}\partial_t H^\varepsilon_m = \partial_t H^\varepsilon_m$, $S_{um}\partial_t \boldsymbol{u}^\varepsilon_m = \partial_t \boldsymbol{u}^\varepsilon_m$, $S_{cm}\partial_t c^\varepsilon_m = \partial_t c^\varepsilon_m$, we obtain

$$\partial_t H^\varepsilon_m = S_{\Theta m}(Q - A_\Theta \Theta^\varepsilon_m - h_{\Theta m}), \tag{12.519}$$

$$\partial_t \boldsymbol{u}^\varepsilon_m = S_{um}(\boldsymbol{F} - \hat{\mu} A_u \boldsymbol{u}^\varepsilon_m - \hat{g} j'_\varepsilon(\boldsymbol{u}^\varepsilon_m) - h_{um}), \tag{12.520}$$

$$\partial_t c^\varepsilon_m = S_{cm}(\varphi - A_c c^\varepsilon_m - h_{cm}). \tag{12.521}$$

Due to Equations 12.506-12.509, $\{A_\Theta \Theta^\varepsilon_m\}$, $\{A_u \boldsymbol{u}^\varepsilon_m\}$, $\{A_c c^\varepsilon_m\}$ are bounded subsets of $L^2(I; {}^1V')$, $L^2(I; V')$, $L^2(I; {}^2V')$, respectively.

Due to the definition

$$(J'_\varepsilon(\boldsymbol{v}), \boldsymbol{w}) = \int_{\Omega\times I} \hat{g} D_{\mathrm{II}}(\boldsymbol{v})^{1/2(\varepsilon-1)} D_{ij}(\boldsymbol{v}) D_{ij}(\boldsymbol{w}) \mathrm{d}\boldsymbol{x}\mathrm{d}t, \ \left\| j'_\varepsilon(\boldsymbol{v}) \right\|_* \le c \left(\int_\Omega (D_{\mathrm{II}}(\boldsymbol{v}))^\varepsilon \mathrm{d}\boldsymbol{x} \right)^{1/2},$$

where $\|\cdot\|_*$ denotes the dual norm, then $j'_\varepsilon(\boldsymbol{u}^\varepsilon_m)$ is a bounded subset of $L^2(I; V')$. Thus Equations 12.519-12.521 indicate that $\partial_t H^\varepsilon_m = S_{\Theta m} p_{\Theta m}$, where $p_{\Theta m} \in \mathscr{P}_{\Theta p} \subset L^2(I; H^{-1}(\Omega))$, and $\mathscr{P}_{\Theta p}$ is a bounded subset of $L^2(I; H^{-1}(\Omega))$. Similarly, $\partial_t \boldsymbol{u}^\varepsilon_m = S_{um} \boldsymbol{p}_{um}$, where $\boldsymbol{p}_{um} \in \mathscr{P}_{up} \subset L^2(I; V')$, and $\mathscr{P}_{up}$ is a bounded subset of $L^2(I; V')$, and finally, $\partial_t c^\varepsilon_m = S_{cm} p_{cm}$, where $p_{cm} \in \mathscr{P}_{cp} \subset L^2(I; L^2(\Omega))$, and $\mathscr{P}_{cp}$ is a bounded subset of $L^2(I; (H^1(\Omega))')$. Moreover, due to the properties of eigenfunctions, then $\|S_{\Theta m} p_{\Theta m}\|_{{}^1V'} \le c \, \|p_{\Theta m}\|_{{}^1V'}$; $\|S_{um} \boldsymbol{p}_{um}\|_{V'} \le c \, \|\boldsymbol{p}_{um}\|_{V'}$; $\|S_{cm} p_{cm}\|_{{}^2V'} \le c \, \|p_{cm}\|_{{}^2V'}$, which completes this part of the proof.

We now turn to the passages to the limit over m. We shall prove the convergence of the finite-dimensional approximation for ε being fixed.

From the a priori estimates I and II, and from Equations 12.507-12.512 the subsequences $\{H^\varepsilon_\mu(t), \mu \in \mathcal{N}\}$, $\{\boldsymbol{u}^\varepsilon_\mu(t), \mu \in \mathcal{N}\}$, and $\{c^\varepsilon_\mu(t), \mu \in \mathcal{N}\}$ of the sequences $\{H^\varepsilon_m(t), m \in \mathcal{N}\}$, $\{\boldsymbol{u}^\varepsilon_m(t), m \in \mathcal{N}\}$ and $\{c^\varepsilon_m(t), m \in \mathcal{N}\}$, respectively, can be taken such that

$$H^\varepsilon_\mu \to H^\varepsilon \text{ in } L^2(I; L^2(\Omega)) \text{ strongly}, \tag{12.522}$$

$$H^\varepsilon_\mu \to H^\varepsilon \text{ in } L^2(I; H^1_o(\Omega)) \text{ weakly}, \tag{12.523}$$

$$H^\varepsilon_\mu \to H^\varepsilon \text{ in } H^1(I; H^{-1}(\Omega)) \text{ weakly}, \tag{12.524}$$

$$\boldsymbol{u}^\varepsilon_\mu \to \boldsymbol{u}^\varepsilon \text{ in } L^\infty(I; H) \text{ *-weakly (weakly star)}, \tag{12.525}$$

$$\boldsymbol{u}^\varepsilon_\mu \to \boldsymbol{u}^\varepsilon \text{ in } L^2(I; V) \text{ weakly}, \tag{12.526}$$

$$\boldsymbol{u}^\varepsilon_\mu \to \boldsymbol{u}^\varepsilon \text{ in } L^2(I; H) \text{ strongly}, \tag{12.527}$$

$$\partial_t \boldsymbol{u}^\varepsilon_\mu \to \partial_t \boldsymbol{u}^\varepsilon \text{ in } L^2(I; V') \text{ weakly}, \tag{12.528}$$

$$c^\varepsilon_\mu \to c^\varepsilon \text{ in } L^2(I; L^2(\Omega)) \text{ strongly}, \tag{12.529}$$

$$c^\varepsilon_\mu \to c^\varepsilon \text{ in } L^2(I; H^1(\Omega)) \text{ weakly}, \tag{12.530}$$

$$c^\varepsilon_\mu \to c^\varepsilon \text{ in } L^2(I; (H^1(\Omega))') \text{ weakly}, \tag{12.531}$$

where $H^{\varepsilon} \in L^2(I; H_o^1(\Omega)) \cap H^1(I; H^{-1}(\Omega))$, $\boldsymbol{u}^{\varepsilon} \in L^2(I; V) \cap L^{\infty}(I; H)$,

$$c^{\varepsilon} \in L^2(I; H^1(\Omega)) \cap H^1(I; (H(\Omega))').$$

For the appropriate components $u_{i\mu}^{\varepsilon}$ of $\boldsymbol{u}_{\mu}^{\varepsilon}$ we have

$$u_{i\mu}^{\varepsilon} \to u_i^{\varepsilon} \text{ weakly a.e. in } \Omega \times I, \tag{12.532}$$

since $\boldsymbol{u}_{\mu}^{\varepsilon} \to \boldsymbol{u}^{\varepsilon}$ in $L^2(I; H)$ strongly. Furthermore, $\{j'_{\varepsilon}(\boldsymbol{u}_m^{\varepsilon})\}$ and $\{u_{i\mu}^{\varepsilon} u_{j\mu}^{\varepsilon}\}$, due to the estimate $\left\| \boldsymbol{u}_{\mu}^{\varepsilon} \right\|_{[L^p(\Omega)]^2} \leq c \left\| \boldsymbol{u}_{\mu}^{\varepsilon} \right\|_{1,2}^{1/2} \left\| \boldsymbol{u}_{\mu}^{\varepsilon} \right\|_{[L^2(\Omega)]^2}^{1/2}$ $\forall \boldsymbol{u} \in [H_o^1(\Omega)]^2$, and Equation 12.508, are bounded subsets of spaces $L^2(I; V')$ and $L^2(I; [L^{p/2}(\Omega)]^2)$, respectively. Then we can also assume that

$$j'_{\varepsilon}(\boldsymbol{u}_{\mu}^{\varepsilon}) \to \chi_o \text{ weakly in } L^2(I; V'), \tag{12.533}$$

$$u_{i\mu}^{\varepsilon} u_{j\mu}^{\varepsilon} \to \Xi_{ij} \text{ weakly in } L^2(I; L^{p/2}(\Omega)). \tag{12.534}$$

From Equation 12.532, $u_{i\mu}^{\varepsilon} u_{j\mu}^{\varepsilon} \to u_i^{\varepsilon} u_j^{\varepsilon}$ in the sense of distributions in $\Omega \times I$, which comparing with Equation 12.534 then gives

$$\Xi_{ij} = u_i^{\varepsilon} u_j^{\varepsilon}.$$

But

$$b(\boldsymbol{u}_{\mu}^{\varepsilon}, \boldsymbol{u}_{\mu}^{\varepsilon}, \boldsymbol{w}_j) = -b(\boldsymbol{u}_{\mu}^{\varepsilon}, \boldsymbol{w}_j, \boldsymbol{u}_{\mu}^{\varepsilon}) \to -b(\boldsymbol{u}_{\mu}^{\varepsilon}, \boldsymbol{w}_j, \boldsymbol{u}_{\mu}^{\varepsilon}) \text{ weakly in } L^2(I) \quad \forall \boldsymbol{w}_j,$$

and similarly for the other trilinear forms.

Then from Equations 12.499–12.501 for $m = \mu$ we obtain

$$\int_{\mathrm{I}} \{(\partial_t H^{\varepsilon}(t), z_j(t)) + b_{\Theta}(\boldsymbol{u}^{\varepsilon}(t), \Theta^{\varepsilon}(t), z_j(t)) + a_{\Theta}(\Theta^{\varepsilon}(t), z_j(t)) + b_p(\boldsymbol{u}^{\varepsilon}(t), z_j(t))$$
$$- (Q(t), z_j(t))\} \mathrm{d}t = 0 \quad \forall j, \tag{12.535}$$

$$\int_{\mathrm{I}} \{(\partial_t \boldsymbol{u}^{\varepsilon}(t), \boldsymbol{v}_j(t)) + b(\boldsymbol{u}^{\varepsilon}(t), \boldsymbol{u}^{\varepsilon}(t), \boldsymbol{v}_j(t)) + \hat{\mu} a(\boldsymbol{u}^{\varepsilon}(t), \boldsymbol{v}_j(t)) + b_s(\Theta^{\varepsilon}(t) - \Theta_o, \boldsymbol{v}_j(t))$$
$$- (\boldsymbol{F}(t), \boldsymbol{v}_j(t))\} \mathrm{d}t + (\chi(t), \boldsymbol{v}_j(t)) = 0 \quad \forall j, \tag{12.536}$$

$$\int_{\mathrm{I}} \{(\partial_t c^{\varepsilon}(t), \mathrm{d}_j(t)) + b_c(\boldsymbol{u}^{\varepsilon}(t), c_L^{\varepsilon}(t), \mathrm{d}_j(t)) + a_c(c^{\varepsilon}(t)), \mathrm{d}_j(t))$$
$$- (\varphi(t), \mathrm{d}_j(t))\} \mathrm{d}t = 0 \quad \forall j, \tag{12.537}$$

where we denote $\chi(t) = J'_{\varepsilon}(\boldsymbol{u}^{\varepsilon}(t))$.

But the systems of functions $\{z_j\}$, $\{\boldsymbol{v}_j\}$, $\{w_j\}$ are complete in ${}^1V$, V, ${}^2V$, respectively, so that from Equations 12.535–12.537 we obtain

$$\int_I \{(\partial_t H^{\varepsilon}(t), z(t)) + b_{\Theta}(\boldsymbol{u}^{\varepsilon}(t), \Theta^{\varepsilon}(t), z(t)) + a_{\Theta}(\Theta^{\varepsilon}(t), z(t)) + b_p(\boldsymbol{u}^{\varepsilon}(t), z(t))$$
$$- (Q(t), z(t))\} \mathrm{d}t = 0 \quad \forall z \in {}^1V,$$

$$\int_I \{(\partial_t \boldsymbol{u}^\varepsilon(t), \boldsymbol{v}(t)) + b(\boldsymbol{u}^\varepsilon(t), \boldsymbol{u}^\varepsilon(t), \boldsymbol{v}(t)) + \hat{\mu} a(\boldsymbol{u}^\varepsilon(t), \boldsymbol{v}(t)) + b_s(\Theta^\varepsilon(t) - \Theta_o, \boldsymbol{v}(t))$$
$$- (\boldsymbol{F}(t), \boldsymbol{v}(t))\}\mathrm{d}t + (\chi(t), \boldsymbol{v}(t)) = 0 \quad \forall \boldsymbol{v} \in V,$$
$$\int_I \{(\partial_t c^\varepsilon(t), \mathrm{d}(t)) + b_c(\boldsymbol{u}^\varepsilon(t), c_L^\varepsilon(t), \mathrm{d}(t)) + a_c(c^\varepsilon(t), \mathrm{d}(t))$$
$$- (\varphi(t), \mathrm{d}(t))\}\mathrm{d}t = 0 \quad \forall \mathrm{d} \in {}^2V.$$

Since Equations 12.493-12.495 and the a priori estimates of Equation 12.506 are satisfied, then it is sufficient to prove that

$$\chi(t) = J'_\varepsilon(\boldsymbol{u}^\varepsilon(t)).$$

To prove this, the property of monotonicity will be used. Then the proof is parallel to that of Nedoma (1995).

Let $\boldsymbol{v} \in L^2(I; V)$ such that $\boldsymbol{v} \in L^2(I; V')$, $\boldsymbol{v}(t_o) = \boldsymbol{u}_o$. Let us put

$$X_\mu = (J'_\varepsilon(\boldsymbol{u}_\mu(t)) - J'_\varepsilon(\boldsymbol{v}(t)), \boldsymbol{u}_\mu(t) - \boldsymbol{v}(t)) + \hat{\mu} \int_\mathrm{I} a(\boldsymbol{u}_\mu(t) - \boldsymbol{v}(t), \boldsymbol{u}_\mu(t) - \boldsymbol{v}(t))\mathrm{d}t$$
$$+ \int_\mathrm{I} (\partial_t \boldsymbol{u}_\mu(t) - \partial_t \boldsymbol{v}(t), \boldsymbol{u}_\mu(t) - \boldsymbol{v}(t))\mathrm{d}t.$$

Using Equation 12.494 we have

$$X_\mu = \int_\mathrm{I} (\boldsymbol{F}(t), \boldsymbol{u}_\mu(t))\mathrm{d}t - (J'_\varepsilon(\boldsymbol{u}_\mu(t)), \boldsymbol{v}(t)) - (J'_\varepsilon(\boldsymbol{v}(t)), \boldsymbol{u}_\mu(t) - \boldsymbol{v}(t))$$
$$- \hat{\mu} \int_\mathrm{I} [a(\boldsymbol{u}_\mu(t), \boldsymbol{v}(t)) + a(\boldsymbol{v}(t), \boldsymbol{u}_\mu(t) - \boldsymbol{v}(t))]\mathrm{d}t$$
$$- \int_\mathrm{I} [(\partial_t \boldsymbol{u}_\mu(t), \boldsymbol{v}(t)) - (\partial_t \boldsymbol{v}(t), \boldsymbol{u}_\mu(t) - \boldsymbol{v}(t))]\mathrm{d}t.$$

Hence $X_\mu \to X$, where

$$X = \int_\mathrm{I} (\boldsymbol{F}(t), \boldsymbol{u}^\varepsilon(t))\mathrm{d}t - [(\chi(t), \boldsymbol{v}(t)) + (J'_\varepsilon(\boldsymbol{v}(t)), \boldsymbol{u}^\varepsilon(t) - \boldsymbol{v}(t))]$$
$$- \hat{\mu} \int_\mathrm{I} [a(\boldsymbol{u}^\varepsilon(t), \boldsymbol{v}(t)) + a(\boldsymbol{v}(t), \boldsymbol{u}^\varepsilon(t) - \boldsymbol{v}(t))]\mathrm{d}t$$
$$- \int_\mathrm{I} (\partial_t \boldsymbol{u}^\varepsilon(t), \boldsymbol{v}(t)) - (\partial_t \boldsymbol{v}(t), \boldsymbol{u}^\varepsilon(t) - \boldsymbol{v}(t))]\mathrm{d}t.$$

Since $X_\mu(t) \geq 0$ for all μ then $X(t) \geq 0$.

Let us put $\boldsymbol{v} = \boldsymbol{u}^\varepsilon - \lambda \boldsymbol{w}$, $\lambda \geq 0$, where $\boldsymbol{w} \in L^2(I; V)$, $\partial_t \boldsymbol{w} \in L^2(I; V')$, $\boldsymbol{w}(t_o) = 0$. Substituting for $\boldsymbol{v}$ and dividing by λ we obtain

$$(\chi(t) - J'_\varepsilon(\boldsymbol{u}^\varepsilon(t) - \lambda \boldsymbol{w}(t)), \boldsymbol{w}(t)) + \lambda \int_\mathrm{I} \{\hat{\mu} a(\boldsymbol{w}(t), \boldsymbol{w}(t)) + (\partial_t \boldsymbol{w}(t), \boldsymbol{w}(t))\}\mathrm{d}t \geq 0.$$

Hence, in the limit $\lambda \to 0$ we obtain

$$(\chi(t) - J'_\varepsilon(\boldsymbol{u}^\varepsilon(t)), \boldsymbol{w}(t)) \geq 0 \quad \forall \boldsymbol{w},$$

from which the assertion $\chi(t) = J'_\varepsilon(\boldsymbol{u}^\varepsilon(t))$ follows.

Therefore, we have proved the existence of H^ε, $\boldsymbol{u}^\varepsilon$, c^ε satisfying Equations 12.493-12.495 and the conditions

$$\begin{aligned}
&H^\varepsilon(t) \text{ is bounded in } L^2(I; {}^1V) \cap H^1(I; H^{-1}(\Omega)),\\
&\partial_t H^\varepsilon(t) \text{ is bounded in } L^2(I; H^{-1}(\Omega)),\\
&\boldsymbol{u}^\varepsilon(t) \text{ is bounded in } L^2(I; V) \cap L^\infty(I; H),\\
&\partial_t \boldsymbol{u}^\varepsilon(t) \text{ is bounded in } L^2(I; V'),\\
&c^\varepsilon(t) \text{ is bounded in } L^2(I; {}^2V) \cap H^1(I; (H^1(\Omega))'),\\
&\partial_t c^\varepsilon(t) \text{ is bounded in } L^2(I; (H^1(\Omega))').
\end{aligned}$$

To prove that $H^\varepsilon(\boldsymbol{x}, t_o) = H_o(\boldsymbol{x})$, $\boldsymbol{u}^\varepsilon(\boldsymbol{x}, t_o) = \boldsymbol{u}_o(\boldsymbol{x})$, $c^\varepsilon(\boldsymbol{x}, t_o) = c_o(\boldsymbol{x})$ the Arzelà-Ascoli's theorem(*) will be used. Due to this theorem, the sequences $\{H^\varepsilon_\mu\}$, $\{\boldsymbol{u}^\varepsilon_\mu\}$, $\{c^\varepsilon_\mu\}$, $\mu \in \mathcal{N}$, converge towards $H^\varepsilon \in C^o(I; H^{-1}(\Omega))$ and/or toward $\boldsymbol{u}^\varepsilon \in C^o(I; V')$, and/or towards $c^\varepsilon \in C^o(I; (H^1(\Omega))')$, respectively, and therefore $\{H^\varepsilon_\mu(.,t_o)\}$, $\{\boldsymbol{u}^\varepsilon_\mu(.,t_o)\}$, $\{\boldsymbol{c}^\varepsilon_\mu(.,t_o)\}$, $\mu \in \mathcal{N}$, converge towards $H^\varepsilon(.,t_o)$ in $H^{-1}(\Omega)$, or towards $\boldsymbol{u}^\varepsilon(.,t_o) \in V'$, or towards $c^\varepsilon(.,t_o) \in (H^1(\Omega))'$, respectively, and at the same time $\{H^\varepsilon_\mu(.,t_o)\}$, $\mu \in \mathcal{N}$, converges to $H_o(\boldsymbol{x})$ in $L^2(\Omega)$ by definition. Similarly, for $\{\boldsymbol{u}^\varepsilon_\mu(.,t_o)\}_{\mu\in\mathcal{N}} \to \boldsymbol{u}_o(\boldsymbol{x})$ and $\{c^\varepsilon_\mu(.,t_o)\}_{\mu\in\mathcal{N}} \to c_o(\boldsymbol{x})$ in $H(\Omega)$ or $L^2(\Omega)$, respectively, by definition. Since $H^\varepsilon \in C^o(I; L^2(\Omega))$, the initial condition holds in $L^2(\Omega)$ and therefore a.e. in Ω. In a similar way we finish the proof for $\boldsymbol{u}_o(\boldsymbol{x})$ and $c_o(\boldsymbol{x})$.

Consider the limitation process $\varepsilon \to 0$. We shall prove that estimates of the regularized enthalpy H^ε, its derivative $\partial_t H^\varepsilon$, and the regularized temperature T^ε do not depend on ε. The proof of that is parallel of that of Lemma 3.4 of Blanc and Gasser (1996) (see also Brezis (1971)).

Firstly, we prove the auxiliary estimate which will be used later. Let assumptions (A2)(iii), (A4)(iii) be satisfied. Let α satisfy (A2), (A4). Let $\varphi = \alpha'(c^\varepsilon + \varepsilon)$ be a test function (this is possible as $c^\varepsilon \geq 0$ in $\Omega \times I$ and α' is uniformly Lipschitzian on $\langle \varepsilon, \infty))$. Then Equation 12.495 yields

$$\int_\Omega \alpha(c^\varepsilon(.,t_1) + \varepsilon)\mathrm{d}\boldsymbol{x} - \int_\Omega \alpha(c_o + \varepsilon)\mathrm{d}\boldsymbol{x} + \int_I\int_\Omega \alpha''(c^\varepsilon + \varepsilon)\left\|\text{grad } c^\varepsilon\right\|^2 \mathrm{d}\boldsymbol{x}\,\mathrm{d}t$$

$$- \int_I\int_\Omega c^\varepsilon_L(c^\varepsilon, T^\varepsilon(c^\varepsilon, H^\varepsilon))\alpha''(c^\varepsilon + \varepsilon)\boldsymbol{u}^\varepsilon \text{grad } c^\varepsilon \mathrm{d}\boldsymbol{x}\,\mathrm{d}t = 0.$$

(*) Let K be a subset of $C^o(I; X)$, where X is a Banach space. Then K is said to be equi-continuous if for every $\varepsilon > 0$ there exists a $\delta = \delta(\varepsilon) > 0$ such that $\|u(t) - u(s)\|_X < \varepsilon$ holds for all $u \in K$ and all $t, s \in I$ for which $|t - s| < \delta$.

Arzelà-Ascoli theorem: A subset K of $C^o(I; K)$ is relatively compact if and only if (i) K is equi-continuous, (ii) the set $K(t)$, $K(t) = \{u(t) \mid u \in K\}$, is relatively compact in Banach space X for every $t \in I$, i.e. every sequence in $K(t)$ contains a subsequence convergent in X.

Hence, due to the Young inequality and the property of α as well as to Lemma 12.18(iv), we then obtain

$$\int_{\mathrm{I}}\int_{\Omega}\alpha''(c^{\varepsilon}+\varepsilon)\left\|\mathrm{grad}\ c^{\varepsilon}\right\|^{2}\mathrm{d}\boldsymbol{x}\,\mathrm{d}t\leq\int_{\Omega}\alpha(c_{o}+\varepsilon)\mathrm{d}\boldsymbol{x}-\inf_{R_{+}}(\alpha)\ \mathrm{meas}\ (\Omega)$$
$$+\int_{\mathrm{I}}\int_{\Omega}\left|c_{L}^{\varepsilon}(c^{\varepsilon},T^{\varepsilon}(c^{\varepsilon},H^{\varepsilon}))\right|\alpha''(c^{\varepsilon}+\varepsilon)\left\|\boldsymbol{u}^{\varepsilon}\right\|\left\|\mathrm{grad}\ c^{\varepsilon}\right\|\mathrm{d}\boldsymbol{x}\,\mathrm{d}t$$
$$\leq c+\frac{1}{2}\int_{I}\int_{\Omega}\alpha''(c^{\varepsilon}+\varepsilon)\left\|\mathrm{grad}\ c^{\varepsilon}\right\|^{2}\mathrm{d}\boldsymbol{x}\,\mathrm{d}t$$
$$+\frac{1}{2}\left\|\boldsymbol{u}^{\varepsilon}\right\|_{L^{\infty}(\Omega)}\int_{\mathrm{I}}\int_{\Omega}\left|c_{L}^{\varepsilon}(c^{\varepsilon},T^{\varepsilon}(c^{\varepsilon},H^{\varepsilon}))\right|^{2}\alpha''(c^{\varepsilon}+\varepsilon)\mathrm{d}\boldsymbol{x}\,\mathrm{d}t\leq c. \quad (12.538)$$

Now, let $t\in I$ be arbitrary. Then Equation 12.493 for $z=\hat{\chi}(t)H^{\varepsilon}$, where $\hat{\chi}(t)$ is a characteristic function of the interval I, yields

$$0=\frac{1}{2}\int_{\Omega}(H^{\varepsilon}(.,t))^{2}\mathrm{d}\boldsymbol{x}-\frac{1}{2}\int_{\Omega}(H_{o})^{2}\mathrm{d}\boldsymbol{x}$$
$$+\int_{\mathrm{I}}\int_{\Omega}\boldsymbol{u}^{\varepsilon}\mathrm{grad}\ T^{\varepsilon}(c^{\varepsilon},H^{\varepsilon})(\partial_{2}T^{\varepsilon}(c^{\varepsilon},H^{\varepsilon}))^{-1/2}(\partial_{2}T^{\varepsilon}(c^{\varepsilon},H^{\varepsilon}))^{1/2}H^{\varepsilon}\mathrm{d}\boldsymbol{x}\,\mathrm{d}t$$
$$+\int_{I}\int_{\Omega}\mathrm{grad}\ T^{\varepsilon}(c^{\varepsilon},H^{\varepsilon})[\mathrm{grad}\ T^{\varepsilon}(c^{\varepsilon},H^{\varepsilon})$$
$$-\partial_{1}T^{\varepsilon}(c^{\varepsilon},H^{\varepsilon})\mathrm{grad}\ c^{\varepsilon}](\partial_{2}T^{\varepsilon}(c^{\varepsilon},H^{\varepsilon}))^{-1}\mathrm{d}\boldsymbol{x}\,\mathrm{d}t.$$

Hence, after some modifications

$$\frac{1}{2}\int_{\Omega}(H^{\varepsilon}(.,t))^{2}\mathrm{d}\boldsymbol{x}+\int_{\mathrm{I}}\int_{\Omega}\mathrm{grad}\ T^{\varepsilon}(c^{\varepsilon},H^{\varepsilon})\mathrm{grad}\ T^{\varepsilon}(c^{\varepsilon},H^{\varepsilon})(\partial_{2}T^{\varepsilon}(c^{\varepsilon},H^{\varepsilon}))^{-1}\mathrm{d}\boldsymbol{x}\,\mathrm{d}t$$
$$=\frac{1}{2}\int_{\Omega}(H_{o})^{2}\mathrm{d}\boldsymbol{x}-\int_{\mathrm{I}}\int_{\Omega}\mathrm{grad}\ T^{\varepsilon}(c^{\varepsilon},H^{\varepsilon})(\partial_{2}T^{\varepsilon}(c^{\varepsilon},H^{\varepsilon}))^{-1}\partial_{1}T^{\varepsilon}(c^{\varepsilon},H^{\varepsilon})\mathrm{grad}\ c^{\varepsilon}\mathrm{d}\boldsymbol{x}\,\mathrm{d}t$$
$$+\int_{\mathrm{I}}\int_{\Omega}\mathrm{grad}\ T^{\varepsilon}(c^{\varepsilon},H^{\varepsilon})(\partial_{2}T^{\varepsilon}(c^{\varepsilon},H^{\varepsilon}))^{-1/2}\boldsymbol{u}^{\varepsilon}(\partial_{2}T^{\varepsilon}(c^{\varepsilon},H^{\varepsilon}))^{1/2}H^{\varepsilon}\mathrm{d}\boldsymbol{x}\,\mathrm{d}t,$$

and after twice applying the Young inequality, we find

$$\frac{1}{2}\int_{\Omega}(H^{\varepsilon}(.,t))^{2}\mathrm{d}\boldsymbol{x}+\int_{\mathrm{I}}\int_{\Omega}\left\|\mathrm{grad}\ T^{\varepsilon}(c^{\varepsilon},H^{\varepsilon})\right\|^{2}(\partial_{2}T^{\varepsilon}(c^{\varepsilon},H^{\varepsilon}))^{-1}\mathrm{d}\boldsymbol{x}\,\mathrm{d}t$$
$$\leq\frac{1}{2}\int_{\Omega}(H_{o})^{2}\mathrm{d}\boldsymbol{x}+\frac{1}{4}\int_{\mathrm{I}}\int_{\Omega}\left\|\mathrm{grad}\ T^{\varepsilon}(c^{\varepsilon},H^{\varepsilon})\right\|^{2}(\partial_{2}T^{\varepsilon}(c^{\varepsilon},H^{\varepsilon}))^{-1}\mathrm{d}\boldsymbol{x}\,\mathrm{d}t$$
$$+\int_{\mathrm{I}}\int_{\Omega}\left|\partial_{1}T^{\varepsilon}(c^{\varepsilon},H^{\varepsilon})\right|^{2}(\partial_{2}T^{\varepsilon}(c^{\varepsilon},H^{\varepsilon}))^{-1}\left\|\mathrm{grad}\ c^{\varepsilon}\right\|^{2}\mathrm{d}\boldsymbol{x}\,\mathrm{d}t$$
$$+\frac{1}{4}\int_{\mathrm{I}}\int_{\Omega}\left\|\mathrm{grad}\ T^{\varepsilon}(c^{\varepsilon},H^{\varepsilon})\right\|^{2}(\partial_{2}T^{\varepsilon}(c^{\varepsilon},H^{\varepsilon}))^{-1}\mathrm{d}\boldsymbol{x}\,\mathrm{d}t$$
$$+\left\|\boldsymbol{u}^{\varepsilon}\right\|_{L^{\infty}(\Omega\times I)}\left\|\partial_{2}T^{\varepsilon}(c^{\varepsilon},H^{\varepsilon})\right\|_{L^{\infty}(\Omega\times I)}\int_{\mathrm{I}}\int_{\Omega}(H^{\varepsilon})^{2}\mathrm{d}\boldsymbol{x}\,\mathrm{d}t.$$

Hence

$$\int_\Omega (H^\varepsilon(.,t))^2 \mathrm{d}\boldsymbol{x} + \int_I \int_\Omega \left\|\mathrm{grad}\ T^\varepsilon(c^\varepsilon, H^\varepsilon)\right\|^2 (\partial_2 T^\varepsilon(c^\varepsilon, H^\varepsilon))^{-1} \mathrm{d}\boldsymbol{x}\,\mathrm{d}t \le \int_\Omega (H_o^\varepsilon)^2 \mathrm{d}\boldsymbol{x}$$

$$+ 2 \int_I \int_\Omega \left|\partial_1 T^\varepsilon(c^\varepsilon, H^\varepsilon)\right|^2 (\partial_2 T^\varepsilon(c^\varepsilon, H^\varepsilon))^{-1} \left\|\mathrm{grad}\ c^\varepsilon\right\|^2 \mathrm{d}\boldsymbol{x}\,\mathrm{d}t + c \int_I \int_\Omega (H^\varepsilon)^2 \mathrm{d}\boldsymbol{x}\,\mathrm{d}t.$$

Hence, using Lemma 12.18(v), Equation 12.538 and the Gronwall lemma, we obtain

$$\left\|H^\varepsilon\right\|_{L^\infty(I;L^2(\Omega))} \le c, \tag{12.539}$$

where $c =$ constant independent of ε. Since $\partial_2 T^\varepsilon$ is bounded from above, then

$$\left\|T^\varepsilon(c^\varepsilon, H^\varepsilon)\right\|_{L^2(I;H_o^1(\Omega))} \le c, \tag{12.540}$$

where $c =$ constant independent of ε.
From Equation 12.493 it follows that

$$\int_I (\partial_t H^\varepsilon, z)\mathrm{d}t = -\int_I (\boldsymbol{u}^\varepsilon \mathrm{grad}\ T^\varepsilon(c^\varepsilon, H^\varepsilon), z)\mathrm{d}t - \int_I a_\Theta(T^\varepsilon(c^\varepsilon, H^\varepsilon), z)\mathrm{d}t$$

$$\le \left\|\boldsymbol{u}^\varepsilon\right\|_{L^\infty(\Omega\times I)} \int_I \left\|\mathrm{grad}\ T^\varepsilon(c^\varepsilon, H^\varepsilon)\right\|_{0,1} \|z\|_{0,1}\,\mathrm{d}t$$

$$+ \int_I \left\|\mathrm{grad}\ T^\varepsilon(c^\varepsilon, H^\varepsilon)\right\|_{0,1} \|\mathrm{grad}\ z\|_{0,1}\,\mathrm{d}t \quad \forall z \in L^2(I; H_o^1(\Omega)).$$

Since $\|z\|^2_{L^2(I;H^{-1}(\Omega))} \underset{\text{def}}{=\!=} \int_I \sup_{w\in H_0^1(\Omega), \|w\|_{H_0^1(\Omega)}\le 1} (z, w))^2 \mathrm{d}t$ and using Equation 12.540, then we have

$$\left\|\partial_t H^\varepsilon\right\|_{L^2(I;H^{-1}(\Omega))} \le c, \tag{12.541}$$

where $c =$ constant independent of ε.

The existence of a weak solution of Problem $(\mathscr{P}_{\mathrm{centh}})_v$ will be established as the limit of a subsequance of weak solutions of the regularized problems. In the previous steps we proved that

$$H^\varepsilon \text{ remains in a bounded set of } L^\infty(I; L^2(\Omega)), \tag{12.542}$$

$$\partial_t H^\varepsilon \text{ remains in a bounded set of } L^\infty(I; H^{-1}(\Omega)), \tag{12.543}$$

$$T^\varepsilon \text{ remains in a bounded set of } L^2(I; H_o^1(\Omega)), \tag{12.544}$$

$$\boldsymbol{u}^\varepsilon \text{ remains in a bounded set of } L^2(I; V) \cap L^\infty(I; H), \tag{12.545}$$

$$\partial_t \boldsymbol{u}^\varepsilon \text{ remains in a bounded set of } L^\infty(I; V'), \tag{12.546}$$

$$c^\varepsilon \text{ remains in a bounded set of } L^\infty(I; H^1(\Omega)), \tag{12.547}$$

$$\partial_t c^\varepsilon \text{ remains in a bounded set of } L^\infty(I; (H^1(\Omega))') \tag{12.548}$$

for any weak regularized solution $\{H^\varepsilon, \Theta^\varepsilon, \boldsymbol{u}^\varepsilon, c^\varepsilon\}$.

In this part of the proof we shall prove that

$$H^\varepsilon \to H \text{ in } C^o(I; H^{-1}(\Omega)),$$

$$T^\varepsilon \to T \text{ strongly in } C^o(I;L^2(\Omega)),$$

$$\boldsymbol{u}^\varepsilon \to \boldsymbol{u}\text{*-weakly in } L^\infty(I;H) \text{ and weakly in } L^2(I;V),$$

$$\partial_t\boldsymbol{u}^\varepsilon \to \partial_t\boldsymbol{u} \text{ weakly in } L^2(I;V'),$$

$$c^\varepsilon \to c \text{ strongly in } C^o(I;L^2(\Omega)).$$

Due to Equations 12.542 and 12.543 the set $\{H^\varepsilon(t), 0 < \varepsilon \leq 1\}$ is relatively compact in $H^{-1}(\Omega)\forall t \in \bar{I}$, and the family $\{H^\varepsilon \mid 0 < \varepsilon \leq 1\}$ is equi-continuous in $H^{-1}(\Omega)$. Then due to the Arzela-Ascoli theorem there exists a subsequence $\{H^{\varepsilon j}\}$, $j \in \mathcal{N}$, converging in $C^o(I;H^{-1}(\Omega))$. Moreover, due to Equations 12.545 and 12.546 there exists a subsequence $\{\boldsymbol{u}^{\varepsilon j}\}$, $j \in \mathcal{N}$, such that

$$\boldsymbol{u}^{\varepsilon j}(t) \to \boldsymbol{u}(t) \text{ in } L^\infty(I;H) \text{ weakly star and weakly in } L^2(I;V), \tag{12.549}$$

$$\partial_t\boldsymbol{u}^{\varepsilon j}(t) \to \partial_t\boldsymbol{u}(t) \text{ weakly in } L^2(I;V'). \tag{12.550}$$

For $\boldsymbol{v} \in W$ let us put (the index j will be omitted)

$$Y_\varepsilon = \int_I \{(\partial_t\boldsymbol{v}(t), \boldsymbol{v}(t) - \boldsymbol{u}^\varepsilon(t)) + \hat{\mu}a(\boldsymbol{u}^\varepsilon(t), \boldsymbol{v}(t) - \boldsymbol{u}^\varepsilon(t)) + b(\boldsymbol{u}^\varepsilon(t), \boldsymbol{u}^\varepsilon(t), \boldsymbol{v}(t) - \boldsymbol{u}^\varepsilon(t))$$
$$+ b_s(\Theta^\varepsilon(t) - \Theta_o, \boldsymbol{v}(t) - \boldsymbol{u}^\varepsilon(t)) - (\boldsymbol{F}(t), v(t) - \boldsymbol{u}^\varepsilon(t))\}\mathrm{d}t + J_\varepsilon(\boldsymbol{v}(t)) - J_\varepsilon(\boldsymbol{u}^\varepsilon(t)).$$

By virtue of Equation 12.494,

$$Y_\varepsilon = \int_I (\partial_t\boldsymbol{v}(t) - \partial_t\boldsymbol{u}^\varepsilon(t), \boldsymbol{v}(t) - \boldsymbol{u}^\varepsilon(t))\mathrm{d}t + J_\varepsilon(\boldsymbol{v}(t)) - J_\varepsilon(\boldsymbol{u}^\varepsilon(t)) - (J'_\varepsilon(\boldsymbol{u}^\varepsilon(t)), \boldsymbol{v}(t) - \boldsymbol{u}^\varepsilon(t)).$$

Due to the initial conditions the first term is equivalent to $(1/2)\,|\boldsymbol{v}(t_1) - \boldsymbol{u}^\varepsilon(t_1)|^2$ and since the functional $\boldsymbol{v} \to j_\varepsilon(\boldsymbol{v})$ is convex, the second term ≥ 0, and thus $Y_\varepsilon \geq 0$. Hence

$$\int_I \{(\partial_t\boldsymbol{v}(t), \boldsymbol{v}(t) - \boldsymbol{u}^\varepsilon(t)) + \hat{\mu}a(\boldsymbol{u}^\varepsilon(t), v(t)) - b(\boldsymbol{u}^\varepsilon(t), \boldsymbol{v}(t), \boldsymbol{u}^\varepsilon(t))$$
$$+ b_s(\Theta^\varepsilon(t) - \Theta_o, \boldsymbol{v}(t) - \boldsymbol{u}^\varepsilon(t)) - (\boldsymbol{F}(t), \boldsymbol{v}(t) - \boldsymbol{u}^\varepsilon(t))\}\mathrm{d}t + J_\varepsilon(\boldsymbol{v}(t))$$
$$\geq \hat{\mu}\int_I a(\boldsymbol{u}^\varepsilon(t), \boldsymbol{u}^\varepsilon(t))\mathrm{d}t + J_\varepsilon(\boldsymbol{u}^\varepsilon(t)).$$

Hence and using Equations 12.549 and 12.550

$$\int_I \{(\partial_t\boldsymbol{u}(t), \boldsymbol{v}(t) - \boldsymbol{u}(t)) + \hat{\mu}a(\boldsymbol{u}(t), \boldsymbol{v}(t)) - b(\boldsymbol{u}(t), \boldsymbol{v}(t), \boldsymbol{u}(t))$$
$$+ b_s(\Theta(t) - \Theta_o, \boldsymbol{v}(t) - \boldsymbol{u}(t)) - (\boldsymbol{F}(t), \boldsymbol{v}(t) - \boldsymbol{u}(t))\}\mathrm{d}t + J(\boldsymbol{v}(t))$$
$$\geq \liminf \int_I [\hat{\mu}a(\boldsymbol{u}^\varepsilon(t), \boldsymbol{u}^\varepsilon(t)) + \hat{g}j_\varepsilon(\boldsymbol{u}^\varepsilon(t))]\mathrm{d}t$$
$$\geq \int_I \hat{\mu}a(\boldsymbol{u}(t), \boldsymbol{u}(t))\mathrm{d}t + J(\boldsymbol{u}(t)), \tag{12.551}$$

as $\liminf \int_I [\hat{\mu}a(\boldsymbol{u}^\varepsilon(t), \boldsymbol{u}^\varepsilon(t))]\mathrm{d}t \geq \int_I [\hat{\mu}a(\boldsymbol{u}(t), \boldsymbol{u}(t))]\mathrm{d}t$ (due to the fact that the function $\boldsymbol{u} \to \int_I \hat{\mu}(\boldsymbol{u}, \boldsymbol{u})\mathrm{d}t$ is lower semi-continuous on $L^2(I;V)$ with the weak topology) and since

$$\int_{\mathrm{I}} j(\boldsymbol{u}(t))\mathrm{d}t \leq \left(\int_{\Omega\times I} D_{\mathrm{II}}(\boldsymbol{u}(t))^{(1/2)(1+\varepsilon)}\mathrm{d}\boldsymbol{x}\,\mathrm{d}t\right)^{1/(1+\varepsilon)} \left(\int_{\Omega\times I} \mathrm{d}\boldsymbol{x}\,\mathrm{d}t\right)^{\varepsilon/(1+\varepsilon)}.$$

Hence

$$\int_{\mathrm{I}} j_{\varepsilon}(\boldsymbol{u}^{\varepsilon}(t))\mathrm{d}t \geq c(\varepsilon)\left(\int_{\mathrm{I}} j(\boldsymbol{u}^{\varepsilon}(t))\mathrm{d}t\right)^{(1+\varepsilon)}, \quad c(\varepsilon) = |\mathrm{meas}\ (\Omega\times I)|^{-\varepsilon},$$

$$\liminf \int_{\mathrm{I}} j_{\varepsilon}(\boldsymbol{u}^{\varepsilon}(t))\mathrm{d}t \geq \liminf \int_{\mathrm{I}} j(\boldsymbol{u}^{\varepsilon}(t))\mathrm{d}t. \tag{12.552}$$

Since the function $\boldsymbol{v} \to \int_{\mathrm{I}} j(\boldsymbol{v}(t))\mathrm{d}t$ is convex and continuous on $L^2(I;V)$, then it is lower semi-continuous in the weak topology of the space $L^2(I;V)$ and thus

$$\liminf \int_{\mathrm{I}} j(\boldsymbol{u}^{\varepsilon}(t))\mathrm{d}t \geq \int_{\mathrm{I}} j(\boldsymbol{u}(t))\mathrm{d}t,$$

which together with Equation 12.552 proves

$$\liminf \int_{\mathrm{I}} j_{\varepsilon}(\boldsymbol{u}^{\varepsilon}(t))\mathrm{d}t \geq \int_{\mathrm{I}} j(\boldsymbol{u}(t))\mathrm{d}t.$$

Then Equations 12.549–12.552 yield that $\boldsymbol{u}$ satisfies Equation 12.487.

Furthermore, due to Equations 12.547 and 12.548 there exists a subsequence $\{c^{\varepsilon j}\}$, $j \in \mathcal{N}$, which strongly converges to $c(t)$ in $C^o(I;L^2(\Omega))$.

Now we shall prove the existence of a subsequence of $\{T^{\varepsilon j}(c^{\varepsilon j}, H^{\varepsilon j})\}$, $j \in \mathcal{N}$, which strongly converges in $L^2(I;L^2(\Omega))$ towards $T(c,H)$. The monotonicity arguments (see Brezis (1971), Cannon, DiBenedetto and Knightly (1980)) will be used. We proved above that $H^{\varepsilon j} \to H$ in $C^o(I;H^{-1}(\Omega))$, $c^{\varepsilon j} \to c$ strongly in $L^2(I;L^2(\Omega))$, $j \to \infty$. Let $\{H^j, c^j\}$, $j \in \mathcal{N}$, be the corresponding sequence and $\{T^j(c^j, H^j)\}$, $j \in \mathcal{N}$, the associated sequence. Further, for simplicity we shall write H^j, c^j instead of $H^{\varepsilon j}$ and $c^{\varepsilon j}$, respectively. Then, due to Lemmata 4.1 of Blanc and Gasser (1996), there exist subsequences, still indexed by j, and null measure $E \subset I$ such that, for any $t \in I\backslash E$

$$\lim_{j\to\infty} \int_{\Omega} T^j(c^j(t), H^j(t))(H^j(t) - H(t))\mathrm{d}\Omega = 0, \tag{12.553}$$

$$\lim_{j\to\infty} \int_{\Omega} \left|c^j(t) - c(t)\right|^2 \mathrm{d}\Omega = 0 \quad \forall t \in I\backslash E. \tag{12.554}$$

Due to assumptions (A2)(i), (ii) and Equations 12.542 and 12.543, then there exists a subsequence $\{T^k(c^k(.,t), H^k(.,t))\}$, $k \in \mathcal{N}$, which is bounded in $L^2(\Omega)$. Hence, for every $t \in I\backslash E$ there exists a subsequence $\{T^l(c^l(.,t), H^l(.,t))\}$, $l \in \mathcal{N}$, dependent on t and such that (see Lemmata 4.2. of Blanc and Gasser (1996)

$$T^l(c^l(.,t), H^l(.,t)) \to T(c(.,t), H(.,t)) \text{ weakly in } L^2(\Omega). \tag{12.555}$$

It remains to prove that $T^j(c^j, H^j) \to T(c,H)$ strongly in $L^2(I;L^2(\Omega))$.

Let $\{T^j(c^j, H^j)\}$, $j \in \mathcal{N}$, be the subsequence resulting from the associated sequence defined above. Then, firstly we prove that

$$f^j \equiv T^j(c^j, H^j) - T(c, H) \to 0 \text{ weakly in } L^2(\Omega), \quad \forall t \in I \backslash E, \tag{12.556}$$

where E is a null measure set defined above.

We prove Equation 12.556 by contradiction. Let $\{f^j\}$, $j \in \mathcal{N}$, be a subsequence, $f \neq 0$, $f \in L^2(\Omega)$ function. Then, there can not be $t \in I \backslash E$, a subsequence $\{f^j\}$ and a function $f \neq 0$, $f \in L^2(\Omega)$ does not exist, such that $\{f^k\}$, $k \in \mathcal{N}$, weakly converges towards f in $L^2(\Omega)$. Thus no subsequence $\{f^k\} \to 0$ weakly, $k \in \mathcal{N}$. But this is in contradiction with Equation 12.555. Hence, we deduce the existence of a subsequence $\{f^j\}$, $j \in \mathcal{N}$, weakly converging towards 0.

Let $\{w_k\}$, $k \in \mathcal{N}$, be a basis of $L^2(\Omega)$. Since for almost every $t \in I \backslash E$, the functions $f^j \in H^1(\Omega)$, then due to Fridrichs' lemma for any $\varepsilon > 0$

$$\left\| f^j(.,t) \right\|^2_{L^2(\Omega)} \leq \varepsilon \left\| f^j(.,t) \right\|^2_{H^1_o(\Omega)} + \sum_{k=0}^{n} \left(\int_\Omega f^j(\boldsymbol{x}, t) w_k(\boldsymbol{x}) \mathrm{d}\boldsymbol{x} \right)^2, n = n(\varepsilon).$$

Hence, by integrating over $t \in I$ and putting $\alpha^{k,j}(t) = (\int_\Omega f^j(\boldsymbol{x}, t) w_k(\boldsymbol{x}) \mathrm{d}\boldsymbol{x})^2$, and due to the estimate in Equation 12.538 and the weak convergence of $\{f^j\}$, $j \in \mathcal{N}$, we firstly find that $\{\alpha^{k,j}(t)\}$, $j \in \mathcal{N}$, converges towards 0 in $L^1(I)$. Secondly, then there exists a constant c, independent of ε, such that

$$\lim_{j \to \infty} \left\| f^j \right\|_{L^2(\Omega \times I)} \equiv \lim_{j \to \infty} \left\| T^j(c^j, H^j) - T(c, H) \right\|_{L^2(\Omega \times I)} \leq c\varepsilon.$$

Since the last estimate is valid for any $\varepsilon > 0$, then

$$T^j(c^j, H^j) \to T(c, H) \text{ strongly in } L^2(I; L^2(\Omega)).$$

Summing all these results we find that if $(H^\varepsilon, \boldsymbol{u}^\varepsilon, c^\varepsilon)$, $\varepsilon \in (0, 1\rangle$, is a weak regularized solution, then there is a sequence indexed by j and a tetrad $\{H, \Theta, \boldsymbol{u}, c\}$ satisfying the regularity conditions in Definition 12.3 such that

$$\begin{aligned}
\partial_t H^j &\to \partial_t H \text{ weakly in } L^2(I; H^{-1}(\Omega)),\\
\partial_t c^j &\to \partial_t c \text{ weakly in } L^2(I; (H^1(\Omega))'),\\
\text{grad } c^j &\to \text{grad } c \text{ weakly in } L^2(I; L^2(\Omega)),\\
\text{grad } T^j(c^j, H^j) &\to \text{grad } \Theta \text{ weakly in } L^2(I; L^2(\Omega)),\\
c_L(c^j, T(c^j, H^j)) &\to c_L(c, T) \text{ strongly in } L^2(I; L^2(\Omega)),
\end{aligned}$$

where $\Theta = T(c, H)$.

Due to Duvaut and Lions (1976) for $\hat{g} = 0$ it follows that the Bingham rheology leads to the Newtonian liquid, and then

$$\boldsymbol{u} \to \boldsymbol{u}_L \text{ weakly in } L^2(I; V), \tag{12.557}$$

$$\partial_t \boldsymbol{u} \to \partial_t \boldsymbol{u}_L \text{ weakly in } L^2(I; V'), \tag{12.558}$$

where $\boldsymbol{u}_L$ denotes the solution corresponding to the liquid state. The proof of Equations 12.557 and 12.558 is parallel to that of Theorem 5.1. of Duvaut and Lions (1976), pp 306–307, where in addition $b_s(\Theta - \Theta_o, \boldsymbol{v}) \in L^2(I; [L^2(\Omega)]^2)$ was assumed to be a part of the body forces.

We now prove this assertion in more detail. We proved above that for $\hat{g} > 0$ there exists a weak solution of a coupled two-Stefan-like problem in a Bingham visco-plastic rheology. It was shown that in the mushy zones both phases, solid (visco-plastic) and liquid (strongly visco-plasic with low $\hat{g}$), are microscopically present. Therefore, we must show that for the threshold of plasticity $\hat{g}$ tends to zero, i.e. if $g \to 0$, the Bingham viscoplastic rheology leads to the Newtonian viscous fluid. It requires us to prove Equations 12.557 and 12.558 if $\hat{g} \to 0$.

We proved above that

$$\|\boldsymbol{u}\|_{L^2(I;V)} + \|\partial_t \boldsymbol{u}\|_{L^2(I;V')} \leq c \text{ for all } \hat{g} > 0,$$

where $\hat{g}$ is bounded. Now, let us assume that $\hat{g} \to 0$. Then we can take out a sequence, again denoted by $\{\boldsymbol{u}\}$, such that

$$\boldsymbol{u} \to \boldsymbol{w} \text{ in } L^2(I; V) \text{ weakly,} \tag{12.559}$$

$$\partial_t \boldsymbol{u} \to \partial_t \boldsymbol{w} \text{ in } L^2(I; V') \text{ weakly,} \tag{12.560}$$

$$u_i \to w_i \text{ in } L^2(\Omega \times I) \text{ strongly in } L^2(I; L^2(\Omega)) \text{ and a.e. in } \Omega \times I, \tag{12.561}$$

$$u_i u_j \to w_i w_j \text{ in } L^2(I; L^2(\Omega)), \tag{12.562}$$

where u_i, w_i are components of $\boldsymbol{u}$ and/or $\boldsymbol{w}$, where $\boldsymbol{w}$ is a solution of the Navier-Stokes problem: $\boldsymbol{u}_L \in L^2(I; V)$, $\partial_t \boldsymbol{u}_L \in L^2(I; V')$

$$(\partial_t \boldsymbol{u}_L(t), \boldsymbol{v}) + \hat{\mu} a(\boldsymbol{u}_L(t), \boldsymbol{v}) + b(\boldsymbol{u}_L(t), \boldsymbol{u}_L(t), \boldsymbol{v}) = (\boldsymbol{F}(t), \boldsymbol{v}) \quad \forall \boldsymbol{v} \in V, \tag{12.563}$$

$$\boldsymbol{u}_L(\boldsymbol{x}, t_o) = \boldsymbol{u}_o(\boldsymbol{x}), \tag{12.564}$$

where the coupled term was included into the body forces as above. Then, similarly as above,

$$b(\boldsymbol{u}, \boldsymbol{u}, \boldsymbol{v}) = -b(\boldsymbol{u}, \boldsymbol{v}, \boldsymbol{u}) \to -b(\boldsymbol{w}, \boldsymbol{v}, \boldsymbol{w}) \text{ in } L^2(I) \text{ weakly,} \quad \forall \boldsymbol{v} \in V. \tag{12.565}$$

Let us put $\boldsymbol{v} = \boldsymbol{v}(t)$, $\boldsymbol{v} \in L^2(I; V)$, then

$$\int_I [(\partial_t \boldsymbol{u}(t), \boldsymbol{v}(t) - \boldsymbol{u}(t)) + \hat{\mu} a(\boldsymbol{u}(t), \boldsymbol{v}(t) - \boldsymbol{u}(t)) + b(\boldsymbol{u}(t), \boldsymbol{u}(t), \boldsymbol{v}(t))$$
$$- (\boldsymbol{F}(t), \boldsymbol{v}(t) - \boldsymbol{u}(t))] \mathrm{d}t + J(\boldsymbol{v}(t)) - J(\boldsymbol{u}(t)) \geq 0,$$

then

$$\int_I [(\partial_t \boldsymbol{u}(t), \boldsymbol{v}(t)) + \hat{\mu} a(\boldsymbol{u}(t), \boldsymbol{v}(t)) - b(\boldsymbol{u}(t), \boldsymbol{v}(t), \boldsymbol{u}(t)) - (\boldsymbol{F}(t), \boldsymbol{v}(t) - \boldsymbol{u}(t))] \mathrm{d}t$$
$$+ J(\boldsymbol{v}(t)) - J(\boldsymbol{u}(t)) \geq \int_I [(\partial_t \boldsymbol{u}(t), \boldsymbol{u}(t)) + \hat{\mu} a(\boldsymbol{u}(t), \boldsymbol{u}(t))] \mathrm{d}t$$
$$\geq \tfrac{1}{2} |\boldsymbol{u}(t_1)|^2 - \tfrac{1}{2} |\boldsymbol{u}_o|^2 + \hat{\mu} \int_I a(\boldsymbol{u}(t), \boldsymbol{u}(t))] \mathrm{d}t.$$

Hence, using Equations 12.559–12.562 and 12.565 and since $\int_I \hat{g} j(\boldsymbol{u}(t))\mathrm{d}t \to 0$, we obtain

$$\begin{aligned}&\int_I [(\partial_t \boldsymbol{w}, \boldsymbol{v}) + \hat{\mu} a(\boldsymbol{w}, \boldsymbol{v}) - b(\boldsymbol{w}, \boldsymbol{v}, \boldsymbol{w}) - (\boldsymbol{F}, \boldsymbol{v} - \boldsymbol{w})]\mathrm{d}t \\ &\geq \liminf \frac{1}{2} |\boldsymbol{u}(t_1)|^2 - \frac{1}{2} |\boldsymbol{u}_o|^2 + \hat{\mu} \liminf \int_I a(\boldsymbol{u}(t), \boldsymbol{u}(t))\mathrm{d}t \\ &\geq \frac{1}{2} |\boldsymbol{w}(t_1)|^2 - \frac{1}{2} |\boldsymbol{u}_o|^2 + \hat{\mu} \int_I a(\boldsymbol{w}(t), \boldsymbol{w}(t))\mathrm{d}t \\ &= \int_I [(\partial_t \boldsymbol{w}, \boldsymbol{w}) + \hat{\mu} a(\boldsymbol{w}(t), \boldsymbol{w}(t))]\mathrm{d}t.\end{aligned}$$

Hence,

$$\begin{aligned}&\int_I [(\partial_t \boldsymbol{w}(t), \boldsymbol{v}(t) - \boldsymbol{w}(t)) + \hat{\mu} a(\boldsymbol{w}(t), \boldsymbol{v}(t) - \boldsymbol{w}(t)) - b(\boldsymbol{w}(t), \boldsymbol{v}(t), \boldsymbol{w}(t)) \\ &\quad - (\boldsymbol{F}(t), \boldsymbol{v}(t) - \boldsymbol{w}(t))]\mathrm{d}t \geq 0 \quad \forall \boldsymbol{v} \in L^2(I; V)\end{aligned}$$

and therefore, for almost everywhere on $\bar{I}$

$$\begin{aligned}&(\partial_t \boldsymbol{w}(t), \boldsymbol{v} - \boldsymbol{w}(t)) + \hat{\mu} a(\boldsymbol{w}(t), \boldsymbol{v} - \boldsymbol{w}(t)) - b(\boldsymbol{w}(t), \boldsymbol{v}, \boldsymbol{w}(t)) \\ &\quad - (\boldsymbol{F}(t), \boldsymbol{v} - \boldsymbol{w}(t)) \geq 0 \quad \forall \boldsymbol{v} \in V.\end{aligned}$$

Hence, putting $\boldsymbol{v} = \boldsymbol{w}(t) \pm \vec{\omega}$, $\vec{\omega} \in V$, it follows that

$$(\partial_t \boldsymbol{w}(t), \vec{\omega}) + \hat{\mu} a(\boldsymbol{w}(t), \vec{\omega}) - b(\boldsymbol{w}(t), \vec{\omega}, \boldsymbol{w}(t)) - (\boldsymbol{F}(t), \vec{\omega})) \geq 0 \quad \forall \vec{\omega} \in V. \tag{12.566}$$

As is known (see, for example, Temam (1977)) the Navier-Stokes problem in Equations 12.563 and 12.564 has a unique solution, and then from Equation 12.566 it follows that

$$\boldsymbol{w}(t) = \boldsymbol{u}_L(t),$$

which proves Equations 12.557 and 12.558. Moreover, due to Equations 12.525 and 12.526 it follows that

$$\partial_t \boldsymbol{u}(t) \to \partial_t \boldsymbol{u}_L(t) \text{ weakly in } L^2(I; V) \text{ and } *\text{-weakly in } L^2(I; H). \tag{12.567}$$

At the end we shall prove that there exist functions $m = \{m_{ij}\}$, the so-called multipliers, satisfying assumptions (A3), which continuously depend on $\hat{g}$ in a suitable topology. For this case from Equation 12.487 we have

$$\begin{aligned}&\int_I \{(\partial_t \boldsymbol{u}(t), \boldsymbol{v} - \boldsymbol{u}(t)) + \hat{\mu} a(\boldsymbol{u}(t), \boldsymbol{v} - \boldsymbol{u}(t)) + b(\boldsymbol{u}(t), \boldsymbol{u}(t), \boldsymbol{v} - \boldsymbol{u}(t)) + b_s(\Theta(t) \\ &\quad - \Theta_o, \boldsymbol{v} - \boldsymbol{u}(t)) - (\boldsymbol{F}(t), \boldsymbol{v} - \boldsymbol{u}(t))\}\mathrm{d}t + J(\boldsymbol{v}) - J(\boldsymbol{u}(t)) \\ &= J(\boldsymbol{v}) - J(\boldsymbol{u}(t)) - 2\hat{g} \int_I [(m_{ij}, D_{ij}(\boldsymbol{v}) - D_{ij}(\boldsymbol{u}))]\mathrm{d}t \\ &\equiv X \geq 0 \quad \forall \boldsymbol{v} \in W,\end{aligned} \tag{12.568}$$

as due to the assumptions (A3)(ii), (iii), and $j(\boldsymbol{u}) = 2(m_{ij}, D_{ij}(\boldsymbol{u}))$, we find

$$2(m_{ij}, D_{ij}(\boldsymbol{v})) \leq 2\int_{\Omega}(D_{ij}(\boldsymbol{v})D_{ij}(\boldsymbol{v}))^{1/2}\mathrm{d}\boldsymbol{x} = j(\boldsymbol{v});$$

and therefore $X \geq 0$.

The existence of the multipliers m_{ij} in parallel follows from the ideas of Duvaut and Lions (1976).

Let us define $B(\boldsymbol{u}, \boldsymbol{v}) \in V$ by

$$b(\boldsymbol{u}, \boldsymbol{v}, \boldsymbol{w}) = (B(\boldsymbol{u}, \boldsymbol{v}), \boldsymbol{w}), \boldsymbol{u}, \boldsymbol{v}, \boldsymbol{w} \in V,$$

and set

$$\underline{\boldsymbol{F}} = \partial_t \boldsymbol{u} - \hat{\mu}\Delta\boldsymbol{u} + B(\boldsymbol{u}, \boldsymbol{u}) - \boldsymbol{F}_o, \tag{12.569}$$

where $\boldsymbol{F}_o = \boldsymbol{F} - b_s(\Theta - \Theta_o, \boldsymbol{u})$, as $b_s(\Theta - \Theta_o, \boldsymbol{u}) \in L^2(\Omega)\,\forall t$. Then $\underline{\boldsymbol{F}} \in L^2(I; V')$ and the variational inequality can be written in the form

$$(\underline{\boldsymbol{F}}(t), \boldsymbol{v}) + \hat{g}j(\boldsymbol{v}) - [(\underline{\boldsymbol{F}}(t), \boldsymbol{u}(t)) + \hat{g}j(\boldsymbol{u})] \geq 0 \quad \forall \boldsymbol{v} \in V. \tag{12.570}$$

Let us put $\boldsymbol{v} = \pm\lambda\boldsymbol{v}$, $\lambda > 0$, then Equation 12.570 leads to

$$\lambda[\pm(\underline{\boldsymbol{F}}(t), \boldsymbol{v}) + \hat{g}j(\boldsymbol{v})] - [(\underline{\boldsymbol{F}}(t), \boldsymbol{u}(t)) + \hat{g}j(\boldsymbol{u}(t))] \geq 0.$$

Hence, putting $\boldsymbol{v} = \boldsymbol{v}(t)$, where $t \to \boldsymbol{v}(t) \in L^2(I; V)$, and then integrating over $t \in I$, we have

$$\lambda\left[\pm\int_{\mathrm{I}}(\underline{\boldsymbol{F}}(t), \boldsymbol{v})\mathrm{d}t + J(\boldsymbol{v})\right] - \left[\int_{\mathrm{I}}(\underline{\boldsymbol{F}}(t), \boldsymbol{u}(t))\mathrm{d}t + J(\boldsymbol{u}(t))\right] \geq 0, \quad \forall \boldsymbol{v} \in L^2(I; V), \quad \forall\lambda \geq 0. \tag{12.571}$$

Hence,

$$\left|\int_{\mathrm{I}}(\underline{\boldsymbol{F}}(t), \boldsymbol{v})\mathrm{d}t\right| \leq J(\boldsymbol{v}) \quad \forall \boldsymbol{v} \in L^2(I; V), \tag{12.572}$$

and

$$\int_{\mathrm{I}}(\underline{\boldsymbol{F}}(t), \boldsymbol{u}(t))\mathrm{d}t + J(\boldsymbol{u}(t)) = 0. \tag{12.573}$$

From Equation 12.571 it follows that $\int_{\mathrm{I}}(\underline{\boldsymbol{F}}(t), \boldsymbol{u}(t)\mathrm{d}t + J(\boldsymbol{u}(t)) \leq 0$ and since, due to Equation 12.572, $\int_{\mathrm{I}}(\underline{\boldsymbol{F}}(t), \boldsymbol{u}(t))\mathrm{d}t + J(\boldsymbol{u}(t)) \geq 0$, then Equation 12.573 is valid.

Let us introduce the space

$$W_o = \{w \mid w = \{w_{ij}\}, w_{ij} = w_{ji}, w_{ij} \in L^1(\Omega \times I)\},$$

with the norm

$$\|w\|_{W_o} = \int_{\Omega\times I}(w_{ij}, w_{ij})^{1/2}\mathrm{d}\boldsymbol{x}\,\mathrm{d}t.$$

Let $P : \boldsymbol{v} \to D_{ij}(\boldsymbol{v}) = \frac{1}{2}(\partial_j v_i + \partial_i v_j)$ be the mapping of $L^2(I; V) \to W_o$. Then Equation 12.572 is equivalent to

$$\left|\int_{\mathrm{I}}(\underline{\boldsymbol{F}}(t), \boldsymbol{v})\mathrm{d}t\right| \leq \hat{g}2^{1/2}\,\|P\boldsymbol{v}\|_{W_o}. \tag{12.574}$$

Applying the Hahn-Banach theorem, then there exists

$$m \in W_o' = \{w \mid w = \{w_{ij}\}, w_{ij} = w_{ji}, w_{ij} \in L^\infty(\Omega \times I)\},$$

such that

$$\int_I (\underline{\boldsymbol{F}}(t), \boldsymbol{v})\mathrm{d}t = -\hat{g}2^{1/2} \int_{\Omega \times I} m_{ij} D_{ij}(\boldsymbol{v})\mathrm{d}\boldsymbol{x}\,\mathrm{d}t \tag{12.575}$$

and such that

$$\|m\|_{W_o'} \le 1. \tag{12.576}$$

It is evident that we can take $\{m_{ij}\}$ in W_o' such that $m_{kk} = 0$ as $D_{kk} = 0$, and W_o' is dual to W_o.

Equation 12.576 is equivalent to $m_{ij}m_{ij} \le 1$ a.e. in $\Omega \times I$. By virtue of the definition of $\underline{\boldsymbol{F}}$, Equation 12.575 is equivalent to

$$\partial_j \tau_{ij} + f_i = \rho \partial_t v_i, \quad \text{with } \tau_{ij} \text{ defined by (A3)(iv).}$$

Using Equation 12.575, then Equation 12.573 is equivalent to

$$\int_{\Omega \times I} m_{ij} D_{ij}(\boldsymbol{u})\mathrm{d}\boldsymbol{x}\,\mathrm{d}t = \int_{\Omega \times I} (D_{ij}(\boldsymbol{u}) D_{ij}(\boldsymbol{u}))^{1/2} \mathrm{d}\boldsymbol{x}\,\mathrm{d}t.$$

Hence, since due to (A3)(ii) $m_{ij}m_{ij} \le 1$ a.e. in $\Omega \times I$, then

$$m_{ij} D_{ij}(\boldsymbol{u}) = (D_{ij}(\boldsymbol{u}) D_{ij}(\boldsymbol{u}))^{1/2} \text{ a.e. in } \Omega \times I.$$

Moreover, Equation 12.573 is equivalent to

$$\partial_t u_i + \partial_j u_i u_j = \partial_j \tau_{ij} + f_i,$$

which completes this part of the proof.

Summing all the above obtained results we have proved the existence of the weak solution of the investigated problem.

Remark: In reality the mushy zone is created not only by the liquid and solid phases, but instead of the pure liquid there exist microscopical elements of strongly visco-plastic rocks, near to the liquid phase and having the consistency of an alloy of both phases. To determine the boundary between this phase and the solid visco-plastic phase is very difficult but it can be achieved through the determination of the threshold of plasticity $\hat{g}$, by the volumic fractions of both phases, the visco-plastic and the strongly visco-plastic with a low threshold of plasticity, so that $f_S + f_L = 1$. The dependence of the pressure and the temperature plays a fundamental role as the phase diagrams show. Therefore, future problems must be studied with physical coefficients depending on temperature and pressure. Solving Problem $(\mathscr{P}_{\text{centh}})_v$ numerically, we find the pressure p directly (as a result of the Uzawa algorithm) and then put this into the physical parameters of the rocks. The temperature will be obtained as a solution of the thermal part of the studied problem. In such a way we can directly study the influences of temperatures and pressures in anomalous zones of the Earth as well as in anomalous conditions of technological problems connected with the construction of high level radioactive waste repositories.

12.4. MATHEMATICAL ANALYSIS OF GEODYNAMIC PROCESSES IN THE EARTH'S AND PLANET'S INTERIOR. GLOBAL GEODYNAMIC, GEOMAGNETIC AND GRAVITY MODELS

12.4.1. Governing Equations of Magnetodynamics of Incompressible Thermo-Bingham's Fluid Under the Gravity Effect

12.4.1.1. Introduction

One of the important problems in present geophysics is the connection of the geodynamic processes and the geomagnetic field in the Earth's interior and the mutual influence of the geomagnetic fields in the core and the mantle of the real Earth. So for the most part, the plastic and fluid motions in the Earth's interior consist of an axi-symmetric non-uniform rotation and an axi-symmetric statistical distribution of cyclonic convention. There may be giant circulation cells in the core and the mantle of the Earth violating this simple generalization, but we know so little about non-symmetric circulation that we cannot include it in any meaningful way.

Contemporary models of the Earth are built in isolation corresponding to individual geophysical fields. The model presented gives a new approach of how to study mutual relations between geodynamic and geomagnetic processes in the Earth's interior as well as in its separate parts (the core, the mantle and the lithosphere). The geodynamic model of the Earth is based on the assumption that the important factors in the course of the development of the Earth's body are gravity, radioactivity, temperature, density differentiation associated with phase transitions of the Earth's material under high temperatures, great pressures and chemical reactions under the same conditions (see Nedoma (1986)). Then the Earth's body can be characterized by a thermo-elasto-visco-plastic state in a relatively weak magnetic field. Consequently, for the derivation of equations describing geodynamic processes in an irregular rotating Earth's body, the thermo-elasto-visco-plastic rheology and the magnetodynamic theory can be used. In the book the Earth's body will be interpreted by a moving and heating incompressible Bingham's medium, which produces the magnetic and gravity fields. Boundary conditions on the surface of the Earth follow from our knowledge of the velocity, temperature and gravity potential fields on the Earth surface and from the presumption that the surface of the Earth is strongly electrically conductive. The model based on the thermo-Bingham rheology represents only a simple model of the real Earth. However, it gives the idea of how to solve the coupling of the geodynamic, geothermal and geomagnetic processes under the gravity effect in the core and the mantle of the Earth, as well as the coupling of the magnetic field in the core and the upper parts of the Earth, i.e. the mantle and the lithosphere.

Here we shall assume that the Earth's body and other planets and stars are approximated by the moving visco-plastic electrically-conducting incompressible thermo-Bingham's medium which is under the gravity effect. For the case of materials processing in the space outside the Earth, similar assumptions can be considered.

The classical equations of magnetodynamics were solved during the last few years in several papers. So an initial boundary value problem for a system of equations of magneto-dynamics of incompressible, electrically-conducting and viscous Newtonian fluid was studied in Stupjalis (1977). The incompressible electrically-conducting Bingham's fluid was first investigated by Duvaut and Lions (1972). The mathematical model of a global gravity field coupled with geodynamic and thermo-magnetodynamic processes in the Earth's interior based on thermo-Bingham's rheology was discussed in Nedoma (1986),(1988). Štědrý and Vejvoda (1983), (1985) prove the existence of time-periodic solutions of equations of magnetodynamics — the incompressible case in Štědrý and Vejvoda (1983),(1986) and the compressible case in Štědrý and Vejvoda (1985). Various existence results have been proved in numerous papers, we mention here only Milani (1980a,b) and Byhovskii (1957).

The aim here is to prove the existence and uniqueness of the time-dependent solutions to a system of equations appearing in generalized thermo-magnetodynamics based on the thermo-Bingham rheology. With sufficient accuracy the solutions simulate geodynamic and geomagnetic processes in the planet's and stellar interiors, as well as processes in materials processing in the cosmic environment.

Subsequently we shall deal with the incompressible moving thermo-Bingham fluid under gravity and magnetic effects.

We shall suppose that $\Omega \subset R^N$, $N = 2, 3$, is a bounded domain with a Lipschitz boundary $\partial\Omega$, occupied by a moving Bingham's fluid. Let $I = (t_o, t_1)$ be a time interval. We shall use the notation $u_{i,j} = \partial u_i/\partial x_j$, $\dot{\boldsymbol{u}} = \partial \boldsymbol{u}/\partial t(\equiv \boldsymbol{v})$ and $\mathrm{d}v_i/\mathrm{d}t = \partial v_i/\partial t + v_j v_{i,j}$. The Einstein summation convention over the range $1 \ldots, N$ will be used. Let $\boldsymbol{n}$ denote the unit outward normal vector at $\boldsymbol{x} \in \partial\Omega$. Let [.,.] denote the vector product in R^N and $\boldsymbol{N} = 2, 3$.

The strain-displacement relations are

$$\varepsilon_{ij}(\boldsymbol{u}) = \tfrac{1}{2}(u_{i,j} + u_{j,i}), \quad i, j = 1 \ldots N. \tag{12.577}$$

We introduce the notation $D_{ij}(\boldsymbol{v}) = e_{ij}(\dot{\boldsymbol{u}})$. The stress-strain relations of the thermo-Bingham medium is defined by

$$\tau_{ij} = \tau_{ij}^B + \tau_{ij}^T + \tau_{ij}^M, \tag{12.578}$$

where τ_{ij}^B is the Bingham stress defined by

$$\tau_{ij}^B = -p\delta_{ij} + \hat{g}D_{ij}/(D_{\text{II}})^{1/2} + 2\hat{\mu}D_{ij}, \tag{12.579}$$

and $\hat{g}$ and $\hat{\mu}$ are thresholds of plasticity and viscosity. Here p represents a spherical part of the stress tensor τ_{ij} (p has a meaning of pressure), δ_{ij} is the well-known Kronecker symbol, and $D_{\text{II}} = (1/2)D_{ij}D_{ij}$ (see, for example, Prager (1961), Duvaut and Lions (1972)). Due to Nowacki (1970), the thermal stress satisfies the relation

$$\tau_{ij}^T = \beta_{ij}(T - T_o), \tag{12.580}$$

where β_{ij} is a coefficient of thermal expansion, and $T_o > 0$ an initial temperature in which the medium is in the initial stress-strain state. Furthermore, due to Landau and

Lifshitz (1984), the Maxwell stress τ_{ij}^M is defined by

$$\tau_{ij}^M = H_i B_j - (\boldsymbol{HB})\delta_{ij}, \tag{12.581}$$

where $\boldsymbol{H}$ and $\boldsymbol{B}$ represent the intensity of the magnetic field and the magnetic induction vector, respectively, and $\boldsymbol{B} = \mu \boldsymbol{H}$, where μ is the magnetic susceptibility.

Without any loss of generality of the problem studied we can solve the following problem.

In dealing with the motion of the visco-plastic electrically-conducting incompressible thermo-Bingham's medium under the gravity effect, the following system of equations for the velocity $\boldsymbol{v} = (v_i)$, the magnetic intensity $\boldsymbol{B}$, the temperature T and the perturbed gravitational potential Φ can be considered in the form

$$\rho \mathrm{d}v_i/\mathrm{d}t = (-p\delta_{ij} + \hat{\mu} D_{ij}/(D_{II})^{1/2} + 2\hat{\mu} D_{ij} + \beta_{ij}(T - T_o) + (H_i B_j - (\boldsymbol{HB})\delta_{ij}))_{,j} + f_i, \boldsymbol{B} = \mu \boldsymbol{H} \quad \text{on } \Omega \times I, \tag{12.582}$$

$$\mathrm{Div}\, \boldsymbol{v} = 0 \quad \text{on } \Omega \times I, \tag{12.583}$$

$$\partial \boldsymbol{B}/\partial t + \mathrm{Rot}\, ((\mu\sigma)^{-1}\, \mathrm{Rot}\, \boldsymbol{B}) - \mathrm{Rot}\, [\boldsymbol{v}, \boldsymbol{B}] = 0 \quad \text{on } \Omega \times I, \tag{12.584}$$

$$\mathrm{Div}\, \boldsymbol{B} = 0 \quad \text{on } \Omega \times I, \tag{12.585}$$

$$\rho \beta_{ij} T_o D_{ij} + \rho c_e \mathrm{d}T/\mathrm{d}t = W - (\kappa_{ij} T_{,j})_{,i} \quad \text{on } \Omega \times I, \tag{12.586}$$

$$\Delta \Phi = -4\pi \rho G_c p K^{-1} + 2\omega^2 \quad \text{on } \Omega \times I, \tag{12.587}$$

with given ρ, $\hat{g}$, $\hat{\mu}$, β_{ij}, T_o, μ, σ, c_e, (κ_{ij}), G_c, K, ω. We define $\Psi = -4\pi\rho G_c p K^{-1} + 2\omega^2$.

Moreover, the following boundary value conditions

$$\boldsymbol{v} = 0 \quad \text{on } \partial\Omega \times I, \tag{12.588}$$

$$\boldsymbol{Bn} = 0 \quad \text{on } \partial\Omega \times I, \tag{12.589}$$

$$[\boldsymbol{n}, \mathrm{Rot}\, \boldsymbol{B}] = 0 \quad \text{on } \partial\Omega \times I, \tag{12.590}$$

$$T = 0 \quad \text{on } \partial\Omega \times I, \tag{12.591}$$

$$\Phi = 0 \quad \text{on } \partial\Omega \times I, \tag{12.592}$$

and the initial conditions

$$\boldsymbol{v}(\boldsymbol{x}, t_o) = 0, \boldsymbol{B}(\boldsymbol{x}, t_o) = 0, T(\boldsymbol{x}, t_o) = 0 \tag{12.593}$$

are considered for the above system of equations.

The generalized model of the real Earth is much more complicated (see Nedoma (1986)). In this model problem the metallized state of the rocks, the volume density of the electromagnetic force, the Joule heat, the viscous force evoked by the viscous stress, the Coriolis force, diffusion processes, etc. are omitted.

12.4.1.2. Variational formulation of the problem

We introduce the following notation of the employed spaces, scalar products, norms and forms.

We denote by $C(\Omega)$ the set of all functions defined and continuous on Ω, and by $C^k(\Omega)$, $1 \leq k < \infty$, the set of all functions defined on Ω which have continuous derivatives up to order k on Ω. Moreover, we denote by $C(\overline{\Omega})$ the space of all functions in $C(\Omega)$ which are bounded and uniformly continuous on $\overline{\Omega}$, and for an integer $1 \leq k < \infty$ we denote by $C^k(\overline{\Omega})$ the space of all functions $u \in C^k(\Omega)$ such that $D^\alpha u \in C(\overline{\Omega})$ for all $\alpha \in M_{N,k}$, where $M_{N,k}$ is the set of all N-dimensional multi-indices of length less than or equal to k. We denote $C^\infty(\Omega) = \cap_{k=0}^{\infty} C^k(\Omega)$ and $C^\infty(\overline{\Omega}) = \cap_{k=0}^{\infty} C^k(\overline{\Omega})$. We introduce the space $D(\Omega)$ as a space of all functions in $C^\infty(\Omega)$ with a compact support in Ω. The space is equipped with the ordinary countable system of semi-norms and as usual $D(\Omega, R^N) = [D(\Omega)]^N$.

We introduce the space $L^p(\Omega)$, $1 \leq p < \infty$, as the space of all measurable functions such that

$$\|f\|_{L^p(\Omega)} = \left(\int_\Omega |f(\boldsymbol{x})|^p \mathrm{d}\boldsymbol{x} \right)^{1/p} < \infty.$$

By $L^\infty(\Omega)$ we denote the set of all measurable functions f defined almost everywhere on Ω and such that $\|f\|_\infty = \text{ ess sup }_\Omega |f(\boldsymbol{x})|$ is finite. The space is endowed with this norm.

We shall employ the Sobolev spaces of functions having generalized derivatives of (possibly fractional) order s of the type $[H^s(\Omega)]^k \equiv H^{s,k}(\Omega)$ (see Lions and Magenes (1970), Nikol'skij (1975)). The norm will be denoted by $\|.\|_{s,k}$, and the scalar product by $(.,.)_s$ (for each integer k). We set $H^{o,k}(\Omega) \equiv L_2^k(\Omega) \equiv L^{2,k}(\Omega)$ and denote by $(.,.)$ the scalar product in $L_2^k(\Omega)$ (as there is no danger of any confusion with k). Denote by $H_o^{s,k}(\Omega)$ the closure of $D(\Omega)$ in the norm of the space $H^{s,k}(\Omega)$. Moreover, we shall denote by $\|.\|_E$ the equivalent norm in $H^{1,N}(\Omega)$ induced by the small strain tensor e_{ij}, i.e. $\|\boldsymbol{u}\|_E^2 = \int_\Omega e_{ij}(\boldsymbol{u}) e_{ij}(\boldsymbol{u}) \mathrm{d}\boldsymbol{x} + \|\boldsymbol{u}\|_{o,N}^2$.

Moreover, for $s \geq 1$ we define the following spaces

$$^1H^{s,N}(\Omega) = \{\boldsymbol{v} \mid \boldsymbol{v} \in H^{s,N}(\Omega),\ \text{Div } \boldsymbol{v} = 0; \boldsymbol{v} = 0 \quad \text{on } \partial\Omega, N = 2, 3\},$$

$$^2H^{s,N}(\Omega) = \{\boldsymbol{C} \mid \boldsymbol{C} \in H^{s,N}(\Omega),\ \text{Div } \boldsymbol{C} = 0; \boldsymbol{n}\boldsymbol{C} = 0 \quad \text{on } \partial\Omega, N = 2, 3\},$$

$$^3H^{s,1}(\Omega) = \{z \mid z \in H^{s,1}(\Omega), z = 0 \quad \text{on } \partial\Omega, N = 2, 3\},$$

$$\underline{H}^s = {}^1H^{s,N}(\Omega) \times {}^2H^{s,N}(\Omega) \times {}^3H^{s,1}(\Omega). \text{ We put } \underline{H}^1 = \underline{H}.$$

Then ${}^1\mathscr{H}(\Omega) = {}^1H^{1,N}(\Omega) \cap D(\Omega, R^N)$. ${}^1H^{s,N}(\Omega)$ is a Hilbert space with the norm $\|.\|_{s,N}$. For the sake of simplicity we put ${}^1H^{1,N}(\Omega) = {}^1H(\Omega)$, $\|.\|_1 = \|.\|_{1,N}$, and in $H^{o,N}(\Omega)\ \|.\|_o = \|.\|_{o,N}$. We shall denote by $\boldsymbol{w}_j$ eigen-functions of a canonical isomorphism ${}^1\Lambda_s : {}^1H^{s,N} \to ({}^1H^{s,N})'$, i.e. $(\boldsymbol{w}, \boldsymbol{v})_s = \lambda_i(\boldsymbol{w}_j, \boldsymbol{v})\, \forall \boldsymbol{v} \in {}^1H^{s,N}$, $\|\boldsymbol{w}_j\|_{o,1} = 1$ (Duvaut and Lions (1972a)). Moreover, ${}^2\mathscr{H}(\Omega) = {}^2H^{1,1}(\Omega) \cap D(\Omega, R^N)$. ${}^2H^{s,N}(\Omega)$, $s \geq 1$, is a Hilbert space with the norm $\|.\|_{s,N}$. We put ${}^2H(\Omega) = {}^2H^{1,N}(\Omega)$. Furthermore, ${}^3\mathscr{H}(\Omega) = {}^3H^{1,1}(\Omega) \cap D(\Omega)$. ${}^3H^{s,1}(\Omega)$, $s \geq 1$, is a Hilbert space with the norm $\|.\|_s$. We put ${}^3H(\Omega) = {}^3H^{1,1}(\Omega)$. Similarly as above, we define $\boldsymbol{C}_j$ and z_j, $\boldsymbol{C}_j \in {}^2H^{s,N}(\Omega)$, $z_j \in {}^3H^{s,1}(\Omega)$, eigen-functions of canonical isomorphisms ${}^2\Lambda_s : {}^2H^{s,N} \to ({}^2H^{s,N})'$ and ${}^3\Lambda_s : {}^3H^{s,1} \to ({}^3H^{s,1})'$. The dual space of ${}^1H^{s,N}(\Omega)$ we shall denote by $({}^1H^{s,N}(\Omega))'$, similarly in all other cases. Furthermore, we define the following spaces

$$
\begin{aligned}
{}^1\boldsymbol{H} &= \{\boldsymbol{v} \mid \boldsymbol{v} \in L^2(I; {}^1H^{s,N}(\Omega)), \boldsymbol{v}' \in L^2(I; {}^1H^{o,N}(\Omega)), \boldsymbol{v}(t_o) = 0\}, \\
{}^2\boldsymbol{H} &= \{\boldsymbol{C} \mid \boldsymbol{C} \in L^2(I; {}^2H^{s,N}(\Omega)), \boldsymbol{C}' \in L^2(I; {}^2H^{o,N}(\Omega)), \boldsymbol{C}(t_o) = 0\}, \\
{}^3\boldsymbol{H} &= \{z \mid z \in L^2(I; {}^3H^{s,1}(\Omega)), z' \in L^2(I; {}^3H^{o,1}(\Omega)), z(t_o) = 0\}, \\
\underline{\boldsymbol{H}} &= {}^1\boldsymbol{H} \times {}^2\boldsymbol{H} \times {}^3\boldsymbol{H}.
\end{aligned}
$$

Throughout this section we shall assume that the body forces $\boldsymbol{f}(\boldsymbol{x}, t) \in L^2(I, H^{1,N}(\overline{\Omega}))$ and thermal sources $W(\boldsymbol{x}, t) \in L^2(I; H^{1,1}(\Omega))$, $\overline{I} = \langle t_o, t_1 \rangle$. We suppose that the density ρ, the thresholds of plasticity and viscosity $\hat{g}$ and $\hat{\mu}$, the bulk modulus K, the specific heat c_e, the electric conductivity σ, and magnetic permeability μ are positive constants. G_c is the gravitational constant. The space-dependent coefficients of thermal expansion (β_{ij}) will be supposed such that $\partial\beta_{ij}/\partial x_j \in L^\infty(\Omega)\, \forall i, j \in \{1, \ldots, N\}$.

Moreover, we suppose that $\kappa_{ij} \in L^\infty(\Omega)$ are Lipschitz on Ω, fulfil the usual symmetry condition $\kappa_{ij} = \kappa_{ji}$ on Ω for every $i, j \in \{1, \ldots, N\}$ and

$$\kappa_{ij}\xi_i\xi_j \geq c_T \|\xi\|^2_{1,N}, \boldsymbol{x} \in \Omega, \xi \in R^N, c_T = \text{constant.} > 0. \tag{12.594}$$

For $\boldsymbol{v}, \boldsymbol{w} \in H^{1,N}(\Omega)$, $\boldsymbol{B}, \boldsymbol{C} \in H^{1,N}(\Omega)$, $T, z \in H^{1,1}(\Omega)$, $\Phi, \varphi \in H^{1,1}(\Omega)$ we put

$$a(\boldsymbol{v}, \boldsymbol{w}) = 2\int_\Omega D_{ij}(\boldsymbol{v})D_{ij}(\boldsymbol{w})\mathrm{d}\boldsymbol{x}, \tag{12.595}$$

$$b_M(\boldsymbol{B}, \boldsymbol{C}) = (\sigma\mu^{-1} \text{ Rot } \boldsymbol{B},\ \text{Rot } \boldsymbol{C}), \tag{12.596}$$

$$a_T(T, z) = \int_\Omega \kappa_{ij}T_{,i}z_{,j}\mathrm{d}\boldsymbol{x}, \tag{12.597}$$

$$a_g(\Phi, \varphi) = \int_\Omega \Phi_{,i}\varphi_{,i}\mathrm{d}\boldsymbol{x}, \tag{12.598}$$

$$j(\boldsymbol{v}) = 2\int_\Omega (D_{II}(\boldsymbol{v}))^{1/2}\mathrm{d}\boldsymbol{x}, \tag{12.599}$$

$$b_o(\boldsymbol{v}, g, z) = \int_\Omega \rho c_e v_k g_{,k} z\mathrm{d}\boldsymbol{x}, \tag{12.600}$$

$$b(\boldsymbol{u}, \boldsymbol{v}, \boldsymbol{w}) = \int_\Omega \rho u_i v_{j,i} w_j\mathrm{d}\boldsymbol{x}, \tag{12.601}$$

$$b_s(T, \boldsymbol{v}) = \int_\Omega (\beta_{ij}T)_{,j}v_i\mathrm{d}\boldsymbol{x}, \tag{12.602}$$

$$b_p(\boldsymbol{v}, g) = \int_\Omega \rho T_o\beta_{ij}v_{i,j}g\mathrm{d}\boldsymbol{x}. \tag{12.603}$$

Thus we introduce the bilinear form A on $\underline{H}$ for $\boldsymbol{a} \equiv (\boldsymbol{v}, \boldsymbol{B}, T)$, $\vec{\alpha} \equiv (\boldsymbol{w}, \boldsymbol{C}, z)$ by

$$A(\boldsymbol{a}, \vec{\alpha}) = \hat{\mu} a(\boldsymbol{v}, \boldsymbol{w}) + b_M(\boldsymbol{B}, \boldsymbol{C}) + a_T(T, z), \tag{12.604}$$

the matrix

$$\left.\begin{aligned} &\mathscr{E} = (\mathscr{E}_{ij}),\ \mathscr{E}_{ij} = \rho\delta_{ij}, && i = 1, \ldots, 2N+1, && j = 1, \ldots, N, \\ &\mathscr{E}_{ij} = \mu^{-1}\delta_{ij}, && i = 1, \ldots, 2N+1, && j = N+1, \ldots, 2N, \\ &\mathscr{E}_{ij} = \rho c_e \delta_{ij}, && i = 1, \ldots, 2N+1, && j = 2N+1, \end{aligned}\right\} \tag{12.605}$$

and the right-hand side

$$\boldsymbol{P} = (\boldsymbol{f}, O, W) \in L^2(I; H^{1,2N+1}(\Omega)). \tag{12.606}$$

Furthermore, we have

$$D_{ij}(\boldsymbol{u})D_{ij}(\boldsymbol{v}) \le 2D_{\mathrm{II}}^{1/2}(\boldsymbol{u})D_{\mathrm{II}}^{1/2}(\boldsymbol{v}), \quad D_{ij} = \frac{1}{2\hat{\mu}}(1 - \hat{g}\tau_{II}^{(-1/2)})\tau_{ij}^D,$$

where τ_{ij}^D are components of the stress tensor deviator and where $\tau_{\mathrm{II}} = (\hat{g} + 2\hat{\mu}D_{\mathrm{II}}^{1/2})^2$.

Let $\boldsymbol{v}, \boldsymbol{w} \in {}^1H^{1,N}(\Omega)$, $\boldsymbol{B}, \boldsymbol{C} \in {}^2H^{1,N}(\Omega)$. Then, according to Equation 12.582 and Maxwell's stresses for incompressible media, we have

(a) $(\partial/\partial x_i(\mu^{-1}\,|\boldsymbol{B}|^2), w_i) = 0$, since $\operatorname{Div} \boldsymbol{w} = 0$,

(b) $(\operatorname{Rot}(\eta_o \operatorname{Rot}\boldsymbol{B}), \mu^{-1}\boldsymbol{C}) = \int_{\partial\Omega} \sigma^{-1}\mu^{-2}[\boldsymbol{n}, \operatorname{Rot}\boldsymbol{B}]\boldsymbol{C}\,\mathrm{d}s + (\eta_o \operatorname{Rot}\boldsymbol{B}, \operatorname{Rot}(\mu^{-1}\boldsymbol{C}))$

$= (\eta_o \operatorname{Rot} \boldsymbol{B}, \operatorname{Rot}(\mu^{-1}\boldsymbol{C}))$ as $[\boldsymbol{n}, \operatorname{Rot} \boldsymbol{B}] = 0$ on $\partial\Omega$, where $\eta_o = (\sigma\mu)^{-1}$,

(c) $(\partial/\partial x_i(\mu^{-1}B_i\boldsymbol{B}), \boldsymbol{w}) = (\mu^{-1}\boldsymbol{w}, B_i\partial\boldsymbol{B}/\partial x_i - \operatorname{grad} \frac{1}{2}\,|\boldsymbol{B}|^2)$

$= (\mu^{-1}w_j, B_i(\partial B_j/\partial x_i - \partial B_i/\partial x_j)) = -(\mu^{-1}[\boldsymbol{w}, \boldsymbol{B}], \operatorname{Rot} \boldsymbol{B})$

$= -(\mu^{-1} \operatorname{Rot}[\boldsymbol{w}, \boldsymbol{B}], \boldsymbol{B})$ as $(\mu^{-1}\boldsymbol{w}, \operatorname{grad} \frac{1}{2}\,|\boldsymbol{B}|^2) = 0$

(for $\operatorname{Div} \boldsymbol{w} = 0$ and $\boldsymbol{w} = 0$ on $\partial\Omega$), hence

$(\partial/\partial x_i(\mu^{-1}B_i\boldsymbol{B}), \boldsymbol{w}) + (\mu^{-1} \operatorname{Rot}[\boldsymbol{w}, \boldsymbol{B}], \boldsymbol{B}) = 0$, (see Duvaut and Lions, (1972*a*, *b*)),

(d) $b(\boldsymbol{v}, \boldsymbol{v}, \boldsymbol{v}) = 0$ (see Temam (1977)).

We multiply Equations 12.582, 12.584, 12.586 and 12.587 by $\boldsymbol{w} - \boldsymbol{v}(t)$, $\boldsymbol{C} - \boldsymbol{B}(t)$, $z - T(t)$, and $\varphi - \Phi(t)$ respectively, and we add the first three equations. We integrate both the sum and the last equation over Ω and we apply the Green theorem, respecting the boundary conditions. According to $(a) - (c)$ the terms $(\partial/\partial x_i(\mu^{-1}\,|\boldsymbol{B}(t)|^2), w_i - v_i(t))$, $(\partial/\partial x_i(\mu^{-1}B_i(t)\boldsymbol{B}(t)), \boldsymbol{v}(t))$ and $(\mu^{-1} \operatorname{Rot}[\boldsymbol{v}(t), \boldsymbol{B}(t)], \boldsymbol{B}(t))$ disappear. Then, after certain modifications (including amongst others the integration in time over the interval I and the substitution $J(\boldsymbol{w}) = \int_I j(\boldsymbol{w}(t))\mathrm{d}t$, we obtain the following variational formulation.

Find a vector function $\boldsymbol{a} \equiv (\boldsymbol{v}, \boldsymbol{B}, T) \in \underline{\boldsymbol{H}}$ and a scalar function $\Phi \in {}^3H^{s,1}(\Omega)$ such that

$$\int_I [(\mathscr{E}\boldsymbol{a}'(t), \vec{\alpha}(t) - \boldsymbol{a}(t)) + A(\boldsymbol{a}(t), \vec{\alpha}(t) - \boldsymbol{a}(t)) + (\Xi(\boldsymbol{a}(t)), \vec{\alpha}(t) - \boldsymbol{a}(t))$$
$$+ (\boldsymbol{P}(t), \vec{\alpha}(t) - \boldsymbol{a}(t)) + \hat{g}j(\boldsymbol{w}(t)) - \hat{g}j(\boldsymbol{v}(t))]\mathrm{d}t \geq 0 \quad \forall\vec{\alpha} \equiv (\boldsymbol{w}, \boldsymbol{C}, z) \in \underline{\boldsymbol{H}}, \tag{12.607}$$

and

$$a_g(\Phi(t), \varphi(t) - \Phi(t)) - (\Psi(t), \varphi(t) - \Phi(t)) \geq 0 \quad \forall\varphi(t) \in {}^3H^{s,1}(\Omega) \quad \forall t \in I, \tag{12.608}$$

where

$$(\Xi(\boldsymbol{a}(t)), \vec{\alpha}(t) - \boldsymbol{a}(t)) = (\Xi_o(\boldsymbol{a}(t)), \boldsymbol{a}(t), \vec{\alpha}(t) - \boldsymbol{a}(t)) + b_s(T(t) - T_o, \boldsymbol{w}(t) - \boldsymbol{v}(t)) + b_p(\boldsymbol{v}(t), z(t) - T(t)), \tag{12.609}$$

$$(\Xi_o(\boldsymbol{a}(t)), \boldsymbol{a}(t), \vec{\alpha}(t) - \boldsymbol{a}(t)) = b(\boldsymbol{v}(t), \boldsymbol{v}(t), \boldsymbol{w}(t) - \boldsymbol{v}(t)) + b_o(\boldsymbol{v}(t), T(t), z(t) - T(t))$$
$$+ (\partial/\partial x_i(\mu^{-1}B_i(t)\boldsymbol{B}(t)), \boldsymbol{w}(t) - \boldsymbol{v}(t))$$
$$- (\mu^{-1} \text{ Rot } [\boldsymbol{v}(t), \boldsymbol{B}(t)], \boldsymbol{C}(t) - \boldsymbol{B}(t)). \tag{12.610}$$

PRELIMINARY RESULTS AND MAIN THEOREM.

Subsequently the following lemmas will be used.

Lemma 12.19: Let $s = \frac{1}{2}N$, then $v_{i,j} \in L^N(\Omega)$, for each $\boldsymbol{v} \in {}^1H^{s,N}$, $i, j = 1 \ldots, N$.
For the proof see Duvaut and Lions (1972a), Lions and Magenes (1970).

From the above given assumption the symmetry

$$a(\boldsymbol{u}, \boldsymbol{v}) = a(\boldsymbol{v}, \boldsymbol{u}), b_M(\boldsymbol{B}, \boldsymbol{C}) = b_M(\boldsymbol{C}, \boldsymbol{B}),$$
$$a_T(T, z) = a_T(z, T), A(\vec{\alpha}, \vec{\gamma}) = A(\vec{\gamma}, \vec{\alpha}),$$
$$a_g(\Phi, \varphi) = a_g(\varphi, \Phi)$$

holds. Moreover, they yield that for $\boldsymbol{v} \in {}^1H^{1,N}(\Omega)$, $\boldsymbol{B} \in {}^2H^{1,N}(\Omega)$, $T, \Phi \in {}^3H^{1,1}(\Omega)$ there exist constants $c_B > 0$, $c_M > 0$, $c_T > 0$, $c_\Phi > 0$ such that

$$a(\boldsymbol{v}, \boldsymbol{v}) \geq c_B \|\boldsymbol{v}\|^2_{1,N} \quad \text{for all } \boldsymbol{v} \in {}^1H^{1,N}(\Omega), \tag{12.611}$$

$$b_M(\boldsymbol{B}, \boldsymbol{B}) \geq c_M \|\boldsymbol{B}\|^2_{1,N} \quad \text{for all } \boldsymbol{B} \in {}^2H^{1,N}(\Omega), \tag{12.612}$$

(which is the corollary of Theorem 6.1, Chapter 7 of Duvaut and Lions (1972a)),

$$a_T(T, T) \geq c_T \|T\|^2_{1,1} \quad \text{for all } T \in {}^3H^{1,1}(\Omega), \tag{12.613}$$

$$a_g(\Phi, \Phi) \geq c_\Phi \|\Phi\|^2_{1,1} \quad \text{for all } \Phi \in {}^3H^{1,1}(\Omega). \tag{12.614}$$

Therefore, Equation 12.611-12.614 yield that for $\boldsymbol{a} = (\boldsymbol{v}, \boldsymbol{B}, T)$ and $\vec{\alpha} = (\boldsymbol{w}, \boldsymbol{C}, z)$ there exists a suitable positive constant c_a independent of $\vec{\alpha}$ such that

$$A(\vec{\alpha}, \vec{\alpha}) \geq c_a \, |\vec{\alpha}|^2_{1,2N+2} \qquad \text{for all } \vec{\alpha} \in \underline{H}. \tag{12.615}$$

From the Hölder inequality applied to $b(\boldsymbol{u}, \boldsymbol{v}, \boldsymbol{w})$ we have

$$|b(\boldsymbol{u}, \boldsymbol{v}, \boldsymbol{w})| \leq c_1 \, \|\boldsymbol{u}\|_{L^{p,N}(\Omega)} \, \|\boldsymbol{w}\|_{L^{p,N}(\Omega)} \sum_{i,j} \left\|D_i v_j\right\|_{L^{N,N}(\Omega)}, \tag{12.616}$$

where $2/p + 1/N = 1$, i.e.

$$p = 2N/(N-1). \tag{12.617}$$

Applying Lemma 12.19 we have

$$|b(\boldsymbol{u}, \boldsymbol{v}, \boldsymbol{w})| \leq c_2 \, \|\boldsymbol{u}\|_{L^{p,N}(\Omega)} \, \|\boldsymbol{w}\|_{L^{p,N}(\Omega)} \, \|\boldsymbol{v}\|_{s,N}, \; s = \tfrac{1}{2}N, \tag{12.618}$$

(see Duvaut and Lions (1972a)). Since

$$\|\boldsymbol{v}\|_{1/2,N} \leq c \, \|\boldsymbol{v}\|^{1/2}_{L^{2,N}(\Omega)} \, \|\boldsymbol{v}\|^{1/2}_{1,N} \text{ and } H^{1/2,N}(\Omega) \subset L^{p,N}(\Omega), \tag{12.619}$$

for p from Equation 12.617, then

$$\|\boldsymbol{v}\|_{L^{p,N}(\Omega)} \leq c_3 \, \|\boldsymbol{v}\|^{1/2}_{1,N} \, \|\boldsymbol{v}\|^{1/2}_{L^{p,N}(\Omega)} \qquad \forall \boldsymbol{v} \in H^{1,N}_o(\Omega), \tag{12.620}$$

where p is defined by Equation 12.617.
From Equation 12.618 and 12.620 we obtain

$$|b(\boldsymbol{u}, \boldsymbol{v}, \boldsymbol{w})| \leq c_4 \, \|\boldsymbol{u}\|^{1/2}_{1,N} \, \|\boldsymbol{u}\|^{1/2}_{o,N} \, \|\boldsymbol{w}\|^{1/2}_{1,N} \, \|\boldsymbol{w}\|^{1/2}_{o,N} \, \|\boldsymbol{v}\|_{s,N}, \qquad s = \tfrac{1}{2}N. \tag{12.621}$$

Furthermore, if $\boldsymbol{u}, \boldsymbol{v}, \boldsymbol{w} \in {}^1\mathscr{H}(\Omega)$ then $b(\boldsymbol{u}, \boldsymbol{v}, \boldsymbol{w}) + b(\boldsymbol{u}, \boldsymbol{w}, \boldsymbol{v}) = 0$. The last equality is also valid for $\boldsymbol{u} \in L^{2,N}(\Omega)$, $\boldsymbol{v}, \boldsymbol{w} \in {}^1H^{s,N}(\Omega)$. It also holds that $b(\boldsymbol{u}, \boldsymbol{u}, \boldsymbol{v}) = -b(\boldsymbol{u}, \boldsymbol{v}, \boldsymbol{u})$ for $\boldsymbol{u}, \boldsymbol{v} \in {}^1\mathscr{H}(\Omega)$.

Similarly $b_o(\boldsymbol{u}, y, z) + b_o(\boldsymbol{u}, z, y) = 0$, $\boldsymbol{u} \in {}^1H^{0,N}(\Omega)$, $y, z \in H^{1,1}_o(\Omega)$ and there exist positive constants c_5, c_6, c_7 independent of $\boldsymbol{u}$, y, z such that

$$\begin{aligned} |b_o(\boldsymbol{u}, y, z)| &\leq c_5 \, \|\boldsymbol{u}\|_{L^{p,N}(\Omega)} \, \|y\|_{L^p(\Omega)} \, \|\{D_i z\}\|_{L^{N,N}(\Omega)} \\ &\leq c_6 \, \|\boldsymbol{u}\|_{L^{p,N}(\Omega)} \, \|y\|_{L^p(\Omega)} \, \|z\|_{s,1} \\ &\leq c_7 \, \|\boldsymbol{u}\|^{1/2}_{1,N} \, \|\boldsymbol{u}\|^{1/2}_{o,N} \, \|y\|^{1/2}_{1,1} \, \|y\|^{1/2}_{o,1} \, \|z\|_{s,1}, \qquad s = \tfrac{1}{2}N, \end{aligned} \tag{12.622}$$

where the second and third inequality hold for $z \in H^{s,1}(\Omega)$ only, and for the third one the additional relation $\boldsymbol{u} \in H^{1,N}(\Omega)$ is needed. Since

$$\begin{aligned} (\partial/\partial x_i(\mu^{-1} B_i \boldsymbol{B}), \boldsymbol{w}) &= \int_\Omega \mu^{-1}(\boldsymbol{B} B_{i,i} + B_i \boldsymbol{B}_{,i}) \boldsymbol{w} \, \mathrm{d}\boldsymbol{x} \\ &= (\mu^{-1} B_i \boldsymbol{B}, \partial \boldsymbol{w}/\partial x_i) \quad (\text{ due to Div } \boldsymbol{B} = 0), \end{aligned} \tag{12.623}$$

and

$$(\mu^{-1}\ \text{Rot}\ [\boldsymbol{v}, \boldsymbol{B}], \boldsymbol{C}) = \int_\Omega \mu^{-1}(v_{i,j}B_j - v_{j,j}B_i + v_iB_{j,j} - v_jB_{i,j})C_i \mathrm{d}\boldsymbol{x}$$
$$= -\int_\Omega \mu^{-1}(v_iB_jC_{j,i} - v_jB_iC_{j,i} - v_iB_jC_{i,j} + v_jB_iC_{i,j})\mathrm{d}\boldsymbol{x}$$
$$= -\int_\Omega \mu^{-1}(v_iB_j - v_jB_i)(C_{j,i} - C_{i,j})\mathrm{d}\boldsymbol{x}$$
$$= -(\mu^{-1}[\boldsymbol{v}, \boldsymbol{B}], \text{Rot}\ \boldsymbol{C}), \tag{12.624}$$

where the Green theorem and the boundary conditions were applied, then (as in Equation 12.615 and 12.617) there exist positive constants c_8, c_9, c_{10} and c_{11} such that

$$\left|(\partial/\partial x_i(\mu^{-1}B_i\boldsymbol{B}), \boldsymbol{w})\right| \le c_8\, \|\boldsymbol{B}\|^2_{L^{p,N}(\Omega)}\, \|D_i\boldsymbol{w}\|_{L^{N,N}(\Omega)}\,,\ p \text{ from Equation 12.616,}$$
$$\le c_9\, \|\boldsymbol{B}\|_{1,N}\, \|\boldsymbol{B}\|_{o,N}\, \|\boldsymbol{w}\|_{s,N}\,, \tag{12.625}$$

and

$$\left|(\mu^{-1}\ \text{Rot}\ [\boldsymbol{v}, \boldsymbol{B}], \vec{\boldsymbol{C}}\right| \le c_{10}\, \|\boldsymbol{v}\|_{L^N_p(\Omega)}\, \|\boldsymbol{B}\|_{L^N_p(\Omega)}\, \|\boldsymbol{C}\|_{s,N}$$
$$\le c_{11}\, \|\boldsymbol{v}\|^{1/2}_{1,N}\, \|\boldsymbol{v}\|^{1/2}_{o,N}\, \|\boldsymbol{B}\|^{1/2}_{1,N}\, \|\boldsymbol{B}\|^{1/2}_{o,N}\, \|\boldsymbol{C}\|_{s,N}\,, \tag{12.626}$$

where the second and third inequality are valid for $\boldsymbol{C} \in H^{s,N}(\Omega)$ and the third one for $\boldsymbol{v} \in H^{1,N}(\Omega)$.

Therefore, due to the estimates in Eqautions 12.616, 12.618, 12.621, 12.622, 12.625 and 12.626 there exist positive constants c_{12}, c_{13} and c_{14} independent of $\vec{\alpha}$, $\vec{\gamma}$ such that

$$|\Xi_o(\vec{\alpha}, \vec{\alpha}, \vec{\gamma})| \le c_{12}\, \|\vec{\alpha}\|^2_{L^{p,2N+1}(\Omega)} \sum_{i,j} \left\|D_i\gamma_j\right\|_{L^{N,2N+1}(\Omega)}$$
$$\le c_{13}\, \|\vec{\alpha}\|^2_{L^{p,2N+1}(\Omega)}\, \|\vec{\gamma}\|_{s,2N+1}$$
$$\le c_{14}\, \|\vec{\alpha}\|_{1,2N+1}\, \|\vec{\alpha}\|_{o,2N+1}\, \|\vec{\gamma}\|^{1/2}_{s,2N+1}\,, \quad s = \tfrac{1}{2}N,$$

where the second and third inequality are valid for $\vec{\gamma} \in H^{s,2N+1}(\Omega)$ and the third for $\vec{\alpha} \in H^{1,2N+1}(\Omega)$.

The main result of this section gives the following theorem.

Theorem 12.18: Let $N \ge 2$, $s = \frac{1}{2}N$. Let

$$\boldsymbol{f} \in L^2(I; ({}^1H^{s,N}(\Omega))'),\ W \in L^2(I; ({}^3H^{s,N}(\Omega))'). \tag{12.628}$$

Then there exist vector function $\boldsymbol{a} \equiv (\boldsymbol{v}, \boldsymbol{B}, T)$ and scalar function Φ such that

$$\boldsymbol{v} \in L(I; {}^1H^{s,N}(\Omega)) \cap L^\infty(I; {}^1H^{o,N}(\Omega)), \tag{12.629}$$

$$v' \in L^2(I; ({}^1H^{s,N}(\Omega))'), \tag{12.630}$$

$$B \in L^2(I; {}^2H^{s,N}(\Omega)) \cap L^\infty(I; {}^2H^{o,N}(\Omega)), \tag{12.631}$$

$$B' \in L^2(I; ({}^2H^{s,N}(\Omega))'), \tag{12.632}$$

$$T \in L^2(I, {}^3H^{s,1}(\Omega)) \cap L^\infty(I; {}^3H^{o,1}(\Omega)), \tag{12.633}$$

$$T' \in L^2(I; ({}^3H^{s,1}(\Omega))'), \tag{12.634}$$

$$\Phi \in L^2(I; {}^3H^{1,1}(\Omega)), \tag{12.635}$$

$$v(t_o) = 0, B(t_o) = 0, T(t_o) = 0, \tag{12.636}$$

and which satisfy the variational inequality and equality conditions of Equations 12.607 and 12.608.

PROOF OF THEOREM 12.18

To prove Theorem 12.18 the double regularization for the problem in Equation 12.607 and the Lax-Milgram theorem for the problem in Equation 12.608 will be used.

Let $J_\varepsilon(w)$ be a regularized functional of the functional $J(w)$ defined by

$$J_\varepsilon(w) = \int_I \hat{g} j_\varepsilon(w(t))\mathrm{d}t, \tag{12.637}$$

where
$$j_\varepsilon(w(t)) = \frac{2}{1+\varepsilon}\int_\Omega (D_{II}(w(t)))^{(1/2)(1+\varepsilon)}\mathrm{d}x, \ \varepsilon > 0,$$

for which

$$(J'_\varepsilon(v), w) = \int_{\Omega\times I} \hat{g} D_{\mathrm{II}}(v)^{(1/2)(\varepsilon-1)} D_{ij}(v) D_{ij}(w)\mathrm{d}x\,\mathrm{d}t. \tag{12.638}$$

Now for the problem in Equation 12.60 we introduce the regularized problem $(\mathscr{P}_r)_v$:

Find a vector function $a_{\varepsilon\eta} \equiv (v_{\varepsilon\eta}, B_{\varepsilon\eta}, T_{\varepsilon\eta}) \in \underline{H}$ such that

$$\begin{aligned}\int_I [(\mathscr{E} a'_{\varepsilon\eta}(t), \vec{\alpha}(t) - a_{\varepsilon\eta}(t)) + A(a_{\varepsilon\eta}(t), \vec{\alpha}(t) - a_{\varepsilon\eta}(t)) + \eta(a_{\varepsilon\eta}(t), \vec{\alpha}(t) - a_{\varepsilon\eta}(t))_s \\ + (\Xi(a_{\varepsilon\eta}(t)), \vec{\alpha}(t) - a_{\varepsilon\eta}(t)) - (P(t), \vec{\alpha}(t) - a_{\varepsilon\eta}(t))]\mathrm{d}t + J_\varepsilon(w) - J_\varepsilon(v_{\varepsilon\eta}(t)) \\ \geq 0 \quad \forall\vec{\alpha} \equiv (w, C, z) \in H,\end{aligned} \tag{12.639}$$

where Ξ is defined in Equation 12.609, $s = \frac{1}{2}N$ and η is a positive number. For $N = 2$ we obtain $s = 1$ and the added term is of the same order as the bilinear forms and therefore can be omitted. We note that the added term has the physical meaning of viscosity.

The method of the proof is as follows:

(a) the existence of the solution of Equation 12.639 based on the Galerkin approximation will be proved,

(b) a priori estimates I and II independent of ε and η will be derived,

(c) limitation processes over m, ε and η will be performed,

(d) we shall prove the uniqueness of the solution of Equation 12.607,

(e) the existence and uniqueness of the solution of Eqaution 12.608 for every $t \in I$ will be proved via the usual elliptic equation technique.

The existence of $\boldsymbol{a}_{\varepsilon\eta}$ will be proved by means of the finite-dimensional approximation. Let $M_H = \{\vec{\alpha}_i\}$ be a countable basis of the space $\underline{H}^s(\Omega)$, i.e. each finite subset of M_H is linearly independent and span $\{\vec{\alpha}_i \mid i = 1, 2, \ldots\}$ is dense in $\underline{H}^s(\Omega)$, as $\underline{H}^s(\Omega)$ is a separable space. Let $\mathscr{W}^m$ be the space spanned by $\{\vec{\alpha}_j \mid 1 \leq j, k \leq m\}$. Then the approximate solution $\boldsymbol{a}_m$ of order m satisfies

$$(\mathscr{E}\boldsymbol{a}'_m(t), \vec{\alpha}_j) + A(\boldsymbol{a}_m(t), \vec{\alpha}_j) + \eta(\boldsymbol{a}_m(t), \vec{\alpha}_j)_s + (\Xi(\boldsymbol{a}_m(t)), \vec{\alpha}_j) + \hat{g}(j'_\varepsilon(\boldsymbol{a}_m(t), \alpha_j)$$
$$- (\boldsymbol{P}(t), \vec{\alpha}_j) = 0, \quad 1 \leq j, k \leq m, \tag{12.640}$$
$$\boldsymbol{a}_m(\boldsymbol{x}, t_o) = 0. \tag{12.641}$$

Since $\{\vec{\alpha}_j\}_{j=1}^m$ are linearly independent, then the system in Equations 12.640 and 12.641 is a regular system of ordinary differential equations of first order and therefore Equations 12.640 and 12.641 uniquely define $\boldsymbol{a}_m$ on the interval $\overline{I}_m = \langle t_o, t_m \rangle$. Therefore, Equation 12.640 is valid for every test function $\vec{\gamma}(t) = \sum_{i=1}^m c_i(t)\alpha_i$, $t \in \overline{I}_m$ where c_i are continuously differentiable functions on $\overline{I}_m$, $i = 1, \ldots, m$. In particularly, it holds for $\vec{\gamma}(t) = \boldsymbol{a}_m(t)$, $t \in I_m$.

A priori estimates I and II show that $t_m = t_1$.

Consider priori estimate I. Using (c) and (d) mentioned above we have

$$(\partial/\partial x_i(\mu^{-1}B_{mi}(t)\boldsymbol{B}_m(t)), \boldsymbol{v}_m(t)) + (\mu^{-1} \operatorname{Rot} [\boldsymbol{v}_m(t), \boldsymbol{B}_m(t)], \boldsymbol{B}_m(t)) = 0, \tag{12.642}$$
$$b(\boldsymbol{v}_m(t), \boldsymbol{v}_m(t), \boldsymbol{v}_m(t)) = 0, \; b_o(\boldsymbol{v}_m(t), T_m(t), T_m(t)) = 0.$$

Let us introduce the notation

$$X_m = b_s(T_m - T_o, \boldsymbol{v}_m) + b_p(\boldsymbol{v}_m, T_m). \tag{12.643}$$

Let $\partial\beta_{ij}/\partial x_j \in L^\infty(\Omega)\,\forall i, j$. Then there exists a positive constant c independent of m (and also of ε and η) such that

$$|X_m| = \left|b_s(T_m - T_o, \boldsymbol{v}_m) + b_p(\boldsymbol{v}_m, T_m)\right|$$
$$= \left|\int_\Omega (\partial/\partial x_j(\beta_{ij}(T_m(t) - T_o))v_i(t) + \rho T_o\beta_{ij}\partial v_i(t)/\partial x_j T_m(t))\mathrm{d}\boldsymbol{x}\right|$$
$$\leq c(1 + \|T_m(t)\|_{1,1}\,\|\boldsymbol{v}_m(t)\|_{o,N} + \|T_m(t)\|_{o,1}\,\|\boldsymbol{v}_m(t)\|_{1,N}). \tag{12.644}$$

Via the integration of Equation 12.640 (with $\vec{\gamma}(t) = \boldsymbol{a}_m(t)$) in time over $I_m = (t_o, t_m)$ we obtain

$$\int_{I_m} [(\mathscr{E}\boldsymbol{a}'_m(t), \boldsymbol{a}_m(t)) + A(\boldsymbol{a}_m(t), \boldsymbol{a}_m(t)) + (\Xi(\boldsymbol{a}_m(t)), \boldsymbol{a}_m(t)) + \eta(\boldsymbol{a}_m(t), \boldsymbol{a}_m(t))_s$$
$$- (\boldsymbol{P}(t), \boldsymbol{a}_m(t))]\mathrm{d}t + (J'_\varepsilon(\boldsymbol{v}_m(t), \boldsymbol{a}_m(t)) = 0. \tag{12.645}$$

Since $(j'_\varepsilon(\boldsymbol{v}), \boldsymbol{v}) \geq 0$, according to the ellipticity (Equations 12.611–12.614) of the bilinear form $A(\boldsymbol{a}, \vec{\alpha})$, and due to Equations 12.642 and 12.644, after some modification we obtain

$$
\begin{aligned}
& e_o \, \|\boldsymbol{a}_m(t)\|_{o,N}^2 + 2c_a \int_{I_m} \|\boldsymbol{a}_m(\tau)\|_{1,N}^2 \, \mathrm{d}\tau + 2\eta \int_{I_m} \|\boldsymbol{a}_m(\tau)\|_s \, \mathrm{d}\tau \\
& \leq 2c \int_{I_m} (1 + \|T_m(\tau)\|_{1,1} \, \|\boldsymbol{v}_m(\tau)\|_{o,N} + \|T_m(\tau)\|_{o,1} \, \|\boldsymbol{v}_m(\tau)\|_{1,N}) \mathrm{d}\tau \\
& \quad + 2 \int_{I_m} \|\boldsymbol{P}(\tau)\|_* \, \|\boldsymbol{a}_m(\tau)\|_{1,N} \, \mathrm{d}\tau, \qquad (12.646)
\end{aligned}
$$

where $e_o = \min\{\rho^{1/2}, \mu^{-1/2}, (\rho c_e)^{1/2}\}$. According to the Gronwall lemma, after some modification we have the following estimates

$$
\|\boldsymbol{a}_m(t)\|_{o,N} \leq c, \, t \in I, \int_I \|\boldsymbol{a}_m(\tau)\|_{1,N}^2 \, \mathrm{d}\tau \leq c, \, \eta \int_I \|\boldsymbol{a}_m(\tau)\|_s^2 \, \mathrm{d}\tau \leq c, \qquad (12.647)
$$

where $c = \text{constant} > 0$ independent of m.

From these estimates we obtain

$$
\{\boldsymbol{a}_m(t); m \in \boldsymbol{N}\} \text{ is a bounded subset in } L^2(I; H), \qquad (12.648)
$$

$$
\{\eta^{1/2}\boldsymbol{a}_m(t); m \in \boldsymbol{N}\} \text{ is a bounded subset in } L^2(I; H^s), \qquad (12.649)
$$

We now consider a priori estimate II. We shall show that

$$
\boldsymbol{a}_m'(t) \text{ is a bounded subset of } L^2(I; (H^s)'). \qquad (12.650)
$$

By virtue of equations 12.627, 12.648 and 12.649 we obtain

$$
\begin{aligned}
& \Xi_o(\boldsymbol{a}_m(t), \boldsymbol{a}_m(t), \vec{\alpha}(t)) = (\boldsymbol{h}_m(t), \vec{\alpha}(t)), \quad \vec{\alpha}(t) \in L^2(I; H^{s,2N+1}(\Omega)), \\
& \|\boldsymbol{h}_m(t)\|_{L^2(I;({}^1H^{s,N}(\Omega))')} \leq c, \qquad (12.651)
\end{aligned}
$$

where c is a positive constant independent of m, ε, η and where the separate terms of $\Xi_o(\boldsymbol{a}_m(t), \boldsymbol{a}_m(t), \vec{\alpha}(t))$ have a sense due to the estimates in Equations 12.617, 12.619, 12.620, 12.623 and 12.624.

Linear form $\boldsymbol{a} \to A(\boldsymbol{a}, \vec{\alpha})$ for fixed $\alpha \in \underline{H}$ is continuous on $\underline{H}$ so that

$$
A(\boldsymbol{a}(t), \vec{\alpha}) = (A_B\boldsymbol{a}(t), \vec{\alpha}), A_B \in \mathscr{L}(\underline{H}, (\underline{H})'), \qquad (12.652)
$$

$$
(\boldsymbol{a}(t), \vec{\alpha})_s = (\Lambda_s\boldsymbol{a}(t), \vec{\alpha}), \Lambda_s \in \mathscr{L}(\underline{H}, (\underline{H})') \qquad (12.653)
$$

Then Equations 12.641 and 12.642 are equivalent to

$$
(\mathscr{E}\boldsymbol{a}_m' + A_B\boldsymbol{a}_m + \eta\Lambda_s\boldsymbol{a}_m + \hat{g}j_\varepsilon'(\boldsymbol{v}_m) + \boldsymbol{h}_m - \boldsymbol{P}, \vec{\alpha}_j) = 0, \quad 1 \leq j \leq m. \qquad (12.654)
$$

Let S_m be orthogonal projections $\underline{H}^o \to \mathscr{W}^m$, where $\mathscr{W}^m$ was defined above. Then

$$
S_m\boldsymbol{h} = \sum_{j=1}^{m} (\boldsymbol{h}, \boldsymbol{w}_j)\boldsymbol{w}_j.
$$

Hence from Equation 12.654 and the fact that $S_m\boldsymbol{a}_m' = \boldsymbol{a}_m$ we obtain

$$
\mathscr{E}\boldsymbol{a}' = S_m(\boldsymbol{P} - A_B\boldsymbol{a}_m - \eta\Lambda_s\boldsymbol{a}_m - \hat{g}j_\varepsilon'(\boldsymbol{v}_m) - \boldsymbol{h}_m). \qquad (12.655)
$$

Then according to Equations 12.648 and 12.653

$$A_B \boldsymbol{a}_m \text{ is a bounded subset of } L^2(I; (\underline{H})') \cap L^2(I; (\underline{H}^s)'),$$

and therefore from Equation 12.649

$$\|\eta \Lambda_s \boldsymbol{a}_m\|_{L^2(I;(\underline{H})')} = O(\eta^{1/2}).$$

But due to Equation 12.638

$$\left\| j'_\varepsilon(\boldsymbol{v}) \right\|_* \leq c \left(\int_\Omega (D_{\mathrm{II}}(\boldsymbol{v}))^\varepsilon \mathrm{d}\boldsymbol{x} \right)^{1/2},$$

where $\|\cdot\|_*$ is the norm in $({}^1H^{s,N}(\Omega))'$, and then

$$j'_\varepsilon(\boldsymbol{v}_m) \text{ is a bounded subset of } L^2(I; ({}^1H^{1,N}(\Omega))').$$

Thus Equation 12.655 indicates that $\boldsymbol{a}'_m = S_m p_m$, where $p_m \in \mathscr{P}_P \subset L^2(I; (\underline{H}^s)')$, and where $\mathscr{P}_P$ is a bounded subset of $L^2(I; (\underline{H}^s)')$.

Since $\lambda^{1/2}\vec{\alpha}_j$ is an orthogonal system in $(\underline{H}^s)'$ (with respect to the norm $\|\boldsymbol{p}\|^2_{\underline{H}^s} = \left\|\Lambda_s^{-1}\boldsymbol{p}\right\|_{\underline{H}^s}$) and since $\|\boldsymbol{p}\|^2_{(\underline{H}^s)'} = \Sigma_{j=1}^m(\boldsymbol{p}, \alpha_j\lambda_j^{1/2})$, $\|S_m\boldsymbol{p}\|^2_{(\underline{H}^s)'} = \Sigma_{j=1}^m(\boldsymbol{p}, \alpha_j\lambda_j^{1/2})$, then

$$\|S_m \boldsymbol{p}\|_{(\underline{H}^s)'} \leq c\, \|\boldsymbol{p}\|_{(\underline{H}^s)'}, \tag{12.656}$$

which proves Equation 12.650.

We now investigate the passages to the limit over m. At first the limitation process with m, i.e. the convergence of the finite-dimensional approximation, will be proved for ε, η being fixed.

From the estimates I and II and from Equations 12.647–12.649 the subsequence $\{\boldsymbol{a}_\mu(t), \mu \in \boldsymbol{N}\}$ of the sequence $\{\boldsymbol{a}_m, m \in \boldsymbol{N}\}$ can be taken such that

$$\begin{aligned}
&\boldsymbol{a}_\mu \to \boldsymbol{a}_{\varepsilon\eta} \text{ in } L^\infty(I; \underline{H}^o) \text{ *-weakly},\\
&\boldsymbol{a}_\mu \to \boldsymbol{a}_{\varepsilon\eta} \text{ in } L^2(I; \underline{H}^1) \text{ weakly},\\
&\boldsymbol{a}'_\mu \to \boldsymbol{a}'_{\varepsilon\eta} \text{ in } L^2(I; (\underline{H}^s)') \text{ weakly},\\
&\boldsymbol{a}_\mu \to \boldsymbol{a}_{\varepsilon\eta} \text{ in } L^2(I; \underline{H}^o) \text{ strongly}.
\end{aligned} \tag{12.657}$$

Thus for the appropriate components $\boldsymbol{v}_\mu$ and $\boldsymbol{B}_\mu$ of $\vec{\alpha}_\mu$ we have

$$\begin{aligned}
&v_{i\mu} \to v_{i\varepsilon\eta} \text{ weakly a.e. in } G \times I,\\
&B_{i\mu} \to B_{i\varepsilon\eta} \text{ weakly a.e. in } G \times I.
\end{aligned} \tag{12.658}$$

(where $v_{i\mu}$, $v_{i\varepsilon\eta}$, $B_{i\mu}$, $B_{i\varepsilon\eta}$ represent the i-th component of $\boldsymbol{v}_\mu$, $\boldsymbol{v}_{\mu\eta}$, $\boldsymbol{B}_\mu$, $\boldsymbol{B}_{\mu\eta}$, respectively), since $\boldsymbol{v}_\mu \to \boldsymbol{v}_{\varepsilon\eta}$, $\boldsymbol{B}_\mu \to \boldsymbol{B}_{\varepsilon\eta}$ in $L^2(I; {}^1H^{o,N}(\Omega))$ strongly or in $L^2(I; {}^2H^{o,N}(\Omega))$ strongly, respectively.

Furthermore, due to Equations 12.620 and 12.648, $\{j'_\varepsilon(\boldsymbol{v}_m)\}$, $\{v_{i\mu}v_{j\mu}\}$, $\{B_{i\mu}B_{j\mu}\}$, are bounded subsets of spaces $L^2(I; ({}^1H^{1,N}(\Omega))')$ and $L^2(I; L^N_{p/2}(\Omega))$, respectively. Then we can also assume that

$$j'_\varepsilon(\vec{v}_\mu) \to \chi \text{ weakly in } L^2(I; ({}^1H^{1,N}(\Omega))'). \tag{12.659}$$

$$v_{i\mu}v_{j\mu} \to \Theta_{ij} \text{ weakly in } L^2(I; L^{p/2}(\Omega)), \tag{12.660}$$

$$v_{i\mu}B_{j\mu} \to {}^1\Theta_{ij} \text{ weakly in } L^2(I; L^{p/2}(\Omega)), \tag{12.661}$$

$$B_{i\mu}B_{j\mu} \to {}^2\Theta_{ij} \text{ weakly in } L^2(I; L^{p/2}(\Omega)). \tag{12.662}$$

From Equation 12.658 $v_{i\mu}v_{j\mu} \to v_{i\varepsilon\eta}v_{j\varepsilon\eta}$, $v_{i\mu}B_{j\mu} \to v_{i\varepsilon\eta}B_{j\varepsilon\eta}$, $B_{i\mu}B_{j\mu} \to B_{i\varepsilon\eta}B_{j\varepsilon\eta}$ in the sense of distributions in $\Omega \times I$, which comparing with equations 12.660-12.662 gives

$$\Theta_{ij} = v_{i\varepsilon\eta}v_{j\varepsilon\eta},\ {}^1\Theta_{ij} = v_{i\varepsilon\eta}B_{j\varepsilon\eta},\ {}^2\Theta_{ij} = B_{i\varepsilon\eta}B_{j\varepsilon\eta}.$$

But

$$b(\boldsymbol{v}_\mu, \boldsymbol{v}_\mu, \boldsymbol{w}_j) = -b(\boldsymbol{v}_\mu, \boldsymbol{w}_j, \boldsymbol{v}_\mu) \to -b(\boldsymbol{v}_{\varepsilon\eta}, \boldsymbol{w}_j, \boldsymbol{v}_{\varepsilon\eta}) \text{ weakly in } L^2(I), \quad \forall \boldsymbol{w}_j.$$

(see Duvaut and Lions (1972*a*)). Similarly for the other trilinear forms, so that

$$\Xi_o(\boldsymbol{a}_\mu, \boldsymbol{a}_\mu, \vec{\alpha}_j) \to -\Xi_o(\boldsymbol{a}_{\varepsilon\eta}, \vec{\alpha}_j, \boldsymbol{a}_{\varepsilon\eta}) \text{ weakly in } L^2(I), \quad \forall \vec{\alpha}_j$$

From Equation 12.640 for $m = \mu$ we obtain

$$\int_I [(\mathscr{E}\boldsymbol{a}'_{\varepsilon\eta}(t), \vec{\alpha}_j(t)) + A(\boldsymbol{a}_{\varepsilon\eta}(t), \vec{\alpha}_j(t)) + \Xi(\boldsymbol{a}_{\varepsilon\eta}(t), \vec{\alpha}_j(t))$$
$$+ \eta(\boldsymbol{a}_{\varepsilon\eta}(t), \vec{\alpha}_j)_s - (\boldsymbol{P}(t), \vec{\alpha}_j(t))]\mathrm{d}t + (\chi(t), \boldsymbol{w}_j(t)) = 0 \quad \forall j, \tag{12.663}$$

where we denote $\chi = J'_\varepsilon(\boldsymbol{v}_{\varepsilon\eta}(t))$.

But the system of functions $\{\vec{\alpha}_j\}$ is complete in $\underline{H}^s$, and hence from Equation 12.663 it follows that

$$\int_\mathrm{I} [(\mathscr{E}\boldsymbol{a}'_{\varepsilon\eta}(t), \vec{\alpha}(t)) + A(t; \boldsymbol{a}_{\varepsilon\eta}(t), \vec{\alpha}(t)) + \Xi(t; \boldsymbol{a}_{\varepsilon\eta}(t), \vec{\alpha}(t)) + \eta(\boldsymbol{a}_{\varepsilon\eta}(t), \vec{\alpha}(t))_s]\mathrm{d}t$$
$$- \int_\mathrm{I} (\boldsymbol{P}(t), \vec{\alpha}(t)) + (\chi(t), \boldsymbol{w}(t)) = 0 \quad \forall \vec{\alpha} \in \underline{H}^s.$$

Since Equations 12.639 and 12.647 are satisfied then it is sufficient to prove that

$$\chi(t) = J'_\varepsilon(\boldsymbol{v}_{\varepsilon\eta}(t)). \tag{12.664}$$

To prove Equation 12.664 then similarly as in Duvaut and Lions (1972a) the property of monotonicity will be used. Let

$$\boldsymbol{a} \in L^2(I; \underline{H}^s) \text{ such that } \boldsymbol{a}' \in L^2(I; (\underline{H}^s)'), \quad \boldsymbol{a}(t_o) = 0.$$

Let us put

$$X_\mu = (J'_\varepsilon(\boldsymbol{a}_\mu(t)) - J'_\varepsilon(\vec{\alpha}(t)), \boldsymbol{a}_\mu(t) - \vec{\alpha}(t)) + \int_\mathrm{I} A(t; \boldsymbol{a}_\mu(t) - \vec{\alpha}(t), \boldsymbol{a}_\mu(t) - \alpha(t))\mathrm{d}t$$
$$+ \eta \int_\mathrm{I} \|\boldsymbol{a}_\mu(t) - \vec{\alpha}(t)\|^2_{s,N}\, \mathrm{d}t + \int_\mathrm{I} (\mathscr{E}\boldsymbol{a}'_\mu(t) - \mathscr{E}\vec{\alpha}'(t), \boldsymbol{a}_\mu(t) - \alpha(t))\mathrm{d}t.$$

From Equation 12.639 we have

$$X_\mu = \int_I (\boldsymbol{P}(t), \boldsymbol{a}_\mu(t))\mathrm{d}t - (J'_\varepsilon(\boldsymbol{a}_\mu(t), \vec{\alpha}(t)) - (J'_\varepsilon(\vec{\alpha}(t), \boldsymbol{a}_\mu(t) - \vec{\alpha}(t))$$
$$- \int_I [A(t; \boldsymbol{a}_\mu(t), \vec{\alpha}(t)) + A(t; \vec{\alpha}(t), \boldsymbol{a}_\mu(t) - \vec{\alpha}(t))]\mathrm{d}t$$
$$- \eta \int_I [(\boldsymbol{a}_\mu(t), \vec{\alpha}(t))_s + (\vec{\alpha}(t), \boldsymbol{a}_\mu(t) - \vec{\alpha}(t))_s]\mathrm{d}t$$
$$- \int_I [(\mathscr{E}\boldsymbol{a}'_\mu(t), \vec{\alpha}(t)) - (\mathscr{E}\vec{\alpha}'(t), \boldsymbol{a}_\mu(t) - \vec{\alpha}(t))]\mathrm{d}t.$$

Hence $X_\mu \to X$, where

$$X = \int_I (\boldsymbol{P}(t), \boldsymbol{a}_{\varepsilon\eta}(t))\mathrm{d}t - [(\chi(t), \vec{\alpha}(t)) + (J'_\varepsilon(\vec{\alpha}(t), \boldsymbol{a}_{\varepsilon\eta}(t) - \vec{\alpha}(t))]$$
$$- \int_I [A(\boldsymbol{a}_{\varepsilon\eta}(t), \vec{\alpha}(t)) + A(\vec{\alpha}(t), \boldsymbol{a}_{\varepsilon\eta}(t) - \vec{\alpha}(t))]\mathrm{d}t$$
$$- \eta \int_I [(\boldsymbol{a}_{\varepsilon\eta}(t), \vec{\alpha}(t))_s + (\vec{\alpha}(t), \boldsymbol{a}_{\varepsilon\eta}(t) - \vec{\alpha}(t))_s]\mathrm{d}t$$
$$- \int_I [(\mathscr{E}\boldsymbol{a}'_{\varepsilon\eta}(t), \vec{\alpha}(t)) - (\mathscr{E}\vec{\alpha}'(t), \boldsymbol{a}_{\varepsilon\eta}(t) - \alpha(t))]\mathrm{d}t.$$

Since $X_\mu(t) \geq 0$ for all μ, then $X(t) \geq 0$.

Let us put $\vec{\alpha} = \boldsymbol{a}_{\varepsilon\eta} - \lambda\vec{u}$, $\lambda \geq 0$, where

$$\boldsymbol{u} \in L^2(I; \underline{H}^s) \text{ such that } \boldsymbol{u}' \in L^2(I; (\underline{H}^s)'), \quad \boldsymbol{u}(t_o) = 0.$$

Substituting for $\vec{\alpha}$ and dividing by λ, then we successively have

$$(\chi(t) - J'_\varepsilon(\boldsymbol{a}_{\varepsilon\eta}(t) - \lambda\boldsymbol{u}(t)), \boldsymbol{u}(t)) + \lambda \int_I \{A(\boldsymbol{u}(t), \boldsymbol{u}(t)) + \eta \|\boldsymbol{u}(t)\|_{s,N} + (\boldsymbol{u}'(t), \boldsymbol{u}(t))\}\mathrm{d}t \geq 0.$$

Hence in the limit $\lambda \to 0$ we obtain

$$(\chi(t) - J'_\varepsilon(\boldsymbol{a}_{\varepsilon\eta}(t)), \boldsymbol{u}(t)) \geq 0 \quad \forall \boldsymbol{u},$$

from which

$$\chi(t) = J'_\varepsilon(\boldsymbol{a}_{\varepsilon\eta}(t)).$$

Therefore, we have proved the existence of the vector function $\boldsymbol{a}_{\varepsilon\eta}$ satisfying Equation 12.639 and the conditions

$$\left.\begin{array}{l} \boldsymbol{a}_{\varepsilon\eta}(t) \text{ is bounded in } L^2(I; \underline{H}^1) \cap L^\infty(I; \underline{H}^o), \\ \boldsymbol{a}'_{\varepsilon\eta}(t) \text{ is bounded in } L^2(I; (\underline{H}^s)'), \\ \eta^{1/2}\boldsymbol{a}_{\varepsilon\eta}(t) \text{ is bounded in } L^2(I; \underline{H}^s). \end{array}\right\} \tag{12.665}$$

Now the limitation process ε, $\eta \to 0$ will be investigated. For $\vec{\alpha} \in \underline{H}$ fixed let us put

$$Y_{\varepsilon\eta} = \int_I [(\mathscr{E}\vec{\alpha}'(t), \vec{\alpha}(t) - \boldsymbol{a}_{\varepsilon\eta}(t)) + A(\boldsymbol{a}_{\varepsilon\eta}(t), \vec{\alpha}(t) - \boldsymbol{a}_{\varepsilon\eta}(t)) + \Xi(\boldsymbol{a}_{\varepsilon\eta}(t), \vec{\alpha}(t) - \boldsymbol{a}_{\varepsilon\eta}(t))$$
$$+ \eta(\boldsymbol{a}_{\varepsilon\eta}(t), \vec{\alpha}(t) - \boldsymbol{a}_{\varepsilon\eta}(t))_s - (\boldsymbol{P}(t), \vec{\alpha}(t) - \boldsymbol{a}_{\varepsilon\eta}(t))]\mathrm{d}t + J_\varepsilon(\boldsymbol{w}) - J_\varepsilon(\boldsymbol{v}_{\varepsilon\eta}).$$

By virtue of Equation 12.639 we have

$$Y_{\varepsilon\eta} = \int_I [\mathscr{E}(\vec{\alpha}'(t) - \boldsymbol{a}'_{\varepsilon\eta}(t)), \vec{\alpha} - \boldsymbol{a}_{\varepsilon\eta}(t)]\mathrm{d}t + J_\varepsilon(\boldsymbol{w}) - J_\varepsilon(\boldsymbol{v}_{\varepsilon\eta}) - (J'_\varepsilon(\boldsymbol{v}_{\varepsilon\eta}), \boldsymbol{w} - \boldsymbol{v}_{\varepsilon\eta}).$$

According to the initial conditions the first term is equivalent to $(1/2)|\mathscr{E}(\vec{\alpha}(t_1) - \boldsymbol{a}_{\varepsilon\eta}(t_1))|^2$ and since the functional $\boldsymbol{w} \to j_\varepsilon(\boldsymbol{w})$ is convex then the second term is ≥ 0. Thus $Y_{\varepsilon\eta} \geq 0$, i.e.

$$\int [(\mathscr{E}\vec{\alpha}'(t), \vec{\alpha}(t) - \boldsymbol{a}_{\varepsilon\eta}(t)) + A(\boldsymbol{a}_{\varepsilon\eta}(t), \vec{\alpha}(t)) - \Xi(\boldsymbol{a}_{\varepsilon\eta}(t), \vec{\alpha}(t) - \boldsymbol{a}_{\varepsilon\eta}(t))$$
$$+ \eta(\boldsymbol{a}_{\varepsilon\eta}(t), \vec{\alpha}(t))_s - (\boldsymbol{P}(t), \vec{\alpha}(t) - \boldsymbol{a}_{\varepsilon\eta}(t))]\mathrm{d}t + J_\varepsilon(\boldsymbol{w})$$
$$\geq \int_I [A(\boldsymbol{a}_{\varepsilon\eta}(t), \boldsymbol{a}_{\varepsilon\eta}(t)) + \eta \left\|\boldsymbol{a}_{\varepsilon\eta}(t)\right\|^2]\mathrm{d}t + J_\varepsilon(\boldsymbol{v}_{\varepsilon\eta}(t))$$
$$\geq \int_I [A(\boldsymbol{a}_{\varepsilon\eta}(t), \boldsymbol{a}_{\varepsilon\eta}(t))]\mathrm{d}t + J_\varepsilon(\boldsymbol{v}_{\varepsilon\eta}(t)). \tag{12.666}$$

Due to Equation 12.665 there exists a subsequence, denoted by $\{\boldsymbol{a}_{\varepsilon\eta}\}$, such that

$$\boldsymbol{a}_{\varepsilon\eta}(t) \to \boldsymbol{a}(t) \text{ in } L^\infty(I, \underline{H}^o) \text{ *-weakly and weakly in } L^2(I; \underline{H}^1),$$
$$\boldsymbol{a}'_{\varepsilon\eta}(t) \to \boldsymbol{a}'(t) \text{ weakly in } L^2(I; (\underline{H}^s)'). \tag{12.667}$$

From Equations 12.666 and 12.667 it follows that

$$\int_I \{(\mathscr{E}\boldsymbol{a}'(t), \vec{\alpha}(t) - \boldsymbol{a}(t)) + A(\boldsymbol{a}(t), \vec{\alpha}(t) - \boldsymbol{a}(t)) - \Xi(\boldsymbol{a}(t), \vec{\alpha}(t), \boldsymbol{a}(t))$$
$$- (\boldsymbol{P}(t), \vec{\alpha}(t) - \boldsymbol{a}(t))\}\mathrm{d}t + J(\boldsymbol{w}) \geq \liminf \int_I [A(\boldsymbol{a}_{\varepsilon\eta}(t), \boldsymbol{a}_{\varepsilon\eta}(t)) + \hat{g} j_\varepsilon(\boldsymbol{a}_{\varepsilon\eta})]\mathrm{d}t$$
$$\geq \int_I [A(\boldsymbol{a}(t), \boldsymbol{a}(t))]\mathrm{d}t + J(\boldsymbol{v}(t))$$

as

$$\liminf \int_I A(\boldsymbol{a}_{\varepsilon\eta}(t), \boldsymbol{a}_{\varepsilon\eta}(t))\mathrm{d}t \geq \int_I A(\boldsymbol{a}(t), \boldsymbol{a}(t))\mathrm{d}t,$$

and since (see Duvaut and Lions (1972a))

$$\int_I j(\boldsymbol{v}(t))\mathrm{d}t \leq \left(\int_{G\times I} D_{\mathrm{II}}(\boldsymbol{v}(t))^{(1+\varepsilon)/2}\mathrm{d}\boldsymbol{x}\,\mathrm{d}t\right)^{1/(1+\varepsilon)} \left(\int_{G\times I} \mathrm{d}\boldsymbol{x}\,\mathrm{d}t\right)^{\varepsilon/(1+\varepsilon)}, \tag{12.667a}$$

then

$$\int_I j_\varepsilon(\boldsymbol{v}_{\varepsilon\eta}(t))\mathrm{d}t \geq c\left(\int_I j(\boldsymbol{v}_{\varepsilon\eta}(t))\mathrm{d}t\right)^{(1+\varepsilon)}, \quad c = c(\varepsilon), \tag{12.668}$$

$$\liminf \int_I j_\varepsilon(\boldsymbol{v}_{\varepsilon\eta}(t))\mathrm{d}t \geq \liminf \int_I j(\boldsymbol{v}_{\varepsilon\eta}(t))\mathrm{d}t. \tag{12.669}$$

Also, since the function $\boldsymbol{w} \to \int_I j(\boldsymbol{w}(t))\mathrm{d}t$ is convex and continuous on $L^2(I; {}^1H^{1,N}(\Omega))$, then it is l.s.c. in the weak topology of the space $L^2(I; {}^1H^{1,N}(\Omega))$ and thus

$$\liminf \int_I j(\boldsymbol{v}_{\varepsilon\eta}(t))\mathrm{d}t \geq \int_I j(\boldsymbol{v}(t))\mathrm{d}t \tag{12.670}$$

which together with Equation 12.669 proves

$$\liminf \int_I j_\varepsilon(\boldsymbol{v}_{\varepsilon\eta}(t))\mathrm{d}t \geq \int_I j(\boldsymbol{v}(t))\mathrm{d}t. \tag{12.671}$$

Then equations (12.667a), (12.668), (12.669) prove that $(\boldsymbol{v}(t), \boldsymbol{B}(t), T(t))$ satisfy Equation 12.607.

The existence of $\Phi(t)$ follows from the Lax-Milgram theorem for every $t \in I$. Thus we have proved that $(\boldsymbol{a}(t), \Phi(t))$ is the solution of the given problem.

Now consider the uniqueness of the solution. We shall assume that $N = 2$ and that $(\boldsymbol{v}(t), \boldsymbol{B}(t), T(t), \Phi(t))$ and $(\boldsymbol{v}_1(t), \boldsymbol{B}_1(t), T_1(t), \Phi_1(t))$ are two solutions of the problem discussed. Then $s = 1$ and ${}^1H^{s,N}(\Omega) = {}^1H^{1,N}(\Omega)$, ${}^2H^{s,N}(\Omega) = {}^2H^{1,N}(\Omega)$ and ${}^3H^{s,1}(\Omega) = {}^3H^{1,1}(\Omega)$, and therefore the terms of viscosity regularization $\eta(\boldsymbol{v}(t), \boldsymbol{w})_s$, $\eta(\boldsymbol{B}(t), \boldsymbol{C})_s$, $\eta(T(t), z)_s$ are of the same order as $\hat{\mu}a(t; \boldsymbol{v}(t), \boldsymbol{w})$, $b_M(t; \boldsymbol{B}(t), \boldsymbol{C})$, $A(t; T(t), z)$ and therefore can be neglected. Let us denote $\boldsymbol{U}(t) = \boldsymbol{v}(t) - \boldsymbol{v}_1(t)$, $\boldsymbol{B}^*(t) = \boldsymbol{B}(t) - \boldsymbol{B}_1(t)$, $T^*(t) = T(t) - T_1(t)$, $\Phi^*(t) = \Phi(t) - \Phi_1(t)$, then $(\boldsymbol{a}^*(t), \Phi^*(t)) = (\boldsymbol{a}(t) - \boldsymbol{a}_1(t), \Phi(t) - \Phi_1(t))$.

Let us put $\vec{\alpha} = \boldsymbol{a}(t)$ or $\vec{\alpha} = \boldsymbol{a}_1(t)$, respectively, into Equation 12.607. Then

$$\int_I \Big\{ -\frac{1}{2}\frac{\mathrm{d}}{\mathrm{d}t}\{|\mathscr{E}\boldsymbol{a}^*(t)|^2\} - A(\boldsymbol{a}^*(t), \boldsymbol{a}^*(t)) - \Xi_o(\boldsymbol{a}(t), \boldsymbol{a}(t), \boldsymbol{a}^*(t))$$
$$+\Xi_o(\boldsymbol{a}_1(t), \boldsymbol{a}_1(t), \boldsymbol{a}^*(t)) - [b_s(T^*(t), \boldsymbol{U}(t)) + b_p(\boldsymbol{U}(t), T^*(t))]\Big\}\,\mathrm{d}t \geq 0.$$

Hence

$$\frac{1}{2}\frac{\mathrm{d}}{\mathrm{d}t}\{\mathscr{E}\,|\boldsymbol{a}^*(t)|^2\} + A(\boldsymbol{a}^*(t), \boldsymbol{a}^*(t)) \leq -b(\boldsymbol{v}(t) - \boldsymbol{U}(t), \boldsymbol{v}(t) - \boldsymbol{U}(t), \boldsymbol{U}(t))$$
$$- b(\boldsymbol{v}(t), \boldsymbol{v}(t), \boldsymbol{U}(t)) + F + G, \tag{12.672}$$

where

$$F = (\partial/\partial x_i(\mu^{-1}B_i(t)\boldsymbol{B}(t)), \boldsymbol{U}(t)) - (\partial/\partial x_i(\mu^{-1}B_{1i}(t)\boldsymbol{B}_1(t)), \boldsymbol{U}(t))$$
$$+ (\mu^{-1}\ \mathrm{Rot}\ [\boldsymbol{v}(t), \boldsymbol{B}(t)], \boldsymbol{B}^*(t)) - (\mu^{-1}\ \mathrm{Rot}\ [\boldsymbol{v}_1(t), \boldsymbol{B}_1(t)], \boldsymbol{B}^*(t)),$$
$$G = b_o(\boldsymbol{U}(t), T^*(t), T^*(t)) + c(\|\boldsymbol{U}(t)\|_{1,1}\,\|T^*(t)\|_{o,N} + \|\boldsymbol{U}(t)\|_{1,N}\,\|T^*(t)\|_{o,1}),$$

and where

$$\left|b_s(T^*(t), \boldsymbol{U}(t)) + b_p(\boldsymbol{U}(t), T^*(t))\right| \le c(\left\|T^*(t)\right\|_{1,1} \|\boldsymbol{U}(t)\|_{o,N} + \left\|T^*(t)\right\|_{o,1} \|\boldsymbol{U}(t)\|_{1,N}).$$

Furthermore, as above

$$(\partial/\partial x_i(\mu^{-1} B_i(t)\boldsymbol{B}(t)), \boldsymbol{U}(t)) + (\mu^{-1} \text{ Rot } [\boldsymbol{U}(t), \boldsymbol{B}(t)], \boldsymbol{B}(t)) = 0,$$

then

$$\begin{aligned} F &= (\partial/\partial x_i(\mu^{-1}(B_i(t)\boldsymbol{B}^*(t) + B_i^*(t)\boldsymbol{B}(t))), \boldsymbol{U}(t)) \\ &\quad + (\mu^{-1} \text{ Rot } [\boldsymbol{v}(t), \boldsymbol{B}(t)], \boldsymbol{B}(t)) + (\mu^{-1} \text{ Rot } [\boldsymbol{U}(t), \boldsymbol{B}(t)], \boldsymbol{B}(t)) \\ &= -(\mu^{-1}(B_i(t)\boldsymbol{B}^*(t) + B_i^*(t)\boldsymbol{B}(t)), \partial \boldsymbol{U}(t)/\partial x_i) \\ &\quad + (\mu^{-1}([\boldsymbol{v}(t), \boldsymbol{B}(t)] + [\boldsymbol{U}(t), \boldsymbol{B}(t)]), \text{ Rot } \boldsymbol{B}^*(t)). \end{aligned}$$

Since as above

$$\begin{aligned} \left|(\partial/\partial x_i(\mu^{-1} B_i(t)\boldsymbol{B}(t)), \boldsymbol{w})\right| &\le c_o \|\boldsymbol{B}(t)\|_{1,N} \|\boldsymbol{B}(t)\|_{o,N} \|\boldsymbol{w}\|_{s,N} \le c_o \|\boldsymbol{B}(t)\|_{1,N} \|\boldsymbol{w}\|_{s,N}, \\ \left|(\mu^{-1} \text{ Rot } [\boldsymbol{v}(t), \boldsymbol{B}(t)], \boldsymbol{C})\right| &\le c_1 \|\boldsymbol{v}(t)\|_{1,N}^{1/2} \|\boldsymbol{v}(t)\|_{o,N}^{1/2} \|\boldsymbol{B}(t)\|_{1,N}^{1/2} \|\boldsymbol{B}(t)\|_{o,N}^{1/2} \|\boldsymbol{C}\|_{s,N} \\ &\le c_1 \|\boldsymbol{v}(t)\|_{1,N}^{1/2} \|\boldsymbol{B}(t)\|_{1,N}^{1/2} \|\boldsymbol{C}\|_{s,N}, \\ \left|b_s(t; \boldsymbol{w}, z) + b_p(t; \boldsymbol{w}, z)\right| &\le c_s(\|z\|_{1,1} \|\boldsymbol{w}\|_{o,N} + \|z\|_{o,1} \|\boldsymbol{w}\|_{1,N}), \\ |b(t; \boldsymbol{w}, \boldsymbol{v}(t), \boldsymbol{w})| &\le c_3 \|\boldsymbol{w}\|_{1,N} \|\boldsymbol{v}(t)\|_{o,N} \|\boldsymbol{w}\|_{s,N} \\ &\le c_1 \|\boldsymbol{w}\|_{1,N}^2 + c_2 \|\boldsymbol{v}(t)\|_{o,N}^2 \|\boldsymbol{w}\|_{s,N}^2 \end{aligned}$$

and therefore

$$\begin{aligned} |F(t)| &\le c_o(\|\boldsymbol{B}(t)\|_{1,N}^{1/2} \|\boldsymbol{B}(t)\|_{o,N}^{1/2} \left\|\boldsymbol{B}^*(t)\right\|_{1,N}^{1/2} \left\|\boldsymbol{B}^*(t)\right\|_{o,N}^{1/2} \|\boldsymbol{U}(t)\|_{1,N} \\ &\quad + \|\boldsymbol{v}(t)\|_{1,N}^{1/2} \|\boldsymbol{v}(t)\|_{o,N}^{1/2} \left\|\boldsymbol{B}^*(t)\right\|_{1,N}^{1/2} \left\|\boldsymbol{B}^*(t)\right\|_{1,N}^{1/2} \\ &\quad + \|\boldsymbol{U}(t)\|_{1,N}^{1/2} \|\boldsymbol{U}(t)\|_{o,N}^{1/2} \|\boldsymbol{B}(t)\|_{1,N}^{1/2} \|\boldsymbol{B}(t)\|_{o,N}^{1/2} \left\|\boldsymbol{B}^*(t)\right\|_{1,N}) \\ &\le c_{p1}(\|\boldsymbol{U}(t)\|_{1,N}^2 + \|\boldsymbol{B}(t)\|_{1,N}^2) + K(c_{p1}) \|\boldsymbol{B}(t)\|_{1,N}^2 \|\boldsymbol{B}(t)\|_{o,N}^2 \|\boldsymbol{B}(t)\|_{o,N}^2 \\ &\quad + c_{p2} \left\|\boldsymbol{B}^*(t)\right\|_{1,N}^2 + K(c_{p2}) \|\boldsymbol{v}(t)\|_{1,N}^2 \|\boldsymbol{v}(t)\|_{o,N}^2 \left\|\boldsymbol{B}^*(t)\right\|_{1,N}^2 \\ &\quad + c_{p3}(\|\boldsymbol{U}(t)\|_{1,N}^2 + \left\|\boldsymbol{B}^*(t)\right\|_{1,N}^2 + K(c_{p3}) \|\boldsymbol{B}(t)\|_{1,N}^2 \left\|\vec{\boldsymbol{B}}(t)\right\|_{o,N}^2 \left\|\vec{\boldsymbol{U}}(t)\right\|_{o,N}^2, \end{aligned}$$

$$\begin{aligned} |b(\boldsymbol{U}(t), \boldsymbol{w}, \boldsymbol{U}(t))| + |F(t)| &\le c_p(\|\boldsymbol{U}(t)\|_{1,N}^2 + \|\boldsymbol{B}(t)\|_{1,N}^2) + K(c_p)(\|\boldsymbol{w}\|_{1,N}^2 + \|\boldsymbol{B}(t)\|_{1,N}^2) \\ &\quad \times (\|\boldsymbol{U}(t)\|_{o,N}^2 + \|\boldsymbol{B}(t)\|_{o,N}^2), \\ A(\boldsymbol{a}^*(t), \boldsymbol{a}^*(t)) &\ge c_\alpha \left\|\boldsymbol{a}^*(t)\right\|_{1,2N+2}^2. \end{aligned}$$

Then Equation 12.672 yields

$$\frac{\mathrm{d}}{\mathrm{d}t}(\mathscr{E}\,|\boldsymbol{a}^*(t)|^2) \le c_\alpha\left(\|\boldsymbol{a}^*(t)\|^2_{1,2N+2}\right)\|\boldsymbol{a}^*(t)\|^2_{o,2N+2} = c_\alpha g(t)\left(\|\boldsymbol{a}^*(t)\|^2_{o,2N+2}\right),$$

where $g(t)$ is an integrable function in variable t and thus

$$\|\boldsymbol{a}^*(t)\|^2_{o,2N+1} \le \int_{t_o}^{t} g(\tau)\left(\|\boldsymbol{a}^*(\tau)\|^2_{o,2N+1}\right)\mathrm{d}\tau.$$

Hence

$$\boldsymbol{a}^*(t) = 0, \ \text{ i.e. } \boldsymbol{U}(t) = 0, \boldsymbol{B}^*(t) = 0, T^*(t) = 0.$$

Let $\Phi(t)$, $\Phi_1(t) \in {}^3H^{1,1}(\Omega) \times I$ be two solutions of Equation 12.608. Then

$$a_g(\Phi(t), \varphi(t) - \Phi(t)) - (\Psi(t), \varphi(t) - \Phi(t)) \ge 0, \tag{12.673}$$

$$a_g(\Phi_1(t), \varphi(t) - \Phi_1(t)) - (\Psi(t), \varphi(t) - \Phi_1(t)) \ge 0$$
$$\forall\varphi(t) \in {}^3H^{1,1}(\Omega), \quad \forall t \in I. \tag{12.674}$$

Putting $\varphi = \Phi_1$ in Equation 12.673 and $\varphi = \Phi$ in Equation 12.674 and adding, we obtain

$$0 \ge a_g(\Phi(t) - \Phi_1(t), \Phi(t) - \Phi_1(t)) \ge c_\phi \|\Phi(t) - \Phi_1(t)\|^2_{1,1},$$

hence $\Phi(t) = \Phi_1(t)$, which completes the proof.

12.5 GENERALIZED THERMO-MAGNETODYNAMIC PROBLEM. A NEW APPROACH TO MANTLE-CORE COUPLING STUDIES. PROBLEMS OF IONIZATION AND METALLIZATION OF THE EARTH ROCKS

12.5.1 Introduction

One of the important problems in the analysis of the magnetic field of the Earth is the analysis of mantle-core coupling. The convection in the Earth's mantle and the thermo-magnetodynamic turbulence in the Earth's core are much studied, but little understood, subjects. The early concept of plate tectonics was based on the idea that the flow is broadly akin to high Rayleigh number convection. Cathles (1975), Kirby (1983), Christensen (1984), Quareni *et al.* (1985), (1986), Fowler (1983), (1993), Nedoma (1984), (1986) showed that the rheologies of the lithosphere and the mantle are temperature-dependent or temperature- and pressure-dependent (namely temperature-dependent or temperature- and pressure-dependent viscosity). Quareni *et al.* (1985), (1986), Fowler (1983), (1985), (1993), Jeanloz and Richter (1979) suggested that convecting fluids with a strongly temperature- and pressure-dependent viscosity are realistic models. But there remains the issue of inhomogeneities such as subduction zones, the GTLV channels and bumps (Hide and Malin (1971), Nedoma (1984), (1986), etc.) in the mantle and the core which create local visco-plastic inhomogeneous regions. In the case of subduction zones, stresses of order 10 kbars and temperature- and pressure-dependent viscosity of the rocks cause plastic yielding, so the rheology is made thermo-visco-plastic. Nevertheless, the real Earth is still some distance away, due to the question of the existence and effects of the phase changes and chemical discontinuities at various depths. We see that the realistic Earth's model will be very complicated. In our contribution we shall discuss the model based on the thermo-visco-plastic Bingham rheology which makes it possible to study the mantle-core coupling as well as inhomogeneities in the Earth's interior, e.g. the Hide's bumps, anomalous regions under subduction zones, the GTLV channels, etc.. The model presented gives a new approach of how to study mutual relations between geodynamic and geomagnetic processes in the Earth's interior as well as in its separate parts, e.g the core, the mantle and the lithosphere. The geodynamic model of the Earth is based on the assumption that in the course of the development of the Earth's body the important factors are gravity, radioactivity, temperature, density differentiation associated with the phase transitions of the Earth's materials under high temperatures, great pressures and chemical reactions under the same conditions (Nedoma (1984), (1986)). Then the Earth can be modelled by a thermo-elasto-visco-plastic rheology in a relatively weak magnetic field and under the gravity field.

12.5.2 The Model: Discussion of Properties of Rocks Inside the Earth

Over long geological periods, that is over time scales of ten and hundred million years, the Earth behaves like a thermo-visco-plastic fluid. Since the Earth is from the outward side cooled, the mantle material removes heat from the inner parts of the Earth by the transport of mass, represented by thermal convection. This convective circulation in the Earth's mantle is directly three dimensional and time dependent, and is manifested by the different combinations of internal and basal heating, which consistently have upwelling concentrations in the form move of cylindrical plumes, called the GTLV channels (great temperature and low viscosity channels) by Nedoma (1984), (1986) and downwelling in planar sheets. No stronger evidence of this exists than in the very surface expression of mantle convection, that is plate tectonics with its diverse, migrating plate boundaries and wide distribution of hot spots. The strength and number of upwelling plumes, the structure of sinking sheets and the nature of time dependence all change with the heating mode. The thermal history of the Earth indicates that the mantle is $\cong$70–80% internally heated. Recent studies on the melting curve of iron under very high pressure ($\cong$250 Gigapascals, see Williams *et al.* (1987)) imply a much higher core/mantle boundary temperature than previously believed and therefore possibly even the greatest part of the basal heating. The inhomogeneities of the mass distribution in the outer, strongly thermo-plastic up to thermo-liquid, core and the thermo-plastic mantle, and non-uniformity in the rotation of the Earth are responsible for the non-uniformity in the action of Coriollis' and Poincaré's forces, which in turn is reflected in the effect on the thermo-magneto-hydrodynamic turbulence in the outer core. In the deeper parts of the outer thermo-visco-liquid core, the molten liquid matter has a tendency to rotate faster on average than its upper parts, which rotate slower due to the viscosity of the materials and because of the stratification of the outer core. Since the viscous friction at the mantle/core boundary is small the mantle is not coupled with the core by large mechanical forces, and thus the mantle probably rotates faster than the core. The mantle/core boundary, the boundary at which the density differentiation of the melted material takes place, is itself uneven due to the non-uniform melting of the mantle rocks, or to possible recrystallization processes in accumulated light substances and the possible consequences of the shape non-uniformity of the initial core/mantle boundary. This may in turn be responsible for temporal changes in the rotation of the mantle and of the whole Earth. The changes in the motion of masses of the outer core and their irregularities may cause inhomogeneities in the regeneration of the magnetic field and, consequently, also an inversion of the Earth's magnetic field. The core/mantle boundary is a very dynamic region, most likely as diverse and heterogeneous as the surface of the Earth. It is a thermal and chemical boundary layer between the convecting crystalline mantle and the strongly thermo-plastic up to thermo-liquid core, and may vary laterally in much the same way as does the top 100 km of the Earth. An understanding of the mantle-core coupling is substantial to understanding the important processes that occur within the Earth, e.g. mantle convection, hot spot plume generation, core cooling, generation of magnetic field, changes of gravity field in time, and plate tectonic processes.

The inhomogeneities in the distribution of masses in the outer core and the mantle, and the non-uniformity in the rotation of the Earth are responsible for the non-uniformity in the action of Coriolis' and Poincaré's forces, which can in turn reflect the effect onto the thermo-magneto-hydrodynamic turbulence in the outer core. The changes in the motion of masses of the outer core and their irregularities, together with strong dependence of fields and velocities in the core, and (weak) core/mantle friction may cause inhomogeneities in the regeneration of the magnetic field and consequently also a reversal of the magnetic field of the Earth. In fact the Coriolis' and magnetic forces, and thermal processes in the core play a substantial role in the thermo-magneto-hydrodynamics. The density differentiation at the core/mantle boundary is continuous and is still present.

The model will be based on the assumption that the inner core is probably formed by a stratified polycrystallic structure in a thermo-visco-plastic state, which is in a non-metallized state in the first investigated case or in a metallized state in the second, characterized by a high density $\rho \cong 13\text{-}14$ gcm^{-3} and kinematic viscosities $\eta \geq 10^{14}$ poise. The outer core is probably formed by a strongly thermo-visco-plastic to thermo-visco-liquid material in a non-metallized state in the first case or in a metallized state in the second, characterized by a density $\rho \cong 9.8\text{-}11.5$ gcm^{-3} and viscosity $\eta \cong 10^{-2}\text{-}10^{9}$poise (Nedoma (1984), (1986), Jeanloz (1990)). Between them the phase transition layer is observed. Moreover, both parts of the core are strongly electrically and heat conducting, and in addition the effective viscosity of the core materials is an extremely strong function of temperature and pressure.

The structure of the mantle depends on the mantle convection. The effects of the heating mode have particular significance for the Earth's mantle, which is heated from below by the hotter outer core and from within by radiogenic heat sources (in the lithosphere) and secular cooling, as well as being heated due to the dissipative energy changing deformation energy into heat and to the friction on the contact boundaries between moving solid parts in the Earth's interior. Quareni *et al.* (1985),(1986), Fowler (1985),(1993), Jeanloz and Richter (1979), Nedoma (1984), (1986) assume a fluid with strongly temperature- and pressure-dependent viscosity in the mantle. Fowler (1985) suggested that convecting fluids of this type will in fact adopt a flow that achieves this, although quite how was not clear. Laboratory experiments are not obviously of any help, and only numerical computations bring several possibilities for illustration of the probable mantle structure (this is one of the reasons why the thermo-Bingham visco-plastic fluid in our model will be taken into account). Best estimates of the variation of mantle parameters with depth to infer density and viscosity variation with depth were found: a viscosity rising from an asthenospheric value of 10^{20} poise to 5.10^{21} poise at 800 km depth and then back down to 10^{20} poise at the core/mantle boundary; and a density rising from $\rho \cong 3.3$-3.8 gcm^{-3} in the upper mantle and 4.65-5.68 gcm^{-3} in the lower mantle. One finds a (probably adiabatic) temperature at the core/mantle boundary of 2300° K. Constraints on the melting point of iron alloy require a basal temperature of at least 3000° K and most probably a good deal more. Therefore, Jeanloz and Richter (1979) and others assume a thermal boundary layer at the base of the mantle, across which the temperature jumps by $\geq 1000°$ K. Other physical parameters such as the threshold of plasticity, the bulk modulus, and the gravitational intensity g can be found from our knowledge of P and S wave distribution and knowledge of the initial gravity field in the Earth's interior (see Bullen (1975), Bercovici *et al.* (1989), Jeanloz (1990)).

The coupling between the core and the mantle is investigated from the geodynamic and electromagnetic point of view. The electric conductivities of the lower mantle and the outer core are characteristic parameters for electromagnetic coupling between the core and the mantle (Roden (1963), Rochester *et al.* (1975)), given by the structure of the Elman-Hartman boundary hydromagnetic layer and magnetic diffusion, and generated as a result of Maxwell's tangential stress, caused by the relative motion of the core and the mantle. Hide (1970) showed the possibility of another type of inertial coupling based on the existence of unevenness at the core/mantle boundary. The viscous friction in the viscous boundary layer transmits the rotational moment between the mantle and the core, and thus mediates a viscous coupling between the core and the mantle of the Earth. The electrical conductivity of the Earth's materials depends mainly on the temperature, pressure and its chemical composition.

Wysession *et al.* (1992) assume that the D'' velocity low under Indonesia, more than 3% slower than P and S wave velocities, cannot be explained just as a thermal anomaly but would suggest an additional increase in iron (here the $Mg/(Mg + Fe)$ ratio is very low-about 0.7). A possible geodynamic explanation that would account for both the slow velocities and the iron temperature variations may involve coupling with the core flow. There is a good correlation between D'' velocities and the geomagnetically determined core flow (Voorhies (1986)). Stevenson (1986), Knittle and Jeanloz (1987), (1989), (1991) proposed that reactions occur between the core liquid iron and the crystalline perovskite, with the resulting products $MgSiO_3$, SiO_2 (but one can assume that SiO_2 will also be metallized) and the iron alloys FeO and FeSi. While the Mg-perovskite and stishovite will have faster than average velocities, the overall effect will be a decrease in velocity due to the presence of the metallic alloy. But it can be predicated that this effect can also occur due to the metallized materials in this region, which behave like the metallic alloy. In such an explanation the constitution of the D'' layer will be enriched by metallized materials (SiO_2). It is probable that the compression of silicates under a constant pressure of about 0.9 Mbar leads to their metallization, and under a load of several Mbar the silicate rocks (Fe_2SiO_4, Mg_2SiO_4) decompose into oxides (SiO_2, FeO, MgO, etc.) (see, for example, Kawai and Mochizuki (1971), Kawai and Nishiyama (1974),(1975)). While the reaction should only occur directly at the core/mantle boundary, it is possible that the strongly thermo-visco-plastic (up to thermo-liquid) iron together with metallized rocks seep up into the mantle through capillary effect (similarly as the water circulates through the crust). It can be noted that this is a typical diffusion process.

Over very long time scales a dynamic coupling between the core and the mantle produces a significant mantle anomaly from the core actions. Many studies of the secular variation of core flow patterns do suggest the necessity of core/mantle coupling, which may be gravitational coupling between density inhomogeneities or topographic-type coupling due to the pressure gradients in the core near the core/mantle boundary topography. Due to lateral variations in the D'' electrical conductivity of more than eleven orders of magnitude, electromagnetic coupling between the core and the mantle occurs. As showen by Zhang and Yuen (1988), Hansen and Yuen (1989), Jeanloz (1990) the metal-rich heterogeneities in the D'' layer would pin the magnetic field lines from the core, either distorting the image of flow patterns or controlling core flow near the core/mantle boundary. The layer D'', rich in the metal materials and in the metallized ones, needed to maintain this electromagnetic coupling could be the FeO and FeSi created from the iron-silicate reactions, locally aggregated through intra-D'' convective sweeping, and would have significantly slower velocities than perovskite. The electromagnetic coupling between the core and the mantle is the significant one. Nevertheless, the layer of light molten materials accumulated on the core/mantle boundary also plays a significant role in dynamic, gravitational and electromagnetic couplings. Nedoma (1984), (1986), (1988) referred to the fact that due to the laboratory experiments of Kawai and Mochizuki (1971), Kawai and Nishiyama (1974), (1975), Shilling and Shelton (1981), Ross (1985), Kamarád (1993), (1996), which proved that under a load of several Mbar silicate rocks (Fe_2SiO_4, Mg_2SiO_4, etc.) decompose into oxides (SiO_2, FeO, MgO, etc.), it is probable that similar processes also take place at the core/mantle boundary. Moreover, they are accompanied by phase transition of oxides into a metallized state. FeO and Fe alloys, and possibly also other metallized oxides and alloys, become part of the core, and the unmetallized and metallized oxides SiO_2, MgO and oxides of other light elements, enriched with oxides of heavier elements, form the light materials, which accumulate below the mantle. These light materials create an unstable layer. Their physical behaviour is similar to that of a strongly thermo-visco-plastic to thermo-visco-liquid molten mass. Light materials display a tendency to create a stable situation by escaping through GTLV channels into the upper parts of the Earth under the aid of convective flow in the mantle and of the Coriolis' force. But firstly these lighter and viscous liquid materials accumulate in a nozzle of the GTLV channels, which creates the bumps (Hide and Malin (1971)), They then flow through the GTLV channels up to the surface, giving the origin of the rift zones in the first case and, in the second, the origin of one probable type of hot spot (the origin of hot spots can also be of another type discussed in connection with the plate tectonic hypothesis, and based on the bulged out bottom part of the lithosphere and on rapid melting of rocks in these areas, see below). By passing through the GTLV channels, the light materials rise to the bottom of the lithosphere. The lithosphere is then heated, melted and strained mechanically in the vertical direction, and after effusing below the lithosphere (where they create the asthenosphere) the light materials also strain it horizontally. But here one must state that this is not the unique mechanism of the hot spots origin; an alternate method can be interpreted by the following mechanism. If the speed of colliding plates or blocks is low and their thickness is of practically the same order, the collision will not be reflected in the subduction of one of them; both plates will be deformed and their thickness will increase and, possibly, folded sediments will be thrust over them (e.g. the African plate). At the same time, the centre of the deformed lithospheric plate will bulge out at the bottom and grow thinner, and will be broken. The bulged out part of the lithosphere is filled by the molten lithospheric and asthenospheric materials and, as the thickness of the lithosphere is smaller, the centre of this region is displayed by the anomalous heat flow (See Figure 12.3). A

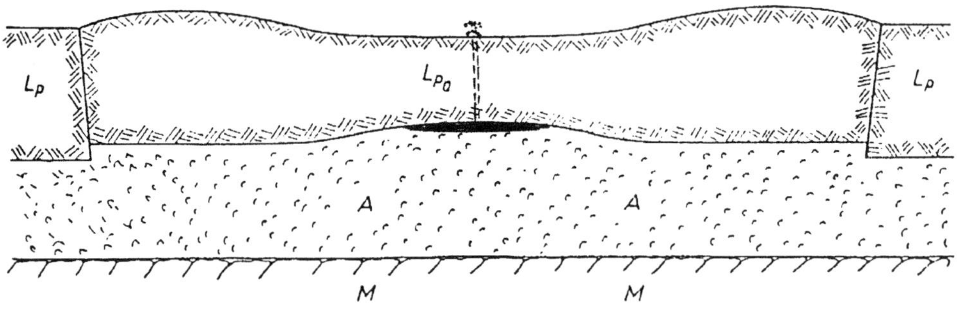

Figure 12.3. Origin and evolution of the thermally anomalous Areas, e.g. African plate

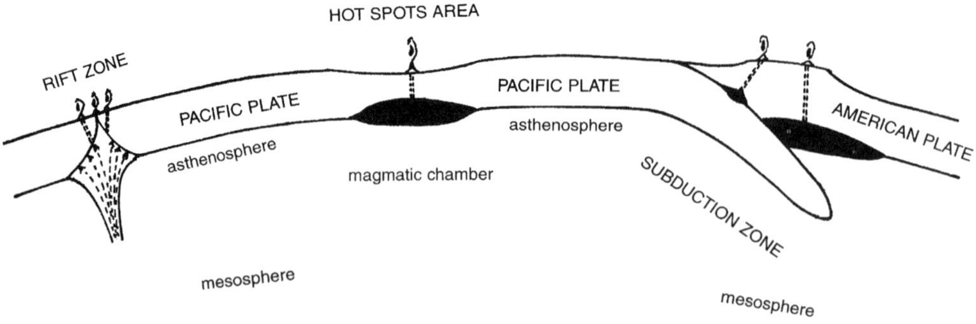

Figure 12.4. Origin and evolution of the hot spots areas

similar mechanism of the origin of hot spots located for example in the Pacific Ocean can be a consequence arising from this fact (See Figure 12.4).

Mathematically this mechanism is described and discussed in detail in Nedoma (1986), (1990) and also in Chapter 7, and therefore, will not be studied in this section.

12.5.3 The Model: Theory

12.5.3.1 Introduction

In the previous section we saw that the thermo-magneto-hydrodynamic processes in the core and the upper parts of the Earth are closely connected with the geodynamic and diffusion processes under gravity in the Earth's interior. The mathematical model of the geomagnetic and geodynamic processes passing in the Earth shows that the self-excitation of the magnetic field under the geodynamic processes in separate parts of the Earth's interior make is possible, in general, to describe the processes for the whole Earth's body by a uniform system of equations. The system of equations for coupled geophysical fields for the whole Earth was given, and the discussion of the limit of their validity was made, in Nedoma (1986), and in the first part of this chapter and in this section we give the fundamental results only.

12.5.3.2 Derivation of the system of equations

Over time scales of ten and hundred million years the Earth behaves as a thermo-visco-plastic and some of its parts as a strongly thermo-visco-plastic (by which is meant rock materials with rheology near a visco-liquid fluid), up to a thermo-visco liquid.

As a continuum representing the Earth's mass moves, it is natural to try to specify the position of each point as a function of time t. To identify a given point, we specify its position $\boldsymbol{x}^o$ at a certain initial time t_o. If $\boldsymbol{x}$ is the position vector of this point at the moment t, the motion of this medium can be described by $\boldsymbol{x} = \boldsymbol{x}(\boldsymbol{x}^o, t)$, $\boldsymbol{x}(\boldsymbol{x}^o, t_o) = \boldsymbol{x}^o$. We denote by $\Omega(t_o) \equiv \Omega$ the region occupied by the designated portion of the Earth's material at time $t = t_o (\equiv 0)$, and by $\Omega(t)$ the region occupied by the designated portion of the Earth's material at time t. We shall assume that the Earth is formed by anisotropic, inhomogeneous, generally compressible thermo-elasto-visco-plastic materials at the initial temperature T_o at $t = t_o$, and that at this time it is in an initial stress-strain state.

That the mass of a portion of the Earth's mass does not vary as time increases is one way to state what is meant by the conservation of mass, i.e.

$$\frac{\mathrm{d}}{\mathrm{d}t} \int_{\Omega} \rho(\boldsymbol{x}, t)\mathrm{d}\boldsymbol{x} = 0, \tag{12.675}$$

where $\rho(\boldsymbol{x}, t)$ is the density of the point $\boldsymbol{x}$ at time t. Since the velocity field is continuous in Ω, then equation 12.675 can be written in the equivalent differential form

$$\partial\rho/\partial t + \mathrm{div}\ (\rho \boldsymbol{v}) = 0. \tag{12.676}$$

Let us assume that $\boldsymbol{x}(t) = \underline{\boldsymbol{x}} + \boldsymbol{u}(\boldsymbol{x}, t)$, where $\underline{\boldsymbol{x}}$ represents co-ordinates in the non-deformed state and $\boldsymbol{u}$ is the displacement vector. Then equation 12.676 is equivalent to $\partial\rho/\partial t + \rho\partial v_i(\boldsymbol{x}, t)/\partial x_i = 0$. As we assume that $\boldsymbol{u}$ is small then $\mathrm{d}\rho/\rho + \partial v_i/\partial x_i \mathrm{d}t = 0$, and therefore

$$\rho(\boldsymbol{x}, t) = \rho_o(\boldsymbol{x})\left(1 - \int_{t_o}^{t} \partial v_i(\boldsymbol{x}, \tau)/\partial x_i \mathrm{d}\tau\right), \quad i = 1, 2, 3, \tag{12.677}$$

where we denote $\rho_o(\boldsymbol{x}) = \rho_o(\boldsymbol{x}, t_o)$. For the case when the Earth is incompressible, $\rho = \rho_o(\boldsymbol{x}) > 0$ (ρ being piecewise constant) and then

$$\operatorname{div} \boldsymbol{v} = 0. \tag{12.678}$$

The fundamental equations for a moving thermo-visco-plastic medium of the inner Earth follow from the law of momentum conservation as

$$\frac{\mathrm{d}}{\mathrm{d}t}\int_{\Omega} \rho(\boldsymbol{x}, t) v_i(\boldsymbol{x}, t)\mathrm{d}\boldsymbol{x} = \int_{\partial\Omega} \tau_{ij}(\boldsymbol{x}, t) n_j \mathrm{d}s + \int_{\Omega} \rho(\boldsymbol{x}, t) f_{oi}(\boldsymbol{x}, t)\mathrm{d}\boldsymbol{x}, \tag{12.679}$$

which represents the rate of change of linear momentum in a material region as equal to the contribution of the body forces plus the contribution of the surface forces, and where ρ is a density, τ_{ij} is a stress tensor, $\boldsymbol{v} = \partial\boldsymbol{u}/\partial t$ is a velocity vector, $\boldsymbol{u}$ is a displacement vector and $\boldsymbol{f}_o$ the body forces. As we shall assume that all functions in Equation 12.679 are sufficiently smooth then Equation 12.679 can be rewritten in its differential form

$$\begin{aligned}\rho(\boldsymbol{x}, t)\frac{\mathrm{d}}{\mathrm{d}t} v_i(\boldsymbol{x}, t) &\equiv \rho(\boldsymbol{x}, t)(\partial v_i(\boldsymbol{x}, t)/\partial t + v_j(\boldsymbol{x}, t)\partial v_i(\boldsymbol{x}, t)/\partial x_i) \\ &= \partial\tau_{ij}(\boldsymbol{x}, t)/\partial x_j + \rho(\boldsymbol{x}, t) f_{oi}(\boldsymbol{x}, t), \quad i, j = 1, 2, 3.\end{aligned} \tag{12.680}$$

Here $\boldsymbol{f} = (\rho f_i)$ represents the body forces of the following types: $\boldsymbol{f}_a = \rho D\boldsymbol{g}$, the gravity force (the Archimedean force), where ρ is a density, $D \leq 1$ is a function representing density inhomogeneities in the Earth's interior, and $\boldsymbol{g}$ is the gravity acceleration vector; $\boldsymbol{f}_b = [\boldsymbol{j}, \boldsymbol{B}]$ represents the volume density of the electromagnetic force (the Lorentz's force), where $\boldsymbol{j}$ the current density and $\boldsymbol{B}$ the magnetic induction vector will be defined below; $\boldsymbol{f}_v \cong ((1/2)\rho\eta_{ijkl}\partial v_i/\partial x_j \partial v_k/\partial x_l)$ is a viscous force, where η_{ijkl} is a coefficient of kinematic viscosity; and $\boldsymbol{f}_c = 2\rho[\vec{\omega}, \boldsymbol{v}]$ is the Coriolis' force. The term on the left-hand side represents the inertia forces.

From the law of angular momentum balance (i.e. moment of momentum) $\int_{\Omega} \varepsilon_{ijk}\tau_{kj}\mathrm{d}\boldsymbol{x} = 0$, it follows that the stress tensor is symmetric, i.e. that $\tau_{kj} = \tau_{jk}$, $k, j = 1, 2, 3$, where ε_{ijk} is the tensor of third order.

An energy balance relation, the general conservation law of energy, is give by

$$\int_{\Omega} \rho\frac{\mathrm{d}e}{\mathrm{d}t}\mathrm{d}\boldsymbol{x} = \int_{\Omega} \tau_{ij} D_{ij}\mathrm{d}\boldsymbol{x} + \int_{\Omega} (\rho W - \partial q_i/\partial x_i)\mathrm{d}\boldsymbol{x}, \quad i = 1, 2, 3, \tag{12.681}$$

where $e = e(F, S, M)$ is a scalar, characterizing an intrinsic energy, F is a free energy, S is an entropy, $S = -\partial F/\partial T$, M is a chemical potential, $M = \partial F/\partial \mathrm{d}$, d is a function characterizing the concentration of a composition of Earth's materials, $\tau_{ij} = \partial F/\partial e_{ij}$

is a stress tensor, e_{ij} is a strain tensor, D_{ij} is a strain rate tensor defined below, w is an influx of energy, and $\boldsymbol{q}$ is a vector of an energy transport, all referred to the unit volume Equation 12.681 is the basic relation for the determination of the thermal distribution in the Earth's interior. Since the domain Ω is arbitrary, we can rewrite equation 12.681 in its differential form

$$\rho\frac{\mathrm{d}e}{\mathrm{d}t}(\boldsymbol{x},t) = \tau_{ij}(\boldsymbol{x},t)D_{ij}(\boldsymbol{x},t) + \rho(\boldsymbol{x},t)W(\boldsymbol{x},t) - \partial q_i(\boldsymbol{x},t)/\partial x_i, \tag{12.682}$$

where $i = 1, 2, 3$. In Equations 12.679–12.682 and hereafter the Einstein's summation convention is used. Subscripts range from 1 to 3.

Since $c(T,\boldsymbol{x},t) = c_e(T,\boldsymbol{x},t)\rho(T,\boldsymbol{x},t)$, where c is a specific heat, and T is a temperature, and since $\boldsymbol{q}(T,\boldsymbol{x},t) = -\kappa(T,\boldsymbol{x},t)$ grad $T(\boldsymbol{x},t)$ is the well-known Fourier's law, where κ is a coefficient of thermal conductivity satisfying $\kappa_{ij}\zeta_i\zeta_j \geq \kappa_o\zeta_i\zeta_i$, $\kappa_o =$ constant > 0, and since $e = F + ST$, $S = \beta_{ij}e_{ij} + c_eT/T_{\boldsymbol{o}}$, where F is the free energy defined by the second law of thermodynamics, then the generalized heat equation can be written in the following form

$$\begin{aligned}\rho(\boldsymbol{x},t)\beta_{ij}(\boldsymbol{x},t)T_o(x)D_{ij}(\boldsymbol{x},t) + \rho(\boldsymbol{x},t)c_e(T,\boldsymbol{x},t)(\partial T(\boldsymbol{x},t)/\partial t + v_k\partial T/\partial x_k)\\ = \rho(\boldsymbol{x},t)W(\boldsymbol{x},t) - \partial/\partial x_i(\kappa_{ij}\partial T/\partial x_j).\end{aligned} \tag{12.683}$$

The first term on the left-hand side represents the energy dissipated in the form of heat in the moving masses of the Earth and generated by their deformation, and W represents the dissipation function representing heat sources as radioactive sources W_1, sources W_2 evoked by the viscous stresses $(1/2)\eta_{ijkl}D_{ij}D_{kl}$, the Joule's heat $W_3 = \sigma_{ij}^{-1}j_ij_j$, where $\sigma = (\sigma_{ij})$ is a coefficient of electric conductivity ($\cong 10^5 - 10^6\Omega^{-1}\mathrm{m}^{-1}$), etc.. Generally, the thermal conductivity consists of three parts with different origins, i.e. the phonon conductivity (decreases with temperature as $T^{-1}{}^\circ K$), the radiative conductivity (increases with temperature but not faster than $(T^oK)^3$) and excited conductivity (becomes effective only at temperatures higher than $\cong 1000^\circ K$).

The behaviour of the magnetic field in the Earth's body can be described by the well-known Maxwell's equations

$$\text{rot}\,\boldsymbol{H} = \boldsymbol{j} + \partial\boldsymbol{D}/\partial t, \tag{12.684}$$

$$\text{rot}\,\boldsymbol{E} = -\mu\partial\boldsymbol{H}/\partial t, \tag{12.685}$$

$$\text{div}\,\boldsymbol{D} = \rho_e, \tag{12.686}$$

$$\text{div}\,\boldsymbol{B} = 0, \tag{12.687}$$

where $\boldsymbol{B} = \mu\boldsymbol{H}$, $\boldsymbol{D} = \varepsilon\boldsymbol{E}$, and $\boldsymbol{E}$ and $\boldsymbol{H}$ are intensities of electric and of magnetic fields, $\boldsymbol{D}$ and $\boldsymbol{B}$ are vectors of electric and magnetic inductivities for moving the Earth's mass, ε is a dielectric tensor and μ is a magnetic permeability tensor which both depend on the temperature and the pressure, $\boldsymbol{j} = \sigma(\boldsymbol{E} + [\boldsymbol{v},\boldsymbol{B}])$ is a current density vector for a moving non-metallized medium, and ρ_e is a volume density of charge.

To describe planetary magnetic processes we shall start from the Maxwell's equations for the quasi-stationary case, in which the Maxwell's current can be omitted. Then from Equation 12.684 we have $\boldsymbol{E} = (\mu\sigma)^{-1}$ rot $\boldsymbol{B} - [\boldsymbol{v}, \boldsymbol{B}]$, where $(\mu\sigma)^{-1} = \eta_o$ denotes a coefficient of magnetic diffusion. Applying the operation rot, then from the second Maxwell's equation we obtain the so-called magnetodynamic equation

$$\partial \boldsymbol{B}(\boldsymbol{x}, t)/\partial t - \text{rot}\,[\boldsymbol{v}(\boldsymbol{x}, t), \boldsymbol{B}(\boldsymbol{x}, t)] + \text{rot}\,(\eta_o(\boldsymbol{x}, t)\,\text{rot}\,\boldsymbol{B}(\boldsymbol{x}, t)) = 0. \qquad (12.688)$$

If a moving thermally and electrically conductive and compressible (or incompressible) thermo-visco-liquid and thermo-visco-plastic media, which contain a relatively high concentration of electrons and ions due to their metallization and ionization, be caught in a magnetic field, then due to their hydrodynamic motions electric currents are originated and an electric field is induced in the Earth's interior. Moreover, the Lorentz's force acting on charge elements (these electric currents) has a great influence on the hydrodynamic motions in these media. On the other hand, the Earth's interior rocks under high temperatures and high pressures change their magnetic properties many times. In planetary physics, anomalous magnetic fields are observed, for example at present the eccentricity of Uranus' and Neptune's magnetism (see Hide (1988), (1993)). Voyager 2 makes measurements of the magnetic field of Uranus. Experts from the NASA Goddard Space Flight Center found that Uranus' magnetic dipole field is apparently inclined to the rotation axis by about 60° and displaced from the centre of the planet by about 0.3 times its radius, which is a much bigger value than the corresponding values for the Earth, Jupiter and Saturn. The effective dipole moment was found to be about 0.23 Gauss R_U^3 (where the radius of Uranus $R_U \cong 25000$ km), corresponding to minimum and maximum fields at the surface of Uranus of about 0.1 to 1.1 Gauss, respectively (Hide (1987)). Moreover, the rotational period of the magnetic field, which should be close to that of the deep interior of the planet, is 17.29 ± 0.1 hours. One probable explanation is the collision of Uranus with a small planet. During this collision the inhomogeneity originates with special magnetic properties (which will be solved in the next section). During the last geological epoch such a situation is also not precluded in the case of the Earth. In a strategy towards an interpretation of the phenomenon, the magnetization under the critical Curie's temperature point represents, in addition to the action of rotation and of the Lorentz' and Coriolis' forces and other effects, one of the relevant roles in the generation of magnetic fields of the planets.

From electro-magnetic theory it is known that the magnetic induction vector $\boldsymbol{B} = \mu_o(1 + \chi)\boldsymbol{H} = \mu\boldsymbol{H}$, where μ is magnetic permeability, χ is magnetic susceptibility which for paramagnetic and diamagnetic rocks is constant, and the vector of magnetization $\boldsymbol{M} = \chi\boldsymbol{H}$. For ferro-magnetic rocks χ is a function of the intensity of the magnetic field $\boldsymbol{H}$, and then the vector of magnetization $\boldsymbol{M} = \chi(\boldsymbol{H})\boldsymbol{H}$. If remaining magnetization is assumed then $\boldsymbol{M} = \boldsymbol{M}_o + \chi(\boldsymbol{M}_o)\boldsymbol{H}$, where $\boldsymbol{M}_o$ is a vector of remaining magnetization. Since some ferro-magnetic rocks under high temperature change into paramagnetic ones, we define such a law as follows.

There exists temperature T_c, the so-called Curie's temperature or Curie's point, magnetization vector $\boldsymbol{M}_o$, magnetic susceptibility χ_o and intensity of magnetic field $\boldsymbol{H}_o$ such that

if $T < T_c$ then $\chi =$ constant and $\chi > 0$, $\mu > 0$ for paramagnetic rocks,
($\chi < 0$, $\mu < 0$ for diamagnetic rocks),
and the vector of magnetization $\boldsymbol{M} = \chi \boldsymbol{H}$, (12.689)
if $T \geq T_c$ then $\chi = \chi(\boldsymbol{H})$ for ferromagnetic rocks and then
the vector of magnetization $\boldsymbol{M} = \chi(\boldsymbol{H})\boldsymbol{H}$, (12.690)

and, moreover,

if $\boldsymbol{H} < \boldsymbol{H}_o$ then $\chi(\boldsymbol{H}) \gg \chi_o (\cong 10^3)$,
if $\boldsymbol{H} \gg H_o$ then $\chi(\boldsymbol{H}) \rightarrow \chi_{\text{sat}}$ and $\boldsymbol{M} \rightarrow M_{\text{sat}}$, i.e. to a certain
maximal value of magnetization corresponding to the state of saturation.

In a simple version of the investigated problem we can assume that χ and therefore also μ are bounded continuous functions. This type of approximation of χ and μ will be discussed in the next section.

Since the moving materials of the Earth are or are not metallized then a current density vector $\boldsymbol{j}$ is defined by the Ohm's law

$$\boldsymbol{j} = \sigma(\boldsymbol{E} + [\boldsymbol{v}, \boldsymbol{B}]) \text{ in a non-metallized case,} \tag{12.691}$$

Let $p_m \gg 0$ be given, where p_m represents the pressure for which a metallized state of material occurs, then

if $p < p_m$ then $\boldsymbol{j} = \sigma(\boldsymbol{E} + [\boldsymbol{v}, \boldsymbol{B}])$ (i.e. a non-metallized state),
if $p \geq p_m$ then there exists $\sigma_m = \overline{\mu}\underline{n}\,\underline{e} > 0$,

where $\overline{\mu}$ represents mobility of an electron, $\underline{n}$ is a density of numbers of carriers and $\underline{e}$ is an electric charge, such that

$$\boldsymbol{j} = (\sigma + \sigma_m)(\boldsymbol{E} + [\boldsymbol{v}, \boldsymbol{B}]) \text{ in a metallized case.} \tag{12.692}$$

Since the moving Earth's materials are under the influence of an electrical field, then they are ionized and in this case the Ohm's law can be written in the following form.

Let there exist a positive constant field $\boldsymbol{E}_o$, the threshold of ionization, and suitable positive scalar σ_i such that

$$\boldsymbol{j} = \sigma(\boldsymbol{E} + [\boldsymbol{v}, \boldsymbol{B}]) \quad \text{if } |\boldsymbol{E}| < E_o \text{ in a non-ionized case,} \tag{12.693}$$

$$\boldsymbol{j} = (\sigma + \sigma_i)(\boldsymbol{E} + [\boldsymbol{v}, \boldsymbol{B}]) \quad \text{if } |\boldsymbol{E}| = \boldsymbol{E}_o \text{ in an ionized case.} \tag{12.694}$$

By combining the Ohm's law for the metallized Earth's materials and the Ohm's law for the ionized Earth's materials, we thus obtain the Ohm's law for the metallized-and-ionized Earth's materials.

Then using the same technique as above we obtain the magnetodynamic equation for the metallized-and-ionized case

$$\partial \boldsymbol{B}(\boldsymbol{x}, t)/\partial t + \operatorname{rot}\,(\underline{\eta}_o(\boldsymbol{x}, t) \operatorname{rot} \boldsymbol{B}(\boldsymbol{x}, t)) - \operatorname{rot}\,([\boldsymbol{v}(\boldsymbol{x}, t), \boldsymbol{B}(\boldsymbol{x}, t)]) = 0, \tag{12.695}$$

where $\underline{\eta}_o = (\mu(\sigma + \sigma_m + \sigma_i))^{-1}$.

In a deforming thermo-elasto-visco-plastic Earth's body thermo-diffusion processes are also observed. Since during the geological period the distribution of a concentration of fluids and gases in the Earth's materials changes, then diffusion processes in the Earth's interior are one of the fundamental sources of dissipative energy. Let d be a function characterizing the concentration of a composition of the Earth's materials, u be the displacement, and M be a chemical potential, then the equation for a concentration d can be derived from the relation between the vector of a flux of mass $\boldsymbol{i}_m$ and the chemical potential M. Since $M = \partial F/\partial \mathrm{d}$, $\boldsymbol{i}_m = L_{dd}/T_p$ grad M, where L_{dd} is the Onsager's constant and F is a free energy, then

$$\boldsymbol{i}_m = D_d \text{ grad d} + K(T) \text{ grad } T + P_p(p) \text{ grad div } \boldsymbol{u}, \tag{12.696}$$

where $D_d = K_d L_{dd} T_p^{-1}$, K_d is a material constant, $K(T)$ is a coefficient of thermodiffusion, $P_p(p)$ is a coefficient of barodiffusion, and p is the pressure. Thus the vector of a flux of mass depends on the concentration d, temperature T and dilatation gradients. Hence the expression for the flux of mass is of the form (see Landau and Lifshitz (1984))

$$\rho(\partial \mathrm{d}/\partial t + \boldsymbol{v} \text{ grad d}) + \text{ div } \boldsymbol{i}_m = h, \tag{12.697}$$

where h is the intensity of the source of diffusion mass.

The effect of gravity has the main importance for all processes acting in the interior of the Earth. Therefore, it cannot be negligible. Thus, it is hardly a restriction to assume a gravity potential Φ (see Nedoma (1986), (1987), (1988)) satisfying the generalized Poisson's equation

$$\Delta\Phi(\boldsymbol{x}, t) = -4\pi G_c p(\boldsymbol{x}, t)\rho(\boldsymbol{x}, t)K^{-1}(\boldsymbol{x}, t) + 2\omega^2, \tag{12.698}$$

where the density $\rho(\boldsymbol{x}, t)$ satisfies Equation 12.676, 12.677 or 12.678, $G_c = 6.6710^{-8}$ cm^{3}g^{-1} sec^{-2} is the gravitational constant, $K(\boldsymbol{x}, \boldsymbol{H}, \boldsymbol{E}, T,.,t) = \rho(\boldsymbol{x}, \boldsymbol{H}, \boldsymbol{E}, T,.,t)(c_p^2(\boldsymbol{x}, \boldsymbol{H}, \boldsymbol{E}, T,.,t) - 4\frac{1}{3}c_s^2(\boldsymbol{x}, \boldsymbol{H}, \boldsymbol{E}, T,.,t)) \cong 5/8\hat{g}^{1/2}e_{kk}^{-1}$ is a bulk modulus, reckoned from the elastic constants, the density or the velocities of seismic P and S waves c_p and c_s, respectively, $\hat{g}$ is a threshold of plasticity defined by the Mises' type relation $(1/2)\tau_{ij}^D\tau_{ij}^D \geq \hat{g}^2$ and represents also the relation for determination of the threshold of plasticity $\hat{g}$ in our model studies. Furthermore,

$$e_{ij} = \tfrac{1}{2}(\partial u_i/\partial x_j + \partial u_j/\partial x_i)(+\partial u_k/\partial x_i \partial u_k/\partial x_j),\ D_{ij}(\boldsymbol{v}) = e_{ij}(\dot{u}) \tag{12.699}$$

are small (finite) strain and strain rate tensors, and the gravity acceleration vector $\boldsymbol{g}$ satisfies

$$\boldsymbol{g} = \text{ grad } \Phi.$$

The constitutional stress-strain and strain rate relations are defined by

$$\tau_{ij} = \tau_{ij}^B + \tau_{ij}^T + \tau_{ij}^M, \tag{12.701}$$

where

$$\tau_{ij}^B = -p\delta_{ij} + \hat{g}D_{ij}D_{\mathrm{II}}^{-1/2} + 2\hat{\mu}D_{ij}, \tag{12.702}$$

represents the stress-strain rate relation for a visco-plastic Bingham medium, $D_{\mathrm{II}} = (1/2)D_{ij}D_{ij}$ is an invariant of the strain rate tensor D_{ij} (for incompressible materials of the Earth $D_{kk} = 0$), $\hat{g}$ and $\hat{\mu}$ are thresholds of plasticity and viscosity, p is the spherical part of the stress tensor τ_{ij}^{B} and has a meaning of pressure, and δ_{ij} is the Kronecker symbol, and

$$\tau_{ij}^{T} = \beta_{ij}(T - T_o) \tag{12.703}$$

represents thermal stresses, T is temperature, T_o initial temperature and β_{ij} a coefficient of thermal expansion. Furthermore,

$$\tau_{ij}^{M} = E_i D_j + H_i B_j - ((\boldsymbol{ED}) + (\boldsymbol{HB}))\delta_{ij} \tag{12.704}$$

is the Maxwell's stress, $\boldsymbol{E}$ and $\boldsymbol{H}$ represent the total intensity vector of the electric and magnetic fields of the moving Earth's mass, and $\boldsymbol{D}$ and $\boldsymbol{B}$ the corresponding vectors of electric and magnetic inductivities. Moreover, the state equation, and the boundary and the initial conditions must be given.

12.5.3.3 Boundary condition

We shall assume that on the surface of the Earth the display of the velocity field $\boldsymbol{v}(\boldsymbol{x}, t)$ or the loading forces $\boldsymbol{P}(\boldsymbol{x}, t)$, the temperature $T_1(\boldsymbol{x}, t)$ or the heat flux $\boldsymbol{q}(\boldsymbol{x}, t)$, the gravity potential $\Phi_1(\boldsymbol{x}, t)$ and the concentration $\mathrm{d}_1(\boldsymbol{x}, t)$ are known, so that

$$\boldsymbol{v}(\boldsymbol{x}, t) = \boldsymbol{v}_1(\boldsymbol{x}, t) \tag{12.705}$$

$$\text{or } \tau_{ij}(\boldsymbol{x}, t)n_j = P_i(\boldsymbol{x}, t), \tag{12.706}$$

$$T(\boldsymbol{x}, t) = T_1(\boldsymbol{x}, t) \tag{12.707}$$

$$\text{or } \kappa_{ij}\partial T(\boldsymbol{x}, t)/\partial x_j n_i = q(\boldsymbol{x}, t), \tag{12.708}$$

$$\Phi(\boldsymbol{x}, t) = \Phi_1(\boldsymbol{x}, t), \tag{12.709}$$

$$\mathrm{d}(\boldsymbol{x}, t) = \mathrm{d}_1(\boldsymbol{x}, t) \tag{12.710}$$

are given, where $\boldsymbol{n}$ is a normal to the Earth's surface.

As the Earth is non-homogeneous then on the interfaces between two neighbouring media the following conditions hold

$$\mathrm{Div}\ \boldsymbol{B} = 0, \tag{12.705}$$

$$\mathrm{Rot}\ \boldsymbol{E} = 0, \tag{12.705}$$

$$\mathrm{Rot}\ \boldsymbol{B} = \mu \boldsymbol{j}, \tag{12.705}$$

$$\mathrm{Div}\ \boldsymbol{D} = \eta_e, \tag{12.705}$$

$$\mathrm{Div}\ \boldsymbol{j} = -\partial\eta_e/\partial t, \tag{12.705}$$

where $\boldsymbol{j}$ is a surface current density, η_e a free surface density, $\mathrm{Div}\ \boldsymbol{A} = \boldsymbol{n}(\boldsymbol{A}^k - \boldsymbol{A}^l)$, $\mathrm{Rot}\ \boldsymbol{A} = [\boldsymbol{n}, \boldsymbol{A}^k - \boldsymbol{A}^l]$, and $\boldsymbol{n}$ is a unit normal to the interface boundary Γ^{kl} between the neighbouring k-th and l-th media, with direction from the k-th to the l-th medium.

Then for the Earth's surface, which we shall assume behaves like the interface between the normal and strongly conductive phases (in which $\boldsymbol{E} = 0$, $\boldsymbol{B} = 0$), we have the following boundary conditions

$$[\boldsymbol{n}, \text{Rot } \boldsymbol{B}] = 0, \tag{12.716}$$

$$\boldsymbol{B} \cdot \boldsymbol{n} = 0. \tag{12.717}$$

Across the interface boundaries Γ^{kl} the velocities and normal stresses are continuous, so that

$$\boldsymbol{v}^k(\boldsymbol{x}, t) = \boldsymbol{v}^l(\boldsymbol{x}, t), \tag{12.718}$$

$$\tau_{ij} n_j n_i \mid_{(k)} = \tau_{ij} n_j n_i \mid_{(l)}, \tag{12.719}$$

and the temperature and the heat flux are continuous, so that

$$T^k(\boldsymbol{x}, t) = T^l(\boldsymbol{x}, t), \tag{12.720}$$

$$\kappa_{ij}(T, \boldsymbol{x}, t) \partial T(\boldsymbol{x}, t)/\partial x_i n_j \Big|_{(k)} = \kappa_{ij}(T, \boldsymbol{x}, t) \partial T(\boldsymbol{x}, t)/\partial x_i n_j \Big|_{(l)}. \tag{12.721}$$

The next conditions follow from Equations 12.711-12.715. We have

$$[\boldsymbol{n}, \boldsymbol{B}^{(k)}(\boldsymbol{x}, t) - \boldsymbol{B}^{(l)}(\boldsymbol{x}, t)] = 0 \tag{12.722}$$

$$\boldsymbol{B}(\boldsymbol{x}, t) \cdot \boldsymbol{n} \mid_{(k)} = \boldsymbol{B}(\boldsymbol{x}, t) \cdot \boldsymbol{n} \mid_{(l)}. \tag{12.723}$$

For the gravity potential Φ, the concentration of a composition of Earth's materials d and flux of mass $\boldsymbol{i}_m$ we shall assume that across the interface boundaries Γ^{kl} they are continuous, so that

$$\Phi^k(\boldsymbol{x}, t) = \Phi^l(\boldsymbol{x}, t), \tag{12.724}$$

$$\mathrm{d}^k(\boldsymbol{x}, t) = \mathrm{d}^l(\boldsymbol{x}, t), \tag{12.725}$$

$$\boldsymbol{i}_m^k(T, \boldsymbol{x}, t) = \boldsymbol{i}_m^l(T, \boldsymbol{x}, t). \tag{12.726}$$

Moreover, at the same time in the Earth's interior the Earth's materials are heated and melted as well as recrystallized or later solidified after losing geodynamic and tectonic motion. Across any change boundary $\Gamma_R(t)$ the conservation law of energy momentum and mass must be fulfilled and the material density is required to be continuous. Generally, the continuum has two or more phases (solid-liquid, etc.) in which some material properties are markedly different owing to non-uniform heating and straining. Most interest is in the investigation of the problem which lies in studies of the properties of the boundary between two phases, e.g. the core/mantle boundary, boundaries between the solid and liquid phases in the case of GTLV channels and the subducted lithosphere, and the boundary between two recrystallized rocks in the mantle and in the lithosphere. If T represents the temperature, then the latent heat must be added in the case of melting and recrystallization, and released in the case of solidification at the phase change boundary. Energy balance arguments show that although the temperature is continuous everywhere, the heat flux has a jump discontinuity across the phase change boundary $\Gamma_R(t)$. The real situation is much more complicated as the phase change boundaries are accompanied by transition layers characterized by mushy zones where both phases co-exist (Nedoma (1986), (1996), and see also Section 12.3.3.3). The investigation of mushy zones is important for detailed studies but not in our global studies. Let $\Gamma_R(t)$ denote the phase change boundary surface, let $\Omega_i(t)$, $i = 1, \ldots, s$, denote the unmelted (unrecrystallized or unsolidified) and melted (recrystallized or solidified) parts of the Earth's interior at time t and let $T_i(\boldsymbol{x}, t)$ denote the temperature in $\Omega_i(t)$. Further let

$$\kappa^k(T, p, \ldots \boldsymbol{x}, t) = \lim_{T \to T_R^-} \kappa(T, p, \ldots \boldsymbol{x}, t),$$

$$\kappa^l(T, p, \ldots \boldsymbol{x}, t) = \lim_{T \to T_R^-} \kappa(T, p, \ldots \boldsymbol{x}, t).$$

Generally, at the phase change boundary surface $R^{kl}(\boldsymbol{x}, t) = 0$ we have

$$T^k(\boldsymbol{x}, t) = T^l(\boldsymbol{x}, t) = T_R^{kl}, \tag{12.727}$$

$$\boldsymbol{q}^k \boldsymbol{n} - \boldsymbol{q}^l \boldsymbol{n} = \rho L^{kl} v_n^{kl}, \tag{12.728}$$

where $v_n^{kl} = \mathrm{d}R^{kl}/\mathrm{d}t$ is the speed of the phase change boundary surface $\Gamma_R^{kl}(t)$ along the normal $\boldsymbol{n}^k$, $R^{kl}(\boldsymbol{x}, t) = 0$ defines the position vector of a point on $\Gamma_R^{kl}(t)$, $L^{kl}(T, p, \ldots \boldsymbol{x}, t)$ is the latent heat, and $q_i = -\kappa_i$ grad T is the heat flux in $\Omega_i(t)$. Thus Equations 12.718 and 12.719 leads to

$$T^k(\boldsymbol{x}, t) = T^l(\boldsymbol{x}, t) = T_R^{kl}, \tag{12.729}$$

$$\kappa^k(T, p, \ldots \boldsymbol{x}, t) \operatorname{grad} T^k \operatorname{grad} R^{kl} - \kappa^l(T, p, \ldots \boldsymbol{x}, t) \operatorname{grad} T^l \operatorname{grad} R^{kl}$$
$$= \rho(\boldsymbol{x}, t) L^{kl}(T, p, \ldots \boldsymbol{x}, t) v_n^{kl}(\boldsymbol{x}, t), \tag{12.730}$$

where T_R^{kl} is the temperature of melting or recrystallization or solidification. In the mathematical literature equations 12.729 and 12.730 are known as the Stefan conditions.

Furthermore, we shall assume that at initial time t_o

$$\boldsymbol{v}(\boldsymbol{x}, t_o) = \boldsymbol{v}_o(\boldsymbol{x}), \tag{12.731}$$

$$T(\boldsymbol{x}, t_o) = T_o(\boldsymbol{x}), \tag{12.732}$$

$$\boldsymbol{B}(\boldsymbol{x}, t_o) = \boldsymbol{B}_o(\boldsymbol{x}), \tag{12.733}$$

$$\Phi(\boldsymbol{x}, t_o) = \Phi_o(\boldsymbol{x}), \tag{12.734}$$

$$\mathrm{d}(\boldsymbol{x}, t_o) = \mathrm{d}_o(\boldsymbol{x}). \tag{12.735}$$

For simplicity in the following we shall assume that $t_o = 0$.

Remark: We will pay attention to the case when the properties of rocks depend discontinuously on $\boldsymbol{x} \in \Omega$, confining ourselves to the case when the medium is piecewise homogeneous and, moreover, depends on all geophysical fields. For the simplest case when the medium is only piecewise homogeneous, we divide the domain Ω into s disjoint parts Ω^ι, $\iota = 1, \ldots, s$, $s \in \mathcal{N}$. Let us assume that the rocks are homogeneous on each Ω^ι. Then $c(\boldsymbol{x}, T) = c^\iota(T)$, $\kappa(\boldsymbol{x}, T) = \kappa^\iota(T)$ for $\boldsymbol{x} \in \Omega^\iota$ and $c^\iota(T) = c^\iota_o(T) + \Sigma_{m=1}^{M-1} L^\iota_m \delta(T - T^\iota_m)$, where c^ι_o is a regular part of the distribution c^ι, δ is the Dirac function, T^ι_m and L^ι_m are the temperature and latent heat, respectively, of the transition between the m-th and $(m+1)$-th phases of rocks that occupy the subdomain Ω^ι. In that case we speak about an M-phase Stefan problem.

The problem in its full version leads to solving the system of non-linear variational inequalities and is, as yet, unsolved. Its particular versions were discussed in the previous sections. The simple version of the second phase change was presented in Nedoma (1995) and will also be investigated in the next section.

12.6. SECOND PHASE-CHANGE TRANSITION AND THE PLANET URANUS MAGNETIC FIELD ANOMALY

12.6.1. Introduction

One of the important problems of present geophysics is the study of geodynamic processes and the geomagnetic field in the planet's interior, and the mutual influence of the geomagnetic fields in the core and mantle of the real Earth. The mathematical analysis of this type of problem is given in Nedoma (1988), (1995). But new investigations of our planetary system indicate an anomalous magnetic field of the planet Uranus, which can be explained by collision with a small space body and second-phase transition. The planet's body can then be characterized by a thermo-visco-plastic state in a relatively weak magnetic field. Consequently, thermo-visco-plastic rheology, magnetodynamic theory and the Landau phase transition theory will be used to derive the equations describing these processes in the planet. In this section the planet will be approximated by a moving and heated incompressible Bingham's medium. The boundary conditions on the surface of the planet follow from our knowledge of the velocity (or loading) and temperature on the planet surface and from the assumption that the surface of the planet is strongly electrically conductive. This model, however, provides an idea of how to investigate the anomalous magnetic field of the planet evoked by the second phase change.

The results reported in this section were obtained in the course of an investigation of a simple model problem of the global geodynamics and magnetic field of the planet Uranus. The problem leads to an initial-boundary value problem for a coupled system of equations of magnetodynamics of an incompressible, electrically conducting, thermo-viscous and ferro-magnetic Bingham's fluid. The variational formulation of the problem and the existence and uniqueness theorem will be given. The variational formulation then represents the initial formulation for numerical solutions.

12.6.2. Mathematical Model

We shall consider an incompressible moving $2D$-Bingham fluid under thermal and magnetic effects.

Let $\Omega \subset R^N$, $N = 2$, be a bounded domain with a Lipschitz boundary $\partial\Omega$, occupied by a homogeneous ferro-magnetic Bingham body. Let $I = (t_o, t_1)$ be a time interval. Furthermore, we shall use the notations $u_{i,j} = \partial u_i/\partial x_j$, $\dot{\boldsymbol{u}} = \partial \boldsymbol{u}/\partial t \equiv \boldsymbol{v}$ and $\mathrm{d}v_i/\mathrm{d}t = \partial v_i/\partial t + v_j v_{i,j}$, and similarly for $\mathrm{d}T/\mathrm{d}t$. The Einstein summation convention over the range of $1, \ldots N$ will be applied. Let $\boldsymbol{n}$ be the vector of the outward unit normal at $\boldsymbol{x} \in \partial\Omega$. By [.,.] we denote the vector product in R^N.

The strain-displacement relation, strain rate relation and stress-strain relation of the thermo-Bingham medium are defined as

$$e_{ij}(\boldsymbol{u}) = \tfrac{1}{2}(u_{i,j} + u_{j,i}), D_{ij}(\boldsymbol{v}) = e_{ij}(\dot{\boldsymbol{u}}) = \tfrac{1}{2}(v_{i,j} + v_{j,i}), \quad i, j = 1, 2,$$

$$\tau_{ij} = \tau_{ij}^B + \tau_{ij}^T + \tau_{ij}^M, \tau_{ij}^B = -p\delta_{ij} + \hat{g}D_{ij}(D_{\mathrm{II}})^{-1/2} + 2\hat{\mu}D_{ij}, \quad \tau_{ij}^T = \beta_{ij}(T - T_o),$$

$$\tau_{ij}^M = H_i B_j - (\boldsymbol{HB})\delta_{ij}, \quad i, j = 1, 2,$$

where $\tau_{ij}^B, \tau_{ij}^T, \tau_{ij}^M$ are the Bingham visco-plastic, thermal and Maxwell's stresses, and where $\hat{g}$ and $\hat{\mu}$ are the thresholds of plasticity and viscosity. Here, p represents the spherical part of the stress tensor τ_{ij} and has the meaning of pressure, δ_{ij} is the Kronecker symbol, $D_{\mathrm{II}} = (1/2)D_{ij}D_{ij}$, $\beta_{ij} = \beta\delta_{ij}$ is the coefficient of thermal expansion, and $T_o\rangle 0$ is the initial temperature at which the medium is in the initial stress-strain state. Furthermore, $\boldsymbol{H}$ and $\boldsymbol{B}$ represent the intensity of the magnetic field and the magnetic induction vector respectively, $\boldsymbol{B} = \mu\boldsymbol{H}$, where $\mu = 1 + 4\pi\chi$, is the magnetic permeability, and χ is the magnetic susceptibility, and $\boldsymbol{M} = \chi\boldsymbol{H}$ represents the magnetization vector.

The state of the body will be characterized by temperature T, velocity $\boldsymbol{v}$, intensity of the magnetic field $\boldsymbol{H}$, and magnetization $\boldsymbol{M}$. The magnetization $\boldsymbol{M}$ is a function of position $\boldsymbol{x}$ and of the velocities of particles $\boldsymbol{v}$, which are small in the neighbourhood of the Curie point, T_c, under weak magnetic field $\boldsymbol{H}$. From the classical Landau phase change theory (see Landau and Lifshitz (1959),(1984), Stanley (1971)) it is known that near the critical point, the Curie point, the magnetization $\boldsymbol{M}$ is small and the spontaneous magnetization in the ferro-magnetic phase varies with temperature, while above the Curie point there is no spontaneous magnetization. The behaviour of magnetic susceptibility χ near the Curie point is characterized by a discontinuity of type $(T - T_c)^{-1}$ and such that $\chi_p = a_1^{-1}(p)(T - T_c)^{-1} > 0$ (for a paramagnetic media), and $\chi_{fo} = 2^{-1}a_1^{-1}(p)(T_c - T)^{-1} > 0$ (for a ferro-magnetic media). Heat capacity c at constant pressure p and $\boldsymbol{H} = 0$ at Curie point T_c has a jump $\Delta c = c_p - c_f = -T(2b(T, p))^{-1}a_1^2(p)$, where a_1 is a function of pressure p, and b is a function of pressure p and temperature T. Moreover, a thermal expansion $\beta_{ij} = \beta\delta_{ij}$, where δ_{ij} is the Kronecker symbol, also has a jump $\Delta\beta = \beta_f - \beta_p$ (where $\beta = v^{-1}(\partial^2 g/\partial T\partial p)$, g is a Gibbs function), and an intensity of magnetic field $\boldsymbol{B}$ has a jump of $\sim -6\pi^{-1}a_1^{-1}(p)(T_c - T)^{-1}\boldsymbol{H}$. Since some ferro-magnetic rocks change to paramagnetic under high temperatures, and since is it practically impossible to measure $T = T_c$, we shall assume that χ, μ, c, β are bounded functions. Then we shall solve the following problem:

Let $\Omega \subset R^N$, $N = 2$, be a bounded domain with a Lipschitzian boundary $\partial\Omega$, occupied by a homogeneous ferro-magnetic Bingham body. Let $\Omega^p \equiv \{\boldsymbol{x} \mid T(\boldsymbol{x}) >$

$T_c\}$, $\Omega^f \equiv \{\boldsymbol{x} \mid T(\boldsymbol{x}) < T_c\}$, $\Omega^c \equiv \{\boldsymbol{x} \mid T(\boldsymbol{x}) = T_c\}$, $\Omega = \Omega^p \cup \Omega^f \cup \Omega^c = \cup_\iota \Omega^\iota$, $\iota = p, f, c$. Let $c = c^p$, $\beta = \beta^p$, $\mu = \mu^p$ for $\boldsymbol{x} \in \Omega^p$, $c = c^f$, $\beta = \beta^f$, $\mu = \mu^f$ for $\boldsymbol{x} \in \Omega^f \cup \Omega^c$. The remaining physical parameters are the same in subdomains Ω^p, Ω^f, Ω^c. In dealing with the motion of the thermo-visco-plastic incompressible Bingham medium under the effect of the magnetic field and heating, the following system of equations for temperature T, velocity $\boldsymbol{v} = (v_i)$ and magnetic intensity $\boldsymbol{B} = (B_i)$ can be considered in the form

$$\rho\beta_{ij}T_o D_{ij} + \rho c \mathrm{d}T/\mathrm{d}t = W - (\kappa_{ij}T_{,j})_{,i} \quad \text{on } \Omega \times I, \tag{12.736}$$

$$\begin{aligned}\rho \mathrm{d}v_i/\mathrm{d}t = (&-p\delta_{ij} + \hat{g}D_{ij}(D_{\mathrm{II}})^{-1/2} + 2\hat{\mu}D_{ij} + \beta_{ij}(T - T_o) \\ &+(H_iB_j - (\boldsymbol{HB})\delta_{ij}))_{,j} + f_i, \boldsymbol{B} = \mu\boldsymbol{H} \quad \text{on } \Omega \times I,\end{aligned} \tag{12.737}$$

$$\mathrm{Div}\ \boldsymbol{v} = 0 \quad \text{on } \Omega \times I, \tag{12.738}$$

$$\partial\boldsymbol{B}/\partial t + \mathrm{Rot}\ ((\mu\sigma)^{-1}\ \mathrm{Rot}\ \boldsymbol{B}) - \mathrm{Rot}\ [\boldsymbol{v}, \boldsymbol{B}] = 0 \quad \text{on } \Omega \times I, \tag{12.739}$$

$$\mathrm{Div}\ \boldsymbol{B} = 0 \text{ on } \Omega\times, I \tag{12.740}$$

with a given density ρ, thresholds of plasticity $\hat{g}$ and viscosity $\hat{\mu}$, a specific heat c, thermal expansion $\beta_{ij} = \beta\delta_{ij}$, thermal conductivity $\kappa_{ij} = \kappa\delta_{ij}$, electric conductivity σ, magnetic permeability μ, and initial temperature T_o. For simplicity we assume that the body occupying the domain Ω is homogeneous and incompressible, so that all physical parameters are constants and Div $\boldsymbol{v} = 0$.

Moreover, the following boundary conditions

$$T(\boldsymbol{x}, t) = 0 \quad (\text{ or } \kappa_{ij}\partial T(\boldsymbol{x}, t)/\partial x_j n_i = q) \quad \text{on } \partial\Omega_1 \times I, \tag{12.741}$$

$$\kappa_{ij}\frac{\partial T(\boldsymbol{x}, t)}{\partial x_j}n_i = 0 \quad \text{on } \partial\Omega_2 \times I, \tag{12.742}$$

$$\boldsymbol{v}(\boldsymbol{x}, t) = 0 \quad \text{on } \partial\Omega \times I, \tag{12.743}$$

$$\boldsymbol{B}(\boldsymbol{x}, t)\boldsymbol{n} = 0, [\boldsymbol{n},\ \mathrm{Rot}\ \boldsymbol{B}] = 0 \quad \text{on } \partial\Omega \times I, \tag{12.744}$$

and initial conditions

$$T(\boldsymbol{x}, t_o) = 0, \tag{12.745}$$

$$\boldsymbol{v}(\boldsymbol{x}, t_o) = 0, \tag{12.746}$$

$$\boldsymbol{B}(\boldsymbol{x}, t_o) = 0 \tag{12.747}$$

are considered in the above system of equations.

12.6.3. Variational Formulation and Weak Solution of the Problem

As in the previous sections (see also Nedoma (1995),(1996)) we shall define the following spaces: we introduce space $D(\Omega)$ as the space of all functions in $C^\infty(\Omega)$ with a compact support in Ω. The space has an ordinary countable system of semi-norms and, as usual, $D(\Omega, R^N) = [D(\Omega)]^N$. The Sobolev spaces of functions having generalized derivatives of order s are denoted $[H^s(\Omega)]^k \equiv H^{s,k}(\Omega)$. The norm will be denoted by $\|.\|_{s,k}$, and the scalar product by $(.,.)_s$ (for each integer k). We put $H^{o,k}(\Omega) \equiv L_2^k(\Omega) \equiv L^{2,k}(\Omega)$ and denote $(.,.)$ the scalar product in $L^{2,k}(\Omega)$ (as there is no danger of confusion with k). Denote by $H_o^{s,k}(\Omega)$ the closure of $D(\Omega)$ in the norm of space $H^{s,k}(\Omega)$. Moreover, we shall denote by $\|.\|_E$ the equivalent norm in $H^{1,N}(\Omega)$ induced by small strain tensor e_{ij}, i.e. $\|\boldsymbol{u}\|_E^2 = \int_\Omega e_{ij}(\boldsymbol{u})e_{ij}(\boldsymbol{u})\mathrm{d}\boldsymbol{x} + \|\boldsymbol{u}\|_{o,N}^2$.

Moreover, for $s \geq 1$ we define the following spaces as

$$^1H^{s,N}(\Omega) = \{\boldsymbol{v} \mid \boldsymbol{v} \in H^{s,N}(\Omega),\ \mathrm{Div}\,\boldsymbol{v} = 0; \boldsymbol{v} = 0 \text{ on } \partial\Omega\} = \prod_\iota {}^1H^{s,N}(\Omega^\iota), \quad N = 2,$$

$$^2H^{s,N}(\Omega) = \{\boldsymbol{C} \mid \boldsymbol{C} \in H^{s,N}(\Omega),\ \mathrm{Div}\,\boldsymbol{C} = 0; \boldsymbol{nC} = 0 \text{ on } \partial\Omega\} = \prod_\iota {}^2H^{s,N}(\Omega^\iota), \quad N = 2,$$

$$^3H^{s,1}(\Omega) = \{z \mid z \in H^{s,1}(\Omega), z = 0 \text{ on } \partial\Omega\} = \prod_\iota {}^3H^{s,N}(\Omega^\iota),$$

$$\overline{H}^s = {}^1H^{s,N}(\Omega) \times {}^2H^{s,N}(\Omega) \times {}^3H^{s,1}(\Omega). \text{ We put } \underline{H}^1 = \underline{H}.$$

Then $^1\mathscr{H}(\Omega) = {}^1H^{1,N}(\Omega) \cap D(\Omega, R^N)$. Note that $^1H^{s,N}(\Omega)$ is a Hilbert space with norm $\|.\|_{s,N}$. For the sake of simplicity, we put $^1H^{1,N}(\Omega) = {}^1H(\Omega)$, $\|.\|_1 = \|.\|_{1,N}$, and in $H^{o,N}(\Omega)$ $\|.\|_o = \|.\|_{o,N}$. We shall denote by w_j the eigen-functions of canonical isomorphism $^1\Lambda_s : {}^1H^{s,N} \to ({}^1H^{s,N})'$, i.e. $(\boldsymbol{w}, \boldsymbol{v})_s = \lambda_i(\boldsymbol{w}_j, \boldsymbol{v})\, \forall \boldsymbol{v} \in {}^1H^{s,N}$, $\left\|\boldsymbol{w}_j\right\|_{o,1} = 1$. Let $^2\mathscr{H}(\Omega) = {}^2H^{1,1}(\Omega) \cap D(\Omega, R^N)$, and $^2H^{s,N}(\Omega)$, $s \geq 1$ be a Hilbert space with norm $\|.\|_{s,N}$. We put $^2H(\Omega) = {}^2H^{1,N}(\Omega)$. Furthermore, $^3\mathscr{H}(\Omega) = {}^3H^{1,1}(\Omega) \cap D(\Omega)$; $^3H^{s,1}(\Omega)$, $s \geq 1$ is a Hilbert space with norm $\|.\|_s$. We put $^3H(\Omega) = {}^3H^{1,1}(\Omega)$. Similarly as above, we define C_j and z_j, $C_j \in {}^2H^{s,N}(\Omega)$, $z_j \in {}^3H^{s,1}(\Omega)$, eigen-functions of canonical isomorphisms $^2\Lambda_s : {}^2H^{s,N} \to ({}^2H^{s,N})'$ and $^3\Lambda_s : {}^3H^{s,1} \to ({}^3H^{s,1})'$. We shall denote by $({}^1H^{s,N}(\Omega))'$ the dual space of $^1H^{s,N}(\Omega)$, and similarly in all other cases. Furthermore, we define the following spaces

$$^1\boldsymbol{H} = \{\boldsymbol{v} \mid \boldsymbol{v} \in L^2(I; {}^1H^{s,N}(\Omega)), \boldsymbol{v}' \in L^2(I; {}^1H^{o,N}(\Omega)), \boldsymbol{v}(t_o) = 0\} = \prod_\iota {}^1\boldsymbol{H}^\iota,$$

$$^2\boldsymbol{H} = \{\boldsymbol{C} \mid \boldsymbol{C} \in L^2(I; {}^2H^{s,N}(\Omega)), \boldsymbol{C}' \in L^2(I; {}^2H^{o,N}(\Omega)), \boldsymbol{C}(t_o) = 0\} = \prod_\iota {}^2\boldsymbol{H}^\iota,$$

$$^3\boldsymbol{H} = \{z \mid z \in L^2(I; {}^3H^{s,1}(\Omega)), z' \in L^2(I; {}^3H^{o,1}(\Omega)), z(t_o) = 0\} = \prod_\iota {}^3\boldsymbol{H}^\iota,$$

$$\overline{\boldsymbol{H}} = \prod_\iota \boldsymbol{H}^\iota = {}^1\boldsymbol{H} \times {}^2\boldsymbol{H} \times {}^3\boldsymbol{H}.$$

Throughout this section we shall assume that the body forces $\boldsymbol{f}(\boldsymbol{x}, t) \in L^2(I, H^{1,2}(\overline{\Omega}))$ and thermal sources $W(\boldsymbol{x}, t) \in L^2(I; H^{1,1}(\Omega))$, and $\overline{I} = \langle t_o, t_1 \rangle$. We assume that density ρ, thresholds of plasticity and viscosity $\hat{g}$ and $\hat{\mu}$, and electric conductivity σ are positive constants. Specific heat c and magnetic permeability μ are constants in Ω_p and Ω_f. The coefficient of thermal expansion (β_{ij}) will be assumed such that $\partial\beta_{ij}/\partial x_j \in L^\infty(\Omega)\, \forall i, j = 1, 2$. We assume that $\kappa_{ij} \in L^\infty(\Omega)$ are Lipschitzian on Ω, that they fulfil the usual symmetry condition $\kappa_{ij} = \kappa_{ji}$ on Ω for every $i, j = 1, 2$, and that

$$\kappa_{ij}\xi_i\xi_j \geq c_T \|\xi\|^2_{1,2}, \tag{12.748}$$

$$\kappa_{i,j} = \kappa_{i,j} \quad \boldsymbol{x} \in \Omega, \quad \xi \in R^2, c_T = \text{constant} > 0, \tag{12.749}$$

For $\boldsymbol{v}, \boldsymbol{w} \in H^{1,2}(\Omega)$, $\boldsymbol{B}, \boldsymbol{C} \in H^{1,2}(\Omega)$, $T, z \in H^{1,1}(\Omega)$, we put

$$a_T(T, z) = \int_\Omega \kappa_{ij} T_{,i} z_{,j} \mathrm{d}\boldsymbol{x}, \tag{12.750}$$

$$a(\boldsymbol{v}, \boldsymbol{w}) = 2\int_\Omega D_{ij}(\boldsymbol{v}) D_{ij}(\boldsymbol{w}) \mathrm{d}\boldsymbol{x}, \tag{12.751}$$

$$b_M(\boldsymbol{B}, \boldsymbol{C}) = (\sigma(\mu)^{-1} \text{ Rot } \boldsymbol{B}, \text{ Rot } \boldsymbol{C}), \tag{12.752}$$

$$j(\boldsymbol{v}) = 2\int_\Omega (D_{\text{II}}(\boldsymbol{v}))^{1/2} \mathrm{d}\boldsymbol{x}, \tag{12.753}$$

$$b_o(\boldsymbol{v}, g, z) = \int_\Omega \rho c v_k g_{,k} z \mathrm{d}\boldsymbol{x}, \tag{12.754}$$

$$b(\boldsymbol{u}, \boldsymbol{v}, \boldsymbol{w}) = \int_\Omega \rho u_i v_{j,i} w_j \mathrm{d}\boldsymbol{x}, \tag{12.755}$$

$$b_s(T, \boldsymbol{v}) = \int_\Omega (\beta_{ij} T)_{,j} v_i \mathrm{d}\boldsymbol{x}, \tag{12.756}$$

$$b_p(\boldsymbol{v}, g) = \int_\Omega \rho T_o \beta_{ij} v_{i,j} g \mathrm{d}\boldsymbol{x}. \tag{12.757}$$

Thus we introduce the bilinear form A on $\overline{H}$ for $\boldsymbol{a} \equiv (T, \boldsymbol{v}, \boldsymbol{B})$, $\vec{\alpha} \equiv (z, \boldsymbol{w}, \boldsymbol{C})$ by

$$A(\boldsymbol{a}, \vec{\alpha}) = a_T(T, z) + \hat{\mu} a(\boldsymbol{v}, \boldsymbol{w}) + b_M(\boldsymbol{B}, \boldsymbol{C}), \tag{12.758}$$

where matrix

$$\begin{aligned} &\mathscr{E} = (\mathscr{E}_{ij}), \mathscr{E}_{ij} = \rho\delta_{ij}, \quad i = 1, \ldots, 5, \; j = 1, 2, \\ &\mathscr{E}_{ij} = (\mu)^{-1}\delta_{ij}, \quad i = 1, \ldots, 5, \; j = 3, 4, \\ &\mathscr{E}_{ij} = \rho c\delta_{ij}, \quad i = 1, \ldots, 5, \; j = 5, \end{aligned} \tag{12.759}$$

and the right-hand side

$$\boldsymbol{P} = (\boldsymbol{f}, O, W) \in L^2(I; H^{1,2N+1}(\Omega)), \quad N = 2. \tag{12.760}$$

As $D_{ij} = (1/2)(\hat{\mu})^{-1}(1 - \hat{g}\tau_{\text{II}}^{-1/2})\tau^D_{ij}$, we have $D_{ij}(\boldsymbol{u})D_{ij}(\boldsymbol{v}) \leq 2D^{1/2}_{\text{II}}(\boldsymbol{u})D^{1/2}_{\text{II}}(\boldsymbol{v})$, where τ^D_{ij} is the stress tensor deviator and $\tau_{\text{II}} = (\hat{g} + 2\hat{\mu}D^{1/2}_{\text{II}})^2$.

Let $\boldsymbol{v}, \boldsymbol{w} \in {}^1H^{1,2}(\Omega)$, $\boldsymbol{B}, \boldsymbol{C} \in {}^2H^{1,2}(\Omega)$. Then, according to Equation 12.737 and Maxwell's stresses for incompressible media, we have

$$(\partial/\partial x_i((\mu)^{-1}|\boldsymbol{B}|^2), w_i) = 0, \text{ since Div } \boldsymbol{w} = 0, \tag{12.761}$$

$$\begin{aligned}(\text{Rot } (\eta_o \text{Rot}\boldsymbol{B}), (\mu)^{-1}\boldsymbol{C}) &= \int_{\partial\Omega} (\sigma)^{-1}(\mu)^{-2}[\boldsymbol{n}, \text{Rot}\boldsymbol{B}]\boldsymbol{C}\,\mathrm{d}s + (\eta_o \text{Rot}\boldsymbol{B}, \text{ Rot }((\mu)^{-1}\boldsymbol{C})) \\ &= (\eta_o \text{ Rot } \boldsymbol{B}, \text{ Rot } ((\mu)^{-1}\boldsymbol{C}))\end{aligned} \tag{12.762}$$

as $[\boldsymbol{n},\ \text{Rot}\ \boldsymbol{B}] = 0$ on $\partial\Omega$, where $\eta_o = (\sigma\mu)^{-1}$.

$$(\partial/\partial x_i((\mu)^{-1}B_i\boldsymbol{B}), \boldsymbol{w}) = ((\mu)^{-1}\boldsymbol{w}, B_i\partial\boldsymbol{B}/\partial x_i - \text{ grad } \tfrac{1}{2}\,|\boldsymbol{B}|^2)$$

$$= -((\mu)^{-1}\ \text{Rot}\ [\boldsymbol{w}, \boldsymbol{B}], \boldsymbol{B}), \qquad (12.763)$$

as $((\mu)^{-1}\boldsymbol{w},\ \text{grad}\ (1/2)\,|\boldsymbol{B}|^2) = 0$ (for Div $\boldsymbol{w} = 0$ and $\boldsymbol{w} = 0$ on $\partial\Omega$),

hence $(\partial/\partial x_i((\mu)^{-1}B_i\boldsymbol{B}), \boldsymbol{w}) + ((\mu)^{-1}\ \text{Rot}\ [\boldsymbol{w}, \boldsymbol{B}], \boldsymbol{B}) = 0,$

$$b(\boldsymbol{v}, \boldsymbol{v}, \boldsymbol{v}) = 0. \qquad (12.764)$$

We multiply Equations 12.761, 12.762 and 12.764 by $z - T(t)$, $\boldsymbol{w} - \boldsymbol{v}(t)$, and $\boldsymbol{C} - \boldsymbol{B}(t)$ respectively, and add the first three equations. We integrate both the sum and the last equation over $\Omega = \cup_\iota \Omega^\iota$. We apply the Green theorem respecting the boundary conditions. According to Equations 12.761-12.763 terms $(\partial/\partial x_i((\mu)^{-1}\,|\boldsymbol{B}(t)|^2), w_i - v_i(t))$, $(\partial/\partial x_i((\mu)^{-1}B_i(t)\boldsymbol{B}(t)), \boldsymbol{v}(t))$, $((\mu)^{-1}\ \text{Rot}\ [\boldsymbol{v}(t), \boldsymbol{B}(t)], \boldsymbol{B}(t))$ vanish. After some modification (including, inter alia, the integration in time over interval I and the substitution $J(\boldsymbol{w}) = \int_I j(\boldsymbol{w}(t))\mathrm{d}t$), we obtain the following variational formulation:

Find vector function $\boldsymbol{a} \equiv (\boldsymbol{v}, \boldsymbol{B}, T) \in \boldsymbol{H}$ such that

$$\int_I [(\mathscr{E}\boldsymbol{a}'(t), \vec{\alpha}(t) - \boldsymbol{a}(t)) + A(\boldsymbol{a}(t), \vec{\alpha}(t) - \boldsymbol{a}(t)) + (\Xi(\boldsymbol{a}(t)), \vec{\alpha}(t) - \boldsymbol{a}(t)) + (\boldsymbol{P}(t), \vec{\alpha}(t) - \boldsymbol{a}(t))$$

$$+ \hat{g}\,j(\boldsymbol{w}(t)) - \hat{g}\,j(\boldsymbol{v}(t))]\mathrm{d}t \geq 0 \quad \forall\vec{\alpha} \equiv (\boldsymbol{w}, \boldsymbol{C}, z) \in \overline{\boldsymbol{H}}, \qquad (12.765)$$

where

$$(\Xi(\boldsymbol{a}(t)), \vec{\alpha}(t) - \boldsymbol{a}(t)) = (\Xi_o(\boldsymbol{a}(t)), \boldsymbol{a}(t), \vec{\alpha}(t) - \boldsymbol{a}(t)) + b_s(T(t) - T_o, \boldsymbol{w}(t) - \boldsymbol{v}(t))$$

$$+ b_p(\boldsymbol{v}(t), z(t) - T(t)), \qquad (12.766)$$

$$(\Xi_o(\boldsymbol{a}(t)), \boldsymbol{a}(t), \vec{\alpha}(t) - \boldsymbol{a}(t)) = b(\boldsymbol{v}(t), \boldsymbol{v}(t), \boldsymbol{w}(t) - \boldsymbol{v}(t)) + b_o(\boldsymbol{v}(t), T(t), z(t) - T(t))$$

$$+ (\partial/\partial x_i((\mu)^{-1}B_i(t)\boldsymbol{B}(t)), \boldsymbol{w}(t) - \boldsymbol{v}(t)) - ((\mu)^{-1}\ \text{Rot}\ [\boldsymbol{v}(t), \boldsymbol{B}(t)], \boldsymbol{C}(t) - \boldsymbol{B}(t)). \qquad (12.767)$$

The motivation for such a definition of the variational (weak) solution arises from the fact that domain $\Omega \times I$ decomposes into domains $\Omega^p \times I$ and $\Omega^f \times I$ corresponding to $T < T_c$ and $T > T_c$, respectively, bounded by smooth phase change boundary $S_2(t)$, the so-called free boundary, on which $T = T_c$, such that Equation 12.748-12.761 hold classically and such that they display a discontinuity across the free boundary $S_2(t)$ μ, β_{ij} and c.

The Gronwall lemma and Lemma 12.19 will be used to prove the main result of this study. Moreover,

$$a_T(T, z) = a_T(z, T), a(\boldsymbol{u}, \boldsymbol{v}) = a(\boldsymbol{v}, \boldsymbol{u}), b_M(\boldsymbol{B}, \boldsymbol{C}) = b_M(\boldsymbol{C}, \boldsymbol{B}), A(\vec{\alpha}, \vec{\gamma}) = A(\vec{\gamma}, \vec{\alpha}).$$

For $T \in {}^3H^{1,1}(\Omega)$, $\boldsymbol{v} \in {}^1H^{1,2}(\Omega)$, $\boldsymbol{B} \in {}^2H^{1,2}(\Omega)$ there exist constants $c_T > 0$, $c_B > 0$, $c_M > 0$, such that

$$a_T(T, T) \geq c_T\,\|T\|^2_{1,1} \quad \forall T \in {}^3H^{1,1}(\Omega), \qquad (12.768)$$

$$a(\boldsymbol{v}, \boldsymbol{v}) \geq c_B\,\|\boldsymbol{v}\|^2_{1,2} \quad \forall \boldsymbol{v} \in {}^1H^{1,2}(\Omega), \qquad (12.769)$$

$$b_M(\boldsymbol{B}, \boldsymbol{B}) \geq c_M\,\|\boldsymbol{B}\|^2_{1,2} \quad \forall \boldsymbol{B} \in {}^2H^{1,2}(\Omega) \qquad (12.770)$$

Thus for $\boldsymbol{a} = (\boldsymbol{v}, \boldsymbol{B}, T)$ and $\vec{\alpha} = (\boldsymbol{w}, \boldsymbol{C}, z)$ there exists a suitable positive constant c_a independent of $\vec{\alpha}$, such that

$$A(\vec{\alpha}, \vec{\alpha}) \geq c_a \|\vec{\alpha}\|^2_{1,2N+2} \text{ for all } \vec{\alpha} \in \underline{H}. \tag{12.771}$$

From the Hölder inequality applied to $b(t; \boldsymbol{u}, \boldsymbol{v}, \boldsymbol{w})$ we have (see Temam (1977))

$$|b(\boldsymbol{u}, \boldsymbol{v}, \boldsymbol{w})| \leq c_1 \|\boldsymbol{u}\|_{L^{2,p}(\Omega)} \|\boldsymbol{w}\|_{L^{2,p}(\Omega)} \sum_{i,j} \left\|D_i v_j\right\|_{L^{2,2}(\Omega)}, \tag{12.772}$$

where

$$2/p + \tfrac{1}{2} = 1. \tag{12.773}$$

Applying Lemma 12.19 we have

$$|b(\boldsymbol{u}, \boldsymbol{v}, \boldsymbol{w})| \leq c_2 \|\boldsymbol{u}\|_{L^{2,p}(\Omega)} \|\boldsymbol{w}\|_{L^{2,p}(\Omega)} \|\boldsymbol{v}\|_{1,2}. \tag{12.774}$$

Since (see Temam (1977),Nedoma (1995))

$$\|\boldsymbol{v}\|_{(1/2),2} \leq c \|\boldsymbol{v}\|^{1/2}_{L^{2,2}(\Omega)} \|\boldsymbol{v}\|^{1/2}_{1,2} \text{ and } H^{1/2,2}(\Omega) \subset L^{2,p}(\Omega), \tag{12.775}$$

for p from Equation 12.773, then

$$\|\boldsymbol{v}\|_{L^{2,p}(\Omega)} \leq c_3 \|\boldsymbol{v}\|^{1/2}_{1,2} \|\boldsymbol{v}\|^{1/2}_{L^{2,2}(\Omega)} \quad \forall \boldsymbol{v} \in H^{1,2}_o(\Omega). \tag{12.776}$$

From Equations 12.774 and 12.776 we obtain

$$|b(\boldsymbol{u}, \boldsymbol{v}, \boldsymbol{w})| \leq c_4 \|\boldsymbol{u}\|^{1/2}_{1,2} \|\boldsymbol{u}\|^{1/2}_{o,2} \|\boldsymbol{w}\|^{1/2}_{1,2} \|\boldsymbol{w}\|^{1/2}_{o,2} \|\boldsymbol{v}\|_{1,2}. \tag{12.777}$$

Furthermore, if $\boldsymbol{u}, \boldsymbol{v}, \boldsymbol{w} \in {}^1\mathscr{H}(\Omega)$, then $b(\boldsymbol{u}, \boldsymbol{v}, \boldsymbol{w}) + b(\boldsymbol{u}, \boldsymbol{w}, \boldsymbol{v}) = 0$ and the last equality is also valid for $\boldsymbol{u} \in L^{2,2}(\Omega)$, $\boldsymbol{v}, \boldsymbol{w} \in {}^1H^{1,2}(\Omega)$. Also $b(\boldsymbol{u}, \boldsymbol{u}, \boldsymbol{v}) = -b(\boldsymbol{u}, \boldsymbol{v}, \boldsymbol{u})$ for $\boldsymbol{u}, \boldsymbol{v} \in {}^1\mathscr{H}(\Omega)$.

Similarly $b_o(\boldsymbol{u}, y, z) + b_o(\boldsymbol{u}, z, y) = 0$, $\boldsymbol{u} \in {}^1H^{0,2}(\Omega)$, $y, z \in H^{1,1}_o(\Omega)$ and there exist positive constants c_5 independent of $z, \boldsymbol{u}, y$ such that

$$|b_o(\boldsymbol{u}, y, z)| \leq c_5 \|\boldsymbol{u}\|^{1/2}_{1,2} \|\boldsymbol{u}\|^{1/2}_{o,2} \|y\|^{1/2}_{1,1} \|y\|^{1/2}_{o,1} \|z\|_{1,1}. \tag{12.778}$$

Since

$$(\partial/\partial x_i((\mu)^{-1} B_i \boldsymbol{B}), \boldsymbol{w}) = \int_\Omega (\mu)^{-1} (\boldsymbol{B} B_{i,i} + B_i \boldsymbol{B}_{,i}) \boldsymbol{w} \, \mathrm{d}\boldsymbol{x} = ((\mu)^{-1} B_i \boldsymbol{B}, \partial \boldsymbol{w}/\partial x_i) \tag{12.779}$$

(due to Div $\boldsymbol{B} = 0$),

$$\begin{aligned} ((\mu)^{-1} \operatorname{Rot}\, [\boldsymbol{v}, \boldsymbol{B}], \boldsymbol{C}) &= \int_\Omega (\mu)^{-1} (v_{i,j} B_j - v_{j,j} B_i + v_i B_{j,j} - v_j B_{i,j}) C_i \mathrm{d}\boldsymbol{x} \\ &= -\int_\Omega (\mu)^{-1} (v_i B_j - v_j B_i)(C_{j,i} - C_{i,j}) \mathrm{d}\boldsymbol{x} = -((\mu)^{-1} [\boldsymbol{v}, \boldsymbol{B}], \operatorname{Rot} \boldsymbol{C}), \end{aligned} \tag{12.780}$$

where the Green theorem and the boundary conditions were applied, there exist positive constants c_6 and c_7 such that

$$\left|(\partial/\partial x_i((\mu)^{-1} B_i \boldsymbol{B}), \boldsymbol{w})\right| \leq c_6 \|\boldsymbol{B}\|_{1,2} \|\boldsymbol{B}\|_{0,2} \|\boldsymbol{w}\|_{1,2}, \tag{12.781}$$

$$\left|((\mu)^{-1} \operatorname{Rot}\, [\boldsymbol{v}, \boldsymbol{B}], \boldsymbol{C})\right| \leq c_7 \|\boldsymbol{v}\|^{1/2}_{1,2} \|\boldsymbol{v}\|^{1/2}_{0,2} \|\boldsymbol{B}\|^{1/2}_{1,2} \|\boldsymbol{B}\|^{1/2}_{0,2} \|\boldsymbol{C}\|_{1,2}. \tag{12.782}$$

Therefore, due to the estimates in Equations 12.772, 12.774, 12.777, 12.778,12.781 and 12.782, there exist positive constants c_8 independent of $\vec{\alpha}$, $\vec{\gamma}$ such that

$$|\Xi_o(\vec{\alpha}, \vec{\alpha}, \vec{\gamma})| \leq c_8 \, \|\vec{\alpha}\|_{1,2N+1} \, \|\vec{\alpha}\|_{0,2N+1} \, \|\vec{\gamma}\|^{1/2}_{1,2N+1}, \quad N = 2. \tag{12.783}$$

Moreover, we have the following estimate

$$|X| = \left|b_s(T - T_o, \boldsymbol{v}) + b_p(\boldsymbol{v}, T)\right| \leq c(1 + \|T(t)\|_{1,1} \, \|\boldsymbol{v}(t)\|_{0,2} + \|T(t)\|_{0,1} \, \|\boldsymbol{v}(t)\|_{1,2}), \tag{12.784}$$

as we have assumed that $\partial\beta_{ij}/\partial x_j \in L^\infty(\Omega)$, $\forall i, j$, and where constant c does not depend on ε.

The main result yields the following theorem.

Theorem 12.19: Let $N = 2$. Let $\Omega = \cup_\iota \Omega^\iota$, $\iota = p, f$, and let

$$W \in L^2(I; ({}^3H^{1,2}(\Omega))'), \boldsymbol{f} \in L^2(I; ({}^1H^{1,2}(\Omega))'). \tag{12.785}$$

Then there exists vector function $\boldsymbol{a} \equiv (\boldsymbol{v}, \boldsymbol{B}, T)$ such that

$$T \in L^2(I, {}^3H^{1,1}(\Omega)) \cap L^\infty(I; {}^3H^{0,1}(\Omega)),\ T' \in L^2(I; ({}^3H^{1,1}(\Omega))'), \tag{12.786}$$

$$\boldsymbol{v} \in L^2(I; {}^1H^{1,2}(\Omega)) \cap L^\infty(I; {}^1H^{0,2}(\Omega)),\ \boldsymbol{v}' \in L^2(I; ({}^1H^{1,2}(\Omega))'), \tag{12.787}$$

$$\boldsymbol{B} \in L^2(I; {}^2H^{1,2}(\Omega)) \cap L^\infty(I; {}^2H^{0,2}(\Omega)),\ \boldsymbol{B}' \in L^2(I; ({}^2H^{1,2}(\Omega))'), \tag{12.788}$$

$$\boldsymbol{v}(t_o) = 0, \boldsymbol{B}(t_o) = 0, T(t_o) = 0 \tag{12.789}$$

and which satisfies Equations 12.765-12.767.

Proof: Regularization will be used to prove the theorem. Let $J_\varepsilon(\boldsymbol{w})$ be a regularized functional of functional $J(\boldsymbol{w})$ defined as

$$J_\varepsilon(\boldsymbol{w}) = \int_I \hat{g} j_\varepsilon(\boldsymbol{w}(t))\mathrm{d}t, \tag{12.790}$$

where $j_\varepsilon(\boldsymbol{w}(t)) = 2(1+\varepsilon)^{-1} \int_\Omega (D_{\mathrm{II}}(\boldsymbol{w}(t)))^{1/2(1+\varepsilon)} \mathrm{d}\boldsymbol{w}$, $\varepsilon > 0$, for which

$$(J'_\varepsilon(\boldsymbol{v}), \boldsymbol{w}) = \int_{\Omega\times I} \hat{g} D_{\mathrm{II}}(\boldsymbol{v})^{1/2(\varepsilon-1)} D_{ij}(\boldsymbol{v}) D_{ij}(\boldsymbol{w}) \mathrm{d}\boldsymbol{x}\, \mathrm{d}t. \tag{12.791}$$

The problem in Equation 12.765 is then replaced by the following regularized

Problem $(\mathscr{P}_r)_v$: Find vector function $\boldsymbol{a}_\varepsilon \equiv (\boldsymbol{v}_\varepsilon, \boldsymbol{B}_\varepsilon, T_\varepsilon) \in \underline{\boldsymbol{H}}$ such that

$$\int_I [(\mathscr{E}\boldsymbol{a}'_\varepsilon(t), \vec{\alpha}(t) - \boldsymbol{a}_\varepsilon(t)) + A(\boldsymbol{a}_\varepsilon(t), \vec{\alpha}(t) - \boldsymbol{a}_\varepsilon(t)) + (\Xi(\boldsymbol{a}_\varepsilon(t)), \vec{\alpha}(t) - \boldsymbol{a}_\varepsilon(t))$$
$$- (\boldsymbol{P}(t), \vec{\alpha}(t) - \boldsymbol{a}_\varepsilon(t))]\mathrm{d}t + J_\varepsilon(\boldsymbol{w}) - J_\varepsilon(\boldsymbol{v}_\varepsilon(t)) \geq 0 \quad \forall\vec{\alpha} \equiv (\boldsymbol{w}, \boldsymbol{C}, z) \in \underline{\boldsymbol{H}}, \tag{12.792}$$

where Ξ is defined by Equations 12.766 and 12.767.

The proof is parallel to that of Nedoma (1995) for the bounded coefficients and depends on the following:

(a) proving the existence of the solution of Equation 12.792, which, as in Section 12.4.1 and Nedoma (1995), is based on the Galerkin approximation,

(b) deriving a priori estimates I and II independent of ε,

(c) performing the limitation procedures over m and ε,

(d) proving the uniqueness of the solution.

Firstly consider the a priori estimates I and II.

Using Equation 12.763 and 12.784, $b(\boldsymbol{v}, \boldsymbol{v}, \boldsymbol{v}) = 0$ we obtain

$$(\partial/\partial x_i((\mu)^{-1}B_{mi}(t)\boldsymbol{B}_m(t)), \boldsymbol{v}_m(t)) + ((\mu)^{-1} \text{ Rot } [\boldsymbol{v}_m(t), \boldsymbol{B}_m(t)], \boldsymbol{B}_m(t)) = 0, \tag{12.793}$$

$$b(\boldsymbol{v}_m(t), \boldsymbol{v}_m(t), \boldsymbol{v}_m(t)) = 0, \tag{12.794}$$

$$b_o(\boldsymbol{v}_m(t), T_m(t), T_m(t)) = 0, \tag{12.795}$$

$$\begin{aligned}|X_m| &= \left|b_s(T_m - T_o, \boldsymbol{v}_m) + b_p(\boldsymbol{v}_m, T_m)\right| \\ &\leq c(1 + \|T_m(t)\|_{1,1}\, \|\boldsymbol{v}_m(t)\|_{o,2} + \|T_m(t)\|_{o,1}\, \|\boldsymbol{v}_m(t)\|_{1,2}),\end{aligned} \tag{12.796}$$

where c is a positive constant independent of m and of ε.

The integration of the Galerkin approximation (with $\vec{\gamma}(t) = \boldsymbol{a}_m(t)$) in time over $I_m = (t_o, t_m)$ yields

$$\int_{I_m} [(\mathscr{E}\boldsymbol{a}'_m(t), \boldsymbol{a}_m(t)) + A(\boldsymbol{a}_m(t), \boldsymbol{a}_m(t)) + (\Xi(\boldsymbol{a}_m(t)), \boldsymbol{a}_m(t)) - (\boldsymbol{P}(t), \boldsymbol{a}_m(t))]\mathrm{d}t + (J'_\varepsilon(\boldsymbol{v}_m(t), \boldsymbol{a}_m(t)) = 0. \tag{12.797}$$

Since $(j'_\varepsilon(\boldsymbol{v}), \boldsymbol{v}) \geq 0$, in view of the ellipticity Equation 121.768-12.770 of the bilinear form $A(\boldsymbol{a}, \vec{\alpha})$, due to Equation 12.975 and 12.796, and after some modification we obtain

$$\begin{aligned}&e_o\, \|\boldsymbol{a}_m(t)\|^2_{0,2} + 2c_a \int_{I_m} \|\boldsymbol{a}_m(\tau)\|^2_{1,2}\, \mathrm{d}\tau \\ &\leq 2c \int_{I_m} (1 + \|T_m(\tau)\|_{1,1}\, \|\boldsymbol{v}_m(\tau)\|_{0,2} + \|T_m(\tau)\|_{0,1}\, \|\boldsymbol{v}_m(\tau)\|_{1,N})\mathrm{d}\tau + 2\int_{I_m} \|\boldsymbol{P}(\tau)\|_*\, \|\boldsymbol{a}_m(\tau)\|_{1,2}\, \mathrm{d}\tau,\end{aligned} \tag{12.798}$$

where $e_o = \min\{(\rho)^{1/2}, (\mu)^{-1/2}, (\rho c)^{1/2}\}$. According to the Gronwall lemma after some more algebra (see Nedoma (1995)) we arrive at the following estimates

$$\|\boldsymbol{a}_m(t)\|_{o,2} \leq c, \quad t \in I, \quad \int_I \|\boldsymbol{a}_m(\tau)\|^2_{1,2}\, \mathrm{d}\tau \leq c, \tag{12.799}$$

where $c =$ constant < 0 independent of m.

The linear form $\boldsymbol{a} \rightarrow A(\boldsymbol{a}, \vec{\alpha})$ for fixed $\vec{\alpha}$ is continuous on H so that

$$A(\boldsymbol{a}(t), \vec{\alpha}) = (A_B\boldsymbol{a}(t), \vec{\alpha}), A_B \in \mathscr{L}(\underline{H}, (\underline{H})'), \tag{12.800}$$

and consequently

$$(\mathscr{E}\boldsymbol{a}'_m + A_B\boldsymbol{a}_m + \hat{g}j'_\varepsilon(\boldsymbol{v}_m) + \boldsymbol{h}_m - \boldsymbol{P}, \vec{\alpha}_j) = 0, \quad 1 \leq j \leq m. \tag{12.801}$$

Hence $\mathscr{E}\boldsymbol{a}' = S_m(\boldsymbol{P} - A_B\boldsymbol{a}_m - \hat{g}j'_\varepsilon(\boldsymbol{v}_m) - \boldsymbol{h}_m)$, where S_m is the orthogonal projection, $\underline{H}_o \rightarrow \mathscr{W}^m$ and where $\mathscr{W}^m$ is the space spanned by $\{\vec{\alpha}_j \mid 1 \leq j \leq m\}$. As in Nedoma (1995) we have

$$\|S_m\boldsymbol{p}\|_{(\underline{H}^1)'} \leq c\, \|\boldsymbol{p}\|_{(H^1)'} \,. \tag{12.802}$$

The estimates in Equation 12.799 and 12.802 prove that

$$\{\boldsymbol{a}_m(t); m \in \boldsymbol{N}\} \text{ is a bounded subset in } L^2(I; \underline{H}), \tag{12.803}$$

$$\boldsymbol{a}'_m(t) \text{ is a bounded subset of } L^2(I; (\underline{H}')').$$

The convergence of the finite-dimensional approximation, i.e. the limitation procedure over m, ε being fixed, follows from Equation 12.803 together with the fact that $\chi(t) = J'_\varepsilon(\boldsymbol{v}_\varepsilon(t))$. Since the system of functions $\{\alpha_j\}$ is complete in $\underline{H}^1$, the regularized problem $(\mathscr{P}_r)_v$ yields

$$\int_I [(\mathscr{E}\boldsymbol{a}'_\varepsilon(t), \vec{\alpha}(t)) + A(\boldsymbol{a}_\varepsilon(t), \vec{\alpha}(t)) + (\Xi(\boldsymbol{a}_\varepsilon(t)), \vec{\alpha}(t)) + (\chi(t), \boldsymbol{w}(t)) = 0 \quad \forall\vec{\alpha} \in \overline{H}^1.$$

The regularized problem $(\mathscr{P}_r)_v$ and Equation 12.799 having been satisfied, it is sufficient to prove that $\chi(t) = J'_\varepsilon(\boldsymbol{v}_\varepsilon(t))$.

To prove the limitation procedure over ε, we define for $\vec{\alpha} \in \overline{H}$ fixed

$$\begin{aligned} Y_\varepsilon = &\int_I [(\mathscr{E}\vec{\alpha}'(t), \vec{\alpha}(t) - \boldsymbol{a}_\varepsilon(t)) + A(\boldsymbol{a}_\varepsilon(t), \vec{\alpha}(t) - \boldsymbol{a}_\varepsilon(t)) + (\Xi(\boldsymbol{a}_\varepsilon(t)), \vec{\alpha}(t) - \boldsymbol{a}_\varepsilon(t)) \\ &- (\boldsymbol{P}(t), \vec{\alpha}(t) - \boldsymbol{a}_\varepsilon(t))]\mathrm{d}t + J_\varepsilon(\boldsymbol{w}) - J_\varepsilon(\boldsymbol{v}_\varepsilon), \quad Y_\varepsilon \geq 0. \end{aligned}$$

Since there exists subsequence $\{\boldsymbol{a}_\varepsilon(t)\}$ such that

$$\begin{aligned} &\boldsymbol{a}_\varepsilon(t) \to \boldsymbol{a}(t) \quad \text{in } L^\infty(I; \underline{H}^o) * - \text{ weakly and weakly in } L^2(I; \overline{H}^1), \\ &\boldsymbol{a}'_\varepsilon(t) \to \boldsymbol{a}'(t) \text{ weakly in } L^2(I; (\overline{H}^1)') \end{aligned}$$

and as $\lim \inf \int_I j_\varepsilon(\boldsymbol{v}_\varepsilon(t))\mathrm{d}t \geq \int_I j(\boldsymbol{v}(t))\mathrm{d}t$, then (see Nedoma (1995)) we have proved that $(\boldsymbol{v}(t), \boldsymbol{B}(t), T(t))$ satisfies Equation 12.765 and that $\boldsymbol{a}'_m(t)$ is a bounded subset of $L^2(I; (\overline{H}^1)')$.

The uniqueness of the solution can be proved by the usual technique as in Nedoma (1995) or in Section 12.4.1.

12.7. GLOBAL GRAVITY MODEL OF THE EARTH AND OF PLANETS

12.7.1. Introduction

Former and present studies show that most major features in the Earth's global gravity field have a correlation with plate tectonic features, with the lower mantle convection and with all geodynamic, thermal and geomagnetic processes in the Earth's interior (see, for example, Kaula (1969), (1972a,b), Bott (1971), Hide and Malin (1971)). However, observed gravity and geoid anomalies are substantially less than the anomalies predicted from the present models. Since it is unikely that the observed gravity data show great errors, this suggests that these models are based on mistaken conceptions or that they are based on conceptions and mathematical models which simplifying too much. The discrepancies between observed and predicted data are caused (after some analysis) by inaccuracies in their mathematical descriptions. Such models assume the influences of the upper parts of the Earth but do not assume all of the much more complicated geodynamic processes in the Earth's interior. These depend greatly on the thermo-magneto-plastodynamic, thermal and diffusion processes, gravity, density differentiation of matters in the Earth's interior, tidal and rotation forces, friction forces, and to a great extent effects occurring on the contact boundaries between different inhomogeneities and the special properties of the Earth's materials. The mass distribution in the Earth's interior (see Arkani-Hamed (1970)) depends on gravity differentiation and geodynamic processes taking place in the

Earth's interior, as well as on the external (the tidal effects, see Melchior (1978)) and internal force fields. It is evident that this changes during geological periods, i.e. that it changes during time. These changes evoke a number of geodynamic effects, e.g. change of the shape of the Earth manifested as recent movements, variation of the gravity field during geological epochs, rate of rotation of the Earth in time, shifting of the principal axes of inertia and centre of gravity of the Earth in time.

Hence, for his analyses Kaula (1972) used solutions and some results of Gaposchkin and Lamberk (1971), Schubert and Turcotte (1971), Turcotte and Oxburgh (1967), which are based on simple models. Nevertheless, he obtains fundamental results as it is impossible to rule out phenomena in the mantle of no direct relationship to the lithosphere as the cause of major features in the gravity field. Most of the major features in the Earth's gravitational field show a fairly systematic correlation with plate tectonic and post-glacial features. This correlation varies in quality from spotty for oceanic rises to consistently close for trench and island arcs. However, the correlation of most gravity features with lithospheric properties encourages this explanation arising from asthenospheric-lithospheric interaction. The more such features can be explained this way the more the remainder become significant indicators of deeper mantle conditions.

The other geodynamic model of the Earth (see Figure 12.5) is based on the results of Nedoma (1983), (1984), (1986), (1994), Nedoma *et al.* (1985), as well as of Section 12.2 of this chapter, where it is shown that the model is based on the following assumptions:

- the origin of the Earth is about 4.5-5 mld years ago and the Earth was probably created by an accumulation of interstellar matter on material condensation nuclei,
- in the course of the development of the Earth's body the fundamental factors were gravity, radioactivity, temperature, diffusion, and density differentiation, together with phase changes of Earth's masses at high temperatures, great pressures and chemical reactions under the same conditions,
- that interior Earth's masses can be characterized by a thermo-elasto-visco-plastic (weak-elastic up to very strong visco-plastic, characterized by welded on and melted rocks) state in a relatively weak magnetic field.

Due to results of Nedoma (1986a,b), Nedoma *et al.* (1985), the theoretical gravity profiles can be determined directly from the best deflection (vertical displacement component) profiles by using the Bouguer relation. The gravity anomaly $\Delta g(\boldsymbol{x}, t, T, p, \boldsymbol{H}, d, \ldots)$ is thus also dependent on all geodynamic processes in the Earth or planets, and thus we speak about a time-dependent gravity field of the Earth and planets.

The aim of this section is to analyse the inner and outer gravity field of the Earth during geological epochs from the point of view of global and local geodynamic processes taking place within the Earth's and planets' interiors. Our investigations will be based on the global geodynamic model of the Earth, discussed in the previous sections.

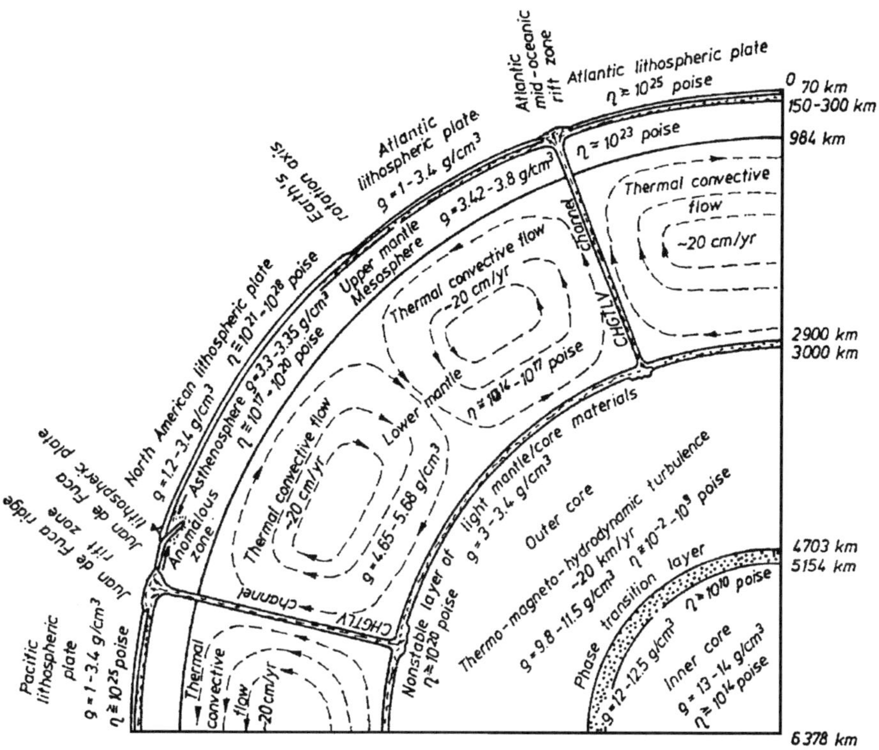

Figure 12.5. Global geodynamic model of the Earth

12.7.2. Generalized Theory of Gravity

12.7.2.1. Introduction

The time-dependent global gravity field during the last geological periods has been closely connected with the geodynamic processes manifested by a transfer of masses in the Earth's interior during the same past geological epochs. Since velocities of mass transfer in the Earth's interior are very slow ($\sim$ several cm yr^{-1} in the mantle or several km yr^{-1} in the core), the gravity field can be studied as a parametrized potential field dependent on the geodynamic processes and all geophysical fields (geothermal, geomagnetic, etc.) and processes (diffusion, melting, metallization) (see Nedoma (1986),(1997)).

Since all geodynamic, geothermal, etc. processes act during the geological epochs we shall denote the investigated geological epoch by the time interval $\bar{I} = \langle t_o, t_1 \rangle$. Then for a moving continuum to identify a given point, we shall specify its position $\boldsymbol{x}_o$ at a certain time t_o and its position $\boldsymbol{x}$ at the moment t by $\boldsymbol{x} = \boldsymbol{x}(\boldsymbol{x}_o, t)$. Then the motion of the medium can be described by $\boldsymbol{x} = \boldsymbol{x}(\boldsymbol{x}_o, t)$, $\boldsymbol{u}(\boldsymbol{x}_o, t_o) = \boldsymbol{u}_o$. We denote by $\Omega(0)$ the region occupied by the Earth at the moment $t = t_o$, and by $\Omega(t)$ the region occupied by the Earth at the moment t. Then the region $\Omega(t)$ is the image of $\Omega(0)$ at time t under the mapping $\boldsymbol{x} = \boldsymbol{x}(\boldsymbol{x}_o, t)$.

To derive the system of equations we shall start from

(a) the law of conservation of mass in its integral or differential form,
(b) the law of momentum conservation in its integral or differential form,
(c) the generalized conservation law of energy in its integral or differential form,
(d) Maxwell's equations,
(e) the equation of thermo-diffusion,
(f) the equation for the potential of gravity and the gravity acceleration vector $\boldsymbol{g}$ (see Section 12.2).

The force acting on a material point at rest on the Earth's surface is the resultant of the gravitational force $\boldsymbol{f}_g(\boldsymbol{x}, t, \ldots)$ and the centrifugal force of the Earth's rotation $\boldsymbol{f}_c(\boldsymbol{x}, t, \ldots)$, and of other forces such as the precession force $\boldsymbol{f}_p(\boldsymbol{x}, t, \ldots)$, the nutation force $\boldsymbol{f}_n(\boldsymbol{x}, t, \ldots)$, the electromagnetic force $\boldsymbol{f}_b(\boldsymbol{x}, t, \ldots)$ and the tidal force $\boldsymbol{f}_t(\boldsymbol{x}, t, \ldots)$, etc. We denote them by $\boldsymbol{f}_r$ (where $r = p, n, b, t$) and they are defined by $\boldsymbol{f}_r = \operatorname{grad} \psi_r$, where ψ_r are the potentials of precession, nutation, electomagnetic and tidal forces, etc., defined in the usual sense. The total force, the resultant of gravitational and centrifugal forces and of all above mentioned forces, is called gravity g. The potential of gravity, denoted by Φ, is defined as the sum of the potential of gravitational force $\boldsymbol{f}_g$, we shall denote it by V, and the centrifugal force $\boldsymbol{f}_c$, we shall denote it by ψ, as well as of the potentials of precession, nutation, electromagnetic, tidal and the other forces, we denote them by ψ_p, ψ_n, ψ_b, ψ_t, etc., i.e.

$$\Phi = \Phi(\boldsymbol{x}, t, \ldots) = V(\boldsymbol{x}, t, \ldots) + \psi(\boldsymbol{x}, t, \ldots) + \sum_r \psi_r(\boldsymbol{x}, t, \ldots). \tag{12.804}$$

For simplicity in the further accounts, the effects of precession, nutation, electromagnetic, tidal and other forces can be omitted.

To derive gravity, let us introduce a rectangular co-ordinate system $0x_1x_2x_3$. Let us denote by x_i, $i = 1, 2, 3$, and by ξ_i, $i = 1, 2, 3$, the co-ordinates of the attracting point or the point mass dm, respectively (see Figure 12.6). The distribution of the density of the Earth's body is defined by

$$\frac{\partial \rho(\boldsymbol{x}, t, \ldots)}{\partial t} + \operatorname{div}\,(\boldsymbol{v}(\boldsymbol{x}, t, \ldots)\rho(\boldsymbol{x}, t, \ldots)) = 0 \text{ at every time } t \in I. \tag{12.805}$$

The potential of gravitation is defined by

$$\begin{aligned} V &= G_c \int_{\Omega(t)} \mathrm{d}m/r = G_c \int_{\Omega(t)} \rho(\boldsymbol{x}, t, \ldots) p(\boldsymbol{x}, t, \ldots) K^{-1}(\boldsymbol{x}, t, \ldots) r^{-1} \mathrm{d}\boldsymbol{x} \\ &= G_c \int_{\Omega(t)} \frac{\rho(\vec{\xi}, t, \ldots)}{\left(\sum_{i=1}^{3} (x_i - \xi_i)^2\right)^{1/2}} \frac{p(\vec{\xi}, t, \ldots)}{K(\vec{\xi}, t, \ldots)} \mathrm{d}\vec{\xi} \text{ at every time } t, \end{aligned} \tag{12.806}$$

where G_c is the gravitational constant which in cgs units has the value $G_c = 6.67\ 10^{-8}\ \mathrm{cm}^3\ \mathrm{g}^{-1}\ \mathrm{sec}^{-2}$, r is the distance between the mass element dm and the

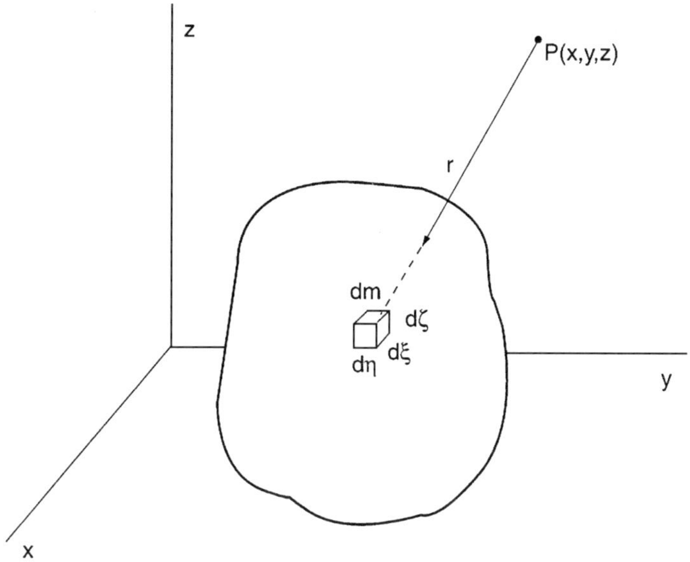

Figure 12.6. Derivation of gravity. Potential of a non-rotating solid body; $\mathrm{d}m = \mathrm{d}m(\xi, \eta, \zeta)$ - the mass element in the local coordinate system (ξ, η, ζ)

point P, $p = p(\boldsymbol{x}, t, \ldots)$ is the excess pressure, $\rho = \rho(\boldsymbol{x}, t, \ldots)$ is the density and $K = K(\boldsymbol{x}, t, \ldots)$ is the bulk modulus reckoned from the elastic constants obtained from analyses of seismic waves propagated inside the Earth. The potential V is continuous through the whole space and time t and vanishes at infinity as r^{-1} for every t. The potential V satisfies Poisson's equation

$$\begin{aligned}\operatorname{div} \boldsymbol{f}_g(\boldsymbol{x}, t, \ldots) &= \Delta V(\boldsymbol{x}, t, \ldots)\\ &= -4\pi G_c p(\boldsymbol{x}, t, \ldots) K^{-1}(\boldsymbol{x}, t, \ldots)\rho(\boldsymbol{x}, t, \ldots) \text{ at every time } t\end{aligned} \tag{12.807}$$

inside the Earth's body, and the Laplace equation

$$\Delta V(\boldsymbol{x}, t, \ldots) = 0 \text{ at every time } t \tag{12.808}$$

outside the Earth's body.

The relation $\operatorname{div} \boldsymbol{f}_g = -4\pi G_c p\rho K^{-1}$ can be interpreted as follows. The masses are the sources of the gravitational field and the changes in geological epochs; the strength of the sources $-\operatorname{div} \boldsymbol{f}_g$ is proportional to the mass density $\rho = \rho(\boldsymbol{x}, t, T, \boldsymbol{H}, , p, d, \ldots)$; and $-4\pi G_c p\rho K^{-1}$ is the gravitational flux in analogy with fluid mechanics. Moreover, it is evident that quantities ρ, p, K, $\boldsymbol{f}_g$ are functions of spatial co-ordinates x_i, $i = 1, 2, 3$, time t, temperature T, concentration of magnetic and electric intensity $\boldsymbol{H}$ and $\boldsymbol{E}$, pressure p, etc., which follows from the fact that the system of equations describing geodynamic processes (in the general point of view) is coupled.

Further, the components f_{gi}, $i = 1, 2, 3$, of the graviational force $\boldsymbol{f}_g$ are then defined by $f_{gi} = \partial V/\partial x_i \equiv V_{,i} \equiv \partial_i V$, $i = 1, 2, 3$, or in the vector symbolism by $\boldsymbol{f}_g = \operatorname{grad} V$, and are also continuous through space and time, but not so the second derivatives. At points where the density at the time moment t has a discontinuity, the second derivatives also have a discontinuity, which is evident as the gravitational potential V satisfies Poisson's equation at every time moment t.

The centrifugal force $\boldsymbol{f}_c$ (see Figure 12.7) is defined by

$$\boldsymbol{f}_c(\boldsymbol{x}, t, \ldots) = \omega^2 \boldsymbol{h} = (\omega^2 x_1, \omega^2 x_2, 0) = \operatorname{grad} \psi(\boldsymbol{x}, t, \ldots) \equiv (\psi_{,1}, \psi_{,2}, \psi_{,3}),$$

where $\psi = (1/2)\omega^2(x_1^2 + x_2^2) = (1/2)\omega^2 h^2$, $h = (x_1^2 + x_2^2)^{1/2}$ and $\vec{\omega} = \vec{\omega}(t)$ is the angular velocity of the rotating Earth.

Thus, the potential of gravity Φ fulfils the generalized Poisson's equation

$$\Delta\Phi(\boldsymbol{x}, t, \ldots) = -4\pi G_c p(\boldsymbol{x}, t, \ldots) K^{-1}(\boldsymbol{x}, t, \ldots)\rho(\boldsymbol{x}, t, \ldots) + 2\omega^2(t) + \sum_r \operatorname{div} \boldsymbol{f}_r \quad \text{for every time } t, \tag{12.809}$$

where $\rho(\boldsymbol{x}, t, \ldots)$ fulfils Equation 12.805.

Since ψ is an analytic function, the discontinuities of Φ are those of V, i.e. second derivatives have jumps at discontinuities of density ρ. The gravity vector $\boldsymbol{g}$ is thus defined by

$$\boldsymbol{g} = \boldsymbol{g}(\boldsymbol{x}, t, \ldots) = \operatorname{grad} \Phi(\boldsymbol{x}, t, \ldots) \text{ for every time } t. \tag{12.810}$$

Generally, we have to solve the following global problem. Let $\boldsymbol{v} = \partial \boldsymbol{u}/\partial t$ be known, then

$$\boldsymbol{u}(\boldsymbol{x}, t, \ldots) = \int_{t_o}^{t_1} \boldsymbol{v}(\boldsymbol{x}, t, \ldots)\mathrm{d}t + \boldsymbol{u}_o(\boldsymbol{x}). \tag{12.811}$$

And thus we have the problem of finding tensor, vector and scalar functions $\{\boldsymbol{v}\}$, $\{\tau_{ij}\}$, $\{\boldsymbol{E}\}$, $\{\boldsymbol{H}\}$, $\{d\}$, $\{T\}$, $\{p\}$, $\{\Phi\}$, $\{\boldsymbol{g}\}$ satisfying

$$\rho\left(\frac{\partial v_i}{\partial t} + v_j \frac{\partial v_j}{\partial x_i}\right) = \frac{\partial \tau_{ij}}{\partial x_j} + F_i, \quad i = 1, 2, 3 \quad \text{in } \Omega(t), \quad \forall t, \tag{12.812}$$

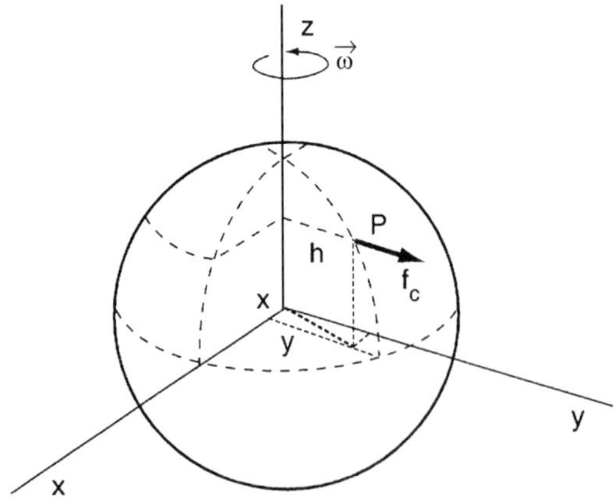

f_c - centrifugal force
h - distance of the point from rotating axis z

Figure 12.7. The centrifugal force at the Earth's surface

$$\frac{\partial \rho}{\partial t} + \operatorname{div}\,(\rho \boldsymbol{v}) = 0 \quad \text{in } \Omega(t), \quad \forall t, \tag{12.813}$$

$$\rho\beta_{ij}T_o\left(D_{ij} + v_k \frac{\partial e_{ij}^{e+v+p}}{\partial x_j}\right) + \rho c_e\left(\frac{\partial T}{\partial t} + v_k \frac{\partial T}{\partial x_k}\right)$$
$$= W - \frac{\partial}{\partial x_i}\left(\kappa_{ij}\frac{\partial T}{\partial x_j}\right) - \sigma_{ij}^{-1} j_i j_j + T j_i \frac{\partial \alpha_{ij}^o}{\partial x_j} \quad \text{in } \Omega(t), \quad \forall t, \tag{12.814}$$

$$\frac{\partial \boldsymbol{B}}{\partial t} - \operatorname{rot}\,[\boldsymbol{v}, \boldsymbol{B}] = (\mu\sigma)^{-1}\Delta \boldsymbol{B} - [\nabla(\mu\sigma)^{-1}, \operatorname{rot} \boldsymbol{B}] + \operatorname{rot} \frac{\partial \boldsymbol{D}}{\partial t} \quad \text{in } \Omega(t), \quad \forall t, \tag{12.815}$$

$$\rho\left(\frac{\partial d}{\partial t} + \boldsymbol{v}\nabla d\right) + \operatorname{div} \boldsymbol{i}_m = h \quad \text{in } \Omega(t), \quad \forall t, \tag{12.816}$$

$$\Delta\Phi = -4\pi G_c p\rho K^{-1} + 2\omega^2 + \sum_r \operatorname{div} \boldsymbol{f}_r \quad \text{in } \Omega(t), \quad \forall t, r = p, n, b, t, \ldots \tag{12.817}$$

Moreover, the functions $\boldsymbol{v}, \rho, \boldsymbol{B}, \boldsymbol{E}, T, d, \Phi, \boldsymbol{g}$ satisfy the boundary conditions on $\partial\Omega(t)$, the interface and contact conditions on interface and contact boundaries between neighbouring geological blocks, the initial conditions on $\Omega(t_o) = \Omega(0)$ for $t = t_o$, the conditions of recrystallization, melting and solidification on the phase change boundaries (see Nedoma (1986),(1995), (1997)) and the boundary conditions on the faces of the cracks and faults.

At present, such problems as a whole are mathematically unsolved. The existence and uniqueness of the solution of such problems are unknown; only the simple models discussed here were proved (see Section 12.3 and Chapters 4–7). Currently, only the algorithms based on their variational formulation for simpler model problems can be given.

12.7.2.2. *Level surfaces and plumb lines*

The surface $\Phi(\boldsymbol{x}, t, \ldots) = \Phi_k(\boldsymbol{x}, t, \ldots)$, $k = 0, 1, 2, \ldots$, are called equipotential surfaces or level surfaces, and for every time moment t are equal to constant Φ_k. The surface $\Phi(\boldsymbol{x}, t, \ldots) = \Phi_o(t)$ is the time-dependent geoid, and it is evident that it changes through geological periods. From this we obtain

$$\mathrm{d}\Phi(\boldsymbol{x}, t, \ldots) = \operatorname{grad}\,\Phi(\boldsymbol{x}, t, \ldots)\mathrm{d}\boldsymbol{x} = \boldsymbol{g}(\boldsymbol{x}, t, \ldots)\mathrm{d}\boldsymbol{x}(t) \text{ for every time moment } t,$$

where t is a parameter.

As t is a parameter and $\Phi(\boldsymbol{x}, t, \ldots) = \text{constant}$ for every t and as the vector $\mathrm{d}\boldsymbol{x}$ is taken along the equi-potential (level) surface, then $\mathrm{d}\Phi = 0$ and then $g(\boldsymbol{x}, t, \ldots)\mathrm{d}\boldsymbol{x}(t) = 0$ for every time moment t. Then we see that at every time moment t the gravity vector $\boldsymbol{g}$ is normal to the equi-potential (level) surface. The lines that intersect all equi-potential surfaces normally at time t are imperceptibly curved and are called lines of force or plumb lines. Moreover, the gravity vector is tangential to them at any point at every time moment t. Thus the plumb lines are defined as a curve whose line element vector $\mathrm{d}\boldsymbol{x}(t) = (\mathrm{d}x_1(t), \mathrm{d}x_2(t), \mathrm{d}x_3(t))$ has the direction of the gravity vector $\boldsymbol{g}(\boldsymbol{x}, t, \ldots) = \text{grad}\ \Phi(\boldsymbol{x}, t, \ldots)$, that is, that $\mathrm{d}\boldsymbol{x}$ and $\boldsymbol{g}$ differ only by a proportionality factor. Further, let $H = H(t)$ be the height of a point above the geoid (sea level), the so-called orthometric height, measured along the curved plumb line starting from the geoid.

It is evident that the geoid and the level surfaces were changed through the geological epochs, so that the classical gravity theory holds for every time moment. From this point of view for the curvature of level surfaces a modification of classical techniques can be used (see Heiskanen and Moritz (1962)).

Thus let P be a point on a level surface S (see Figure 12.8). Let $x_1x_2x_3 \equiv xyz$ be a local co-ordinate system with the origin at P, whose x_3-axis is vertical and therefore identical to the normal level surface. Let us intersect the surface $\Phi(\boldsymbol{x}, t, \ldots) = \Phi_k(t)$ and the plane x_1x_3 by setting $x_2 = 0$.

From analytical geometry it is known that for the curvature of the intersection of the level surface with the x_1x_3-plane, $K_1 = \mathrm{d}^2x_3/\mathrm{d}x_1^2$ holds for every t. From $\Phi(\boldsymbol{x}, t, \ldots) = \Phi_k(t)$ for all t, differentiating with respect to x_1 we obtain

$$\frac{\partial \Phi}{\partial x_1} + \frac{\partial \phi}{\partial x_3}\frac{\mathrm{d}x_3}{\mathrm{d}x_1} = 0,$$

$$\frac{\partial^2 \Phi}{\partial^2 x_1} + 2\frac{\partial^2 \Phi}{\partial x_1 \partial x_3}\frac{\mathrm{d}x_3 \mathrm{d}x_3}{\mathrm{d}x_1 \mathrm{d}x_1} + \frac{\partial \Phi}{\partial x_3}\frac{\mathrm{d}^2 x_3}{\mathrm{d}x_1^2} = 0 \text{ for all } t,$$

as $x_2 = 0$ and x_3 is a function of x_1. From Figure 12.8 the x_3-axis is perpendicular to the level surface and then $\partial\Phi/\partial x_3 = -g$, and the x_1-axis is a tangent at P to the level surface, and then $\mathrm{d}x_3/\mathrm{d}x_1 = 0$ for all t. Thus

$$\frac{\mathrm{d}^2 x_3}{\mathrm{d}x_1^2} = -\frac{\partial^2 \Phi}{\partial^2 x_1}\left(\frac{\partial \Phi}{\partial x_3}\right)^{-1}, \quad K_1 = g^{-1}\frac{\partial^2 \Phi}{\partial^2 x_1} \text{ for every } t.$$

Replacing x_1 by x_2 we immediately obtain the expression of the curvature of the intersection of the level surface with the x_2x_3-plane in the form $K_2 = g^{-1}\partial^2\Phi/\partial x_2^2$ for every t.

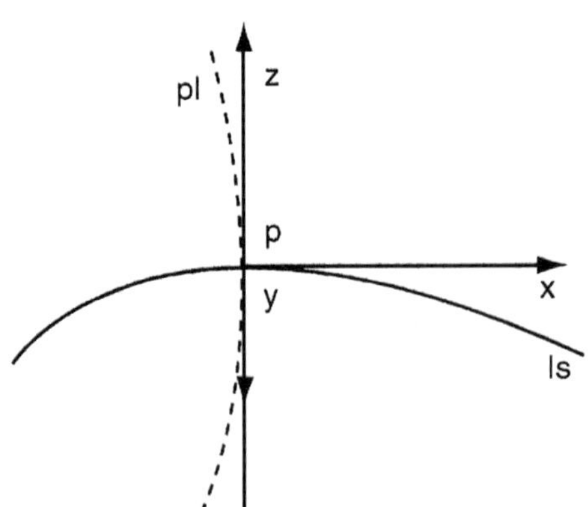

Figure 12.8. Definition of the local coordinate system $(x, y, z) \equiv (x_1, x_2, x_3)$ - the local coordinate system; pl = the plumbline and ls = the level surface

We shall now define the mean curvature J of a surface at a point P as the arithmetical mean of the curvature of the curves in which two mutually perpendicular planes through the surface normal intersect the surface at the time moment t. Thus we have

$$J = -\tfrac{1}{2}(K_1 + K_2) = -\tfrac{1}{2}g^{-1}(\partial^2\Phi/\partial x_1^2 + \partial^2\Phi/\partial x_2^2) \text{ at every time moment } t. \tag{12.818}$$

From Equations 12.817 and 12.818 we obtain

$$\begin{aligned} &-gJ(\boldsymbol{x}, t, \ldots) + \frac{\partial^2\Phi(\boldsymbol{u}, t, \ldots)}{\partial^2 x_3} \\ &\quad = -4\pi G_c p(\boldsymbol{x}, t, \ldots)K^{-1}(\boldsymbol{x}, t, \ldots)\rho(\boldsymbol{x}, t, \ldots) + 2\omega^2(t) + \sum_r \operatorname{div} \boldsymbol{f}_r(\boldsymbol{x}, t). \end{aligned}$$

Since $\partial\Phi/\partial x_3 = -g$, $\partial^2\Phi/\partial x_3^2 = -\partial g/\partial x_3 \doteq -\partial g/\partial H$, where H is the ortometric height, we obtain the generalized Brun's relation

$$\begin{aligned} \frac{\partial g(\boldsymbol{x}, t, \ldots)}{\partial H} &= -2g(\boldsymbol{x}, t, \ldots)J(\boldsymbol{x}, t, \ldots) + 4\pi G_c p(\boldsymbol{x}, t, \ldots)K^{-1}(\boldsymbol{x}, t, \ldots)\rho(\boldsymbol{x}, t, \ldots) \\ &\quad - 2\omega^2(t) - \sum_r \operatorname{div} \boldsymbol{f}_r(\boldsymbol{x}, t) \text{ at every time moment } t. \end{aligned} \tag{12.819}$$

This equation gives the relation between the geometrical and dynamic concepts of geodesy through geological epochs. It relates the vertical gradient of gravity $\partial g/\partial H$ to the mean curvature of the level surface J.

12.7.2.3. Integral equation for the gravity potential

Gauss' formula gives

$$\int_{\Omega(t)} \Delta\Phi \mathrm{d}\boldsymbol{x} = \int_{\partial\Omega(t)} \frac{\partial\Phi}{\partial n}\mathrm{d}s \text{ for every time moment } t \in I,$$

where $\partial\Omega(t)$ represents the physical surface of the Earth at moment t, and $\Omega(t)$ the domain occupied by the real Earth at moment t. Then

$$\begin{aligned} \int_{\partial\Omega(t)} \frac{\partial\Phi}{\partial n}\mathrm{d}s &= \int_{\Omega(t)} \Delta\Phi \mathrm{d}\boldsymbol{x} \\ &= \int_{\Omega(t)} \left[-4\pi G_c p(\boldsymbol{x}, t, \ldots)K^{-1}(\boldsymbol{x}, t, \ldots)\rho(\boldsymbol{x}, t, \ldots) + 2\omega^2(t) + \sum_r \operatorname{div} \boldsymbol{f}_r \right] \mathrm{d}\boldsymbol{x} \\ &= -\int_{\partial\Omega(t)} g_n \mathrm{d}s, \end{aligned} \tag{12.820}$$

where $\partial\Phi/\partial n = -g_n$, and g_n is the component of the gravity normal to the real Earth's surface $\partial\Omega(t)$ at the time moment t. Since $M_o = \int_{\Omega(t)} \rho(\boldsymbol{x}, t, \ldots)\mathrm{d}\boldsymbol{x}$ is the mass of the Earth, then we obtain

$$M_o = (4\pi G_c)^{-1} \int_{\Omega(t)} g_n \mathrm{d}s + (2\pi G_c)^{-1}\omega^2(\xi) \int_{\Omega(t)} \mathrm{d}\boldsymbol{x}$$
$$+ (4\pi G_c)^{-1} \int_{\Omega(t)} \sum_r \operatorname{div} \boldsymbol{f}_r \mathrm{d}\boldsymbol{x} \text{ for every time moment } t \in I. \quad (12.821)$$

The last equation permits the determination of the mass of the Earth at an arbitrary time moment $t \in I$ from the measured gravity onto the Earth's surface at the same time moment t. By using the third Green's identity $u = r^{-1}$ we gradually obtain

$$\int_{\Omega(t)} r^{-1}\Delta\Phi \mathrm{d}\boldsymbol{x} + 2\pi\Phi - \int_{\partial\Omega(t)} \left[r^{-1}\frac{\partial\Phi}{\partial n} - \Phi\frac{\partial}{\partial n}(r^{-1}) \right] \mathrm{d}s = 0$$

and

$$\int_{\Omega(t)} r^{-1} \Bigg[-4\pi G_c p(\boldsymbol{x}, t, \ldots) K^{-1}(\boldsymbol{x}, t, \ldots)\rho(\boldsymbol{x}, t, \ldots) + 2\omega^2(t)$$
$$+ \sum_r \operatorname{div} \boldsymbol{f}_r(t) \Bigg] \mathrm{d}\boldsymbol{x} + 2\pi\Phi(\boldsymbol{x}, t, \ldots) + \int_{\partial\Omega(t)} [\Phi(\boldsymbol{x}, t, \ldots)\partial/\partial n(r^{-1})$$
$$+ g_n(\boldsymbol{x}, t, \ldots) r^{-1}]\mathrm{d}s = 0.$$

Since $\Phi = G_c \int_{\Omega(t)} r^{-1} p(\boldsymbol{x}, t, \ldots) K^{-1}(\boldsymbol{x}, t, \ldots)\rho(\boldsymbol{x}, t, \ldots)\mathrm{d}\boldsymbol{x} + (1/2)\omega^2(t)[x_1^2(t) + x_2^2(t)]$, and omitting the term containing $\boldsymbol{f}_r(t)$ we obtain

$$-2\pi\Phi(\boldsymbol{x}, t, \ldots) + \int_{\partial\Omega(t)} [\Phi(\boldsymbol{x}, t, \ldots)\partial/\partial n(r^{-1}) + g_n(\boldsymbol{x}, t, \ldots)r^{-1}]\mathrm{d}s + 2\pi\omega^2(t)$$
$$\times [x_1^2(t) + x_2^2(t)] + 2\omega^2(t) \int_{\Omega(t)} r^{-1}\mathrm{d}\boldsymbol{x} = 0 \text{ for every } t \in I. \quad (12.822)$$

All quantities in this equation are referred to the physical surface of the Earth $\partial\Omega(t)$ at the moment t, and where r denotes the distance between a fixed point P and the variable surface element ds of the physical Earth's surface $\partial\Omega$, $\boldsymbol{n}$ denotes the outward normal to the physical Earth's surface $\partial\Omega$, g_n denotes the normal component of the gravity vector, ω denotes the angular velocity of the rotating Earth, and $\bar{r}$ denotes the distance between the point P and the volume element $\mathrm{d}\boldsymbol{x}$. Equation 12.822 relates the physical surface of the Earth $\partial\Omega(t)$ at the moment t to the gravity potential $\Phi(\boldsymbol{x}, t, \ldots)$ and the gravity $\boldsymbol{g}(\boldsymbol{x}, t, \ldots)$ at the same moment t. On the other hand, if we assume that $\Phi(\boldsymbol{x}, t, \ldots)$ and $\boldsymbol{g}(\boldsymbol{x}, t, \ldots)$ are known then in the same way this equation determines the surface $\partial\Omega(t)$ at the time moment t. This is a way to obtain the generalized time-dependent Molodensky's theory for the determination of the surface $\partial\Omega(t)$ of the real Earth from the measured gravity potential and gravity $\boldsymbol{g}(\boldsymbol{x}, t, \ldots)$ at an arbitrary time moment t during the geological evolution of the Earth.

12.7.2.4 The anomalous gravity field

Let $U = U(\boldsymbol{x}, t)$ be the gravity potential of a normal body, the so-called normal gravity potential, which is the same for every time moment t. The difference between the real gravity potential of the Earth $\Phi(\boldsymbol{x}, t, \ldots)$ and the normal gravity potential $U = U(\boldsymbol{x}, t) \equiv U(\boldsymbol{x})$ is the so-called anomalous or disturbing potential $T = T(\boldsymbol{x}, t, \ldots) = \Phi(\boldsymbol{x}, t, \ldots) - U(\boldsymbol{x}, t, \ldots)$ for every t. We compare the geoid $\Phi(\boldsymbol{x}, t, \ldots) = \Phi_o(\boldsymbol{x}, t)$ and the reference ellipsoid $U(\boldsymbol{x}, t) = U_o = \Phi_o$. From Figure 12.9 we denote $N = \overline{PQ}$ the geoidal height (geoidal undulation), $\boldsymbol{g}_P$ the gravity vector at point P, $\vec{\gamma}_P$ the normal gravity vector at point Q, and $\Delta\boldsymbol{g}(\boldsymbol{x}, t, \ldots) = \boldsymbol{g}_P(\boldsymbol{x}, t, \ldots) - \vec{\gamma}_Q(\boldsymbol{x}, t)$ the gravity anomaly vector, whose magnitude is the gravity anomaly $\Delta g = g_P(\boldsymbol{x}, t, \ldots) - \gamma_Q(\boldsymbol{x}, t)$ at the time moment t, and whose difference in direction is the deflection of the vertical at time moment t. It is evident that both of these change during geological epochs. Further, since $\boldsymbol{g}(\boldsymbol{x}, t, \ldots) = \text{grad } \Phi(\mathbf{x}, t, \ldots)$, $\vec{\gamma}(\boldsymbol{x}, t) = \text{grad } U(\boldsymbol{x}, t)$ then

$$\delta(t) = \boldsymbol{g}_P(\boldsymbol{x}, t, \ldots) - \vec{\gamma}_P(\boldsymbol{x}, t) = \text{grad } (\Phi(\boldsymbol{x}, t, \ldots) - U(\boldsymbol{x}, t)) = \text{grad } T(\boldsymbol{x}, t, \ldots).$$

Since $g(\boldsymbol{x}, t, \ldots) = -\partial\Phi(\boldsymbol{x}, t, \ldots)/\partial n$, $\gamma(\boldsymbol{x}, t) = -\partial U(\boldsymbol{x}, t)/\partial n_e = -\partial U(\boldsymbol{x}, t)/\partial n$, where $\boldsymbol{n}, \boldsymbol{n}_e$ are the normal to the geoid and to the normal body (ellipsoid), then

$$\begin{aligned}\delta g &= g_P(\boldsymbol{x}, t, \ldots) - \gamma_P(\boldsymbol{x}, t) = -(\partial\Phi(\boldsymbol{x}, t, \ldots)/\partial n - \partial U(\boldsymbol{x}, t)/\partial n_e) \\ &\doteq -(\partial\Phi(\boldsymbol{x}, t, \ldots)/\partial n - \partial U(\boldsymbol{x}, t)/\partial n) = -\partial T(\boldsymbol{x}, t, \ldots)/\partial n,\end{aligned}$$

and we see that the gravity disturbance δg is the difference in magnitude of the actual and the normal gravity vector, the normal component of the gravity disturbance vector $\vec{\delta}$ for every time t.

Since $\gamma_P(\boldsymbol{x}, t) = \gamma_Q(\boldsymbol{x}, t) + \partial\gamma(\boldsymbol{x}, t)/\partial n N(t)$ for every time moment t, we have

$$\begin{aligned}-\frac{\partial T(\boldsymbol{x}, t, \ldots)}{\partial n} &= \delta g = g_P(\boldsymbol{x}, t, \ldots) - \gamma_P(\boldsymbol{x}, t) \\ &= g_P(\boldsymbol{x}, t, \ldots) - \gamma_Q(\boldsymbol{x}, t) - \frac{\partial\gamma(\boldsymbol{x}, t)}{\partial n} N(t) \text{ for every } t. \quad (12.823)\end{aligned}$$

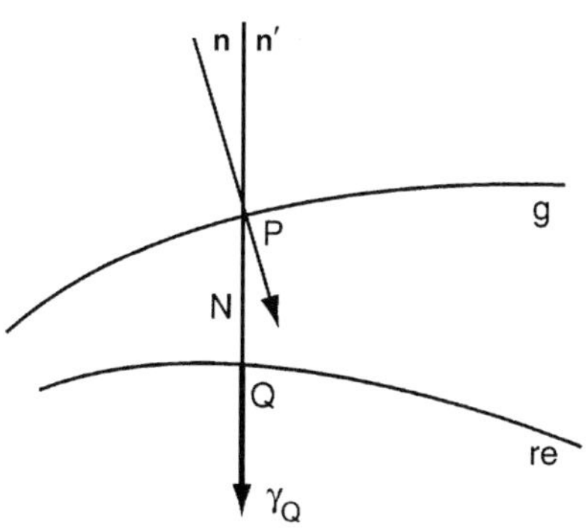

Figure 12.9. Geoid and reference ellipsoid; $g =$ geoid, $re =$ the reference ellipsoid, $N = \overline{PQ}$-the geoidal height $\boldsymbol{n}, \boldsymbol{n}' =$ normals to the geoid and the reference ellipsoid, $\vec{\gamma}_Q =$ the normal gravity vector at Q

Hence

$$\frac{\partial T(\boldsymbol{x}, t, \ldots)}{\partial n} - (\gamma(\boldsymbol{x}, t))^{-1}\frac{\partial \gamma(\boldsymbol{x}, t)}{\partial n} T(\boldsymbol{x}, t, \ldots) = -\Delta g(\boldsymbol{x}, t, \ldots)$$

$$\text{for every time moment } t \in I, \quad (12.824)$$

which is the so-called generalized fundamental equation of physical geodesy. Here $\Delta g(\boldsymbol{x}, t, \ldots)$ is assumed to be known at every point of the geoid at the time moment t, and $\partial T/\partial n$, and T are given upon the surface of the geoid at the same time moment t. Thus the problem of determination of the disturbing potential $T(\boldsymbol{x}, t, \ldots)$ leads to a third boundary value problem of the potential theory with time t as a parameter. If this is solved for $T(\boldsymbol{x}, t, \ldots)$, then the geoidal height can be computed by the generalized Bruns' relation $N(t) = T(\boldsymbol{x}, t, \ldots)/\gamma(\boldsymbol{x}, t)$ for every time moment $t \in I$.

12.7.3 Gravity Field Outside the Earth

Outside the physical Earth's surface the potential of gravity $\Phi(\boldsymbol{x}, t, \ldots)$ fulfils the Laplace equation

$$\Delta\Phi(\boldsymbol{x}, t, \ldots) = 0 \text{ in the outer space } \Omega'(t), \text{ for every time } t. \quad (12.825)$$

On the physical surface of the Earth $\partial\Omega(t)$ the gravity vector can be given, so that

$$\partial\Phi(\boldsymbol{x}, t, \ldots)/\partial n = -g(\boldsymbol{x}, t, \ldots) \text{ for every time } t \quad (12.826)$$

is given. At infinity it fulfils the conditions of regularity

$$\lim_{r\to\infty} r\Phi(\boldsymbol{x}, t, \ldots) = 0. \quad (12.827)$$

The same problem can be investigated from the point of view of another formulation in which any satellite's measurements can be used. Then we construct a surface of known shape, denoted by S, lying at a height H_o above the physical surface of the Earth. We shall assume that it turns around the same axis passing through the same centre of gravity with the same angular velocity $\vec{\omega}(t)$. Then the potential of gravity can be formed on S from the satellite's measurements, i.e. (see Figure 12.10)

$$\Phi(\boldsymbol{x}, t, \ldots) = \Phi_k(t) \text{ on } S(t). \quad (12.828)$$

The problem investigated can be formulated as a boundary value problem for the Laplace equation in $3D$, and then analysed and solved by methods of functional and numerical analyses such as the finite element method or the boundary element method (see, for example, Nedelec and Nedoma (1980)). We shall give an idea for solving this problem by the second method.

It is well known (see Theorem 1.1 and Theorem 1.2 in Nedelec and Nedoma (1980)) that such problems, i.e. the problems in Equations 12.825, 12.826 and 12.828 and in Equations 12.825-12.827 have unique solutions.

By using Green's formula and by the technique used in Nedelec and Nedoma (1980) we obtain the following integral equations

$$\Phi(\boldsymbol{x}, t, \ldots) + \int_{\partial\Omega\cup S} \Phi\frac{\partial v}{\partial n}\mathrm{d}s - (4\pi)^{-1}\int_{\partial\Omega\cup S} \frac{\partial\Phi(\boldsymbol{x}, t, \ldots)}{\partial n_i}|\boldsymbol{y} - \boldsymbol{x}|^{-1}\,\mathrm{d}s = 0, \quad (12.829)$$

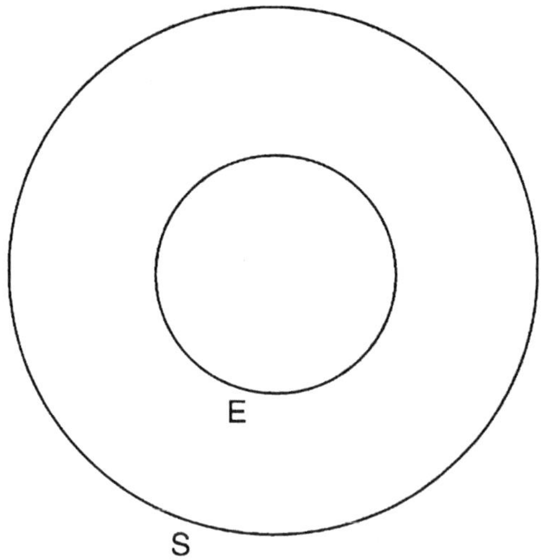

Figure 12.10. The bounded domain outside the Earth; $E =$ the Earth's surface, $S =$ the spherical surface rotating with the same angular velocity like the Earth

where $\boldsymbol{y} \in \Omega'(t)$, $\partial\Phi/\partial n_i$ is the derivative along the normal of the restriction of Φ to Ω, $|\cdot|$ is the Euklidean norm in R^3, $\boldsymbol{y} = (y_i)^T$, $\boldsymbol{x} = (x_i)^T$, $i = 1, 2, 3$, and $v \equiv v(\boldsymbol{x}) = (4\pi)^{-1} |\boldsymbol{x} - \boldsymbol{y}|^{-1}$, and

$$\Phi(\boldsymbol{x}, t, \ldots) = (4\pi)^{-1} \int_{\partial\Omega(t)} \kappa(\boldsymbol{y}, t, \ldots) |\boldsymbol{y} - \boldsymbol{x}|^{-1} \, \mathrm{d}s, \tag{12.830}$$

where $S(t) \to \infty$, $\partial\Phi/\partial n_e - \partial\Phi/\partial n_i = -4\pi\kappa$, $\kappa = \mathrm{d}m/\mathrm{d}\partial\Omega$ is the surface density, $\boldsymbol{y} \in \Omega$, and $\partial\Phi/\partial n_e$, $\partial\Phi/\partial n_i$ are the normal derivative of Φ in $\Omega'(t)$ and the derivative along the normal of the restriction of Φ to Ω'. The problem in Equation 12.830 is equivalent to its variational formulation

$$b(\kappa, \hat{\kappa}) = (4\pi)^{-1} \int_{\partial\Omega'(t)} \int_{\partial\Omega'(t)} \kappa(\boldsymbol{y}, t)\hat{\kappa}(\boldsymbol{x}, t) |\boldsymbol{y} - \boldsymbol{x}|^{-1} \, \mathrm{d}s(\boldsymbol{y}, t)\mathrm{d}s(\boldsymbol{x}, t), \tag{12.831}$$

where $b(\kappa, \hat{\kappa})$ is a symmetric bilinear coercive form. Both problems can be solved by applying the boundary element technique (see, for example, Nedelec and Nedoma (1980) and, for more details, Brebbia (1978), Banerjee and Butterfield (1981), Venturini (1983), Brebbia, Telles and Wrobel (1984), Brebbia and Walker (1980), Beer and Watson (1992)).

REFERENCES

Anufriev, A.P. and Braginsky, S.I. (1975), "About the Influence of Boundary Unevenness of the Earth Nucleus on Speed of Liquid and Magnetic Field", *Geomagn. i Aeronomiya*, **15**, 1075-1082 (in Russian).

Anufriev, A.P., Cupal, I. and Hejda, P. (1993), "Time Evolution of the Solution in Model Z", in *The Cosmic Dynamo* (Krause, F. *et al.* (Eds.)), 447-452.

Arkani-Hamed, J. (1970), "Lateral Variations of Density in the Mantle", *Geophys. J.*, **20**, 431-455.

Baker, H.B. (1914), "The Origin of Continental Forms", *S.*, 1-111.

Banerjee, P.K. and Butterfield, R. (1981), *Boundary Element Methods in Engineering Science*, McGraw-Hill, London.

Beer, G. and Watson, J.O. (1992), *Introduction to Finite and Boundary Element Methods for Engineers*, J. Wiley & Sons Ltd., Chichester, New York, Brisbane, Toronto, Singapore.

Benfield, A.E. (1950), The Temperature of an Accreating Earth. *Trans. Aen. Geophys. Union*, **31**, 53-68.

Bercovici, D., Schubert, G. and Glatzmaier, G.A. (1989a), "Influence of Heating Mode on Three-Dimensional Mantle Convection", *Geophys. Res. Lett.* **16**, 617-620.

Bercovici, D., Schubert, G., Glatzmaier, G.A. and Zebib, A.I. (1989b), "Three-Dimensional Thermal Convection in a Spherical Shell", *J. Fluid Mech.*, **206**, 75-104.

Blanc, P. and Gasser, L. (1996), "Existence of a Weak Solution for Binary Alloy Problem", Submitted to *Mathematische Annalen.*

Bott, M.H.P. (1971), "The Mantle Transition Zone as Possible Source of Global Gravity Anomalies", *Earth Planet. Sci. Lett.*, **11**, 28-34.

Braginsky, S.I. (1967), "Magnetic Waves in the Earth's Core", *Geomagn. i Aeronomiya*, **7**, 1050-1060.

Braginsky, S.I. (1975), "Almost an Axially-Symmetrical Model of the Hydromagnetic Dynamo of the Earth", *Geomagn. i Aeronomiya*, **15**, 149-158.

Braginsky, S.I. and Roberts, P.H. (1987), "Model-Z Geodynamo", *Geophys. Astrophys. Fluid Dynam.*, **38**, 327-349.

Brebbia, C.A. (1978), *The Boundary Element Method For Engineers*, Pentach Press, Plymouth.

Brebbia, C.A. and Walker, S. (1980). *Boundary Element Techniques in Engineering*, Newnes-Butterworths, London.

Brebbia, C.A., Telles, J.C.F. and Wrobel, L.C. (1983), *Boundary Element Technique*, Springer Verlag, Berlin.

Brezis, H. (1971), "Monotonicity Methods in Hilbert Spaces and Some Applications to Non-linear Partial Differential Equations", in *Contributions to Non-linear Functional Analysis* (Zarantonello E.H. (Ed.)), Academic Press, New York, 101-156.

Bucha, V. (Ed.) (1983), *Magnetic Field and the Processes in the Earth's Interior*, Academia, Prague.

Bullen, K.E. (1975), *The Earth's Density*, Chapman and Hall, London.

Byhovskii, N.B. (1957), "A Solution of a Mixed Problem for a System of Maxwell's Equations in the Case of Ideally Conducting Boundary", *Vestnik Leningrad. Univ.*, **13**, 50-66 (in Russian).

Cannon, J.R., DiBenedetto, E. and Knightly, G.H. (1983), "The Bi-dimensional Stefan Problem With Convection: The Time Dependent Case", *Partial Differential Equations*, **14**, 1549-1604.

Cathles, L.M. (1975), *The Viscosity of the Earth's Mantle*, Princeton University Press. Princeton, NJ.

Céa, J. (1971), *Optimisation, théorie et algorithmes*, Dunod, Paris.

Chandrasekhar, S. (1961), *Hydrodynamic and Hydromagnetic Stability*, Clarendon Press, Oxford.

Christensen, U.R. (1984), "Convection With Pressure- and Temperature-Dependent Non-Newtonian Rheology", *Geophys. J. R. Astron. Soc.*, **77**, 343-384.

Christensen, U.R. and Yuen, D.A. (1984), "An Interaction of a Subducting Lithospheric Slab with a Chemical or Phase Boundary", *J. Geophys. Res*, **89**, **B6**, 4389-4402.

Ciavaldini, J.F. (1975), "Analyse Numérique d'un Probléme de Stefan á Deux Par un Méthod D'éléments Finis", *SIAM J. Numer. Anal.*, **12**, 464-489.

Cupal, I. (1983), "The Magnetic Flux of an Oscillatory Nearly Symmetric Dynamo", in *Stellar and Planetary Magnetism* (Soward, A.M. (Ed.)), Gordon and Breach Science Publishers, New York, London, Paris.

Cupal, I. and Hejda, P. (1989), "On the Computation of a Model-Z with Electromagnetic Core-Mantle Coupling", *Geophys. & Astrophys. Fluid Dynam.*, **49**, 161-172.

Dietz, R.S. (1961), "Continent and Ocean Basin Evolution by Spreading of the Sea Floor", *Nature*, **190**, 854-857.

Du Toit, A. (1937), *Our Wandering Continents*, Oliver and Boyd, Edinburgh.

Duvant, G. (1975). Two phases Stefan Problem with ranging Specific Head Coefficients. *An. Acad. Brasil. Cienc.*, **47**, 314.

Duvaut, G. and Lions, J.L. (1972a), *Les Inéquations en Mécanique et en Physique*, Dunod, Paris.

Duvaut, G. and Lions, J.L. (1972b), "Inéquations en Thermoélasticité et Magnétohydrodynamique", *Arch. Rational. Mech. Anal.*, **46**, 241-279.

Duvaut, G. and Lions, J.L. (1976), *Inequalities in Mechanics and Physics*, Springer Verlag, Berlin, Heidelberg, New York.

Dziewonski, A.M., Hager, B.H. and O'Connell, R.J. (1977), "Large Scale Heterogeneities in the Lower Mantle", *J. Geophys. Res.*, **82**, 239-235.

Dziewonski, A.M., Hales, A.M. and Lapwood, E.R. (1975), "Parametrically Simple Earth Models consistent with Geophysical Data", *Physics of the Earth and Planetary Interiors*, **10**, 12-48.

Ekeland, I. and Temam, R. (1976), *Convex Analysis and Variational Problems*, North-Holland Publ. Co., Amsterdam, Oxford.

Elliot, C.M. (1981), "On the Finite Element Approximation of an Elliptic Variational Inequality arising from an Implicit Time Discretization of the Stefan Problem", *IMA J. Numer. Anal.*, **1**, 115-125.

Fortin, M. (1972), *Calcul Numérique des Écoulements des Fluides de Bingham et des Fluides Newtoniens Incompressible par la Methode des Elements Finis*, Thése d'Etat Université Pierre et Marie Curie, Paris.

Fowler, A.C. (1983), "On the Thermal State of the Earth's Mantle", *J. Geophys.*, **53**, 42-51.

Fowler, A.C. (1985), "Fast Thermo-viscous Convection", *Stud. Appl. Math.*, **72**, 189-219.

Fowler, A.C. (1993), "Toward a Description of Convection with Temperature- and Pressure-Dependent Viscosity", *Stud. Appl. Math.*, **88**, 113-139.

Fox, L. (1975). "What Are the Best Methods?", in *Moving Boundary Problems in Heat Flow and Diffusion* (Ockendon, J.R. and Hodkins, W.R. (Eds)), Clarendon Press, 210-240.

Friedman, A. (1968), "The Stefan Problem in Several Space Variables", *Trans. Amer. Math. Soc.*, **133**, 51-87.

Friedman, A. (1982), *Variational Principles and Free Boundary Problems*, Wiley-Interscience, New York.

Fučík, S. and Kufner, A. (1980), *Non-linear Differential Equations*, Elsevier, Amsterdam, Oxford, New York.

Gaposchkin, E.M. and Lamberk, K. (1971), "Earth's Gravity Field to 16th Degree and Station Co-ordinates From Satellite and Terrestrial Data", *J. Geophys. Res.*, **76**, 4855-4883.

Gasser, L. (1995), *Existence Analysis and Numerical Schemes for Models of Binary Alloy Solidification*, EPFL Thesis, Lausanne.

Girault, V. and Raviart, P.A. (1986), *Finite Element Methods for Navier-Stokes Equations*, Springer Verlag, Berlin.

Glatzmaier, G.A. (1988), "Numerical Simulations of Mantle Convection: Time-Dependent, Three-Dimensional, Compressible, Spherical Shell", *Geophys. Astrophys. Fluid Dyn.*, **43**, 223-264.

Hansen, U. and Yuen, D.A. (1989), "Dynamical Influences from Thermal-Chemical Instabilities at the Core-Mantle Boundary", *Geophys. Res. Lett.*, **16**, 629-632.

Heiskanen, W.A. and Moritz, H. (1962), *Physical Geodesy*, W.H. Freeman Co, San Francisco.

Hejda, P.(1982), "Boundary Conditions in the Hydromagnetic Dynamo Problem", *Studia Geoph. et Geod.*, **26**, 361-372.

Hejda, P. and Anufriev, A.P. (1993), "A New Numerical Scheme in the Solution of the Geodynamo Z-Model", in *The Cosmic Dynamo* (Krause, F. *et al.* (Eds.)), 441-445.

Hess, H.H. (1962), "History of the Ocean Basins", *Geol. Soc. Amer. Bull., Buddington Memorial Volume: Petrologic Studies*, 599-600.

Hide, R. (1966), "Free Hydromagnetic Oscillations of the Earth's Core and the Geomagnetic Theory of the Geomagnetic Secular Variation", *Phil. Trans. Roy. Soc.*, **259A**, 615-647.
Hide, R. (1970), "On the Earth's Core-Mantle Interface", *Quart. J. Roy. Meteorolog. Soc.*, **96**, 579.
Hide, R. (1977), "Towards a Theory of Irregular Variations in the Length of the Day and Core-Mantle Coupling", *Phil. Trans. Roy. Soc.*, **284a**, 547-553.
Hide, R. (1979), "Dynamo Theorems", *Geophys. Astrophys. Fluid Dyn.*, **14**, 183-186.
Hide, R. (1982), "On the Role of Rotation in the Generation of Magnetic Fields by Fluid Motions", *Phil. Trans. Roy. Soc.*, **306A**, 223-234.
Hide, R. (1985), "Thermal Convection in a Rotating Fluid Subject to a Horizontal Temperature Gradient", *Societa Italiana di Fisica*, Bologna, Italy, 159-171.
Hide, R. (1988), "Towards an Interpretation of Uranus' Eccentric Magnetic Field", *Geophys. Astrophys. Fluid Dyn.*, **44**, 207-209.
Hide, R. (1993). Personal Communication.
Hide, R. and Malin, S.R.C. (1971), "Bumps on the Core-Mantle Boundary", *Comments on Earth's Sci Geophys.*, **2**, 1-13.
Hodkins, V.R. (Ed.) (1975), "Moving Boundary Problems in Heat Flow and Diffusion", *Proc. Conf. Univ. of Oxford*, 25-27 March 1974, Clarendon Press, Oxford.
Holmes, A. (1931), "Radioactivity and Earth' Movements", *Trans. Geol. Soc. Glasgow*, **18**, 559-606.
Holmes, A. (1965), *Principles of Physical Geology* (6th Revised Edition), Ronald Press, New York.
Jarvis, G.T. and Peltier, W.R. (1982), "Mantle Convection as a Boundary Layer Phenomenon", *Geophys. J. R. Astron. Soc.*, **68**, 389-427.
Jeanloz, R. (1990), "The Nature of the Earth's Core", *Annu. Rev. Earth Planet Sci.*, **18**, 357-386.
Jeanloz, R. and Richter, F.M. (1979), "Convection, Composition and the Thermal State of the Lower Mantle", *J. Geophys. Res.*, **84**, 5497-5504.
Jerome, J.W. (1977), "Non-linear Equations of Evolution and a Generalizad Stefan Problem", *J. Diff. Equations*, **26**, 240-261.
Jerome, J.W. (1983), *Approximation of Non-linear Evolution Systems*, Academic Press, New York, London.
Jerome, J.W. and Rose, M.E. (1977), "Error Estimates for Multi-Dimenmensional Two-Phase Stefan Problem", *Math. Comp.*, **39**, **160**, 377-414.
Kamarád, J. (1993), (1996), Personal Communication.
Kamenomotskaya, S. (1958), "On the Stefan Problem", *Naučnye Dokl. Vysšei Školy*, **1**, 60-62 (in Russian),
Kamenomotskaya, S. (1961), "On the Stefan Problem", *Matem. Sbornik*, **53**, **(95)4**, 489-514 (in Russian),
Kaula, W.A. (1969), "A Tectonic Classification of the Main Features of the Earth's Gravitational Field", *J. Geophys. Res.*, **74**, 4807-4826.
Kaula, W.A. (1972a), "Global Gravity and Tectonics", in *The Nature of the Solid Earth* (Robertson, E.C. (Ed.)). McGraw-Hill, New York, 385-405.
Kaula, W.A. (1972b), "Global Gravity and Mantle Convection", in *The Upper Mantle* (Ritsema, A.R. (Ed.)). Elsevier, Amsterdam, London, New York.
Kawai, N. and Mochizuki, S. (1971), "Metallic States in the Three 3rd Transition Metal Oxides FeO_3, CrO_3 and TiO_3 under Static High Pressure," *Phys. Rev. Lett.*, **A36**, 1.
Kawai, N. and Nishiyama, A. (1974), "Conductive SiO_2 at High Pressure", *Proc. Jap. Acad.*, 50.
Kawai, N. and Nishiyama, A. (1975), "Insulator-Metal Transition SiO_2, FeO and MgO", *Rev. Phys. and Chem. Japan* (Special Issue).
Kirby, S.H. (1983), "Rheology of the Lithosphere", *Revs. Geophys. Space Phys.*, **21**, 1458-1487.
Knittle, E. and Jeanloz, R. (1987), "Synthesis and Equation of State of $(Mg,Fe)SiO_3$ Perovskite to Over 100 GPa", *Science*, **235**, 668-670.
Knittle, E. and Jeanloz, R. (1989), "Simulating the Core-Mantle Boundary: An Experimental Study of High-Pressure Reactions between Silicates and Liquid Iron", *Geophys. Res. Lett.*, **16**, 609-612.

Knittle, E. and Jeanloz, R. (1991), "The High Pressure Phase Diagram of $Fe_{0.94}O$: A Possible Constituent of the Earth's Core", *J. Geophys. Res.*, **96**, **16**, 16169-16180.

Knopoff, L. (1964), "The Convection Current Hypothesis", *Rev. Geophys.*, **2**, 1, 89-122.

Knopoff, L. (1967), "On Convection in the Upper Mantle", *Geophys. J.*, **14**, 341-346.

Ladyzenskaya, O.A. (1969), *The Mathematical Theory of Viscous Incompressible Flow*, Gordon and Breach, New York.

Ladyzenskaya, O.A., Solonnikov, V.A. and Ural'ceva, N. (1968), *Linear and Quasi-linear Equations of Parabolic Type*, American Mathematical Society, Providence, RI.

Landau, L.D and Lifshitz, E.M. (1954), *Mechanics of Continuous Media*, GITTL, Moscow (in Russian),

Landau, L.D and Lifshitz, E.M. (1959), *Course of Theoretical Physics: Fluid Mechanics*, Pergamon Press, New York.

Landau, L.D and Lifshitz, E.M. (1984), *Electrodynamics of Continuous Media*, Pergamon Press, Oxford, New York, Beijing, Frankfurt, Sao Paulo, Sydney, Tokyo, Toronto.

Le Pichon, X., Francheteau, J. and Bonnin, J. (1973), *Plate Tectonics* (Developments in Geophysics, 6), Elsevier, Amsterdam.

Lions, J.L. (1969), *Quelques Méthodes des Résolution des Problémes aux Limites Non Linéaires*, Dunod, Paris.

Lions, J.L. and Magenes, E. (1970), *Problémes aux Limites Non Homogénes et Applications, 103*, Dunod, Paris.

Ljubimova, E.A. (1967), "Theory of Thermal State of the Earth's Mantle", in *The Earth's Mantle* (Gaskel, T.F. (Ed.)), Academic Press, London, New York.

Martinec, Z., Matyska, C., Čadek, O. and Hrdina, P. (1993), "The Stokes Problem with *3D* Newtonian Rheology in a Spherical Shell", *Computer Physics Communications*, **76**, 63-79.

Mauersberger, (1956), "Zur Bestimmung des Potentiallosen Anteiles des Geomagnetischen Hauptfeldes", *Gerlands Beitr. Geophysik*, **65**, 295-308.

McKenzie, D.P. (1983), "The Earth's Mantle", *Sci. Am.*, **249**, 67-78.

McKenzie, D.P. and Parker, R.L. (1967), "The North Pacific: An Example of Tectonics on a Sphere", *Nature*, **216**, N5122.

McKenzie, D.P. *et al.* (1974), "Convection in the Earth's Mantle: Towards a Numerical Simulation", *J. Fluid. Mech.*, **62**, **3**, 465-538.

Melchior, P. (1978), *The Tides of the Planet Earth*, Oxford.

Milani, A. (1980), "On a Singular Perturbation Problem for the Maxwell Equations in a Multiply Connected Domain", *Rend. Sem. Mat. Univ. Politech. Torino*, **38**, **1**, 123-132.

Moffatt, H.K. (1977), "Topographic Coupling of the Core-Mantle Interface", *Geophys. Astrophys. Fluid Dyn.*, **9**, 279-288.

Moffatt, H.K. (1978), *Magnetic Field Generation in Electrical Conducting Fluids*, Cambridge University Press.

Moffatt, H.K. and Dillon, R.I. (1976), "The Correlation between Gravitational and Geomagnetic Fields Caused by Interaction of Core Fluid Motion with a Bumpy Core-Mantle Interface", *Phys. Earth. Planet. Int.*, **13**, 67-78.

Morgan, W.J. (1968), "Rises, Trenches, Great Faults and Crustal Blocks", *J. Geophys. Res.*, **73**, 1959-1982.

Nec̄as, J. (1967). *Jes méthods directes en théorie des équations elliptiques*. Academia, Prague.

Nedelec, J.C. and Nedoma, J. (1980), "Numerical Approximation of Potential Problems", *Geod. Geop. Veröff.*, **III**, **H45**, 96-110.

Nedoma, J. (1983), "On One Type of Signorini Problem without Friction in Linear Thermoelasticity", *Appl. Mat.*, **28**, **6**, 393-407.

Nedoma, J. (1984), "On the Global Geodynamic Model of the Earth," *Gerlands Beitr. Geophysik*, **93**, **4**, 241-260.

Nedoma, J. (1986), "On the Geodynamic Model of the Earth and the Plate Tectonic Hypothesis: I-III", *Gerlands Beitr. Geophysik*, **95**, **1**, 36-62 (I,II), **95**, **2**, 89-105 (III),

Nedoma, J. (1987), "On the Signorini Problem with Friction in Linear Thermoelasticity: The Quasi-Coupled 2D-Case", *Appl. Mat.*, **32**, **3**, 186-199.

Nedoma, J. (1988a), "On the New Problem in the Potential Gravity Theory", in *Potential Theory*, (Král J. *et al.* (Eds.)) Plenum Press, New York and London, 255-262.

Nedoma, J. (1988b), "Global Gravity and the Geodynamic Model of the Earth", *Gerlands Beitr. Geophysik*, **97**, **6**, 495-513.

Nedoma, J. (1990), "Tectonic Evolution of Collision Zones: I-II", *Gerlands Beitr. Geophysik*, **99**, **2**, 97-108, 109-121.

Nedoma, J. (1993), "On a Generalized Thermo-Magnetohydrodynamic Problem: A New Approach to Mantle-Core Coupling", *7th Scientific Assembly of IAGA, Section 1.1*, Buenos Aires, 8-20 August, 1993, Argentina. (See also in *Technical Report V-578*, January 1994, ICS AS CR, Prague, Czech Republic.)

Nedoma, J. (1994), "Finite Element Analysis of Contact Problems in Thermoelasticity: The Semi-coercive Case", *J. Comput. Appl. Math.*, **50**, 411-423.

Nedoma, J.(1995), "Equations of Magnetodynamics of Incompressible Thermo-Bingham's Fluid under the Gravity Effect", *J. Comput. Appl. Math.*, **59(1)**, 109-128.

Nedoma, J. (1996a), "Finite Element Approximation of a Coupled Stefan-Contact Problem in Linear Thermoelasticity arising from the Implicit Discretization in Time: The Semi-coercive Case", *Tech. Report 676*, ICS AS *CR*, Prague.

Nedoma, J. (1996b), "Finite Element Approximation of a Coupled Contact-Stefan-Like Problem arising from the Time Discretization in Deformation Theory of Thermo-Plasticity", *7th Int. Congress on Computational and Applied Mathematics*, Leuven, July, 1996.

Nedoma, J. (1997a), "On a coupled Stephan-Like Problem in Thermo-Visco-Plastic Rhedogy", *J. Comput. Appl. Math.*, **84**(1997) 45-80.

Nedoma, J. (1997b), "Finite Element Approximation of a Coupled Contact-Stefan-Like Problem Arising from the Time Discretization in Deformation Theory of Thermo-Plasticity", *J. Comput. Appl. Math.*, **82**(1997) 313-331.

Nedoma, J. and Dvořák, J. (1995), "On the FEM Solution of a Coupled Contact-Two-Phase Stefan Problem in Thermo-Elasticity", *J. Comput. Appl. Math.*, **63**, 411-420.

Nedoma, J., Haslinger, J. and Hlaváček, I. (1989), "The Problem of an Obducting Lithospheric Plate in the Aleutian Arc System. A Finite Element Analysis in the Frictionless Case", *Math. Comput. Modelling*, **12**, **1**, 61-75.

Nikol'skij, S.M. (1975), *Approximation of Function of Several Variables and Embedding Theorems*, Springer Verlag, Berlin, Heidelberg, New York.

Nochetto, R.H., Paollini, M. and Verdi, C. (1991a), "An Adaptive Finite Element Method for Two-Phase Stefan Problems in Two Space Variables", *Math. Comp.*, **57**, 73-108.

Nochetto, R.H., Paollini, M. and Verdi, C. (1991b), "Local Mesh Refinements for Two-Phase Stefan Problem in Two Space Variables", *SIAM J. Sc. Statistical Comput.*, **12**, 1207-1244.

Nowacki, W. (1970), *Dynamical Problems of Thermoelasticity*, Mir, Moscow (in Russian).

Ockendon, J.R. and Hodgkins, W.R. (Eds.) (1975), *Moving Boundary Problems in Heat Flow and Diffusion*, Clarendon Press, Oxford.

Oleinik, O.A. (1960), "A Method of Solution of the General Stefan Problem", *Dokl. AN SSSR*, **135**, 1054-1057.

Parker, E.N. (1979), *Cosmical Magnetic Fields*, Clarendon Press, Oxford.

Pawlow, I. (1982), "A Variational Inequality Approach to Generalized Two-Phase Stefan Problem in Several Space Variables", *Annali di Matematica Pura ed Applicata*, **IV**, **CXXXI**, 333-373.

Peltier, W.R. (1974), "The Inpulse Response of a Maxwell Earth", *Rev. Geophys. Space Phys.*, **12**, 649-669.

Peltier, W.R. (1984), "The Thickness of the Continental Lithosphere", *J. Geophys. Res.*, **89**, 11303-11316.

Peltier, W.R. (Ed.) (1989), *Mantle Convection, Plate Tectonics and Global Dynamics*, Gordon and Breach Science Publisher, New York, London, Paris, Montreux, Tokyo, Melbourne.

Picasso, M. (1995), "An Adaptive Finite Element Algorithm for a Two-Dimensional Stationary Stefan-Like Problem", *Comput. Methods Appl. Mech. Engng.*, **124**, 213-230.

Prager, W. (1961), *Introduction to Continuum Mechanics*, Izd. IL., Moscow (in Russian); Ginn and Co. (1963).

Quareni, D.A., Yuen, D.A., Sewell, G. and Christensen, U. (1985), "High Rayleigh Number Convection with Strongly Variable Viscosity: A Comparison between Mean Field and Two-Dimensional Solutions", *J. Geophys. Res.*, **90**, 12633-12644.

Quareni, D.A. *et al.* (1986), "Adiabaticity and Viscosity in Deep Mantle Convection", *Geophys. Res. Letts.*, **13**, 38-41.

Quarteroni, A. and Valli, A. (1994), *Numerical Approximation of Partial Differential Equations*, Springer Verlag, Berlin, Heidelberg, New York, London, Paris, Tokyo, Hong Kong, Barcelona, Budapest.

Ranalli, G. (1987), *Rheology of the Earth*, Allen & Unwin, London, Sydney, Wellington.

Ritsema, A.R. (Ed.) (1972), *The Upper Mantle*, Elsevier, Amsterdam, London, New York.

Roberts, P.H. and Soward, B.M. (1972), "Magnetohydrodynamics of the Earth's Care", *Annual Review of Fluid Mechanics*, **4**, 117-154.

Roberts, P.H. and Soward, B.M. (1975), "A United Approach to Mean-Field Electrodynamics", *Astron. Nachr.*, **296**, 49-64.

Roberts, P.H. and Stix, M. (1971), "The Turbulent Dynamo. A Translation of a Series of Papers by Krause, F., Radler, K.H. and Steenbeck, M., *Tech. Note 60*, NCAR, Boulder, Colorado.

Roberts, P.H. and Stix, M. (1972), "Alfa-Effect Dynamos, by the Bullard-Gellman Formalism", *Astron. Astrophys.*, **18**, 453-466.

Rochester, M.G. *et al.* (1975), "Can Precession Power the Geomagnetic Dynamo?" *Geophys. J. Roy. Astr. Soc.*, **43**, 661.

Roden, R.B. (1963), "Electromagnetic Core-Mantle Coupling", *Geophys. J. Roy. Astr. Soc.*, **7**, 361.

Runcorn, S.K. (1962), "Convection Currents in the Earth's Mantle", *Nature*, **195**, N4848.

Runcorn, S.K. (1967), "Flow in the Mantle Inferred from the Low Degree Harmonico of the Geopotential", *Geophys. J. R. Astr. Soc.*, **42**, 375-384.

Runcorn, S.K. (1972), "Dynamical Processes in the Deeper Mantle", in "The Upper Mantle (Ritsema, A.R. (Ed.))", *Tectonophysics*, **13**, **1**-**4**, 623-637.

Schubert, G. and Turcotte, D.L. (1971), "One Dimensional Model of Shallow Mantle Convection", *J. Geophys. Res.*, **77**, 5.

Sclater, J.G., Jaupart, C. and Galson, D. (1980), "The Heat Flow through the Oceanic and Continental Crust and the Heat Loss of the Earth", *Rev. Geophys. Space Phys.*, **18**, 269-311.

Shape, H.N. and Peltier, W.R. (1979), "A Thermal History Model of the Earth based on Parametrized Convection", *Geophys. J. R. Astr. Soc.*, **50**, 171-205.

Shilling, J.S. and Shelton, R.N. (Eds.) (1981), *Physics of Solids Under High Pressure*, North-Holland Publ. Co, Amsterdam.

Soward, A.M. (Ed.) (1983), *Stellar and Planetary Magnetism*, Gordon and Breach Science Publishers, New York, London, Paris.

Stacey, F.D. (1977), *Physics of the Earth*, J. Wiley & Sons, New York.

Stanley, H.E. (1971), *Introduction to Phase Transition and Critical Phenomena*, Clarendon Press, Oxford.

Štědrý, M. and Vejvoda, O. (1983), "Small Time-Periodic Solutions of Equations of Magneto-hydrodynamics as a Singularity Perturbed Problem", *Apl. Mat.*, **28**, **5**, 344-356.

Štědrý, M. and Vejvoda, O. (1985). Equations of Magneto-hydrodynamics of compressible fluid: Periodic solutions, *Apl. Mat*, **30**, 2, 72-91.

Štědrý, M. and Vejvoda, O. (1986), "Equations of Magneto-hydrodynamics: Periodic Solutions", *Časopis Pěst. Mat.*, **111**, 177-184.

Steenbeck, M. and Krause, F. (1966), "Erklarung Stellar und Planetarer Magnetfelder durch einen Turbulenzbedingten Dynamomechanisms", *Z. Naturforch.*, **21a**, 1285-1296.

Stephan, J. (1889), "Über einige Probleme der Theorie der Wärmeleitung. *Akad. Mat. Natur.*, **98**, S.-B., Wien.

Stevenson, D.J. (1986), "On the Role of Surface Tension in the Migration of Melts and Fluids", *Geophys. Res. Lett.*, **13**, 1149-1152.

Stupjalis, L. (1977), "On Solvability of an Initial-Boundary Value Problem of Magneto-hydrodynamics", *Zapiski Naučnych Seminarov LOMI*, **69**, 219-239.

Taylor, F.B. (1910), "Bearing of the Tertiary Mountain Belt on the Origin of the Earth's Plan", *Bull. Geol. Soc. Am.*, **21**, 179-226.

Temam, R. (1977), (1984) *Navier-Stokes Equations. Theory and Numerical Analysis*, North-Holland Publ. Co., Amsterdam.

Temam, R. (1983), *Navier-Stokes Equations and Non-linear Functional Analysis*, SIAM, Philadelphia.

Thomasset, F. (1981), *Implementation of Finite Element Methods for Navier-Stokes Equations*, Springer Verlag, Berlin.

Tough, J.G. (1967), "Nearly Symmetric Dynamos", *Geophys. J.*, **13**, 393-396.

Tough, J.G. and Roberts, P.H. (1968), "Nearly Symmetric Hydromagnetic Dynamos", *Phys. Earth. Planet. Inter.*, **1**, 288-296.

Turcotte, D.L. (1975), "Geophysical Problems with Moving Phase Change Boundaries and Heat Flow", in *Moving Boundary Problems in Heat Flow and Diffusion* (Hodkins, V.R. (Ed.)), Proc. Conf. Univ. of Oxford, 25-27 March 1974, Clarendon Press, Oxford.

Turcotte D.L., Oxburgh, E.R. (1967), "Finite Amplitude Convective Cells and Continental Drift", *J. Fluid. Mech.*, **28**, 29-42.

Turcotte, D.L. and Schubert, G. (1971), "Structure of the Olivine-Spinel Phase Boundary in the Descending Lithosphere", *J. Geophys. Res.*, **76**, 7980-7987.

Turcotte, D.L. and Schubert, G. (1982), *Geodynamics*, J.Wiley & Sons, New York.

Venturini, W.S. (1983), *Boundary Element Method in Geomechanics* (Lecture Notes in Engineering), Springer Verlag, Berlin.

Voorhies, C.V. (1986), "Steady Flows at the Top of Earth's Core Derived from Geomagnetic Field Model", *J. Geophys. Res.*, **91**, **12**, 12444-12466.

Wegener, A. (1912), "Die Entstehung der Kontinente", *Peterm Mitt.*, 185-195, 253-256, 305-309, Jena.

Wegener, A. (1915), "Die Entstehung der Kontinente und Ozeane", *S.*, 1-94, Braunschweig.

Wegener, A. (1929), "Die Entstehung der Kontinente und Ozeane", *Wisenschaften und Technik*, **66**, 1-231, Braunschweig. [Translated in 1929 as *The Origin of Continents and Oceans*, Methuen, London, then reprinted in 1966, by Dover.]

Whitaker, S. (1969), "Advances in Theory of Fluid Motion in Porous Media", *Industrial and Engineering Chemistry*, **61**, 14-28.

White, R.E. (1982a), "An Enthalpy Formulation of the Stefan Problem", *SIAM J. Numer. Anal.*, **19**, 1129-1157.

White, R.E. (1982b), "A Numerical Solution of the Enthalpy Formulation of the Stefan Problem", *SIAM J. Numer. Anal.*, **19**, 1158-1172.

Williams, Q.R. *et al.* (1987), "The Melting Curve of Iron to 250 Gigapascals: Constraint on the Temperature at the Earth's Centre", *Science*, **236**, 181-182.

Wu, P. and Peltier, W.R. (1983), "Glacial Isostatic Adjustment and the Free Air Gravity Anomaly as a Constraint on Deep Viscosity", *Geophys. J. Roy. Astr. Soc.*, **74**, 377-449.

Wysession, M.E. *et al.* (1992), "The Structure of the Core-Mantle Boundary from Diffracted Waves", *J. Geophys. Res.*, **97**, **B6**, 8749-8764.

Zhang, S. and Yuen, D.A. (1988), "Dynamical Effects on the Core-Mantle Boundary From Depth-Dependent Thermodynamical Properties of the Lower Mantle", *Geophys. Res. Lett.*, **15**, 451-454.

Index